Engineering Drawing and Design

Fifth Edition

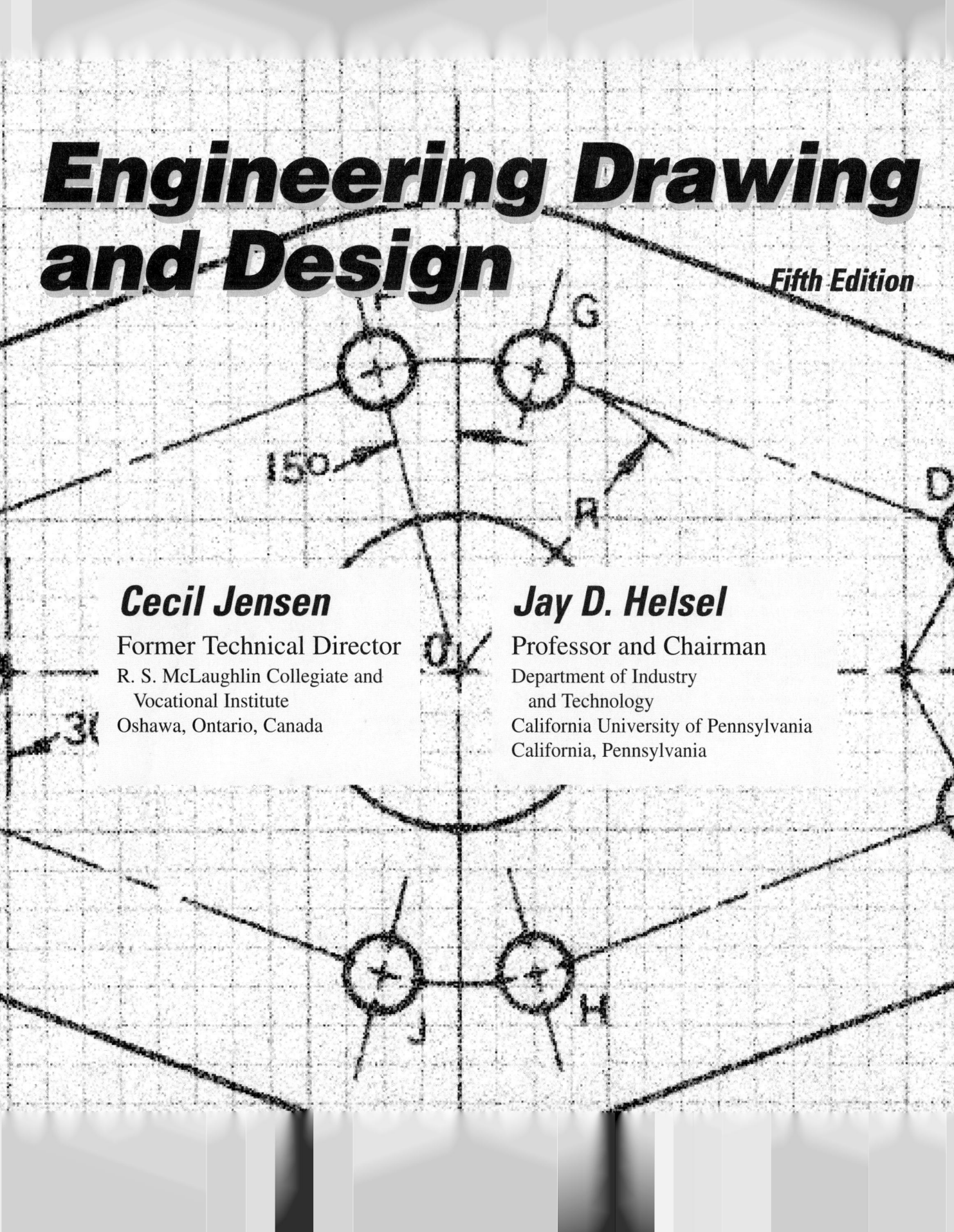

Cecil Jensen

Former Technical Director
R. S. McLaughlin Collegiate and
    Vocational Institute
Oshawa, Ontario, Canada

Jay D. Helsel

Professor and Chairman
Department of Industry
    and Technology
California University of Pennsylvania
California, Pennsylvania

Photo Credit: Cover photo courtesy of Wayne Eastep

Jensen, Cecil Howard
　　[Engineering drawing and design / Cecil Jensen, Jay D. Helsel.
　　5th ed.
　　　　p.　　　cm.
　　Includes index.
　　ISBN 0-02-801795-1
　　1. Mechanical drawing.　　2. Engineering design.　　I. Helsel, Jay D.
　　II. Title.
　　T353.J47　1996
　　604.2—dc20　　　　　　　　　　　　　　　　94–11373
　　　　　　　　　　　　　　　　　　　　　　　　　CIP

Photo credits for part and chapter openers: Part One, Comstock, Inc. Part Two,
Charly Franklin/FPG International. Part Three, Andrew Sacks/Tony Stone Images.
Part Four, Earl Zubkoff Photography. Part Five, Martin Rogers/Tony Stone Images.
Chapters 1–27, Aaron Haupt. Other credits: Photos 27–7–1 and 27–7–5 by OrCAD, Inc.

This printing includes references to ASME Y14.5M-1994.

Printed in the United States of America.

Send all inquiries to:
Glencoe/McGraw-Hill
936 Eastwind Drive
Westerville, OH 43081

ISBN 0-02-801795-1
3 4 5 6 7 8 9 10 11 12 13 14 15　QPK/LP　03 02 01 00 99 98 97 96

CONTENTS

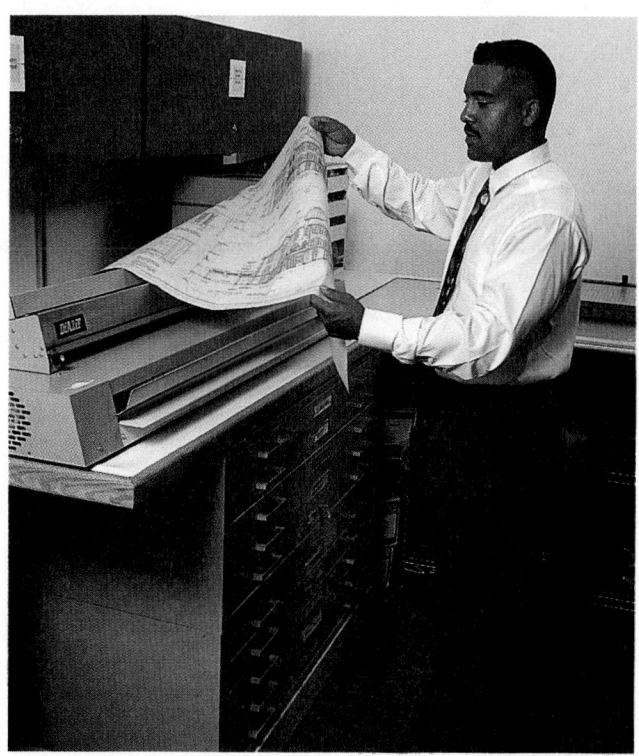

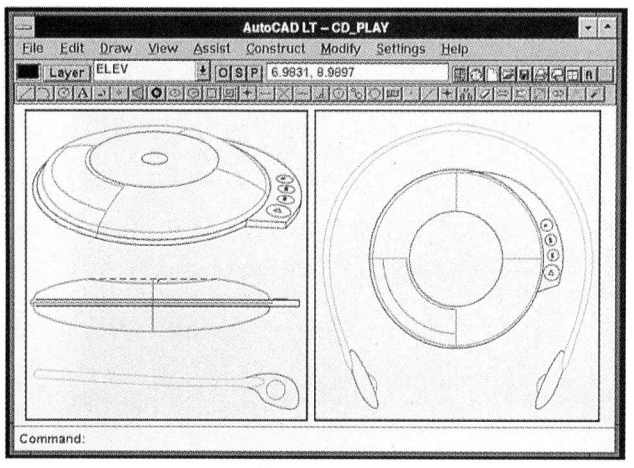

PART 2 Fasteners, Materials, and Forming Processes **259**

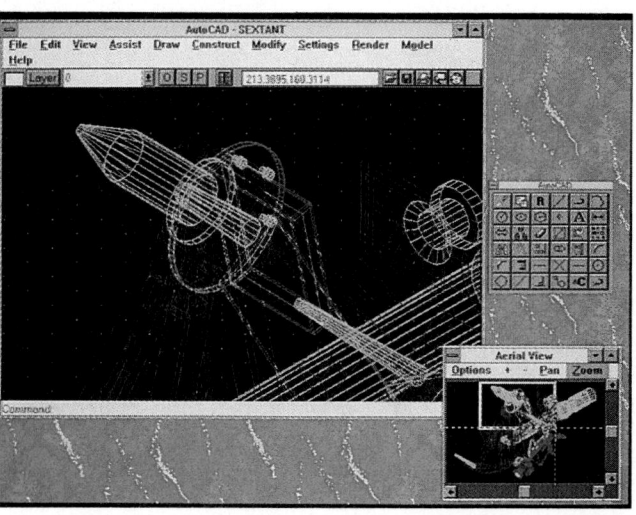

ABOUT THE AUTHORS

CECIL JENSEN is the author or coauthor of many successful technical books, including *Engineering Drawing and Design, Fundamentals of Engineering Drawing, Fundamentals of Engineering Graphics* (formerly called *Drafting Fundamentals), Interpreting Engineering Drawings, Geometric Dimensioning and Tolerancing for Engineering and Manufacturing Technology, Architectural Drawing and Design for Residential Construction, Home Planning and Design,* and *Interior Design.* Some of these books are printed in three languages and are used in many countries.

He has 27 years of teaching experience in mechanical and architectural drafting, and was a technical director for a large vocational school in Canada. He has also been responsible for the supervision of the teaching of technical courses for General Motors apprentices in Oshawa, Canada. Before entering the teaching profession, Mr. Jensen gained several years of design experience in the industry.

Mr. Jensen is a member of the Canadian Standards Committee (CSA) on Technical Drawings (which includes both mechanical and architectural drawing), and is chairman of the Committee on Dimensioning and Tolerancing. Mr. Jensen is Canada's representative on the American (ANSI) Standards for Dimensioning and Tolerancing, and has represented Canada at two world (ISO) conferences in Oslo (Norway) and Paris on the standardization of technical drawings.

He took an early retirement from the teaching profession in order to devote his full attention to writing.

JAY D. HELSEL is a professor of industry and technology at California University of Pennsylvania. He completed his undergraduate work at California State College and was awarded a master's degree from Pennsylvania State University. He has done advanced graduate work at West Virginia and at the University of Pittsburgh, where he completed a doctoral degree in educational communications and technology. In addition, Dr. Helsel holds a certificate in airbrush techniques and technical illustration from the Pittsburgh Art Institute.

He has worked in industry and has taught drafting, metalworking, woodworking, and a variety of laboratory and professional courses at both the secondary and college levels. During the past 25 years, he has also worked as a free-lance artist and illustrator. His work appears in many technical publications.

Dr. Helsel is coauthor of *Engineering Drawing and Design, Fundamentals of Engineering Drawing, Programmed Blueprint Reading, Mechanical Drawing,* and *Computer-Aided Engineering Drawing.*

DRAWING STANDARDS UPDATE

The recently published standard *ASME Y14.5M—1994 Dimensioning and Tolerancing,* a revision of *ANSI Y14.5M—1982,* contains additions and modifications to improve national and international drawing communications. This was accomplished by adopting many of the International Organization for Standardization (ISO) drawing standards.

The publication of this standard came at a time when *Engineering Drawing and Design, 5th edition* was ready for publication. In order that the fifth edition of *Engineering Drawing and Design* reflect the current standards adopted by the ISO, the United States, and Canada, this section of the text was introduced to provide current drawing standards to the user. This second printing includes changes in Chapter 16 to be consistent with *ASME Y14.5M—1994.* This will ensure that the students and teaching staff using this text will be informed of the latest drawing requirements.

Significant changes found in *ASME Y14.5M—1994* are:

- Adoption of the universal (ISO) datum feature symbol.

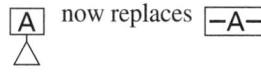

The drawing and placement of this symbol is explained in Unit 16–5.

- Discontinuance of the use of the RFS symbol Ⓢ. The "regardless of feature size" condition now applies where the symbol for MMC and LMC are not stated on size features. This now conforms with ISO practices and is explained in Unit 16–4.

- Placement of the projected tolerance zone symbol and its height in the feature control frame, following the stated tolerance and any modifier. The dimension indicating the minimum height of the projected tolerance zone is placed after the projected tolerance zone symbol. See Unit 16–10.

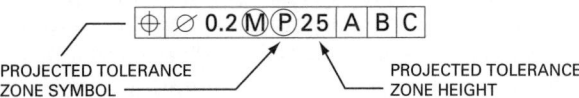

PROJECTED TOLERANCE
ZONE SYMBOL

PROJECTED TOLERANCE
ZONE HEIGHT

- Revision of the designation from ANSI to ASME to reflect The American Society of Mechanical Engineers as the preparing organization.
- Reinstatement of the symmetry symbol =, to be applied only on an RFS basis. See Unit 16–14.

ACKNOWLEDGMENTS

The authors are indebted to the members of ASME Y14.5M –1994 Dimensioning and Tolerancing, and the members of the CAN/CSA–B78.2–M91 Dimensioning and Tolerancing of Technical Drawings for the countless hours they have contributed to making such a successful standard.

We also wish to thank Donald Voisinet, Professor of Engineering Technology and Coordinator of Design and Drafting at the Niagara County Community College, for his assistance in developing Chapter 2 Computer-Aided Drafting (CAD); Sherwood Davis, Professor at Salt Lake Community College, for his revision of Chapter 27 Electrical and Electronic Drawings; and Dennis Short, Associate Professor in the Technical Graphics Department of the School of Technology at Purdue University, for his development of the *Instructor's Wraparound Edition* and the instructional material to be included in *Instructor's Management System.*

The authors thank the many people who helped in the preparation of this edition including John Beck, the executive editor, Freida O'Neil-Robinson, the editor, and Jennifer King, the senior production editor.

The authors and staff of Glencoe/McGraw-Hill wish to express their appreciation to the following individuals for their professional review of the text and sample design.

Kenneth Arnold
Tulsa Equipment Manufacturing Co.

Tom Brennan
ITT Technical Institute

Judith Dalton
ITT Technical Institute

Thomas Eddins
Clayton State College

Melvin Freeman
Houston Community College

Joseph Greenfield
Suny College of Technology

Mel Hartley
Bessmer State Technical

Michael Holler
Paragon Films, Inc.

Stanley Hopkins
New England Institute of Technology

Tommy Justice
John Patterson State Technical College

Hamid Khan
Ball State University

Harold Lott
Calhoun Community College

Walter Reed
Oregon Polytechnic Institute

Deb Rosenweig
York Technical Institute

M. Peter Saxon, III
Porter and Chester Institute

Dan Steinke
ITT Technical Institute

Mostafa Tossi
Penn State Worthington Scranton

George Voll
ITT Technical Institute

David Webb
Salt Lake Community College

PREFACE

Engineering Drawing and Design is written as an introductory course to prepare students for drafting careers in a modern, technology-intensive industry. Technical drafting, like all technical areas, is constantly changing. The computer has revolutionized the way in which drawings and parts are made. Thus, in this new edition, the authors have made every effort to translate the most current technical information available into the most useful form from the standpoint of both instructor and student. The latest developments and current practices in all areas of graphic communication, CAD, functional drafting, material representation, shop processes, geometric tolerancing, true positioning, numerical control, electronic drafting, and metrication have been incorporated into this edition. The approach used synthesizes, simplifies, and converts complex drafting standards and procedures into understandable instructional units.

Before beginning to work on this edition, a questionnaire was mailed to a number of users and non-users of the text requesting their input on text material and format. In response to the reviewers' suggestions and recommendations, we have incorporated major changes in the fifth edition. The suggestions and changes are:

1. Using a two-column format, rather than a three-column format, which is easier for students to read.
2. Updating the photographs of drafting and CAD equipment. The first three chapters of this edition provide up-to-date color photographs of drafting equipment. In addition, two eight-page color inserts are included as photo essays covering various engineering areas.
3. A greater selection of drawing and design projects. For this edition, we have added over three hundred projects throughout the text.
4. More information on geometric tolerancing and how to apply it to various drawings. This edition has three to four times more information and assignments on this subject than any other current text.
5. Deleting CAD icons and applications throughout the existing text. The reviewers felt that the CAD reference manuals which accompany CAD software packages provide adequate coverage of CAD usage.
6. Continuing to use the unit approach to teach the subject matter. Reviewers find this approach to be a real bonus. It allows them to readily put together a customized program that suits the needs of their students and local industry by choosing the appropriate units. This edition continues to divide chapters into mini-teaching units.
7. Supporting the design concepts covered in the text through drawing practice. While time constraints often limit the use of design units in the instructional program, graduates often find these concepts to be extremely useful in drafting and design experience. Instructors can choose those units they feel appropriate to their program.

8. Continuing to cover current ANSI and ISO drawing practices better than every other text. It is a basic requirement of any engineering drawing text to keep the instructors aware of the latest drawing standards and practices. In this edition, we again included the latest drawing standards on:

- Methods of representation (Unit 6–1)
- Symbols representation for installed rivets used in aerospace equipment (Unit 11–5)
- Simplified representation on drawing (Unit 14–1), a new standard being prepared by ISO and ANSI

These simplified representations have been used for years by American industries as a cost-saving feature. They are covered in this edition of *Engineering Drawing and Design* as they have been since the second edition.

In addition to the major changes incorporated from the reviewers' recommendations, this edition also provides other new features including key terms and a new two-color design. Each chapter begins with the important terms and their definitions. The new two-color design is used to highlight the text's features and enhance its appearance.

Engineering Drawing and Design, 5th edition is supported by ancillary products including some which are new to this edition. The following summarizes the major changes in the instructor and student support material:

- *Instructor's Wraparound Edition of the Textbook.* For the first time, the drafting instructor has a comprehensive teaching guide. This unique and innovative two-color teaching guide combines the student edition with the instructor's edition to provide a wealth of teaching support. Chapter and part overviews, unit objectives, teaching tips, and critical thinking exercises specifically focus on major concepts. Real world engineering drawing facts and technical references provide an interesting learning link. Pop quizzes, evaluation standards, and reteaching strategies are combined into this unprecedented publication.
- *Instructor's Management System.* This comprehensive solutions manual details solutions to many of the end-of-chapter drawing problems. It also provides instructors with course objectives, instructional tips, teaching transparencies, and chapter tests.
- *Instructor's CD-ROM Program.* The CD-ROM includes all of the drawing solutions from the Instructor's Management System. It also provides tutorial drawings which are excellent for instructional use.

- *Problems Workbook for Engineering Drawing and Design.* Many reviewers suggested that we include a problems workbook to enable their students to have more practice. A correlated problems workbook is now available.
- *Text-Workbook for Geometric Tolerancing.* Many of the reviewers also suggested that we include additional coverage of geometric tolerancing in a separate book. This up-to-date text is intended for those instructors who emphasize geometric tolerancing in their program.

Comments and suggestions concerning this and future editions of the text are most welcomed.

Cecil Jensen and Jay Helsel

BASIC DRAWING AND DESIGN

CHAPTER 1

ENGINEERING GRAPHICS AS A LANGUAGE

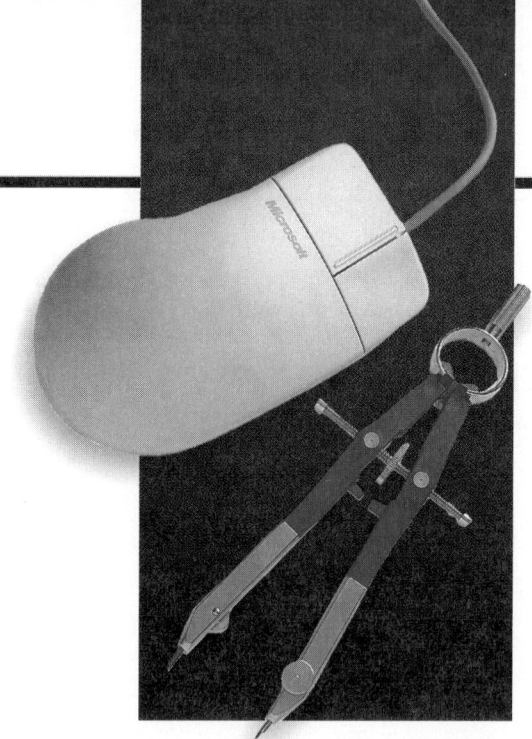

Definitions

Artistic drawing   The expression of real or imagined ideas of a cultural nature.

CADD   An acronym for computer-aided drafting and design.

Computer-aided drafting (CAD)   Drafting and design done with the use of a computer.

Drafting   A language using pictures to communicate thoughts and ideas.

End-product drawings   Drawings which consist of detail or part drawings and assembly or subassembly drawings, but do not include supplementary drawings.

Engineering drawing   The main method of communication between all people concerned with the design and manufacture of parts.

Graphic representation   The expression of ideas through lines or marks impressed on a surface.

Instrument or **manual drawings**   Drawings made with the use of instruments.

Layouts   Drawings made to scale of the object to be built.

Sketches   Drawings made with a pencil or pen but without the assistance of instruments or computers.

Technical drawing   The expression of technical ideas or ideas of a practical nature.

1-1 THE LANGUAGE OF INDUSTRY

Since earliest times people have used drawings to communicate and record ideas so that they would not be forgotten. Figure 1-1-1 shows builders in an early civilization following technical drawings to construct a building.

Graphic representation means dealing with the expression of ideas by lines or marks impressed on a surface. A drawing is a graphic representation of a real thing. Drafting, therefore, is a graphic language, because it uses pictures to communicate thoughts and ideas. Because these pictures are understood by people of different nations, drafting is referred to as a "universal language."

Drawing has developed along two distinct lines, with each form having a different purpose. On the one hand, artistic drawing is concerned mainly with the expression of real or imagined ideas of a cultural nature. Technical drawing, on the

FIG. 1-1-1   Early use of drawing in constructing a building.

TYPICAL BRANCHES OF ENGINEERING GRAPHICS	ACTIVITIES	PRODUCTS	SPECIALIZED AREAS
MECHANICAL 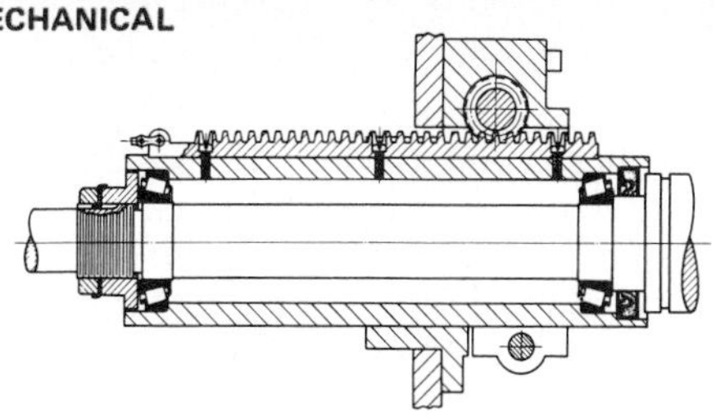	DESIGNING TESTING MANUFACTURING MAINTENANCE CONSTRUCTION	MATERIALS MACHINES DEVICES	POWER GENERATION TRANSPORTATION MANUFACTURING POWER SERVICES ATOMIC ENERGY MARINE VESSELS
ARCHITECTURAL 	PLANNING DESIGNING SUPERVISING	BUILDINGS ENVIRONMENT LANDSCAPE	COMMERCIAL BUILDINGS RESIDENTIAL BUILDINGS INSTITUTIONAL BUILDINGS ENVIRONMENTAL SPACE FORMS
ELECTRICAL 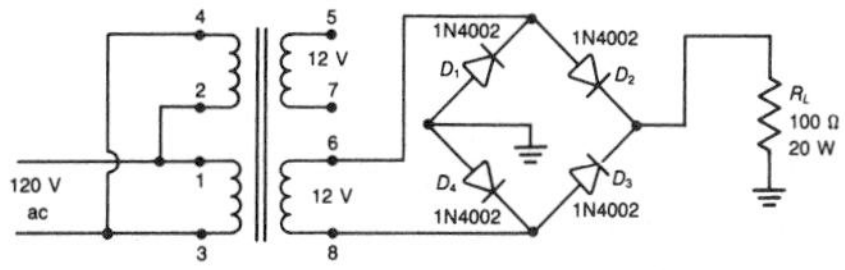	DESIGNING DEVELOPING SUPERVISING PROGRAMMING	COMPUTERS ELECTRONICS POWER ELECTRICAL	POWER GENERATION POWER APPLICATION TRANSPORTATION ILLUMINATION INDUSTRIAL ELECTRONICS COMMUNICATIONS INSTRUMENTATION MILITARY ELECTRONICS
AEROSPACE 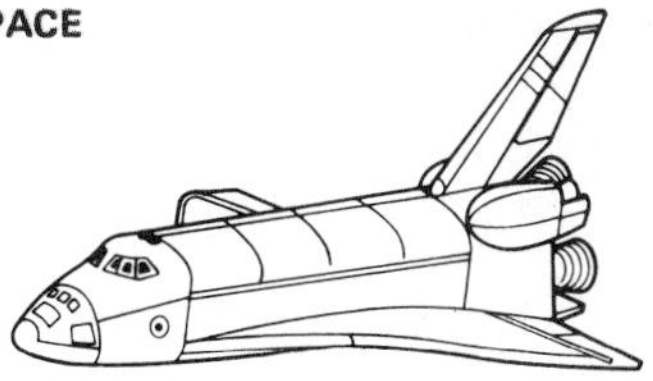	PLANNING DESIGNING TESTING	MISSILES PLANES SATELLITES ROCKETS	AERODYNAMICS STRUCTURAL DESIGN INSTRUMENTATION PROPULSION SYSTEMS MATERIALS RELIABILITY TESTING PRODUCTION METHODS
PIPING 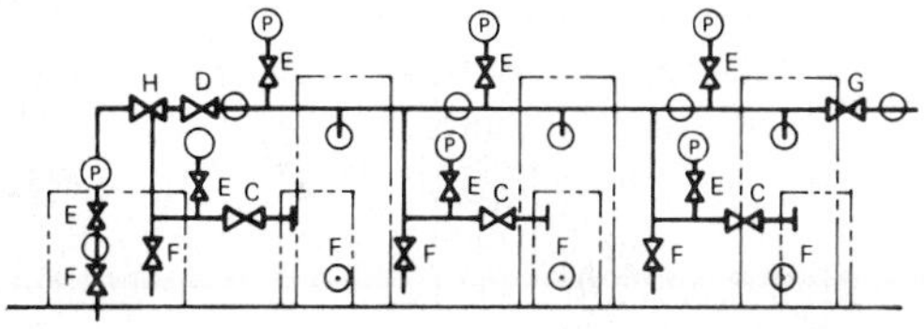	DESIGNING TESTING MANUFACTURING MAINTENANCE CONSTRUCTION	BUILDINGS HYDRAULICS PNEUMATICS PIPE LINES	LIQUID TRANSPORTATION MANUFACTURING POWER SERVICES HYDRAULICS PNEUMATICS
STRUCTURAL 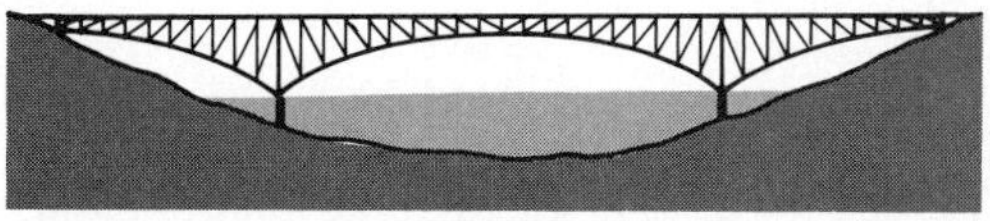	PLANNING DESIGNING MANUFACTURING CONSTRUCTION	MATERIALS BUILDINGS MACHINES VEHICLES BRIDGES	STRUCTURAL DESIGNS BUILDINGS PLANES SHIPS AUTOMOBILES BRIDGES
TECHNICAL ILLUSTRATING 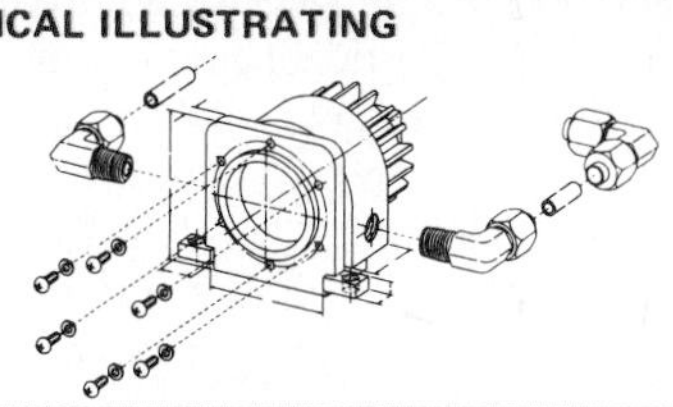	PROMOTION DESIGNING ILLUSTRATING	CATALOGS MAGAZINES DISPLAYS	NEW PRODUCTS ASSEMBLY INSTRUCTIONS PRESENTATIONS COMMUNITY PROJECTS RENEWAL PROGRAMS

FIG. 1-1-2   Various fields of drafting.

other hand, is concerned with the expression of technical ideas or ideas of a practical nature, and it is the method used in all branches of technical industry.

Even highly developed word languages are inadequate for describing the size, shape, and relationship of physical objects. For every manufactured object there are drawings that describe its physical shape completely and accurately, communicating engineering concepts to manufacturing. For this reason, drafting is called the "language of industry."

Drafters translate the ideas, rough sketches, specifications, and calculations of engineers, architects, and designers into working plans that are used in making a product (Fig. 1-1-2, pg. 3). Drafters may calculate the strength, reliability, and cost of materials. In their drawings and specifications, they describe exactly what materials workers are to use on a particular job. To prepare their drawings, drafters use either computer-aided drafting and design (CADD) systems or manual drafting instruments, such as compasses, dividers, protractors, templates, and triangles, as well as drafting machines that combine the functions of several devices. They also may use engineering handbooks, tables, and calculators to assist in solving technical problems.

Drafters are often classified according to their type of work or their level of responsibility. Senior drafters (designers) use the preliminary information provided by engineers and architects to prepare design "layouts" (drawings made to scale of the object to be built). Detailers (junior drafters) make drawings of each part shown on the layout, giving dimensions, material, and any other information necessary to make the detailed drawing clear and complete. Checkers carefully examine drawings for errors in computing or recording sizes and specifications.

Drafters may also specialize in a particular area, such as mechanical, electrical, electronic, aeronautic, structural design, piping, or architectural drafting.

Drawing Standards

Throughout the long history of drafting, many drawing conventions, terms, abbreviations, and practices have come into common use. It is essential that different drafters use the same practices if drafting is to serve as a reliable means of communicating technical theories and ideas.

In the interest of worldwide communication, the International Organization of Standardization (ISO) was established in 1946. One of its committees, ISO TCIO, was formed to deal with the subject of technical drawings. Its goal was to come up with a universally accepted set of drawing standards. Today most countries have adopted, either in full or with minor changes, the standards set up by this committee, making drafting a truly universal language.

The American Society of Mechanical Engineers (ASME) is the governing body that establishes the standards for the United States through its ASME Y14.5 committee, made up of selected personnel from industry, technical organizations, and education. Members from the ASME Y14.5 also serve on the ISO TCIO subcommittee.

The standards used throughout this text reflect the current thinking of the ANSI committee. These standards apply primarily to end-product drawings. *End-product drawings* usually consist of detail or part drawings and assembly or subassembly drawings, and are not intended to fully cover other supplementary drawings, such as checklists, item lists, schematic diagrams, electrical wiring diagrams, flowcharts, installation drawings, process drawings, architectural drawings, and pictorial drawings.

The information and illustrations presented here have been revised to reflect current industrial practices in the preparation and handling of engineering documents. The increased use of reduced-size copies of engineering drawings made from microfilm and the reading of microfilm require the proper preparation of the original engineering document regardless of whether the drawing was made manually or by computer (CAD). All future drawings should be prepared for eventual photographic reduction or reproduction. The observance of the drafting practices described in this text will contribute substantially to the improved quality of photographically reproduced engineering drawings.

1-2 CAREERS IN ENGINEERING GRAPHICS

The Student

While students are learning basic drafting skills (Fig. 1-2-1), they will also be increasing their general technical knowledge, learning about some of the engineering and manufacturing processes involved in production. Not all students will choose a drafting career. However, an understanding of this graphic language is necessary for anyone who works in any of the fields of technology, and is essential for those who plan to enter the skilled trades or become a technician, technologist, or engineer.

Because a drawing is a set of instructions that the worker will follow, it must be accurate, clean, correct, and complete. When drawings are made with the use of instruments, they are called *instrument* (or *manual*) *drawings*. When they are developed with the use of a computer, they are known as *computer-aided drawings*. When made without instruments or the aid of a computer, drawings are referred to as *sketches*. The ability to sketch ideas and designs and to produce accurate drawings is a basic part of drafting skills.

In everyday life, a knowledge of engineering graphics is helpful in understanding house plans and assembly, maintenance, and operating instructions for many manufactured or hobby products.

Places of Employment

There are well over 300,000 people working in drafting positions in the United States. A significant number are women. About 9 out of 10 drafters are employed in private industry. Manufacturing industries that employ a large number of drafters are those making machinery, electrical equipment, transportation equipment, and fabricated metal products. Nonmanufacturing industries employing a large number of drafters are engineering and architectural consulting firms, construction companies, and public utilities.

(A)

(B)

FIG. 1-2-1   College drafting room. *(Left—Glencoe file photo; right—Doug Martin.)*

Drafters also work for the government; the majority work for the armed services. Drafters employed by state and local governments work chiefly for highway and public works departments. Several thousand drafters are employed by colleges and universities and by other nonprofit organizations.

Training, Qualifications, and Advancement

There are many design careers available at different technical levels of performance. Most companies are in need of design and drafting services for growth in technical development, construction, or production. Any person interested in becoming a drafter can acquire the necessary training from a number of sources, including junior and community colleges, extension divisions of universities, vocational/technical schools, and correspondence schools. Others may qualify for drafting positions through on-the-job training programs combined with part-time schooling.

The prospective drafter's training in post-high school drafting programs should include courses in mathematics and physical sciences, as well as in CAD and CADD. Studying fabrication practices and learning some trade skills are also helpful, since many higher-level drafting jobs require knowledge of manufacturing or construction methods. This is especially true in the mechanical discipline due to the implementation of CAD/CAM (computer-aided design/computer-aided manufacturing). Many technical schools offer courses in structural design, strength of materials, physical metallurgy, CAM, and robotics.

As drafters gain skill and experience, they may advance to higher-level positions such as checkers, senior drafters, designers, supervisors, and managers (Fig. 1-2-2). Drafters who take additional courses in engineering and mathematics are often able to qualify for engineering positions.

Qualifications for success as a drafter include the ability to visualize objects in three dimensions and the development of

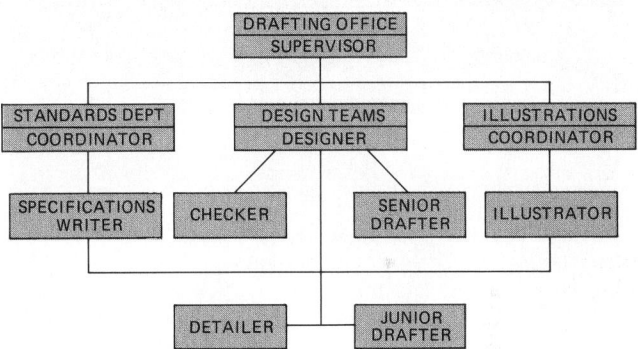

FIG. 1-2-2   Positions within the drafting office.

problem-solving design techniques. Since the drafter is the one who finalizes the details on drawings, attentiveness to detail is a valuable asset.

Employment Outlook

Employment opportunities for drafters are expected to remain stable as a result of the complex design problems of modern products and processes. The need for drafters will, however, fluctuate with local and national economics. Since drafting is a part of manufacturing, job opportunities in this field will also rise or drop in accordance with various manufacturing industries. The demand for drafters will be high in some areas and low in others as a result of high-tech expansion or a slump in sales. In addition, computerization is creating many new products, and support and design occupations, including drafters, will continue to grow. On the other hand, photoreproduction of drawings and expanding use of CAD have eliminated many routine tasks done by drafters. This development will probably reduce the need for some less skilled drafters.

REFERENCES AND SOURCE MATERIALS

1. Charles Bruning Co.
2. *Occupational Outlook Handbook.*

1-3 THE DRAFTING OFFICE

Drafting room technology has progressed at the same rapid pace as the economy of the country. Many changes have taken place in the modern drafting room compared to the typical drafting room scene before CAD, as shown in Fig. 1-3-1. Not only is there far more equipment, but it is of much higher quality. Noteworthy progress has been and continues to be made.

The drafting office is the starting point for all engineering work. Its product, the *engineering drawing,* is the main method of communication among all people concerned with the design and manufacture of parts. Therefore the drafting office must provide accommodations and equipment for the drafters, from designer and checker to detailer or tracer; for the personnel who make copies of the drawings and file the originals; and for the secretarial staff who assist in the preparation of the drawings. Typical drafting workstations are shown in Figs. 1-3-2 and 1-3-3.

Fewer engineering departments now rely on manual drafting methods. Computers are replacing drafting boards at a steady pace because of increased productivity. However, where a high volume of finished or repetitive work is not necessary, manual drafting does the job adequately and inexpensively. CAD and manual drafting can serve as full partners in the design process, enabling the designer to do jobs that are simply not possible or feasible with manual equipment alone.

Besides increasing the speed with which a job is done, a CAD system can perform many of the tedious and repetitive

(A) THE DRAFTING OFFICE AT THE TURN OF THE CENTURY. (Bettman Archives, Inc.)

(B) MANUAL DRAFTING. (Vemco Corp.)

(C) CAD DRAFTING. (Computervision Corp.)

FIG. 1-3-1   Evolution of the drafting office. *(A—Bettman Archive, B—Vemos Corp., C—courtesy of International Business Machines Corporation.)*

FIG. 1-3-2   Manual drafting office. *(Doug Martin)*

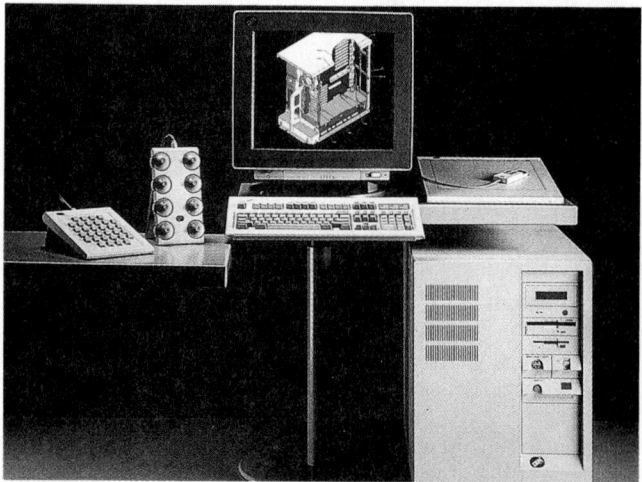

FIG. 1-3-3   CAD drafting office. *(Courtesy of International Business Machines Corporation)*

skills ordinarily required of a drafter, such as lettering and differentiating line weights. CAD thus frees the drafter to be more creative while it quickly performs the mundane tasks of drafting. It is conservatively estimated that CAD has been responsible for an improvement of at least 30 percent in production in terms of time spent on drawing.

A CAD system by itself cannot create. A drafter must create the drawing, and thus a strong design and drafting background remains essential.

It may not be practical to handle all the workload in a design or drafting office on a CAD system. Although most design and drafting work most certainly can benefit from it, some functions will continue to be performed by traditional means. Thus some companies will use CAD for only a portion of the workload. Still others use CAD almost exclusively. Whatever the percentage of CAD use, one fact is certain: It has had, and will continue to have, a dramatic effect on design and drafting careers.

Once a CAD system has been installed, the required personnel must be hired or trained. Trained personnel generally originate from one of three popular sources: educational institutions, CAD equipment manufacturer training courses, and individual company programs.

1-4 MANUAL DRAFTING

Over the years, the designer's chair and drafting table have evolved into a drafting station that provides a comfortable, integrated work area. Yet much of the equipment and supplies employed years ago are still in use today, although vastly improved.

Drafting Furniture

Special tables and desks are manufactured for use in single-station or multi-station design offices. Typical are desks with attached drafting boards (Fig. 1-4-1). The boards may be used by the occupant of the desk to which it is attached, in which case it may swing out of the way when not in use, or may be reversed for use by the person in the adjoining station.

In addition to such special workstations, a variety of individual desks, chairs, tracing tables, filing cabinets, and special storage devices for equipment are available (Fig. 1-4-2).

The simplest manually adjustable tables typically consist of a hinged surface riding on a vertical rod secured by a hand knob screw. The hand knob screw is loosened, the top is set at the desired angle, and the hand knob screw is retightened.

Drafting Equipment

See Fig. 1-4-3 (pg. 8) for a variety of drafting equipment.

FIG. 1-4-2   Drafting office furniture. *(The Mayline Company)*

Drawing Boards

The drawing sheet is attached directly to the surface of a drafting table or a portable drawing board (Fig. 1-4-4, pg. 8). Drafting boards are used in schools and at home and generally have a smaller work surface than that found on drafting tables. They are designed to stay flat and have straight guiding edges. Most professional drafting tables have a special overlay drawing surface material that "recovers" from minor pinholes and dents.

FIG. 1-4-1   Manual drafting workstations.

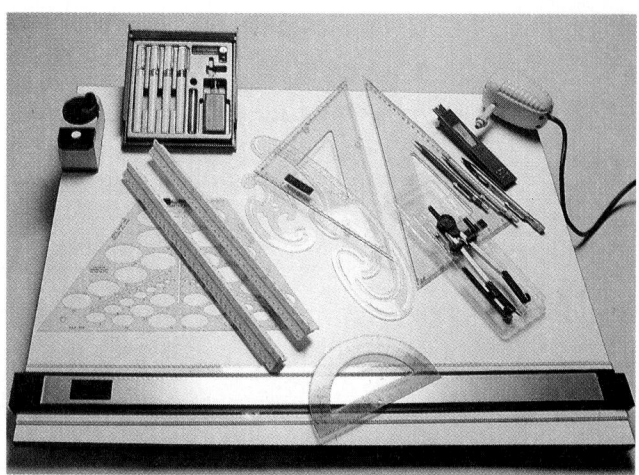

FIG. 1-4-3   Drafting equipment. *(STUDIOHIO)*

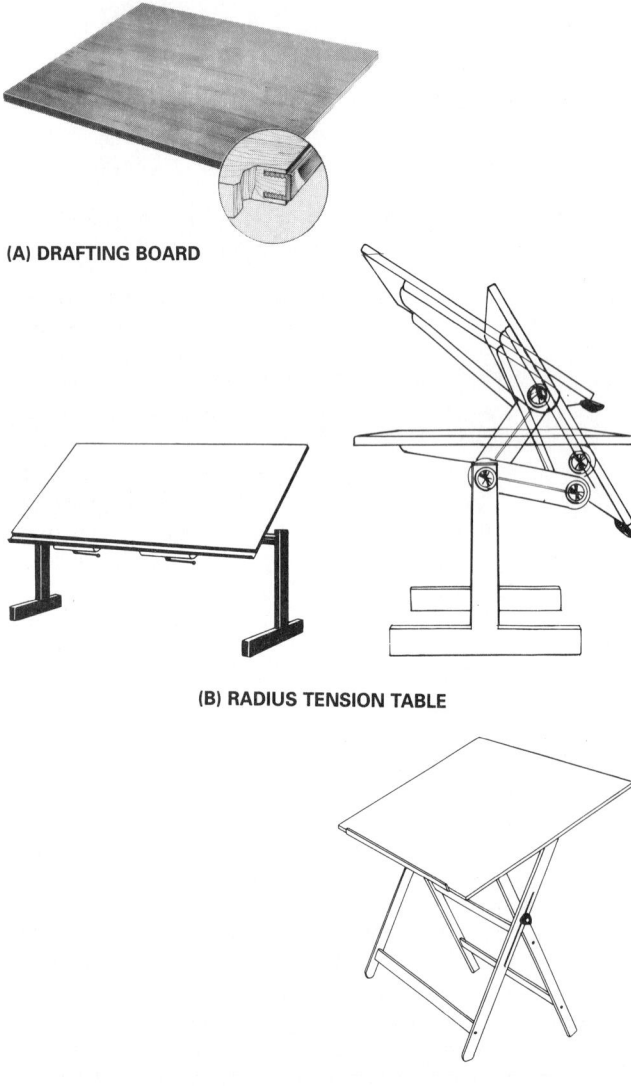

(A) DRAFTING BOARD

(B) RADIUS TENSION TABLE

(C) STUDENT TABLE

FIG. 1-4-4   Drafting tables and boards. *(Norman Wade Co. Ltd.)*

Drafting Machines

In the well-equipped engineering department, where the designer is expected to do accurate drafting, the T square has largely been replaced by the drafting machine. This device, which combines the functions of T square, triangles, scale, and protractor, is estimated to save up to 50 percent of the user's time. All positioning is done with one hand, while the other hand is free to draw.

Drafting machines may be attached to any drafting board or table. Two types are currently available (Fig. 1-4-5). In the track type, a vertical beam carrying the drafting instruments rides along a horizontal beam fastened to the top of the table. In

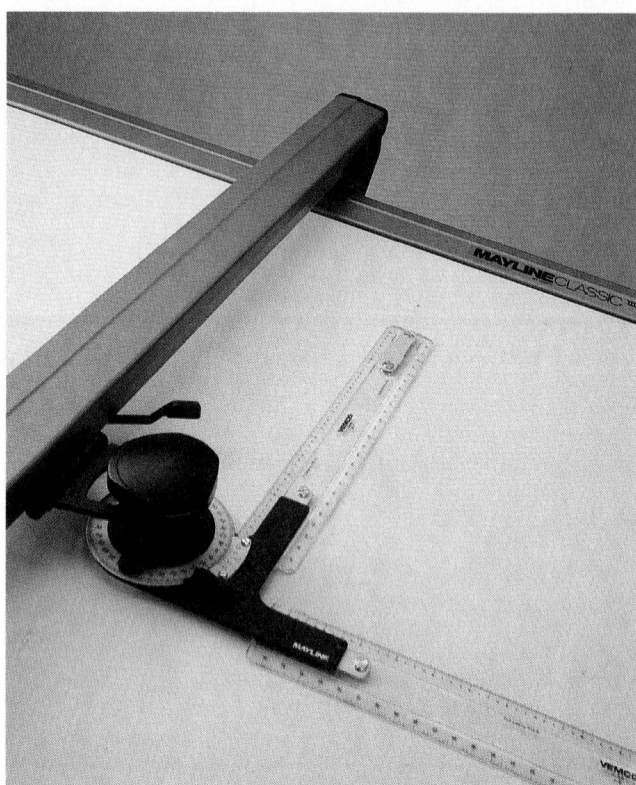

(A) TRACK TYPE

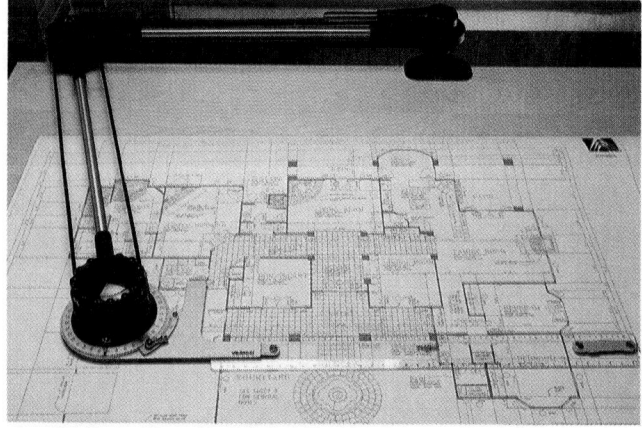

(B) ARM-TYPE

FIG. 1-4-5   Drafting machines. *(A—The Mayline Company, B—Doug Martin)*

the arm (or elbow) type, two arms pivot from the top of the machine and are relative to each other.

The track-type machine has several advantages over the arm-type. It is better suited for large drawings and is normally more stable and accurate. The track type also allows the drafting table to be positioned at a steeper angle and permits locking in the vertical and horizontal positions.

Some track-type drafting machines provide a digital display of angles, the *X-Y* coordinates, and a memory function.

Parallel Slide

The parallel slide, also called the parallel bar, is used in drawing horizontal lines and for supporting triangles when vertical and sloping lines are being drawn (Fig. 1-4-6). It is fastened on each end to cords, which pass over pulleys. This arrangement permits movement up and down the board while maintaining the parallel slide in a horizontal position.

T Squares

Although the T square has been replaced in drafting offices and schools by drafting machines, parallel slides, and computers, it is still used by some students for drafting at home. For this reason it is included in this text.

The T square (Fig. 1-4-7) performs the same function as the parallel slide. T squares are made of various materials, the

FIG. 1-4-6   Drafting table with parallel slide. *(Doug Martin)*

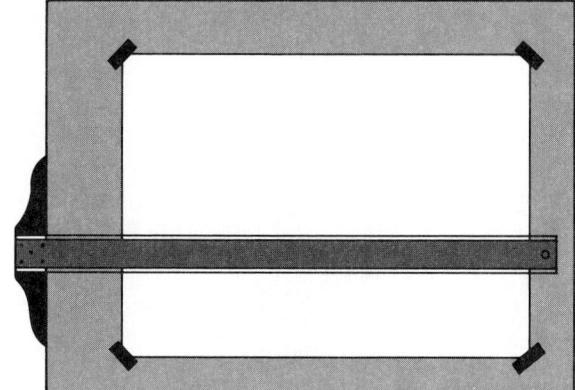

FIG. 1-4-7   T square.

more popular being plastic-edged wood blades with heads made from wood or plastic.

The head of the T square is placed on the left side of a drawing board for use by right-handed people and on the right side of the drawing board for use by left-handed people.

Triangles

Triangles are used together with the parallel slide or T square when you are drawing vertical and sloping lines (Fig. 1-4-8, pg. 10). The triangles most commonly used are the 30/60° and the 45° triangles. Singly or in combination, these triangles can be used to form angles in multiples of 15°. For other angles, the protractor (Fig. 1-4-9, pg. 10) is used. All angles can be drawn with the adjustable triangle (Fig. 1-4-10, pg. 11); this instrument replaces the two common triangles and the protractor.

Scales

Scale may refer to the measuring instrument or the size to which a drawing is to be made.

Measuring Instrument   Shown in Fig. 1-4-11 (pg. 11) are the common shapes of scales used by drafters to make measurements on their drawings. Scales are used only for measuring and are not to be used as a straightedge for drawing lines. It is important that drafters draw accurately to scale. The scale to which the drawing is made must be given in the title block or strip which is part of the drawing.

Sizes to Which Drawings Are Made   When an object is drawn at its actual size, the drawing is called *full scale* or *scale 1:1*. Many objects, however, such as buildings, ships, or airplanes, are too large to be drawn full scale, so they must be drawn to a reduced scale. An example would be the drawing of a house to a scale of ¼ in. = 1 ft or 1:48.

Frequently, objects such as small wristwatch parts are drawn larger than their actual size so that their shape can be seen clearly. Such a drawing has been drawn to an enlarged scale. The minute hand of a wristwatch, for example, could be drawn to a scale of 5:1.

Many mechanical parts are drawn to half scale, 1:2, and quarter scale, 1:4, or nearest metric scale, 1:5. Notice that the scale is expressed as an equation. With reference to the 1:5 scale, the left side of the equation represents a unit of the size drawn; the right side represents the equivalent five units of measurement of the actual object.

Scales are made with a variety of combined scales marked on their surfaces. This combination of scales spares the drafter the necessity of calculating the sizes to be drawn when working to a scale other than full size.

Metric Scales   The linear unit of measurement for mechanical drawings is the millimeter. Scale multipliers and divisors of 2 and 5 are recommended (Fig. 1-4-12, pg. 11).

The numbers shown indicate the difference in size between the drawing and the actual part. For example, the ratio 10:1 shown on the drawing means that the drawing is 10 times the actual size of the part, whereas a ratio of 1:5 on the drawing means the object is 5 times as large as it is shown on the drawing.

FIG. 1-4-8   Triangles.

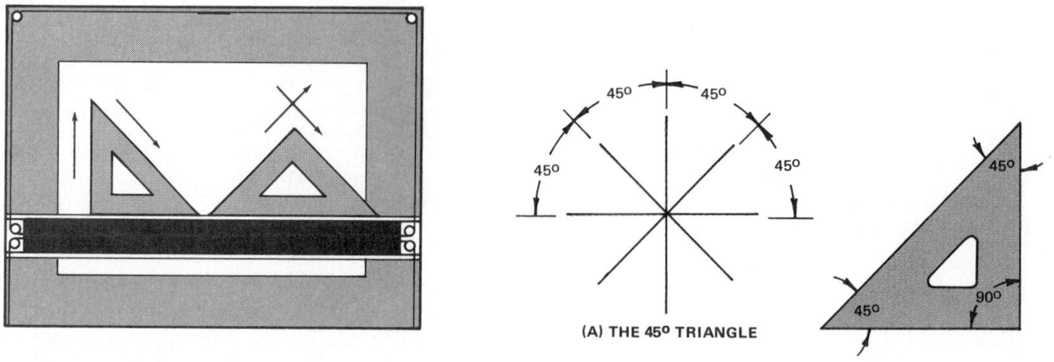

(A) THE 45° TRIANGLE

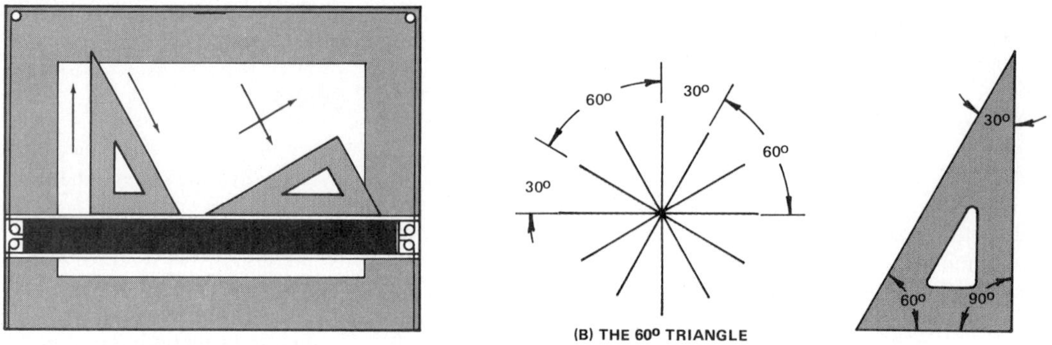

(B) THE 60° TRIANGLE

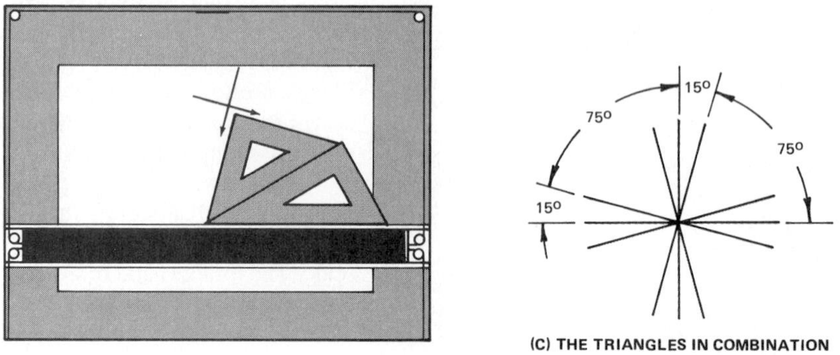

(C) THE TRIANGLES IN COMBINATION

FIG. 1-4-9   Protractors.
(Staedtler Mars Ltd)

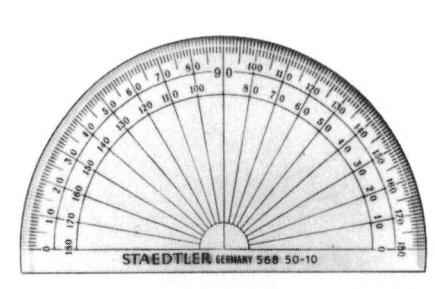

(A) HALF-CIRCLE PROTRACTOR

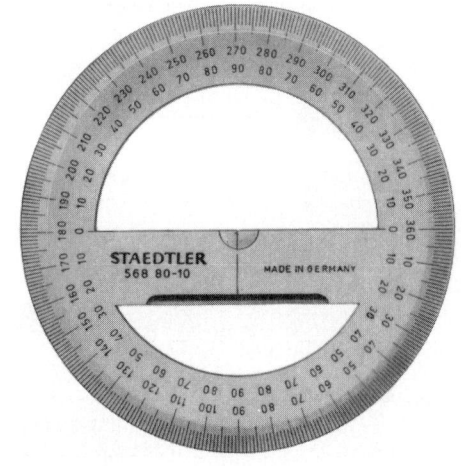

(B) FULL-CIRCLE PROTRACTOR

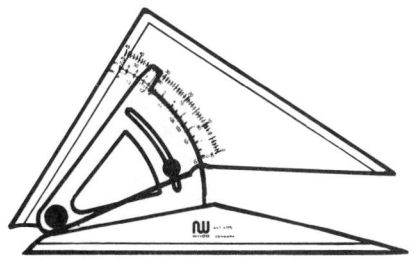

FIG. 1-4-10   Adjustable triangle. *(Norman Wade Co. Ltd.)*

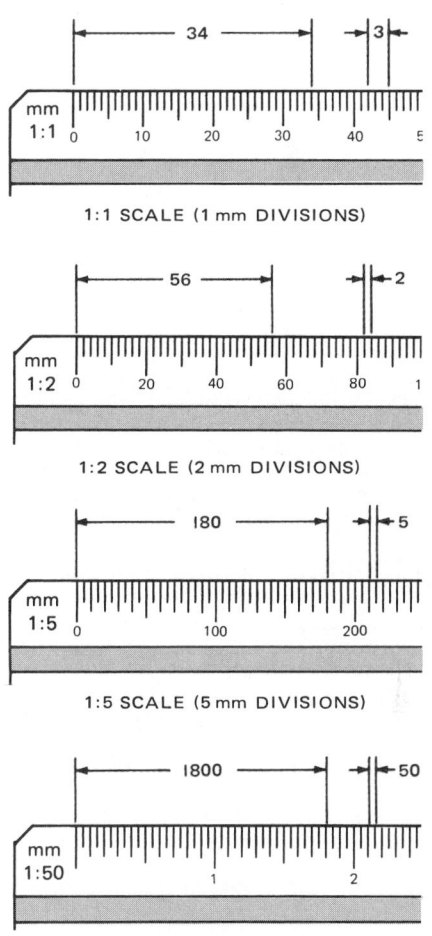

1:1 SCALE (1 mm DIVISIONS)

1:2 SCALE (2 mm DIVISIONS)

1:5 SCALE (5 mm DIVISIONS)

1:50 SCALE (50 mm DIVISIONS)

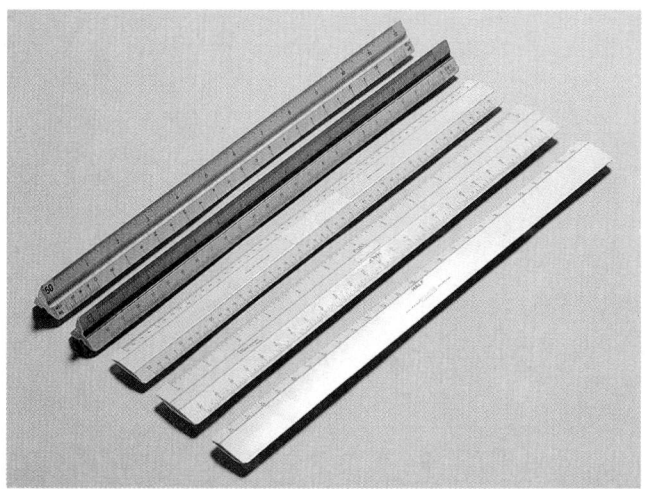

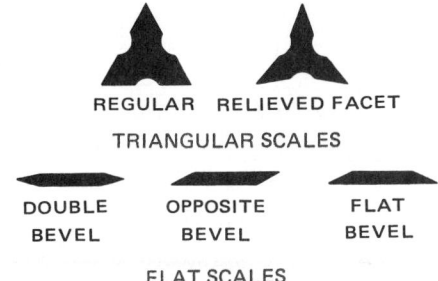

REGULAR   RELIEVED FACET

TRIANGULAR SCALES

DOUBLE BEVEL   OPPOSITE BEVEL   FLAT BEVEL

FLAT SCALES

FIG. 1-4-11   Drafting scales. *(STUDIOHIO)*

ENLARGED	SIZE AS	REDUCED
1000 : 1	1 : 1	1 : 2
500 : 1		1 : 5
200 : 1		1 : 10
100 : 1		1 : 20
50 : 1		1 : 50
20 : 1		1 : 100
10 : 1		1 : 200
5 : 1		1 : 500
2 : 1		1 : 1000

FIG. 1-4-12   Metric scales.

The units of measurement for architectural drawings are the meter and millimeter. The same scale multipliers and divisors used for mechanical drawings are used for architectural drawings.

Inch (U.S. Customary) Scales

Inch Scales   There are three types of scales that show various values that are equal to 1 inch (in.) (Fig. 1-4-13, pg. 12). They are the decimal inch scale, the fractional inch scale, and the scale that has divisions of 10, 20, 30, 40, 50, 60, and 80 parts to the inch. The last scale is known as the civil engineer's scale. It is used for making maps and charts. The divisions, or parts of an inch, can be used to represent feet, yards, rods, or miles. This scale is also useful in mechanical drawing when the drafter is dealing with decimal dimensions.

On fractional inch scales, multipliers or divisors of 2, 4, 8, and 16 are used, offering such scales as full size, half size, and quarter size.

Foot Scales   These scales are used mostly in architectural work (Fig. 1-4-14, pg. 12). They differ from the inch scales in

11

that each major division represents a foot, not an inch, and end units are subdivided into inches or parts of an inch. The more common scales are ⅛ in. = 1 ft, ¼ in. = 1 ft, 1 in. = 1 ft, and 3 in. = 1 ft. The most commonly used inch and foot scales are shown in Fig. 1-4-15.

Compasses

The compass is used for drawing circles and arcs. Several basic types and sizes are available (Fig. 1-4-16).

- *Friction head compass,* standard in most drafting sets.
- *Bow compass,* which operates on the jackscrew or ratchet principle by turning a large knurled nut.
- *Drop bow compass,* used mostly for drawing small circles. The center rod contains the needle point and remains stationary while the pencil or pen leg revolves around it (Fig. 1-4-17).

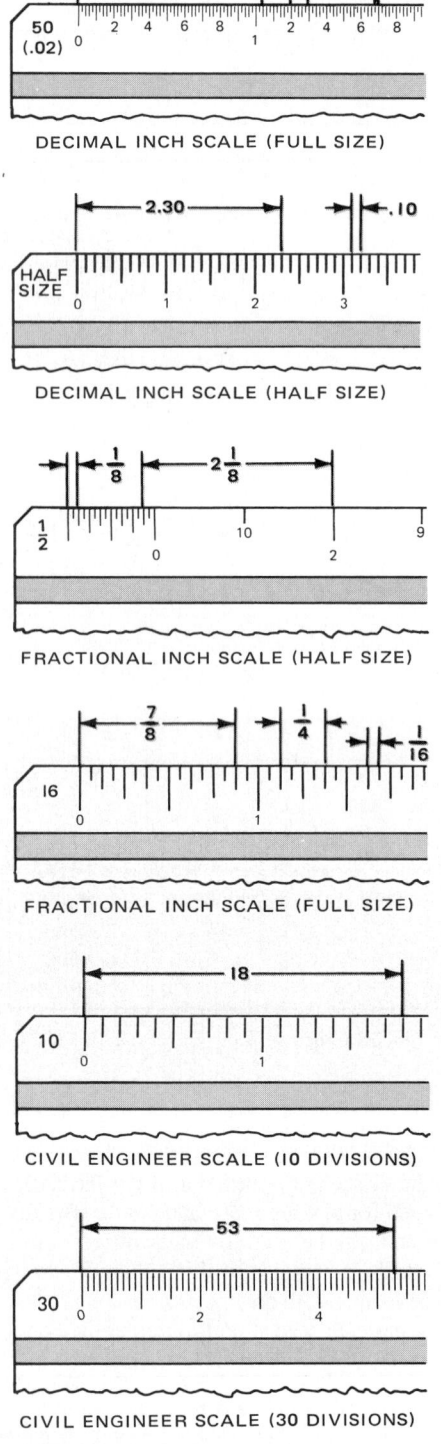

FIG. 1-4-13   Inch scales.

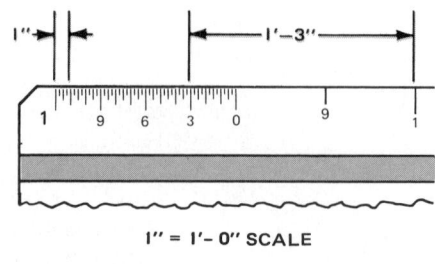

1″ = 1′–0″ SCALE

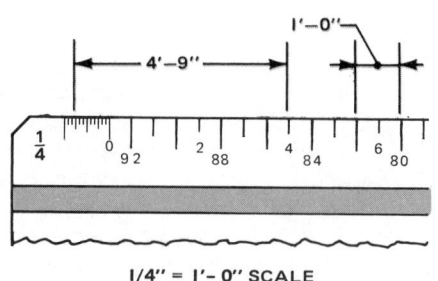

1/4″ = 1′–0″ SCALE

FIG. 1-4-14   Foot and inch scales.

DECIMALLY DIMENSIONED DRAWINGS	FRACTIONALLY DIMENSIONED DRAWINGS	DIMENSIONED IN FEET AND INCHES	
		EQUIVALENT	
		SCALE	RATIO
10:1	8:1	6 IN. = 1 FT	1:2
5:1	4:1	3 IN. = 1 FT	1:4
2:1	2:1	1½ IN. = 1 FT	1:8
1:1	1:1	1 IN. = 1 FT	1:12
1:2	1:2	¾ IN. = 1 FT	1:16
1:5	1:4	½ IN. = 1 FT	1:24
1:10	1:8	⅜ IN. = 1 FT	1:32
1:20	1:16	¼ IN. = 1 FT	1:48
ETC.	ETC.	³⁄₁₆ IN. = 1 FT	1:64
		⅛ IN. = 1 FT	1:96
		¹⁄₁₆ IN. = 1 FT	1:192

FIG. 1-4-15   Recommended drawing scales.

- *Beam compass,* a bar with an adjustable needle and pencil-and-pen attachment for drawing large arcs or circles.
- *Adjustable arc,* a device used to accurately draw any radius from 7 to 20 in. (200 to 5000 mm).

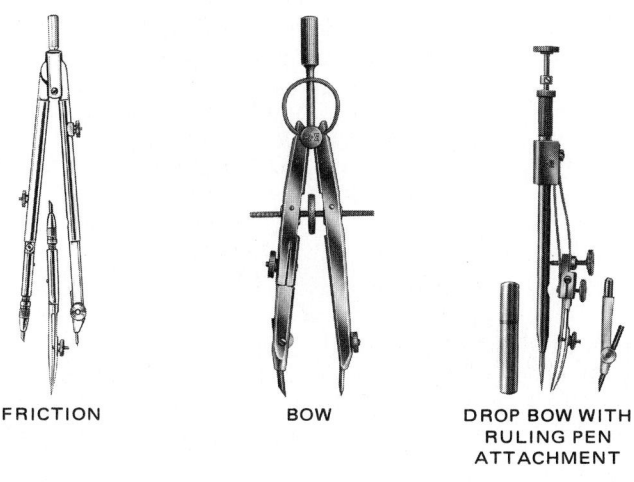

FRICTION   BOW   DROP BOW WITH RULING PEN ATTACHMENT

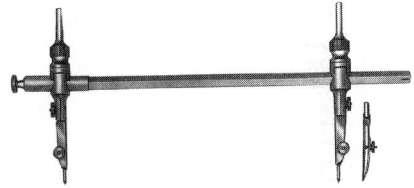

BEAM WITH RULING PEN ATTACHMENT

FIG. 1-4-16   Compasses. *(Keuffel & Esser Co.)*

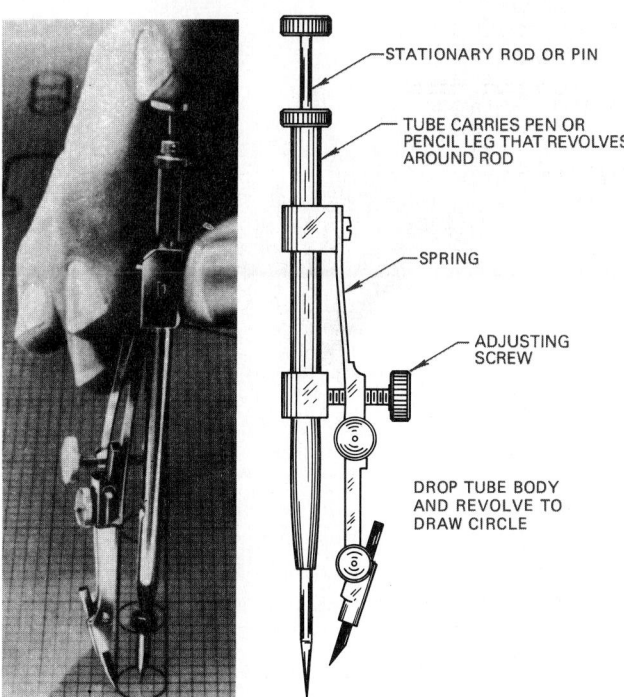

STATIONARY ROD OR PIN

TUBE CARRIES PEN OR PENCIL LEG THAT REVOLVES AROUND ROD

SPRING

ADJUSTING SCREW

DROP TUBE BODY AND REVOLVE TO DRAW CIRCLE

FIG. 1-4-17   The drop-bow compass is used for drawing very small circles.

The bow compass is adjusted by turning a screw whose knurled head is located either in the center or to one side. The bow compass can be used and adjusted with one hand as shown in Fig. 1-4-18 (pg. 14). The proper technique is:

1. Adjust the compass to the correct radius.
2. Hold the compass between the thumb and finger.
3. With greater pressure on the leg with the needle located on the intersection of the center lines, rotate the compass in a clockwise direction. The compass should be slightly tipped in the direction of motion.

Dividers

Lines are divided and distances transferred (moved from one place to another) with dividers. The basic types of dividers are shown in Fig. 1-4-19 (pg. 14).

Dividers have a steel pin insert in each leg and come in a variety of sizes and designs, similar to the compasses. A compass can be used as a divider by replacing its lead point with a steel pin.

One type of divider, known as a proportional divider, is used to enlarge or reduce line segments or an object without the need of mathematical calculations or the use of a scale. An adjustable center point permits you to set the desired proportion. Setting the divider points on one side of the dividers to the original line length or distance automatically determines the reduced or enlarged size on the opposite divider points.

Pencils and Leads

Pencils   As with all other equipment, advances in pencil design have made drawing lines and lettering easier. The new automatic pencils are designed to hold leads of one width, thus eliminating the need to sharpen the lead. These pencils (Fig. 1-4-20, pg. 15) are available in several different lead sizes (color-coded for easy identification) and hardnesses. Leads are designed for use on paper or drafting film, or both. Thus a drafter will have several automatic pencils, each having a selected line width, lead hardness, and make, for performing particular line or lettering tasks on film or paper.

Another type of drafting pencil, often referred to as a mechanical pencil or lead holder, advances a uniformly sized lead that periodically requires sharpening. The leads for the mechanical pencils are usually sharpened in an electric lead pointer, which produces a tapered point. A sandpaper block is used to sharpen compass leads.

Leads   Because of the drawing media used and the type of reproduction required, pencil manufacturers have marketed three types of lead for the preparation of engineering drawings.

Graphite Lead   This is the conventional type of lead, which has been used for years. It is available in a variety of grades or hardnesses—9H, 8H, 7H, 6H (hard); 5H and 4H (medium hard); 3H and 2H (medium); H and F (medium soft); and HB, B, 2B, 3B, 4B, 5B, and 6B (very soft), the very soft grades not being recommended for drafting. The selection of the proper grade of lead is important. A hard lead might penetrate the drawing paper while a soft lead will smear.

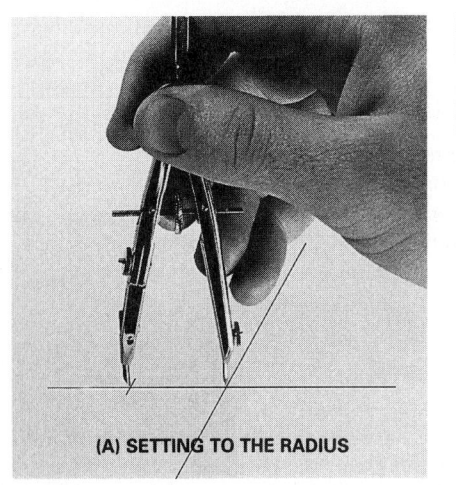
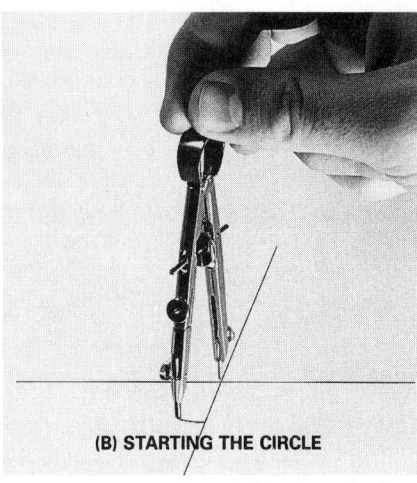
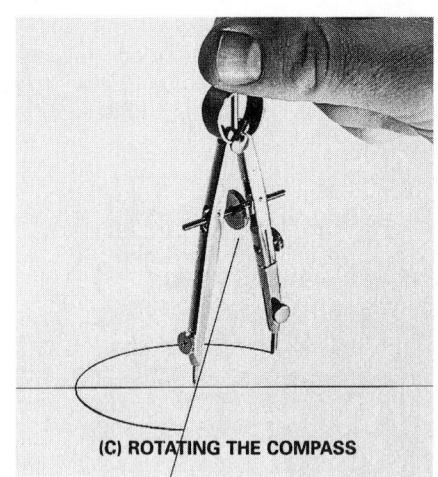

(A) SETTING TO THE RADIUS   **(B) STARTING THE CIRCLE**   **(C) ROTATING THE COMPASS**

FIG. 1-4-18   Adjusting the radius and drawing a circle with the bow compass.

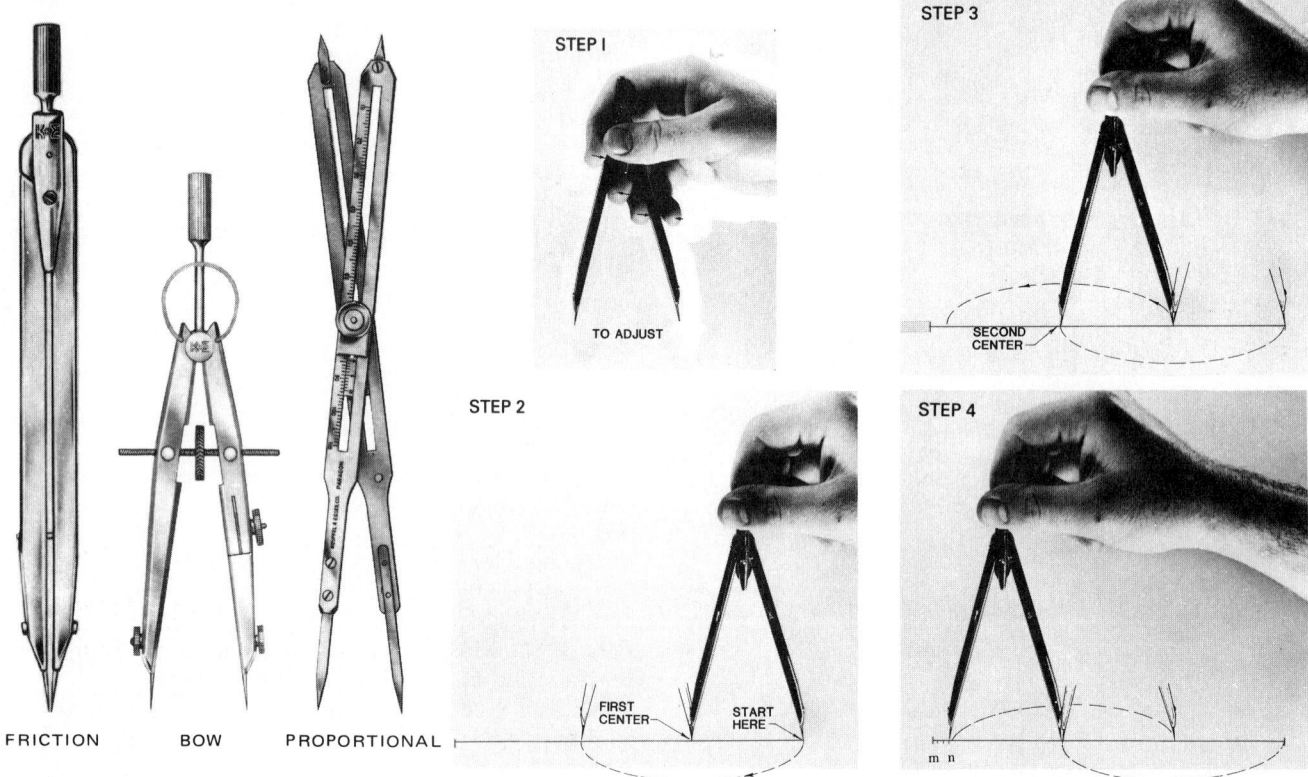

FRICTION   BOW   PROPORTIONAL

(A) TYPES OF DIVIDERS   **(B) DIVIDERS ARE USED TO DIVIDE AND TO TRANSFER DISTANCES**

FIG. 1-4-19   Dividers. *(Keuffel & Esser Co.)*

The next two types of drafting leads were developed as a result of the introduction of film as a drawing medium. A limited number of grades are available in these leads, and they do not correspond to the grades used for graphite lead.

Plastic Lead   This type of lead is designed for use only on film. It has good microform reproduction characteristics.

Plastic-Graphite Lead   As the name implies, this lead is made of plastic and graphite. There are two basic types: fired and

extruded. They are designed for use only on film, erase well, do not readily smear, and produce a good opaque line that is suitable for microform reproduction. The main drawback with this type of lead is that it does not hold a point well.

Erasers and Cleaners

Erasers   A variety of erasers have been designed to do special jobs—remove surface dirt, minimize surface damage on film or vellum, and remove ink or pencil lines (Fig. 1-4-21).

Erasing Machines   Very little pressure is required when using the electrically powered erasing machine because the high-speed rotation of the shaft actually does the clean-up job rapidly and flawlessly. These machines make erasures with pinpoint accuracy (Fig. 1-4-22).

Cleaners   An easy way to clean tracings is to sprinkle them lightly with gum eraser particles while working. Then triangles, scales, etc., stay spotless and clean the surface automatically as they are moved back and forth. The particles contain no grit or abrasive, and will actually improve the lead- or ink-taking quality of the drafting surface.

Erasing Shields   These thin pieces of metal or plastic (Fig. 1-4-23) have a variety of openings to permit the erasure of fine detail lines or lettering without disturbing nearby work that is to be left on the drawing. With this device, erasures can be made quickly and accurately.

Brushes

A light brush (Fig. 1-4-24) is used to keep the drawing area clean. By using a brush to remove eraser particles and any accumulated dirt, the drafter avoids smudging the drawing.

Templates

To save time, drafters use templates (Fig. 1-4-25, pg. 16) for drawing circles and arcs. Templates are available with standard hole sizes ranging from small to 6.00 in. (150 mm) in diameter. Templates are also used for drawing standard square, hexagonal, triangular, and elliptical shapes and standard electrical and architectural symbols.

(A) AUTOMATIC PENCILS

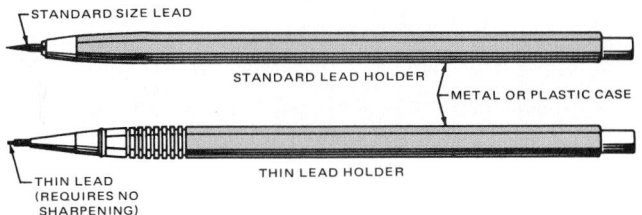

(B) MECHANICAL PENCILS

FIG. 1-4-20   Drafting pencils. *(Koh-I-Noor)*

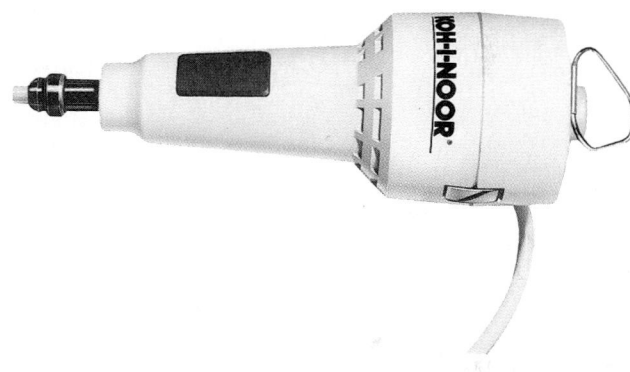

FIG. 1-4-22   Erasing machine. *(Koh-I-Noor)*

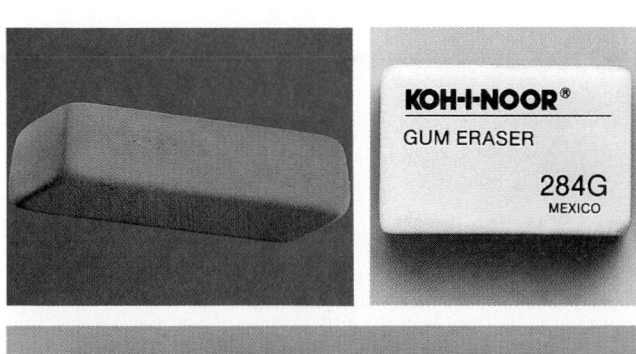

FIG. 1-4-21   Erasers. *(A—STUDIOHIO, B & C—Koh-I-Noor)*

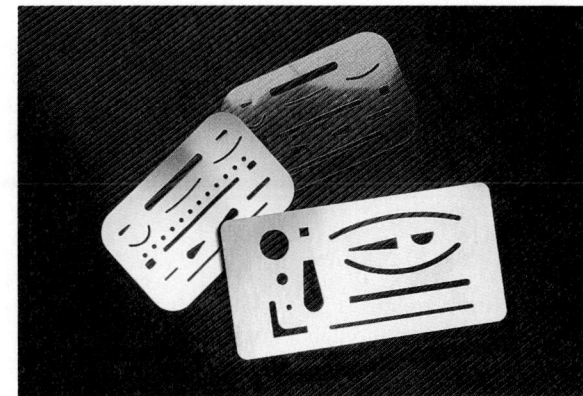

FIG. 1-4-23   Erasing shields. *(STUDIOHIO)*

FIG. 1-4-24   Drafter's brush. *(J.S. Staedtler, Inc.)*

FIG. 1-4-25 Templates. *(Teledyne Post)*

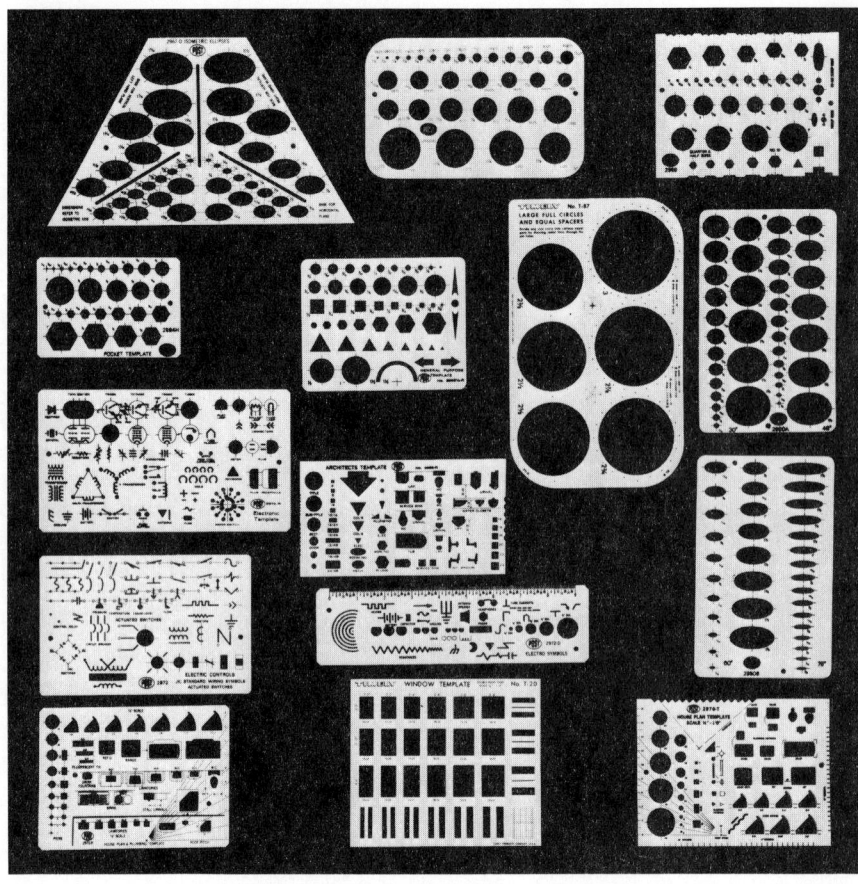

Irregular Curves

For drawing curved lines in which, unlike circular arcs, the radius of curvature is not constant, a tool known as an irregular or French curve (Fig. 1-4-26) is used. The patterns for these curves are based on various combinations of ellipses, spirals, and other mathematical curves. The curves are available in a variety of shapes and sizes. Normally, the drafter plots a series of points of intersection along the desired path and then uses the French curve to join these points so that a smooth-flowing curve results.

Curved Rules and Splines

Curved rules and splines (Fig. 1-4-27) solve the problem of ruling a smooth curve through a given set of points. They lie flat on the board and are as easy to use as a triangle; yet they can be bent to fit any contour to a 3 in. (75 mm) minimum radius and will hold the position without support. A clear, plastic ruling edge stands away from the board just far enough to prevent ink lines from smearing.

Inking Equipment

Although most production drawings are drawn with pencil, in the last few years the number of ink drawings has been on the increase. The use of this type of drawing for technical illustrations and the demand for good, clear drawings for microform

FIG. 1-4-26 Irregular curves.

reproduction have brought about the introduction of new and improved inking methods and techniques.

The three types of inking pens used by drafters today are the ruling pen and the refillable and disposable technical pens. The ruling pen has been around for years and is still used by many drafters. Technical pens are newer and are more commonly used.

Refillable Technical Pens Technical pens have made the inking of technical drawings relatively easy. These pens have points of different sizes for drawing different line widths

FIG. 1-4-27   Curved rule and spline. *(Doug Martin)*

(Fig. 1-4-28). Technical pen points produce uniform line widths because their design provides a steady flow of ink that does not usually clog.

Some technical pens have a refillable cartridge for storing ink. Others have a cartridge that is used once and then replaced.

Technical pen points are made of different materials for use on different surfaces. There are three main kinds of points: hard-chrome stainless steel (for paper or vellum), tungsten-carbide (for the longest wear on film, vellum, and paper), and jewel point for drafting on any media. Technical pen points have a shoulder to prevent smudging. The shoulder is barrel-shaped for use with curved or straight guiding instruments.

The refillable technical pen has some definite advantages over the ruling pen:

1. It is easy to produce lines of uniform width since no line adjustments are required.
2. The barrel-shaped shoulder prevents smudging because point and shoulder are offset.
3. The cartridge makes it easy to load ink. Also, this pen needs refilling less often than a ruling pen.
4. The ink laid down by a technical pen can dry more uniformly because it flows more evenly.
5. The technical pen does not often need to be cleaned because it has a weighted cleaning wire inside the capillary tube of the point.
6. A syringe can be used to flush out the pen point with a cleaning solution.

The technical pen can be stored in a humidified container that prevents the ink from drying out. In some technical pens, a seal forms in the cap when the cap is tightened with a firm twist. If you are careful in filling and cleaning pens, your inking will be better.

00	0	1	2	
0.25	0.35	0.40	0.50	mm
.010	.014	.017	.020	in.

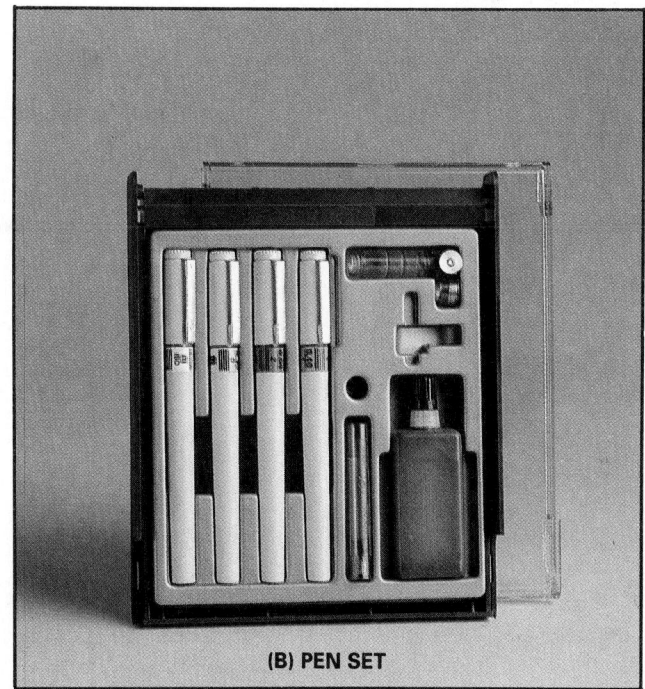

(A) LINE WIDTHS

(B) PEN SET

FIG. 1-4-28   Examples of technical pen nibs and technical pens. *(STUDIOHIO)*

Disposable Technical Pens   Disposable technical pens are similar to refillable pens except that the pen is discarded once the ink within the pen is used. They have the advantage over the other pens because their fast-drying ink prevents smearing. They are available in a variety of line widths and the drawing of lines, circles, and arcs is done in the same way as refillable technical pens.

Ruling Pens   The ruling pen is steadily being replaced by the technical pens. Where technical pens may be used for line work, drawing arrowheads, and lettering, the ruling pen, as the name implies, is used only for ruling lines.

The ruling pen (Fig. 1-4-29) has blades that can be adjusted to draw lines of different widths. It can be used to draw both straight and curved lines. Ruling pens are sometimes called all-purpose pens. They can be filled from a cartridge tube, a squeeze bottle, or a dropper cap.

Calculators

Calculators, such as the one shown in Fig. 1-4-30, are used by drafters to make fast mathematical calculations using division, multiplication, and extractions of square roots, and to solve problems involving areas, volumes, masses, strengths of materials, pressures, etc.

Basic Equipment

The following is a list of items (Fig. 1-4-31) commonly used in drafting.

    Drawing board
    T square, parallel-ruling straightedge (parallel slide), or drafting machine

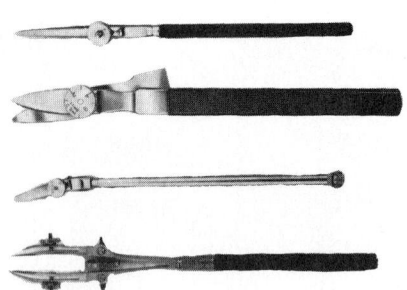

FIG. 1-4-29   A variety of ruling pens.

    Drawing sheets (paper or film)
    Drafting tape
    Drafting pencils
    Erasers
    Erasing shield
    Triangles, 45° and 30/60° (not required with drafting machine)
    Scales
    Templates
    Irregular curve
    Inking pen
    Brush
    Protractor
    Cleaning powder
    Calculator

Your drafting instructor can tell you exactly what equipment will be needed for your course.

ASSIGNMENTS

See Assignments 1 through 4 for Unit 1-4 on pages 20–21.

FIG. 1-4-30   Calculator.
    (Doug Martin)

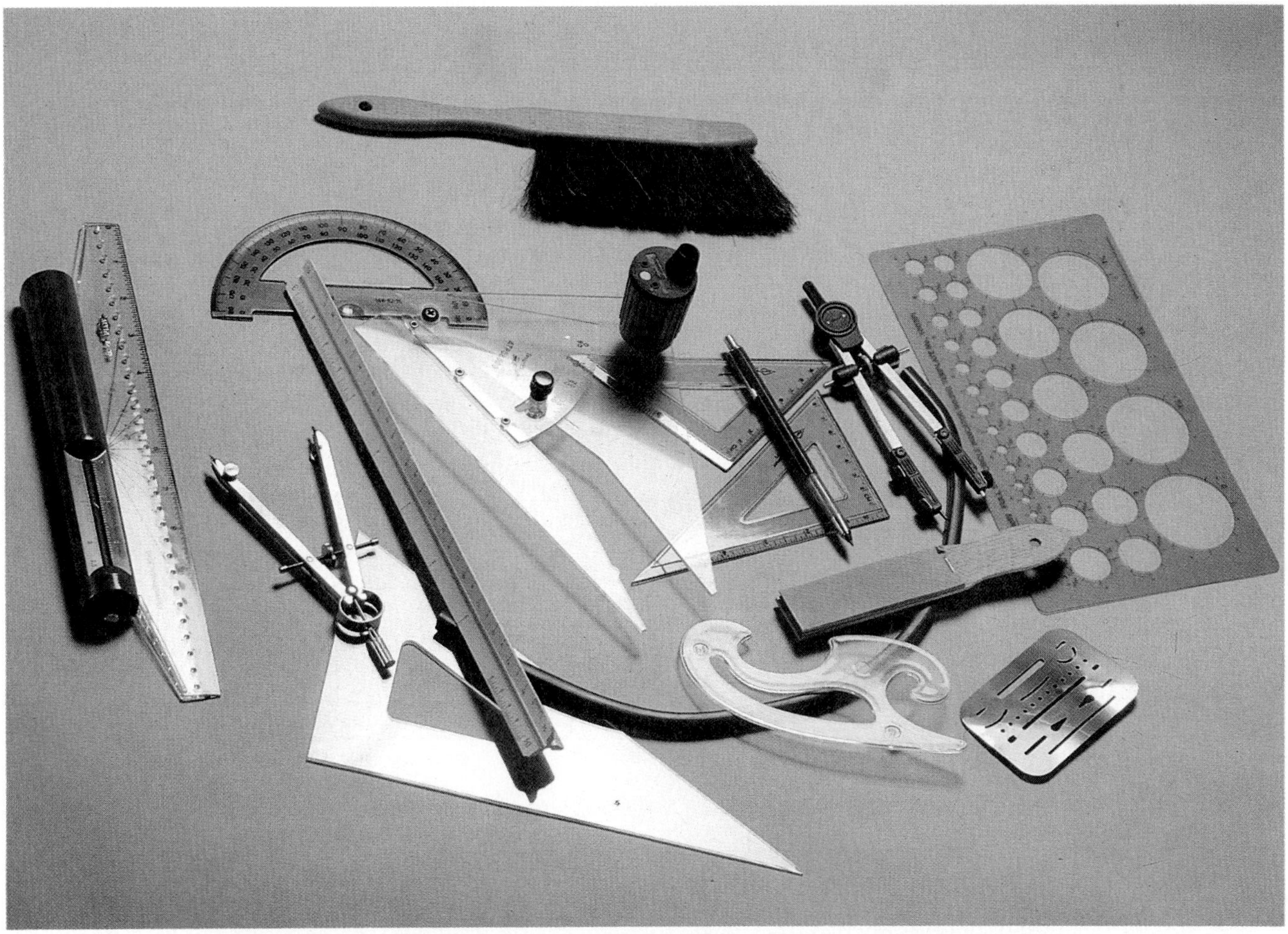

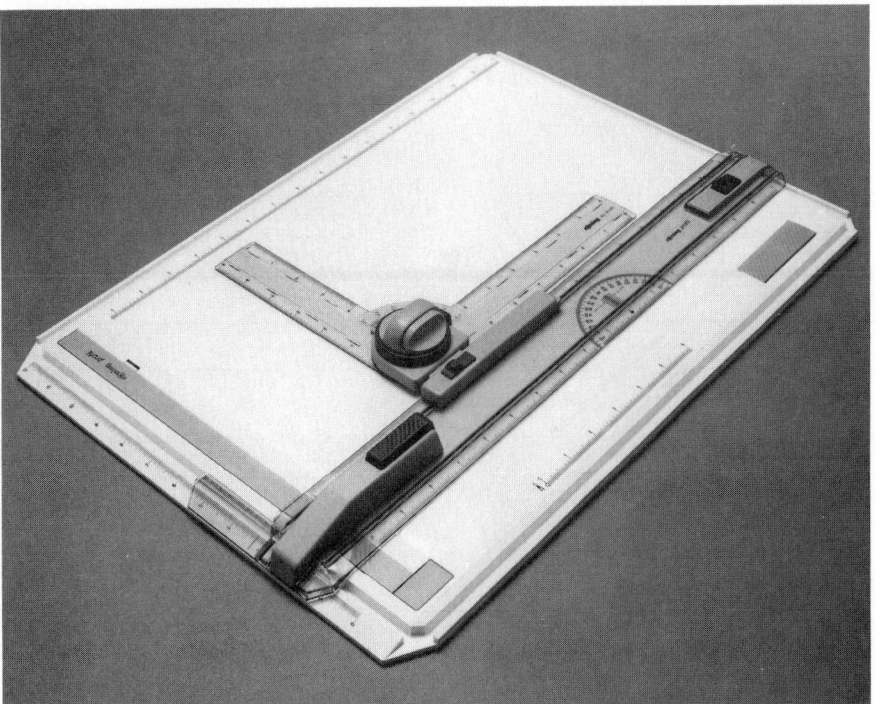

(A) INSTRUMENTS

(B) PORTABLE DRAWING BOARD AND DRAFTING MACHINE

FIG. 1-4-31   Basic drafting equipment. *(A—STUDIOHIO, B—Norman Wade Co. Ltd.)*

ASSIGNMENTS FOR CHAPTER 1

ASSIGNMENTS FOR UNIT 1-4, MANUAL DRAFTING

1. Using the scales shown in Fig. 1-4-A determine lengths A through K.
2. Metric measurements assignment. With reference to Fig. 1-4-B and using the scale:

1:1	measure distances A through E
1:2	measure distances F through K
1:5	measure distances L through P
1:10	measure distances Q through U
1:50	measure distances V through Z

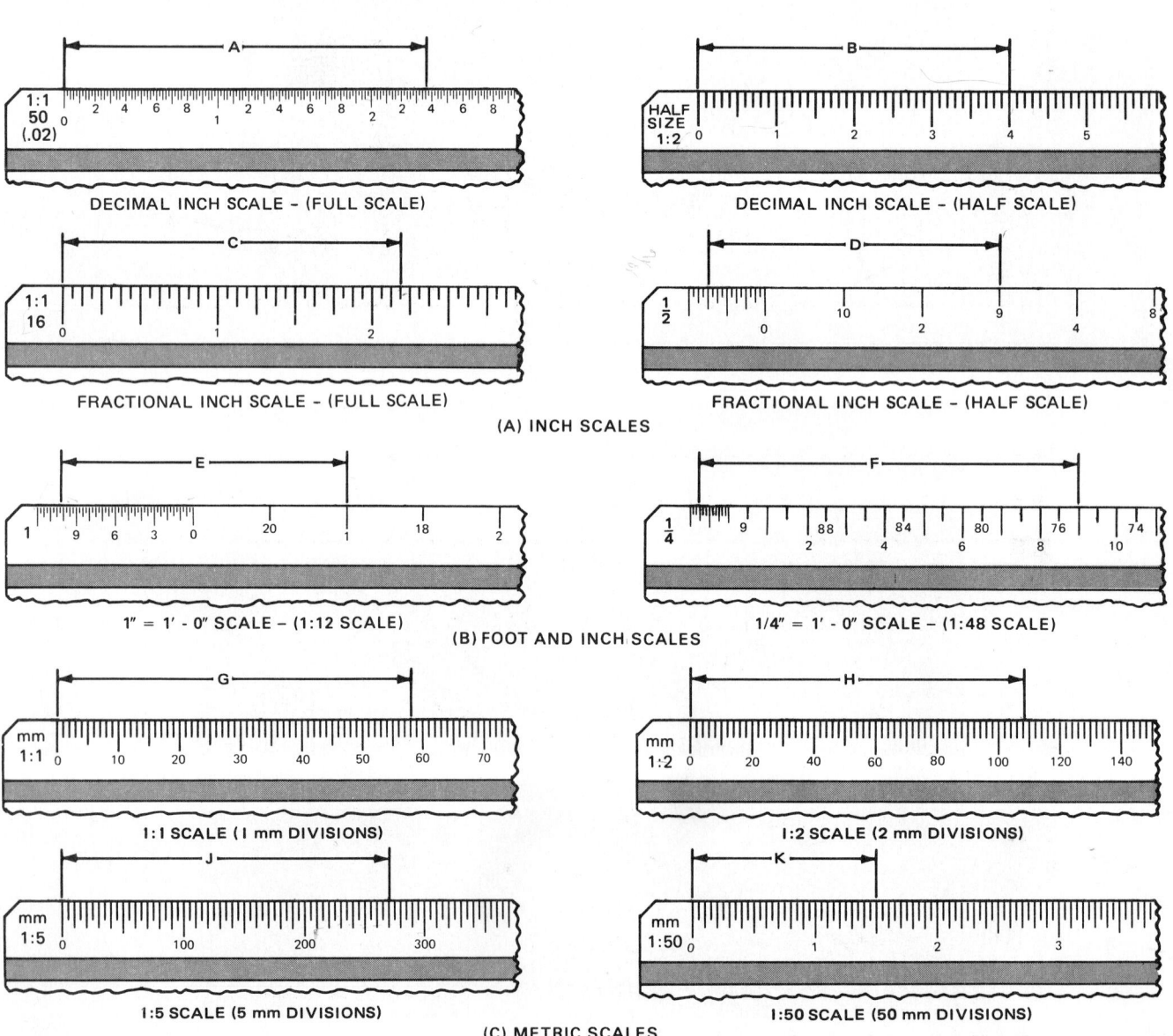

(A) INCH SCALES

(B) FOOT AND INCH SCALES

(C) METRIC SCALES

FIG. 1-4-A   Reading drafting scales.

3. Inch measurement assignment. With reference to Fig. 1-4-B and using the scale:
   1:1 decimal inch scale; measure distances A through F
   1:1 fractional inch scale; measure distances G through M
   1:2 decimal inch scale; measure distances N through T
   1:2 fractional inch scale; measure distances U through Z

4. Foot and inch measurement assignment. With reference to Fig. 1-4-B and using the scale:
   1" = 1' – 0", measure distances A through F
   3" = 1' – 0", measure distances G through M
   ¼" = 1' – 0", measure distances N through T
   ⅜" = 1' – 0", measure distances U through Z

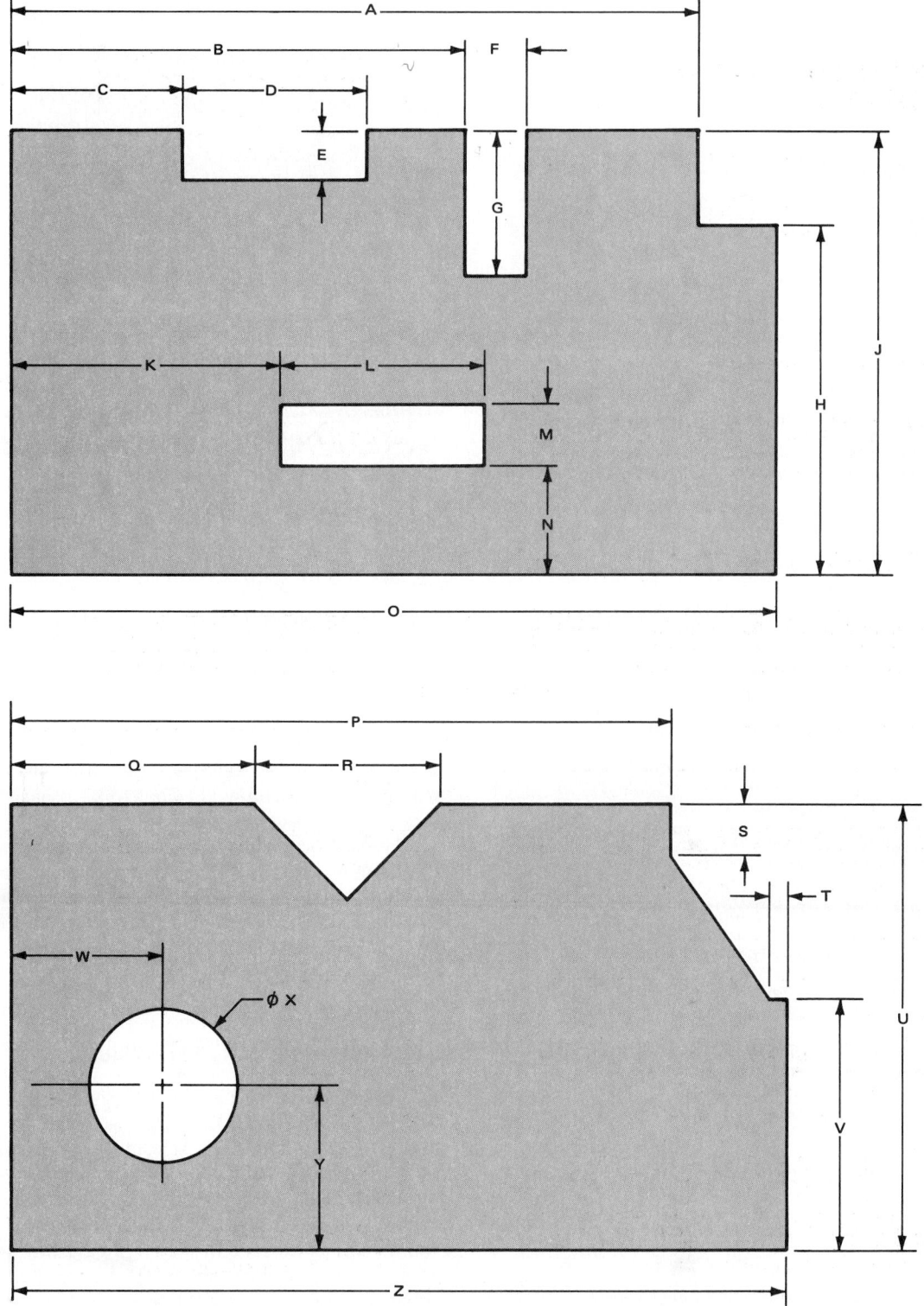

FIG. 1-4-B   Scale measurement assignment.

CHAPTER 2

COMPUTER-AIDED DRAFTING (CAD)

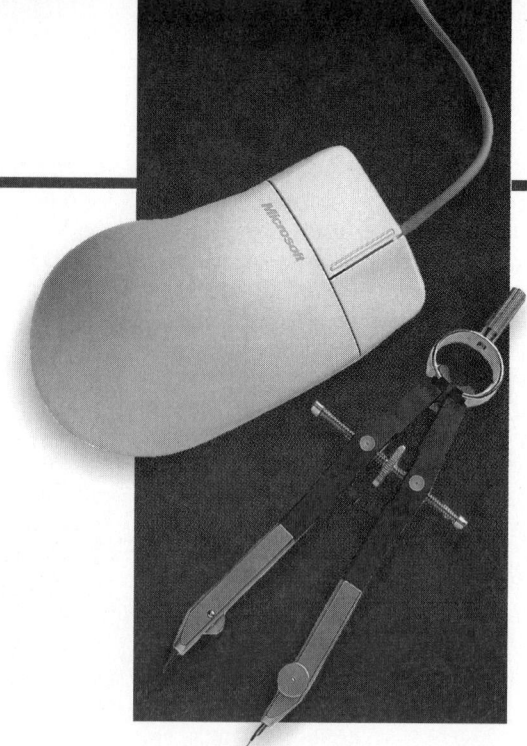

Definitions

Bit   A binary digit; the smallest unit of information recognized by a computer.

Byte   A unit of computer memory consisting of 8 bits.

Central processing unit (CPU)   Part of the computer processing equipment.

CIM   The acronym for computer-integrated manufacturing.

Computer   The combination of a CPU and a terminal.

File server   The central unit holding both the programs and files for a network of computers or terminals.

Hard copy   A paper copy of a finished drawing created on a computer.

Networked system   A group of computers that can communicate with each other.

Personal computer   A microcomputer unit usually consisting of a CPU, a keyboard, and a monitor.

Pixel   A "dot" of color that appears on a computer screen.

Program   A group of written instructions for a computer, logically arranged to perform a task.

Random-access memory (RAM)   Temporary storage locations for data entered into the computer.

Read-only memory (ROM)   The permanent memory in a computer.

Terminal   The combination of a keyboard and a graphics display monitor.

Write protect   A safety feature which prevents the loss of data from storage disks.

X-Y linear coordinate system   A system for specifying locations in a drawing based on horizontal and vertical axes.

2-1 INTRODUCTION TO COMPUTER-AIDED DRAFTING (CAD)

The last 20 years have brought great changes to the drafting room. Its physical appearance, furnishings, even its drafters and engineers have moved quickly from their battered domain of old into the information age. This era is often referred to as the "technical revolution."

These changes were brought about largely by the integrated-circuit (IC) chip. It has, in fact, revolutionized the way we work and play. This era has seen dramatic changes in worldwide communications at all levels—personal, professional, industrial—and in every facet of modern-day life. The microchip is on our wrists (quartz digital watches). It is used to help solve math problems (hand-held calculators). It entertains (video games) and helps to run businesses (computers). The technical changes it has launched have affected many careers, and retraining to upgrade job skills has become commonplace. Drafting and design have been at the forefront of the changes. CAD and CADD are familiar acronyms that have swept through the profession.

Components of a CAD System

A CAD system is made up of various combinations of devices. This holds true for small-, medium-, and large-system applications. The specific package selected largely depends on the needs of the user. Various types of drawings, such as check prints or finished drawings, referred to as *hard copy*, may be preferred by certain companies. Other companies that are fully automated may not require any drawing whatsoever. This means that one company will choose a piece of equipment that prepares a drawing one way. Another company will select equipment that uses another method for preparation of

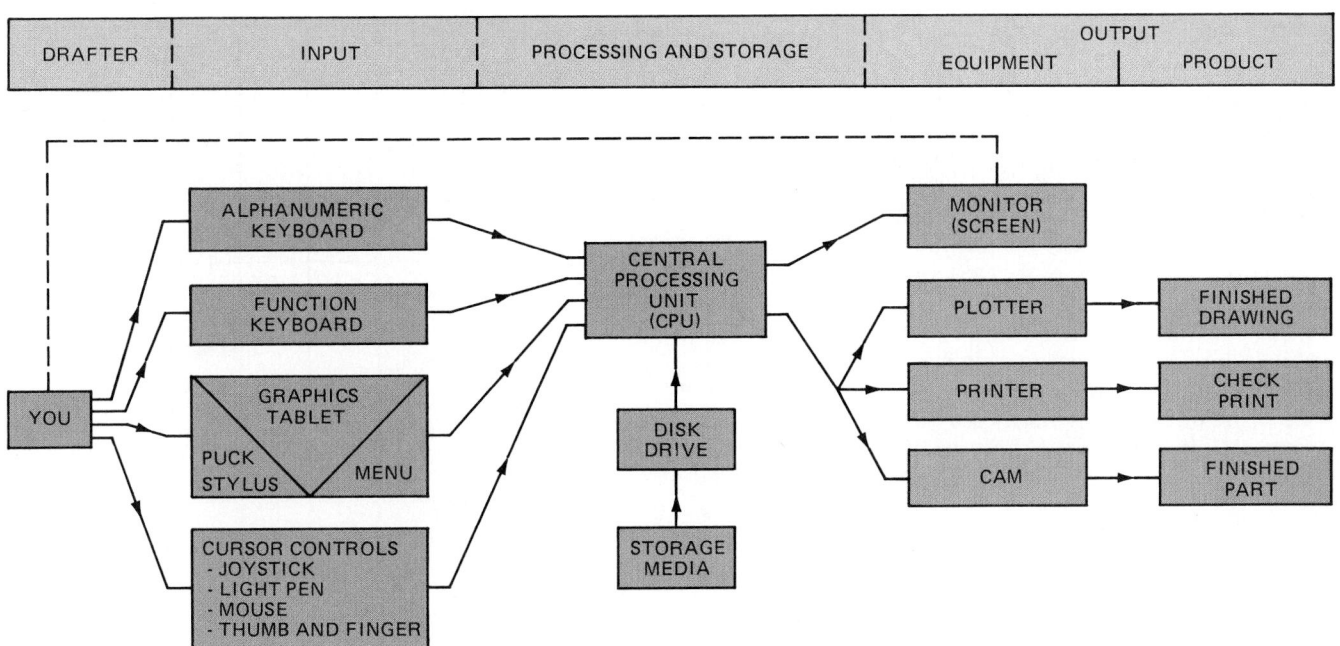

DRAFTER	INPUT	PROCESSING AND STORAGE	OUTPUT	
			EQUIPMENT	PRODUCT

FIG. 2-1-1   Operational flowchart of a CAD system.

drawings. Still a third company will not utilize any equipment to produce hard copies.

Generally, each system element can be categorized as one of the following types:

- Processing and storage
- Input
- Output

This unit analyzes each major component within these categories. Figure 2-1-1 shows an overall diagram of a complete micro-CAD system, including the various input and output devices. The purpose and function of each piece, as well as its relationship to the complete system, will be described. The items shown are typical components found in any system. It would be unlikely, however, to find all of these items in one particular system. The central processing unit (CPU) is considered part of the processing equipment. An alphanumeric (letters and numbers) keyboard is used to manually input data and is normally found with the graphics display monitor as one unit. In combination, a keyboard and graphics display monitor are commonly referred to as a *terminal*. A CPU is normally with the terminal; this combination is commonly referred to as a *computer*. Thus, a computer is composed of a CPU, alphanumeric keyboard, and graphics display monitor.

The typical system arrangement is *interactive*. This means that a person must cause the interaction between the CPU and the graphics display shown on the monitor screen. An alphanumeric keyboard or other input device will aid this process. After the design and/or drawing on the CAD unit has been completed, the information may be transferred to various output devices.

Figure 2-1-2 illustrates one type of system arrangement. To the right is a workstation that includes a monitor, a keyboard and a mouse for inputting data, and a CPU with disk drives. In the background to the left is a plotter.

FIG. 2-1-2   CAD system equipment. *(Courtesy of International Business Machines Corporation)*

2-2 PROCESSING AND STORAGE EQUIPMENT

The CAD program, drawings, and symbols are stored on disk. The CPU collects input information and places the lines and letters in such a way as to produce the required drawings and data.

Central Processing Unit

Bits and Bytes

The CPU is the computing portion of the system. A large number of integrated-circuit (IC) chips are combined into a microprocessor. This is where the number crunching occurs, i.e., the performance of fundamental computations. The number of computations, or the capacity of the unit, is designated by the number of bytes. *Byte* is the base term used by the system manufacturers, describing a character of memory containing 8 bits. A *bit* is a binary digit, the smallest unit of information recognized by a computer.

The size of the CPU normally determines the type of CAD system. *Micro* is the term used for small systems; *networked systems* and *mainframe*, for large systems.

Micro System

The micro unit has a typical user (or dynamic) memory capacity up to 32 megabytes (MB) or greater. This means there is enough space for about 32 million characters. Due to binary numbering limitations, 1K is not exactly 1K in the computer. It is the closest (1024) multiple. To get a feel for memory size, keep in mind that 640K roughly equals 400 typewritten pages. The term "larger and smaller" has been used to describe micro units. This means that larger and larger capacity is put into smaller and smaller units. The chart shown in Fig. 2-2-1 illustrates the rapid advancement in microcomputing technology. The associated cost per bit (or byte) has fallen at a rate inversely proportional to this increased capacity. As a result, the microcomputer revolution was launched.

A single or dual monitor and alphanumeric keyboard are normally part of a micro unit, as shown in Fig. 2-2-2. These units are known as *personal computers* or *PCs*. PC systems are economically priced, are readily available, and are used for a multitude of drafting applications. With their greatly increased capacity, they can accomplish a large amount of work.

Complex, expensive systems need not be tied up with drafting requirements. The data gathered on the PC (or system) can later be "translated" to a larger system. CAD has thus become accessible to a growing number of designers and drafters.

Networked System

AutoCAD may be utilized on multiple systems that are tied together (networked). A *file server* is the central unit holding both the program and drawing files. The network provides the full power of the system to each terminal. All of the software and drawing information may be shared resulting in reduced cost and standardization.

Mainframe

A mainframe system has a huge CPU (or host computer). CAD is only one of a multitude of functions that it can execute. Work such as the creation of payroll spreadsheets may be part of its operation. It offers more capability than the PC and networked systems. Mainframes were available long before CAD

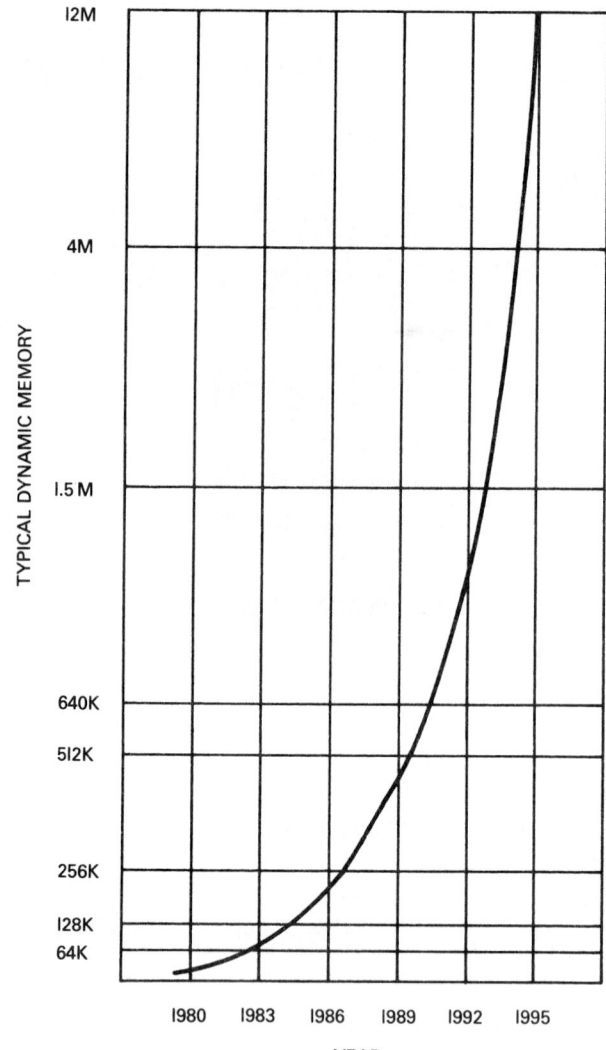

FIG. 2-2-1   Microcomputer dynamic memory (RAM).

became popular; consequently, in the early days of CAD, the mainframe was the most commonly used type of system for computer graphics. Mainframe terminals are normally found in a remote location of the workplace and are not combined into a single unit, as is the PC. For example, several terminals, each of which may be in a remote location, are still connected to the same processing unit, as shown in Fig. 2-2-3.

Software

Program Language

Software is the set of instructions that cause hardware (machines) to function. A *program* is synonymous with software and can refer to the original source code or to the executable (machine language) version. The instructions tell the computer what to do and when to do it. They are used to input information into the system. CAD programs are written in a variety of languages. Common ones are LISP (List Processing

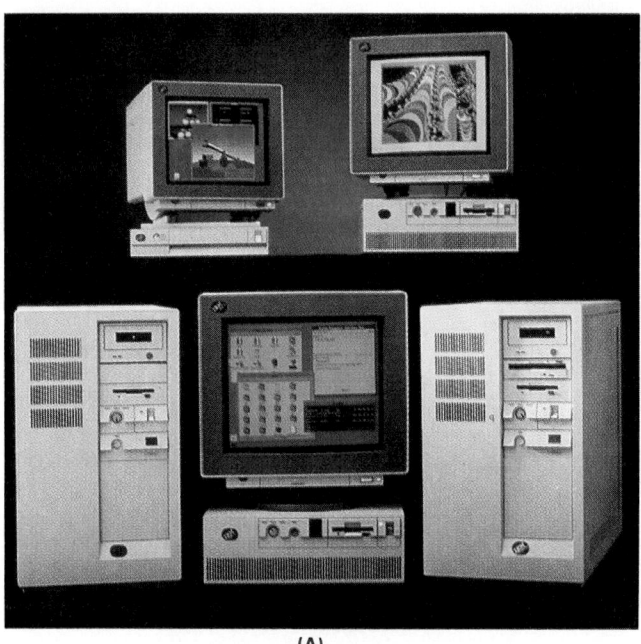

FIG. 2-2-2   A variety of microcomputers. *(Courtesy of International Business Machines Corporation)*

(A)

(B)

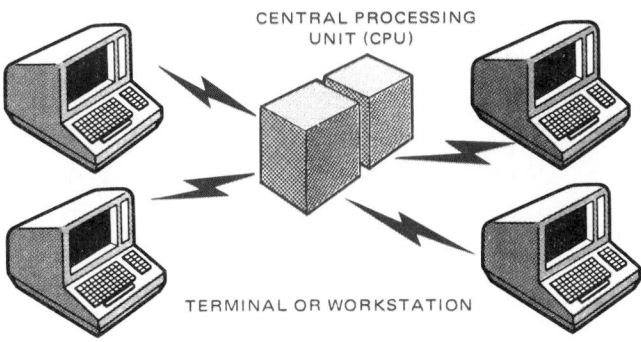

FIG. 2-2-3   Mainframe or networked system.

Language for Symbolic Computing), BASIC (*Beginners' All-purpose Symbolic Instruction Code*), and C. BASIC is popular since it uses Englishlike and mathlike "easy-to-use" language. The notations are made up of statements rather than sentences.

No matter which language is used, drafters and designers need not necessarily develop a knowledge of it. A drafter or designer will normally serve as the user of programs rather than the developer. Some proficiency, however, may be desired, and an introductory programming course or LISP workshop may be taken for this purpose.

Software has been extensively developed over the years. A computer programmer will spend thousands of work hours to develop a program. The computer explosion has dropped hundreds of CAD software and hardware products onto the markets and a multitude of complete programs are available. The recommendation with software is, "Buy—don't build." If the desired software to do the job can be purchased, why invest thousands of work hours to develop a new program?

Storage Medium

CAD systems will magnetically store programs and drawings on disk. A disk economically stores drawings and programs and allows their speedy retrieval.

Two types of disks are available: *hard* (fixed) and *floppy* (removable). Disks are also single or dual. This means that data may be stored on one side (single) or on both sides (dual) of the disk.

Two standard diameters for floppy disks are 5.25 and 3.50 in. (Fig. 2-2-4, pg. 26). Note the protective covering over both disks. Disks must be handled gently. Even a small scratch can ruin the contents of the disk. The cover will protect the disk from dust, dirt, and accidental scratching. It cannot, however, prevent mishandling. A disk, for example, should not be exposed to heat or magnetic fields. The contents will immediately become damaged.

Notice the small tabs shown in Fig. 2-2-4D. They are used to protect disk contents. It is possible to accidentally write over the data on the disk with new data. If this occurs, the original data is destroyed. The small tab covers the slot in order to prevent the loss of data. This safety feature is known as *write protect.*

Areas of the disk are divided into tracks and sectors. *Tracks* are circles that surround the disk. *Sectors* are arc-shaped sections of tracks. Tracks and sectors divide the disk so that information can be stored in findable places (Fig. 2-2-5, pg. 26).

The number of sides and density of a disk are given as:

- DSDD—Double sided, double density.
- DSHD—Double sided, high density.

25

(A) 5.25 IN. COVERED FLOPPY DISK   (B) 5.25 IN. UNCOVERED FLOPPY DISK

(C) 3.50 IN. FLOPPY DISK

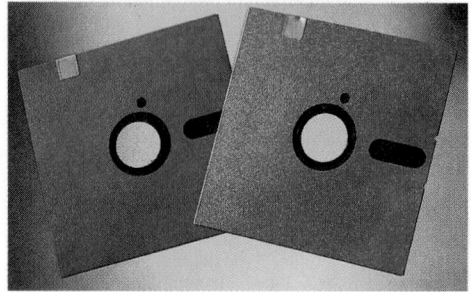

(D) WRITE-PROTECT TABS FOR 5.25 IN. FLOPPY DISKS

FIG. 2-2-4   Floppy disks. *(A & C—courtesy of International Business Machines Corporation; B, C, & D—STUDIOHIO)*

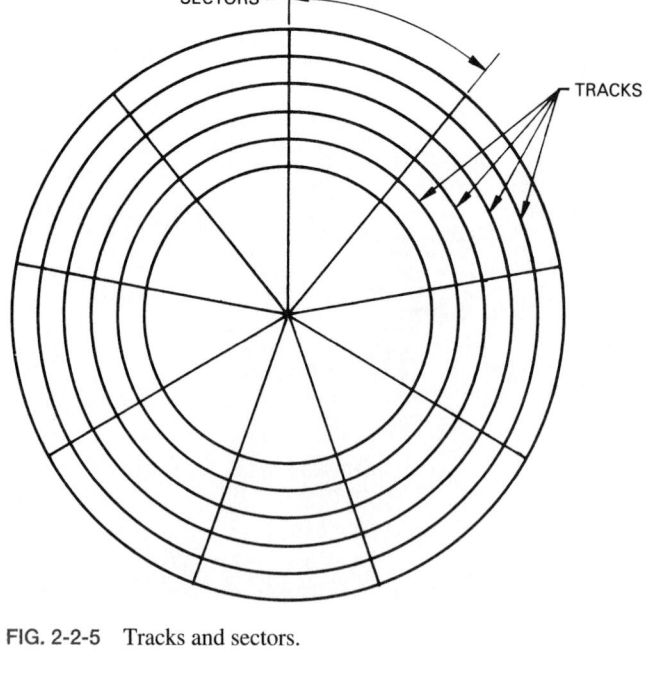

FIG. 2-2-5   Tracks and sectors.

Choose floppy disks for which your computer's disk drive was designed. A lower quality diskette or higher quality diskette may not function properly. High-density disks are the most common.

The compact disc (CD) distributes a larger amount of data than a floppy disk (Fig. 2-2-6). This 12 cm optical disc can store numerical data in excess of 600 Megabytes on a single disc. This is equivalent to approximately 500 times the storage capacity of a standard 5¼" floppy diskette. As a mass storage device for the computer, CD-ROM (Read-Only Memory) uses an optical system for reading data. This read-only system allows only prerecorded disc to be read out. The user cannot directly store or modify the data. The CD-ROM has become an integral part of the new multimedia systems that combine the best features of both television and computer. The CD-ROM features include animation, video, sound, and user interaction. CD-ROM is finding its greatest use wherever the access and storage of large files are required. Applications in this area include CAD/CAM, imaging, and graphics. It is this growing list of applications that makes the CD-ROM a major success.

Storage Units

Disk Drives

Additional devices are required to allow input to, or output from, the processing equipment. Among these are disk drives and memories. Systems that use the floppy disk require equipment to drive the software. A disk drive (Fig. 2-2-7) receives the floppy disk directly and may be used as permanent storage. The information on the disk may be a program, symbols, or drawings. When loading a disk into a drive, be certain to insert it correctly. The label is to be up, the slotted portion is placed in first, and the disk carefully slid into the drive.

FIG. 2-2-6   Compact discs.
    (Mak-I Photo Design)

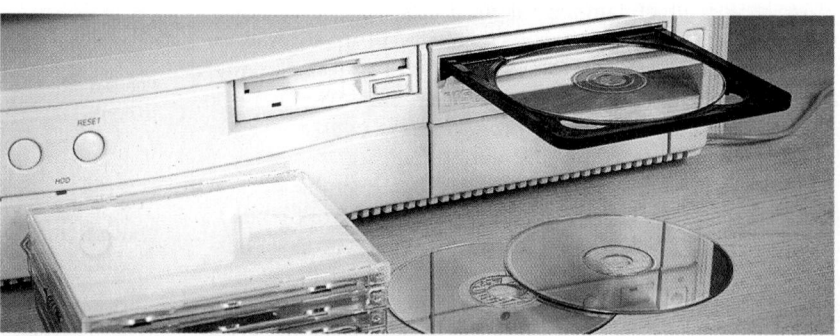

FIG. 2-2-7   Floppy disk drive. *(Mak-I Photo Design)*

Memory Systems

Computers have memory systems to store programs and data. The memory may be either permanent or temporary. Permanent memory is referred to as *read-only memory* (ROM). ROM may also be referred to as *firmware*. Firmware is software inside of hardware. Temporary memory is referred to as *dynamic*, *user*, or *random-access memory* (RAM). RAM provides temporary storage locations for entries made by any input device, allowing a drawing to be developed. A complete set of statements (a program), up to system capacity, is stored in RAM. Computers are generally designated by system capacity. A 640K unit, for example, has 640,000 bytes of dynamic memory. The contents in dynamic memory are destroyed when power is lost or when a power surge occurs. Thus power surge protectors should be used on all systems. In the long run, this will prevent many hours of work from being lost.

The contents stored on a disk are not lost as a result of a power interruption. Remember, though, not to remove a floppy disk while placing information onto it (writing) or taking information from it (reading). A glitch (discontinuity) may result. The drive busy light on the disk drive will be lit when information is being read from, or written to, the disk.

Computer systems allow you to see what data are held on the floppy disk. Data related to an individual program or drawing make up a *file*. All the files on the disk are listed in a *directory*. To see what files are on the disk, insert the disk in the drive and type in the directory command.

2-3 INPUT DEVICES

Skills in using and operating input devices are replacing the skills of manual drafting. This equipment instructs the computer to draw lines and circles of various widths and sizes and

to apply such graphics as symbols, dimensions, and notes to a drawing.

Alphanumeric Keyboard

An alphanumeric keyboard allows you to communicate directly with the CPU. It may be used to manually input data for:

- Nongraphic work.
- Text and notes additions.
- Exact coordinate input.
- Command selection.

Every CAD system provides a means to select the part of a program that allows the use of a particular command. One way to effect this is to key in the appropriate letter(s) after the COMMAND prompt. For example, LINE is the command for line creation. To select that part of the program, key in LINE or L and press the ENTER key.

The keyboard is an extended version of a standard typewriter keyboard (Fig. 2-3-1). The alphanumeric keys are the same. Usually, though, there are additional keys that allow some specialized command options to be quickly accomplished. "Alpha" refers to the keys that input letters of the alphabet. "Numeric" refers to the other keys, each of which inputs a number. You may type in an alphanumeric instruction, and complete it by pressing the ENTER key. Knowledge of the keyboard is indispensable for anyone involved with CAD. The speed of inputting data with the keyboard is a function of the ability to use it.

Function Keypad

The function keypad is an input device that is used to retrieve a program or part of a program. (The term *program function keypad* is probably more descriptive, but for simplicity, the term *function keypad* will be used to describe this input device.) A function keypad contains several buttons or keys. A part of a program is electronically connected to one of the buttons or keys, and this is operated during the execution of a particular function.

Graphics Tablet

The graphics tablet is a flat surface area electronically sensitized beneath the surface. The tablet is available in a wide range of

FIG. 2-3-1   Alphanumeric keyboard. *(STUDIOHIO)*

sizes. It may vary from a small surface [11 × 11 in. or 279 × 279 millimeters (mm)] to one that exceeds an E-size drawing (36 in. × 48 in. or 910 × 1220 mm). Beneath the surface lies a grid pattern of many horizontal and vertical sensors. When proper contact is made on the surface, electrical impulses are transmitted to the computer. The information that is transmitted provides the programmed instructions to the CPU.

Digitizing

The graphics tablet is another type of input device. In terms of graphics creation, it is more important than the keyboard. Also, it has many purposes. One use is quick and accurate graphic conversion. A rough sketch can be converted to a finished drawing by the transferring of point and line locations to the screen. Information is based on the *X-Y* (horizontal-vertical) linear coordinate system. These are entered quickly and efficiently into the computer, using this so-called "electronic drawing board" (the graphics tablet). The result of the input data is graphically displayed on a monitor screen. This method is commonly known as *digitizing*. Consequently, a graphics tablet is also referred to as a *digitizer*. Other functions, such as

symbol input, may be performed with the aid of a tablet. Figure 2-3-2 shows a graphics tablet menu.

Tablet Menu

CAD systems are menu-driven. This means that a selection is made from a preprogrammed menu (tablet or screen) by the user to call up a particular part of a program. Graphics tablets are provided with menus having a variety of options. A menu may be placed on the tablet surface. A graphics tablet menu is shown in Fig. 2-3-3. Each of the small boxes, or cells, illustrates the list of available choices. Many of the terms are abbreviated to fit into the box. To make a selection, place and activate the stylus or puck over the desired item, such as a polygon, as shown in Fig. 2-3-3. Notice both a name and an icon within each cell.

Any selection on the tablet will retrieve a portion of the program corresponding to the desired title or symbol. For example, if you wish to draw lines, select the menu labeled LINE with the line *icon* (graphical representation), as shown in Fig. 2-3-2. This will call up the part of the program that allows the creation of lines.

FIG. 2-3-2   Graphics tablet menu. *(Autodesk Inc.)*

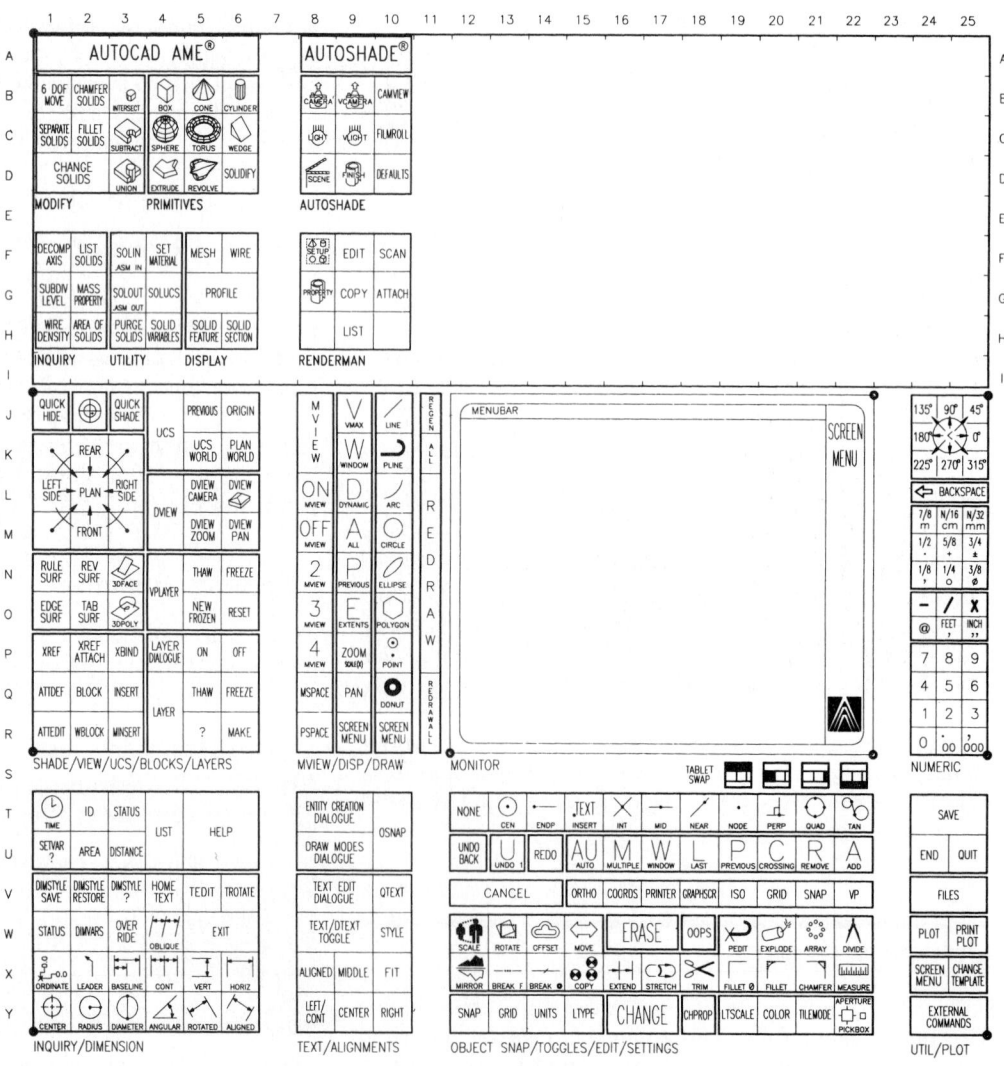

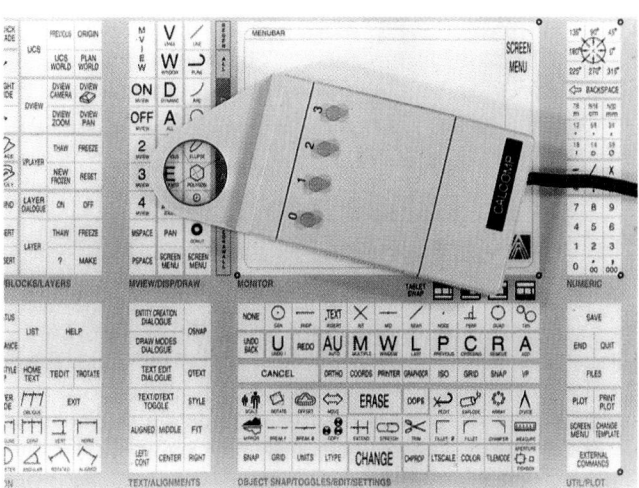

FIG. 2-3-3   Graphics tablet menu and puck.

A menu limits the design drafter to certain types of commands. Both the software and the menu, however, can be changed on many of these systems. One way is to insert a symbol disk or a menu disk, or enter LOAD LISP commands. Completely different applications are thus programmed into the unit. A different mask, or overlay, corresponding to a particular program, such as the partial landscape menu shown in Fig. 2-3-4, can be used. The option to switch menus is particularly useful when one wishes to create various schematic diagrams or other specialized graphics.

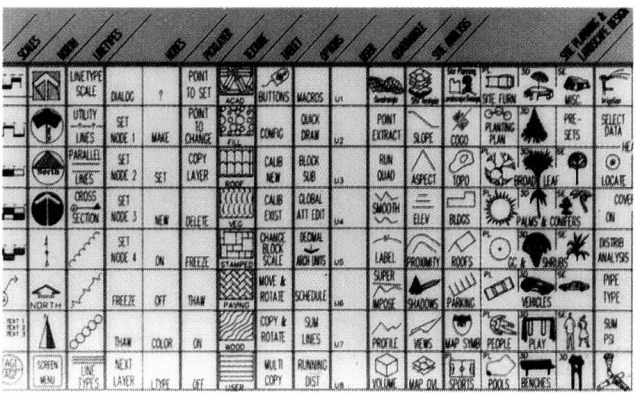

FIG. 2-3-4   Partial landscape design menu. *(LandCADD)*

Cursor Controls

Several pieces of equipment are used in conjunction with the graphics tablet. These may include a stylus or pen, a push-button cursor or puck, and a power module or console.

Puck and Stylus

A *stylus* (electric pen) or a puck is used to select the desired horizontal-vertical coordinate locations on the graphics tablet. These locations are transferred to the computer. Figure 2-3-3 illustrates a tablet with a push-button puck. Several styles of pucks and styluses are available. Figure 2-3-5 illustrates these cursor controls.

FIG. 2-3-5   Puck and styluses.
(Doug Martin)

Each puck has fine black cross hairs that are used for positioning. The buttons on a puck may be used for various purposes, depending on the manner in which the software has been created. Pressing the appropriate button causes horizontal (X) and vertical (Y) data to be accurately sent to the CPU by an electrical signal. The result is displayed on the monitor screen as a bright mark, or *cursor*. This selection process is repeated as often as necessary to complete the drawing.

A stylus essentially operates the same way as a single-button puck. The position is selected by the tip of the pen. As you move the pen tip across the tablet surface, the position will change correspondingly on the monitor. It will appear as a cursor on the monitor. After the desired position has been located, it is digitized by activating the stylus, usually by pressing down on it.

Other Cursor Controls

The Mouse   Figure 2-3-6A shows another example of a puck or *mouse*, which is the most popular type of cursor control. The mouse rolls across a rectangularly shaped pad which relates to the shape of the monitor. A mouse generally has one or more buttons that are used to "click" the cursor into place on the screen.

Light Pen   A light pen is used as a direct-entry input pointing device. It is also considered a digitizer since it can change displayed points and select menu options on the screen. The pen is electronic and contains a photocell sensory element to detect the presence of light. Hence the term *light pen*. Attached to one end is a cable through which the signal is transmitted. The other end of the pen may be positioned by hand to a desired screen location. After positioning, touch the screen with the tip of the pen. Depressing it causes the pen to become activated. Light spots are sensed. A signal is sent to the system, indicating the position. By this method, any element of the graphics display may be identified to the computer. One disadvantage of the light pen is that after several hours of prolonged use a drafter may tire from holding it.

Other Devices   Several other types of input devices may be used to position the cursor. They include a joystick (Fig. 2-3-6B), touch pad, and Tracball. While these devices are normally not used, it is possible for one to be found on a particular system.

2-4 OUTPUT EQUIPMENT

The drawings that are being produced are observed on the monitor screen. At any time, a print of what is seen on the screen can readily be obtained from a printer or plotter. When the drawing or design is completed, a finished drawing with near perfect lines and lettering is made on a plotter. This is equivalent to the drawing made manually (only better). Prints or microfilm are then produced from this drawing.

If a company has a developed CAD/CAM operation, a drawing may not be required. Instead, a set of instructions in coded form is directly transmitted to the fabricating equipment. The instructions include all location and size information necessary to produce the product.

Monitors

As a drawing is being made, its image can be produced on the screen of a monitor (or graphics display station). The *monitor* displays data in both alphanumeric (written) and graphical (pictorial) form. The user can view a picture of the design as the design is being entered into the system. This display can be "called up" in a variety of ways, depending on the type of monitor used.

Cathode-Ray Tube

The most popular monitor is the cathode-ray tube (CRT). The display method is similar to a television screen. A sample screen display is shown in Fig. 2-4-1. There are several types of CRTs.

FIG. 2-3-6   Other cursor controls. *(A—Glencoe file photo; B—STUDIOHIO.)*

FIG. 2-4-1   Monitor display. *(Autocad/Macintosh)*

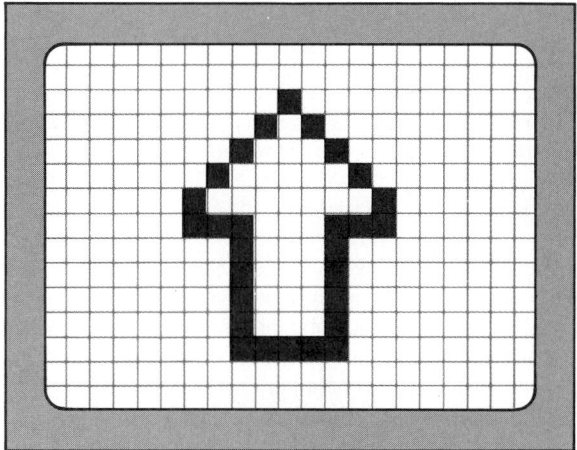

FIG. 2-4-2   Raster display on a CRT.

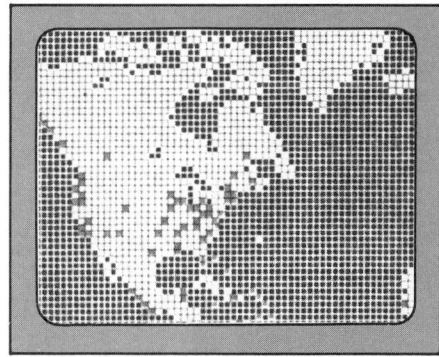

FIG. 2-4-3   Pixel representation.

Vector   The vector-writing CRT is drawn on by an *X-Y* direction coordinate system. This is similar in principle to the popular Etch-a-Sketch toy. The computer first locates points and then connects the points.

Raster   Raster screens have become the dominant type of CRT displays. The raster uses a grid network to display the image. Each grid is either a dark or a light image that falls within a square area that appears on the screen as a dot (dark) or undot (light) in many color choices. Each dot is known as a *pixel* (picture element). An analogy for how this works is a placard pattern used by fans in the stands of a football game. The cards are used to show a graphical message, and each fan within the pattern holds up either a dark or a light card. At a sufficient distance, the combinations of dark and light spots produce a recognizable image. This phenomenon is shown by the arrow in Fig. 2-4-2.

Another example of pixel representation is shown by the map in Fig. 2-4-3. The horizontal and vertical scan lines are shown so that you can clearly see each pixel. Since the pixels are so large, this would be considered a low-resolution image. Inclined lines will always appear jagged, a phenomenon known as "jaggies" or "stairstepping" (Fig. 2-4-4).

The *resolution* (clearness) of an image depends on the closeness of lines forming the grid pattern. The smaller the pixels, the more resolution the image has. The greater the number of dots per unit area, the greater the resolution. The greater number of dots improves picture quality. For example, an enlarged photograph has a loss in resolution, or clearness, because of the reduced number of dots per unit area.

Significant advancement in raster technology has been achieved in recent years. New techniques enhance resolution. Jaggies appearing on the lines and polygon boundaries have been significantly reduced. Consequently, low resolution is virtually eliminated. The improved resolution will more

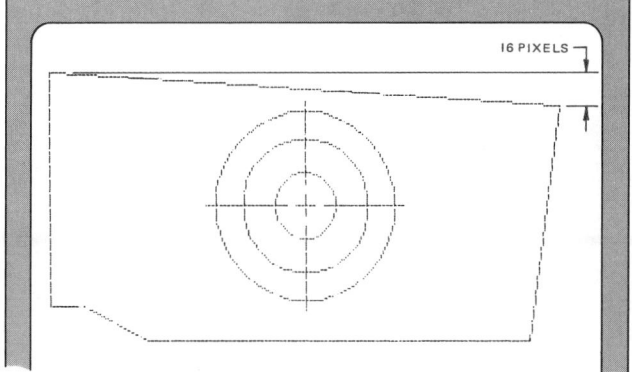

FIG. 2-4-4   Low-resolution display.

likely be seen in the medium-resolution to "hi-res" (high-resolution) range, as illustrated in the art throughout the text and in Fig. 2-4-1.

Color   The types of CRTs mentioned are color-enhanced. This display screen is similar to the color television screen. Each of three electron guns emits one of the primary television colors: red, green, or blue. From a combination of these, any pattern of colors may be utilized.

Other Monitors

There are other less-popular types of monitors. One type is *plasma*. Plasma technique uses a flat, thin panel and the glow from an inert gas to display the image. Neon is the gas commonly used. The gas is ionized to emit visible light in a dot matrix pattern. The plasma technique is potentially popular since it is better for your eyes than the CRT. The CRT has low-level radiation emission and a flicker associated with it. This is eliminated with the plasma panel display, which is virtually flicker-free. The primary disadvantage of plasma display is that it produces poor resolution and the size of the panel is small.

Other types of monitors include electroilluminescent, liquid crystal, and projection techniques. Each has its associated problems, and their use is minimal.

Plotters

There are three types of plotters commonly used by industry to produce a finished drawing:

1. Pen plotter
2. Electrostatic plotter
3. Laser plotter

Pen Plotters

A pen plotter is an electromechanical graphics output device. Pen plotters (shown in Fig. 2-4-5) are the most popular type used. They move a pen in one direction across the paper medium. The second direction to produce the lines is accomplished by the paper movement. The plotter is used to produce a finished drawing. This can be any combination of lines and alphanumerics. If a CAD system is thought of as an automated drafting machine, the plotter is the part replacing the activity of "laying lead." It produces the finished original drawing that was previously developed and displayed on the monitor.

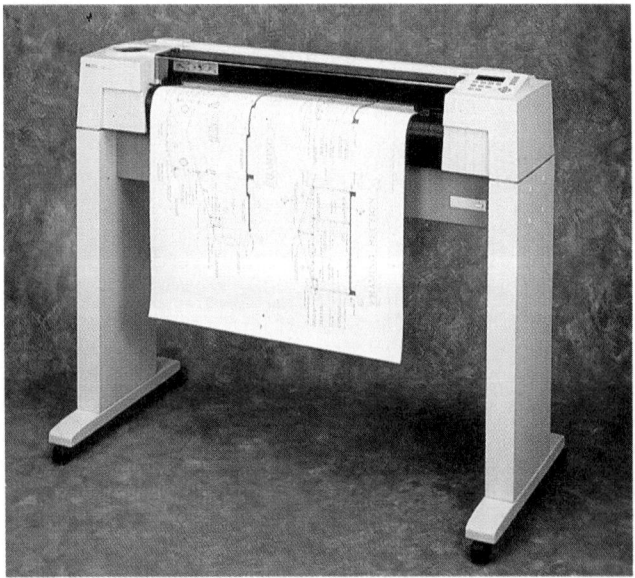

FIG. 2-4-5   Pen plotter. *(Courtesy of Hewlett-Packard Co.)*

FIG. 2-4-6   Pen holder for multiple pen plotter. *(Doug Martin)*

Regardless of screen resolution, a plotter will produce quality lines. There are more addressable dots per unit area. Thus line quality is excellent and virtually jaggie-free.

Various types of ink pens, such as wet ink, felt tip, or liquid ball, can be used. They may be a single color or multicolor. More importantly, plotter pens offer a variable line width (weight) option, usually 0.3 and 0.7 mm widths, with modern models now offering a pencil option. Different pens are inserted or removed rather quickly. The desired pen (different width or color) is inserted and the holder returned to the plotter (Fig. 2-4-6).

The surface area of a pen plot drawing may be as small as an A size (8.5 × 11 in.) and larger than an E size (34 × 44 in.) sheet or drawing format. The size of the drawing to be placed on the paper can also vary. You are able to vary the scale of the lines and characters by manually setting the plotting surface area and/or scale.

Pens will draw on various types of media. Most popular are opaque drawing paper, vellum, and polyester film. Being able to match the medium to its purpose is a distinct advantage of the pen plotter. Another advantage is that the drawing produced is of high quality and is uniform and precise. On the other hand, an average D size (22 × 34 in.) pen plot will tie up the plotter for long periods of time. Thus the plotter should not be used as a print machine. Once the drawing has been finalized, prints may be produced by whiteprint photocopy, or microfilm equipment.

The pen plotter is slow compared with other output devices. It will take from several seconds (simple drawing) to several minutes (complex drawing) to produce a drawing—a disadvantage for a user requiring large-scale production. Yet due to cost factors pen plotters will remain the most common output device for low- to medium-volume applications. This still represents a major reduction of time from manual methods of producing these drawings. In fact, a rather complicated D size drawing, which may have taken a week to create on the screen, can be plotted in less than 20 minutes.

Pen plotter technology includes the drum, microgrip, and flatbed types.

Drum A drum plotter consists of a long cylinder and a pen carriage. The surface area is curved rather than flat and is in the shape of a cylinder. Hence the term *drum*. The drum rotates to provide one axis of movement. The carriage moves the pen(s) to provide the other axis of movement.

Microgrip The medium is gripped at the edges with a microgrip plotter. The paper is moved back and forth. High performance is attained at a low cost.

Flatbed Pen movement occurs in both axes with the flatbed plotter. The pen carriage is controlled in both the *X* and the *Y* axes. Motors and cable are used for control. Short digital steps, normally less than .010 in. long, produce the line. The vellum, polyester film, or other medium is held on the bed surface by electrostatic attraction. Flatbed plotters are no longer popular.

Electrostatic Plotters

Electrostatic plotters will replace pen plotters in applications requiring high production. See Fig. 2-4-7. Plots are made using an ion deposition process toner at a speed exceeding one inch of paper length per second. This means that a plot can be prepared as much as 20 times quicker. A D size drawing is plotted in less than a minute with crisp, black lines. The disadvantage of the electrostatic plotter is that it is expensive. As the cost of equipment decreases, however, a wider use is anticipated.

Laser Plotters

A laser plotter uses a moving laser beam to alter a point-by-point electrical charge on the surface of a rotating drum. The drum is exposed to dry ink, which adheres to the charged areas of the drum. This is then transferred to a hot roller where the ink is fused to the medium. Laser plotters, like electrostatic plotters, produce drawings at high speed.

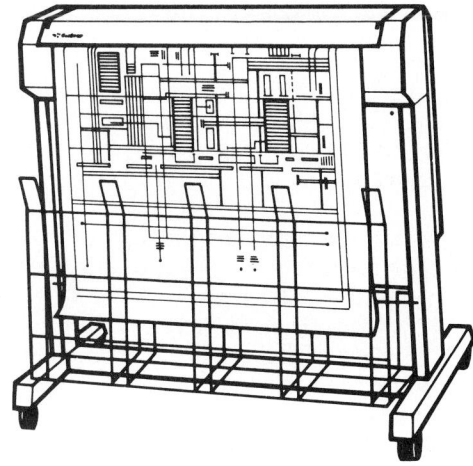

FIG. 2-4-7 Electrostatic plotter. *(Norman Wade Co. Ltd.)*

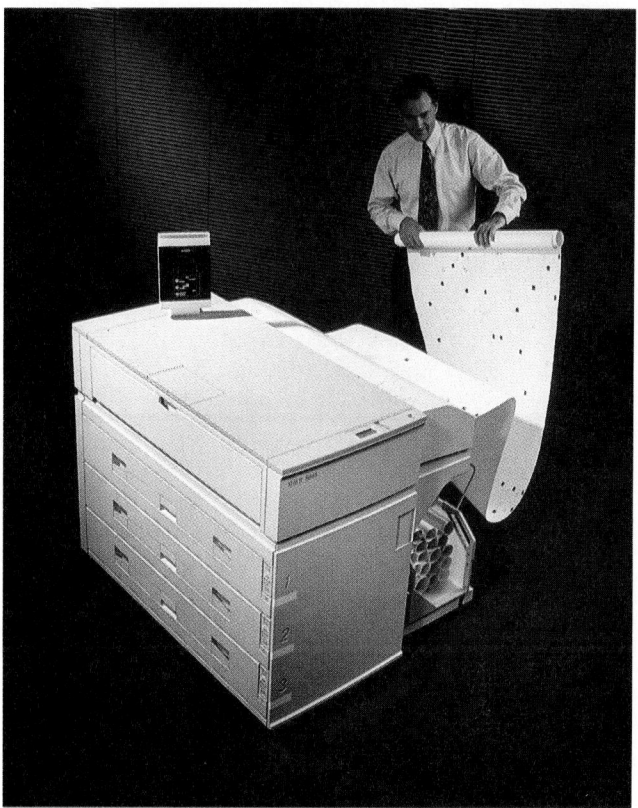

FIG. 2-4-8 Laser printer/plotter. *(Courtesy of Xerox Corporation)*

Printers and Printer/Plotters

A *print* is a preliminary drawing that is produced by a printer. Older printer models produce an image of what is seen on the monitor, complete with all jagged lines and circles. All lines are shown as one thickness. The print is used to check preliminary stages of the design. Recent technical advances, however, have vastly improved the resolution and led to the development of printer/plotters (Fig. 2-4-8).

The newer printers duplicate the screen display quickly and conveniently. The primary advantage is speed. They produce output much more quickly and less expensively than pen plotting. Complex graphic screen displays may be copied by the PRINT command. Whatever is on the screen will be copied, including any combination of graphic and nongraphic (text) displays. This procedure is called a *screen dump*. The entire process requires seconds. The copy, except for the dot matrix type, approaches the level of quality produced by plotters.

Laser Printer/Plotters

Laser technology is developed to the point where it will meet plotter quality. The primary disadvantage is that it cannot produce the large drawings required by industry.

Dot Matrix Printer

The newer printers have a much higher resolution than the old style. They do not, however, approach the quality of a plotter. Their advantage is low cost.

Ink-jet Printer

The ink-jet process deposits variously colored ink droplets on the medium. The result is a multiple-color copy. They are now capable of producing large drawings at a fairly reasonable cost.

Photoprinter

A typical photoprinter involves the use of fiberoptics technology to reproduce the image on dry silver paper. The unit produces a small A size copy that is satisfactory for quick preview during intermediate work steps.

2-5 COMPUTER-AIDED MANUFACTURING

Computer-aided manufacturing (CAM) uses the result of a computer-aided design. Combining CAD and CAM has had the effect of radically increasing productivity and accuracy. CAD/CAM means that an engineering drawing is no longer produced, since there is a direct, hard-wired connection between design and production. The output of the CAD system is a drawing stored in a geometric data base. This drawing is transmitted directly into the CAM equipment.

Numerical Control

One method of transmitting the design information is known as *numerical control* (NC) or *computer numerical control* (CNC). CNC can store the designs that are used with a variety of production-related processes. The result is a finished manufactured part. This means that errors will no longer occur, since a machinist will not have to interpret an engineering drawing.

True, CAD/CAM is the ultimate goal of industry. Many have CAD; many have CAM; few, however, have CAD/CAM. CAD/CAM and *computer-integrated manufacturing* (CIM) have been implemented at a slow rate. There are, however, software packages available, such as NC CODE and Smart CAM, that operate quite well in conjunction with AutoCAD.

Robotics

The other part of CAM is known as robotics. Robot machinery differs from CNC machinery in that movement is now the prime duty. Automatic manipulators are used to perform a variety of material-handling functions. The robot manipulators are arms and hands (Fig. 2-5-1). They will grasp, operate, assemble, and handle with great consistency and dependability. They are able to perform tasks that are considered too difficult, dangerous, or monotonous for human workers. This is especially true in environments that are intolerable to human beings. The range of tasks includes:

- Working with metals at extremely high temperatures (e.g., spot welding).
- Working in rooms filled with toxic fumes (e.g., spray painting).
- Exerting great forces (e.g., lifting and moving heavy, awkward products).
- Working with delicate parts (e.g., adjusting electronic systems).

FIG. 2-5-1   Robot display on a monitor screen.

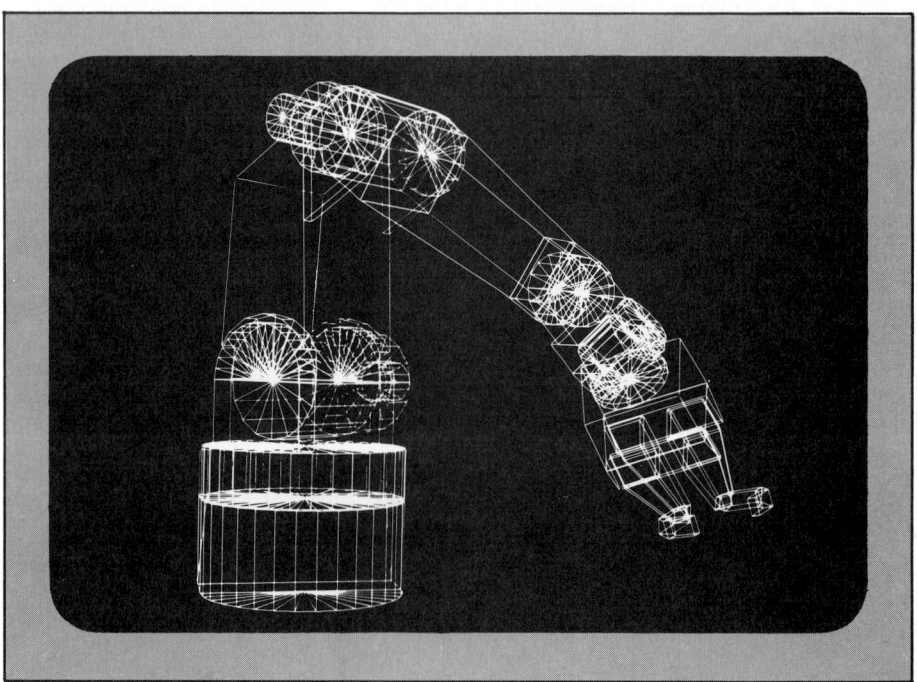

CHAPTER 3

DRAWING MEDIA, FILING, STORAGE, AND REPRODUCTION

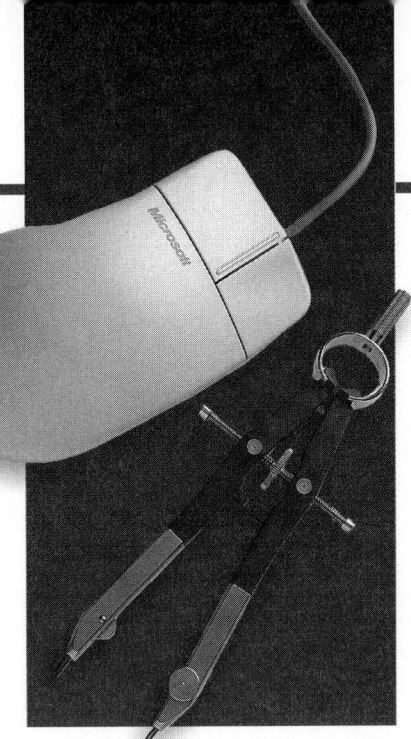

Definitions

Boot   To start up the computer.

CD-ROM (Compact Disc Read-Only Memory)   A high-capacity (approximately 600 MB) mass storage device that uses an optical rather than a magnetic system for reading data. CD-ROMs are frequently used to distribute large amounts of data.

Diazo (whiteprint) process   A drawing reproduction method involving photosensitive paper and chemicals.

Drawing media   The material on which the original drawing is made.

Drawing paper   Primarily opaque paper used for drawing originals.

Magneto-optic drive   An optical mass storage device similar to the CD-ROM that is erasable and rewritable. These drives allow for the recording of large amounts of data on compact, reusable, high-capacity, removable media.

Margin marks   Marks made in the margins of a drawing to convey information such as folding marks or graphical scale.

Photoreproduction   A drawing reproduction method using an engineering plain-paper copier.

Title block   Margin notes containing the drawing number, the name of the firm or organization, the title or description, and the scale; located in the lower right-hand corner.

Tracing paper   Primarily translucent paper used for drawing originals or tracing original drawings.

Zoning   A method of dividing large drawings into segments for easy reference.

3-1 DRAWING MEDIA AND FORMAT

Drawing Media

The term *drawing media* in this text refers only to the material on which the original drawing is made. Most engineering drawings that are to be reproduced are prepared on a transparent material such as vellum (a drafting paper having translucent properties) or polyester film. These drawing media differ sufficiently between each other and within themselves to provide a wide choice of qualities and characteristics for selection of the perfect material for specific drawing requirements.

Paper

Drafting papers come with a wide range of qualities—strength, erasability, permanence, translucency, etc. The distinguishing feature between drawing and tracing paper is translucency.

Opaque papers are used primarily as drawing papers. Because of a change in drafting practices, the need for drawing papers today is limited to the two extremes in the scale of quality—the very highest-grade, permanent drawing papers with the best possible erasing quality for maps or master drawings, which are later photographed, and the inexpensive school type of papers for the educational market. Since master drawings often must be revised and corrected, the major consideration in high-grade drawing papers is erasability.

Vellum is an inexpensive drawing paper that has good transparency qualities. However, it is not recommended if the drawing will be subject to excessive handling or rough usage. It is suitable for ink or pencil and permits good-quality reproduction by any of the reproduction processes. Generally, the

35

thinner papers are more translucent, but the heavier papers possess more strength, withstand repeated erasures more satisfactorily, and have greater durability in handling and filing.

Formerly, tracing papers were used almost exclusively for the ink tracing of pencil drawings made on opaque papers. Translucency was the prime requirement. Today the usual drafting practice is to develop the master drawing directly on translucent paper from which reproductions can be made, thereby saving the time, expense, and checking involved in the tracing process. This practice means that, in addition to good translucency, modern high-grade tracing paper must be able to withstand considerable handling. The paper should retain these qualities for a long time to avoid the eventual necessity of redrawing, either manually or photographically.

If electrostatic reproduction or microfilming is used to produce prints or microfilm of the CAD- or manually-produced drawing, either opaque or translucent paper can be used as the drawing medium.

Polyester Film

The advantages of polyester film as a drawing material are many. Raw polyester has natural dimensional stability, strong resistance to tearing, high transparency, age and heat resistance, nonsolubility, and waterproofness. The outstanding virtue of film over any other drafting medium is that film is almost indestructible. Its amazing permanence safeguards the important investment in engineering drawings and records. It is permanently translucent, waterproof, unaffected by aging, superb for pencil drafting, ink work, and typewriting, and has unequaled erasability. The working, or drawing, side of the film has a matte surface for accepting pencil or ink and permitting erasures.

Polyester materials, however, present some problems. The material must be sufficiently dense to avoid reflection from the copyboard in microforming, but translucent enough for backlighting or contact printing.

Should a diazo (whiteprint) machine be used to make prints of the CAD or manually produced drawing, translucent film or paper is used as the drawing medium. However, reproducible prints (diazo intermediates) can be made from the original drawing and finished prints then made from the intermediates.

Preprinted Grid Sheets

This type of drawing paper makes the job of making manually prepared drawings easier and quicker. Cross-sectional lines printed directly on the paper or used as a liner beneath the drawing paper provide an accurate guide for all drawing work. These cross-sectional lines are available in several grid sizes. The squared and pictorial styles (isometric, perspective, and oblique) are the more common preprinted grid papers used by drafters. These grids, when applied directly on the paper, as shown in Fig. 3-1-1, or on film with special nonreproducible ink, will not appear on the prints when the drawing is reproduced by diazo or photographic methods.

For freehand work, whether rough sketches in the field or in the drafting room, whether these are preliminary or to be used as the finished drawings, gridline papers are an invaluable aid. The cross-sectional patterns serve as ready-made guides for base lines, dimensioning, and angles (Fig. 3-1-2).

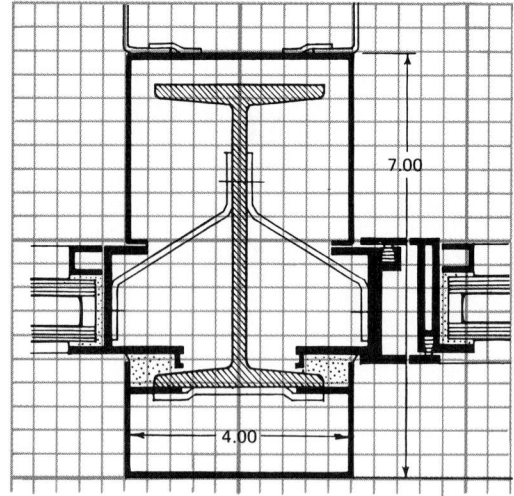

(A) DRAWING ON GRID SHEET

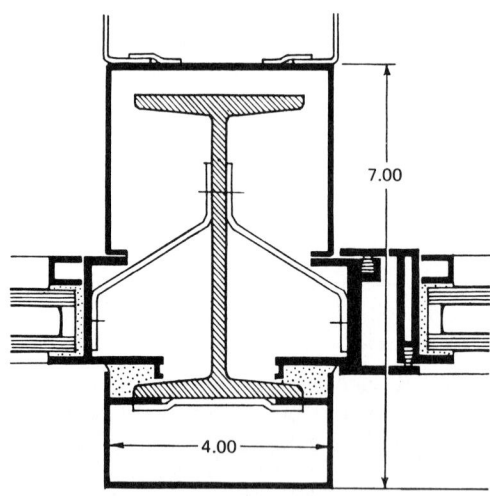

(B) PRINT OF DRAWING SHOWN IN (A)

FIG. 3-1-1   Drawing directly on preprinted grid sheet.

FIG. 3-1-2   Preprinted grid sheet being used as an underlay.
(Doug Martin)

Standard Drawing Sizes

Inches   Drawing sizes in the inch system are based on dimensions of commercial letterheads, 8.5 × 11 in., and standard rolls

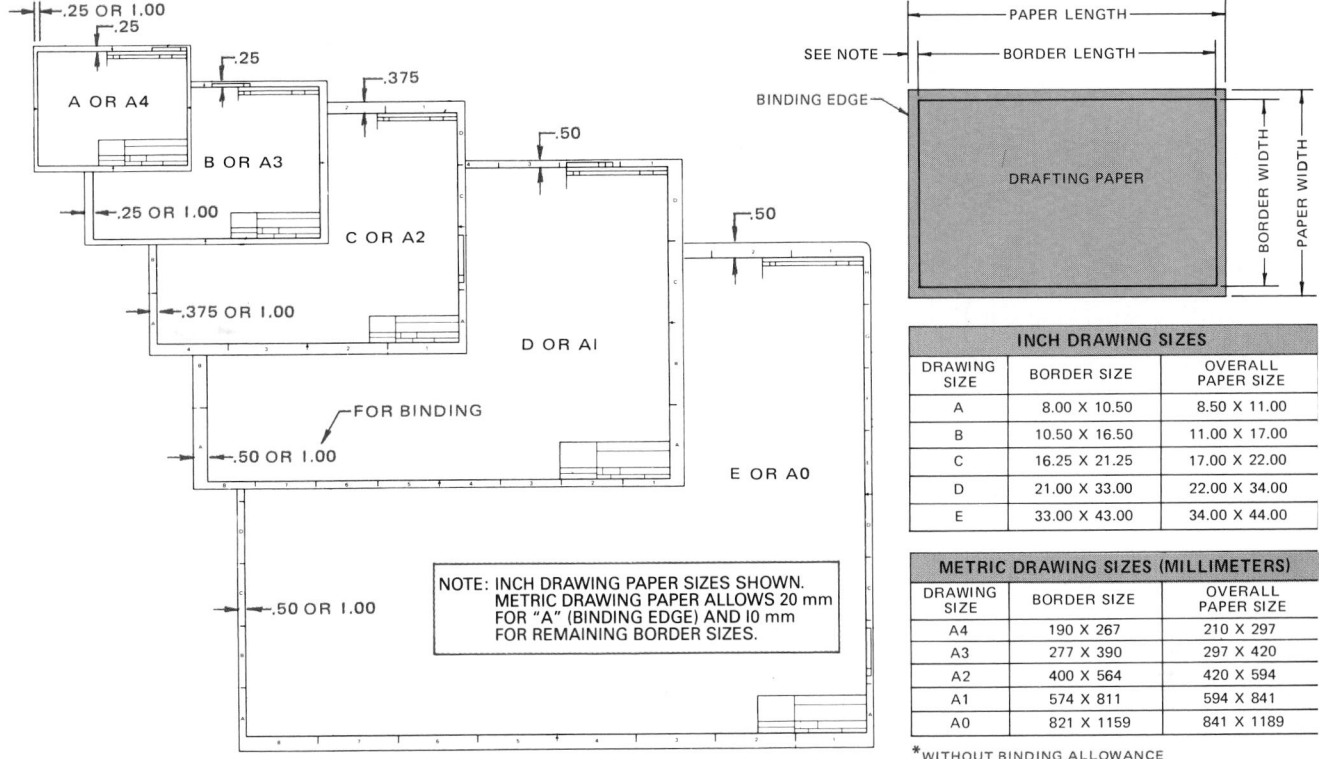

INCH DRAWING SIZES		
DRAWING SIZE	BORDER SIZE	OVERALL PAPER SIZE
A	8.00 X 10.50	8.50 X 11.00
B	10.50 X 16.50	11.00 X 17.00
C	16.25 X 21.25	17.00 X 22.00
D	21.00 X 33.00	22.00 X 34.00
E	33.00 X 43.00	34.00 X 44.00

METRIC DRAWING SIZES (MILLIMETERS)		
DRAWING SIZE	BORDER SIZE	OVERALL PAPER SIZE
A4	190 X 267	210 X 297
A3	277 X 390	297 X 420
A2	400 X 564	420 X 594
A1	574 X 811	594 X 841
A0	821 X 1159	841 X 1189

*WITHOUT BINDING ALLOWANCE

NOTE: INCH DRAWING PAPER SIZES SHOWN. METRIC DRAWING PAPER ALLOWS 20 mm FOR "A" (BINDING EDGE) AND 10 mm FOR REMAINING BORDER SIZES.

FIG. 3-1-3   Standard drawing sizes.

of paper or film, 36 and 42 in. wide. They can be cut from these standard rolls with a minimum of waste (Fig. 3-1-3).

Metric   Metric drawing sizes are based on the A0 size, having an area of 1 square meter (m^2) and a length-to-width ratio of $1:\sqrt{2}$. Each smaller size has an area half of the preceding size, and the length-to-width ratio remains constant (Fig. 3-1-4).

CAD   After *booting* the CAD system (preparing it to run an application program), the drawing size limits must be set prior to starting the drawing. These limits will be determined by the space that the object to be drawn will require. For example, a full scale single-view drawing of a small object will require only a small drawing size. A larger drawing size will be required to prepare a full-scale multiview drawing of a larger object. There are several standard drawing sizes from which to choose.

An alternate to this procedure would be to draw to full scale and then scale down the finished plot and insert it into an appropriate paper size.

A third alternative would be to draw the paper and border at whatever scale will fit the drawing and adjust the scale when plotting to the corrected plotted size.

Drawing Format

A general format for drawings is shown in Fig. 3-1-5 (pg. 38), which illustrates a drawing trimmed to size. It is recommended that preprinted drawing forms be made to the trimmed size and have rounded corners, as shown, to minimize dog-ears and tears.

Zoning System

Drawings larger than B size may be *zoned* for easy reference by dividing the space between the trimmed size and the inside border into zones measuring 4.25×5.50 in. These zones are numbered horizontally and lettered vertically, with uppercase letters, from the lower RH (right-hand) corner, as in Fig. 3-1-5, so that any area of the drawing can be identified by a letter and

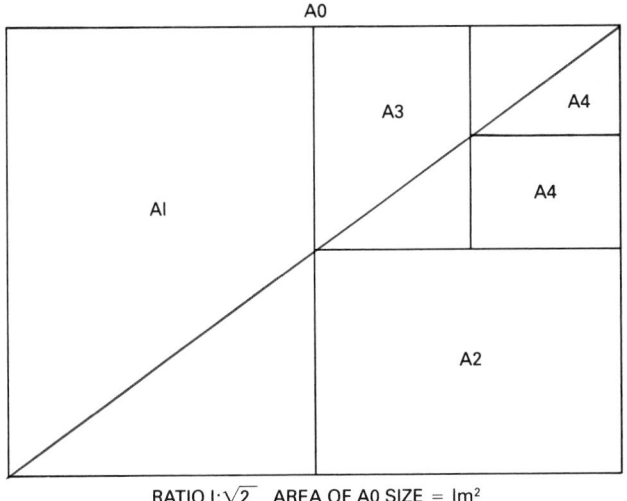

RATIO $1:\sqrt{2}$ AREA OF A0 SIZE = $1m^2$

FIG. 3-1-4   Metric drawing sizes.

FIG. 3-1-5   Drawing format.

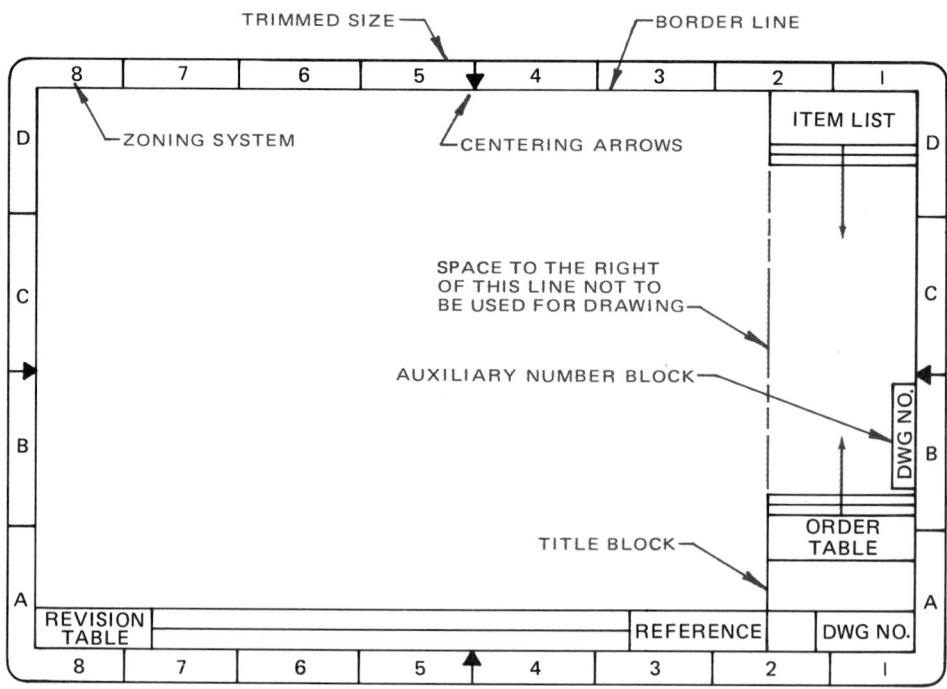

a number, such as B3, similar to reading a road map. Just as with maps, zoning is useful to locate fine detail on complex drawings.

Marginal Marking

In addition to zone identification, the margin may also carry fold marks to enable folding and graphical scale to facilitate reproduction to a specific size. In the process of microforming, it is necessary to center the drawing within rather close limits in order to meet standards. To facilitate this operation, it has become common practice to put a centering arrow or mark on at least three sides of the drawing. Most practices include the arrows on each of the four sides. If three sides are used, the arrows should be on the two sides and on the bottom. This helps the camera operator align the drawing properly since the copyboard usually contains cross hairs through the center of the board at right angles. With any three arrows aligned on the cross hairs, centering is automatic. The arrows should be on the center of the border that outlines the information area of the drawing, not at the edge of the sheet on which the drawing is made.

Title Block

Title blocks vary greatly and are usually preprinted. Drafters are rarely required to make their own.

The title block is located in the lower right-hand corner. The arrangement and size of the title block are optional, but the following information should be included:

1. Drawing number
2. Name of firm or organization
3. Title or description
4. Scale

Provision may also be made within the title block for the date of issue, signatures, approvals, sheet number, drawing size; job, order, or contract number; references to this or other documents; and standard notes, such as tolerances or finishes. An example of a typical title block is shown in Fig. 3-1-6. In classrooms, a title strip (Fig. 3-1-7) is often used on A and B size drawings.

Item (Material) List

The whole space above the title block, with the exception of the auxiliary number block, should be reserved for tabulating materials, change of order, and revision. Drawing in this space should be avoided. On preprinted forms, the right-hand inner border may be graduated to facilitate ruling for an item list. Fig. 3-1-8 shows a combined item list, order table, and title block.

Change or Revision Table

All drawings should carry a change or revision table, either down the right-hand side or across the bottom of the drawing.

NORDALE MACHINE COMPANY		
PITTSBURGH, PENNSYLVANIA		
COVER PLATE		
MATERIAL- MS		NO. REQD-4
SCALE- 1:2	DN BY *D Scott*	
DATE- 3/6/94	CH BY *B Jensen*	**A - 7628**

FIG. 3-1-6   Title block.

DRAFTING TECHNOLOGY	NAME:		DWG NAME:	DWG NO.
CALIFORNIA UNIVERSITY OF PENNSYLVANIA	COURSE:			
CALIFORNIA, PENNSYLVANIA	DATE:	APPD:	SCALE:	

FIG. 3-1-7   Title strip.

AMT	DET	STOCK SIZE	MAT.

NORDALE MACHINE COMPANY

PITTSBURGH, PENNSYLVANIA

MODEL _____

PART NAME _____

_____

PART NO. _____

OPERATION _____

FOR USE ON _____

METAL _____ DIE CLEARANCE _____

TOLERANCE: ±0.5mm UNLESS OTHERWISE SPECIFIED   METRIC

LAYOUT _____ DRAWN _____

CHECKED _____ APPROVED _____

SCALE _____ DATE _____

FROM B. P. _____ DATED _____

SHEETS	SHEET	**NO.**

(A) TYPICAL SET-UP

QTY	ITEM	MATL	DESCRIPTION	PT NO.
1	BASE	G1	PATTERN #A3154	1
1	CAP	G1	PATTERN # B7156	2
1	SUPPORT	AISI-1212	.38 × 2.00 × 4.38	3
1	BRACE	AISI-1212	.25 × 1.00 × 2.00	4
1	COVER	AISI-1035	.1345 (# 10 GA USS) × 6.00 × 7.50	5
1	SHAFT	AISI-1212	Ø1.00 × 6.50	6
2	BEARINGS	SKF	RADIAL BALL # 6200Z	7
2	RETAINING CLIP	TRUARC	N5000-725	8
1	KEY	STL	WOODRUFF # 608	9
1	SET SCREW	CUP POINT	HEX SOCKET .25UNC × 1.50	10
4	BOLT-HEX HD-REG	SEMI-FIN	.38UNC × 1.50LG	11
4	NUT-REG HEX	STL	.38UNC	12
4	LOCK WASHER-SPRING	STL	.38-MED	13

NOTE: Parts 7 to 13 are purchased items

(B) MATERIAL LIST APPLICATION

FIG. 3-1-8   Combined order table, item list, and title block.

In addition to the description of drawing changes, provision may be made for recording a revision symbol zone location, issue number, date, and approval of the change. Typical revision tables are shown in Fig. 3-1-9 (pg. 40).

Auxiliary Number Blocks

An auxiliary number block, approximately 2 × .25 in. (50 × 10 mm), is placed above the title block so that after prints are folded, the number will appear close to the top RH corner of the print, as in Fig. 3-1-5. This is done to facilitate identification when the folded prints are filed on edge.

Auxiliary number blocks are usually placed within the inside border, but they may be placed in the margin outside the border line if space permits.

REFERENCES AND SOURCE MATERIAL

1. Keuffel and Esser Co.
2. *Machine Design* and National Microfilm
3. Eastman Kodak Co.

3-2 FILING AND STORAGE

One of the most common and difficult problems facing an engineering department is how to set up and maintain an efficient engineering filing area. Normal office filing methods are not considered satisfactory for engineering drawings. To properly serve its function, an engineering filing area must meet

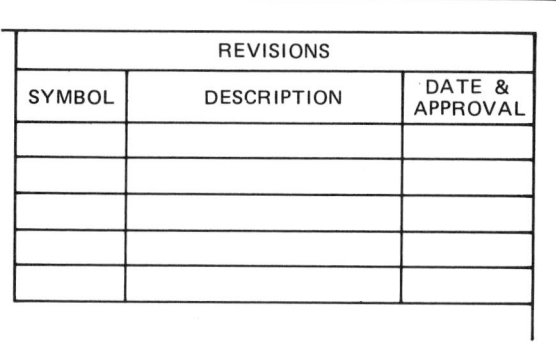

(A) VERTICAL REVISION TABLE

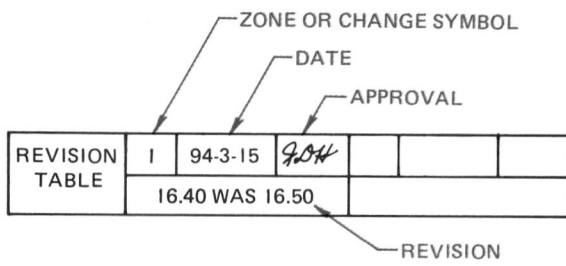

(B) HORIZONTAL REVISION TABLE

FIG. 3-1-9   Revision tables.

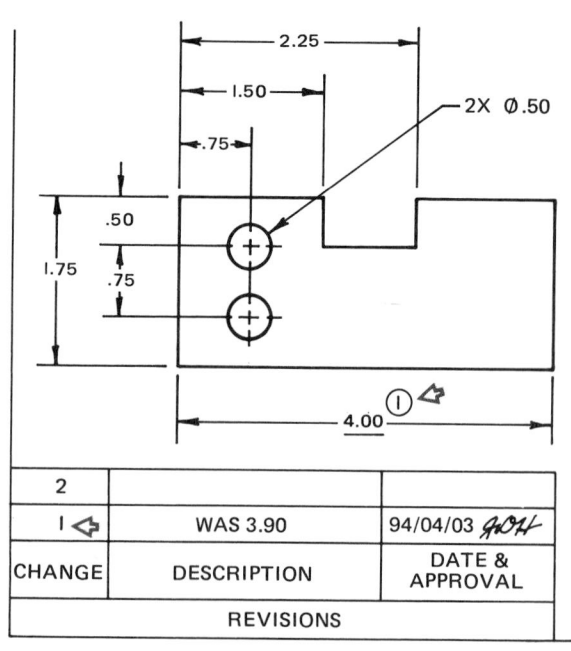

(C) APPLICATION

two important criteria: accessibility of information and protection of valuable documentation.

For this kind of system to be effective, drawings must be readily accessible. The degree of accessibility depends on whether drawings are considered active, semiactive, or inactive.

Filing Systems

Original Drawings

Unless a company has developed a full microforming system, the drafter's original drawings, produced either manually or by CAD, are kept and filed for future use or reference, and prints are made as required. To avoid crease lines, the originals, unlike prints, must *not* be folded. They are filed in either a flat or rolled position (Fig. 3-2-1).

In determining the type of equipment to use for engineering files, remember that different types of drawings require different kinds of files. Also, in planning a filing system, keep in mind that filing requirements are always increasing; unlike normal office files that can be purged each year, the more drawings produced, the more need to be stored. Therefore, any filing systems must have the flexibility of being easily expanded—usually in a minimum of space.

Microfilm Filing Systems

Although microfilming (Fig. 3-2-2) has been an established practice in many engineering offices for some time, the advent of CAD and new high-speed reproduction methods has made it less significant. It seems logical that reducing drawings to tiny images on film would make them more difficult to locate.

However, this is not the case, for while they are reduced in size, they are made more uniform. This results in improved file arrangements.

Forms of Film   One way to classify microfilm is according to the physical form in which it is used.

Roll Film   This is the form of the film after it has been removed from the camera and developed. Microfilm comes in four different widths—16, 35, 70, and 105 mm—and is stored in magazines.

Aperture Cards   Perhaps the simplest of the flat microforms is finished roll film cut into separate frames, each mounted on a card having a rectangular hole. Aperture cards are available in many sizes.

Jackets   Jackets are made of thin, clear plastic and have channels into which short strips of microfilms are inserted. They come in a variety of film-channel combinations for 16 and/or 35 mm microfilm. Like aperture cards, jackets can be viewed easily.

Microfiche   A microfiche is a sheet of clear film containing a number of microimages arranged in rows (Fig. 3-2-3). A common size is 100×150 mm, frequently arranged to contain 98 images. Microfiches are especially well suited for quantity distribution of standard information, such as parts and service lists.

CAD Storage

Original drawings in CAD are digital information and stored on magnetic media, such as tape, floppy disks, and hard disks, or

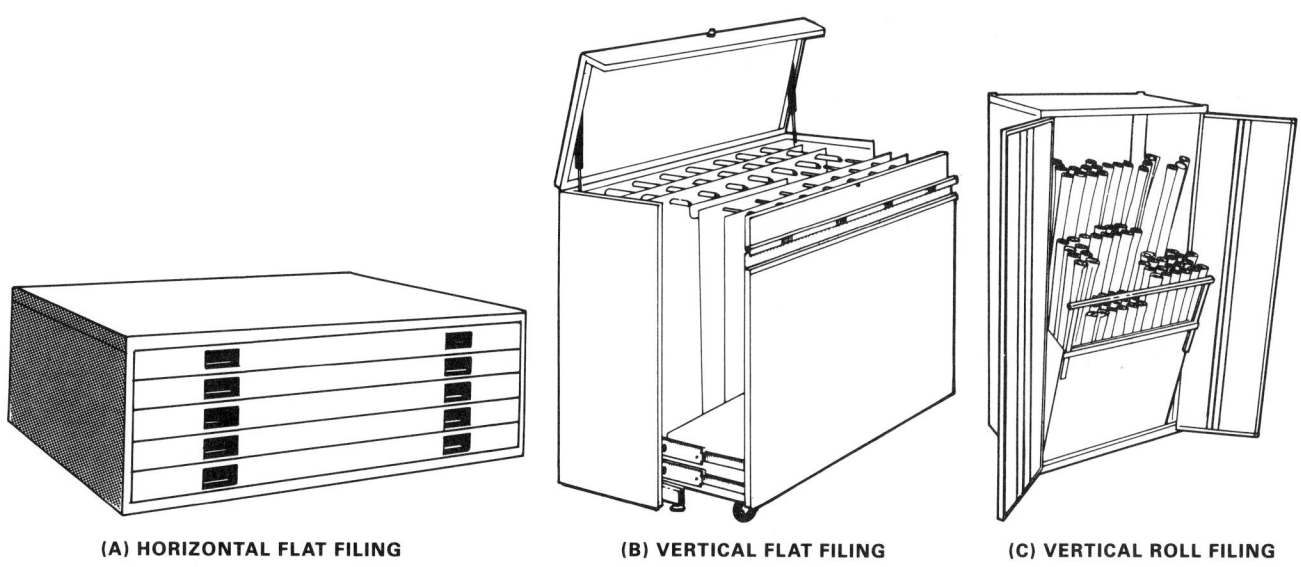

(A) HORIZONTAL FLAT FILING   **(B) VERTICAL FLAT FILING**   **(C) VERTICAL ROLL FILING**

FIG. 3-2-1   Filing systems for original drawings. *(Norman Wade Co. Ltd.)*

FIG. 3-2-2   Microfilming.
(Doug Martin)

FIG. 3-2-3   Microfiche.
(STUDIOHIO)

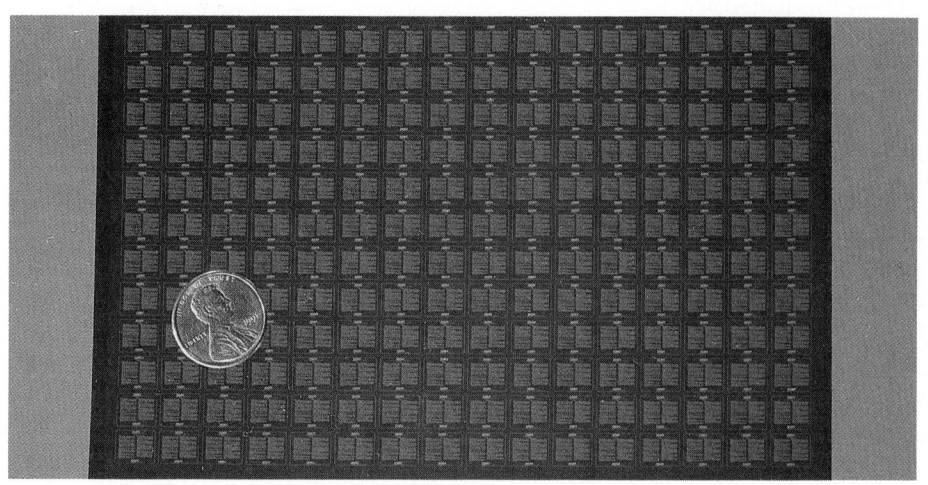

FIG. 3-2-4   Floppy disk with protective cover. *(STUDIOHIO)*

on optical media such as laser disks, CD-ROMs, or magneto-optical diskettes. Since magnetic media can be easily damaged, special procedures are needed to protect original drawings. One finished drawing plot or printer plot (often referred to as "hard copy") made on film, vellum, or paper may be stored as a permanent record in the same way that manually drawn originals are preserved. Optical disks are not easily damaged and make excellent permanent records.

Handling Diskettes

The diskettes used in personal computers and workstations for removable storage should be properly handled, labeled, and stored. The 3.5 in., double sided, high density diskette (Fig. 3-2-4) is the most common type of diskette and stores approximately 1.44 MB of data. All diskettes should be clearly labeled as to their content and ownership. Diskettes not in use should be properly stored in a diskette storage unit such as a file box or binder page. Magnetic media are not permanent, and multiple copies of diskettes with important data should be maintained.

The following are some rules for handling and storing diskettes or any other similar computer media:

1. Always properly label and store the diskette.
2. Never use a pencil to write on the diskette label. The loose graphite can damage the disk drive's read/write head.
3. Do not touch the disk surface with your fingers or anything else. This can ruin the surface of the diskette and make the data unreadable.
4. Keep diskettes away from magnetic fields such as motors, speakers, and some desk lamps.
5. Do not expose diskettes to temperature extremes. The safe temperature range for diskettes is between 50° and 140°F (10–60°C). Diskettes left in the direct sun will be destroyed by the high temperature.
6. Keep diskettes away from liquids and loose debris. Do not use a diskette that has been wet or dirty as the diskette may damage the drive.
7. Always make a backup diskette of important data (like your assignments).

Folding of Prints

To facilitate handling, mailing, and filing, prints should be folded to letter size, 8.5 × 11 in. (210 × 297 mm), in such a

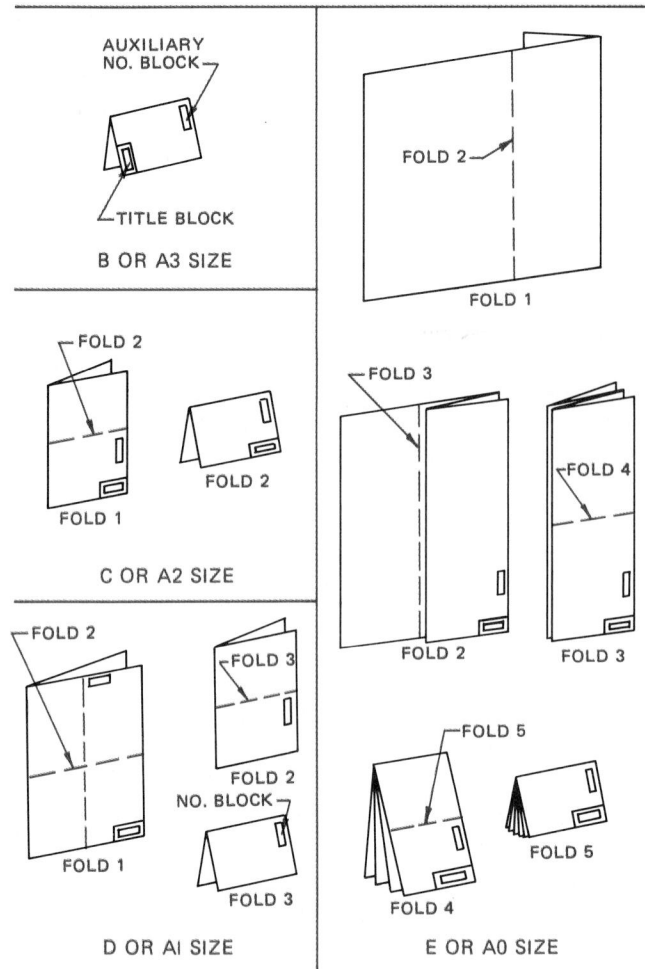

FIG. 3-2-5   Folding of prints.

way that the title block and auxiliary number always appear on the front face and the last fold is always at the top. In filing, this prevents other drawings from being pushed into the folds of filed prints. Recommended methods of folding standard-size prints are illustrated in Fig. 3-2-5.

On preprinted forms, it is recommended that fold marks be included in the margin of size B and larger drawings and be identified by number, for example, "fold 1," "fold 2." In zoned prints, the fold lines will coincide with zone boundaries, but they should nevertheless be identified.

To avoid loss of clarity by frequent folding, important details should not be placed close to fold areas. As a time-saver, some copiers are equipped to automatically fold prints.

REFERENCES AND SOURCE MATERIAL

1. Eastman Kodak Co. and "Setting Up and Maintaining an Effective Drafting Filing System," *Reprographics.*

3-3 DRAWING REPRODUCTION

A revolution in reproduction technologies and methods began several decades ago. It brought with it new equipment and supplies that have made quick copying commonplace. The introduction of high-quality, moderate-cost CAD printers and plotters in the 1990s has also influenced the choice of reproduction equipment. The new technologies make it possible to apply improved systems approaches and new information-handling techniques to all types of files, ranging from small documents to large engineering drawings (Fig. 3-3-1).

The pressures on business and government for greater efficiency, space savings, cost reductions, lower investment costs, and equally important factors provided a fertile field for the new reproduction technologies. There is no reason to believe that such pressures will diminish. In fact, as the years go by, it is certain that more and more improvements will occur, newer and better reproduction and information-handling equipment and methods will be discovered, and the advantages they offer will find ever-widening application.

The following reproduction methods apply whether the original drawing was prepared manually or by CAD (plotter).

Reproduction Equipment

Studies of reproduction facilities, existing or proposed, should first consider the nature of the demand for this service, then the processes which best satisfy the demand, and finally the particular machines which employ the processes. Factors to consider at these stages of study include:

- *Input originals*—sizes, paper mass, color, artwork
- *Quality of output copies*—depending on expected use and degree of legibility required
- *Size of copies*—same size, enlarged, reduced
- *Color*—copy paper and ink or pen
- *Registration*—in multiple-color work
- *Volume*—numbers of orders and copies per order
- *Speed*—machine productivity, convenient start-stop and load-unload
- *Cost*—direct labor, direct material, overhead, service
- *Future requirements*

Copiers

The methods used to produce copies are the diazo process, photoreproduction, copying with printer/plotters, and microfilming.

Diazo Process (Whiteprint)   In this process (Fig. 3-3-2) paper or film coated with a photosensitive diazonium salt is exposed to light passing through the original drawing made on a translucent paper or film. The exposed coated sheet is then developed by an ammonia vapor or agent. Where the light passes through the clear areas of the original drawing, it decomposes the diazonium salt, leaving a clear area on the copy (print). Where markings on the original block the light, the ammonia and the unexposed coating produce an opaque dye image of the original markings. A positive original makes a positive copy, and a negative original makes a negative copy. The three diazo processes currently used differ mainly in the way the developing agent is introduced to the diazo coat. These are ammonia vapor developing, moist developing, and pressure developing (PD). The most significant characteristic of the diazo process is that it is the most economical method of making prints. The main disadvantages of this process are that

FIG. 3-3-1   Drawing reproduction. *(Glencoe file photo)*

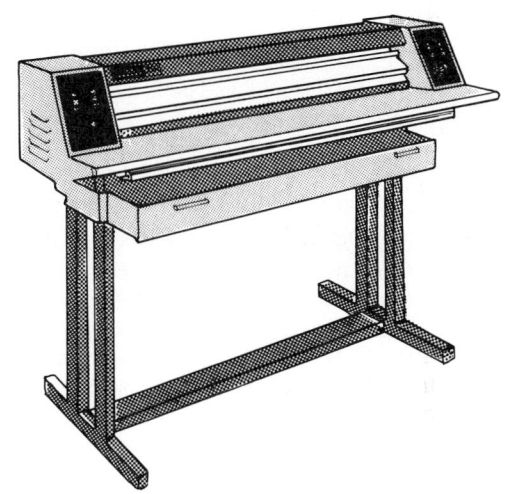

FIG. 3-3-2   Diazo (whiteprint) machine. *(Norman Wade Co. Ltd.)*

only full-size prints can be made and the original drawing being copied must be made on a translucent material.

Photoreproduction   Photoreproduction using an engineering plain-paper copier (Fig. 3-3-3, pg. 44) has become popular because there is no need to use transparent/translucent originals. It prints on bond paper, vellum, and drafting film, in sizes from 8.5 × 11 in. up to 36 in. wide by any manageable length. It combines high-speed productivity with many efficient, often automatic, features to rapidly deliver high-quality, large-size copies. The many advantages include:

1. No need for translucent originals. Copies are made equally well from opaque or translucent drawings.
2. No ammonia or developing agent necessary. Consequently, this method is environmentally safe.
3. High-volume copying. A standard office copier will produce clear multiple copies at the rate of several D or E size drawings per minute.

FIG. 3-3-3   Photoreproduction (plain-paper) copier. *(OCE-USA, Inc.)*

4. Large-size copies. In addition to same-size copies, there are reduction (less than 50 percent) and enlargement (up to 200 percent) options.
5. Time-saving functions. Some plain-paper copies will reduce, automatically fold, and collate the prints.
6. Ideal for cut-and-paste drafting.

The disadvantage of an engineering plain paper copier is the cost of service.

Digital Plain Paper Printer/Plotters   One of the newest and most versatile methods for reproducing engineering drawings is to plot and copy directly from CAD (Fig. 3-3-4). Using a Digital Laser system with CAD plotting allows you to bypass the plotter bottleneck. High-resolution, 600 dpi (dots per inch) engineering drawings are plotted directly from the CAD system with multiple copies produced in just seconds. A laser system, for example, is able to create a complex D size CAD plot in less than 15 seconds and identical copies at the rate of 12 prints per minute. This technology is an important step toward a fully integrated engineering/reprographics department. The disadvantage of this method is the cost of equipment. Plotters or printer/plotters should be used only when the number of reproduction copies is low and color is important.

Microfilm Equipment   Microfilm equipment includes readers and viewers and reader-printers.

Readers and Viewers   Microfilm readers magnify film images large enough to be read and project the images onto a translucent or opaque screen. Some readers accommodate only one microfilm (roll, jacket, microfiche, or aperture card). Others can be used with two or more. Scanning-type readers, having a variable-type magnification, are used when frames containing a large drawing are viewed. Generally, only parts of the drawing can be viewed at one time.

Reader-Printers   Two kinds of equipment are used to make enlarged prints from microfilm: reader-printers and

enlarger-printers. The *reader-printer* (Fig. 3-3-5) is a reader that incorporates a means of making hard copy from the projected image. The *enlarger-printer* is designed only for copying and does not include the means for reading.

Scanners   Many engineering offices have converted their manual drawings into CAD drawings. Completely redrawing each one is a tedious and time-consuming process. Even using a large graphics tablet and retracing the drawing by digitizing takes time and is expensive. The most effective means to accomplish this conversion is to use a scanner. Scanning technology allows you to feed in a manual drawing and convert the vectors (line lengths, circles, etc.) into the computer data base. The conversion is not perfect, so you will have to call up each drawing and edit it as needed.

REFERENCES AND SOURCE MATERIALS

1. Oce Bruning

FIG. 3-3-4   Laser printer/plotter. *(Courtesy of Xerox Corporation)*

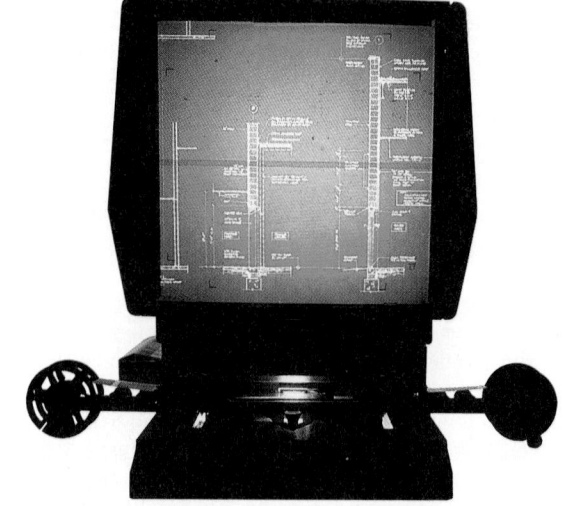

FIG. 3-3-5   Microfilm reader/printer. *(Doug Martin)*

CHAPTER 4

BASIC DRAFTING SKILLS

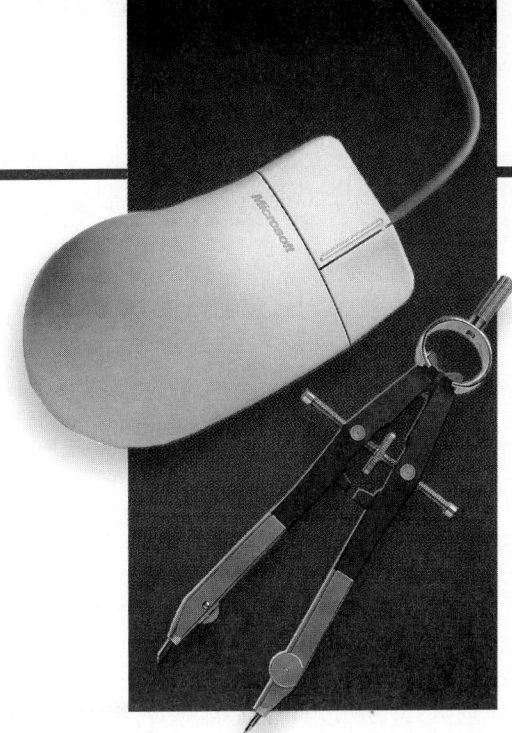

Definitions

Arcs   Parts of a circle.

Center lines   Lines consisting of alternating long and short dashes used to represent the axis of symmetrical parts and features, bolt circles, and paths of motion.

Concentric circles   Circles of different diameters that have the same center.

Construction lines   Lightly drawn lines used during the preliminary stages of a drawing.

Coordinates   Numerical information used to describe a specific point in an *X-Y* linear coordinate system.

Ellipses   Ovals.

Gothic-style lettering   The preferred style of lettering used in drafting.

Guidelines   Lines used to ensure uniform lettering.

Irregular curves   Nonconcentric, nonstraight lines drawn smoothly through a series of points.

Line   The most important single entity on a technical drawing.

Multiview or **orthographic projection**   A type of drawing in which an object is usually shown in more than one view.

Overlay   A piece of tracing paper used to redraw a portion of a sketch or drawing.

Pictorial drawing   A drawing which shows the width, height, and depth of an object in one view.

Radius   The distance from the center of a circle to its edge.

Refined sketch   A neatly drawn, finished-looking freehand sketch.

Tangent arcs   Parts of two circles that touch.

Unrefined or **rough sketch**   A quickly drawn, freehand sketch.

Visible lines   Clearly drawn lines used to represent visible edges or contours of objects.

4-1 STRAIGHT LINE WORK, LETTERING, AND ERASING

Manual Drafting

Line Work

A line is the fundamental, and perhaps the most important, single entity on a technical drawing. Lines are used to help illustrate and describe the shape of objects that will later become real parts. The various lines used in drawing form the "alphabet" of the drafting language. Like letters of the alphabet, they are different in appearance (see Figs. 4-1-1 and 4-1-2, pp. 46–48). The distinctive features of all lines that form a permanent part of the drawing are the differences in their width and construction. Lines must be clearly visible and stand out in sharp contrast to one another. This line contrast is necessary if the drawing is to be clear and easily understood.

The drafter first draws very light construction lines, setting out the main shape of the object in various views. Since these first lines are very light, they can be erased easily should changes or corrections be necessary. When the drafter is satisfied that the layout is accurate, the construction lines are then changed to their proper type, according to the alphabet of lines. Guidelines, used to ensure uniform lettering, are also drawn very lightly.

TYPE OF LINE	APPLICATION	DESCRIPTION
HIDDEN LINE — — — — THIN — — — —		THE HIDDEN OBJECT LINE IS USED TO SHOW SURFACES, EDGES, OR CORNERS OF AN OBJECT THAT ARE HIDDEN FROM VIEW.
CENTER LINE THIN ALTERNATE LINE AND SHORT DASHES	CENTER LINE	CENTER LINES ARE USED TO SHOW THE CENTER OF HOLES AND SYMMETRICAL FEATURES.
SYMMETRY LINE CENTER LINE THICK SHORT LINES	SYMMETRY LINE	SYMMETRY LINES ARE USED WHEN PARTIAL VIEWS OF SYMMETRICAL PARTS ARE DRAWN. IT IS A CENTER LINE WITH TWO THICK SHORT PARALLEL LINES DRAWN AT RIGHT ANGLES TO IT AT BOTH ENDS.
EXTENSION AND DIMENSION LINES THIN DIMENSION LINE EXTENSION LINE		EXTENSION AND DIMENSION LINES ARE USED WHEN DIMENSIONING AN OBJECT.
LEADERS ARROW · DOT THIN		LEADERS ARE USED TO INDICATE THE PART OF THE DRAWING TO WHICH A NOTE REFERS. ARROWHEADS TOUCH THE OBJECT LINES WHILE THE DOT RESTS ON A SURFACE.
BREAK LINES THIN LONG BREAK THICK SHORT BREAK		BREAK LINES ARE USED WHEN IT IS DESIRABLE TO SHORTEN THE VIEW OF A LONG PART.
CUTTING-PLANE LINE THICK OR		THE CUTTING-PLANE LINE IS USED TO DESIGNATE WHERE AN IMAGINARY CUTTING TOOK PLACE.

FIG. 4-1-1   Types of lines.

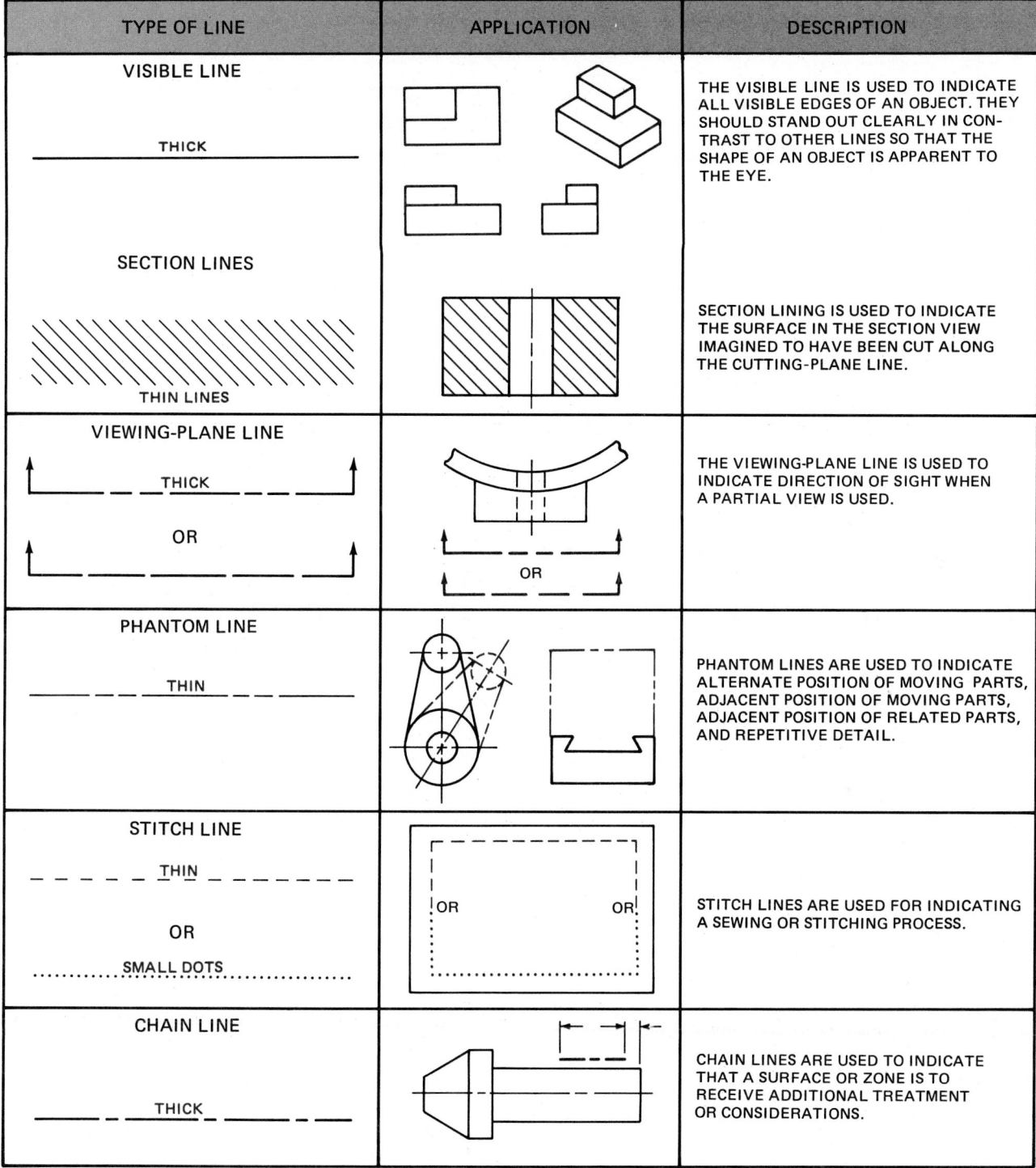

TYPE OF LINE	APPLICATION	DESCRIPTION
VISIBLE LINE THICK		THE VISIBLE LINE IS USED TO INDICATE ALL VISIBLE EDGES OF AN OBJECT. THEY SHOULD STAND OUT CLEARLY IN CONTRAST TO OTHER LINES SO THAT THE SHAPE OF AN OBJECT IS APPARENT TO THE EYE.
SECTION LINES THIN LINES		SECTION LINING IS USED TO INDICATE THE SURFACE IN THE SECTION VIEW IMAGINED TO HAVE BEEN CUT ALONG THE CUTTING-PLANE LINE.
VIEWING-PLANE LINE THICK OR	OR	THE VIEWING-PLANE LINE IS USED TO INDICATE DIRECTION OF SIGHT WHEN A PARTIAL VIEW IS USED.
PHANTOM LINE THIN		PHANTOM LINES ARE USED TO INDICATE ALTERNATE POSITION OF MOVING PARTS, ADJACENT POSITION OF MOVING PARTS, ADJACENT POSITION OF RELATED PARTS, AND REPETITIVE DETAIL.
STITCH LINE THIN OR SMALL DOTS	OR OR	STITCH LINES ARE USED FOR INDICATING A SEWING OR STITCHING PROCESS.
CHAIN LINE THICK		CHAIN LINES ARE USED TO INDICATE THAT A SURFACE OR ZONE IS TO RECEIVE ADDITIONAL TREATMENT OR CONSIDERATIONS.

FIG. 4-1-1   Types of lines. (continued)

Line Widths   Two widths of lines, thick and thin, as shown in Fig. 4-1-3 (pg. 49), are recommended for use on drawings. Thick lines are .030 to .038 in. (0.5 to 0.8 mm) wide, thin lines between .015 and .022 in. (0.3 to 0.5 mm) wide. The actual width of each line is governed by the size and style of the drawing and the smallest size to which it is to be reduced. All lines of the same type should be uniform throughout the drawing.

Spacing between parallel lines should be such that there is no "fill-in" when the copy is reproduced by available photographic methods. Spacing of no less than .12 in. (3 mm) normally meets reproduction requirements.

All lines should be sharp, clean-cut, opaque, uniform, and properly spaced for legible reproduction by all commonly used methods, including microforming, in accordance with industry

FIG. 4-1-2   Application of lines. *(ANSI Y14.2M, 1982)*

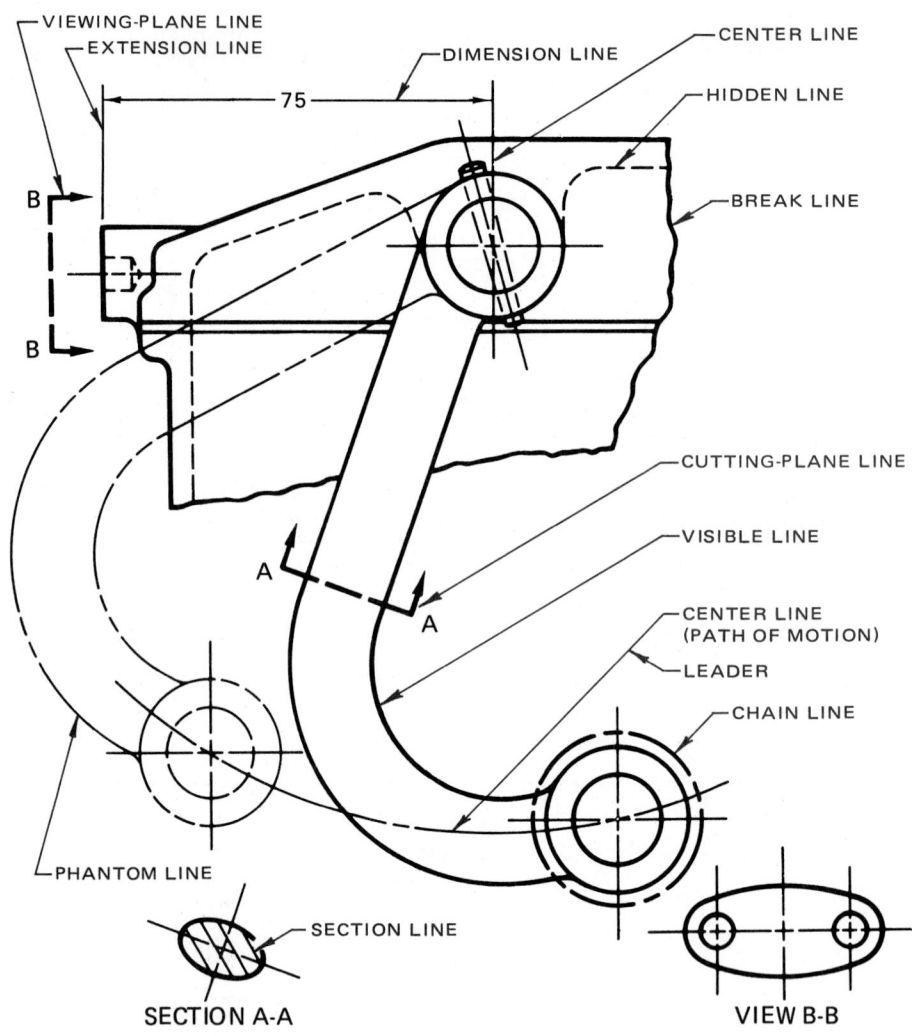

VIEWING-PLANE LINE
EXTENSION LINE
DIMENSION LINE
CENTER LINE
HIDDEN LINE
BREAK LINE
CUTTING-PLANE LINE
VISIBLE LINE
CENTER LINE (PATH OF MOTION)
LEADER
CHAIN LINE
PHANTOM LINE
SECTION LINE

SECTION A-A   VIEW B-B

and government requirements. There should be a distinct contrast between the two widths of lines.

Visible Lines   The "visible lines" should be used for representing visible edges or contours of objects. Visible lines should be drawn so that the views they outline clearly stand out on the drawing with a definite contrast between these lines and secondary lines.

The applications of the other types of lines are explained in detail throughout this text.

Drawing Straight Lines   When using a T square to draw horizontal lines (Fig. 4-1-4), hold the head of the T square against the edge of the drawing board and slide the T square either up or down to the desired position. Firmly press down on the blade of the T square to prevent it from moving, then proceed to draw the line. When drawing vertical lines, a triangle, which rests on the top side of the T square, is moved to the desired position and both the blade of the T square and the triangle are held firmly to the drawing board with the hand not holding the pencil.

When a parallel slide is used, as in Fig. 4-1-5, it will always be in a horizontal position as the wire and rollers in the slide move both ends of the slide simultaneously and at the same speed.

A general rule to follow when drawing straight lines is to lean the pencil in the direction of the line you are about to draw. A right-handed person would lean the pencil to the right and draw horizontal lines from left to right. The left-handed person would reverse this procedure. When drawing vertical lines, lean the pencil away from yourself, toward the top of the drafting board, and draw lines from bottom to top. Lines sloping from the bottom to the top right are drawn from bottom to top; lines sloping from the bottom to the top left are drawn from top to bottom. This procedure for drawing sloping lines would be reversed for a left-handed person.

When using a conical-shaped lead, rotate the pencil slowly between your thumb and your forefinger when drawing lines (Fig. 4-1-6). This keeps the lines uniform in width and the pencil sharp. Do not rotate a pencil having a bevel or wedge-shaped lead.

Many drafters today use the 0.5 mm mechanical pencil. Holding the pencil perpendicular to the paper, the drafter can produce uniform lines easily. The pencil is not rotated for this procedure. Pencils and leads are available from 0.2 to 0.9 mm in diameter for creating different line widths.

THICK

WIDTH .032 IN. (0.7 mm)

THIN

WIDTH .016 IN. (0.35 mm)

FIG. 4-1-3   Line widths.

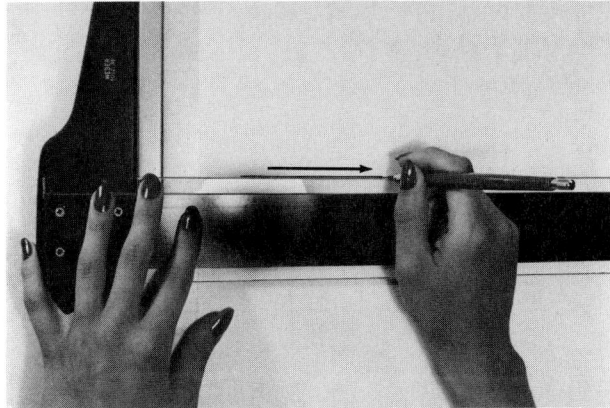

(A) DRAWING A HORIZONTAL LINE

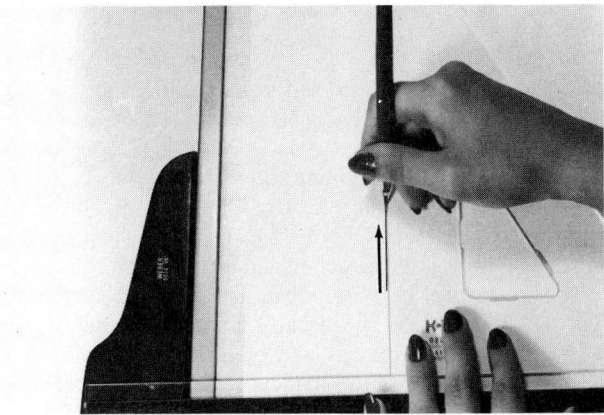

(B) DRAWING A VERTICAL LINE

FIG. 4-1-4   Drawing horizontal and vertical lines.

FIG. 4-1-5   Drawing horizontal and sloping lines with the aid of a parallel slide.

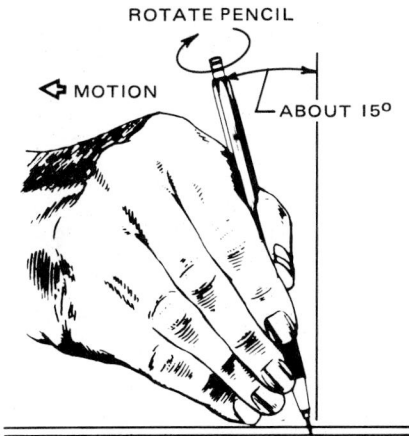

FIG. 4-1-6   Rotate the pencil when drawing lines.

Lettering

Single-Stroke Gothic Lettering   The most important requirements for lettering are legibility, reproducibility, and ease of execution. These are particularly important because of the increased use of microforming, which requires optimum clarity and adequate size of all details and lettering. It is recommended that all drawings be made to conform to these requirements and that particular attention be paid to avoid the following common faults:

1. Unnecessarily fine detail
2. Poor spacing of details
3. Carelessly drawn figures and letters
4. Inconsistent delineation
5. Incomplete erasures that leave ghost images
6. Use of differing densities, such as pencil, ink, and typescript on the same drawing

These requirements are met in the recommended single-stroke Gothic characters shown in Fig 4-1-7 (pg. 50) or adaptations thereof, which improve reproduction legibility. One such adaptation by the National Microfilm Association is the vertical Gothic style Microfont alphabet (Fig. 4-1-8, pg. 50) intended for general use.

Either inclined or vertical lettering is permissible, but only one style of lettering should be used throughout a drawing. The preferred slope for the inclined characters is 2 in 5, or approximately 68° with the horizontal.

Uppercase letters should be used for all lettering on drawings unless lowercase letters are required to conform with other established standards, equipment nomenclature, or marking.

Lettering for titles, subtitles, drawing numbers, and other uses may be made freehand, by typewriter, or with the aid of mechanical lettering devices, such as templates and lettering machines. Regardless of the method used, all characters are to conform, in general, with the recommended Gothic style and must be legible in full- or reduced-size copy by any accepted method of reproduction.

Letters in words should be spaced so that the background areas between the letters are approximately equal, and words are to be clearly separated by a space equal to the height of the lettering (Fig. 4-1-9, pg. 50). The vertical space between lines

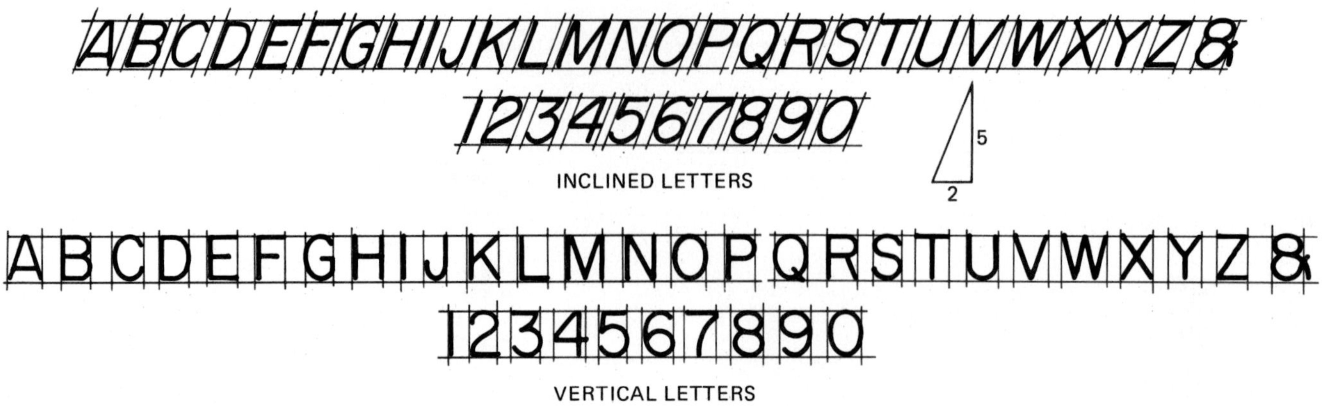

INCLINED LETTERS

VERTICAL LETTERS

FIG. 4-1-7   Approved Gothic lettering for engineering drawings.

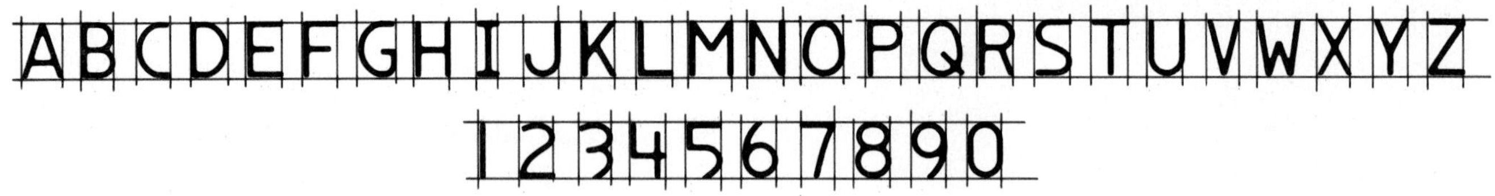

FIG. 4-1-8   Microfont letters. *(National Microfilm Assoc.)*

of lettering should be no more than the height of the lettering and no less than half the height of the lettering.

The recommended minimum freehand and mechanical letter heights for various applications are given in Fig 4-1-10. So that lettering will be uniform and of proper height, light guidelines, properly spaced, are drawn first and then the lettering is drawn between these lines.

Notes should be placed horizontally on drawings and separated vertically by spaces at least equal to double the height of the character size used, to maintain the identity of each note.

Decimal points must be uniform, dense, and large enough to be clearly visible on approved types of reduced copy. Decimal points should be placed in line with the bottom of the associated digits and be given adequate space.

Lettering should not be underlined except when special emphasis is required. The underlining should not be less than .06 in. (1.5 mm) below the lettering.

When drawings are being made for microforming, the size of the lettering is an important consideration. A drawing may be reduced to half size when microformed at 30X reduction and blown back at 15X magnification. (Most microform engineering readers and blowback equipment have a magnification

of 15X. If a drawing is microformed at 30X reduction, the enlarged blown-back image is 50 percent; at 24X, it is 62 percent of its original size.)

Standards generally do not allow characters smaller than .12 in. (3 mm) for drawings to be reduced 30X, and the trend is toward larger characters. Figure 4-1-11 shows the proportionate size of letters after reduction and enlargement.

The lettering heights, spacing, and proportions in Figs. 4-1-10 and 4-1-11 normally provide acceptable reproduction or camera reduction and blowback. However, manually, mechanically, optimechanically, or electromechanically applied lettering (typewriter, etc.) with heights, spacing, and proportions less than those recommended are acceptable when the reproducibility requirements of the accepted industry or military reproduction specifications are met.

Erasing Techniques

Revision or change practice is inherent in the method of making engineering drawings. It is much more economical to introduce changes or additions on an original drawing than to redraw the entire drawing. Consequently, erasing has become a science all

FIG. 4-1-9   Spacing of lettering. *(National Microfilm Assoc.)*

GOOD SPACING OF CHARACTERS AND EVEN *LINE WEIGHT PRODUCE CONSISTENTLY GOOD RESULTS ON MICROFILM*

PREFERRED
(OPEN-TYPE LETTERING)

POORLY SPACED AND FORMED, OR CRAMPED *LETTERING MEANS POOR RESULTS IN MICROFILMING*

UNDESIRABLE
(CRAMPED LETTERING)

USE	INCH		METRIC mm		DRAWING SIZE
	FREEHAND	MECHANICAL	FREEHAND	MECHANICAL	
DRAWING NUMBER IN TITLE BLOCK	0.250	0.240	7	7	UP TO AND INCLUDING 17 x 22 INCHES
	0.312	0.290			LARGER THAN 17 x 22 INCHES
DRAWING TITLE	0.250	0.240	7	7	ALL
SECTION AND TABULATION LETTERS	0.250	0.240	7	7	
ZONE LETTERS AND NUMERALS IN BORDER	0.188	0.175	5	5	
DIMENSION, TOLERANCE, LIMITS, NOTES, SUBTITLES FOR SPECIAL VIEWS, TABLES, REVISIONS, AND ZONE LETTERS FOR THE BODY OF THE DRAWING	0.125	0.120	3.5	3.5	UP TO AND INCLUDING 17 x 22 INCHES
	0.156	0.140	5	5	LARGER THAN 17 x 22 INCHES

THIS IS AN EXAMPLE OF .125 IN. LETTERING

THIS IS AN EXAMPLE OF .188 IN. LETTERING

THIS IS AN EXAMPLE OF .250 IN. LETTERING

FIG. 4-1-10   Recommended lettering heights. *(ANSI Y14.2M, 1982)*

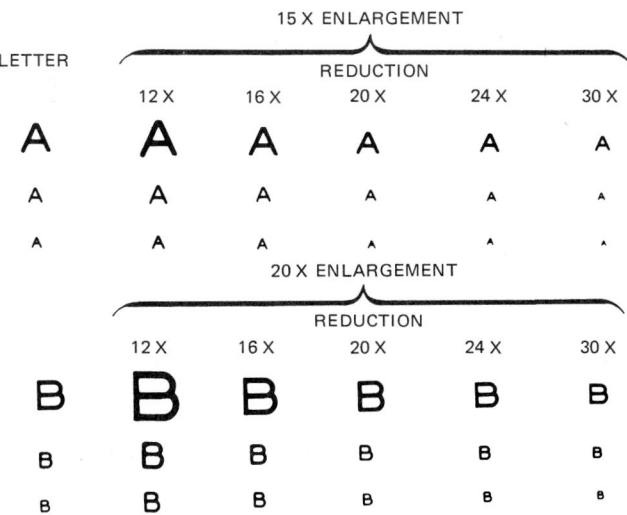

FIG. 4-1-11   Proportionate size of letters after reduction and enlargement. *(National Microfilm Assoc.)*

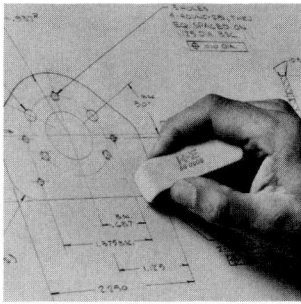

(A) USE OF ERASING SHIELD   (B) CLEANING UP AN AREA

FIG. 4-1-12   Erasing. *(Keuffel & Esser Co.)*

its own. Proper erasing is extremely important since some drawings are revised a great number of times. Consequently, good techniques and materials must be used that permit repeated erasures on the same area. Some recommendations follow.

1. Avoid damaging the surface of the drawing medium by selecting the proper eraser.
2. Lines not thoroughly erased produce ghostlike images on prints, resulting in reduced legibility.
3. A hard, smooth surface such as a triangle, placed under the lines being removed makes erasing easier.
4. Using an erasing shield protects the adjacent lines and lettering and also eliminates wrinkling (Fig. 4-1-12).

5. Also erase on the back side of the paper. Lines frequently pick up dirt or graphite on the underside, and if not erased, will still produce lines on the print.
6. Be sure to completely remove erasure debris from the drawing surface.
7. When extensive changes are required, it may be more economical to cut and paste or make an intermediate drawing.
8. When erasing, use no more pressure than necessary.
9. If the drawing quality of the paper or other medium has been damaged by erasing, it may be improved by sprinkling an inking powder on the surface and rubbing it with a cloth.

In addition, it is necessary to match the density of the surrounding background when erasures are made. Often, the erased area is much cleaner than the rest of the drawing. If the change is made on this clean area, the contrast between line and background is different and that area presents a problem in reproduction. It is usual practice to "smudge" the erased area so that it looks about the same shade as the surrounding area.

Removing Lines on Film   Lines on photoreproduction film fall into two classifications: photographic lines and pencil-and-ink lines. All these lines can be removed easily so that the erased area can be used for further drafting. Here are some tips for removing lines.

Erasers   There are three basic types of erasers: rubber, plastic, and liquid. Rubber and plastic erasers may tend to cause a shine on the drafting surface. When plastic erasers are used, the shiny appearance may actually be transparentizing, because of the plasticizers used in their manufacture, rather than wearing on the drafting surface. The transparentizing effect is not detrimental since it does not reduce the ability of the surface to take a pencil line. Good drafting lines can easily be drawn over areas from which lines have been erased many times. A good general rule to follow is to use a soft, nonabrasive eraser and only enough pressure to remove the line. Liquid erasers do not put a shine on the drafting surface and can be used to make many erasures in the same place.

Erasing machines require special techniques. Many different types of erasing inserts are available. To avoid burning holes in paper or melting the matte surface on film, you must use a light touch and keep the eraser moving.

When the drafting surface is affected by excessive erasures, it can be repaired by rolling a regular typewriter eraser across the smooth area or by rubbing a small amount of drafting powder into the area with a finger.

Pencil Lines   Pencil lines can be removed from all film with a soft, non-abrasive rubber or plastic eraser or with liquid eraser. To keep the drafting surface from becoming too shiny, avoid excessive pressure.

Fastening Paper to the Board

The most common method of holding the drawing paper to the drafting board is with drafting tape.

When fastening the paper to the board, line up the bottom or top edge of the paper with the top horizontal edge of the T square, parallel slide, or horizontal scale of the drafting machine (Fig. 4-1-13). When refastening a partially completed drawing, use lines on the drawing rather than the edge of the paper for alignment.

 CAD

Skills in drawing lines and lettering are not required with CAD. Rather, skill in operating the CAD equipment becomes essential.

Before learning the fundamentals of drafting, you should become familiar with the basics of line creation.

Coordinate Input

One of the most common ways to create lines using CAD is by coordinate input (keying in distances). There are three methods of coordinate input:

1. Absolute coordinate
2. Relative coordinate
3. Polar coordinate (line length and angle)

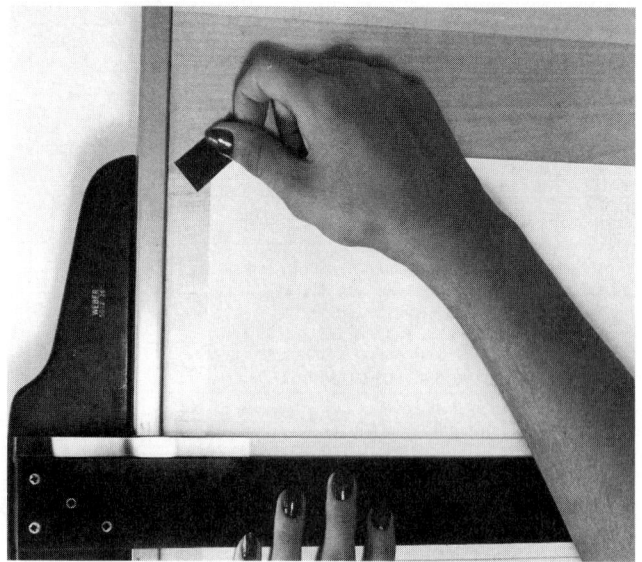

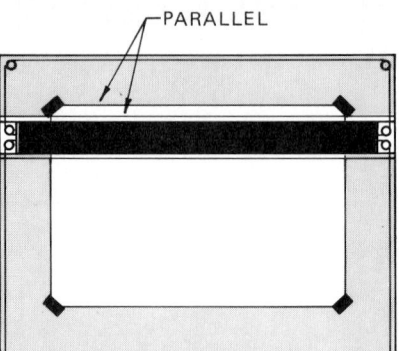

FIG. 4-1-13   Positioning the paper on the board.

Absolute Coordinates   Coordinate input is based on the rectangular (horizontal and vertical) measurement system. All absolute distances are described in terms of their distance from the drawing origin. Relative and polar coordinates may be described with respect to a particular point (position) on the drawing. This position can be at any location on the drawing and is described by two-dimensional coordinates, horizontal (X) and vertical (Y). The X axis is horizontal and is considered the first and basic reference axis. The Y axis is vertical and is 90° to the X axis. Any distance to the right of the drawing origin is considered a positive X value and any distance to the left of the drawing origin a negative X value. Distances above the drawing origin are positive Y values and distances below the drawing origin are negative Y values.

For example, four points lie in a plane, as shown in Fig. 4-1-14. The plane is divided into four quadrants. Point A lies in quadrant 1 and is located at position (6, 5), with the X coordinate first, followed by the Y coordinate. Point B lies in quadrant 2 and is located at position (−4, 3). Point C lies in quadrant 3 and is located at position (−5, −4). Point D lies in quadrant 4 and is located at position (3, −2).

Using the coordinate system to locate a point on a drawing would be greatly simplified if all parts of the drawing were located in the first quadrant because all the values would be positive and the plus and minus signs would not be required.

For this reason, on CAD systems the drawing origin is normally located at the lower left of the monitor. Thus the origin is a base reference point from which all positions on the drawing are measured. Its dimensions are $X = 0$ and $Y = 0$, referred to as 0,0. Figure 4-1-15 shows two points located on a monitor screen. Point 1 (shown as a cross) has an X coordinate of 2.50 and a Y coordinate of 6.00 with reference to the origin (0,0). Point 2 has an X coordinate of 8.35 and a Y coordinate of 2.20 with reference to the origin.

Relative Coordinates   A relative coordinate is located with respect to the current access location (last cursor position selected) rather than the origin (0,0). In other words, it is located with respect to another point on your drawing. Often, both absolute and relative coordinate input are used during drawing preparation. A line may be created by combining these two methods.

An example of this type of dimensioning is shown in Fig. 4-1-16. Point 1 was located in the identical way in which point 1 in Fig. 4-1-15 was located, that is, using absolute coordinates. In Fig. 4-1-16, point 2 is located relative to point 1. Thus the relative coordinates of point 2 would be 5.85,−3.80 (5.85 to the right and 3.80 below the last position). Note the minus sign in front of the 3.80 dimension. Point 2 in both figures is in the same position relative to the origin.

Polar Coordinates   A polar coordinate is similar to a relative coordinate since it is positioned with respect to the current access location. A line, however, will be specified according to its actual length and a direction rather than as X, Y coordinate distance. The direction is measured angularly in a counterclockwise direction from a horizontal line. Zero degrees is located horizontally to the right (at 3 o'clock) as shown in Fig. 4-1-17.

For example, a line 6.50 in. long to be drawn from point E to point F, as shown in Fig. 4-1-17A, would have polar coordinates of 6.50 and 45°. The line drawn from point G to point H (Fig. 4-1-17B) would have polar coordinates of 8.00 and 210°.

Line Styles, Text, and Erasing

All CAD systems have the options to create or apply line styles and text. The LINETYPE command allows you to select the desired line style. The TEXT command allows you to add

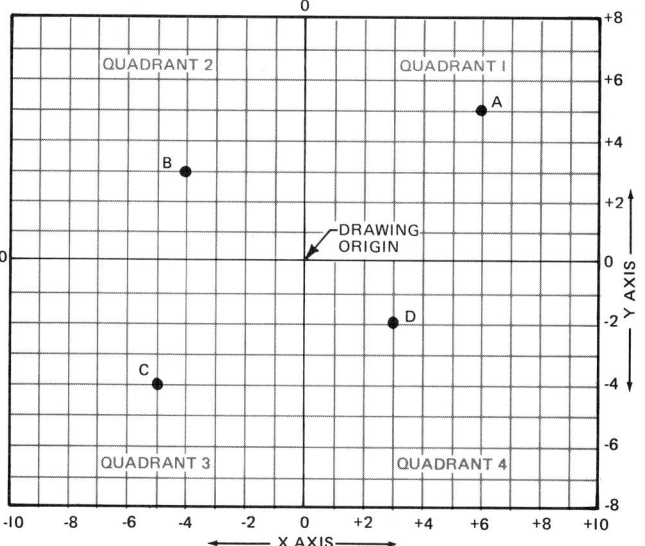

FIG. 4-1-14   Two-dimensional coordinates (X and Y).

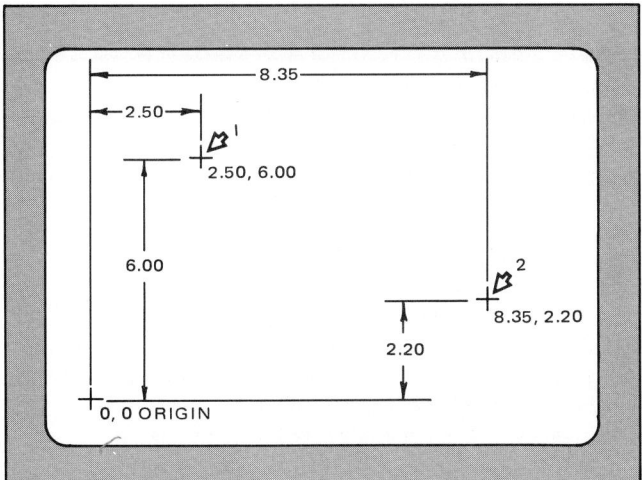

FIG. 4-1-15   Point locations shown on CAD monitor using absolute coordinates.

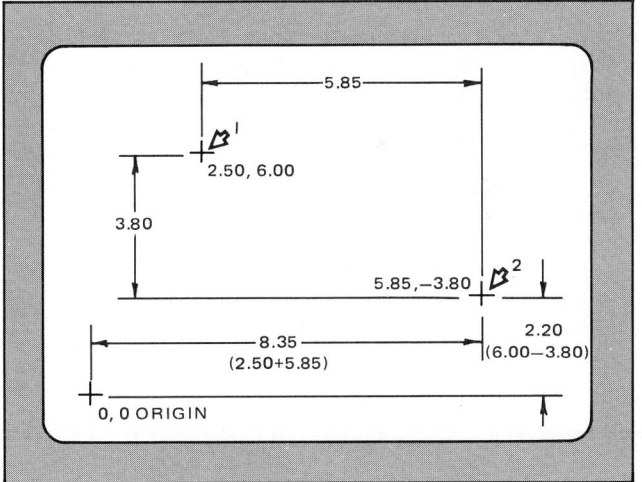

FIG. 4-1-16   Locating point 2 from point 1 using the relative coordinate option.

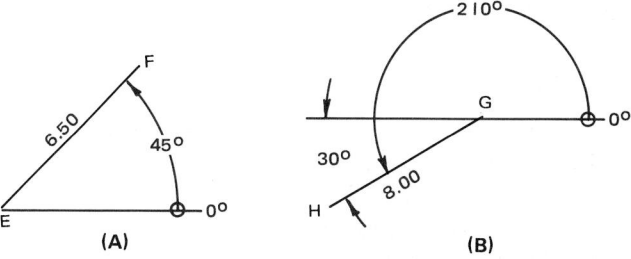

FIG. 4-1-17   Polar coordinates.

53

letters, numbers, words, notes, symbols, and messages to the drawing. Additionally, it can be used for size description (lengths, diameters, etc.).

As noted earlier, the need for revision or change is inherent in preparing technical drawings. Consequently, a variety of editing commands for removing unwanted lines and lettering from the drawing are available on all CAD systems.

REFERENCES AND SOURCE MATERIAL

1. National Microfilm Association
2. Keuffel and Esser Co.
3. Eastman Kodak Co.

ASSIGNMENTS

See Assignments 1 through 14 for Unit 4-1 on pages 65–70.

4-2 CIRCLES AND ARCS

Center Lines

Center lines consist of alternating long and short dashes (Fig. 4-2-1). They are used to represent the axis of symmetrical parts and features, bolt circles, and paths of motion. The long dashes of the center lines may vary in length, depending upon the size of the drawing. Center lines should start and end with long dashes and should not intersect at the spaces between dashes. Center lines should extend uniformly and distinctly a short distance beyond the object or feature of the drawing unless a longer extension is required for dimensioning or for some other purpose. They should not terminate at other lines of the drawing, nor should they extend through the space between views.

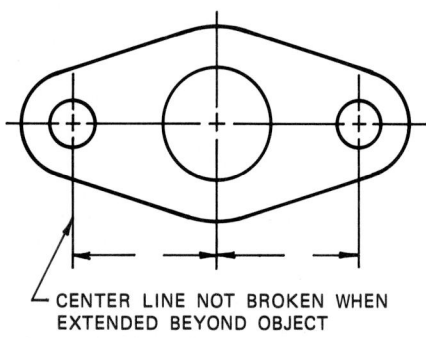

— CENTER LINE NOT BROKEN WHEN EXTENDED BEYOND OBJECT

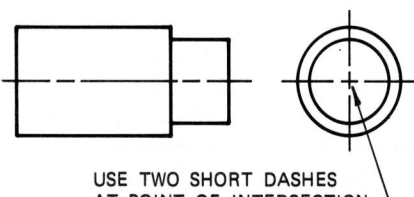

USE TWO SHORT DASHES AT POINT OF INTERSECTION ⌐

FIG. 4-2-1   Center line technique.

Very short center lines may be unbroken if no confusion results with other lines.

Center lines are used to locate the center of circles and arcs. They are first drawn as light construction lines, then finished as alternate long and short dashes, with the short dashes intersecting at the center of the circle.

CAD

Center lines may be drawn by following the line commands explained in your CAD manual. The procedure by which you construct center lines in CAD varies greatly among different software packages.

Drawing Circles and Arcs

Circles and arcs are drawn with the aid of a compass or template. When circles and arcs are drawn with a compass, it is recommended that the compass lead be softer and blacker than the pencil lead being used on the same drawing. For example, if you are drawing with a 2H or 3H pencil, use an H compass lead. This will produce a drawing having similar line work since it is necessary to compensate for the weaker impression left on the drawing medium by the compass lead as compared with the stronger direct pressure of the pencil point. For drawing circles and arcs, see Figs. 4-2-2 and 4-2-3.

It is essential that the compass lead be reasonably sharp at all times in order to ensure proper line width. The compass lead should be sharpened to a bevel point, with the top rounded off, as shown in Fig. 4-2-4. The lead is slightly shorter than the needle point.

For drawing large circles and arcs, a beam compass, as shown in Fig. 4-2-5A (pg. 56), is used. For drawing very large arcs, the adjustable arc may be used (Fig. 4-2-5B). It accurately draws any radius from 7 to 200 m.

Drafters find it much easier and faster to use circle templates. There are sets that contain all common sizes and shapes of holes that most drafters are ever called upon to draw. When using a circle template, choose the correct diameter, line up the marks on the template with the center lines, and trace a dark thick line.

The drawing of arcs should be done before the tangent lines are made heavy. Draw light construction lines to establish the compass point location and check to make certain that the compass lead meets properly with both tangent lines before drawing the arc.

CAD

Circles   There are several common methods used to construct circles. The common methods used to draw circles include: (1) center and radius; (2) center and diameter; (3) three-point circle; and (4) two-point circle.

Arcs   Among the 10 most common methods used to draw arcs and fillets are: (1) three points; (2) start, center, and end; (3) start, end, and radius; (4) start, center, and angle; and (5) fillet.

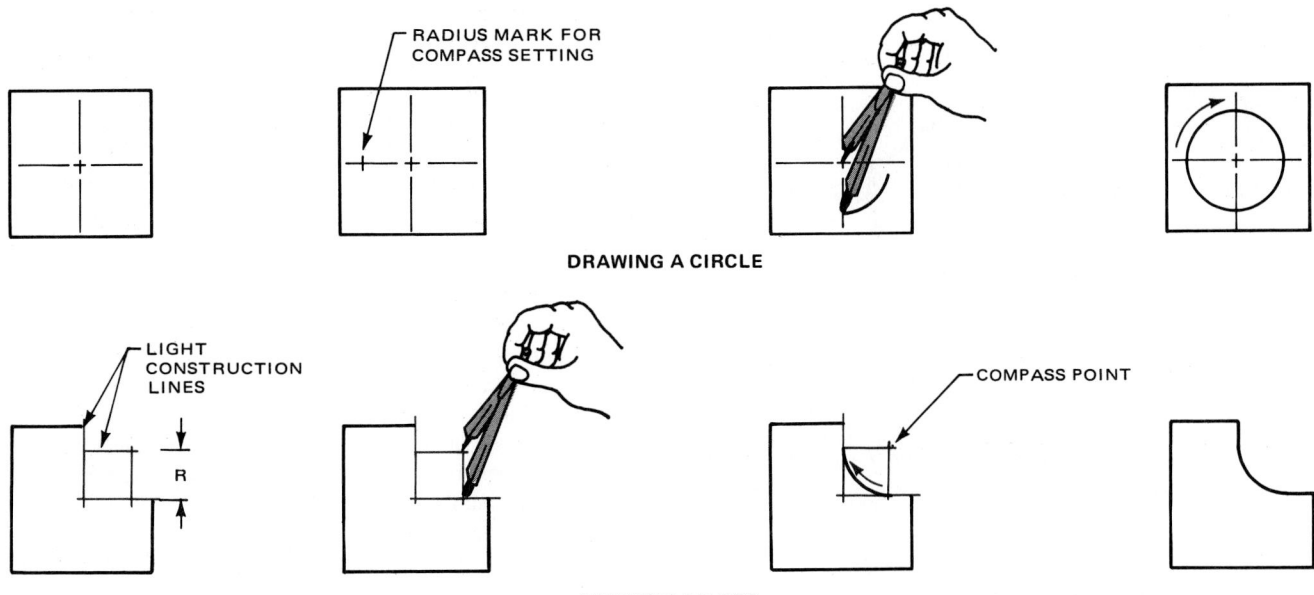

FIG. 4-2-2   Drawing circles and arcs.

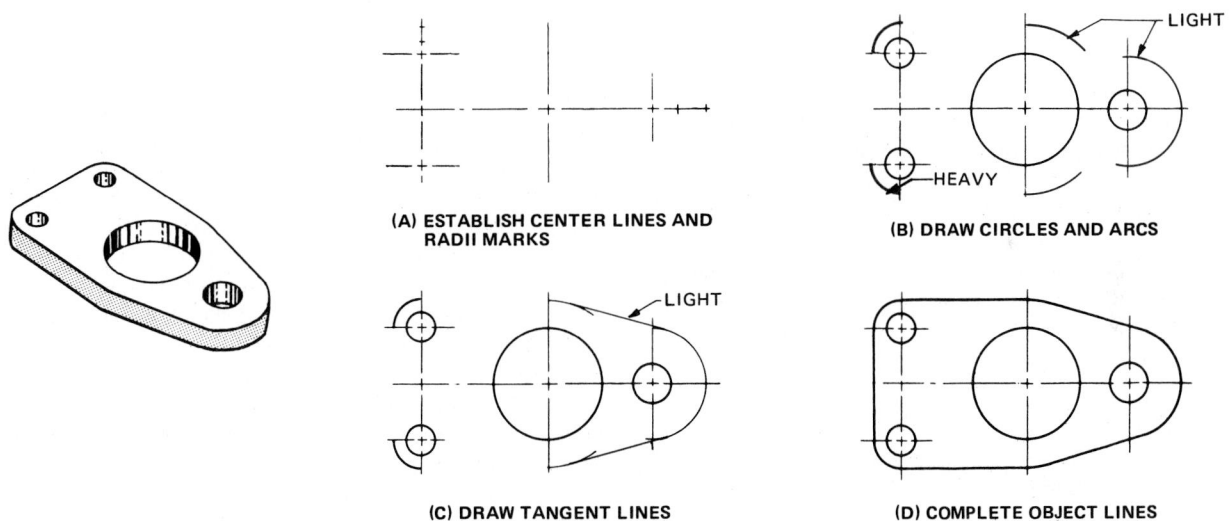

(A) ESTABLISH CENTER LINES AND RADII MARKS

(B) DRAW CIRCLES AND ARCS

(C) DRAW TANGENT LINES

(D) COMPLETE OBJECT LINES

FIG. 4-2-3   Sequence of steps for drawing a view having circles and arcs.

FIG. 4-2-4   Sharpening and setting the compass lead.

ASSIGNMENTS

See Assignments 15 through 22 for Unit 4-2 on pages 70–73.

4-3 DRAWING IRREGULAR CURVES

Curved lines may be drawn with the aid of irregular curves, flexible curves, and elliptical templates. Using an irregular curve, establish the points through which the curved line passes (Fig. 4-3-1, pg. 56), and draw a light freehand line through these points. Next, fit the irregular curve or other instrument by trial against a part of the curved line and draw a portion of line. Move the curve to match the next portion, and so forth. Each new position should fit enough of the part just drawn (overlap) to ensure continuing a smooth line. It is very important to notice whether the radius of the curved line is increasing or decreasing and to place the irregular curve in the same way.

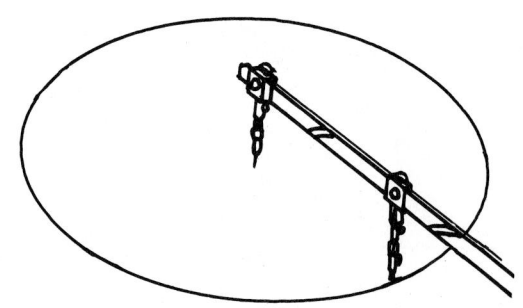

(A) USING A BEAM COMPASS

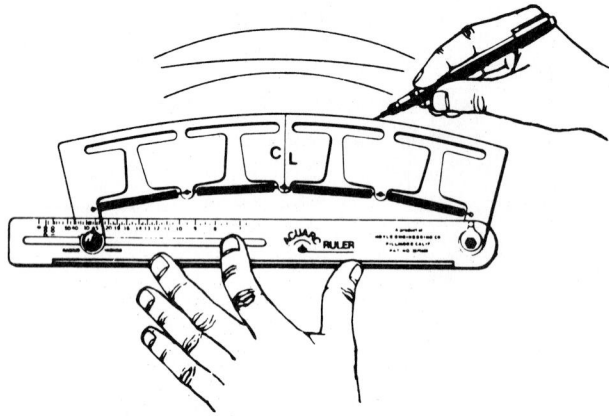

(B) USING AN ADJUSTABLE ARC

FIG. 4-2-5   Beam compass and adjustable arc.

If the curved line is symmetrical about the axis, the position of the axis may be marked on the irregular curve with a pencil for one side and then reversed to match and draw the other side.

 **CAD**

An irregular curve is a nonconcentric, nonstraight line drawn smoothly through a series of points. In CAD systems, it is commonly referred to as a *spline*.

ASSIGNMENTS

See Assignments 23 through 25 for Unit 4-3 on page 73.

4-4 SKETCHING

Freehand sketching is the simplest form of drawing. It is one of the quickest ways to express ideas. The drafter may use sketches to help simplify and explain thoughts and concepts to other people in discussions of mechanical parts and mechanisms. Sketching, therefore, is an important method of communication.

Freehand sketching is also a necessary part of drafting because the drafter in industry frequently sketches ideas and designs prior to making drawings with instruments. Practice in sketching helps the student develop a good sense of proportion and accuracy of observation.

Freehand drawings generally need some freehand lettering to explain features of a new idea or how a new product works. A drawing well planned may be worth a thousand words, as an old saying goes, but a few choice words well organized can explain some details.

The notes lettered in the rough sketch in Fig. 4-4-1 describe some functional features that are important to operation. Simple freehand lettering will complement an idea that is captured in a sketch, especially if the lettering is neat and carefully placed on a drawing.

Types of Sketches

Any image drawn on paper freehand (without a straightedge or other instrument) may be called a *sketch*. The sketching may be *rough*, or *unrefined*. That is, the sketch may be drawn quickly with jagged lines. The rough sketch can quickly express thoughts. Instruments are not used in the preparation of rough sketches. Figure 4-4-2 shows rough sketches that were used to develop preliminary (early) designs of a two-position automobile mirror.

The other way of sketching is *refined*. That is, the sketch is neat and looks finished. A refined sketch is carefully drawn. Templates and straightedges may be used to complement the freehand lines. It shows good proportion and excellent line values. It may be more persuasive than an unrefined sketch. Many refined sketches are based on a rough sketch that has captured the general idea.

Overlays

A very good way to *refine* (improve) a sketch is to use an overlay. Sketches are often drawn on paper that can be seen through. This paper is called translucent paper or tracing paper.

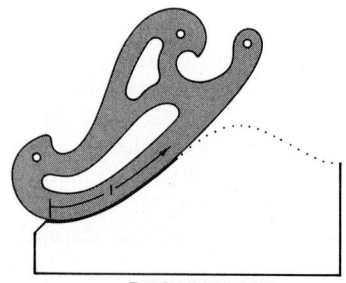

FIRST POSITION

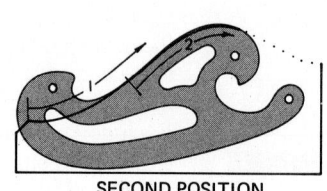

SECOND POSITION

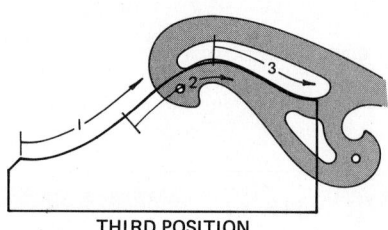

THIRD POSITION

FIG. 4-3-1   Drawing a curved line.

FIG. 4-4-1 A rough sketch with notes about important features.

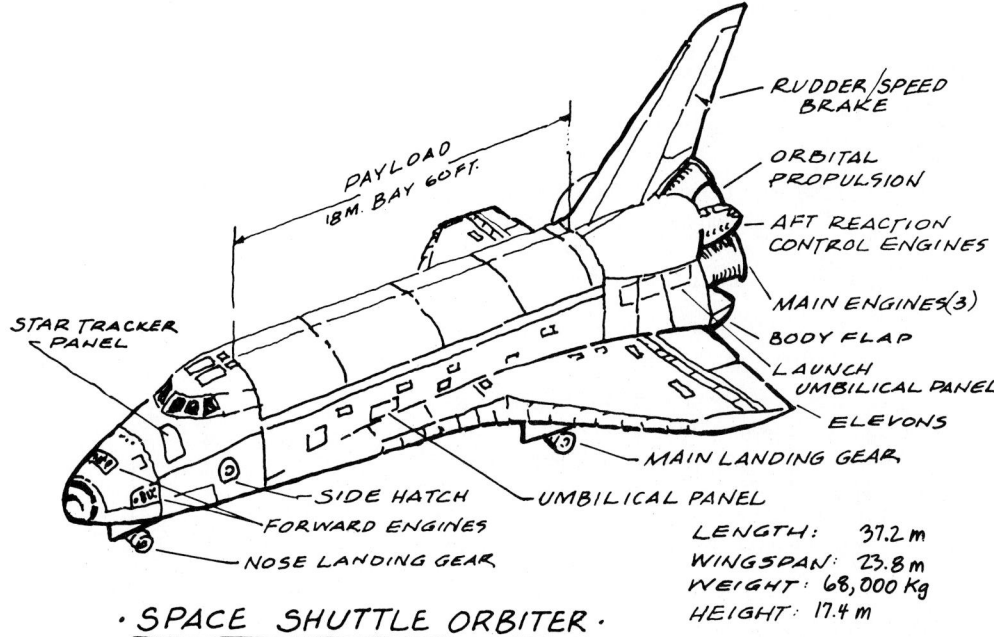

SPACE SHUTTLE ORBITER

LENGTH: 37.2 m
WINGSPAN: 23.8 m
WEIGHT: 68,000 Kg
HEIGHT: 17.4 m

FIG. 4-4-2 Study of a two-position mirror for a racing car.

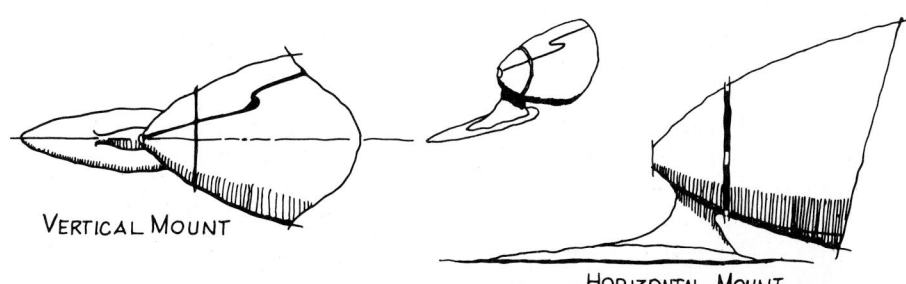

VERTICAL MOUNT

HORIZONTAL MOUNT

The best parts of a sketch may be quickly traced by putting a new piece of this paper, called the *overlay*, on top (Fig. 4-4-3). Thus, refining ideas means sketching over and over again on tracing paper until the design is right.

The overlay is used in two important ways that may seem very much alike. The first use is for reshaping an idea. This might include refining the proportions of the parts of an object or changing its shape entirely. Second, an overlay can be used to refine, or improve, the drawing itself without really changing the design. These two operations can be, and usually are, done at the same time.

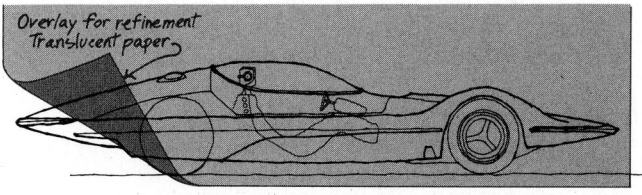

FIG. 4-4-3 The overlay can speed up the drawing process.

Nature of a Sketch

A sketch is an important form of graphic (drawn) communication. There are many levels of sketches for different uses (Fig. 4-4-4). If you know the language of mechanical drawing well, you will know how to sketch your ideas quickly, clearly, and accurately.

Temporary Sketch

Many technical sketches have short lives. They are done merely to solve an immediate problem, then thrown away.

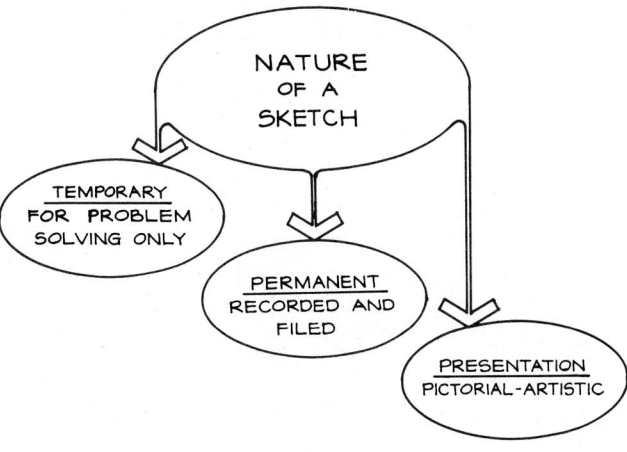

FIG. 4-4-4 Sketches are classified in three ways.

Other technical sketches are kept longer. It may take weeks or even months to study some sketches and make mechanical drawings from them. However, these sketches too may be thrown away some day.

Permanent Sketch

Sometimes the engineering department or the management of a company will include a sketch in a notice to other employees. Such a sketch is an important record and should be kept. Therefore, some sketches are filed as part of a company's permanent records.

Presentation Sketch

The sketch that is refined for presentation is usually *pictorial* (picturelike). It is used to convince a client or manager to accept and approve the ideas presented. The pictorial sketch has a three-dimensional view that can be understood easily by nontechnical people. Such sketches are usually drawn so that they appeal to the eye.

Views Needed For a Sketch

In Chap. 6, multiview projection will be discussed in full. However, you need to know some basic things about views and how they are placed in order to make sketches.

There are two types of drawings that you can sketch easily. The first is a pictorial drawing. In this type of drawing, the width, height, and depth of an object are shown in one view (Fig. 4-4-5). This type of sketching is covered in Chap. 15. The other type is called *multiview projection* or *orthographic projection*. This type of projection is fully explained in Chap. 6. In this type of drawing an object is usually shown in more than one view. You do this by drawing sides of the object and relating them to each other, as shown in Fig. 4-4-6.

If an object can be fully shown in only one view, this means that its shape can be described well in two dimensions (sizes). These are height and width. Things shown in one-view drawings generally have a depth or thickness that is uniform (the same throughout). This third dimension can be given as a note rather than being drawn. A typical one-view drawing is shown in Fig. 4-4-7A. The thickness of the stamping is shown by a note on the sketch. Many parts that are cylindrical in shape can also be shown in single views if the diameters are noted (Fig. 4-4-7B).

Materials for Sketching

Paper

You can use plain paper for sketching. If you need to refine the sketch, use tracing paper. Control proportions while sketching by using cross-section paper, also called graph paper or squared paper (Fig. 4-4-8). Graph paper should be used whenever possible because it cuts down on drawing time. The graph paper most often used has heavily ruled 1 in. squares. The 1 in. squares are then subdivided into lightly ruled $\frac{1}{10}$, $\frac{1}{8}$, $\frac{1}{4}$, or $\frac{1}{2}$ in. squares. This paper is called 10 to the inch, 8 to the inch, and so on. Graph paper is also ruled in millimeters. There are many

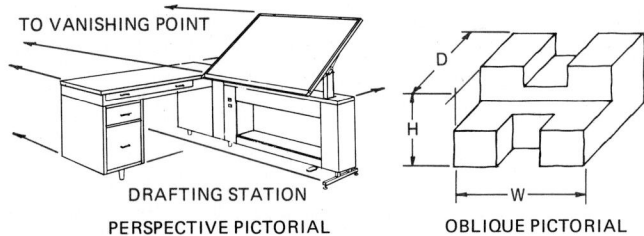

FIG. 4-4-5 Pictorial sketches.

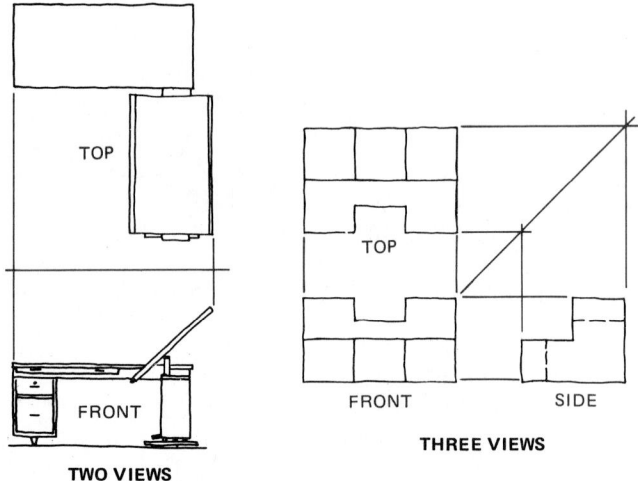

FIG. 4-4-6 Multiview sketches.

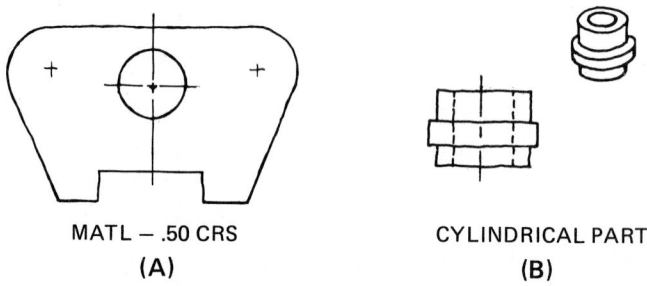

FIG. 4-4-7 One-view sketches.

specially ruled graph papers for particular kinds of drawing, such as isometric and perspective. These kinds of drawings are explained in Chap. 15.

Pencils and Erasers

Most drafters like to use soft lead pencils (grades F, H, or HB), properly sharpened. They also use an eraser that is good for soft leads, such as a plastic eraser or a kneaded-rubber eraser.

Freehand Lines

Lines drawn freehand have a natural look. They show freedom of movement because of their slight changes in direction. Hold the pencil far enough from the point that you can move your fingers easily and yet put enough pressure on the point to make dense, black lines when you need to. Draw light construction lines with very little pressure on the point. They should be light enough that they need not be erased.

Straight Lines

You can sketch lines in the following ways:

1. Draw them continuously.
2. Draw short dashes where the line should start and end. Then place the pencil point on the starting dash, draw toward it.
3. Draw a series of strokes that touch each other or are separated by very small spaces.
4. Draw a series of overlapping strokes.

Before you try to draw objects, practice sketching straight lines to improve your line technique. Draw horizontal lines from left to right if you're right-handed, and right to left if you're left-handed. Draw vertical lines from the top down.

Inclined Lines

Sketch sloped, or inclined, lines from left to right. It might be easiest to turn the paper and draw an inclined line the same way as a horizontal one. When trying to sketch a specific angle, first draw a vertical line and a horizontal line to form a right angle (90°). Divide the right angle in half to form two 45° angles. The sketched line should pass through the corners of the squares on the grid (Fig. 4-4-9A). Use an isometric grid

when angles of 30° and 60° are required (Fig. 4-4-9B). By starting with these simple angles, you can estimate (guess) other angles more exactly. Note the direction for drawing the inclined lines.

Proportions for Sketching

Sketches are not usually made to scale (exact measure). Nonetheless, it is important to keep sketches in proportion (similar to exact measure). In preparing the layout, look at the largest overall dimension, usually width, and estimate the size. Next, determine the proportion of the height to the width. Then, as the front view takes shape with the width and height, compare the smaller details with the larger ones and fill them in too (Fig. 4-4-10, pg. 60).

It is important that the design drafter, in sketching an object, have a good sense of how distances relate to each other. This ability will allow the drafter to show the width, height, and depth of an object in the right proportions.

For example, suppose that the design drafter plans a contemporary stereo cabinet to be 4 units wide. The height of the cabinet is 2 units. The depth of the cabinet is 1 unit. This is a proportion of 2:1. If you were designing this cabinet in customary units, you might choose a 15 in. width, height, and depth unit. The width overall would be 60 in. The height would be 30 in., and the depth 15 in. The proportions are developed in 2:1 units.

As previously mentioned, using a grid sheet will cut down on drafting time and result in a more accurate drawing. In the

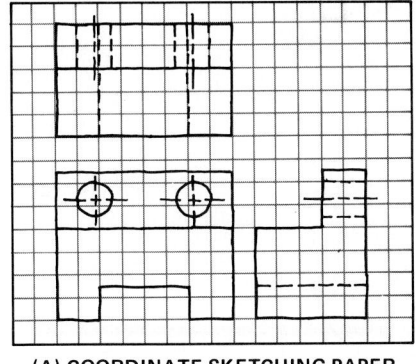

(A) COORDINATE SKETCHING PAPER

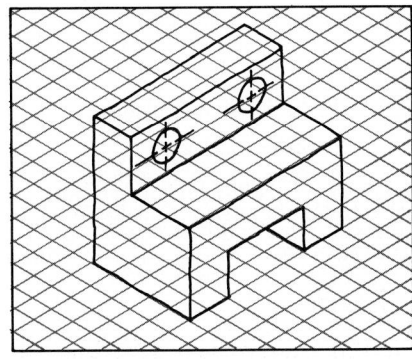

(B) ISOMETRIC SKETCHING PAPER

FIG. 4-4-8   Graph paper.

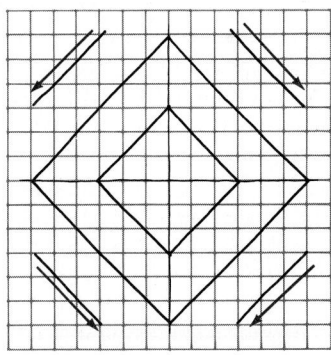

(A) SQUARE GRID

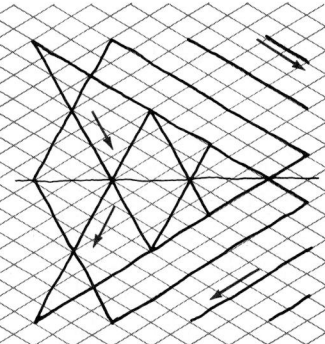

(B) ISOMETRIC GRID

FIG. 4-4-9   Using graph paper to help in sketching sloped lines.

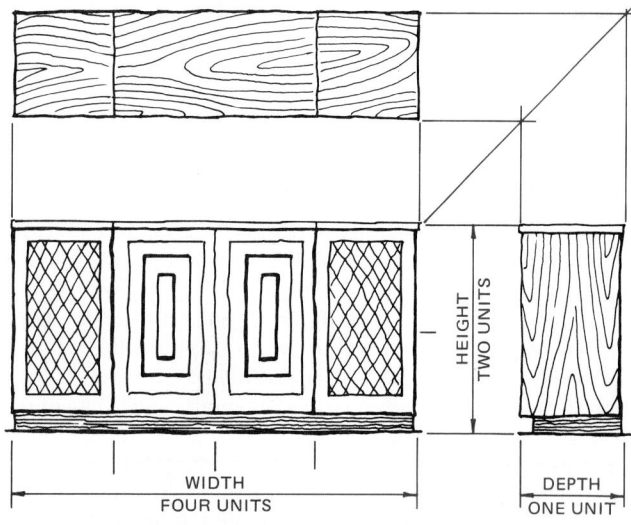

FIG. 4-4-10   Sketching a contemporary stereo with proportional units.

example given in Fig. 4-4-10, using a 4-to-the-inch, or .25 grid, and making each space equal to 3 in., a unit would be equal to five spaces on the grid sheet.

Circles and Arcs

There are two ways to sketch circles. For the first way (Fig. 4-4-11A), draw very light horizontal and vertical lines. Next, estimate the length of the *radius* (the distance from the center of the circle to its edge; plural, *radii*) and mark it off. Then draw a square in which you can sketch the circle. For the second way (Fig. 4-4-11B), first draw very light center lines. Then draw *bisecting* (halving) lines through the center at convenient angles. Next, mark off estimated radii of the same length on all lines. The bottom of the curve is generally easier to form, so draw it first. Then turn the paper so that the rest of the circle to be drawn is on the bottom. Complete the circle. You can sketch *arcs* (parts of a circle); *tangent arcs* (parts of two circles that touch); and *concentric circles* (circles of different diameters that have the same center) in the same way you sketch circles (Fig. 4-4-12). Use light, straight, construction lines to block in the area of the figure.

For large circles and arcs and for *ellipses* (ovals), use a scrap of paper with the radius marked off along one edge (Fig. 4-4-13). Put one end of the marked-off radius on the center of the circle. Draw the arc by placing a pencil at the other end of the radius and turning the scrap paper. Note that two radii are needed for an ellipse. To draw the ellipse, sketch both center lines. Keep both radius points on the center lines as you draw with the pencil at the other end.

Getting Started

When you are starting to sketch a view (or views), first lightly sketch the overall size as a rectangular or square shape, carefully estimating its proportions. Then add lines for the details of the shape and thicken all lines forming the view (Fig. 4-4-14).

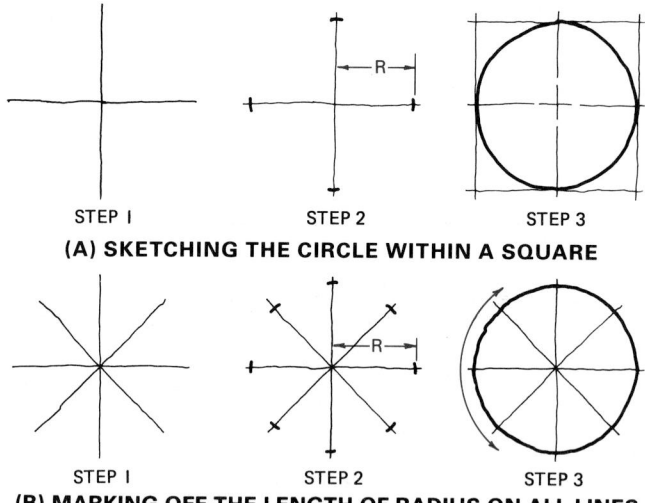

(A) SKETCHING THE CIRCLE WITHIN A SQUARE

(B) MARKING OFF THE LENGTH OF RADIUS ON ALL LINES

FIG. 4-4-11   Sketching circles.

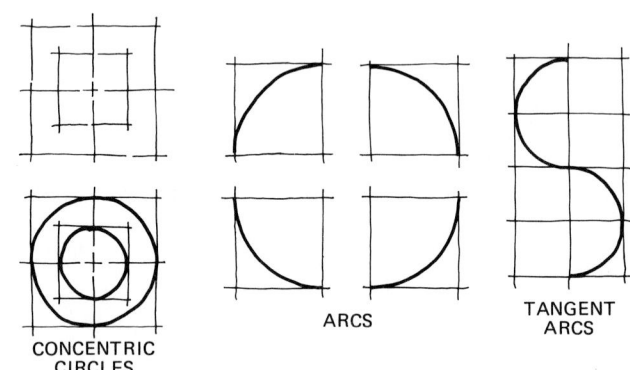

CONCENTRIC CIRCLES   ARCS   TANGENT ARCS

FIG. 4-4-12   Sketching arcs and concentric circles.

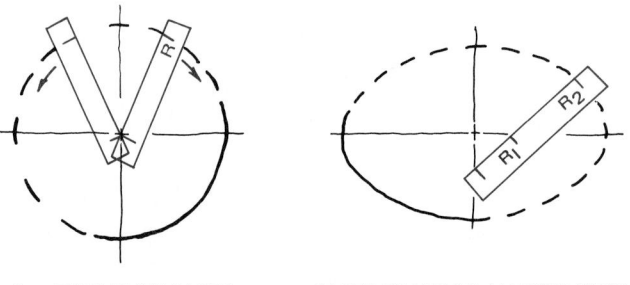

R = RADIUS ON PAPER STRIP FOR LARGE CIRCLES

LARGE CIRCLE

PAPER TRAMMEL MARKED WITH MAJOR (R2) AND MINOR (R1) RADII

ELLIPSE

FIG. 4-4-13   Sketching large circles and ellipses.

Figure 4-4-15 shows the two methods of sketching circles. Figure 4-4-16 illustrates, both pictorially and orthographically, the use of graph paper for sketching a machine part.

 CAD

Sketching is still a necessary part of drafting because the drafter in industry frequently sketches ideas and designs prior

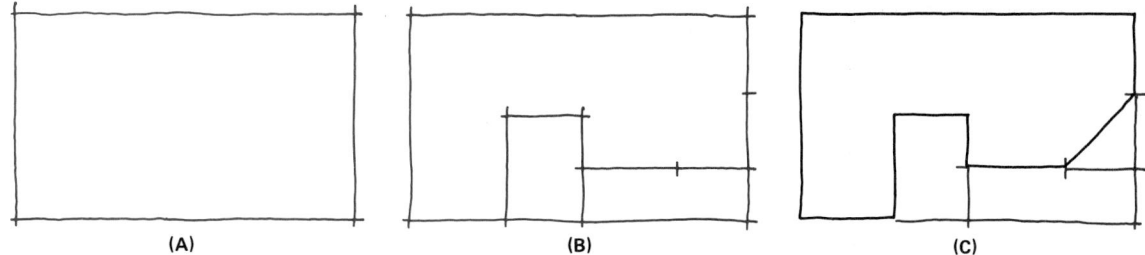

FIG. 4-4-14   Sketching a view having straight lines.

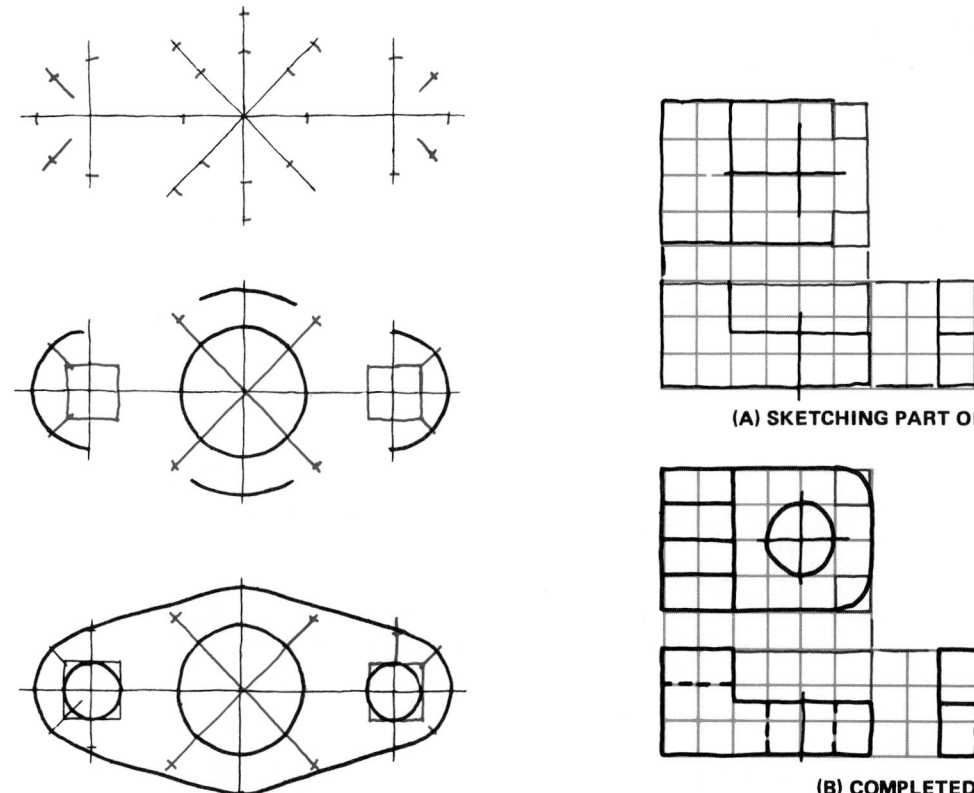

FIG. 4-4-15   Sketching a figure having circles and arcs.

(A) SKETCHING PART OF THE OUTLINE

(B) COMPLETED VIEWS

FIG. 4-4-16   Usual procedure for sketching three views.

to making a CAD drawing. Sketching may be accomplished manually or on a CAD monitor using the SKETCH command.

ASSIGNMENTS

See Assignments 26 through 31 for Unit 4-4 on pages 74–75.

4-5 INKING

Most technical drawings are made with pencil on a good-quality tracing paper (vellum) or on drafting film. However, ink drawings have seen an increase in the past few decades. Ink is often used to make high-quality tracings that can be copied (whiteprinted, photographed, or microfilmed). The need for

highly efficient ink plotters for CAD brought about improvements in ink and inking equipment for manual drafting. For example, the technical pen with refillable cartridges, cartridge ink supplies, erasable ink, accessories, and attachments has made inking more efficient and improved the quality of the finished drawing.

Drawing Ink

Ink used for technical drawings is also called *india ink*. It must have some special characteristics. For example, it must flow freely and yet dry quickly. It must adhere (stick) well and not chip, crack, peel, or smear after drying. It must also be completely opaque in order to produce good uniform line tone and yet be erasable on all drafting media.

Both waterproof and nonwaterproof ink can be used on high-quality paper, cloth, polyester film, and illustration board.

The waterproof ink is best suited for mechanical drawings. Nonwaterproof ink is best for fine-line pictorial drawings. Both types of ink produce opaque lines that reproduce well when they are photographed or copied.

Using Ink Equipment

Figure 4-5-1 shows the correct position for drawing straight lines with a technical pen. Note the direction of the stroke and the angle of the pen. Hold the technical pen in a nearly vertical position (perpendicularly to the medium) to get the most uniform line.

For making circles and arcs, templates and compasses can be used with technical pens, as shown in Fig. 4-5-2. When inking with a ruling-pen type of compass, allow the compass to lean about 15° in the direction of the rotating motion.

Templates having built-in risers for inking are available (Fig. 4-5-3). Regular templates may also be used by building up one side of the template with small pieces of masking tape. These act as spacers between the inking surface of the drawing medium and the bottom side of the template.

Irregular (French) curves can be used to guide the pen when curves that are not circular arcs are being inked.

The ruling pen has blades that you can adjust to draw lines of different widths. It can be used to draw both straight and curved lines. Ruling pens are sometimes called all-purpose pens. They can be filled from a cartridge tube, a squeeze bottle, or a dropper cap.

The simple rules listed below will help you produce high-quality ink drawings using the ruling pen. Study them carefully.

1. Do not hold the pen over the drawing when you are filling it.
2. Do not fill the pen too full; about .10 to .20 in. (3 to 6 mm) is usually enough.
3. Never dip the pen into the ink bottle or allow ink to get on the outside of the blades.
4. Keep the blades clean by wiping them often with a soft towel, tissue, or pen cleaner.
5. While inking a line, keep both nibs (points) of the blades in contact with the drawing surface.
6. Keep the blades slanted in the direction of the pen stroke (Fig. 4-5-4).
7. Do not press the nibs hard against the straightedge. Pressing too hard will make line widths uneven.
8. Do not press the nibs too hard against the drawing surface. The blades may get dull or damage the drawing surface.

Many compasses have ruling-pen attachments that are used for inking arcs and circles. To use a compass for inking, remove

FIG. 4-5-1   The position of a technical pen is important when drawing lines. *(Teledyne Post)*

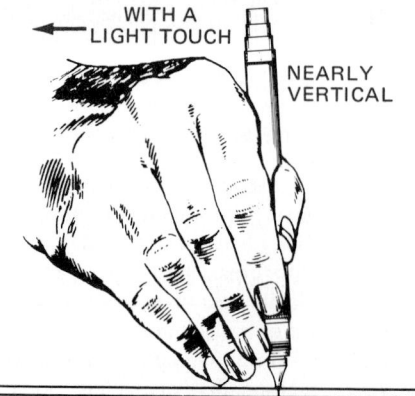

WITH A
LIGHT TOUCH

NEARLY
VERTICAL

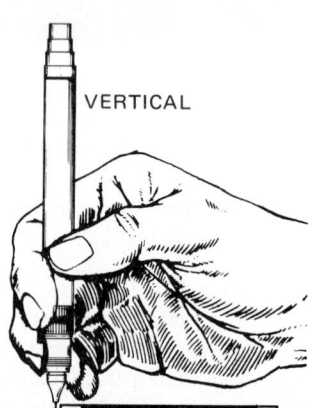

VERTICAL

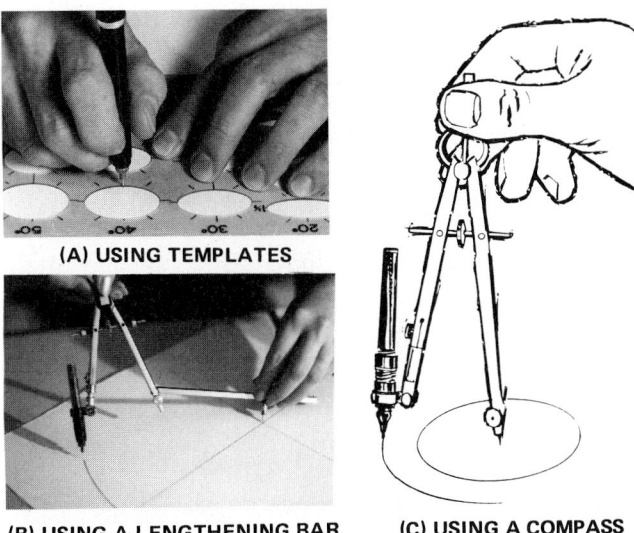

(A) USING TEMPLATES

(B) USING A LENGTHENING BAR FOR LARGE CIRCLES

(C) USING A COMPASS

FIG. 4-5-2   Drawing circles and arcs with a technical pen.

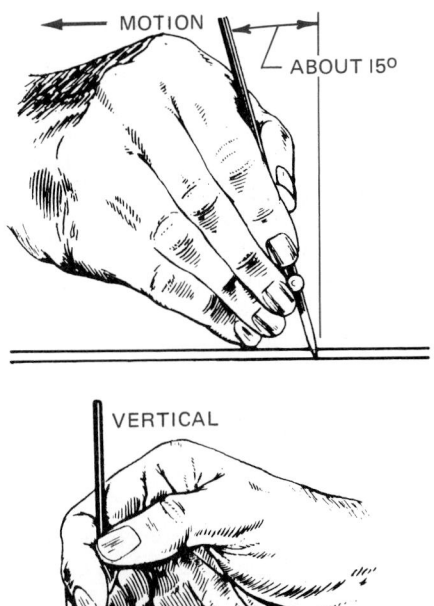

FIG. 4-5-4   Inking straight lines with a ruling pen.

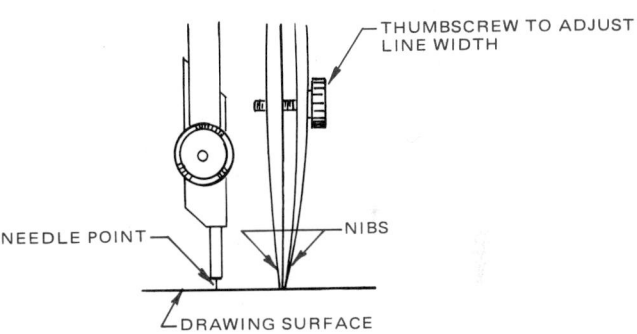

FIG. 4-5-5   Ruling nibs and needle point should be perpendicular to drawing surface.

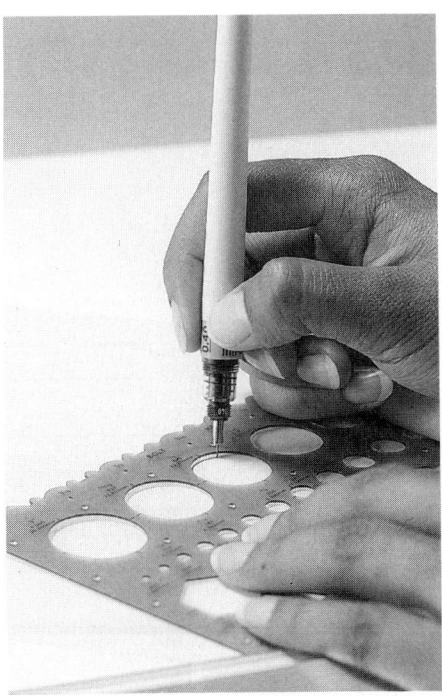

FIG. 4-5-3   A template with built-in risers for inking. *(STUDIOHIO)*

the pencil leg from the compass and replace it with the ruling-pen leg. Adjust the needle point carefully, as shown in Fig. 4-5-5, so that the legs are perpendicular to the drawing surface. Always draw the circle in one stroke. The compass should be slanted slightly when a ruling pen is being used. The lengthening bar, or beam compass, can be used for large circles.

Pen Cleaning

The technical pen and ruling pen sometimes need cleaning. The ultrasonic cleaner is very useful because it works so rapidly. The cleaning usually takes from 20 to 30 seconds, and

the pen is ready to use again. To use an ultrasonic cleaner, just dip the pen point into a well containing cleaning fluid. From 60,000 to 80,000 cycles per second of sound energy act to loosen the dried ink on the point. The normal high range of human hearing is 15,000 cycles per second.

To clean a technical pen thoroughly, you may have to take it apart. Remove the cap and ink cartridge and soak the point in a special cleaning fluid. After removing the dried ink from the point, flush the parts in a stream of cold water. Dry all the parts before using the pen again. If the parts are not dry, the ink will be diluted (mixed with water) and lines will not be dark enough. If care is taken in handling and storing inking equipment, it will work well and need very little repair.

Lettering, Arrowheads, and Symbols

Lettering and drawing of arrowheads and symbols are done in the same manner as with a pencil except that either a dry or wet technical pen is used.

63

Erasing Techniques

The ink used on polyester drafting film is waterproof. However, you can easily remove ink from the film by rubbing it with a moistened plastic eraser. Do not use any pressure in rubbing. The polyester film does not absorb ink, and therefore all the ink dries on top of its highly finished surface. Remove ink from other surfaces, such as tracing vellum or illustration board, with regular ink erasers or chemically treated ink erasers that absorb ink. But be very careful. Press lightly with strokes in the direction of the line to remove ink caked on the surface. Too much pressure damages the surface and makes it hard to revise the drawing. Different erasing techniques work better on different surfaces. Try different techniques to find the best one for each surface.

Order of Inking

Smooth joints and tangents, sharp corners, and neat fillets give a drawing a professional look and make it easy to read. Good inking requires careful practice and a definite order of working procedures.

The usual order of inking or tracing is the same order as for pencil drawings (Fig. 4-5-6). First, ink the arcs, centered over the pencil lines, as in (A). Ink the horizontal lines next, as in (B). Complete the drawing with the vertical lines, as in (C). Then add the dimension lines, arrowheads, finish marks, and so on, and fill in the dimensions, as in (D).

Inking continues to be a prime professional procedure for preparing engineering documents. Inked drawings will produce excellent reproductions and, of course, enhance the corporate image for presentation drawings. The drafting technician with manual inking skills will be able to complement those drawings prepared with CAD systems using multipen plotters.

ASSIGNMENTS

See Assignments 32 and 33 for Unit 4-5 on page 75.

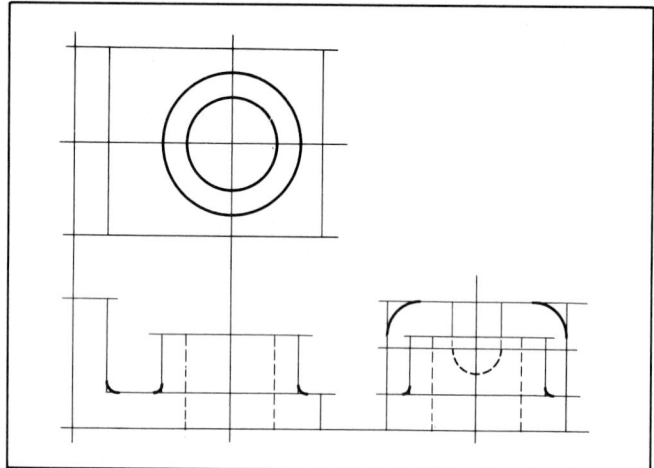

(A) DRAW CIRCLES AND ARCS

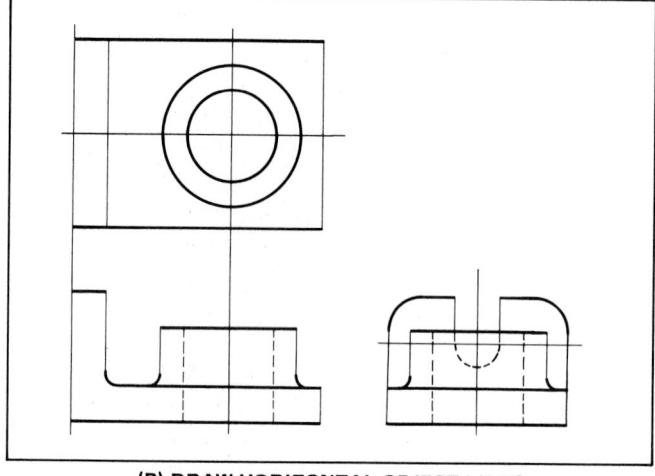

(B) DRAW HORIZONTAL OBJECT LINES

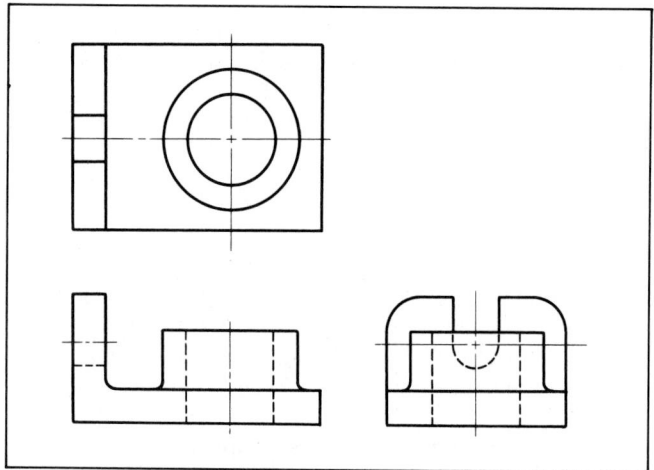

(C) DRAW VERTICAL OBJECT LINES

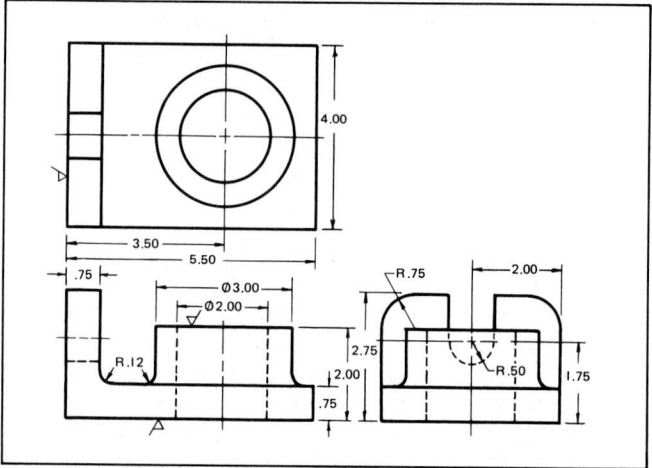

(D) ADD CENTER LINES, DIMENSION LINES, LETTERING, AND SYMBOLS

FIG. 4-5-6   Sequence of steps in developing a finished ink drawing.

ASSIGNMENTS FOR CHAPTER 4

NOTE ABOUT DRAWING ASSIGNMENTS

The assignments throughout this text are designed to teach the student certain aspects of technical drawings. It does not matter whether the drawing is made manually or by CAD. However, in order to simplify set-up instructions, the *X*, *Y*, and *Z* axes are shown in lieu of an arrow to indicate the viewing direction of the front view on all pictorial drawing assignments.

NOTE ABOUT DUAL DIMENSIONING

The dual dimensions shown in this book, especially in the assignment sections, are neither hard nor soft conversions. Instead, the sizes are those that would be most commonly used in the particular dimensioning units and so are only approximately equal. Dual dimensioning this way avoids awkward sizes and allows instructor and student to be confident when using either set of dimensions. Where dual dimensioning is shown, the dimensions given first or placed above are inch units of measurement. The other values shown are in millimeters.

ASSIGNMENTS FOR UNIT 4-1, STRAIGHT LINE WORK, LETTERING, AND ERASING

1. Lettering assignment. Set up a B (A3) size sheet similar to that shown in Fig. 4-1-A. Using uppercase Gothic lettering shown in Fig. 4-1-7, complete each line. Each letter and number is to be drawn several times to the three recommended lettering heights shown. Very light guidelines must be drawn first.

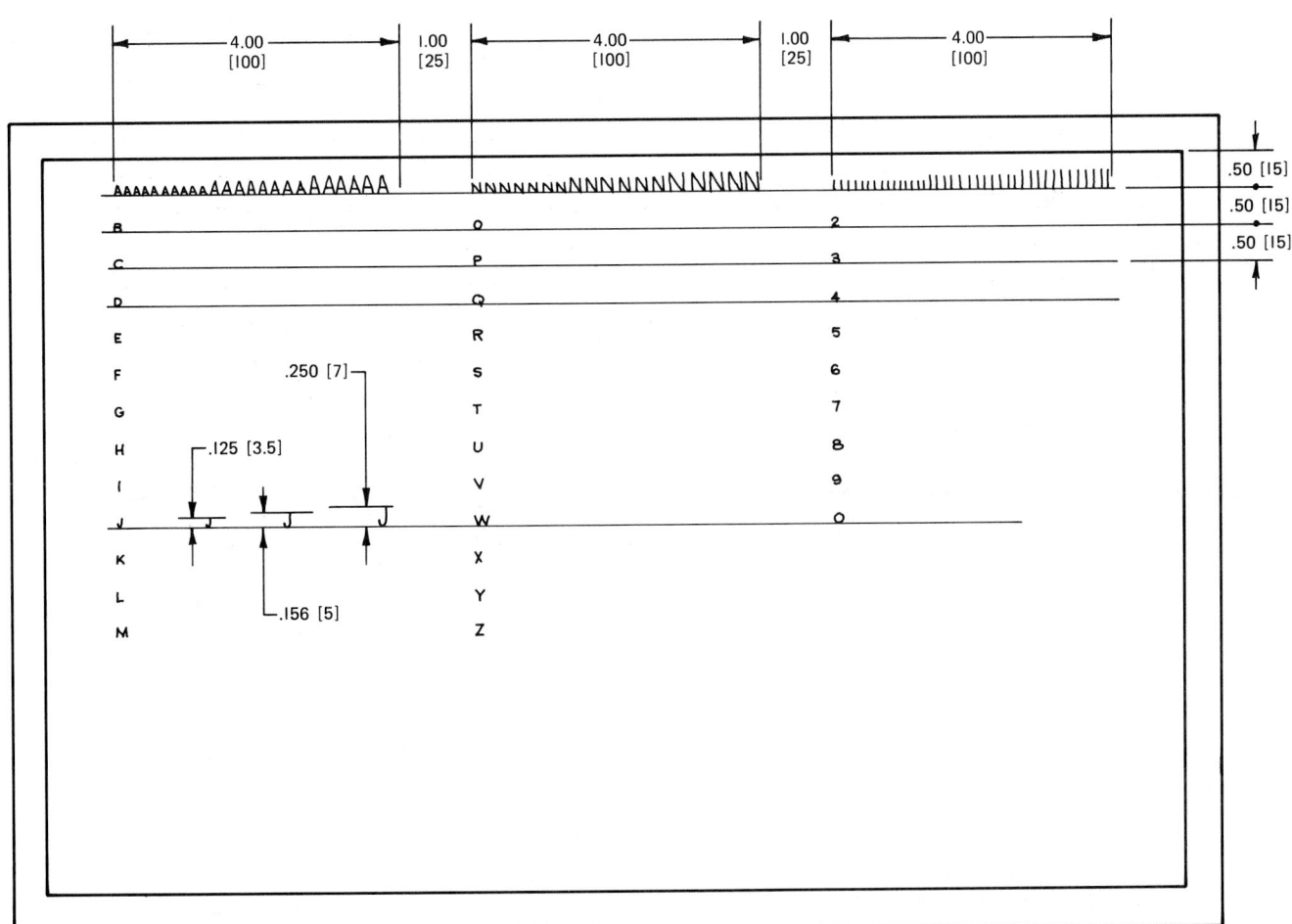

FIG. 4-1-A   Lettering assignment.

2. On a B (A3) size sheet draw one of the templates shown in Figs. 4-1-B or 4-1-C. Scale 1:1. Do not dimension.

3. On a B (A3) size sheet draw the shearing blank shown in Fig. 4-1-D. Scale 1:1. Do not dimension.

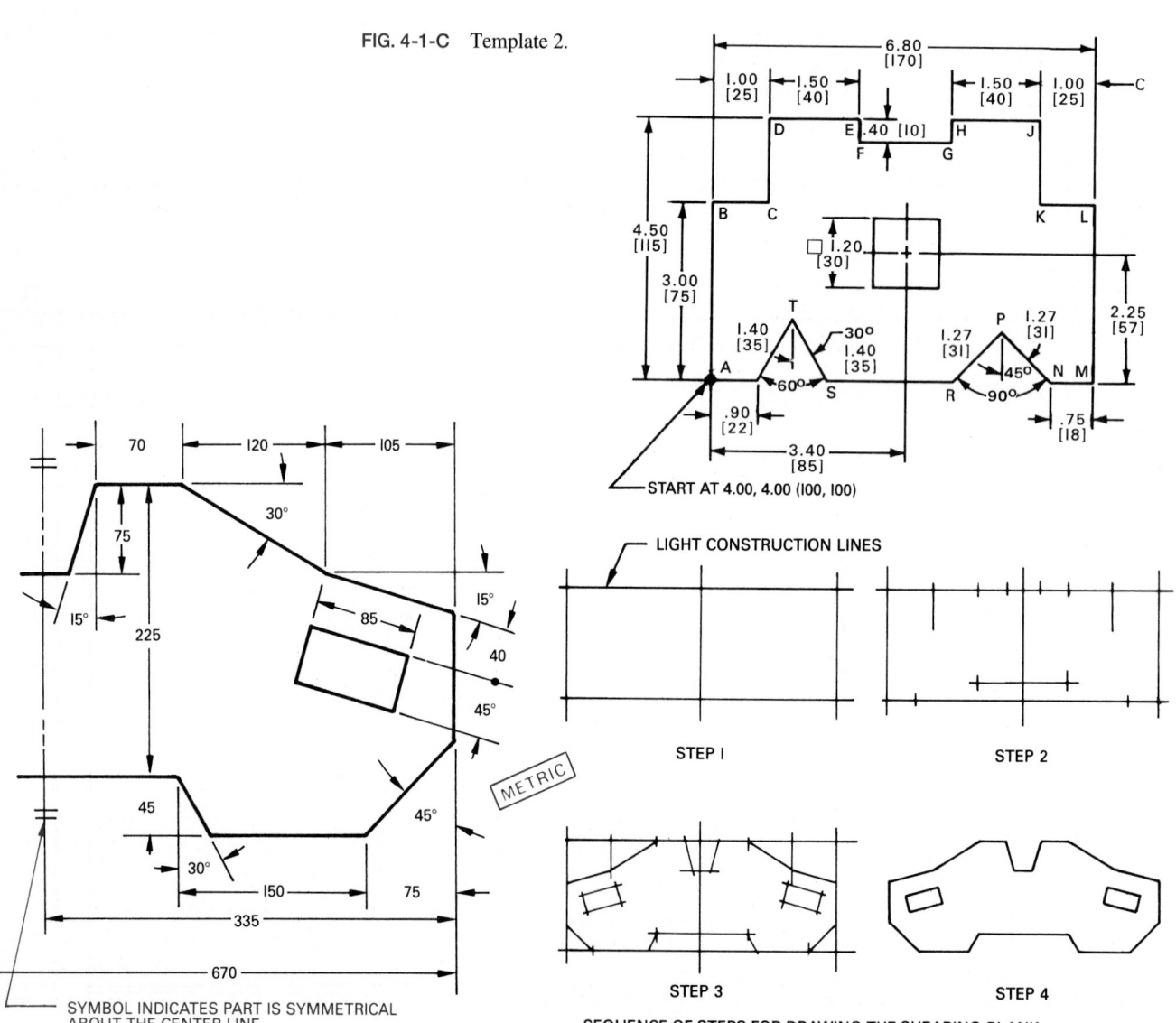

FIG. 4-1-B   Template 1.

FIG. 4-1-C   Template 2.

FIG. 4-1-D   Shearing blank.

SEQUENCE OF STEPS FOR DRAWING THE SHEARING BLANK

4. On a B (A3) size sheet draw the three parts shown in Fig. 4-1-E. Scale 1:1. Do not dimension.

5. On a B (A3) size sheet draw the three parts shown in Fig. 4-1-F. Use light construction lines as parts of each line will not be required. Scale 1:1. Do not dimension.

6. On a B (A3) size sheet draw any two of the three parts shown in Fig. 4-1-G. Scale 1:1. Do not dimension.

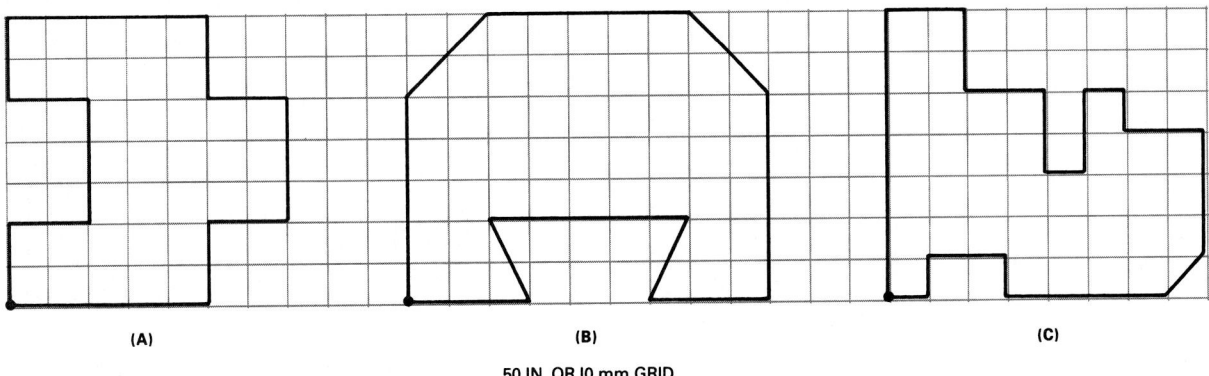

(A) (B) (C)

.50 IN. OR 10 mm GRID

FIG. 4-1-E   Line drawing assignment.

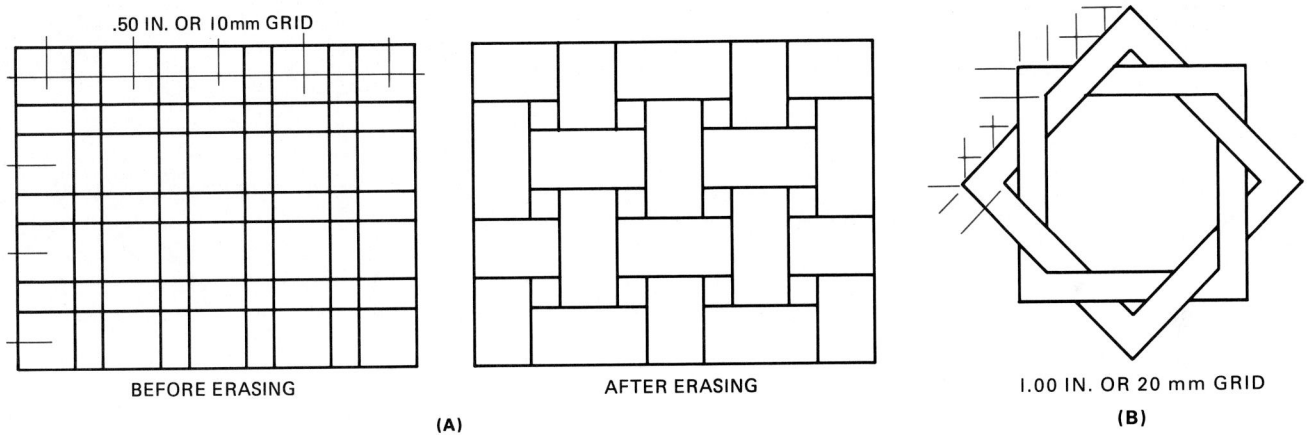

BEFORE ERASING AFTER ERASING 1.00 IN. OR 20 mm GRID

(A) (B)

FIG. 4-1-F   Line drawing assignment.

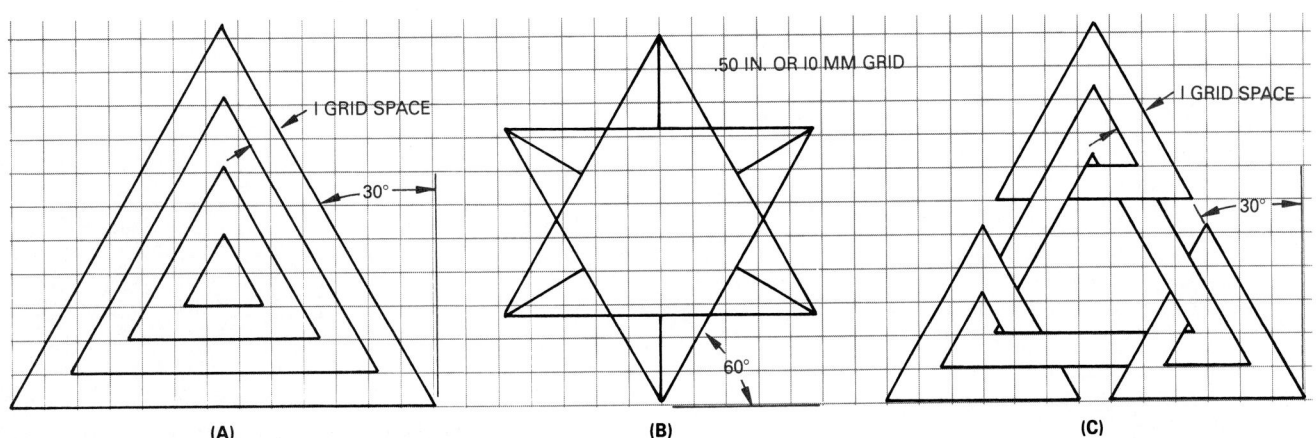

(A) (B) (C)

FIG. 4-1-G   Line drawing assignment.

7. On a B (A3) size sheet draw the three patterns shown in Fig. 4-1-H. Scale 1:1. Do not dimension.
8. On a B (A3) size sheet draw the designs shown in Fig. 4-1-J. Scale 1:1. Do not dimension.
9. On a B (A3) size sheet draw the designs shown in Fig. 4-1-K. Scale 1:1. Do not dimension.
10. Using absolute coordinates draw Figs. 4-1-L and 4-1-M on a B (A3) size format. The bottom left corner of the drawing (point 1) is the starting point. Scale 1:1.
11. Using relative coordinates, draw Fig. 4-1-N on a B (A3) size format. The bottom left corner of the drawing (point 1) is the starting point. Scale 1:1.

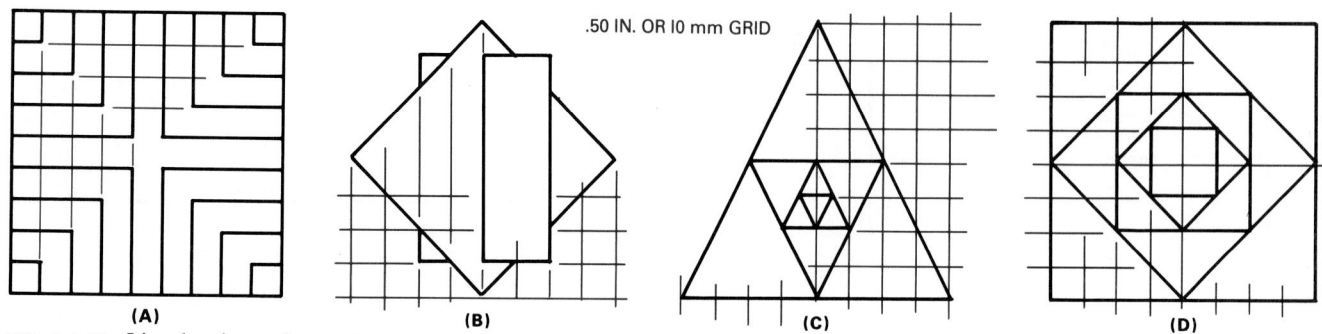

FIG. 4-1-H   Line drawing assignment.

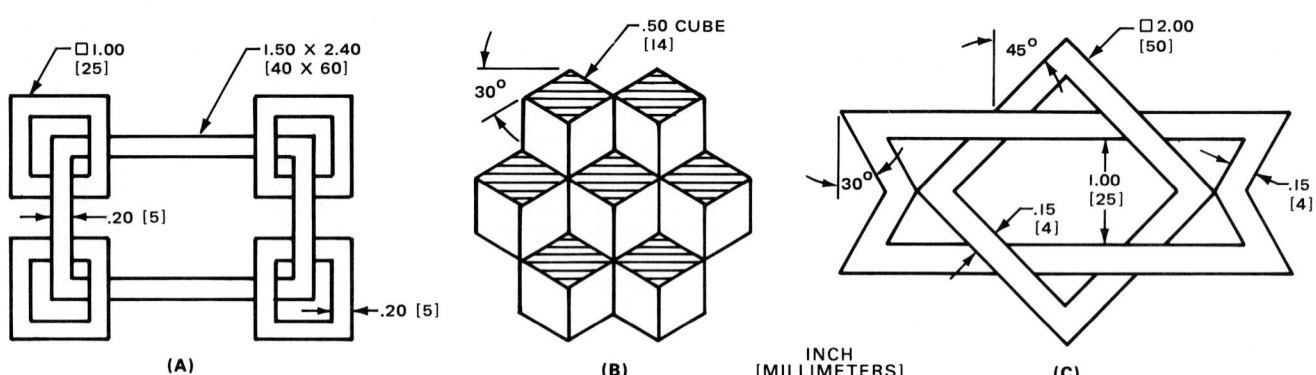

FIG. 4-1-J   Inlay designs.

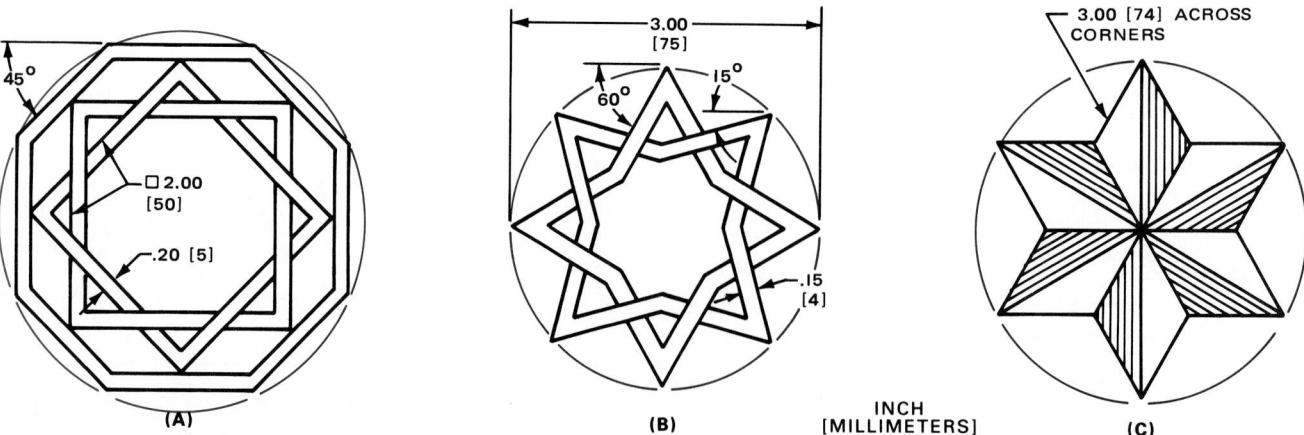

FIG. 4-1-K   Inlay designs.

68

ABSOLUTE COORDINATES (INCHES)		
Point	X Axis	Y Axis
1	.25	.25
2	7.00	.25
3	8.50	1.00
4	8.50	2.25
5	7.00	3.00
6	8.50	3.75
7	8.50	5.25
8	7.00	6.25
9	5.50	6.25
10	5.50	4.50
11	4.75	4.50
12	4.75	6.25
13	3.50	6.25
14	3.50	5.50
15	1.50	5.50
16	1.50	6.25
17	.75	6.25
18	.25	5.50
19	.25	.25
New Start		
20	.75	1.50
21	2.25	3.50
22	.75	3.50
23	.75	1.50
New Start		
24	3.00	2.25
25	5.75	2.25
26	5.75	3.75
27	3.00	3.75
28	3.00	2.25
New Start		
29	2.75	.75
30	6.00	.75
31	5.25	1.50
32	3.50	1.50
33	2.75	.75

FIG. 4-1-L   Absolute coordinates.

ABSOLUTE COORDINATES (MILLIMETERS)		
Point	X Axis	Y Axis
1	10	10
2	50	10
3	50	20
4	120	20
5	120	10
6	150	10
7	180	30
8	220	30
9	220	100
10	160	100
11	160	130
12	140	130
13	140	160
14	110	160
15	120	140
16	90	140
17	70	100
18	40	120
19	60	160
20	40	160
21	10	140
22	10	80
23	20	40
24	10	40
25	10	10
New Start		
26	40	50
27	160	50
28	120	90
29	50	70
30	40	50

FIG. 4-1-M   Absolute coordinates.

RELATIVE COORDINATES (INCHES)		
Point	X Axis	Y Axis
1	0	0
2	4.50	0
3	0	.75
4	−.75	0
5	0	.75
6	−.75	0
7	0	.75
8	−3.00	0
9	0	−2.25
New Start—Solid		
10	0	.75
11	3.75	0
New Start		
12	−.75	.75
13	−3.00	0
New Start—Solid		
14	0	1.50
15	4.50	0
16	0	2.25
17	−4.50	0
18	0	−2.25
New Start—Solid		
19	0	.75
20	3.75	0
21	0	1.50
New Start—Solid		
22	−.75	0
23	0	−.75
24	−3.00	0
New Start—Solid		
25	5.25	−4.50
26	2.25	0
27	0	2.25
28	−.75	0
29	0	−.75
30	−.75	0
31	0	−.75
32	−.75	0
33	0	−.75
New Start—Solid		
34	.75	.75
35	1.50	0
New Start—Solid		
36	0	.75
37	−.75	0

FIG. 4-1-N   Relative coordinates.

RELATIVE COORDINATES (MILLIMETERS)		
Point	X Axis	Y Axis
1	0	0
2	30	0
3	0	10
4	10	0
5	0	−10
6	30	0
7	0	50
8	−10	0
9	0	−15
10	−50	0
11	0	15
12	−10	0
13	0	−50
New Start—Solid		
14	5	10
15	15	0
16	0	20
17	−15	0
18	0	−20
New Start—Solid		
19	45	0
20	15	0
21	0	20
22	−15	0
23	0	−20

FIG. 4-1-P   Relative coordinates.

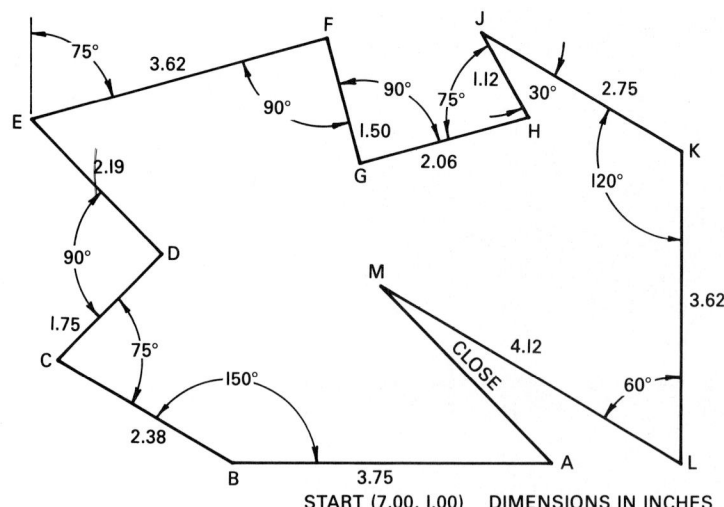

FIG. 4-1-R   Template—polar coordinates.

12. Using relative coordinates, draw Fig. 4-1-P on a B (A3) size format. The bottom left corner of the drawing (point 1) is the starting point. Scale 1:1.
13. On a B (A3) size format draw the template shown in Fig. 4-1-R. Start at point A. Scale 1:1.
14. On a chart list show the polar coordinates for the templates shown in Figs. 4-1-R, 4-1-B, and 4-1-C. Move in a clockwise direction starting at point A.

ASSIGNMENTS FOR UNIT 4-2, CIRCLES AND ARCS

15. On a B (A3) size format, draw the dial indicator shown in Fig. 4-2-A. Scale 2:1. Do not dimension but add the word DEGREES and the degree numbers shown.
16. On a B (A3) size format, draw the dart board shown in Fig. 4-2-B. Scale 1:2. Use diagonal line shading and add the numbers. Do not dimension.

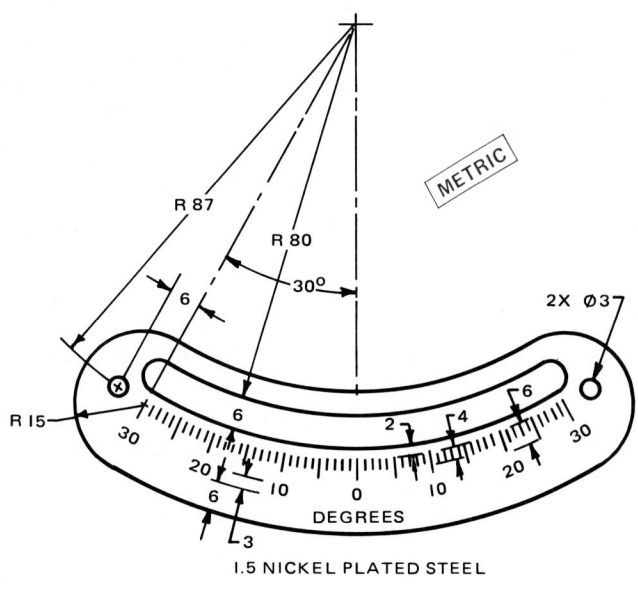

FIG. 4-2-A   Dial indicator.

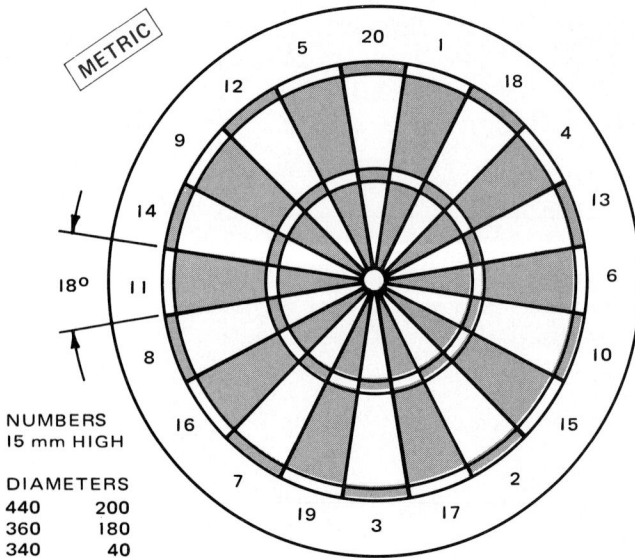

FIG. 4-2-B   Dart board.

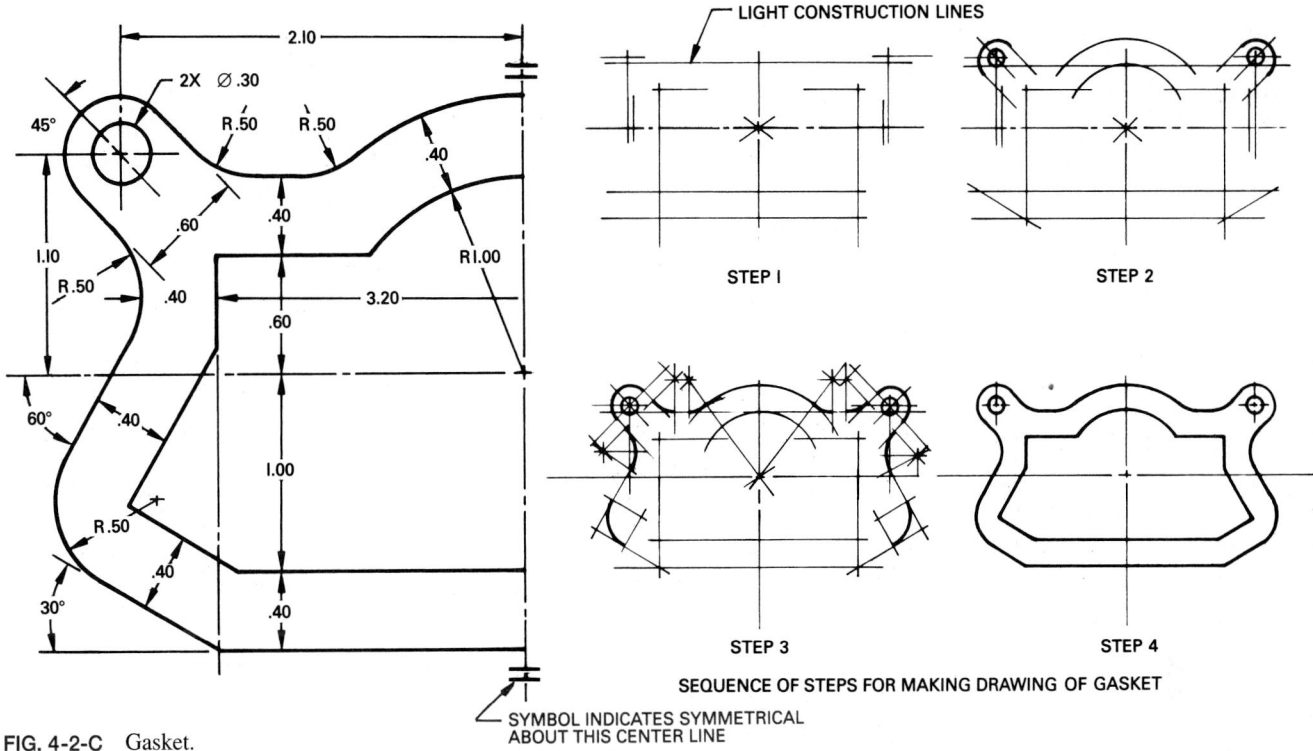

FIG. 4-2-C   Gasket.

SEQUENCE OF STEPS FOR MAKING DRAWING OF GASKET

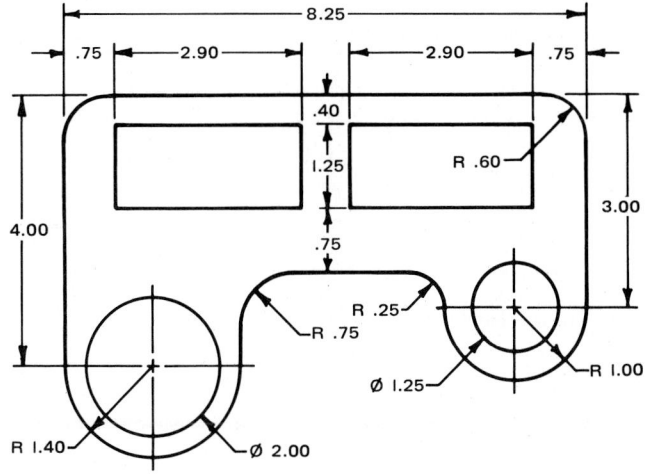

FIG. 4-2-D   Template.

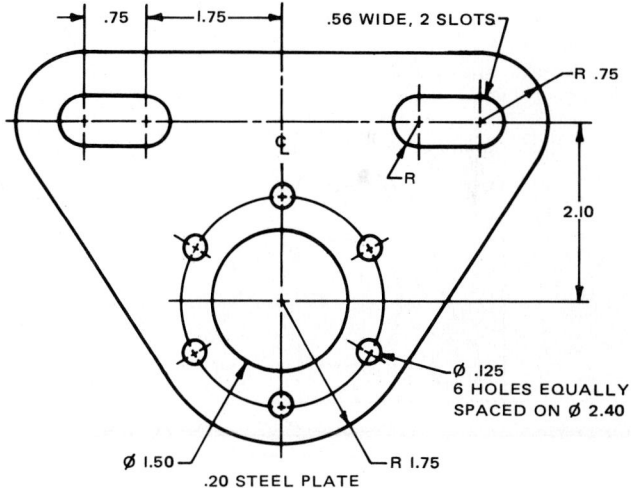

FIG. 4-2-F   Shaft support.

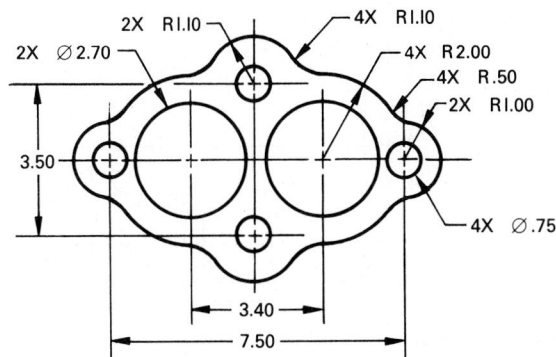

FIG. 4-2-E   Carburetor gasket.

17. On an A (A4) size format, draw the gasket shown in Fig. 4-2-C. Scale 1:1. Do not dimension.
18. On a B (A3) size format, draw the template shown in Fig. 4-2-D. Scale 1:1. Do not dimension.
19. On a B (A3) size format draw one of the parts shown in Figs. 4-2-E and 4-2-F. Scale 1:1. Do not dimension.

20. On an A (A4) size format draw one of the parts shown in Figs. 4-2-G to 4-2-J. Scale 1:1. Do not dimension.
21. On an A (A4) size format draw one of the parts shown in Figs. 4-2-K to 4-2-M. Scale 1:1. Do not dimension.
22. On an A (A4) size format draw the reel shown in Fig. 4-2-N. Scale 1:1. Do not dimension.

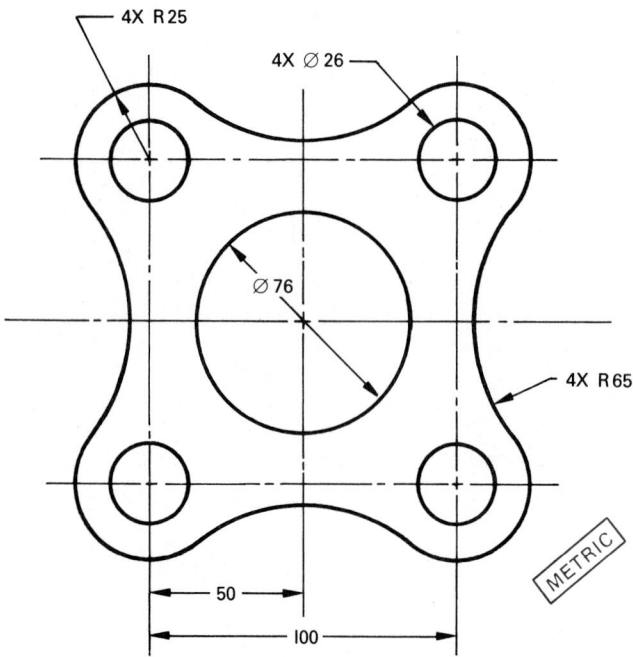

FIG. 4-2-G   Anchor plate.

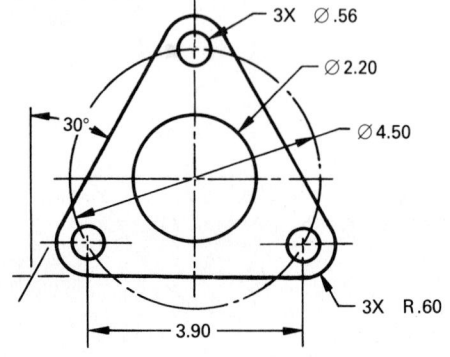

FIG. 4-2-H   Base plate.

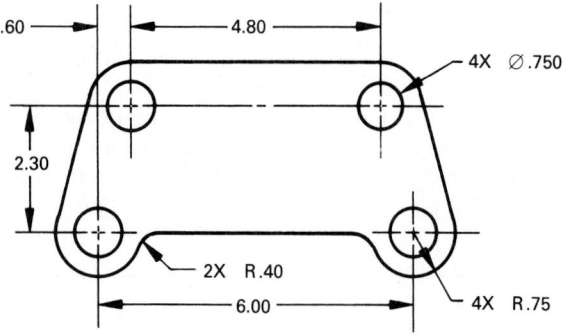

FIG. 4-2-J   Cover plate.

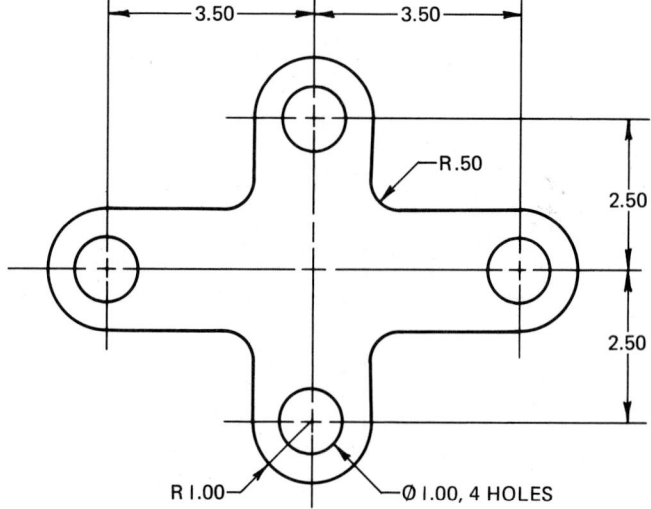

FIG. 4-2-K   Offset link.

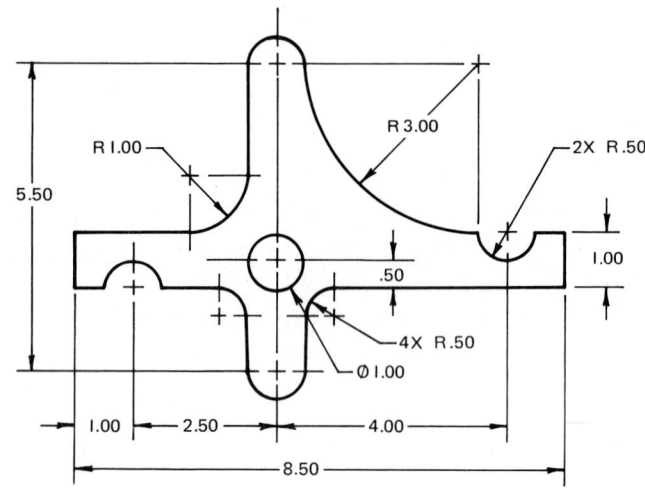

FIG. 4-2-L   Pawl.

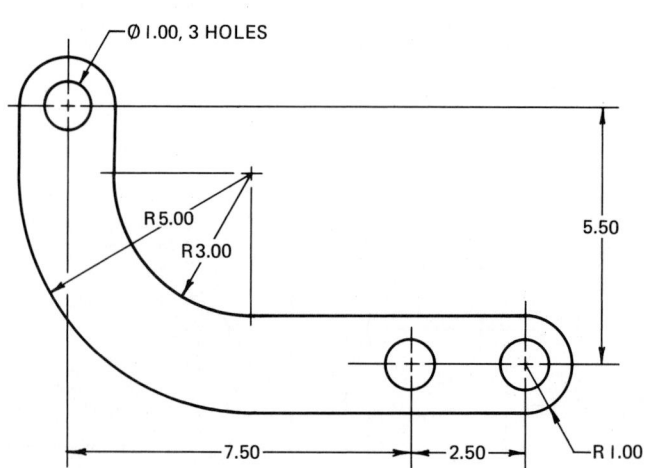

FIG. 4-2-M   Rod support.

FIG. 4-2-N   Reel side.

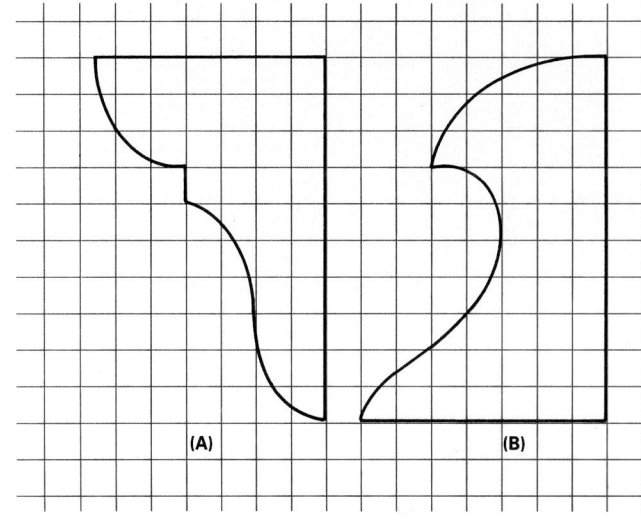

ASSIGNMENTS FOR UNIT 4-3, IRREGULAR CURVES

23. On a B (A3) size format, lay out the pattern for the table leg shown in Fig. 4-3-A to the scale 1:2.
24. Using grid paper or creating a grid on the monitor, draw the furniture patterns shown in Fig. 4-3-B. Use .50 in. or 10 mm grid.
25. Using grid paper or creating a grid on the monitor, draw the line graph shown in Fig. 4-3-C. Use .25 in. or 5 mm grid.

FIG. 4-3-B   Furniture patterns.

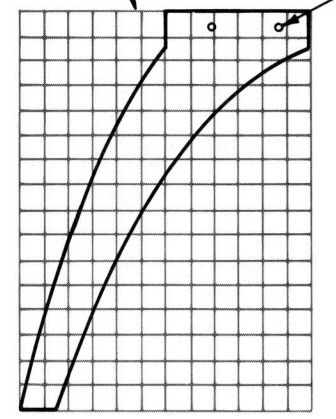

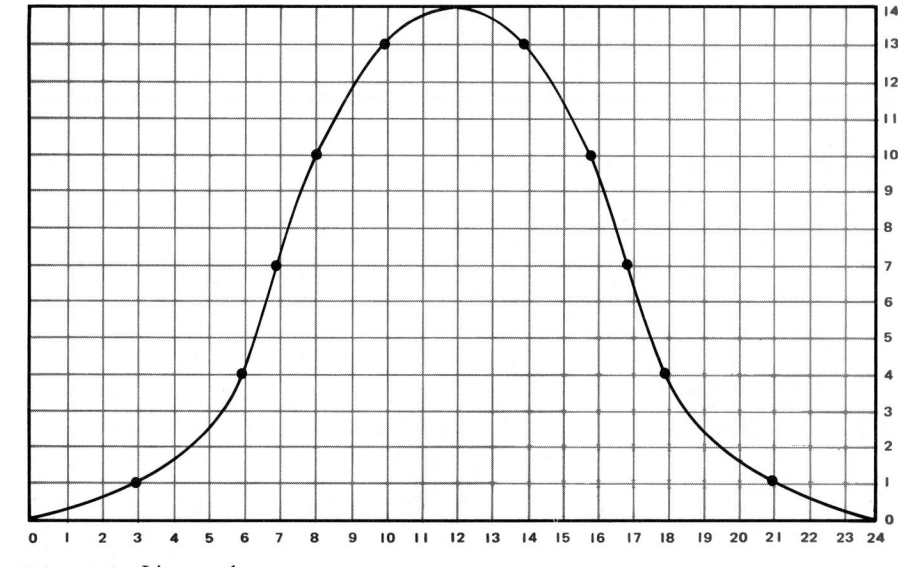

FIG. 4-3-A   Table leg.

FIG. 4-3-C   Line graph.

ASSIGNMENTS FOR UNIT 4-4, SKETCHING

26. Using grid paper, sketch the template shown in Fig. 4-2-D.

27. Using grid paper, sketch the shaft support shown in Fig. 4-2-F.
28. Using grid paper, sketch the patterns shown in Fig. 4-4-A.
29. Using grid paper, sketch the patterns shown in Fig. 4-4-B.

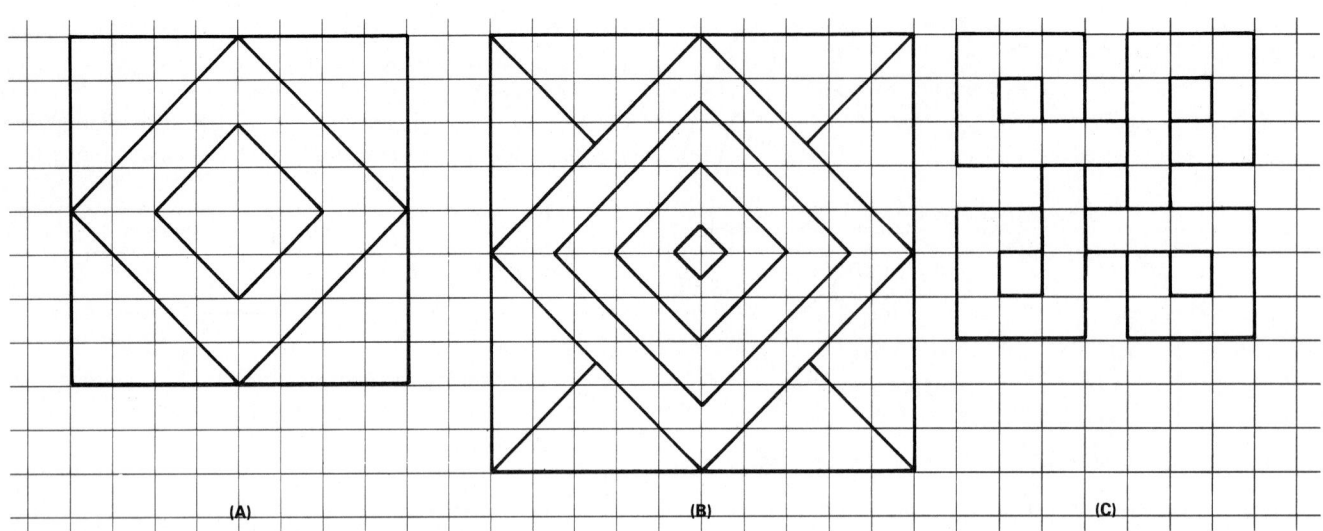

FIG. 4-4-A   Sketching assignment.

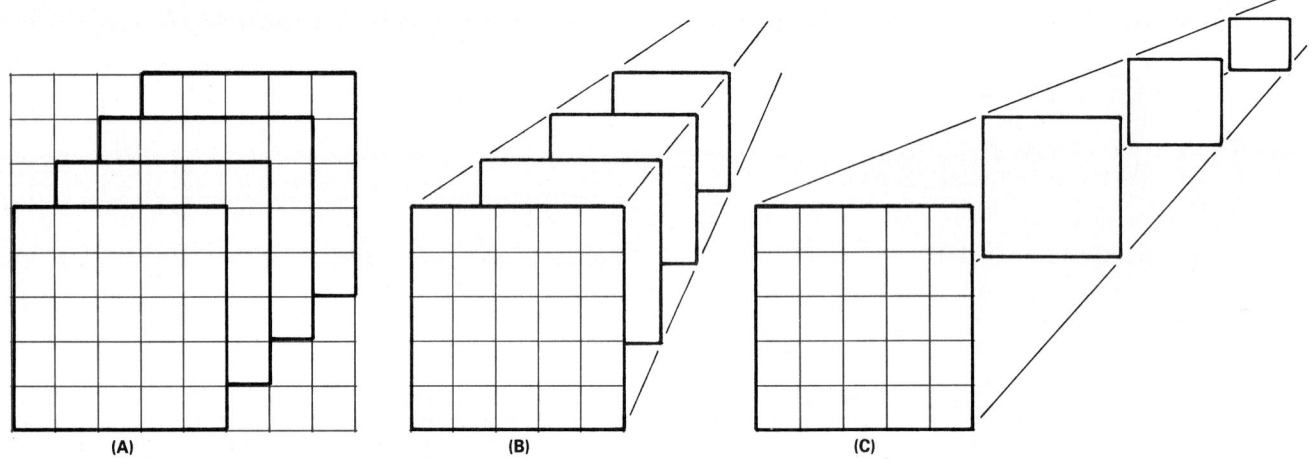

FIG. 4-4-B   Sketching assignment.

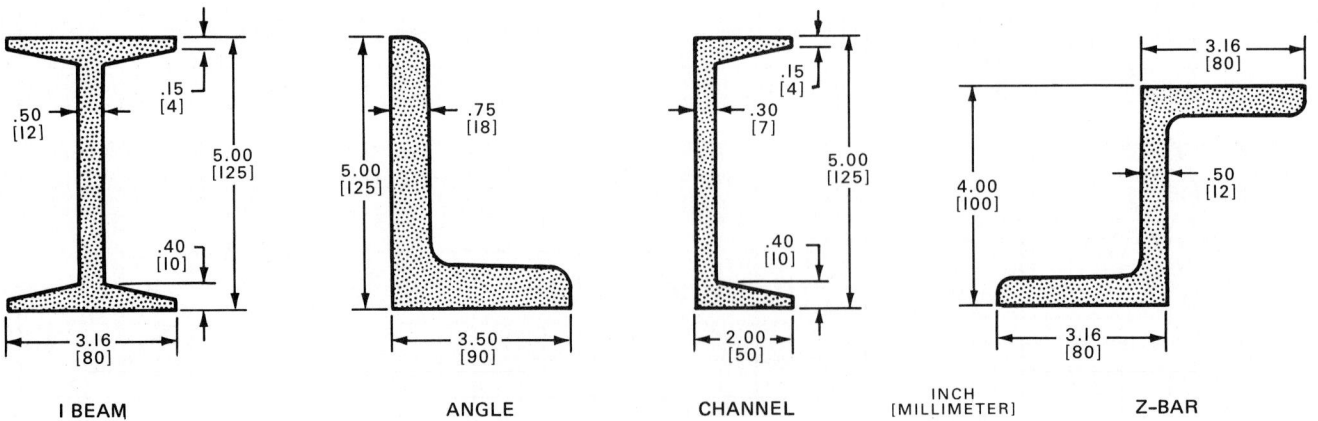

FIG. 4-4-C   Structural steel shapes.

FIG. 4-4-D   Sketching lines, circles, and arcs.

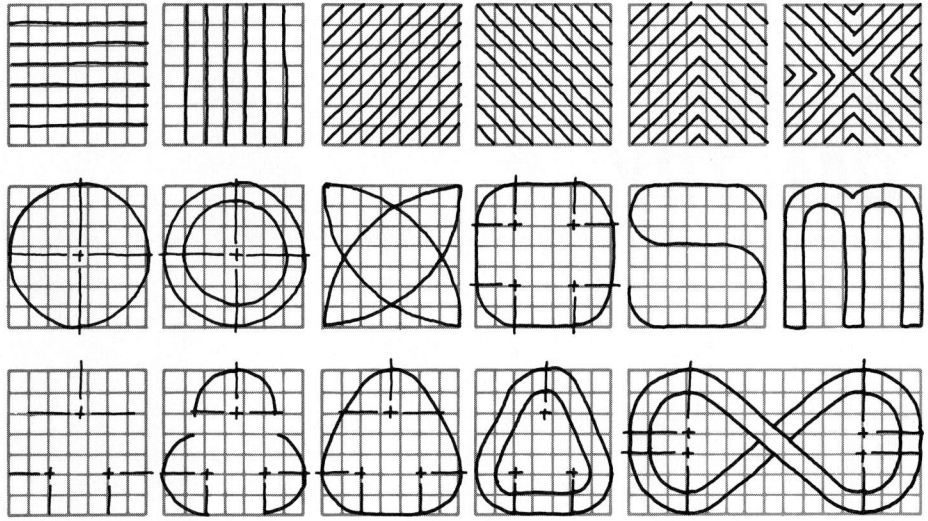

30. Using grid paper, sketch the structural steel shapes shown in Fig. 4-4-C. The shapes do not have to be drawn to scale, but should be drawn in proportion.
31. Using grid paper, sketch the patterns shown in Fig. 4-4-D.

ASSIGNMENTS FOR UNIT 4-5, INKING

32. As assigned by your instructor, make an ink drawing of any of the assignments for Units 4-1 through 4-4.
33. Make an ink drawing of one of the parts shown in Figs. 4-5-A to 4-5-C. Scale 1:1. Do not dimension.

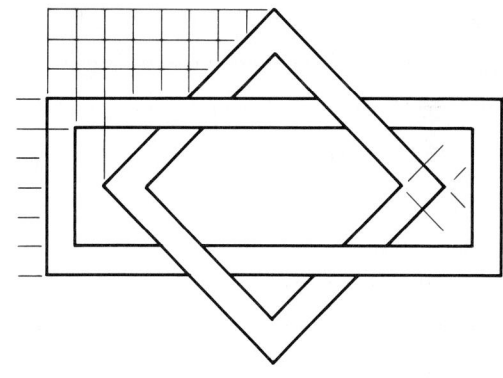

.50 IN. OR 10 mm GRID

FIG. 4-5-B   Inlay design.

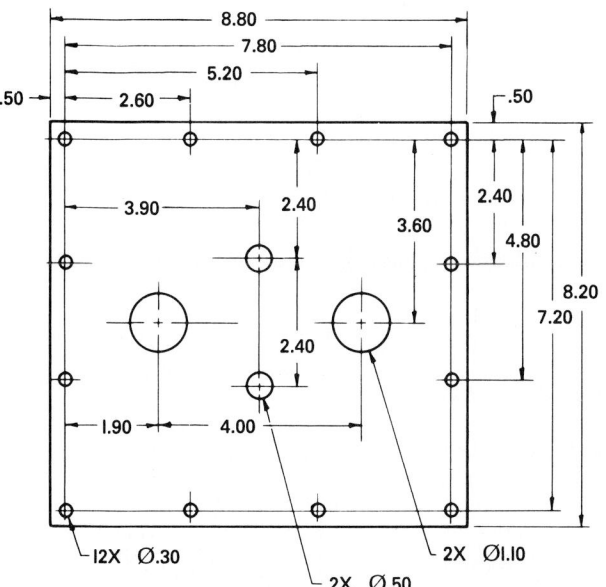

FIG. 4-5-A   Cover plate.

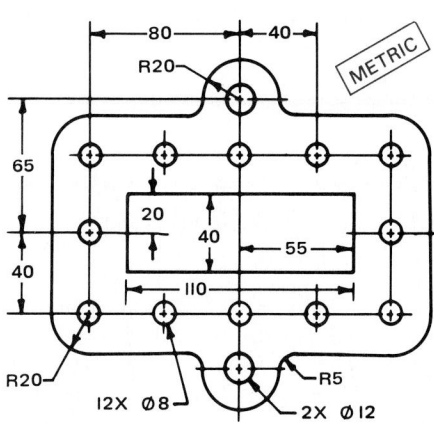

FIG. 4-5-C   Gasket.

CHAPTER 5

APPLIED GEOMETRY

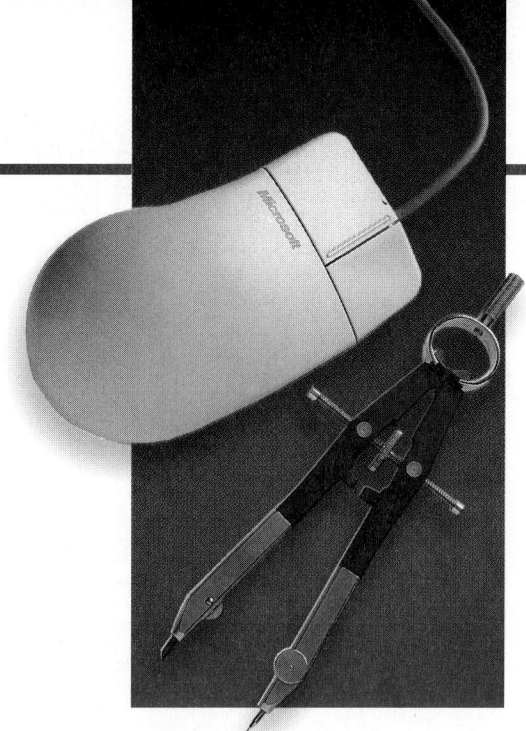

Definitions

Angle   The shape formed by two lines that extend from the same point.

Bisect   To cut or divide into two.

Circumscribe   To place a figure around another, touching it at points but not cutting it.

Geometry   The study of the size, shape, and relationship of objects.

Helix   The curve generated by a point that revolves uniformly around and up or down the surface of a cylinder.

Inscribe   To place a figure within another so that all angular points of it lie on the boundary (circumference).

Parabola   A plane curve generated by a point that moves along a path equidistant from a fixed line and fixed point.

Parallel   Continuously equidistant lines or surfaces.

Perpendicular   At right angles (90°) to a line or surface.

Polygon   A figure bounded by five or more straight lines, possibly of uneven length.

Tangent   Meeting a line or surface at a point but not intersecting it.

5-1 BEGINNING GEOMETRY: STRAIGHT LINES

Geometry is the study of the size and shape of objects. The relationship of straight and curved lines in drawing shapes is also a part of geometry. Some geometric figures used in drafting include circles, squares, triangles, hexagons, and octagons (Fig. 5-1-1).

Geometric constructions are made of individual lines and points drawn in proper relationship to one another. Accuracy is extremely critical.

Geometric constructions are very important to drafters, surveyors, engineers, architects, scientists, mathematicians, and designers. Geometric constructions have important uses, both in making drawings and in solving problems with graphs and diagrams. Sometimes it is necessary to use geometric constructions, particularly if, when doing manual drafting, the drafter does not have the advantages afforded by a drafting machine, an adjustable triangle, or templates for drawing hexagonal and elliptical shapes. Therefore, nearly everyone in all technical fields needs to know the constructions explained in this chapter.

All of the lines and shapes shown in this chapter can be drawn using CAD commands. This chapter covers manual drafting using the instruments and equipment described in Chap. 4. The following exercises provide practice in geometric constructions.

To Draw a Line or Lines Parallel to and at a Given Distance from an Oblique Line

1. Given line *AB* (Fig. 5-1-2), erect a perpendicular *CD* to *AB*.
2. Space the given distance from the line *AB* by scale measurement or by an arc along line *CD*.
3. Position a triangle, using a second triangle or a T square as base, so that one side of the triangle is parallel with the given line.
4. Slide this triangle along the base to the point at the desired distance from the given line, and draw the required line.

To Draw a Straight Line Tangent to Two Circles

Place a T square or straightedge so that the top edge just touches the edges of the circles, and draw the tangent line (Fig. 5-1-3). Perpendiculars to this line from the centers of the circles give the tangent points T_1 and T_2.

To Bisect a Straight Line

1. Given line *AB* (Fig. 5-1-4, pg. 78), set the compass to a radius greater than ½ *AB*.

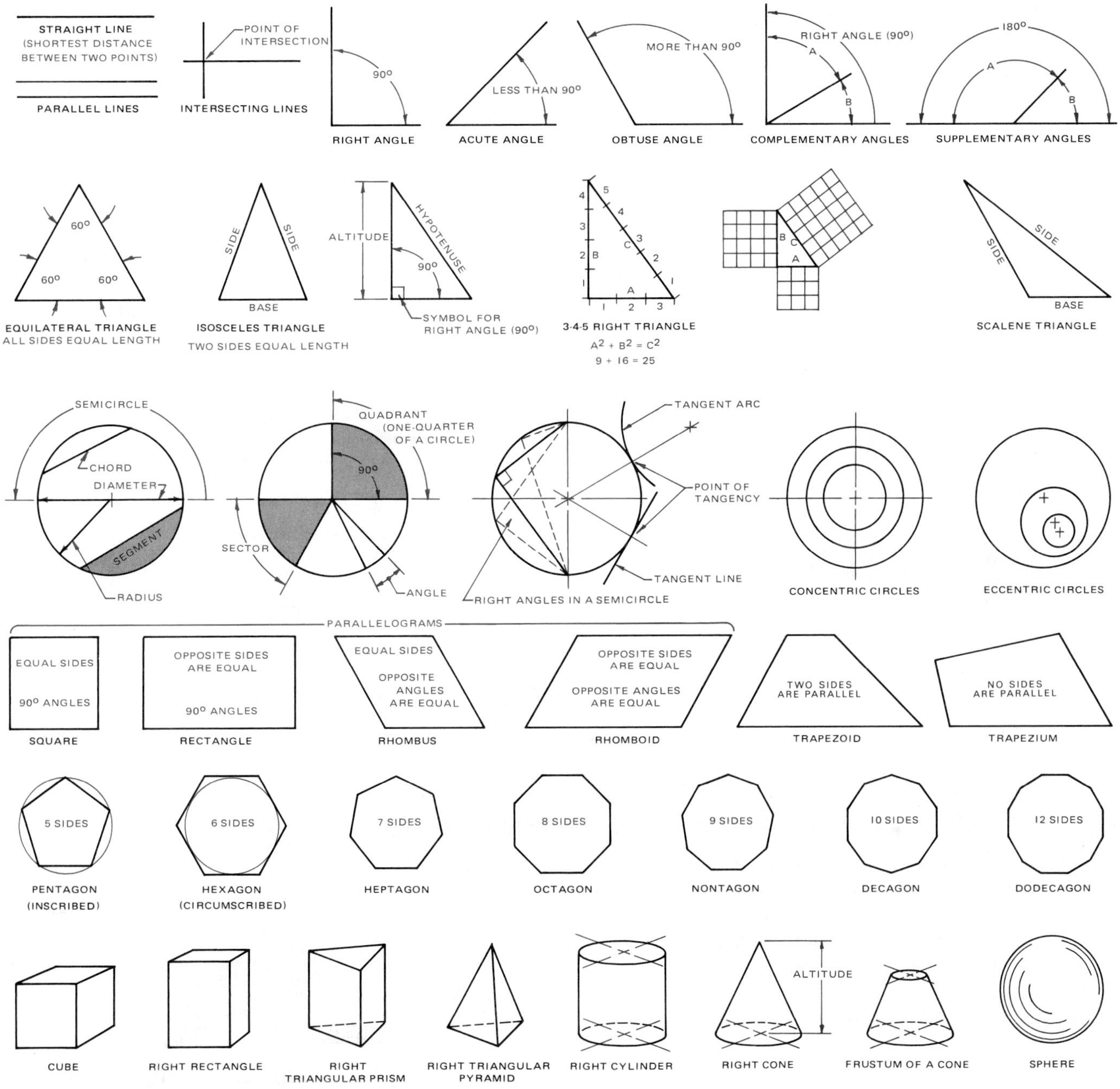

STRAIGHT LINE
(SHORTEST DISTANCE
BETWEEN TWO POINTS)

PARALLEL LINES

POINT OF
INTERSECTION

INTERSECTING LINES

90°

RIGHT ANGLE

LESS THAN 90°

ACUTE ANGLE

MORE THAN 90°

OBTUSE ANGLE

RIGHT ANGLE (90°)

A

B

COMPLEMENTARY ANGLES

180°

A

B

SUPPLEMENTARY ANGLES

60°

60°   60°

EQUILATERAL TRIANGLE
ALL SIDES EQUAL LENGTH

SIDE   SIDE

BASE

ISOSCELES TRIANGLE
TWO SIDES EQUAL LENGTH

ALTITUDE

HYPOTENUSE

90°

SYMBOL FOR
RIGHT ANGLE (90°)

5
4
3
2
1

B

C

A

1 2 3

3-4-5 RIGHT TRIANGLE
$A^2 + B^2 = C^2$
$9 + 16 = 25$

B
C
A

SIDE   SIDE

BASE

SCALENE TRIANGLE

SEMICIRCLE

CHORD

DIAMETER

SEGMENT

RADIUS

QUADRANT
(ONE-QUARTER
OF A CIRCLE)

90°

SECTOR

ANGLE

TANGENT ARC

POINT OF
TANGENCY

TANGENT LINE

RIGHT ANGLES IN A SEMICIRCLE

CONCENTRIC CIRCLES

ECCENTRIC CIRCLES

PARALLELOGRAMS

EQUAL SIDES

90° ANGLES

SQUARE

OPPOSITE SIDES
ARE EQUAL

90° ANGLES

RECTANGLE

EQUAL SIDES

OPPOSITE
ANGLES
ARE EQUAL

RHOMBUS

OPPOSITE SIDES
ARE EQUAL

OPPOSITE ANGLES
ARE EQUAL

RHOMBOID

TWO SIDES
ARE PARALLEL

TRAPEZOID

NO SIDES
ARE PARALLEL

TRAPEZIUM

5 SIDES

PENTAGON
(INSCRIBED)

6 SIDES

HEXAGON
(CIRCUMSCRIBED)

7 SIDES

HEPTAGON

8 SIDES

OCTAGON

9 SIDES

NONTAGON

10 SIDES

DECAGON

12 SIDES

DODECAGON

CUBE

RIGHT RECTANGLE

RIGHT
TRIANGULAR PRISM

RIGHT TRIANGULAR
PYRAMID

RIGHT CYLINDER

ALTITUDE

RIGHT CONE

FRUSTUM OF A CONE

SPHERE

FIG. 5-1-1   Dictionary of drafting geometry.

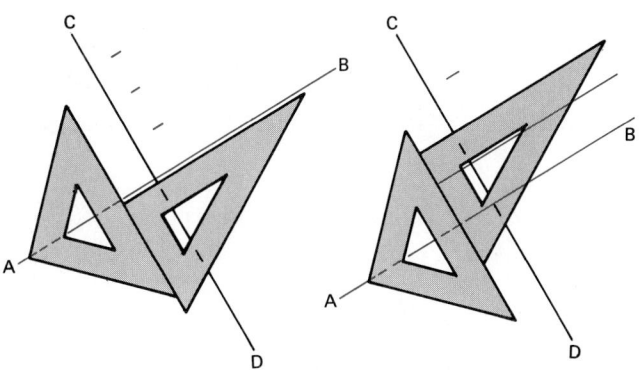

FIG. 5-1-2   Drawing parallel lines with the use of triangles.

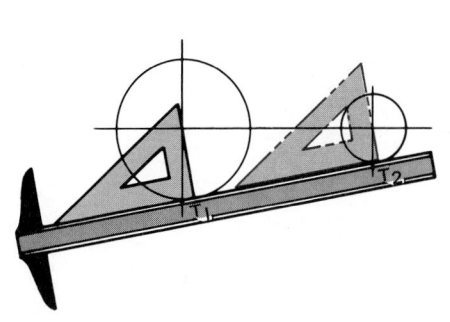

FIG. 5-1-3   Drawing a straight line tangent to two circles.

2. Using centers at *A* and *B*, draw intersecting arcs above and below line *AB*. A line *CD* drawn through the intersections will divide *AB* into two equal parts and will be perpendicular to line *AB*.

To Bisect an Arc

1. Given arc *AB* (Fig. 5-1-5), set the compass to a radius greater than ½ *AB*.
2. Using points *A* and *B* as centers, draw intersecting arcs above and below arc *AB*. A line drawn through the intersections *C* and *D* will divide the arc *AB* into two equal parts.

To Bisect an Angle

1. Given angle *ABC*, with center *B* and a suitable radius (Fig. 5-1-6), draw an arc to cut *BC* at *D* and *BA* at *E*.
2. With centers *D* and *E* and equal radii, draw arcs to intersect at *F*.
3. Join *B* and *F* and extend to *G*. Line *BG* is the required bisector.

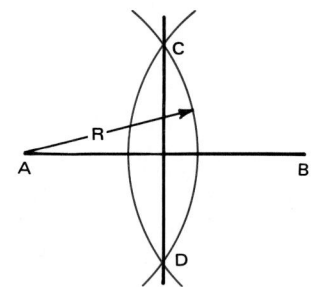

FIG. 5-1-4   Bisecting a line.

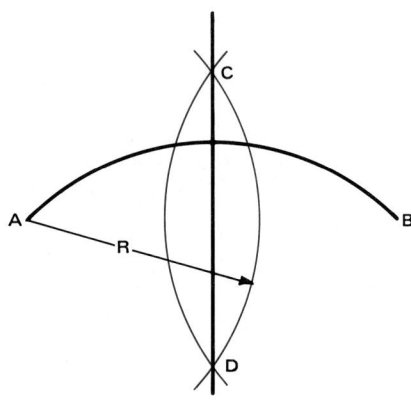

FIG. 5-1-5   Bisecting an arc.

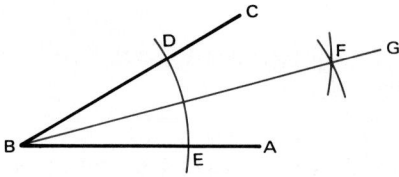

FIG. 5-1-6   Bisecting an angle.

To Divide a Line into a Given Number of Equal Parts

1. Given line *AB* and the number of equal divisions desired (12, for example), draw a perpendicular from *A*.
2. Place the scale so that the desired number of equal divisions is conveniently included between *B* and the perpendicular. Then mark these divisions, using short vertical marks from the scale divisions, as in Fig. 5-1-7.
3. Draw perpendiculars to line *AB* through the points marked, dividing the line *AB* as required.

ASSIGNMENTS

See Assignments 1 through 3 for Unit 5-1 on pages 84–85.

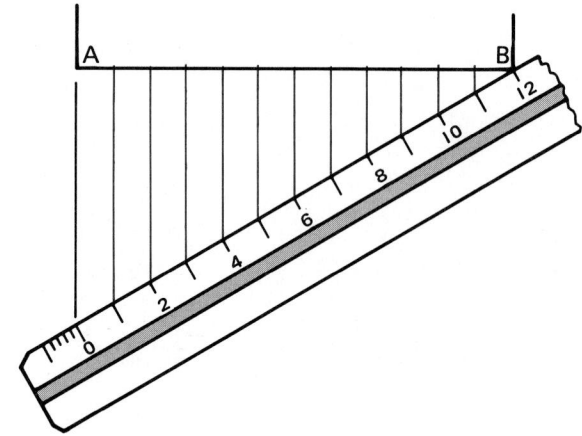

FIG. 5-1-7   Dividing a straight line into equal parts.

5-2 ARCS AND CIRCLES

To Draw an Arc Tangent to Two Lines at Right Angles to Each Other

Given radius R of the arc (Fig. 5-2-1):

1. Draw an arc having radius *R* with center at *B*, cutting the lines *AB* and *BC* at *D* and *E*, respectively.
2. With *D* and *E* as centers and with the same radius *R*, draw arcs intersecting at *O*.
3. With center *O*, draw the required arc. The tangent points are *D* and *E*.

To Draw an Arc Tangent to the Sides of an Acute Angle

Given radius *R* of the arc (Fig. 5-2-2):

1. Draw lines inside the angle, parallel to the given lines, at distance *R* away from the given lines. The center of the arc will be at *C*.

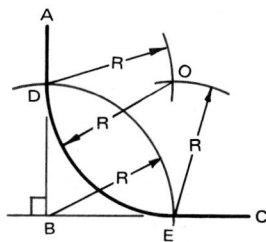

FIG. 5-2-1   Arc tangent to two lines at right angles to each other.

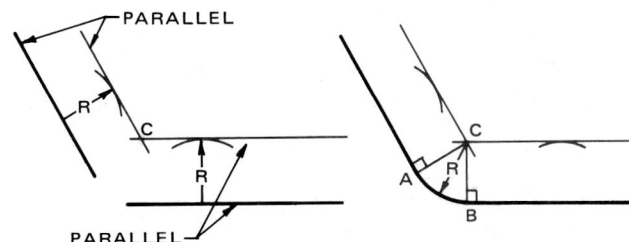

FIG. 5-2-3   Drawing an arc tangent to the sides of an obtuse angle.

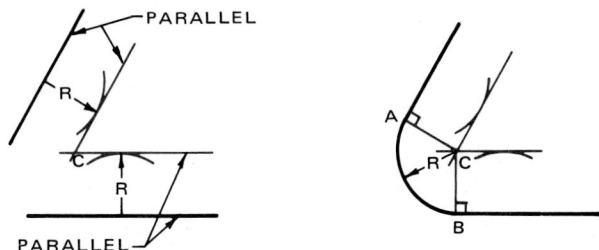

FIG. 5-2-2   Drawing an arc tangent to the sides of an acute angle.

2. Set the compass to radius *R*, and with center *C* draw the arc tangent to the given sides. The tangent points *A* and *B* are found by drawing perpendiculars through point *C* to the given lines.

To Draw an Arc Tangent to the Sides of an Obtuse Angle

Given radius *R* of the arc (Fig. 5-2-3):

1. Draw lines inside the angle, parallel to the given lines at distance *R* away from the given lines. The center of the arc will be at *C*.
2. Set the compass to radius *R*, and with center *C* draw the arc tangent to the given sides. The tangent points *A* and *B* are found by drawing perpendiculars through point *C* to the given lines.

To Draw a Circle on a Regular Polygon

1. Given the size of the polygon (Fig. 5-2-4), bisect any two sides, for example, *BC* and *DE*. The center of the polygon is where bisectors *FO* and *GO* intersect at point *O*.
2. The inner circle radius is *OH*, and the outer circle radius is *OA*.

To Draw a Reverse, or Ogee, Curve Connecting Two Parallel Lines

1. Given two parallel lines *AB* and *CD* and distances *X* and *Y* (Fig. 5-2-5), join points *B* and *C* with a line.
2. Erect a perpendicular to *AB* and *CD* from points *B* and *C*, respectively.
3. Select point *E* on line *BC* where the curves are to meet.
4. Bisect *BE* and *EC*.
5. Points *F* and *G* where the perpendiculars and bisectors meet are the centers for the arcs forming the ogee curve.

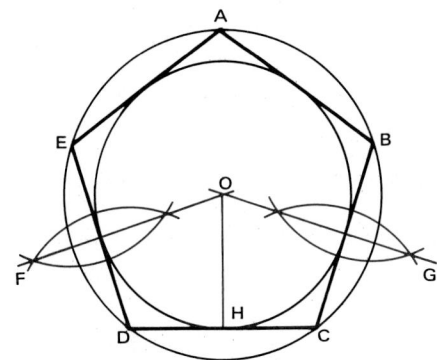

FIG. 5-2-4   Drawing a circle on a regular polygon.

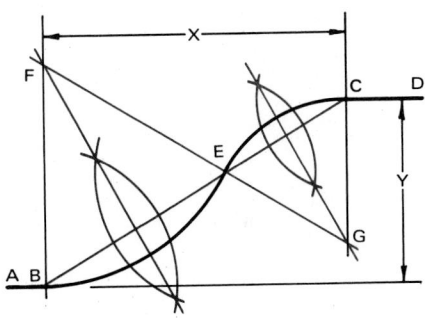

FIG. 5-2-5   Drawing a reverse (ogee) curve connecting two parallel lines.

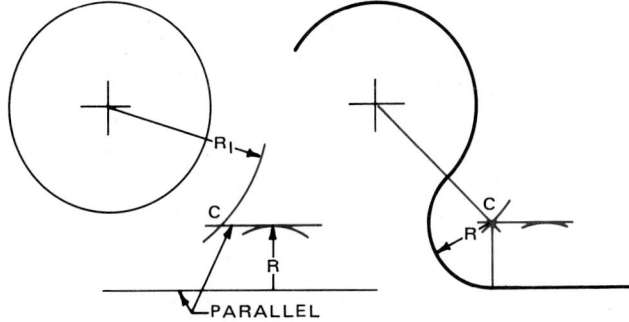

FIG. 5-2-6   Drawing an arc tangent to a circle and a straight line.

To Draw an Arc Tangent to a Given Circle and Straight Line

1. Given *R*, the radius of the arc (Fig. 5-2-6), draw a line parallel to the given straight line between the circle and the line at distance *R* away from the given line.

2. With the center of the circle as center and radius R_1 (radius of the circle plus R), draw an arc to cut the parallel straight line at C.

3. With center C and radius R, draw the required arc tangent to the circle and the straight line.

To Draw an Arc Tangent to Two Circles

1. Given the radius of arc R (Fig. 5-2-7A), with the center of circle A as center and radius R_2 (radius of circle A plus R), draw an arc in the area between the circles.

2. With the center of circle B as center and radius R_3 (radius of circle B plus R), draw an arc to cut the other arc at C.

3. With center C and radius R, draw the required arc tangent to the given circles.

As an alternative:

1. Given radius of arc R (Fig. 5-2-7B), with the center of circle A as center and radius $R - R_2$, draw an arc in the area between the circles.

2. With the center of circle B as center and radius $R - R_3$, draw an arc to cut the other arc at C.

3. With center C and radius R, draw the required arc tangent to the given circles.

To Draw an Arc or Circle Through Three Points Not in a Straight Line

1. Given points A, B, and C (Fig. 5-2-8), join points A, B, and C as shown.

2. Bisect lines AB and BC and extend bisecting lines to intersect at O. Point O is the center of the required circle or arc.

3. With center O and radius OA draw an arc.

ASSIGNMENTS

See Assignments 4 through 8 for Unit 5-2 on pages 86–87.

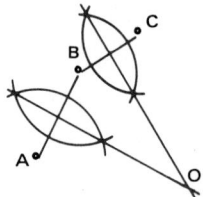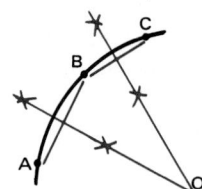

FIG. 5-2-8   Drawing an arc or circle through three points not in a straight line.

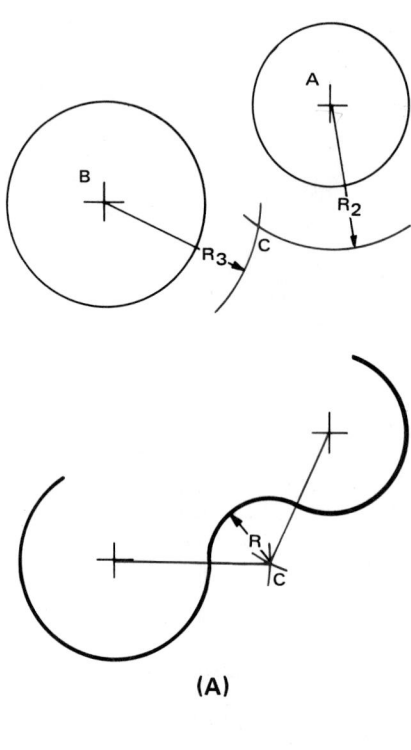

(A)

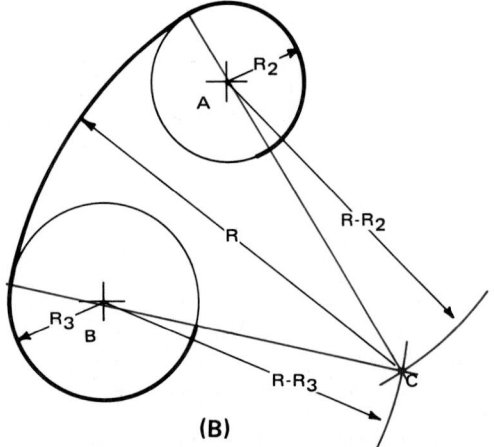

(B)

FIG. 5-2-7   Drawing an arc tangent to two circles.

5-3 POLYGONS

A *polygon* is a plane figure bounded by five or more straight lines not necessarily of equal length. A *regular polygon* is a plane figure bounded by five or more straight lines of equal length and containing angles of equal size.

To Draw a Hexagon, Given the Distance Across the Flats

1. Establish horizontal and vertical center lines for the hexagon (Fig. 5-3-1).

2. Using the intersection of these lines as center, with radius one-half the distance across the flats, draw a light construction circle.

3. Using the 60° triangle, draw six straight lines, equally spaced, passing through the center of the circle.

4. Draw tangents to these lines at their intersection with the circle.

To Draw a Hexagon, Given the Distance Across the Corners

1. Establish horizontal and vertical center lines, and draw a light construction circle with radius one-half the distance across the corners (Fig. 5-3-2).
2. With a 60° triangle, establish points on the circumference 60° apart.
3. Draw straight lines connecting these points.

To Draw an Octagon, Given the Distance Across the Flats

1. Establish horizontal and vertical center lines and draw a light construction circle with radius one-half the distance across the flats (Fig. 5-3-3).
2. Draw horizontal and vertical lines tangent to the circle.
3. Using the 45° triangle, draw lines tangent to the circle at a 45° angle from the horizontal.

To Draw an Octagon, Given the Distance Across the Corners

1. Establish horizontal and vertical center lines and draw a light construction circle with radius one-half the distance across the corners (Fig. 5-3-4).
2. With the 45° triangle, establish points on the circumference between the horizontal and vertical center lines.

3. Draw straight lines connecting these points to the points where the center lines cross the circumference.

To Draw a Regular Polygon, Given the Length of the Sides

As an example, let a polygon have seven sides.
1. Given the length of side AB (Fig. 5-3-5), with radius AB and A as center, draw a semicircle and divide it into seven equal parts using a protractor.

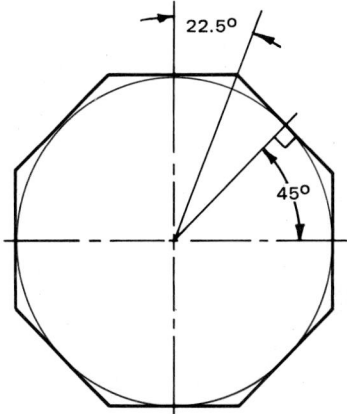

FIG. 5-3-3   Constructing an octagon, given the distance across flats.

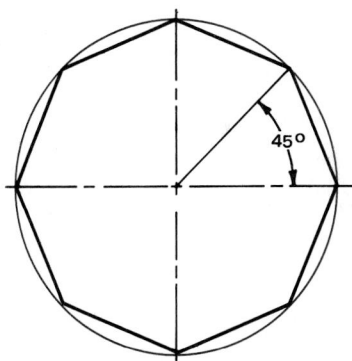

FIG. 5-3-4   Constructing an octagon, given the distance across corners.

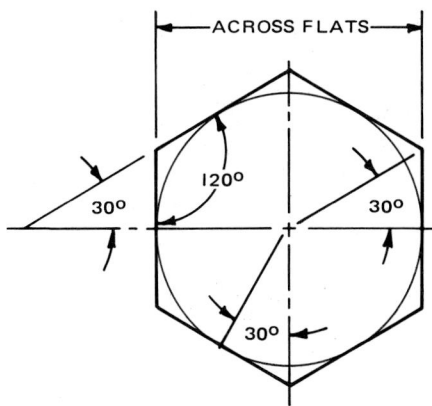

FIG. 5-3-1   Constructing a hexagon, given the distance across flats.

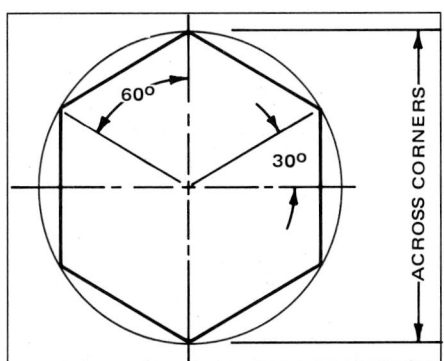

FIG. 5-3-2   Constructing a hexagon, given the distance across corners.

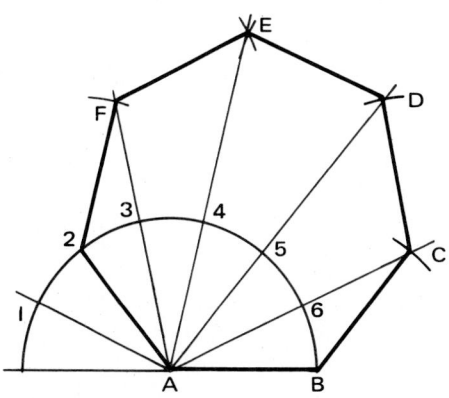

FIG. 5-3-5   Constructing a regular polygon, given the length of one side.

2. Through the second division from the left, draw radial line *A*2.
3. Through points 3, 4, 5, and 6 extend radial lines as shown.
4. With *AB* as radius and *B* as center, cut line *A*6 at *C*. With the same radius and *C* as center, cut line *A*5 at *D*. Repeat at *E* and *F*.
5. Connect these points with straight lines.

These steps can be followed in drawing a regular polygon with any number of sides.

To Inscribe a Regular Pentagon in a Given Circle

1. Given circle with center *O* (Fig. 5-3-6), draw the circle with diameter *AB*.
2. Bisect line *OB* at *D*.
3. With center *D* and radius *DC*, draw arc *CE* to cut the diameter at *E*.
4. With *C* as center and radius *CE*, draw arc *CF* to cut the circumference at *F*. Distance *CF* is one side of the pentagon.
5. With radius *CF* as a chord, mark off the remaining points on the circle. Connect the points with straight lines.

ASSIGNMENTS

See Assignments 9 through 11 for Unit 5-3 on pages 88–89.

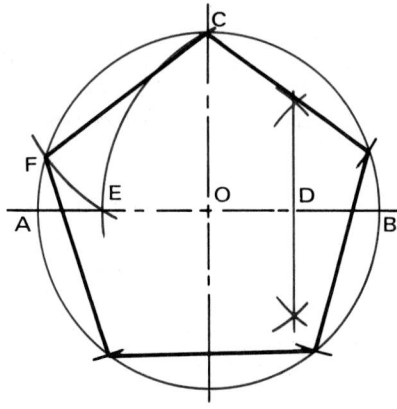

FIG. 5-3-6   Inscribing a regular pentagon in a given circle.

5-4 ELLIPSE

The *ellipse* is the plane curve generated by a point moving so that the sum of the distances from any point on a curve to two fixed points, called *foci*, is a constant.

Often a drafter is called upon to draw oblique and inclined holes and surfaces that take the approximate form of an ellipse. Several methods, true and approximate, are used for its construction. The terms *major diameter* and *minor diameter* will be used in place of *major axis* and *minor axis* so the reader won't become confused with the mathematical *X* and *Y* axes.

To Draw an Ellipse—Two-Circle Method

1. Given the major and minor diameters (Fig. 5-4-1), construct two concentric circles with diameters equal to *AB* and *CD*.
2. Divide the circles into a convenient number of equal parts. Figure 5-4-1 shows 12.
3. Where the radial lines intersect the outer circle, as at 1, draw lines parallel to line *CD* inside the outer circle.
4. Where the same radial line intersects the inner circle, as at 2, draw a line parallel to axis *AB* away from the inner circle. The intersection of these lines, as at 3, gives points on the ellipse.
5. Draw a smooth curve through these points.

To Draw an Ellipse—Four-Center Method

1. Given the major diameter *CD* and the minor diameter *AB* (Fig. 5-4-2), join points *A* and *C* with a line.
2. Draw an arc with point *O* as the center and radius *OC* and extend line *OA* to locate point *E*.
3. Draw an arc with point *A* as the center and radius *AE* to locate point *F*.

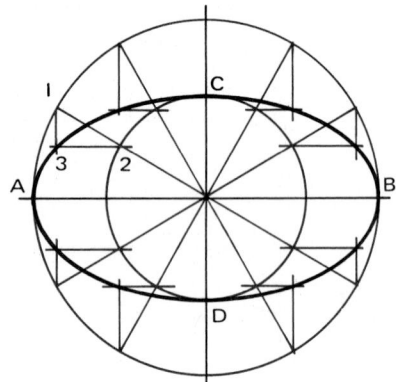

FIG. 5-4-1   Drawing an ellipse—two-circle method.

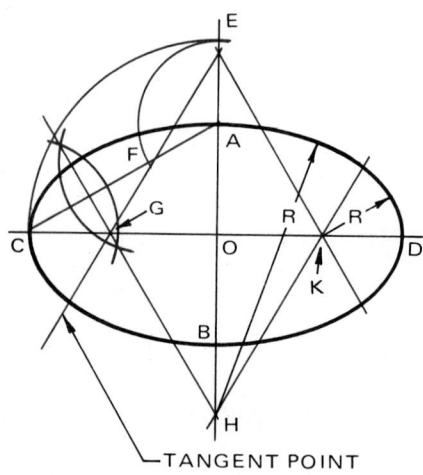

FIG. 5-4-2   Drawing an ellipse—four-center method.

4. Draw the perpendicular bisector of line *CF* to locate points *G* and *H*.
5. Draw arcs with *G* and *K* as centers and radii *HA* and *EB* to complete the ellipse.

To Draw an Ellipse—Parallelogram Method

1. Given the major diameter *CD* and minor diameter *AB* (Fig. 5-4-3), construct a parallelogram.
2. Divide *CO* into a number of equal parts. Divide *CE* into the same number of equal parts. Number the points from *C*.
3. Draw a line from *B* to point 1 on line *CE*. Draw a line from *A* through point 1 on *CO*, intersecting the previous line. The point of intersection will be one point on the ellipse.
4. Proceed in the same manner to find other points on the ellipse.
5. Draw a smooth curve through these points.

ASSIGNMENTS

See Assignments 12 and 13 for Unit 5-4 on page 90.

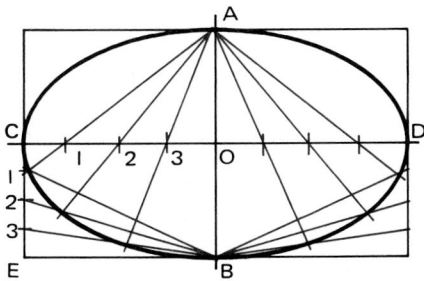

FIG. 5-4-3   Drawing an ellipse—parallelogram method.

FIG. 5-5-1   Drawing a cylindrical helix.

5-5 HELIX AND PARABOLA

Helix

The *helix* is the curve generated by a point that revolves uniformly around and up or down the surface of a cylinder. The *lead* is the vertical distance that the point rises or drops in one complete revolution.

To Draw a Helix

1. Given the diameter of the cylinder and the lead (Fig. 5-5-1), draw the top and front views.
2. Divide the circumference (top view) into a convenient number of parts (use 12) and label them.
3. Project lines down to the front view.
4. Divide the lead into the same number of equal parts and label them as shown in Fig. 5-5-1.
5. The points of intersection of lines with corresponding numbers lie on the helix. *Note*: Since points 8 to 12 lie on the back portion of the cylinder, the helix curve starting at point 7 and passing through points 8, 9, 10, 11, 12 to point 1 will appear as a hidden line.
6. If the development of the cylinder is drawn, the helix will appear as a straight line on the development.

Parabola

The *parabola* is a plane curve generated by a point that moves along a path equidistant from a fixed line (*directrix*) and a fixed point (*focus*). Again, these methods produce an approximation of the true conic section.

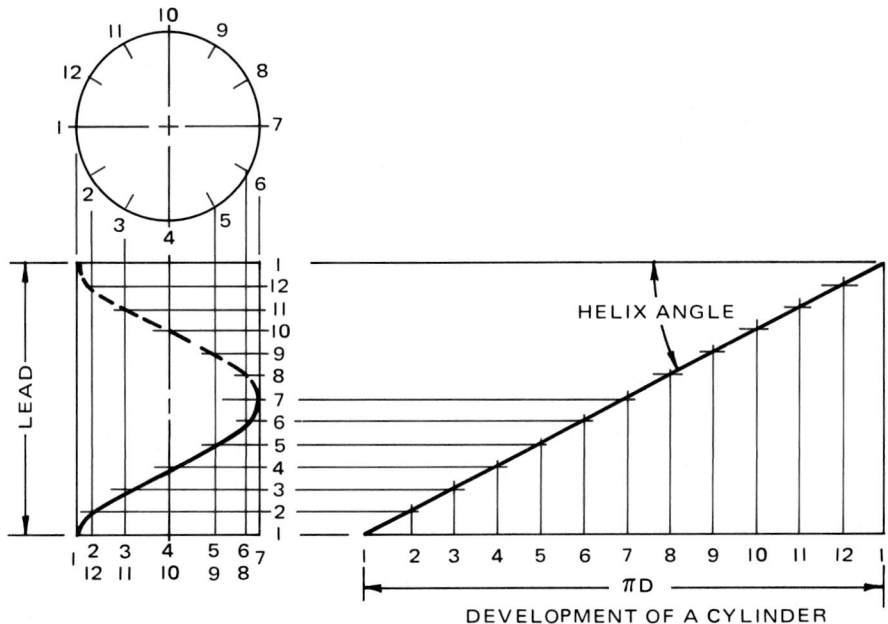

To Construct a Parabola—Parallelogram Method

1. Given the sizes of the enclosing rectangle, distances *AB* and *AC* (Fig. 5-5-2A), construct a parallelogram.
2. Divide *AC* into a number of equal parts. Number the points as shown. Divide distance A–O into the same number of equal parts.
3. Draw a line from *O* to point 1 on line *AC*. Draw a line parallel to the axis through point 1 on line *AO*, intersecting the previous line *O1*. The point of intersection will be one point on the parabola.
4. Proceed in the same manner to find other points on the parabola.
5. Connect the points using an irregular curve.

To Construct a Parabola—Offset Method

1. Given the sizes of the enclosing rectangle, distances *AB* and *AC* (Fig. 5-5-2B), construct a parallelogram.
2. Divide *OA* into four equal parts.
3. The offsets vary in length as the square of their distances from *O*. Since *OA* is divided into four equal parts, distance *AC* will be divided into 4^2, or 16, equal divisions. Thus since *O1* is one-fourth the length of *OA*, the length of line $1\text{-}1_1$ will be $(\frac{1}{4})^2$, or $\frac{1}{16}$, the length of *AC*.
4. Since distance *O2* is one-half the length of *OA*, the length of line $2\text{-}2_1$ will be $(\frac{1}{2})^2$, or $\frac{1}{4}$, the length of *AC*.
5. Since distance *O3* is three-fourths the length of *OA*, the length of line $3\text{-}3_1$ will be $(\frac{3}{4})^2$, or $\frac{9}{16}$, the length of *AC*.
6. Complete the parabola by joining the points with an irregular curve.

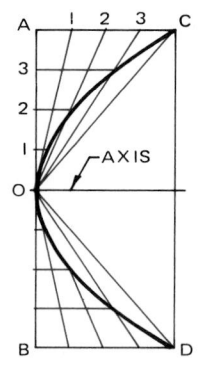

PARALLELOGRAM METHOD

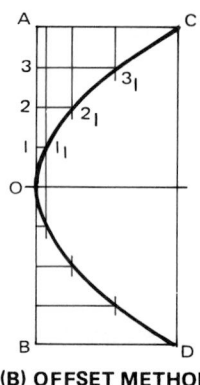

(B) OFFSET METHOD

FIG. 5-5-2   Common methods used to construct a parabola.

ASSIGNMENTS

See Assignments 14 through 16 for Unit 5-5 on pages 90–91.

ASSIGNMENTS FOR CHAPTER 5

Note: In the preparation of the following drawings, it is advisable to begin by using light construction lines. This will permit overruns and miscalculations without damage to the worksheet. After the drawing has been roughed out, make final lines of the appropriate type and thickness.

ASSIGNMENTS FOR UNIT 5-1, STRAIGHT LINES

1. Divide a B (A3) size sheet as in Fig. 5-1-A. In the designated areas draw the geometric constructions. Scale 1:1.

2. On an A (A4) size sheet, draw the geometric constructions in Fig. 5-1-B. For part C use light construction circles to develop the squares. Scale 1:1. Do not dimension.

3. On an A (A4) size sheet, draw the geometric constructions in Fig. 5-1-C. Use light construction circles to develop the figures. Scale 1:1. Do not dimension.

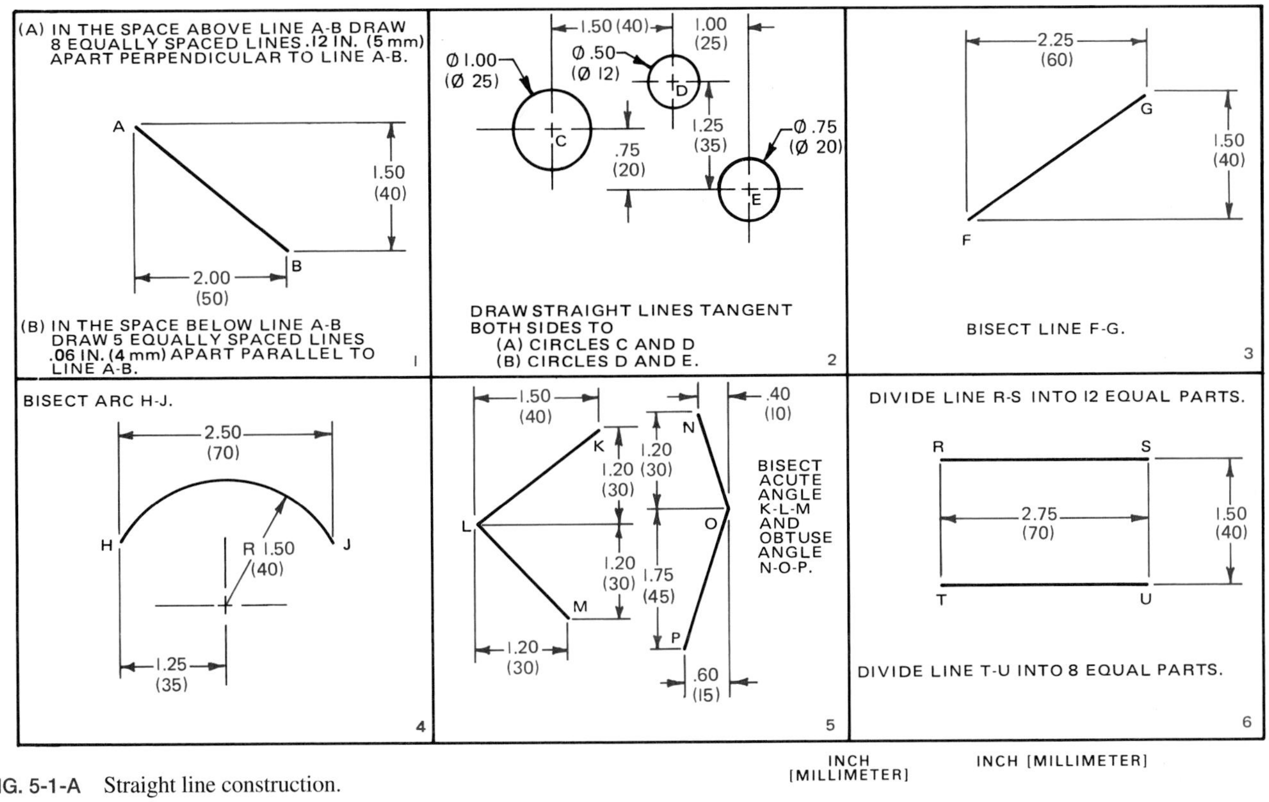

(A) IN THE SPACE ABOVE LINE A-B DRAW
8 EQUALLY SPACED LINES .12 IN. (5mm)
APART PERPENDICULAR TO LINE A-B.

A

1.50
(40)

2.00
(50)

B

(B) IN THE SPACE BELOW LINE A-B
DRAW 5 EQUALLY SPACED LINES
.06 IN. (4mm) APART PARALLEL TO
LINE A-B.

1

Ø 1.00
(Ø 25)

Ø .50
(Ø 12)

D

C

.75
(20)

1.50 (40)

1.00
(25)

1.25
(35)

E

Ø .75
(Ø 20)

DRAW STRAIGHT LINES TANGENT
BOTH SIDES TO
(A) CIRCLES C AND D
(B) CIRCLES D AND E.

2

2.25
(60)

G

1.50
(40)

F

BISECT LINE F-G.

3

BISECT ARC H-J.

2.50
(70)

H

R 1.50
(40)

J

1.25
(35)

4

1.50
(40)

K

N

.40
(10)

1.20
(30)

1.20
(30)

L

O

1.20
(30)

M

1.75
(45)

1.20
(30)

P

.60
(15)

BISECT
ACUTE
ANGLE
K-L-M
AND
OBTUSE
ANGLE
N-O-P.

INCH
[MILLIMETER]

5

DIVIDE LINE R-S INTO 12 EQUAL PARTS.

R

S

2.75
(70)

1.50
(40)

T

U

DIVIDE LINE T-U INTO 8 EQUAL PARTS.

INCH [MILLIMETER]

6

FIG. 5-1-A   Straight line construction.

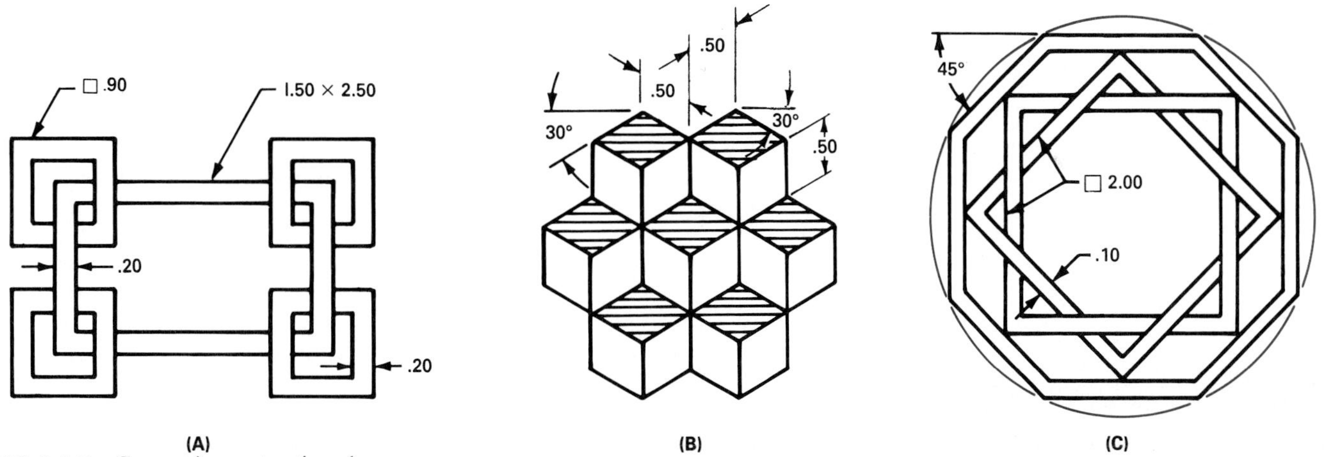

□ .90

1.50 × 2.50

.20

.20

(A)

.50

.50

30°

30°

.50

(B)

45°

□ 2.00

.10

(C)

FIG. 5-1-B   Geometric constructions 1.

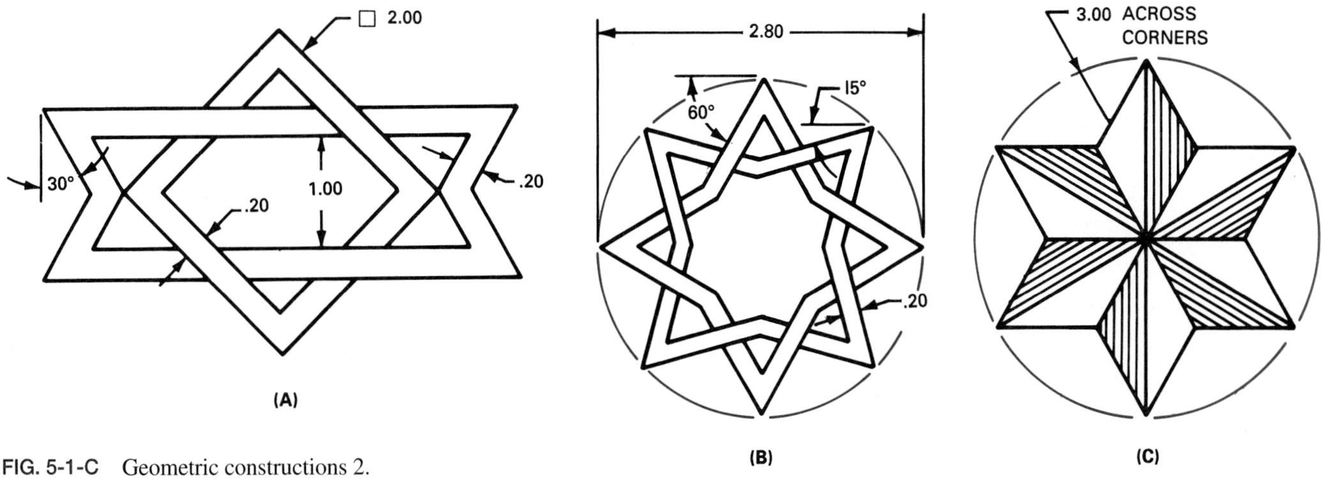

□ 2.00

30°

.20

.20

1.00

(A)

2.80

60°

15°

.20

(B)

3.00 ACROSS
CORNERS

(C)

FIG. 5-1-C   Geometric constructions 2.

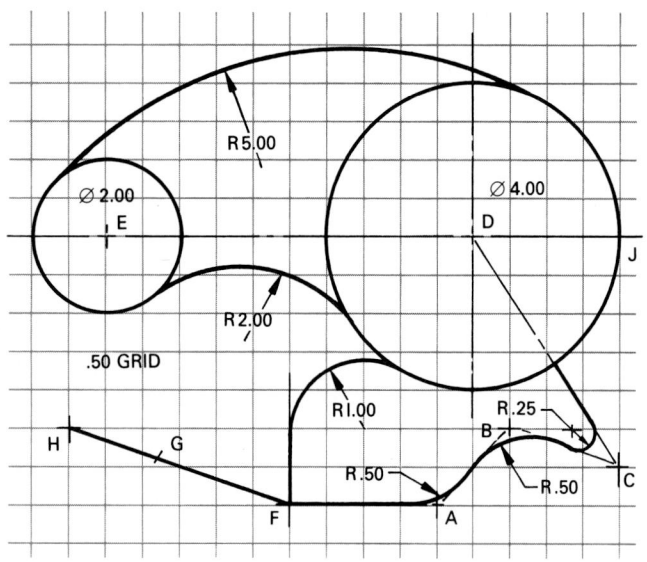

FIG. 5-2-A   Geometric constructions 3.

ASSIGNMENTS FOR UNIT 5-2, ARCS AND CIRCLES

4. On an A (A4) size sheet, complete the drawing shown in Fig. 5-2-A, given the following additional information. Points A to J are located on the .50 in. grid.
   - Draw straight lines between the points shown, then draw the arcs.
   - Divide the upper right-hand quadrant of the Ø4.00 into six 15° segments.
   - Divide the upper left-hand quadrant of the Ø4.00 into nine equal segments.
   - Bisect the arc JDC.
   - Line GF is twice as long as line HG. Draw a reverse ogee curve passing through points H, G, and F. Scale 1:1. Do not dimension.
5. Make a drawing of one of the parts shown in Fig. 5-2-B or 5-2-C. Scale 1:1. Do not dimension.

FIG. 5-2-B   Adjustable fork.

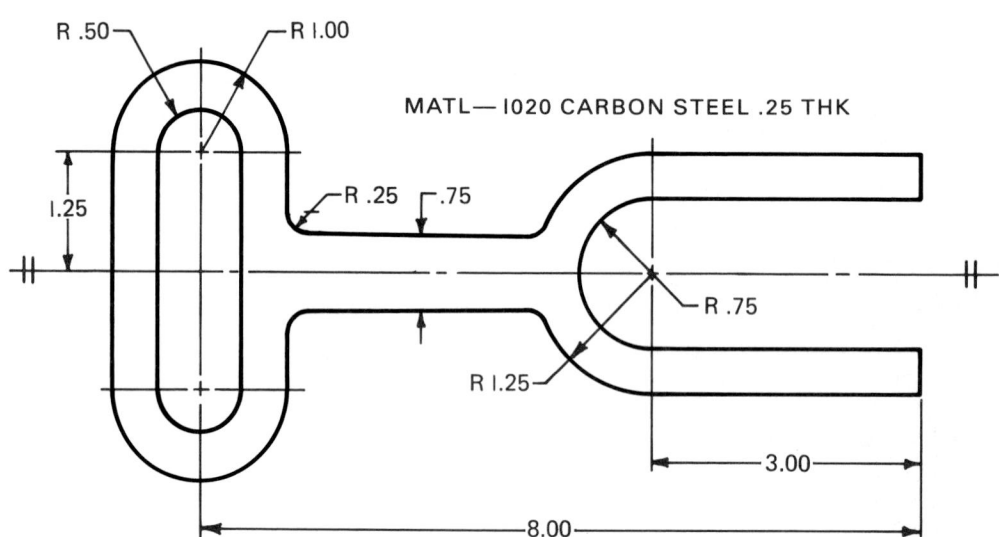

FIG. 5-2-C   Rocker arm.

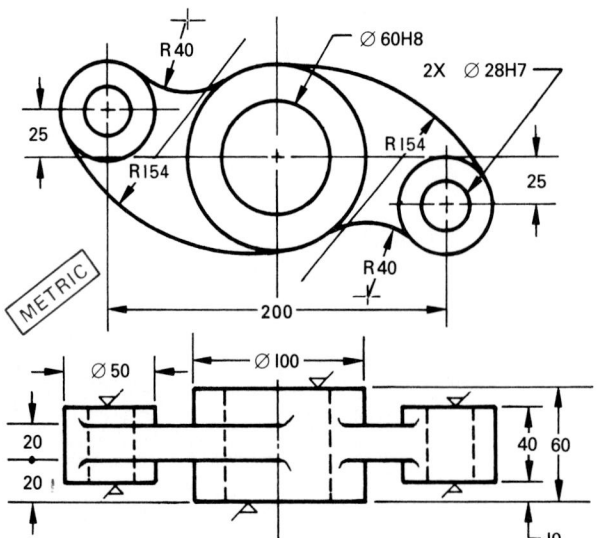

6. Divide a B (A3) size sheet as shown in Fig. 5-2-D. In the designated areas draw the geometric constructions. Scale 1:1.

7. Make a drawing of the belt drive shown in Fig. 5-2-E. The idler pulley is located midway between the B and driver

shafts. The diameters of the pulley are shown in inches on the drawing. Given the RPM of the driver pulley, calculate the RPM of the other pulleys A to D. Scale: 1 in. = 1 ft.

8. Make a drawing of PT 2 shown in Fig. 5-2-F. Scale 1:1.

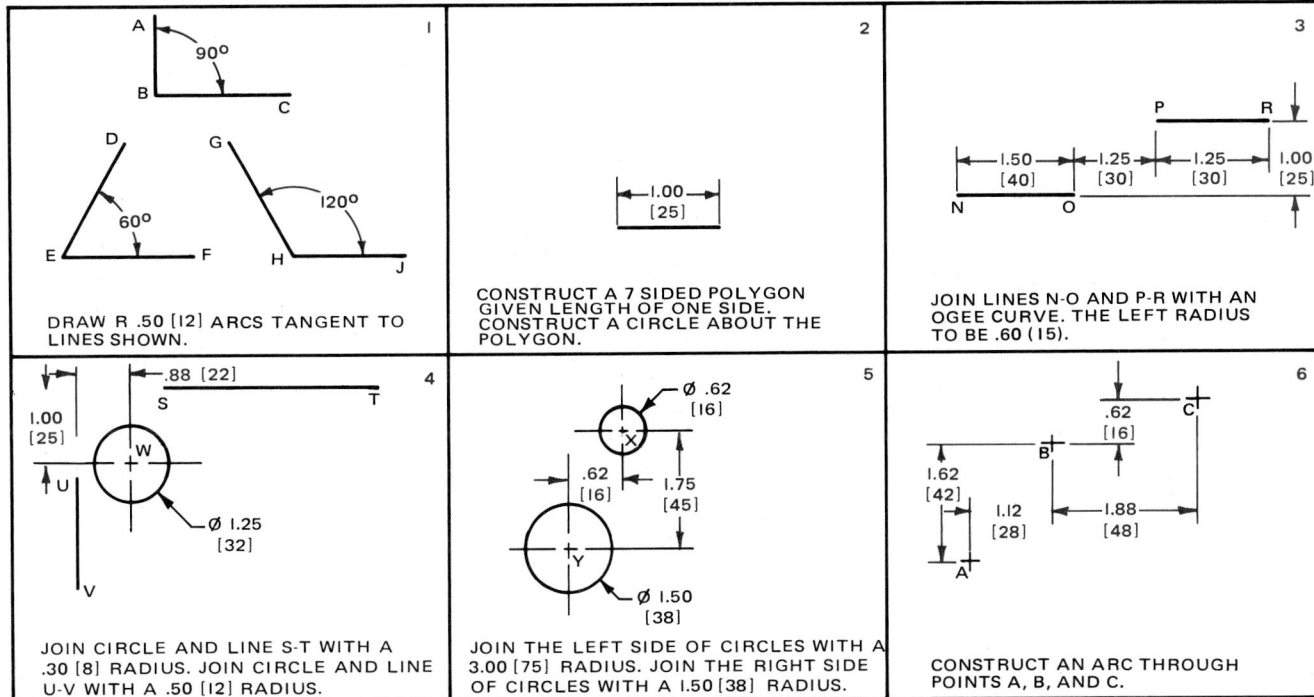

FIG. 5-2-D   Curved-line construction.

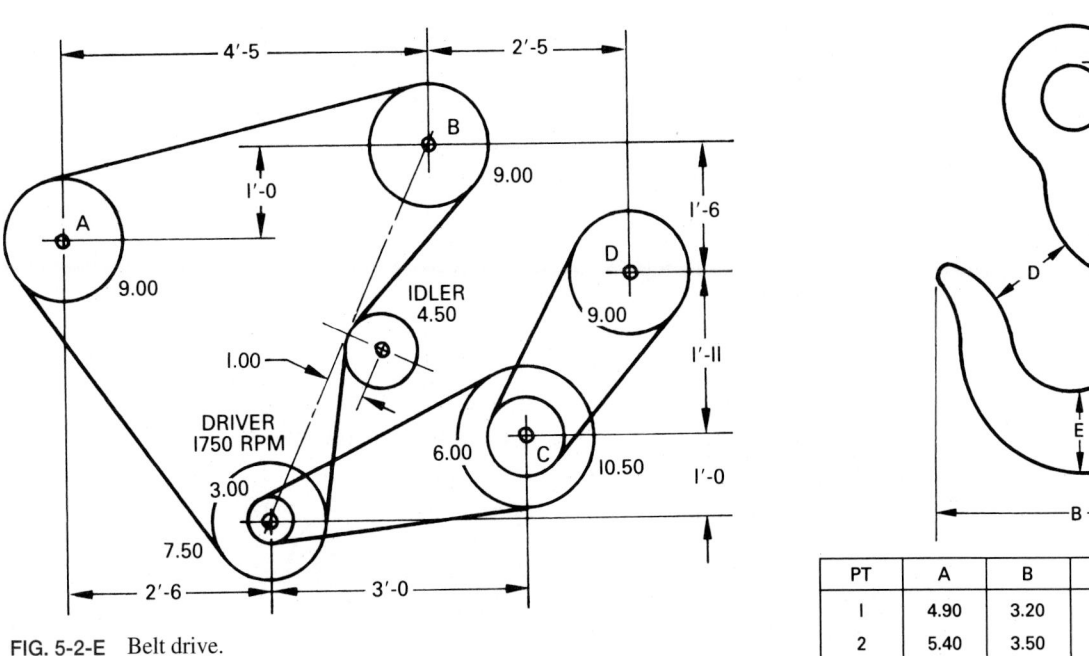

FIG. 5-2-E   Belt drive.

PT	A	B	C	D	E
1	4.90	3.20	.90	1.06	.84
2	5.40	3.50	1.00	1.10	.90
3	6.25	4.10	1.10	1.24	1.10
4	6.90	4.54	1.24	1.40	1.30

FIG. 5-2-F   Wire rope hook.

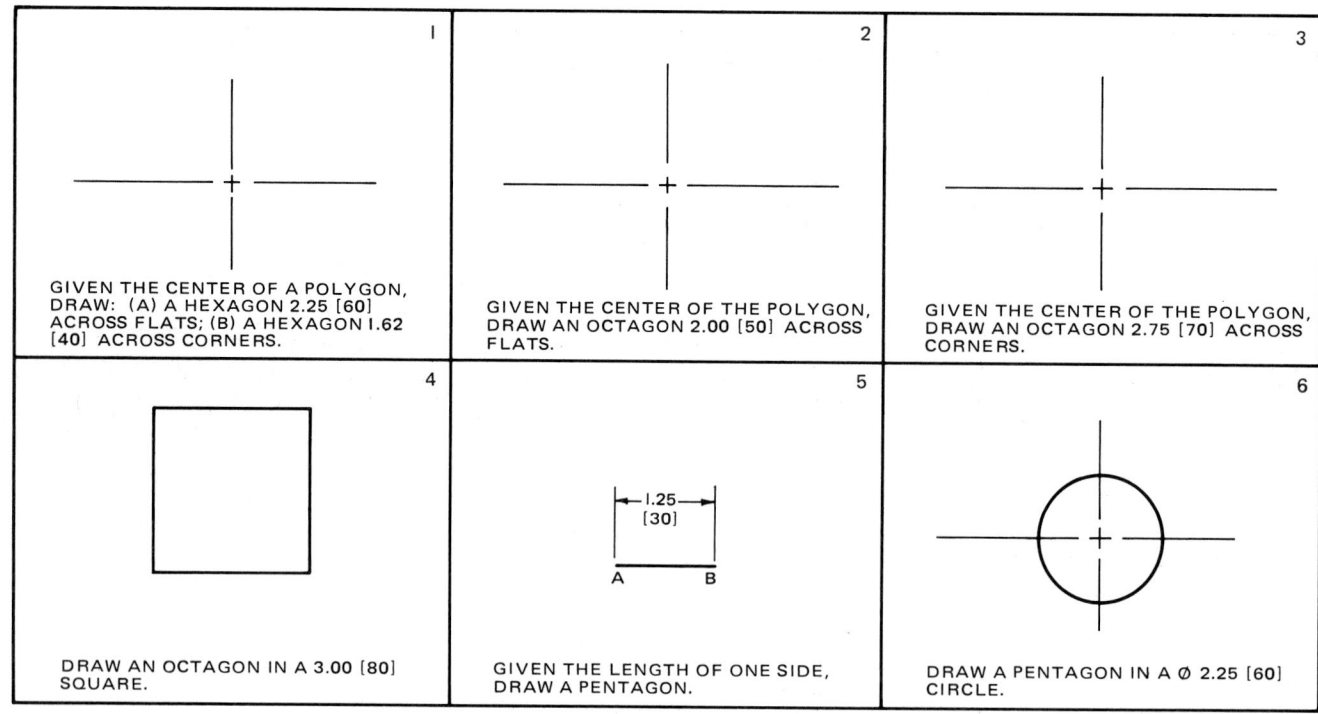

1 GIVEN THE CENTER OF A POLYGON, DRAW: (A) A HEXAGON 2.25 [60] ACROSS FLATS; (B) A HEXAGON 1.62 [40] ACROSS CORNERS.	2 GIVEN THE CENTER OF THE POLYGON, DRAW AN OCTAGON 2.00 [50] ACROSS FLATS.	3 GIVEN THE CENTER OF THE POLYGON, DRAW AN OCTAGON 2.75 [70] ACROSS CORNERS.
4 DRAW AN OCTAGON IN A 3.00 [80] SQUARE.	5 GIVEN THE LENGTH OF ONE SIDE, DRAW A PENTAGON.	6 DRAW A PENTAGON IN A Ø 2.25 [60] CIRCLE.

INCH [MILLIMETER]

INCH [MILLIMETER]

FIG. 5-3-A   Polygon constructions.

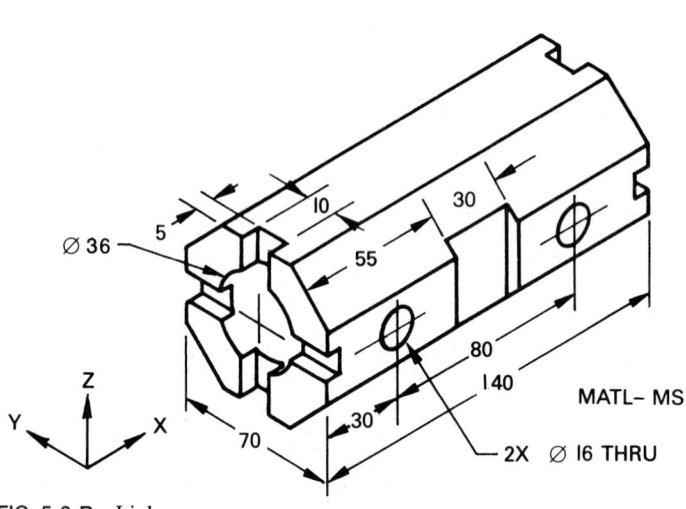

FIG. 5-3-B   Link.

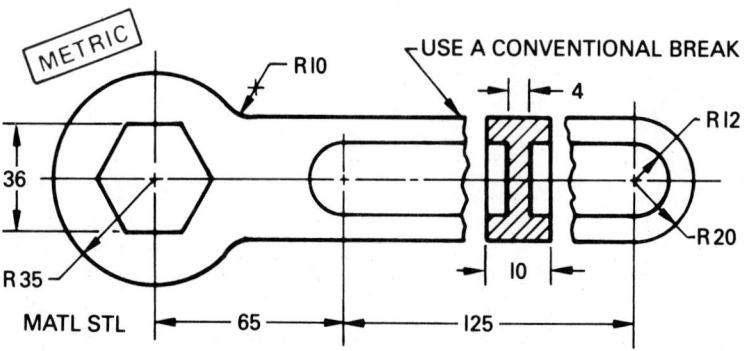

FIG. 5-3-C   Hex wrench.

ASSIGNMENTS FOR UNIT 5-3, POLYGONS

9. Divide a B (A3) size sheet as shown in Fig. 5-3-A. In the designated areas, draw the geometric constructions.
10. Make a working drawing of one of the parts shown in Figs. 5-3-B to 5-3-D. Scale 1:1.
11. Make a working drawing of one of the parts shown in Figs. 5-3-E and 5-3-F. Scale 1:1.

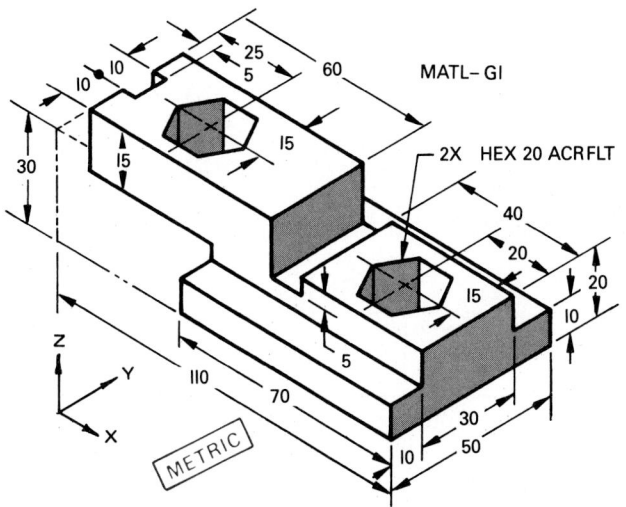

FIG. 5-3-D   Control guide.

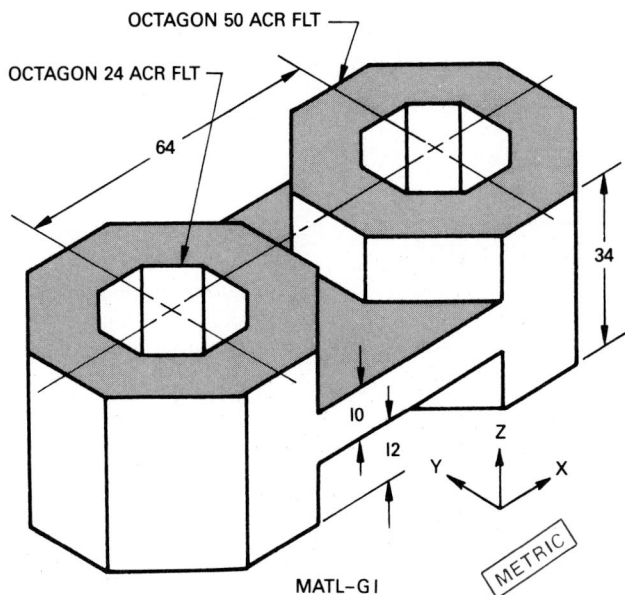

FIG. 5-3-E   Spacer.

FIG. 5-3-F   Template.

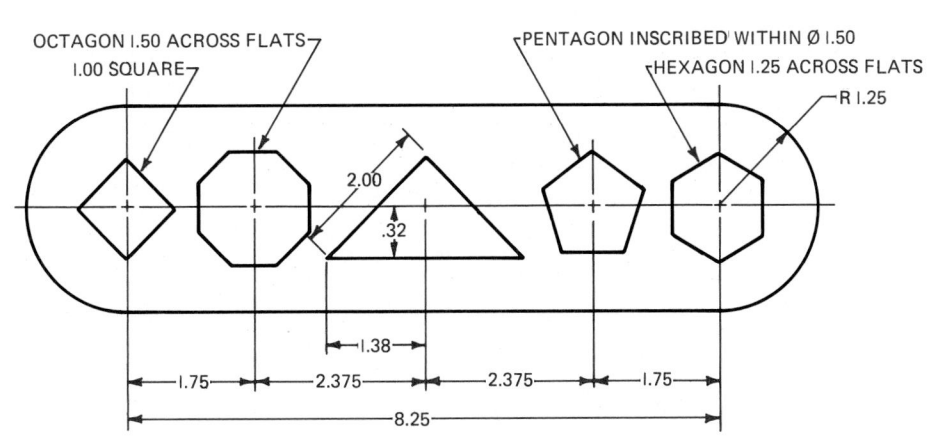

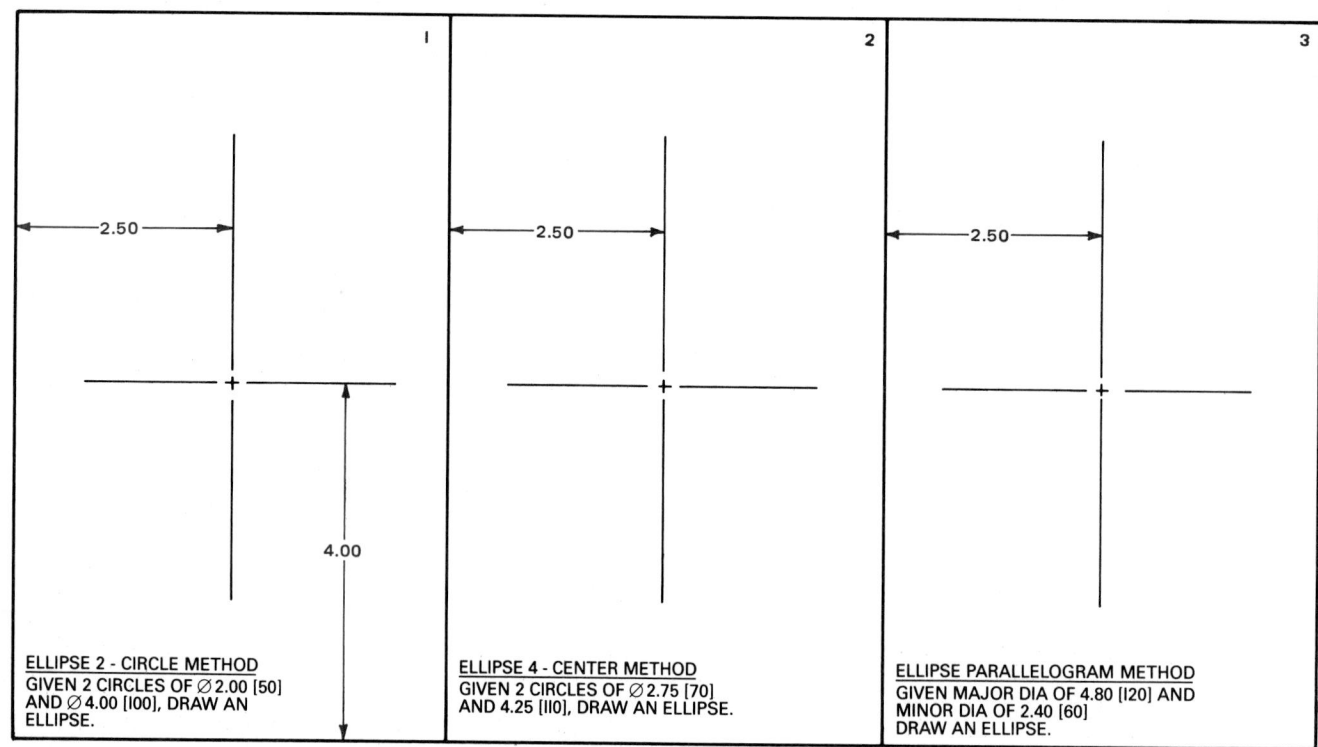

FIG. 5-4-A Ellipse constructions.

INCH [MILLIMETERS]

ASSIGNMENTS FOR UNIT 5-4, ELLIPSE

12. Divide a B (A3) size sheet as shown in Fig. 5-4-A. In the designated areas draw the geometric constructions.
13. Draw two concentric circles of 4.00 and 6.00 in. diameter. Using these circles, construct an ellipse using these two diameters. The smaller circle represents the minor diameter and the larger circle the major diameter of the ellipse. Inscribe a regular pentagon within the 4.00 in. diameter. Within the pentagon, draw a circle with the diameter tangent to the sides of the pentagon. Scale 1:1.

ASSIGNMENTS FOR UNIT 5-5, HELIX AND PARABOLA

14. On an A (A4) size sheet make a working drawing of the fan base shown in Fig. 5-5-A. Leave on the construction lines for developing the parabolic curves. Use the parallelogram method. Scale 1:1.

15. On a B (A3) size sheet lay out the three angles shown in Fig. 5-5-B and develop a parabolic curve using the parallelogram method on each. Each line has ten equal divisions of 10 mm. Scale 1:1.
16. Divide a B (A3) size sheet, as shown in Fig. 5-5-C. In the designated areas, draw the geometric constructions. Scale 1:1.

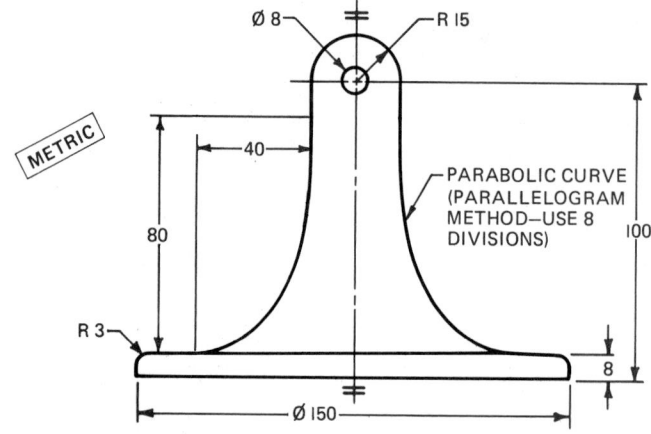

FIG. 5-5-A Fan base.

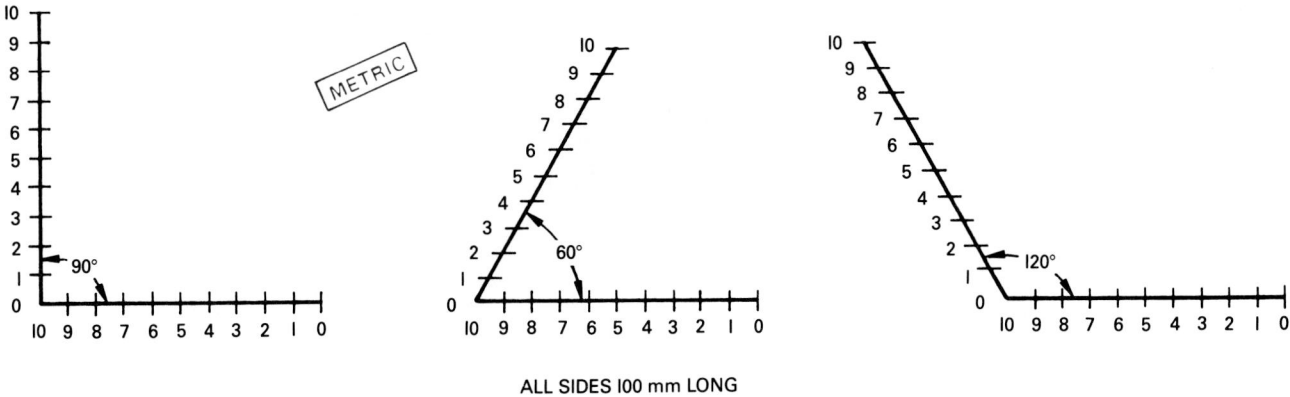

ALL SIDES 100 mm LONG

FIG. 5-5-B   Parabolic curves.

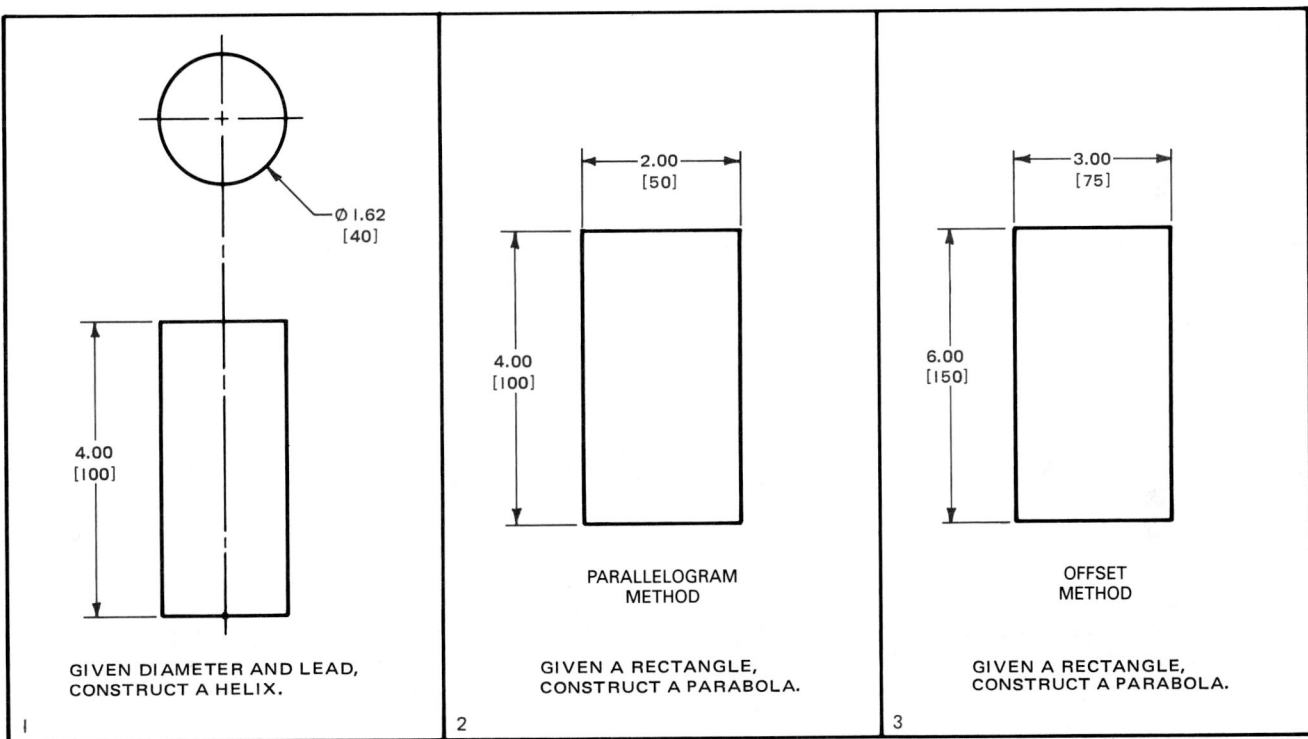

GIVEN DIAMETER AND LEAD,
CONSTRUCT A HELIX.

PARALLELOGRAM
METHOD

GIVEN A RECTANGLE,
CONSTRUCT A PARABOLA.

OFFSET
METHOD

GIVEN A RECTANGLE,
CONSTRUCT A PARABOLA.

INCH
[MILLIMETER]

FIG. 5-5-C   Helix and parabola constructions.

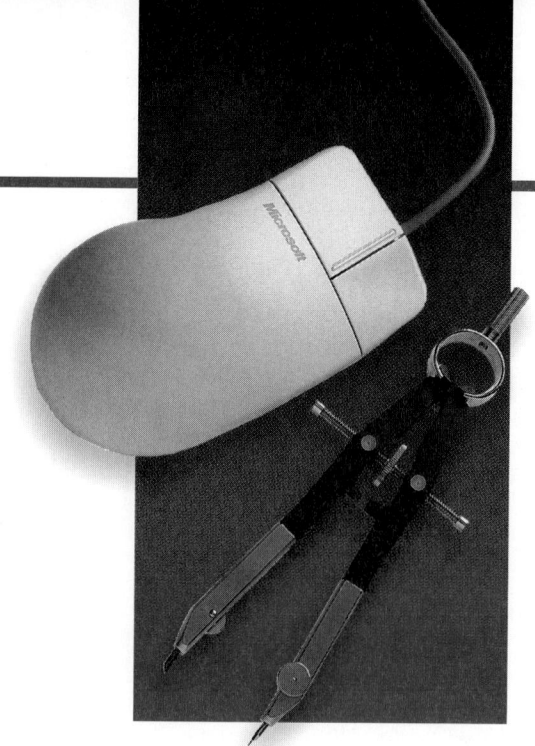

CHAPTER 6

THEORY OF SHAPE DESCRIPTION

Definitions

Auxiliary view   An additional view used to depict an inclined surface that must be shown clearly and without distortion.

First-angle projection   A projection method in which the object to be represented appears between the observer and the coordinate viewing planes on which the object is orthogonally projected.

Hidden lines   Lines used to indicate the hidden edges of an object in a drawing.

Knurling   An operation that puts patterned indentations in the surface of a metal part.

Mirrored orthographic representation   The projection method in which the object to be represented is a reproduction of the image in a mirror (face up).

Miter line   A line used to quickly and accurately construct the third view of an object after two views are established.

Orthogonal projection   A projection method in which more than one view is used to define an object.

Reference arrows layout   A projection method which permits the various views to be freely positioned.

Runout   A type of curve that describes the point at which one feature blends into another feature.

Side or **end view**   An additional view used to depict cylindrically shaped surfaces containing special features.

Third-angle projection   A projection method in which the object appears behind the coordinate viewing planes on which the object is orthographically projected.

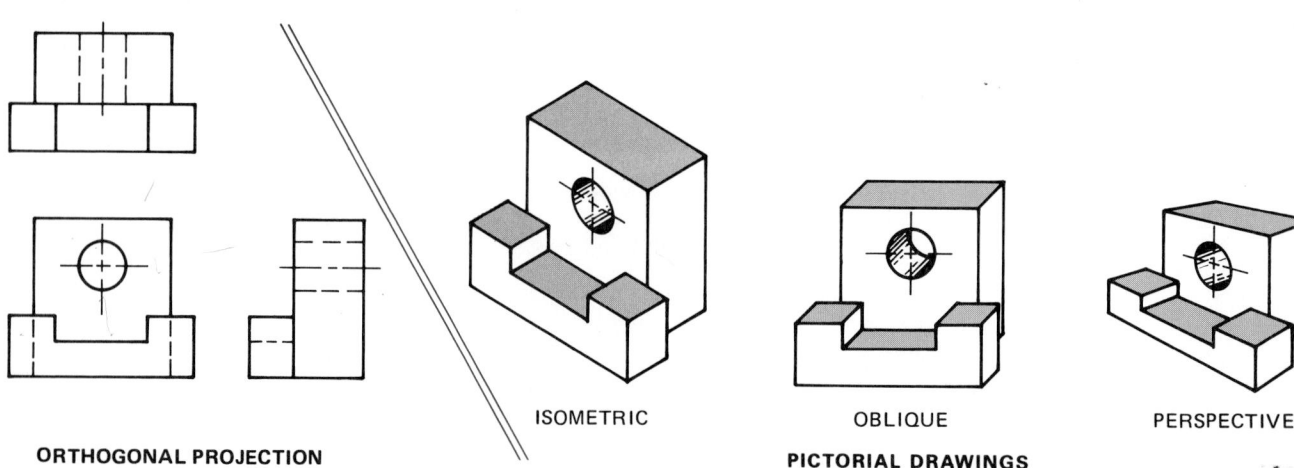

ORTHOGONAL PROJECTION   ISOMETRIC   OBLIQUE   PERSPECTIVE

PICTORIAL DRAWINGS

FIG. 6-1-1   Types of projection used in drafting.

6-1 ORTHOGRAPHIC REPRESENTATIONS

Theory of Shape Description

In the broad field of technical drawings, various projection methods are used to represent objects. Each method has its advantages and disadvantages.

The normal technical drawing is often shown in *orthogonal projection*, in which more than one view is used to draw and completely define an object (Fig. 6-1-1). The drawing of two-dimensional representations, however, requires an understanding of both the projection method and its interpretation so that the reader of the drawing, looking at two-dimensional views, will be able to visualize a three-dimensional object.

For many technical fields and their stages of development, the drafter must supply the viewer with an easily understood drawing. Pictorial drawings provide a three-dimensional view of an object as it would appear to the observer, as shown in Fig. 6-1-1. These pictorial drawings are described in detail in Chap. 15.

The steady increase in global technical intercommunication, and the interchange of drawings with different countries, as well as the evolution of methods of computer-aided design and drafting with their various types of three-dimensional representation, require that today's drafters have a knowledge of all methods of representation.

Orthographic Representations

Orthographic representation is obtained by means of parallel orthogonal projections and results in flat, two-dimensional views systematically positioned relative to each other. To show the object completely, the six views in the directions *A*, *B*, *C*, *D*, *E*, and *F* may be necessary (Fig. 6-1-2).

The most informative view of the object to be represented is normally chosen as the principal view (front view). This is view *A* according to the direction of viewing *a* and usually shows the object in the functioning, manufacturing, or mounting position. The position of the other views relative to the principal view in the drawing depends on the projection method (third-angle, first-angle, reference arrows). In practice, not all six views (*A* to *F*) are needed. When views other than the principal view are necessary, these should be selected in order to:

- Limit the number of views and sections to the minimum necessary to fully represent the object without ambiguity.
- Avoid unnecessary repetition of detail.

Methods of Representation

There are four methods of orthographic representation: third-angle projection, first-angle projection, reference arrows layout, and mirrored orthographic representation. Third-angle projection is used in the United States, Canada, and many other countries throughout the world. First-angle projection is used mainly in European and Asiatic countries.

Third-Angle Projection

The *third-angle projection* method is an orthographic representation in which the object to be represented and seen by the observer appears behind the coordinate viewing planes on which the object is orthographically projected (Fig. 6-1-3B, pg. 94). On each projection plane, the object is represented as if seen orthogonally from in front of each plane.

The positions of the various views relative to the principal (front) view are then rotated or positioned so that they lie on the same plane (drawing surface) on which the front view *A* is projected.

FIG. 6-1-2   Designation of views.

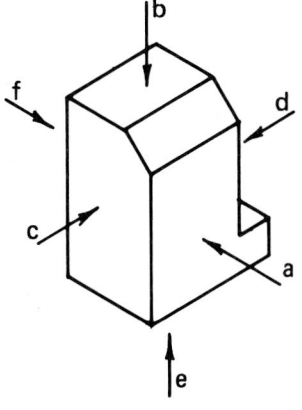

DIRECTION OF OBSERVATION		DESIGNATION OF VIEW
VIEW IN DIRECTION	VIEW FROM	
a	THE FRONT	A
b	ABOVE	B
c	THE LEFT	C
d	THE RIGHT	D
e	BELOW	E
f	THE REAR	F

Therefore in Fig. 6-1-3C with reference to the principal view *A*, the other views are arranged as follows:

- View *B*—The view from above is placed above.
- View *E*—The view from below is placed underneath.
- View *C*—The view from the left is placed on the left.
- View *D*—The view from the right is placed on the right.
- View *F*—The view from the rear may be placed on the left or on the right, as convenient.

The letters *A* to *F* are shown here only to identify the location of views when third-angle projection is used. On working drawings these letters would not be shown.

The identifying symbol for this method of representation is shown in Fig. 6-1-3D.

First-Angle Projection

The *first-angle projection* method is an orthographic representation in which the object to be represented appears between the observer and the coordinate viewing planes on which the object is orthogonally projected (Fig. 6-1-4B).

The position of the various views relative to the principal (front) view *A* are then rotated or positioned so that they lie on the same plane (drawing surface) on which the front view *A* is projected.

Therefore, in Fig. 6-1-4C with reference to the principal view *A*, the other views are arranged as follows:

- View *B*—The view from above is placed underneath.
- View *E*—The view from below is placed above.
- View *C*—The view from the left is placed on the right.
- View *D*—The view from the right is placed on the left.
- View *F*—The view from the rear is placed on the right or on the left, as convenient.

The letters *A* to *F* are shown here only to identify the location of views when first-angle projection is used. On working drawings these letters would not be shown.

The identifying symbol of this method of representation is shown in Fig. 6-1-4D.

Reference Arrows Layout

In those cases where it is advantageous not to position the views according to the strict pattern of the third- or the first-angle projection method, *reference arrows layout* permits the various views to be freely positioned.

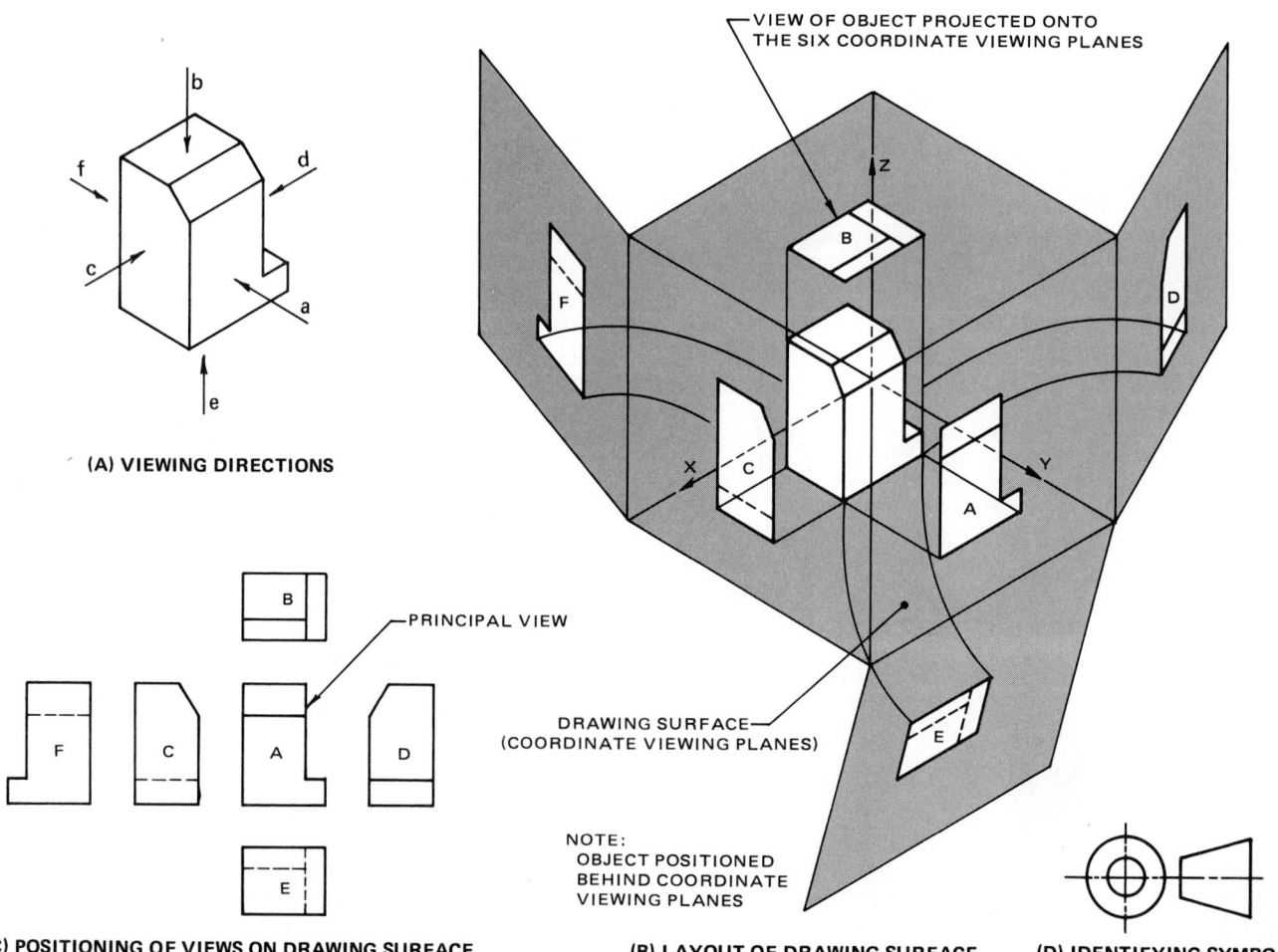

(A) VIEWING DIRECTIONS

(C) POSITIONING OF VIEWS ON DRAWING SURFACE

VIEW OF OBJECT PROJECTED ONTO THE SIX COORDINATE VIEWING PLANES

PRINCIPAL VIEW

DRAWING SURFACE (COORDINATE VIEWING PLANES)

NOTE: OBJECT POSITIONED BEHIND COORDINATE VIEWING PLANES

(B) LAYOUT OF DRAWING SURFACE

(D) IDENTIFYING SYMBOL

FIG. 6-1-3   Third-angle projection.

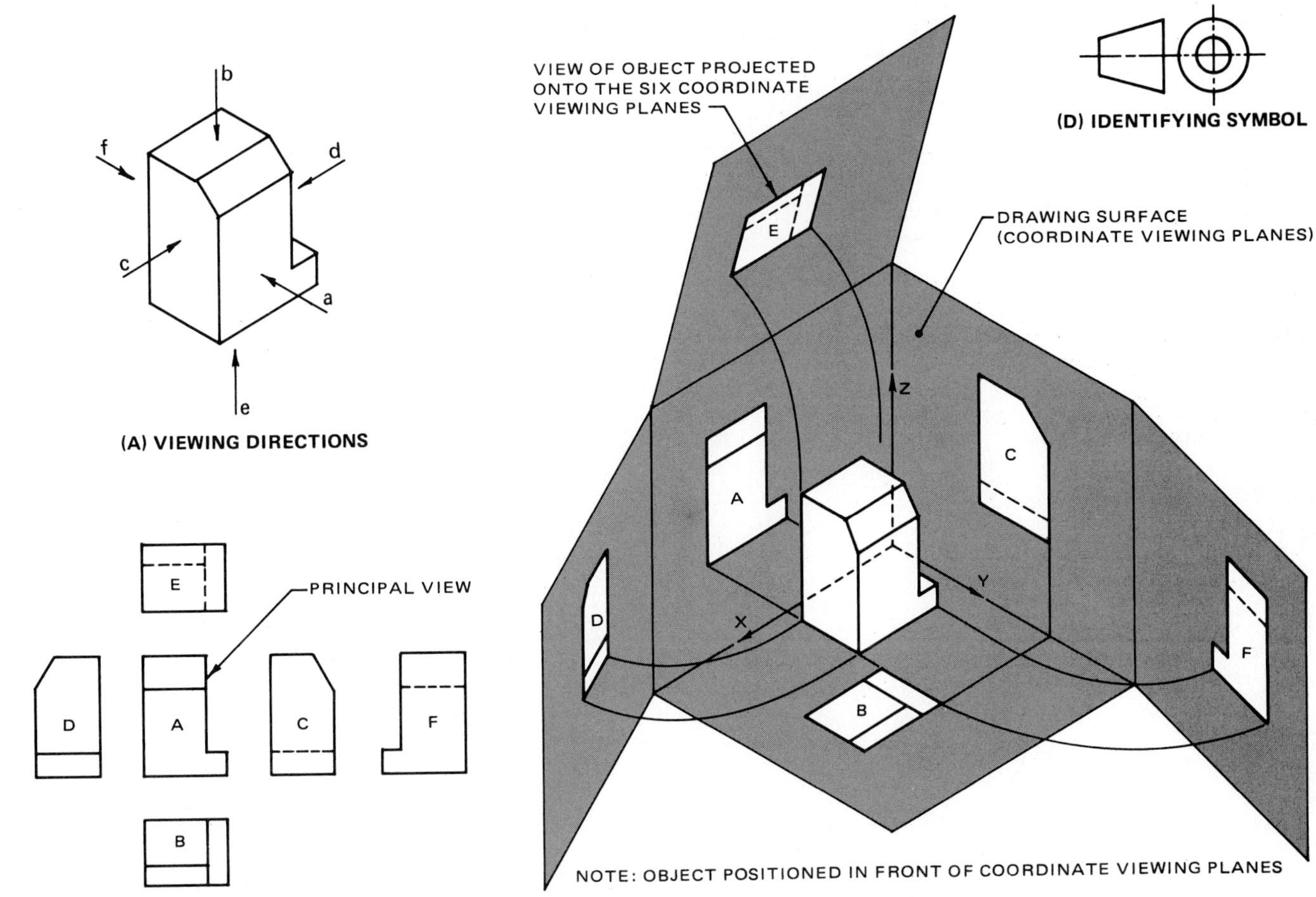

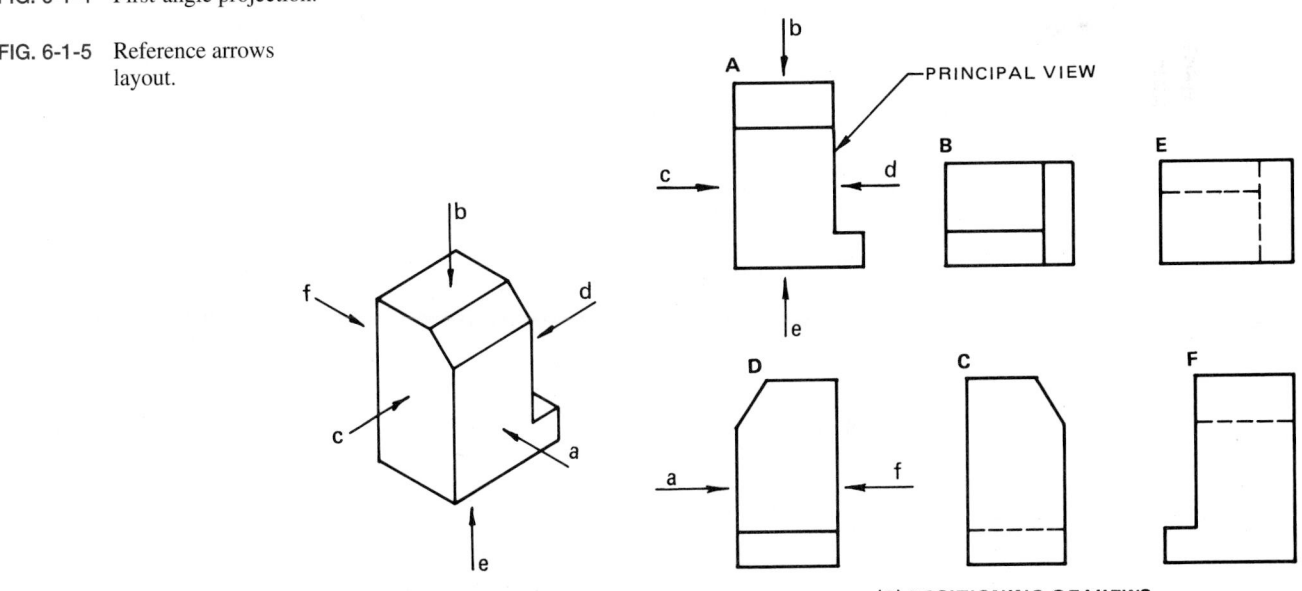

(A) VIEWING DIRECTIONS

VIEW OF OBJECT PROJECTED
ONTO THE SIX COORDINATE
VIEWING PLANES

(D) IDENTIFYING SYMBOL

DRAWING SURFACE
(COORDINATE VIEWING PLANES)

(C) POSITIONING OF VIEWS ON DRAWING SURFACE

PRINCIPAL VIEW

NOTE: OBJECT POSITIONED IN FRONT OF COORDINATE VIEWING PLANES

(B) LAYOUT OF DRAWING SURFACE

FIG. 6-1-4   First-angle projection.

FIG. 6-1-5   Reference arrows
layout.

(A) VIEWING DIRECTIONS

PRINCIPAL VIEW

(B) POSITIONING OF VIEWS

With the exception of the principal view, each view is identified by a letter (Fig. 6-1-5B). A lowercase letter on the principal view, and where required on one of the side views, indicates the direction of observation of the other views, which are identified by the corresponding capital letter placed immediately above the view on the left.

The identified views may be located irrespective of the principal view. Whatever the direction of the observer, the capital letters identifying the views should always be positioned to be read from the direction from which the drawing is normally viewed. No symbol is needed on the drawing to identify this method.

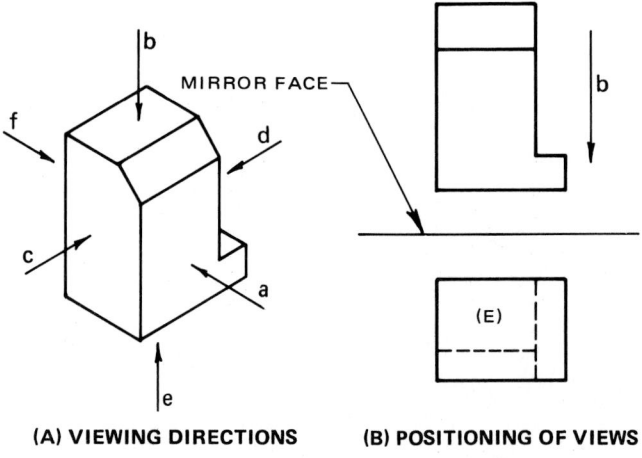

(A) VIEWING DIRECTIONS

(B) POSITIONING OF VIEWS

Mirrored Orthographic Representation

Mirrored orthographic representation is the method preferred for use in construction drawings. In this method the object to be represented is a reproduction of the image in a mirror (face up), positioned parallel to the horizontal planes of the object (Fig. 6-1-6).

The identifying symbol for this method is shown in Fig. 6-1-6C.

Identifying Symbols

The symbol used to identify the method of representation should be shown on all drawings, preferably in the lower right-hand corner of the drawing, adjacent to the title block (Fig. 6-1-7).

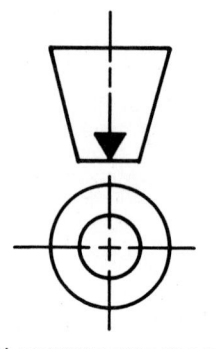

(C) IDENTIFYING SYMBOL

FIG. 6-1-6   Mirrored orthographic projection.

CAD Coordinate Input for Orthographic Representation

In Unit 4-1 you learned how to locate points and lines using coordinate input. The positions were described on the drawing

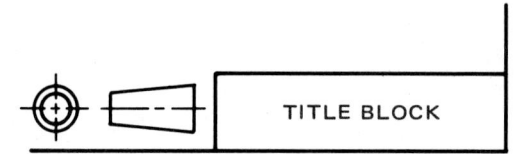

FIG. 6-1-7   Location of identifying symbol for method of representation.

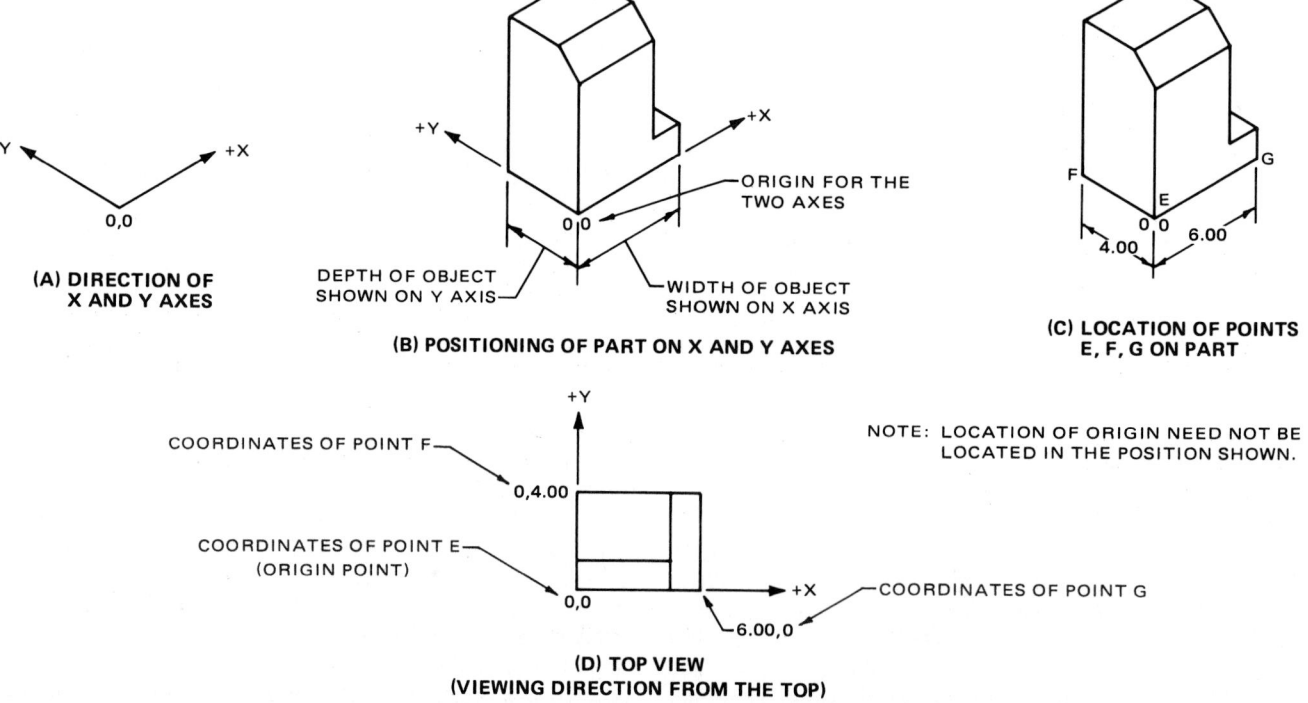

(A) DIRECTION OF X AND Y AXES

(B) POSITIONING OF PART ON X AND Y AXES

(C) LOCATION OF POINTS E, F, G ON PART

(D) TOP VIEW
(VIEWING DIRECTION FROM THE TOP)

NOTE: LOCATION OF ORIGIN NEED NOT BE LOCATED IN THE POSITION SHOWN.

FIG. 6-1-8   Locating points on a part by two-axis (X and Y) coordinate input.

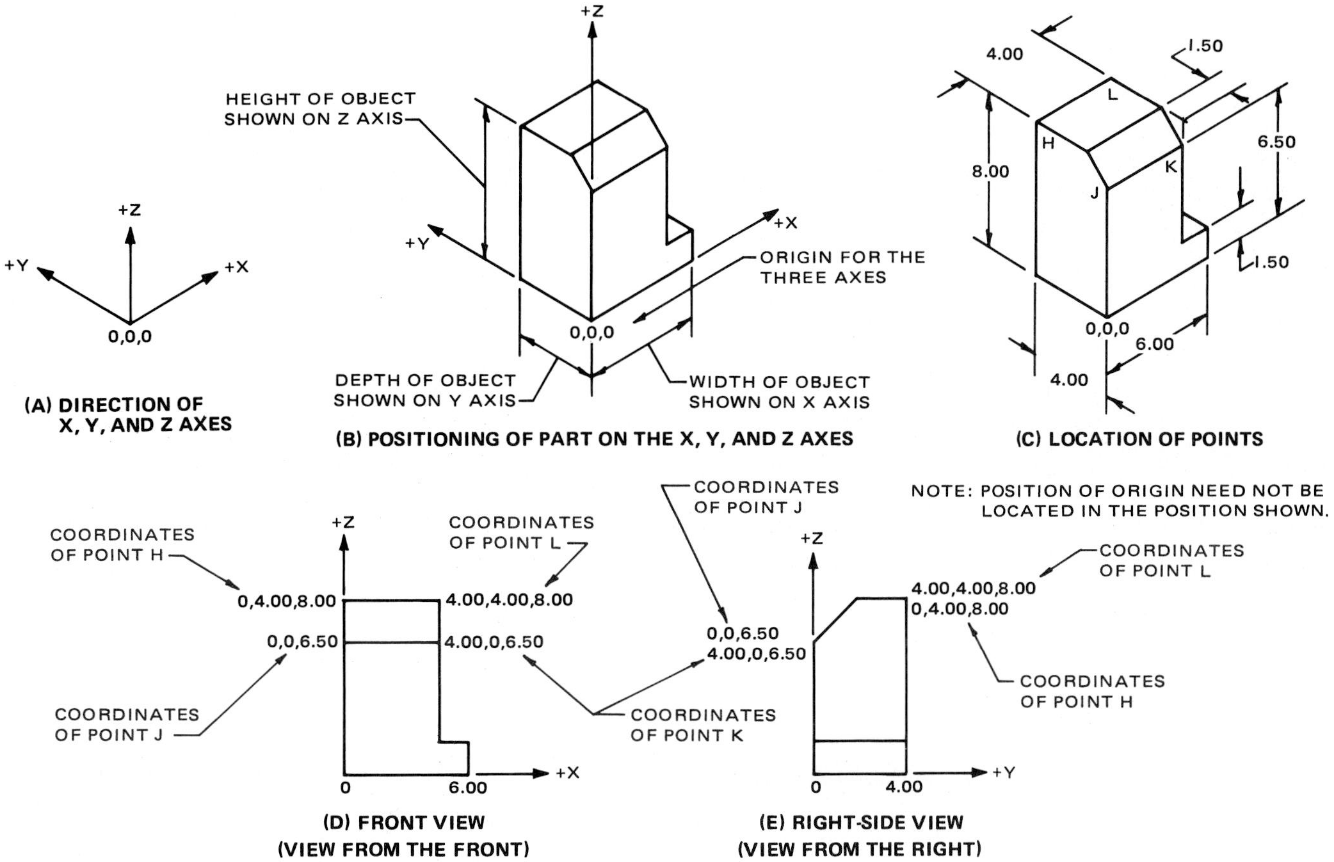

HEIGHT OF OBJECT
SHOWN ON Z AXIS

+Z

+Z

+Y   +X

0,0,0

**(A) DIRECTION OF
X, Y, AND Z AXES**

+Y

+X

ORIGIN FOR THE
THREE AXES

0,0,0

DEPTH OF OBJECT
SHOWN ON Y AXIS

WIDTH OF OBJECT
SHOWN ON X AXIS

(B) POSITIONING OF PART ON THE X, Y, AND Z AXES

4.00   1.50

L

H   6.50

K

J

0,0,0
6.00

4.00   1.50

8.00

(C) LOCATION OF POINTS

COORDINATES
OF POINT H

+Z

COORDINATES
OF POINT L

0,4.00,8.00   4.00,4.00,8.00

0,0,6.50   4.00,0,6.50

COORDINATES
OF POINT J

COORDINATES
OF POINT K

0   6.00   +X

**(D) FRONT VIEW
(VIEW FROM THE FRONT)**

COORDINATES
OF POINT J

+Z

0,0,6.50
4.00,0,6.50

4.00,4.00,8.00
0,4.00,8.00

0   4.00   +Y

NOTE: POSITION OF ORIGIN NEED NOT BE
LOCATED IN THE POSITION SHOWN.

COORDINATES
OF POINT L

COORDINATES
OF POINT H

**(E) RIGHT-SIDE VIEW
(VIEW FROM THE RIGHT)**

FIG. 6-1-9   Locating points on a part by three-axis (X, Y, and Z) coordinate input.

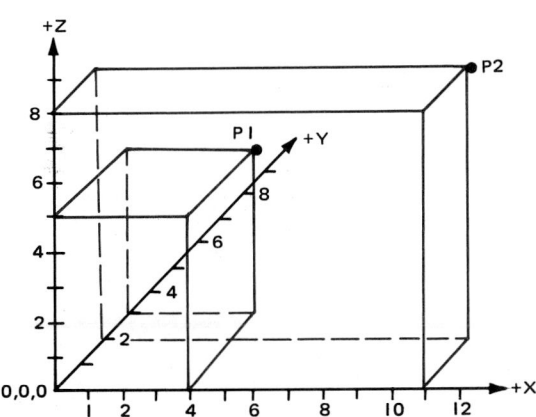

+Z

P2

8

P1   +Y

6   8

4   6

2   4

2

0,0,0   +X

2   4   6   8   10   12

FIG. 6-1-10   Points in space.

by two-dimensional coordinates, horizontal (X) and vertical (Y). The X axis is horizontal and is considered the first and basic reference axis. The Y axis is vertical and is 90° to the X axis.

When third-angle orthographic representation is used to show a part, the view from above is known as the top view. The X and Y axes and coordinates are identified with this view, and width and depth features are shown (Fig. 6-1-8).

With the exceptions of the views from the top and below, all the other views require information regarding the height of features. This is provided by introducing a third axis, called the

Z axis, to the system. The coordinates for the origin of the three axes are then identified by the numbers 0, 0, 0, the last coordinate representing distances on the Z axis (Fig. 6-1-9A). As previously mentioned, the point of origin may be located at any convenient location on the drawing. The coordinates for points H, J, K, and L shown in Fig. 6-1-9C would be 0, 4.00, 8.00 (point H); 0, 0, 6.50 (point J); 4.00, 0, 6.50 (point K); and 4.00, 4.00, 8.00 (point L). Note that the coordinates for a point remain the same, regardless of the view on which they are shown.

Coordinate Input to Locate Points in Space

A point in space can be described by its X, Y, and Z coordinates. For example, $P1$ in Fig. 6-1-10 can be described by its (X, Y, Z) coordinates as (4, 3, 5), and $P2$ as (11, 2, 8).

A pictorial drawing of a part can be described as lines joining a series of points in space (Fig. 6-1-11, pg. 98). The 0, 0, 0 reference indicates the absolute X, Y, Z coordinate origin. It has been designated to be the lower left-front corner position of the front view. The lower right-front position is labeled 12, 0, 0. This means that the coordinate location for that point is 12 units (in.) to the right and has the same elevation (height) and depth as the coordinate origin. All other positions are interpreted in the same manner.

A symbol similar to that shown in Fig. 6-1-9A is used in the pictorial drawings that follow to designate the direction of the X, Y, and Z axes for that particular part.

97

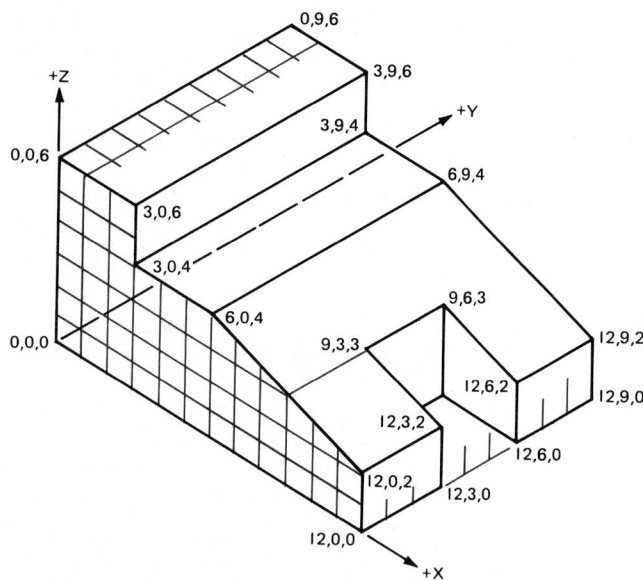

FIG. 6-1-11   Three-dimensional coordinates.

REFERENCES AND SOURCE MATERIAL

1. ISO 5456 *Technical Drawings—Projection Methods*

ASSIGNMENTS

See Assignments 1 through 4 for Unit 6-1 on pages 110–114.

6-2 ARRANGEMENT AND CONSTRUCTION OF VIEWS

Spacing the Views

For clarity and good appearance the views should be well balanced on the drawing paper, whether the drawing shows one view, two views, three views, or more. The drafter must anticipate the approximate space required. This is determined from the size of the object to be drawn, the number of views, the scale used, and the space between views. Ample space should be provided between views to permit placement of dimensions on the drawing without crowding. Space should also be allotted so that notes can be added. However, space between views should not be excessive.

Figure 6-2-1 shows how to balance the views for a three-view drawing. For a drawing with two or more views, follow these guidelines:

1. Decide on the views to be drawn and the scale to be used, e.g., 1:1 or 1:2.
2. Make a sketch of the space required for each of the views to be drawn, showing these views in their correct location. A simple rectangle for each view will be adequate (Fig. 6-2-1B).
3. Put on the overall drawing sizes for each view. (These sizes are shown as *W*, *D*, and *H*.)
4. Decide upon the space to be left between views. These spaces should be sufficient for the parallel dimension

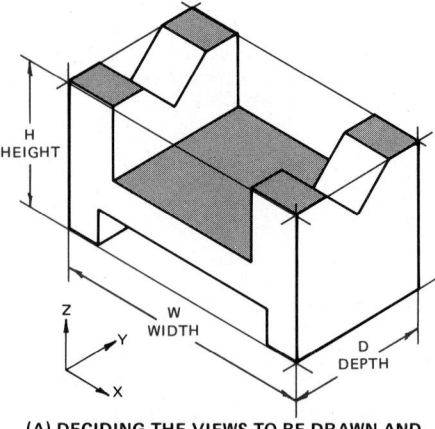

(A) DECIDING THE VIEWS TO BE DRAWN AND THE SCALE TO BE USED

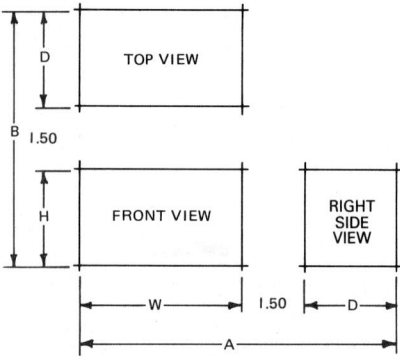

(B) CALCULATING DISTANCES A AND B

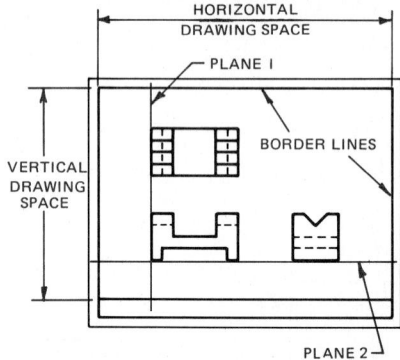

(C) ESTABLISHING LOCATION OF PLANES 1 AND 2 ON DRAWING PAPER OR CRT MONITOR

FIG. 6-2-1   Balancing the drawing on the drawing paper or monitor.

lines to be placed between views. For most drawing projects, 1.50 in. (40 mm) is sufficient.
5. Total these dimensions to get the overall horizontal distance (*A*) and overall vertical distance (*B*).
6. Select the drawing sheet that best accommodates the overall size of the drawing with suitable open space around the views.
7. Measure the "drawing space" remaining after all border lines, title strip or title block, etc., are in place (Fig. 6-2-1C).
8. Take one-half of the difference between distance *A* and the horizontal drawing space to establish plane 1.
9. Take one-half of the difference between distance *B* and the vertical drawing space to establish plane 2.

Use of a Miter Line

The use of a miter line provides a fast and accurate method of constructing the third view once two views are established (Fig. 6-2-2).

Using a Miter Line to Construct the Right Side View

1. Given the top and front views, project lines to the right of the top view.
2. Establish how far from the front view the side view is to be drawn (distance *D*).
3. Construct the miter line at 45° to the horizon.
4. Where the horizontal projection lines of the top view intersect the miter line, drop vertical projection lines.
5. Project horizontal lines to the right of the front view and complete the side view.

Using a Miter Line to Construct the Top View

1. Given the front and side views, project vertical lines up from the side view.
2. Establish how far away from the front view the top view is to be drawn (distance *D*).
3. Construct the miter line at 45° to the horizon.
4. Where the vertical projection lines of the side view intersect the miter line, project horizontal lines to the left.
5. Project vertical lines up from the front view and complete the top view.

CAD

In a CAD environment, construction lines are usually placed on a separate working layer, and the geometry on that layer is given an identifying color. This layer can be echoed off when a plot is generated, leaving only the finished drawing. In more complicated drawings, several different construction layers may be required.

The working area of a drawing is established by a LIMITS command. The limits of the working drawing area are normally expressed as the lower left and upper right corners of the drawing and correspond to the size of the drawing form. When the drawing is plotted, the limits are used to determine the overall size of the sheet that is required.

ASSIGNMENTS

See Assignments 5 through 8 for Unit 6-2 on pages 114–115.

6-3 ALL SURFACES PARALLEL AND ALL EDGES AND LINES VISIBLE

To help you fully appreciate the shape and detail of views drawn in third-angle orthographic projection, the units for this chapter have been designed according to the types of surfaces generally found on objects. These surfaces can be divided into flat surfaces parallel to the viewing planes with and without hidden features; flat surfaces that appear inclined in one plane and parallel to the other two principal reference planes (called *inclined surfaces*); flat surfaces that are inclined in all three reference planes (called *oblique surfaces*); and surfaces that have diameters or radii. These drawings are so designed that only the top, front, and right side views are required.

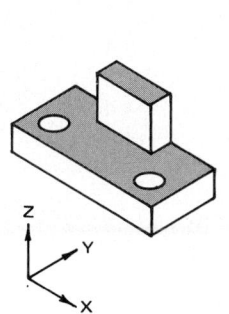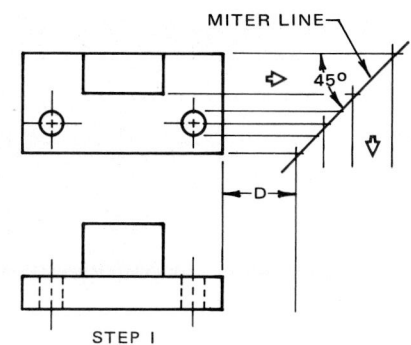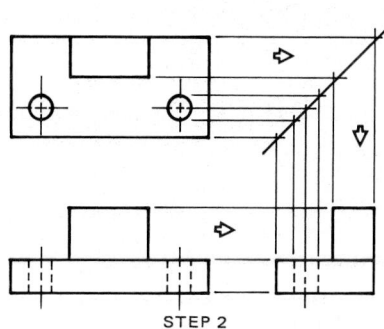

(A) ESTABLISHING WIDTH LINES ON SIDE VIEW

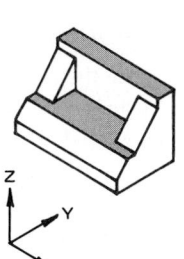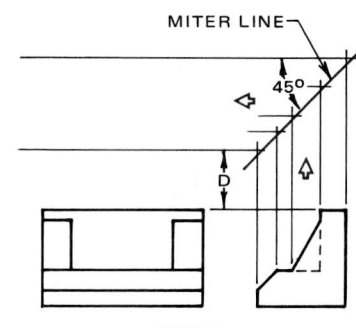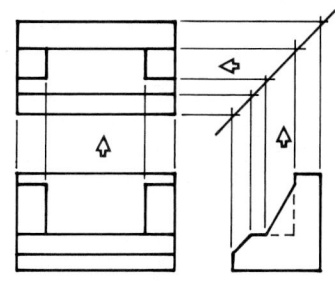

(B) ESTABLISHING WIDTH LINES ON TOP VIEW

FIG. 6-2-2   Use of a miter line.

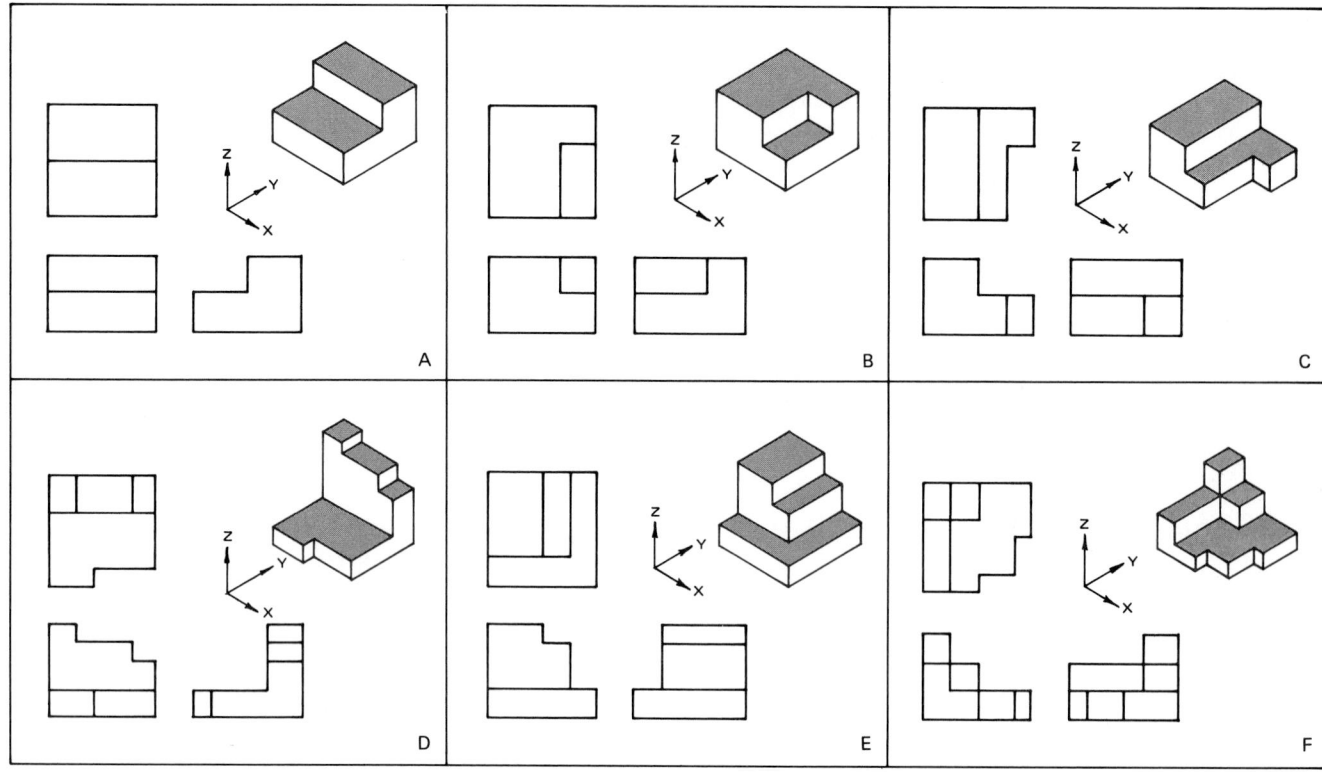

NOTE: ARROWS INDICATE DIRECTION OF SIGHT WHEN LOOKING AT THE FRONT VIEW.

FIG. 6-3-1   Illustrations of objects drawn in third-angle orthographic projection.

All Surfaces Parallel to the Viewing Planes and All Edges and Lines Visible   When a surface is parallel to the viewing planes, that surface will show as a surface on one view and a line on the other views. The lengths of these lines are the same as the lines shown on the surface view. Figure 6-3-1 shows examples.

ASSIGNMENTS

See Assignments 9 and 10 for Unit 6-3 on pages 115–116.

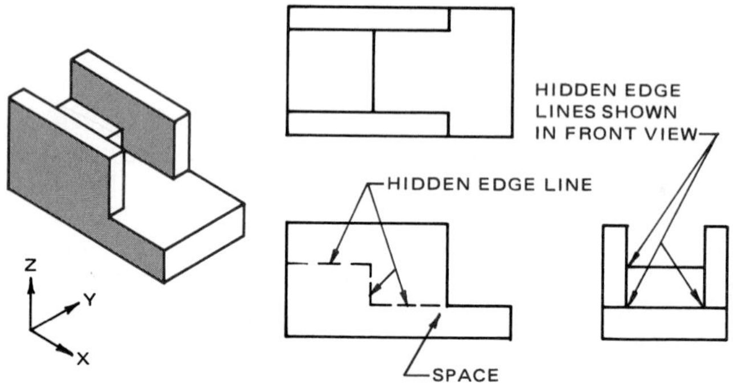

FIG. 6-4-1   Hidden lines.

6-4   HIDDEN SURFACES AND EDGES

Most objects drawn in engineering offices are more complicated than the one shown in Fig. 6-4-1. Many features (lines, holes, etc.) cannot be seen when viewed from outside the piece. These hidden edges are shown with hidden lines and are normally required on the drawing to show the true shape of the object.

Hidden lines consist of short, evenly spaced dashes. They should be omitted when not required to preserve the clarity of the drawing. The length of dashes may vary slightly in relation to the size of the drawing.

Lines depicting hidden features and phantom details should always begin and end with a dash in contact with the line at which they start and end, except when such a dash would form

a continuation of a visible detail line. Dashes should join at corners. Arcs should start with dashes at the tangent points (Fig. 6-4-2). Figure 6-4-3 shows additional examples of objects requiring hidden lines.

All CAD systems have the option to create different line styles. On large systems, these options are found on the auxiliary menu. On smaller systems, the line style selection is made directly from the tablet menu. Any style of line may be set using the LINETYPE command.

ASSIGNMENTS

See Assignments 11 through 15 for Unit 6-4 on pages 116 through 119.

FIG. 6-4-2 Application of hidden lines.

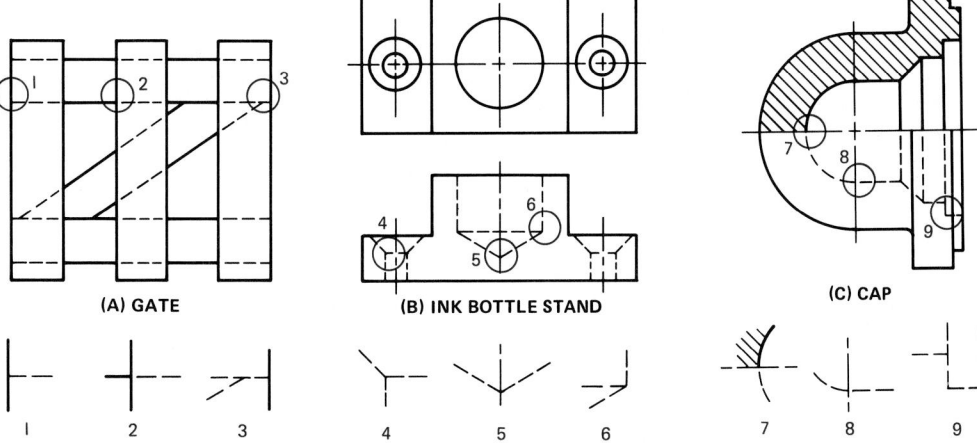

(A) GATE   (B) INK BOTTLE STAND   (C) CAP

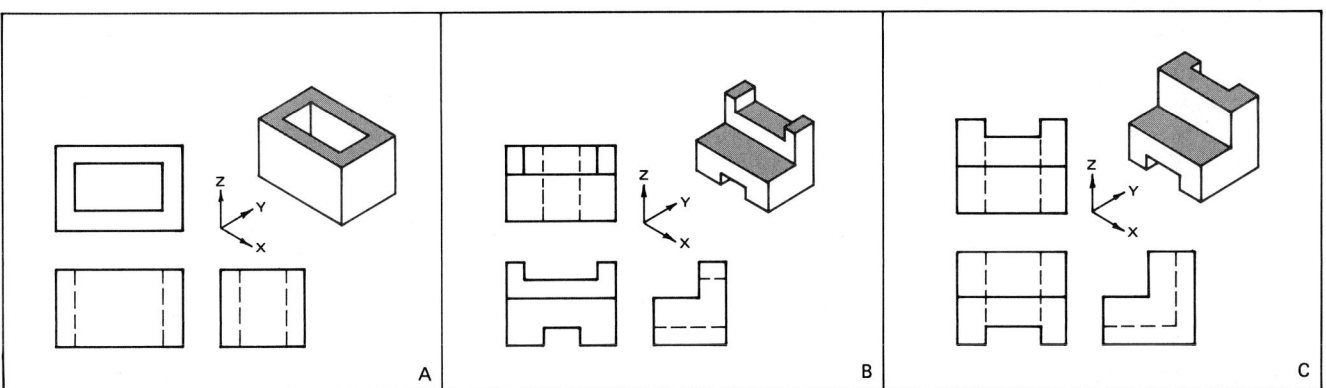

FIG. 6-4-3   Illustrations of objects having hidden features.

6-5 INCLINED SURFACES

If the surfaces of an object lie in either a horizontal or a vertical position, the surfaces appear in their true shapes in one of the three views, and these surfaces appear as a line in the other two views.

When a surface is inclined or sloped in only one direction, that surface is not seen in its true shape in the top, front, or side view. It is, however, seen in two views as a distorted surface. On the third view it appears as a line.

The true length of surfaces *A* and *B* in Fig. 6-5-1 is seen in the front view only. In the top and side views, only the width of surfaces *A* and *B* appears in its true size. The length of these surfaces is foreshortened. Figure 6-5-2 (pg. 102) shows additional examples.

Where an inclined surface has important features that must be shown clearly and without distortion, an *auxiliary*, or helper, view must be used. This type of view will be discussed in detail in Chap. 7.

ASSIGNMENTS

See Assignments 16 through 21 for Unit 6-5 on pages 120 through 124.

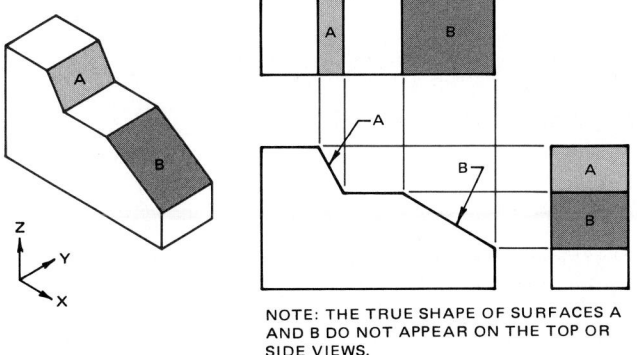

NOTE: THE TRUE SHAPE OF SURFACES A AND B DO NOT APPEAR ON THE TOP OR SIDE VIEWS.

FIG. 6-5-1 Sloping surfaces.

6-6 CIRCULAR FEATURES

Typical parts with circular features are illustrated in Fig. 6-6-1 (pg. 102). Note that the circular feature appears circular in one view only and that no line is used to show where a curved surface joins a flat surface. Hidden circles, like hidden flat surfaces, are represented on drawings by a hidden line.

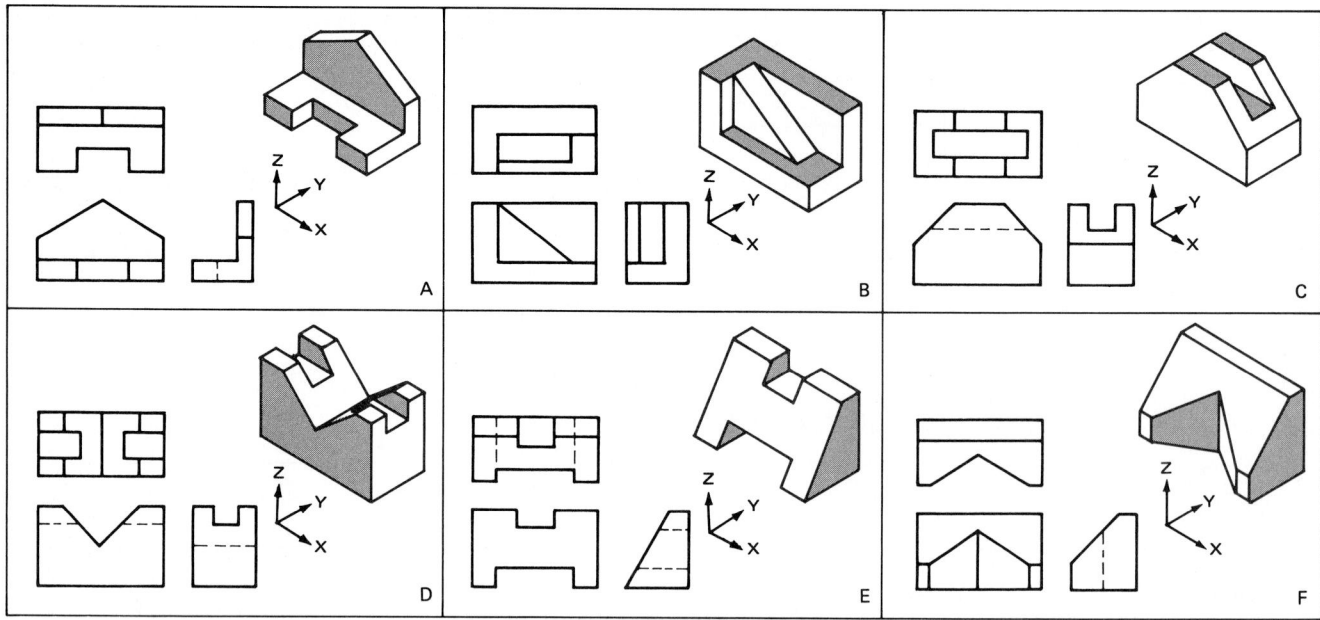

FIG. 6-5-2   Illustrations of objects having sloping surfaces.

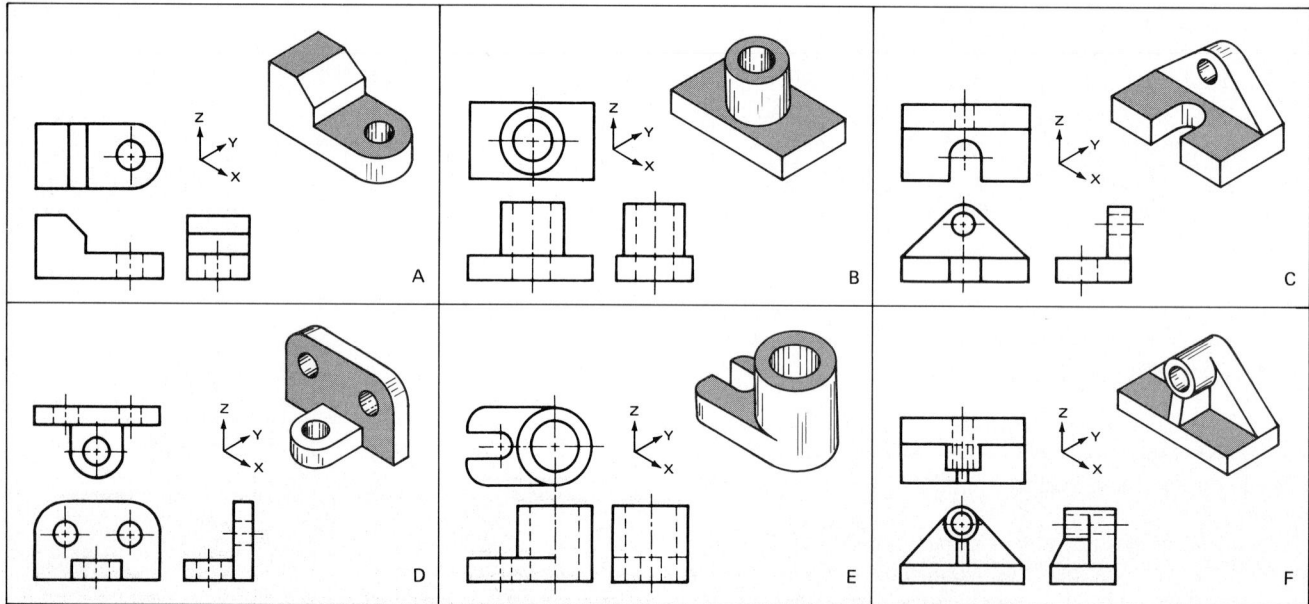

FIG. 6-6-1   Illustrations of objects having circular features.

The intersection of unfinished surfaces, such as found on cast parts, that are rounded or filleted at the point of theoretical intersection may be indicated conventionally by a line (see Unit 6-15).

Center Lines

A center line is drawn as a thin, broken line of long and short dashes, spaced alternately. Such lines may be used to locate center points, axes of cylindrical parts, and axes of symmetry, as shown in Fig. 6-6-2. Solid center lines are often used when the circular features are small. Center lines should project for a short distance beyond the outline of the part or feature to which they refer. They must be extended for use as extension lines for dimensioning purposes, but in this case the extended portion is not broken.

On views showing the circular features, the point of intersection of the two center lines is shown by the two intersecting short dashes.

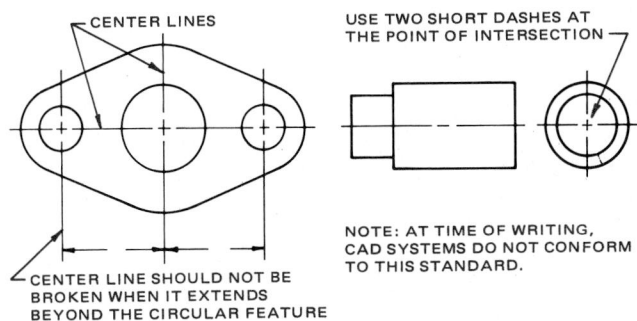

FIG. 6-6-2   Center line applications.

See Assignments 22 through 26 for Unit 6-6 on pages 125 through 128.

6-7 OBLIQUE SURFACES

When a surface is sloped so that it is not perpendicular to any of the three viewing planes, it will appear as a surface in all three views but never in its true shape. This is referred to as an *oblique surface* (Fig 6-7-1). Since the oblique surface is not perpendicular to the viewing planes, it cannot be parallel to them and consequently appears foreshortened. If a true view is required for this surface, two auxiliary views—a primary and a secondary view—need to be drawn. This is discussed in detail in Unit 7-4, "Secondary Auxiliary Views." Figure 6-7-2 shows additional examples of objects having oblique surfaces.

See Assignments 27 through 29 for Unit 6-7 on pages 129 to 130.

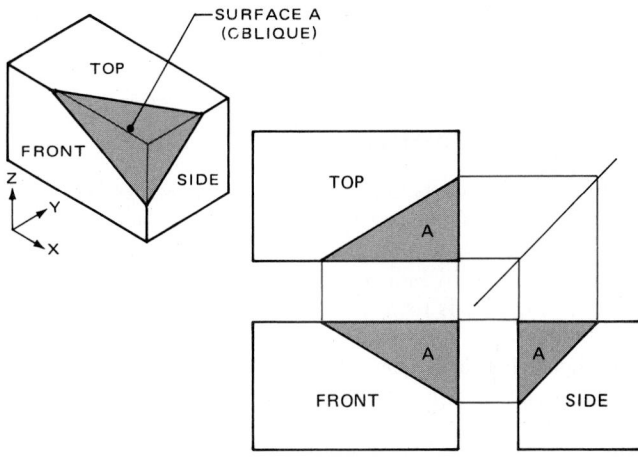

FIG. 6-7-1   Oblique surface is not its true shape in any of the three views.

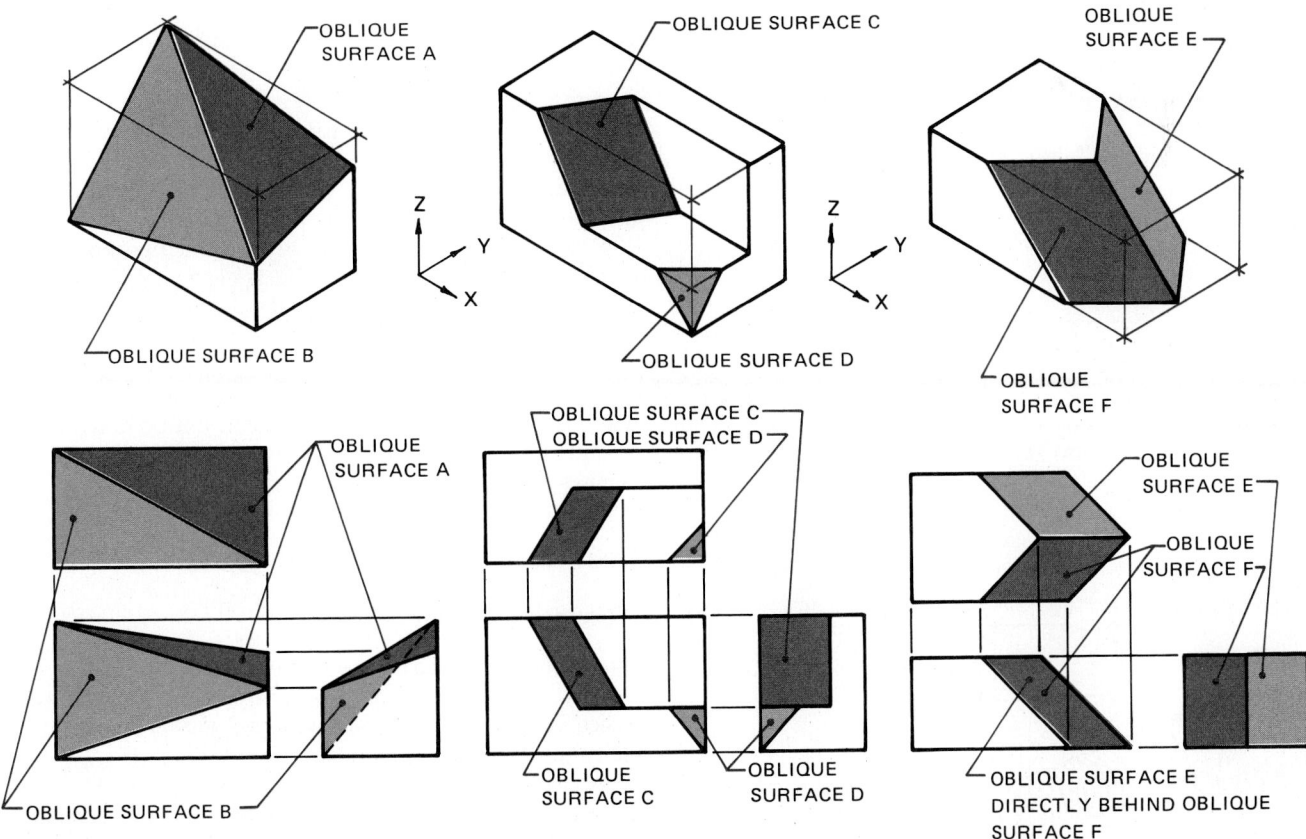

FIG. 6-7-2   Illustrations of objects having oblique surfaces.

6-8 ONE- AND TWO-VIEW DRAWINGS

View Selection

Views should be chosen that will best describe the object to be shown. Only the minimum number of views that will completely portray the size and shape of the part should be used. They should also be chosen to avoid hidden feature lines whenever possible, as shown in Fig. 6-8-1.

Except for complex objects of irregular shape, it is seldom necessary to draw more than three views. For representing simple parts, one- or two-view drawings will often be adequate.

One-View Drawings

In one-view drawings, the third dimension, such as thickness, may be expressed by a note or by descriptive words or abbreviations, such as DIA, Ø, or HEX ACRFLT. Square sections may be indicated by light, crossed, diagonal lines. This applies whether the face is parallel or inclined to the drawing plane (Fig. 6-8-2).

When cylindrically shaped surfaces include special features, such as a keyseat, a side view (often called an *end view*) is required.

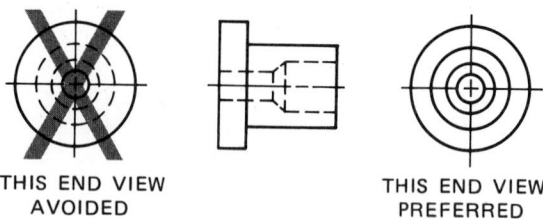

THIS END VIEW AVOIDED   THIS END VIEW PREFERRED

FIG. 6-8-1   Avoidance of hidden-line features.

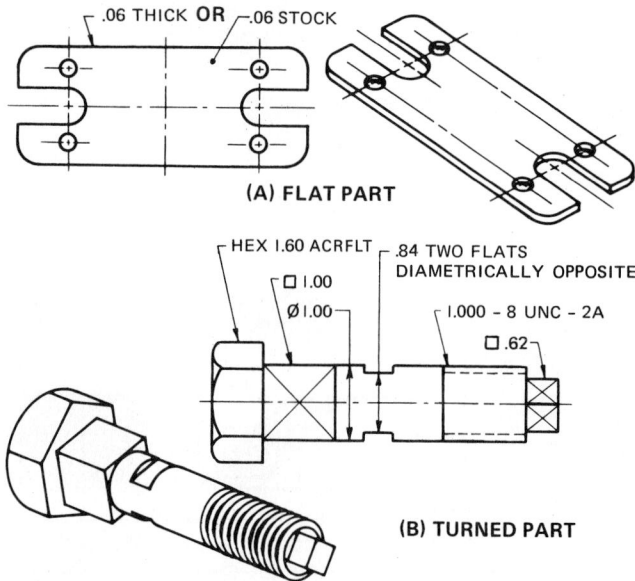

.06 THICK **OR** .06 STOCK

(A) FLAT PART

HEX 1.60 ACRFLT   .84 TWO FLATS DIAMETRICALLY OPPOSITE
□ 1.00
Ø1.00   1.000 - 8 UNC - 2A
□ .62

(B) TURNED PART

FIG. 6-8-2   One-view drawings.

Two-View Drawings

Frequently the drafter will decide that only two views are necessary to explain fully the shape of an object (Fig. 6-8-3). For this reason, some drawings consist of two adjacent views such as the top and front views only or front and right side views only. Two views are usually sufficient to explain fully the shape of cylindrical objects; if three views were used, two of them would be identical, depending on the detail structure of the part.

ASSIGNMENT

See Assignment 30 for Unit 6-8 on pages 130–131.

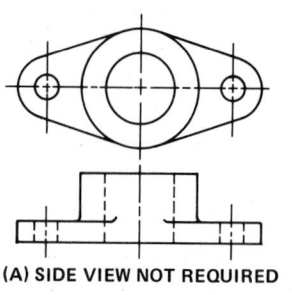
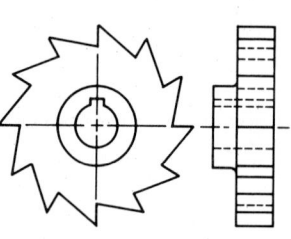

(A) SIDE VIEW NOT REQUIRED   **(B) TOP VIEW NOT REQUIRED**

FIG. 6-8-3   Two-view drawings.

6-9 SPECIAL VIEWS

Partial Views

Symmetrical objects may often be adequately portrayed by half views (Fig. 6-9-1A). A center line is used to show the axis of symmetry. Two short thick lines, above and below the view of the object, are drawn at right angles to, and on, the center line to indicate the line of symmetry.

Partial views, which show only a limited portion of the object with remote details omitted, should be used, when necessary, to clarify the meaning of the drawing (Fig. 6-9-1B). Such views are used to avoid the necessity of drawing many hidden features.

On drawings of objects where two side views can be used to better advantage than one, each need not be complete if together they depict the shape. Show only the hidden lines of features immediately behind the view (Fig. 6-9-1C).

Rear Views and Enlarged Views

Placement of Views

When views are placed in the relative positions shown in Fig. 6-1-3, it is rarely necessary to identify them. When they are placed in other than the regular projected position, the removed view must be clearly identified.

Whenever appropriate, the orientation of the main view on a detail drawing should be the same as on the assembly drawing. To avoid the crowding of dimensions and notes, ample space must be provided between views.

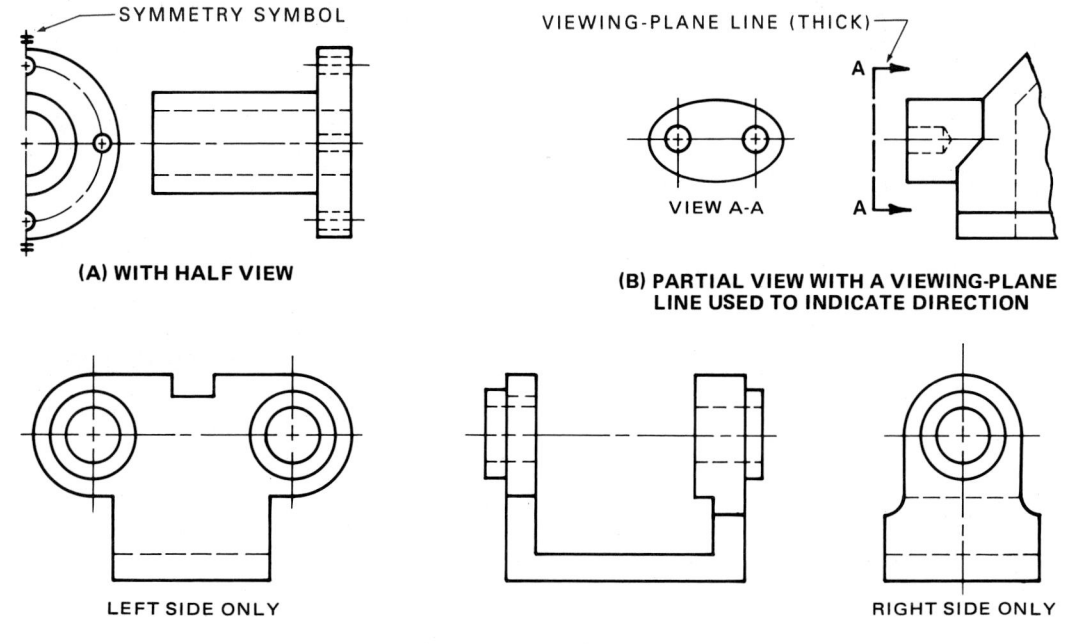

(A) WITH HALF VIEW

(B) PARTIAL VIEW WITH A VIEWING-PLANE LINE USED TO INDICATE DIRECTION

LEFT SIDE ONLY   RIGHT SIDE ONLY

(C) PARTIAL SIDE VIEWS

FIG. 6-9-1   Partial views.

Rear Views

Rear views are normally projected to the right or left. When this projection is not practical because of the length of the part, particularly for panels and mounting plates, the rear view must not be projected up or down. Doing so would result in the part being shown upside down. Instead, the view should be drawn as if it were projected sideways but located in some other position, and it should be clearly labeled REAR VIEW REMOVED (Fig. 6-9-2). Alternately the reference arrows layout method of representation may be used as explained in Unit 6-1.

Enlarged Views

Enlarged views are used when it is desirable to show a feature in greater detail or to eliminate the crowding of details or dimensions (Fig. 6-9-3). The enlarged view should be oriented

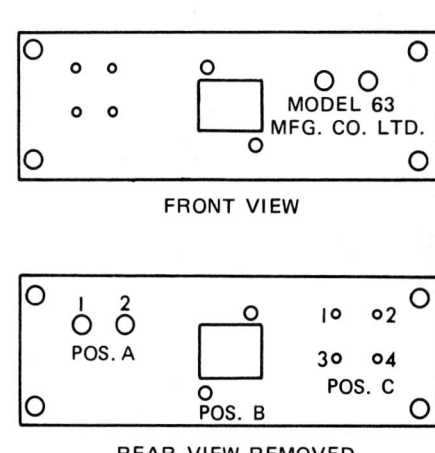

FRONT VIEW

REAR VIEW REMOVED

FIG. 6-9-2   Removed rear views.

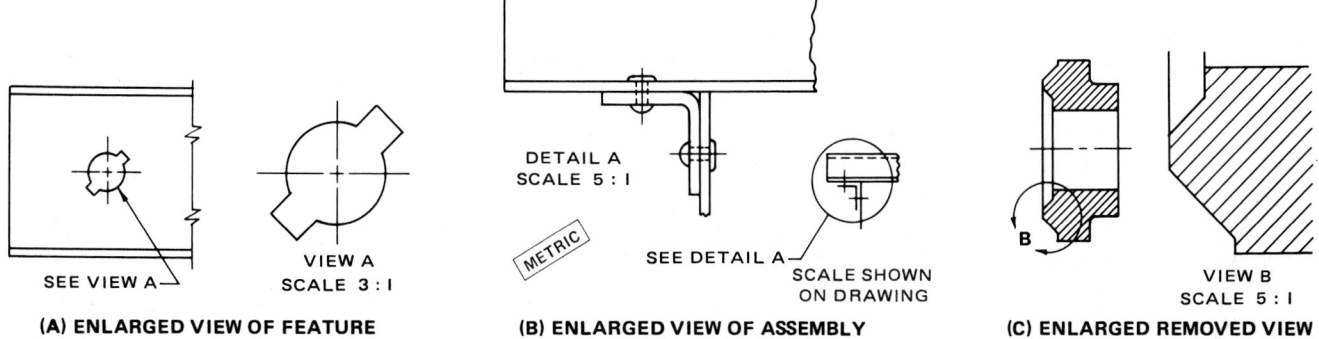

(A) ENLARGED VIEW OF FEATURE

(B) ENLARGED VIEW OF ASSEMBLY

(C) ENLARGED REMOVED VIEW

FIG. 6-9-3   Enlarged views.

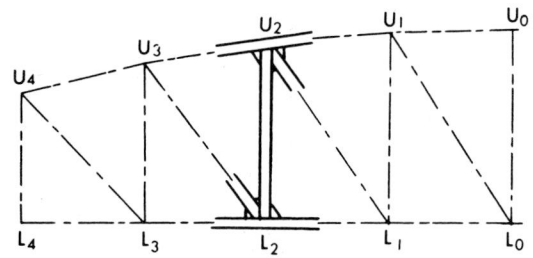

FIG. 6-9-4   Key plan.

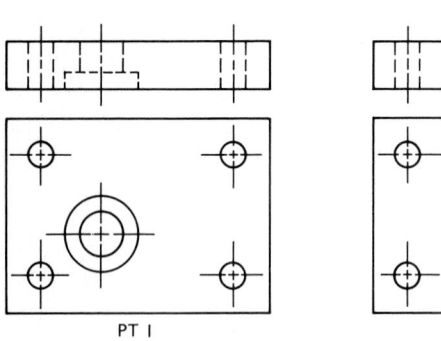

(A) TWO DRAWINGS

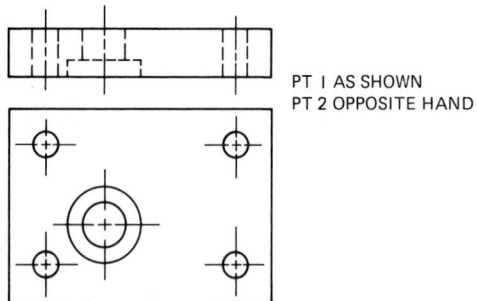

PT I AS SHOWN
PT 2 OPPOSITE HAND

(B) ONE DRAWING REPLACES TWO VIEWS

FIG. 6-9-5   Opposite-hand views.

in the same manner as the main view. However, if an enlarged view is rotated, state the direction and the amount of rotation of the detail. The scale of enlargement must be shown, and both views should be identified by one of the three methods shown.

Key Plans

A method particularly applicable to structural work is to include a small *key plan*, using bold lines, on each sheet of a drawing series that shows the relationship of the detail on that sheet to the whole work, as in Fig. 6-9-4.

Opposite-Hand Views

Where parts are symmetrically opposite, such as for right- and left-hand usage, one part is drawn in detail and the other is described by a note such as PART B SAME EXCEPT

OPPOSITE HAND. It is preferable to show both part numbers on the same drawing (Fig. 6-9-5).

ASSIGNMENTS

See Assignments 31 through 33 for Unit 6-9 on pages 131 to 132.

6-10 CONVENTIONAL REPRESENTATION OF COMMON FEATURES

To simplify the representation of common features, a number of conventional drafting practices are used. Many conventions are deviations from the true projection for the purpose of clarity; others are used to save drafting time. These conventions must be executed carefully, for clarity is even more important than speed.

Many drafting conventions, such as those used on thread, gear, and spring drawings, appear in various chapters throughout the text. Only the conventions not described in those chapters appear here.

Repetitive Details

Repetitive features, such as gear and spline teeth, are depicted by drawing a partial view, showing two or three of these features, with a phantom line or lines to indicate the extent of the remaining features (Figs. 6-10-1A and B). Alternatively, gears and splines may be shown with a solid thick line representing the basic outline of the part and a thin line representing the root of the teeth. This is essentially the same convention that is used for screw threads. The pitch line may be added by using the standard center line.

Knurls

Knurling is an operation that puts patterned indentations in the surface of a metal part to provide a good finger grip (Figs. 6-10-1C and D). Commonly used types of knurls are straight, diagonal, spiral, convex, raised diamond, depressed diamond, and radial. The pitch refers to the distance between corresponding indentations, and it may be a straight pitch, a circular pitch, or a diametral pitch. For cylindrical surfaces, the latter is preferred. The pitch of the teeth for coarse knurls (measured parallel to the axis of the work) is 14 teeth per inch (TPI) or about 2 mm; for medium knurls, 21 TPI or about 1.2 mm; and for fine knurls, 33 TPI or 0.8 mm. The medium-pitch knurl is the most commonly used.

As a time-saver, the knurl symbol is shown on only a part of the surface being knurled.

Holes

A series of similar holes is indicated by drawing one or two holes and showing only the center for the others (Figs. 6-10-1E and F).

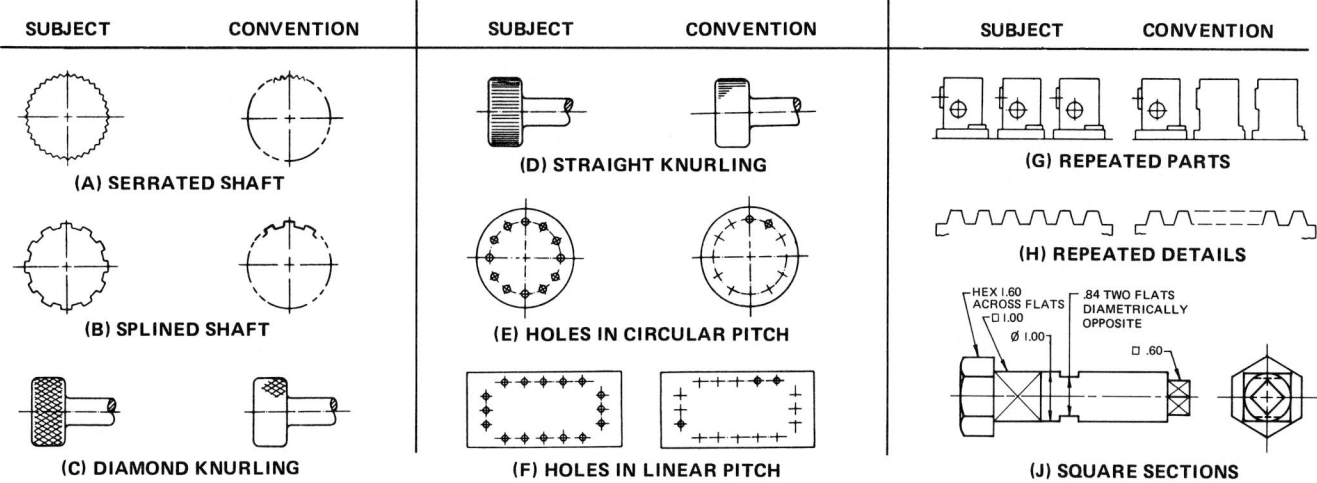

FIG. 6-10-1   Conventional representation of common features.

Repetitive Parts

Repetitive parts, or intricate features, are shown by drawing one in detail and the others in simple outline only. A covering note is added to the drawing (Figs. 6-10-1G and H).

Square Sections

Square sections on shafts and similar parts may be illustrated by thin, crossed, diagonal lines, as shown in Fig. 6-10-1J.

ASSIGNMENT

See Assignment 34 for Unit 6-10 on page 133.

6-11 CONVENTIONAL BREAKS

Long, simple parts, such as shafts, bars, tubes, and arms, need not be drawn to their entire length. Conventional breaks located at a convenient position may be used and the true length indicated by a dimension. Often a part can be drawn to a larger scale to produce a clearer drawing if a conventional break is used. Two types of break lines are generally used (Fig. 6-11-1A). Thick freehand lines are used for short breaks. The thin line with freehand zig-zag is recommended for long breaks and may be used for solid details or for assemblies containing open spaces.

Special break lines, shown in Fig. 6-11-1B, are used when it is desirable to indicate the shape of the features.

ASSIGNMENT

See Assignment 35 for Unit 6-11 on page 133.

6-12 MATERIALS OF CONSTRUCTION

Symbols used to indicate materials in sectional views are shown in Fig. 9-1-6 (pg. 229). Those shown for concrete, wood, and transparent materials are also suitable for outside

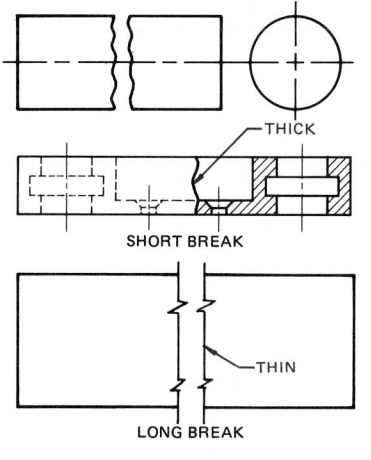

(A) GENERAL-USE BREAK LINES

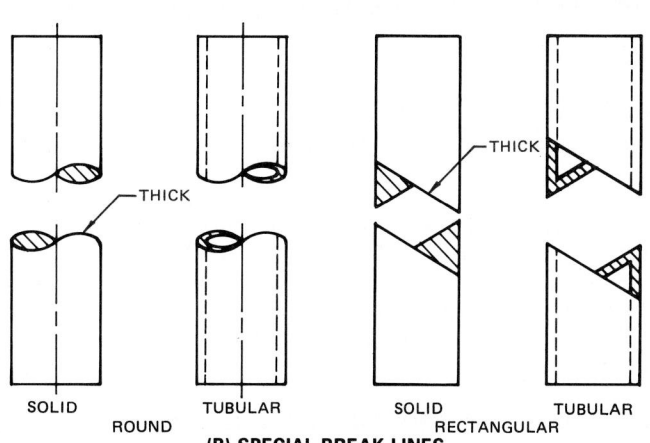

(B) SPECIAL BREAK LINES

FIG. 6-11-1   Conventional breaks.

views. Other symbols that may be used to indicate areas of different materials are shown in Fig. 6-12-1. It is not necessary to cover the entire area affected with such symbolic lining as long as the extent of the area is shown on the drawing.

Transparent Materials

These should generally be treated in the same manner as opaque materials; i.e., details behind them are shown with hidden lines if such details are necessary.

ASSIGNMENT

See Assignment 36 for Unit 6-12 on page 134.

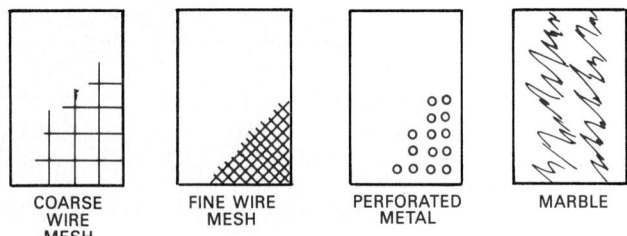

COARSE WIRE MESH · FINE WIRE MESH · PERFORATED METAL · MARBLE

PERFORATED METALS

FIG. 6-12-1   Symbols to indicate materials of construction.

6-13 CYLINDRICAL INTERSECTIONS

The intersections of rectangular and circular contours, unless they are very large, are shown conventionally, as in Figs. 6-13-1 and 6-13-2. The same convention may be used to show the intersection of two cylindrical contours, or the curve of intersection may be shown as a circular arc.

ASSIGNMENT

See Assignment 37 for Unit 6-13 on page 134.

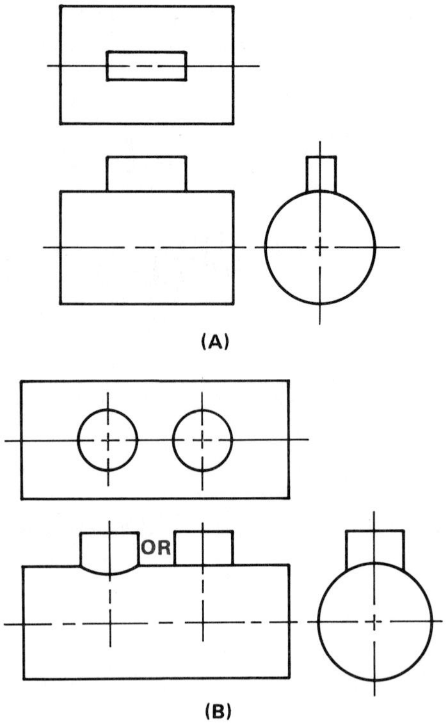

6-14 FORESHORTENED PROJECTION

When the true projection of a feature would result in confusing foreshortening, it should be rotated until it is parallel to the line of the section or projection (Fig. 6-14-1).

Holes Revolved to Show True Distance from Center

Drilled flanges in elevation or section should show the holes at their true distance from the center, rather than the true projection.

(A)

(B)

FIG. 6-13-1   Conventional representation of external intersections.

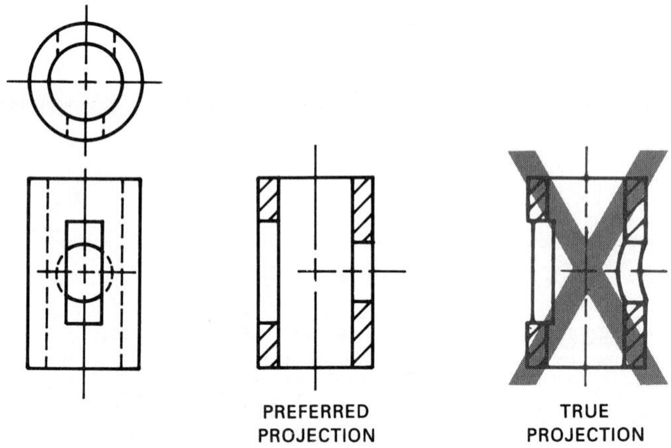

PREFERRED PROJECTION · TRUE PROJECTION

FIG. 6-13-2   Conventional representation of holes in cylinders.

ASSIGNMENT

See Assignment 38 for Unit 6-14 on page 135.

6-15 INTERSECTIONS OF UNFINISHED SURFACES

The intersections of unfinished surfaces that are rounded or filleted may be indicated conventionally by a line coinciding with the theoretical line of intersection. The need for this convention is demonstrated by the examples shown in Fig. 6-15-1, where the upper top views are shown in true projection. Note that in each example the true projection would be misleading. In the case of the large radius, such as shown in Fig. 6-15-1C, no line is drawn. Members such as ribs and arms that blend into other features terminate in curves called

runouts. With manual drafting small runouts are usually drawn freehand. Large runouts are drawn with an irregular curve, template, or compass (Fig. 6-15-2, pg. 110).

ASSIGNMENT

See Assignment 39 for Unit 6-15 on page 135.

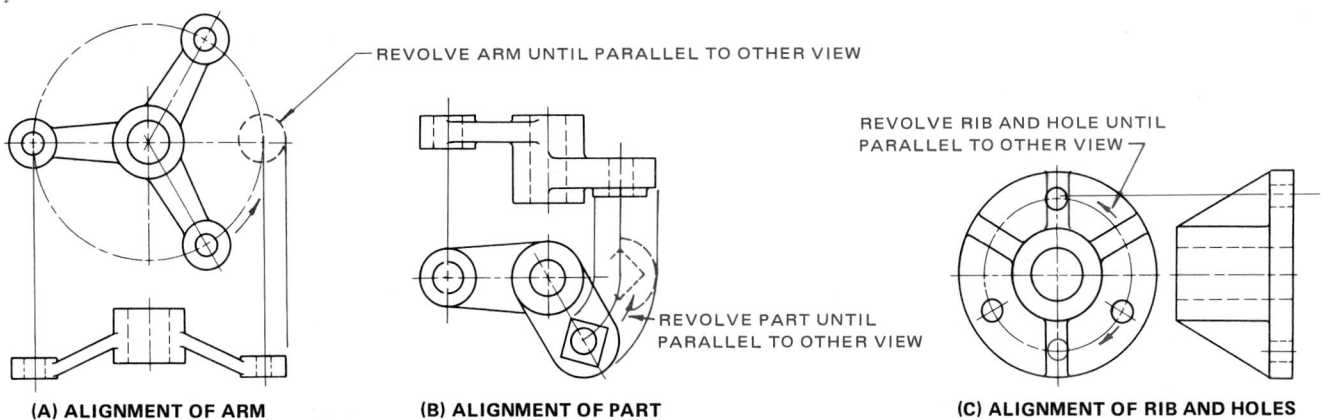

(A) ALIGNMENT OF ARM **(B) ALIGNMENT OF PART** **(C) ALIGNMENT OF RIB AND HOLES**

FIG. 6-14-1   Alignment of parts and holes to show true relationship.

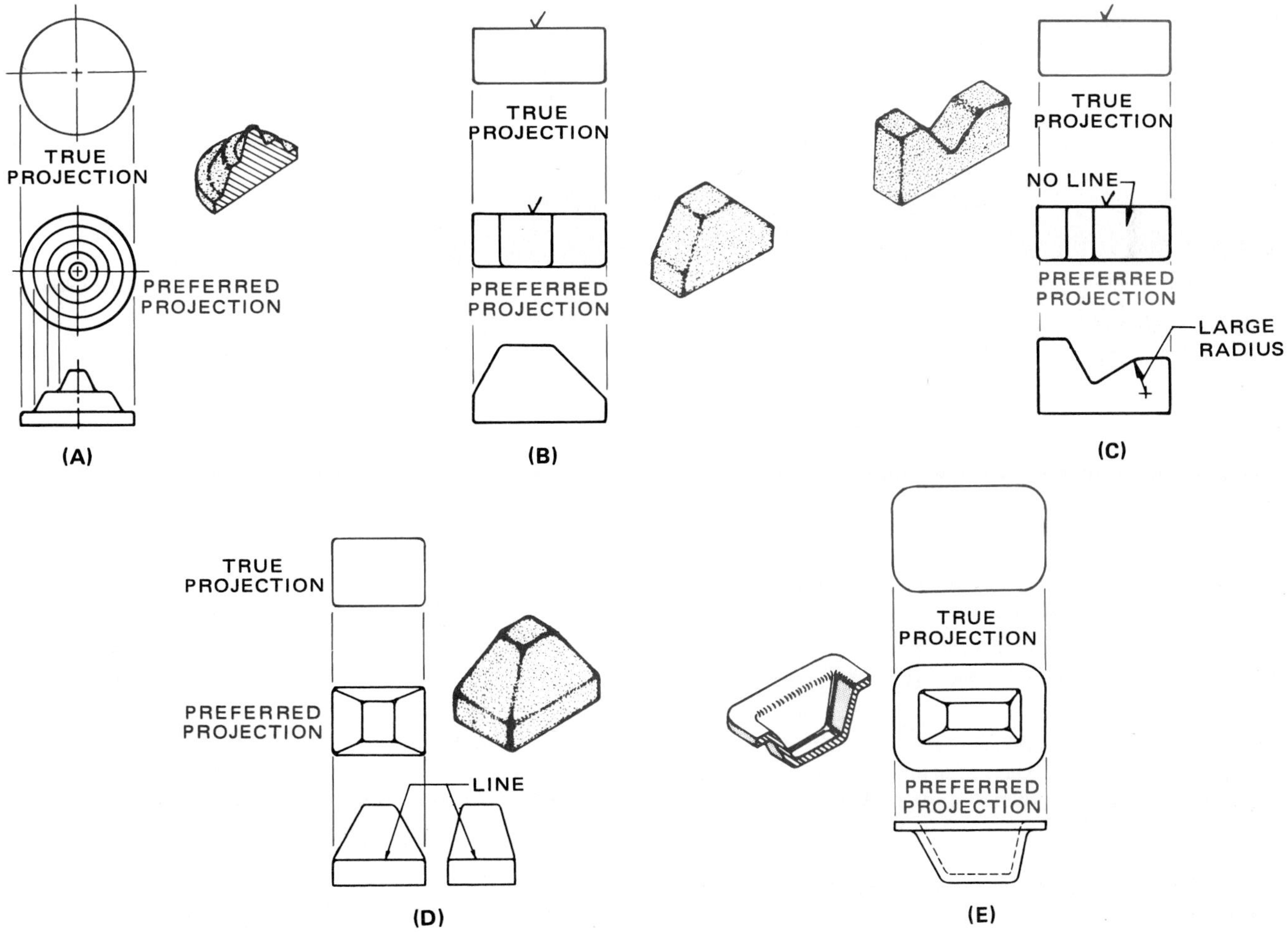

FIG. 6-15-1   Conventional representation of rounds and fillets.

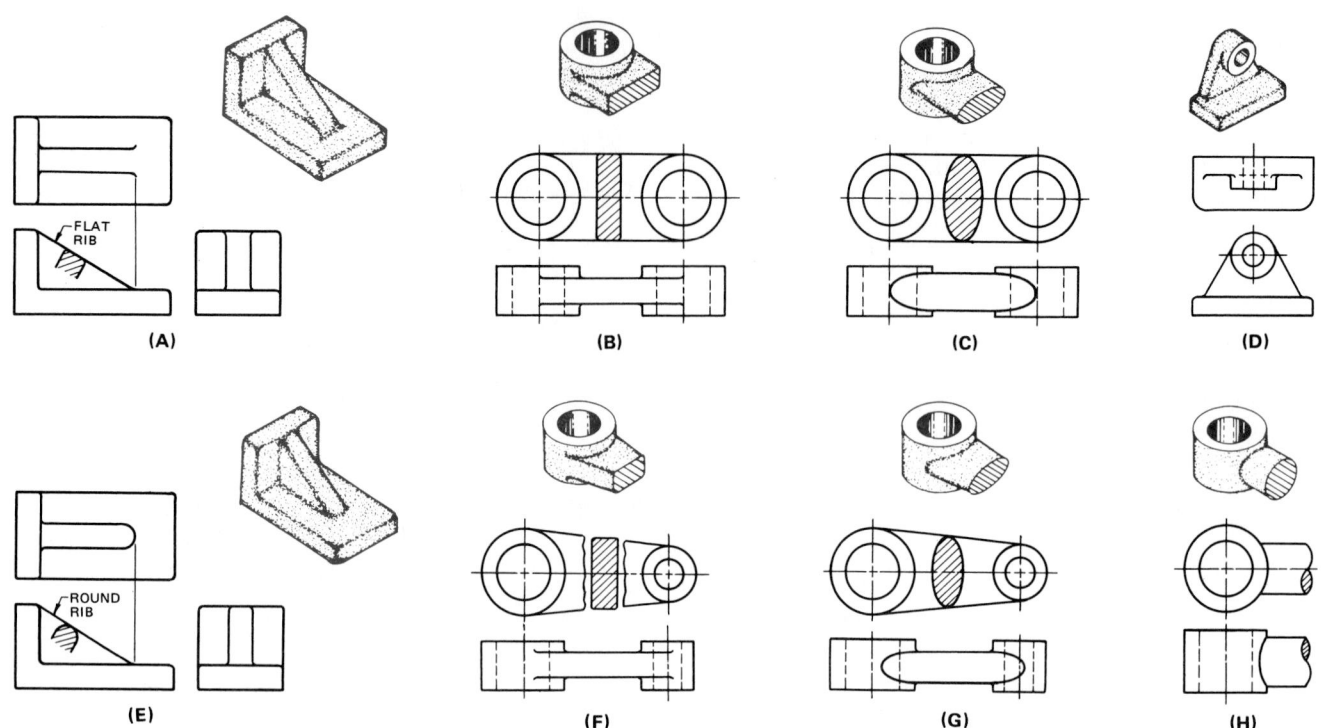

FIG. 6-15-2   Conventional representation of runouts.

ASSIGNMENTS FOR CHAPTER 6

Notes: (1) CAD may be substituted for manual drafting for any assignments in this chapter. (2) Unless otherwise specified all drawings are to be drawn in third-angle projection.

ASSIGNMENTS FOR UNIT 6-1, ORTHOGRAHIC REPRESENTATIONS

1. Draw the six views for any two parts shown in Figs. 6-1-A through 6-1-E using the following methods of representation: (A) third-angle projection; (B) first-angle projection; (C) reference arrows layout. Show only what can be seen when viewing the object. Do not attempt to show any hidden features. Viewing in the direction of axis *Y* will represent the principal view. Identify the views as shown in Figs. 6-1-3 through 6-1-5.

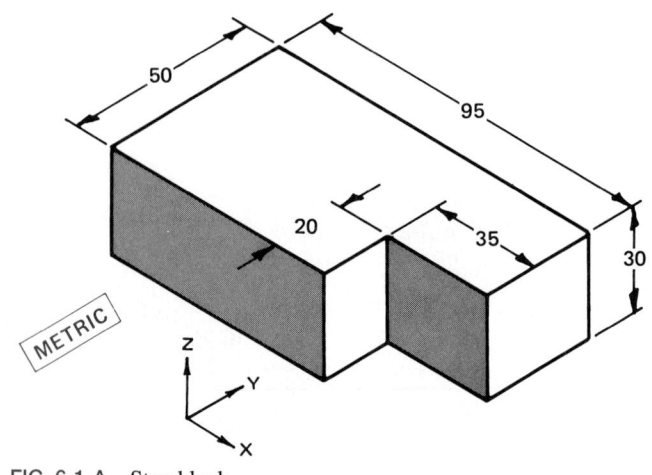

FIG. 6-1-A   Stop block.

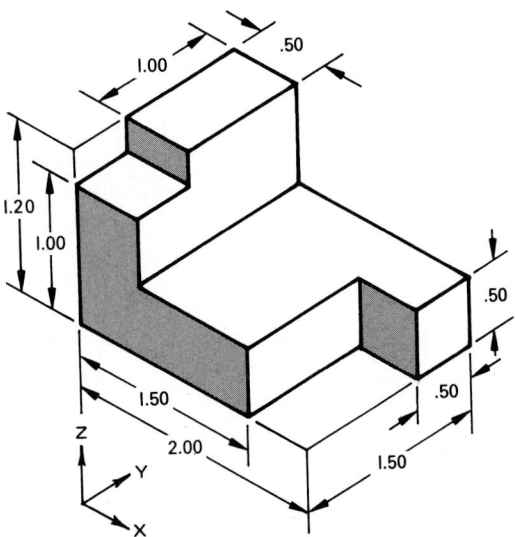

FIG. 6-1-B   Angle bracket.

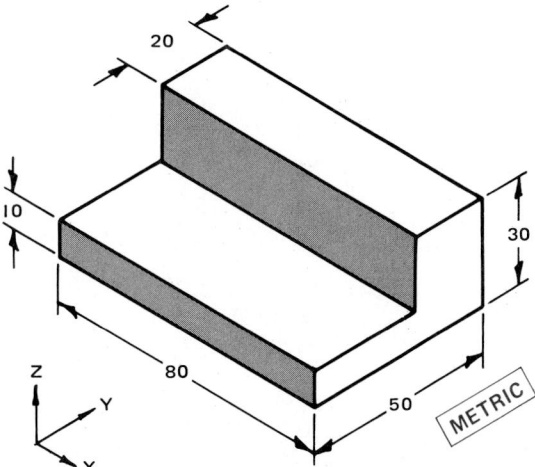

FIG. 6-1-C   Step block.

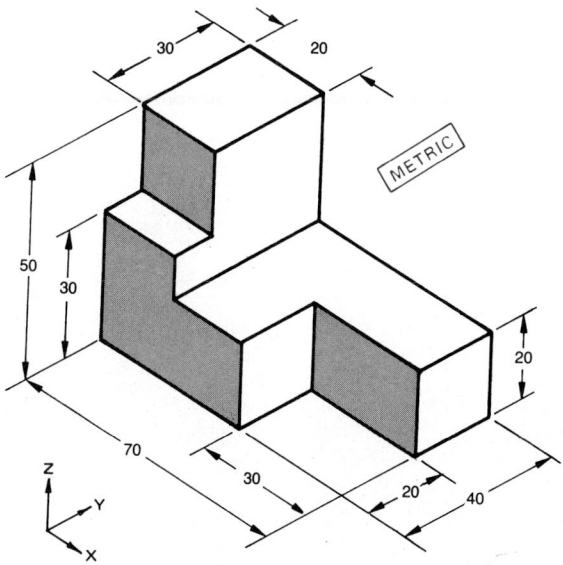

FIG. 6-1-D   Corner bracket.

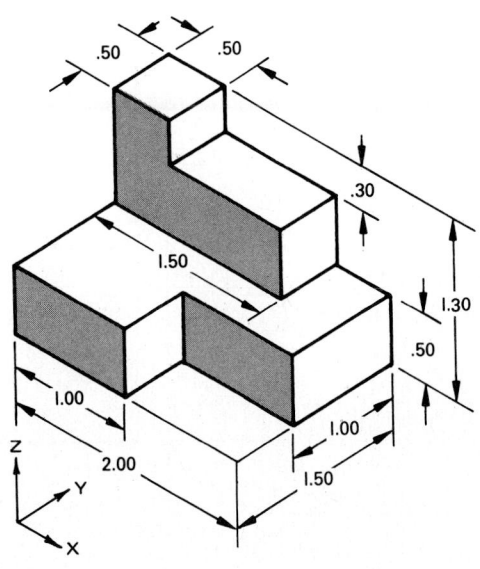

FIG. 6-1-E   Locating block.

2. Using squared graph paper of 4 or 5 squares to the inch (one square representing 1.00 in.) or 10 mm squares (one square representing 10 mm) or the grid on the CAD monitor, sketch or plot the views using the two-dimensional absolute coordinates shown in Figs. 6-1-F through 6-1-H. Scale 1:1.

ABSOLUTE COORDINATES (IN.)		
Point	X Axis	Y Axis
1	0	0
2	3.50	0
3	3.50	1.00
4	2.00	1.00
5	2.00	2.00
6	0	2.00
7	0	0
NEW START		
8	0	.50
9	3.50	.50
NEW START		
10	1.50	0
11	1.50	.50
NEW START		
12	0	2.50
13	1.50	2.50
14	1.50	3.00
15	3.50	3.00
16	3.50	4.50
17	0	4.50
18	0	2.50
NEW START		
19	0	3.50
20	3.50	3.50
NEW START		
21	2.00	3.50
22	2.00	4.50
NEW START		
23	4.00	0
24	6.00	0
25	6.00	2.00
26	5.00	2.00
27	5.00	.50
28	4.00	.50
29	4.00	0
NEW START		
30	5.00	1.00
31	6.00	1.00
NEW START		
32	4.50	0
33	4.50	.50

FIG. 6-1-F   Absolute coordinate (inches) assignment.

ABSOLUTE COORDINATES (IN.)		
Point	X Axis	Y Axis
1	0	0
2	2.50	0
3	2.50	.50
4	0	.50
NEW START		
5	2.00	.50
6	2.00	1.00
7	0	1.00
NEW START		
8	1.50	1.00
9	1.50	1.50
10	0	1.50
11	0	0
NEW START		
12	0	2.00
13	2.50	2.00
14	2.50	3.50
15	0	3.50
16	0	2.00
NEW START		
17	0	2.50
18	2.00	2.50
19	2.00	3.50
NEW START		
20	0	3.00
21	1.50	3.00
22	1.50	3.50
NEW START		
23	4.50	.50
24	3.00	.50
25	3.00	0
26	4.50	0
27	4.50	1.50
28	4.00	1.50
29	4.00	1.00
NEW START		
30	4.50	1.00
31	3.50	1.00
32	3.50	.50

FIG. 6-1-G   Absolute coordinate (inches) assignment.

ABSOLUTE COORDINATES (mm)		
Point	X Axis	Y Axis
1	0	0
2	90	0
3	90	10
4	0	10
NEW START		
5	0	0
6	0	40
7	50	40
8	50	10
NEW START		
9	70	0
10	70	10
NEW START		
11	0	50
12	70	50
13	70	70
14	90	70
15	90	90
16	0	90
17	0	50
NEW START		
18	0	80
19	50	80
20	50	90
NEW START		
21	140	10
22	100	10
23	100	0
24	140	0
25	140	40
26	130	40
27	130	10
NEW START		
28	120	0
29	120	10

FIG. 6-1-H   Absolute coordinate (metric) assignment.

3. Using squared graph paper of 4 or 5 squares to the inch (one square representing 1.00 in.) or 10 mm squares (one square representing 10 mm) or the grid on the CAD monitor, sketch or plot the views using the two-dimensional relative coordinates shown in Figs. 6-1-J through 6-1-L. Scale 1:1.

RELATIVE COORDINATES (IN.)		
Point	X Axis	Y Axis
1	0	0
2	3.00	0
3	0	.50
4	−2.00	0
5	0	1.50
6	−1.00	0
7	0	−2.00
NEW START		
8	0	1.50
9	1.00	0
NEW START		
10	1.50	0
11	0	.50
NEW START		
12	0	2.50
13	1.50	0
14	0	.50
15	1.50	0
16	0	1.50
17	−3.00	0
18	0	−2.00
NEW START		
19	1.00	2.50
20	0	2.00
NEW START		
21	0	3.50
22	1.00	0
NEW START		
23	3.50	0
24	2.00	0
25	0	2.00
26	−1.00	0
27	0	−.50
28	−1.00	0
29	0	−1.50
NEW START		
30	3.50	.50
31	2.00	0
NEW START		
32	4.00	0
33	0	.50

FIG. 6-1-J   Relative coordinate (inches) assignment.

RELATIVE COORDINATES (IN.)		
Point	X Axis	Y Axis
1	0	0
2	3.00	0
3	0	1.00
4	−1.00	0
5	0	.50
6	−2.00	0
7	0	−1.50
NEW START		
8	1.50	0
9	0	.50
10	−1.00	0
11	0	1.00
NEW START		
12	0	2.00
13	0	1.50
14	3.00	0
15	0	−.50
16	−2.50	0
17	0	−1.00
NEW START		
18	0	2.00
19	1.50	0
20	0	1.00
NEW START		
21	2.00	3.00
22	0	.50
NEW START		
23	3.50	0
24	1.50	0
25	0	1.50
26	−1.50	0
27	0	−1.50
NEW START		
28	4.50	0
29	0	1.50
NEW START		
30	3.50	.50
31	1.00	0
NEW START		
32	4.50	1.00
33	.50	0

FIG. 6-1-K   Relative coordinate (inches) assignment.

RELATIVE COORDINATES (mm)		
Point	X Axis	Y Axis
1	0	0
2	70	0
3	0	20
4	−70	0
NEW START		
5	0	0
6	0	30
7	30	0
8	0	−10
NEW START		
9	20	30
10	0	−10
NEW START		
11	40	20
12	0	−20
NEW START		
13	0	40
14	40	0
15	0	20
16	30	0
17	0	20
18	−70	0
19	0	−40
NEW START		
20	0	60
21	20	0
22	0	10
23	10	0
24	0	10
NEW START		
25	120	20
26	−40	0
27	0	−20
28	40	0
29	0	30
30	−20	0
31	0	−30
NEW START		
32	110	20
33	0	10

FIG. 6-1-L   Relative coordinate (metric) assignment.

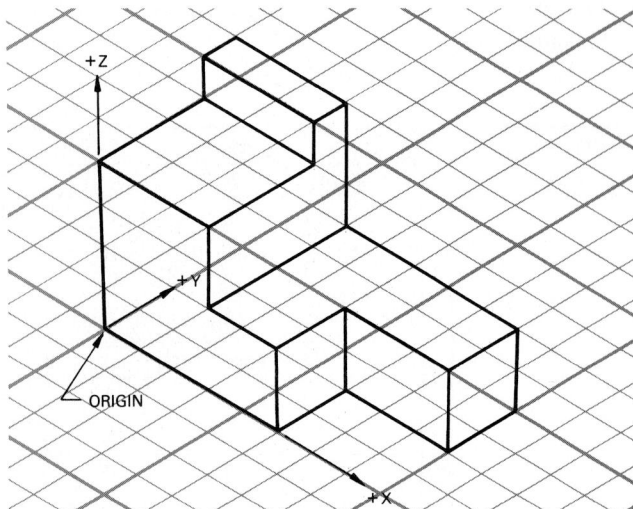

FIG. 6-1-M   Stand.

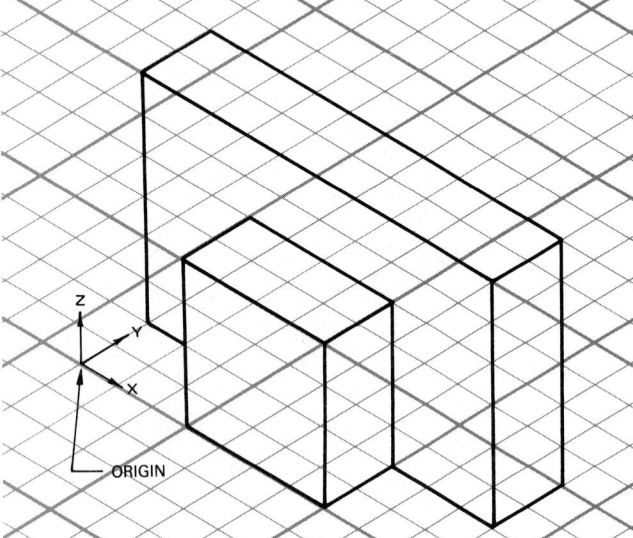

FIG. 6-1-N   Spacer.

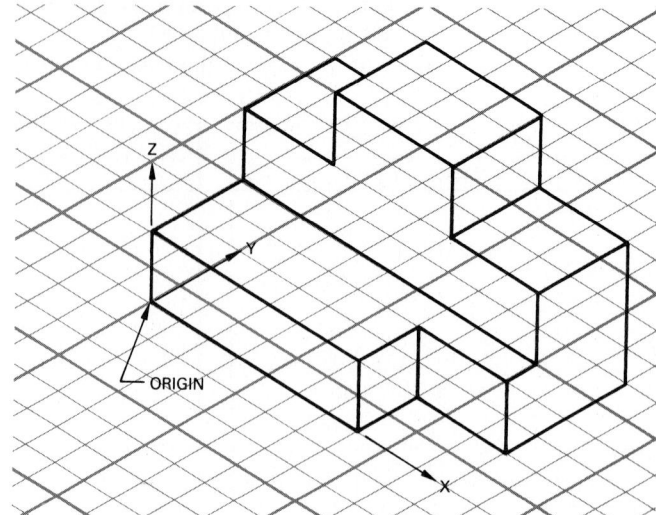

FIG. 6-1-P   Slide bracket.

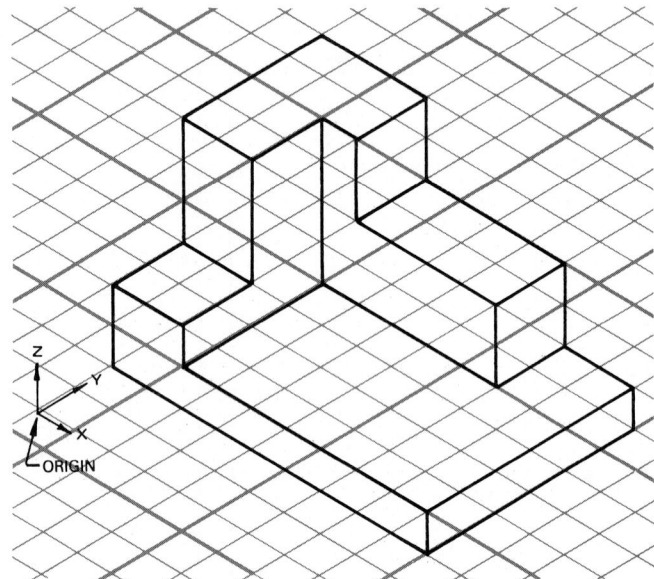

FIG. 6-1-R   Corner stand.

4. Using isometric graph paper sketch (copy) any three of the parts shown in Figs. 6-1-M through 6-1-S. Each square on the grid should represent .50 in. or 10 mm. After completing the views, add the *X*, *Y*, and *Z* coordinates where the lines intersect one another. Identify only those intersections that can be seen. Note the location of the origin for each part.

ASSIGNMENTS FOR UNIT 6-2, ARRANGEMENT AND CONSTRUCTION OF VIEWS

5. Make a three-view sketch similar to Figs. 6-2-1B and C and establish the distance between plane 1 and the left border line and between plane 2 and the bottom border line, given the following: scale 1:1; drawing space

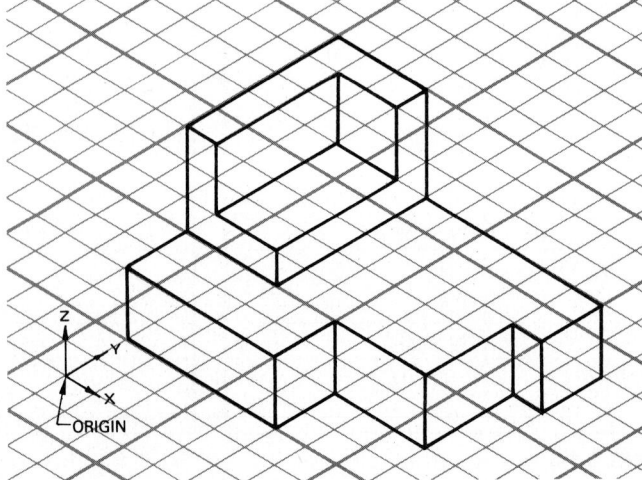

FIG. 6-1-S   Guide bracket.

8.00 × 10.50 in.; part size: W = 4.10, H = 1.40, D = 2.10; space between views to be 1.00 in.

6. Make a three-view sketch similar to Figs. 6-2-1B and C and establish the distance between plane 1 and the left border line and between plane 2 and the bottom border line, given the following: scale 1:2; drawing space 8.00 × 10.50 in.: part size: W = 8.50, H = 4.90, D = 4.50; space between views to be 1.00 in.

7. Angle bracket, Fig. 6-1-B, sheet size A (A4), scale 1:1. Make a three-view drawing using a miter line to complete the right side view. Space between views to be 1.00 in.

8. Locating block, Fig. 6-1-E, sheet size A (A4), scale 1:1. Make a three-view drawing using a miter line to complete the top view. Space between views to be 1.00 in.

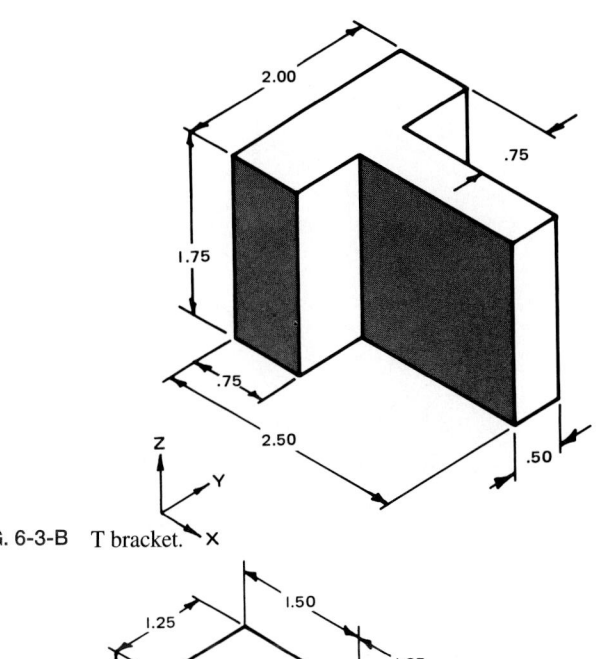

FIG. 6-3-B   T bracket.

ASSIGNMENTS FOR UNIT 6-3, ALL SURFACES PARALLEL AND ALL EDGES AND LINES VISIBLE

9. On preprinted grid paper (.25 in. or 10 mm grid) sketch three views of each of the objects shown in Fig. 6-3-A. Each square shown on the objects represents one square on the grid paper. Allow one grid space between views and a minimum of two grid spaces between objects. Identify the type of projection used by placing the identifying symbol at the bottom of the drawing.

10. Draw three views of one of the parts shown in Figs. 6-3-B through 6-3-E (here and on pg. 116). Allow 1.00 in. or 25 mm between views. Scale full or 1:1. Do not dimension.

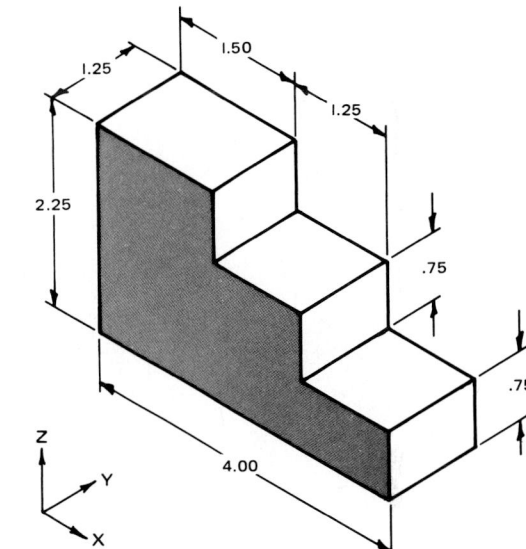

FIG. 6-3-C   Step support.

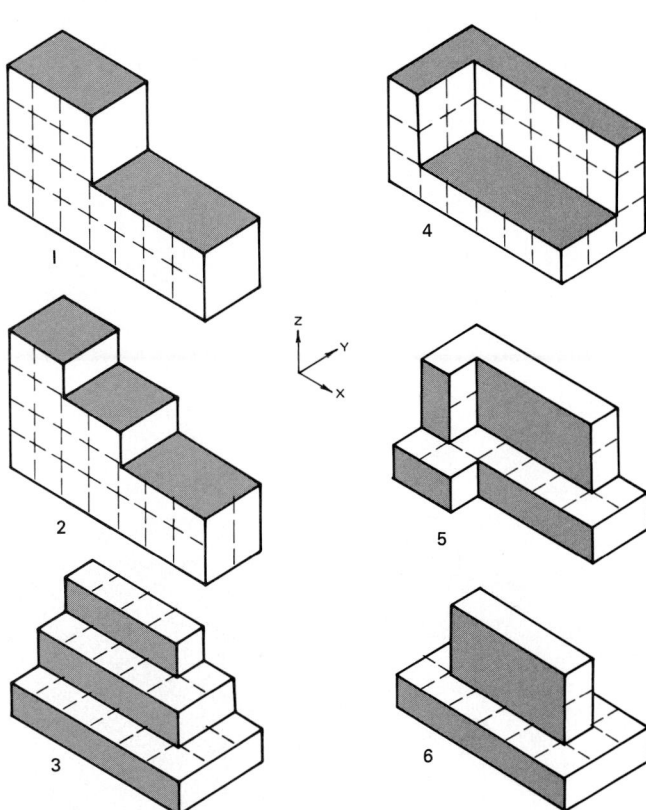

FIG. 6-3-A   Sketching assignment.

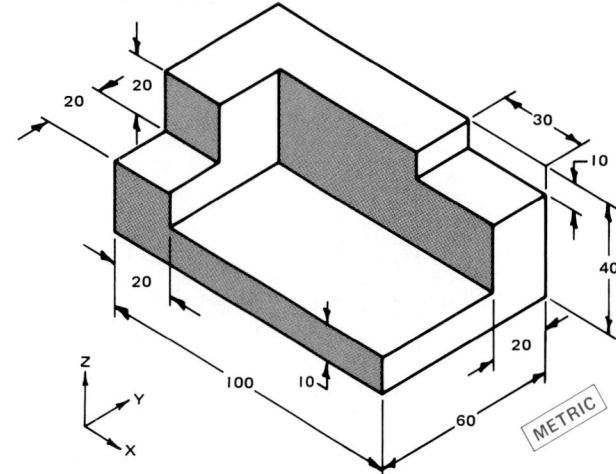

FIG. 6-3-D   Corner block.

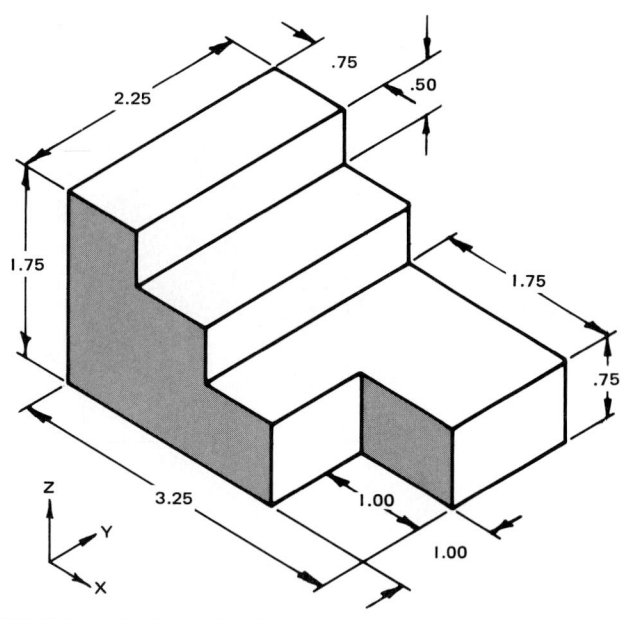

FIG. 6-3-E　Angle step bracket.

ASSIGNMENTS FOR UNIT 6-4, HIDDEN SURFACES AND EDGES

11. On preprinted grid paper (.25 in. or 10 mm grid) sketch three views of each of the objects shown in Figs. 6-4-A and 6-4-B. Each square shown on the objects represents one square on the grid paper. Allow one grid space between views and a minimum of two spaces between objects. Identify the type of projection by placing the identifying symbol at the bottom of the drawing.
12. Sketch the views needed for a multiview drawing of the parts shown in Fig. 6-4-C. Choose your own sizes and estimate proportions.
13. Match the pictorial drawings to the orthographic drawings shown in Fig. 6-4-D.

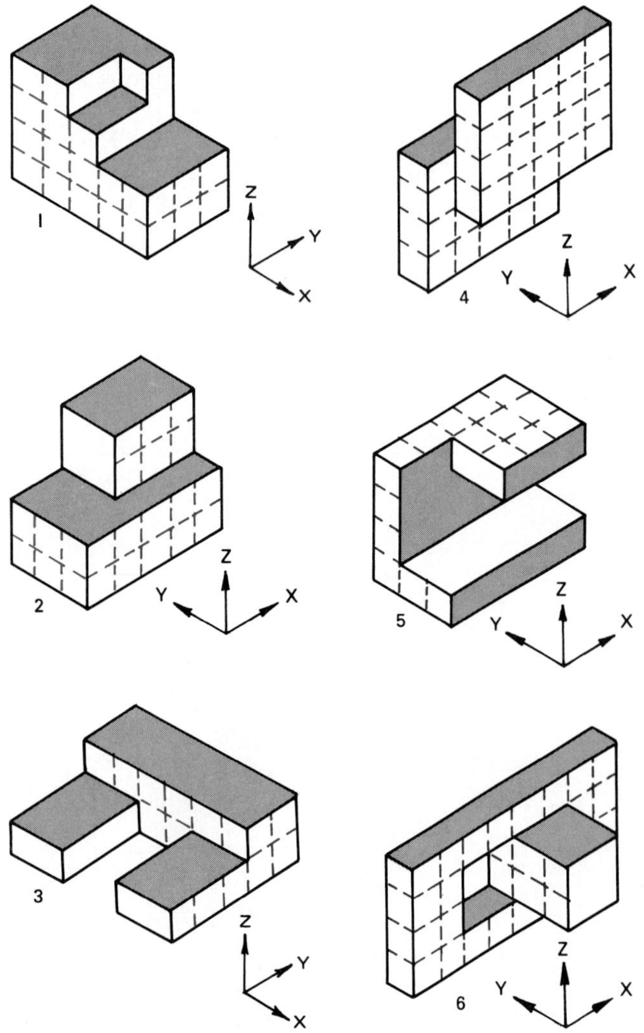

FIG. 6-4-A　Sketching assignment.

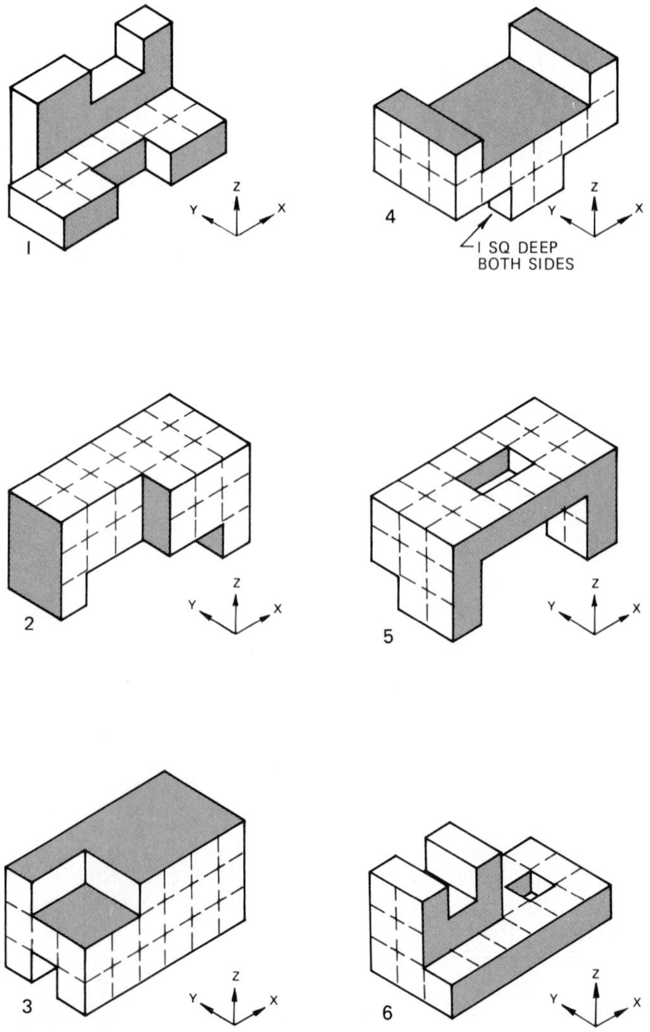

FIG. 6-4-B　Sketching assignment.

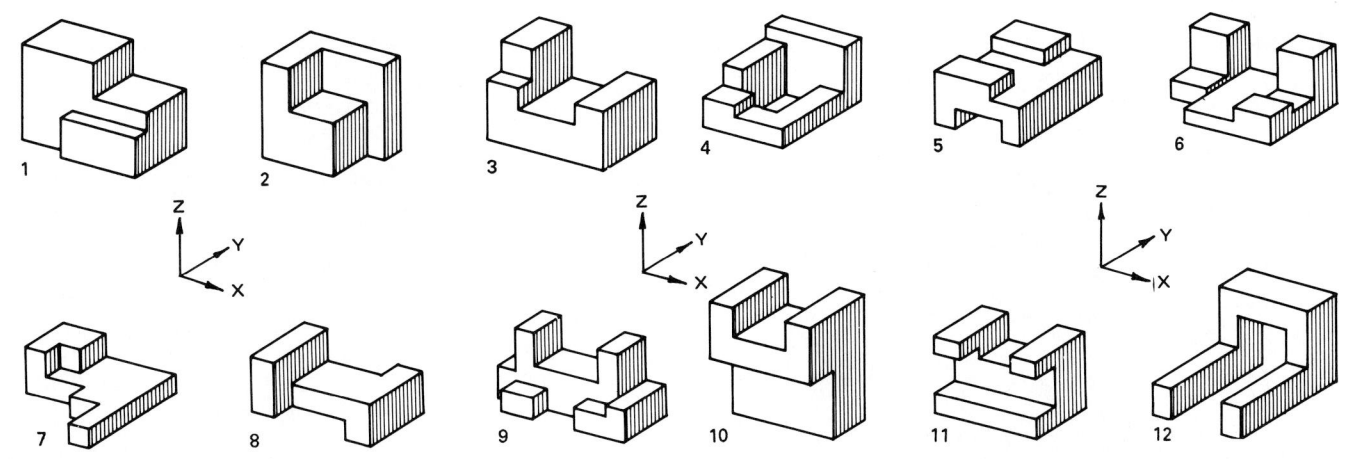

FIG. 6-4-C   Sketching assignment.

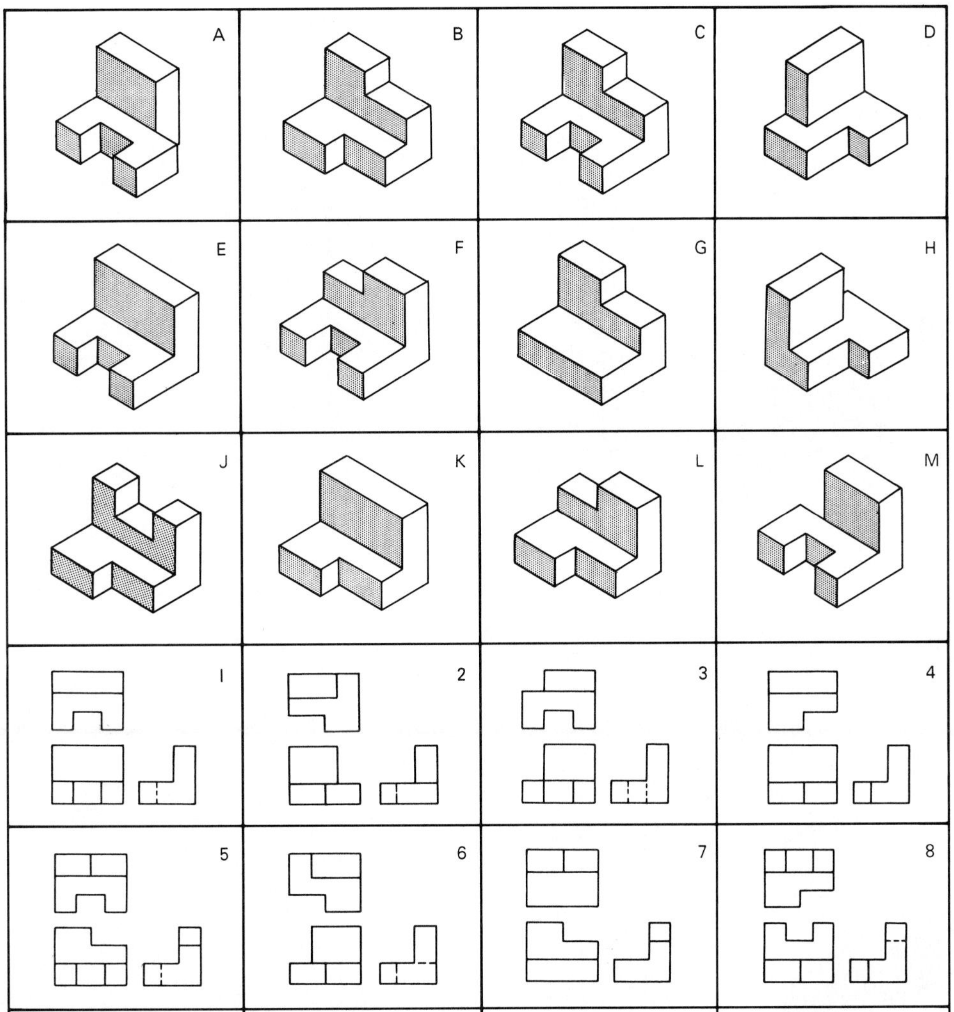

FIG. 6-4-D   Matching test.

14. Make a three-view drawing of one of the parts shown in Figs. 6-4-E through 6-4-K. Allow 1.00 in. or 25 mm between views. Do not dimension.

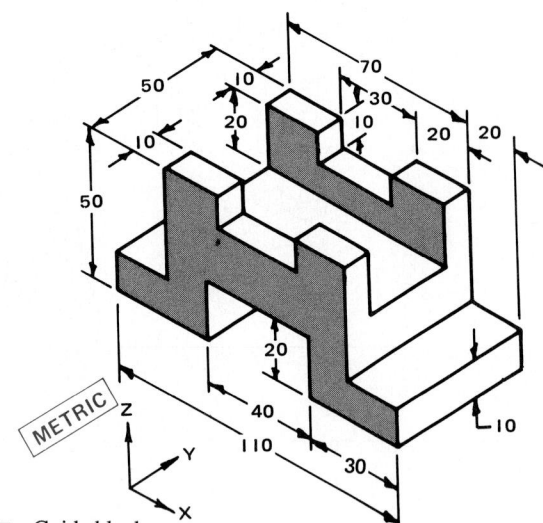

FIG. 6-4-E   Guide block.

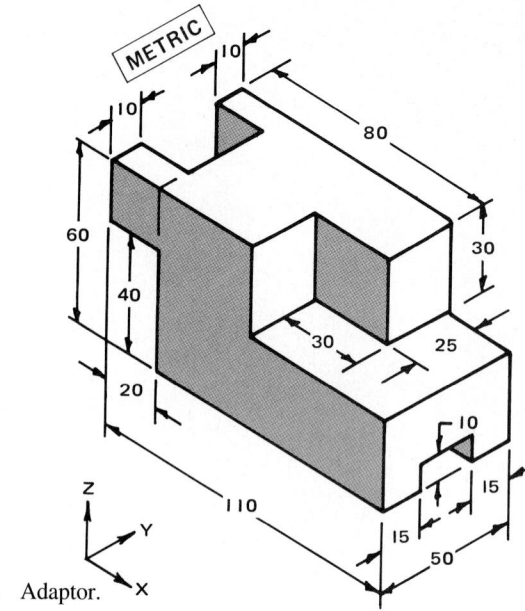

FIG. 6-4-H   Adaptor.

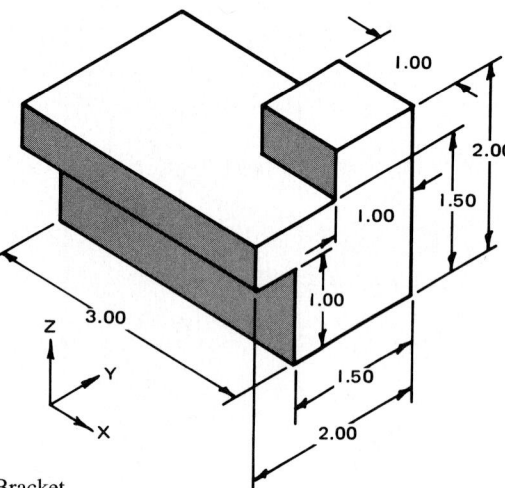

FIG. 6-4-F   Bracket.

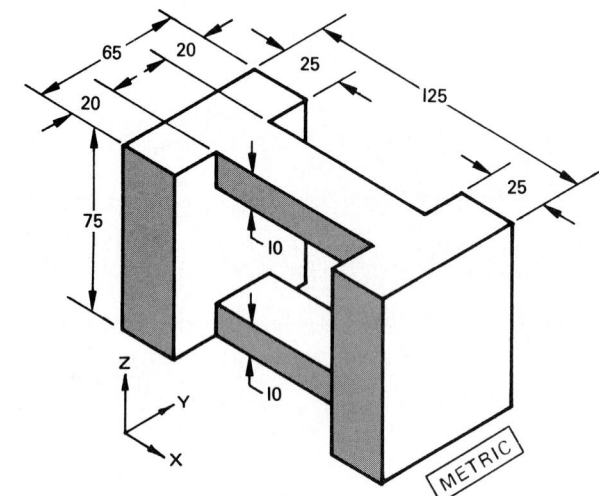

FIG. 6-4-J   Bracket.

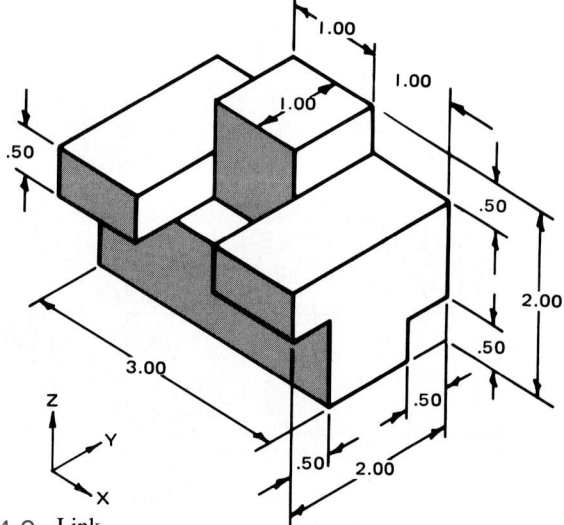

FIG. 6-4-G   Link.

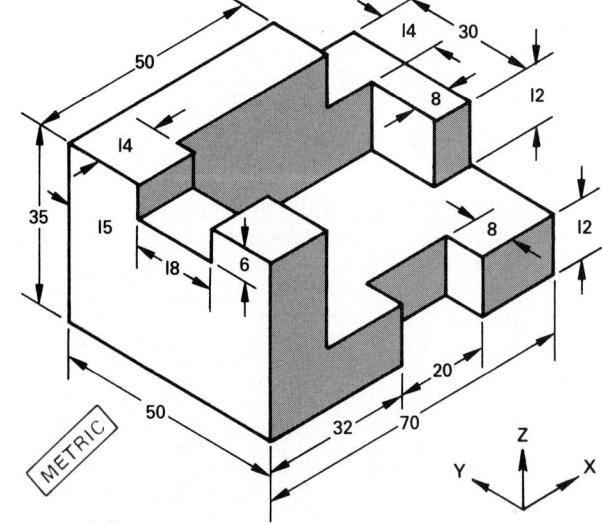

FIG. 6-4-K   Adjusting guide.

15. Make a three-view drawing of one of the parts shown in Fig. 6-4-L through 6-4-S. Allow 1.00 in. or 25 mm between views. Do not dimension.

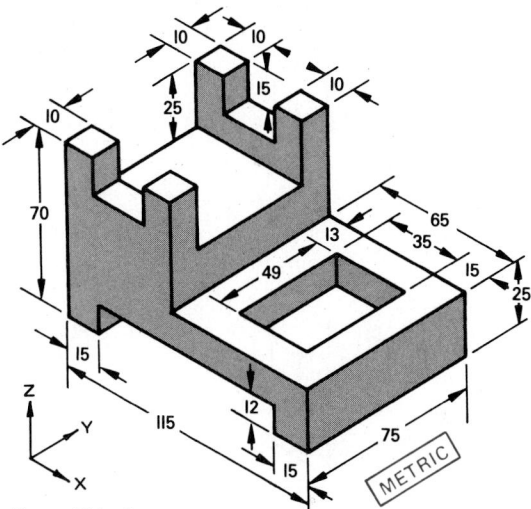

FIG. 6-4-L   Control block.

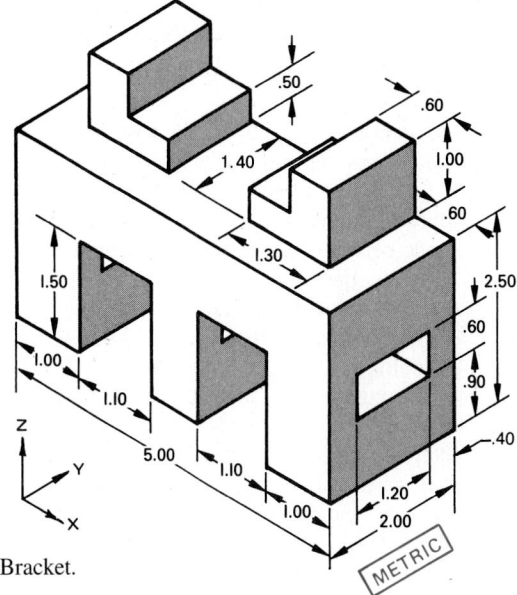

FIG. 6-4-P   Bracket.

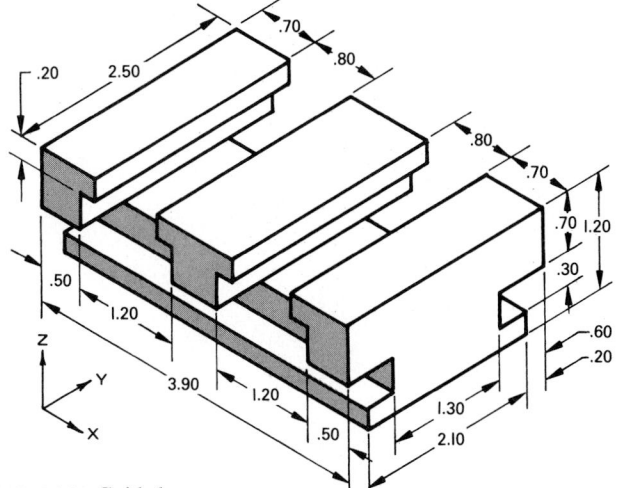

FIG. 6-4-M   Guide bar.

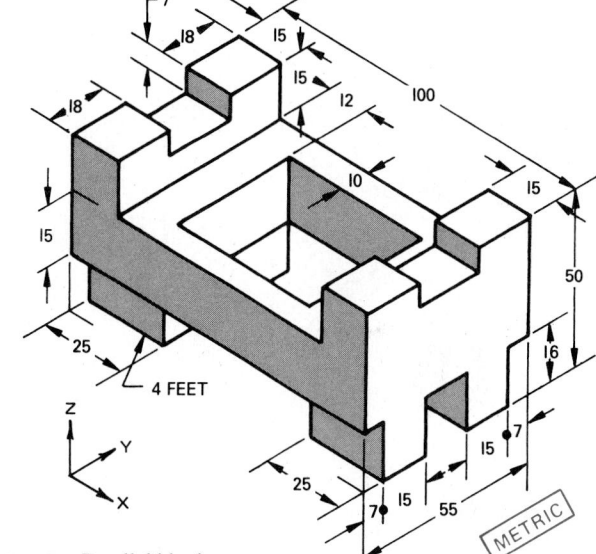

FIG. 6-4-R   Parallel block.

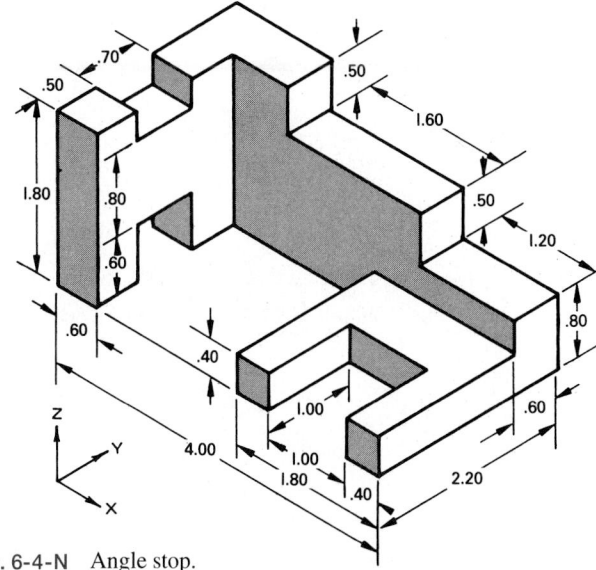

FIG. 6-4-N   Angle stop.

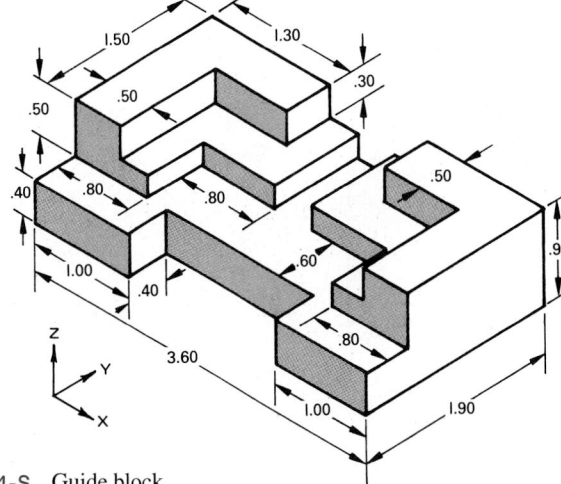

FIG. 6-4-S   Guide block.

ASSIGNMENTS FOR UNIT 6-5, INCLINED SURFACES

16. On preprinted grid paper (.25 in or 10 mm grid using 5 mm squares) sketch three views of each of the objects shown in Figs. 6-5-A or 6-5-B. Each square shown on the objects represents one square on the grid paper. Allow one grid space between views and a minimum of two grid spaces between objects. The sloped (inclined) surfaces on each of the three objects are identified by a letter. Identify the sloped surfaces on each of the three views with a corresponding letter. Also identify the type of projection used by placing the appropriate identifying symbol at the bottom of the drawing.
17. Sketch the views needed for a multiview drawing of the parts shown in Fig. 6-5-C. Choose your own sizes and estimate proportions.
18. Make three-view sketches of the parts shown in Figs. 6-5-D through 6-5-G. Follow the same instructions shown for Assignment 16.

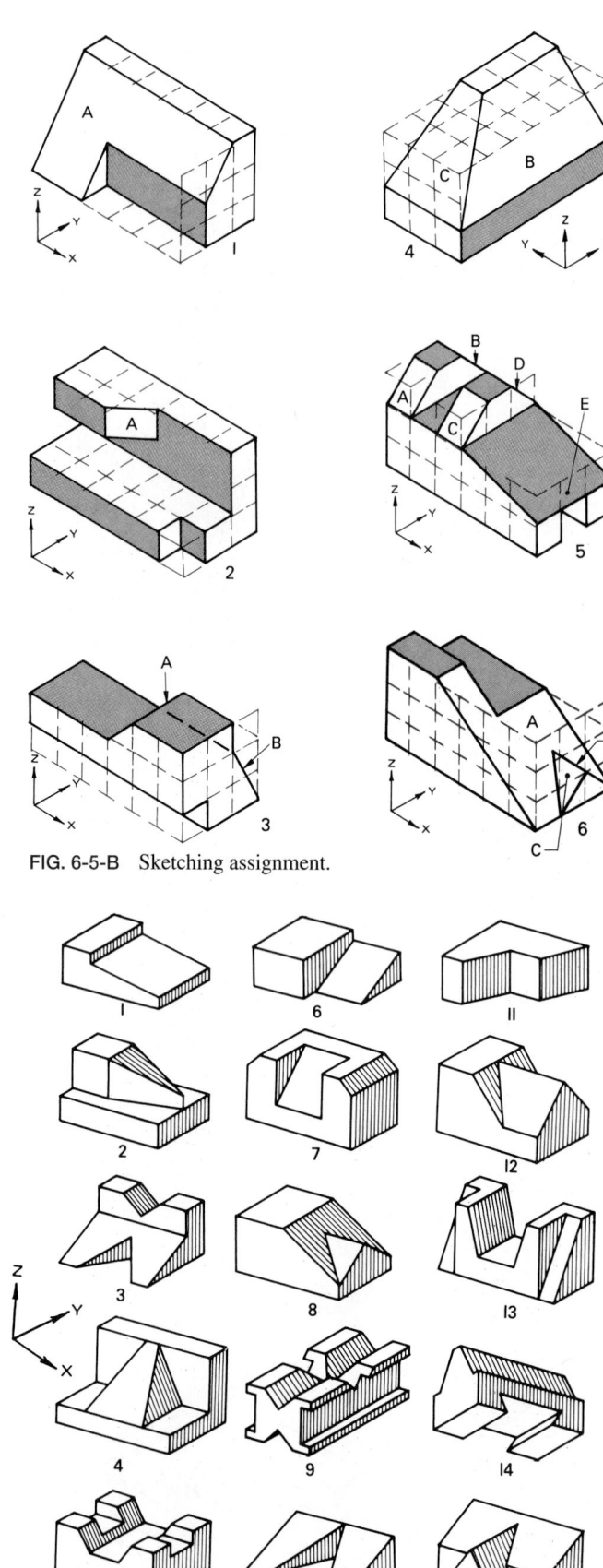

FIG. 6-5-B   Sketching assignment.

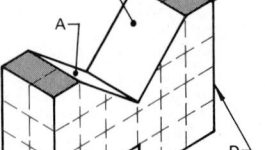

FIG. 6-5-A   Sketching assignment.

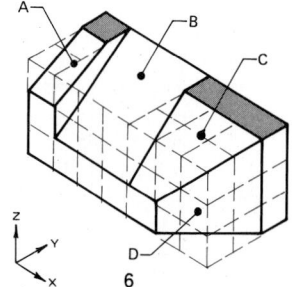

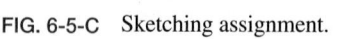

FIG. 6-5-C   Sketching assignment.

120

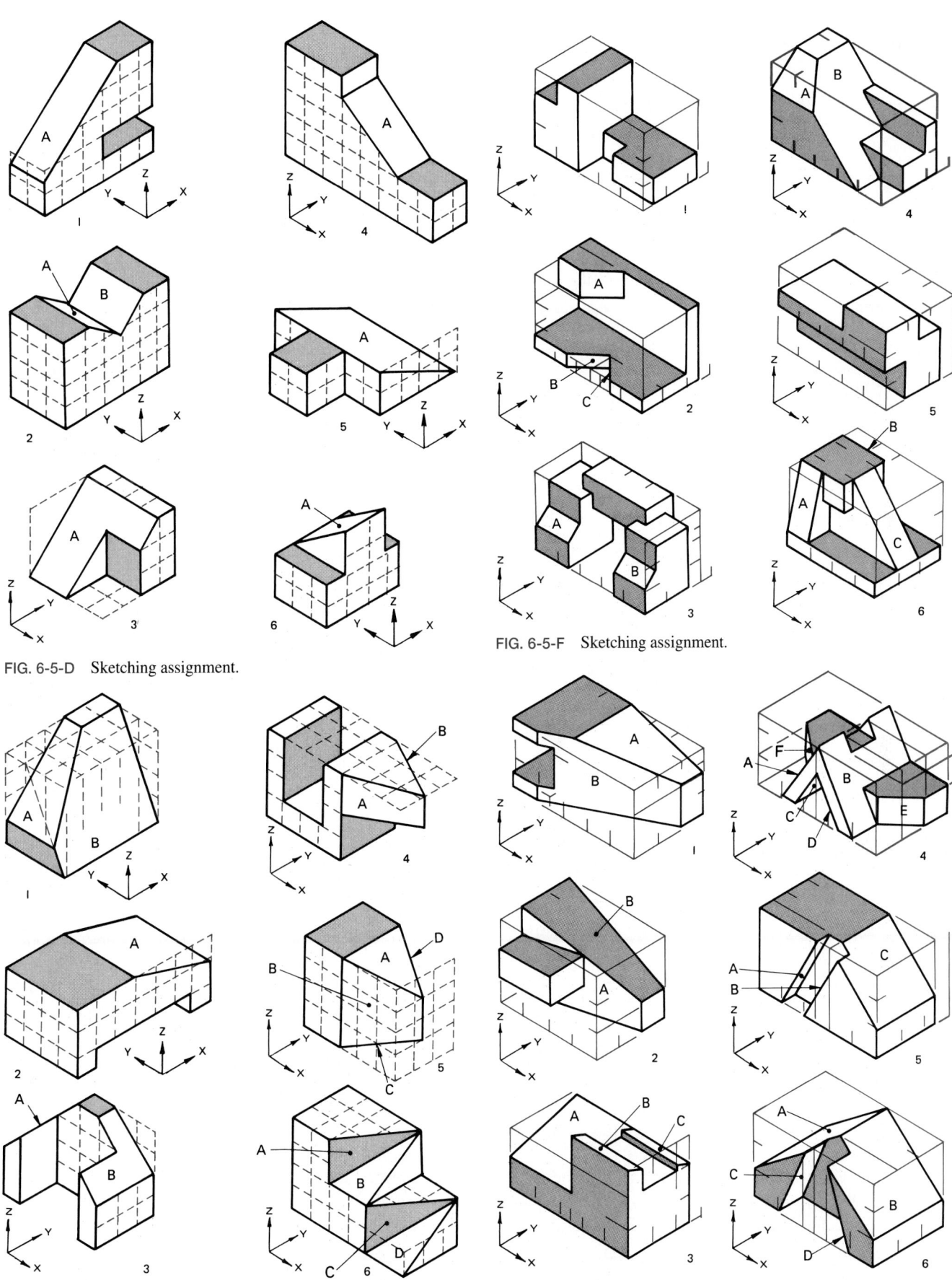

FIG. 6-5-D   Sketching assignment.

FIG. 6-5-E   Sketching assignment.

FIG. 6-5-F   Sketching assignment.

FIG. 6-5-G   Sketching assignment.

19. Match the pictorial drawings to the orthographic drawings
    shown in Fig. 6-5-H.

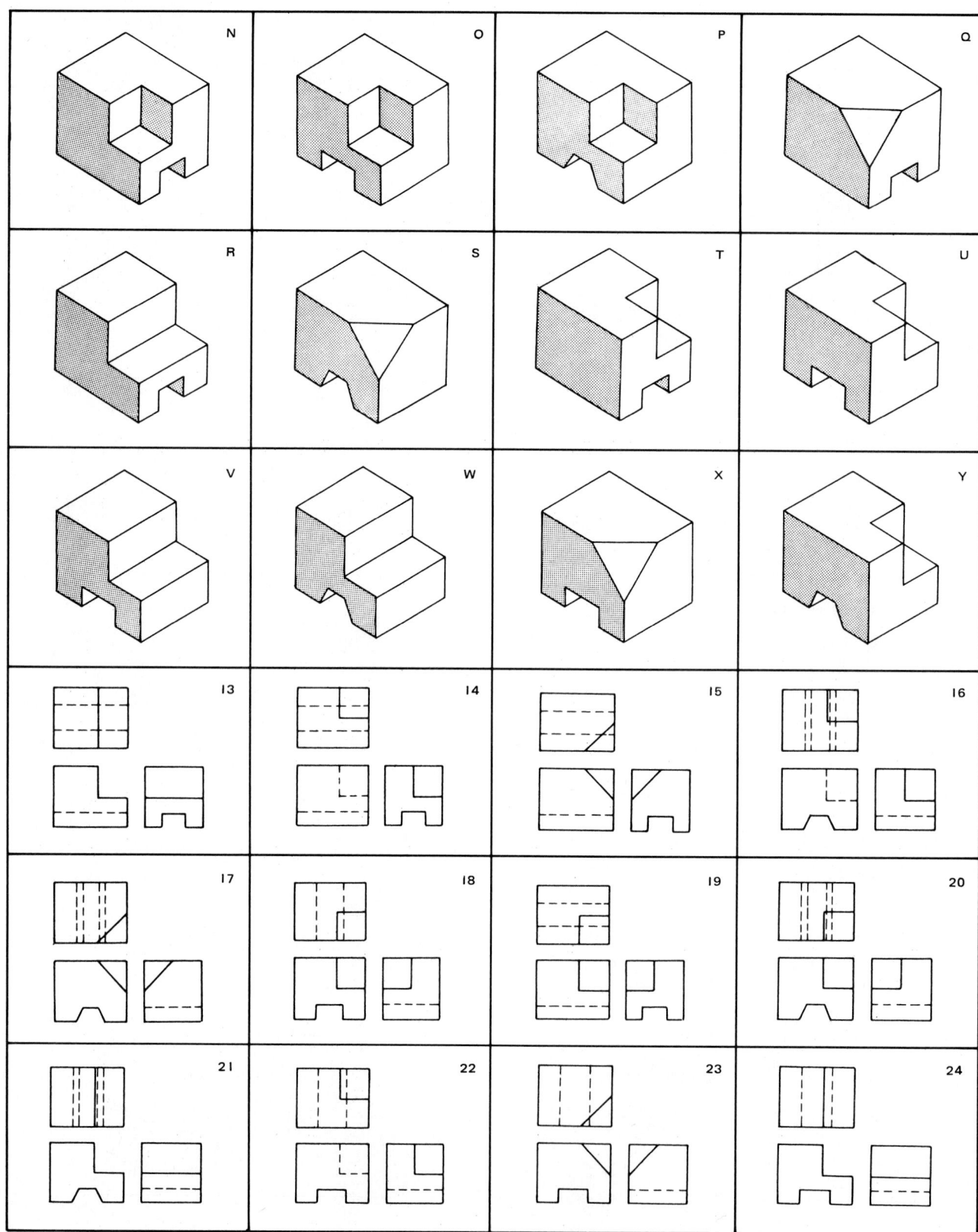

FIG. 6-5-H   Matching test.

20. Make a three-view drawing of one of the parts shown in Figs. 6-5-J through 6-5-P. Allow 1.00 in. or 25 mm between views. Do not dimension.

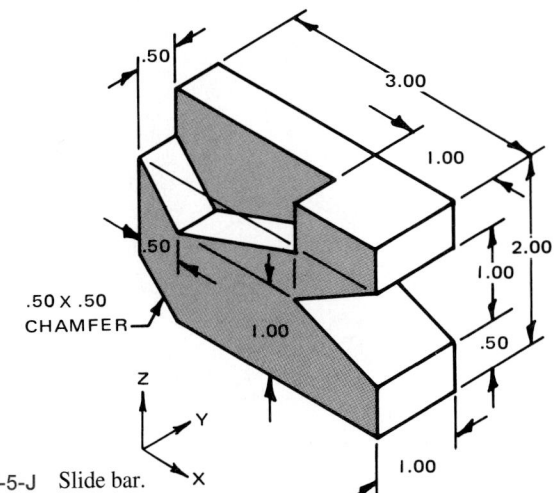

FIG. 6-5-J   Slide bar.

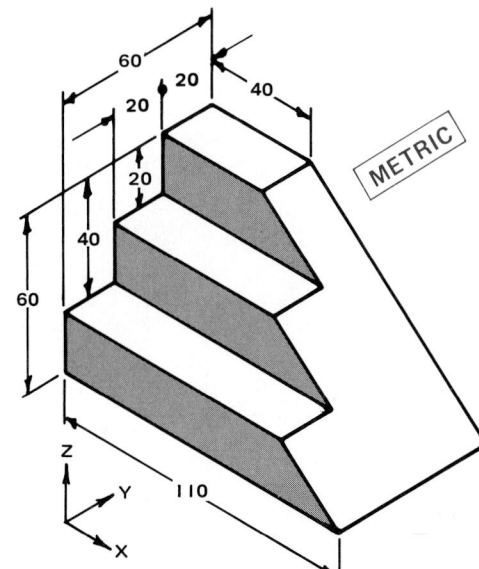

FIG. 6-5-K   Adjusting guide.

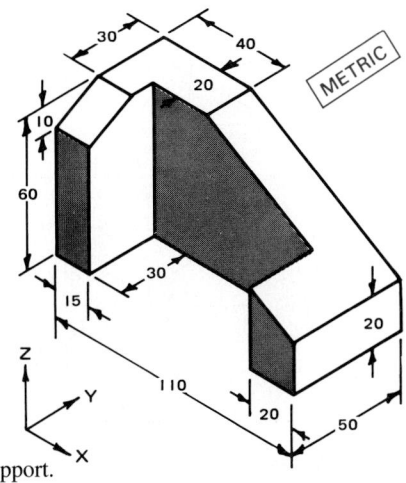

FIG. 6-5-L   Flanged support.

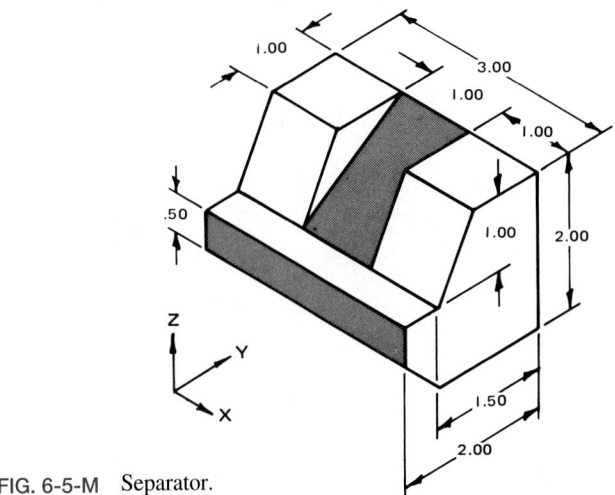

FIG. 6-5-M   Separator.

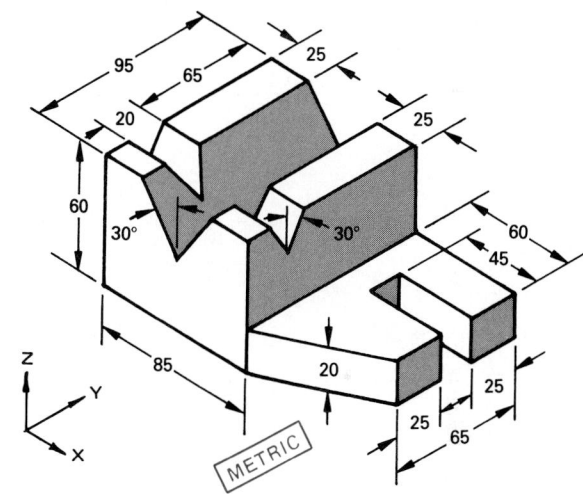

FIG. 6-5-N   Guide block.

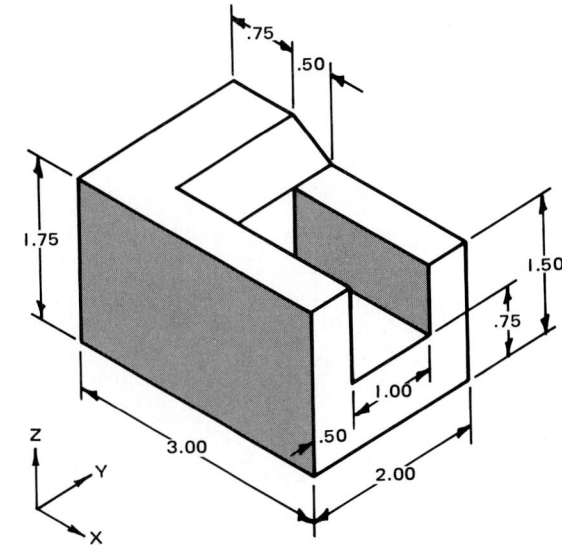

FIG. 6-5-P   Spacer.

123

21. Make a three-view drawing of one of the parts shown in Figs. 6-5-R through 6-5-W. Allow 1.00 in. or 25 mm between views.

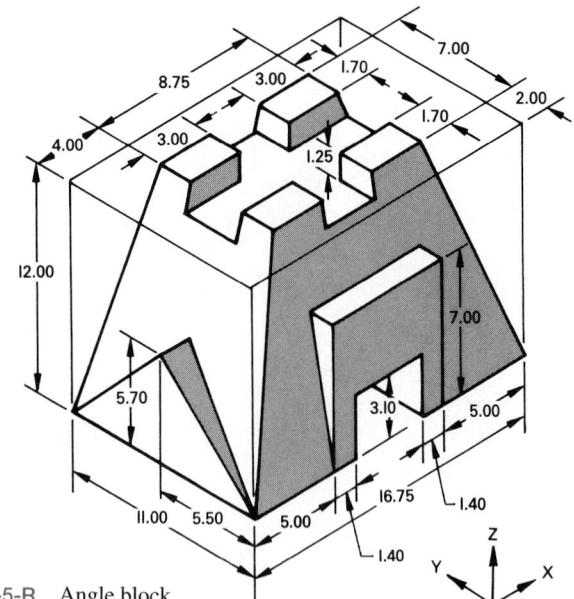

FIG. 6-5-R   Angle block.

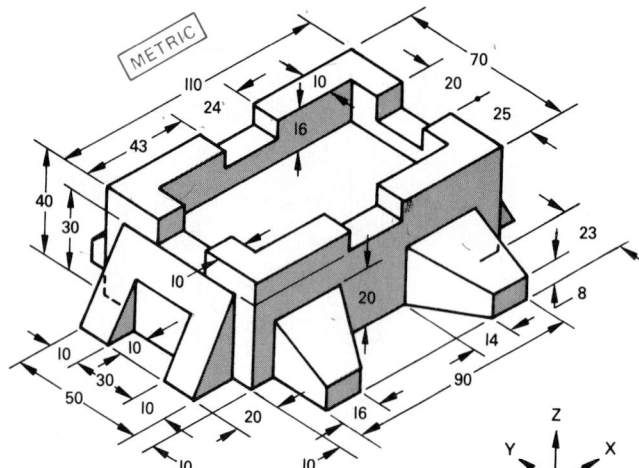

FIG. 6-5-S   Base plate.

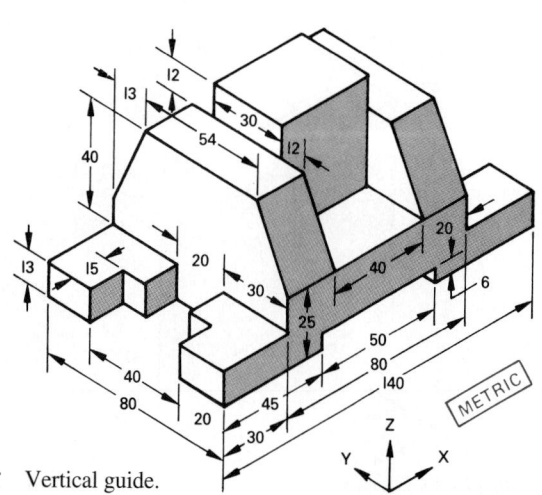

FIG. 6-5-T   Vertical guide.

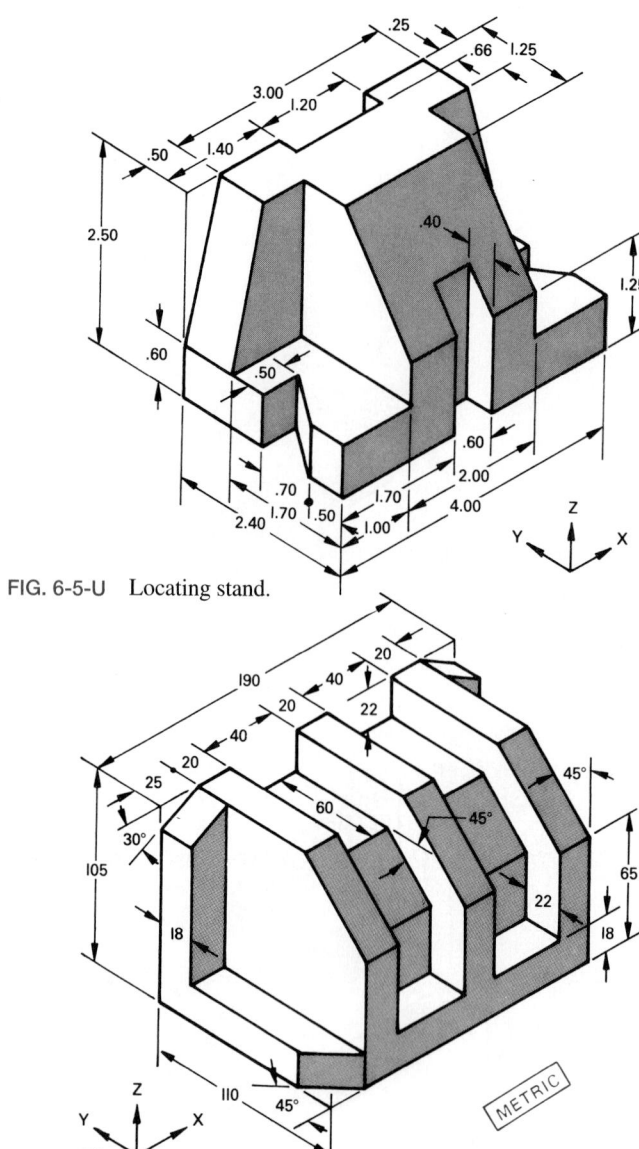

FIG. 6-5-U   Locating stand.

FIG. 6-5-V   Base.

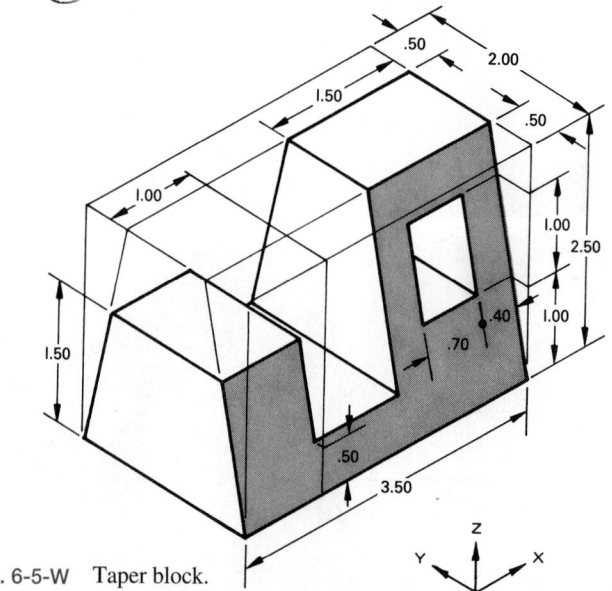

FIG. 6-5-W   Taper block.

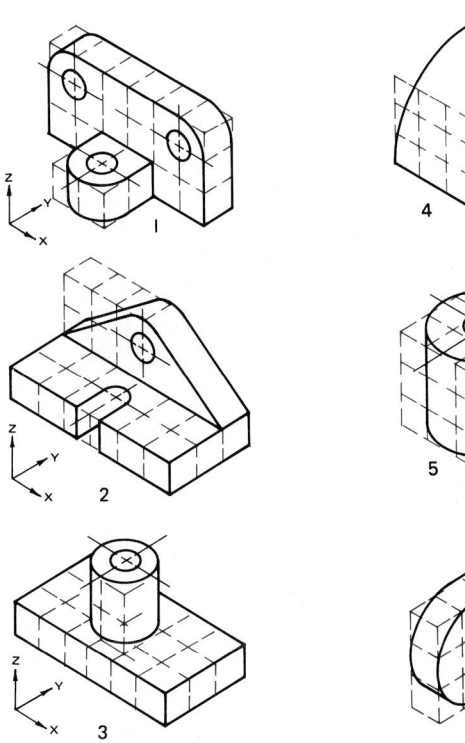

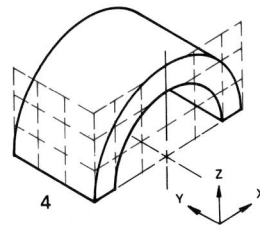

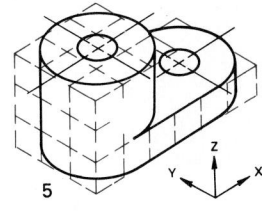

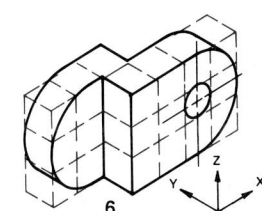

FIG. 6-6-A   Sketching assignment.

ASSIGNMENTS FOR UNIT 6-6, CIRCULAR FEATURES

22. On preprinted grid paper (.25 in. or 10 mm grid) sketch three views of each of the objects shown in Fig. 6-6-A or 6-6-B. Each square shown on the objects represents one square on the grid paper. Allow one grid space between views and a minimum of two grid spaces between objects. Identify the type of projection used by placing the appropriate identifying symbol at the bottom of the drawing.

23. Sketch the views needed for a multiview drawing of the parts shown in Fig. 6-6-C. Choose your own sizes and estimate proportions.

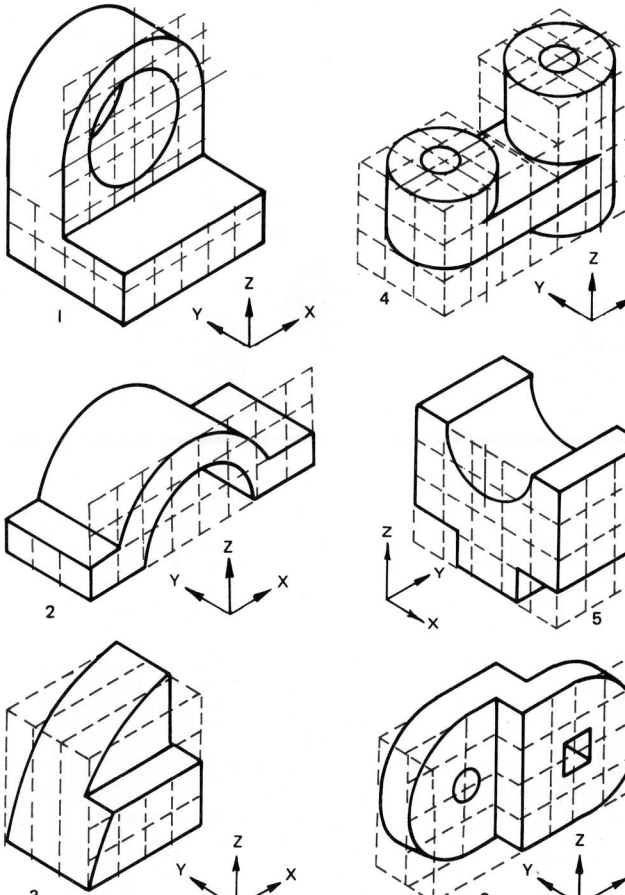

FIG. 6-6-B   Sketching assignment.

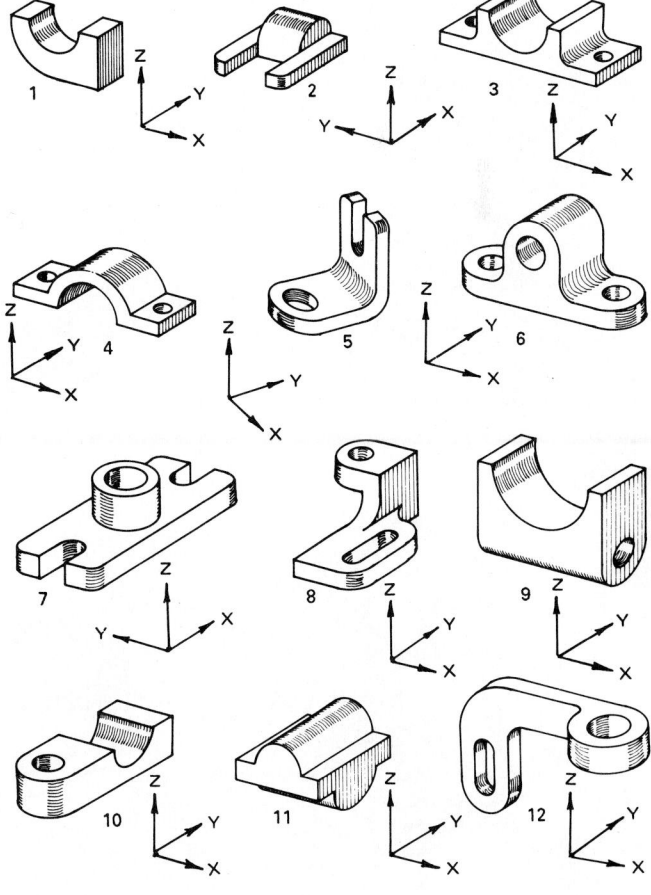

FIG. 6-6-C   Sketching assignent.

125

24. Sketch the views needed for a multiview drawing of the parts shown in Fig. 6-6-D or 6-6-E. Choose your own sizes and estimate proportions.

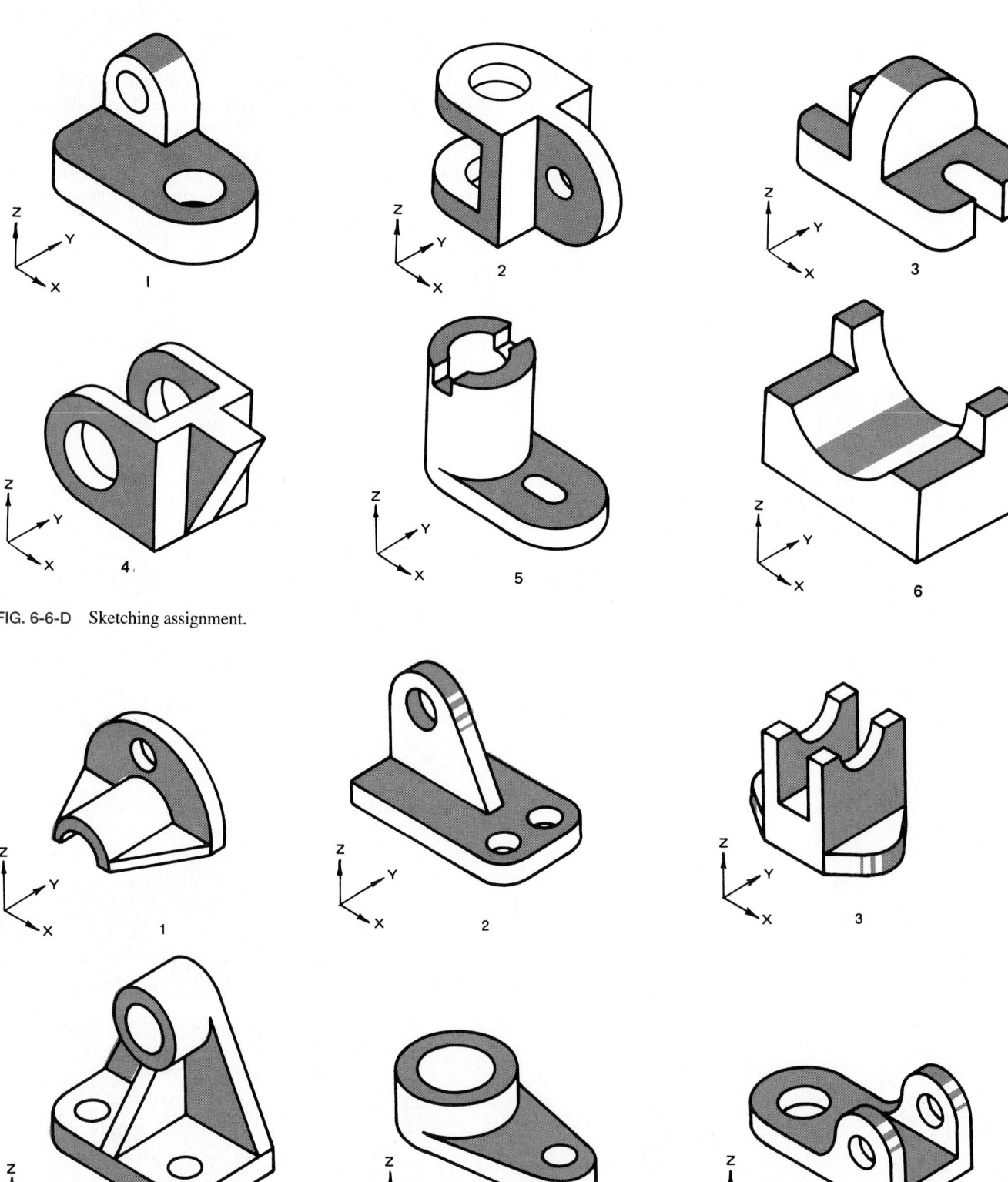

FIG. 6-6-D   Sketching assignment.

FIG. 6-6-E   Sketching assignment.

25. Make a three-view drawing of one of the parts shown in Figs. 6-6-F through 6-6-L. Allow 1.00 in. or 25 mm between views. Scale 1:1. Do not dimension.

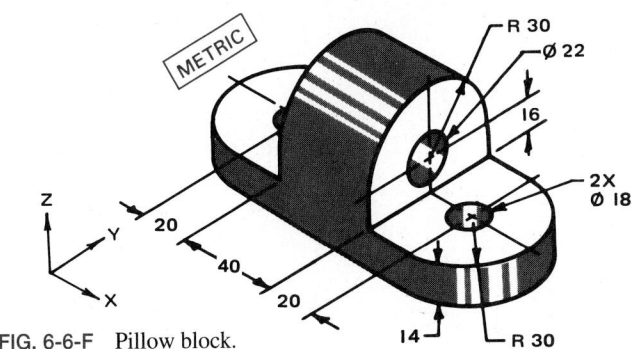

FIG. 6-6-F   Pillow block.

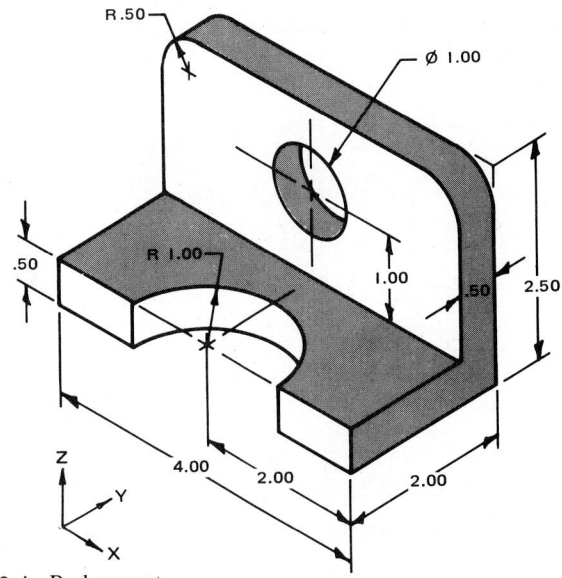

FIG. 6-6-J   Rod support.

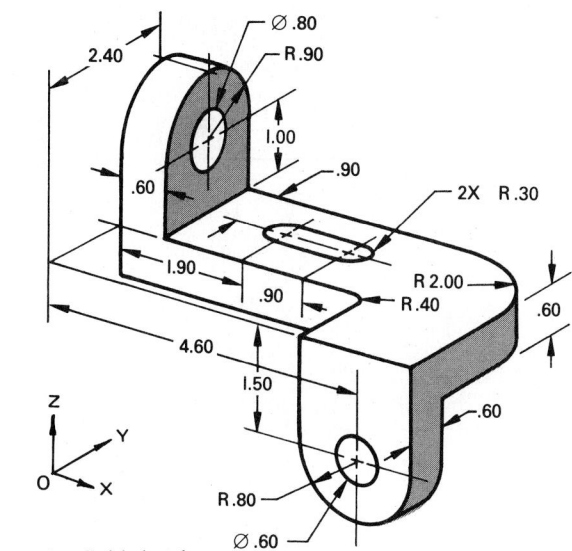

FIG. 6-6-G   Guide bracket.

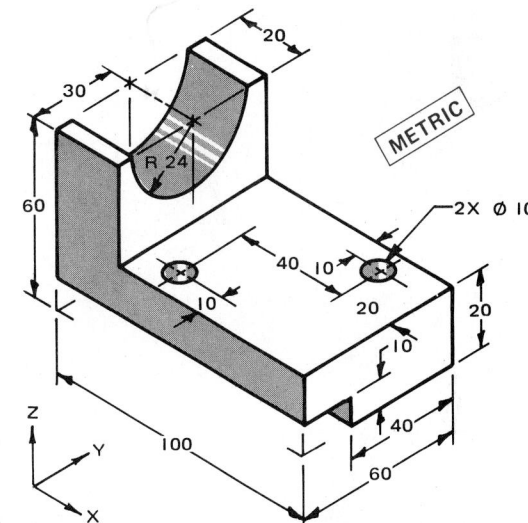

FIG. 6-6-K   Cradle support.

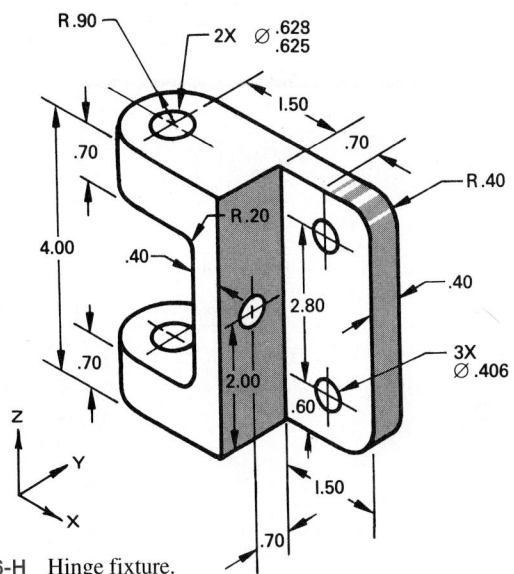

FIG. 6-6-H   Hinge fixture.

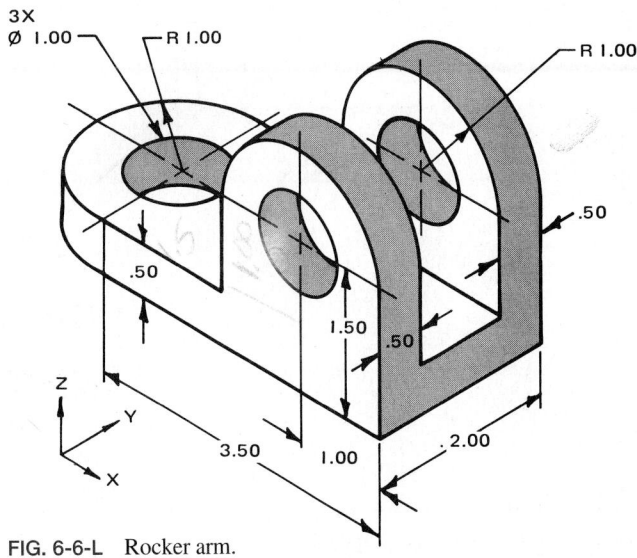

FIG. 6-6-L   Rocker arm.

26. Sketch the missing views for the parts shown in Fig.
6-6-M.

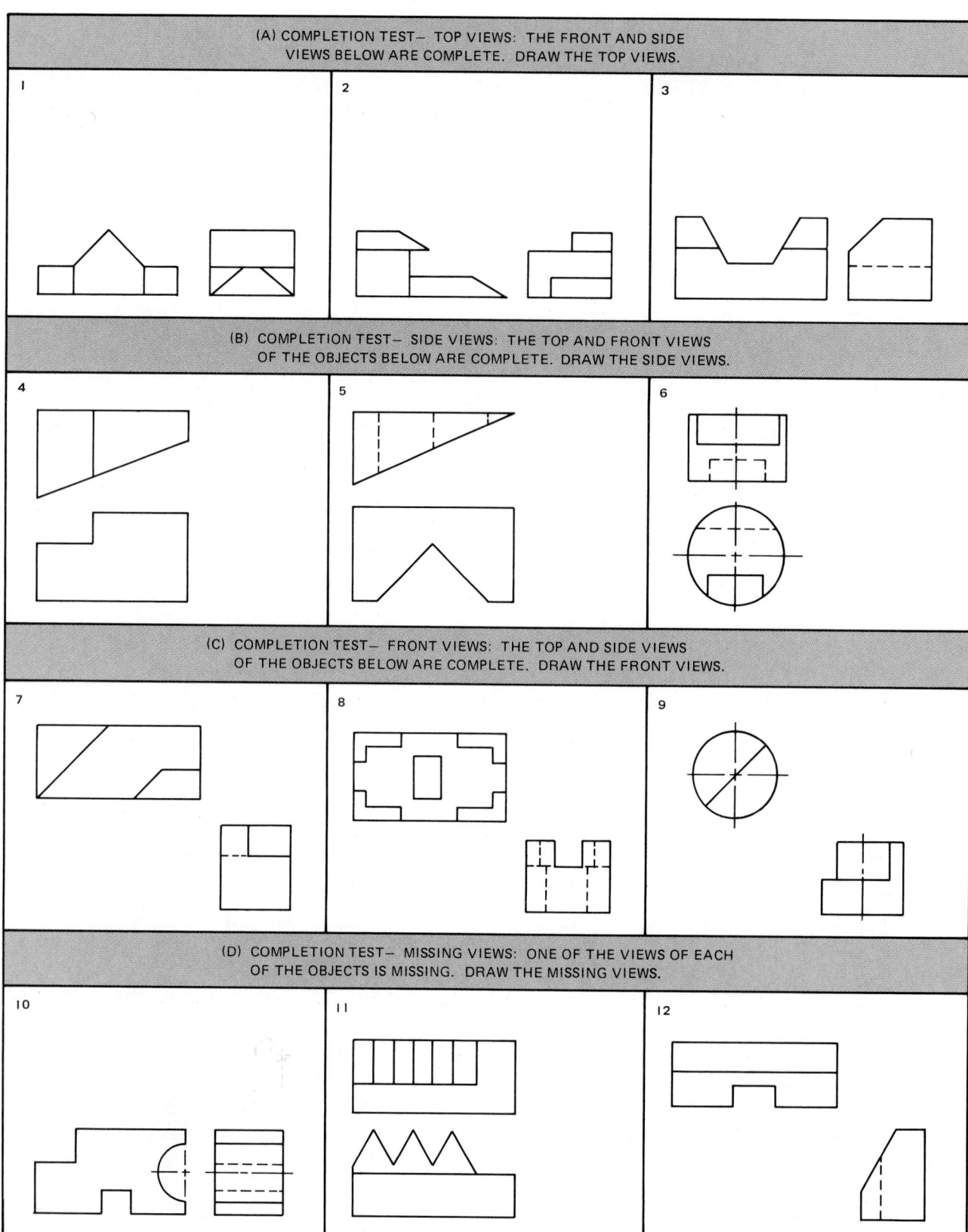

FIG. 6-6-M   Completion tests.

ASSIGNMENTS FOR UNIT 6-7, OBLIQUE SURFACES

27. On preprinted grid paper (.25 in. or 10 mm grid) sketch three views of each of the objects shown in Fig. 6-7-A or 6-7-B. Each square on the objects represents one square on the grid paper. Allow one grid space between views and a minimum of two grid spaces between objects. The oblique surfaces on the objects are identified by a letter. Identify the oblique surfaces on each of the three views with a corresponding letter. Also identify the type of projection used by placing the appropriate identifying symbol on the drawing.

28. Make a three-view drawing of one of the parts shown in Figs. 6-7-C and 6-7-D. Allow 1.20 in. (30 mm) between views. Do not dimension. The oblique surfaces on the objects are identified by a letter. Identify the oblique surfaces on each of the three views with a corresponding letter.

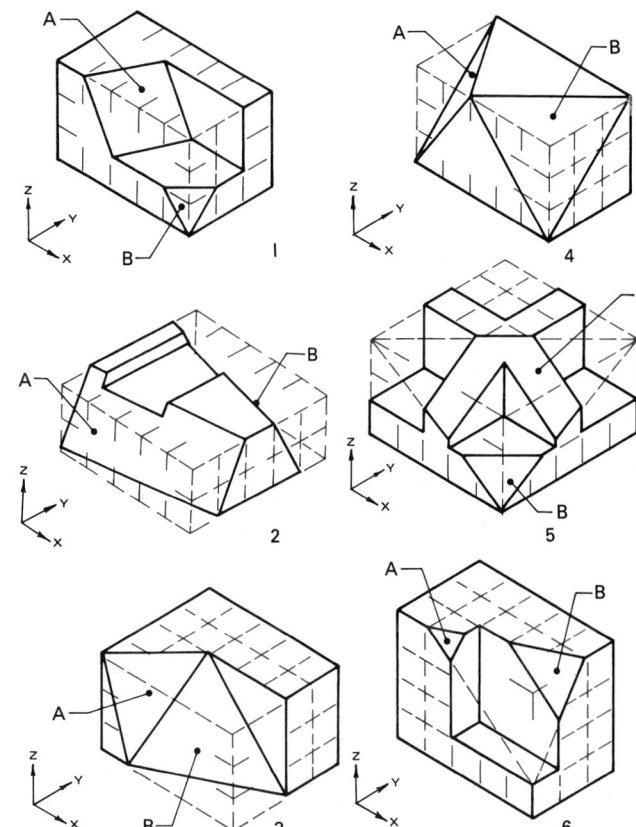

FIG. 6-7-B   Sketching assignment.

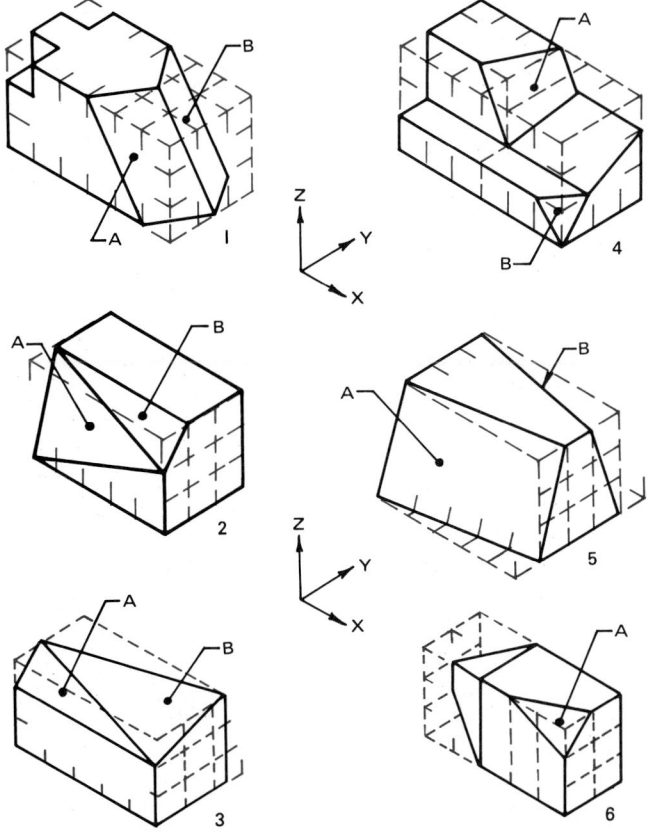

FIG. 6-7-A   Sketching assignment.

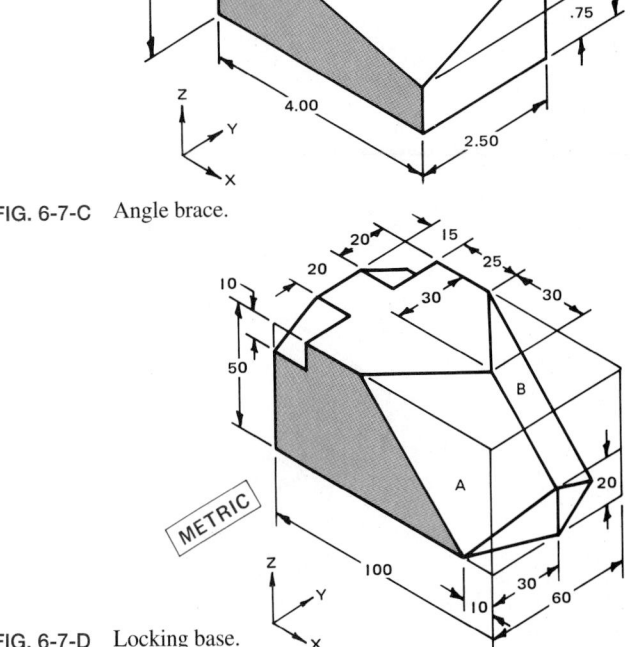

FIG. 6-7-C   Angle brace.

FIG. 6-7-D   Locking base.

129

29. Make a three-view drawing of one of the parts shown in Figs. 6-7-E through 6-7-G. Allow 1.20 in. (30 mm) between views. Do not dimension. The oblique surfaces on each part are identified by a letter. Identify the oblique surfaces on each of the three views with a corresponding letter.

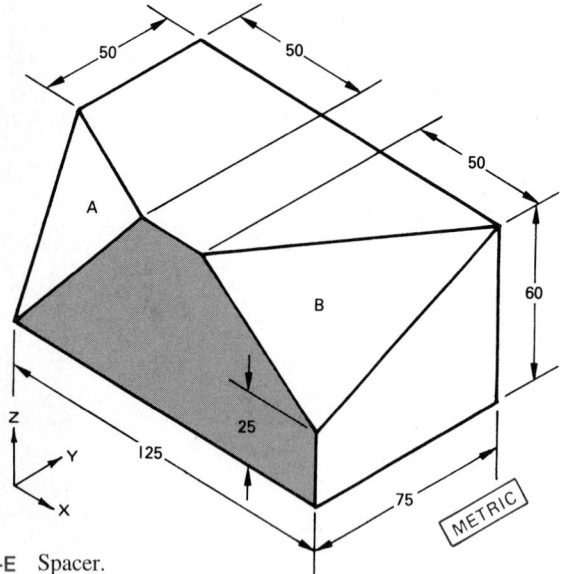

FIG. 6-7-E   Spacer.

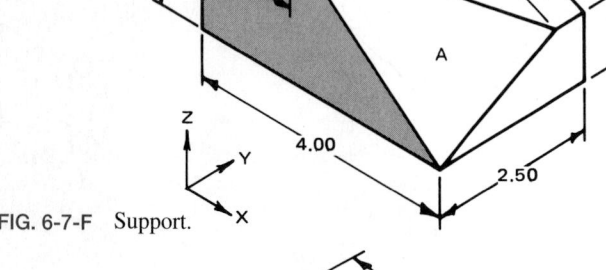

FIG. 6-7-F   Support.

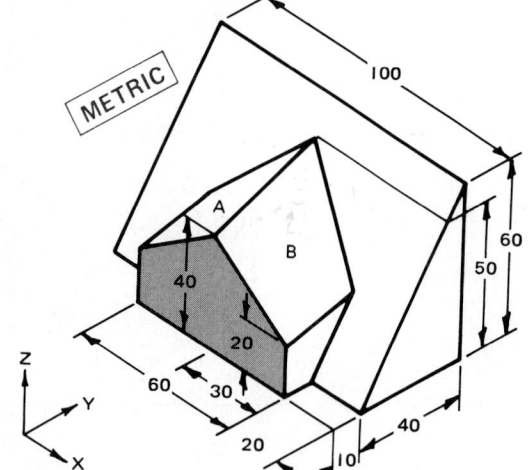

FIG. 6-7-G   Base plate.

ASSIGNMENT FOR UNIT 6-8, ONE- AND TWO-VIEW DRAWINGS

30. Select any six of the objects shown in Fig. 6-8-A and draw only the necessary views that will completely describe

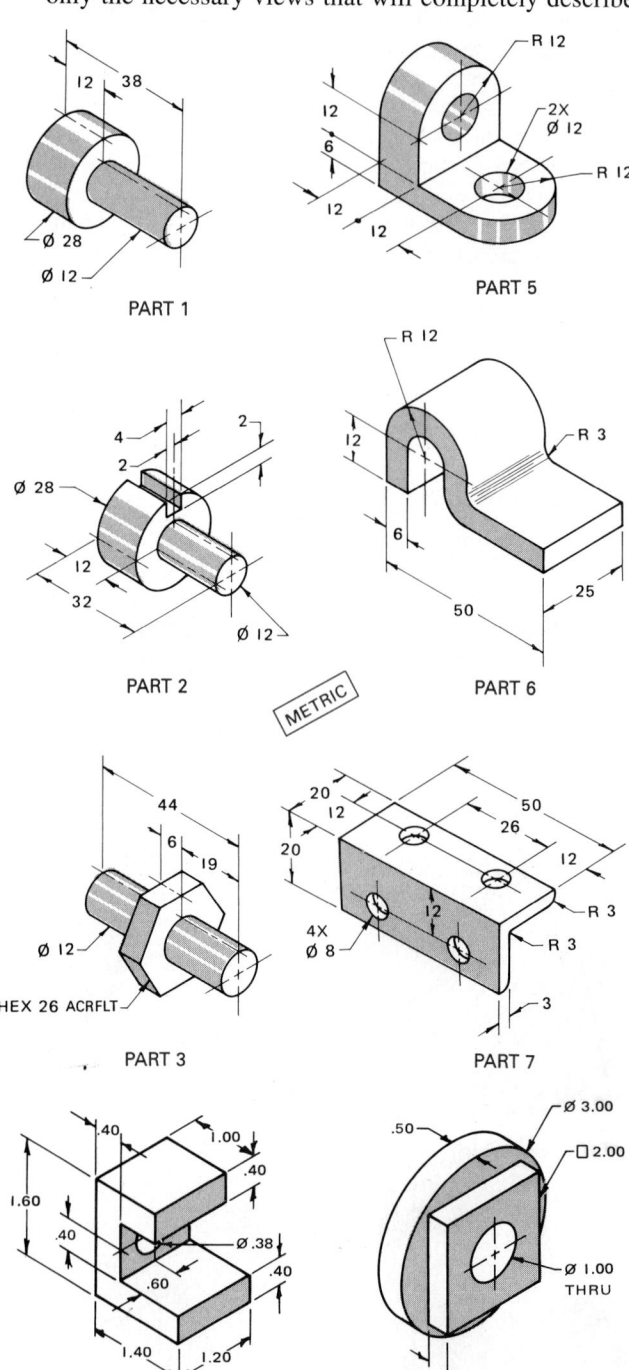

FIG. 6-8-A   Drawing assignment.

each part. Use symbols or abbreviations where possible. The drawings need not be to scale but should be drawn in proportion to the illustrations shown.

ASSIGNMENTS FOR UNIT 6-9, SPECIAL VIEWS

31. Select one of the objects shown in Figs. 6-9-A through 6-9-D and draw only the necessary views (full and partial) that will completely describe each part. Add dimensions and machining symbols where required. Scale 1:1.

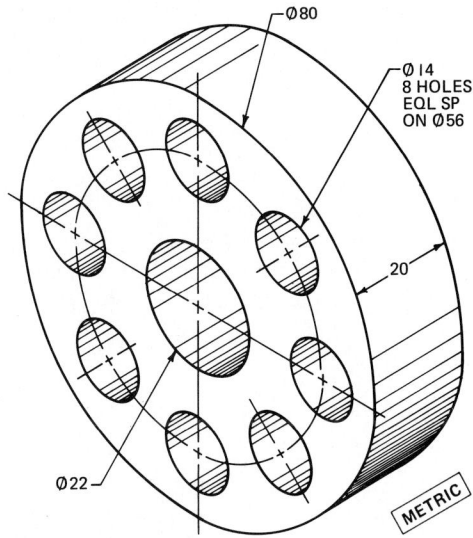

FIG. 6-9-A   Round flange.

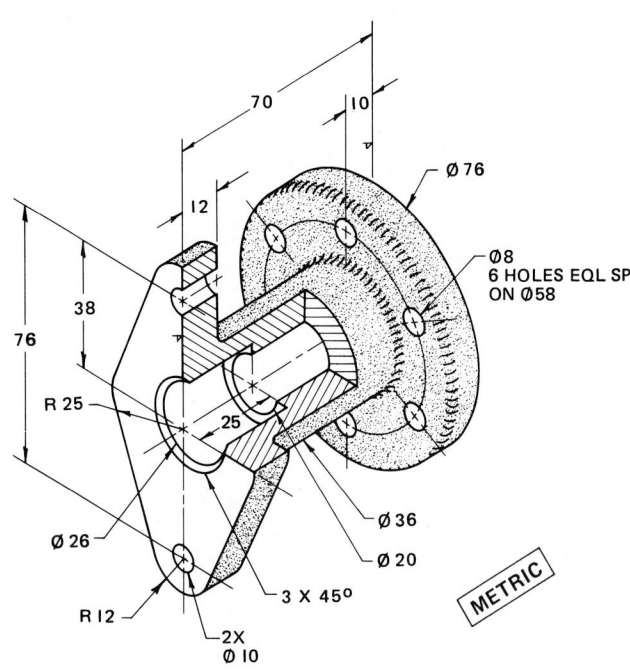

FIG. 6-9-C   Flanged coupling.

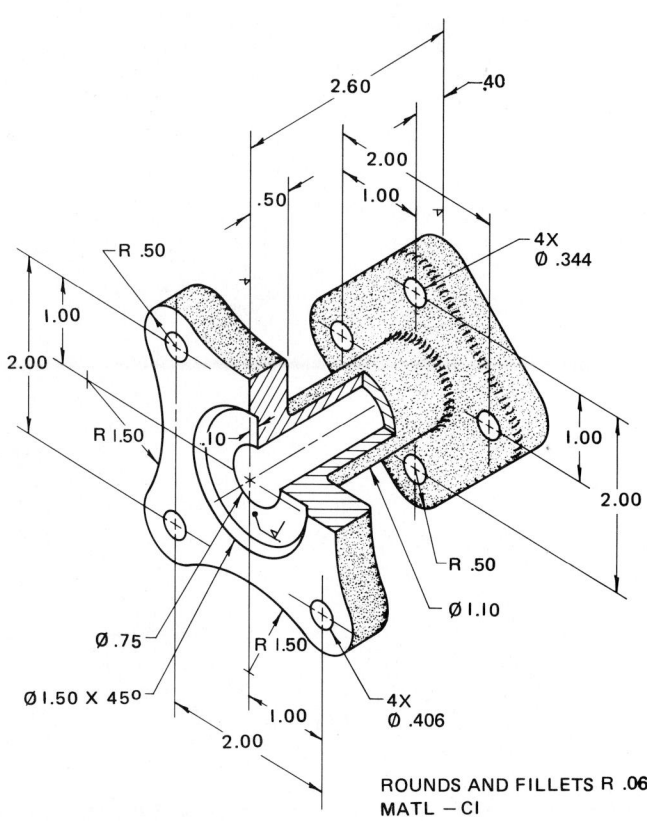

FIG. 6-9-B   Flanged adaptor.

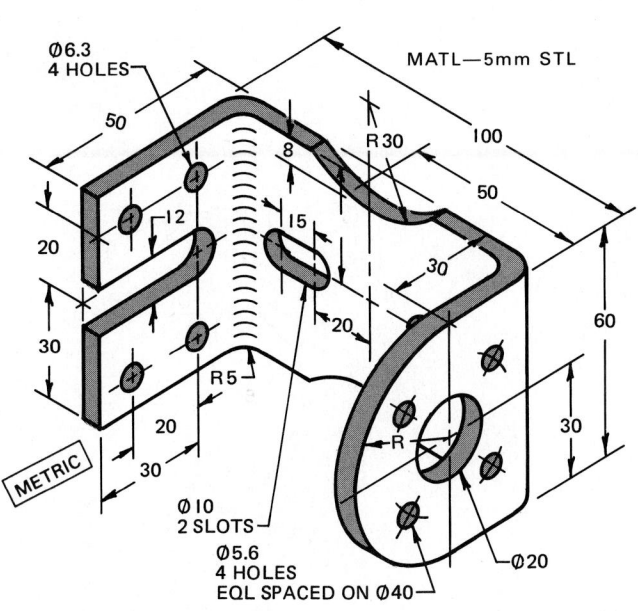

FIG. 6-9-D   Connector.

32. Select one of the panels shown in Figs. 6-9-E and 6-9-F and make a detail drawing of the part. Enlarged views are recommended. Panels such as these, where labeling is used to identify the terminals, are used extensively in the electrical and electronics industry.

33. With most truss drawings, the scale used on the overall assembly is such that intricate detail cannot be clearly shown. As a result, enlarged detail views are added. With this type of assembly, many parts are opposite-hand to their counterparts.

Bolting Data: All members to be bolted together with five .375 high-strength bolts. Spacing is 1.50 in. from end and 3.00 in. center to center.

On a B size sheet, draw the enlarged views of the gusset assemblies shown in Fig. 6-9-G. Scale 1:12.

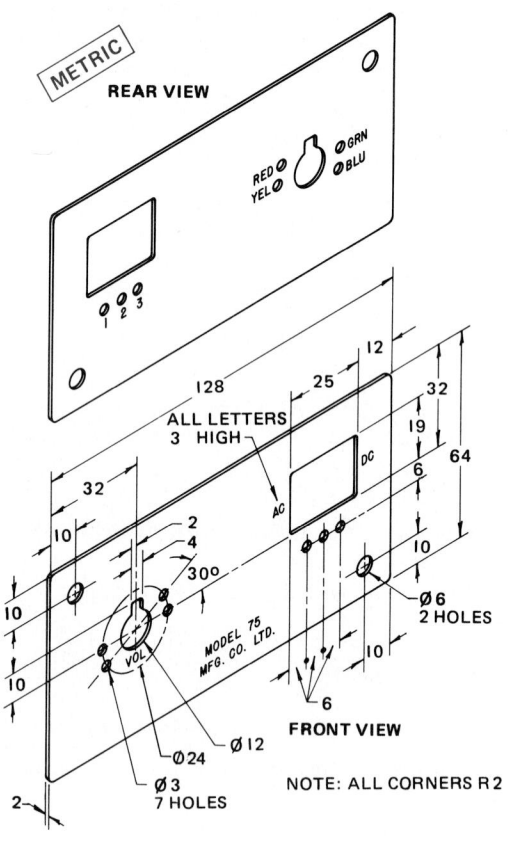

FIG. 6-9-E   Radio cover plate.

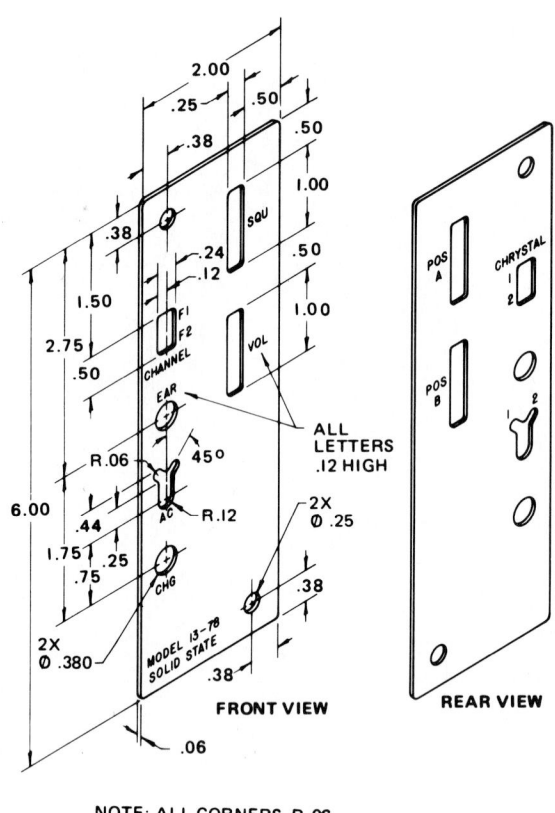

FIG. 6-9-F   Transceiver cover plate.

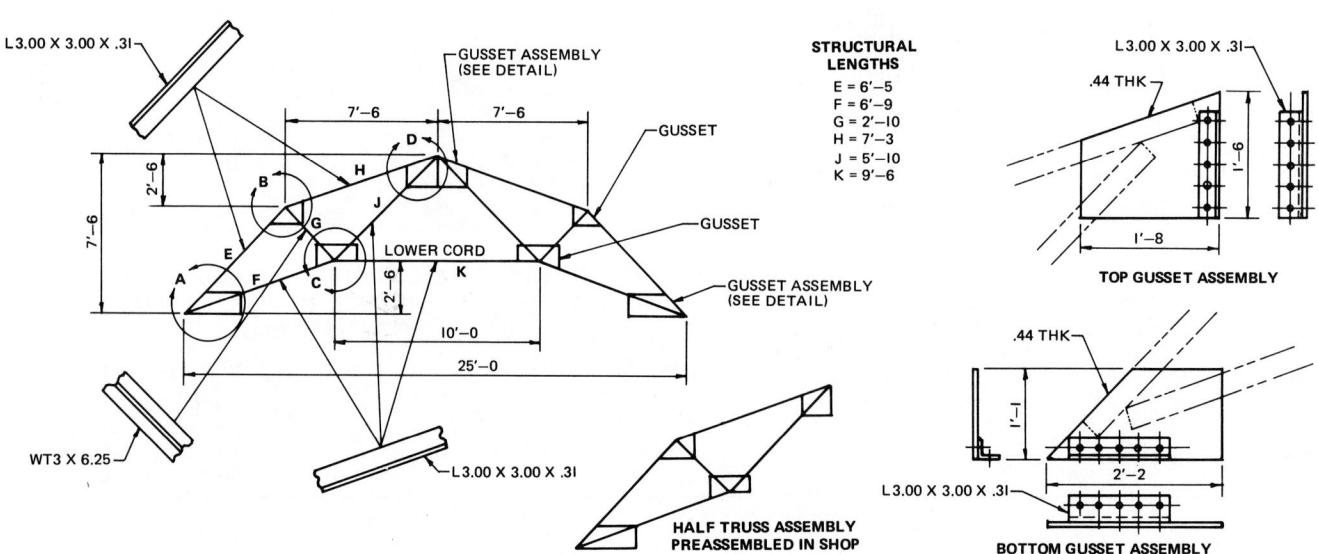

FIG. 6-9-G   Crescent truss.

ASSIGNMENT FOR UNIT 6-10, CONVENTIONAL REPRESENTATION OF COMMON FEATURES

34. Make a working drawing of one of the parts shown in Fig. 6-10-A or 6-10-B. Wherever possible, simplify the

drawing by using conventional representation of features and symbolic dimensioning (including symmetry). Scale 10:1.

ASSIGNMENT FOR UNIT 6-11, CONVENTIONAL BREAKS

35. Make a working drawing of one of the parts shown in Fig. 6-11-A or 6-11-B. Use conventional breaks to shorten the length of the part. An enlarged view is also recommended where the detail cannot be clearly shown at full scale. Apply the symmetry symbol and use symbolic dimensioning wherever possible. Scale 1:1.

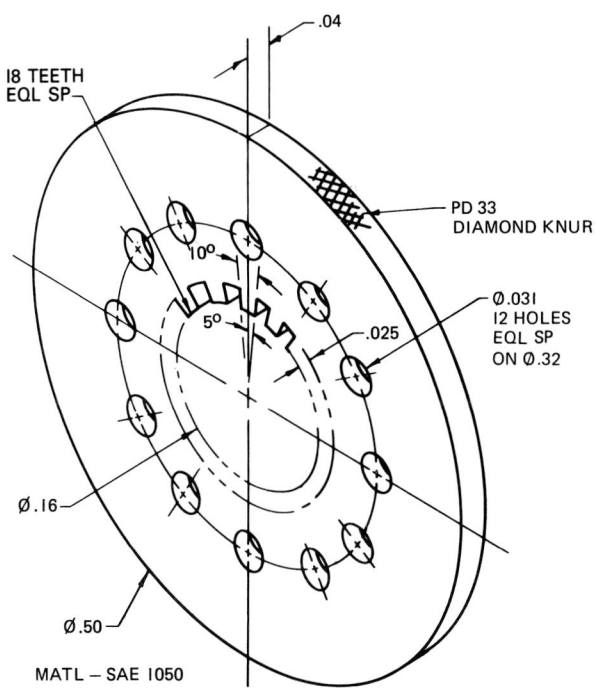

FIG. 6-10-A Adjustable locking plate.

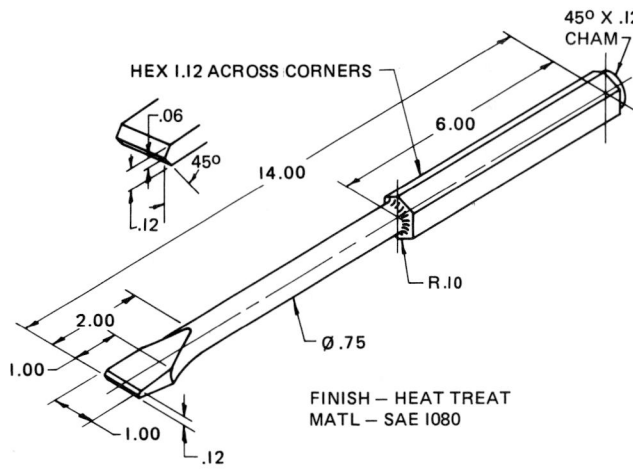

FIG. 6-11-A Hand chisel.

FIG. 6-10-B Clock stem.

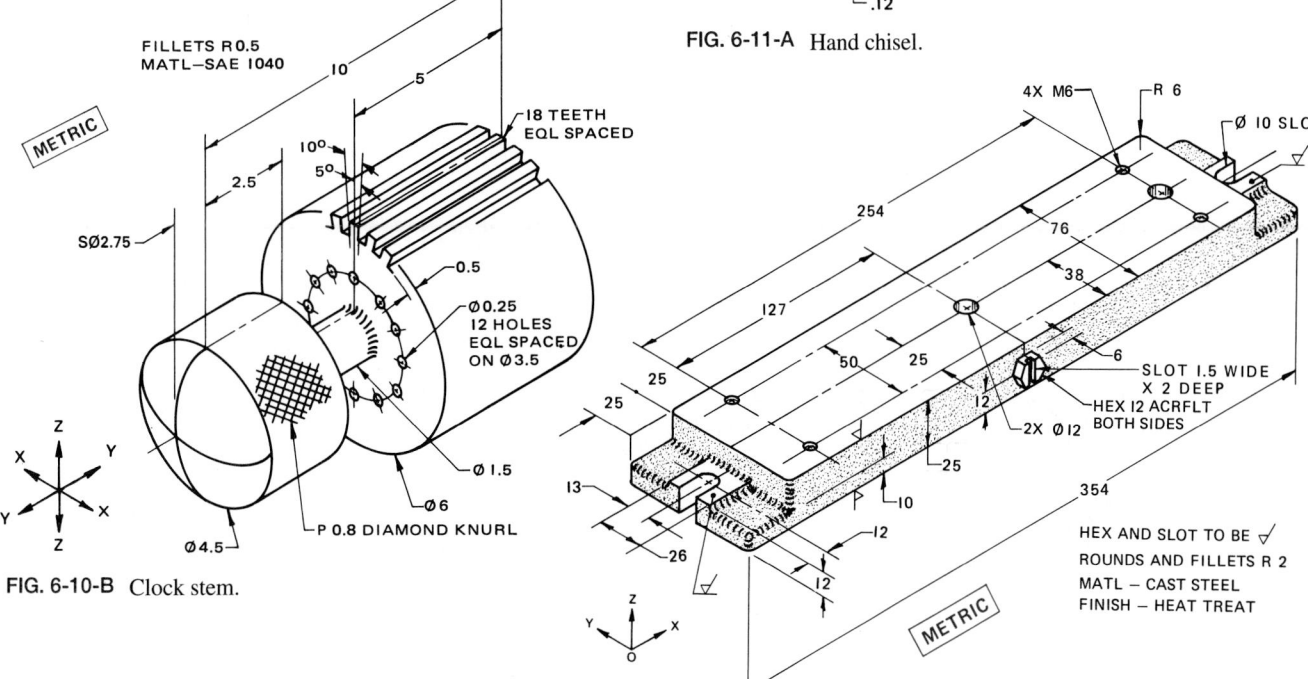

FIG. 6-11-B Fixture base.

ASSIGNMENT FOR UNIT 6-12, MATERIALS OF CONSTRUCTION

36. Make a detailed assembly drawing of one of the assemblies shown in Fig. 6-12-A or 6-12-B. Enlarged details are recommended for the steel mesh and joints. Use conventional breaks to shorten the length. Scale 1:5.

ASSIGNMENT FOR UNIT 6-13, CYLINDRICAL INTERSECTIONS

37. Make a working drawing of one of the parts shown in Fig. 6-13-A or 6-13-B. For 6-13-A a bushing is to be pressed (H7/s6) into the large hole and the stepped smaller hole is to have a running fit (H8/f7) with its respective shaft. These sizes are to be given as limit dimensions. All other finished surfaces are to have a 3.2 μm (micrometers) finish. For Fig. 6-13-B an LN3 fit is required for the two large holes. Finished surfaces are to have a 63 μin. (microinches) finish with a .06 in. material-removal allowance. Use your judgment in selecting the number of views required and deciding whether some form of sectional view would be desirable to improve the readability of the drawing. Scale 1:1.

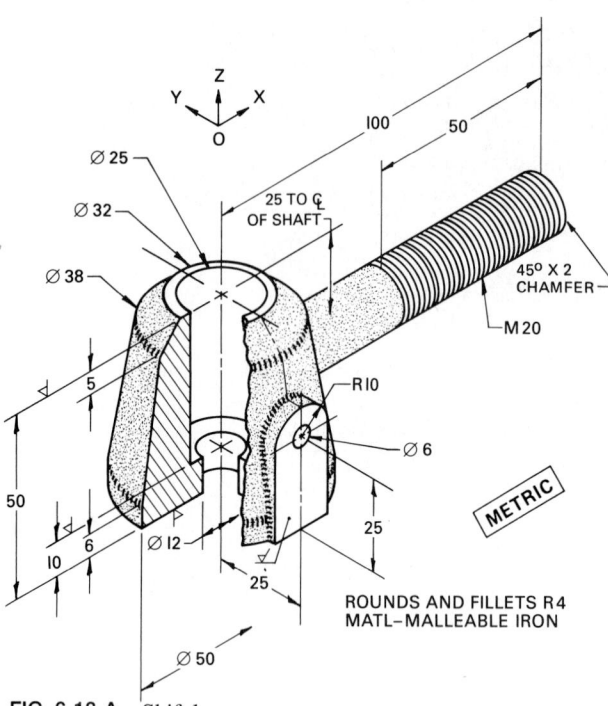

FIG. 6-13-A   Shift lever.

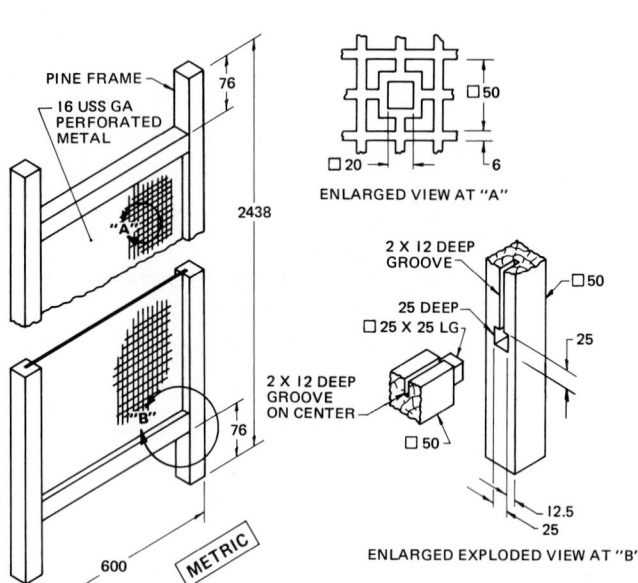

FIG. 6-12-A   Room divider.

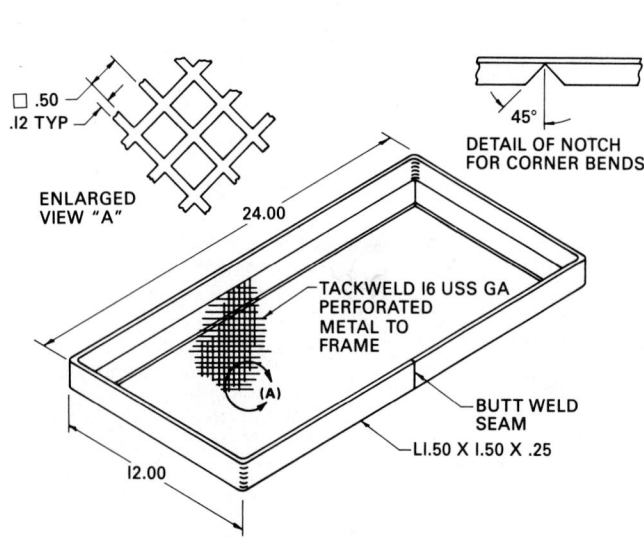

FIG. 6-12-B   Barbecue grill.

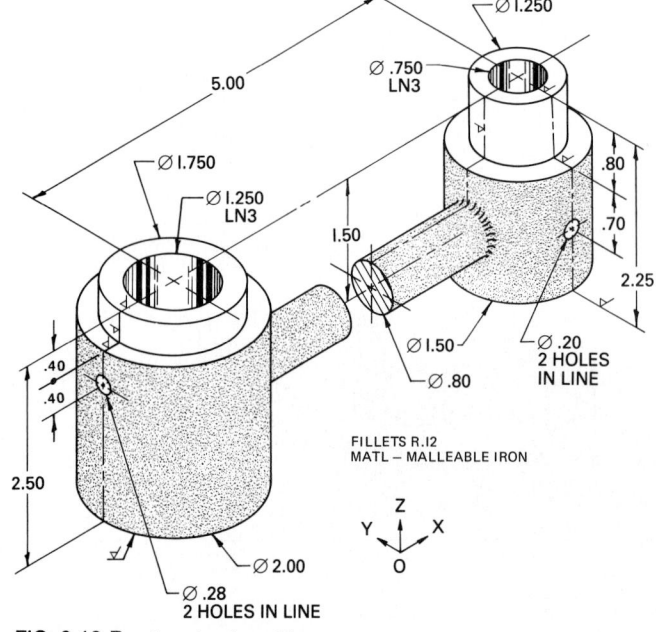

FIG. 6-13-B   Steering knuckle.

134

ASSIGNMENT FOR UNIT 6-14, FORESHORTENED PROJECTION

38. Make a working drawing of one of the parts shown in Fig. 6-14-A or 6-14-B. All surface finishes are to be 1.6 μm or 63 μin. Keyed holes will have H9/d9 or RC6 fits with shafts. Where required, rotate the features to show their true distances from the centers and edges. To show the true shape of the ribs or arms, a revolved section is recommended. Dimension the keyseat as per Chap. 11 and the Appendix. Scale 1:1.

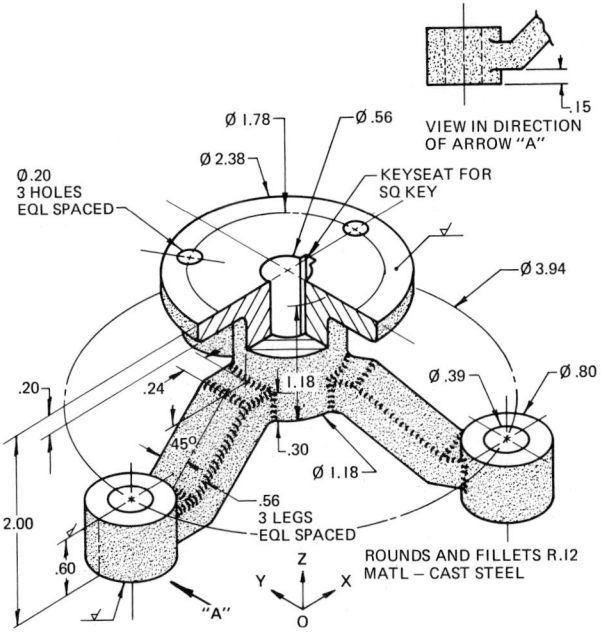

FIG. 6-14-A   Clutch.

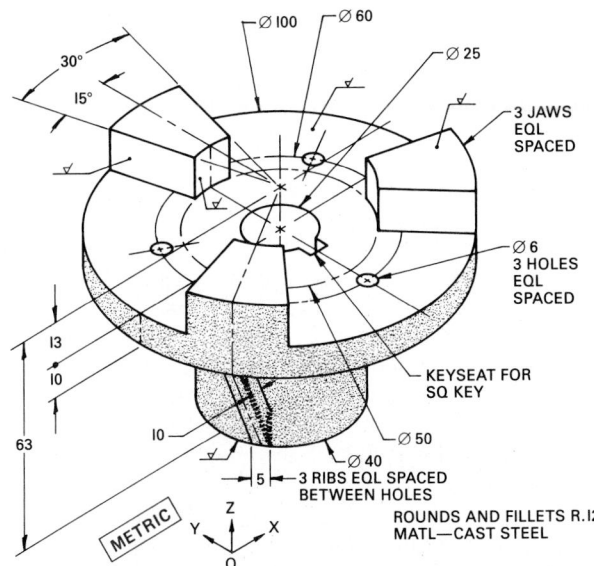

FIG. 6-14-B   Mounting bracket.

ASSIGNMENT FOR UNIT 6-15, INTERSECTIONS OF UNFINISHED SURFACES

39. Make a three-view detail drawing of one of the problems shown in Fig. 6-15-A or 6-15-B. Scale 1:1. Surface finish requirements are essential for all parts. Use symbolic dimensioning wherever possible. For Fig. 6-15-A the T slot surfaces should have a maximum roughness of 0.8 μm and a maximum waviness of 0.05 mm for a 25 mm length. The back surface should have a maximum roughness of 3.2 μm with no restrictions on waviness. For Fig. 6-15-B the back surface and notch should have an equivalent control as the T slot in Fig. 6-15-A. The faces on the boss should have a maximum roughness of 125 μin. with no restrictions on waviness.

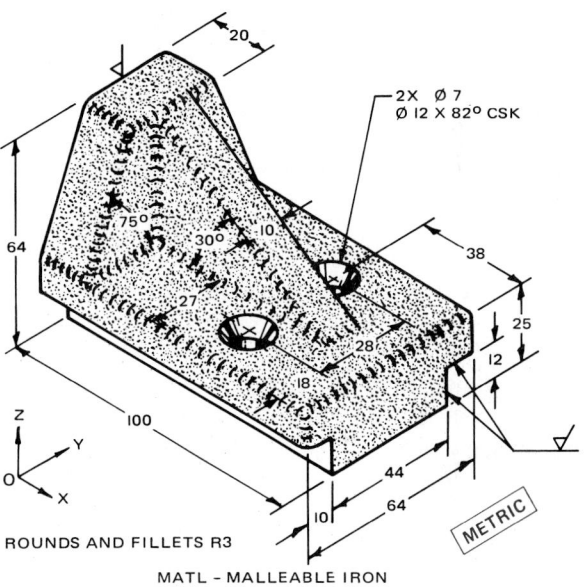

FIG. 6-15-A   Cutoff stop.

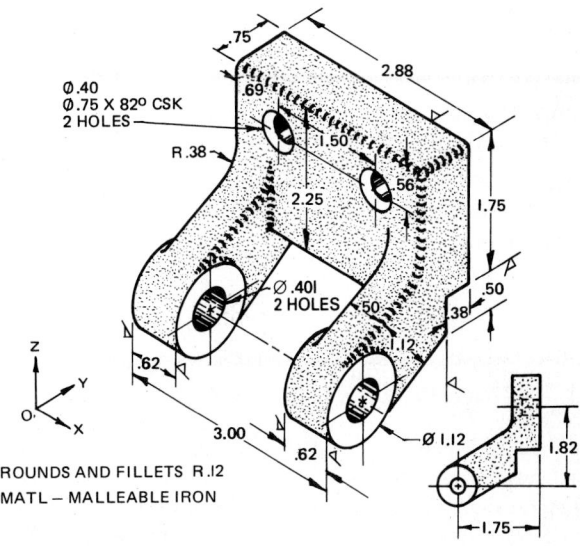

FIG. 6-15-B   Sparker bracket.

135

CHAPTER 7

AUXILIARY VIEWS AND REVOLUTIONS

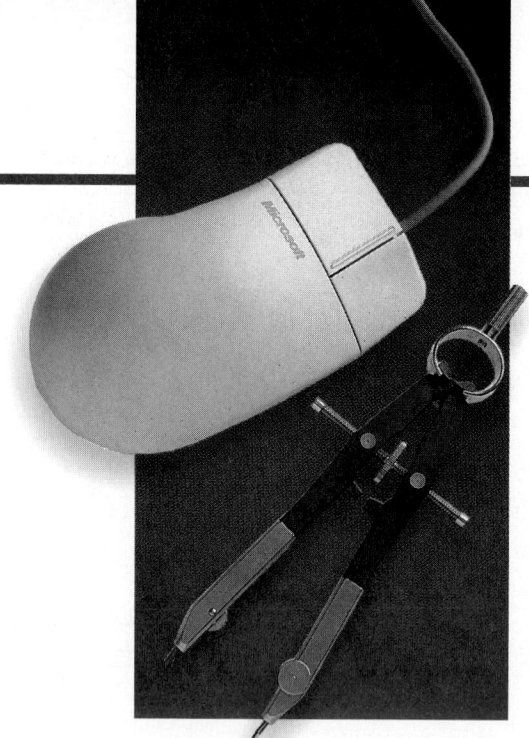

Definitions

Descriptive geometry   The use of graphic representations to solve mathematical problems.

Inclined line   In descriptive geometry, a line which appears as a true-length line in one view but is foreshortened in the other two views.

Inclined plane   In descriptive geometry, a plane which appears distorted in two views and as a line in the third view.

Multi-auxiliary view   Several additional views used to depict surfaces that must be shown clearly and without distortion.

Normal line   In descriptive geometry, a line which appears as a point in one view and as a true-length line in the other two views.

Normal plane   In descriptive geometry, a plane which appears in its true shape in one view and as a line in the other two views.

Oblique line   In descriptive geometry, a line which appears inclined in all views.

Oblique plane   In descriptive geometry, a plane which appears distorted in all views.

Revolutions   A method of representing an object through a series of views that have been "revolved" or turned on an imaginary axis.

7-1   PRIMARY AUXILIARY VIEWS

Many machine parts have surfaces that are not perpendicular, or at right angles, to the plane of projection. These are referred to as *sloping* or *inclined surfaces*. In the regular orthographic views, such surfaces appear to be distorted and their true shape is not shown. When an inclined surface has important characteristics that should be shown clearly and without distortion, an auxiliary (additional or helper) view is used so that the drawing completely and clearly explains the shape of the object. In many cases, the auxiliary view will replace one of the regular views on the drawing, as illustrated in Fig. 7-1-1.

One of the regular orthographic views will have a line representing the edge of the inclined surface. The auxiliary view is projected from this edge line, at right angles, and is drawn parallel to the edge line.

Only the true-shape features on the views need be drawn, as shown in Fig. 7-1-2. Since the auxiliary view shows only the true shape and detail of the inclined surface or features, a partial auxiliary view is all that is necessary. Likewise, the distorted features on the regular views may be omitted. Hidden lines are usually omitted unless required for clarity. This procedure is recommended for functional and production drafting where drafting costs are an important consideration. However, the drafter may be called upon to draw the complete views of the part. This type of drawing is often used for catalog and standard parts drawing.

Additional examples of auxiliary view drawings are shown in Fig. 7-1-3 (pg. 138).

Figure 7-1-4 (pg. 138) shows how to make an auxiliary view of a symmetrical object. Figure 7-1-4A shows the object in a pictorial view. In this illustration the center plane is used as the reference plane. In Fig. 7-1-4B the center plane is drawn parallel to the inclined surface shown in the front view. The edge view of this plane appears as a center line, line XY, on the top view. Number the points of intersection between the inclined surface and the vertical lines on the top view. Then transfer these numbers to the edge view of the inclined surface on the front view, as shown. Parallel to this edge view and at a convenient distance from it, draw the line $X'Y'$, as in Fig. 7-1-4C. Now, in the top view, find the distances D_1 and D_2 from the numbered points to the center line. These are the depth measurements. Transfer them onto the corresponding construction lines that you have just drawn, measuring them off on either

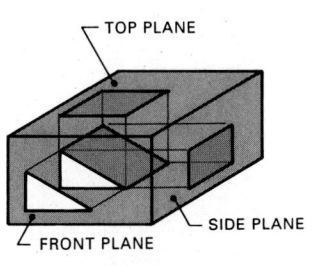

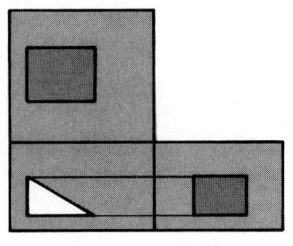

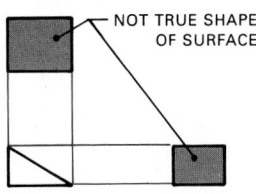
— NOT TRUE SHAPE OF SURFACE

THREE PRINCIPAL PLANES OF PROJECTION HINGED TOGETHER

PLANES UNFOLDED

PLANES REMOVED SHOWING THREE REGULAR (TOP, FRONT, SIDE) VIEWS

NOTE: IN NONE OF THESE VIEWS DOES THE SLANTED (COLORED) SURFACE APPEAR IN ITS TRUE SHAPE.

(A) WEDGED BLOCK SHOWN IN THREE REGULAR VIEWS

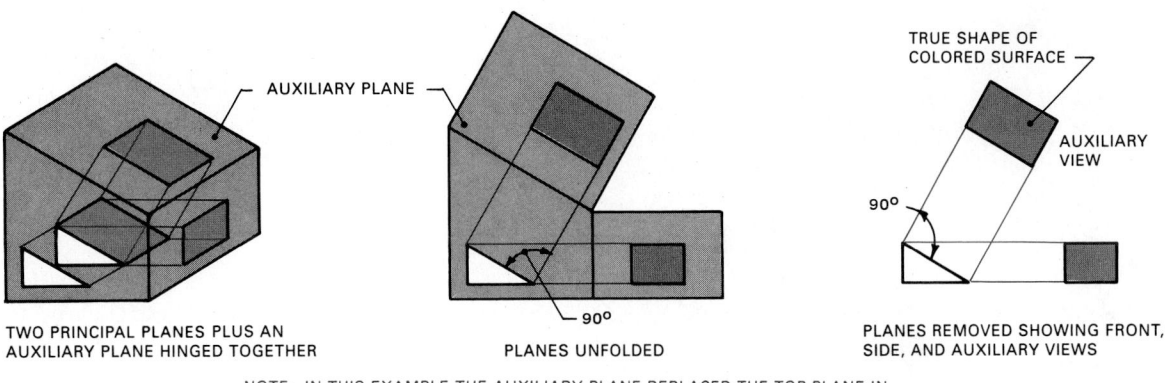

TWO PRINCIPAL PLANES PLUS AN AUXILIARY PLANE HINGED TOGETHER

PLANES UNFOLDED

PLANES REMOVED SHOWING FRONT, SIDE, AND AUXILIARY VIEWS

NOTE: IN THIS EXAMPLE THE AUXILIARY PLANE REPLACED THE TOP PLANE IN ORDER THAT THE SLANTED (COLORED) SURFACE MAY BE SHOWN IN ITS TRUE SHAPE.

(B) REPLACING THE TOP PLANE WITH AN AUXILIARY PLANE

FIG. 7-1-1   Relationship of the auxiliary plane to the three principal planes.

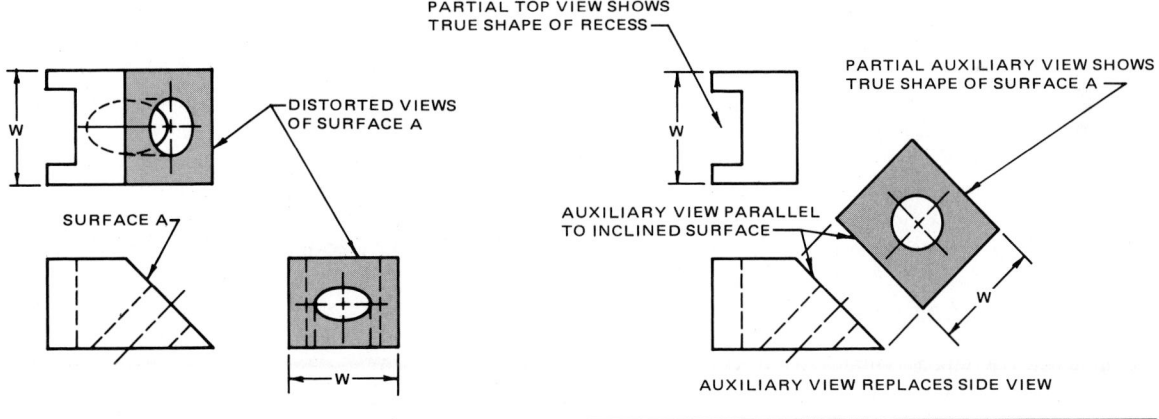

PARTIAL TOP VIEW SHOWS TRUE SHAPE OF RECESS

DISTORTED VIEWS OF SURFACE A

SURFACE A

PARTIAL AUXILIARY VIEW SHOWS TRUE SHAPE OF SURFACE A

AUXILIARY VIEW PARALLEL TO INCLINED SURFACE

AUXILIARY VIEW REPLACES SIDE VIEW

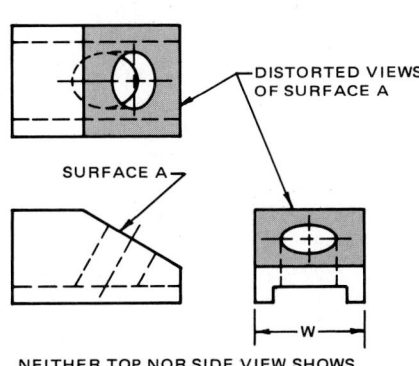

DISTORTED VIEWS OF SURFACE A

SURFACE A

NEITHER TOP NOR SIDE VIEW SHOWS TRUE SHAPE OF SURFACE A

PARTIAL VIEWS SHOWING ONLY THE NECESSARY DETAILS ARE RECOMMENDED

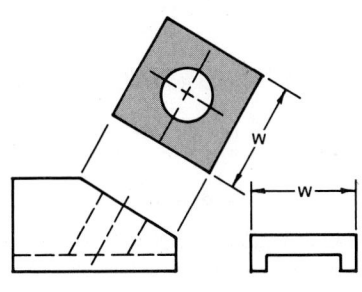

AUXILIARY VIEW REPLACES TOP VIEW

FIG. 7-1-2   Auxiliary views replacing regular views.

137

NOTE: ONLY CONVENTIONAL BREAK ON PROJECTED
      SURFACE NEED BE SHOWN ON PARTIAL VIEWS.

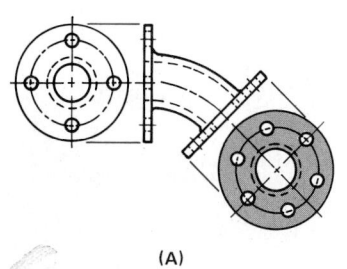
(A)

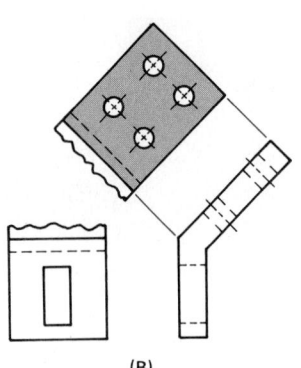
(B)

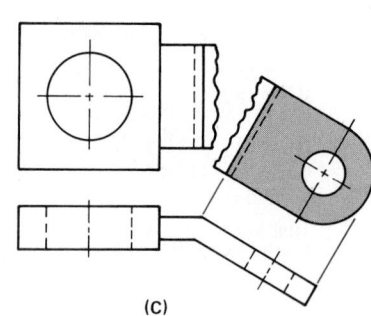
(C)

FIG. 7-1-3   Examples of auxiliary view drawings.

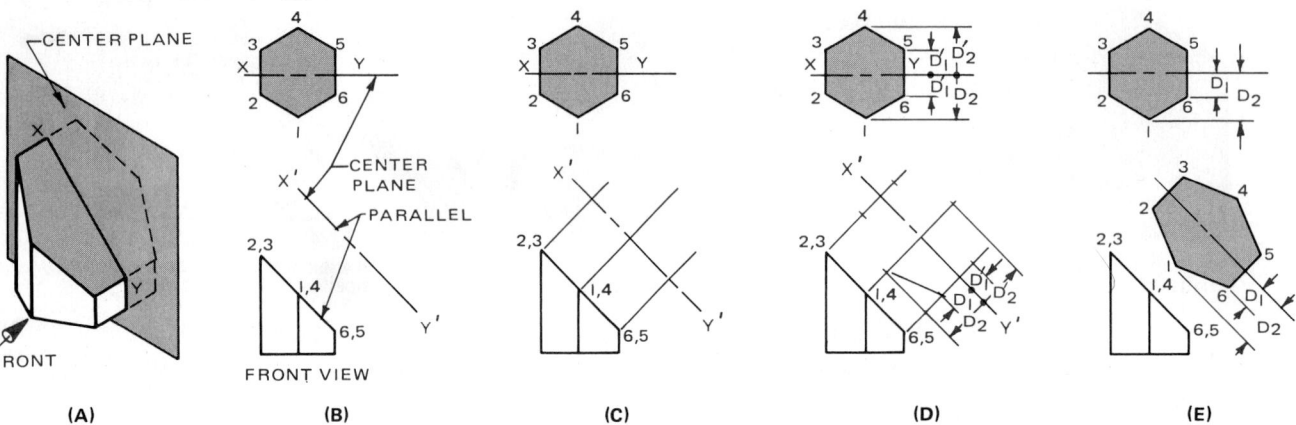

(A)   **(B)**   **(C)**   **(D)**   **(E)**

FIG. 7-1-4   Drawing an auxiliary view using the center plane reference.

side of line $X'Y'$, as shown in Fig. 7-1-4D. The result will be a set of points on the construction lines. Connect and number these points, as shown in Fig. 7-1-4E, and the front auxiliary view of the inclined surface results. The remaining portions of the object may also be projected from the center reference plane.

Dimensioning Auxiliary Views

One of the basic rules of dimensioning is to dimension the feature where it can be seen in its true shape and size. Thus the auxiliary view will show only the dimensions pertaining to those features for which the auxiliary view was drawn. The recommended dimensioning method for engineering drawings is the *unidirectional system* (Fig. 7-1-5).

ASSIGNMENT ▓▓▓▓▓▓▓▓▓▓▓▓▓▓▓▓▓

See Assignment 1 for Unit 7-1 on pages 158–159.

7-2 CIRCULAR FEATURES IN AUXILIARY PROJECTION

As mentioned in Unit 7-1, at times it is necessary to show the complete views of an object. If circular features are involved

in auxiliary projection, the surfaces appear elliptical, not circular, in one of the views.

The method most commonly used to draw the true-shape projection of the curved surface is the plotting of a series of points on the line, the number of points being governed by the accuracy of the curved line required.

Figure 7-2-1 illustrates an auxiliary view of a truncated cylinder. The shape seen in the auxiliary view is an ellipse.

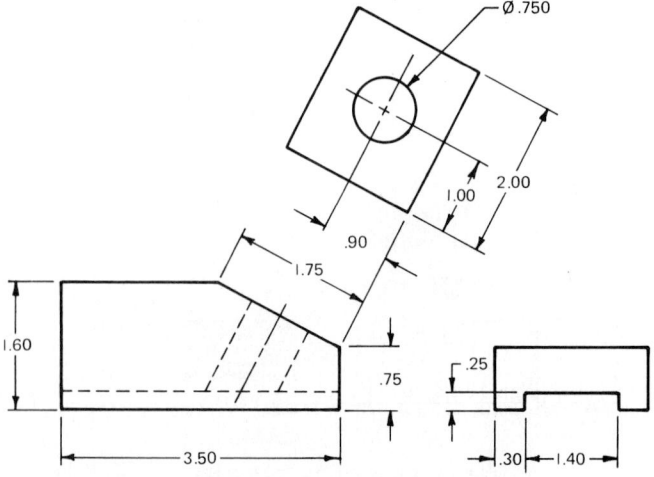

FIG. 7-1-5   Dimensioning auxiliary view drawings.

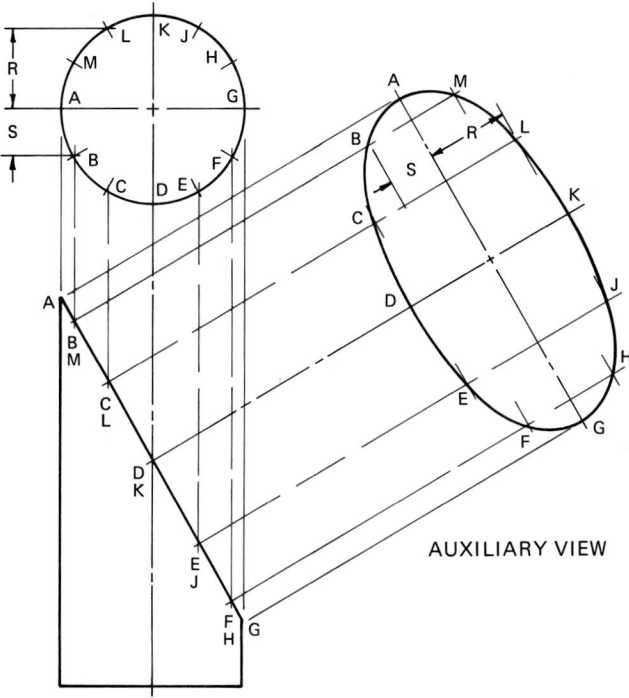

FIG. 7-2-1   Establishing true shape of truncated cylinder.

This shape is drawn by plotting lines of intersection. The perimeter of the circle in the top view is divided to give a number of equally spaced points—in this case, 12 points, *A* to *M*, spaced 30° apart (360°/12 = 30°). These points are projected down to the edge line on the front view, then at right angles to the edge line to the area where the auxiliary view will be drawn. A center line for the auxiliary view is drawn parallel

FIG. 7-2-2   Constructing the true shape of a curved surface by plotting method.

to the edge line, and width settings taken from the top view are transferred to the auxiliary view. Note width setting *R* for point *L*. Because the illustration shows a true cylinder and the point divisions in the top view are all equal, the width setting *R* taken at *L* is also the correct width setting for *C*, *E*, and *J*. Width setting *S* for *B* is also the correct width setting for *F*, *H*, and *M*. When all the width settings have been transferred to the auxiliary view, the resulting points of intersection are connected by means of an irregular curve to give the desired elliptical shape.

It is often necessary to construct the auxiliary view first in order to complete the regular views, shown in Fig. 7-2-2.

ASSIGNMENT

See Assignment 2 for Unit 7-2 on pages 159–160.

7-3 MULTI-AUXILIARY-VIEW DRAWINGS

Some objects have more than one surface not perpendicular to the plane of projection. In working drawings of these objects, an auxiliary view may be required for each surface. Naturally, this would depend upon the amount and type of detail lying on these surfaces. This type of drawing is often referred to as the *multi-auxiliary-view* drawing (Fig. 7-3-1, pg. 140).

One can readily see the advantage of using the unidirectional system of dimensioning for dimensioning an object such as the one shown in Fig. 7-3-2 (pg. 140).

ASSIGNMENTS

See Assignments 3 and 4 for Unit 7-3 on pages 161–162.

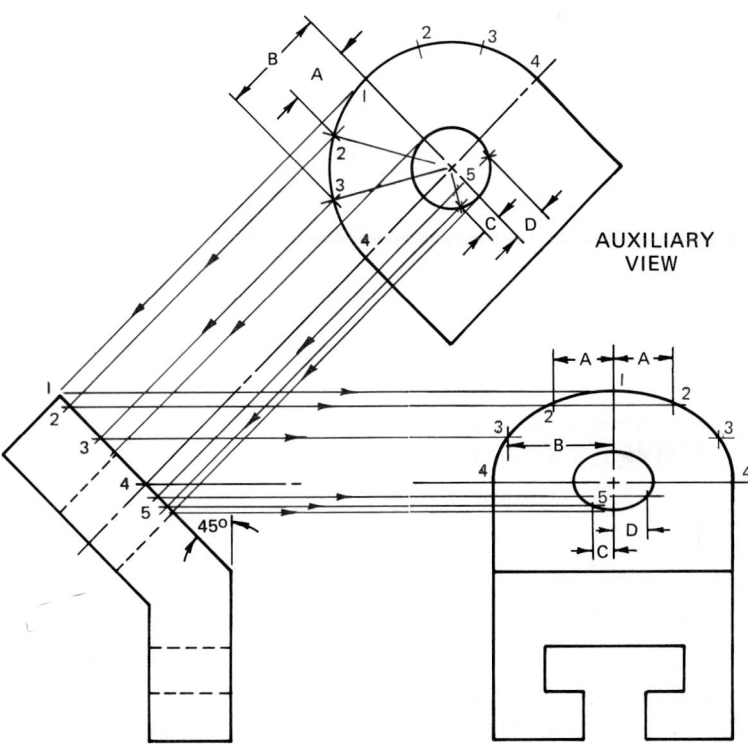

FIG. 7-3-1   Auxiliary views added to regular views to show true shape of features.

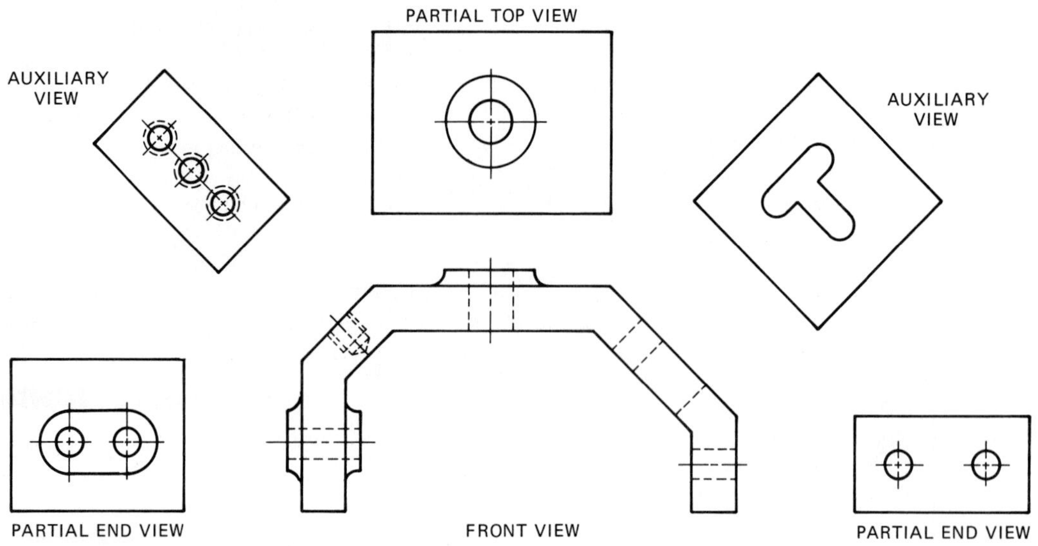

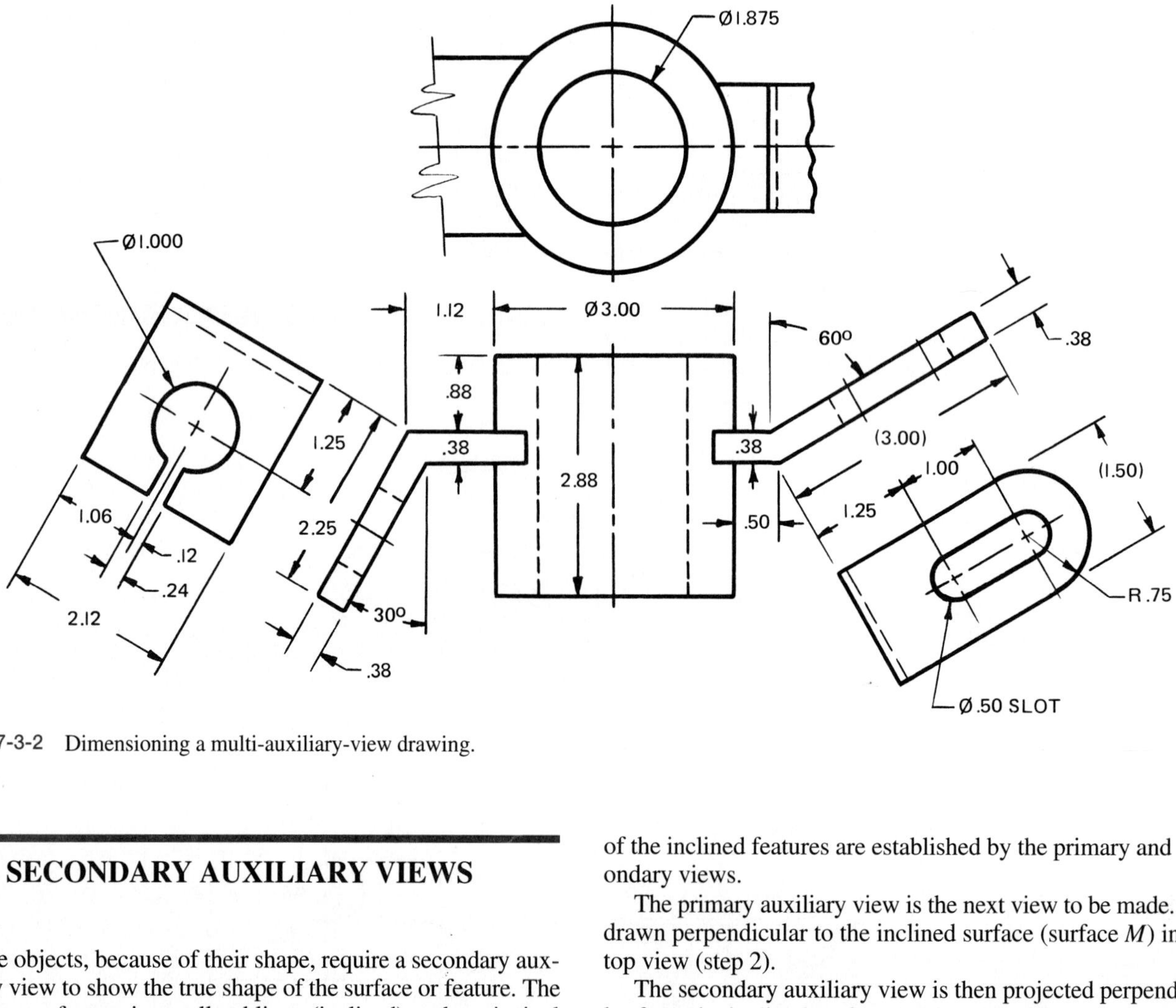

FIG. 7-3-2   Dimensioning a multi-auxiliary-view drawing.

7-4 SECONDARY AUXILIARY VIEWS

Some objects, because of their shape, require a secondary auxiliary view to show the true shape of the surface or feature. The surface or feature is usually oblique (inclined) to the principal planes of projection. In order to draw a secondary auxiliary view, such as the one shown in Fig. 7-4-1, only portions of the front and top views are drawn first (step 1). The remainder of these two views can be drawn only after the location and size of the inclined features are established by the primary and secondary views.

The primary auxiliary view is the next view to be made. It is drawn perpendicular to the inclined surface (surface M) in the top view (step 2).

The secondary auxiliary view is then projected perpendicular from the inclined surface (surface N) of the primary auxiliary view (step 3). Once the secondary auxiliary view is drawn, the size and location of the features on the secondary auxiliary view, in this case the hexagon hole, can be located on the primary auxiliary view.

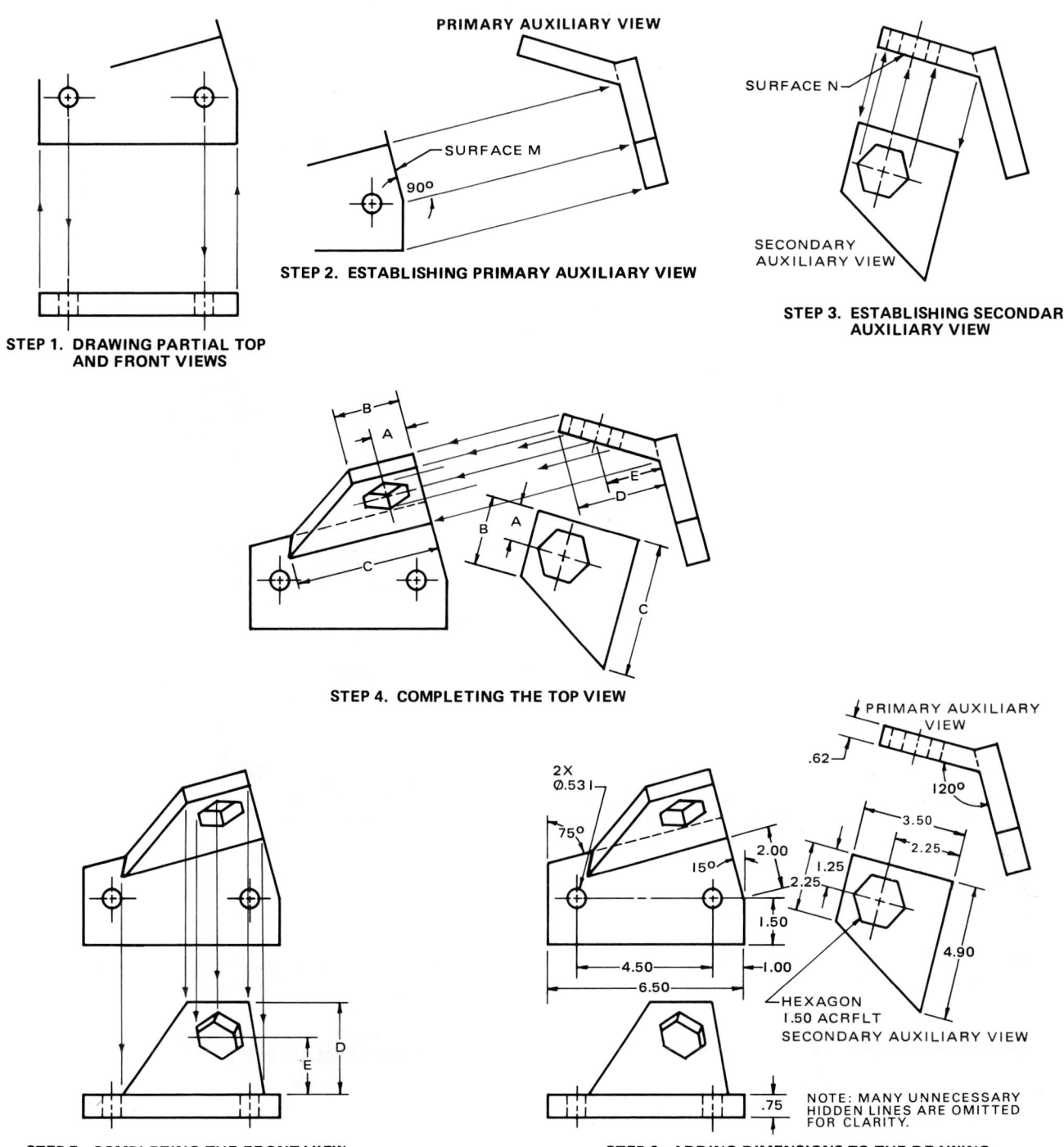

FIG. 7-4-1 Steps in drawing a secondary auxiliary view.

Next the top view is completed (step 4) by projecting lines and points from the primary auxiliary view and distances, such as *A*, *B*, and *C*, from the secondary auxiliary view.

Last, the front view is completed by projecting lines and points from the top view and distances, such as *D* and *E*, from the primary auxiliary view (step 5), then adding dimensions (step 6).

Another example of using auxiliary views in establishing the true shape and size of an oblique surface is shown

in Fig. 7-4-2 (pg. 142). Surface 1-2-3-4 is inclined to the three regular planes (front, top, and side) of projection. In this example, the three regular views are drawn first. The primary auxiliary view is drawn next by projecting lines parallel to line 1-2 in the top view. Note that in this view, points 1, 2, 3, and 4 appear as a line or edge view of the surface 1-2-3-4. The secondary auxiliary view is drawn next by projecting lines perpendicular to line 1-2-3-4 in the primary auxiliary view in order to show the true shape and size of surface 1-2-3-4.

FIG. 7-4-2 Secondary
auxiliary view
required to find the
true shape of
surface 1-2-3-4.

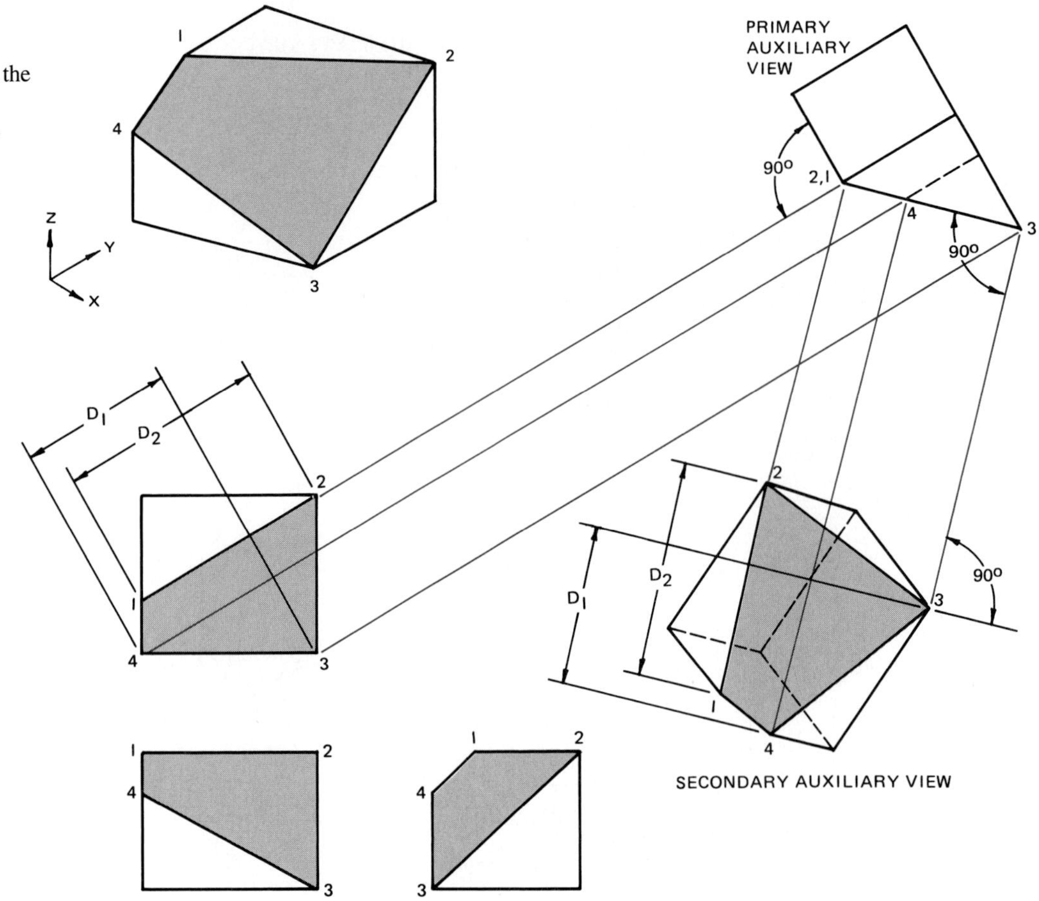

SECONDARY AUXILIARY VIEW

When preparing auxiliary view drawings, be sure to allow sufficient space between views to ensure that the views and dimensions that have to be drawn later will fit in the allotted space.

ASSIGNMENT

See Assignment 5 for Unit 7-4 on page 163.

7-5 REVOLUTIONS

A major problem in technical drawing and design is the creation of projections for finding the true views of lines and planes. The following is a brief review of the principles of descriptive geometry involved in the solution of such problems. The designer, working along with an engineering team, can solve problems graphically with geometric elements. Structures that occupy space have three-dimensional forms made up of a combination of geometric elements (Fig. 7-5-1).

The graphic solutions of three-dimensional forms require an understanding of the space relations that points, lines, and planes share in forming any given shape. Problems that many times require mathematical solutions can often be solved graphically with an accuracy that will allow manufacturing and construction. *Basic descriptive geometry* is one of the designer's methods of thinking through and solving problems.

Reference Planes

In Unit 6-1 reference planes were used to show how the six basic views of an object were positioned on a flat surface. Unfolding these reference planes forms a two-dimensional surface that a drafter uses to construct views and solve problems.

To identify the different planes being used on a drawing, an identification code is needed for these planes. One such system is to identify the top or horizontal reference plane by the letter *T*; identify the front or vertical reference plane by the letter *F*; and identify the side or profile reference plane by the letter *S*. Thus point 1 on a part, line, or plane would be identified as 1*F* on the front reference plane, 1*T* on the top reference plane, and 1*S* on the side reference plane.

The folding lines shown on the box are referred to as reference lines on the drawing, as shown in Fig. 7-5-2. Other reference planes and reference lines are drawn and labeled as required.

Revolutions

As we have seen, when the true size and shape of an inclined surface do not show in a drawing, one solution is to make an auxiliary view. Another, however, is to keep using the regular reference planes while imagining that the object has been revolved (turned) as shown in Fig. 7-5-3. Remember, in auxiliary views, you set up new reference planes in order to look at objects from new directions. Understanding *revolutions*

FIG. 7-5-1   Geometric space-frame structure.

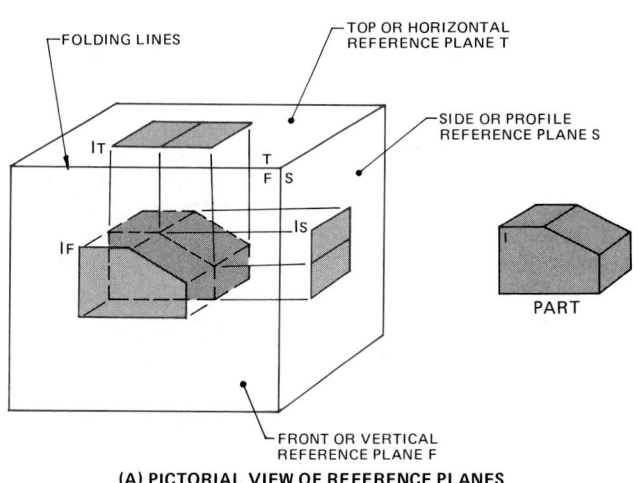

(A) PICTORIAL VIEW OF REFERENCE PLANES

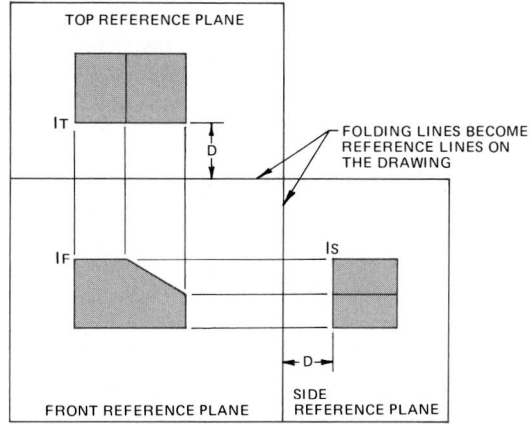

(B) UNFOLDING OF THE THREE REFERENCE PLANES

FIG. 7-5-2   Reference lines.

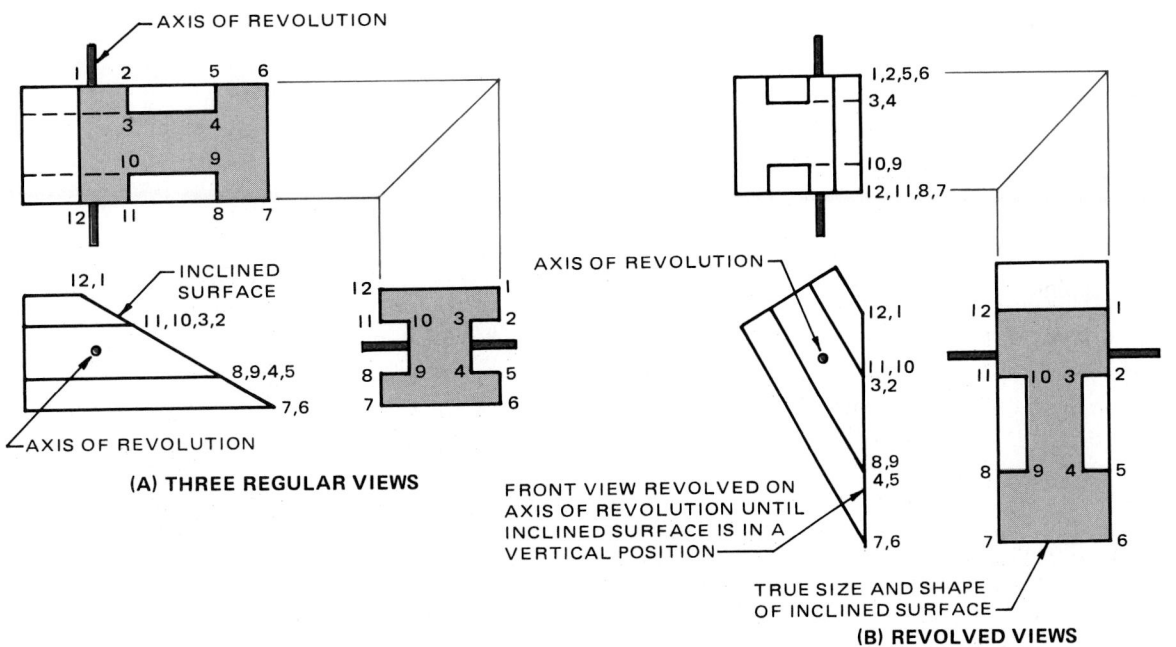

(A) THREE REGULAR VIEWS

FRONT VIEW REVOLVED ON AXIS OF REVOLUTION UNTIL INCLINED SURFACE IS IN A VERTICAL POSITION

TRUE SIZE AND SHAPE OF INCLINED SURFACE

(B) REVOLVED VIEWS

FIG. 7-5-3   Revolving front view to obtain true size and shape of inclined surface.

(ways of revolving objects) should help you better understand auxiliary views.

Axis of Revolution

An easy way to picture an object being revolved is to imagine that a shaft or an axis has been passed through it. Imagine, also, that this axis is perpendicular to one of the principal planes. In Fig. 7-5-4 the three principal planes are shown with an axis passing through each one and through the object beyond.

An object can be revolved to the right (clockwise) or to the left (counterclockwise) about an axis perpendicular to either the vertical or the horizontal plane. The object can be revolved forward (counterclockwise) or backward (clockwise) about an axis perpendicular to the profile plane. As we have seen, an axis of revolution can be perpendicular to the vertical, horizontal, or profile plane. In Fig. 7-5-5A the usual front and top views of an object are shown at the left. To the right the same views of the object are shown after the object has been revolved 45° counterclockwise about an axis perpendicular to the vertical plane. Notice that the front view is still the same in

size and shape as before except that it has a new position. The new top view has been made by projecting up from the new front view and across from the old top view. Note that the depth remains the same from one top view to the other.

In Fig. 7-5-5B, a second object is shown at the left in the usual top and front views. To the right the same views of the object are shown after it has been revolved 60° clockwise about an axis perpendicular to the horizontal plane. The new top view is the same in size and shape as before. The new front view has been made by projecting down from the new top view and across from the old front view. Note that the height remains the same from the original front view to the revolved front view.

In Fig. 7-5-5C a third object is shown at the top in the usual front and side views. Below, the same views of the object are shown after it has been revolved forward (counterclockwise) 30° about an axis perpendicular to the profile plane. The new front view has been made by projecting across from the new side view and down from the old front view in space 1. Note that the width remains the same from one front view to the other. Revolution can be clockwise, as in Fig. 7-5-5B, or it can be counterclockwise, as in A and C.

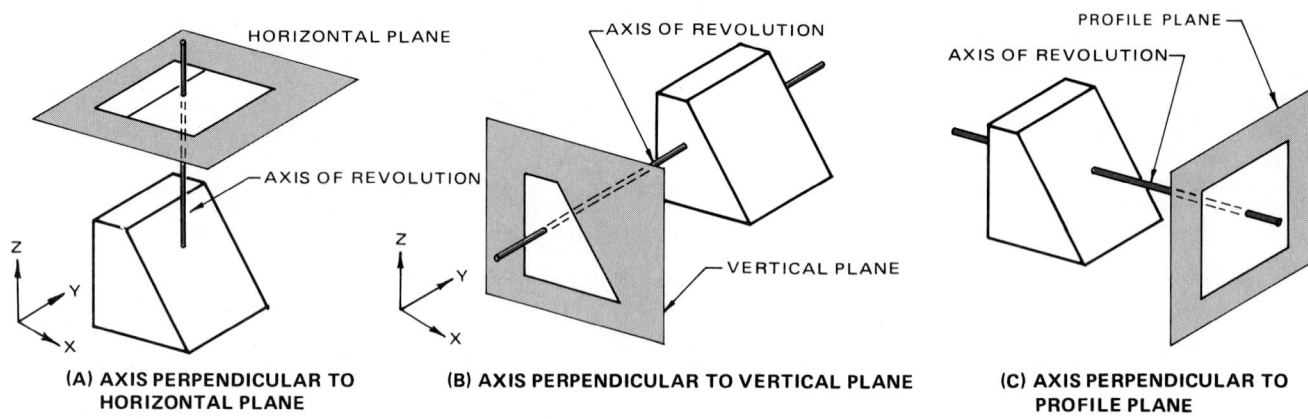

(A) AXIS PERPENDICULAR TO HORIZONTAL PLANE

(B) AXIS PERPENDICULAR TO VERTICAL PLANE

(C) AXIS PERPENDICULAR TO PROFILE PLANE

FIG. 7-5-4   The axis of revolution is perpendicular to the principal planes.

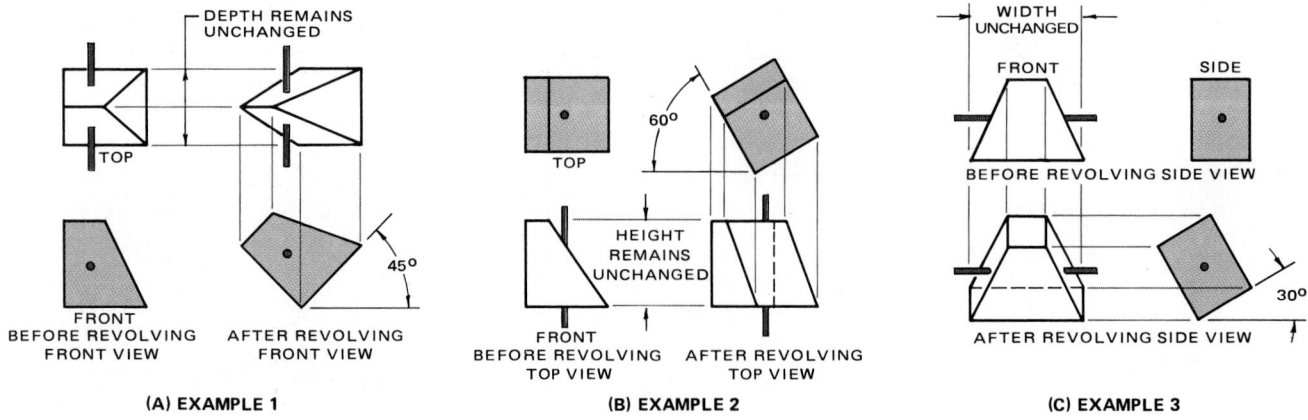

(A) EXAMPLE 1

(B) EXAMPLE 2

(C) EXAMPLE 3

FIG. 7-5-5   Single revolutions about the three axes.

The Rule of Revolution

The rule of revolution has two parts:

1. The view that is perpendicular to the axis of revolution stays the same except in position. (This is true because the axis is perpendicular to the plane on which it is projected.)
2. Distances parallel to the axis of revolution stay the same. (This is true because they are parallel to the plane or planes on which they are projected.)

Figure 7-5-6 illustrates the two parts of the rule of revolution.

True Shape of an Oblique Surface Found by Successive Revolutions

A surface shows its true shape when it is parallel to one of the principal planes. In Fig. 7-5-7A an object is shown pictorially and in orthographic projection. Note that surface 1-2-3-4 is an oblique surface because it is inclined in all three of the normal views. To find the true shape and size of this surface by revolutions, the following rotations must be made.

First Revolution   Rotate the top view until line 1-2 is in a vertical position (Fig. 7-5-7B). By projection, complete the front and side views. Note that surface 1-2-3-4 now appears as line 1-3 in the front view (frontal plane).

Second Revolution   Next rotate the front view until line 1-3 is in a vertical position (Fig. 7-5-7C). The true shape of surface 1-2-3-4 can now be found by revolving the part about an axis perpendicular to the frontal plane until the surface is parallel to the profile plane. The depth distances in the side and top views are identical to the depth distances in the side view shown in Fig. 7-5-7B. Note that this is the same part shown in Fig. 7-4-2, in which a secondary auxiliary view was used to establish the true size and shape of the surface.

Auxiliary Views and Revolved Views

You can show the true size of an inclined surface by either an auxiliary view (Fig. 7-5-8A, pg. 146) or a revolved view (Fig. 7-5-8B). In a revolved view, the inclined surface is turned until it is parallel to one of the principal planes. The revolved view in B is similar to the auxiliary view in A.

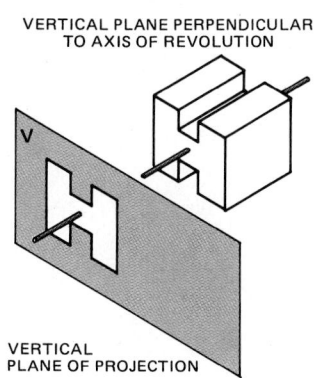

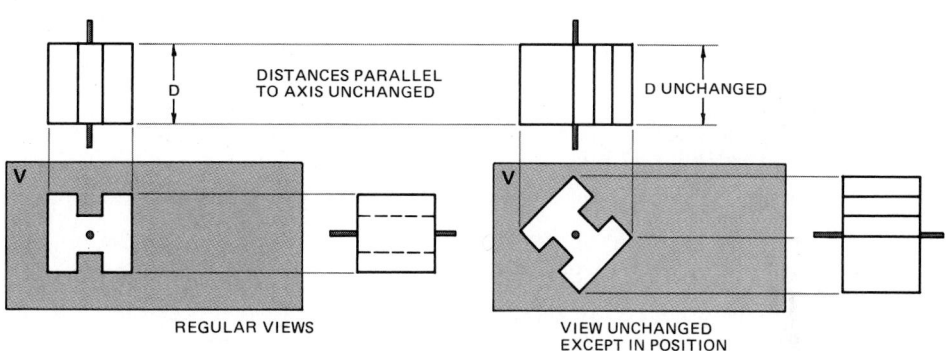

FIG. 7-5-6   The rule of revolution.

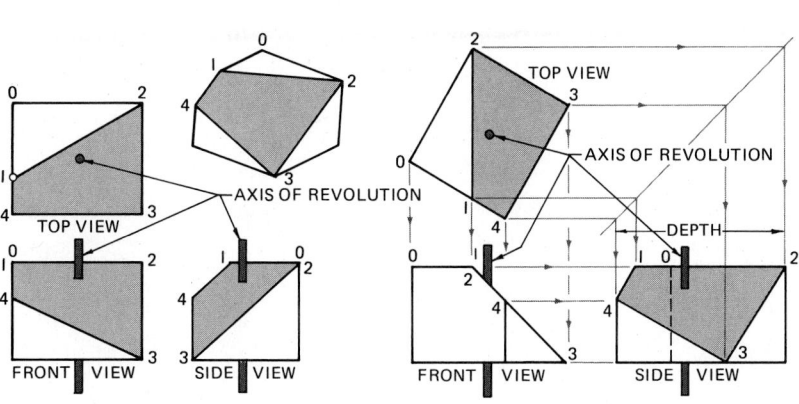

(A) THREE REGULAR VIEWS AND PICTORIAL OF PART

(B) FIRST REVOLUTION – ROTATE TOP VIEW UNTIL LINE 1-2 IS VERTICAL

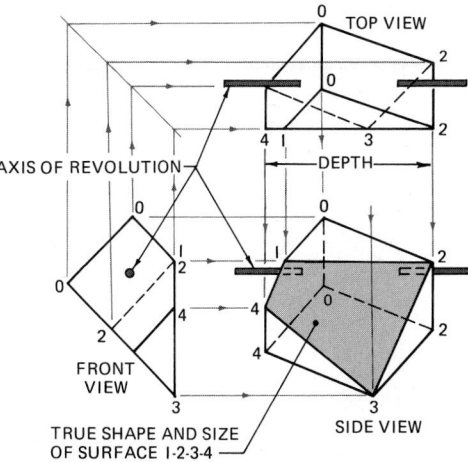

(C) SECOND REVOLUTION – ROTATE FRONT VIEW UNTIL LINE REPRESENTING SURFACE 1-2-3-4 IS VERTICAL

FIG. 7-5-7   The true shape of surface 1-2-3-4 is obtained by successive revolutions.

In the auxiliary view, it is as if the observer has changed position to look at the object from a new direction. Conversely, in the revolved view, it is as if the object has changed position. Both revolutions and auxiliaries improve your ability to visualize objects. They also work equally well in solving problems.

True Length of a Line

Since an auxiliary view shows the true size and shape of an inclined surface, it can also be used to find the true length of a line. In Fig. 7-5-9A the line *OA* does not show its true length in the top, front, or side view because it is inclined to all three of these planes of projection. In the auxiliary view in Fig. 7-5-9B, however, it does show its true length (*TL*), because the auxiliary plane is parallel to the surface *OAB*.

Figure 7-5-9C shows another way to show the true length (*TL*) of line *OA*. In this case, revolve the object about an axis perpendicular to the vertical plane until surface *OAB* is parallel to the profile plane. The side view will then show the true size of surface *OAB* and also the true length of line *OA*. A shorter method of showing the true length of *OA* is to revolve only the surface *OAB*, as shown in Fig. 7-5-9D.

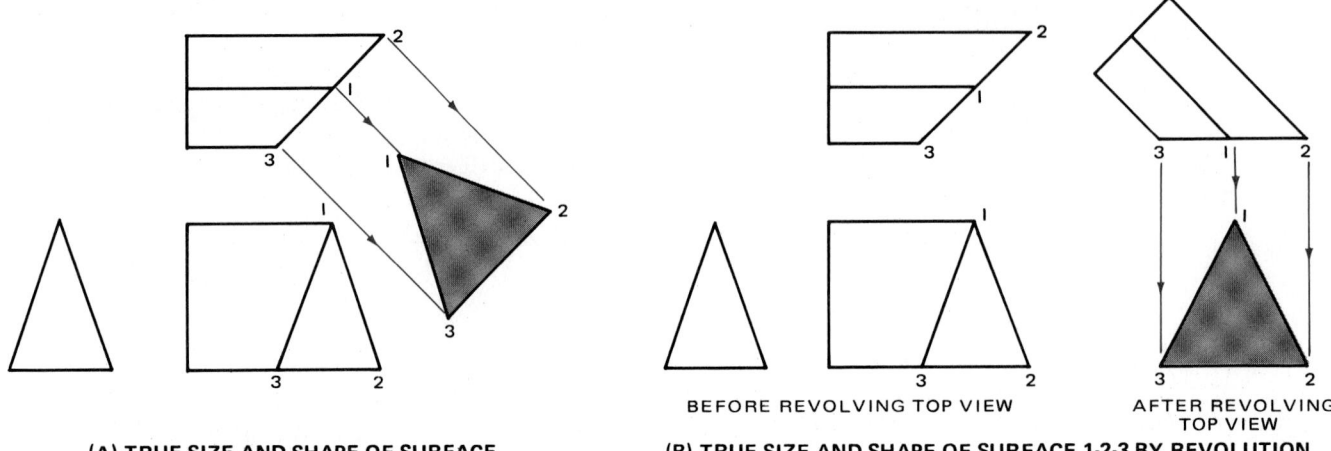

(A) TRUE SIZE AND SHAPE OF SURFACE 1-2-3 BY AUXILIARY VIEW

(B) TRUE SIZE AND SHAPE OF SURFACE 1-2-3 BY REVOLUTION

BEFORE REVOLVING TOP VIEW

AFTER REVOLVING TOP VIEW

FIG. 7-5-8   True size of a surface obtained by using auxiliary and revolved views.

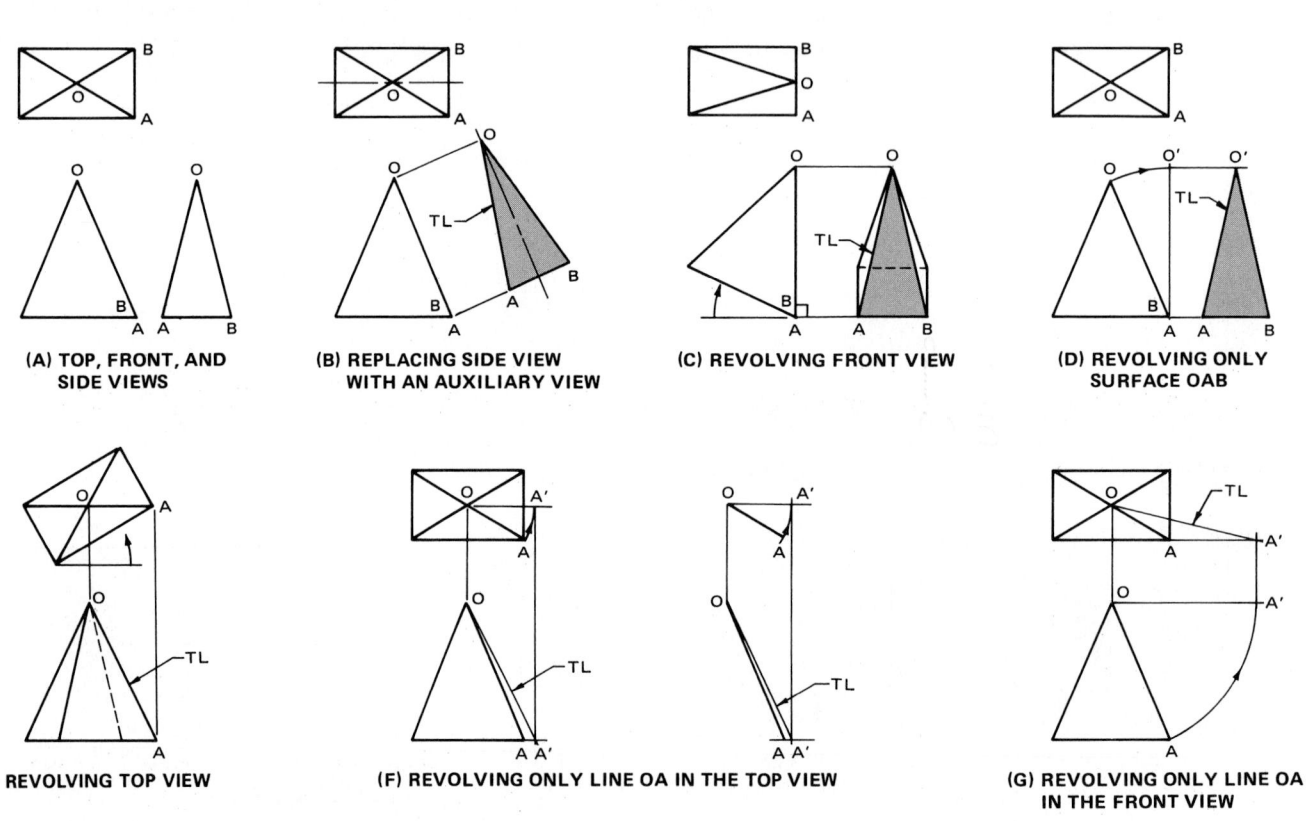

(A) TOP, FRONT, AND SIDE VIEWS

(B) REPLACING SIDE VIEW WITH AN AUXILIARY VIEW

(C) REVOLVING FRONT VIEW

(D) REVOLVING ONLY SURFACE OAB

(E) REVOLVING TOP VIEW

(F) REVOLVING ONLY LINE OA IN THE TOP VIEW

(G) REVOLVING ONLY LINE OA IN THE FRONT VIEW

FIG. 7-5-9   Typical true-length problems examined and solved.

In Fig. 7-5-9E the object is revolved in the top view until line OA in that view is horizontal. The front view now shows line OA in its true length because this line is now parallel to the vertical plane.

In Fig. 7-5-9F still another method is shown. In this case, instead of the whole object being revolved, just line OA is turned in the top view until it is horizontal at OA. The point A₁ can then be projected to the front view. There, OA₁ will show line OA at its true length.

You can revolve a line in any view to make it parallel to any one of the three principal planes. Projecting the line on the plane to which it is parallel will show its true length. In Fig. 7-5-9G the line has been revolved parallel to the horizontal plane. The true length of line OA then shows in the top view.

Fig. 7-5-10 shows a simple part with one view revolved in each of the examples. Space 1 shows a three-view drawing of a block in its simplest position. In space 2 (upper right) the block is shown after being revolved from the position in space 1 through 45° about an axis perpendicular to the frontal plane. The front view was drawn first, copying the front view in space 1. The top view was obtained by projecting up from the front view and across from the top view of space 1.

In space 3 (lower left) the block has been revolved from position 1 through 30° about an axis perpendicular to the horizontal plane. The top view was drawn first, copied from the top view of space 1.

In space 4 the block has been tilted from position 2 about an axis perpendicular to the side plane 30°. The side view was drawn first, copied from the side view in space 2. The widths of the front and top views were projected from the front view of space 2.

ASSIGNMENT

See Assignment 6 for Unit 7-5 on page 164.

7-6 LOCATING POINTS AND LINES IN SPACE

Points in Space

A point can be considered physically real and can be located by a small dot or a small cross. It is normally identified by two or more projections. In Fig. 7-6-1A point A and B are located on all three reference planes. Notice that the unfolding of the three planes forms a two-dimensional surface with the fold lines remaining. The fold lines are labeled as shown to indicate that F represents the front view, T represents the top view, and S represents the profile or right-side view. In Fig. 7-6-1B the

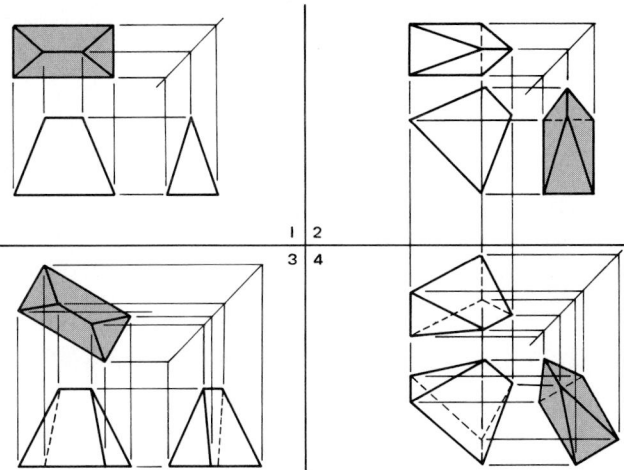

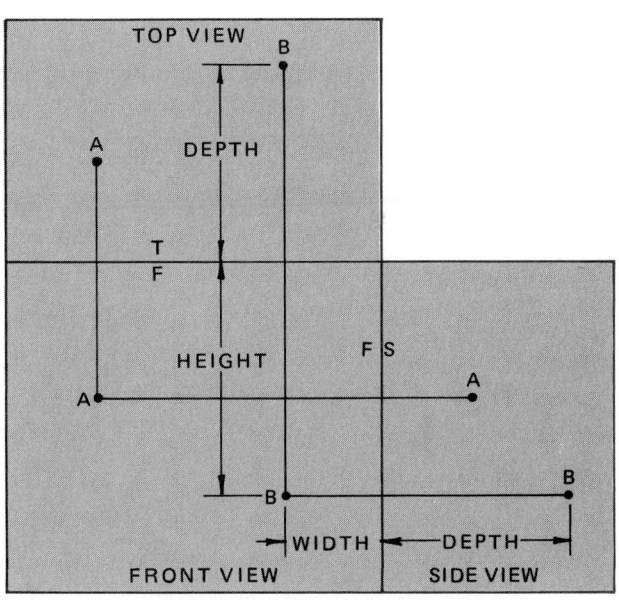

FIG. 7-5-10   Revolving a view of a part.

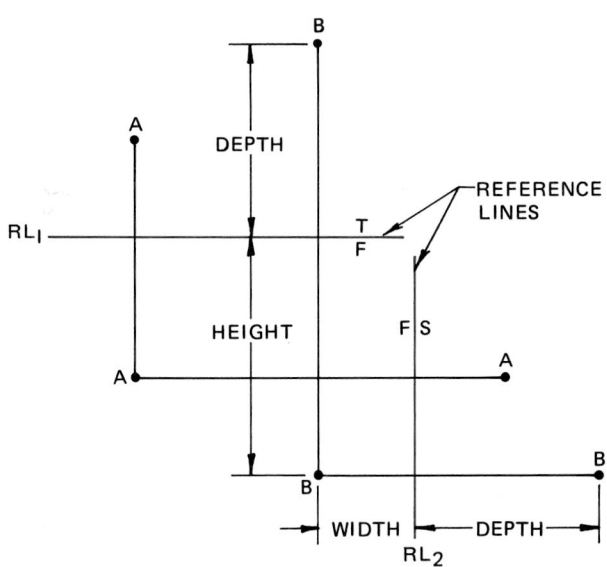

(A) POINTS A AND B IDENTIFIED ON UNFOLDED REFERENCE PLANE

(B) POINTS A AND B IDENTIFIED BY REFERENCE LINES

FIG. 7-6-1   Points in space.

planes are replaced with reference lines RL_1 and RL_2, which are placed in the same position as the fold lines in Fig. 7-6-1A.

Lines in Space

Lines in descriptive geometry are grouped into three classes depending on how they are positioned in relation to the reference lines.

Normal Lines   A line that is perpendicular to the reference plane will project as a point on that plane. In Fig. 7-6-2A line AB is perpendicular to the front reference plane. As such, it is shown as a point ($A_F B_F$) in the front view and as a true-length line in the top and side views (lines $A_T B_T$ and $A_S B_S$, respectively).

Inclined Lines   A line that appears inclined in one plane, as shown in Fig. 7-6-2B, and is parallel to the other two principal views which will appear foreshortened in the other two views. The inclined line shown in the front view will be the true length of line AB.

Oblique Lines   A line that appears inclined in all three views is an oblique line. It is neither parallel nor perpendicular to any of the three planes. The true length of the line is not shown in any of these views (Fig. 7-6-2C).

True Length of an Oblique Line by Auxiliary Projection

Since a normal line and an inclined line have projections parallel to a principal plane, the true length of each can be seen in that projection. Since an oblique line is not parallel to any of the three principal reference planes, an auxiliary reference line RL_3 can be placed parallel to any one of the oblique lines, as shown in Fig. 7-6-3. Transfer distances M and N shown in the regular views to the auxiliary view, locating points A_1 and B_1, respectively. Join points A_1 and B_1 with a line, obtaining the true length of line AB.

Point on a Line

The line $A_F B_F$ in the front view of Fig. 7-6-4A contains a point C. To place point C on the line in the other two views, it is necessary to project construction lines perpendicular to the reference lines RL_1 and RL_2, as shown in Fig. 7-6-4B. The construction lines are projected to line $A_T B_T$ in the top view and line $A_S B_S$ in the side view, locating point C on the line in these views.

If point C is to be located on the true length of line AB, another reference line, such as RL_3, is required, and the distances N and M in the front view are then used to locate the true length of line $A_1 B_1$ in the auxiliary view. Position C is then projected perpendicular to line $A_S B_S$ in the side view to locate C on the true-length line.

Point-On-Point View of a Line

When the front view $A_F B_F$ and top view $A_T B_T$ are given, as in Fig. 7-6-5, and the point-on-point view of line AB is required, the procedure is as follows:

- Draw a true-length view of AB by the method described in "True Length of an Oblique Line by Auxiliary Projection" and shown in Fig. 7-6-3.
- Draw a reference line perpendicular to the true length of line $A_1 B_1$, and label it RL_3.
- The next adjacent, second auxiliary view $A_2 B_2$ will be a point-on-point view of line AB.

ASSIGNMENTS

See Assignments 7 and 8 for Unit 7-6 on page 165.

7-7 PLANES IN SPACE

Planes for practical studies are considered to be without thickness and can be extended without limit. A plane may be represented or determined by intersecting lines, two parallel lines, a line and a point, three points, or a triangle.

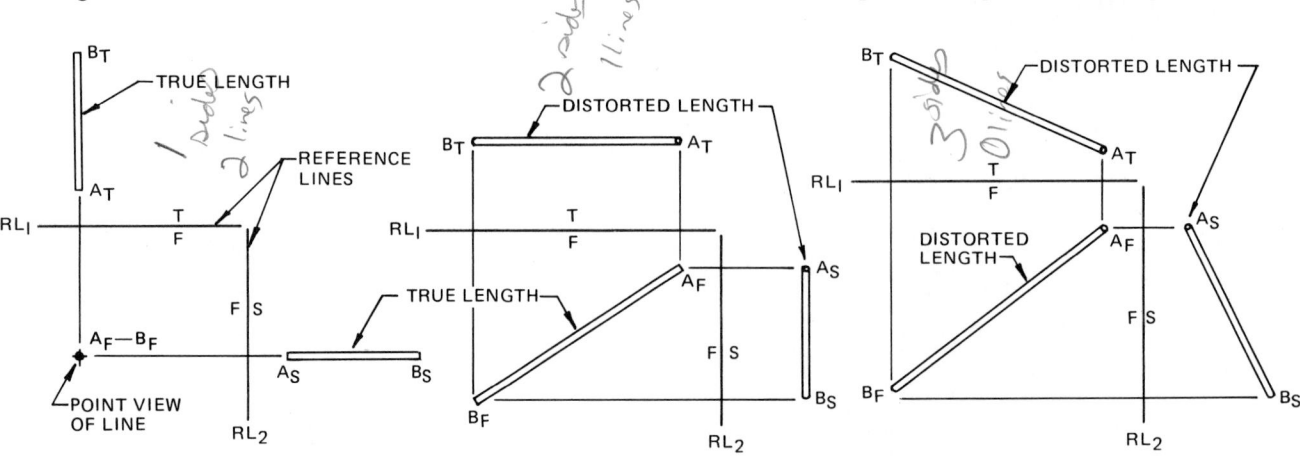

FIG. 7-6-2   Lines in space.

(A) A NORMAL LINE   **(B) INCLINED LINE**   **(C) OBLIQUE LINES**

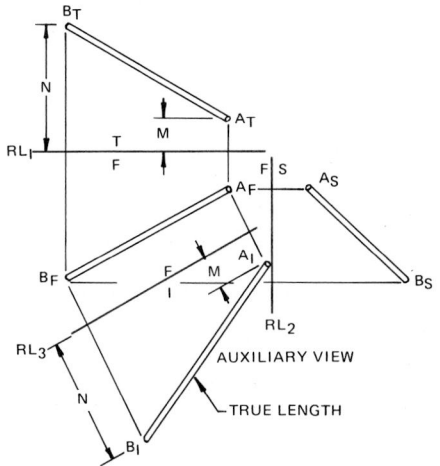

(A) REFERENCE LINE RL₃ PLACED PARALLEL TO FRONT VIEW

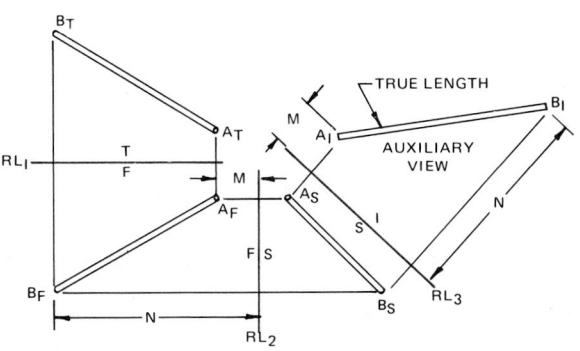

(B) REFERENCE LINE RL₃ PLACED PARALLEL TO SIDE VIEW

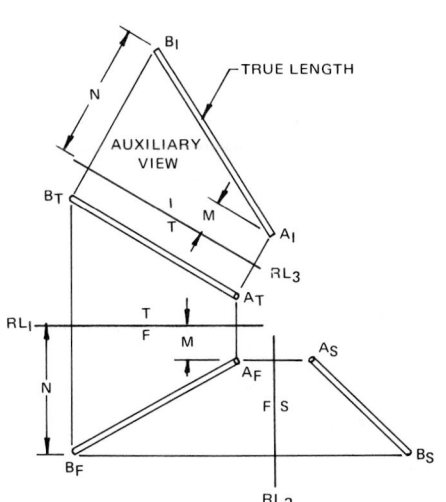

(C) REFERENCE LINE RL₃ PLACED PARALLEL TO TOP VIEW

FIG. 7-6-3   The length of an oblique line by auxiliary view projection.

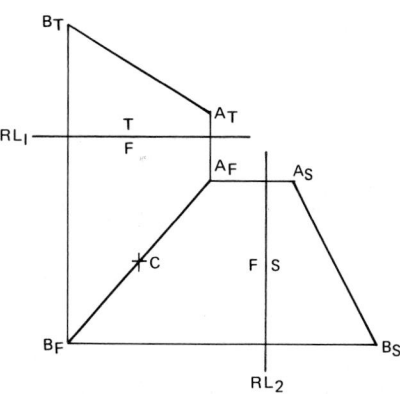

(A) PROBLEM—TO LOCATE POINT C ON LINE A-B IN OTHER VIEWS

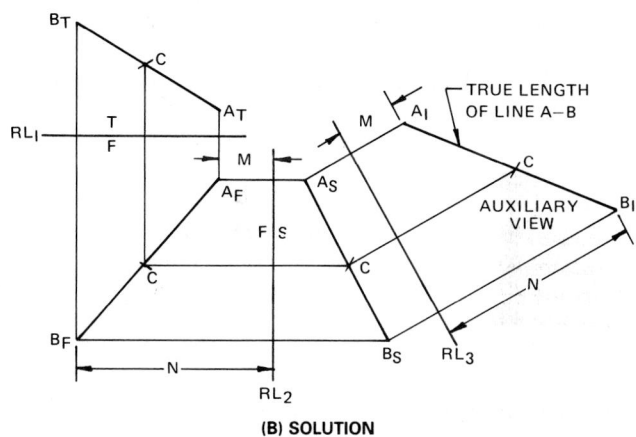

(B) SOLUTION

FIG. 7-6-4   Point on a line.

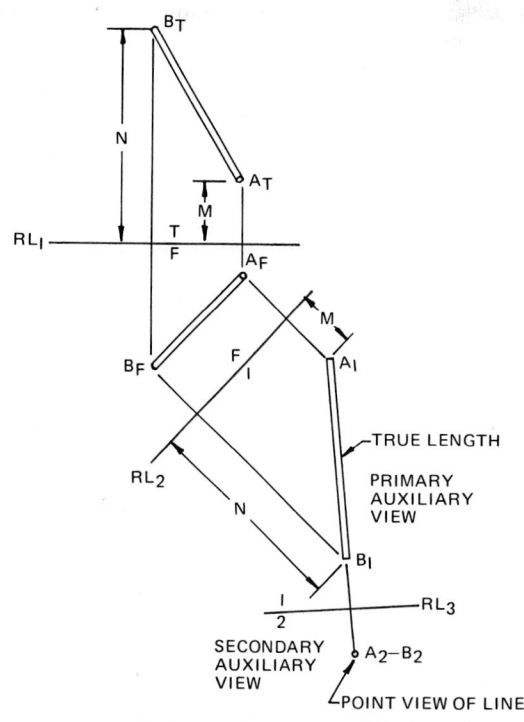

FIG. 7-6-5   Point-on-point view of a line.

The three basic planes, referred to as the *normal plane*, *inclined plane*, and *oblique plane*, are identified by their relationship to the three principal reference planes. Figure 7-7-1 illustrates the three basic planes, each plane being triangular in shape.

Normal Plane   A plane whose surface, in this case a triangular surface, appears in its true shape in the front view and as a line in the other two views.

Inclined Plane   This results when the shape of the triangular plane appears distorted in two views and as a line in the other view.

Oblique Plane   A plane whose shape appears distorted in all three views.

Locating a Line in a Plane

The top and front views in Fig. 7-7-2A show a triangular plane *ABC* and lines *RS* and *MN*, each located in one of the views.

To find their location in the other views, refer to Fig. 7-7-2B and the following procedures.

To locate line *RS* in the front view:

- Line $R_T S_T$ crosses over lines $A_T B_T$ and $A_T C_T$ at points D_T and E_T, respectively.
- Project points D_T and E_T to front view, locating points D_F and E_F.
- Extend a line through points D_F and E_F.
- The length of the line can be found by projecting points R_T and S_T to the front view, locating the end points R_F and S_F.

To locate line *MN* in the top view:

- Extend line $M_F N_F$ in the front view, locating points H_F and G_F on lines $A_F B_F$ and $A_F C_F$, respectively.
- Project points H_F and G_F to the top view, locating points H_T and G_T.
- Draw a line through points H_T and G_T.
- Project points M_F and N_F to top view, locating points on line $M_T N_T$.

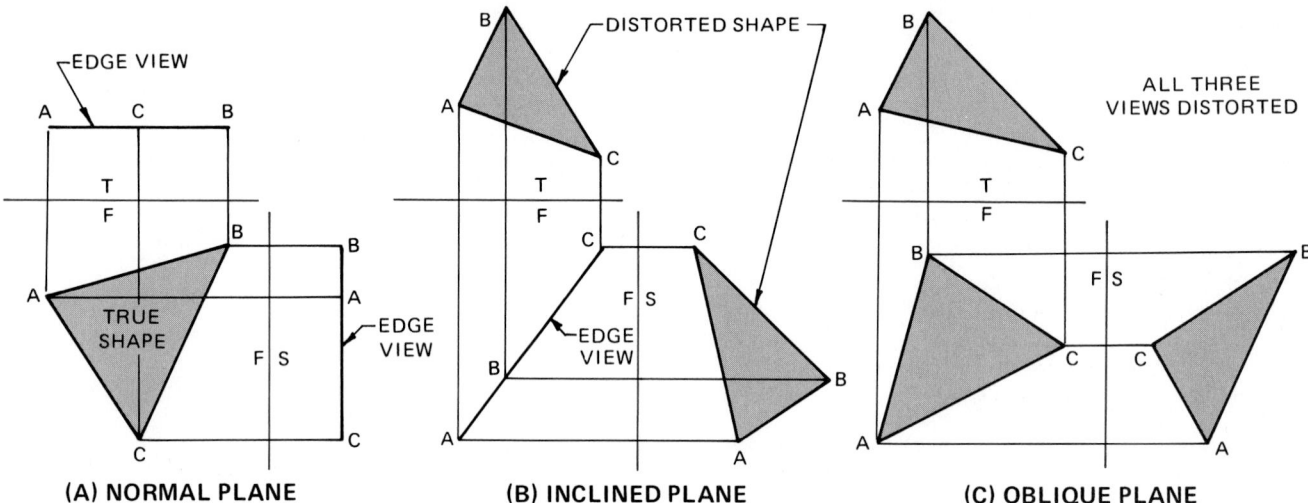

(A) NORMAL PLANE   **(B) INCLINED PLANE**   **(C) OBLIQUE PLANE**

FIG. 7-7-1   Planes in space.

FIG. 7-7-2   Locating a line on a plane.

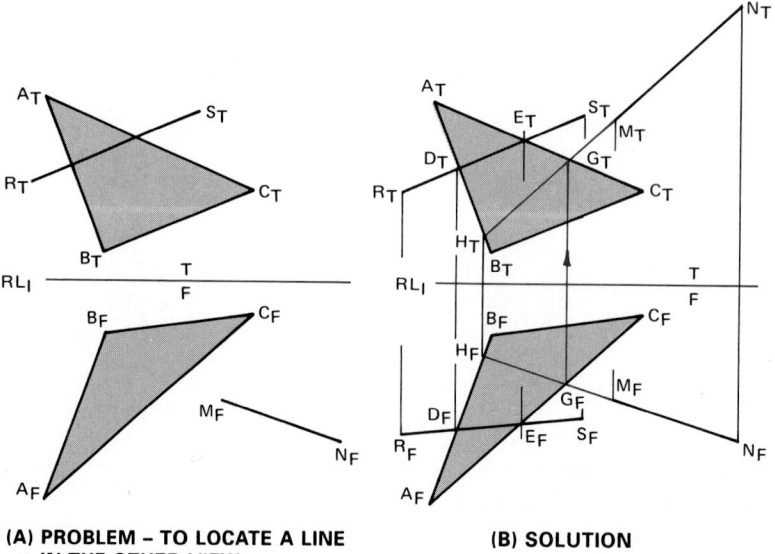

(A) PROBLEM – TO LOCATE A LINE IN THE OTHER VIEW   **(B) SOLUTION**

Locating a Point on a Plane

The top and front views shown in Fig. 7-7-3A show a triangular plane ABC and points R and S each located in one of the views. To find their location in the other views, refer to Fig. 7-7-3B and the following procedures.

To locate point R in the front view:

- Draw a line from A_T passing through point R_T to a point M_T on line B_TC_T.
- Project point M_T to front view, locating point M_F.
- Join points A_F and M_F with a line.
- Project point R_T to front view, locating point R_F.

To locate point S in the top view:

- Draw a line between points B_F and S_F, locating point N_F on line A_FC_F.
- Project point N_F to top view, locating point N_T.
- Draw a line through points B_T and N_T.
- Project point S_F to top view, locating point S_T.

Locating the Piercing Point of a Line and a Plane—Cutting-Plane Method

The top and front views shown in Fig. 7-7-4 show a line UV passing somewhere through plane ABC. The piercing point of the line through the plane is found as follows:

- Locate points D_T and E_T in the top view.
- Project points D_T and E_T to the front view, locating points D_F and E_F.
- The intersection of lines D_FE_F and U_FV_F is the piercing point, labeled O_F.
- Project point O_F to top view, locating point O_T.

Locating the Piercing Point of a Line and a Plane—Auxiliary View Method

The top and front views in Fig. 7-7-5 show a line UV passing somewhere through plane ABC. The piercing point of the line through the plane is found as follows:

- Draw line A_TD_T in the top view parallel to reference line RL_1.
- Project point D_T to front view, locating point D_F.

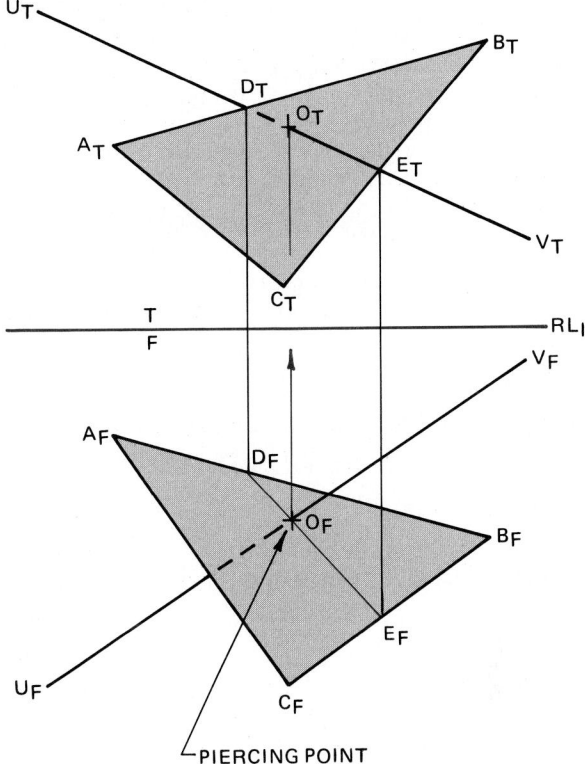

FIG. 7-7-4   Locating the piercing point of a line and a plane, cutting-plane method.

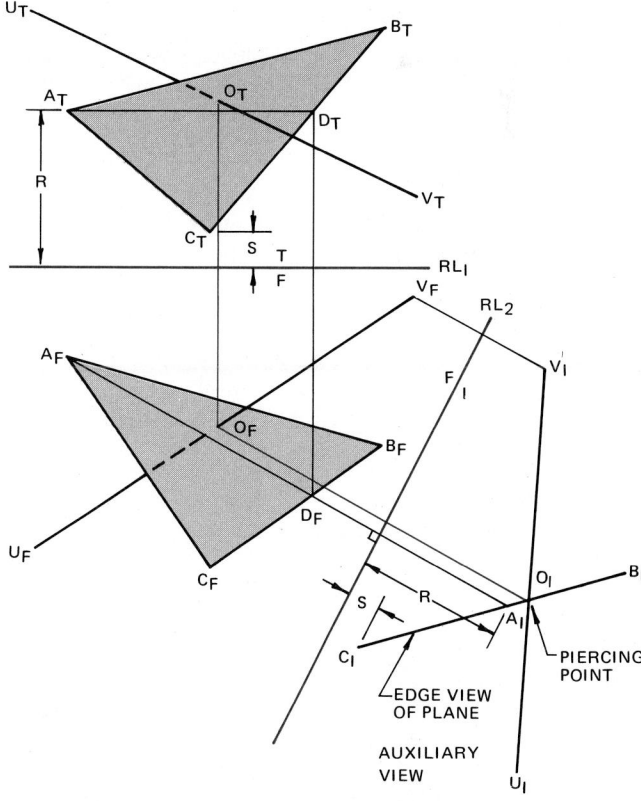

FIG. 7-7-5   Locating the piercing point of a line and a plane, auxiliary view method.

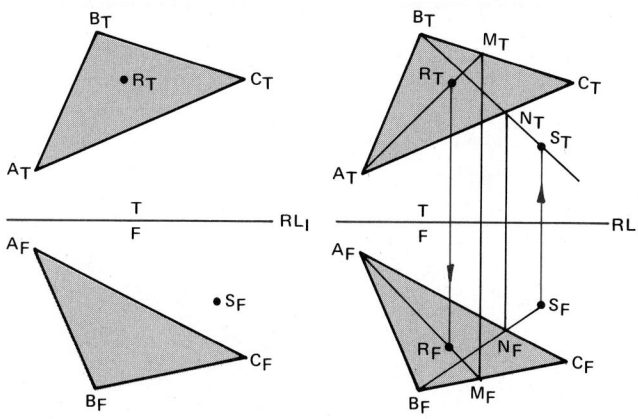

(A) PROBLEM – TO LOCATE A POINT IN THE OTHER VIEW

(B) SOLUTION

FIG. 7-7-3   Locating a point on a plane.

151

- Draw reference line RL_2 perpendicular to a line intersecting points A_F and D_F in the front view.
- Draw an auxiliary view which shows the edge view of the plane. The intersection of the plane and line locates the piercing point O_1.
- Project the piercing point to front and top views.

ASSIGNMENTS ▰▰▰▰▰▰▰▰▰▰

See Assignments 9 and 10 for Unit 7-7 on page 166.

7-8 ESTABLISHING VISIBILITY OF LINES IN SPACE

Visibility of Oblique Lines by Testing

In the example of the two nonintersecting pipes shown in Fig. 7-8-1A, it is not apparent which pipe is nearest the viewer at the crossing points in the two views. To establish which of the pipes lies in front of the other, the following procedure is used.

To establish the visible pipe at the crossing shown in the top view (Fig. 7-8-1B):

- Label the crossing of lines $A_T B_T$ and $C_T D_T$ as ①, ②.
- Project the crossing point to the front view, establishing points ① and ②.
- Point ① is closer to reference line RL_1, which means that line $A_T B_T$ is nearer when the top view is being observed and thus is visible.

To establish the visible pipe at the crossing shown in the front view:

- Label the crossing of lines $A_F B_F$ and $C_F D_F$ as ③, ④.
- Project the crossing point to top view, establishing points ③ and ④.

- Point ④ is closer to reference line RL_1, which means that line $C_F D_F$ is nearer when the front view is being observed and thus is visible. Figure 7-8-1C shows the correct crossings of the pipes.

Visibility of Lines and Surfaces by Testing

In cases where points or lines are approximately the same distance away from the viewer, it may be necessary to graphically check the visibility of lines and points, as in Fig. 7-8-2.

To check visibility of lines $A_T C_T$ and $B_T D_T$ in the top view:

- Label intersection of lines $A_T C_T$ and $B_T D_T$ as ①, ②.
- Project the point of intersection to front view, establishing point ① on line $A_F C_F$ and point ② on line $B_F D_F$.
- Point ① is closer to reference line RL_1, which means that line $A_T C_T$ is nearer when the top view is being observed and thus is visible.
- Point ② is farther away from reference line RL_1, which means that line $B_T D_T$ would not be seen when one is viewing from the top.

To check visibility of lines $A_F C_F$ and $B_F D_F$ in the front view:

- Label the intersection of lines $A_F C_F$ and $B_F D_F$ as ③, ④.
- Project the point of intersection to top view, establishing point ③ on line $A_T C_T$ and point ④ on line $B_T D_T$.
- Point ③ is closer to reference line RL_1, which means that line $A_F C_F$ is nearer when the front view is being observed and thus is visible.
- Point ④ is farther away from reference line RL_1, which means that line $B_F D_F$ would not be seen when one is viewing from the front. Figure 7-8-2C shows the completed top and front views of the part.

Visibility of Lines and Surfaces by Observation

In order to fully understand the shape of an object, it is necessary to know which lines and surfaces are visible in each of the views. Determining their visibility can, in most cases, be done

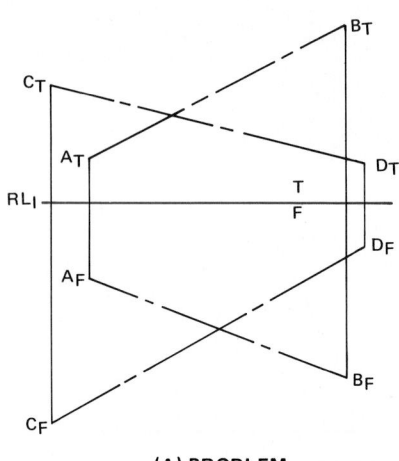

(A) PROBLEM

(B) ESTABLISHING LINES WHICH ARE CLOSER TO OBSERVER

(C) SOLUTION

FIG. 7-8-1   Visibility of oblique lines by testing.

by inspection. With reference to Fig. 7-8-3A, the outline of the part is obviously visible. However, the visibility of lines and surfaces within the outline must be determined. This is accomplished by determining the position of O_F in the front view. Since position O is the closest point to the reference line RL_1 in the front view, it must be the point which is closest to the observer when viewing the top view. Thus it can be seen, and the lines converging to point O_T are visible.

With reference to determining the visibility of the lines in the front view, see the top view. Plane $O_T C_T D_T$ is closest to reference line RL_1. Therefore it must be the closest surface when the observer is looking at the front view, and it must be visible. Since point B_T in the top view is farthest away from

reference line RL_1, it is the point which is farthest away when the observer is looking at the front view. Since it lies behind surface $O_F C_F D_F$, it cannot be seen.

From this example it may be stated that lines or points closest to the observer will be visible, and lines and points farthest away from the viewer but lying within the outline of the view will be hidden.

ASSIGNMENTS

See Assignments 11 and 12 for Unit 7-8 on pages 166–167.

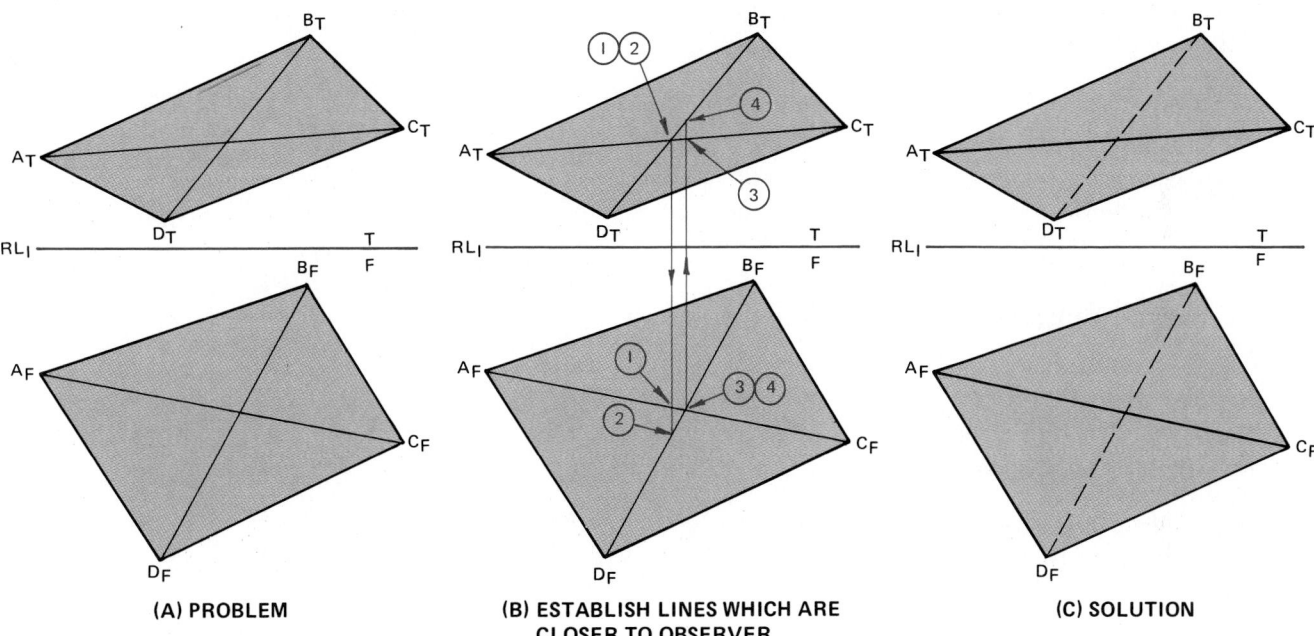

(A) PROBLEM

(B) ESTABLISH LINES WHICH ARE CLOSER TO OBSERVER

(C) SOLUTION

FIG. 7-8-2   Establishing visibility of lines and surfaces by testing.

FIG. 7-8-3   Visibility of lines and surface by observation.

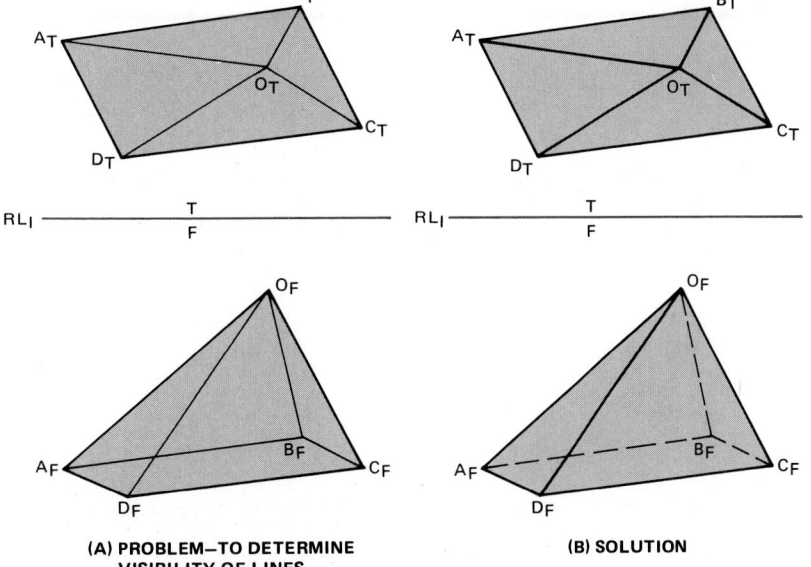

(A) PROBLEM—TO DETERMINE VISIBILITY OF LINES

(B) SOLUTION

7-9 DISTANCES BETWEEN LINES AND POINTS

Distance from a Point to a Line

When the front and side views are given, as in Fig. 7-9-1, and the shortest distance between line AB and point P is required, the procedure is as follows:

- Draw reference line RL_2 parallel to line A_SB_S in the side view.
- Transfer distances designated as R, S, and U in the front view to the primary auxiliary view. The resulting line A_1B_1 in the auxiliary view is the true length of line AB.
- Next draw reference line RL_3 perpendicular to line A_1B_1.
- Transfer distances designated as V and W in the side view to the secondary auxiliary view, establishing points P_2 and A_2B_2, the latter being the point view of line AB.
- The shortest distance between point P and line AB is shown in the secondary auxiliary view.

Design Application Figure 7-9-2 illustrates the application of the point-on-point view of a line to determine the clearance between a hydraulic cylinder and a clip on the wheel housing.

Shortest Distance Between Two Oblique Lines

When the front and top views are given, as in Fig. 7-9-3, and the shortest distance between the two lines AB and CD is required, the procedure is as follows:

- Draw reference line RL_2 parallel to line A_FB_F in the front view.
- Transfer the distances designated as R, S, U, and V in the top view to the primary auxiliary view. The resulting line A_1B_1 in the auxiliary view is the true length of line AB.

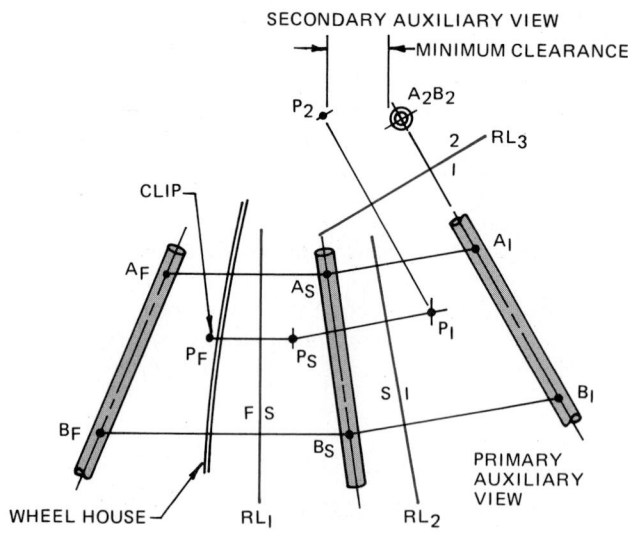

FIG. 7-9-2 Design application of distance from a point to a line.

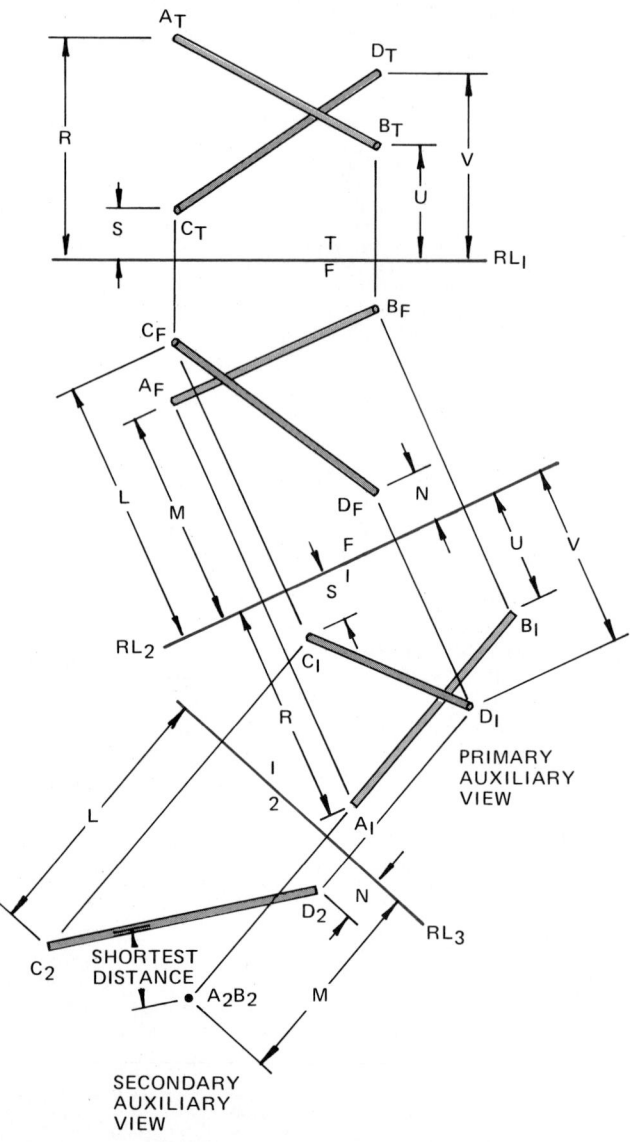

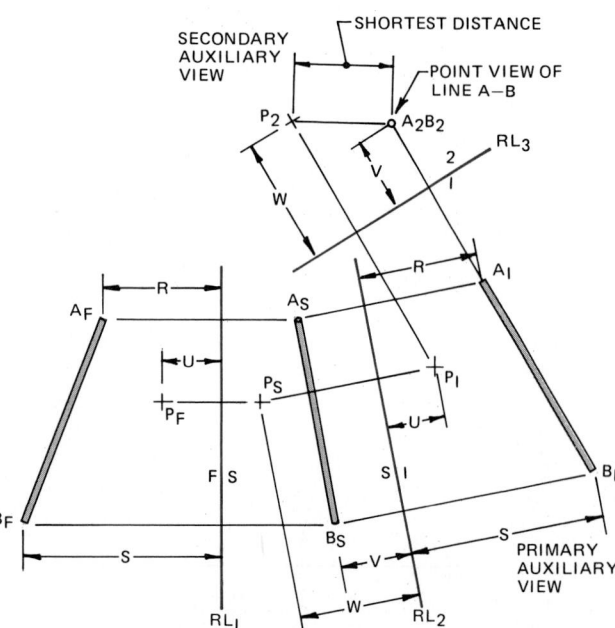

FIG. 7-9-1 Distance from a point to a line.

FIG. 7-9-3 Shortest distance between oblique lines.

- Next draw reference line RL_3 perpendicular to line A_1B_1.
- Transfer the distances designated as L, M, and N in the front view to the secondary auxiliary view. Point A_2B_2 is a point view of line AB.
- The shortest distance between the two lines AB and CD is shown in the secondary auxiliary view.

ASSIGNMENTS

See Assignments 13 and 14 for Unit 7-9 on page 168.

7-10 EDGE AND TRUE VIEW OF PLANES

The three primary planes of projection are horizontal, vertical (or frontal), and profile.

A plane that is not parallel to a primary plane is not visible in its true dimensions. To show a plane in true view, it must be revolved until it is parallel to a projection plane. Figure 7-10-1 shows an oblique plane ABC in the top and front views. The object is to find the true view of this plane. When the top and front views are examined carefully, no line is parallel to the reference line in either view. However, a line C_TD_T can be drawn on the plane parallel to the reference line RL_1 and projected to the front view for its true length D_FC_F. Next:

- Draw reference line RL_2 perpendicular to line D_FC_F in the front view.
- Transfer the distances designated R, S, and U in the top view to the primary auxiliary view. The resulting line A_1B_1 is the edge view of the plane.

- Draw reference line RL_3 parallel to line A_1B_1.
- Transfer the distances L, M, and N in the front view to construct the true shape of plane ABC in the secondary auxiliary view.

Design Application   Figure 7-10-2 shows the application of the procedure followed for Fig. 7-10-1. Points A, B, C, and D

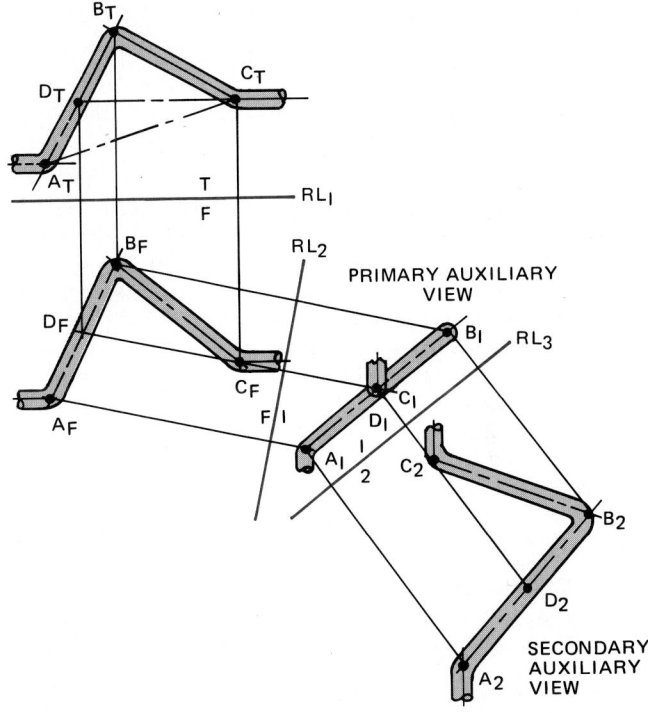

FIG. 7-10-2   Design application of true view of a plane for Fig. 7-10-1.

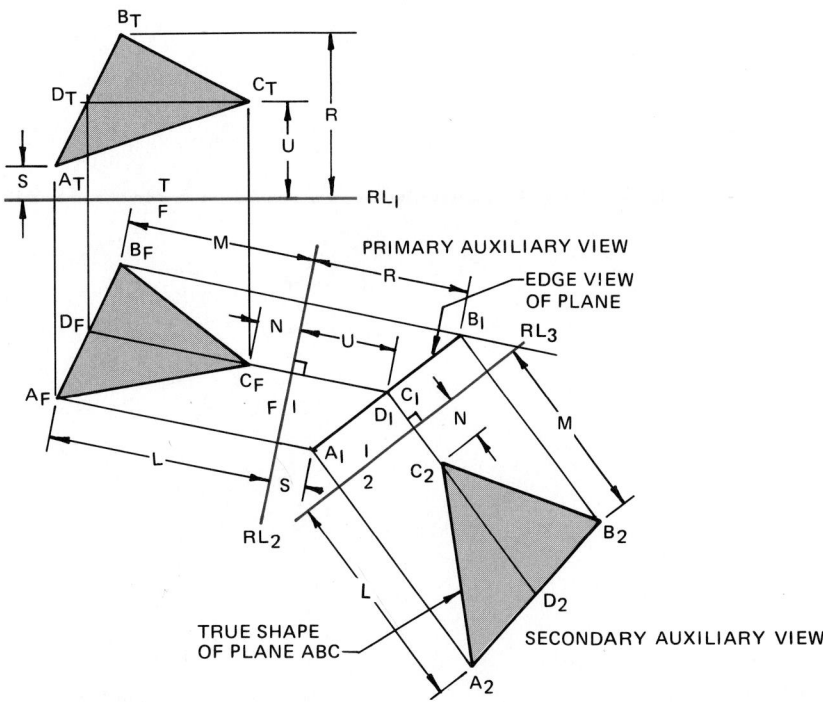

FIG. 7-10-1   True view of a plane.

correspond in both drawings, but line *AC* is omitted in Fig. 7-10-2 since it serves no practical purpose in the design.

Planes in Combination

Figure 7-10-3 demonstrates a solution where a combination of planes is involved. Note that $A_T B_T C_T$ and $A_F B_F C_F$ form one plane while $B_T C_T D_T$ and $B_F C_F D_F$ form another. Also line *BC* is common to both planes. The objective in the problem is to find the true bends at the angles *ABC* and *BCD*. The procedure is as follows:

- Construct the primary auxiliary view, which shows the true length of line *BC*.
- Construct a secondary auxiliary view, which shows *BC* as a point-on-point view. The result is the edge view of both planes *ABC* and *BCD* in the secondary auxiliary view.
- Since any view adjacent to a point-on-point view of a line must show the line in its true length, *BC* will be in true length in secondary auxiliary views 2 and 3. Therefore, projecting perpendicularly from the edge views in the

secondary auxiliary view 1 to the secondary auxiliary views 2 and 3 gives not only *BC* in true length in auxiliary views 2 and 3, but also the true angles *ABC* and *BCD*.

ASSIGNMENTS

See Assignments 15 and 16 for Unit 7-10 on pages 168–169.

7-11 ANGLES BETWEEN LINES AND PLANES

The Angle a Line Makes with a Plane

The top and front views in Fig. 7-11-1 show a line *UV* passing somewhere through plane *ABC*. The true angle between the line and the plane will be shown in the view that shows the

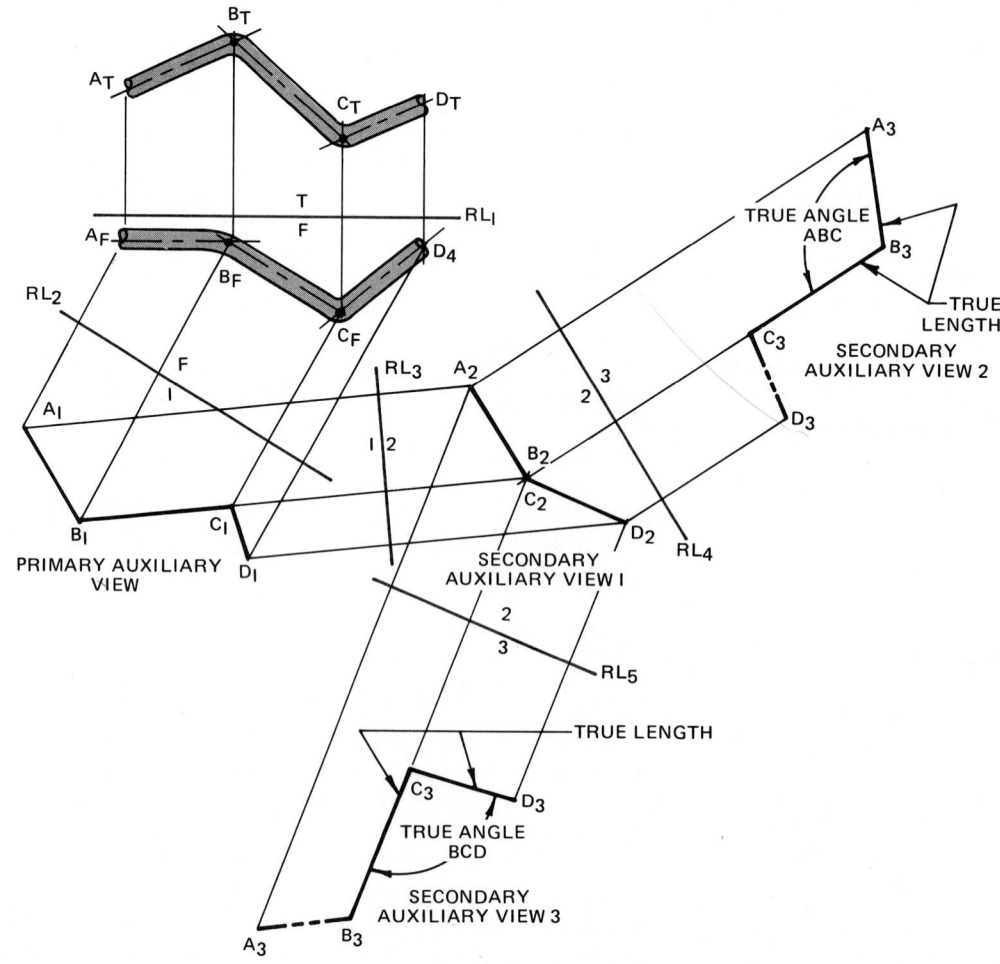

FIG. 7-10-3   Use of planes in combination.

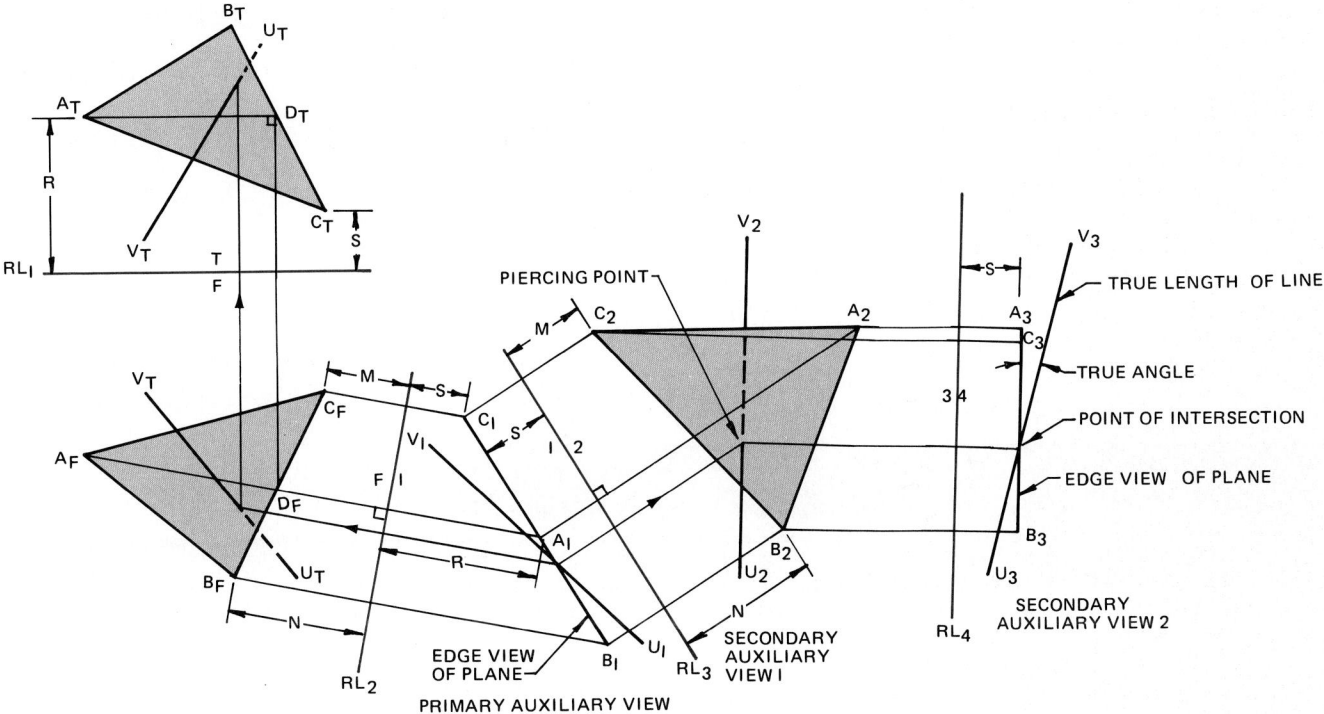

FIG. 7-11-1   The angle a line makes with a plane.

edge view of the plane and the true length of the line. This view is found as follows:

- Draw line $A_T D_T$ in the top view parallel to reference line RL_1.
- Project point D_T to front view, locating point D_F.
- Draw reference line RL_2 perpendicular to a line intersecting points A_F and D_F in the front view.
- Draw the primary auxiliary view, which shows the edge view of the plane but distorted length of line UV. The point of intersection between the line and edge view of the plane is established.
- Draw reference line RL_3 parallel to edge view of the plane shown in the auxiliary view.
- Draw the secondary auxiliary view 1, which shows the true view of the plane and location of the piercing point.
- Draw reference line RL_4 parallel to line $V_2 U_2$.
- Draw the secondary auxiliary view 2, which shows the true length of line UV and the true angle between line and edge view of plane.

Edge Lines of Two Planes

Figure 7-11-2 shows a line of intersection AB made by two planes, triangles ABC and ABD. When the top and front views are given, the point-on-point view of line AB and the true angle between the planes are found as follows:

- Draw reference line RL_2 parallel to line $A_F B_F$ in the front view.
- Transfer the distances designated R, S, and U in the top view to the primary auxiliary view. The resulting line $A_1 B_1$ in the auxiliary view is the true length of line AB.

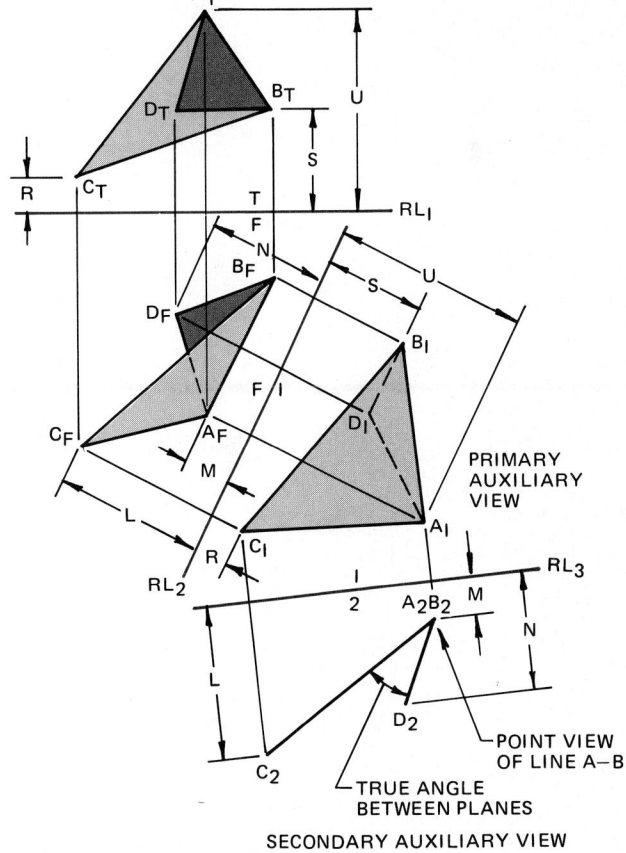

FIG. 7-11-2   Edge lines of two planes.

157

- Next draw reference line RL_3 perpendicular to line A_1B_1.
- Transfer the distances designated as L, M, and N in the front view to the secondary auxiliary view.
- Point A_2B_2 is a point-on-point view of line AB. The true angle between the two planes is seen in the secondary auxiliary view.

ASSIGNMENTS

See Assignments 17 through 19 for Unit 7-11 on pages 169 to 170.

ASSIGNMENTS FOR CHAPTER 7

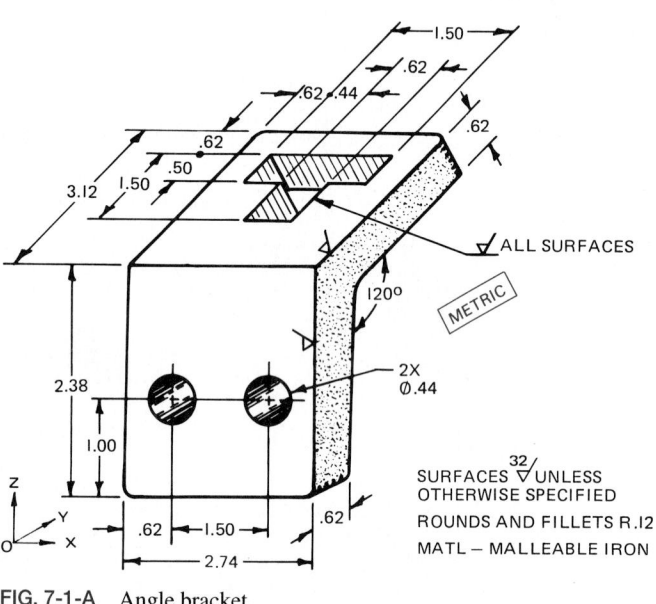

FIG. 7-1-A   Angle bracket.

ASSIGNMENT FOR UNIT 7-1, PRIMARY AUXILIARY VIEWS

1. Make a working drawing of one of the parts shown in Figs. 7-1-A through 7-1-E. For Fig. 7-1-A draw the front, side, and auxiliary view. For all others draw the top, front, and auxiliary view. Partial views are to be used unless otherwise directed by your instructor. Hidden lines may be added for clarity. Scale 1:1.

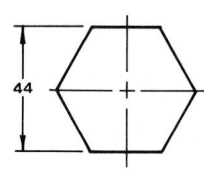
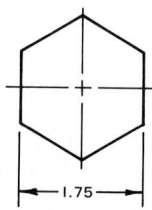
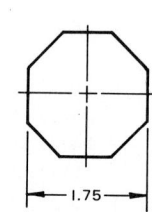

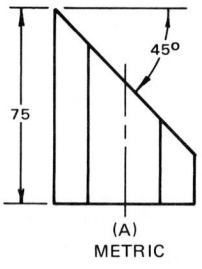
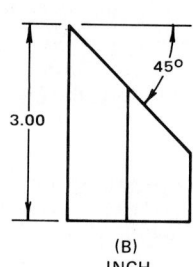
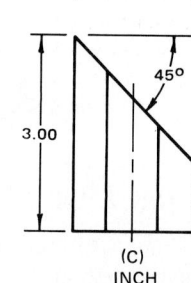

(A) METRIC	(B) INCH	(C) INCH

FIG. 7-1-C   Truncated prisms.

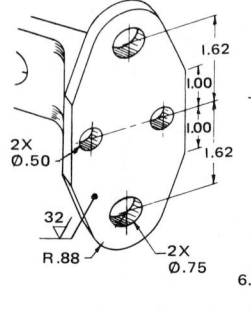
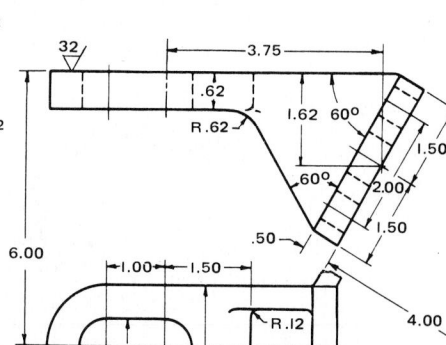

FIG. 7-1-B   Angle plate.

FIG. 7-1-D   Cross-slide bracket.

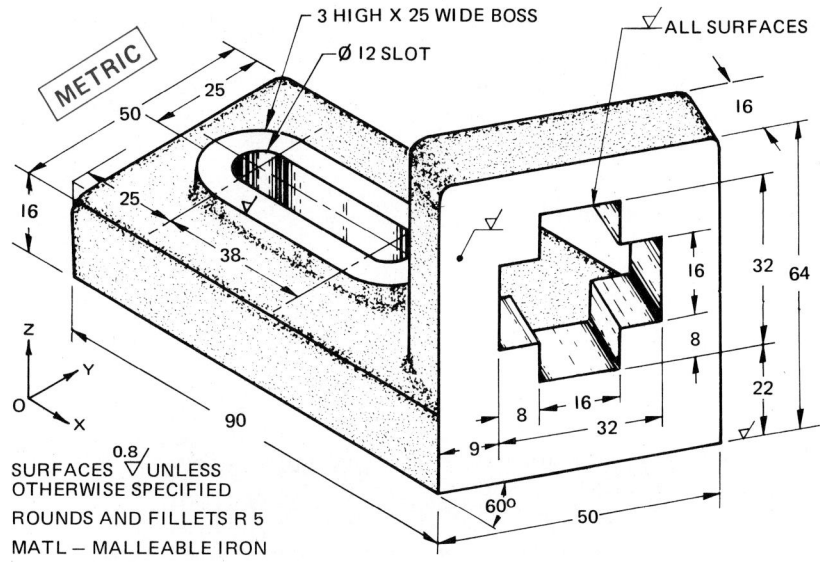

3 HIGH X 25 WIDE BOSS
Ø 12 SLOT
ALL SURFACES

METRIC

SURFACES ▽ UNLESS OTHERWISE SPECIFIED
ROUNDS AND FILLETS R 5
MATL — MALLEABLE IRON

FIG. 7-1-E   Statue bases.

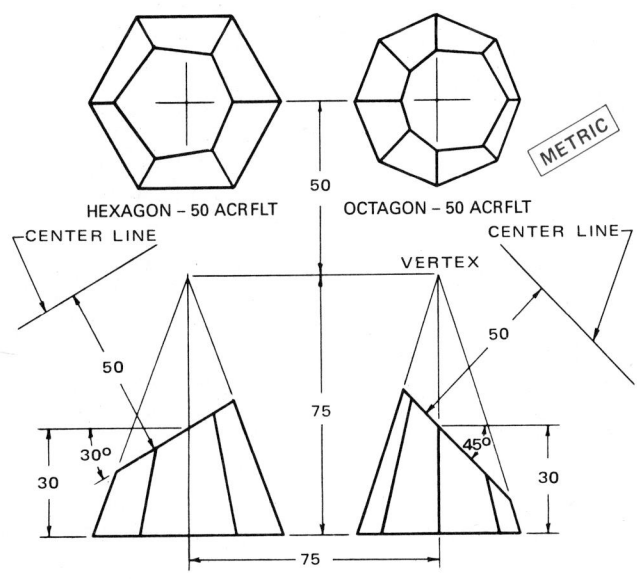

HEXAGON – 50 ACRFLT   OCTAGON – 50 ACRFLT

METRIC

CENTER LINE   CENTER LINE

VERTEX

ASSIGNMENT FOR UNIT 7-2, CIRCULAR FEATURES IN AUXILIARY PROJECTION

2. Make a working drawing of one of the parts shown in Figs. 7-2-A through 7-2-D (here and on pg. 160). Refer to the drawing for set-up of views. Draw complete top and front views and a partial auxiliary view. Hidden lines may be added for clarity.

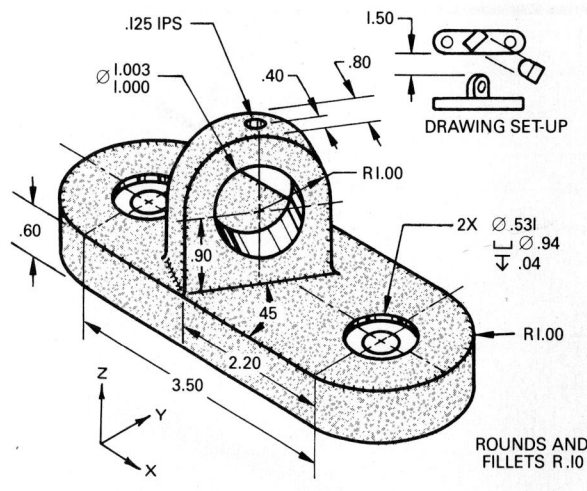

.125 IPS
1.50
Ø 1.003 / 1.000
.40
.80
DRAWING SET-UP
R1.00
2X Ø.531
⌴ Ø.94
↧ .04
R1.00
ROUNDS AND FILLETS R.10

FIG. 7-2-A   Shaft support.

FIG. 7-2-B   Link.

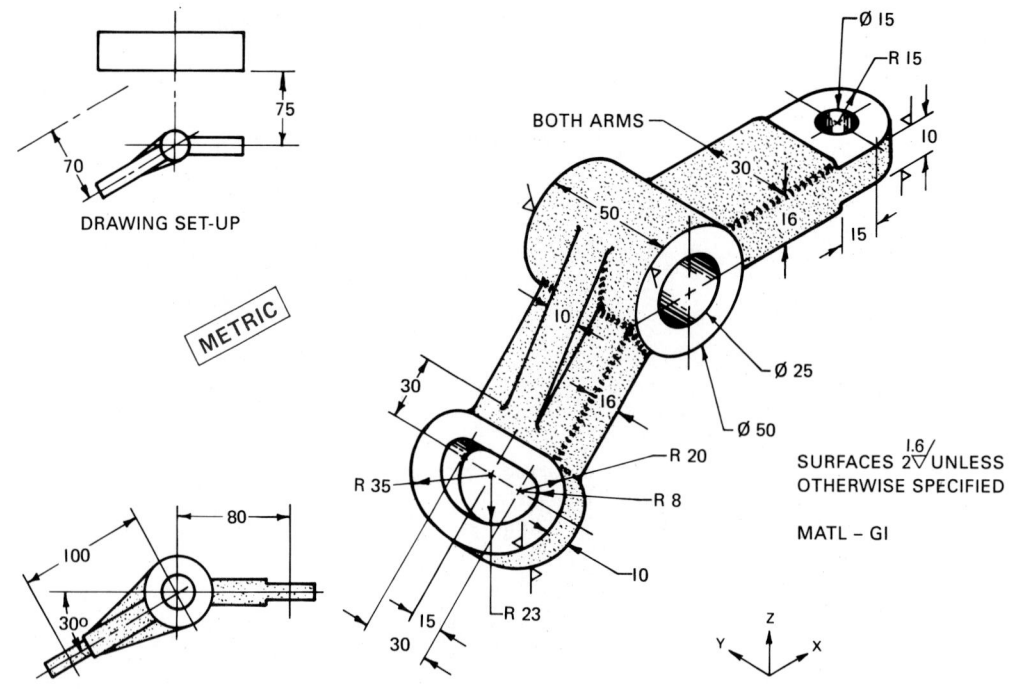

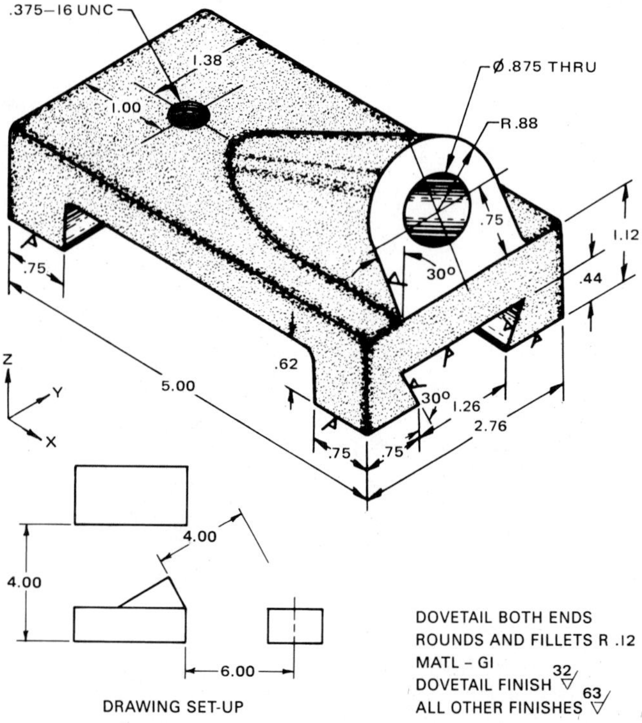

DRAWING SET-UP

METRIC

BOTH ARMS

Ø 15
R 15
10
30
16
15
50
10
16
30
R 20
R 8
R 35
Ø 25
Ø 50

SURFACES 2∇ UNLESS OTHERWISE SPECIFIED
1.6

MATL – GI

100
80
300
30°

R 23
15
30
10
75
70

Z
Y   X

.375–16 UNC

1.38
1.00
Ø .875 THRU
R .88

.75
.75
30°
1.12
.44
.62
.75   .75
1.26
2.76

Z
Y
X

5.00

4.00
4.00
6.00

DRAWING SET-UP

FIG. 7-2-C   Control block.

DOVETAIL BOTH ENDS
ROUNDS AND FILLETS R .12
MATL – GI
DOVETAIL FINISH ∇ 32
ALL OTHER FINISHES ∇ 63

10.50
3.00

DRAWING SET-UP

.70
2.20
.30
30°
.60 SQUARE HOLE,
Ø I.10 C BORE X I.20 DEEP

HEX
3.00 ACRFLT

1.50

Ø 2.10
Ø1.00

4 RIBS

4.90

15°

4X Ø .406
EQL SP ABOUT
RIBS ON Ø 2.40

.20   1.40

.40

Ø 4.00

ROUNDS & FILLETS R .10

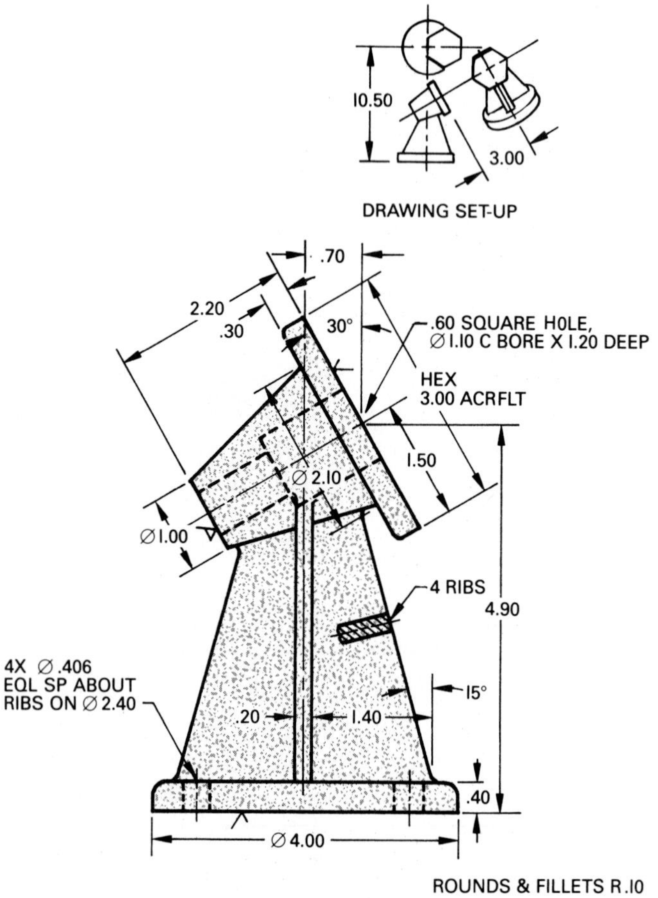

FIG. 7-2-D   Pedestal.

ASSIGNMENTS FOR UNIT 7-3, MULTI-AUXILIARY-VIEW DRAWINGS

3. Make a working drawing of one of the parts shown in Figs. 7-3-A through 7-3-C. Refer to the drawings for set-up of views. Draw complete top and front views and partial auxiliary views.

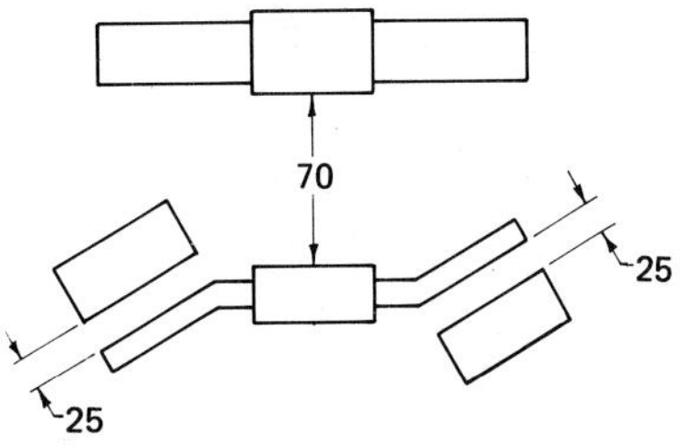

70
25
25

DRAWING SET-UP

MATL-.12 ALUMINUM

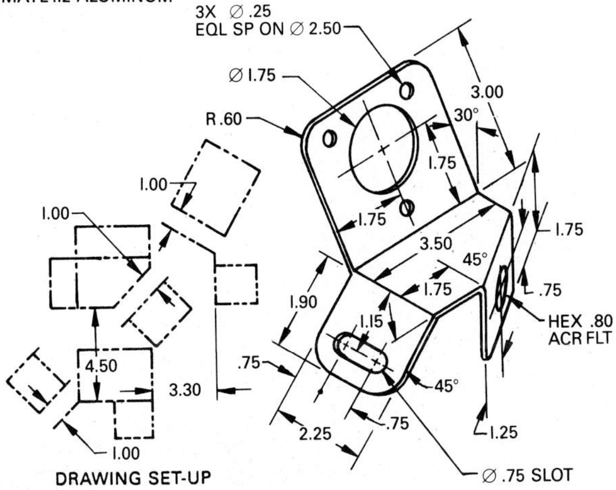

3X ∅ .25
EQL SP ON ∅ 2.50
∅ 1.75
R .60
3.00
30°
1.75
1.75
1.00
1.00
3.50
45°
1.75
1.75
.75
HEX .80
ACR FLT
1.90
1.15
4.50
.75
3.30
45°
2.25
.75
1.25
1.00
∅ .75 SLOT

DRAWING SET-UP

FIG. 7-3-A Mounting plate.

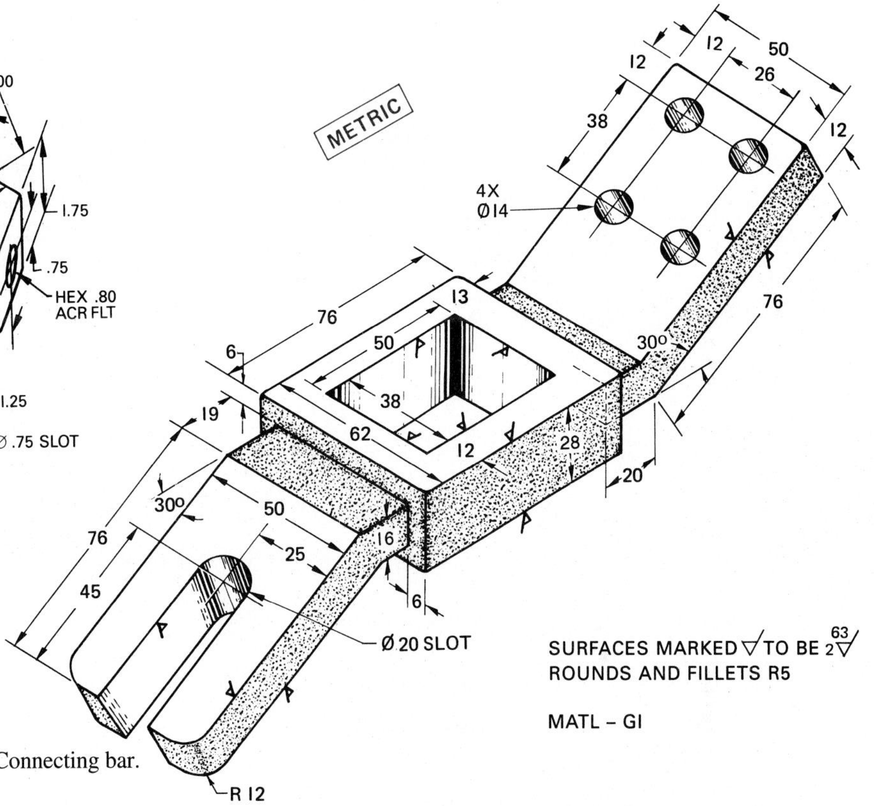

METRIC
12
12
50
26
12
38
4X
Ø14
13
76
76
6
50
38
30°
19
62
12
28
20
30°
50
76
25
16
45
6
Ø 20 SLOT
R 12

SURFACES MARKED ∀ TO BE $2\overset{63}{\nabla}$
ROUNDS AND FILLETS R5

MATL – GI

FIG. 7-3-B Connecting bar.

MATERIAL-GRAY IRON
SURFACES MARKED ∀ TO BE .06 $\overset{1.6}{\forall}$
ROUNDS AND FILLETS R.12

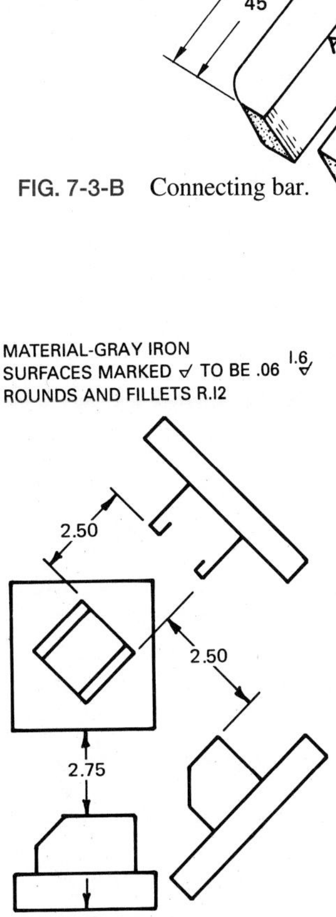

2.50
2.50
2.75

DRAWING SET-UP

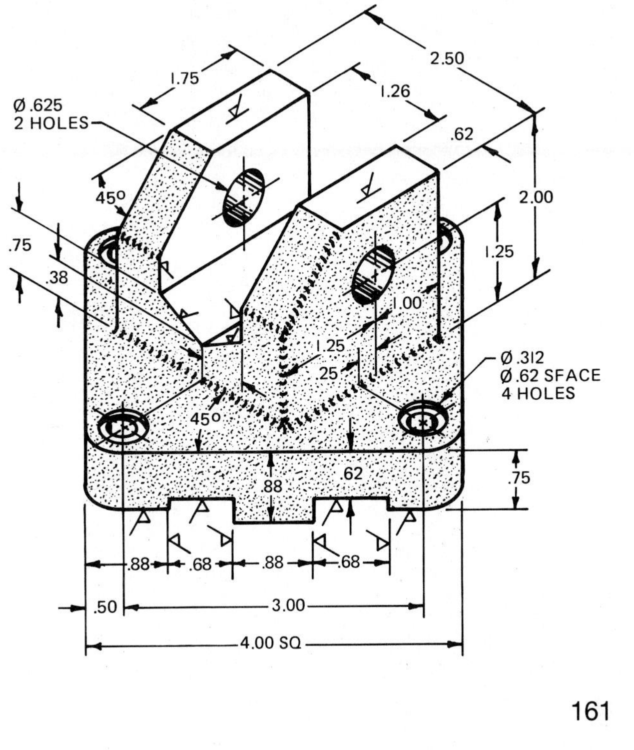

2.50
1.75
1.26
Ø .625
2 HOLES
.62
45°
2.00
.75
1.25
.38
1.00
1.25
.25
Ø .312
Ø .62 SFACE
4 HOLES
45°
.62
.88
.75
.88
.68
.88
.68
.50
3.00
4.00 SQ

FIG. 7-3-C Angle slide.

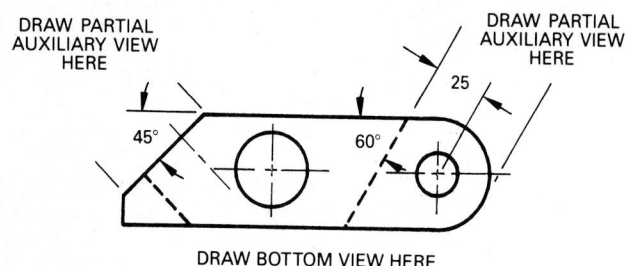

DRAW PARTIAL
AUXILIARY VIEW
HERE

DRAW PARTIAL
AUXILIARY VIEW
HERE

45° 60° 25

DRAW BOTTOM VIEW HERE

4. Make a working drawing of one of the parts shown in Figs. 7-3-D through 7-3-F. Draw partial auxiliary views for Fig. 7-3-D; partial auxiliary and side views for Fig. 7-3-E; and full top and front views, and partial auxiliary views for Fig. 7-3-F.

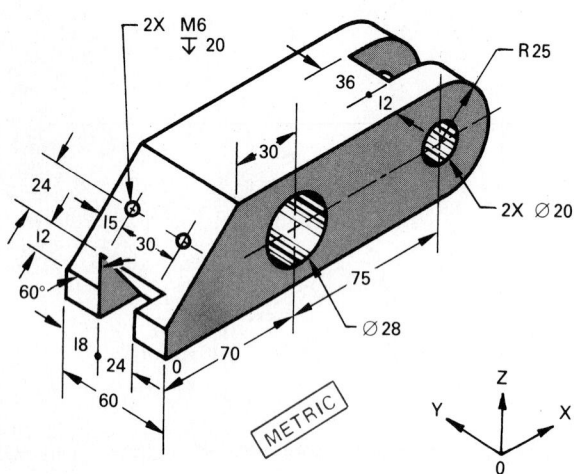

FIG. 7-3-D Dovetail bracket.

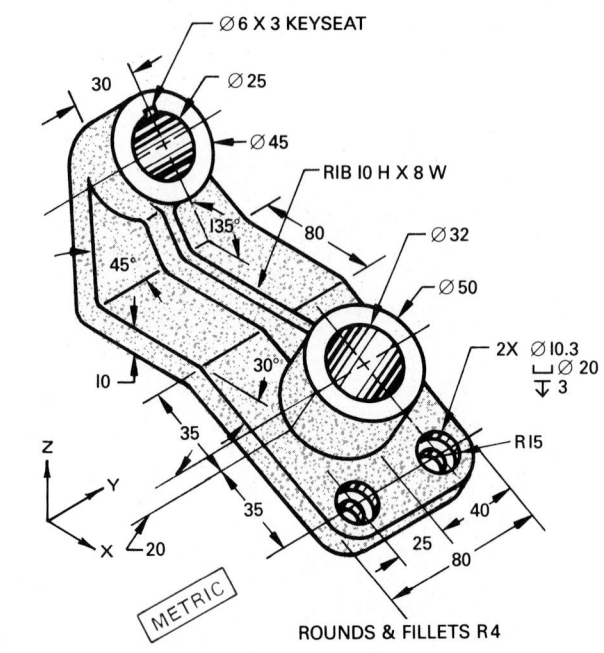

ROUNDS & FILLETS R 4

FIG. 7-3-F Offset guide.

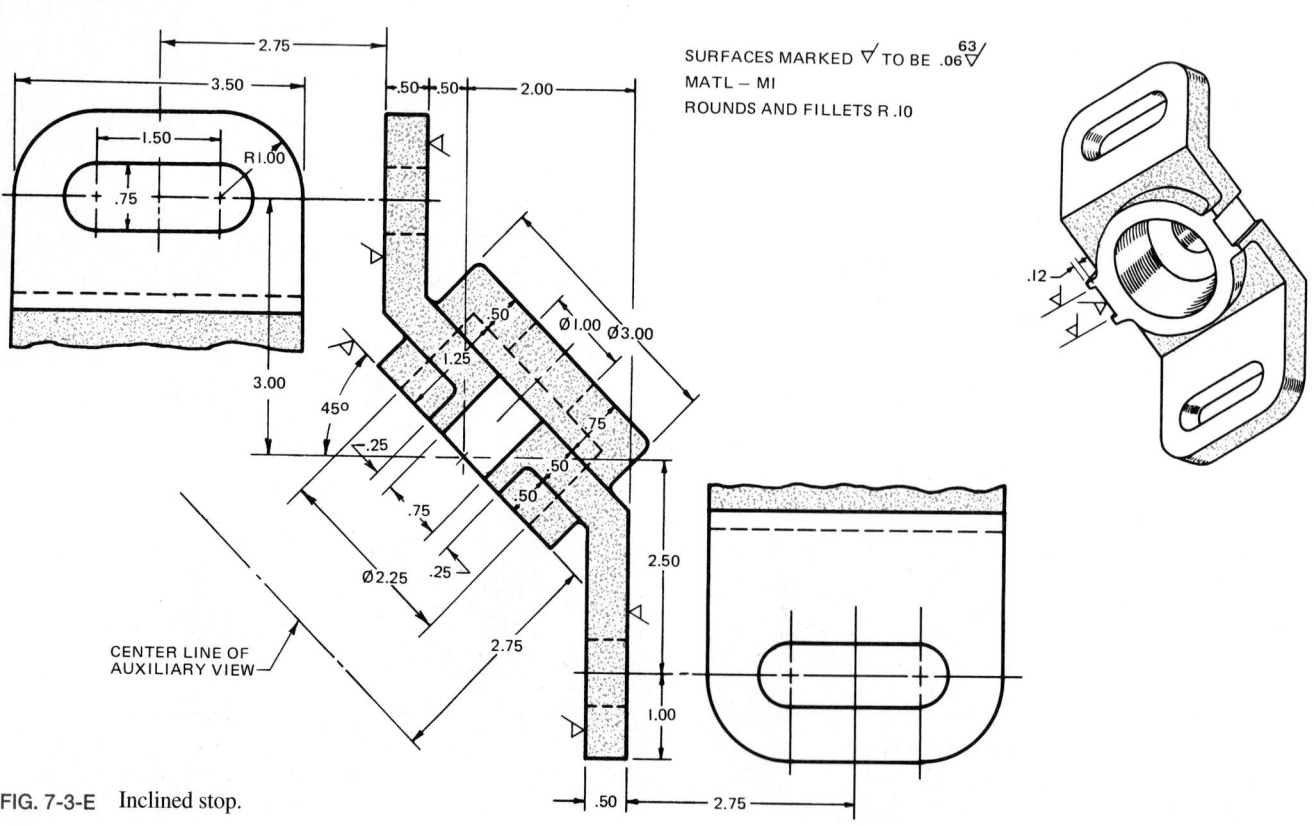

SURFACES MARKED ∇ TO BE .06 ⁶³∇
MATL — MI
ROUNDS AND FILLETS R .10

FIG. 7-3-E Inclined stop.

ASSIGNMENT FOR UNIT 7-4, SECONDARY AUXILIARY VIEWS

5. Make a working drawing of one of the parts shown in Figs. 7-4-A through 7-4-C. The selection and placement of views are shown with the drawing. Only partial auxiliary views need be drawn, and hidden lines may be added to improve clarity.

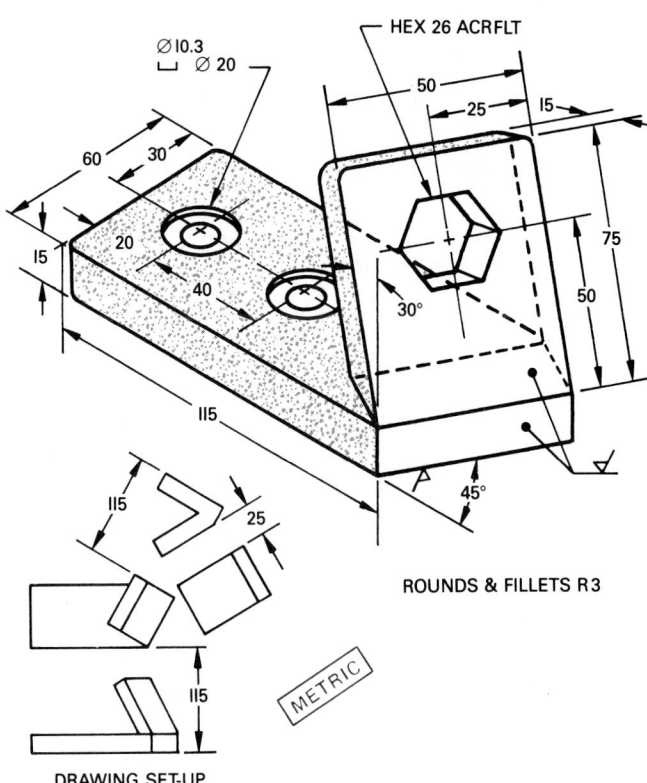

ROUNDS & FILLETS R3

DRAWING SET-UP

FIG. 7-4-A   Hexagon slot support.

FIG. 7-4-B   Dovetail bracket.

LOCATION OF Ø 20 HOLE
ON AUXILIARY VIEW

DRAWING SET-UP

FIG. 7-4-C   Pivot arm.

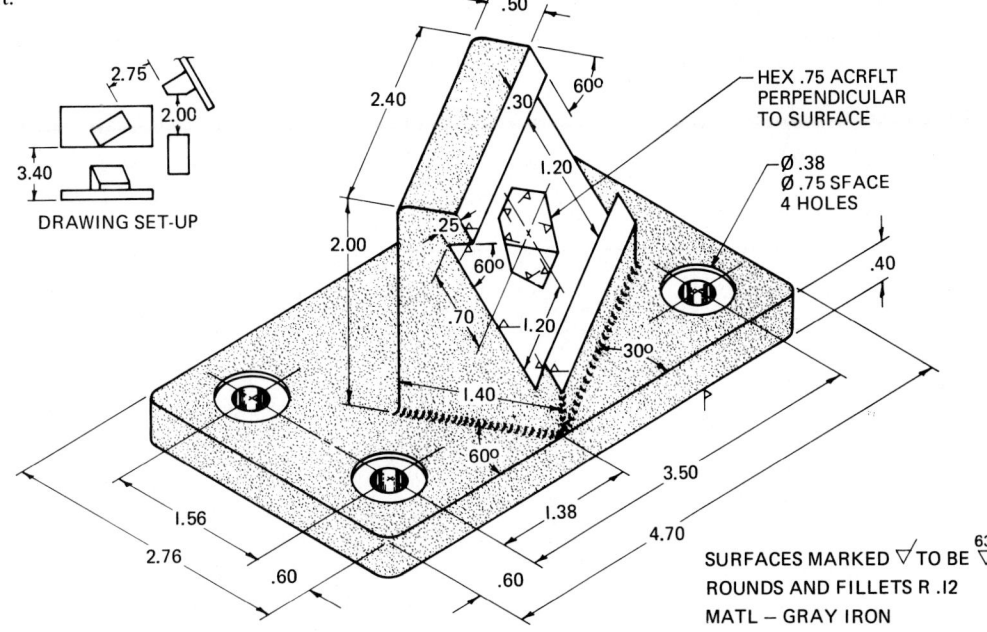

HEX .75 ACRFLT
PERPENDICULAR
TO SURFACE

Ø .38
Ø .75 SFACE
4 HOLES

SURFACES MARKED ▽ TO BE ▽ 63
ROUNDS AND FILLETS R .12
MATL – GRAY IRON

163

ASSIGNMENT FOR UNIT 7-5, REVOLUTIONS

6. Select one of the parts shown in Fig. 7-5-A and draw the views as positioned in the example given.

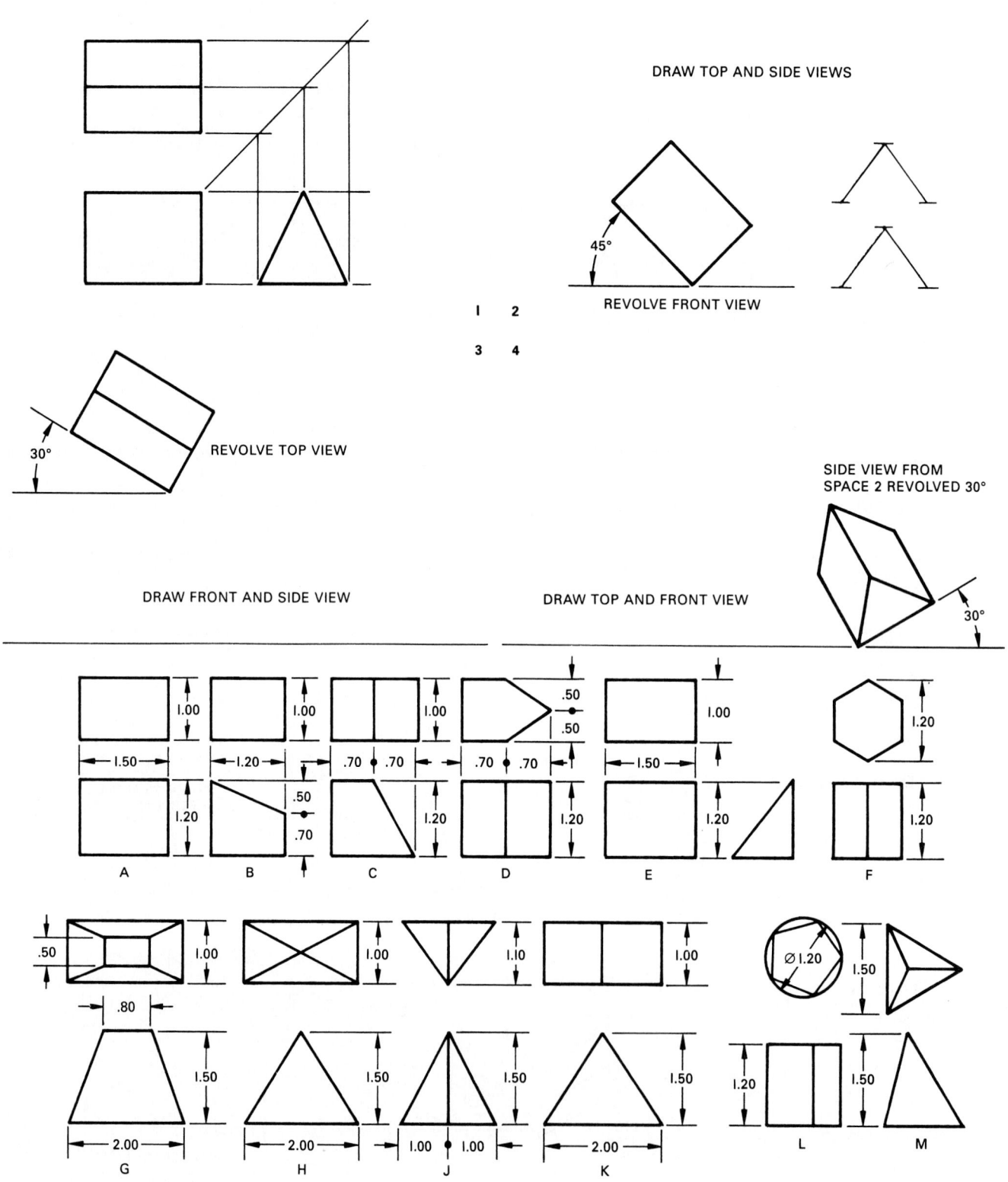

FIG. 7-5-A Revolution assignments.

ASSIGNMENTS FOR UNIT 7-6, LOCATING POINTS AND LINES IN SPACE

7. With the use of a grid and reference lines, locate the points and/or lines in the third view for the drawings shown in Fig. 7-6-A. Scale to suit.

8. With the use of a grid and reference lines, locate the lines in the other views for the drawings shown in Fig. 7-6-B. Scale to suit.

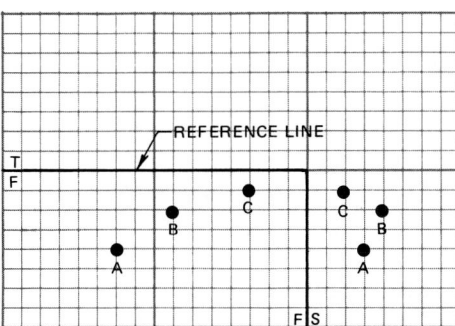

(1) LOCATE POINTS IN THE TOP VIEW

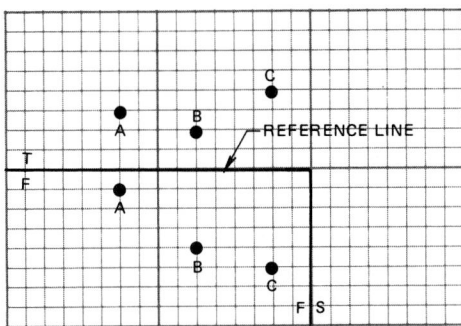

(2) LOCATE POINTS IN THE SIDE VIEW

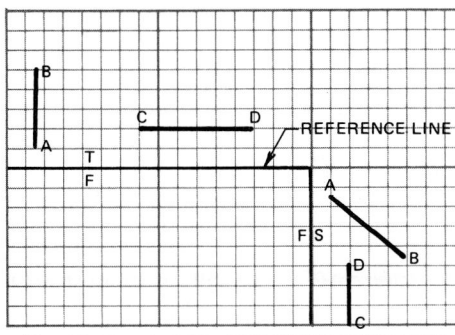

(3) LOCATE INCLINED LINES IN THE OTHER VIEW

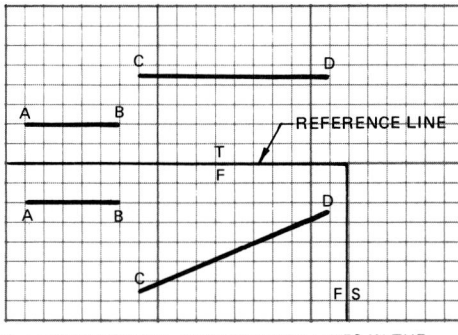

(4) LOCATE NORMAL AND INCLINED LINES IN THE OTHER VIEW

FIG. 7-6-A Assignments.

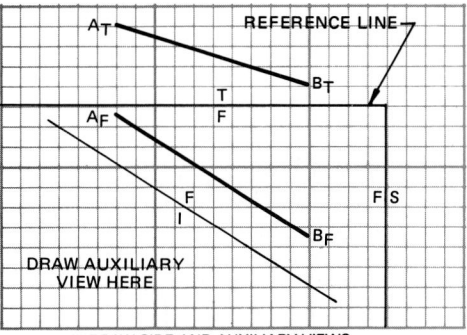

(1) DRAW SIDE AND AUXILIARY VIEWS
(INDICATE TRUE LENGTH OF LINE A–B)

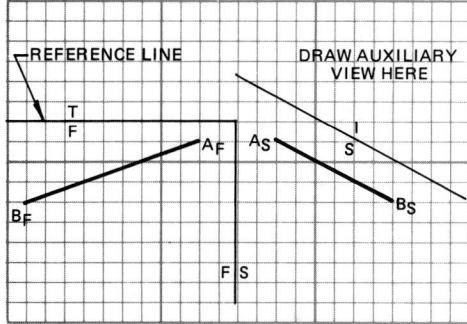

(2) DRAW TOP AND AUXILIARY VIEWS
(INDICATE TRUE LENGTH OF LINE A–B)

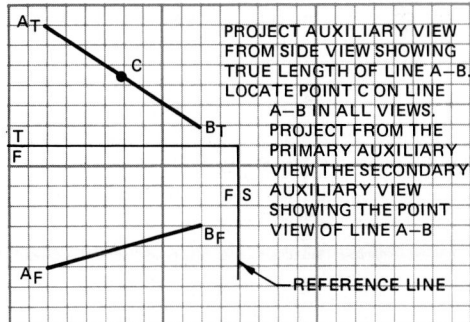

(3) DRAW SIDE AND AUXILIARY VIEWS

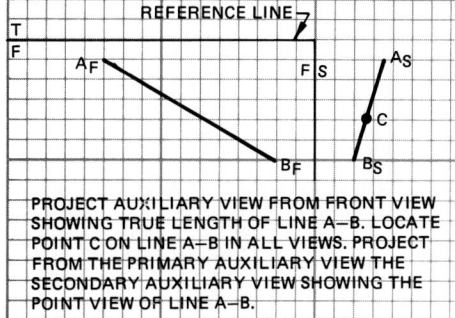

(4) DRAW TOP AND AUXILIARY VIEWS

FIG. 7-6-B Assignments.

ASSIGNMENTS FOR UNIT 7-7, PLANES IN SPACE

9. *Locating a Plane or a Line in Space.* With the use of a grid and reference lines, complete the three drawings shown in Fig. 7-7-A. Scale to suit.
10. *Locating a Point in Space and the Piercing Point of a Line and a Plane.* With the use of a grid and reference lines, complete the drawings shown in Fig. 7-7-B. Scale to suit.

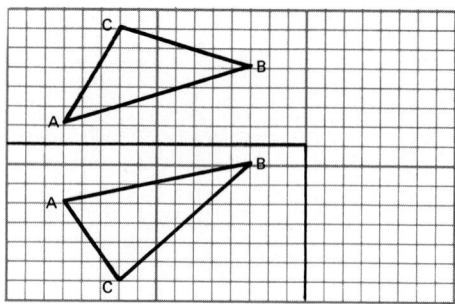

(1) DRAW THE SIDE VIEW OF TRIANGLE ABC

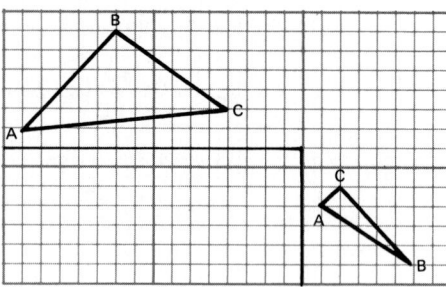

(2) DRAW THE FRONT VIEW OF TRIANGLE ABC

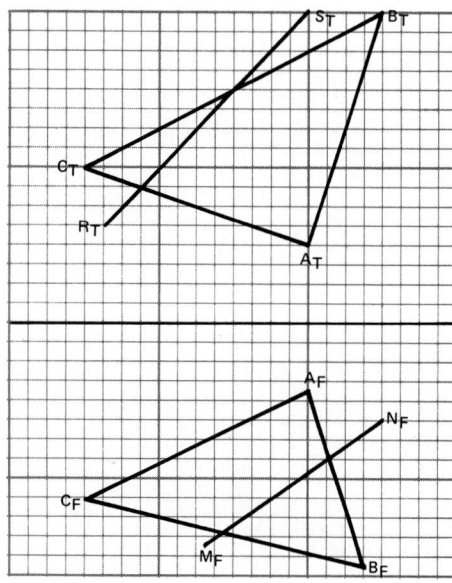

(3) LOCATING A LINE IN THE OTHER VIEW

FIG. 7-7-A   Assignments.

ASSIGNMENTS FOR UNIT 7-8, ESTABLISHING VISIBILITY OF LINES IN SPACE

11. *Visibility of Lines and Surfaces by Observation and Testing.* With the use of a grid and reference lines, lay out the drawings shown in Fig. 7-8-A. By observation, sketch the circular pipes (drawings 1 and 2) in a manner similar to that shown in Fig. 7-8-1, showing the direction in which the pipes are sloping and which pipe is closer to the

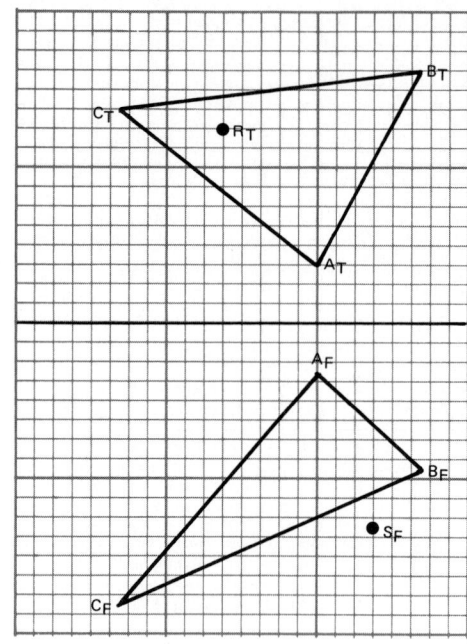

(1) LOCATE THE POINTS IN SPACE

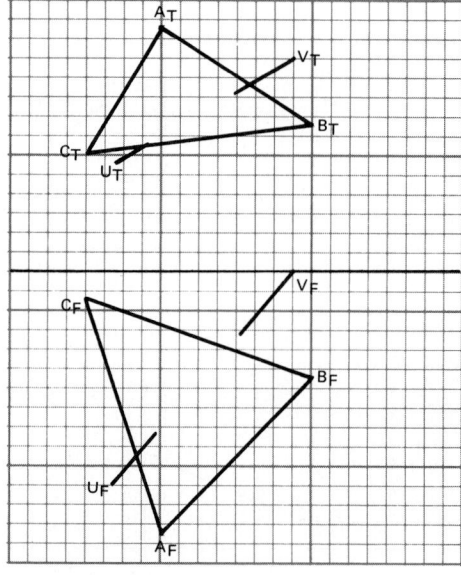

(2) LOCATE THE PIERCING POINT OF THE LINE AND PLANE AND COMPLETE LINE UV.

FIG. 7-7-B   Assignments.

observer in the two views. By testing in a manner similar to that shown in Fig. 7-8-2, establish the visibility of lines and surfaces of drawings 3 and 4. Scale to suit.

12. *Establishing Visibility of Lines and Surfaces by Testing.* With the use of a grid and reference lines, draw the four

parts shown in Fig. 7-8-B. By testing in the manner shown in Fig. 7-8-2, complete the two-view drawings, showing the visible and hidden lines. Scale to suit.

FIG. 7-8-A   Assignments.

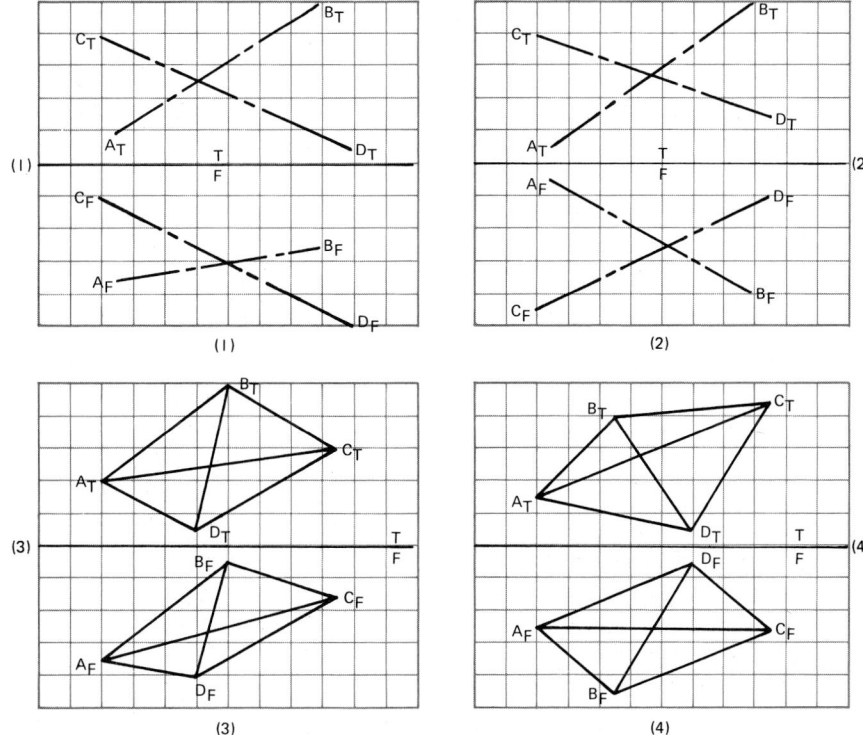

FIG. 7-8-B   Assignments.

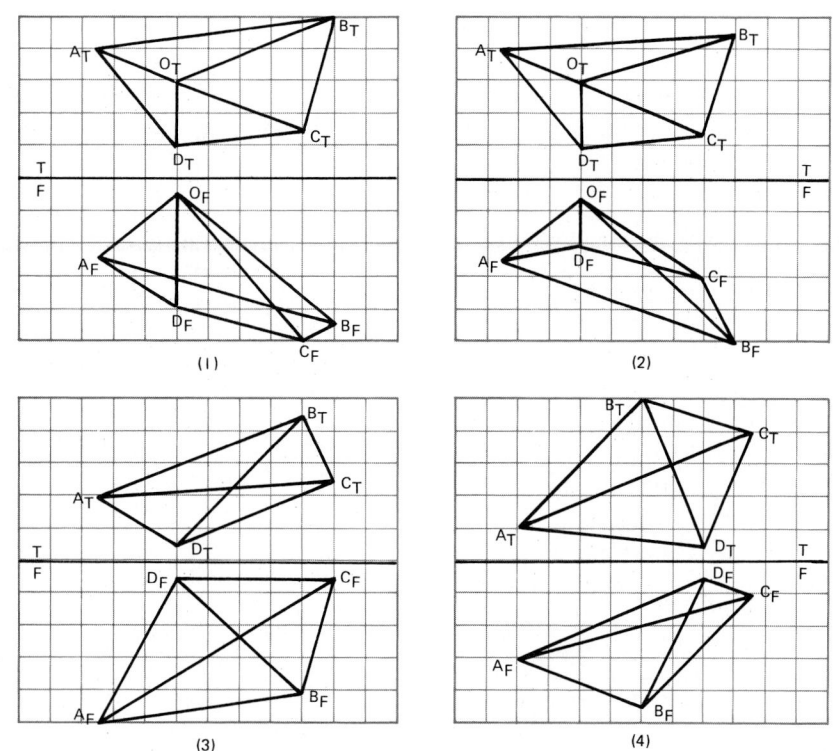

ASSIGNMENTS FOR UNIT 7-9, DISTANCES BETWEEN LINES AND POINTS

13. *Finding Distance from a Point to a Line.* With the use of a grid, reference lines and auxiliary views, find the distance from a point to a line in the two problems shown in Fig. 7-9-A. Scale to suit.

14. *Finding the Shortest Distance Between Oblique Lines.* With the use of a grid, reference lines, and auxiliary views, find the shortest distance between the oblique lines for the two problems shown in Fig. 7-9-B. Scale to suit.

ASSIGNMENTS FOR UNIT 7-10, EDGE AND TRUE VIEW OF PLANES

15. *Finding True Angles of Intersecting Planes.* With the use of grid, reference lines, and auxiliary views, find the true angles of the intersecting planes for the two problems shown in Fig. 7-10-A. Scale to suit.

16. *True View of a Plane.* With the use of a grid, reference lines, and auxiliary views, find the true view of the plane for the two problems shown in Fig. 7-10-B. Scale to suit.

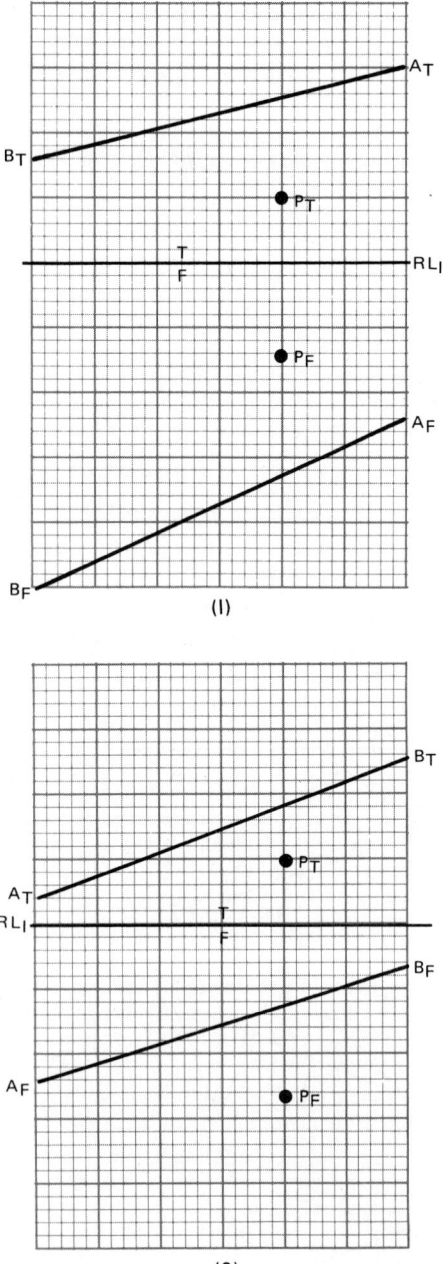

(1)

(2)

FIG. 7-9-A   Assignments.

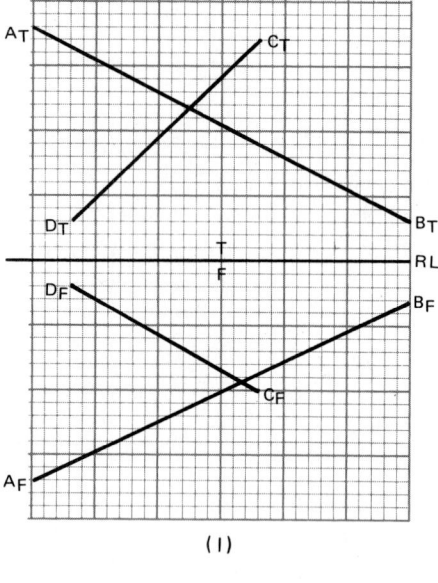

(1)

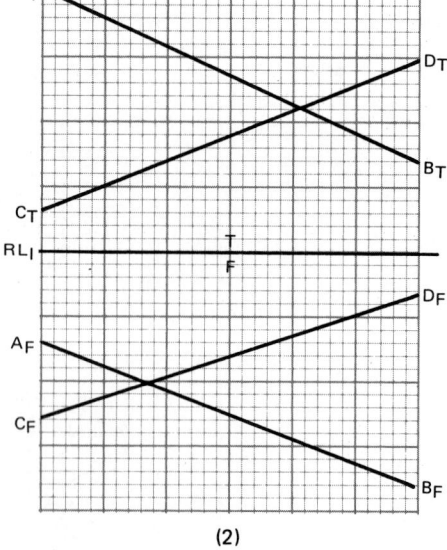

(2)

FIG. 7-9-B   Assignments.

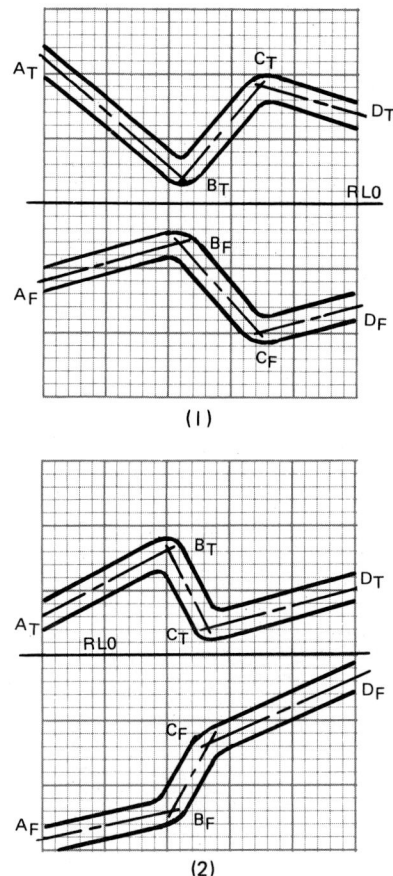

FIG. 7-10-A   Assignments.

ASSIGNMENTS FOR UNIT 7-11, ANGLES BETWEEN LINES AND PLANES

17. *The Angle a Line Makes with a Plane.* With the use of a grid, reference lines, and auxiliary views, find the angle the line makes with a plane for the layout shown in Fig. 7-11-A. Scale to suit.

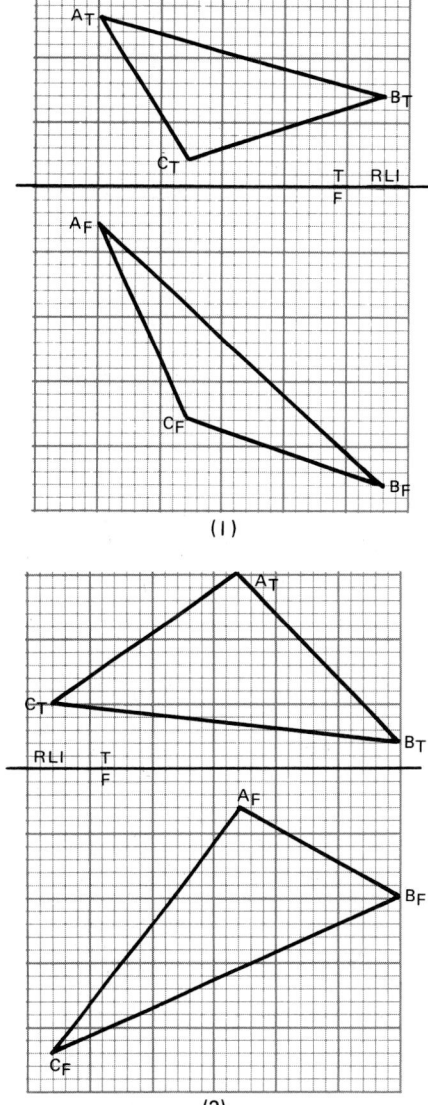

FIG. 7-10-B   Assignments.

FIG. 7-11-A   Assignments.

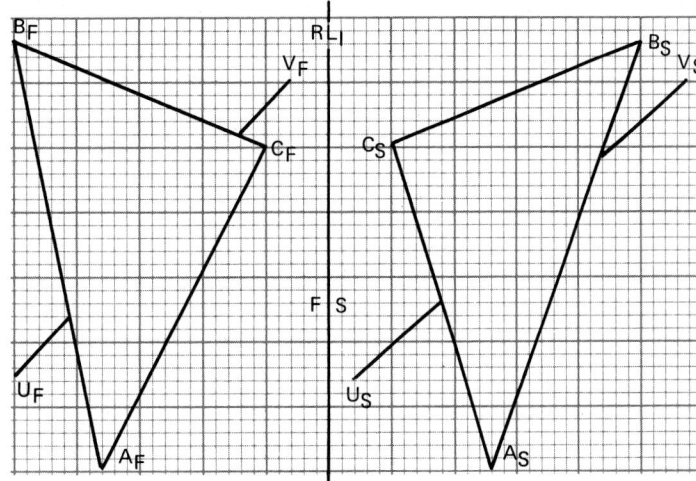

18. *The Angle a Line Makes with a Plane.* With the use of a grid, reference lines, and auxiliary views, find the angle the line makes with a plane for the layout shown in Fig. 7-11-B. Scale to suit.

19. *Edge Line of Two Planes.* With the use of a grid, reference lines, and auxiliary views, find the edge line of the two planes shown in the two problems in Fig. 7-11-C. Scale to suit.

FIG. 7-11-B   Assignments.

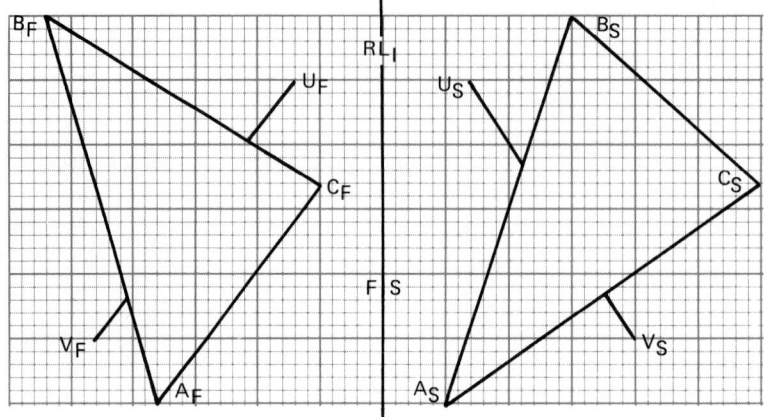

FIG. 7-11-C   Assignments.

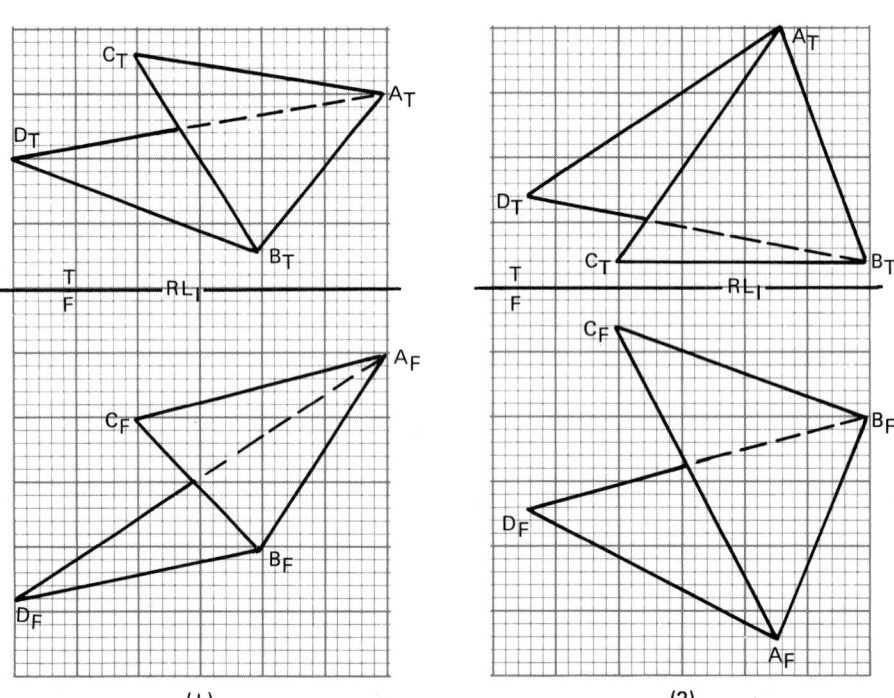

(1)   (2)

CHAPTER 8

BASIC DIMENSIONING

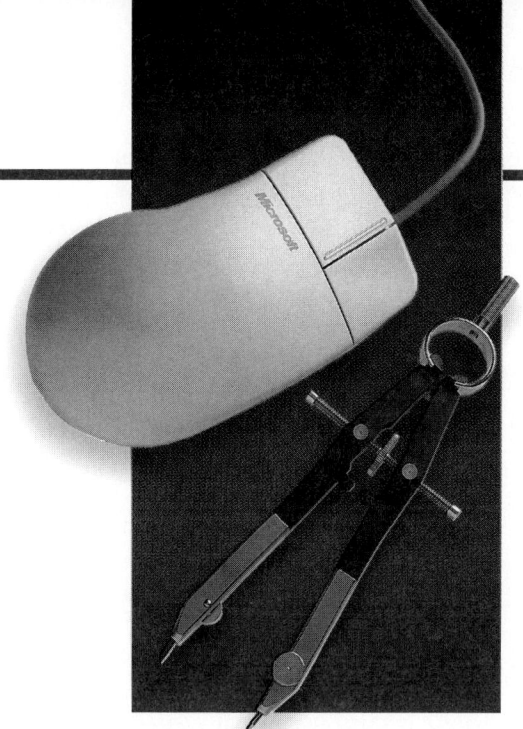

Definitions

Chamfer   To cut away the inside or outside piece to facilitate assembly.

Counterbore   A flat-bottomed, cylindrical recess that permits the head of a fastening device to lie recessed into the part.

Countersink   An angular-sided recess to accommodate the head of flathead screws, rivets, and similar items.

Datum dimensioning   The method in which several dimensions emanate from a common reference point or line.

Dimensions   Lines and numerical values used to define geometric characteristics, such as lengths, diameters, angles, and locations.

Extension (projection) lines   Lines used to indicate the point or line on the drawing to which the dimension applies.

Fits   The clearance or interference between two mating parts.

Leaders   A straight, inclined line used to direct notes, dimensions, symbols, item numbers, or part numbers to features on the drawing.

Mass production   Production in which parts are produced in quantity, requiring special tools and gages.

Notes   Written information used to simplify or complement dimensions; they may be general or local (specific).

Slope   The slant of a line representing an inclined surface.

Spotface   An area where the surface is machined just enough to provide smooth, level seating for a bolt head, nut, or washer.

Symmetrical   The quality in which features on each side of the center line or median plane are identical in size, shape, and location.

Taper   The ratio of the difference in the diameters of two sections.

Tolerances   The permissible variations in the specified form, size, or location of individual features of a part.

Undercutting or **necking**   The process of cutting a recess in a diameter to permit two parts to come together.

Unit production   Production in which each part is made separately using general-purpose tools and machines.

8-1 BASIC DIMENSIONING

A working drawing is one from which a part can be produced. The drawing must be a complete set of instructions, so that it will not be necessary to give further information to the people fabricating the object. A *working drawing,* then, consists of the views necessary to explain the shape, the dimensions needed for manufacture, and required specifications, such as material and quantity needed. The latter information may be found in the notes on the drawing, or it may be located in the title block.

Dimensioning

Dimensions are given on drawings by extension lines, dimension lines, leaders, arrowheads, figures, notes, and symbols. They define geometric characteristics, such as lengths, diameters, angles, and locations (Fig. 8-1-1, pg. 172). The lines used in dimensioning are thin in contrast to the outline of the object. The dimensions must be clear and concise and permit only one interpretation. In general, each surface, line, or point is located by only one set of dimensions. These dimensions are not duplicated in other views. Deviations from the approved rules for dimensioning should be made only in exceptional cases, when they will improve the clarity of the dimensions. An exception to these rules is for arrowless and tabular dimensioning, which is discussed in Unit 8-4.

Drawings for industry require some form of tolerancing on dimensions so that components can be properly assembled and manufacturing and production requirements can be met. This

FIG. 8-1-1   Basic dimensioning elements.

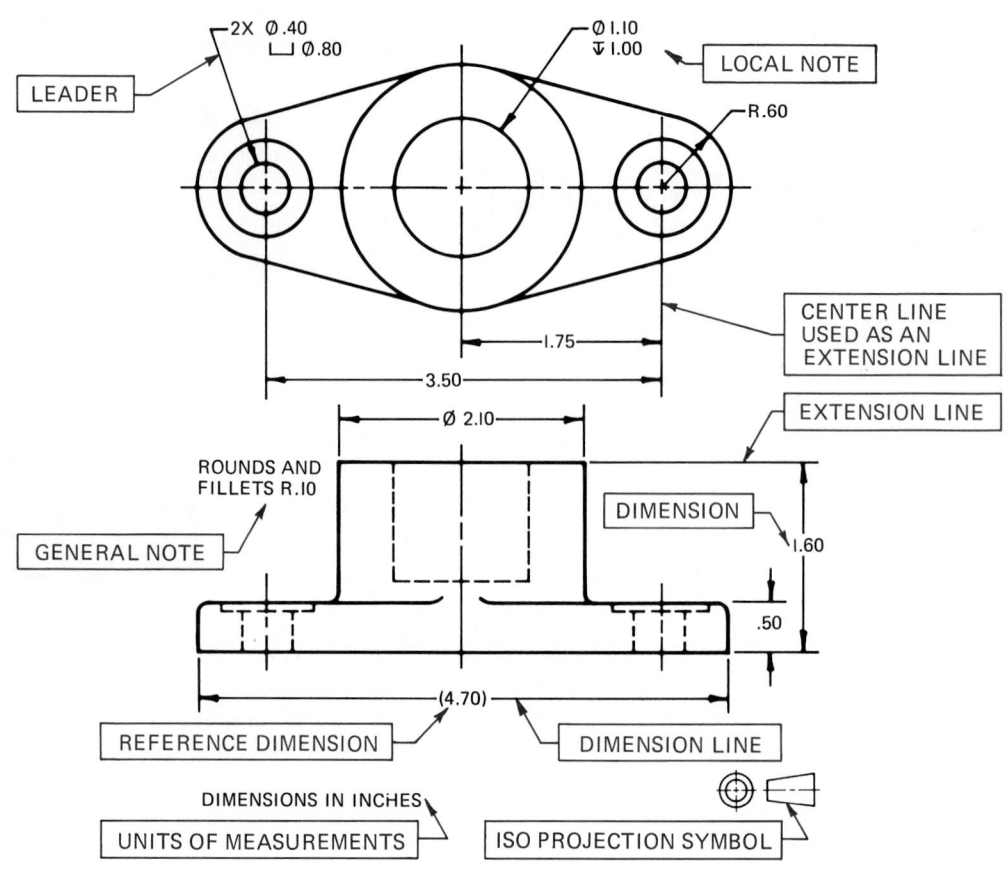

chapter deals only with basic dimensioning and tolerancing techniques. Geometric tolerancing, such as true positioning and tolerance of form, is covered in detail in Chap. 16.

Dimension and Extension Lines

Dimension lines are used to determine the extent and direction of dimensions, and they are normally terminated by uniform arrowheads, as shown in Fig. 8-1-2. Using an oblique line in lieu of an arrowhead is a common method used in architectural drafting. The recommended length and width of arrowheads should be in a ratio of 3:1 (Fig. 8-1-3B). The length of the arrowhead should be equal to the height of the dimension numerals.

A single style of arrowhead should be used throughout the drawing. An industrial practice is to use a small filled-in circle in lieu of two arrowheads where space is limited (Fig. 8-1-3D).

Preferably, dimension lines should be broken for insertion of the dimension that indicates the distance between the extension lines. Where dimension lines are not broken, the dimension is placed above the dimension line.

When several dimension lines are directly above or next to one another, it is good practice to stagger the dimensions in order to improve the clarity of the drawing. The spacing suitable for most drawings between parallel dimension lines is .38 in. (8mm), and the spacing between the outline of the object and the nearest dimension line should be approximately

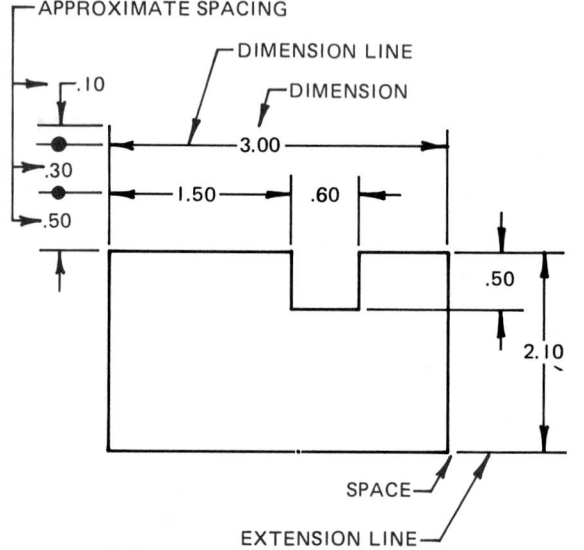

FIG. 8-1-2   Dimension and extension lines.

.50 in. (10mm). When the space between the extension lines is too small to permit the placing of the dimension line complete with arrowheads and dimension, the alternate method of placing the dimension line, dimension, or both outside the extension lines is used (Fig. 8-1-3D). Center lines should never be

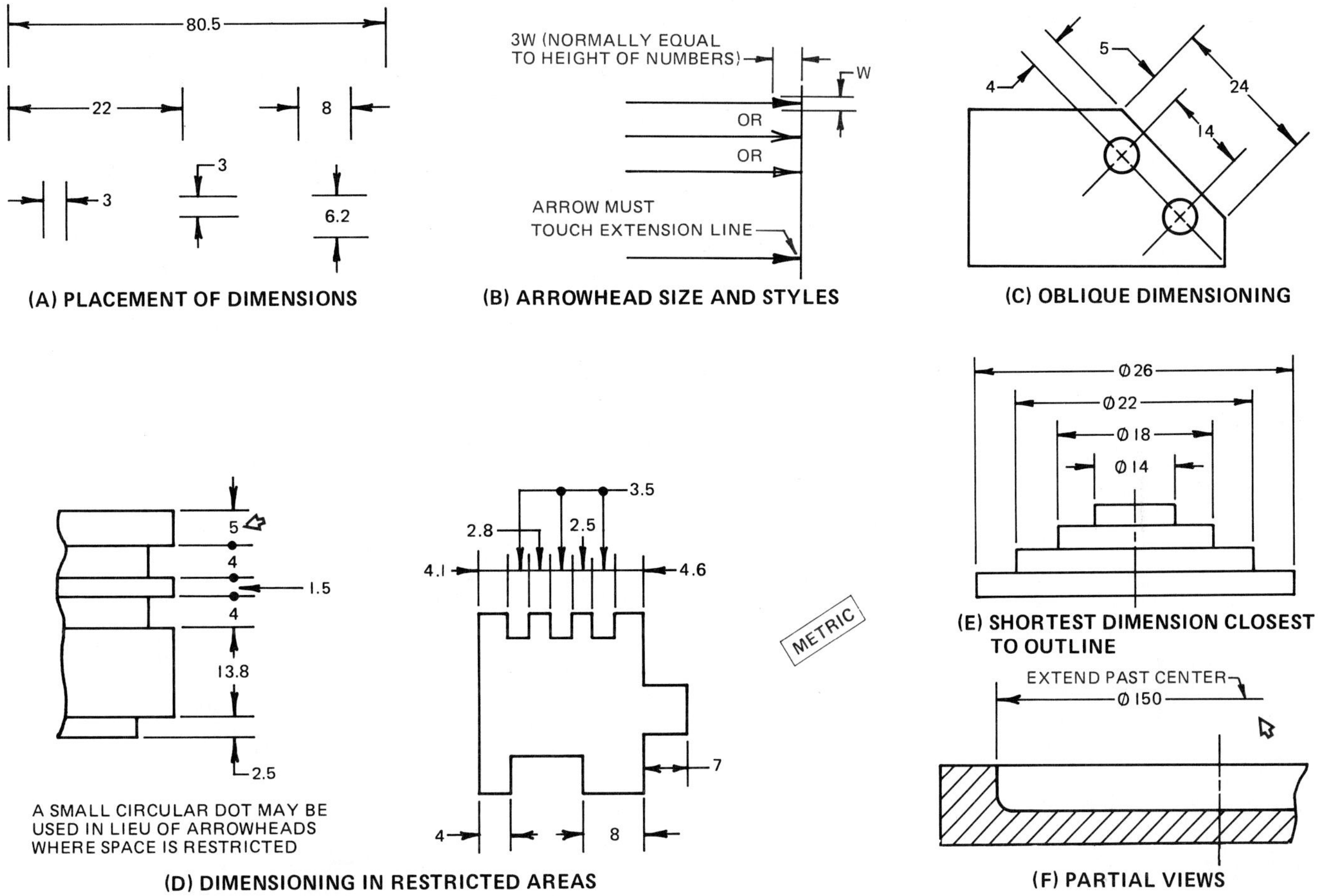

(A) PLACEMENT OF DIMENSIONS

(B) ARROWHEAD SIZE AND STYLES

3W (NORMALLY EQUAL TO HEIGHT OF NUMBERS)

OR

OR

ARROW MUST TOUCH EXTENSION LINE

(C) OBLIQUE DIMENSIONING

(E) SHORTEST DIMENSION CLOSEST TO OUTLINE

METRIC

A SMALL CIRCULAR DOT MAY BE USED IN LIEU OF ARROWHEADS WHERE SPACE IS RESTRICTED

(D) DIMENSIONING IN RESTRICTED AREAS

EXTEND PAST CENTER

(F) PARTIAL VIEWS

FIG. 8-1-3   Dimensioning linear features.

used for dimension lines. Every effort should be made to avoid crossing dimension lines by placing the shortest dimension closest to the outline (Fig. 8-1-3E).

Avoid dimensioning to hidden lines. In order to do so, it may be necessary to use a sectional view or a broken-out section. When the termination for a dimension is not included, as when used on partial or sectional views, the dimension line should extend beyond the center of the feature being dimensioned and shown with only one arrowhead (Fig. 8-1-3F).

Dimension lines should be placed outside the view where possible and should extend to extension lines rather than visible lines. However, when readability is improved by avoiding either extra long extension lines (Fig. 8-1-4) or the crowding of dimensions, placing of dimensions on views is permissible.

Extension (projection) lines are used to indicate the point or line on the drawing to which the dimension applies (Fig. 8-1-5, pg. 174). A small gap is left between the extension line and the outline to which it refers, and the extension line extends about .12 in. (3mm) beyond the outermost dimension line. However, when extension lines refer to points, as in Fig. 8-1-5E, they should extend through the points. Extension lines are usually drawn perpendicular to dimension lines. However, to improve clarity or when there is overcrowding, extension lines may be drawn at an oblique angle as long as clarity is maintained.

Center lines may be used as extension lines in dimensioning. The portion of the center line extending past the circle is not broken, as in Fig. 8-1-5B.

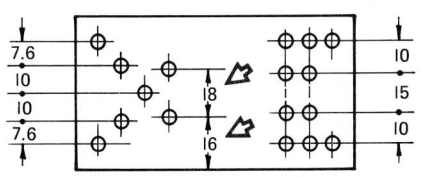

(A) IMPROVING READABILITY OF DRAWING

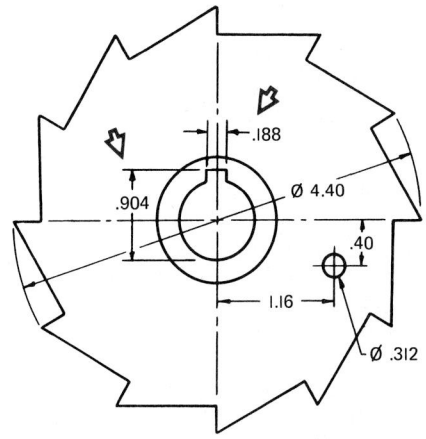

(B) AVOIDING LONG EXTENSION LINES

FIG. 8-1-4   Placing dimensions on views.

Where extension lines cross other extension lines, dimension lines, or visible lines, they are not broken. However, if extension lines cross arrowheads or dimension lines close to arrowheads, a break in the extension line is recommended (Fig. 8-1-5C).

Leaders

Leaders are used to direct notes, dimensions, symbols, item numbers, or part numbers to features on the drawing. See Fig. 8-1-6. A leader should generally be a single straight inclined line (not vertical or horizontal) except for a short horizontal portion extending to the center of the height of the first or last letter or digit of the note. The leader is terminated by an arrowhead or a dot of at least .06 in. (1.5 mm) in diameter. Arrowheads should always terminate on a line; dots should be used within the outline of the object and rest on a surface. Leaders should not be bent in any way unless it is unavoidable. Leaders should not cross one another, and two or more leaders adjacent to one another should be drawn parallel if practicable. It is better to repeat dimensions or references than to use long leaders.

Where a leader is directed to a circle or circular arc, its direction should point to the center of the arc or circle. Regardless of the reading direction used, aligned or unidirectional, all notes and dimensions used with leaders are placed in a horizontal position.

Notes

Notes are used to simplify or complement dimensioning by giving information on a drawing in a condensed and systematic manner. They may be general or local notes, and should be in the present or future tense.

General Notes   These refer to the part or the drawing as a whole. They should be shown in a central position below the view to which they apply or placed in a general note column. Typical examples of this type of note are:

- FINISH ALL OVER
- ROUNDS AND FILLETS R .06
- REMOVE ALL SHARP EDGES

Local Notes   These apply to local requirements only and are connected by a leader to the point to which the note applies.

Repetitive features and dimensions may be specified in the local note by the use of an X in conjunction with the numeral to indicate the "number of times" or "places" they are required (see Figs. 8-1-1 and 8-1-6). A full space is left between the X and the feature dimension. For additional information refer to Unit 8-3.

Typical examples are:

- 4 × Ø6
- 2 × 45°
- Ø3
  ∨ Ø 11.5 × 86°
- M12 × 1.25

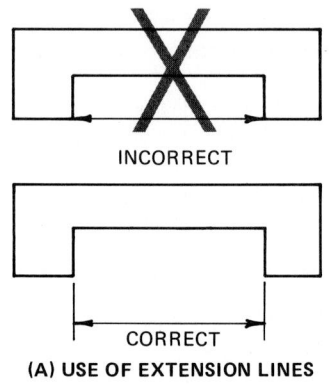

(A) USE OF EXTENSION LINES

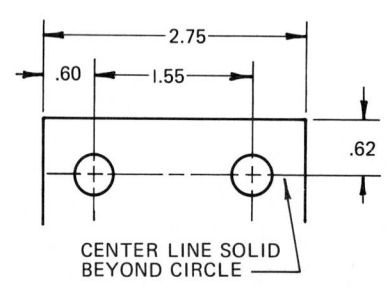

CENTER LINE SOLID BEYOND CIRCLE

(B) CENTER LINE USED AS EXTENSION LINE

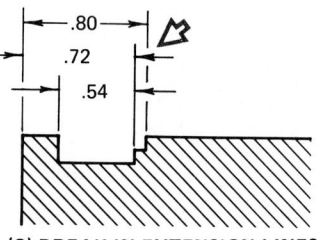

(C) BREAK IN EXTENSION LINES

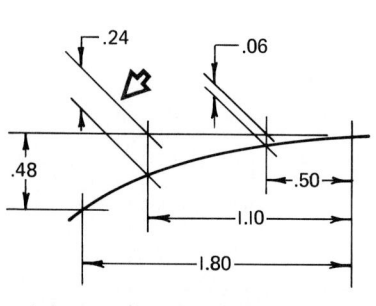

(D) OBLIQUE EXTENSION LINES

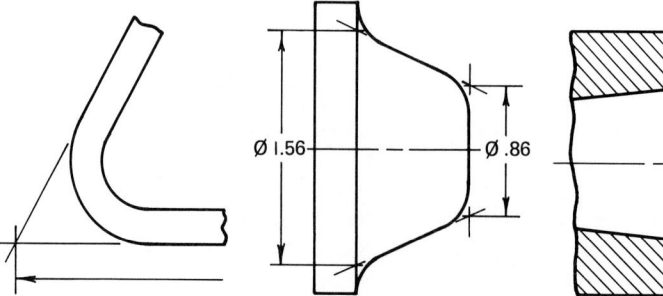

(E) EXTENSION LINE FROM POINTS

FIG. 8-1-5   Extension (projection) lines.

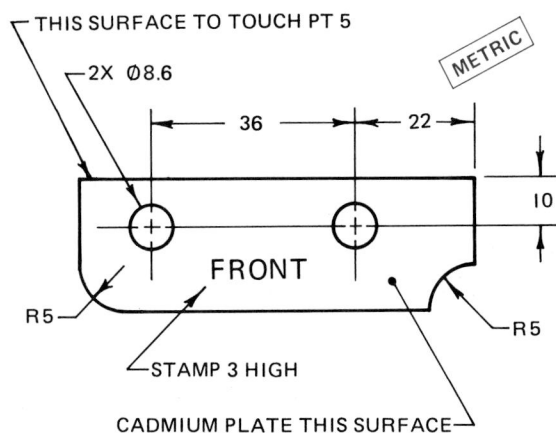

FIG. 8-1-6   Leaders.

Units of Measurement

Although the metric system of dimensioning is becoming the official international standard of measurement, most drawings in the United States are still dimensioned in inches or feet and inches. For this reason, drafters should be familiar with all the dimensioning systems that they may encounter. The dimensions used in this book are primarily decimal inch. However, metric and dual dimensions shown in the problems in this text are also used.

On drawings where all dimensions are either inches or millimeters, individual identification of linear units is not required. However, the drawing should contain a note stating the units of measurement.

Where some inch dimensions, such as nominal pipe sizes, are shown on a millimeter-dimensioned drawing, the abbreviation IN. must follow the inch values.

Inch Units of Measurement

Decimal-Inch System   (U.S. customary linear units)   Parts are designed in basic decimal increments, preferably .02 in., and are expressed with a minimum of two figures to the right of the decimal point (Fig. 8-1-7). Using the .02 in. module, the second decimal place (hundredths) is an even number or zero. By using the design modules having an even number for the last digit, dimensions can be halved for center distances without

increasing the number of decimal places. Decimal dimensions that are not multiples of .02, such as .01, .03, and .15, should be used only when it is essential to meet design requirements, such as to provide clearance, strength, smooth curves, etc. When greater accuracy is required, sizes are expressed as three- or four-place decimal numbers, for example, 1.875.

Whole dimensions will show a minimum of two zeros to the right of the decimal point.

$$24.00 \quad not \quad 24$$

An inch value of less than 1 is shown without a zero to the left of the decimal point.

$$.44 \quad not \quad 0.44$$

In cases where parts have to be aligned with existing parts or commercial products, which are dimensioned in fractions, it may be necessary to use decimal equivalents of fractional dimensions.

Fractional-Inch System   This system of dimensioning has not been recommended for use by ANSI for many years. It is shown in this text only for reference purposes or when making changes to drawings that may still be in use. In this system, parts are designed in basic units of common fractions down to 1/64 in. Decimal dimensions are used when finer divisions than 1/64 in. must be made. Common fractions may be used for specifying the size of holes that are produced by drills ordinarily stocked in fraction sizes and for the sizes of standard screw threads.

When common fractions are used on drawings, the fraction bar must not be omitted and should be horizontal except when applied with a typewriting machine that does not have a horizontal fraction bar.

When a dimension intermediate between 1/64 increments is necessary, it is expressed in decimals, such as .30, .257, or .2575 in.

The inch marks (") are not shown with dimensions. A note such as

DIMENSIONS ARE IN INCHES

should be clearly shown on the drawing. The exception is when the dimension "1 in." is shown on the drawing. The 1 should then be followed by the inch marks—1", *not* 1.

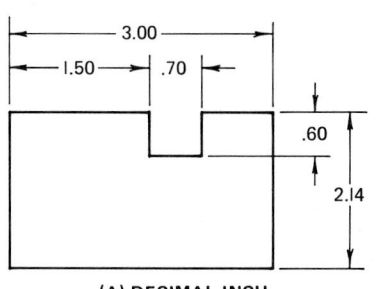

(A) DECIMAL INCH

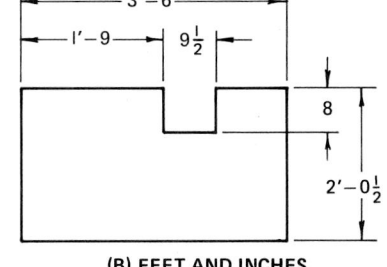

(B) FEET AND INCHES

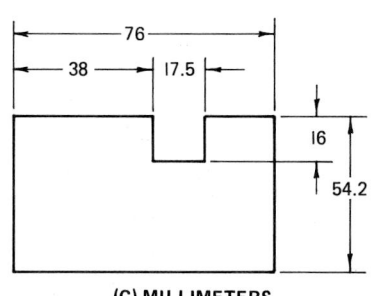

(C) MILLIMETERS

FIG. 8-1-7   Dimensioning units.

Foot-and-Inch System   Feet and inches are often used for installation drawings, drawings of large objects, and floor plans associated with architectural work. In this system all dimensions 12 in. or greater are specified in feet and inches. For example, 24 in. is expressed as 2'–0, and 27 in. is expressed as 2'–3. Parts of an inch are usually expressed as common fractions, rather than as decimals.

The inch marks (") are not shown on drawings. The drawing should carry a note such as

DIMENSIONS ARE IN FEET AND INCHES UNLESS OTHERWISE SPECIFIED

A dash should be placed between the foot and inch values. For example, 1'–3, not 1'3.

SI Metric Units of Measurement

The standard metric units on engineering drawings are the millimeter (mm) for linear measure and micrometer (μm) for surface roughness (Fig. 8-1-7C). For architectural drawings, meter and millimeter units are used.

Whole numbers from 1 to 9 are shown without a zero to the left of the number or a zero to the right of the decimal point.

2   *not*   02 or 2.0

A millimeter value of less than 1 is shown with a zero to the left of the decimal point.

0.2   *not*   .2 or .20

0.26   *not*   .26

Decimal points should be uniform and large enough to be clearly visible on reduced-size prints. They should be placed in line with the bottom of the associated digits and be given adequate space.

Neither commas nor spaces are used to separate digits into groups in specifying millimeter dimensions on drawings.

32545   *not*   32 545

Identification   A metric drawing should include a general note, such as

UNLESS OTHERWISE SPECIFIED DIMENSIONS ARE IN MILLIMETERS

and be identified by the word METRIC prominently displayed near the title block.

Units Common to Either System

Some measurements can be stated so that the callout will satisfy the units of both systems. For example, tapers such as .006 in. per inch and 0.006 mm per millimeter can both be expressed simply as the ratio 0.006:1 or in a note such as TAPER 0.006:1. Angular dimensions are also specified the same in both inch and metric systems.

Standard Items

Fasteners and Threads   Either inch or metric fasteners and threads may be used. Refer to the Appendix and Chap. 10 for additional information.

Hole Sizes   Tables showing standard inch and metric drill sizes are shown in the Appendix.

Dual Dimensioning

With the great exchange of drawings taking place between the United States and the rest of the world, at one time it became advantageous to show drawings in both inches and millimeters. As a result, many internationally operated companies adopted a dual system of dimensioning. Today, however, this type of dimensioning should be avoided if possible. When it may be necessary or desirable to give dimensions in both inches and millimeters on the same drawing, as shown in Figs. 8-1-8 and 8-1-9, a note or illustration should be located near the title block or strip to identify the inch and millimeter dimensions. Examples are

$$\frac{\text{MILLIMETER}}{\text{INCH}} \quad \text{and/or} \quad \text{MILLIMETER [INCH]}$$

Angular Units

Angles are measured in degrees. The decimal degree is now preferred over the use of degrees, minutes, and seconds. For example, the use of 60.5° is preferred to the use of 60° 30'. Where only minutes or seconds are specified, the number of minutes or seconds is preceded by 0°, or 0°0', as applicable. Some examples follow.

Decimal Degree	Degrees, Minutes, and Seconds
10° ± 0.5°	10° ± 0°30'
0.75°	0°45'
0.004°	0°0'15"
90° ± 1.0°	90° ± 1°
25.6 ± 0.2°	25°36' ± 0°12'
25.51°	25° 30'36"

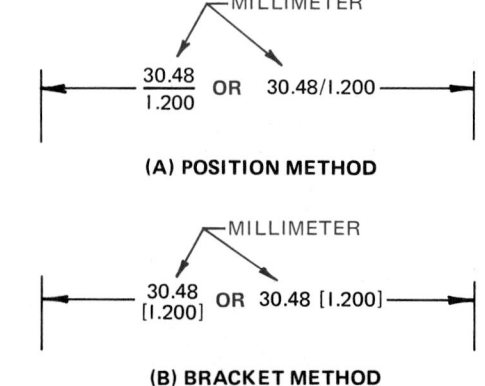

(A) POSITION METHOD

(B) BRACKET METHOD

FIG. 8-1-8   Dual dimensioning.

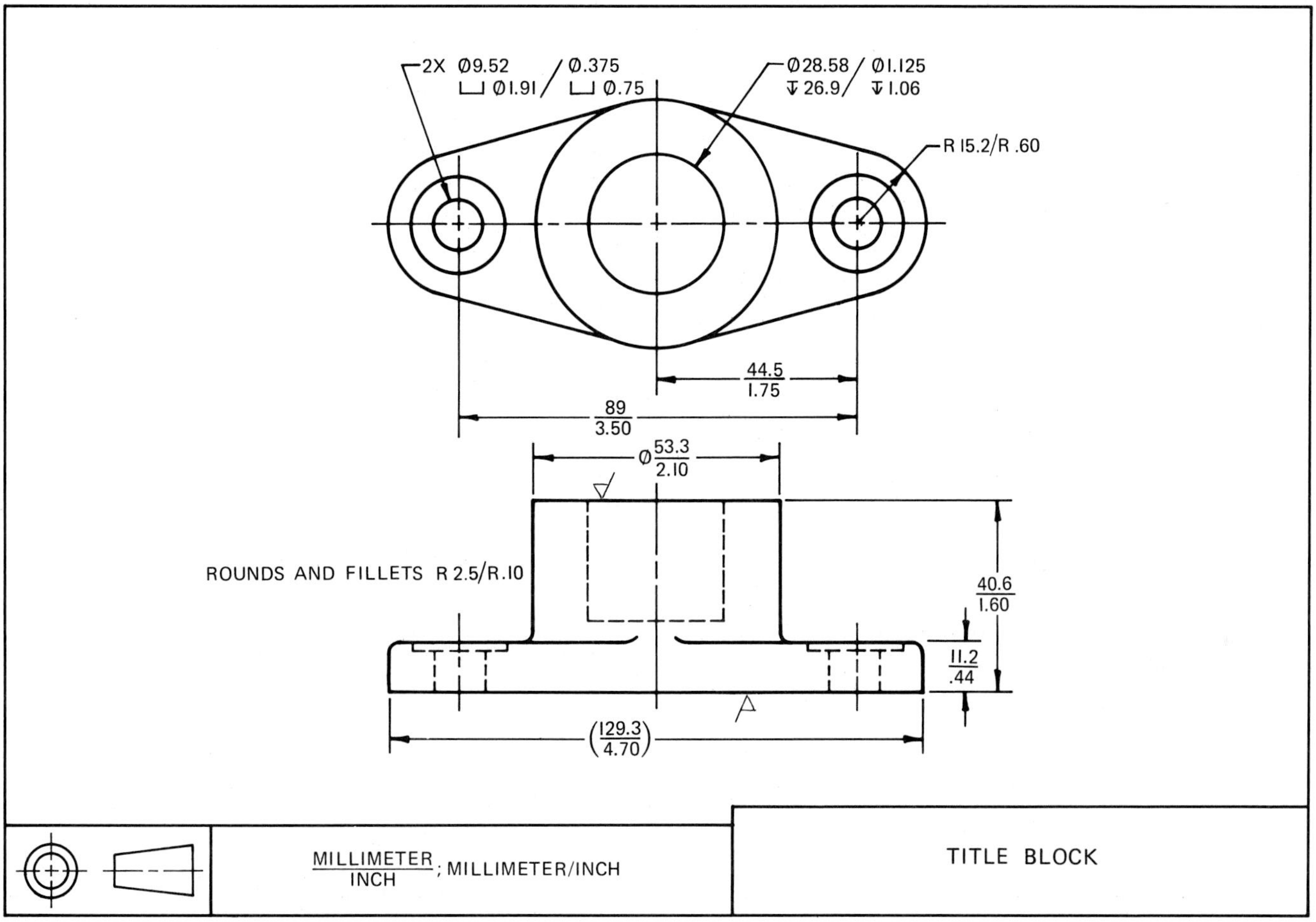

2X Ø9.52 / Ø.375
⊔ Ø1.91 / ⊔ Ø.75

Ø28.58 / Ø1.125
⊽26.9 / ⊽1.06

R15.2/R.60

44.5
1.75

89
3.50

Ø 53.3
2.10

ROUNDS AND FILLETS R 2.5/R.10

40.6
1.60

11.2
.44

(129.3)
(4.70)

MILLIMETER
INCH ; MILLIMETER/INCH

TITLE BLOCK

FIG. 8-1-9   Dual-dimensioned drawing.

FIG. 8-1-10   Angular units.

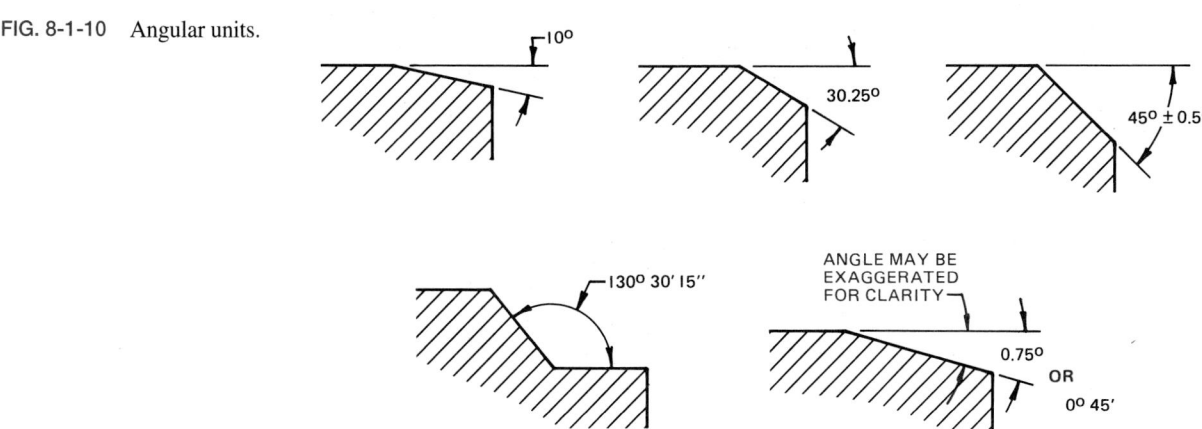

10°

30.25°

45° ± 0.5°

130° 30′ 15″

ANGLE MAY BE
EXAGGERATED
FOR CLARITY

0.75°
OR
0° 45′

The dimension line of an angle is an arc drawn with the apex of the angle as the center point for the arc, wherever practicable. The position of the dimension varies according to the size of the angle and appears in a horizontal position. Refer to the recommended arrangements as shown in Fig. 8-1-10.

Reading Direction

Dimensions and notes should be placed to be read from the bottom of the drawing for engineering drawings (unidirectional system). For architectural and structural drawings the aligned system of dimensioning is used (Fig. 8-1-11, pg. 178).

In both methods angular dimensions and dimensions and notes shown with leaders should be aligned with the bottom of the drawing.

Basic Rules for Dimensioning

Refer to Fig. 8-1-12.

- Place dimensions between the views when possible.
- Place the dimension line for the shortest length, width, or height nearest the outline of the object. Parallel dimension lines are placed in order of their size, making the longest dimension line the outermost.

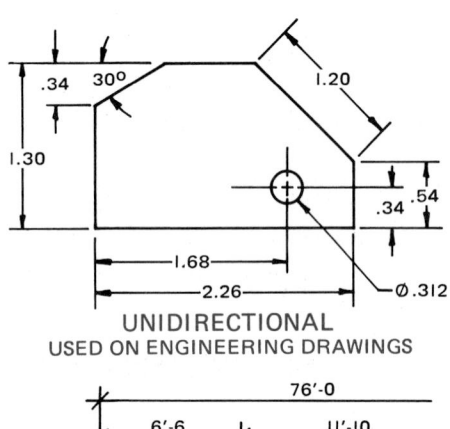

UNIDIRECTIONAL
USED ON ENGINEERING DRAWINGS

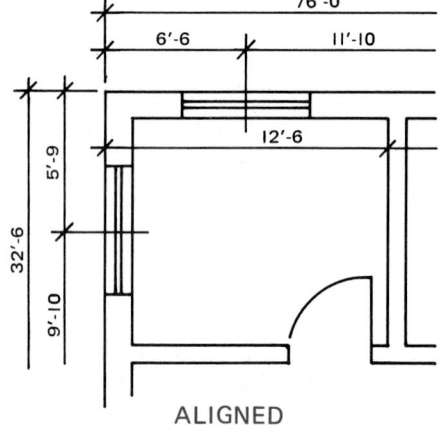

ALIGNED
USED ON ARCHITECTURAL AND STRUCTURAL DRAWINGS

FIG. 8-1-11   Reading direction of dimensions.

- Place dimensions with the view that best shows the characteristic contour or shape of the object. When this rule is applied, dimensions will not always be between views.
- On large views, dimensions can be placed on the view to improve clarity.
- Use only one system of dimensions, either the unidirectional or the aligned, on any one drawing.
- Dimensions should not be duplicated in other views.
- Dimensions should be selected so that it will not be necessary to add or subtract dimensions in order to define or locate a feature.

Symmetrical Outlines

A part is said to be *symmetrical* when the features on each side of the center line or median plane are identical in size, shape, and location. Partial views are often drawn for the sake of economy or space. With CAD duplicating, the other half of the view requires little effort. However, space constraints may preclude this possibility. When only one-half the outline of a symmetrically shaped part is drawn, symmetry is indicated by applying the symmetry symbol to the center line on both sides of the part (Fig. 8-1-13). In such cases, the outline of the part should extend slightly beyond the center line and terminate with a break line. Note the dimensioning method of extending the dimension lines to act as extension lines for the perpendicular dimensions.

Reference Dimensions

A reference dimension is shown for information only, and it is not required for manufacturing or inspection purposes. It is enclosed in parentheses, as shown in Fig. 8-1-14. Formerly the abbreviation REF was used to indicate a reference dimension.

Not-to-Scale Dimensions

When a dimension on a drawing is altered, making it not to scale, it should be underlined (underscored) with a straight, thick line (Fig. 8-1-15), except when the condition is clearly shown by break lines.

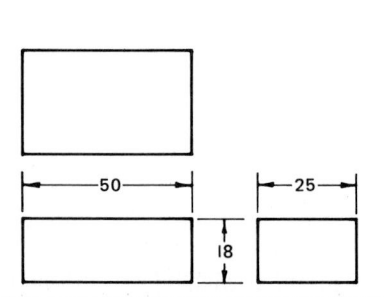

(A) PLACE DIMENSIONS BETWEEN VIEWS

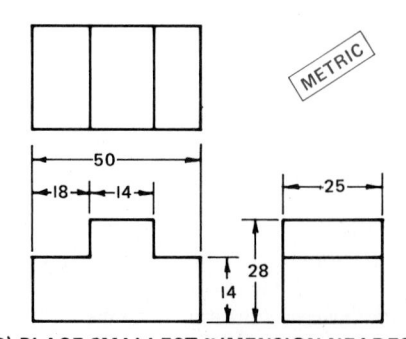

(B) PLACE SMALLEST DIMENSION NEAREST THE VIEW BEING DIMENSIONED

(C) DIMENSION THE VIEW THAT BEST SHOWS THE SHAPE

FIG. 8-1-12   Basic dimensioning rules.

FIG. 8-1-13   Dimensioning symmetrical outlines or features.

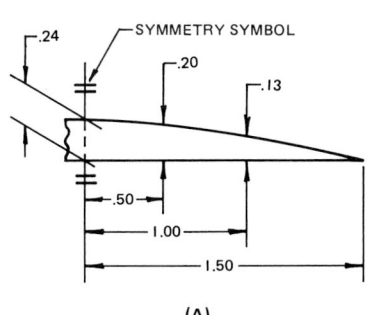

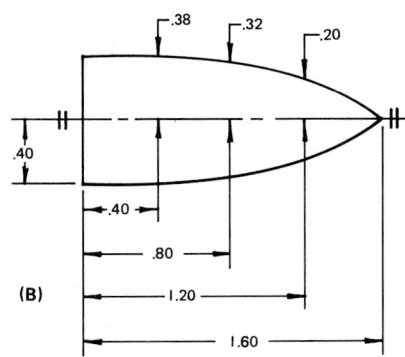

(A)

(B)

Operational Names

The use of operational names, such as turn, bore, grind, ream, tap, and thread, with dimensions should be avoided. While the drafter should be aware of the methods by which a part can be produced, the method of manufacture is better left to the producer. If the completed part is adequately dimensioned and has surface texture symbols showing finish quality desired, it remains a shop problem to meet the drawing specifications.

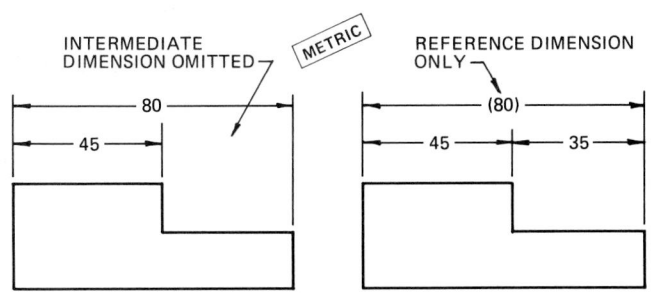

FIG. 8-1-14   Reference dimensions.

FIG. 8-1-15   Not-to-scale dimensions.

FIG. 8-2-1   Dimensioning diameters.

Abbreviations

Abbreviations and symbols are used on drawings to conserve space and time, but only where their meanings are quite clear. See the Appendix for commonly accepted abbreviations.

REFERENCES AND SOURCE MATERIAL

1. ASME Y14.5M–1994, *Dimensioning and Tolerancing.*
2. CAN/CSA B78.2–M91, *Dimensioning and Tolerancing of Technical Drawings.*

ASSIGNMENTS

See Assignments 1 through 3 for Unit 8-1 on pages 209–211.

8-2 DIMENSIONING CIRCULAR FEATURES

Diameters

Where the diameter of a single feature or the diameters of a number of concentric cylindrical features are to be specified, it is recommended that they be shown on the longitudinal view (Fig. 8-2-1).

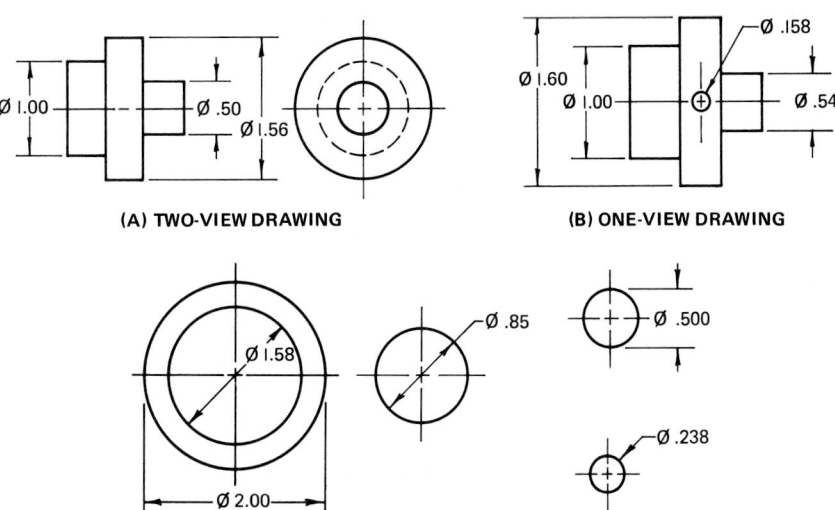

(A) TWO-VIEW DRAWING

(B) ONE-VIEW DRAWING

(C) DIMENSIONING DIAMETERS ON END VIEW

Where space is restricted or when only a partial view is used, diameters may be dimensioned as illustrated in Fig. 8-2-2. Regardless of where the diameter dimension is shown, the numerical value is preceded by the diameter symbol Ø for both customary and metric dimensions.

Radii

The general method of dimensioning a circular arc is by giving its radius. A radius dimension line passes through, or is in line with, the radius center and terminates with an arrowhead touching the arc (Fig. 8-2-3). An arrowhead is never used at the radius center. The size of the dimension is preceded by the abbreviation R for both customary and metric dimensioning. Where space is limited, as for a small radius, the radial dimension line may extend through the radius center. Where it is inconvenient to place the arrowhead between the radius center and the arc, it may be placed outside the arc, or a leader may be used (Fig. 8-2-3A).

Where a dimension is given to the center of the radius, a small cross should be drawn at the center (Fig. 8-2-3B). Extension lines and dimension lines are used to locate the center. Where the location of the center is unimportant, a radial arc may be located by tangent lines (Fig. 8-2-3E).

Where the center of a radius is outside the drawing or interferes with another view, the radius dimension line may be foreshortened (Fig. 8-2-3D). The portion of the dimension line next to the arrowhead should be radial relative to the curved line. Where the radius dimension line is foreshortened and the center is located by coordinate dimensions, the dimensions locating the center should be shown as foreshortened or the dimension shown as not to scale.

Simple fillet and corner radii may also be dimensioned by a general note, such as

ALL ROUNDS AND FILLETS UNLESS OTHERWISE SPECIFIED R.20
or
ALL RADII R5

Where a radius is dimensioned in a view that does not show the true shape of the radius, TRUE R is added before the radius dimension, as illustrated in Fig. 8-2-4.

Rounded Ends

Overall dimensions should be used for parts or features having rounded ends. For fully rounded ends, the radius (R) is shown but not dimensioned (Fig. 8-2-5A). For parts with partially rounded ends, the radius is dimensioned (Fig. 8-2-5B). Where a hole and radius have the same center and the hole location is more critical than the location of a radius, either the radius or the overall length should be shown as a reference dimension (Fig. 8-2-5C).

FIG. 8-2-2   Dimensioning diameters where space is restricted.

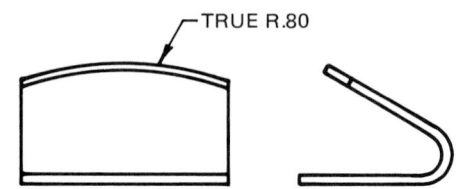

FIG. 8-2-4   Indicating true radius.

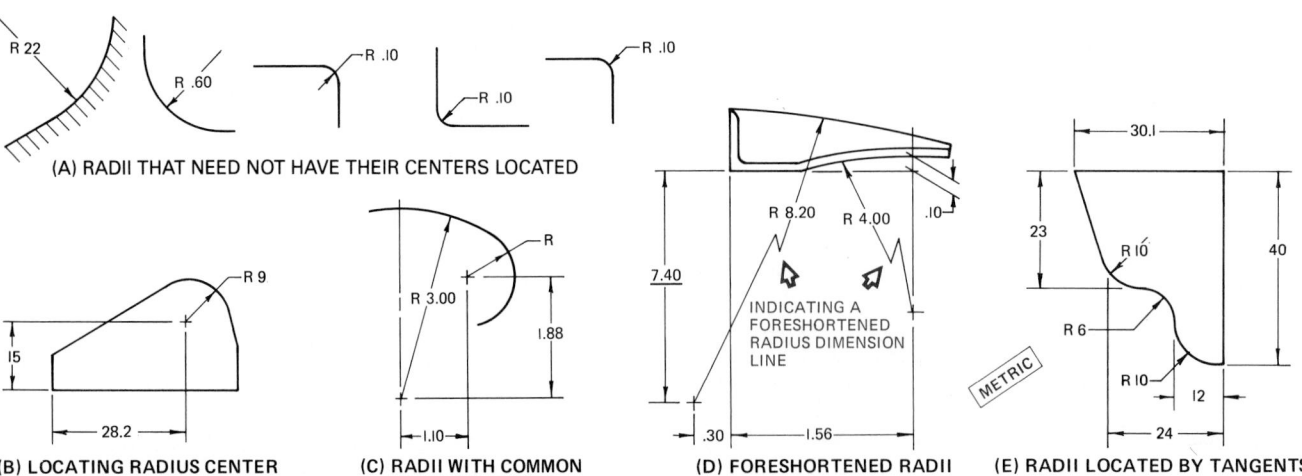

(A) RADII THAT NEED NOT HAVE THEIR CENTERS LOCATED

(B) LOCATING RADIUS CENTER

(C) RADII WITH COMMON TANGENT POINTS

(D) FORESHORTENED RADII

(E) RADII LOCATED BY TANGENTS

FIG. 8-2-3   Dimensioning radii.

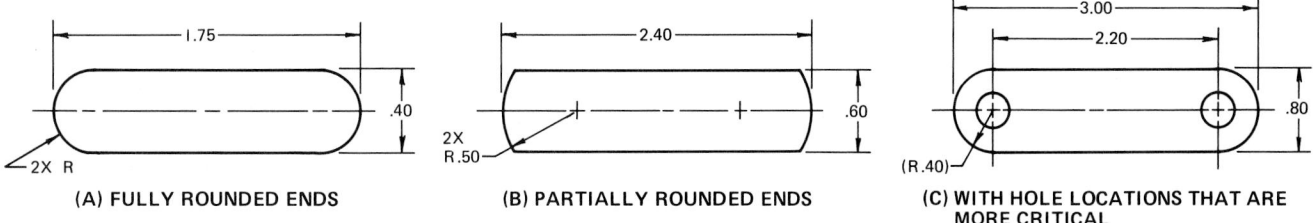

FIG. 8-2-5   Dimensioning external surfaces with rounded ends.

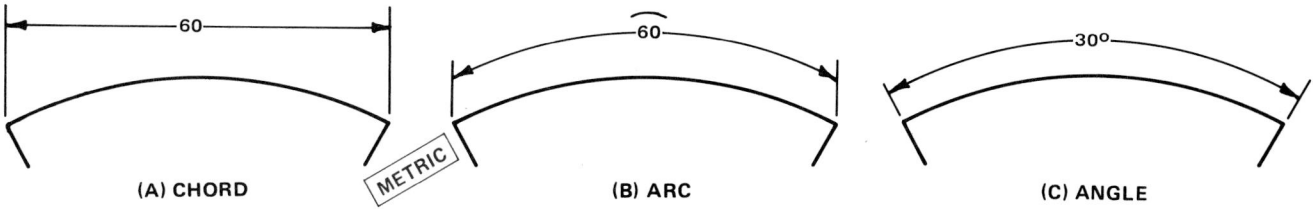

FIG. 8-2-6   Dimensioning chords, arcs, and angles.

Dimensioning Chords, Arcs, and Angles

The difference in dimensioning chords, arcs, and angles is shown in Fig. 8-2-6.

Spherical Features

Spherical surfaces may be dimensioned as diameters or radii, but the dimension should be used with the abbreviations SR or SØ (Fig. 8-2-7).

Cylindrical Holes

Plain, round holes are dimensioned in various ways, depending upon design and manufacturing requirements (Fig. 8-2-8). However, the leader is the method most commonly used. When a *leader* is used to specify diameter sizes, as with small holes, the dimension is identified as a diameter by preceding the numerical value with the diameter symbol Ø.

The size, quantity, and depth may be shown on a single line, or on several lines if preferable. For through holes, the abbreviation THRU should follow the dimension if the drawing does not make this clear. The depth dimension of a blind hole is the depth of the full diameter and is normally included as part of the dimensioning note.

When more than one hole of a size is required, the number of holes should be specified. However, care must be taken to avoid placing the hole size and quantity values together without adequate spacing. It may be better to show the note on two or more lines than to use a single line note that might be misread (Fig. 8-2-8D).

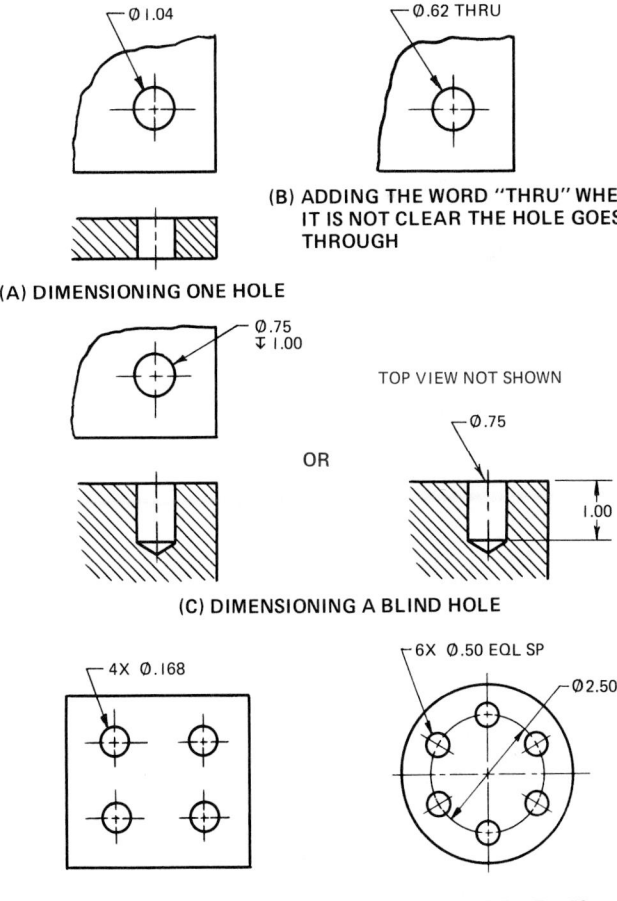

(B) ADDING THE WORD "THRU" WHEN IT IS NOT CLEAR THE HOLE GOES THROUGH

(A) DIMENSIONING ONE HOLE

TOP VIEW NOT SHOWN

OR

(C) DIMENSIONING A BLIND HOLE

NOTE: SEE UNIT 8–3 FOR DIMENSIONING REPETITIVE FEATURES

(D) DIMENSIONING A GROUP OF HOLES

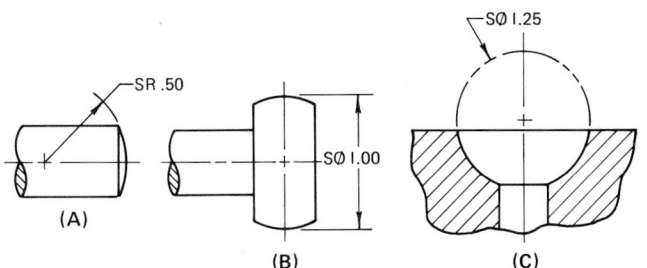

FIG. 8-2-7   Dimensioning spherical surfaces.

FIG. 8-2-8   Dimensioning cylindrical holes.

Minimizing Leaders   If too many leaders would impair the legibility of a drawing, letters or symbols as shown in Fig. 8-2-9 should be used to identify the features.

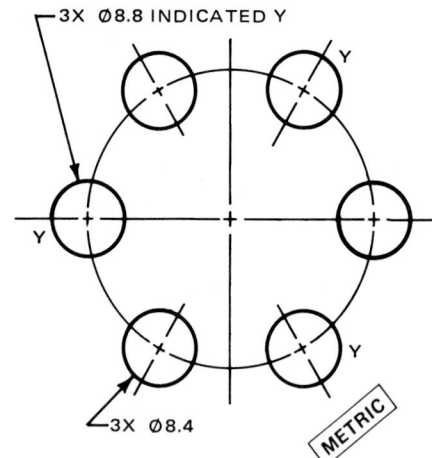

FIG. 8-2-9   Minimizing leaders.

Slotted Holes

Elongated holes and slots are used to compensate for inaccuracies in manufacturing and to provide for adjustment. See Fig. 8-2-10. The method selected to locate the slot would depend on how the slot was made. The method shown in Fig. 8-2-10B is used when the slot is punched out and the location of the punch is given. Figure 8-2-10A shows the dimensioning method used when the slot is machined out.

Countersinks, Counterbores, and Spotfaces

Counterbores, spotfaces, and countersinks are specified on drawings by means of dimension symbols or abbreviations, the symbols being preferred. The symbols or abbreviations indicate the form of the surface only and do not restrict the methods used to produce that form. The dimensions for them are usually given as a note, preceded by the size of the through hole (Figs. 8-2-11 and 8-2-12).

A *countersink* is an angular-sided recess to accommodate the head of flathead screws, rivets, and similar items. The diameter at the surface and the included angle are given. When the depth of the tapered section of the countersink is critical, it

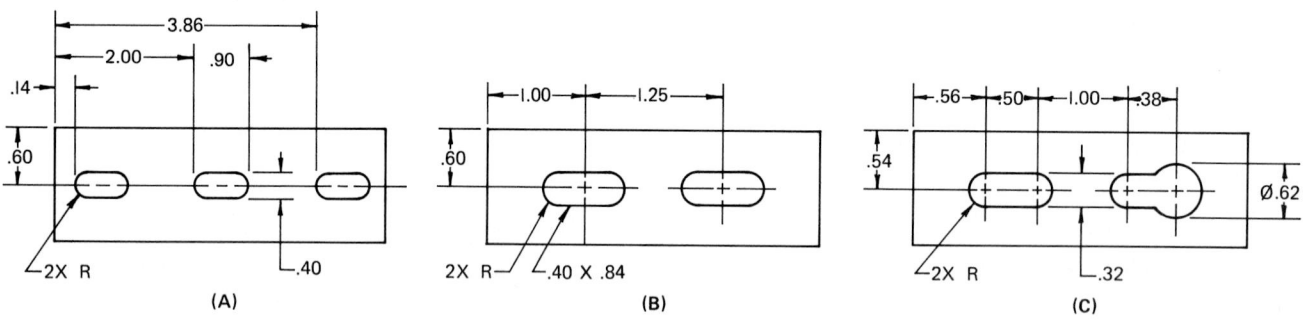

FIG. 8-2-10   Slotted holes.

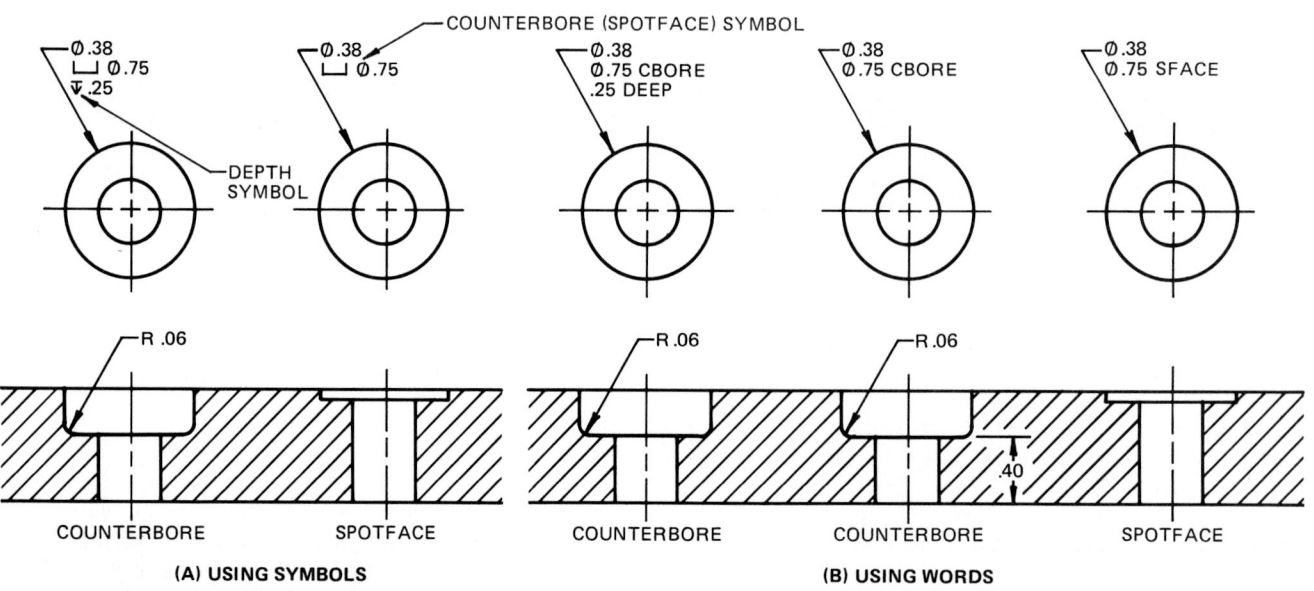

FIG. 8-2-11   Counterbored and spotfaced holes.

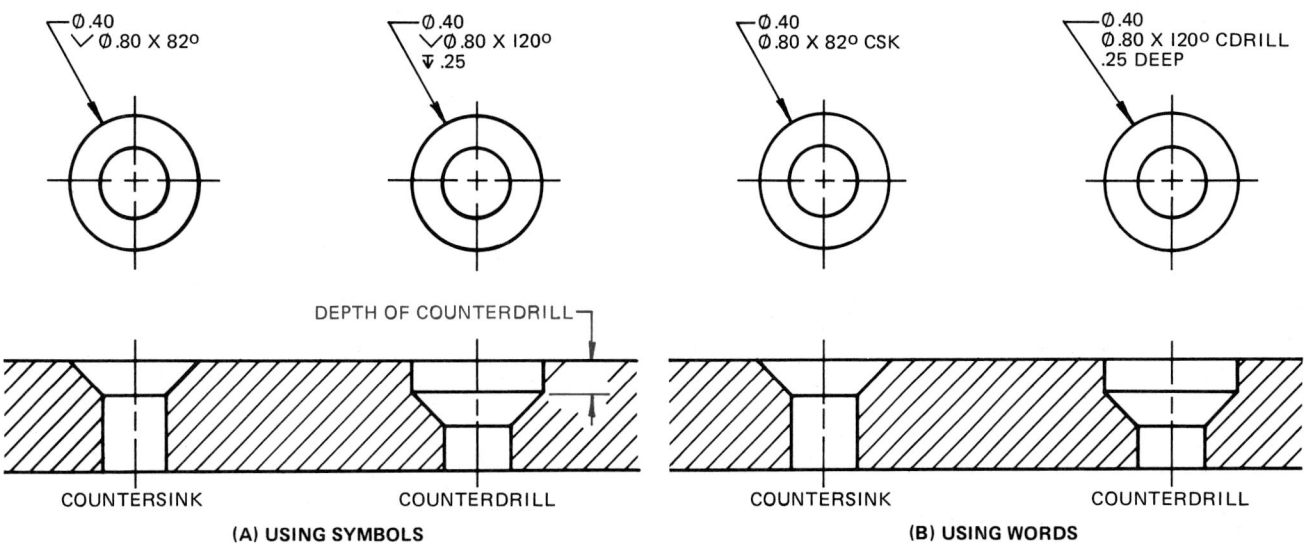

FIG. 8-2-12   Countersunk and counterdrilled holes.

is specified in the note or by a dimension. For counterdrilled holes, the diameter, depth, and included angle of the counterdrill are given.

A *counterbore* is a flat-bottomed, cylindrical recess that permits the head of a fastening device, such as a bolt, to lie recessed into the part. The diameter, depth, and corner radius are specified in a note. In some cases, the thickness of the remaining stock may be dimensioned rather than the depth of the counterbore.

A *spotface* is an area where the surface is machined just enough to provide smooth, level seating for a bolt head, nut, or washer. The diameter of the faced area and either the depth or the remaining thickness are given. A spotface may be specified by a general note and not delineated on the drawing. If no depth or remaining thickness is specified, it is implied that the spotfacing is the minimum depth necessary to clean up the surface to the specified diameter.

The symbols for counterbore or spotface, countersink, and depth are shown in Figs. 8-2-11 and 8-2-12. In each case the symbol precedes the dimension.

REFERENCE AND SOURCE MATERIAL

1. ASME Y14.5M–1994, *Dimensioning and Tolerancing.*
2. CAN/CSA B78.2–M91, *Dimensioning and Tolerancing of Technical Drawings.*

ASSIGNMENTS

See Assignments 4 through 6 for Unit 8-2 on pages 212–214.

8-3 DIMENSIONING COMMON FEATURES

Repetitive Features and Dimensions

Repetitive features and dimensions may be specified on a drawing by the use of an X in conjunction with the numeral to indicate the "number of times" or "places" they are required. A space is shown between the X and the dimension.

An X that means "by" is often used between coordinate dimensions specified in note form. Where both are used on a drawing, care must be taken to ensure each is clear (Fig. 8-3-1, pg. 184).

Chamfers

The process of *chamfering,* that is, cutting away the inside or outside piece, is done to facilitate assembly. Chamfers are normally dimensioned by giving their angle and linear length (Fig. 8-3-2, pg. 184). When the chamfer is 45°, it may be specified as a note.

When a very small chamfer is permissible, primarily to break a sharp corner, it may be dimensioned but not drawn, as in Fig. 8-3-2C. If not otherwise specified, an angle of 45° is understood.

Internal chamfers may be dimensioned in the same manner, but it is often desirable to give the diameter over the chamfer. The angle may also be given as the included angle if this is a design requirement. This type of dimensioning is generally necessary for larger diameters, especially those over 2 in. (50mm), whereas chamfers on small holes are usually expressed as countersinks. Chamfers are never measured along the angular surface.

Slopes and Tapers

Slope

A *slope* is the slant of a line representing an inclined surface. It is expressed as a ratio of the difference in the heights at right angles to the base line, at a specified distance apart (Fig. 8-3-3, pg. 185). Figure 8-3-3D is the preferred method of dimensioning slopes on architectural and structural drawings.

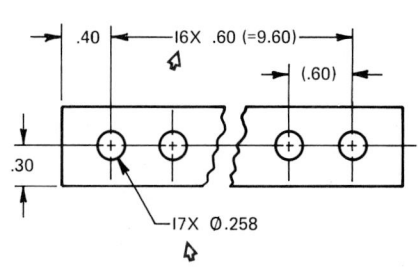

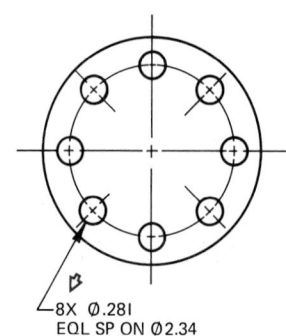

(A) USING "NUMBER OF TIMES" SYMBOL

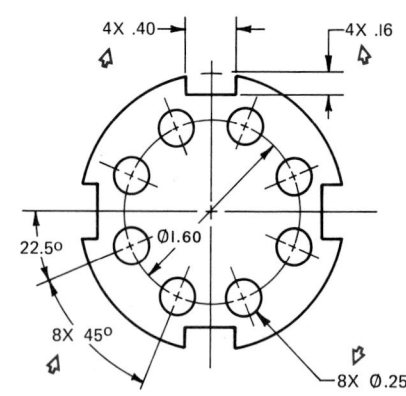

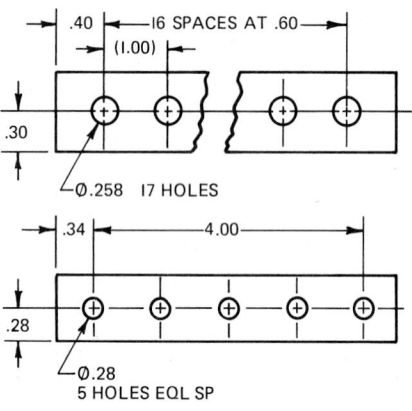

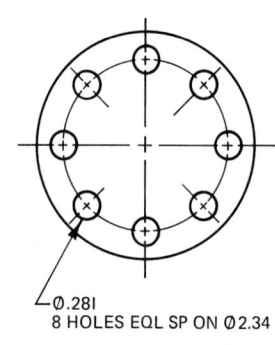

(B) USING DESCRIPTIVE NOTES

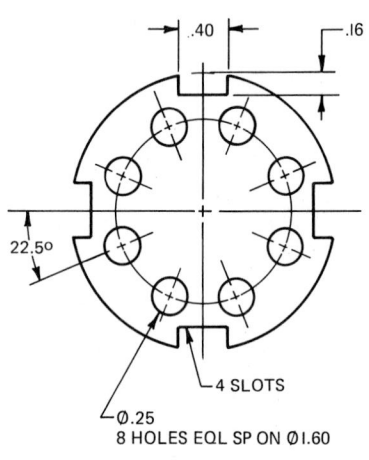

FIG. 8-3-1   Dimensioning repetitive detail.

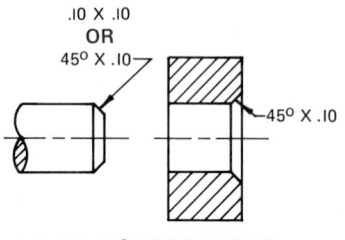

(A) FOR 45° CHAMFERS ONLY

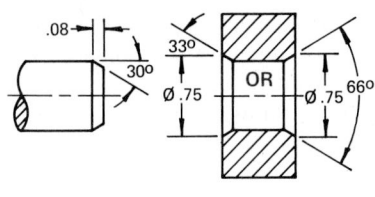

(B) FOR ALL CHAMFERS

(C) SMALL CHAMFERS

(D) CHAMFERS BETWEEN SURFACES AT OTHER THAN 90°

FIG. 8-3-2   Dimensioning chamfers.

The following dimensions and symbol may be used, in different combinations, to define the slope of a line or flat surface:

- The slope specified as a ratio combined with the slope symbol (Fig. 8-3-3A).
- The slope specified by an angle (Fig. 8-3-3B).

- The dimensions showing the difference in the heights of two points from the base line and the distance between them (Fig. 8-3-3C).

Taper

A *taper* is the ratio of the difference in the diameters of two sections (perpendicular to the axis of a cone to the distance between these two sections), as shown in Fig. 8-3-4. When the taper symbol is used, the vertical leg is always shown to the left and precedes the ratio figures. The following dimensions may be used, in suitable combinations, to define the size and form of tapered features:

- The diameter (or width) at one end of the tapered feature
- The length of the tapered feature
- The rate of taper
- The included angle
- The taper ratio
- The diameter at a selected cross section
- The dimension locating the cross section

Knurls

Knurling is specified in terms of type, pitch, and diameter before and after knurling (Fig. 8-3-5). The letter P precedes the pitch number. Where control is not required, the diameter after knurling is omitted. Where only portions of a feature require

184

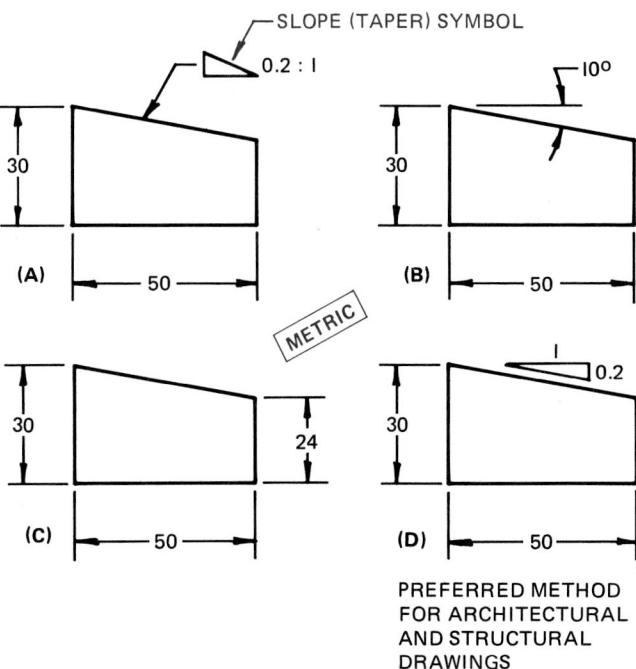

FIG. 8-3-3   Dimensioning slopes.

PREFERRED METHOD
FOR ARCHITECTURAL
AND STRUCTURAL
DRAWINGS

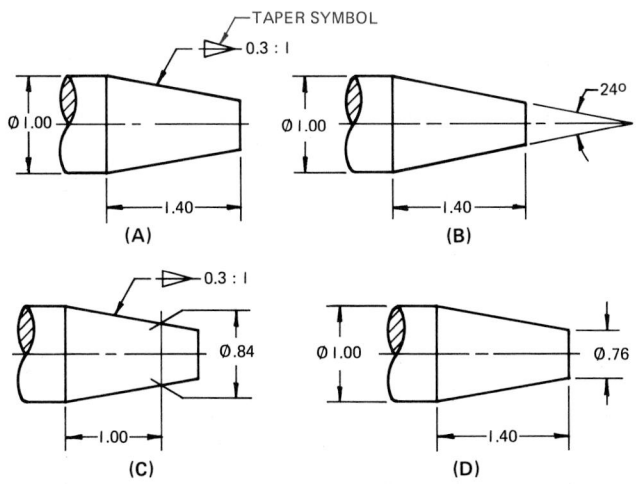

FIG. 8-3-4   Dimensioning tapers.

knurling, axial dimensions must be provided. Where required to provide a press fit between parts, knurling is specified by a note on the drawing that includes the type of knurl required, the pitch, the toleranced diameter of the feature prior to knurling, and the minimum acceptable diameter after knurling. Commonly used types are straight, diagonal, spiral, convex, raised diamond, depressed diamond, and radial. The pitch is usually expressed in terms of teeth per inch or millimeter and may be the straight pitch, circular pitch, or diametral pitch. For cylindrical surfaces, the latter is preferred.

The knurling symbol is optional and is used only to improve clarity on working drawings.

Formed Parts

In dimensioning formed parts, the inside radius is usually specified, rather than the outside radius, but all forming dimensions should be shown on the same side if possible. Dimensions apply to the side on which the dimensions are shown unless otherwise specified (Fig. 8-3-6).

Undercuts

The operation of *undercutting* or *necking,* that is, cutting a recess in a diameter, is done to permit two parts to come together, as illustrated in Fig. 8-3-7A (pg. 186). It is indicated on the drawing by a note listing the width first and then the diameter. If the radius is shown at the bottom of the undercut, it will be assumed that the radius is equal to one-half the width

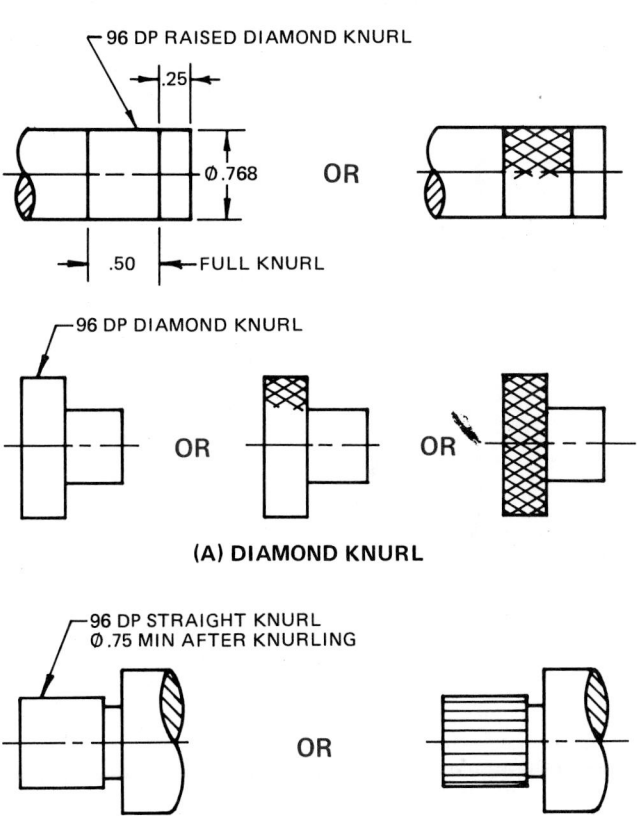

FIG. 8-3-5   Dimensioning knurls.

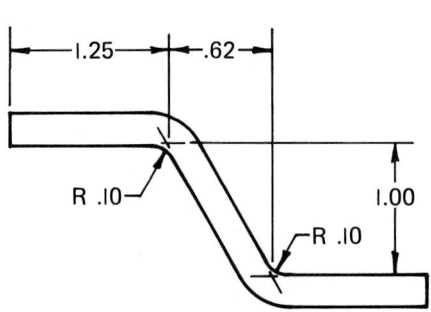

FIG. 8-3-6   Dimensioning theoretical points of intersection.

185

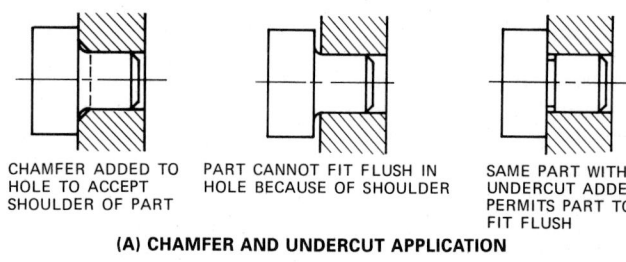

| CHAMFER ADDED TO HOLE TO ACCEPT SHOULDER OF PART | PART CANNOT FIT FLUSH IN HOLE BECAUSE OF SHOULDER | SAME PART WITH UNDERCUT ADDED PERMITS PART TO FIT FLUSH |

(A) CHAMFER AND UNDERCUT APPLICATION

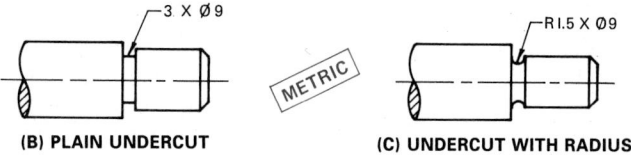

(B) PLAIN UNDERCUT   **(C) UNDERCUT WITH RADIUS**

FIG. 8-3-7   Dimensioning undercuts.

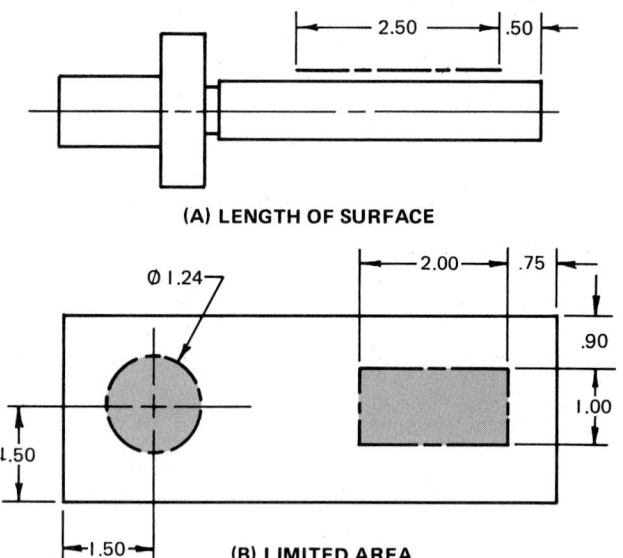

(A) LENGTH OF SURFACE

(B) LIMITED AREA

FIG. 8-3-8   Dimensioning limited lengths and areas.

unless otherwise specified, and the diameter will apply to the center of the undercut. When the size of the undercut is unimportant, the dimension may be left off the drawing.

Limited Lengths and Areas

Sometimes it is necessary to dimension a limited length or area of a surface to indicate a special condition. In such instances, the area or length is indicated by a chain line (Fig. 8-3-8A). When indicating a length of surface, the chain line is drawn parallel and adjacent to the surface. When indicating an area of surface, the area is cross-hatched within the chain line boundary (Fig. 8-3-8B).

Wire, Sheet Metal, and Drill Rod

Wire, sheet metal, and drill rod, which are manufactured to gage or code sizes, should be shown by their decimal dimensions; but gage numbers, drill letters, etc., may be shown in parentheses following those dimensions.

EXAMPLES

Sheet —.141 (NO. 10 USS GA)
      —.081 (NO. 12 B & S GA)

REFERENCES AND SOURCE MATERIAL

1. ASME Y14.5M–1994, *Dimensioning and Tolerancing.*
2. CAN/CSA B78.2–M91, *Dimensioning and Tolerancing of Technical Drawings.*

ASSIGNMENTS

See Assignments 7 through 12 for Unit 8-3 on pages 215 to 216.

8-4 DIMENSIONING METHODS

The choice of the most suitable dimensions and dimensioning methods will depend, to some extent, on how the part will be produced and whether the drawings are intended for unit or mass production. *Unit production* refers to cases where each part is to be made separately, using general-purpose tools and machines. *Mass production* refers to parts produced in quantity, where special tools and gages are usually provided.

Either linear or angular dimensions may locate features with respect to one another (point-to-point) or from a datum. Point-to-point dimensions may be adequate for describing simple parts. Dimensions from a datum may be necessary if a part with more than one critical dimension must mate with another part.

The following systems of dimensioning are used more commonly for engineering drawings.

Rectangular Coordinate Dimensioning

This is a method for indicating distance, location, and size by means of linear dimensions measured parallel or perpendicular to reference axes or datum planes that are perpendicular to one another. Coordinate dimensioning with dimension lines must clearly identify the datum features from which the dimensions originate (Fig. 8-4-1).

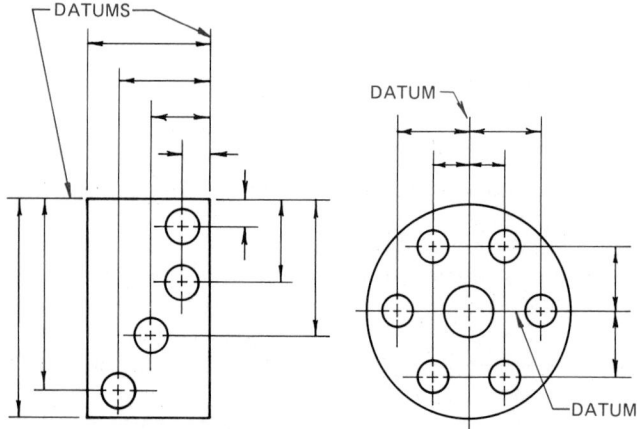

FIG. 8-4-1   Rectangular coordinate dimensioning.

Rectangular Coordinates for Arbitrary Points   Coordinates for arbitrary points of reference without a grid appear adjacent to each point (Fig. 8-4-2) or in tabular form (Fig. 8-4-3). CAD systems will automatically display any point coordinate when it is picked.

Rectangular Coordinate Dimensioning Without Dimension Lines   Dimensions may be shown on extension lines without the use of dimension lines or arrowheads. The base lines may be zero coordinates or they may be labeled as X, Y, and Z (Fig. 8-4-4).

Tabular Dimensioning   Tabular dimensioning is a type of coordinate dimensioning in which dimensions from mutually perpendicular planes are listed in a table on the drawing rather than on the pictorial delineation. This method is used on drawings that require the location of a large number of similarly shaped features when dimensioning parts for numerical control (Fig. 8-4-5).

It may be advantageous to have a part dimensioned symmetrically about its center, as shown in Fig. 8-4-6 (pg. 188). When the center lines are designated as the base (zero) lines, positive and negative values will occur. These values are shown with the dimensions locating the holes.

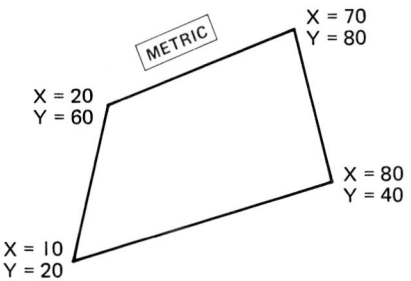

FIG. 8-4-2   Coordinates for arbitrary points.

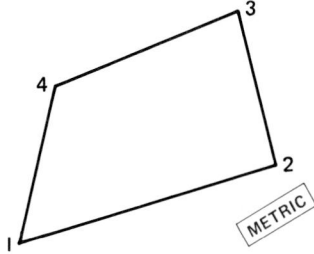

POINT	X	Y
1	10	20
2	80	40
3	70	80
4	20	60

FIG. 8-4-3   Coordinates for arbitrary points in tabular form.

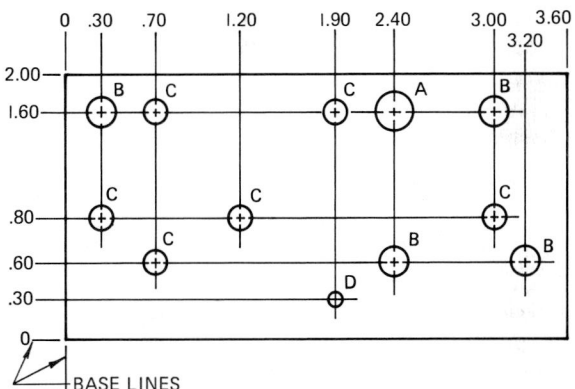

HOLE SYMBOL	HOLE SIZE
A	.246
B	.189
C	.154
D	.125

FIG. 8-4-4   Rectangular coordinate dimensioning without dimension lines (arrowless dimensioning).

FIG. 8-4-5   Tabular dimensioning.

HOLE DIA	HOLE SYMBOL	LOCATION		
		X	Y	Z
5.6	A₁	60	40	18
4.8	B₁	10	40	THRU
	B₂	75	40	THRU
	B₃	60	16	THRU
	B₄	80	16	THRU
4	C₁	18	40	THRU
	C₂	55	40	THRU
	C₃	10	20	THRU
	C₄	30	20	THRU
	C₅	75	20	THRU
	C₆	18	16	THRU
3.2	D₁	55	8	12
8.1	E₁	42	20	12

FIG. 8-4-6   Tabular dimensioning with origin (0, 0) for X and Y axes located at the center of the part.

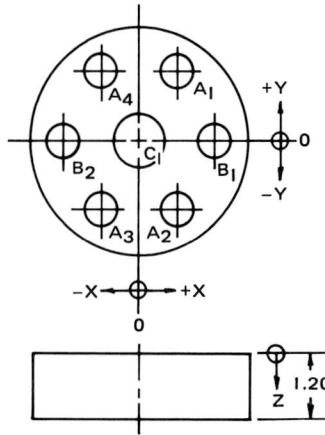

HOLE DIA	HOLE SYMBOL	LOCATION		
		X	Y	Z
.375	A₁	.50	.75	THRU
	A₂	.50	−.75	THRU
	A₃	−.50	−.75	THRU
	A₄	−.50	.75	THRU
.250	B₁	1.00	0	.60
	B₂	−1.00	0	.60
.500	C₁	0	0	THRU

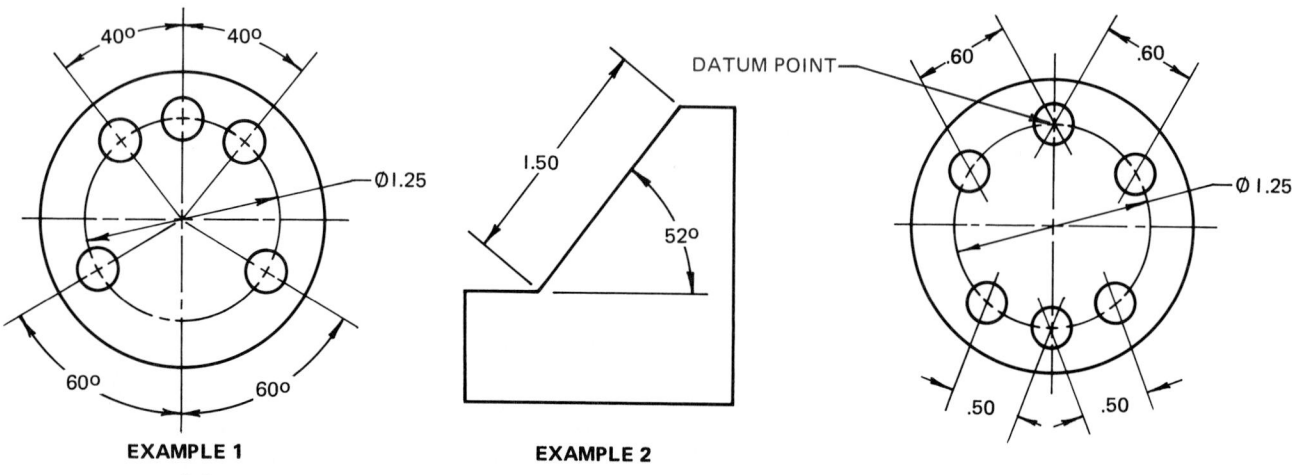

(A) POLAR COORDINATE DIMENSIONING

(B) CHORDAL DIMENSIONING

FIG. 8-4-7   Polar coordinate and chordal dimensioning.

Polar Coordinate Dimensioning

Polar coordinate dimensioning is commonly used in circular planes or circular configurations of features. It is a method of indicating the position of a point, line, or surface by means of a linear dimension and an angle, other than 90°, that is implied by the vertical and horizontal center lines (Fig. 8-4-7A).

Chordal Dimensioning

The chordal dimensioning system may also be used for the spacing of points on the circumference of a circle relative to a datum, where manufacturing methods indicate that this will be convenient (Fig. 8-4-7B).

True-Position Dimensioning

True-position dimensioning has many advantages over the coordinate dimensioning system (Fig. 8-4-8). Because of its scope, it is covered as a complete topic in Chap. 16.

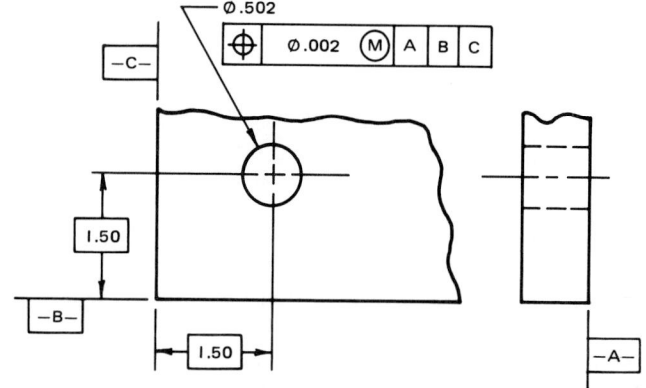

FIG. 8-4-8   True-position dimensioning.

Chain Dimensioning

When a series of dimensions is applied on a point-to-point basis, it is called chain dimensioning (Fig. 8-4-9). A possible disadvantage of this system is that it may result in an undesirable accumulation of tolerances between individual features. See Unit 8-5.

188

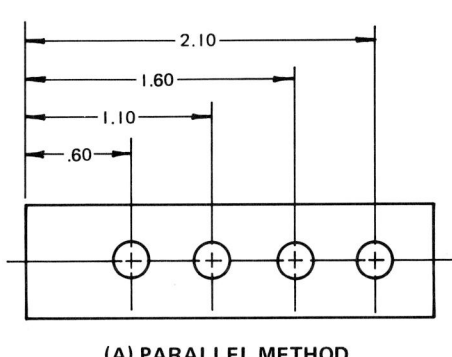

FIG. 8-4-9   Chain dimensioning.

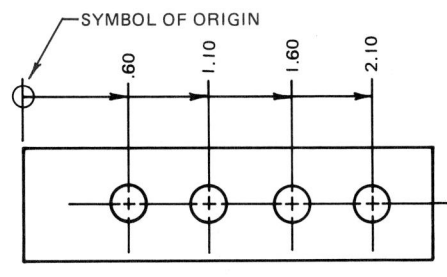

(A) PARALLEL METHOD

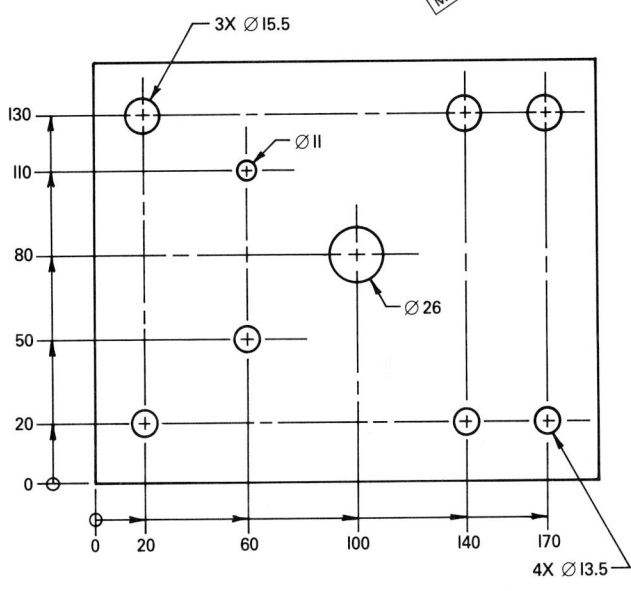

(B) SUPERIMPOSED METHOD

FIG. 8-4-10   Common-point (baseline) dimensioning.

Datum or Common-Point Dimensioning

When several dimensions emanate from a common reference point or line, the method is called *common-point* or *datum dimensioning*. Dimensioning from reference lines may be executed as parallel dimensioning or as a superimposed running dimensioning (Fig. 8-4-10).

Superimposed running dimensioning is simplified parallel dimensioning and may be used where there are space problems. Dimensions should be placed near the arrowhead, in line with the corresponding extension line, as shown in Fig. 8-4-10B. The origin is indicated by a circle and the opposite end of each dimension is terminated with an arrowhead.

It may be advantageous to use superimposed running dimensions in two directions. In such cases, the origins may be shown as in Fig. 8-4-11, or at the center of a hole or other feature.

REFERENCES AND SOURCE MATERIAL

1. ASME Y14.5M–1994, *Dimensioning and Tolerancing.*
2. CAN/CSA B78.2–M91, *Dimensioning and Tolerancing of Technical Drawings.*

ASSIGNMENTS

See Assignments 13 through 18 for Unit 8-4 on pages 217 through 219.

FIG. 8-4-11   Superimposed running dimensions in two directions.

8-5 LIMITS AND TOLERANCES

In the 6000 years of the history of technical drawing as a means for the communication of engineering information, it seems inconceivable that such an elementary practice as the tolerancing of dimensions, which we take so much for granted today, was introduced for the first time about 80 years ago.

Apparently, engineers and fabricators came in a very gradual manner to the realization that exact dimensions and shapes could not be attained in the manufacture of materials and products.

The skilled tradespeople of old prided themselves on being able to work to exact dimensions. What they really meant was that they dimensioned objects with a degree of accuracy greater than that with which they could measure. The use of modern measuring instruments would easily have shown the deviations from the sizes that they called exact.

As soon as it was realized that variations in the sizes of parts had always been present, that such variations could be restricted but not avoided, and also that slight variations in the size that a part was originally intended to have could be tolerated without its correct functioning being impaired, it was evident that interchangeable parts need not be identical parts, but that it would be sufficient if the significant sizes that controlled

189

FIG. 8-5-1   A working drawing.

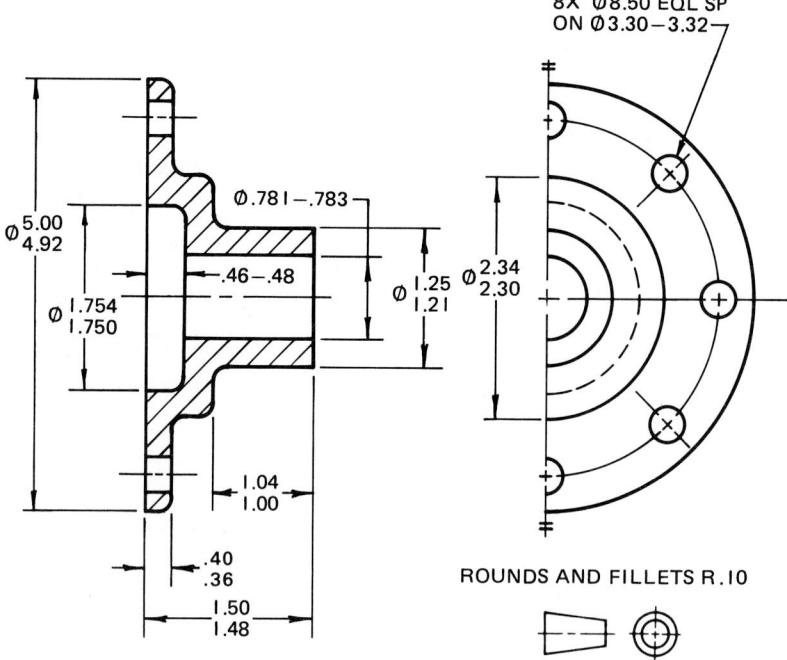

8X Ø8.50 EQL SP
ON Ø3.30−3.32

ROUNDS AND FILLETS R.10

their fits lay between definite limits. Accordingly, the problem of interchangeable manufacture evolved from the making of parts to a would-be exact size, to the holding of parts between two limiting sizes lying so closely together that any intermediate size would be acceptable.

Tolerances are the permissible variations in the specified form, size, or location of individual features of a part from that shown on the drawing. The finished form and size into which material is to be fabricated are defined on a drawing by various geometric shapes and dimensions.

As mentioned previously, the manufacturer cannot be expected to produce the exact size of parts as indicated by the dimensions on a drawing, so a certain amount of variation on each dimension must be tolerated. For example, a dimension given as 1.500 ± .004 in. means that the manufactured part can be anywhere between 1.496 and 1.504 in. and that the tolerance permitted on this dimension is .008 in. The largest and smallest permissible sizes (1.504 and 1.496 in., respectively) are known as the *limits*.

Greater accuracy costs more money, and since economy in manufacturing would not permit all dimensions to be held to the same accuracy, a system for dimensioning must be used (Fig. 8-5-1). Generally, most parts require only a few features to be held to high accuracy.

In order that assembled parts may function properly and to allow for interchangeable manufacturing, it is necessary to permit only a certain amount of tolerance on each of the mating parts and a certain amount of allowance between them.

Key Concepts

In order to calculate limit dimensions, the following concepts should be clearly understood (refer to Fig. 8-5-2).

Actual Size   The actual size is the measured size.

Basic Size   The basic size of a dimension is the theoretical size from which the limits for that dimension are derived by the application of the allowance and tolerance.

Design Size   Design size refers to the size from which the limits of size are derived by the application of tolerances.

Limits of Size   These limits are the maximum and minimum sizes permissible for a specific dimension.

Nominal Size   The nominal size is the designation used for the purpose of general identification.

Tolerance   The tolerance on a dimension is the total permissible variation in the size of a dimension. The tolerance is the difference between the limits of size.

Bilateral Tolerance   With bilateral tolerance, variation is permitted in both directions from the specified dimension.

TERMINOLOGY	EXAMPLE	EXPLANATION
BASIC SIZE	1.500	
BASIC SIZE WITH TOLERANCE ADDED	1.500 ±.004 ←	HALF OF TOTAL TOLERANCE
LIMITS OF SIZE	1.504 1.496	LARGEST AND SMALLEST SIZES PERMITTED
TOLERANCE	.008	DIFFERENCE BETWEEN LIMITS OF SIZE

FIG. 8-5-2   Limit and tolerance terminology.

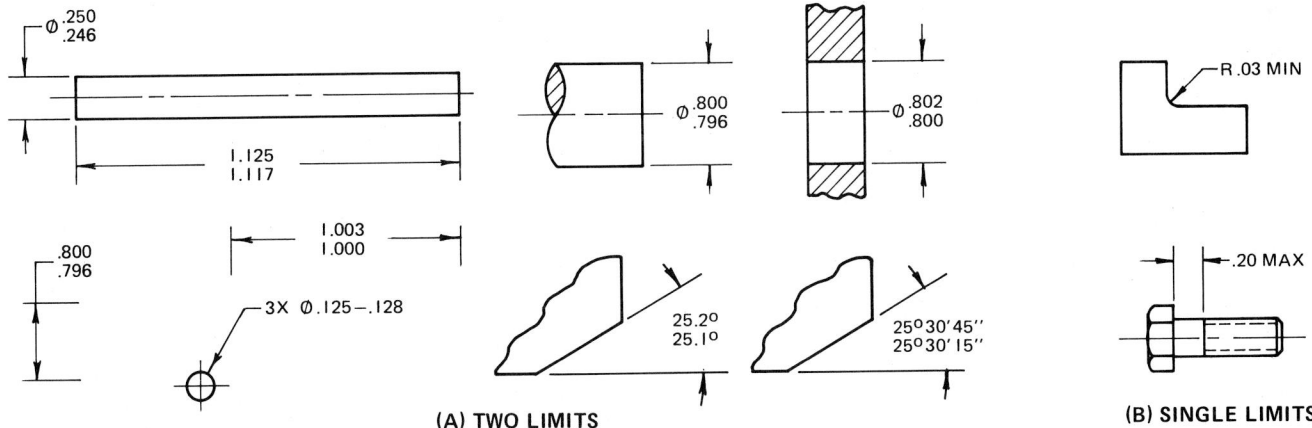

FIG. 8-5-3   Methods of indicating tolerances on drawings.

Unilateral Tolerance   With unilateral tolerance, variation is permitted in only one direction from the specified dimension.

Maximum Material Size   The maximum material size is that limit of size of a feature that results in the part containing the maximum amount of material. Thus it is the maximum limit of size for a shaft or an external feature, or the minimum limit of size for a hole or internal feature.

Tolerancing

All dimensions required in the manufacture of a product have a tolerance, except those identified as reference, maximum, minimum, or stock. Tolerances may be expressed in one of the following ways:

- As specified limits of tolerances shown directly on the drawing for a specified dimension (Fig. 8-5-3).
- As plus-and-minus tolerancing.
- Combining a dimension with a tolerance symbol. (See "Symbols" in Unit 8-6.)
- In a general tolerance note, referring to all dimensions on the drawing for which tolerances are not otherwise specified.
- In the form of a note referring to specific dimensions.
- Tolerances on dimensions that locate features may be applied directly to the locating dimensions or by the positional tolerancing method described in Chap. 16.
- Tolerancing applicable to the control of form and runout, referred to as *geometric tolerancing,* is also covered in detail in Chap. 16.

Direct Tolerancing Methods

A tolerance applied directly to a dimension may be expressed in two ways.

Limit Dimensioning   For this method, the high limit (maximum value) is placed above the low limit (minimum value). When it is expressed in a single line, the low limit precedes the high limit and they are separated by a dash (Figs. 8-5-3 and 8-5-4).

Where limit dimensions are used and where either the maximum or minimum dimension has digits to the right of the decimal point, the other value should have the zeros added so that both the limits of size are expressed to the same number of decimal places. This applies to both U.S. customary and metric drawings. For example:

$$\begin{matrix} 30.75 \\ 30.00 \end{matrix} \quad not \quad \begin{matrix} 30.75 \\ 30 \end{matrix} \quad and \quad \begin{matrix} .750 \\ .748 \end{matrix} \quad not \quad \begin{matrix} .75 \\ .748 \end{matrix}$$

Plus-and-Minus Tolerancing   (Refer to Fig. 8-5-5, pg. 192.) For this method the dimension of the specified size is given first and is followed by a plus or minus expression of tolerancing. The plus value should be placed above the minus value. This type of tolerancing can be broken down into bilateral and

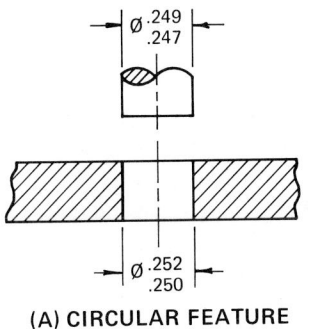

(A) CIRCULAR FEATURE

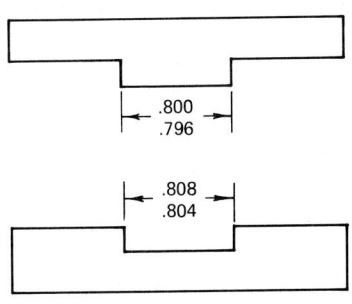

(B) FLAT FEATURE

FIG. 8-5-4   Limit dimensioning application.

unilateral tolerancing. In a bilateral tolerance, the plus-and-minus tolerances should normally be equal, but special design considerations may sometimes dictate unequal values (Fig. 8-5-6). The specified size is the design size, and the tolerance represents the desired control of quality and appearance.

Metric Tolerancing   In the metric system the dimension need not be shown to the same number of decimal places as its tolerance. For example:

$$1.5 \pm 0.04 \quad not \quad 1.50 \pm 0.04$$

$$10 \pm 0.1 \quad not \quad 10.0 \pm 0.1$$

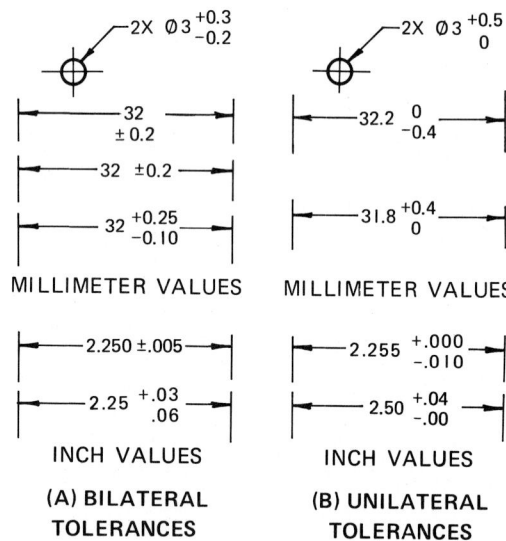

MILLIMETER VALUES   MILLIMETER VALUES

INCH VALUES   INCH VALUES

(A) BILATERAL TOLERANCES   **(B) UNILATERAL TOLERANCES**

FIG. 8-5-5   Plus-and-minus tolerancing.

Where bilateral tolerancing is used, both the plus and minus values have the same number of decimal places, using zeros where necessary. For example:

$$30^{+0.15}_{-0.10} \quad not \quad 30^{+0.15}_{-0.1}$$

Where unilateral tolerancing is used and either the plus or minus value is nil, a single zero is shown without a plus or minus sign. For example:

$$40^{0}_{-0.15} \quad or \quad 40^{+0.15}_{0}$$

An application of unilateral tolerancing is shown in Fig. 8-5-6B.

Inch Tolerancing   In the inch system the dimension is given to the same number or decimal places as its tolerance. For example:

Bilateral:

$$.500 \pm .004 \quad not \quad .50 \pm .004$$

Unilateral:

$$.750^{+.005}_{-.000} \quad not \quad .750^{+.005}_{-0}$$

$$30.0° \pm .2° \quad not \quad 30° \pm .2°$$

Conversion charts for tolerances are shown in Fig. 8-5-7.

General Tolerance Notes   The use of general tolerance notes greatly simplifies the drawing and saves considerable layout in

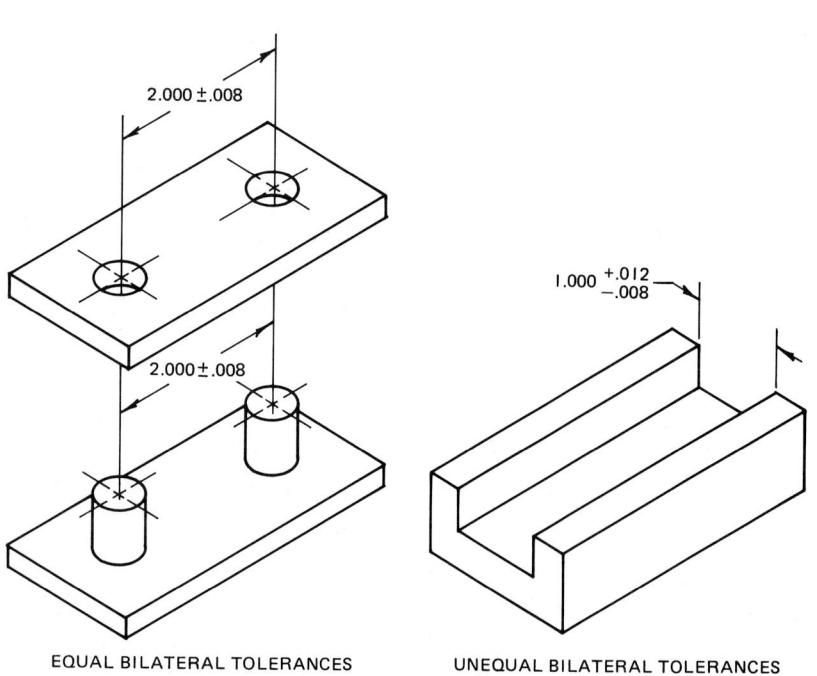

EQUAL BILATERAL TOLERANCES   UNEQUAL BILATERAL TOLERANCES

(A) BILATERAL TOLERANCES

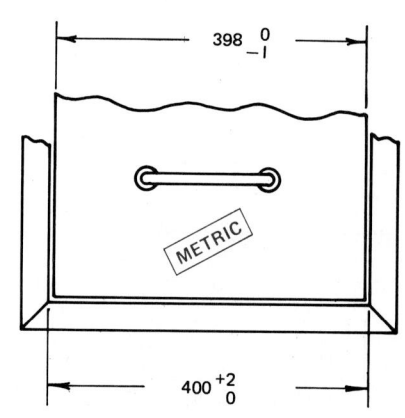

(B) UNILATERAL TOLERANCES

FIG. 8-5-6   Application of tolerances.

its preparation. The following examples illustrate the wide field of application of this system. The values given in the examples are typical.

EXAMPLE 1

EXCEPT WHERE STATED OTHERWISE, TOLERANCES ON FINISHED DECIMAL DIMENSIONS ±0.1.

TOTAL TOLERANCE IN INCHES		MILLIMETER CONVERSION ROUNDED TO
AT LEAST	LESS THAN	
.00004	.0004	4 DECIMAL PLACES
.0004	.004	3 DECIMAL PLACES
.004	.04	2 DECIMAL PLACES
.04	.4	1 DECIMAL PLACE
.4 AND OVER		WHOLE mm

TOTAL TOLERANCE IN MILLIMETERS		INCH CONVERSION ROUNDED TO
AT LEAST	LESS THAN	
0.002	0.02	5 DECIMAL PLACES
0.02	0.2	4 DECIMAL PLACES
0.2	2	3 DECIMAL PLACES
2 AND OVER		2 DECIMAL PLACES

FIG. 8-5-7 Conversion charts for tolerances.

EXAMPLE 2

EXCEPT WHERE STATED OTHERWISE, TOLERANCES ON FINISHED DIMENSIONS TO BE AS FOLLOWS:

Dimension (in.)	Tolerance
UP TO 4.00	± .004
FROM 4.01 TO 12.00	± .003
FROM 12.01 TO 24.00	± .02
OVER 24.00	± .04

A comparison between the tolerancing methods described is shown in Fig. 8-5-8.

Tolerance Accumulation

It is necessary also to consider the effect of each tolerance with respect to other tolerances, and not to permit a chain of tolerances to build up a cumulative tolerance between surfaces or points that have an important relation to one another. Where the position of a surface in any one direction is controlled by more than one tolerance, the tolerances are cumulative. Figure 8-5-9 (pg. 194) compares the tolerance accumulation resulting from three different methods of dimensioning.

Chain Dimensioning The maximum variation between any two features is equal to the sum of the tolerances on the intermediate distances. This results in the greatest tolerance accumulation, as illustrated by the ± .08 variation between holes X and Y, as shown in Fig. 8-5-9A.

Datum Dimensioning The maximum variation between any two features is equal to the sum of the tolerances on the two dimensions from the datum to the feature. This reduces the

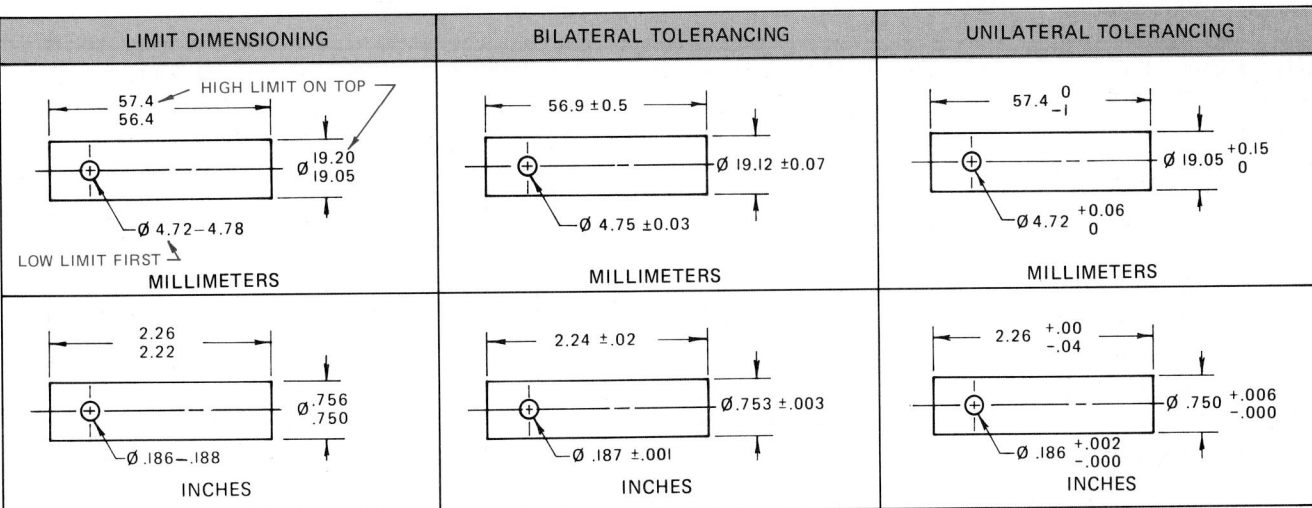

FIG. 8-5-8 A comparison of the tolerance methods.

FIG. 8-5-9   Dimensioning method comparison.

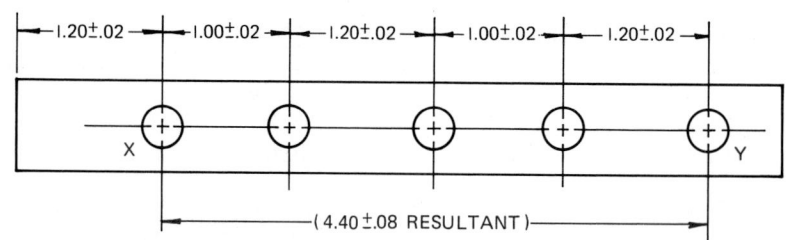

(A) CHAIN DIMENSIONING (GREATEST TOLERANCE ACCUMULATION)

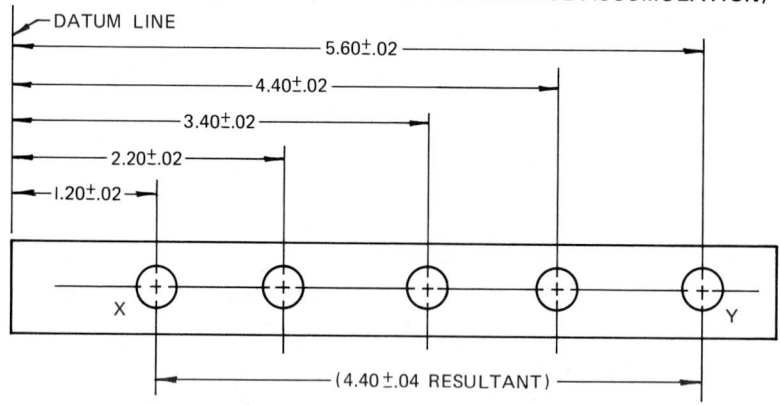

(B) DATUM DIMENSIONING (LESSER TOLERANCE ACCUMULATION)

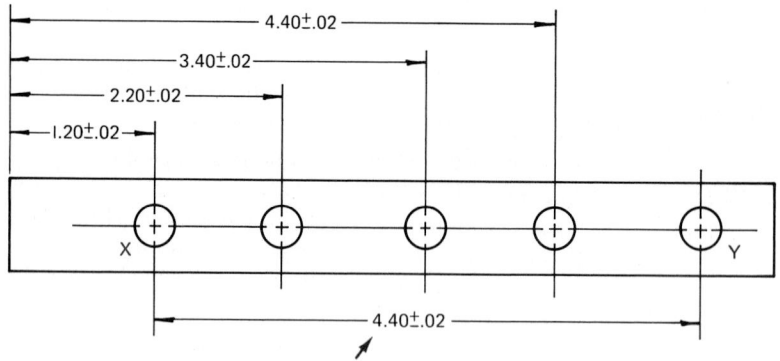

(C) DIRECT DIMENSIONING (LEAST TOLERANCE ACCUMULATION)

tolerance accumulation, as illustrated by the ±.04 variation between holes *X* and *Y* in Fig. 8-5-9B.

Direct Dimensioning   The maximum variation between any two features is controlled by the tolerance on the dimension between the features. This results in the least tolerance accumulation, as illustrated by the ± .02 variation between holes *X* and *Y* in Fig. 8-5-9C.

Additional Rules for Dimensioning

- The engineering intent must be clearly defined.
- Dimensions must be complete enough to describe the total geometry of each feature. Determining a shape by measuring its size on a drawing or by assuming a distance or size is not acceptable.
- Dimensions should be selected and arranged to avoid unsatisfactory accumulation of tolerances to preclude

more than one interpretation, and to ensure a proper fit between mating parts.
- The finished part should be defined without specifying manufacturing methods. Thus only the diameter of a hole is given, without indicating if it is to be drilled, reamed, punched, or made by any other operation.
- Dimensions must be selected to give required information directly. Dimensions should preferably be shown in true profile views and refer to visible outlines rather than to hidden lines. A common exception to this general rule is a diametral dimension on a section view.
- Drawings that illustrate part surfaces or center lines at right angles to each other, but without an angular dimension, are interpreted as being 90° between these surfaces or center lines. Actual surfaces, axes, and center planes may vary within their specified tolerance of perpendicularity.
- Dimension lines are placed outside the outline of the part and between the views unless the drawing may be simplified or clarified by doing otherwise.

- Dimension lines should be aligned, if practicable, and should be grouped for uniform appearance.

REFERENCES AND SOURCE MATERIAL

1. ASME Y14.5M–1994, *Dimensioning and Tolerancing.*
2. CAN/CSA B78.2–M91, *Dimensioning and Tolerancing of Technical Drawings.*

ASSIGNMENT

See Assignment 19 for Unit 8-5 on page 220.

8-6 FITS AND ALLOWANCES

In order that assembled parts may function properly and to allow for interchangeable manufacturing, it is necessary to permit only a certain amount of tolerance on each of the mating parts and a certain amount of allowance between them.

Fits

The fit between two mating parts is the relationship between them with respect to the amount of clearance or interference present when they are assembled. There are three basic types of fits: clearance, interference, and transition.

Clearance Fit   A fit between mating parts having limits of size so prescribed that a clearance always results in assembly.

Interference Fit   A fit between mating parts having limits of size so prescribed that an interference always results in assembly.

Transition Fit   A fit between mating parts having limits of size so prescribed as to partially or wholly overlap, so that either a clearance or interference may result in assembly.

Allowance

An *allowance* is an intentional difference between the maximum material limits of mating parts. It is the minimum clearance (positive allowance) or maximum interference (negative allowance) between such parts.

The most important terms relating to limits and fits are shown in Fig. 8-6-1. The terms are defined as follows:

Basic Size   The size to which limits or deviations are assigned. The basic size is the same for both members of a fit.

Deviation   The algebraic difference between a size and the corresponding basic size.

Upper Deviation   The algebraic difference between the maximum limit of size and the corresponding basic size.

Lower Deviation   The algebraic difference between the minimum limit of size and the corresponding basic size.

Tolerance   The difference between the maximum and minimum size limits on a part.

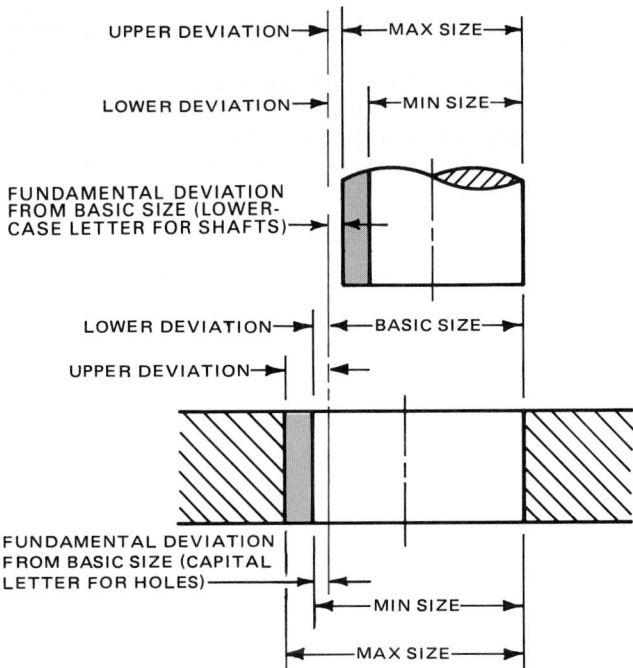

FIG. 8-6-1   Illustration of definitions.

Tolerance Zone   A zone representing the tolerance and its position in relation to the basic size.

Fundamental Deviation   The deviation closest to the basic size.

Description of Fits

Running and Sliding Fits

Running and sliding fits, for which tolerances and clearances are given in the Appendix, represent a special type of clearance fit. These are intended to provide a similar running performance, with suitable lubrication allowance, throughout the range of sizes.

Locational Fits

Locational fits are intended to determine only the location of the mating parts; they may provide rigid or accurate location, as with interference fits, or some freedom of location, as with clearance fits. Accordingly, they are divided into three groups: clearance fits, transition fits, and interference fits.

Locational clearance fits are intended for parts that are normally stationary but that can be freely assembled or disassembled. They run from snug fits for parts requiring accuracy of location, through the medium clearance fits for parts such as ball, race, and housing, to the looser fastener fits where freedom of assembly is of prime importance.

Locational transition fits are a compromise between clearance and interference fits for application where accuracy of location is important but a small amount of either clearance or interference is permissible.

Locational interference fits are used where accuracy of location is of prime importance and for parts requiring rigidity and alignment with no special requirements for bore pressure. Such fits are not intended for parts designed to transmit frictional loads from one part to another by virtue of the tightness of fit; these conditions are covered by force fits.

Drive and Force Fits

Drive and force fits constitute a special type of interference fit, normally characterized by maintenance of constant bore pressures throughout the range of sizes. The interference therefore varies almost directly with diameter, and the difference between its minimum and maximum values is small, to maintain the resulting pressures within reasonable limits.

Interchangeability of Parts

Increased demand for manufactured products led to the development of new production techniques. Interchangeability of parts became the basis for mass-production, low-cost manufacturing, and it brought about the refinement of machinery, machine tools, and measuring devices. Today it is possible and generally practical to design for 100 percent interchangeability.

No part can be manufactured to exact dimensions. Tool wear, machine variations, and the human factor all contribute to some degree of deviation from perfection. It is therefore necessary to determine the deviation and permissible clearance, or interference, to produce the desired fit between parts.

Modern industry has adopted three basic approaches to manufacturing:

1. *The completely interchangeable assembly.* Any and all mating parts of a design are toleranced to permit them to assemble and function properly without the need for machining or fitting at assembly.
2. *The fitted assembly.* Mating features of a design are fabricated either simultaneously or with respect to one another. Individual members of mating features are not interchangeable.
3. *The selected assembly.* All parts are mass-produced, but members of mating features are individually selected to provide the required relationship with one another.

Standard Inch Fits

Standard fits are designated for design purposes in specifications and on design sketches by means of symbols, as shown in Fig. 8-6-2. These symbols, however, are not intended to be shown directly on shop drawings; instead the actual limits of size are determined and specified on the drawings.

The letter symbols used are as follows:

RC   Running and sliding fit
LC   Locational clearance fit
LT   Locational transition fit
LN   Locational interference fit
FN   Force or shrink fit

These letter symbols are used in conjunction with numbers representing the class of fit; for example, FN4 represents a class 4 force fit.

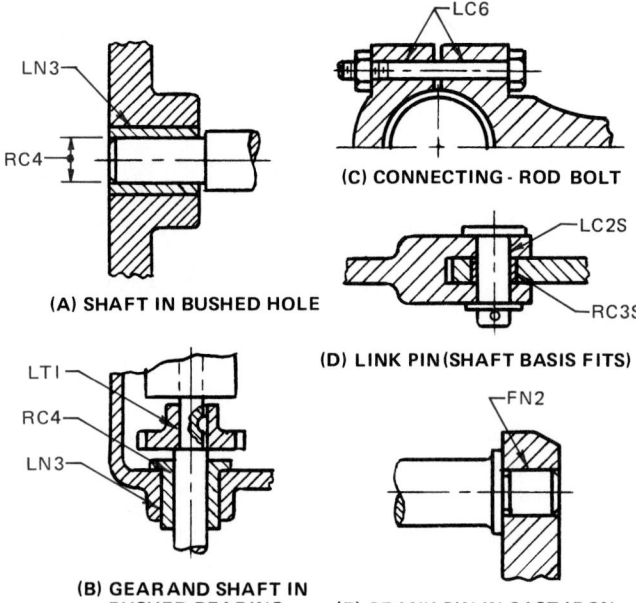

FIG. 8-6-2   Typical design sketches showing classes of fits.

Each of these symbols (two letters and a number) represents a complete fit, for which the minimum and maximum clearance or interference, and the limits of size for the mating parts, are given directly in Appendix tables 43 through 47. See Fig. 8-6-3.

Running and Sliding Fits

RC1 Precision Sliding Fit   This fit is intended for the accurate location of parts that must assemble without perceptible play, for high-precision work such as gages.

RC2 Sliding Fit   This fit is intended for accurate location, but with greater maximum clearance than class RC1. Parts made to this fit move and turn easily but are not intended to run freely, and in the larger sizes may seize with small temperature changes.

Note: LC1 and LC2 locational clearance fits may also be used as sliding fits with greater tolerances.

RC3 Precision Running Fit   This fit is about the closest fit that can be expected to run freely and is intended for precision work for oil-lubricated bearings at slow speeds and light journal pressures, but is not suitable where appreciable temperature differences are likely to be encountered.

RC4 Close Running Fit   This fit is intended chiefly as a running fit for grease- or oil-lubricated bearings on accurate machinery with moderate surface speeds and journal pressures, where accurate location and minimum play are desired.

RC5 and RC6 Medium Running Fits   These fits are intended for higher running speeds and/or where temperature variations are likely to be encountered.

RC7 Free Running Fit   This fit is intended for use where accuracy is not essential and/or where large temperature variations are likely to be encountered.

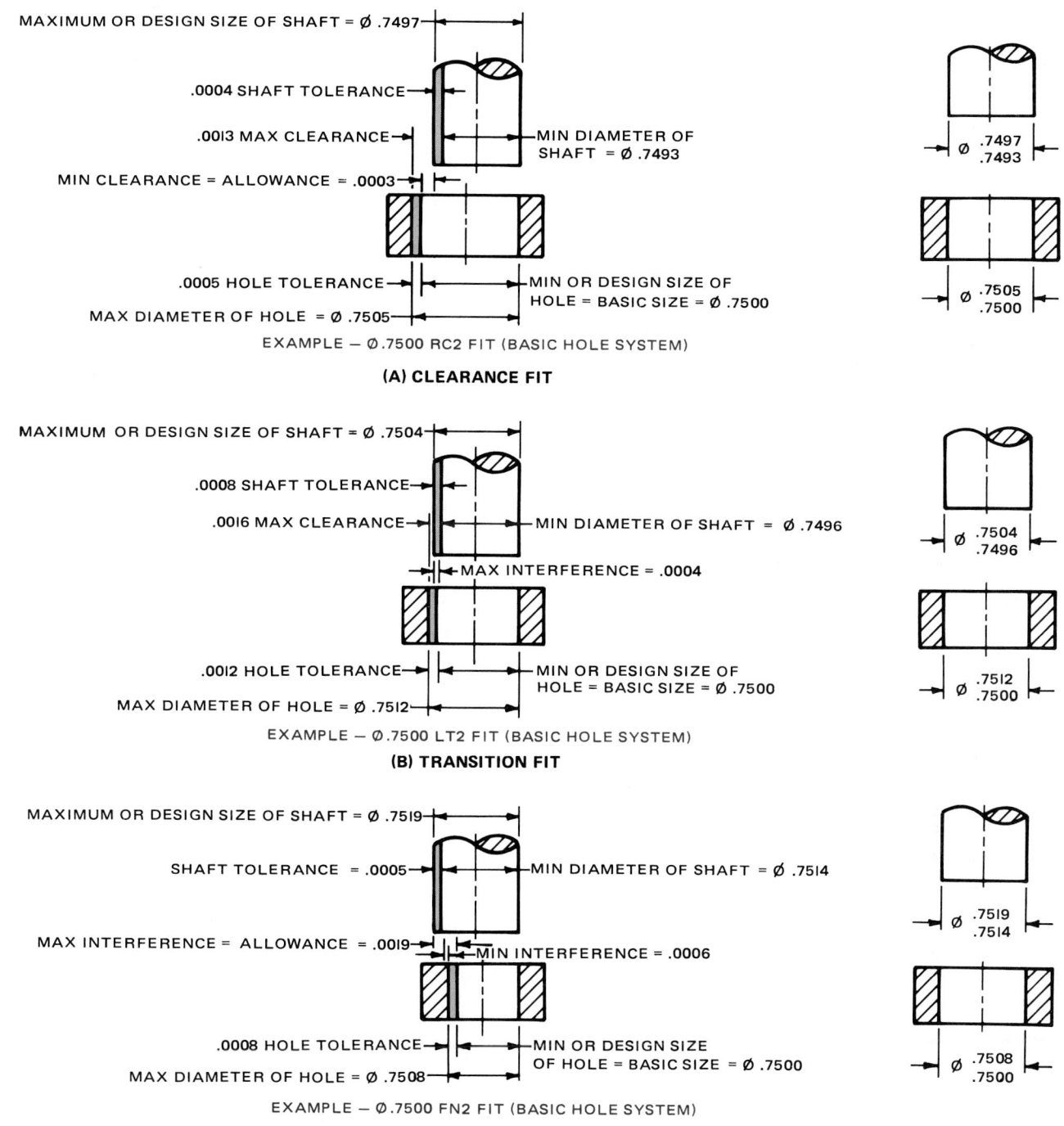

MAXIMUM OR DESIGN SIZE OF SHAFT = Ø .7497

.0004 SHAFT TOLERANCE

.0013 MAX CLEARANCE

MIN DIAMETER OF SHAFT = Ø .7493

MIN CLEARANCE = ALLOWANCE = .0003

.0005 HOLE TOLERANCE

MIN OR DESIGN SIZE OF HOLE = BASIC SIZE = Ø .7500

MAX DIAMETER OF HOLE = Ø .7505

Ø .7497/.7493

Ø .7505/.7500

EXAMPLE — Ø.7500 RC2 FIT (BASIC HOLE SYSTEM)

(A) CLEARANCE FIT

MAXIMUM OR DESIGN SIZE OF SHAFT = Ø .7504

.0008 SHAFT TOLERANCE

.0016 MAX CLEARANCE

MIN DIAMETER OF SHAFT = Ø .7496

MAX INTERFERENCE = .0004

.0012 HOLE TOLERANCE

MIN OR DESIGN SIZE OF HOLE = BASIC SIZE = Ø .7500

MAX DIAMETER OF HOLE = Ø .7512

Ø .7504/.7496

Ø .7512/.7500

EXAMPLE — Ø.7500 LT2 FIT (BASIC HOLE SYSTEM)

(B) TRANSITION FIT

MAXIMUM OR DESIGN SIZE OF SHAFT = Ø .7519

SHAFT TOLERANCE = .0005

MIN DIAMETER OF SHAFT = Ø .7514

MAX INTERFERENCE = ALLOWANCE = .0019

MIN INTERFERENCE = .0006

.0008 HOLE TOLERANCE

MIN OR DESIGN SIZE OF HOLE = BASIC SIZE = Ø .7500

MAX DIAMETER OF HOLE = Ø .7508

Ø .7519/.7514

Ø .7508/.7500

EXAMPLE — Ø.7500 FN2 FIT (BASIC HOLE SYSTEM)

(C) INTERFERENCE FIT

FIG. 8-6-3   Types of inch fits.

RC8 and RC9 Loose Running Fits   These fits are intended for use where materials made to commercial tolerances, such as cold-rolled shafting, tubing, etc., are involved.

Locational Clearance Fits

Locational clearance fits are intended for parts that are normally stationary but that can be freely assembled or disassembled. They run from snug fits for parts requiring accuracy of location, through the medium-clearance fits for parts such as spigots, etc., to the looser fastener fits where freedom of assembly is of prime importance.

These are classified as follows:

LC1 to LC4   These fits have a minimum zero clearance, but in practice the probability is that the fit will always have a clearance. These fits are suitable for location of nonrunning parts and spigots, although classes LC1 and LC2 may also be used for sliding fits.

197

LC5 and LC6   These fits have a small minimum clearance, intended for close location fits for nonrunning parts. LC5 can also be used in place of RC2 as a free-slide fit, and LC6 may be used as a medium running fit having greater tolerances than RC5 and RC6.

LC7 and LC11   These fits have progressively larger clearances and tolerances and are useful for various loose clearances for assembly of bolts and similar parts.

Locational Transition Fits

Locational transition fits are a compromise between clearance and interference fits for application where accuracy of location is important, but either a small amount of clearance or interference is permissible.

These are classified as follows:

LT1 and LT2   These fits average a slight clearance, giving a light push fit, and are intended for use where the maximum clearance must be less than for the LC1 to LC3 fits, and where slight interference can be tolerated for assembly by pressure or light hammer blows.

LT3 and LT4   These fits average virtually no clearance and are for use where some interference can be tolerated, for example, to eliminate vibration. These are sometimes referred to as an *easy keying fit* and are used for shaft keys and ball race fits. Assembly is generally by pressure or hammer blows.

LT5 and LT6   These fits average a slight interference, although appreciable assembly force will be required when extreme limits are encountered, and selective assembly may be desirable. These fits are useful for heavy keying, for ball race fits subject to heavy duty and vibration, and as light press fits for steel parts.

Locational Interference Fits

Locational interference fits are used where accuracy of location is of prime importance, and for parts requiring rigidity and alignment with no special requirements for bore pressure. Such fits are not intended for parts designed to transmit frictional loads from one part to another by virtue of the tightness of fit, as these conditions are covered by force fits.

These are classified as follows:

LN1 and LN2   These are light press fits, with very small minimum interference, suitable for parts such as dowel pins, which are assembled with an arbor press in steel, cast iron, or brass. Parts can normally be dismantled and reassembled, as the interference is not likely to overstrain the parts, but the interference is too small for satisfactory fits in elastic materials or light alloys.

LN3   This is suitable as a heavy press fit in steel and brass, or a light press fit in more elastic materials and light alloys.

LN4 to LN6   While LN4 can be used for permanent assembly of steel parts, these fits are primarily intended as press fits for more elastic or soft materials, such as light alloys and the more rigid plastics.

Force or Shrink Fits

Force or shrink fits constitute a special type of interference fit, normally characterized by maintenance of constant bore pressures throughout the range of sizes. The interference therefore varies almost directly with diameter, and the difference between its minimum and maximum values is small to maintain the resulting pressures within reasonable limits.

These fits may be described briefly as follows:

FN1 Light Drive Fit   Requires light assembly pressure and produces more or less permanent assemblies. It is suitable for thin sections or long fits, or in cast iron external members.

FN2 Medium Drive Fit   Suitable for ordinary steel parts or as a shrink fit on light sections. It is about the tightest fit that can be used with high-grade cast iron external members.

FN3 Heavy Drive Fit   Suitable for heavier steel parts or as a shrink fit in medium sections.

FN4 and FN5 Force Fits   Suitable for parts that can be highly stressed and/or for shrink fits where the heavy pressing forces required are impractical.

Basic Hole System

In the basic hole system, which is recommended for general use, the basic size will be the design size for the hole, and the tolerance will be plus. The design size for the shaft will be the basic size minus the minimum clearance, or plus the maximum interference, and the tolerance will be minus, as given in the tables in the Appendix. For example (see Table 43 in the Appendix), for a 1 in. RC7 fit, values of +.0020, .0025, and −.0012 are given; hence limits will be:

$$\text{Hole } \varnothing \; 1.0000 \; {}^{+.0020}_{-.0000}$$

$$\text{Shaft } \varnothing \; .9975 \; {}^{+.0000}_{-.0012}$$

Basic Shaft System

Fits are sometimes required on a basic shaft system, especially in cases where two or more fits are required on the same shaft. This is designated for design purposes by a letter S following the fit symbol; for example, RC7S.

Tolerances for holes and shaft are identical with those for a basic hole system, but the basic size becomes the design size for the shaft and the design size for the hole is found by adding the minimum clearance or subtracting the maximum interference from the basic size.

For example, for a 1 in. RC7S fit, values of +.0020, .0025, and −.0012 are given; therefore, limits will be:

$$\text{Hole } \varnothing \; 1.0025 \; {}^{+.0020}_{-.0000}$$

$$\text{Shaft } \varnothing \; 1.0000 \; {}^{+.0000}_{-.0012}$$

Preferred Metric Limits and Fits

The ISO system of limits and fits for mating parts is approved and adopted for general use in the United States. It establishes the designation symbols used to define specific dimensional limits on drawings.

The general terms "hole" and "shaft" can also be taken as referring to the space containing or contained by two parallel faces of any part, such as the width of a slot, or the thickness of a key.

An "International Tolerance grade" establishes the magnitude of the tolerance zone or the amount of part size variation allowed for internal and external dimensions alike (Table 40, Appendix). There are 18 tolerance grades, which are identified by the prefix IT, such a IT6, IT11, etc. The smaller the grade number, the smaller the tolerance zone. For general applications of IT grades, see Fig. 8-6-4.

Grades 1 to 4 are very precise grades intended primarily for gage making and similar precision work, although grade 4 can also be used for very precise production work.

Grades 5 to 16 represent a progressive series suitable for cutting operations, such as turning, boring, grinding, milling, and sawing. Grade 5 is the most precise grade, obtainable by fine grinding and lapping, while 16 is the coarsest grade for rough sawing and machining.

Grades 12 to 16 are intended for manufacturing operations such as cold heading, pressing, rolling, and other forming operations.

As a guide to the selection of tolerances, Fig. 8-6-4B has been prepared to show grades that may be expected to be held by various manufacturing processes for work in metals. For work in other materials, such as plastics, it may be necessary to use coarser tolerance grades for the same process.

A fundamental deviation establishes the position of the tolerance zone with respect to the basic size. Fundamental deviations are expressed by *tolerance position letters*. Capital letters are used for internal dimensions, and lowercase letters for external dimensions.

Tolerance Symbol

For metric application of limits and fits, the tolerance may be indicated by a basic size and tolerance symbol. By combining the IT grade number and the tolerance position letter, the tolerance symbol is established that identifies the actual maximum and minimum limits of the part. The toleranced sizes are thus defined by the basic size of the part followed by the symbol composed of a letter and a number (Fig. 8-6-5, pg. 200).

Preferred Tolerance Grades

The preferred tolerance grades are shown in Table 40 of the Appendix. The encircled tolerance grades (13 each) are first choice, the framed tolerance grades are second choice, and the open tolerance grades are third choice.

Hole Basis Fits System

In the hole basis fits system (see Table 48 of the Appendix) the basic size will be the minimum size of the hole. For example,

FIG. 8-6-4 International Tolerance (IT) grades.

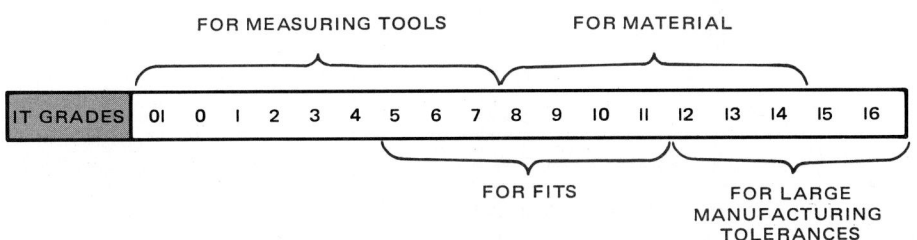

(A) APPLICATIONS

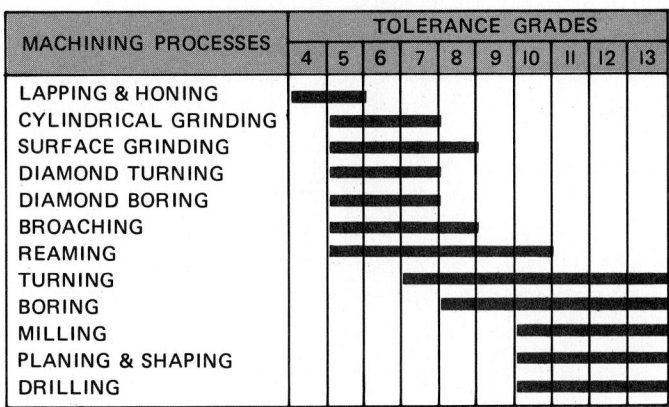

(B) APPLICATIONS FOR MACHINING PROCESSES

TOLERANCE ZONE SYMBOL ─┐

40 H 8

BASIC SIZE ─┘

FUNDAMENTAL DEVIATION ─
(POSITION LETTER – CAPITAL
LETTER FOR INTERNAL DIMENSION)

└ INTERNATIONAL
  TOLERANCE GRADE
  (IT NUMBER)

(A) INTERNAL DIMENSION (HOLES)

TOLERANCE ZONE SYMBOL ─┐

40 f 7

BASIC SIZE ─┘

FUNDAMENTAL DEVIATION ─
(POSITION LETTER – LOWERCASE
LETTER FOR EXTERNAL DIMENSION)

└ INTERNATIONAL
  TOLERANCE GRADE
  (IT NUMBER)

(B) EXTERNAL DIMENSION (SHAFTS)

FIG. 8-6-5   Metric tolerance symbol.

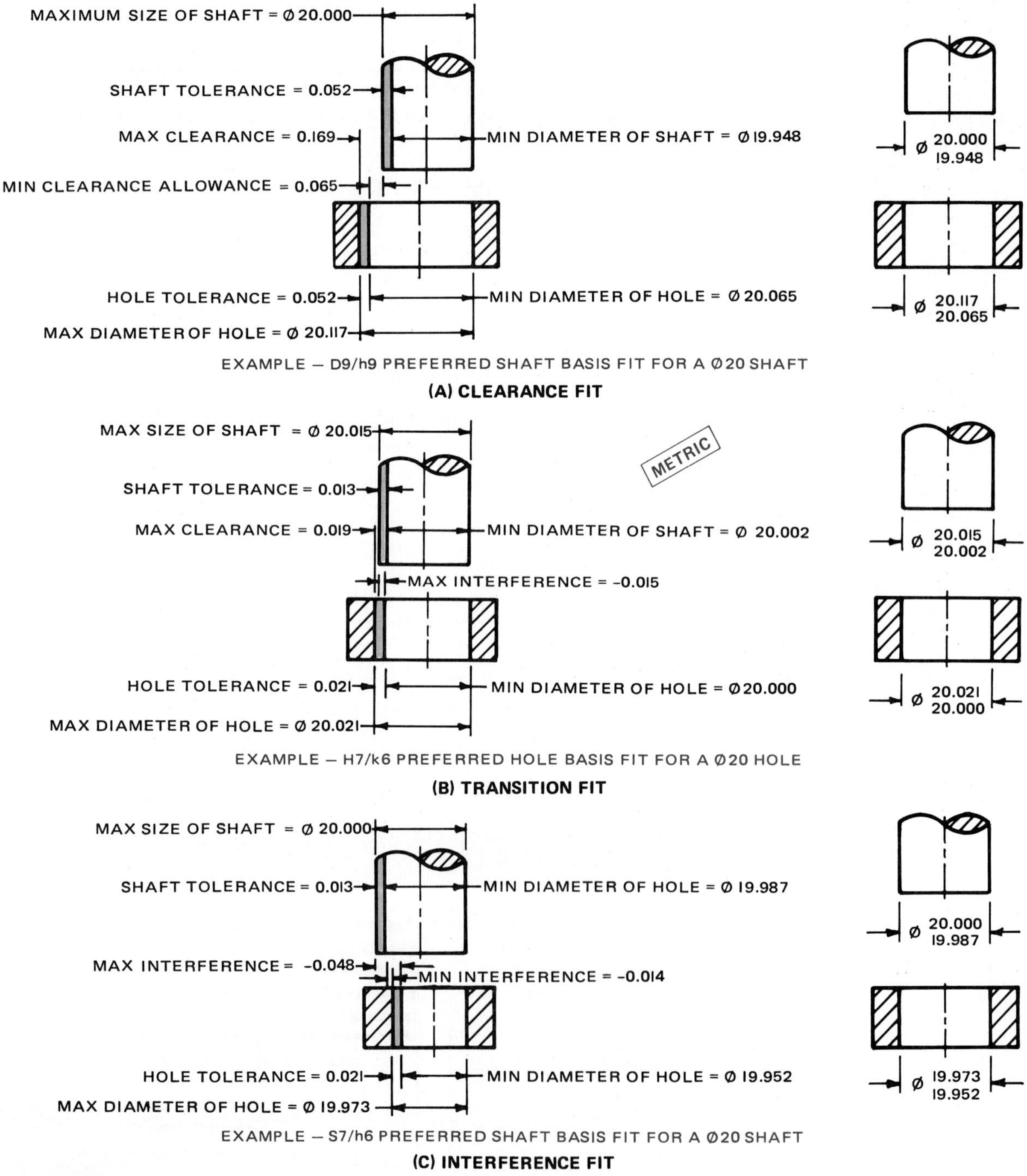

MAXIMUM SIZE OF SHAFT = Ø 20.000

SHAFT TOLERANCE = 0.052

MAX CLEARANCE = 0.169 ── MIN DIAMETER OF SHAFT = Ø 19.948

MIN CLEARANCE ALLOWANCE = 0.065

HOLE TOLERANCE = 0.052 ── MIN DIAMETER OF HOLE = Ø 20.065
MAX DIAMETER OF HOLE = Ø 20.117

Ø 20.000 / 19.948

Ø 20.117 / 20.065

EXAMPLE — D9/h9 PREFERRED SHAFT BASIS FIT FOR A Ø 20 SHAFT

(A) CLEARANCE FIT

MAX SIZE OF SHAFT = Ø 20.015

SHAFT TOLERANCE = 0.013

MAX CLEARANCE = 0.019 ── MIN DIAMETER OF SHAFT = Ø 20.002

MAX INTERFERENCE = –0.015

HOLE TOLERANCE = 0.021 ── MIN DIAMETER OF HOLE = Ø 20.000
MAX DIAMETER OF HOLE = Ø 20.021

METRIC

Ø 20.015 / 20.002

Ø 20.021 / 20.000

EXAMPLE — H7/k6 PREFERRED HOLE BASIS FIT FOR A Ø 20 HOLE

(B) TRANSITION FIT

MAX SIZE OF SHAFT = Ø 20.000

SHAFT TOLERANCE = 0.013 ── MIN DIAMETER OF HOLE = Ø 19.987

MAX INTERFERENCE = –0.048 ── MIN INTERFERENCE = –0.014

HOLE TOLERANCE = 0.021 ── MIN DIAMETER OF HOLE = Ø 19.952
MAX DIAMETER OF HOLE = Ø 19.973

Ø 20.000 / 19.987

Ø 19.973 / 19.952

EXAMPLE — S7/h6 PREFERRED SHAFT BASIS FIT FOR A Ø 20 SHAFT

(C) INTERFERENCE FIT

FIG. 8-6-6   Types of metric fits.

for a Ø25H8/f7 fit, which is a preferred hole basis clearance fit, the limits for the hole and shaft will be as follows:

Hole limits = Ø 25.000 – Ø 25.033
Shaft limits = Ø 24.959 – Ø 24.980
Minimum interference = - 0.020
Maximum interference = - 0.074

If a Ø25H7/s6 preferred hole basis interference fit is required, the limits for the hole and shaft will be as follows:

Hole limits = Ø 25.000 – Ø 25.021
Shaft limits = Ø 25.035 – Ø 25.048
Minimum interference = –0.014
Maximum interference = –0.048

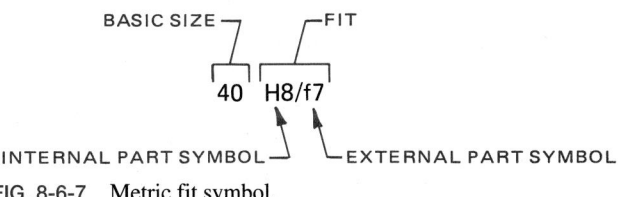

FIG. 8-6-7   Metric fit symbol.

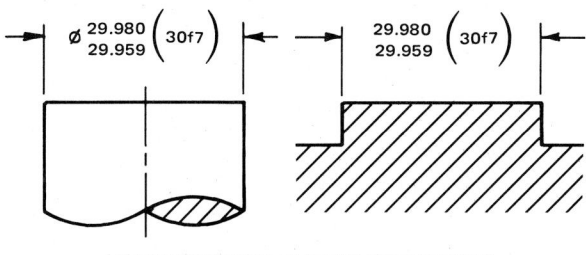

(A) WHEN SYSTEM IS FIRST INTRODUCED

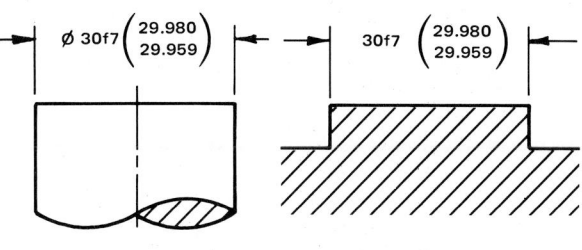

(B) AS EXPERIENCE IS GAINED

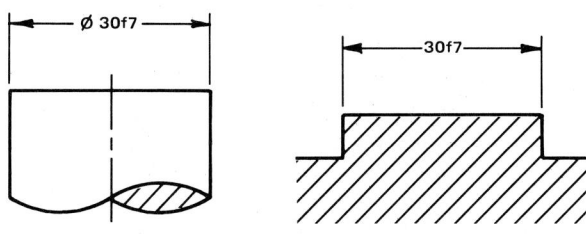

(C) WHEN SYSTEM IS ESTABLISHED

FIG. 8-6-8   Metric tolerance symbol application.

Shaft Basis Fits System

Where more than two fits are required on the same shaft, the shaft basis fits system is recommended. Tolerances for holes and shaft are identical with those for a basic hole system. However, the basic size becomes the maximum shaft size. For example, for a Ø16 C11/h11 fit, which is a preferred shaft basis clearance fit, the limits for the hole and shaft will be as follows (refer to Table 49 of the Appendix):

Hole limits = Ø 16.095 – Ø 16.205
Shaft limits = Ø 15.890 – Ø 16.000
Minimum clearance = 0.095
Maximum clearance = 0.315

Preferred Fits

First-choice tolerance zones are shown to relative scale in Tables 41 and 42 of the Appendix. Hole basis fits have a fundamental deviation of "H" on the hole, and shaft basis fits have a fundamental deviation of "h" on the shaft. Normally, the hole basis system is preferred. Figure 8-6-6 shows examples of three common fits.

Fit Symbol   A fit is indicated by the basic size common to both components, followed by a symbol corresponding to each component, with the internal part symbol preceding the external part symbol (Fig. 8-6-7).

The limits of size for a hole having a tolerance symbol 40H8 (see Table 41) is:

Ø 40.039 Maximum limit
Ø 40.000 Minimum limit

The limits of size for the shaft having a tolerance symbol 40f7 (see Table 42) is:

Ø 39.975 Maximum limit
Ø 39.950 Minimum limit

The method shown in Fig. 8-6-8A is recommended when the system is first introduced. In this case limit dimensions are specified, and the basic size and tolerance symbol are identified as reference.

As experience is gained, the method shown in Fig. 8-6-8B can be used. When the system is established and standard tools, gages, and stock materials are available with size and symbol identification, the method shown in Fig. 8-6-8C may be used.

This would result in a clearance fit of 0.025 to 0.089 mm. A description of the preferred metric fits is shown in Fig. 8-6-9 (pg. 202).

REFERENCES AND SOURCE MATERIAL

1. ANSI B4.2, *Preferred Metric Limits and Fits.*

ASSIGNMENTS

See Assignments 20 through 23 for Unit 8-6 on pages 221 through 224.

FIG. 8-6-9 Description of preferred metric fits.

	ISO SYMBOL			
	HOLE BASIS	SHAFT BASIS	DESCRIPTION	
CLEARANCE FITS	H11/c11	C11/h11	LOOSE RUNNING FIT FOR WIDE COMMERCIAL TOLERANCES OR ALLOWANCES ON EXTERNAL MEMBERS.	MORE CLEARANCE
CLEARANCE FITS	H9/d9	D9/h9	FREE RUNNING FIT NOT FOR USE WHERE ACCURACY IS ESSENTIAL, BUT GOOD FOR LARGE TEMPERATURE VARIATIONS, HIGH RUNNING SPEEDS, OR HEAVY JOURNAL PRESSURES.	MORE CLEARANCE
CLEARANCE FITS	H8/f7	F8/h7	CLOSE RUNNING FIT FOR RUNNING ON ACCURATE MACHINES AND FOR ACCURATE LOCATION AT MODERATE SPEEDS AND JOURNAL PRESSURES.	MORE CLEARANCE
CLEARANCE FITS	H7/g6	G7/h6	SLIDING FIT NOT INTENDED TO RUN FREELY, BUT TO MOVE AND TURN FREELY AND LOCATE ACCURATELY.	MORE CLEARANCE
CLEARANCE FITS	H7/h6	H7/h6	LOCATIONAL CLEARANCE FIT PROVIDES SNUG FIT FOR LOCATING STATIONARY PARTS; BUT CAN BE FREELY ASSEMBLED AND DISASSEMBLED.	MORE CLEARANCE
TRANSITION FITS	H7/k6	K7/h6	LOCATIONAL TRANSITION FIT FOR ACCURATE LOCATION, A COMPROMISE BETWEEN CLEARANCE AND INTERFERENCE.	MORE INTERFERENCE
TRANSITION FITS	H7/n6	N7/h6	LOCATIONAL TRANSITION FIT FOR MORE ACCURATE LOCATION WHERE GREATER INTERFERENCE IS PERMISSIBLE.	MORE INTERFERENCE
INTERFERENCE FITS	H7/p6	P7/h6	LOCATIONAL INTERFERENCE FIT FOR PARTS REQUIRING RIGIDITY AND ALIGNMENT WITH PRIME ACCURACY OF LOCATION BUT WITHOUT SPECIAL BORE PRESSURE REQUIREMENTS.	MORE INTERFERENCE
INTERFERENCE FITS	H7/s6	S7/h6	MEDIUM DRIVE FIT FOR ORDINARY STEEL PARTS OR SHRINK FITS ON LIGHT SECTIONS, THE TIGHTEST FIT USABLE WITH CAST IRON.	MORE INTERFERENCE
INTERFERENCE FITS	H7/u6	U7/h6	FORCE FIT SUITABLE FOR PARTS WHICH CAN BE HIGHLY STRESSED OR FOR SHRINK FITS WHERE THE HEAVY PRESSING FORCES REQUIRED ARE IMPRACTICAL.	MORE INTERFERENCE

8-7 SURFACE TEXTURE

Modern development of high-speed machines has resulted in higher loadings and increased speeds of moving parts. To withstand these more severe operating conditions with minimum friction and wear, a particular surface finish is often essential, making it necessary for the designer to accurately describe the required finish to the persons who are actually making the parts.

For accurate machines it is no longer sufficient to indicate the surface finish by various grind marks, such as "g," "f," or "fg." It becomes necessary to define surface finish and take it out of the opinion or guesswork class.

All surface finish control starts in the drafting room. The designer has the responsibility of specifying the right surface to give maximum performance and service life at the lowest cost. In selecting the required surface finish for any particular part, designers base decisions on experience with similar parts, on field service data, or on engineering tests. Such factors as size and function of the parts, type of loading, speed and direction of movement, operating conditions, physical characteristics of both materials on contact, whether they are subjected to stress reversals, type and amount of lubricant, contaminants, temperature, etc., influence the choice.

There are two principal reasons for surface finish control:

1. To reduce friction
2. To control wear

Whenever a film of lubricant must be maintained between two moving parts, the surface irregularities must be small enough so they will not penetrate the oil film under the most severe operating conditions. Bearings, journals, cylinder boxes, piston pins, bushings, pad bearings, helical and worm gears, seal surfaces, machine ways, and so forth, are examples where this condition must be fulfilled.

Surface finish is also important to the wear of certain pieces that are subject to dry friction, such as machine tool bits, threading dies, stamping dies, rolls, clutch plates, brake drums, etc.

Smooth finishes are essential on certain high-precision pieces. In mechanisms such as injectors and high-pressure cylinders, smoothness and lack of waviness are essential to accuracy and pressure-retaining ability.

Surfaces, in general, are very complex in character. Only the height, width, and direction of surface irregularities will be covered in this section since these are of practical importance in specific applications.

Surface Texture Characteristics

Refer to Fig. 8-7-1.

Microinch A microinch is one-millionth of an inch (.000 001 in.). For written specifications or reference to surface roughness requirements, microinch may be abbreviated as μin.

Micrometer A micrometer is one-millionth of a meter (0.000 001 m). For written specifications or reference to surface roughness requirements, micrometer may be abbreviated as μm.

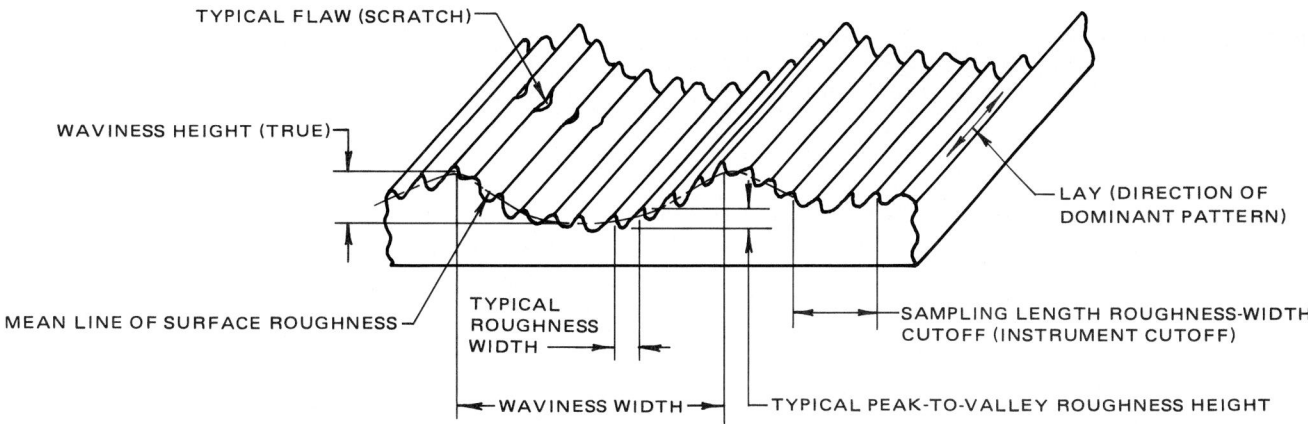

FIG. 8-7-1   Surface texture characteristics.

Roughness   Roughness consists of the finer irregularities in the surface texture, usually including those that result from the inherent action of the production process. These include traverse feed marks and other irregularities within the limits of the roughness-width cutoff.

Roughness-Height Value   Roughness-height value is rated as the arithmetic average (AA) deviation expressed in microinches or micrometers measured normal to the center line. ISO and many European countries use the term CLA (center line average) in lieu of AA. Both have the same meaning.

Roughness Spacing   Roughness spacing is the distance parallel to the nominal surface between successive peaks or ridges that constitute the predominant pattern of the roughness. Roughness spacing is rated in inches or millimeters.

Roughness-Width Cutoff   The greatest spacing of repetitive surface irregularities is included in the measurement of average roughness height. Roughness-width cutoff is rated in inches or millimeters and must always be greater than the roughness width in order to obtain the total roughness-height rating.

Waviness   Waviness is usually the most widely spaced of the surface texture components and normally is wider than the roughness-width cutoff. Waviness may result from such factors as machine or work deflections, vibration, chatter, heat treatment, or warping strains. Roughness may be considered as

superimposed on a "wavy" surface. Although waviness is not currently in ISO Standards, it is included as part of the surface texture symbol to follow present industrial practices in the United States.

Lay   The direction of the predominant surface pattern, ordinarily determined by the production method used, is the lay.

Flaws   Flaws are irregularities that occur at one place or at relatively infrequent or widely varying intervals in a surface. Flaws include such defects as cracks, blow holes, checks, ridges, and scratches. Unless otherwise specified, the effect of flaws is not included in the roughness-height measurements.

Surface Texture Symbol

Surface characteristics of roughness, waviness, and lay may be controlled by applying the desired values to the surface texture symbol, shown in Figs. 8-7-2 and 8-7-3 (pg. 204), in a general note, or both. Where only the roughness value is indicated, the horizontal extension line on the symbol may be omitted. The horizontal bar is used whenever any surface characteristics are placed above the bar or to the right of the symbol. The point of the symbol should be located on the line indicating the surface, on an extension line from the surface, or on a leader pointing to the surface or extension line. If necessary, the symbol may be connected to the surface by a leader line terminating in an arrow. The symbol applies to the entire surface, unless otherwise specified. The symbol for the same surface should not be duplicated on other views.

When numerical values accompany the symbol, the symbol should be in an upright position in order to be readable from the bottom. This means that the long leg and extension line are always on the right. When no numerical values are shown on the symbol, the symbol may also be positioned to be readable from the right side.

Application

Plain (Unplated or Uncoated) Surfaces   Surface texture values specified on plain surfaces apply to the completed surface unless otherwise noted.

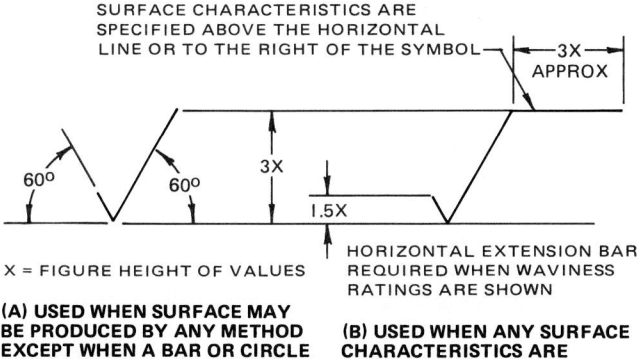

SURFACE CHARACTERISTICS ARE SPECIFIED ABOVE THE HORIZONTAL LINE OR TO THE RIGHT OF THE SYMBOL

X = FIGURE HEIGHT OF VALUES

HORIZONTAL EXTENSION BAR REQUIRED WHEN WAVINESS RATINGS ARE SHOWN

(A) USED WHEN SURFACE MAY BE PRODUCED BY ANY METHOD EXCEPT WHEN A BAR OR CIRCLE IS SPECIFIED

(B) USED WHEN ANY SURFACE CHARACTERISTICS ARE SPECIFIED

FIG. 8-7-2   Basic surface texture symbol.

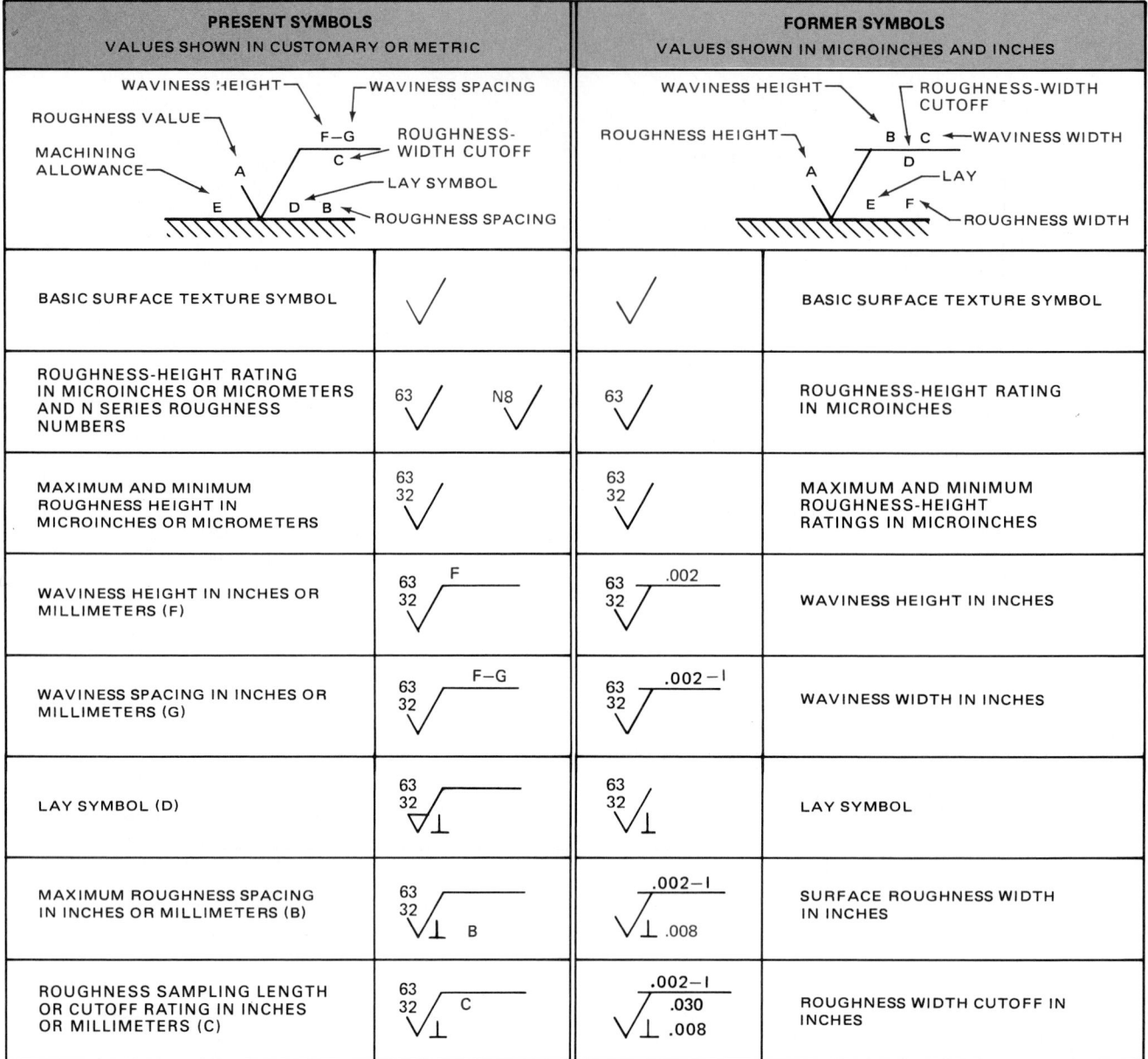

FIG. 8-7-3   Location of notes and symbols on surface texture symbol.

Plated or Coated Surfaces   Drawings or specifications for plated or coated parts must indicate whether the surface texture value applies before, after, or both before and after plating or coating.

Surface Texture Ratings   The roughness value rating is indicated at the left of the long leg of the symbol (Fig. 8-7-3). The specification of only one rating indicates the maximum value, and any lesser value is acceptable. The specification of two ratings indicates the minimum and maximum values, and anything lying within that range is acceptable. The maximum value is placed over the minimum.

Waviness-height rating is specified in inches or millimeters and is located above the horizontal extension of the symbol. Any lesser value is acceptable.

Waviness spacing is indicated in inches or millimeters and is located above the horizontal extension and to the right, separated from the waviness-height rating by a dash. Any lesser value is acceptable. If the waviness value is a minimum, the abbreviation MIN should be placed after the value.

The surface roughness range for common production methods is shown in Fig. 8-7-4.

Typical surface roughness-height applications are shown in Figs. 8-7-5 and 8-7-6 (pp. 206–207).

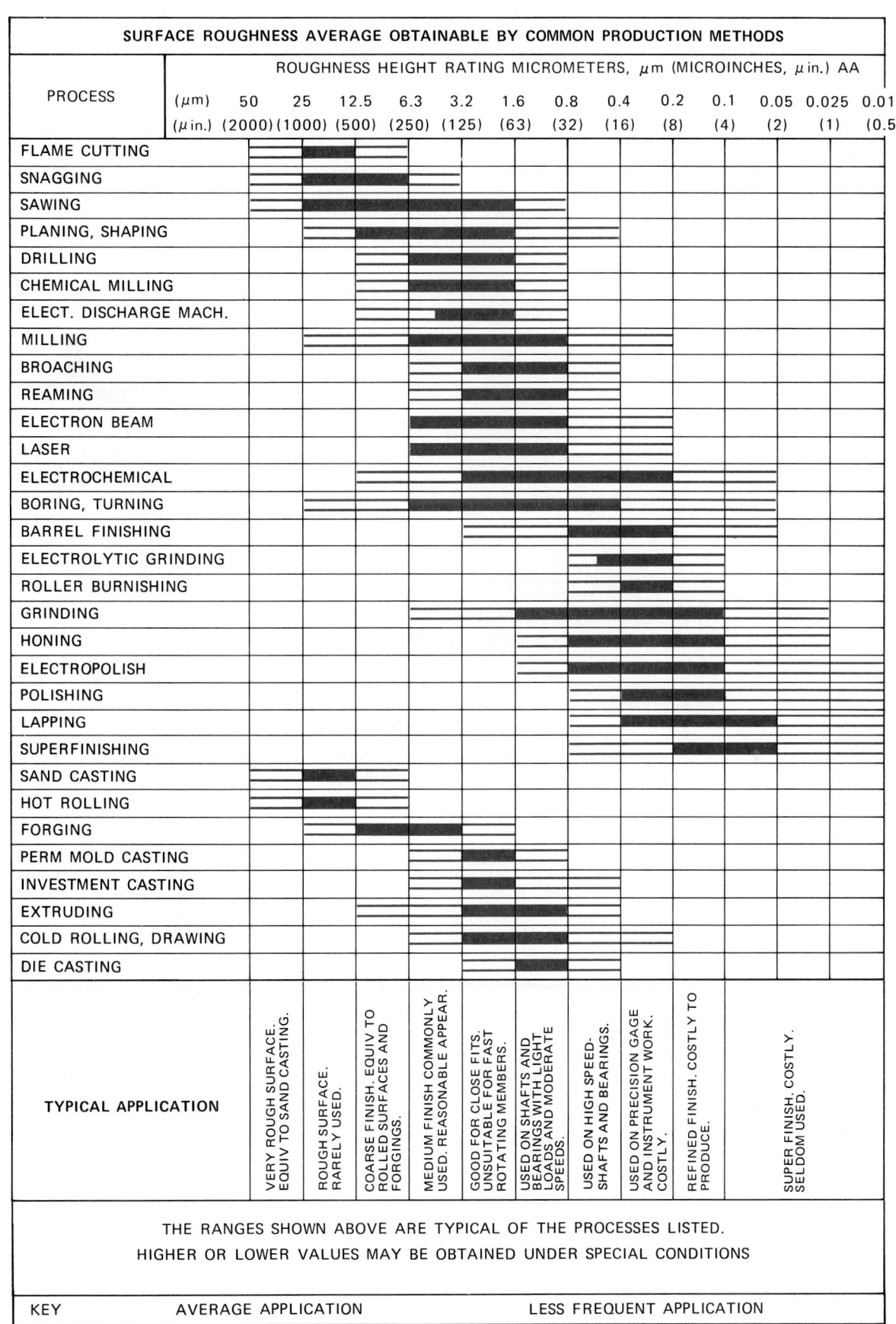

FIG. 8-7-4   Surface roughness range for common production.

MICROINCHES AA RATING	MICROMETERS AA RATING	APPLICATION
1000	25.2	ROUGH, LOW-GRADE SURFACE RESULTING FROM SAND CASTING, TORCH OR SAW CUTTING, CHIPPING, OR ROUGH FORGING. MACHINE OPERATIONS ARE NOT REQUIRED AS APPEARANCE IS NOT OBJECTIONABLE. THIS SURFACE, RARELY SPECIFIED, IS SUITABLE FOR UNMACHINED CLEARANCE AREAS ON ROUGH CONSTRUCTION ITEMS.
500	12.5	ROUGH, LOW-GRADE SURFACE RESULTING FROM HEAVY CUTS AND COARSE FEEDS IN MILLING, TURNING, SHAPING, BORING, AND ROUGH FILING, DISC GRINDING, AND SNAGGING. IT IS SUITABLE FOR CLEARANCE AREAS ON MACHINERY, JIGS, AND FIXTURES. SAND CASTING OR ROUGH FORGING PRODUCES THIS SURFACE.
250	6.3	COARSE PRODUCTION SURFACES, FOR UNIMPORTANT CLEARANCE AND CLEAN-UP OPERATIONS, RESULTING FROM COARSE SURFACE GRIND, ROUGH FILE, DISC GRIND, RAPID FEEDS IN TURNING, MILLING, SHAPING, DRILLING, BORING, GRINDING, ETC., WHERE TOOL MARKS ARE NOT OBJECTIONABLE. THE NATURAL SURFACES OF FORGINGS, PERMANENT MOLD CASTINGS, EXTRUSIONS, AND ROLLED SURFACES ALSO PRODUCE THIS ROUGHNESS. IT CAN BE PRODUCED ECONOMICALLY AND IS USED ON PARTS WHERE STRESS REQUIREMENTS, APPEARANCE, AND CONDITIONS OF OPERATIONS AND DESIGN PERMIT.
125	3.2	THE ROUGHEST SURFACE RECOMMENDED FOR PARTS SUBJECT TO LOADS, VIBRATION, AND HIGH STRESS. IT IS ALSO PERMITTED FOR BEARING SURFACES WHEN MOTION IS SLOW AND LOADS LIGHT OR INFREQUENT. IT IS A MEDIUM COMMERCIAL MACHINE FINISH PRODUCED BY RELATIVELY HIGH SPEEDS AND FINE FEEDS TAKING LIGHT CUTS WITH SHARP TOOLS. IT MAY BE ECONOMICALLY PRODUCED ON LATHES, MILLING MACHINES, SHAPERS, GRINDERS, ETC., OR ON PERMANENT MOLD CASTINGS, DIE CASTINGS, EXTRUSIONS, AND ROLLED SURFACES.
63	1.6	A GOOD MACHINE FINISH PRODUCED UNDER CONTROLLED CONDITIONS USING RELATIVELY HIGH SPEEDS AND FINE FEEDS TO TAKE LIGHT CUTS WITH SHARP CUTTERS. IT MAY BE SPECIFIED FOR CLOSE FITS AND USED FOR ALL STRESSED PARTS, EXCEPT FAST-ROTATING SHAFTS, AXLES, AND PARTS SUBJECT TO SEVERE VIBRATION OR EXTREME TENSION. IT IS SATISFACTORY FOR BEARING SURFACES WHEN MOTION IS SLOW AND LOADS LIGHT OR INFREQUENT. IT MAY ALSO BE OBTAINED ON EXTRUSIONS, ROLLED SURFACES, DIE CASTINGS, AND PERMANENT MOLD CASTINGS WHEN RIGIDLY CONTROLLED.
32	0.8	A HIGH-GRADE MACHINE FINISH REQUIRING CLOSE CONTROL WHEN PRODUCED BY LATHES, SHAPERS, MILLING MACHINES, ETC., BUT RELATIVELY EASY TO PRODUCE BY CENTERLESS, CYLINDRICAL, OR SURFACE GRINDERS. ALSO, EXTRUDING, ROLLING, OR DIE CASTING MAY PRODUCE A COMPARABLE SURFACE WHEN RIGIDLY CONTROLLED. THIS SURFACE MAY BE SPECIFIED IN PARTS WHERE STRESS CONCENTRATION IS PRESENT. IT IS USED FOR BEARINGS WHEN MOTION IS NOT CONTINUOUS AND LOADS ARE LIGHT. WHEN FINER FINISHES ARE SPECIFIED, PRODUCTION COSTS RISE RAPIDLY; THEREFORE, SUCH FINISHES MUST BE ANALYZED CAREFULLY.
16	0.4	A HIGH-QUALITY SURFACE PRODUCED BY FINE CYLINDRICAL GRINDING, EMERY BUFFING, COARSE HONING, OR LAPPING. IT IS SPECIFIED WHERE SMOOTHNESS IS OF PRIMARY IMPORTANCE, SUCH AS RAPIDLY ROTATING SHAFT BEARINGS, HEAVILY LOADED BEARINGS, AND EXTREME TENSION MEMBERS.
8	0.2	A FINE SURFACE PRODUCED BY HONING, LAPPING, OR BUFFING. IT IS SPECIFIED WHERE PACKINGS AND RINGS MUST SLIDE ACROSS THE DIRECTION OF THE SURFACE GRAIN, MAINTAINING OR WITHSTANDING PRESSURES, OR FOR INTERIOR HONED SURFACES OF HYDRAULIC CYLINDERS. IT MAY ALSO BE REQUIRED IN PRECISION GAGES AND INSTRUMENT WORK, OR SENSITIVE-VALUE SURFACES, OR ON RAPIDLY ROTATING SHAFTS AND ON BEARINGS WHERE LUBRICATION IS NOT DEPENDABLE.
4	0.1	A COSTLY REFINED SURFACE PRODUCED BY HONING, LAPPING, AND BUFFING. IT IS SPECIFIED ONLY WHEN THE REQUIREMENTS OF DESIGN MAKE IT MANDATORY. IT IS REQUIRED IN INSTRUMENT WORK, GAGE WORK, AND WHERE PACKINGS AND RINGS MUST SLIDE ACROSS THE DIRECTION OF SURFACE GRAIN, SUCH AS ON CHROME-PLATED PISTON RODS, ETC., WHERE LUBRICATION IS NOT DEPENDABLE.
2 1	0.05 0.025	COSTLY REFINED SURFACES PRODUCED ONLY BY THE FINEST OF MODERN HONING, LAPPING, BUFFING, AND SUPERFINISHING EQUIPMENT. THESE SURFACES MAY HAVE A SATIN OR HIGHLY POLISHED APPEARANCE DEPENDING ON THE FINISHING OPERATION AND MATERIAL. THESE SURFACES ARE SPECIFIED ONLY WHEN DESIGN REQUIREMENTS MAKE IT MANDATORY. THEY ARE SPECIFIED ON FINE OR SENSITIVE INSTRUMENT PARTS OR OTHER LABORATORY ITEMS, AND CERTAIN GAGE SURFACES, SUCH AS PRECISION GAGE BLOCKS.

FIG. 8-7-5   Typical surface roughness-height applications.

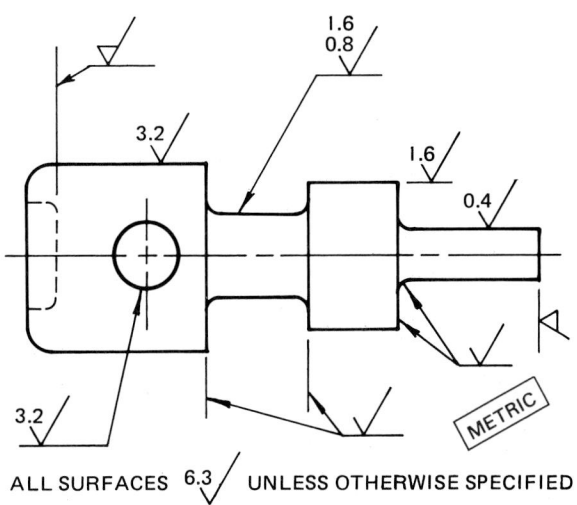

ALL SURFACES 6.3/ UNLESS OTHERWISE SPECIFIED

NOTE: VALUES SHOWN ARE IN MICROMETERS.

FIG. 8-7-6   Application of surface texture symbols and notes.

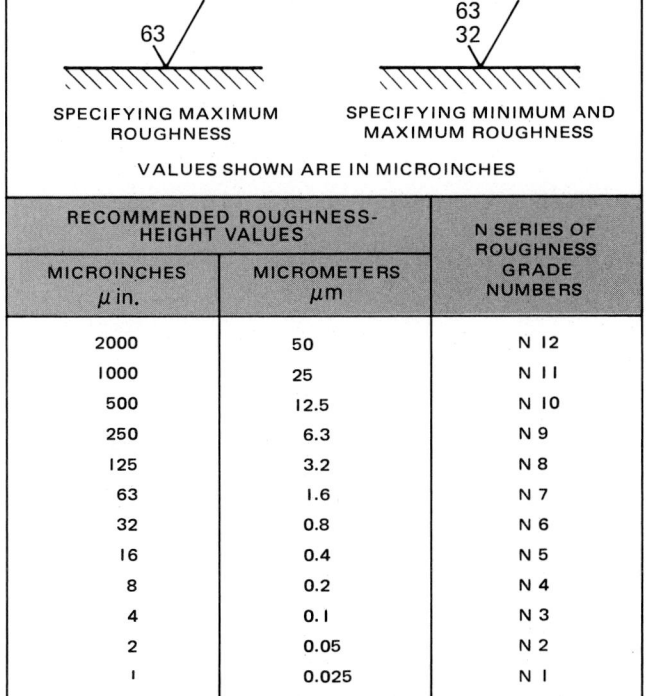

SPECIFYING MAXIMUM ROUGHNESS   SPECIFYING MINIMUM AND MAXIMUM ROUGHNESS

VALUES SHOWN ARE IN MICROINCHES

RECOMMENDED ROUGHNESS-HEIGHT VALUES		N SERIES OF ROUGHNESS GRADE NUMBERS
MICROINCHES μ in.	MICROMETERS μm	
2000	50	N 12
1000	25	N 11
500	12.5	N 10
250	6.3	N 9
125	3.2	N 8
63	1.6	N 7
32	0.8	N 6
16	0.4	N 5
8	0.2	N 4
4	0.1	N 3
2	0.05	N 2
1	0.025	N 1

FIG. 8-7-7   Roughness-height ratings and their equivalent N series grade numbers.

Roughness height ratings and their equivalent N series grade numbers are shown in Fig. 8-7-7.

Lay symbols, which indicate the directional pattern of the surface texture, are shown in Fig. 8-7-8. The symbol is located to the right of the long leg of the symbol. On surfaces having parallel or perpendicular lay designated, the lead resulting from machine feeds may be objectionable. In these cases, the symbol should be supplemented by the words NO LEAD.

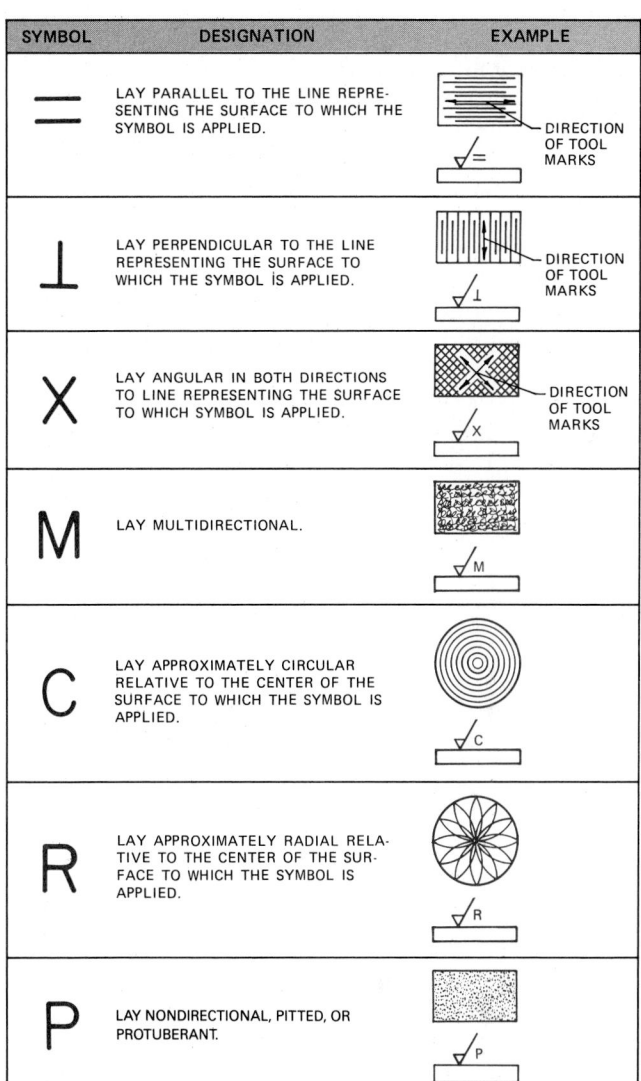

SYMBOL	DESIGNATION	EXAMPLE
=	LAY PARALLEL TO THE LINE REPRESENTING THE SURFACE TO WHICH THE SYMBOL IS APPLIED.	DIRECTION OF TOOL MARKS
⊥	LAY PERPENDICULAR TO THE LINE REPRESENTING THE SURFACE TO WHICH THE SYMBOL IS APPLIED.	DIRECTION OF TOOL MARKS
X	LAY ANGULAR IN BOTH DIRECTIONS TO LINE REPRESENTING THE SURFACE TO WHICH SYMBOL IS APPLIED.	DIRECTION OF TOOL MARKS
M	LAY MULTIDIRECTIONAL.	
C	LAY APPROXIMATELY CIRCULAR RELATIVE TO THE CENTER OF THE SURFACE TO WHICH THE SYMBOL IS APPLIED.	
R	LAY APPROXIMATELY RADIAL RELATIVE TO THE CENTER OF THE SURFACE TO WHICH THE SYMBOL IS APPLIED.	
P	LAY NONDIRECTIONAL, PITTED, OR PROTUBERANT.	

FIG. 8-7-8   Lay symbols.

PRESENT SYMBOL	FORMER SYMBOL
LAY SYMBOLS	
STANDARD ROUGHNESS SAMPLING LENGTH VALUES	
INCHES	MILLIMETERS
.003	0.08
.010	0.25
.030	0.8
.100	2.54
.300	8
1.000	25.4

FIG. 8-7-9   Lay and roughness sampling length applications.

Roughness sampling length or cutoff rating is in inches or millimeters and is located below the horizontal extension (Fig. 8-7-3). Unless otherwise specified, roughness sampling length is .03 in. (0.08 mm). See Fig. 8-7-9.

Notes

Notes relating to surface roughness can be local or general. Normally, a general note is used where a given roughness requirement applies to the whole part or the major portion. Any exceptions to the general note are given in a local note (Fig. 8-7-10).

Machined Surfaces

In preparing working drawings or parts to be cast, molded, or forged, the drafter must indicate the surfaces on the drawing that will require machining or finishing. The symbol √ identifies those surfaces that are produced by machining operations (Fig. 8-7-11). It indicates that material is to be provided for removal by machining. Where all the surfaces are to be machined, a general note, such as FINISH ALL OVER, may

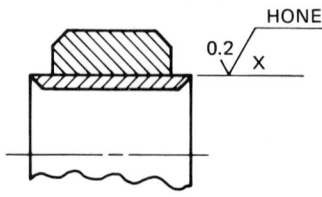

ROUGHNESS VALUE SHOWN IN MICROMETERS.

(A) LOCAL NOTE   **(B) GENERAL NOTE**

FIG. 8-7-10   Surface texture notes.

REMOVAL OF MATERIAL BY MACHINING IS

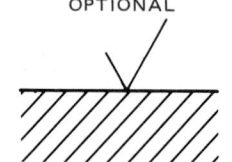

FIG. 8-7-11   Indicating the removal of material on the surface texture symbol.

FIG. 8-7-12   Application of surface texture symbol when machining of surface is required.

be used, and the symbols on the drawing may be omitted. Where space is restricted, the machining symbol may be placed on an extension line.

Machining symbols, like dimensions, are not normally duplicated. They should be used on the same view as the dimensions that give the size or location of the surfaces concerned. The symbol is placed on the line representing the surface or, where desirable, on the extension line locating the surface. Figures 8-7-12 and 8-7-13 show examples of the use of machining symbols.

Material Removal Allowance

When it is desirable to indicate the amount of material to be removed, the amount of material in inches or millimeters is shown to the left of the symbol. Illustrations showing material removal allowance are shown in Figs. 8-7-14 and 8-7-15.

Material Removal Prohibited

When it is necessary to indicate that a surface must be produced without material removal, the machining prohibited symbol shown in Fig. 8-7-16 must be used.

Former Machining Symbols

Former machining symbols, as shown in Fig. 8-7-17, may be found on many drawings in use today. When called upon to make changes or revisions to a drawing already in existence, a drafter must adhere to the drawing conventions shown on that drawing.

REFERENCES AND SOURCE MATERIALS

1. ANSI Y14.36, *Surface Texture Symbols.*
2. GAR.
3. General Motors.

ASSIGNMENTS

See Assignments 24 through 28 for Unit 8-7 on pages 225 to 226.

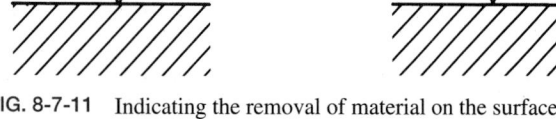

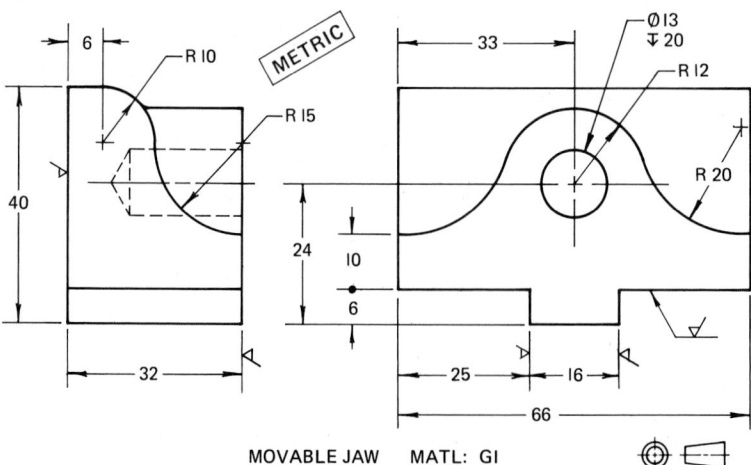

MOVABLE JAW   MATL: GI

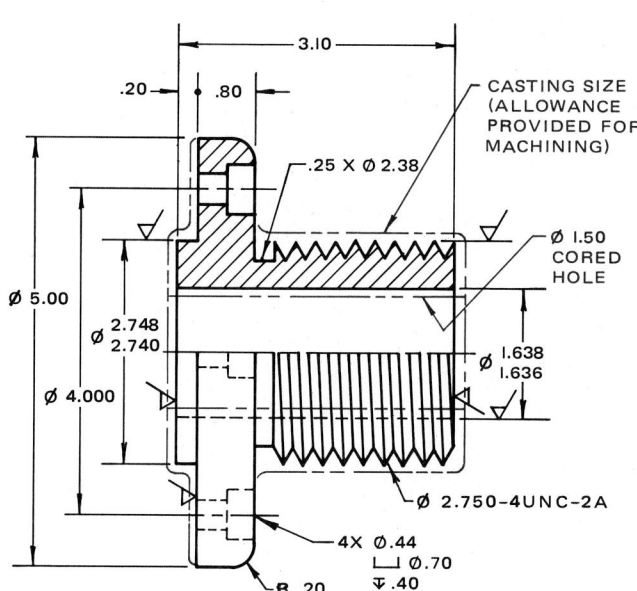

FIG. 8-7-13   Extra metal allowance for machined surfaces.

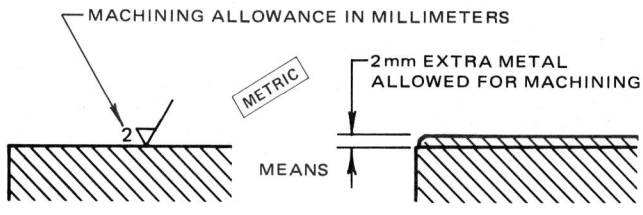

FIG. 8-7-14   Indication of machining allowance.

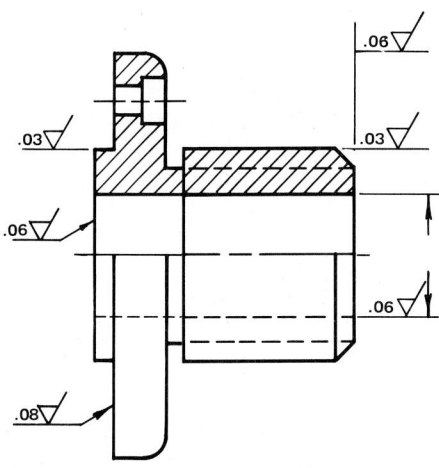

FIG. 8-7-15   Indicating machining allowance on drawings.

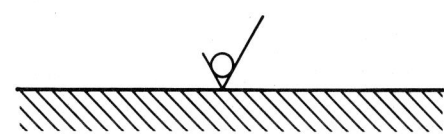

FIG. 8-7-16   Symbol to indicate the removal of material is not permitted.

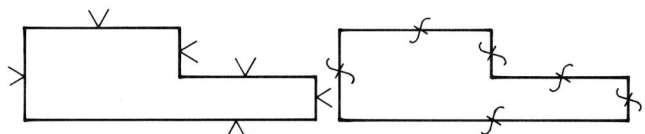

FIG. 8-7-17   Former machining symbols.

ASSIGNMENTS FOR CHAPTER 8

ASSIGNMENTS FOR UNIT 8-1, BASIC DIMENSIONING

1. Select one of the template drawings (Fig. 8-1-A or 8-1-B, pg. 210) and make a one-view drawing, complete with dimensions, of the part.

2. Select one of the parts shown in Figs. 8-1-C through 8-1-F (pg. 210) and make a three-view drawing, complete with dimensions, of the part.

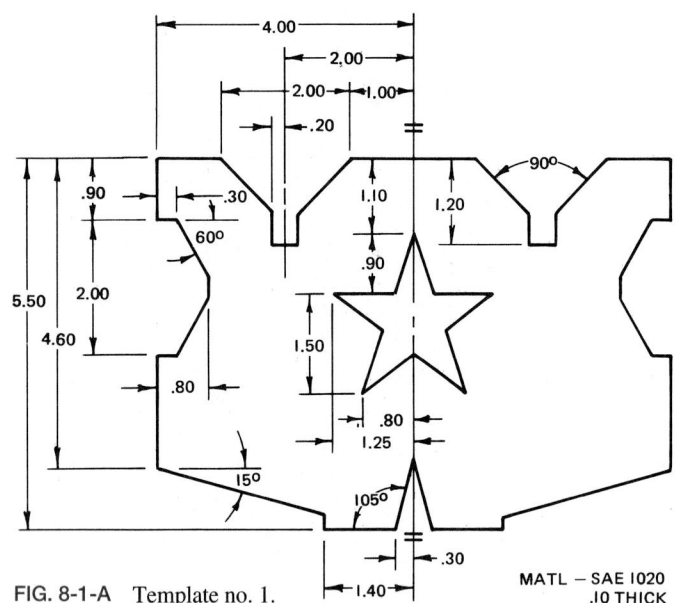

FIG. 8-1-A   Template no. 1.

MATL — SAE 1020
.10 THICK

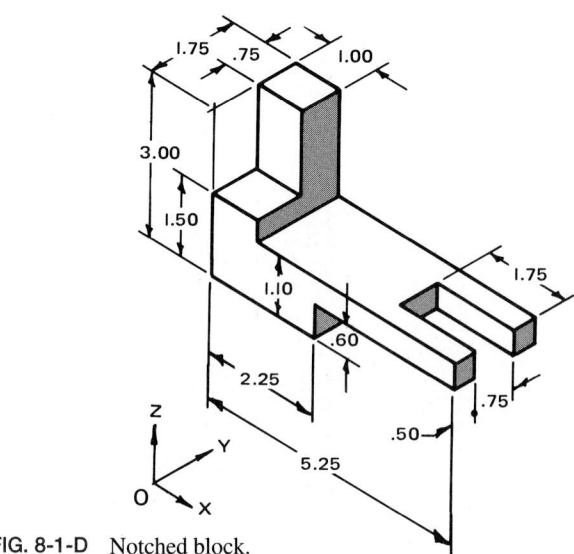

FIG. 8-1-D   Notched block.

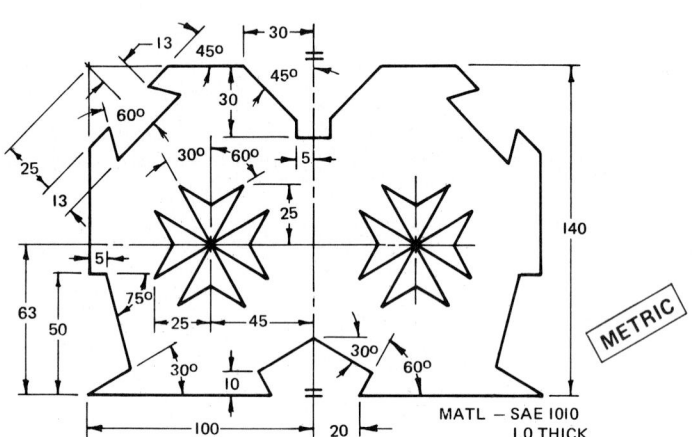
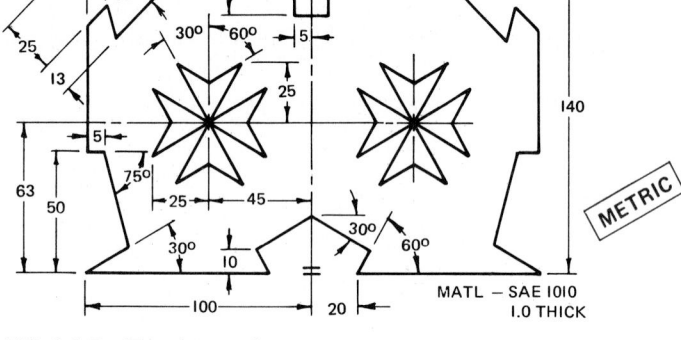

MATL — SAE 1010
1.0 THICK

FIG. 8-1-B   Template no. 2.

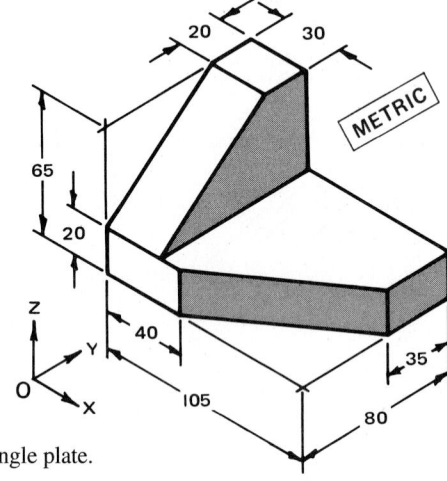

FIG. 8-1-E   Angle plate.

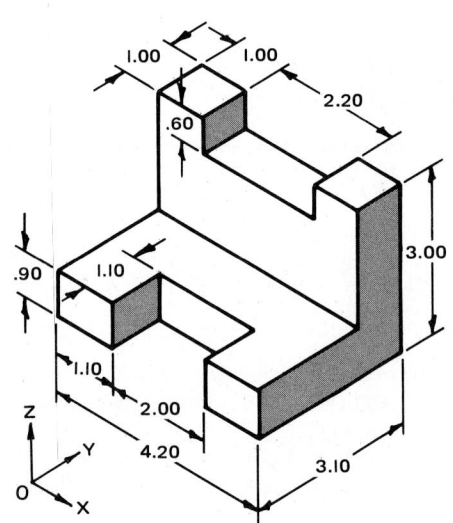

FIG. 8-1-C   Cross slide.

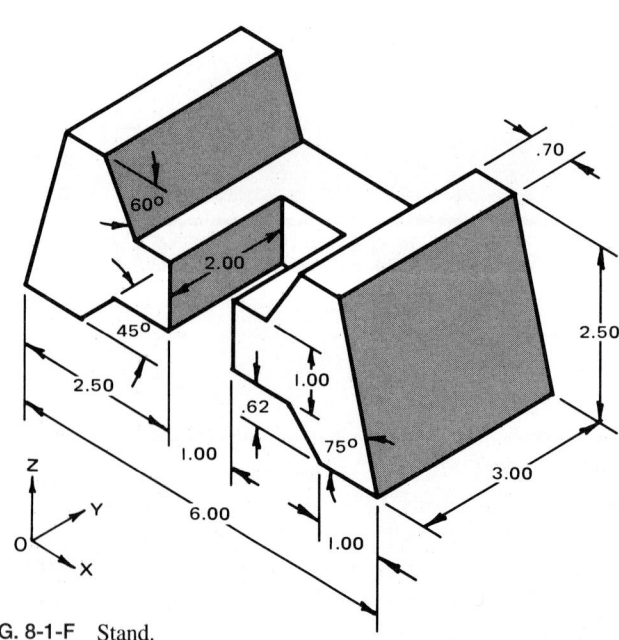

FIG. 8-1-F   Stand.

3. Select one of the parts shown in Figs. 8-1-G through 8-1-L and make a three-view drawing, complete with dimensions, of the part. Show the dimensions with the view that best shows the shape of the part or feature.

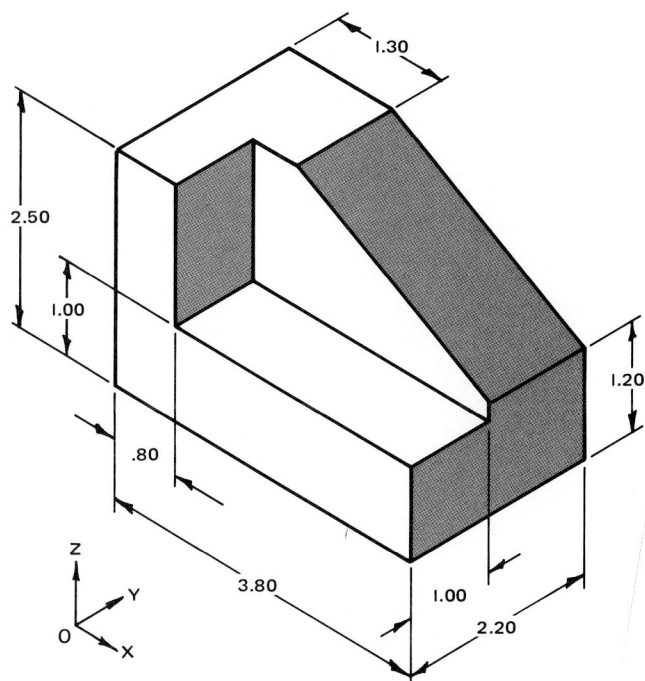

FIG. 8-1-G   Guide stop.

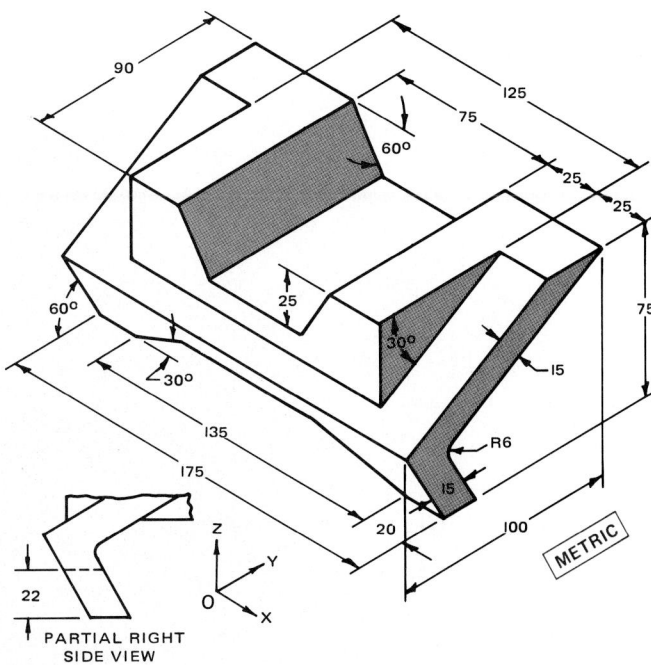

FIG. 8-1-H   Separator.

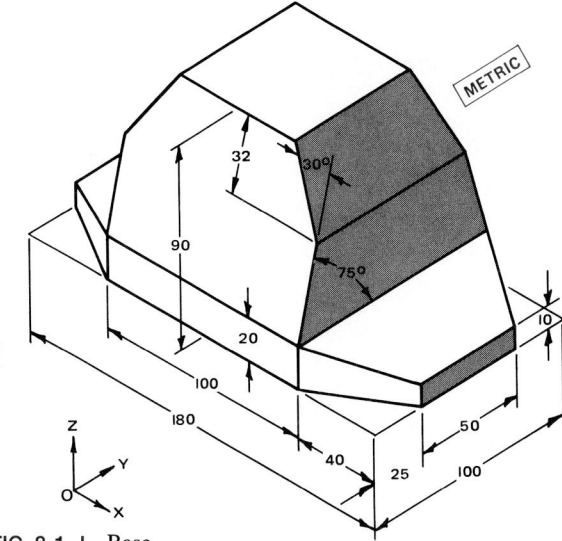

FIG. 8-1-J   Base.

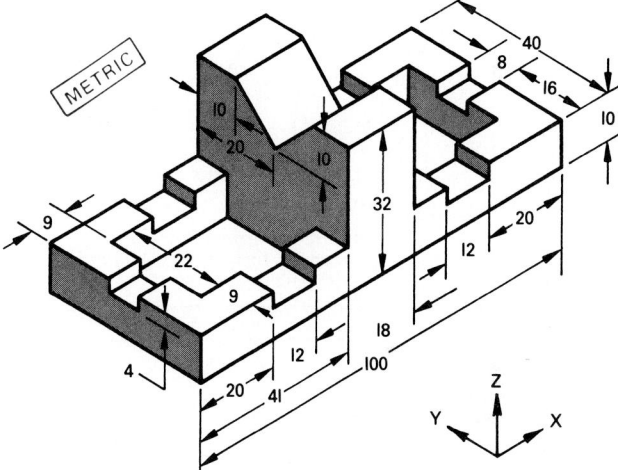

FIG. 8-1-K   Guide support.

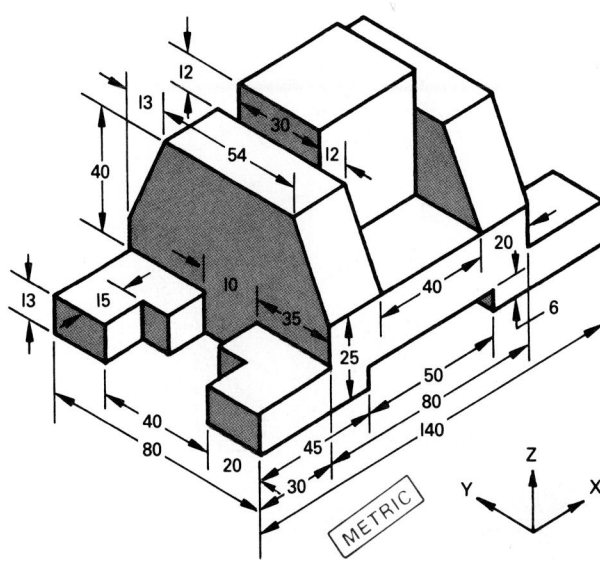

FIG. 8-1-L   Vertical guide.

ASSIGNMENTS FOR UNIT 8-2, DIMENSIONING CIRCULAR FEATURES

4. Select one of the problems shown in Figs. 8-2-A through 8-2-E and make a one-view drawing, complete with dimensions, of the part.

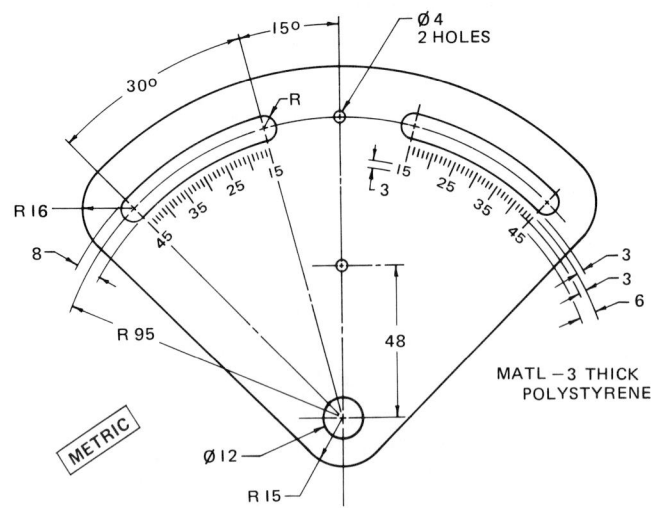

FIG. 8-2-C   Dial indicator.

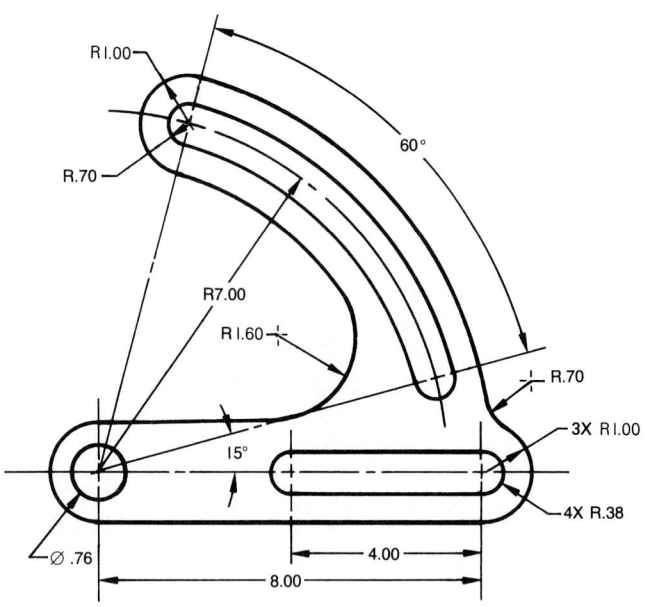

FIG. 8-2-A   Adjustable table support.

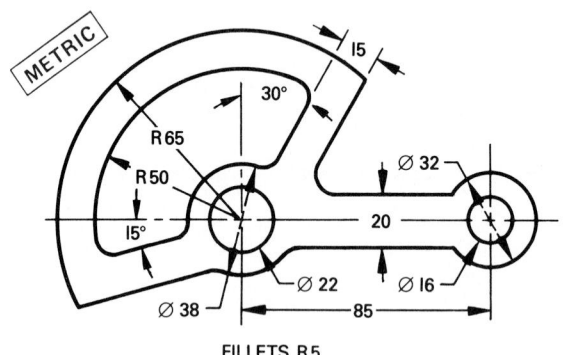

FIG. 8-2-D   Adjustable sector.

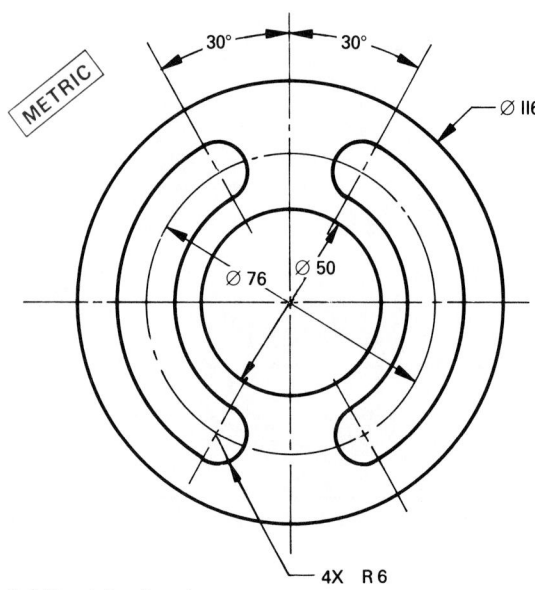

FIG. 8-2-B   Adjusting ring.

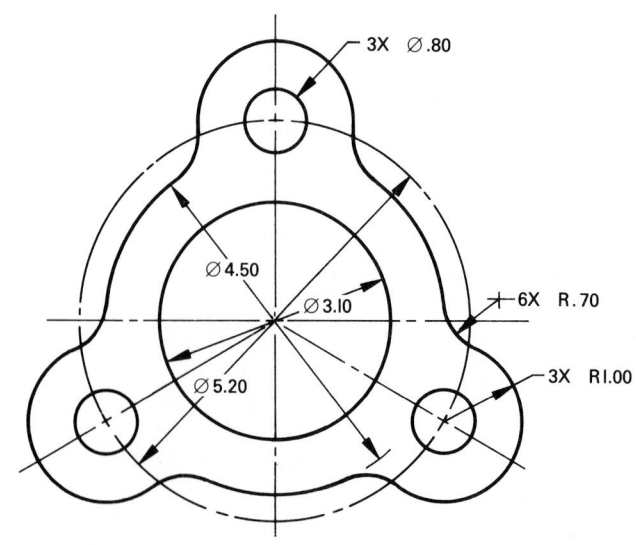

FIG. 8-2-E   Gasket.

5. Select one of the problems shown in Figs. 8-2-F through 8-2-K and make a three-view drawing, complete with dimensions, of the part.

ROUNDS & FILLETS R.10

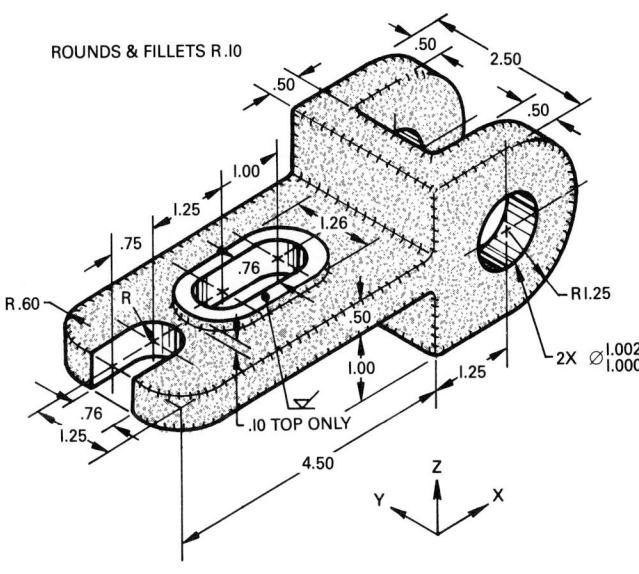

FIG. 8-2-F   Swing bracket.

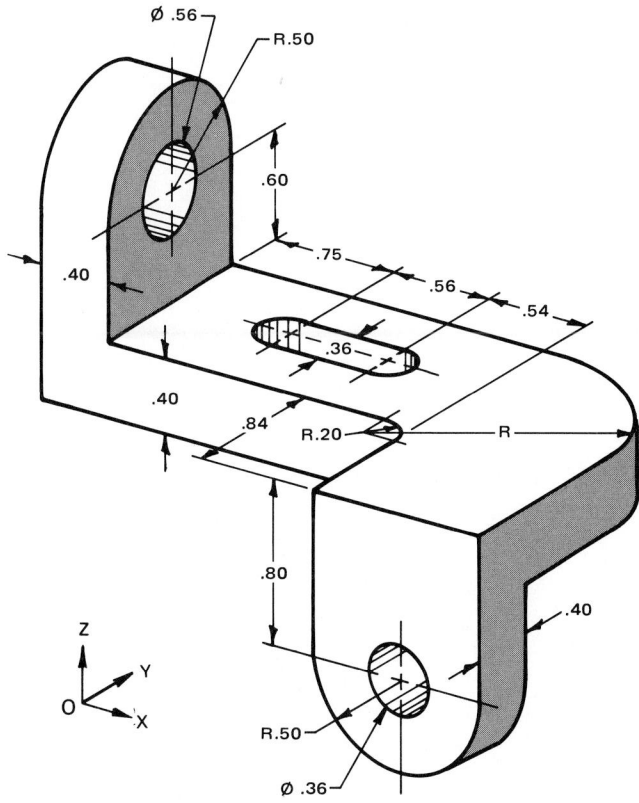

FIG. 8-2-G   Bracket.

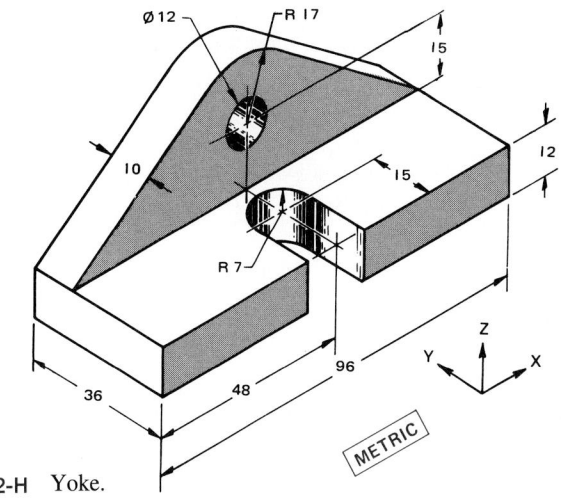

FIG. 8-2-H   Yoke.

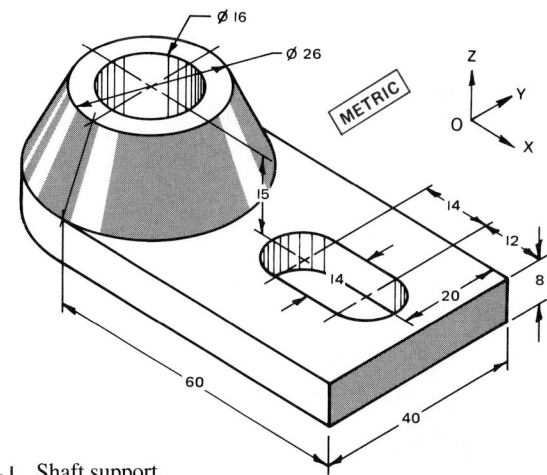

FIG. 8-2-J   Shaft support.

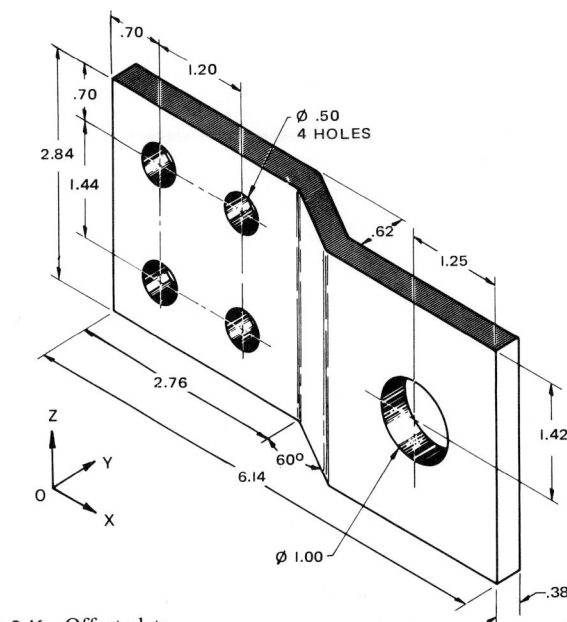

FIG. 8-2-K   Offset plate.

6. Select one of the parts shown in Fig. 8-2-L, and using one
   of the scales shown, redraw the part and add dimensions.

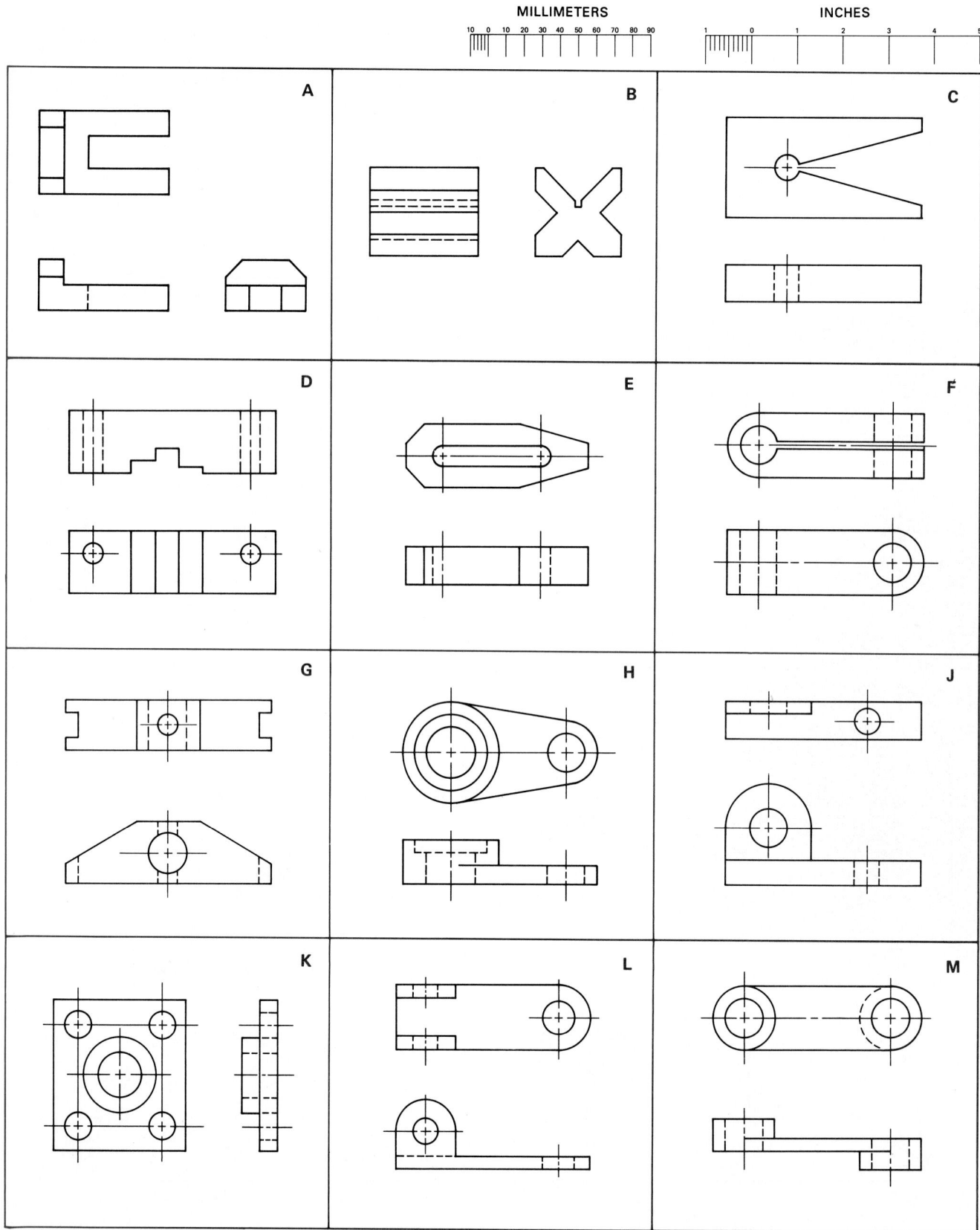

MILLIMETERS

INCHES

FIG. 8-2-L   Problems in dimensioning practice.

ASSIGNMENTS FOR UNIT 8-3, DIMENSIONING COMMON FEATURES

7. Redraw the handle shown in Fig. 8-3-A. The following features are to be added and dimensioned:
   (a) 45° × .10 chamfer
   (b) 33DP diamond knurl for 1.20 in. starting .80 in. from left end
   (c) 1:8 circular taper for 1.20 in. length on right end of Ø1.25

(d) .16 × Ø.54 in. undercut on Ø.75
(e) Ø.189 × .25 in. deep, 4 holes equally spaced
(f) 30° × .10 chamfer, the .10 in. dimension taken horizontally along the shaft.

8. Redraw the selector shaft shown in Fig. 8-3-B and dimension. Scale the drawing for sizes.
9. Make a one-view drawing (plus a partial view of the blade), with dimensions, of the screwdriver shown in Fig. 8-3-C.
10. Make a one-view drawing with dimensions of the indicator rod shown in Fig. 8-3-D.

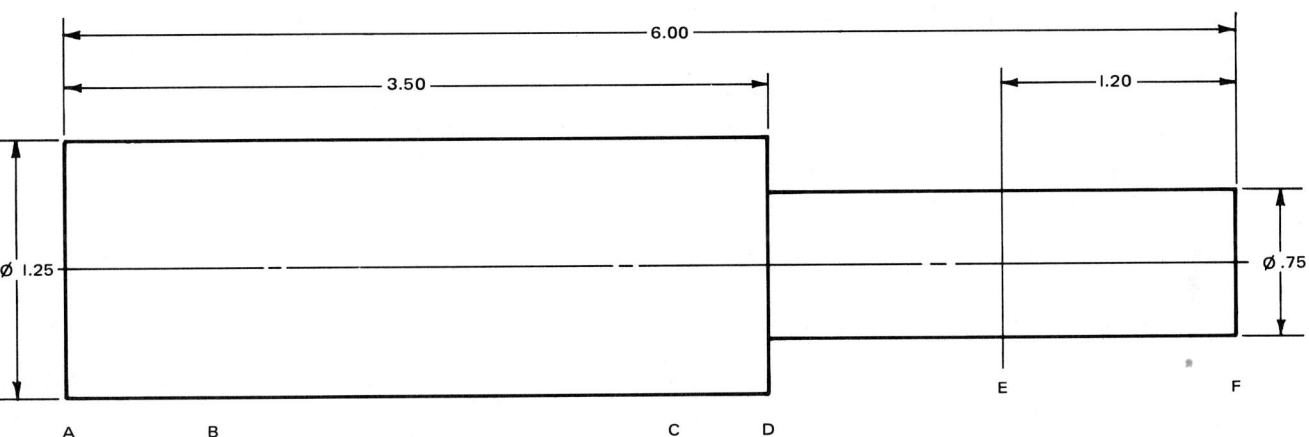

FIG. 8-3-A Handle.

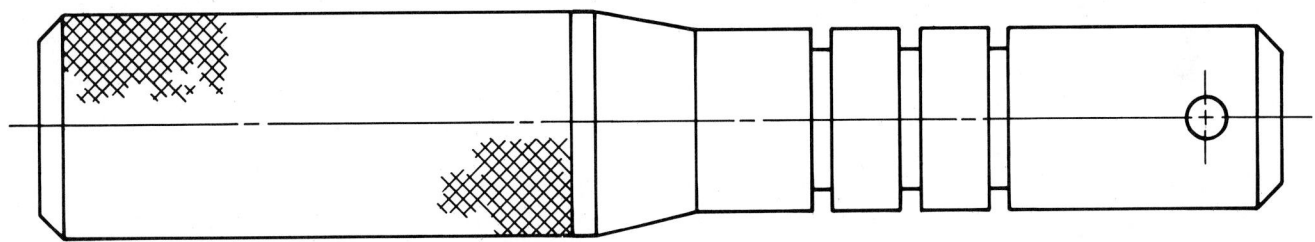

FIG. 8-3-B Selector shaft.

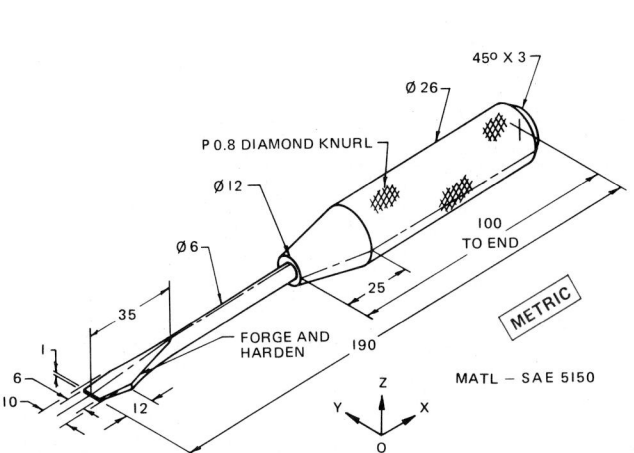

FIG. 8-3-C Screwdriver.

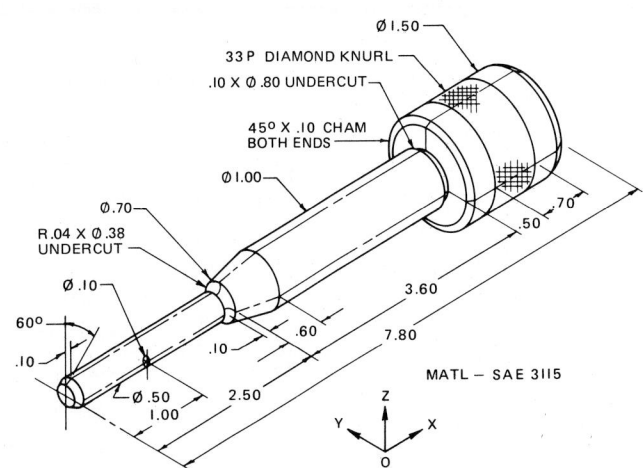

FIG. 8-3-D Indicator rod.

215

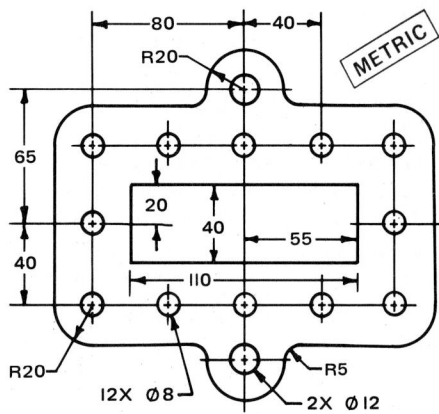

MATL—2mm FLUOROCARBON PLASTIC (MYLAR)

FIG. 8-3-E Gasket.

11. Make a half-view drawing of one of the parts shown in Figs. 8-3-E through 8-3-G. Add the symmetry symbol to the drawing and dimension using symbolic dimensioning wherever possible. Use the MIRROR command to create the view if CAD is used.

12. Make a one-view drawing of the adjusting locking plate shown in Fig. 8-3-H. If manually drawn, show only two or three holes and teeth. Scale 10:1.

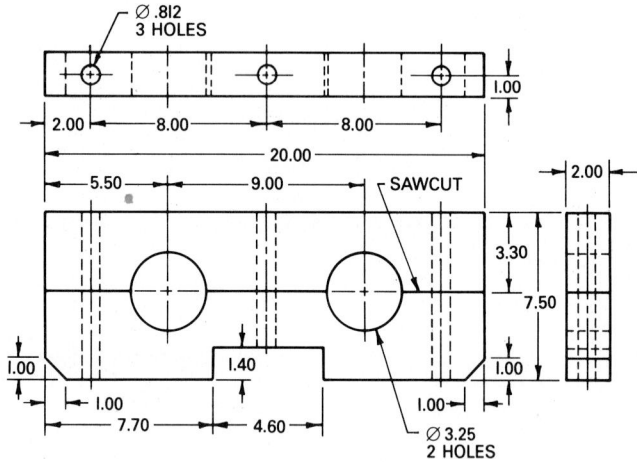

FIG. 8-3-F Tube support.

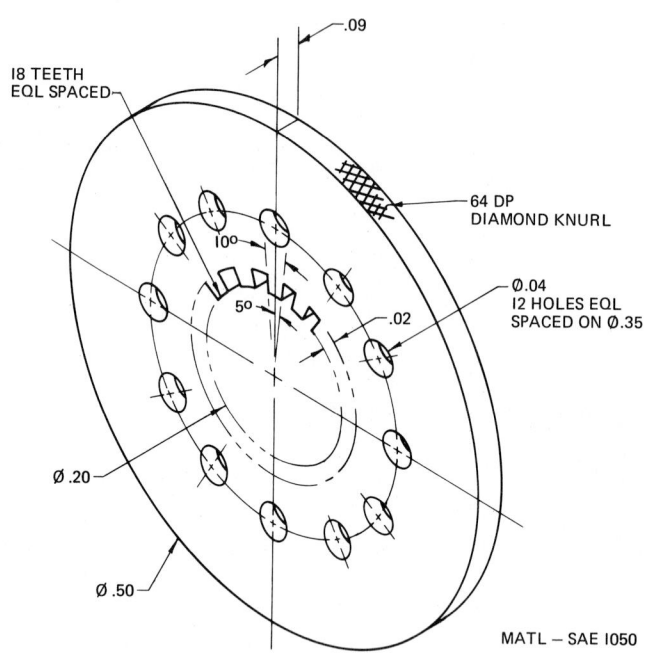

FIG. 8-3-H Adjusting locking plate.

FIG. 8-3-G Gasket.

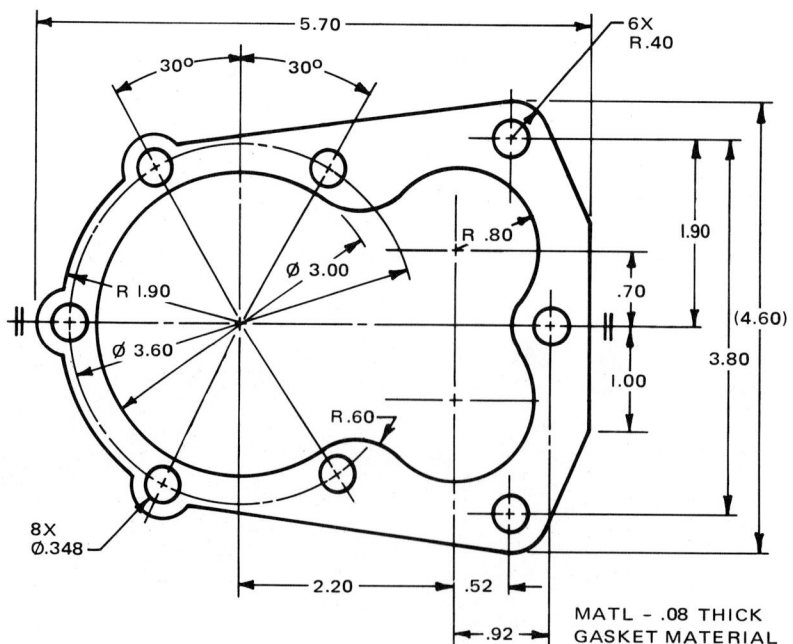

MATL – .08 THICK
GASKET MATERIAL

ASSIGNMENTS FOR UNIT 8-4, DIMENSIONING METHODS

13. Select one of the problems shown in Figs. 8-4-A and 8-4-B, and make a working drawing of the part. The arrowless dimensioning shown is to be replaced with rectangular coordinate dimensioning and has the following dimensioning changes.
   For Fig. 8-4-A:
   - Holes A, E, and D are located from the zero coordinates.
   - Holes B are located from center of hole E.
   - Holes C are located from center of hole D.
   For Fig. 8-4-B:
   - Holes E and D are located from left and bottom edges.
   - Holes A and C are located from center of hole D.
   - Holes B are located from center of hole E.

14. Redraw the terminal board shown in Fig. 8-4-C using tabular dimensioning. Use the bottom and left-hand edge for the datum surfaces to locate the holes.

15. Divide a sheet into four quadrants by bisecting the vertical and horizontal sides. In each quadrant draw the adaptor plate shown in Fig. 8-4-D. Different methods of dimensioning are to be used for each drawing. The methods are rectangular coordinate, chordal, arrowless, and tabular.

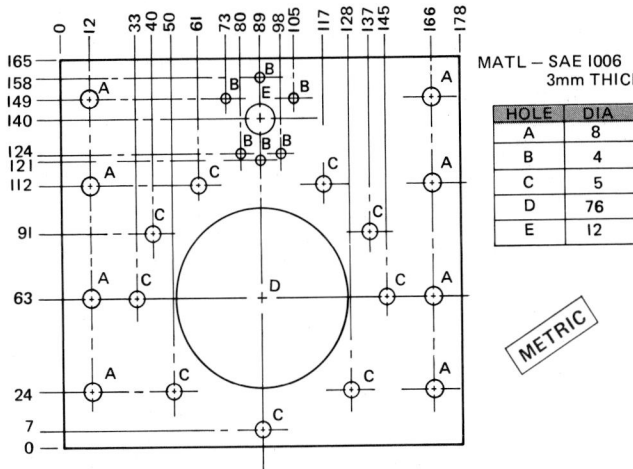

FIG. 8-4-A   Cover plate.

HOLE	DIA
A	8
B	4
C	5
D	76
E	12

MATL – SAE 1006
3mm THICK

METRIC

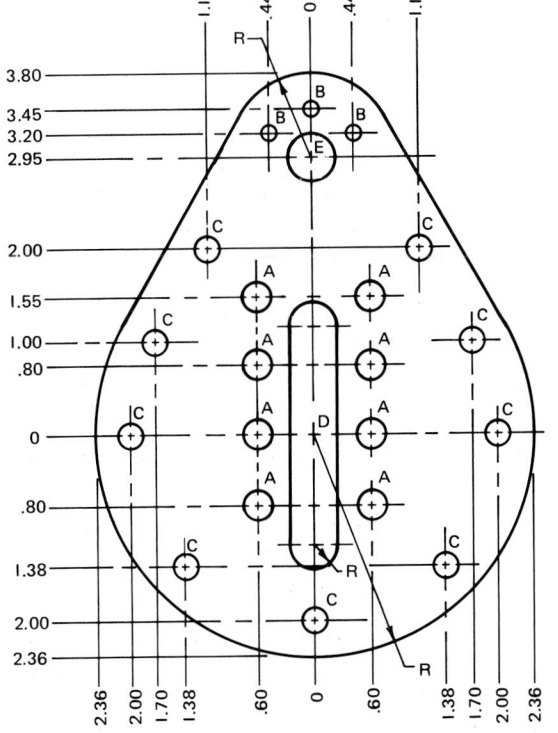

FIG. 8-4-B   Transmission cover.

MATL—SAE 1008
12 THICK

HOLE	SIZE
A	.30
B	.16
C	.24
D	.40 X 2.80
E	.50

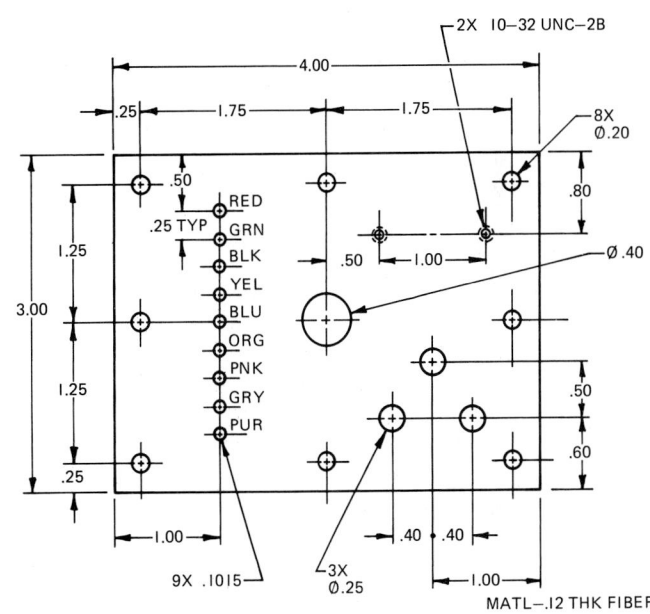

FIG. 8-4-C   Terminal board.

MATL—.12 THK FIBER

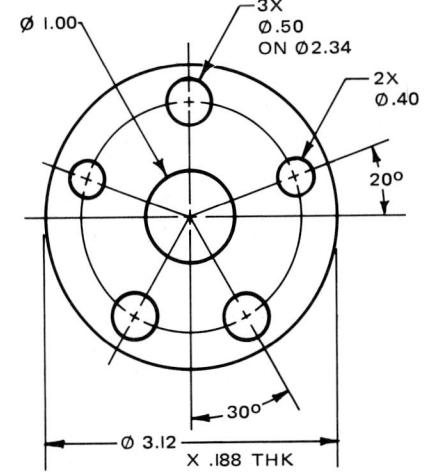

FIG. 8-4-D   Adaptor plate.

16. Redraw the oil chute shown in Fig. 8-4-E using tabular dimensioning to locate the holes from datum surfaces X, Y, and Z.

17. Redraw one of the parts shown in Figs. 8-4-F and 8-4-G. Use arrowless or tabular dimensioning. For Fig. 8-4-F use the bottom and left-hand edge for the datum surfaces to locate the holes. For Fig. 8-4-G use the bottom and the center of the part to locate the features. Use the MIRROR command to create the views if CAD is used.

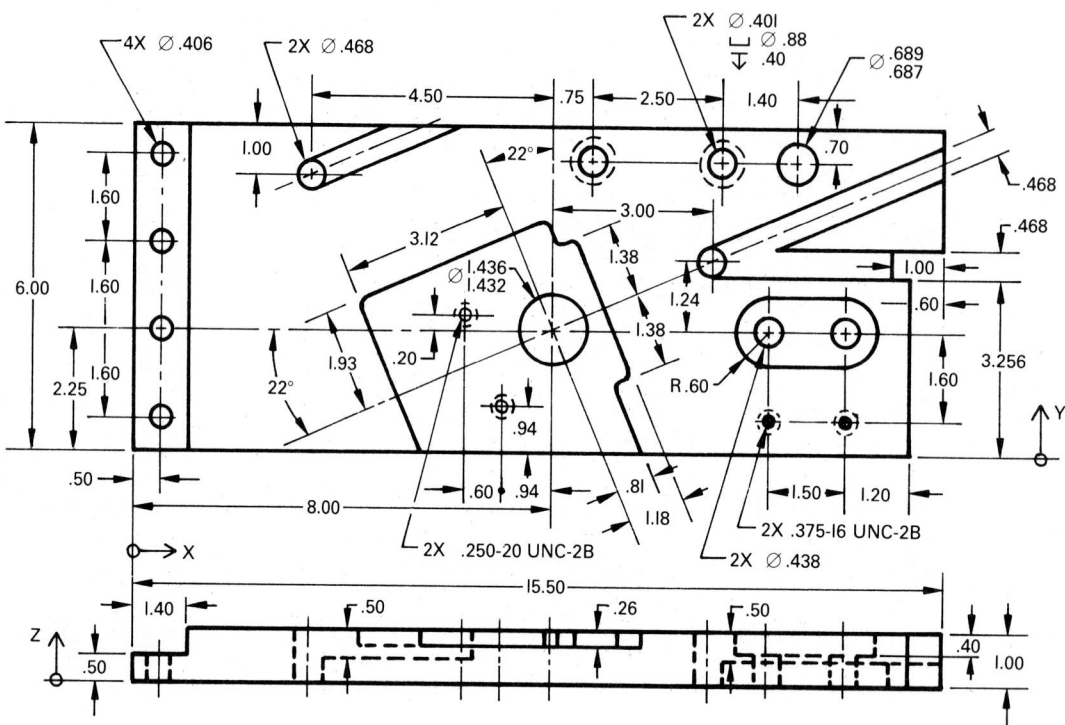

FIG. 8-4-E   Oil chute.

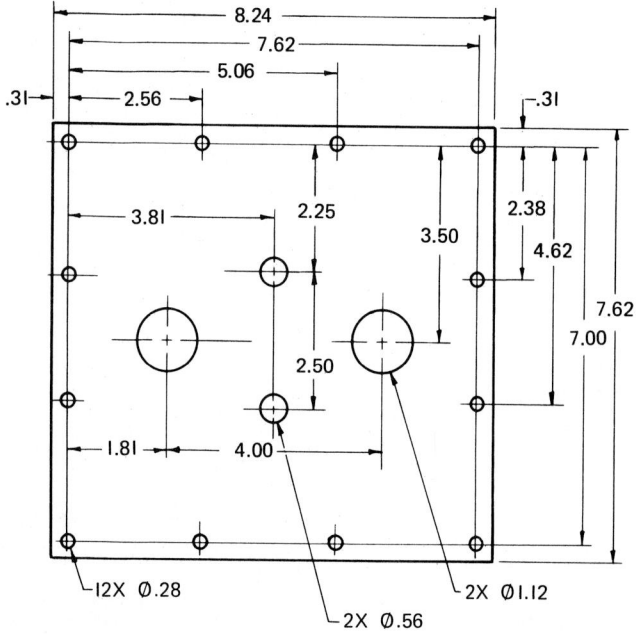

FIG. 8-4-F   Cover plate.

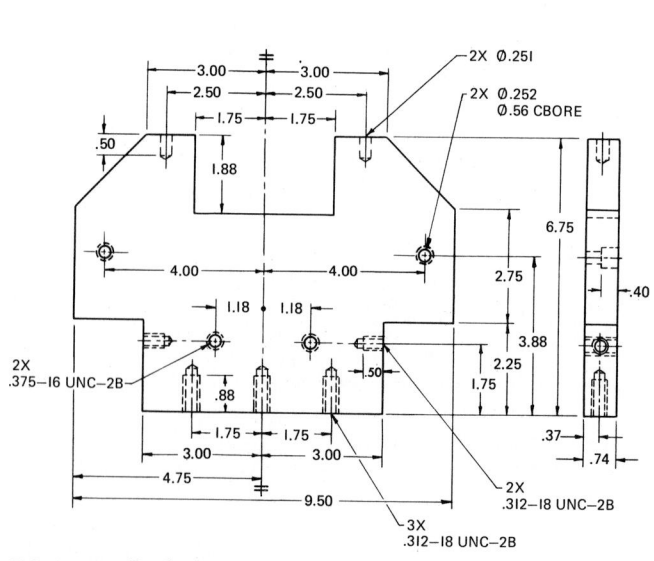

FIG. 8-4-G   Back plate.

18. Redraw the part shown in Fig. 8-4-H and use tabular dimensioning for locating the holes. Use the X, Y, and Z datums to locate the holes.

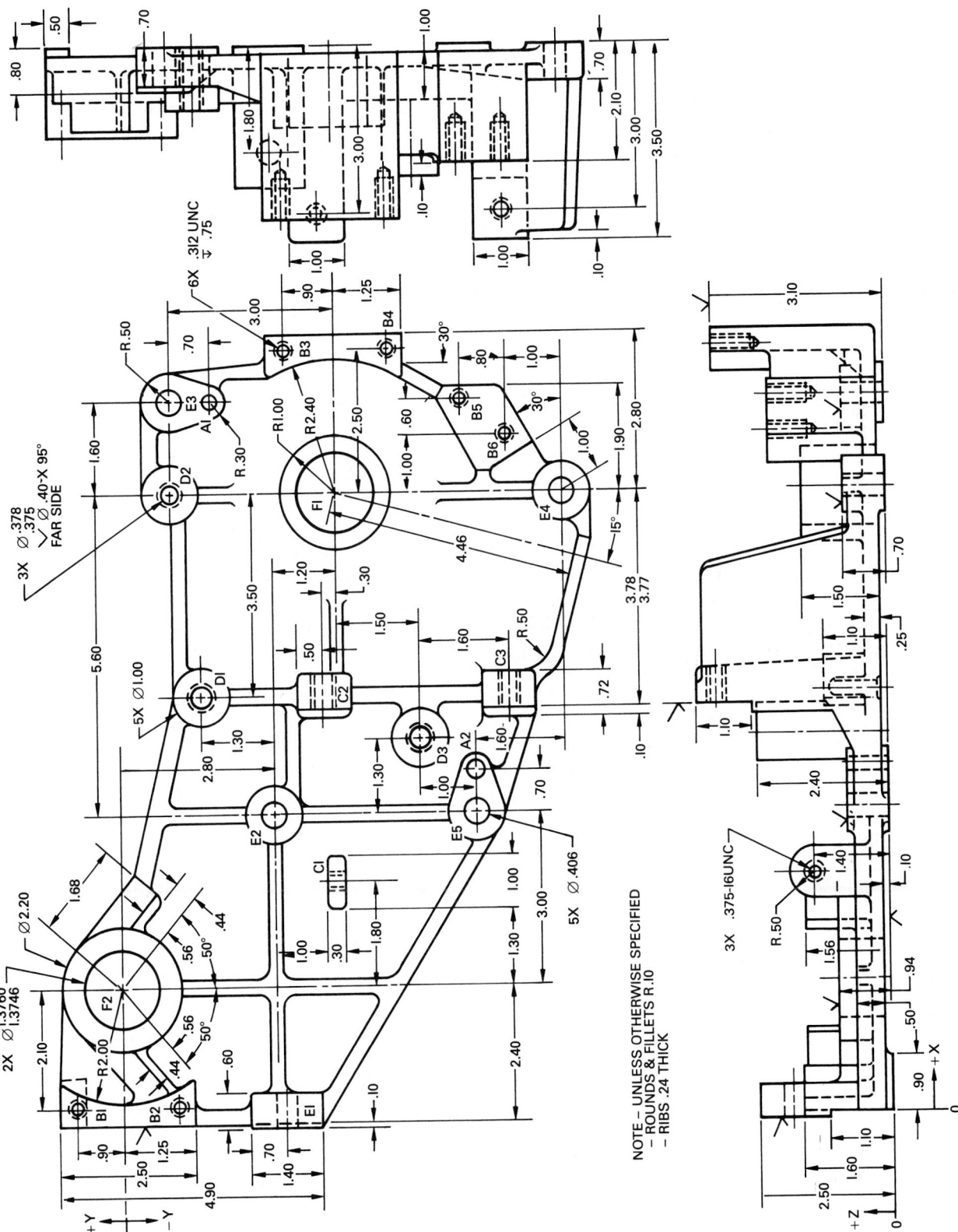

FIG. 8-4-H Interlocking base.

ASSIGNMENT FOR UNIT 8-5, LIMITS AND TOLERANCES

19. Calculate the sizes and tolerances for one of the drawings shown in Fig. 8-5-A or 8-5-B.

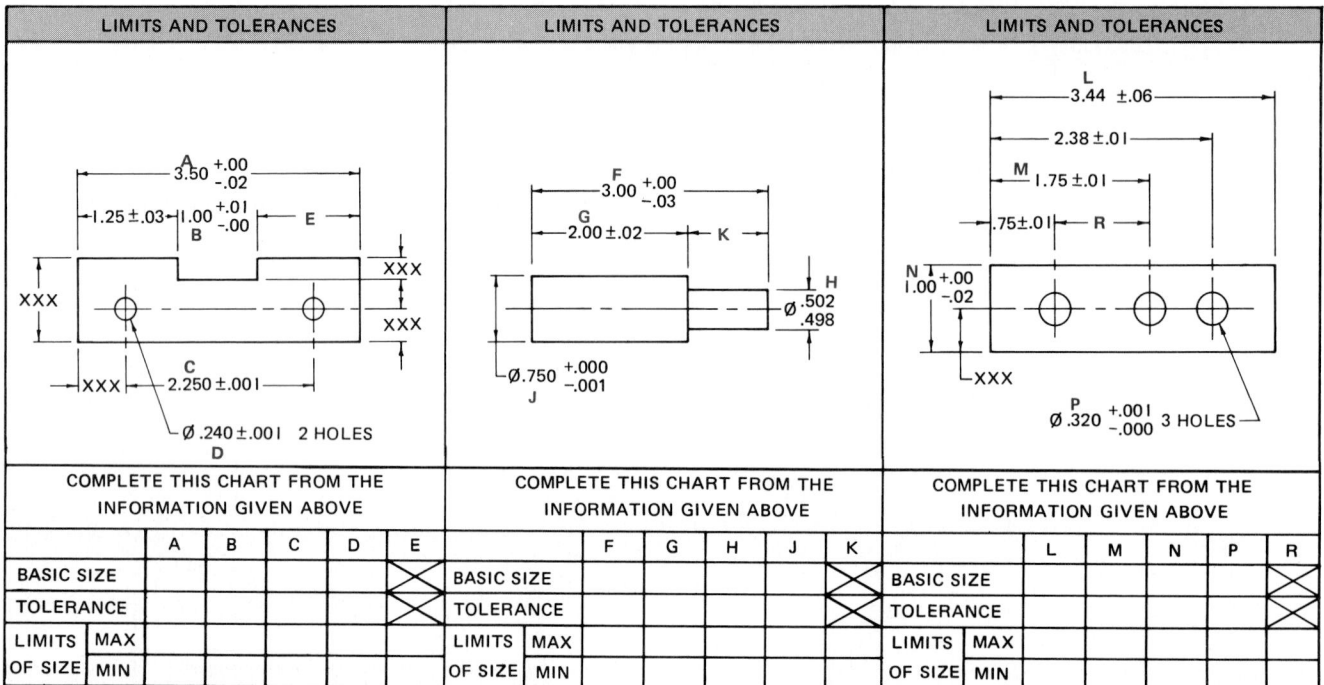

FIG. 8-5-A   Inch limits and tolerances.

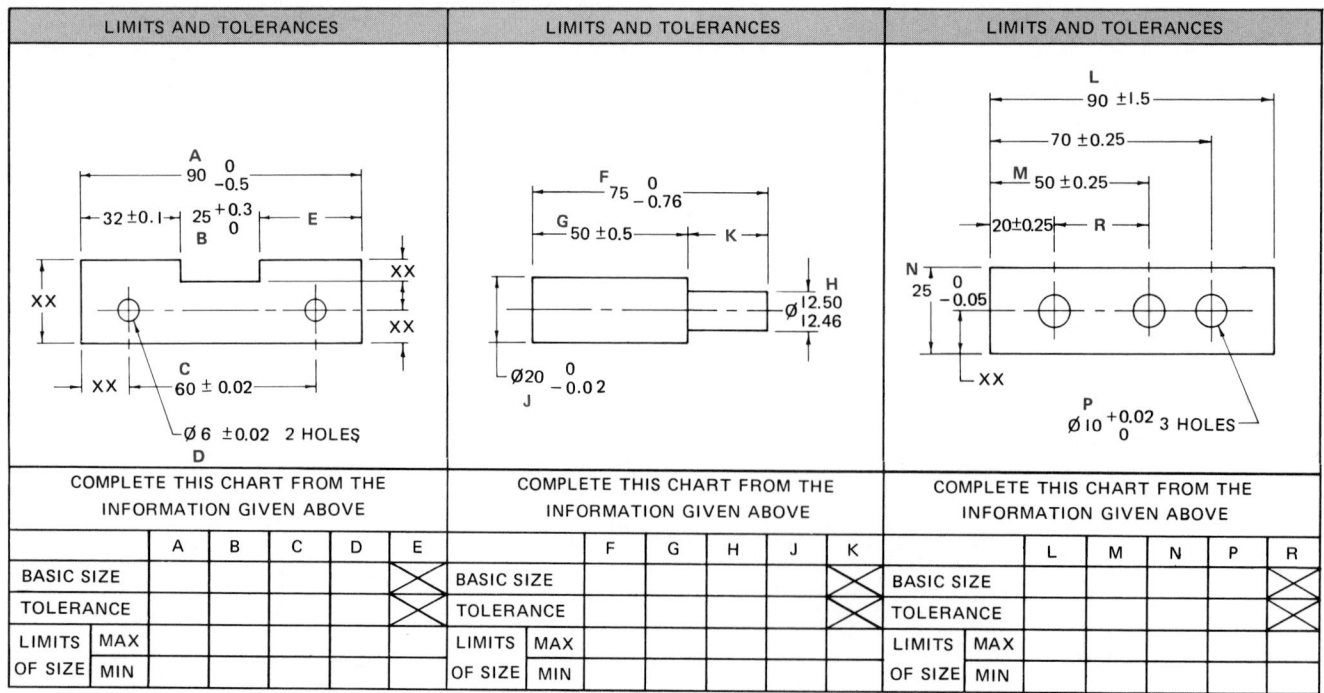

FIG. 8-5-B   Metric limits and tolerances.

ASSIGNMENTS FOR UNIT 8-6, FITS AND ALLOWANCES

20. Using the tables of fits located in the Appendix, calculate the missing dimensions in any of the four charts shown in Figs. 8-6-A through 8-6-D (here and on pg. 222).

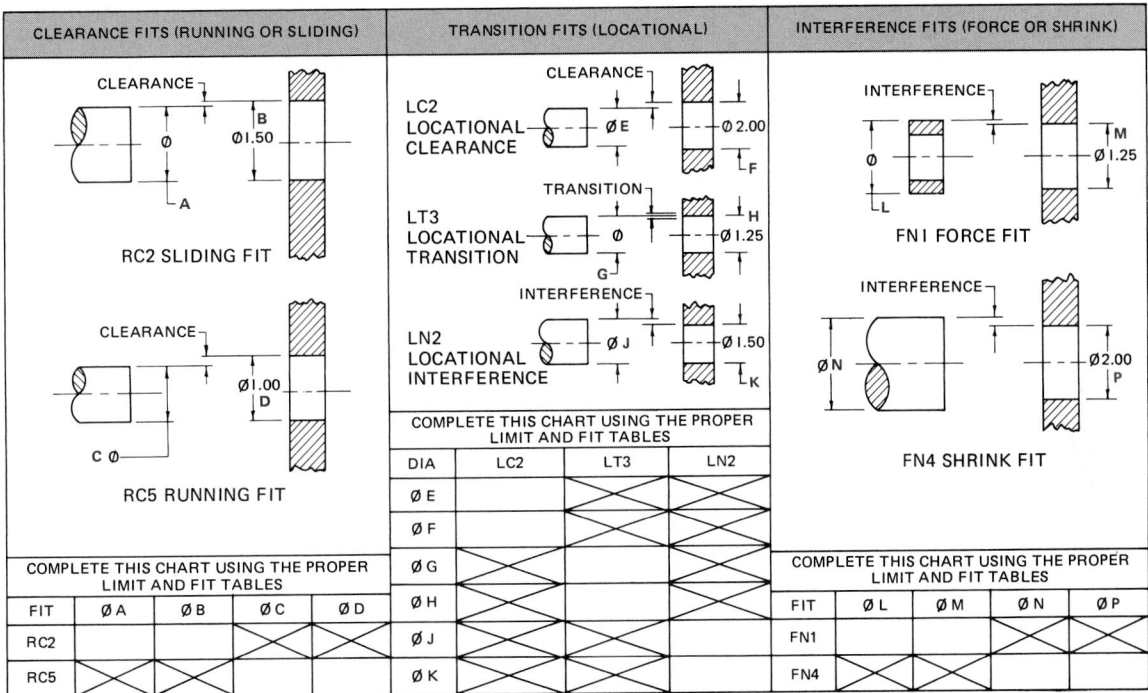

FIG. 8-6-A   Inch fits.

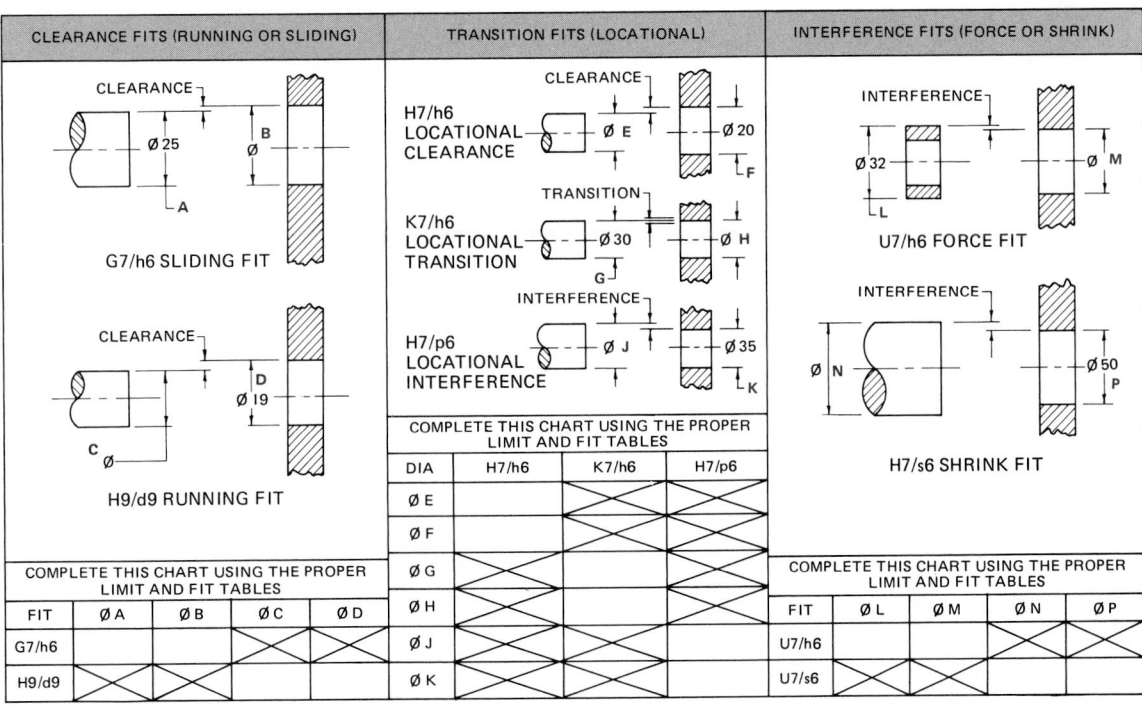

FIG. 8-6-B   Metric fits.

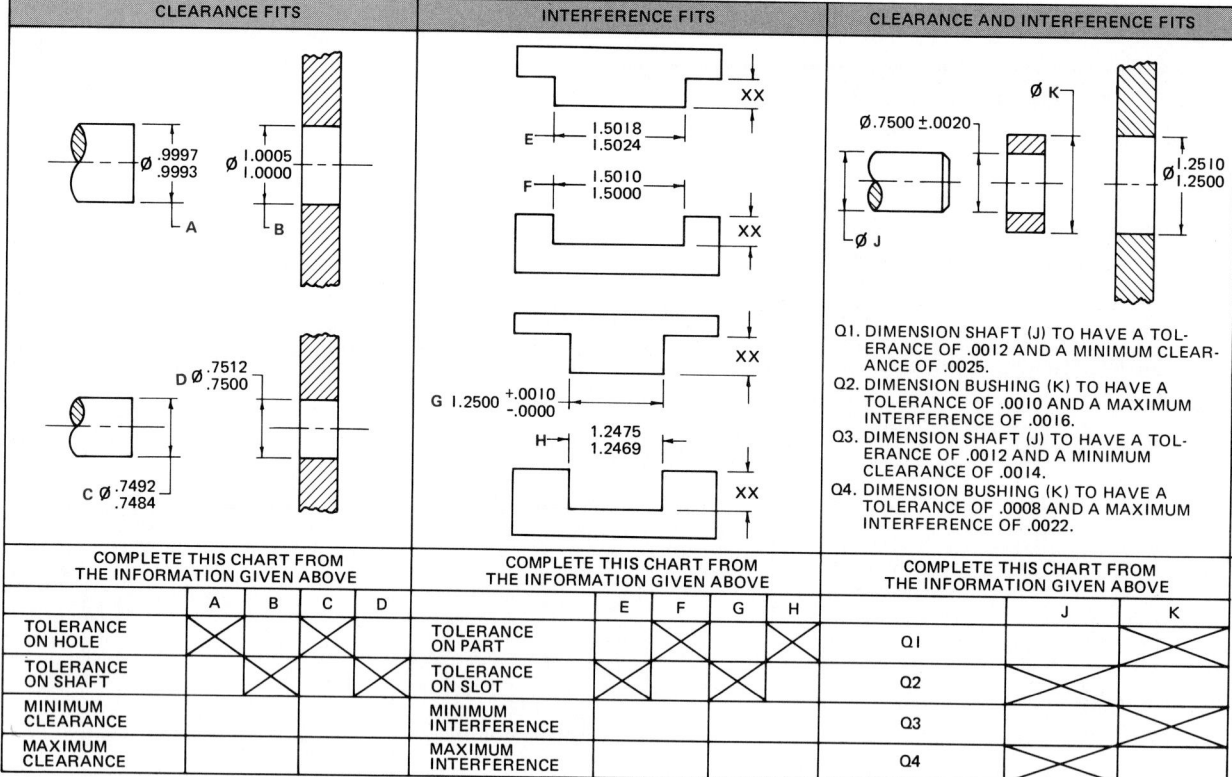

FIG. 8-6-C   Inch fits.

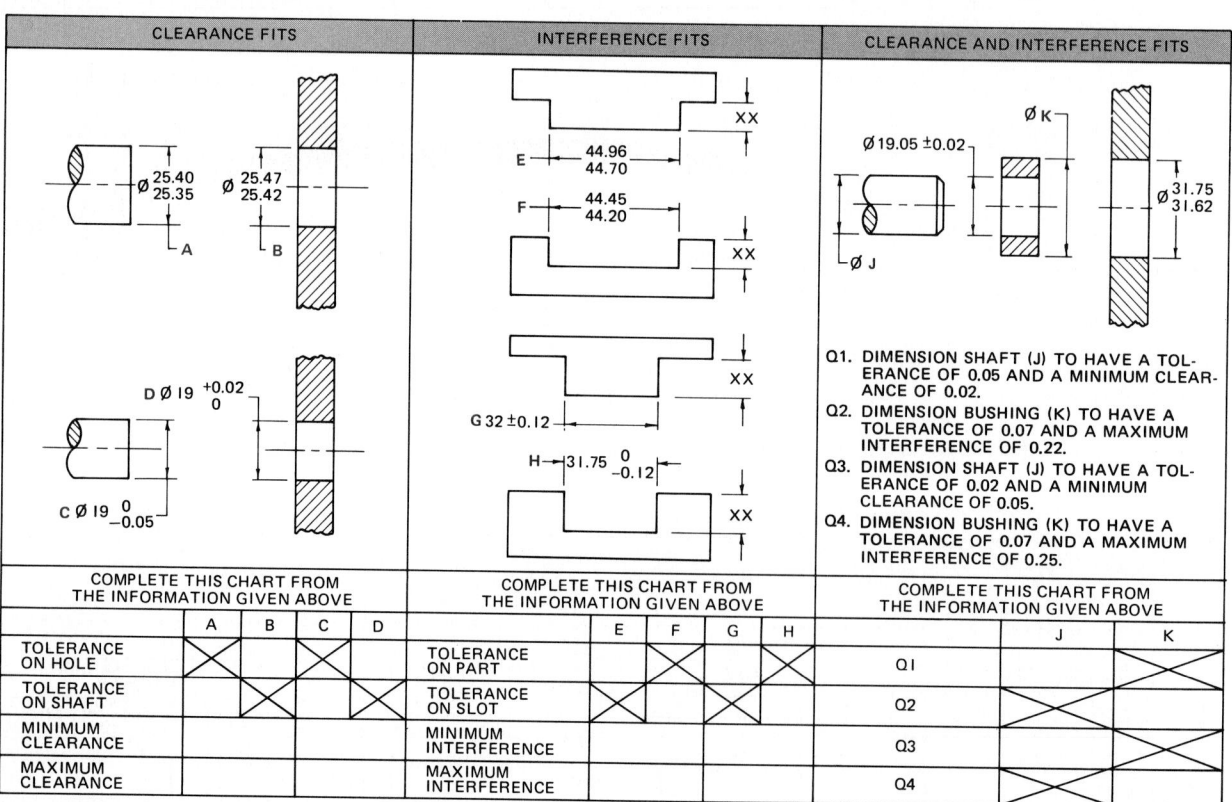

FIG. 8-6-D   Metric fits.

21. Using the fit tables in the Appendix, complete the table shown in Fig. 8-6-E using either U.S. customary or metric sizes.

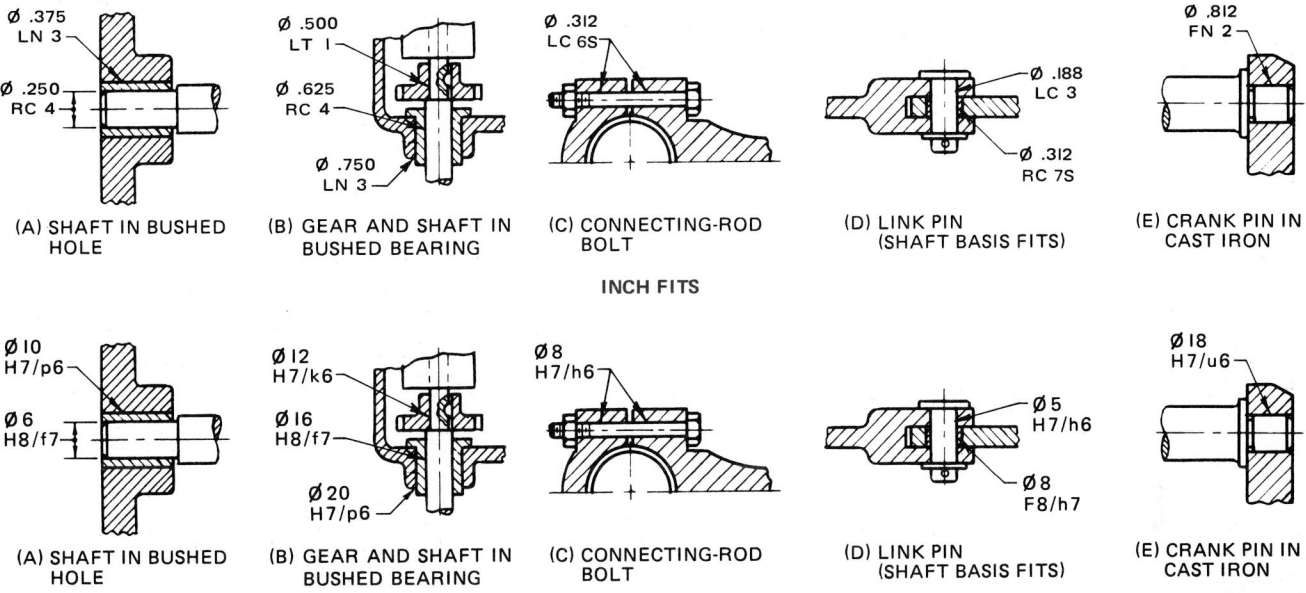

(A) SHAFT IN BUSHED HOLE

(B) GEAR AND SHAFT IN BUSHED BEARING

(C) CONNECTING-ROD BOLT

(D) LINK PIN (SHAFT BASIS FITS)

(E) CRANK PIN IN CAST IRON

INCH FITS

(A) SHAFT IN BUSHED HOLE

(B) GEAR AND SHAFT IN BUSHED BEARING

(C) CONNECTING-ROD BOLT

(D) LINK PIN (SHAFT BASIS FITS)

(E) CRANK PIN IN CAST IRON

METRIC FITS

DESIGN SKETCH	BASIC DIAMETER SIZE IN. [mm]	SYMBOL	BASIS	FEATURE	LIMITS OF SIZE		CLEARANCE OR INTERFERENCE	
					MAX	MIN	MAX	MIN
A	.375 [10]		HOLE	HOLE				
				SHAFT				
A	.250 [6]		HOLE	HOLE				
				SHAFT				
B	.500 [12]		HOLE	HOLE				
				SHAFT				
B	.625 [16]		HOLE	HOLE				
				SHAFT				
B	.750 [20]		HOLE	HOLE				
				SHAFT				
C	.312 [8]		SHAFT	HOLE				
				SHAFT				
D	.188 [5]		HOLE	HOLE				
				SHAFT				
D	.312 [8]		SHAFT	HOLE				
				SHAFT				
E	.812 [18]		HOLE	HOLE				
				SHAFT				

FIG. 8-6-E   Fit problems.

22. Make a detail drawing of the spindle shown in Fig. 8-6-F. Scale the part for sizes using one of the scales shown with the drawing. Other considerations are:
    (a) "A" diameter to have an LC3 (inch) or H7/h6 (metric) fit.
    (b) "B" diameter requires a 96 diamond knurl or its equivalent.
    (c) "C" diameter to have an LT3 (inch) or H7/k6 (metric) fit.
    (d) "D" diameter to be a minimum relief (undercut).
    (e) "E" to be a standard No. 807 Woodruff key in center of segment and the diameter to be controlled by an RC3 (inch) or H7/g6 (metric) fit.

(f) "F" to be undercut for a standard external retaining ring and dimensioned to manufacturer's specifications.
(g) Dimension in decimal inch or metric.

23. Make a detail drawing of the roller guide base shown in Fig. 8-6-G. Use one of the scales shown on the drawing to scale the part for sizes. Other considerations are:
    (a) Keyseat for a standard square key and limits on the hole controlled by either an H9/d9 (metric) or RC6 (inch) fit.
    (b) Control critical machine surfaces to 0.8 μm or 32 μin.
    (c) Dimension in metric or decimal inch.

FIG. 8-6-F   Spindle.

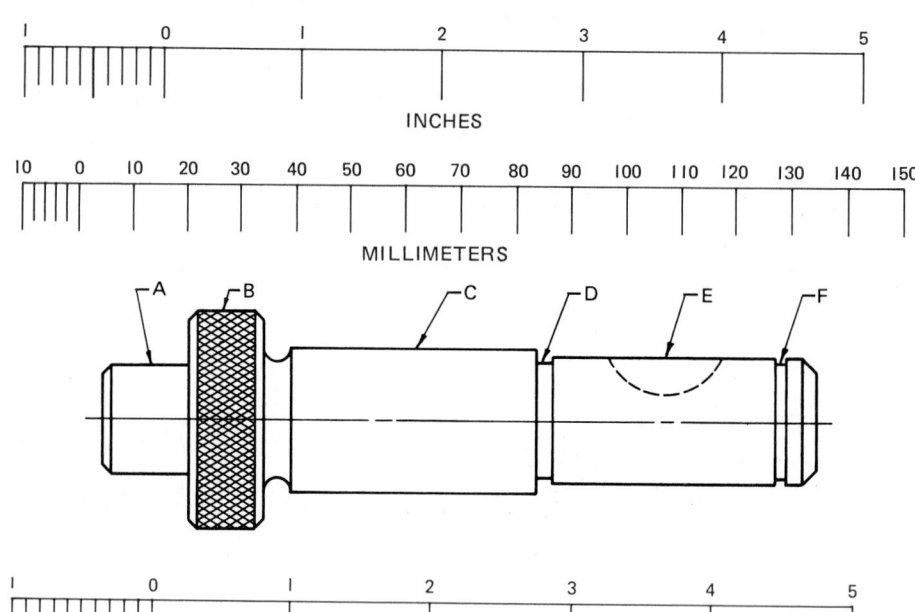

FIG. 8-6-G   Roller guide base.

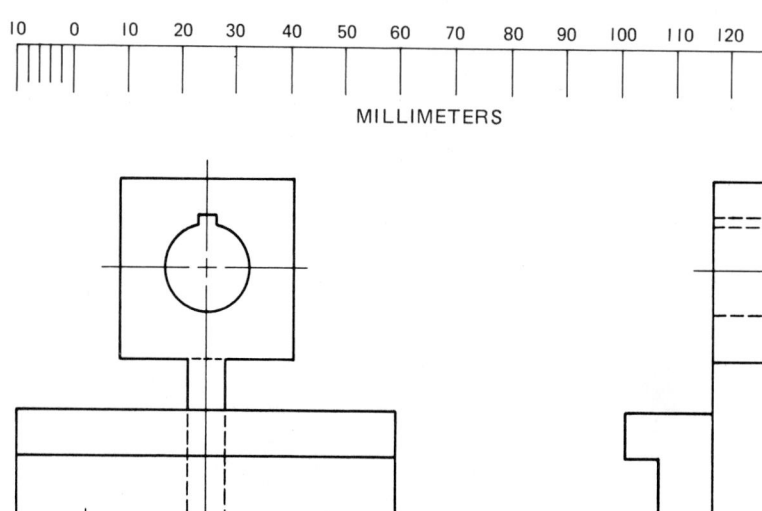

ASSIGNMENTS FOR UNIT 8-7, SURFACE TEXTURE

24. Make a working drawing of the link shown in Fig. 8-7-A. The amount of material to be removed from the end surfaces of the hub is .09 in. and .06 in. on the bosses and bottom of the vertical hub. The two large holes are to have an LN3 fit for journal bearings. Scale 1:1.

25. Make a working drawing of the cross slide shown in Fig. 8-7-B. Scale 1:1. The following information is to be added to the drawing:
    - The dovetail slot is to have a maximum roughness value of 3.2 μm and a machining allowance of 2mm.
    - The ends of the shaft support are to have maximum and minimum roughness values of 1.6 and 0.8 μm and a machining allowance of 2 mm.
    - The hole is to have an H8 tolerance.

26. Make a working drawing of the column bracket shown in Fig. 8-7-C. Scale 1:1. The following information is to be added to the drawing:
    - The bottom of the base is to have a maximum roughness value of 125 μin. and a machining allowance of .06 in.
    - The tops of the bosses are to have a maximum roughness value of 250 min. and a machining allowance of .04 in.
    - The end surfaces of the hubs supporting the shafts are to have maximum and minimum roughness values of 63 and 32 μin. and a machining allowance of .04 in.
    - The large hole is to be dimensioned for an RC4 fit. The small hole is to be dimensioned for an LN3 fit for plain bearings.

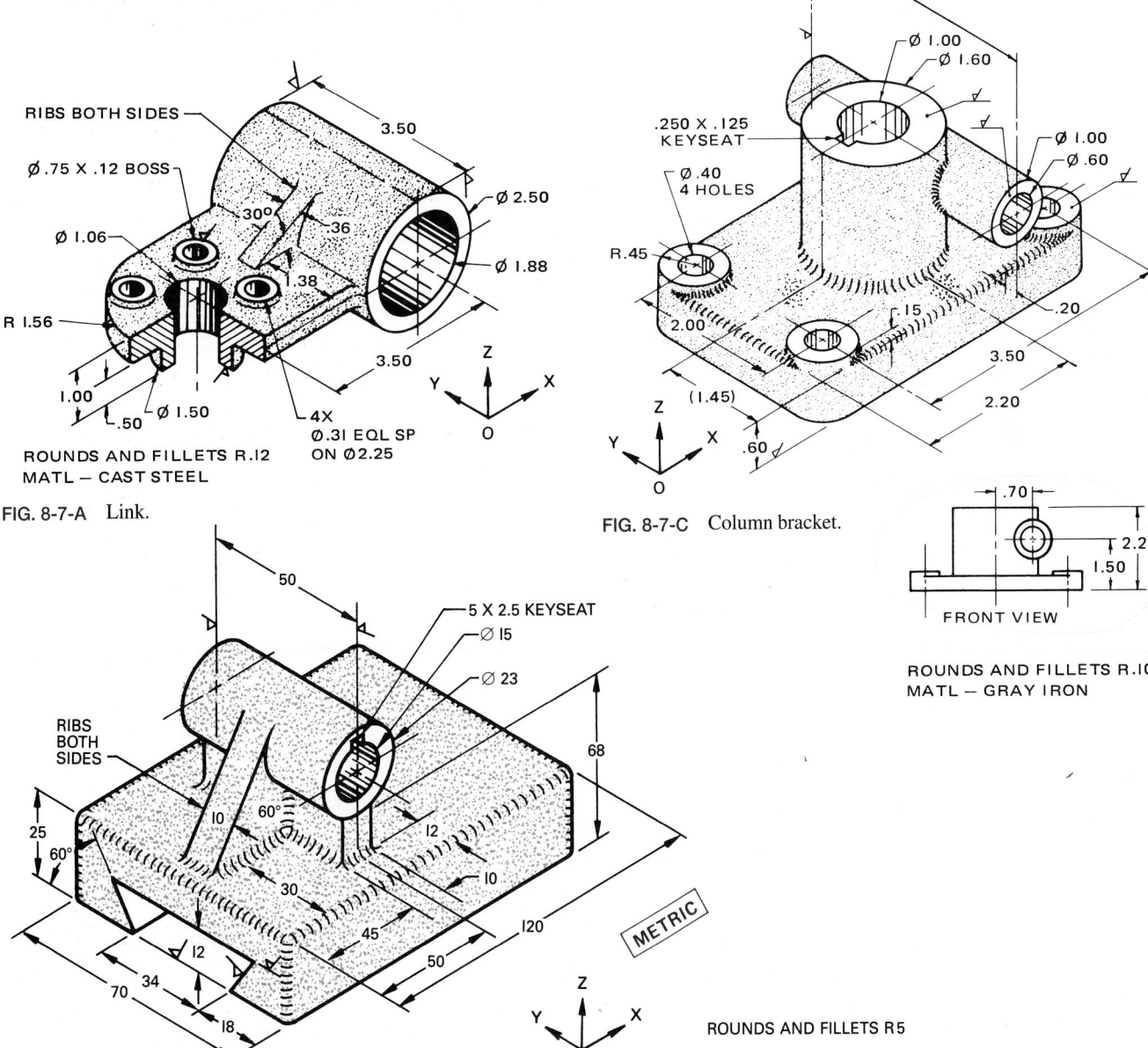

FIG. 8-7-A   Link.

FIG. 8-7-B   Cross slide.

FIG. 8-7-C   Column bracket.

225

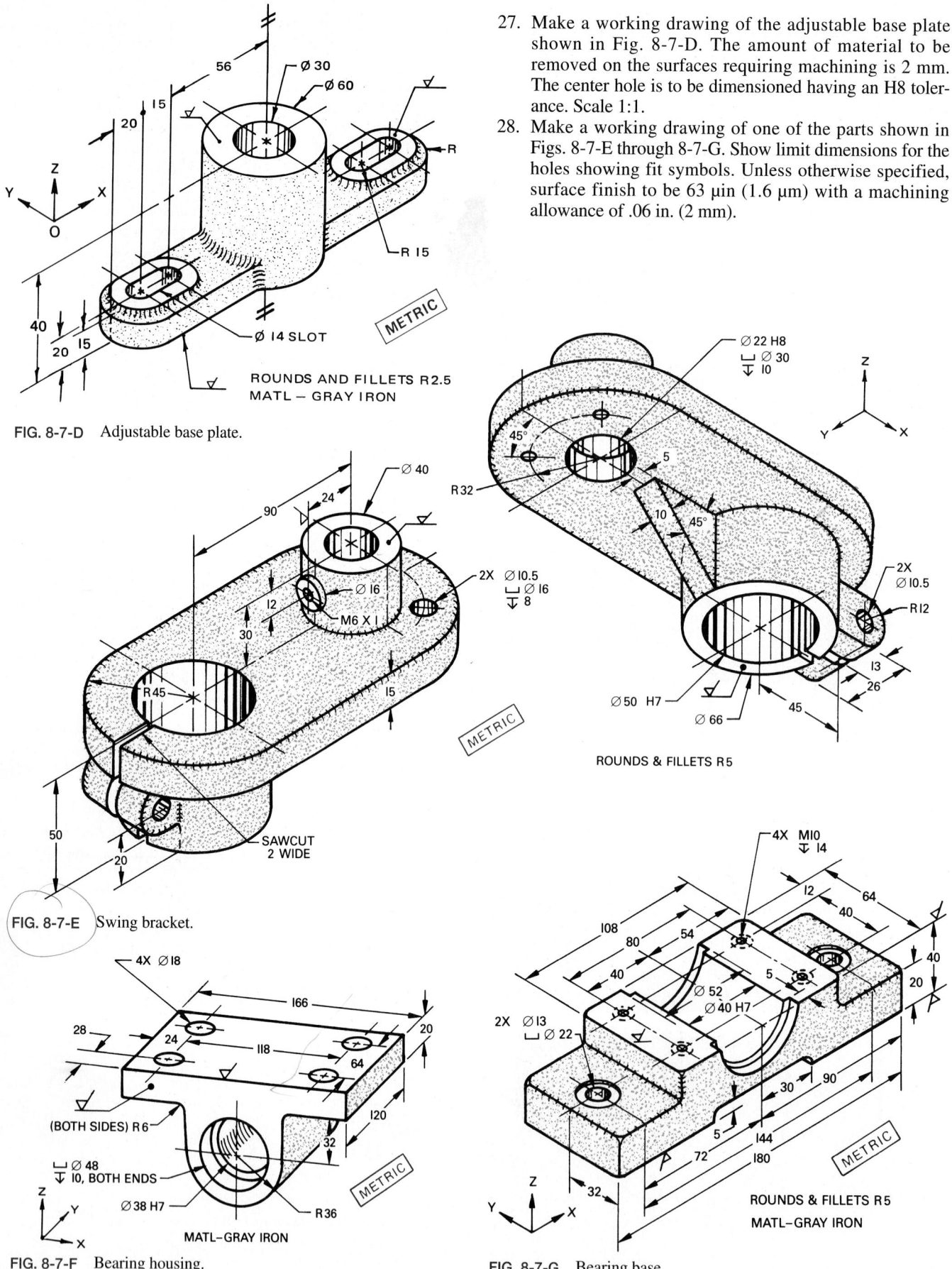

27. Make a working drawing of the adjustable base plate shown in Fig. 8-7-D. The amount of material to be removed on the surfaces requiring machining is 2 mm. The center hole is to be dimensioned having an H8 tolerance. Scale 1:1.

28. Make a working drawing of one of the parts shown in Figs. 8-7-E through 8-7-G. Show limit dimensions for the holes showing fit symbols. Unless otherwise specified, surface finish to be 63 μin (1.6 μm) with a machining allowance of .06 in. (2 mm).

FIG. 8-7-D   Adjustable base plate.

FIG. 8-7-E   Swing bracket.

FIG. 8-7-F   Bearing housing.

FIG. 8-7-G   Bearing base.

CHAPTER 9

SECTIONS

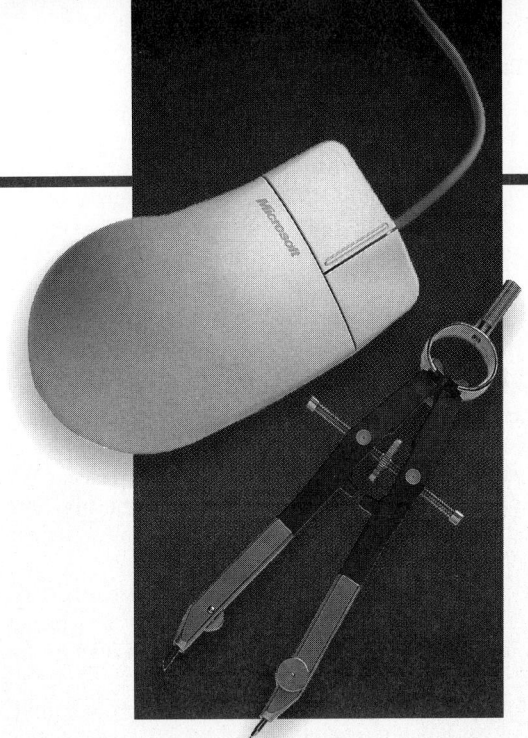

Definitions

Cutting plane   The plane at which the exterior view is cut away to reveal the interior view.

Cutting-plane line   A line to indicate the location of the cutting plane.

Half-section   A view of an assembly or object, usually symmetrical, that shows one-half of the full view in section.

Phantom section   A sectional view superimposed on the regular view, used to show the typical interior shape and/or mating part within the object.

Section lining or **cross-hatching**   Lines used to indicate either the surface that has been theoretically cut or the material from which the object is to be made.

Sectional views or sections   Drawings used to show interior details of objects that are too complicated to be shown clearly in regular views.

9-1 SECTIONAL VIEWS

Sectional views, commonly called *sections,* are used to show interior detail that is too complicated to be shown clearly by regular views containing many hidden lines. For some assembly drawings, they show a difference in materials. A sectional view is obtained by supposing that the nearest part of the object to be cut or broken away is on an imaginary cutting plane. The exposed or cut surfaces are identified by section lining or cross-hatching. Hidden lines and details behind the cutting-plane line are usually omitted unless they are required for clarity or dimensioning. It should be understood that only in the sectional view is any part of the object shown as having been removed.

A sectional view frequently replaces one of the regular views. For example, a regular front view is replaced by a front view in section, as shown in Fig. 9-1-1.

Whenever practical, except for revolved sections, sectional views should be projected perpendicularly to the cutting plane and be placed in the normal position for third-angle projection.

FIG. 9-1-1   A full-section drawing.

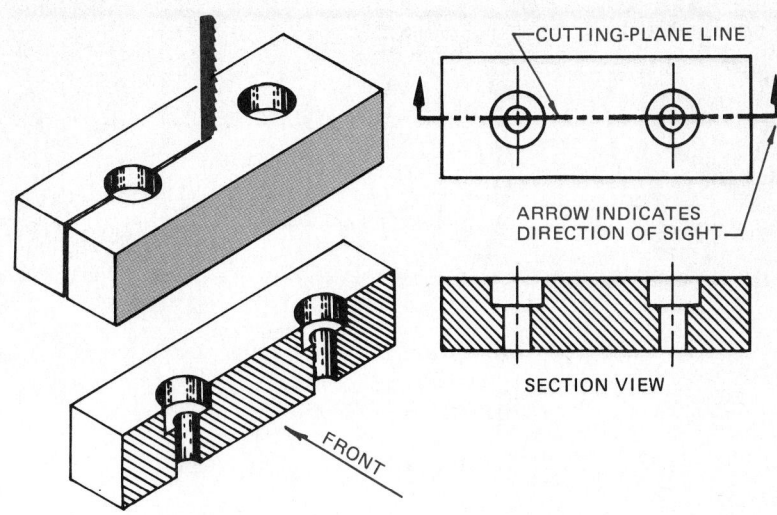

When the preferred placement is not practical, the sectional view may be moved to some other convenient position on the drawing, but it must be clearly identified, usually by two upper-case letters, and labeled.

Cutting-Plane Lines

Cutting-plane lines (Fig. 9-1-2) are used to show the location of cutting planes for sectional views. Two forms of cutting-plane lines are approved for general use.

The first form consists of evenly spaced, thick dashes with arrowheads. The second form consists of alternating long dashes and pairs of short dashes. The long dashes may vary in length, depending on the size of the drawing.

Both forms of lines should be drawn to stand out clearly on the drawing. The ends of the lines are bent at 90° and terminated by bold arrowheads to indicate the direction of sight for viewing the section.

The cutting-plane line can be omitted when it corresponds to the center line of the part and it is obvious where the cutting

plane lies. On drawings with a high density of line work and for offset sections (see Unit 9-6), cutting-plane lines may be modified by omitting the dashes between the line ends for the purpose of obtaining clarity, as shown in Fig. 9-1-2B.

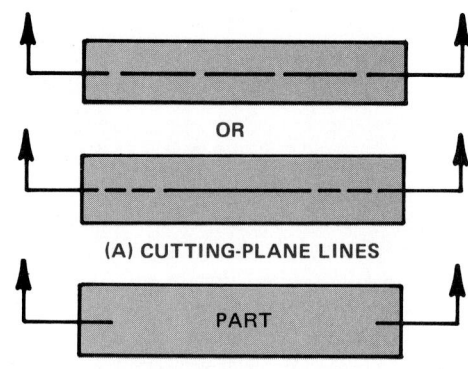

FIG. 9-1-2   Cutting-plane lines.

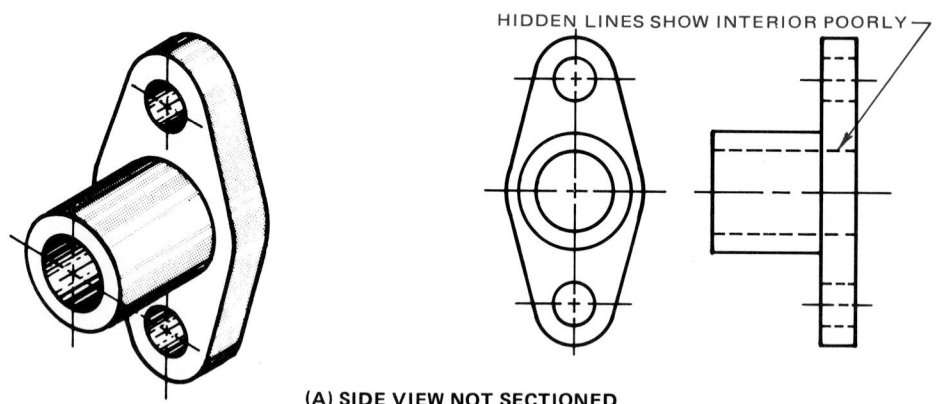

(A) SIDE VIEW NOT SECTIONED

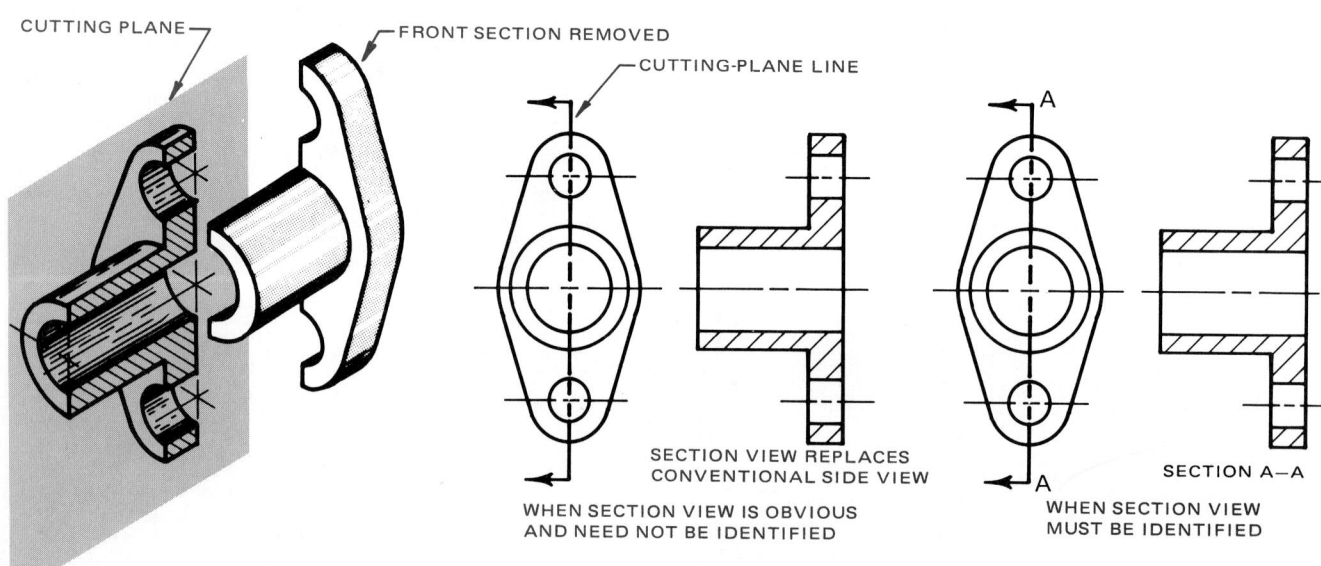

(B) SIDE VIEW IN FULL SECTION

FIG. 9-1-3   Full-section view.

228

FIG. 9-1-4   Visible and hidden lines on section views.

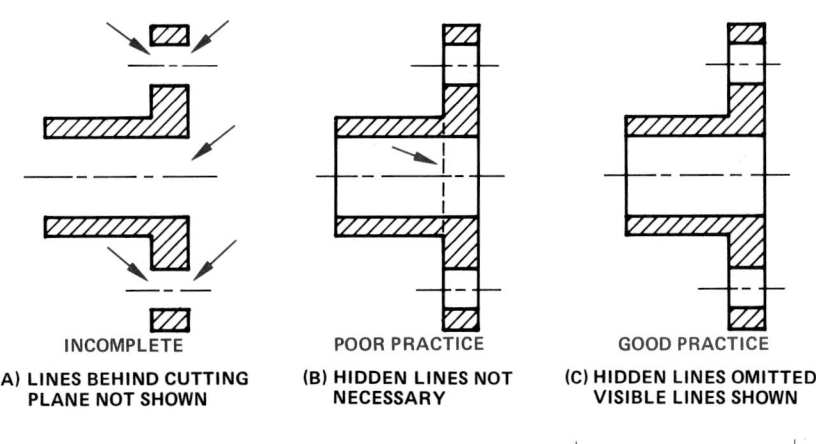

| (A) LINES BEHIND CUTTING PLANE NOT SHOWN | (B) HIDDEN LINES NOT NECESSARY | (C) HIDDEN LINES OMITTED, VISIBLE LINES SHOWN |
| INCOMPLETE | POOR PRACTICE | GOOD PRACTICE |

Full Sections

When the cutting plane extends entirely through the object in a straight line and the front half of the object is theoretically removed, a *full section* is obtained (Figs. 9-1-3 and 9-1-4). This type of section is used for both detail and assembly drawings. When the section is on an axis of symmetry, it is not necessary to indicate its location (Fig. 9-1-5). However, it may be identified and indicated in the normal manner to increase clarity, if so desired.

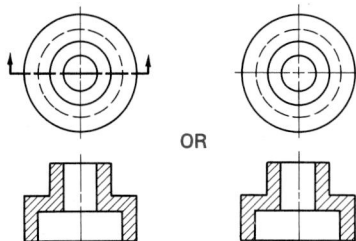

FIG. 9-1-5   Cutting-plane line may be omitted when it corresponds with a center line.

Section Lining

Section lining, sometimes referred to as *cross-hatching,* can serve a double purpose. It indicates the surface that has been theoretically cut and makes it stand out clearly, thus helping the observer to understand the shape of the object. Section lining may also indicate the material from which the object is to be made, when the lining symbols shown in Fig. 9-1-6 are used.

Section Lining for Detail Drawings

Since the exact material specifications for a part are usually given elsewhere on the drawing, the general-purpose section lining symbol is recommended for most detail drawings. An exception may be made for wood when it is desirable to show the direction of the grain.

The lines for section lining are thin and are usually drawn at an angle of 45° to the major outline of the object. The same angle is used for the whole "cut" surface of the object. If the part shape would cause section lines to be parallel, or nearly so, to one of the sides of the part, some angle other than 45° should be chosen. See Fig. 9-1-7 (pg. 230). The spacing of the hatching lines should be reasonably uniform to give a good appearance to the drawing. The pitch, or distance between lines, normally varies between .03 and .12 in. (1 and 3 mm), depending on the size of the area to be sectioned.

As a cost reduction in manual drafting, large areas need not be entirely section-lined (Fig. 9-1-8, pg. 230). Section lining around the outline will usually be sufficient, providing clarity is not sacrificed.

Dimensions or other lettering should not be placed in sectioned areas. When this is unavoidable, the section lining should be omitted for the numerals or lettering (Fig. 9-1-9, pg. 230).

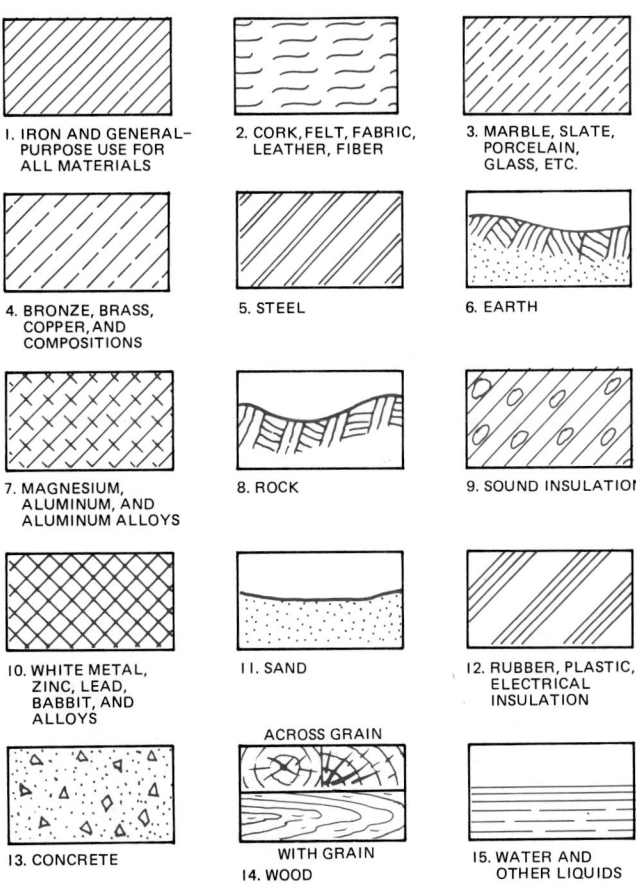

1. IRON AND GENERAL-PURPOSE USE FOR ALL MATERIALS

2. CORK, FELT, FABRIC, LEATHER, FIBER

3. MARBLE, SLATE, PORCELAIN, GLASS, ETC.

4. BRONZE, BRASS, COPPER, AND COMPOSITIONS

5. STEEL

6. EARTH

7. MAGNESIUM, ALUMINUM, AND ALUMINUM ALLOYS

8. ROCK

9. SOUND INSULATION

10. WHITE METAL, ZINC, LEAD, BABBIT, AND ALLOYS

11. SAND

12. RUBBER, PLASTIC, ELECTRICAL INSULATION

13. CONCRETE

ACROSS GRAIN
WITH GRAIN
14. WOOD

15. WATER AND OTHER LIQUIDS

FIG. 9-1-6   Symbolic section lining.

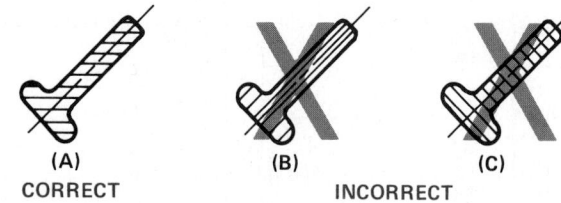

FIG. 9-1-7   Direction of section lining.

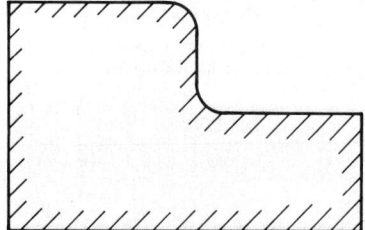

FIG. 9-1-8   Outline section lining.

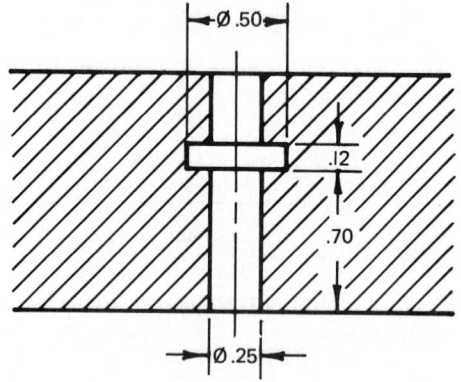

FIG. 9-1-9   Section lining omitted to accommodate dimensions.

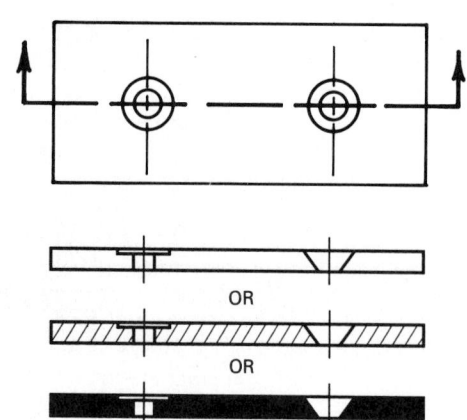

FIG. 9-1-10   Thin parts in section.

Sections that are too thin for effective section lining, such as sheet-metal items, packing, and gaskets, may be shown without section lining, or the area may be filled in completely (Fig. 9-1-10).

ASSIGNMENTS

See Assignments 1 and 2 for Unit 9-1 on pages 240–241.

9-2 TWO OR MORE SECTIONAL VIEWS ON ONE DRAWING

If two or more sections appear on the same drawing, the cutting-plane lines are identified by two identical large Gothic letters, one at each end of the line, placed behind the arrowhead so that the arrow points away from the letter. Normally, begin alphabetically with A-A, then B-B, and so on (Fig. 9-2-1). The identification letters should not include I, O, Q, or Z.

FIG. 9-2-1   Detail drawing
having two
sectional views.

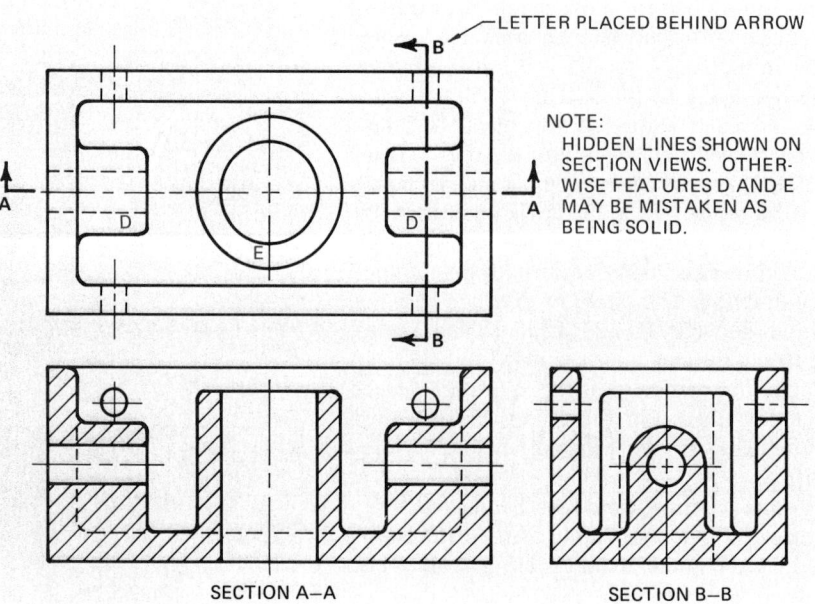

LETTER PLACED BEHIND ARROW

NOTE:
HIDDEN LINES SHOWN ON SECTION VIEWS. OTHER-WISE FEATURES D AND E MAY BE MISTAKEN AS BEING SOLID.

SECTION A—A   SECTION B—B

Sectional-view subtitles are given when identification letters are used and appear directly below the view, incorporating the letters at each end of the cutting-plane line thus: SECTION A-A, or abbreviated, SECT. B-B. When the scale is different from the main view, it is stated below the subtitle thus:

SECTION A-A
SCALE 1:10

ASSIGNMENTS

See Assignments 3 and 4 for Unit 9-2 on pages 241–242.

9-3 HALF-SECTIONS

A *half-section* is a view of an assembly or object, usually symmetrical, showing one-half of the view in section (Figs. 9-3-1 and 9-3-2). Two cutting-plane lines, perpendicular to each other, extend halfway through the view, and one-quarter of the view is considered removed with the interior exposed to view.

Similar to the practice followed for full-section drawings, the cutting-plane line need not be drawn for half-sections when it is obvious where the cutting took place. Instead, center lines may be used. When a cutting-plane is used, the common

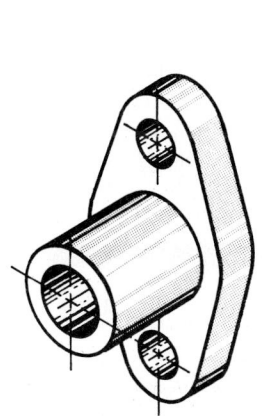
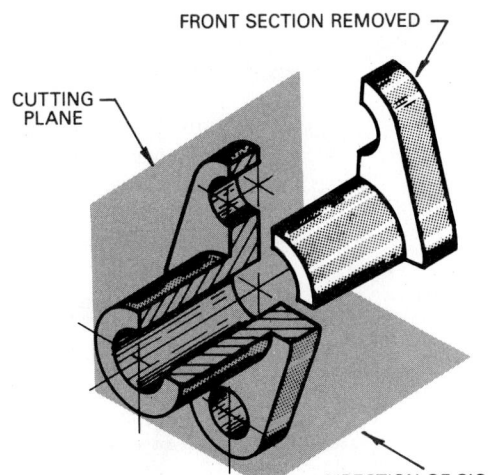

FIG. 9-3-1 Half-section drawing.

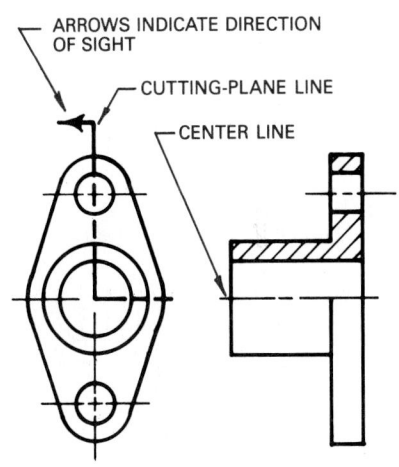

FIG. 9-3-2 Half-section views.

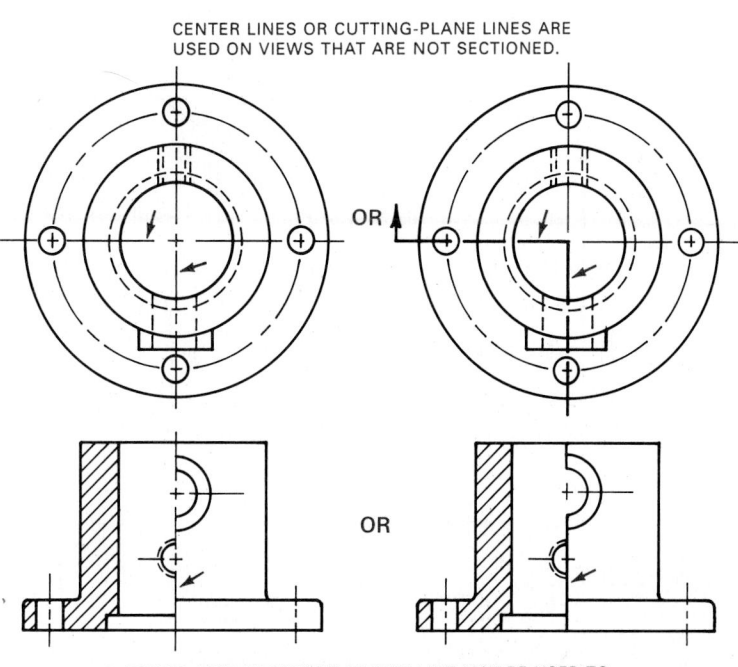

A CENTER LINE OR VISIBLE OBJECT LINE MAY BE USED TO DIVIDE THE SECTIONED HALF FROM THE UNSECTIONED HALF.

practice is to show only one end of the cutting-plane line, terminating with an arrow to show the direction of sight for viewing the section.

On the sectional view a center line or a visible object line may be used to divide the sectioned half from the unsectioned half of the drawing. This type of sectional drawing is best suited for assembly drawings where both internal and external construction is shown on one view and where only overall and center-to-center dimensions are required. The main disadvantage of using this type of sectional drawing for detail drawings is the difficulty in dimensioning internal features without adding hidden lines. However, hidden lines may be added for dimensioning, as shown in Fig. 9-3-3.

ASSIGNMENT

See Assignment 5 for Unit 9-3 on pages 242–243.

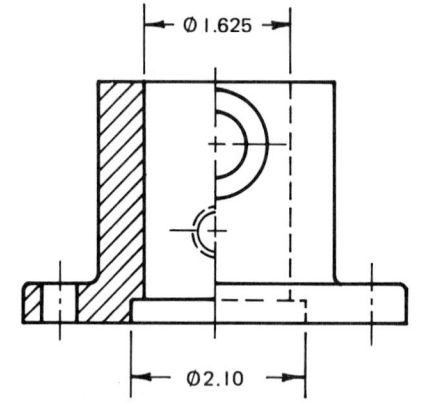

HIDDEN LINES ADDED FOR DIMENSIONING

FIG. 9-3-3 Dimensioning a half-section view.

9-4 THREADS IN SECTION

True representation of a screw thread is seldom provided on working drawings because it would require very laborious and accurate drawing involving repetitious development of the helix curve of the thread. A symbolic representation of threads is now standard practice.

Three types of conventions are in general use for screw-thread representation (Fig. 9-4-1). These are known as detailed, schematic, and simplified representations. Simplified representation should be used whenever it will clearly portray the requirements. Schematic and detailed representations require more drafting time but are sometimes necessary to avoid confusion with other parallel lines or to more clearly portray particular aspects of the threads.

Threaded Assemblies

Any of the thread conventions shown here may be used for assemblies of threaded parts, and two or more methods may be used on the same drawing, as shown in Fig. 9-4-2. In sectional views, the externally threaded part is always shown covering the internally threaded part (Fig. 9-4-3).

ASSIGNMENTS

See Assignments 6 and 7 for Unit 9-4 on pages 243–244.

9-5 ASSEMBLIES IN SECTION

Section Lining on Assembly Drawings

General-purpose section lining is recommended for most assembly drawings, especially if the detail is small. Symbolic

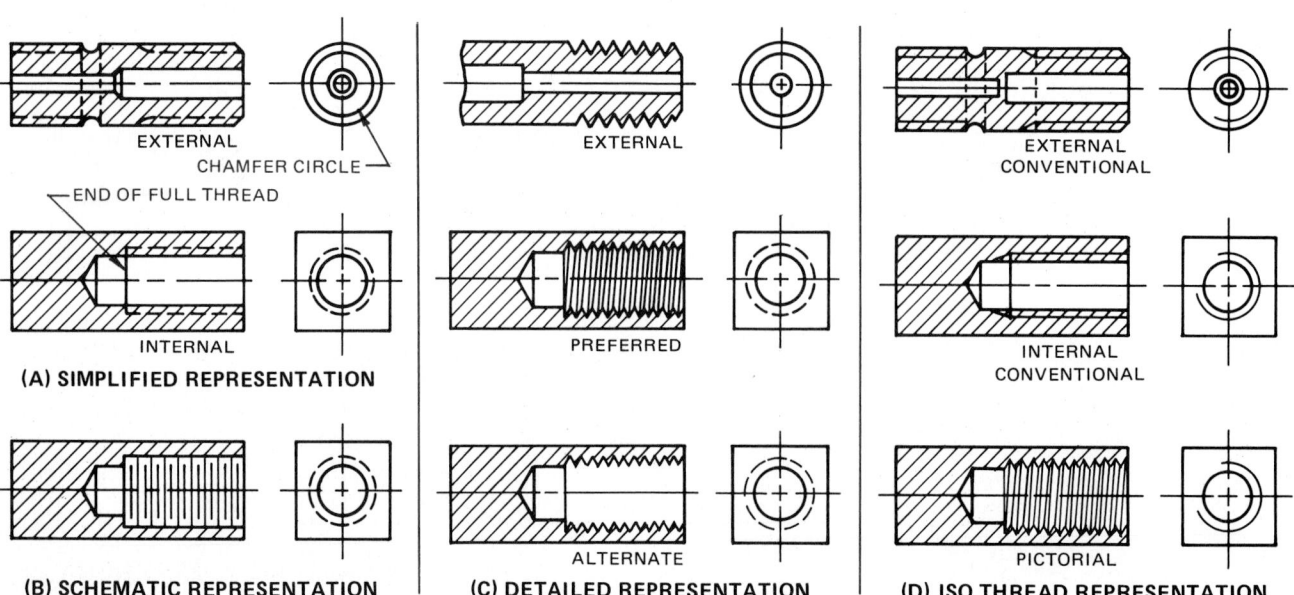

(A) SIMPLIFIED REPRESENTATION

(B) SCHEMATIC REPRESENTATION

(C) DETAILED REPRESENTATION

(D) ISO THREAD REPRESENTATION

FIG. 9-4-1 Threads in section.

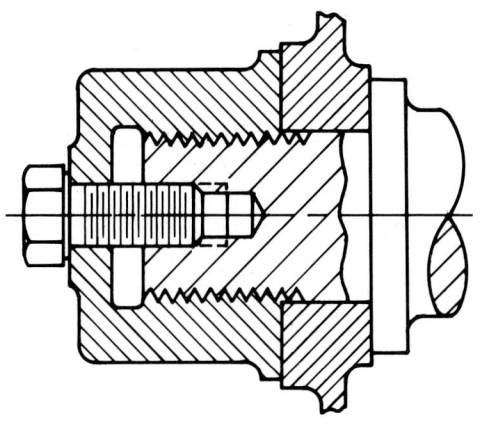

FIG. 9-4-2　Threaded assembly.

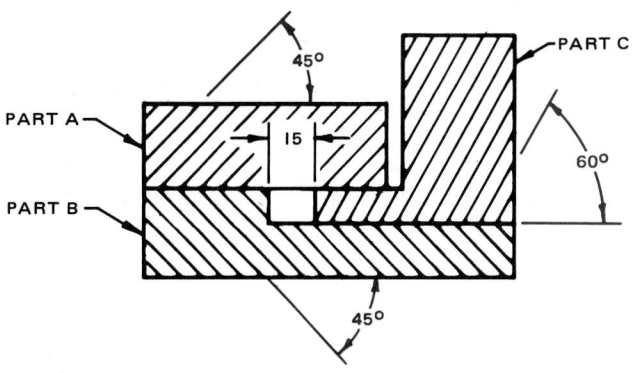

FIG. 9-5-1　Direction of section lining.

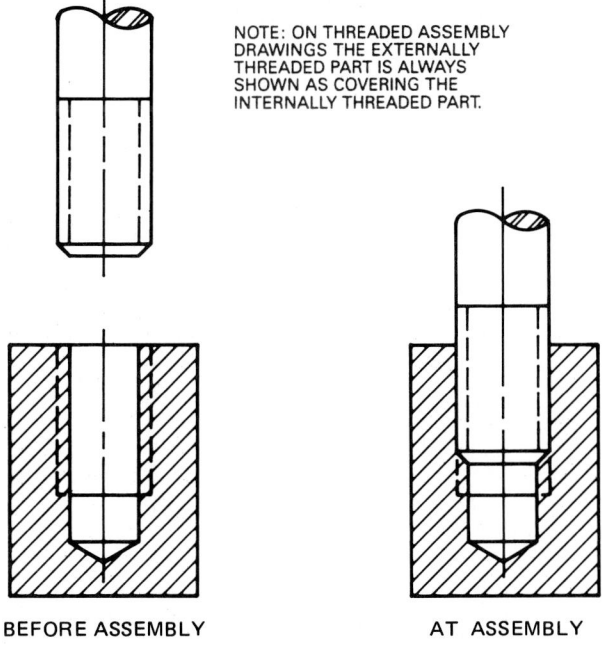

NOTE: ON THREADED ASSEMBLY
DRAWINGS THE EXTERNALLY
THREADED PART IS ALWAYS
SHOWN AS COVERING THE
INTERNALLY THREADED PART.

BEFORE ASSEMBLY　　　AT ASSEMBLY

FIG. 9-4-3　Drawing threads in assembly drawings.

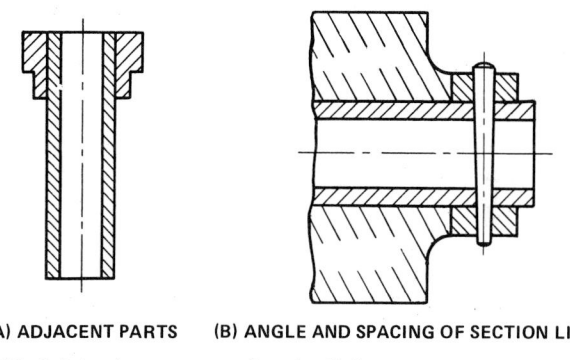

(A) ADJACENT PARTS　(B) ANGLE AND SPACING OF SECTION LINING

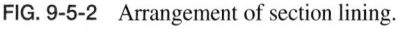

FIG. 9-5-2　Arrangement of section lining.

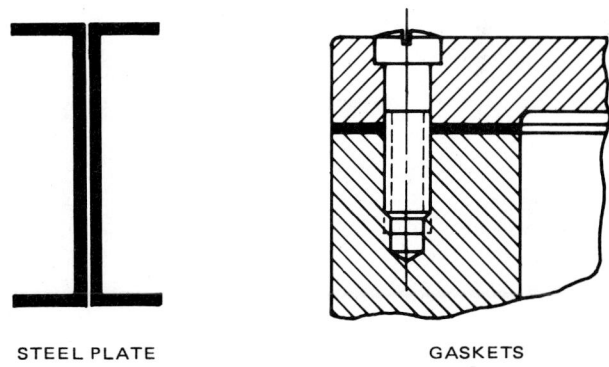

STEEL PLATE　　　　GASKETS

FIG. 9-5-3　Assembly of thin parts in section.

section lining is generally not recommended for drawings that will be microformed.

General-purpose section lining should be drawn at an angle of 45° with the main outlines of the view. On adjacent parts, the section lines should be drawn in the opposite direction, as shown in Figs. 9-5-1 and 9-5-2.

For additional adjacent parts, any suitable angle may be used to make each part stand out separately and clearly. Section lines should not be purposely drawn to meet at common boundaries.

When two or more thin adjacent parts are filled in, a space is left between them, as shown in Fig. 9-5-3.

Symbolic section lining is used on special-purpose assembly drawings, such as illustrations for parts catalogs, display assemblies, and promotional materials, when it is desirable to distinguish between different materials (Fig. 9-1-6).

All assemblies and subassemblies pertaining to one particular set of drawings should use the same symbolic conventions.

Shafts, Bolts, Pins, Keyseats, and Similar Solid Parts, in Section　Shafts, bolts, nuts, rods, rivets, keys, pins, and similar solid parts, the axes of which lie in the cutting plane, should not be sectioned except that a broken-out section of the shaft may be used to describe more clearly the key, keyseat, or pin (Fig. 9-5-4, pg. 234).

ASSIGNMENTS

See Assignments 8 through 10 for Unit 9-5 on pages 244–246.

233

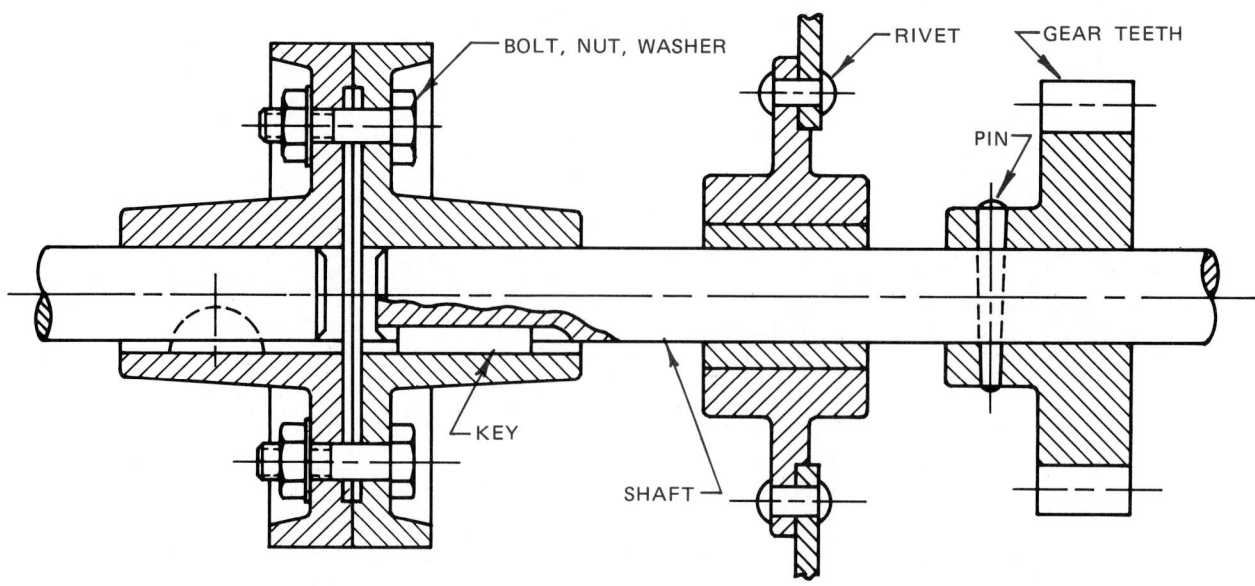

FIG. 9-5-4   Parts that are not section-lined even though the cutting plane passes through them.

9-6 OFFSET SECTIONS

In order to include features that are not in a straight line, the cutting plane may be offset or bent, so as to include several planes or curved surfaces (Figs. 9-6-1 and 9-6-2).

An *offset section* is similar to a full section in that the cutting-plane line extends through the object from one side to the other. The change in direction of the cutting-plane line is not shown in the sectional view.

ASSIGNMENT

See Assignment 11 for Unit 9-6 on pages 246–247.

9-7 RIBS, HOLES, AND LUGS IN SECTION

Ribs in Sections

A true-projection sectional view of a part, such as shown in Fig. 9-7-1, would be misleading when the cutting plane passes longitudinally through the center of the rib. To avoid this impression of solidity, a section not showing the ribs section-lined is preferred. When there is an odd number of ribs, such as those shown in Fig. 9-7-1B, the top rib is aligned with the bottom rib to show its true relationship with the hub and flange. If the rib is not aligned or revolved, it would appear distorted on the sectional view and would therefore be misleading.

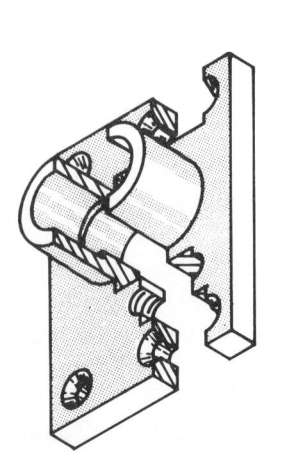
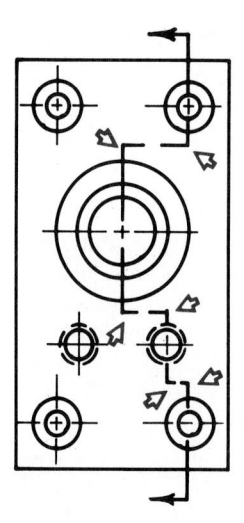
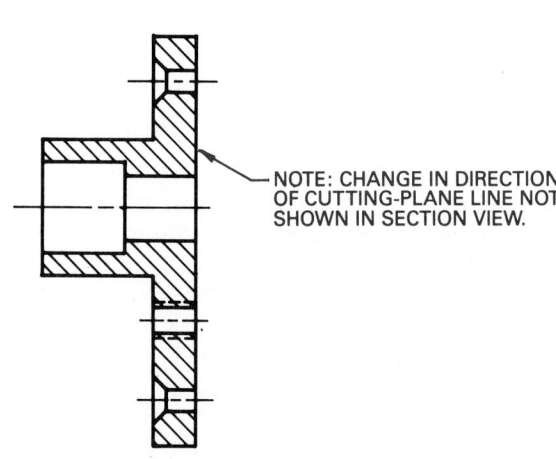

NOTE: CHANGE IN DIRECTION OF CUTTING-PLANE LINE NOT SHOWN IN SECTION VIEW.

FIG. 9-6-1   An offset section.

FIG. 9-6-2   Positioning offset
sections.

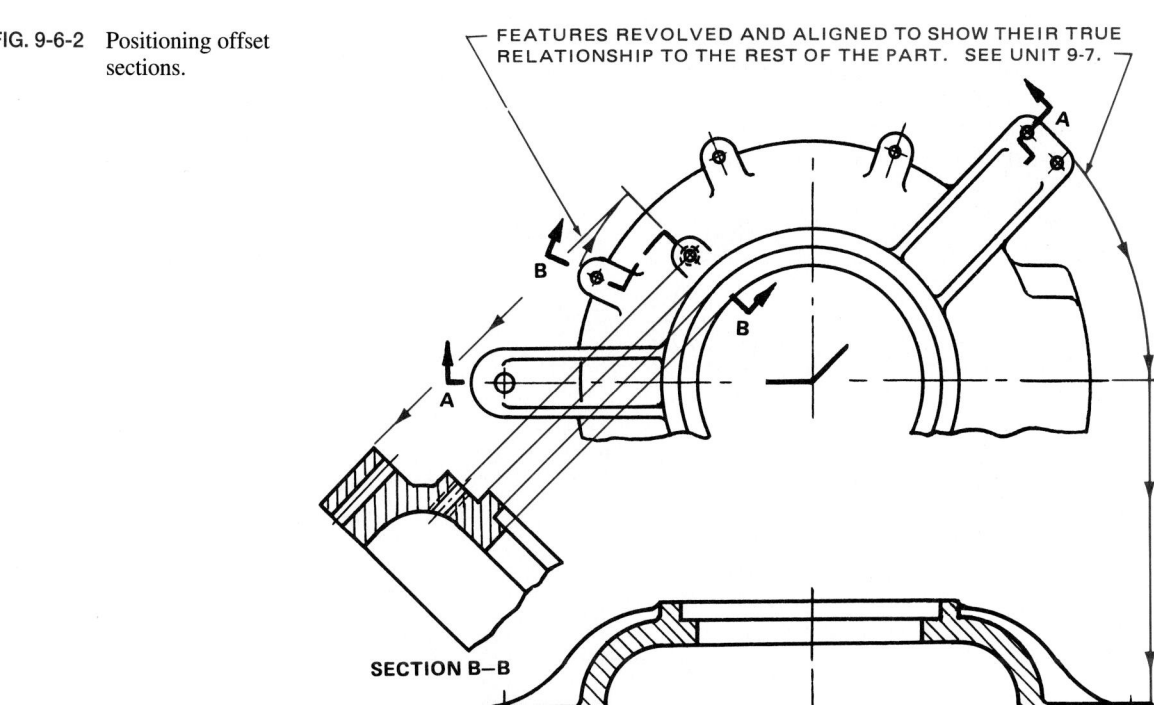

FEATURES REVOLVED AND ALIGNED TO SHOW THEIR TRUE
RELATIONSHIP TO THE REST OF THE PART.   SEE UNIT 9-7.

SECTION B–B

SECTION A–A

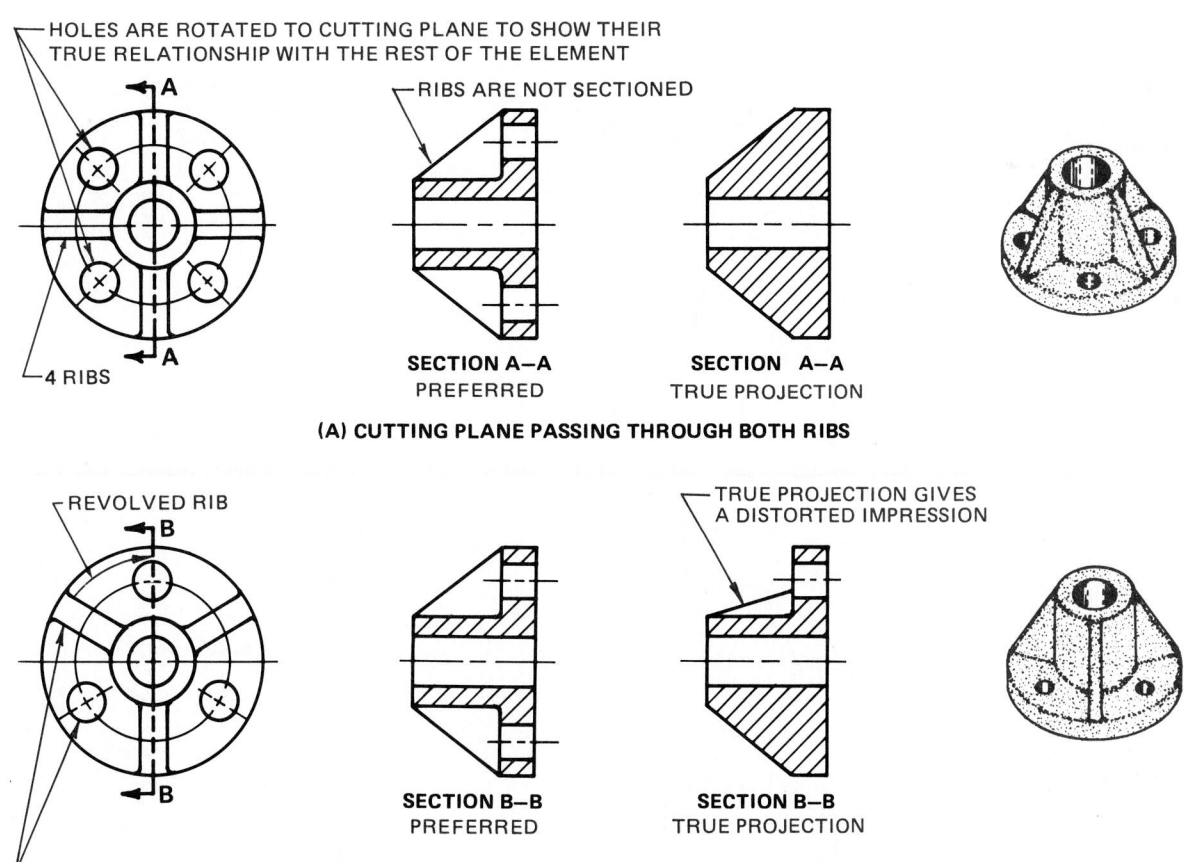

HOLES ARE ROTATED TO CUTTING PLANE TO SHOW THEIR
TRUE RELATIONSHIP WITH THE REST OF THE ELEMENT

RIBS ARE NOT SECTIONED

4 RIBS

SECTION A–A
PREFERRED

SECTION A–A
TRUE PROJECTION

(A) CUTTING PLANE PASSING THROUGH BOTH RIBS

REVOLVED RIB

TRUE PROJECTION GIVES
A DISTORTED IMPRESSION

HOLES AND RIB ARE ROTATED TO CUTTING PLANE

SECTION B–B
PREFERRED

SECTION B–B
TRUE PROJECTION

(B) CUTTING PLANE PASSING THROUGH ONE RIB AND ONE HOLE

FIG. 9-7-1   Preferred and true projection through ribs and holes.

At times it may be necessary to use an alternative method of identifying ribs in a sectional view. Figure 9-7-2 shows a base and a pulley in section. If rib A of the base was not sectioned as previously mentioned, it would appear exactly like rib B in the sectional view and would be misleading. Similarly, ribs C shown on the pulley may be overlooked. To clearly show the relationship of the ribs with the other solid features on the base and pulley, alternate section lining on the ribs is used. The line between the rib and solid portions is shown as a broken line.

FIG. 9-7-2   Alternate method of showing ribs in section.

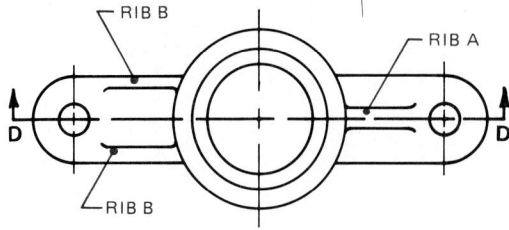
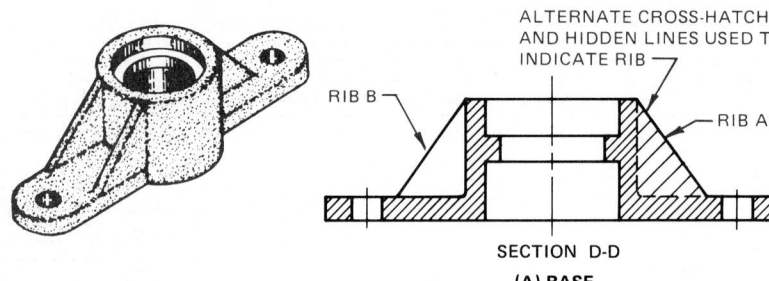
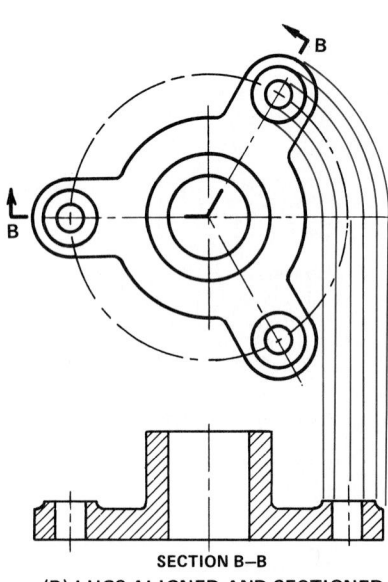

SECTION D-D

(A) BASE

(B) PULLEY

FIG. 9-7-3   Lugs in section.

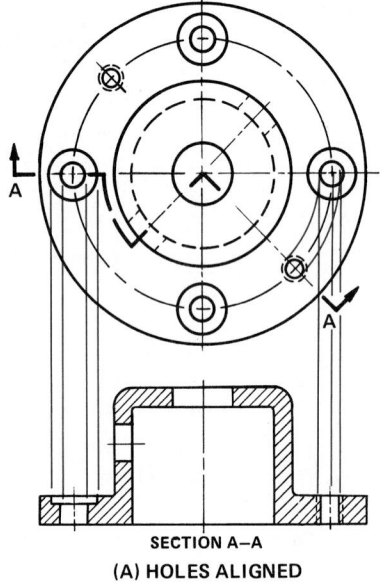

SECTION A–A

(A) HOLES ALIGNED

SECTION B–B

(B) LUGS ALIGNED AND SECTIONED

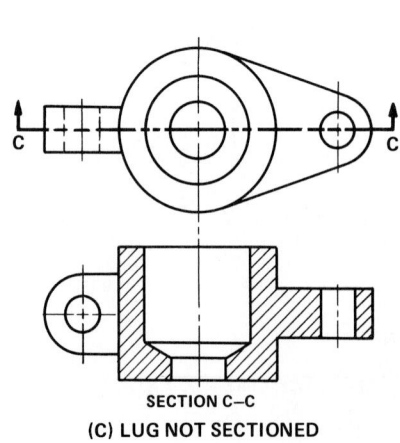

SECTION C–C

(C) LUG NOT SECTIONED

SECTION D–D

(D) LUGS ALIGNED AND SECTIONED

Holes in Sections

Holes, like ribs, are aligned as shown in Fig. 9-7-1 to show their true relationship to the rest of the part.

Lugs in Section

Lugs, like ribs and spokes, are also aligned to show their true relationship to the rest of the part, because true projection may be misleading. Figure 9-7-3 shows several examples of lugs in section. Note how the cutting-plane line is bent or offset so that the features may be clearly shown in the sectional view.

Some lugs are shown in section, and some are not. When the cutting plane passes through the lug crosswise, the lug is sectioned; otherwise, the lugs are treated in the same manner as ribs.

ASSIGNMENT

See Assignment 12 for Unit 9-7 on pages 247–248.

9-8 REVOLVED AND REMOVED SECTIONS

Revolved and removed sections are used to show the cross-sectional shape of ribs, spokes, or arms when the shape is not obvious in the regular views (Figs. 9-8-1 through 9-8-3, here and on pg. 238). Often end views are not needed when a revolved section is used. For a revolved section, draw a center line through the shape on the plane to be described, imagine the part to be rotated 90°, and superimposed on the view of the

shape that would be seen when rotated (Figs. 9-8-1 and 9-8-2). If the revolved section does not interfere with the view on which it is revolved, the view is not broken unless it would provide for clearer dimensioning. When the revolved section interferes or passes through lines on the view on which it is revolved, the general practice is to break the view (Fig. 9-8-2). Often the break is used to shorten the length of the object. In no circumstances should the lines on the view pass through the section. When superimposed on the view, the outline of the revolved section is a thin, continuous line.

The removed section differs in that the section, instead of being drawn right on the view, is removed to an open area on the drawing (Fig. 9-8-3). Frequently the removed section is drawn to an enlarged scale for clarification and easier dimensioning. Removed sections of symmetrical parts should be placed, whenever possible, on the extension of the center line (Fig. 9-8-3B).

On complicated drawings where the placement of the removed view may be some distance from the cutting plane, auxiliary information, such as the reference zone location (Fig. 9-8-4, pg. 238), may be helpful.

Placement of Sectional Views

Whenever practical, except for revolved sections, sectional views should be projected perpendicularly to the cutting plane and be placed in the normal position for third-angle projection (Fig. 9-8-5, pg. 238).

When the preferred placement is not practical, the sectional view may be removed to some other convenient position on the drawing, but it must be clearly identified, usually by two capital letters, and be labeled.

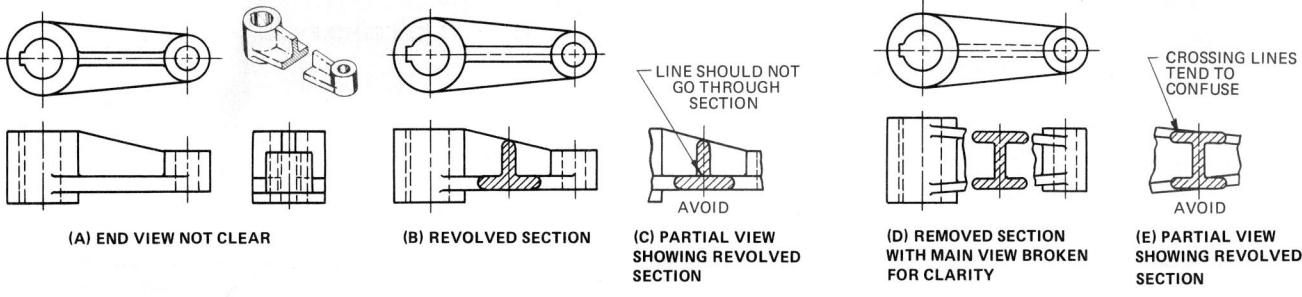

(A) END VIEW NOT CLEAR

(B) REVOLVED SECTION

(C) PARTIAL VIEW SHOWING REVOLVED SECTION

(D) REMOVED SECTION WITH MAIN VIEW BROKEN FOR CLARITY

(E) PARTIAL VIEW SHOWING REVOLVED SECTION

FIG. 9-8-1   Revolved sections.

FIG. 9-8-2   Revolved (superimposed) sections.

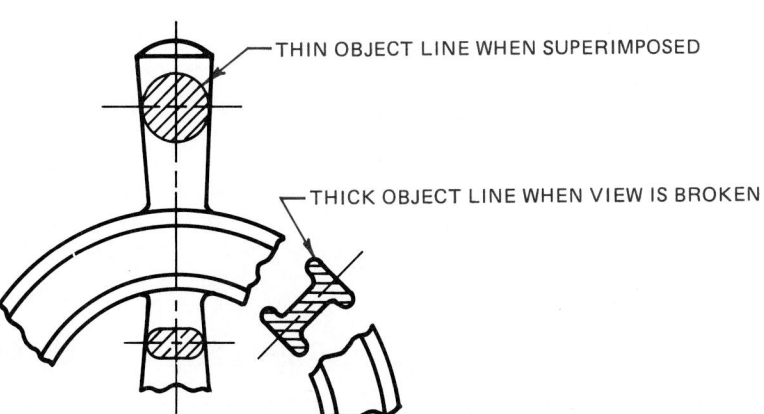

THIN OBJECT LINE WHEN SUPERIMPOSED

THICK OBJECT LINE WHEN VIEW IS BROKEN

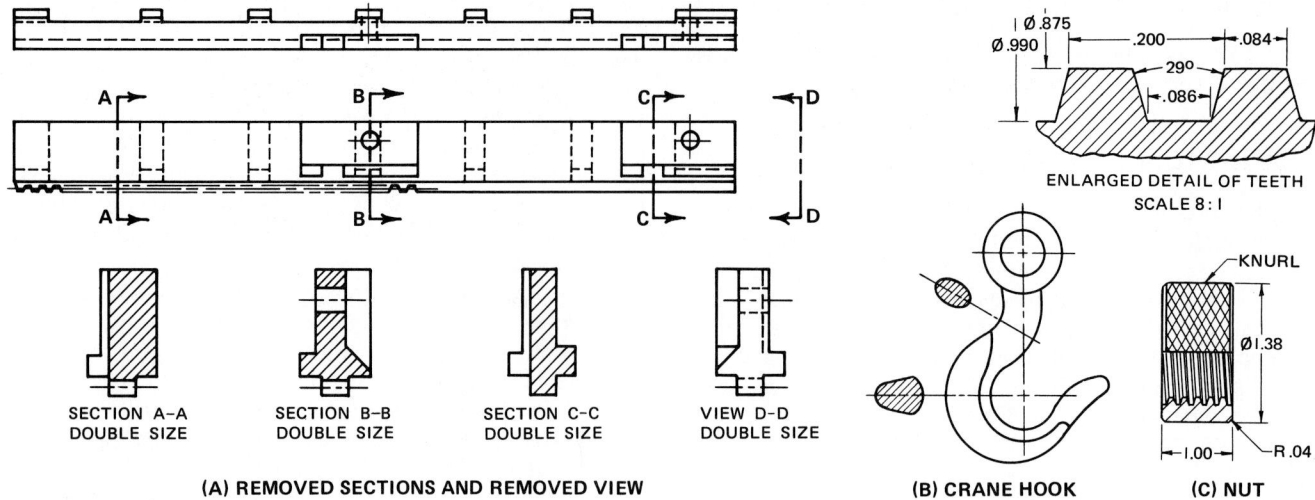

SECTION A-A
DOUBLE SIZE

SECTION B-B
DOUBLE SIZE

SECTION C-C
DOUBLE SIZE

VIEW D-D
DOUBLE SIZE

(A) REMOVED SECTIONS AND REMOVED VIEW

ENLARGED DETAIL OF TEETH
SCALE 8 : 1

(B) CRANE HOOK   **(C) NUT**

FIG. 9-8-3   Removed sections.

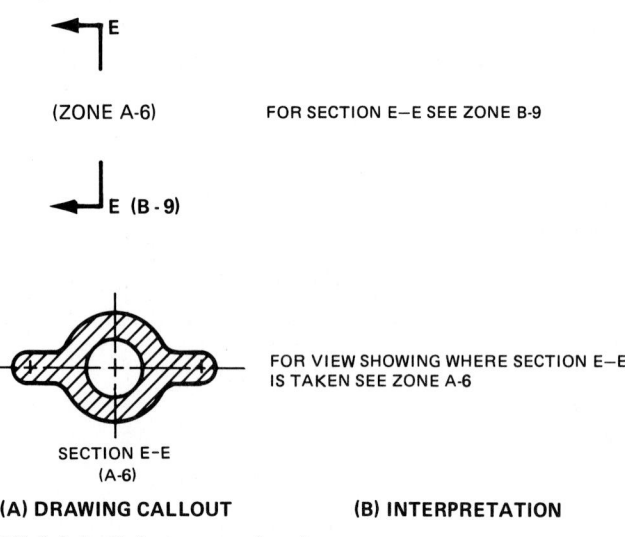

(ZONE A-6)   FOR SECTION E–E SEE ZONE B-9

E (B - 9)

FOR VIEW SHOWING WHERE SECTION E–E
IS TAKEN SEE ZONE A-6

SECTION E-E
(A-6)

(A) DRAWING CALLOUT   **(B) INTERPRETATION**

FIG. 9-8-4   Reference zone location.

INCORRECT

SECTION A-A
REMOVED

SECTION A-A
REMOVED AND
REVOLVED 60°
CLOCKWISE

ACCEPTABLE

CORRECT

FIG. 9-8-5   Placement of sectional views.

ASSIGNMENTS

See Assignments 13 through 15 for Unit 9-8 on pages 249 to 250.

9-9 SPOKES AND ARMS IN SECTION

A comparison of the true projection of a wheel with spokes and a wheel with a web is made in Figs. 9-9-1A and B. This comparison shows that a preferred section for the wheel and spoke is desirable so that it will not appear to be a wheel with a solid web. In preferred sectioning, any part that is not solid or continuous around the hub is drawn without the section lining, even though the cutting plane passes through it. When there is an odd number of spokes, as shown in Fig. 9-9-1C, the bottom spoke is aligned with the top spoke to show its true relationship to the wheel and to the hub. If the spoke was not revolved or aligned, it would appear distorted in the sectional view.

ASSIGNMENT

See Assignment 16 for Unit 9-9 on page 250.

9-10 PARTIAL OR BROKEN-OUT SECTIONS

Where a sectional view of only a portion of the object is needed, partial sections may be used (Fig. 9-10-1). An irregular break line is used to show the extent of the section. With this type of section, a cutting-plane line is not required.

ASSIGNMENT

See Assignment 17 for Unit 9-10 on pages 250–251.

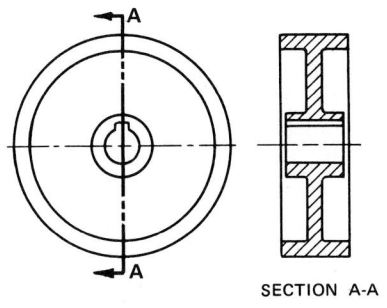

(A) FLAT PULLEY WITH WEB

SECTION A-A

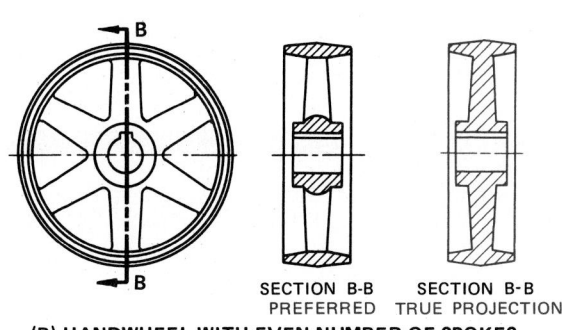

(B) HANDWHEEL WITH EVEN NUMBER OF SPOKES

SECTION B-B
PREFERRED

SECTION B-B
TRUE PROJECTION

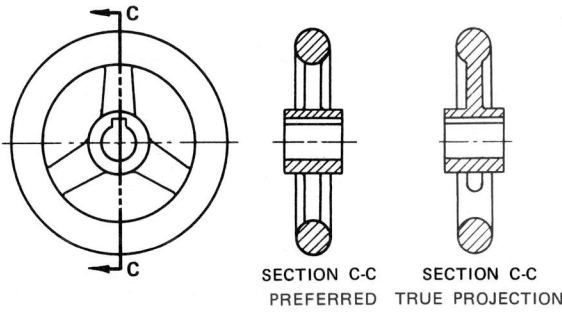

SECTION C-C
PREFERRED

SECTION C-C
TRUE PROJECTION

(C) HANDWHEEL WITH ODD NUMBER OF SPOKES

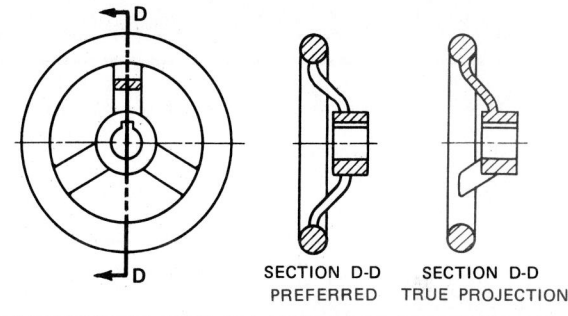

SECTION D-D
PREFERRED

SECTION D-D
TRUE PROJECTION

(D) HANDWHEEL WITH ODD NUMBER OF OFFSET SPOKES

FIG. 9-9-1   Preferred and true projection of spokes.

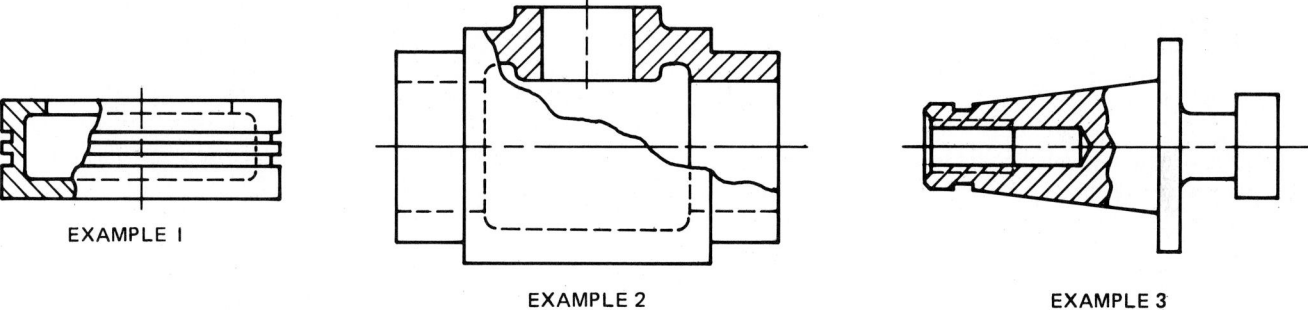

EXAMPLE I

EXAMPLE 2

EXAMPLE 3

FIG. 9-10-1   Broken-out or partial sections.

9-11 PHANTOM OR HIDDEN SECTIONS

A *phantom section* is used to show the typical interior shapes of an object in one view when the part is not truly symmetrical in shape, as well as to show mating parts in an assembly drawing (Fig. 9-11-1). It is a sectional view superimposed on the regular view without the removal of the front portion of the object. The section lining used for phantom sections consists of thin, evenly spaced, broken lines.

ASSIGNMENT

See Assignment 18 for Unit 9-11 on pages 251–252.

9-12 SECTIONAL DRAWING REVIEW

In Units 9-1 through 9-11 the different types of sectional views have been explained and drawing problems have been assigned with each type of section drawing.

In the drafting office it is the drafter who must decide which views are required to fully explain the part to be made. In addition, the drafter must select the proper scale(s) which will show the features clearly.

FIG. 9-11-1   Phantom or hidden sections.

This unit has been designed to review the sectional-view options open to the drafter.

ASSIGNMENT

See Assignment 19 for Unit 9-12 on pages 253–257.

ASSIGNMENTS FOR CHAPTER 9

ASSIGNMENTS FOR UNIT 9-1, SECTIONAL VIEWS

1. Select one of the problems shown in Fig. 9-1-A or 9-1-B and make a working drawing of the part. Surfaces shown with √ should have a surface texture rating of 125 μin. or 3.2 μm and a machining allowance of .06 in. or 1.5 mm. Use symbolic dimensioning wherever possible. For Fig. 9-1-A draw the top and a full-section front view. For Fig. 9-1-B draw the right side and a full-section front view.

2. Select one of the problems shown in Fig. 9-1-C or 9-1-D and make a three-view working drawing of the part. Surfaces shown with √ should have a surface texture rating of 63 μin. or 1.6 μm and a machining allowance of .06 in. or 2 mm. Use limit dimensions for the holes showing a fit. For Fig. 9-1-C draw the front view in full section. For Fig. 9-1-D draw the right side view in full section through the Ø16 hole.

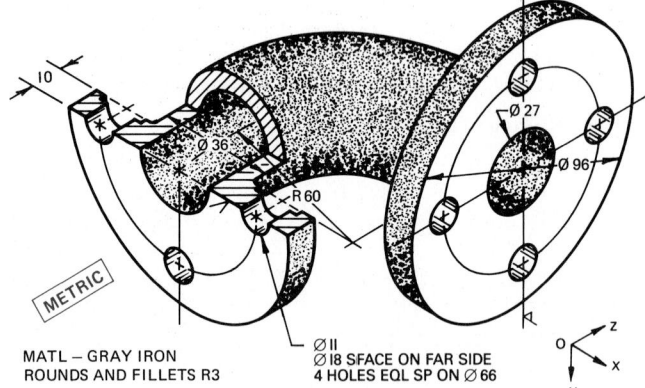

MATL – GRAY IRON
ROUNDS AND FILLETS R3

Ø 11
Ø 18 SFACE ON FAR SIDE
4 HOLES EQL SP ON Ø 66

FIG. 9-1-B   Flanged elbow.

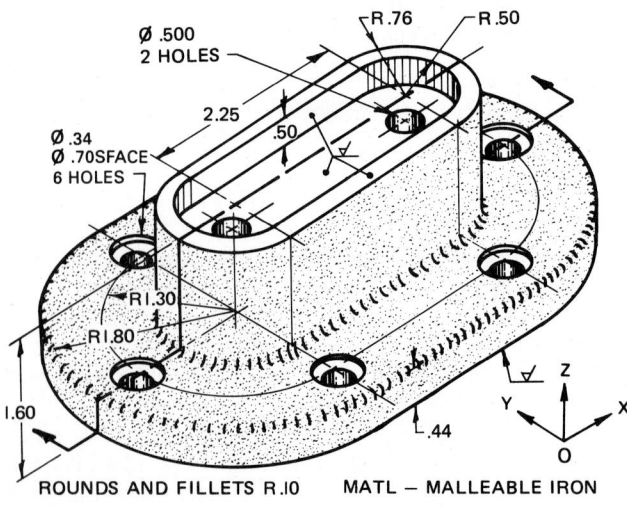

FIG. 9-1-A   Shaft base.

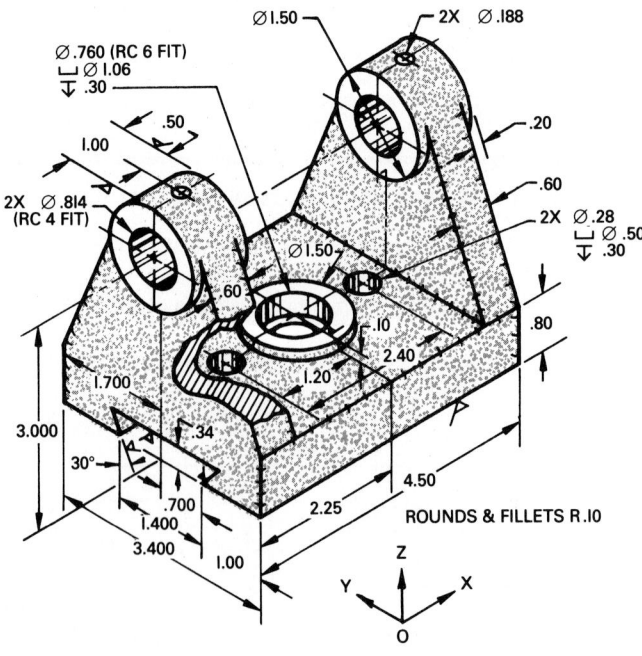

FIG. 9-1-C   Slide bracket.

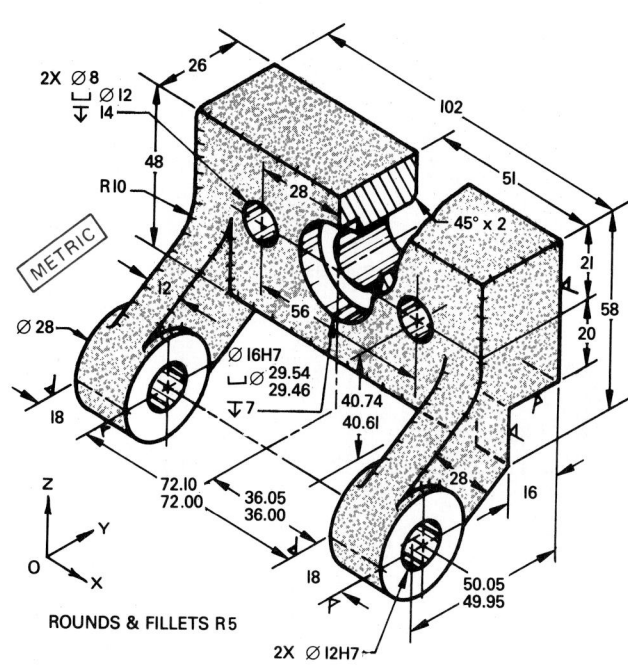

FIG. 9-1-D   Bracket.

ROUNDS & FILLETS R5

2X Ø8
Ø12
14
26
102
48
51
R10
METRIC
28
45° x 2
21
58
20
12
56
Ø 28
Ø 16H7
29.54
29.46
7
40.74
40.61
18
72.10
72.00
36.05
36.00
16
28
18
50.05
49.95
2X Ø12H7
Z
Y
O
X

ASSIGNMENTS FOR UNIT 9-2, TWO OR MORE SECTIONAL VIEWS ON ONE DRAWING

3. Select one of the problems shown in Fig. 9-2-A or 9-2-B and make a working drawing of the part showing the appropriate views in sections. Refer to the Appendix for taper sizes. Use symbolic dimensioning wherever possible.

FIG. 9-2-A   Casing.

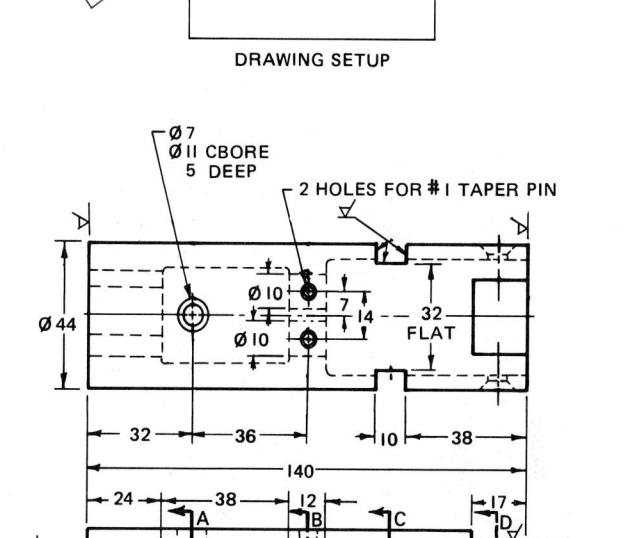

FILLETS R 3
MATL – GRAY IRON

METRIC

DRAWING SETUP

Ø 7
Ø 11 CBORE
5 DEEP
2 HOLES FOR #1 TAPER PIN
Ø 44
Ø 10
7
14
Ø 10
32
FLAT
32
36
10
38
140
24
38
12
17
HEX 22
ACRFLT
Ø 28
Ø 34
10
38
FLAT
Ø 7
A
B
C
D
A
B
C
D
Ø 7
Ø 11 X 82° CSK
2 HOLES

FIG. 9-2-B   Housing.

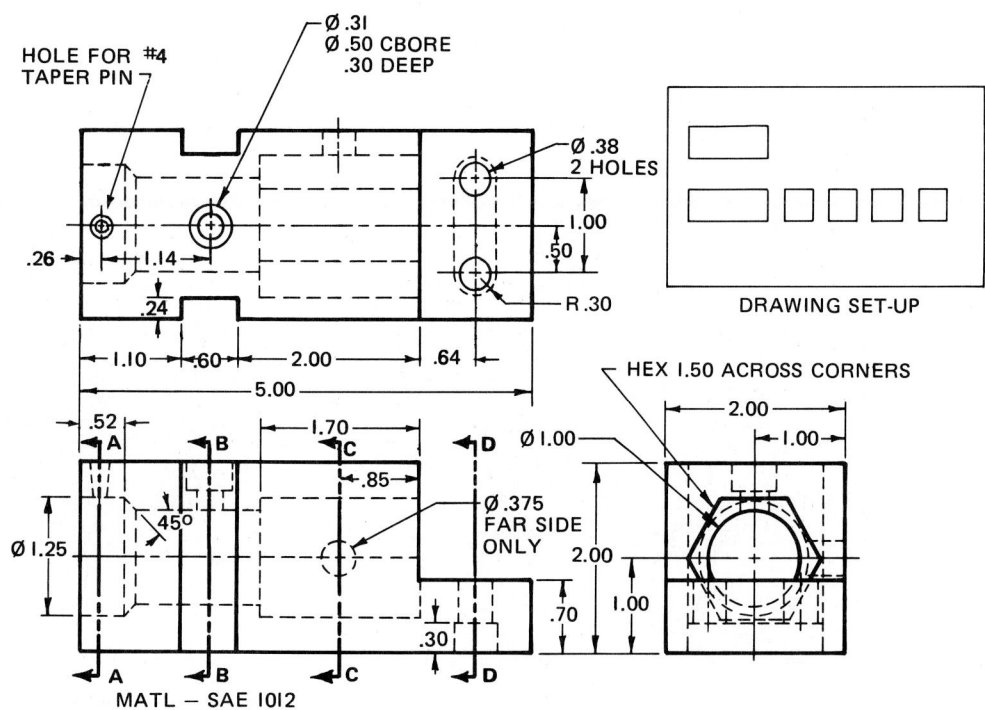

HOLE FOR #4
TAPER PIN
Ø .31
Ø .50 CBORE
.30 DEEP
Ø .38
2 HOLES
1.00
.26
1.14
.50
.24
R .30
1.10
.60
2.00
.64
5.00
.52
A
B
1.70
C
D
Ø 1.00
45°
Ø 1.25
.85
Ø .375
FAR SIDE
ONLY
.30
.70
A
B
C
D
MATL – SAE 1012

DRAWING SET-UP

HEX 1.50 ACROSS CORNERS
2.00
1.00
2.00
1.00

241

FIG. 9-2-C Guide block.

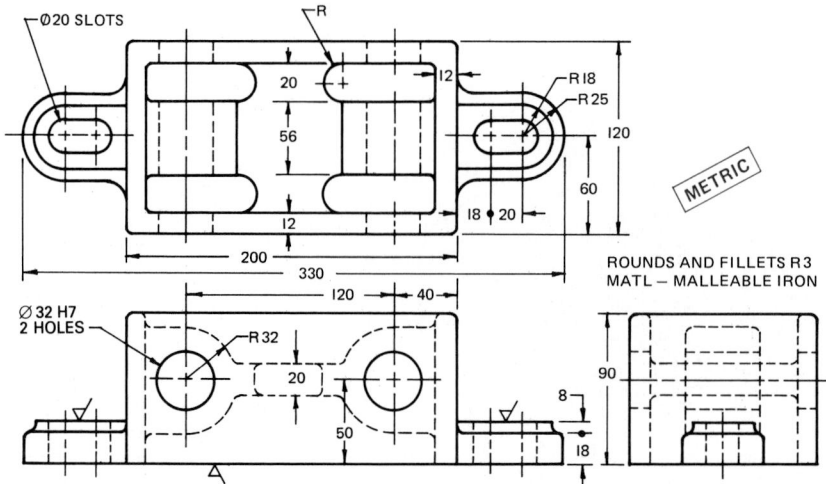

ROUNDS AND FILLETS R3
MATL — MALLEABLE IRON

4. Make a three-view working drawing of the guide block shown in Fig. 9-2-C showing the front and side views in full section. The cutting plane for the side view should be taken through the right Ø32 hole. The surface finish for the bottom should have a surface texture rating of 1.6 and a machining allowance of 2 mm. The surface finish for the two bosses should have a surface texture rating of 0.8 and a machining allowance of 1 mm. Show the limits for the Ø32 holes.

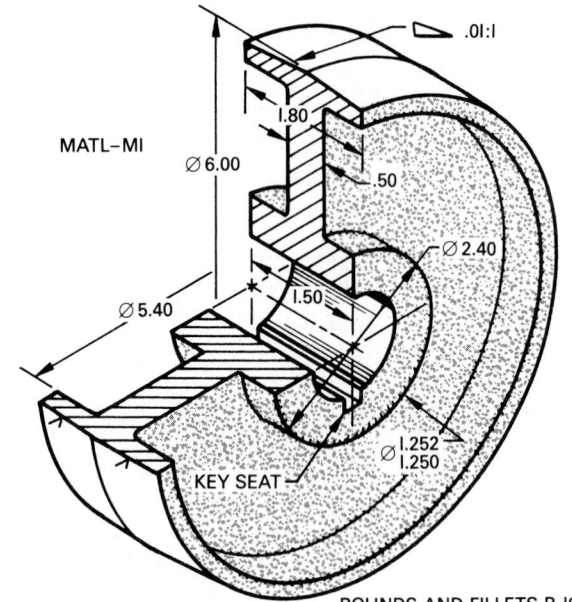

FIG. 9-3-B Flat belt pulley.

ASSIGNMENT FOR UNIT 9-3, HALF-SECTIONS

5. Select one of the parts shown in Figs. 9-3-A through 9-3-D and make a two-view working drawing of the part showing the side view in half-section. Dimension the keyseat as per Chap. 11.

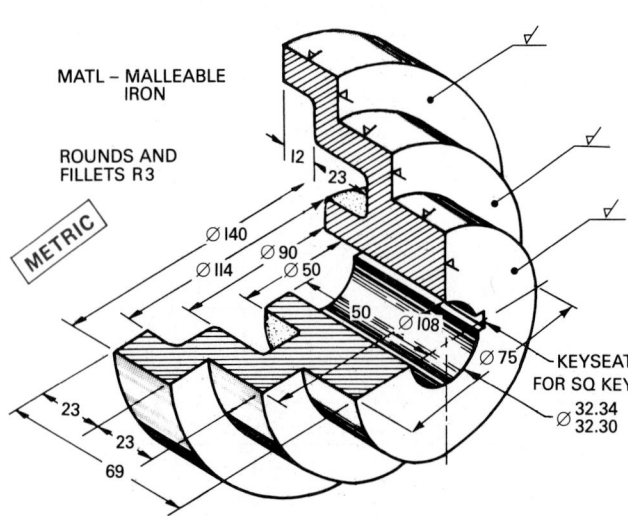

FIG. 9-3-A Step pulley.

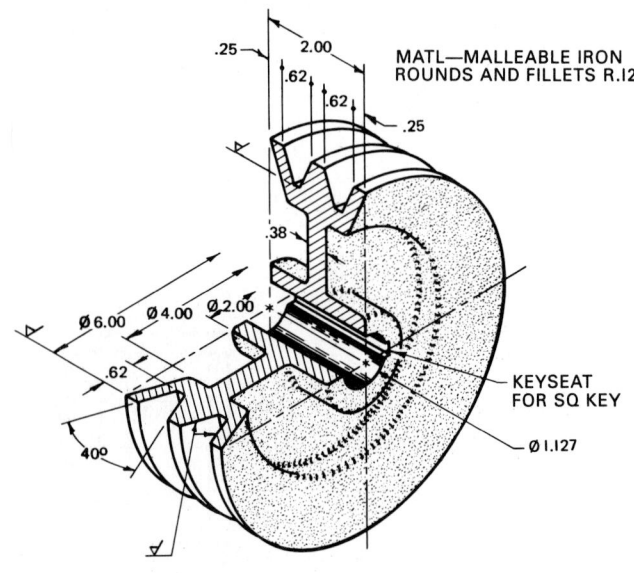

FIG. 9-3-C Double -V pulley.

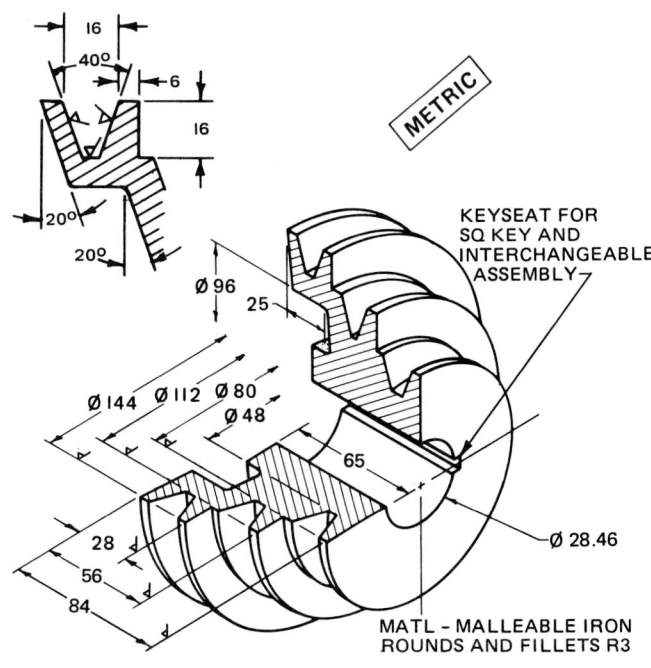

FIG. 9-3-D   Step-V pulley.

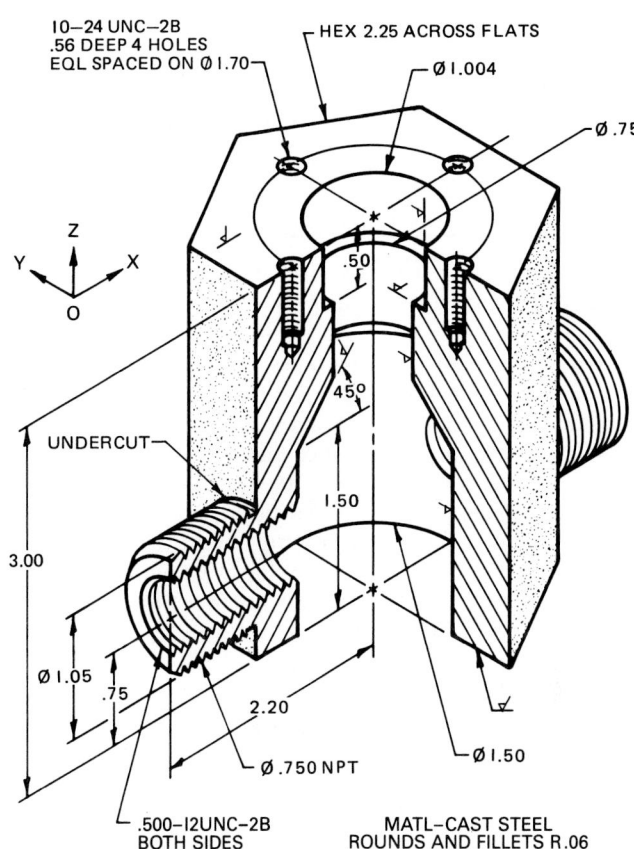

FIG. 9-4-B   Valve body.

ASSIGNMENTS FOR UNIT 9-4, THREADS IN SECTION

6. Select one of the problems shown in Figs. 9-4-A through 9-4-C and make a working drawing of the part. Determine the number of views and the best type of section that will clearly describe the part. Use symbolic dimensioning wherever possible and add undercut sizes.

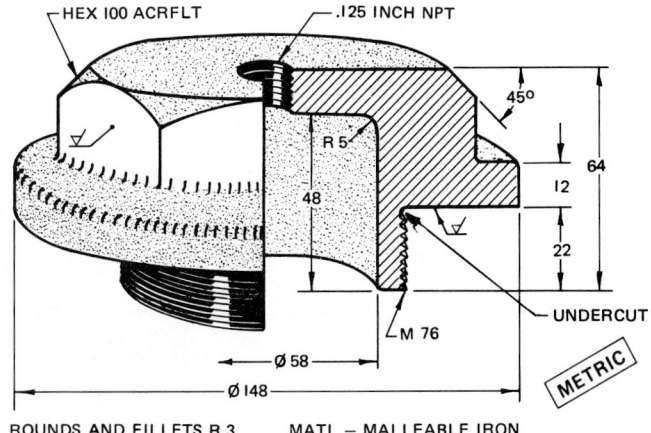

FIG. 9-4-A   Pipe plug.

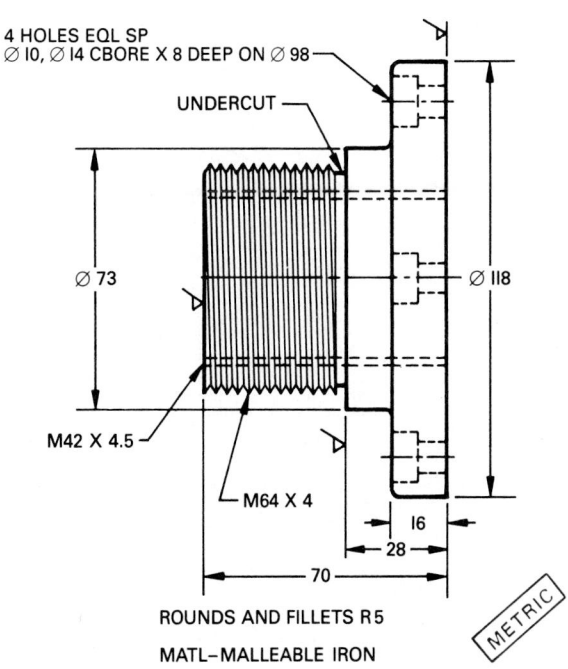

FIG. 9-4-C   End plate.

7. Make a three-view working drawing of the control arm shown in Fig. 9-4-D. Draw the front view in full section. Tabular dimensioning is to be used to locate the holes. The origin is located at the bottom left side of the part. Locate the two centers for each of the slotted holes. The two Ø6.5-6.6 holes will be drilled as one operation.

ASSIGNMENTS FOR UNIT 9-5, ASSEMBLIES IN SECTION

8. Make a one-view section assembly drawing of one of the problems shown in Figs. 9-5-A through 9-5-C. Include on your drawing an item list and identify the parts on the assembly. Assuming that this drawing will be used in a catalog, place on the drawing the dimensions and information required by the potential buyer. Scale 1:1.
9. Make a two-view assembly drawing of the cam slide shown in Fig. 9-5-D. Show the top view with the cover plate removed and the front view in full section. Use your judgment for dimensions not shown.

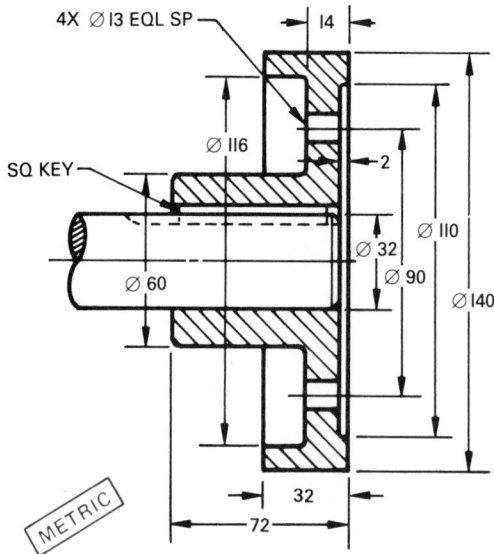

FLANGES HELD TOGETHER BY MI2 X 1.75 X 50 LG HEX HD BOLTS WITH LOCKWASHERS

2mm NEOPRENE GASKET BETWEEN FLANGES

FIG. 9-5-A   Flanged connection.

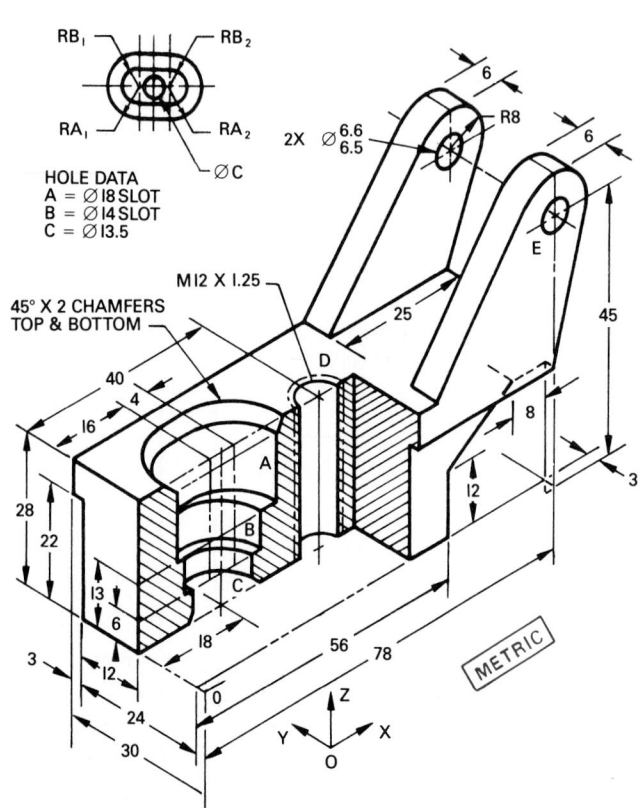

FIG. 9-4-D   Control arm.

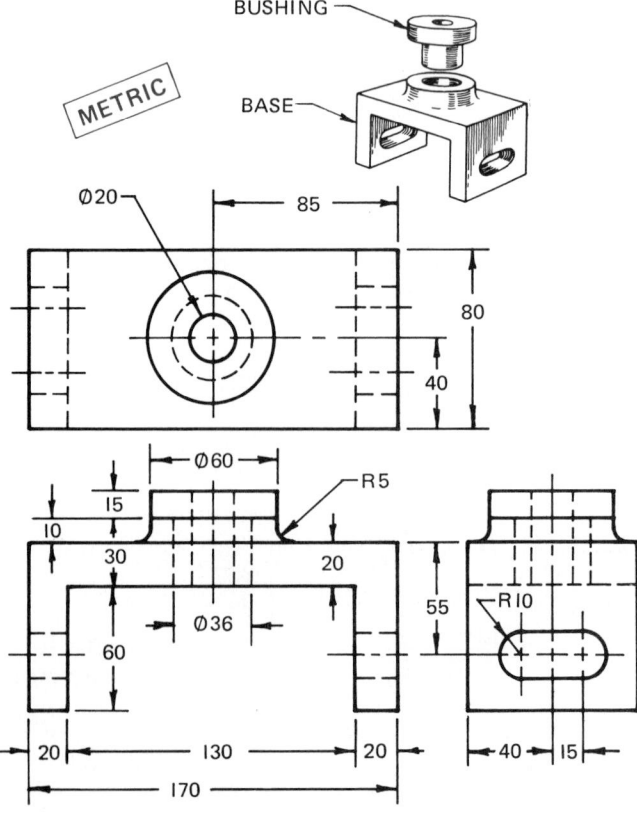

FIG. 9-5-B   Bushing holder.

244

FIG. 9-5-C   Caster.

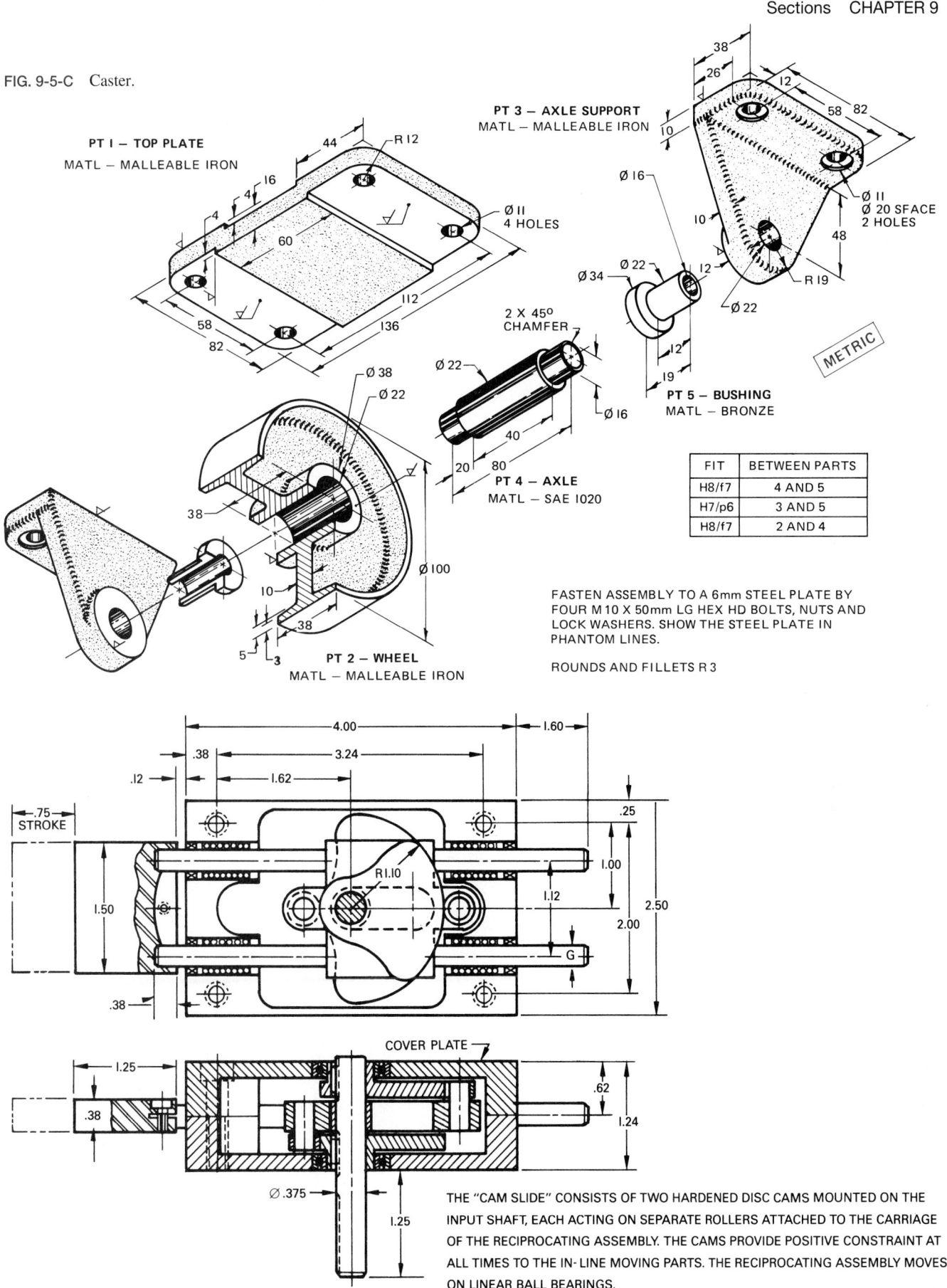

PT I — TOP PLATE
MATL — MALLEABLE IRON

R 12

44

16

4

4

60

112

58

136

82

Ø II
4 HOLES

PT 3 — AXLE SUPPORT
MATL — MALLEABLE IRON

38

26

12

58   82

10

Ø 16

10

12

Ø 22

Ø 34

Ø II
Ø 20 SFACE
2 HOLES

48

R 19

Ø 22

12

19

METRIC

PT 5 — BUSHING
MATL — BRONZE

2 X 45°
CHAMFER

Ø 22

Ø 16

20   80

40

PT 4 — AXLE
MATL — SAE 1020

Ø 38

Ø 22

38

Ø 100

10

38

5   3

PT 2 — WHEEL
MATL — MALLEABLE IRON

FIT	BETWEEN PARTS
H8/f7	4 AND 5
H7/p6	3 AND 5
H8/f7	2 AND 4

FASTEN ASSEMBLY TO A 6mm STEEL PLATE BY
FOUR M 10 X 50mm LG HEX HD BOLTS, NUTS AND
LOCK WASHERS. SHOW THE STEEL PLATE IN
PHANTOM LINES.

ROUNDS AND FILLETS R 3

.75
STROKE

4.00

1.60

.38

3.24

.12

1.62

.25

1.50

R1.10

1.00

1.12

2.50

2.00

.38

G

1.25

COVER PLATE

.62

.38

1.24

Ø .375

1.25

THE "CAM SLIDE" CONSISTS OF TWO HARDENED DISC CAMS MOUNTED ON THE
INPUT SHAFT, EACH ACTING ON SEPARATE ROLLERS ATTACHED TO THE CARRIAGE
OF THE RECIPROCATING ASSEMBLY. THE CAMS PROVIDE POSITIVE CONSTRAINT AT
ALL TIMES TO THE IN- LINE MOVING PARTS. THE RECIPROCATING ASSEMBLY MOVES
ON LINEAR BALL BEARINGS.

FIG. 9-5-D   Cam slide (protected by patent) courtesy Stelron Cam Co.

10. Make a two-view assembly drawing of the connecting link shown in Fig. 9-5-E. Show the top and front views with the front view in full section.

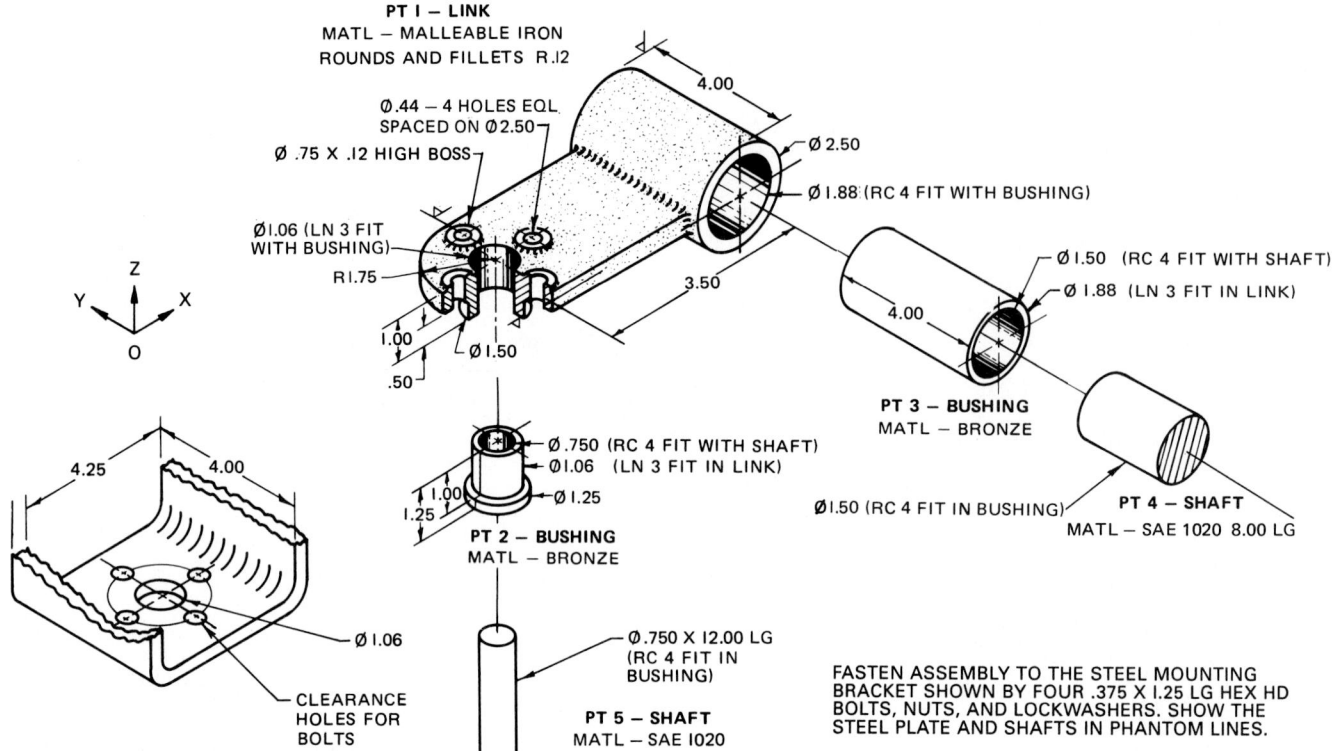

PT I – LINK
MATL – MALLEABLE IRON
ROUNDS AND FILLETS R.12

Ø.44 – 4 HOLES EQL SPACED ON Ø2.50
Ø.75 X .12 HIGH BOSS
Ø1.06 (LN 3 FIT WITH BUSHING)
R1.75
1.00
.50
Ø1.50
4.00
Ø2.50
Ø1.88 (RC 4 FIT WITH BUSHING)
3.50

Z
Y   X
O

Ø.750 (RC 4 FIT WITH SHAFT)
Ø1.06 (LN 3 FIT IN LINK)
1.00
1.25
Ø1.25
PT 2 – BUSHING
MATL – BRONZE

Ø1.50 (RC 4 FIT WITH SHAFT)
Ø1.88 (LN 3 FIT IN LINK)
4.00
PT 3 – BUSHING
MATL – BRONZE

Ø1.50 (RC 4 FIT IN BUSHING)
PT 4 – SHAFT
MATL – SAE 1020 8.00 LG

Ø.750 X 12.00 LG (RC 4 FIT IN BUSHING)
PT 5 – SHAFT
MATL – SAE 1020

4.25   4.00
Ø1.06
CLEARANCE HOLES FOR BOLTS
PARTIAL DETAIL OF MOUNTING BRACKET

FASTEN ASSEMBLY TO THE STEEL MOUNTING BRACKET SHOWN BY FOUR .375 X 1.25 LG HEX HD BOLTS, NUTS, AND LOCKWASHERS. SHOW THE STEEL PLATE AND SHAFTS IN PHANTOM LINES.

FIG. 9-5-E   Connecting link.

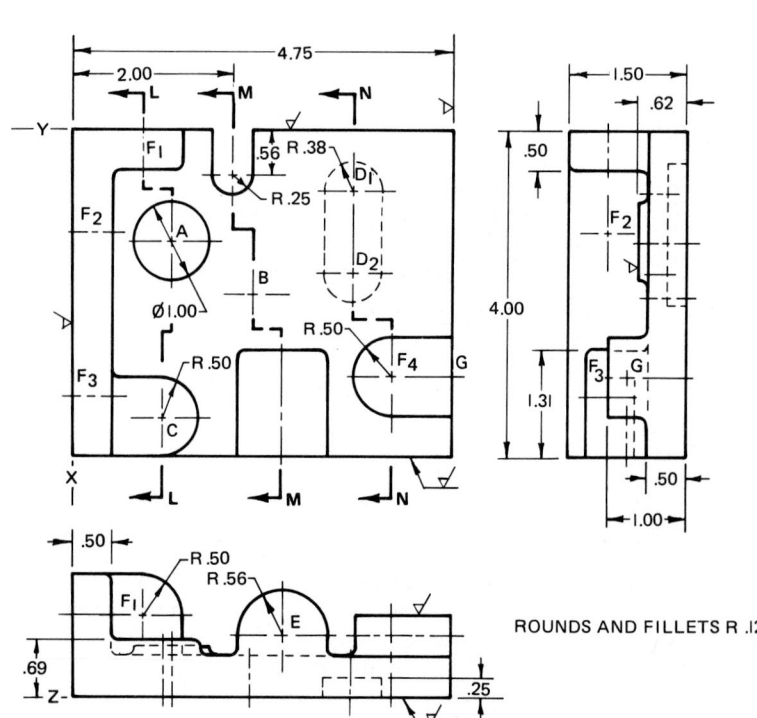

4.75
2.00
L   M   N
Y
F1
R.38
.56
R.25
D1
F2
A
B
D2
Ø1.00
R.50
F3
R.50
F4   G
C
X
L   M   N

1.50
.62
.50
F2
4.00
F3   G
1.31
.50
1.00

ROUNDS AND FILLETS R.12

.50
R.50
R.56
F1
E
.69
.25
Z

FIG. 9-6-A   Base plate.

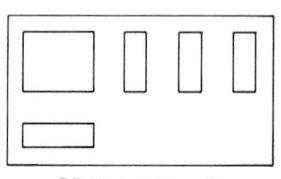

DRAWING SET-UP

REPLACE RIGHT SIDE VIEW WITH SECTIONS G-G, H-H, AND J-J.

HOLE	HOLE SIZE	LOCATION		
		X	Y	Z
A	.500–13UNC–2B	1.25	1.38	
B	Ø.281 CSK Ø.50 X 82°	2.25	1.94	
C	Ø.281 CBORE Ø.50 X .25 DEEP	1.12	3.50	
D1	Ø.31	3.50	.75	
D2	Ø.31	3.50	1.75	
E	.500–13UNC–2B X .75 DEEP	2.62		.75
F1	Ø.50	.88		1.00
F2	Ø.50		1.25	1.00
F3	Ø.50		3.25	1.00
F4	Ø.50	4.00	3.00	
G	Ø.12 THROUGH		3.00	.75

DRAW TOP, FRONT AND 3 SECTION VIEWS
MATL – MALLEABLE IRON

ASSIGNMENT FOR UNIT 9-6, OFFSET SECTIONS

11. Select one of the problems shown in Fig. 9-6-A or 9-6-B and make a working drawing of the part. Scale 1:1.

ASSIGNMENT FOR UNIT 9-7, RIBS, HOLES, AND LUGS IN SECTION

12. Select one of the problems shown in Figs. 9-7-A through 9-7-E (here and on pg. 248) and make a three-view working drawing of the part showing the front and side views in section. For Fig. 9-7-C draw only the front and side views.

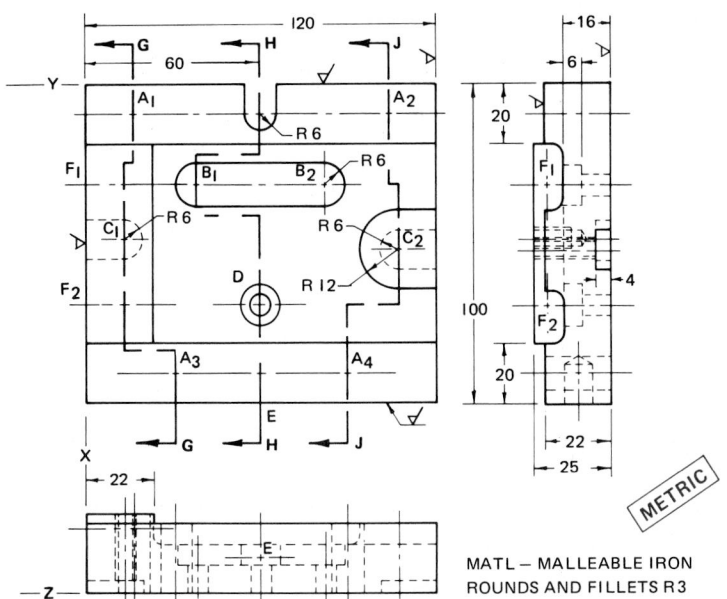

FIG. 9-6-B   Mounting plate.

HOLE	HOLE SIZE	LOCATION		
		X	Y	Z
A1	Ø 12	16	9	
A2	Ø 12	100	9	
A3	Ø 12	30	92	
A4	Ø 12	87	92	
B1	Ø 8	38	32	
B2	Ø 8	80	32	
C1	M6 X 12 DEEP	12	50	
C2	M6	104	52	
D	Ø 6 CBORE Ø 12 X 6 DEEP	58	70	
E	Ø 10 X 12 DEEP	58		11
F1	Ø 6		32	20
F2	Ø 6		70	20

REPLACES RIGHT SIDE VIEW WITH
SECTIONS L-L, M-M, AND N-N

MATL – MALLEABLE IRON
ROUNDS AND FILLETS R3

METRIC

FIG. 9-7-A   Shaft support.

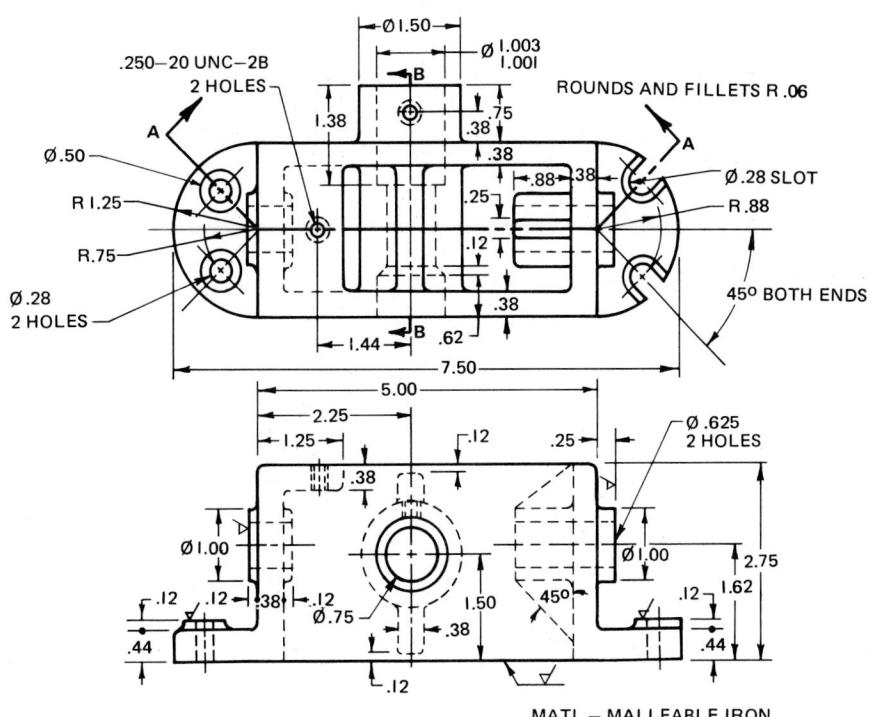

MATL – MALLEABLE IRON

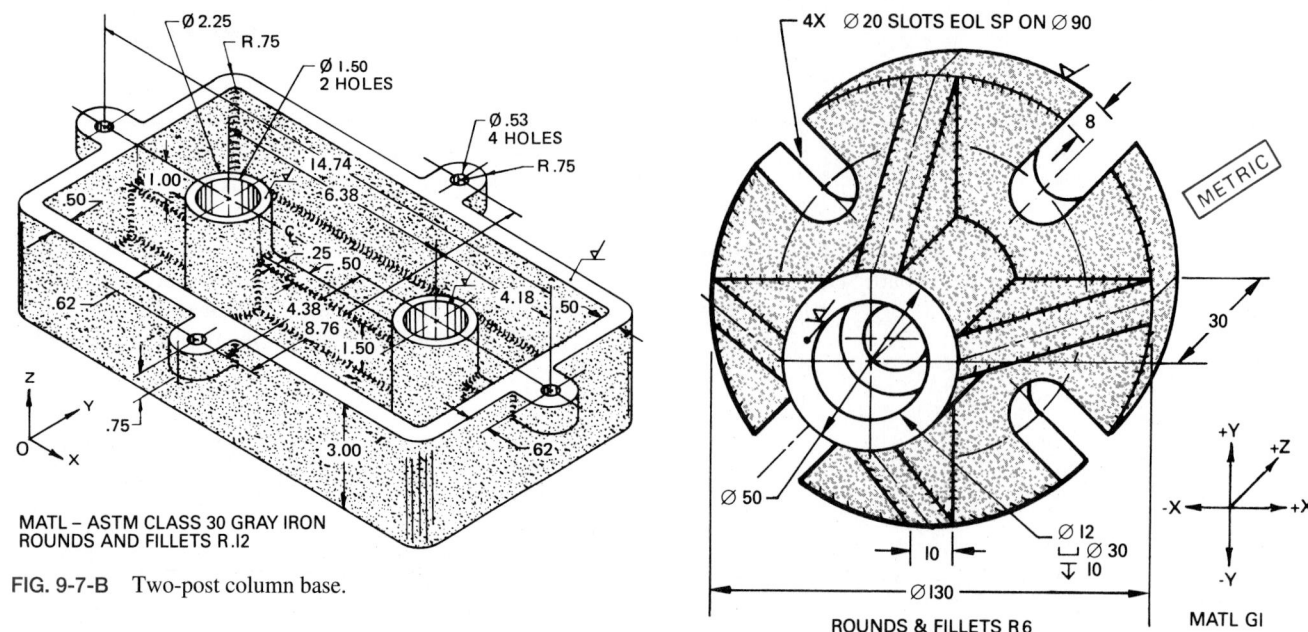

MATL – ASTM CLASS 30 GRAY IRON
ROUNDS AND FILLETS R.I2

FIG. 9-7-B   Two-post column base.

4X   ∅ 20 SLOTS EOL SP ON ∅ 90

METRIC

∅ 50

∅ I2
⌴ ∅ 30
↧ I0

∅ I30

ROUNDS & FILLETS R 6

+Y
+Z
-X
+X
-Y

MATL GI

FIG. 9-7-C   Flanged support.

FIG. 9-7-D   Bracket bearing.

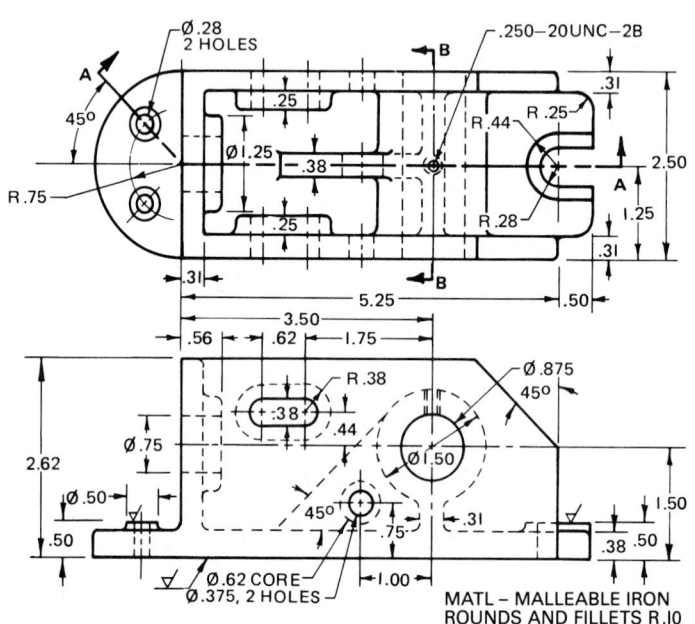

∅.28
2 HOLES

.250–20UNC–2B

R.75

∅1.25

2.50

1.25

R.44   R.25

R.28

5.25   .50

3.50

.56   .62   1.75

R.38

∅.875   45°

2.62

∅.75   .44

∅.50   ∅1.50

45°   .31

∅.62 CORE   .75   1.00   .38   .50
∅.375, 2 HOLES   1.50

MATL – MALLEABLE IRON
ROUNDS AND FILLETS R.I0

FIG. 9-7-E   Shaft support base.

ROUNDS AND FILLETS R3
MATL – MALLEABLE IRON

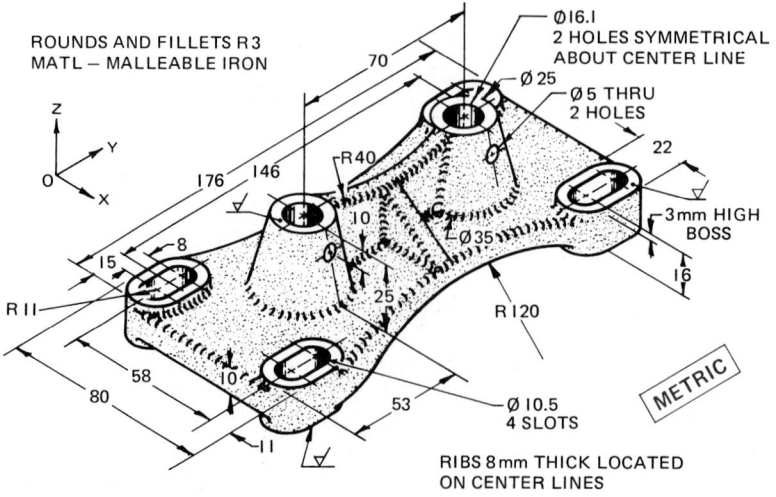

∅16.1
2 HOLES SYMMETRICAL
ABOUT CENTER LINE

∅ 25
∅ 5 THRU
2 HOLES

70

22

R40

146

176

⌴ ∅ 35

I0

3mm HIGH
BOSS

I5   8

25   I6

R II

R I20

80   58

I0

53

∅ 10.5
4 SLOTS

I1

METRIC

RIBS 8mm THICK LOCATED
ON CENTER LINES

248

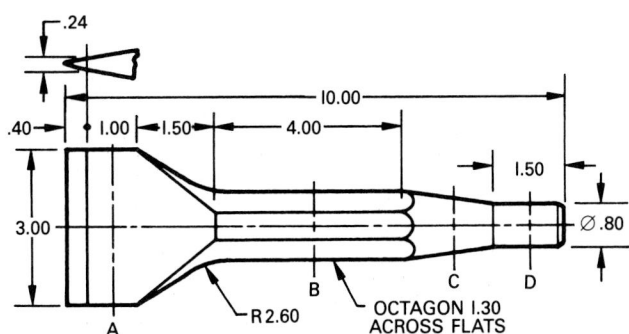

FIG. 9-8-A   Connector.

ASSIGNMENTS FOR UNIT 9-8,
REVOLVED AND REMOVED SECTIONS

13. Make a two-view working drawing of the connector shown in Fig. 9-8-A. Show a revolved section of the arm on the top view. The machined surfaces are to have a surface texture rating of 1.6 and a machining allowance of 2 mm.
14. Make a working drawing of the chisel shown in Fig. 9-8-B. Show either revolved or removed sections taken at planes A through D.
15. Select one of the problems shown in Fig. 9-8-C or 9-8-D (pg. 250) and make a working drawing of the part. For clarity it is recommended that an enlarged removed view be used to show the detail of the inclined hole. Use symbolic dimensioning wherever possible. Scale 1:1.

FIG. 9-8-B   Chisel.

FIG. 9-8-C   Shaft support.

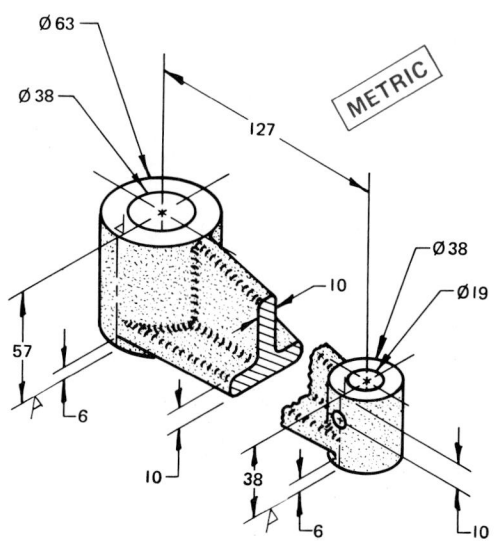

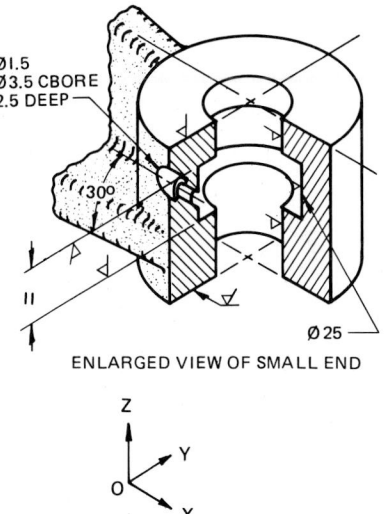

ENLARGED VIEW OF SMALL END

MATL — MALLEABLE IRON
ROUNDS AND FILLETS R3

249

FIG. 9-8-D   Idler support.

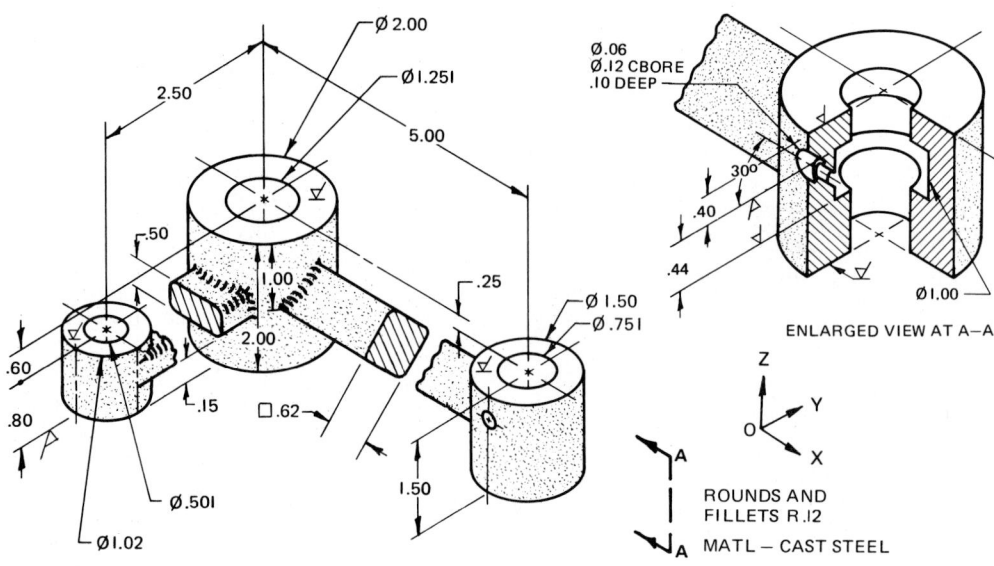

ASSIGNMENT FOR UNIT 9-9, SPOKES AND ARMS IN SECTION

16. Select one of the problems shown in Fig. 9-9-A or 9-9-B and make a two-view working drawing of the part. Draw the side view in full section, and show a revolved section of the spoke in the front view. Scale 1:1.

ASSIGNMENT FOR UNIT 9-10, PARTIAL OR BROKEN-OUT SECTIONS

17. Select one of the problems shown in Fig. 9-10-A or 9-10-B and make a two-view working drawing of the part. Use partial sections where clarity of drawing can be achieved. Scale 1:1.

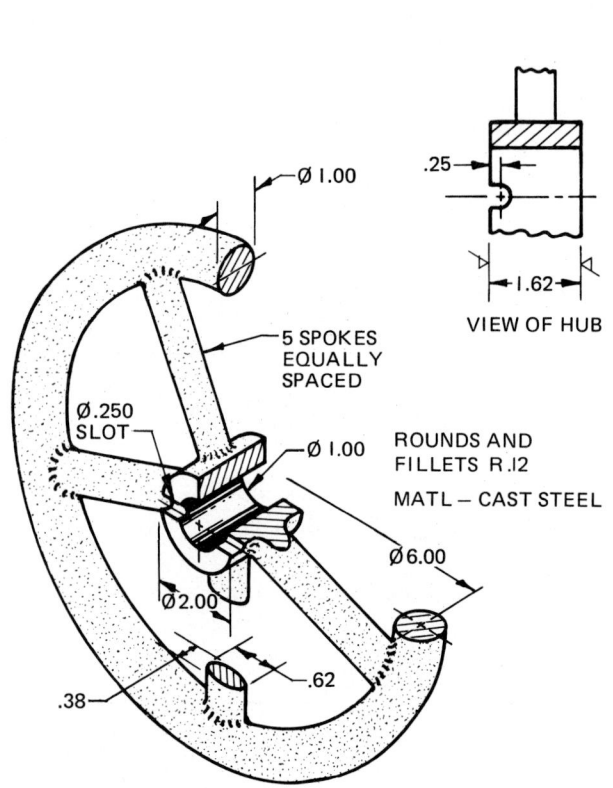

FIG. 9-9-A   Handwheel.

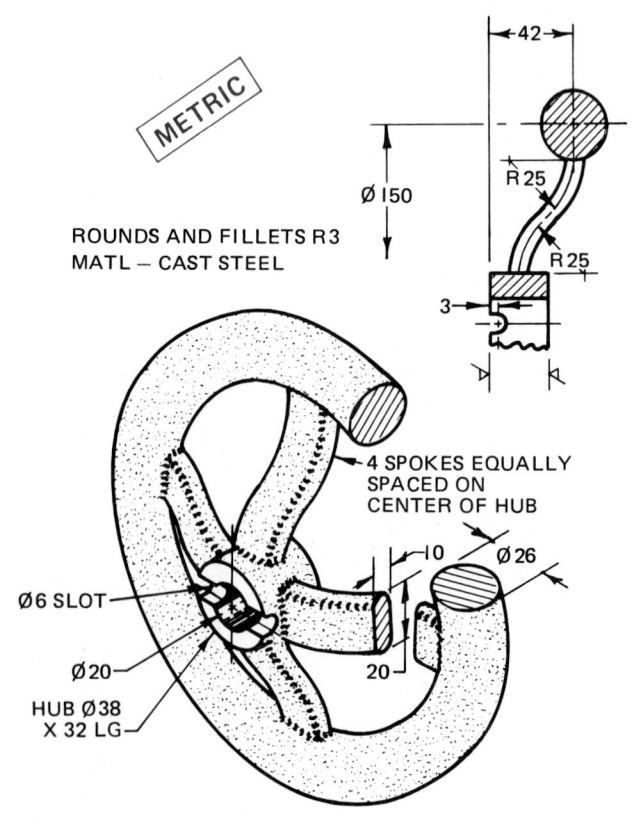

FIG. 9-9-B   Offset handwheel.

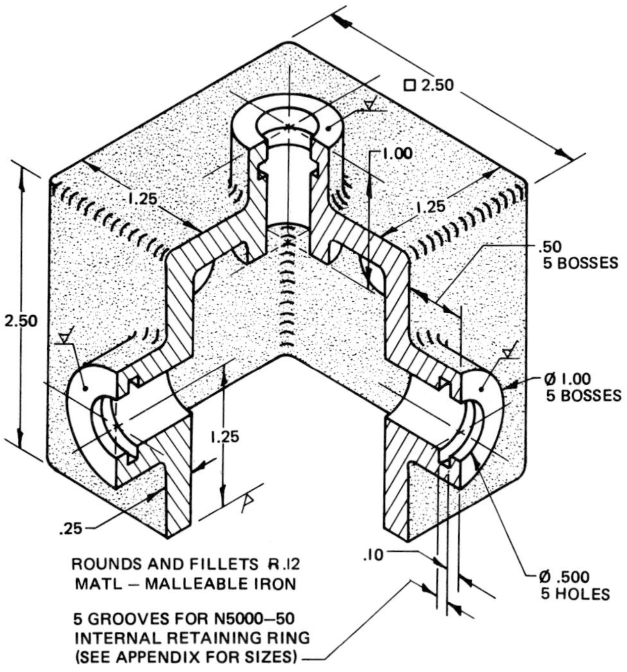

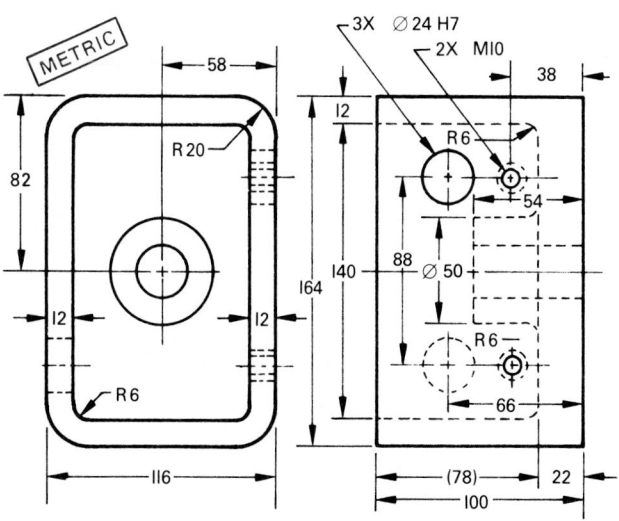

FIG. 9-10-A   Tumble box.

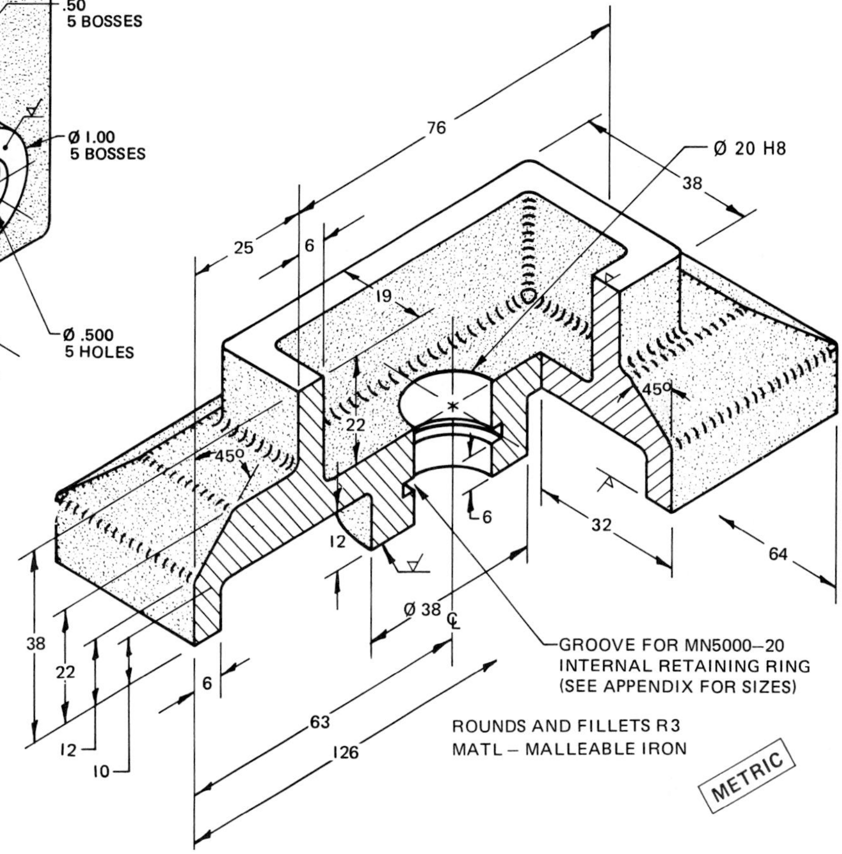

ROUNDS AND FILLETS R.12
MATL – MALLEABLE IRON

5 GROOVES FOR N5000–50
INTERNAL RETAINING RING
(SEE APPENDIX FOR SIZES)

□ 2.50
1.00
.50 5 BOSSES
Ø 1.00 5 BOSSES
2.50
1.25
1.25
1.25
.25
.10
Ø .500 5 HOLES

FIG. 9-10-B   Hold-down bracket.

Ø 20 H8
76
38
25
6
19
45°
22
6
32
64
45°
12
38
22
6
12
10
63
126
Ø 38
GROOVE FOR MN5000–20
INTERNAL RETAINING RING
(SEE APPENDIX FOR SIZES)
ROUNDS AND FILLETS R3
MATL – MALLEABLE IRON
METRIC

ASSIGNMENT FOR UNIT 9-11, PHANTOM OR HIDDEN SECTIONS

18. Make a two-view drawing of the part or one of the assemblies shown in Figs. 9-11-A through 9-11-C (here and on pg. 252). One view is to be drawn as a phantom section drawing. Show only the hole and shaft sizes for the fits shown. Scale 1:1.

FIG. 9-11-A   Bearing housing.

METRIC
58
R 20
82
12
12
164
116
R 6
3X Ø 24 H7
2X M10
38
12
R 6
54
140
88
Ø 50
R 6
66
(78)
22
100

251

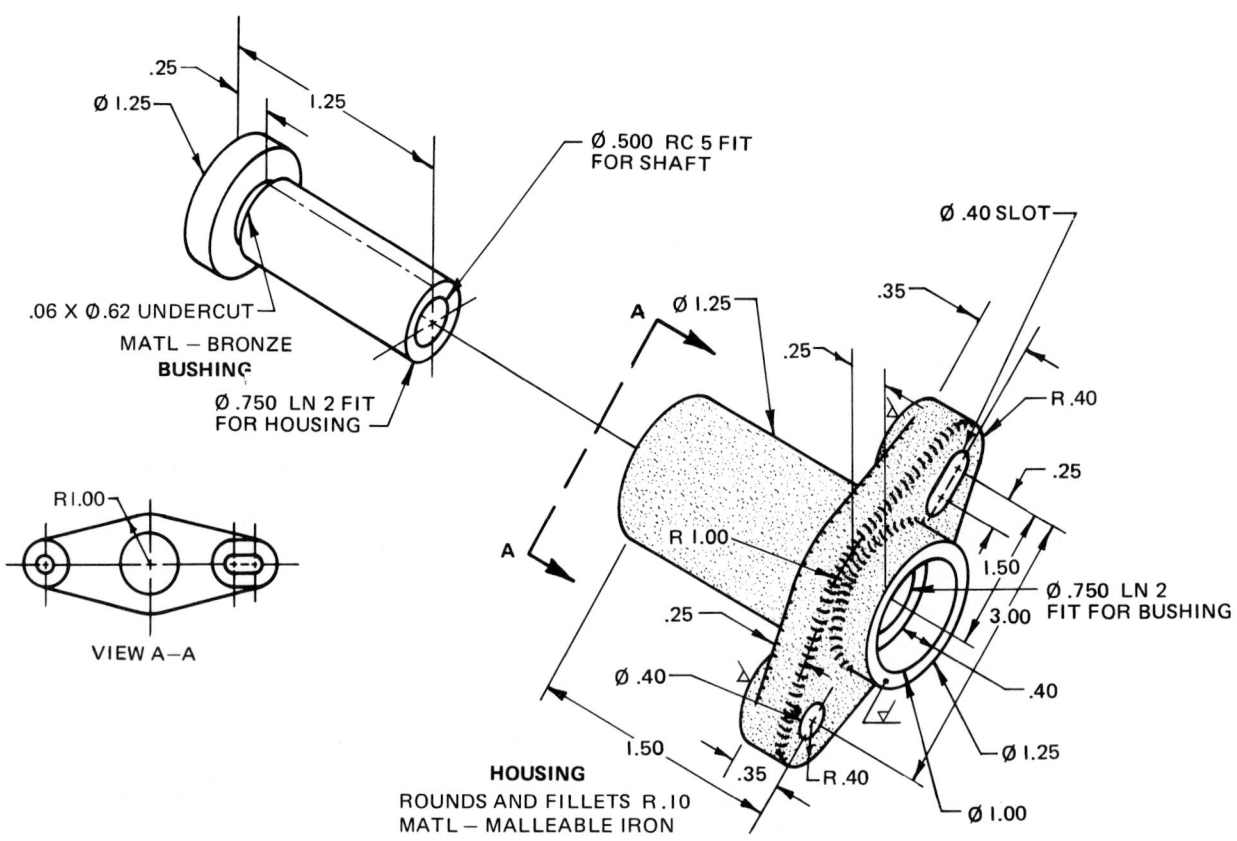

.25

Ø 1.25

1.25

Ø .500 RC 5 FIT
FOR SHAFT

Ø .40 SLOT

.35

.06 X Ø.62 UNDERCUT
MATL — BRONZE
BUSHING

Ø .750 LN 2 FIT
FOR HOUSING

Ø 1.25

.25

R .40

.25

R 1.00

R 1.00

1.50

Ø .750 LN 2
FIT FOR BUSHING

3.00

.25

Ø .40

.40

A

A

R1.00

VIEW A—A

1.50

.35

R .40

Ø 1.25

Ø 1.00

HOUSING
ROUNDS AND FILLETS R.10
MATL — MALLEABLE IRON

FIG. 9-11-B   Drill jig assembly.

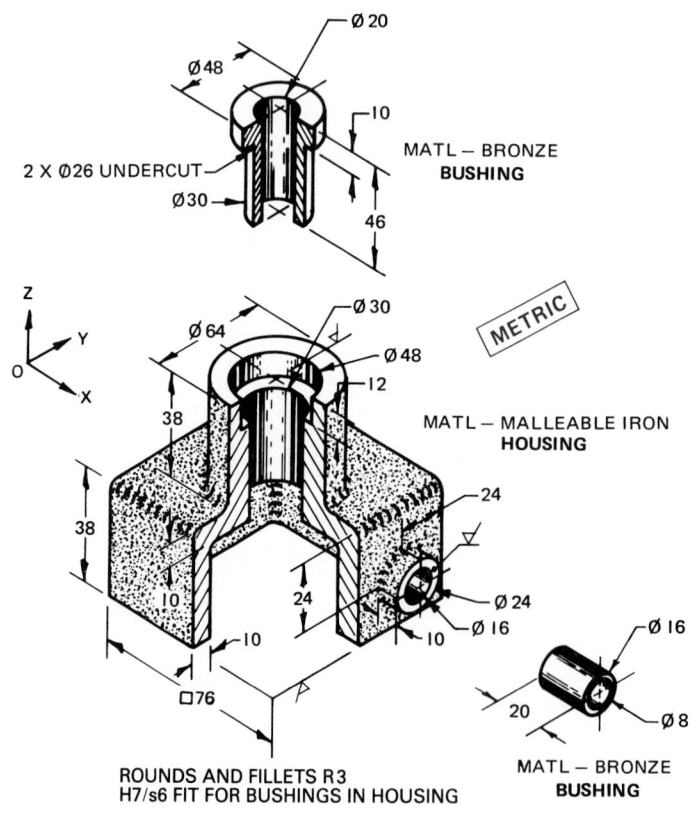

Ø 20

Ø 48

10

MATL — BRONZE
BUSHING

2 X Ø26 UNDERCUT

Ø 30

46

Z

Y

O

X

Ø 30

Ø 64

Ø 48

12

METRIC

MATL — MALLEABLE IRON
HOUSING

38

38

10

24

24

10

Ø 24

Ø 16

10

□ 76

Ø 16

20

Ø 8

ROUNDS AND FILLETS R3
H7/s6 FIT FOR BUSHINGS IN HOUSING

MATL — BRONZE
BUSHING

FIG. 9-11-C   Housing.

ASSIGNMENT FOR UNIT 9-12, SECTIONAL DRAWING REVIEW

19. Make a working drawing of one of the parts shown in Figs. 9-12-A through 9-12-H (pp. 253–257). From the information on section drawings found in Units 9-1 through 9-11, select appropriate sectional views that will improve the clarity of the drawing.

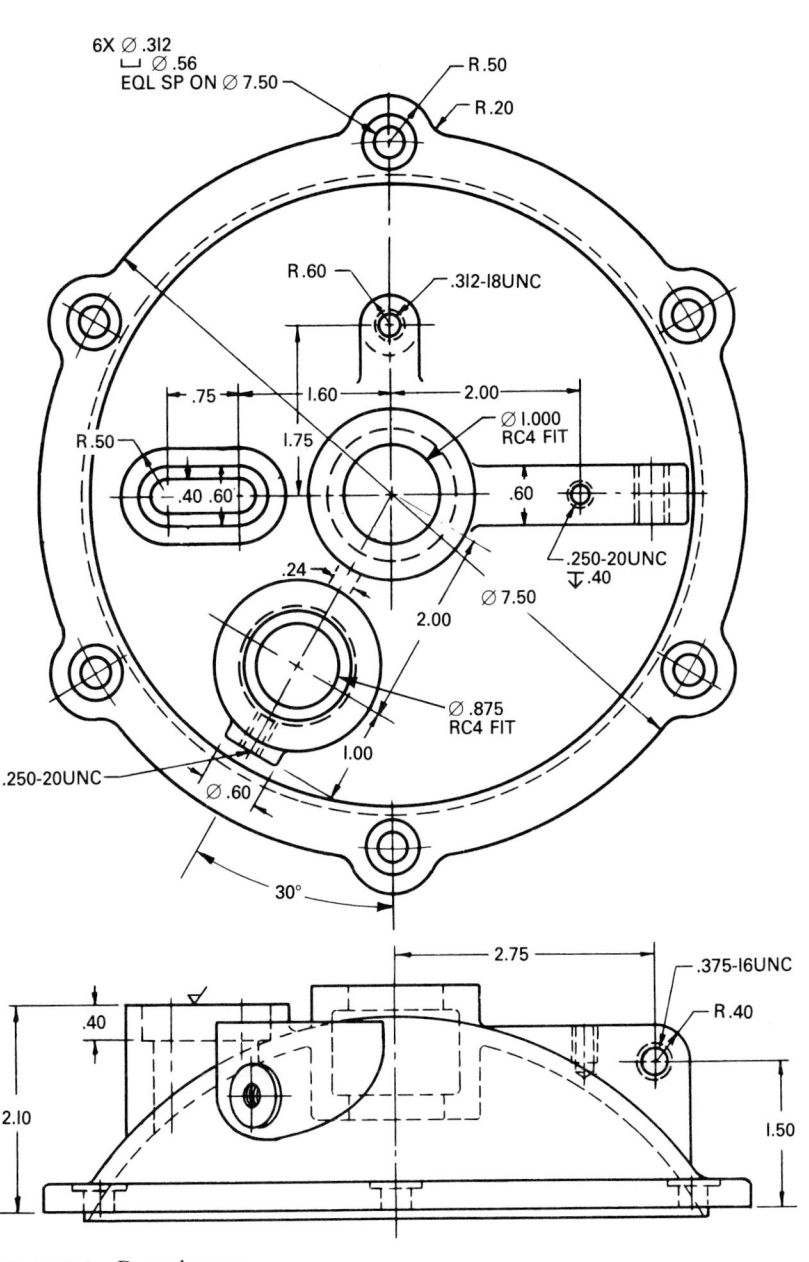

FIG. 9-12-A   Domed cover.

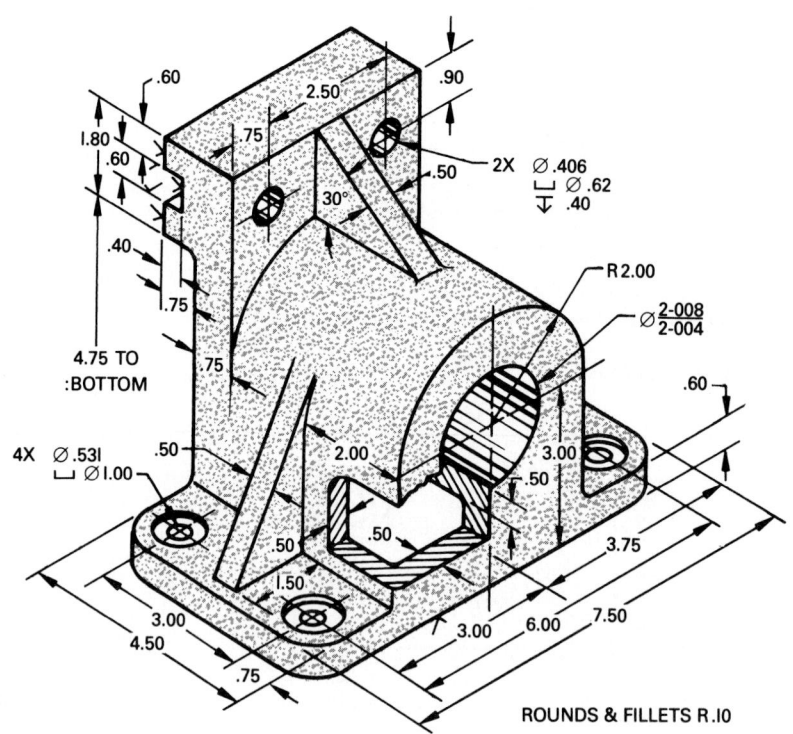

FIG. 9-12-B   Slide support.

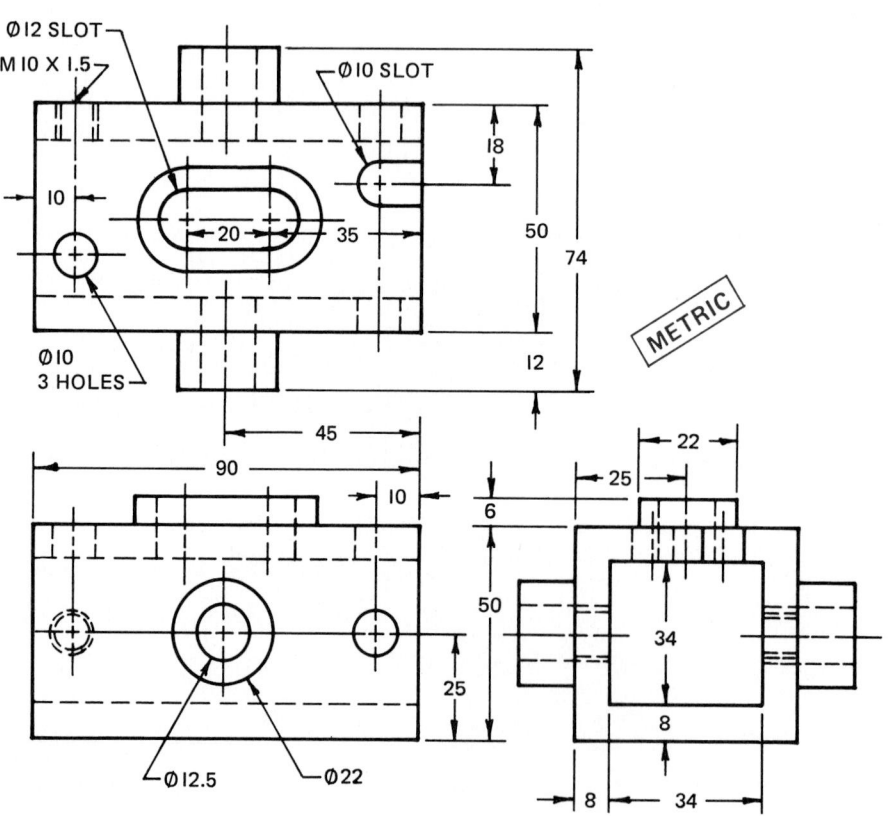

FIG. 9-12-C   Jacket.

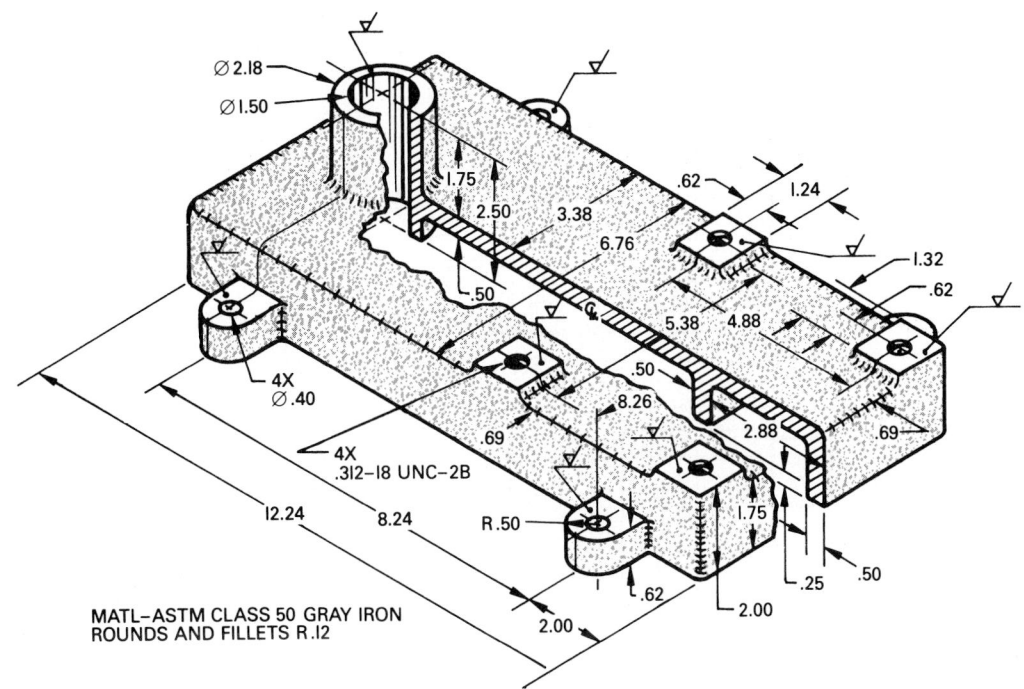

FIG. 9-12-D   Drill press base.

Ø 2.18
Ø 1.50
1.75
2.50
3.38
6.76
.62
1.24
1.32
5.38
4.88
.62
.50
4X
Ø .40
.50
.69
8.26
2.88
.69
4X
.312–18 UNC–2B
R.50
1.75
12.24
8.24
.62
.25
.50
2.00
2.00

MATL–ASTM CLASS 50 GRAY IRON
ROUNDS AND FILLETS R.12

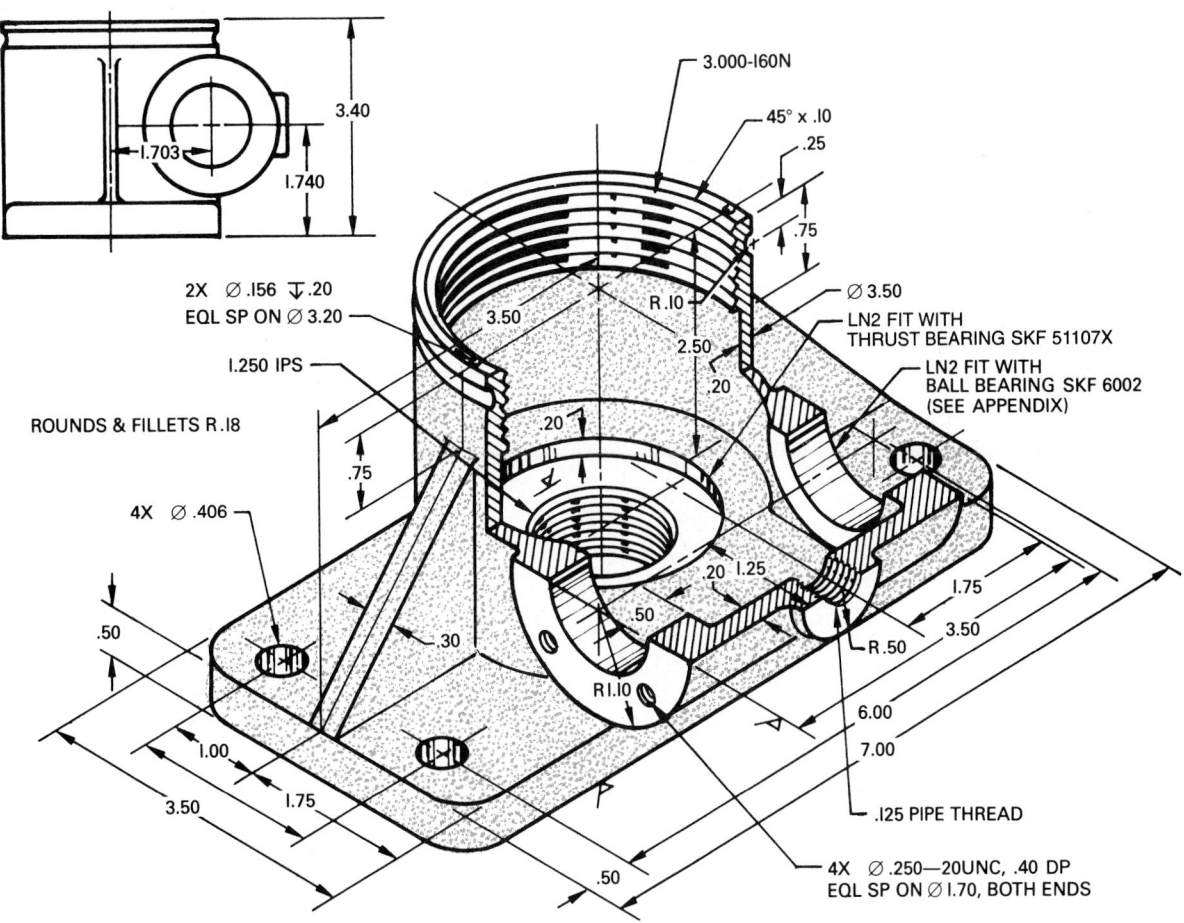

FIG. 9-12-E   Base.

3.40
1.703
1.740

3.000-160N
45° x .10
.25
.75

2X   Ø .156 ▽.20
EQL SP ON Ø 3.20

1.250 IPS

ROUNDS & FILLETS R.18

4X   Ø .406

3.50
R.10
2.50
.20

Ø 3.50
LN2 FIT WITH
THRUST BEARING SKF 51107X

LN2 FIT WITH
BALL BEARING SKF 6002
(SEE APPENDIX)

.75
.20

.50
4X   Ø .250—20UNC, .40 DP
EQL SP ON Ø 1.70, BOTH ENDS

.20   1.25

1.75
3.50

R.50

R1.10

6.00
7.00

.50

1.00
1.75

3.50

.50

.30

.125 PIPE THREAD

FIG. 9-12-F   Swivel base.

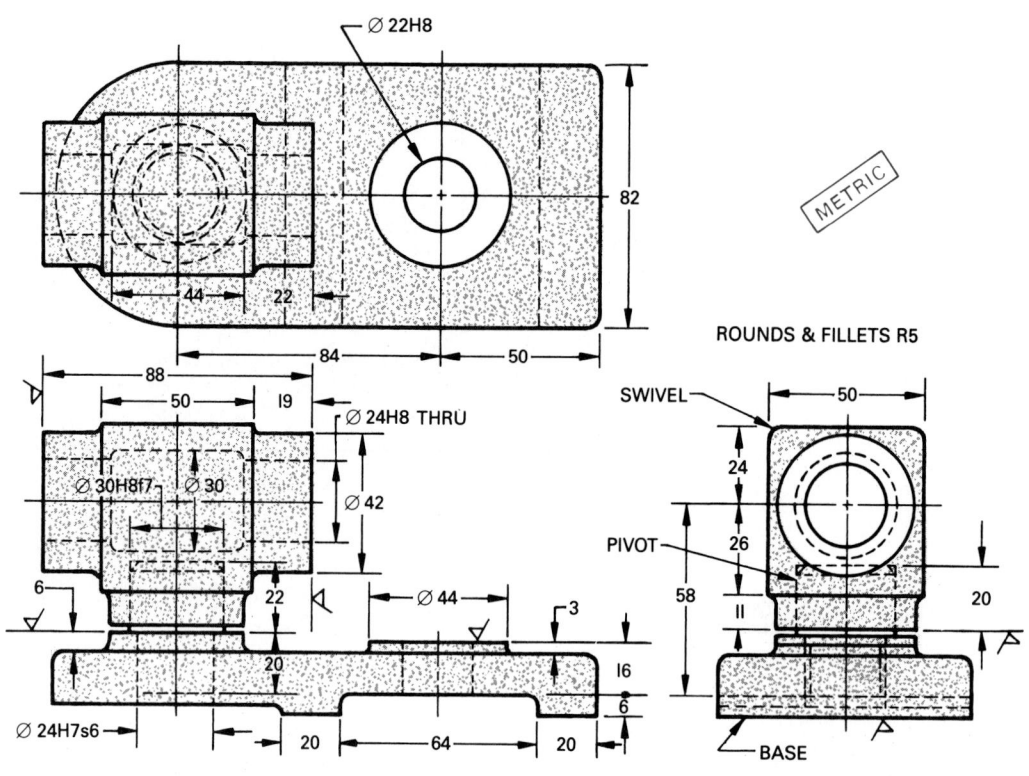

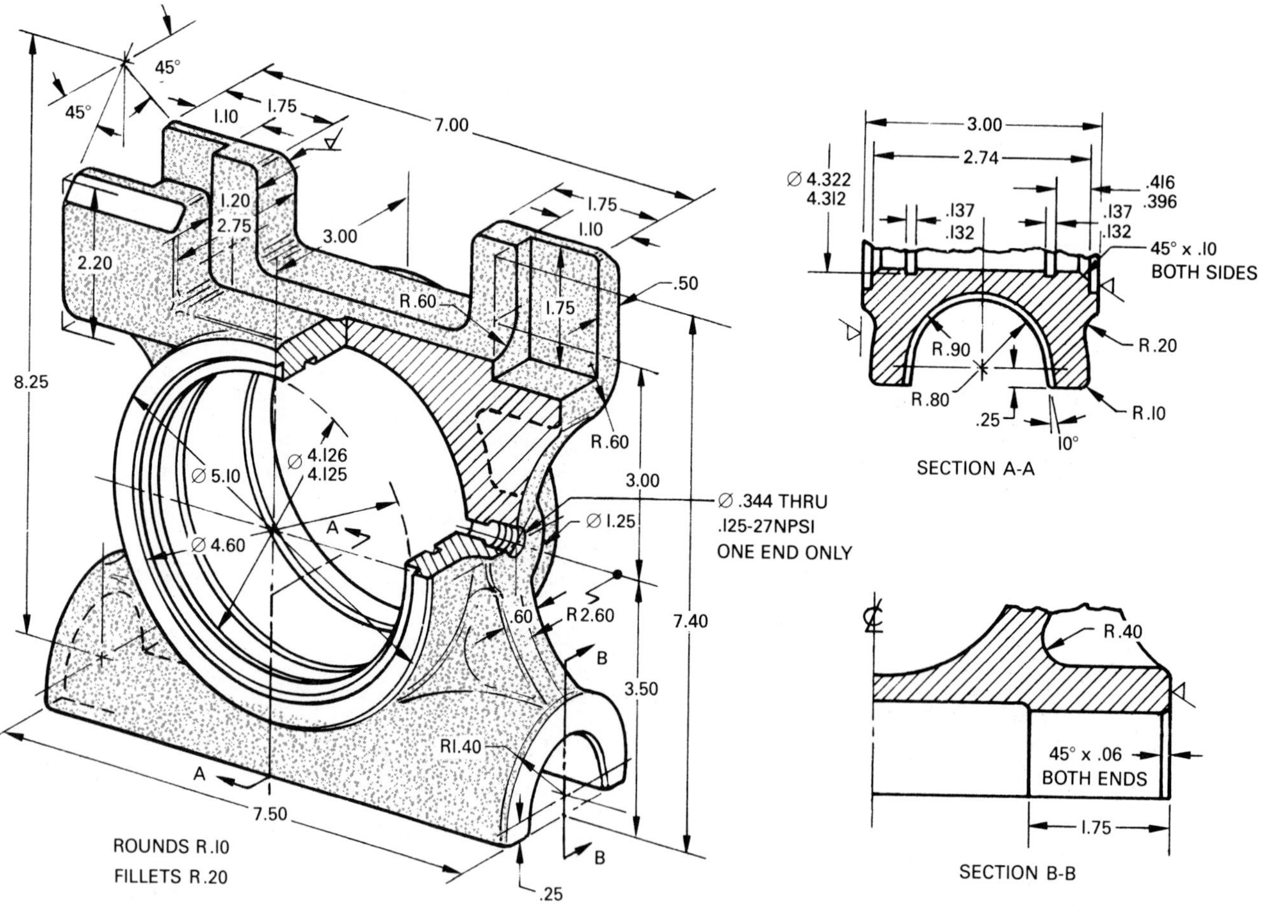

ROUNDS & FILLETS R5

∅ 22H8

METRIC

FIG. 9-12-G   Housing.

ROUNDS R.10
FILLETS R.20

SECTION A-A

SECTION B-B

FIG. 9-12-H   Pump base.

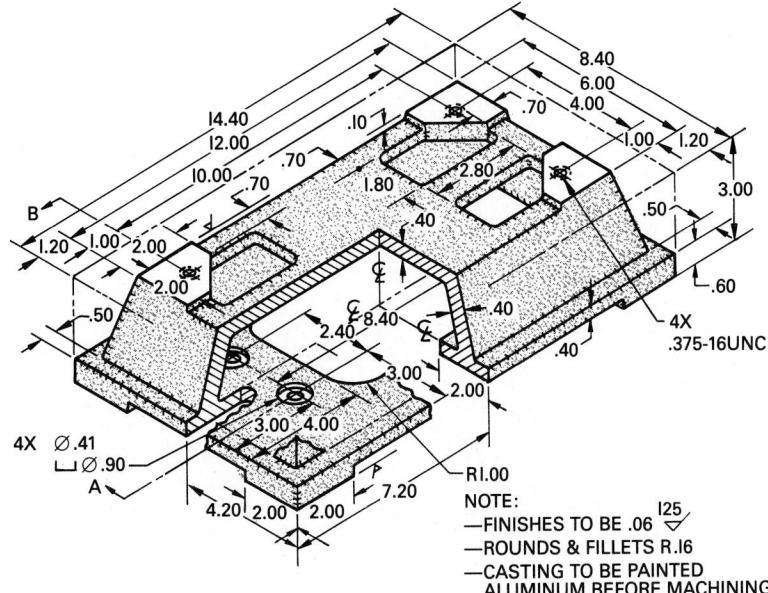

4X ∅.41
⊔∅.90

4X
.375-16UNC

R1.00

NOTE:
—FINISHES TO BE .06 125
—ROUNDS & FILLETS R.16
—CASTING TO BE PAINTED
   ALUMINUM BEFORE MACHINING

FASTENERS, MATERIALS, AND FORMING PROCESSES

CHAPTER 10

THREADED FASTENERS

Definitions

Chemical locking Fastening achieved by means of an adhesive.

Clearance drill size A diameter slightly larger than the major diameter of the bolt which permits the free passage of the bolt.

Counterbored hole A circular, flat-bottomed recess that permits the head of a bolt or cap screw to rest below the surface of the part.

Countersunk hole An angular-sided recess that accommodates the shape of a flat-head cap screw or machine screw or an oval-head machine screw.

Detailed representation A method of representation used to show the detail of a screw thread.

Free-spinning devices Fasteners that spin free in the clamping direction, which makes them easy to assemble, and have break-loose torque greater than the seating torque.

Lead The distance the threaded part would move parallel to the axis during one complete rotation in relation to a fixed mating part.

Pitch The distance from a point on the thread form to the corresponding point on the next form.

Prevailing-torque methods Fasteners that make use of increased friction between nut and bolt.

Screw thread A ridge of uniform section in the form of a helix on the external or internal surface of a cylinder.

Series The number of threads per inch, set for different diameters.

Simplified representation A method of representation used to clearly portray the requirements of a thread.

Spotfacing A machine operation that provides a smooth, flat surface where a bolt head or a nut will rest.

Standardization The manufacture and use of parts of similar types and sizes to reduce cost and simplify inventory and quality control.

Tap drill size A diameter equal to the minor diameter of the thread for a tapped hole.

10-1 SIMPLIFIED THREAD REPRESENTATION

Fastening devices are important in the construction of manufactured products, in the machines and devices used in manufacturing processes, and in the construction of all types of buildings. Fastening devices are used in the smallest watch to the largest ocean liner (Fig. 10-1-1).

FIG. 10-1-1 Fasteners. *(Industrial Fasteners Institute)*

There are two basic kinds of fasteners: permanent and removable. Rivets and welds are permanent fasteners. Bolts, screws, studs, nuts, pins, rings, and keys are removable fasteners. As industry progressed, fastening devices became standardized, and they developed definite characteristics and names. A thorough knowledge of the design and graphic representation of the more common fasteners is an essential part of drafting.

The cost of fastening, once considered only incidental, is fast becoming recognized as a critical factor in total product cost. "It's the in-place cost that counts, not the fastener cost" is an old saying of fastener design. The art of holding down fastener cost is not learned simply by scanning a parts catalog. More subtly, it entails weighing such factors as standardization, automatic assembly, tailored fasteners, and joint preparation.

A favorite cost-reducing method, *standardization,* not only cuts the cost of parts but reduces paperwork and simplifies inventory and quality control. By standardizing on type and size, it may be possible to reach the level of usage required to make power tools or automatic assembly feasible.

Screw Threads

A *screw thread* is a ridge of uniform section in the form of a helix on the external or internal surface of a cylinder (Fig. 10-1-2). The helix of a square thread is shown in Fig. 10-1-3.

The *pitch* of a thread, *P,* is the distance from a point on the thread form to the corresponding point on the next form, measured parallel to the axis (Fig. 10-1-4, pg. 262). The *lead, L,* is the distance the threaded part would move parallel to the axis during one complete rotation in relation to a fixed mating part (the distance a screw would enter a threaded hole in one turn).

Thread Forms

Figure 10-1-5 (pg. 262) shows some of the more common thread forms in use today. The ISO metric thread will eventually replace all the V-shaped metric and inch threads. As for the other thread forms shown, the proportions will be the same for both metric- and inch-size threads.

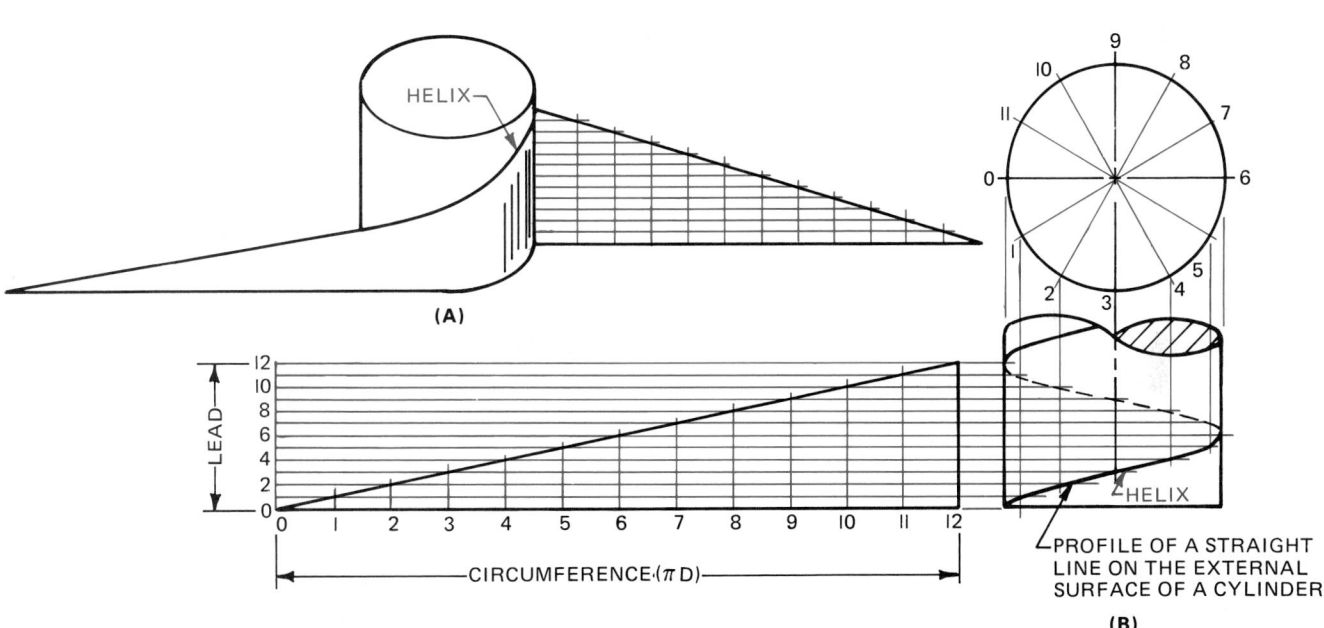

FIG. 10-1-2   The helix.

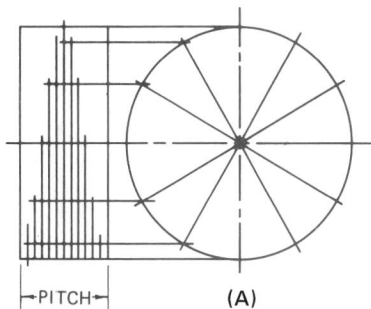
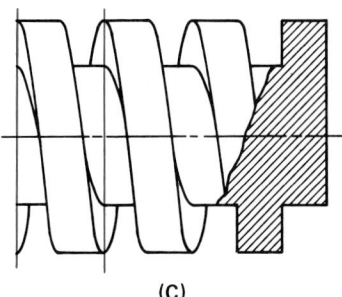

FIG. 10-1-3   The helix of a square thread.

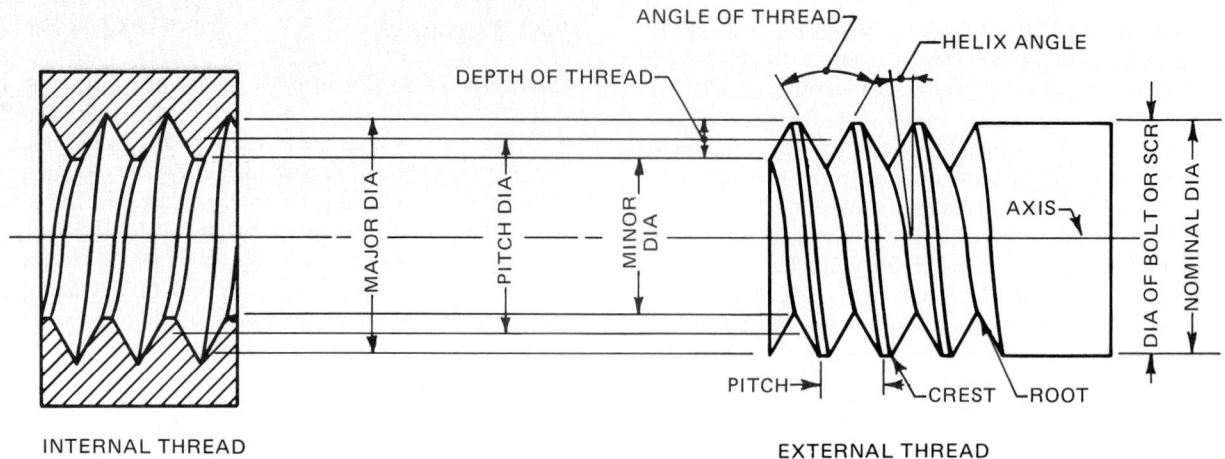

FIG. 10-1-4   Screw thread terms.

FIG. 10-1-5   Common thread forms and proportions.

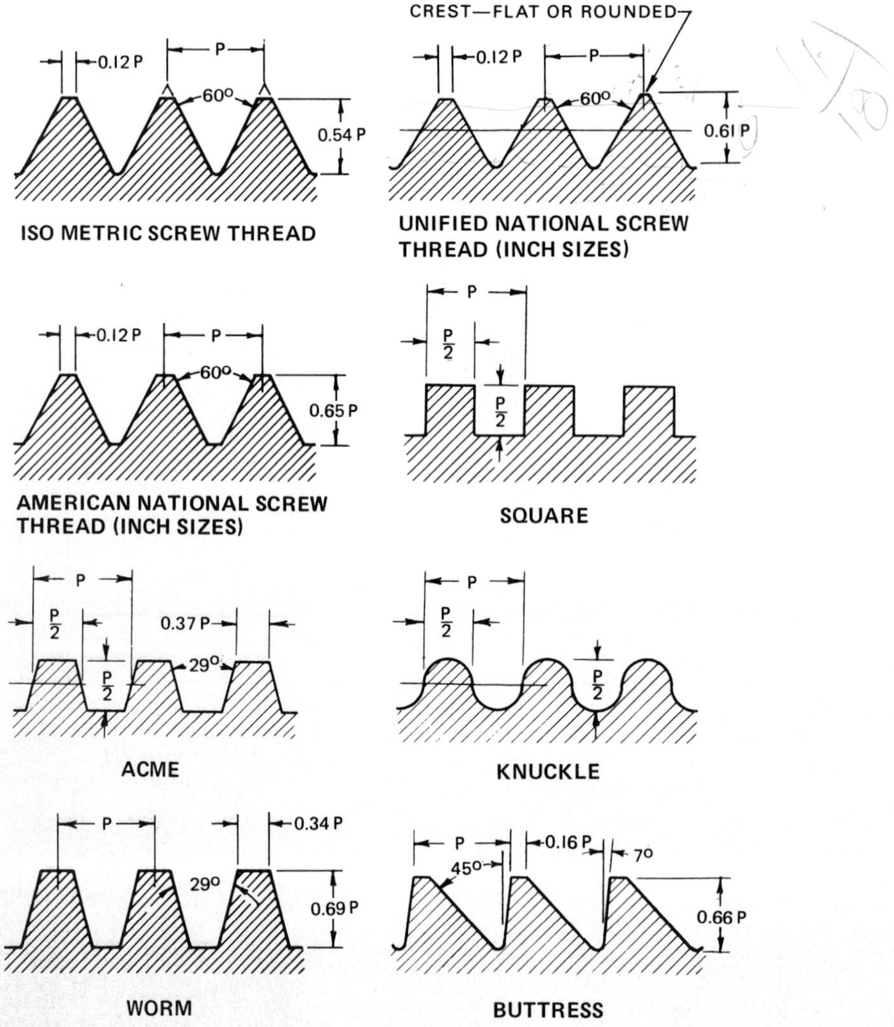

ISO METRIC SCREW THREAD

UNIFIED NATIONAL SCREW THREAD (INCH SIZES)

AMERICAN NATIONAL SCREW THREAD (INCH SIZES)

SQUARE

ACME

KNUCKLE

WORM

BUTTRESS

The knuckle thread is usually rolled or cast. A familiar example of this form is seen on electric light bulbs and sockets (Fig. 10-1-6). The square and acme forms are designed to transmit motion or power, as on the lead screw of a lathe. The buttress thread takes pressure in only one direction—against the surface perpendicular to the axis.

Thread Representation

True representation of a screw thread is seldom used on working drawings. Symbolic representation of threads is now standard practice. There are three types of conventions in general use for screw thread representation. These are known as

FIG. 10-1-6   Application of a knuckle thread. *(STUDIOHIO)*

simplified, detailed, and *schematic* (Fig. 10-1-7). *Simplified* representation should be used whenever it will clearly portray the requirements. *Detailed* representation is used to show the detail of a screw thread, especially for dimensioning in enlarged views, layouts, and assemblies. The *schematic* representation is nearly as effective as the detailed representation and is much easier to draw when manual drafting is used. This representation has given way to the simplified representation,

and as such, has been discarded as a thread symbol by most countries.

Right- and Left-Hand Threads

Unless designated otherwise, threads are assumed to be right-hand. A bolt being threaded into a tapped hole would be turned in a right-hand (clockwise) direction (Fig. 10-1-8). For some special applications, such as turnbuckles, left-hand threads are required. When such a thread is necessary, the letters LH are added after the thread designation.

Single and Multiple Threads

Most screws have single threads. It is understood that unless the thread is designated otherwise, it is a single thread. The single thread has a single ridge in the form of a helix (Fig. 10-1-9). The lead of a thread is the distance traveled parallel to the axis in one rotation of a part in relation to a fixed mating part (the distance a nut would travel along the axis of a bolt with one rotation of the nut). In single threads, the lead is equal to the pitch. A double thread has two ridges, started 180° apart, in the form of helices, and the lead is twice the pitch. A triple thread has three ridges, started 120° apart, in the form of helices, and the lead is three times the pitch. Multiple threads are used where fast movement is desired with a minimum

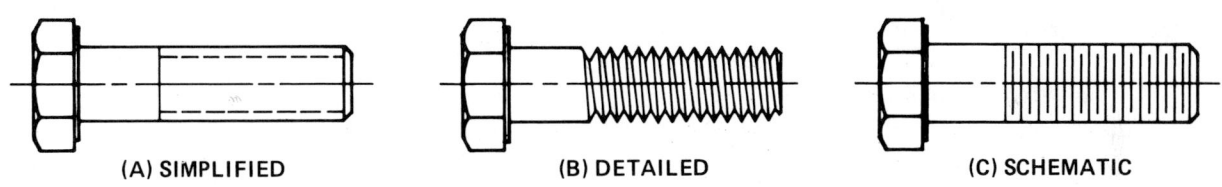

(A) SIMPLIFIED   **(B) DETAILED**   **(C) SCHEMATIC**

FIG. 10-1-7   Symbolic thread representation.

FIG. 10-1-8   Right- and left-hand threads.

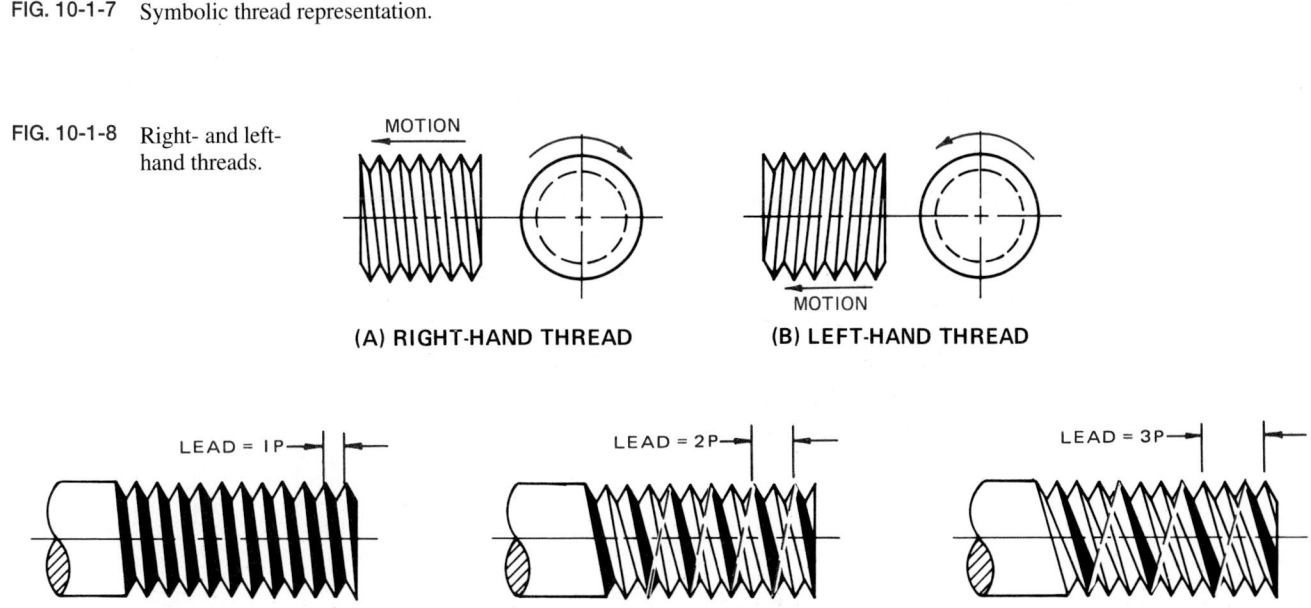

(A) RIGHT-HAND THREAD   **(B) LEFT-HAND THREAD**

(A) SINGLE THREAD   **(B) DOUBLE THREAD**   **(C) TRIPLE THREAD**

FIG. 10-1-9   Single and multiple threads.

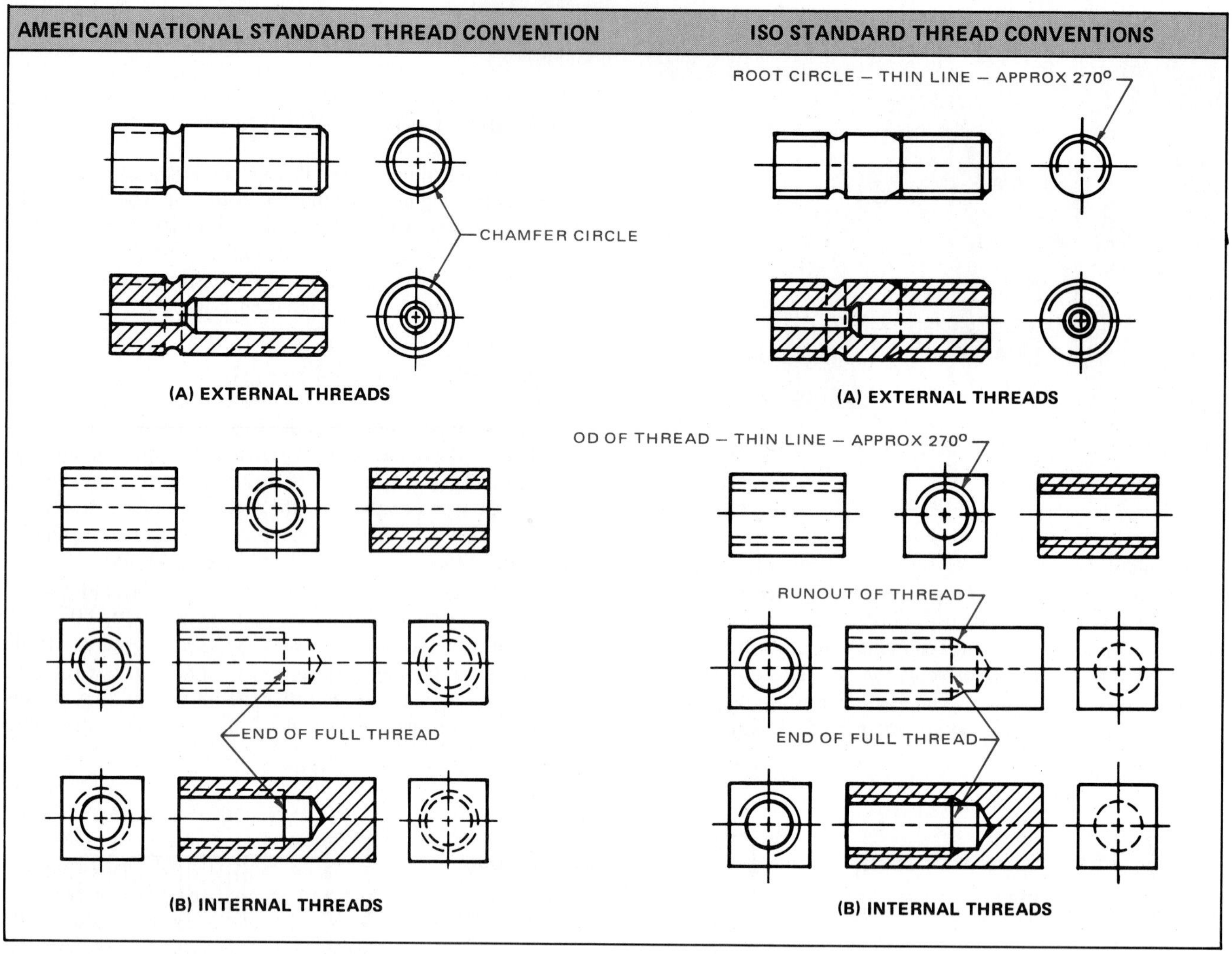

AMERICAN NATIONAL STANDARD THREAD CONVENTION	ISO STANDARD THREAD CONVENTIONS

ROOT CIRCLE — THIN LINE — APPROX 270°

CHAMFER CIRCLE

(A) EXTERNAL THREADS

(A) EXTERNAL THREADS

OD OF THREAD — THIN LINE — APPROX 270°

RUNOUT OF THREAD

END OF FULL THREAD

END OF FULL THREAD

(B) INTERNAL THREADS

(B) INTERNAL THREADS

FIG. 10-1-10   Simplified thread representation.

number of rotations, such as on threaded mechanisms for opening and closing windows.

Simplified Thread Representation

In this system the thread crests, except in hidden views, are represented by a thick outline and the thread roots by a thin broken line (Fig. 10-1-10). The end of the full-form thread is indicated by a thick line across the part, and imperfect or runout threads are shown beyond this line by running the root line at an angle to meet the crest line. If the length of runout threads is unimportant, this portion of the convention may be omitted.

Threaded Assemblies

For general use, the simplified representation of threaded parts is recommended for assemblies (Fig. 10-1-11). In sectional views, the externally threaded part is always shown covering the internally threaded part.

Inch Threads

In the United States and Canada a great number of threaded assemblies are still designed using inch-sized threads. In this system the pitch is equal to

$$\frac{1}{\text{Number of threads per inch}}$$

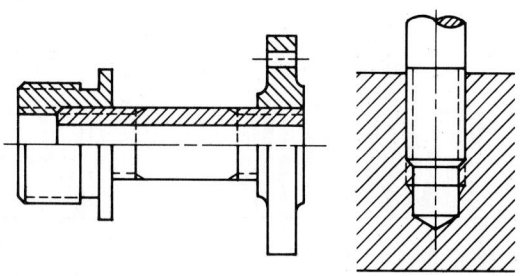

FIG. 10-1-11   Simplified representation of threads in assembly drawings.

The number of threads per inch is set for different diameters in what is called a thread *series*. For the Unified National system, there is the coarse-thread series (UNC) and the fine-thread series (UNF). See Table 8 of the Appendix.

In addition, there is an extra-fine-thread series (UNEF) for use where a small pitch is desirable, such as on thin-walled tubing. For special work and for diameters larger than those specified in the coarse and fine series, the Unified National thread system has three series that provide for the same number of threads per inch regardless of the diameter. These are the 8-thread series, the 12-thread series, and the 16-thread series. These are called *constant-pitch* threads.

Thread Class

Three classes of external thread (classes 1A, 2A, and 3A) and three classes of internal thread (classes 1B, 2B, and 3B) are provided. These classes differ in the amount of allowances and tolerances provided in each class.

The general characteristics and uses of the various classes are as follows.

Classes 1A and 1B   These classes produce the loosest fit, that is, the greatest amount of play (free motion) in assembly. They are useful for work where ease of assembly and disassembly is essential, such as for stove bolts and other rough bolts and nuts.

Classes 2A and 2B   These classes are designed for the ordinary good grade of commercial products, such as machine screws and fasteners, and for most interchangeable parts.

Classes 3A and 3B   These classes are intended for exceptionally high-grade commercial products, where a particularly close or snug fit is essential and the high cost of precision tools and machines is warranted.

Thread Designation

Thread designation for inch threads, whether external or internal, is expressed in this order: diameter (nominal or major diameter in decimal form with a minimum of three or maximum of four decimal places), number of threads per inch, thread form and series, and class of fit (number and letter). See Fig. 10-1-12.

Metric Threads

Metric threads are grouped into diameter-pitch combinations distinguished from one another by the pitch applied to specific diameters (Fig. 10-1-13, pg. 266).

The pitch for metric threads is the distance between corresponding points on adjacent teeth. In addition to a coarse- and fine-pitch series, a series of constant pitches is available. See Table 9 of the Appendix.

Coarse-Thread Series   This series is intended for use in general engineering work and commercial applications.

Fine-Thread Series   The fine-thread series is for general use where a finer thread than the coarse-thread series is desirable. In comparison with a coarse-thread screw, the fine-thread screw is stronger in both tensile and torsional strength and is less likely to loosen under vibration.

Thread Grades and Classes

The *fit* of a screw thread is the amount of clearance between the internal and external threads when they are assembled.

For each of the two main thread elements—pitch diameter and crest diameter—a number of tolerance grades have been established. The number of the tolerance grades reflects the size of the tolerance. For example, grade 4 tolerances are smaller than grade 6 tolerances, and grade 8 tolerances are larger than grade 6 tolerances.

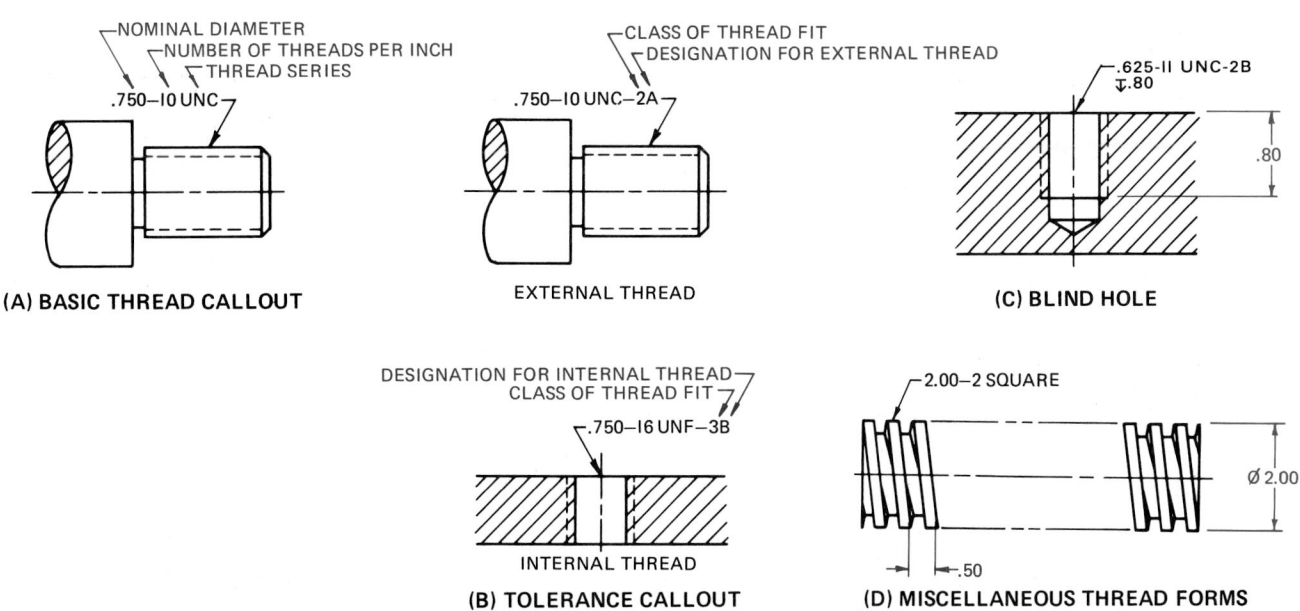

(A) BASIC THREAD CALLOUT

EXTERNAL THREAD

(C) BLIND HOLE

(B) TOLERANCE CALLOUT

(D) MISCELLANEOUS THREAD FORMS

FIG. 10-1-12   Thread specifications for inch-size threads.

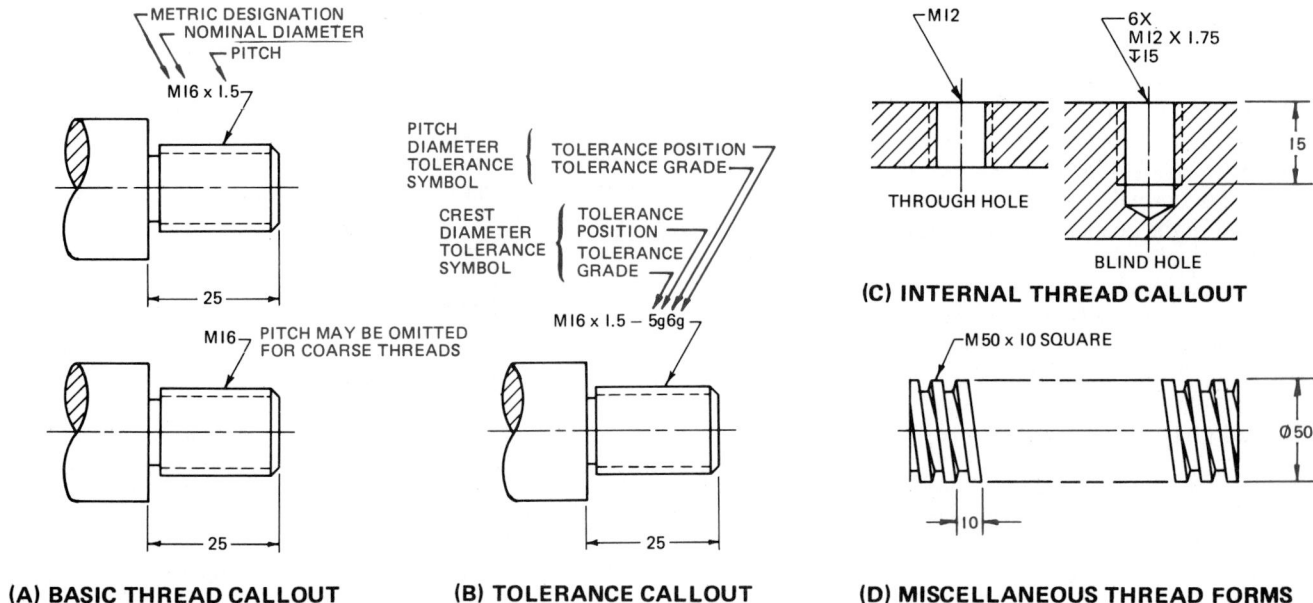

FIG. 10-1-13 Thread specifications for metric threads.

Grade 6 tolerances should be used for medium-quality length-of-engagement applications. The tolerance grades below grade 6 are intended for applications involving fine quality and/or short lengths of engagement. Tolerance grades above grade 6 are intended for coarse quality and/or long lengths of engagement.

In addition to the tolerance grade, a positional tolerance is required. This tolerance defines the maximum-material limits of the pitch and crest diameters of the external and internal threads and indicates their relationship to the basic profile.

In conformance with current coating (or plating) thickness requirements and the demand for ease of assembly, a series of tolerance positions reflecting the application of varying amounts of allowance has been established as follows.

For external threads:

- Tolerance position e (large allowance)
- Tolerance position g (small allowance)
- Tolerance position h (no allowance)

For internal threads:

- Tolerance position G (small allowance)
- Tolerance position H (no allowance)

ISO Metric Screw Thread Designation

ISO metric screw threads are defined by the nominal size (basic major diameter) and pitch, both expressed in millimeters. An M specifying an ISO metric screw thread precedes the nominal size, and an × separates the nominal size from the pitch (Fig. 10-1-13). For the coarse-thread series only, the pitch is not shown unless the dimension for the length of the thread is required. In specifying the length of thread, an × is used to separate the length of thread from the rest of the designations.

For external threads, the length or depth of thread may be given as a dimension on the drawing.

For example, a 10 mm diameter, 1.25 pitch, fine-thread series is expressed as M10 × 1.25. A 10 mm diameter, 1.5 pitch, coarse-thread series is expressed as M10; the pitch is not shown unless the length of thread is required. If the latter thread were 25 mm long and this information was required on the drawing, the thread callout would be M10 × 1.5 × 25.

A complete designation for an ISO metric screw thread comprises, in addition to the basic designation, an identification for the tolerance class. The tolerance class designation is separated from the basic designation by a dash and includes the symbol for the pitch diameter tolerance followed immediately by the symbol for crest diameter tolerance. Each of these symbols consists of a numeral indicating the grade tolerance followed by a letter representing the tolerance position (a capital letter for internal threads and a lowercase letter for external threads). Where the pitch and crest diameter symbols are identical, the symbol needs to be given only once. The complete designation for an ISO metric screw is used only when design requirements warrant it.

For external threads, the length of thread may be given as a dimension on the drawing. The length given is to be the minimum length of full thread. For threaded holes that go all the way through the part, the term THRU is sometimes added to the note. If no depth is given, the hole is assumed to go all the way through. For threaded holes that do not go all the way through, the depth (in conjunction with the depth symbol or word) is given in the note, for example, M12 × 1.75 × 20 DEEP. The depth given is the minimum depth of full thread.

Neither the chamfer shown at the beginning of a thread, nor the undercut at the end of a thread where a small diameter meets a larger diameter is required to be dimensioned, as shown in Fig. 10-1-14. A comparison of customary and metric thread sizes is shown in Fig. 10-1-15.

Pipe Threads

The pipe universally used is the inch-sized pipe. When pipe is ordered, the nominal diameter and wall thickness (in inches or millimeters) are given. In calling for the size of thread, the note used is similar to that for screw threads. When calling for a pipe thread on a metric drawing, the abbreviation IN follows the pipe size (Fig. 10-1-16).

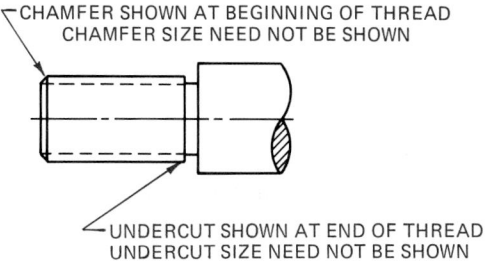

FIG. 10-1-14   Omission of thread information on detail drawings.

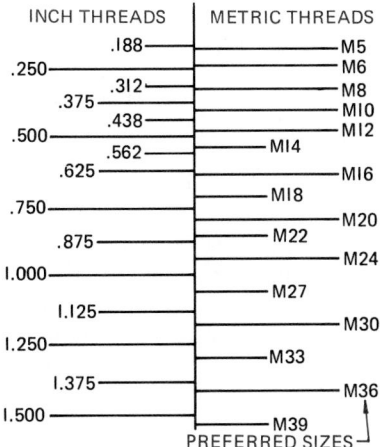

FIG. 10-1-15   Comparison of thread sizes.

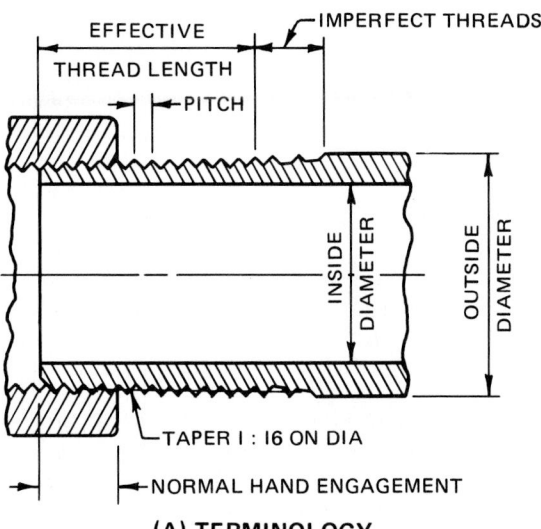

(A) TERMINOLOGY

FIG. 10-1-16   Pipe thread terminology and conventions.

EXAMPLE 1

4×8NPT

EXAMPLE 2

4×8NPS

where   4 = nominal diameter of pipe, in inches
       8 = number of threads per inch
      N = American Standard
      P = pipe
      S = straight pipe thread
      T = taper pipe thread

REFERENCES AND SOURCE MATERIAL

1. ANSI Y14.6, *Screw Thread Representation.*

ASSIGNMENTS

See Assignments 1 through 8 for Unit 10-1 on pages 283–285.

10-2 DETAILED AND SCHEMATIC THREAD REPRESENTATION

Detailed Thread Representation

Detailed representation of threads is a close approximation of the actual appearance of a screw thread. The form of the thread is simplified by showing the helices as straight lines and the truncated crests and roots as sharp Vs. It is used when a more realistic thread representation is required (Fig. 10-2-1, pg. 268).

Detailed Representation of V Threads   The detailed representation for V-shaped threads uses the sharp-V profile.

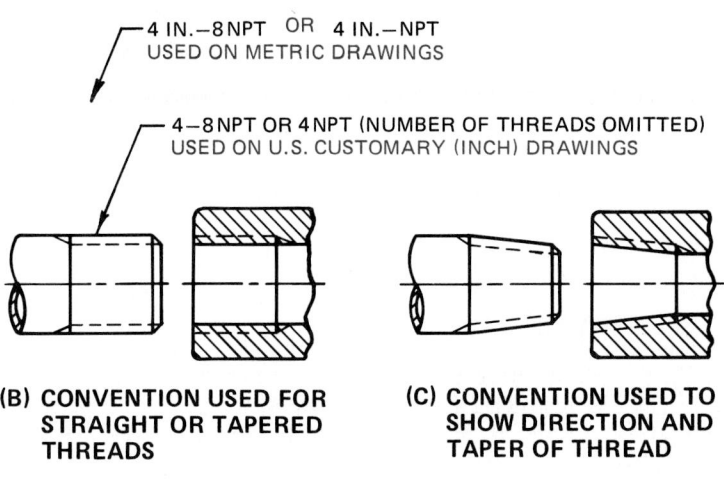

(B) CONVENTION USED FOR STRAIGHT OR TAPERED THREADS

(C) CONVENTION USED TO SHOW DIRECTION AND TAPER OF THREAD

The order of drawing the screw threads is shown in Fig. 10-2-1. The pitch is seldom drawn to scale; generally it is approximated. Lay off (establish) the pitch P and the half-pitch $P/2$, as shown in step 1. Add the crest lines. In step 2 add the V profile for one thread, top and bottom, locating the root diameter. Add construction lines for the root diameter. In step 3 add one side of the remaining Vs (thread profile), then add the other side of the Vs, completing the thread profile. In step 4, add the root lines which complete the detailed representation of the threads.

Detailed Representation of Square Threads   The depth of the square thread is one-half the pitch. In Fig. 10-2-2A, lay off spaces equal to $P/2$ along the diameter and add construction lines to locate the depth (root dia.) of thread. At B add the crest lines. At C add the root lines, as shown. At D the internal square thread is shown in section. Note the reverse direction of the crest and root lines.

Detailed Representation of Acme Threads   The depth of the acme thread is one-half the pitch (Fig. 10-2-2E through G). The stages in drawing acme threads are shown at E. For drawing purposes locate the pitch diameter midway between the outside diameter and the root diameter. The pitch diameter locates the pitch line. On the pitch line, lay off half-pitch spaces and the root lines to complete the view. The construction shown at F is enlarged.

Sectional views of an internal acme thread are shown at G. Showing the root and crest lines beyond the cutting plane on sectional views is optional.

Threaded Assemblies

It is often desirable to show threaded assembly drawings in detailed form, e.g., in presentation or catalog drawings. Hidden lines are normally omitted on these drawings, as they do nothing

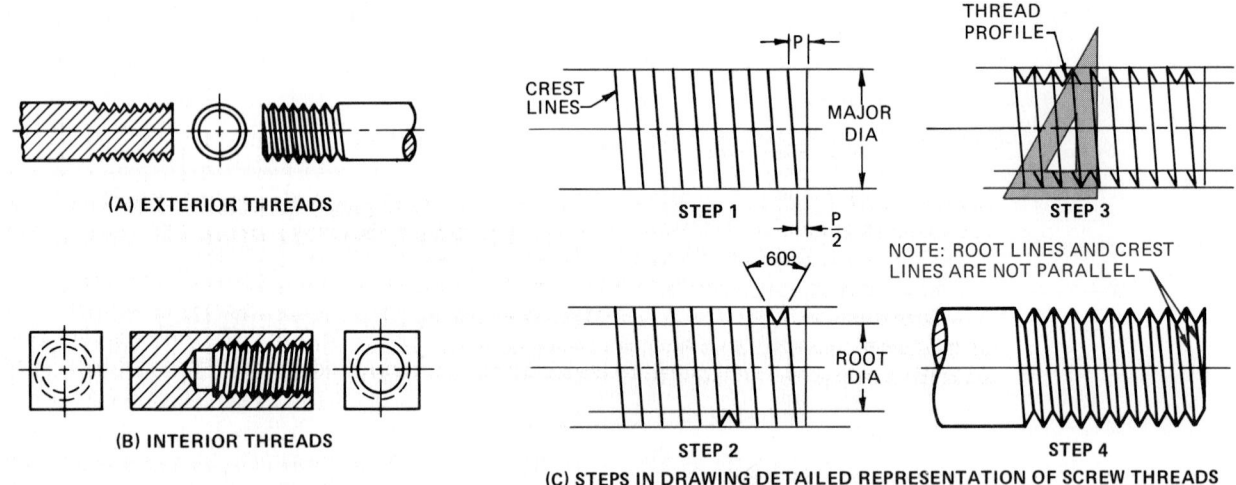

FIG. 10-2-1   Detailed representation of threads.

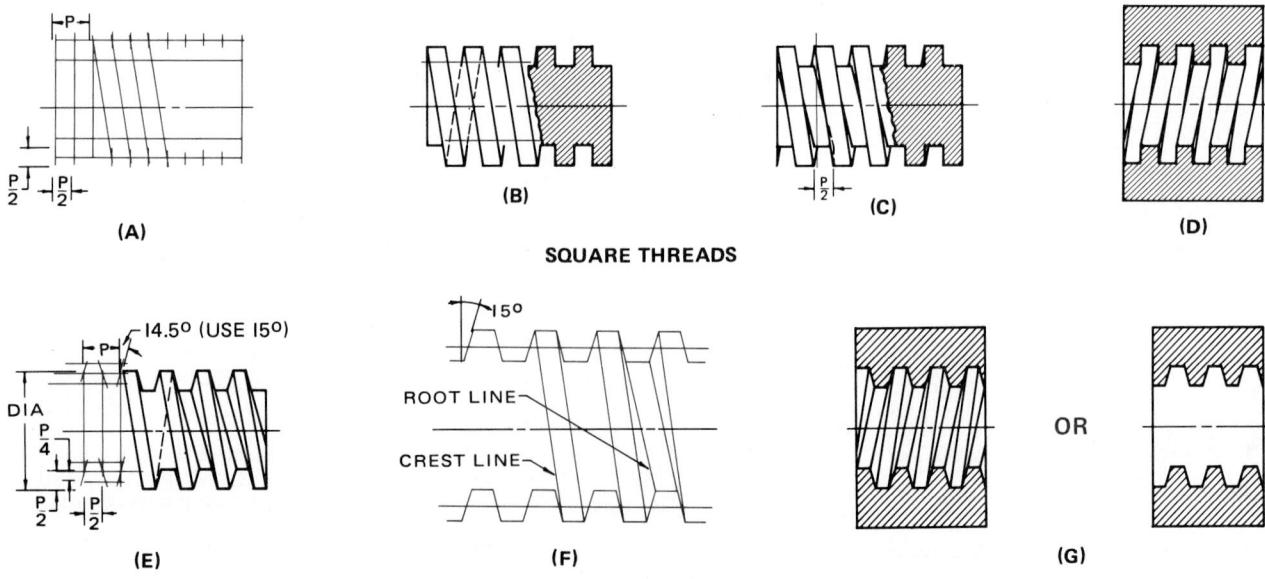

FIG. 10-2-2   Steps in drawing detailed representation of square and acme threads.

FIG. 10-2-3   Detailed threaded assembly.

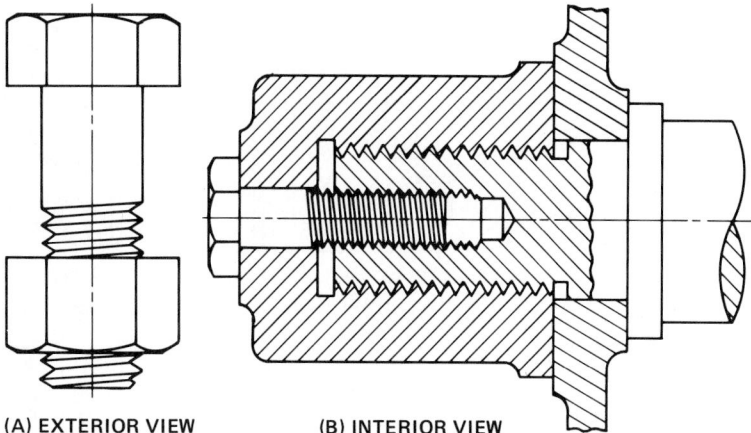

(A) EXTERIOR VIEW   (B) INTERIOR VIEW

to add to the clarity of the drawing (Fig. 10-2-3). One type of thread representation is generally used within any one drawing. When required, however, all three methods may be used.

Schematic Thread Representation

The staggered lines, symbolic of the thread root and crests, normally are perpendicular to the axis of the thread. The spacing between the root and crest lines and the length of the root lines are drawn to any convenient size (Fig. 10-2-4). At one time the root line was shown as a thick line.

REFERENCES AND SOURCE MATERIAL

1. ANSI Y14.6, *Screw Threads Representation.*

ASSIGNMENTS

See Assignments 9 through 11 for Unit 10-2 on pages 286 to 287.

10-3 COMMON THREADED FASTENERS

Fastener Selection

Fastener manufacturers agree that product selection must begin at the design stage. For it is here, when a product is still a figment of someone's imagination, that the best interests of the designer, production manager, and purchasing agent can be served. Designers, naturally, want optimum performance; production people are interested in the ease and economics of assembly; purchasing agents attempt to minimize initial costs and stocking costs.

The answer, pure and simple, is to determine the objectives of the particular fastening job and then consult fastener suppliers. These technical experts can often shed light on the situation and then recommend the right item at the best in-place cost.

Machine screws are among the most common fasteners in industry (Figs. 10-3-1 and 10-3-2, pg. 270). They are the easiest

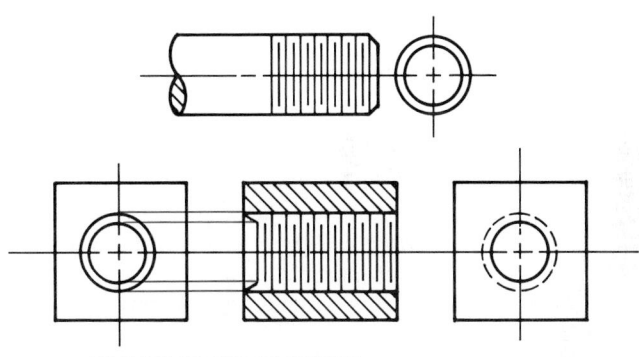

CHAMFERED END OF THREAD

FIG. 10-2-4   Schematic representation of threads.

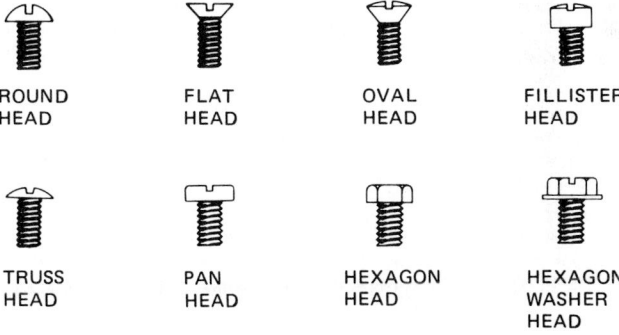

ROUND HEAD	FLAT HEAD	OVAL HEAD	FILLISTER HEAD
TRUSS HEAD	PAN HEAD	HEXAGON HEAD	HEXAGON WASHER HEAD

(A) SCREWS

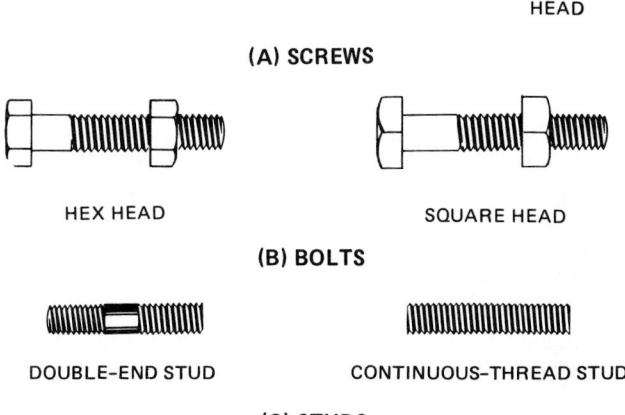

HEX HEAD   SQUARE HEAD

(B) BOLTS

DOUBLE-END STUD   CONTINUOUS-THREAD STUD

(C) STUDS

FIG. 10-3-1   Common threaded fasteners.

FIG. 10-3-2   Fastener applications.

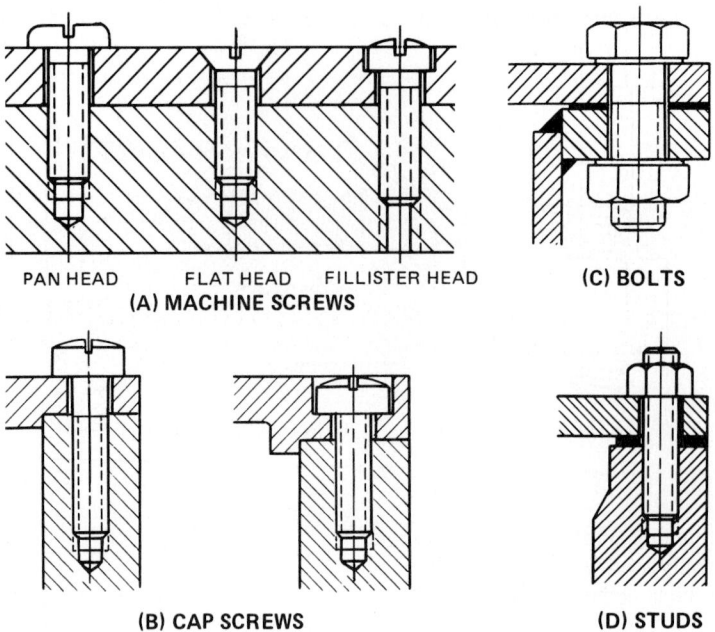

PAN HEAD   FLAT HEAD   FILLISTER HEAD

(A) MACHINE SCREWS

(B) CAP SCREWS

(C) BOLTS

(D) STUDS

to install and remove. They are also among the least understood. To obtain maximum machine-screw efficiency, thorough knowledge of the properties of both the screw and the materials to be fastened together is required.

For a given application, a designer should know the load that the screw must withstand, whether the load is one of tension or shear, and whether the assembly will be subject to impact shock or vibration. Once these factors have been determined, the size, strength, head shape, and thread type can be selected.

Fastener Definitions

Machine Screws   Machine screws have either fine or coarse threads and are available in a variety of heads. They may be used in tapped holes as shown in Fig. 10-3-2A, or with nuts.

Cap Screws   A cap screw is a threaded fastener that joins two or more parts by passing through a clearance hole in one part and screwing into a tapped hole in the other, as in Fig. 10-3-2B. A cap screw is tightened or released by torquing the head. Cap screws range in size from .25 in. (6 mm) in diameter and are available in five basic types of head.

Captive Screws   Captive screws are those that remain attached to the panel or parent material even when disengaged from the mating part. They are used to meet military requirements, to prevent screws from being lost, to speed assembly and disassembly operations, and to prevent damage from loose screws falling into moving parts or electrical circuits.

Tapping Screws   Tapping screws cut or form a mating thread when driven into preformed holes.

Bolts   A bolt is a threaded fastener that passes through clearance holes in assembled parts and threads into a nut (Fig. 10-3-2C).

Bolts and nuts are available in a variety of shapes and sizes. The square and hexagon head are the two most popular designs.

Studs   Studs are shafts threaded at both ends, and they are used in assemblies. One end of the stud is threaded into one of the parts being assembled; and the other assembly parts, such as washers and covers, are guided over the studs through clearance holes and are held together by means of a nut that is threaded over the exposed end of the stud (Fig. 10-3-2D).

Explanatory Data

A bolt is designed for assembly with a nut. A screw is designed to be used in a tapped or other preformed hole in the work. However, because of basic design, it is possible to use certain types of screws in combination with a nut.

The Change to Metric Fasteners

In the United States, the Industrial Fasteners Institute (IFI) has undertaken a major compilation of standards in its *Metric Fastener Standards* book.

Fastener Configuration

Head Styles

Which of the various head configurations to specify depends on the type of driving equipment used (screwdriver, socket wrench, etc.), the type of joint load, and the external appearance desired. The head styles shown in Fig. 10-3-3 can be used for both bolts and screws but are most commonly identified with the fastener category called machine screw or cap screw.

Hex and Square   The hex head is the most commonly used head style. The hex head design offers greater strength, ease of torque input, and area than the square head.

Pan   This head combines the qualities of the truss, binding, and round head types.

Binding   This type of head is commonly used in electrical connections because its undercut prevents fraying of stranded wire.

Washer (flanged)   This configuration eliminates the need for a separate assembly step when a washer is required, increases the bearing areas of the head, and protects the material finish during assembly.

Oval   Characteristics of this head type are similar to those of the flat head but it is sometimes preferred because of its neat appearance.

Flat   Available with various head angles, this fastener centers well and provides a flush surface.

Fillister   The deep slot and small head allow a high torque to be applied during assembly.

Truss   This head covers a large area. It is used where extra holding power is required, holes are oversize, or the material is soft.

12-Point   This 12-sided head is normally used on aircraft-grade fasteners. Multiple sides allow for a very sure grip and high torque during assembly.

Drive Configurations

Figure 10-3-4 shows 16 different driving designs.

Shoulders and Necks

The shoulder of a fastener is the enlarged portion of the body of a threaded fastener or the shank of an unthreaded fastener (Fig. 10-3-5, pg. 272).

Point Styles

The *point* of a fastener is the configuration of the end of the shank of a headed or headless fastener. Standard point styles are shown in Fig. 10-3-6 (pg. 272).

Cup   Most widely used when the cutting-in action of point is not objectionable.

Flat   Used when frequent resetting of a part is required. Particularly suited for use against hardened steel shafts. This point is preferred where walls are thin or the threaded member is a soft material.

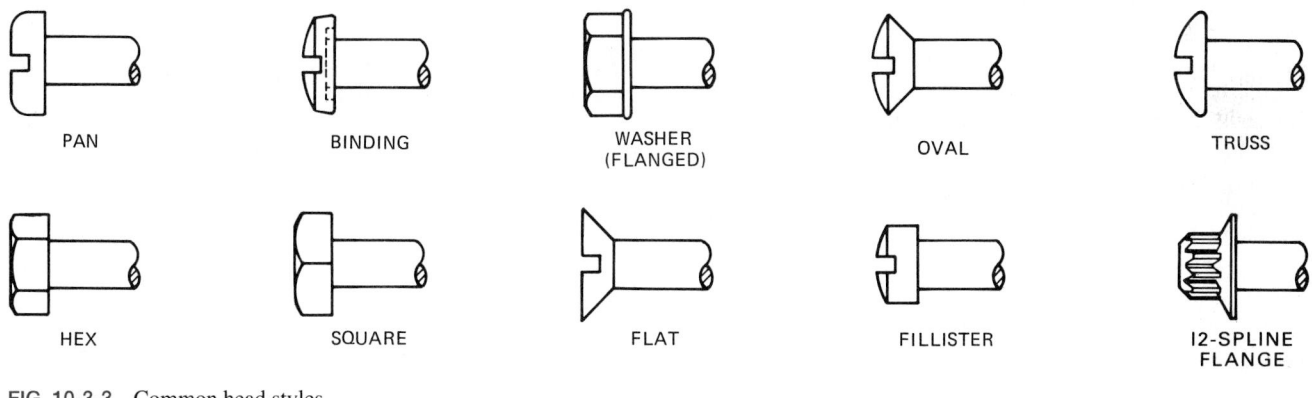

FIG. 10-3-3   Common head styles.

FIG. 10-3-4   Drive configurations.

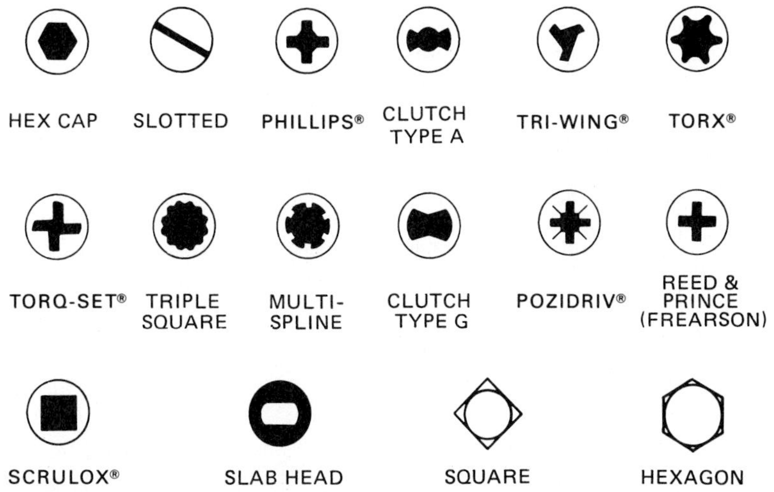

FIG. 10-3-5   Shoulder and necks.

OVAL SHOULDER

ROUND SHOULDER

FIN NECK

SQUARE (CARRIAGE) NECK

CUP

FLAT

CONE

HALF DOG

OVAL

FIG. 10-3-6   Point styles.

Cone   Used for permanent location of parts. Usually spotted in a hole to half its length.

Oval   Used when frequent adjustment is necessary or for seating against angular surfaces.

Half Dog   Normally applied where permanent location of one part in relation to another is desired.

Property Classes of Fasteners

Inch Fasteners

The strength of customary fasteners for most common uses is determined by the size of the fastener and the material from which it is made. Property classes are defined by the Society of Automotive Engineers (SAE) or the American Society for Testing and Materials (ASTM).

Figure 10-3-7 lists the mechanical requirements of inch-sized fasteners and their identification patterns.

Metric Fasteners

For mechanical and material requirements, metric fasteners are classified under a number of property classes. Bolts, screws, and studs have seven property classes of steel suitable for general engineering applications. The property classes are designated by numbers where increasing numbers generally represent increasing tensile strengths. The designation symbol consists of two parts: the first numeral of a two-digit symbol or the first two numerals of a three-digit symbol approximates one-hundredth of the minimum tensile strength in megapascals (MPa) and the last numeral approximates one-tenth of the ratio expressed as a percentage of minimum yield strength and minimum tensile strength.

EXAMPLE 1

A property class 4.8 fastener (see Fig. 10-3-8) has a minimum tensile strength of 420 MPa and a minimum yield strength of 340 MPa. One percent of 420 is 4.2. The first digit is 4. The minimum yield strength of 340 MPa is equal to approximately 80 percent of the minimum tensile strength of 420 MPa. One-tenth of 80 percent is 8. The last digit of the property class is 8.

Property Class (Equal to or Less Than)	Nominal Diameter	Minimum Tensile Strength MPa	Minimum Yield Strength MPa
4.6	M5 thru M36	400	240
4.8	M1.6 thru M16	420	340
5.8	M5 thru M24	520	420
8.8	M16 thru M36	830	660
9.8	M1.6 thru M16	900	720
10.9	M5 thru M36	1040	940
12.9	M1.6 thru M36	1220	1100

FIG. 10-3-8   Mechanical requirements for metric bolts, screws, and studs.

FIG. 10-3-7   Mechanical requirements for inch-size threaded fasteners.

HEAD DESIGNATION						
GRADE	GRADES 0, 1, 2	GRADE 3	GRADE 5	GRADE 7	GRADE 8
MINIMUM TENSILE STRENGTH KIPS	0—NO REQUIRE-MENTS 1—55 2—69 64 55	110 100	120 115 105	133	150

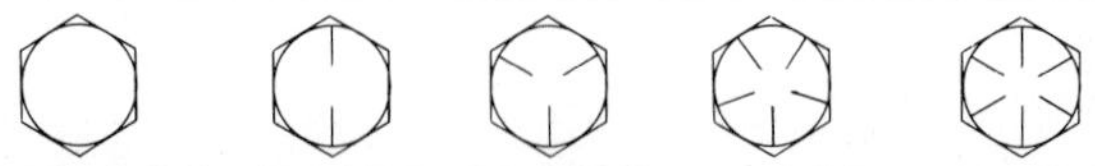

EXAMPLE 2

A property class 10.9 fastener (see Fig. 10-3-8) has a minimum tensile strength of 1040 MPa and a minimum yield strength of 940 MPa. One percent of 1040 is 10.4. The first two numerals of the three-digit symbol are 10. The minimum yield strength of 940 MPa is equal to approximately 90 percent of the minimum tensile strength of 1040 MPa. One-tenth of 90 percent is 9. The last digit of the property class is 9.

Machine screws are normally available only in classes 4.8 and 9.8; other bolts, screws, and studs are available in all classes within the specified product size limitations given in Fig. 10-3-7.

For guidance purposes only, to assist designers in selecting a property class, the following information may be used:

- Class 4.6 is approximately equivalent to SAE grade 1 and ASTM A 307, grade A.
- Class 5.8 is approximately equivalent to SAE grade 2.
- Class 8.8 is approximately equivalent to SAE grade 5 and ASTM A 449.
- Class 9.8 has properties approximately 9 percent stronger than SAE grade 5 and ASTM A 449.
- Class 10.9 is approximately equivalent to SAE grade 8 and ASTM A 354 grade BD.

Fastener Markings

Slotted and crossed recessed screws of all sizes and other screws and bolts of sizes .25 in. or M4 and smaller need not be marked. All other bolts and screws of sizes .25 in. or M5 and larger are marked to identify their strength. The property class symbols for metric fasteners are shown in Fig. 10-3-9. The symbol is located on the top of the bolt head or screw. Alternatively, for hex-head products, the markings may be indented on the side of the head.

All studs of size .25 in. or M5 and larger are identified by the property class symbol. The marking is located on the

extreme end of the stud. For studs with an interference-fit thread the markings are located at the nut end. Studs smaller than .50 in. or M12 use different identification symbols.

Nuts

The customary terms *regular* and *thick* for describing nut thicknesses have been replaced by the terms *style 1* and *style 2* for metric nuts. The design of style 1 and 2 steel nuts shown in Fig. 10-3-10 is based on providing sufficient nut strength to reduce the possibility of thread stripping. There are three property classes of steel nuts available (Fig. 10-3-11).

Hex-Flanged Nuts   These nuts are intended for general use in applications requiring a large bearing contact area. The two styles of flanged hex nuts differ dimensionally in thickness

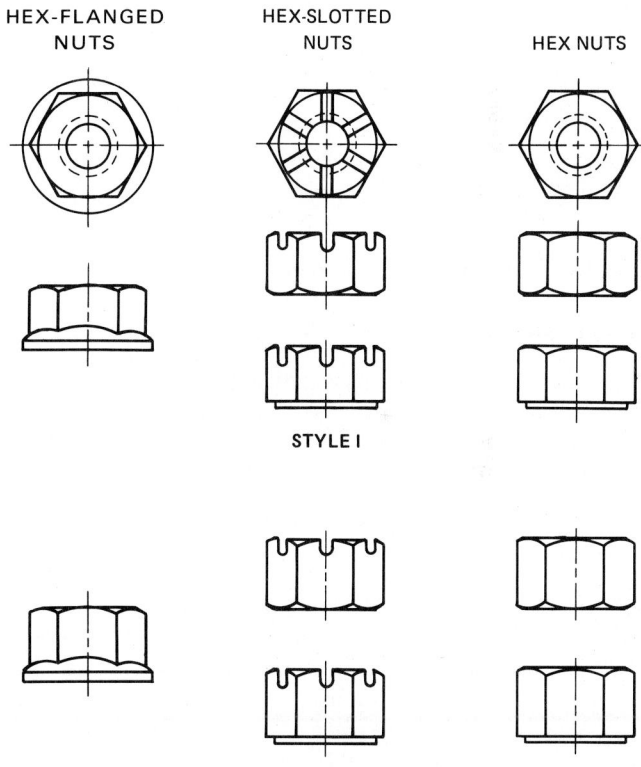

FIG. 10-3-10   Hex-nut styles.

Property Class	Identification Symbol	
	Bolts, Screws and Studs	Studs Smaller Than M12
4.6	4.6	—
4.8	4.8	—
5.8	5.8	—
8.8 (I)	8.8	○
9.8 (I)	9.8	+
10.9 (I)	10.9	□
12.9	12.9	△

Note I: Products made of low-carbon martensite steel shall be additionally identified by underlining the numerals.

FIG. 10-3-9   Metric property class identification symbols for bolts, screws, and studs.

Property Class	Nominal Nut Size	Suggested Property Class of Mating Bolt, Screw, or Stud
5	M5 thru M36	4.6, 4.8, 5.8
9	M5 thru M16 M20 thru M36	5.8, 9.8 5.8, 8.8
10	M6.3 thru M36	10.9

FIG. 10-3-11   Metric nut selection for bolts, screws, and studs.

FIG. 10-3-12 Approximate head proportions for hex-head cap screws, bolts, and nuts.

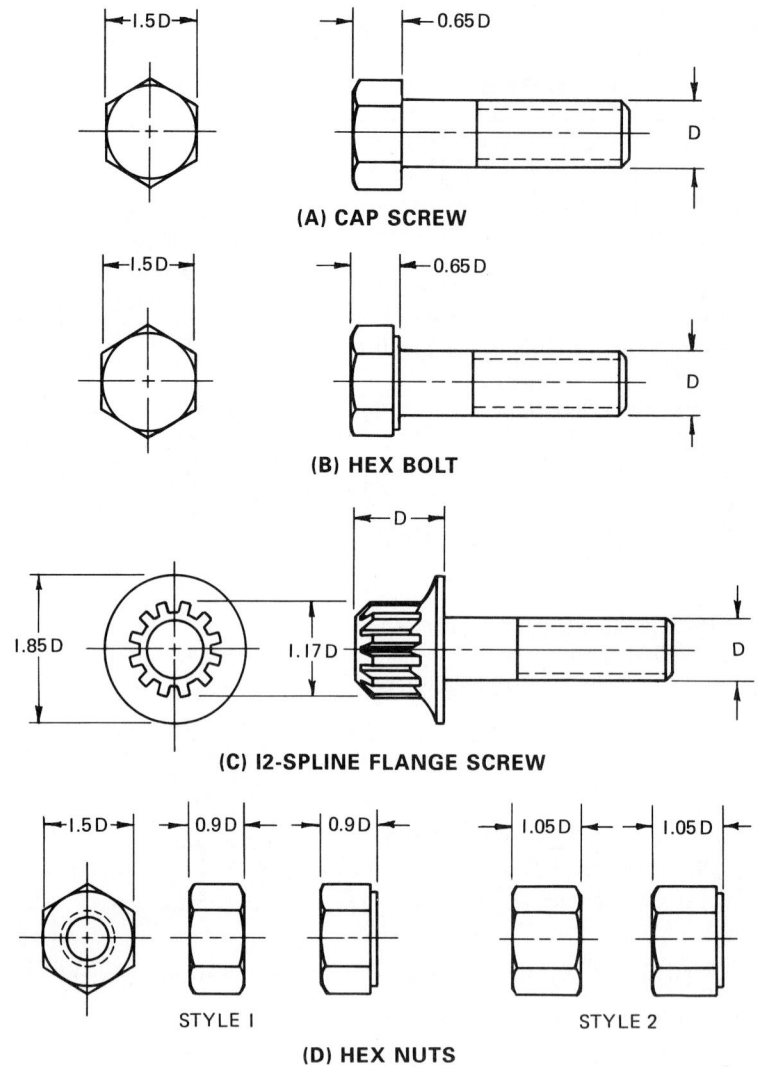

(A) CAP SCREW

(B) HEX BOLT

(C) I2-SPLINE FLANGE SCREW

STYLE I

STYLE 2

(D) HEX NUTS

only. The standard property classes for hex-flanged nuts are identical to the hex nuts. All metric nuts are marked to identify their property class.

Drawing a Bolt and Nut

Bolts and nuts are not normally drawn on detail drawings unless they are of a special size or have been modified. On some assembly drawings it may be necessary to show a nut and bolt. Approximate nut and bolt sizes are shown in Fig. 10-3-12. Actual sizes are found in Table 11 of the Appendix. Nut and bolt templates are also available and are recommended as a cost-saving device for manual drafting. Conventional drawing practice is to show the nuts and bolt heads in the across-corners position in all views.

Studs

Studs, as shown in Fig. 10-3-13, are still used in large quantities to best fulfill the needs of certain design functions and for overall economy.

Double-End Studs These studs are designated in the following sequence: type and name; nominal size; thread information; stud length; material, including grade identification; and finish (plating or coating) if required.

EXAMPLE

TYPE 2 DOUBLE-END STUD
.500—13 UNC—2A × 4.00
CADMIUM PLATED

Continuous-Thread Studs These studs are designated in the following sequence: product name, nominal size, thread

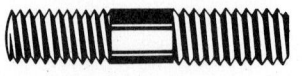

(A) DOUBLE-END

(B) CONTINUOUS-THREAD

FIG. 10-3-13 Studs.

information, stud length, material, and finish (plating or coating) if required.

EXAMPLE

TYPE 3 CONTINUOUS-THREAD STUD, M24 × 3 × 200, STEEL CLASS 8.8, ZINC PHOSPHATE AND OIL

Washers

Washers are one of the most common forms of hardware and perform many varied functions in mechanically fastened assemblies. They may only be required to span an oversize clearance hole, to give better bearing for nuts or screw faces, or to distribute loads over a greater area. Often, they serve as locking devices for threaded fasteners. They are also used to maintain a spring-resistance pressure, to guard surfaces against marring, and to provide a seal.

Classification of Washers

Washers are commonly the elements that are added to screw systems to keep them tight, but not all washers are locking types. Many washers serve other functions, such as surface protection, insulation, sealing, electrical connection, and spring-tension take-up devices.

Flat Washers   Plain, or flat, washers are used primarily to provide a bearing surface for a nut or a screw head, to cover large clearance holes, and to distribute fastener loads over a large area—particularly on soft materials such as aluminum or wood (Fig. 10-3-14).

Conical Washers   These washers are used with screws to effectively add spring take-up to the screw elongation.

Helical Spring Washers   These washers are made of slightly trapezoidal wire formed into a helix of one coil so that the free height is approximately twice the thickness of the washer section (Fig. 10-3-15).

Tooth Lock Washers   Made of hardened carbon steel, a tooth lock washer has teeth that are twisted or bent out of the plane of the washer face so that sharp cutting edges are presented to both the workpiece and the bearing face of the screw head or nut (Fig. 10-3-16).

Spring Washers   There are no standard designs for spring washers (Fig. 10-3-17). They are made in a great variety of sizes and shapes and are usually selected from a manufacturer's catalog for some specific purpose.

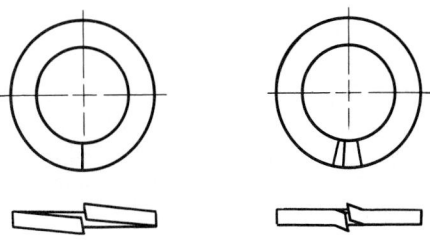

(A) PLAIN   (B) NONLINK POSITIVE

FIG. 10-3-15   Helical spring washers.

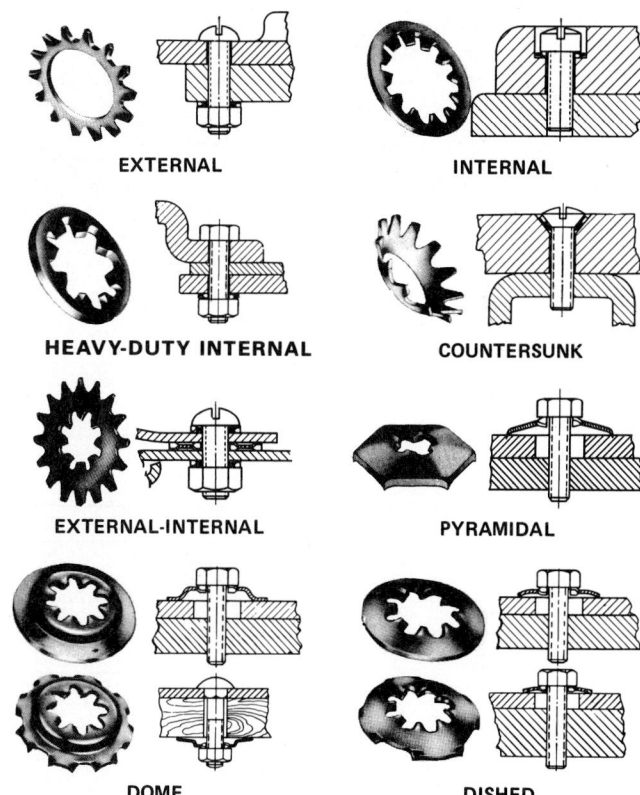

EXTERNAL   INTERNAL

HEAVY-DUTY INTERNAL   COUNTERSUNK

EXTERNAL-INTERNAL   PYRAMIDAL

DOME   DISHED

FIG. 10-3-16   Tooth lock washers.

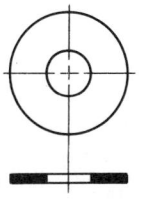 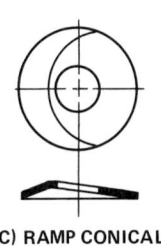

(A) FLAT   (B) CONICAL   (C) RAMP CONICAL

FIG. 10-3-14   Flat and conical washers.

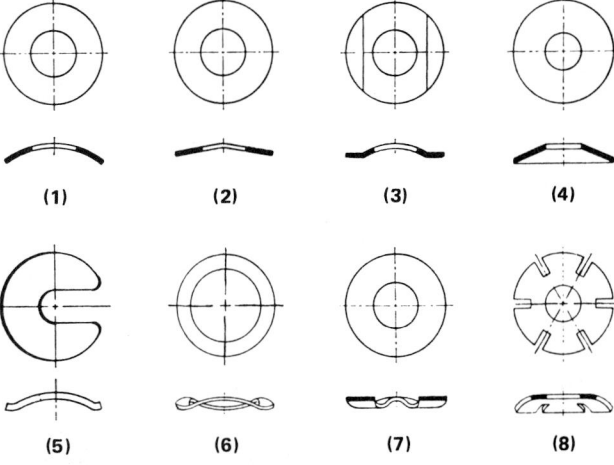

(1)   (2)   (3)   (4)

(5)   (6)   (7)   (8)

FIG. 10-3-17   Typical spring washers.

275

Special-Purpose Washers   Molded or stamped nonmetallic washers are available in many materials and may be used as seals, electrical insulators, or for protection of the surface of assembled parts.

Many plain, cone, or tooth washers are available with special mastic sealing compounds firmly attached to the washer. These washers are used for sealing and vibration isolation in high-production industries.

Terms Related to Threaded Fasteners

The *tap drill size* for a threaded (tapped) hole is a diameter equal to the minor diameter of the thread. The *clearance drill size,* which permits the free passage of a bolt, is a diameter slightly greater than the major diameter of the bolt (Fig. 10-3-18). A *counterbored hole* is a circular, flat-bottomed recess that permits the head of a bolt or cap screw to rest below the surface of

FIG. 10-3-18   Specifying threaded fasteners and holes.

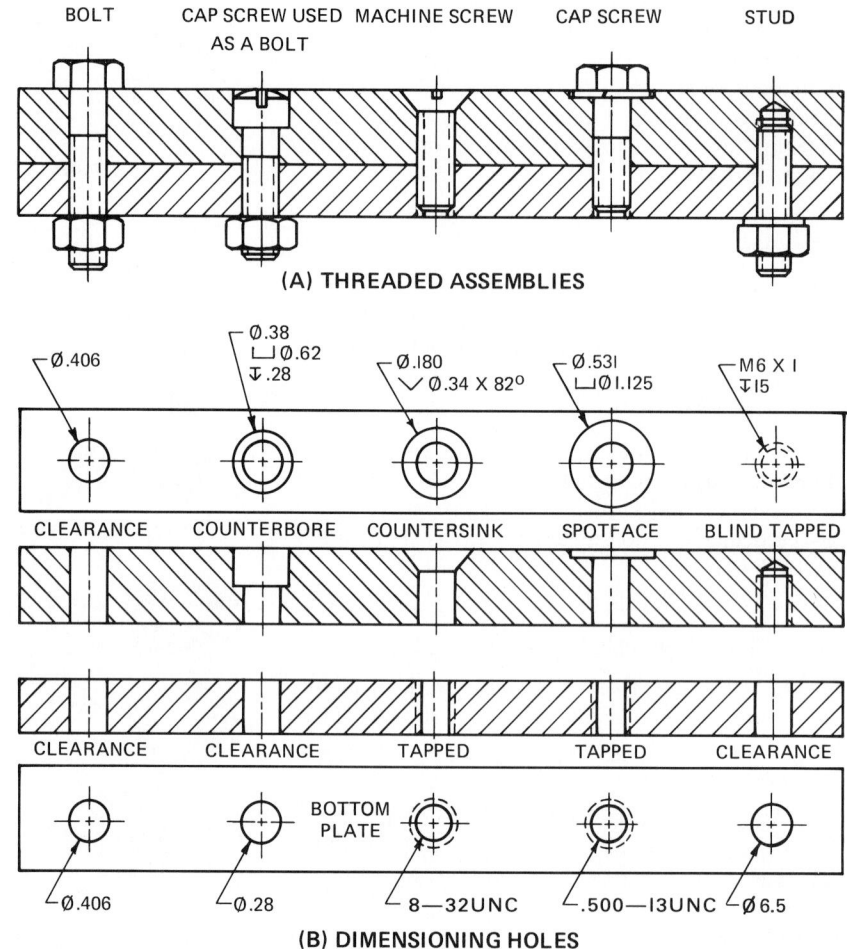

(A) THREADED ASSEMBLIES

(B) DIMENSIONING HOLES

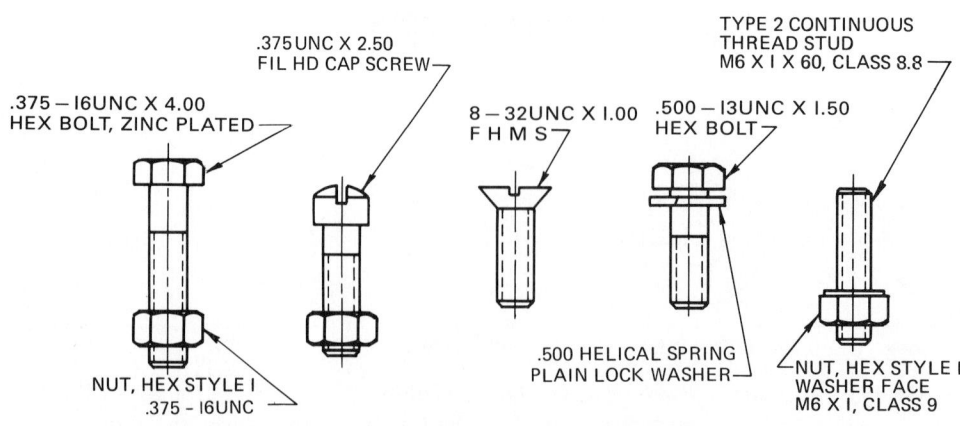

(C) DESCRIPTION OF FASTENERS

the part. A *countersunk hole* is an angular-sided recess that accommodates the shape of a flat-head cap screw or machine screw or an oval-head machine screw. *Spotfacing* is a machine operation that provides a smooth, flat surface where a bolt head or a nut will rest.

Specifying Fasteners

In order for the purchasing department to properly order the fastening device that has been selected in the design, the following information is required. (*Note:* The information listed will not apply to all types of fasteners.)

1. Type of fastener
2. Thread specifications
3. Fastener length
4. Material
5. Head style
6. Type of driving recess
7. Point type (setscrews only)
8. Property class
9. Finish

EXAMPLES

.375—16 UNC—2A × 4.00 HEX BOLT, ZINC PLATED

M10 × 1.5 × 50, 9.8 12-SPLINE FLANGE SCREW, CADMIUM PLATED

TYPE 2 DOUBLE-END STUD, M10 × 1.5 × 100, STEEL CLASS 9.8, CADMIUM PLATED

NUT, HEX, STYLE 1, .500 UNC STEEL

MACH SCREW, PHILLIPS ROUND HD, 8—32 UNC × 1.00, BRASS

WASHER, FLAT 8.4 ID × 17 OD × 2 THK, STEEL HELICAL SPRING

REFERENCES AND SOURCE MATERIAL

1. *Machine Design*, Fastening and joining reference issue.
2. *Design Engineering* and Staff of Stelco's "Fastener Facts."
3. ANSI Y14.6, *Screw Thread Representation.*
4. "Metric Fastener Standards Handbook" by Industrial Fasteners Institute.

ASSIGNMENTS

See Assignments 12 through 17 for Unit 10-3 on pages 287 through 289.

10-4 SPECIAL FASTENERS

Setscrews

Setscrews are used as semipermanent fasteners to hold a collar, sheave, or gear on a shaft against rotational or translational forces. In contrast to most fastening devices, the setscrew is essentially a compression device. Forces developed by the screw point on tightening produce a strong clamping action that resists relative motion between assembled parts. The basic problem in setscrew selection is to find the best combination of setscrew form, size, and point style that provides the required holding power.

Setscrews can be categorized in two ways: by their head style and by the point style desired (Fig. 10-4-1). Each setscrew style is available in any one of five point styles.

STANDARD POINTS	
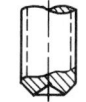	**CUP** Most generally used. Suitable for quick and semi-permanent location of parts on soft shafts, where cutting in of edges of cup shape on shaft is not objectionable.
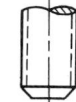	**FLAT** Used where frequent resetting is required, on hard steel shafts, and where minimum damage to shafts is necessary. Flat is usually ground on shaft for better contact.
	CONICAL For setting machine parts permanently on shaft, which should be spotted to receive cone point. Also used as a pivot or hanger.
	SPHERICAL Should be used against shafts spotted, splined, or grooved to receive it. Sometimes substituted for cup point.
	HALF DOG For permanent location of machine parts, although cone point is usually preferred for this purpose. Point should fit closely to diameter of drilled hole in shaft. Sometimes used in place of a dowel pin.
STANDARD HEADS	
	HEXAGON SOCKET Standard size range: No. 0 to 1.0 in. (2 to 24mm), threaded entire length of screw in .06 in. (2mm) increments from .25 to .62 in. (6 to 16mm), .12 in. (3mm) increments from .62 to 1.0 in. (16 to 24 mm). Coarse or fine thread series.
	SLOTTED Standard size range: No. 5 to .75 in. (3 to 20mm) threaded entire length of screw. Coarse or fine thread series.
	FLUTED SOCKET Same as hexagon socket. No. 0 and 1 (2 and 3mm) have four flutes. All others have six flutes.
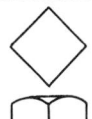	**SQUARE HEAD** Standard size range: No. 10 to 1.50 in. (5 to 36mm). Entire body is threaded. Coarse or fine thread series. Sizes .25 in. (6mm) and larger are normally available in coarse threads only.

FIG. 10-4-1   Setscrews.

The conventional approach to selecting the setscrew diameter is to make it roughly equal to one-half the shaft diameter. This rule of thumb often gives satisfactory results, but its range of usefulness is limited.

Setscrews and Keyseats

When a setscrew is used in combination with a key, the screw diameter should be equal to the width of the key. In this combination the setscrew is locating the parts in an axial direction only. The torsional load on the parts is carried by the key.

The key should be tight-fitting, so that no motion is transmitted to the screw. Key design is covered in Chap. 11.

Keeping Fasteners Tight

Fasteners are inexpensive, but the cost of installing them can be substantial. Probably the simplest way to cut assembly costs is to make sure that, once installed, fasteners stay tight.

The American National Standards Institute has identified three basic locking methods: free-spinning, prevailing-torque, and chemical locking. Each has its own advantages and disadvantages (Fig. 10-4-2).

Free-spinning devices include toothed and spring lockwashers and screws and bolts with washerlike heads. With these arrangements, the fasteners spin free in the clamping direction, which makes them easy to assemble, and the break-loose torque is greater than the seating torque. However, once break-loose torque is exceeded, free-spinning washers have no prevailing torque to prevent further loosening.

Prevailing torque methods make use of increased friction between nut and bolt. Metallic types usually have deformed threads or contoured thread profiles that jam the threads on assembly. Non-metallic types make use of nylon or polyester insert elements that produce interference fits on assembly.

Chemical locking is achieved by coating the fastener with an adhesive.

Locknuts

A *locknut* is a nut with special internal means for gripping a threaded fastener to prevent rotation. It usually has the dimensions, mechanical requirements, and other specifications of a standard nut, but with a locking feature added.

Locknuts are divided into three general classifications: prevailing-torque, free-spinning, and other types. These are shown in Figs. 10-4-3 and 10-4-4.

Prevailing-Torque Locknuts

Prevailing-torque locknuts spin freely for a few turns, and then must be wrenched to final position. The maximum holding and locking power is reached as soon as the threads and the locking feature are engaged. Locking action is maintained until the nut is removed. Prevailing-torque locknuts are classified by basic design principles:

1. Thread deflection causes friction to develop when the threads are mated; thus the nut resists loosening.
2. The out-of-round top portion of the tapped nut grips the bolt threads and resists rotation.

 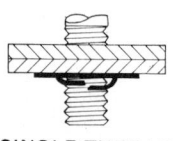

TOOTHED WASHER   SINGLE-THREAD LOCKNUT GRIP SCREW

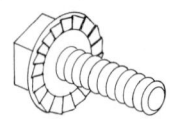 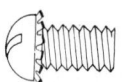

SERRATED TOOTH   PREASSEMBLED WASHER AND SCREW

(A) FREE-SPINNING

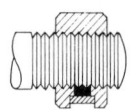 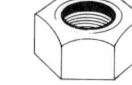

NYLON PLUG FOR WEDGING ACTION   NONMETALLIC PLUG GRIPS BOLT THREADS

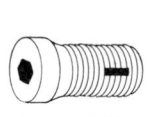 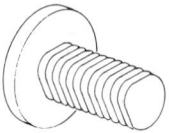

STRIP INSERT   THREAD DEFORMATION

(B) PREVAILING-TORQUE

FIG. 10-4-2   Basic locking methods for setscrews.

3. The slotted section of the locknut is pressed inward to provide a spring frictional grip on the bolt.
4. Inserts, either nonmetallic or of soft metal, are plastically deformed by the bolt threads to produce a frictional interference fit.
5. A spring wire or pin engages the bolt threads to produce a wedging or ratchet-locking action.

Free-Spinning Locknuts

Free-spinning locknuts are free to spin on the bolt until seated. Additional tightening locks the nut.

Since most free-spinning locknuts depend on clamping force for their locking action, they are usually not recommended for joints that might relax through plastic deformation or for fastening materials that might crack or crumble.

Other Locknut Types

Jam nuts are thin nuts used under full-sized nuts to develop locking action. The large nut has sufficient strength to elastically deform the lead threads of the bolt and jam nut. Thus, a considerable resistance against loosening is built up. The use of jam nuts is decreasing; a one-piece, prevailing-torque locknut usually is used instead at a savings in assembled cost.

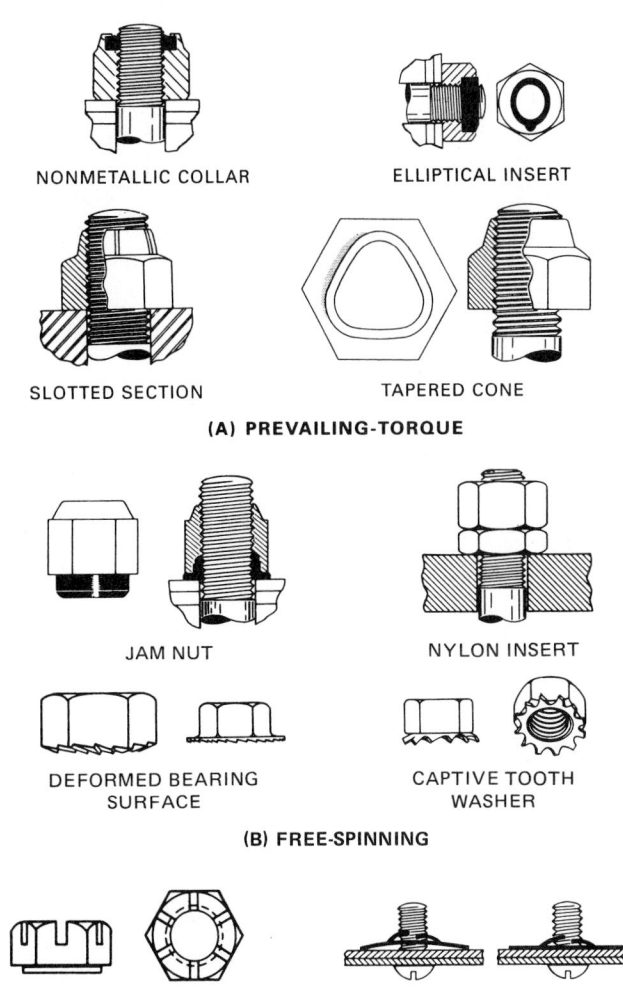

NONMETALLIC COLLAR

ELLIPTICAL INSERT

SLOTTED SECTION

TAPERED CONE

(A) PREVAILING-TORQUE

JAM NUT

NYLON INSERT

DEFORMED BEARING SURFACE

CAPTIVE TOOTH WASHER

(B) FREE-SPINNING

SLOTTED NUT AND COTTER PIN

SINGLE THREAD

(C) OTHER TYPES

FIG. 10-4-3   Locknuts.

Slotted and castle nuts have slots that receive a cotter pin that passes through a drilled hole in the bolt and thus serves as the locking member. Castle nuts differ from slotted nuts in that they have a circular crown of a reduced diameter.

Single-thread locknuts are spring steel fasteners which may be speedily applied. Locking action is provided by the grip of the thread-engaging prongs and the reaction of the arched base. Their use is limited to nonstructural assemblies and usually to screw sizes below 6 mm in diameter (Fig. 10-4-5, pg. 280).

Captive or Self-Retaining Nuts

Captive or self-retaining nuts provide a permanent, strong, multiple-thread fastener for use on thin materials (Fig. 10-4-6, pg. 280). They are especially good where there are blind locations, and they can normally be attached without damaging finishes. Methods of attaching these types of nuts vary and tools required for assembly are generally uncomplicated and inexpensive. The self-retained nuts are grouped according to four means of attachment.

1. Plate or anchor nuts: These nuts have mounting lugs that can be riveted, welded, or screwed to the part.
2. Caged nuts: A spring-steel cage retains a standard nut. The cage snaps into a hole or clips over an edge to hold the nut in position.
3. Clinch nuts: They are specially designed nuts with pilot collars that are clinched or staked into the parent part through a precut hole.
4. Self-piercing nuts: A form of clinch nut that cuts its own hole.

Inserts

Inserts are a special form of nut designed to serve the function of a tapped hole in blind or through-hole locations (Fig. 10-4-7, pg. 280).

FIG. 10-4-4   Single-thread engaging nuts.

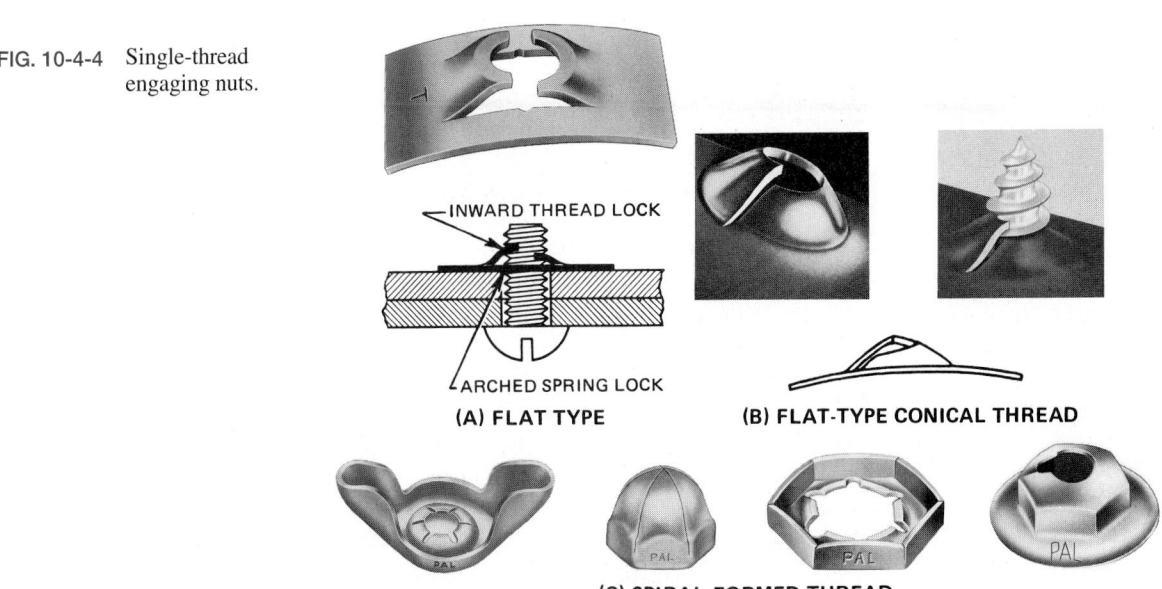

INWARD THREAD LOCK

ARCHED SPRING LOCK

(A) FLAT TYPE

(B) FLAT-TYPE CONICAL THREAD

(C) SPIRAL-FORMED THREAD

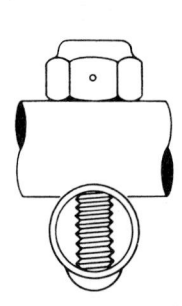

USE OF LOCKNUT FOR
TUBULAR FASTENING

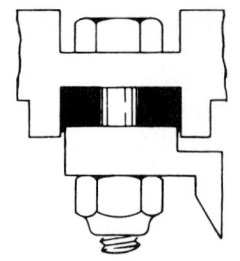

FOR RUBBER-INSULATED
AND CUSHION MOUNT-
INGS WHERE THE NUT
MUST REMAIN STATION-
ARY

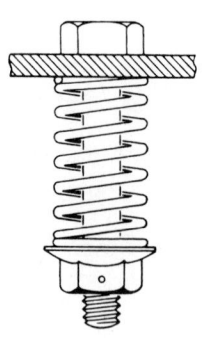

USE OF LOCKNUT ON
A SPRING CLAMP

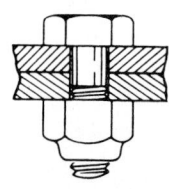

USE OF LOCKNUT WHERE
ASSEMBLY IS SUBJECTED
TO VIBRATORY OR CYCLIC
MOTIONS THAT COULD
CAUSE LOOSENING

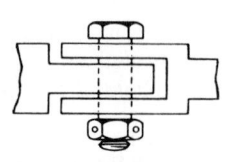

USE OF LOCKNUT ON A
BOLTED CONNECTION
THAT REQUIRES PRE-
DETERMINED PLAY

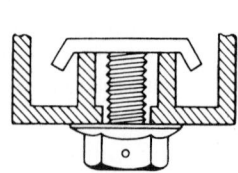

FOR AN EXTRUDED PART
ASSEMBLY

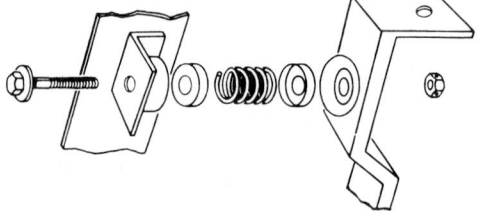

FOR HOLDING A MOTOR MOUNTING
SECURELY IN POSITION

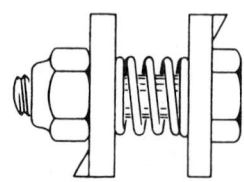

FOR SPRING-MOUNTED
CONNECTIONS WHERE
THE NUT MUST REMAIN
STATIONARY OR IS SUB-
JECT TO ADJUSTMENT

FIG. 10-4-5 Typical locknut applications.

(A) PLATE NUT

(B) CAGED NUT

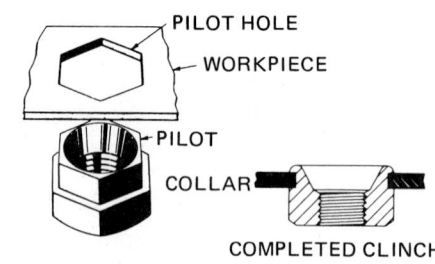

(C) CLINCH NUT

(I) UNIVERSAL PIERCE NUT

(2) HIGH-STRESS PIERCE NUT

(D) PIERCE NUTS

FIG. 10-4-6 Captive or self-retaining nuts.

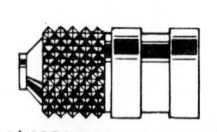

(A) MOLDED-IN INSERT

(B) SELF-TAPPING INSERT

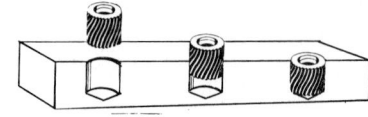

(C) PRESSED-IN INSERT

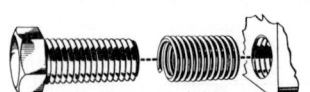

**(D) EXTERNAL-INTERNAL
THREADED INSERT**

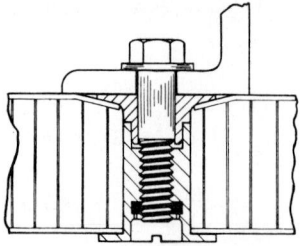

(E) SANDWICH PANEL INSERT

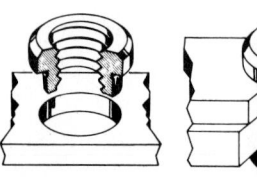

(F) THIN MATERIAL INSERT

FIG. 10-4-7 Inserts.

Sealing Fasteners

Fasteners hold two or more parts together, but they can perform other functions as well. One important auxiliary function is that of sealing gases and liquids against leakage.

Two types of sealed-joint construction are possible with fasteners (Fig. 10-4-8). In one approach, the fasteners enter the sealed medium and are separately sealed. The second approach uses a separate sealing element that is held in place by the clamping forces produced by conventional fasteners, such as rivets or bolts.

There are many methods of obtaining a seal using sealing fasteners, as shown in Fig. 10-4-9.

REFERENCES AND SOURCE MATERIAL

1. *Machine Design,* Fastening and joining reference issue.

ASSIGNMENTS

See Assignments 18 through 20 for Unit 10-4 on page 290.

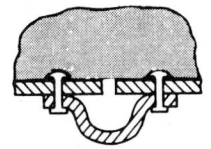 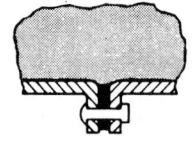

(A) FASTENERS SEPARATELY SEALED   **(B) SEALING ELEMENT CLAMPED IN PLACE**

FIG. 10-4-8   Types of sealed-joint construction.

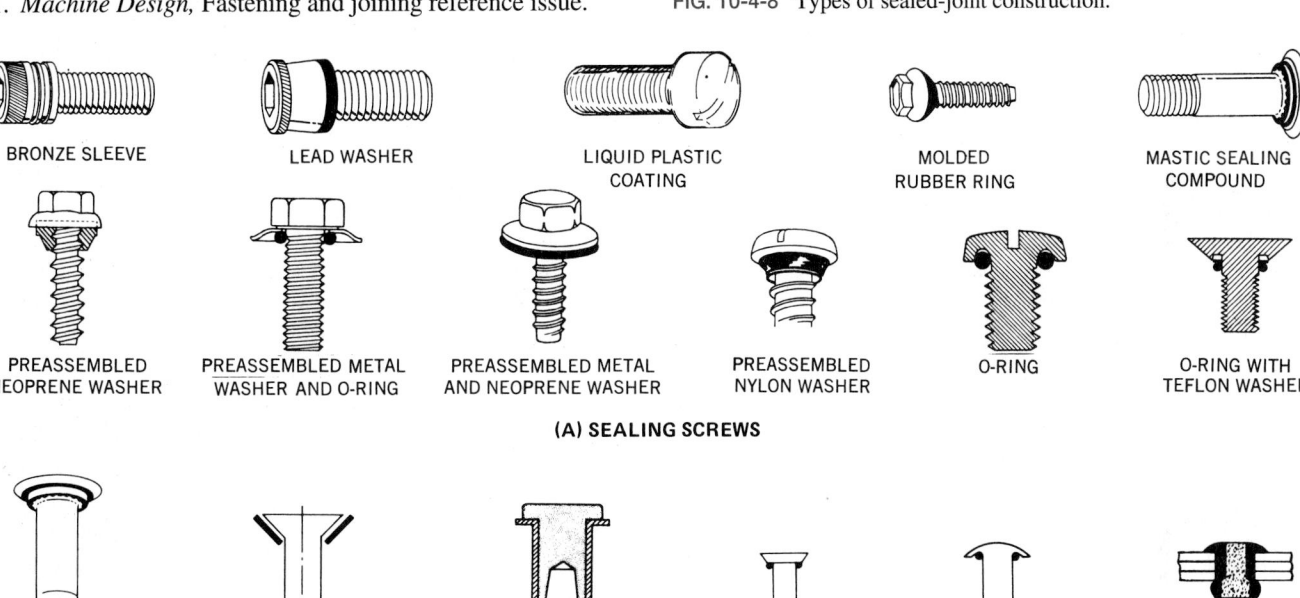

BRONZE SLEEVE LEAD WASHER LIQUID PLASTIC COATING MOLDED RUBBER RING MASTIC SEALING COMPOUND

PREASSEMBLED NEOPRENE WASHER PREASSEMBLED METAL WASHER AND O-RING PREASSEMBLED METAL AND NEOPRENE WASHER PREASSEMBLED NYLON WASHER O-RING O-RING WITH TEFLON WASHER

(A) SEALING SCREWS

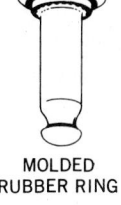  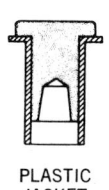 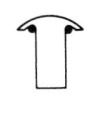 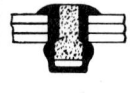

MOLDED RUBBER RING SOFT–ALUMINUM WASHER PLASTIC JACKET O-RING O-RING INTERFERENCE FIT

(B) SEALING RIVETS

 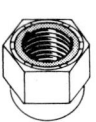  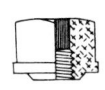  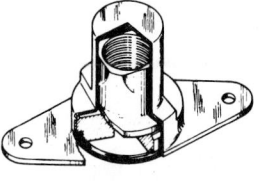

NYLON PELLET COPPER INSERT NYLON COLLAR NYLON BODY FLOWED-IN SEALANT MOLDED RUBBER GASKET OR O-RING

(C) SEALING NUTS

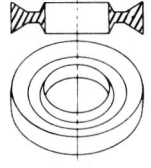   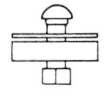  

MOLDED NYLON– SEAL RING MOLDED RUBBER TOROID LAMINATED NEOPRENE TO METAL NYLON SLEEVE O-RING FLOWED-IN SEALANT

(D) SEALING WASHERS

FIG. 10-4-9   Sealing fasteners.

10-5 FASTENERS FOR LIGHT-GAGE METAL, PLASTIC, AND WOOD

Tapping Screws

Tapping screws cut or form a mating thread when driven into drilled or cored holes. These one-piece fasteners permit rapid installation, since nuts are not used and access is required from only one side of the joint. The mating thread produced by the tapping screw fits the screw threads closely, and no clearance is necessary. This close fit usually keeps the screws tight, even under vibrating conditions (Fig. 10-5-1).

Tapping screws are practically all case-hardened and, therefore, can be driven tight and have a relatively high ultimate torsional strength. The screws are used in steel, aluminum (cast, extruded, rolled, or die-formed) die castings, cast iron, forgings, plastics, reinforced plastics, asbestos, and resin-impregnated plywood (Fig. 10-5-2). Coarse threads should be used with weak materials.

Self-drilling tapping screws have special points for drilling and then tapping their own holes (Fig. 10-5-3). These eliminate drilling or punching, but they must be driven by a power screwdriver.

Special Tapping Screws

Typical special tapping screws are the self-captive screws and double-thread combinations for limited drive. Self-captive screws combine a coarse-pitched starting thread (similar to type B) with a finer pitch (machine-screw thread) farther along the screw shank.

Sealing tapping screws, with pre-assembled washers or O-rings (Fig. 10-5-4B) are available in a variety of styles.

TYPE AB   TYPE B   TYPE F   TYPE U

FIG. 10-5-1 Self-tapping screws.

HEAVY GAGE SHEET METAL AND STRUCTURAL STEEL USE TYPES B, U, F.	LIGHT GAGE SHEET METAL USE TYPES AB, B.
  Holes may be drilled or clean-punched. Two parts may have pierced holes to nest burrs. This results in a stronger joint. 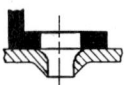 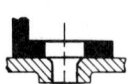 Use a pierced hole in workpiece if clearance hole is needed in part to be fastened. Extruded hole may also be used in workpiece if clearance hole is needed in fastened part.	  Holes may be drilled or clean-punched the same size in both sheet metal parts. For thicker sheet metal and structural steel, a clearance hole should be provided in the part to be fastened. Hole size depends on thickness of the workpiece. Notes: I. Use hex-head on type B screws. 2. With type U screws, material should be thick enough to permit sufficient thread engagement—at least one screw diameter.
PLASTICS USE TYPES B, U, F.	CASTINGS AND FORGINGS USE TYPES B, U, F.
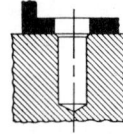 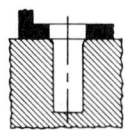 Screw holes may be molded or drilled. If material is brittle or friable, molded holes should be formed with a rounded chamfer, and drilled holes should be machine-chamfered. Provide a clearance in the part to be fastened. Depth of penetration should be held within the "minimum and maximum" limits recommended. The hole should be deeper than the screw penetration to allow for chip clearance.	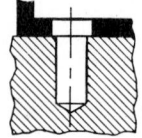 Holes may be cored if it is practical to maintain close tolerances. Otherwise blind-drill holes to recommended hole size. Provide a clearance hole for screw in the part to be fastened. The hole in the casting, if it is a blind hole, should be deeper than the screw penetration to allow for chip clearance. Notes: I. Hole in fastened part may be the same size as workpiece hole for type U screws. 2. Type B is suitable for use only in nonferrous castings.

FIG. 10-5-2 Tapping-screw application chart.

REFERENCES AND SOURCE MATERIAL

1. *Machine Design,* Fastening and joining reference issue.

ASSIGNMENTS ▨▨▨▨▨▨▨▨▨▨▨▨▨▨▨

See Assignments 21 through 23 for Unit 10-5 on page 291.

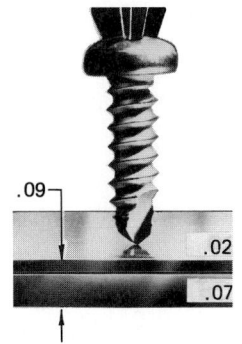
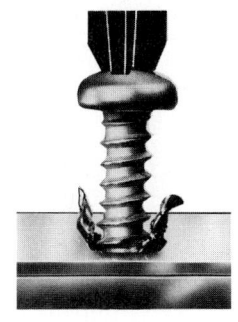

FIG. 10-5-3   Self-drilling tapping screws.

FIG. 10-5-4   Special tapping
screws.

(A) TAPPING SCREWS WITH PREASSEMBLED WASHERS

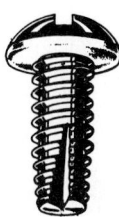

(B) TAPPING SCREWS WITH PREASSEMBLED SEALING WASHERS OR COMPOUNDS

ASSIGNMENTS FOR CHAPTER 10

ASSIGNMENTS FOR UNIT 10-1, SIMPLIFIED THREAD REPRESENTATION

1. Make a working drawing of the gear box shown in Fig. 10-1-A. Draw the top, front, a full section right-side view, and an auxiliary view showing the surface with the tapped holes. Scale 1:1.

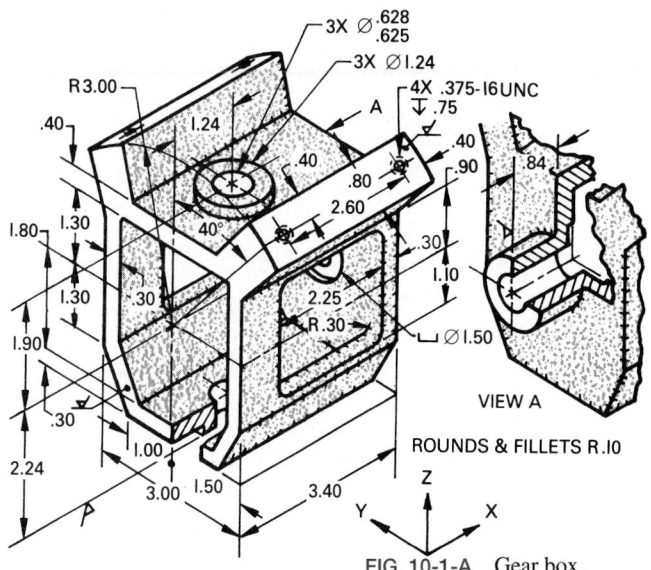

FIG. 10-1-A   Gear box.

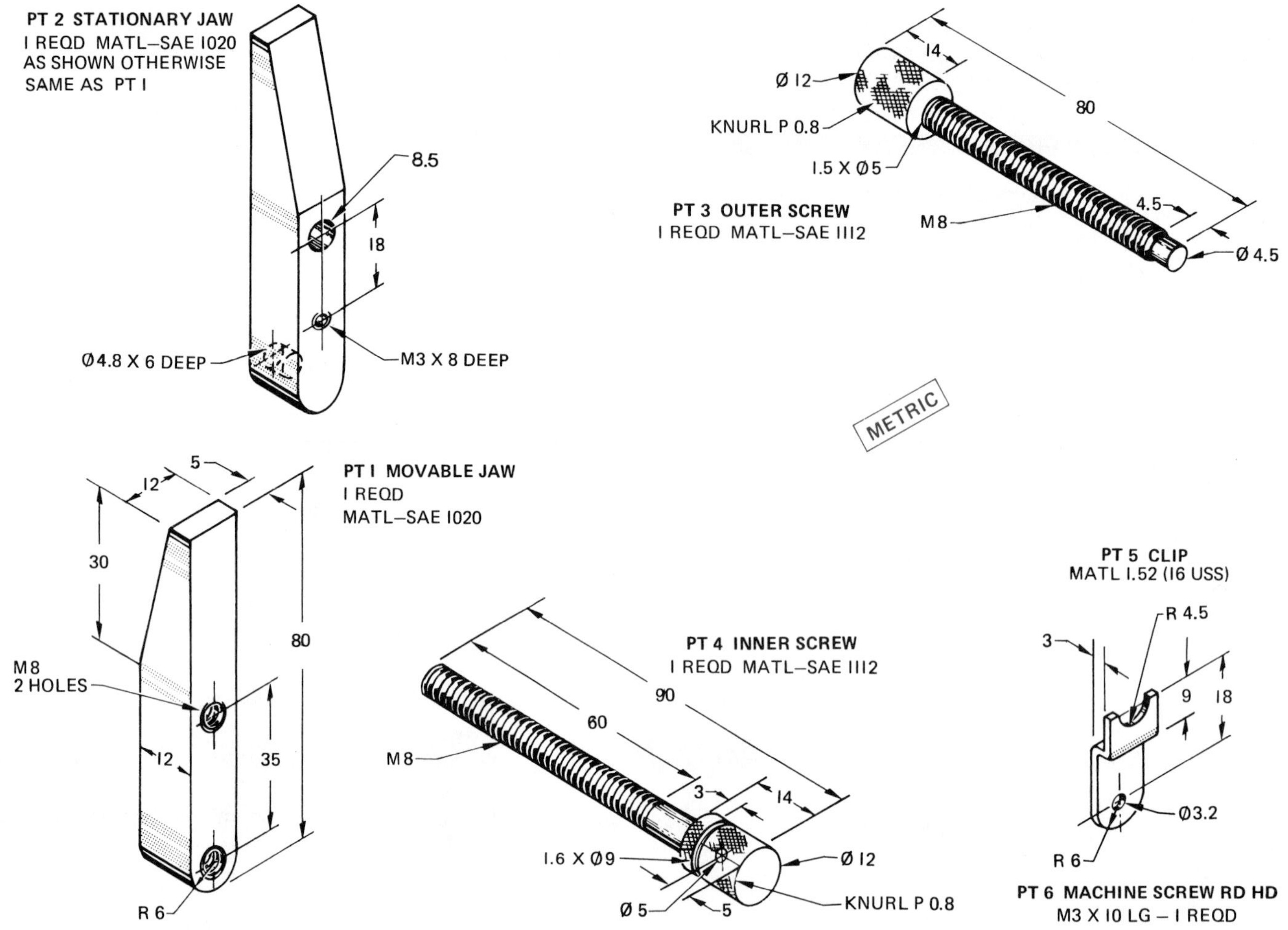

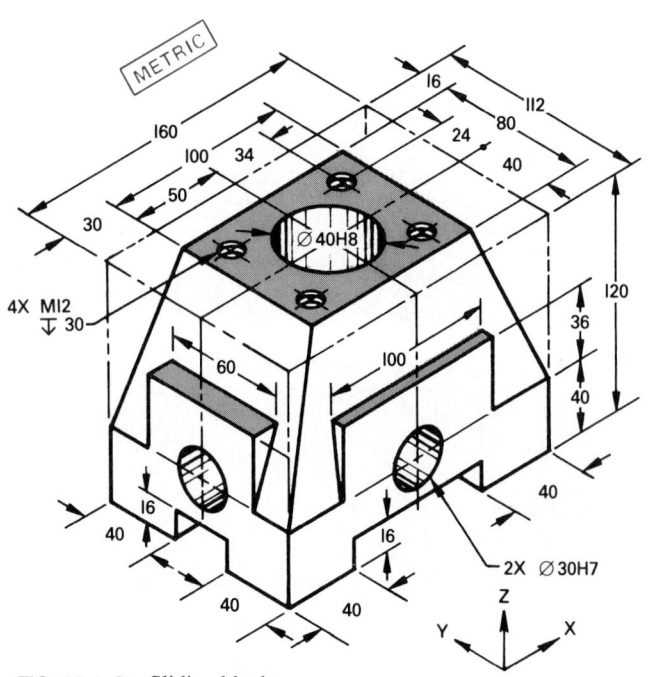

PT 2 STATIONARY JAW
I REQD MATL—SAE 1020
AS SHOWN OTHERWISE
SAME AS PT I

8.5

18

Ø4.8 X 6 DEEP

M3 X 8 DEEP

PT 3 OUTER SCREW
I REQD MATL—SAE 1112

Ø 12

KNURL P 0.8

1.5 X Ø 5

14

80

4.5

M8

Ø 4.5

METRIC

PT I MOVABLE JAW
I REQD
MATL—SAE 1020

5

12

30

80

M8
2 HOLES

12

35

R 6

PT 5 CLIP
MATL 1.52 (16 USS)

R 4.5

3

9   18

Ø3.2

R 6

PT 6 MACHINE SCREW RD HD
M3 X 10 LG — I REQD

PT 4 INNER SCREW
I REQD MATL—SAE 1112

90

60

M8

3

14

1.6 X Ø9

Ø 12

Ø 5

5

KNURL P 0.8

FIG. 10-1-B   Parallel clamps.

METRIC

16

112

160

100   34

80

24

50

30

40

Ø 40H8

120

36

4X M12
⤓ 30

60

100

40

16

40

16

2X Ø30H7

40

40

40

Z

Y   X

FIG. 10-1-C   Sliding block.

2. Make a two-view assembly drawing of the parallel clamps shown in Fig. 10-1-B. Use simplified thread conventions and include an item list calling for all the parts. The only dimension required on the drawing is the maximum opening of the jaws. Identify the parts on the assembly. Scale 1:1.

3. Make detail drawings of the parts shown in Fig. 10-1-B. Scale 1:1. Use your judgment for the number of views required for each part.

4. Make a working drawing of the sliding block shown in Fig. 10-1-C. Show the top, front, and right-side views. Use simplified thread representation. Use limit dimensions where fit symbols are shown. Scale 1:2.

5. Make a three-view detail drawing of the terminal block shown in Fig. 10-1-D. Use tabular dimensioning for locating the holes from the X, Y, and Z axes. The origin for the X and Y axes will be the center of the Ø4.80 hole. The origin for the Z axis will be the bottom of the part. Scale 1:2.

6. Make a detail drawing of the guide block shown in Fig. 10-1-E. Draw the top, front, and left-side views. Scale 1:1.
7. Make a one-view assembly drawing of the turnbuckle shown in Fig. 10-1-F. Show the assembly in its shortest length and also include the maximum position shown in phantom lines. The only dimensions required are the minimum and maximum distances between the eye centers. Scale 1:1.
8. Make detail drawings of the parts shown in Fig. 10-1-F. Scale 1:1.

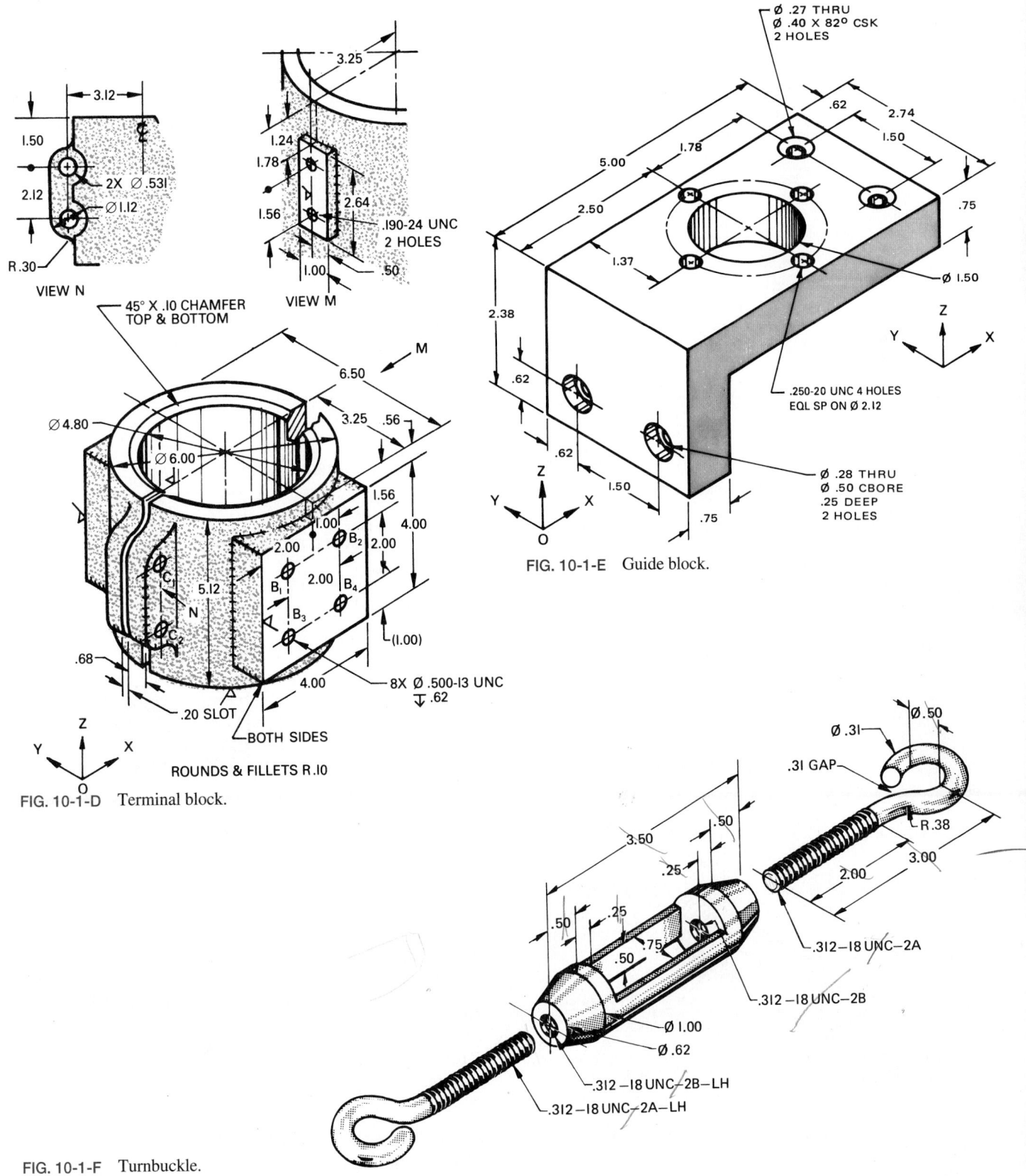

VIEW N

VIEW M

FIG. 10-1-D   Terminal block.

FIG. 10-1-E   Guide block.

FIG. 10-1-F   Turnbuckle.

ASSIGNMENTS FOR UNIT 10-2, DETAILED AND SCHEMATIC THREAD REPRESENTATION

9. Lay out the parts as shown in either Fig. 10-2-A or 10-2-B and draw the threads in detailed representation. The end rods are to be drawn in section. Scale 1:1.

10. Make one-view drawings of the parts shown in Figs. 10-2-C and 10-2-D using detailed thread representation. Use a conventional break to shorten the length of each part. Scale 1:1.

11. Make a one-view drawing of one of the parts shown in Fig. 10-2-E or 10-2-F using detailed thread representation. Scale 1:1 for Fig. 10-2-E and scale 2:1 for Fig. 10-2-F.

FIG. 10-2-A   Connector and supports.

FIG. 10-2-B   Connector and supports.

FIG. 10-2-C   Guide rod.

FIG. 10-2-D   Jack screw.

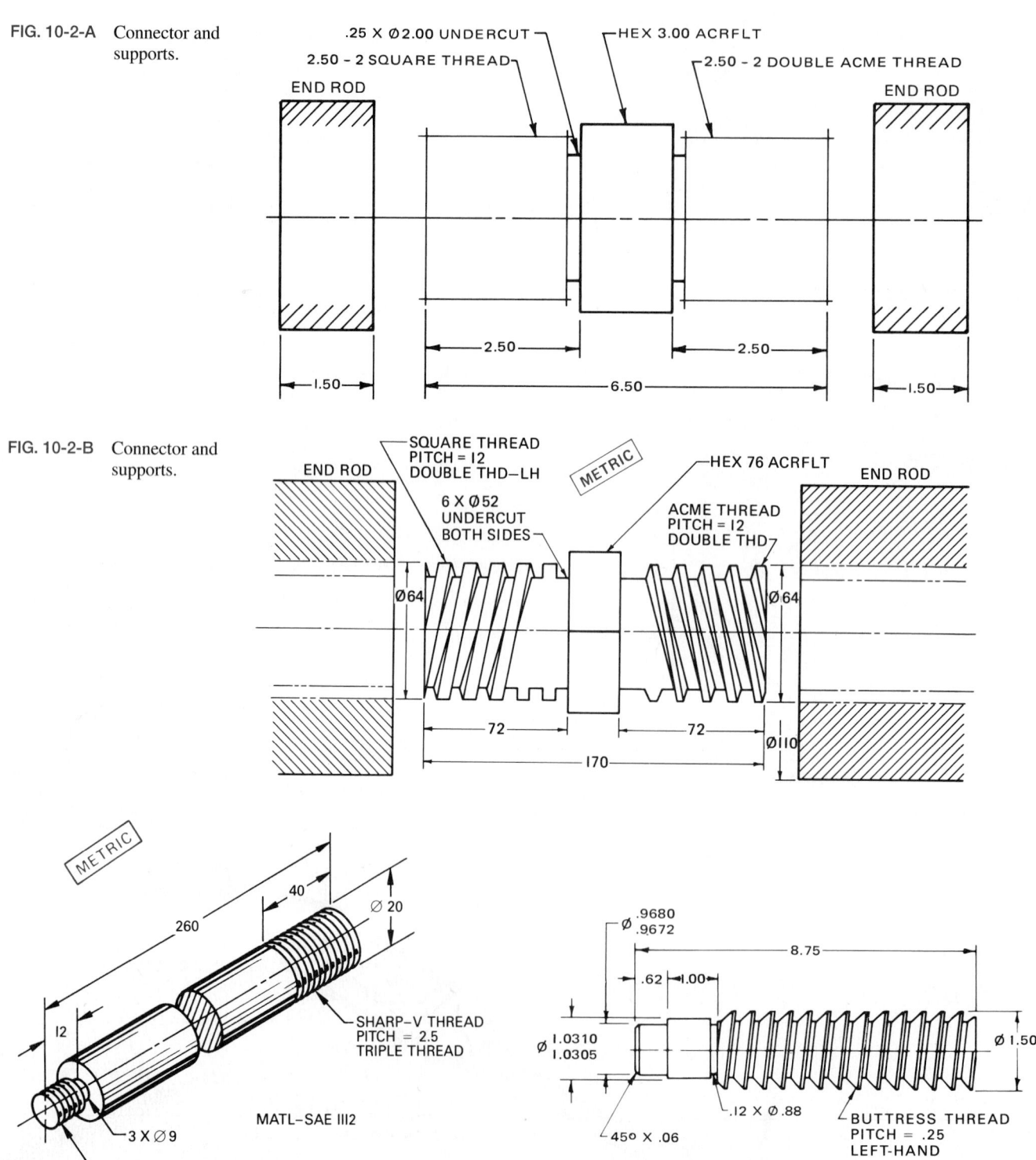

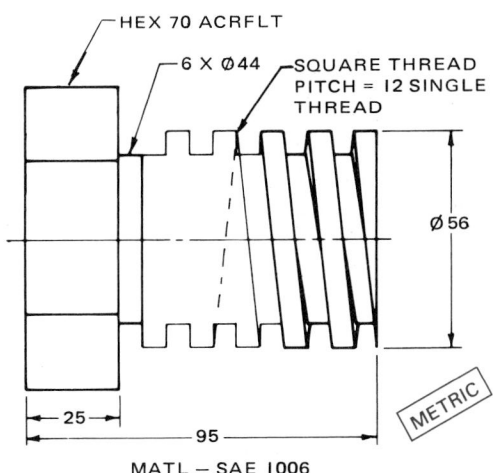

FIG. 10-2-E Plug.

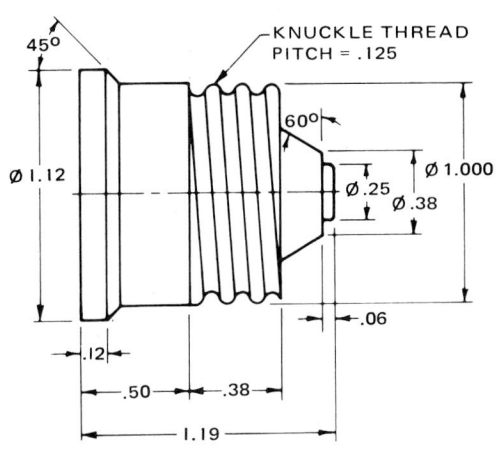

FIG. 10-2-F Fuse.

ASSIGNMENTS FOR UNIT 10-3, COMMON THREADED FASTENERS

12. Prepare a full-section assembly drawing of the four fastener assemblies shown in Fig. 10-3-A. Dimension both the clearance and threaded holes. A top view may be shown if required. Scale 1:1.

13. Prepare full-section assembly drawings of the four fastener assemblies shown in Fig. 10-3-B. Dimension both the clearance and threaded holes. A top view of the fastener may be shown if required. Scale 1:1.

FIG. 10-3-A Threaded fasteners.

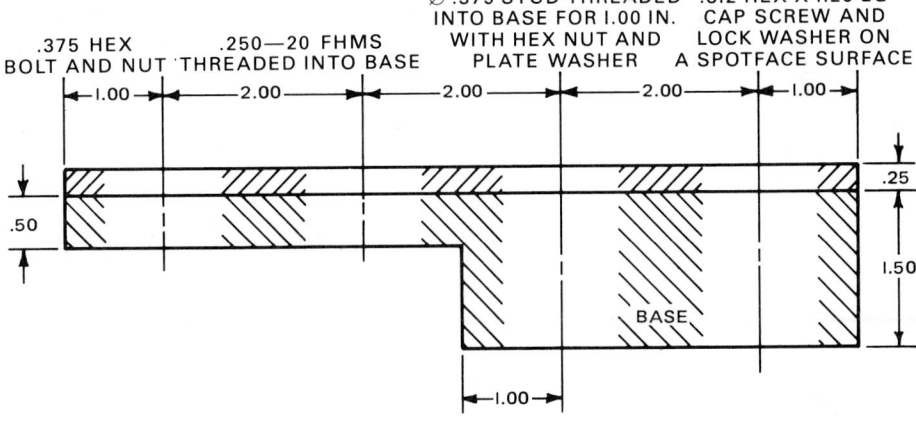

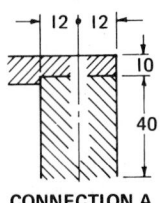

CONNECTION A
M 10 X 30 LG
HEX HD CAP SCREW

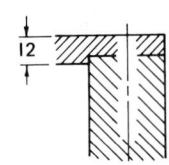

CONNECTION B
M 10 X 40 LG STUD
THREAD EACH END 20 LG
HEX NUT STYLE I AND
SPRING LOCK WASHER

CONNECTION C
M 10 X 30 LG
FL HD CAP SCREW

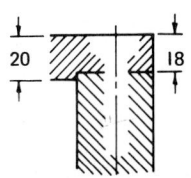

CONNECTION D
M 10 X 1.25 X 25 LG
SOCKET HEAD CAP SCREW
AND SPRING LOCK WASHER

FIG. 10-3-B Threaded fasteners.

287

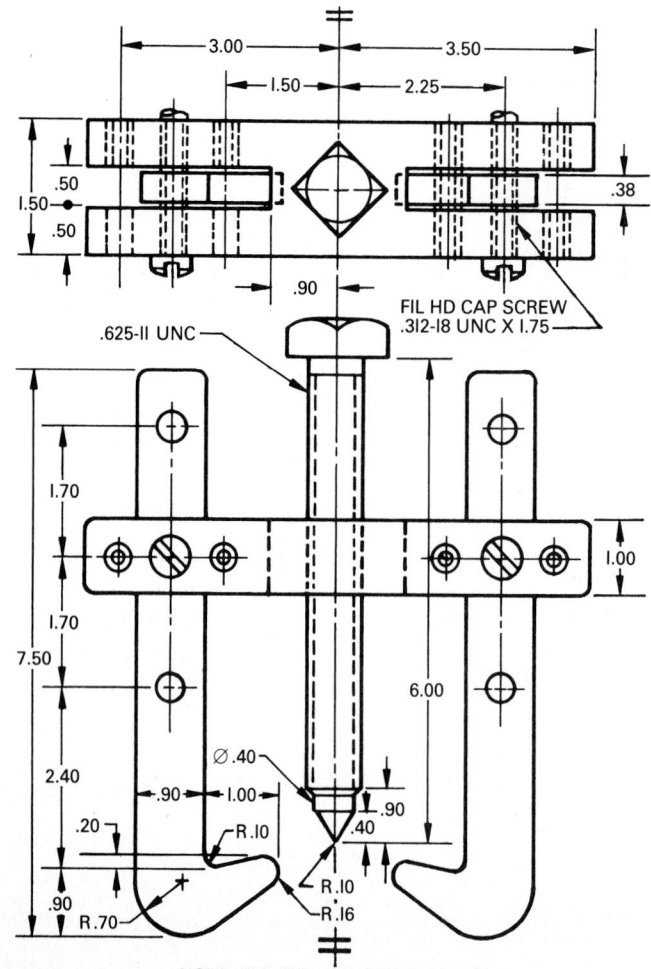

FIG. 10-3-C Wheel-puller assembly.

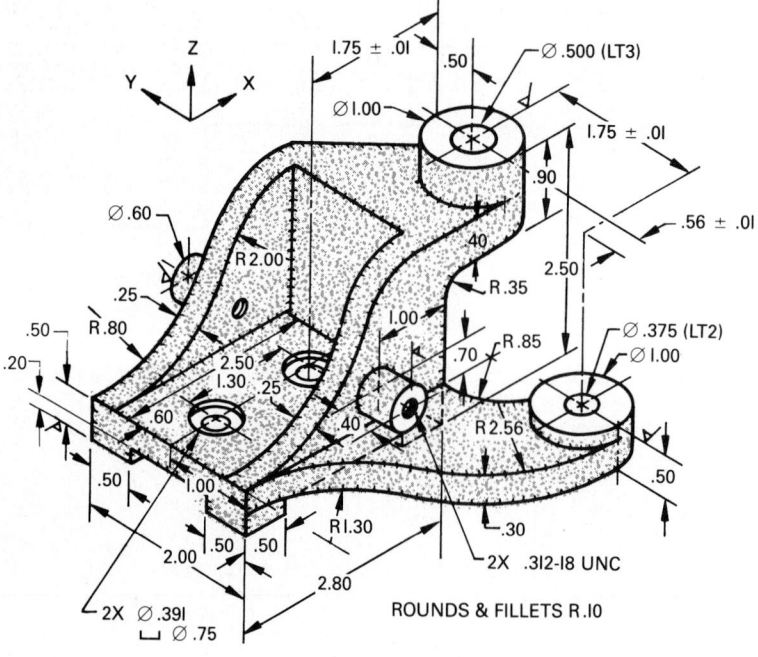

FIG. 10-3-D Shaft intermediate support.

14. Make a two-view assembly drawing of the wheel-puller shown in Fig. 10-3-C. Use simplified thread representation and include an item list calling for all the parts. Scale 1:1.

15. Make detail drawings of the parts shown in Fig. 10-3-C. Use your judgment for the selection and number of views required for each part. Scale 1:1.

16. Make a working drawing of the shaft intermediate support shown in Fig. 10-3-D. Show the top, front, and left-side views. Surfaces shown √ to have a maximum roughness value of 250 μin. and a machining allowance of .04 in. Show the limits of size for the Ø.500 and Ø.375 holes. Scale 1:1.

17. Make a working drawing of the base in Fig. 10-3-E. Show the limits of size for the Ø15 and Ø18 holes. Surfaces shown √ to have a maximum roughness value of 3.2 μm and a machining allowance of 2 mm. Scale 1:1.

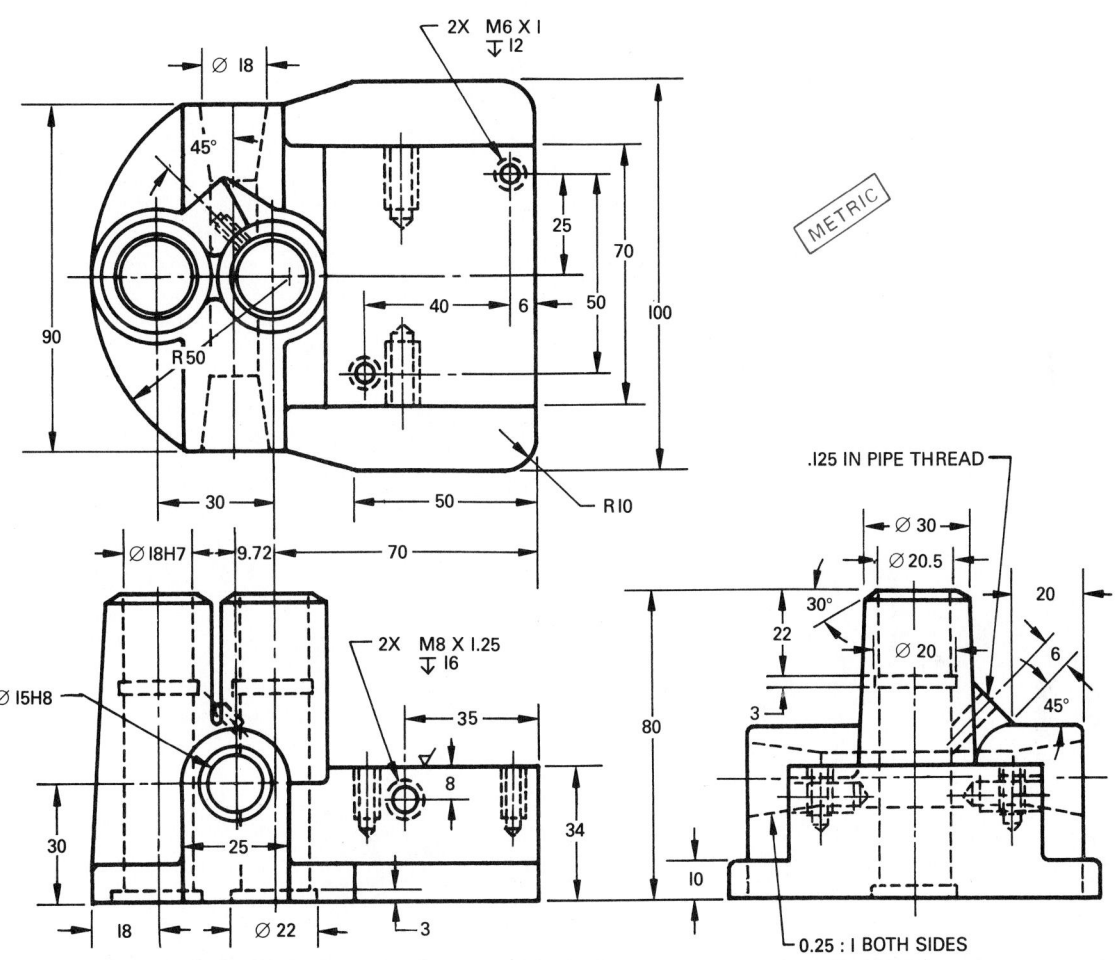

FIG. 10-3-E   Base.

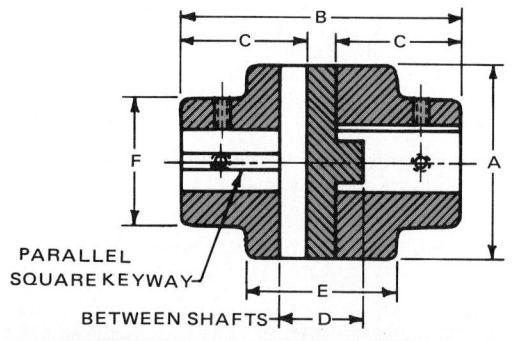

PARALLEL
SQUARE KEYWAY

BETWEEN SHAFTS

MAXIMUM BORE	A	B	C	D	E	F
.9375	3.00	3.75	1.75	.88	1.50	2.38
1.1875	3.50	4.69	2.19	1.06	1.81	2.75
1.4375	4.00	5.62	2.62	1.25	2.12	3.12
1.6875	5.00	6.56	3.06	1.44	2.44	3.50
1.9375	5.50	7.50	3.50	1.50	2.50	4.00
2.1875	6.00	8.44	3.94	1.81	3.06	4.38

DIMENSIONS SHOWN ARE IN INCHES

ASSIGNMENTS FOR UNIT 10-4, SPECIAL FASTENERS

18. Make a one-view assembly drawing of the flexible coupling shown in Fig. 10-4-A. The shafts, which are coupled, are 1.50 in. in diameter and are to be shown in the assembly. They are to extend beyond the coupling for approximately 2.00 in. and end with a conventional break. Show the setscrews and keys in position. Scale 1:1.

FIG. 10-4-A   Flexible coupling.

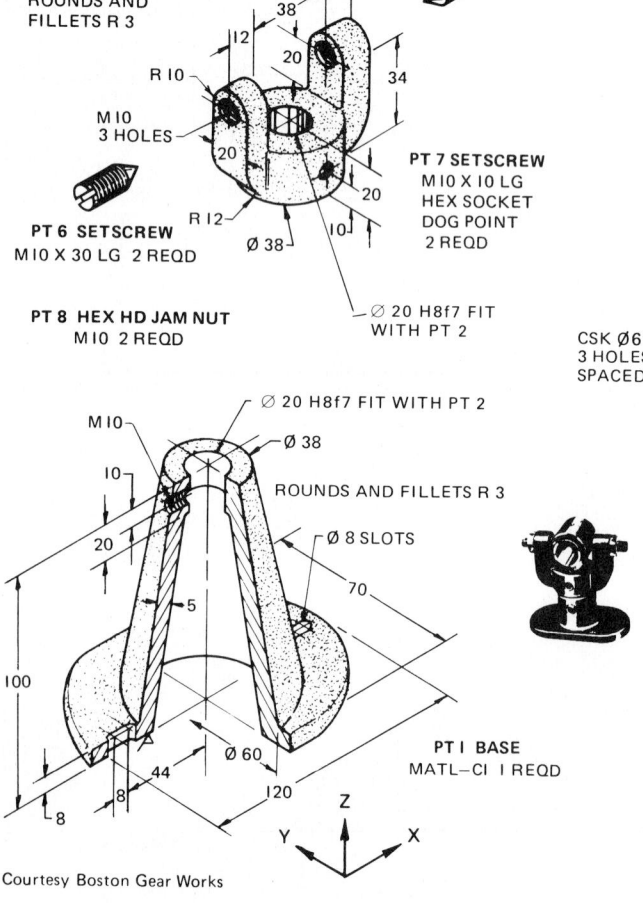

PT 3 YOKE
MATL—CI I REQD
ROUNDS AND
FILLETS R 3

R 10

M 10
3 HOLES

PT 6 SETSCREW
M10 X 30 LG 2 REQD

PT 8 HEX HD JAM NUT
M10 2 REQD

PT 7 SETSCREW
M10 X 10 LG
HEX SOCKET
DOG POINT
2 REQD

Ø 20 H8f7 FIT
WITH PT 2

M10

Ø 20 H8f7 FIT WITH PT 2

Ø 38

ROUNDS AND FILLETS R 3

Ø 8 SLOTS

PT I BASE
MATL—CI I REQD

Courtesy Boston Gear Works

FIG. 10-4-B   Adjustable shaft support. (Boston Gear Works)

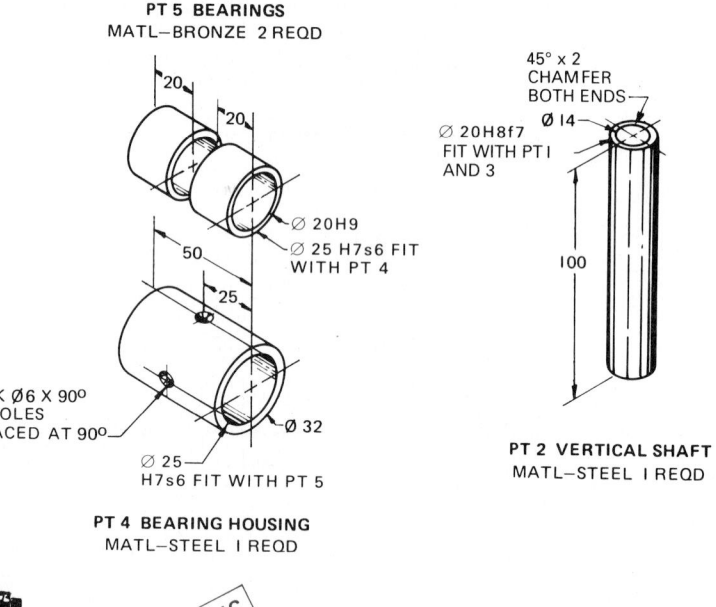

PT 5 BEARINGS
MATL—BRONZE 2 REQD

Ø 20H9

Ø 25 H7s6 FIT
WITH PT 4

CSK Ø6 X 90°
3 HOLES
SPACED AT 90°

Ø 25
H7s6 FIT WITH PT 5

Ø 32

PT 4 BEARING HOUSING
MATL—STEEL I REQD

METRIC

45° x 2
CHAMFER
BOTH ENDS

Ø 14

Ø 20H8f7
FIT WITH PT I
AND 3

PT 2 VERTICAL SHAFT
MATL—STEEL I REQD

19. Make a one-view assembly drawing of the adjustable shaft support shown in Fig. 10-4-B. Show the base in full section. A broken-out section is recommended to clearly show the setscrews in the yoke. Add part numbers to the assembly drawing and include an item list. Do not dimension. Scale 1:1.

20. Make detail drawings for the parts shown in Fig. 10-4-B. Use your judgment for the number of views required for each part. Show the limits of size for the holes. Scale 1:1.

ASSIGNMENTS FOR UNIT 10-5, FASTENERS FOR LIGHT-GAGE METAL, PLASTIC, AND WOOD

21. Make a drawing of the assemblies shown in Fig. 10-5-A. Either inch or metric fasteners may be used. The steel post is fastened to the panel by two rows of tapping screws. The steel strap is held to the post by a single tapping screw which has the equivalent strength (body area) of at least three of the other tapping screws. Dimension the holes and fastener sizes. Scale to suit.

22. Make a two-view assembly drawing of the woodworking vise shown in Fig. 10-5-B. Have the opening between jaws 1.50 in. Include on the drawing an item list calling for all the parts. Scale 1:2.

23. Make detail drawings of the parts shown in Fig. 10-5-B. Use your judgment for the selection and number of views required for each part. Show the limits of size where fits are given. Scale 1:2.

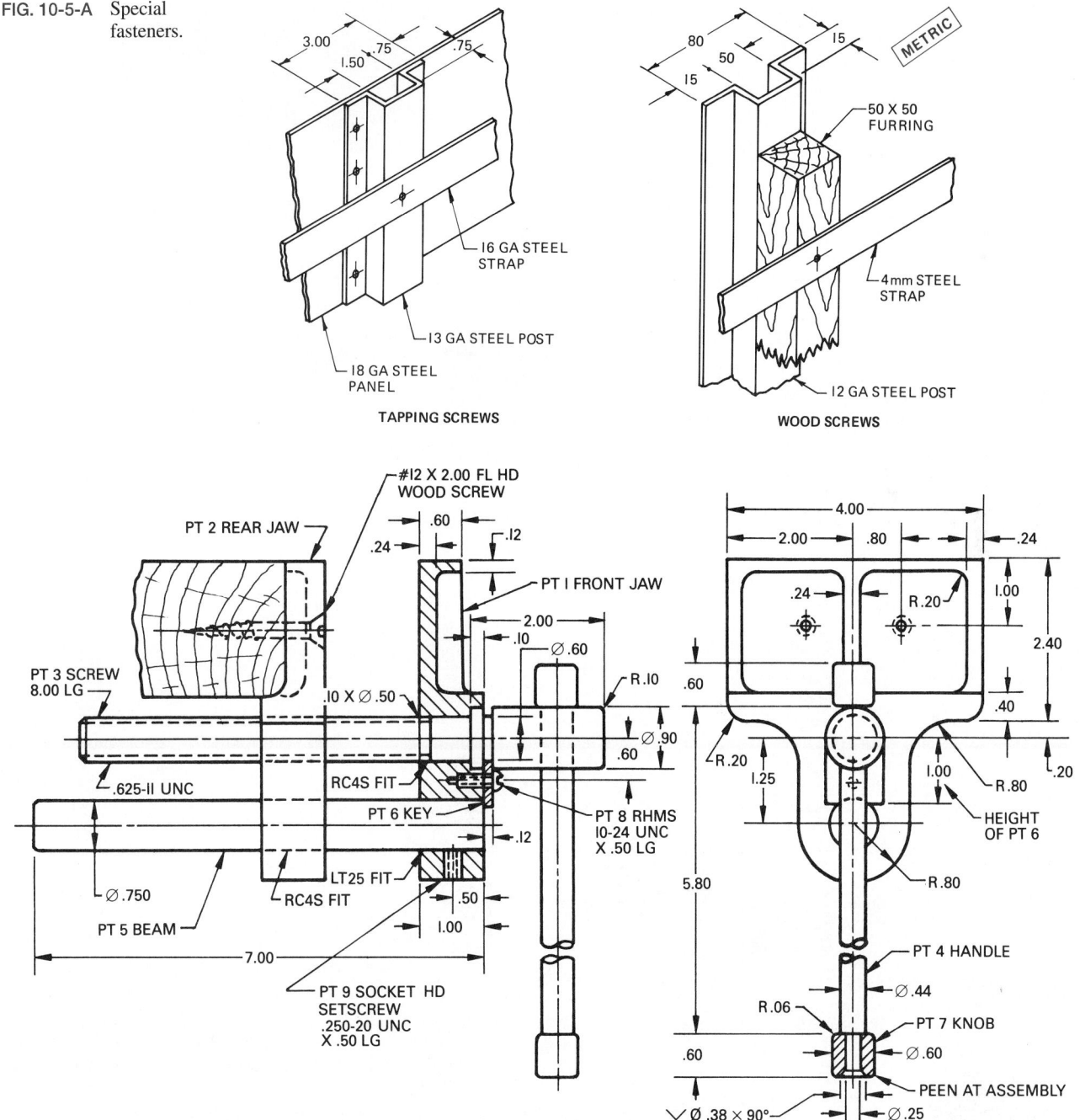

FIG. 10-5-A   Special fasteners.

FIG. 10-5-B   Woodworking vise.

CHAPTER 11

MISCELLANEOUS TYPES OF FASTENERS

Definitions

Adhesion   The force that holds materials together.

Compression spring   An open-coiled, helical spring that offers resistance to a compressive force.

Edge distance   The interval between the edge of the part and the center line of the fastener (rivet, bolt, etc.).

Extension spring   A close-coiled, helical spring that offers resistance to a pulling force.

Key   A piece of steel lying partly in a groove in the shaft and extending into another groove in the hub.

Keyseat   The groove that holds the key in the shaft.

Keyway   The groove in the hub or surrounding part that holds the key.

Pitch distance   The interval between center lines of adjacent fasteners (rivets, bolts, etc.).

Quick-release pins   Fasteners designed to release quickly and easily.

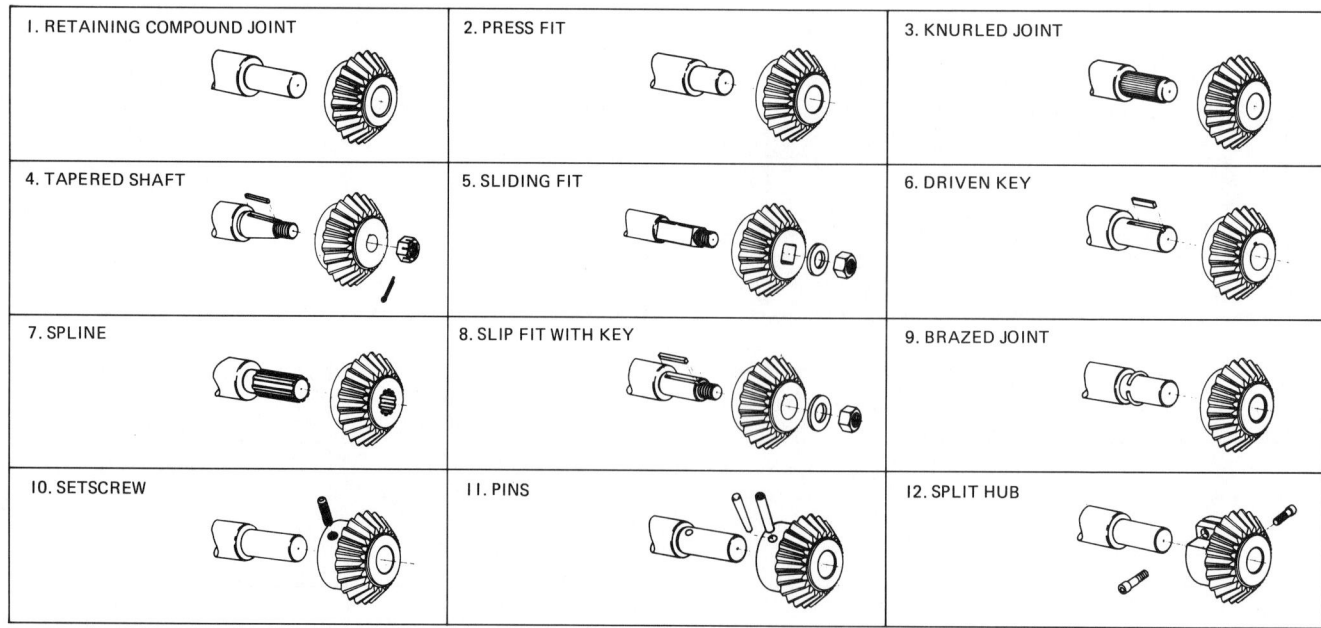

1. RETAINING COMPOUND JOINT	2. PRESS FIT	3. KNURLED JOINT
4. TAPERED SHAFT	5. SLIDING FIT	6. DRIVEN KEY
7. SPLINE	8. SLIP FIT WITH KEY	9. BRAZED JOINT
10. SETSCREW	11. PINS	12. SPLIT HUB

FIG. 11-1-1   Miscellaneous types of fasteners.

Resistance-welded fastener   A threaded metal part designed to be fused permanently in place by standard production welding equipment.

Retaining or **snap rings**   Fasteners used to provide a removable shoulder to accurately locate, retain, or lock components on shafts and in bores of housings.

Rivet   A ductile metal pin that is inserted through holes in two or more parts, and having the ends formed over to securely hold the parts.

Semipermanent pins   Fasteners that require application of pressure or the aid of tools for installation or removal.

Serrations   Shallow, involute splines with 45° pressure angles.

Splined shaft   A shaft having multiple grooves, or keyseats, cut around its circumference.

Spring clip   A self-retaining clip, requiring only a flange, panel edge, or mounting hole to clip to.

Stress   The force pulling materials apart.

Torsion spring, flat coil spring, and **flat spring**   Types of springs that exert pressure in a circular arc.

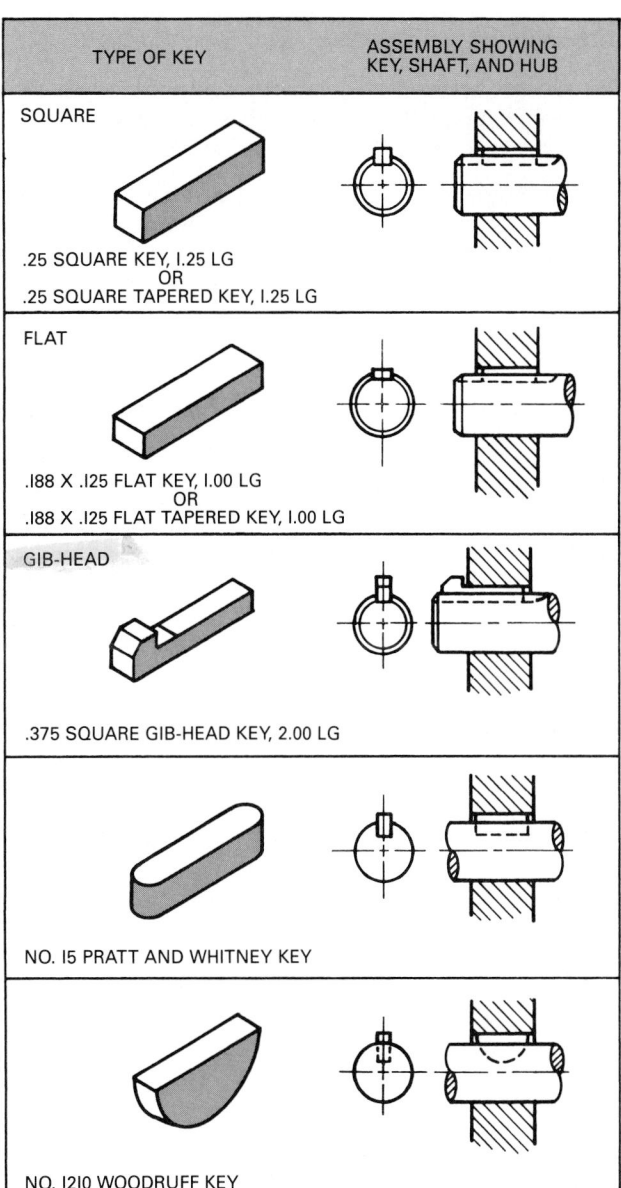

TYPE OF KEY	ASSEMBLY SHOWING KEY, SHAFT, AND HUB
SQUARE .25 SQUARE KEY, 1.25 LG OR .25 SQUARE TAPERED KEY, 1.25 LG	
FLAT .188 X .125 FLAT KEY, 1.00 LG OR .188 X .125 FLAT TAPERED KEY, 1.00 LG	
GIB-HEAD .375 SQUARE GIB-HEAD KEY, 2.00 LG	
NO. 15 PRATT AND WHITNEY KEY	
NO. 1210 WOODRUFF KEY	

FIG. 11-1-2   Common keys.

11-1 KEYS, SPLINES, AND SERRATIONS

Keys

A *key* is a piece of steel lying partly in a groove in the shaft and extending into another groove in the hub. The groove in the shaft is referred to as a *keyseat,* while the groove in the hub or surrounding part is referred to as a *keyway* (see boxes 6 and 8 in Fig. 11-1-1). A key is used to secure gears, pulleys, cranks, handles, and similar machine parts to shafts, so that the motion of the part is transmitted to the shaft, or the motion of the shaft to the part, without slippage. The key may also act in a safety capacity; its size is generally calculated so that when overloading takes place, the key will shear or break before the part or shaft breaks or deforms.

There are many kinds of keys. The most common types are shown in Fig. 11-1-2. Square and flat keys are widely used in industry. The width of the square and flat key should be approximately one-quarter the shaft diameter, but for proper key selection refer to Table 21 in the Appendix. These keys are also available with a 1:100 taper on their top surfaces and are then known as *squared-taper* or *flat-tapered* keys. The keyway in the hub is tapered to accommodate the taper on the key.

The gib-head key is the same as the squared or flat-tapered key but has a head added for easy removal.

The Pratt and Whitney key is rectangular with rounded ends. Two-thirds of this key sits in the shaft, one-third sits in the hub.

The Woodruff key is semicircular and fits into a semicircular keyseat in the shaft and a rectangular keyway in the hub. The width of the key should be approximately one-quarter

the diameter of the shaft, and its diameter should approximate the diameter of the shaft. Half the width of the key extends above the shaft and into the hub. Refer to the Appendix for exact sizes. Woodruff keys are identified by a number that gives the nominal dimensions of the key. The numbering system, which originated many years ago, is identified with the fractional-inch system of measurement. The last two digits of the number give the normal diameter in eighths of an inch, and the digits preceding the last two give the nominal width in thirty-seconds of an inch. For example, a No. 1210 Woodruff key describes a key $\frac{12}{32} \times \frac{10}{8}$ in., or a $\frac{3}{8} \times 1\frac{1}{4}$ in. key.

In calling up keys on the item list, only the information shown in the callout in Fig. 11-1-2 need be given.

Dimensioning of Keyseats

Keyseats and keyways are dimensioned by width, depth, location, and if required, length. The depth is dimensioned from the opposite side of the shaft or hole (Fig. 11-1-3).

Tapered Keyseats   The depth of tapered keyways in hubs, which is shown on the drawing, is the nominal depth $H/2$ minus an allowance. This is always the depth at the large end of the tapered keyseat and is indicated on the drawing by the abbreviation LE.

The radii of fillets, when required, must be dimensioned on the drawing.

Since standard milling cutters for Woodruff keys have the same appropriate number, it is possible to call for a Woodruff keyseat by the number only. Where it is desirable to detail Woodruff keyseats on a drawing, all dimensions are given in the form of a note in the following order: width, depth, and radius of cutter. Woodruff keyseats may alternately be dimensioned in the same manner as for square and flat keys, specifying first the width and then the depth (Fig. 11-1-4).

Splines and Serrations

A *splined shaft* is a shaft having multiple grooves, or keyseats, cut around its circumference for a portion of its length, in order that a sliding engagement may be made with corresponding internal grooves of a mating part.

Splines are capable of carrying heavier loads than keys, permit lateral movement of a part while maintaining positive rotation, and allow the attached part to be indexed or changed to another angular position.

Splines have either straight-sided teeth or curved-sided teeth. The latter type is known as an *involute spline*.

Involute Splines   The splines are similar in shape to involute gear teeth but have pressure angles of 30, 37.5 or 45°. There are two types of fits, the *side fit* and the *major-diameter fit* (Fig. 11-1-5).

Straight-Side Splines   The most popular are the SAE straight-side splines, as shown in Fig. 11-1-6. They have been

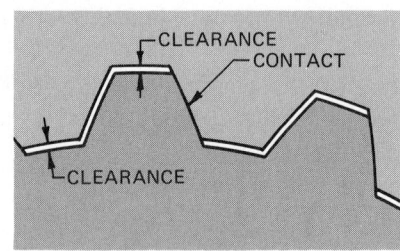

(A) SIDE FIT

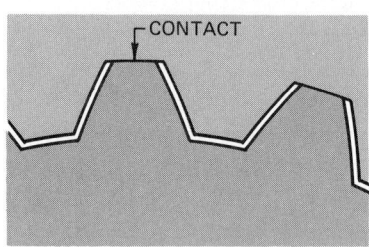

(B) MAJOR DIAMETER FIT

FIG. 11-1-5   Involute splines.

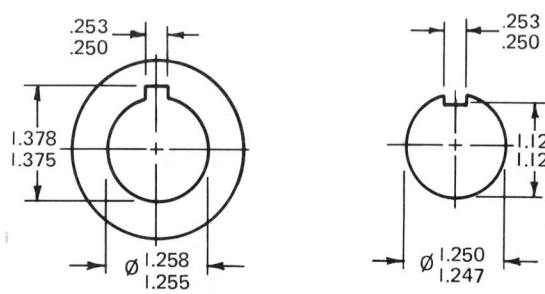

FIG. 11-1-3   Dimensioning keyseats.

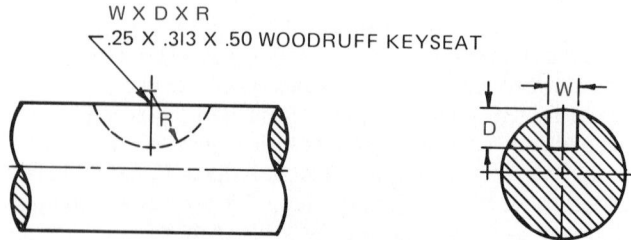

FIG. 11-1-4   Alternate method of detailing a Woodruff keyseat.

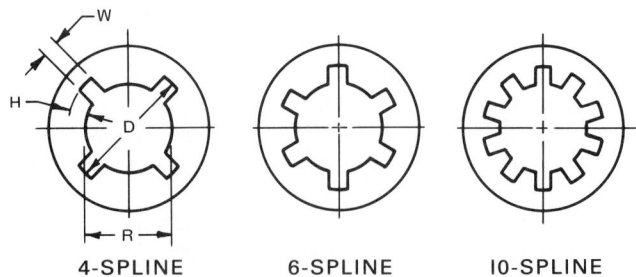

4-SPLINE   6-SPLINE   I0-SPLINE

NUMBER OF SPLINES	W FOR ALL FITS	PERMANENT FIT		TO SLIDE WITHOUT LOAD		TO SLIDE UNDER LOAD	
		H	R	H	R	H	R
4	0.241 D	0.075 D	0.85 D	0.125 D	0.75 D		
6	0.250 D	0.050 D	0.90 D	0.075 D	0.85 D	0.100 D	0.80 D
10	0.156 D	0.045 D	0.91 D	0.070 D	0.86 D	0.095 D	0.81 D
16	0.098 D	0.045 D	0.91 D	0.070 D	0.86 D	0.095 D	0.81 D

FIG. 11-1-6   Sizes of SAE parallel-side splines.

used in many applications in the automotive and machine industries.

Serrations *Serrations* are shallow, involute splines with 45° pressure angles. They are primarily used for holding parts, such as plastic knobs, on steel shafts.

Drawing Data

It is essential that a uniform system of drawing and specifying splines and serrations be used on drawings. The conventional method of showing and calling out splines on a drawing is shown in Fig. 11-1-7. Distance *L* does not include the cutter

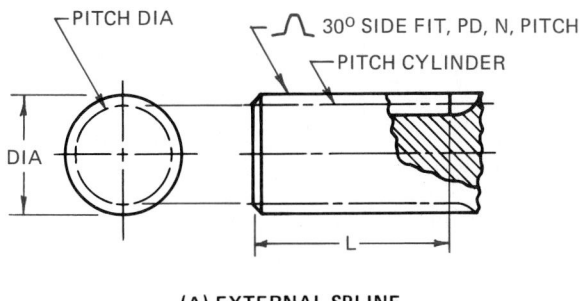

(A) EXTERNAL SPLINE

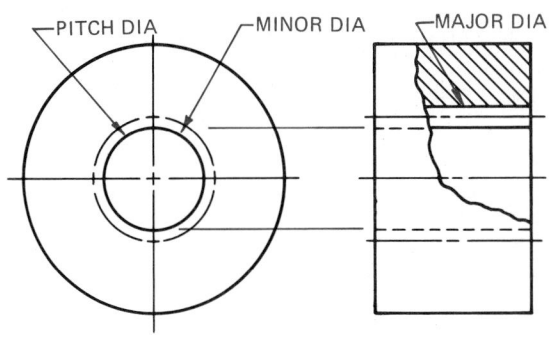

(B) INTERNAL SPLINE

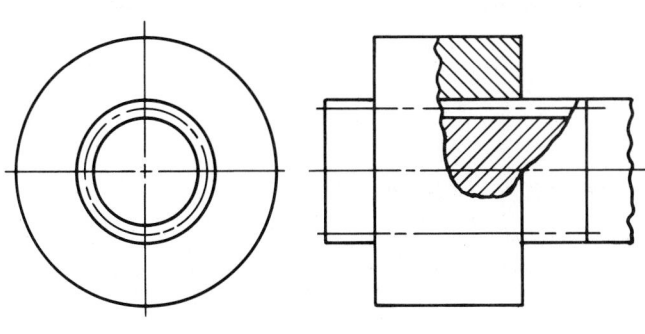

(C) ASSEMBLY DRAWING

INVOLUTE SPLINES

FIG. 11-1-7   Callout and representation of splines.

(A) EXTERNAL SPLINE

(B) INTERNAL SPLINE

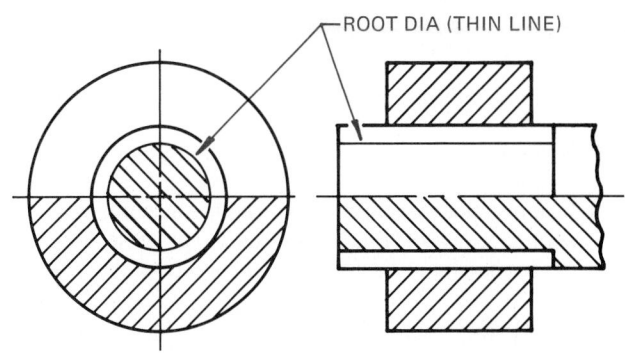

(C) ASSEMBLY DRAWING

STRAIGHT-SIDED SPLINES

HARDENED AND GROUND DOWEL PIN	TAPER PIN	CLEVIS PIN	COTTER PIN
Standardized in nominal diameters ranging from .12 to .88 (3 to 22mm). 1. Holding laminated sections together with surfaces either drawn up tight or separated in some fixed relationship. 2. Fastening machine parts where accuracy of alignment is a primary consideration. 3. Locking components on shafts, in the form of transverse pin key.	Standard pins have a taper of 1:48 measured on the diameter. Basic dimension is the diameter of the large end. Used for light-duty service in the attachment of wheels, levers, and similar components to shafts. Torque capacity is determined on the basis of double shear, using the average diameter along the tapered section in the shaft for area calculations.	Standard nominal diameters for clevis pins range from .19 to 1.00 (5 to 25mm). Basic function of the clevis pin is to connect mating yoke, or fork, and eye members in knuckle-joint assemblies. Held in place by a small cotter pin or other fastening means, it provides a mobile joint construction, which can be readily disconnected for adjustment or maintenance.	Sizes have been standardized in nominal diameters ranging from .03 to .75 (1 to 20mm). Locking device for other fasteners. Used with a castle or slotted nut on bolts, screws, or studs, it provides a convenient, low-cost locknut assembly. Hold standard clevis pins in place. Can be used with or without a plain washer as an artificial shoulder to lock parts in position on shafts.

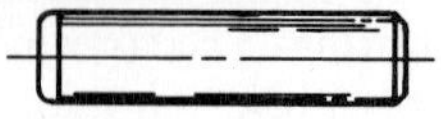

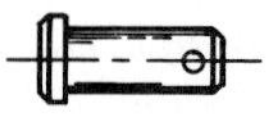

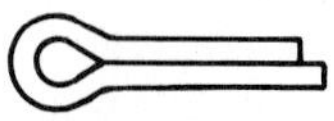

FIG. 11-2-1 Machine pins.

runout. The drawing callout shows the symbol indicating the type of spline followed by the type of fit, the pitch diameter, number of teeth and pitch for involute splines, and number of teeth and outside diameter for straight-sided teeth.

ASSIGNMENTS

See Assignments 1 and 2 for Unit 11-1 on pages 315–316.

11-2 PIN FASTENERS

Pin fasteners are an inexpensive and effective method of assembly where loading is primarily in shear. They can be separated into two groups: semipermanent and quick-release.

Semipermanent Pins

Semipermanent pin fasteners require application of pressure or the aid of tools for installation or removal. The two basic types are machine pins and radial locking pins.

The following general design rules apply to all types of semipermanent pins:

■ Avoid conditions where the direction of vibration parallels the axis of the pin.
■ Keep the shear plane of the pin a minimum distance of one diameter from the end of the pin.
■ In applications where engaged length is at a minimum and appearance is not critical, allow pins to protrude the length of the chamfer at each end for maximum locking effect.

Machine Pins

Four types are generally considered to be most important: hardened and ground dowel pins and commercial straight pins, taper pins, clevis pins, and standard cotter pins. Descriptive data and recommended assembly practices for these four traditional types of machine pins are presented in Fig. 11-2-1. For proper size selection of cotter pins, refer to Fig. 11-2-2.

Radial Locking Pins

Two basic pin forms are employed: solid with grooved surfaces and hollow spring pins, which may be either slotted or spiral-wrapped.

Grooved Straight Pins Locking action of the grooved pin is provided by parallel, longitudinal grooves uniformly spaced around the pin surface. Rolled or pressed into solid pin stock, the grooves expand the effective pin diameter. When the pin is driven into a drilled hole corresponding in size to nominal pin diameter, elastic deformation of the raised groove edges produces a secure force-fit with the hole wall. Figure 11-2-3 shows

NOMINAL THREAD SIZE	NOMINAL COTTER PIN SIZE	COTTER PIN HOLE	END CLEARANCE*
.250 (6)	.062 (1.5)	.078 (1.9)	.11 (3)
.312 (8)	.078 (2)	.094 (2.4)	.11 (3)
.375 (10)	.094 (2.5)	.109 (2.8)	.14 (4)
.500 (12)	.125 (3)	.141 (3.4)	.17 (6)
.625 (14)	.156 (3)	.172 (3.4)	.23 (5)
.750 (20)	.156 (4)	.172 (4.5)	.27 (7)
1.000 (24)	.188 (5)	.203 (5.6)	.31 (8)
1.125 (27)	.188 (5)	.203 (5.6)	.39 (8)
1.250 (30)	.219 (6)	.234 (6.3)	.41 (10)
1.375 (36)	.219 (6)	.234 (6.3)	.44 (11)
1.500 (42)	.250 (6)	.266 (6.3)	.48 (12)
1.750 (48)	.312 (8)	.328 (8.5)	.55 (14)

*DISTANCE FROM EXTREME POINT OF BOLT OR SCREW TO CENTER OF COTTER PIN HOLE.   Inch (mm)

FIG. 11-2-2 Recommended cotter pin sizes.

FIG. 11-2-3   Grooved radial locking pins.

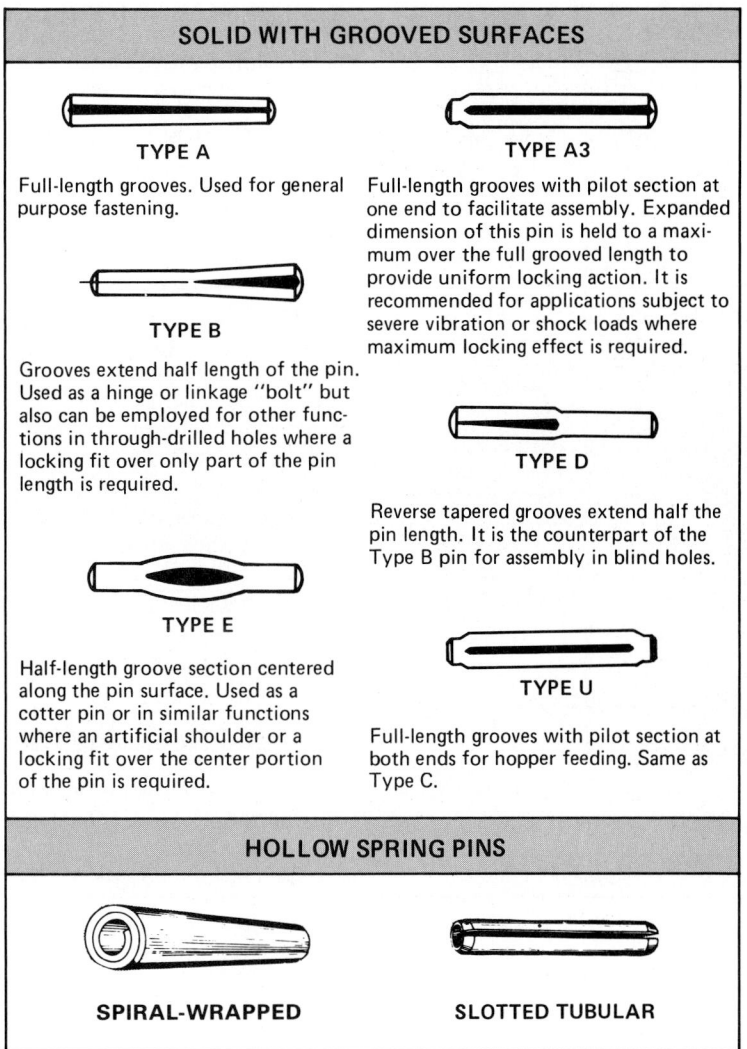

SOLID WITH GROOVED SURFACES

TYPE A

Full-length grooves. Used for general purpose fastening.

TYPE B

Grooves extend half length of the pin. Used as a hinge or linkage "bolt" but also can be employed for other functions in through-drilled holes where a locking fit over only part of the pin length is required.

TYPE E

Half-length groove section centered along the pin surface. Used as a cotter pin or in similar functions where an artificial shoulder or a locking fit over the center portion of the pin is required.

TYPE A3

Full-length grooves with pilot section at one end to facilitate assembly. Expanded dimension of this pin is held to a maximum over the full grooved length to provide uniform locking action. It is recommended for applications subject to severe vibration or shock loads where maximum locking effect is required.

TYPE D

Reverse tapered grooves extend half the pin length. It is the counterpart of the Type B pin for assembly in blind holes.

TYPE U

Full-length grooves with pilot section at both ends for hopper feeding. Same as Type C.

HOLLOW SPRING PINS

SPIRAL-WRAPPED

SLOTTED TUBULAR

six of the grooved-pin constructions that have been standardized. For typical grooved pin applications and size selection, refer to Figs. 11-2-4 and 11-2-5 (pg. 298).

Hollow Spring Pins   Resilience of hollow cylinder walls under radial compression forces is the principle of spiral-wrapped and slotted tubular pins (Fig. 11-2-3). Both pin forms are made to controlled diameters greater than the holes into which they are pressed. Compressed when driven into the hole, the pins exert spring pressure against the hole wall along their entire engaged length to develop locking action.

Standard slotted tubular pins are designed so that several sizes can be used inside one another. In such combinations, shear strengths of the individual pins are additive. For spring pin applications, refer to Fig. 11-2-6 (pg. 299).

Quick-Release Pins

Commercially available quick-release pins vary widely in head styles, types of locking and release mechanisms, and range of pin lengths (Fig. 11-2-7, pg. 299).

Quick-release pins may be divided into two basic types: push-pull and positive-locking pins. The positive-locking pins can be further divided into three categories: heavy-duty cotter pins, single-acting pins, and double-acting pins.

Push-Pull Pins

These pins are made with either a solid or a hollow shank, containing a detent assembly in the form of a locking lug, button, or ball, backed up by some type of resilient core, plug, or spring. The detent member projects from the surface of the pin body until sufficient force is applied in assembly or removal to cause it to retract against the spring action of the resilient core and release the pin for movement.

Positive-Locking Pins

For some quick-release fasteners, the locking action is independent of insertion and removal forces. As in the case of push-pull pins, these pins are primarily suited for shear-load applications. However, some degree of tension loading usually can be tolerated without affecting the pin function.

SHAFT DIA		TRANSVERSE KEY			LONGITUDINAL KEY	
		PIN DIA		TAPER PIN NO.	PIN DIA	
IN.	(MM)	IN.	(MM)		IN.	(MM)
.375	(10)	.125	(3)	3/0	.094	(2.5)
.438	(12)	.156	(4)	0	.125	(3)
.500	(14)	.156	(5)	0	.125	(4)
.562	(16)	.188	(5)	2	.156	(4)
.625	(18)	.188	(6)	2	.156	(5)
.750	(20)	.250	(6)	4	.156	(5)
.875	(22)	.250	(6)	4	.219	(6)
1.000	(24)	.312	(8)	6	.250	(6)
1.062	(26)	.312	(8)	6	—	—
1.125	(28)	.375	(10)	7	—	—
1.188	(30)	.375	(10)	7	—	—
1.250	(32)	.375	(10)	7	.312	(8)
1.375	(34)	.438	(11)	7	.375	(10)
1.438	(36)	.438	(11)	7	—	—
1.500	(38)	.500	(12)	8	.438	(11)

FIG. 11-2-4   Recommended groove pin sizes.

REFERENCES AND SOURCE MATERIAL

1. *Machine Design*, Fastening and joining reference issue.

ASSIGNMENTS

See Assignments 3 through 5 for Unit 11-2 on pages 316 through 318.

11-3   RETAINING RINGS

Retaining rings, or *snap rings*, are used to provide a removable shoulder to accurately locate, retain, or lock components on shafts and in bores of housings (see Fig. 11-3-1, pg. 300). They are easily installed and removed, and since they are usually made of spring steel, retaining rings have a high shear strength and impact capacity. In addition to fastening and positioning, a number of rings are designed for taking up end play caused by accumulated tolerances or wear in the parts being retained. In general, these devices can be placed into three categories, which describe the type and method of fabrication: stamped retaining rings, wire-formed rings, and spiral-wound retaining rings.

Stamped Retaining Rings

Stamped retaining rings, in contrast to wire-formed rings with their uniform cross-sectional area, have a tapered radial width

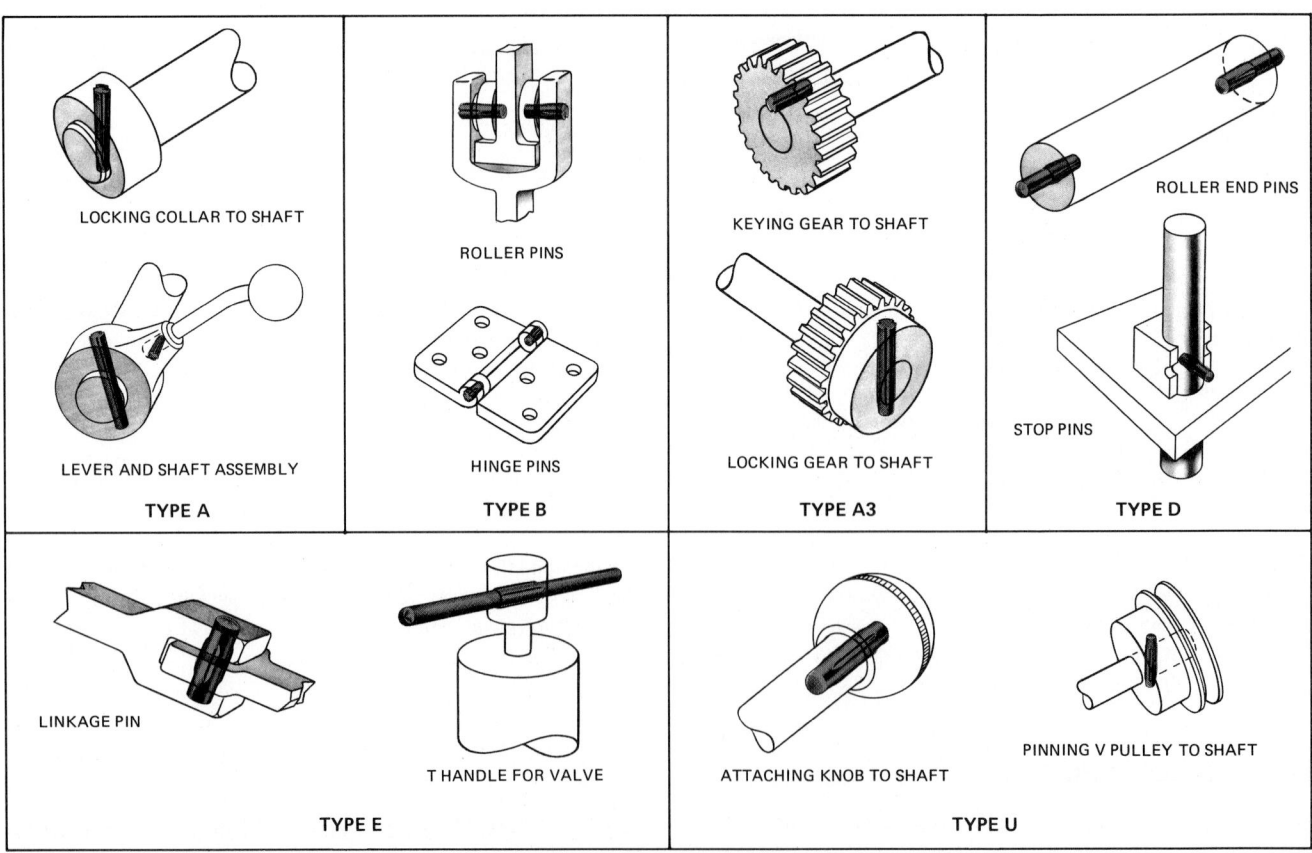

FIG. 11-2-5   Groove pin applications.

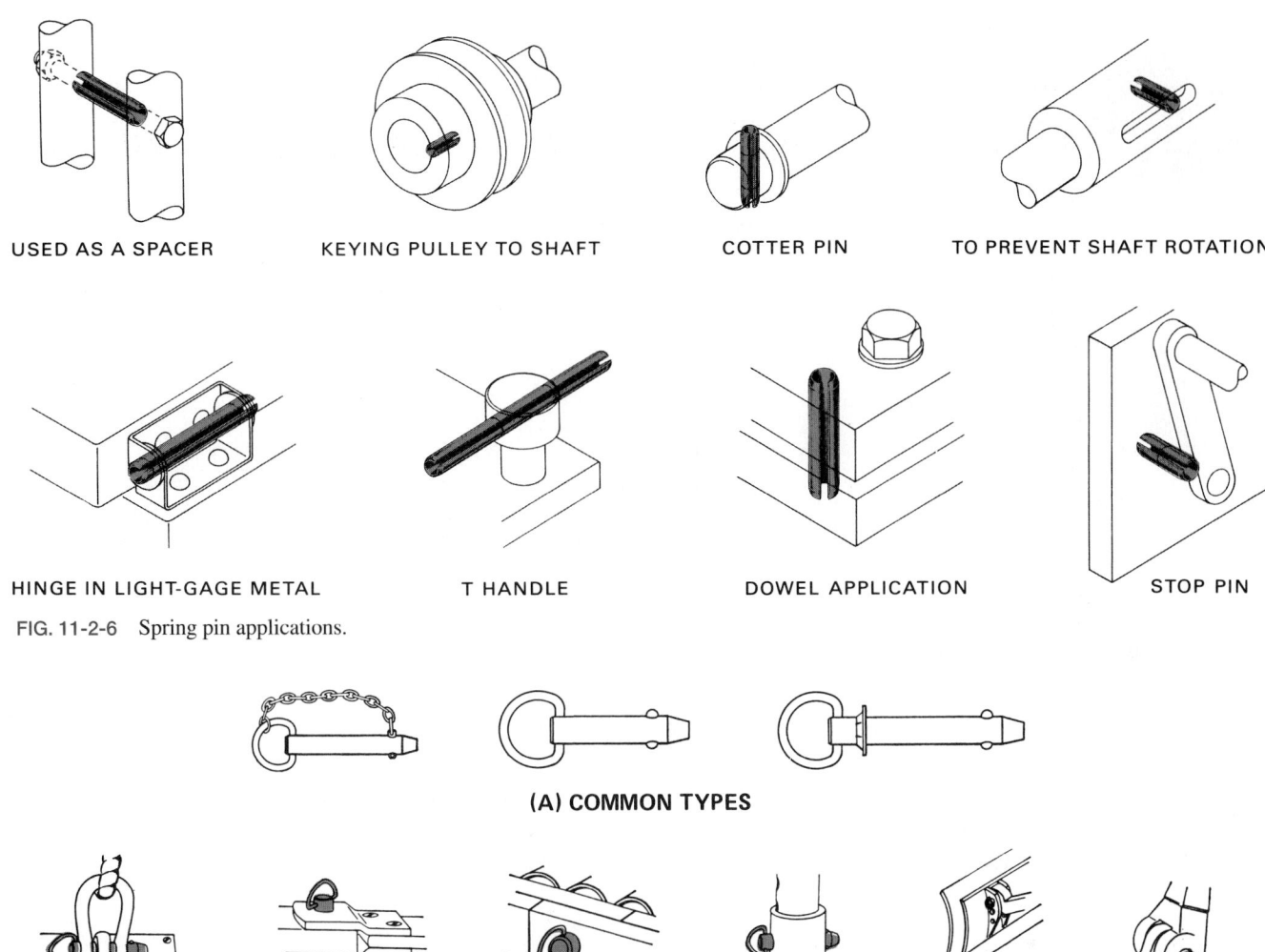

USED AS A SPACER   KEYING PULLEY TO SHAFT   COTTER PIN   TO PREVENT SHAFT ROTATION

HINGE IN LIGHT-GAGE METAL   T HANDLE   DOWEL APPLICATION   STOP PIN

FIG. 11-2-6   Spring pin applications.

(A) COMMON TYPES

CLEVIS-SHACKLE PIN   DRAW–BAR HITCH PIN   RIGID COUPLING PIN   TUBING LOCKPIN   ADJUSTMENT PIN   SWIVEL HINGE PIN

(B) APPLICATIONS

FIG. 11-2-7   Quick-release pins.

that decreases symmetrically from the center section to the free ends. The tapered construction permits the rings to remain circular when they are expanded for assembly over a shaft or contracted for insertion into a bore or housing. This constant circularity ensures maximum contact surface with the bottom of the groove.

Stamped retaining rings can be classified into three groups: axially assembled rings, radially assembled rings, and self-locking rings which do not require grooves. *Axially assembled rings* slip over the ends of shafts or down into bores, while *radially assembled rings* have side openings that permit the rings to be snapped directly into grooves on a shaft.

Commonly used types of stamped retaining rings are illustrated and compared in Fig. 11-3-2 (pg. 300).

Wire-Formed Retaining Rings

The *wire-formed retaining ring* is a split ring formed and cut from spring wire of uniform cross-sectional size and shape. The wire is cold-drawn or rolled into shape from a continuous coil or bar. Then the gap ends are cut into various configurations for ease of application and removal.

Rings are available in many cross-sectional shapes, but the most commonly used are the rectangular and circular cross sections.

Spiral-Wound Retaining Rings

Spiral-wound retaining rings consist of two or more turns of rectangular material, wound on edge to provide continuous crimped or uncrimped coil.

REFERENCES AND SOURCE MATERIAL

1. *Machine Design*, Fastening and joining reference issue.

ASSIGNMENTS ▓▓▓▓▓▓▓▓▓▓▓▓▓▓▓▓▓▓▓▓

See Assignments 6 through 8 for Unit 11-3 on pages 318–319.

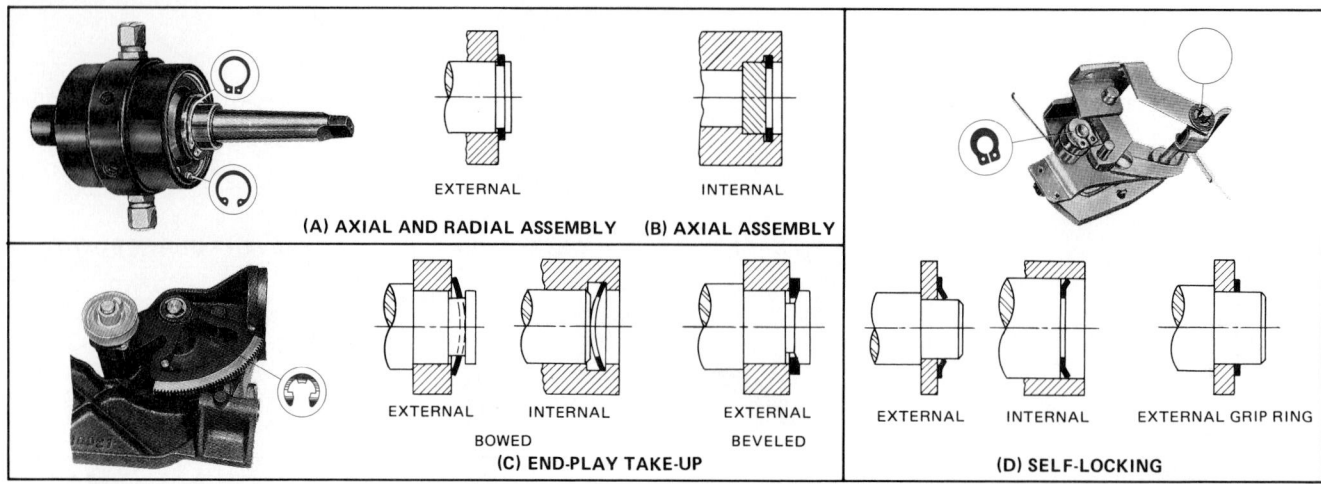

FIG. 11-3-1   Retaining ring applications.

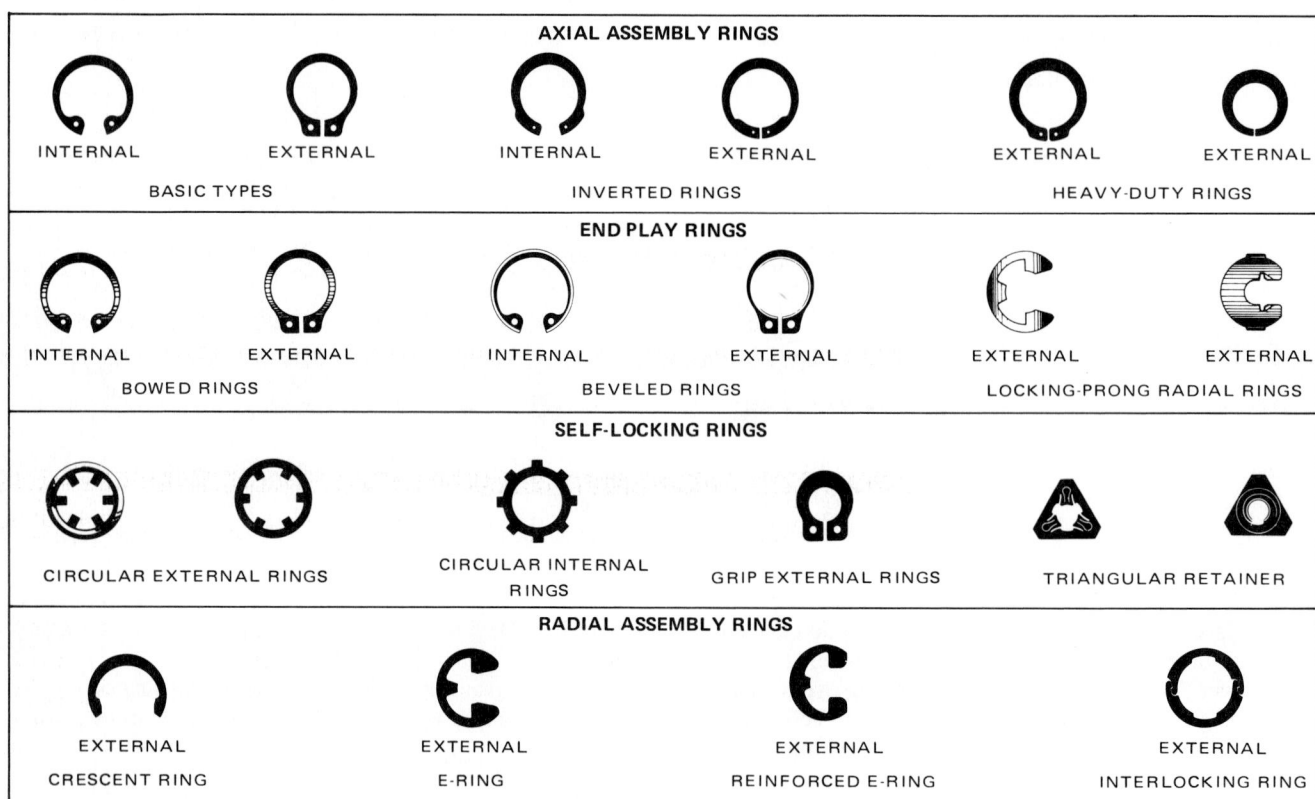

FIG. 11-3-2   Stamped retaining rings.

11-4 SPRINGS

Springs may be classified into three general groups according to their application.

Controlled Action Springs   Controlled action springs have a well-defined function, or a constant range of action for each cycle of operation. Examples are valve, die, and switch springs.

Variable-Action Springs   Variable-action springs have a changing range of action because of the variable conditions imposed upon them. Examples are suspension, clutch, and cushion springs.

Static Springs   Static springs exert a comparatively constant pressure or tension between parts. Examples are packing or bearing pressure, antirattle, and seal springs.

Types of Springs

The type or name of a spring is determined by characteristics such as function, shape of material, application, or design.

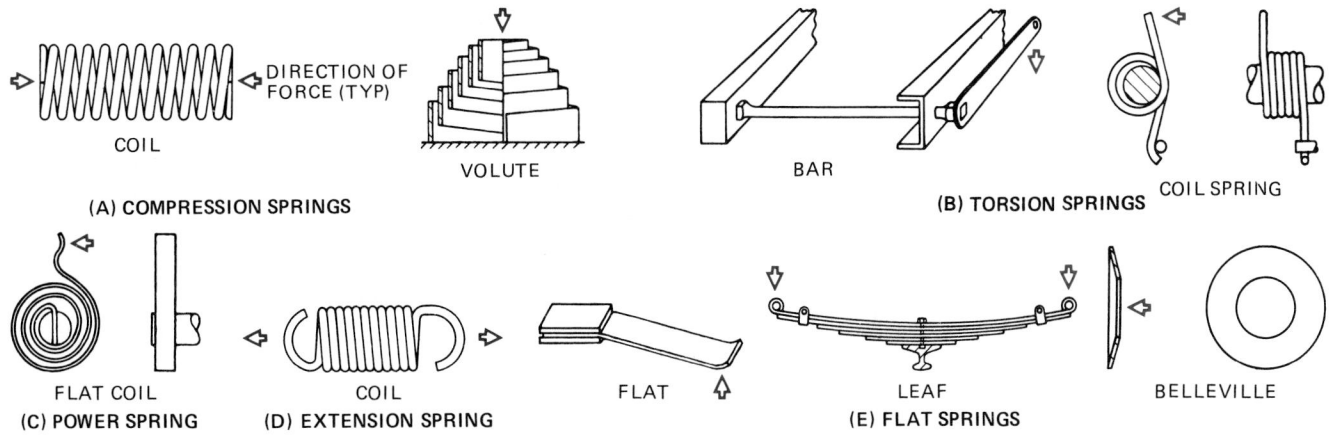

FIG. 11-4-1   Types of springs.

Figure 11-4-1 illustrates common springs in use. Figure 11-4-2 designates spring nomenclature.

Compression Springs

A *compression spring* is an open-coiled helical spring that offers resistance to a compressive force.

Compression Spring Ends   Figure 11-4-3A (pg. 302) shows the ends commonly used on compression springs.

Plain open ends are produced by straight cutoff with no reduction of helix angle (Fig. 11-4-4, pg. 302). The spring should be guided on a rod or in a hole to operate satisfactorily.

Ground open ends are produced by parallel grinding of open-end coil springs. Advantages of this type of end are improved stability and a larger number of total coils.

Plain closed ends are produced with a straight cutoff and with reduction of helix angle to obtain closed-end coils, resulting in a more stable spring.

Ground closed ends are produced by parallel grinding of closed-end coil springs, resulting in maximum stability.

Extension Springs

An *extension spring* is a close-coiled, helical spring that offers resistance to a pulling force. It is made from round or square wire.

Extension Spring Ends   The end of an extension spring is usually the most highly stressed part. Thus, proper consideration should be given to its selection. The types of ends shown in Fig. 11-4-3B are most commonly used on extension springs. Different types of ends can be used on the same spring.

Torsion Springs

Springs exerting pressure along a path that is a circular arc, or in other words, providing a torque, are called torsion springs, motor springs, power springs, etc. The term *torsion spring* is usually applied to a helical spring of round, square, or rectangular wire, loaded by torque.

The variation in ends used is almost limitless, but a few of the more common types are illustrated in Fig. 11-4-3C.

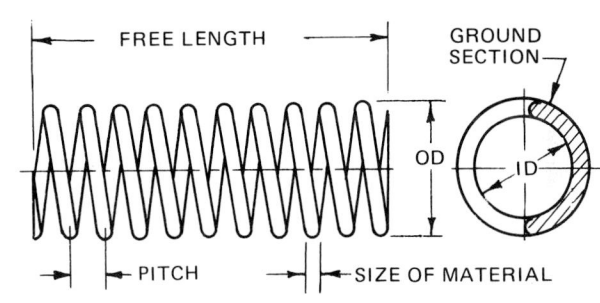

FIG. 11-4-2   Spring nomenclature.

Torsion Bar Springs   A *torsion bar spring* is a relatively straight bar anchored at one end, on which a torque may be exerted at the other end, thus tending to twist it about its axis.

Power Springs

Clock or Motor Type   A *flat coil spring*, also known as a clock or motor spring, consists of a strip of tempered steel wound on an arbor and usually confined in a case or drum.

Flat Springs

Flat springs are made of flat material formed in such a manner as to apply force in the desired direction when deflected in the opposite direction.

Leaf Springs   A *leaf spring* is composed of a series of flat springs nested together and arranged to provide approximately uniform distribution of stress throughout its length. Springs may be used in multiple arrangements, as shown in Fig. 11-4-5A (pg. 302).

Belleville Springs   *Belleville springs* are washer-shaped, made in the form of a short, truncated cone.

Belleville washers may be assembled in series to accommodate greater deflections, in parallel to resist greater forces, or in combination of series and parallel, as shown in Fig. 11-4-5B.

Spring Drawings

On working drawings, a schematic drawing of a helical spring is recommended to save drafting time (Fig. 11-4-6, pg. 303).

FIG. 11-4-3 End styles for helical springs.

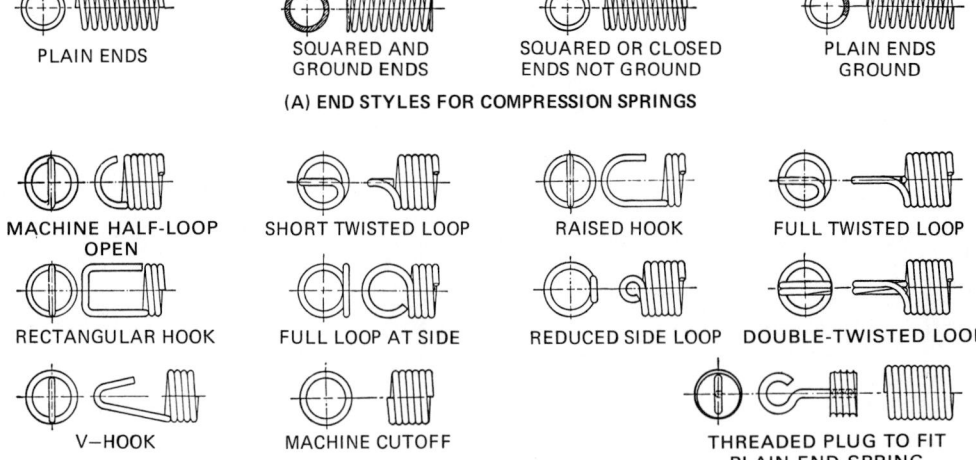

PLAIN ENDS   SQUARED AND GROUND ENDS   SQUARED OR CLOSED ENDS NOT GROUND   PLAIN ENDS GROUND

(A) END STYLES FOR COMPRESSION SPRINGS

MACHINE HALF-LOOP OPEN   SHORT TWISTED LOOP   RAISED HOOK   FULL TWISTED LOOP

RECTANGULAR HOOK   FULL LOOP AT SIDE   REDUCED SIDE LOOP   DOUBLE-TWISTED LOOP

V–HOOK   MACHINE CUTOFF   THREADED PLUG TO FIT PLAIN-END SPRING

(B) END STYLES FOR EXTENSION SPRINGS

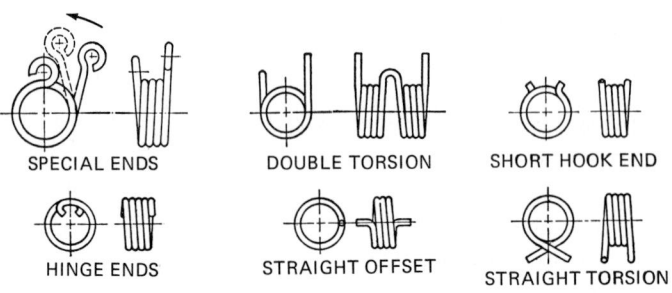

SPECIAL ENDS   DOUBLE TORSION   SHORT HOOK END

HINGE ENDS   STRAIGHT OFFSET   STRAIGHT TORSION

(C) END STYLES FOR TORSION SPRINGS

FIG. 11-4-4 Coil definitions.

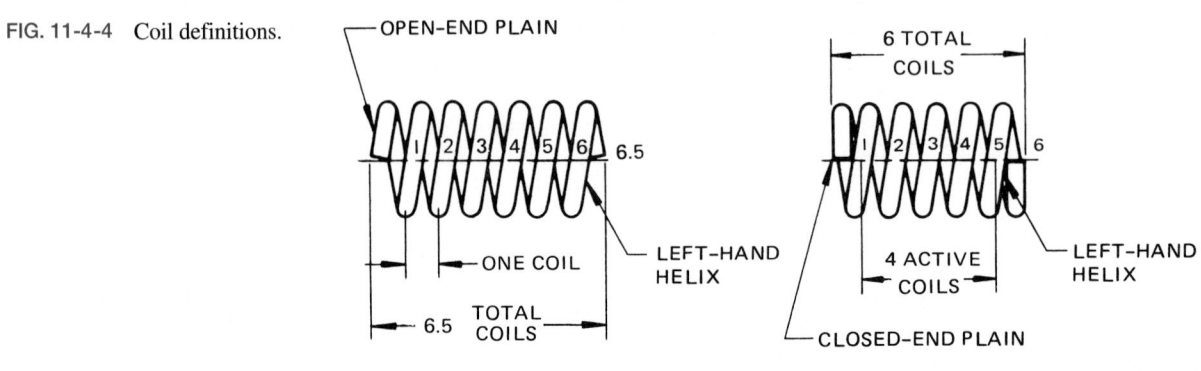

As in screw-thread representation, straight lines are used in place of the helical curves. On assembly drawings, springs are normally shown in section, and either cross-hatching lines or solid black shading is recommended, depending on the size of the wire's diameter (Fig. 11-4-7).

Dimensioning Springs

The following information should be given on a drawing of a spring:

- Size, shape, and kind of material used in the spring
- Diameter (outside or inside)
- Pitch or number of coils

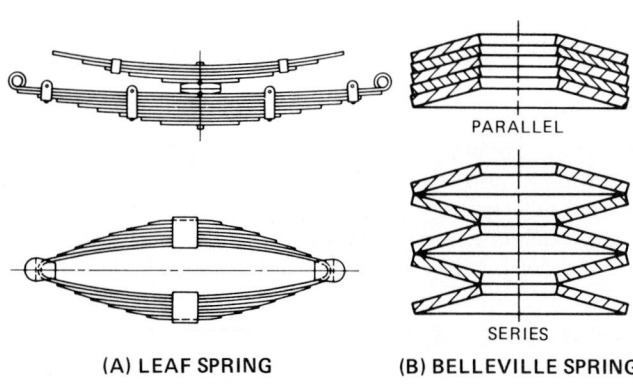

(A) LEAF SPRING   (B) BELLEVILLE SPRING

FIG. 11-4-5 Spring arrangements.

FIG. 11-4-6   Schematic drawing of springs.

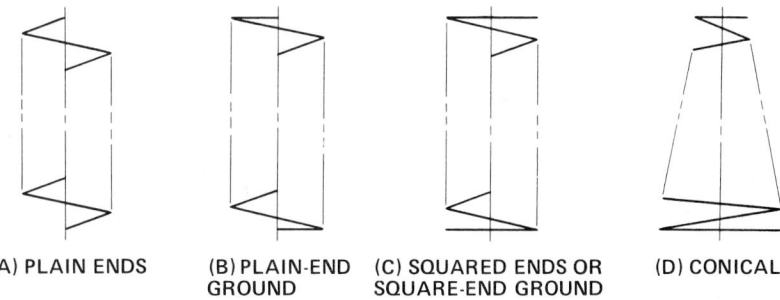

(A) PLAIN ENDS   (B) PLAIN-END GROUND   (C) SQUARED ENDS OR SQUARE-END GROUND   (D) CONICAL

FIG. 11-4-7   Showing helical springs on assembly drawings.

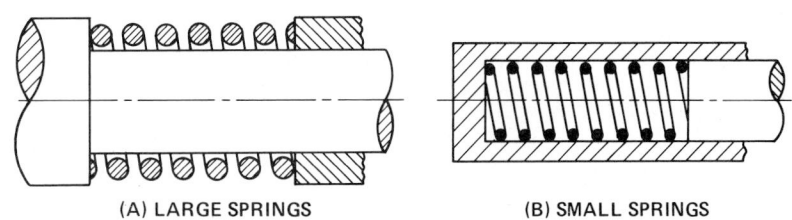

(A) LARGE SPRINGS   (B) SMALL SPRINGS

- Shape of ends
- Length
- Load and rate (not covered in this text)

EXAMPLE

ONE HELICAL TENSION SPRING 3.00 LG (OR NUMBER OF COILS), .50 ID, PITCH .25, 18 B & S GA SPRING BRASS WIRE

In using single-line representation, the dimensions should state the applicable size of material required to ensure correct interpretation on such features as inside diameter, outside diameter, and end loops (Fig. 11-4-8).

Spring Clips

Spring clips are a relatively new class of industrial fasteners. They perform multiple functions, eliminate the handling of several small parts, and thus reduce assembly costs (Fig. 11-4-9, pg. 304).

The *spring clip* is generally self-retaining, requiring only a flange, panel edge, or mounting hole to clip to. Basically, spring clips are light-duty fasteners and serve the same function as small bolts and nuts, self-tapping screws, clamps, spot welding, and formed retaining plates.

Dart-type Spring Clips   Dart-shaped panel retaining elements have hips to engage within panel or component holes. The top of arms of the fastener can be formed in any shape to perform unlimited fastening functions.

Stud Receiver Clips   There are three basic types of stud receivers: push-ons, tubular types, and self-threading fasteners. All are designed to make attachments to unthreaded studs, rivets, pins, or rods of metal or plastic.

Cable, Wire, and Tube Clips   These fasteners incorporate self-retaining elements for engaging panel holes or mounting on panel edges and flanges.

Spring-clip cable, wire, and tubing fasteners are front-mounting devices, requiring no access to the back of the panel.

Spring Molding Clips   Molding retaining clips are formed with legs that hold the clips to a panel and arms that positively engage the flanges of various sizes and shapes of trim molding and pull the molding tightly to the attaching panel.

U-, S-, and C-Shaped Spring Clips   These spring clips get their names from their shapes. The fastening function is accomplished by using inward compressive spring force to secure assembly components or provide self-retention after installation.

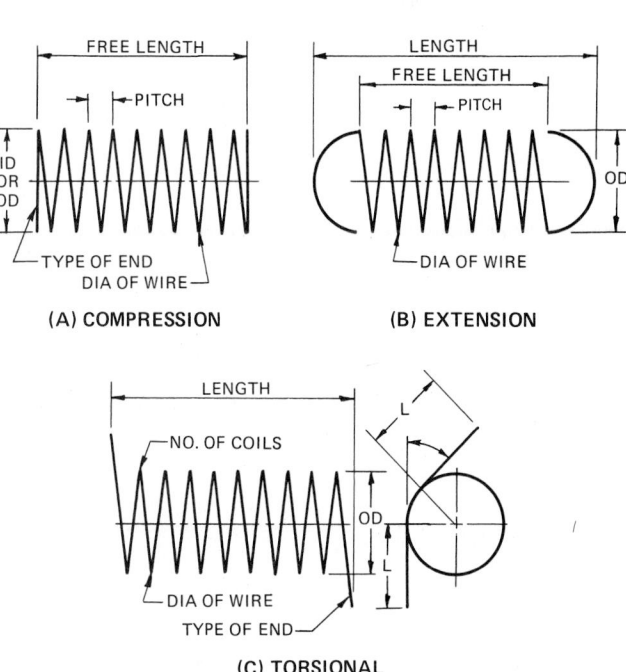

FIG. 11-4-8   Dimensioning springs.

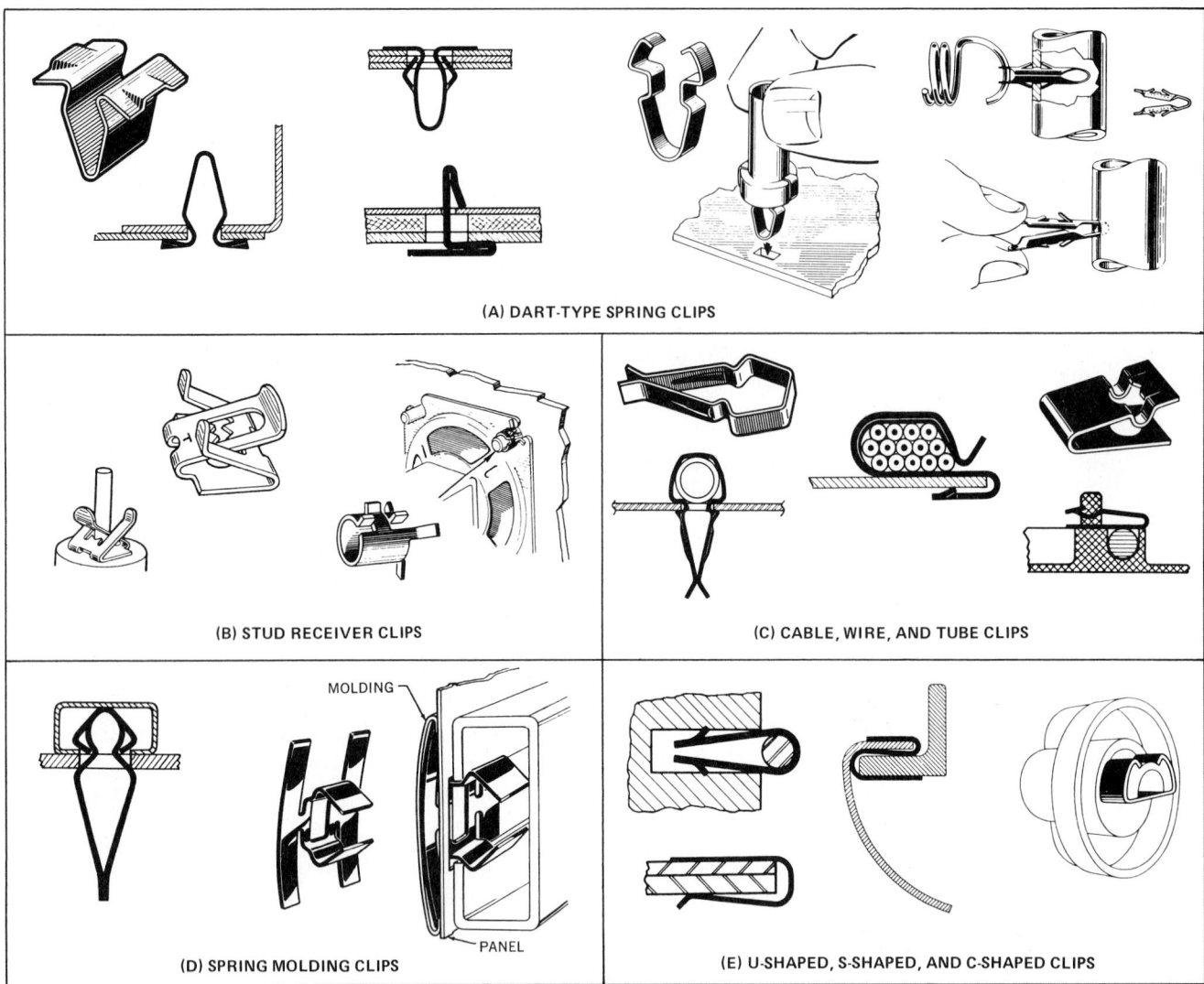

FIG. 11-4-9   Spring clips.

REFERENCES AND SOURCE MATERIAL

1. General Motors Corp.
2. The Wallace Barnes Co. Ltd.
3. *Machine Design*, Fastening and joining reference issue.

ASSIGNMENTS

See Assignments 9 through 11 for Unit 11-4 on pages 320 to 321.

11-5   RIVETS

Standard Rivets

Riveting is a popular method of fastening and joining, primarily because of its simplicity, dependability, and low cost.

A myriad of manufactured products and structures, both small and large, are held together by these fasteners. Rivets are classified as permanent fastenings, as distinguished from removable fasteners, such as bolts and screws.

Basically, a *rivet* is a ductile metal pin that is inserted through holes in two or more parts, and having the ends formed over to securely hold the parts.

Another important reason for riveting is versatility, with respect to both the properties of rivets as fasteners and the method of clinching.

- *Part materials:* Rivets can be used to join dissimilar materials, metallic or nonmetallic, in various thicknesses.
- *Multiple functions:* Rivets can serve as fasteners, pivot shafts, spacers, electric contacts, stops, or inserts.
- *Fastening finished parts:* Rivets can be used to fasten parts that have already received a final painting or other finishing.

Riveted joints are neither watertight nor airtight, although such a joint may be attained at some added cost by using a sealing compound. The riveted parts cannot be disassembled for maintenance or replacement without knocking the rivet out and clinching a new one in place for reassembly. Common riveted joints are shown in Fig. 11-5-1.

Large Rivets

Large rivets are used in structural work of buildings and bridges. Today, however, high-strength bolts have almost completely replaced rivets in field connections because of cost, strength, and the noise factor. Rivet joints are of two types: butt and lapped. The more common types of large rivets are shown in Fig. 11-5-2. In order to show the difference between *shop rivets* (rivets that are put in the structure at the shop) and *field rivets* (rivets that are used on the site), two types of symbols are used. In drawing shop rivets, the diameter of the rivet head is shown on the drawings. For field rivets, the shaft diameter is used. Figure 11-5-3 shows the conventional rivet symbols adopted by the American and Canadian Institutes of Steel Construction.

Rivets for Aerospace Equipment

The following representation of rivets on drawings for aerospace equipment is also recommended for other fields of work involving rivets.

The symbolic representation for a set (installed) rivet consists of a cross indicating its position. This representation is

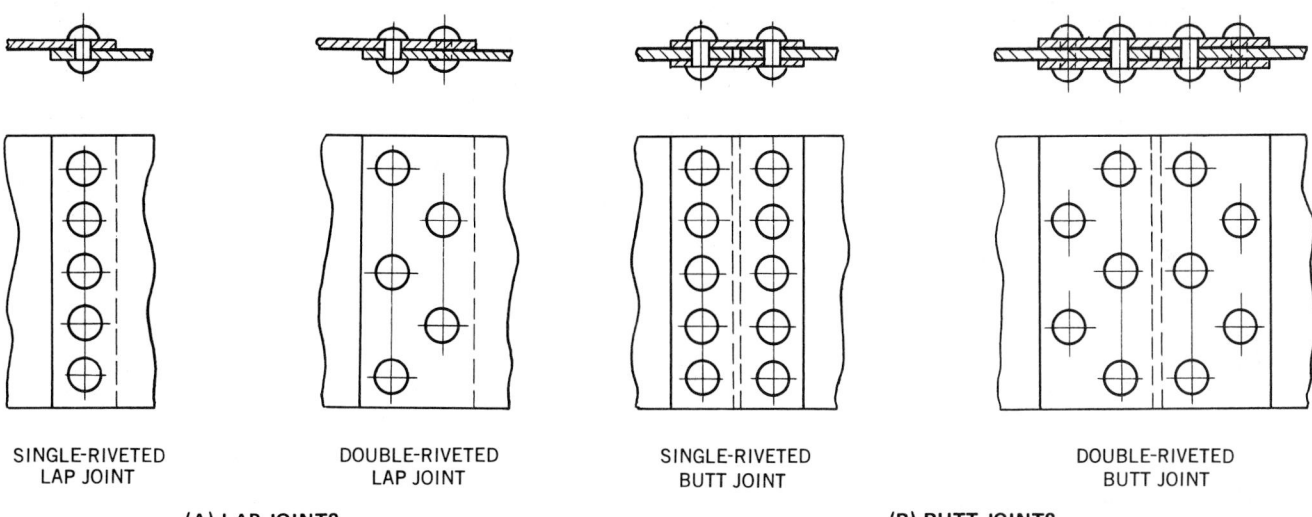

SINGLE-RIVETED
LAP JOINT

DOUBLE-RIVETED
LAP JOINT

SINGLE-RIVETED
BUTT JOINT

DOUBLE-RIVETED
BUTT JOINT

(A) LAP JOINTS

(B) BUTT JOINTS

FIG. 11-5-1   Common riveted joints.

FIG. 11-5-2   Approximate sizes and types of large rivets .50 in. (12 mm) and up.

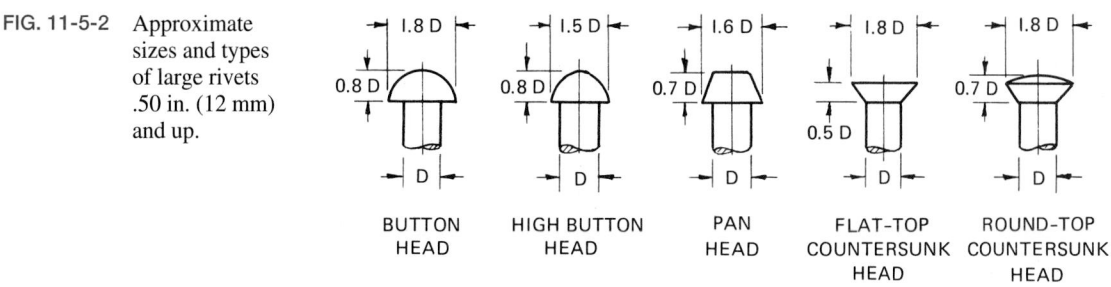

BUTTON
HEAD

HIGH BUTTON
HEAD

PAN
HEAD

FLAT-TOP
COUNTERSUNK
HEAD

ROUND-TOP
COUNTERSUNK
HEAD

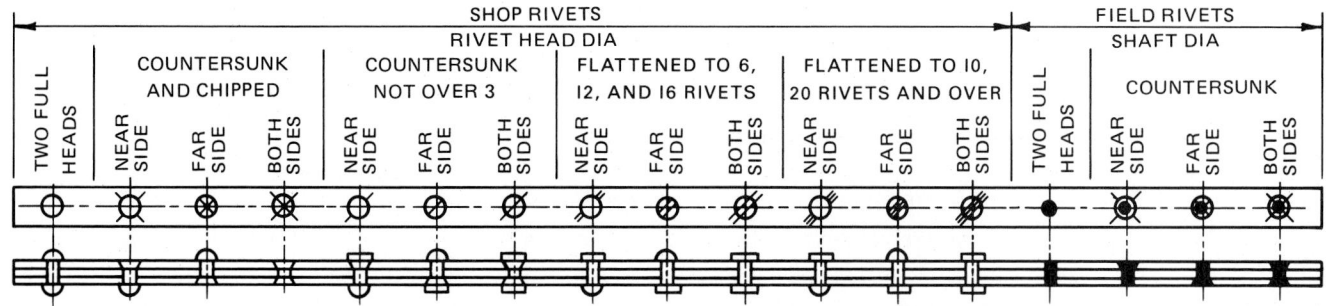

FIG. 11-5-3   Conventional rivet symbols.

supplemented by the relevant information regarding rivet and rivet assembly (Fig. 11-5-4).

The upper left-hand quadrant of the symbol shows the part number for the rivet used in the item list on the drawing or in a table on the drawing that clearly defines the part. This number is preceded by the capital letter R (Fig. 11-5-4B). Where a composite rivet is used (rivet plus sleeve), the item reference numbers for both rivet and sleeve are shown (Fig. 11-5-4C).

SYMBOLIC REPRESENTATION	DESCRIPTION AND MEANING
A	POSITION OF RIVET
B R 17	SOLID RIVET R 17 = RIVET, ITEM REFERENCE 17 SHOWN ON ITEM LIST OR TABLE ON THE DRAWING 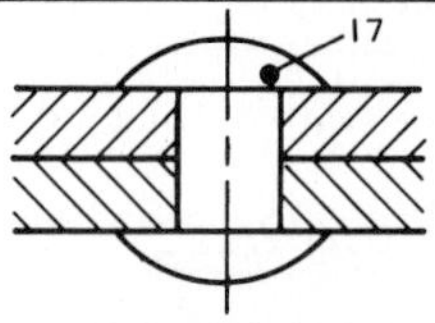
C R 32 35	COMPOSITE RIVET R 32 = RIVET, ITEM REFERENCE 32 SHOWN ON ITEM LIST OR TABLE ON THE DRAWING 35 = SLEEVE, ITEM REFERENCE 35 SHOWN ON ITEM LIST OR TABLE ON THE DRAWING 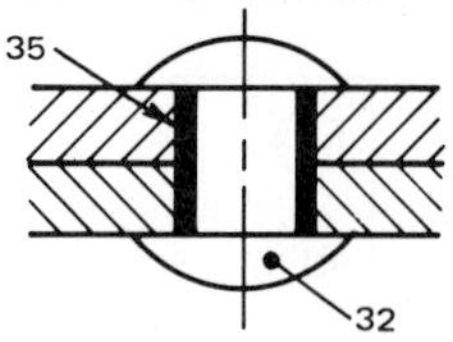
D N OR F	N = PREFORMED HEAD OF THE RIVET ON NEAR SIDE F = PREFORMED HEAD OF THE RIVET ON FAR SIDE
E	100° COUNTERSINK ON NEAR SIDE 82° COUNTERSINK ON FAR SIDE 100° COUNTERSINK ON BOTH SIDES 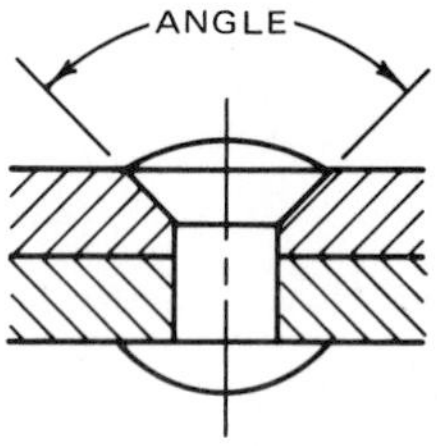
F	100° DIMPLING ON NEAR SIDE 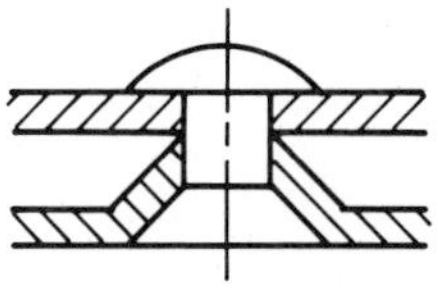 TWO SHEETS 82° DIMPLED ON FAR SIDE 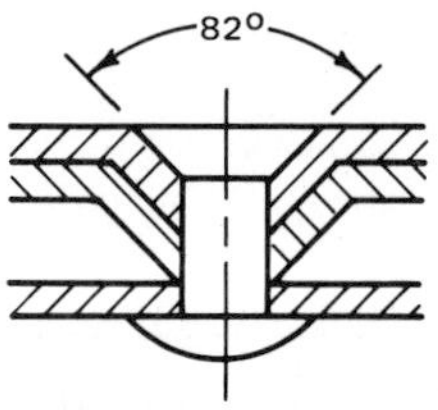
G	{ FIRST SHEET DIMPLED 100° ON NEAR SIDE { SECOND SHEET COUNTERSINK 100° ON FAR SIDE { FIRST SHEET DIMPLED 82° ON NEAR SIDE { SECOND SHEET DIMPLED 82° ON FAR SIDE

FIG. 11-5-4   Symbolic representation for a set (installed) rivet used on aerospace equipment.

The upper right-hand quadrant of the symbol contains a capital letter giving the position of the preformed head (Fig. 11-5-4D).

The lower left-hand quadrant of the symbol contains information on the position of either a countersink or a dimpling, or a combination of both. The countersink to be made on the parts to be riveted is indicated by an equilateral triangle oriented to indicate either near or far side (Fig. 11-5-4E). If other than 100°, the value of the angle in degrees is placed on the right of the countersink symbol. Where dimpling of the sheets to be riveted is required, it is indicated by an open isosceles triangle oriented to indicate either near or far side (Fig. 11-5-4F). If other than 100°, the value of the angle in degrees is placed on the right of the dimpling symbol. Where the combination of a countersink on one part and a dimpling on the other part is required, it is indicated by showing both the countersink and dimpling symbols. If other than 100°, the value of the angle in degrees is placed to the right of the countersink and dimpling symbol (Fig. 11-5-4G). The lower right-hand quadrant of the symbol is left blank.

Symbolic Representation of a Line of Rivets   The crosses (symbol representing the fixed rivet) are aligned along the axes of the drawing, and the number of places for rivets are indicated.

The supplemental information is placed directly on the drawing if space is available or with a leader line indicating the corresponding rivet assembly (Fig. 11-5-5A).

When the rivets are aligned, identical, and equidistant, the symbols should be shown in the first and last positions, together with the total number of pitches and distance (Fig. 11-5-5B).

Small Rivets

Design of small rivet assemblies is influenced by two major considerations:

1. The joint itself, its strength, appearance, and configuration
2. The final riveting operation, in terms of equipment capabilities and production sequence

Types of Small Rivets

Four types of small rivets are illustrated in Fig. 11-5-6 and described as follows.

Semitubular   This is the most widely used type of small rivet. The depth of the hole in the rivet, measured along the wall, does not exceed 112 percent of the mean shank diameter.

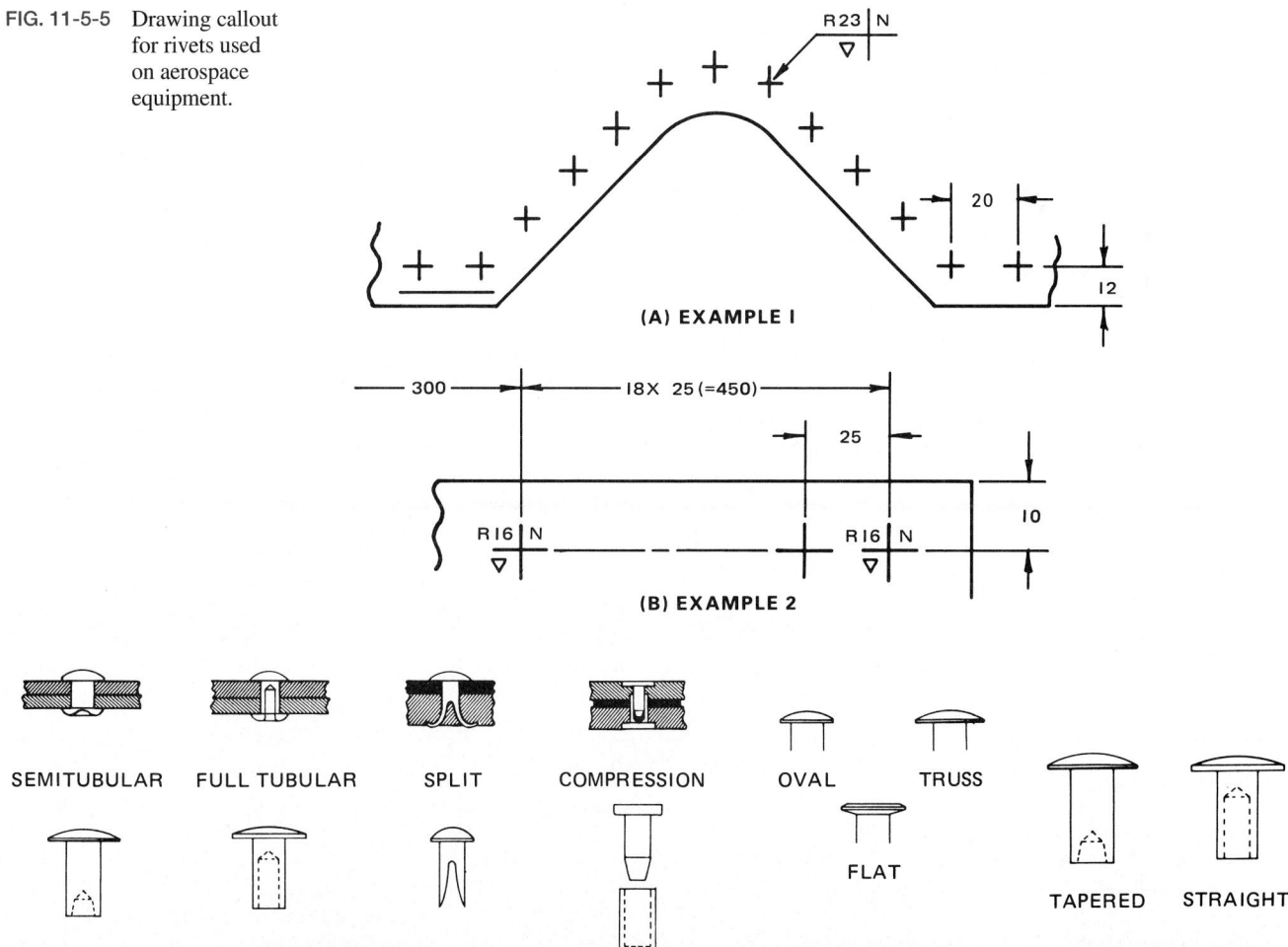

FIG. 11-5-5 Drawing callout for rivets used on aerospace equipment.

FIG. 11-5-6 Basic types of small rivets.

The hole may be extruded (straight or tapered) or drilled (straight), depending on the manufacturer and/or rivet size.

Full Tubular   This rivet has a drilled shank with a hole depth more than 112 percent of the mean shank diameter. It can be used to punch its own hole in fabric, some plastic sheets, and other soft materials, eliminating a preliminary punching or drilling operation.

Bifurcated (Split)   The rivet body is sawed or punched to produce a pronged shank that punches its own hole through fiber, wood, or plastic.

Compression   This rivet consists of two elements: the solid or blank rivet and the deep-drilled tubular member. Pressed together, these form an interference fit.

Design Recommendations

Figure 11-5-7 shows the preferred methods of drawing small-rivet connections for various types of joints, materials, clearances, and so forth. The following are considerations that should be taken into account when small rivets are to be used as fasteners.

Select the Right Rivets   Basic types are covered in Fig. 11-5-6. Rivet standards for all types but compression rivets have been published by the Tubular and Split Rivet Council.

Rivet Diameters   The optimum rivet diameter is determined, not by performance requirements, but by economics—the costs of the rivet and the labor to install it. The rivet length-to-diameter ratio should not exceed 6:1.

Rivet Positioning   The location of the rivet in the assembled product influences both joint strength and clinching requirements. The important dimensions are edge distance and pitch distance.
   Edge distance is the interval between the edge of the part and the center line of the rivet. The recommended edge distance for plastic materials, either solid or laminated, is between two and three diameters, depending on the thickness and inherent strength of the material.
   Pitch distance—the interval between center lines of adjacent rivets—should not be too small. Unnecessarily high stress concentrations in the riveted material and buckling at adjacent empty holes can result if the pitch distance is less than three times the diameter of the largest rivet in the assembly (metal parts) or five times the diameter (plastic parts).

Blind Rivets

Blind riveting is a technique for setting a rivet without access to the reverse side of the joint. However, blind rivets may also be used in applications in which both sides of the joint are actually accessible.
   Blind rivets are classified according to the methods with which they are set: pull-mandrel, drive-pin, and chemically expanded (Fig. 11-5-8).

Design Considerations

Blind-rivet design data are illustrated in Fig. 11-5-9 (pg. 310).

Type of Rivet   Selection depends on a number of factors, such as speed of assembly, clamping capacity, available sizes, adaptability to the assembly, ease of removal, cost, and structural integrity of the joint.

Joint Design   Factors that must be known include allowable tolerances of rivet length versus assembly thickness, hole clearance, joint configuration, and type of loading.

Speed of Installation   The fastest, most efficient installation is done with power tools—air, hydraulic, or electric. Manual tools, such as special pliers, can be used efficiently with practically no training.

In-Place Costs   Blind rivets often have lower in-place costs than solid rivets or tapping screws.

Loading   A blind-rivet joint is usually in compression or shear.

Material Thickness   Some rivets can be set in materials as thin as .02 in. (0.5 mm). Also, if one component is of compressible material, rivets with extra-large head diameter should be used.

Edge Distance   The average recommended edge distance is twice the diameter of the rivet.

Spacing   Rivet pitch should be three times the diameter of the rivet.

Length   The amount of length needed for clinching action varies greatly. Most rivet manufacturers provide data on grip ranges of their rivets.

Backup Clearance   Full entry of the rivet is essential for tightly clinched joints. Sufficient backup clearance must be provided to accommodate the full length of the unclinched rivet.

Blind Holes or Slots   A useful application of a blind rivet is in fastening members in a blind hole. At A in Fig. 11-5-9, the formed head bears against the side of the hole only. This joint is not as strong as the other two (B and C).

Riveted Joints   Riveted cleat or batten holds a butt joint, A. The simple lap joint, B, must have sufficient material beyond the hole for strength. Excessive material beyond rivet hole C may curl up or vibrate or cause interference problems, depending on the installation.

Flush Joints   Generally, flush joints are made by countersinking one of the sections and using a rivet with a countersunk head, A.

Weatherproof Joints   A hollow-core rivet can be sealed by capping it, A; by plugging it, B; or by using both a cap and a plug, C. To obtain a true seal, however, a gasket or mastic should be used between the sections and perhaps under the rivet head. An ideal solution is to use a closed-end rivet.

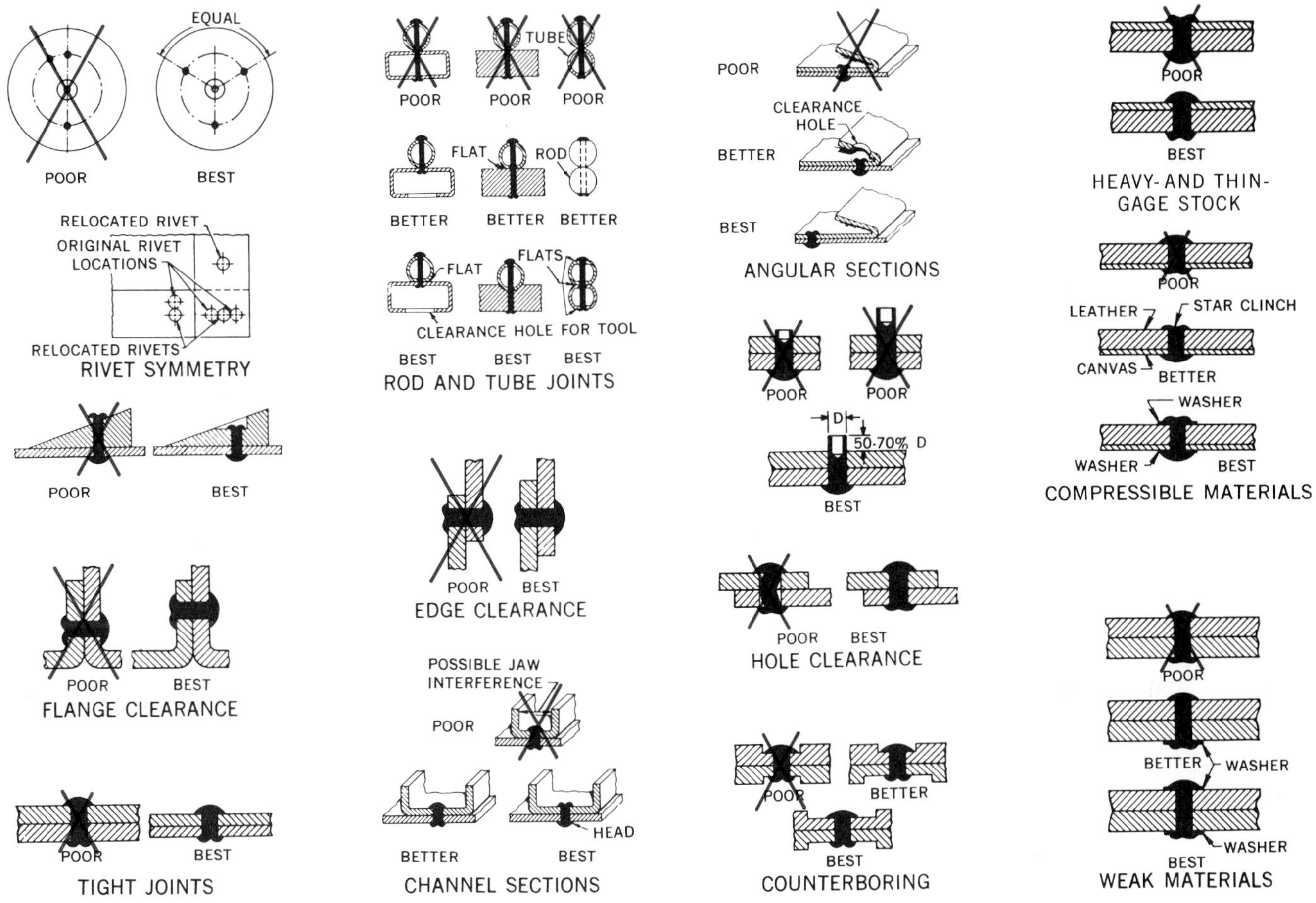

FIG. 11-5-7 Small rivet design data.

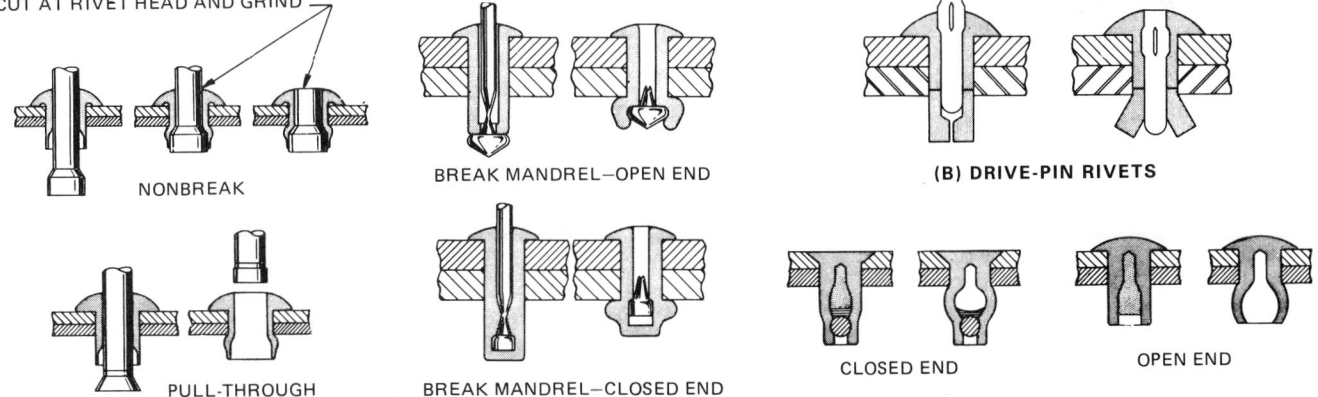

FIG. 11-5-8 Basic types of blind rivets and methods of setting.
(*Machine Design,* Vol. 53, No. 26)

Rubber, Plastic, and Fabric Joints Some plastics, such as reinforced molded fiberglass or polystyrene, which are reasonably rigid, present no problem for most small rivets. However, when the material is very flexible or is a fabric, set the rivet as shown at A or B, with the upset head against the solid member. If this practice is not possible, use a backup strip as shown at C.

Pivoted Joints There are a number of ways of producing a pivoted assembly. Three are shown.

Attaching Solid Rod When attaching a rod to other members, the usual practice is to pass the rivet completely through the rod.

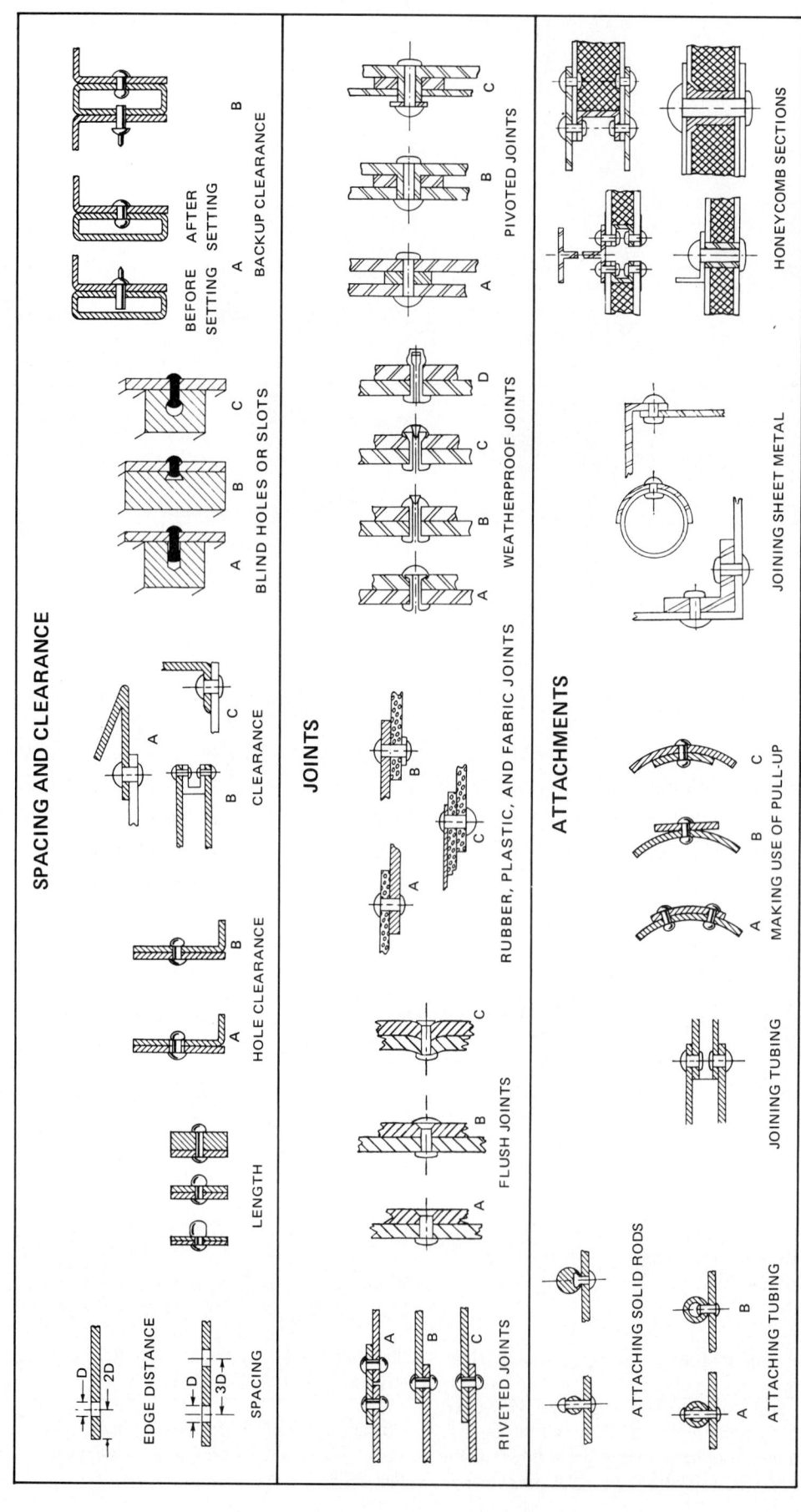

FIG. 11-5-9   Blind rivet design data.

Attaching Tubing Attaching tubing is an application for which the blind rivet is ideally suited.

Joining Tubing This tubing joint is a common form of blind riveting, used for both structural and low-cost power transmission assemblies.

Making Use of Pull-up By judicious positioning of rivets and parts that are to be assembled with rivets, the setting force can sometimes be used to pull together unlike parts.

Honeycomb Sections Inserts should be employed to strengthen the section and provide a strong joint.

REFERENCES AND SOURCE MATERIAL

1. *Machine Design,* Fastening and joining reference issue.

ASSIGNMENTS

See Assignments 12 and 13 for Unit 11-5 on pages 322–323.

11-6 WELDED FASTENERS

The most common forms of welded fasteners are screws and nuts. In this unit, welded fasteners are grouped into resistance-welded threaded fasteners and arc-welded studs.

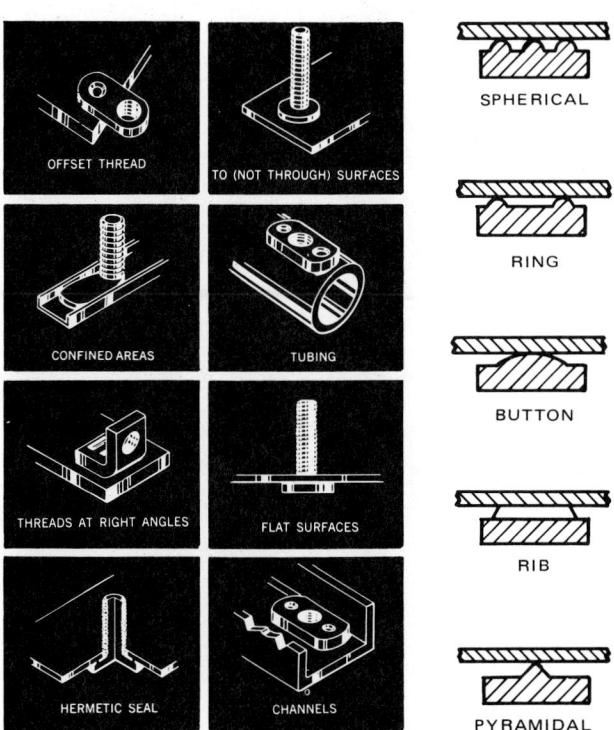

| (A) APPLICATION | (B) WELD PROJECTION |

OFFSET THREAD · TO (NOT THROUGH) SURFACES · SPHERICAL · CONFINED AREAS · TUBING · RING · THREADS AT RIGHT ANGLES · FLAT SURFACES · BUTTON · HERMETIC SEAL · CHANNELS · RIB · PYRAMIDAL

FIG. 11-6-1 Resistance-welded fasteners.

Resistance-Welded Fasteners

Simply defined, a *resistance-welded fastener* is an externally or internally threaded metal part designed to be fused permanently in place by standard production welding equipment. Two methods of resistance welding are used to attach these fasteners: projection welding and spot welding.

Design Considerations

Before fasteners can be used, three basic requirements must be met (Figs. 11-6-1 to 11-6-3, here and on pg. 312).

1. The materials to be joined, both part and fastener, must be suitable for resistance welding.
2. The parts to be welded must be portable enough to be carried to the welder.
3. Production volume should be great enough to justify tooling costs.

Figure 11-6-1 shows typical resistance-welded fasteners.

Arc-Welded Studs

There are two basic stud welding processes: electric-arc and capacitor-discharge.

Use Projection-Weld Fasteners When
■ Suitable projection welding equipment is available.
■ Appearance is an important consideration. Projection welding does not mark the surface on the opposite side of the weld.
■ Simultaneous welding of multiple fasteners is required.
■ Spacing between fasteners must be kept close.
■ Fasteners must be welded to part sections of varying thicknesses.
■ Fasteners must be welded to parts of unusual shape or a watertight weld joint is required.
■ Welding fixtures can be used for easier locating or automatic feeding.
■ Length of production run without maintenance is critical.

Use Spot-Weld Fasteners When
■ Suitable rocker-arm welding equipment is available.
■ Appearance of the part surface opposite the weld is not critical. Spot welding leaves a slight indentation from the electrode tips.
■ Other spot welds are being performed on parts of the assembly.
■ Length of production run without maintenance is not too important. Spot-welding electrode tips will mushroom to some extent in production welding. Shorter runs before refacing or redressing must be expected.
■ Dissimilar materials, such as aluminum, copper, or magnesium, are being welded.
■ Shape, size, or space requirements do not permit use of projection-welded fasteners.

FIG. 11-6-2 Guide to weld fastener selection.

311

Spot-Weld Nuts — columns A, B, C, D

Projection-Weld Nuts — columns E, F, G, H, I, J, K, L, M

Spot-Weld Screws and Pins — columns N, O, P, Q, R, S

Projection-Weld Screws and Pins — columns T, U, V, W, X, Y

Application Factors	A	B	C	D	E	F	G	H	I	J	K	L	M
Flat Surfaces	▲	▼	▼	▲	▲	▼	▲	▲	▲	▲	▲	▲	▲
Curved Surfaces (concave)	▲	▼	▼	▲	▲	▼	▲	▲	▲		▼	▲	
Round Surfaces (convex)	▲	▼	▼	▼	▲	▼	▲	▲	▲	▼	▼	▲	
Tubing	▲	▼	▼		▲	▼	▲	▲	▲		▼		
Channels	▲	▼	▲	▲	▲	▼	▲	▲	▲			▲	
Narrow Flanges	▲	▼	▲	▲	▲	▼	▲	▲	▲		▼	▲	
Offset	▲	▼	▼		▲	▼		▲	▲				
Wall Corners					▲	▲		▲	▲	▲		▲	▲
Blind Hole								▲					▲
Wire	▲	▼		▼	▲			▲	▲				▲
Through Hole	▲	▼		▼	▲		▲	▲	▼	▼	▲	▲	▲
Tension Against Weld		▲			▲	▼	▲	▲	▲		▲		▲
Hermetic Seal								▼					▲
Right Angle	▼		▼	▼					▲				▲
Extra Thread					▲		▲			▲		▲	
Bridging			▲	▲			▲						
Dual Tapped			▲	▲	▲			▲	▲	▲		▲	
Self-Locating Pilot	▲	▲		▼	▲	▲	▼	▲	▲	▲	▲	▲	▲
No Hole Required in Sheet	▼			▼									
Used with Keyhole Slot				▼									
Pilotless						▼	▼	▲		▲		▲	▲

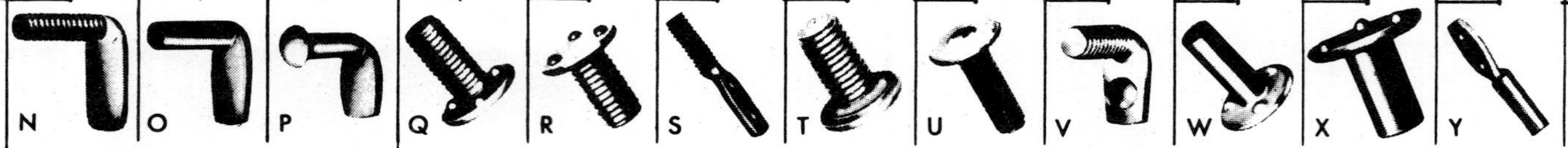

Spot-Weld Screws and Pins — N, O, P, Q, R, S

Projection-Weld Screws and Pins — T, U, V, W, X, Y

LEGEND

Spot-Weld Screws and Pins
A Single Tab
B Targeted
C Double Tab
D Dual Tapped
E Dual Projection
F Button Projection
G Four-Button Projection
H Pilotless
I Right-Angle Bracket

J Blind-Hole Flange
K Through Hole
L Tee-Shape
M Hermetic Seal
N Right-Angle Spade

Projection-Weld Screws and Pins
O Right-Angle Spade Pin
P Keyhole-Slot, Right-Angle Spade Pin
Q Through Hole
R Blind Hole
S Spade
T Hermetic Seal
U Button-Projection, Blind Hole

V Button, Right-Angle Spade
W Through-Hole Pin
X Blind-Hole Pin
Y Spade Pin

FIG. 11-6-3 Resistance-welded fastener guide.

Electric-Arc Stud Welding   The more widely used stud welding process is a semiautomatic electric-arc process. To avoid burn-through, the plate thickness should be at least one-fifth the weld base diameter.

Capacitor-Discharge Stud Welding   This stud welding process derives its heat from an arc produced by a rapid discharge of stored electrical energy.

Design Considerations

In most instances, the thickness of the plate for stud attachment will determine the stud welding process. Electric-arc stud welding is generally used for fasteners .32 in. (8 mm) and larger.

REFERENCES AND SOURCE MATERIAL

1. *Machine Design,* Fastening and joining reference issue.

ASSIGNMENT

See Assignment 14 for Unit 11-6 on page 324.

11-7 ADHESIVE FASTENINGS

Industrial designers and manufacturers are relying on adhesives more than ever before. They allow greater versatility in design, styling, and materials. They can also cut costs. However, as with any engineering tool, there are limitations as well as advantages. For physical properties and application data of typical adhesives, refer to Table 51 of the Appendix.

Adhesion Versus Stress

Adhesion is the force that holds materials together. *Stress* on the other hand, is the force pulling materials apart (Fig. 11-7-1). The basic types of stress in adhesives are:

1. *Tensile.* Pull is exerted equally over the entire joint. Pull direction is straight and away from the adhesive bond. All adhesive contributes to bond strength.

2. *Shear.* Pull direction is across the adhesive bond. The bonded materials are being forced to slide over one another.

3. *Clearance.* Pull is concentrated at one edge of the joint and exerts a prying force on the bond. The other edge of the joint is theoretically under zero stress.

4. *Peel.* One surface must be flexible. Stress is concentrated along a thin line at the edge of the bond.

Resistance to stress is one reason for the rapid increase in the use of adhesives for product assembly. The following points elaborate on stress resistance and the other advantages of adhesives.

Advantages

1. Adhesives allow uniform distribution of stress over the entire bond area. (Fig. 11-7-1). This eliminates stress concentration caused by rivets, bolts, spot welds, and similar fastening techniques. Lighter, thinner materials can be used without sacrificing strength.
2. Adhesives can effectively bond dissimilar materials.
3. Continuous contact between mating surfaces effectively bonds and seals against many environmental conditions.
4. Adhesives eliminate holes needed for mechanical fasteners and surface marks resulting from spot welding, brazing, etc.

Limitations

1. Adhesive bonding can be slow or require critical processing. This is particularly true in mass production. Some adhesives require heat and pressure or special jigs and fixtures to establish the bond.
2. Adhesives are sensitive to surface conditions. Special surface preparation may be required.
3. Some adhesive solvents present hazards. Special ventilation may be required to protect employees from toxic vapors.
4. Environmental conditions can reduce bond strength of some adhesives. Some do not hold well when exposed to low temperatures, high humidity, severe heat, chemicals, water, etc.

FIG. 11-7-1   Stresses in bonded joints.

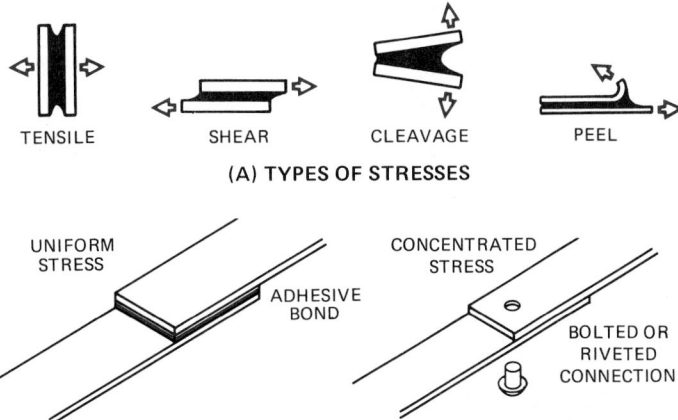

TENSILE   SHEAR   CLEAVAGE   PEEL

(A) TYPES OF STRESSES

UNIFORM STRESS

ADHESIVE BOND

CONCENTRATED STRESS

BOLTED OR RIVETED CONNECTION

(B) STRESSES CAUSED BY FASTENERS

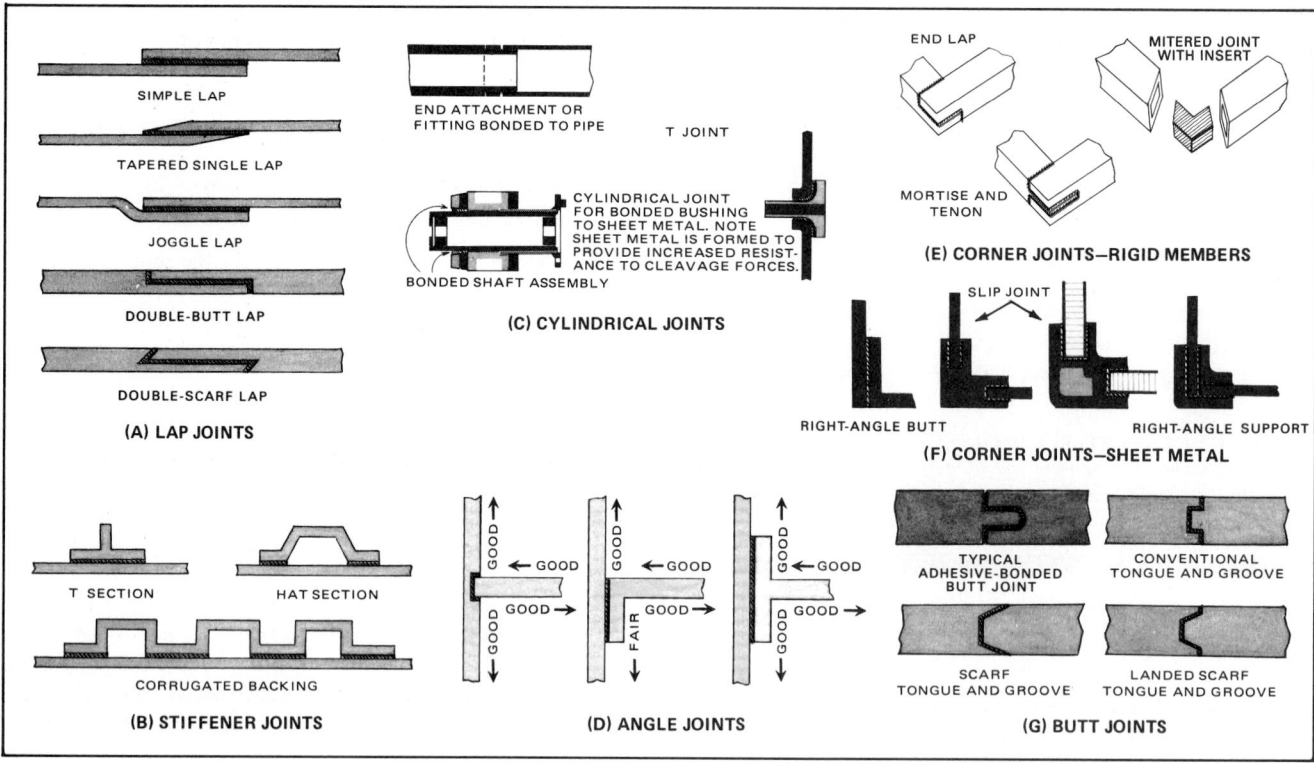

FIG. 11-7-2   Adhesive joint design guide.

Joint Design

Joints should be specifically designed for use with structural adhesives. First, the joint should be designed so that all the bonded area shares the load equally. Second, the joint configuration should be designed so that basic stress is primarily in shear or tensile, with cleavage and peel minimized or eliminated.

The following structural joints and their advantages and disadvantages illustrate some typical design alternatives (Fig. 11-7-2).

Lap Joints   Lap joints are most practical and applicable in bonding thin materials. The simple lap joint is offset. This can result in cleavage and peel stress under load when thin materials are used. A tapered single lap joint is more efficient than a simple lap joint. The tapered edge allows bending of the joint edge under stress. The joggle lap joint gives more uniform stress distribution than either the simple or tapered lap joint.

The double-butt lap joint gives more uniform stress distribution in the load-bearing area than the above joints. This type of joint, however, requires machining, which is not always feasible with thinner-gage metals. Double-scarf lap joints have better resistance to bending forces than double-butt joints.

Angle Joints   Angle joints give rise to either peel or cleavage stress depending on the gage of the metal. Typical approaches to the reduction of cleavage are illustrated.

Butt Joints   The following recessed butt joints are recommended: landed scarf tongue and groove, conventional tongue and groove, and scarf tongue and groove.

Cylindrical Joints   The T joint and overlap slip joint are typical for bonding cylindrical parts such as tubing, bushings, and shafts.

Corner Joints—Sheet Metal   Corner joints can be assembled with adhesives by using simple supplementary attachments. This permits joining and sealing in a single operation. Typical designs are right-angle butt joints, slip joints, and right-angle support joints.

Corner Joints—Rigid Members   Corner joints, as in storm doors or decorative frames, can be adhesive-bonded. End lap joints are the simplest design type, although they require machining. Mortise and tenon joints are excellent from a design standpoint, but they also require machining. The mitered joint with an insert is best if both members are hollow extrusions.

Stiffener Joints   Deflection and flutter of thin metal sheets can be minimized with adhesive-bonded stiffeners.

REFERENCES AND SOURCE MATERIAL

1. 3M Co.

ASSIGNMENT

See Assignment 15 for Unit 11-7 on page 325.

11-8 FASTENER REVIEW FOR CHAPTERS 10 AND 11

In chapters 10 and 11 the more common types of fasteners were explained and drawing problems were assigned for each type of fastener. In this unit selected assignments, which incorporate a variety of fasteners, were chosen to provide a thorough review of the numerous types of fasteners available to the designer.

ASSIGNMENTS

See Assignments 16 and 17 for Unit 11-8 on pages 325–326.

ASSIGNMENTS FOR CHAPTER 11

ASSIGNMENTS FOR UNIT 11-1, KEYS, SPLINES, AND SERRATIONS

1. Lay out the two fastener assemblies shown in Fig. 11-1-A or 11-1-B (pg. 316). The following fasteners are used:

   For 11-1-A
   - *Assembly A:* flat key
   - *Assembly B:* serrations

   For 11-1-B
   - *Assembly A:* square key
   - *Assembly B:* Woodruff key

   Refer to the Appendix and manufacturers' catalogs for sizes and use your judgment for dimensions not shown. Show the dimensions for the keyseats and serrations. Scale 1:1.

FIG. 11-1-A   Key and serration fasteners.

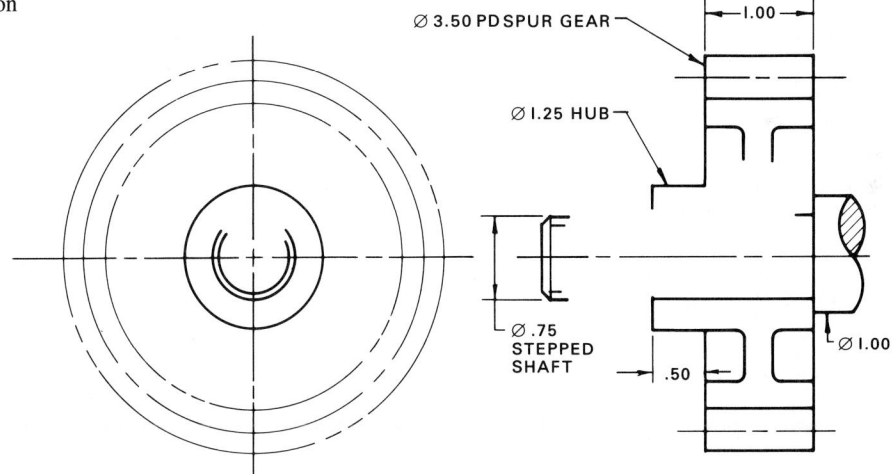

ASSEMBLY A (FLAT KEY)

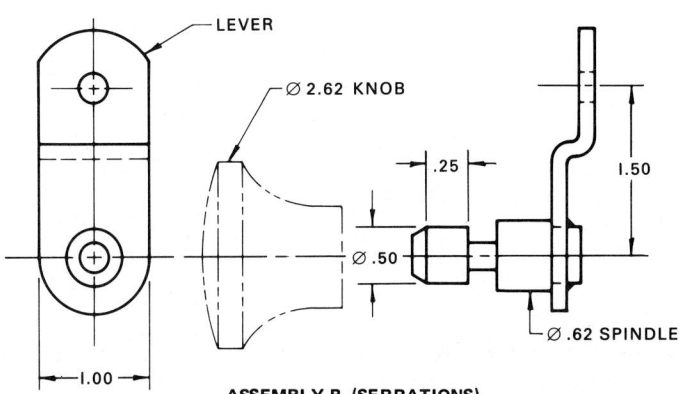

ASSEMBLY B (SERRATIONS)

2. Make a working drawing of the axle shown in Fig. 11-1-C. Dimension the keyseats according to Fig. 11-1-3. Refer to the Appendix.

ASSIGNMENTS FOR UNIT 11-2, PIN FASTENERS

3. Complete the pin assemblies shown in Fig. 11-2-A or 11-2-B, given the following information:

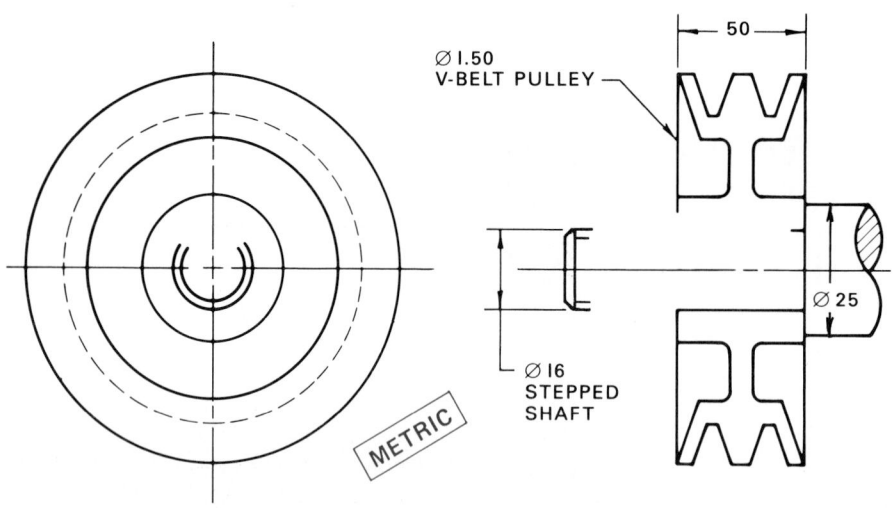

Ø 1.50
V-BELT PULLEY

50

Ø 25

Ø 16
STEPPED
SHAFT

METRIC

ASSEMBLY A (SQUARE KEY)

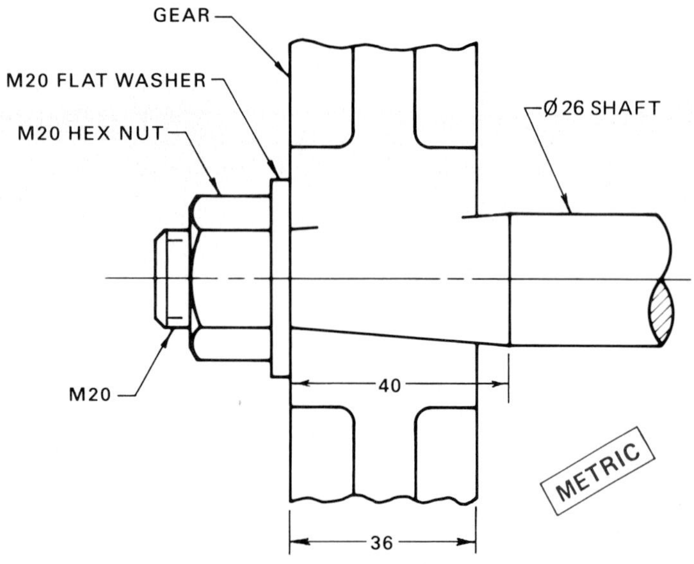

GEAR

M20 FLAT WASHER

M20 HEX NUT

Ø 26 SHAFT

M20

40

METRIC

36

ASSEMBLY B (WOODRUFF KEY)

FIG. 11-1-B   Key fasteners.

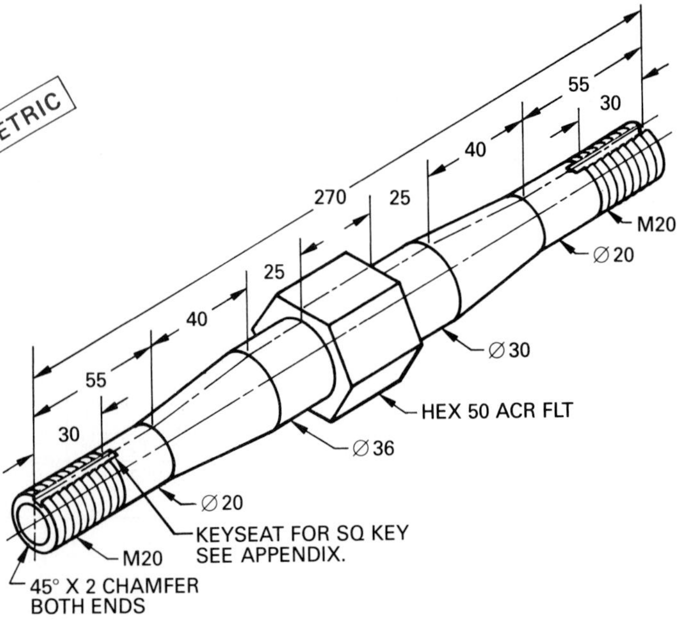

55
30
M20
Ø 20
270
25
40
25
Ø 30
40
HEX 50 ACR FLT
55
Ø 36
30
Ø 20
KEYSEAT FOR SQ KEY
SEE APPENDIX.
M20
45° X 2 CHAMFER
BOTH ENDS

FIG. 11-1-C   Axle.

For Fig. 11-2-A
- *Assembly A.* Slotted tubular spring pins are used to fasten the cap and handle to the shaft. Scale 1:2.
- *Assembly B.* A clevis pin whose area is equal to the four rivets is used to fasten the trailer hitch to the tractor draw bar. Scale 1:2.

For Fig. 11-2-B
- *Assembly A.* A type E grooved pin holds the roller to the bracket. A washer and cotter pin are used to fasten the bracket to the push rod. Scale 1:1.
- *Assembly B.* A type A3 grooved pin holds the V-belt pulley to the shaft. Scale 1:1.

Refer to manufacturers' catalogs for pin sizes and provide the complete information to order each fastener.

FIG. 11-2-A   Pin fasteners.

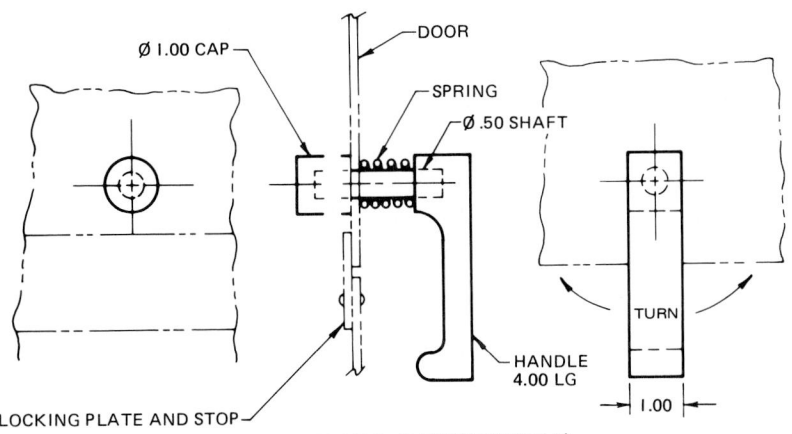

ASSEMBLY A (CABINET HANDLE)

ASSEMBLY B (DRAW BAR HITCH)

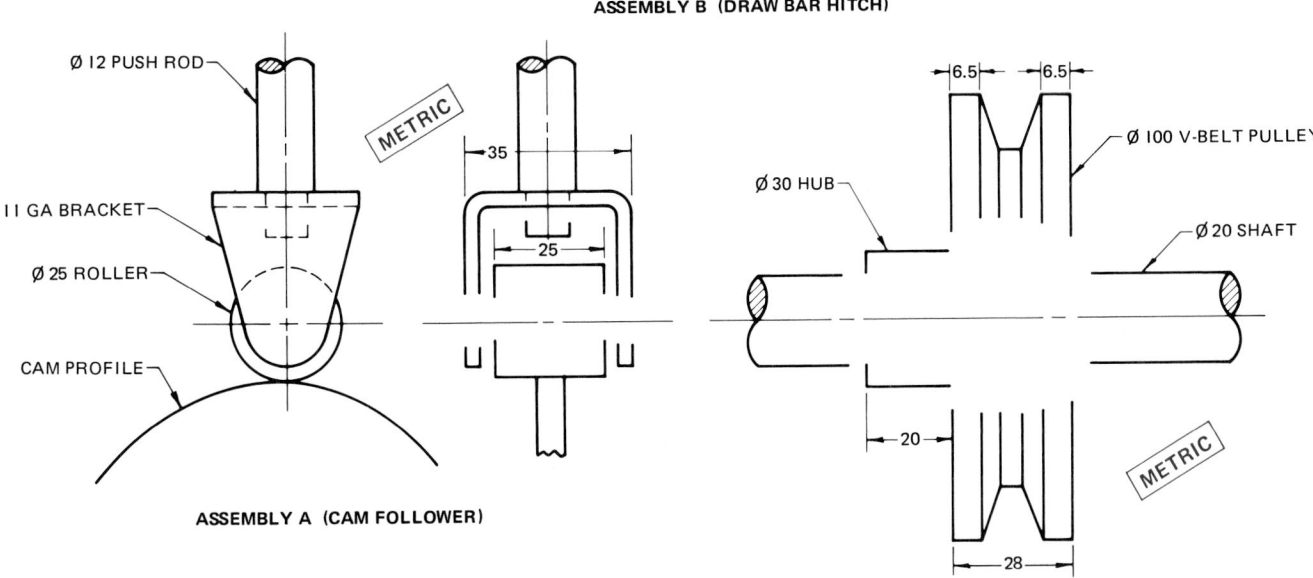

ASSEMBLY A (CAM FOLLOWER)

ASSEMBLY B (V-BELT PULLEY)

FIG. 11-2-B   Pin fasteners.

4. Make a two-view assembly drawing of the crane hook shown in Fig. 11-2-C. The hook is to be held to the U-frame with a slotted locknut. A spring pin is inserted through the locknut slots to prevent the nut from turning. A clevis pin with washer and cotter pin holds the pulley to the frame. Include on the drawing an item list. Scale 1:1.

5. Prepare detail drawings of the parts in Assignment 4. Use your judgment for the scale and selection of views.

ASSIGNMENTS FOR UNIT 11-3, RETAINING RINGS

6. Complete the assemblies shown in Fig. 11-3-A or 11-3-B by adding suitable retaining rings as per the information supplied below. Refer to the Appendix and manufacturers'

FIG. 11-2-C   Crane hook.

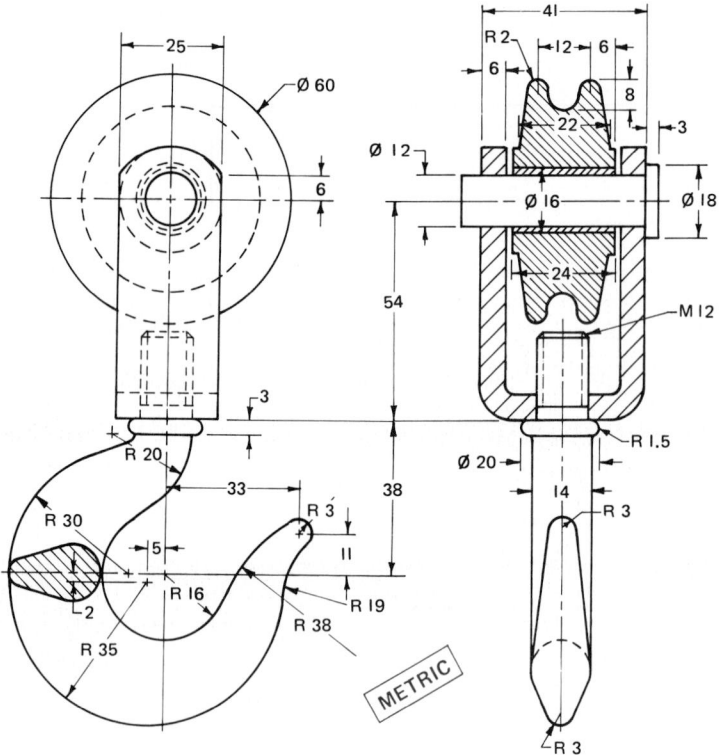

FIG. 11-3-A   Retaining ring fasteners.

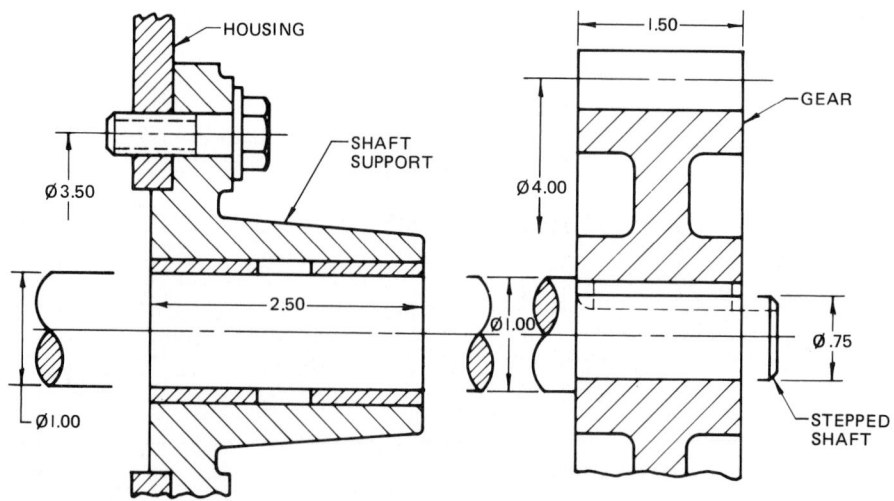

ASSEMBLY A (EXTERNAL RETAINING RINGS)

catalogs and show on the drawing the catalog number for the retaining ring. Add ring and groove sizes. Scale 1:1. Use your judgment for dimensions not shown.

For Fig. 11-3-A an external radial retaining ring mounted on the shaft is to act as a shoulder for the shaft support. An external axial retaining ring is required to hold the gear on the shaft.

For Fig. 11-3-B

- *Assembly A.* External self-locking retaining rings hold the roller shaft in position on the bracket.
- *Assembly B.* An external self-locking ring holds the plastic housing to the viewer case. An internal self-locking ring holds the lens in position.

7. Complete the power drive assembly shown in Fig. 11-3-C given the following information. The shaft is positioned in the housing by an SKF #6005 bearing. The end cap and a retaining ring hold the bearing in place and a retaining ring holds the cap in the housing. Two retaining rings position the bearing on the shaft. The gear is positioned on the clutch by a retaining ring and a square key. The clutch is locked to the shaft by a square key held in position by a setscrew. The pulley drive is positioned and held to the shaft by a square key and two retaining rings. Include an item list on the drawing calling out the purchased parts.

8. Make detail drawings of the end cap and the partial view of the shaft in Assignment 7.

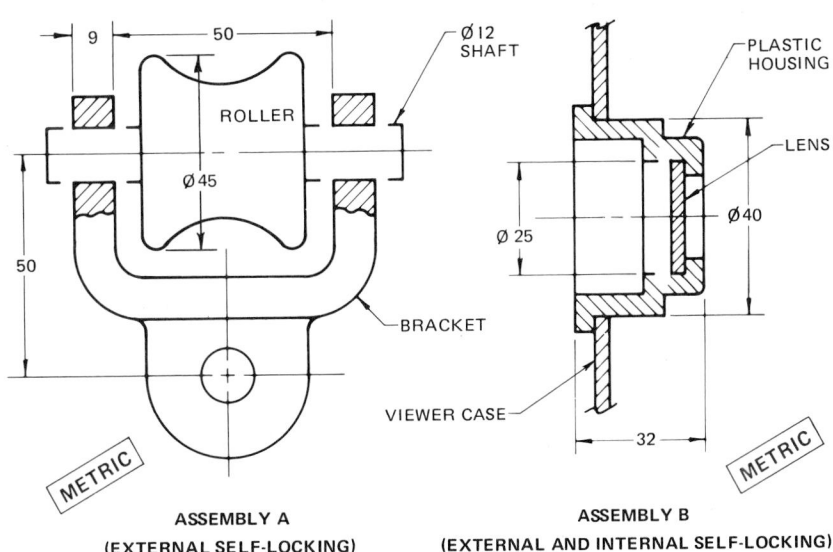

FIG. 11-3-B Retaining ring fasteners.

ASSEMBLY A
(EXTERNAL SELF-LOCKING)

ASSEMBLY B
(EXTERNAL AND INTERNAL SELF-LOCKING)

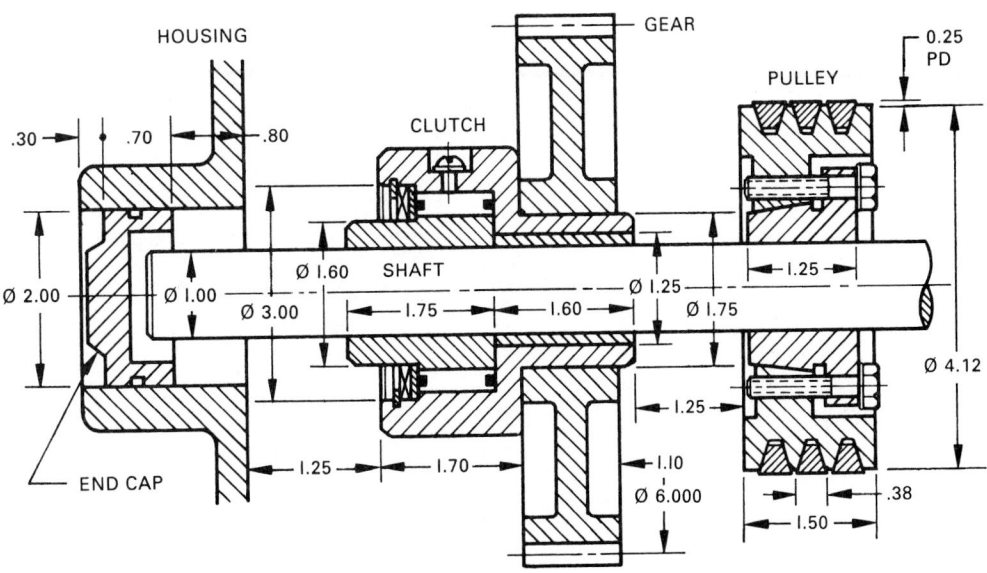

FIG. 11-3-C Power drive assembly.

ASSIGNMENTS FOR UNIT 11-4, SPRINGS

9. Lay out the two assembly drawings as shown in Figs. 11-4-A or 11-4-B. Complete the drawings from the information supplied below, and make detail drawings of the springs. Use your judgment for sizes not given.
For Fig. 11-4-A
- *Assembly A.* The license plate holder is held to the frame of the car by a hinge. A torsion spring is required to keep the plate holder in position. The torsion spring is slipped over the hinge pin during assembly, and one end of the spring passes through the hole in the bumper. The other end of the spring is locked into the spring-retaining notch in the license plate holder. Scale 1:2.
- *Assembly B.* Flat springs are positioned in openings C and D in the tape deck player. These springs hold the cassette against the bottom and the locating pin positioned in the left side of the tape deck. Scale 1:2.

For Fig. 11-4-B
- *Assembly A.* An extension spring controls the lever. The spring is fastened to the neck in the pin and through the hole in the lever. Scale 1:1.

FIG. 11-4-A   Spring fasteners.

- *Assembly B.* A compression spring mounted on the shaft of the handle provides sufficient pressure to hold the lever in position, thus maintaining the door against the panel. To open the door, the handle is pushed in and turned. This action compresses the spring and forces the lever away from the notch in the panel edge, thus permitting the lever to turn. Scale 1:1.
10. Complete the punch holder assembly shown in Fig. 11-4-C given the following information. The two helical springs have plain closed ends and are Ø.06 and have a pitch of .10. The plunger and punch are held in the punch holder by retaining rings. An RC3 fit is required for the Ø.30 shaft. Include on the drawing an item list.
11. Make working drawings of the parts from the completed assembly drawing in Assignment 10. Use your judgment for dimensions not shown.

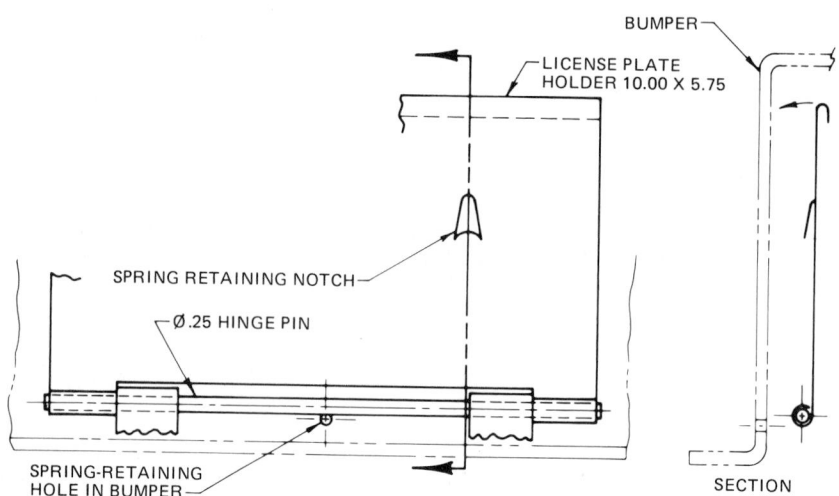

ASSEMBLY A (TORSION SPRING)

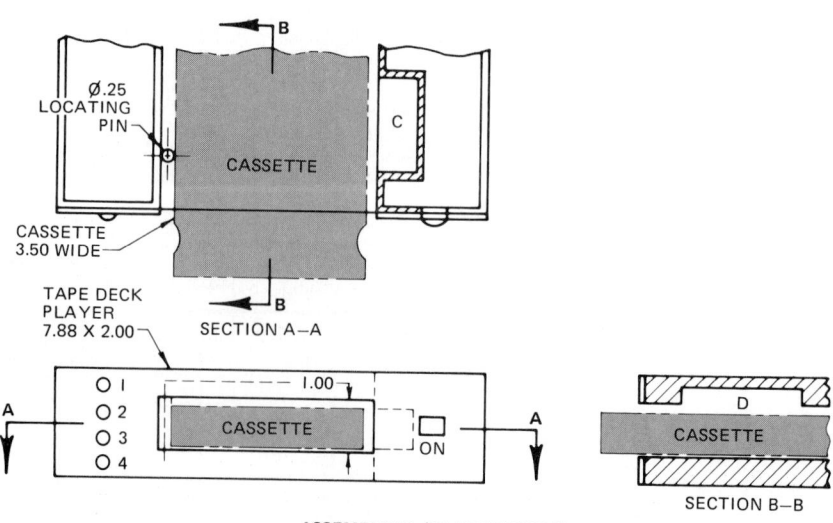

ASSEMBLY B (FLAT SPRINGS)

FIG. 11-4-B Spring fasteners.

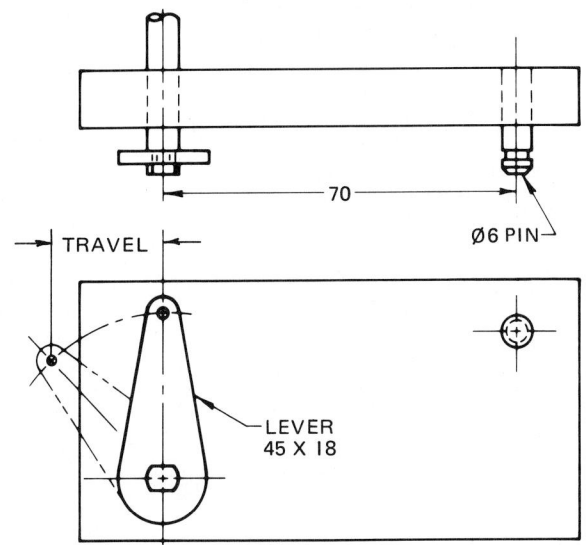

ASSEMBLY A (EXTENSION SPRING)

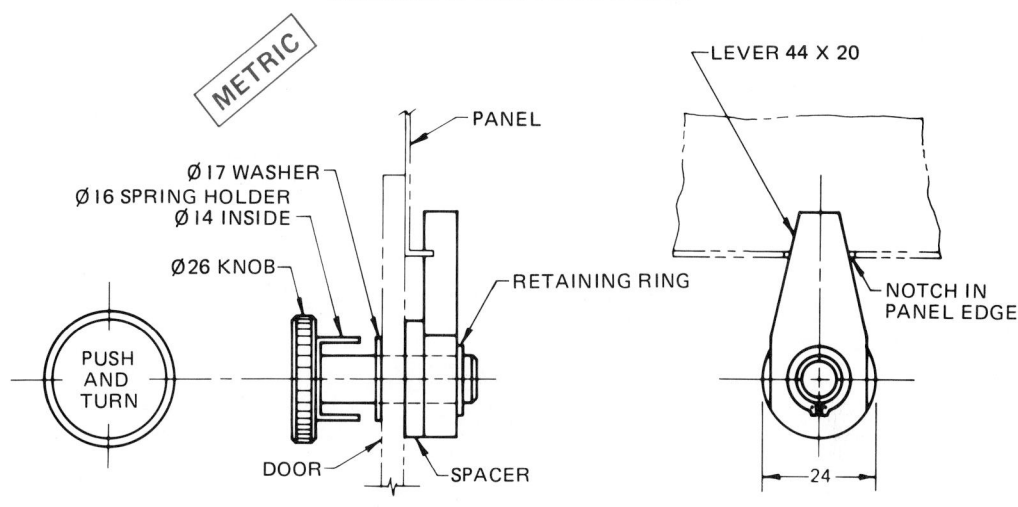

ASSEMBLY B (COMPRESSION SPRING)

FIG. 11-4-C Punch holder assembly.

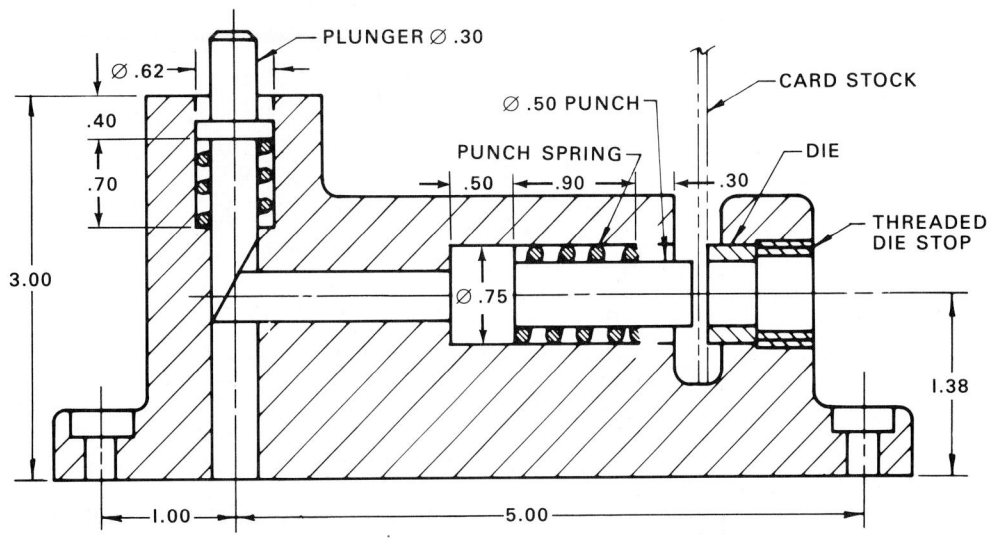

321

ASSIGNMENTS FOR UNIT 11-5, RIVETS

12. Complete the two assembly drawings shown in Fig. 11-5-A or 11-5-B from the information supplied below. Refer to manufacturers' catalogs for rivet type and sizes, and on each assembly show the callout for the rivets. Use your judgment for sizes not given.
    For Fig. 11-5-A
    - *Assembly A.* Padlock brackets are riveted to the locker door and door frame with two blind rivets in each bracket. Scale 1:1.
    - *Assembly B.* The roof truss is assembled in the shop with five evenly spaced Ø.50 in. (12 mm) rivets in each angle. Scale 1:4.
    For Fig. 11-5-B
    - *Assembly A.* The grill is held to the panel by four truss-head full tubular rivets. Scale 1:1.

- *Assembly B.* The support is held to the plywood panel by drive rivets uniformly spaced on the gage lines. Two rivets hold the bracket to the support. Scale 1:1.

13. Complete the assembly shown in Fig. 11-5-C using the graphical symbols of rivets for aerospace equipment and given the following information:
    - *Assembly A.* Ø8 rivets equally spaced at 55 OC; item reference 22; 100° countersunk both sides; preformed head near side.
    - *Assembly B.* Ø6 combined rivets equally spaced at 50 OC; item reference 19, sleeve item reference 21; preformed head far side.
    - *Assembly C.* Ø4 rivets equally spaced at 40 OC (4 sides); item reference 16; preformed head far side; 82° dimple near side.

FIG. 11-5-A   Rivet fasteners.

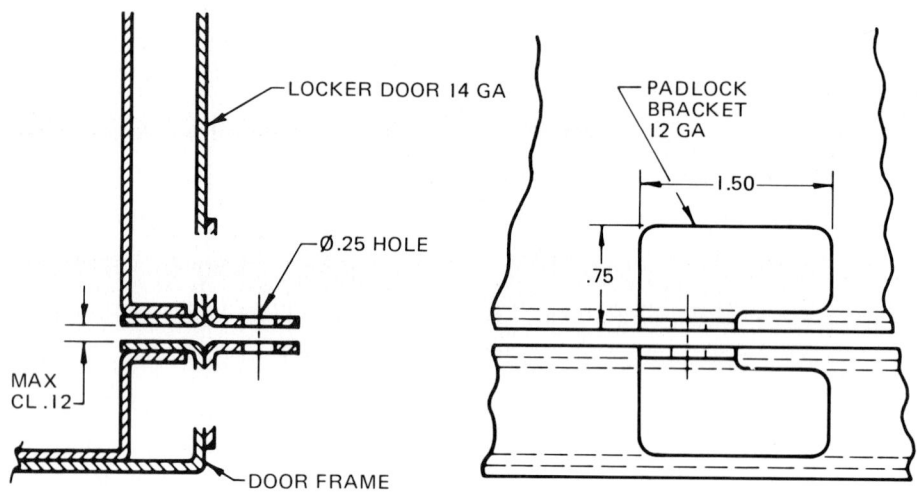

LOCKER DOOR 14 GA

PADLOCK BRACKET 12 GA

Ø.25 HOLE

1.50

.75

MAX CL .12

DOOR FRAME

ASSEMBLY A (BLIND RIVETS)

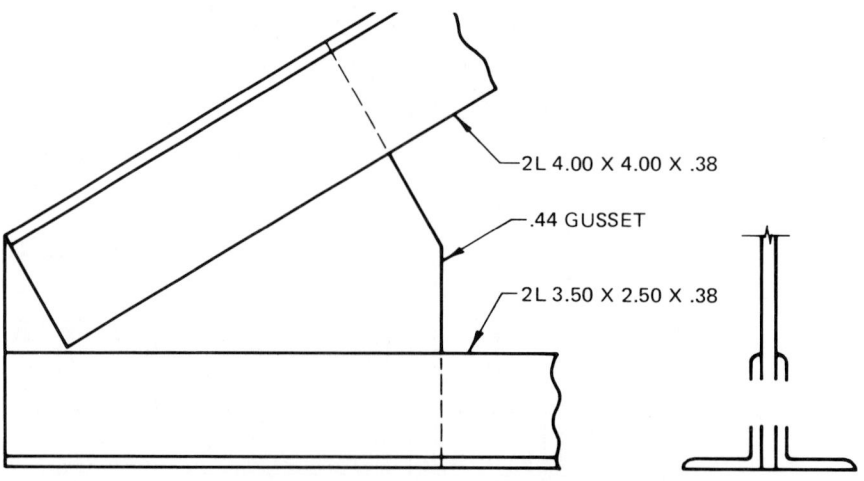

2L 4.00 X 4.00 X .38

.44 GUSSET

2L 3.50 X 2.50 X .38

ASSEMBLY B (LARGE STRUCTURAL RIVETS)

FIG. 11-5-B   Rivet fasteners.

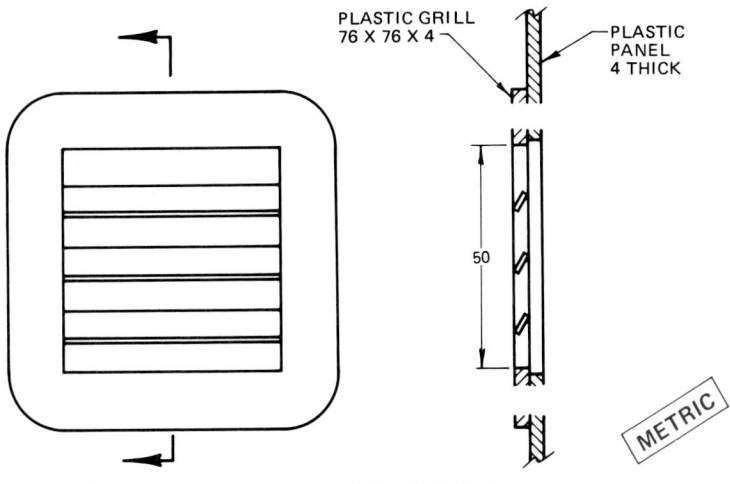

ASSEMBLY A (SMALL RIVETS)

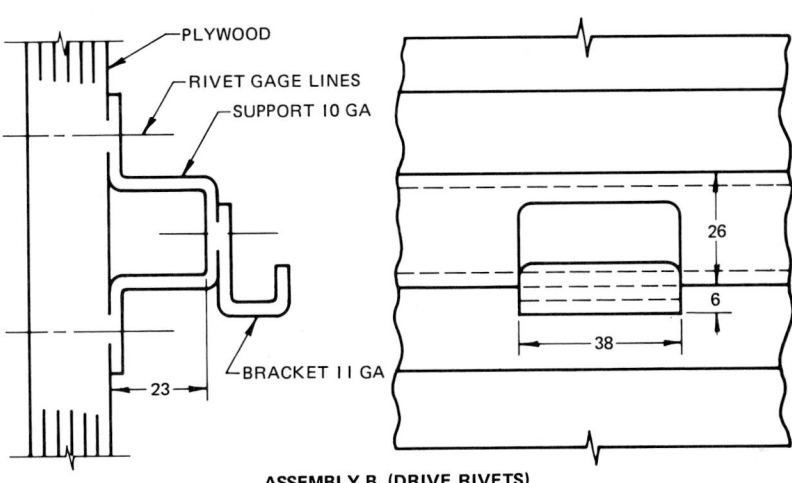

ASSEMBLY B (DRIVE RIVETS)

FIG. 11-5-C   Rivets for aerospace equipment.

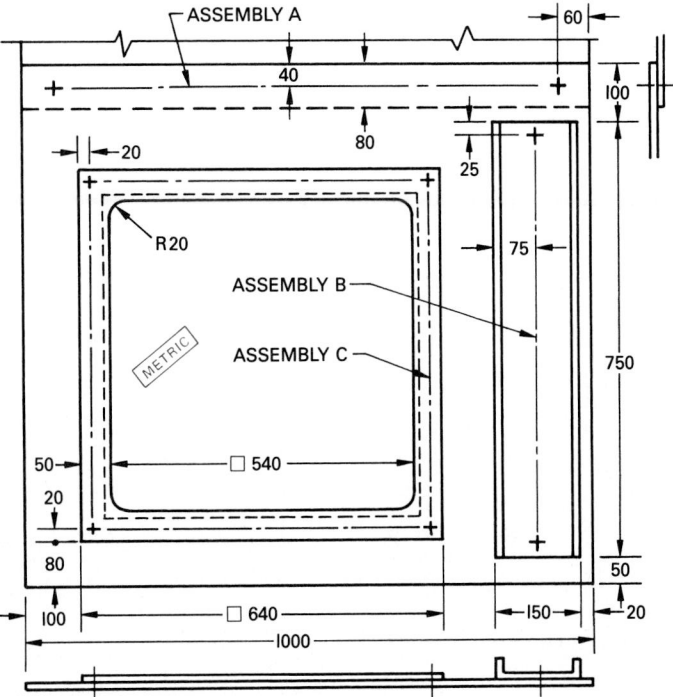

ASSIGNMENT FOR UNIT 11-6, WELDED FASTENERS

14. Complete the two assemblies shown in Fig. 11-6-A or 11-6-B. Refer to manufacturers' catalogs and the Appendix for standard fastener components. Complete the drawings from the information supplied below. Use your judgment for sizes not given. Scale 1:1.

    For Fig. 11-6-A

    ■ *Assembly A.* Two resistance-welded threaded fasteners, one on each side of the pipe, are required. The bracket drops over the fasteners, and lock washers and nuts secure the bracket to the pipe.

    ■ *Assembly B.* A leakproof attaching method (stud welding) is required to hold the adaptor to the panel.

    For Fig. 11-6-B

    ■ *Assembly A.* A spot-weld nut is to be attached to the panel. A hole in the clamp permits a machine screw to fasten the pipe clamp to the nut.

    ■ *Assembly B.* A right-angle bracket is to be fastened (projection welding) to the bottom plate. The vertical plate is secured to the bracket by a machine screw and lock washer.

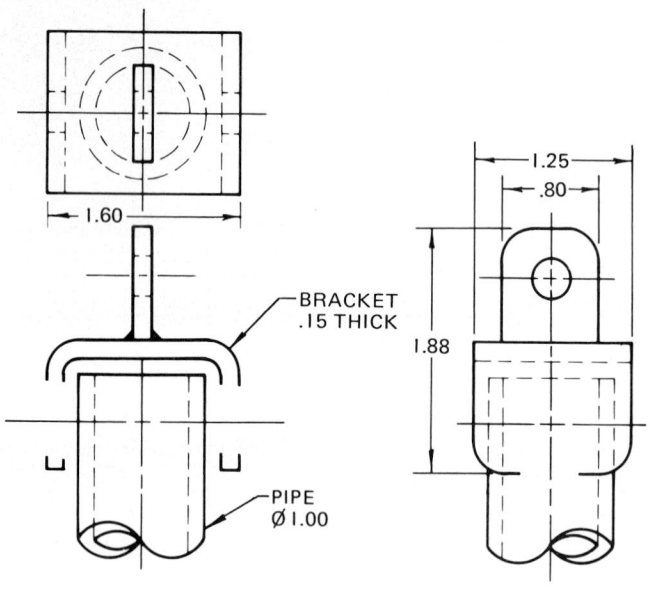

ASSEMBLY A (PIPE ATTACHMENT)

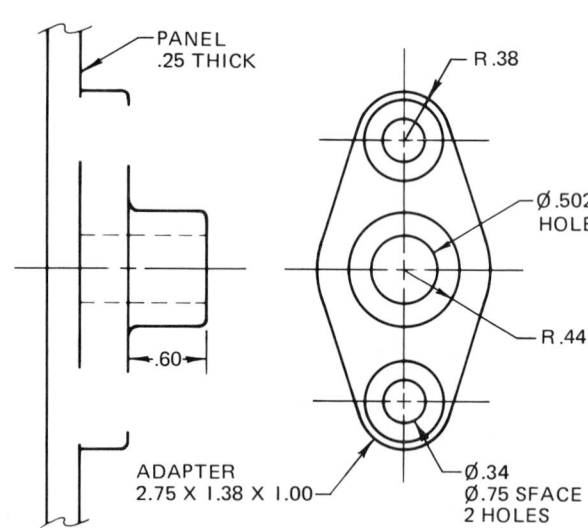

ASSEMBLY B (LEAKPROOF ATTACHMENT)

FIG. 11-6-A   Welded fasteners.

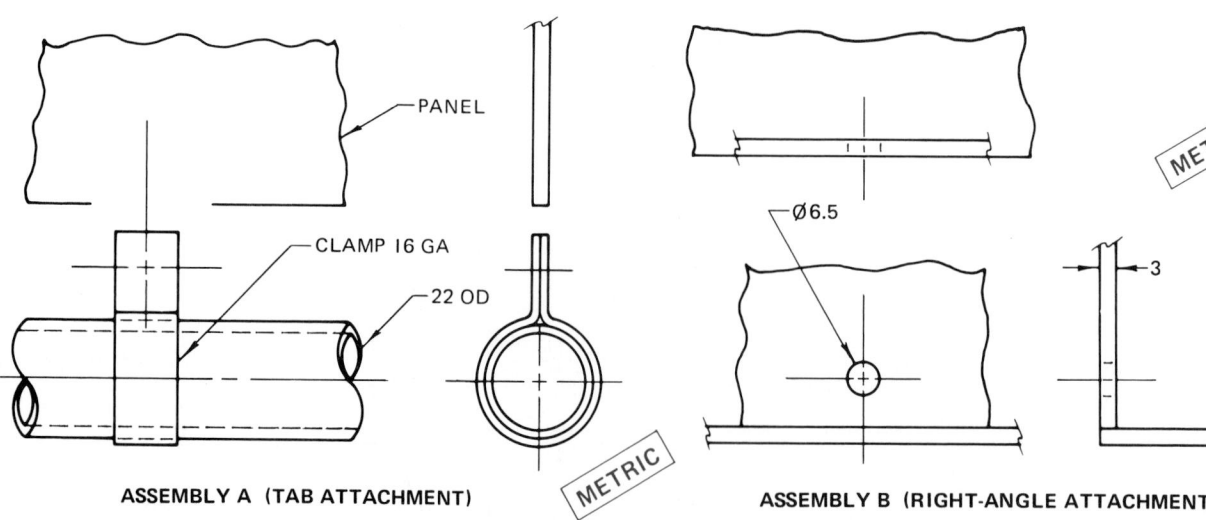

ASSEMBLY A (TAB ATTACHMENT)

ASSEMBLY B (RIGHT-ANGLE ATTACHMENT)

FIG. 11-6-B   Welded fasteners.

ASSIGNMENT FOR UNIT 11-7, ADHESIVE FASTENINGS

15. Complete the two adhesive-bonded assemblies shown in Fig. 11-7-A or 11-7-B from the information supplied

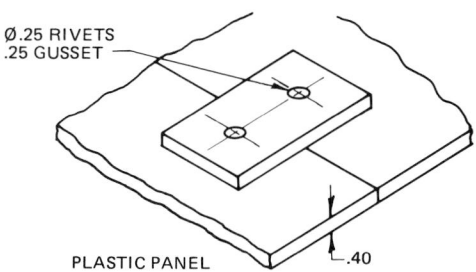

ASSEMBLY A (BUTT JOINT)

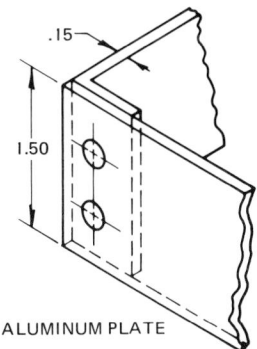

ASSEMBLY B (SLIP JOINT)

FIG. 11-7-A   Adhesive fastenings.

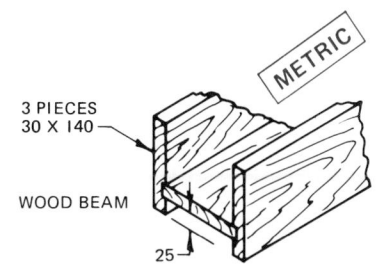

ASSEMBLY A (ANGLE JOINT)

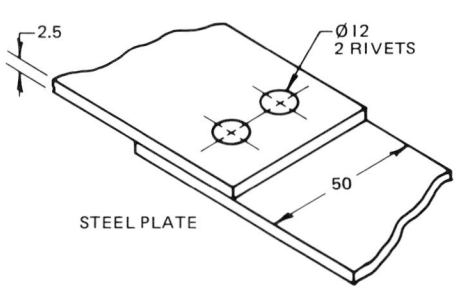

ASSEMBLY B (LAP JOINT)

FIG. 11-7-B   Adhesive fastenings.

below and the adhesive chart in the Appendix. List the adhesive product number and state the method of application you would recommend. Use your judgment for sizes not shown, and dimension the joint. Scale is to suit.

For Fig. 11-7-A
- *Assembly A.* The riveted joint shown is to be replaced by a joggle lap joint. It must be fast-drying.
- *Assembly B.* The sheet-metal corner joint shown is to be replaced by a slip joint. It must be water-resistant.

For Fig. 11-7-B
- *Assembly A.* Three pieces of wood are to be assembled into the shape shown. Joint design has not been shown.
- *Assembly B.* The riveted joint shown is to be replaced by a joggle lap joint. See Table 51 of the Appendix for more information on military (MMM) specs.

ASSIGNMENTS FOR UNIT 11-8, FASTENER REVIEW FOR CHAPTERS 10 AND 11

16. Prepare detail drawings of the parts shown in Fig. 11-8-A. Include on the drawing an item list. The shaft is to have an RC4 fit with the bushing and the bushing an LN3 fit in the body. Use your judgment for the selection and number of views for each part.

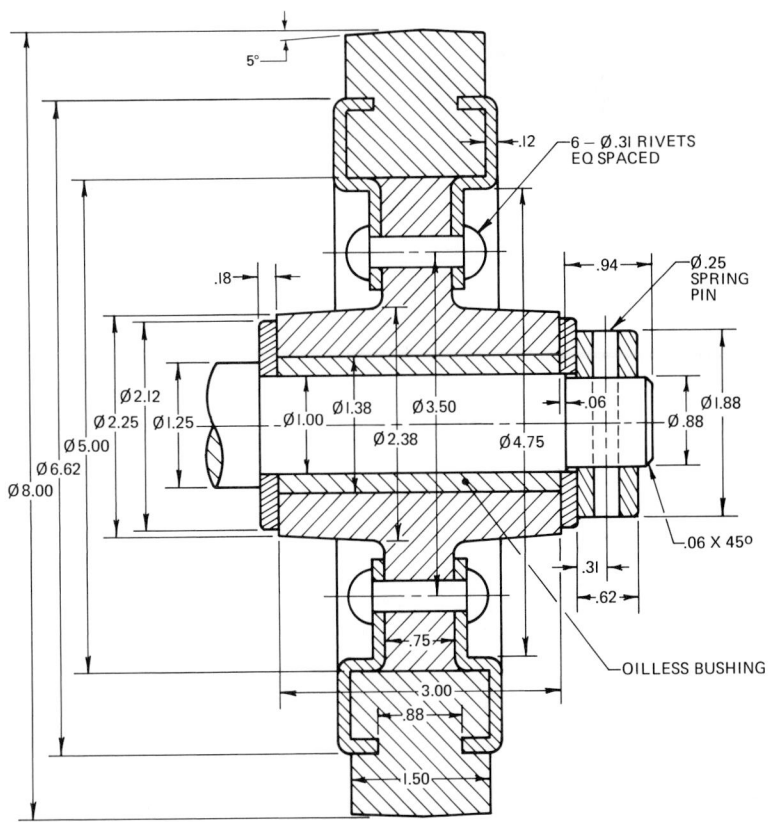

FIG. 11-8-A   Wheel assembly.

17. Make a one-view assembly drawing of the universal joint
    shown in Fig. 11-8-B. Include on the drawing an item list.

FIG. 11-8-B   Universal joint.

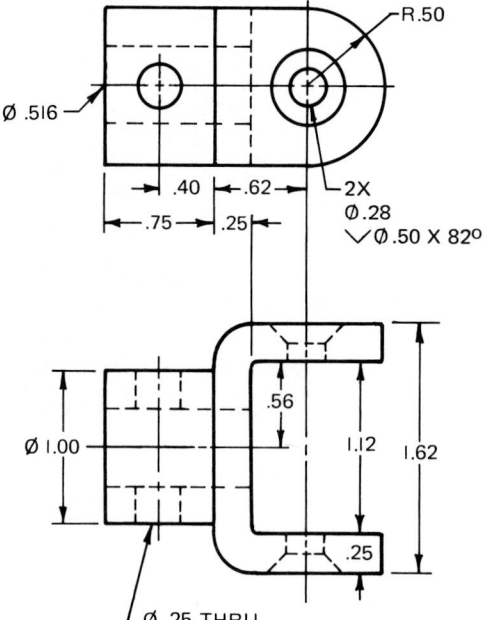

PT I - FORK - 2 REQD

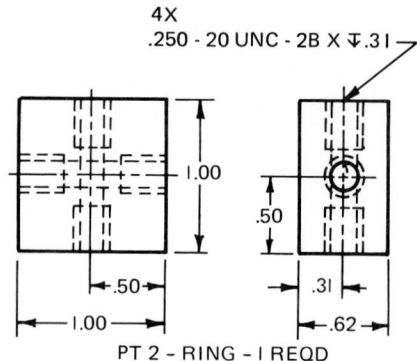

PT 2 - RING - I REQD

PT 3 - Ø .25 SPRING PIN - 2 REQD
PT 4 - .250 - 20 FHMS - .62 LG - 4 REQD

CHAPTER 12

MANUFACTURING MATERIALS

Definitions

Chilling   A process that produces white iron.

High alloy   Steel castings that contain a minimum of eight percent nickel and/or chromium.

Plastics   Nonmetallic materials capable of being formed or molded with the aid of heat, pressure, chemical reactions, or a combination of these.

Thermoplastics   Materials that soften, or liquefy, and flow when heat is applied.

Thermosetting plastics   Materials which undergo an irreversible chemical change when heat is applied or when a catalyst or reactant is added.

12-1 CAST IRONS AND FERROUS METALS

This chapter is an up-to-date reference on manufacturing materials. It provides the drafter and designer with basic information on materials and their properties to ensure the proper selection of the product material.

Ferrous Metals

Iron and the large family of iron alloys called steel are the most frequently specified metals. Iron is abundant (iron ore constitutes about five percent of the earth's crust), easy to convert from ore to a useful form, and iron and steel are sufficiently strong and stable for most engineering applications.

All commercial forms of iron and steel contain carbon, which is an integral part of the metallurgy of iron and steel.

Cast Iron

Because of its low cost, cast iron is often considered a simple metal to produce and to specify. Actually, the metallurgy of cast iron is more complex than that of steel and other familiar design materials. Whereas most other metals are usually specified by a standard chemical analysis, the same analysis of cast iron can produce several entirely different types of iron, depending upon rate of cooling, thickness of the casting, and how long the casting remains in the mold. By controlling these variables, the foundry can produce a variety of irons for heat- or wear-resistant uses, or for high-strength components (Fig. 12-1-1, pg. 328).

Types of Cast Iron

Ductile (Nodular) Iron   Ductile iron, sometimes called nodular iron, is not as available as gray iron, and it is more difficult to control in production. However, ductile iron can be used where higher ductility or strength is required than is available in gray iron (Fig. 12-1-2, pg. 328).

Ductile iron is used in applications such as crankshafts because of its good machinability, fatigue strength, and high modulus of elasticity; heavy-duty gears because of its high yield strength and wear resistance; and automobile door hinges because of its ductility.

Gray Iron   Gray iron is a supersaturated solution of carbon in an iron matrix. The excess carbon precipitates out in the form of graphite flakes. Typical applications of gray iron include automotive blocks, flywheels, brake disks and drums, machine bases, and gears. Gray iron normally serves well in any machinery application because of its fatigue resistance.

White Iron   White iron is produced by a process called *chilling,* which prevents graphite carbon from precipitating out. Either gray or ductile iron can be chilled to produce a surface of white iron. In castings that are white iron throughout, however,

FIG. 12-1-1   Schematic diagram of a blast furnace, hot blast stone, and skiploader. *(American Iron and Steel Institute)*

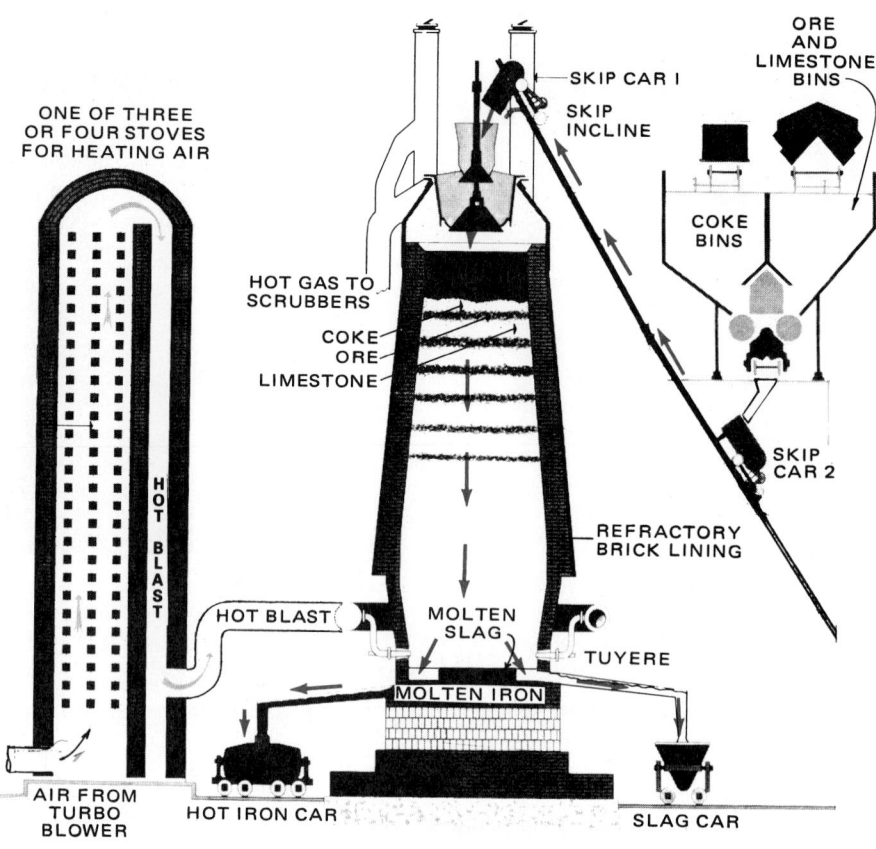

FIG. 12-1-2   Mechanical properties of cast iron.

MECHANICAL PROPERTY		DUCTILE				WHITE	GRAY						MALLEABLE						
		50-55-06	60-40-18	100-70-03	120-90-02		20	25	30	40	50	60	32510	35018	40010	45006	50005	70003	90001
Yield strength	10^3 lb/in.^{2}	60–75	45–60	75–90	90–125		*	*	*	*	*	*	32	35	40	45	50	70	90
	MPa	410–520	310–410	520–620	620–860								220	240	275	310	345	485	620
Tensile strength	10^3 lb/in.^{2}	90–110	60–80	100–120	125–150	20–50	20–25	25–30	30–35	40–48	50–57	60–66	50	53	60	65	70	85	105
	MPa	620–760	410–550	690–825	860–1035	140–345	140–170	170–205	205–240	275–330	345–390	415–455	345	365	415	450	480	585	725
Elongation in 2.00 in. (50 mm)	%	3–10	10–25	6–10	2–7	—	1	1	1	0.8	0.5	0.5	10	18	10	6	5	3	1
Modulus of elasticity	10^3 lb/in.^{2}	22–25	22–25	22–25	22–25	—	12	13	15	17	19	20	25	25	26	26	26	26–28	26–28
	10^3 MPa	150–170	150–170	150–170	150–170	8 —	83	90	103	117	131	138	172	172	180	180	180	180–193	180–193

*Yield strength usually about 65–80% of tensile strength.

the composition of iron is selected according to part size to ensure that the volume of metal involved can chill rapidly enough to produce white iron.

Because of their extreme hardness, white irons are used primarily for applications requiring wear and abrasion resistance, such as mill liners and shot-blasting nozzles. Other uses include railroad brake shoes, rolling-mill rolls, clay-mixing and brick-making equipment, and crushers and pulverizers. Plain (unalloyed) white iron usually costs less than other cast irons.

The principal disadvantage of white iron is that it is very brittle.

High-Alloy Irons   High-alloy irons are ductile, gray, or white irons that contain over three percent alloy content. These irons have properties that are significantly different from the unalloyed irons and are usually produced by specialized foundries.

Malleable Iron   Malleable iron is white iron that has been converted to a malleable condition by a two-stage heat-treating process.

It is a commercial cast material that is similar to steel in many respects. It is strong and ductile, has good impact and fatigue properties, and has excellent machining characteristics.

The two basic types of malleable iron are ferritic and pearlitic. Ferritic grades are more machinable and ductile, whereas the pearlite grades are stronger and harder.

Forming Process

For design information on the preparation of metal castings, see Chap. 13, Units 13-1 and 13-3.

REFERENCES AND SOURCE MATERIAL

1. *Machine Design,* Materials reference issue.

ASSIGNMENT

See Assignment 1 for Unit 12-1 on page 344.

12-2 CARBON STEEL

Carbon steel is essentially an iron-carbon alloy with small amounts of other elements (either intentionally added or unavoidably present), such as silicon, magnesium, copper, and sulfur. Steels can be either cast to shape or wrought into various mill forms from which finished parts can be machined, forged, formed, stamped, or otherwise generated.

Wrought steel is either poured into ingots or is sand-cast. After solidification the metal is reheated and hot-rolled—often in several steps—into the finished wrought form. Hot-rolled steel is characterized by a scaled surface and a decarburized skin.

Carbon and Low-Alloy Cast Steels

Carbon and low-alloy cast steels lend themselves to the formation of streamlined, intricate parts with high strength and rigidity. A number of advantages favor steel casting as a method of construction:

1. The metallographic structure of steel castings is uniform in all directions. It is free from the directional variations in properties of wrought-steel products.
2. Cast steels are available in a wide range of mechanical properties depending on the compositions and heat treatments.
3. Steel castings can be annealed, normalized, tempered, hardened, or carburized.
4. Steel castings are as easy to machine as wrought steels.
5. Most compositions of carbon and low-alloy cast steels are easily welded because their carbon content is under 0.45 percent.

The making of steel is illustrated in Fig. 12-2-1 (pg. 330).

High-Alloy Cast Steels

The term *high alloy* is applied arbitrarily to steel castings containing a minimum of eight percent nickel and/or chromium. Such castings are used mostly to resist corrosion or provide strength at temperatures above 1200°F (560°C).

Carbon Steels

Carbon steels are the workhorse of product design. They account for over 90 percent of total steel production. More carbon steels are used in product manufacturing than all other metals combined.

A thorough understanding of the selection and specification criteria for all types of steel requires knowledge of what is implied by carbon-steel mill forms, qualities, grades, tempers, finishes, edges, and heat treatments; also how and where these terms relate to dimensions, tolerances, physical and mechanical properties, and manufacturing requirements.

The designer's specification job really begins the instant that molten steel hits the mold. The conditions under which steel solidifies have a significant effect on production and on performance of subsequent mill products.

Steel Specification

Several ways are used to identify a specific steel: by chemical or mechanical properties, by its ability to meet a standard specification or industry-accepted practice, or by its ability to be fabricated into an identified part.

Chemical Composition

The steel producer can be instructed to produce a desired composition in one of three ways:

1. By a maximum limit
2. By a minimum limit
3. By an acceptable range

The following are some commonly specified elements.

Carbon   Carbon is the principal hardening element in steel. As carbon content is increased to about 0.85 percent, hardness and tensile strength increase, but ductility and weldability decrease.

Manganese   Manganese is a lesser contributor to hardness and strength. Properties depend on carbon content. Increasing manganese increases the rate of carbon penetration during carburizing but decreases weldability.

Phosphorus   Large amounts of phosphorus increase strength and hardness but reduce ductility and impact toughness, particularly in the higher-carbon grades. Phosphorus in low-carbon, free-machining steels improves machinability.

Silicon   A principal deoxidizer in the steel industry, silicon increases strength and hardness but to a lesser extent than manganese. However, it reduces machinability.

329

FIG. 12-2-1 Flowchart for steelmaking.

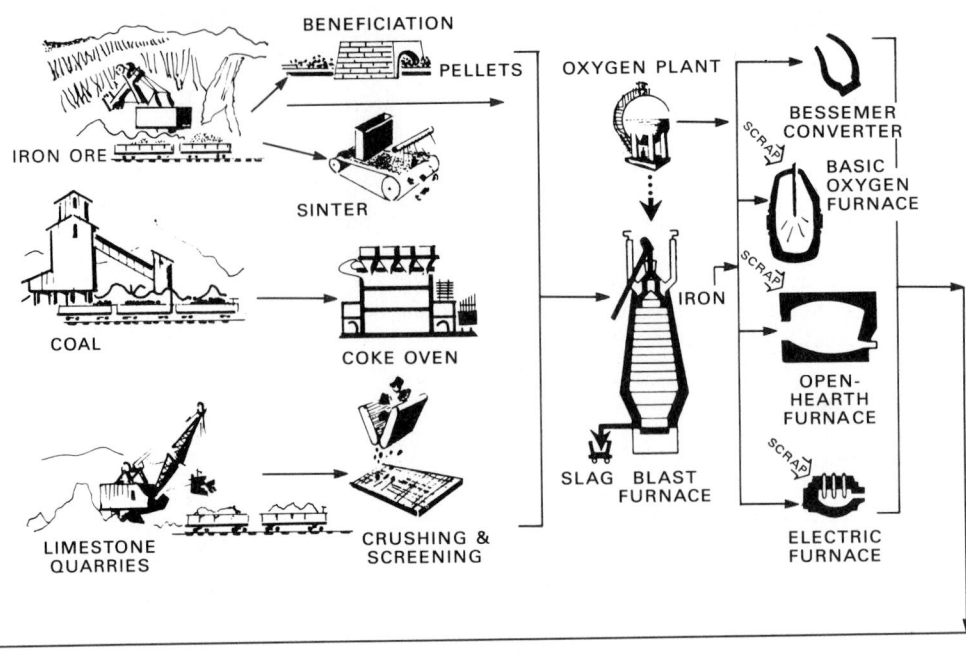

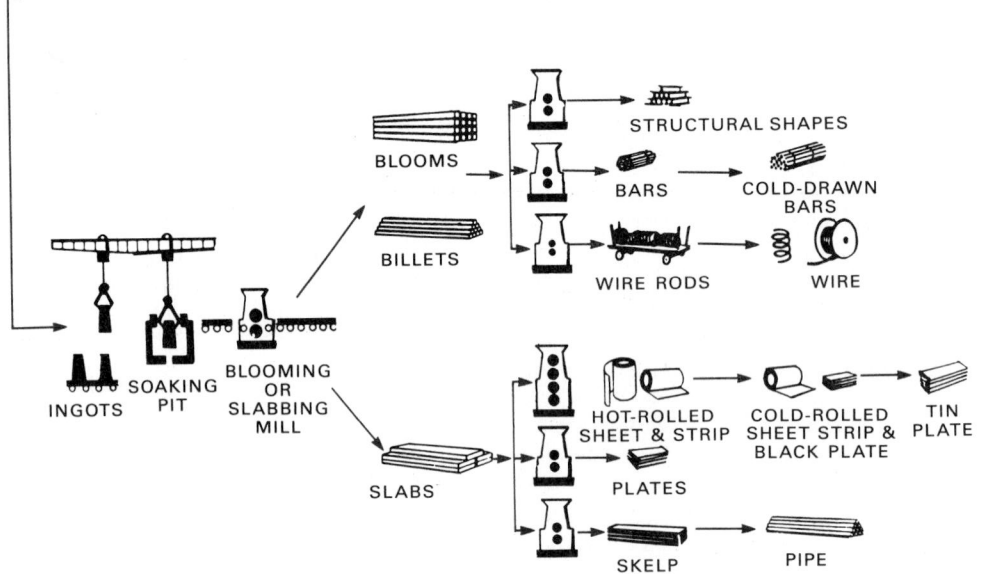

Sulfur Increased sulfur content reduces transverse ductility, notch-impact toughness, and weldability. Sulfur is added to improve machinability of steel.

Copper Copper improves atmospheric corrosion resistance when present in excess of 0.15 percent.

Lead Lead improves the machinability of steel.

Classification Bodies

The specifications covering the composition of iron and steel have been issued by various classification bodies. These specifications serve as a selection guide and provide a means for the buyer to conveniently specify certain known and recognized requirements. The main classification bodies are:

SAE—Society of Automotive Engineers

AISI—American Iron and Steel Institute This is an association of steel producers that issues steel specifications for the steelmaking industry and cooperates with the SAE in using the same numbers for the same steel.

ASTM—American Society for Testing and Materials This group is interested in materials of all kinds and writes specifications. The ASTM steel specifications for steel plate and structural shapes are used by all steelmakers in North America.

The ASTM has several specifications covering structural steel. Both the AISA and AISC (American Institute of Steel Construction) refer to ASTM specifications.

ASME—American Society of Mechanical Engineers This group is interested in the steel used in pressure vessels and other mechanical equipment.

SAE and AISI—Systems of Steel Identification

The specifications for steel bar are based on a code that specifies the composition of each type of steel covered. They include both plain carbon and alloy steels. The code is a four-number system (Figs. 12-2-2 and 12-2-3). Each figure in the number has the following specific function: the first or left-side figure represents the major class of steel, the second figure represents a subdivision of the major class. For example, the series having *one* (1) as the left-hand figure covers the carbon steels. The second figure breaks this class up into normal low-sulfur steels, the high-sulfur free-machining grades, and another grade having higher than normal manganese.

Originally the second figure represented the percentage of the major alloying element present, and this is true of many of the alloy steels. However, this had to be varied in order to account for all the steels that are available.

The third and fourth figures represent carbon content in hundredths of one percent, thus the figure xx15 means 0.15 of one percent carbon.

EXAMPLE

SAE 2335 is a nickel steel containing 3.5 percent nickel and 0.35 of one percent carbon.

Carbon-Steel Sheets

Flat-rolled carbon-steel sheets are made from heated slabs that are progressively reduced in size as they move through a series of rolls. Typical properties of rolled carbon steels are shown in Fig. 12-2-4 (pg. 332).

Hot-Rolled Sheets   Hot-rolled sheets are produced in three principal qualities: commercial, drawing, and physical.

Cold-Rolled Sheets   Cold-rolled sheets are made from hot-rolled coils that are pickled, then cold-reduced to the desired thickness. The commercial quality of cold-rolled sheets is normally produced with a matte finish suitable for painting or enameling but not suitable for electroplating.

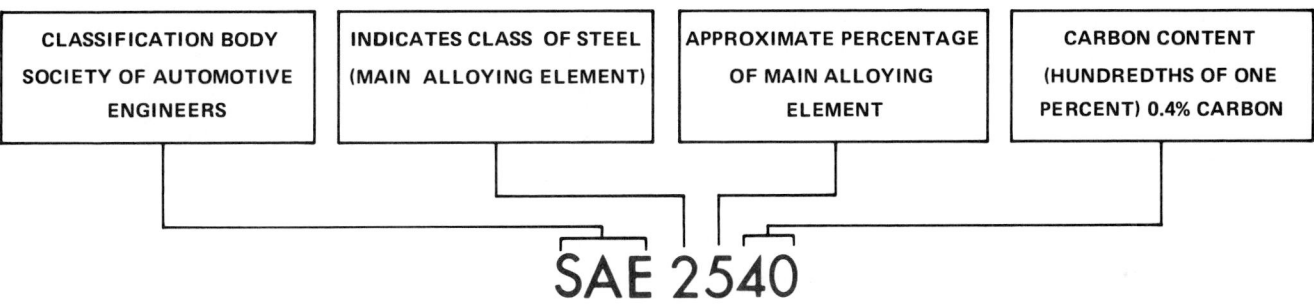

FIG. 12-2-2   Steel designation system.

FIG. 12-2-3   Carbon steel designations, properties and uses.

TYPE OF CARBON STEEL	NUMBER SYMBOL	PRINCIPAL PROPERTIES	COMMON USES
Plain carbon	10XX		
Low-carbon steel (0.06 to 0.20% carbon)	1006 to 1020	Toughness and less strength	Chains, rivets, shafts, and pressed steel products
Medium-carbon steel (0.20 to 0.50% carbon)	1020 to 1050	Toughness and strength	Gears, axles, machine parts, forgings, bolts, and nuts
High-carbon steel (over 0.50% carbon)	1050 and over	Less toughness and greater hardness	Saws, drills, knives, razors, finishing tools, and music wire
Sulfurized (free-cutting)	11XX	Improves machinability	Threads, splines, and machined parts
Phosphorized	12XX	Increases strength and hardness but reduces ductility	
Manganese steels	13XX	Improves surface finish	

MECHANICAL PROPERTY		AISI STEEL											
		1015/1020/1022			1035/1040			1045/1050			1095		
		HOT-ROLLED	COLD DRAWN	ANNEALED	HOT-ROLLED	COLD-DRAWN	QUENCHED AND TEMPERED	HOT-ROLLED	COLD-DRAWN	QUENCHED AND TEMPERED	HOT-ROLLED	COLD-DRAWN AND ANNEALED	QUENCHED AND TEMPERED
Yield strength	10^3 lb/in.^{2}	40	51	42	42	71	63–96	49	84	68–117	66	76	80–152
	MPa	270	350	295	290	440	435–660	335	580	470–800	455	525	580–1050
Tensile strength	10^3 lb/in.^{2}	65	61	60	76	85	96–130	90	100	105–137	130	99	130–216
	MPa	450	420	415	525	585	660–895	620	690	725–945	895	680	895–1490
% Elongation in 2.00 in. (50 mm)		25	15	38	18	12	17–24	15	10	25–15	9	13	10–84

FIG. 12-2-4   Typical mechanical properties of rolled carbon steel.

Carbon-Steel Plates

Carbon-steel plates are produced (in rectangular plates or in coils) by hot rolling directly from the ingot or slab. Plate thickness ranges from .19 in. (4 mm) and thicker for plates up to 48 in. (1200 mm) wide, and from .25 in. (6 mm) and thicker for plates wider than 48 in. (1200 mm). Thickness is specified in millimeters or inches. It can also be specified by weight (lb/ft^2) or mass (kg/m^2).

Carbon-Steel Bars

Hot-Rolled Bars   Hot-rolled carbon-steel bars are produced from blooms or billets in a variety of cross sections and sizes (Figs. 12-2-5 and 12-2-6).

Cold-Finished Bars   Cold-finished carbon-steel bars are produced from hot-rolled steel by a cold-finishing process which improves surface finish, dimensional accuracy, and alignment. Cold drawing and cold rolling also increase the yield and tensile strength. For machinability ratings of cold-drawn carbon steel, see Fig. 12-2-7.

Steel Wire

Steel wire is made from hot-rolled rods produced in continuous-length coils. Most wire is drawn, but some special shapes are rolled.

Pipe and Tubing

Pipe and tubing range from the familiar plumber's black pipe to high-precision mechanical tubing for bearing races. Pipe and tubing may contain fluids, support structures, or be a primary shape from which products are fabricated.

Welded Tubular Products   Welded tubular products are made from hot-rolled or cold-rolled flat steel coils.

Pipe   Pipe is produced from carbon or alloy steel to nominal dimensions. Nominal pipe sizes are expressed in inch sizes, but in the metric system the outside diameter and the wall thickness are expressed in millimeters. The outside diameter is often much larger than the nominal size. For example, a .75-in. standard-weight pipe has an outside diameter of 1.050 in. (26.7 mm).

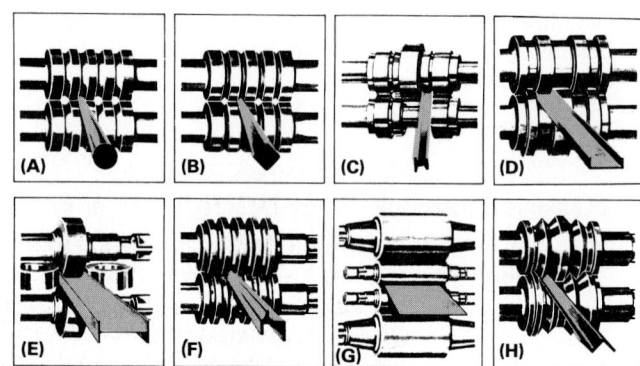

FIG. 12-2-5   Standard steel stock. *(American Iron and Steel Institute)*

The outside diameter of nominal-size pipe always remains the same and the mass or wall thickness changes. ANSI B36 has developed 10 different wall thicknesses (schedules) of pipe. (See Table 57 in the Appendix.)

Nominal pipe-size designation stops at 12 in. Pipe 14 in. and over is listed on the basis of outside diameter and wall thickness.

Tubing   Tubing is usually specified by a combination of either outside diameter, inside diameter, or wall thickness. Sizes range from approximately .25 to 5.00 in. (6 to 125 mm), in increments of .12 in. (3 mm). Wall thickness is usually specified in inches, millimeters, or by gage numbers.

Structural-Steel Shapes

A large tonnage of structural-steel shapes goes into manufactured products rather than buildings. The frame of a truck, railroad car, or earth-moving equipment is a structural design problem, just as is a high-rise building.

Size Designation   Several ways are used to describe a structural section in a specification, depending primarily on its shape.

1. Beams and channels are measured by the depth of the section in inches (millimeters) and by weight (lb/ft) or mass (kg/m).

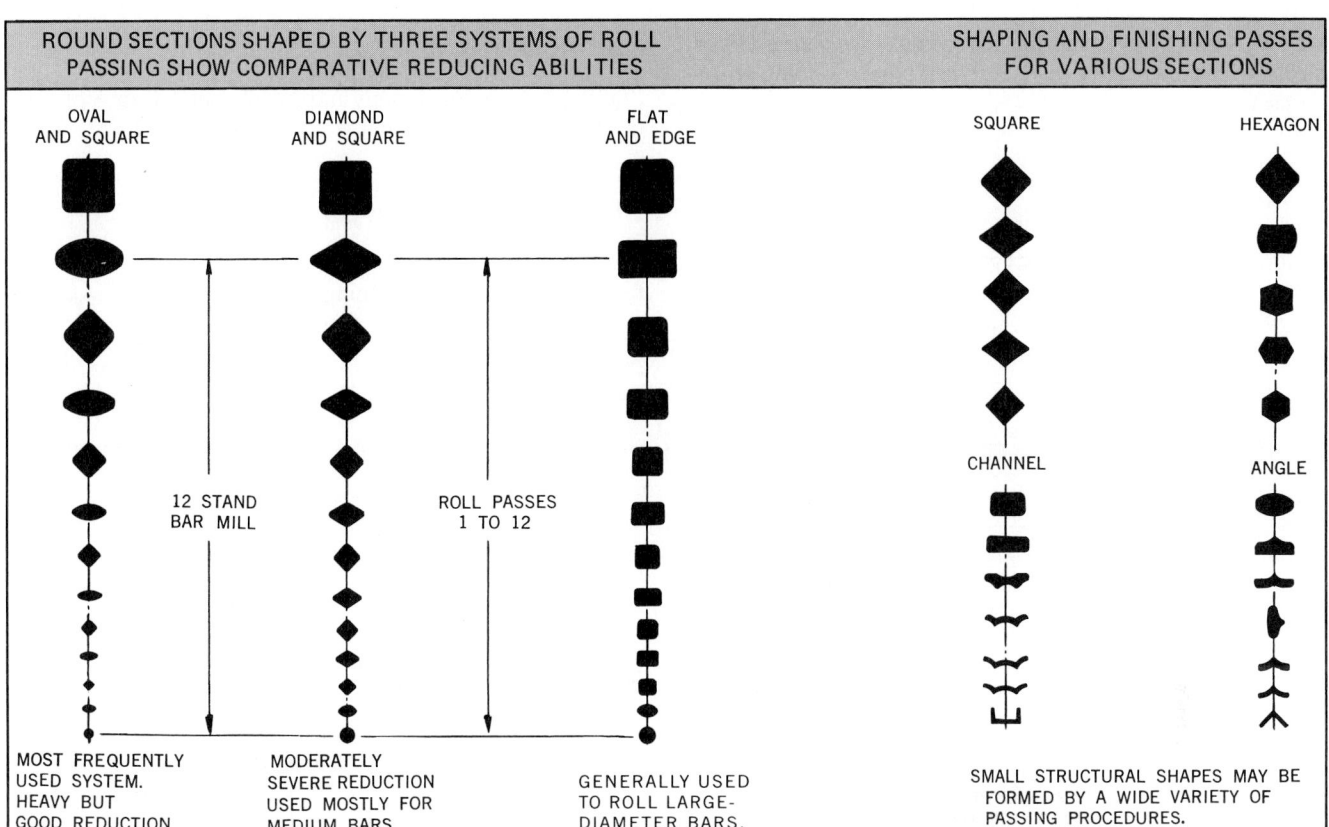

| ROUND SECTIONS SHAPED BY THREE SYSTEMS OF ROLL PASSING SHOW COMPARATIVE REDUCING ABILITIES | | | SHAPING AND FINISHING PASSES FOR VARIOUS SECTIONS | |

FIG. 12-2-6   Bar-mill roll passes. *(American Iron and Steel Institute)*

AISI NO.	RATING*	AISI NO.	RATING*
12L14	195	1114	85
1213	137	1137	72
1215	137	1141	69
1212	100	1018	66
1211	91	1045	55
1117	89		

*Based on 1212 = 100%

FIG. 12-2-7   Machinability rating of cold-drawn carbon steel.

2. Angles are described by length of legs and thickness in inches (millimeters), or more commonly, by length of legs and weight (lb/ft) or mass (kg/m). The longest leg is always stated first.
3. Tees are specified by width of flange, overall depth of stem, and pounds per foot (kilograms per meter) in that order.
4. Zees are specified by width of flange and thickness in inches (millimeters), or by depth, width across flange, and pounds per foot (kilograms per meter).
5. Wide-flange sections are described by depth, width across flange, and pounds per foot (kilograms per meter).

High-Strength Low-Alloy Steels

The properties of high-strength low-alloy (HSLA) steels generally exceed those of conventional carbon structural steels. These low-alloy steels are usually chosen for their high ratios of yield to tensile strength, resistance to puncturing, abrasion resistance, corrosion resistance, and toughness.

ASTM Specifications

ASTM has six specifications covering high-strength low-alloy steels. These are:

ASTM A94   Used primarily for riveted and bolted structures and for special structural purposes.

ASTM A242   Used primarily for structural members where light weight or low mass and durability are important.

ASTM A374   Used where high strength is required and where resistance to atmospheric corrosion must be at least equal to that of plain copper-bearing steel.

ASTM A375   This specification differs slightly from ASTM A374 in that material can be specified in the annealed or normalized condition.

ASTM A440   This covers high-strength intermediate-manganese steels for nonwelded applications.

ASTM A441   This covers the intermediate-manganese HSLA steels, which are readily weldable when proper welding procedures are used.

Low- and Medium-Alloy Steels

There are two basic types of alloy steel: *through hardenable* and *surface hardenable*. Each type contains a broad family of steels whose chemical, physical, and mechanical properties make them suitable for specific product applications (Fig. 12-2-8).

Stainless Steels

Stainless steels have many industrial uses because of their desirable corrosion-resistance and strength properties.

Free-Machining Steels

A whole family of free-machining steels has been developed for fast and economical machining (Fig. 12-2-9). These steels are available in bar stock in various compositions, some standard

TYPE OF STEEL	ALLOY SERIES	APPROXIMATE ALLOY CONTENT (%)	PRINCIPAL PROPERTIES	COMMON USES
Manganese steel	13xx	Mn 1.6–1.9	Improve surface finish	
Molybdenum steels	40xx 41xx 43xx 44xx 46xx 47xx 48xx	Mo 0.15–0.3 Cr 0.4–1.1; Mo 0.08–0.35 Ni 1.65–2; Cr 0.4–0.9; Mo 0.2–0.3 Mo 0.45–0.6 Ni 0.7–2; Mo 0.15–0.3 Ni 0.9–1.2; Cr 0.35–0.55; Mo 0.15–0.4 Ni 3.25–3.75; Mo 0.2–0.3	High strength	Axles, forgings, gears, cams, mechanical parts
Chromium steels	50xx 51xx E51100 E52100	Cr 0.3–0.5 Cr 0.7–1.15 C 1.0; Cr 0.9–1.15 C 1.0; Cr 0.9–1.15	Hardness, great strength and toughness	Gears, shafts, bearings, springs, connecting rods
Chromium-vanadium steel	61xx	Cr 0.5–1.1; V 0.1–0.15	Hardness and strength	Punches and dies, piston rods, gears, axles
Nickel-chromium-molybdenum steels	86xx 87xx 88xx	Ni 0.4–0.7; Cr 0.4–0.6; Mo 0.15–0.25 Ni 0.4–0.7; Cr 0.4–0.6; Mo 0.2–0.3 Ni 0.4–0.7; Cr 0.4–0.6; Mo 0.3–0.4	Rust resistance, hardness, and strength	Food containers, surgical equipment
Silicon-manganese steel	92xx	Si 1.8–2.2	Springiness and elasticity	Springs

FIG. 12-2-8   AISI designation system for alloy steel.

DESIGNATION	PHOSPHORIZED				SULFURIZED								
	12L13/12L14/ 12L15		1211/1212/ 1213		1117/1118/1119			1137			1141/1144		
MECHANICAL PROPERTY	HOT-ROLLED	COLD-DRAWN	HOT-ROLLED	COLD-DRAWN	HOT-ROLLED	COLD-DRAWN	QUENCHED AND TEMPERED	HOT-ROLLED	COLD-DRAWN	QUENCHED AND TEMPERED	HOT-ROLLED	COLD-DRAWN	QUENCHED AND TEMPERED
Yield strength 10^3 lb/in.^{2}	34	60-80	33	58	34-46	51-68	50-76	48	82	136	51	90	163
MPa	235	416-550	225	400	235-315	350-470	345-525	330	565	335	350	620	1120
Tensile strength 10^3 lb/in.^{2}	57	70-90	55	75	62-76	69-78	89-113	88	98	157	94	100	190
MPa	390	480-620	380	517	425-525	475-535	615-780	605	675	1080	650	690	1310
% Elongation in 2.00 in. (50 mm)	22	10-18	25	10	23-33	15-20	17-22	15	10	5	15	10	9
Machinability (B1212 = 100)	195–296		91–137		89–100			71			69		

FIG. 12-2-9   Typical mechanical properties of free-machining carbon steels.

and some proprietary. When utilized properly, they lower the cost of machining by reducing metal removal time.

REFERENCES AND SOURCE MATERIAL

1. *Machine Design,* Materials reference issue.

ASSIGNMENT

See Assignment 2 for Unit 12-2 on page 344.

12-3 NONFERROUS METALS

Although ferrous alloys are specified for more engineering applications than all nonferrous metals combined, the large family of nonferrous metals offers a wider variety of characteristics and mechanical properties. For example, the lightest metal is lithium, .02 lb/in.^{3} (0.53 g/cm^3); the heaviest is osmium with a weight of .81 lb/in.^{3} (mass of 22.5 g/cm^3)—nearly twice the weight of lead. Mercury melts at around −38°F (−30°C), while tungsten, the highest-melting metal, liquefies at 6170°F (3410°C).

Availability, abundance, and the cost to convert the metal into useful forms all play an important part in selecting a nonferrous metal. Although nearly 80 percent of all elements are called "metals," only about two dozen of these are used as structural engineering materials. Of the balance, however, many are used as coatings, in electronic devices, as nuclear materials, and as minor constituents in other systems.

One of the most important aspects in selecting a material for a mechanical or structural application is how easily the material can be shaped into the finished part—and how its properties can be either intentionally or inadvertently altered in the process (Fig. 12-3-1). Frequently, metals are simply cast into the finished part. In other cases, metals are cast into an intermediate form (such as an ingot), then worked or "wrought" by rolling, forging, extruding, or other deformation processes.

Manufacturing with Metals

Machining   Most metals can be machined. Machinability is best for metals that allow easy chip removal with minimum tool wear.

Powder Metallurgy (PM) Compacting   Parts can be made from most metals and alloys by PM compacting, although only a few are economically justified. Iron and iron-copper alloys are most commonly used.

Casting   Theoretically, any metal that can be melted and poured can be cast. However, economic limitations usually narrow down the number of ways metals are cast commercially.

Extruding and Forging   Metals to be forged or extruded must be ductile and not work-harden at working temperature. Some metals show these characteristics at room temperature and can be cold-worked; others must be heated.

Stamping and Forming   Most metals, except brittle alloys, can be press-worked.

FORMING METHOD	ALUMINUM	COPPER	IRON	LEAD	MAGNESIUM	NICKEL	SILVER, GOLD, PLATINUM	MOLYBDENUM, COPPER TANTALUM, TUNGSTEN	STEEL	TIN	TITANIUM	ZINC
Casting												
Centrifugal	✔	✔	✔			✔			✔			
Continuous	✔	✔	✔	✔		✔			✔			✔
Ceramic mold	✔	✔	✔		✔	✔	✔		✔			
Investment	✔	✔			✔	✔			✔			✔
Permanent mold	✔	✔	✔	✔	✔	✔			✔	✔		✔
Sand	✔	✔	✔	✔	✔	✔			✔	✔		
Shell mold	✔	✔	✔		✔	✔					✔	✔
Die casting	✔	✔			✔							✔
Cold heading	✔	✔		✔		✔	✔		✔			
Deep drawing	✔	✔			✔	✔			✔		✔	✔
Extruding	✔	✔	✔		✔	✔		✔	✔	✔	✔	
Forging	✔	✔			✔	✔		✔	✔		✔	
Machining	✔	✔	✔		✔	✔	✔	✔	✔		✔	✔
PM compacting	✔	✔	✔			✔	✔	✔	✔		✔	
Stamping and forming	✔	✔			✔	✔	✔	✔	✔		✔	✔

FIG. 12-3-1   Common methods of forming metals.

Cold Heading   Metals must be ductile and should not work-harden rapidly. Annealing should restore ductility and softness in cold-heading alloys.

Deep Drawing   Deep drawing involves severe deformation and the metal is usually stretched over the die.

Aluminum

The density of aluminum is about one-third that of steel, brass, nickel, or copper. Yet some alloys of aluminum are stronger than structural steel. Under most service conditions, aluminum has high resistance to corrosion and forms no colored salts that might stain or discolor adjacent components (Fig. 12-3-2).

Copper

Copper alloys, approximately 250 of them, are fabricated in rod, sheet, tube, and wire form. Each of these alloys has some property or combination of properties that makes it unique. They can be grouped into several general headings, such as coppers, brasses, leaded brasses, phosphor bronzes, aluminum bronzes, silicon bronzes, beryllium coppers, cupronickels, and nickel silvers (Fig. 12-3-3).

Copper alloys are used where one or more of the following properties is needed: thermal or electrical conductivity, corrosion resistance, strength, ease of forming, ease of joining, and color.

The major alloy usages are:

1. Copper in pure form as a conductor in the electrical industry
2. Copper or alloy tubing for water, drainage, air conditioning, and refrigeration lines
3. Brasses, phosphor bronzes, and nickel silvers as springs or in construction of equipment if corrosive conditions are too severe for iron or steel

An advantage of copper and its alloys, offered by no other metals, is the wide range of colors available.

Nickel

Commercially pure wrought nickel is a grayish-white metal capable of taking a high polish. Because of its combination of attractive mechanical properties, corrosion resistance, and formability, nickel or its alloys is used in a variety of structural applications usually requiring specific corrosion resistance.

Magnesium

Magnesium, with density of only .06 lb/in.^{3} (1.74 g/cm^3), is the world's lightest structural metal. The combination of low density and good mechanical strength makes possible alloys with a high strength-to-weight ratio.

Zinc

Zinc is a relatively inexpensive metal that has moderate strength and toughness and outstanding corrosion resistance in many types of service.

The principal characteristics that influence the selection of zinc alloys for die castings include the dimensional accuracy obtainable, castability of thin sections, smooth surface, dimensional stability, and adaptability to a wide variety of finishes.

Titanium

Titanium is a light metal at .16 lb/in.^{3} (4.43 g/cm^3); it is 60 percent heavier than aluminum but 45 percent lighter than alloy steel. It is the fourth most abundant metallic element in the earth's crust and the ninth most common element.

Titanium-based alloys are much stronger than aluminum alloys and superior in many respects to most alloy steels.

Beryllium

Beryllium has a strength-to-weight ratio comparable to high-strength steel, yet it is lighter than aluminum. Its melting point

MAJOR ALLOYING ELEMENT	DESIGNATION
Aluminum (99% or more)	1xxx
Copper	2xxx
Manganese	3xxx
Silicon	4xxx
Magnesium	5xxx
Magnesium and silicon	6xxx
Zinc	7xxx
Other elements	8xxx
Unused series	9xxx

FIG. 12-3-2   Wrought aluminum alloy designations.

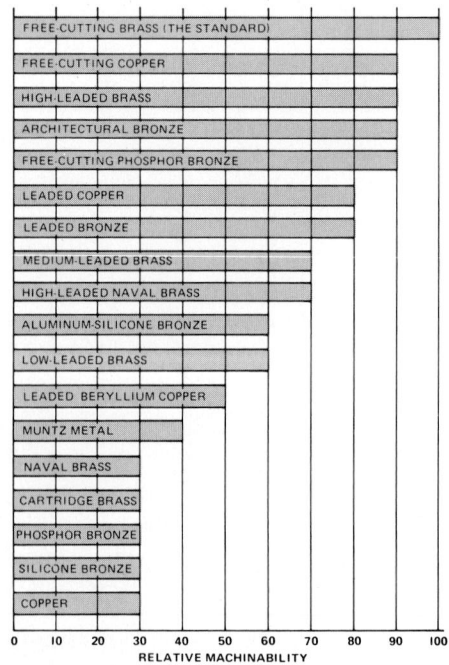

FIG. 12-3-3   Free-machining copper alloys.

is 2345°F (1285°C) and it has excellent thermal conductivity. It is nonmagnetic and a good conductor of electricity.

Refractory Metals

Refractory metals are those metals with melting points above 3600°F (2000°C). Among these, the best known and most extensively used are tungsten, tantalum, molybdenum and niobium. Refractory metals are characterized by high-temperature strength, corrosion resistance, and high melting points.

Tantalum and Niobium

Tantalum and niobium are usually discussed together, since most of their working operations are identical. Unlike molybdenum and tungsten, tantalum and niobium can be worked at room temperatures. The major differences between tantalum and niobium are in density, nuclear cross section, and corrosion resistance. The density of tantalum is almost twice that of niobium.

Molybdenum

Molybdenum is widely used in missiles, aircraft, industrial furnaces, and nuclear projects. Its melting point is lower than that of tantalum and tungsten. Molybdenum has a high strength-to-weight ratio and a low vapor pressure, is a good conductor of heat and electricity, and has a high modulus of elasticity and a low coefficient of expansion.

Tungsten

Tungsten is the only refractory metal that has the combination of excellent corrosion resistance, good electrical and thermal conductivity, a low coefficient of expansion, and high strength at elevated temperatures.

Precious Metals

Gold costs over 8,000 times more than an equal amount of iron; rhodium costs nearly 32,000 times more than copper. With prices such as these, why are precious metals ever specified?

In some cases, precious metals are used for their unique surface characteristics. They reflect light better than other metals. Gold, for example, is specified as a surface for heat reflectors, insulators, and collectors because of its outstanding ability to reflect ultraviolet radiation.

The family of metals called precious metals can be divided into three sub-groups: silver and silver alloys; gold and gold alloys; and the so-called platinum metals, which are platinum, palladium, rhodium, ruthenium, iridium, and osmium.

REFERENCES AND SOURCE MATERIAL

1. *Machine Design,* Materials reference issue.

ASSIGNMENTS

See Assignments 3 and 4 for Unit 12-3 on pages 344–345.

12-4 PLASTICS

This unit will acquaint drafters with the general characteristics of commercially available plastics so that they can make proper use of plastics in products.

Plastics may be defined as nonmetallic materials capable of being formed or molded with the aid of heat, pressure, chemical reactions, or a combination of these (Fig. 12-4-1).

Plastics are strong, tough, durable materials that solve many problems in machine and equipment design. Metals, it is true, are hard and rigid. This means that they can be machined, to very close tolerances, into cams, bearings, bushings, and gears, which will work smoothly under heavy loads for long periods. Although some come close, no plastic has the hardness and creep resistance of steel, for example. However, metals have many weaknesses that engineering plastics do not. Metals corrode or rust, they must be lubricated, their working surfaces wear readily, they cannot be used as electrical or thermal insulators, they are opaque and noisy, and where they must flex, metals fatigue rapidly.

Plastics can resolve these weaknesses, though not necessarily all with one material. The engineering plastics are resistant to most chemicals; fluorocarbon is one of the most chemically inert substances known. None of the engineering plastics corrode or rust; acetal resin and fluorocarbon are unaffected even when continuously immersed in water. Engineering plastics can be run at low speeds and loads and, without lubrication, are among the world's slipperiest solids, being comparable to ice. Engineering plastics are resilient; therefore, they run more quietly and smoothly than equivalent metal products, and they are able to stand periodic overloads without harmful effects.

Plastics are a family of materials, not a single material. Each material has its special advantages. Being manufactured, plastics raw materials are capable of being variously combined to give almost any property desired in an end product. But these are controlled variations unlike those of nature's products. Some thermoplastics can be sterilized.

The widespread and growing use of plastics in almost every phase of modern living can be credited in large part to their unique combinations of advantages. These advantages are light

FIG. 12-4-1   A variety of plastics parts.

THERMOPLASTICS			
NAME OF PLASTIC	**PROPERTIES**	**FORMS AND METHODS OF FORMING**	**USES**
ABS (Acrylonitrile Butadiene-Styrene)	Strong, tough, good electrical properties.	Available in powder or granules for injection molding, extrusion, and calendering and as sheet for vacuum forming.	Pipe, wheels, football helmets, battery cases, radio cases, children's skates, tote boxes.
Acetal Resin	Rigid without being brittle, tough, resistant to extreme temperatures, good electrical properties.	Produced in powder form for molding and extrusion, available in rod, bar, tube, strip, slab.	Automobile instrument clusters, gears, bearings, bushings, door handles, plumbing fixtures, threaded fasteners, cams.
Acrylics	Exceptional clarity and good light transmission. Strong, rigid, and resistant to sharp blows. Excellent insulator. Colorless or full range of transparent, translucent, or opaque colors.	Available in sheet, rod, tube, and molding powders. Plastic products can be produced by fabricating of sheets, rods, and tubes, hot forming of sheets, injection and compression molding of powder, extrusion, casting.	Airplane canopies and windows, television and camera viewing lenses, combs, costume jewelry, salad bowls, trays, lamp bases, scale models, automobile tail lights, outdoor signs.
Cellulosics (A) Cellulose Acetate		Available in pellets, sheets, film, rods, tubes, strips, coated cord. Can be made into products by injection, compression molding, extrusion, blow molding, and vacuum forming, or sheets and coating.	Eyeglass frames, toys, lamp shades, combs, shoe heels.
(B) Cellulose Acetate Butyrate	Among the toughest of plastics. Retains a lustrous finish under normal wear. Transparent, translucent, or opaque in wide variety of colors and in clear transparent. Good insulators.	Available in pellets, sheets, rods, tubes, strips and as a coating. Can be made into products by injection, compression molding, extrusion, blowing and drawing of sheet, laminating, coating.	Steering wheels, radio cases, pipe and tubing, tool handles, playing cards.
(C) Cellulose Propionate		Available in pellets for injection extrusion or compression molding.	Appliance housing, telephone handsets, pens and pencils.
(D) Ethyl Cellulose		Available in granules, flake, sheet, rod, tube, film, or foil. Can be made into finished products by injection, compression molding, extrusion, drawing.	Edge moldings, flashlights, electrical parts.
(E) Cellulose Nitrate		Available in rods, tubes, sheets for machining and as a coating.	Shoe heel covers, fabric coating.
Fluorocarbons	Low coefficient of friction, resistant to extreme heat and cold. Strong, hard, and good insulators.	Available as powder and granules in resin form. Sheet, rod, tube, film, tape, and dispersions. Molded, extruded, and machined.	Valve seats, gaskets, coatings, linings, tubings.

FIG. 12-4-2   Thermoplastics. *(The Society of Plastics Industry)*

weight, range of color, good physical properties, adaptability to mass-production methods, and often, lower cost.

Aside from the range of uses attributable to the special qualities of different plastics, these materials achieve still greater variety through the many forms in which they can be produced.

They may be made into definite shapes like dinnerware and electric switchboxes. They may be made into flexible film and sheeting such as shower curtains and upholstery. Plastics may be made into sheets, rods, and tubes that are later shaped or machined into internally lighted signs or airplane blisters. They may be made into filaments for use in household screening, industrial strainers, and sieves. Plastics may be used as a coating on textiles and paper. They may be used to bind such materials as fibers of glass and sheets of paper or wood to form boat hulls, airplane wing tips, and tabletops.

Plastics are usually classified as either thermoplastic or thermosetting.

Thermoplastics

Thermoplastics soften, or liquefy, and flow when heat is applied. Removal of the heat causes these materials to set or solidify. They may be reheated and reformed or reused. In this group fall the acrylics, the cellulosics, nylons (polyamides),

THERMOPLASTICS			
NAME OF PLASTIC	PROPERTIES	FORMS AND METHODS OF FORMING	USES
Nylon (Polyamides)	Resistant to extreme temperatures. Strong and long-wearing range of soft colors.	Available as a molding powder, in sheets, rods, tubes, and filaments. Injection, compression, blow molding, and extrusion.	Tumblers, faucet washers, gears. As a filament, it is used as brush bristles, fishing line.
Polycarbonate	High impact strength, resistant to weather, transparent.	Primarily a molding material, may take form of film, extrusion, coatings, fibers, or elastomers.	Parts for aircraft, automobiles, business machines, gages, safety-glass lenses.
Polyethylene	Excellent insulating properties, moisture proof, clear transparent, translucent.	Available in pellet, powder, sheet, film, filament, rod, tube, and foamed. Injection, compression, blow molding, extrusion, coating, and casting.	Ice cube trays, tumblers, dishes, bottles, bags, balloons, toys, moisture barriers.
Polystyrene	Clear, transparent, translucent, or opaque. All colors. Water and weather resistant, resistance to heat or cold.	Available in molding powders or granules, sheets, rods, foamed blocks, liquid solution, coatings, and adhesives. Injection, compression molding, extrusion, laminating, machining.	Kitchen items, food containers, wall tile, toys, instrument panels.
Polypropylenes	Good heat resistance. High resistance to cracking. Wide range of colors.	Processed by injection molding, blow molding, and extrusion.	Thermal dishware, washing machine agitators, pipe and pipe fittings, wire and cable insulation, battery boxes, packaging film and sheets.
Urethanes	Tough and shock-resistant for solid materials. Flexible for foamed material, can be foamed in place.	Solid type—starting two reactants, final article can be extruded, molded, calendered, or cast. Foamed type—can be made by either a prepolymer or one-shot process. In either slab stock or molded form.	Mattresses, cushioning, padding, toys, rug underlays, crash-pads, sponges, mats, adhesion, thermal insulation, industrial tires.
Vinyls	Strong and abrasion-resisting. Resistant to heat and cold. Wide color range.	Available in molding powder, sheet, rod, tube, granules, powder. It can be formed by extrusion, casting, calendering, compression, and injection molding.	Raincoats, garment bags, inflatable toys, hose, records, floor and wall tile, shower curtains, draperies, pipe, paneling.

FIG. 12-4-2 Thermoplastics. (continued)

polyethylene, polystyrene, polyfluorocarbons, the vinyls, polyvinylidene, ABS, acetal resin, polypropylene, and polycarbonates (Fig. 12-4-2).

Thermosetting Plastics

Thermosetting plastics undergo an irreversible chemical change when heat is applied or when a catalyst or reactant is added. They become hard, insoluble, and infusible, and they do not soften upon reapplication of heat. Thermosetting plastics include phenolics, amino plastics (melamine and urea), cold-molded polyesters, epoxies, silicones, alkyds, allylics, and casein (Fig. 12-4-3, pg. 340).

Machining

Practically all thermoplastics and thermosets can be satisfactorily machined on standard equipment with adequate tooling.

The nature of the plastic will determine whether heat should be applied, as in some laminates, or avoided, as in buffing some thermoplastics. Standard machining operations can be used, such as turning, drilling, tapping, milling, blanking, and punching.

Material Selection

One of the first decisions a designer makes is the choice of materials. The choice is influenced by many factors, such as the end use of the product and the properties of the selected material (see Fig. 12-4-4, pg. 341). No attempt is made at this point to discuss the engineering approach to selection of materials.

However, a basic examination and selection of a plastic material at this time will help acquaint the drafter with the wide range of plastics available.

THERMOSETTING PLASTICS			
NAME OF PLASTIC	PROPERTIES	FORMS AND METHODS OF FORMING	USES
Alkyds	Excellent dielectric strength, heat resistance, and resistance to moisture.	Available in molding powder and liquid resin. Finished molded products are produced by compression molding.	Light switches, electric motor insulator and mounting cases, television tuning devices and tube supports. Enamels and lacquers for automobiles, refrigerators, and stoves are typical uses for the liquid form.
Allylics	Excellent dielectric strength and insulation resistance. No moisture absorption; stain resistance. Full range of opaque and transparent colors.	Available in the form of monomers, prepolymers, and powders. Finished articles may be made by transfer or compression molding, lamination, coating, or impregnation.	Electrical connectors, appliance handles, knobs, etc. Laminated overlays or coatings for plywood, hardboard, and other laminated materials needing protection from moisture.
Amino (Melamine and Urea)	Full range of translucent and opaque colors. Very hard, strong, but not unbreakable. Good electrical qualities.	Available as molding powder or granules, as a foamed material in solution, and as resins. Finished products can be made by compression, transfer, plunger molding, and laminating with wood, paper, etc.	Melamine—Tablewear, buttons, distributor cases, tabletops, plywood adhesive, and as a paper and textile treatment. Urea—Scale housing, radio cabinets, electrical devices, appliance housings, stove knobs in resin form as baking enamel coatings, plywood adhesive and as a paper and textile treatment.
Casein	Excellent surface polish. Wide range of near transparent and opaque colors. Strong, rigid, affected by humidity and temperature changes.	Available in rigid sheets, rods, and tubes, as a powder and liquid. Finished products are made by machining of the sheets, rods, and tubes.	Buttons, buckles, beads, game counters, knitting needles, toys, and adhesives.
Cold-Molded 3 Types: Bitumin Phenolic Cement-Asbestos	Resistance to high heat, solvents, water, and oil.	Available in compounds. Finished articles produced by molding and curing.	Switch bases and plugs, insulators, small gears, handles and knobs, tiles, jigs and dies, toy building blocks.
Epoxy	Good electrical properties; water and weather resistance.	Available as molding compounds, resins, foamed blocks, liquid solutions, adhesives, coatings, sealants.	Protective coating for appliances, cans, drums, gymnasium floors, and other hard-to-protect surfaces. They firmly bond metals, glass, ceramics, hard rubber and plastics, printed circuits, laminated tools and jigs, and liquid storage tanks.
Phenolics	Strong and hard. Heat and cold resistant; excellent insulators.	Cast and molded.	Radio and tv cabinets, washing machine agitators, jukebox housings, jewelry, pulleys, electrical insulation.
Polyesters (Fiberglass)	Strong and tough, bright and pastel colors. High dielectric qualities.	Produced as liquids, dry powders, premix molding compounds, and as cast sheets, rods, and tubes. They are formed by molding, casting, impregnating, and premixing.	Used to impregnate cloth or mats of glass fibers, paper, cotton, and other fibers in the making of reinforced plastic for use in boats, automobile bodies, luggage.
Silicones	Heat resistant, good dielectric properties.	Available as molding compounds, resins, coatings, greases, fluids, and silicon rubber. Finished by compression and transfer molding, extrusion, coating, calendering, casting, foaming, and impregnating.	Coil forms, switch parts, insulation for motors, and generator coils.

FIG. 12-4-3   Thermosetting plastics. *(The Society of Plastics Industry)*

MATERIALS \ PROPERTY	Thermoplastics											Thermosets				
	ABS	ACETAL	ACRYLIC	CELLULOSIC	FLUOROCARBON	POLYAMIDE	POLYCARBONATE	POLYETHYLENE	POLYPROPYLENE	POLYVINYL CHLORIDE	POLYVINYL	EPOXY	PHENOLICS	POLYESTER	SILICONE	UREA & MELAMINE
Tensile Strength	–	2	2	4	–	1	1	–	–	3	–	3	3	2	1	2
Flexural Strength	4	4	3	4	1	2	3	–	1	3	3	3	2	1	3	3
Impact Strength	2	–	–	2	1	–	2	–	3	2	2	–	1	2	3	3
Hardness	2	2	1	–	–	2	2	–	–	3	–	3	1	3	4	2
Continuous Heat Resistance	–	4	–	–	1	2	–	–	3	–	–	4	2	3	1	4
Weather Resistance	–	2	1	–	1	2	–	2	2	–	2	3	3	1	2	4
Resistance to Heat Expansion	2	–	3	–	–	1	2	–	–	2	3	4	2	2	1	3
Electrical Properties	–	–	4	–	2	–	1	3	3	2	1	1	1	1	1	1
Maximum Volume per Kilogram	3	–	–	4	–	–	–	2	1	3	–	4	1	1	3	2
Chemical Resistance	–	2	–	–	1	3	–	4	4	–	–	1	1	1	1	1
Transparency	–	–	1	3	–	–	4	–	–	2	4	–	–	–	–	–
Resistance to Cold Flow	2	–	3	–	–	–	3	–	–	2	1	–	–	–	–	–
Dimensional Stability to Moisture	2	2	3	–	1	–	2	1	1	1	2	2	–	1	3	2
Colorability	1	3	1	2	4	3	2	3	3	1	3	–	–	1	–	2

Note: Materials rated number 1 are best of those listed for property indicated. Dashes indicate material is not considered for that particular property. Do not compare properties between thermoplastic and thermosetting materials.

FIG. 12-4-4   Property comparison chart for plastics *(General Motors Corp.)*

PRODUCTION REPORT

Part Name	Material 1st Choice	Material 2nd Choice	Reason for Selection	Machining Required	Color
Telephone Case	ABS	Cellulosics	Good impact strength Good range of colors Excellent surface finish Good electrical properties Variety of forming methods	None	Green Blue White Tan Red Black

FIG. 12-4-5   Selection of material. *(STUDIOHIO)*

For instance, the preliminary production report for the material selection of the telephone case shown in Fig. 12-4-5 is an example of the type of research required in selecting a material.

Forming Processes

For design information on the preparation of molded plastics, see Chap. 13, Unit 13-4.

REFERENCES AND SOURCE MATERIAL

1. The Society of the Plastics Industry, Inc.
2. Crystaplex Plastics.
3. General Motors Corp.

ASSIGNMENTS ▨▨▨▨▨▨▨▨▨▨▨▨▨▨▨▨▨▨▨▨▨▨▨▨

See Assignments 5 through 8 for Unit 12-4 on page 345–346.

12-5 RUBBER

This unit is to acquaint the drafter with the general characteristics of rubber, both natural and synthetic.

The use of rubber is advantageous when design considerations involve one or more of the following factors:

- Electrical insulation
- Vibration isolation
- Sealing surfaces
- Chemical resistance
- Flexibility

Material and Characteristics

Elastomers (rubber-like substances) are derived from either natural or synthetic sources. Rubber can be formed into useful rigid or flexible shapes, usually with the aid of heat, pressure, or both. The most outstanding characteristics of vulcanized rubber are its low modulus of elasticity and its ability to withstand large deformations and to quickly recover its shape when released. Vulcanized rubber is most compressible. In general, natural rubber has good flex life and low temperature flexibility. Certain synthetic rubbers have characteristics that offer improved performance under conditions that involve such deteriorating effects as heat, oil, and weather.

The cost of each type of rubber and ease of processing are factors to be considered when selecting materials for any application. The use of rubber varies from cements and coatings to soft or hard mechanical goods. Typical formed part are tires, tubes, battery cases, drive belts, machinery mounts, hoses, seals, floormats, gaskets, and weather strips.

Kinds of Rubber

Rubber parts are produced in either mechanical (solid) or cellular form, depending upon the desired performance of the part.They are categorized into two kinds of rubber, natural and synthetic. The synthetic rubbers are divided into several kinds.

Mechanical Rubber

Mechanical rubber is used in either pressure-molded, cast, or extruded forms. Typical parts produced by these methods are tires, belts, and bumpers. Mechanical rubber should be used in preference to sponge rubber because of its superior physical properties.

Cellular Rubber

Cellular rubber can be produced with "open" or "closed" cells. Open-cell sponge rubber is made by the inclusion of a gas-forming chemical compound in the mixture before vulcanization. The heat of the vulcanizing process causes a gas to form in the rubber, making a cellular structure. If compressed, the air is expelled from the cells. When the pressure is released, air is absorbed, allowing the part to quickly recover its shape. Typical applications are pads and weather stripping. Foam rubber is a specialized type of open cell.

Closed-cell sponge rubber is made by an inert gas solution method which produces innumerable ball-shaped cells with continuous walls. When closed-cell sponge rubber is deformed, the cells are displaced rather than deflated. Closed-cell rubber is very springy when squeezed.

Both open- and closed-cell sponge rubber is available in block or sheet form that can be cut to size and shape. This characteristic can sometimes provide a low-cost method of producing relatively simple parts.

Assembly Methods

Several methods of fastening rubber parts to other components of an assembly can be used. When selecting the method of attachment, the designer should consider the hardness of the rubber, operating conditions, and disassembly requirements.

Fastener Inserts

Rubber can be molded to various metallic inserts, as illustrated in Fig. 12-5-1 (pg. 343). Some of the advantages of this practice are the elimination of loose attaching parts, simplification of assembly operations, and reduction of assembly equipment.

Inserts should be designed with holes, undercuts, or such shape that the rubber can overhang an edge. This design provides a mechanical anchor and additional adhesive bond of the rubber to the metal. Sharp edges that cause stress concentrations should be avoided.

Grip Fit

Many molded and extruded soft rubber parts and shapes are designed to take advantage of their gripping action to hold them in place at assembly. This action derives from the characteristic of rubber that permits it to be stretched or extended (Fig. 12-5-2). The grip fit can also serve as a seal for most elements. In applications containing liquids under pressure, additional fasteners should be used to ensure retention and a positive seal.

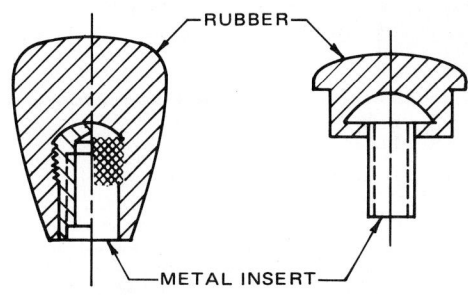

FIG. 12-5-1   Fastener inserts.

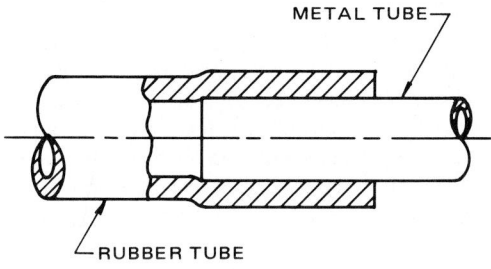

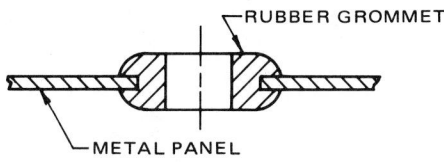

FIG. 12-5-2   Grip fits.

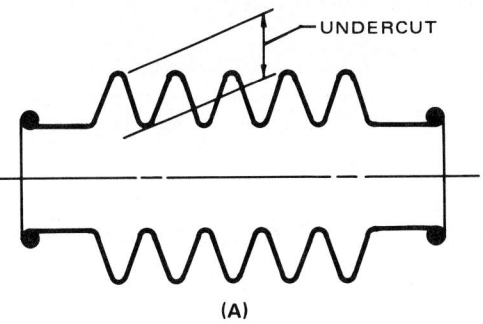

(A)

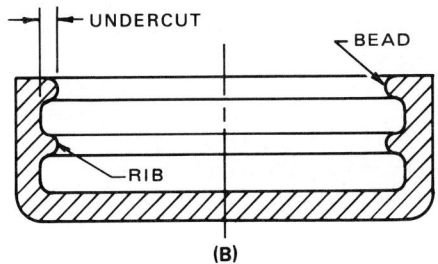

(B)

FIG. 12-5-3   Ribs, undercuts, and beads.

Design Considerations

Hard rubber molded parts present problems similar to those of plastics, which are described in Unit 12-4. The following points should be considered in the designing of soft rubber molded parts.

- Reenforcing ribs generally do not represent molding problems. When the inside size is relatively large and the undercut is not too deep, the part may easily be stripped from the mold because of its elasticity (Fig. 12-5-3).
- The thickness of walls and sections depends upon the loading requirements and the hardness of the rubber. Because of the resilience of soft rubber, sections should be of uniform cross section (Fig. 12-5-4).
- Due to the flexibility of rubber and the size and shape of the part, many items do not require draft. However, draft, or taper, usually facilitates molding. The amount of draft depends upon the hardness of the rubber, length of surface, and types of inserts. Generally, at least 0.5° draft per side should be provided.
- Fillets and radii improve the flow of rubber to the various sections. Where rubber is bonded to inserts, the bond will be less likely to fail if a certain amount of rubber is permitted to overhang the edges of the inserts (Fig. 12-5-4).

Specifying Rubber on Drawings

Rubber specifications should always be determined in consultation with a material engineer or rubber parts supplier. Since the broad spectrum of properties of rubber are easily varied by

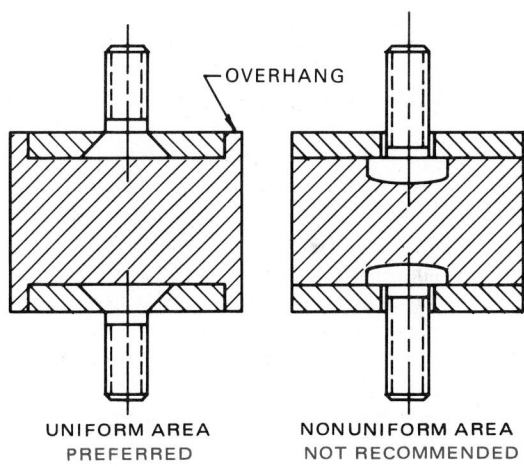

UNIFORM AREA
PREFERRED

NONUNIFORM AREA
NOT RECOMMENDED

FIG. 12-5-4   Wall and section thickness.

the ingredients and conditioning of processing, rubber materials should be specified on the basis of performance rather than chemical composition. Specifications normally cover tensile strength, elongation, hardness, compression, set, and various aging and weather tests.

In parts formed of soft rubber compounds, the hardness is usually specified because it is quickly and easily measured and is related to modulus.

REFERENCES AND SOURCE MATERIAL

1. General Motors Corp.

ASSIGNMENTS

See Assignments 9 through 11 for Unit 12-5 on page 346.

ASSIGNMENTS FOR CHAPTER 12

ASSIGNMENT FOR UNIT 12-1, CAST IRONS

1. Make a two-view working drawing of one of the parts shown in Fig. 12-1-A or 12-1-B. Use a revolved section to show the center section of the arm. Select a suitable cast iron for the part. Scale 1:1.

ASSIGNMENT FOR UNIT 12-2, CARBON STEEL

2. Make a working drawing of one of the parts shown in Fig. 12-2-A or 12-2-B. Show the worm threads in pictorial form. Scale 2:1. Select a suitable steel for the part. Conventional breaks may be used to shorten the length of the view.

ASSIGNMENTS FOR UNIT 12-3, NONFERROUS METALS

3. Make a three-view working drawing of the outboard motor clamp shown in Fig. 12-3-A or 12-3-B. Add a full section top view having the cutting plane located at line *AD*. Use lines or surfaces marked *A*, *B*, and *C* as the zero lines and use arrowless dimensioning. Scale 1:2. Select a suitable material noting that the part must be water-resistant, have a painted finish, have moderate strength, and have a light weight or mass.

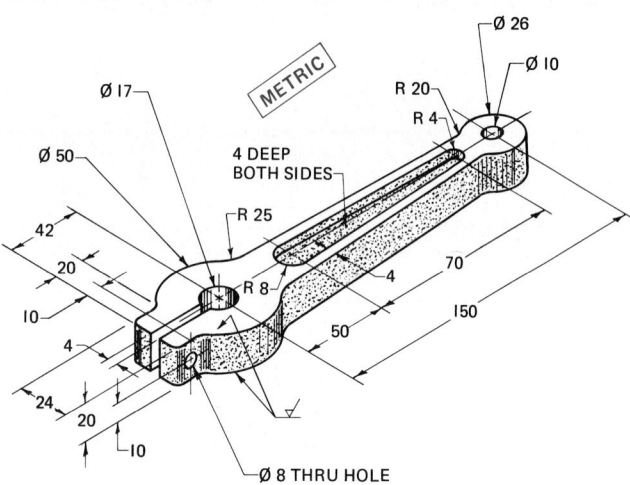

FIG. 12-1-A   Plug wrench.

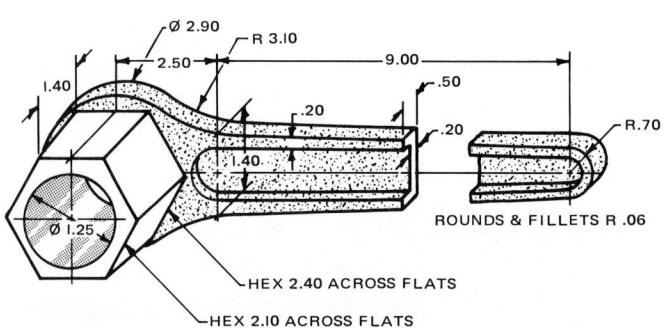

FIG. 12-1-B   Door closer arm.

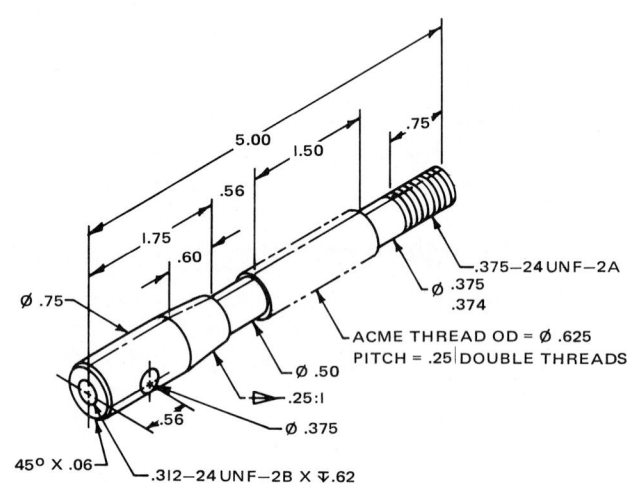

FIG. 12-2-A   Raising bar.

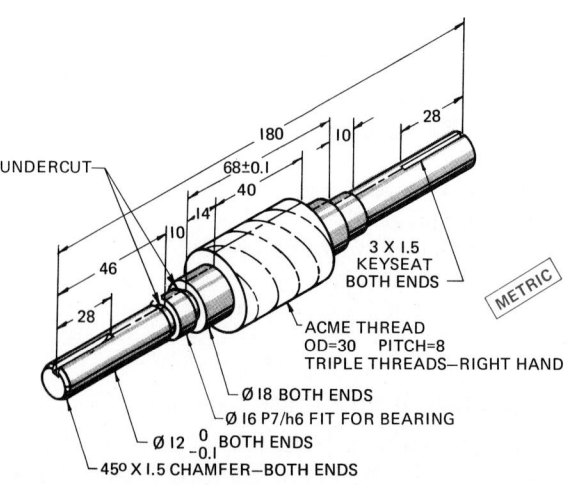

FIG. 12-2-B   Worm for gear jack.

4. Make a two-view assembly drawing, complete with an item list, of the coupling assembly shown in Fig. 12-3-C. The coupling is bolted to an 8 mm steel plate. Using phantom lines, show the plate and the shafts extending a short distance beyond the parts. Select suitable material for the parts. Scale 1:1.

ASSIGNMENTS FOR UNIT 12-4, PLASTICS

5. Design and prepare working drawings for a plastic tee and rubber grommet used for golfing. The design can be standard or novel. The tee is held on to a #20 Am. St. gage aluminum plate by the grommet.

6. Design and prepare a working drawing of a gearshift handle to screw on to a Ø.375 in. (or Ø10 mm) shaft. A threaded insert (see Unit 13-4) is recommended. Selection of material and color to be included on the drawing.

7. Make a one-view detailed assembly drawing of the shaft coupling shown in Fig. 12-4-A. The metal hubs are joined by an elastomer. Assembly sizes: overall length 2.90; shaft diameter .750 (RC4 hole basis fit); hub Ø1.50. Use your judgment for dimensions not shown. Include on the drawing an item list. Scale 1:1.

8. Make a two-view exploded orthographic assembly drawing of the connecting link shown in Fig. 12-4-B (pg. 346). Include on the drawing a material list. The student is to select the material. Show only overall dimensions and shaft sizes.

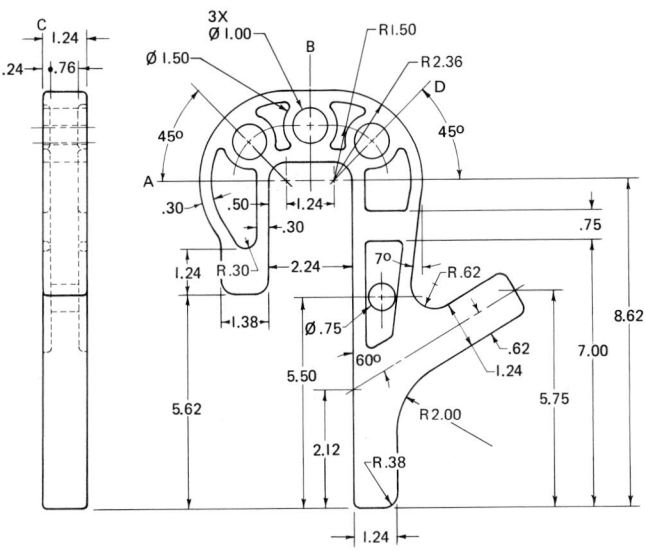

ROUNDS AND FILLETS R.12

FIG. 12-3-A   Outboard motor clamp.

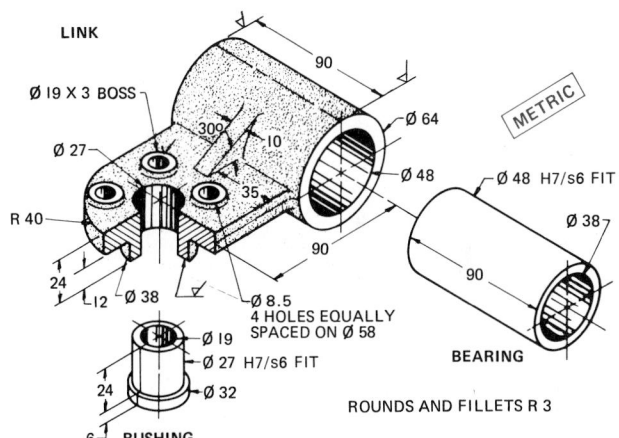

FIG. 12-3-C   Coupling.

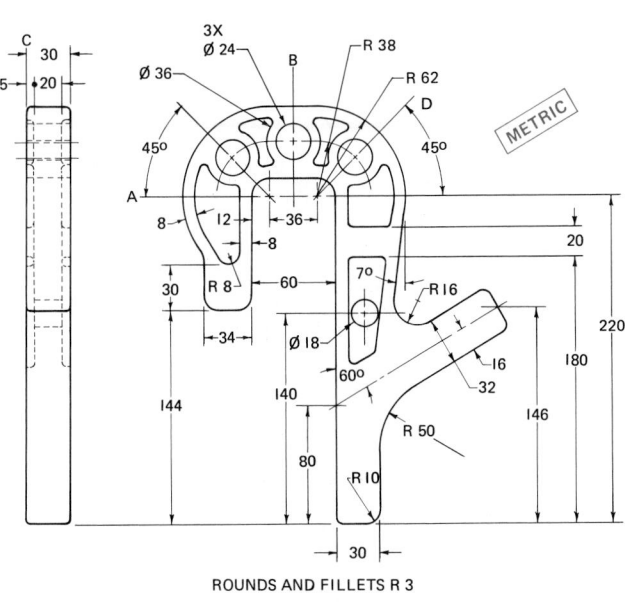

ROUNDS AND FILLETS R 3

FIG. 12-3-B   Outboard motor clamp.

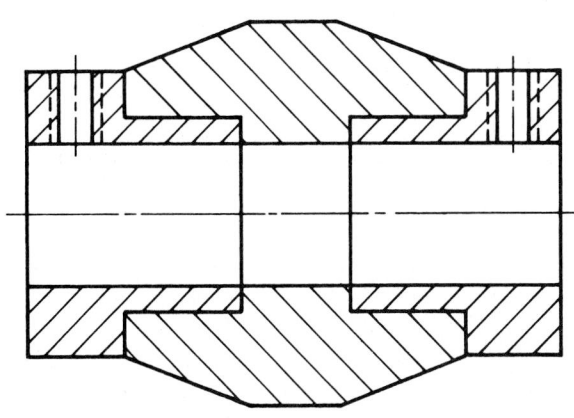

FIG. 12-4-A   Shaft coupling.

ASSIGNMENTS FOR UNIT 12-5, RUBBER

9. Design a rubber boot, similar to that shown in Fig. 12-5-3, to fit on the universal joint shown in Fig. 12-5-A. The purpose of the boot is to prevent dirt and other contaminants from forming around the joint.

10. Make a detail drawing of the boot in Assignment 9.
11. Make a one-view full section assembly drawing of the caster assembly shown in Fig. 12-5-B. Include on the drawing an item list and overall assembly sizes. Select a suitable material for part 1. Scale 1:1.

FIG. 12-4-B   Connecting link.

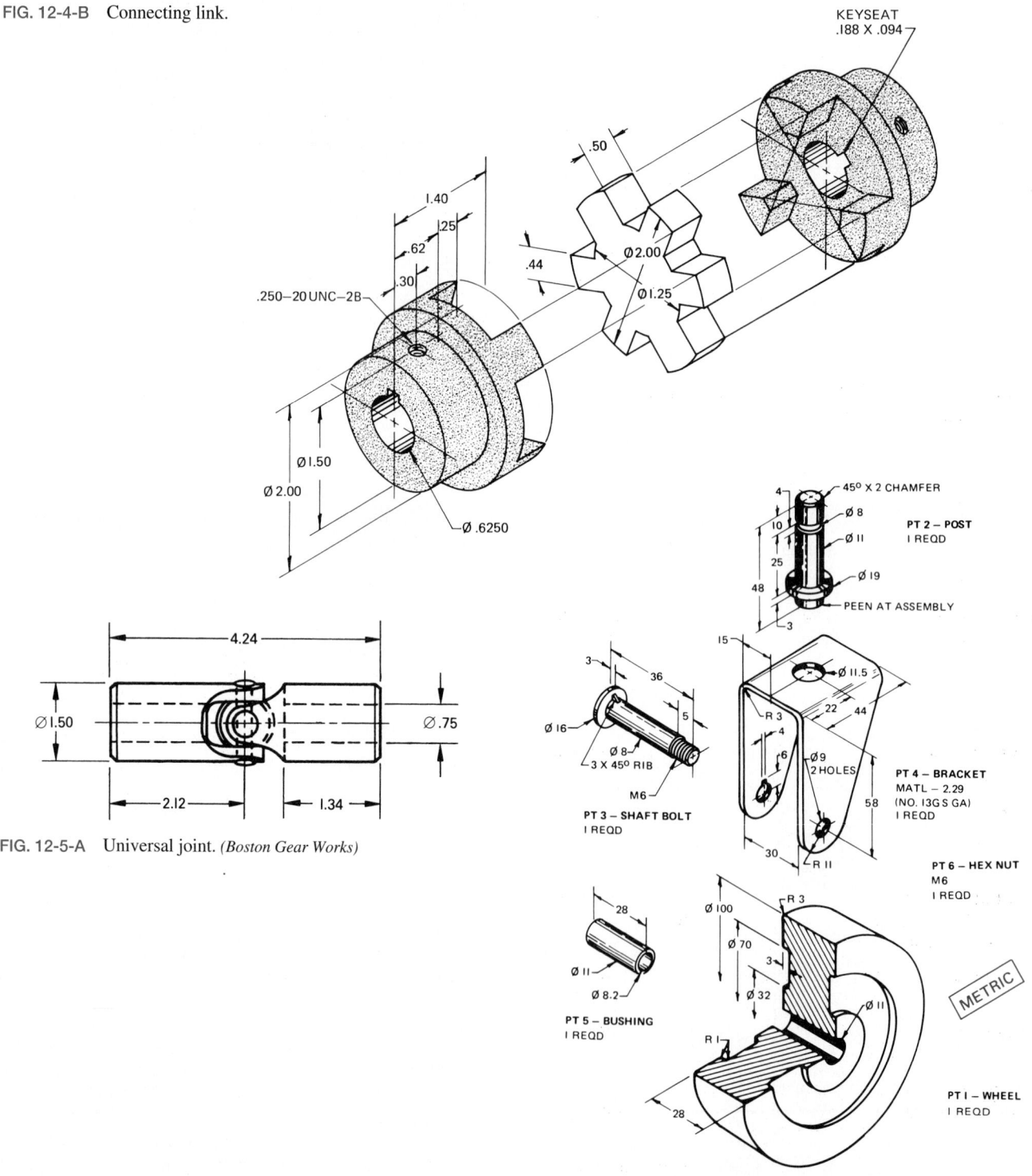

FIG. 12-5-A   Universal joint. *(Boston Gear Works)*

FIG. 12-5-B   Caster assembly.

CHAPTER 13

FORMING PROCESSES

Definitions

Briquetting machines   Machines used to compress powders into finished shapes.

Casting   The process whereby parts are produced by pouring molten metal into a mold.

Closed-die forging   All forging operations involving three-dimensional control.

Cold chamber   A type of die-casting machine used for high-melting nonferrous alloys.

Cold shut   A lap where two surfaces of metal have folded against each other.

Datum surfaces   The surfaces that provide a common reference for measuring, machining, and assembly.

Datum-locating dimension   The dimension between each casting datum surface and the corresponding machining datum surface.

Draft   The slope given to the side walls of a pattern to facilitate easy removal from a mold, or a casting from a die.

Ferrous   Metals that contain iron.

Flash space   The space provided between the die surfaces for the escape of the excess metal, called flash.

Forging   Plastically deforming a cast or sintered ingot, a wrought bar or billet, or a powder-metal shape to produce a desired shape with good mechanical properties.

Investment   The refractory material used to encase a wax pattern.

Nonferrous   Alloys that contain metals such as aluminum, magnesium, and copper but contain no iron.

Parting line   A line along which the pattern is divided for molding, or along which the sections of a mold separate.

Powder metallurgy   The process of making parts by compressing and sintering various metallic and nonmetallic powders into shape.

Shrinkage   The difference between dimensions of the mold and the corresponding dimensions of the molded part.

Slurry   A mixture of plaster of paris and fillers with water and setting-control agents.

Submerged plunger   A type of die-casting machine used for low-melting alloys.

13-1 METAL CASTINGS

Forming Processes

When a component of a machine takes shape on the drawing board or CAD monitor of the designer, the method of its manufacture may still be entirely open. The number of possible manufacturing processes is increasing day by day, and the optimum process is found only by carefully weighing technological advantages and drawbacks in relation to the economics of production.

The choice of the manufacturing process depends on the size, shape, and quantity of the component. Manufacturing processes are therefore important to the engineer and drafter in order to properly design a part. They must be familiar with the advantages, disadvantages, costs, and machines necessary for manufacturing. Since the cost of the part is influenced by the production method, such as welding or casting, the designer must be able to choose wisely the method that will reduce the cost. In some cases it may be necessary to recommend the purchase of a new or different machine in order to produce the part at a competitive price. This means the designer should design the part for the process as well as for the function. Most of all, unnecessarily close tolerances on nonfunctional dimensions should be avoided.

This chapter covers the following manufacturing processes: casting, forging, and powder metallurgy. Forming by means of welding is covered in Chap. 18.

Casting Processes

Casting is the process whereby parts are produced by pouring molten metal into a mold. A typical cast part is shown in Fig. 13-1-1. Casting processes for metals can be classified by either the type of mold or pattern or the pressure or force used to fill the mold. Conventional sand, shell, and plaster molds use a permanent pattern, but the mold is used only once. Permanent molds and die-casting dies are machined in metal or graphite sections and are employed for a large number of castings. Investment casting and the relatively new full mold process involve both an expendable mold and an expendable pattern.

Casting metals are usually alloys or compounds of two or more metals. They are generally classed as ferrous or nonferrous metals. *Ferrous metals* are those that contain iron, the most common being gray iron, steel, and malleable iron. *Nonferrous alloys,* which contain no iron, are those containing metals such as aluminum, magnesium, and copper.

Sand Mold Casting

The most widely employed casting process for metals uses a permanent pattern of metal or wood that shapes the mold cavity when loose molding material is compacted around the pattern. This material consists of a relatively fine sand, which serves as the refractory aggregate, plus a binder.

A typical sand mold, with the various provisions for pouring the molten metal and compensating for contraction of the solidifying metal, and a sand core for forming a cavity in the casting are shown in Fig. 13-1-2. Sand molds consist of two or more sections: bottom (*drag*), top (*cope*), and intermediate sections (*cheeks*) when required. The sand is contained in flasks equipped with pins and plates to ensure the alignment of the cope and drag.

Molten metal is poured into the sprue, and connecting runners provide flow channels for the metal to enter the mold cavity through gates. Riser cavities are located over the heavier sections of the casting. A vent is usually added to permit the escape of gases that are formed during the pouring of metal.

When a hollow casting is required, a form called a *core* is usually used. Cores occupy that part of the mold that is intended to be hollow in the casting. Cores, like molds, are formed of sand and placed in the supporting impressions, or *core prints,* in the molds. The core prints ensure positive location of the core in the mold and, as such, should be placed so that they support the mass of the core uniformly to prevent shifting or sagging. Metal core supports called *chaplets*, which

FIG. 13-1-1   Typical cast part.

are used in the mold cavity and which fuse into the casting, are sometimes used by the foundry in addition to core prints (Fig. 13-1-3A, pg. 350). Chaplets and their locations are not usually specified on drawings.

In producing sand molds, a metal or wooden pattern must first be made. The pattern, normally made in two parts, is slightly larger in every dimension than the part to be cast, to allow for shrinkage when the casting cools. This is known as *shrinkage allowance*, and the pattern maker allows for it by using a shrink rule for each of the cast metals.

Drafts, or slight tapers, are also placed on the pattern to allow for easy withdrawal from the sand mold. The parting line location and amount of draft are very important considerations in the design process.

In the construction of patterns for castings in which various points on the surface of the casting must be machined, sufficient excess metal should be provided for all machined surfaces. Allowance depends on the metal used, the shape and size of the part, the tendency to warp, the machining method, and set-up.

After a sand mold has been used, the sand is broken and the casting removed. Next the excess metal, gates, and risers are removed and remelted.

Shell Mold Casting

The refractory sand used in shell molding is bonded by a thermostable resin that forms a relatively thin shell mold. A heated, reusable metal pattern plate (Fig. 13-1-3B) is used to form each half of the mold by either dumping a sand-resin mixture on top of the heated pattern or by blowing resin-coated sand under air pressure against the pattern.

Plaster Mold Casting

Plaster of paris and fillers are mixed with water and setting-control agents to form a *slurry*. This slurry is poured around a reusable metal or rubber pattern and sets to form a gypsum mold (Fig. 13-1-3C). The molds are then dried, assembled, and filled with molten (nonferrous) metals. Plaster mold casting is ideal for producing thin, sound walls. As in sand mold casting, a new mold is required for each casting. Castings made by this process have smoother finish, finer detail, and greater dimensional accuracy than sand castings.

Permanent Mold Casting

Permanent mold casting makes use of a metal mold, similar to a die, which is utilized to produce many castings from each mold (Fig. 13-1-4, pg. 350). It is used to produce some ferrous alloy castings, but due to rapid deterioration of the mold caused by the high pouring temperatures of these alloys and the high mold cost, the process is confined largely to production of nonferrous alloy castings.

Investment Mold Casting

Investment castings have been better known in the past by the term *lost wax castings*. The term *investment* refers to the refractory material used to encase the wax patterns.

This process uses both an expendable pattern and an expendable mold. Patterns of wax, plaster, or frozen mercury

FIG. 13-1-2   Sequence in preparing a sand casting.

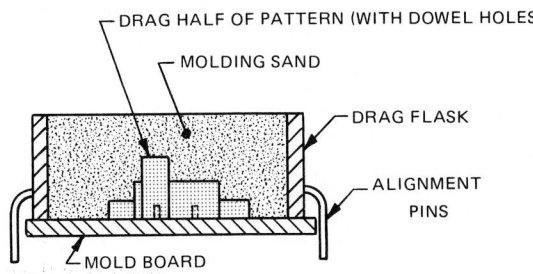

DRAG HALF OF PATTERN (WITH DOWEL HOLES)
MOLDING SAND
DRAG FLASK
ALIGNMENT PINS
MOLD BOARD

(A) STARTING TO MAKE THE SAND MOLD

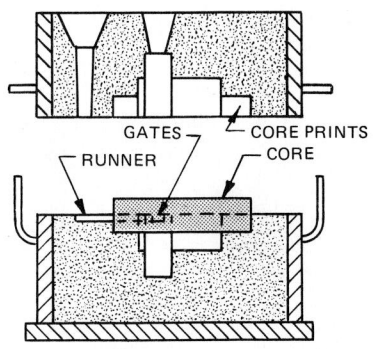

GATES
CORE PRINTS
RUNNER
CORE

(E) PARTING FLASKS TO REMOVE PATTERN AND TO ADD CORE AND RUNNER

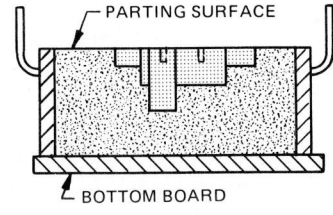

PARTING SURFACE
BOTTOM BOARD

(B) AFTER ROLLING OVER THE DRAG

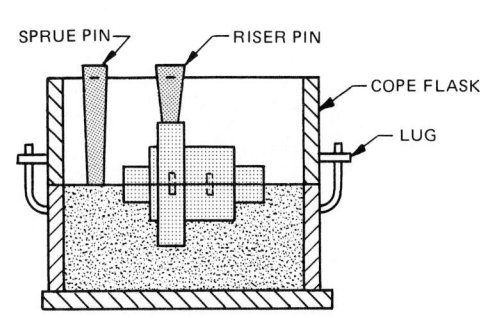

SPRUE PIN
RISER PIN
COPE FLASK
LUG

(C) PREPARING TO RAM MOLDING SAND IN COPE

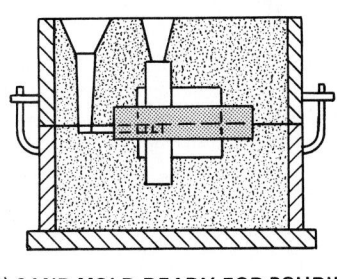

(F) SAND MOLD READY FOR POURING

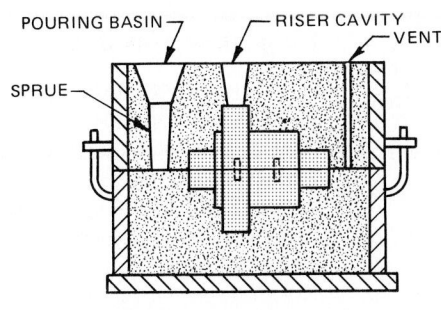

POURING BASIN
RISER CAVITY
VENT
SPRUE

(D) REMOVING RISER AND SPRUE PINS AND ADDING POURING BASIN

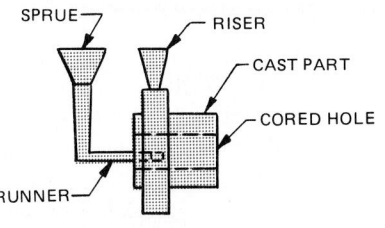

SPRUE
RISER
CAST PART
CORED HOLE
RUNNER

SPRUE, RISER, AND RUNNER TO BE REMOVED FROM CASTING.

(G) CASTING AS REMOVED FROM THE MOLD

FIG. 13-1-3   Mold casting techniques.

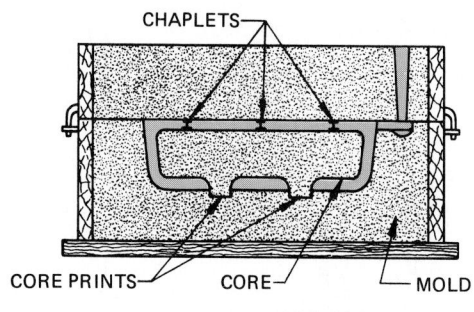

(A) CHAPLETS FOR SAND MOLD

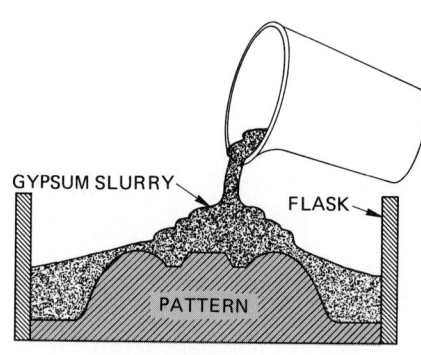

(B) SHELL MOLD BEING STRIPPED FROM PATTERN

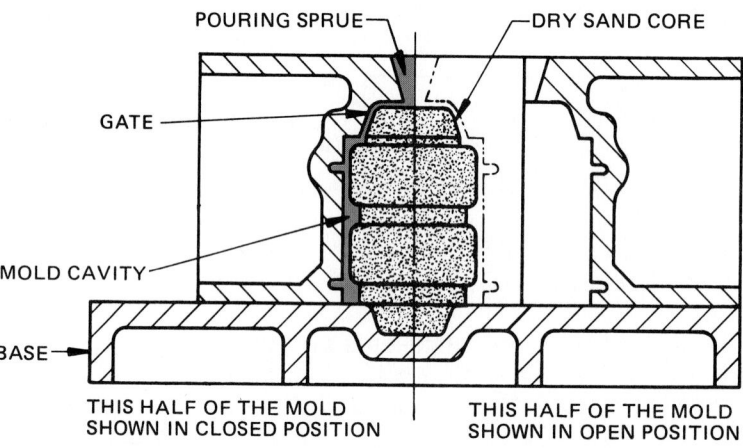

(C) POURING SLURRY OVER A PLASTER MOLD PATTERN

FIG. 13-1-4   Permanent mold casting. *(General Motors Corp.)*

THIS HALF OF THE MOLD SHOWN IN CLOSED POSITION   THIS HALF OF THE MOLD SHOWN IN OPEN POSITION

are cast in metal dies. The molds are formed either by pouring a slurry of a refractory material around the pattern positioned in a flask or by building a thick layer of shell refractory on the pattern by repeated dipping into slurries and drying. The arrangement of the wax patterns in the flask method is shown in Fig. 13-1-5.

Full Mold Casting

The characteristic feature of the full mold process is the use of consumable patterns made of foamed plastic. These are not extracted from the mold but are vaporized by the molten metal.

The full mold process is suitable for individual castings. The advantages it offers are obvious: it is very economical and reduces the delivery time required for prototypes, articles urgently needed for repair jobs, or individual large machine parts.

Centrifugal Casting

In the centrifugal casting process, commonly applied to cylindrical casting of either ferrous or nonferrous alloys, a permanent mold is rotated rapidly about the axis of the casting while a measured amount of molten metal is poured into the mold cavity (Fig. 13-1-6A). The centrifugal force is used to hold the metal against the outer walls of the mold with the volume of metal poured determining the wall thickness of the casting. Rotation speed is rapid enough to form the central hole without a core. Castings made by this method are smooth, sound, and clean on the outside.

Continuous Casting

Continuous casting produces semifinished shapes, such as uniform section rounds, ovals, squares, rectangles, and plates.

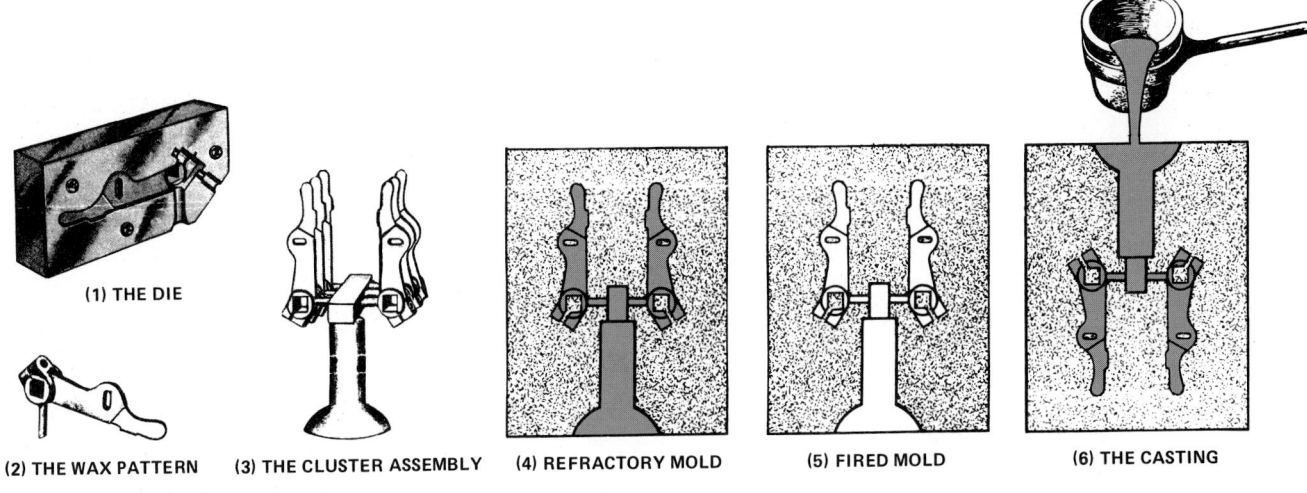

FIG. 13-1-5   Investment mold casting.

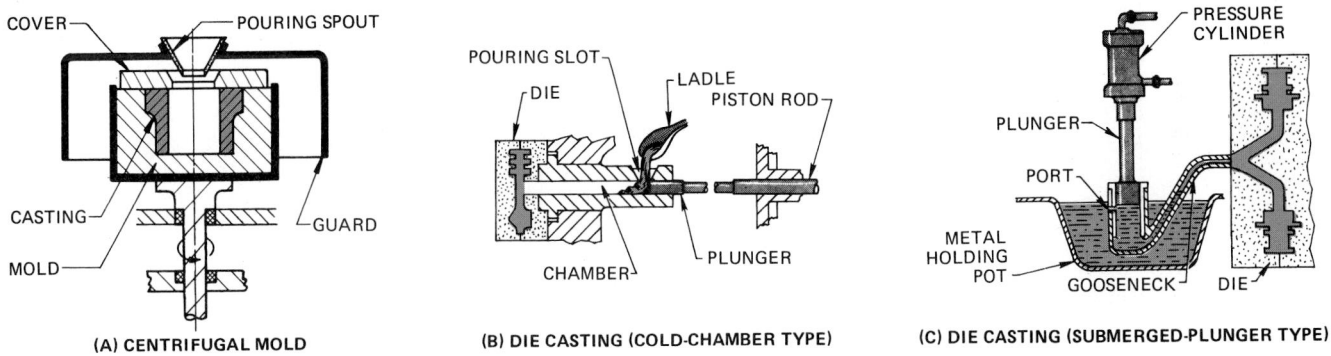

FIG. 13-1-6   Casting equipment and processes.

These shapes are cast from nearly all ferrous and nonferrous metals by continuously pouring the molten metal into a water-jacketed mold. The metal solidifies in the mold, and the solid billet exits continuously into a water spray. These sections are processed further by rolling, drawing, or extruding into smaller, more intricate shapes. Iron bars cast by this process are finished by machining.

Die Casting

One of the least expensive, fastest, and most efficient processes used in the production of metal parts is die casting. Die castings are made by forcing molten metal into a die or mold. Large quantities, accurately cast, can be produced with a die-casting die, thus eliminating or reducing machining costs. Many parts are completely finished when taken from a die. Since die castings can be accurate to within .001 in. (0.02 mm) of size, internal and external threads, gear teeth, and lugs can readily be cast.

Die casting has its limitations. Only nonferrous alloys can be die-cast economically because of the lack of a suitable die material to withstand the higher temperatures required for steel and iron.

Die-casting machines are of two types: the *submerged-plunger* type for low-melting alloys containing zinc, tin, lead, etc., and the *cold-chamber* type for high-melting nonferrous

alloys containing aluminum and magnesium (Figs. 13-1-6B and C).

Selection of Process

Selection of the most feasible casting process for a given part requires an evaluation of the type of metal, the number of castings required, their shape and size, the dimensional accuracy required, and the casting finish required. When the casting can be produced by a number of methods, selection of the process is based on the most economical production of the total requirement. Since final cost of the part, rather than price of the rough casting, is the significant factor, the number of finishing operations necessary on the casting is also considered. Those processes that provide the closest dimensions, the best surface finish, and the most intricate detail usually require the smallest number of finishing operations.

A direct comparison of the capabilities, production characteristics, and limitations of several processes is provided in Fig. 13-1-7 (pg. 352).

Design Considerations

The advantages of using castings for engineering components are well appreciated by designers. Of major importance is the

PROCESS	METALS CAST	USUAL WEIGHT (MASS) RANGE	MINIMUM PRODUCTION QUANTITIES	RELATIVE SET-UP COST	CASTING DETAIL FEASIBLE	MINIMUM THICKNESS IN. (MM)	DIMENSIONAL TOLERANCES IN. (MM)	SURFACE FINISH, RMS (μIN.)
SAND (Green, Dry, and Core) CO₂ Sand	All ferrous and nonferrous	Less than 1 lb. (0.5 kg) to several tons	3, without mechanization	Very low to high depending on mechanization	Fair	.12 to .25 (3 to 6) .10 to .25 (2.5 to 6)	± .03 (0.8) ± .02 (0.5)	350 250
SHELL	All ferrous and nonferrous	0.5 to 30 lb. (0.2 to 15 kg)	50	Moderate to high depending on mechanization	Fair to good	.03 to .10 (0.8 to 2.5)	± .015 (0.4)	200
PLASTER	Al, Mg, Cu, and Zn alloys	Less than 1 lb. to 3000 lb. (0.5 to 1350 kg)	1	Moderate	Excellent	.03 to .08 (0.8 to 2)	± .01 (0.2)	100
INVESTMENT	All ferrous and nonferrous	Less than 1 oz. to 50 lb. (30 g to 25 kg)	25	Moderate	Excellent	0.2 to .06 (0.5 to 1.5)	± .005 (0.1)	80
PERMANENT MOLD Metal Mold Graphite Mold	Nonferrous and cast iron Steel	1 to 40 lb. (0.5 to 20 kg) 5 to 300 lb. (2 to 150 kg)	100 100	Moderate to high	Poor	.18 to .25 (4.5 to 6) .25 (6)	± .02 (0.5) ± .03 (0.8)	200 200
DIE	Sn, Pb, Zn, Al, Mg, and Cu alloys	Less than 1 lb. to 20 lb. (0.5 to 10 kg)	1000	High	Excellent	.05 to .08 (1.2 to 2)	± .002 (0.5)	60

* Values listed are primarily for aluminum alloys, but data applies generally to other metals also.

‡ Depends on surface area. Double if dimension is across parting line.

FIG. 13-1-7   General characteristics of casting processes.

fact that they can produce shapes of any degree of complexity and of virtually any size.

Solidification of Metal in a Mold

While this is not the first step in the sequence of events, it is of such fundamental importance that it forms the most logical point to begin understanding the making of a casting. Consider a few simple shapes transformed into mold cavities and filled with molten metal.

In a sphere, heat dissipates from the surface through the mold while solidification commences from the outside and proceeds progressively inward, in a series of layers (Fig. 13-1-8A). As liquid metal solidifies, it contracts in volume, and unless feed metal is supplied, a shrinkage cavity may form in the center.

The designer must realize that a shrinkage problem exists and that the foundry worker must attach risers to the casting or resort to other means to overcome it.

When the simple sphere has solidified further, it continues to contract in volume, so that the final casting is smaller than the mold cavity.

Consider a shape with a square cross section, such as the one shown in Fig. 13-1-8B. Here again, cooling proceeds at

right angles to the surface and is necessarily faster at the corners of the casting. Thus solidification proceeds more rapidly at the corners.

The resulting hot spot prolongs solidification, promoting solidification shrinkage and lack of density in this area. The only logical solution, from the designer's viewpoint, is the provision of very generous fillets or radii at the corners. Additionally, the relative size or shape of the two sections forming the corner is of importance. If they are materially different, as in Fig. 13-1-8E, contraction in the lighter member will occur at a different rate from that in the heavier member. Differential contraction is the major cause of casting stress, warping, and cracking.

General Design Rules

Design for Casting Soundness   Most metals and alloys shrink when they solidify. Therefore, the design must be such that all members of the parts increase in dimension progressively to one or more suitable locations where feeder heads can be placed to offset liquid shrinkage (Fig. 13-1-9).

Fillet or Round All Sharp Angles   Fillets have three functional purposes: to reduce stress concentration in the casting in service; to eliminate cracks, tears, and draws at reentry angles;

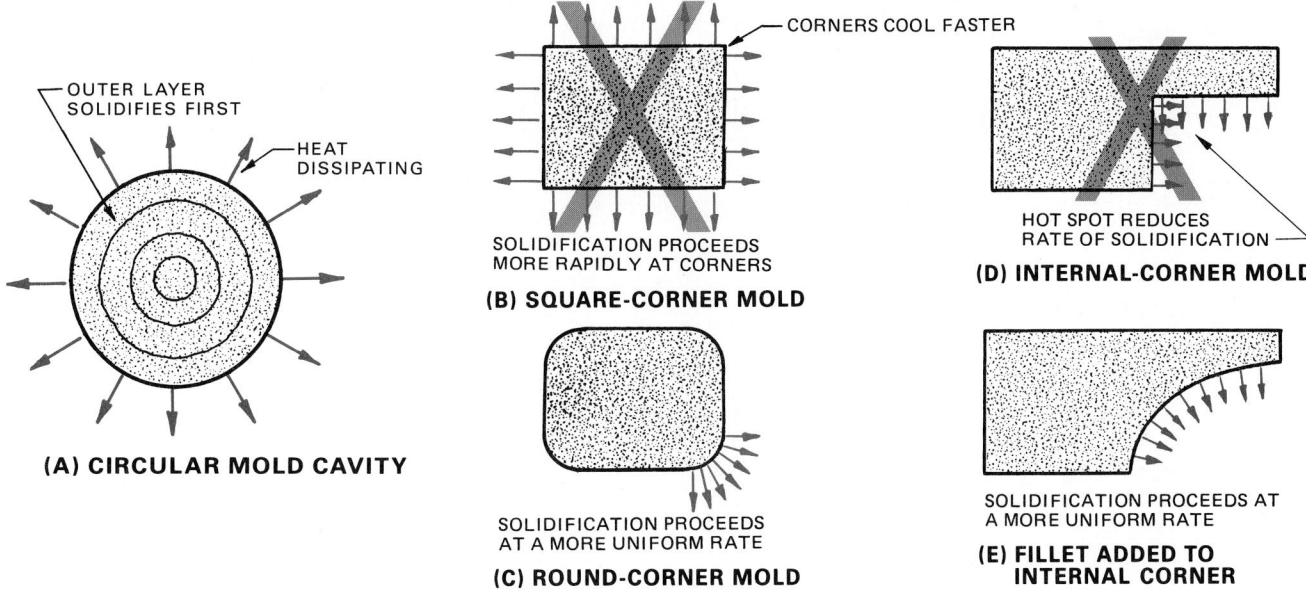

FIG. 13-1-8   Cooling effect on mold cavities filled with molten metal.

(A) CIRCULAR MOLD CAVITY

OUTER LAYER
SOLIDIFIES FIRST

HEAT
DISSIPATING

CORNERS COOL FASTER

SOLIDIFICATION PROCEEDS
MORE RAPIDLY AT CORNERS

(B) SQUARE-CORNER MOLD

HOT SPOT REDUCES
RATE OF SOLIDIFICATION

(D) INTERNAL-CORNER MOLD

SOLIDIFICATION PROCEEDS
AT A MORE UNIFORM RATE

(C) ROUND-CORNER MOLD

SOLIDIFICATION PROCEEDS AT
A MORE UNIFORM RATE

(E) FILLET ADDED TO
INTERNAL CORNER

FIG. 13-1-9   Design members so that all parts increase progressively to feeder risers. *(Meehanite Metal Corp.)*

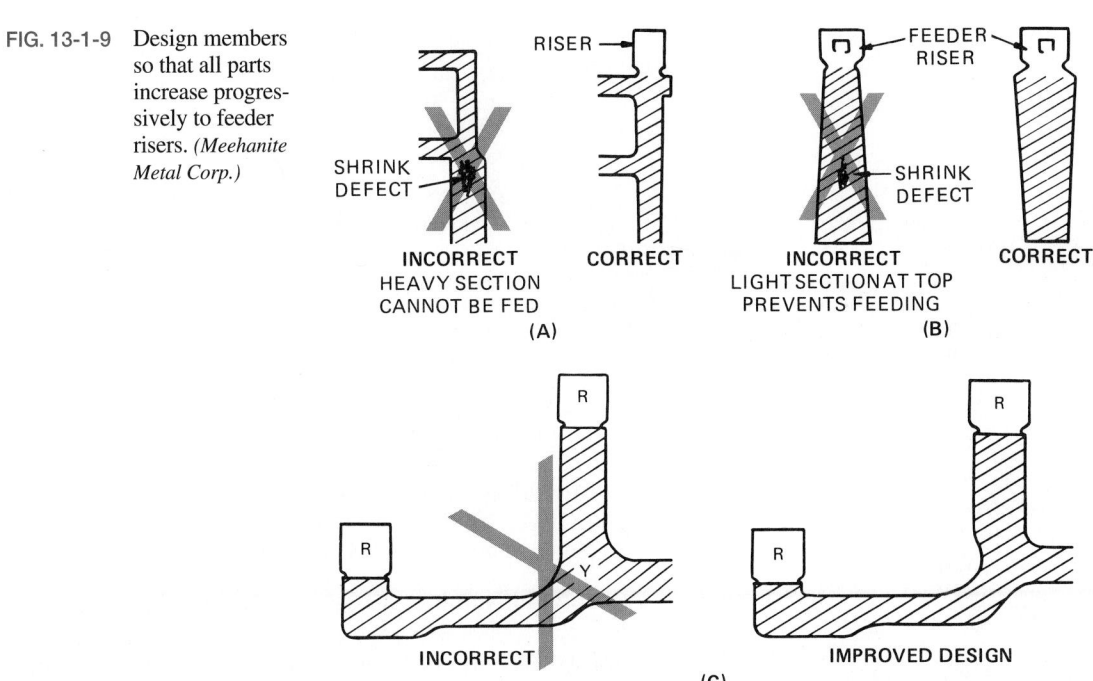

RISER

SHRINK
DEFECT

INCORRECT
HEAVY SECTION
CANNOT BE FED

CORRECT

(A)

FEEDER
RISER

SHRINK
DEFECT

INCORRECT
LIGHT SECTION AT TOP
PREVENTS FEEDING

CORRECT

(B)

INCORRECT

IMPROVED DESIGN

(C)

and to make corners more moldable to eliminate hot spots (Fig. 13-1-10, pg. 354).

Bring the Minimum Number of Adjoining Sections Together
A well-designed casting brings the minimum number of sections together and avoids acute angles (Fig. 13-1-11, pg. 354).

Design All Sections as Nearly Uniform in Thickness as Possible   Shrink defects and casting strains existed in the casting illustrated in Fig. 13-1-12 (pg. 354). Redesigning eliminated excessive metal and resulted in a casting that was free from defects, was lighter in weight (mass), and prevented the development of casting strains in the light radial veins.

Avoid Abrupt Section Changes—Eliminate Sharp Corners at Adjoining Sections   The difference in the relative thickness of adjoining sections should be minimum and not exceed a 2:1 ratio (Fig. 13-1-13, pg. 355).

When a change of thickness must be less than 2:1, it may take the form of a fillet; where the difference must be greater, the form recommended is that of a wedge.

Wedge-shaped changes in wall thickness are to be designated with a taper not exceeding 1 in 4.

Design Ribs for Maximum Effectiveness   Ribs have two functions: to increase stiffness and to reduce the mass. If too

FIG. 13-1-10   Fillet all sharp angles.

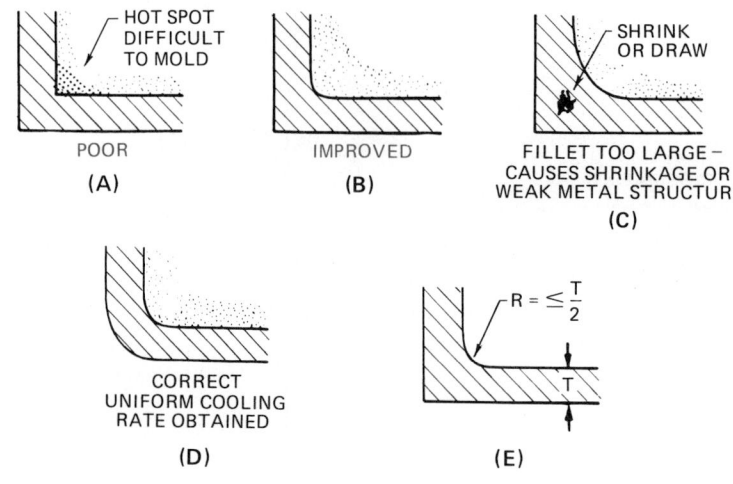

FIG. 13-1-11   Bringing the minimum number of adjoining sections together. *(Meehanite Metal Corp.)*

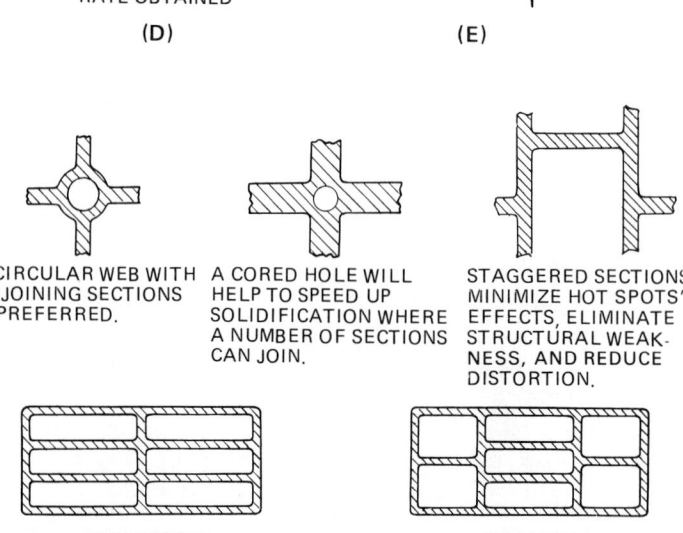

A CIRCULAR WEB WITH ADJOINING SECTIONS IS PREFERRED.

A CORED HOLE WILL HELP TO SPEED UP SOLIDIFICATION WHERE A NUMBER OF SECTIONS CAN JOIN.

STAGGERED SECTIONS MINIMIZE HOT SPOTS' EFFECTS, ELIMINATE STRUCTURAL WEAKNESS, AND REDUCE DISTORTION.

INCORRECT

CORRECT

TO PREVENT UNEVEN COOLING, BRING THE MINIMUM NUMBER OF SECTIONS TOGETHER OR STAGGER SO THAT NO MORE THAN TWO SECTIONS CAN JOIN.

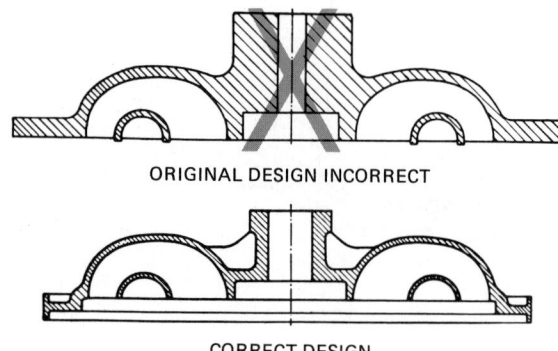

ORIGINAL DESIGN INCORRECT

CORRECT DESIGN

FIG. 13-1-12   Design all sections as nearly uniform in thickness as possible. *(Meehanite Metal Corp.)*

shallow in depth or too widely spaced, they are ineffectual (Fig. 13-1-14).

Avoid Bosses and Pads Unless Absolutely Necessary
Bosses and pads increase metal thickness, create hot spots, and cause open grain or draws. Blend these into the casting by tapering or flattening the fillets. Bosses should not be included

in casting design when the surface to support bolts, etc., may be obtained by milling or countersinking.

Use Curved Spokes   In spoked wheels, a curved spoke is preferred to a straight one. It will tend to straighten slightly, thereby offsetting the dangers of cracking (Fig. 13-1-15).

Use an Odd Number of Spokes   A wheel having an odd number of spokes will not have the same direct tensile stress along the arms as one having an even number and will have more resiliency to casting stresses.

Consider Wall Thicknesses   Walls should be of minimum thickness, consistent with good foundry practice, and should provide adequate strength and stiffness. Wall thicknesses for different materials are as follows:

1. Walls of gray-iron castings and aluminum sand castings should not be less than .16 in. (4 mm) thick.
2. Walls of malleable iron and steel castings should not be less than .18 in. (5 mm) thick.
3. Walls of bronze, brass, or magnesium castings should not be less than .10 in. (2.4 mm) thick.

Select Parting Lines   A *parting line* is a line along which the pattern is divided for molding, or along which the sections of a

FIG. 13-1-13   Avoid abrupt changes. *(Meehanite Metal Corp.)*

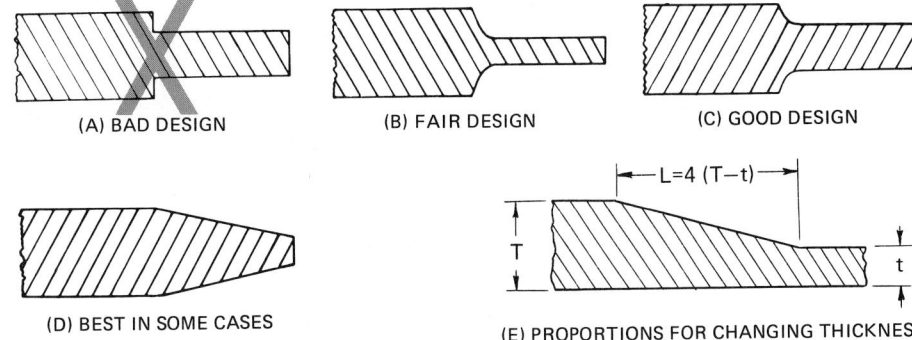

(A) BAD DESIGN   (B) FAIR DESIGN   (C) GOOD DESIGN

(D) BEST IN SOME CASES

(E) PROPORTIONS FOR CHANGING THICKNESS

$L = 4\,(T-t)$

FIG. 13-1-14   Design ribs for maximum effectiveness. *(Meehanite Metal Corp.)*

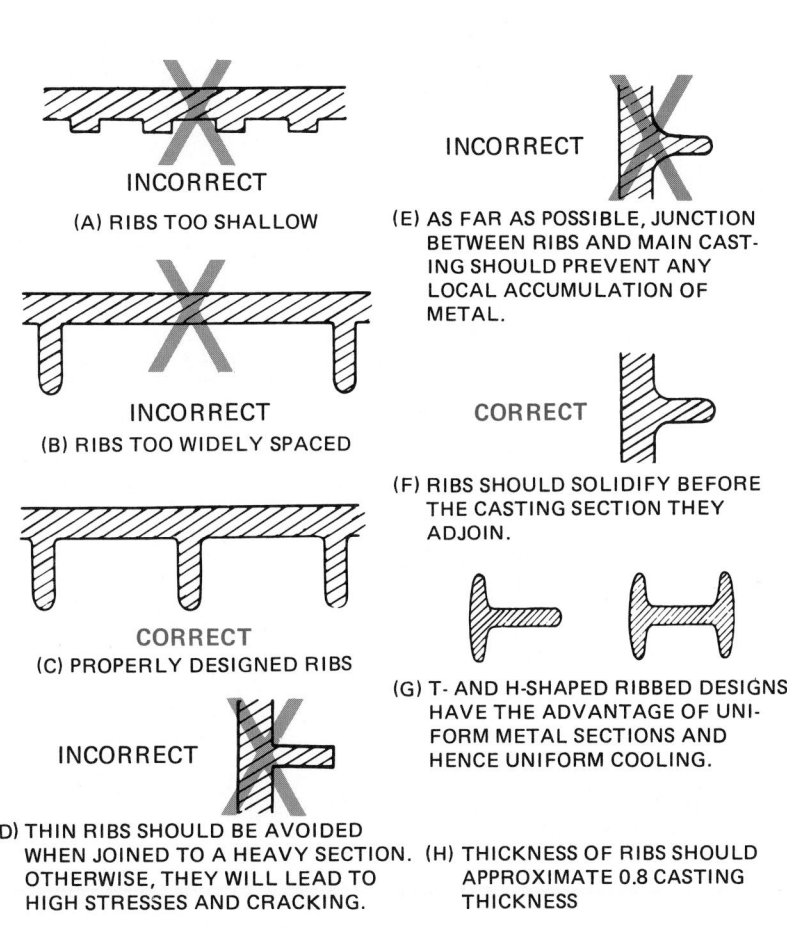

INCORRECT

(A) RIBS TOO SHALLOW

INCORRECT

(B) RIBS TOO WIDELY SPACED

CORRECT

(C) PROPERLY DESIGNED RIBS

INCORRECT

(D) THIN RIBS SHOULD BE AVOIDED WHEN JOINED TO A HEAVY SECTION. OTHERWISE, THEY WILL LEAD TO HIGH STRESSES AND CRACKING.

INCORRECT

(E) AS FAR AS POSSIBLE, JUNCTION BETWEEN RIBS AND MAIN CASTING SHOULD PREVENT ANY LOCAL ACCUMULATION OF METAL.

CORRECT

(F) RIBS SHOULD SOLIDIFY BEFORE THE CASTING SECTION THEY ADJOIN.

(G) T- AND H-SHAPED RIBBED DESIGNS HAVE THE ADVANTAGE OF UNIFORM METAL SECTIONS AND HENCE UNIFORM COOLING.

(H) THICKNESS OF RIBS SHOULD APPROXIMATE 0.8 CASTING THICKNESS

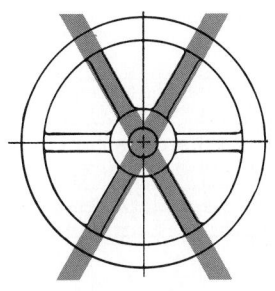

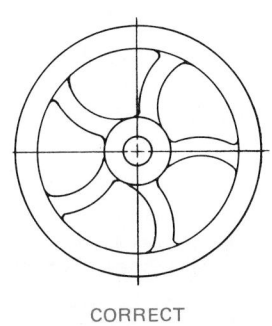

INCORRECT   CORRECT

(A) USE AN ODD NUMBER OF CURVED SPOKES

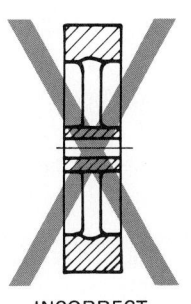

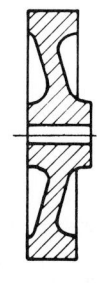

INCORRECT   CORRECT
CAREFULLY BLEND SECTIONS
(B) AVOID EXCESSIVE SECTION VARIATION

FIG. 13-1-15   Spoked-wheel design. *(Meehanite Metal Corp.)*

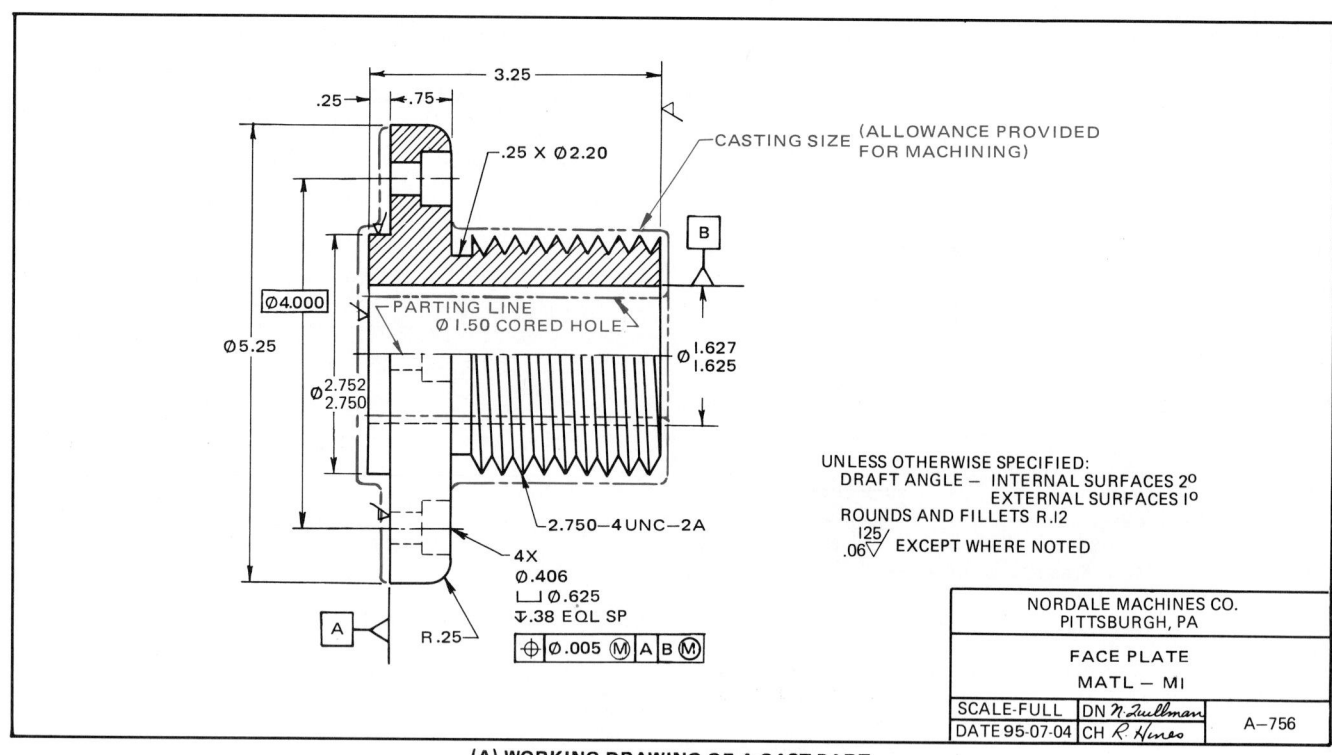

(A) WORKING DRAWING OF A CAST PART

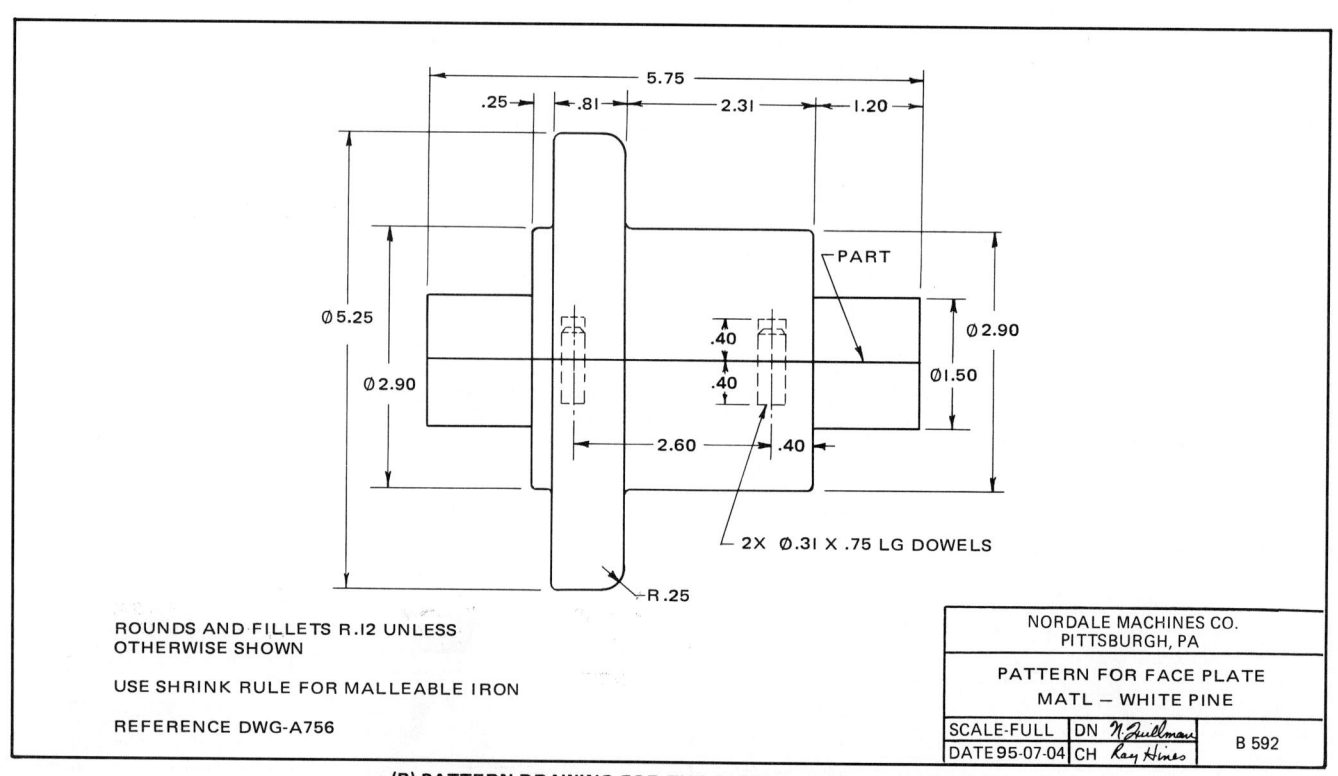

(B) PATTERN DRAWING FOR THE CAST PART SHOWN IN (A)

FIG. 13-1-16   Cast-part drawings.

mold separate. Selection of a parting line depends on a number of factors:

- Shape of the casting
- Elimination of machining on draft surfaces
- Method of supporting cores
- Location of gates and feeders

Drill Holes in Castings   Small holes usually are drilled and not cored.

Drafting Practices

It is important that a detail drawing give complete information on all cast parts, e.g.:

- Machining allowances
- Surface texture
- Draft angles
- Limits on cast surfaces that must be controlled
- Locating points
- Parting lines

On small, simple parts all casting information is included on the finished drawing (Fig. 13-1-16). On more complicated parts, it may be necessary to show additional casting views and sections to completely illustrate the construction of the casting. These additional views should show the rough casting outline in phantom lines and the finished contour in solid lines.

Material   In the selection of material for any particular application, the designer is influenced primarily by the physical characteristics, such as strength, hardness, density, resistance to wear, mass, antifrictional properties, conductivity, corrosion resistance, shrinkage, and melting point.

CASTING ALLOY	DIMENSIONS WITHIN THIS RANGE	CASTING ALLOWANCE	STANDARD DRAWING TOLERANCE (±)
Cast Iron, Aluminum, Bronze, Etc. Sand Castings	Up to 8.00	.06	.03
	8.00 to 16.00	.09	.06
	16.00 to 24.00	.12	.07
	24.00 to 32.00	.18	.09
	Over 32.00	.25	.12
Pearlitic, Malleable, and Steel Sand Castings	Up to 8.00	.06	.03
	8.00 to 16.00	.09	.06
	16.00 to 24.00	.18	.09
	Over 24.00	.25	.12
Permanent and Semipermanent Mold Castings	Up to 12.00	.06	.03
	12.00 to 24.00	.09	.06
	Over 24.00	.18	.09
Plaster Mold Castings	Up to 8.00	.03	.02
	8.00 to 12.00	.06	.03
	Over 12.00	.10	.06

FIG. 13-1-17   Guide to machining and tolerance allowance in inches for castings.

Machining Allowance   In the construction of patterns for castings in which various points on the surface of the casting must be machined, sufficient excess metal should be provided for all machined surfaces. Unless otherwise specified, Fig. 13-1-17 may be used as a guide to machine finish allowance.

Fillets and Radii   Generous fillets and radii (rounds) should be provided on cast corners and specified on the drawing.

Casting Tolerances   A great many factors contribute to the dimensional variations of castings. However, the standard drawing tolerances specified in Fig. 13-1-17 can be satisfactorily attained in the production of castings.

Draft   All casting methods require a draft or taper on all surfaces perpendicular to the parting line to facilitate removal of the pattern and ejection of the casting. The permissible draft must be specified on the drawing, in either degrees of taper for each surface, inches of taper per inch of length, or millimeters of taper per millimeter of length.

Suitable draft angles in general use for both sand and die castings are 1° for external surfaces and 2° for internal surfaces, as shown in Fig. 13-1-18.

The drawing must always clearly indicate whether the draft should be added to, or subtracted from, the casting dimensions.

Casting Datums

It is recognized that in many cases a drawing is made of the fully machined end product, and casting dimensions, draft, and machining allowances are left entirely to the pattern maker or foundry worker. However, for mass-production purposes it is generally advisable to make a separate casting drawing, with carefully selected datums, to ensure that parts will fit into machining jigs and fixtures and will meet final requirements after machining. Under these circumstances, dimensioning requires the selection of two sets of datum surfaces, lines, or points—one for the casting and one for the machining—to provide common reference points for measuring, machining, and assembly. To select suitable datums, the designer must know how the casting is to be made, where the parting line or lines are to be, and how the part is going to fit into machining jigs and fixtures.

The first step in dimensioning is to select a primary datum surface, sometimes referred to as the *base surface* for the casting, and to identify it as datum *A* (Fig. 13-1-19, pg. 358).

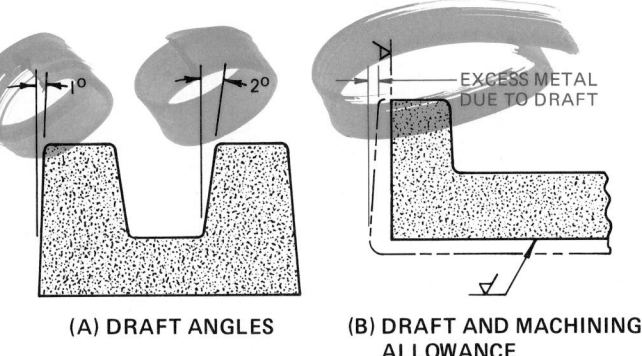

(A) DRAFT ANGLES   (B) DRAFT AND MACHINING ALLOWANCE

FIG. 13-1-18   Draft for removing of casting from mold.

357

This primary datum should be a surface that meets the following criteria as closely as possible:

1. It must be a surface, or datum targets on a surface (see Fig. 13-1-20), that can be used as the basis for measuring the casting and that can later be used for mounting and locating the part in a jig or fixture, for the purpose of machining the finished part.
2. It should be a surface that will not be removed by machining, so that control of material to be removed is not lost, and can be checked at final inspection.
3. It should be parallel with the top of the mold, or parting line, that is, a surface that has no draft or taper.
4. It should be integral with the main body of the casting, so that measurements from it to the main surfaces of the casting will be least affected by cored surfaces, parting lines, or gated surfaces.
5. It should be a surface, or target areas on a surface, on which the part can be clamped without causing any distortion, so that the casting will not be under a distortional stress for the first machining operation.

6. It should be a surface that will provide locating points as far apart as possible, so that the effect of any flatness error will be minimized.

The second step is to select two other planes to serve as secondary and tertiary surfaces. These planes should be at right angles to one another and to the primary datum surface. They probably will not coincide with actual surfaces, because of taper or draft, except at one point, usually a point adjacent to the primary datum surface. These are identified as datum *B* and datum *C*, respectively, as shown in Fig. 13-1-21.

In the case of a circular part, the endview center lines may be selected as secondary and tertiary datums, as shown in Fig. 13-1-22B. In this case, unless otherwise specified, the center lines represent the center of the outside or overall diameter of the part.

Machining Datums

The first step in dimensioning the machined or finished part is to select a primary datum surface for machining and to identify it as datum *D* (Fig. 13-1-22A). This surface is the first surface on the casting to be machined and is thereafter used as the datum surface for all other machining operations. It should be selected to meet the following criteria:

1. It is generally preferable, though not essential, that it be a surface that is parallel to the primary casting datum surface.
2. It may be a large, flat, machined surface or several small areas of surfaces in the same or parallel planes.

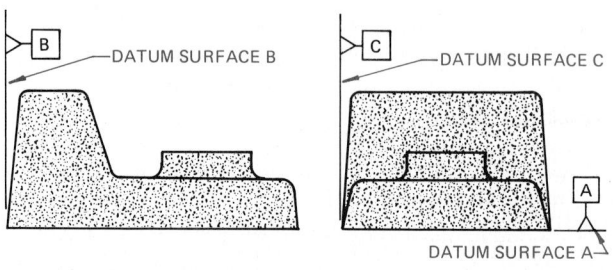

FIG. 13-1-19   Casting datums.

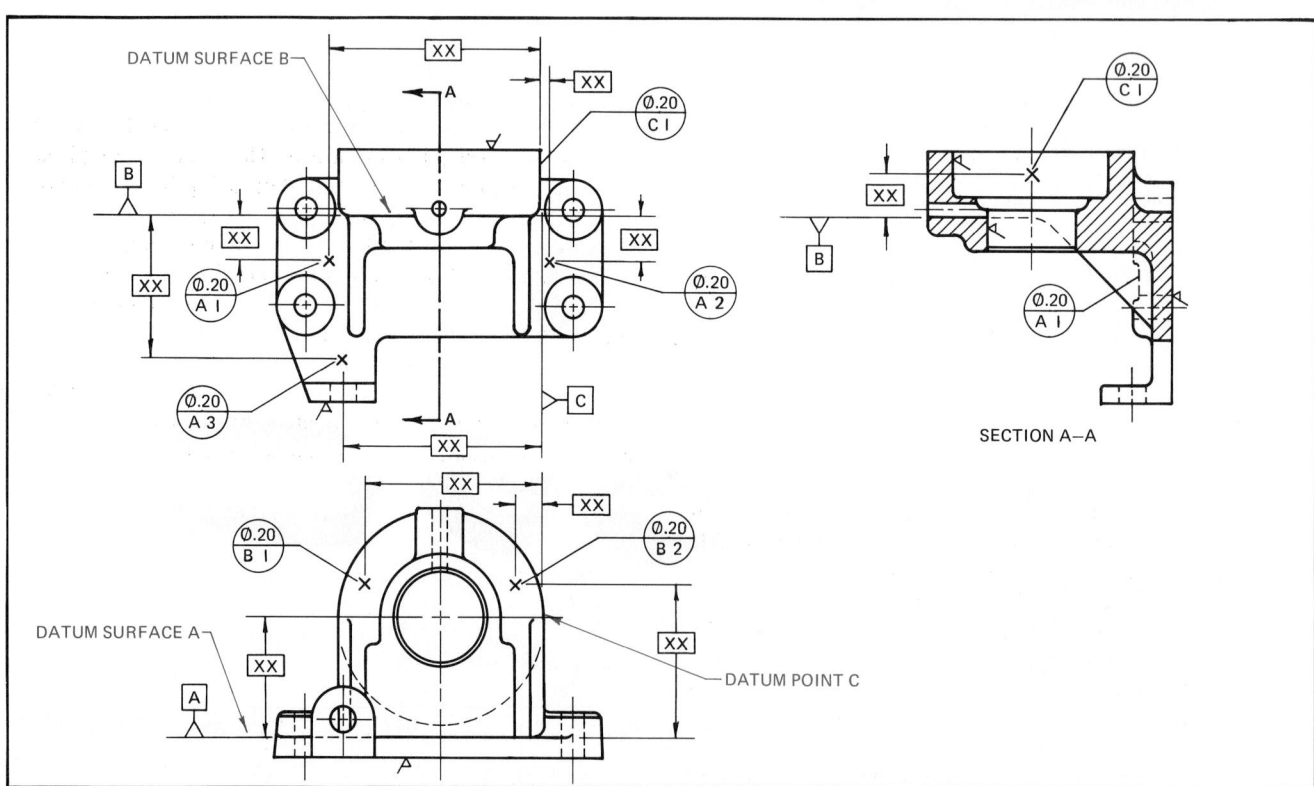

FIG. 13-1-20   Machined cast drawing illustrating datum lines, set-up points, and surface finish.

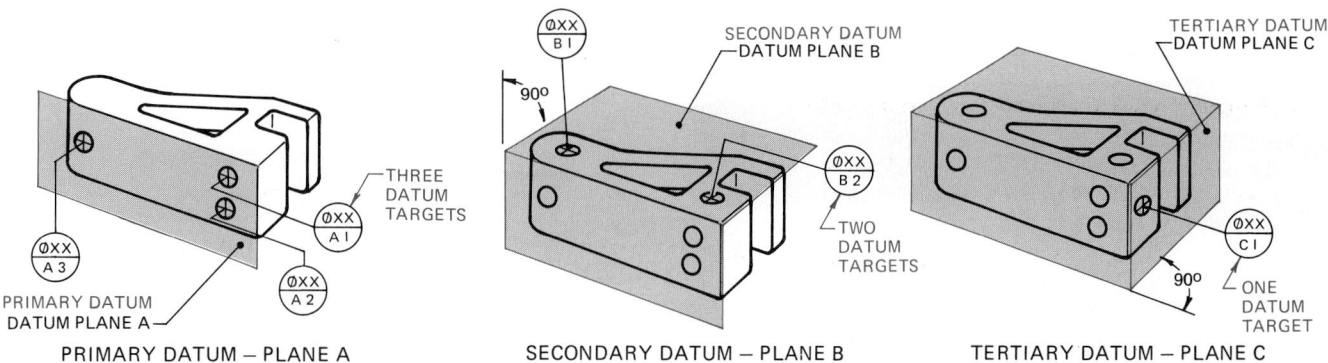

FIG. 13-1-21   Datum planes and datum targets.

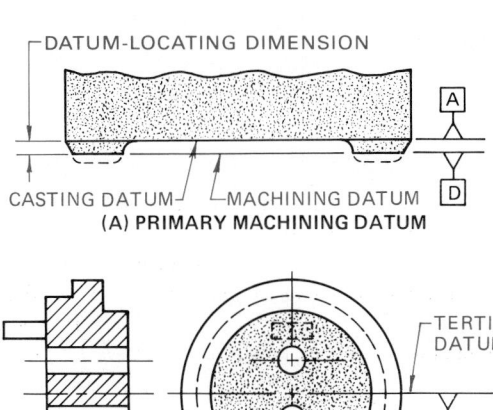

(A) PRIMARY MACHINING DATUM

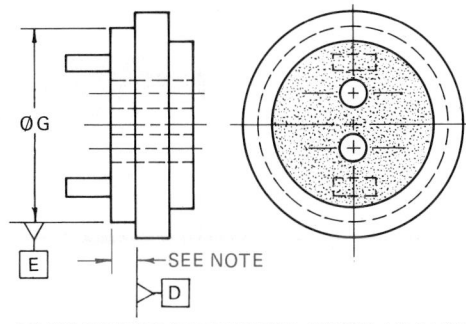

(B) CASTING DATUM FOR CIRCULAR PART

(C) MACHINING DATUMS FOR CIRCULAR PART
SHOWN IN (B)

FIG. 13-1-22   Casting and machining datums.

3. If the primary casting datum surface is smooth and does not require machining, as in die castings, or if suitable target areas have been selected, the same surface may be used as the machining datum surface.
4. If the primary casting datum surface of sand castings appears to be the only suitable surface, it is recommended that three or four pads be provided, which can be machined to form the machining datum surface, as shown in Fig. 13-1-22A.
5. When pads or small target areas are selected, they should be placed as far apart as possible and located where the part can be readily clamped in jigs or fixtures without distorting it or interfering with other machining operations.

The second step is to select two other surfaces to serve as secondary and tertiary datums. If these datum surfaces are required only for locating and dimensioning purposes, and not for clamping in a jig or fixture, some suitable datums other than flat, machined surfaces may be chosen. These could be the same datums as used for casting, if the locating point in each case is clearly defined and is not removed in machining. For circular parts, a hole drilled in the center, or a turned diameter other than the outside diameter, may provide a suitable center line for use as secondary datum surfaces (Fig. 13-1-22C).

The third step is to specify the *datum-locating dimension*, that is, the dimension between each casting datum surface and the corresponding machining datum surface (Fig. 13-1-22A). There is never more than one such dimension from each casting datum surface.

Dimensions

When suitable datum surfaces have been selected, with datum-locating dimensions for the machined-casting drawing, dimensioning may proceed, with dimensions being specified directly from the datums to all main surfaces. However, where it is necessary to maintain a particular relationship between two or more surfaces or features, regular point-to-point dimensioning is usually the preferred method. This will normally include all such items as thickness of ribs, height of bosses, projections, depth of grooves, most diameters and radii, and center distances between holes and similar features. Whenever possible, specify dimensions to surfaces or surface intersections, rather than to radii centers or nonexistent center lines. Dimensions given on the casting drawing should not be repeated, except as reference dimensions, on the machined-part drawing.

REFERENCES AND SOURCE MATERIAL

1. American Iron and Steel Institute, *Principles of Forging Design*.
2. General Motors Corp.
3. Meehanite Metal Corp.

ASSIGNMENTS

See Assignments 1 through 6 for Unit 13-1 on pages 371 through 373.

13-2 FORGINGS

Forging consists of plastically deforming, either by a squeezing pressure or sharp blows, a cast or sintered ingot, a wrought bar or billet, or a powder-metal shape, to produce a desired shape with good mechanical properties. Practically all ductile metals can be forged (Fig. 13-2-1).

Closed-Die Forging

Impression Dies

Closed-die forgings are made by hammering or pressing metal until it conforms closely to the shape of the enclosing dies. Grain flow in the closed-die-forged parts can be oriented in the direction requiring greatest strength. In practice, *closed-die forging* has become the term applied to all forging operations involving three-dimensional control.

Three-dimensional control of the material to be forged requires a closed die, a simple and common form of which is the impression die.

In the simplest example of impression-die forging (Fig. 13-2-2) the workpiece is cylindrical and is placed in the bottom-half die. On closing of the top-half die, the cylinder undergoes elastic compression until its enlarged sides touch the side walls of the die impression. At this point, a small amount of excess material begins to form the flash between the two die faces.

The forging impression die gives control over all three directions, except when the die is similar to that shown in Fig. 13-2-2, and the deforming forging machine tool has an unlimited stroke (e.g., a hammer or hydraulic press). In the latter case, the die must be shaped to allow complete closing of the striking faces at the end of the stroke.

Forging dies can be divided into three main classes: single-impression, double-impression, and interlocking (Fig. 13-2-3). Single-impression dies have the impression of the desired forging entirely in one half of the die. Double-impression dies have part of the impression of the desired forging sunk in each die in such a manner that no part of the die projects past the parting line into the other die. This type is the most common class of forging.

Trimming Dies

Because the quantity of forging metal is generally in excess of the space in the die cavity, space is provided between the die surfaces for the escape of the excess metal. This space is called the *flash space*, and the excess metal that flows into it is called *flash*. The flash thickness is proportionate to the mass of the forging.

The flash is removed from forgings by trimming dies, which are formed to the outline of the part (Fig. 13-2-4).

General Design Rules

Corner and Fillet Radii   It is important in forging design to use correct radii where two surfaces meet. Corner and fillet radii on forgings should be sufficient to facilitate the flow of metal.

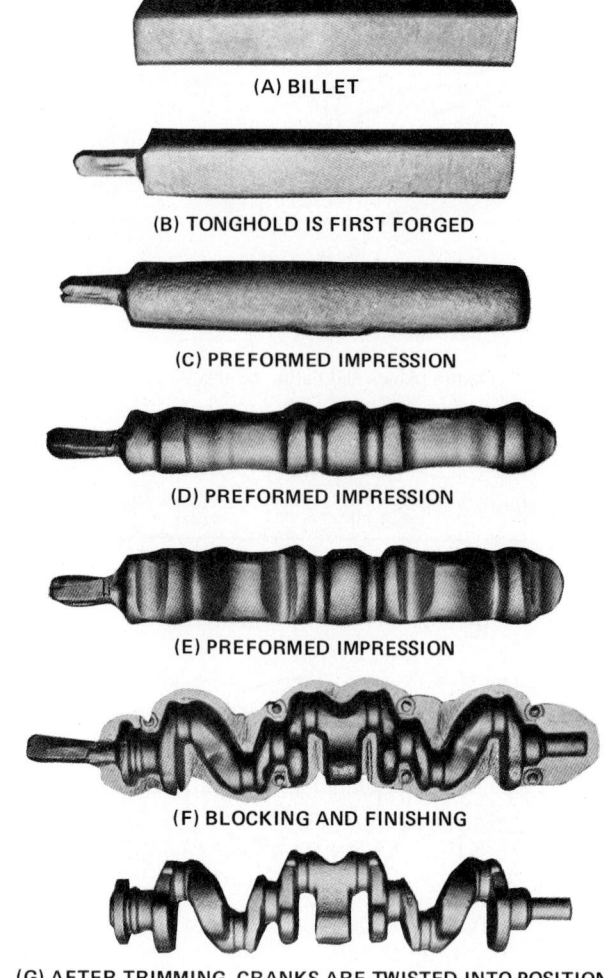

(A) BILLET

(B) TONGHOLD IS FIRST FORGED

(C) PREFORMED IMPRESSION

(D) PREFORMED IMPRESSION

(E) PREFORMED IMPRESSION

(F) BLOCKING AND FINISHING

(G) AFTER TRIMMING, CRANKS ARE TWISTED INTO POSITION

FIG. 13-2-1   The forging of a crankshaft. *(Wyman-Gordon Co.)*

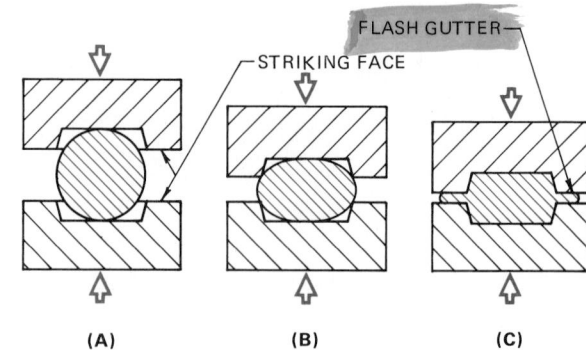

FIG. 13-2-2   Compression in impression dies.

Stress concentrations resulting from abrupt changes in section thickness or direction are minimized by corner and fillet radii of correct size. Any radius larger than recommended will increase die life. Any radius smaller than recommended will decrease die life. See Fig. 13-2-5 for recommendations.

Sharp fillets cause the formation of cold shuts. In a forging, a *cold shut* is a lap where two surfaces of metal have folded against each other, forming an undesirable flow of metal. A cold shut causes a weak spot that may be opened into a crack

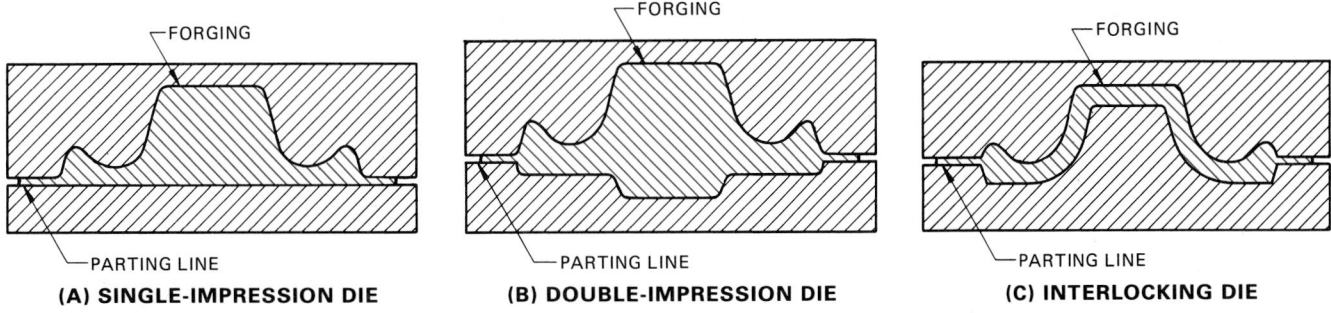

(A) SINGLE-IMPRESSION DIE   **(B) DOUBLE-IMPRESSION DIE**   **(C) INTERLOCKING DIE**

FIG. 13-2-3   Forging dies.

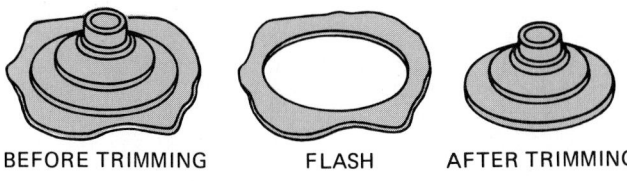

BEFORE TRIMMING   FLASH   AFTER TRIMMING

FIG. 13-2-4   Flash trimming.

by heat treatment. Cold shuts are most likely to form at fillets in deep depressions or in deep sections, especially where the metal is confined (Fig. 13-2-6, pg. 362). In these cases larger fillets are required, as shown in Fig. 13-2-5.

Draft Angle   Draft is one of the first factors to be considered in designing a forged part (Fig. 13-2-7, pg. 362). *Draft* is defined as the slope given to the side walls of the die in order to facilitate removal of the forging. Where little or no draft is allowed, stripper or ejection mechanisms must be used. The usual amount of draft for exterior contours is 7° and for interior contours, 10°.

Die draft equivalent is the amount of offset that results from draft. Figure 13-2-8 (pg. 362) shows the draft equivalents for varying angles and depth of draft.

Parting Line   The surfaces of dies that meet in forgings are the striking surfaces. The line of meeting is the parting line. The parting line of the forging must be established in order to determine the amount of draft and its location.

The location and the type of parting as applied to simple forgings are shown in Fig. 13-2-9 (pg. 362).

Drafting Practices

In preparing forging drawings, it is important to consider drafting practices that may be peculiar to forgings, such as:

- Draft angles and parting lines
- Corner and fillet radii
- Forging tolerances
- Machining allowances

FIG. 13-2-5   Corner and fillet radii.

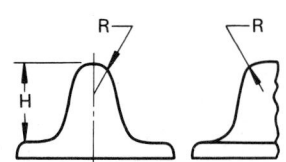

H		R
OVER	TO AND INCL	
0	1.00 (25)	.06 (1.5)
1.00 (25)	1.50 (35)	.09 (2.5)
1.50 (35)	2.00 (50)	.12 (3)
2.00 (50)	3.00 (80)	.18 (4.5)

MIN CORNER RADII

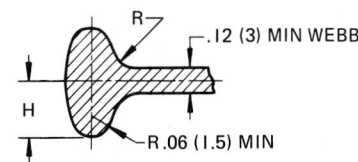

H		R
OVER	TO AND INCL	
0	.30 (8)	R = H
.30 (8)	.50 (13)	$R = \dfrac{3H}{4}$

FILLET RADII FOR SMALL RIBS

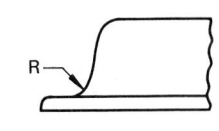

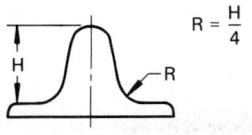

$R = \dfrac{H}{4}$

FILLET RADII WHEN
METAL IS CONFINED

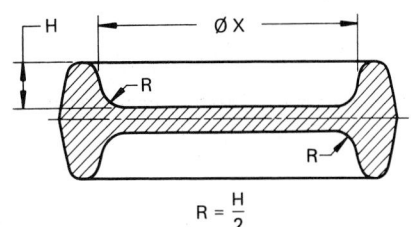

$R = \dfrac{H}{2}$

DEPTH OF A FORGED RECESS SHOULD
NOT EXCEED 0.67 X DIA.

FILLET RADII WHEN METAL
IS NOT CONFINED

FIG. 13-2-6   Cold shut.

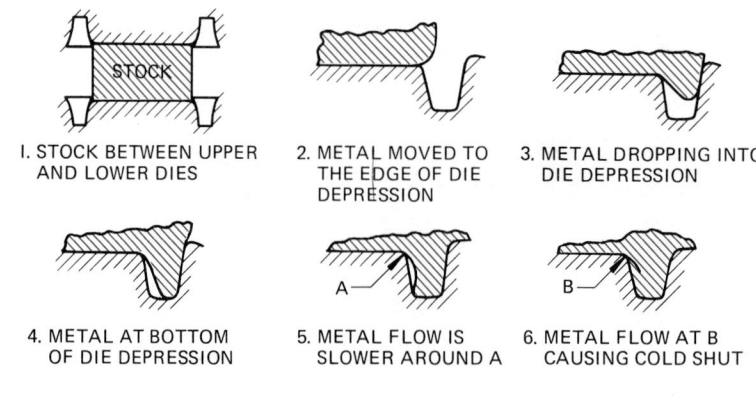

1. STOCK BETWEEN UPPER AND LOWER DIES

2. METAL MOVED TO THE EDGE OF DIE DEPRESSION

3. METAL DROPPING INTO DIE DEPRESSION

4. METAL AT BOTTOM OF DIE DEPRESSION

5. METAL FLOW IS SLOWER AROUND A

6. METAL FLOW AT B CAUSING COLD SHUT

FIG. 13-2-7   Draft application.

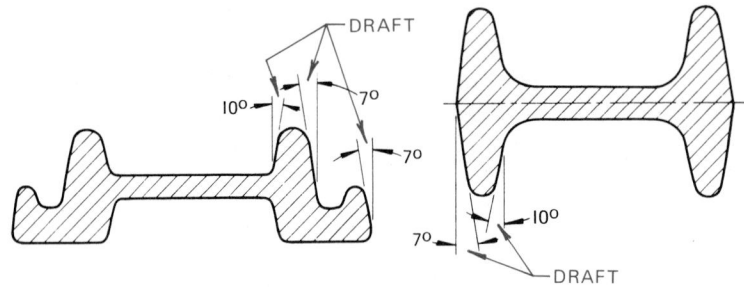

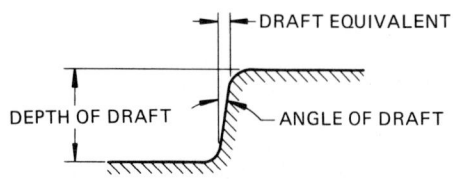

DEPTH OF DRAFT	DRAFT EQUIV FOR ANGLE OF		
	5°	7°	10°
.20 (5)	.018 (0.437)	.024 (0.614)	.035 (0.882)
.40 (10)	.035 (0.875)	.050 (1.228)	.070 (1.763)
.60 (15)	.052 (1.312)	.074 (1.842)	.106 (2.645)
.80 (20)	.070 (1.750)	.100 (2.456)	.140 (3.527)
1.00 (25)	.088 (2.187)	.123 (3.070)	.176 (4.408)

FIG. 13-2-8   Die draft equivalent.

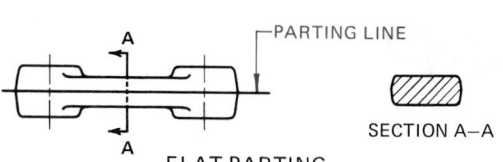

FLAT PARTING

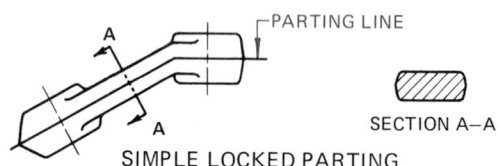

SIMPLE LOCKED PARTING

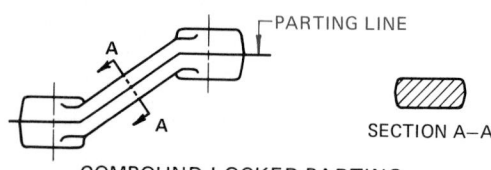

COMPOUND LOCKED PARTING

FIG. 13-2-9   Parting line application.

- Heat treatment
- Location of trademark, part number, and vendor specification

Dimensioning   It is usually desirable to apply dimensions of the depths of the die impressions to the forged part. Draft is additive to these dimensions and should be expressed in degrees or linear dimensions.

When the depth of the die impression is located, only one dimension should originate from the parting line. This surface should then be used to establish other dimensions, as shown in Fig. 13-2-10A.

Allowance for Machining   When a forging is to be machined, allowance must be made for metal to be removed.

Composite Drawings   Generally, a forged part should be shown on one drawing with the forging outline shown in phantom lines, as in Fig. 13-2-10B. Forging outlines for machining allowance should not be dimensioned unless the amount of finish cannot be controlled by the machining symbol.

FIG. 13-2-10   Forged-part drawings.

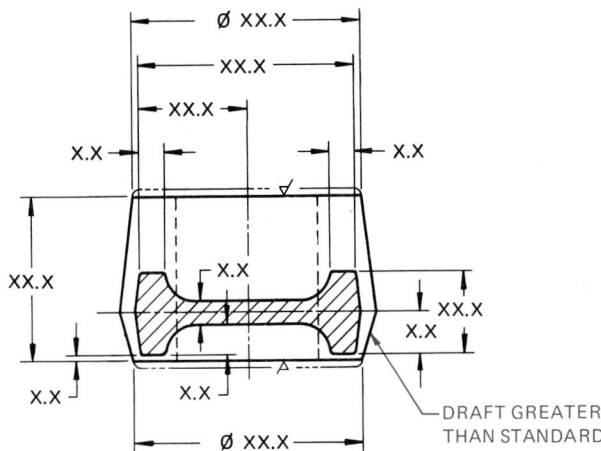

(A) DIMENSIONING A FORGED DRAWING

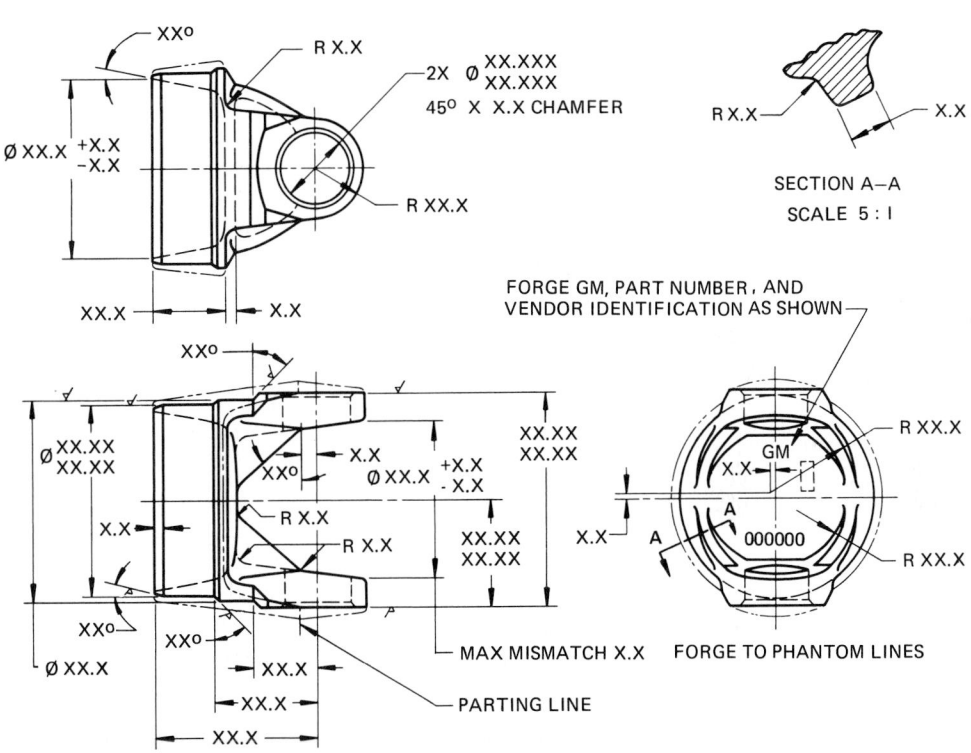

(B) A COMPOSITE FORGED DRAWING

Separate drawings for rough forgings should be made only when the part is complicated and the outline of the rough forging cannot be clearly visualized, or where the outline of the rough forging must be maintained for tooling purposes.

Where both the forging and machining drawings are shown on the same sheet, as in Fig. 13-2-11 (pg. 364), place the headings FORGING DRAWING and MACHINING DRAWING directly under the corresponding views.

REFERENCES AND SOURCE MATERIAL

1. Frank Burbank, "Forging," *Machine Design*, Vol. 37, No. 21.

ASSIGNMENTS

See Assignments 7 and 8 for Unit 13-2 on page 374.

13-3 POWDER METALLURGY

Powder metallurgy is the process of making parts by compressing and sintering various metallic and nonmetallic powders into shape (Fig. 13-3-1, pg. 364).

Dies and presses known as *briquetting machines* are used to compress the powders into shape. These briquets or compacts

FIG. 13-2-11 Separate forging and machining drawings.

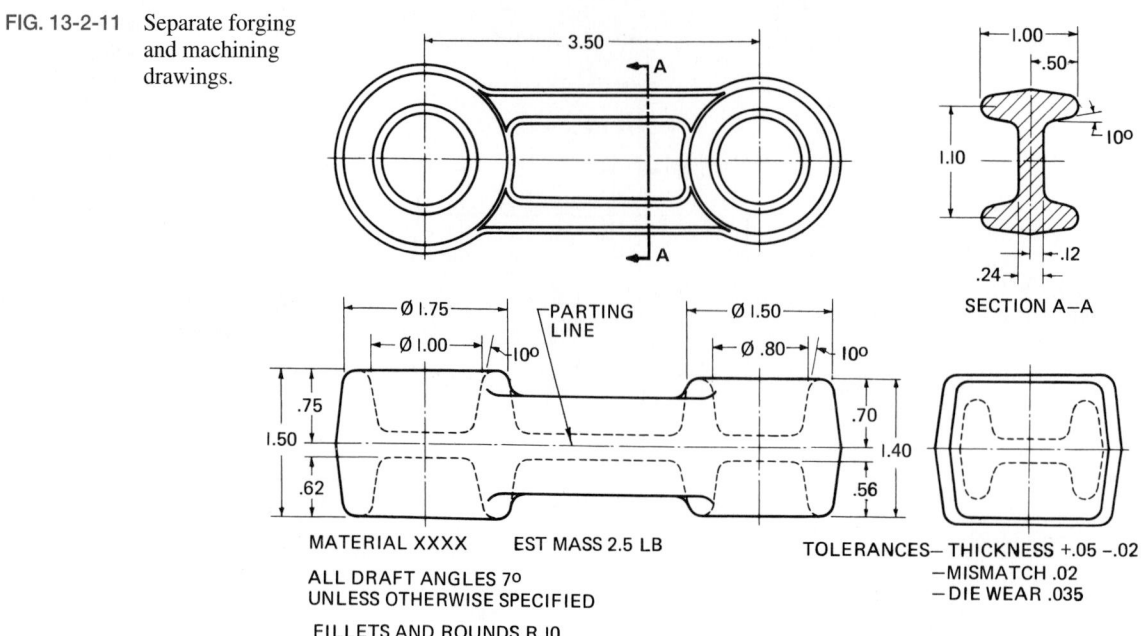

PARTING LINE

MATERIAL XXXX   EST MASS 2.5 LB

ALL DRAFT ANGLES 7°
UNLESS OTHERWISE SPECIFIED

FILLETS AND ROUNDS R.10

TOLERANCES— THICKNESS +.05 −.02
—MISMATCH .02
—DIE WEAR .035

SECTION A–A

(A) FORGING DRAWING

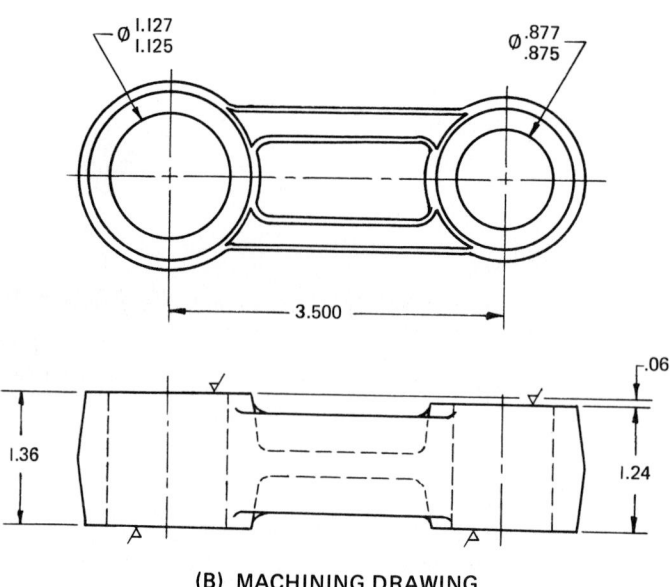

(B) MACHINING DRAWING

FIG. 13-3-1 Compacting sequence for powder metallurgy.

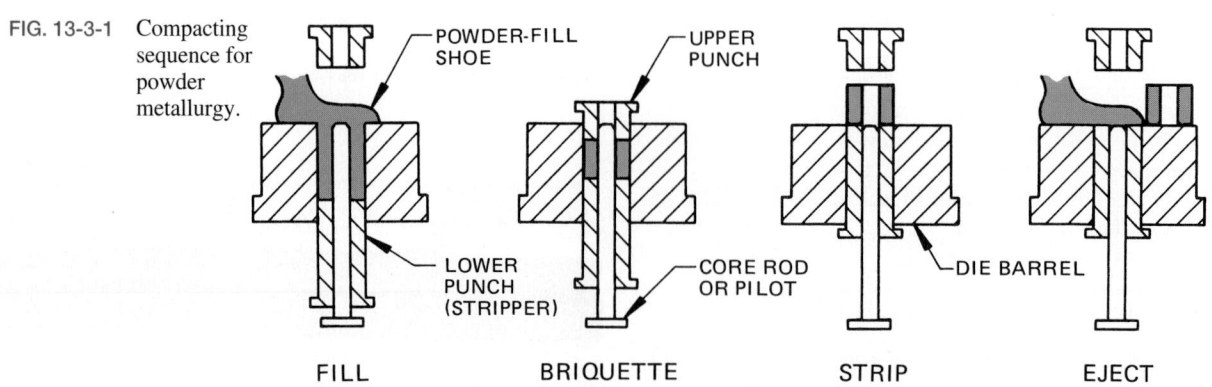

FILL   BRIQUETTE   STRIP   EJECT

are then sintered, or heated, in an atmosphere-controlled furnace, bonding the powdered materials.

Design Considerations

The following considerations should be taken into account when powder-metal parts are designed in order to realize the maximum benefits from the powder metallurgy process (Fig. 13-3-2). This process is most applicable to the production of cylindrical, rectangular, or irregular shapes that do not have large variations in cross-sectional dimensions. Splines, gear teeth, axial holes, counterbores, straight knurls, serrations, slots, and keyseats present few problems.

Ejection from the Die   The shape of the part must permit ejection from the die. The design requirements for some parts can be achieved only by subsequent machining, as in some corner relief designs, reverse tapers, holes at right angles to the direction of pressing, diamond knurls, and undercuts.

Axial Variations   Slots having a depth greater than one-fourth the axial length of the part require multiple-punch action and result in high production costs.

Corner Reliefs   Corner reliefs can be molded or machined. A molded corner relief will save machining.

Reverse Tapers   Reverse tapers cannot be molded. They must be machined.

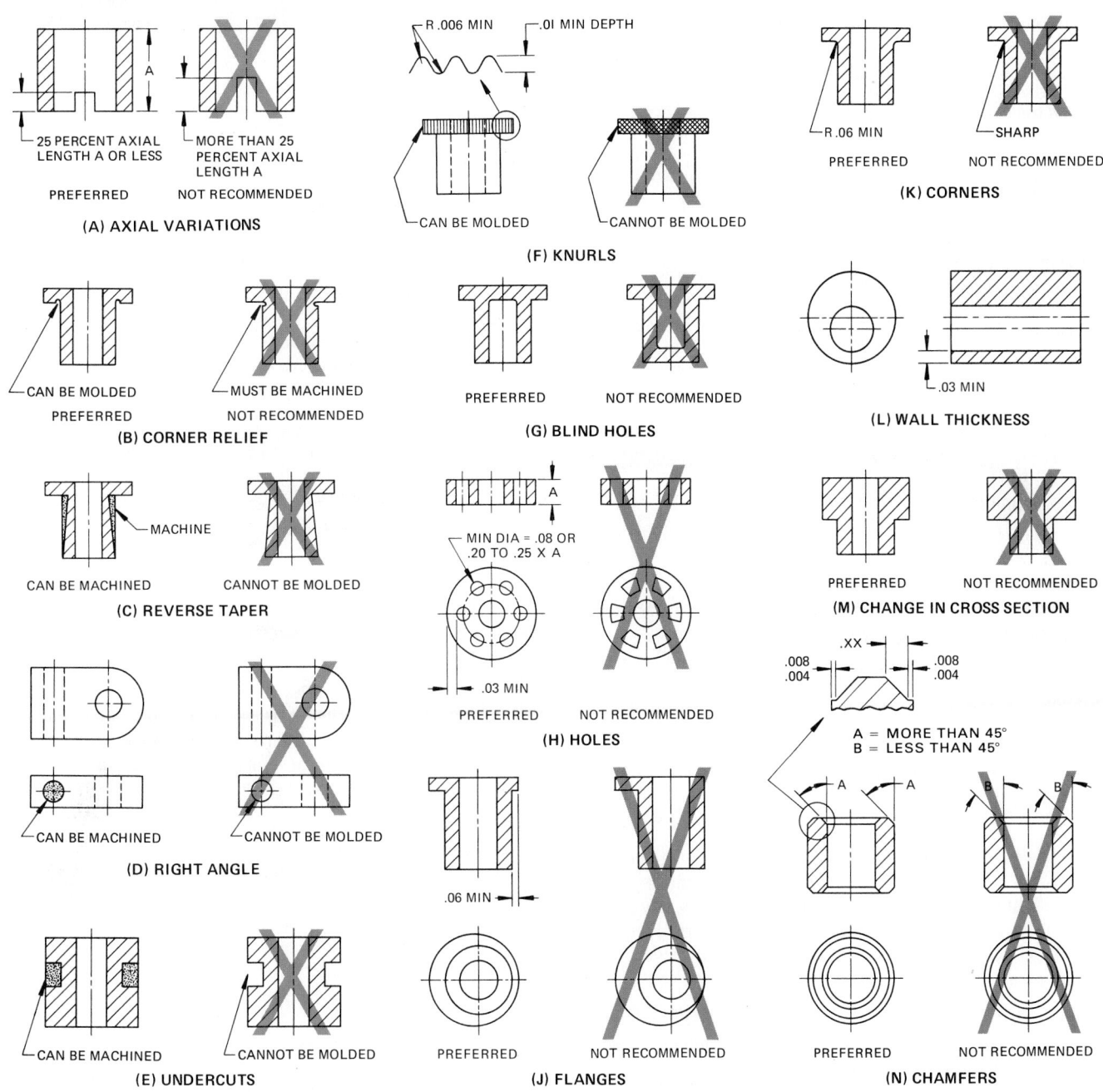

FIG. 13-3-2   Design considerations for powder metallurgy.

Holes at Right Angles to the Direction of Pressing   Right-angle holes must be machined.

Undercuts   Undercuts must be machined.

Knurls   Straight knurls can be molded; diamond knurls cannot.

Blind Holes   If a flange is opposite the blind end of the hole, the part must be modified to allow powder to fill in the die.

Holes   The use of round holes, instead of odd-shaped holes, will simplify tooling, strengthen the part and reduce costs.

Flanges   A .06 in. (1.5 mm) minimum flange overhang is desired to provide longer tool life.

Corners   A fillet radius must be provided under the flange on a flanged part. It allows uniform powder flow in the die and produces a high-strength part.

Wall Thickness   In general, sidewalls bordering a depression or hole should be a minimum of .03 in. (0.8 mm) thick.

Changes in Cross Section   Large changes in cross section should be avoided because they cause density variation. Warping and cracking are likely to occur during sintering.

Chamfers   Care in the design of chamfers minimizes sharp edges on tools and improves tool life.

REFERENCES AND SOURCE MATERIALS

1. General Motors Corp.

ASSIGNMENTS ▐▐▐▐▐▐▐▐▐▐▐▐▐▐▐▐▐▐▐▐▐▐▐▐▐▐▐▐▐

See Assignments 9 and 10 for Unit 13-3 on pages 375–376.

13-4 PLASTIC MOLDED PARTS

Single Parts

The design of molded parts involves several factors not normally encountered with machine-fabricated and assembled parts. It is important that designers take these factors into consideration.

Shrinkage   *Shrinkage* is defined as the difference between dimensions of the mold and the corresponding dimensions of the molded part. Normally the mold designer is more concerned with shrinkage than the molded-part designer. Shrinkage does, however, affect dimensions, warpage, residual stress, and moldability.

Section Thickness   Solidification is a function of heat transfer from or to the mold for both thermoplastics and thermosets. Each material has a fixed rate of heat transfer. Therefore, where section thickness varies, areas within a molded part will solidify at different rates. The varying rates will cause irregular

shrinkage, sink marks, additional strain, and warpage. For these reasons, uniform section thickness is important and may be maintained by adding holes or depressions, as shown in Fig. 13-4-1A and B.

Gates   Gate location should be anticipated during the design stage. Avoid gating into areas subjected to high stress levels, fatigue, or impact. To optimize molding, locate gates in the heaviest section of the part (Fig. 13-4-1C).

Parting or Flash Line   As described earlier, flash is that portion of the molding material that flows or exudes from the mold parting line during molding. Any mold that is made of two or more parts may produce flash at the line of junction of the mold parts. The thickness of flash usually varies between .002 and .016 in. (0.05 and 0.40 mm), depending upon the accuracy of the mold, type of material, and the process used (Fig. 13-4-1D).

Fillets and Radii   The principal functions of fillets and radii (rounds) are to ease the flow of plastic within the mold, to facilitate ejection of the part, and to distribute stress in the part in service. During molding, the material is liquefied, but it is a heavy, viscous liquid that does not easily flow around sharp corners. The liquid tends to bend around corners; therefore, rounded corners permit the liquid plastic to flow smoothly and easily through the mold. For recommended radii, see Fig. 13-4-1E.

Molded Holes   A through hole is more advantageous than a blind hole since it is more accurate and economical. Blind holes should not be more than twice as deep as their diameter, as shown in Fig. 13-4-1F. Avoid placing holes at angles other than perpendicular to the flash line. If such holes are necessary, consider using a drilled hole to maintain simple molding.

Internal and External Draft   Draft is necessary on all rigid molded articles to facilitate removal of the part from the mold. Draft may vary from 0.25 to 4° per side, depending upon the length of the vertical wall, surface area, finish, kind of material, and the mold or method of ejection used.

Threads   External and internal threads can be easily molded by means of loose-piece inserts and rotating core pins. External threads may be formed by placing the cavity so that the threads are formed in the mold pattern.

Ribs and Bosses   Ribs increase rigidity of a molded part without increasing wall thickness and sometimes facilitate flow during molding. Bosses reinforce small, stressed areas, providing sufficient strength for assembly with inserts or screws. Recommended proportions for ribs and bosses are shown in Fig. 13-4-1G.

Undercuts   Parts with undercuts should be avoided. Normally, parts with external undercuts cannot be withdrawn from a one-piece mold. Internal undercuts are considered impractical and should be avoided. If an internal undercut is essential, it may be achieved by machining or by use of a flexible mold core material (Fig. 13-4-1H).

Assemblies

The design of molded parts that are to be assembled with typical fastening methods involves factors different from those normally encountered with metal.

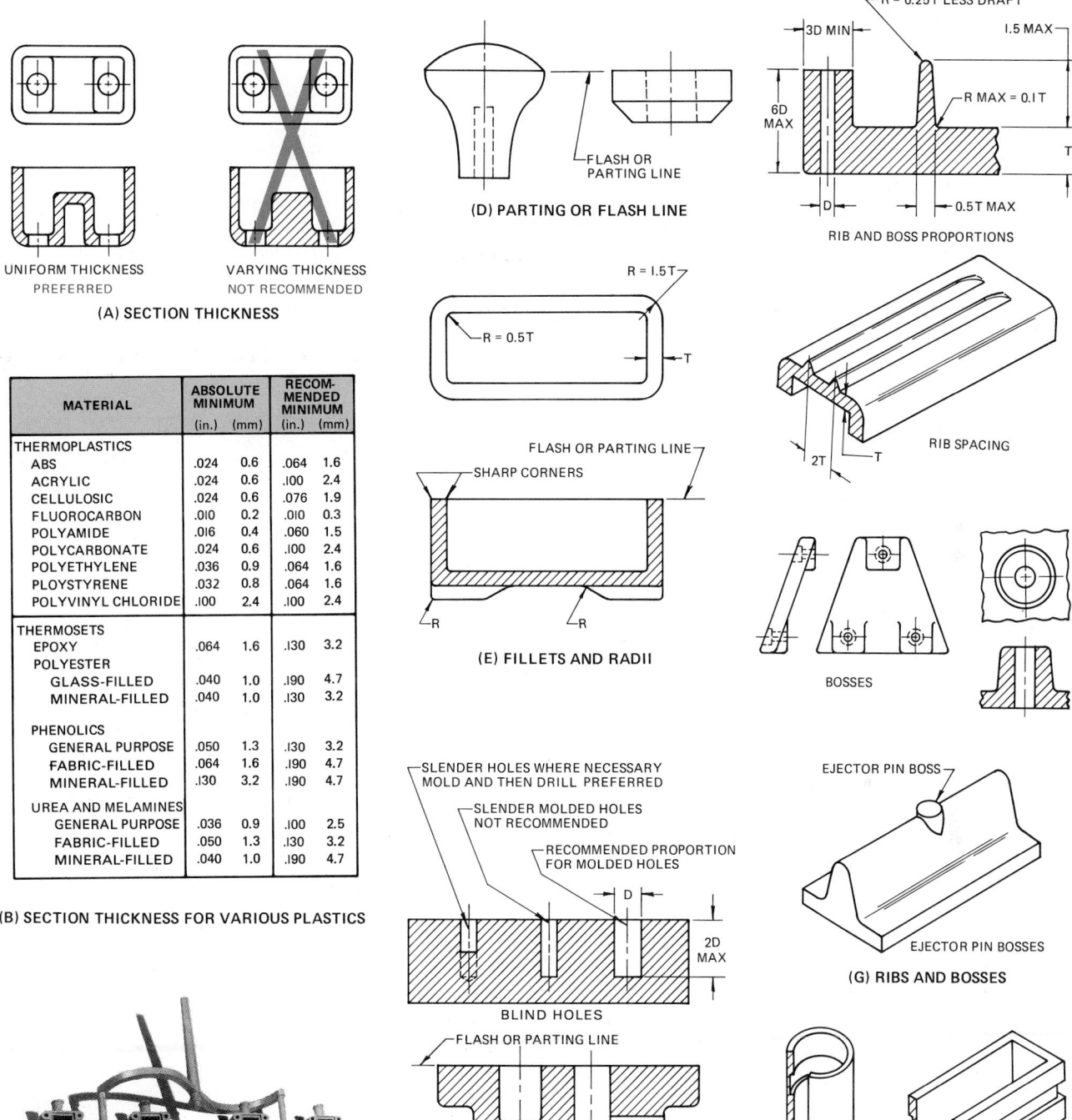

FIG. 13-4-1   Design considerations for plastic molded parts.

Holes and Threads   Mechanical fasteners, in general, depend upon a hole of some type. Holes should be designed and located to provide maximum strength and minimum molding problems. Any straight hole, molded or machined, should have between it and an adjacent hole, or side wall, an amount of material equal to or greater than the diameter or width of the hole. Any threaded hole, molded or tapped, should have between it and an adjacent hole, or side wall, an amount of material at least three times the outside diameter of the thread. Spacing may be reduced, however, by proper use of bosses.

Drilled holes are often more accurate and easier to produce than molded holes, even though they require a second operation.

Tapped holes provide an economical means of joining a molded part to its assembly. The designer should avoid threads

367

with a pitch of less than .03 in. (0.8 mm). Holes that are to be tapped should be countersunk to prevent chipping when the tap is inserted.

External and internal threads can be molded integrally with the part. Molded threads are usually more expensive to form than other threads because either a method of unscrewing the part from the mold must be provided or a split mold must be used.

Inserts   After the molding material has been determined, the insert should be designed. The molded part should be designed around the insert.

Inserts of round rod stock, coarse diamond-knurled and grooved, provide the strongest anchorage under torque and tension. A large single groove with knurling on each end is superior to two or more grooves with smaller knurled surface areas. See examples of inserts in Fig. 13-4-2.

Press and Shrink Fits   Inserts may be secured by a press fit, or the plastic molding material may be assembled to a larger part by a shrink fit, as shown in Fig. 13-4-3. Both methods rely on shrinkage of the material, which is greatest immediately after removal from the mold.

Heat Forming and Heat Sealing   Most thermoplastics can be re-formed by the application of heat and pressure, as shown in Fig. 13-4-4. This re-forming often eliminates the need for other assembly methods, such as adhesive bonding and mechanical fasteners. This method cannot be used with thermosetting materials.

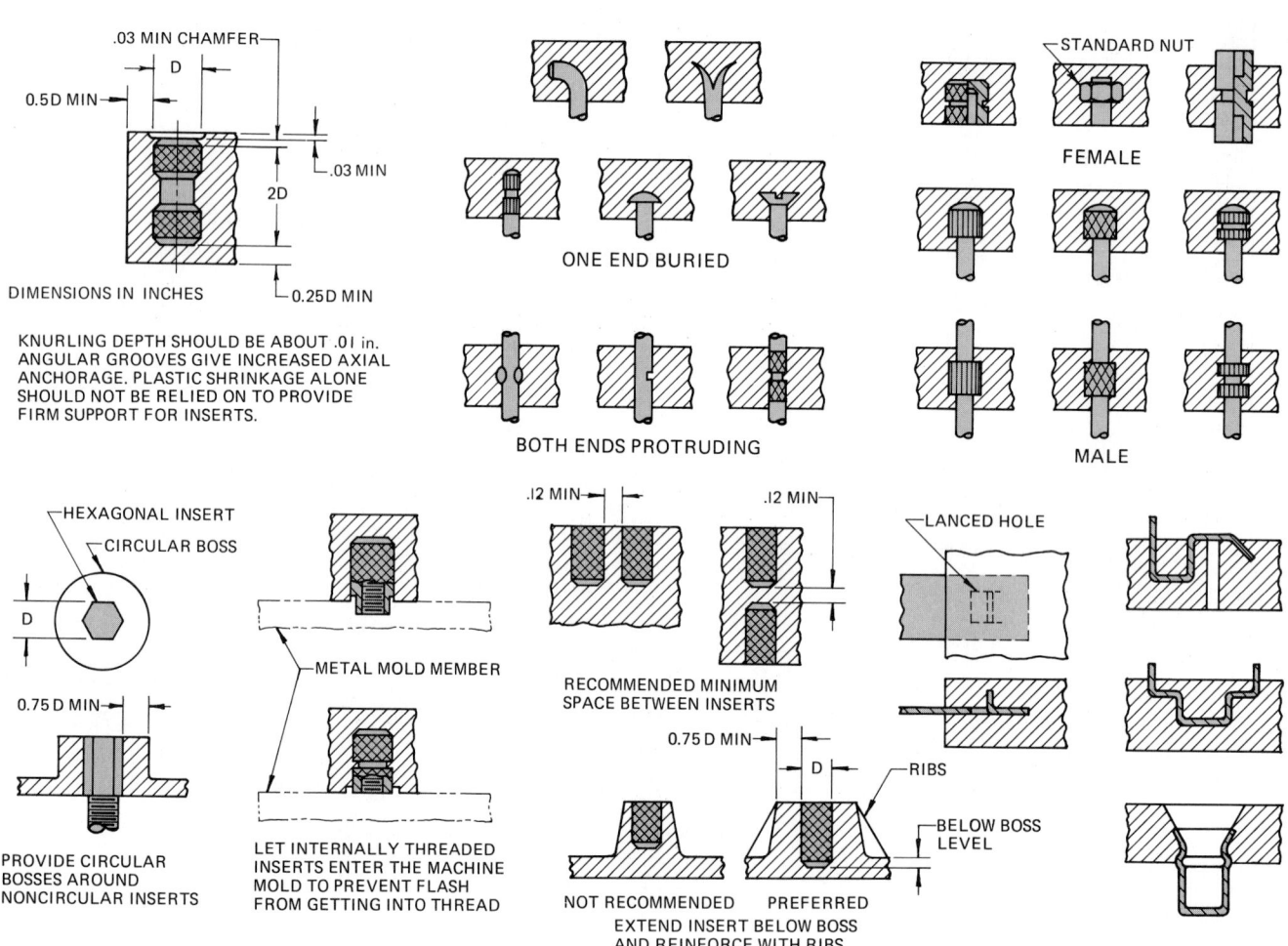

FIG. 13-4-2   Insert applications.

FIG. 13-4-3   Press and shrink fits.

Mechanical Fastening Various designs of mechanical fasteners are commercially available. Spring-type metal hinges and clips, speed clips or nuts, and expanding rivets are a few of these designs. Design of the parts for assembly requires that molded parts have sufficient sectional strength to withstand the stresses that will be encountered with fasteners. A strengthening of the area that will receive the brunt of these applied stresses is usually required (Fig. 13-4-5).

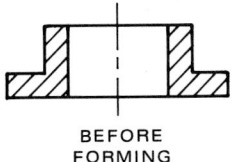 

BEFORE FORMING | AFTER FORMING IN ASSEMBLY

FIG. 13-4-4 Heat forming.

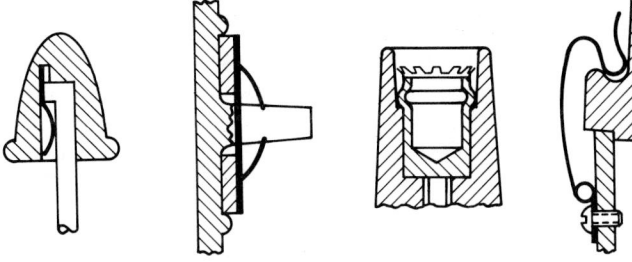

FIG. 13-4-5 Mechanical fasteners.

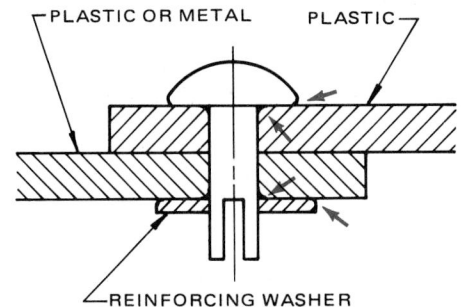

NOTE: BREAK ALL SHARP EDGES ON RIVET, WASHER, AND HOLES.

FIG. 13-4-6 Recommended riveting procedures.

FIG. 13-4-7 Boss cap design.

Rivets Conventional riveting equipment and procedures can be used with plastics. Care must be exercised to minimize stresses induced during the fastening operation. To do this, the rivet head should be 2.5 to 3 times the shank diameter. Also, rivets should be backed with either plates or washers to avoid high localized stresses (Fig. 13-4-6).

Drilled holes rather than punched holes are preferred for fasteners. If possible, fastener clearance in the hole should be at least .01 in. (0.3 mm) to maintain a plane stress condition at the fastener.

Boss Caps A *boss cap* is a cup-shaped metal ring that is pressed onto the boss by hand, with an air cylinder, or with a light-duty press. It is designed to reinforce the boss against the expansion force exerted by tapping screws (Fig. 13-4-7).

Adhesive Bonding When two or more parts are to be joined into an assembly, adhesives permit a strong, durable fastening between similar materials and often are the only fastening method available for joining dissimilar materials. Structural adhesives are made from the same basic resins as many plastics and thus react to their operating environment in a similar manner. In order to provide maximum strength, adhesives must be applied as a liquid to thoroughly wet the surface of the part. The bonding surface must be chemically clean to permit complete wetting. Basic plastics vary in physical properties, so adhesives made from these materials also vary. Fig. 13-4-8 (pg. 370) shows adhesively bonded joints.

Ultrasonic Bonding Ultrasonic bonding often is used instead of solvent cementing to bond plastic parts. By using this technique, irregularly shaped parts can be bonded in two seconds or less. The bonded parts may be handled and used at reasonable temperatures within minutes after joining.

Only one of the mating parts comes in contact with the horn (Fig. 13-4-9, pg. 370). The part transmits the ultrasonic vibration to small, hidden bonding areas, resulting in fast, perfect welds. Both mating halves remain cool except at the seam, where the energy is quickly dissipated.

This technique is not recommended where high impact strength is required in the bond area.

Ultrasonic Staking Ultrasonic staking frequently involves the assembly of metal parts. In this technique, a stud molded

THREAD SIZE	DIMENSIONS				
	A	B	C	D	R
# 6(.138)	.14	.21	.10	.28	.02
# 8(.164)	.16	.25	.13	.34	.02
# 10(.190)	.18	.29	.17	.40	.02

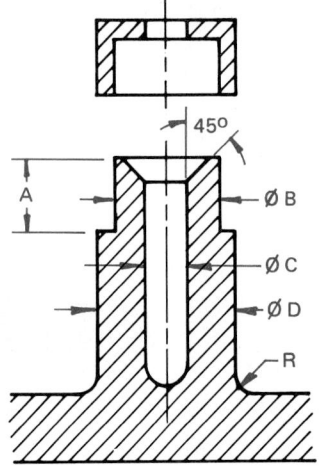

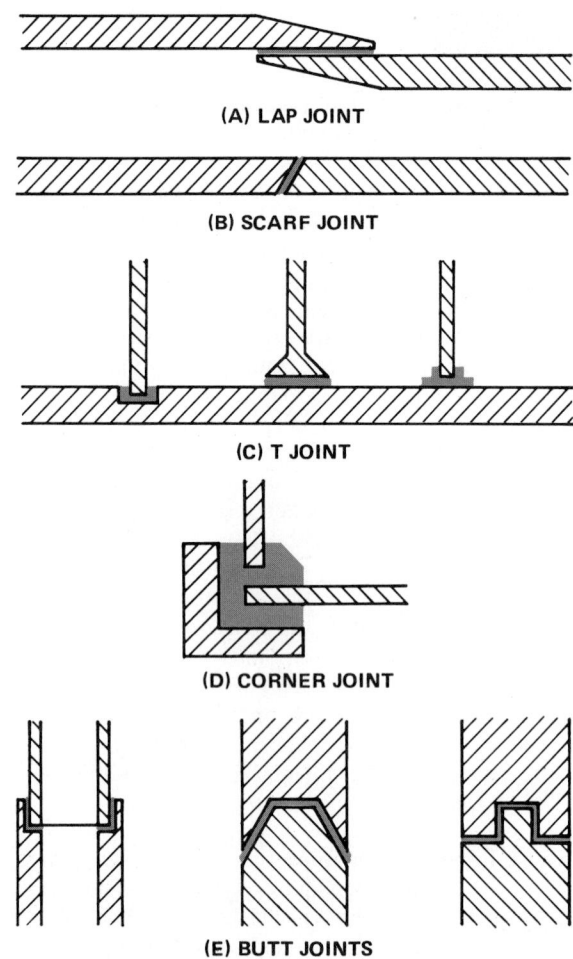

(A) LAP JOINT

(B) SCARF JOINT

(C) T JOINT

(D) CORNER JOINT

(E) BUTT JOINTS

FIG. 13-4-8   Adhesive bonding.

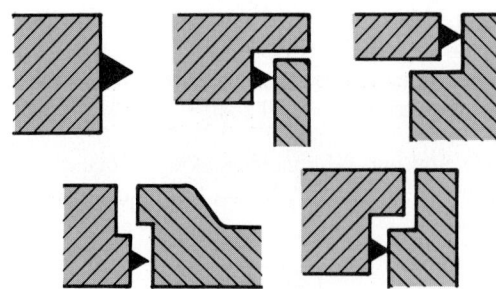

FIG. 13-4-9   Design joints for ultrasonic bonding.

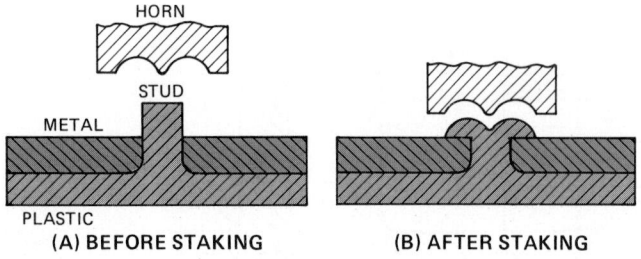

HORN

STUD

METAL

PLASTIC

(A) BEFORE STAKING   **(B) AFTER STAKING**

FIG. 13-4-10   Typical ultrasonic staking operation.

into the plastic part protrudes through a hole in the metal part. The surface of the stud is vibrated with a horn having high amplitude and a relatively small contact area. The vibration causes the stud to melt and re-form in the configuration of the horn tip (Fig. 13-4-10).

Friction or Spin Welding   This welding technique is limited to parts with circular joints. It is especially useful for large parts where ultrasonic welding or chemical bonding is impractical.

In friction or spin welding, the faces to be joined are pressed together while one part is spun and the other is held fixed. Frictional heat produces a molten zone that becomes a weld when spinning stops (Fig. 13-4-11).

Drawings

In addition to the usual considerations, the following points should be taken into account when a detail drawing of a plastic part is made:

1. Can the part be removed from the mold?
2. Is location of flash line consistent with design requirements?
3. Is section thickness consistent? Are there thick sections? Thin sections? Could greater uniformity of section thickness be maintained?
4. Has the material been correctly specified?
5. Is each feature in accordance with the thinking of competent materials engineers and molders?
6. Have close tolerance requirements been reviewed with responsible engineers?
7. Have marking requirements been specified to inform field service people of the material from which the part is fabricated?

REFERENCES AND SOURCE MATERIAL

1. General Motors Corp.
2. General Electric Co.

ASSIGNMENTS

See Assignments 11 through 13 for Unit 13-4 on pages 376 to 377.

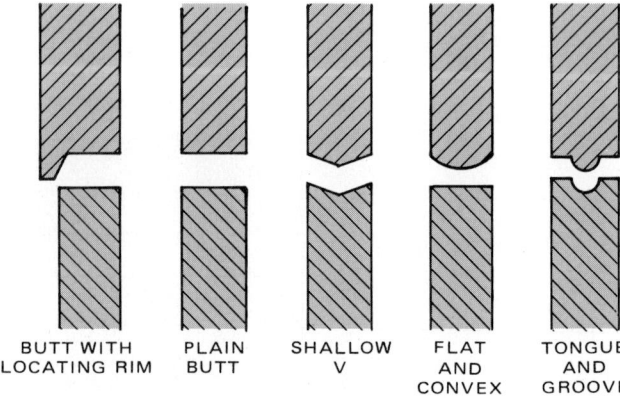

| BUTT WITH LOCATING RIM | PLAIN BUTT | SHALLOW V | FLAT AND CONVEX | TONGUE AND GROOVE |

FIG. 13-4-11   Joint shapes for spin welding.

ASSIGNMENTS FOR CHAPTER 13

ASSIGNMENTS FOR UNIT 13-1, CASTINGS

1. Complete the assembly drawing of the fork for the hinged pipe vise shown in Fig. 13-1-A. Use your judgment for dimensions not given. Scale 1:1.

2. Complete the detail drawing of the base for the adjustable shaft support assembly shown in Fig. 13-1-B. Cored holes are to be used for the shaft holes. Scale 1:1.

FIG. 13-1-A   Pipe vise.

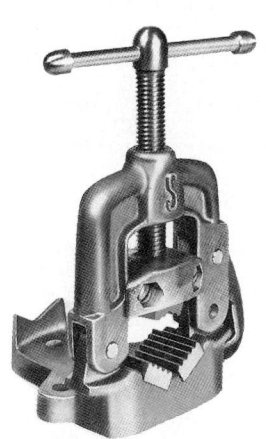

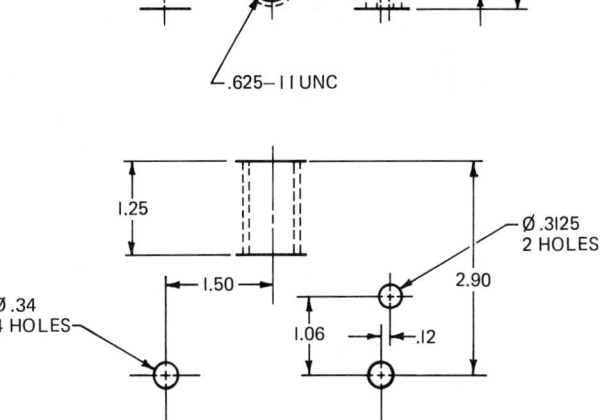

FIG. 13-1-B   Adjustable shaft support.

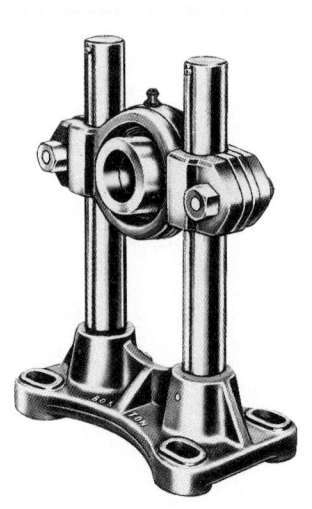

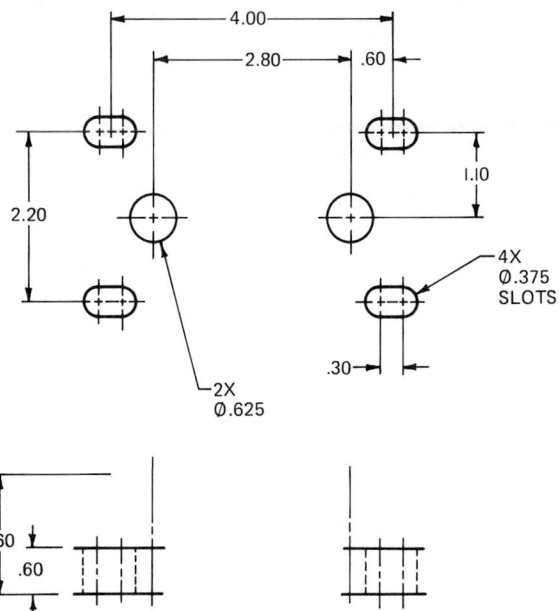

3. Prepare both the casting and the machining drawings for the connector shown in Fig. 13-1-C. Draw a one-view full section, complete with the necessary dimensions for each drawing. Scale 1:1.

4. Prepare both the casting and the machining drawing for the pump bracket shown in Fig. 13-1-D. Cored holes of Ø20 and Ø9 are to be used for the Ø24 and Ø12 holes. The machined surfaces are to have a maximum roughness

FIG. 13-1-C   Connector.

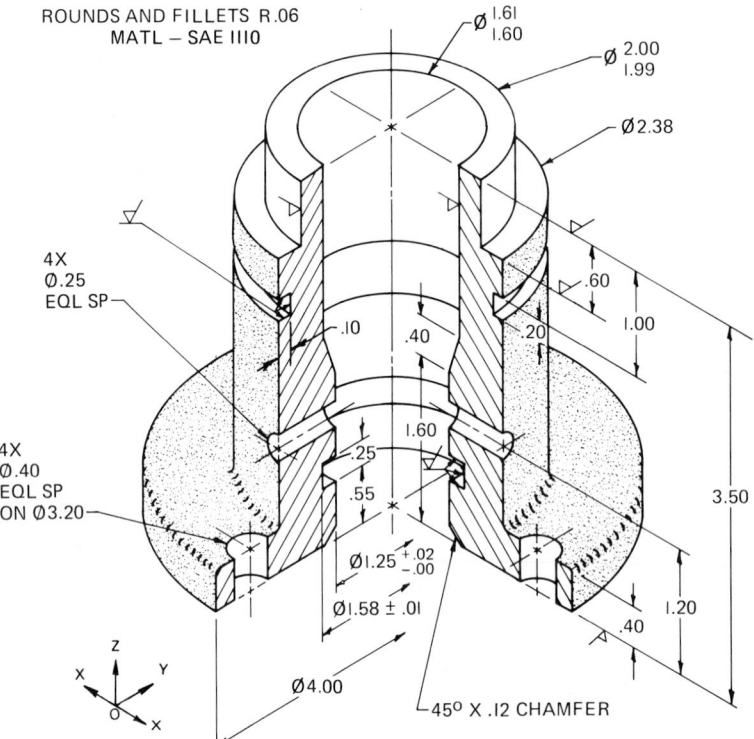

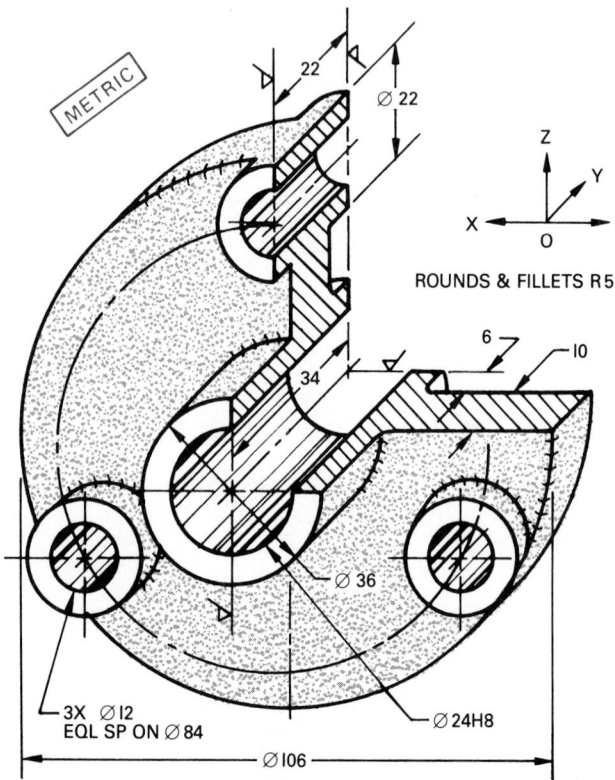

FIG. 13-1-D   Pump bracket.

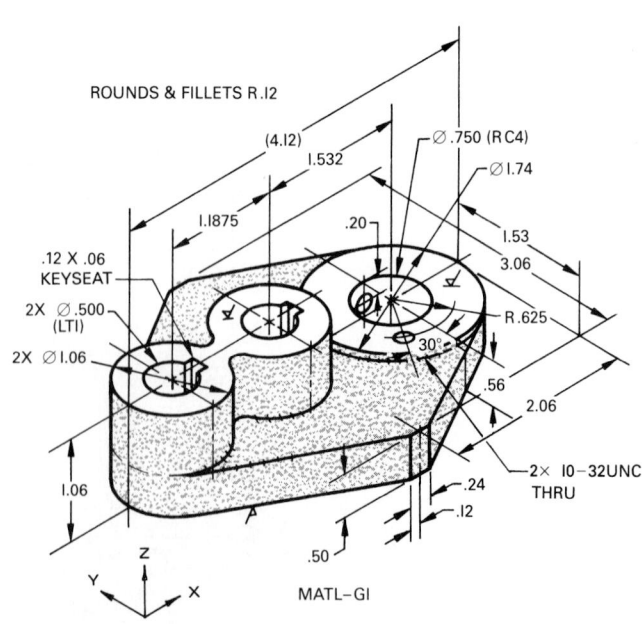

FIG. 13-1-E   Top plate.

372

value of 3.2 μm and a machining allowance of 2 mm. Show the limit dimensions for the Ø24 hole.

5. Prepare both the casting and machining drawings of the top plate shown in Fig. 13-1-E. Cored holes are to be used for the three vertical holes. The machined surfaces are to have a maximum roughness value of 63 μin and a machining allowance of .06 in. Show the limit dimensions where fits are indicated. Scale 1:1.

6. Redesign one of the welded parts shown in Figs. 13-1-F through 13-1-H into a cast part. Make a machine drawing given the following information. Show the limit sizes where fits are indicated. Surfaces shown with the letter A are to have a maximum roughness value of 1.6 μm and a machining allowance of 2 mm or its equivalent. Use symbolic dimensioning. For Fig. 13-1-F add a spotface to the Ø.78 hole and increase the top and bottom thickness.

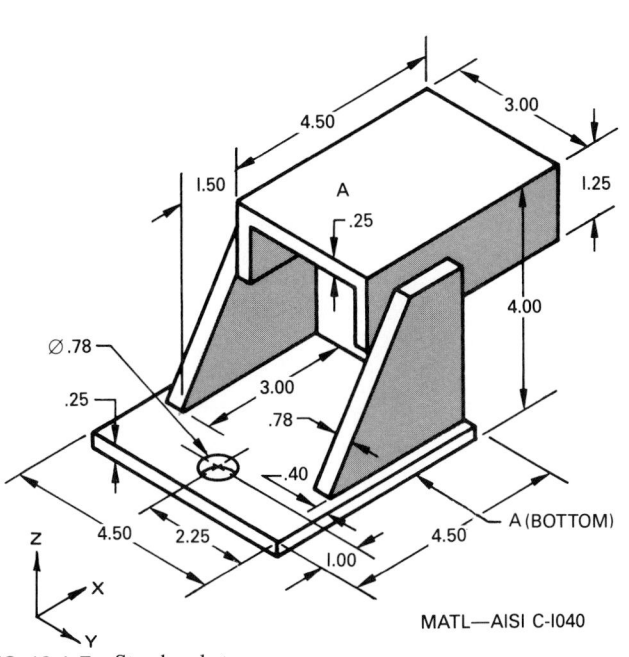

FIG. 13-1-F   Step bracket.

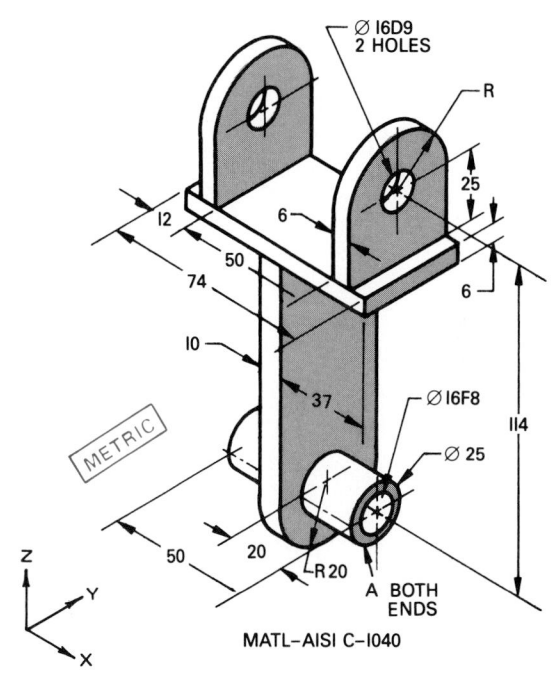

FIG. 13-1-G   Swing bracket.

FIG. 13-1-H   Shaft support.

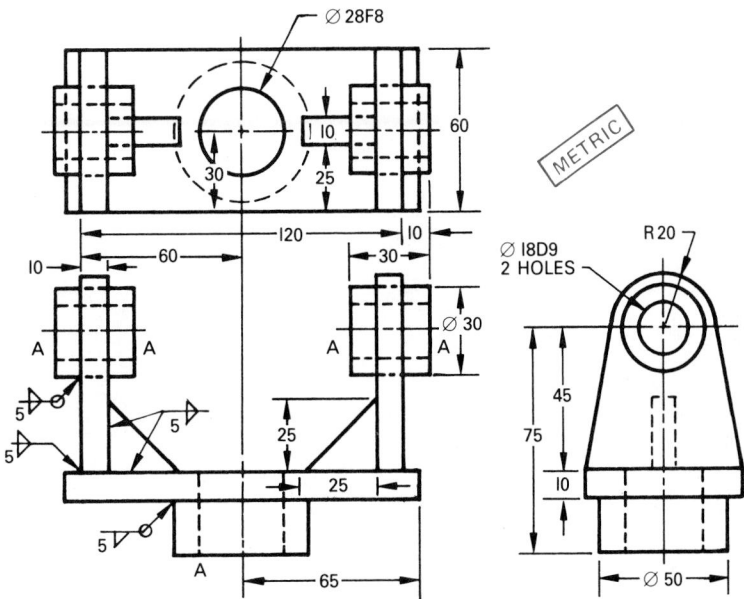

373

ASSIGNMENTS FOR UNIT 13-2, FORGINGS

7. Prepare a forging drawing of one of the parts shown in Fig. 13-2-A or 13-2-B. Scale 1:1.

8. Prepare a forging drawing for the wrench handle shown in Fig. 13-2-C. Scale 1:1.

FIG. 13-2-A   Open-end wrench.

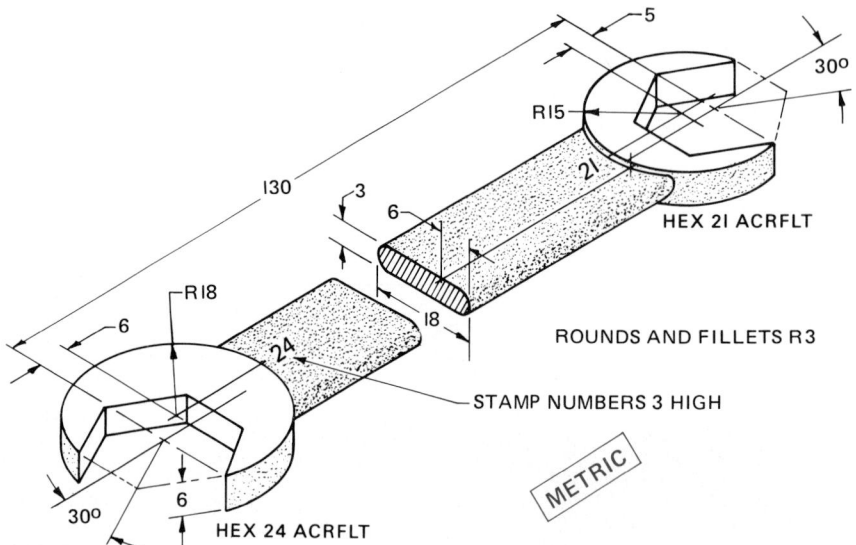

FIG. 13-2-B   Bracket.

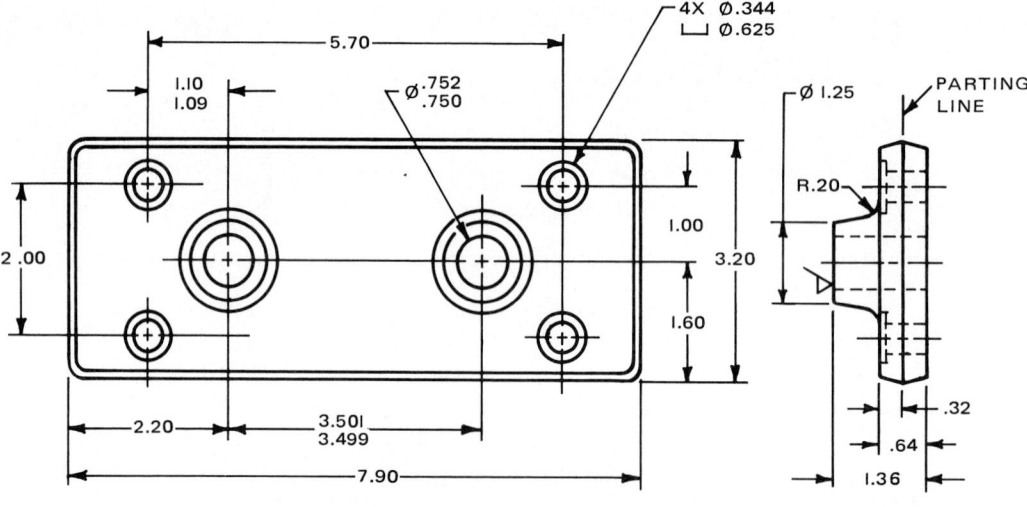

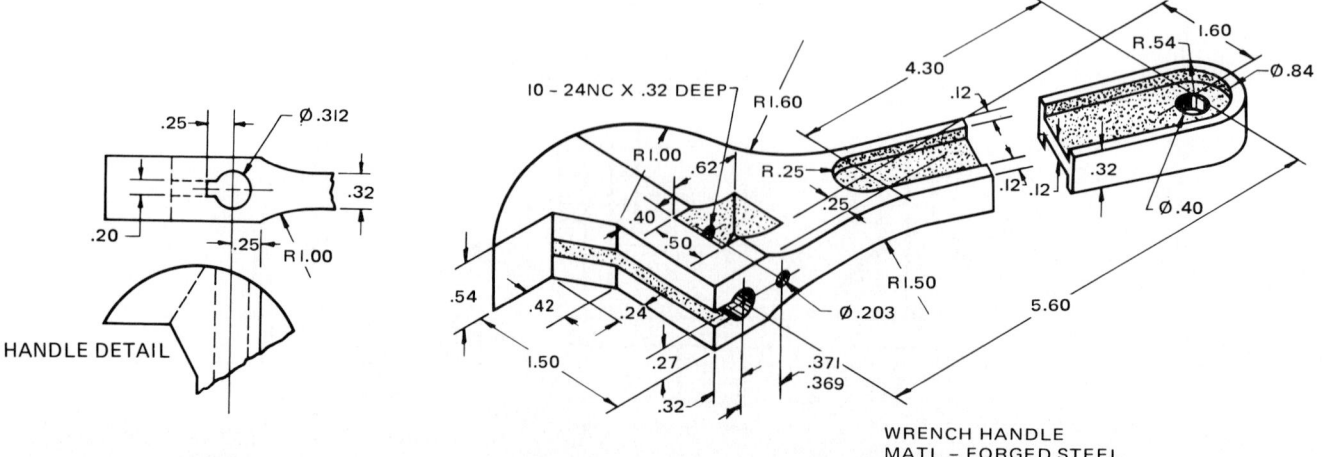

FIG. 13-2-C   Wrench handle.

ASSIGNMENTS FOR UNIT 13-3, POWDER METALLURGY

9. Prepare two drawings, one for machining the part, the second for the making of the briquet (powder metallurgy) for one of the parts shown in Fig. 13-1-C, 13-3-A, or 13-3-B. Scale 1:1.

10. Prepare one or two drawings as required, one for machining the part, the second for making the briquet (powder metallurgy) for one of the prefabricated (welded) parts in Figs. 13-3-C through 13-3-E (here and on pg. 376). Scale 1:1.

FIG. 13-3-A   Bracket.

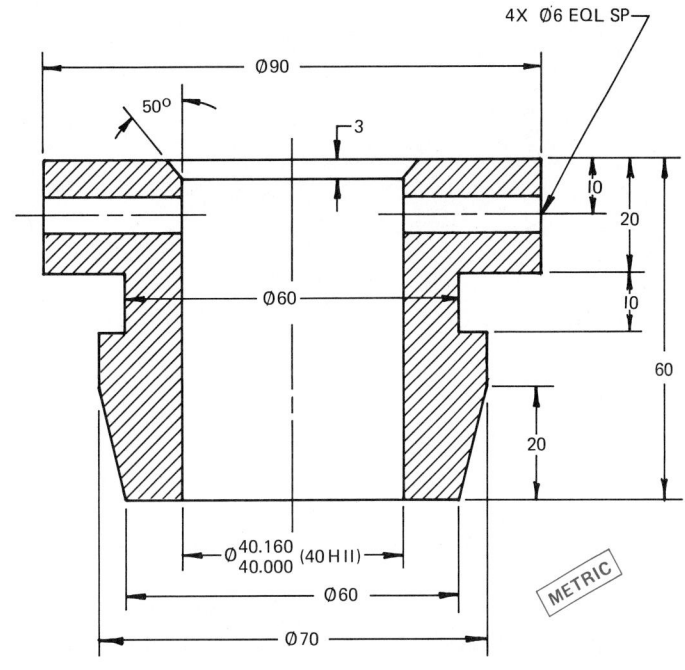

FIG. 13-3-B   Tool holder.

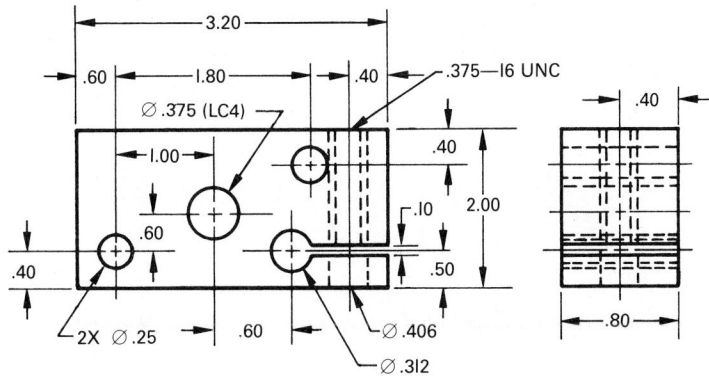

FIG. 13-3-C   Shaft base.

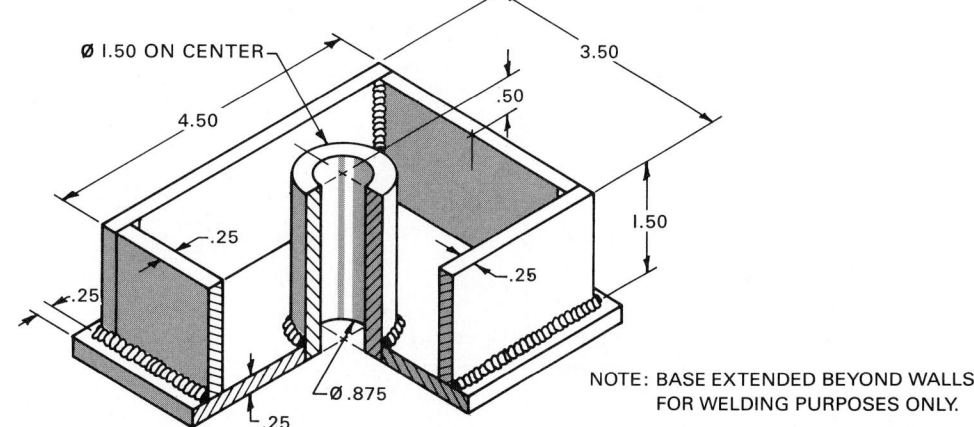

NOTE: BASE EXTENDED BEYOND WALLS
FOR WELDING PURPOSES ONLY.

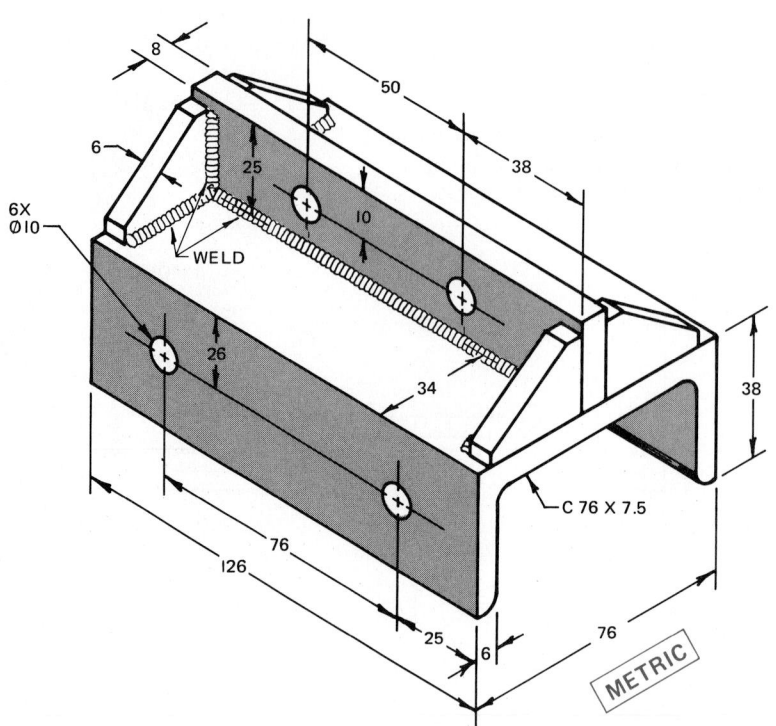

FIG. 13-3-D　Caster frame.

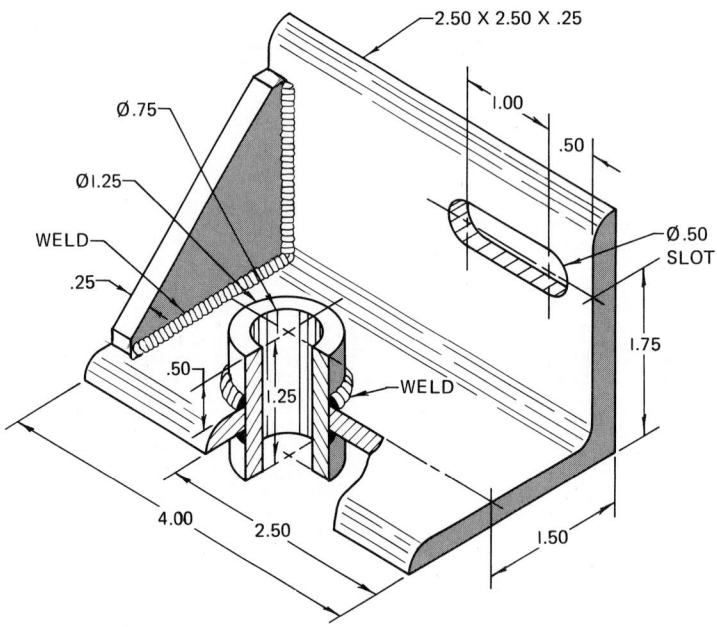

FIG. 13-3-E　Slide bracket

ASSIGNMENTS FOR UNITS 13-4, PLASTIC MOLDED PARTS

11. Redesign for plastic molding one of the parts shown in Figs. 13-3-C through 13-3-E. Where required, prepare two drawings—one for the mold, the other for machining.

Refer to the molding recommendations shown in this unit and indicate the parting line on the drawing. Use your judgment for dimensions not given. Scale 1:1.

12. Using a plastic molding design, add threaded inserts to one of the parts shown in Figs. 13-4-A through 13-4-D. Use your judgment for dimensions not shown and the type and number of views required. Scale 2:1.

13. Make a plastic molding assembly drawing of the parts shown in Fig. 13-4-E. The retaining ring is to be positioned in the center of the part and molded into position. Modification to the retaining ring may be required to prevent the ring from turning in the wheel. Scale 5:1. Show a top view and a full-section view. Dimension the finished assembly.

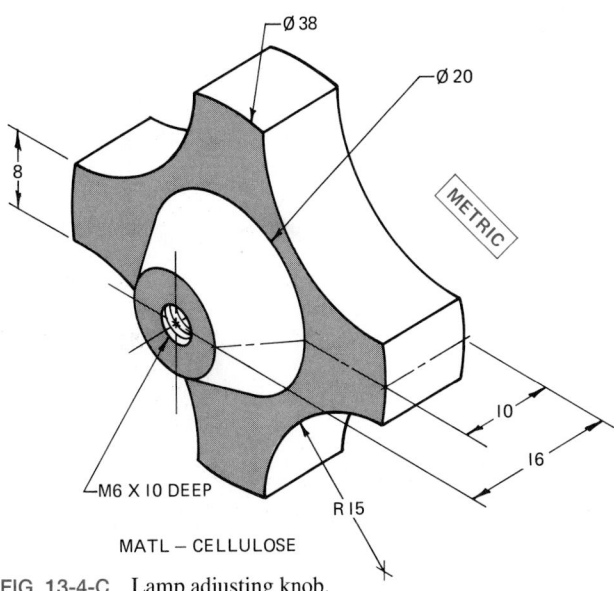

MATL – CELLULOSE

FIG. 13-4-C   Lamp adjusting knob.

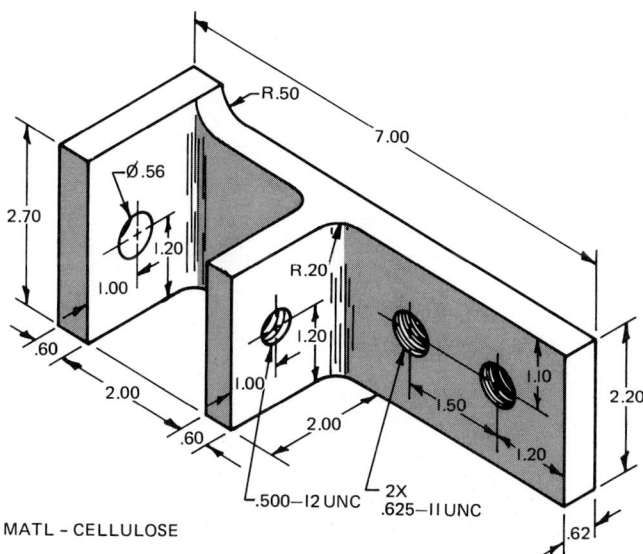

MATL – CELLULOSE

FIG. 13-4-A   Connector.

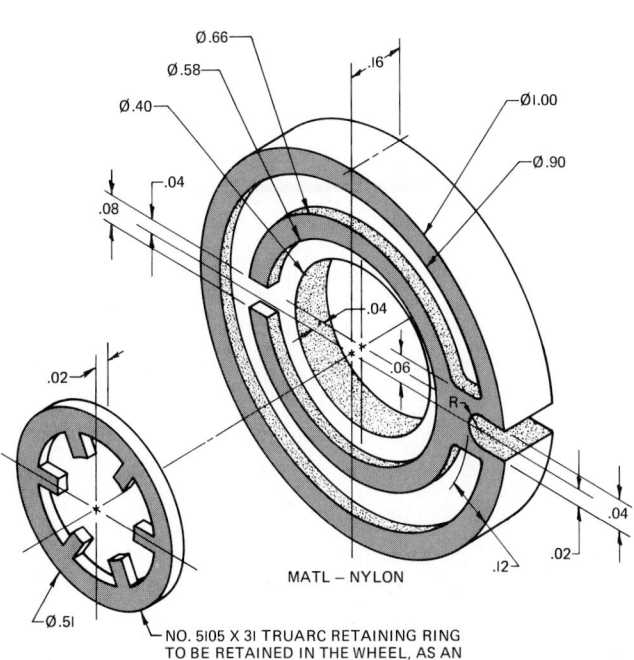

FIG. 13-4-D   Gear clamp.

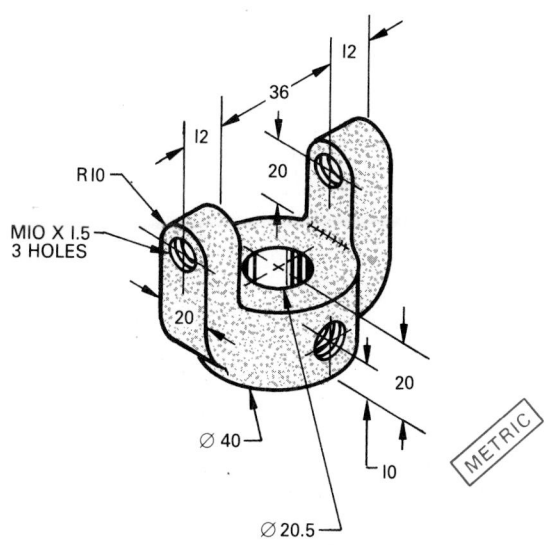

FIG. 13-4-B   Pivot arm.

FIG. 13-4-E   Cassette-tape drive wheel.

MATL – NYLON

NO. 5105 X 31 TRUARC RETAINING RING TO BE RETAINED IN THE WHEEL, AS AN INSERT, AT THE CENTER OF THE THICKNESS.

WORKING DRAWINGS AND DESIGN

CHAPTER 14

DETAIL AND ASSEMBLY DRAWINGS

Definitions

Appliqués Pressure-sensitive overlays used to depict common parts, shapes, symbols, surface textures, or notes.

Assembly drawing A drawing showing the product in its completed state.

Detailer The draftsperson (drafter) who works from a complete set of instructions and drawings, or who makes working drawings of parts that involve the design of the part.

Item list or **bill of material** An itemized list of all the components shown on an assembly drawing or a detail drawing.

Subassemblies Preassembled components and individual parts used to make up a finished product.

Working drawing A drawing that supplies information and instructions for the manufacture or construction of machines or structures.

14-1 DRAWING QUALITY ASSURANCE

The requirement that engineering drawings be complete, clear, accurate, conform to standards, and ensure proper functional operation is the responsibility of the drafter, checker, and other specialists assigned to review the drawing prior to release. Use of the following recommendations and specific considerations, as applicable, when reviewing drawings is advised as a means of promoting the preparation of quality drawings. The term *drawings* refers to the depiction of details, assemblies, installations, or other types of graphical representations.

Knowledge of the design requirements, the manufacturing process involved, and drafting practices on the part of the drafter, checker, and other reviewers has a definite influence on the accuracy and cost of parts and assemblies. Layouts should be carefully studied and, where necessary, discussed with the designer and responsible engineer to ensure full understanding of the function and application of the design. Suggestions for improvement in design or manufacture should be discussed with accountable personnel. Finished drawings should reflect the objective findings of all responsible reviewers.

Although the review procedure may vary, it is recommended that reference to layouts, proven similar designs, and other pertinent design data be used by the reviewers. Throughout the review it is vital to be constantly on the alert for omitted or incomplete information.

Review Considerations

The following items are typical of those that need to be considered, as applicable, in the preparation and review of drawings.

Applied Surface Finish Any applied surface finish requirements must be completely defined.

Expansion Dimensions and tolerances should be adjusted for thermal expansion or contraction during operation. Differences in expansion coefficients of various materials should be kept in mind.

Grain Flow A part made from a forging or from sheet metal must have direction of grain flow indicated where it is important to the durability of the part.

Inspection Processes Inspection processes, such as magnetic particle, fluorescent penetrants, and X rays, must be noted on the drawing where required.

Interchangeability Requirements for interchangeability must be considered.

Locking Feature Locking features for the retention of parts, such as lockwire holes and tab washer slots, should be shown where required.

Material   Proper material and heat treatment must be specified.

Procurability   Where an item is vendor-supplied, or includes vendor-controlled features, such as material, process, or operational devices, its availability should be considered.

Protective Finish   Protective finish specifications, such as painting or plating, should be specified.

Seizure   Where parts come in contact, material and surface treatments subject to "seizing," galvanic action, or similar effects should normally be avoided.

Service   Accessibility must be provided for servicing, assembling, inspection, and adjustment.

Standard Parts   Standard parts should be used wherever applicable.

Standard Practices   Standards pertaining to design, materials, processes, etc. should be used.

Strength   Design must adequately meet all stress requirements, such as thermal, dynamic, and fatigue stresses. Deterioration (embrittlement, corrosion, and wear) must be considered.

Surface Texture (Roughness)   Surface texture values must be specified for all surfaces requiring control. The values shown should be compatible with overall design requirements.

Tolerances   The tolerances indicated by the linear and angular dimensions and by local, general, or title block notes must ensure the proper assembly and functioning of the parts. Tolerances should be as liberal as the design will permit.

Drawing Considerations

Abbreviations   Abbreviations should conform to the country's, or the individual company's, drawing standards.

Conformance to Drawing Standards   Drawings should conform to the country's, or the individual company's, drawing standards in regard to size of sheet, format, zone marking, microfilm alignment arrowheads, arrangement of views, line characteristics, scale, letter and dimension heights, notes, and general appearance. Lines and lettering must be distinct and dark enough to ensure legible reproduction, including microfilm reduction. Letter (and number) form and size must be compatible with microfilming and reduced-size prints.

Dimensions   The part must be fully dimensioned and the dimensions clearly positioned. True-position relationship should be shown where applicable. Dimensions should not be repeated or shown in a manner that constitutes double dimensioning. Dimensions should not result in objectionable tolerance accumulation. Dimensions should emphasize function of design in preference to production operations or processes and should be such as to minimize shop calculations. Developed lengths and stock size should be specified as applicable.

Draft Angle and Radii   Proper draft angles, fillets, and corner radii should be specified (see Chap. 13).

Geometric Surface Relationship   All requirements covering necessary geometric relationships, such as straightness, runout, squareness, and parallelism, must be shown (see Chap. 16).

Revisions   All revisions must be properly recorded and all lines damaged by erasing during the making of revisions must be restored. All related drawings should be revised to conform.

Scale   The drawing should be to scale and the scale should be identified. Where drawings are to no scale, they should be identified as such.

Surface Texture Symbols   Surface texture symbols and values must be specified for all surfaces requiring control. The values should be compatible with overall design requirements.

Tolerances   The tolerances specified by the linear and angular dimensions and by local, general, or title block notes must ensure the proper assembly and functioning of the part. The selection of positional tolerancing or coordinate tolerancing should be carefully considered. Tolerances should be as liberal as the design will permit.

Symbols   Wherever possible, symbols should be used in lieu of words. The placement and use of symbols should reflect the latest standards.

Symmetrical Opposite Parts   An AS SHOWN and OPPOSITE HAND note with proper identification numbers must be shown for all such parts, unless a separate drawing is made for each hand.

Views   Sufficient full and sectional views must be shown and must be in proper relation to each other if third-angle projection is used, or properly identified if the reference arrow layout method is used (see Chap. 6).

Fabrication Considerations

Adhesives   The drawing must clearly identify the type of joint and adhesive used.

Brazing, Soldering, and Welding   The drawing must include local or general notes or symbols, as applicable, for the method of fabrication used.

Casting   Where the part is made as a casting, sufficient tolerances must be provided for draft, warpage, core shifting, or crossing of the parting line. Can coring be simplified or eliminated? Is the cast part number located in a practical position?

Centers   Where manufacturing can be facilitated by providing machining centers, they should be specified on the drawing.

Economy   Is the design the most economical approach; or would redesign result in a more economical approach without sacrificing quality?

Forged and Molded Parts   For parts made by forging or molding, sufficient tolerances must be allowed for warping, die shift, and die closure.

Holes   Are tolerances adequate to permit economical drilling or reaming? Blind holes must be sufficiently deep to permit threading and reaming.

Machining Lugs   Where a part is cast or forged, manufacturing can often be facilitated by providing clamping lugs and locating pads. Removal of such lugs after machining, where required, should be noted on the drawing.

Numerical Control Machining   Parts to be machined on numerically controlled equipment may be dimensioned to facilitate programming.

Processing Clearance   Design must allow sufficient clearance for drills, cutters, grinding wheels, as well as welding, riveting, and other processing tools.

Special Considerations   Notes for sandblast, vapor blast, and any other special operations should be included where required.

Stamping   All dimensions should be given to the same side of metal, where practical.

Tooling   Dimensions on drawings should reflect the use of standard tooling, such as reamers, cutters, and drills, wherever possible, without specifically calling out the type of tooling to be used, e.g. Ø6.30, not 6.30 DRILL.

Installation Considerations

Assembly   Parts should be designed so there is no possibility of misassembly. Often a dowel, offset bolt hole, or similar feature can be provided to ensure correct, one-way assembly. The design should permit servicing without unreasonable complications.

Clearance   The part must have sufficient clearance with surrounding parts to permit assembly and operation.

Driving Feature   Threaded parts require a slot, hex, or other driving feature.

Puller Feature   Where a part has a tight fit, it may require a puller lip, a jackscrew thread, a knockout hole, or some similar extraction feature.

Tool Clearance   Adequate clearance must be provided for wrenches or other assembly tools.

Torque Values   Required wrench torque values should be specified where items are assembled by means of bolts, cap screws, nuts, or similar features.

REFERENCES AND SOURCE MATERIAL

1. General Motors Corp.

14-2   FUNCTIONAL DRAFTING

Since the basic function of the drafting department is to provide sufficient information to produce or assemble parts, functional drafting must embrace every possible means to communicate this information in the least expensive manner. Functional drafting also applies to any method that would lower the cost of producing the part. New technological developments have provided many new ways of producing drawings at lower costs and/or in less time. This means that drafters must be prepared to discard some of the old, traditional methods in favor of these newer means of communication.

There are many ways to reduce drafting time in preparing a drawing. These drawing shortcuts, when collectively used, are of prime importance in an effective drafting system. These newer techniques cannot be blindly applied, however, but must be carefully evaluated to make certain that the benefits outweigh the potential disadvantages. This evaluation should answer the following questions:

- What is its purpose?
- Is it a personal preference disguised as a project requirement?
- Does it meet contractual requirements?
- Will the shortcut increase costs in other areas, such as manufacturing, purchasing, or inspection?
- Is it an effective communication link?
- How much training or education is required to make effective use of it?
- Are facilities available to implement it?
- Does the shortcut bypass a real bottleneck?

As each of these categories is examined, the advantages of the shortcuts will become apparent.

Procedural Shortcuts

There are a number of procedural shortcuts which, if properly applied and carefully managed, can shorten the drawing preparation cycle and result in savings.

Streamlined Approval Requirements   It is obvious that the more signatures required on a drawing, the greater the delays in releasing data. The decision as to who will approve drawings and drawing changes must be carefully considered to make certain that all necessary functions have been taken into account (checkers, responsible engineers, important technical specialists, etc.) without imposing undue restrictions. Project ground rules and contractual requirements also play an important part in this decision.

Eliminating the Drawing Check from the Preparation Cycle   One of the most common suggested shortcuts, usually proposed when a project is behind schedule or exceeding its budget or when experienced personnel are involved, is to eliminate checking from the drawing preparation cycle.

Using Standard and Existing Drawings   Drawings of parts are constantly being prepared that are repetitions of existing drawings. If the drafter were to incorporate the existing design parts that were already drawn with the new drawings, many drawing hours would be saved. Good drawing application records and an efficient multiple-use drawing system can eliminate a great deal of duplication. Standard tabulated drawings may be used to eliminate hundreds of drawings (Figs. 14-2-1 and 14-2-2).

Standard Drafting Practices   Standard drafting practices are obviously the backbone of efficient drafting room operations. The best way to establish and implement these practices is through a good drafting room manual, with requirements that must be strictly observed by all personnel.

The drafting room manual should contain data on the use and preparation of specific types of drawings, drawing and part number requirements, standard and special drafting practices,

FIG. 14-2-1   Standard tabulated
drawings.

QTY	PART	MATL	DESCRIPTION	PT NO.
2	CABLE SUPPORT	MAPLE	A—5374 PT 1	1
2	CABLE SUPPORT	MAPLE	A—5374 PT 2	2
3	CABLE SUPPORT	MAPLE	A—5374 PT 4	3

(A) DRAWING CALLOUT

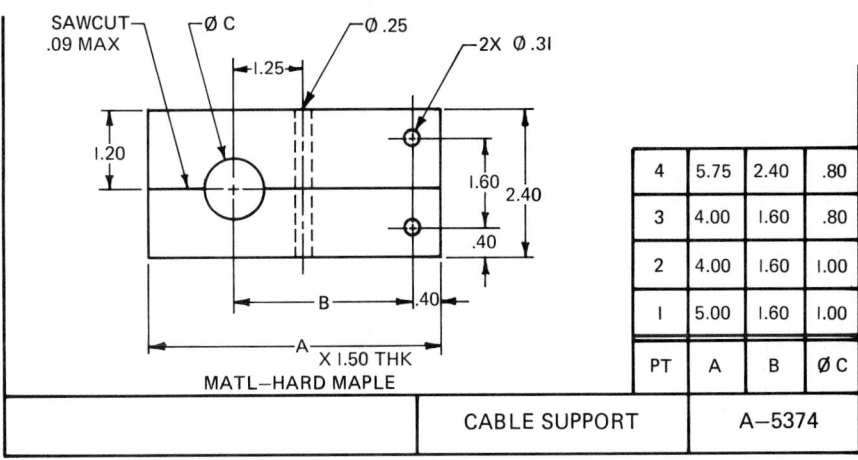

4	5.75	2.40	.80
3	4.00	1.60	.80
2	4.00	1.60	1.00
1	5.00	1.60	1.00
PT	A	B	Ø C

CABLE SUPPORT	A—5374

(B) STANDARD PART

FIG. 14-2-2   Standard parts drawings stored on microfilm. *(Eastman Kodak Co.)*

rules for dimensioning and tolerancing, specifications for associated lists, and company procedures for the preparation, handling, release, and control of drawings.

Team Drafting   Many engineering departments have turned out drawings by the method of one drafter to one drawing. Team drafting involves a number of people producing one drawing. While this may seem uneconomical, it is an expeditious approach, with visible cost savings over the traditional method.

Some firms are using team drafting because it is a better utilization of skill levels. It is a training program through which drafting skills are taught and semiskilled people are given an opportunity to gain experience.

Data Retrieval   The use of microform reader-printers in the drafting room provides quick and ready access to standard drawings and parts. The use of microfiche cards is becoming popular, because they can hold up to 70 pages of information. However, for this method to be effective, a full-time librarian is needed.

Standard Parts and Design-Standard Information Encouraging the use of standard parts and standard approaches to design will not only result in drafting time saved but will cut costs in areas such as purchasing, material control, manufacturing, etc. The odd-size cutout that requires special tooling, the design that calls for nonstandard hardware, and the equipment that uses a wide variety of fasteners when only one or two would suffice are typical cases where properly applied standards would reduce both time and cost.

Copying Machines   One of the most important time-saving devices, which should be available in every drafting area, is a copying machine for reference copies, checking prints of work in preparation, and similar uses (Fig. 14-2-3, pg. 384). When a drafter needs a copy, work is delayed until the copy is made available. Therefore, a good copying machine will soon pay for itself in drawing hours saved.

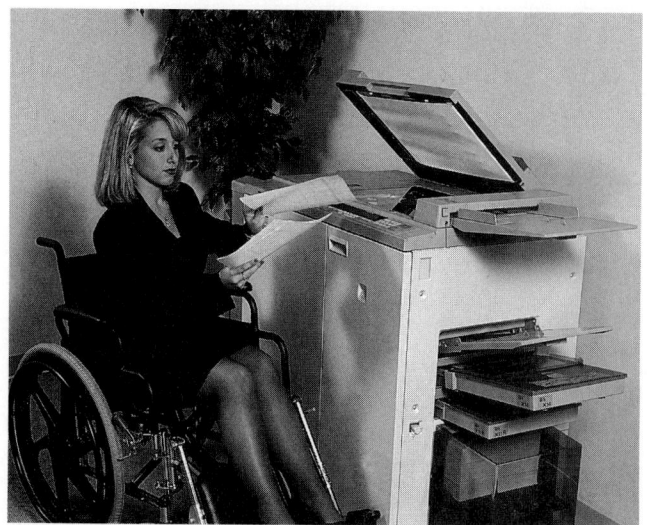

FIG. 14-2-3   Copying machine. *(Doug Martin)*

Training Programs   To provide drafters with standard procedures and technical information is not enough; they must be trained in their use. New drafters are frequently overwhelmed by a strange environment, while old employees fail to keep up with new requirements or properly use the services available. Training programs for the indoctrination of new personnel and the updating of long-service employees are rewarded by more efficient and versatile operation.

Manual Drafting Equipment and Materials

The quality of the material and supplies used in the preparation of drawings is as important as the quality of the instruments used in fabrication.

Numerous timesaving devices are available for manual drafting: templates for every application, "pens" for easier line work, and more application of tape to artwork, transfer-type lettering, etc. Since drafting applications vary so widely, only the drafting supervisor can determine which devices will increase the drafting production.

Drafting aids are designed to facilitate the making of drawings by removing or reducing some of the more tedious aspects of drafting.

Templates, such as shown in Fig. 14-2-4, play an important part in functional drafting, for they save a great deal of time in drawing common shapes of details such as rounds, squares, hexagons, and ellipses. In addition to common shapes, templates have been made for standard parts, such as nuts and bolt heads, for electrical symbols, outlines of tools and equipment, and many other outlines that are often repeated.

Reducing the Number of Drawings Required

The cost of a project is, to some extent, directly related to the number of drawings that must be prepared. Therefore, careful planning to reduce the number of drawings required can result in significant savings. Some ways to reduce the number of drawings are explained below.

FIG. 14-2-4   Templates are made for many different uses and save a lot of time.

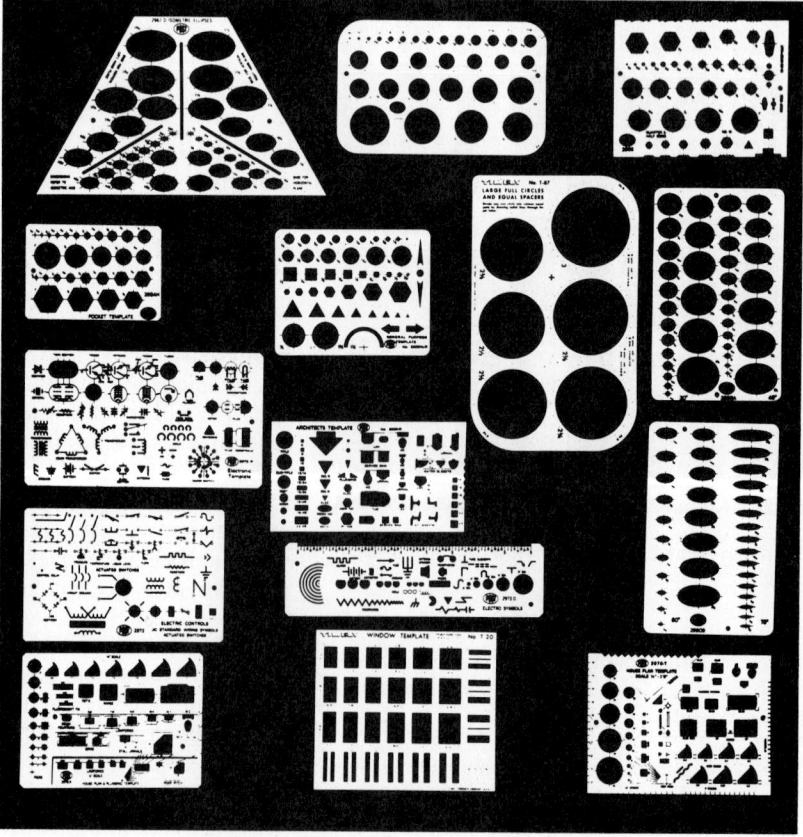

Detail Assembly Drawings   Detail assembly drawings, in which parts are detailed in place on the assembly (Fig. 14-2-5), and multidetail assembly drawings, in which there are separate detail views for the assembly and each of its parts, will reduce the number of drawings required. However, they must be used with extreme care. They can easily become too complicated and confusing to be an effective means of communication (see Unit 14-8).

Selecting the Most Suitable Type of Projection to Describe the Part   The selection of the type of projection (orthographic, isometric, or oblique) can greatly increase the ease with which some drawings can be read and, in many cases, reduce drafting time. For example, a single-line piping drawing drawn in isometric projection simplifies an otherwise difficult drawing problem in orthographic projection (Fig. 14-2-6).

Simplified Representations in Drawings

The steady rise of simplified representation in drawings by various industries has prompted ISO to prepare an international standard that lists together the various methods of simplified representation in general use for detail and assembly drawings.

Simplified representation in drawings is not new. Simplified thread and pipe symbols are two simplified representations that have been used for many years. Promoting and using simplified representation has many advantages; simplified representation:

- Raises the design efficiency.
- Accelerates the course of design.
- Reduces the workload in the drafting office.
- Enhances legibility of the drawing, so as to meet the requirements for drawings in computer graphics and microcopying.

In addition to the following recommendations, simplified representation of features is shown throughout this text where the appropriate drawing practices are explained.

1. Avoid unnecessary views. In many cases one or two views are sufficient to explain the part fully.
2. Use simplified drawing practices, as described throughout this text, especially on threads and common features.
3. The use of the symmetry symbol means that all dimensions are symmetrical about that line.

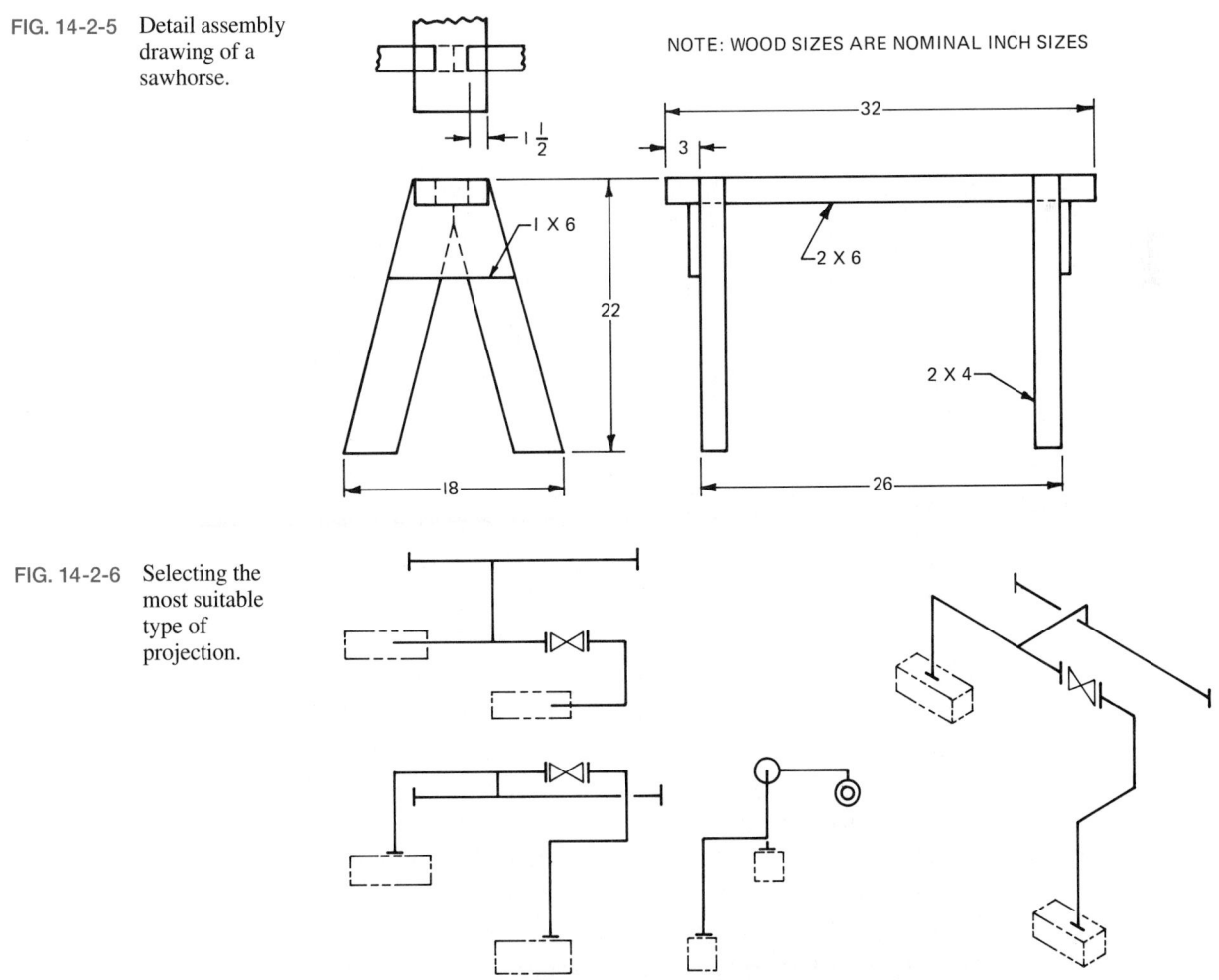

FIG. 14-2-5   Detail assembly drawing of a sawhorse.

NOTE: WOOD SIZES ARE NOMINAL INCH SIZES

FIG. 14-2-6   Selecting the most suitable type of projection.

(A) ORTHOGRAPHIC PROJECTION   (B) ISOMETRIC PROJECTION

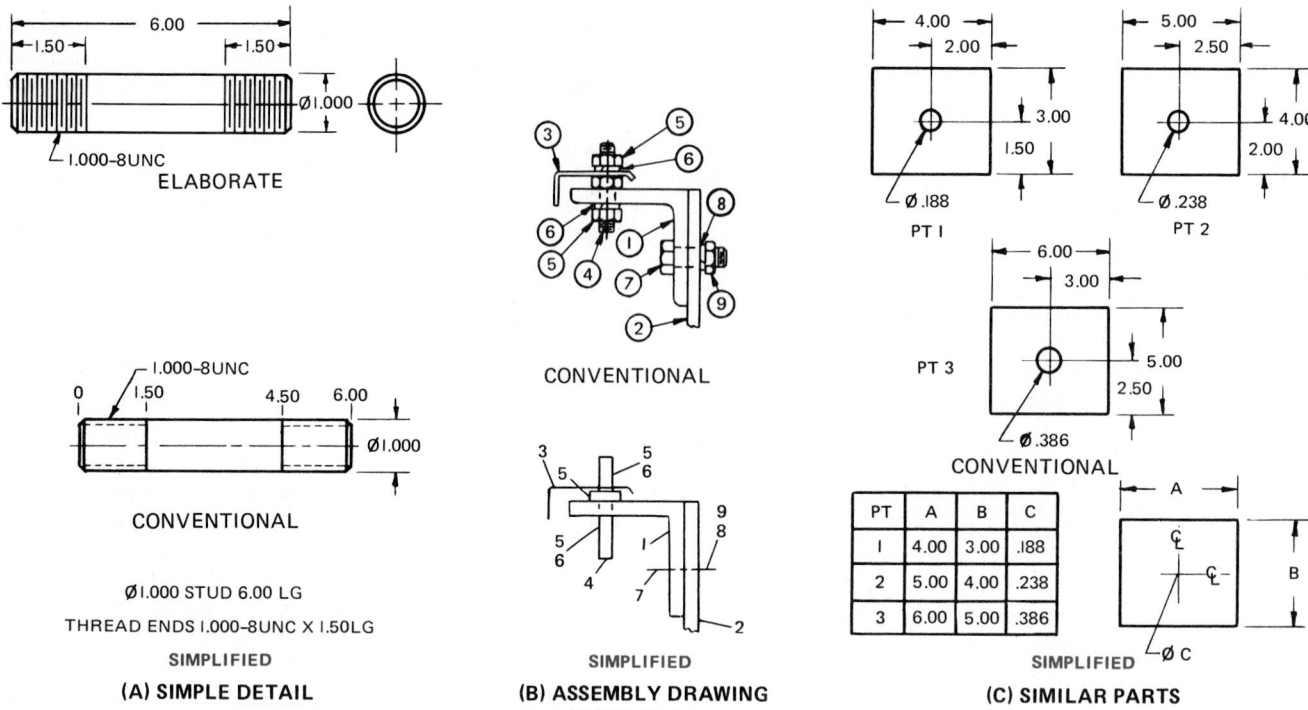

FIG. 14-2-7 Comparison between conventional and simplified representation.

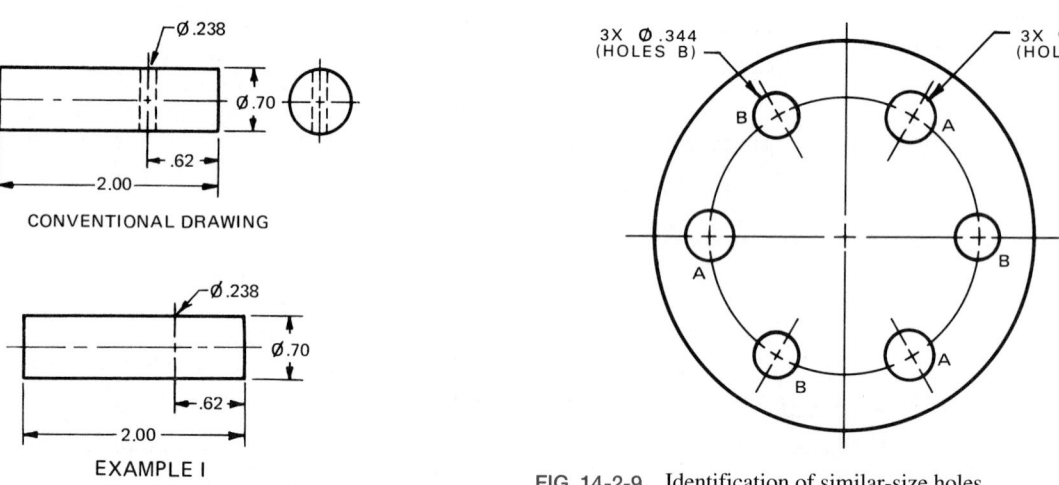

CONVENTIONAL DRAWING

EXAMPLE I

NOTE PT 2 Ø.70 X 2.00LG
Ø.238 HOLE – .62 FROM END

PART DESCRIBED BY A NOTE

EXAMPLE 2

FIG. 14-2-8 Simplified representation for detailed parts.

FIG. 14-2-9 Identification of similar-size holes.

4. Complicated parts are best described by means of a drawing. However, explanatory notes can complement the drawing, thereby eliminating views that are time-consuming to draw (Figs. 14-2-7 and 14-2-8).

5. When a number of holes of similar size are to be made in a part, there is a chance that the person producing the part may misinterpret the diameter of some of the holes. In such cases, the identification of similar-size holes should be made clear (Fig. 14-2-9).

6. A simplified assembly drawing should be used for assembly purposes only. Some means of simplification are:
   - Standard parts, such as nuts, bolts, and washers, need not be drawn.
   - Small fillets and rounds on cast parts need not be shown.
   - Phantom outlines of complicated details can often be used.

7. Use templates or symbol libraries.

8. Within limits, a small drawing is made more quickly than a large drawing when manual drafting is used.

9. Eliminate hidden lines that do not add clarification.

10. Show only partial views of symmetrical objects (Fig. 14-2-10).
11. Avoid the use of elaborate pictorial and repetitive detail when manual drafting is used.
12. Eliminate repetitive data by use of general notes or phantom lines when manual drafting is used.
13. Eliminate views where the shape or dimension can be given by description, for example, Ø, □, HEX, THK, etc.

Freehand Sketching

Many shops care little whether the drawing is freehand, whether one view is shown, or whether the drawing is to scale, as long as the proportions are approximate. They want the necessary information clearly shown. Freehand sketches and drawings made with instruments can be shown on one sheet. However, it must be clearly understood that the use of freehand sketching does not give the drafter a license to turn out sloppy work.

Savings as high as 30 percent in the preparation of working drawings have been attributed to the use of freehand sketches as opposed to instrument-produced drawings. However, freehand sketching has its limitations. It is highly effective on simple detailed parts, for small radii, such as rounds and fillets, and for small holes. In many cases the term *freehand* is not entirely correct. For instance, templates may be used to draw circles, resistors, or other common features, or a straightedge may be used to produce long lines since it is faster and more accurate than freehand sketching. But short lines are drawn more quickly freehand.

Drawing paper with nonreproducible grid lines is ideal for freehand sketching (Fig. 14-2-11). For this reason many companies have their drawing paper made with nonreproducible grid lines over the entire drawing area. Other advantages of having the grid lines on the paper are that (1) they may serve as guidelines in lettering notes and dimensions and (2) they may be used for measuring distances, thereby reducing the number of times the scale is used for measuring.

Reproduction Shortcuts

In the past few years, a number of reproduction techniques have been developed which, if properly used, can greatly reduce drawing preparation time. An understanding of available techniques and their limitations, supported by the close cooperation of a reproduction group familiar with drafting operations, can help the drafting supervisor make significant cost savings.

New Drawings Made From Existing Drawings

When a new drawing is to be made from an existing drawing with few changes, CAD makes this task easy by simply removing the unwanted material and drawing in the new. With manual drafting, reproducibles will save a great deal of preparation time. This procedure involves making a translucent or transparent print from the original drawing, removing unwanted material from this print, and adding the new information to the drawing. The main drawback to this method is that the existing drawing may not conform to the latest standard drawing practice.

Cut-and-Paste Drafting

No matter how original a design may be, a great number of part features are repetitive. With the aid of modern reproduction methods, drawings can be created by using unchanged portions of existing drawings. When manual drafting is used, transferring them from one drawing to the next is accomplished by scissors and paste-up drafting. It provides a way of using all or parts of existing drawings, notes, charts, and drawing forms to revise existing drawings and to create new drawings. Through the utilization of existing drawings much valuable drafting time is freed for creative design drafting rather than hand copying.

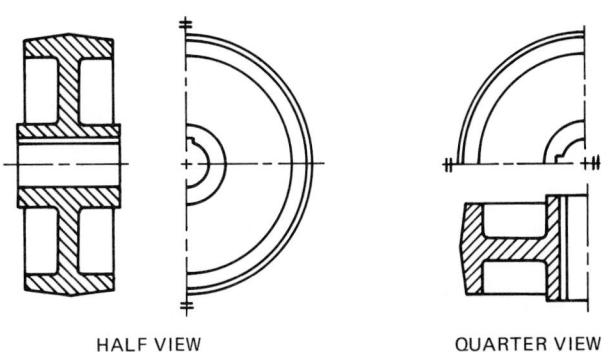

HALF VIEW   QUARTER VIEW

FIG. 14-2-10   Partial views.

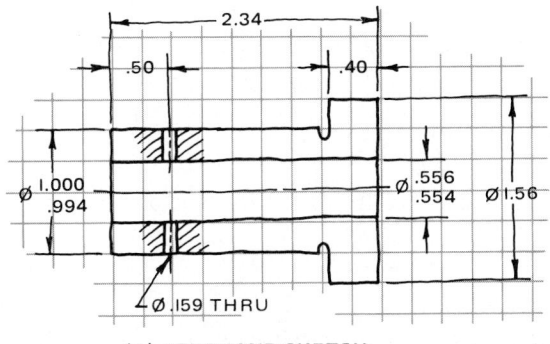

(A) FREEHAND SKETCH

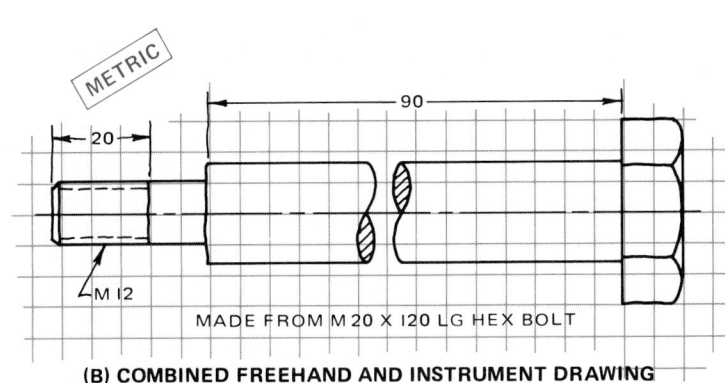

(B) COMBINED FREEHAND AND INSTRUMENT DRAWING

FIG. 14-2-11   Sketching of parts on coordinate paper.

Finished prints can be made on paper, acetate, or vellum. They can be the same size or reduced to different sizes, depending on the reproduction equipment being used.

Another important advantage of cut-and-paste drafting is that materials copied from existing drawings do not have to be minutely rechecked, as must be done with new drawings. Rechecking time is reduced (Fig. 14-2-12).

Appliqués

One of the most successful methods of reducing drawing time, when manual drafting is used, is the use of appliqués. When parts, shapes, symbols, or notes are used repeatedly, *appliqués* should be considered. These pressure-sensitive overlays may be printed on opaque, transparent, or translucent sheets with an adhesive backing.

Appliqués are available in a great variety of standard symbols or patterns (Fig. 14-2-13) and in blank (unprinted) sheets. A matte surface on the blank sheet will accept typewriter copy as well as pencil or ink lines. This material is often used for making corrections on drawings and for adding materials lists or detailed notes that can be typed faster than they can be lettered. Figure 14-2-14 shows a drawing that used many of the appliqués shown in Fig. 14-2-13.

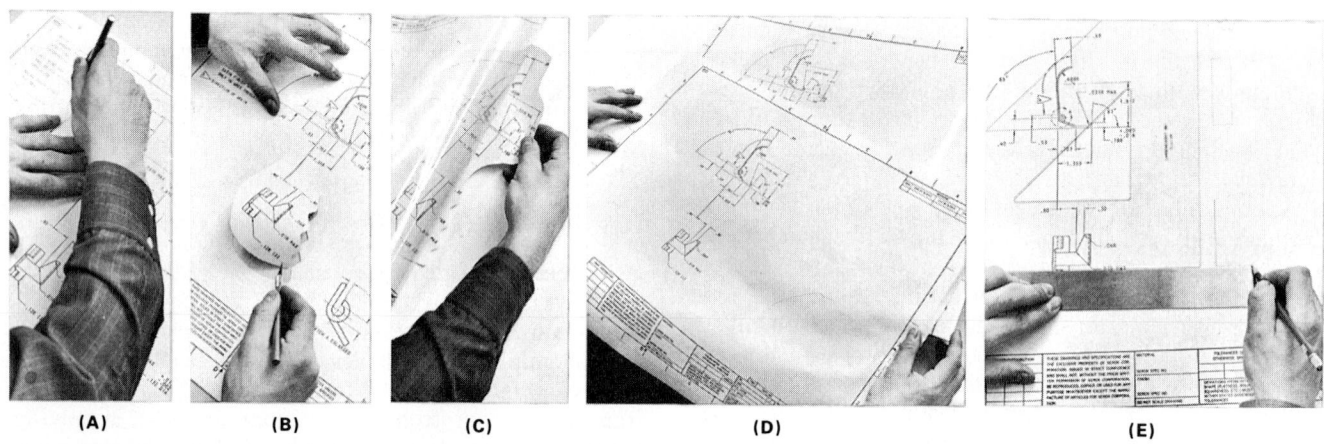

FIG. 14-2-12   Cut-and-paste drafting. *(Xerox Corp.)*

FIG. 14-2-13   A variety of shapes and sizes of appliqués. *(Graphic Standard Instruments Co.)*

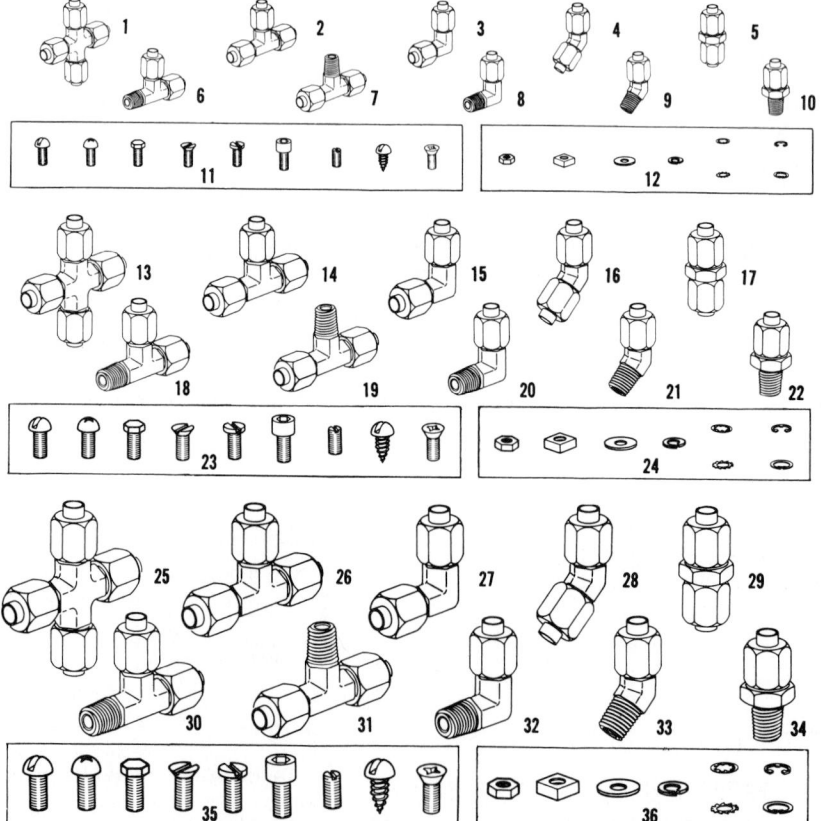

FIG. 14-2-14   Application of appliqués shown in Fig. 14-2-13. *(Graphic Standard Instruments Co.)*

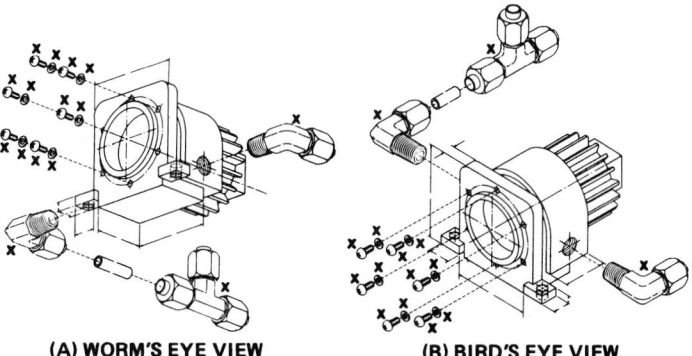

(A) WORM'S EYE VIEW   **(B) BIRD'S EYE VIEW**

Appliqués are available in two basic types: cutout and transfer. *Cutout appliqués* are applied by positioning the desired image in the correct position on the drawing, burnishing (rubbing) the image area, and cutting around it to remove the portion not wanted. The *transfer-type* pressure-sensitive appliqué works on a somewhat different principle. The carrier is removed from the translucent image sheet, and the area to be transferred is placed in position on the drawing. The image to be transferred is then rubbed over the top surface of the transfer sheet with a burnishing stick.

The combined use of cut-and-paste drafting and appliqués for new drawings is found extensively in industry, especially in the electronics and piping fields (Fig. 14-2-15).

Photodrawings

Photodrawings, that is, engineering drawings into which one photograph, or more, is incorporated, have increased in popularity because they can sometimes present a subject even more clearly than conventional drawings. Photodrawings supplement rather than replace conventional engineering drawings by eliminating much of the tedious and time-consuming effort involved when the subject is difficult to draw. They are particularly useful for assembly drawings, piping diagrams, large machine installations, switchboards, etc., provided, of course, that the subject of the drawings exists so that it may be photographed.

Photodrawings are also a comprehensive means of clearly transmitting technical information; they free the drafter from having to draw things that already exist. See Fig. 14-2-16 (pg. 390).

Photodrawings have other advantages. They are easy to make and usually take much less time to prepare than an equivalent amount of conventional drafting.

Background   Any photodrawing must begin with a photograph of an object, a part or assembly, a building, a model, or whatever else may be the subject of the drawing.

Photography   The best photographic angle usually is one that shows the subject in a flat view with as little perspective as possible. (If the situation calls for a perspective, select the angle that best describes the object.) Make certain that all the parts important to the photodrawing are in view of the camera.

ASSIGNMENTS ▰▰▰▰▰▰▰▰▰▰▰▰▰▰

See Assignments 1 through 9 for Unit 14-2 on pages 401–406.

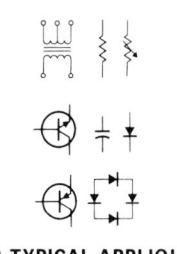

(A) TYPICAL APPLIQUÉS

(B) APPLICATION

FIG. 14-2-15   Electronics appliqués. *(Bishop Industries Corp.)*

14-3 DETAIL DRAWINGS

A *working drawing* is a drawing that supplies information and instructions for the manufacture or construction of machines or structures. Generally, working drawings may be classified into two groups: *detail drawings,* which provide the necessary information for the manufacture of the parts, and *assembly drawings,* which supply the necessary information for their assembly.

Since working drawings may be sent to other companies to make or assemble the parts, the drawings should conform with the drawing standards of that company. For this reason, most companies follow the drawing standards of their country. The drawing standards recommended by ANSI and ASME have been adopted by the majority of industries in the United States.

FIG. 14-2-16   Photodrawings.
(Eastman Kodak Co.)

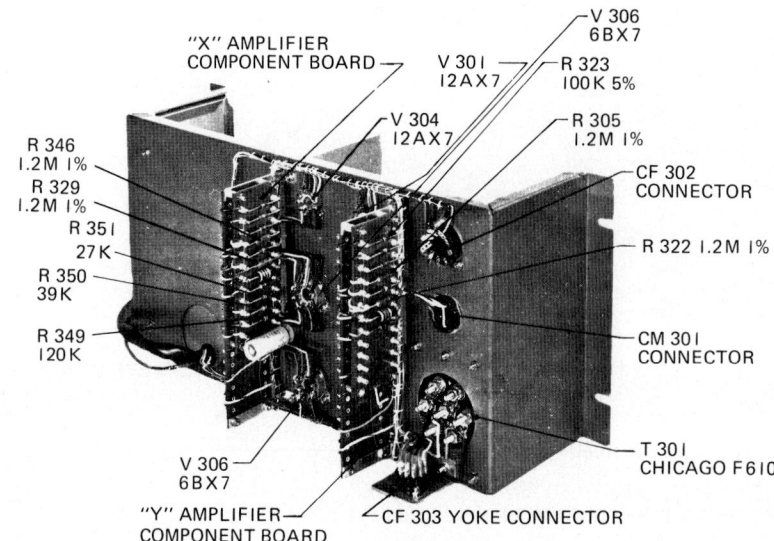

Detail Drawing Requirements

A detail drawing (Fig. 14-3-1) must supply the complete information for the construction of a part. This information may be classified under three headings: shape description, size description, and specifications.

Shape Description   This term refers to the selection and number of views to show or describe the shape of the part. The part may be shown in either pictorial or orthographic projection, the latter being used more frequently. Sectional views, auxiliary views, and enlarged detail views may be added to the drawing in order to provide a clearer image of the part.

Size Description   Dimensions that show the size and location of the shape features are then added to the drawing. The manufacturing process will influence the selection of some dimensions, such as datum features. Tolerances are then selected for each dimension.

Specifications   This term refers to general notes, material, heat treatment, finish, general tolerances, and number required. This information is located on or near the title block or strip.

Additional Drawing Information   In addition to the information pertaining to the part, a detail drawing includes additional information such as drawing number, scale, method of projection, date, name of part or parts, and the drafter's name.

The selection of paper or finished plot size is determined by the number of views selected, the number of general notes required, and the drawing scale used. If the drawing is to be microformed, the lettering size would be another factor to consider. The drawing number may carry a prefix or suffix number or letter to indicate the sheet size, such as A-571 or 4-571; the letter A indicates that it is made on an 8.50 × 11.00 in. sheet, and the number 4 indicates that the drawing is made on a 210 × 297 mm sheet.

Drawing Checklist

As an added precaution against errors occurring on a drawing, many companies have provided checklists for drafters to follow before a drawing is issued to the shop. A typical checklist may be as follows:

1. *Dimensions.* Is the part fully dimensioned, and are the dimensions clearly positioned? Is the drawing dimensioned to avoid unnecessary shop calculations?
2. *Scale.* Is the drawing to scale? Is the scale shown? What will the plot scale be?
3. *Tolerances.* Are the clearances and tolerances specified by the linear and angular dimensions and by local, general, or title block notes suitable for proper functioning? Are they realistic? Can they be liberalized?
4. *Standards.* Have standard parts, design, materials, processes, or other items been used where possible?
5. *Surface texture.* Have surface roughness values been shown where required? Are the values shown compatible with overall design requirements?
6. *Material.* Have proper material and heat treatment been specified?

Qualifications of a Detailer

The detailer should have a thorough understanding of materials, shop processes, and operations in order to properly dimension the part and call for the correct finish and material. In addition, the detailer must have a thorough knowledge of how the part functions in order to provide the correct data and tolerances for each dimension.

The detailer may be called upon to work from a complete set of instructions and drawings, or he or she may be required to make working drawings of parts that involve the design of the part. Design considerations are limited in this unit but are covered in detail in Chap. 19.

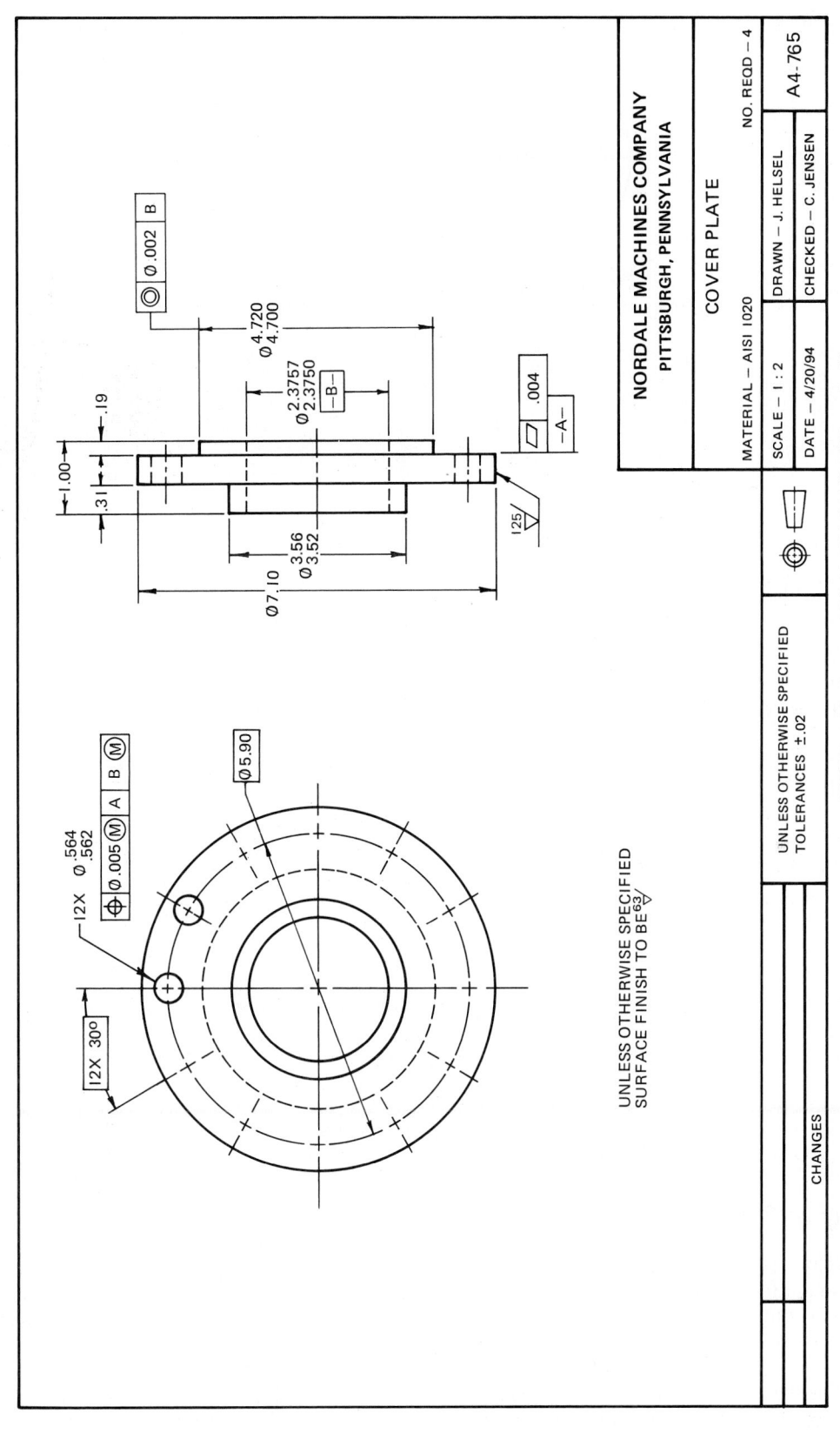

FIG. 14-3-1   A simple detail drawing.   [NOTE: This drawing, created prior to ASME Y14.5M-1994, uses ANSI Y14.5M-1982 (R 1988) standard.]

FIG. 14-3-2   Manufacturing process influences the shape of the part.

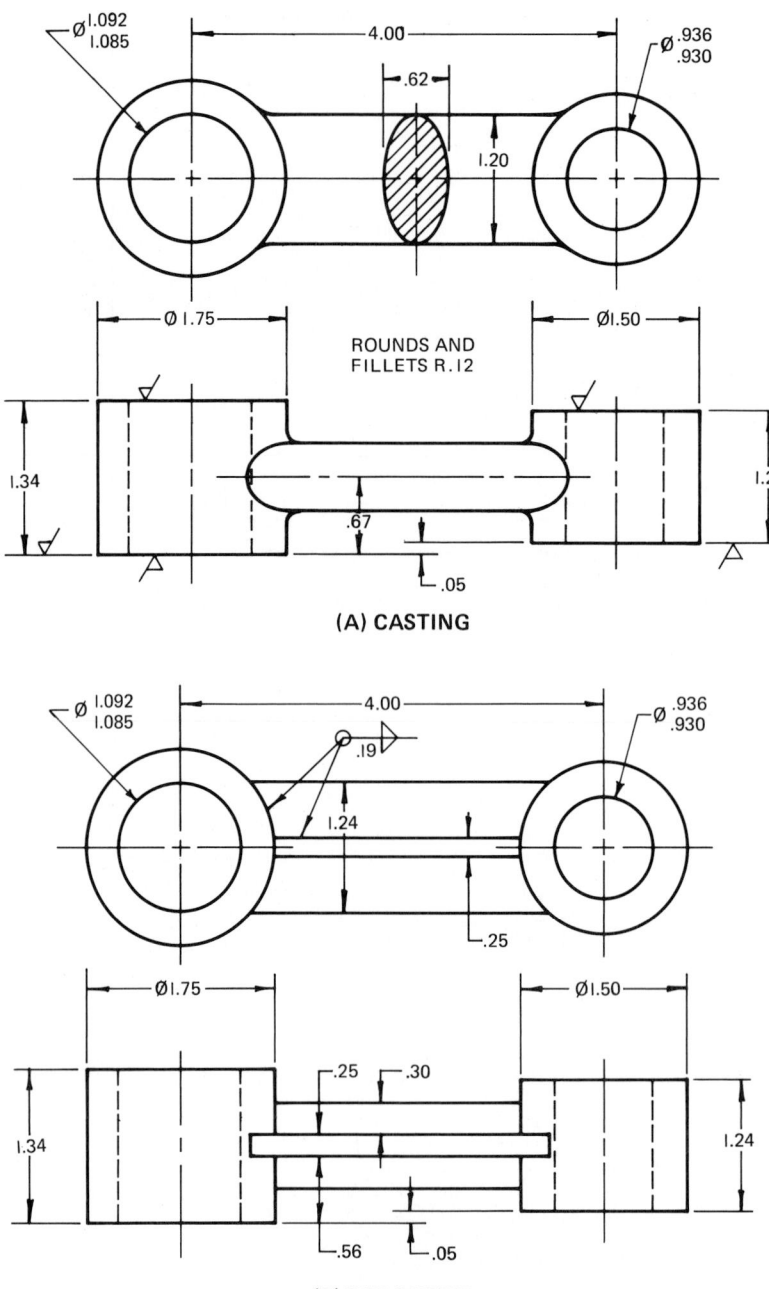

(A) CASTING

(B) WELDMENT

Manufacturing Methods

The type of manufacturing process will influence the selection of material and detailed feature of a part (Fig. 14-3-2). For example, if the part is to be cast, rounds and fillets will be added. Additional material will also be required where surfaces are to be finished.

The more common manufacturing processes are machining from standard stock; prefabrication, which includes welding, riveting, soldering, brazing, and gluing; forming from sheet stock; casting; and forging. The latter two processes can be justified only when large quantities are required for specially designed parts. All these processes are described in detail in other chapters.

Several drawings may be made for the same part, each one giving only the information necessary for a particular step in the manufacture of the part. A part that is to be produced by forging, for example, may have one drawing showing the original rough forged part and one detail of the finished forged part. (Figs. 14-3-2C and 14-3-2D).

ASSIGNMENTS

See Assignments 10 through 20 for Unit 14-3 on pages 407 through 414.

FIG. 14-3-2 Manufacturing process influences the shape of the part. (continued)

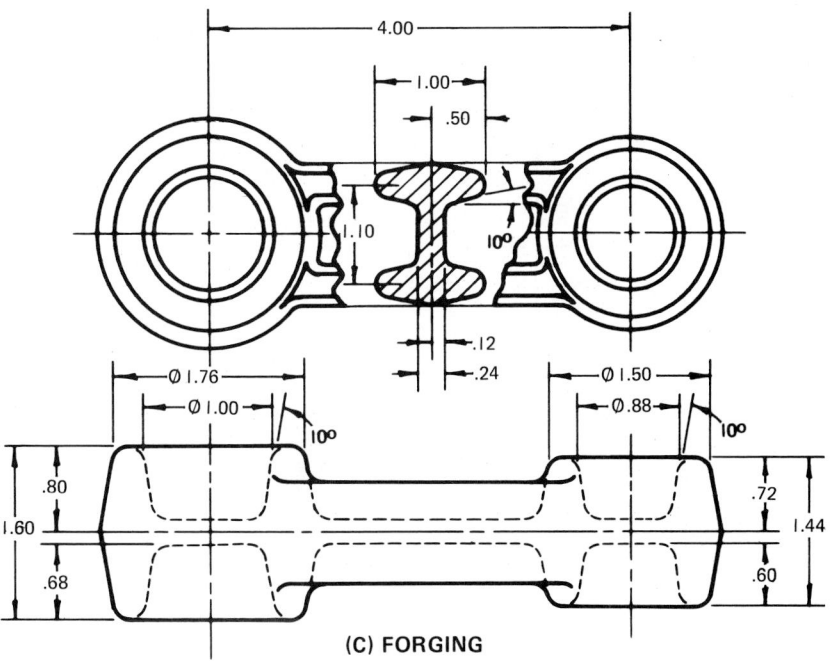

(C) FORGING

UNLESS OTHERWISE SPECIFIED FINISH IS $\frac{32}{}$
TOLERANCE ON DIMENSIONS ±.02

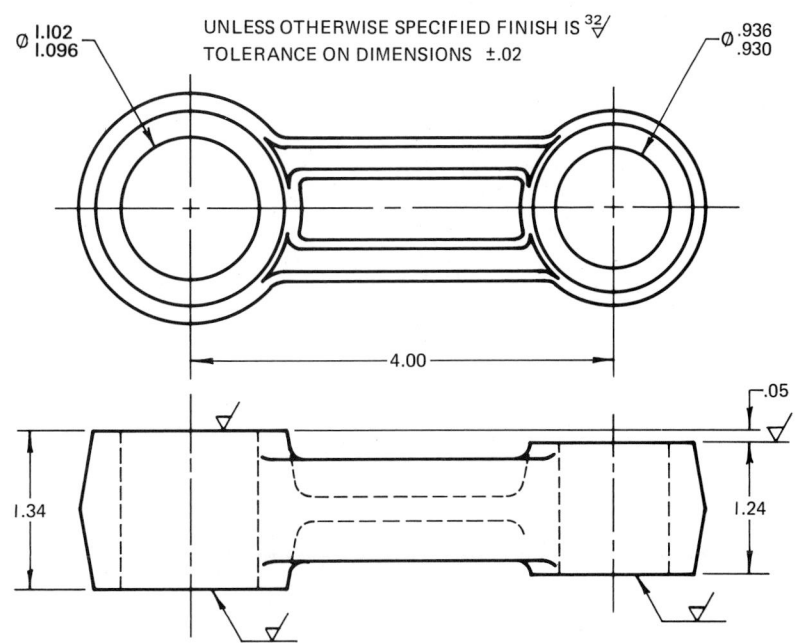

(D) MACHINING DRAWING FOR FORGED PART SHOWN IN (C)

14-4 MULTIPLE DETAIL DRAWINGS

Detail drawings may be shown on separate sheets, or they may be grouped on one or more large sheets.

Often the detailing of parts is grouped according to the department in which they are made. For example, wood, fiber, and metal parts are used in the assembly of a transformer. Three separate detail sheets—one for wood parts, one for fiber parts, and the third for the metal parts—may be drawn. These parts would be made in the different shops and sent to another area for assembly. In order to facilitate assembly, each part is given an identification part number, which is shown on the assembly drawing. A typical detail drawing showing multiple parts is illustrated in Fig. 14-4-1 (pg. 394).

If the details are few, the assembly drawing may appear on the same sheet or sheets.

ASSIGNMENTS

See Assignments 21 through 27 for Unit 14-4 on pages 415 through 421.

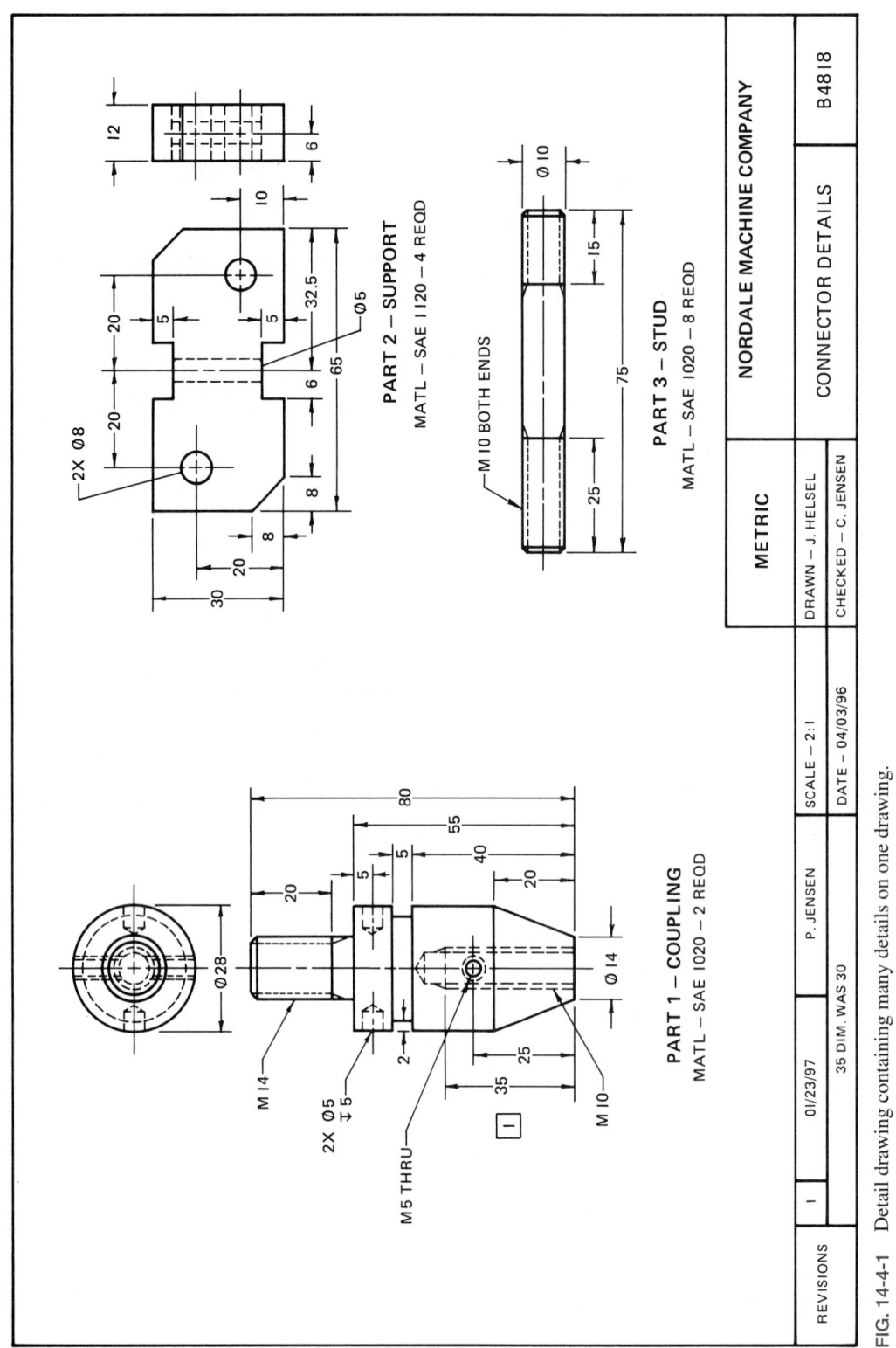

FIG. 14-4-1   Detail drawing containing many details on one drawing.

14-5 DRAWING REVISIONS

Revisions are made to an existing drawing when manufacturing methods are improved, to reduce cost, to correct errors, and to improve design. A clear record of these revisions must be registered on the drawing.

All drawings must carry a change or revision table, either down the right-hand side or across the bottom of the drawing. In addition to a description of drawing changes, provision may be made for recording a revision symbol, a zone location, an issue number, a date, and the approval of the change. Should the drawing revision cause a dimension or dimensions to be other than the scale indicated on the drawing, the dimensions that are not to scale should be indicated by the method shown in Fig. 8-1-15. Typical revision tables are shown in Fig. 14-5-1.

At times, when there are a large number of revisions to be made, it may be more economical to make a new drawing. When this is done, the words REDRAWN and REVISED should appear in the revision column of the new drawing. A new date is also shown for updating old prints.

REFERENCES AND SOURCE MATERIAL

1. ANSI Y14.5M, 1982, *Dimensions and Tolerancing*.
2. CAN/CSA-B78.2, *Dimensioning and Tolerancing of Technical Drawings*.

ASSIGNMENT

See Assignment 28 for Unit 14-5 on page 421.

14-6 ASSEMBLY DRAWINGS

All machines and mechanisms are composed of numerous parts. A drawing showing the product in its completed state is called an *assembly drawing* (Fig. 14-6-1).

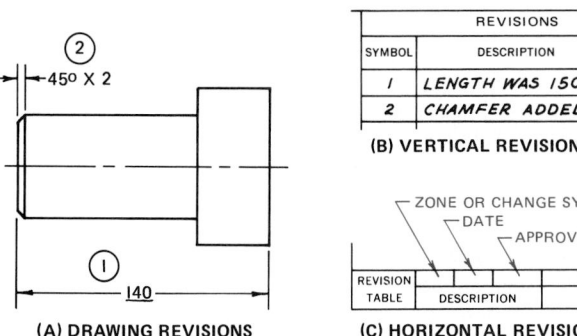

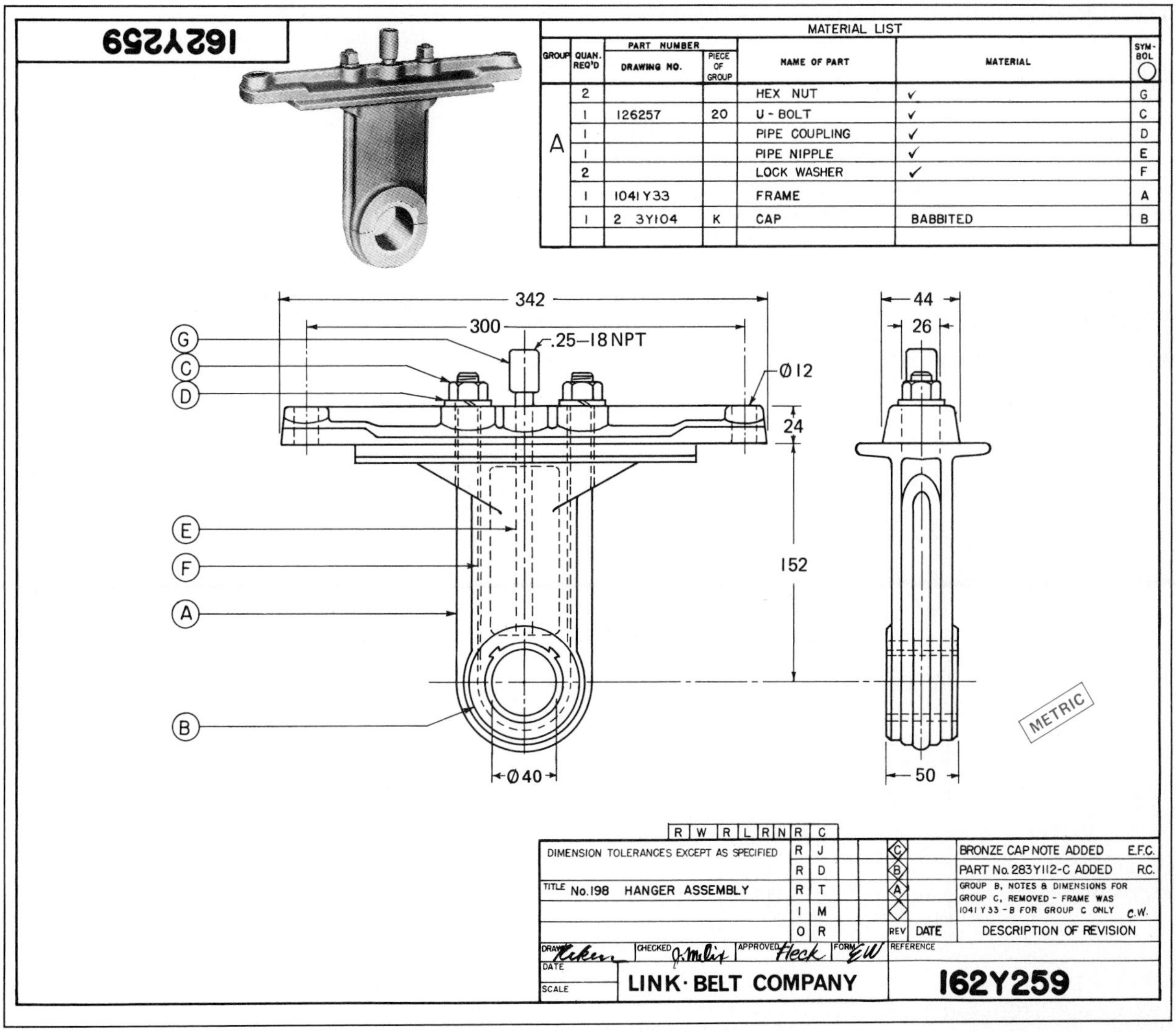

(A) DRAWING REVISIONS

REVISIONS		
SYMBOL	DESCRIPTION	DATE & APPROVAL
1	LENGTH WAS 150	J. Helsel 3-4-96
2	CHAMFER ADDED	F. Newman 2-2-97

(B) VERTICAL REVISION BLOCK

ZONE OR CHANGE SYMBOL
DATE
APPROVAL

REVISION TABLE	DESCRIPTION		

(C) HORIZONTAL REVISION BLOCK

FIG. 14-5-1   Drawing revisions.

162Y259

MATERIAL LIST						
GROUP	QUAN. REQ'D	PART NUMBER DRAWING NO.	PIECE OF GROUP	NAME OF PART	MATERIAL	SYMBOL
A	2			HEX NUT	✓	G
	1	126257	20	U-BOLT	✓	C
	1			PIPE COUPLING	✓	D
	1			PIPE NIPPLE	✓	E
	2			LOCK WASHER	✓	F
	1	1041Y33		FRAME		A
	1	2 3Y104	K	CAP	BABBITED	B

342
300
.25-18 NPT
Ø12
24
152
44
26
50
Ø40
METRIC

R	W	R	L	R	N	R	C				
		DIMENSION TOLERANCES EXCEPT AS SPECIFIED		R	J		◇		BRONZE CAP NOTE ADDED		E.F.C.
				R	D		◇B		PART No. 283Y112-C ADDED		R.C.
		TITLE No.198 HANGER ASSEMBLY		R	T		◇A		GROUP B, NOTES & DIMENSIONS FOR GROUP C, REMOVED - FRAME WAS 1041Y33 -B FOR GROUP C ONLY		C.W.
				I	M						
				O	R				REV	DATE	DESCRIPTION OF REVISION

DRAWN *Riken*   CHECKED *J. Miles*   APPROVED *Heck*   FORM *EW*   REFERENCE
DATE
SCALE

LINK·BELT COMPANY   **162Y259**

FIG. 14-6-1   Assembly drawing. *(Link-Belt Co.)*

Assembly drawings vary greatly in the amount and type of information they give, depending on the nature of the machine or mechanism they depict. The primary functions of the assembly drawing are to show the product in its completed shape, to indicate the relationship of its various components, and to designate these components by a part or detail number. Other information that might be given includes overall dimensions, capacity dimensions, relationship dimensions between parts (necessary information for assembly), operating instructions, and data on design characteristics.

Design Assembly Drawings

When a machine is designed, an assembly drawing or a design layout is first drawn to clearly visualize the performance, shape, and clearances of the various parts. From this assembly drawing, the detail drawings are made and each part is given a part number. To assist in the assembling of the machine, part numbers of the various details are placed on the assembly drawing. The part number is attached to the corresponding part with a leader, as illustrated in Fig. 14-6-2. It is important that the detail drawings not use identical numbering schemes when several item lists are used. Circling the part number is optional.

Installation Assembly Drawings

This type of assembly drawing is used when many unskilled people are employed to mass-assemble parts. Since these people are not normally trained to read technical drawings, simplified pictorial assembly drawings similar to the one shown in Fig. 14-6-3 are used.

Assembly Drawings for Catalogs

Special assembly drawings are prepared for company catalogs. These assembly drawings show only pertinent details and dimensions that would interest the potential buyer. Often one drawing, having letter dimensions accompanied by a chart, is used to cover a range of sizes, such as the pillow block shown in Fig. 14-6-4B.

Item List

An *item list,* often referred to as a *bill of material* (BOM), is an itemized list of all the components shown on an assembly drawing or a detail drawing (Fig. 14-6-5, pg. 398). Often an item list is placed on a separate sheet for ease of handling and duplicating. Since the item list is used by the purchasing department to order the necessary material for the design, the item list should show the raw material size rather than the finished size of the part.

For castings a pattern number should appear in the size column in lieu of the physical size of the part.

Standard components, which are purchased rather than fabricated, such as bolts, nuts, and bearings, should have a part number and appear on the item list. Information in the descriptive column should be sufficient for the purchasing agent to order these parts.

Item lists placed on the bottom of the drawing should read from bottom to top, while item lists placed on the top of the drawing should read from top to bottom. This practice allows additions to be made at a later date.

ASSIGNMENTS

See Assignments 29 through 40 for Unit 14-6 on pages 421 through 434.

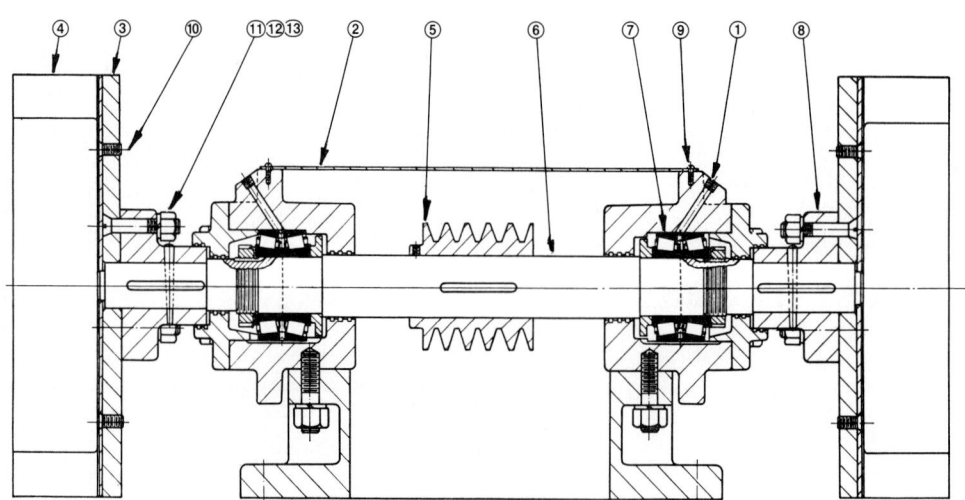

FLOOR STAND GRINDING MACHINE

FIG. 14-6-2   Design assembly drawing. *(Timken Roller Bearing Co.)*

FIG. 14-6-3   Installation assembly drawing.

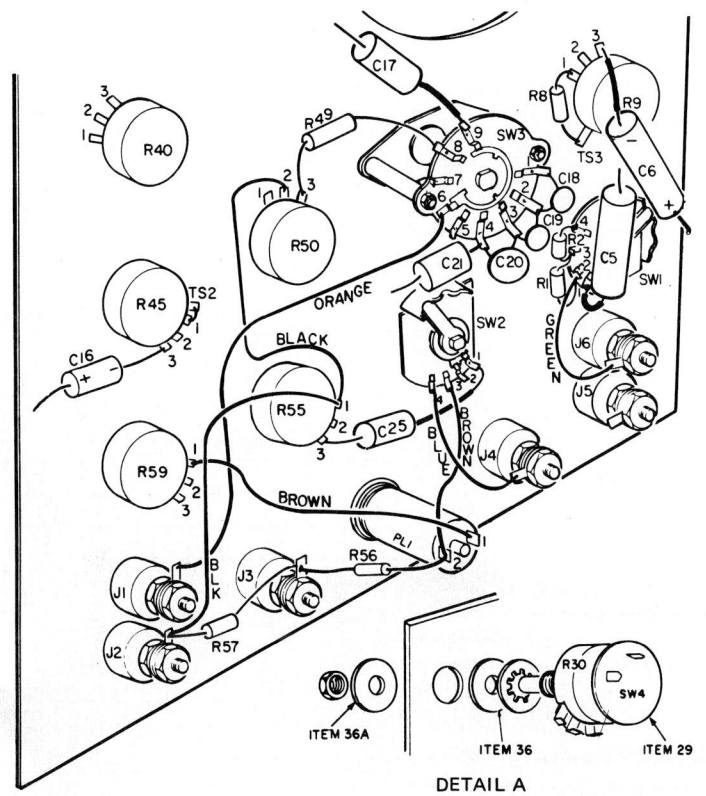

DETAIL A

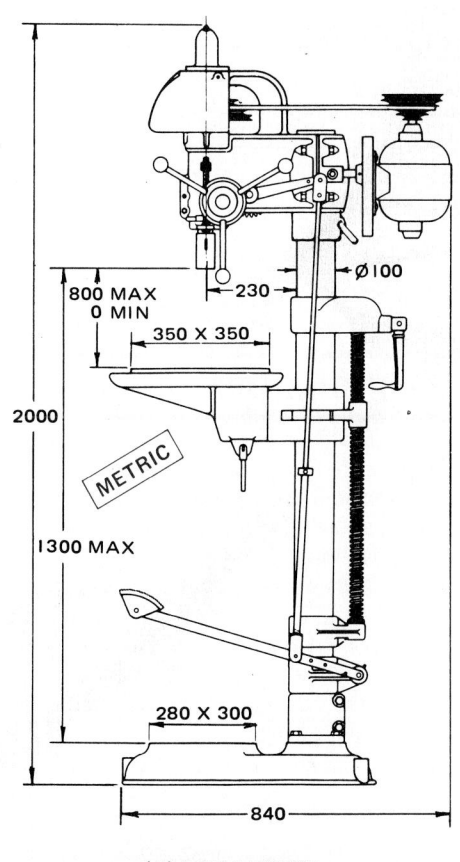

(A) DRILL PRESS

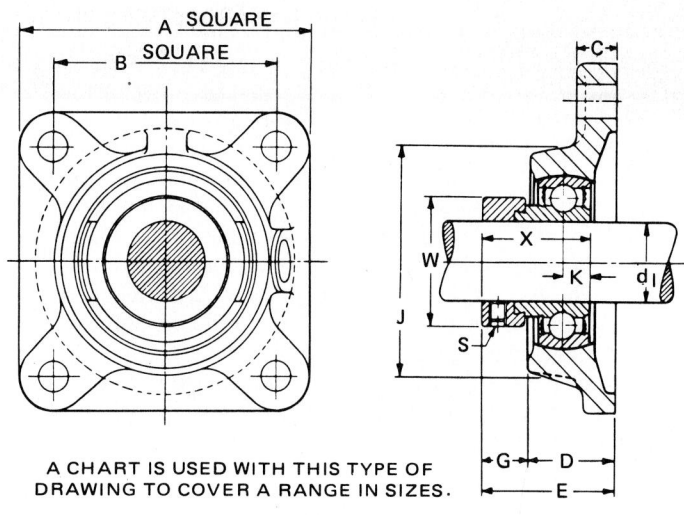

A CHART IS USED WITH THIS TYPE OF DRAWING TO COVER A RANGE IN SIZES.

(B) PILLOW BLOCK

FIG. 14-6-4   Assembly drawings used in catalogs.

FIG. 14-6-5   Item list.

QTY	ITEM	MATL	DESCRIPTION	PT NO.
I	BASE	GI	PATTERN # A3I54	I
I	CAP	GI	PATTERN # B7I56	2
I	SUPPORT	AISI-I2I2	.38 X 2.00 X 4.38	3
I	BRACE	AISI-I2I2	.25 X I.00 X 2.00	4
I	COVER	AISI-I035	.I345 (#I0 GA USS) X 6.00 X 7.50	5
I	SHAFT	AISI-I2I2	ØI.00 X 6.50	6
2	BEARINGS	SKF	RADIAL BALL # 6200Z	7
2	RETAINING CLIP	TRUARC	N5000-725	8
I	KEY	STL	WOODRUFF # 608	9
I	SETSCREW	CUP POINT	HEX SOCKET .25UNC X I.50	I0
4	BOLT—HEX HD—REG	SEMI-FIN	.38UNC X I.50LG	II
4	NUT—REG HEX	STL	.38UNC	I2
4	LOCK WASHER—SPRING	STL	.38 - MED	I3
				I4

NOTE: PARTS 7 TO I3 ARE PURCHASED ITEMS.

(A) TYPICAL ITEM LIST.

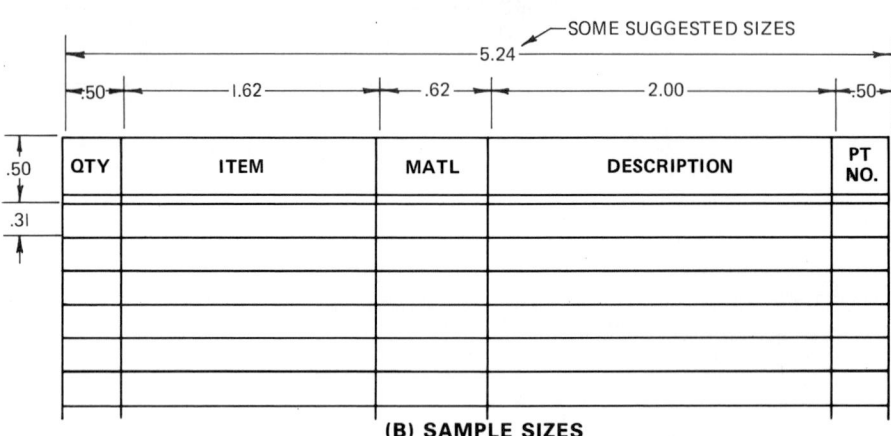

(B) SAMPLE SIZES

14-7 EXPLODED ASSEMBLY DRAWINGS

In many instances parts must be identified or assembled by persons unskilled in the reading of engineering drawings. Examples are found in the appliance-repair industry, which relies on assembly drawings for repair work and for reordering parts. Exploded assembly drawings, like that shown in Fig. 14-7-1, are used extensively in these cases, for they are easier to read. This type of assembly drawing is also used frequently by companies that manufacture do-it-yourself assembly kits, such as model-making kits.

For this type of drawing, the parts are aligned in position. Frequently, shading techniques are used to make the drawings appear more realistic.

14-8 DETAIL ASSEMBLY DRAWINGS

Often these are made for fairly simple objects, such as pieces of furniture, when the parts are few in number and are not intricate in shape. All the dimensions and information necessary for the construction of each part and for the assembly of the parts are given directly on the assembly drawing. Separate views of specific parts, in enlargements showing the fitting together of parts, may also be drawn in addition to the regular assembly drawing. Note that in Fig. 14-8-1 (pg. 400) the enlarged views are drawn in picture form, not as regular orthographic views. This method is peculiar to the cabinet-making trade and is not normally used in mechanical drawing.

ASSIGNMENTS

See Assignments 41 and 42 for Unit 14-7 on pages 435–436.

ASSIGNMENT

See Assignment 43 for Unit 14-8 on pages 436–437.

FIG. 14-7-1   Exploded assembly drawings.

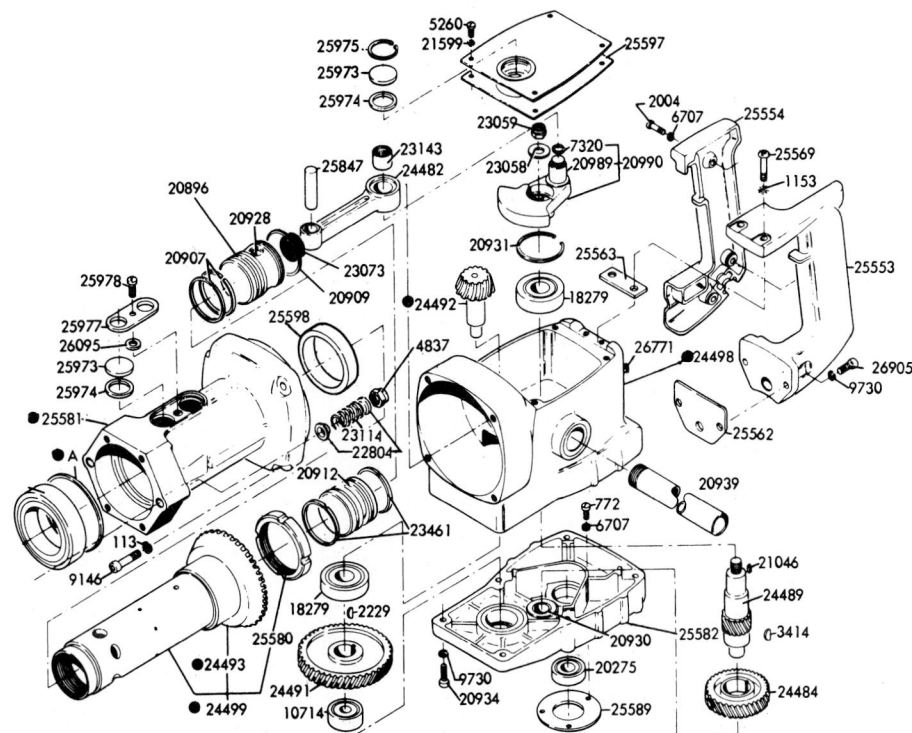

(A) PICTORIAL EXPLODED ASSEMBLY

NOTE:
FRICTION PLATE USES 3 CLUTCH DISC UNITS WITH 4 CLUTCH DISCS ON FRICTION PLATE.

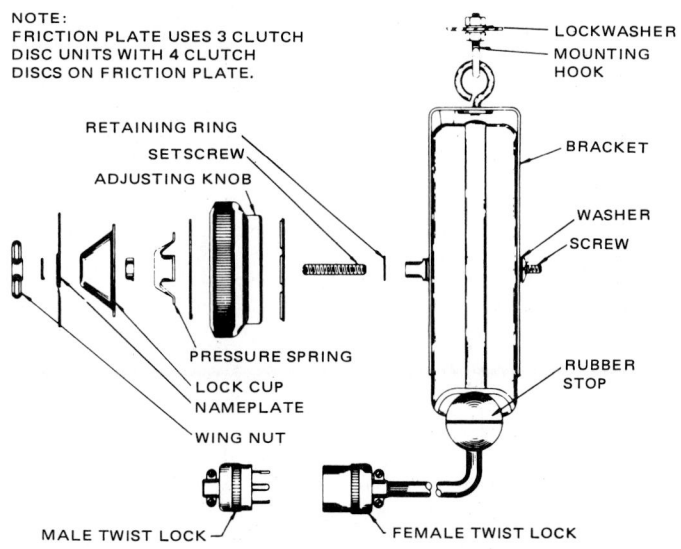

(B) ORTHOGRAPHIC EXPLODED ASSEMBLY

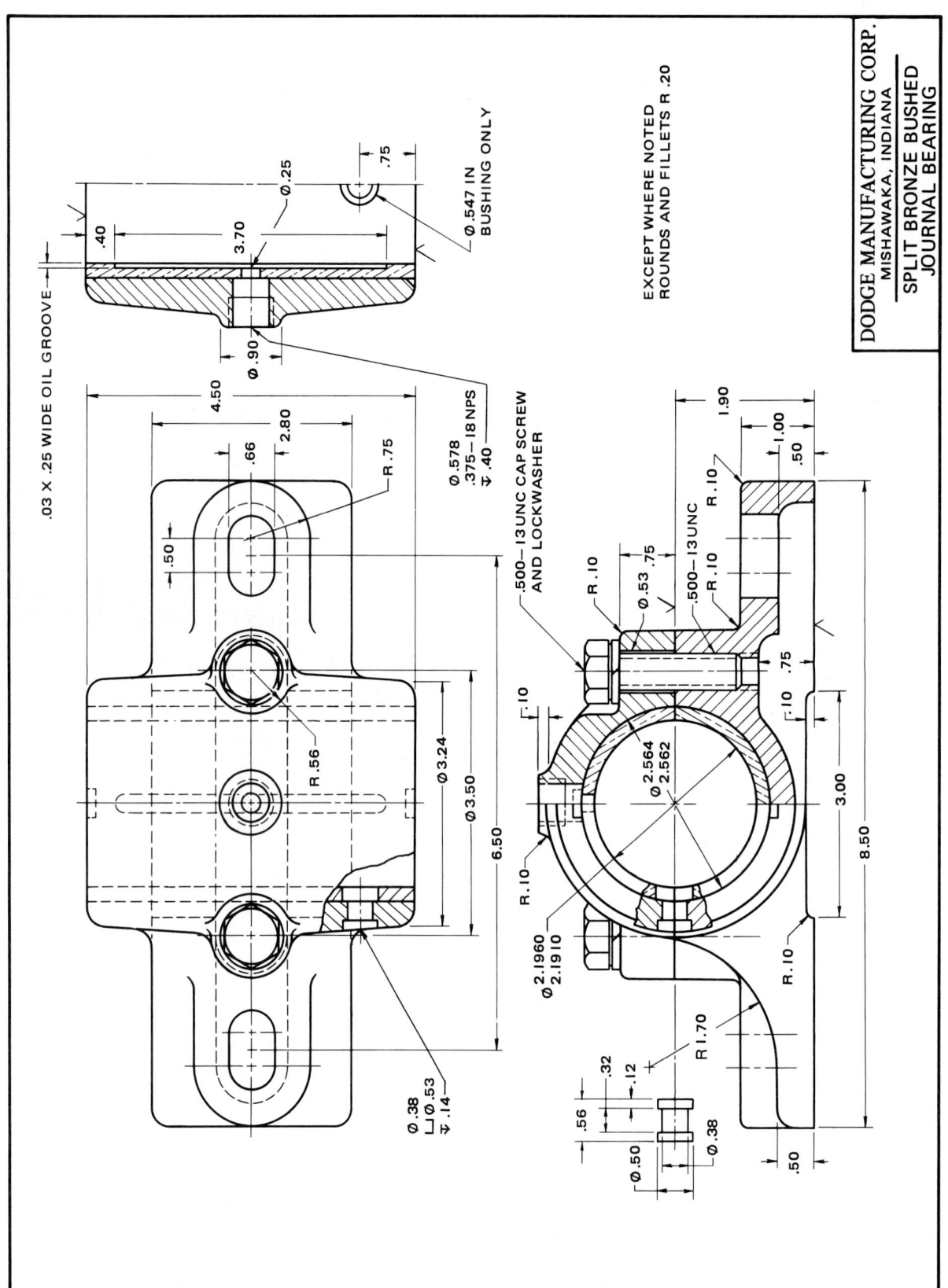

FIG. 14-8-1   Detail assembly drawing.

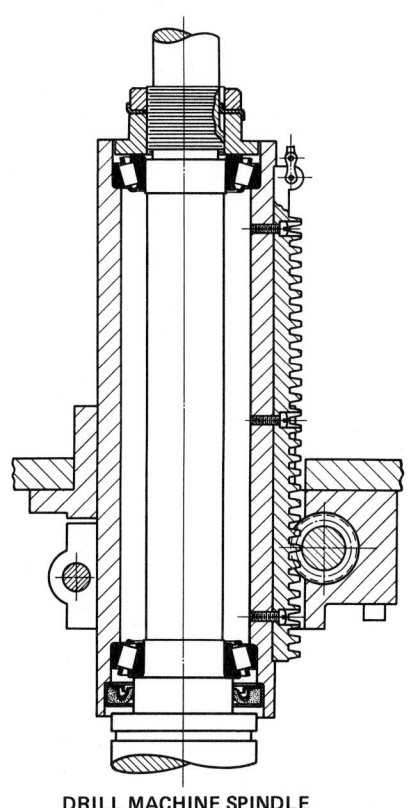

DRILL MACHINE SPINDLE

FIG. 14-9-1   Subassembly drawing. *(Timken Roller Bearing Co.)*

14-9 SUBASSEMBLY DRAWINGS

Many completely assembled items, such as a car and a television set, are assembled with many preassembled components as well as individual parts. These preassembled units are referred to as *subassemblies* (Fig. 14-9-1). The assembly drawings of a transmission for an automobile and the transformer for a television set are typical examples of subassembly drawings.

Subassemblies are designed to simplify final assembly as well as permit the item to be either assembled in a more suitable area or purchased from an outside source. This type of drawing shows only those dimensions that would be required for the completed assembly. Examples are size of the mounting holes and their location, shaft locations, and overall sizes. This type of drawing is found frequently in catalogs. The pillow block shown in Fig. 14-6-4B is a typical subassembly drawing.

ASSIGNMENTS

See Assignments 44 through 47 for Unit 14-9 on pages 438 through 440.

ASSIGNMENTS FOR CHAPTER 14

Note: Convert to symbolic and limit dimensioning, wherever practical, for all drawing assignments in this chapter.

ASSIGNMENTS FOR UNIT 14-2, FUNCTIONAL DRAFTING

1. After the number of drawings made over the last 6 months was reviewed, it was discovered that a great number of cable straps, shown in Fig. 14-2-A (pg. 402), were being made that were similar in design. Prepare a standard tabulated drawing similar to Fig. 14-1-1, reducing the number of standard parts to 4. Scale 1:1.
2. The rod guide shown in Fig. 14-2-B (pg. 402) is to be drawn twice and the drawing time for each recorded. First, on plain paper make an isometric drawing of the part, using a compass to draw the circles and arcs. Next, repeat the drawing, only this time use isometric grid paper and a template for drawing the circles and arcs. From the drawing times recorded, state in percentage the time saved by the use of grid paper and templates. Scale 1:1. Do not dimension.
3. The book rack shown in Fig. 14-2-C (pg. 402) is to be drawn twice and the drawing time for each drawing recorded. The first drawing is to show a three-view orthographic projection of the book rack assembly showing only those dimensions and instructions pertinent to the assembly. On the same drawing prepare detail drawings for the parts required. Scale to suit. On the second drawing make an orthographic detailed assembly drawing of the book rack showing the dimensions and instructions necessary to completely make and assemble the parts. Scale to suit. From the drawing times recorded, state a percentage of time saved by the use of detailed assembly drawings.

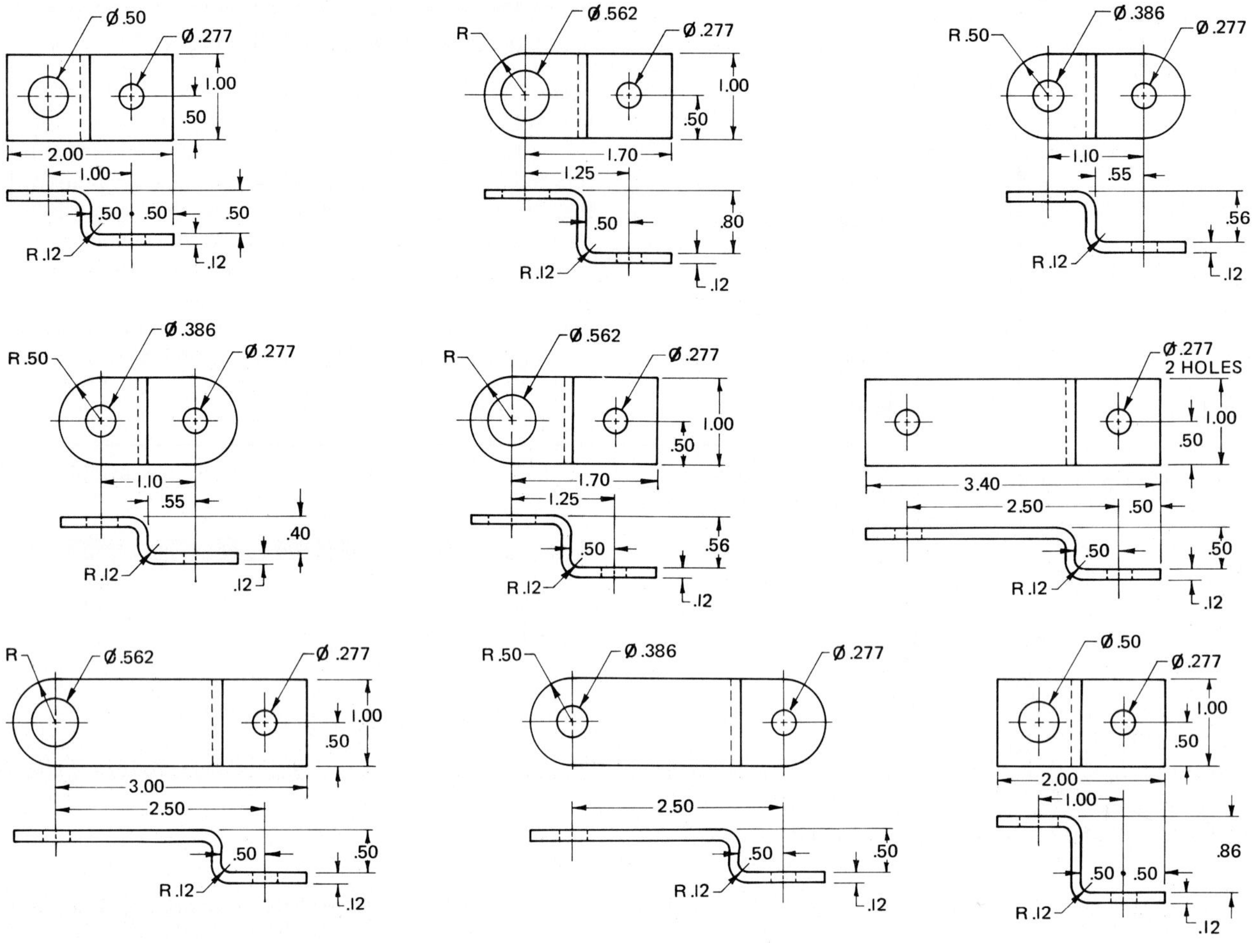

FIG. 14-2-A   Cable straps.

FIG. 14-2-B   Rod guide.

FIG. 14-2-C   Book rack.

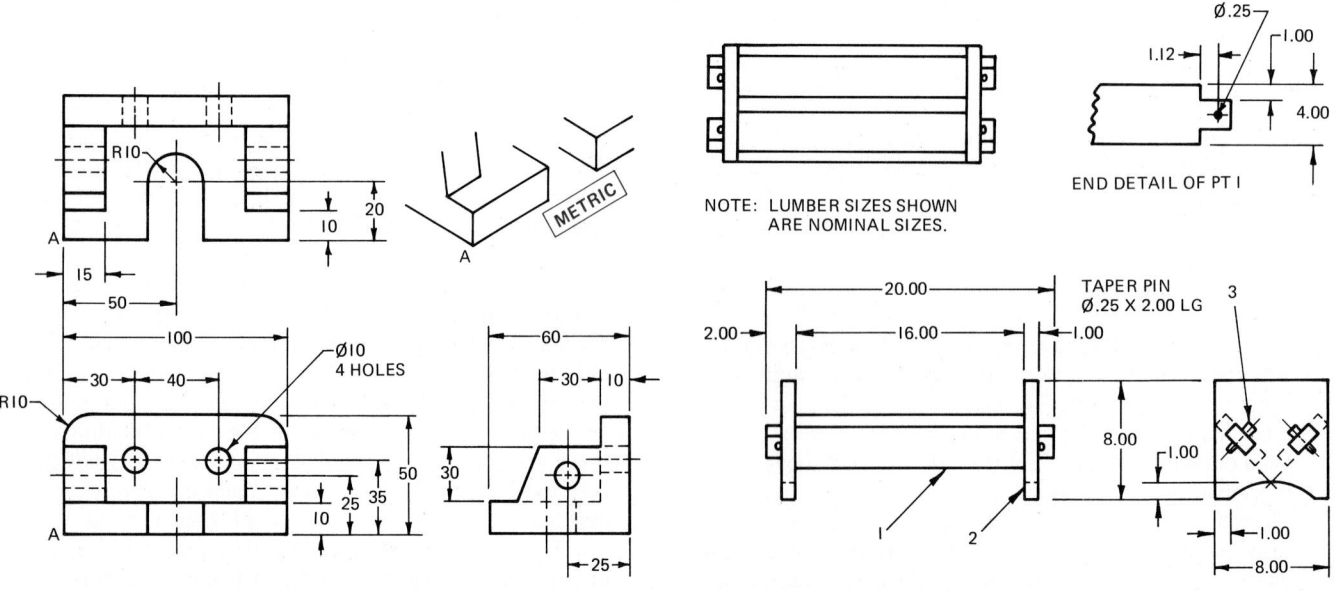

NOTE: LUMBER SIZES SHOWN
ARE NOMINAL SIZES.

END DETAIL OF PT 1

TAPER PIN
Ø.25 X 2.00 LG

4. Redraw the part shown in Fig. 14-2-D using arrowless dimensioning and simplified drawing practices. Scale 1:12. Use the bottom and left-hand edge for the datum surfaces.

5. Redraw the two parts shown in Figs. 14-2-E and 14-2-F using partial views and the symmetry symbol. Scale to suit.

FIG. 14-2-D   Cover plate.

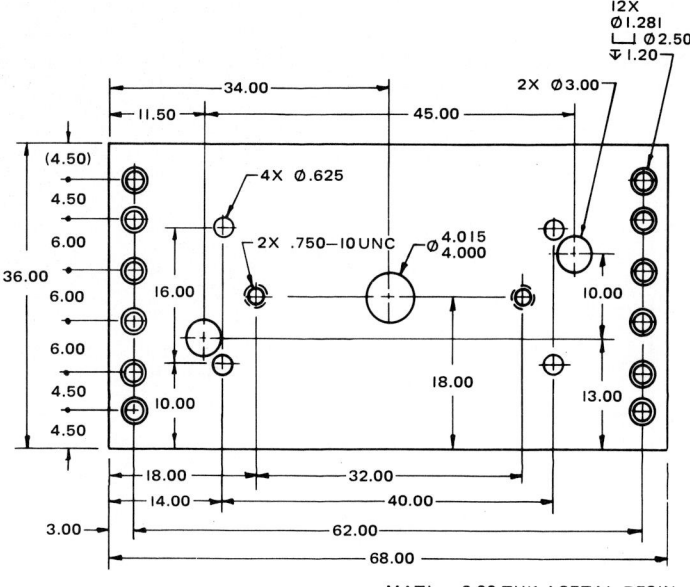

MATL — 3.00 THK ACETAL RESIN

FIG. 14-2-E   Tube support.

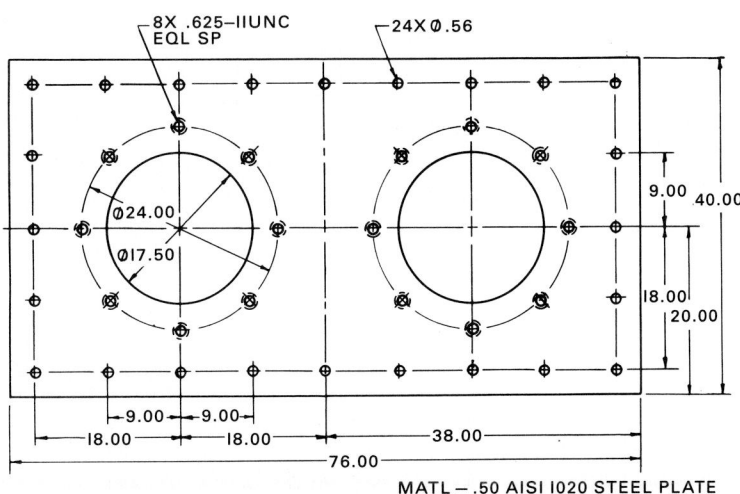

MATL — .50 AISI 1020 STEEL PLATE

FIG. 14-2-F   Gasket.

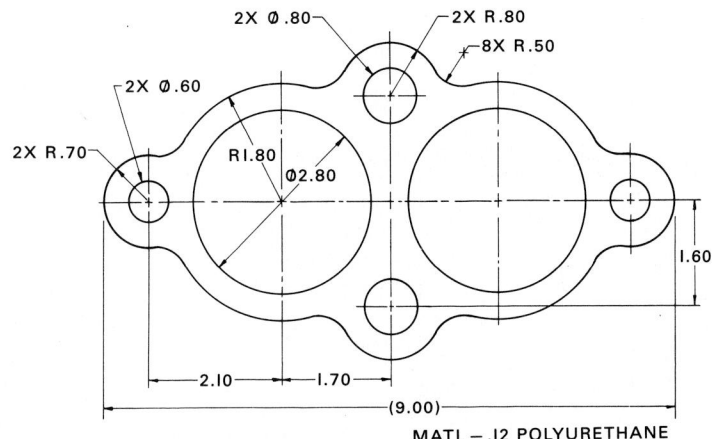

MATL — .12 POLYURETHANE

6. Make simplified drawings of the parts shown in Figs. 14-2-G and 14-2-H. Refer to Fig. 14-2-7. Scale to suit.

7. An exploded orthographic assembly drawing of the wheel-puller shown in Fig. 14-2-J is urgently required. Time does not permit one drafter to do the entire drawing; thus, three drafters will be required to draw the parts. Scale 1:2. Draw all the parts. When all the parts have been completed, make prints of them. Cut out the parts and assemble them in the exploded position on a B (A3) size sheet. Make a suitable print of the exploded assembly on a copying machine.

8. Draw the electronics diagram shown in Fig. 14-2-K using the CAD library (you may have to make your own) or use appliqués and templates if manual drafting is to be used. If appliqués of the electronic components are not available, make your own by photostating Fig. 14-2-15, then cut them out and glue them to your drawing. There is no scale.

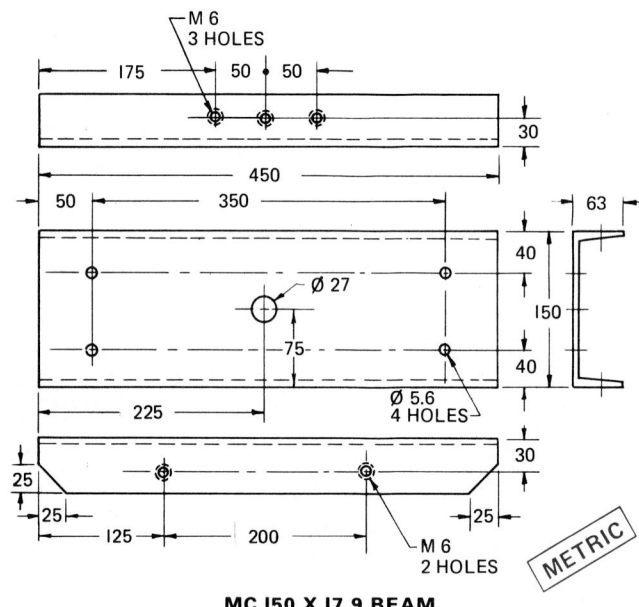

MC 150 X 17.9 BEAM

FIG. 14-2-G   Clamp.

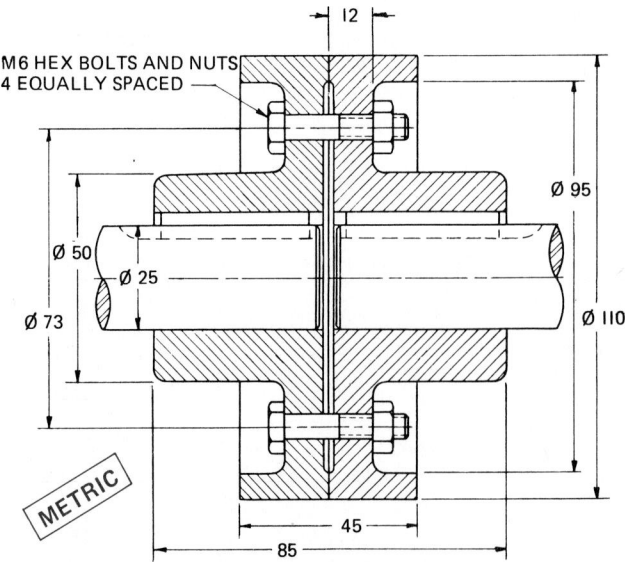

FIG. 14-2-H   Flanged coupling.

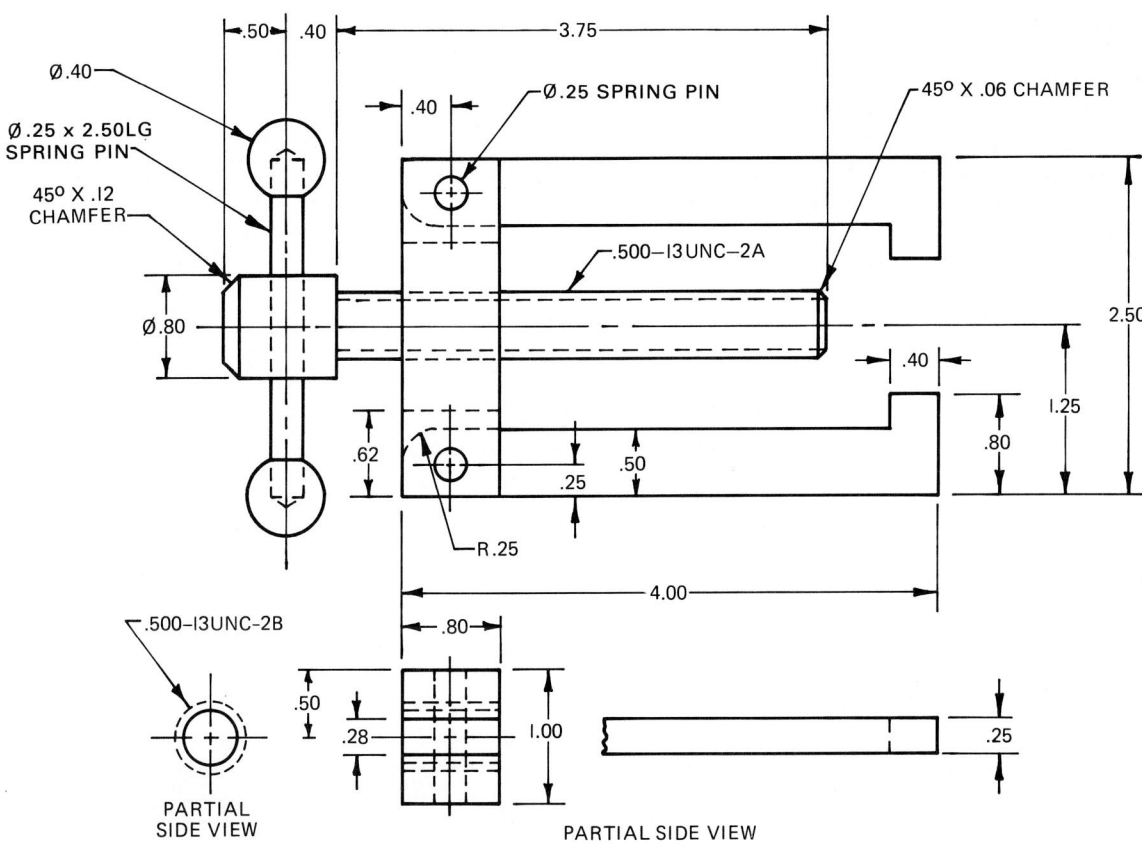

FIG. 14-2-J   Wheel-puller.

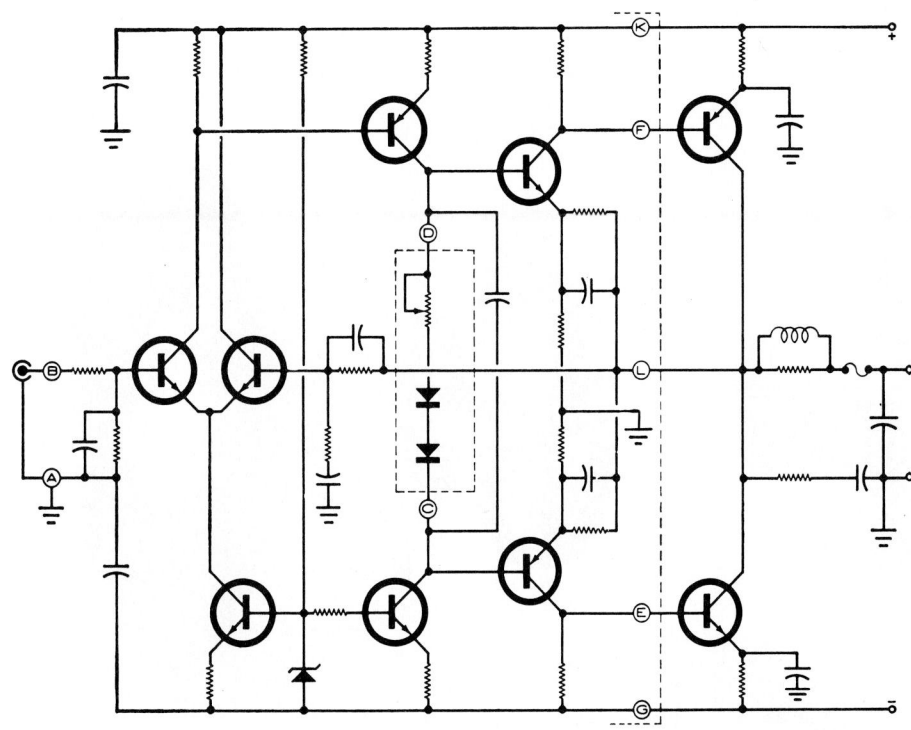

FIG. 14-2-K   Electronics diagram.

9. Make a photostat of the exploded sprocket assembly shown in Fig. 14-2-L. The photostat, which is to serve as a photodrawing, is to replace the two-view drawing. Make a new chart listing metric sizes to replace the existing chart. Leaders, dimensions, and dimension lines are to be added to the photodrawing.

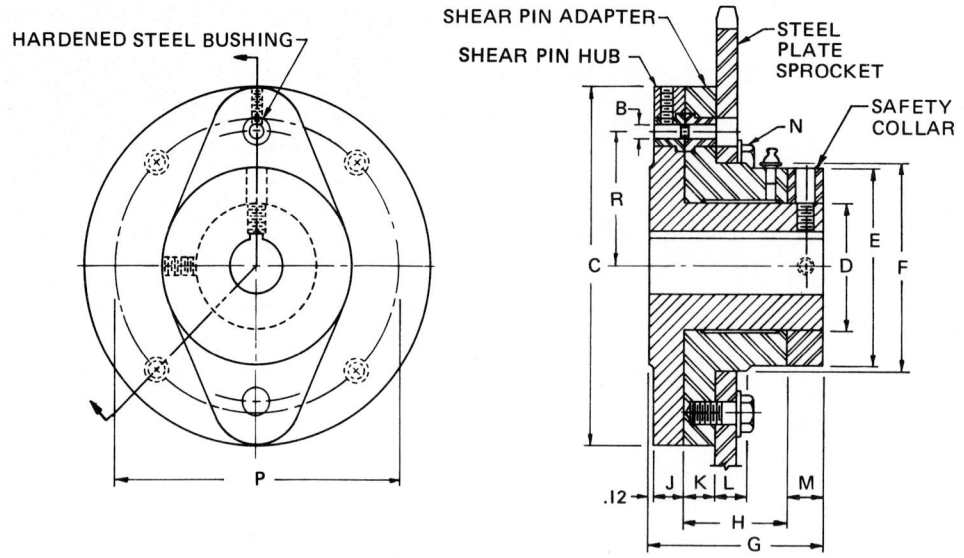

Hub Bore Range	Shear Pin Assembly Number	Shear Pin		Diameters				Length Thru			Hub Flange Thickness	Adapt Flange Thickness	Sprocket Seat Width	Bolts	
		Radius	Pin Dia.	Flange	Shear Pin Hub	Adapt Hub & Collar	Sprocket Seat	Shear Pin Hub	Adapt	Collar				Number & Size	Bolt Circle
		R	B	C	D	E	F	G	H	M	J	K	L	N	P
1.00 & under	SP-17	1.80	.25	5.25	1.75	2.50	2.62	2.44	1.38	.38	.56	.56	.44	4-.38	4.00
1.06-1.25	SP-18	2.18	.25	6.00	2.25	3.25	3.38	2.94	1.75	.50	.56	.56	.56	4-.38	4.75
1.30-1.50	SP-19	2.56	.30	6.75	2.75	4.00	4.12	3.56	2.12	.62	.68	.68	.68	4-.50	5.50
1.56-1.75	SP-20	3.00	.38	7.75	3.25	4.75	4.88	4.18	2.50	.75	.80	.80	.68	4-.50	6.25
1.80-2.00	SP-21	3.30	.45	8.75	3.75	5.25	5.38	4.80	2.88	.88	.94	.94	.94	4-.62	7.00
2.06-2.25	SP-22	3.80	.50	9.75	4.25	6.25	6.38	5.18	3.00	1.00	1.06	1.06	1.18	4-.62	8.00
2.30-2.50	SP-23	4.00	.50	10.00	4.50	6.50	6.62	5.68	3.50	1.00	1.06	1.06	1.38	4-.62	8.25
2.56-2.75	SP-24	4.40	.55	11.50	5.00	7.00	7.12	6.30	3.88	1.12	1.18	1.18	1.38	4-.62	9.25
2.80-3.00	SP-25	4.90	.62	12.50	5.50	8.00	8.12	6.94	4.25	1.25	1.30	1.30	1.38	6-.62	10.25

FIG. 14-2-L   Sprocket assembly.

ASSIGNMENTS FOR UNIT 14-3, DETAIL DRAWINGS

10. Make a detail drawing of one of the parts shown in Figs. 14-3-A through 14-3-G. Select the scale and the number of views required.

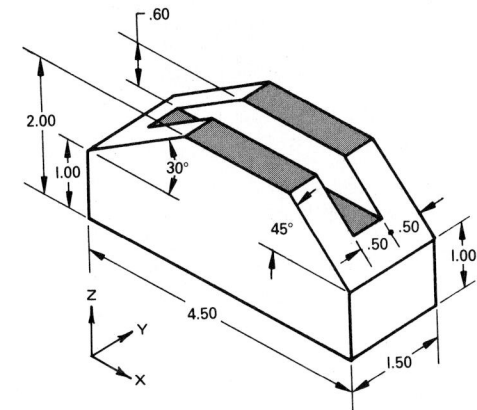

FIG. 14-3-A   Guide block.

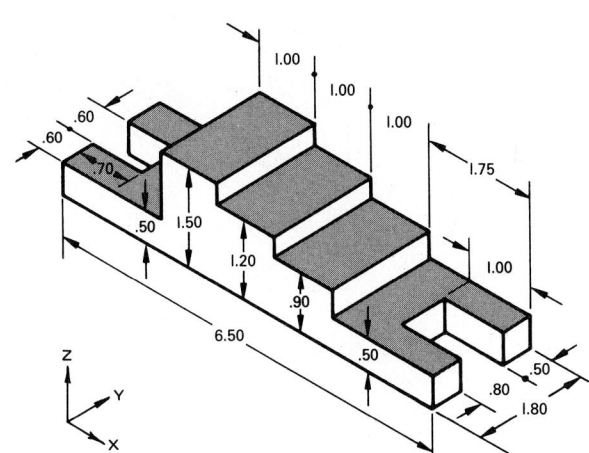

FIG. 14-3-B   Step block.

METRIC

FIG. 14-3-D   Angle block.

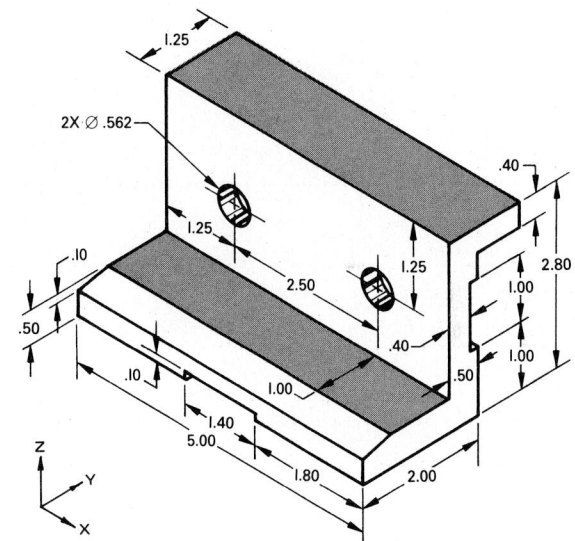

FIG. 14-3-E   Guide bracket.

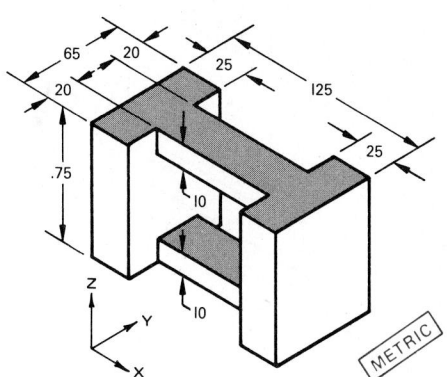

FIG. 14-3-C   Hanger.

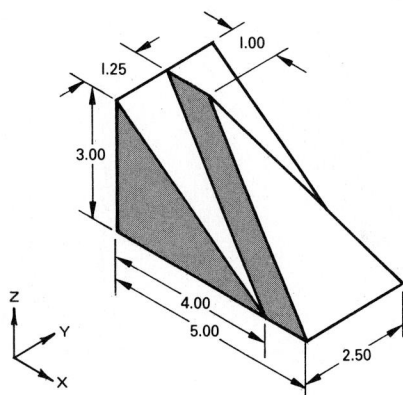

FIG. 14-3-F   Control link.

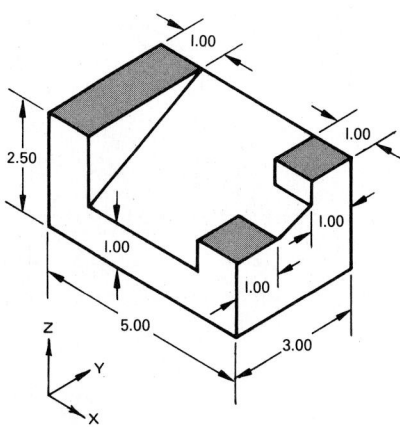

FIG. 14-3-G   End bracket.

11. Make a detail drawing of one of the parts shown in Figs. 14-3-H through 14-3-N. Select the scale and the number of views required.

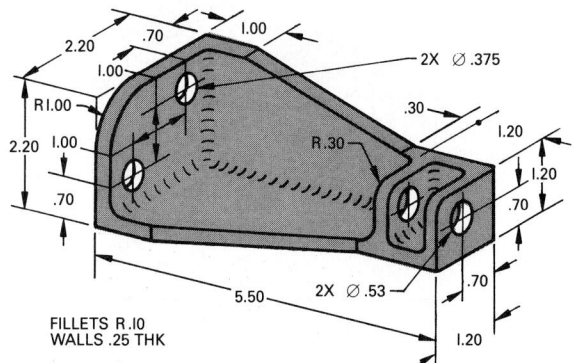

FILLETS R.10
WALLS .25 THK

FIG. 14-3-H Caster leg.

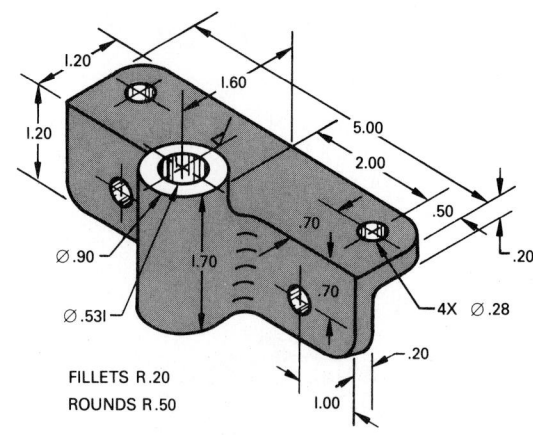

FILLETS R.20
ROUNDS R.50

FIG. 14-3-L Oarlock socket.

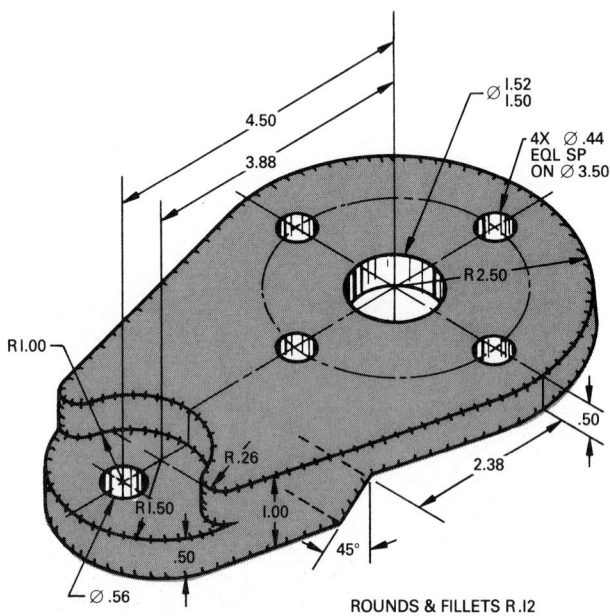

ROUNDS & FILLETS R.12

FIG. 14-3-J Spacer.

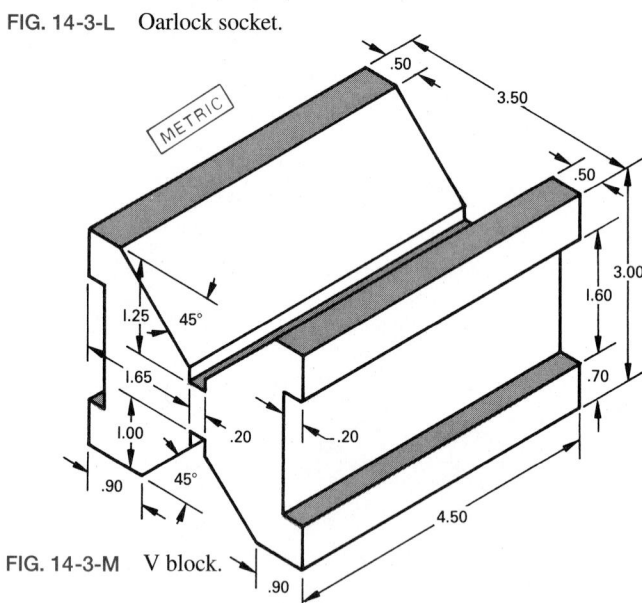

FIG. 14-3-M V block.

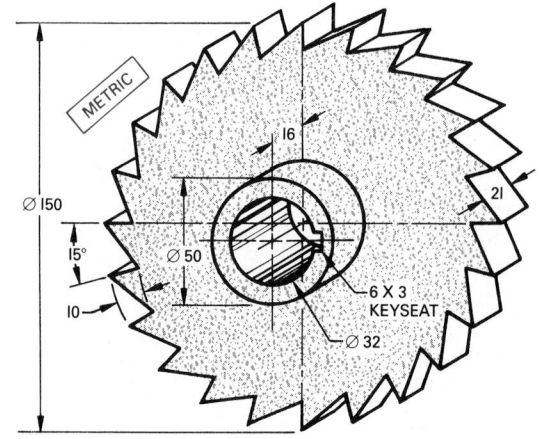

FIG. 14-3-K Ratchet wheel.

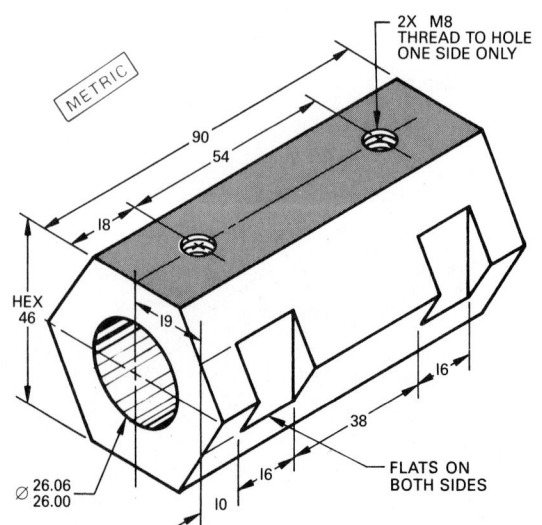

FIG. 14-3-N Guide rack.

12. Make a detail drawing of one of the parts shown in Figs. 14-3-P through 14-3-T. Select the scale and the number of views required.

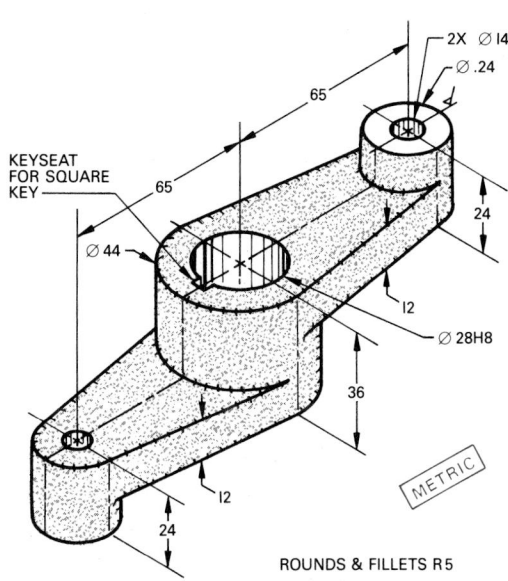

FIG. 14-3-P   Control arm.

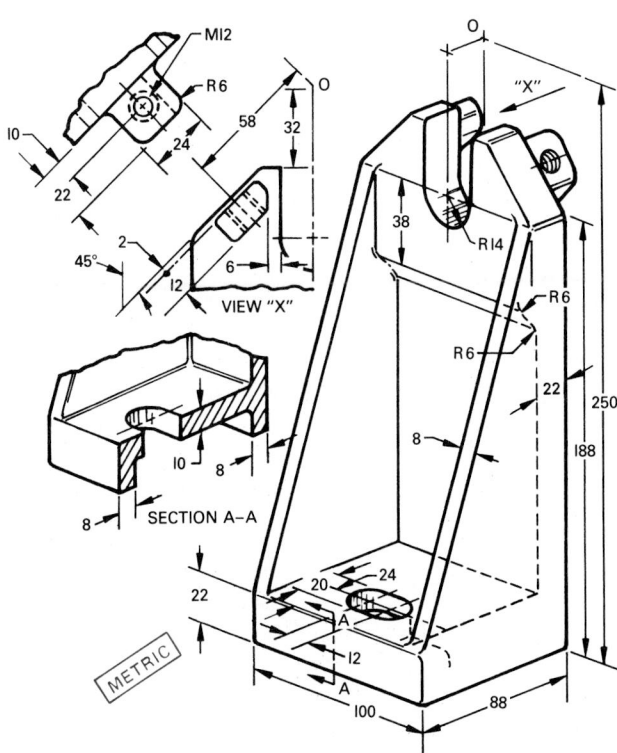

FIG. 14-3-Q   End base.

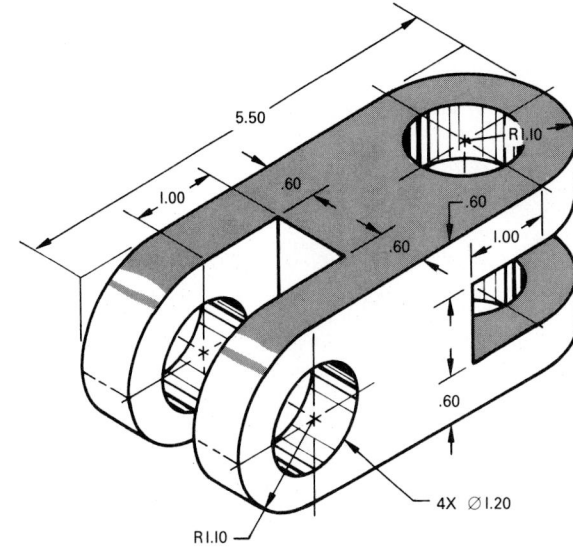

FIG. 14-3-R   Coupling.

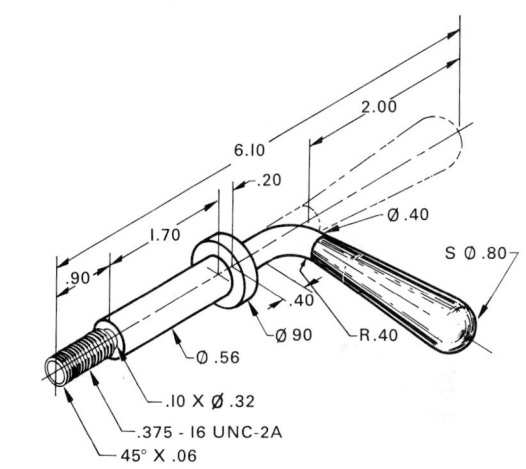

FIG. 14-3-S   Handle.

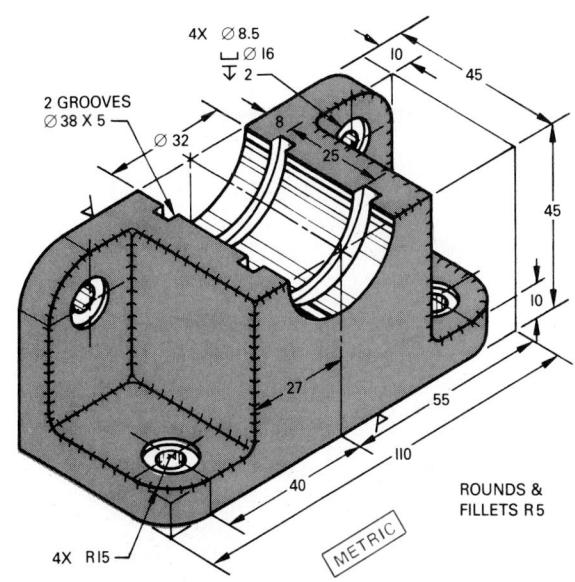

FIG. 14-3-T   Column support.

13. Make a detail drawing of one of the parts shown in Figs. 14-3-U through 14-3-Y. Select the scale and the number of views required.

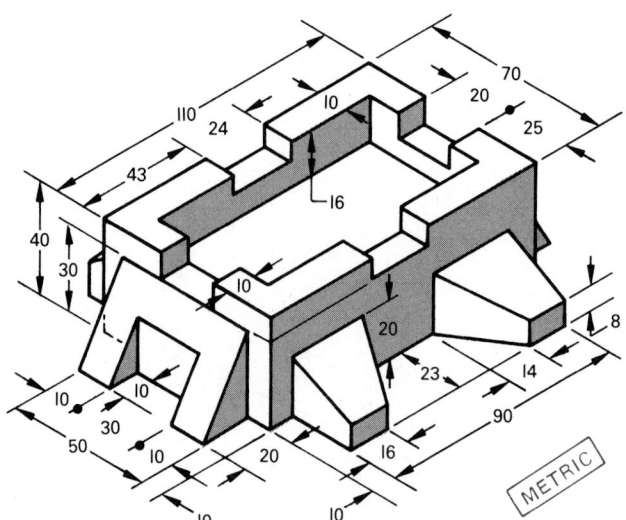

FIG. 14-3-W   Base plate.

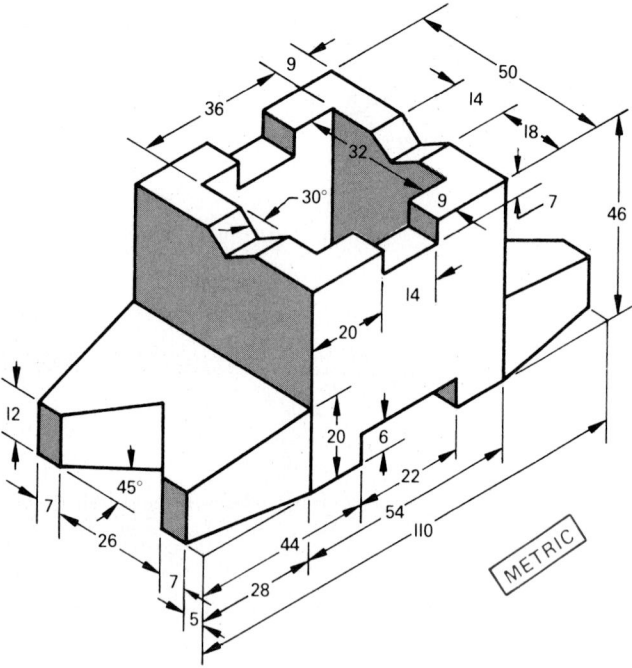

FIG. 14-3-U   Trunion.

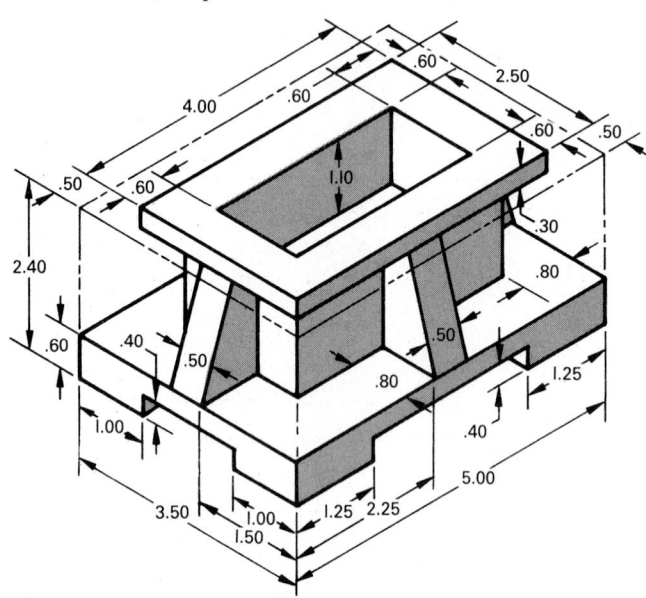

FIG. 14-3-X   Cradle bracket.

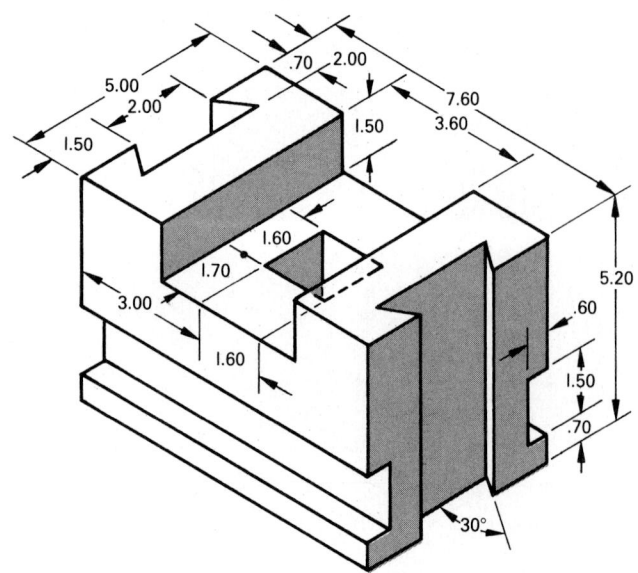

FIG. 14-3-V   Base.

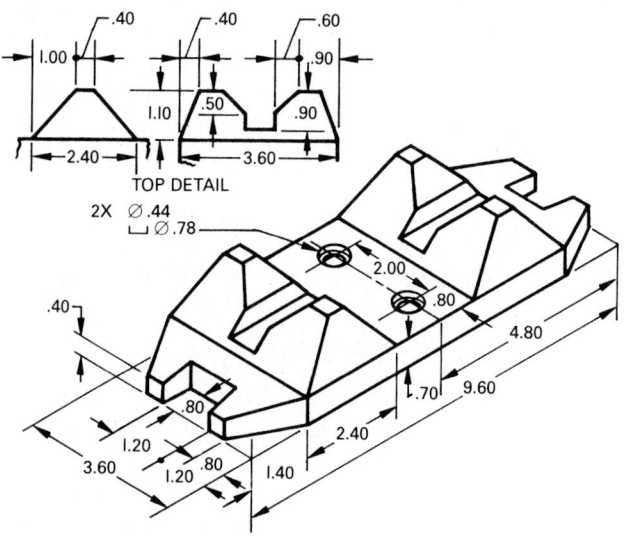

FIG. 14-3-Y   Sliding block.

14. Select one of the parts shown in Figs. 14-3-Z and 14-3-AA and make a three-view working drawing. Dimensions are to be converted to millimeters. Only the dovetail and T slot dimensions are critical and must be taken to an accuracy of two points beyond the decimal point. All other dimensions are to be rounded off to whole numbers.

FIG. 14-3-Z   Locating stand.

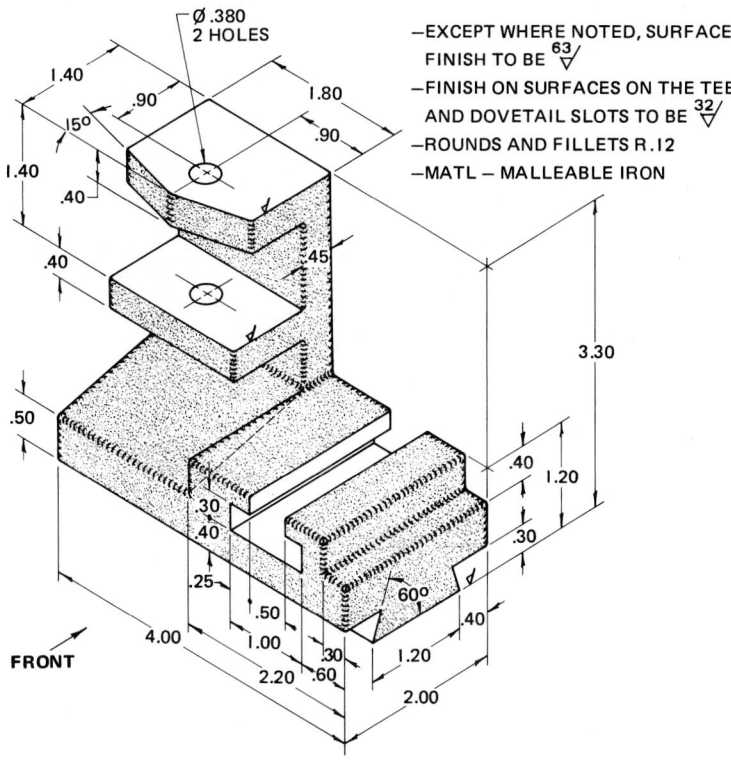

FIG. 14-3-AA   Cross slide.

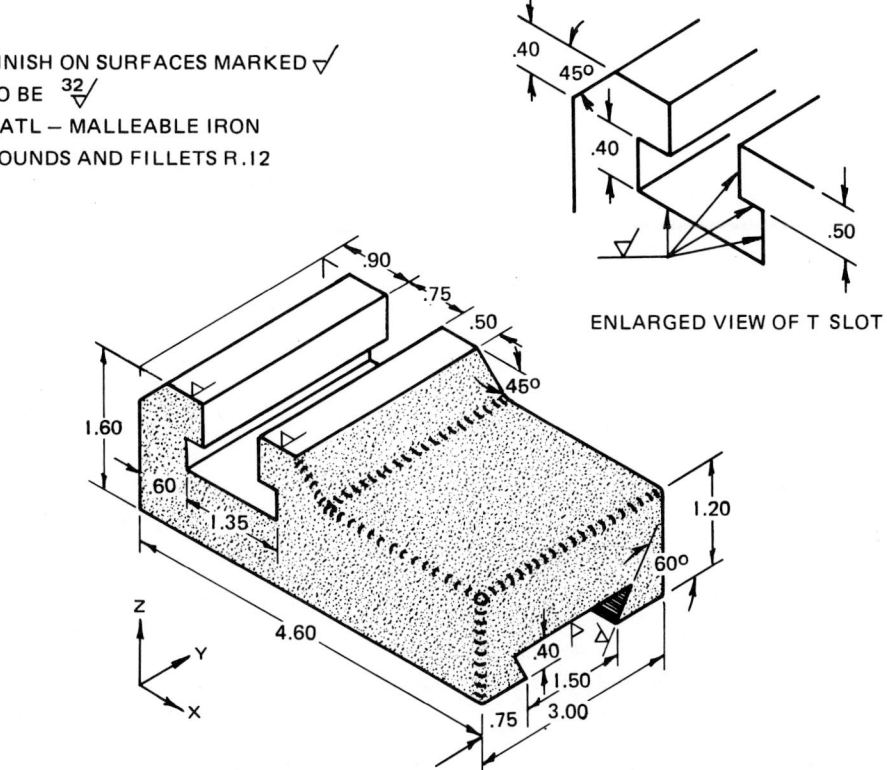

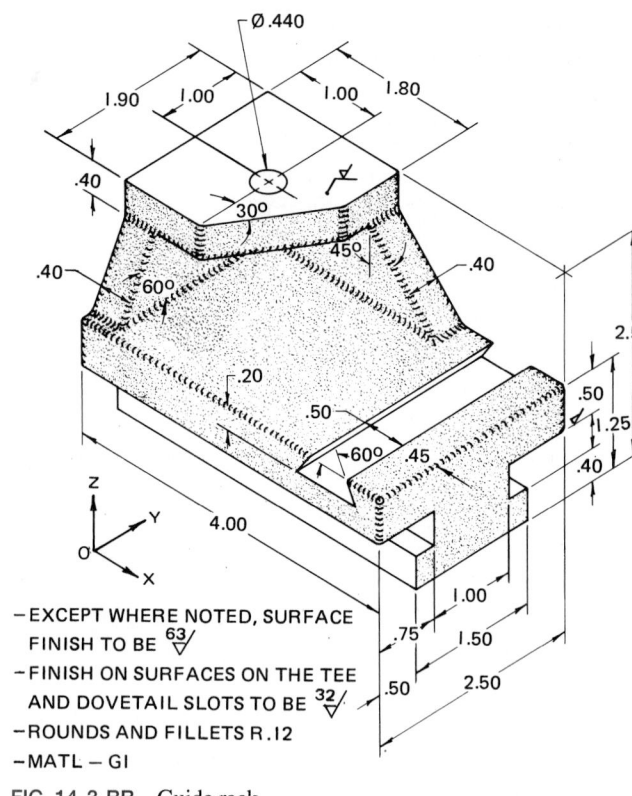

–EXCEPT WHERE NOTED, SURFACE
FINISH TO BE $\overset{63}{\triangledown}$

–FINISH ON SURFACES ON THE TEE
AND DOVETAIL SLOTS TO BE $\overset{32}{\triangledown}$

–ROUNDS AND FILLETS R.12

–MATL – GI

FIG. 14-3-BB   Guide rack.

15. Make a detail drawing of one of the parts shown in Figs. 14-3-BB and 14-3-CC. Select the scale and the number of views required.

16. On a C (A2) size sheet, make a detail drawing of the part shown in Fig. 14-3-DD. To clearly show the features, a section and bottom view should also be drawn. The recommended drawing layout is shown. The slots are to have

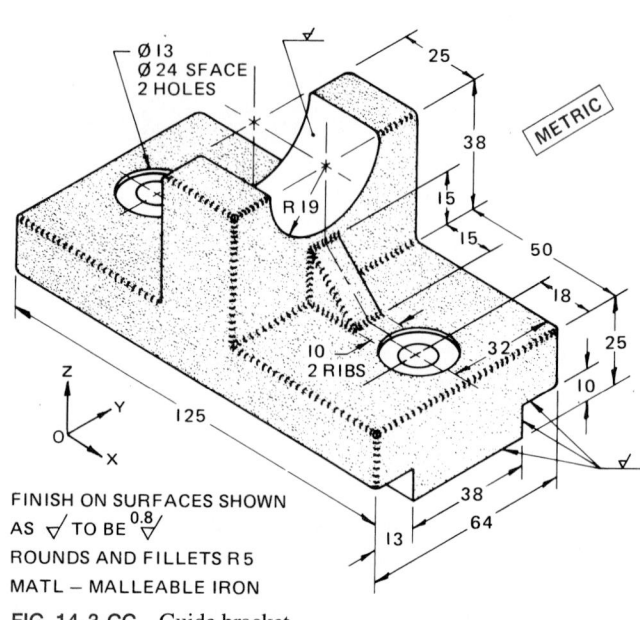

FINISH ON SURFACES SHOWN
AS $\triangledown$ TO BE $\overset{0.8}{\triangledown}$
ROUNDS AND FILLETS R 5
MATL – MALLEABLE IRON

FIG. 14-3-CC   Guide bracket.

FIG. 14-3-DD   Pipe vise base.

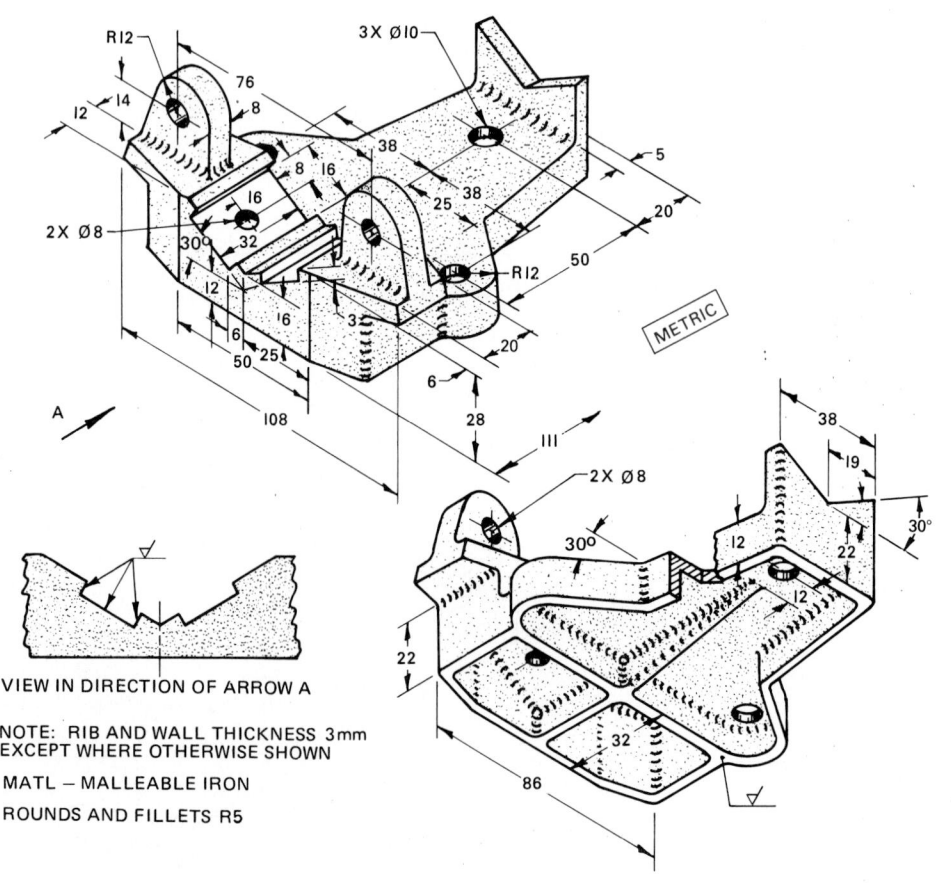

VIEW IN DIRECTION OF ARROW A

NOTE: RIB AND WALL THICKNESS 3mm
EXCEPT WHERE OTHERWISE SHOWN

MATL – MALLEABLE IRON

ROUNDS AND FILLETS R5

412

a surface finish of 3.2 mm and a machining allowance of 2 mm. The base is to have the same surface finish but with a 3 mm machining allowance.

17. Make a detail drawing of one of the parts shown in Figs. 14-3-EE and 14-3-FF. Select the scale and the number of views required.

18. Make a detail drawing of the part shown in Fig. 14-3-GG. The surfaces shown with a √ are to have a surface finish of 63 μin and a machining allowance of .06 in. Select the scale and the number of views required.

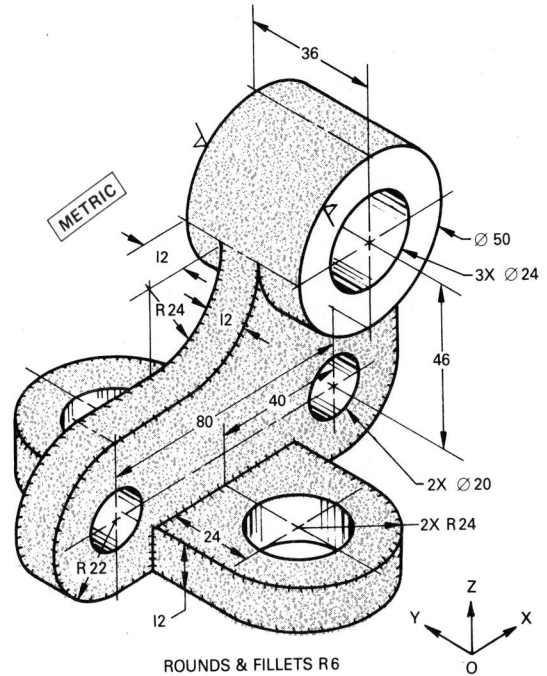

ROUNDS & FILLETS R 6

FIG. 14-3-EE   Swivel hanger.

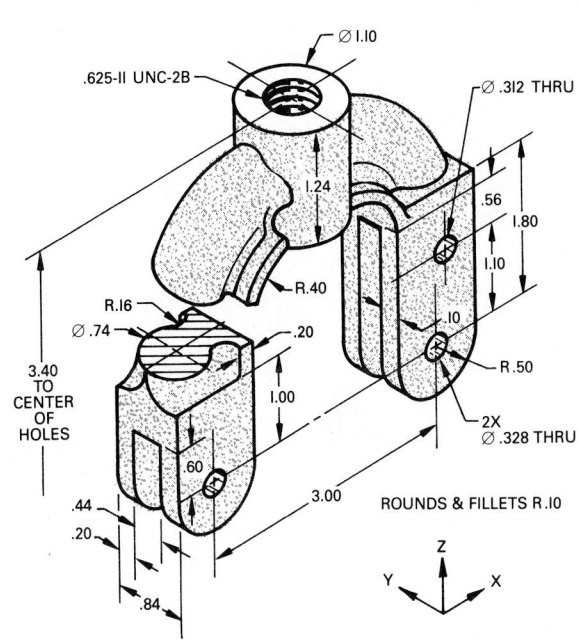

ROUNDS & FILLETS R.10

FIG. 14-3-FF   Fork.

FIG. 14-3-GG   Offset bracket.

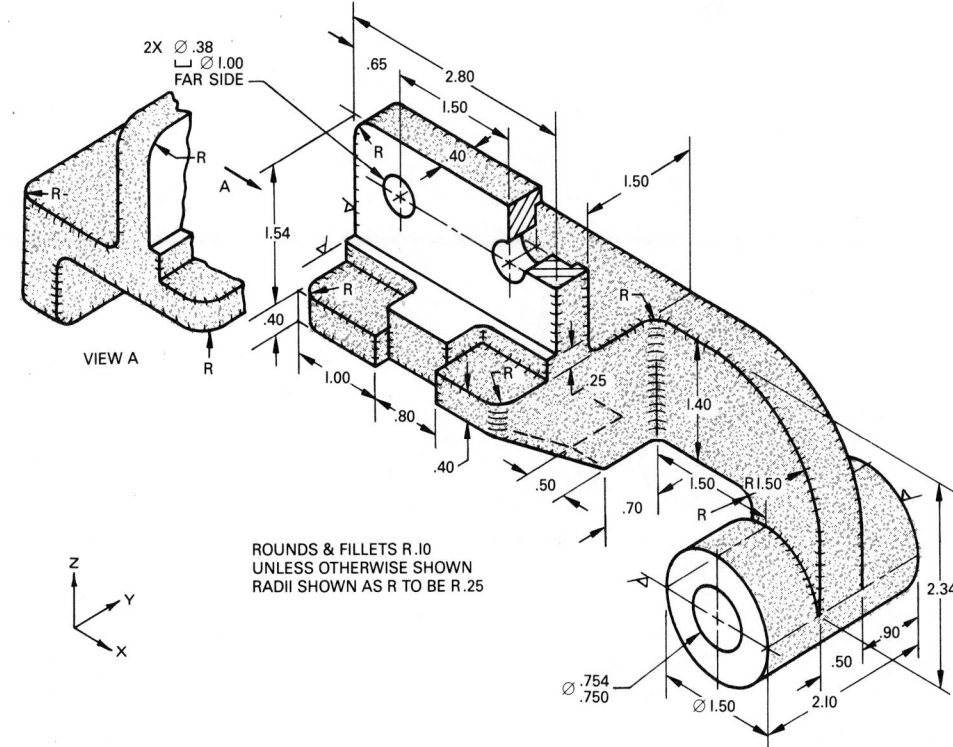

VIEW A

ROUNDS & FILLETS R.10
UNLESS OTHERWISE SHOWN
RADII SHOWN AS R TO BE R.25

19. On a C (A2) size sheet draw the part shown in Fig. 14-3-HH. Draw all six views plus a partial auxiliary view for the .250 threaded hole. Scale 1:2.

20. Make a detail drawing of the part shown in Fig. 14-3-JJ. The surfaces shown with a ∛ are to have a surface finish of 63 μin and a machining allowance of .06 in. Select the scale and the number of views required.

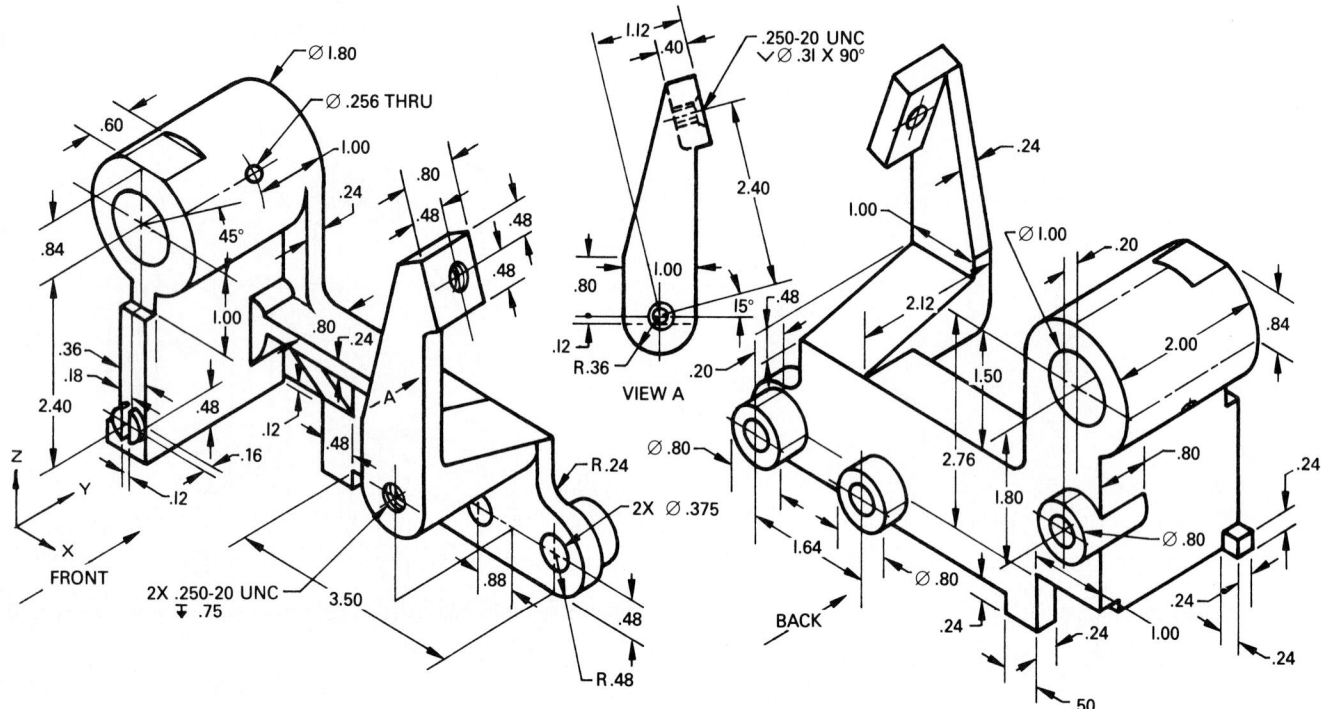

FIG. 14-3-HH   Control bracket.

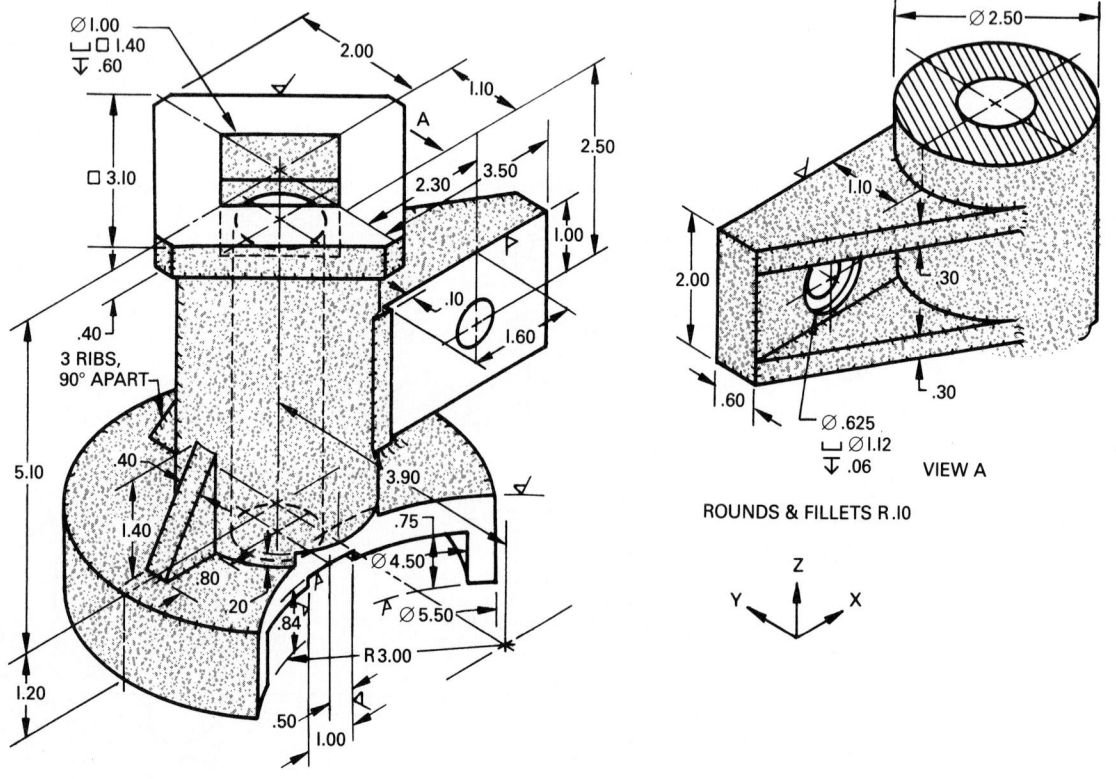

FIG. 14-3-JJ   Pedestal.

ASSIGNMENTS FOR UNIT 14-4, MULTIPLE DETAIL DRAWINGS

21. Make detail drawings of all the parts shown of one of the assemblies in Figs. 14-4-A and 14-4-B. Since time is money, select only the views necessary to describe each part. Below each part show the following information: part number, name of part, material, number required. Scale 1:1.

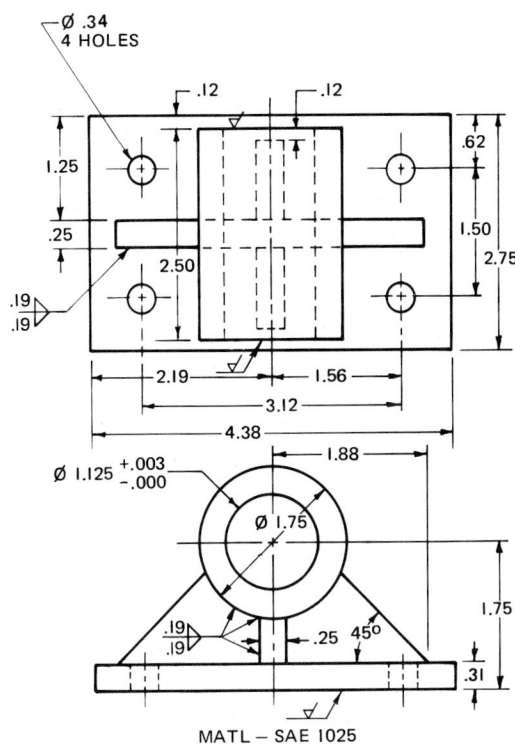

FIG. 14-4-A   Shaft support.

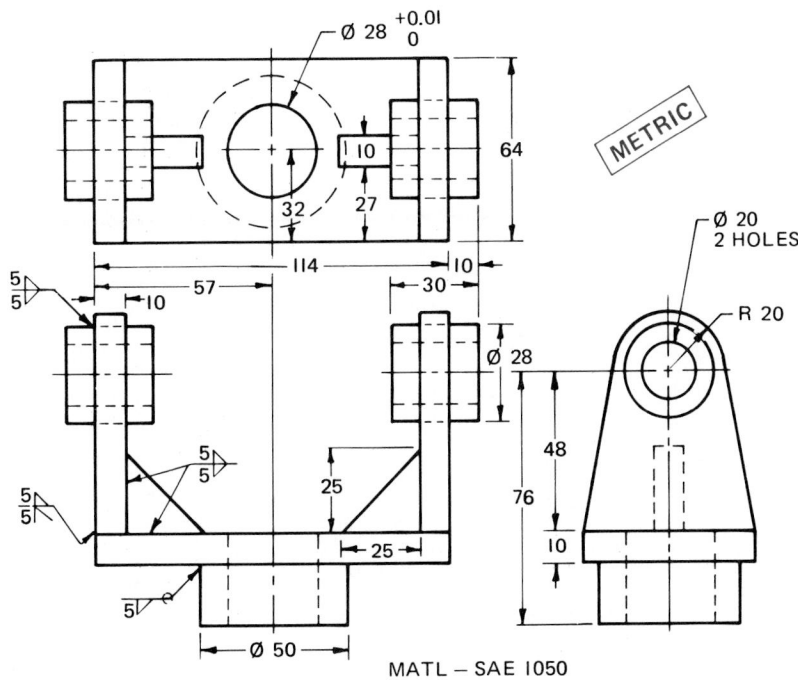

FIG. 14-4-B   Shaft pivot support.

22. Make detail drawings of the nonstandard parts shown on one of the assemblies in Figs. 14-4-C and 14-4-D. Select the scale and the number of views required. Add to the drawing an item list.

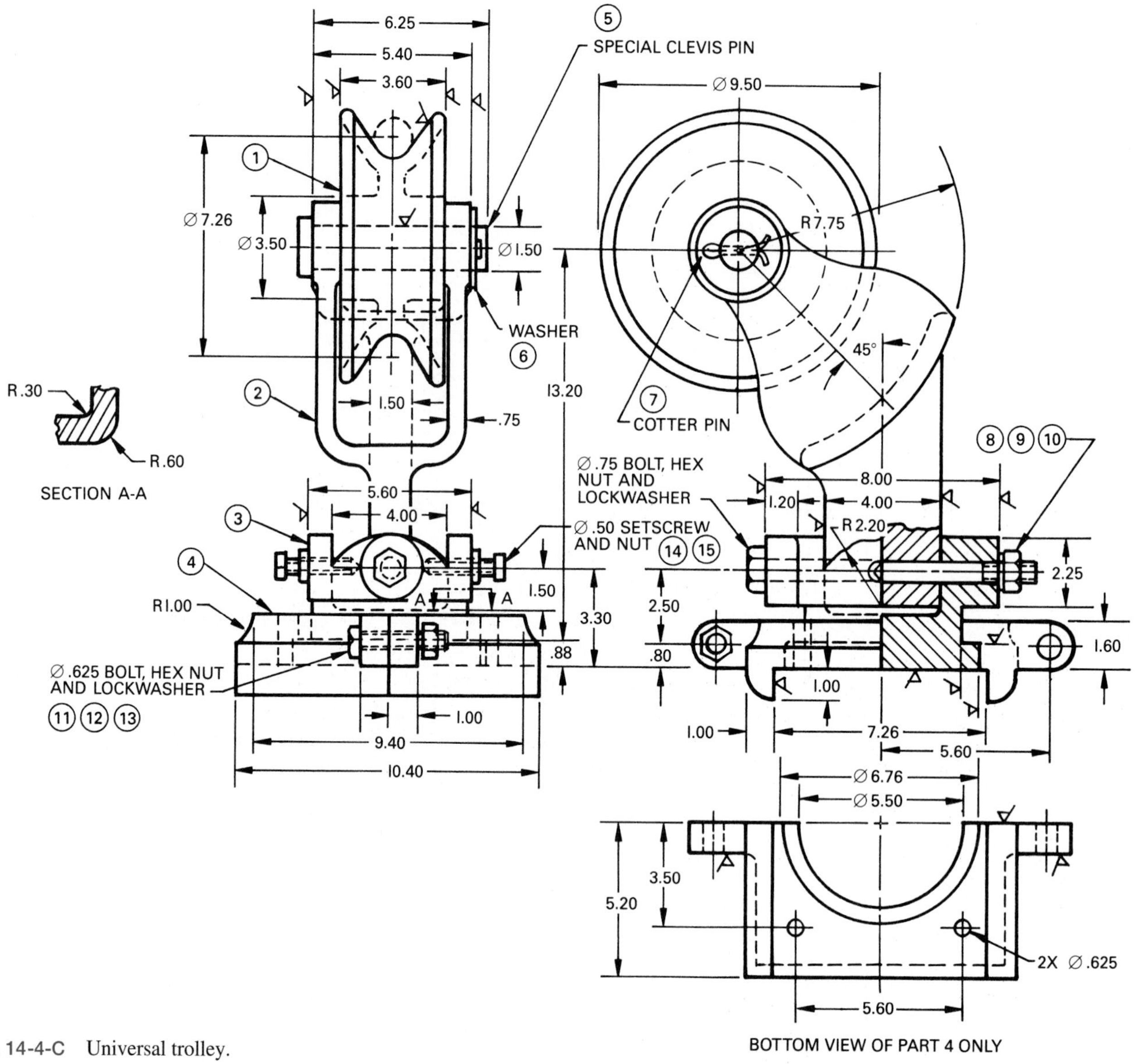

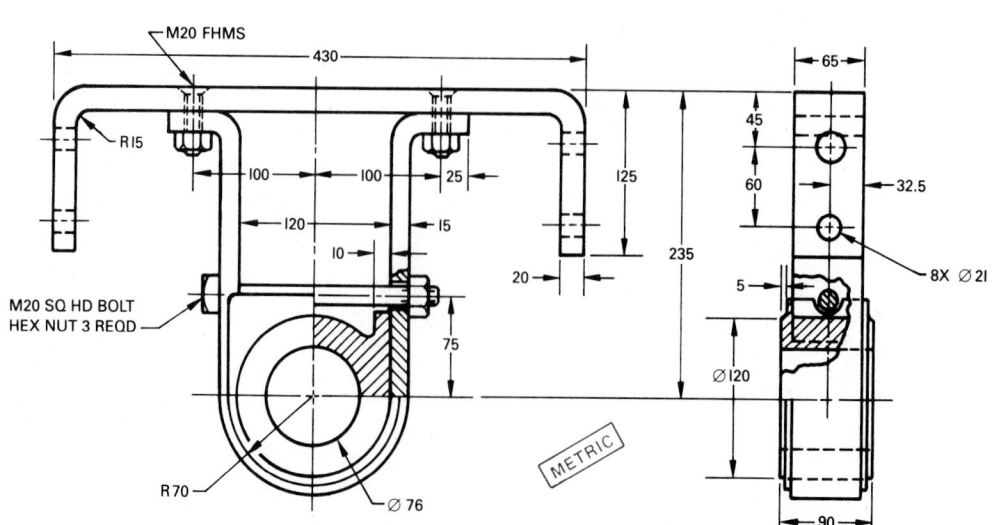

FIG. 14-4-C   Universal trolley.

FIG. 14-4-D   Bearing bracket.

23. Make detail drawings of the nonstandard parts shown in Fig. 14-4-E. Select the scale and the number of views required. Add to the drawing an item list. The surfaces shown with a √ are to have a surface finish of 63 μin and a machining allowance of .06 in.

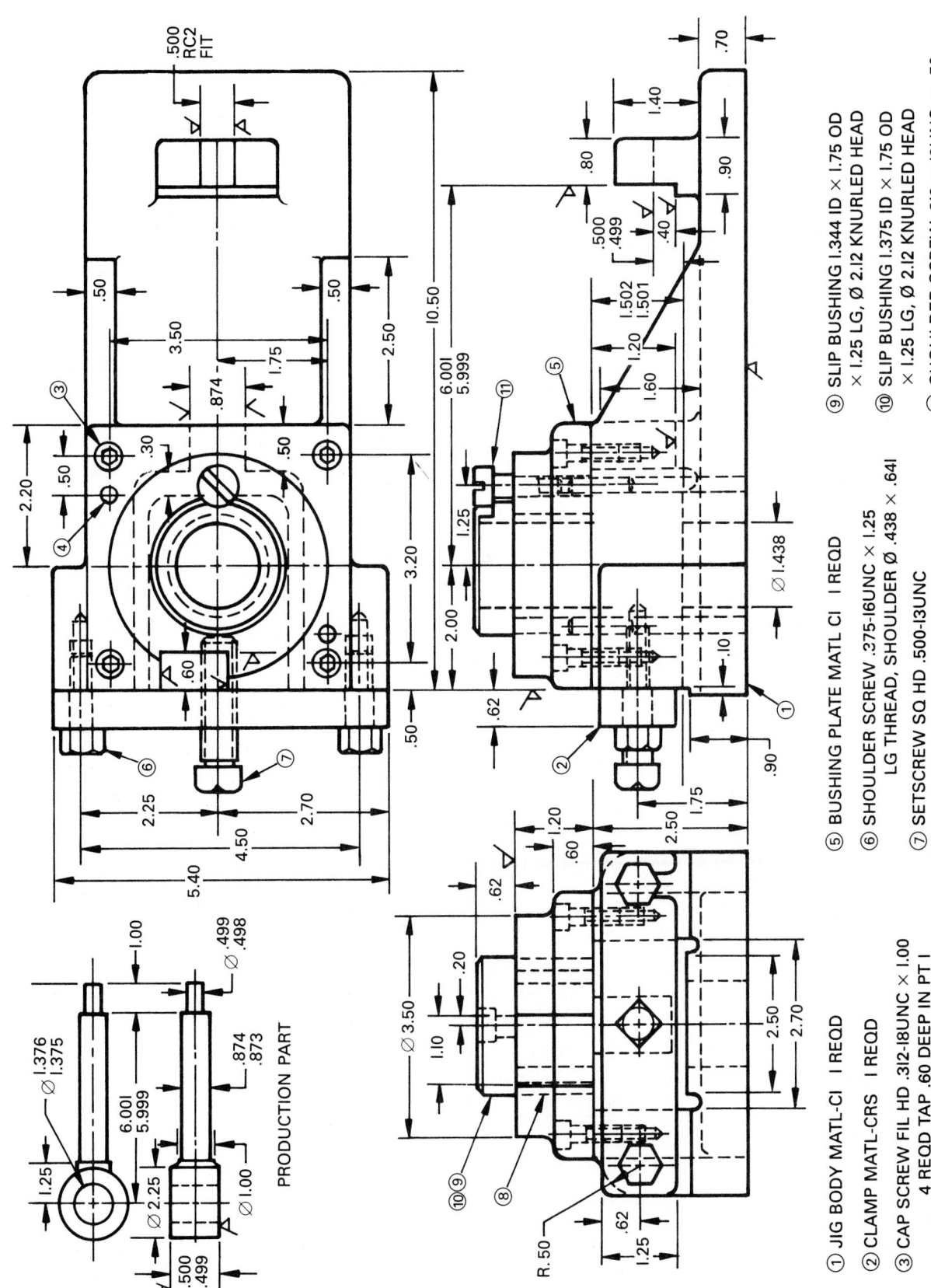

⑨ SLIP BUSHING 1.344 ID × 1.75 OD
    × 1.25 LG, Ø 2.12 KNURLED HEAD

⑩ SLIP BUSHING 1.375 ID × 1.75 OD
    × 1.25 LG, Ø 2.12 KNURLED HEAD

⑪ SHOULDER SCREW .312 × 18UNC × .50
    LG THREAD, SHOULDER Ø .375
    × .312 LG, HEAD Ø .62 × .31 HIGH

⑤ BUSHING PLATE MATL CI I REQD

⑥ SHOULDER SCREW .375-16UNC × 1.25
    LG THREAD, SHOULDER Ø .438 × .641

⑦ SETSCREW SQ HD .500-13UNC
    × 2.00 LG, FLAT POINT, I REQD

⑧ LINER BUSHING 1.75 ID × 2.12 OD
    × 1.25 LG I REQD

① JIG BODY MATL-CI I REQD

② CLAMP MATL-CRS I REQD

③ CAP SCREW FIL HD .312-18UNC × 1.00
    4 REQD TAP .60 DEEP IN PT I

④ SPRING PIN Ø .25 × 1.00 2 REQD

PRODUCTION PART

FIG. 14-4-E   Drill jig.

417

24. On C (A2) size sheets make detail drawings of the parts shown in Fig. 14-4-F. Select the scale and the number of views required. An LN3 fit is required between the bushings and housing, and an RC4 fit between the shafts and bushing. Include an item list for the parts.

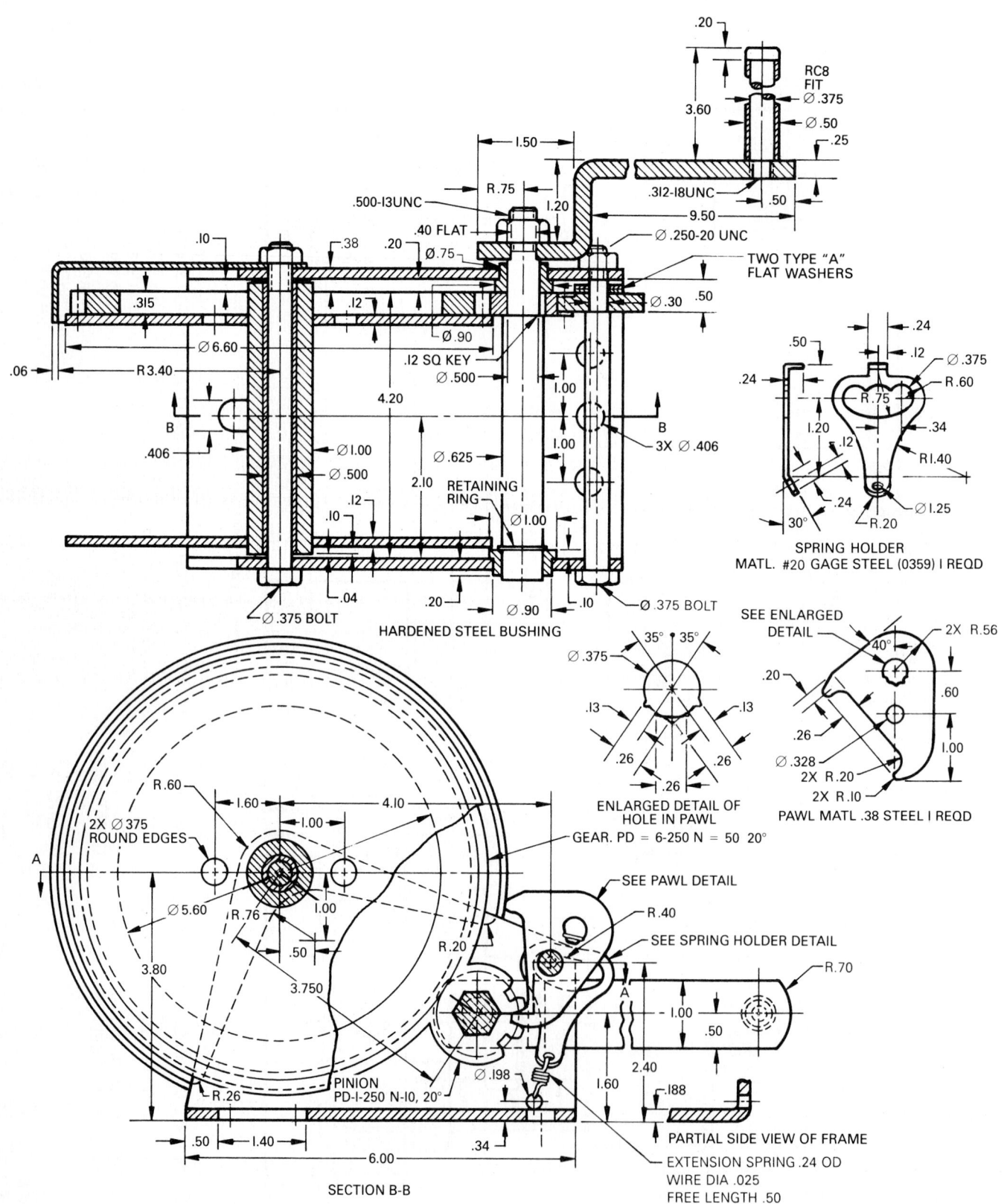

FIG. 14-4-F Winch. *(Fulton Co.)*

25. Prepare detail drawings of any of the parts assigned by your instructor from the assembly drawings shown in Figs. 14-4-G and 14-4-H. For Fig. 14-4-G the following fits are to be used: Ø28H9/d9; Ø45H7/s6; and Ø35H8/f7.

For Fig. 14-4-H an RC4 fit is required for the Ø1.20 shaft. The scale and selection of views are to be decided by the student. Include an item list for the parts.

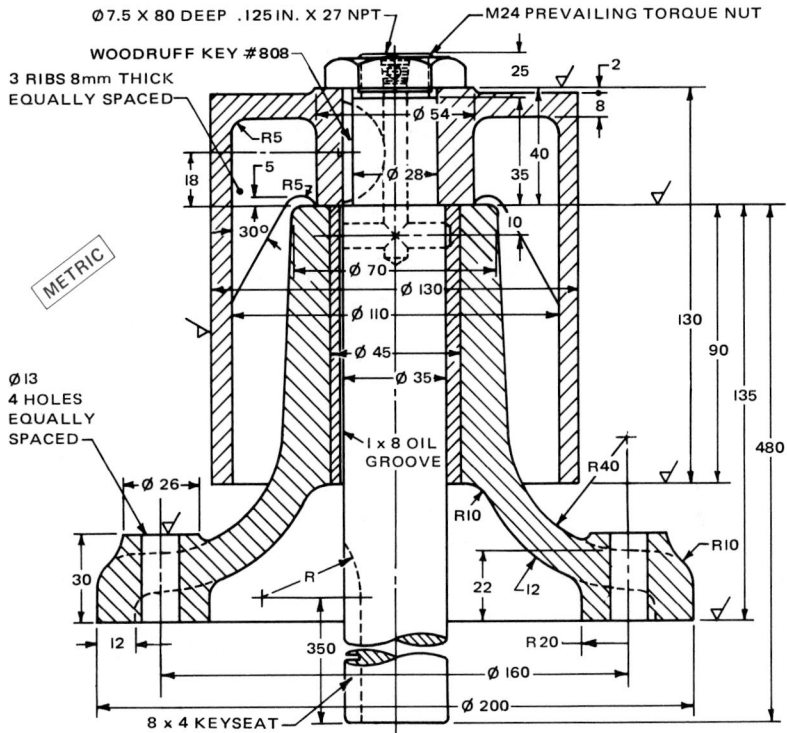

FIG. 14-4-G Pulley assembly.

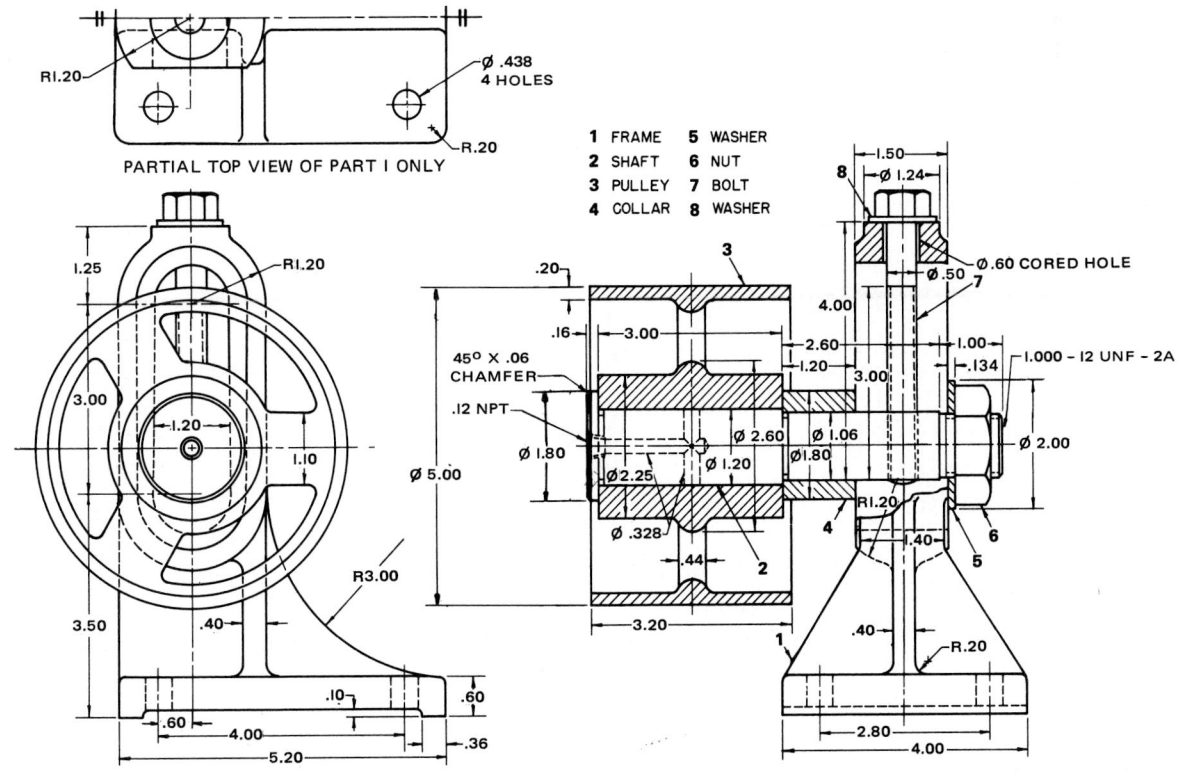

FIG. 14-4-H Adjustable pulley.

26. Prepare detail drawings for the nonstandard parts for the assemblies shown in Figs. 14-4-J and 14-4-K. Select the scale and number of views required. Include an item list for the parts.

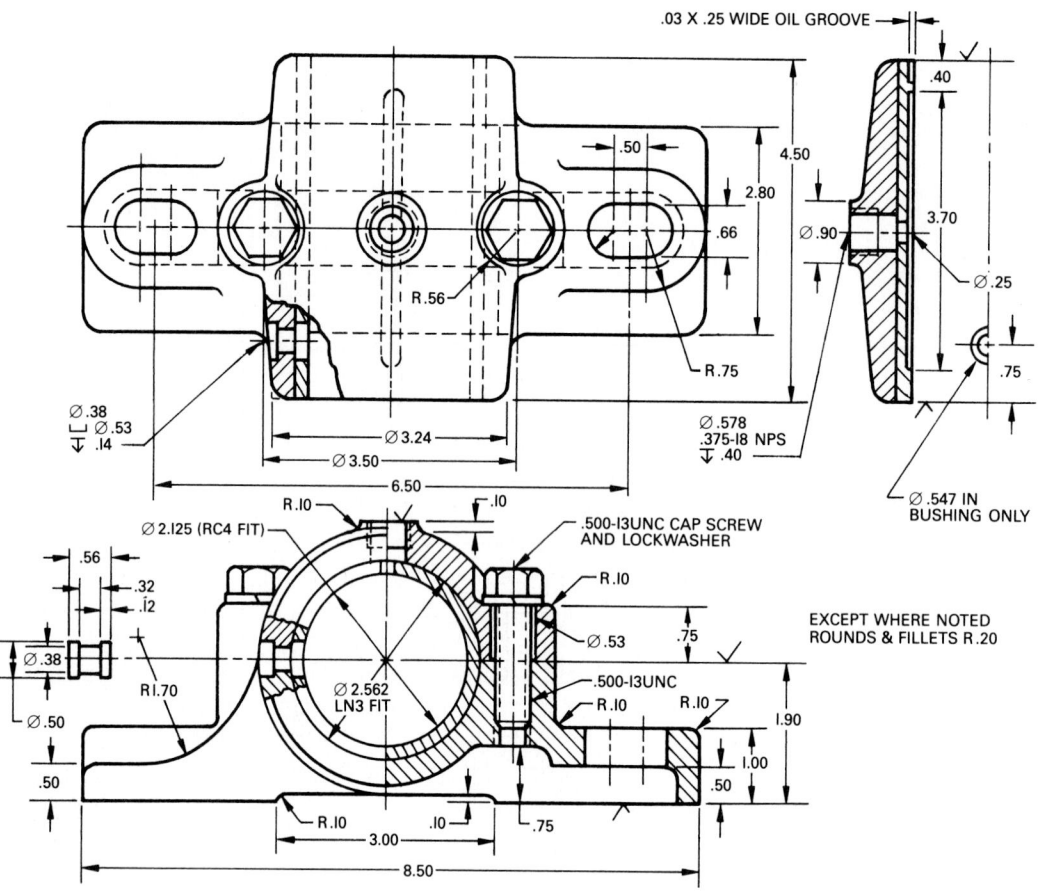

FIG. 14-4-J   Split bushing.

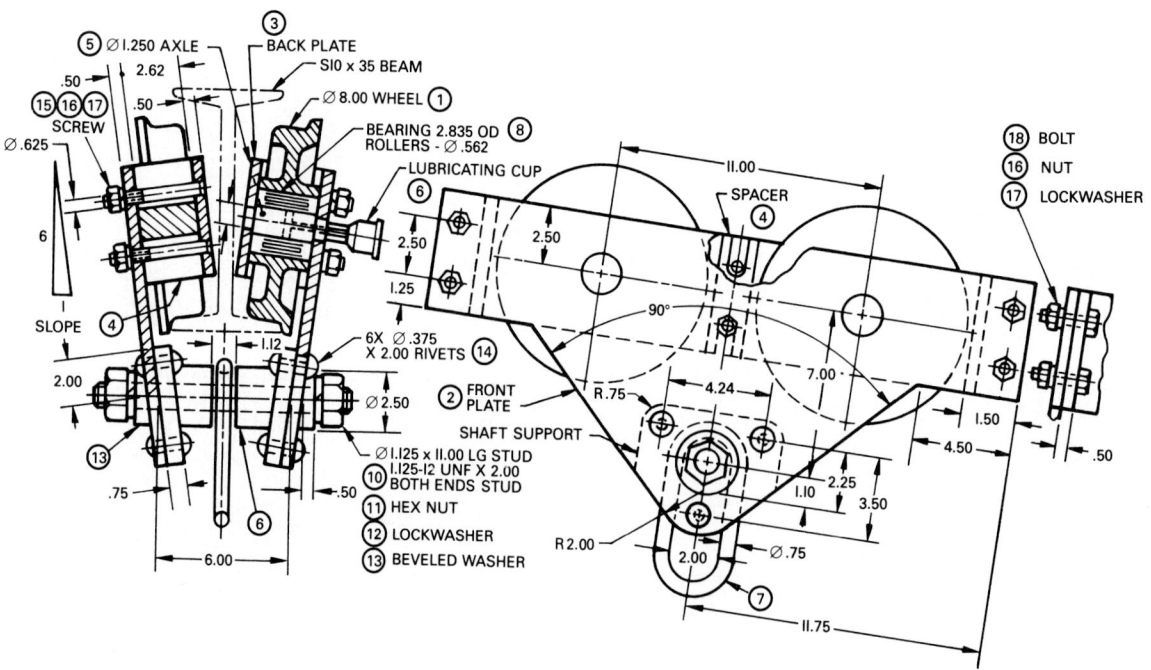

FIG. 14-4-K   Four-wheel trolley.

27. Prepare detail drawings for the nonstandard parts for any of the assemblies shown in Figs. 14-6-A through 14-6-G (here and through pg. 427). The student is to select the scale and the number of views required. Include an item list for the parts.

ASSIGNMENT FOR UNIT 14-5, DRAWING REVISIONS

28. Select one of the drawings shown in Figs. 14-5-A and 14-5-B and make appropriate revisions to these drawings, recording the changes in a drawing revision column and indicating which dimensions are not to scale.

ASSIGNMENTS FOR UNIT 14-6, ASSEMBLY DRAWINGS

29. Make a one-view assembly drawing of one of the assemblies shown in Figs. 14-6-A and 14-6-B. For Fig. 14-6-B show a round bar Ø24 mm in phantom being held in position. Include on the drawing an item list and identification part numbers. Scale 1:1.

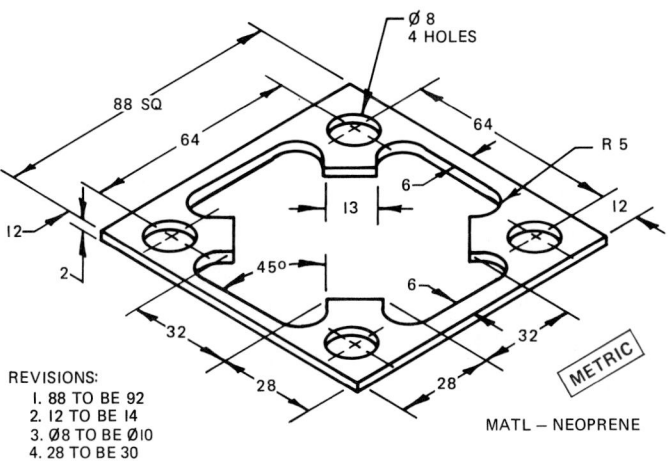

REVISIONS:
1. 88 TO BE 92
2. 12 TO BE 14
3. Ø8 TO BE Ø10
4. 28 TO BE 30

MATL — NEOPRENE

FIG. 14-5-A Gasket.

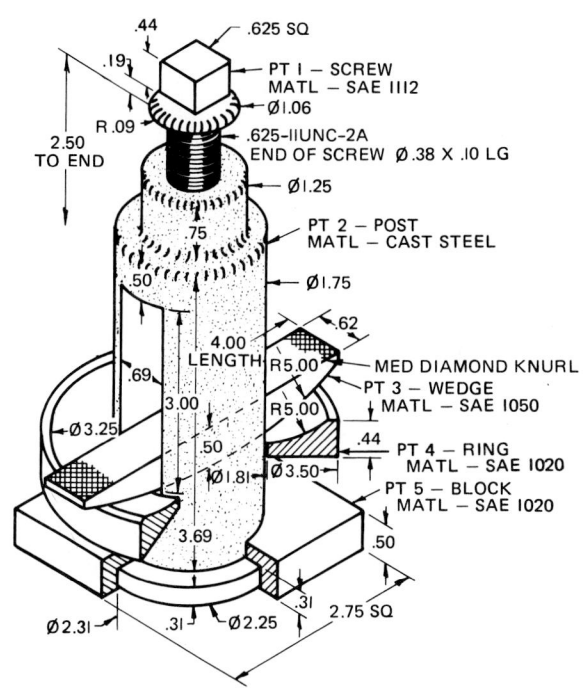

FIG. 14-6-A Tool post holder.

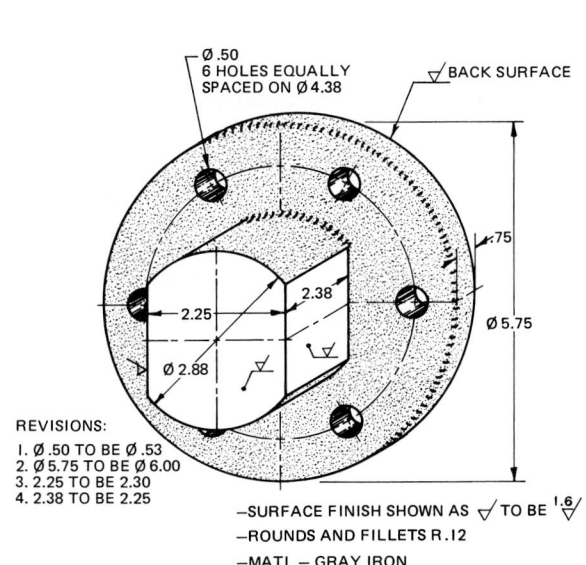

REVISIONS:
1. Ø.50 TO BE Ø.53
2. Ø5.75 TO BE Ø6.00
3. 2.25 TO BE 2.30
4. 2.38 TO BE 2.25

—SURFACE FINISH SHOWN AS ∇ TO BE 1.6∇
—ROUNDS AND FILLETS R.12
—MATL — GRAY IRON

FIG. 14-5-B Axle cap.

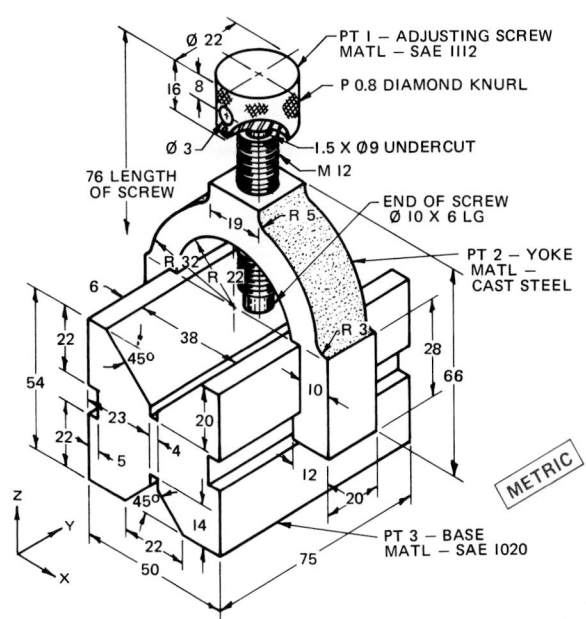

FIG. 14-6-B V-block clamp.

30. Make a one-view assembly drawing of the bench vise shown in Fig. 14-6-C. Show the vise jaws open 50 mm and place on the drawing only pertinent dimensions. Include on the drawing an item list and identification part numbers. Scale 1:1.

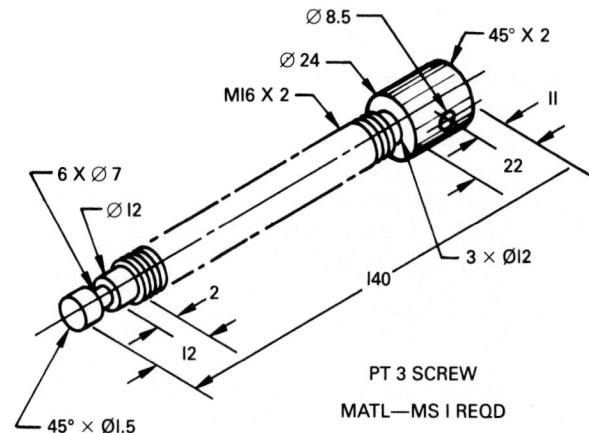

Ø 8.5
45° X 2
Ø 24
MI6 X 2
II
22
6 X Ø 7
Ø I2
3 × ØI2
2
I40
I2
45° × ØI.5

PT 3 SCREW
MATL—MS I REQD

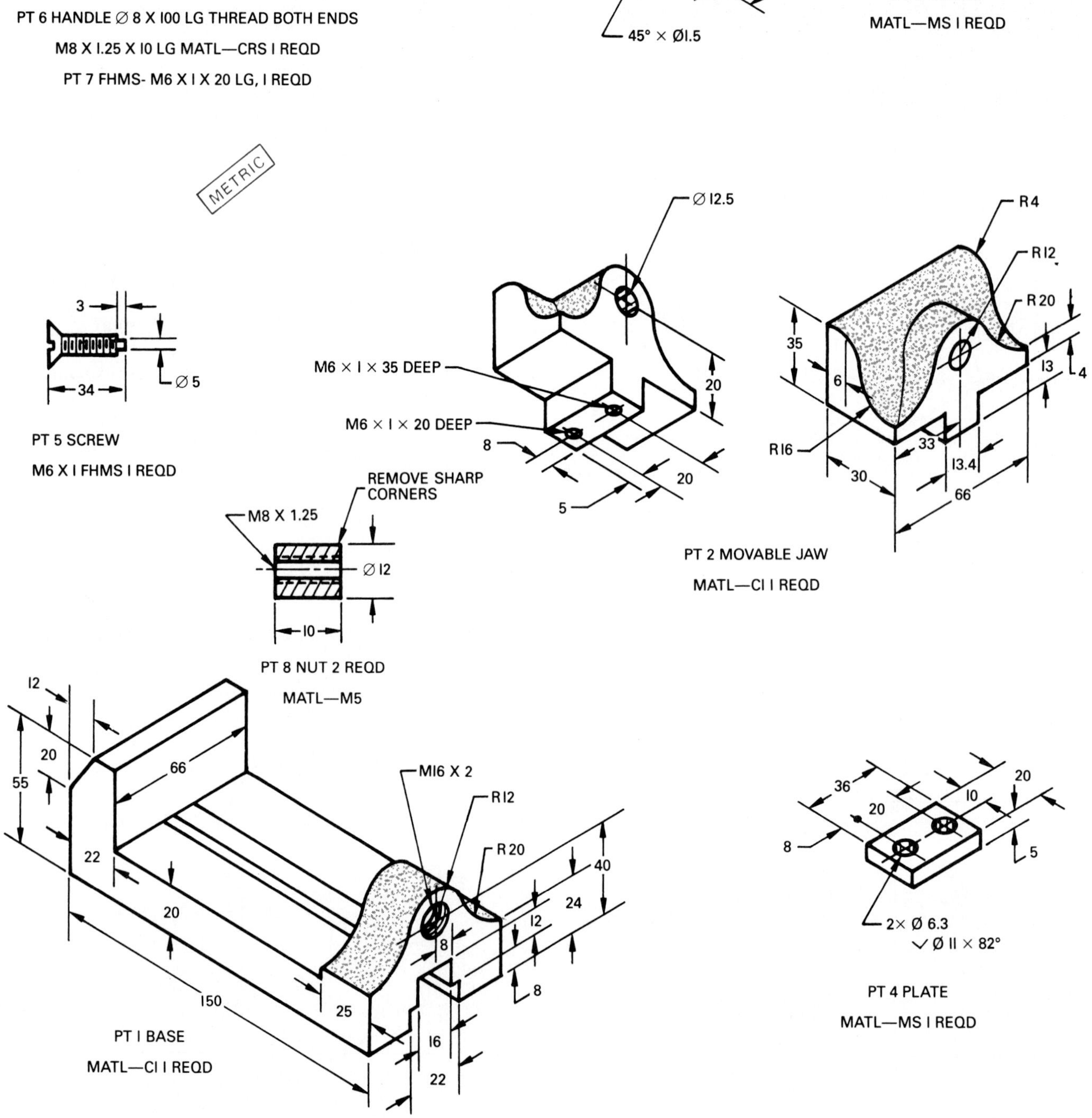

PT 6 HANDLE Ø 8 X I00 LG THREAD BOTH ENDS
M8 X I.25 X I0 LG MATL—CRS I REQD
PT 7 FHMS- M6 X I X 20 LG, I REQD

METRIC

Ø I2.5
R 4
R I2
R 20
35
6
I3
4
M6 × I × 35 DEEP
M6 × I × 20 DEEP
20
R I6
8
20
33
I3.4
30
66
5

PT 2 MOVABLE JAW
MATL—CI I REQD

3
Ø 5
34

PT 5 SCREW
M6 X I FHMS I REQD

REMOVE SHARP
CORNERS
M8 X I.25
Ø I2
I0

PT 8 NUT 2 REQD
MATL—M5

I2
20
66
55
22
20
I50
25
I6
22

MI6 X 2
R I2
R 20
40
24
I2
8
8

PT I BASE
MATL—CI I REQD

36
20
I0
20
8
5
2× Ø 6.3
∨ Ø II × 82°

PT 4 PLATE
MATL—MS I REQD

FIG. 14-6-C   Bench vise.

31. Make a one-view assembly drawing of the wrench shown in Fig. 14-6-D. Include on the drawing an item list and identification part numbers. Scale 1:1.

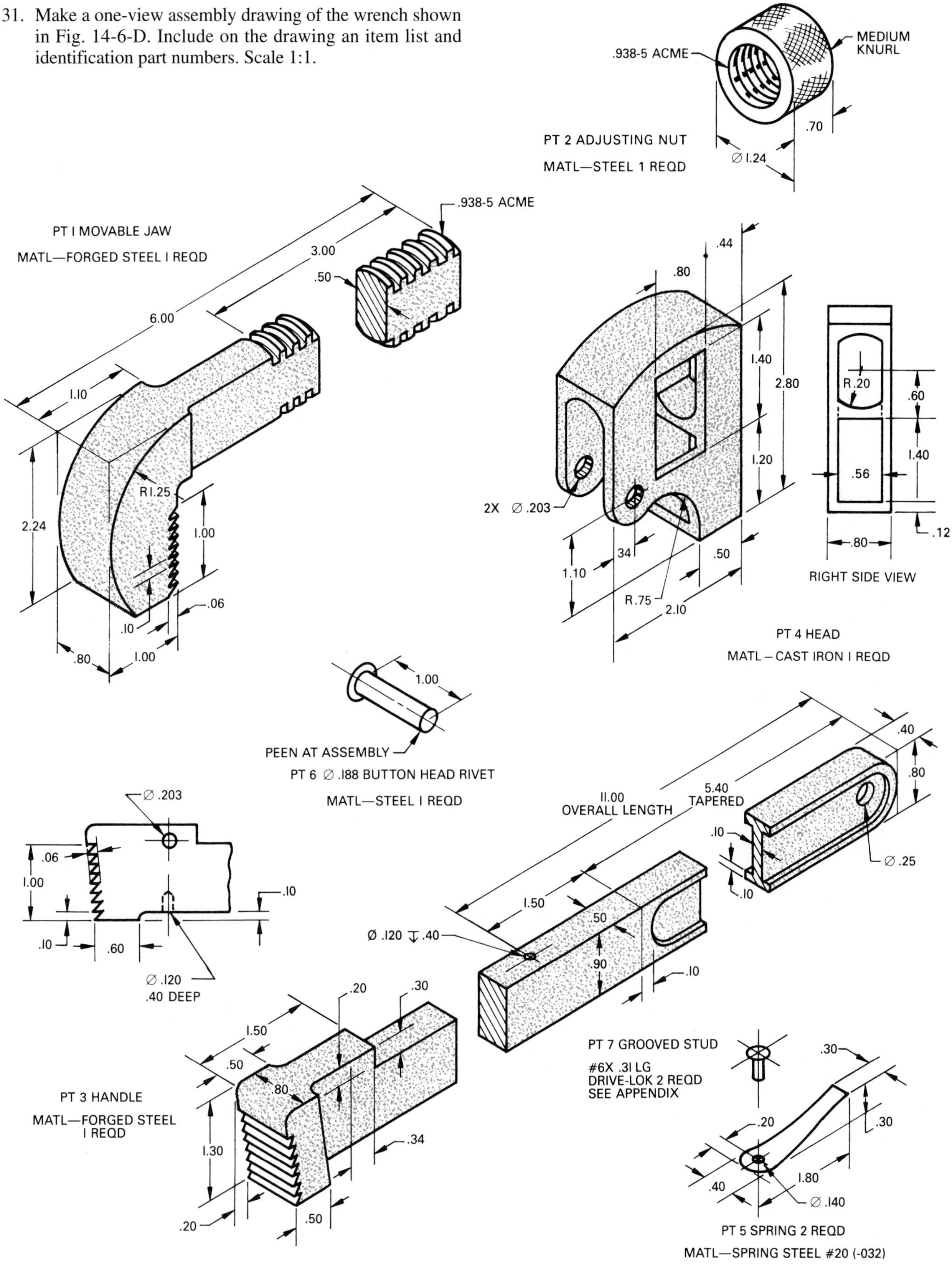

.938-5 ACME

MEDIUM KNURL

PT 2 ADJUSTING NUT
MATL—STEEL 1 REQD

Ø 1.24

.70

.938-5 ACME

PT 1 MOVABLE JAW
MATL—FORGED STEEL 1 REQD

3.00

6.00

.50

1.10

R1.25

2.24

1.00

.06

.10

.80

1.00

.44

.80

1.40

2.80

1.20

R.20

.60

1.40

.56

.80

.12

RIGHT SIDE VIEW

2X   Ø .203

1.10

34

.50

R.75

2.10

PT 4 HEAD
MATL – CAST IRON 1 REQD

1.00

PEEN AT ASSEMBLY
PT 6 Ø .188 BUTTON HEAD RIVET
MATL—STEEL 1 REQD

.40

.80

.10

Ø .25

.10

11.00
OVERALL LENGTH

5.40
TAPERED

Ø .203

.06

1.00

.10

.60

Ø .120
.40 DEEP

Ø .120 ⊤ .40

1.50

.50

.90

.10

.20

.30

1.50

.50

80

1.30

.34

.20

.50

PT 3 HANDLE
MATL—FORGED STEEL
1 REQD

PT 7 GROOVED STUD

#6X .31 LG
DRIVE-LOK 2 REQD
SEE APPENDIX

.30

.20

.30

.40

1.80

Ø .140

PT 5 SPRING 2 REQD
MATL—SPRING STEEL #20 (-032)

FIG. 14-6-D   Stillson wrench.

32. On a C (A2) sheet make a three-view assembly drawing of the wood vise shown in Fig. 14-6-E with the jaws open one inch. Draw the front view in full section, a bottom view, and a half end view. Include on the drawing an item list and identification part numbers. Scale 1:1.

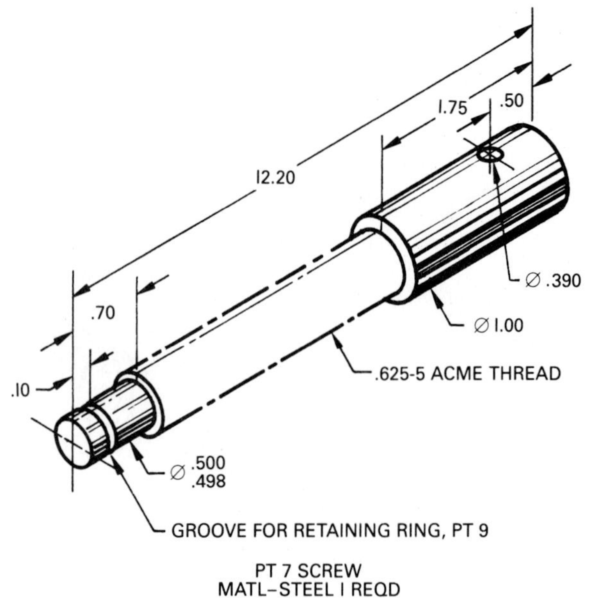

12.20
1.75
.50
Ø .390
Ø 1.00
.70
.10
.625-5 ACME THREAD
Ø .500 .498
GROOVE FOR RETAINING RING, PT 9
PT 7 SCREW
MATL–STEEL 1 REQD

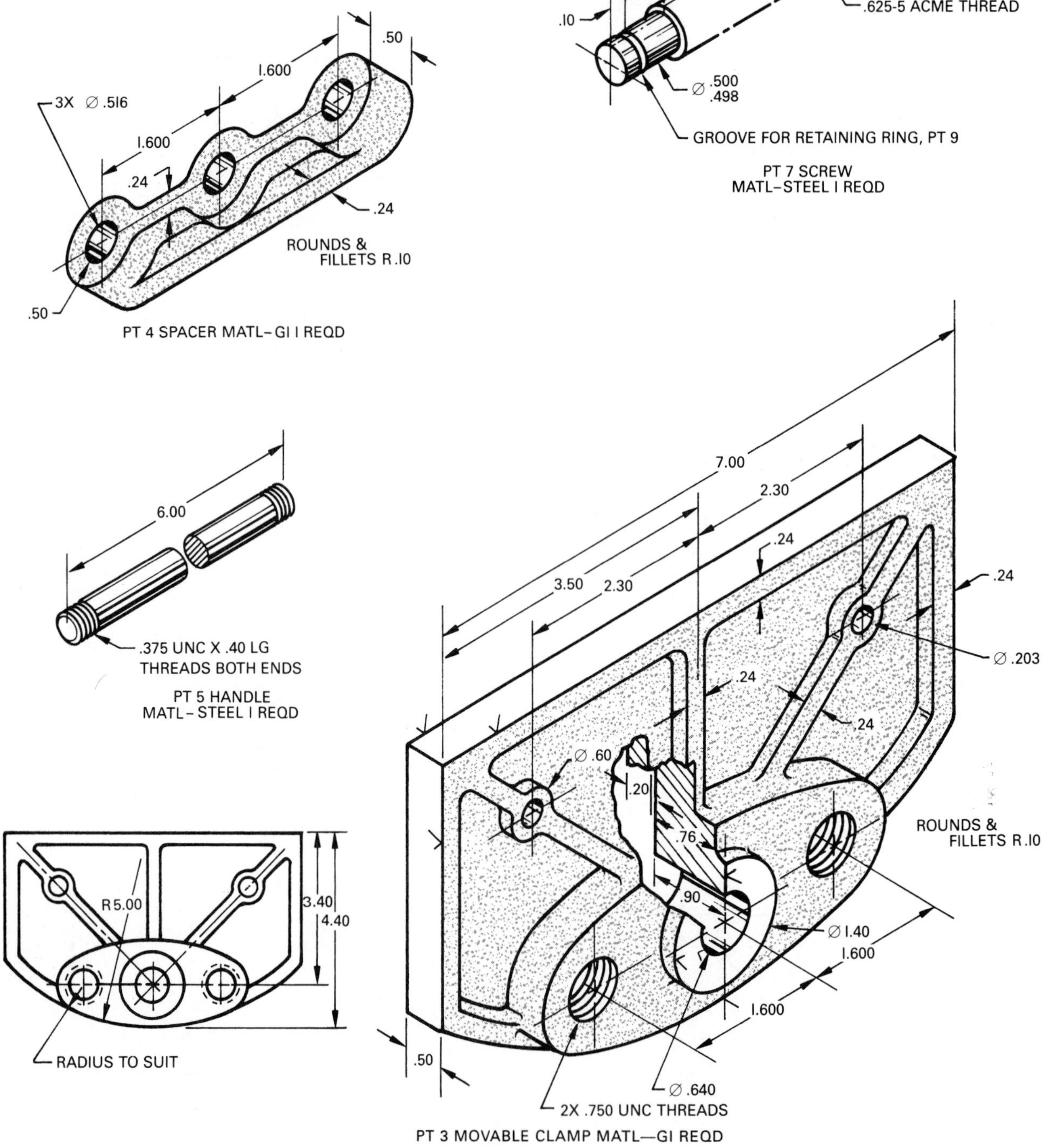

.50
1.600
3X Ø .516
1.600
.24
.24
ROUNDS & FILLETS R.10
.50
PT 4 SPACER MATL–GI 1 REQD
6.00
.375 UNC X .40 LG
THREADS BOTH ENDS
PT 5 HANDLE
MATL–STEEL 1 REQD
R5.00
3.40
4.40
RADIUS TO SUIT
7.00
2.30
.24
3.50
2.30
.24
Ø .203
.24
.24
Ø .60
.20
.76
.90
ROUNDS & FILLETS R.10
Ø 1.40
1.600
1.600
.50
Ø .640
2X .750 UNC THREADS
PT 3 MOVABLE CLAMP MATL—GI REQD

FIG. 14-6-E Wood vise. *(Woden Tools)*

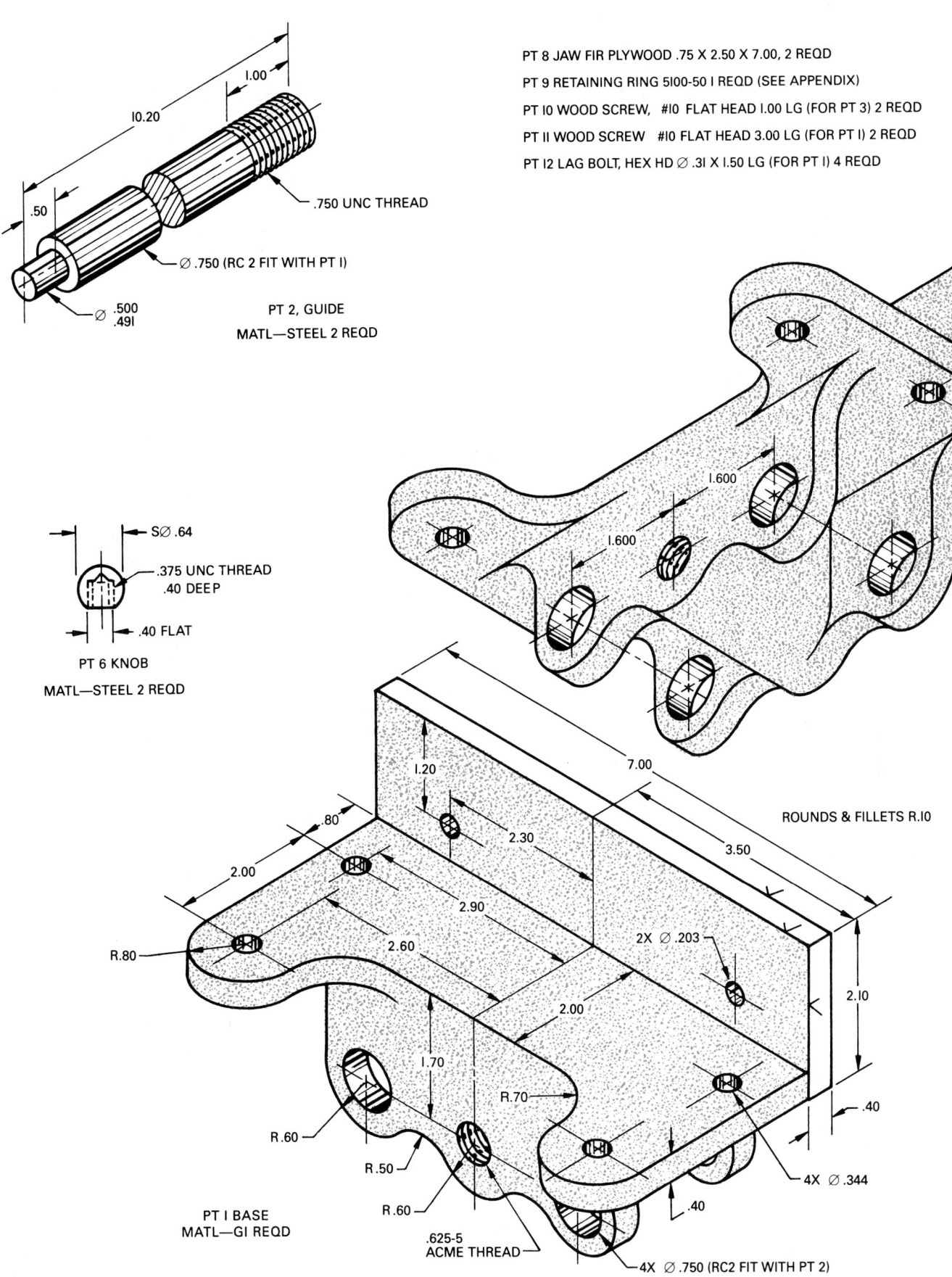

PT 8 JAW FIR PLYWOOD .75 X 2.50 X 7.00, 2 REQD
PT 9 RETAINING RING 5100-50 I REQD (SEE APPENDIX)
PT I0 WOOD SCREW,   #I0 FLAT HEAD I.00 LG (FOR PT 3) 2 REQD
PT II WOOD SCREW   #I0 FLAT HEAD 3.00 LG (FOR PT I) 2 REQD
PT I2 LAG BOLT, HEX HD ∅ .3I X I.50 LG (FOR PT I) 4 REQD

I.00
I0.20
.50
.750 UNC THREAD
∅ .750 (RC 2 FIT WITH PT I)
∅ .500 / .49I
PT 2, GUIDE
MATL—STEEL 2 REQD

S∅ .64
.375 UNC THREAD
.40 DEEP
.40 FLAT
PT 6 KNOB
MATL—STEEL 2 REQD

I.600
I.600

I.20
7.00
ROUNDS & FILLETS R.I0
3.50
.80
2.00
2.30
2.90
2.60
2.00
2X ∅ .203
2.I0
R.80
I.70
R.70
.40
R.60
R.50
R.60
4X ∅ .344
.625-5 ACME THREAD
.40
4X ∅ .750 (RC2 FIT WITH PT 2)
PT I BASE
MATL—GI REQD

FIG. 14-6-E   Wood vise. (continued)

33. Make a two-view assembly drawing of the check valve
shown in Fig. 14-6-F. Show the front view in full section.
Include on the drawing an item list and identification part
numbers. Scale 1:1.

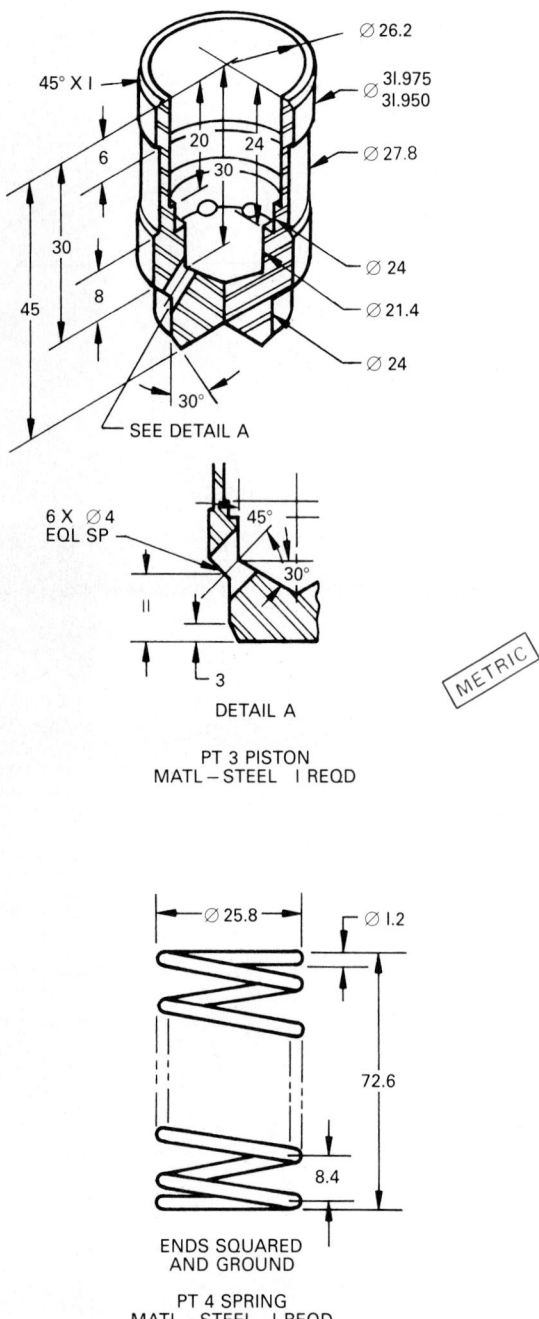

DETAIL A

PT 3 PISTON
MATL — STEEL   I REQD

ENDS SQUARED
AND GROUND

PT 4 SPRING
MATL — STEEL   I REQD

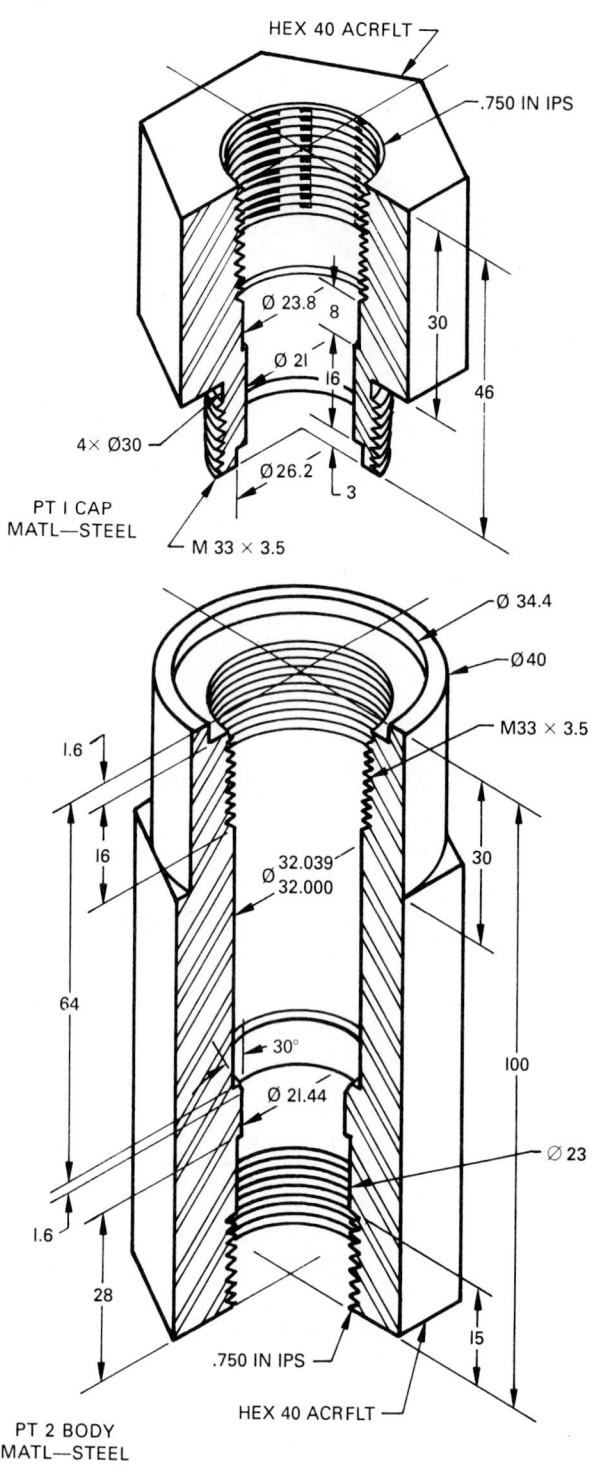

PT I CAP
MATL—STEEL

PT 2 BODY
MATL—STEEL

PT 5 O-RING Ø 2 × 30 ID

FIG. 14-6-F   Check valve. *(Bellows-Valvair)*

34. Make a two-view assembly drawing of the check valve shown in Fig. 14-6-G. Show the front view in full section and the top view through part 5. Include on the drawing an item list and identification part numbers. Scale 1:1.

PT 7 CAP SCREW, SOCKET HD
10-24 UNC X 1.75 LONG 4 REQD

PT 8 LOCKWASHER, #10, 4 REQD

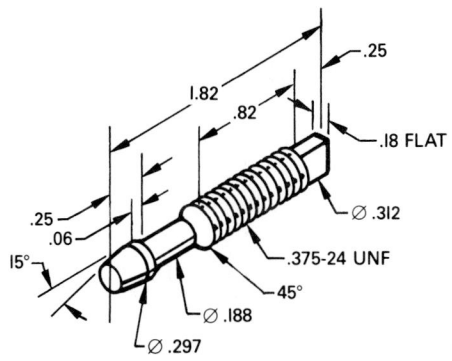

PT 5 VALVE
MATL—STEEL I REQD.

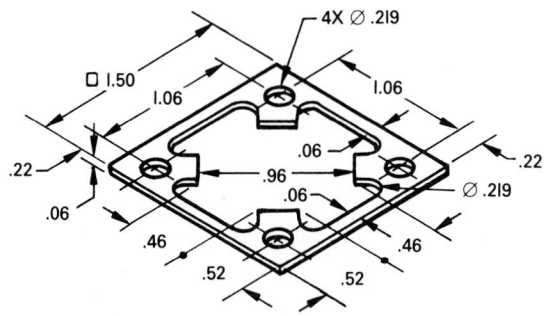

PT 6 GASKET MATL—NEOPRENE I REQD

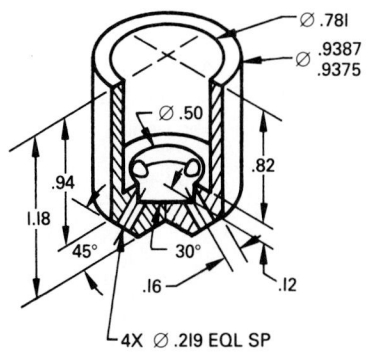

PT 3 PISTON
MATL—ALUMINUM I REQD

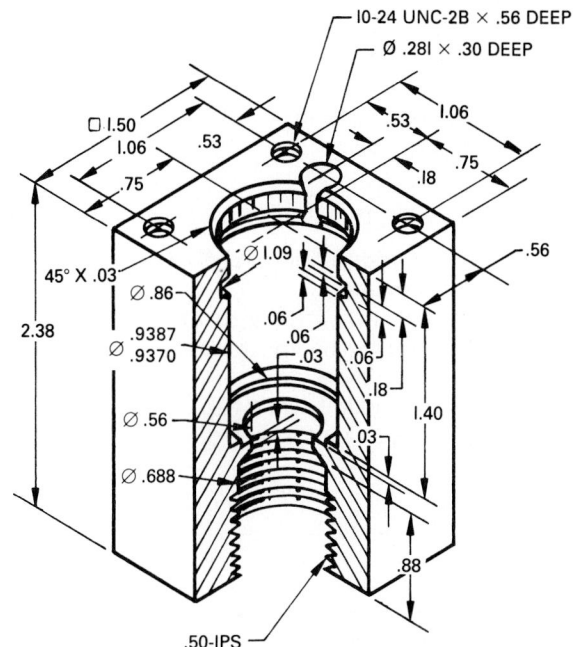

10-24 UNC-2B × .56 DEEP
Ø .281 × .30 DEEP

PT 2 BODY MATL—STEEL I REQD

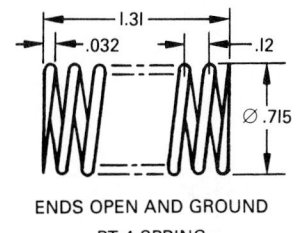

ENDS OPEN AND GROUND
PT 4 SPRING
MATL—STEEL I REQD

Ø .234 ⊤ .56
DRILL FROM
BOTTOM
SURFACE

Ø .234 ⊤ 1.31
Ø .332 ⊤ 1.12
.375-24, UNF ⊤ .62

SECTION A

4X Ø .203
⊔ Ø .348
⊤ 25

.50 iPS

SEE
SECTION A

45° X .03

PT I CAP MATL— STEEL I REQD

FIG. 14-6-G   Check valve. *(Bellows-Valvair)*

427

35. On a B (A3) size sheet make a two-view assembly drawing of the trolley shown in Fig. 14-6-H mounted on an S200 × 34 beam. Show the side view in half section and place on the drawing the dimensions suitable for a catalog. Scale 1:2. Refer to Figs. 25-1-7 and 25-1-8 for dimensions of the beam.

PT 8 ADJUSTING WASHER 26 ID × 44 OD × 4 THK
    12 REQD, MATL—STL
PT 9 RIVET, BUTTON HEAD, Ø10 × 60 LG, 4 REQD
PT 10 WASHER 26 ID × 65 OD × 3 THK, 4 REQD
PT 11 LOCKNUT M16 × 2, 4 REQD
    PREVAILING TORQUE INSERT-TYPE
PT 12 COTTER PIN Ø6 × 40 LG, 6 REQD

METRIC

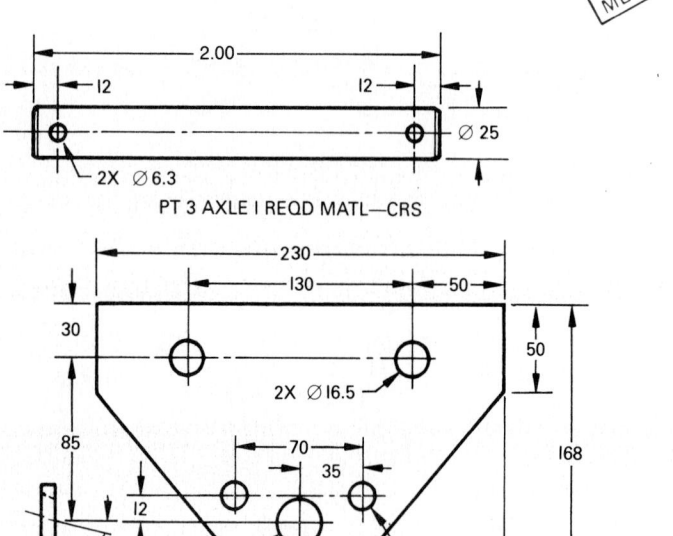

PT 3 AXLE 1 REQD MATL—CRS

2.00
12   12
Ø 25
2X Ø 6.3

PT 1 SIDE PLATE 2 REQD MATL—MST

230
130   50
30
50
85
70
35
12
8.5°
Ø 26
2X Ø16.5
2X Ø10.3
50   90
168

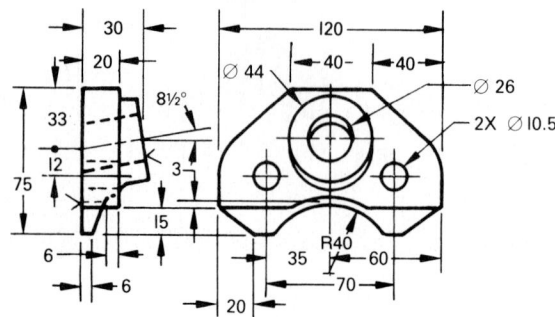

PT 2 GUIDE 2 REQD MATL—CI

30
20
33
75
12
6
6
8½°
3
15
20
120
40   40
Ø 44
Ø 26
2X Ø10.5
R40
35   60
70

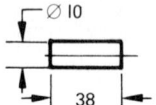

PT 5 ROLLER BEARING

Ø 10
38

44 REQD MATL—CRS
CASE HARDEN!

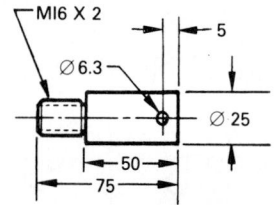

PT 7 AXLE 4 REQD MAT—CRS

M16 X 2
Ø 6.3
Ø 25
5
50
75

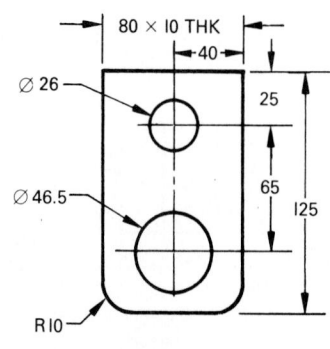

PT 4 HOOK 1 REQD MATL—MST

80 × 10 THK
40
Ø 26
Ø 46.5
R 10
25
65
125

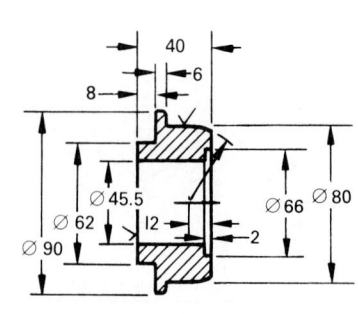

PT 6 WHEEL 4 REQD MAT—CI

40
6
8
Ø 45.5
Ø 62
Ø 90
12
Ø 66
Ø 80
2

FIG. 14-6-H   Trolley.

428

36. On a C (A2) size sheet make a two-view (one view can be a partial view) assembly drawing of the pipe cutter shown in Fig. 14-6-J. Sizes shown are nominal sizes. Include an item list and identification part numbers on the drawing. Scale 1:1.

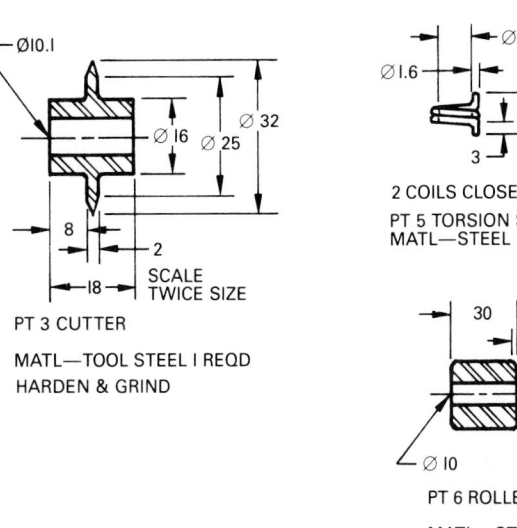

PT 3 CUTTER

MATL—TOOL STEEL I REQD
HARDEN & GRIND

SCALE TWICE SIZE

2 COILS CLOSED
PT 5 TORSION SPRING
MATL—STEEL I REQD

PT 6 ROLLER

MATL—STEEL
CASE HARDEN 2 REQD

METRIC

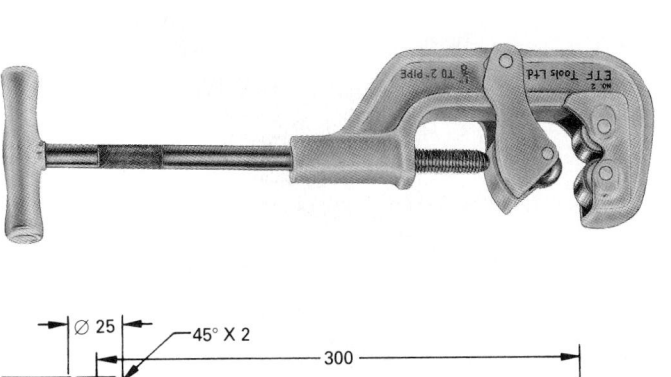

PT 2 HANDLE MATL—CRS I REQD

PT 4 ∅ 10 SPRING PIN

(SEE APPENDIX) 4 REQD

NOTCH IN ONE LEG ONLY FOR PT 5

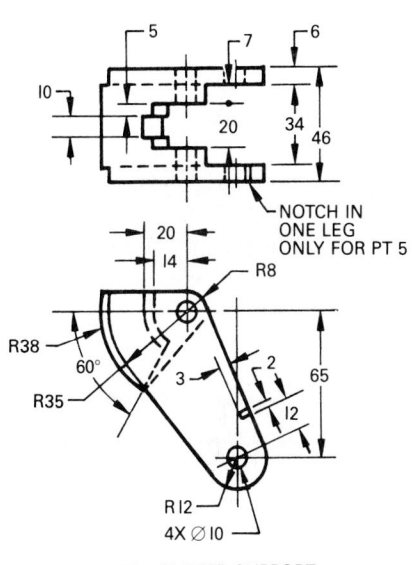

PT 7 CUTTER SUPPORT

MATL—I REQD

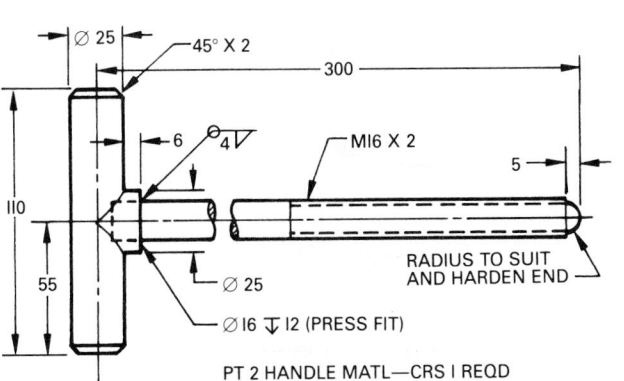
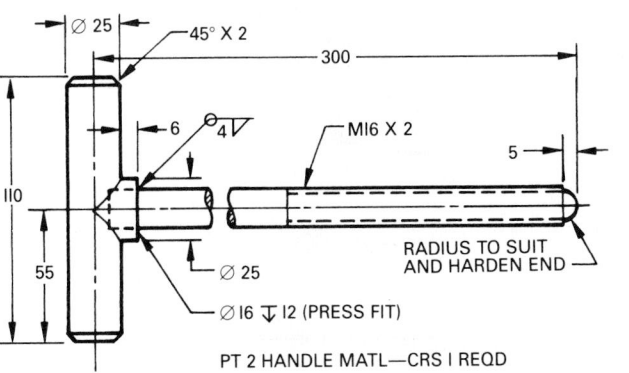

ROUNDS AND FILLETS R2

PT I FRAME MATL—GI I REQD

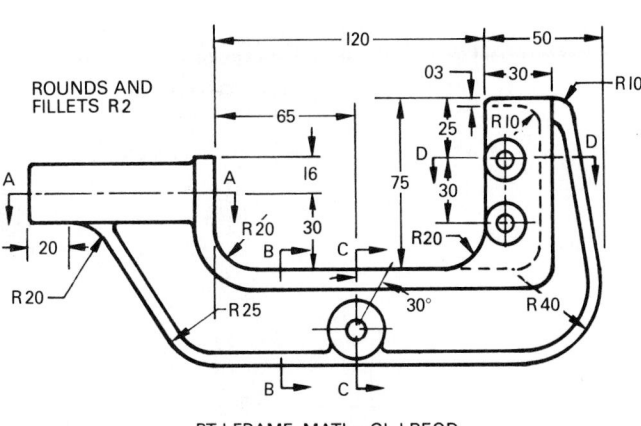

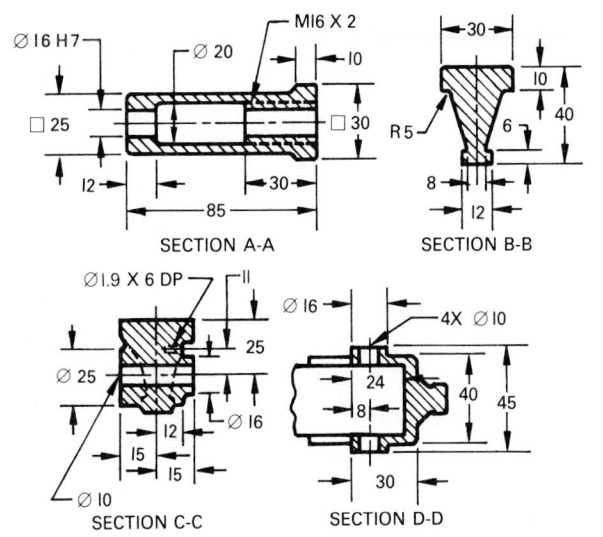

SECTION A-A

SECTION B-B

SECTION C-C

SECTION D-D

FIG. 14-6-J   Pipe cutter.

37. On a C (A2) size sheet make a one-view assembly drawing of the two-arm parallel puller shown in Fig. 14-6-K
removing the tapered roller bearing from the shaft shown
in assembly A. Include on the drawing an item list and
identification part numbers. Scale 1:1.

PT 19 BALL BEARING ⌀ .375 MATL—STEEL, 1 REQD

PT 20 GREASE CUP

PT 21 BOLT—HEX HD .312 UNF X 1.50 LG, 6 REQD

PT 22 BOLT—HEX HD .312 UNF X 1.75 LG, 5 REQD

PT 23 BOLT—HEX HD .312 UNF X 2.50 LG, 4 REQD

PT 24 MACH SCREW—HEX HD 8-32 X 1.25 LG, 4 REQD

PT 25 NUT—HEX HD .312 UNF, 10 REQD

PT 26 NUT—HEX HD 8-32, 4 REQD

PT 27 SETSCREW—HEADLESS .375 UNF X .50 LG
                                    CUP POINT, 2 REQD

PT 28 SETSCREW—HEADLESS 8-32 X .25 LG
                                    FULL DOG, 2 REQD

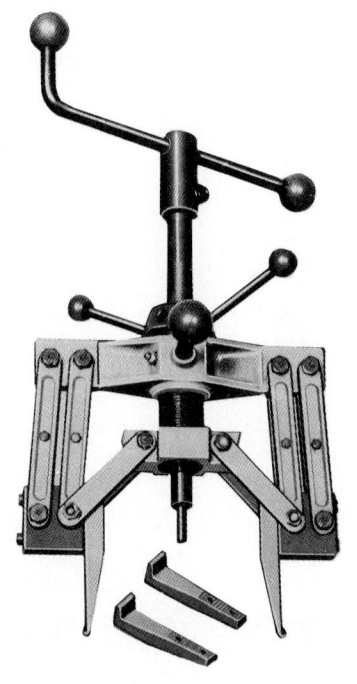

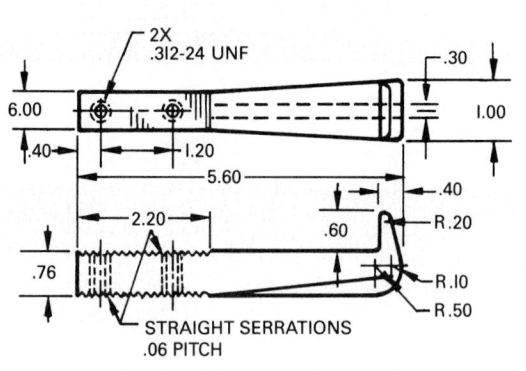

PT 11   FINGERS MATL—CI 2 REQD

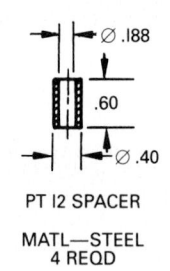

PT 12 SPACER

MATL—STEEL
4 REQD

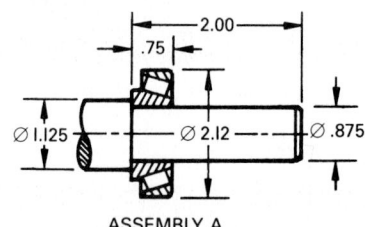

ASSEMBLY A

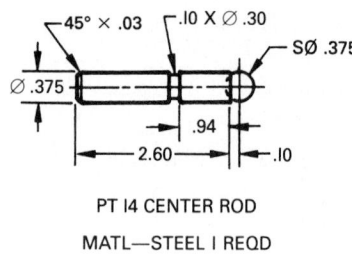

PT 14 CENTER ROD

MATL—STEEL 1 REQD

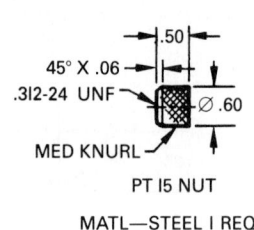

PT 15 NUT

MATL—STEEL 1 REQD

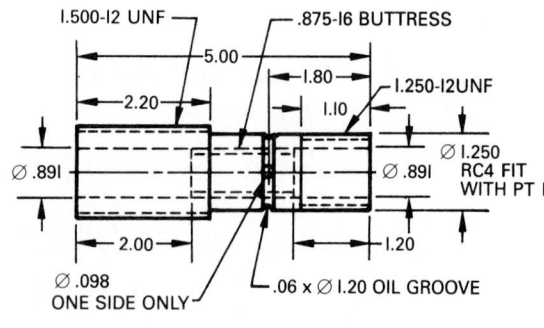

PT 13 ADJUSTING SCREW MATL—STEEL 1 REQD

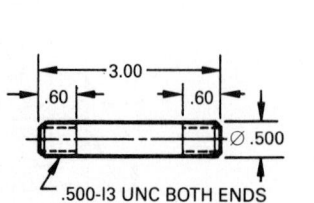

PT 16 HANDLE MATL—STEEL 2 REQD

PT 17 HANDLE MATL—STEEL 1 REQD

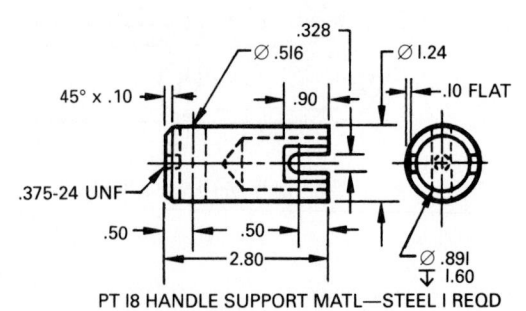

PT 18 HANDLE SUPPORT MATL—STEEL 1 REQD

FIG. 14-6-K   Two-arm parallel puller. *(Delro Industries)*

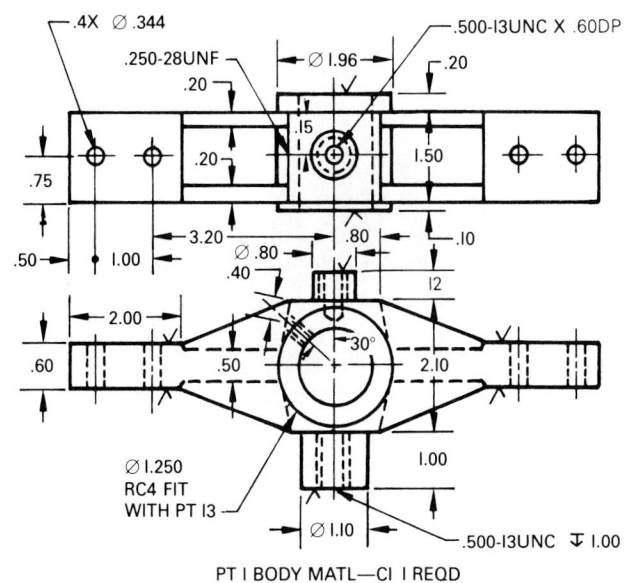

PT I BODY MATL—CI I REQD

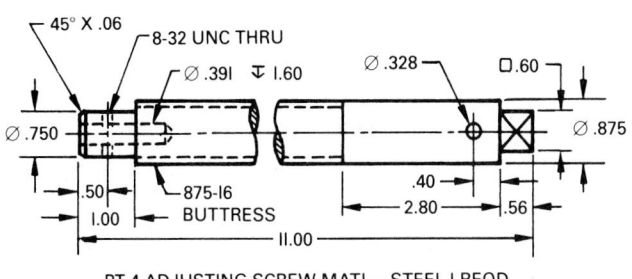

PT 4 ADJUSTING SCREW MATL—STEEL I REQD

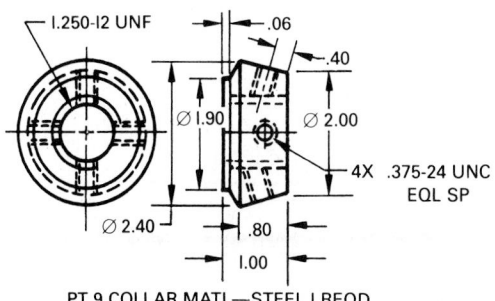

PT 9 COLLAR MATL—STEEL I REQD

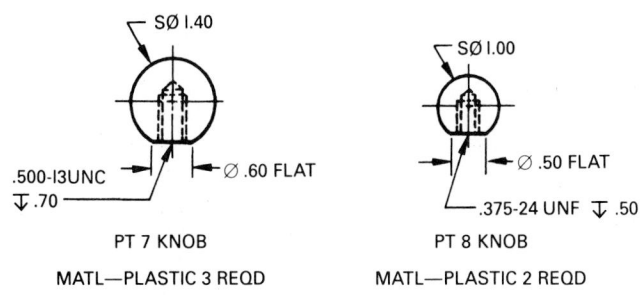

PT 7 KNOB

MATL—PLASTIC 3 REQD

PT 8 KNOB

MATL—PLASTIC 2 REQD

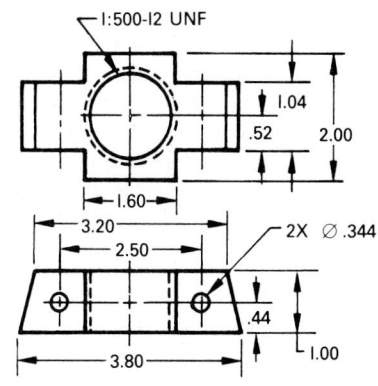

PT 2 SLIDE BAR MATL—STEEL I REQD

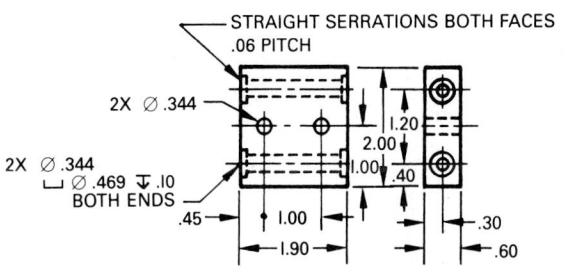

PT 3 JAW MATL—STEEL 2 REQD

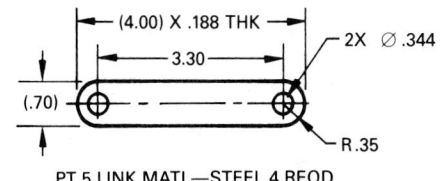

PT 5 LINK MATL—STEEL 4 REQD

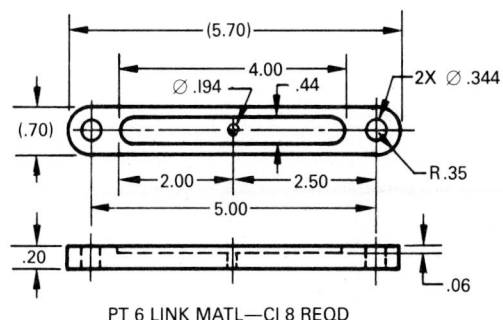

PT 6 LINK MATL—CI 8 REQD

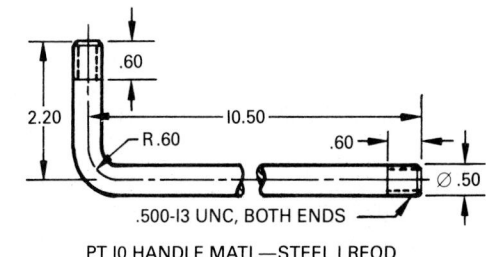

PT I0 HANDLE MATL—STEEL I REQD

FIG. 14-6-K   Two-arm parallel puller. (continued)

431

38. On a C (A2) size sheet make a front-view, full-section assembly drawing plus a partial side view of the journal jack shown in Fig. 14-6-L. Show the jack in its lowest position and a phantom view of part three 3.00 in. higher. Use this phantom view to indicate the maximum jack height and the regular view to indicate the minimum height. Include an item list and identification part numbers. Scale 1:1.

PT I9 FLAT WASHER MATL—STEEL

    1.19 ID × 2.25 OD × .180 I REQD

PT 20 PIN MATL—STEEL Ø.188 × 1.00 LG I REQD

PT 2I BALL BEARING SØ .625 MATL—STEEL I2 REQD

PT 22 KEY—608 WOODRUFF, I REQD

PT 23 PIN MATL—STEEL Ø.25 × .40 LG I REQD

PT 24 COTTER PIN Ø.I25 × I.25 LG I REQD

PT 25 COTTER PIN Ø.094 × .75 LG I REQD

PT 26 HANDLE .875 ID × 1.00 OD × I8.00 LG STL I REQD

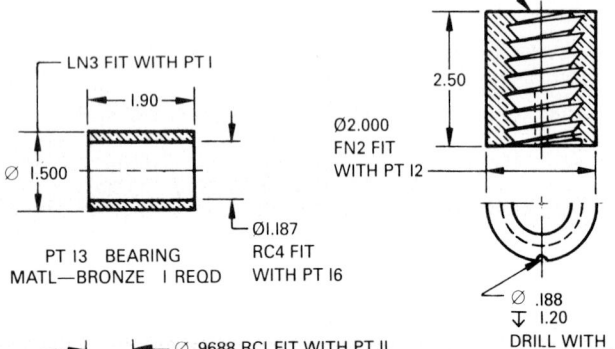

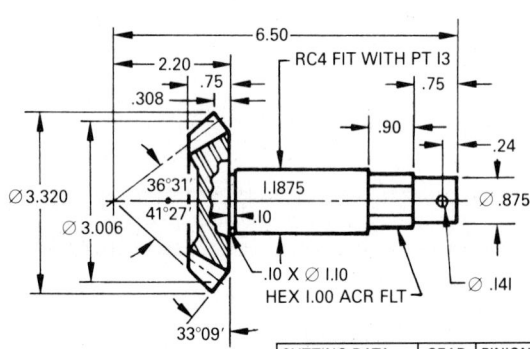

CUTTING DATA	GEAR	PINION
NO. OF TEETH	20	I5
DIAMETRAL PITCH	5	5
TOOTH FORM	I4-1/2°	I4-1/2°
CUTTING ANGLE	47°-50'	3I°-46'
WHOLE DEPTH	.43I	.43I
CHORDAL ADD	.203	.203
CHORDAL THICK	.3I4	.3I4

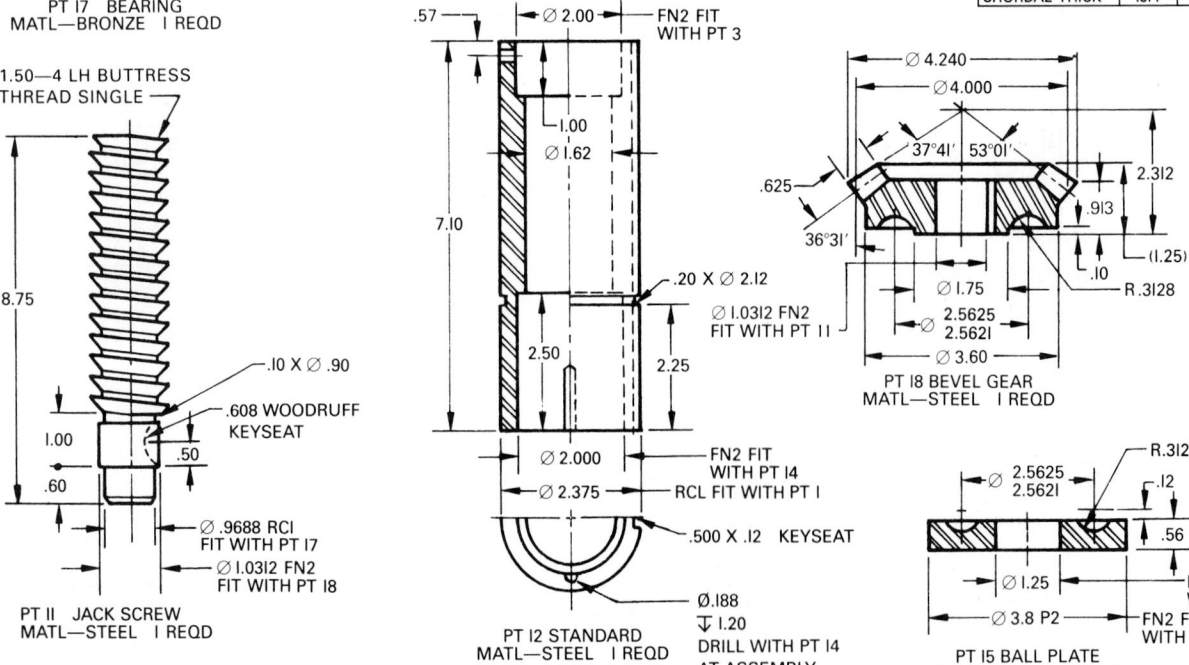

FIG. 14-6-L   Journal jack. *(Duff-Norton Co.)*

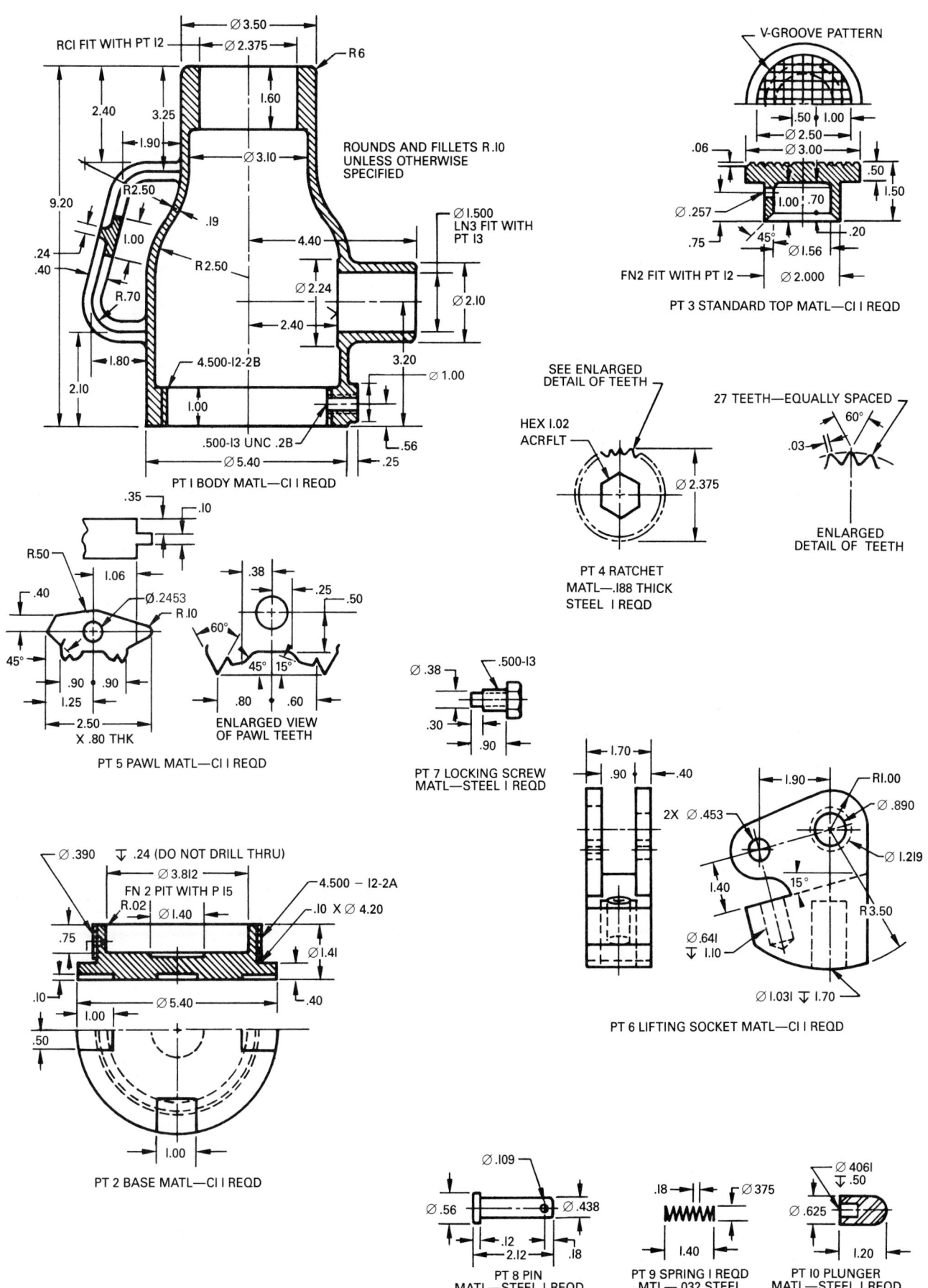

FIG. 14-6-L   Journal jack. (continued)

39. On a C (A2) size sheet make a two-view assembly drawing (front and side) of the grinder shown in Fig. 14-6-M. Show the front view in half section. Include on the drawing an item list and identification part numbers. Scale 1:1.

40. Make detail drawings of parts 1 and 4 shown in Fig. 14-6-M, replacing the descriptive fit terms with appropriate dimensions.

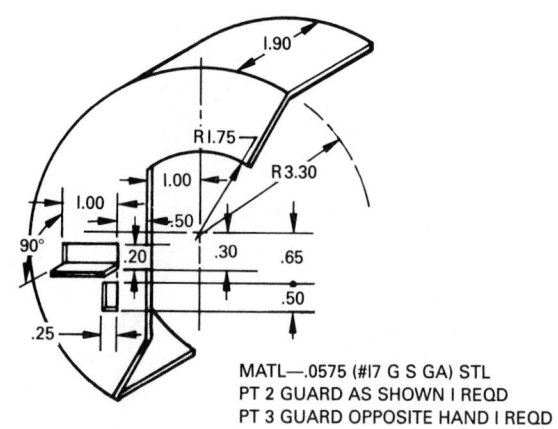

MATL—.0575 (#17 G S GA) STL
PT 2 GUARD AS SHOWN 1 REQD
PT 3 GUARD OPPOSITE HAND 1 REQD

PT 9 BEARING SKF 600Z-2Z (SEE APPENDIX) 2 REQD

PT 10 GRINDING WHEEL—FINE 6.00 OD × 1.00 THK × Ø .501 BORE, 1 REQD

PT 11 GRINDING WHEEL—MED. 6.00 OD × 1.00 THK × Ø .501 BORE, 1 REQD

PT 12 CARRIAGE BOLT .250-20 UNC × 1.00 LG, 4 REQD

PT 13 WASHER PLAIN TYPE A .281 ID × .625 OD × .065, 4 REQD

PT 14 SETSCREW—SLOTTED HEAD CUP POINT .250—20UNC × .31 LG, 1 REQD

PT 15 WING NUT .250-20 UNC, 4 REQD

PT 16 NUT—REG HEX .500 UNC, 1 REQD

PT 17 NUT—REG HEX .500 UNC-LH, 1 REQD

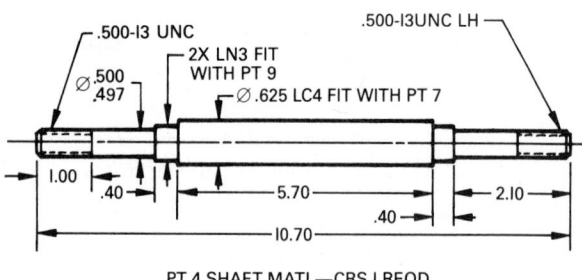

PT 4 SHAFT MATL—CRS 1 REQD

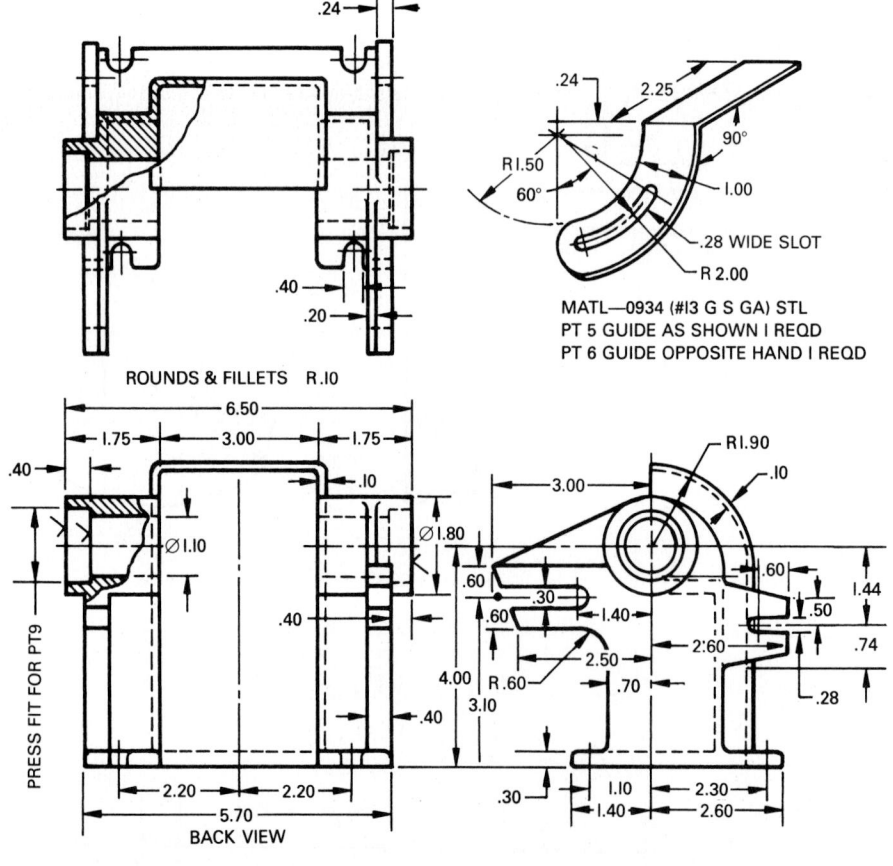

ROUNDS & FILLETS   R.10

MATL—0934 (#13 G S GA) STL
PT 5 GUIDE AS SHOWN 1 REQD
PT 6 GUIDE OPPOSITE HAND 1 REQD

PT 7 PULLEY
MATL—STL 1 REQD

PT 1 BASE MATERIAL—CI 1 REQD

BACK VIEW

PT 8 SPACER
MATL—.0934 (#13 G S GA)
4 REQD

FIG. 14-6-M   Bench grinder.

ASSIGNMENTS FOR UNIT 14-7, EXPLODED ASSEMBLY DRAWINGS

Use center lines to align parts. Include on the drawing an item list and identification part numbers. Scale 1:1.

41. Make an exploded isometric assembly drawing of one of the assemblies shown in Figs. 14-2-J, 14-7-A, and 14-7-B.

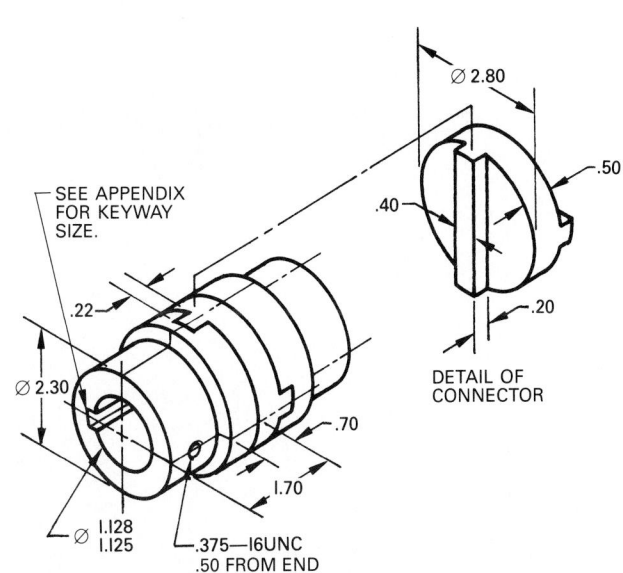

SEE APPENDIX FOR KEYWAY SIZE.

DETAIL OF CONNECTOR

FIG. 14-7-A   Coupling.

FIG. 14-7-B   Universal joint.

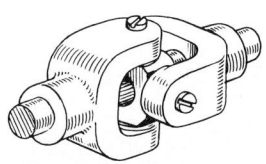

ITEM LIST			
PT	ITEM	QTY	MATL	DESCRIPTION
1	FORK	2	WI	
2	RING	1	STL	
3	STUD	4	STL	
4	NO.4 TAPER PIN	2	STL	PURCHASED

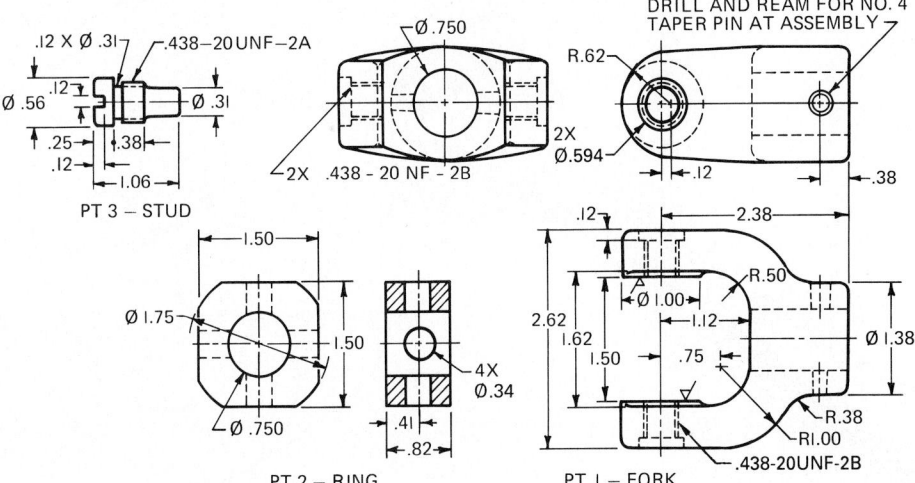

PT 3 – STUD

PT 2 – RING

PT 1 – FORK

42. Make an exploded assembly drawing in orthographic projection of one of the assemblies shown in Figs. 14-7-B and 14-7-C. Use center lines to align the parts. Include on the drawing an item list and identification part numbers.

ASSIGNMENT FOR UNIT 14-8, DETAILED ASSEMBLY DRAWINGS

43. Make a three-view detailed assembly drawing of any one of the assemblies shown in Figs. 14-8-A, 14-8-C, and

FIG. 14-7-C   Caster.

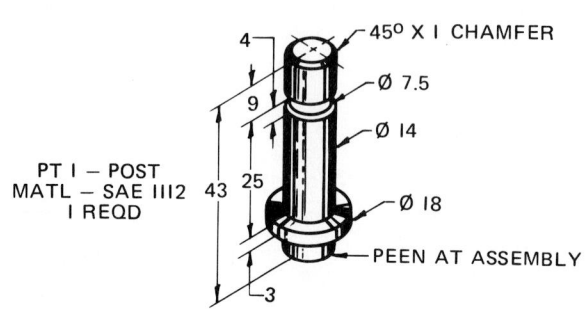

PT I – POST
MATL – SAE III2
I REQD

45° X I CHAMFER
Ø 7.5
Ø 14
Ø 18
PEEN AT ASSEMBLY

4
9
43
25
3

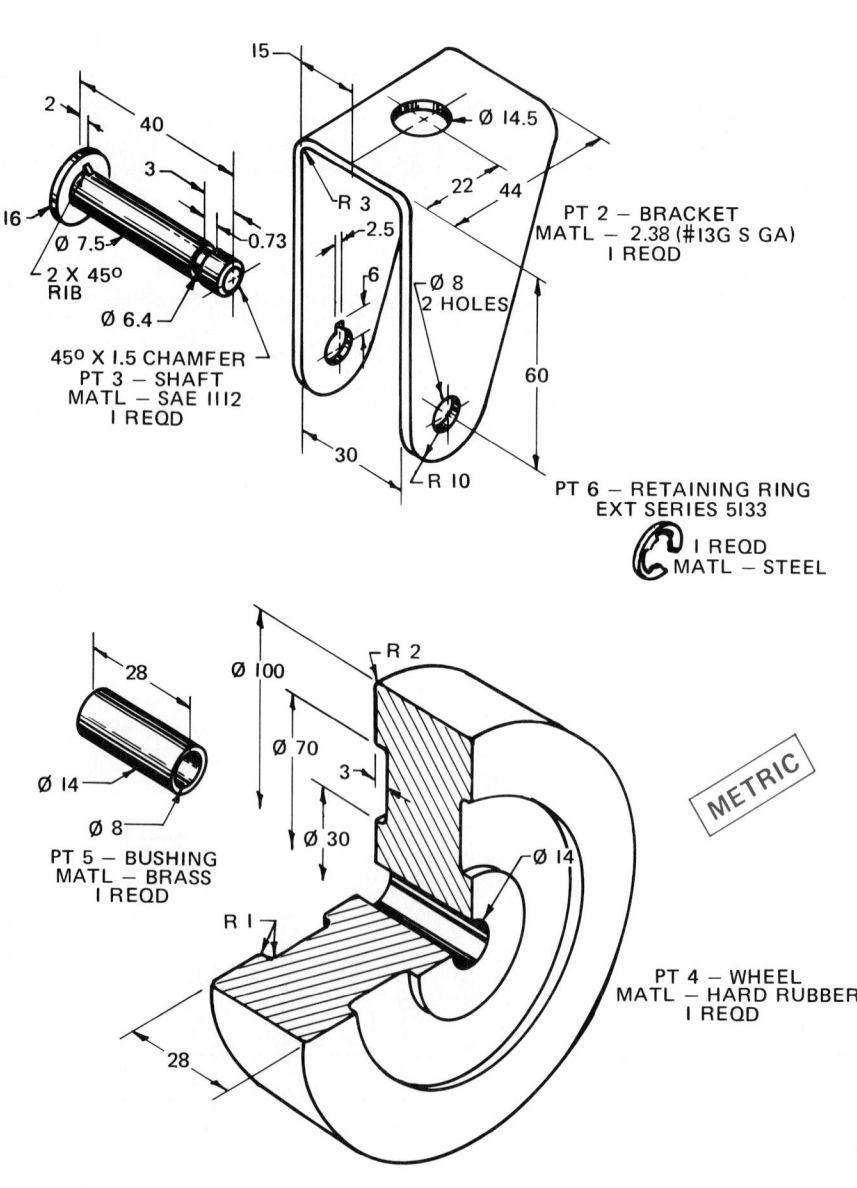

2
40
Ø 16
Ø 7.5
3
0.73
2 X 45°
RIB
Ø 6.4
45° X 1.5 CHAMFER
PT 3 – SHAFT
MATL – SAE III2
I REQD

15
Ø 14.5
R 3
2.5
6
Ø 8
2 HOLES
22   44
60
30
R 10
PT 2 – BRACKET
MATL – 2.38 (#13G S GA)
I REQD

PT 6 – RETAINING RING
EXT SERIES 5133
I REQD
MATL – STEEL

28
Ø 14
Ø 8
PT 5 – BUSHING
MATL – BRASS
I REQD

Ø 100
Ø 70
Ø 30
R 2
3
R I
Ø 14
28
METRIC
PT 4 – WHEEL
MATL – HARD RUBBER
I REQD

14-8-D. For Fig. 14-8-B only a partial assembly drawing is required. Include on the drawing the method of assembly (i.e., nails, wood screws, dowels, etc.) and an item list and identification part numbers. Include in the item list the assembly parts. The student is to select scale and drawing paper size. For Fig. 14-8-C the basic sizes are given. Design a table of your choice. Show on the drawing how the sides and feet are designed and fastened.

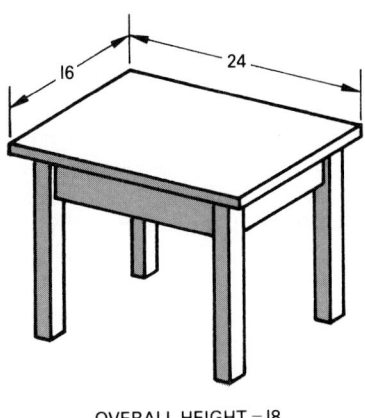

OVERALL HEIGHT = 18

DIMENSIONS SHOWN ARE IN INCHES.

FIG. 14-8-C   Night table.

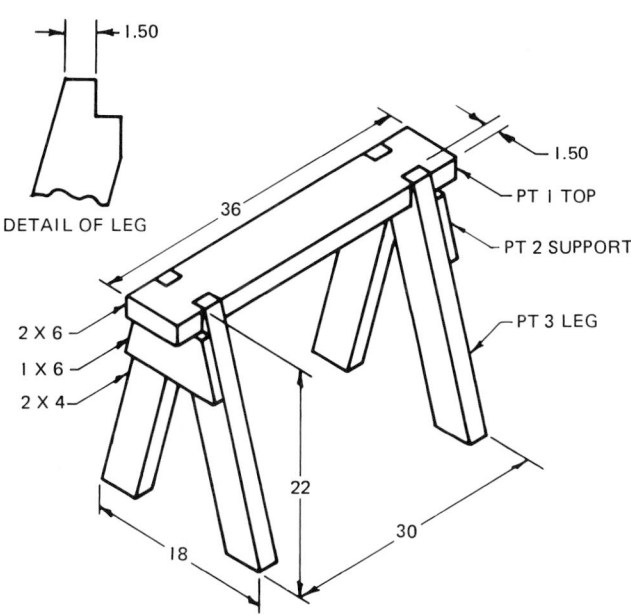

DETAIL OF LEG

PT 1 TOP
PT 2 SUPPORT
PT 3 LEG

2 X 6
1 X 6
2 X 4

MATL — CONSTRUCTION GRADE SPRUCE

NOTE: WOOD SIZES (THICKNESS AND WIDTH) ARE NOMINAL INCH SIZES.

FIG. 14-8-A   Sawhorse.

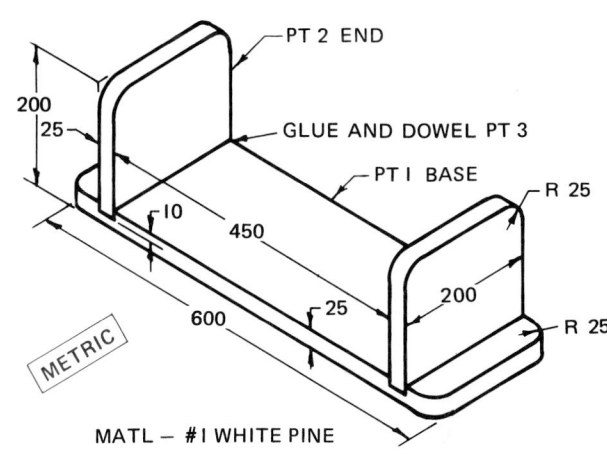

PT 2 END
GLUE AND DOWEL PT 3
PT 1 BASE
R 25
R 25
METRIC

MATL — #1 WHITE PINE

FIG. 14-8-D   Book rack.

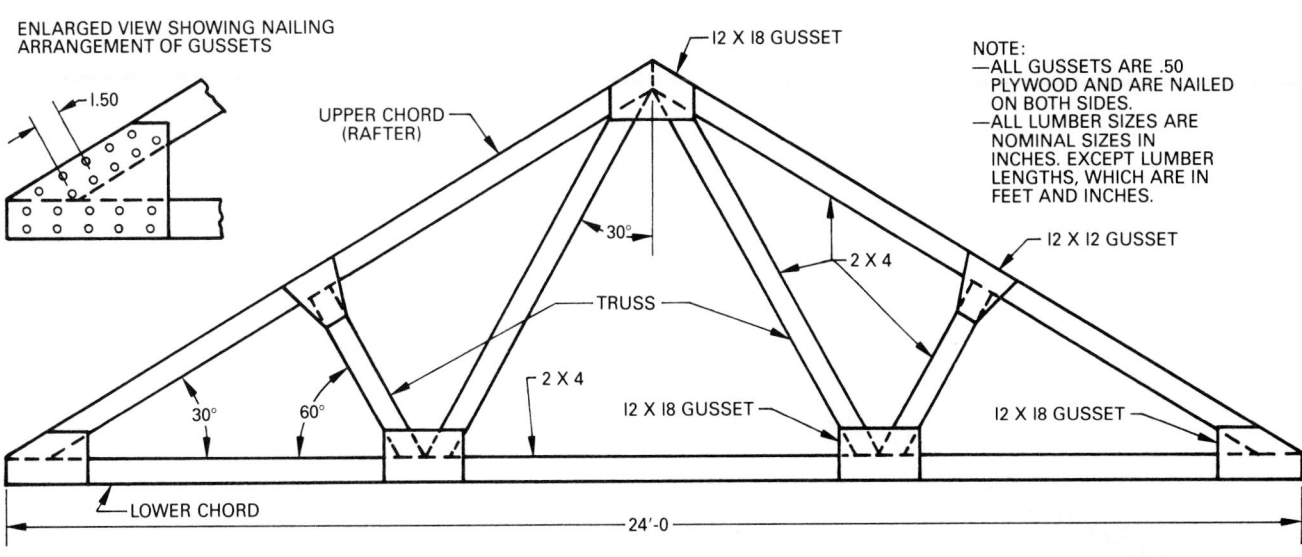

ENLARGED VIEW SHOWING NAILING ARRANGEMENT OF GUSSETS

12 X 18 GUSSET

UPPER CHORD (RAFTER)

NOTE:
—ALL GUSSETS ARE .50 PLYWOOD AND ARE NAILED ON BOTH SIDES.
—ALL LUMBER SIZES ARE NOMINAL SIZES IN INCHES. EXCEPT LUMBER LENGTHS, WHICH ARE IN FEET AND INCHES.

12 X 12 GUSSET

30°

2 X 4

TRUSS

30°   60°

2 X 4

12 X 18 GUSSET

12 X 18 GUSSET

LOWER CHORD

24'-0

FIG. 14-8-B   Roof truss.

ASSIGNMENTS FOR UNIT 14-9, SUBASSEMBLY DRAWINGS

44. For Fig. 14-9-A make a two-view subassembly drawing of die set B1-361 with dimensions, identification part numbers, and an item list. K = 7.50. Select a plain (journal) bearing from the Appendix. Convert to decimal inch dimensions. Scale 1:2. Identify the hole and shaft sizes for the fits shown.

45. For Fig. 14-9-B make a one-view subassembly drawing of the wheel. A broken-out or partial section view is recommended to show the interior features. Include on the drawing pertinent dimensions, identification part numbers, and an item list. Four Ø10 mm bolts fasten the wheel to an 8 mm plate. Scale 1:1.

RC4 FIT BETWEEN BUSHING AND SHAFT J

LN2 FIT BETWEEN BUSHING AND PUNCH HOLDER

LN2 FIT BETWEEN SHAFT AND DIE SHOE

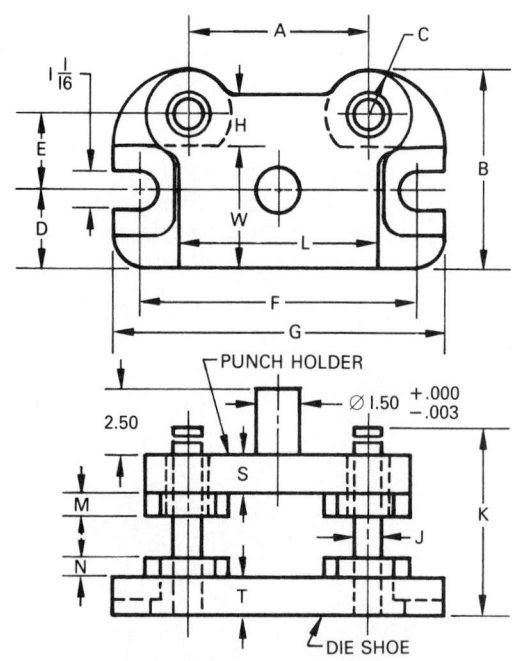

Width Inches W	Length Inches L	Thickness Die T	Thickness Punch S	Catalog Number	A	B	C	D	E	F	G	H	J	K	M	N
3	4	1-1/4	1	BI-341	3	5	1-1/8	2	1-7/8	6-1/2	7-1/2	1-1/2	3/4	SPECIFY	5/8	1/2
		1-1/2	1	BI-342												
3	6	1-1/2	1-1/4	BI-361	5	5-1/4	1-1/4	2	2	8-1/2	10	1-11/16	7/8	SPECIFY	3/4	5/8
		1-1/2	1-3/4	BI-362												
		2	1-1/4	BI-363												
		2	1-3/4	BI-364												

FIG. 14-9-A   Die sets. *(E.A. Baumbach Mfg. Co.)*

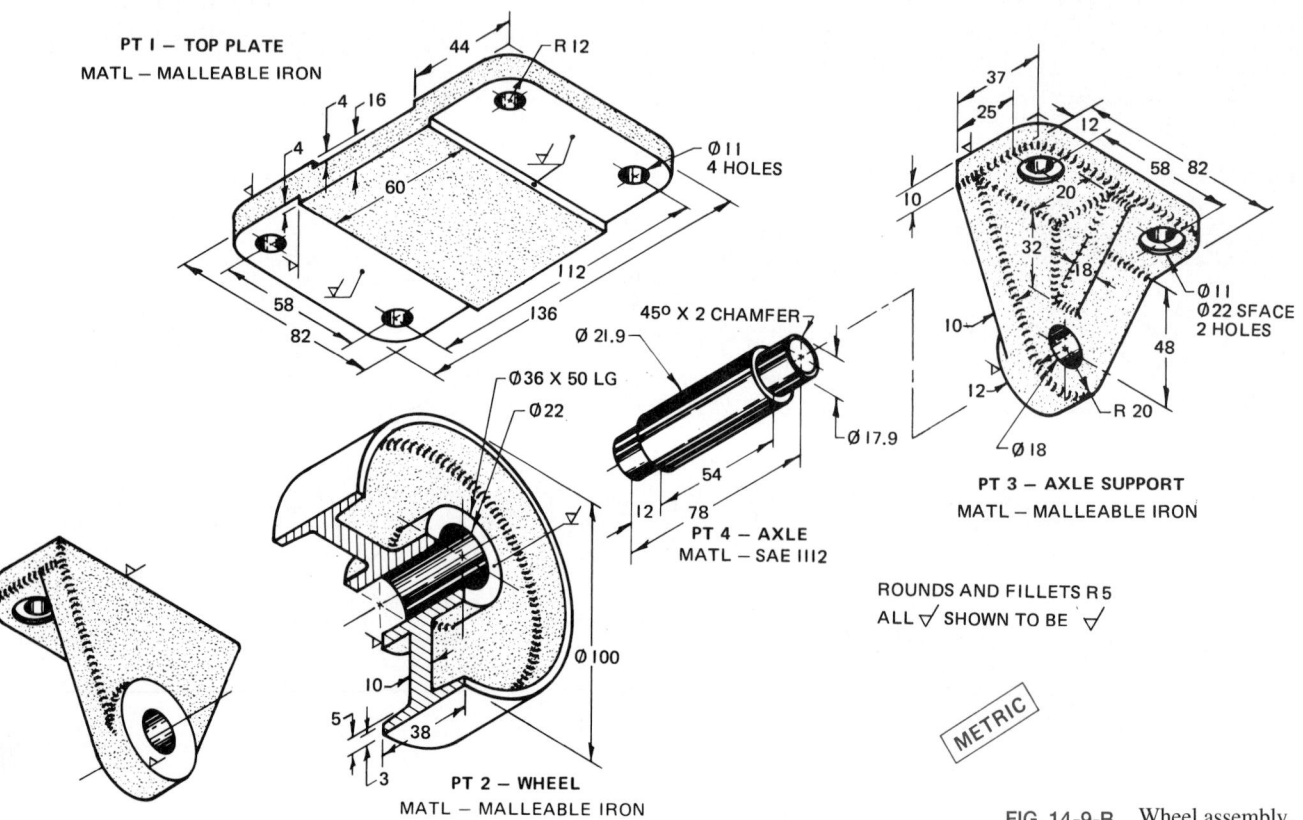

FIG. 14-9-B   Wheel assembly.

46. On a B (A3) size sheet make a partial-section assembly drawing of the idler pulley shown in Fig. 14-9-C. Place on the drawing dimensions suitable for a catalog. Add to the drawing an item list and identification part numbers. Scale 1:1.

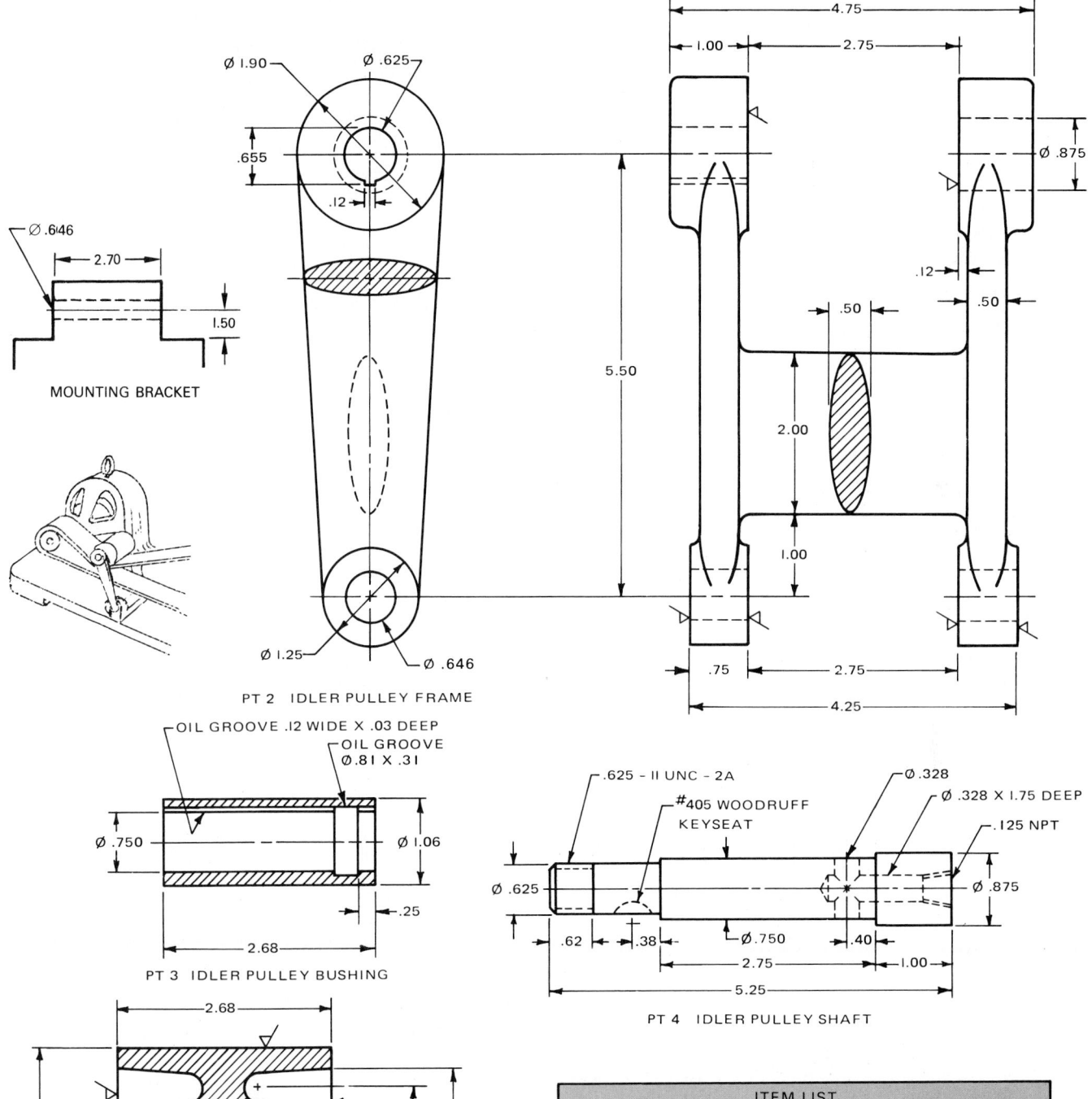

MOUNTING BRACKET

PT 2   IDLER PULLEY FRAME

OIL GROOVE .12 WIDE X .03 DEEP
OIL GROOVE Ø.81 X .31

PT 3   IDLER PULLEY BUSHING

PT 4   IDLER PULLEY SHAFT

.625 - 11 UNC - 2A
#405 WOODRUFF KEYSEAT
Ø .328
Ø .328 X 1.75 DEEP
.125 NPT

PT 1   IDLER PULLEY

ITEM LIST				
PT	ITEM	MATL	DESCRIPTION	QTY
1	IDLER PULLEY	GI	A - 5432	1
2	IDLER PULLEY FRAME	GI	A - 1734	1
3	IDLER PULLEY BUSHING	BRZ		1
4	IDLER PULLEY SHAFT	SAE 1120		1
5	HEX NUT	STEEL	.625 UNC	1
6	WOODRUFF KEY	STD	NO. 405	1
7	OILER	STD	.125 NPT	1

FIG. 14-9-C   Idler pulley.

439

47. On a B (A3) size sheet make a one-view, full-section drawing of the clutch shown in Fig. 14-9-D. A gear is mounted on the hub of a Formsprag overrunning clutch, Model 12. Shaft dia. 1.375 in. Use your judgment for dimensions not shown. Gear data: 20° spur gear; 6.000 PD; DP = 4; 1.00 tooth face; hub Ø3.50 × 1.340 wide; hub projection one side. Scale 1:1.

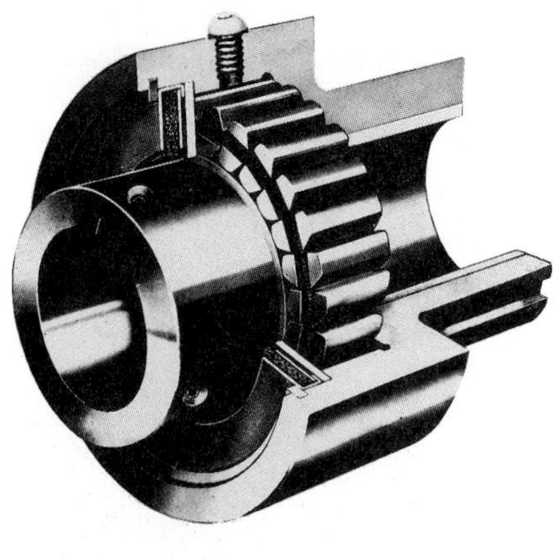

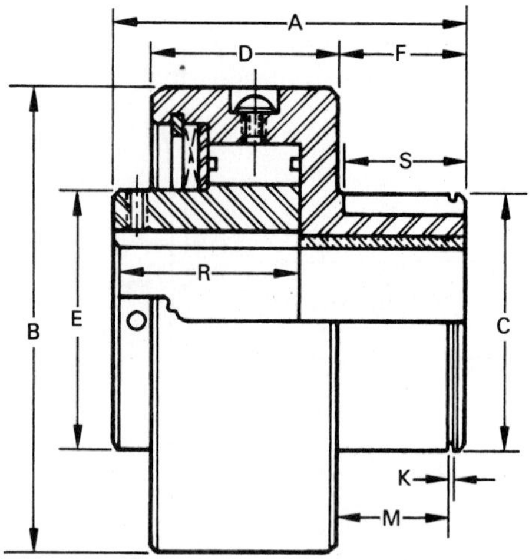

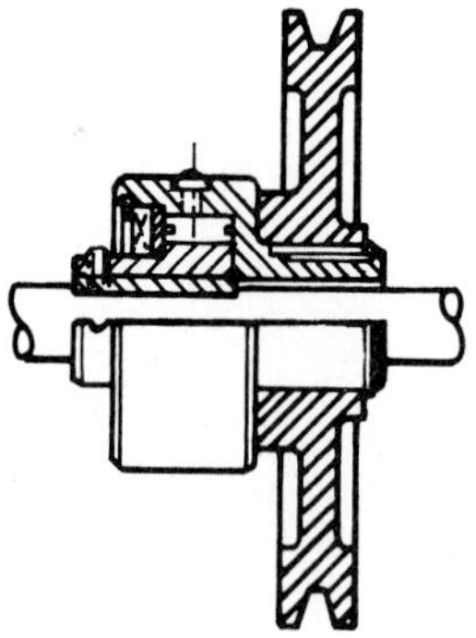

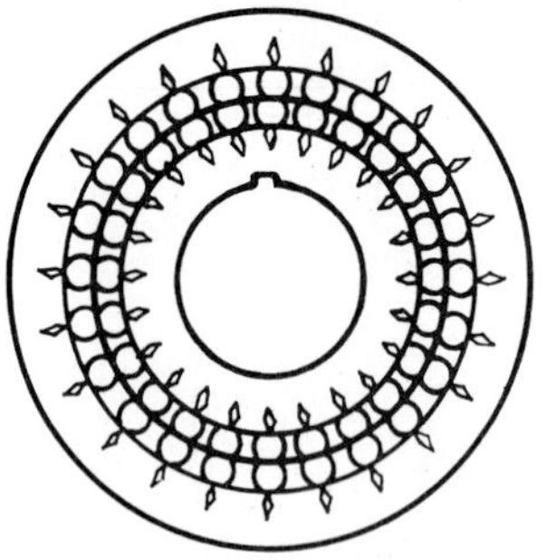

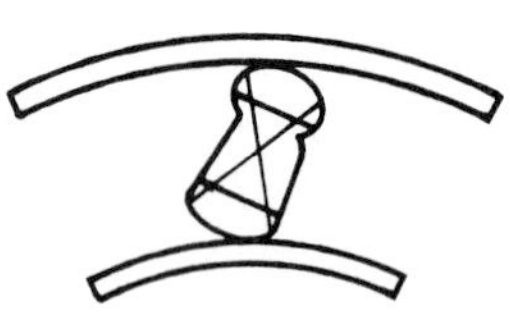

F-S SERIES MODEL NUMBER	3	5	6	8	10	12	14	16
Torque Capacity (Lb.Ft.)	15	40	130	190	420	600	1100	1300
Standard Bore Size	.375	.500 .625	.750	.875 1.000	1.125 1.250	1.375 1.500	1.625 1.750	1.875 2.000
Keyway		1/8 × 1/16 3/16 × 3/32	3/16 × 3/32	1/4 × 1/8	5/16 × 5/32	5/16 × 5/32 3/8 × 3/16	7/16 × 7/32	1/2 × 1/4
Standard Hub Keyway	1/8 × 1/16	3/16 × 3/32	3/16 × 3/32	1/4 × 1/8	5/16 × 5/32	3/8 × 3/16	7/16 × 7/32	1/2 × 1/4
A	2	2 3/4	3 3/16	3 5/16	3 5/8	3 7/8	4 3/8	4.75
B	1 5/8	2 3/16	2 7/8	3 1/4	3 3/4	4 7/16	5 1/4	5.90
C	.875 .874	1.250 1.249	1.375 1.374	1.750 1.749	2.250 2.249	2.500 2.499	2.875 2.874	3.250 3.249
D	13/16	1 1/4	1 15/32	1 5/8	1 3/4	1 15/16	2 1/4	2 1/2
E	1	1 1/16	1 1/4	1 5/8	2 1/32	2 3/8	2 7/8	3 1/4
F	13/16	1	1 5/16	1 5/16	1 7/16	1 7/16	1 5/8	1 3/4
K	.056 .036	.068 .048	.068 .048	.076 .056	.076 .056	.076 .056	.076 .056	.088 .068
M	.720 .715	.905 .900	1.220 1.215	1.220 1.215	1.345 1.340	1.345 1.340	1.530 1.525	1.655 1.650
R	1	1 15/32	1 9/16	1 11/16	1 51/64	2 1/8	2 11/32	2 1/2
S	1/2	9/16	15/16	7/8	15/16	1 3/8	1 9/16	1 11/16
Oil Hole		1/4 – 28	1/4 – 28	1/4 – 28	1/4 – 28	1/4 – 28	1/4 – 28	1/4 – 28

A full complement of sprags between concentric inner and outer races transmits power from one race to the other by wedging action of the sprags when either race is rotated in the driving direction. Rotation in the opposite direction frees the sprags and clutch is disengaged or "overruns."

FIG. 14-9-D Overrunning clutch. *(Formsprag)*

CHAPTER 15

PICTORIAL DRAWINGS

Definitions

Axonometric projection Projection in which the lines of sight are perpendicular to the plane of projection, but in which the three faces of a rectangular object are all inclined to the plane of projection.

Isometric drawing Projection based on the procedure of revolving the object at an angle of 45° to the horizontal, so that the front corner is toward the viewer, then tipping the object up or down at an angle of 35°16'.

Non-isometric lines Lines used to depict sloping surfaces in isometric drawings.

Oblique projection Projection based on the procedure of placing the object with one face parallel to the frontal plane and placing the other two faces on oblique planes.

Perspective projection A method of drawing that depicts a three-dimensional object on a flat plane as it appears to the eye.

Unidirectional dimensioning The preferred method of dimensioning isometric drawings.

15-1 PICTORIAL DRAWINGS

Pictorial drawing is the oldest written method of communication known, but the character of pictorial drawing has continually changed with the advance of civilization. In this text only those kinds of pictorial drawings commonly used by the engineer, designer, and drafter are considered. Pictorial drawings are useful in design, construction or production, erection or assembly, service or repairs, and sales.

There are three general types into which pictorial drawings may be divided: axonometric, oblique, and perspective. These three differ from one another in the fundamental scheme of projection, as shown in Fig. 15-1-1. The type of pictorial drawing used depends on the purpose for which it is drawn. They are used to explain complicated engineering drawings to people who do not have the training or ability to read the conventional multiview drawings; to help the designer work out problems in space, including clearances and interferences; to train new employees in the shop; to speed up and clarify the

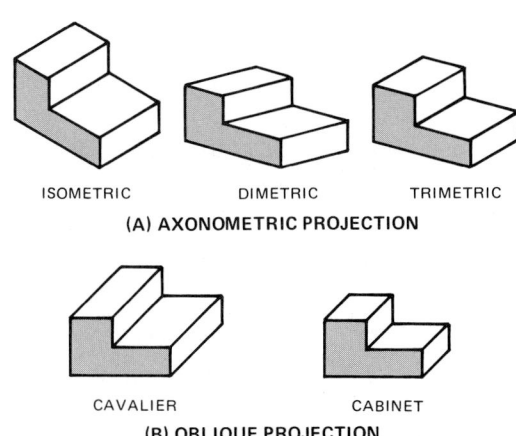

ISOMETRIC   DIMETRIC   TRIMETRIC
(A) AXONOMETRIC PROJECTION

CAVALIER   CABINET
(B) OBLIQUE PROJECTION

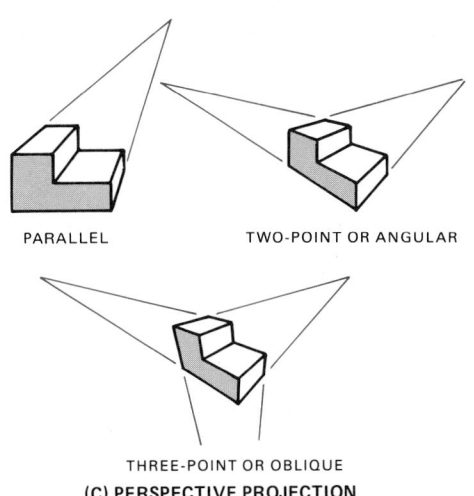

PARALLEL   TWO-POINT OR ANGULAR

THREE-POINT OR OBLIQUE
(C) PERSPECTIVE PROJECTION

FIG. 15-1-1 Types of pictorial drawings.

441

assembly of a machine or the ordering of new parts; to transmit ideas from one person to another, from shop to shop, or from salesperson to purchaser; and as an aid in developing the power of visualization.

Axonometric Projection

A projected view in which the lines of sight are perpendicular to the plane of projection, but in which the three faces of a rectangular object are all inclined to the plane of projection, is called an *axonometric projection* (Fig. 15-1-2). The projections of the three principal axes may make any angle with one another except 90°. Axonometric drawings, as shown in Figs. 15-1-3 and 15-1-4, are classified into three forms: *isometric drawings,* where the three principal faces and axes of the object are equally inclined to the plane of projection; *dimetric drawings,* where two of the three principal faces and axes of the object are equally inclined to the plane of projection; and *trimetric drawings,* where all three faces and axes of the object make different angles with the plane of projection. The most popular form of axonometric projection is the isometric.

Figure 15-1-5 illustrates the three types of axonometric projection, showing the compatible ellipse selection and the percentage the lines are foreshortened. The 15° angles for dimetric projection and the 11.5° and 30° angles for trimetric projection are shown as these angles are used extensively by industry.

Isometric Drawings

This method is based on a procedure of revolving the object at an angle of 45° to the horizontal, so that the front corner is toward the viewer, then tipping the object up or down at an angle of 35° 16' (Fig. 15-1-6, pg. 444). When this is done to a cube, the three faces visible to the viewer appear equal in shape and size, and the side faces are at an angle of 30° to the horizontal. If the isometric view were actually projected from a view of the object in the tipped position, the lines in the isometric view would be foreshortened and would, therefore, not be seen in their true length. To simplify the drawing of an isometric view, the actual measurements of the object are used. Although the object appears slightly larger without the allowance for shortening, the proportions are not affected.

All isometric drawings are started by constructing the isometric axes, which are a vertical line for height and isometric lines to left and right, at an angle of 30° from the horizontal, for length and width. The three faces seen in the isometric view are the same faces that would be seen in the normal orthographic views: top, front, and side. Figure 15-1-6B illustrates the selection of the front corner *(A)*, the construction of the isometric axes, and the completed isometric view. Note that all lines are drawn to their true length, measured along the isometric axes, and that hidden lines are usually omitted. Vertical

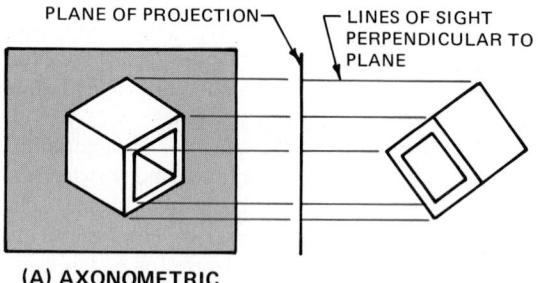

(A) AXONOMETRIC

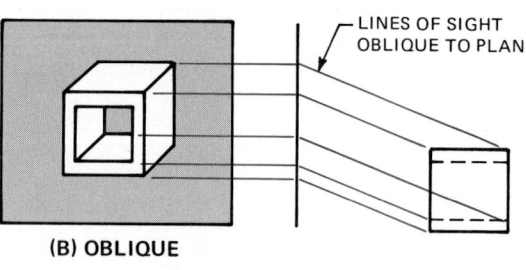

(B) OBLIQUE

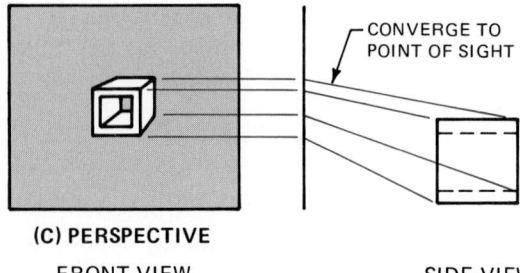

(C) PERSPECTIVE

FRONT VIEW   SIDE VIEW

FIG. 15-1-2   Kinds of projections.

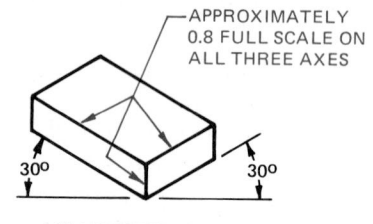

(A) ISOMETRIC PROJECTION

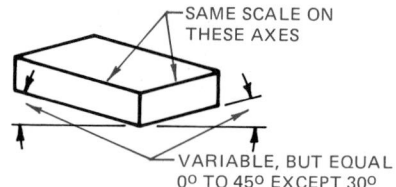

(B) DIMETRIC PROJECTION

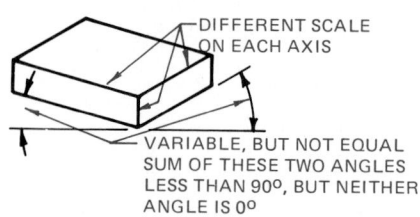

(C) TRIMETRIC PROJECTION

FIG. 15-1-3   Types of axonometric drawings.

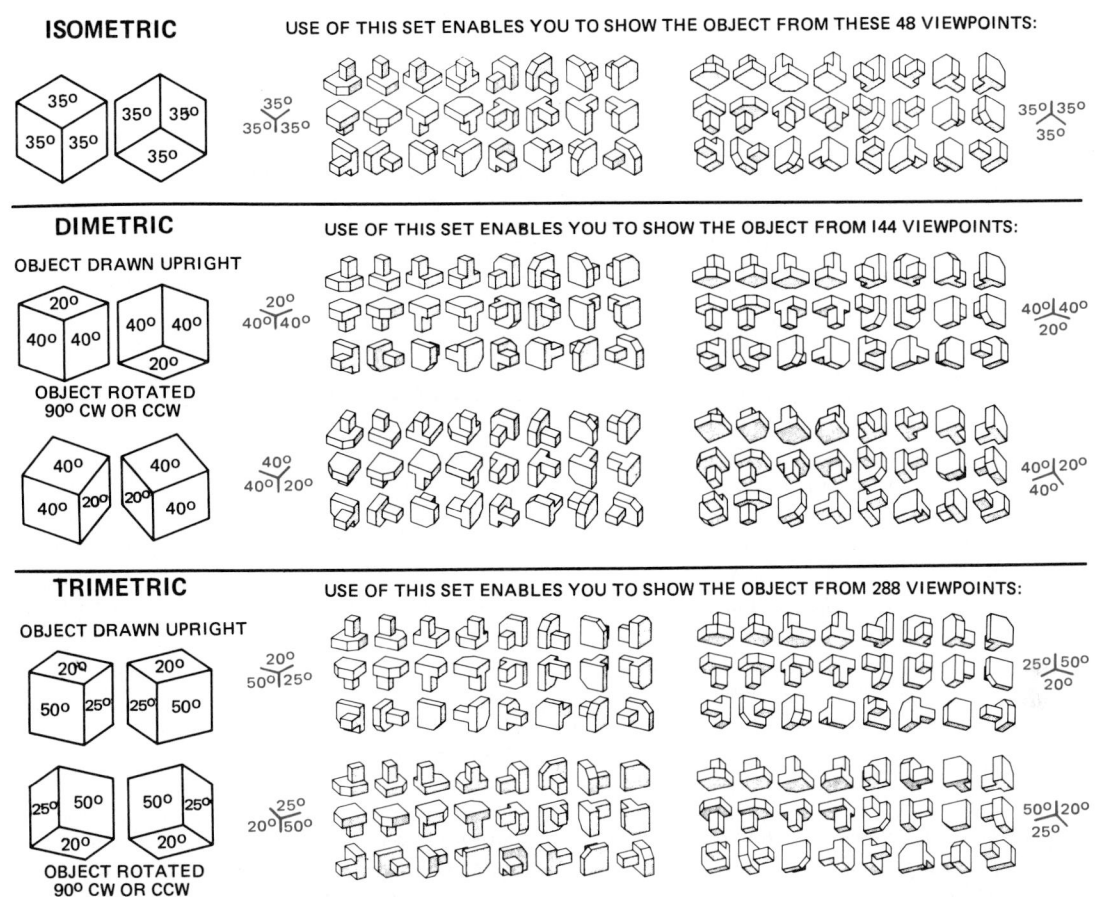

FIG. 15-1-4   Axonometric projection. *(Graphic Standard Instrument Co.)*

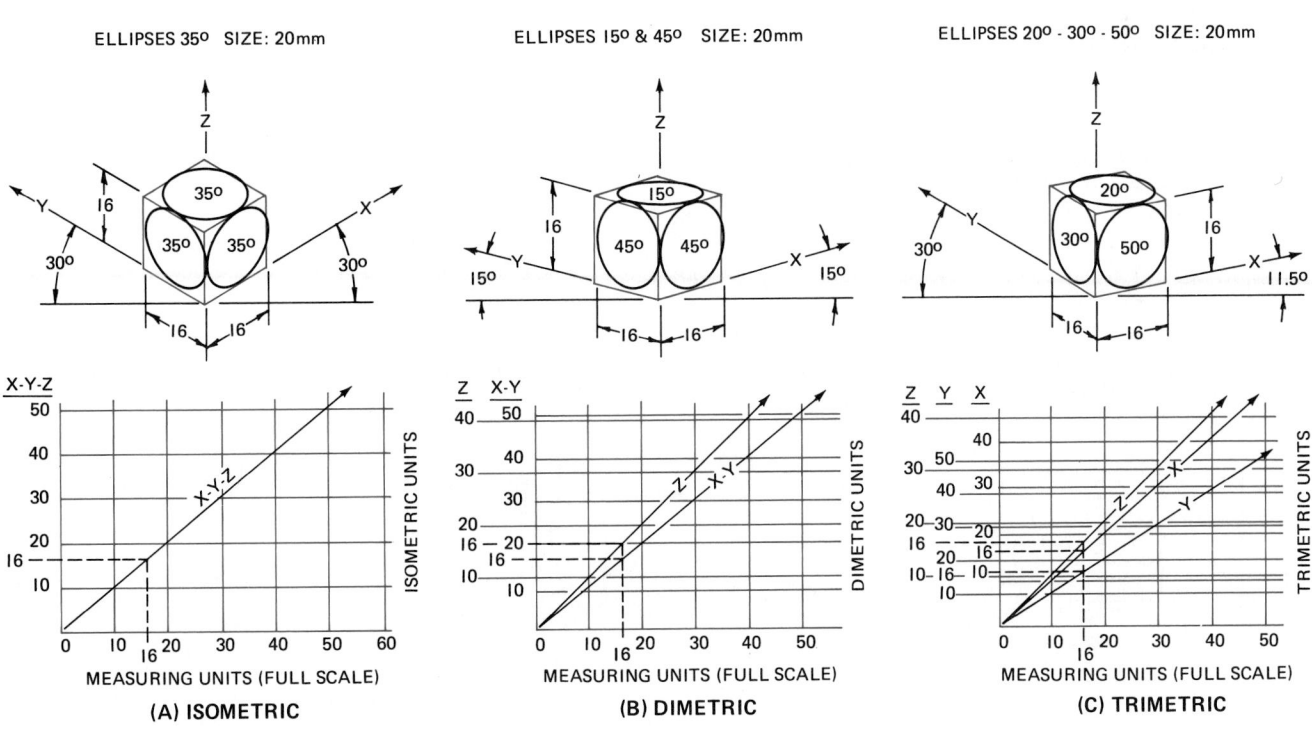

FIG. 15-1-5   Ellipse and line reduction sizes for axonometric projection.
(General Motors Corp.)

443

FIG. 15-1-6   Isometric axes and projection.

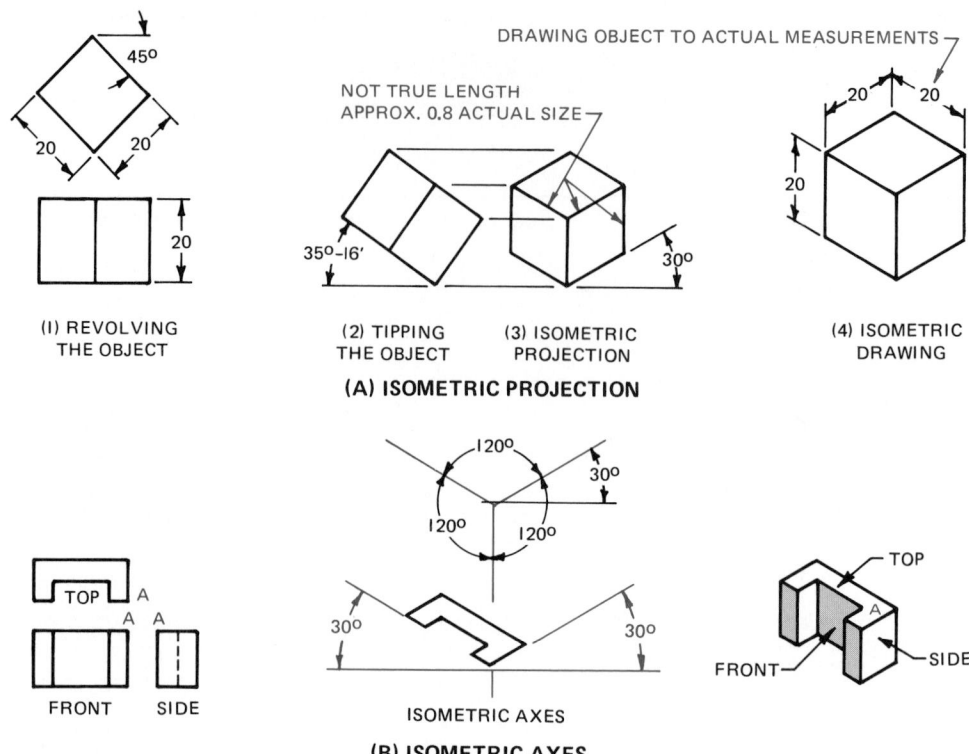

DRAWING OBJECT TO ACTUAL MEASUREMENTS

45°

20   20

20

(I) REVOLVING THE OBJECT

NOT TRUE LENGTH
APPROX. 0.8 ACTUAL SIZE

35°-16'   30°

(2) TIPPING THE OBJECT   **(3) ISOMETRIC PROJECTION**

20   20

20

(4) ISOMETRIC DRAWING

(A) ISOMETRIC PROJECTION

TOP   A

A   A

FRONT   SIDE

120°
30°
120°   120°

30°   30°

ISOMETRIC AXES

(B) ISOMETRIC AXES

TOP

FRONT   SIDE

A

edges are represented by vertical lines, and horizontal edges by lines at 30° to the horizontal.

Two techniques can be used for making an isometric drawing of an irregularly shaped object, as illustrated in Fig. 15-1-7. In one method, the object is divided mentally into a number of sections and the sections are created one at a time in their proper relationship to one another. In the second method, a box is created with the maximum height, width, and depth of the object; then the parts of the box that are not part of the object are removed, leaving the pieces that form the total object.

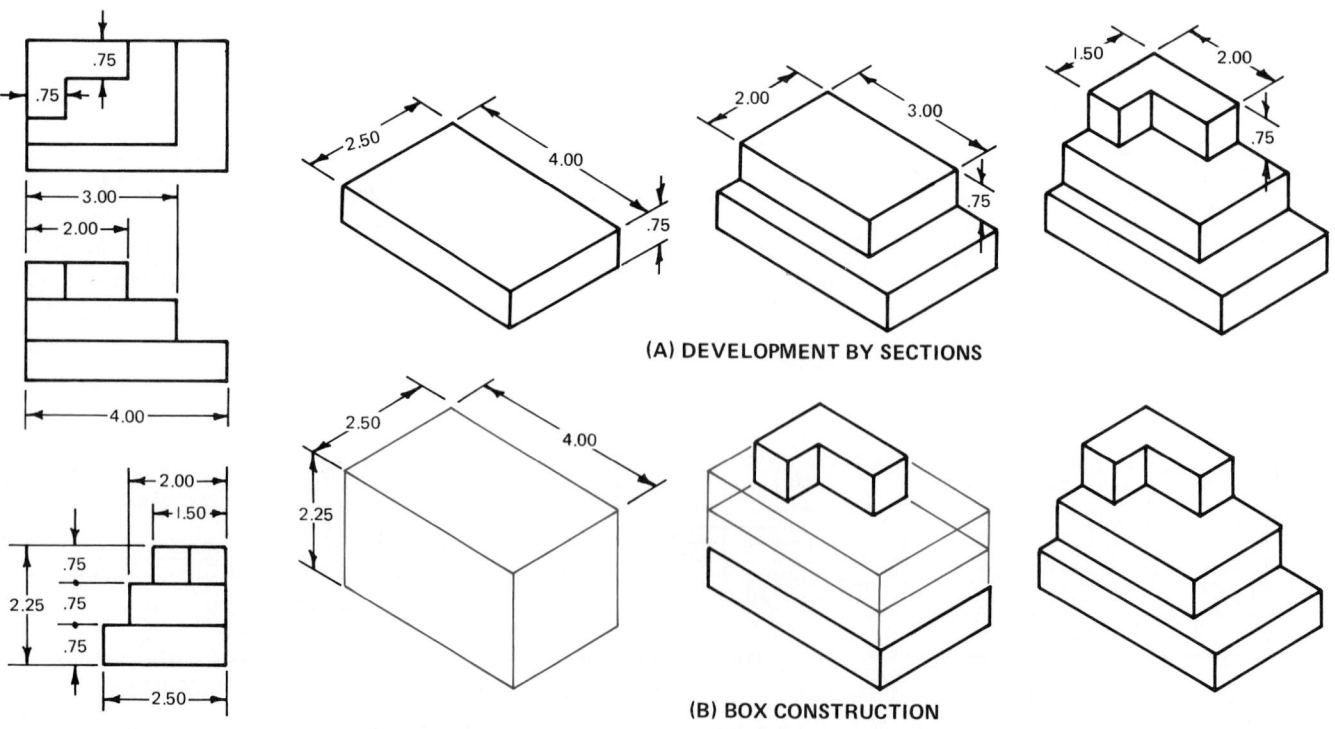

.75
.75
3.00
2.00
4.00

2.50   4.00

.75

2.00   3.00

.75

1.50   2.00

.75

(A) DEVELOPMENT BY SECTIONS

2.00
1.50
.75
2.25   .75
.75
2.50

2.50   4.00

2.25

(B) BOX CONSTRUCTION

FIG. 15-1-7   Developing an isometric drawing.

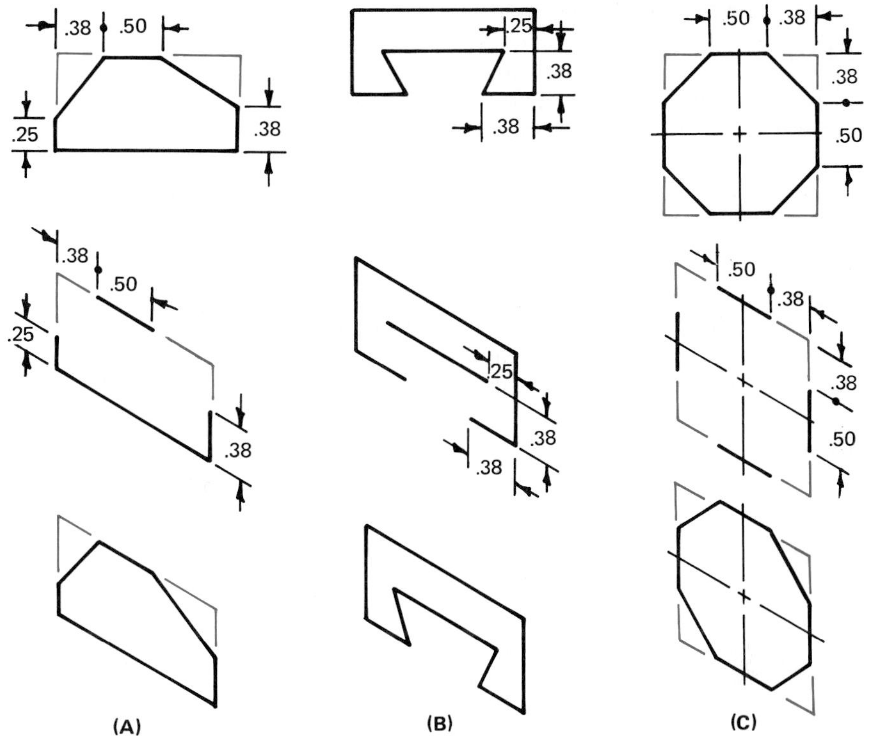

FIG. 15-1-8   Examples in the construction of nonisometric lines.

(A)   (B)   (C)

(A) PART   (B) BLOCK IN FEATURES   (C) DARKEN ISOMETRIC LINES   (D) COMPLETE NONISOMETRIC LINES

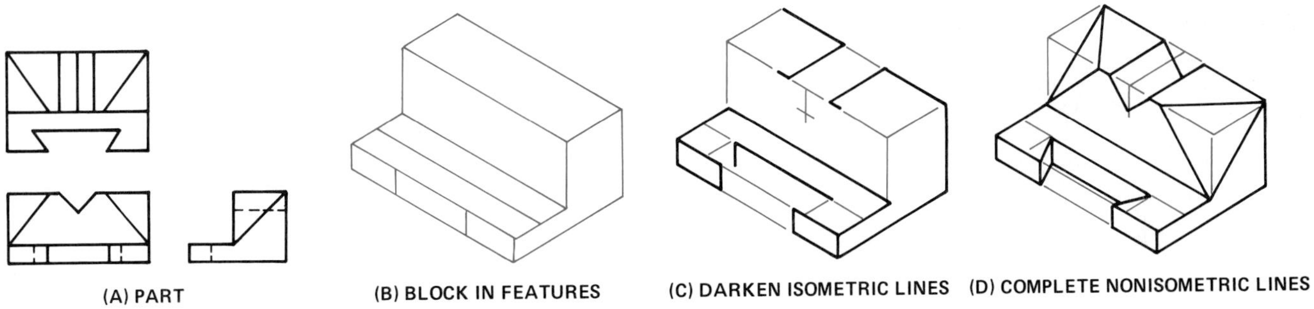

FIG. 15-1-9   Sequence in drawing an object having nonisometric lines.

Nonisometric Lines

Many objects have sloping surfaces that are represented by sloping lines in the orthographic views. In isometric drawing, sloping surfaces appear as *nonisometric* lines. To create them, locate their endpoints, found on the ends of isometric lines, and join them with a straight line. Figures 15-1-8 and 15-1-9 illustrate examples in the construction of nonisometric lines.

Dimensioning Isometric Drawings

At times, an isometric drawing of a simple object may serve as a working drawing. In such cases, the necessary dimensions and specifications are placed on the drawing.

Dimension lines, extension lines, and the line being dimensioned are shown in the same plane. Arrowheads should be in the plane of the dimension and extension lines. (Fig. 15-1-10).

Unidirectional dimensioning is the preferred method of dimensioning isometric drawings. The letters and numbers are

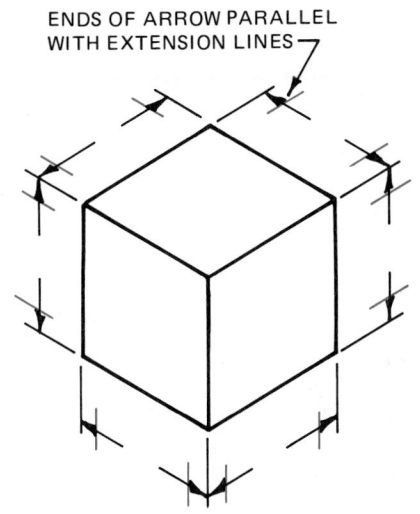

ENDS OF ARROW PARALLEL WITH EXTENSION LINES

FIG. 15-1-10   Orienting the dimension line, arrowhead, and extension line.

vertical and read from the bottom of the sheet. An example of this type of dimensioning is shown in Fig. 15-1-11.

Since the isometric is a one-view drawing, it is not usually possible to avoid placing dimensions on the view or across dimension lines. However, this practice should be avoided whenever possible.

Isometric Grid Paper

Isometric grid sheets are another timesaving device. Designers and engineers frequently use isometric grid paper on which they sketch their ideas and designs (Fig. 15-1-12). Many companies, such as those which prepare pipe drawings, have large drawing sheets made with nonreproducible isometric grid lines.

 CAD

A pictorial drawing may be created by using any CAD two-dimensional (2D) system. Normally, a grid pattern peculiar to the type of pictorial will be employed. Since isometrics are the most popular, an ISOMETRIC GRID option is found on virtually every system.

Automatic generation of a pictorial is common to the axonometric and perspective types only.

CAD systems provide a MODELING option, often referred to as 3D MODELING. With this option the model, isometric

or otherwise, can be automatically generated from the multiview drawing.

REFERENCES AND SOURCE MATERIAL

1. ANSI Y14.4, *American Drafting Standards Manual, Pictorial Drawing.*
2. General Motors Corp.

ASSIGNMENTS ▮▮▮▮▮▮▮▮▮▮▮▮▮▮▮▮▮▮▮▮▮▮▮▮▮▮▮▮

See Assignments 1 through 6 for Unit 15-1 on pages 465 through 468.

15-2 CURVED SURFACES IN ISOMETRIC

Circles and Arcs in Isometric

A circle on any of the three faces of an object drawn in isometric has the shape of an ellipse (Fig. 15-2-1). Figure 15-2-2 illustrates the steps in drawing circular features on isometric drawings.

1. Draw the center lines and a square, with sides equal to the circle diameter, in isometric.
2. Using the obtuse-angled (120°) corners as centers, draw arcs tangent to the sides forming the obtuse-angled corners, stopping at the points where the center lines cross the sides of the square.
3. Draw construction lines from these same points to the opposite obtuse-angled corners. The points at which these construction lines intersect are the centers for arcs drawn tangent to the sides forming the acute-angled corners, meeting the first arcs.

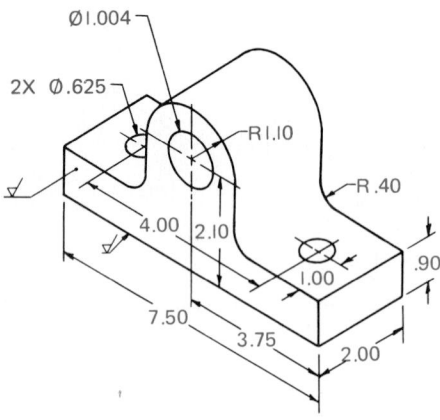

FIG. 15-1-11   Isometric dimensioning.

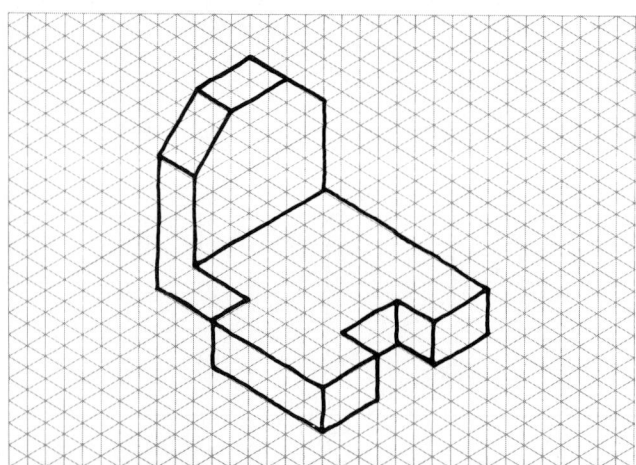

FIG. 15-1-12   Isometric grid paper.

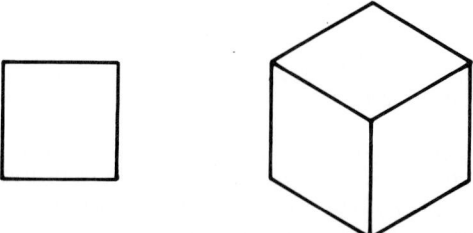

(A) A SQUARE DRAWN IN THE THREE ISOMETRIC POSITIONS

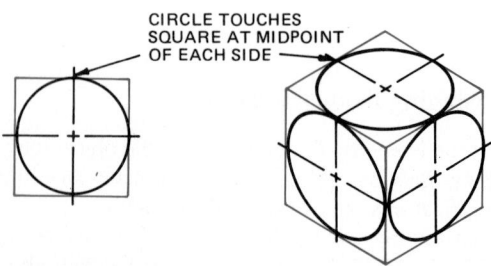

(B) A CIRCLE PLACED INSIDE A SQUARE AND DRAWN IN THE THREE ISOMETRIC POSITIONS

FIG. 15-2-1   Circles in isometric.

FIG. 15-2-2 Sequence in drawing isometric circles.

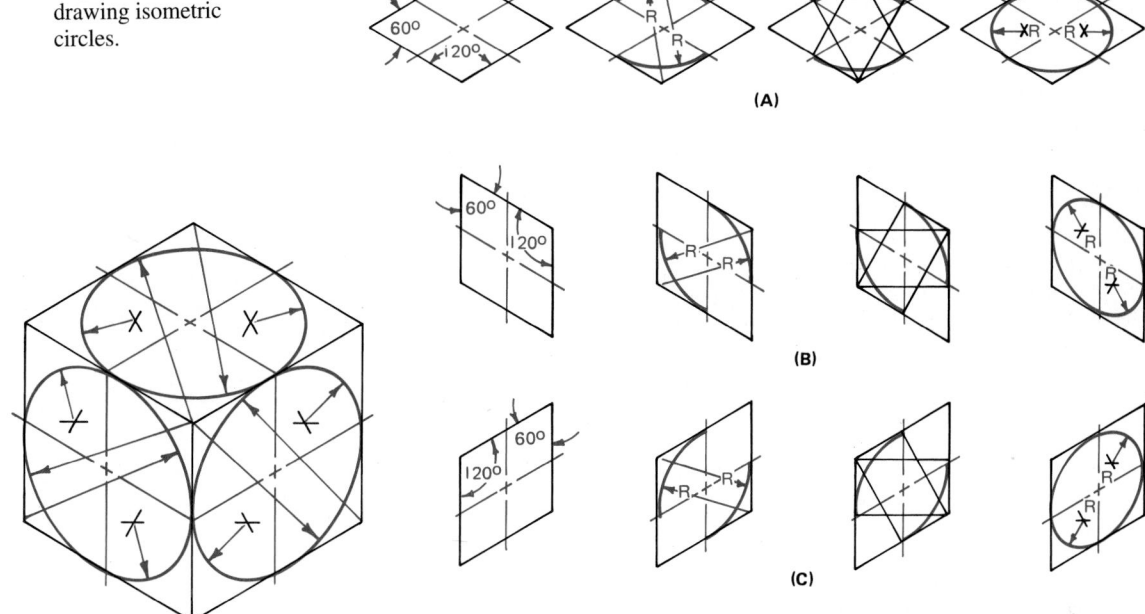

(A)

(B)

(C)

When concentric circles are drawn, each circle must have its own set of centers for the arcs, as shown in Fig. 15-2-3.

The same technique is used for drawing part-circles (arcs), as shown in Fig. 15-2-4. Construct an isometric square with sides equal to twice the radius, and draw that portion of the ellipse necessary to join the two faces. When these faces are

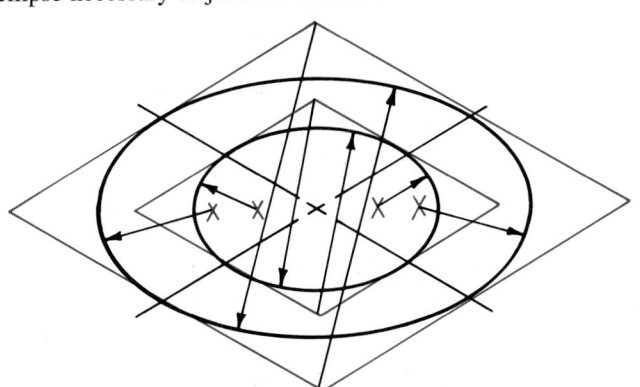

FIG. 15-2-3 Drawing concentric isometric circles.

FIG. 15-2-4 Drawing isometric arcs and circles.

parallel, draw half of an ellipse (one long radius and one short radius); when they are at an obtuse angle (120°), draw one long radius; and when they are at an acute angle (60°), draw one short radius.

Isometric Templates

For convenience and time saving, isometric ellipse templates should be used whenever possible. A wide variety of elliptical templates are available. The template shown in Fig. 15-2-5 (pg. 448) combines ellipses, scales, and angles. Markings on the ellipses coincide with the center lines of the holes, speeding up the drawing of circles and arcs. Figure 15-2-6 (pg. 448) shows the same part that appears in Fig. 15-2-4 but with the arcs and circles being constructed with a template.

Sketching Circles and Arcs

In sketching circles and arcs on isometric grid paper, locate the center lines first, then lightly sketch in construction boxes

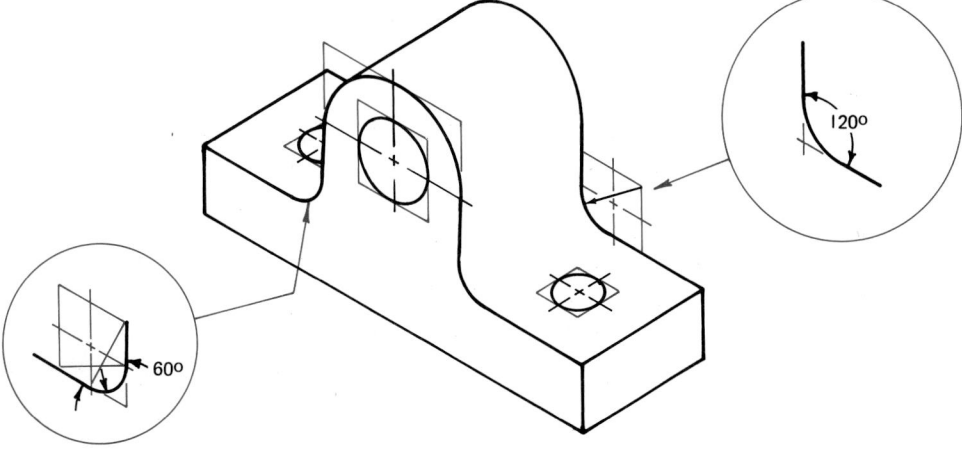

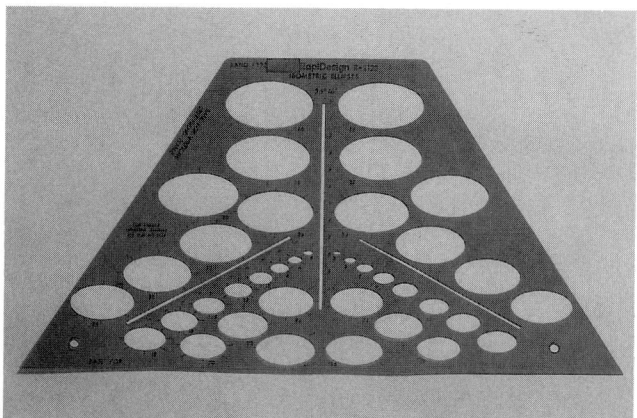

FIG. 15-2-5   Isometric ellipse template. *(STUDIOHIO)*

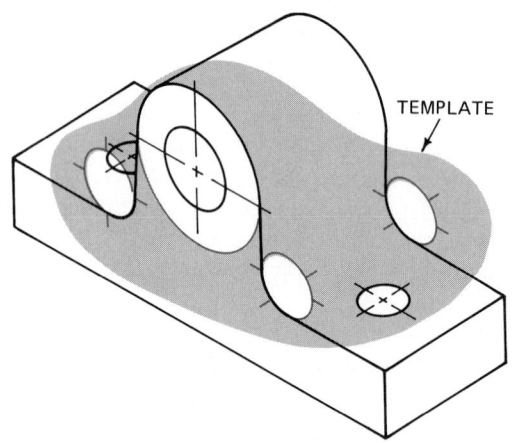

FIG. 15-2-6   Circles and arcs drawn with isometric ellipse template.

(isometric squares) where the circles and arcs should be (Fig. 15-2-7). Sketch the ellipse (isometric circle) just touching the center of each of the four sides of the square.

Drawing Irregular Curves in Isometric

To draw curves other than circles or arcs, the plotting method shown in Fig. 15-2-8 is used.

1. Draw an orthographic view, and divide the area enclosing the curved line into equal squares.
2. Produce an equivalent area on the isometric drawing, showing the offset squares.
3. Take positions relative to the squares from the orthographic view, and plot them on the corresponding squares on the isometric view.
4. Draw a smooth curve through the established points with the aid of an irregular curve.

🖥 CAD

The ELLIPSE command in CAD provides two basic methods for constructing an ellipse. The first method allows for the specification of the major and minor axes or diameters. While this is the most common method of ellipse construction,

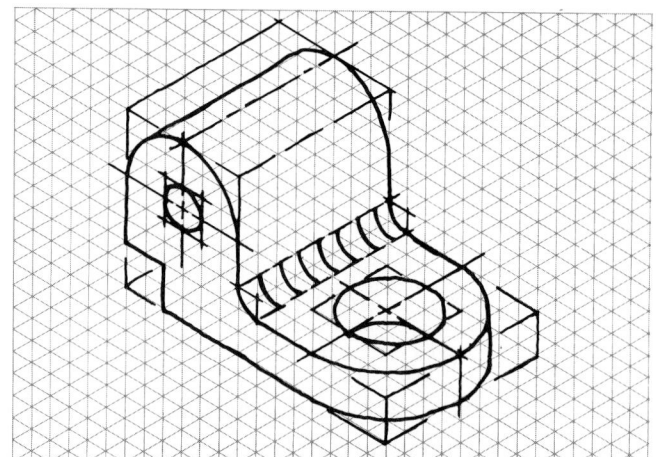

FIG. 15-2-7   Sketching isometric circles and arcs.

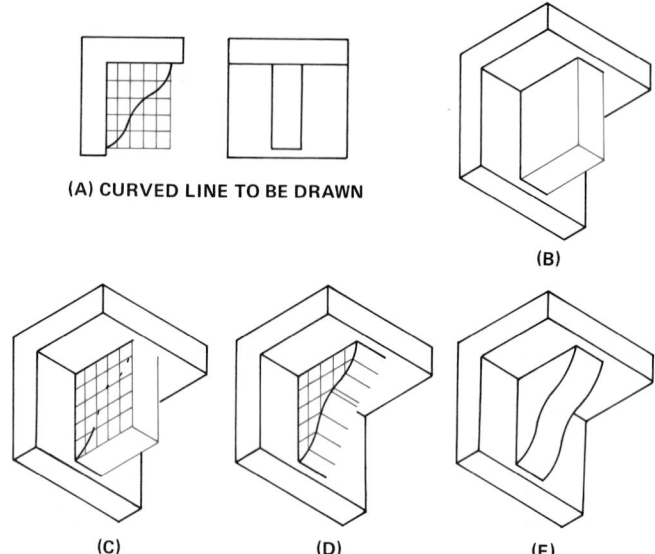

(A) CURVED LINE TO BE DRAWN

(B)

(C)   **(D)**   **(E)**

FIG. 15-2-8   Curves drawn in isometric by means of offset measurements.

specification of the major and minor axis values does not allow for the construction of an ellipse at a known degree of exposure or eccentricity. The second method of ellipse construction in CAD is used to construct an ellipse with a known angle of exposure. The center and major diameter of the ellipse are identified, and then the angle of exposure or rotation in degrees is specified. For curves other than circles or arcs, a grid pattern is employed to allow for offset construction. A series of points are located on the curve using offset construction techniques. A polyline or spline is then generated through those points to create the curve. The larger the number of points used to generate the curve, the more accurate the construction of the curve will be.

ASSIGNMENTS

See Assignments 7 through 9 for Unit 15-2 on pages 468 to 469.

15-3 COMMON FEATURES IN ISOMETRIC

Isometric Sectioning

Isometric drawings are usually made showing exterior views, but sometimes a sectional view is needed. The section is taken on an isometric plane, that is, on a plane parallel to one of the faces of the cube. Figure 15-3-1 shows isometric full sections taken on a different plane for each of three objects. Note the construction lines representing the part that has been cut away. Isometric half-sections are illustrated in Fig. 15-3-2.

When an isometric drawing is sectioned, the section lines are shown at an angle of 60° with the horizontal or in a horizontal position, depending on where the cutting-plane line is located. In half sections, the section lines are sloped in opposite directions, as shown in Fig. 15-3-2.

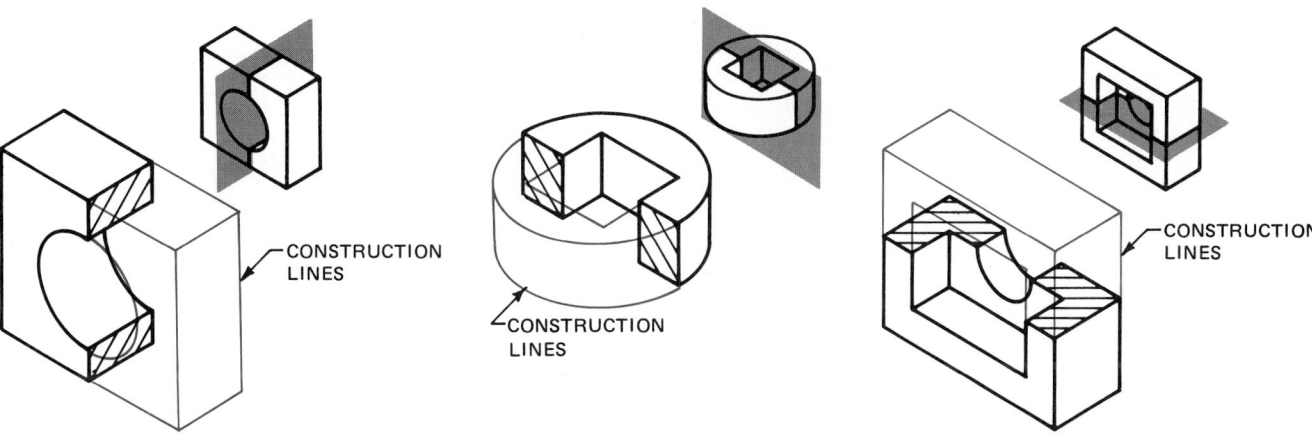

FIG. 15-3-1   Examples of isometric full sections.

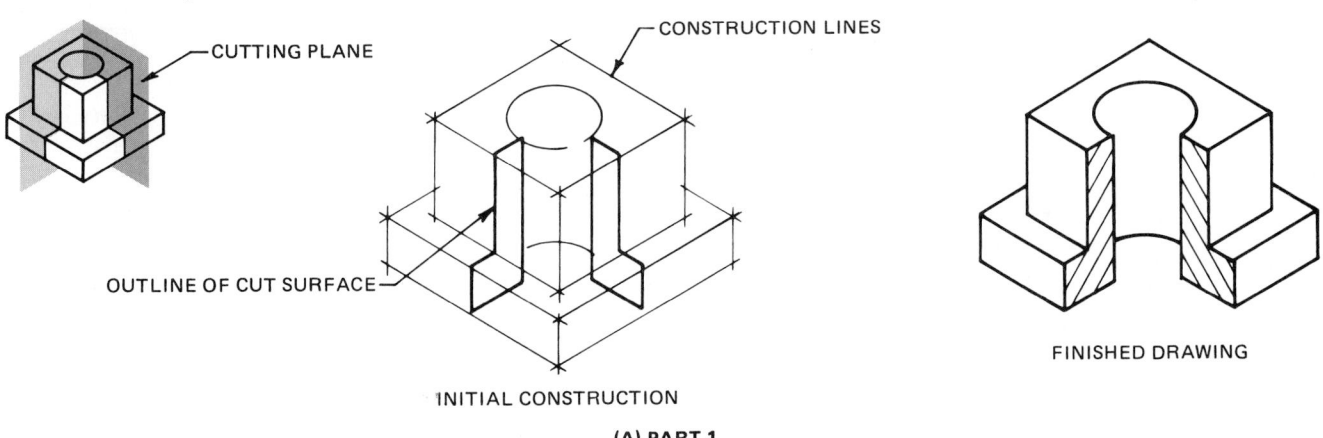

(A) PART 1

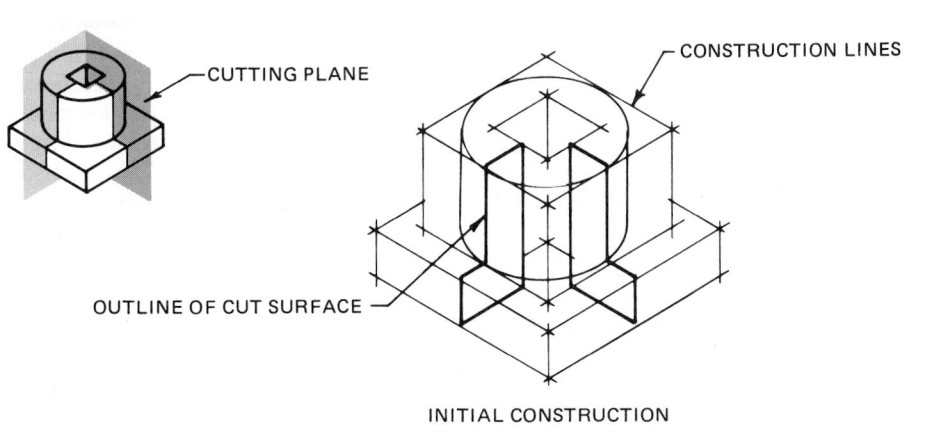
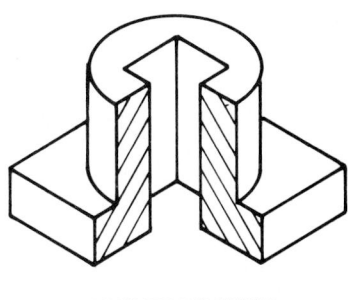

(B) PART 2

FIG. 15-3-2   Examples of isometric half-sections.

449

Fillets and Rounds

For most isometric drawings of parts having small fillets and rounds, the adopted practice is to draw the corners as sharp features. However, when it is desirable to represent the part, normally a casting, as having a more realistic appearance, either of the methods shown in Fig. 15-3-3 may be used.

Threads

The conventional method for showing threads in isometric is shown in Fig. 15-3-4. The threads are represented by a series of ellipses uniformly spaced along the center line of the thread. The spacing of the ellipses need not be the spacing of the actual pitch.

Break Lines

For long parts, break lines should be used to shorten the length of the drawing. Freehand breaks are preferred, as shown in Fig. 15-3-5.

Isometric Assembly Drawings

Regular or exploded assembly drawings are frequently used in catalogs and sales literature, as illustrated by Fig. 15-3-6.

ASSIGNMENTS ▰▰▰▰▰▰▰▰▰▰▰▰▰▰▰▰▰▰

See Assignments 10 through 16 for Unit 15-3 on pages 469 through 472.

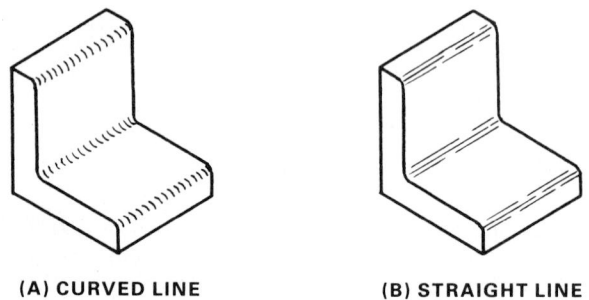

(A) CURVED LINE   **(B) STRAIGHT LINE**

FIG. 15-3-3   Representation of fillets and rounds in isometric.

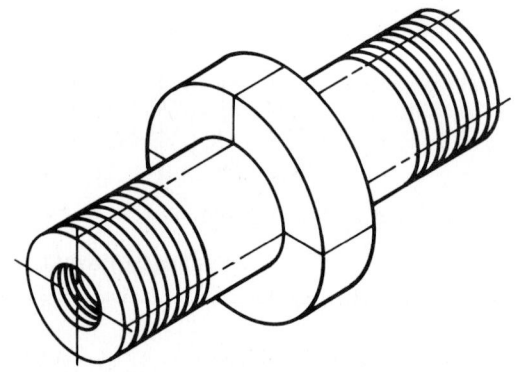

FIG. 15-3-4   Representation of threads in isometric.

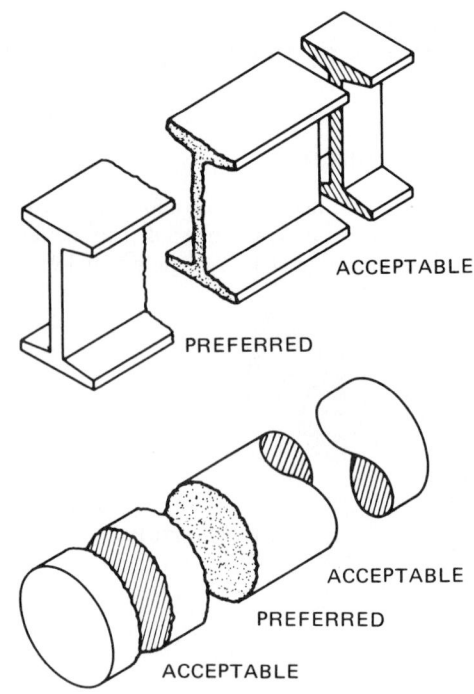

FIG. 15-3-5   Conventional breaks in isometric.

15-4 OBLIQUE PROJECTION

This method of pictorial drawing is based on the procedure of placing the object with one face parallel to the frontal plane and placing the other two faces on oblique (or receding) planes, to left or right, top or bottom, at a convenient angle. The three axes of projection are vertical, horizontal, and receding. Figure 15-4-1 illustrates a cube drawn in typical positions with the receding axis at 60°, 45°, and 30°. This form of projection has the advantage of showing one face of the object without distortion. The face with the greatest irregularity of outline or contour, or the face with the greatest number of circular features, or the face with the longest dimension faces the front (Fig. 15-4-2).

Two types of oblique projection are used extensively. In *cavalier oblique,* all lines are made to their true length, measured on the axes of the projection. In *cabinet oblique,* the lines on the receding axis are shortened by one-half their true length to compensate for distortion and to approximate more closely what the human eye would see. For this reason, and because of the simplicity of projection, cabinet oblique is a commonly used form of pictorial representation, especially when circles and arcs are to be drawn. Figure 15-4-3 (pg. 452) shows a comparison of cavalier and cabinet oblique. Note that hidden lines are omitted unless required for clarity. Many of the drawing techniques for isometric projection apply to oblique projection. Figure 15-4-4 (pg. 452) illustrates the construction of an irregularly shaped object by the box method.

Inclined Surfaces

Angles that are parallel to the picture plane are drawn as their true size. Other angles can be laid off by locating the ends of the inclined line.

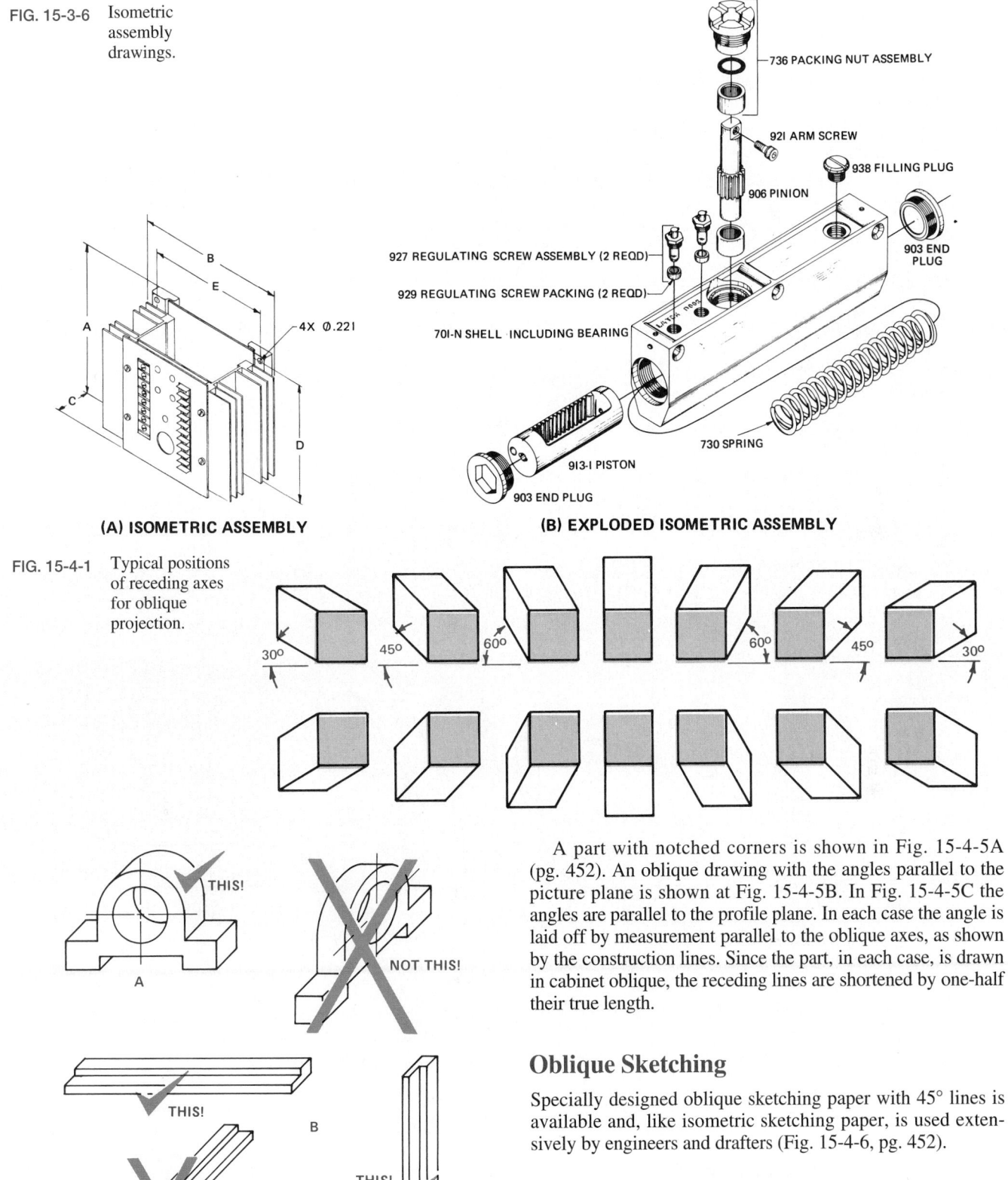

FIG. 15-3-6   Isometric assembly drawings.

(A) ISOMETRIC ASSEMBLY

736 PACKING NUT ASSEMBLY

921 ARM SCREW

938 FILLING PLUG

906 PINION

927 REGULATING SCREW ASSEMBLY (2 REQD)

929 REGULATING SCREW PACKING (2 REQD)

701-N SHELL ·INCLUDING BEARING

903 END PLUG

913-1 PISTON

903 END PLUG

730 SPRING

(B) EXPLODED ISOMETRIC ASSEMBLY

FIG. 15-4-1   Typical positions of receding axes for oblique projection.

FIG. 15-4-2   Two general rules for oblique drawings.

A part with notched corners is shown in Fig. 15-4-5A (pg. 452). An oblique drawing with the angles parallel to the picture plane is shown at Fig. 15-4-5B. In Fig. 15-4-5C the angles are parallel to the profile plane. In each case the angle is laid off by measurement parallel to the oblique axes, as shown by the construction lines. Since the part, in each case, is drawn in cabinet oblique, the receding lines are shortened by one-half their true length.

Oblique Sketching

Specially designed oblique sketching paper with 45° lines is available and, like isometric sketching paper, is used extensively by engineers and drafters (Fig. 15-4-6, pg. 452).

Dimensioning Oblique Drawings

Dimension lines are drawn parallel to the axes of projection. Extension lines are projected from the horizontal and vertical object lines whenever possible.

The dimensioning of an oblique drawing is similar to that of an isometric drawing. The recommended method is

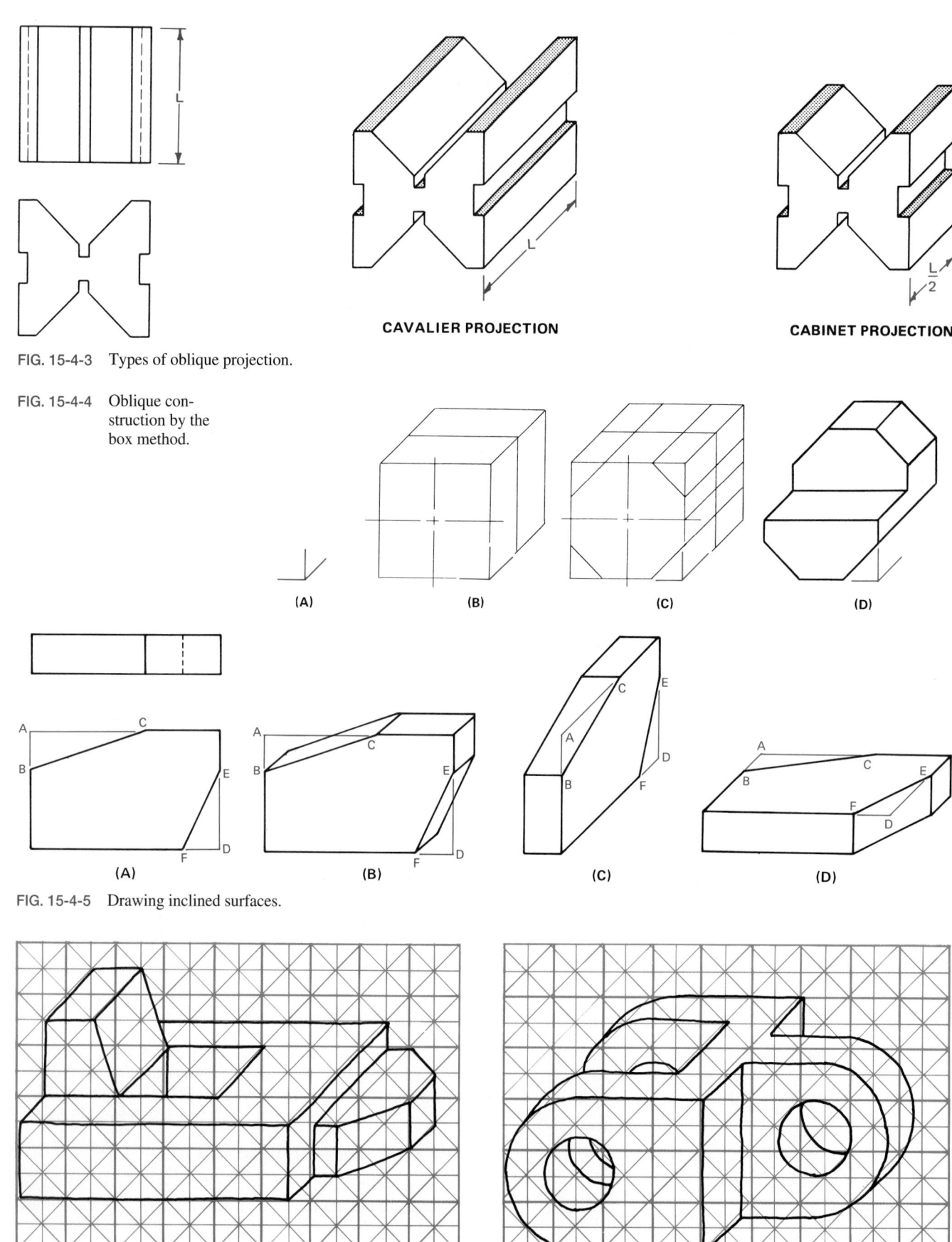

FIG. 15-4-3 Types of oblique projection.

CAVALIER PROJECTION

CABINET PROJECTION

FIG. 15-4-4 Oblique construction by the box method.

(A) (B) (C) (D)

FIG. 15-4-5 Drawing inclined surfaces.

(A) (B) (C) (D)

FIG. 15-4-6 Oblique sketching paper.

FIG. 15-4-7 Dimensioning an
oblique drawing.

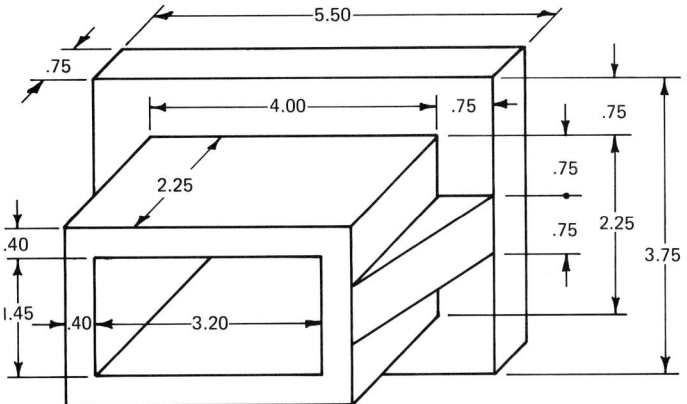

unidirectional dimensioning, which is shown in Fig. 15-4-7. As in isometric dimensioning, some dimensions must be placed directly on the view.

CAD

Specially designed oblique lines or grids with horizontal, vertical, and 45° reference lines are available on some CAD systems. If not available, a rectangular pattern may be used to develop the true-size front face. For the two oblique faces a 45° or 60° line pattern is used.

Oblique modeling is not a CAD option.

ASSIGNMENTS

See Assignments 17 through 21 for Unit 15-4 on pages 472 through 474.

15-5 COMMON FEATURES IN OBLIQUE

Circles and Arcs

Whenever possible, the face of the object having circles or arcs should be selected as the front face, so that such circles or arcs can be easily drawn in their true shape (Fig. 15-5-1). When circles or arcs must be drawn on one of the oblique faces, the *offset measurement method* illustrated in Fig. 15-5-2 (pg. 454) may be used.

1. Draw an oblique square about the center lines, with sides equal to the diameter.
2. Draw a true circle within the oblique square, and establish equally spaced points about its circumference.
3. Project these point positions to the edge of the oblique square, and draw lines on the oblique axis from these

FIG. 15-5-1 Application of an
oblique drawing.

SCHEMATIC OF A COMPLETELY AUTOMATIC REGISTRATION CONTROL SYSTEM
MAINTAINING THE LOCATION OF CUTOFF ON A CONTINUOUS PRINTED WEB.

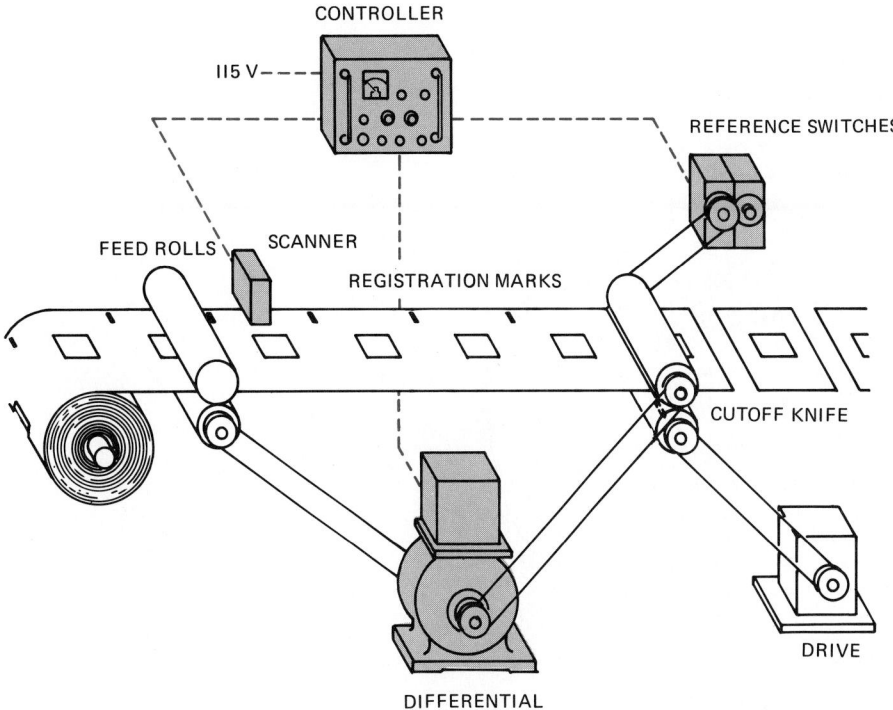

positions. Similarly spaced lines are drawn on the other axis for a cavalier oblique drawing and the spaces halved for a cabinet oblique drawing, forming offset squares and giving intersection points for the oval shape.

Another method used when circles or arcs must be drawn on one of the oblique surfaces is the *four-center method*. In Fig. 15-5-3 a circle is shown as it would be drawn on a front plane, a side plane, and a top plane.

Circles not parallel to the picture plane when drawn by the approximate method are not pleasing but are satisfactory for some purposes. Ellipse templates, when available, should be used because they reduce drawing time and give much better results. If a template is used, the oblique circle should first be blocked in as an oblique square in order to locate the proper

position of the circle. Blocking in the circle first also helps the drafter select the proper size and shape of the ellipse. The construction and dimensioning of an oblique part are shown in Fig. 15-5-4.

Oblique Sectioning

Oblique drawings are usually made as outside views, but sometimes a sectional view is necessary. The section is taken on a plane parallel to one of the faces of an oblique cube. Figure 15-5-5 shows an oblique full section and an oblique half-section. Construction lines show the part that has been cut away.

Treatment of Conventional Features

Fillets and Rounds   Small fillets and rounds normally are drawn as sharp corners. When it is desirable to show the corners rounded, either of the methods shown in Fig. 15-5-6 is recommended.

Threads   The conventional method of showing threads in oblique is shown in Fig. 15-5-7. The threads are represented by a series of circles uniformly spaced along the center line of the thread. The spacing of the circles need not be the spacing of the pitch.

Breaks   Figure 15-5-8 shows the conventional method for representing breaks.

🖳 CAD

The CIRCLE, ARC, and FILLET commands are used to create circles and arcs on the front face of an oblique drawing. For the oblique faces use the ELLIPSE command, and specify the degree of exposure (i.e., 45°) to construct circles and arcs.

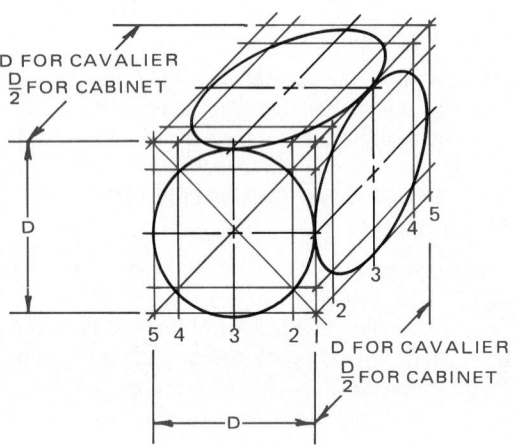

FIG. 15-5-2   Drawing oblique circles by means of offset measurements.

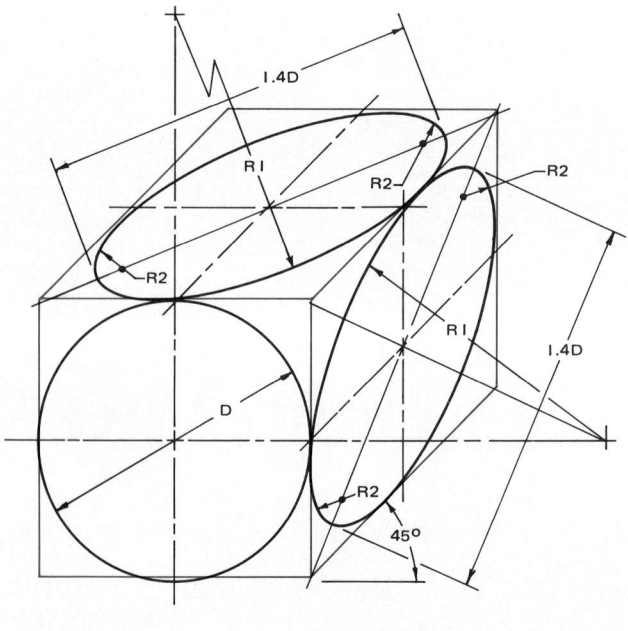

(A) CABINET OBLIQUE

(B) CAVALIER OBLIQUE

FIG. 15-5-3   Approximate ellipse construction for oblique drawings with 45° axis.

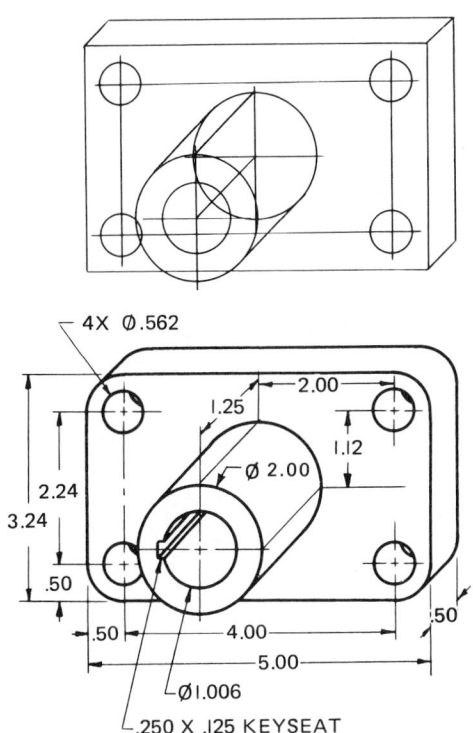

4X Ø.562

2.00

1.25

1.12

Ø 2.00

2.24

3.24

.50

.50

.50

4.00

5.00

Ø1.006

.250 X .125 KEYSEAT

FIG. 15-5-4   Construction and dimensioning of an oblique object.

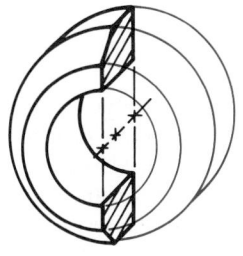

(A) FULL SECTION

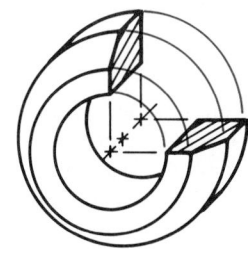

(B) HALF-SECTION

FIG. 15-5-5   Oblique full and half-sections.

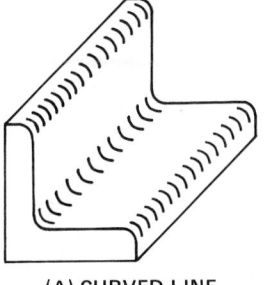

(A) CURVED LINE

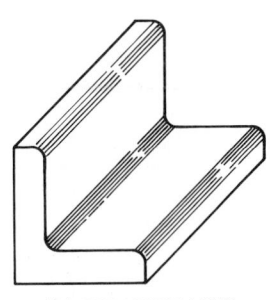

(B) STRAIGHT LINE

FIG. 15-5-6   Representing rounds and fillets.

Thus far, all discussion has been confined to two-dimensional (2D) software. For the vast majority of engineering drawing applications, two-dimensional drafting will suffice. Multiview projection, which uses two or more two-dimensional views to describe a three-dimensional object, is considered the standard operating procedure for many design offices. Occasionally, however, three-dimensional (axonometric)

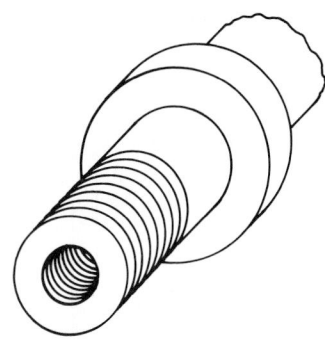

FIG. 15-5-7   Representation of threads in oblique.

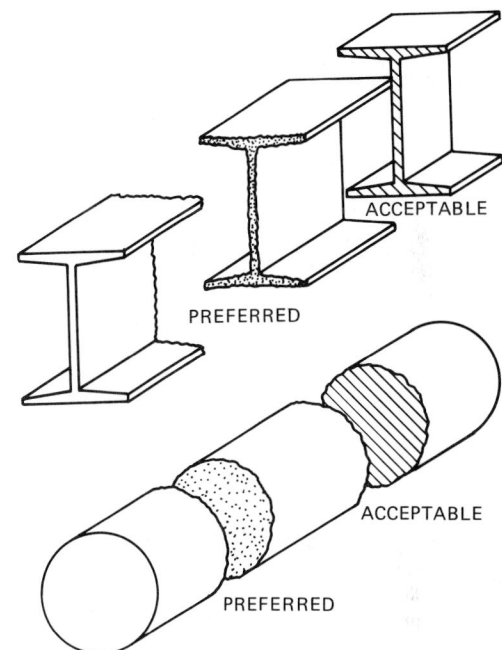

ACCEPTABLE

PREFERRED

ACCEPTABLE

PREFERRED

FIG. 15-5-8   Conventional breaks.

drawing capability is desirable. This can be accomplished as previously shown using a basic two-dimensional system. Lines are drawn inclined both to the right or left at 30° from the horizontal. These, with a vertical line, establish the three major axes of an isometric drawing. This method, however, may become quite cumbersome, requiring a considerable time expenditure. CAD systems provide a modeling option often referred to as *3D modeling*. With this option the model, isometric or otherwise, is automatically generated from the drawing. The method commonly used in modeling is to draw one view (often the plan or top view) and key in the third dimension (thickness or height). All *X, Y,* and *Z* coordinate information is provided to the system. The single-view (plan or top) method to create the object is shown in Fig. 15-5-9 (pg. 456).

ASSIGNMENTS

See Assignments 22 through 25 for Unit 15-5 on pages 474 through 477.

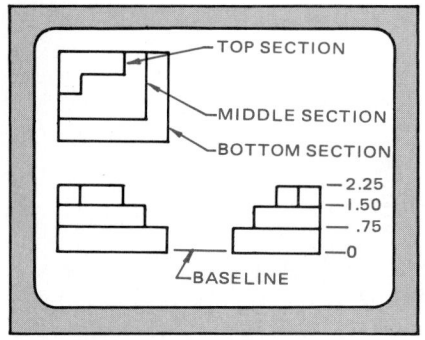

(A) THE PART TO BE MODELED

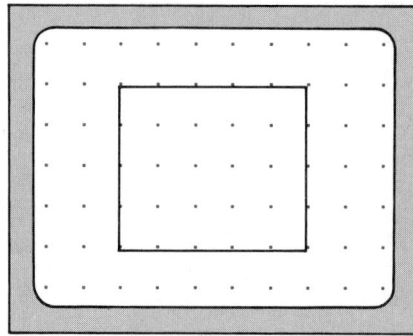

(B) PLAN VIEW OF BOTTOM SECTION

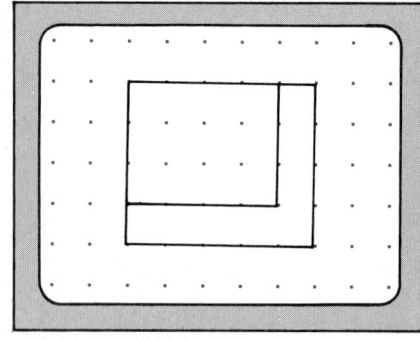

(C) PLAN VIEW WITH CENTER SECTION ADDED

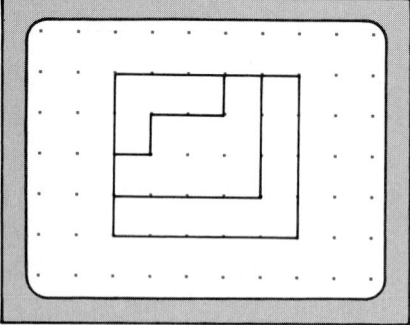

(D) TOP SECTION ADDED

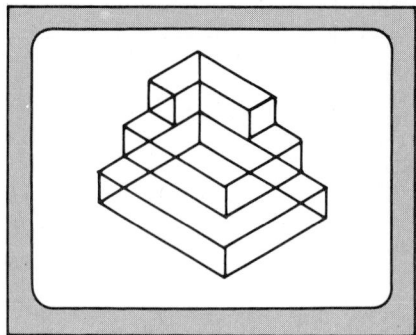

(E) WIRE-FRAME ISOMETRIC ILLUSTRATED WITH ALL LINES

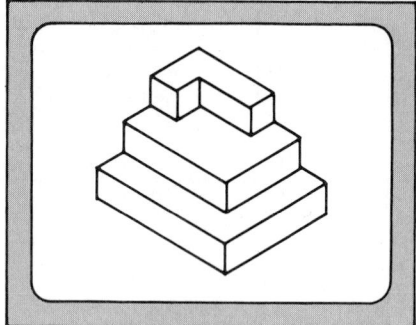

(F) WIRE FRAME ISOMETRIC ILLUSTRATED WITH HIDDEN LINES REMOVED

FIG. 15-5-9   CAD modeling.

15-6 PARALLEL, OR ONE-POINT, PERSPECTIVE

Perspective Projection

Perspective is a method of drawing that depicts a three-dimensional object on a flat plane as it appears to the eye (Fig. 15-6-1). A pictorial drawing made by the intersection of the picture plane with lines of sight converging from points on the object to the point of sight, which is located at a finite distance from the picture plane, is called a perspective (Fig. 15-6-2).

Perspective drawings are more realistic than axonometric or oblique drawings because the object is shown as the eye would see it. Since they are far more difficult to draw than the other types of pictorial drawings, their use in drafting is limited mainly to production or presentation illustrations and illustrations of proposed structures by architects.

FIG. 15-6-1   Application of parallel and angular perspective drawings. *(General Motors Corp.)*

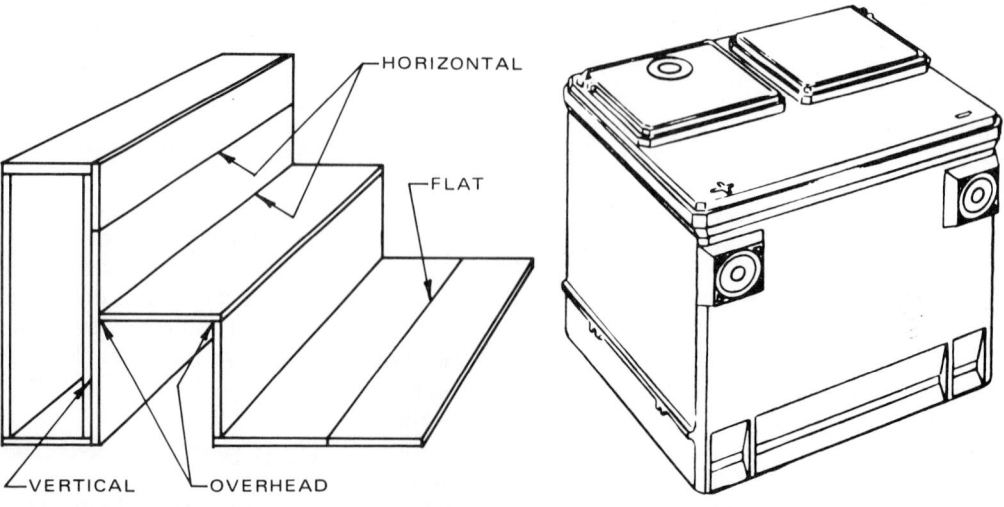

(A) PARALLEL PERSPECTIVE

(B) ANGULAR PERSPECTIVE

The main elements of a perspective drawing are the *picture plane* (plane of projection), the *station point* (the position of the observer's eye when viewing the object), the *horizon* (an imaginary horizontal line taken at eye level), the *vanishing point* or *points* (a point or points on the horizon where all the receding lines converge), and the *ground line* (the base line of the picture plane and object).

To avoid undue distortion in perspective, the point of sight (station point) should be located so that the cone of rays from the observer's eye has an angle at the apex not greater than 30°. This would place the station point a distance away from the outside portion of the object of approximately 2 to 2 ½ times the width of the object being viewed (see Figs. 15-6-2 and 15-6-3).

Types of Perspective Drawings

There are three types of perspective drawings:

1. *Parallel:* one vanishing point
2. *Angular:* two vanishing points
3. *Oblique:* three vanishing points

In industry they are normally referred to as one-point, two-point, and three-point perspectives, respectively (Fig. 15-6-4, pg. 458). Only parallel and angular perspectives are covered in this text.

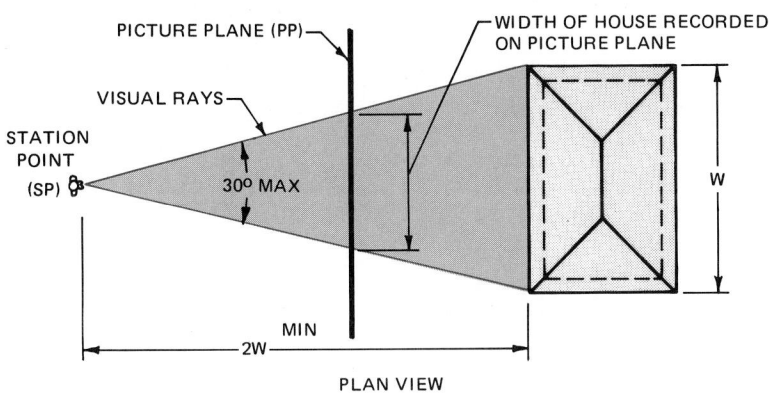

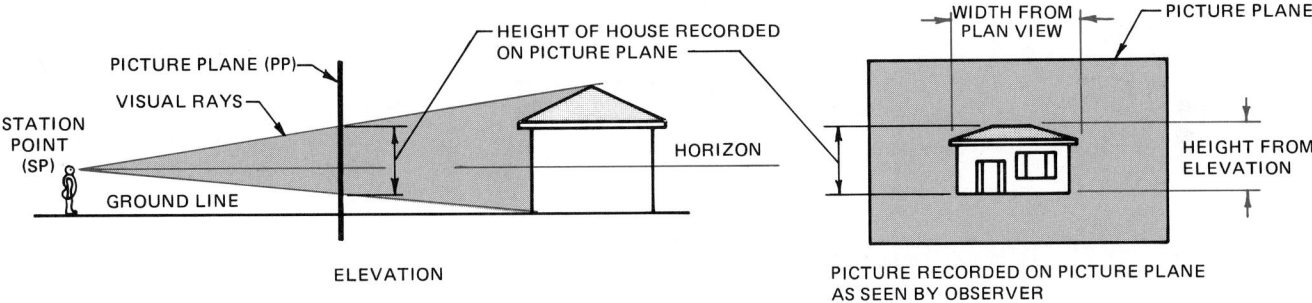

FIG. 15-6-2 Recording the picture on the picture plane.

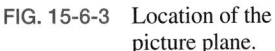

FIG. 15-6-3 Location of the picture plane.

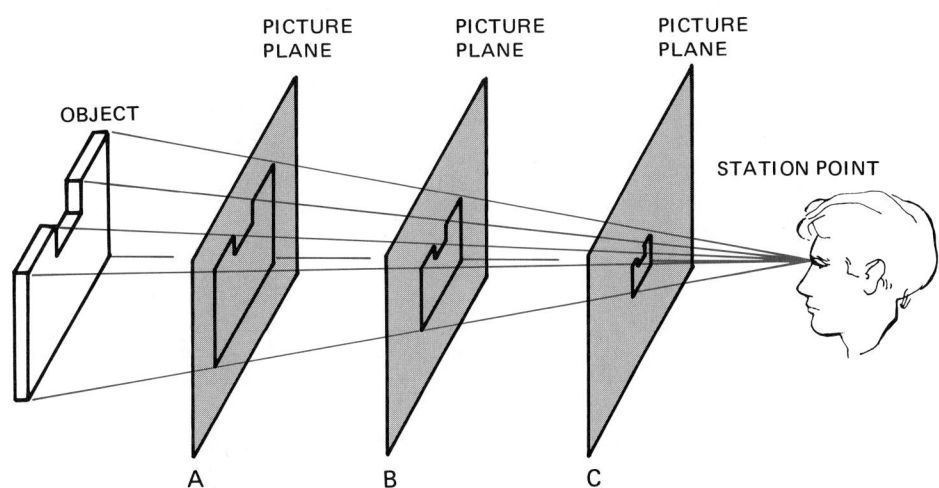

Parallel, or One-Point, Perspective

Parallel-perspective drawings are similar to oblique drawings, except the receding lines all converge at one point on the horizon. In drawing a *parallel-perspective drawing,* one face of the object is placed on the picture-plane line so that it will be drawn in its true size and shape, as shown in Fig. 15-6-5. The *PP* line shown in the top view represents the picture-plane line, and

point *SP* (station point) is the position of the observer. The lines of the object, which are not on the picture-plane, are found by projecting lines down from the top view from the point of intersection of the visual ray and the picture plane, as shown by point *N* in Fig. 15-6-5A(1).

Where the true height of a line or a point does not lie on the picture plane, such as point *P* in Fig. 15-6-5A(2), the true height may be found by extending line *PR* to point *S* on the

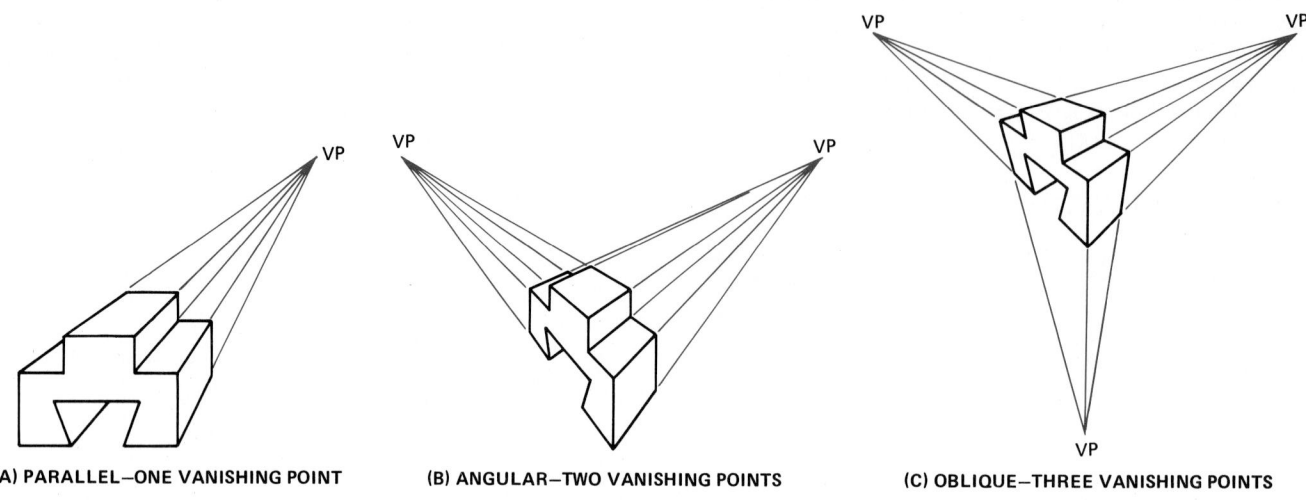

(A) PARALLEL—ONE VANISHING POINT   (B) ANGULAR—TWO VANISHING POINTS   (C) OBLIQUE—THREE VANISHING POINTS

FIG. 15-6-4   Types of perspective drawings.

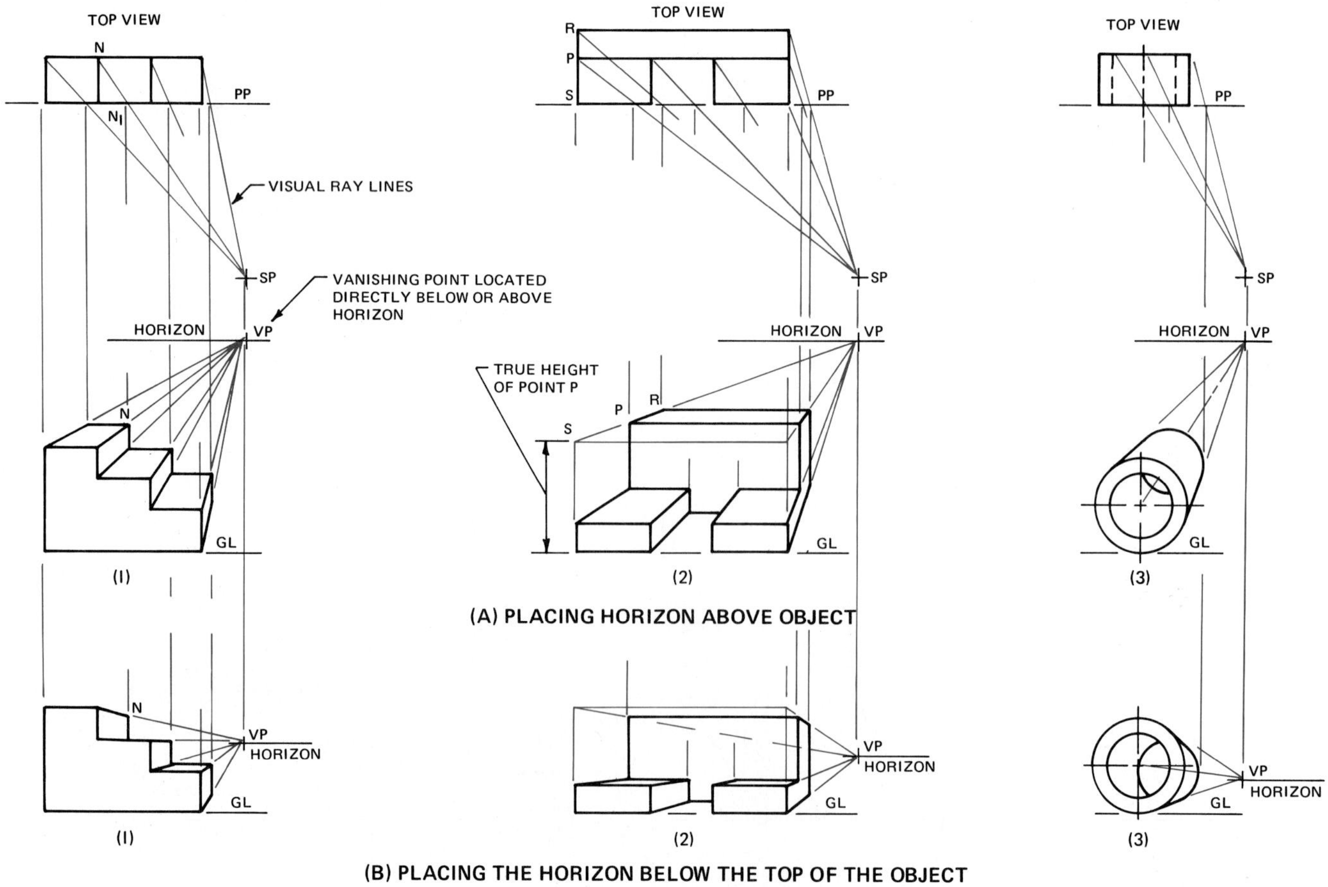

(A) PLACING HORIZON ABOVE OBJECT

(B) PLACING THE HORIZON BELOW THE TOP OF THE OBJECT

FIG. 15-6-5   Parallel, or one-point, perspective.

picture plane. Since point *S* lies on the picture plane and is the same height as point *P,* it may readily be found on the perspective drawing. Point *P* will lie on the receding line joining point *S* to the line *VP.*

In drawing a one-point perspective, a side or front view and a top view are normally drawn first—the top view to locate the part with respect to the picture plane and the side or front view to obtain the height of the various features. Figure 15-6-6 shows a simple, one-point perspective drawing with construction lines.

One of the most common uses of a parallel-perspective drawing is for representing the interior of a building. With this type of drawing, the vanishing point is located inside the room and is normally at eye level (Fig. 15-6-7).

Parallel-Perspective Grids

A variety of perspective grid sheets are available which enable the drafter to produce perspective drawings in less time than the conventional manner. Using a grid eliminates the tedious

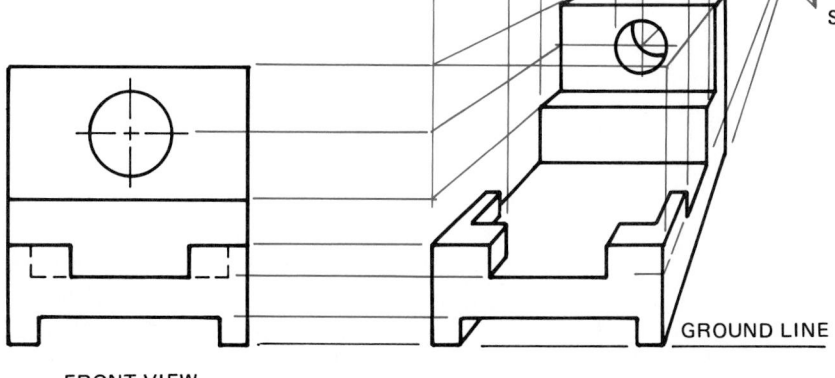

FIG. 15-6-6   Construction of a one-point perspective.

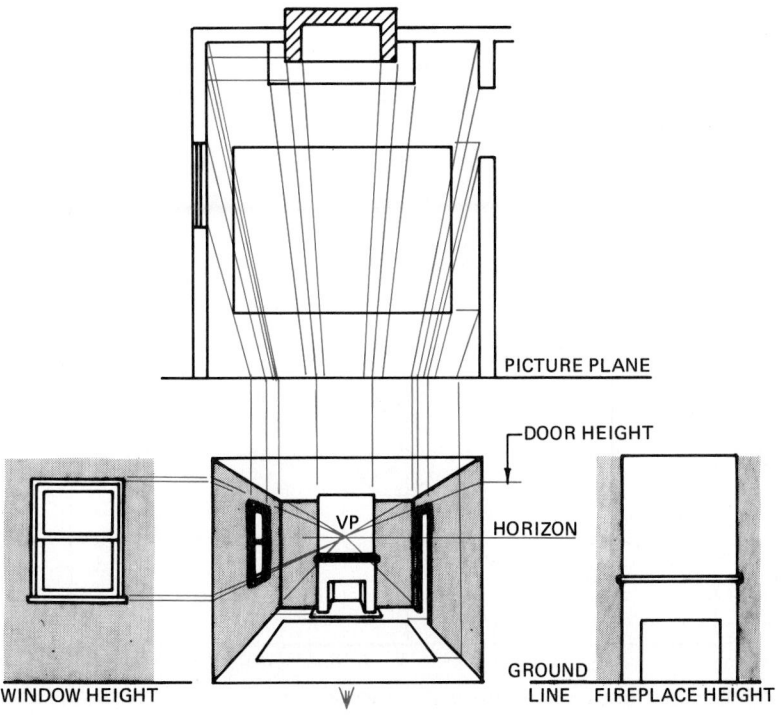

FIG. 15-6-7   Parallel-perspective drawing of an interior of a house.

459

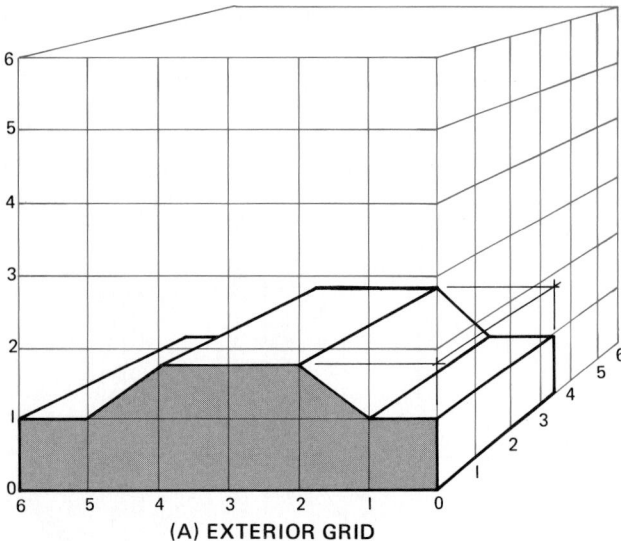

(A) EXTERIOR GRID

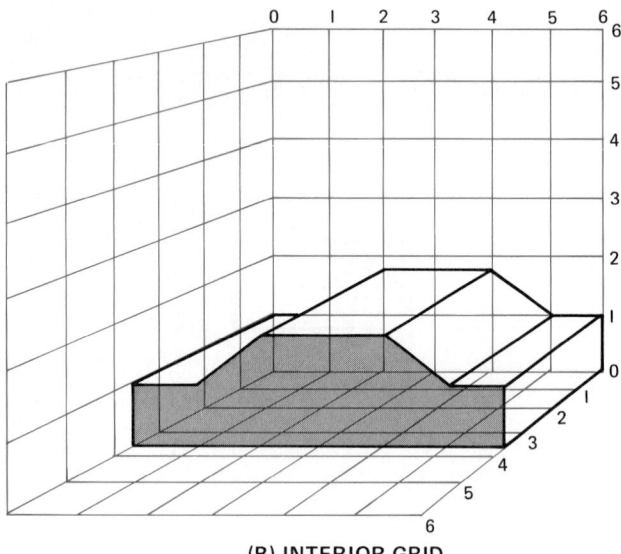

(B) INTERIOR GRID

FIG. 15-6-8   Parallel-perspective grids.

effort of establishing and projecting from the vanishing points for each individual feature. It also eliminates the problem of having the vanishing points located, in many instances, beyond the drawing area.

The cube grid, which is most widely used, has two basic variations: an exterior grid and an interior grid (Fig. 15-6-8). The grid sizes are dependent upon the desired scale of the parts to be drawn. The height and width planes are subdivided into identical increments, each increment representing any convenient size, such as 1.00 in., 1 ft or 10, 100, or 1000 mm. The plane or surface representing the depth is subdivided into increments that are proportionately foreshortened as they recede from the picture plane and thus create the perspective illusion (Fig. 15-6-9).

REFERENCES AND SOURCE MATERIAL

1. ANSI Y14.4, *American Drafting Standards Manual, Pictorial Drawing.*
2. General Motors Corp.

ASSIGNMENTS

See Assignments 26 through 28 for Unit 15-6 on pages 478 to 479.

15-7 ANGULAR, OR TWO-POINT, PERSPECTIVE

Two-point perspective is used quite extensively for architectural and product illustration, as shown in Fig. 15-7-1. *Angular-perspective drawings* are similar to axonometric drawings except that the receding lines converge at two vanishing points located on the horizon. Normally the height, or vertical, lines are parallel to the picture plane, and the length and width lines recede.

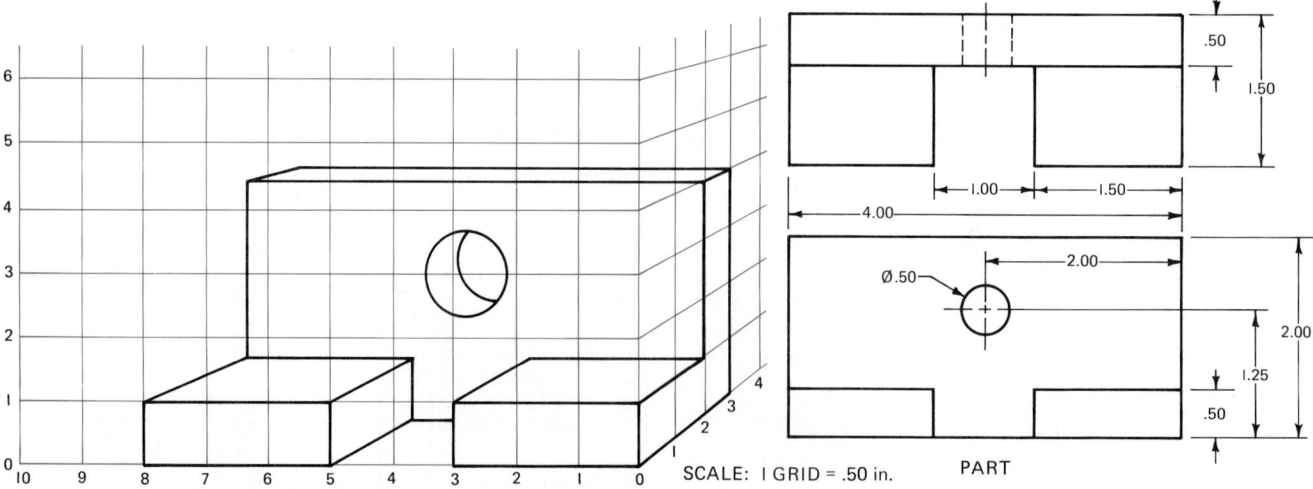

SCALE: 1 GRID = .50 in.   PART

FIG. 15-6-9   Part drawn on a parallel-perspective exterior grid.

The construction for a simple prism is shown in Fig. 15-7-2. Since line 1-2 rests on the picture plane, it will appear as its true height on the perspective drawing and will be located directly below line 1-2 on the top view. The next step is to join points 1 and 2 with light receding lines to both vanishing points. These receding lines represent the width and length lines of the prism; the width lines recede to *VPL* and the length lines recede to *VPR*. Since line 3-4 on the top view does not rest on the picture plane, it will not appear in its true height nor as its true distance from line 1-2 in the perspective. To find its position on the perspective drawing, join line 3-4, which appears as a point in the top view, to *SP* with a visual ray line. Where this visual ray line intersects the picture plane at *C*, project a vertical line down to the perspective view until it intersects the receding lines 1-*VPR* and 2-*VPR* at points 3 and 4, respectively. Line 5-6 may be found in the same manner. Next join point 3 to *VPL* and point 5 to *VPR* with light receding lines. The intersection of these lines is point 7.

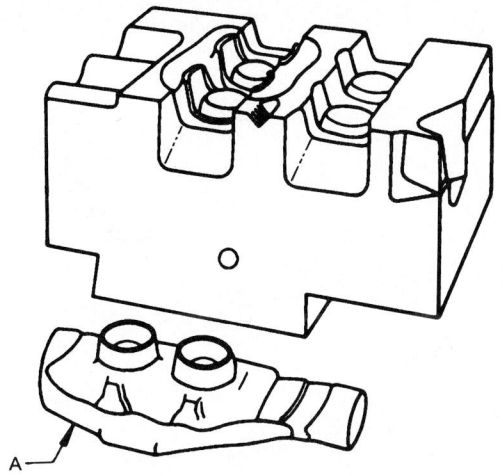

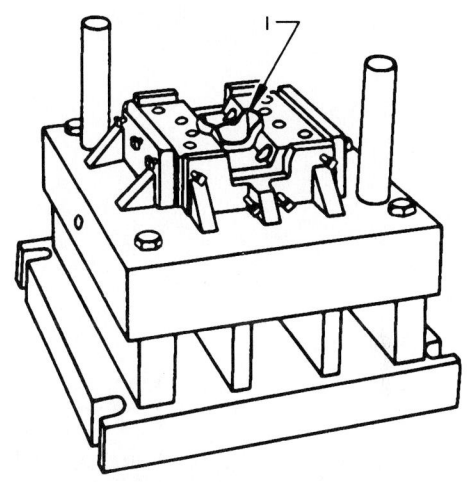

FIG. 15-7-1   Angular-, or two-point, perspective drawings.
(General Motors Corp.)

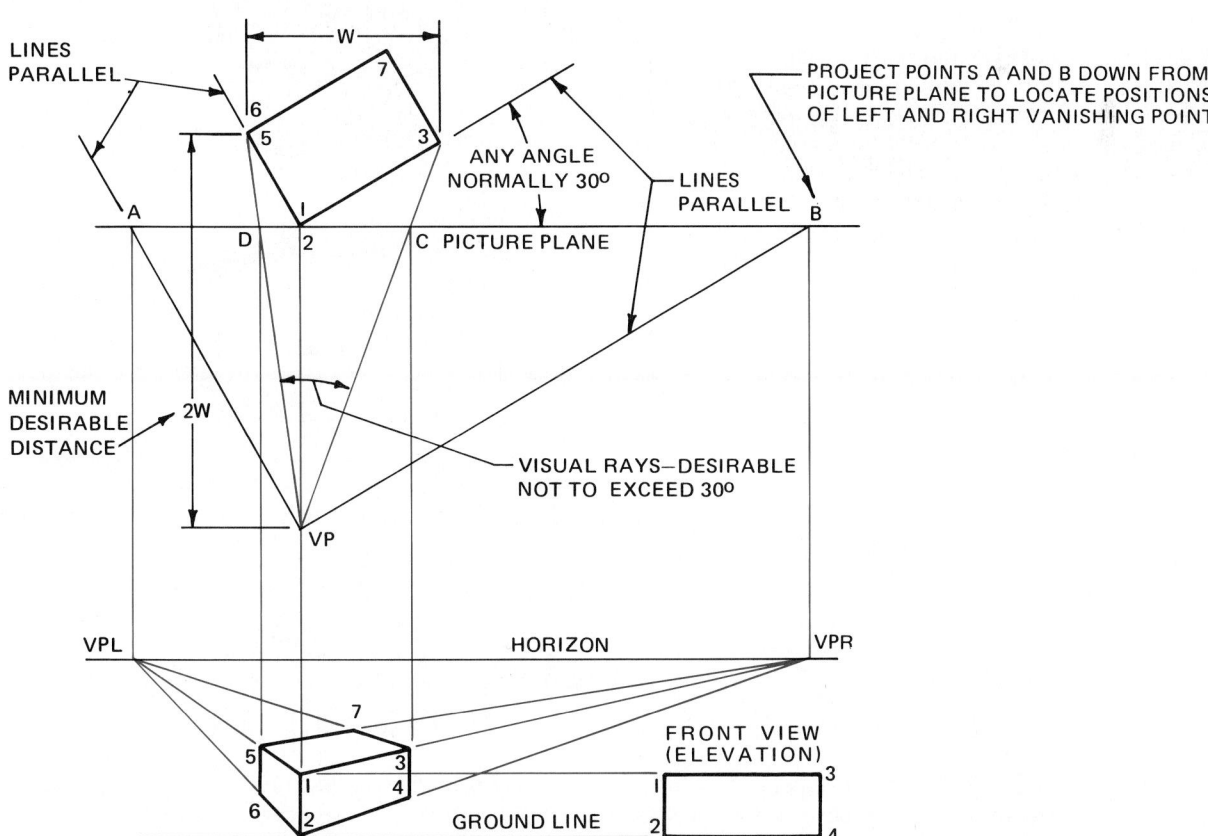

FIG. 15-7-2   Angular-perspective drawing of a prism.

461

FIG. 15-7-3   Angular-perspective
drawing of an object
that does not touch
the picture plane.

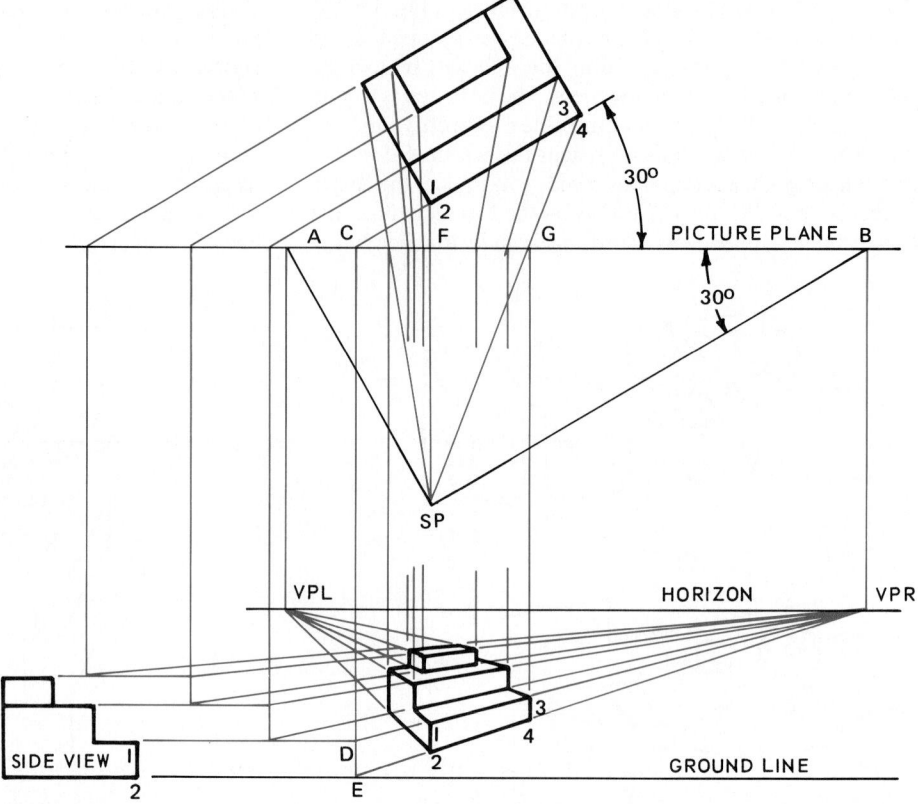

Lines Not Touching on the Picture Plane   Figure 15-7-3 illustrates the construction of a perspective drawing where none of the object lines touch the picture plane.

All these lines can be constructed by using the following procedure, which locates the position and size of lines 1-2 and 3-4. Extend line 1-3 (and 2-4) in the top view to intersect the picture plane at point *C*. Project a line down from *C* to intersect horizontal lines 1-*D* and 2-*E* at *D* and *E,* respectively. Had line 1-2 been located at *C* in the top view, it would have appeared at its true height and at *D-E* on the perspective. Join points *D* to *VPR* and *E* to *VPR* with light receding lines. Somewhere along these lines are points 1, 2, 3, and 4. Next join lines 1-2 and 3-4 in the top view to *SP* with visual ray lines. Where these visual ray lines intersect the picture plane at *F* and *G,* respectively, project vertical lines down to the perspective view intersecting line *D-VPR* at 1 and 3 and *E-VPR* at 2 and 4.

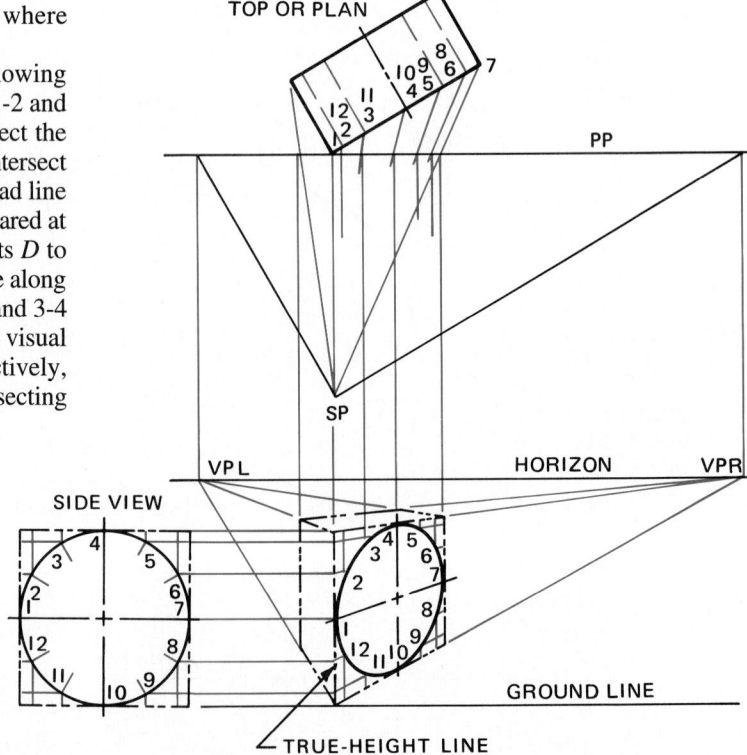

FIG. 15-7-4   Construction of
a circle in
angular
perspective.

Construction of Circles and Curves in Perspective   Circles and curves may be constructed in perspective, as illustrated in Fig. 15-7-4. Using orthographic projections, oriented with respect to the subject in the plan and side views, plot and label the desired points (using numbers) on the curved surfaces.

From the plan view project these points to the picture plane, then vertically down to the perspective view. Project the height of the plotting numbers horizontally from the side view to the true-height line in the perspective view. The position of the plotting numbers may now be located on the perspective view.

Locate the points of intersection of the lines projected down from the picture plane with the visual ray lines receding to the right vanishing point from the appropriate numbers on the true-height line.

Horizon Line   Figure 15-7-5 illustrates different effects produced by repositioning the object with respect to the horizon.

Angular-Perspective Grids

Exterior Grid   When the three adjacent exterior planes of the cube are developed, the resultant image is referred to as an *exterior grid.* When this grid is used, the points are projected from the top plane downward and from the picture planes away from the observer (Figs. 15-7-6 and 15-7-7).

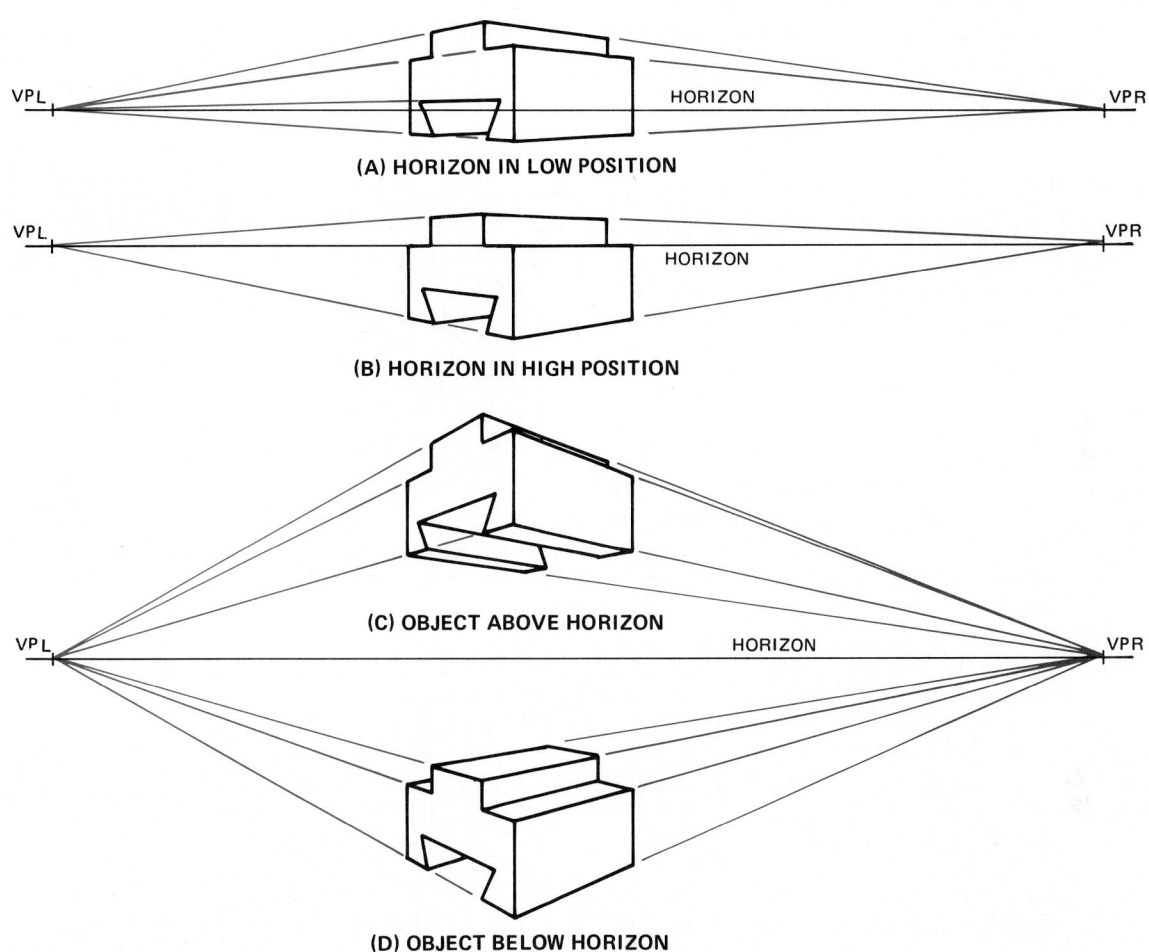

FIG. 15-7-5   Horizon lines.

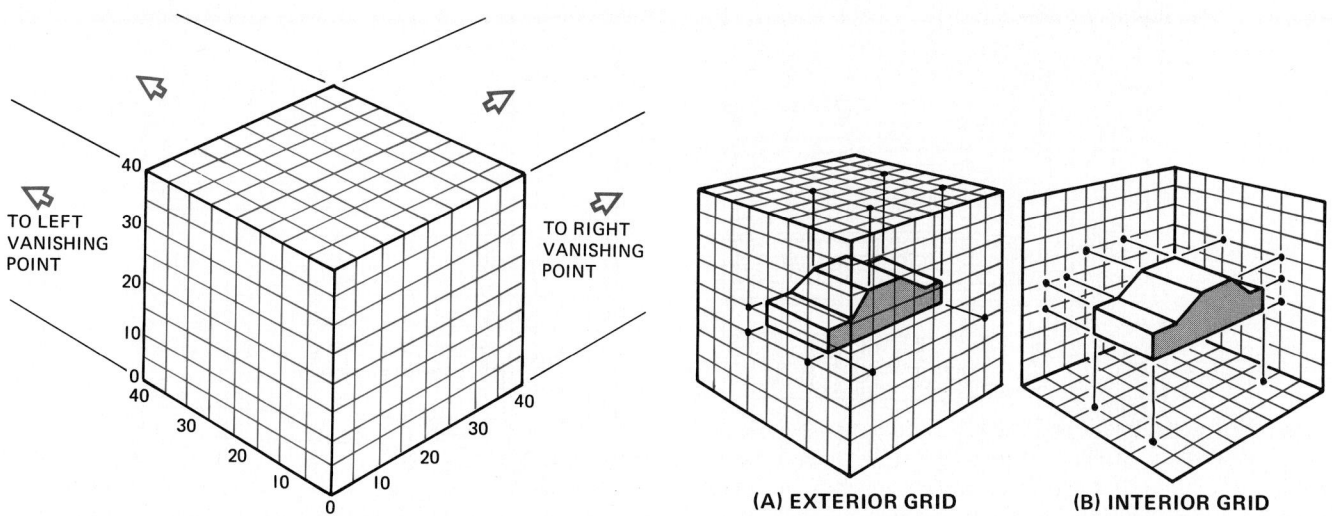

FIG. 15-7-6   Angular-perspective grid.

FIG. 15-7-7   Types of angular-perspective grids.

463

Interior Grid   When the three adjacent interior planes of the cube are exposed and developed, the resultant image is referred to as an *interior grid.* When this grid is used, the points are projected from the base plane upward and from the picture planes toward the observer. Each produces the same results.

Two further variations of both the exterior and interior grids are known as the *bird's-eye* and *worm's-eye grids.* These effects are achieved by rotating the vertical plane of the grid about the horizon line. Objects drawn in the bird's-eye grid appear as if they were being viewed from above the horizon line, as seen in Fig. 15-7-8. Objects drawn in the worm's-eye grid appear as if they were being viewed from below the horizon line.

Grid Increments

The three surfaces or planes of the grid are subdivided into multiple vertical and horizontal increments. Each increment is proportionately foreshortened as it recedes from the picture plane and thus creates the perspective illusion. The grid increments can be any size desired (Fig. 15-7-9).

 CAD

CAD systems that support 3-D modeling usually support the generation perspective drawings. In the construction of the perspective drawing, the position of the "camera" represents the person viewing the object. Several parameters such as zoom, distance, and view clip are then manipulated to create the required perspective view of the object. Only the view is manipulated as the original model of the object remains unaffected.

REFERENCES AND SOURCE MATERIAL

1. General Motors Corp.

ASSIGNMENT

See Assignment 29 for Unit 15-7 on pages 479–480.

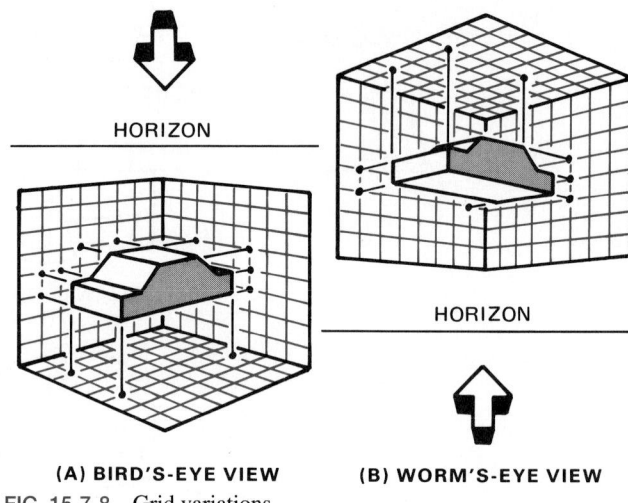

(A) BIRD'S-EYE VIEW   **(B) WORM'S-EYE VIEW**

FIG. 15-7-8   Grid variations.

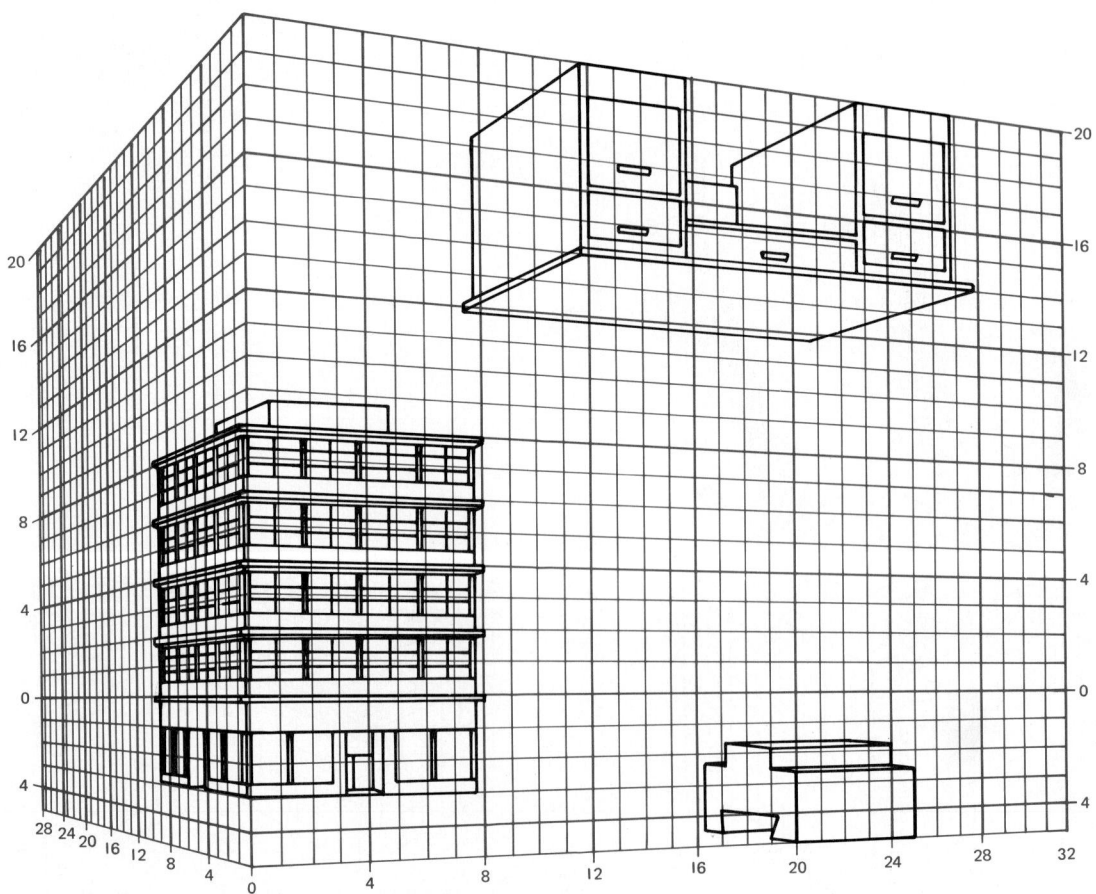

FIG. 15-7-9   Angular-perspective grid applications.

ASSIGNMENTS FOR CHAPTER 15

ASSIGNMENTS FOR UNIT 15-1, PICTORIAL DRAWINGS

1. On isometric grid paper or using the CAD isometric grid, draw the parts shown in Fig. 15-1-A. Do not show hidden lines. Each square shown on the drawing represents one isometric square on the grid.

2. On isometric grid paper or using the CAD isometric grid, draw the parts shown in Fig. 15-1-B. Do not show hidden lines. Each square shown on the drawing represents one isometric square on the grid.

3. On isometric grid paper or using the CAD isometric grid, sketch the parts shown in Fig. 15-1-C. Do not show hidden lines.

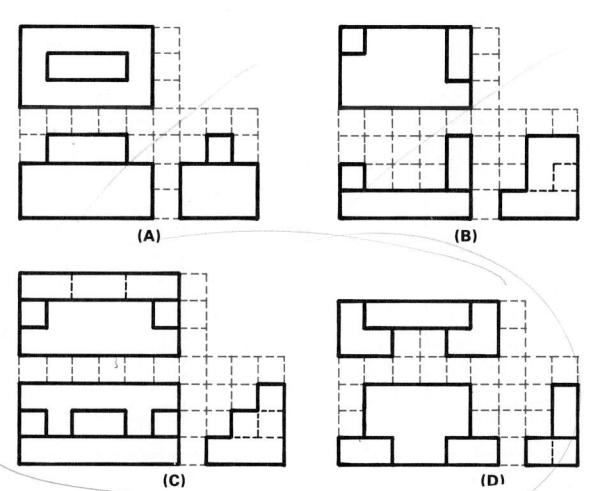

FIG. 15-1-A   Isometric flat-surface assignments.

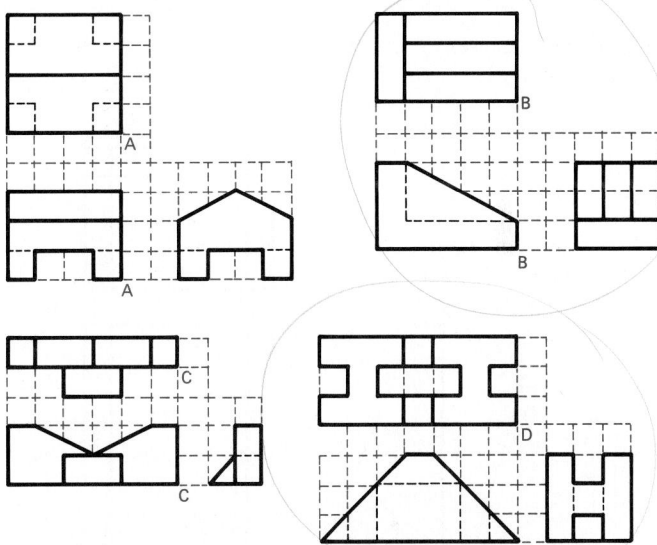

FIG. 15-1-B   Isometric flat-surface assignments.

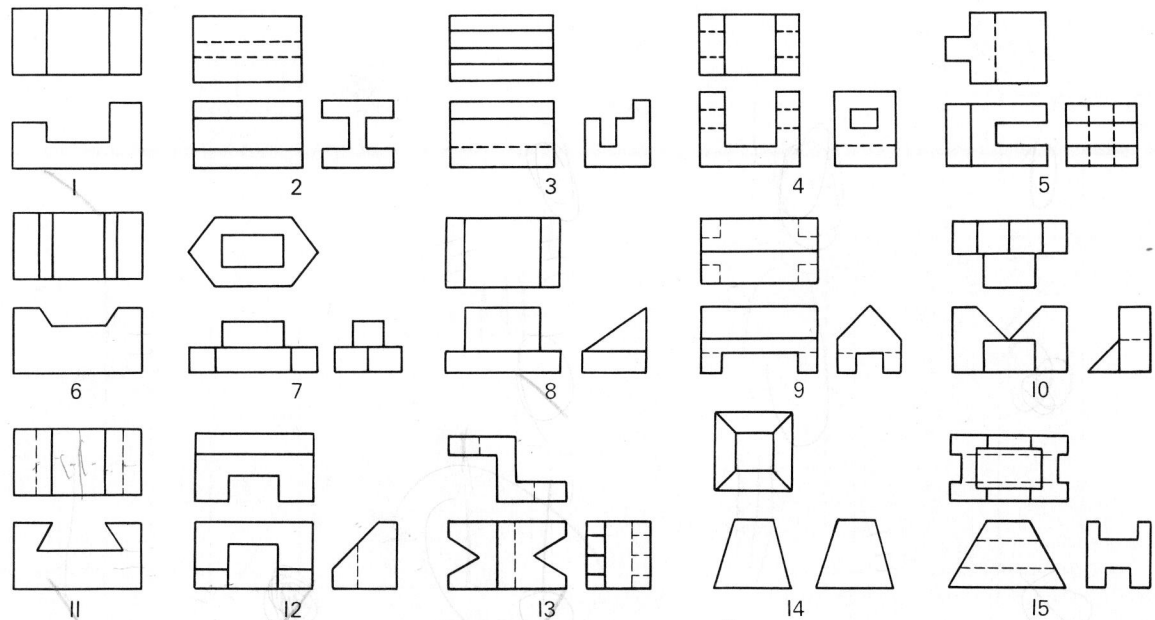

FIG. 15-1-C   Sketching assignments.

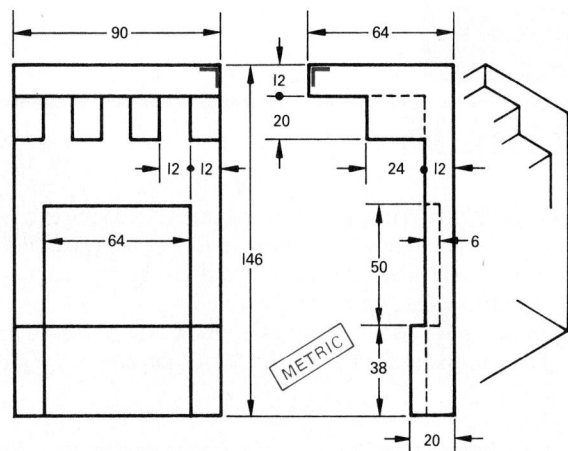

FIG. 15-1-D   Tablet.

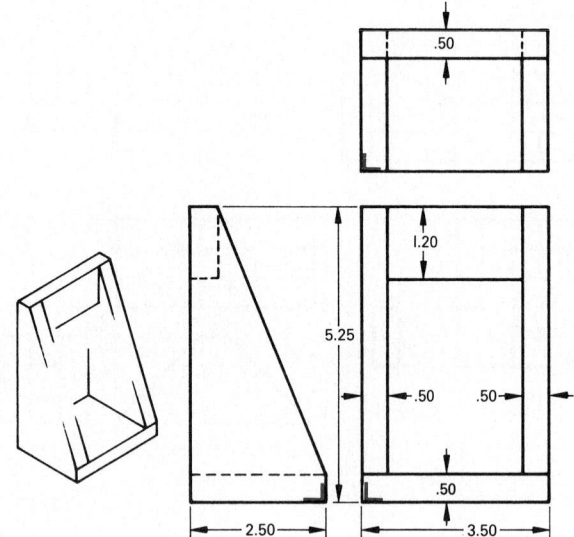

FIG. 15-1-E   Stirrup.

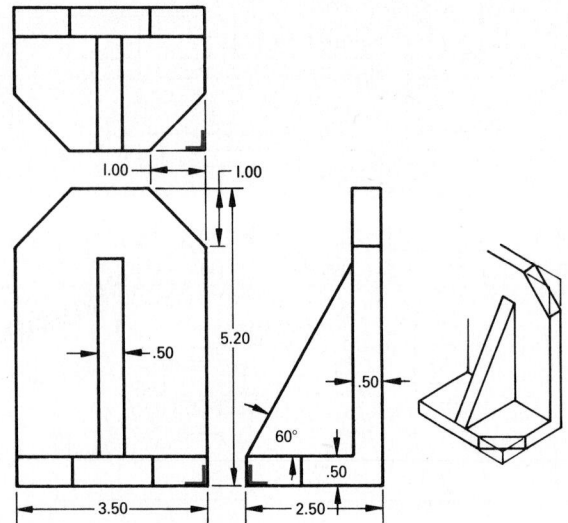

FIG. 15-1-F   Brace.

4. On isometric grid paper or using the CAD isometric grid, make an isometric drawing of one of the parts shown in Figs. 15-1-D through 15-1-F. A partial starting layout is provided for each of the parts shown. Start at the corner indicated by thick lines. Scale 1:1.

5. On isometric grid paper or using the CAD isometric grid, make an isometric drawing of one of the parts shown in Figs. 15-1-G through 15-1-J. For Fig. 15-1-G use the layout shown in Fig. 15-1-E; for Fig. 15-1-H use the layout shown

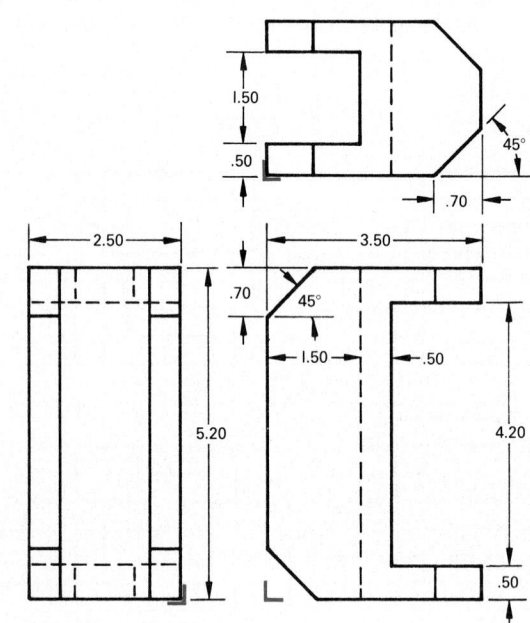

FIG. 15-1-G   Cross slide.

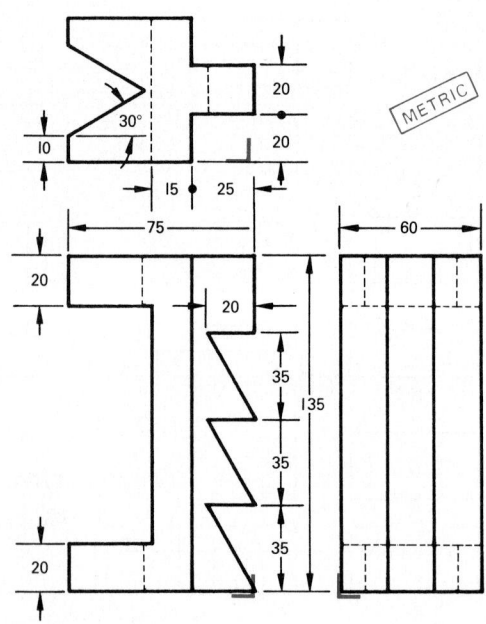

FIG. 15-1-H   Ratchet.

in Fig. 15-1-F; and for Fig. 15-1-J use the layout shown in Fig. 15-1-D. Scale 1:1.

6. Make an isometric drawing, complete with dimensions, of one of the parts shown in Figs. 15-1-K through 15-1-Q (here and on pg. 468). Scale 1:1.

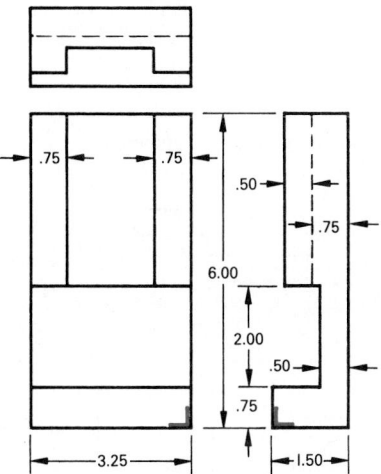

FIG. 15-1-J Stop.

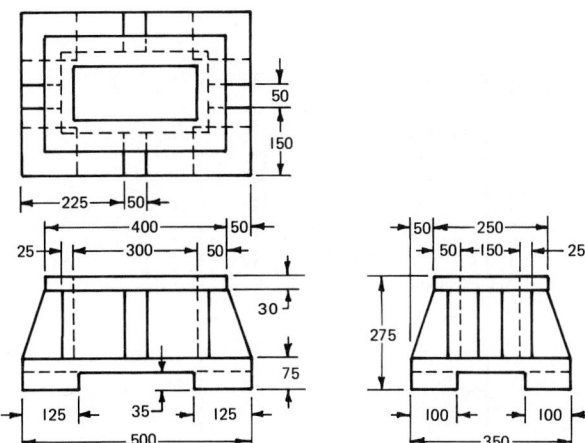

FIG. 15-1-K Planter box.

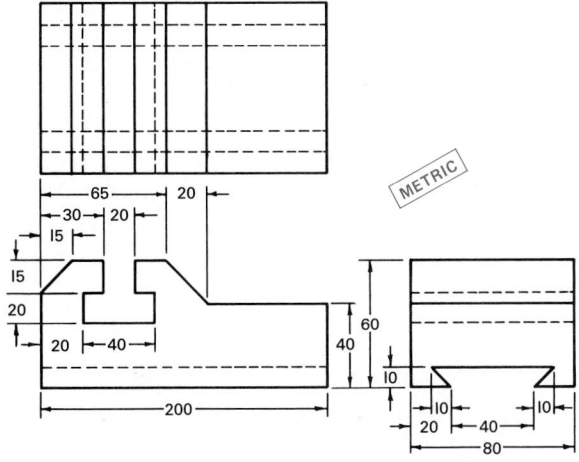

FIG. 15-1-L Cross slide.

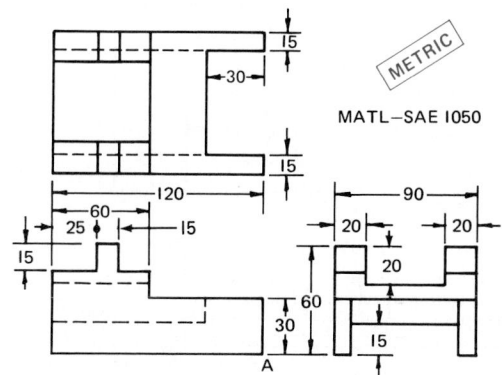

FIG. 15-1-M Step block.

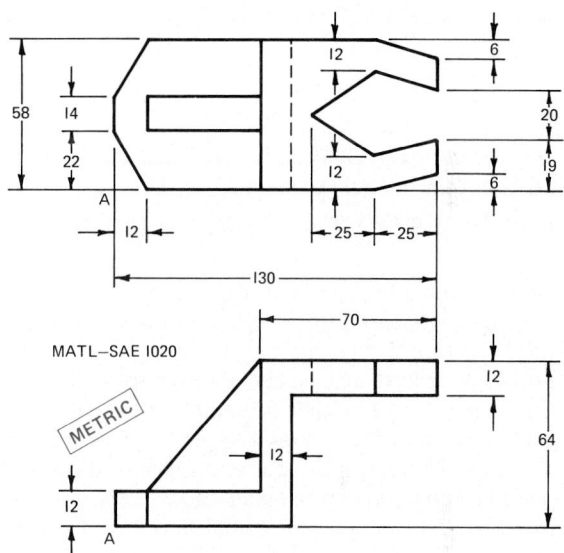

FIG. 15-1-N Support bracket.

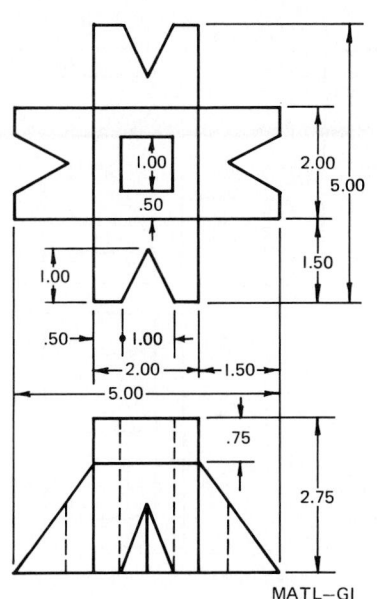

FIG. 15-1-P Stand.

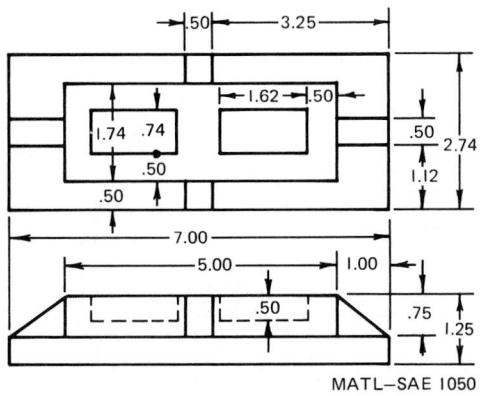

FIG. 15-1-Q   Base plate.

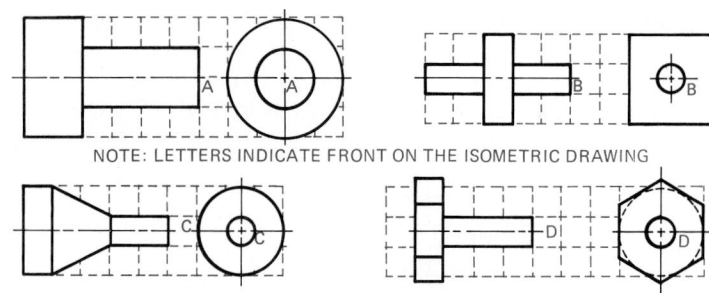

FIG. 15-2-B   Isometric curved-surface assignments.

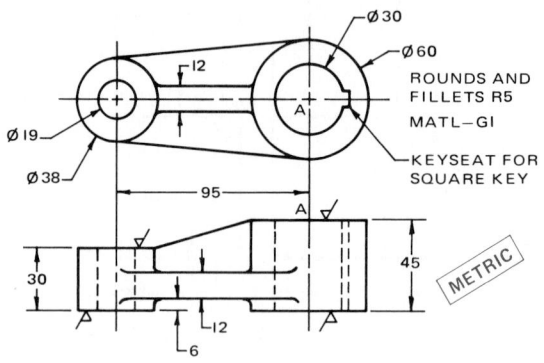

FIG. 15-2-C   Link.

ASSIGNMENTS FOR UNIT 15-2, CURVED SURFACES IN ISOMETRIC

7. On isometric grid paper or using the CAD isometric grid, draw the parts shown in Fig. 15-2-A. Each square shown on the figure represents one square on the isometric grid. Hidden lines may be omitted for clarity.

8. On isometric grid paper or using the CAD isometric grid, draw the parts shown in Fig. 15-2-B. Each square shown on the figure represents one square on the isometric grid. Hidden lines may be omitted for clarity.

9. Make an isometric drawing, complete with dimensions, of one of the parts shown in Figs. 15-2-C through 15-2-F. Scale 1:2 for Fig. 15-2-F. For all others the scale is 1:1.

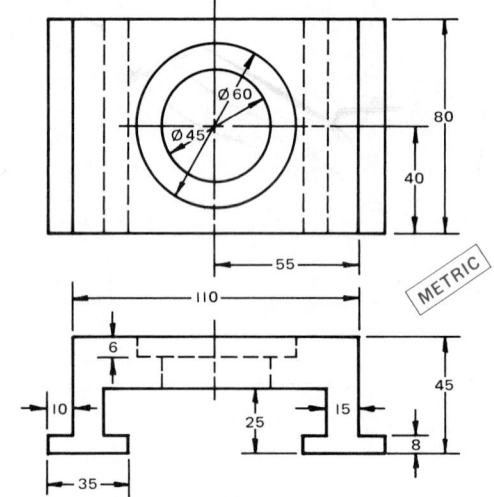

FIG. 15-2-D   T-guide.

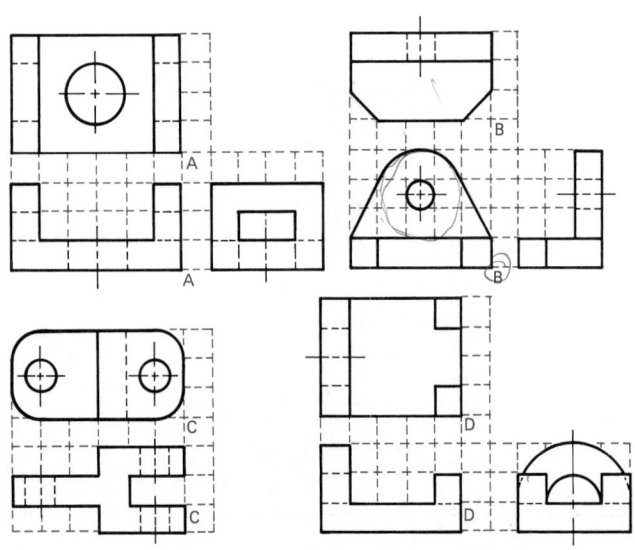

NOTE: LETTERS INDICATE POSITION OF LOWER FRONT CORNER ON THE ISOMETRIC DRAWING

FIG. 15-2-A   Isometric curved-surface assignments.

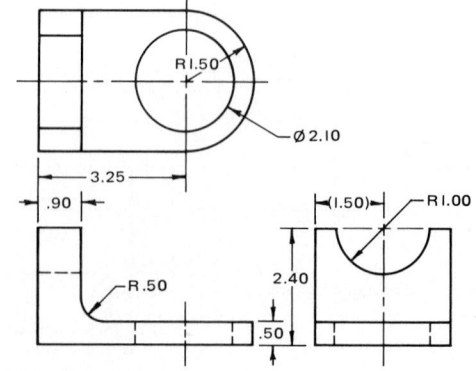

FIG. 15-2-E   Cradle bracket.

FIG. 15-2-F   Base.

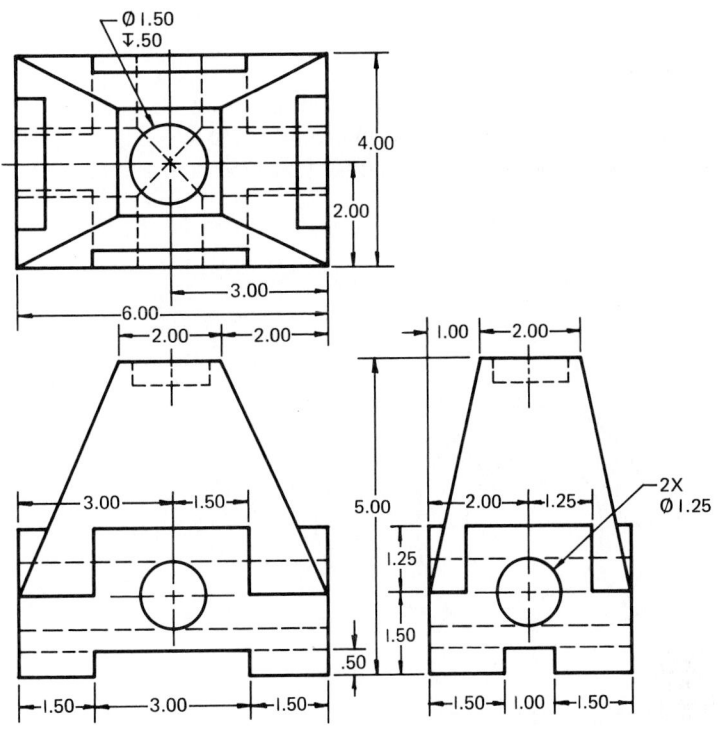

ASSIGNMENTS FOR UNIT 15-3, COMMON FEATURES IN ISOMETRIC

10. Make an isometric half-section view, complete with dimensions, of one of the parts shown in Figs. 15-3-A and 15-3-B. Scale 1:1.

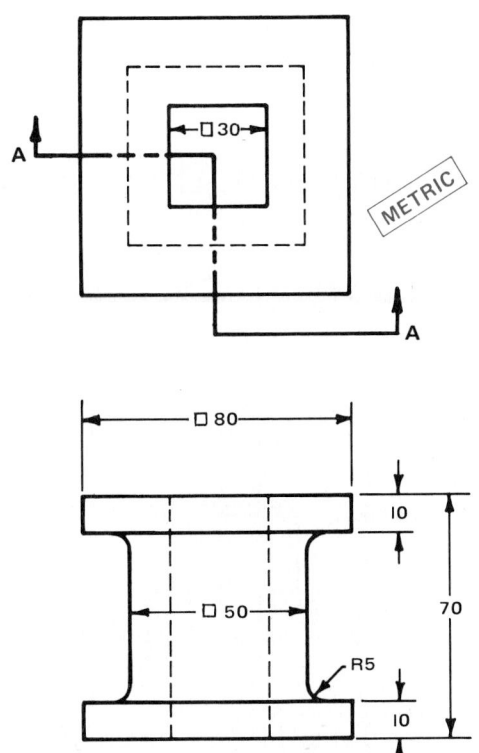

FIG. 15-3-A   Guide block.

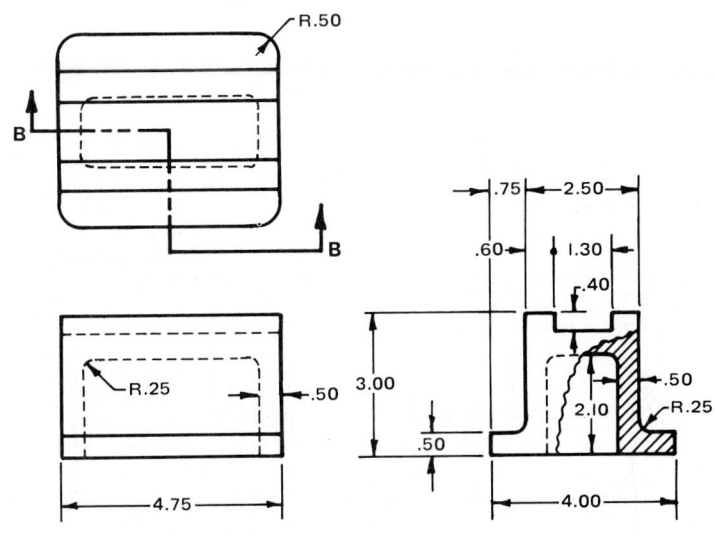

FIG. 15-3-B   Base.

11. Make an isometric full-section drawing, complete with dimensions, of one of the parts shown in Figs. 15-3-C through 15-3-E. Scale 1:1.
12. Make an isometric drawing, complete with dimensions, of one of the parts shown in Figs. 15-3-F and 15-3-G. Use a conventional break to shorten the length for Fig. 15-3-F. Scale 1:1.

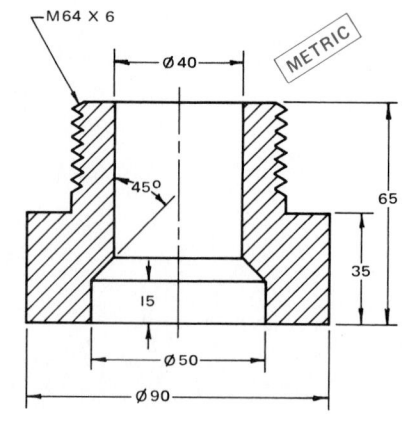

FIG. 15-3-E   Adapter.

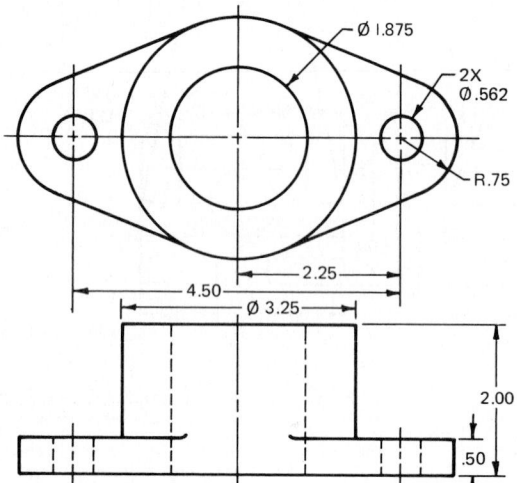

FIG. 15-3-C   Bearing support.

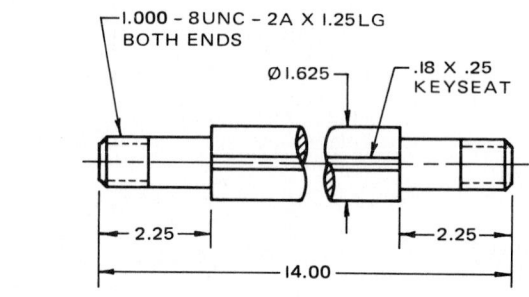

FIG. 15-3-F   Shaft.

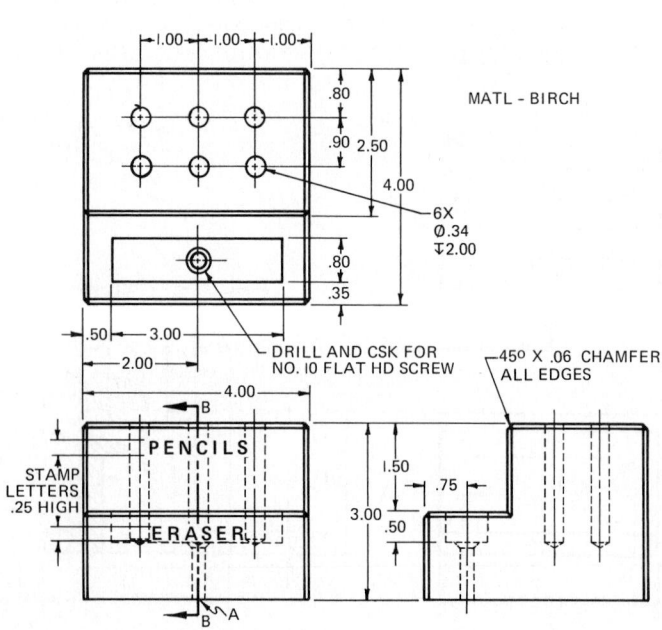

FIG. 15-3-D   Pencil holder.

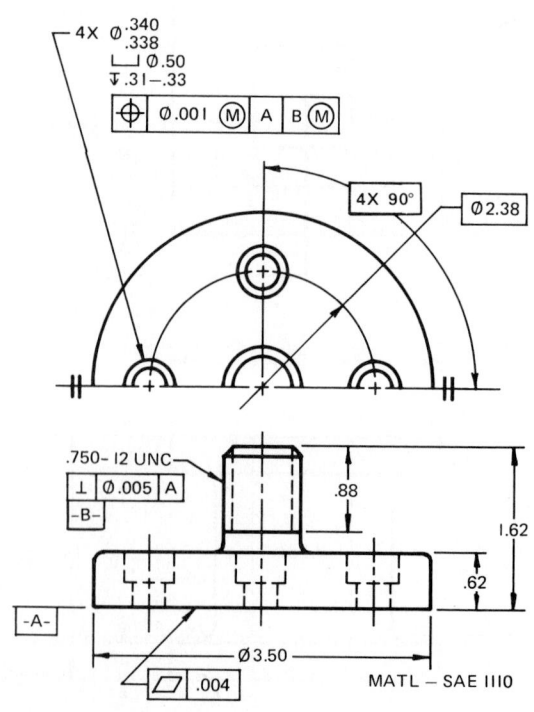

FIG. 15-3-G   End plate.

13. Make an isometric assembly drawing of the two-post die set, Model 302, shown in Fig. 15-3-H. Allow 2 in. between the top and base. Scale 1:2. Do not dimension. Include on the drawing an item list. Using part numbers, identify the parts on the assembly.

14. Make an isometric exploded assembly drawing of the book rack shown in Fig. 15-3-J. Use a B (A3) sheet. Scale 1:2. Do not dimension. Include on the drawing an item list. Using part numbers, identify the parts on the assembly.

FIG. 15-3-H   Two-post die set.

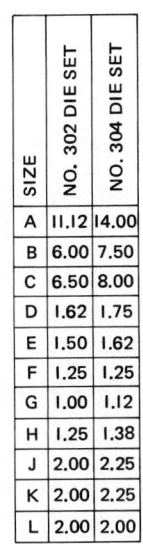

SIZE	NO. 302 DIE SET	NO. 304 DIE SET
A	11.12	14.00
B	6.00	7.50
C	6.50	8.00
D	1.62	1.75
E	1.50	1.62
F	1.25	1.25
G	1.00	1.12
H	1.25	1.38
J	2.00	2.25
K	2.00	2.25
L	2.00	2.00

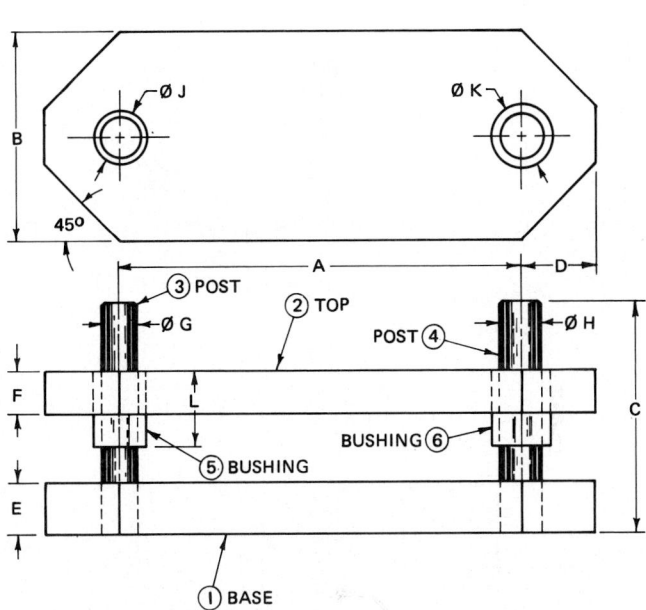

FIG. 15-3-J   Book rack.

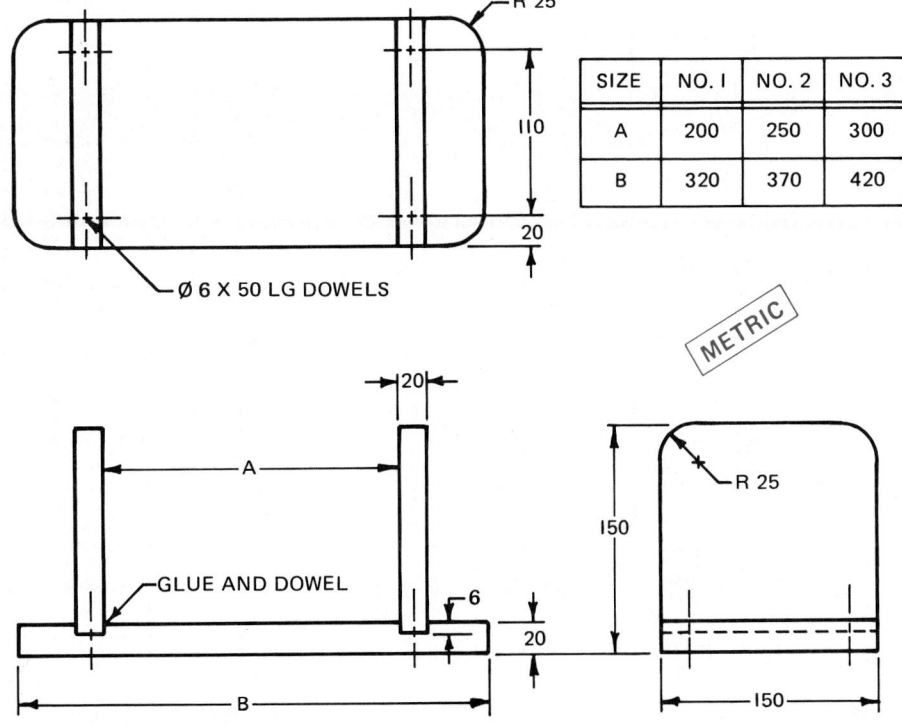

SIZE	NO. I	NO. 2	NO. 3
A	200	250	300
B	320	370	420

METRIC

471

15. Make an isometric assembly drawing of the gear clamp shown in Fig. 15-3-K. Add dimensions, identification part numbers, and an item list. Scale 1:1.

16. Make an isometric exploded assembly drawing of the universal joint shown in Fig. 15-3-L. Scale 1:1. Do not dimension. Include on the drawing an item list. Using part numbers, identify the parts on the assembly.

ASSIGNMENTS FOR UNIT 15-4, OBLIQUE PROJECTION

17. On coordinate grid paper or using the CAD grid, make oblique drawings of the three parts shown in Fig. 15-4-A. Each square shown on the figure represents one square on the grid. Hidden lines may be omitted to improve clarity.

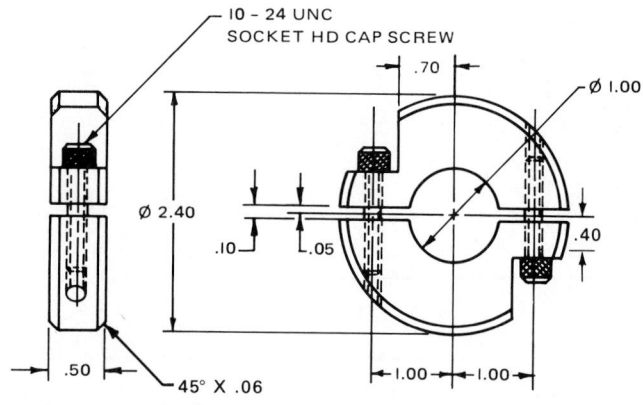

FIG. 15-3-K   Gear clamp.

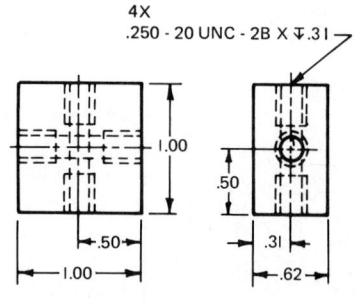

PT 1 - FORK - 2 REQD

PT 2 - RING - 1 REQD
PT 3 - Ø .25 SPRING PIN - 2 REQD
PT 4 - .250 - 20 FHMS - .62 LG - 4 REQD

FIG. 15-3-L   Universal joint.

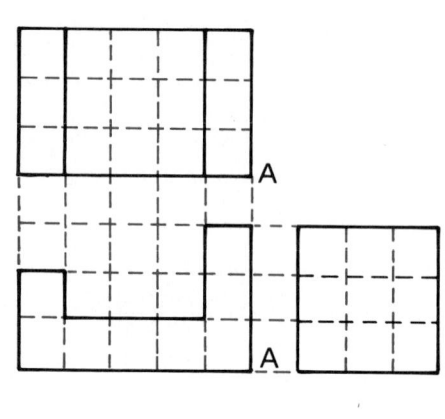

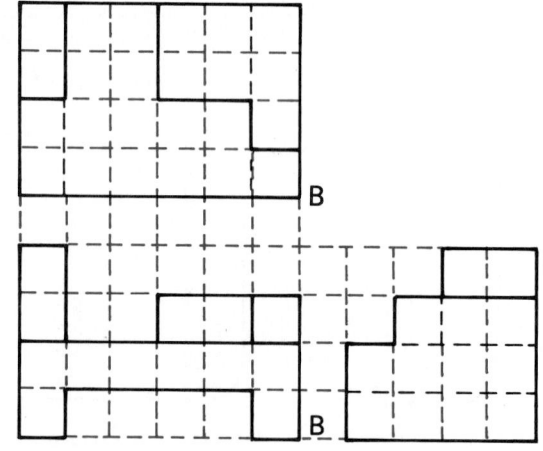

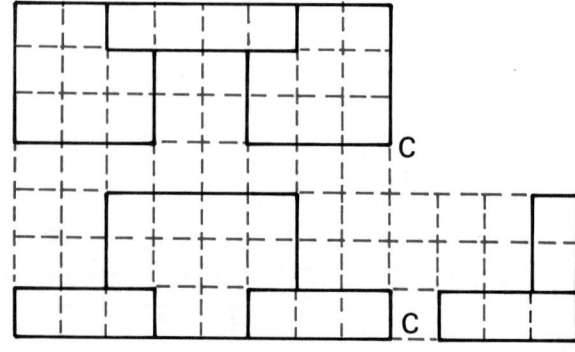

FIG. 15-4-A   Oblique flat-surface assignments.

18. On coordinate grid paper or using the CAD grid, make oblique drawings of the three parts shown in Fig. 15-4-B. Each square shown on the figure represents one square on the grid. Hidden lines may be omitted to improve clarity.

19. On oblique grid paper or using the CAD grid, sketch the parts shown in Fig. 15-4-C. Do not dimension or show hidden lines.

20. Make an oblique drawing of one of the parts shown in Figs. 15-4-D and 15-4-E. A partial starting layout is provided for each of the parts shown. Start at the corner indicated by thick lines. Scale 1:1.

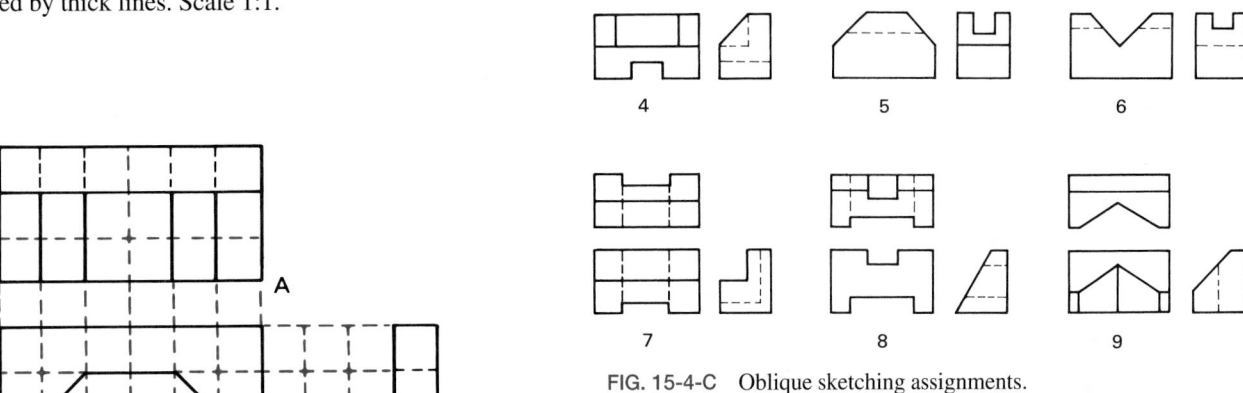

FIG. 15-4-C Oblique sketching assignments.

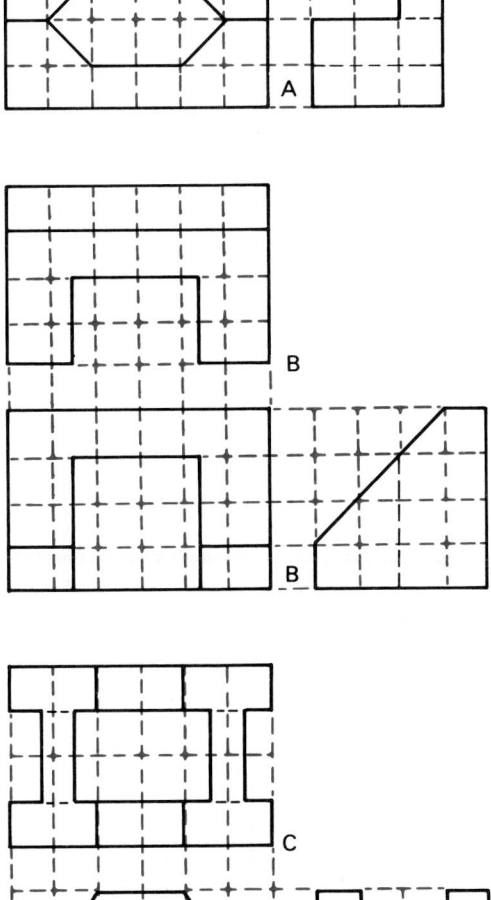

FIG. 15-4-B Oblique flat-surface assignments.

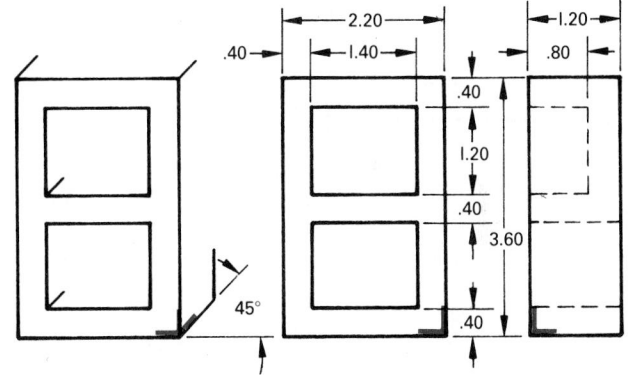

FIG. 15-4-D Stand.

FIG. 15-4-E V-block.

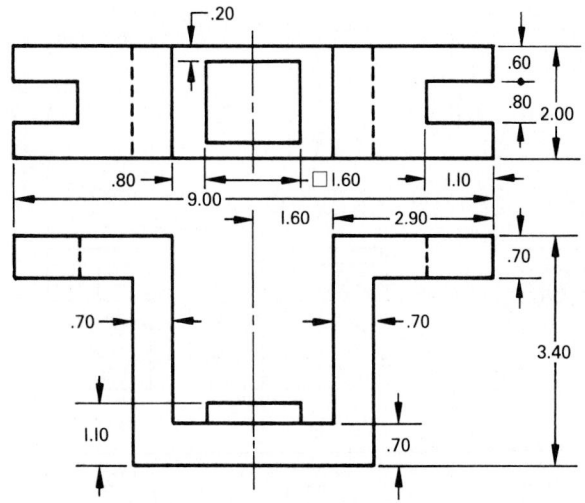

FIG. 15-4-F   Control arm.

21. Make an oblique drawing, complete with dimensions, of one of the parts shown in Figs. 15-4-F through 15-4-J. Scale 1:1.

ASSIGNMENTS FOR UNIT 15-5, COMMON FEATURES IN OBLIQUE

22. Make an oblique drawing of one of the parts shown in Figs. 15-5-A and 15-5-B. A partial starting layout is provided for each of the parts shown. Add dimensions. Scale 1:1.
23. Make an oblique drawing, complete with dimensions, of one of the parts shown in Figs. 15-5-C through 15-5-G. Scale 1:1. For Fig. 15-5-G show either a broken-out or phantom section to show the Ø.406 hole.

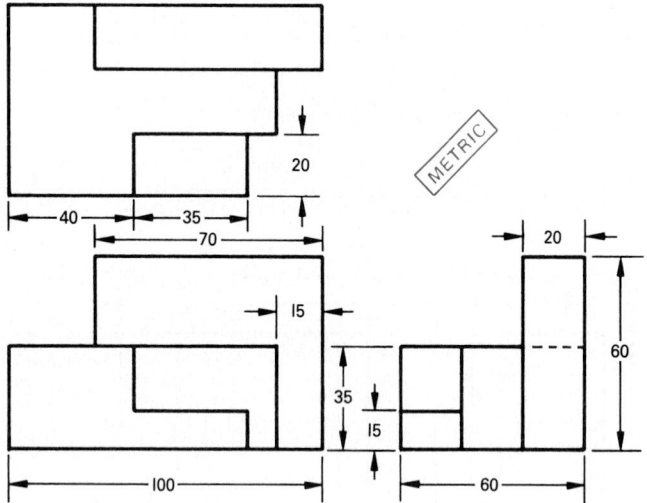

FIG. 15-4-G   Spacer block.

FIG. 15-4-H   V-block rest.

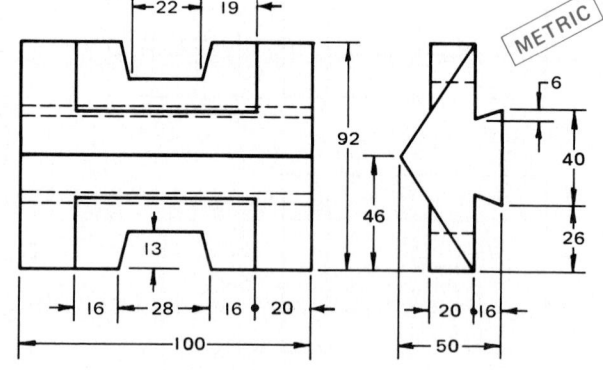

FIG. 15-4-J   Dovetail guide.

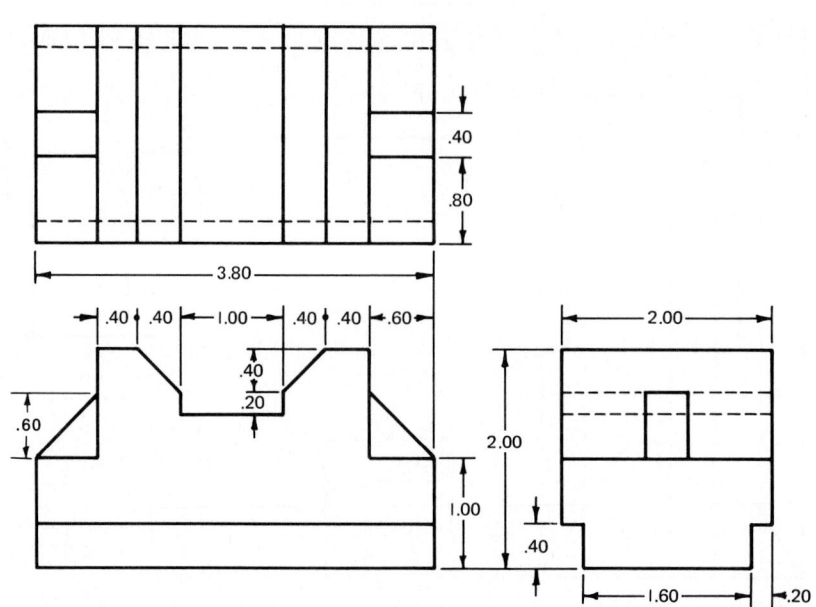

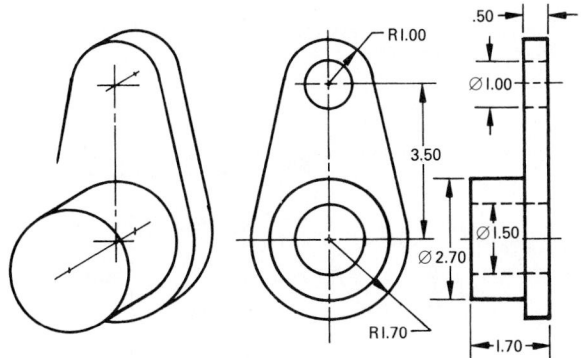

FIG. 15-5-A   Crank.

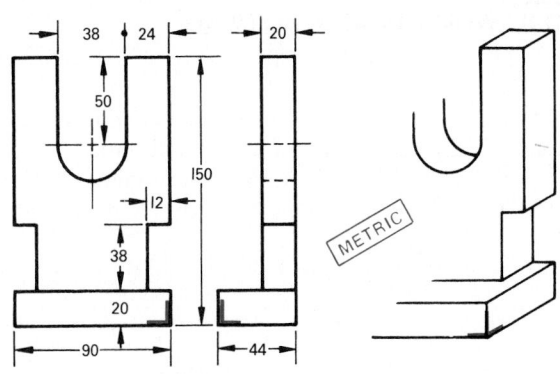

FIG. 15-5-B   Shaft support.

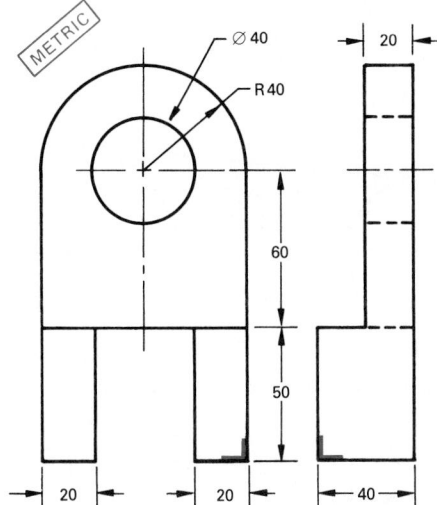

FIG. 15-5-C   Forked guide.

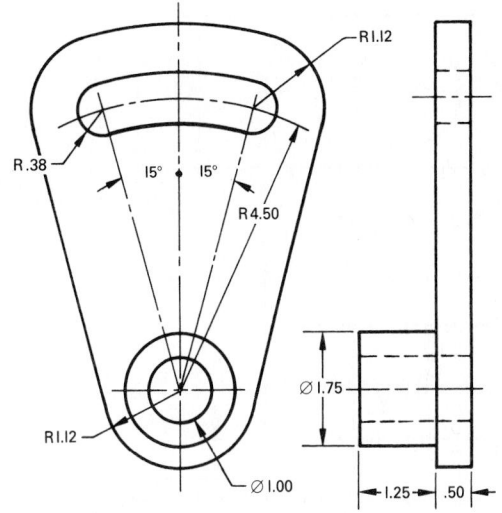

FIG. 15-5-D   Slotted sector.

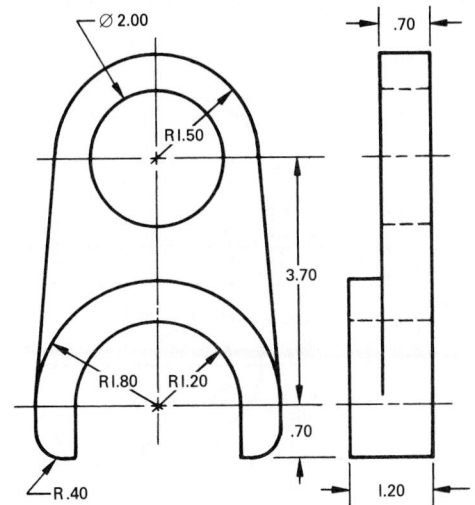

FIG. 15-5-E   Guide link.

FIG. 15-5-G   Tool holder.

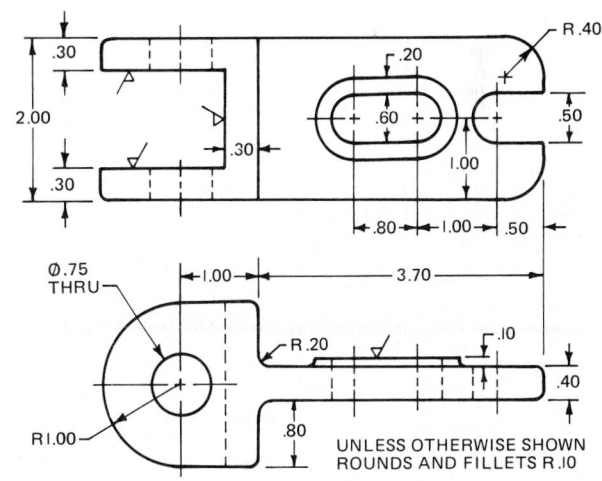

FIG. 15-5-F   Swing bracket.

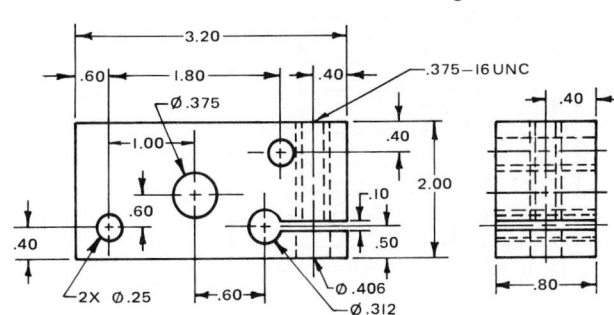

24. Make an oblique drawing, complete with dimensions, of one of the parts shown in Figs. 15-5-H through 15-5-M. For Figs. 15-5-J through 15-5-L use the straight line method of showing the rounds and fillets. Scale to suit.

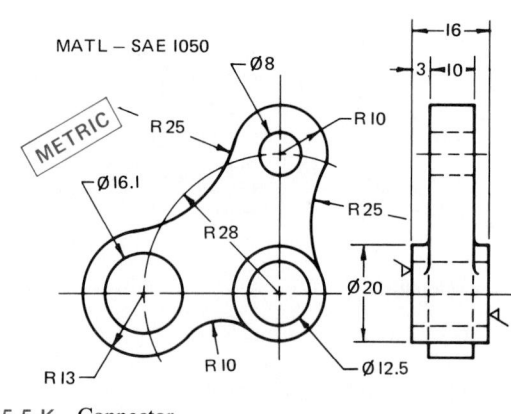

FIG. 15-5-K   Connector.

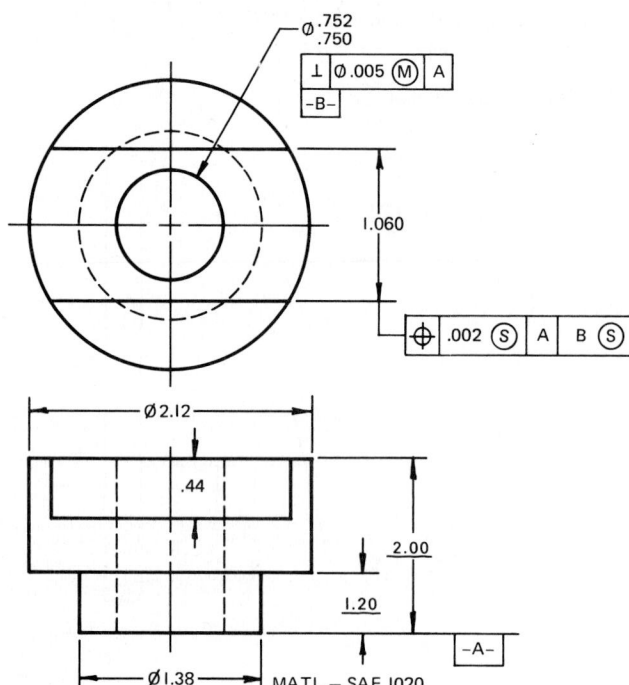

FIG. 15-5-H   Drive link.

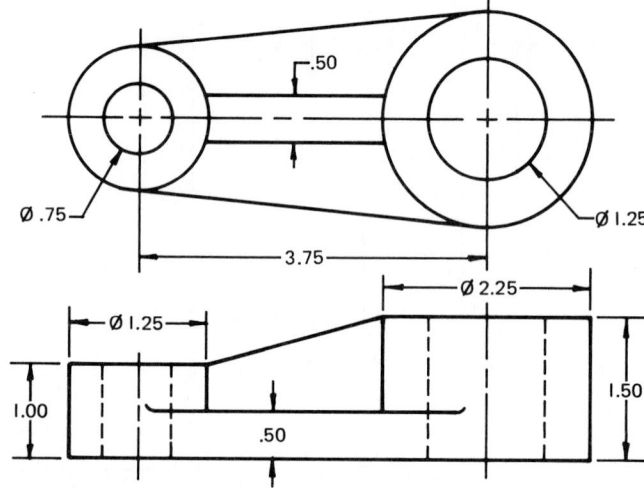

FIG. 15-5-L   Link.

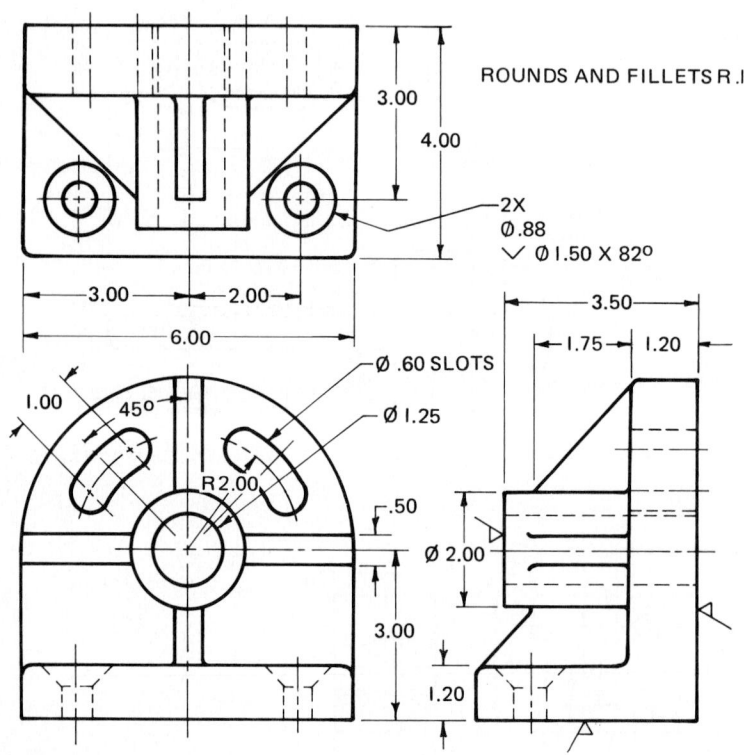

FIG. 15-5-J   End bracket.

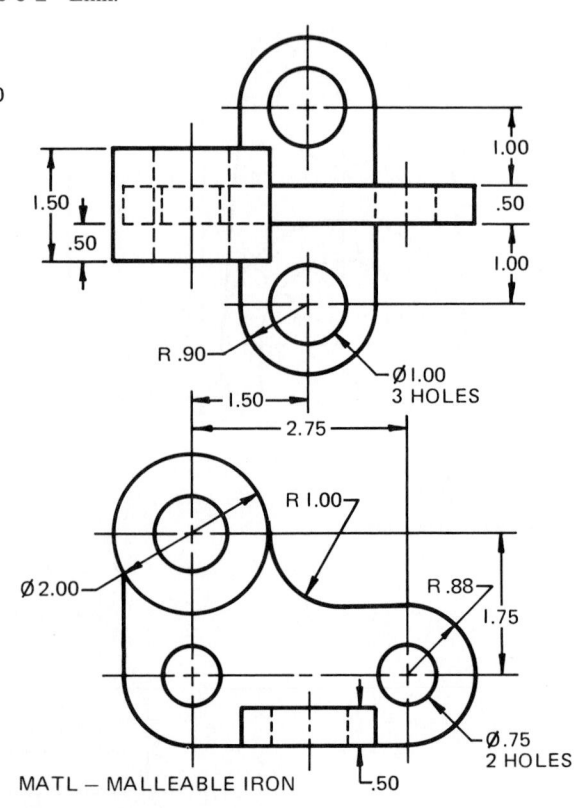

MATL — MALLEABLE IRON

FIG. 15-5-M   Swivel hanger.

25. Make an oblique drawing, complete with dimensions, of one of the parts shown in Figs. 15-5-N through 15-5-U. For Fig. 15-5-P use a conventional break to shorten the length. Show a half-section for Figs. 15-5-S and 15-5-T. Scale to suit.

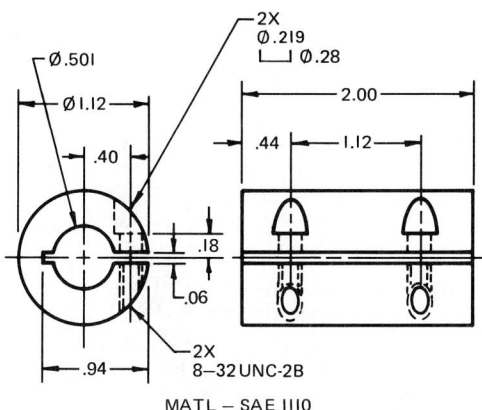

FIG. 15-5-N   Coupling.

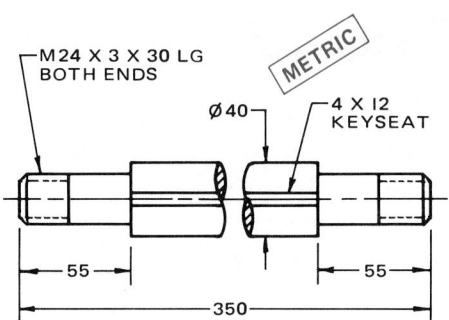

FIG. 15-5-P   Shaft.

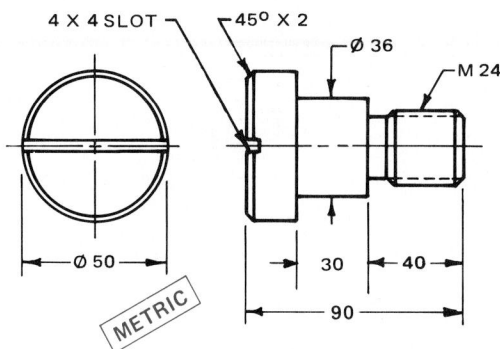

FIG. 15-5-R   Stop button.

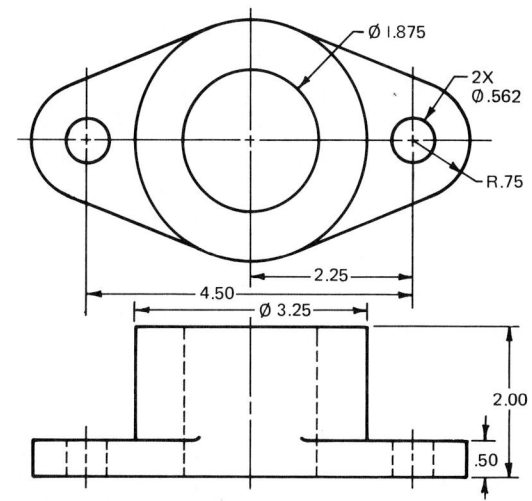

FIG. 15-5-S   Bearing support.

FIG. 15-5-T   Bushing holder.

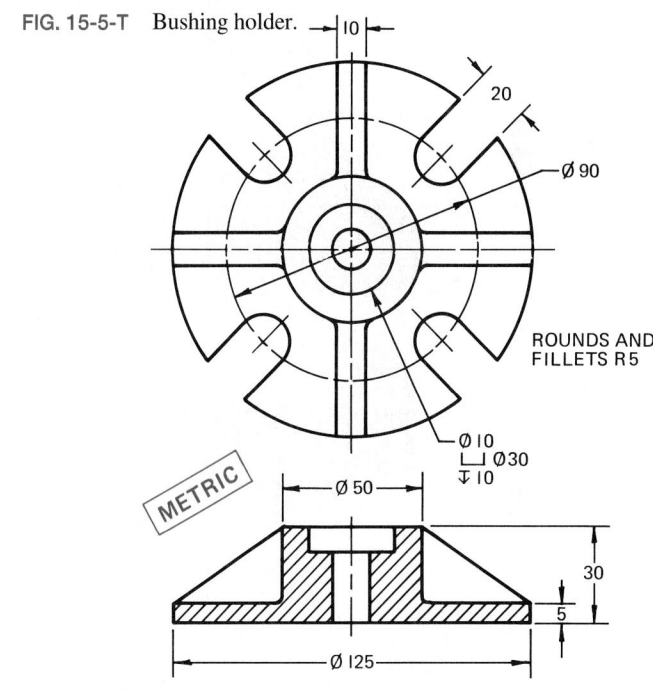

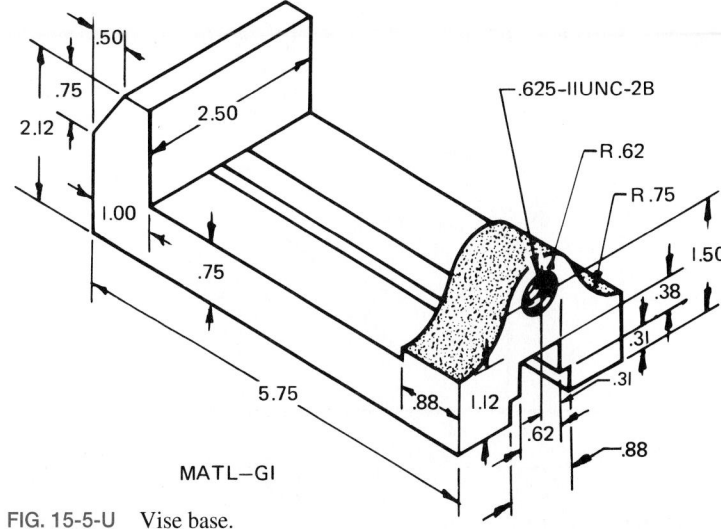

FIG. 15-5-U   Vise base.

ASSIGNMENTS FOR UNIT 15-6, PARALLEL, OR ONE-POINT, PERSPECTIVE

Note: If a one-point perspective grid is not available, copy the grid shown in Fig. 15-6-9. For a worm's-eye view rotate the grid 180°. Position the part on the grid in order to best show the part.

26. Using a one-point perspective grid, make a drawing of one of the parts shown in Figs. 15-6-A through 15-6-C. Add dimensions. Scale to suit.

27. Using a one-point perspective grid, make a drawing of one of the parts shown in Figs. 15-6-D through 15-6-F. Add dimensions. Scale to suit.

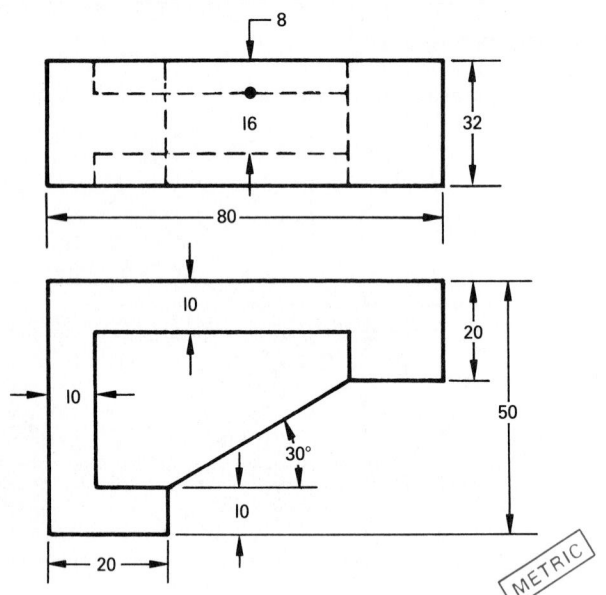

FIG. 15-6-A   Bracket.

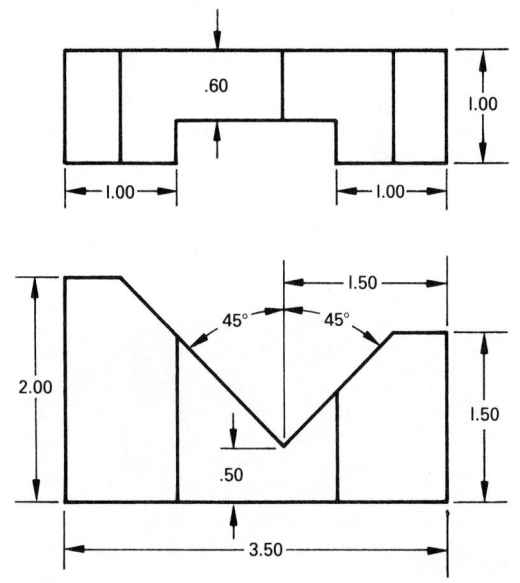

FIG. 15-6-B   V-slide.

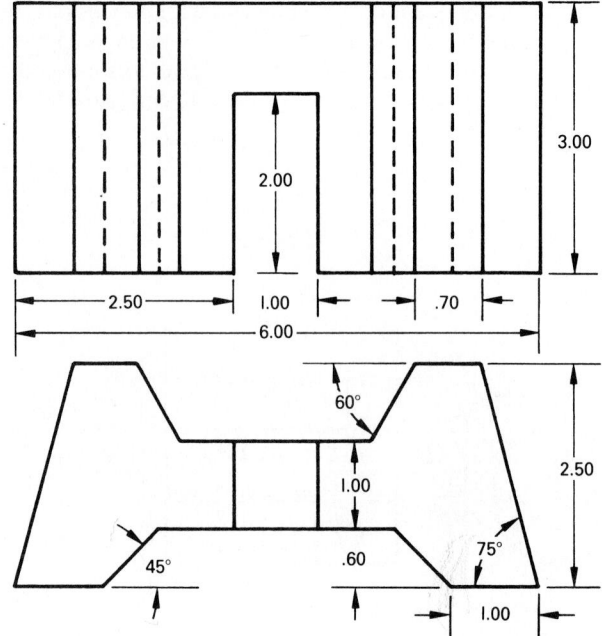

FIG. 15-6-C   Base.

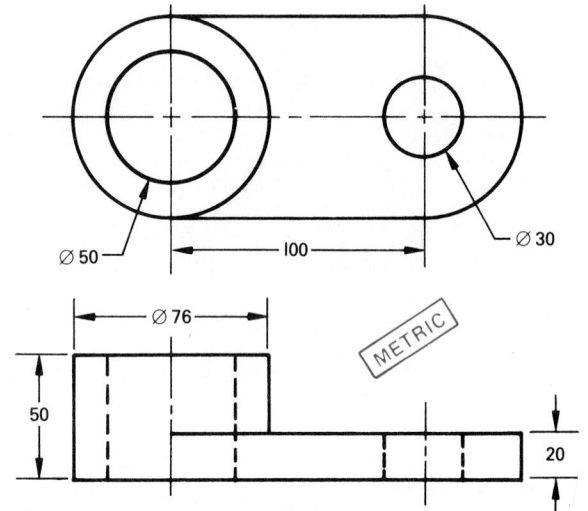

FIG. 15-6-D   Bearing.

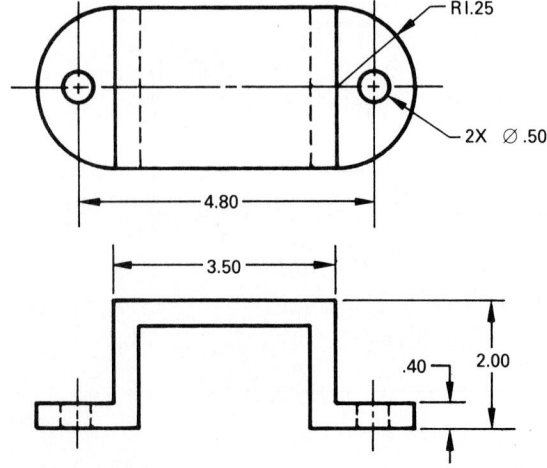

FIG. 15-6-E   Clamp.

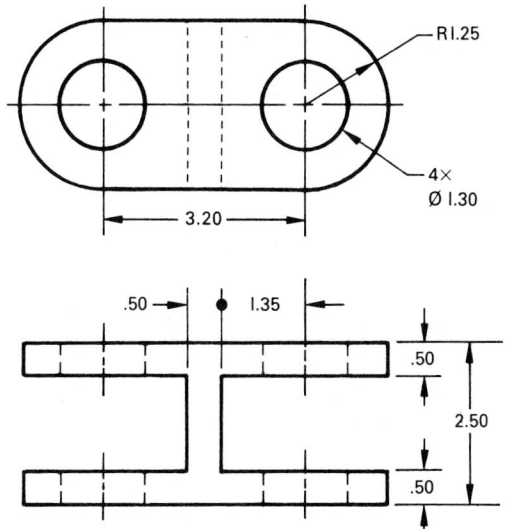

FIG. 15-6-F   Rod spacer.

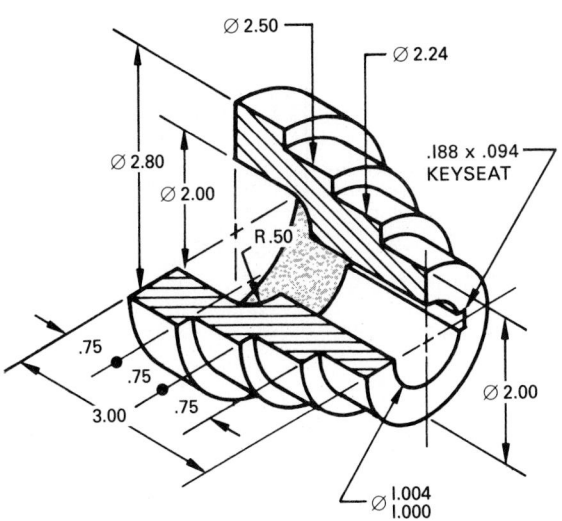

FIG. 15-6-G   Step pulley.

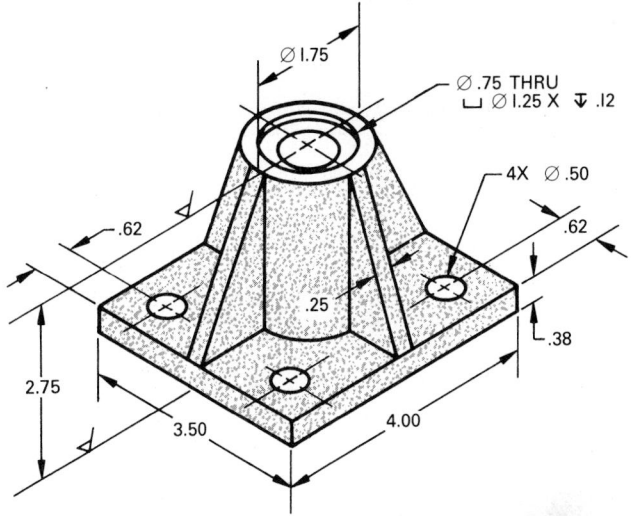

FIG. 15-6-J   Base plate.

28. Using a one-point perspective grid, make a half-section drawing of one of the parts shown in Figs. 15-6-G through 15-6-J. Add dimensions. Scale to suit.

ASSIGNMENT FOR UNIT 15-7, ANGULAR, OR TWO-POINT, PERSPECTIVE

Note: If a two-point perspective grid is not available, copy the grid shown in Fig. 15-7-9. For a bird's-eye view rotate the grid 180°. Position the part on the grid in order to best show the part.

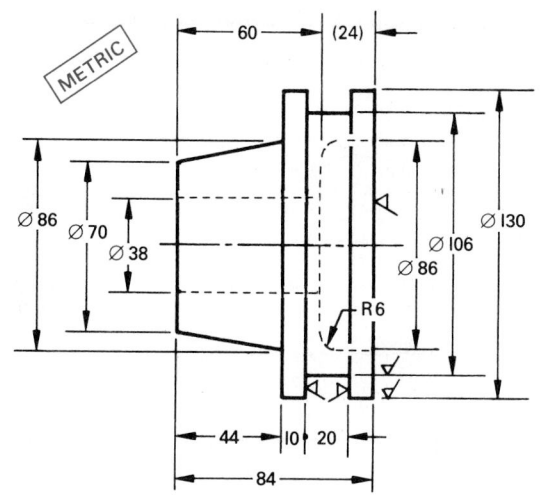

FIG. 15-6-H   Cone spacer.

479

29. Using a two-point perspective grid make a drawing of one of the parts shown in Figs. 15-7-A through 15-7-G. Add dimensions. Scale to suit.

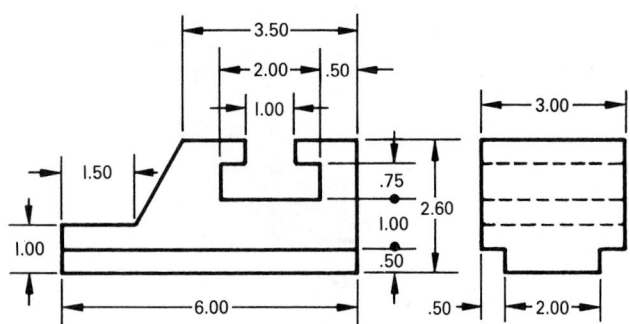

FIG. 15-7-A   Tool support.

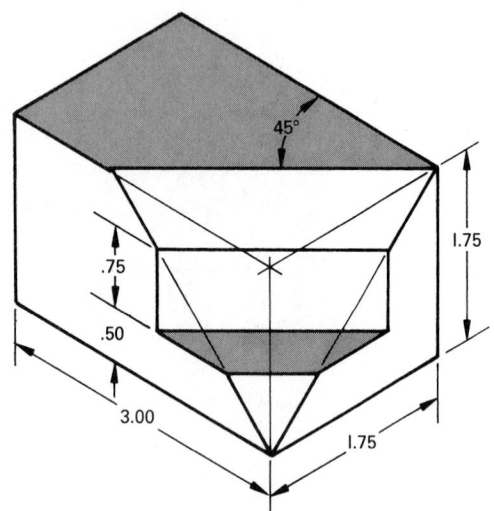

FIG. 15-7-B   Corner block.

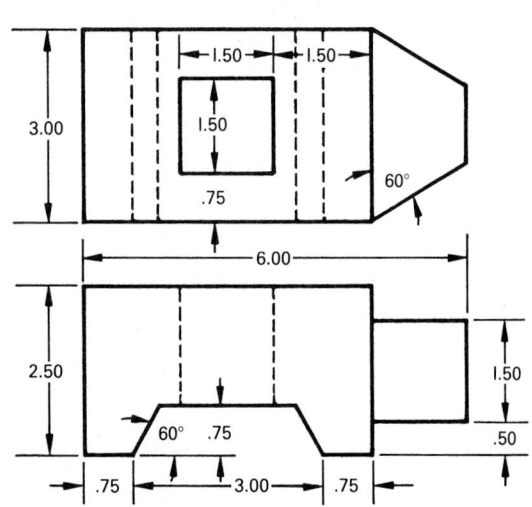

FIG. 15-7-C   Locating support.

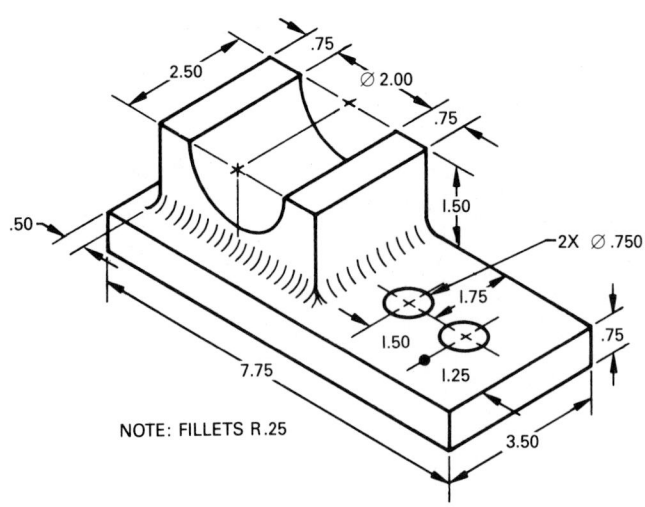

NOTE: FILLETS R.25

FIG. 15-7-D   Horizontal guide.

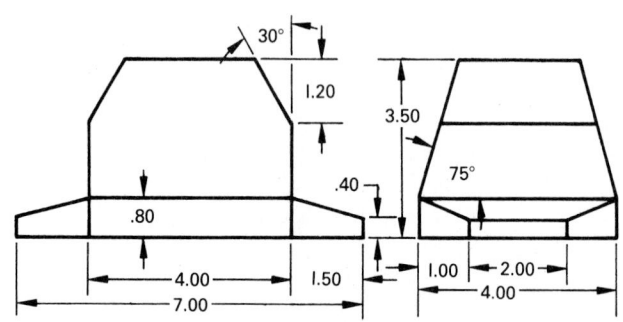

FIG. 15-7-E   Base.

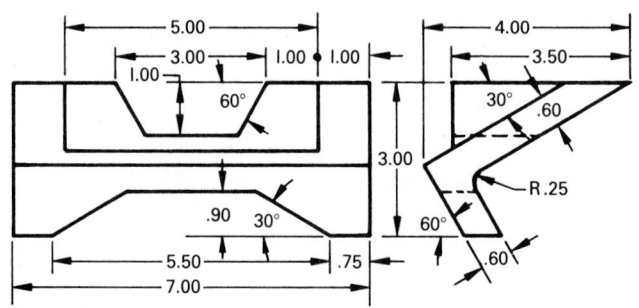

FIG. 15-7-F   Separator.

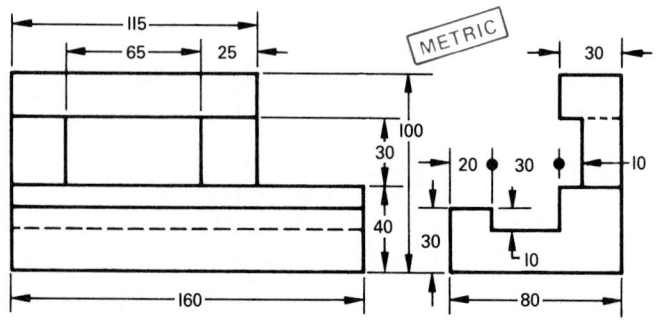

FIG. 15-7-G   Support guide.

CHAPTER 16

GEOMETRIC DIMENSIONING AND TOLERANCING

Definitions

Angularity   The condition of a surface or axis at a specified angle from a datum plane or axis.

Axis   A theoretical straight line about which a part or circular feature revolves.

Basic dimension   The theoretical exact size, profile, orientation, or location of a feature or datum target.

Bilateral tolerance zone   A form in which the profile tolerance zone is equally disposed about the basic profile.

Circular runout   Tolerance which provides control of circular elements of a surface.

Circularity   A condition of a circular line or the surface of a circular feature where all points on the line or on the circumference of a plane cross section of the feature are the same distance from a common axis or center point.

Coaxiality   A condition in which two or more circular or similar features are arranged with their axes in the same straight line.

Concentricity   A condition in which two or more features, such as circles, spheres, cylinders, cones, or hexagons share a common center or axis.

Coordinate tolerancing   Tolerances applied directly to the coordinate dimensions or to applicable tolerances specified in a general tolerance note.

Coplanarity   The condition of two or more surfaces having all elements in one plane.

Correlative geometric tolerancing   Tolerancing for the control of two or more features intended to be correlated in position or attitude.

Cylindricity   A condition of a surface in which all points of the surface are the same distance from a common axis.

Datum   A point, line, plane, or other geometric surface from which dimensions are measured when so specified or to which geometric tolerances are referenced.

Datum feature   A feature of a part, such as an edge or a surface, that forms the basis for a datum or is used to establish its location.

Dimension   A geometric characteristic, the size of which is specified.

Feature   A specific, characteristic portion of a part, such as a surface, hole, slot, screw, thread, or profile.

Feature control frame   A method of specifying geometric tolerances; a rectangular box divided into compartments containing the geometric characteristic symbol followed by the tolerance reference datums where applicable.

Fixed-fastener case   The condition in which one of the parts to be assembled has restrained fasteners.

Floating-fastener case   The condition in which two or more parts are assembled with fasteners, such as bolts and nuts, and all parts have clearance holes for the bolts.

Geometric tolerance   The maximum permissible variation of form, profile, orientation, location, and runout from that indicated or specified on the drawing.

Line profile   The outline of a part or feature as depicted in a view on a drawing.

Lobing   A circular feature where the diametral values may be constant or nearly so.

Orientation   The angular relationship between two or more lines, surfaces, or other features.

Ovality   A circular feature where differences appear between the major and minor axes.

Parallelism   The condition of a surface equidistant at all points from a datum plane.

Perpendicularity   The condition of a surface at 90° to a datum plane or axis.

Positional tolerancing   Normally a circular tolerance zone within which the center line of the hole or shaft is permitted to vary from its true position.

481

Profile   The outline form or shape of a line or surface.

Runout   A composite tolerance used to control the functional relationship of one or more features of a part to a datum axis.

Surface profile   The form or shape of a complete surface in three dimensions.

Symmetry   A condition in which a feature or features are positioned about the center plane of a datum feature.

Three-plane datum system or **datum reference frame**   A system used to indicate datums of mutually perpendicular plane surfaces in positional relationships.

Tolerance   The total permissible variation in the size of a dimension, which is equal to the difference between the limits of size.

Total runout   The runout of a complete surface, not merely the runout of each circular element.

Unilateral tolerance zone   A form in which the tolerance zone is wholly on one side of the basic profile instead of equally divided on both sides.

Upper and lower deviations   The differences between the basic, or zero, line and the maximum and minimum sizes.

Virtual condition   The overall envelope of perfect form within which the feature would just fit.

16-1 MODERN ENGINEERING TOLERANCING

An engineering drawing of a manufactured part is intended to convey information from the designer to the manufacturer and inspector. It must contain all information necessary for the part to be correctly manufactured. It must also enable an inspector to make a precise determination of whether the part is acceptable.

Therefore each drawing must convey three essential types of information:

1. The material to be used
2. The size or dimensions of the part
3. The shape or geometric characteristics

The drawing must also specify permissible variations for each of these aspects in the form of tolerance or limits.

Materials are usually covered by separate specifications or supplementary documents, and the drawings need only make reference to these.

Size is specified by linear and angular dimensions. Tolerances may be applied directly to these dimensions or may be specified by means of a general tolerance note.

Shape and geometric characteristics, such as orientation and position, are described by views on the drawing, supplemented to some extent by dimensions.

In the past tolerances were often shown for which no precise interpretation existed, for example, on dimensions that originated at nonexistent center lines. The specification of datum features was often omitted, resulting in measurements being made from actual surfaces when, in fact, datums were intended. There was confusion concerning the precise effect of various methods of expressing tolerances and of the number of decimal places used. While tolerancing of geometric characteristics was sometimes specified in the form of notes, no precise methods or

interpretations were established. Straight or circular lines were drawn without specifying how straight or round they should be. Square corners were drawn without specifying by how much the 90° angle could vary.

Modern systems of tolerancing, which include geometric and positional tolerancing, use of datum and datum targets, and more precise interpretations of linear and angular tolerances, provide designers and drafters with a means of expressing permissible variations in a very precise manner. Furthermore, the methods and symbols are international in scope and therefore help break down language barriers.

It is not necessary to use geometric tolerances for every feature on a part drawing. In most cases it is to be expected that if each feature meets all dimensional tolerances, form variations will be adequately controlled by the accuracy of the manufacturing process and equipment used.

The chapter covers the application of modern tolerancing methods to drawing.

National and International Standards   References are made to technical drawing standards published by United States and international standardizing bodies. These bodies are usually referred to by their acronyms, as shown in Fig. 16-1-1.

Most of the symbols in all these standards are identical, but there are some variations. These are chiefly in the methods of indicating datum features and of applying the symbols to drawings. In view of the exchange of drawings among the United States, Canada, and other countries, it is advantageous for drafters and designers to be acquainted with these symbols.

For this reason whenever differences between United States and ISO standards occur, two methods are shown in some of the illustrations, and each is labeled with the acronym of the appropriate standardizing body, ANSI or ISO. However, differences in symbols or methods of application do not affect the principles or interpretation of tolerances, unless noted.

Illustrations   Most of the drawings in this chapter are not complete working drawings. They are intended only to illustrate a principle. Therefore, to avoid distraction from the information being presented, most of the details that are not essential to explain the principle have been omitted.

Basic Concepts

Some of the basic concepts used in dimensioning and tolerancing of drawings follow. While these are not new, their exact

ACRONYM	STANDARDIZING BODY	STANDARD FOR DIMENSIONING AND TOLERANCING
ANSI	American National Standards Institute American Society of Mechanical Engineers	ASME Y14.5M–1994
ISO	International Organization for Standardization	ISO R1101

FIG. 16-1-1   Standardizing bodies.

meanings warrant special attention in order that there be no ambiguity in the precise interpretation of tolerancing methods described in this chapter.

Dimension

A *dimension* is a geometric characteristic the size of which is specified, such as diameter, length, angle, location, or center distance. The term is also used for convenience to indicate the magnitude or value of a dimension, as specified on a drawing (Fig. 16-1-2).

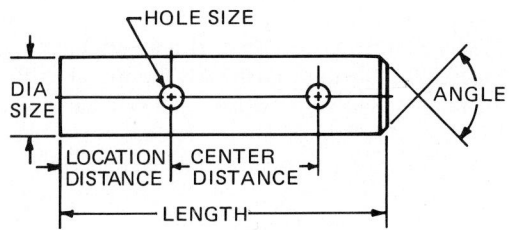

FIG. 16-1-2   Dimensions of a part.

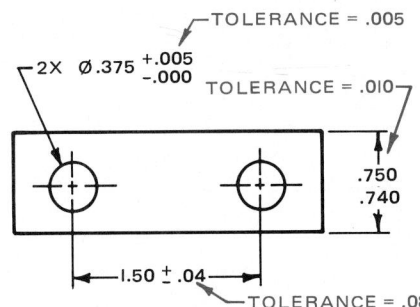

(A) TOLERANCE SIZE

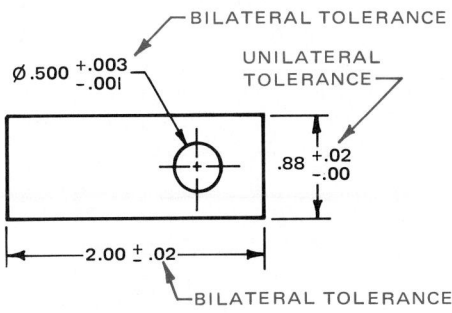

(B) TYPE OF TOLERANCE

FIG. 16-1-3   Tolerances.

FIG. 16-1-4   Sizes of mating parts.

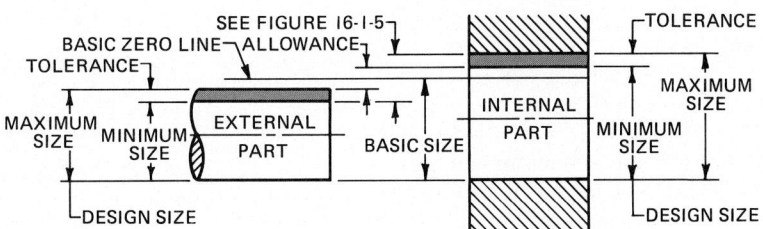

Tolerance

The *tolerance* on a dimension is the total permissible variation in its size, which is equal to the difference between the limits of size. The plural *tolerances* is sometimes used to denote the permissible variations from the specified size when the tolerance is expressed bilaterally.

For example, in Fig. 16-1-3A the tolerance on the center distance dimension $1.50 \pm .04$ is .08 in., but in common practice the values $+.04$ and $-.04$ are often referred to as the tolerances.

Size of Dimensions

In theory it is impossible to produce a part to an exact size, because every part, if measured with sufficient accuracy, would be found to be a slightly different size. However, for purposes of discussion and interpretation, a number of distinct sizes for each dimension have to be recognized: actual size, nominal size, specified size, and design size.

Actual size   Actual size is the measured size of an individual part.

Nominal size   The nominal size is the designation of size used for purposes of general identification.

The nominal size is used in referring to a part in an assembly drawing stocklist, in a specification, or in other such documents. It is very often identical to the basic size but in many instances may differ widely; for example, the external diameter of a .50 in. steel pipe is .84 in. (21.34 mm). The nominal size is .50 in.

Specified size   This is the size specified on the drawing when the size is associated with a tolerance. The specified size is usually identical to the design size or, if no allowance is involved, to the basic size.

Figure 16-1-4 shows two mating features with the tolerance and allowance zones exaggerated, to illustrate the sizes, tolerances, and allowances. This figure also illustrates the origin of tolerance block diagrams, as shown in Fig. 16-1-5 (pg. 484), which are commonly used to show the relationships among part limits, gage or inspection limits, and gage tolerances.

Design size   The design size of a dimension is the size in relation to which the tolerance for that dimension is assigned.

Theoretically, it is the size on which the design of the individual feature is based, and therefore it is the size that should be specified on the drawing. For dimensions of mating features it is derived from the basic size by the application of the allowance, but when there is no allowance, it is identical to the basic size.

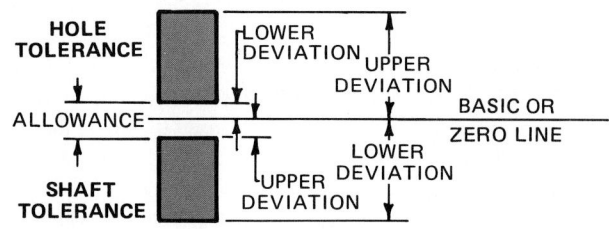

FIG. 16-1-5   Tolerance block diagram.

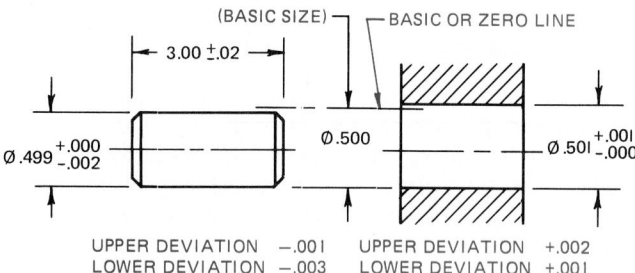

UPPER DEVIATION   −.001   UPPER DEVIATION   +.002
LOWER DEVIATION   −.003   LOWER DEVIATION   +.001

FIG. 16-1-6   Deviations.

Deviations

The differences between the basic, or zero, line and the maximum and minimum sizes are called the *upper* and *lower deviations,* respectively.

Thus in Fig. 16-1-6 the upper deviation of the external part is −.001, and the lower deviation is −.003. For the hole diameter, the upper deviation is +.002, and the lower deviation is +.001, whereas for the length of the pin the upper and lower deviations are +.02 and −.02, respectively.

Basic (Exact) Dimensions

A *basic dimension* represents the theoretical exact size, profile, orientation, or location of a feature or datum target. It is the basis from which permissible variations are established by tolerances or other dimensions, in notes, or in feature control frames (Fig. 16-1-7). They are shown without tolerances, and each basic dimension is enclosed in a rectangular frame to indicate that the tolerances in the general tolerance note do not apply.

Feature

A *feature* is a specific, characteristic portion of a part, such as a surface, hole, slot, screw thread, or profile.

While a feature may include one or more surfaces, the term is generally used in geometric tolerancing in a more restricted sense, to indicate a specific point, line, or surface. Some examples are the axis of a hole, the edge of a part, or a single flat or curved surface, to which reference is being made or which forms the basis for a datum.

Axis

An *axis* is a theoretical straight line about which a part or circular feature revolves or could be considered to revolve (Fig. 16-1-8).

Interpretation of Drawings and Dimensions

It should not be necessary to specify the geometric shape of a feature unless some particular precision is required. Lines that appear to be straight imply straightness; those that appear to be round imply circularity; those that appear to be parallel imply parallelism; those that appear to be square imply perpendicularity; center lines imply symmetry; and features that appear to be concentric about a common center line imply concentricity.

Therefore it is not necessary to add angular dimensions of 90° to corners of rectangular parts nor to specify that opposite sides are parallel.

However, if a particular departure from the illustrated form is permissible, or if a certain degree of precision of form is required, these must be specified. If a slight departure from the true geometric form or position is permissible, it should be exaggerated pictorially in order to show clearly where the dimensions apply. Figure 16-1-9 shows some examples. Dimensions that are not to scale should be underlined.

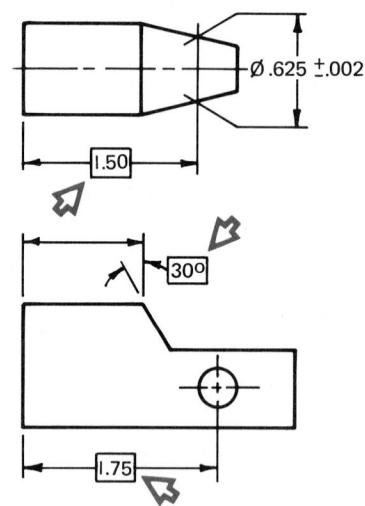

FIG. 16-1-7   Basic (exact) dimensions.

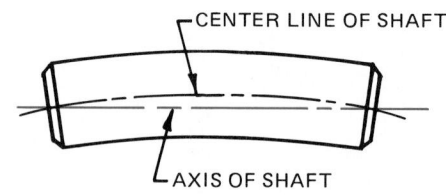

FIG. 16-1-8   Divergence of axis and center line when part is deformed.

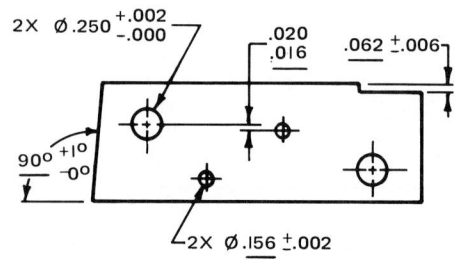

FIG. 16-1-9   Exaggeration of small dimensions.

Point-to-Point Dimensions

When datums are not specified, linear dimensions are intended to apply on a point-to-point basis, either between opposing points on the indicated surfaces or directly between the points marked on the drawing.

The examples shown in Fig. 16-1-10 should help to clarify this principle of point-to-point dimensions.

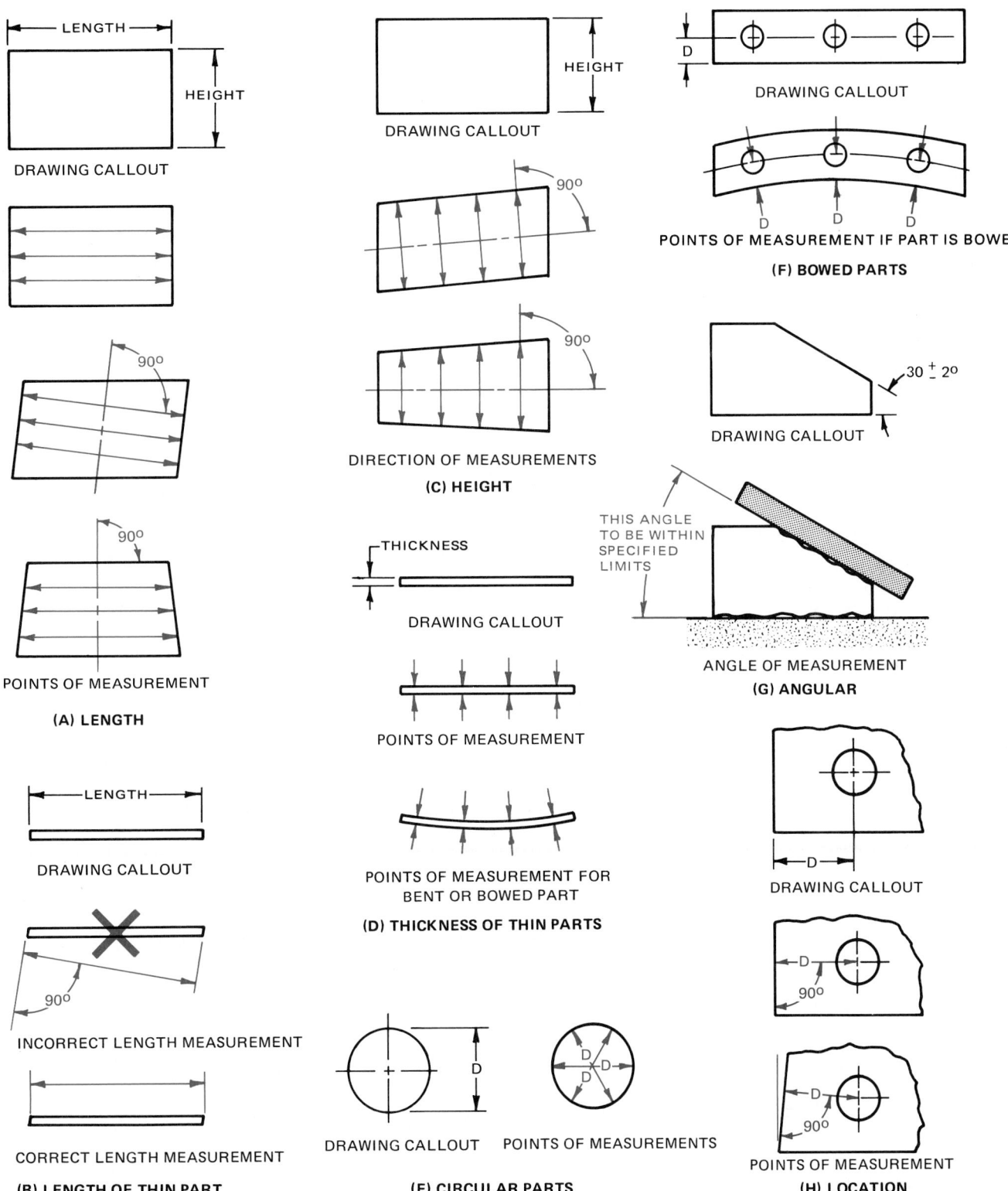

FIG. 16-1-10   Point-to-point dimensions when datums are not used.

485

Location Dimensions with Datums

When location dimensions originate from a feature or surface specified as a datum, measurement is made from the theoretical datum, not from the actual feature or surface of the part.

There will be many cases where a curved center line, as shown in Fig. 16-1-10F, would not meet functional requirements or where the position of the hole in Fig. 16-1-10H would be required to be measured parallel to the base. This can easily be specified by referring the dimension to a datum feature, as shown in Fig. 16-1-11. This will be more fully explained in Unit 16-9, where the interpretation of coordinate tolerances is compared with positional tolerances.

Assumed Datums

There are often cases where the basic rules for measurements on a point-to-point basis cannot be applied, because the originating points, lines, or surfaces are offset in relation to the features located by the dimensions (Fig. 16-1-12). It is then necessary to assume a suitable datum, which is usually the theoretical extension of one of the lines or surfaces involved.

The following general rules cover three types of dimensioning procedures commonly encountered.

1. If a dimension refers to two parallel edges or planes, the longer edge or larger surface, which has the greatest influence in the measurement, is assumed to be the datum feature. For example, if the surfaces of the part shown in Fig. 16-1-12A were not quite parallel, as shown in the lower view, dimension D would be acceptable if the top surface was within limits when measured at a and b, but need not be within limits if measured at c.
2. If only one of the extension lines refers to a straight edge or surface, the extension of that edge or surface is assumed to be the datum. Thus in Fig. 16-1-12B measurement of dimension A is made to a datum surface as shown at a in the bottom view.
3. If both extension lines refer to offset points rather than to edges or surfaces, generally it should be assumed that the datum is a line running through one of these points and parallel to the line or surface to which it is dimensionally related. Thus in Fig. 16-1-12C dimension A is measured from the center of hole D to a line through the center of hole C that is parallel to the datum.

Permissible Form Variations

The actual size of a feature must be within the limits of size, as specified on the drawing, at all points of measurement. This means that each measurement, made at any cross section of the feature, must not be greater than the maximum limit of size nor smaller than the minimum limit of size (Fig. 16-1-13).

In the case of mating parts, such as holes and shafts, it is usually necessary to ensure that they do not deviate from perfect form at the maximum material size (envelope principle), by reason of being bent or otherwise deformed. This condition is shown in Fig. 16-1-14 (pg. 488), where features conform to perfect form at the maximum material condition, but are permitted to deviate from perfect form at the minimum material condition.

If only size tolerances or limits of size are specified for an individual feature, no element of the feature would be permitted to extend beyond the maximum material boundary. Examples are shown in Fig. 16-1-15 (pg. 488).

ANSI DATUM SYMBOL

D

-R-

DATUM FEATURE

NOTE: DATUM PLANE R APPLIES TO ALL DIMENSIONS ORIGINATING FROM THIS SURFACE.

(A) DRAWING CALLOUT

DATUM PLANE R

D   D   D

POINTS OF MEASUREMENT TO DATUM

(B) INTERPRETATION IF PART IS BOWED

FIG. 16-1-11   Dimensions referenced to a datum.

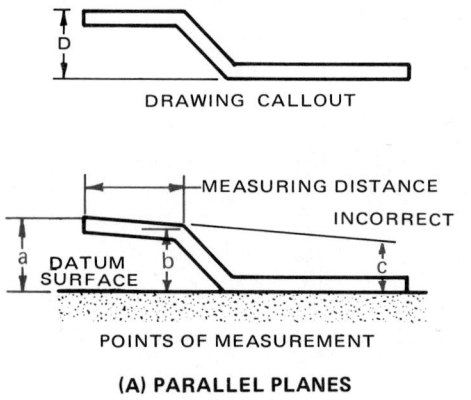

DRAWING CALLOUT

MEASURING DISTANCE
INCORRECT

a   DATUM   b   c
SURFACE

POINTS OF MEASUREMENT

(A) PARALLEL PLANES

FIG. 16-1-12   Assumed datums.

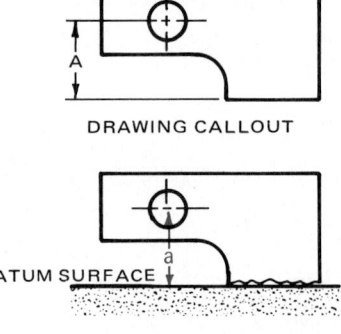

DRAWING CALLOUT

DATUM SURFACE   a

POINT OF MEASUREMENT

(B) SINGLE PLANE

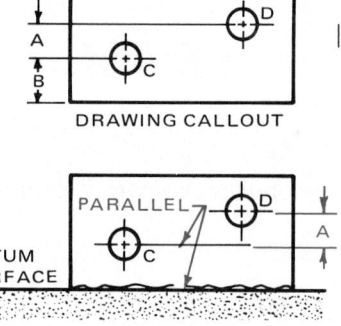

DRAWING CALLOUT

PARALLEL   D
DATUM   C   A
SURFACE

POINT OF MEASUREMENT

(C) OFFSET POINTS

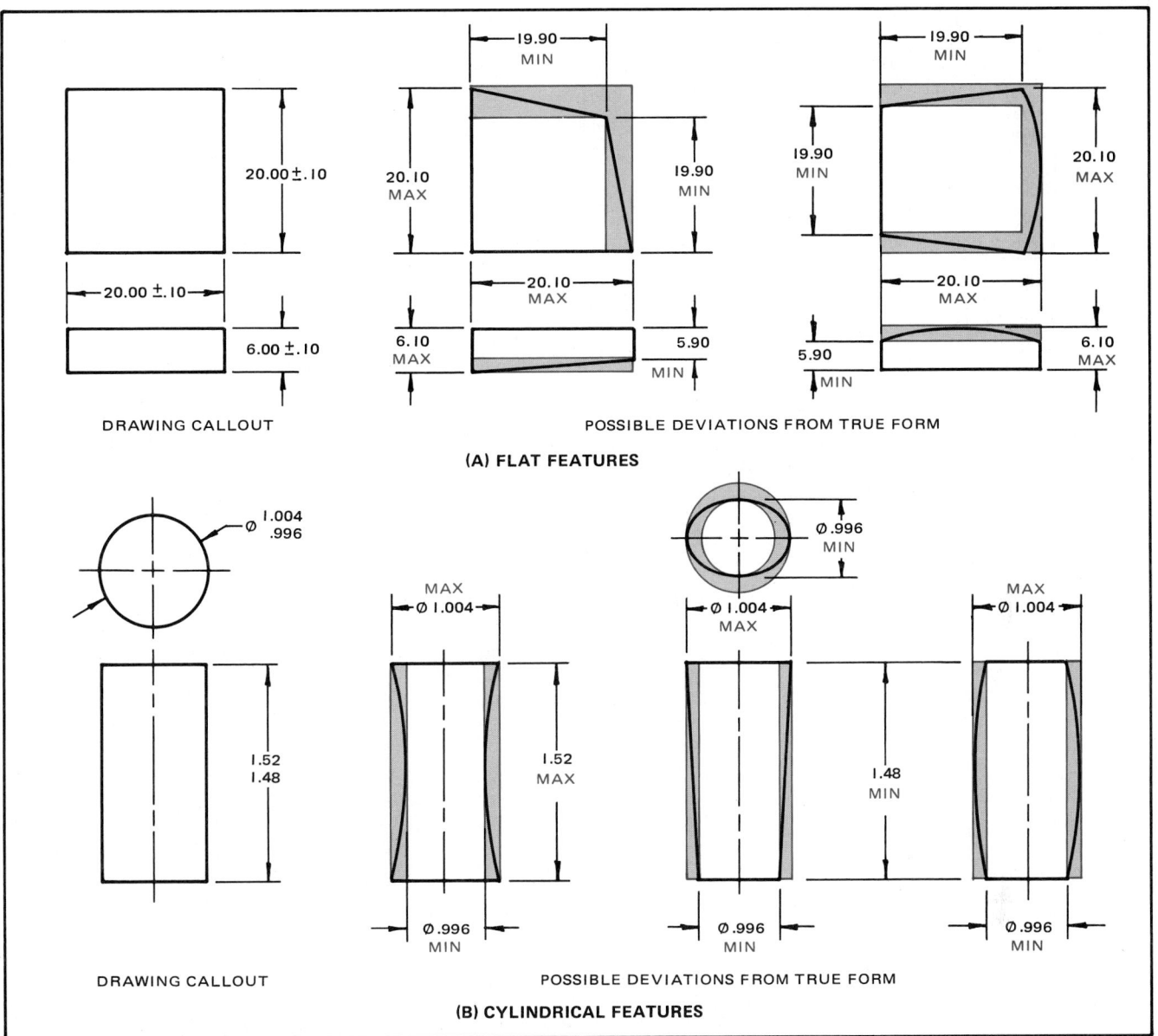

DRAWING CALLOUT   POSSIBLE DEVIATIONS FROM TRUE FORM

(A) FLAT FEATURES

DRAWING CALLOUT   POSSIBLE DEVIATIONS FROM TRUE FORM

(B) CYLINDRICAL FEATURES

FIG. 16-1-13   Deviations in form permitted by toleranced dimensions.

REFERENCES AND SOURCE MATERIAL

1. ASME Y14.5M–1994, *Dimensioning and Tolerancing.*
2. CAN/CSA B78.2-M91, *Dimensioning and Tolerancing of Technical Drawings.*
3. ISO drawing standards.

ASSIGNMENTS

See Assignments 1 and 2 for Unit 16-1 on pages 569–570.

16-2 GEOMETRIC TOLERANCING

By themselves, toleranced linear dimensions, or limits of size, do not give specific control over many other variations of form, orientation, and to some extent, position. These variations could be errors of squareness, perpendicularity, or deviations caused by bending of parts, lobing, and eccentricity.

In order to meet functional requirements, it is often necessary to control such deviations. Geometric tolerances are added to ensure that parts are not only within their limits of size but are also within specified limits of geometric form, orientation, and position.

The most commonly used geometric tolerances are the simple form tolerances of straightness and flatness, the orientation tolerances of perpendicularity and parallelism, and positional tolerances for location of small holes. These geometric tolerances will be explained, together with their rules, symbols, and methods for their application to engineering drawings.

A *geometric tolerance* is the maximum permissible variation of form, profile, orientation, location, and runout from that indicated or specified on the drawing. The tolerance value represents the width or diameter of the tolerance zone, within which the point, line, or surface of the feature shall lie.

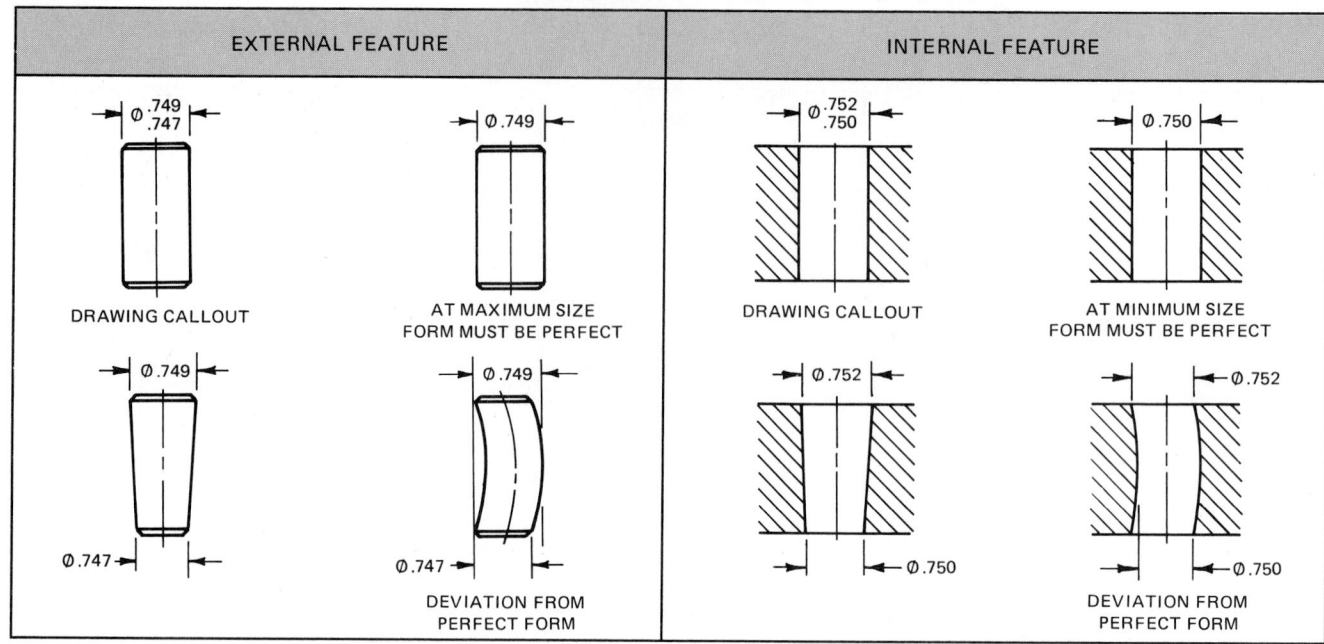

FIG. 16-1-14   Examples of deviation of form when perfect form at the maximum material condition is required.

FIG. 16-1-15   Form variations accepted by gage limits.

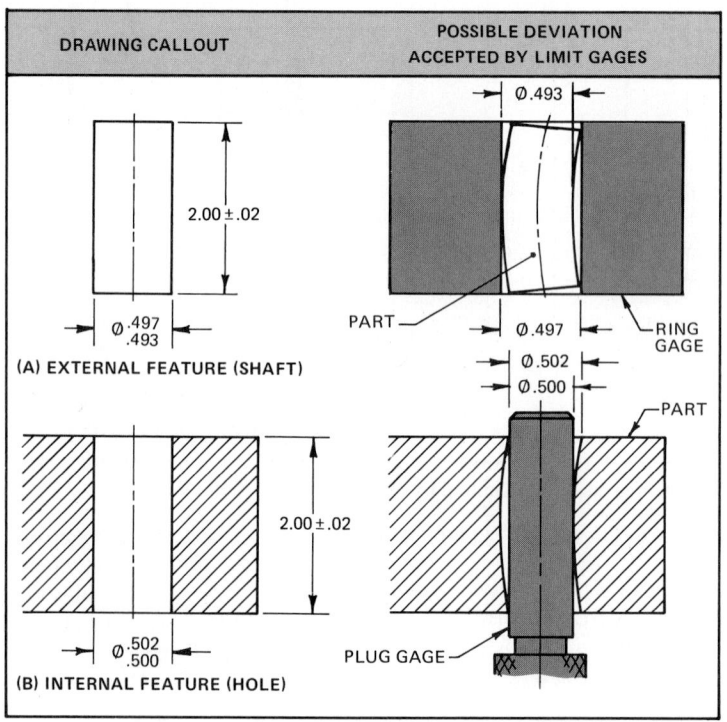

From this definition it follows that a feature would be permitted to have any variation of form, or take up any position, within the specified geometric tolerance zone.

For example, a line controlled in a single plane by a straightness tolerance of .006 in. must be contained within a tolerance zone .006 in. wide (Fig. 16-2-1).

Points, Lines, and Surfaces   The production and measurement of engineering parts deals, in most cases, with surfaces

of objects. These surfaces may be flat, cylindrical, conical, or spherical or have some more or less irregular shape or contour. Measurement, however, usually has to take place at specific points. A line or surface is evaluated dimensionally by making a series of measurements at various points along its length.

Geometric tolerances are chiefly concerned with points and lines, while surfaces are considered to be composed of a series of line elements running in two or more directions.

Points have position but no size, so position is the only characteristic that requires control. Lines and surfaces have to be

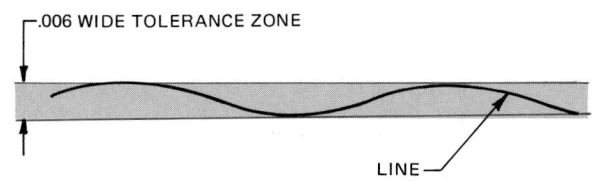

.006 WIDE TOLERANCE ZONE

LINE

FIG. 16-2-1   Tolerance zone for straightness of a line.

controlled for form, orientation, and location. Therefore geometric tolerances provide for control of these characteristics, as shown in Fig. 16-2-2.

Feature Control Frame

Some geometric tolerances have been used for many years in the form of notes, such as PARALLEL WITH SURFACE *A*

WITHIN .001 and STRAIGHT WITHIN .12. While such notes are now obsolete, the reader should be prepared to recognize them on older drawings.

The current method is to specify geometric tolerances by means of the feature control frame. A *feature control frame* for an individual feature is divided into compartments containing the geometric characteristic symbol followed by the tolerance (Fig. 16-2-3, pg. 490). Where applicable, the tolerance is preceded by the diameter symbol and may be followed by a material condition symbol (see Unit 16-4).

When necessary, other compartments are added to contain datum references, as explained in Unit 16-5.

Geometric characteristic symbols relating to lines (straightness, angularity, perpendicularity, profile of a line, parallelism, position) are shown in Fig. 16-2-2. These and other symbols will be introduced as required, but all are shown in the figure for reference purposes.

FIG. 16-2-2   Geometric characteristic symbols.

FEATURE	TYPE OF TOLERANCE	CHARACTERISTIC	SYMBOL	SEE UNIT
INDIVIDUAL FEATURES	FORM	STRAIGHTNESS	——	16-2, 16-5
		FLATNESS	▱	16-3
		CIRCULARITY (ROUNDNESS)	◯	16-12
		CYLINDRICITY	⌯	
INDIVIDUAL OR RELATED FEATURES	PROFILE	PROFILE OF A LINE	⌒	16-13
		PROFILE OF A SURFACE	⌓	
RELATED FEATURES	ORIENTATION	ANGULARITY	∠	16-7, 16-8
		PERPENDICULARITY	⊥	
		PARALLELISM	//	
	LOCATION	POSITION	⌖	16-9
		CONCENTRICITY	◎	16-14
		SYMMETRY	⌰	16-14
	RUNOUT	CIRCULAR RUNOUT	*↗	16-14
		TOTAL RUNOUT	*↗↗	
SUPPLEMENTARY SYMBOLS		MAXIMUM MATERIAL CONDITION	Ⓜ	16-4
		LEAST MATERIAL CONDITION	Ⓛ	
		PROJECTED TOLERANCE ZONE	Ⓟ	16-9
		BASIC DIMENSION	XX	16-9, 16-11
		DATUM FEATURE	A◁	16-5
		DATUM TARGET	⌀.50 / A2	16-11

* MAY BE FILLED IN

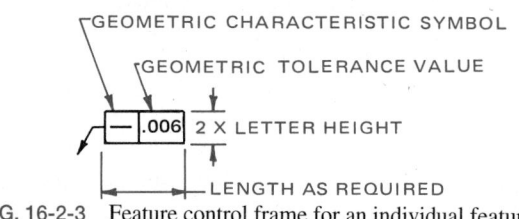

FIG. 16-2-3   Feature control frame for an individual feature.

Application to Drawings

The feature control frame is related to the feature by one of the following methods (shown in Fig. 16-2-4):

1. Running a leader from the frame to the feature, Fig. 16-2-4A. This method is used when control of the surface element is required.
2. Running a leader from the frame to an extension line of the surface but not in line with the dimension, Fig. 16-2-4A. This method is also used when control of the surface elements is required.
3. Attaching the side or end of the frame to an extension line extending from a plane-surface feature, Fig. 16-2-4A.
4. Attaching the frame to the dimension line pertaining to the feature, Figure 16-2-4B. (See Unit 16-4.)
5. Locating the frame below the size dimension to control the center line, axis, or center plane of the feature, Fig. 16-2-4B. (See Unit 16-4.)

Application to Surfaces

The arrowhead of the leader from the feature control frame should touch the surface of the feature or the extension line of the surface.

The leader from the feature control frame should be directed at the feature in its characteristic profile. Thus in Fig. 16-2-5 the straightness tolerance is directed to the side view, and the circularity tolerance to the end view. This may not always be possible, and a tolerance connected to an alternative view, such as a circularity tolerance connected to a side view, is acceptable. When it is more convenient, or when space is limited, the arrowhead may be directed to an extension line, but not in line with the dimension line.

When two or more feature control frames apply to the same feature, they are drawn together with a single leader and arrowhead, as shown in Fig. 16-2-6.

Form Tolerances

Form tolerances control straightness, flatness, circularity, and cylindricity. They are applicable to single (individual) features or elements of single features and, as such, do not require locating dimensions. Orientation tolerances control angularity, parallelism, and perpendicularity.

Form and orientation tolerances critical to function and interchangeability are specified where the tolerances of size and location do not provide sufficient control. A tolerance of form or orientation may be specified where no tolerance of size is given, e.g., the control of flatness.

Form tolerances specify the maximum permissible variation from the desired form and apply to all points on the considered surface.

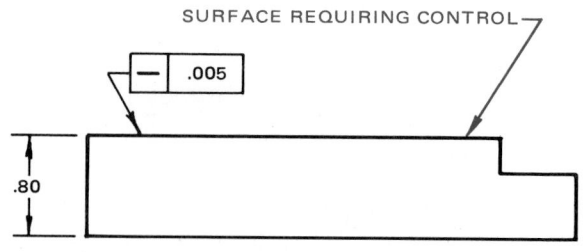

RUNNING A LEADER FROM THE FRAME TO THE FEATURE

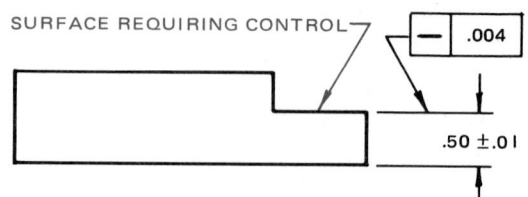

ATTACHED TO AN EXTENSION LINE USING A LEADER

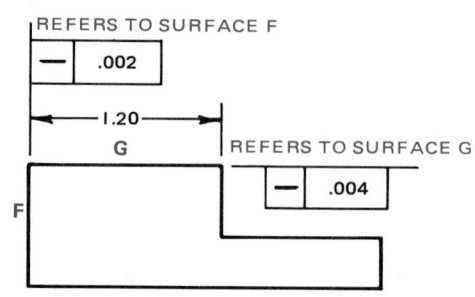

ATTACHED DIRECTLY TO AN EXTENSION LINE

(A) CONTROL OF SURFACE OR SURFACE ELEMENTS

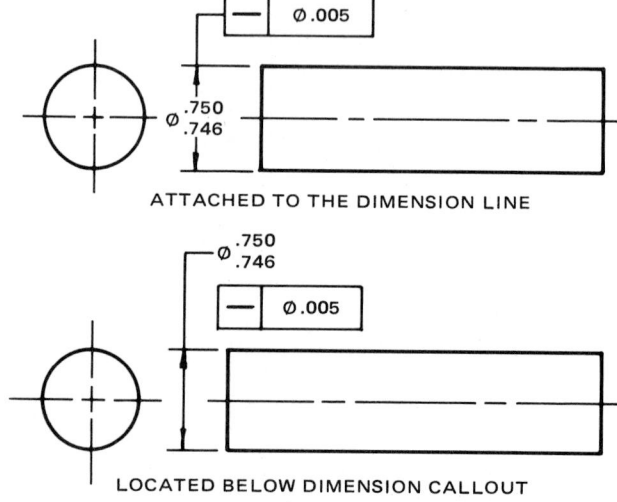

ATTACHED TO THE DIMENSION LINE

LOCATED BELOW DIMENSION CALLOUT

(B) CONTROL OF FEATURE OF SIZE

FIG. 16-2-4   Application of feature control frame.

Straightness

Straightness is a condition where the element of a surface or a center line is a straight line. The geometric characteristic symbol for straightness is a horizontal line (Fig. 16-2-5).

A straightness tolerance specifies a tolerance zone within which the considered element of the surface or center line must lie. A straightness tolerance is applied to the view where the elements to be controlled are represented by a straight line.

Straightness Controlling Surface Elements

Lines   Straightness is fundamentally a characteristic of a line, such as the edge of a part or a line scribed on a surface. A straightness tolerance is specified on a drawing by means of a feature control frame, which is directed by a leader to the line requiring control, as shown in Fig. 16-2-7. It states in symbolic form that the line shall be straight within .006 in. This means that the line shall be contained within a tolerance zone

consisting of the area between two parallel straight lines in the same plane, separated by the specified tolerance.

Theoretically, straightness could be measured by bringing a straightedge into contact with the line and determining that any space between the straightedge and the line does not exceed the specified tolerance. The straightness error will be the maximum space between the feature and the straightedge. For example, in Fig. 16-2-8, the measured straightness error of the top edge of the part is that shown at $H1$, not $H2$.

Cylindrical Surfaces   For cylindrical parts, or curved surfaces that are straight in one direction, the feature control frame should be directed to the side view, where line elements appear as a straight line, as shown in Fig. 16-2-9 and 16-2-10 (pg. 492).

A straightness tolerance thus applied to the surface controls surface elements only. Therefore it would control bending or a wavy condition of the surface or a barrel-shaped part, but it would not necessarily control the straightness of the center line or the conicity of the cylinder.

Straightness of a cylindrical surface is interpreted to mean that each line element of the surface shall be contained within a

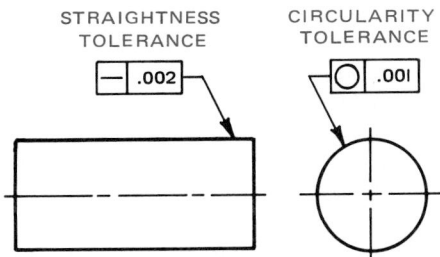

FIG. 16-2-5   Preferred location of feature control frame.

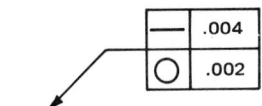

FIG. 16-2-6   Combined feature control frames directed to one surface.

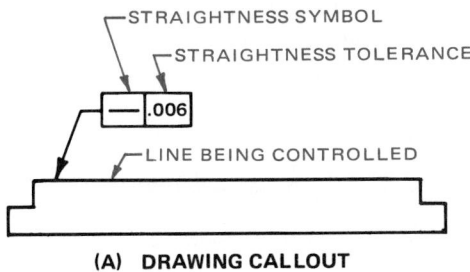

(A) DRAWING CALLOUT

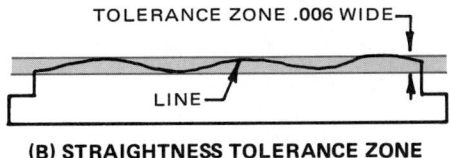

(B) STRAIGHTNESS TOLERANCE ZONE

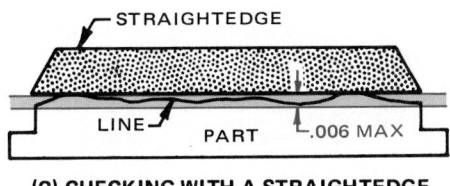

(C) CHECKING WITH A STRAIGHTEDGE

FIG. 16-2-7   Straightness symbol and application.

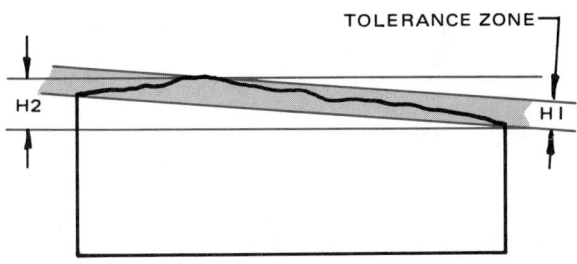

FIG. 16-2-8   Evaluating an uneven surface.

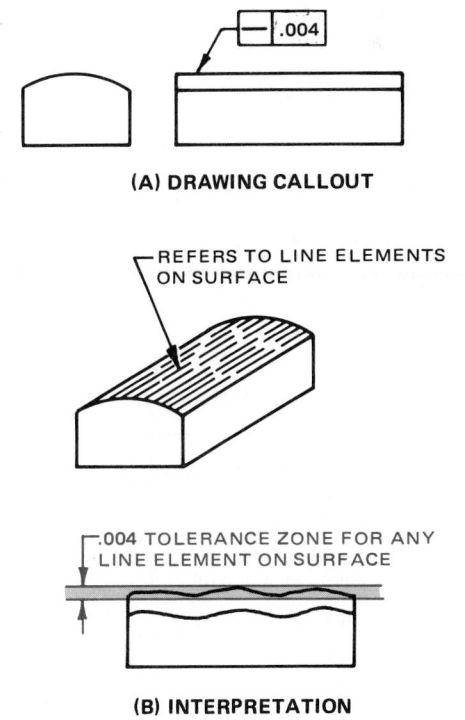

FIG. 16-2-9   Straightness of surface line elements.

491

tolerance zone consisting of the space between two parallel lines, separated by the width of the specified tolerance, when the part is rolled along one of the planes. All circular elements of the surface must be within the specified size tolerance. When only limits of size are specified, no error in straightness would be permitted if the diameter were at its maximum material size.

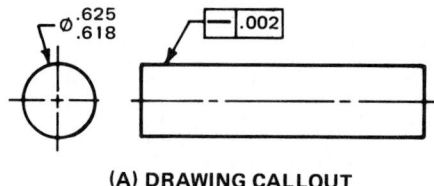

(A) DRAWING CALLOUT

REFERS TO LINE
ELEMENTS ON SURFACE

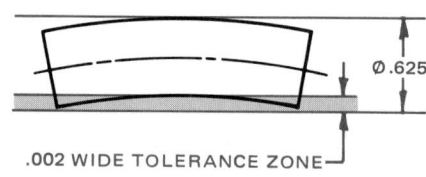

.002 WIDE TOLERANCE ZONE

BENDING ERROR

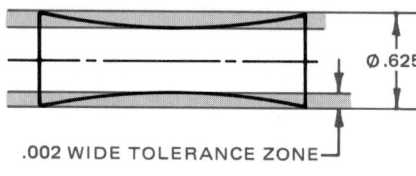

.002 WIDE TOLERANCE ZONE

CONCAVE ERROR

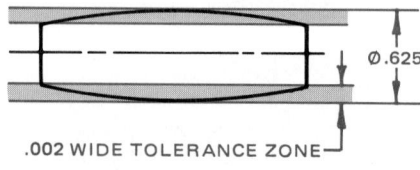

.002 WIDE TOLERANCE ZONE

CONVEX ERROR

(B) INTERPRETATION

NOTE: NO PART OF THE CYLINDRICAL SURFACE
MAY LIE OUTSIDE THE LIMITS OF SIZE.

FIG. 16-2-10   Straightness errors in surface elements of a cylindrical part.

The straightness tolerance must be less than the size tolerance. Since the limits of size must be respected, the full straightness tolerance may not be available for opposite elements in the case of waisting or barreling of the surface (Fig. 16-2-9).

Conical Surfaces   A straightness tolerance can be applied to a conical surface in the same manner as for a cylindrical surface, as shown in Fig. 16-2-11, and will ensure that the rate of taper is uniform. The actual rate of taper, or the taper angle, must be separately toleranced.

Flat Surfaces   A straightness tolerance applied to a flat surface indicates straightness control in one direction only and must be directed to the line on the drawing representing the surface to be controlled and the direction in which control is required, as shown in Fig. 16-2-12. It is then interpreted to mean that each line element on the surface in the indicated direction shall lie within a tolerance zone.

Different straightness tolerances may be specified in two or more directions when required, as shown in Fig. 16-2-13. However, if the same straightness tolerance is required in two coordinate directions on the same surface, a flatness tolerance rather than a straightness tolerance is used.

If it is not otherwise necessary to draw all three views, the straightness tolerances may all be shown on a single view by indicating the direction with short lines terminated by arrowheads, as shown in Fig. 16-2-13C.

Straightness of center lines and planes is covered in Unit 16-4.

REFERENCES AND SOURCE MATERIAL

1. ASME Y14.5M–1994, *Dimensioning and Tolerancing.*
2. CAN/CSA B78.2-M91, *Dimensioning and Tolerancing of Technical Drawings.*
3. ISO drawing standards.

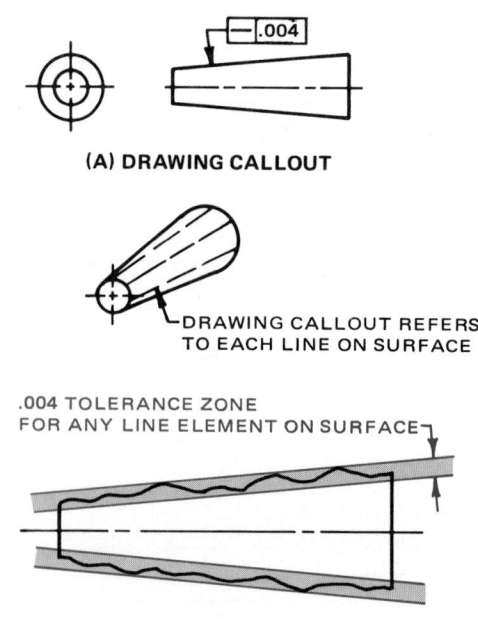

(A) DRAWING CALLOUT

DRAWING CALLOUT REFERS
TO EACH LINE ON SURFACE

.004 TOLERANCE ZONE
FOR ANY LINE ELEMENT ON SURFACE

(B) INTERPRETATION

FIG. 16-2-11   Straightness of a conical surface.

ASSIGNMENT

See Assignment 3 for Unit 16-2 on pages 570–571.

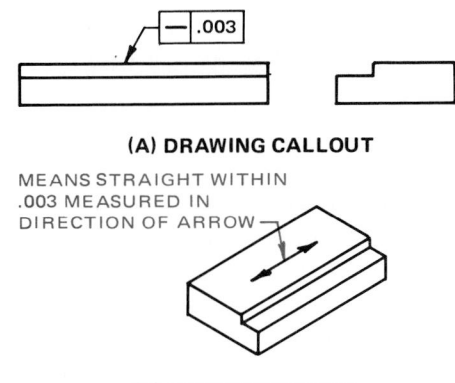

(A) DRAWING CALLOUT

MEANS STRAIGHT WITHIN
.003 MEASURED IN
DIRECTION OF ARROW

(B) INTERPRETATION

FIG. 16-2-12   Straightness in one direction of a flat surface.

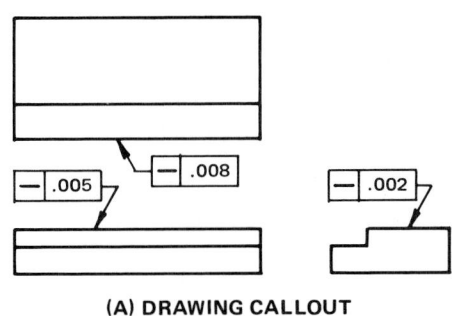

(A) DRAWING CALLOUT

STRAIGHT WITHIN .002 MEASURED
IN DIRECTION OF ARROWS

STRAIGHT WITHIN .005 MEASURED
IN DIRECTION OF ARROWS

STRAIGHT WITHIN .008 MEASURED
IN DIRECTION OF ARROWS

(B) INTERPRETATION

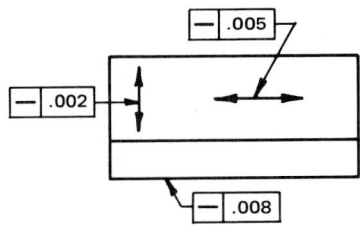

(C) THREE STRAIGHTNESS TOLERANCES ON ONE VIEW

FIG. 16-2-13   Straightness tolerances in different directions.

16-3 FLATNESS

The symbol for flatness is a parallelogram, with angles 60°, as shown in Fig. 16-3-1. The length and height are based on a percentage of the height of the lettering used on the drawing.

Flatness of a Surface

Flatness of a surface is a condition in which all surface elements are in one plane.

A flatness tolerance is applied to a line representing the surface of a part by means of a feature control frame, as shown in Fig. 16-3-2. A flatness tolerance means that all points on the surface shall be contained within a tolerance zone consisting of the space between two parallel planes that are separated by the specified tolerance. These planes may be oriented in any manner to contain the surface, that is, they are not necessarily parallel to the base. The flatness tolerance must be less than the size tolerance.

If the same control is desired on two or more surfaces, a suitable note indicating the number of surfaces may be added instead of repeating the symbol (Fig. 16-3-3, pg. 494).

Flatness Per Unit Area

Flatness may be applied, as in the case of straightness, on a unit basis as a means of preventing an abrupt surface variation

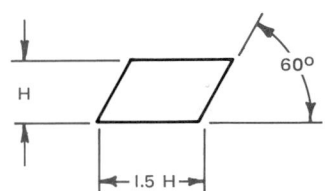

H = RECOMMENDED LETTER HEIGHT

FIG. 16-3-1   Flatness symbol.

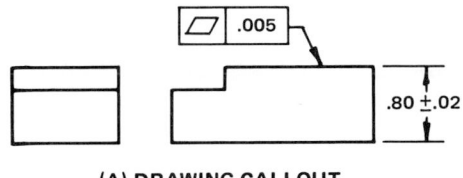

(A) DRAWING CALLOUT

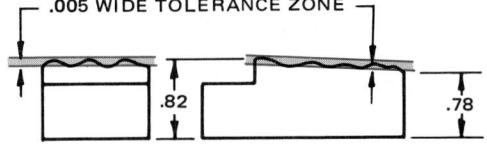

THE SURFACE MUST LIE BETWEEN TWO PARALLEL PLANES
.005 IN. APART. ADDITIONALLY, THE SURFACE MUST BE
LOCATED WITHIN ANY SPECIFIED LIMITS OF SIZE.

(B) INTERPRETATION

FIG. 16-3-2   Specifying flatness of a surface.

493

within a relatively small area of the feature. The unit variation is used either in combination with a specified total variation or alone. Caution should be exercised when using unit control without specifying a maximum limit for the total length because of the relatively large variations that may result if no such restriction is applied. If the feature has a uniformly continuous bow throughout its length that just conforms to the tolerance applicable to the unit length, the overall tolerance may result in an unsatisfactory part.

Since flatness involves surface area, the size of the unit area, for example, 1.00 × 1.00 in., is specified to the right of the flatness tolerance, separated by a slash line (Fig. 16-3-4).

Two or More Flat Surfaces in One Plane

Coplanarity is the condition of two or more surfaces having all elements in one plane. Coplanarity may be controlled by form, orientation, or locational tolerancing, depending on the functional requirements. See Unit 16-14.

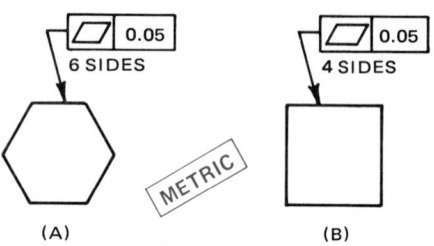

6 SIDES   4 SIDES

METRIC

(A)   (B)

FIG. 16-3-3   Controlling flatness on two or more surfaces.

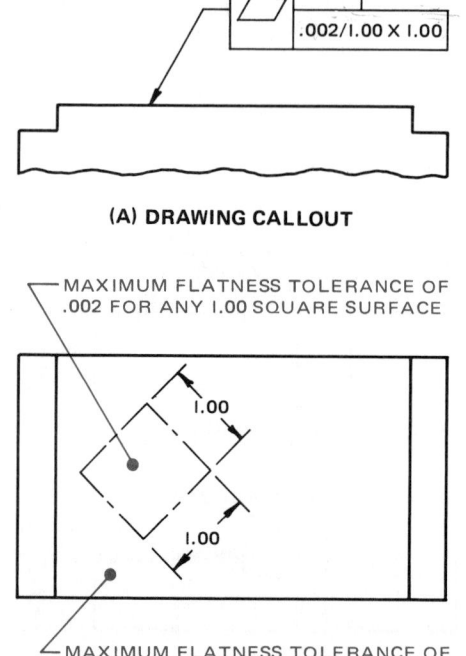

.010

.002/1.00 × 1.00

(A) DRAWING CALLOUT

MAXIMUM FLATNESS TOLERANCE OF .002 FOR ANY 1.00 SQUARE SURFACE

1.00

1.00

MAXIMUM FLATNESS TOLERANCE OF .010 FOR ENTIRE SURFACE AREA

(B) INTERPRETATION

FIG. 16-3-4   Overall flatness tolerance combined with a flatness tolerance of a unit area.

REFERENCES AND SOURCE MATERIAL

1. ASME Y14.5M–1994, *Dimensioning and Tolerancing.*
2. CAN/CSA B78.2-M91, *Dimensioning and Tolerancing of Technical Drawings.*
3. ISO drawing standards.

ASSIGNMENTS

See Assignments 4 through 7 for Unit 16-3 on page 571.

16-4 STRAIGHTNESS OF A FEATURE OF SIZE

Features of Size

So far, only lines, line elements, and single surfaces have been considered. These are features having no diameter or thickness, and geometric tolerances applied to them cannot be affected by feature size.

Features of size are features that do have diameter or thickness. These may be cylinders, such as shafts and holes. They may be slots, tabs, or rectangular or flat parts, where two parallel, flat surfaces are considered to form a single feature. With features of size, the feature control frame is associated with the size dimension (Fig. 16-4-1).

Before proceeding with examples of features of size, it is essential to understand certain terms.

Circular Tolerance Zones   When the resulting tolerance zone is cylindrical, such as when straightness of the center line of a cylindrical feature is specified, a diameter symbol precedes the tolerance value in the feature control frame and the feature control frame is located below the dimension pertaining to the feature (Fig. 16-4-2).

Maximum Material Condition (MMC)   When a feature or part is at the limit of size, which results in its containing the maximum amount of material, it is said to be at MMC. Thus it is the maximum limit of size for an external feature, such as a shaft, or the minimum limit of size for an internal feature, such as a hole (Fig. 16-4-1).

Virtual Condition (Size)   *Virtual condition* refers to the overall envelope of perfect form within which the feature would just fit. For an external feature such as a shaft, it is the maximum measured size plus the effect of permissible form variations, such as straightness, flatness, roundness, cylindricity, and orientation tolerances. For an internal feature such as a hole, it is the minimum measured size minus the effect of such form variations (Fig. 16-4-1).

Parts are generally toleranced so they will assemble when mating features are at MMC. Additional tolerance on form or location is permitted when features depart from their MMC size.

Least Material Condition (LMC)   This term refers to that size of a feature that results in the part containing the minimum

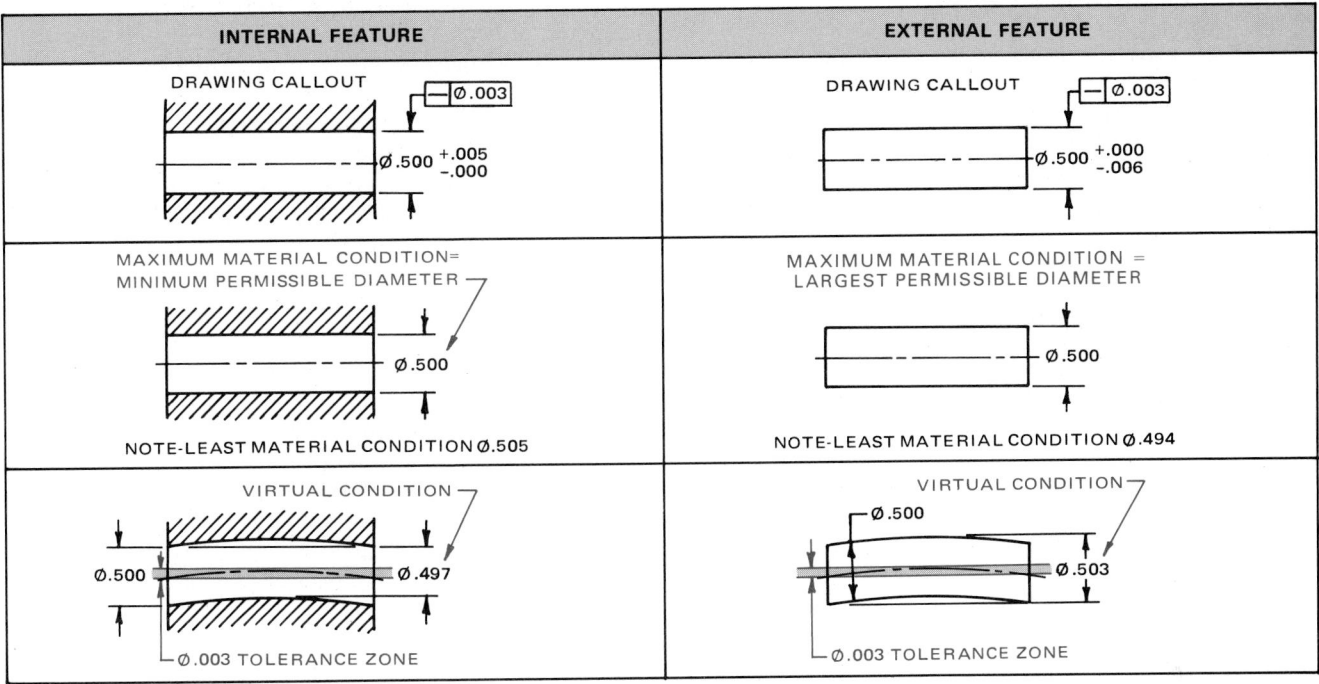

FIG. 16-4-1 Maximum material and virtual conditions.

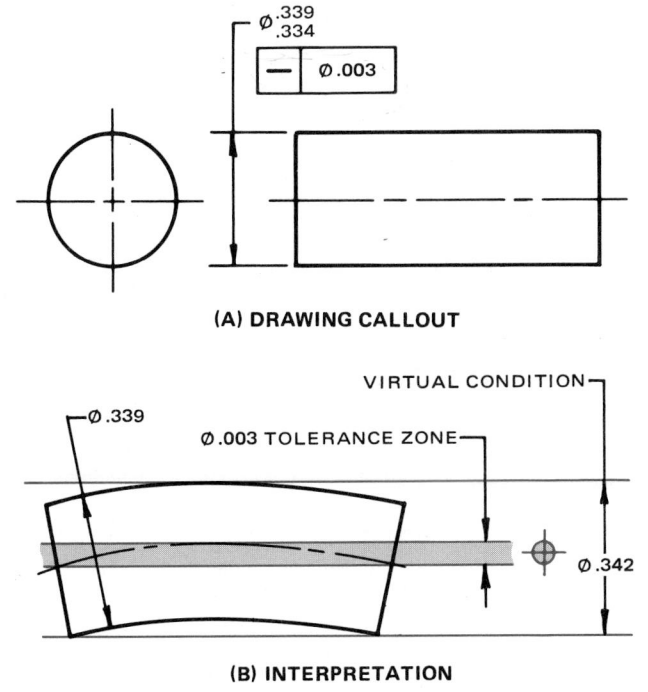

(A) DRAWING CALLOUT

(B) INTERPRETATION

FIG. 16-4-2 Circular tolerance zone.

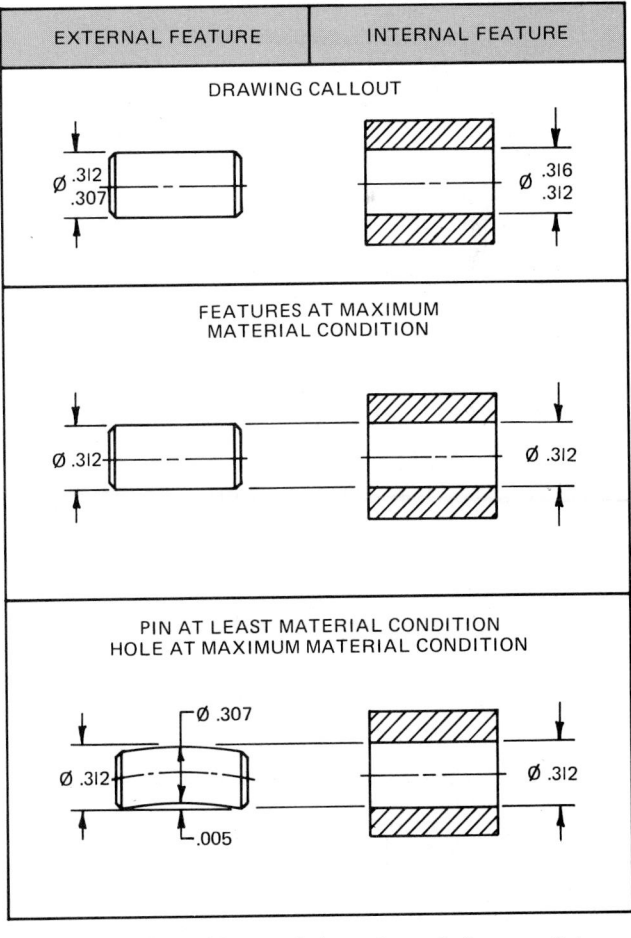

FIG. 16-4-3 Effect of form variations when only features of size are specified.

amount of material. Thus it is the minimum limit of size for an external feature, for example, a shaft, and the maximum limit of size for an internal feature, such as a hole (Fig. 16-4-3).

Regardless of Feature Size (RFS) This term indicates that a geometric tolerance applies to any size of a feature that lies within its size tolerance.

Material Condition Symbols

The symbols used to indicate "at maximum material condition," and "least material condition" are shown in Fig. 16-4-4. The "regardless of feature size" symbol was shown in ANSI standards prior to the implementing of ASME 14.5M–1994. It is shown here as many drawings currently in use show this symbol.

Applicability of RFS, MMC, and LMC

Applicability of RFS, MMC, or LMC is limited to features subject to variations in size. They may be datum features or other features whose axes or center planes are controlled by geometric tolerances. In such cases, the following practices apply:

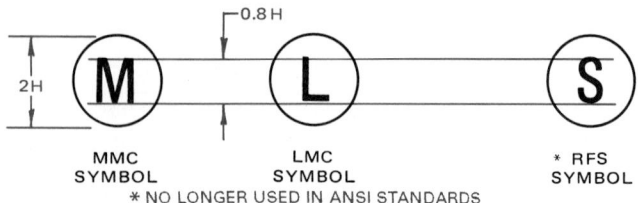

FIG. 16-4-4   Material condition symbols.

RFS applies, with respect to the individual tolerance, datum reference, or both. Where no modifying symbol is specified. MMC or LMC must be specified on the drawing where it is required.

If freedom of assembly of mating parts is the chief criterion for establishing a geometric tolerance for a feature of size, the least favorable assembly condition exists when the parts are made to the maximum material condition. Further geometric variations can then be permitted, without jeopardizing assembly, as the features approach their least material condition.

EXAMPLE 1

The effect of a form tolerance is shown in Fig. 16-4-3, where a cylindrical pin of Ø.307–.312 in. is intended to assemble into a round hole of Ø.312–.316 in. If both parts are at their maximum material condition of Ø.312 in., it is evident that both would have to be perfectly round and straight in order to assemble. However, if the pin was at its least material condition of Ø.307 in., it could be bent up to .005 in. and still assemble in the smallest permissible hole.

EXAMPLE 2

Another example, based on the location of features, is shown in Fig. 16-4-5. This shows a part with two projecting pins required to assemble into a mating part having two holes at the same center distance.

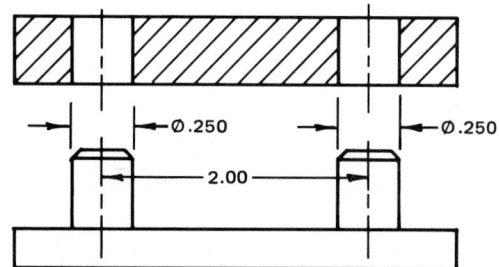

(A) DRAWING CALLOUT

CENTER DISTANCE MUST BE PERFECT IN ORDER TO ASSEMBLE

(B) PINS AND HOLES AT MAXIMUM MATERIAL CONDITION

FIG. 16-4-5   Effect of location.

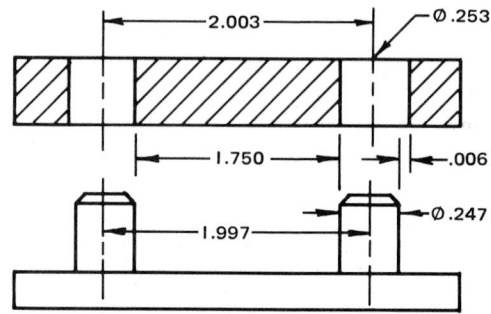

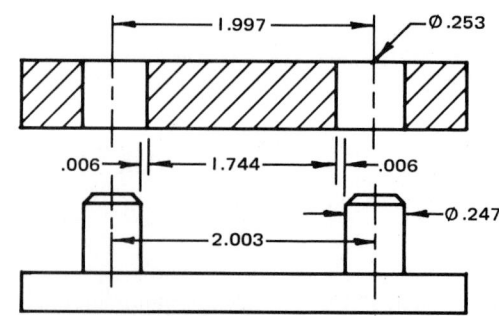

EACH CENTER DISTANCE MAY BE INCREASED OR DECREASED BY .003

(C) PINS AND HOLES AT LEAST MATERIAL CONDITION

The worst assembly condition exists when the pins and holes are at their maximum material condition, which is Ø.250 in. Theoretically, these parts would just assemble if their form, orientation (squareness to the surface), and center distances were perfect. However, if the pins and holes were at their least material condition of Ø.247 and Ø.253 in., respectively, it would be evident that one center distance could be increased and the other decreased by .003 in. without jeopardizing the assembly condition.

Maximum Material Condition (MMC)

If a geometric tolerance is required to be modified on an MMC basis, it is specified on the drawing by including the symbol Ⓜ immediately after the tolerance value in the feature control frame as shown in Fig. 16-4-6.

A form tolerance modified in this way can be applied only to a feature of size; it cannot be applied to a single surface. It controls the boundary of the feature, such as a complete cylindrical surface, or two parallel surfaces of a flat feature. This permits the feature surface or surfaces to cross the maximum material boundary by the amount of the form tolerance unless it is required that the virtual condition be kept within the maximum material boundary, in which case the form tolerance must be specified as zero at MMC, as shown in Fig. 16-4-7.

However, a note such as PERFECT FORM AT MMC NOT REQD must be specified on the drawing (Figs. 16-4-12 and 16-4-13).

Application of MMC to geometric symbols is shown in Fig. 16-4-8.

Application with Maximum Value   It is sometimes necessary to ensure that the geometric tolerance does not vary over the full range permitted by the size variations. For such applications a maximum limit may be set to the geometric tolerance and this is shown in addition to that permitted at MMC, as shown in Fig. 16-4-9.

Regardless of Feature Size (RFS)

When MMC or LMC is not specified with a geometric tolerance for a feature of size, no relationship is intended to exist between the feature size and the geometric tolerance. In other words, the tolerance applies regardless of feature size.

In this case, the geometric tolerance controls the form, orientation, or location of the center line, axis, or median plane of the feature.

The regardless of feature size symbol shown in Fig. 16-4-4 was used only with a tolerance of position. See Unit 16-9 and Fig. 16-4-10.

Least Material Condition (LMC)

The symbol for LMC is shown in Fig. 16-4-4. It is the condition in which a feature of size contains the least amount of material within the stated limits of size.

Specifying LMC is limited to positional tolerance applications where MMC does not provide the desired control and RFS is too restrictive. LMC is used to maintain a desired

CHARACTERISTIC TOLERANCE		FEATURE BEING CONTROLLED
STRAIGHTNESS	—	**NO** FOR A PLANE SURFACE OR A LINE ON A SURFACE **YES** FOR A FEATURE THE SIZE OF WHICH IS SPECIFIED BY A TOLERANCED DIMENSION, SUCH AS A HOLE, SHAFT OR A SLOT
PARALLELISM	//	
PERPENDICULARITY	⊥	
ANGULARITY	∠	
POSITION	⊕	
FLATNESS	▱	**NO** FOR ALL FEATURES
CIRCULARITY (ROUNDNESS)	○	
CYLINDRICITY	⌭	
CONCENTRICITY	◎	
SYMMETRY	≡	
PROFILE OF A LINE	⌒	
PROFILE OF A SURFACE	⌓	
CIRCULAR RUNOUT	↗	
TOTAL RUNOUT	↗↗	

FIG. 16-4-8   Application of MMC to geometric symbols.

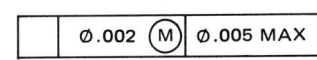

FIG. 16-4-9   Tolerance with a maximum specified value.

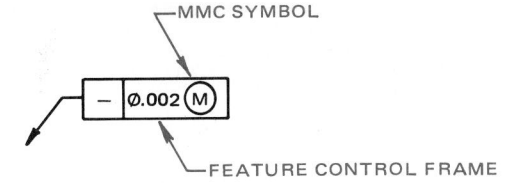

FIG. 16-4-6   Application of MMC symbol.

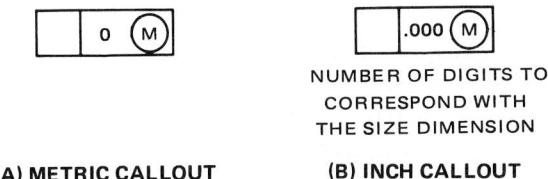

(A) METRIC CALLOUT   **(B) INCH CALLOUT**

FIG. 16-4-7   MMC callout with zero tolerance.

FIG. 16-4-10   Application of RFS symbol prior to publication of ASME Y14.5–1994.

relationship between the surface of a feature and its true position at tolerance extremes. See Unit 16-9 and Fig. 16-4-11.

The symbols for RFS and LMC are used only in ANSI standards and have not been adopted internationally.

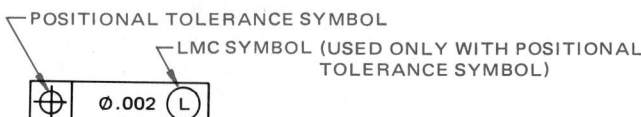

FIG. 16-4-11   Application of LMC symbol.

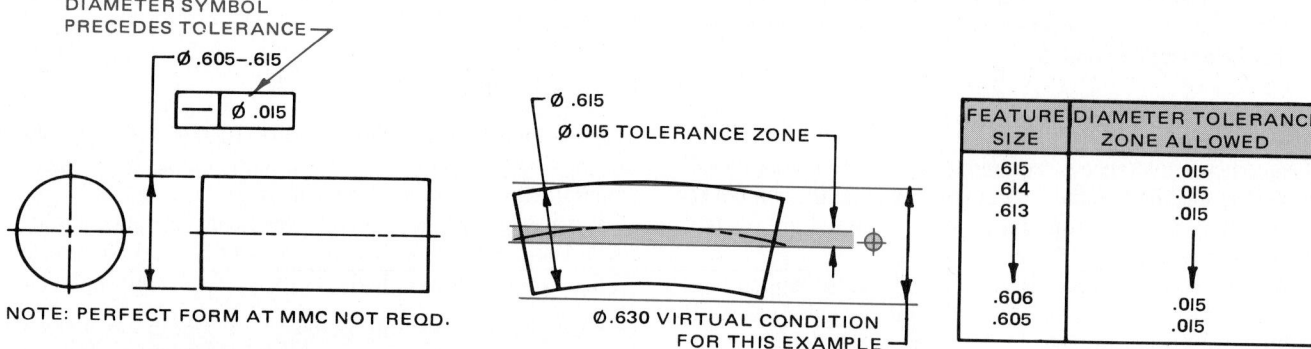

(A) DRAWING CALLOUT

NOTE: PERFECT FORM AT MMC NOT REQD.

(B) INTERPRETATION

FIG. 16-4-12   Specifying straightness—RFS.

FIG. 16-4-13   Specifying straightness—MMC.

FEATURE SIZE	DIAMETER TOLERANCE ZONE ALLOWED
.615	.015
.614	.015
.613	.015
↓	↓
.606	.015
.605	.015

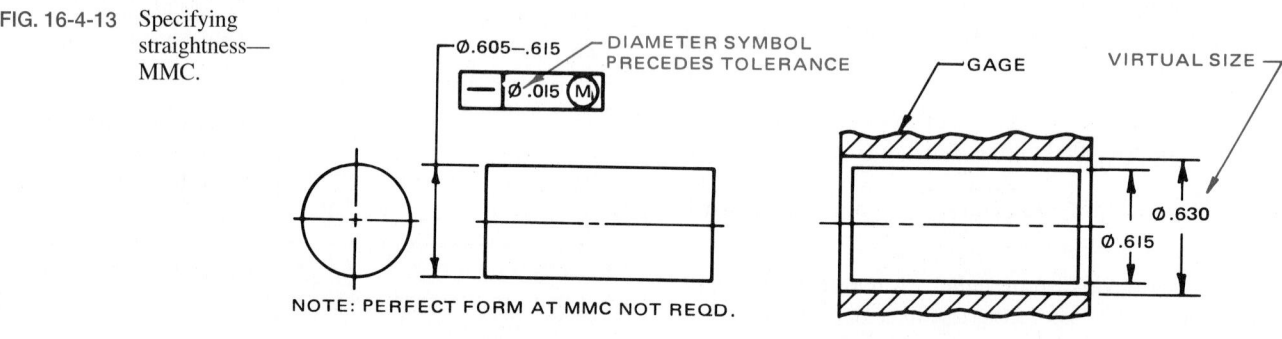

(A) DRAWING CALLOUT

NOTE: PERFECT FORM AT MMC NOT REQD.

THE MAXIMUM DIAMETER OF THE PIN WITH PERFECT FORM IN A GAGE

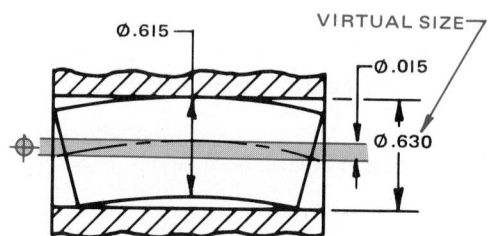

WITH PIN AT MAXIMUM DIAMETER (.615), THE GAGE WILL ACCEPT THE PIN WITH UP TO .015 IN. VARIATION IN STRAIGHTNESS

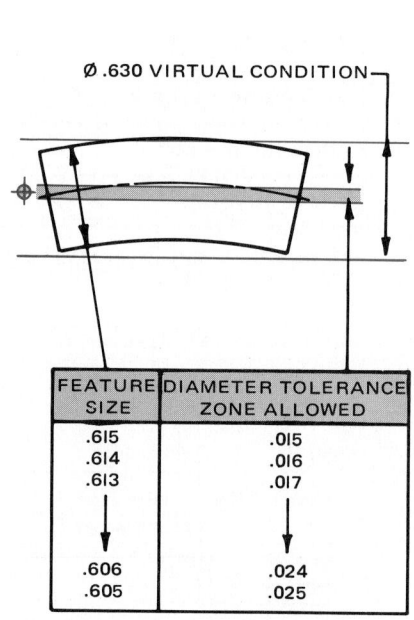

FEATURE SIZE	DIAMETER TOLERANCE ZONE ALLOWED
.615	.015
.614	.016
.613	.017
↓	↓
.606	.024
.605	.025

(B) INTERPRETATION

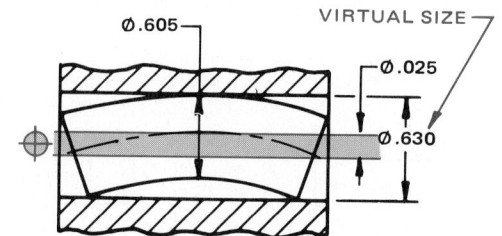

WITH PIN AT MINIMUM DIAMETER (.605), THE GAGE WILL ACCEPT THE PIN WITH UP TO .025 IN. VARIATION IN STRAIGHTNESS

(C) ACCEPTANCE BOUNDARY

498

Straightness of a Feature of Size

Figures 16-4-12 and 16-4-13 show examples of cylindrical parts where all circular elements of the surface are to be within the specified size tolerance; however, the boundary of perfect form at MMC may be violated. This violation is permissible when the feature control frame is associated with the size dimension, or attached to an extension of the dimension line. In the two figures a diameter symbol precedes the tolerance value and the tolerance is applied on an RFS and an MMC basis, respectively. Normally the straightness tolerance is smaller than the size tolerance, but a specific design may allow the situation depicted in the figures. The collective effect of size and form variation can produce a virtual condition equal to the MMC size plus the straightness tolerance (Fig. 16-4-13). The derived center line of the feature must lie within a cylindrical tolerance zone as specified.

Straightness—RFS

When applied on an RFS basis, as in Fig. 16-4-12, the maximum permissible deviation from straightness is .015 in. regardless of the feature size. Note that the absence of a modifying symbol indicates that RFS applies.

Straightness—MMC

If the straightness tolerance of .015 in. is required only at MMC, further straightness error can be permitted without jeopardizing assembly, as the feature approaches its least material size (Fig. 16-4-13). The maximum straightness tolerance is the specified tolerance plus the amount the feature departs from its MMC size. The center line of the actual feature must lie within the derived cylindrical tolerance zone such as given in the table of Fig. 16-4-13.

Straightness—Zero MMC

It is quite permissible to specify a geometric tolerance of zero MMC, which means that the virtual condition coincides with the maximum material size (Fig. 16-4-14). Therefore, if a feature is at its maximum material limit everywhere, no errors of straightness are permitted.

Straightness on the MMC basis can be applied to any part or feature having straight line elements in a plane that includes the diameter or thickness. This includes practically all the parts already shown on an RFS basis. However, it should not be used for features that do not have a uniform cross section.

Straightness with a Maximum Value

If it is desired to ensure that the straightness error does not become too great when the part approaches the least material condition, a maximum value may be added, as shown in Fig. 16-4-15 (pg. 500).

Control in Specific Directions

As already stated, straightness of a center line applies only to center lines that run in the direction of the line or line elements to which the straightness tolerance is directed. If there could be some ambiguity, a note should be added, such as THIS DIRECTION ONLY, as shown in Fig. 16-4-16A (pg. 500). If the part is circular and it is intended that the tolerance apply in all directions, a diameter symbol precedes the tolerance value, as shown in Fig. 16-4-16B.

If different tolerances apply in two directions, the tolerance zone is then a parallelepiped (Fig. 16-4-17, pg. 501).

FIG. 16-4-14   Specifying straightness— zero MMC.

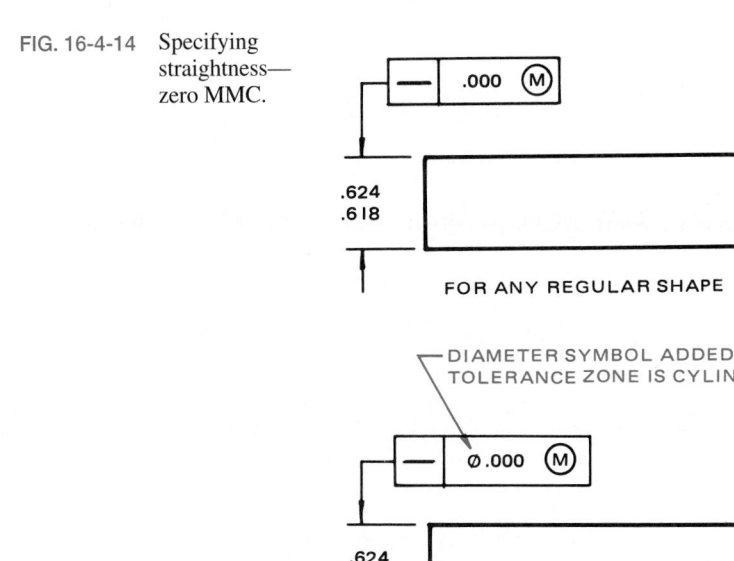

(A) DRAWING CALLOUT

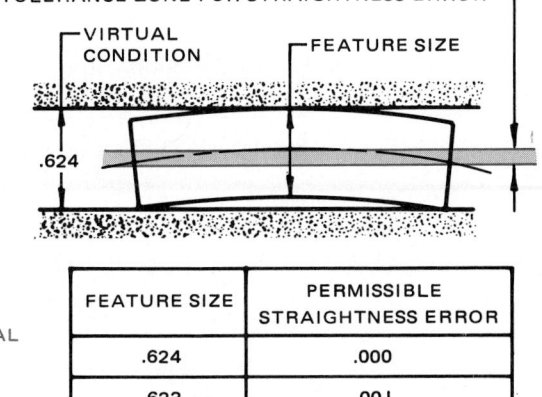

FEATURE SIZE	PERMISSIBLE STRAIGHTNESS ERROR
.624	.000
.623	.001
.622	.002
.621	.003
.620	.004
.619	.005
.618	.006

(B) PERMISSIBLE VARIATIONS

499

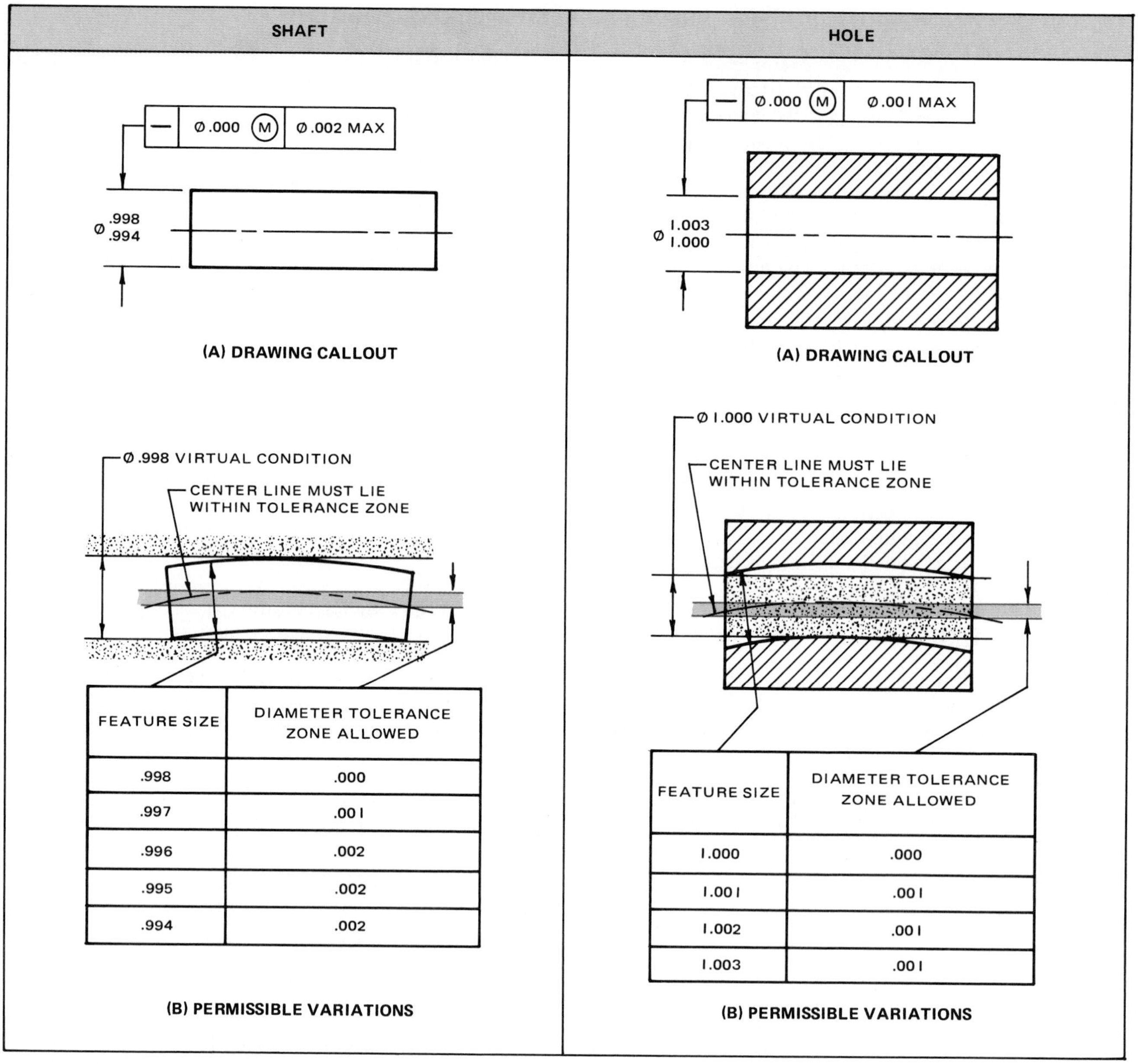

SHAFT	HOLE

— | Ø.000 (M) | Ø.002 MAX

Ø .998 / .994

(A) DRAWING CALLOUT

— | Ø.000 (M) | Ø.001 MAX

Ø 1.003 / 1.000

(A) DRAWING CALLOUT

Ø .998 VIRTUAL CONDITION
CENTER LINE MUST LIE WITHIN TOLERANCE ZONE

FEATURE SIZE	DIAMETER TOLERANCE ZONE ALLOWED
.998	.000
.997	.001
.996	.002
.995	.002
.994	.002

(B) PERMISSIBLE VARIATIONS

Ø 1.000 VIRTUAL CONDITION
CENTER LINE MUST LIE WITHIN TOLERANCE ZONE

FEATURE SIZE	DIAMETER TOLERANCE ZONE ALLOWED
1.000	.000
1.001	.001
1.002	.001
1.003	.001

(B) PERMISSIBLE VARIATIONS

FIG. 16-4-15   Specifying straightness of a shaft and hole with a maximum value.

FIG. 16-4-16   Direction and application of straightness.

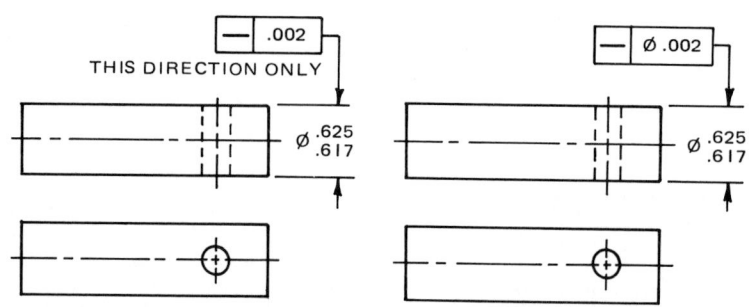

— | .002
THIS DIRECTION ONLY
Ø .625 / .617

— | Ø .002
Ø .625 / .617

(A) APPLIES IN ONE DIRECTION ONLY   **(B) APPLIES IN ALL DIRECTIONS**

Shapes Other Than Round

A straightness tolerance, not modified by MMC, may be applied to parts or features of any size or shape, provided they have a center plane, as in Fig. 16-4-18, which is intended to be straight in the direction indicated. Examples of parts or features with center planes are those having a hexagonal, square, or rectangular cross section.

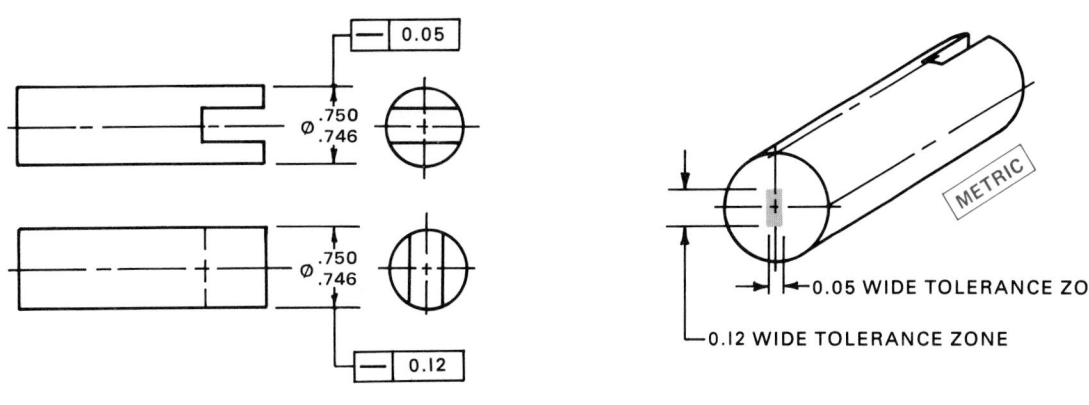

(A) DRAWING CALLOUT

(B) TOLERANCE ZONE

FIG. 16-4-17   Straightness in two directions.

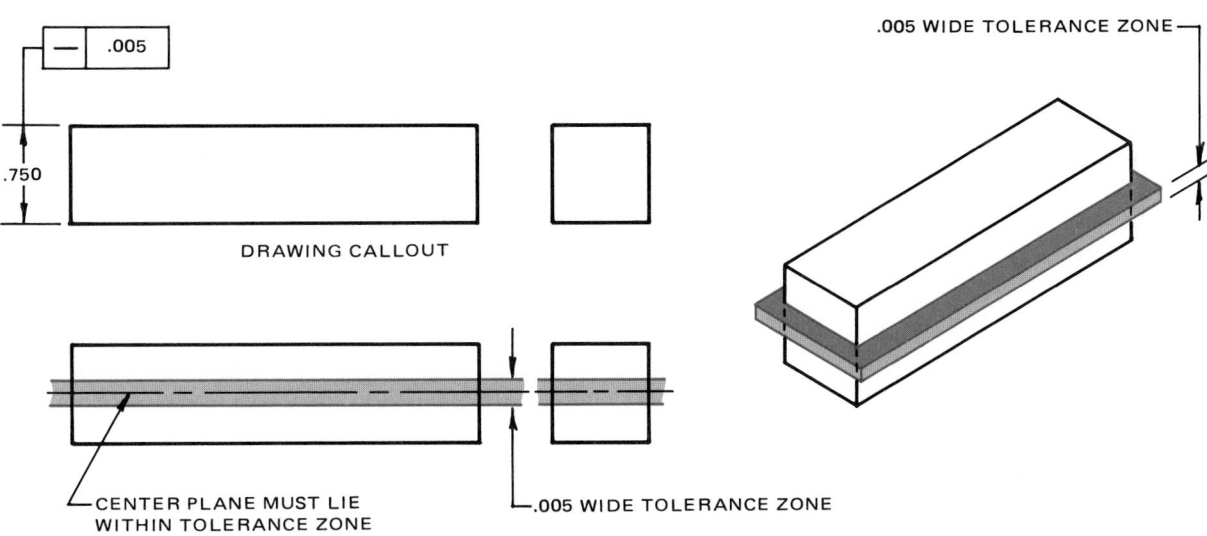

(A) SQUARE AND RECTANGULAR PARTS

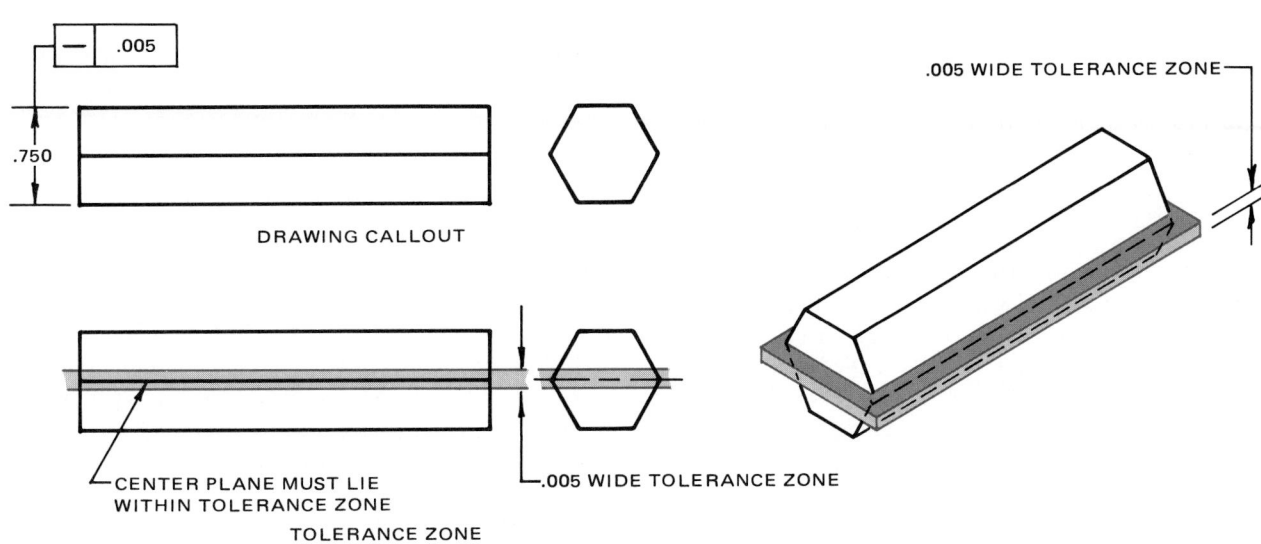

(B) REGULAR POLYGONS

FIG. 16-4-18   Straightness of a center plane—RFS.

501

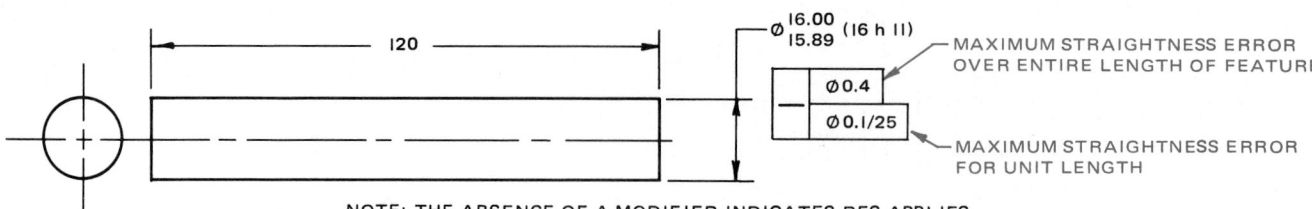

NOTE: THE ABSENCE OF A MODIFIER INDICATES RFS APPLIES.

(A) DRAWING CALLOUT

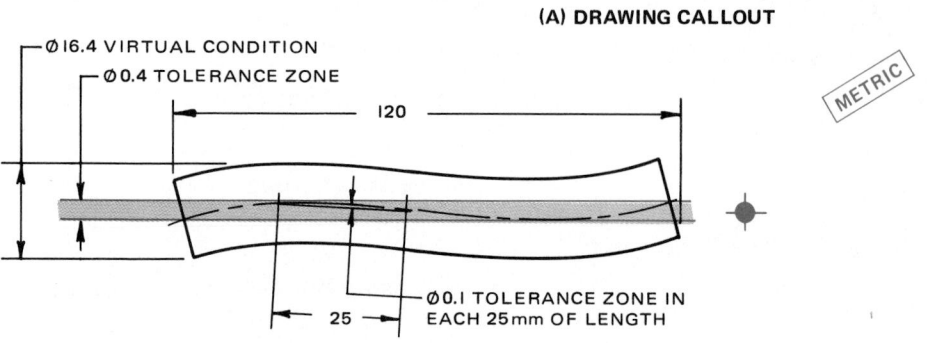

(B) INTERPRETATION

FIG. 16-4-19   Specifying straightness per unit length with specified
total straightness, both RFS.

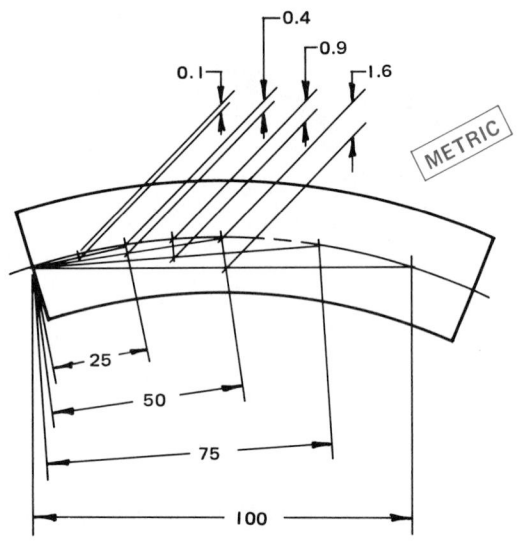

FIG. 16-4-20   Possible results of specifying straightness per unit
length RFS with no maximum specified.

Tolerances directed in this manner apply to straightness of
the center plane between all opposing line elements of the surfaces in the direction to which the control is directed. The
width of the tolerance zone is in the direction of the arrowhead.
If the cross section forms a regular polygon, such as a hexagon
or square, the tolerance applies to the center plane, between
each pair of sides, without it being necessary to so state on the
drawing.

Straightness Per Unit Length

Straightness may be applied on a unit length basis as a means of
preventing an abrupt surface variation within a relatively short
length of the feature (Fig. 16- 4-19). Caution should be exercised when using unit control without specifying a maximum

limit for the total length because of the relatively large variations that may result if no such restriction is applied. If the feature has a uniformly continuous bow throughout its length that
just conforms to the tolerance applicable to the unit length, the
overall tolerance may result in an unsatisfactory part. Figure
16-4-20 illustrates the possible condition if the straightness per
unit length given in Fig. 16-4-19 is used alone, that is, if
straightness for the total length is not specified.

REFERENCES AND SOURCE MATERIAL

1. ASME Y14.5M–1994, *Dimensioning and Tolerancing.*
2. CAN/CSA B78.2-M91, *Dimensioning and Tolerancing of
   Technical Drawings.*
3. ISO drawing standards.

ASSIGNMENTS

See Assignments 8 through 12 for Unit 16-4 on page 572.

16-5 DATUMS AND THE THREE-PLANE CONCEPT

Datums

Datum   A *datum* is a point, line, plane, or other geometric
surface from which dimensions are measured when so specified or to which geometric tolerances are referenced. A datum
has an exact form and represents an exact or fixed location for
purposes of manufacture or measurement.

Datum Feature   A *datum feature* is a feature of a part, such
as an edge or a surface, that forms the basis for a datum or is
used to establish its location.

Datums for Geometric Tolerancing

Datums are exact geometric points, lines, or surfaces, each based on one or more datum features of the part. Surfaces are usually either flat or cylindrical, but other shapes are used when necessary. The datum features, being physical surfaces of the part, are subject to manufacturing errors and variations. For example, a flat surface of a part, if greatly magnified, will show some irregularity. If brought into contact with a perfect plane, it will touch only at the highest points, as shown in Fig. 16-5-1. The true datums are theoretical but are considered to exist, or to be simulated, by locating surfaces of machines, fixtures, and gaging equipment on which the part rests or with which it makes contact during manufacture and measurement.

Three-Plane System

Geometric tolerances, such as straightness and flatness, refer to unrelated lines and surfaces and do not require the use of datums.

Orientation and locational tolerances refer to related features, that is, they control the relationship of features to one another or to a datum or datum system. Such datum features must be properly identified on the drawing.

Usually only one datum is required for orientation purposes, but positional relationships may require a datum system consisting of two or three datums. These datums are designated as primary, secondary, and tertiary. When these datums are plane surfaces that are mutually perpendicular, they are commonly referred to as a *three-plane datum system,* or a *datum reference frame.*

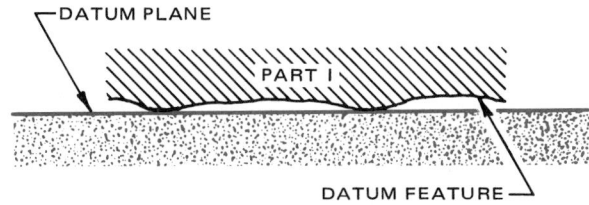

FIG. 16-5-1   Magnified section of a flat surface.

Primary Datum   If the primary datum feature is a flat surface, it could lie on a suitable plane surface, such as the surface of a gage, which would then become a primary datum, as shown in Fig. 16-5-2. Theoretically, there will be a minimum of three high spots on the flat surface that will come in contact with the surface of the gage.

Secondary Datum   If the part, while lying on this primary plane, is brought into contact with a secondary plane, it will theoretically touch at a minimum of two points.

Tertiary Datum   The part can now be slid along, while maintaining contact with the primary and secondary planes, until it contacts a third plane. This plane then becomes the tertiary datum, and the part will theoretically touch it at only one point.

These three planes constitute a datum system from which measurements can be taken. They will appear on the drawing, as shown in Fig. 16-5-3 (pg. 504), except that the datum features will be identified in their correct sequence by the methods described later in this unit.

It must be remembered that the majority of parts are not of the simple rectangular shape, and considerably more ingenuity may be required to establish suitable datums for more complex shapes.

Identification of Datums

Datum symbols are required to serve two purposes:

1. To identify the datum surface or feature on the drawing
2. To identify, for reference purposes, the datum feature

There are two methods of datum symbolization in general use for such purposes: one is shown and used in ANSI standards; the other, the ISO method, is used in most other countries of the world.

Former ANSI Datum Feature Symbol

In the former ANSI system, every datum feature was identified by a capital letter, enclosed in a rectangular box. A dash was placed before and after the letter, to identify it as applying to a datum feature, as shown in Fig. 16-5-4 (pg. 504). This

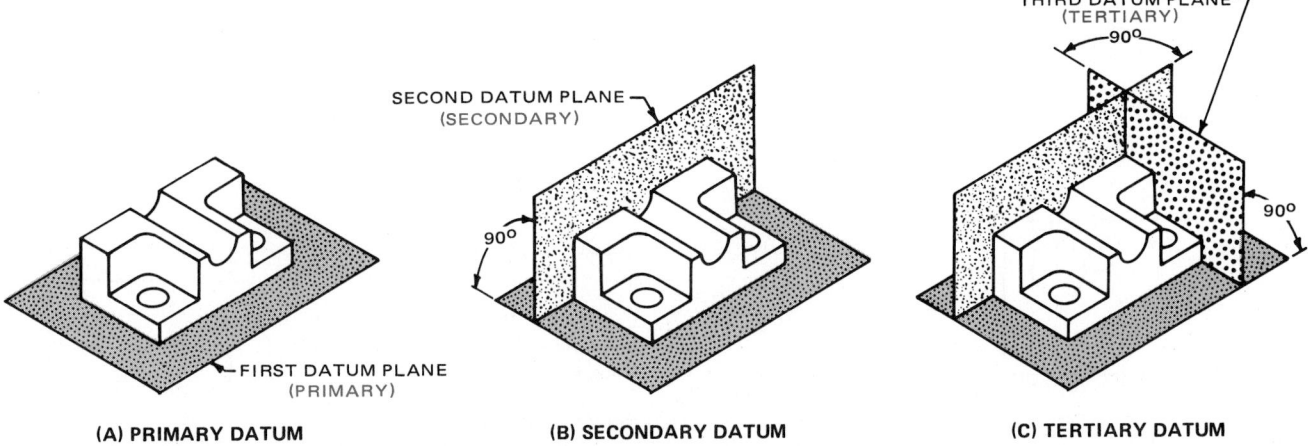

(A) PRIMARY DATUM   **(B) SECONDARY DATUM**   **(C) TERTIARY DATUM**

FIG. 16-5-2   The datum planes.

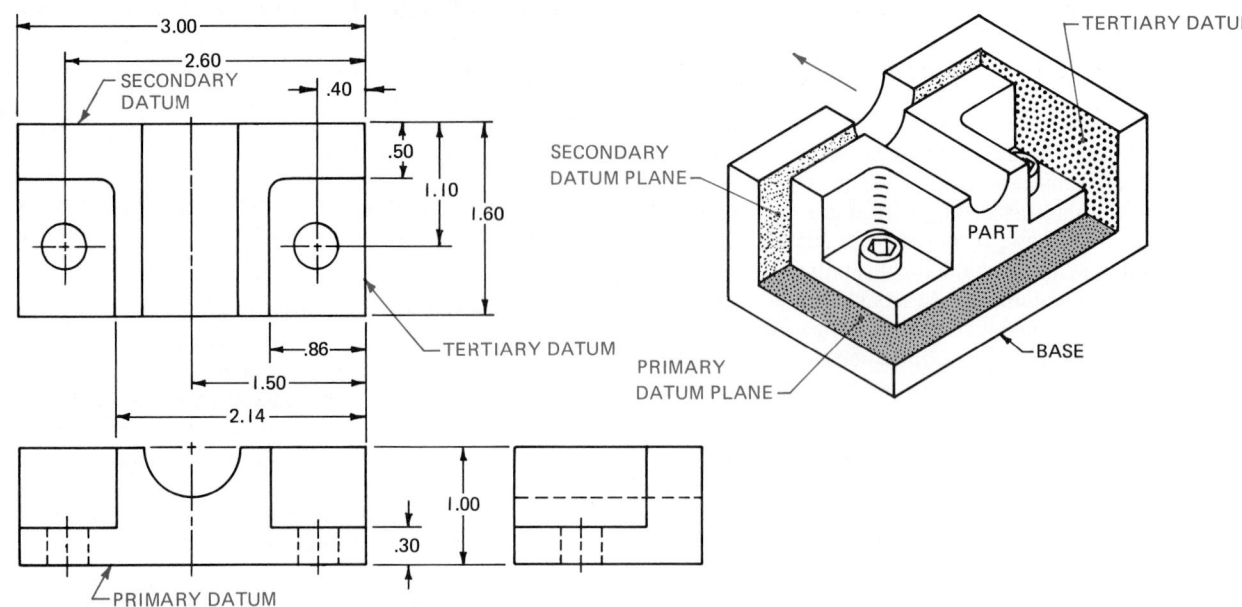

FIG. 16-5-3   Three-plane datum system.

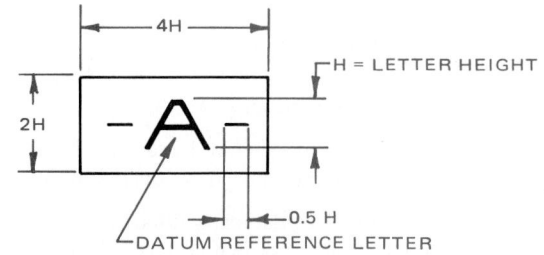

FIG. 16-5-4   Former ANSI datum feature symbol.

identifying symbol was directed to the datum feature in any one of the following ways.

For datum features not subject to size variation:

■ By attaching a side or end of the frame to an extension line from the feature, providing it is a plane surface
■ By running a leader with arrowhead from the frame to the feature
■ By adding the symbol to the feature control frame pertaining to the feature

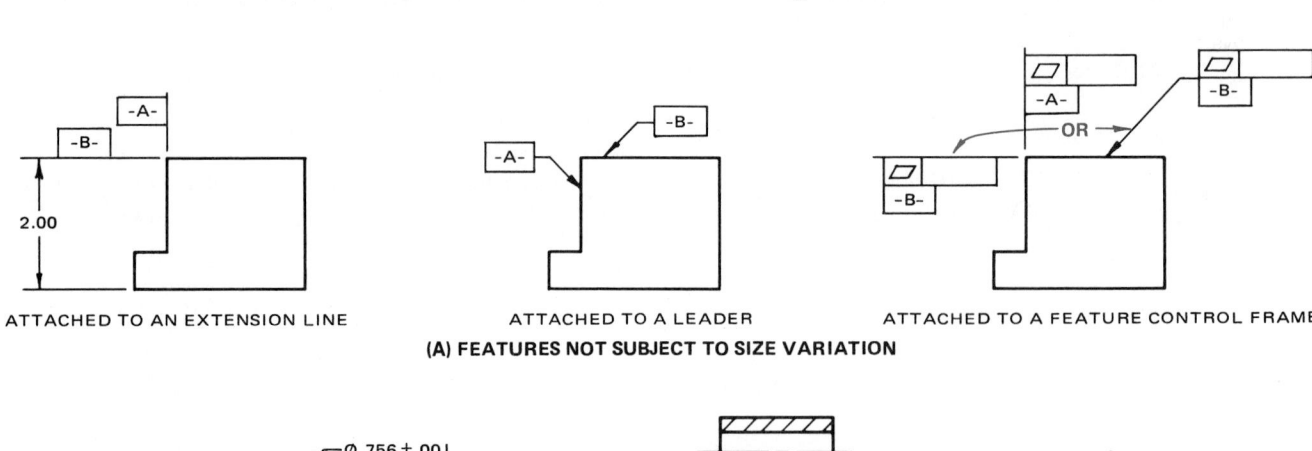

ATTACHED TO AN EXTENSION LINE   ATTACHED TO A LEADER   ATTACHED TO A FEATURE CONTROL FRAME

(A) FEATURES NOT SUBJECT TO SIZE VARIATION

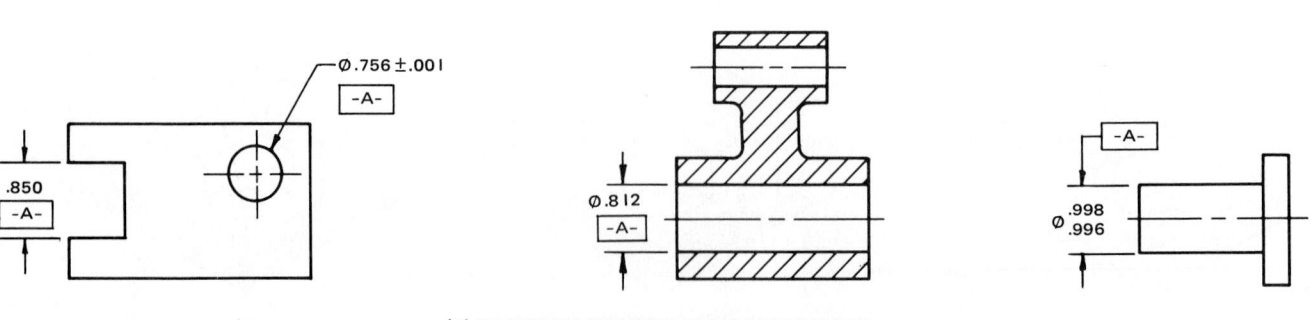

(B) FEATURES SUBJECT TO SIZE VARIATION

FIG. 16-5-5   Former placement of ANSI datum feature symbol.

For datum features subject to size variation (see Unit 16-7):

- By attaching a side or end of the frame to an extension of the dimension line pertaining to a feature of size
- Associating the datum symbol with the size dimension

These methods are illustrated in Fig. 16-5-5.

ISO Datum Feature Symbol

The ISO datum feature symbol, recently adopted by ANSI, is used by all other countries. The ISO datum feature symbol is a right-angle triangle, with a leader projecting from the 90° apex, as shown in Fig. 16-5-6. The base of the triangle should be slightly greater than the height of the lettering used on the drawing. The triangle may be filled in.

The datum is identified by a capital letter placed in a square frame and connected to the leader.

The ISO datum feature symbol may be directed to the datum feature in one of the following ways:

- Placed on the outline of the feature or an extension of the outline (but clearly separated from the dimension line) when the datum feature is the line or surface itself
- Shown as an extension of the dimension line when the datum feature is the axis or median plane
- Attached to the feature control frame when the datum feature is the axis or median (center) plane.
- For small features, where extension lines are not used, the symbol may be placed on the leader line

These methods are illustrated in Fig. 16-5-7.

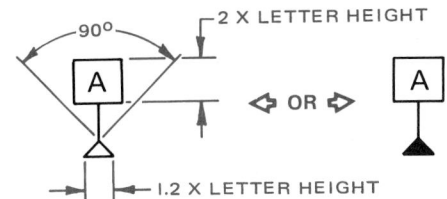

FIG. 16-5-6   ISO datum feature symbol.

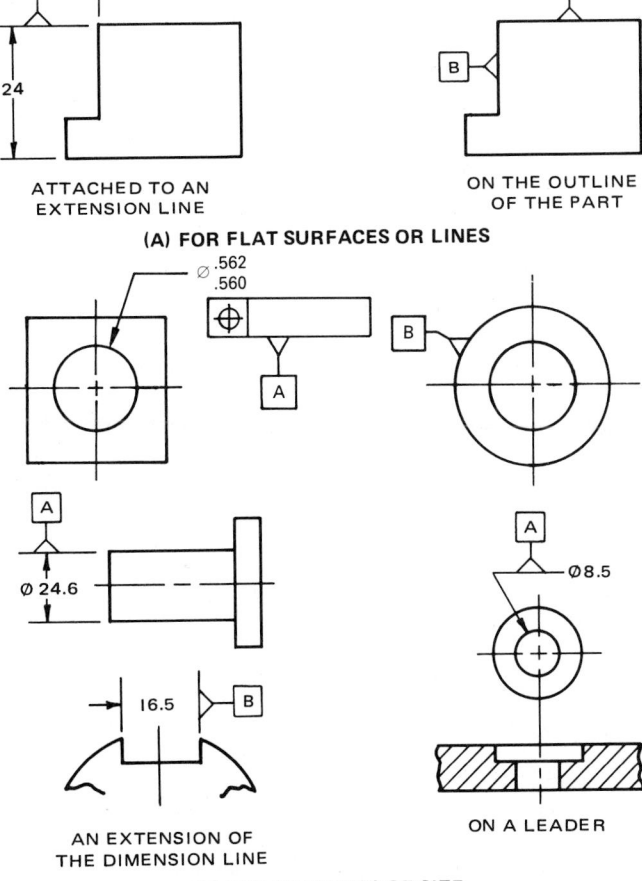

(A) FOR FLAT SURFACES OR LINES

(B) FOR FEATURES OF SIZE

FIG. 16-5-7   Placement of ISO datum feature symbol.

Association with Geometric Tolerances

The datum letter is placed in the feature control frame by adding an extra compartment for the datum reference, as shown in Fig. 16-5-8.

If two or more datum references are involved, additional frames are added and the datum references are placed in these frames in the correct order, that is, primary, secondary, and tertiary datums, as shown in Fig. 16-5-9.

Multiple Datum Features

If a single datum is established by two datum features, such as two ends of a shaft, each feature is identified by a separate letter. Both letters are then placed in the same compartment of the feature control frame, with a dash between them, as shown in Fig. 16-5-10 (pg. 506). The datum, in this case, is the common line between the two datum features.

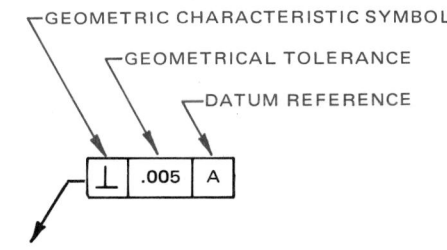

FIG. 16-5-8   Feature control frame referenced to a datum.

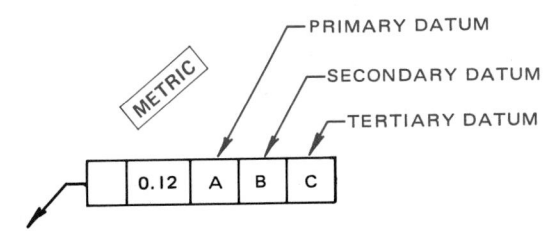

FIG. 16-5-9   Multiple datum references.

505

REFERENCES AND SOURCE MATERIAL

1. ASME Y14.5M–1994, *Dimensioning and Tolerancing.*
2. CAN/CSA B78.2-M91, *Dimensioning and Tolerancing of Technical Drawings.*
3. ISO drawing standards.

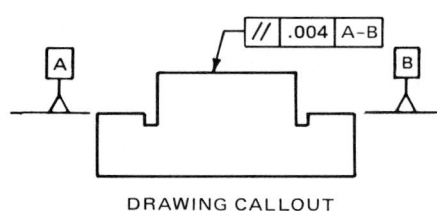

DRAWING CALLOUT

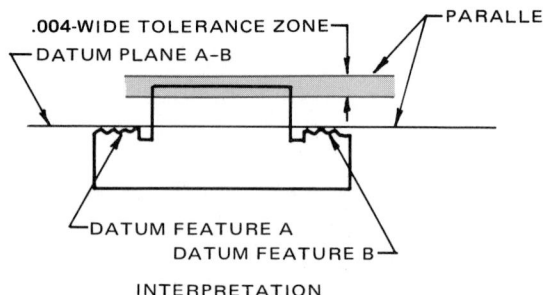

INTERPRETATION

(A) COPLANAR DATUM FEATURES

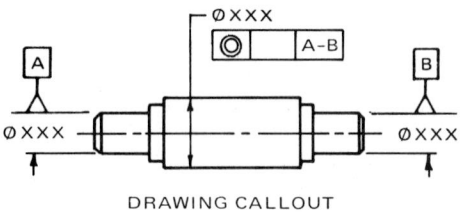

DRAWING CALLOUT

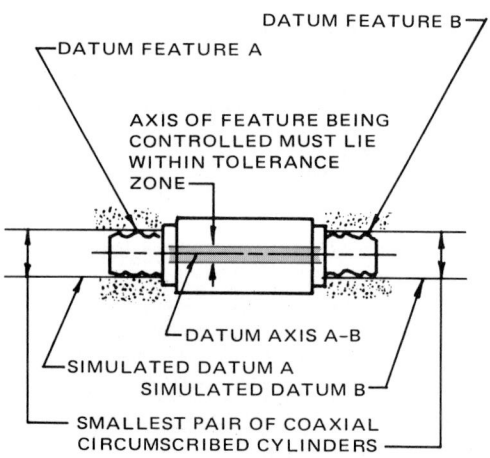

INTERPRETATION

(B) COAXIAL DATUM FEATURES

FIG. 16-5-10   Two datum features for one datum.

ASSIGNMENTS

See Assignments 13 through 16 for Unit 16-5 on pages 573 to 574.

16-6 ORIENTATION TOLERANCING OF FLAT SURFACES

Orientation refers to the angular relationship that exists between two or more lines, surfaces, or other features. Orientation tolerances control angularity, parallelism, and perpendicularity. Since to a certain degree the limits of size control form and parallelism, and tolerances of location (see Unit 16-9) control orientation, the extent of this control should be considered before specifying form or orientation tolerances.

A tolerance of form or orientation may be specified where the tolerances of size and location do not provide sufficient control.

Orientation tolerances, when applied to plane surfaces, control flatness if a flatness tolerance is not specified.

The general geometric characteristic for orientation is termed angularity. This term may be used to describe angular relationships, of any angle, between straight lines or surfaces with straight line elements, such as flat or cylindrical surfaces. For two particular types of angularity special terms are used. These are perpendicularity, or squareness, for features related to each other by a 90° angle and parallelism for features related to one another by an angle of 0°.

An orientation tolerance specifies a zone within which the considered feature, its line elements, its axis, or its center plane must be contained.

Reference to a Datum

An orientation tolerance indicates a relationship between two or more features. Whenever possible, the feature to which the controlled feature is related should be designated as a datum. Sometimes this does not seem possible, e.g., where two surfaces are equal and cannot be distinguished from one another. The geometric tolerance could theoretically be applied to both surfaces without a datum, but it is generally preferable to specify two similar requirements, using each surface in turn as the datum.

Angularity, parallelism, and perpendicularity are orientation tolerances applicable to related features. Relation to more than one datum feature should be considered if required to stabilize the tolerance zone in more than one direction.

There are three geometric symbols for orientation tolerances (Fig. 16-6-1). The proportions are based on the height of the lettering used on the drawing.

Angularity Tolerance

Angularity is the condition of a surface or axis at a specified angle (other than 90° or 0°) from a datum plane or axis. An angularity tolerance for a flat surface specifies a tolerance zone, the width of which is defined by two parallel planes at a specified basic angle from a datum plane or axis. The surface of the considered feature must lie within this tolerance zone (Fig. 16-6-2).

FIG. 16-6-1   Orientation symbols.

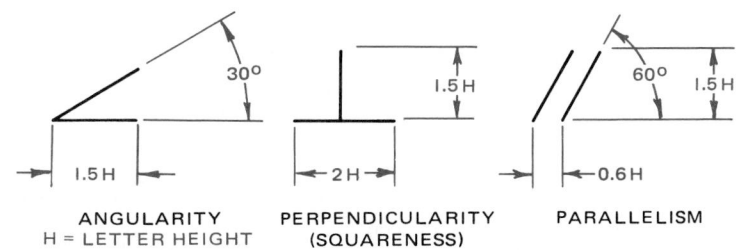

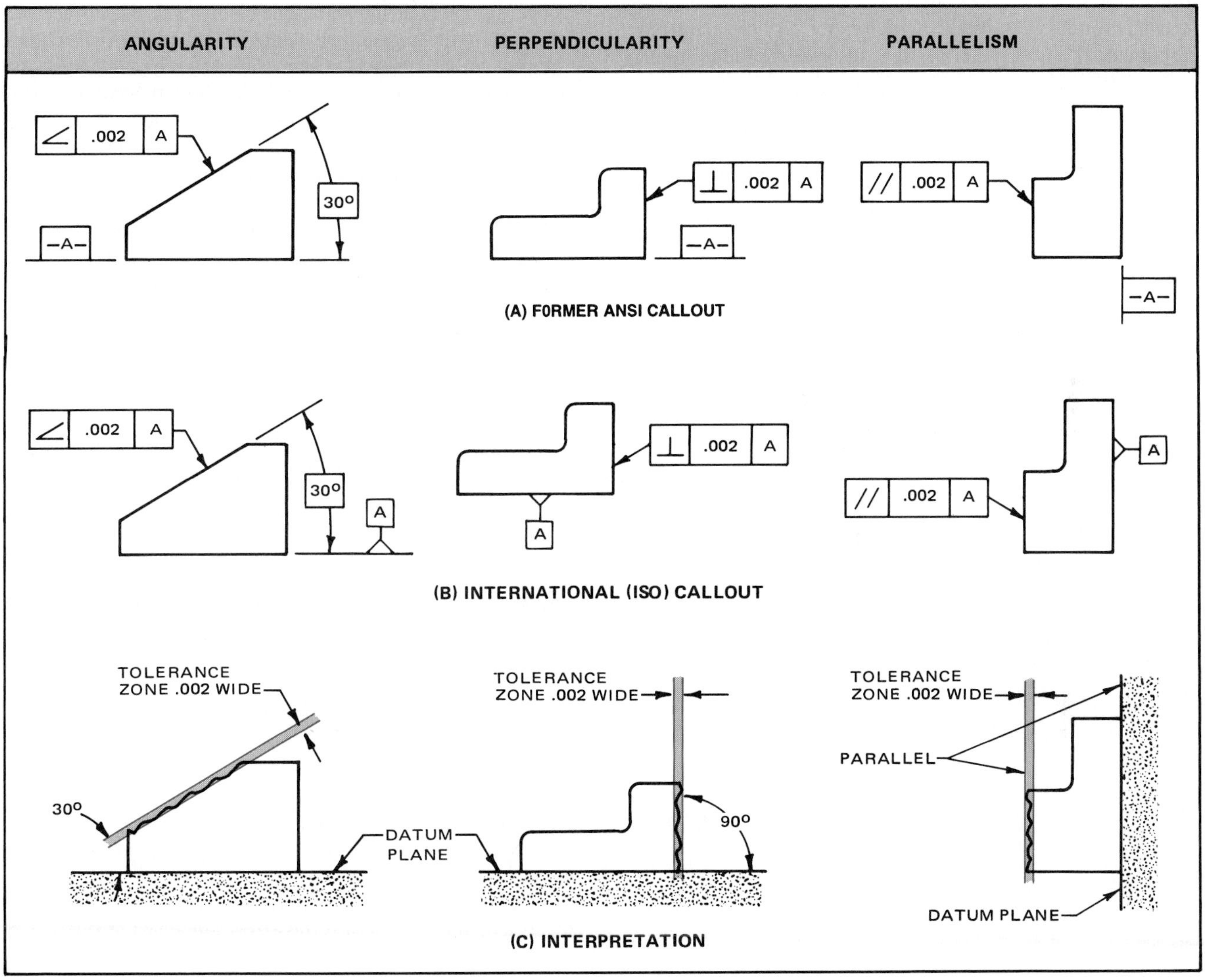

ANGULARITY	PERPENDICULARITY	PARALLELISM
1.5 H	2H	0.6 H
H = LETTER HEIGHT	(SQUARENESS)	

(A) FORMER ANSI CALLOUT

(B) INTERNATIONAL (ISO) CALLOUT

(C) INTERPRETATION

FIG. 16-6-2   Orientation tolerancing for flat surfaces.

For geometric tolerancing of angularity, the angle between the datum and the controlled feature should be stated as a basic angle. Therefore it should be enclosed in a rectangular frame, as shown in Fig. 16-6-2, to show that the general tolerance note does not apply. However, the angle need not be stated for either perpendicularity (90°) or parallelism (0°).

Perpendicularity Tolerance

Perpendicularity is the condition of a surface at 90° to a datum plane or axis. A perpendicularity tolerance for a flat surface specifies a tolerance zone defined by two parallel planes perpendicular to a datum plane or axis. The surface of the considered feature must lie within this tolerance zone (Fig. 16-6-2).

Parallelism Tolerance

Parallelism is the condition of a surface equidistant at all points from a datum plane. A parallelism tolerance for a flat surface specifies a tolerance zone defined by two planes or lines parallel to a datum plane or axis. The line elements of the surface must lie within this tolerance zone.

Figure 16-6-2 shows three simple parts in which one flat surface is designated as a datum feature and another flat surface is

507

related to it by one of the orientation tolerances. Each of these tolerances is interpreted to mean that the designated surface shall be contained within a tolerance zone consisting of the space between two parallel planes, separated by the specified tolerance (.002 in.) and related to the datum by the basic angle specified (30°, 90°, or 0°).

When an orientation tolerance is specified, there is no need to specify a form tolerance for the same feature unless a smaller tolerance is necessary.

Under certain circumstances, such as for very thin parts (e.g., parts made of sheet material), it is often desirable to control

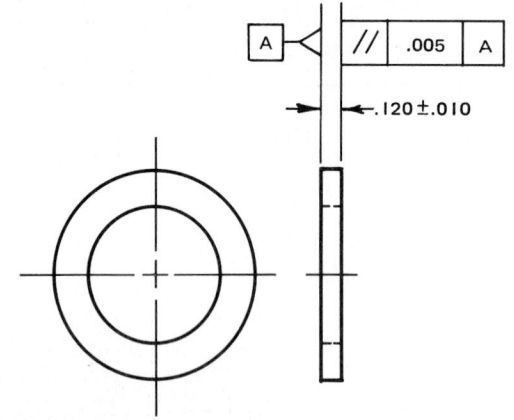

FIG. 16-6-3   Controlling parallelism of a flat part.

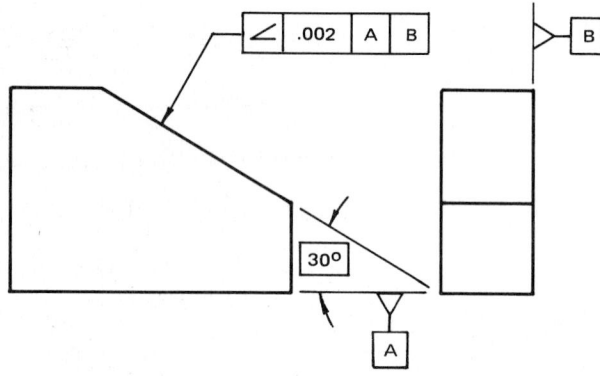

FIG. 16-6-4   Angularity referenced to a datum system.

flatness on both sides. This is accomplished by applying a parallelism tolerance to one surface of the feature and making the opposite side the datum feature (Fig. 16-6-3).

Control in Two Directions

The measuring principles for angularity describe the method of aligning the part prior to making angularity measurements. Proper alignment ensures that line elements of the surface perpendicular to the angular line elements are parallel to the datum.

For example, the part in Fig. 16-6-4 will be aligned so that line elements running horizontally in the left-hand view will be parallel to datum A. However, these line elements will bear a proper relationship with the sides, ends, and top faces only if these surfaces are true and square with datum B.

When both form and orientation tolerances are applied to a single feature, the form tolerance must be less than the orientation tolerance. An example of this is shown in Fig. 16-6-5 where the flatness of the surface must be controlled to a greater degree than its orientation. The flatness tolerance must lie within the angularity tolerance zone.

REFERENCES AND SOURCE MATERIAL

1. ASME Y14.5M–1994, *Dimensioning and Tolerancing.*
2. CAN/CSA B78.2-M91, *Dimensioning and Tolerancing of Technical Drawings.*
3. ISO drawing standards.

ASSIGNMENTS

See Assignments 17 and 18 for Unit 16-6 on page 575.

16-7 DATUM FEATURES SUBJECT TO SIZE VARIATION

In Unit 16-5, we saw how single features, such as flat surfaces or line elements of a surface, were used to establish datums for measuring purposes.

When a feature of size, such as diameter or width, is specified as a datum feature, it differs from singular flat surfaces in

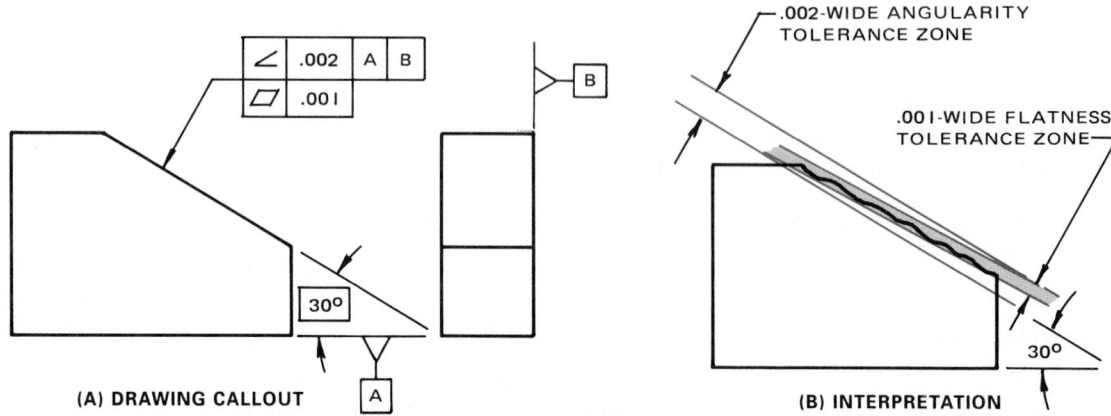

FIG. 16-6-5   Applying both an angularity and a flatness tolerance to a flat surface.

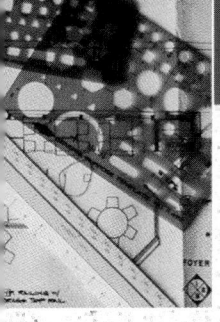

MANUAL DRAFTING TECHNIQUES

This drafter is using various instruments in the preparation of the drawing. These include a drafting machine, scale, templates, and a calculator.

ROBOTICS AND LASERS

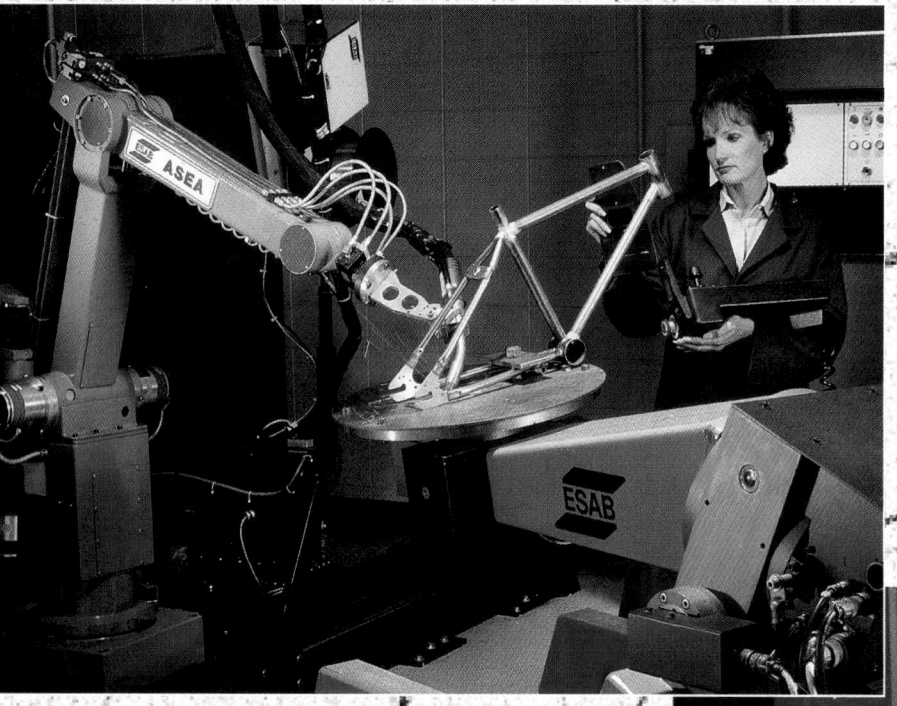

Here are some examples of modern equipment including robotic arms, robotic welding, and laser welding equipment. These products were developed as a result of the combination of computer-aided manufacturing (CAM) and computer-aided design (CAD).

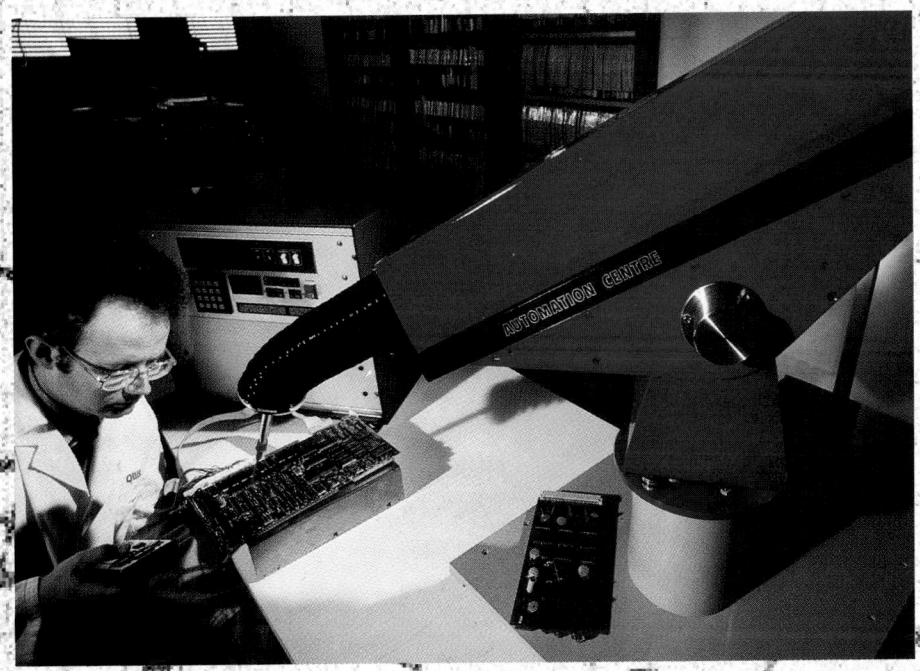

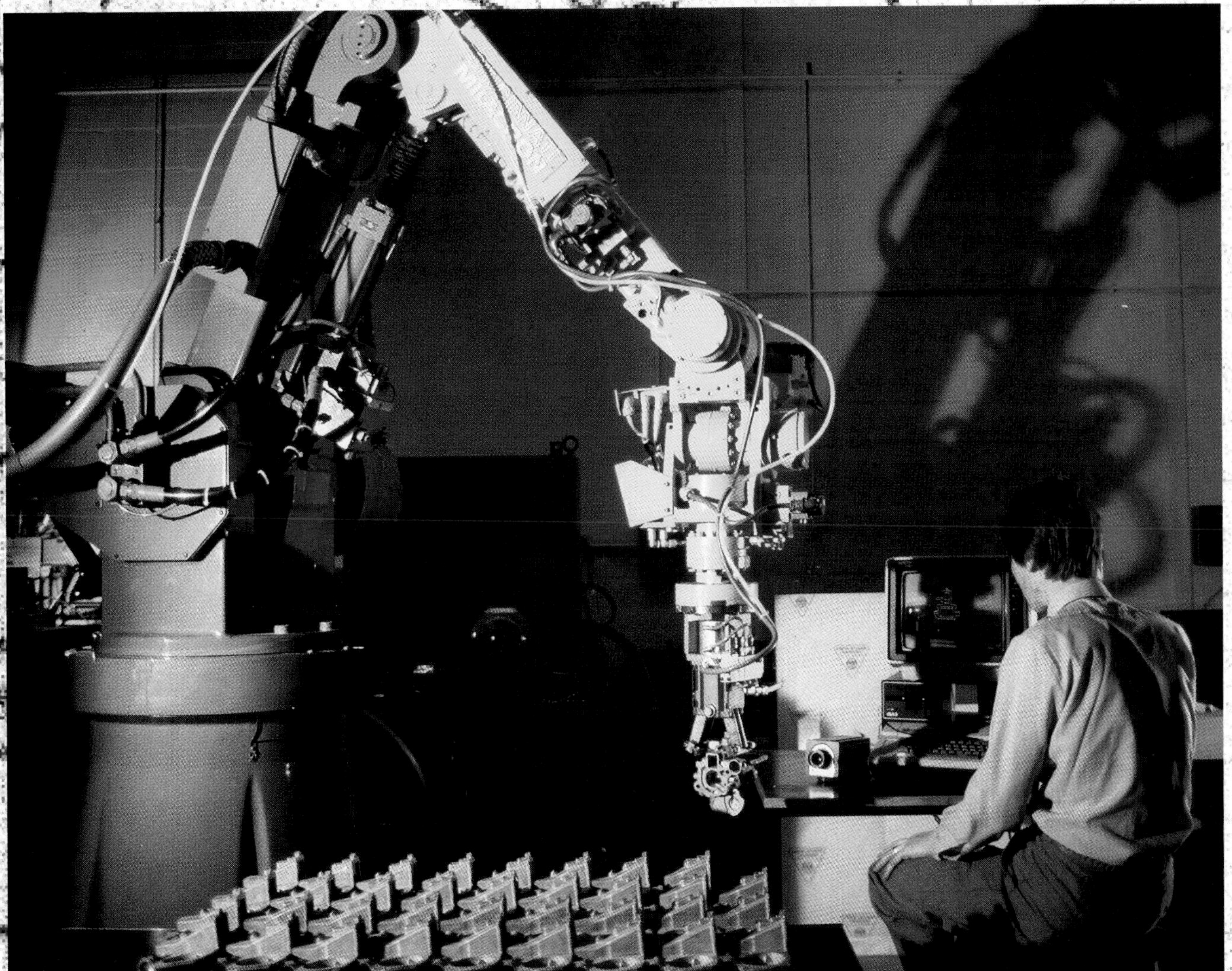

MANUFACTURING PROCESSES

4

0,0

ORIGIN

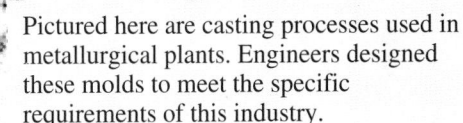

Pictured here are casting processes used in metallurgical plants. Engineers designed these molds to meet the specific requirements of this industry.

AUTOMOBILE PRODUCTION

The automotive industry has changed drastically in recent years. Assembly line equipment designed using computer-aided manufacturing techniques are pictured here and on the next page. Computer-aided design has also become an integral part of the automotive industry.

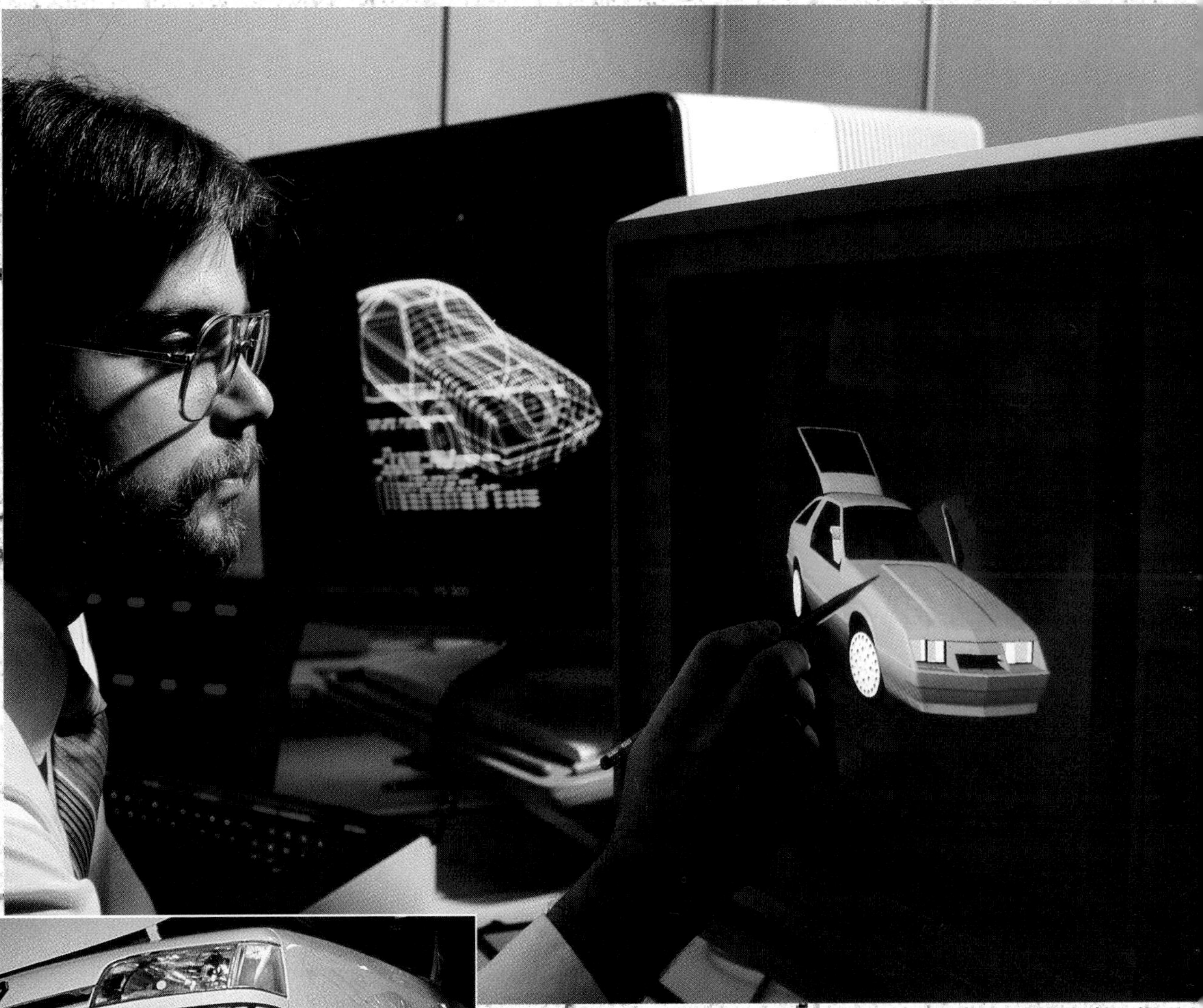

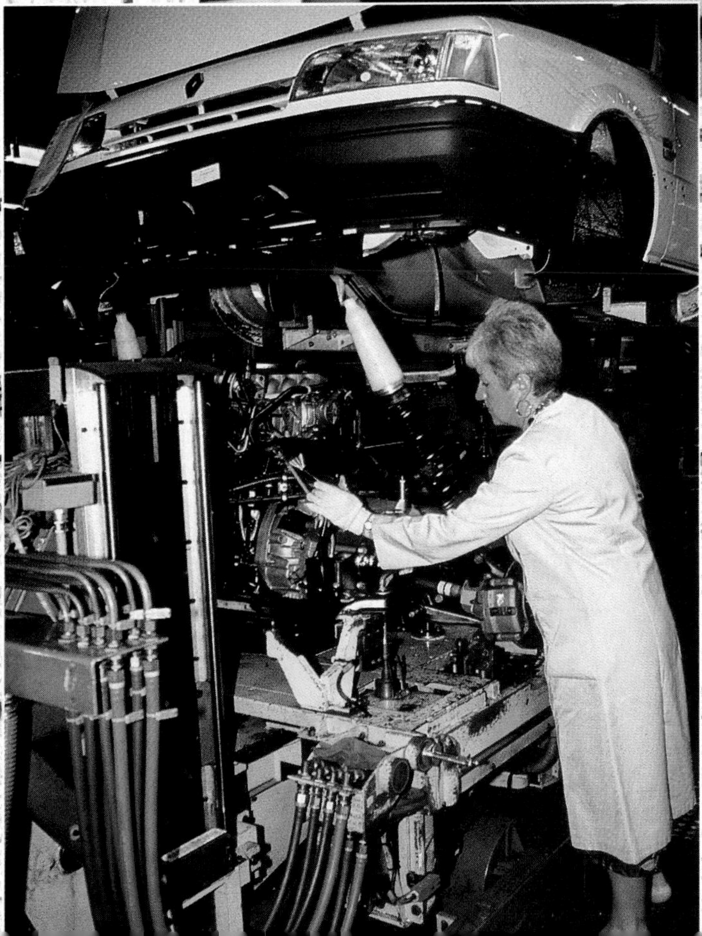

AUTOMOBILE PRODUCTION

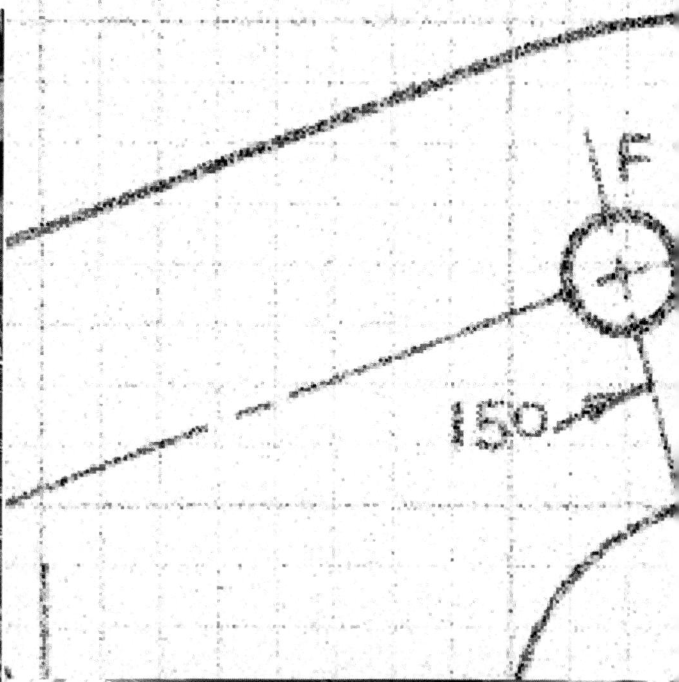

that it is subject to variations in size as well as form. The datum has to be established from the full surface of a cylindrical feature or from two opposing surfaces of other features of size. However, the true datum is a datum axis, center line, or center plane of the feature.

The datum identifying symbol is directed to the datum feature of size by the methods shown in Figs. 16-5-5 and 16-5-7.

Parts with Cylindrical Datum Features

Figure 16-7-1 illustrates a part having a cylindrical datum feature. Primary datum feature *A* relates the part to the first datum plane. Since secondary datum feature *B* is cylindrical, it is associated with two theoretical planes—the secondary and tertiary planes in the three-plane relationship.

These two theoretical planes are represented on the drawing by center lines crossing at right angles. The intersection of these planes coincides with the datum axis. Once established,

the datum axis becomes the origin for related dimensions while the two planes *X* and *Y* indicate the direction of measurements. In such cases, only two datum features are referenced in the feature control frame.

Figure 16-7-2 is another example where the cylindrical feature is used as the secondary datum. The primary datum is then a perfect plane on which the part would normally rest. The secondary datum is still the axis of an imaginary perfect cylinder, but also one that is perpendicular to the primary datum. This cylinder would theoretically touch the feature at only two points. The part has been purposely drawn out of square to show the effect of such deviations.

RFS and MMC Applications

Because variations are allowed by the size dimension, it becomes necessary to determine whether RFS or MMC applies in each case. For a tolerance of position, the datum reference

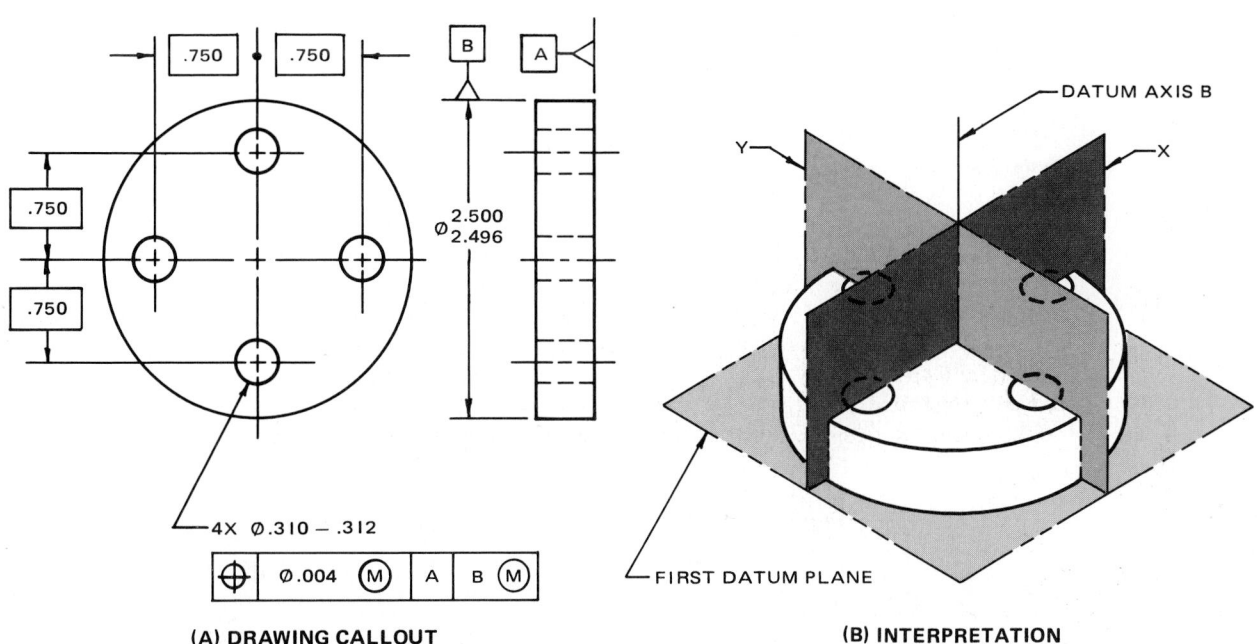

FIG. 16-7-1   Part with cylindrical datum feature.

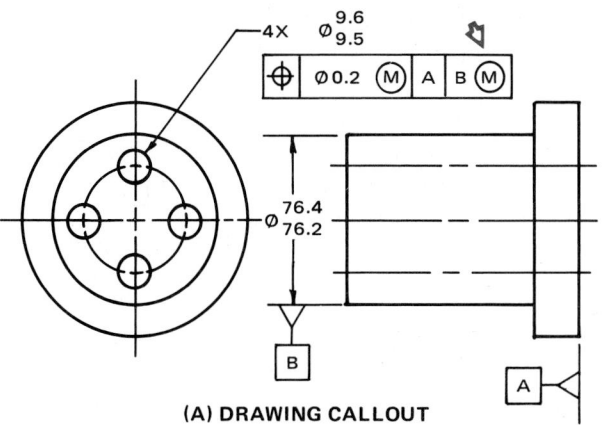

FIG. 16-7-2   Cylindrical feature as secondary datum.

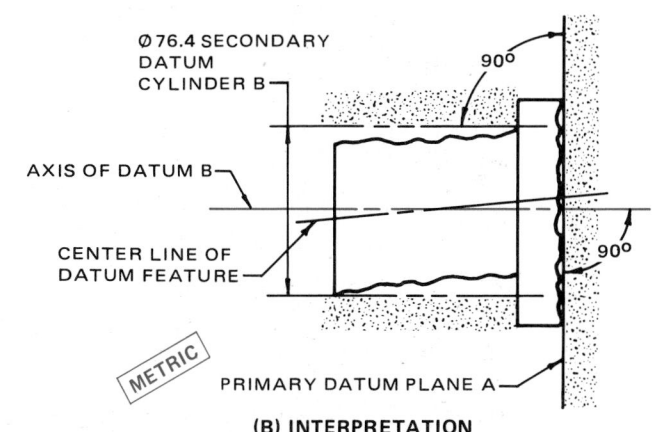

509

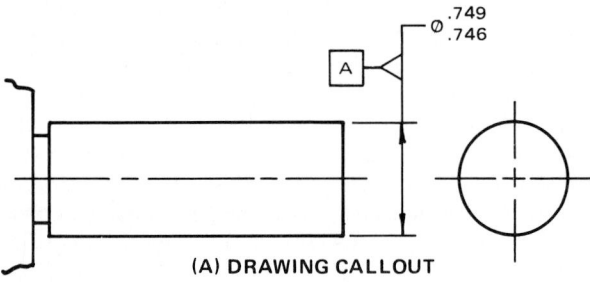

(A) DRAWING CALLOUT

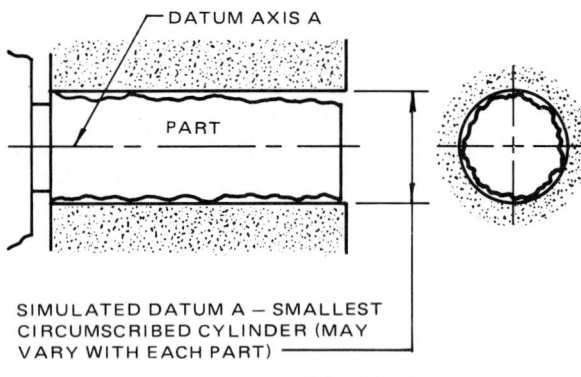

DATUM AXIS A

PART

SIMULATED DATUM A – SMALLEST
CIRCUMSCRIBED CYLINDER (MAY
VARY WITH EACH PART)

(B) INTERPRETATION

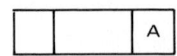

A

**(C) PRIMARY DATUM CALLOUT IN
FEATURE CONTROL FRAME**

FIG. 16-7-3   External primary datum cylinder—RFS.

letter must always be followed by the appropriate modifying symbol in the feature control frame. For all other geometric tolerances, RFS is implied unless otherwise specified.

Datum Features—RFS

Where a datum feature of size is applied on an RFS basis, the datum is established by physical contact between the feature surface, or surfaces, and surfaces of the measuring equipment.

Primary Datum Feature—Cylindrical   If an external feature, such as the shaft shown in Fig. 16-7-3, is specified as a primary datum feature, the datum is the axis of the smallest circumscribed cylinder that contacts the feature surface.

If an internal cylindrical feature, such as the hole shown in Fig. 16-7-4, is specified as a datum feature, the datum is the axis of the largest inscribed cylinder that contacts the feature surface.

Primary Datum Feature—Noncircular   The rules given for cylindrical features also apply to features of other shapes having a uniform cross section, i.e., those whose cross section has the form of a regular polygon. The simulated datum will always be the same shape as the datum feature.

Primary Datum Feature—Parallel Surfaces   The simulated datum features consist of two flat surfaces, such as two opposite faces of a rectangular part or two sides of a slot. For an internal feature, the datum is the center plane between two simulated parallel planes that, at maximum separation, contact the corresponding surfaces of the feature. For an external feature, the datum is the center plane between two simulated parallel planes that, at minimum separation, contact the corresponding surfaces of the feature (Fig. 16-7-5).

For both external and internal features, the secondary datum (axis or center plane) is established in the same manner as indicated above with an additional requirement: The contacting

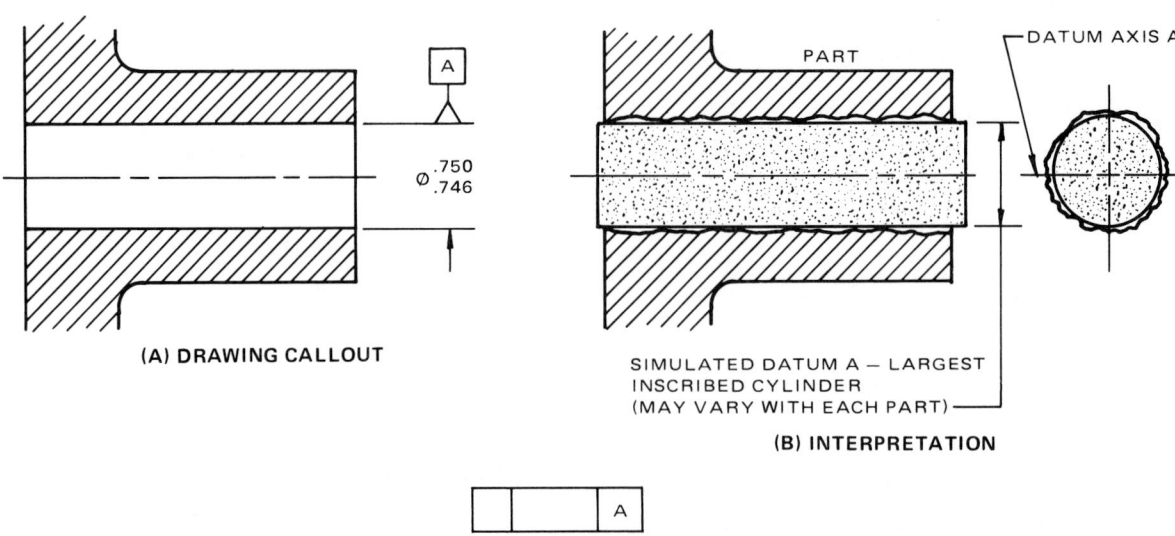

(A) DRAWING CALLOUT

PART

DATUM AXIS A

SIMULATED DATUM A – LARGEST
INSCRIBED CYLINDER
(MAY VARY WITH EACH PART)

(B) INTERPRETATION

A

(C) PRIMARY DATUM CALLOUT IN FEATURE CONTROL FRAME

FIG. 16-7-4   Internal primary datum cylinder—RFS.

cylinder or parallel planes must be oriented perpendicular to the primary datum.

Datum Features—MMC

Where a datum feature of size is applied on an MMC basis, machine and gaging elements in the measuring equipment, which remain constant in size, may be used to simulate a true geometric counterpart of the feature and to establish the datum.

In this case, the size of the simulated datum is established by the specified MMC limit of size of the datum feature or its virtual condition, where applicable. In Fig. 16-7-6, because no form tolerance is specified, the simulated datum is made to the specified MMC limit of size of .565 in.

Where a datum feature of size is controlled by a specified tolerance of form, the size of the simulated datum is the MMC limit of size. There is an exception: Where a straightness tolerance is applied on an RFS or MMC basis, the size of

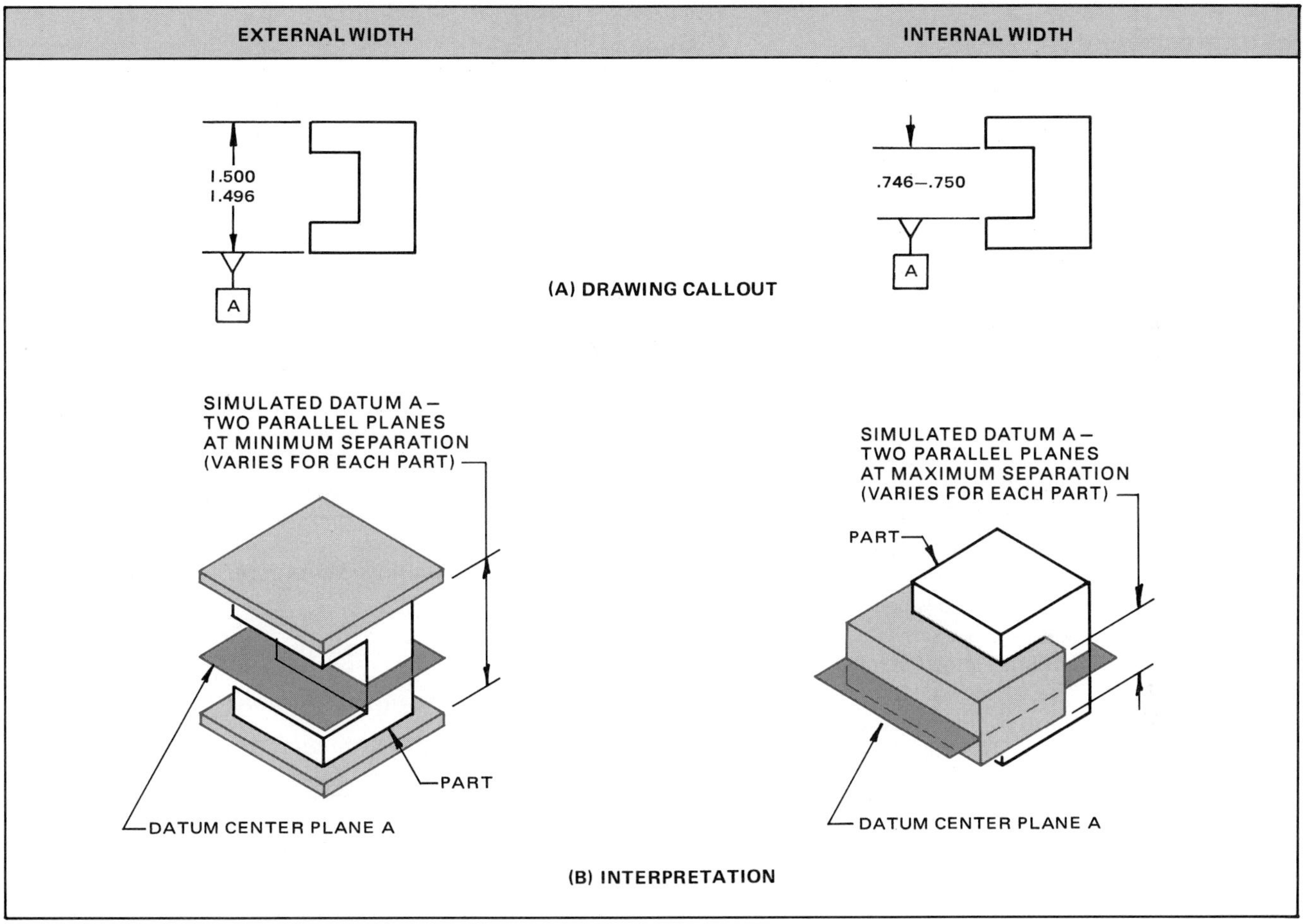

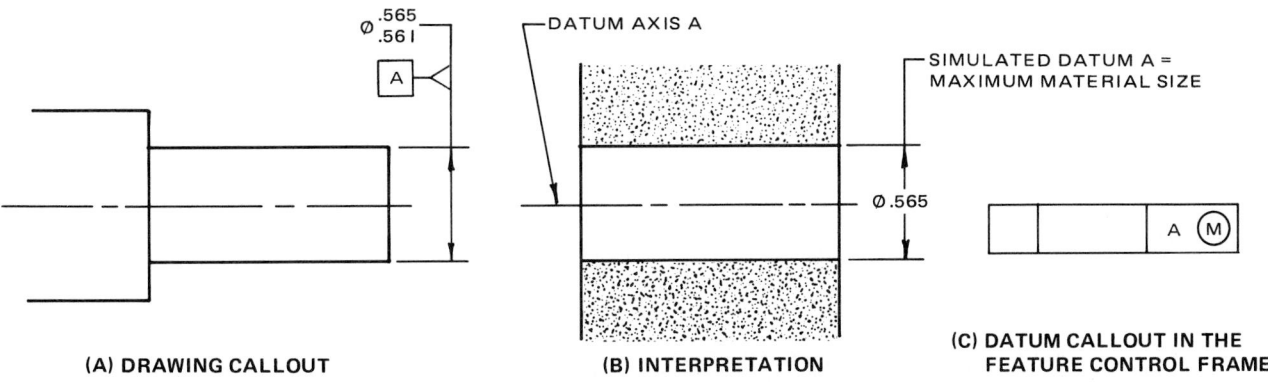

FIG. 16-7-5   Width as primary datum—RFS.

FIG. 16-7-6   External primary datum without form tolerances—MMC.

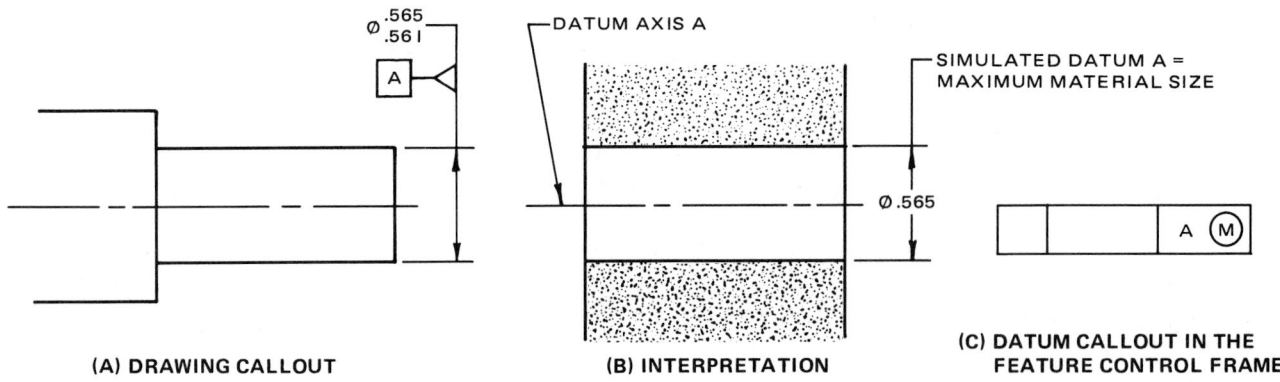

FIG. 16-7-7   External primary datum with straightness tolerance—MMC.

the simulated datum is the virtual condition of the datum feature (Figs. 16-7-7 and 16-7-8).

Figure 16-7-8 shows a flatness tolerance that applies to both datum surfaces. The simulated datum consists of two parallel planes separated by a distance equal to the maximum material condition, since the flatness tolerances must lie within the specified limits of size. It should be noted that the gage does not check the flatness requirement.

Where secondary or tertiary datum features of size in the same datum reference frame are controlled by a specified tolerance of location or orientation with respect to each other, the size of the simulated datum is the virtual condition of the datum feature. Figure 16-7-9 illustrates both secondary and tertiary datums specified at MMC but simulated at virtual condition. Virtual condition refers to the potential boundary of a feature, as specified on a drawing, derived from the collective effect of the maximum material limit of size and the specified form or orientation tolerance. These are added for external features, such as shafts, and subtracted for internal features, such as holes and slots.

Where design requirements disallow a virtual condition, or if no form or orientation tolerance is specified, it is assumed, for datum reference purposes, that the tolerance is zero at MMC.

The fact that a datum applies on an MMC basis is given in the feature control frame by the addition of the MMC symbol Ⓜ immediately following the datum reference (Fig. 16-7-9C). When there is more than one datum reference, the MMC symbol must be added for each datum where the modification is required.

REFERENCES AND SOURCE MATERIAL

1. ASME Y14.5M–1994, *Dimensioning and Tolerancing*.
2. CAN/CSA B78.2-M91, *Dimensioning and Tolerancing of Technical Drawings*.
3. ISO drawing standards.

ASSIGNMENT

See Assignment 19 for Unit 16-7 on page 576.

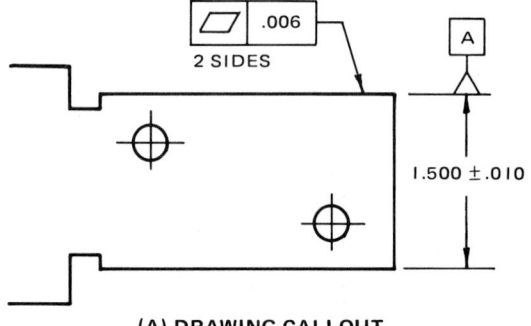

(A) DRAWING CALLOUT

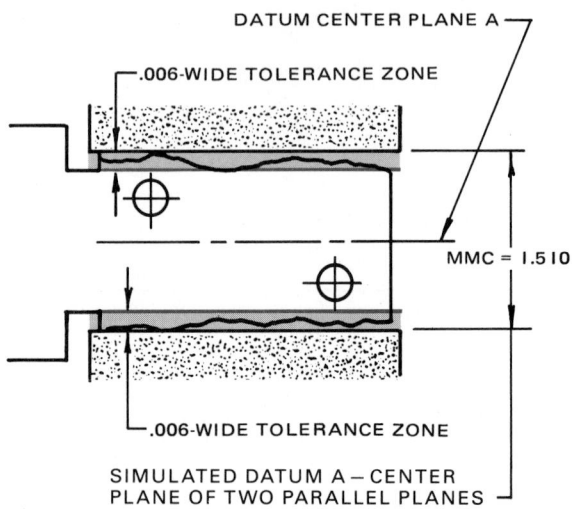

NOTE: FLATNESS TOLERANCE MUST BE CHECKED SEPARATELY AND MUST LIE WITHIN THE LIMITS OF SIZE.

(B) INTERPRETATION

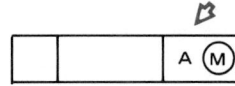

(C) DATUM CALLOUT IN THE FEATURE CONTROL FRAME

FIG. 16-7-8   External primary datum—MMC.

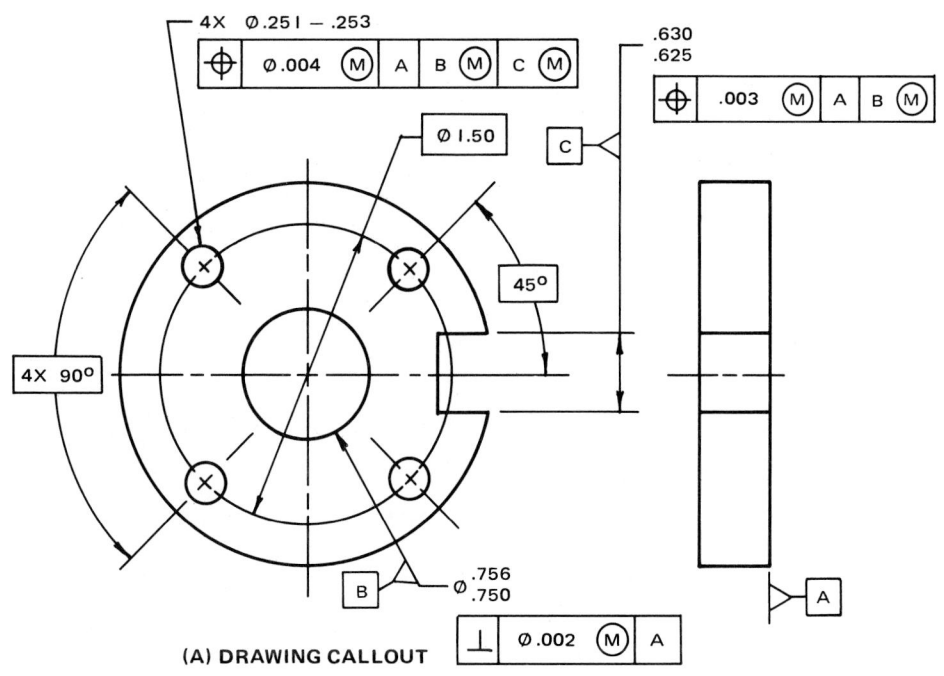

(A) DRAWING CALLOUT

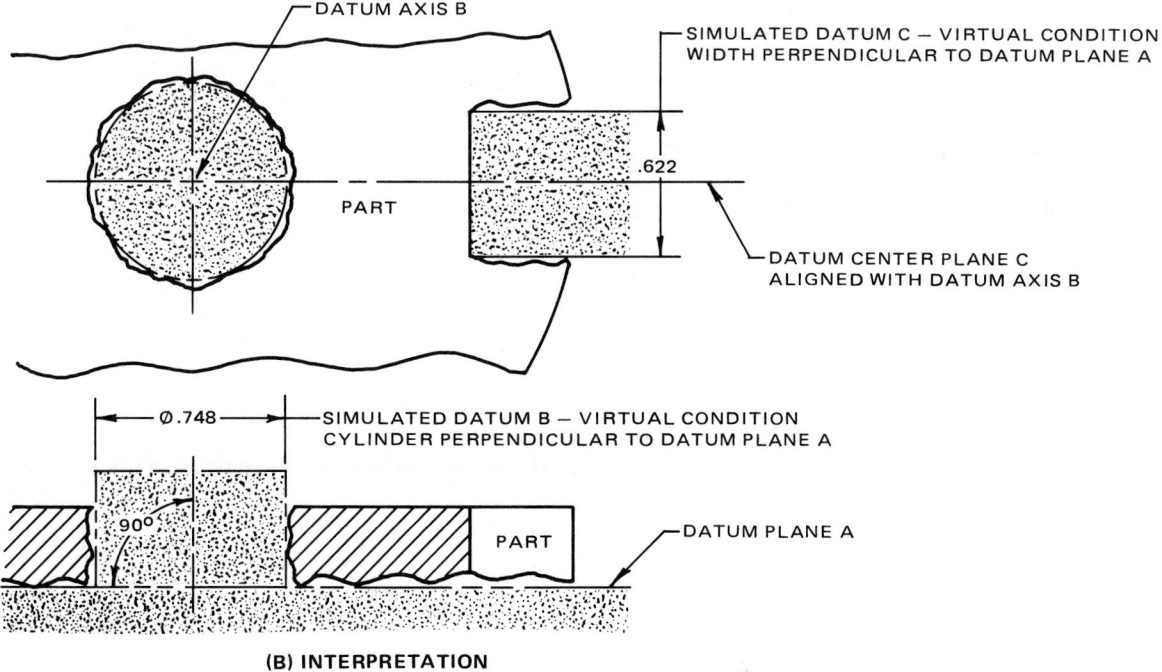

(B) INTERPRETATION

(C) SECONDARY AND TERTIARY DATUM CALLOUTS IN FEATURE CONTROL FRAME

FIG. 16-7-9   Secondary and tertiary datum features—MMC.

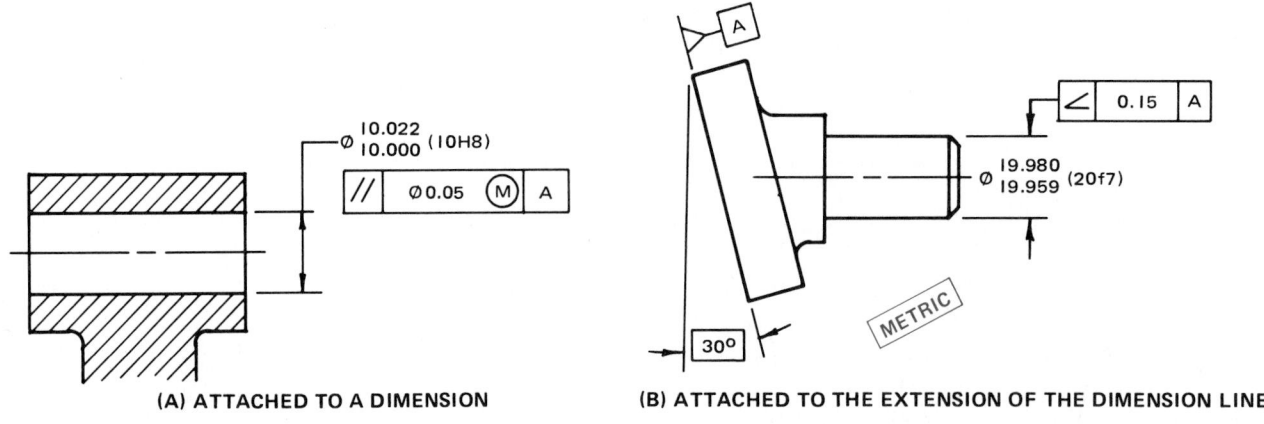

(A) ATTACHED TO A DIMENSION

(B) ATTACHED TO THE EXTENSION OF THE DIMENSION LINE

FIG. 16-8-1   Feature control frame associated with size dimension.

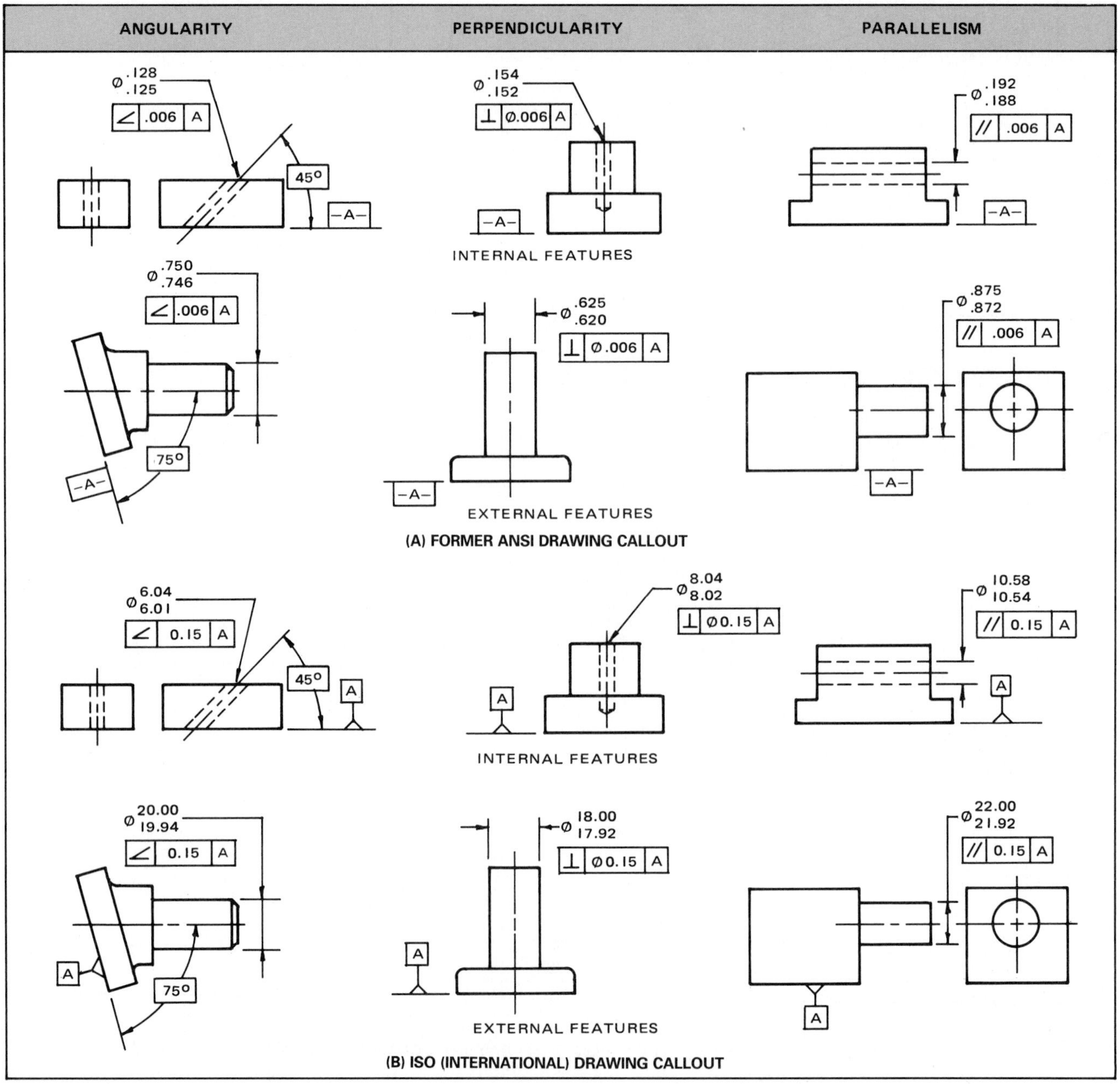

(A) FORMER ANSI DRAWING CALLOUT

(B) ISO (INTERNATIONAL) DRAWING CALLOUT

FIG. 16-8-2   Orientation tolerances for cylindrical features—RFS.

16-8 ORIENTATION TOLERANCING FOR FEATURES OF SIZE

Unit 16-6 outlined how to apply orientation tolerances to flat surfaces. In these instances the feature control frames were directed to the surfaces requiring orientation. When orientation tolerances apply to the axis of cylindrical features or to the datum planes of two flat surfaces, the feature control frame is associated with the size dimension of the feature requiring control (Fig. 16-8-1).

Tolerances intended to control orientation of the axis of a feature are applied to drawings as shown in Fig. 16-8-2. Although this unit deals mostly with cylindrical features, methods similar to those given here can be applied to noncircular features, such as square and hexagonal shapes.

The axis of the cylindrical feature must be contained within a tolerance zone consisting of the space between two parallel planes separated by the specified tolerance. The parallel planes are related to the datum by the basic angles of 45°, 90°, or 0° in Fig. 16-8-2.

The absence of a modifying symbol in the tolerance compartment of the feature control frame indicates that RFS applies.

Angularity Tolerance

The tolerance zone is defined by two parallel planes at the specified basic angle from a datum plane or axis within which the axis of the considered feature must lie. Figure 16-8-3 illustrates the tolerance zone for angularity.

Parallelism Tolerance

Parallelism is the condition of a surface equidistant at all points from a datum plane or an axis equidistant along its length from a datum axis or plane. A parallelism tolerance specifies a tolerance zone defined by two planes or lines parallel to a datum plane or axis within which the axis of the considered feature must lie (Fig. 16-8-4), or a cylindrical tolerance zone, the axis of which is parallel to the datum axis within which the axis of the considered feature must lie (see Fig. 16-8-14, pg. 519).

Perpendicularity Tolerance

A perpendicularity tolerance specifies one of the following:

1. A cylindrical tolerance zone perpendicular to a datum plane or axis within which the center line of the considered feature must lie (Fig. 16-8-2).
2. A tolerance zone defined by two parallel planes perpendicular to a datum axis within which the axis of the considered feature must lie (Fig. 16-8-14, pg. 519).

When the tolerance is one of perpendicularity, the tolerance zone planes can be revolved around the feature axis without

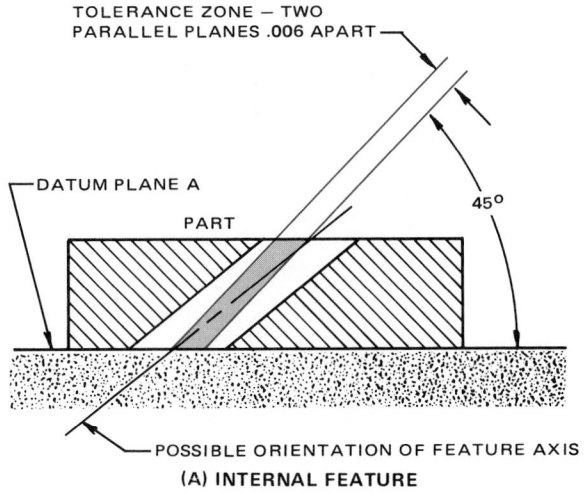

(A) INTERNAL FEATURE

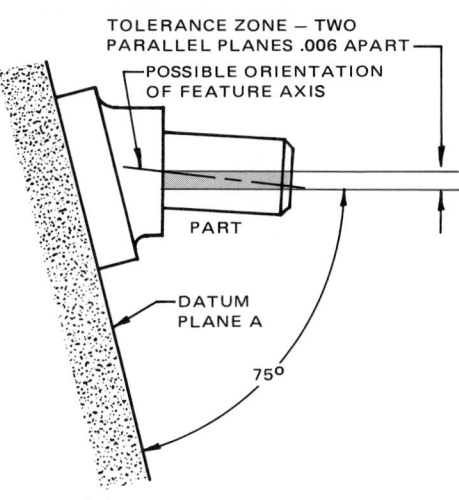

(B) EXTERNAL FEATURE

FIG. 16-8-3   Tolerance zones for angularity shown in Fig. 16-8-2.

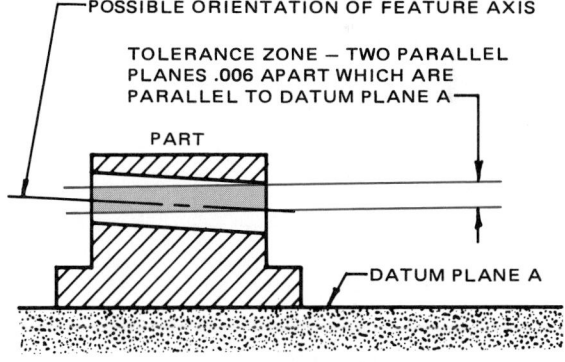

(A) INTERNAL FEATURE

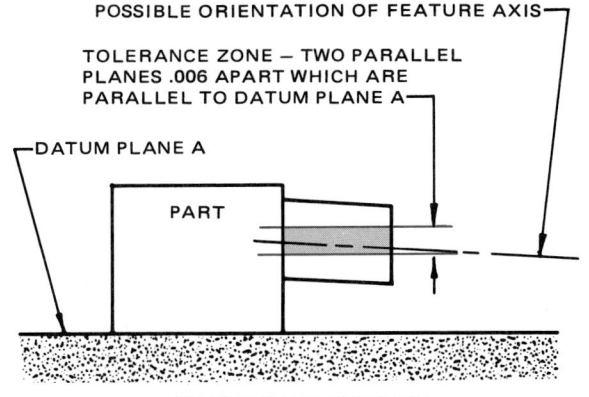

(B) EXTERNAL FEATURE

FIG. 16-8-4   Tolerance zones for parallelism shown in Fig. 16-8-2.

515

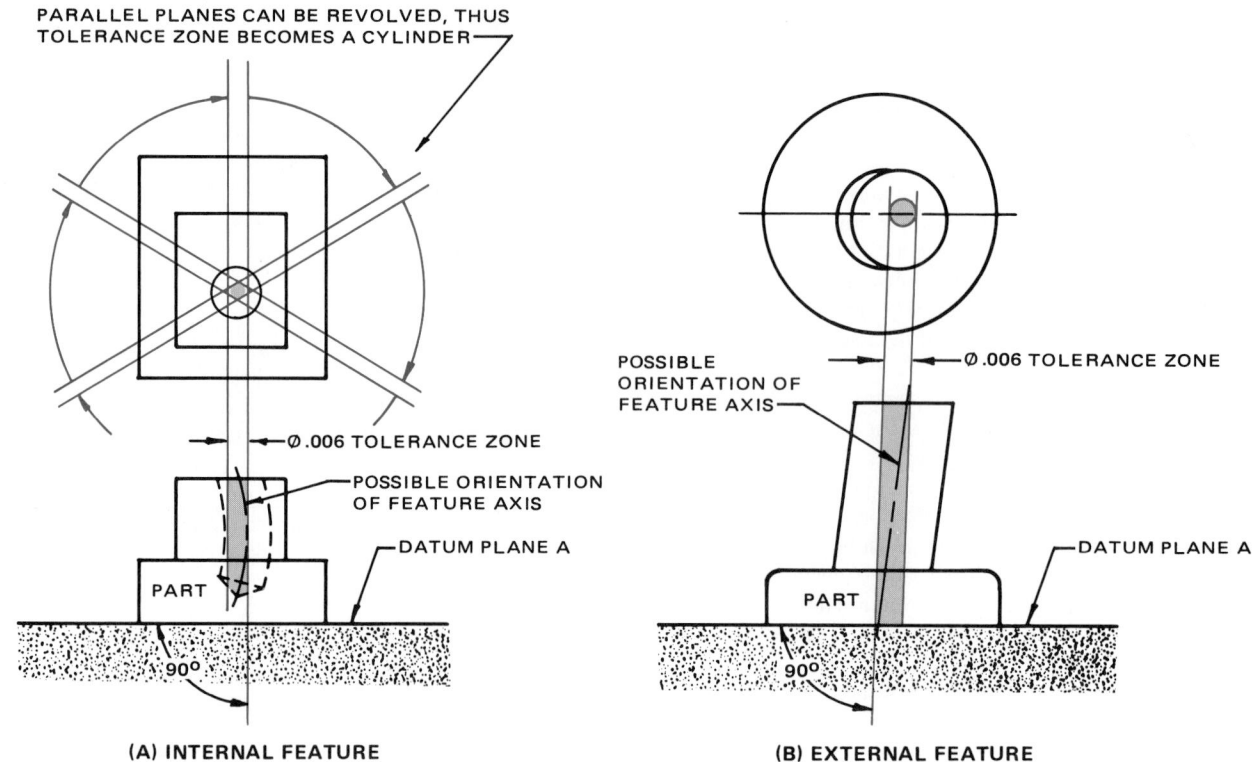

(A) INTERNAL FEATURE

(B) EXTERNAL FEATURE

FIG. 16-8-5 Tolerance zones for perpendicularity shown in Fig. 16-8-2.

FIG. 16-8-6 Orientation tolerances referenced to two datums.

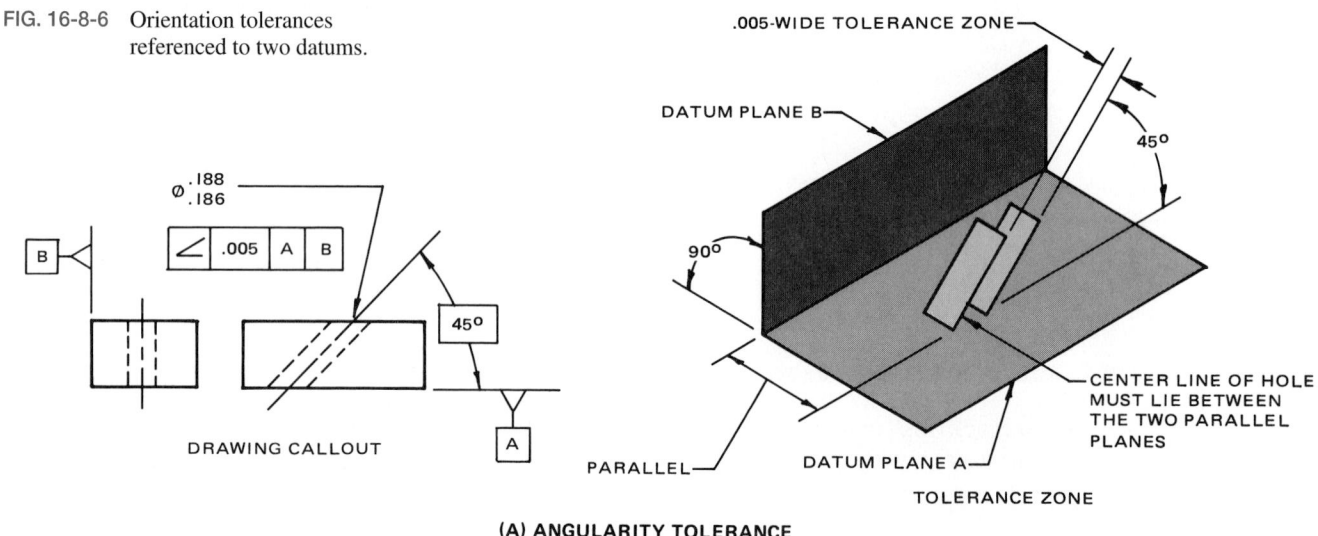

(A) ANGULARITY TOLERANCE

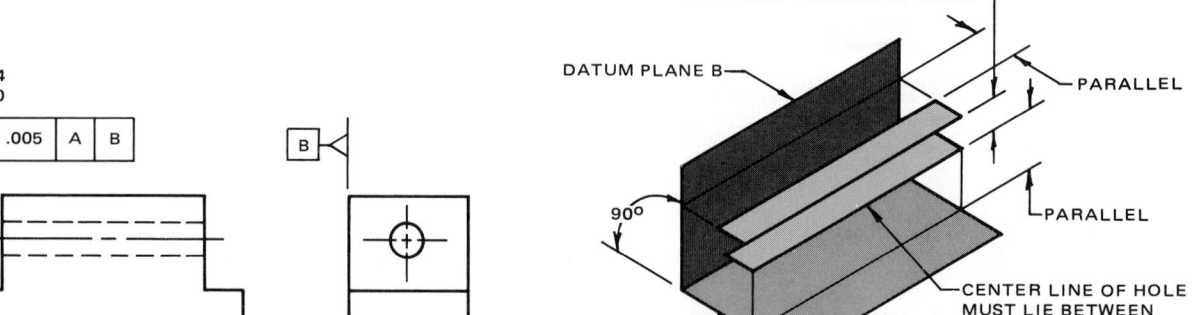

(B) PARALLELISM TOLERANCE

affecting the angle. The tolerance zone therefore becomes a cylinder. This cylindrical zone is perpendicular to the datum and has a diameter equal to the specified tolerance (Fig. 16-8-5). A diameter symbol precedes the perpendicularity tolerance.

Control in Two Directions

The feature control frames shown in Fig. 16-8-2 control angularity and parallelism with the base (datum *A*) only. If control with a side is also required, the side should be designated as the secondary datum (Fig. 16-8-6). The center line of the hole must lie within the two parallel planes.

Control on an MMC Basis

EXAMPLE 1

As a hole is a feature of size, any of the tolerances shown in Fig. 16-8-2 can be modified on an MMC basis. This is specified by adding the symbol Ⓜ after the tolerance. Figure 16-8-7 shows an example.

EXAMPLES 2 AND 3

Because the cylindrical features represent features of size, orientation tolerances may be applied on an MMC basis. This is specified by adding the modifying symbol after the tolerance, as shown in Figs. 16-8-8 and 16-8-9.

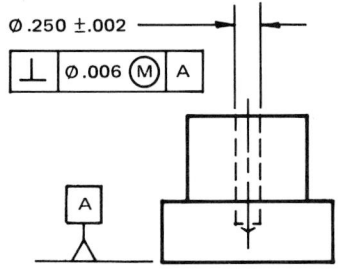

FIG. 16-8-7   Perpendicularity tolerance for a hole on an MMC basis.

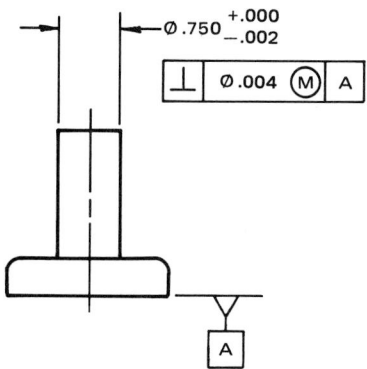

FIG. 16-8-8   Perpendicularity tolerance for a shaft on an MMC basis.

Internal Cylindrical Features

Figure 16-8-2 shows some simple parts in which the axis or center line of a hole is related by an orientation tolerance to a flat surface. The flat surface is designated as the datum feature.

The axis of each hole must be contained within a tolerance zone consisting of the space between two parallel planes. These planes are separated by a specified tolerance of .006 in. for the parts shown in Fig. 16-8-2A, and by a specified tolerance of 0.15 mm for the parts shown in Fig. 16-8-2B.

Specifying Parallelism for an Axis   Regardless of feature size, the feature axis shown in Fig. 16-8-10 must lie between

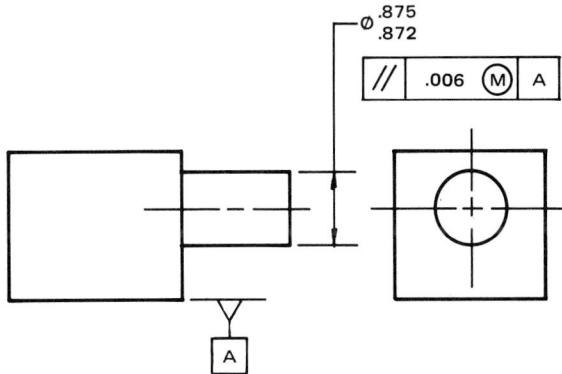

FIG. 16-8-9   Parallelism tolerance for a shaft on an MMC basis.

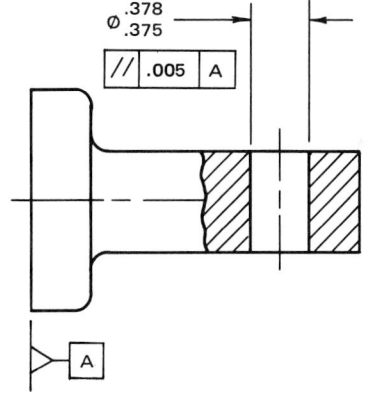

(A) DRAWING CALLOUT

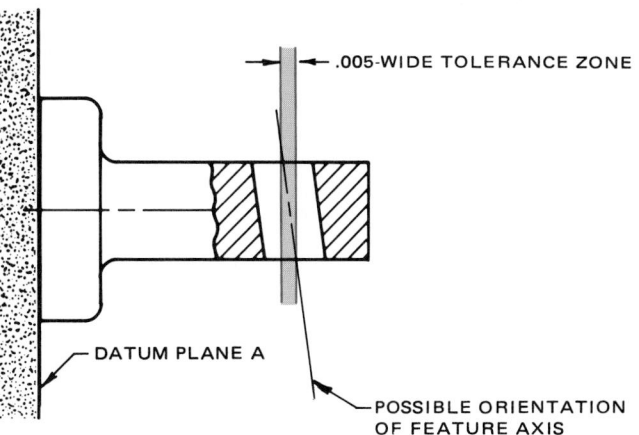

(B) INTERPRETATION

FIG. 16-8-10   Specifying parallelism for an axis (feature RFS).

517

two parallel planes, .005 in. apart, that are parallel to datum plane *A*. Additionally, the feature axis must be within any specified tolerance of location.

Figure 16-8-11 specifies parallelism for an axis when both the feature and the datum feature are shown on an RFS basis. Regardless of feature size, the feature axis must lie within a

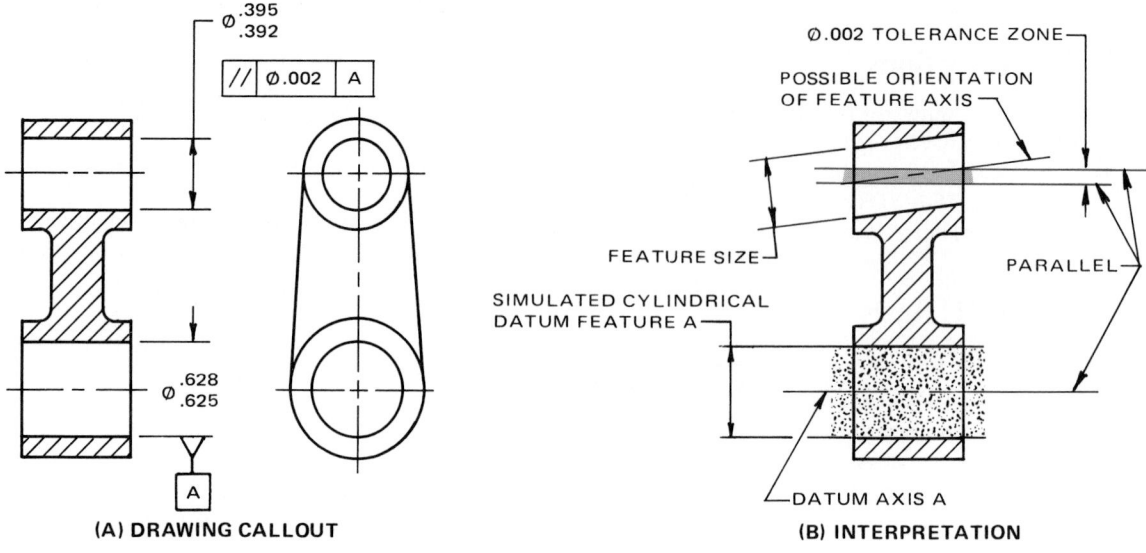

FIG. 16-8-11   Specifying parallelism for an axis (both feature and datum feature RFS).

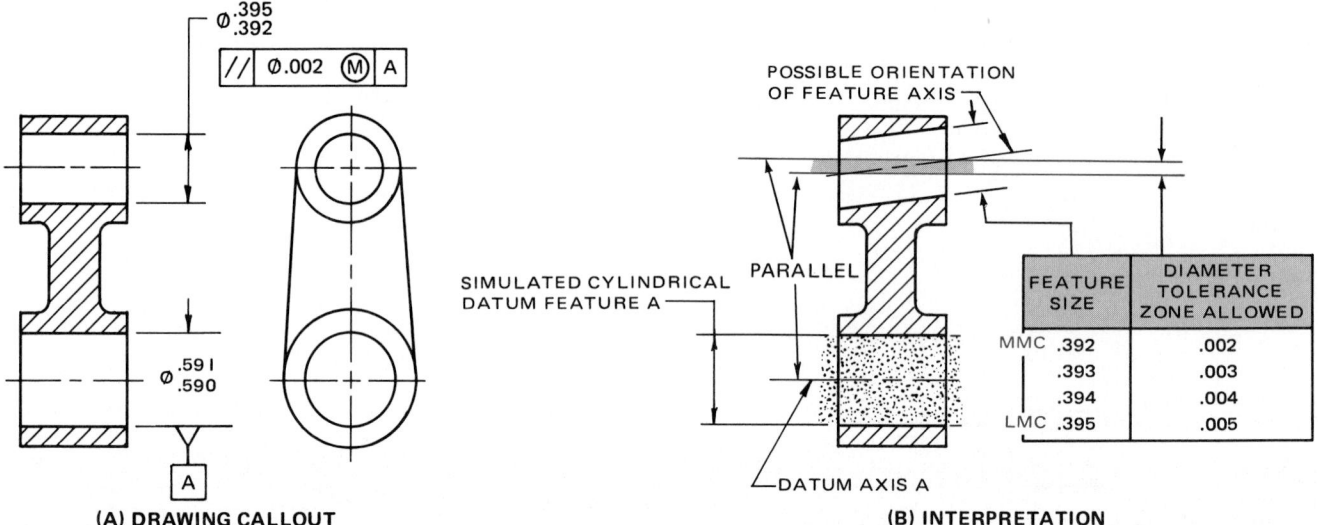

FIG. 16-8-12   Specifying parallelism for an axis (feature at MMC and datum feature RFS).

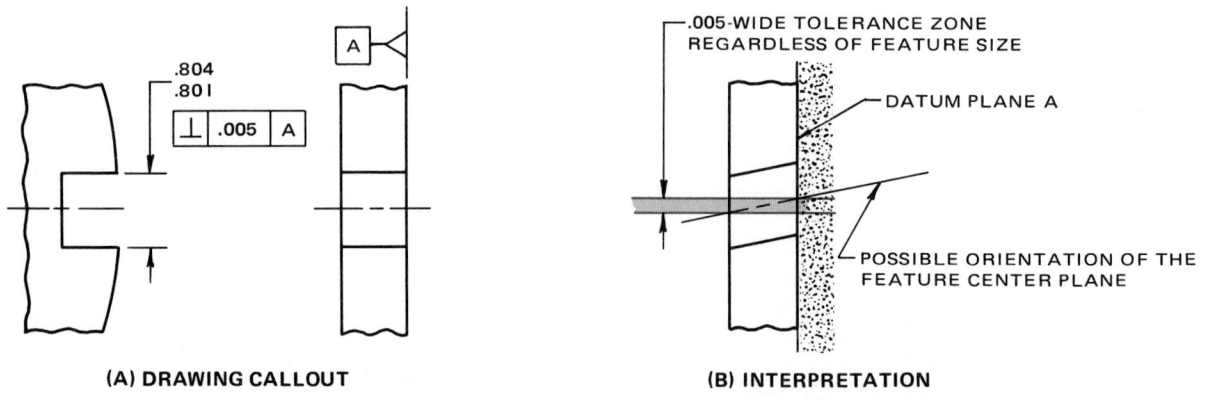

FIG. 16-8-13   Specifying perpendicularity for a median plane (feature RFS).

cylindrical tolerance zone of .002 in. diameter whose axis is parallel to datum axis *A*. Additionally, the feature axis must be within any specified tolerance of location.

Figure 16-8-12 specifies parallelism for an axis when the feature is shown on an MMC basis and the datum feature is shown on an RFS basis. Where the feature is at the maximum material condition (.392 in.), the maximum parallelism tolerance is .002 in. diameter. Where the feature departs from its MMC size, an increase in the parallelism tolerance is allowed equal to the amount of such departure. Additionally, the feature axis must be within any specified tolerance of location.

Perpendicularity for a Median Plane   Regardless of feature size, the center plane of the feature shown in Fig. 16-8-13 must

lie between two parallel planes, .005 in. apart, that are perpendicular to datum plane *A*. Additionally, the feature center plane must be within any specified tolerance of location.

Perpendicularity for an Axis (Both Feature and Datum RFS)   Regardless of feature size, the feature axis shown in Fig. 16-8-14 must lie between two parallel planes, .005 in. apart, that are perpendicular to datum axis *A*. Additionally, the feature axis must be within any specified tolerance of location.

Perpendicularity for an Axis (Tolerance at MMC)   Where the feature shown in Fig. 16-8-15 is at the MMC (Ø 2.000), its axis must be perpendicular within .002 in. to datum plane *A*. Where the feature departs from MMC, an increase in the

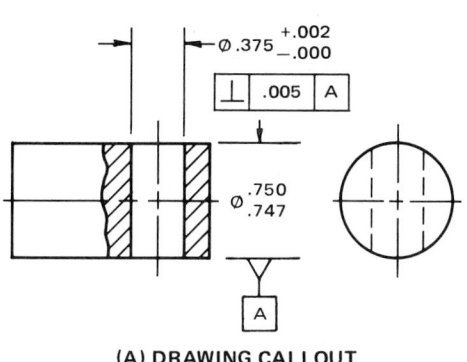

(A) DRAWING CALLOUT

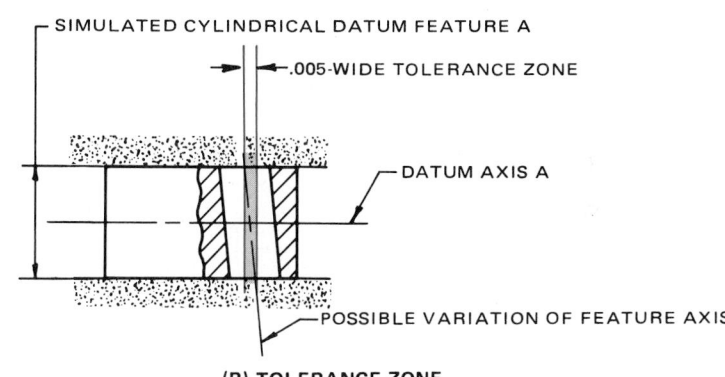

(B) TOLERANCE ZONE

FIG. 16-8-14   Specifying perpendicularity for an axis (both feature and datum feature RFS).

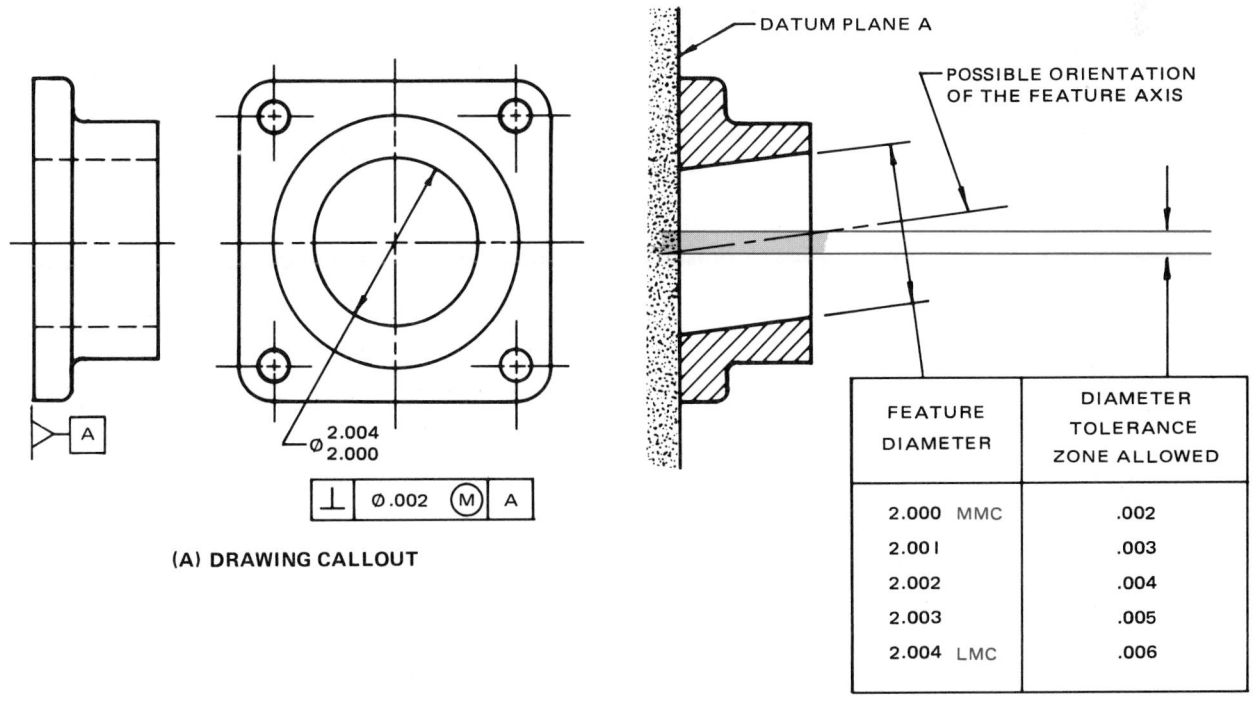

(A) DRAWING CALLOUT

FEATURE DIAMETER		DIAMETER TOLERANCE ZONE ALLOWED
2.000	MMC	.002
2.001		.003
2.002		.004
2.003		.005
2.004	LMC	.006

(B) TOLERANCE ZONE

FIG. 16-8-15   Specifying perpendicularity for an axis (tolerance at MMC).

perpendicularity tolerance is allowed equal to the amount of such departure. Additionally, the feature axis must be within the specified tolerance of location.

Perpendicularity for an Axis (Zero Tolerance at MMC)
Where the feature shown in Fig. 16-8-16 is at the MMC (Ø50.00), its axis must be perpendicular to datum plane *A*. Where the feature departs from MMC, an increase in the perpendicularity tolerance is allowed equal to the amount of such departure. Additionally, the feature axis must be within any specified tolerance of location.

Perpendicularity with a Maximum Tolerance Specified
Where the feature shown in Fig. 16-8-17 is at MMC (Ø50.00), its axis must be perpendicular to datum plane *A*. Where the feature departs from MMC, an increase in the perpendicularity tolerance is allowed equal to the amount of such departure, up to

the 0.1 mm maximum. Additionally, the feature axis must be within any specified tolerance of location.

External Cylindrical Features

Perpendicularity for an Axis (Pin or Boss RFS)   Regardless of feature size, the feature axis shown in Fig. 16-8-18 must lie within a cylindrical zone (Ø 0.4 mm) that is perpendicular to, and projects from, datum plane *A* for the feature height. Additionally, the feature axis must be within any specified tolerance of location.

Perpendicularity for an Axis (Pin or Boss at MMC)   Where the feature shown in Fig. 16-8-19 is at MMC (Ø 15.984 mm), the maximum perpendicularity tolerance is Ø 0.05 mm. Where the feature departs from its MMC size, an increase in the

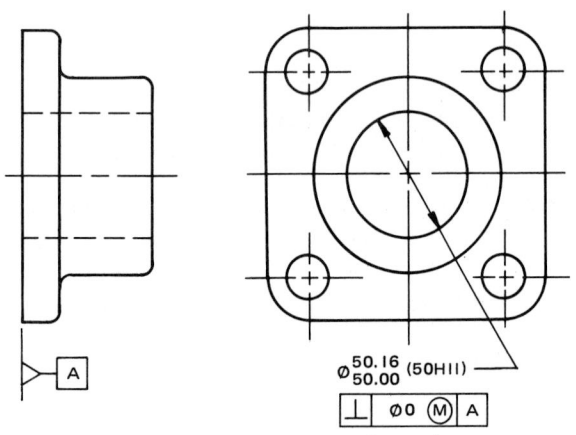

(A) DRAWING CALLOUT

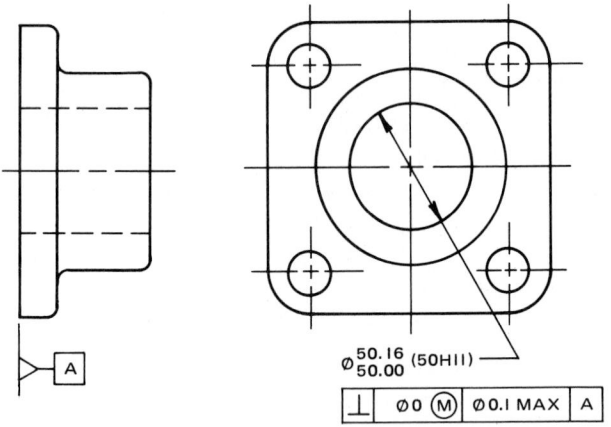

(A) DRAWING CALLOUT

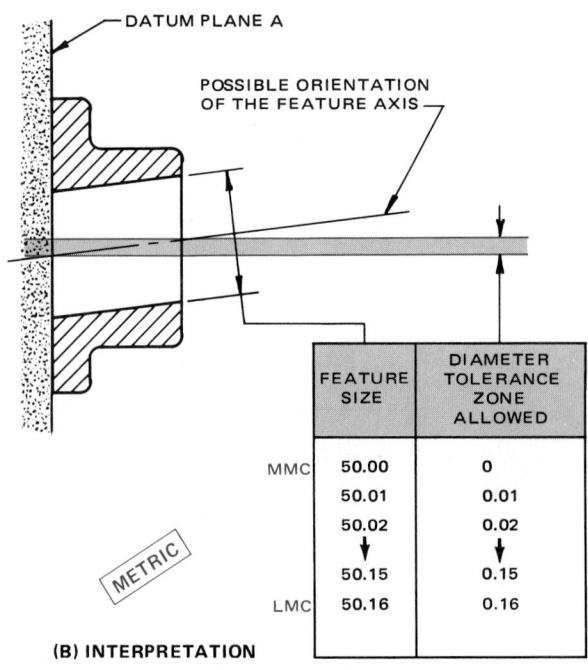

FIG. 16-8-16   Specifying perpendicularity for an axis (zero tolerance at MMC).

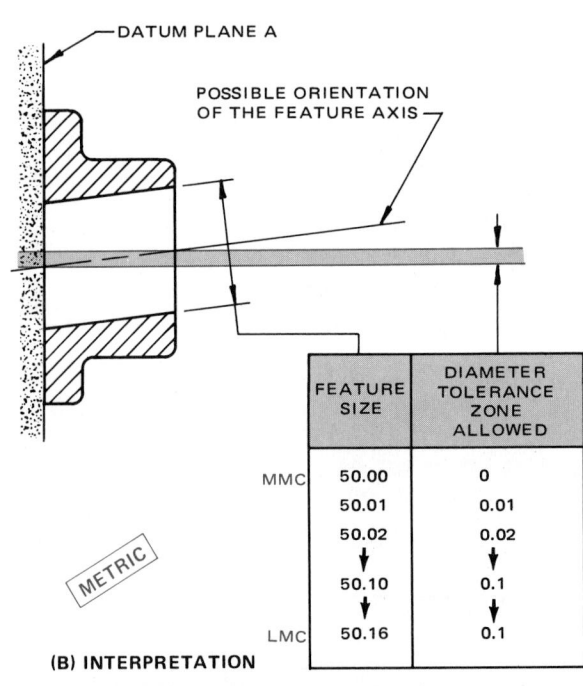

FIG. 16-8-17   Specifying perpendicularity for an axis (zero tolerance at MMC with a maximum specified).

perpendicularity tolerance is allowed equal to the amount of such departure. Additionally, the feature axis must be within any specified tolerance of location.

REFERENCES AND SOURCE MATERIAL

1. ASME Y14.5M–1994, *Dimensioning and Tolerancing.*
2. CAN/CSA B78.2-M91, *Dimensioning and Tolerancing of Technical Drawings.*
3. ISO drawing standards.

ASSIGNMENTS

See Assignments 20 through 22 for Unit 16-8 on pages 576 to 577.

16-9 POSITIONAL TOLERANCING

The location of features is one of the most frequently used applications of dimensions on technical drawings. Tolerancing may be accomplished either by coordinate tolerances applied to the dimensions or by geometric (positional) tolerancing.

Positional tolerancing is especially useful when applied on an MMC basis to groups or patterns of holes or other small features in the mass production of parts. This method meets functional requirements in most cases and permits assessment with simple gaging procedures.

Most examples in this unit are devoted to the principles involved in the location of small, round holes because they represent the most commonly used applications. The same principles apply, however, to the location of other features, such as slots, tabs, bosses, and noncircular holes.

Tolerancing Methods

A single hole is usually located by means of rectangular coordinate dimensions, extending from suitable edges or other features of the part to the axis of the hole. Other dimensioning methods, such as polar coordinates, may be used when circumstances warrant.

There are two standard methods of tolerancing the location of holes, as illustrated in Fig. 16-9-1 (pg. 522).

1. *Coordinate tolerancing* refers to tolerances applied directly to the coordinate dimensions or to applicable tolerances specified in a general tolerance note.
2. *Positional tolerancing* refers to a tolerance zone within which the center line of the hole or shaft is permitted to vary from its true position. Positional tolerancing can be further classified according to the type of modifying symbol associated with the tolerance. These are:

   - Positional tolerancing, regardless of feature size (RFS)
   - Positional tolerancing, maximum material condition basis (MMC)
   - Positional tolerancing, least material condition basis (LMC)

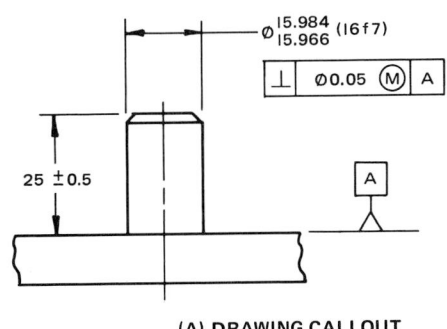

(A) DRAWING CALLOUT

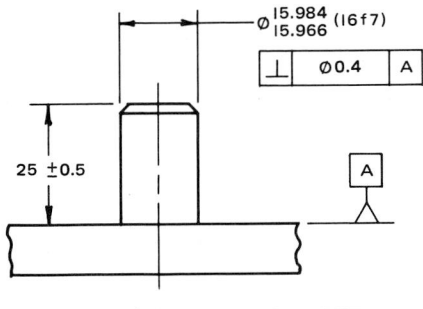

(A) DRAWING CALLOUT

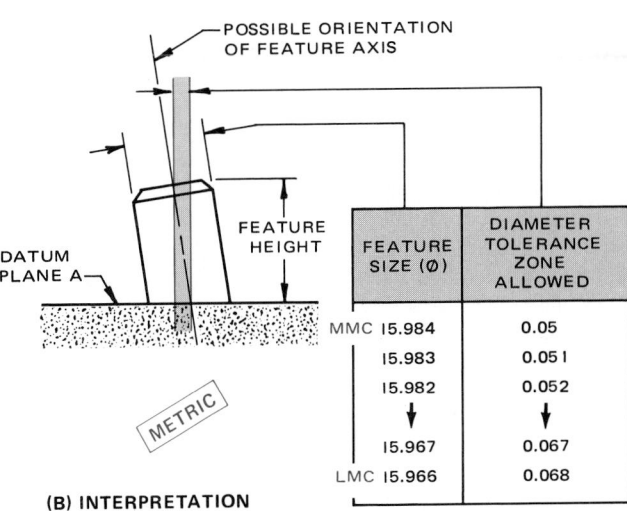

FEATURE SIZE (Ø)	DIAMETER TOLERANCE ZONE ALLOWED
MMC 15.984	0.05
15.983	0.051
15.982	0.052
↓	↓
15.967	0.067
LMC 15.966	0.068

(B) INTERPRETATION

FIG. 16-8-19   Specifying perpendicularity for an axis (pin or boss at MMC).

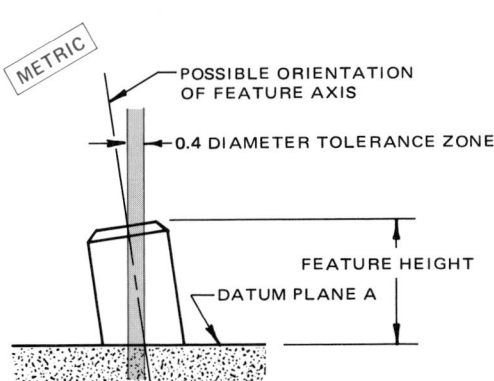

(B) INTERPRETATION

FIG. 16-8-18   Specifying perpendicularity for an axis (pin or boss RFS).

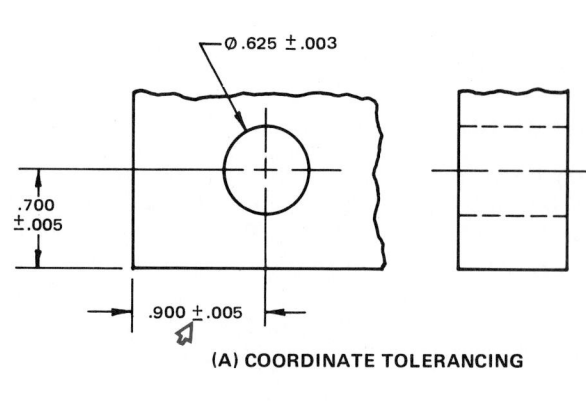

(A) COORDINATE TOLERANCING

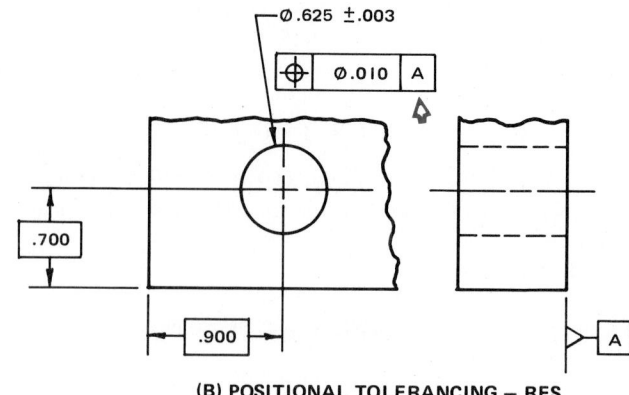

(B) POSITIONAL TOLERANCING – RFS

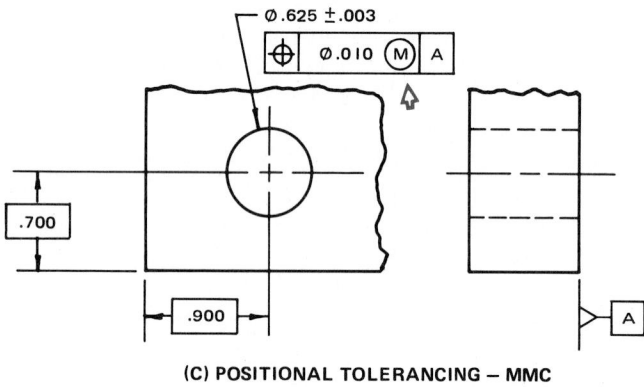

(C) POSITIONAL TOLERANCING – MMC

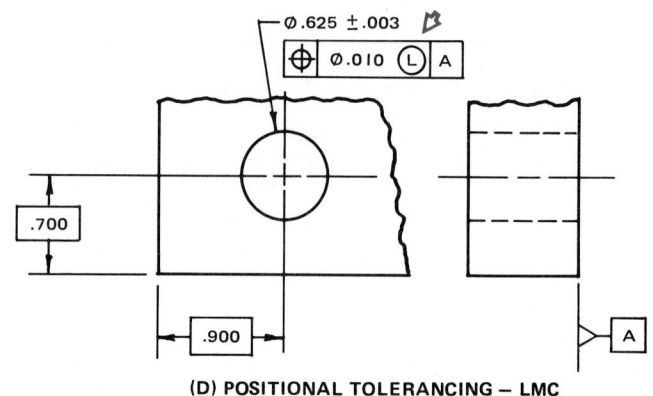

(D) POSITIONAL TOLERANCING – LMC

FIG. 16-9-1   Comparison of tolerancing methods.

These positional tolerancing methods are part of the system of geometric tolerancing.

Any of these tolerancing methods can be substituted one for the other, although with differing results. It is necessary, however, to first analyze the widely used method of coordinate tolerancing in order to explain and understand the advantages and disadvantages of the positional tolerancing methods.

Coordinate Tolerancing

Coordinate dimensions and tolerances may be applied to the location of a single hole, as shown in Fig. 16-9-2. They locate the hole axis and result in either a rectangular or a wedge-shaped tolerance zone within which the axis of the hole must lie.

If the two coordinate tolerances are equal, the tolerance zone formed will be a square. Unequal tolerances result in a rectangular tolerance zone. Where one of the locating dimensions is a radius, polar dimensioning gives a circular ring section tolerance zone. For simplicity, square tolerance zones are used in the analysis of most of the examples in this section.

It should be noted that the tolerance zone extends for the full depth of the hole, that is, the whole length of the axis. This is illustrated in Fig. 16-9-3 and explained in more detail in a later unit. In most of the illustrations, tolerances will be analyzed as they apply at the surface of the part, where the axis is represented by a point.

Maximum Permissible Error

The actual position of the feature axis may be anywhere within the tolerance zone. For square tolerance zones, the maximum allowable variation from the desired position occurs in a direction of 45° from the direction of the coordinate dimensions (Fig. 16-9-4).

For rectangular tolerance zones this maximum tolerance is the square root of the sum of the squares of the individual tolerances, or expressed mathematically.

$$\sqrt{X^2 + Y^2}$$

For the examples shown in Fig. 16-9-2, the tolerance zones are shown in Fig. 16-9-5, and the maximum tolerance values are as shown in the following examples.

EXAMPLE A

$$\sqrt{.010^2 + .010^2} = .014$$

EXAMPLE B

$$\sqrt{.010^2 + .020^2} = .0224$$

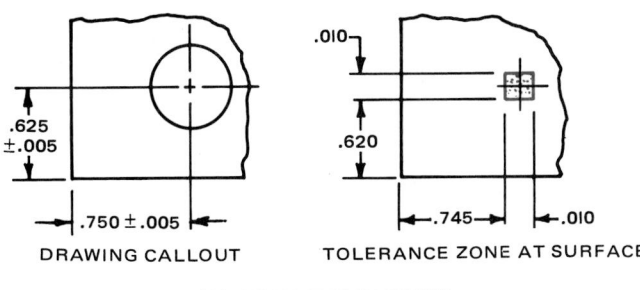

(A) EQUAL TOLERANCES

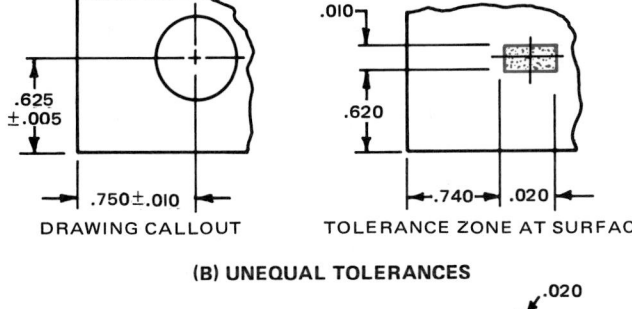

(B) UNEQUAL TOLERANCES

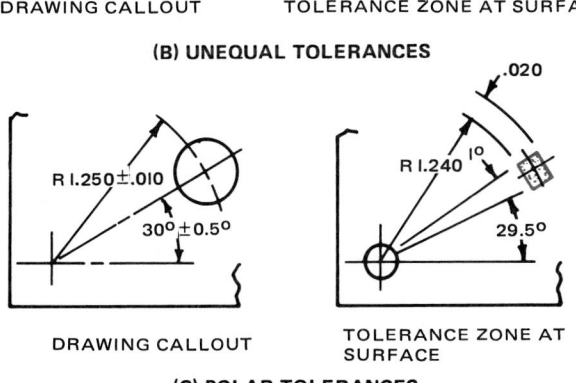

(C) POLAR TOLERANCES

FIG. 16-9-2   Tolerance zones for coordinate tolerancing.

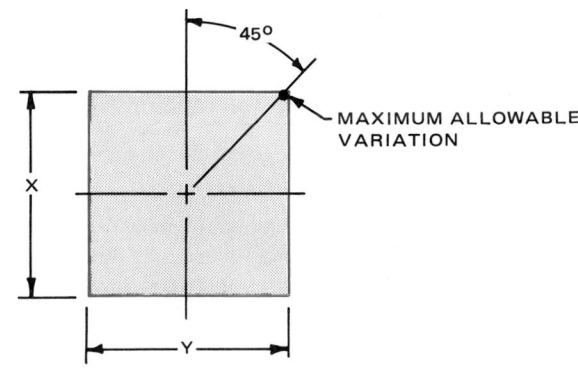

FIG. 16-9-3   Tolerance zone extending through part.

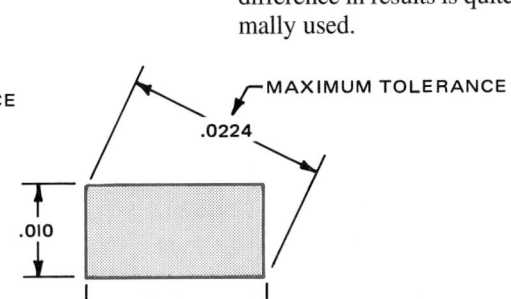

FIG. 16-9-4   Square tolerance zone.

EXAMPLE C

For polar coordinates the extreme variation is

$$\sqrt{A^2 + T^2}$$

where $A = R \tan a$
    T = tolerance on radius

R = mean radius
a = angular tolerance
Thus, the extreme variation in example C is

$$\sqrt{(1.25 \times .017\ 45)^2 + .020^2} = .03$$

Note: Mathematically, A in the above formula should be $2R$ $\tan a/2$, instead of $R \tan a$, and T should be $T \cos A/2$, but the difference in results is quite insignificant for the tolerances normally used.

EXAMPLE A

$$\sqrt{.010^2 + .010^2} = .014$$

EXAMPLE B

$$\sqrt{.010^2 + .020^2} = .0224$$

EXAMPLE C

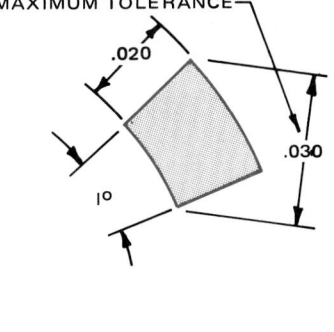

FIG. 16-9-5   Tolerance zones for parts shown in Fig. 16-9-2.

523

Use of Charts

A quick and easy method of finding the maximum positional error permitted with coordinate tolerancing, without having to calculate squares and square roots, is by use of a chart like that shown in Fig. 16-9-6.

In the example shown in Fig. 16-9-2A, the tolerance in both directions is .010 in. The extensions of the horizontal and vertical lines of .010 in the chart intersect at point A, which lies between the radii of .013 and .014 in. When interpolated and rounded to three dimensional places, the maximum permissible variation of position is .014 in.

In the example shown in Fig. 16-9-2B, the tolerances are .010 in. in one direction and .020 in. in the other. The extensions of the vertical and horizontal lines at .010 and .020 in., respectively, in the chart intersect at point B, which lies between the radii of .022 and .023 in. When interpolated and rounded to three decimal places, the maximum variation of position is .022 in. Figure 19-9-6 also shows a chart for use with tolerances in millimeters.

Disadvantages of Coordinate Tolerancing

There are a number of disadvantages of the direct tolerancing method:

1. It results in a square or rectangular tolerance zone within which the axis must lie. For a square zone this permits a variation in a 45° direction of approximately 1.4 times the specified tolerance. This amount of variation may necessitate the specification of tolerances that are only 70 percent of those that are functionally acceptable.
2. It may result in an undesirable accumulation of tolerances when several features are involved, especially when chain dimensioning is used.

3. It is more difficult to assess clearances between mating features and components than when positional tolerancing is used, especially when a group or a pattern of features is involved.
4. It does not correspond to the control exercised by fixed functional "go" gages, often desirable in mass production of parts. This becomes particularly important in dealing with a group of holes. With direct coordinate tolerancing, the location of each hole has to be measured separately in two directions, whereas with positional tolerancing on an MMC basis, one functional gage checks all holes in one operation.

Advantages of Coordinate Tolerancing

The advantages claimed for direct coordinate tolerancing are as follows:

1. It is simple and easily understood and, therefore, it is commonly used.
2. It permits direct measurements to be made with standard instruments and does not require the use of special-purpose functional gages or other calculations.

Positional Tolerancing

Positional tolerancing is part of the system of geometric tolerancing. It defines a zone within which the center, axis, or center plane of a feature of size is permitted to vary from true (theoretically exact) position. A positional tolerance is indicated by the position symbol, a tolerance, and appropriate datum references placed in a feature control frame.

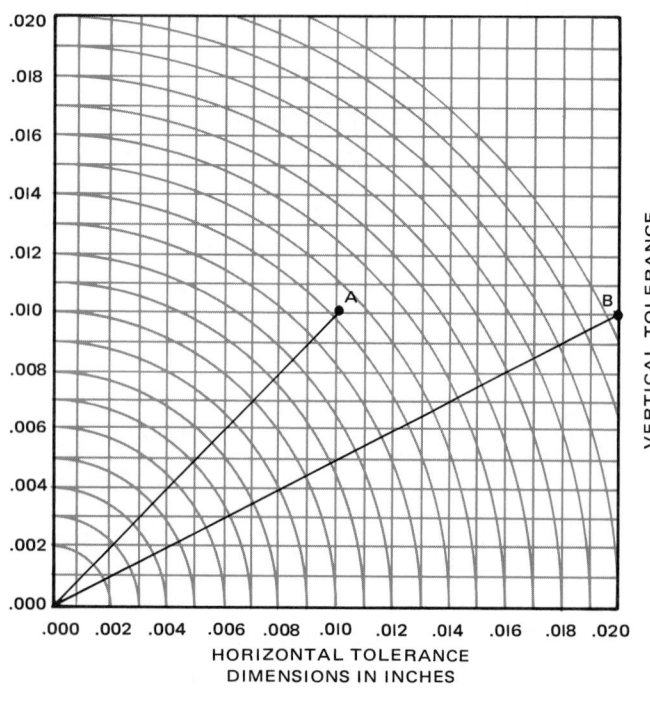

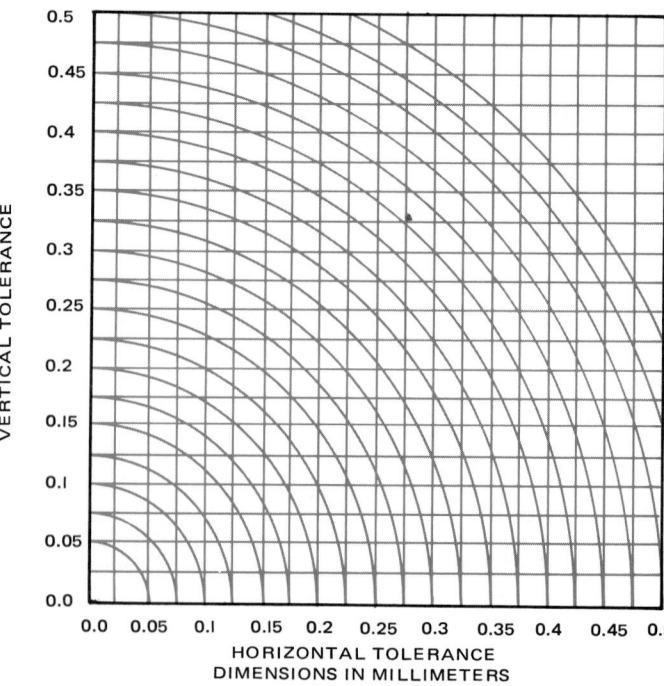

FIG. 16-9-6 Chart for calculating maximum tolerance using coordinate tolerancing.

Basic dimensions represent the exact values to which geometrical positional tolerances are applied elsewhere by symbols or notes on the drawing. They are enclosed in a rectangular frame (basic dimension symbol) as shown in Fig. 16-9-7. Where the dimension represents a diameter or a radius, the symbol Ø or R is included in the rectangular frame. General tolerance notes do not apply to basic dimensions. The frame size need not be any larger than that necessary to enclose the dimension. Permissible deviations from the basic dimension are then given by a positional tolerance as described in this unit.

Formerly the word BASIC or the abbreviation TP was used to indicate such dimensions.

Symbol for Position   The geometric characteristic symbol for position is a circle with two solid center lines, as shown in Fig. 16-9-8. This symbol is used in the feature control frame in the same manner as for other geometric tolerances.

Material Condition Basis

Positional tolerancing is applied on an MMC, RFS, or LMC basis. The appropriate symbol for the above follows the specified tolerance and where required the applicable datum reference in the feature control frame.

As positional tolerance controls the position of the center, axis, or center plane of a feature, the feature control frame is normally attached to the size of the feature, as shown in Fig. 16-9-9.

Positional Tolerancing for Circular Features   The positional tolerance represents the diameter of a cylindrical tolerance zone, located at true position as determined by the basic dimensions on the drawing, within which the axis or center line of the feature must lie.

Except for the fact that the tolerance zone is circular instead of square, a positional tolerance on this basis has exactly the same meaning as direct coordinate tolerancing but with equal tolerances in all directions.

It has already been shown that with rectangular coordinate tolerancing the maximum permissible error in location is not the value indicated by the horizontal and vertical tolerances, but rather is equivalent to the length of the diagonal between the two tolerances. For square tolerance zones this is 1.4 times the specified tolerance values. The specified tolerance can therefore be increased to an amount equal to the diagonal of the coordinate tolerance zone without affecting the clearance between the hole and its mating part.

This does not affect the clearance between the hole and its mating part, yet it offers 57 percent more tolerance area, as

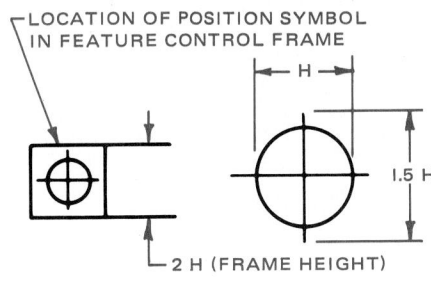

FIG. 16-9-8   Position symbol.

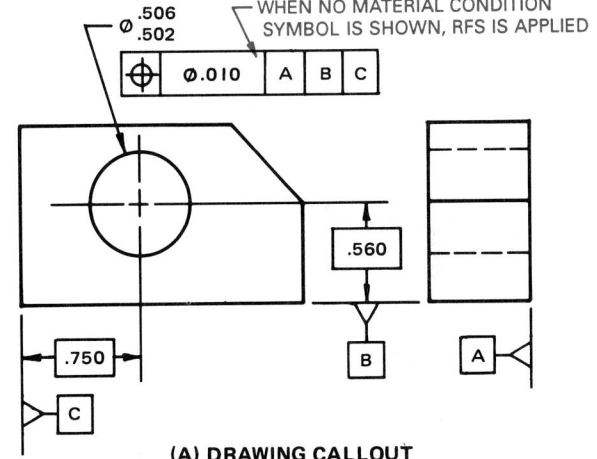

(A) DRAWING CALLOUT

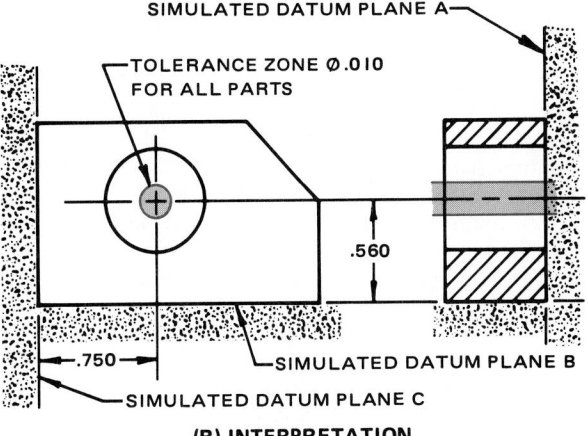

(B) INTERPRETATION

FIG. 16-9-9   Positional tolerancing—RFS.

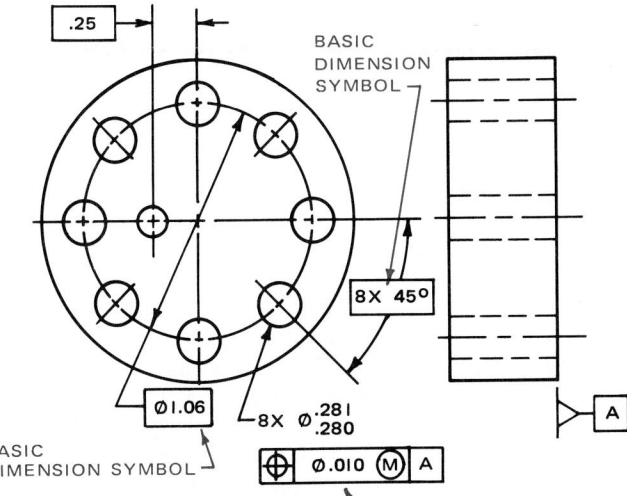

FIG. 16-9-7   Identifying basic dimensions.

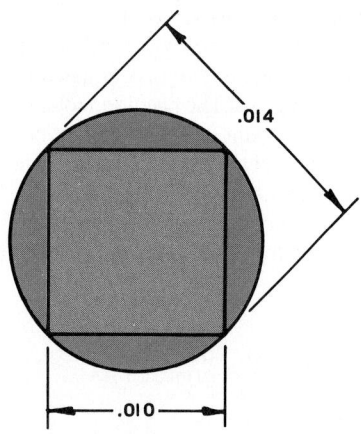

AREA OF CIRCUMSCRIBED CIRCULAR
ZONE = 157% OF SQUARE TOLERANCE ZONE

FIG. 16-9-10   Relationship of tolerance zones.

shown in Fig. 16-9-10. Such a change would most likely result in a reduction in the number of parts rejected for positional errors.

A simpler method is to make coordinate measurements and evaluate them on a chart. For example, if measurements of four parts, made according to Fig. 16-9-9, are as shown in the table below, only two are acceptable: the parts lying in the .010 diameter tolerance zone. These positions are shown on the chart, Fig. 16-9-11.

| PART | LOCATING MEASUREMENTS | | ACCEPTABILITY |
	Y	X	
A	.565	.752	Rejected
B	.562	.754	Accepted
C	.557	.753	Accepted
D	.556	.754	Rejected

Positional Tolerancing — MMC   The positional tolerance and MMC of mating features are considered in relation to each other. MMC by itself means a feature of a finished product contains the maximum amount of material permitted by the toleranced size dimension of that feature. Thus for holes, slots, and other internal features, maximum material is the condition where these features are at their minimum allowable sizes. For shafts, as well as for bosses, lugs, tabs, and other external features, maximum material is the condition where these features are at their maximum allowable sizes.

A positional tolerance applied on an MMC basis may be explained in either of the following ways:

1. *In terms of the surface of a hole.* While maintaining the specified size limits of the hole, no element of the hole surface shall be inside a theoretical boundary having a diameter equal to the minimum limit of size minus the positional tolerance located at true position (Fig. 16-9-12).

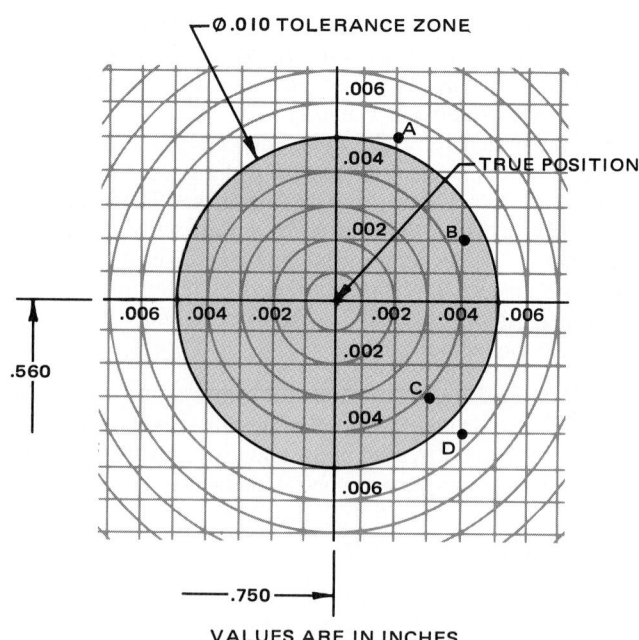

VALUES ARE IN INCHES

FIG. 16-9-11   Chart for evaluating the positional tolerance shown in Fig. 16-9-9.

THEORETICAL BOUNDARY – MINIMUM DIAMETER OF HOLE (MMC) MINUS THE POSITIONAL TOLERANCE

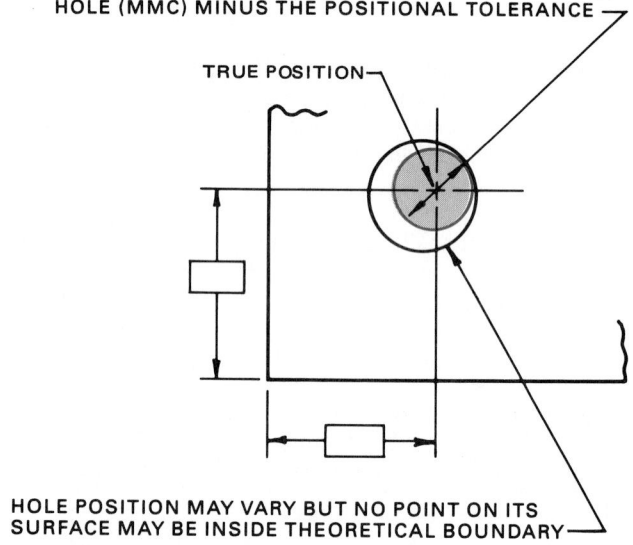

HOLE POSITION MAY VARY BUT NO POINT ON ITS SURFACE MAY BE INSIDE THEORETICAL BOUNDARY

FIG. 16-9-12   Boundary for surface of a hole at MMC.

2. *In terms of the axis of the hole.* Where a hole is at MMC (minimum diameter), its axis must fall within a cylindrical tolerance zone whose axis is located at true position. The diameter of this zone is equal to the positional tolerance (Figs. 16-9-13A and B). This tolerance zone also defines the limits of variation in the attitude of the axis of the hole in relation to the datum surface (Fig. 16-9-13C). It is only when the feature is at MMC that the specified positional tolerance applies. Where the actual size of the feature is larger than MMC, additional positional tolerance results (Fig. 16-9-15, pg. 528). This increase of positional tolerance is equal to the difference between the specified maximum material limit of size (MMC) and the

actual size of the feature. The specified positional tolerance for a feature may be exceeded where the actual size of the hole is larger than MMC and still satisfy function and interchangeability requirements.

The problems of tolerancing for the position of holes are simplified when positional tolerancing is applied on an MMC basis. Positional tolerancing simplifies measuring procedures by using functional "go" gages. It also permits an increase in positional variations as the size departs from the maximum material size without jeopardizing free assembly of mating features.

A positional tolerance on an MMC basis is specified on a drawing, on either the front or the side view, as shown in Fig. 16-9-14. The MMC symbol Ⓜ is added in the feature control frame immediately after the tolerance.

A positional tolerance applied to a hole on an MMC basis means that the boundary of the hole must fall outside a perfect cylinder having a diameter equal to the minimum limit of size minus the positional tolerance. This cylinder is located with its

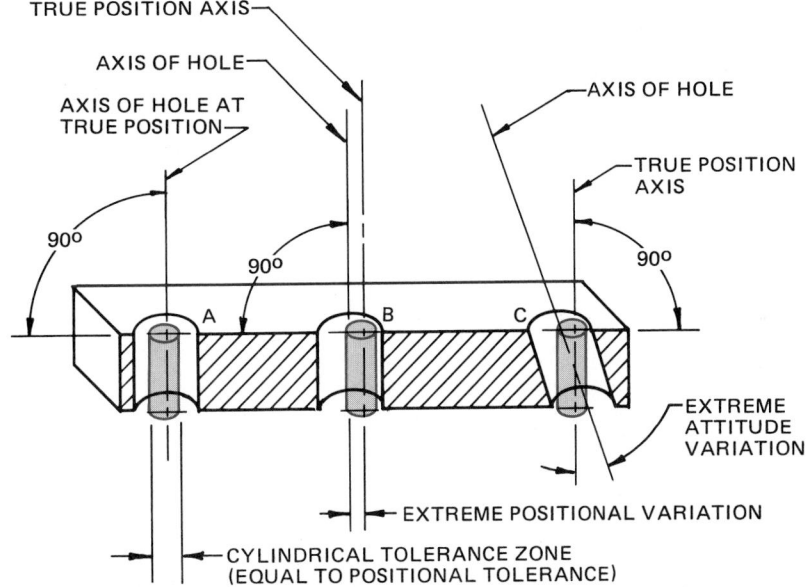

FIG. 16-9-13   Hole axes in relationship to positional tolerance zones.

HOLE A – AXIS OF HOLE IS COINCIDENT WITH TRUE POSITION AXIS
HOLE B – AXIS OF HOLE IS LOCATED AT EXTREME POSITION TO THE LEFT
OF TRUE POSITION AXIS (BUT WITHIN TOLERANCE ZONE)
HOLE C – AXIS OF HOLE IS INCLINED TO EXTREME ATTITUDE WITHIN
TOLERANCE ZONE

NOTE: THE LENGTH OF THE TOLERANCE ZONE IS EQUAL TO THE LENGTH
OF THE FEATURE, UNLESS OTHERWISE SPECIFIED ON THE DRAWING.

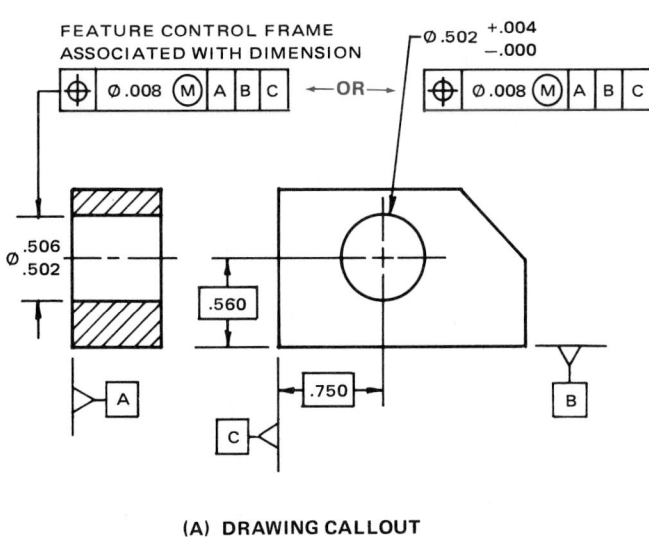

(A) DRAWING CALLOUT

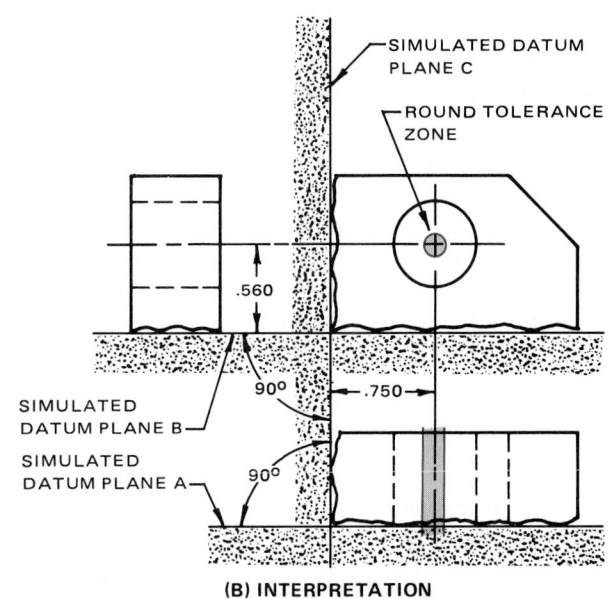

(B) INTERPRETATION

FIG. 16-9-14   Positional tolerancing—MMC.

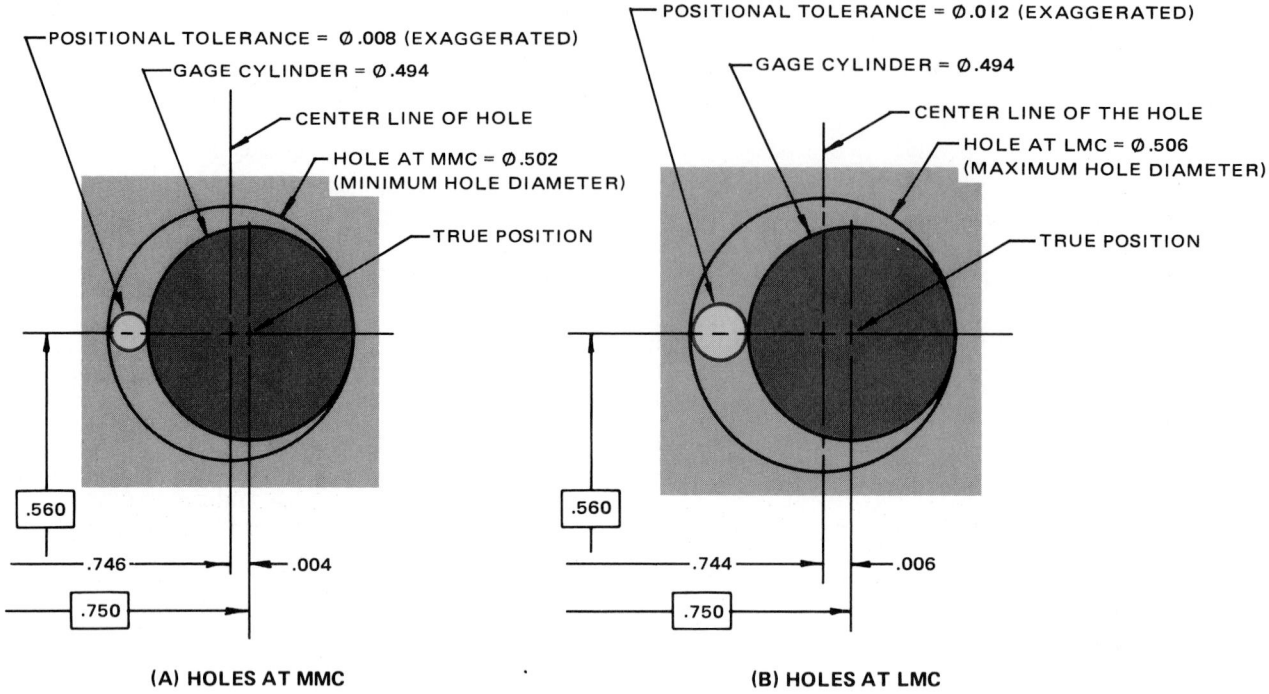

FIG. 16-9-15 Positional variations for tolerancing for Fig. 16-9-14.

axis at true position. The hole must, of course, meet its diameter limits.

The effect is illustrated in Fig. 16-9-15, where the gage cylinder is shown at true position and the minimum and maximum diameter holes are drawn to show the extreme permissible variations in position in one direction.

Therefore if a hole is at its maximum material condition (minimum diameter), the position of its axis must lie within a circular tolerance zone having a diameter equal to the specified tolerance. If the hole is at its maximum diameter (least material condition), the diameter of the tolerance zone for the axis is increased by the amount of the feature tolerance. The greatest deviation of the axis in one direction from true position is therefore:

$$\frac{H + P}{2} = \frac{.004 + .008}{2} = .006$$

where H = hole diameter tolerance
P = positional tolerance

It must be emphasized that positional tolerancing, even on an MMC basis, is not a cure-all for positional tolerancing problems; each method of tolerancing has its own area of usefulness. In each application a method must be selected that best suits that particular case.

Positional tolerancing on an MMC basis is preferred when production quantities warrant the provision of functional "go" gages, because gaging is then limited to one simple operation, even when a group of holes is involved. This method also facilitates manufacture by permitting larger variations in position when the diameter departs from the maximum material

condition. It cannot be used when it is essential that variations in location of the axis be observed regardless of feature size.

Positional Tolerancing at Zero MMC The application of MMC permits the tolerance to exceed the value specified, provided features are within size limits and parts are acceptable. This is accomplished by adjusting the minimum size limit of a hole to the absolute minimum required for the insertion of an applicable fastener located precisely at true position, and specifying a zero tolerance at MMC (Fig. 16-9-16). In this case, the positional tolerance allowed is totally dependent on the actual size of the considered feature.

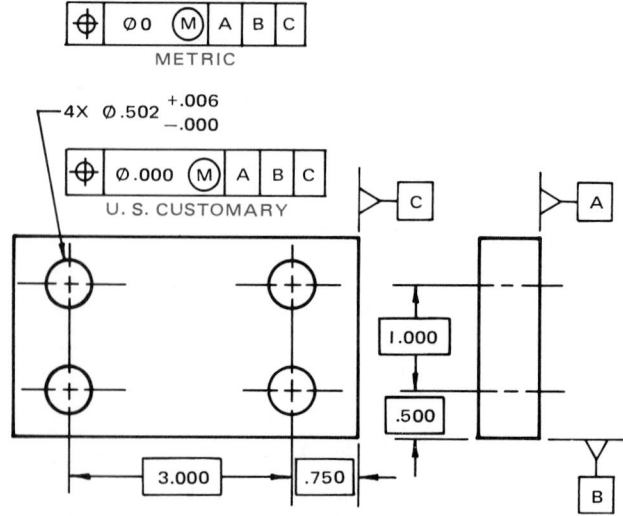

FIG. 16-9-16 Positional tolerancing—zero MMC.

Positional Tolerancing—RFS   In certain cases, the design or function of a part may require the positional tolerance or datum reference, or both, to be maintained regardless of actual feature sizes. RFS, where applied to the positional tolerance of circular features, requires the axis of each feature to be located within the specified positional tolerance regardless of the size of the feature. This requirement imposes a closer control of the features involved and introduces complexities in verification.

Positional Tolerancing—LMC   Where positional tolerancing at LMC is specified, the stated positional tolerance applies when the feature contains the least amount of material permitted by its toleranced size dimension. Specification of LMC further requires perfect form at LMC. Perfect form at MMC is not required. Where the feature departs from its LMC size, an increase in positional tolerance is allowed, equal to the amount of such departure (Fig. 16-9-17). Specifying LMC is limited to

positional tolerancing applications where MMC does not provide the desired control and RFS is too restrictive.

Advantages of Positional Tolerancing

It is practical to replace coordinate tolerances with a positional tolerance having a value equal to the diagonal of the coordinate tolerance zone. This provides 57 percent more tolerance area and would probably result in the rejection of fewer parts for positional errors.

A simple method for checking positional tolerance errors is to evaluate them on a chart, as shown in Fig. 16-9-18 (pg. 530). For example, the four parts listed in Fig. 16-9-19 (pg. 531) were rejected when the coordinate tolerances were applied to them.

If the parts had been toleranced using the positional tolerance RFS method shown in Fig. 16-9-9 and given a tolerance

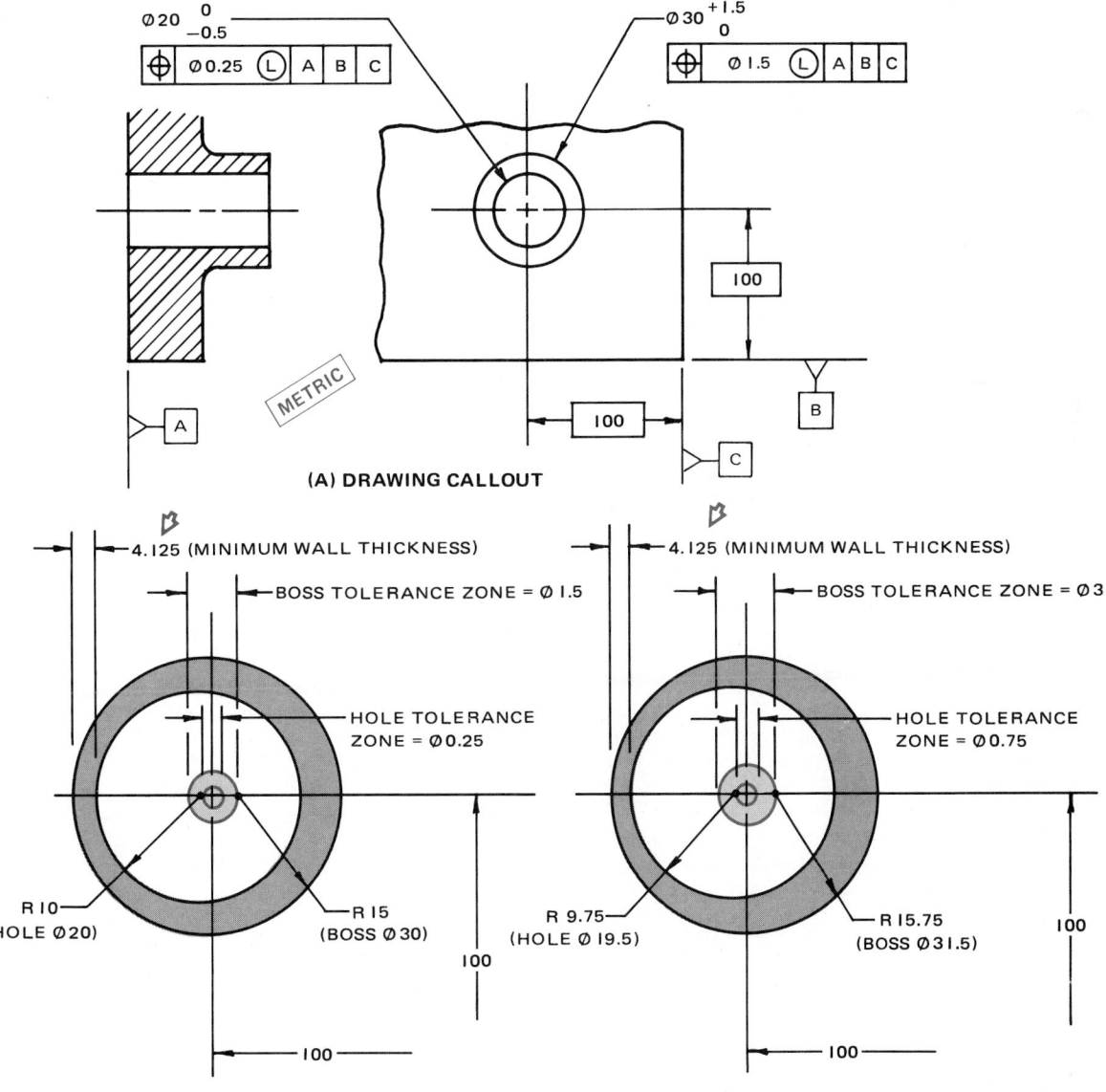

FIG. 16-9-17   LMC applied to a boss and a hole.

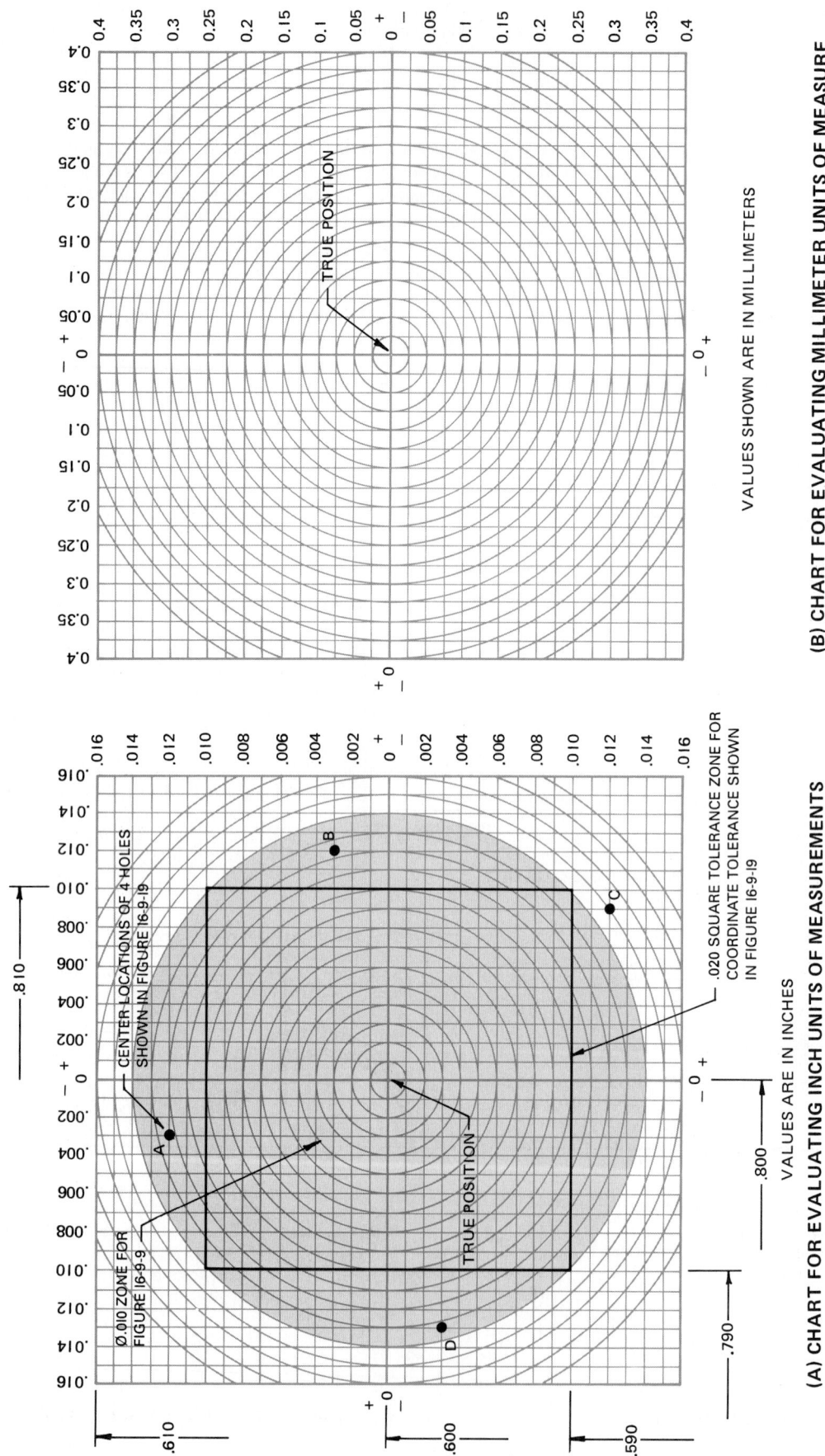

(B) CHART FOR EVALUATING MILLIMETER UNITS OF MEASURE

VALUES SHOWN ARE IN MILLIMETERS

TRUE POSITION

(A) CHART FOR EVALUATING INCH UNITS OF MEASUREMENTS

VALUES ARE IN INCHES

TRUE POSITION

.020 SQUARE TOLERANCE ZONE FOR COORDINATE TOLERANCE SHOWN IN FIGURE 16-9-19

CENTER LOCATIONS OF 4 HOLES SHOWN IN FIGURE 16-9-19

Ø 0.010 ZONE FOR FIGURE 16-9-9

FIG. 16-9-18   Charts for evaluating positional tolerancing.

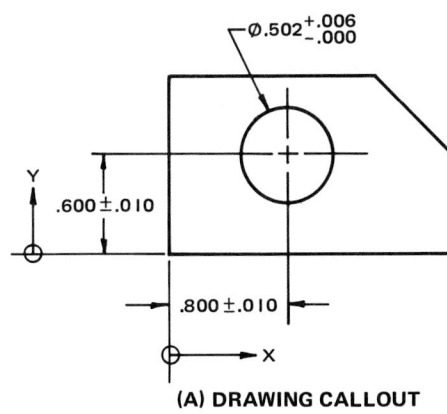

(A) DRAWING CALLOUT

PART	HOLE DIA	HOLE LOCATION		COMMENT
		X	Y	
A	.503	.797	.612	REJECTED
B	.504	.812	.603	REJECTED
C	.508	.809	.588	REJECTED
D	.506	.787	.597	REJECTED

REFER TO FIGURE 16—9—18 FOR LOCATION ON CHART

(B) LOCATION AND SIZE OF REJECTED PARTS

FIG. 16-9-19   Parts A to D rejected because hole centers do not lie within the coordinate tolerance zone.

of Ø.028 in. (equal to the diagonal of the coordinate tolerance zone), three of the parts—*A, B,* and *D*—would not have been rejected.

If the parts shown in Fig. 16-9-19 had been toleranced using the positional tolerance MMC method and given a tolerance of Ø.028 in. at MMC (Fig. 16-9-14), part *C,* which was rejected using the RFS tolerancing method, would not have been rejected if it had been straight. The positional tolerance can be increased to Ø.034 in. for a part having a hole diameter of .508 in. (LMC) without jeopardizing the function of the part (Fig. 16-9-15).

REFERENCES AND SOURCE MATERIAL

1. ASME Y14.5M–1994, *Dimensioning and Tolerancing.*
2. CAN/CSA B78.2-M91, *Dimensioning and Tolerancing of Technical Drawings.*
3. ISO drawing standards.

ASSIGNMENTS

See Assignments 23 through 27 for Unit 16-9 on page 578.

16-10 PROJECTED TOLERANCE ZONE

The application of the projected tolerance zone concept is recommended where the variation in perpendicularity of threaded or press-fit holes could cause fasteners such as screws, studs, or pins to interfere with mating parts (Fig. 16-10-1). An interference can occur where a positional tolerance is applied to the depth of threaded or press-fit holes and the hole axes are inclined within allowable limits. Unlike the floating-fastener application involving clearance holes only, the attitude of a fixed fastener is restrained by the inclination of the produced hole into which it assembles. Figure 16-10-2 (pg. 532) illustrates how the projected tolerance zone concept realistically treats the condition shown in

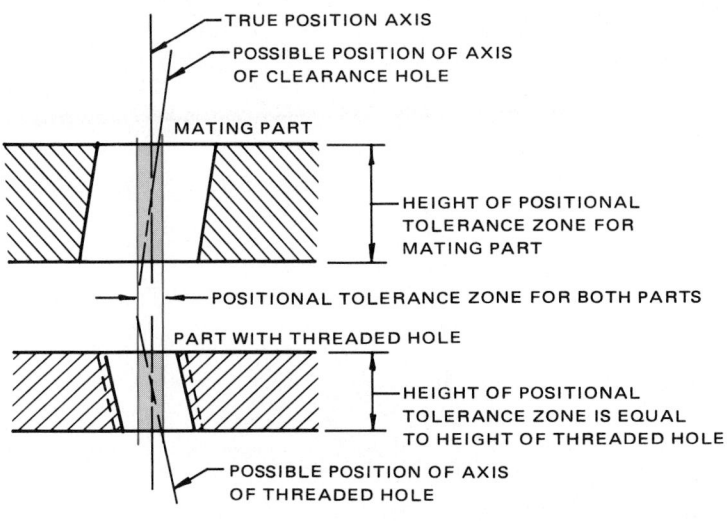

(A) PARTS TO BE ASSEMBLED

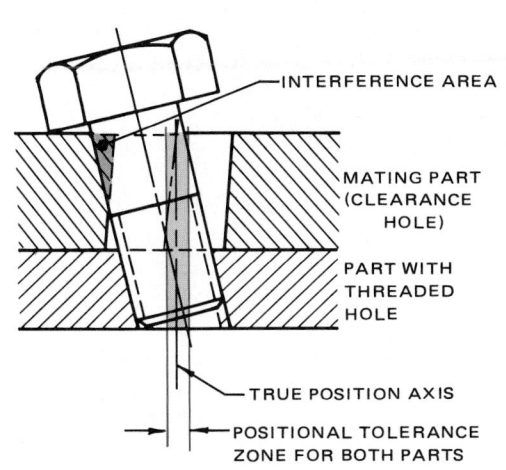

(B) PARTS SHOWN IN ASSEMBLED POSITION

FIG. 16-10-1   Illustrating how a fastener can interfere with a mating part.

531

Fig. 16-10-1. Note that it is the variation in perpendicularity of the portion of the fastener passing through the mating part that is significant. The location and the perpendicularity of the threaded holes are only of importance insofar as they affect the extended portion of the engaging fastener.

Clearance Holes in Mating Parts   Specifying a projected tolerance zone will ensure that fixed fasteners do not interfere with mating parts having clearance-hole sizes determined by the formulas recommended in Unit 16-17.

Symbol   The projected tolerance zone symbol is shown in Fig. 16-10-3. The symbol dimensions are based on percentages of the recommended letter-height dimensions.

Application   Figure 16-10-4 illustrates the application of a positional tolerance using a projected tolerance zone. The projected tolerance zone symbol followed by the minimum projected tolerance zone height is placed after the geometric tolerance in the feature control frame. Prior to the publication of ASME Y14.5M–1994 the minimum projected tolerance zone height followed by the symbol was enclosed in a rectangular frame and attached to the bottom of the feature control frame. The specified value for the projected tolerance zone is a mini-

mum and represents the maximum permissible mating part thickness or the maximum installed length or height of components such as studs or dowel pins. For through holes or in more complex or unusual situations, the direction of the projection from the datum surface may need further clarification. In such instances, the projected tolerance zone may be indicated as illustrated in Fig. 16-10-5. The minimum extent and direction of the projected tolerance zone is shown on the drawing as a dimensioned value with a heavy chain line drawn adjacent to an extension of the center line of the hole.

Where studs or press-fit pins are located on an assembly drawing, the specified positional tolerance applies only to the height of the projected portion of the pin or stud after installation. The specification of a projected tolerance zone is unnecessary. However, a projected tolerance zone is applicable where threaded or plain holes for studs or pins are located on a detail drawing. In these cases the specified projected height should equal the maximum permissible height of the stud or pin after installation, not the mating part thickness (Fig. 16-10-6).

Where design considerations require a closer control in the perpendicularity of a threaded hole than that allowed by the positional tolerance, a perpendicularity tolerance applied as

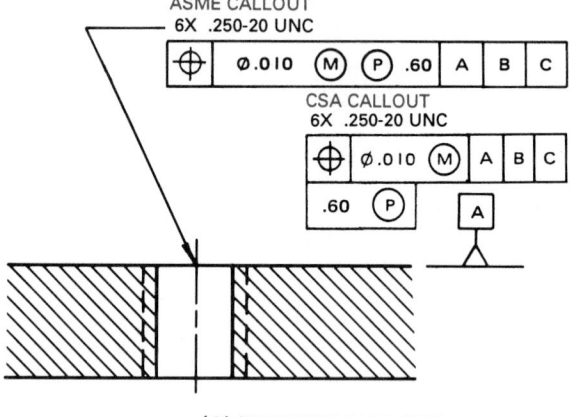

(A) DRAWING CALLOUT

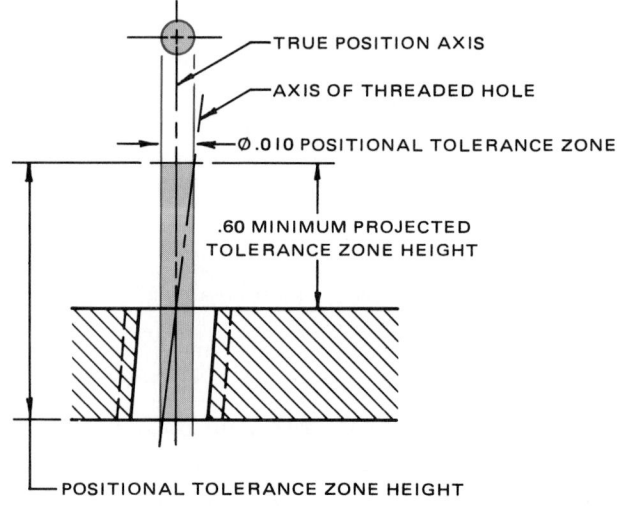

(B) TOLERANCE ZONE

FIG. 16-10-4   Specifying a projected tolerance zone.

FIG. 16-10-2   Basis for projected tolerance zone.

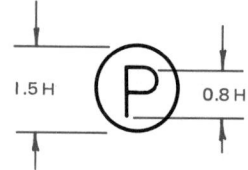

H = LETTERING HEIGHT OF DIMENSIONS

FIG. 16-10-3   Projected tolerance zone symbol.

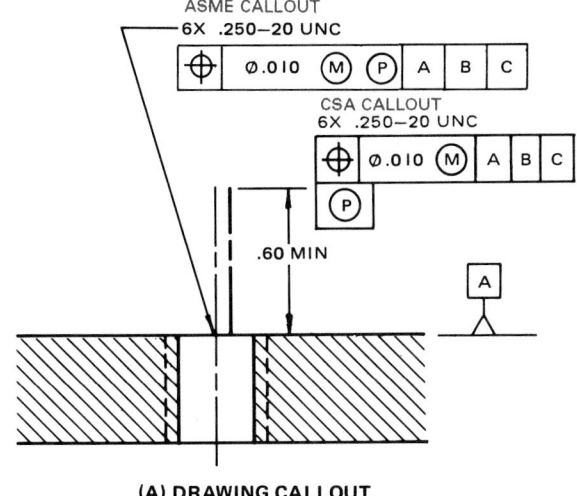

(A) DRAWING CALLOUT

AXES OF CLEARANCE AND THREADED HOLES
MUST LIE WITHIN Ø.010 TOLERANCE ZONE

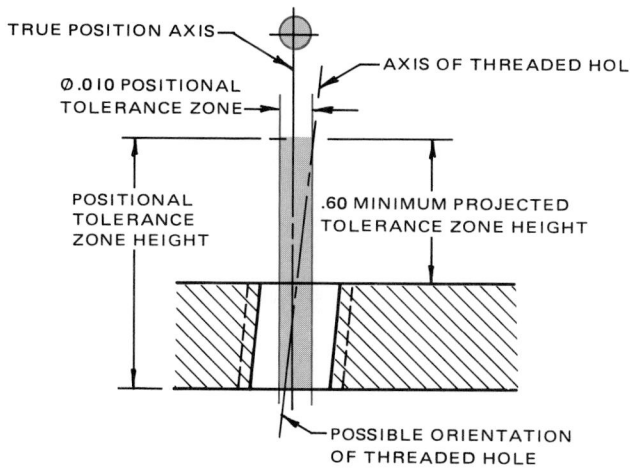

(B) TOLERANCE ZONE

FIG. 16-10-5   Projected tolerance zone indicated by a chain line.

a projected tolerance zone may be specified (Fig. 16-10-7, pg. 534).

REFERENCES AND SOURCE MATERIAL

1. ASME Y14.5M–1994, *Dimensioning and Tolerancing.*
2. CAN/CSA B78.2-M91, *Dimensioning and Tolerancing of Technical Drawings.*
3. ISO drawing standards.

ASSIGNMENTS ▮▮▮▮▮▮▮▮▮▮▮▮▮▮▮▮▮▮▮▮▮▮▮▮

See Assignments 28 and 29 for Unit 16-10 on page 579.

16-11 DATUM TARGETS

The full feature surface was used to establish a datum for the features so far designated as datum features. This may not always be practical for the following reasons:

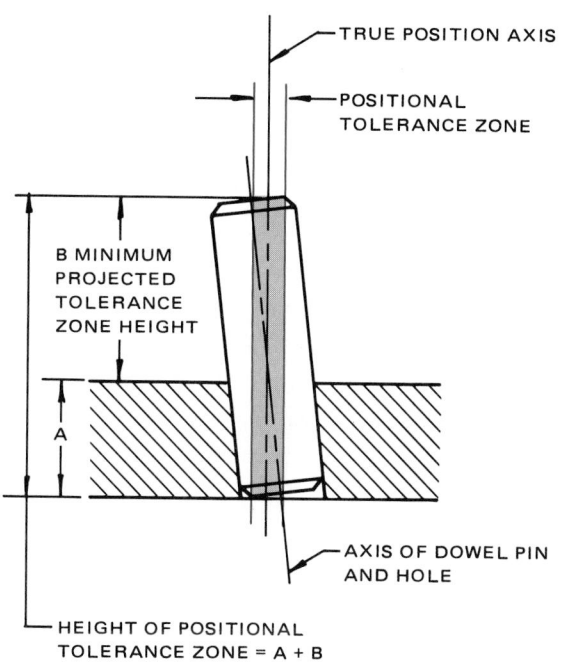

FIG. 16-10-6   Projected tolerance zone applied to studs and dowel pins.

1. The surface of a feature may be so large that a gage designed to make contact with the full surface may be too expensive or too cumbersome to use.
2. Functional requirements of the part may necessitate the use of only a portion of a surface as a datum feature, for example, the portion which contacts a mating part in assembly.
3. A surface selected as a datum feature may not be sufficiently true, and a flat datum feature may rock when placed on a datum plane, so that accurate and repeatable measurements from the surface would not be possible. This is particularly so for surfaces of castings, forgings, weldments, and some sheet-metal and formed parts.

A useful technique to overcome such problems is the datum target method. In this method certain points, lines, or small areas on the surfaces are selected as the bases for establishment of datums. For flat surfaces, this usually requires three target points or areas for a primary datum, two for a secondary datum, and one for a tertiary datum.

It is not necessary to use targets for all datums. It is quite logical, for example, to use targets for the primary datum and other surfaces or features for secondary and tertiary datums if required, or to use a flat surface of a part as the primary datum and to locate fixed points or lines on the edges as secondary and tertiary datums.

Datum targets should be spaced as far apart as possible to provide maximum stability for making measurements.

Datum Target Symbol

Points, lines, and areas on datum features are designated on the drawing by means of a datum target symbol. The symbol

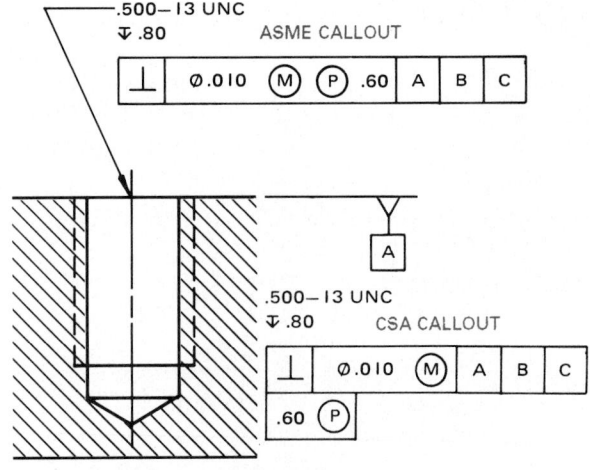

.500–13 UNC
⇩ .80   ASME CALLOUT

⊥ | ⌀ .010 Ⓜ Ⓟ .60 | A | B | C

A

.500–13 UNC
⇩ .80   CSA CALLOUT

⊥ | ⌀ .010 Ⓜ | A | B | C
.60 Ⓟ

(A) DRAWING CALLOUT

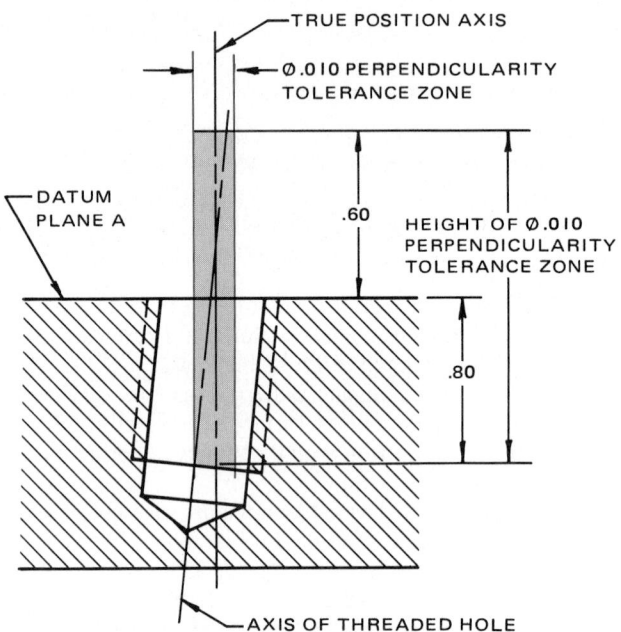

TRUE POSITION AXIS

⌀ .010 PERPENDICULARITY
TOLERANCE ZONE

DATUM
PLANE A

.60

HEIGHT OF ⌀ .010
PERPENDICULARITY
TOLERANCE ZONE

.80

AXIS OF THREADED HOLE

AXIS OF THREADED HOLE AND PROJECTED TOLERANCE
ZONE MUST LIE WITHIN THE PERPENDICULARITY
TOLERANCE ZONE

(B) TOLERANCE ZONE

FIG. 16-10-7   Specifying perpendicularity for a projected tolerance
zone.

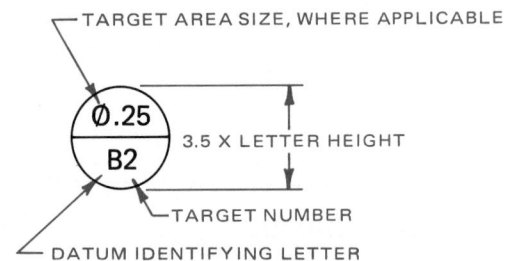

TARGET AREA SIZE, WHERE APPLICABLE

⌀ .25
B2

3.5 X LETTER HEIGHT

TARGET NUMBER

DATUM IDENTIFYING LETTER

FIG. 16-11-1   Datum target symbol.

(Fig. 16-11-1) is placed outside the part outline with a radial
(leader) line directed to the target point (indicated by an "X"),
target line, or target area, as applicable (Fig. 16-11-2). The use
of a solid radial (leader) line indicates that the datum target is
on the near (visible) surface. The use of a dashed radial (leader)
line, as in Fig. 16-11-8B (pg. 536), indicates that the datum
target is on the far (hidden) surface. The leader, shown without an arrowhead, should not be shown in either a horizontal
or vertical position. The datum feature itself is identified in the
usual manner with a datum feature symbol.

The datum target symbol is a circle having a diameter
approximately 3.5 times the height of the lettering used on the
drawing. The circle is divided horizontally into two halves. The
lower half contains a letter identifying the associated datum,
followed by the target number assigned sequentially starting
with 1 for each datum. For example, in a three-plane, six-point
datum system, if the datums are A, B, and C, the datum target
would be A_1, A_2, A_3, B_1, B_2, and C_1 (Fig. 16-11-3). Where the
datum target is an area, the area size may be entered in the
upper half of the symbol; otherwise, the upper half is left blank.

Identification Targets

Datum Target Points   Each target point is shown on the surface, in its desired location, by means of a cross, drawn at

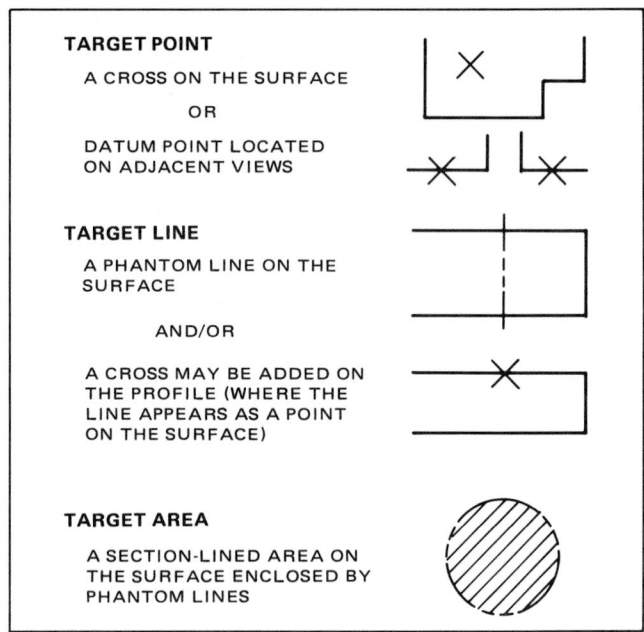

TARGET POINT

A CROSS ON THE SURFACE

OR

DATUM POINT LOCATED
ON ADJACENT VIEWS

TARGET LINE

A PHANTOM LINE ON THE
SURFACE

AND/OR

A CROSS MAY BE ADDED ON
THE PROFILE (WHERE THE
LINE APPEARS AS A POINT
ON THE SURFACE)

TARGET AREA

A SECTION-LINED AREA ON
THE SURFACE ENCLOSED BY
PHANTOM LINES

FIG. 16-11-2   Identification of datum targets.

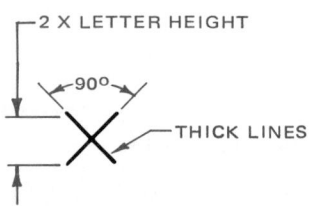

2 X LETTER HEIGHT

90°

THICK LINES

FIG. 16-11-3   Symbol for a datum target point.

approximately 45° to the coordinate dimensions. The cross is twice the height of the lettering used, as shown in Figs. 16-11-3 and 16-11-4A. Where there is no direct view, the point location is dimensioned on two adjacent views (Fig. 16-11-4B).

Target points may be represented on tools, fixtures, and gages by spherically ended pins, as shown in Fig. 16-11-5.

Datum Target Lines   A datum target line is indicated by the symbol X on an edge view of a surface, a phantom line on the direct view, or both (Fig. 16-11-6). Where the length of the datum target line must be controlled, its length and location are dimensioned.

It should be noted that if a line is designated as a tertiary datum feature, it will theoretically touch the gage pin at only one point. If it is a secondary datum feature, it will touch at two points.

The application and use of a surface and three lines as datum features are shown in Fig. 16-11-7 (pg. 536).

Datum Target Areas   Where it is determined that an area or areas of flat contact are necessary to ensure establishment of the datum (where spherical or pointed pins would be inadequate), a target area of the desired shape is specified. The datum target area is indicated by section lines inside a phantom outline of the desired shape, with controlling dimensions added. The diameter of circular areas is given in the upper half of the datum target symbol (Fig. 16-11-8A, pg. 536). Where it becomes impractical to delineate a circular target area, the method of indication shown in Fig. 16-11-8B may be used.

Datum target areas may have any desired shape, a few of which are shown in Fig. 16-11-9 (pg. 536). Target areas should be kept as small as possible, consistent with functional requirements, to avoid having large, cross-hatched areas on the drawing.

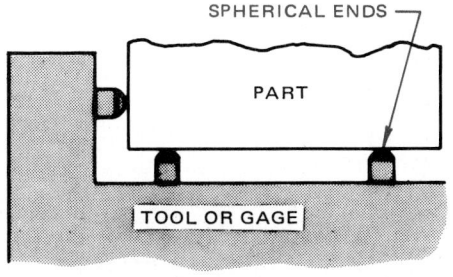

(A)

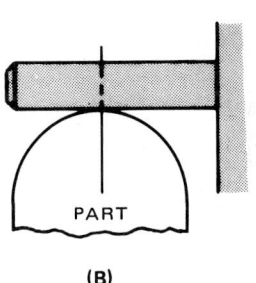

(B)

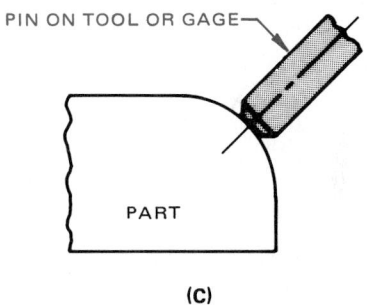

(C)

FIG. 16-11-5   Location of part on datum target points.

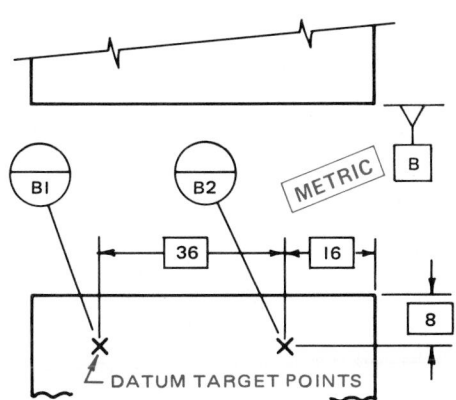

(A) DATUM POINTS SHOWN ON A SURFACE

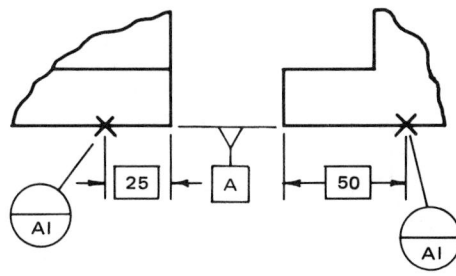

(B) DATUM POINT LOCATED BY TWO VIEWS

FIG. 16-11-4   Datum target points.

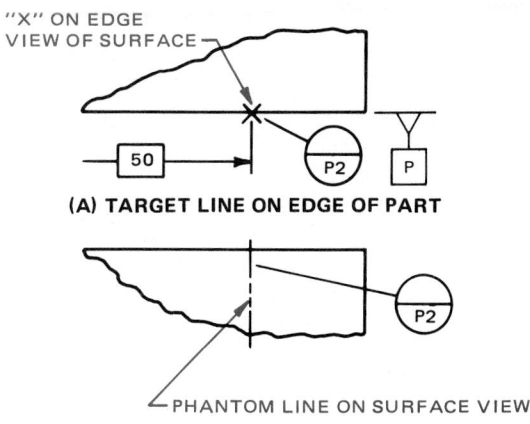

FIG. 16-11-6   Datum target line.

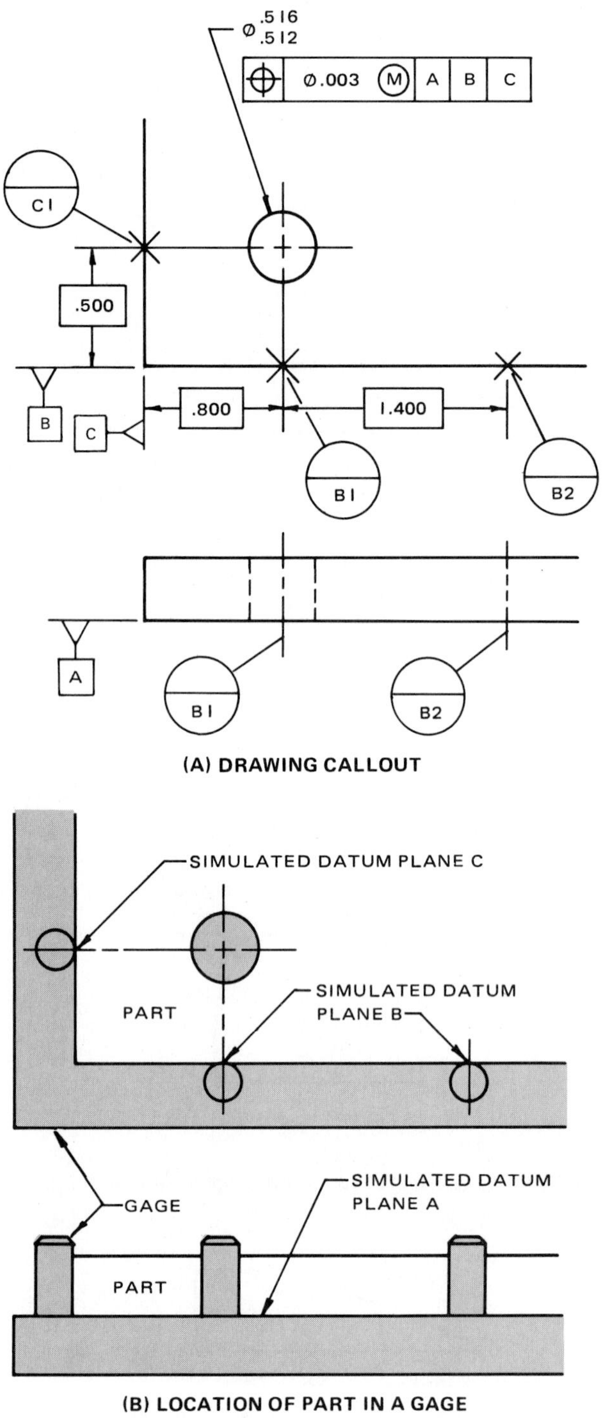

.516
⌀ .512

⊕ | ⌀.003 | Ⓜ | A | B | C

C1

.500

B   C

.800   1.400

B1   B2

A

B1   B2

(A) DRAWING CALLOUT

SIMULATED DATUM PLANE C

SIMULATED DATUM PLANE B

PART

GAGE

SIMULATED DATUM PLANE A

PART

(B) LOCATION OF PART IN A GAGE

FIG. 16-11-7   Part with a surface and three target lines used as datum features.

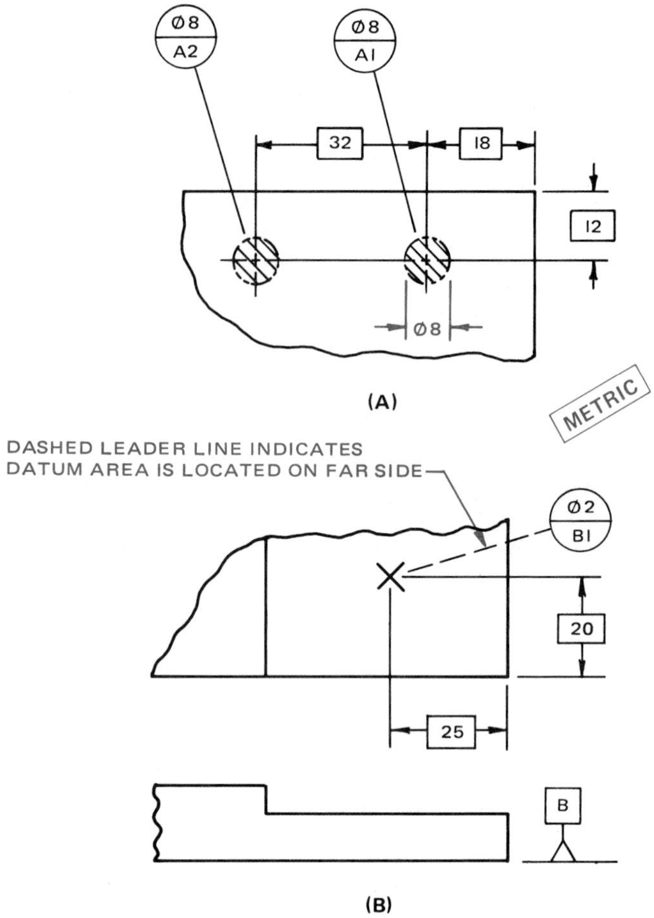

Targets Not in the Same Plane

In most applications datum target points that form a single datum are all located on the same surface, as shown in Fig. 16-11-4A. However, this is not essential. They may be located on different surfaces, to meet functional requirements as shown, for example, in Fig. 16-11-10. In some cases the datum plane may be located in space, that is, not actually touching the part, as shown

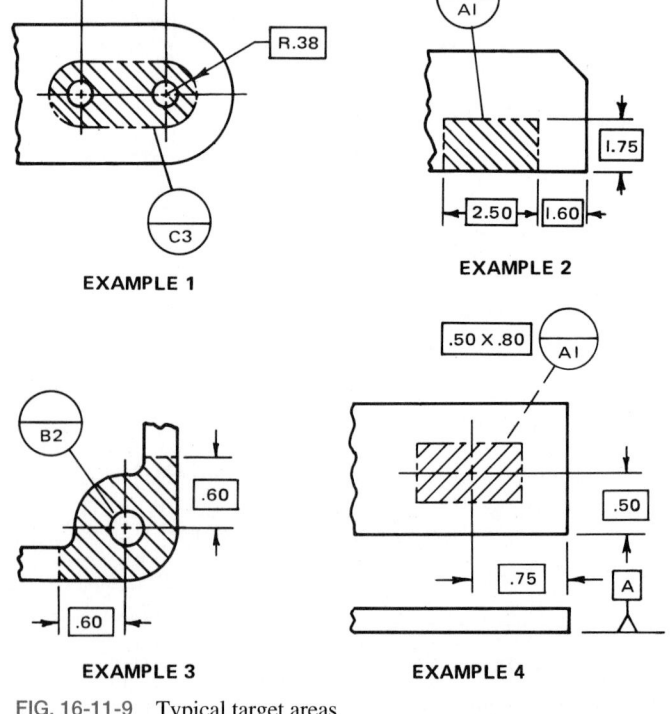

⌀8 / A2   ⌀8 / A1

32   18

12

⌀8

(A)

DASHED LEADER LINE INDICATES
DATUM AREA IS LOCATED ON FAR SIDE

⌀2 / B1

20

25

B

(B)

FIG. 16-11-8   Datum target areas.

1.00

R.38

C3

EXAMPLE 1

A1

1.75

2.50   1.60

EXAMPLE 2

B2

.60

.60

EXAMPLE 3

.50 X .80 | A1

.50

.75   A

EXAMPLE 4

FIG. 16-11-9   Typical target areas.

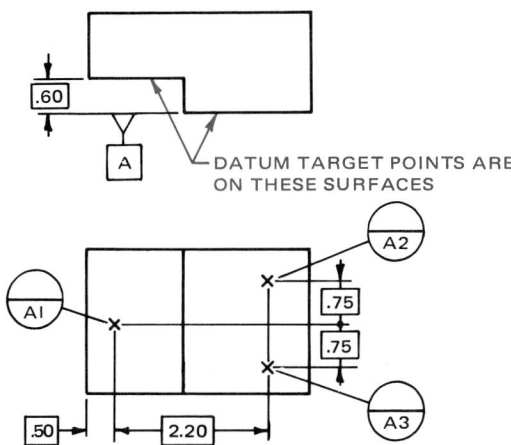

FIG. 16-11-10   Datum target points on different planes.

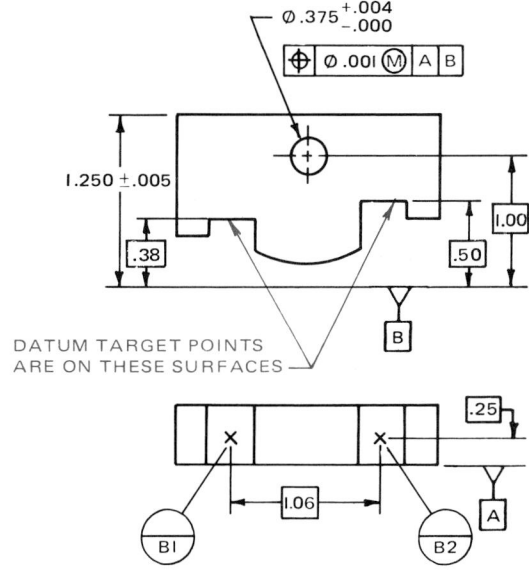

FIG. 16-11-11   Datum outside the part profile.

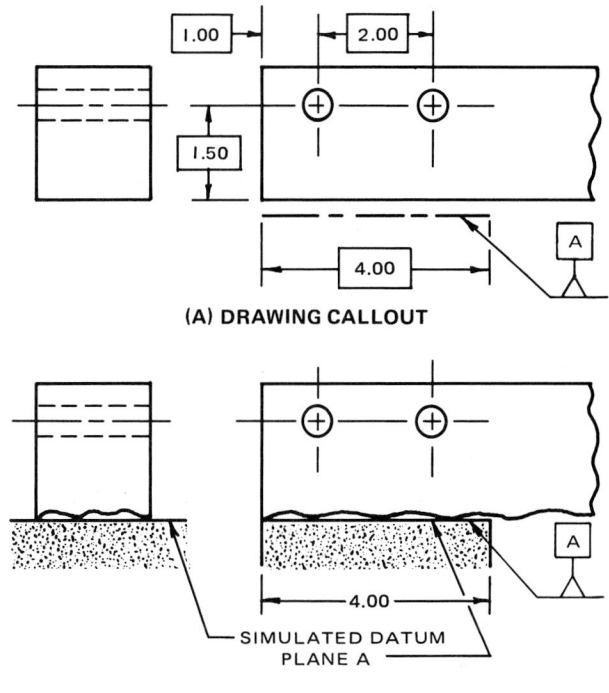

(A) DRAWING CALLOUT

(B) INTERPRETATION

FIG. 16-11-12   Partial datum.

in Fig. 16-11-11. In such applications the controlled features must be dimensioned from the specified datum, and the position of the datum from the datum targets must be shown by means of exact datum dimensions. For example, in Fig. 16-11-11, datum *B* is positioned by means of datum dimensions .38, .50, and 1.06. The top surface is controlled from this datum by means of a toleranced dimension, and the hole is positioned by means of the basic dimension 1.00 and a positional tolerance.

Partial Surfaces as Datums

It is often desirable to specify only part of a surface, instead of the entire surface, to serve as a datum feature. This may be indicated by means of a thick chain line drawn parallel to the surface profile (dimension for length and location) or by a datum target area. Figure 16-11-12 illustrates a long part where holes are located only at one end.

Dimensioning for Target Location

The location of datum targets is shown by means of basic dimensions. Each dimension is shown, without tolerances, enclosed in a rectangular frame, indicating that the general tolerance does not apply. Dimensions locating a set of datum targets should be dimensionally related or have a common origin.

Application of datum targets and datum dimensioning is shown in Fig. 16-11-13 (pg. 538).

REFERENCES AND SOURCE MATERIAL

1. ASME Y14.5M–1994, *Dimensioning and Tolerancing.*
2. CAN/CSA B78.2-M91, *Dimensioning and Tolerancing of Technical Drawings.*
3. ISO drawing standards.

ASSIGNMENTS

See Assignments 30 and 31 for Unit 16-11 on page 580.

16-12 CIRCULARITY (ROUNDNESS) AND CYLINDRICITY

Circularity

Circularity refers to a condition of a circular line or the surface of a circular feature where all points on the line or on the circumference of a plane cross section of the feature are the same

537

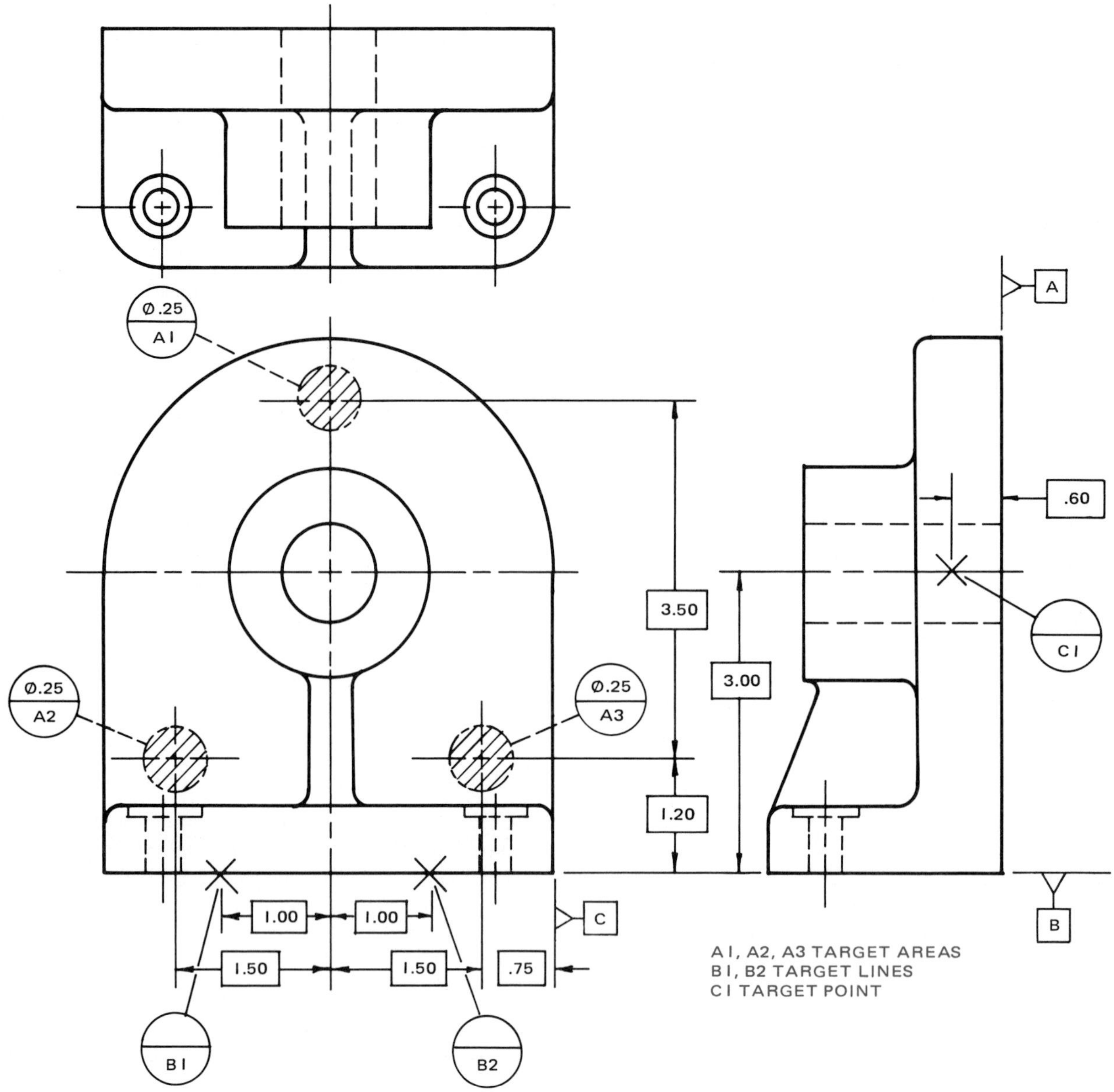

Ø.25
A1

Ø.25
A2

Ø.25
A3

3.50

3.00

1.20

1.00   1.00

1.50   1.50   .75

B1

B2

A

.60

C1

C

B

A1, A2, A3 TARGET AREAS
B1, B2 TARGET LINES
C1 TARGET POINT

FIG. 16-11-13   Part with a surface and three target lines used as datum features.

distance from a common axis or center point. It is similar to straightness except that it is wrapped around a circular cross section.

Examples of circular features would include disks, spheres, cylinders, and cones. The measurement plane for a sphere is any plane that passes through a section of maximum diameter. For a cylinder, cone, or other nonspherical feature, the measurement plane is any plane perpendicular to the axis or center line.

Errors of circularity (out-of-roundness) of a circular line or the periphery of a cross section of a circular feature may occur as *ovality,* where differences appear between the major and minor axes; as *lobing,* where in some instances the diametral values may be constant or nearly so; or as random irregularities from a true circle. All these errors are illustrated in Fig. 16-12-1.

The geometric characteristic symbol for circularity is a circle, having a diameter equal to 1.5 times the height of letters on the drawing, as shown in Fig. 16-12-2.

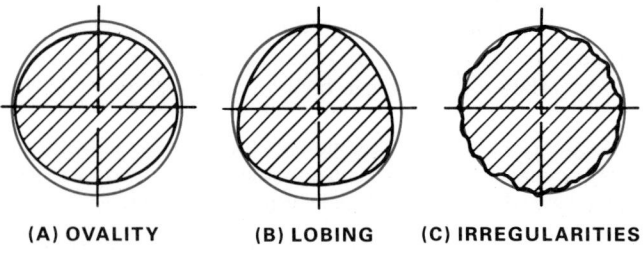

(A) OVALITY   **(B) LOBING**   **(C) IRREGULARITIES**

FIG. 16-12-1   Common types of circularity errors.

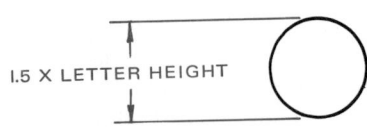

1.5 X LETTER HEIGHT

FIG. 16-12-2   Circularity symbol.

538

Circularity Tolerance

A circularity tolerance is measured radially and specifies the width between two circular rings for a particular cross section within which the circular line or the circumference of the feature in that plane shall lie, as shown in Figs. 16-12-3 and 16-12-4. Additionally, each circular element of the surface must be within the specified limits of size.

As circularity is a form tolerance, it is not related to datums. A circularity tolerance may be specified by using the circularity

symbol in the feature control frame. It is expressed on an RFS basis. The absence of a modifying symbol in the feature control frame means that RFS applies to the circularity tolerance.

A circularity tolerance cannot be modified on an MMC basis since it controls surface elements only. The circularity tolerance must be less than half the size tolerance as it must lie in a space equal to half the size tolerance.

Circularity of Noncylindrical Parts   Noncylindrical parts refer to conical parts and other features that are circular in cross section but that have variable diameters, such as those shown in Fig. 16-12-5. Since many sizes of circles may be shown in the end view, it is usually best to direct the circularity tolerance to the longitudinal surfaces, as shown.

Cylindricity

Cylindricity is a condition of a surface in which all points of the surface are the same distance from a common axis. The cylindricity tolerance is a composite control of form that includes circularity, straightness, and parallelism of the surface elements of a cylindrical feature. It is like a flatness tolerance wrapped around a cylinder.

The geometric characteristic symbol for cylindricity consists of a circle with two tangent lines at 60°, as shown in Fig. 16-12-6.

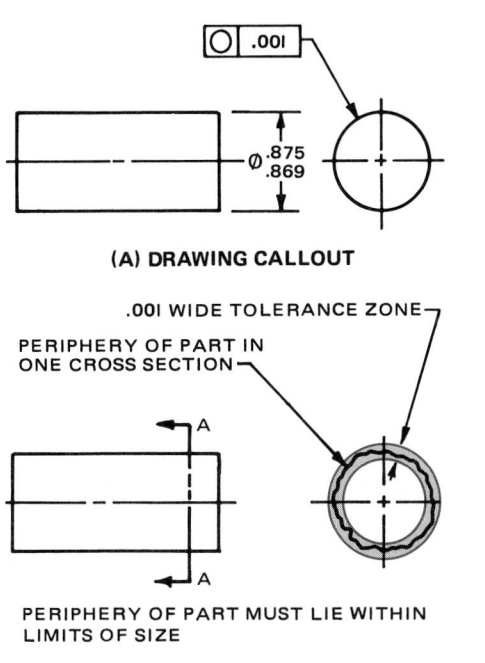

FIG. 16-12-3   Circularity tolerance applied to a cylindrical part.

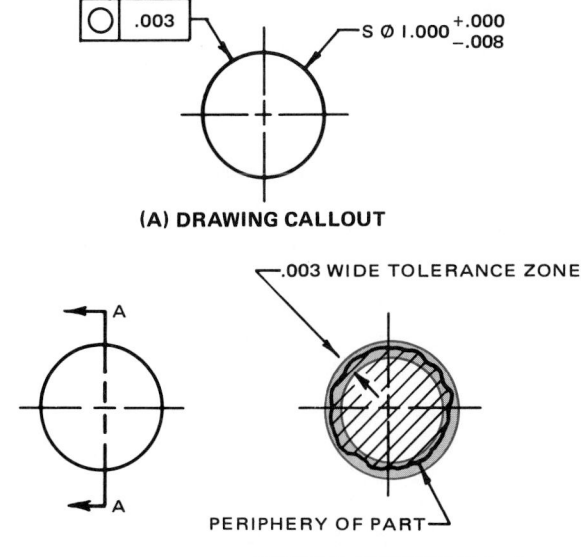

FIG. 16-12-4   Circularity tolerance applied to a sphere.

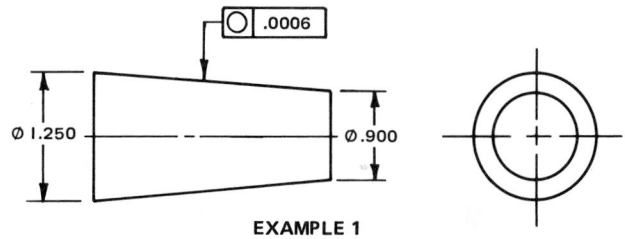

EXAMPLE 1

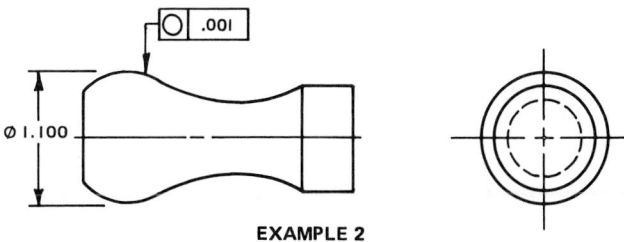

EXAMPLE 2

FIG. 16-12-5   Circularity tolerance applied to noncylindrical features.

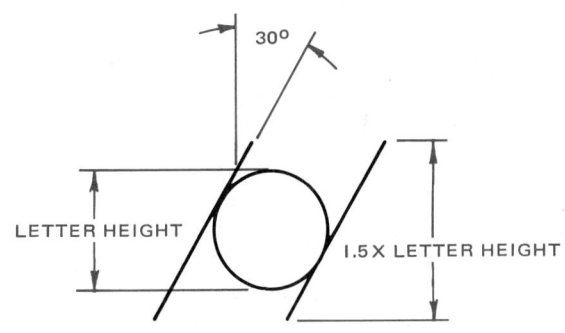

FIG. 16-12-6   Cylindricity symbol.

539

Cylindricity Tolerance

A cylindricity tolerance that is measured radially specifies a tolerance zone bounded by two concentric cylinders within which the surface must lie. The cylindricity tolerance must be within the specified limits of size. In the case of cylindricity, unlike that of circularity, the tolerance applies simultaneously to both circular and longitudinal elements of the surface (Fig. 16-12-7). The leader from the feature control symbol may be directed to either view. The cylindricity tolerance must be less than the size tolerance.

Since each part is measured for form deviation, it becomes obvious that the total range of the specified cylindricity tolerance will not always be available.

The cylindricity tolerance zone is controlled by the measured size of the actual part. The part size is first determined, then the cylindricity tolerance is added as a refinement to the actual size of the part. If, in the example shown in Fig. 16-12-7, the largest measurement of the produced part is Ø.748 in., which is near the high limit of size (.750), the largest diameter of the two concentric cylinders for the cylindricity tolerance would be Ø.748 in. The smaller of the concentric cylinders would be .748 minus twice the cylindricity tolerance (2 × .002) = Ø.744 in. The cylindricity tolerance zone must also lie between the limits of size, and the entire cylindrical surface of the part must lie between these two concentric circles to be acceptable.

If, on the other hand, the largest diameter measured for a part is Ø.743 in., which is near the lower limit of size (.740 in.), the cylindricity deviation of that part cannot be greater than .0015 in. since it would exceed the lower limit of size.

Likewise, if the smallest measured diameter of a part is .748 in., which is near the high limit of size, the largest diameter of the two concentric cylinders for the cylindricity tolerance would be Ø.750 in., which is the maximum permissible

diameter of the part. In this case the cylindricity tolerance could not be greater than (.750 − .748)/2, or .001 in.

Cylindricity tolerances can be applied only to cylindrical surfaces, such as round holes and shafts. No specific geometric tolerances have been devised for other circular forms, which require the use of several geometric tolerances. A conical surface, for example, must be controlled by a combination of tolerances for circularity, straightness, and angularity.

Errors of cylindricity may be caused by out-of-roundness, like ovality or lobing, by errors of straightness caused by bending or by diametral variation, by errors of parallelism, like conicity or taper, and by random irregularities from a true cylindrical form (Fig. 16-12-8).

Since cylindricity is a form tolerance much like that of a flatness tolerance in that it controls surface elements only, it cannot be modified on an MMC basis. The absence of a modifying symbol in the feature control frame indicates that RFS applies.

REFERENCES AND SOURCE MATERIAL

1. ASME Y14.5M–1994, *Dimensioning and Tolerancing.*
2. CAN/CSA B78.2-M91, *Dimensioning and Tolerancing of Technical Drawings.*
3. ISO drawing standards.

ASSIGNMENTS

See Assignments 32 through 36 for Unit 16-12 on page 581.

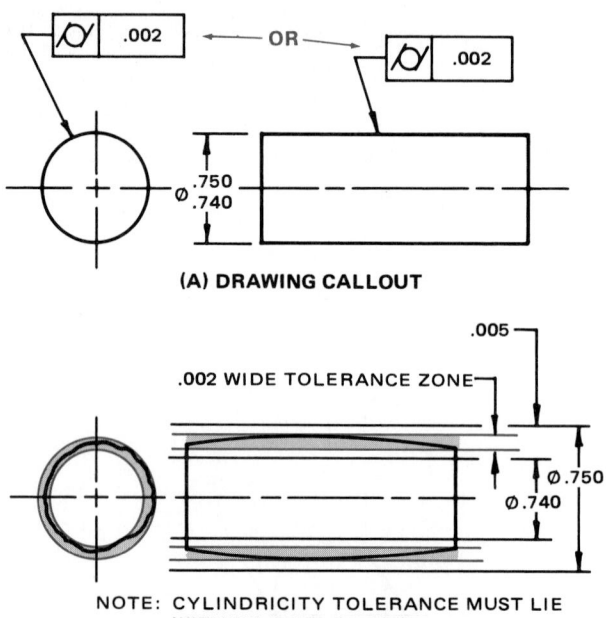

NOTE: CYLINDRICITY TOLERANCE MUST LIE WITHIN LIMITS OF SIZE.

(B) TOLERANCE ZONE

FIG. 16-12-7   Cylindricity tolerance may be directed to either view.

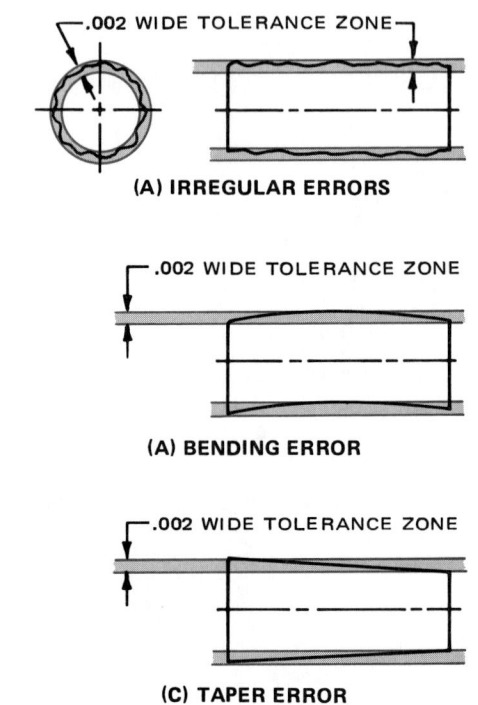

NOTE: CYLINDRICITY TOLERANCE MUST LIE WITHIN LIMITS OF SIZE.

FIG. 16-12-8   Permissible form errors for part shown in Fig. 16-12-7.

16-13 PROFILE TOLERANCING

Profiles

A *profile* is the outline form or shape of a line or surface. A *line profile* may be the outline of a part or feature as depicted in a view on a drawing. It may represent the edge of a part, or it may refer to line elements of a surface in a single direction, such as the outline of cross sections through the part. In contrast, a *surface profile* outlines the form or shape of a complete surface in three dimensions.

The elements of a line profile may be straight lines, arcs, or other curved lines. The elements of a surface profile may be flat surfaces, spherical surfaces, cylindrical surfaces, or surfaces composed of various line profiles in two or more directions.

A profile tolerance specifies a uniform boundary along the true profile within which the elements of a surface must lie. MMC is not applicable to profile tolerances. When used as a refinement of size, the profile tolerance must be contained within the limits of size.

Profile Symbols

There are two geometric characteristic symbols for profiles, one for lines and one for surfaces. Separate symbols are required, because it is often necessary to distinguish between line elements of a surface and the complete surface itself. The symbol for profile of a line consists of a semicircle with a diameter equal to twice the lettering size used on the drawing. The symbol for profile of a surface is identical except that the semicircle is closed by a straight line at the bottom, as shown in Fig. 16-13-1. All other geometric tolerances of form and orientation are merely special cases of profile tolerancing.

Profile tolerances are used to control the position of lines and surfaces that are neither flat nor cylindrical.

Profile-of-a-Line Tolerance

A profile-of-a-line tolerance may be directed to a line of any length or shape. With profile-of-a-line tolerance, datums may be used in some circumstances but would not be used when the only requirement is the profile shape taken cross section by cross section. Profile-of-a-line tolerancing is used where it is not desirable to control the entire surface of the feature as a single entity.

A profile-of-a-line tolerance is specified in the usual manner, by including the symbol and tolerance in a feature control frame directed to the line to be controlled, as shown in Fig. 16-13-2.

If the line on the drawing to which the tolerance is directed represents a surface, the tolerance applies to all line elements of the surface parallel to the plane of the view on the drawing, unless otherwise specified.

The tolerance indicates a tolerance zone consisting of the area between two parallel lines, separated by the specified tolerance, which are themselves parallel to the basic form of the line being toleranced. The tolerance zone established by a profile-of-a-line tolerance is two-dimensional, extending along the length of the considered feature.

Bilateral and Unilateral Tolerances

The profile tolerance zone, unless otherwise specified, is equally disposed about the basic profile in a form known as a *bilateral tolerance zone*. The width of this zone is always measured perpendicular to the profile surface. The tolerance zone may be considered to be bounded by two lines enveloping a series of circles, each having a diameter equal to the specified profile tolerance, with their centers on the theoretical, basic profile, as shown in Fig. 16-13-2.

Occasionally it is desirable to have the tolerance zone wholly on one side of the basic profile instead of equally divided on both sides. Such zones are called *unilateral tolerance zones*. They are specified by showing a phantom line drawn parallel and close to the profile surface. The tolerance is directed to this line, as shown in Fig. 16-13-3. The zone line need extend only a sufficient distance to make its application clear.

All-Around Profile Tolerance

Where a profile tolerance applies all around the profile of a part, the symbol used to designate "all around" is placed on the leader from the feature control frame (Figs. 16-13-4 and 16-13-5, pg. 542).

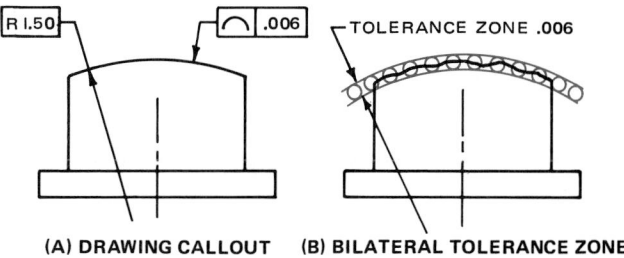

(A) DRAWING CALLOUT   **(B) BILATERAL TOLERANCE ZONE**

FIG. 16-13-2   Simple profile with a bilateral profile tolerance zone.

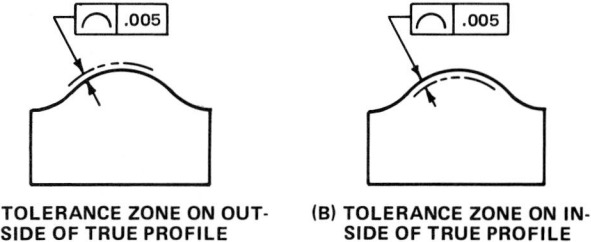

(A) TOLERANCE ZONE ON OUT-SIDE OF TRUE PROFILE   **(B) TOLERANCE ZONE ON IN-SIDE OF TRUE PROFILE**

FIG. 16-13-3   Unilateral tolerance zones.

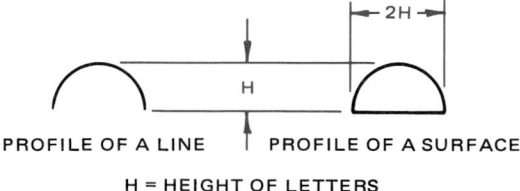

PROFILE OF A LINE   PROFILE OF A SURFACE

H = HEIGHT OF LETTERS

FIG. 16-13-1   Profile symbols.

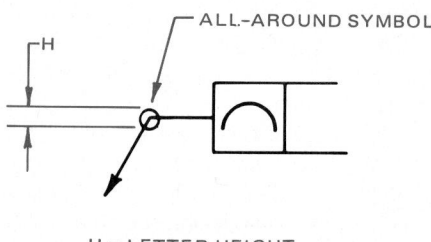

FIG. 16-13-4   All-around symbol.

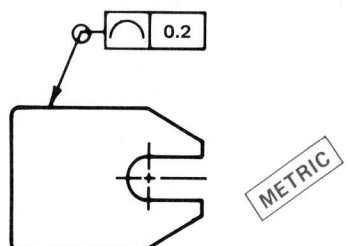

FIG. 16-13-5   Profile tolerance required for all around the surface.

Method of Dimensioning

The true profile is established by means of basic dimensions, each of which is enclosed in a rectangular frame to indicate that the tolerance in the general tolerance note does not apply.

When the profile tolerance is not intended to control the position of the profile, there must be a clear distinction between dimensions that control the position of the profile and those that control the form or shape of the profile.

Any convenient method of dimensioning may be used to establish the basic profile. Examples are chain or common-point dimensions, dimensioning to points on a surface or to the intersection of lines, dimensioning located on tangent radii, and angles.

To illustrate, the simple part in Fig. 16-13-6 shows a dimension of .90 ±.01 controlling the height of the profile. This dimension must be separately measured. The radius of 1.500 in. is a basic dimension, and it becomes part of the profile. Therefore, the profile tolerance zone has radii of 1.497 and 1.503, but is free to float in any direction within the limits of the positional tolerance zone in order to enclose the curved profile.

If the radius were shown as a toleranced dimension, without the rectangular frame, as in Fig. 16-13-7, it would become a separate measurement.

Figure 16-13-8 shows a more complex profile, where the profile is located by a single toleranced dimension. There are, however, five basic dimensions defining the true profile.

In this case, the tolerance on the height indicates a tolerance zone .06 in. wide extending the full length of the profile, because the profile is established by basic dimensions. No other dimension exists to affect the orientation or height. The profile tolerance specifies a .008 in. wide tolerance zone, which may lie anywhere within the .06 in. tolerance zone.

Extent of Controlled Profile   The profile is generally intended to extend to the first abrupt change or sharp corner.

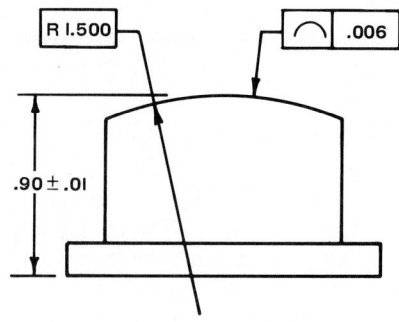

FIG. 16-13-6   Position and form as separate requirements.

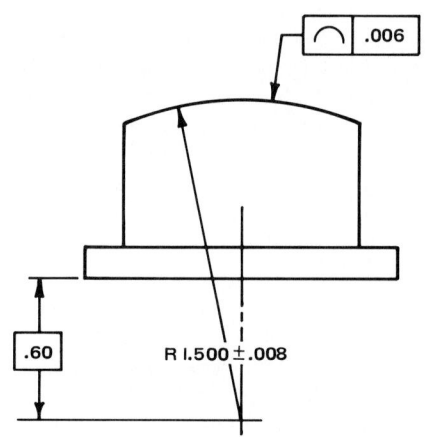

FIG. 16-13-7   Position and radius separate from form.

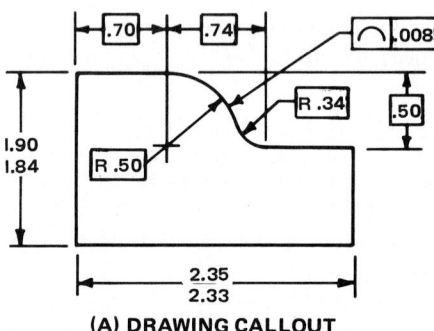

(A) DRAWING CALLOUT

TOLERANCE ZONE FOR FORM OF PROFILE
.008 WIDE SHOWN IN AN EXTREME POSITION

TOLERANCE ZONE FOR POSITION OF PROFILE

ACTUAL PROFILE

(B) PROFILE TOLERANCE ZONE

FIG. 16-13-8   Profile defined by basic dimensions.

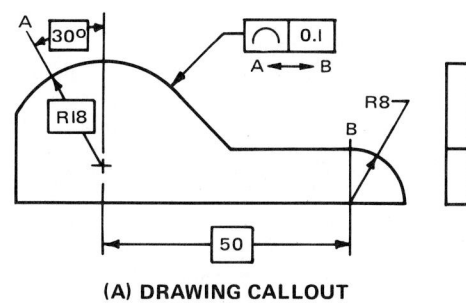

(A) DRAWING CALLOUT

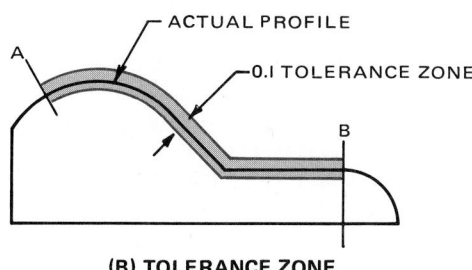

(B) TOLERANCE ZONE

FIG. 16-13-9   Specifying extent of profile.

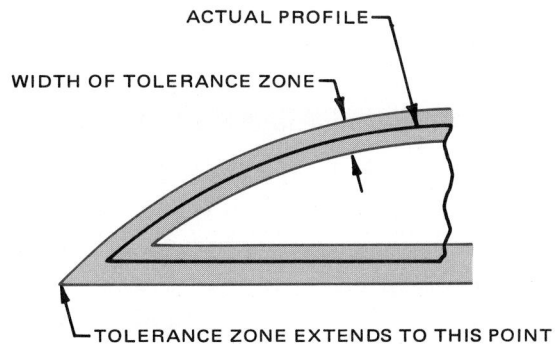

FIG. 16-13-10   Tolerance zone at a sharp corner.

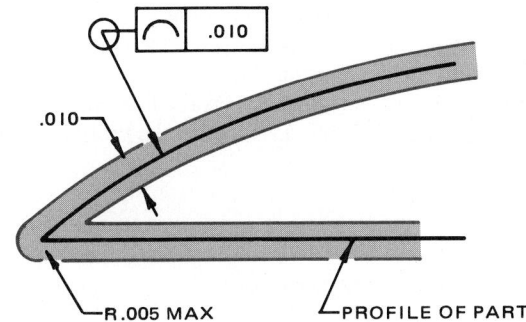

FIG. 16-13-11   Controlling the profile of a sharp corner.

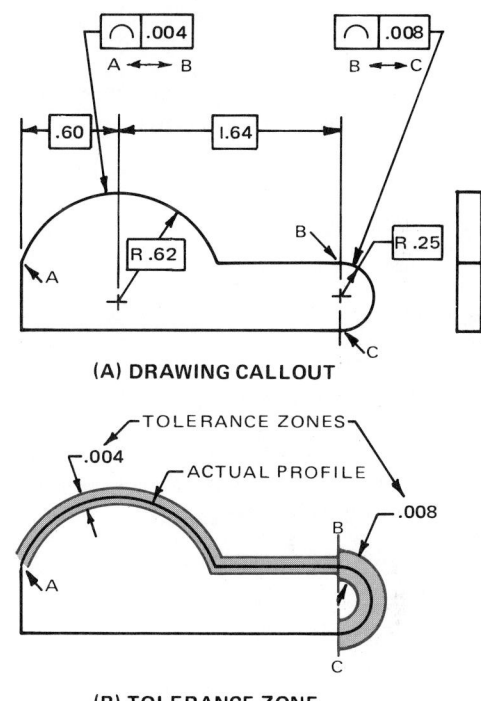

(A) DRAWING CALLOUT

(B) TOLERANCE ZONE

FIG. 16-13-12   Dual tolerance zones.

For example, in Fig. 16-13-8 it extends from the upper left-hand corner to the upper right-hand corner, unless otherwise specified. If the extent of the profile is not clearly identified by sharp corners or by basic profile dimensions, it must be indicated by a note under the feature control symbol, such as A↔B, as shown in Fig. 16-13-9.

If the controlled profile includes a sharp corner, the corner represents a continuity of the tolerance boundary, and the boundary is considered to extend to the intersection of the boundary lines, as shown in Fig. 16-13-10. Since the intersecting surfaces may lie anywhere within the tolerance zone, the actual part contour could conceivably be round. If this is undesirable, the drawing must indicate the design requirements, for example, by specifying the maximum radius as shown in Fig. 16-13-11.

If different profile tolerances are required on different segments of a surface, the extent of each profile tolerance is indicated by the use of reference letters to identify the extremities (Fig. 16-13-12).

Profile-of-a-Surface Tolerance

If the same tolerance is intended to apply over the whole surface, instead of to lines or line elements in specific directions, the profile-of-a-surface tolerance is used (Fig. 16-13-13, pg. 544). While the profile tolerance may be directed to the surface in either view, it is usually directed to the view showing the shape of the profile.

The profile-of-a-surface tolerance indicates a tolerance zone having the same form as the basic surface, with a uniform width equal to the specified tolerance within which the entire surface must lie. It is used to control form or combinations of size, form, and orientation. Where used as a refinement of size, the profile tolerance must be contained within the size limits. The symbol for a profile of a surface is shown in Fig. 16-13-1.

The basic rules for profile-of-a-line tolerancing apply to profile-of-a-surface tolerancing except that in most cases profile-of-a-surface tolerances require references to datums in order to provide proper orientation of the profile. This reference is specified simply by indicating suitable datums. Figures 16-13-13 and 16-13-14 show simple parts where one and two datums are designated.

The criterion that distinguishes a profile tolerance as applying to position or to orientation is whether the profile is related to the datum by a basic dimension or by a toleranced dimension (Fig. 16-13-15).

Profile tolerances controlling position are very useful for parts that can be revolved around a cylindrical datum feature (Fig. 16-13-16).

Profile tolerancing may be used to control the form and orientation of plane surfaces. In Fig. 16-13-17, a profile-of-a-surface tolerance is used to control a plane surface inclined to a datum feature.

A profile tolerance may be specified to control the conicity of a surface in either of two ways: as an independent control of form or as a combined control of form and orientation. Figure 16-13-18 (pg. 546) illustrates a conical feature controlled by a profile-of-a-surface tolerance where conicity of the surface is a refinement of size. In Fig. 16-13-19 (pg. 546) the same control is applied but is oriented to a datum axis. In each case the feature must lie within the size limits.

Where a profile-of-a-surface tolerance applies all around the profile of a part, the symbol used to designate "all around" is placed on the leader from the feature control frame (Fig. 16-13-20, pg. 546).

REFERENCES AND SOURCE MATERIAL

1. ASME Y14.5M–1994, *Dimensioning and Tolerancing.*
2. CAN/CSA B78.2-M91, *Dimensioning and Tolerancing of Technical Drawings.*
3. ISO drawing standards.

ASSIGNMENTS

See Assignments 37 through 41 for Unit 16-13 on pages 582 to 583.

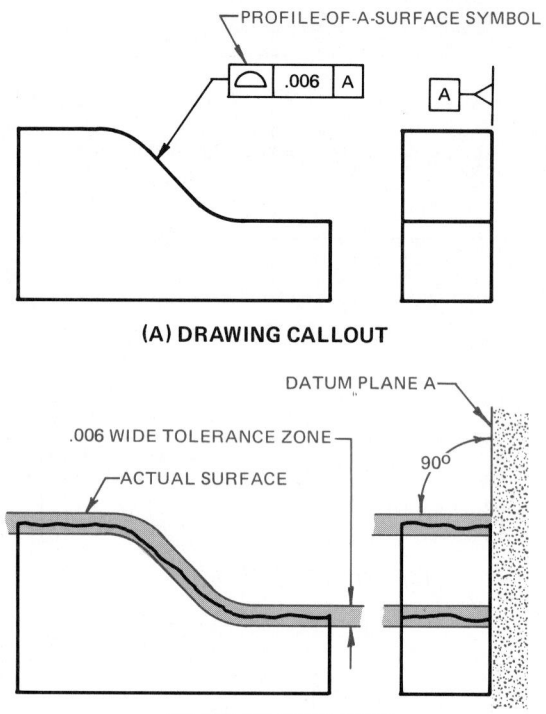

FIG. 16-13-13   Profile-of-a-surface tolerance referenced to a datum.

FIG. 16-13-14   Profile-of-a-surface tolerance referenced to two datums.

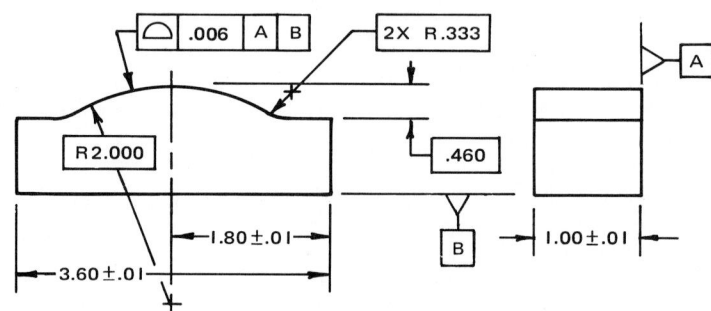

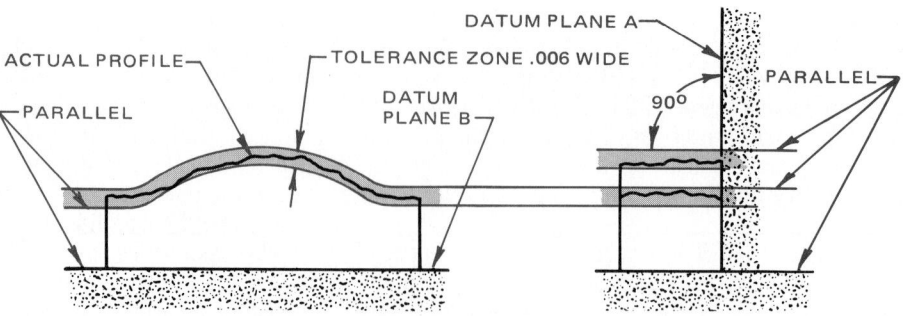

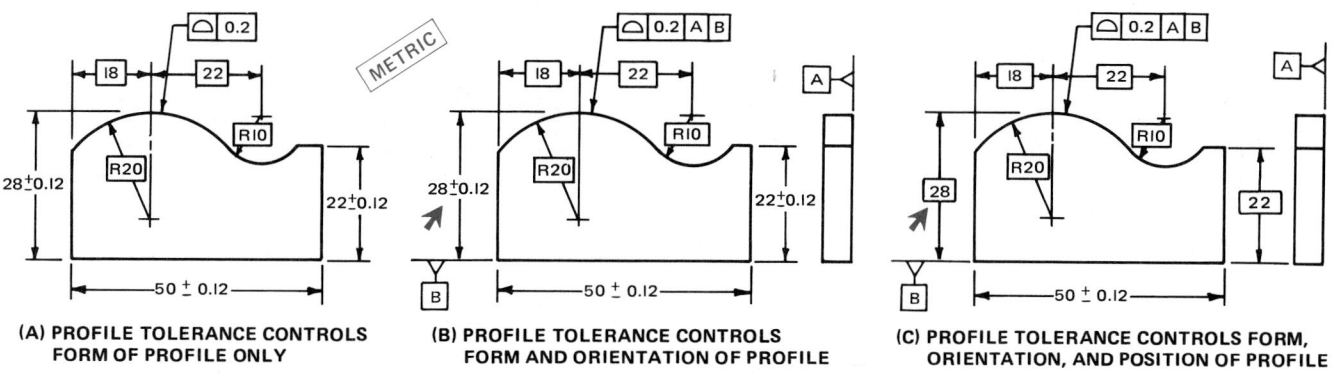

FIG. 16-13-15 Comparison of profile tolerances.

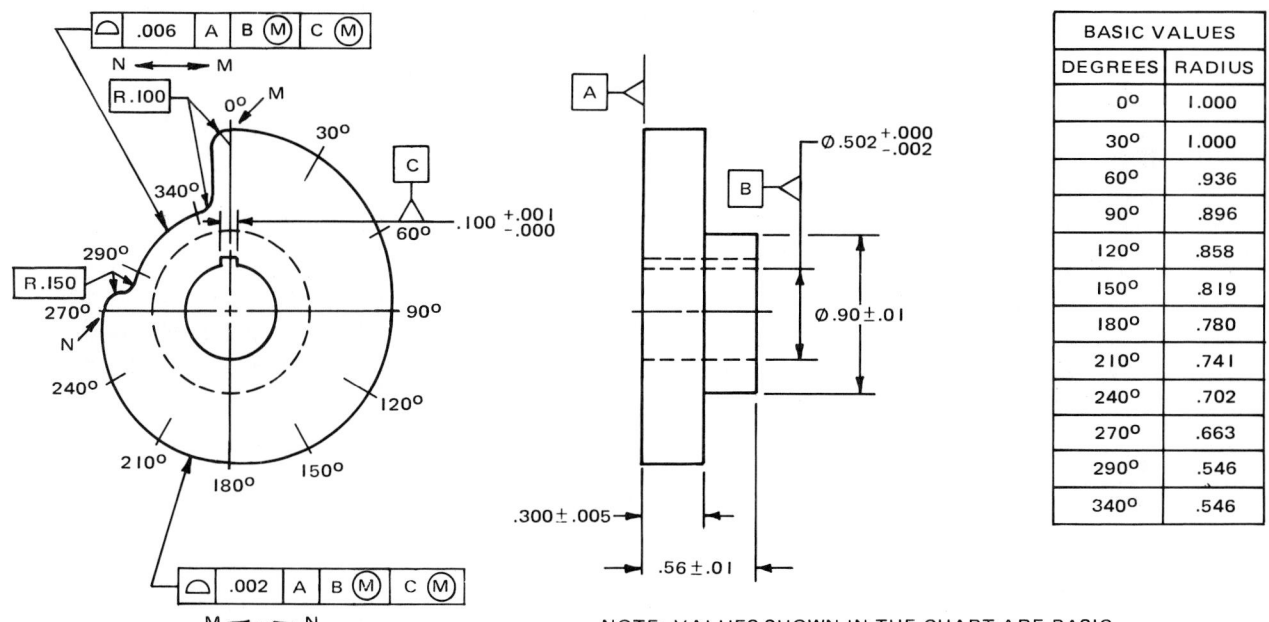

NOTE: VALUES SHOWN IN THE CHART ARE BASIC.

FIG. 16-13-16 Profile-of-a-surface tolerance controls form and size of cam profile.

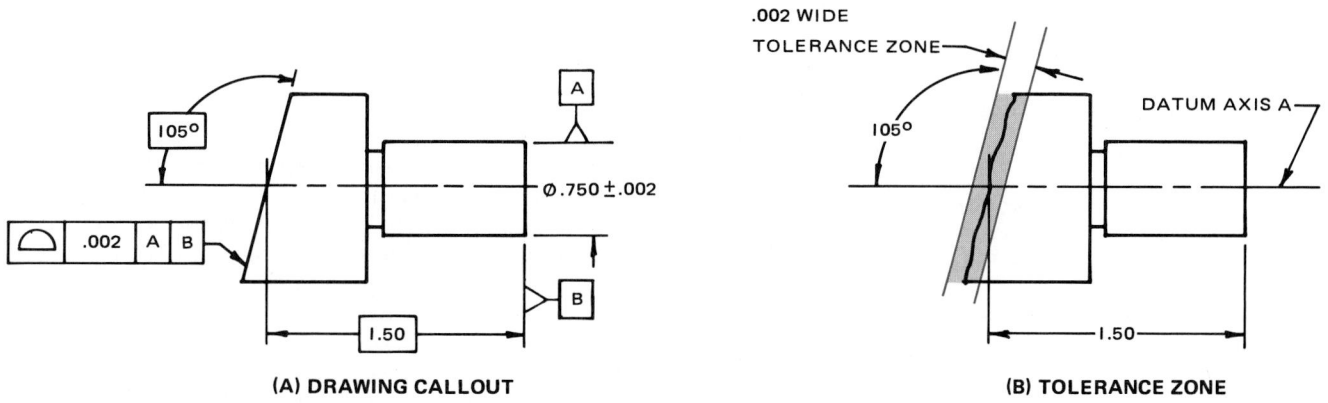

FIG. 16-13-17 Specifying profile-of-a-surface tolerance for a plane surface.

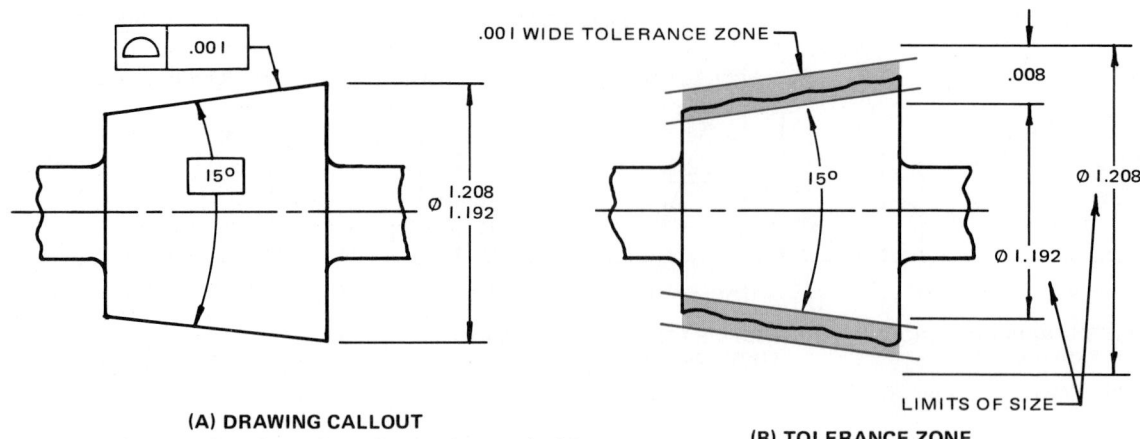

(A) DRAWING CALLOUT

(B) TOLERANCE ZONE

FIG. 16-13-18   Specifying profile-of-a-surface tolerance for a conical feature.

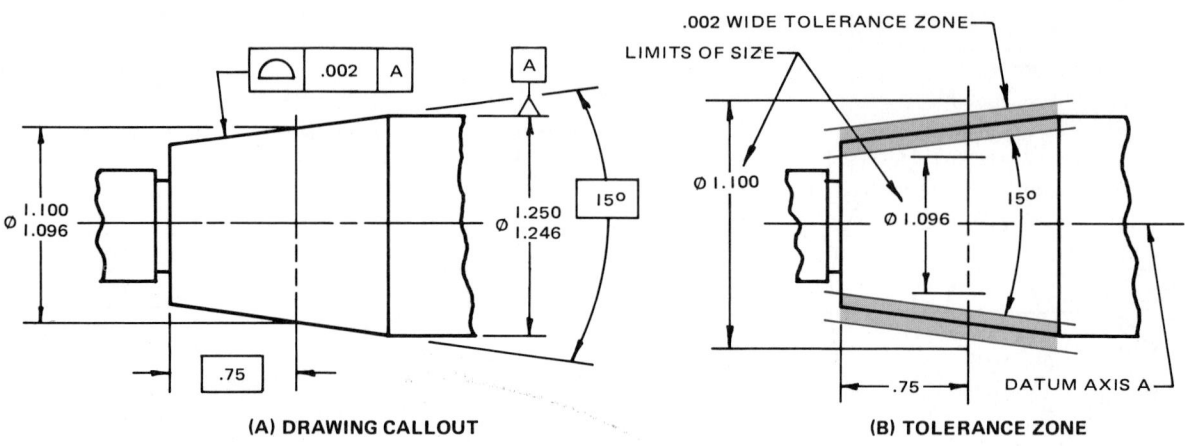

(A) DRAWING CALLOUT

(B) TOLERANCE ZONE

FIG. 16-13-19   Specifying profile-of-a-surface tolerance for a conical feature referenced to a datum.

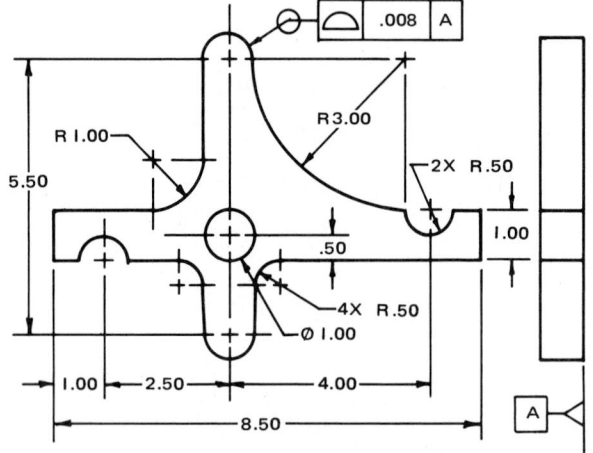

FIG. 16-13-20   Profile-of-a-surface tolerance required for all around the surface.

16-14 CORRELATIVE TOLERANCES

Correlative geometric tolerancing refers to tolerancing for the control of two or more features intended to be correlated in position or attitude. Examples of such correlated tolerancing include coplanarity, for control of two or more flat surfaces; positional tolerance at MMC, for symmetrical relationships, such as control of features equally disposed about a center line; concentricity and coaxiality, for control of features having common axes or center lines; and runout, for control of surfaces related to an axis. These are all tolerances of location; special symbols have been provided for some of them to clarify and simplify drawing callout requirements.

When position is to be separately controlled, other form or orientation tolerances may be applied to control the correlation of features.

Coplanarity

As defined earlier, coplanarity refers to the relative position of two or more flat surfaces that are intended to lie in the same geometric plane. A profile-of-a-surface tolerance may be used where it is desirable to treat two or more surfaces as a single interrupted or noncontinuous surface (Fig. 16-14-1). Each surface must lie between two parallel planes .003 in. apart. Additionally, both surfaces must be within the specified limits of size. No datum reference is stated in Fig. 16-14-1, as in the case of flatness. Since the orientation of the tolerance zone is established from contact of the part against a reference standard, the datum is established by the surfaces themselves.

Where more than two surfaces are involved, it is desirable to identify the surfaces used in contacting the reference standard to establish the tolerance zone. A datum-identifying symbol is applied and the datum reference letter is added to the feature control frame (Fig. 16-14-2). The two designated surfaces must lie between parallel planes equally disposed about datum the plane

Figure 16-14-3 shows a case where the coplanar surfaces are required to be perpendicular to the axis of a hole.

Concentricity

Concentricity is a condition in which two or more features, such as circles, spheres, cylinders, cones, or hexagons, share a

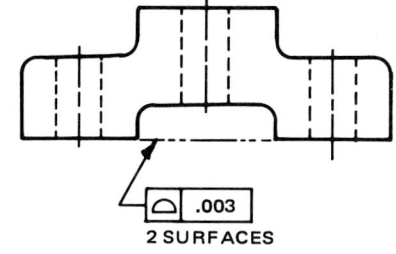

(A) DRAWING CALLOUT

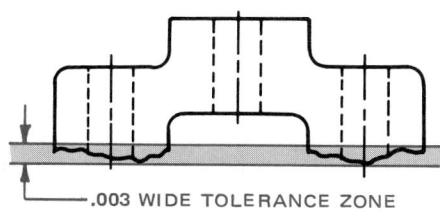

(B) TOLERANCE ZONE

FIG. 16-14-1   Specifying profile-of-a-surface tolerance for coplanar surfaces.

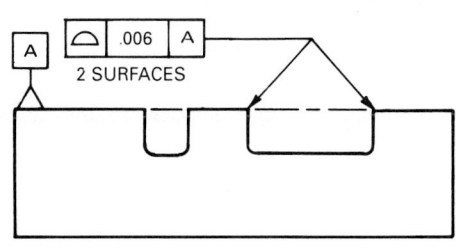

(A) DRAWING CALLOUT

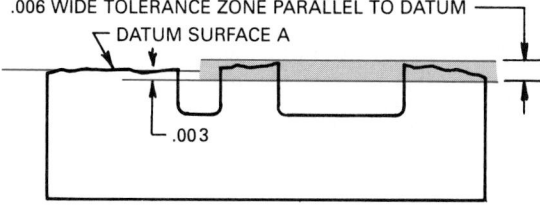

(B) TOLERANCE ZONE

FIG. 16-14-2   Coplanar surfaces with one surface designated as a datum.

common center or axis. An example would be a round hole through the center of a cylindrical part.

A concentricity tolerance is a particular case of a positional tolerance. It controls the permissible variation in position, or eccentricity, of the center line of the controlled feature in relation to the axis of the datum feature. A concentricity tolerance specifies a cylindrical tolerance zone having a diameter equal to the specified tolerance whose axis coincides with a datum axis. The feature control frame is located below, or attached to, a leader-directed callout or dimension pertaining to the feature. The center of all cross sections normal to the axis of the controlled feature must lie within this tolerance zone.

The geometric characteristic symbol used for concentricity consists of two concentric circles, having diameters equal to the actual height (1:1) and 1.5 times the height of lettering used on the drawing (Fig. 16-14-4).

Concentricity—RFS   Concentricity tolerance and the datum reference, because of their unique characteristics, are always used on an RFS basis. (ISO permits concentricity tolerances to be used on an MMC basis.)

Figure 16-14-5 (pg. 548) shows a common type of part where the outer diameter is required to be concentric with the center hole, which is designated as a datum feature.

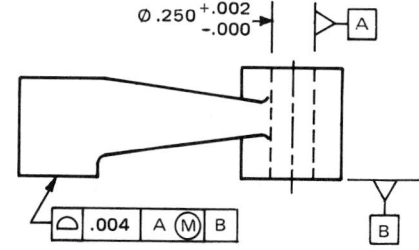

(A) DRAWING CALLOUT

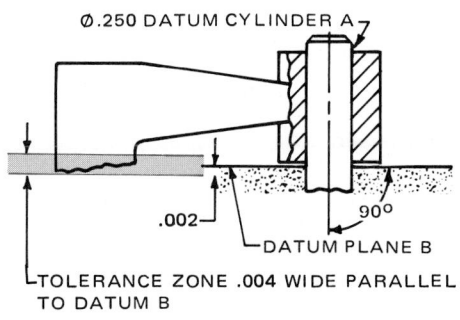

(B) TOLERANCE ZONE

FIG. 16-14-3   Surface referenced to a datum system.

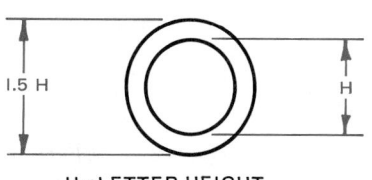

H =LETTER HEIGHT

FIG. 16-14-4   Concentricity symbol.

547

Figure 16-14-6 shows an example of a part where two cylindrical portions are intended to be coaxial. This figure also illustrates the extreme errors of eccentricity and parallelism that the concentricity tolerance would permit. The absence of a modifying symbol after the tolerance indicates that RFS applies.

A concentricity tolerance may be referenced to a datum system, instead of to a single datum, to meet certain functional requirements. Figure 16-14-7 gives an example in which the tolerance zone is perpendicular to datum *A* and also concentric with the axis of datum *B* in the plane of datum *A*.

Coaxiality

Coaxiality is very similar to concentricity in which two or more circular or similar features are arranged with their axes in the

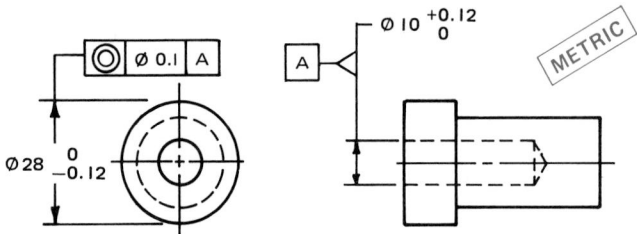

FIG. 16-14-5   Cylindrical part with concentricity tolerance.

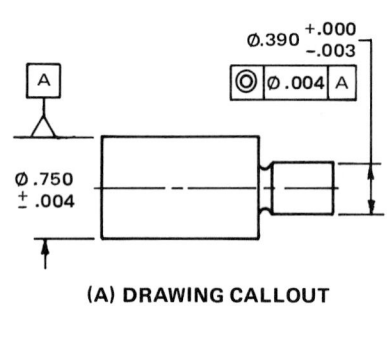

(A) DRAWING CALLOUT

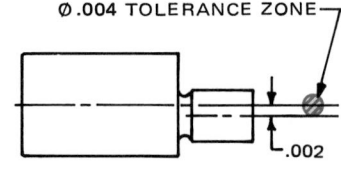

(B) EXTREME ECCENTRICITY

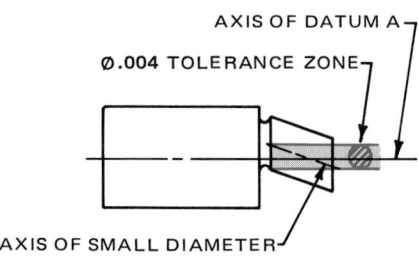

(C) EXTREME ANGULAR VARIATION

FIG. 16-14-6   Concentricity of cylindrical features.

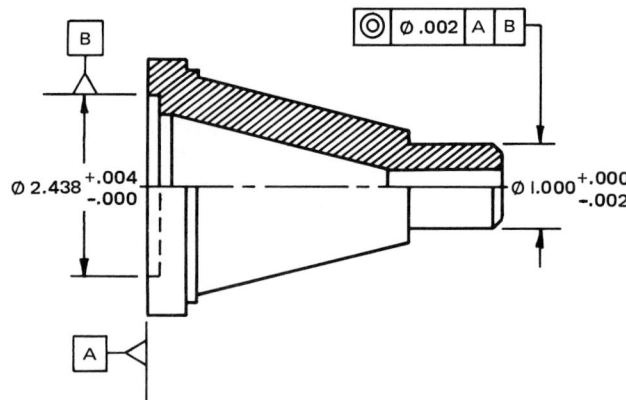

FIG. 16-14-7   Concentricity referenced to a datum system.

same straight line. Examples might be a counterbored hole or a shaft having parts along its length turned to different diameters.

There are three methods of controlling coaxial features. They are positional tolerancing, runout tolerancing, and concentricity tolerancing. Selection of the proper control depends upon the functional requirements of the part.

Positional Tolerance Control   Where the surfaces of revolution are cylindrical and the control of the axes can be applied on an MMC basis, positional tolerancing is recommended (Fig. 16-14-8). This type of tolerancing permits the use of a simple receiver gage for inspection.

Where it is necessary to control coaxiality of related features within their limits of size, a zero tolerance is specified. The datum feature is normally specified on an MMC basis. Where both features are at MMC, boundaries of perfect form are thereby established that are truly coaxial. Variations in coaxiality are permitted only where the features depart from their MMC size.

Runout Tolerance Control   Where a combination of surfaces of revolution is cylindrical, conical, or spherical relative to a common datum axis, a runout tolerance is recommended. MMC is not applicable to runout tolerances as it controls elements of the surface.

Concentricity Tolerance Control   Unlike the controls covered above, where measurements taken along a surface of revolution are cumulative variations of form and displacement (eccentricity), a concentricity tolerance requires the establishment and verification of axes irrespective of surface conditions.

Alignment of Coaxial Holes   A positional tolerance is used to control the alignment of two or more holes on a common axis. It is used where a tolerance of location alone does not provide the necessary control of alignment of these holes and a separate requirement must be specified. Figure 16-14-9 shows an example of four coaxial holes of the same size. Where holes are of different specified sizes and the same requirements apply to all holes, a single feature control system is used, supplemented by a note, such as 2 COAXIAL HOLES (Fig. 16-14-10, pg. 550).

Symmetry

Symmetry is a condition in which a feature or features are positioned about the center plane of a datum feature. Where it is

FIG. 16-14-8 Positional tolerancing for coaxiality.

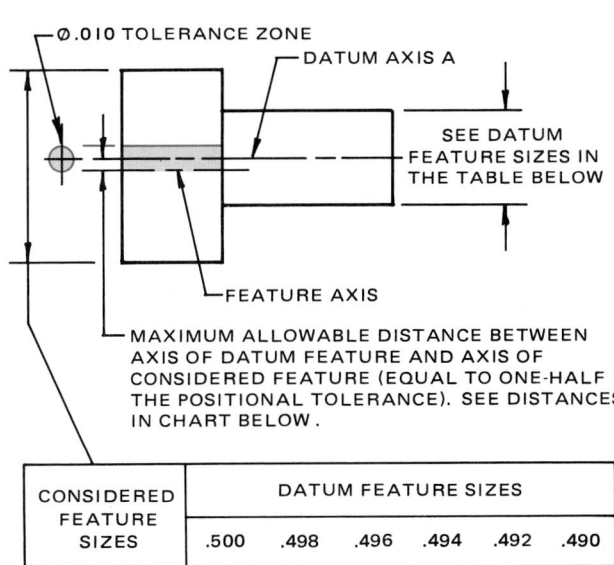

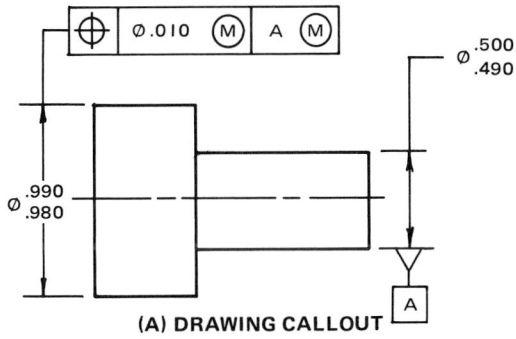

(A) DRAWING CALLOUT

CONSIDERED FEATURE SIZES	DATUM FEATURE SIZES					
	.500	.498	.496	.494	.492	.490
.990	.005	.006	.007	.008	.009	.010
.988	.006	.007	.008	.009	.010	.011
.986	.007	.008	.009	.010	.011	.012
.984	.008	.009	.010	.011	.012	.013
.982	.009	.010	.011	.012	.013	.014
.980	.010	.011	.012	.013	.014	.015

(B) ALLOWABLE DISTANCES BETWEEN AXES

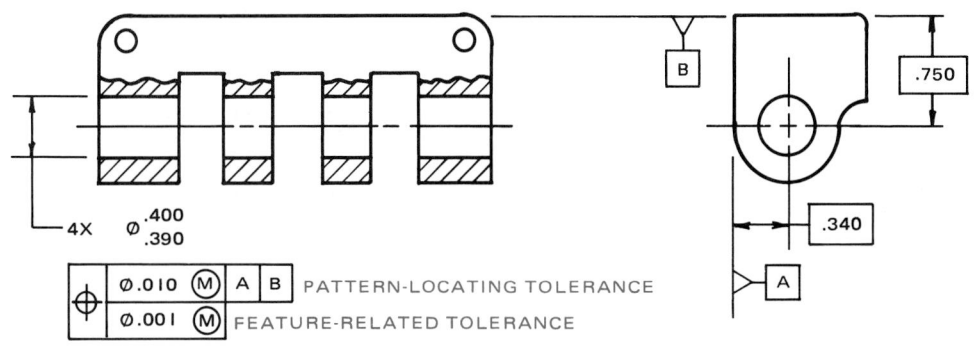

(A) DRAWING CALLOUT

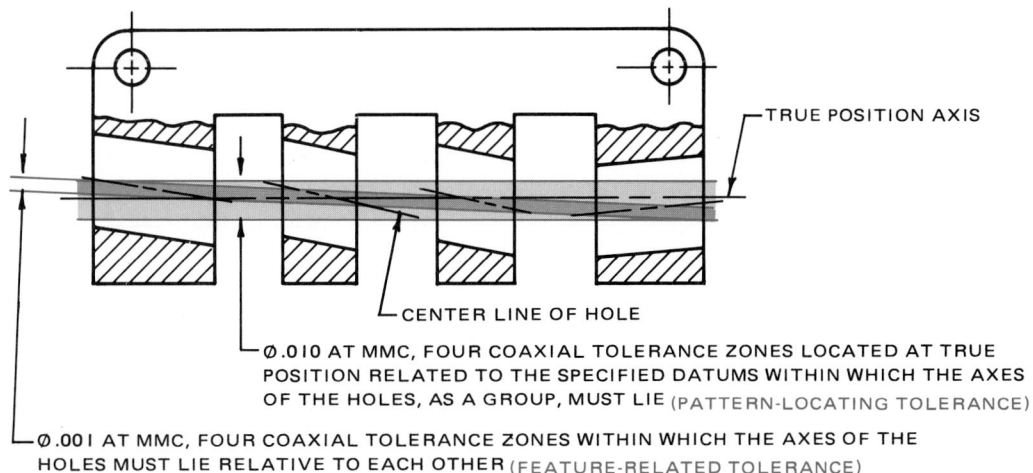

(B) TOLERANCE ZONES

FIG. 16-14-9 Positional tolerancing for coaxial holes of the same size.

FIG. 16-14-10   Positional tolerancing for coaxial holes of different size.

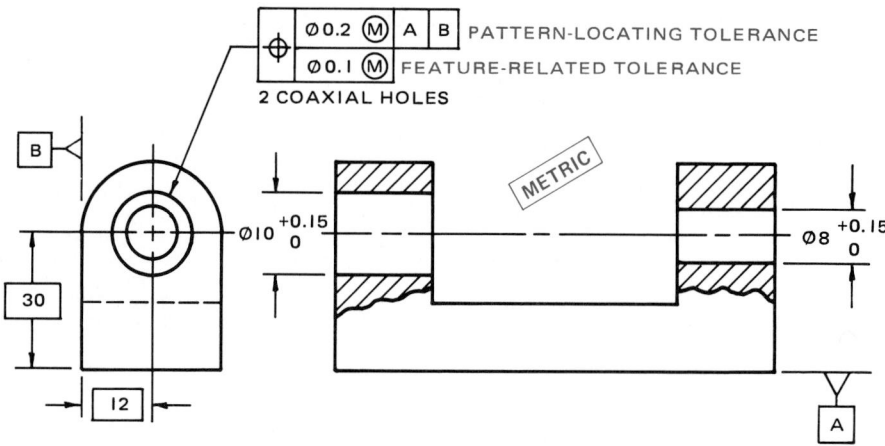

required that a feature be located symmetrically with respect to the center plane of a datum feature, positional tolerancing is used. As the tolerance zone is not cylindrical, the diameter symbol in the feature control frame is not shown.

The symbol for symmetry is shown in Fig. 16-13-8. All other countries, including ISO, use this symbol to define symmetry.

A symmetrical relationship may be controlled by specifying a positional tolerance at MMC as illustrated in Figs. 16-14-12 and 16-14-13. The datum feature may be specified either on an MMC or an RFS basis, depending upon the design requirements.

Where it is necessary to control the symmetry of related features within their limits of size, a zero positional tolerance at MMC is specified and the datum feature is normally specified on an MMC basis. Boundaries of perfect form are thereby established that are truly symmetrical when both features depart from their MMC size toward LMC.

Some designs may require a control of the symmetrical relationship between features regardless of their actual sizes. In such cases, both the specified positional tolerance and the datum reference apply on an RFS basis (Fig. 16-14-14). The center plane of the slot must lie between two parallel planes, .0025 in. apart regardless of the sizes of both datum *B* and the feature, which are equally disposed about the center plane of datum *B*.

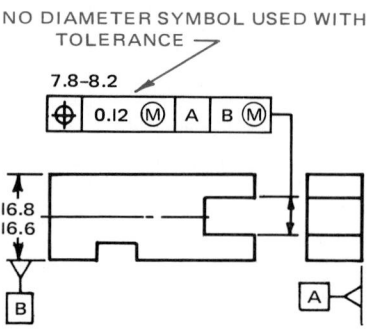

(A) DRAWING CALLOUT

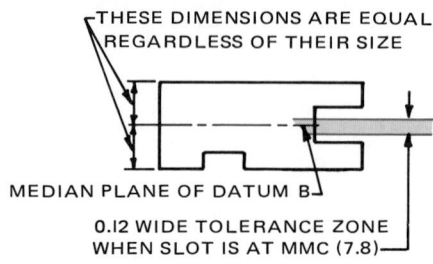

(B) TOLERANCE ZONE

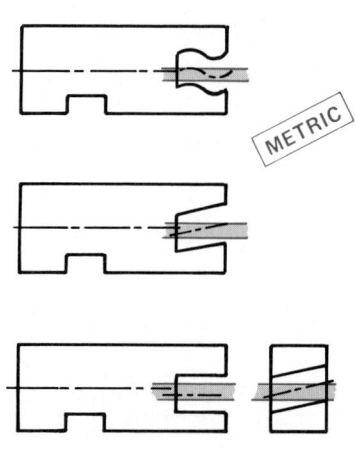

(C) PERMISSIBLE VARIATIONS

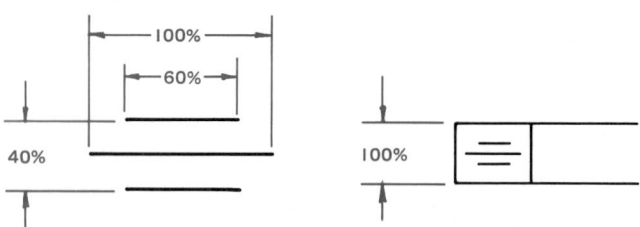

NOTE: DIMENSIONS REFER TO PERCENTAGES OF THE FEATURE CONTROL FRAME HEIGHT.

(A) SYMMETRY SYMBOL   **(B) SHOWN IN THE FEATURE CONTROL FRAME**

FIG. 16-14-11   Symmetry symbol.

FIG. 16-14-12   Positional tolerancing at MMC for symmetry.

FIG. 16-14-13   Examples in tolerancing symmetrically located features.

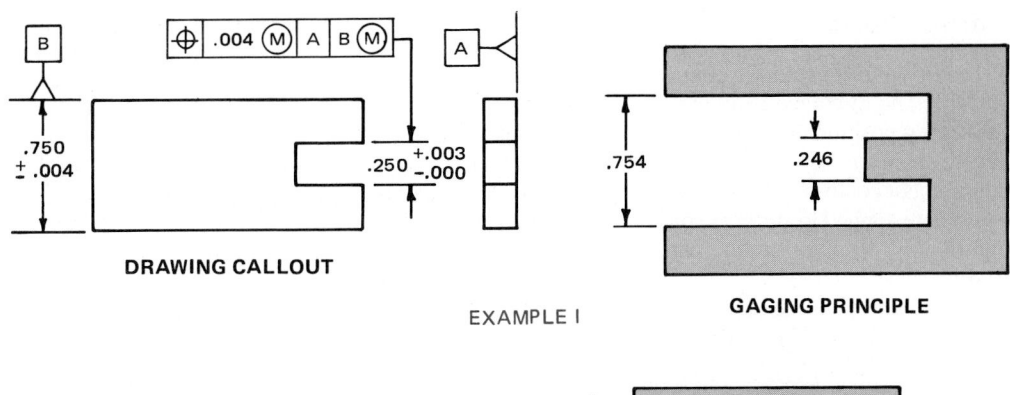

DRAWING CALLOUT

EXAMPLE I

GAGING PRINCIPLE

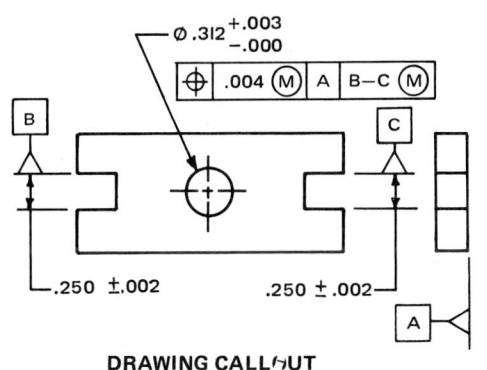

DRAWING CALLOUT

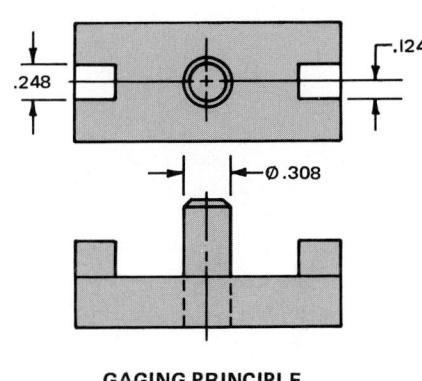

GAGING PRINCIPLE

EXAMPLE 2

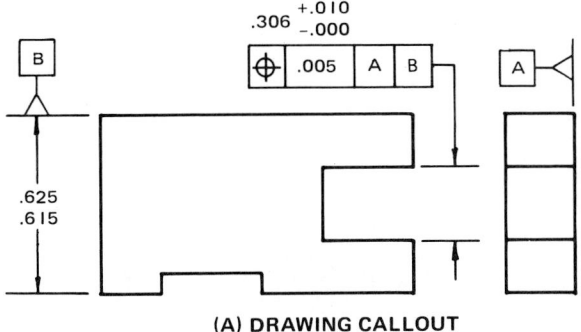

(A) DRAWING CALLOUT

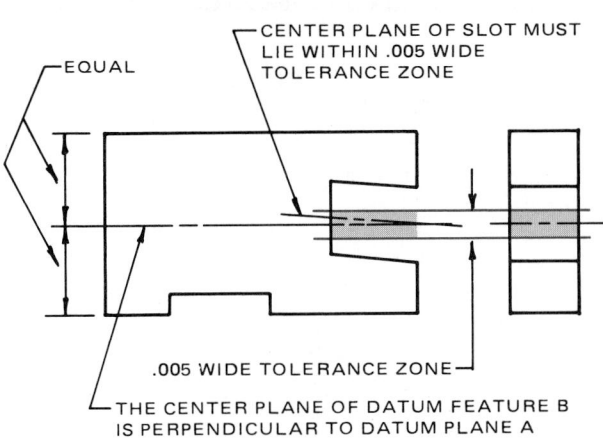

CENTER PLANE OF SLOT MUST LIE WITHIN .005 WIDE TOLERANCE ZONE

EQUAL

.005 WIDE TOLERANCE ZONE

THE CENTER PLANE OF DATUM FEATURE B IS PERPENDICULAR TO DATUM PLANE A

(B) INTERPRETATION

FIG. 16-14-14   Positional tolerancing for symmetry—RFS.

Runout

Runout is a composite tolerance used to control the functional relationship of one or more features of a part to a datum axis. The types of features controlled by runout tolerances include those surfaces constructed around a datum axis and those constructed at right angles to a datum axis (Fig. 16-14-15).

Each feature must be within its runout tolerance when rotated about the datum axis. The tolerance specified for a controlled surface is the total tolerance or full indicator movement (FIM) in inspection and international terminology.

There are two types of runout control, circular runout and total runout. The type used is dependent upon design requirements and manufacturing considerations. See the geometric characteristic symbols for runout in Fig. 16-14-16 (pg. 552).

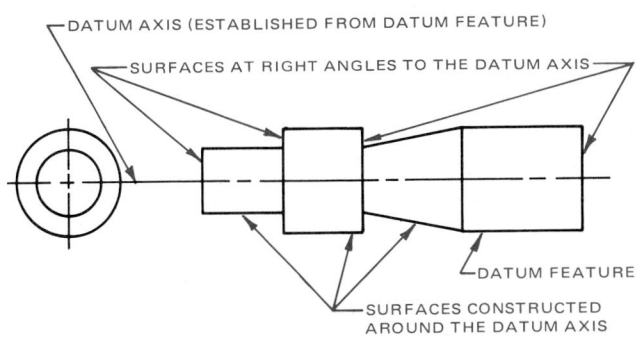

DATUM AXIS (ESTABLISHED FROM DATUM FEATURE)

SURFACES AT RIGHT ANGLES TO THE DATUM AXIS

DATUM FEATURE

SURFACES CONSTRUCTED AROUND THE DATUM AXIS

FIG. 16-14-15   Features applicable to runout tolerancing.

Circular Runout

Circular runout provides control of circular elements of a surface. The tolerance is applied independently at any usual measuring position as the part is rotated 360° (Fig. 16-14-17). Where applied to surfaces constructed around a datum axis, circular runout controls variations such as circularity and coaxiality. Where applied to surfaces constructed at right angles to the datum axis, circular runout controls wobble at all diametral positions.

Where a runout tolerance applies to a specific portion of a surface, a thick chain line is drawn adjacent to the surface profile to show the desired length. Basic dimensions are used to define the extent of the portion so indicated (Fig. 16-14-17).

Total Runout

Total runout concerns the runout of a complete surface, not merely the runout of each circular element. For measurement purposes the checking indicator must traverse the full length or extent of the surface while the part is revolved about its datum axis. Measurements are made over the whole surface without resetting the indicator. Total runout is the difference between the lowest indicator reading in any position and the highest reading in that or in any other position on the same surface. Thus in Fig. 16-14-18 the tolerance zone is the space between two concentric cylinders separated by the specified tolerance and coaxial with the datum axis.

Establishing Datums

In many examples the datum axis has been established from centers drilled in the two ends of the part, in which case the part is mounted between centers for measurement purposes. This is an ideal method of mounting and revolving the part. When centers are not provided, any cylindrical or conical surface may be used to establish the datum axis if it is chosen on the basis of the functional requirements of the part.

Figure 16-14-19 shows a simple, external cylindrical feature specified as the datum feature.

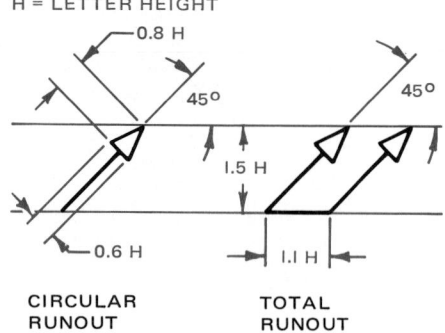

FIG. 16-14-16 Runout symbols.

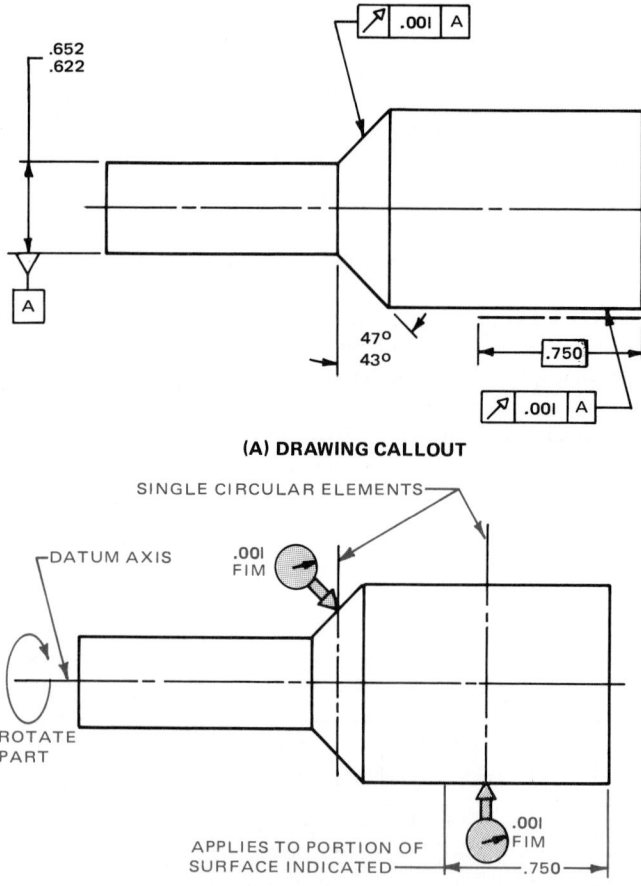

FIG. 16-14-17 Specifying circular runout relative to a datum diameter.

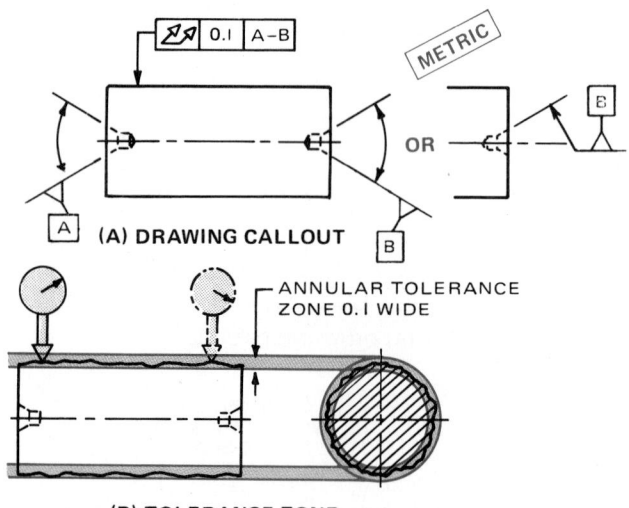

FIG. 16-14-18 Tolerance zone for total runout.

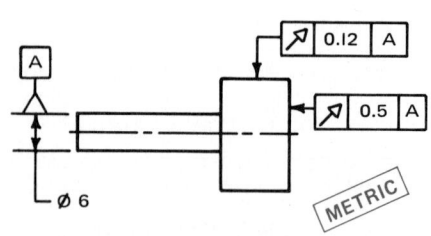

FIG. 16-14-19 Cylindrical datum feature.

FIG. 16-14-20 Specifying runout relative to two datum diameters.

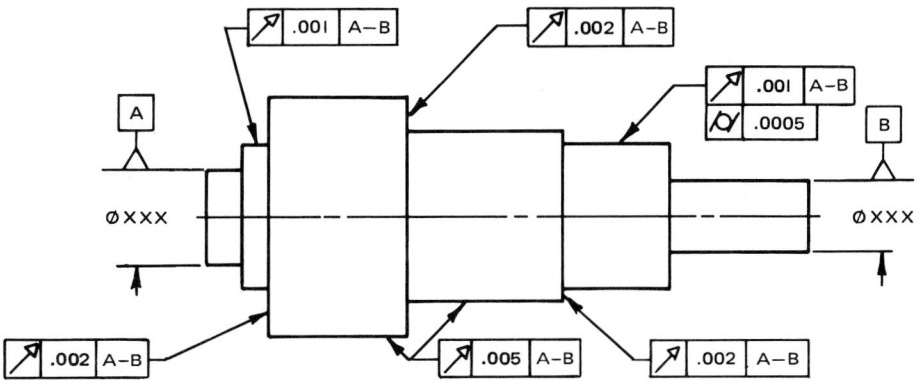

Figure 16-14-20 illustrates the application of runout tolerances where two datum diameters act as a single datum axis to which the features are related.

REFERENCES AND SOURCE MATERIAL

1. ASME Y14.5M–1994, *Dimensioning and Tolerancing.*
2. CAN/CSA B78.2-M91, *Dimensioning and Tolerancing of Technical Drawings.*
3. ISO drawing standards.

ASSIGNMENTS

See Assignments 42 through 48 for Unit 16-14 on pages 584 through 586.

16-15 POSITIONAL TOLERANCING FOR NONCYLINDRICAL FEATURES

Positional tolerancing undoubtedly finds its greatest usefulness in controlling the position of holes, but the same method of tolerancing is equally useful for the control of many other features, such as slots, grooves, tabs, bosses, and studs.

As with holes, positional tolerancing for these miscellaneous features should be specified on an MMC basis wherever possible because of the difficulty of measurement and evaluation when specified on an RFS basis. For this reason most of the examples in this unit are on an MMC basis.

Noncircular Features at MMC

Where a positional tolerance of a noncircular feature (e.g., a slot) is specified at MMC, the following conditions apply.

1. In terms of the surface of a slot (Fig. 16-15-1):
   - The slot must be within the limits of size.
   - A gage having a width equal to the virtual condition of the slot (MMC minus the positional tolerance) and located at true position must be capable of entering the slot.

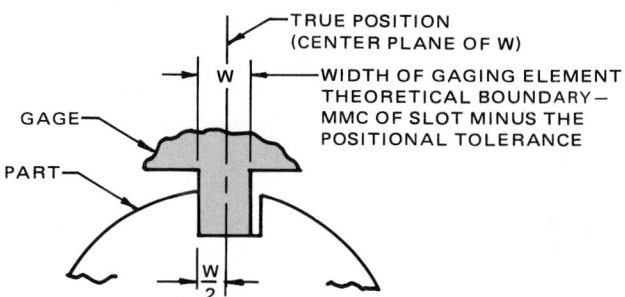

SLOT POSITION MAY VARY AS SHOWN, BUT NO POINT OF EITHER SIDE SURFACE SHALL BE INSIDE OF W

(A) SLOT SHOWN IN EXTREME RIGHT POSITION

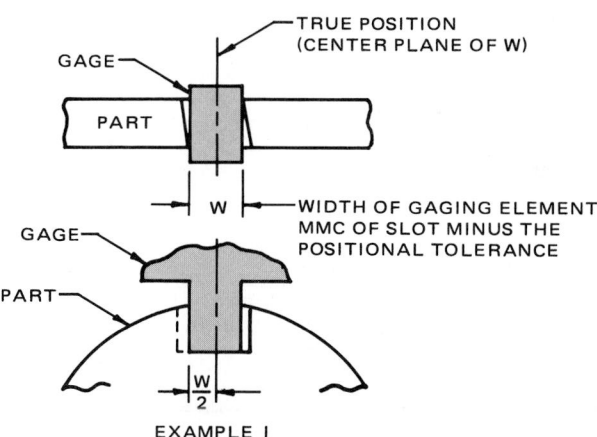

EXAMPLE I

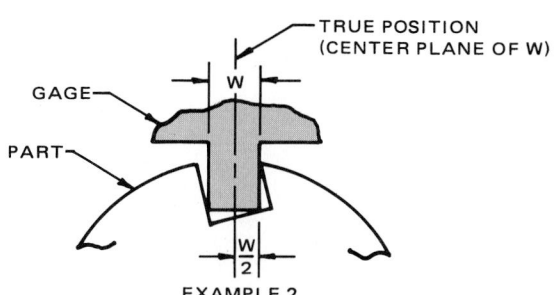

EXAMPLE 2

SIDE SURFACES OF SLOT MAY VARY IN ATTITUDE, PROVIDED W IS NOT VIOLATED AND SLOT WIDTH IS WITHIN LIMITS OF SIZE

(B) EXTREME ANGLE VARIATIONS OF SLOT

FIG. 16-15-1 Boundaries for the surface of a slot at MMC.

553

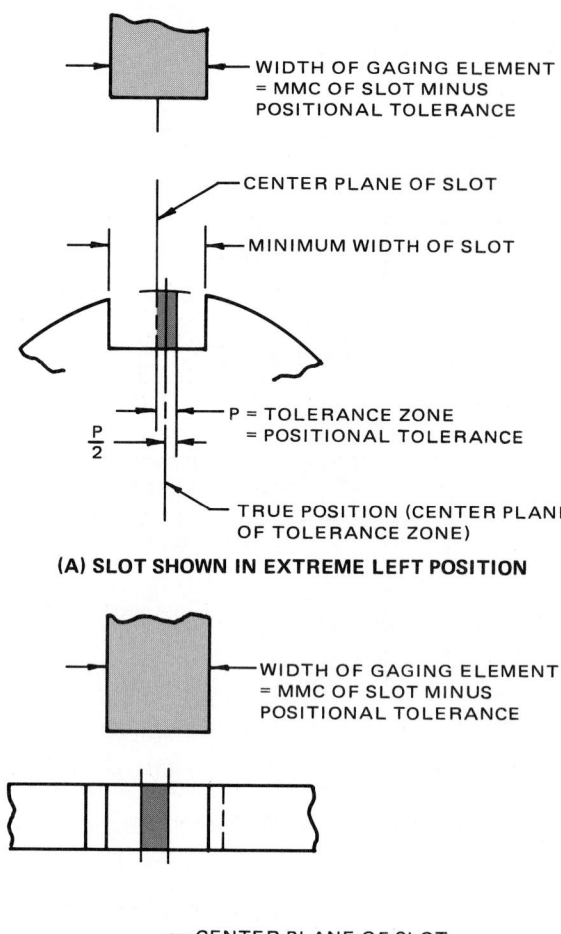

(A) SLOT SHOWN IN EXTREME LEFT POSITION

WIDTH OF GAGING ELEMENT
= MMC OF SLOT MINUS
POSITIONAL TOLERANCE

CENTER PLANE OF SLOT

MINIMUM WIDTH OF SLOT

P = TOLERANCE ZONE
= POSITIONAL TOLERANCE

$\dfrac{P}{2}$

TRUE POSITION (CENTER PLANE
OF TOLERANCE ZONE)

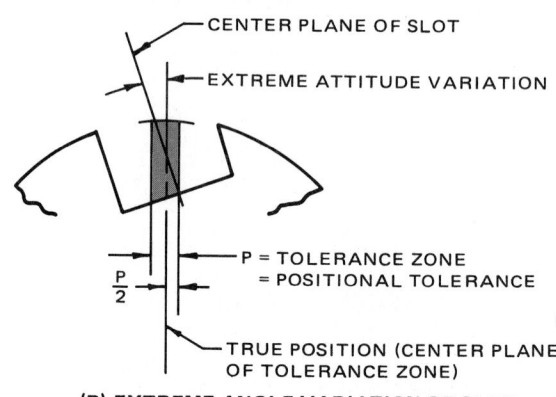

WIDTH OF GAGING ELEMENT
= MMC OF SLOT MINUS
POSITIONAL TOLERANCE

CENTER PLANE OF SLOT

EXTREME ATTITUDE VARIATION

P = TOLERANCE ZONE
= POSITIONAL TOLERANCE

$\dfrac{P}{2}$

TRUE POSITION (CENTER PLANE
OF TOLERANCE ZONE)

(B) EXTREME ANGLE VARIATION OF SLOT

FIG. 16-15-2   Tolerance zone for the center plane of a slot at MMC.

FIG. 16-15-3   Positional tolerancing applied to slots at MMC.

2. In terms of the center plane of a slot (Fig. 16-15-2) the gage must be capable of entering the slot when:
- The slot is at MMC.
- The gage width is equal to the virtual condition of the slot (MMC minus the positional tolerance).
- The center plane of the slot lies within the positional tolerance zone.

Slots—Straight Line Configuration   Slots and grooves arranged in a straight line are dimensioned by specifying the width of the slots as a toleranced dimension and the center distance or distances between them as basic dimensions. The positional tolerance is associated with the slot size dimension, as shown in Fig. 16-15-3.

Tabs or Projections   The same tolerancing methods used for slots are applicable to a series of tabs or projections, as shown in Fig. 16-15-4A. The gage for such parts must have slots of a width equal to the maximum material size plus the positional tolerance, as shown in Fig. 16-15-4B.

If the requirement applies to only one side or face of the tabs or slots, as shown in Fig. 16-15-5, the features cannot be treated as features of size, and the maximum material principle cannot be applied. Thus the tolerance is interpreted to mean that the entire face of each tab or tooth must lie within a tolerance zone bounded by two parallel planes. These planes are separated by the specified tolerance and are located at true position perpendicular to datum A.

Note that the advantage of this method of tolerancing over coordinate dimensioning and tolerancing is that the positional tolerance applies from every tooth to every other tooth. There is no accumulation of tolerances. Errors of form (flatness) and orientation are included within the positional tolerance.

If the positional tolerance in Fig. 16-15-5 applied only to the position of the teeth relative to one another, that is, if datum C was not specified, the position of each tooth would be measured similarly but evaluation would be different. The variations might be calculated in the same manner, except that the distance from the surface of the first tooth would be used instead of datum C.

Slots—Circular Configuration   Figure 16-15-6 (pg. 556) shows a circular configuration of slots in which the positional tolerance controls their position relative to the center hole and one flat face. As both the tolerance and datum B are specified on an MMC basis, inspection can be performed with a functional "go" gage. The diameter of the central plug in this gage

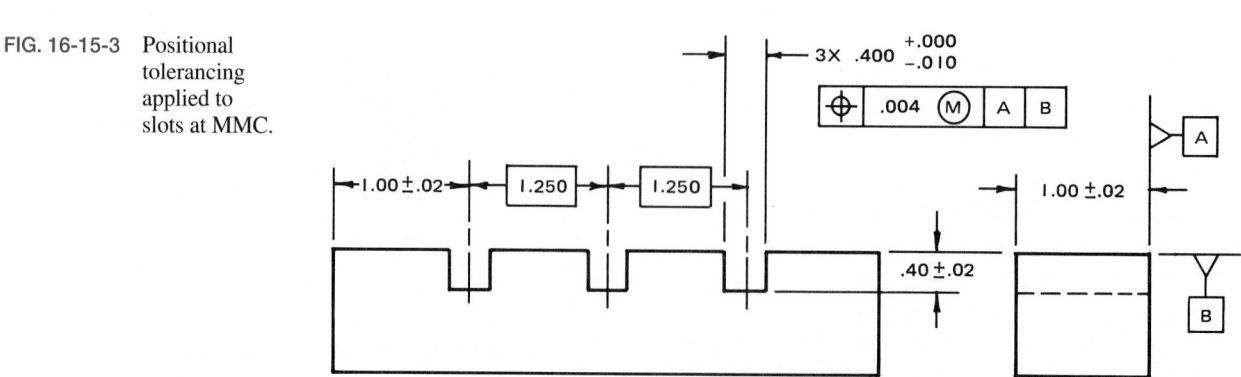

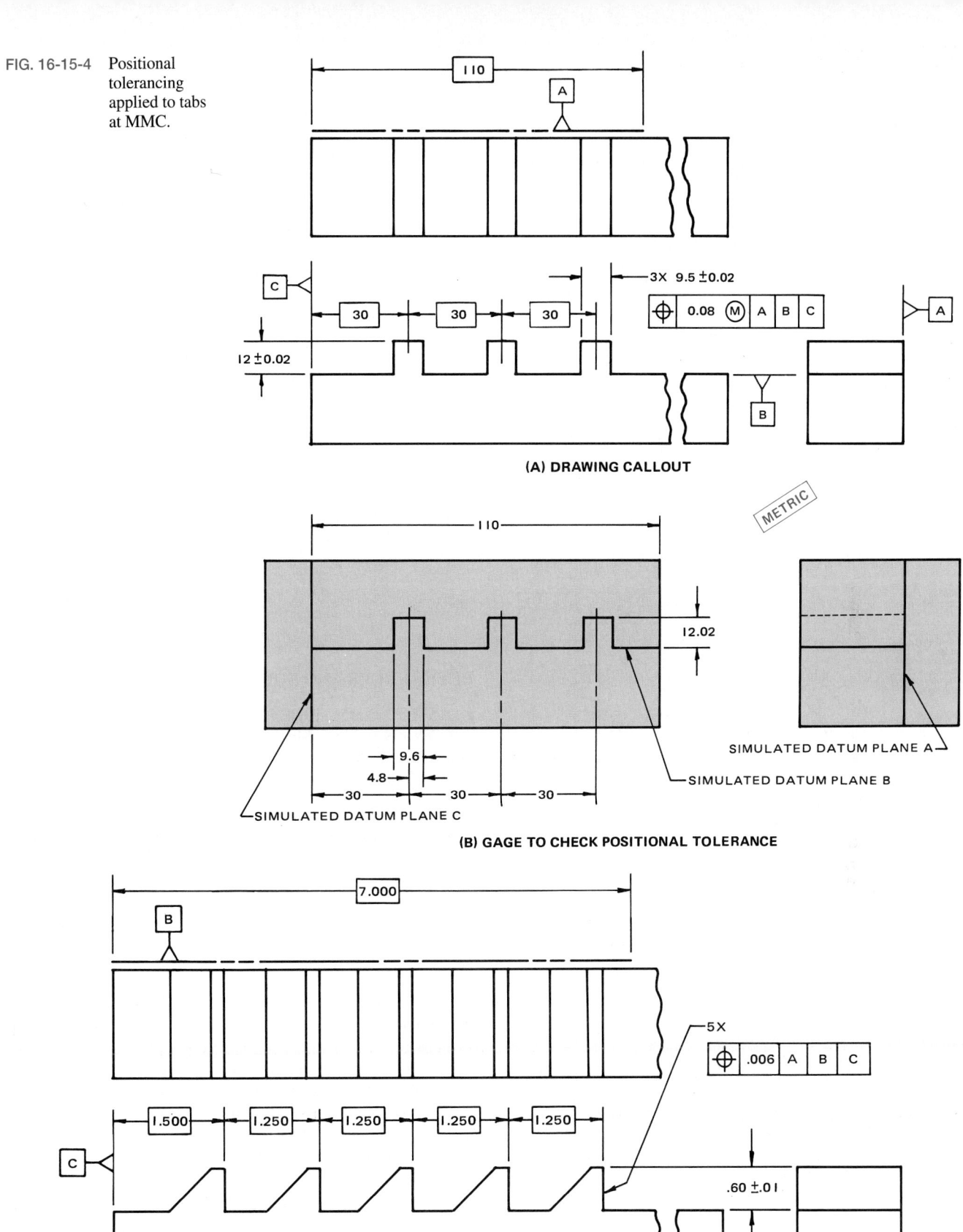

FIG. 16-15-4 Positional tolerancing applied to tabs at MMC.

(A) DRAWING CALLOUT

METRIC

(B) GAGE TO CHECK POSITIONAL TOLERANCE

FIG. 16-15-5 Positional tolerancing applied to tabs at RFS.

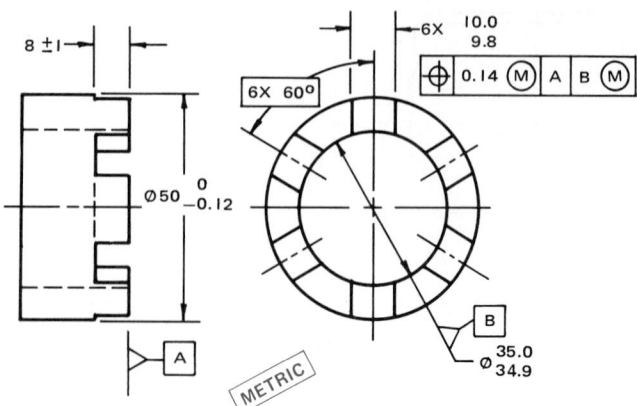

FIG. 16-15-6   Positional tolerancing applied to a configuration of slots at MMC.

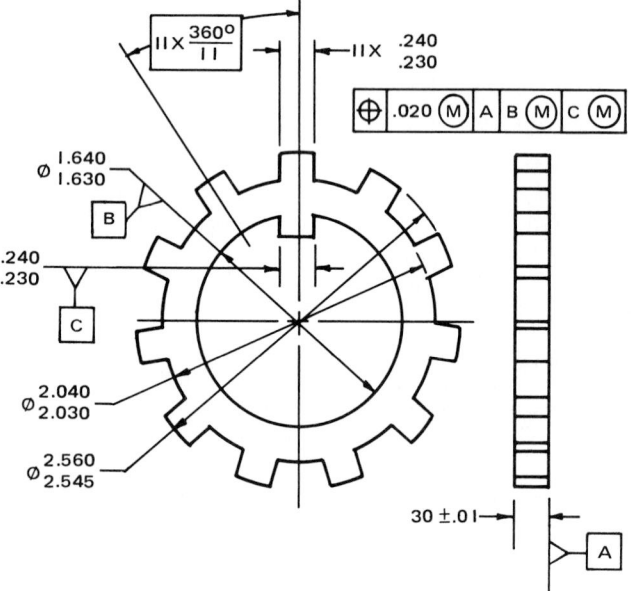

FIG. 16-15-7   Positional tolerancing applied to tabs at MMC.

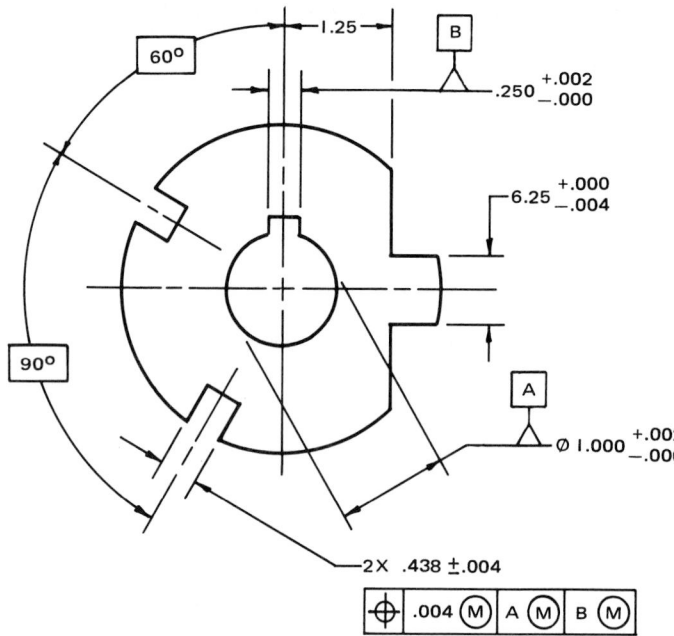

FIG. 16-15-8   Circular configuration of tab and slots at MMC.

REFERENCES AND SOURCE MATERIAL

1. ASME Y14.5M–1994, *Dimensioning and Tolerancing.*
2. CAN/CSA B78.2-M91, *Dimensioning and Tolerancing of Technical Drawings.*
3. ISO drawing standards.

ASSIGNMENTS

See Assignments 49 and 50 for Unit 16-15 on page 587.

16-16 POSITIONAL TOLERANCING FOR MULTIPLE PATTERNS OF FEATURES

Positional tolerancing may also be used to locate multiple patterns of features if each pattern is referenced to common datums and referenced in the same order of precedence.

In Figure 16-16-1 each pattern of feature is located relative to common datum features not subject to size tolerances. Since all locating dimensions are basic and all measurements are from a common datum reference frame, verification of positional tolerance requirements for the part can be collectively accomplished in a single set-up or gage. Figure 16-16-2 shows the tolerance zones for Fig. 16-16-1. The actual centers of all holes must lie on or within their respective tolerance zones when measured from datums *A, B,* and *C.*

Multiple patterns of features, located by basic dimensions from common datum features that are subject to size tolerances, are also considered a single composite pattern if their respective feature control frames contain the same datums in the same order of precedence with the same modifying symbols. If such interrelationship is not required between one pattern and any other pattern or patterns, a notation such as SEP

is equal to the maximum material size, and the gaging elements surrounding the plug have a width equal to the maximum material size of the slots less the positional tolerance.

In these examples of location of slots and tabs, note that the positional tolerance controls orientation and form (straightness) as well as position as illustrated in Fig. 16-15-7.

Tabs and slots can be used as datum features as well as being controlled by positional tolerances. If a form or orientation tolerance is not specified for such datum features, MMC is used. The width or diameter of the corresponding gaging element is then equal to the maximum material size. As the positional tolerance on the 11 tabs and the 2 datum features *B* and *C* are all on an MMC basis, inspection requires the use of a functional "go" gage, made to the same shape as the part except in reverse, that is, a plug and keyway for datums *B* and *C* and 11 slots to gage the positions of the outer teeth.

Figure 16-15-8 shows another example, where a positional tolerance is specified.

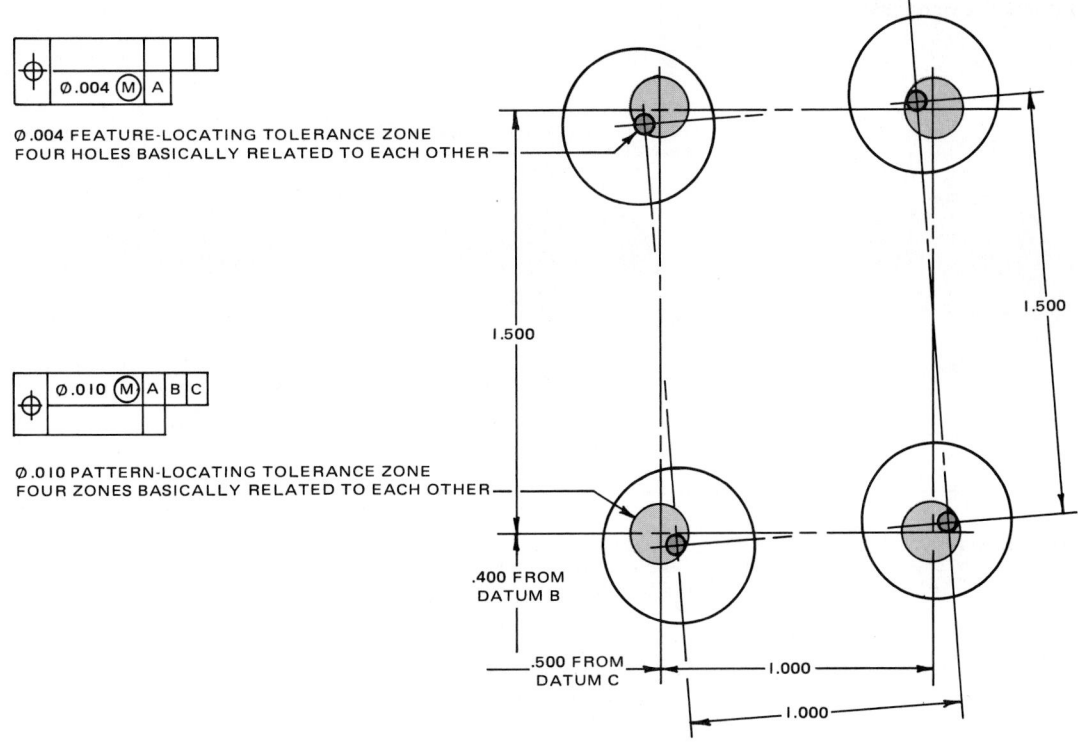

Ø.004 FEATURE-LOCATING TOLERANCE ZONE
FOUR HOLES BASICALLY RELATED TO EACH OTHER

Ø.010 PATTERN-LOCATING TOLERANCE ZONE
FOUR ZONES BASICALLY RELATED TO EACH OTHER

NOTE: THE AXES OF HOLES MUST SIMULTANEOUSLY LIE WITHIN BOTH TOLERANCE ZONES.
VERIFICATION OF PATTERN-LOCATING TOLERANCE AND FEATURE-LOCATING
TOLERANCE IS MADE INDEPENDENTLY OF EACH OTHER.

FIG. 16-16-7   Tolerance zones for the four-hole pattern shown in Fig. 16-16-5.

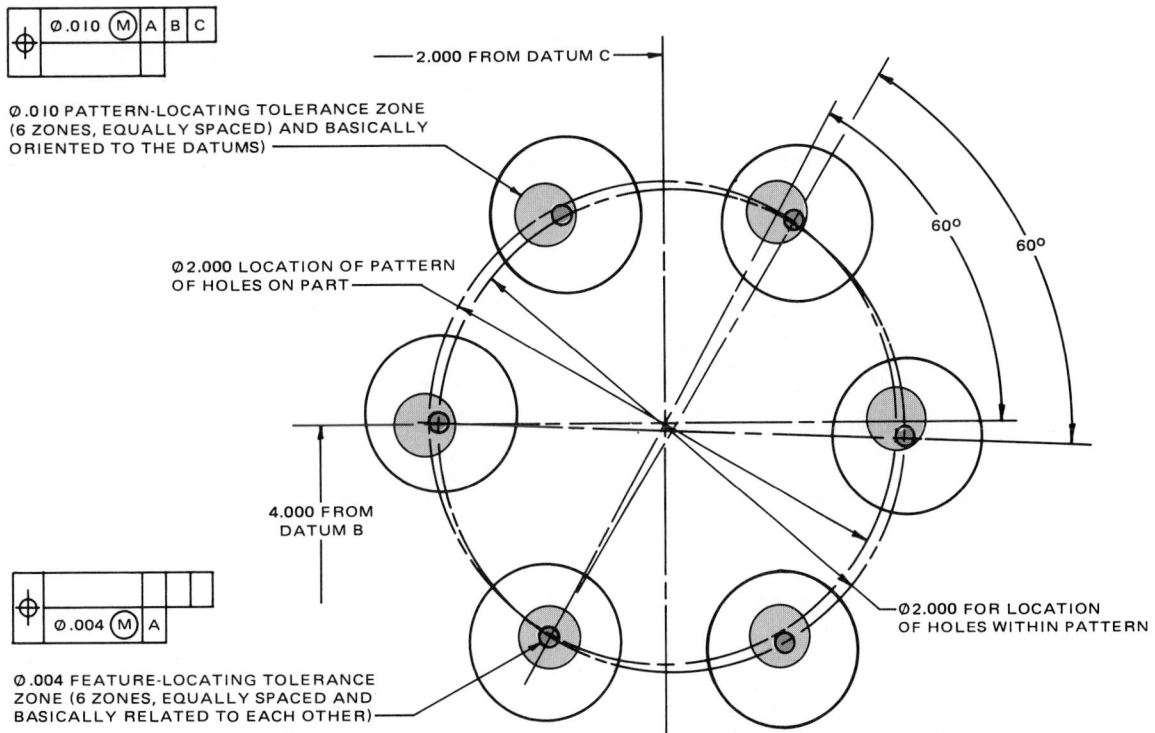

Ø.010 PATTERN-LOCATING TOLERANCE ZONE
(6 ZONES, EQUALLY SPACED) AND BASICALLY
ORIENTED TO THE DATUMS)

Ø2.000 LOCATION OF PATTERN
OF HOLES ON PART

4.000 FROM
DATUM B

Ø.004 FEATURE-LOCATING TOLERANCE
ZONE (6 ZONES, EQUALLY SPACED AND
BASICALLY RELATED TO EACH OTHER)

Ø2.000 FOR LOCATION
OF HOLES WITHIN PATTERN

NOTE: VERIFICATIONS OF FEATURE-LOCATING AND PATTERN-LOCATING TOLERANCE
ZONES ARE MADE INDEPENDENTLY OF EACH OTHER.

AXES OF HOLES MUST SIMULTANEOUSLY LIE WITHIN BOTH TOLERANCE ZONES.

FIG. 16-16-8   Tolerance zones for the six-hole pattern shown in Fig. 16-16-5.

561

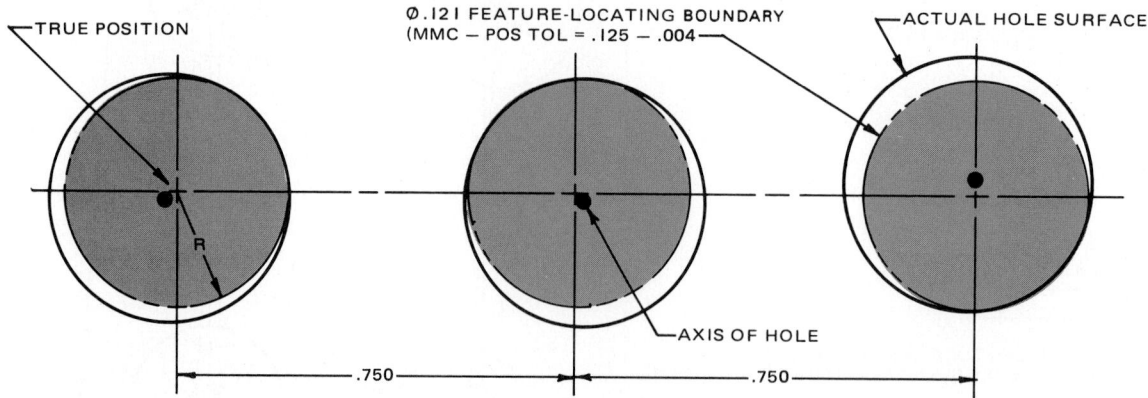

NOTE: — NO PORTION OF THE SURFACE OF ANY HOLE IS PERMITTED TO BE INSIDE ITS RESPECTIVE (Ø.121) FEATURE-LOCATING BOUNDARY, EACH BOUNDARY BEING BASICALLY RELATED TO THE OTHER AND BASICALLY ORIENTED TO DATUM PLANE A.

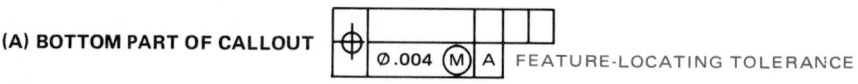

(A) BOTTOM PART OF CALLOUT   FEATURE-LOCATING TOLERANCE

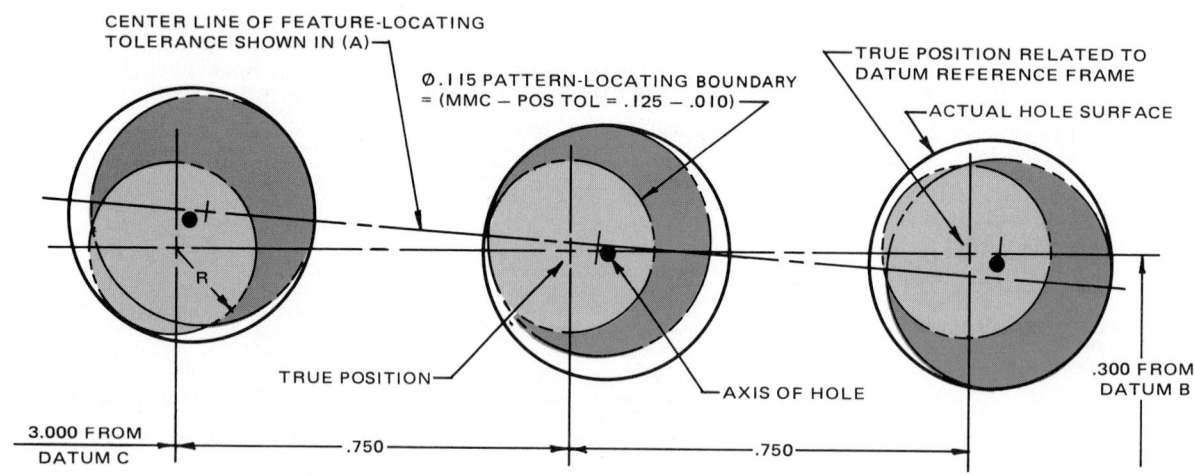

NOTE: NO PORTION OF THE SURFACE OF ANY HOLE IS PERMITTED TO BE INSIDE ITS RESPECTIVE (Ø .115) PATTERN-LOCATING BOUNDARY, EACH BOUNDARY BEING BASICALLY LOCATED IN RELATION TO THE SPECIFIED DATUM REFERENCE FRAME.

VERIFICATIONS OF (A) AND (B) ARE MADE INDEPENDENTLY OF EACH OTHER.

AXES OF HOLES MUST SIMULTANEOUSLY LIE WITHIN BOTH TOLERANCE ZONES.

(B) TOP PART OF CALLOUT   PATTERN-LOCATING TOLERANCE

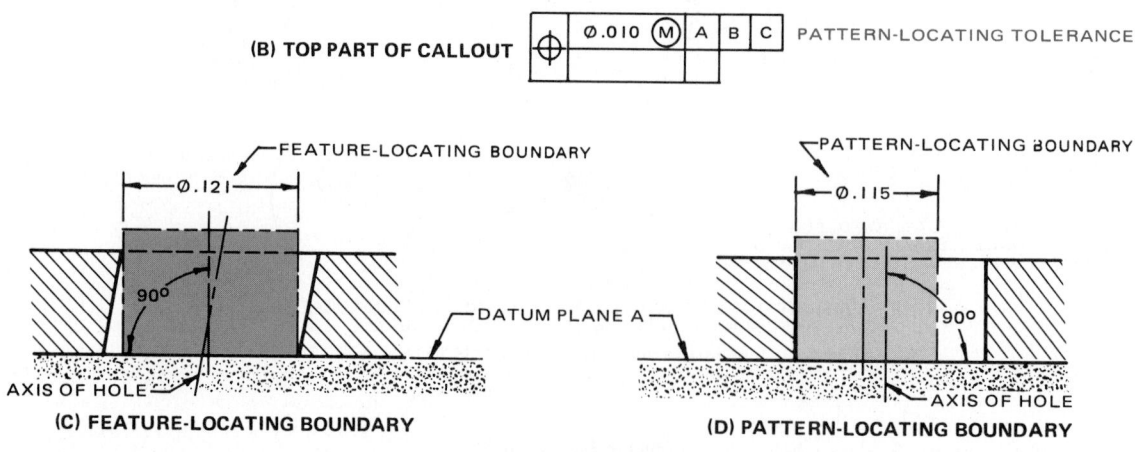

(C) FEATURE-LOCATING BOUNDARY   **(D) PATTERN-LOCATING BOUNDARY**

FIG. 16-16-9   Acceptance boundaries for the three-hole pattern shown in Fig. 16-16-5.

inside its respective Ø.121 feature-locating boundary, each boundary being basically related to the other and basically oriented to datum plane *A*.

This is the preferred tolerancing system for hole groups in mass-produced parts, where it becomes economical to provide location gages. In this system it is recommended that tolerances always be specified on an MMC basis to facilitate gaging procedures. Patterns of features such as these, where composite positional tolerances are specified, may be gaged with two separate gages. For the three-pattern holes described in Fig. 16-16-9, the smaller tolerance requires a simple "go" gage with three gaging cylinders of Ø.121 in. to simulate the condition shown in Fig. 16-16-9A. Note that in the drawing callout the tolerance is related only to datum *A*. It controls the position of the holes relative to one another, but it also controls their perpendicularity with the datum *A* surface within the same tolerance.

The pattern-locating tolerance (the larger tolerance) requires a gage having gaging cylinders of Ø.115 in. to simulate the conditions shown in Fig. 16-16-9B.

REFERENCES AND SOURCE MATERIAL

1. ASME Y14.5M–1994, *Dimensioning and Tolerancing*.
2. CAN/CSA B78.2-M91, *Dimensioning and Tolerancing of Technical Drawings*.
3. ISO drawing standards.

ASSIGNMENTS

See Assignments 51 through 55 for Unit 16-16 on pages 588 to 589.

16-17 FORMULAS FOR POSITIONAL TOLERANCING

The purpose of this unit is to present formulas for determining the required positional tolerances or the required sizes of mating features to ensure that parts will assemble. The formulas are valid for all types of features or patterns of features and will give a "no interference, no clearance" fit when features are at maximum material condition and are located at the extreme of their positional tolerance.

Formulas given in this unit use the following three symbols, as illustrated in Fig. 16-17-1:

- F = maximum diameter of fastener (MMC limit)
- H = minimum diameter of clearance hole (MMC limit)
- T = positional tolerance diameter

Subscripts are used when more than one feature of size or tolerance is involved. For example:

- H_1 = minimum diameter of hole in part 1
- H_2 = minimum diameter of hole in part 2

There are two separate conditions under which fasteners are used, described here as the "floating-fastener case" and the "fixed-fastener case." Each of these conditions will be treated separately.

Floating Fasteners

Where two or more parts are assembled with fasteners such as bolts and nuts and all parts have clearance holes for the bolts, the condition is termed a *floating-fastener case,* as shown in Fig. 16-17-2. Where the fasteners are the same diameter and it is desired to use the same clearance hole diameters and the same positional tolerances for the parts to be assembled, the following formula applies.

$$T = H - F$$

EXAMPLE 1

If the fasteners shown in Fig. 16-17-2 are .312 in. diameter maximum and the clearance holes are .339 in. diameter minimum, the required positional tolerance is

$$T = .339 - .312$$
$$= Ø.027 \text{ in. for each part.}$$

Any number of parts with different hole sizes and positional tolerances may be mated, provided the formula $T = H - F$ is applied to each part individually.

EXAMPLE 2

Figure 16-17-3 (pg. 564) shows two parts with positional tolerances of Ø.030 in. on an RFS basis. Figure 16-17-4 (pg. 564) shows these parts with holes in an extreme position.

In this example the minimum hole diameter for both parts is

$$H = F + T$$
$$= .500 + .030$$
$$= Ø.530 \text{ in.}$$

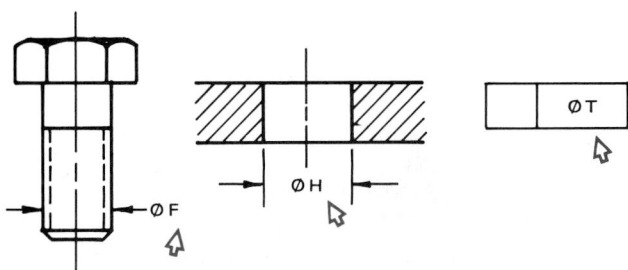

FIG. 16-17-1   Formula symbols.

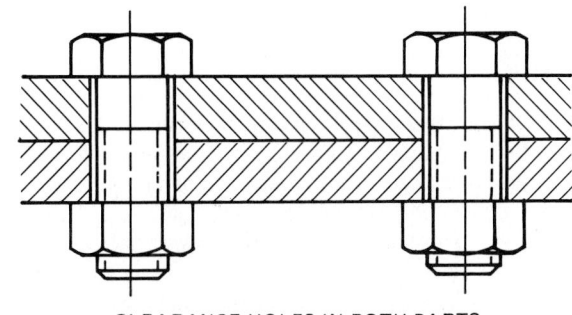

CLEARANCE HOLES IN BOTH PARTS

FIG. 16-17-2   Floating fasteners.

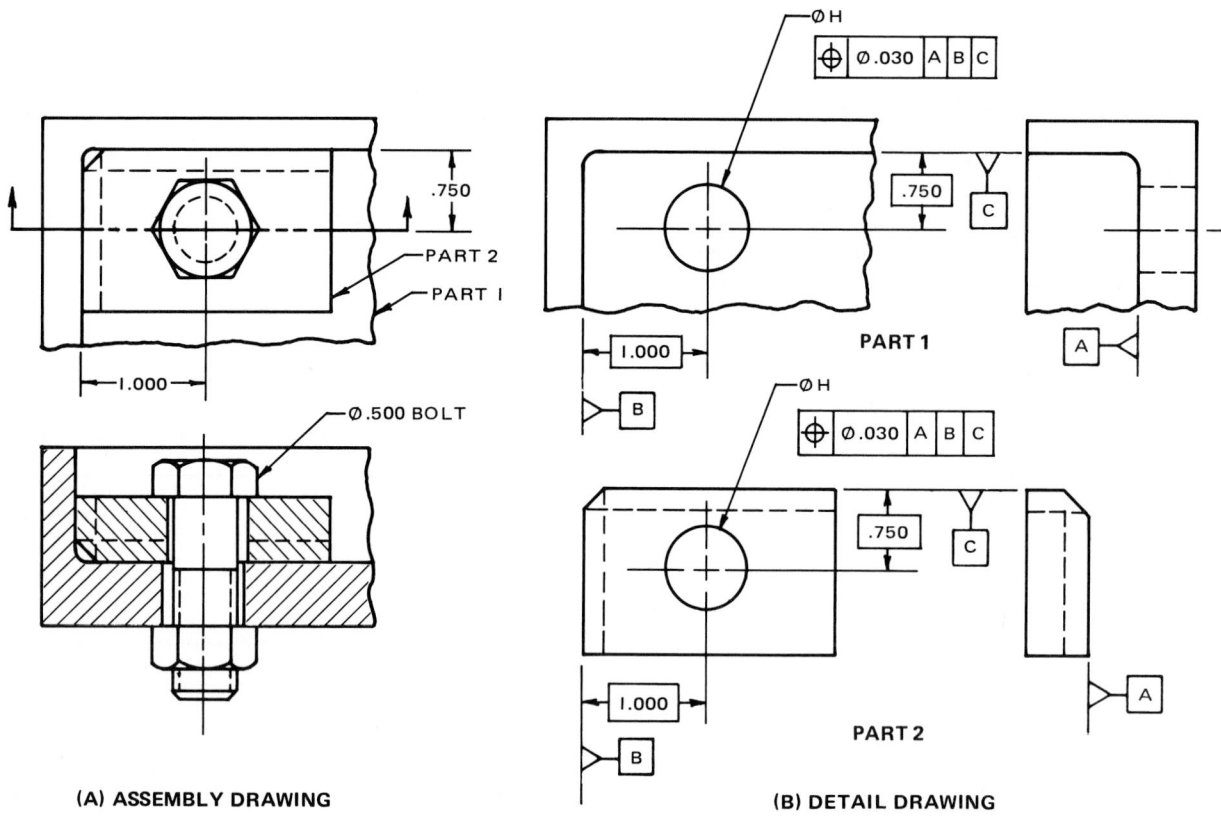

(A) ASSEMBLY DRAWING

(B) DETAIL DRAWING

FIG. 16-17-3   Bolted assembly with floating fastener.

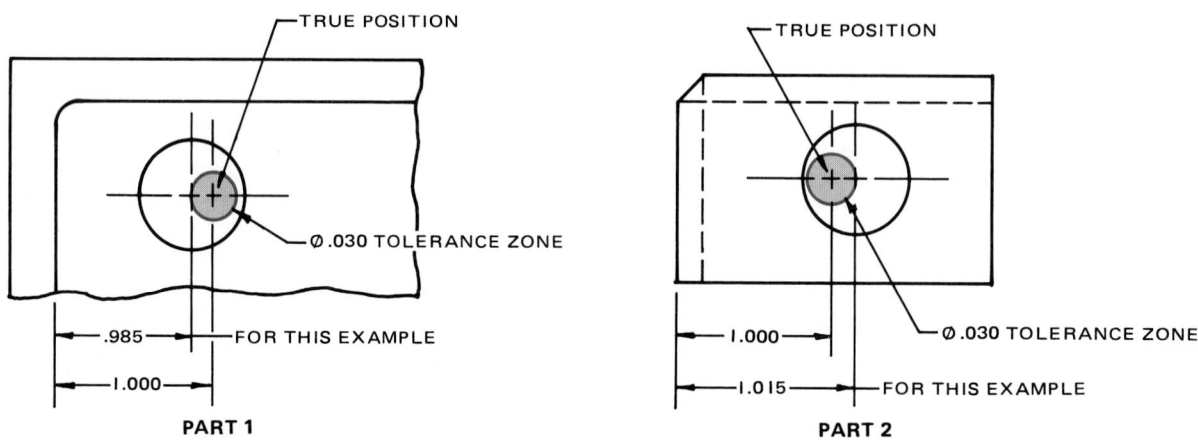

FIG. 16-17-4   Part shown in Fig. 16-17-3 with holes shown in extreme position.

EXAMPLE 3

If MMC is specified with the positional tolerances, as shown in Fig. 16-17-5, the calculations for the minimum hole size are exactly the same as for the RFS condition. The difference in the requirement is that, on an RFS basis, as the size of the hole approaches its maximum diameter, more clearance is provided all around the fastener without the position of the hole changing. When MMC is specified this increase in clearance may be utilized to permit a greater variation in the position of the holes,

as illustrated in Fig. 16-17-6. If the hole in Fig. 16-17-5 was at LMC (Ø.536), a positional tolerance of Ø.036 (.030+.006) could be used.

Calculating Clearance

The formulas given so far have been based on determining the minimum hole diameter or the maximum permissible tolerance for location that would just permit the parts to assemble without any clearance under extreme conditions.

Clearance is usually expressed in terms of the difference between diameters, i.e., the difference between the diameter of a hole and the diameter of the mating part that assembles into it.

The same formulas can be used to determine the minimum clearance for any given drawing specifications. For example, in example 2 we saw how, with a positional tolerance of Ø.030 in., the minimum hole diameter had to be .530 in. If a positional tolerance of Ø.020 in. is substituted, the minimum hole required would be $H = F + T = Ø.520$ in. Therefore a minimum Ø.530 in. hole would provide an extra .010 in. clearance on diameter, or an extra .005 in. all around.

Fixed Fasteners

Where one of the parts to be assembled has restrained fasteners, such as screws or studs in tapped holes, the condition is termed a *fixed-fastener case* (Fig. 16-17-7, pg. 566). Where the fasteners are of the same diameter and it is desired to use the same positional tolerances in the parts to be assembled, the following formula applies, subject to the provisions of perpendicularity errors described later in this unit

$$T = \frac{H - F}{2} \text{ or } H = F + 2T$$

Note that the allowable positional tolerance for fixed fasteners is one-half that for comparable floating fasteners.

EXAMPLE 4

If the fasteners shown in Fig. 16-17-7 have a maximum diameter of 1.00 in. and the clearance holes have a minimum diameter of 1.06 in., the required positional tolerance is

$$T = \frac{1.06 - 1.00}{2}$$

$$= Ø.03 \text{ in. positional tolerance for each part}$$

Unequal Tolerances and Hole Sizes

It is sometimes desirable to have different tolerances for location or different hole sizes in each of the assembled parts. One reason may be because one part already exists and the other must be designed to mate with it. In such cases the hole sizes and the positional tolerances must be separated, and the previous formula, $H = F + T$, becomes

$$H_1 + H_2 = 2F + T_1 + T_2$$
$$\text{or}$$
$$T_1 + T_2 = H_1 + H_2 - 2F$$

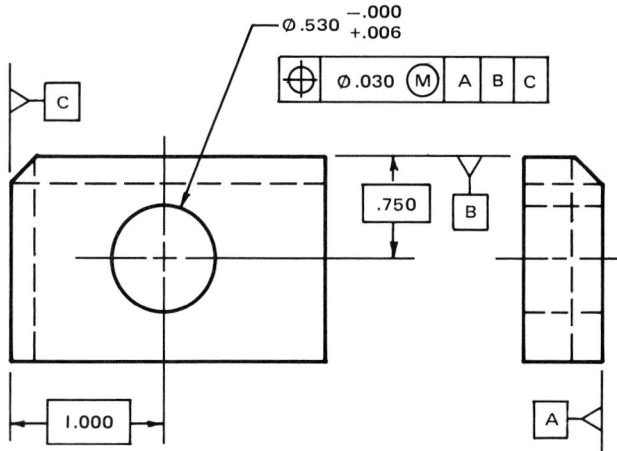

FIG. 16-17-5 Positional tolerance on an MMC basis.

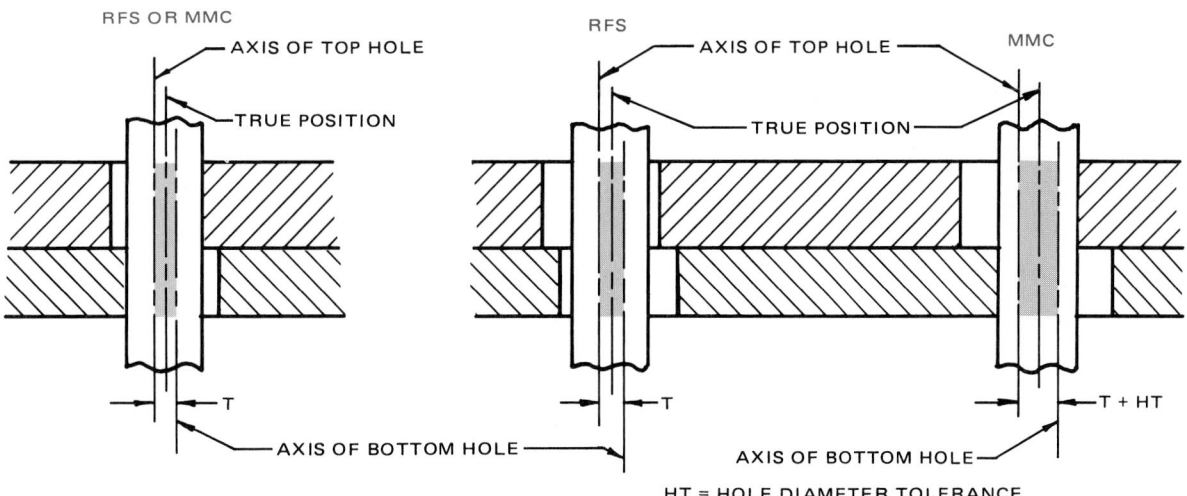

(A) MINIMUM SIZE HOLES **(B) MAXIMUM SIZE HOLES**

FIG. 16-17-6 Extreme position of holes on an RFS and an MMC basis.

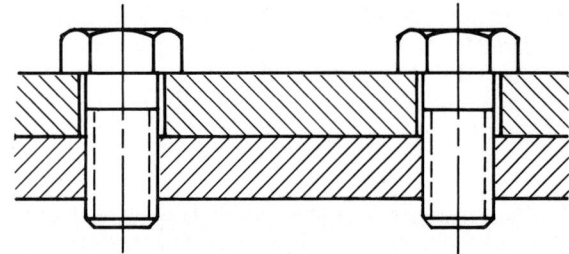

FIG. 16-17-7   Fixed fasteners.

For example, if it is desirable that the part with tapped holes in example 4 have a larger positional tolerance than the part with clearance holes, T can be separated into T_1 and T_2 in any appropriate manner such that

$$T = \frac{T_1 + T_2}{2}$$

EXAMPLE 5

The fasteners shown in Fig. 16-17-7 have a maximum diameter of .500 in. and the clearance holes have a minimum diameter of .560 in. Two different positional tolerances are required, the larger tolerance being for the clearance holes.

$$T = \frac{.560 - .500}{2}$$
$$= .030 \text{ in.}$$

In this example T_1 could be .024 in. instead of .030 in., and then T_2 would be .036 in.

The general formula for fixed fasteners where two mating parts have different positional tolerances is $T_1 + T_2 = H - F$.

Coaxial Features

The formula given for floating fasteners also applies to mating parts having two coaxial features, where one of these features is a datum for the other (Fig. 16-17-8). Where it is desired to divide the available tolerance unequally between the two parts, the following formula can be used

$$(T_1 + T_2) = (H_1 + H_2) - (F_1 + F_2)$$

This formula is valid only for simple two-feature parts as illustrated. By applying the formula above to the example shown in Fig. 16-17-8, the following will result

$$T_1 + T_2 = (H_1 + H_2) - (F_1 + F_2)$$
$$= (1.002 + .503) - (1.000 + .500)$$
$$= .005 \text{ in. total available tolerance}$$

If T_1 is .003 in., then T_2 is .002 in.

Perpendicularity Errors

The formulas do not provide sufficient clearance for fixed fasteners when threaded holes or holes for tight-fitting members, such as dowels, deviate from the perpendicular. To provide

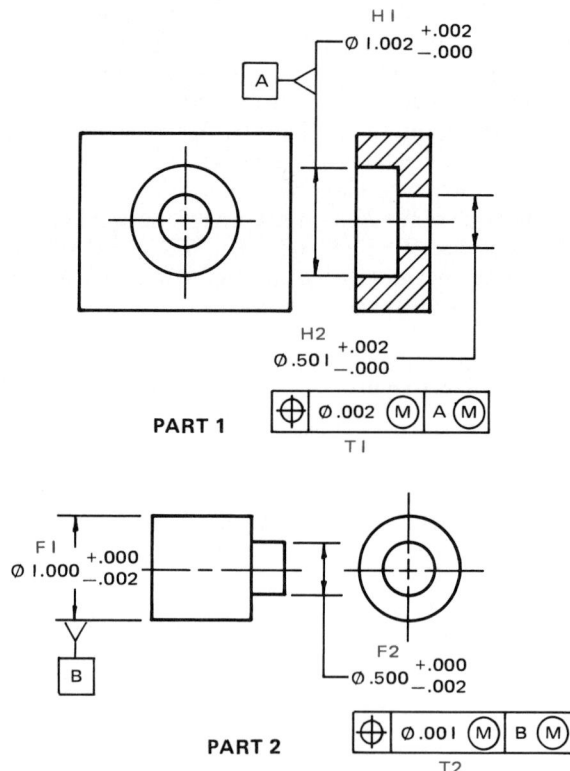

FIG. 16-17-8   Coaxial features—mating parts.

for this condition, the projected tolerance zone method of positional tolerancing should be applied to threaded or tight-fitting holes.

REFERENCES AND SOURCE MATERIAL

1. ASME Y14.5M–1994, *Dimensioning and Tolerancing*.
2. CAN/CSA B78.2-M91, *Dimensioning and Tolerancing of Technical Drawings*.
3. ISO drawing standards.

ASSIGNMENTS ▨▨▨▨▨▨▨▨▨▨▨▨▨▨▨▨▨▨▨▨▨▨▨▨

See Assignments 56 through 60 for Unit 16-17 on page 590.

16-18 SUMMARY OF RULES FOR GEOMETRIC TOLERANCING

When to Use Geometric Tolerancing

It is not necessary to use geometric tolerances for every feature on a part drawing. In most cases it is to be expected that, if each feature meets all dimensional tolerances, form variations will be adequately controlled by the accuracy of the manufacturing process and equipment used. This is supplemented by the partial degree of control exercised by the measuring and gaging procedure used.

If there is any doubt about the adequacy of such control, a geometric tolerance of form, orientation, or position must be specified, as described in this text. This is often necessary when parts are of such size or shape that bending or other distortion is likely to occur. It is also necessary when errors of shape or form must be held within limits other than those that might ordinarily be expected from the manufacturing process, and as a means of meeting functional or interchangeability requirements.

It will perhaps be necessary to specify the most complete and explicit manufacturing requirements (dimensions/tolerances) on drawings prepared for subcontracting to workshops of widely varying equipment and experience, where possible manufacturing process variations are not known. On the other hand, if the same parts are to be manufactured and assembled in a workshop where the method of production has been proved to produce parts and assemblies of satisfactory quality, the same degree of tolerancing may not be necessary.

Basic Rules

The basic rules for geometric tolerancing are summarized here for convenience but are explained in greater detail in the units throughout this text. The summary does not refer to points, which can have location only, nor to runout, which is a composite tolerance requiring separate treatment.

Geometric tolerances can be categorized into three basic groups:

- *Form,* to control the form and shape of features
- *Angularity,* to control orientation of features
- *Position,* to control location of features

Any of these tolerances can be applied to lines and surfaces of any size or shape. Two separate profile symbols have been provided to distinguish between profile of a line and profile of a surface. No such distinction has been made for tolerances of angularity or position, but if ambiguity may result in any particular application, a suitable note should be added.

Since straight lines and circular lines, as well as flat and cylindrical surfaces, occur so frequently in practice, special names and symbols have been established for their control. These special designations should be used for such lines and surfaces instead of the categorized names given above:

- *Form of a line,* straightness and circularity
- *Form of a surface,* flatness and cylindricity
- *Orientation of a line, surface, or feature,* angularity, parallelism, and perpendicularity
- *Location of features,* (true) position and concentricity

Lines usually represent the edges of geometric shapes or line elements in a single direction on a surface. All lines that consist of curves (except complete circles) or a combination of straight and curved lines can be controlled for form by the profile-of-a-line tolerance. Examples are outlines of rectangles, hexagons, ellipses, semicircles, and various curved forms.

Surfaces, other than flat and cylindrical, can be controlled for form by the profile-of-a-surface tolerance. Examples are spherical surfaces, bars of hexagonal, square, or other shapes, and holes of various shapes, such as hexagonal, elongated, or oval.

Positional Tolerancing

The locational tolerance of position may be applied to a line, axis, center plane, surface, or any feature regardless of its shape.

All positional tolerances, when applied to a feature of size that incorporates a dimension, such as a diameter or thickness, may be modified by RFS, MMC, and LMC (Fig. 16-18-1, pg. 568). They may be datum features or other features whose axes or center planes require control. In such cases the following practices apply:

Tolerance of Position   RFS, MMC, or LMC must be specified for tolerances of true position on the drawing with respect to the individual tolerance, datum reference, or both, as applicable.

All Other Geometric Tolerances   RFS applies, with respect to the individual tolerance, datum reference, or both, where no modifying symbol is specified. MMC or LMC is specified on the drawing where required.

Limits of Size

Unless otherwise specified, the limits of size of a feature prescribe the extent within which variations of geometric form, as well as size, are allowed. This control applies solely to individual features of size.

Where only a tolerance of size is specified, the limits of size of an individual feature prescribe the extent to which variations in its geometric form, as well as size, are allowed.

The form of an individual feature is controlled by its limits of size to the extent prescribed as follows:

1. The surface or surfaces of a feature shall not extend beyond a boundary (envelope) of perfect form at MMC. This boundary is the true geometric form represented by the drawing. No variation of form is permitted if the feature is produced at its MMC limit of size.
2. Where it is desired to permit a surface or surfaces of a feature to exceed the boundary of perfect form at MMC, a note such as PERFECT FORM AT MMC NOT REQD is specified, exempting the pertinent size dimension from the provision described above.
3. The limits of size do not control the orientation or location relationship between individual features. Features shown perpendicular, coaxial, or symmetrical to each other must be controlled for location or orientation to avoid incomplete drawing requirements. These controls may be specified by the methods shown in the text.

If it is necessary to establish a boundary of perfect form at MMC to control the relationship between features, the following methods may be used:

- Specify a zero tolerance of orientation at MMC, including a datum reference (at MMC, if applicable), to control angularity, perpendicularity, or parallelism of the feature.
- Specify a zero positional tolerance at MMC, including a datum reference at MMC, to control axial or symmetrical features.
- Indicate this control for the features involved by a note such as PERFECT ORIENTATION (or COAXIALITY or SYMMETRY) AT MMC REQUIRED FOR RELATED FEATURES.
- Relate dimensions to a datum reference plane.

GEOMETRIC CHARACTERISTIC	SYMBOL	APPLICABLE TO FEATURE BEING CONTROLLED	APPLICABLE TO DATUM REFERENCE
STRAIGHTNESS	—	NOT APPLICABLE FOR A PLANE SURFACE OR A LINE ON A SURFACE MMC OR RFS APPLICABLE IF TOLERANCE APPLIES TO AN AXIS OR CENTER PLANE OF A FEATURE OF SIZE, EG., A HOLE, SHAFT, OR SLOT	NO DATUM REFERENCE
FLATNESS	▱	NOT APPLICABLE	NO DATUM REFERENCE
CIRCULARITY	○		
CYLINDRICITY	⌭		
PROFILE OF A LINE	⌒	NOT APPLICABLE	MMC NOT APPLICABLE RFS APPLICABLE ONLY TO DATUM FEATURES OF SIZE HAVING AN AXIS OR CENTER PLANE
PROFILE OF A SURFACE	◠		
PERPENDICULARITY	⊥	NOT APPLICABLE FOR A PLANE SURFACE MMC, LMC, AND RFS APPLICABLE IF TOLERANCE APPLIES TO AN AXIS, OR CENTER PLANE OF A FEATURE OF SIZE	NOT APPLICABLE TO A SINGLE-PLANE SURFACE MMC, LMC, AND RFS APPLICABLE ONLY TO DATUM FEATURES OF SIZE HAVING AN AXIS OR CENTER
PARALLELISM	∥		
ANGULARITY	∠		
POSITION	⌖	MMC, LMC, AND RFS APPLICABLE IF TOLERANCE APPLIES TO AN AXIS OR CENTER PLANE OF A FEATURE OF SIZE.	NOT APPLICABLE TO A SINGLE-PLANE SURFACE MMC, LMC, AND RFS APPLICABLE ONLY TO DATUM FEATURES OF SIZE HAVING AN AXIS OR CENTER PLANE
CONCENTRICITY*	◎	APPLICABLE ONLY TO RFS	APPLICABLE ONLY TO RFS
SYMMETRY	≡		
CIRCULAR RUNOUT	↗ **		
TOTAL RUNOUT	⌰↗ **		

* ISO PERMITS CONCENTRICITY TO BE USED ON AN MMC BASIS.

** ARROWS MAY BE FILLED IN.

FIG. 16-18-1   Application of MMC, LMC, and RFS.

Form and Orientation

Form tolerances control straightness, flatness, circularity, and cylindricity. Orientation tolerances control angularity, parallelism, and perpendicularity. A profile tolerance may control form, orientation, and size, depending on how it is applied. Since, to a certain degree, the limits of size control orientation, the extent of these limits must be considered before specifying form and orientation tolerances.

Form and orientation tolerances critical to function and interchangeability are specified where the tolerance of size and location do not provide sufficient control. As such, straightness when applied to surface elements, flatness, circularity, and cylindricity tolerances must always be less than the size tolerance.

In specifying orientation tolerances to control angularity, perpendicularity, parallelism, and in some cases, profile, the considered feature is related to one or more datum features. Note that angularity, perpendicularity, and parallelism when applied to flat surfaces control flatness if a flatness tolerance is not specified.

When no variations of orientation are permitted at the MMC size limit of a feature, the feature control frame contains a zero for the tolerance, modified by the symbol for MMC. Deviation from perfect orientation can exist only as the feature departs from MMC.

Profile Tolerancing

The profile tolerance specifies a uniform boundary along the true profile within which the elements of the surface must lie. It is used to control form or combinations of size, form, and orientation.

Coaxiality Control

Coaxiality is the condition where the axes of two or more surfaces of revolution are coincident. The amount of permissible variation may be expressed by a positional tolerance, a runout tolerance, or a concentricity tolerance.

Where the surfaces of revolution are cylindrical and the control of the axes can be applied on a material condition basis, positional tolerancing is recommended as it permits the use of simple receiver gages for inspection.

Where a combination of surfaces of revolution are cylindrical, conical, or spherical relative to a common datum axis, a runout tolerance is recommended. MMC is not applicable for runout.

REFERENCES AND SOURCE MATERIAL

1. ASME Y14.5M–1994, *Dimensioning and Tolerancing.*
2. CAN/CSA B78.2-M91, *Dimensioning and Tolerancing of Technical Drawings.*
3. ISO drawing standards.

ASSIGNMENTS

See Assignments 61 through 73 for Unit 16-18 on pages 591 through 597.

ASSIGNMENTS FOR CHAPTER 16

ASSIGNMENTS FOR UNIT 16-1, MODERN ENGINEERING TOLERANCING

1. Parts may deviate from true form and still be acceptable provided the measurements lie within the limits of size.

Show by means of a sketch with dimensions two acceptable form variations for each part shown in Fig. 16-1-A.

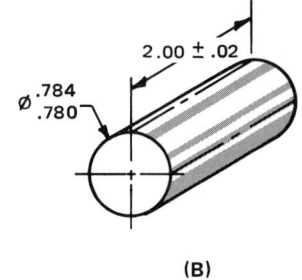

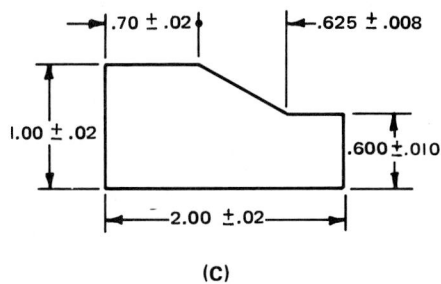

FIG. 16-1-A   Assignment.

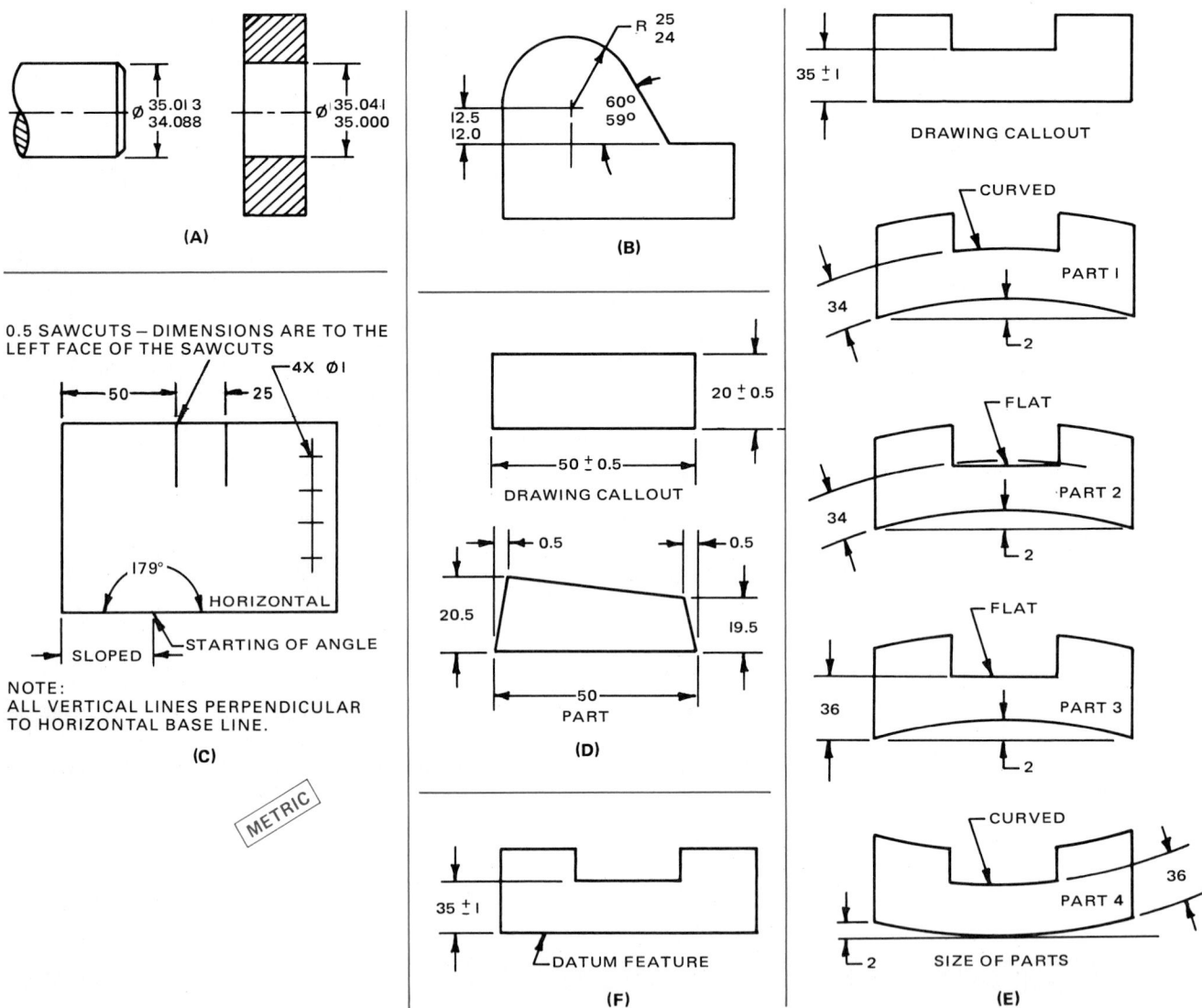

FIG. 16-1-B   Assignment.

2. Prepare sketches from the drawings shown in Fig. 16-1-B and the following information:
   (a) Using illustration (A) make a tolerance block diagram similar to Fig. 16-1-5. Show the deviations and limits of size.
   (b) Draw illustration (B) and shade in and dimension the tolerance zone.
   (c) The exaggeration of sizes is used when it improves the clarity of the drawing. Draw illustration (C) and exaggerate the sizes which would improve the readability of the drawing. Dimension the exaggerated features.
   (d) With reference to illustration (D), is the part acceptable? State your reason.
   (e) With reference to the drawing callout shown in illustration (E), what parts would pass inspection?
   (f) In the drawing callout in illustration (F), what parts in illustration (E) would pass inspection?

ASSIGNMENT FOR UNIT 16-2, GEOMETRIC TOLERANCING

3. With reference to Fig. 16-2-A and the information given below, add the feature control frames to the following parts:
   Part 1. Surface A to have a straightness tolerance of .004 in.
   Part 2. Surface M to have a straightness tolerance of .006 in. Surface N to have a straightness tolerance of .008 in.
   Part 3. Surface R to be straight within .006 in. for direction *A* and straight within .002 in. for direction *B*.
   Part 4. With straightness specified as shown, what is the maximum permissible deviation from straightness of the line elements if the radius is (a) .496 in., (b) .501 in., (c) .504 in.?
   Part 5. Eliminate the top view and place the feature control frames on the front and side views.

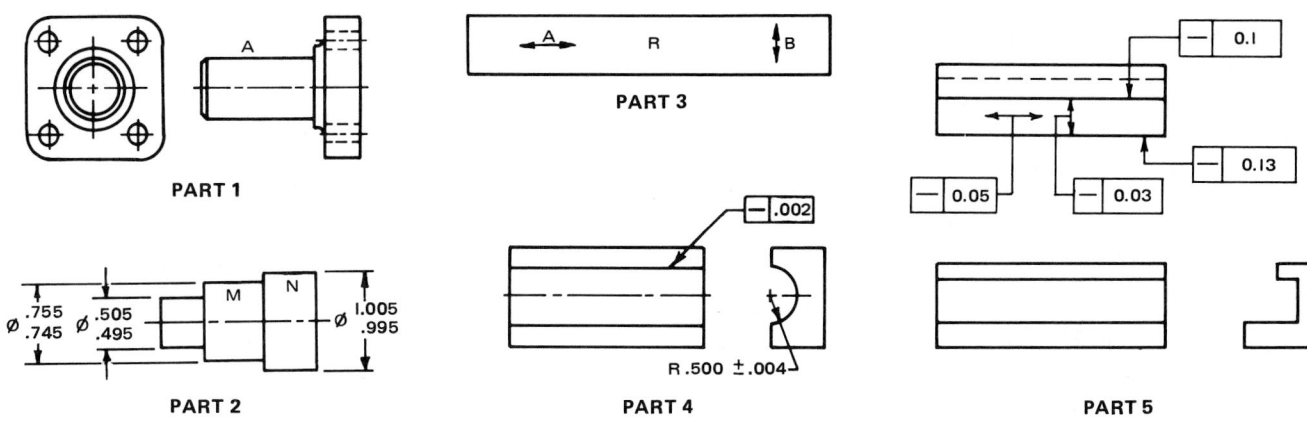

FIG. 16-2-A   Assignment.

ASSIGNMENTS FOR UNIT 16-3, FLATNESS

4. Add a flatness tolerance of 0.03 to the base of the flange shown in Fig. 16-3-A.
5. Add the following tolerances to surface *B* of the base shown in Fig. 16-3-B: (a) Maximum flatness tolerance of .010 in. for entire surface; (b) Limited area flatness tolerance of .005 for any 2.00 × 2.00 in. area.

6. In Fig. 16-3-C part 1 is required to fit into part 2 so that there will not be any interference and the maximum clearance will never exceed .005 in. Add the maximum limits of size to part 2. Flatness tolerances of .001 in. are to be added to the two surfaces of each part.
7. Show the tolerance zones and limits of size dimensions for the two parts shown in Fig. 16-3-D.

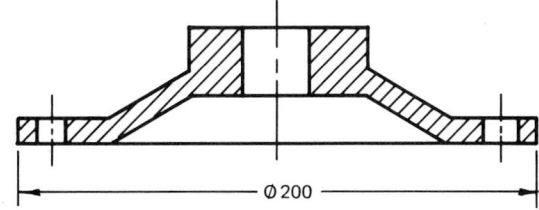

FIG. 16-3-A   Flange.

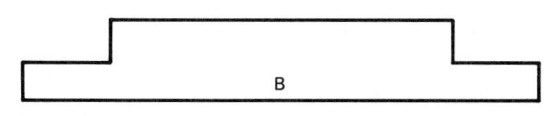

FIG. 16-3-B   Base.

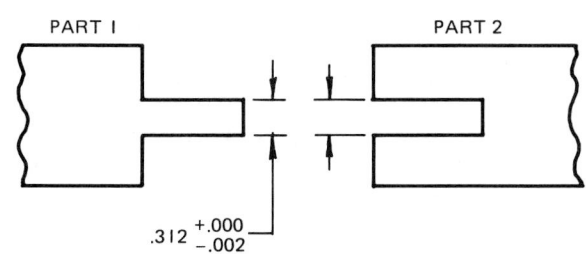

FIG. 16-3-C   Slot assembly.

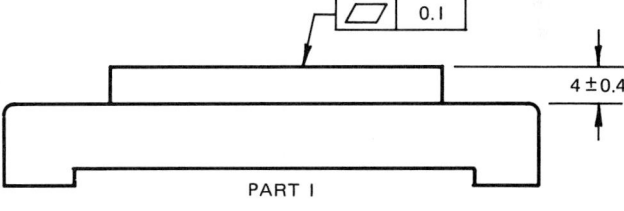

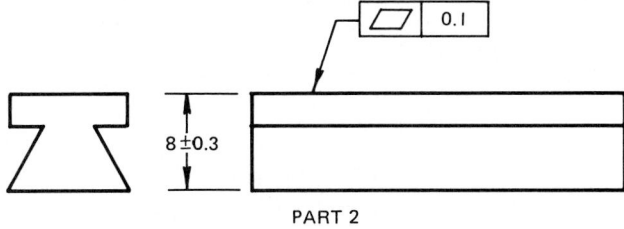

FIG. 16-3-D   Flatness tolerance.

ASSIGNMENTS FOR UNIT 16-4, STRAIGHTNESS OF A FEATURE OF SIZE

8. What is the virtual condition for each of the parts shown in Fig. 16-4-A?
9. The hole shown in Fig. 16-4-B does not have a straightness tolerance. What is the maximum permissible deviation from straightness if perfect form at the maximum material size is required?
10. Complete the charts shown in Figs. 16-4-C and 16-4-D showing the largest permissible straightness error for the feature sizes shown.

11. If the maximum straightness allowance was not added to the straightness tolerance in Fig. 16-4-E, what parts would be acceptable? State your reasons if the part is not acceptable.
12. With reference to Fig. 16-4-F, what is the maximum deviation permitted from straightness for the shaft if it was (a) at MMC? (b) at LMC? (c) Ø.623?

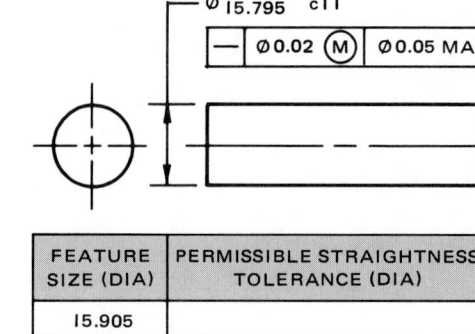

FIG. 16-4-A   Assignment.

FIG. 16-4-B   Assignment.

FIG. 16-4-C   Assignment.

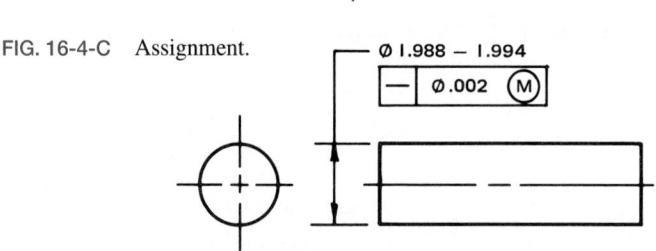

FEATURE SIZE (DIA)	PERMISSIBLE STRAIGHTNESS TOLERANCE (DIA)
15.905	
15.895	
15.885	
15.875	
15.865	
15.845	
15.795	

FIG. 16-4-D   Assignment.

FEATURE SIZE (DIA)	PERMISSIBLE STRAIGHTNESS TOLERANCE (DIA)
1.994	
1.993	
1.992	
1.991	
1.990	
1.989	
1.988	

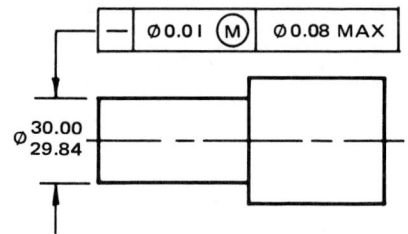

PART	FEATURE SIZE Ø	STRAIGHTNESS DEVIATION
A	29.94	0.08
B	30.02	0.008
C	29.80	0.06
D	29.86	0.10
E	29.97	0.04

FIG. 16-4-E   Assignment.

FIG. 16-4-F   Assignment.

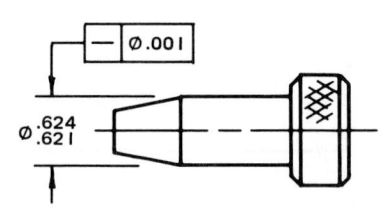

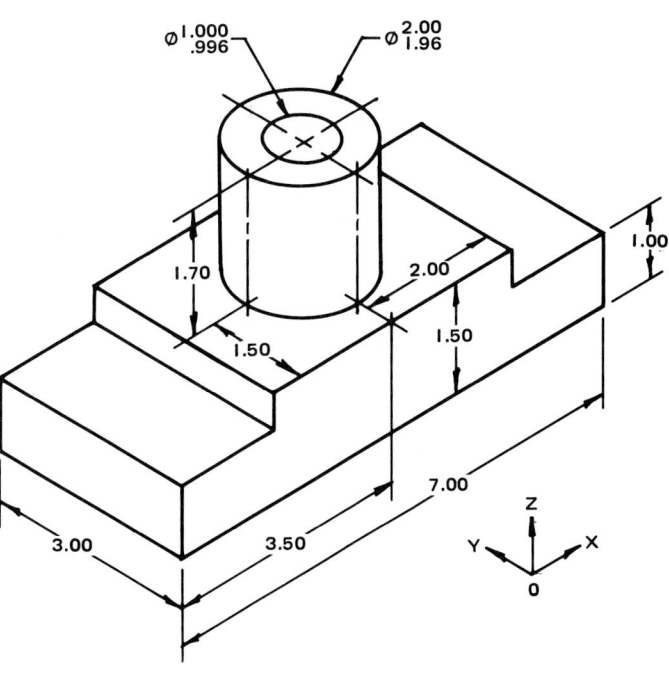

ASSIGNMENTS FOR UNIT 16-5, DATUMS AND THE THREE-PLANE CONCEPT

13. Draw the front and top views of the stand and add the information shown in Fig. 16-5-A to the drawing.
14. Draw the front and top views of the part shown in Fig. 16-5-B and add the information shown to the drawing. With reference to the slot and locational dimensions shown in Fig. 16-5-B (B), are the three parts shown acceptable?

GEOMETRIC TOLERANCING REQUIREMENTS TO BE ADDED TO DRAWING
• THE BOTTOM IS PRIMARY DATUM A
• THE FRONT IS SECONDARY DATUM B
• THE RIGHT SIDE IS TERTIARY DATUM C
• THE BOTTOM IS TO HAVE A FLATNESS TOLERANCE OF .003
• THE Ø 1.000 TO HAVE A STRAIGHTNESS TOLERANCE OF .002 AT MMC

FIG. 16-5-A   Stand.

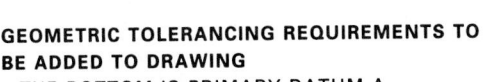

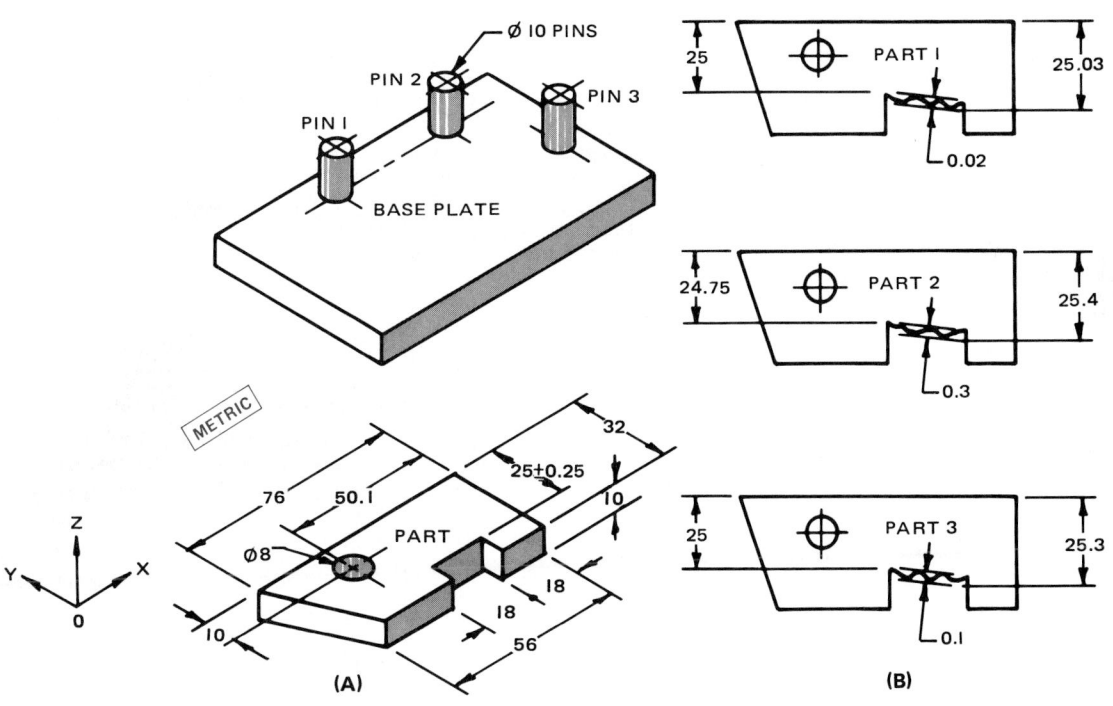

(A)

(B)

GEOMETRIC TOLERANCING REQUIREMENTS TO BE ADDED TO DRAWING
• ALL DATUMS AND TOLERANCES TO BE ON AN MMC BASIS UNLESS OTHERWISE SPECIFIED
• THE TOP SURFACES OF THE BASE PLATE TO BE DATUM A
• PINS I, 2, AND 3 ARE USED TO ESTABLISH THE SECONDARY AND TERTIARY DATUMS FOR THE PART SHOWN
• A FLATNESS TOLERANCE OF 0.2 TO BE ADDED TO THE BACK SURFACE OF THE PART

FIG. 16-5-B   Assignment.

15. Anyone involved with the use of technical drawings must be capable of interpreting drawings from other countries as well as their own. From the information given in Fig. 16-5-C prepare two drawings, one using the former ANSI symbols, the other ISO symbols to show these differences.

16. Make a three-view drawing of the shaft support and add the information shown in Fig. 16-5-D to the drawing.

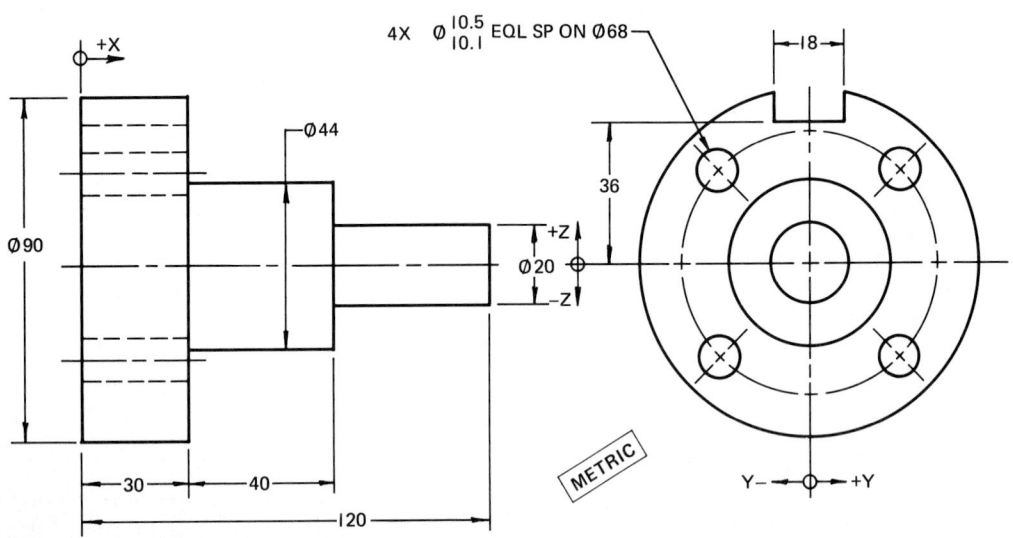

4X Ø |10.5 EQL SP ON Ø68
     |10.1

GEOMETRIC TOLERANCING REQUIREMENTS TO BE ADDED TO DRAWING
- ALL DATUMS AND TOLERANCES TO BE ON AN MMC BASIS UNLESS OTHERWISE SPECIFIED
- THE Ø 44 TO BE DATUM A
- THE END FACE OF Ø 90 TO BE DATUM B
- THE WIDTH OF THE SLOT TO BE DATUM C
- THE END FACE TO BE FLAT WITHIN 0.25
- THE CENTER LINE OF THE Ø 44 MUST BE STRAIGHT WITHIN 0.1 RFS
- THE SURFACE OF Ø 20 MUST BE STRAIGHT WITHIN 0.2

FIG. 16-5-C Stepped shaft.

GEOMETRIC TOLERANCING REQUIREMENTS TO BE ADDED TO DRAWING
- ALL DATUMS AND TOLERANCES TO BE ON AN MMC BASIS UNLESS OTHERWISE SPECIFIED
- SURFACES MARKED A, B, AND C TO BE DATUMS A, B, AND C RESPECTIVELY
- THE SHAFT TO HAVE A STRAIGHTNESS TOLERANCE OF .003 AT MMC AND TO BE DATUM E
- THE BOTTOM TO BE FLAT WITHIN .005 FOR THE ENTIRE SURFACE BUT THE FLATNESS ERROR MUST NOT EXCEED .002 FOR ANY 1.00 × 1.00 AREA
- BOTH SIDES OF THE NOTCH TO BE FLAT WITHIN .001
- THE HOLE TO HAVE A STRAIGHTNESS TOLERANCE OF .002 AT MMC AND TO BE DATUM D

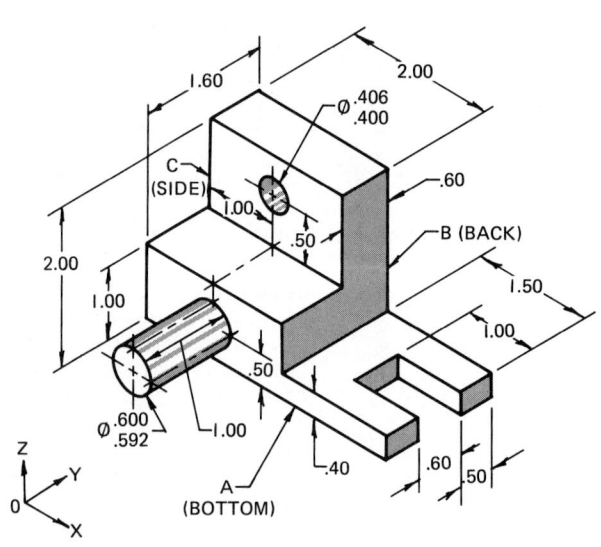

FIG. 16-5-D Shaft support.

574

ASSIGNMENTS FOR UNIT 16-6, ORIENTATION TOLERANCING OF FLAT SURFACES

17. From the information shown in Fig. 16-6-A make a three-view drawing of the stand.

18. From the information shown in Fig. 16-6-B make a three-view drawing of the cutoff stop.

GEOMETRIC TOLERANCING REQUIREMENTS TO BE ADDED TO DRAWING

• SURFACES A, B, AND D TO BE DATUMS A, B, AND D RESPECTIVELY
• THE BACK TO BE PERPENDICULAR TO BOTTOM WITHIN .0I AND BE FLAT WITHIN .006
• THE TOP TO BE PARALLEL TO BOTTOM WITHIN .005
• SURFACE C TO HAVE AN ANGULARITY TOLERANCE OF .008 WITH THE BOTTOM. SURFACE D TO BE THE SECONDARY DATUM FOR THIS REQUIREMENT
• THE BOTTOM TO BE FLAT WITHIN .002
• THE SIDES OF THE SLOT TO BE PARALLEL TO EACH OTHER WITHIN .002 AND PERPENDICULAR WITHIN .004 WITH THE BACK. ONE SIDE OF THE SLOT IS TO BE DATUM E.

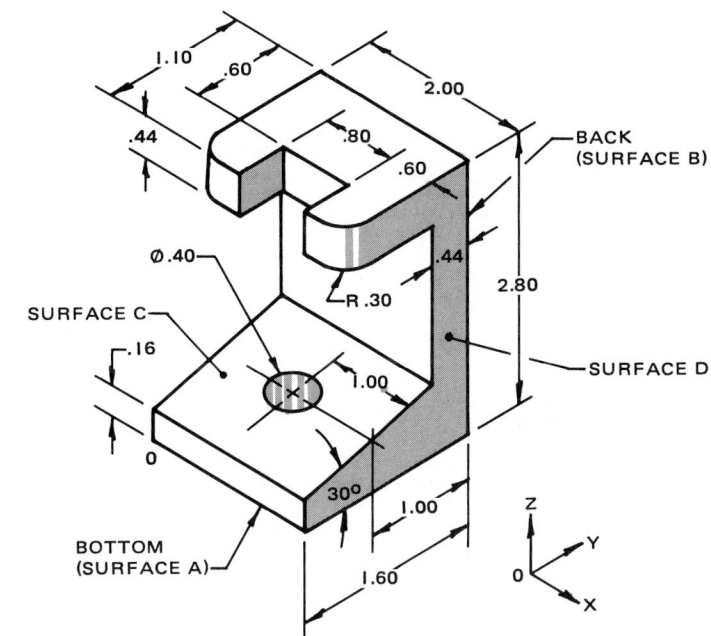

FIG. 16-6-A   Stand.

GEOMETRIC TOLERANCING REQUIREMENTS TO BE ADDED TO DRAWING

• SURFACES A, B, C, D AND E TO BE DATUMS A, B, C, D, AND E RESPECTIVELY
• SURFACE C TO HAVE A FLATNESS TOLERANCE OF 0.2
• SURFACES F AND G OF THE DOVETAIL ARE TO HAVE AN ANGULARITY TOLERANCE OF 0.05 WITH A SINGLE DATUM ESTABLISHED BY THE TWO DATUM FEATURES D AND E. F AND G SURFACES ARE TO BE FLAT WITHIN 0.02
• SURFACE H TO BE PARALLEL TO SURFACE B WITHIN 0.05
• SURFACE C TO BE PERPENDICULAR TO SURFACES D AND E WITHIN 0.04

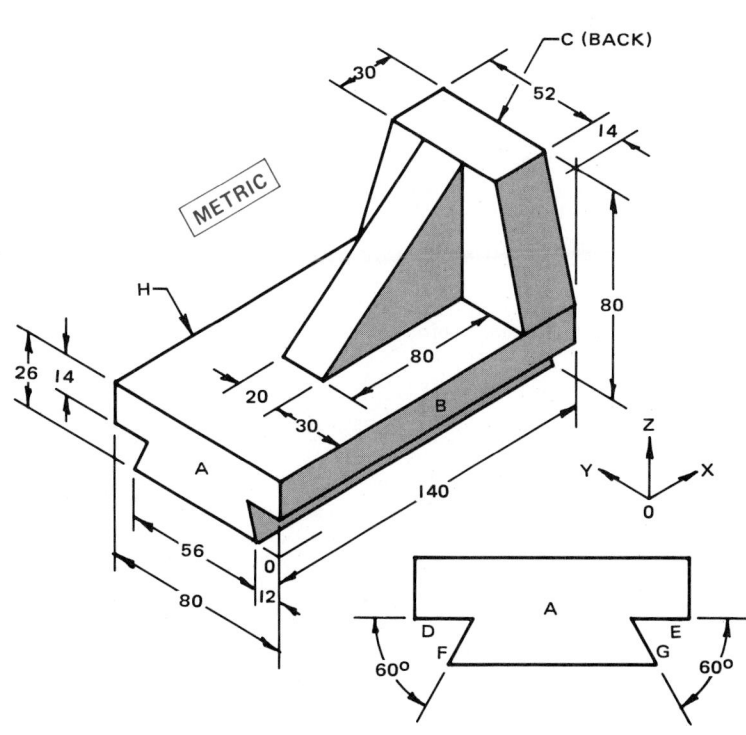

FIG. 16-6-B   Cutoff stop.

575

ASSIGNMENT FOR UNIT 16-7, DATUM FEATURES SUBJECT TO SIZE VARIATION

19. What would be the size of the gaging element to evaluate datum *A* given the drawing callouts and the measured size of the related datum features shown in Fig. 16-7-A?

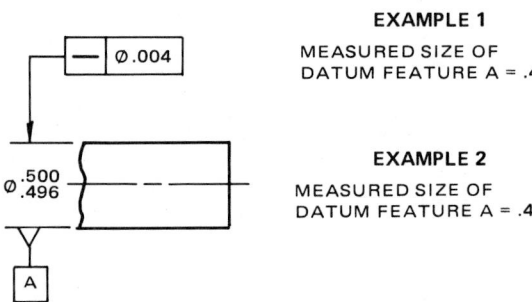

EXAMPLE 1

MEASURED SIZE OF
DATUM FEATURE A = .499

EXAMPLE 2

MEASURED SIZE OF
DATUM FEATURE A = .496

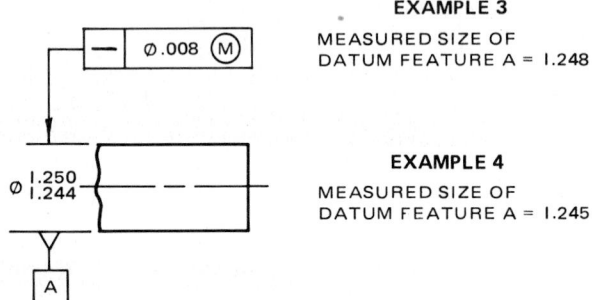

EXAMPLE 3

MEASURED SIZE OF
DATUM FEATURE A = 1.248

EXAMPLE 4

MEASURED SIZE OF
DATUM FEATURE A = 1.245

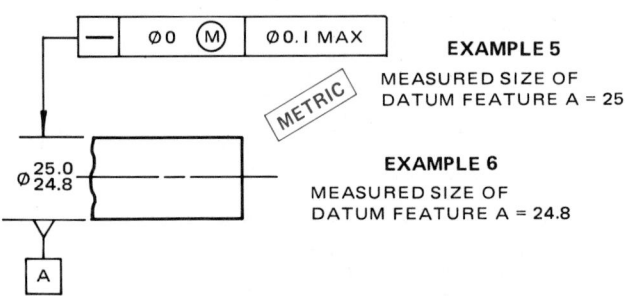

EXAMPLE 5

MEASURED SIZE OF
DATUM FEATURE A = 25

EXAMPLE 6

MEASURED SIZE OF
DATUM FEATURE A = 24.8

FIG. 16-7-A   Datum features subject to size variation.

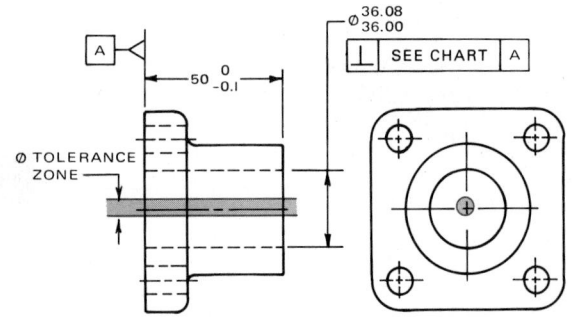

ASSIGNMENTS FOR UNIT 16-8, ORIENTATION TOLERANCING FOR FEATURES OF SIZE

20. Complete the tables shown in Fig. 16-8-A showing the maximum permissible tolerance zone for the three perpendicularity tolerances.

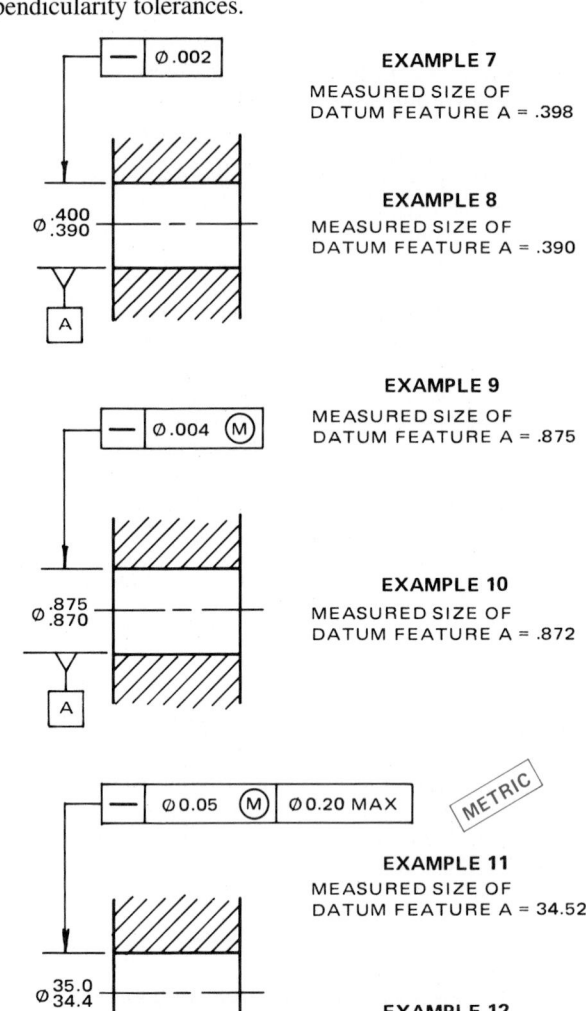

EXAMPLE 7

MEASURED SIZE OF
DATUM FEATURE A = .398

EXAMPLE 8

MEASURED SIZE OF
DATUM FEATURE A = .390

EXAMPLE 9

MEASURED SIZE OF
DATUM FEATURE A = .875

EXAMPLE 10

MEASURED SIZE OF
DATUM FEATURE A = .872

EXAMPLE 11

MEASURED SIZE OF
DATUM FEATURE A = 34.52

EXAMPLE 12

MEASURED SIZE OF
DATUM FEATURE A = 34.94

FIG. 16-8-A   Pillow block.

⊥ ∅0.03 A		⊥ ∅0 Ⓜ A		⊥ ∅0 Ⓜ ∅0.06 MAX A	
FEATURE SIZE ∅	DIAMETER TOLERANCE ZONE ALLOWED	FEATURE SIZE ∅	DIAMETER TOLERANCE ZONE ALLOWED	FEATURE SIZE ∅	DIAMETER TOLERANCE ZONE ALLOWED
36.00		36.00		36.00	
36.01		36.01		36.01	
36.02		36.02		36.02	
36.03		36.03		36.03	
36.04		36.04		36.04	
36.05		36.05		36.05	
36.06		36.06		36.06	
36.07		36.07		36.07	
36.08		36.08		36.08	

21. From the information shown in Fig. 16-8-B make a two-view drawing of the spacer.

22. From the information shown in Fig. 16-8-C make a two-view drawing of the support.

GEOMETRIC TOLERANCING REQUIREMENTS TO BE ADDED TO DRAWING

- ALL DATUMS AND TOLERANCES TO BE ON AN MMC BASIS UNLESS OTHERWISE SPECIFIED
- SURFACES MARKED A, B, AND C ARE DATUMS A, B, AND C RESPECTIVELY
- SURFACE A IS PERPENDICULAR WITHIN .01 TO DATUMS B AND C IN THAT ORDER
- SURFACE D IS PARALLEL WITHIN .004 OF DATUM B
- THE SLOT IS PARALLEL WITHIN .002 TO DATUM C AND PERPENDICULAR WITHIN .001 TO DATUM A
- THE 1.750 HOLE HAS AN RC7 FIT (SHOW THE SIZE OF THE HOLE AS LIMITS) AND IS PERPENDICULAR WITHIN .002 TO DATUM A
- SURFACE E HAS AN ANGULARITY TOLERANCE OF .010 WITH DATUM C
- SURFACE A IS TO BE FLAT WITHIN .002 FOR ANY ONE-INCH SQUARE SURFACE WITH A MAXIMUM FLATNESS TOLERANCE OF .005

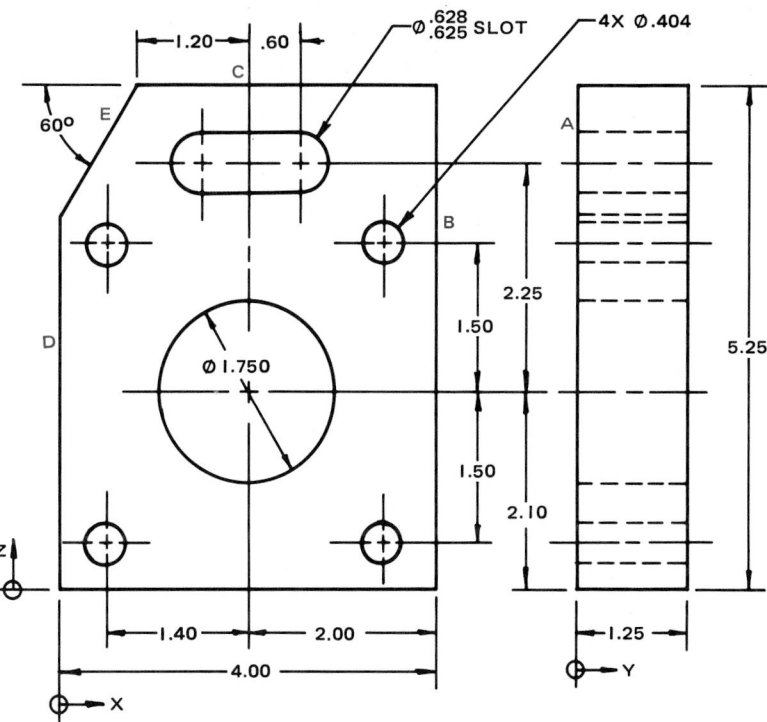

FIG. 16-8-B   Spacer.

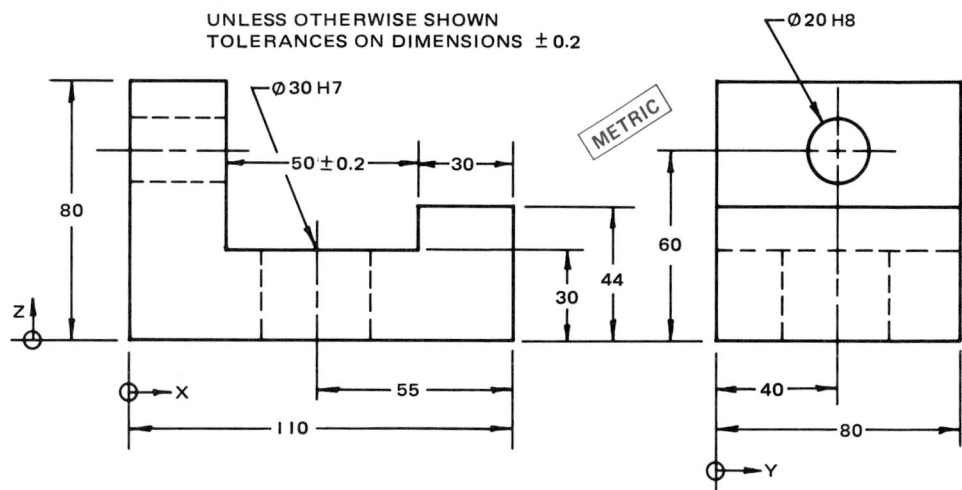

GEOMETRIC TOLERANCING REQUIREMENTS TO BE ADDED TO DRAWING

- ALL DATUMS AND TOLERANCES TO BE ON AN MMC BASIS UNLESS OTHERWISE SPECIFIED
- THE BOTTOM IS DATUM A AND HAS A FLATNESS TOLERANCE OF 0.01 FOR ANY 25 MM SQUARE SURFACE WITH A MAXIMUM FLATNESS TOLERANCE OF 0.03
- THE HORIZONTAL HOLE (Ø 20H8) IS TO BE PARALLEL WITHIN 0.02 WITH DATUM A, RFS. SHOW THE LIMITS OF SIZE FOR THE HOLE.

- THE VERTICAL HOLE (Ø 30H7) IS PERPENDICULAR WITHIN 0.03 TO DATUM A. SHOW THE LIMITS OF SIZE FOR THE HOLE
- THE SLOT WIDTH (50 ± 0.2) IS TO BE PERPENDICULAR WITHIN 0.15 TO DATUM A, RFS

FIG. 16-8-C   Support.

ASSIGNMENTS FOR UNIT 16-9, POSITIONAL TOLERANCING

23. If coordinate tolerances as shown in Fig. 16-9-A are given, what are the shapes of the tolerance zones and the distance between extreme permissible positions of the holes?

24. In order to assemble correctly, the hole shown in Fig. 16-9-B must not vary more than .0014 in. in any direction from its true position when the hole is at its smallest size. Prepare sketches showing suitable toleranc-ing, dimensioning, and datums, where required, to achieve this by using:
    (a) Coordinate tolerancing
    (b) Positional tolerancing—RFS
    (c) Positional tolerancing—MMC
    (d) Positional tolerancing—LMC

25. With reference to Assignment 24, what would be the max-imum permissible deviation from true position when the hole was at its largest size in each of the four examples?

26. In Fig. 16-9-C add the largest equal coordinate tolerances so that if two such parts are assembled with the edges aligned, the distance between their hole centers could never be more than that shown.

27. The part shown in Fig. 16-9-D (A) is set on a revolving table, so adjusted that the part revolves about the true posi-tion center of the 20 mm hole.
    (a) If both indicators give identical readings and the results in Fig. 16-9-D (B) are obtained, which parts are acceptable?
    (b) What is the positional error for each part in Fig. 16-9-D (B)?
    (c) If MMC instead of RFS had been shown in the fea-ture control frame in Fig. 16-9-D (A), what is the diameter of the mandrel that would be required to check the parts?
    (d) What is the maximum permissible tolerance error for each part shown in Fig. 16-9-D (B) if MMC was used?

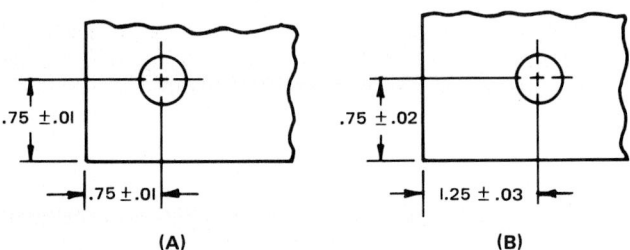

FIG. 16-9-A   Assignment.

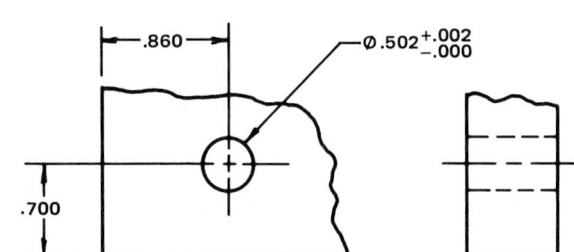

FIG. 16-9-B   Assignment.

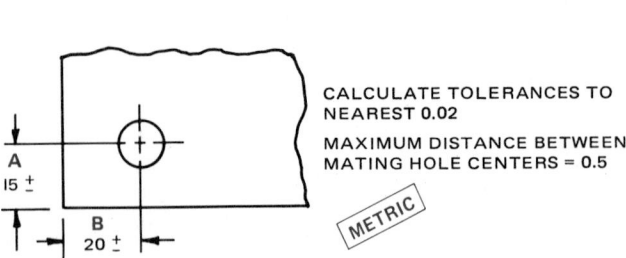

CALCULATE TOLERANCES TO NEAREST 0.02

MAXIMUM DISTANCE BETWEEN MATING HOLE CENTERS = 0.5

FIG. 16-9-C   Assignment.

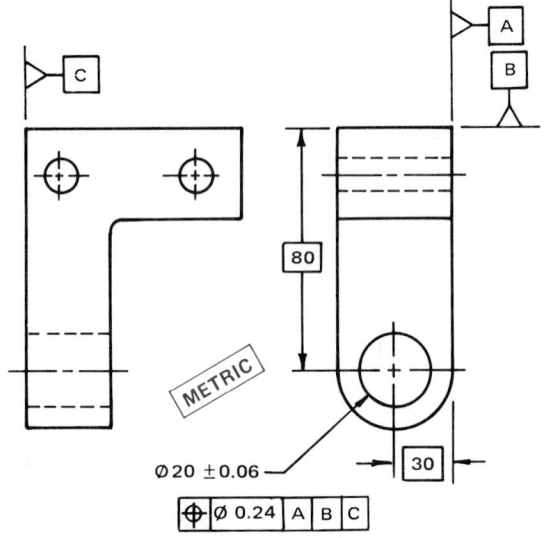

(A) DRAWING CALLOUT

PART NO.	SIZE OF MANDREL	HIGHEST READING	LOWEST READING
1	20.00	1.54	1.32
2	20.06	0.18	-0.07
3	19.96	1.87	1.61
4	19.94	1.72	1.48
5	20.00	1.95	1.85
6	20.05	1.24	1.02

(B) READINGS FOR PARTS

FIG. 16-9-D   Assignment.

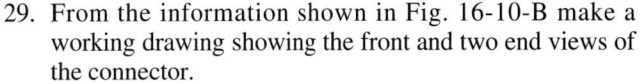

ASSIGNMENTS FOR UNIT 16-10, PROJECTED TOLERANCE ZONE

28. From the information shown in Fig. 16-10-A make a working drawing showing the top and front views of the cover plate.

GEOMETRIC TOLERANCING REQUIREMENTS TO BE ADDED TO DRAWING

- ALL DATUMS AND TOLERANCES TO BE ON AN MMC BASIS UNLESS OTHERWISE SPECIFIED
- SURFACES AND FEATURES MARKED A, B, C, AND D ARE DATUMS A, B, C, AND D RESPECTIVELY
- THE 4 HOLES MUST NOT VARY FROM TRUE POSITION BY MORE THAN .002 IN ANY DIRECTION WHEN THE HOLES ARE AT MMC AND ARE RELATED TO DATUMS A, D, AND B IN THAT ORDER
- A FLATNESS TOLERANCE OF .010 IS REQUIRED FOR THE UNDERSIDE OF THE COVER PLATE
- THE Ø 2.000 HOLE HAS A POSITIONAL TOLERANCE OF .008 AND IS REFERENCED TO DATUMS A, B, AND C IN THAT ORDER
- A PROJECTED TOLERANCE ZONE OF .60 IS REQUIRED FOR THE 4 HOLES, THE PROJECTION BEING DIRECTED AWAY FROM THE TOP OF THE COVER

FIG. 16-10-A   Cover plate.

FIG. 16-10-B   Connector.

29. From the information shown in Fig. 16-10-B make a working drawing showing the front and two end views of the connector.

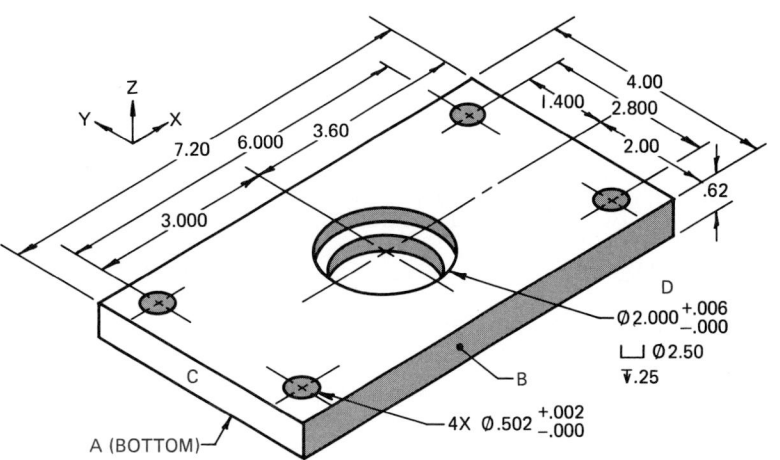

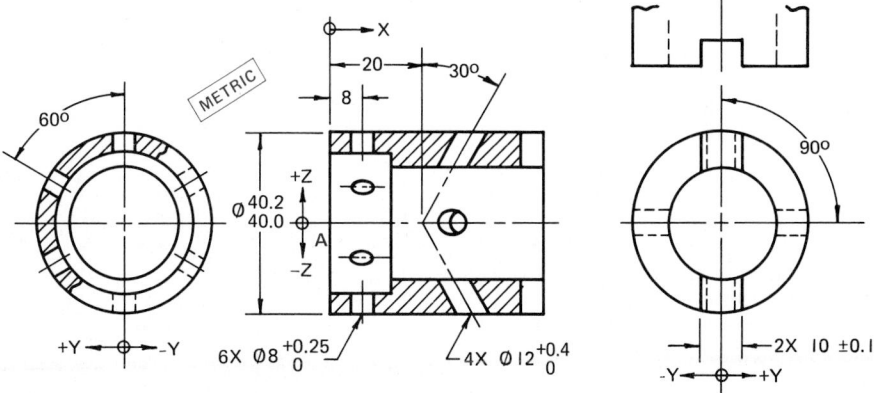

GEOMETRIC TOLERANCING REQUIREMENTS TO BE ADDED TO DRAWING

- ALL DATUMS AND TOLERANCES TO BE ON AN MMC BASIS UNLESS OTHERWISE SPECIFIED
- SURFACE A IS DATUM A AND IS THE PRIMARY DATUM
- Ø 40.0–40.2 IS DATUM B AND IS THE SECONDARY DATUM
- THE 10 MM WIDE SLOTS ARE DATUM C AND FORM THE TERTIARY DATUM
- THE Ø 8 HOLES HAVE A POSITIONAL TOLERANCE OF 0.1 AND ARE REFERENCED TO DATUMS A, B, AND C
- THE Ø 12 HOLES HAVE A POSITIONAL TOLERANCE OF 0.2 WITH A MAXIMUM POSITIONAL TOLERANCE

OF 0.5 AND ARE REFERENCED TO DATUMS A, B, AND C
- BOTH GROUPS OF HOLES ARE TO HAVE A PROJECTED TOLERANCE ZONE OF 10 MM MEASURED PERPENDICULAR TO THE Ø 40.0–40.2 CYLINDRICAL SURFACE

ASSIGNMENTS FOR UNIT 16-11, DATUM TARGETS

30. Make a three-view working drawing of the bearing housing shown in Fig. 16-11-A showing the datum features. Only the dimensions related to the datums need to be shown.

FIG. 16-11-A Bearing housing.

31. From the information shown in Fig. 16-11-B make a three-view working drawing of the bracket guide.

DATUM AND LOCATION				
DATUM DESCRIPTION		LOCATION FROM		
		PRIMARY DATUM PLANE	SECONDARY DATUM PLANE	TERTIARY DATUM PLANE
DATUM A TARGET AREAS Ø .50	AI		.60	I.40
	A2		.60	5.00
	A3		3.60	3.20
DATUM B TARGET LINES	BI			.80
	B2			5.60
DATUM C TARGET POINT	CI	.60	I.80	

ROUNDS & FILLETS R.20

SECONDARY DATUM PLANE

TERTIARY DATUM PLANE

2.00   2.00

6.40

RI.00

Ø 1.200 +.003 −.000

1.80

.60

2.20

4.40

Z

Y

PRIMARY DATUM PLANE

X

GEOMETRIC TOLERANCING REQUIREMENTS TO BE ADDED TO DRAWING
- ALL DATUMS AND TOLERANCES TO BE ON AN MMC BASIS UNLESS OTHERWISE SPECIFIED

- THE HOLE HAS A POSITIONAL TOLERANCE OF .004 REFERENCED TO DATUMS A, B, AND C
- DATUM TARGET INFORMATION SHOWN IN THE CHART BELOW

GEOMETRIC TOLERANCING REQUIREMENTS TO BE ADDED TO DRAWING
- ALL DATUMS AND TOLERANCES TO BE ON AN MMC BASIS UNLESS OTHERWISE SPECIFIED
- PRIMARY DATUM A HAS THREE Ø 3 TARGET AREAS. AI AND A2 ARE LOCATED ON CENTER OF SURFACE M, ONE-FIFTH THE DEPTH DISTANCE FROM THE FRONT AND BACK, RESPECTIVELY. A3 IS LOCATED ON SURFACE N MIDWAY BETWEEN THE CENTER OF THE HOLE AND THE RIGHT END
- SECONDARY DATUM B IS A DATUM LINE LOCATED AT MID-HEIGHT OF SURFACE D
- TERTIARY DATUM C IS A DATUM POINT LOCATED AT THE CENTER OF SURFACE E
- THE Ø 10 HOLE HAS A POSITIONAL TOLERANCE OF 0.2 REFERENCED TO DATUMS A, B, AND C

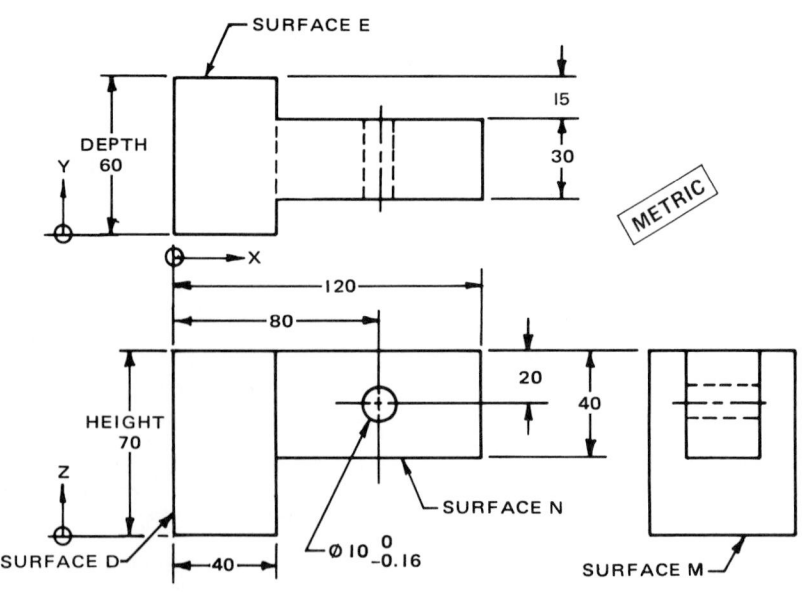

SURFACE E

15

30

DEPTH 60

Y

X

METRIC

120

80

20

40

HEIGHT 70

Z

SURFACE N

SURFACE D

40

Ø 10 0 −0.16

SURFACE M

FIG. 16-11-B Bracket guide.

ASSIGNMENTS FOR UNIT 16-12, CIRCULARITY (ROUNDNESS) AND CYLINDRICITY

32. Add circularity tolerances to the diameters shown in Fig. 16-12-A. The circularity tolerances are to be one-fifth of the size tolerances for each diameter.
33. Show on each part in Fig. 16-12-B a cylindricity tolerance. The size of the cylindricity tolerance is to equal one-quarter the size tolerance for each diameter.
34. Measurements for circularity for the part shown in Fig. 16-12-C were made at the cross sections A-A to C-C. All points on the periphery fell within the two rings. The outer ring was the smallest that could be circumscribed about the profile, and the inner ring the largest that could be inscribed within the profile. State which sections meet drawing requirements.
35. Apply cylindricity tolerances to the three features dimensioned in Fig. 16-12-D. The cylindricity tolerances are to be 25 percent of the size tolerances.
36. Readings were taken at intervals along the shaft shown in Fig. 16-12-E to check the cylindricity tolerance. All points on the periphery fell within the two rings. The outer ring was the smallest that could be circumscribed about the profile, and the inner ring was the largest that could be inscribed within the profile.
    (a) Does the part meet drawing requirements?
    (b) Sketch the cylindricity tolerance zone complete with dimensions.

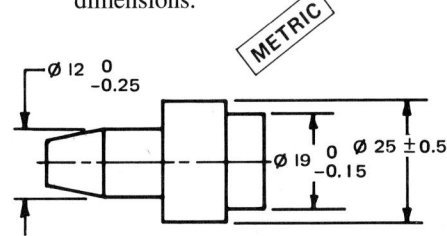

FIG. 16-12-A   Assignment.

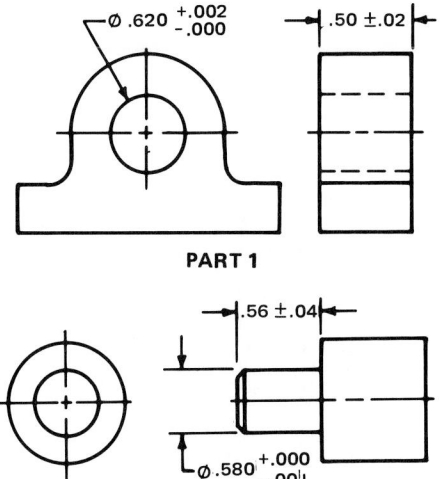

PART 1

PART 2

FIG. 16-12-B   Assignment.

(c) If the cylindricity tolerance was changed to a circularity tolerance, would the part pass inspection?

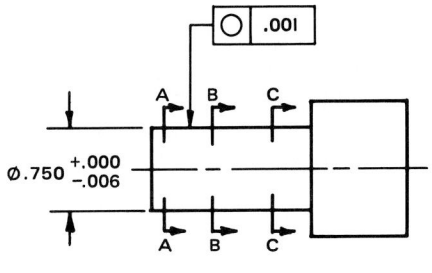

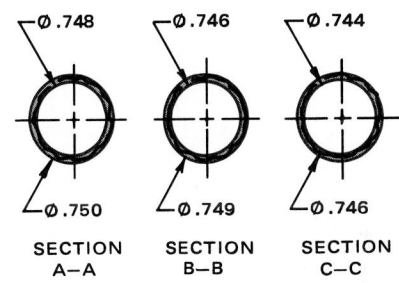

SECTION A–A   SECTION B–B   SECTION C–C

FIG. 16-12-C   Assignment.

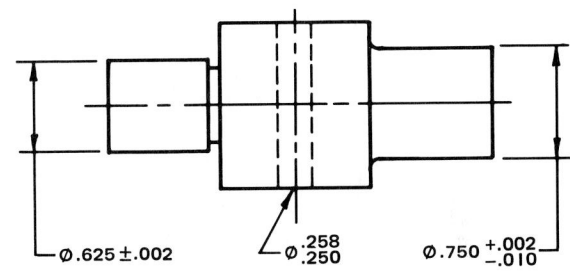

FIG. 16-12-D   Assignment.

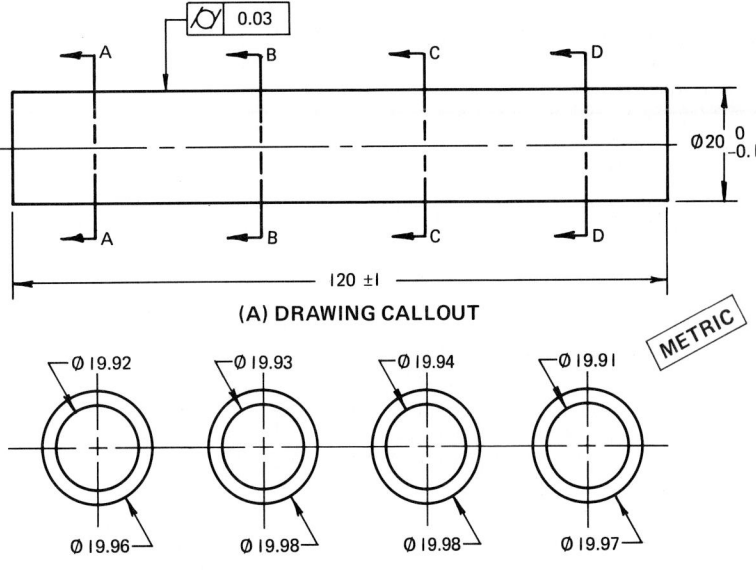

(A) DRAWING CALLOUT

(B) TOLERANCE ZONE

FIG. 16-12-E   Assignment.

ASSIGNMENTS FOR UNIT 16-13, PROFILE TOLERANCING

37. A cam is dimensioned as shown in Fig. 16-13-A. If parts were measured with an indicator that was set to zero and measurements were obtained as shown, which parts shown in the chart would not be acceptable? Of the nonacceptable parts, which could be made acceptable by regrinding?

38. In Fig. 16-13-B the form of the indented portion is to be controlled by the profile-of-a-line (bilateral) tolerance of .006 in. Show the tolerance and sketch the resulting tolerance zone on the drawing and indicate which dimensions are basic.

39. It is required to control the profile in Fig. 16-13-C with the tolerance described on the drawing. Add the profile-of-a-line tolerance to the drawing and sketch the resulting tolerance zone.

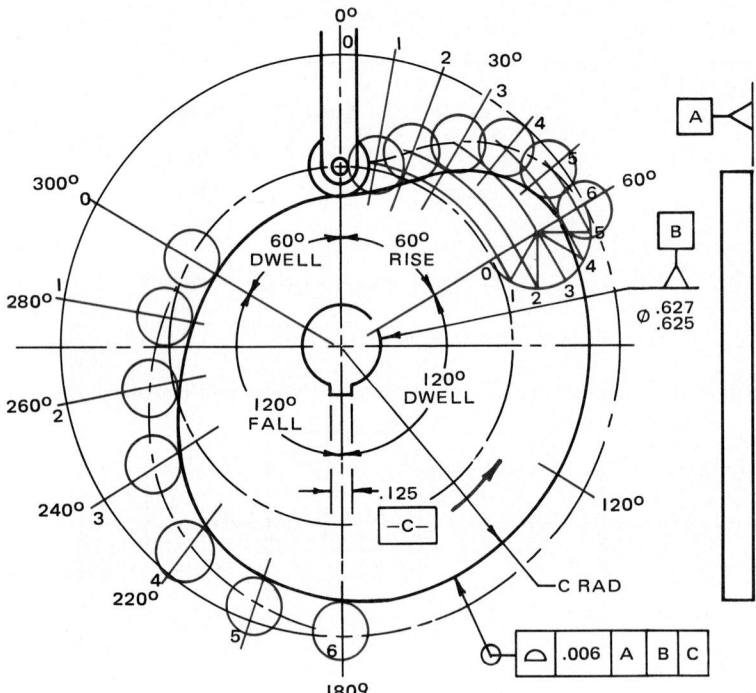

DRAWING CALLOUT FOR C RAD	BASIC DIMENSION OF CAMS				
	ANGLE	CAM 1	CAM 2	CAM 3	CAM 4
1.000	0	.997	1.002	.999	1.003
1.300	30	1.302	1.303	1.297	1.299
1.600	60	1.602	1.600	1.601	1.596
1.600	120	1.598	1.597	1.603	1.601
1.600	180	1.596	1.601	1.604	1.598
1.450	220	1.447	1.453	1.449	1.451
1.300	240	1.303	1.302	1.297	1.299
1.150	260	1.153	1.147	1.151	1.148
1.000	300	1.005	1.001	.997	.999
1.000	330	1.003	.997	1.002	.998

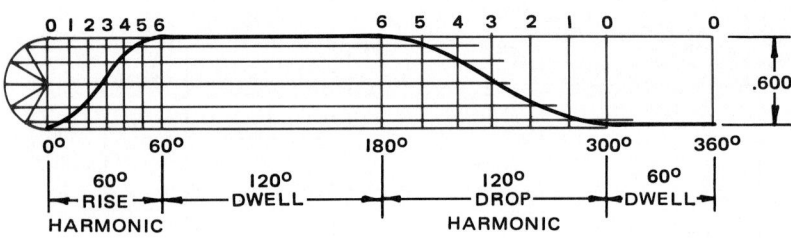

FIG. 16-13-A   Cam.

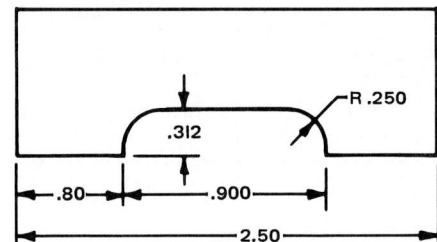

FIG. 16-13-B   Assignment.

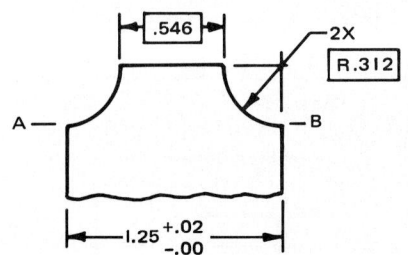

CONTROL THE PROFILE A TO B WITH A LINE BILATERAL PROFILE TOLERANCE OF .003 EXCEPT THAT THE .546 STRAIGHT PORTION CAN BE PERMITTED TO VARY VERTICALLY BY ±.01.

FIG. 16-13-C   Assignment.

40. From the information shown in Fig. 16-13-D make a two-view working drawing of the slide.

41. From the information shown in Fig. 16-13-E make a two-view working drawing of the indicator.

FIG. 16-13-D   Slide.

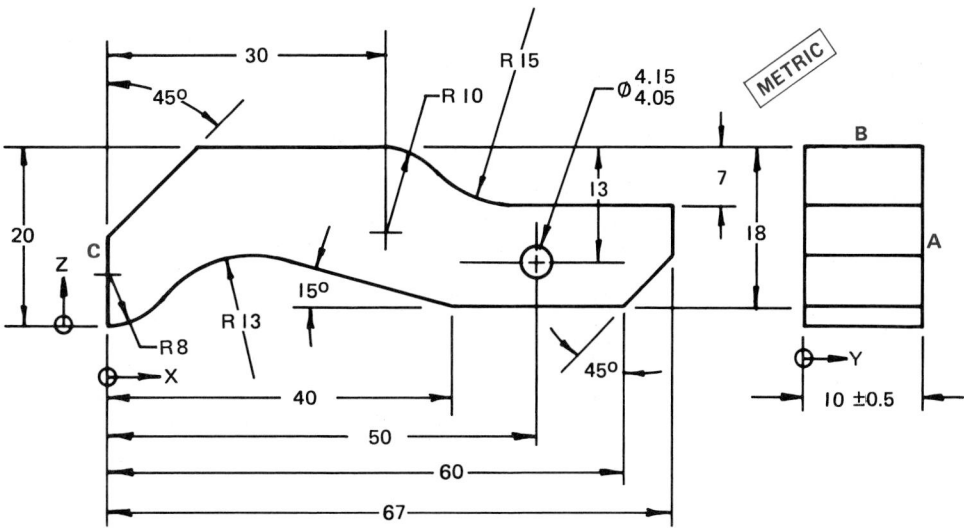

GEOMETRIC TOLERANCING REQUIREMENTS TO BE ADDED TO DRAWING
- ALL DATUMS AND TOLERANCES TO BE ON AN MMC BASIS UNLESS OTHERWISE SPECIFIED
- SURFACES A, B, AND C ARE DATUMS A, B, AND C IN THAT ORDER
- ALL CORNERS ON THE PROFILE ARE TO HAVE A MAXIMUM 0.1 RADIUS
- A PROFILE-OF-A-SURFACE TOLERANCE OF 0.2 IS TO BE ADDED ALL AROUND THE PROFILE OF THE

PART AND THE PART CANNOT EXCEED THE BOUNDARY OF THE DIMENSIONS SHOWN
- THE PROFILE TOLERANCE TO BE REFERENCED TO DATUMS A AND B IN THAT ORDER
- THE HOLE IS TO HAVE A POSITIONAL TOLERANCE OF 0.12 AND BE REFERENCED TO DATUMS A, B, AND C IN THAT ORDER

FIG. 16-13-E   Indicator.

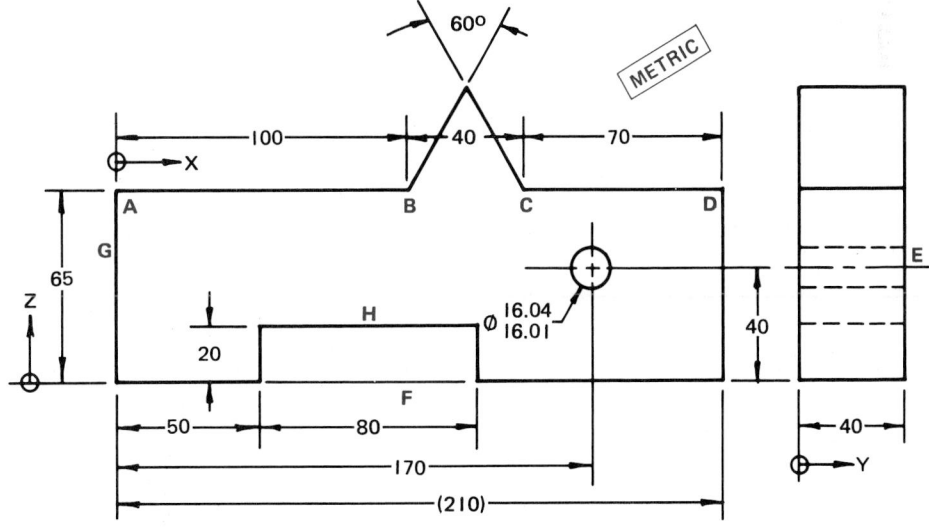

GEOMETRIC TOLERANCING REQUIREMENTS TO BE ADDED TO DRAWING
- THE TRIANGULAR FEATURE (FROM B TO C) HAS A PROFILE-OF-A-LINE TOLERANCE OF 0.1
- THE STRAIGHT FEATURES A TO B AND C TO D HAVE A PROFILE-OF-A-LINE TOLERANCE OF 0.3
- THE TOLERANCE ZONE IS LOCATED ON THE OUTSIDE OF THE TRUE PROFILE
- MAXIMUM RADIUS ON SHARP POINT TO BE 0.1 RADIUS

- INDICATE WHICH DIMENSIONS ARE BASIC
- SURFACES E, F, AND G ARE DATUMS E, F, AND G RESPECTIVELY
- SURFACE H IS PARALLEL WITHIN 0.06 WITH SURFACE F
- THE HOLE HAS A POSITIONAL TOLERANCE OF 0.4 AND IS REFERENCED TO DATUMS E, F, AND G IN THAT ORDER

583

ASSIGNMENTS FOR UNIT 16-14, CORRELATIVE TOLERANCES

42. From the information shown in Fig. 16-14-A make a two-view working drawing of the control arm.

43. From the information shown in Fig. 16-14-B make a two-view working drawing (top and side views) of the adjustable base.

FIG. 16-14-A   Control arm.

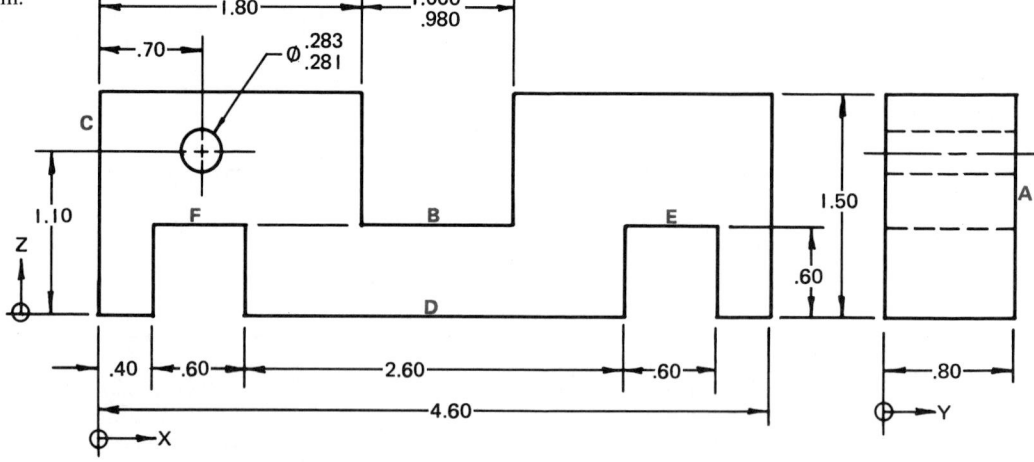

GEOMETRIC TOLERANCING REQUIREMENTS TO BE ADDED TO DRAWING
- ALL DATUMS AND TOLERANCES TO BE ON AN MMC BASIS UNLESS OTHERWISE SPECIFIED
- SURFACES IDENTIFIED AS A, B, C, AND D TO BE DATUMS A, B, C, AND D RESPECTIVELY
- SURFACE A TO BE FLAT WITHIN .010
- A PROFILE-OF-A-SURFACE TOLERANCE OF .010 IS TO BE APPLIED TO THE THREE COPLANAR

SURFACES IDENTIFIED BY THE LETTERS F, B, AND E
- THE MEDIAN PLANE OF THE TOP SLOT IS TO BE PERPENDICULAR WITHIN .005 TO DATUM B
- THE HOLE IS TO HAVE A POSITIONAL TOLERANCE OF .008 AND BE REFERENCED TO DATUMS A, D, AND C IN THAT ORDER

GEOMETRIC TOLERANCING REQUIREMENTS TO BE ADDED TO DRAWING
- ALL DATUMS AND TOLERANCES TO BE ON AN MMC BASIS UNLESS OTHERWISE SPECIFIED
- SURFACES INDICATED BY LETTERS A, B, AND C ARE DATUMS A, B, AND C RESPECTIVELY
- SURFACE A TO BE FLAT WITHIN 0.15
- A PROFILE-OF-A-SURFACE TOLERANCE OF 0.2 IS TO BE APPLIED TO THE THREE COPLANAR SURFACES ON THE LEFT SIDE OF THE PART WITH THE LOWER OF THE THREE SURFACES BEING DESIGNATED AS DATUM D
- A PROFILE-OF-A SURFACE TOLERANCE OF 0.4 IS TO BE APPLIED TO THE BOTTOM FOUR COPLANAR SURFACES WITH THE SURFACE ON THE LEFT BEING DESIGNATED AS DATUM E.
- THE Ø 10 PORTION OF THE COUNTERBORED HOLES IS DATUM N AND IS TO HAVE A POSITIONAL TOLERANCE OF 0.12 RELATED TO DATUMS A, B, AND C IN THAT ORDER.
- A COAXIAL RELATIONSHIP BETWEEN THE Ø 10 AND Ø 16 PORTIONS OF THE COUNTERBORED HOLES IS CONTROLLED

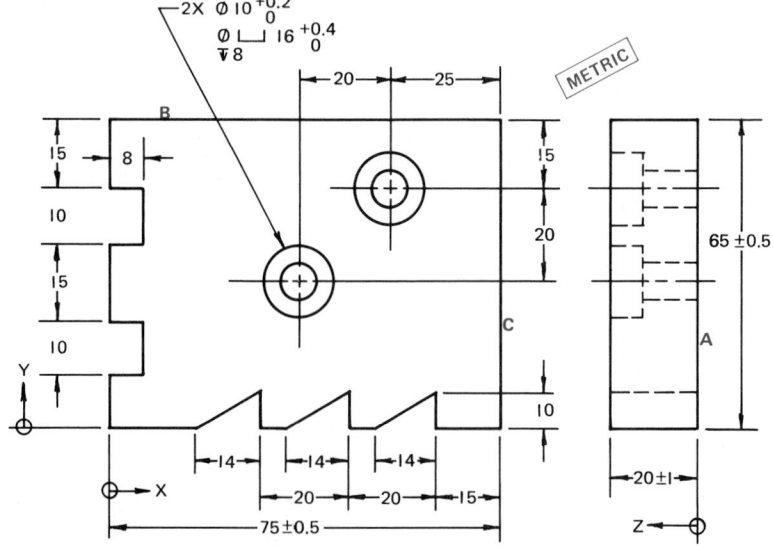

BY APPLYING A POSITIONAL TOLERANCE OF ZERO TO THE COUNTERBORED DIAMETER

FIG. 16-14-B   Adjustable base.

584

44. Complete the chart shown in Fig. 16-14-C showing the maximum allowable distance between the axes of the two coaxial features.

45. From the information shown in Fig. 16-14-D make a two-view working drawing (front and end views) of the axle.

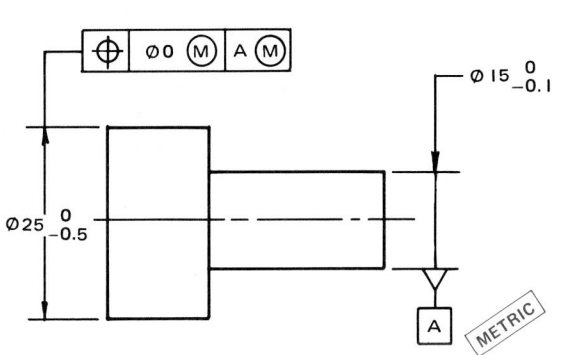

(A) DRAWING CALLOUT

CONSIDERED FEATURE SIZES	DATUM FEATURE SIZES					
	15	14.98	14.96	14.94	14.92	14.9
25						
24.9						
24.8						
24.7						
24.6						
24.5						

(B) ALLOWABLE DISTANCES BETWEEN AXES

FIG. 16-14-C   Stepped shaft.

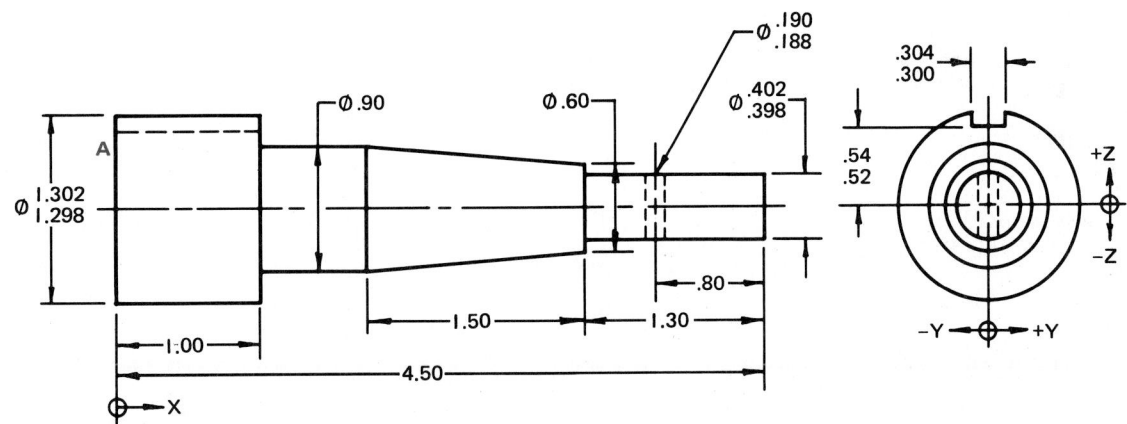

GEOMETRIC TOLERANCING REQUIREMENTS TO BE ADDED TO DRAWING

- ALL DATUMS AND TOLERANCES TO BE ON AN MMC BASIS UNLESS OTHERWISE SPECIFIED
- THE Ø .398–402 SHAFT IS TO BE CONCENTRIC WITHIN .003 WITH THE Ø 1.298–1.302 SHAFT
- SURFACE A IS DATUM A
- THE Ø 1.298–1.302 SHAFT IS DATUM B
- THE Ø .398–402 SHAFT IS DATUM C
- THE SLOT IS TO BE SYMMETRICALLY LOCATED WITHIN ZERO MMC ON THE SHAFT AND REFERENCED TO DATUMS A AND B IN THAT ORDER
- THE Ø .188–.190 HOLE IS TO BE PERPENDICULAR WITHIN .002 WITH THE SHAFT.

FIG. 16-14-D   Axle.

46. From the information shown in Fig. 16-14-E make a one-view working drawing of the stepped shaft.
47. From the information shown in Fig. 16-14-F make a one-view working drawing of the tapered shaft.

48. Show geometric tolerances for the parts shown in Fig. 16-14-G which will ensure that the features are symmetrical with their datum features.

GEOMETRIC TOLERANCING REQUIREMENTS TO BE ADDED TO DRAWING
- Ø 1.183–1.187 TO BE DATUM C
- A 1.200 LENGTH OF SHAFT STARTING. 40 FROM THE RIGHT TO BE DATUM D.
- RUNOUT TOLERANCES ARE REFERENCED TO THE AXIS ESTABLISHED BY DATUMS C AND D
- A TOTAL RUNOUT TOLERANCE OF .002 BETWEEN POSITIONS A AND B
- A CIRCULAR RUNOUT TOLERANCE OF .002 FOR DIAMETERS E AND F
- A CIRCULAR RUNOUT TOLERANCE OF .005 FOR DIAMETER G
- A CIRCULAR RUNOUT TOLERANCE OF .004 FOR SURFACE H
- A CIRCULAR RUNOUT TOLERANCE OF .003 FOR SURFACES J AND K

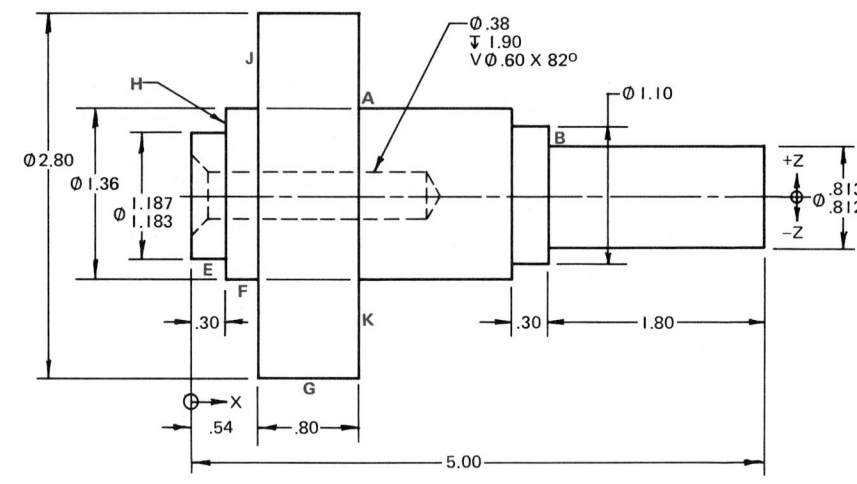

FIG. 16-14-E   Stepped shaft.

GEOMETRIC TOLERANCING REQUIREMENTS TO BE ADDED TO DRAWING
- RUNOUT TOLERANCES ARE REFERENCED TO THE AXIS OF THE CENTERING HOLES A AND B WHICH COLLECTIVELY ACTS AS A COMPOUND DATUM
- CIRCULAR RUNOUT TOLERANCES OF 0.03 FOR SURFACES C, D, E, AND F
- CIRCULAR RUNOUT TOLERANCE OF 0.08 FOR THE TWO CURVED SURFACES
- TOTAL RUNOUT TOLERANCE OF 0.05 FOR THE Ø 28 AND Ø 38 SECTIONS
- TOTAL RUNOUT TOLERANCE OF 0.04 FOR SURFACE G

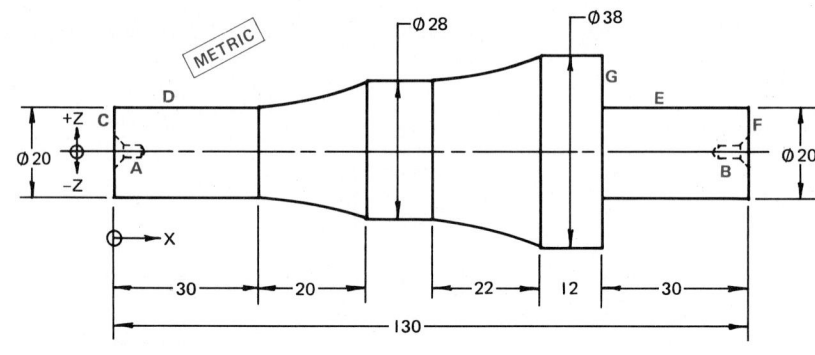

FIG. 16-14-F   Tapered shaft.

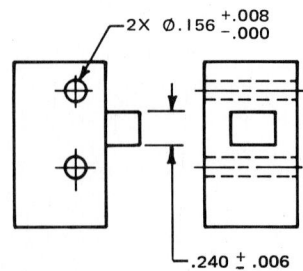

RECTANGULAR PROJECTION TO BE SYMMETRICAL WITHIN .002 AT MMC WITH THE HOLES.

PART 1

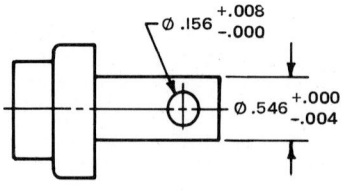

Ø.156 HOLE TO BE SYMMETRICAL WITH Ø.546 WITHIN .001 REGARDLESS OF FEATURE SIZE.

PART 2

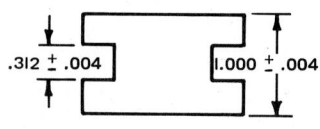

THE 2 SLOTS TO BE SIMULTANEOUSLY SYMMETRICAL WITH THE 1.000 WIDTH WITHIN ZERO TOLERANCE WHEN BOTH THE SLOTS AND WIDTH ARE AT MMC.

PART 3

FIG. 16-14-G   Assignments.

ASSIGNMENTS FOR UNIT 16-15, POSITIONAL TOLERANCING OF NONCYLINDRICAL FEATURES

49. From the information given in Fig. 16-15-A make a working drawing showing the top and right side view of the guide block.

50. From the information given in Fig. 16-15-B make a working drawing showing the top and right side view of the locating block.

FIG. 16-15-A   Guide block.

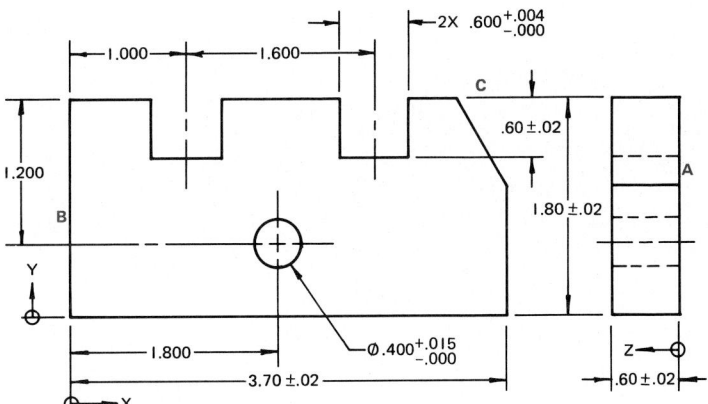

GEOMETRIC TOLERANCING REQUIREMENTS TO BE ADDED TO DRAWING
- ALL DATUMS AND TOLERANCES TO BE ON AN MMC BASIS UNLESS OTHERWISE SPECIFIED
- SURFACES MARKED A, B, AND C ARE THE PRIMARY, SECONDARY, AND TERTIARY DATUMS IN THAT ORDER
- THE TWO SLOTS (VERTICAL SIDES ONLY) ARE TO BE LOCATED BY MEANS OF A

POSITIONAL TOLERANCE OF .006 AND REFERENCED TO THE THREE DATUMS
- THE ⌀ .400 HOLE IS LOCATED BY A POSITIONAL TOLERANCE OF .010 REFERENCED TO DATUMS A, B, AND C
- SURFACE C IS TO BE PERPENDICULAR TO SURFACE B WITHIN .005
- SURFACE A IS TO BE FLAT WITHIN .008

GEOMETRIC TOLERANCING REQUIREMENTS TO BE ADDED TO DRAWING
- ALL DATUMS AND TOLERANCES TO BE ON AN MMC BASIS UNLESS OTHERWISE SPECIFIED
- THE PRIMARY DATUM IS SURFACE A (DATUM A)
- THE SECONDARY DATUM IS THE ⌀ 12 HOLE (DATUM B)
- THE TERTIARY DATUM IS THE KEYWAY (DATUM C)
- THE TWO SLOTS ARE LOCATED BY A POSITIONAL TOLERANCE OF 0.4. THEY ARE LOCATED ON THE HORIZONTAL CENTER LINE OF THE ⌀ 12 HOLE AND REFERENCED TO DATUMS A, B, AND C.
- THE AXES OF THE TWO SMALL HOLES HAVE A POSITIONAL TOLERANCE OF 0.25 AND ARE REFERENCED TO DATUMS A, B, AND C IN THAT ORDER
- DATUM A HAS A FLATNESS TOLERANCE OF 0.2
- THE ⌀ 12 HOLE HAS A POSITIONAL TOLERANCE OF 0.08 AND IS REFERENCED TO DATUM A
- THE KEYWAY IS TO BE SYMMETRICALLY LOCATED ON THE ⌀ 12 HOLE BY A ZERO POSITIONAL TOLERANCE

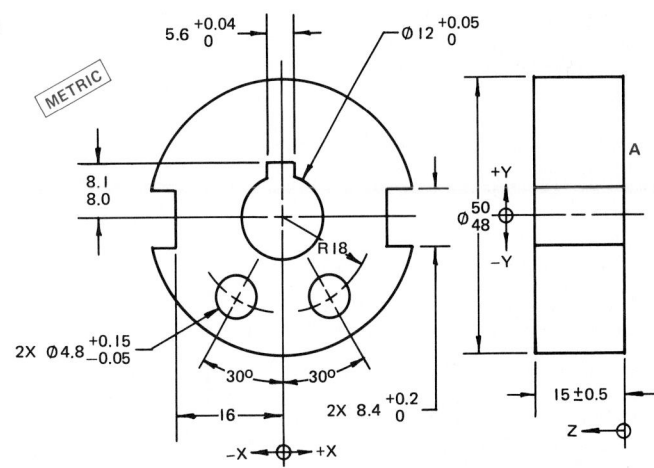

FIG. 16-15-B   Locating block.

ASSIGNMENTS FOR UNIT 16-16, POSITIONAL TOLERANCING FOR MULTIPLE PATTERNS OF FEATURES

51. Prepare a working drawing showing the top and right side view of the locating plate from the information shown in Fig. 16-16-A.

52. Prepare a working drawing showing the top and right side view of the guide from the information shown in Fig. 16-16-B.

53. Figure 16-16-C shows the actual size of the holes produced on one part from the drawing made in Assignment 52. Prepare a chart showing the maximum diameter tolerance permitted for each of the holes shown.

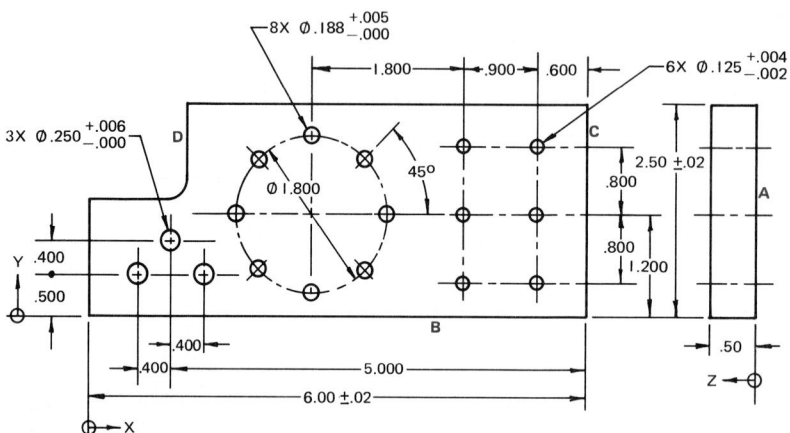

GEOMETRIC TOLERANCING REQUIREMENTS TO BE ADDED TO DRAWING

• ALL DATUMS AND TOLERANCES TO BE ON AN MMC BASIS UNLESS OTHERWISE SPECIFIED

• SURFACES MARKED A, B, AND C ARE THE PRIMARY, SECONDARY, AND TERTIARY DATUMS RESPECTIVELY

• COMPOSITE POSITIONAL TOLERANCING IS REQUIRED FOR THE THREE PATTERNS OF HOLES AND IS REFERENCED TO DATUMS A, B, AND C IN THAT ORDER

• THE POSITIONAL TOLERANCE FOR LOCATION OF HOLE PATTERNS ARE .010, .008, AND .016 FOR THE ∅ .250, ∅ .188, AND ∅ .125 HOLES RESPECTIVELY

• THE POSITIONAL TOLERANCE FOR HOLES WITHIN THE PATTERN ARE .004, .003, AND .006 FOR THE ∅ .250, ∅ .188, AND ∅ .125 HOLES RESPECTIVELY

• A FLATNESS TOLERANCE OF .004 IS REQUIRED FOR DATUM B

• THE TOP SURFACE IS TO BE PARALLEL WITHIN .008 WITH THE BOTTOM SURFACE

• THE RIGHT SIDE SURFACE IS TO BE PERPENDICULAR WITHIN .005 WITH THE BOTTOM SURFACE

• SURFACE D TO HAVE A PERPENDICULARITY TOLERANCE OF .008 WITH THE BOTTOM SURFACE

• DATUM A TO BE FLAT WITHIN .005

FIG. 16-16-A   Locating plate.

GEOMETRIC TOLERANCE REQUIREMENTS TO BE ADDED TO DRAWING

• ALL DATUMS AND TOLERANCES TO BE ON AN MMC BASIS UNLESS OTHERWISE SPECIFIED

• BASIC DIMENSIONS AND DATUMS ARE TO BE SHOWN

• SURFACES MARKED A, B, AND C ARE THE PRIMARY, SECONDARY, AND TERTIARY DATUMS RESPECTIVELY

• ALL OF THE HOLES ARE TO BE TREATED AS A COMPOSITE PATTERN

• THE COORDINATE TOLERANCING IS TO BE CONVERTED TO POSITIONAL TOLERANCING OF EQUIVALENT VALUES

• DATUM A IS TO BE FLAT WITHIN 0.2

• DATUM C TO BE PERPENDICULAR WITHIN 0.3 TO DATUM B

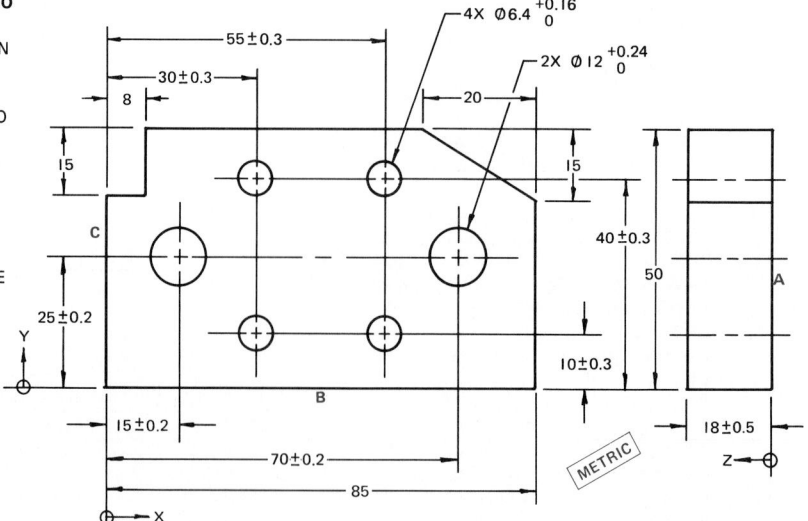

FIG. 16-16-B   Guide.

54. Prepare a working drawing showing the top and right side view of the bracket from the information shown in Fig. 16-16-D.

55. Prepare a working drawing showing the top and right side view of the spacer from the information shown in Fig. 16-16-E.

FIG. 16-16-C Part made from the drawing callout shown in Fig. 16-16-B.

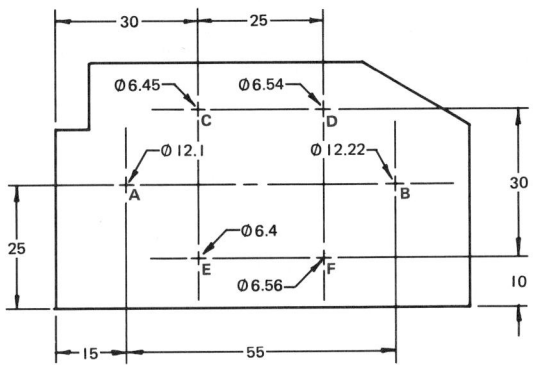

HOLE	MAXIMUM DIAMETER TOLERANCE ZONE PERMITTED
A	
B	
C	
D	
E	
F	

GEOMETRIC TOLERANCING REQUIREMENTS TO BE ADDED TO DRAWING
- ALL DATUMS AND TOLERANCES TO BE ON AN MMC BASIS UNLESS OTHERWISE SPECIFIED
- SURFACES MARKED A, B, AND C ARE THE PRIMARY, SECONDARY, AND TERTIARY DATUMS RESPECTIVELY
- THE DIFFERENT SIZED HOLES ARE TO BE DIMENSIONED HAVING SEPARATE REQUIREMENTS AND REFERENCED TO THE THREE DATUMS
- THE COORDINATE TOLERANCING IS TO BE CONVERTED TO POSITIONAL TOLERANCING OF EQUIVALENT VALUES
- BASIC DIMENSIONS AND DATUMS ARE TO BE SHOWN
- DATUM A IS TO BE FLAT WITHIN .002
- DATUM C TO BE PERPENDICULAR WITHIN .004 TO DATUM B

FIG. 16-16-D Bracket.

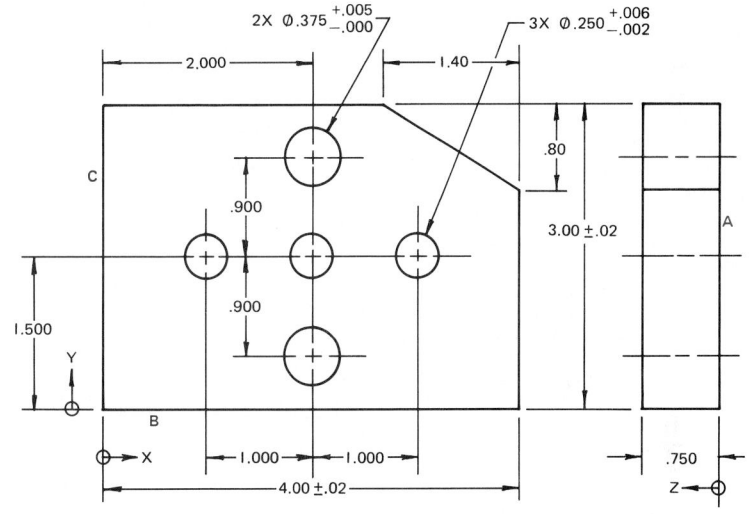

UNLESS OTHERWISE SPECIFIED TOLERANCE ON DIMENSIONS ±.005

GEOMETRIC TOLERANCING REQUIREMENTS TO BE ADDED TO DRAWING
- ALL DATUMS AND TOLERANCES TO BE ON AN MMC BASIS UNLESS OTHERWISE SPECIFIED
- SURFACES MARKED A, B, AND C ARE THE PRIMARY, SECONDARY, AND TERTIARY DATUMS RESPECTIVELY
- FOR THE PRIMARY DATUM THE ENTIRE SURFACE IS TO BE USED
- FOR THE SECONDARY DATUM USE TWO DATUM TARGET LINES LOCATED ONE-FIFTH OF THE WIDTH OF THE PART FROM EACH SIDE
- THE TERTIARY DATUM IS A TARGET LINE LOCATED ON THE CENTER LINE OF THE TWO Ø 12 HOLES
- COMPOSITE POSITIONAL TOLERANCING IS REQUIRED FOR THE TWO PATTERNS OF HOLES
- THE POSITIONAL TOLERANCES OF LOCATION OF PATTERNS ARE 0.8 AND 1.2 FOR THE Ø 8 AND Ø 12 HOLES RESPECTIVELY
- THE POSITIONAL TOLERANCING FOR THE HOLES WITHIN THE PATTERN ARE 0.5 AND 0.8 FOR THE Ø 8 AND Ø 12 HOLES RESPECTIVELY

- ADD A FLATNESS TOLERANCE OF 0.5 TO THE BOTTOM SURFACE
- THE TOP SURFACE IS TO BE PARALLEL WITHIN 0.4 WITH THE BOTTOM SURFACE
- THE LEFT SIDE IS TO BE PERPENDICULAR WITHIN 0.6 WITH THE BOTTOM SURFACE

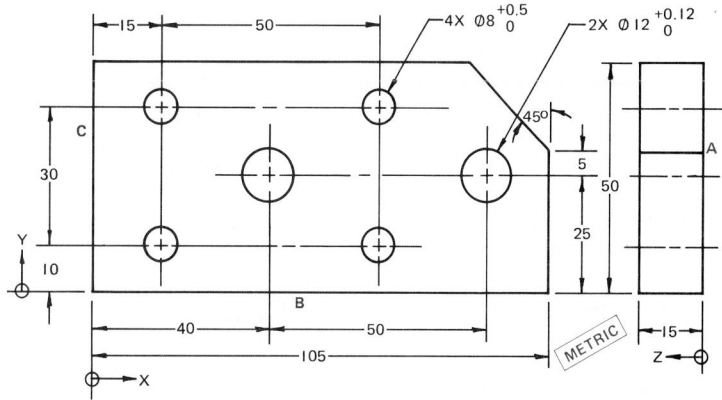

FIG. 16-16-E Spacer.

ASSIGNMENTS FOR UNIT 16-17, FORMULAS FOR POSITIONAL TOLERANCING

56. What is the size of the positional tolerance required for the cover and flange shown in Fig. 16-17-A, detail 1?

57. If the positional tolerance for the holes in the flange in Assignment 56 is to be twice the size as that of the holes in the cover, what size would it be?

58. What is the size of the positional tolerance required for the cover and flange holes in Fig. 16-17-A, detail 2?

59. The bracket is to be assembled to the plate shown in Fig. 16-17-B with two M5 screws (max Ø5 mm). What is the smallest size permissible for the two holes on the plate?

60. Apply equal positional tolerance to the coaxial features shown in Fig. 16-17-C.

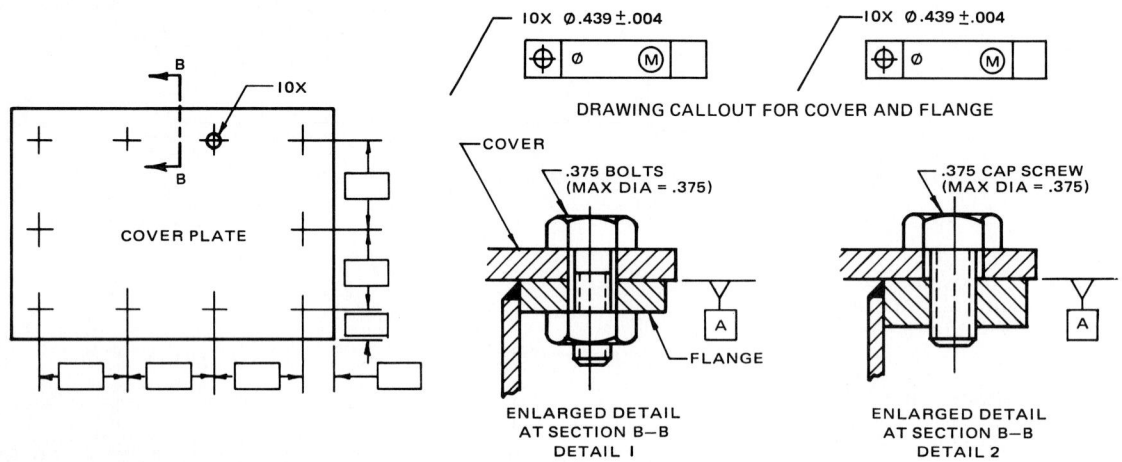

FIG. 16-17-A   Cover plate assembly.

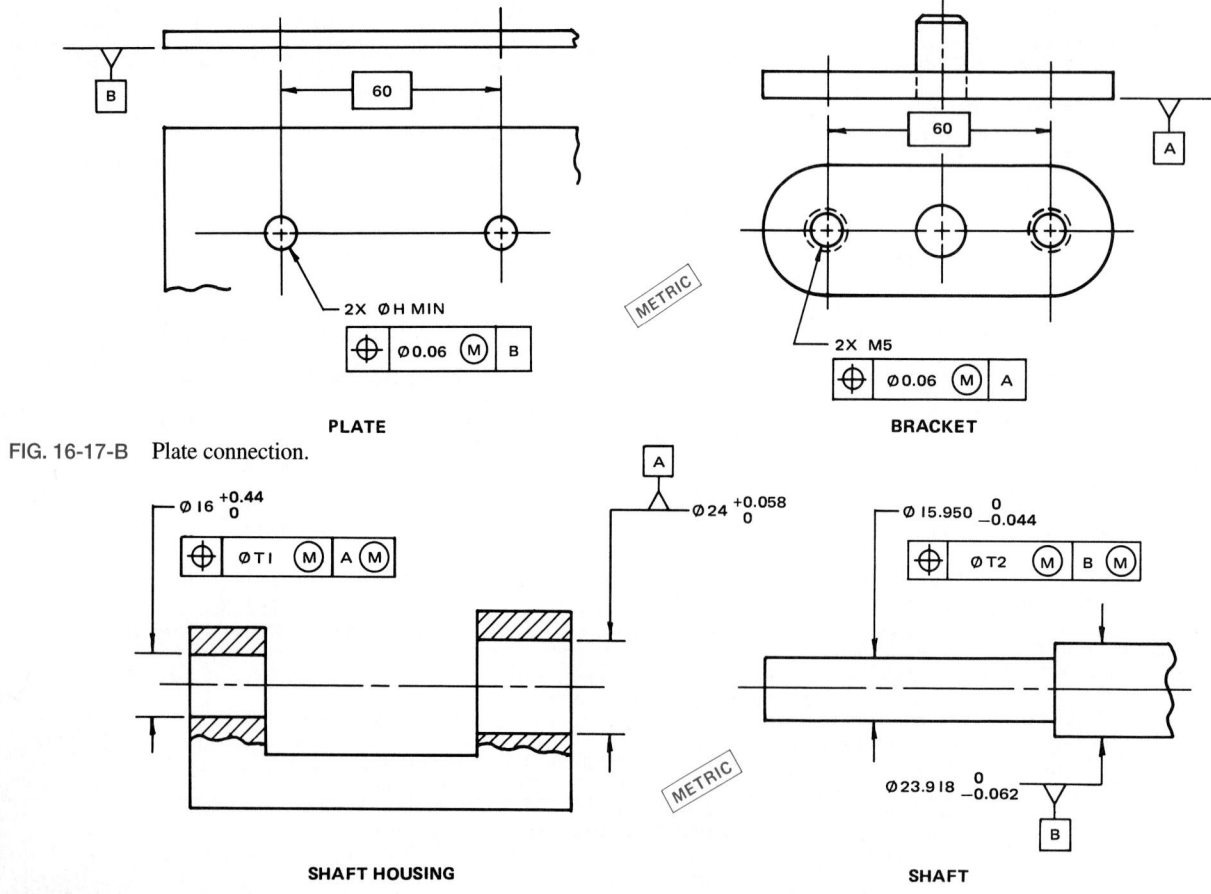

FIG. 16-17-B   Plate connection.

FIG. 16-17-C   Shaft seat.

ASSIGNMENTS FOR UNIT 16-18, SUMMARY OF RULES FOR GEOMETRIC TOLERANCING

61. From the information given in Fig. 16-18-A prepare a working drawing showing the front and right side view of the collar.

62. From the information given in Fig. 16-18-B prepare a working drawing showing the top and right side view of the locating plate.

GEOMETRIC TOLERANCING REQUIREMENTS TO BE ADDED TO DRAWING
- ALL DATUMS AND TOLERANCES TO BE ON AN MMC BASIS UNLESS OTHERWISE SPECIFIED
- THE LARGE HOLE IS DATUM A. THREE EQUALLY SPACED TARGET POINTS LOCATED .50 FROM EACH END ESTABLISH DATUM A. THEY ARE LOCATED FROM DATUM B. THE AXIS OF THE HOLE HAS A PERPENDICULARITY TOLERANCE OF ZERO REFERENCED TO DATUM B
- DATUM B IS SURFACE B. IT IS TO BE FLAT WITHIN .003
- THE EIGHT HOLES ARE TO HAVE A POSITIONAL TOLERANCE OF .004 AND REFERENCED TO DATUMS B AND A IN THAT ORDER
- THE LEFT END OF THE PART IS TO BE PARALLEL WITH DATUM B WITHIN .005
- SURFACES MARKED E AND F ARE TO HAVE TOTAL RUNOUT TOLERANCES OF .002 REFERENCED TO DATUMS A AND B IN THAT ORDER

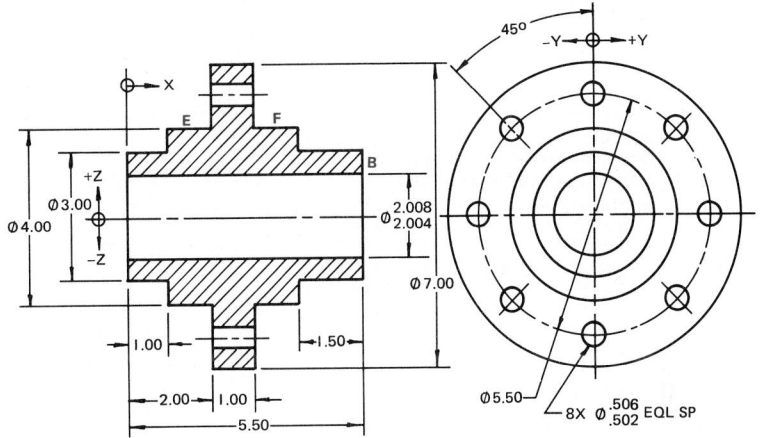

FIG. 16-18-A   Collar.

FIG. 16-18-B   Locating plate.

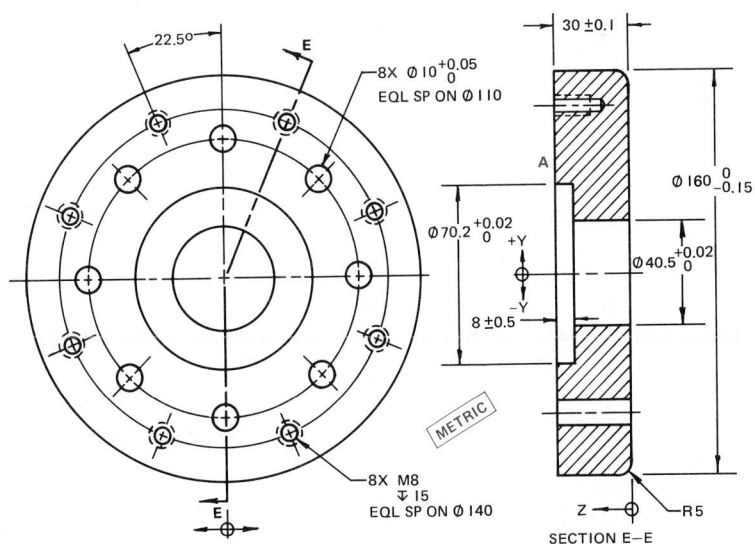

GEOMETRIC TOLERANCING REQUIREMENTS TO BE ADDED TO DRAWING
- ALL DATUMS AND TOLERANCES TO BE ON AN MMC BASIS UNLESS OTHERWISE SPECIFIED
- SURFACE A IS DATUM A AND HAS A FLATNESS TOLERANCE OF 0.04
- THE Ø 70.2 IS DATUM B
- THE Ø 10 HOLES ARE DATUM C
- THE Ø 40.5 IS TO HAVE A POSITIONAL TOLERANCE OF 0.02 REFERENCED TO DATUMS A AND B IN THAT ORDER

- THE M8 TAPPED HOLES ARE TO HAVE A POSITIONAL TOLERANCE OF 0.3 REFERENCED TO DATUMS A, B, AND C IN THAT ORDER
- THE Ø 10 HOLES ARE TO HAVE A POSITIONAL TOLERANCE OF 0.1 REFERENCED TO DATUMS A AND B IN THAT ORDER

63. From the information given in Fig. 16-18-C prepare a working drawing showing the top and right side view of the transmission cover.

64. From the information given in Fig. 16-18-D prepare a working drawing showing the top and right side view of the cover plate.

GEOMETRIC TOLERANCING REQUIREMENTS TO BE ADDED TO DRAWING

- ALL DATUMS AND TOLERANCES TO BE ON AN MMC BASIS UNLESS OTHERWISE SPECIFIED
- DIMENSIONS NOT SHOWN TOLERANCED OR BASIC WILL HAVE A TOLERANCE OF ±.01
- SURFACES OR PLANES F, G, AND H ARE PRIMARY, SECONDARY, AND TERTIARY DATUMS F, G, AND H RESPECTIVELY. NOTE DATUM PLANE G IS A STEPPED DATUM FEATURE, CONSISTING OF TWO DATUM LINES GI AND G2. DATUM H IS A DATUM LINE LOCATED ON THE BOTTOM OF THE PART
- SLOT D IS DATUM D AND HAS A POSITIONAL TOLERANCE OF .01 AND IS REFERENCED TO DATUMS F, G, AND H IN THAT ORDER
- SURFACE F TO BE FLAT WITHIN .005
- HOLE E IS DATUM E AND HAS A POSITIONAL TOLERANCE OF .002 AND IS REFERENCED TO DATUMS F AND D IN THAT ORDER
- HOLES B HAVE A POSITIONAL TOLERANCE OF .003 AND ARE REFERENCED TO DATUMS F, E, AND D IN THAT ORDER
- HOLES C HAVE A POSITIONAL TOLERANCE OF .005 AND ARE REFERENCED TO DATUMS F, D, AND E IN THAT ORDER

HOLE	DIA
A	.300 ±.004
B	.165 +.003 −.000
C	.256 +.002 −.000
D	.40 X 2.80
E	.500 +.004 −.000

MATL – SAE 1008

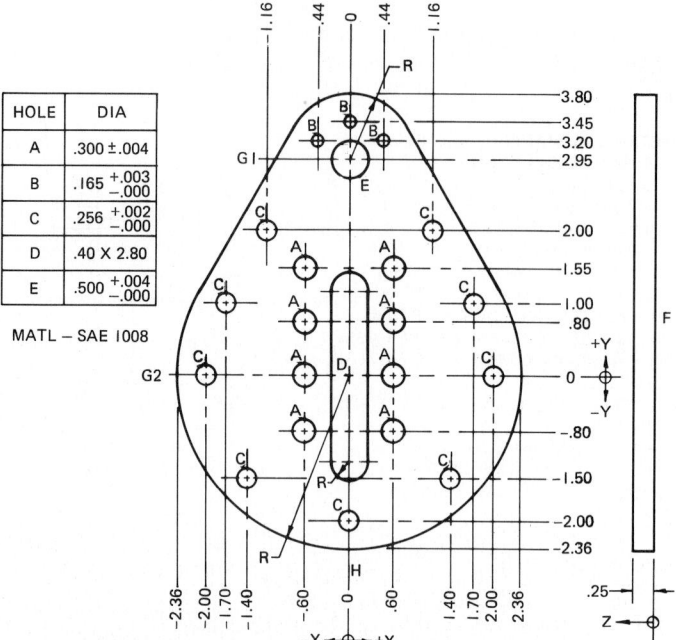

- HOLES A HAVE A POSITIONAL TOLERANCE OF .004 AND ARE REFERENCED TO DATUMS F, D, AND E IN THAT ORDER

FIG. 16-18-C   Transmission cover.

GEOMETRIC TOLERANCING REQUIREMENTS TO BE ADDED TO DRAWING

- ALL DATUMS AND TOLERANCES TO BE ON AN MMC BASIS UNLESS OTHERWISE SPECIFIED
- DIMENSIONS NOT SHOWN TOLERANCED OR BASIC WILL HAVE A TOLERANCE OF ±0.5
- SURFACES F, G, AND H ARE DATUMS F, G, AND H RESPECTIVELY
- SURFACE F TO BE FLAT WITHIN 0.4
- HOLE D IS DATUM D AND HAS A POSITIONAL TOLERANCE OF 0.5 AND IS REFERENCED TO DATUMS F, G, AND H IN THAT ORDER
- HOLE E IS DATUM E AND HAS A POSITIONAL TOLERANCE OF 0.2 AND IS REFERENCED TO DATUMS F, D, AND G IN THAT ORDER
- HOLES A HAVE COMPOSITE POSITIONAL TOLERANCING AND ARE REFERENCED TO DATUMS F, D, AND E IN THAT ORDER. THE PATTERN-LOCATING TOLERANCE IS 0.6 AND THE FEATURE-RELATING TOLERANCE IS 0.3
- HOLES C HAVE COMPOSITE POSITIONAL TOLERANCING AND ARE REFERENCED TO DATUMS F, D, AND E IN THAT ORDER. THE PATTERN-LOCATING TOLERANCE IS 0.8 AND THE FEATURE-RELATING TOLERANCE IS 0.4

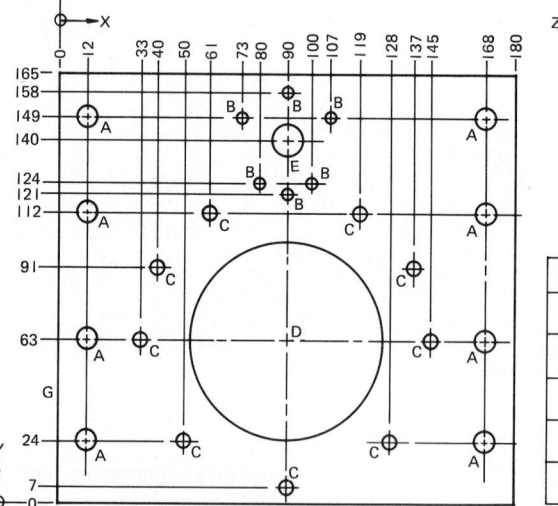

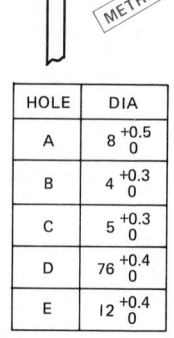

METRIC

HOLE	DIA
A	8 +0.5 0
B	4 +0.3 0
C	5 +0.3 0
D	76 +0.4 0
E	12 +0.4 0

- HOLES B HAVE COMPOSITE POSITIONAL TOLERANCING AND ARE REFERENCED TO DATUMS F, E, AND D IN THAT ORDER. THE PATTERN-LOCATING TOLERANCE IS 0.6 AND THE FEATURE-RELATING TOLERANCE IS 0.2

FIG. 16-18-D   Cover plate.

65. From the information given in Fig. 16-18-E prepare a working drawing showing the front and right side view of the housing.

66. From the information given in Fig. 16-18-F prepare a working drawing showing the front and end view of the end plate.

FIG. 16-18-E   Housing.

G NOTES

GEOMETRIC TOLERANCING REQUIREMENTS TO BE ADDED TO DRAWING

- ALL DATUMS AND TOLERANCES TO BE ON AN MMC BASIS UNLESS OTHERWISE SPECIFIED
- DIAMETER A IS DATUM A. IT HAS A PERPENDICULARITY TOLERANCE OF .020 REFERENCED TO DATUM D
- DIAMETER B IS DATUM B. IT HAS A PERPENDICULARITY OF .010 REFERENCED TO DATUM D. IT HAS A CIRCULAR RUNOUT TOLERANCE OF .030 REFERENCED TO DATUM A
- HOLE C IS DATUM C. IT HAS A POSITIONAL TOLERANCE OF .005 AND IS REFERENCED TO DATUMS D AND B IN THAT ORDER
- SURFACE D IS DATUM D. IT HAS A FLATNESS TOLERANCE OF .005
- DIAMETER E IS DATUM E. IT HAS A CIRCULAR RUNOUT TOLERANCE OF .008 REFERENCED TO DATUMS D AND B IN THAT ORDER
- DIAMETER F IS DATUM F. IT HAS A POSITIONAL TOLERANCE OF .004 AND IS REFERENCED TO DATUMS R, B, AND C IN THAT ORDER
- DIAMETER G IS DATUM G. IT HAS A POSITIONAL TOLERANCE OF .005 AND IS REFERENCED TO DATUMS R AND F IN THAT ORDER
- SURFACE N IS DATUM N. IT HAS A PARALLELISM TOLERANCE OF .005 REFERENCED TO DATUM D
- SURFACE R IS DATUM R
- THE Ø 3.275–3.280 HAS A

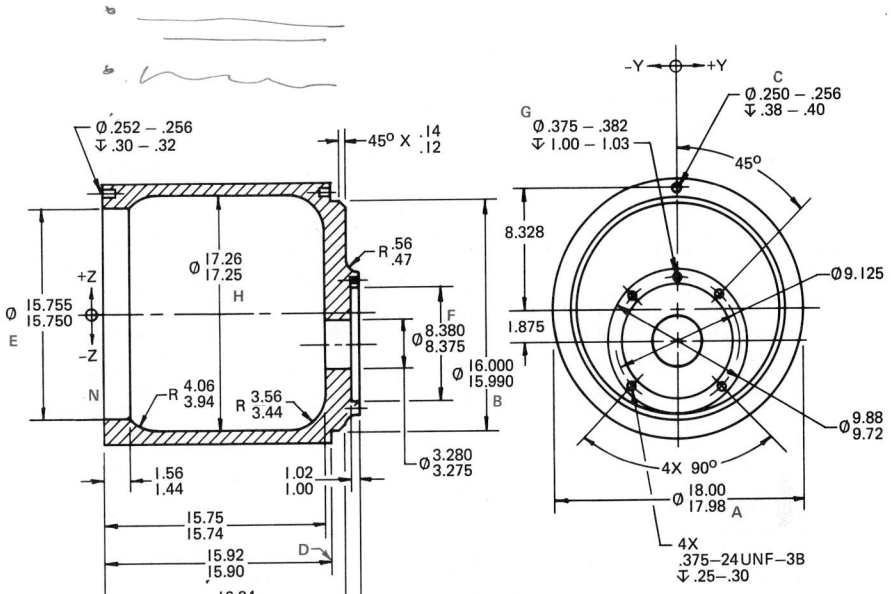

POSITIONAL TOLERANCE OF .010 AND IS REFERENCED TO DATUMS R AND F IN THAT ORDER
- THE FOUR TAPPED HOLES HAVE A POSITIONAL TOLERANCE OF .010 RFS

AND ARE REFERENCED TO DATUMS R, F, AND G IN THAT ORDER
- THE Ø 17.25–17.26 DIAMETER HAS A CIRCULAR RUNOUT OF .025 AND IS REFERENCED TO DATUM A

GEOMETRIC TOLERANCING REQUIREMENTS TO BE ADDED TO DRAWING

- ALL DATUMS AND TOLERANCES TO BE ON AN MMC BASIS UNLESS OTHERWISE SPECIFIED
- SURFACE A IS DATUM A. IT HAS A FLATNESS TOLERANCE OF 0.02
- DIAMETER B IS DATUM B. IT HAS A CIRCULAR RUNOUT OF 0.01 AND IS REFERENCED TO DATUMS A AND C IN THAT ORDER
- DIAMETER C IS DATUM C. IT HAS A PERPENDICULARITY TOLERANCE OF 0.08 AND IS REFERENCED TO DATUM A
- SURFACE D IS DATUM D. IT HAS A FLATNESS TOLERANCE OF 0.02 AND A PARALLELISM TOLERANCE OF 0.05 REFERENCED TO DATUM A
- DIAMETER E HAS A CIRCULAR RUNOUT TOLERANCE OF 0.1 REFERENCED TO DATUM B
- SURFACE F IS PARALLEL TO DATUM A WITHIN 0.1
- THE Ø 10.5–10.8 HOLES HAVE A POSITIONAL TOLERANCE OF 0.2 AND ARE REFERENCED TO DATUMS A AND C IN THAT ORDER
- DATUM A HAS THREE Ø 6 DATUM TARGET AREAS EQUALLY SPACED ON

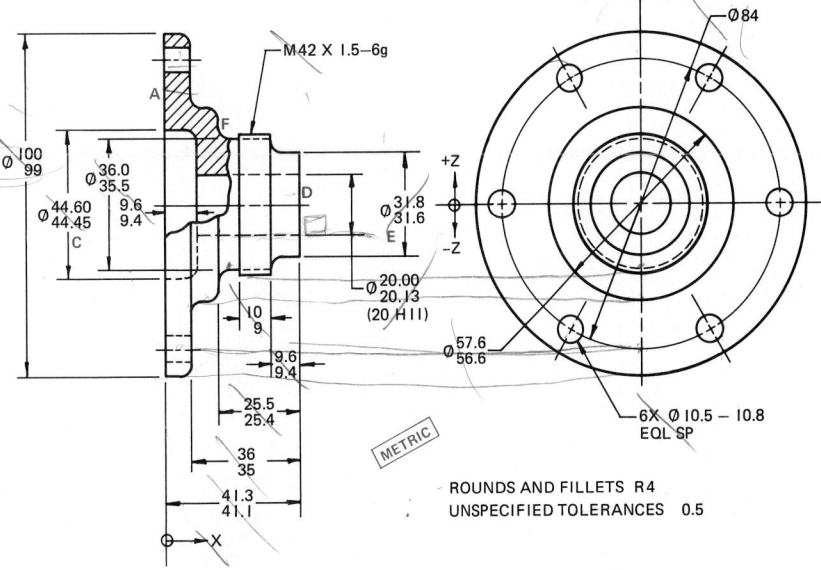

Ø 74 LOCATED RADIALLY MIDWAY BETWEEN THE Ø 10.5-10.8 HOLES

ROUNDS AND FILLETS R4
UNSPECIFIED TOLERANCES 0.5

FIG. 16-18-F   End plate.

67. From the information given in Fig. 16-18-G prepare a working drawing showing the top and right side view of the adjusting plate.

68. From the information given in Fig. 16-18-H prepare a working drawing showing the front and left side view of the bearing housing.

GEOMETRIC TOLERANCING REQUIREMENTS TO BE ADDED TO DRAWING

- ALL DATUMS AND TOLERANCES TO BE ON AN MMC BASIS UNLESS OTHERWISE SPECIFIED
- THE BACK SURFACE OF THE PLATE IS DATUM A
- THE HOLE (Ø .500-.503) WHICH POSITIONS THE ADJUSTING PLATE TO THE BRACKET IS DATUM B. IT IS TO BE PERPENDICULAR WITHIN .00I TO DATUM A
- THE TWO SLOTS (DATUM C) ARE LOCATED BY A POSITIONAL TOLERANCE OF .010 AND REFERENCED TO DATUMS A AND B IN THAT ORDER
- THE HOLES (Ø .378-.39I) WHICH ACCOMMODATES THE BOLT HOLDING THE ROD TO THE PLATE IS LOCATED BY A POSITIONAL TOLERANCE OF .020 AND REFERENCED TO DATUMS A, B, AND C IN THAT ORDER

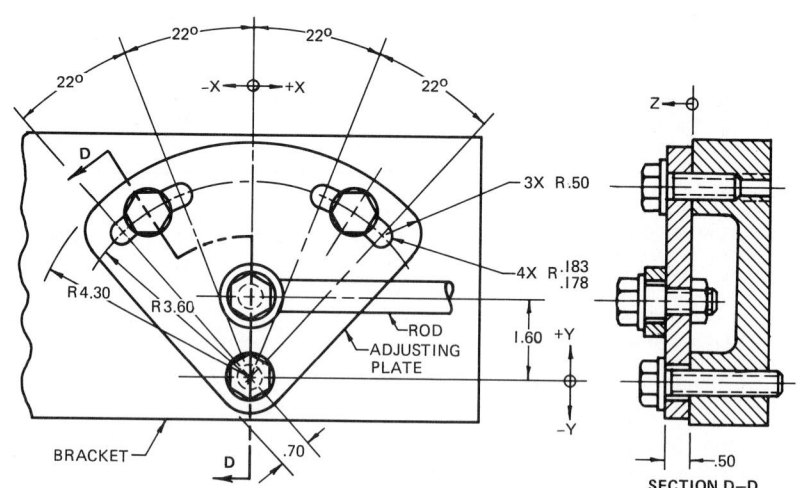

FIG. 16-18-G   Adjusting plate.

GEOMETRIC TOLERANCING REQUIREMENTS TO BE ADDED TO DRAWING

- ALL DATUMS AND TOLERANCES TO BE ON AN MMC BASIS UNLESS OTHERWISE SPECIFIED
- THE BEARING HOUSING SURFACE WHICH IS ATTACHED BY .375 CAP SCREWS TO THE MAIN HOUSING IS DATUM A AND HAS A FLATNESS TOLERANCE OF .003
- THE BEARING SEAT IS DATUM D AND HAS A PARALLEL TOLERANCE OF .004 WITH DATUM A
- AN LN3 FIT IS REQUIRED BETWEEN THE HOUSING AND THE BEARING. THIS DIAMETER IS DATUM E AND HAS A PERPENDICULARITY TOLERANCE OF .00I WITH DATUM D, AND HAS A POSITIONAL TOLERANCE OF .002 REFERENCED TO DATUMS A AND C IN THAT ORDER
- THE SURFACE CONTACTING THE COVER PLATE IS DATUM F
- THE HOLES FOR THE SPRING PINS ARE Ø .375-.376 AND ARE DESIGNATED AS DATUM C. THEY ARE LOCATED BY A POSITIONAL TOLERANCE OF .00I AND REFERENCED TO DATUM A
- THE HOLES FOR THE .375 CAP SCREWS ARE Ø .390-.402 AND ARE

LOCATED BY A POSITIONAL TOLERANCE OF .008 AND REFERENCED TO DATUMS A AND C IN THAT ORDER

- THE FOUR TAPPED HOLES HAVE A POSITIONAL TOLERANCE OF .010 AND ARE REFERENCED TO DATUMS D AND E IN THAT ORDER

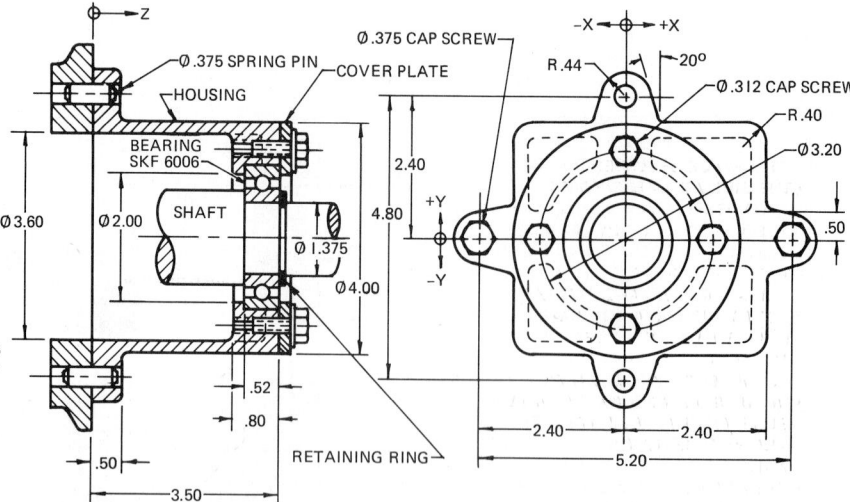

FIG. 16-18-H   Bearing housing.

594

69. From the information given in Fig. 16-18-J prepare a working drawing showing the front and both end views of the adapter.

70. From the information given in Fig. 16-18-K prepare a working drawing showing the front and left side views of the shift lever support.

GEOMETRIC TOLERANCING REQUIREMENTS TO BE ADDED TO DRAWING
- ALL DATUMS AND TOLERANCES TO BE ON AN MMC BASIS UNLESS OTHERWISE SPECIFIED
- THE VERTICAL SURFACE ADJACENT TO THE COVER PLATE IS DATUM A. IT HAS A FLATNESS TOLERANCE OF .002
- THE Ø 5.00 SECTION WHICH POSITIONS THE ADAPTOR ON THE COVER PLATE IS DATUM B AND HAS AN LC2 FIT. IT HAS A PERPENDICULARITY TOLERANCE OF .001 REFERENCED TO DATUM A
- THE Ø 2.40 WHICH POSITIONS THE HYDRAULIC PUMP ON THE ADAPTOR IS DATUM E AND HAS AN LC2 FIT. THIS DIAMETER HAS TWO GEOMETRIC TOLERANCES: A PERPENDICULARITY TOLERANCE OF .002 REFERENCED TO DATUM D, AND A CIRCULAR RUNOUT TOLERANCE OF .004 REFERENCED TO DATUMS A AND B IN THAT ORDER
- THE GROUP OF FOUR TAPPED HOLES HAS A POSITIONAL TOLERANCE OF .012 AND IS REFERENCED TO DATUMS D AND E IN THAT ORDER
- THE VERTICAL FLAT SURFACE CONTACTING THE HYDRAULIC PUMP IS DATUM D AND HAS TWO GEOMETRIC TOLERANCES: A FLATNESS TOLERANCE OF .002 AND A CIRCULAR RUNOUT TOLERANCE OF

.004 REFERENCED TO DATUMS A AND B IN THAT ORDER
- THE GROUP OF SIX CLEARANCE HOLES (Ø .391-.401) HAS A POSITIONAL TOLERANCE OF .012 REFERENCED TO DATUMS A AND B IN THAT ORDER

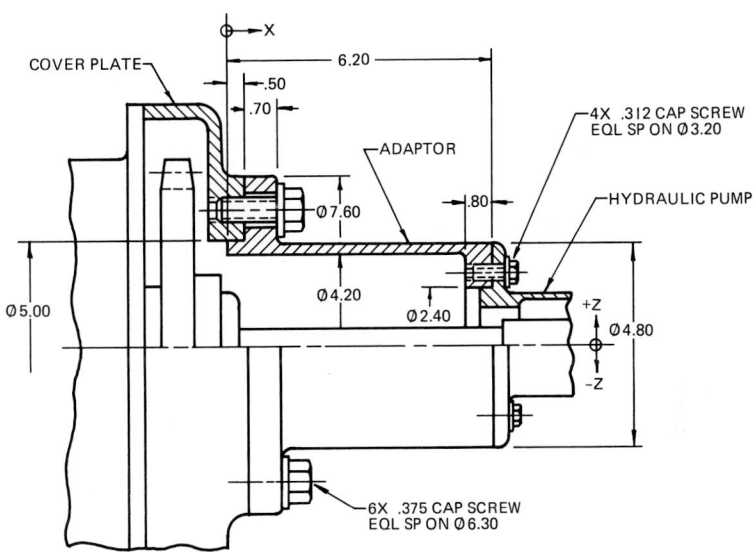

FIG. 16-18-J   Adapter.

GEOMETRIC TOLERANCING REQUIREMENTS TO BE ADDED TO DRAWING
- ALL DATUMS AND TOLERANCES TO BE ON AN MMC BASIS UNLESS OTHERWISE SPECIFIED
- THE SIDE ATTACHED TO THE CYLINDER BLOCK IS DATUM A. IT HAS A FLATNESS TOLERANCE OF 0.15
- THE RIGHT SURFACE OF THE TWO COPLANAR SURFACES AT THE BOTTOM IS DESIGNATED AS DATUM B. A PROFILE-OF-A-SURFACE TOLERANCE OF 0.08 IS TO BE ADDED TO THE BOTTOM SURFACES
- DATUMS A AND B ARE TO HAVE SURFACE TEXTURE RATINGS OF 1.6 AND A MACHINING ALLOWANCE OF 2 MM
- THE FIVE CLEARANCE HOLES (Ø 10.8-11.0) FOR THE CAP SCREWS REQUIRE A POSITIONAL TOLERANCE OF 0.4 AND ARE TO BE REFERENCED TO DATUMS A AND B IN THAT ORDER
- THE Ø 25 HOLE IS TO HAVE AN H9d9 FIT WITH THE SHAFT. A POSITIONAL TOLERANCE OF 0.2 LOCATES THE HOLE WITH REFERENCE TO DATUMS A AND B IN THAT ORDER

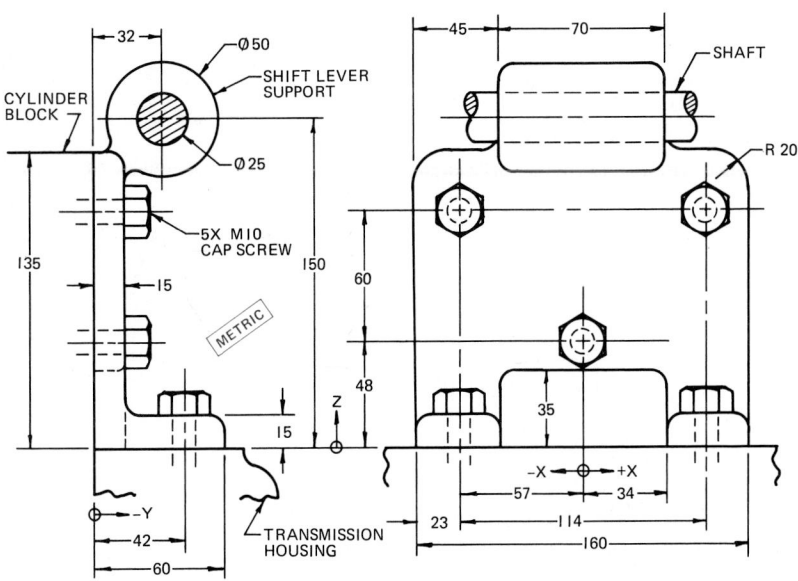

FIG. 16-18-K   Shift lever support.

71. From the information given in Fig. 16-18-L prepare a working drawing showing the front, right side, and partial top view of the motor mount.

72. From the information given in Fig. 16-18-M prepare a working drawing showing the front and right end view of the pulley bracket.

GEOMETRIC TOLERANCING REQUIREMENTS TO BE ADDED TO DRAWING

- ALL DATUMS AND TOLERANCES TO BE ON AN MMC BASIS UNLESS OTHERWISE SPECIFIED
- THE VERTICAL MOUNTING SURFACE FOR THE MOTOR IS DATUM B. IT HAS A PERPENDICULARITY TOLERANCE OF 0.05 WITH DATUM A
- THE BOTTOM IS DATUM A
- DATUMS A AND B ARE TO HAVE SURFACE TEXTURE RATINGS OF 1.6 AND A MACHINING ALLOWANCE OF 2MM
- THE Ø 50 HOLE IS DATUM C AND HAS A PERPENDICULARITY TOLERANCE OF ZERO WITH DATUM B. THE HOLE IS ALSO TO HAVE A POSITIONAL TOLERANCE OF 0.6 REFERENCED TO DATUMS B, A, AND D IN THAT ORDER AND BE DIMENSIONED FOR AN H8/f7 FIT
- THE THREE MOUNTING HOLES TO HAVE A POSITIONAL TOLERANCE OF 0.3 AND BE REFERENCED TO DATUMS B AND C IN THAT ORDER. A CHAMFER IS TO BE ADDED TO THESE HOLES FOR EASE OF ASSEMBLY
- THE TWO MOUNTING HOLES ARE DATUM D AND ARE Ø 8.04-8.07. THEY HAVE A POSITIONAL TOLERANCE OF 0.06 AND ARE REFERENCED TO DATUMS A AND B IN THAT ORDER

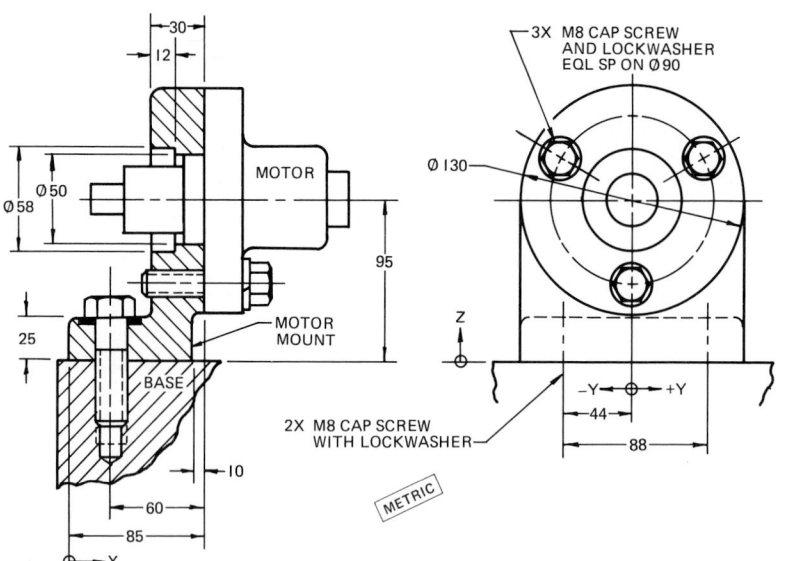

FIG. 16-18-L   Motor mount.

GEOMETRIC TOLERANCING REQUIREMENTS TO BE ADDED TO DRAWING

- ALL DATUMS AND TOLERANCES TO BE ON AN MMC BASIS UNLESS OTHERWISE SPECIFIED
- THE BOLTING SURFACE WITH THE PULLEY IS TO BE DATUM C. IT IS TO HAVE A FLATNESS TOLERANCE OF 0.4 AND BE PARALLEL TO DATUM A WITHIN 0.2
- THE BOLTING SURFACE WITH THE CRANKSHAFT IS TO BE DATUM A. IT HAS A FLATNESS TOLERANCE OF 0.4
- THE Ø 50 IS TO BE DATUM B. IT HAS A PERPENDICULARITY TOLERANCE OF ZERO WITH DATUM A. THE OUTSIDE CORNER REQUIRES A SMALL RADIUS FOR EASE OF ASSEMBLY. IT IS TO BE DIMENSIONED FOR AN H7h6 FIT
- THE FOUR MOUNTING HOLES (Ø 8.66-8.90) FOR THE BRACKET HAVE A POSITIONAL TOLERANCE OF 0.4 AND ARE REFERENCED TO DATUMS A AND B IN THAT ORDER
- THE FOUR TAPPED HOLES FOR MOUNTING THE PULLEY HAVE A POSITIONAL TOLERANCE OF 0.3 AND ARE REFERENCED TO DATUM C. A CHAMFER IS TO BE ADDED TO THESE HOLES FOR EASE OF ASSEMBLY
- DATUMS A AND C ARE TO HAVE SURFACE TEXTURE RATINGS OF 1.6 AND A MACHINING ALLOWANCE OF 2MM

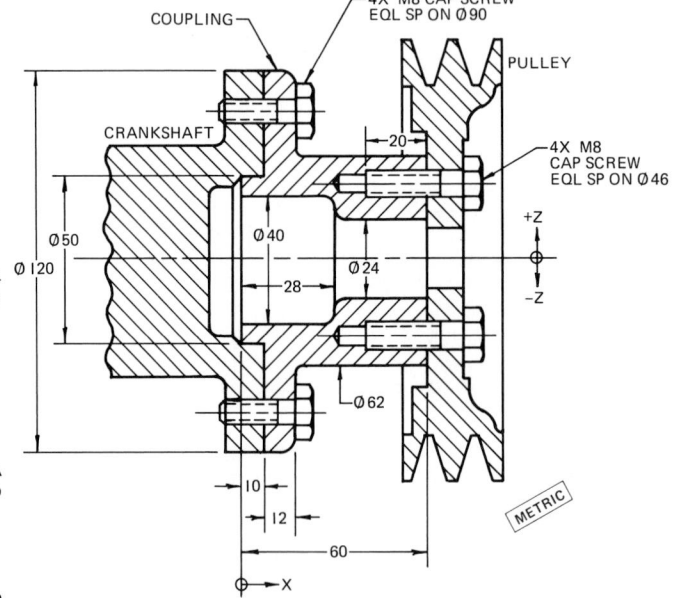

FIG. 16-18-M   Pulley bracket.

73. From the information given in Fig. 16-18-N prepare a working drawing showing the front and right end view of the drive coupling.

GEOMETRIC TOLERANCING REQUIREMENTS TO BE ADDED TO DRAWING

- ALL DATUMS AND TOLERANCES TO BE ON AN MMC BASIS UNLESS OTHERWISE SPECIFIED
- THE COPLANAR SURFACES OF THE ∅ 1.50 BOSSES ARE DATUM B. A PROFILE OF A SURFACE TOLERANCE OF .004 IS APPLIED TO THE COPLANAR SURFACES AND REFERENCED TO DATUM A.
- THE ∅ 1.375 HOLE HAS AN RC4 FIT WITH THE SHAFT AND IS DATUM A
- THE ROOT DIAMETER (∅ .422-.426) OF THE ∅ .500 SHOULDER SCREW IS DATUM C. IT HAS A POSITIONAL TOLERANCE OF .006 AND IS REFERENCED TO DATUMS B AND A IN THAT ORDER
- THE COUNTERBORED HOLE FOR THE SHOULDER OF THE SCREW MUST BE CONCENTRIC WITH DATUM C WITHIN .016.

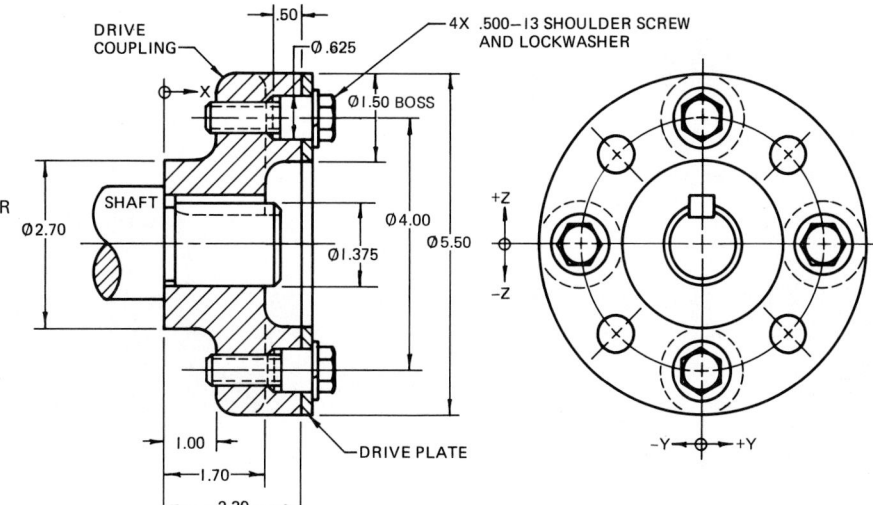

FIG. 16-18-N   Drive coupling.

CHAPTER 17

DRAWINGS FOR NUMERICAL CONTROL

Definitions

Absolute coordinate programming   An NC system in which each position is described in relation to the origin.

Numerical control (NC)   A means of automatically directing some or all of the functions of a machine from instructions.

Origin or **zero point**   The position where the X and Y axes intersect.

Relative coordinate or **point-to-point programming**   An NC system in which each new position is described in relation to the previous position.

Two-axis machine   An NC machine designed to locate points in only the X and Y directions.

Three-axis machine   An NC machine designed to locate points in the X, Y, and Z directions.

17-1 TWO-AXIS CONTROL SYSTEMS

Interactive graphics programs are used to develop assemblies, item lists, 3-D models, and mathematical results. The results are considered to be working engineering drawings, yet the design may never be produced on paper. The data generated by a CAD system can be directly used by a CAM (computer-aided manufacturing) system, thus the term CAD/CAM (Fig. 17-1-1).

In the CAD/CAM system, designers and engineers interact with the computer by means of a graphics terminal. They design and manufacture a part from start to finish. The design and drafting are accomplished electronically. A number and/or security system ensures that the latest copy of the design is available to all departments. Each design activity, as well as all models, is stored in an integrated CAD/CAM data base. On the factory floor, the manufacturing staff is on the same network

as the designers. Programs take the design information and automatically convert it into other programs that run milling machines, lathes, multiple drill presses, assembly lines, and so forth. This ideal system as illustrated in Fig. 17-1-2, helps to epitomize the "total automated factory."

Computer Numerical Control (CNC)

Numerical control (NC) is a means of automatically directing some or all of the functions of a machine from instructions. The instructions are stored on tapes or floppy disks. The controller interprets the coded instructions and directs the machine through the required operations.

It has been established that because of the consistently high accuracy of numerically controlled machines, and because human errors have been almost entirely eliminated, the number of rejected parts has been considerably reduced.

Because both set-up and tape preparation times are short, numerically controlled machines produce a part faster than manually controlled machines. When changes become necessary on a part, they can easily be implemented by changing the original tapes. The process takes very little time and expense in comparison to the alteration of a jig or fixture.

Another area where numerically controlled machines are better is in the quality or accuracy of the work. In many cases a numerically controlled machine can produce parts more accurately at no additional cost, resulting in reduced assembly time and improved interchangeability of parts. The latter is especially important when spare parts are required.

Dimensioning for Numerical Control

Common guidelines have been established that enable dimensioning and tolerancing practices to be used effectively in delineating parts for both NC and conventional fabrication. Each object is prepared using baseline (or coordinate) dimensioning

FIG. 17-1-1   Fully automated
industrial
CAD/CAM.
(Computervision)

Courtesy Computervision

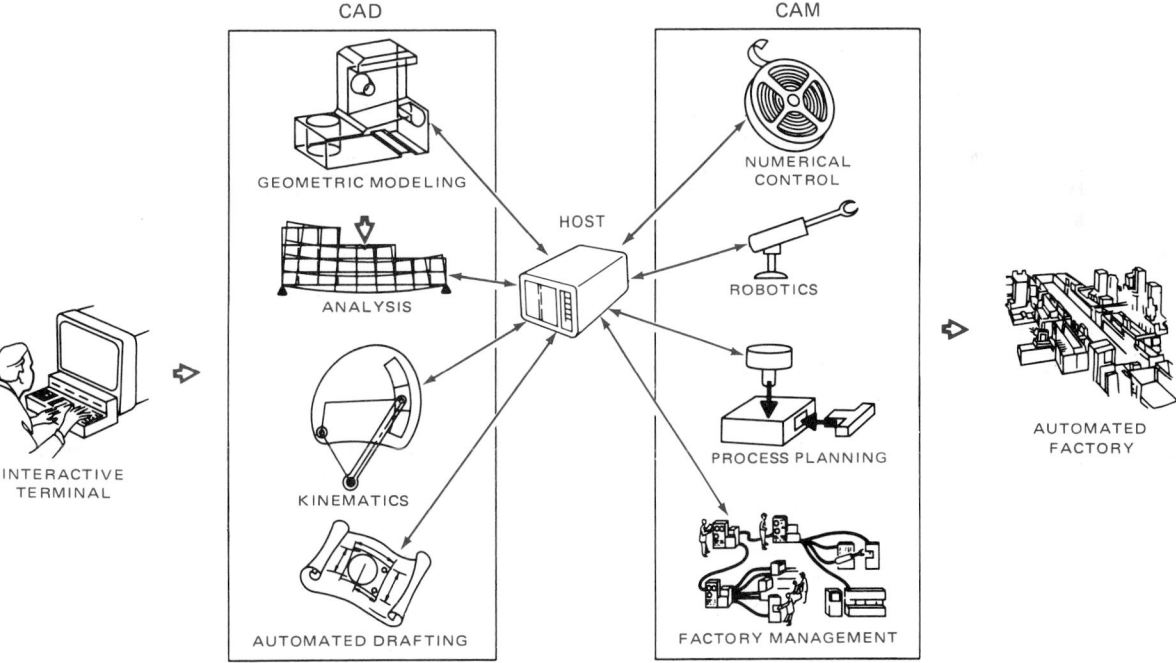

FIG. 17-1-2   Integrated computer-aided engineering (CAE) systems.

methods. First, the selection of an absolute (0, 0, 0) or (0, 0) coordinate origin is made depending on whether the control is three-axis or two-axis. All part dimensions would be referenced from that origin.

After a multiview working drawing is produced, the information is transferred to manufacturing equipment. Most manufacturers use some sort of interactive (human) interface. This approach does not produce fully automated CAD/CAM. Partial information (partial programming) in the form of manufacturing data is generated. This allows the NC computer to compile instructions from programs stored within its memory. The result is a detailed program plan for tool-path generation. While not as automatic as hardwired CAD/CAM, much of the tedium has been removed. Easy-to-use instructions in partial programming require much less effort than completely creating the program.

The complete procedure is illustrated in Fig. 17-1-3. The partial-programming portion of the process is shown within the unshaded box. The machine tool automatically produces the finished part only after this programming has been accomplished. Fully automated CAD/CAM replaces the portion of the process within the unshaded box by a hard wire. The procedures outlined in this chapter are valid for either process.

Dimensioning for a Two-Axis Coordinate System

The CNC concept is based on the system of rectangular or Cartesian coordinates in which any position can be described in terms of distance from an origin point along either two or three mutually perpendicular axes. Two-dimensional coordinates (X,Y) define points in a plane (Fig. 17-1-4).

The X axis is horizontal and is considered the first and basic reference axis. Distances to the right of the Y axis are considered positive X values and to the left of the Y axis as negative X values.

The Y axis is vertical and perpendicular to the X axis in the plane of a drawing showing XY relationships. Distances above the X axis are considered positive Y values and below the X axis as negative Y values. The position where the X and Y axes cross is called the *origin,* or *zero point.*

For example, four points lie in a plane, as shown in Fig. 17-1-4. The plane is divided into four quadrants. Point A lies in

quadrant 1 and is located at position (6, 5), with the X coordinate first, followed by the Y coordinate. Point B lies in quadrant 2 and is located at position (−4, 3). Point C lies in quadrant 3 and is located at position (−5, −4). Point D lies in quadrant 4 and is located at position (3, −2).

Designing for NC would be greatly simplified if all work were done in the first quadrant because all the values would be positive and the plus and minus signs would not be required. For that reason many CAD/CAM systems place the origin (0, 0) to the lower left of the video display. This way only positive values apply. However, any of the four quadrants may be used, and as such, programming in any of the quadrants should be understood.

Some NC machines, called *two-axis machines,* are designed for locating points in only the X and Y directions. The function of these machines is to move the machine table or tool to a specified position in order to perform work, as shown in Fig. 17-1-5. With the fixed spindle and movable table as shown in Fig. 17-1-5B, hole A is drilled; then the table moves to the left, positioning point B below the drill. This is the most frequently used method. With the fixed table and movable spindle as shown in Fig. 17-1-5C, hole A is drilled; then the spindle moves to the right, positioning the drill above point B. This changes the direction of the motion, but the movement of the cutter as related to the work remains the same.

Origin (Zero Point)

As previously mentioned, the origin is the point where the X and Y axes intersect. It is the point from which all coordinate

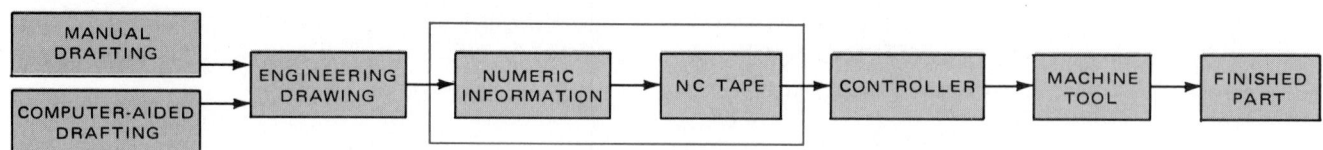

FIG. 17-1-3   Computer numerical control sequence.

FIG. 17-1-4   Two-dimensional coordinates (*X* and *Y*).

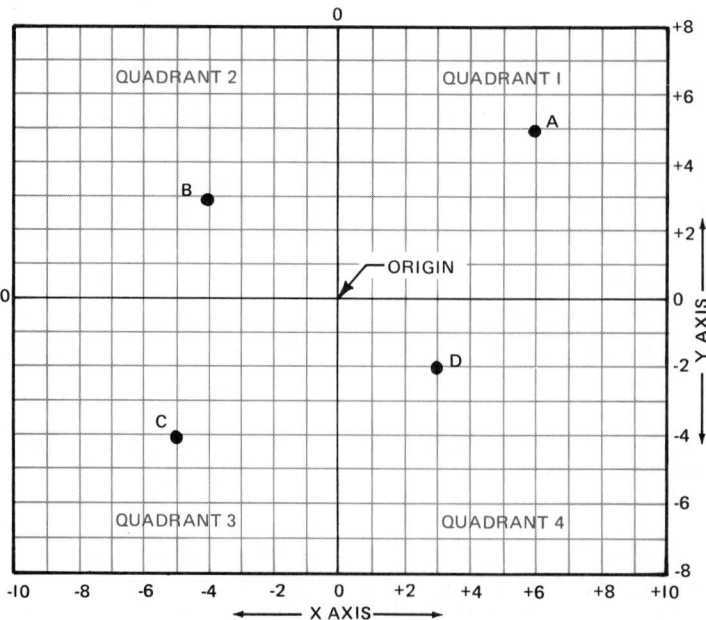

dimensions are measured. Many systems have a fixed origin point built in.

Two examples of fixed origins on machine tables are shown in Fig. 17-1-6. In Fig. 17-1-6A all points are located in the first quadrant, resulting in positive X and Y values. This is the most frequently used method. In Fig. 17-1-6B all points are located in the third quadrant, resulting in negative X and Y values.

Set-Up Point

The set-up point is located on the part or the fixture holding the part. It may be the intersection of two finished surfaces, the center of a previously machined hole in the part, or a feature of the fixture. It must be accurately located in relation to the origin, as shown in Fig. 17-1-7 (pg. 602).

Relative Coordinate (Point-to-Point) Programming

With point-to-point programming, each new position is given from the last position. To compute the next position wanted, it is necessary to establish the sequence in which the work is to be done.

An example of this type of dimensioning is shown in Fig. 17-1-7A. The distance between the left edge of the part and hole 1 is given as .75 in. From hole 1 to hole 4 the dimension shown is 4.50 in. (X axis), and from hole 1 to hole 2 the dimension shown is 1.50 in. (Y axis). These dimensions give the distance from the last drilled hole to the next drilled hole. Assume the holes are to be drilled in the sequence shown in the figure. Hole 1 is located (2.75, 2.75) from the origin or zero point. After hole 1 has been drilled, the drill spindle is positioned above the center of hole 2. Hole 2 has the same

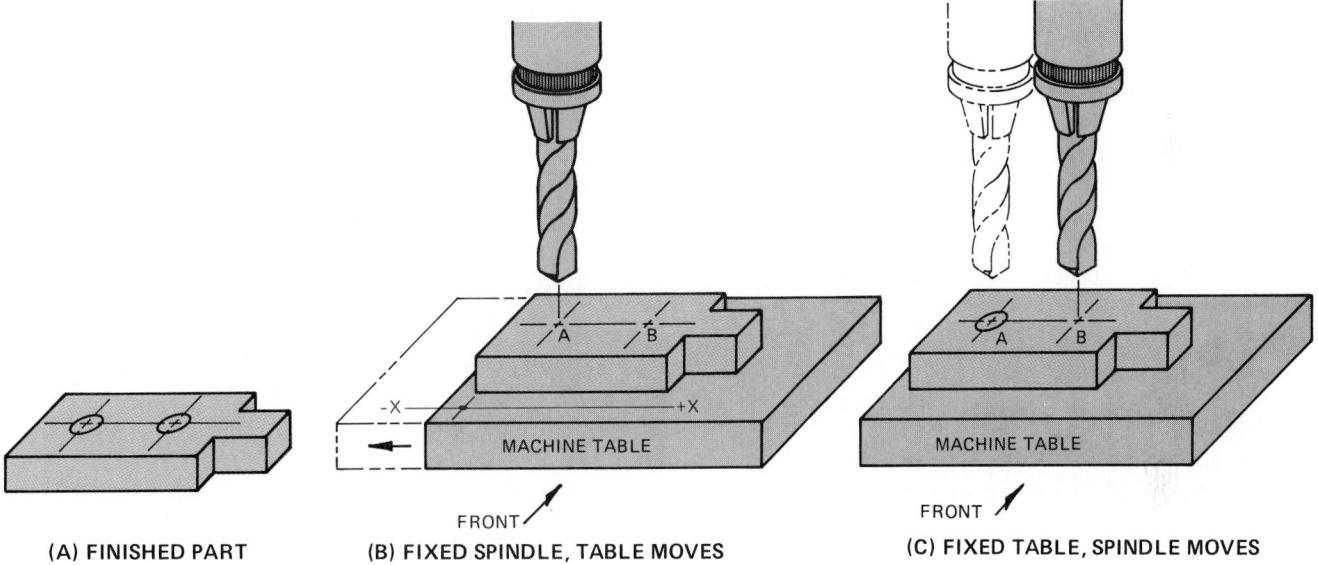

(A) FINISHED PART   (B) FIXED SPINDLE, TABLE MOVES   (C) FIXED TABLE, SPINDLE MOVES

FIG. 17-1-5   Positioning the work.

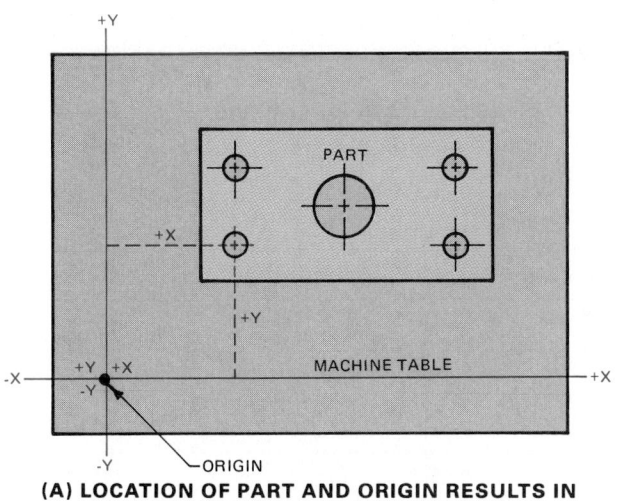

(A) LOCATION OF PART AND ORIGIN RESULTS IN
FIRST QUADRANT CNC DIMENSIONING

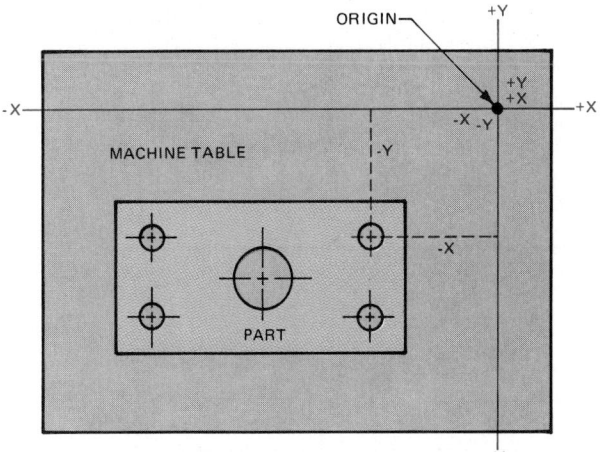

(B) LOCATION OF PART AND ORIGIN RESULTS IN
THIRD QUADRANT CNC DIMENSIONING

FIG. 17-1-6   Origin location.

X-coordinate dimension as hole 1, making the *X* increment zero. Since the vertical distance between holes 1 and 2 is 1.50 in., the *Y* increment becomes +1.50.

After hole 2 is drilled, the drill spindle is positioned above the center of hole 3. Since the horizontal distance between holes 2 and 3 is 4.50 in., the *X* increment is +4.50. Hole 3 has the same *Y*-coordinate dimension as hole 2, making the *Y* increment zero.

From hole 3 the drill spindle is positioned above the center of hole 4. Hole 4 has the same *X*-coordinate dimension as hole 3, making the *X* increment zero. Since the vertical distance between hole 3 and hole 4 is 1.50 in., the *Y* increment is −1.50.

Figure 17-1-8 lists the distance between holes and indicates the direction of motion by plus and minus signs. It can be seen that each pair of coordinates shows the distance between the two locations.

Absolute Coordinate Programming

Many systems use absolute coordinate programming instead of the point-to-point method of dimensioning. With this type of dimensioning, all dimensions are taken from the origin; as such, baseline or datum dimensioning, as shown in Fig. 17-1-7B, is used. For example: after hole 1 is drilled, the drill spindle has to be positioned above the center of hole 2. The coordinates for

hole 2 are (2.75, 4.25). Figure 17-1-9 shows the absolute coordinate dimensions of the holes shown in Fig. 17-1-7B.

ASSIGNMENTS

See Assignments 1 through 4 for Unit 17-1 on pages 605 through 607.

17-2   THREE-AXIS CONTROL SYSTEMS

Many NC machines operate in three directions with the table and carriage moving the *X* and *Y* directions, as explained in Unit 17-1, and the tool spindle, such as a turret drill, traveling in an up-and-down direction. A vertical line taken through the center of the machine spindle is referred to as the *Z* axis and is perpendicular to the plane formed by the *X* and *Y* axes (Fig. 17-2-1).

Thus a point in space can be described by its *X*, *Y*, and *Z* coordinates. For example, P_1 in Fig. 17-2-2 can be described by its (*X*, *Y*, *Z*) coordinates as (4, 3, 5) and P_2 as (11, 2, 8).

An isometric drawing of a part can be described as lines joining a series of points in space (Fig. 17-2-3, pg. 604). The 0, 0, 0 reference indicates the absolute *X*, *Y*, *Z* coordinate origin. It has been designated to be the lower-left front corner position of the

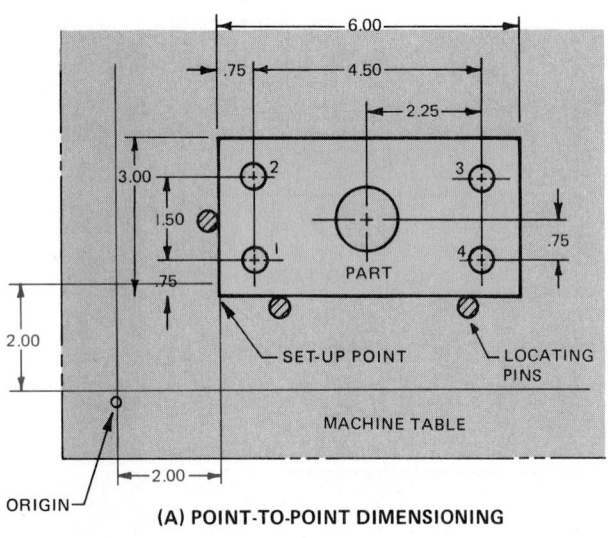

(A) POINT-TO-POINT DIMENSIONING

(B) BASE LINE DIMENSIONING

FIG. 17-1-7   Dimensioning for NC.

HOLE	X	Y
1	+2.75	+2.75
2	0	+1.50
3	+4.50	0
4	0	−1.50

FIG. 17-1-8   Relative coordinate (point-to-point) dimensioning of holes shown in Fig. 17-1-7A.

HOLE	X	Y
1	+2.75	+2.75
2	+2.75	+4.25
3	+7.25	+4.25
4	+7.25	+2.75

FIG. 17-1-9   Absolute coordinate dimensioning of holes shown in Fig. 17-1-7B.

object. The lower-right front position is labeled 12, 0, 0. This means that the coordinate location for that point is 12 units (inches) to the right having the same elevation and depth as the coordinate origin. All other positions are interpreted in the same manner.

A popular system used on many NC machines, such as the turret drill, is to establish the Z zero reference plane above the work piece. Each tool is then adjusted and calibrated to the Z zero reference plane.

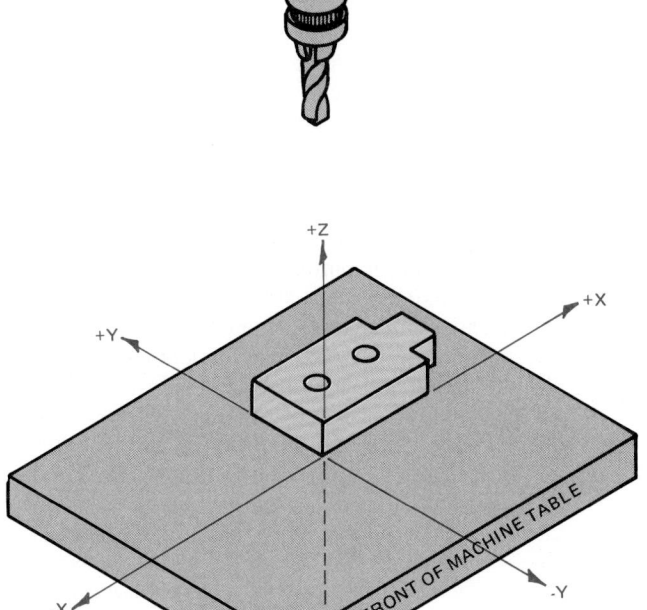

FIG. 17-2-1   *X*, *Y*, and *Z* axes.

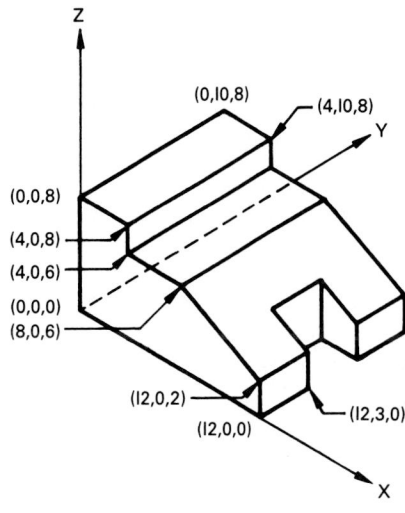

FIG. 17-2-3   Three-dimensional coordinates.

FIG. 17-2-2   Points in space.

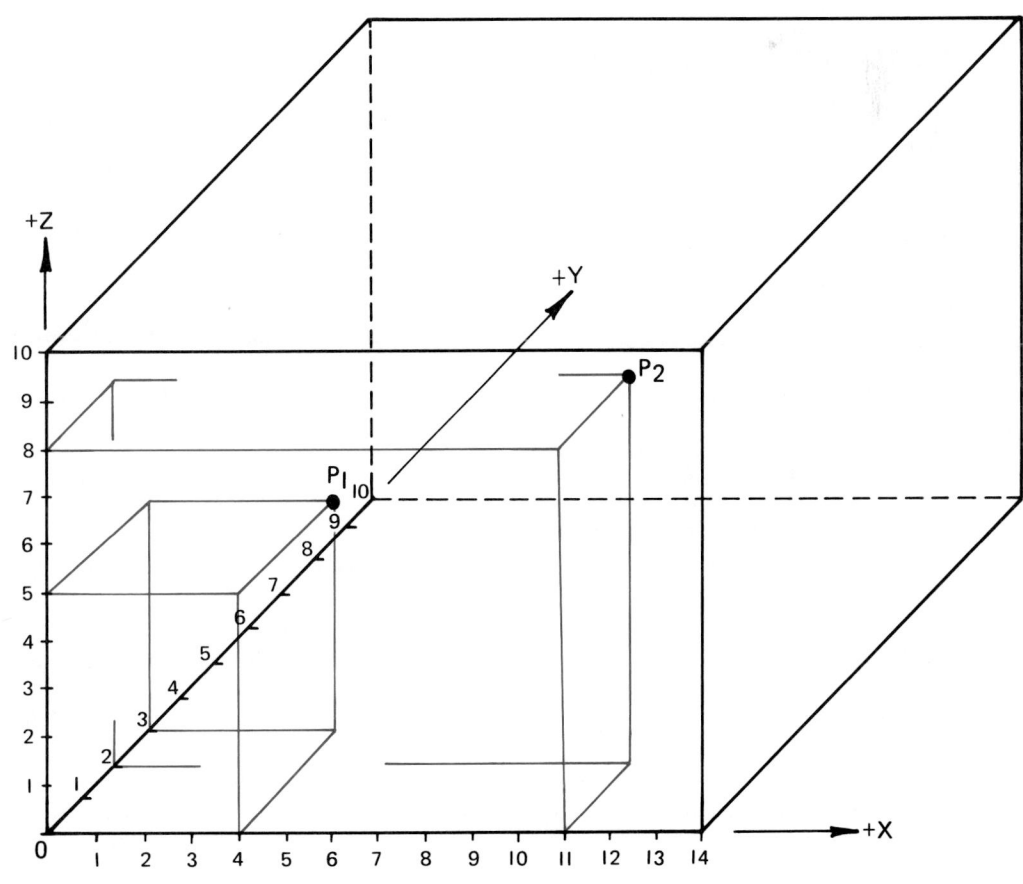

For example, Fig. 17-2-4 shows a part requiring three drilled holes. As the center hole is drilled through, the part is raised by gage blocks so that the drill does not touch the machine table. The height of the gage blocks is determined by the distance the drill passes through the workpiece plus clearance, or .06 in. + 0.3D + .12 in. (Fig. 17-2-5). If a .75 drill was used, the gage block height would be .06 + .23 + .12 = .41 in.

If the distance from the top of the workpiece to the Z zero reference plane is set at .75 in., the Z coordinates for the three holes shown are −(.75 + A), −(.75 + B), and −(.75 + C).

Dimensioning and Tolerancing

Recommended guidelines for dimensioning and tolerancing practices for use in defining parts for NC fabrication are:

1. When the basic coordinate system is established, the set-up point should be placed at an appropriate location on the part itself.

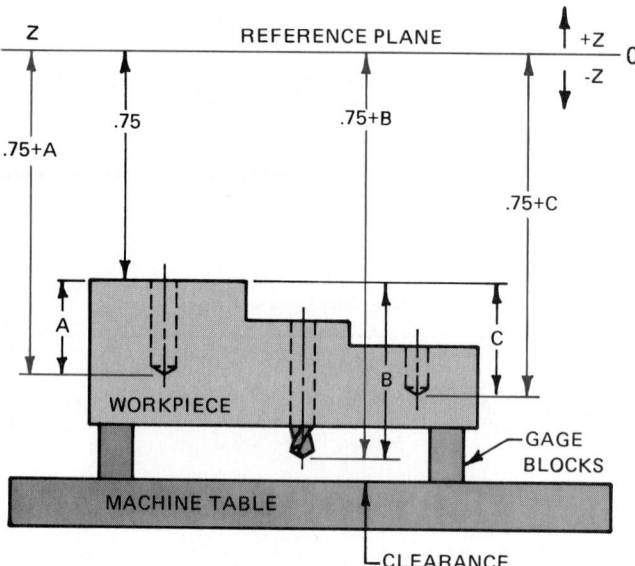

FIG. 17-2-4   Calculating Z distance.

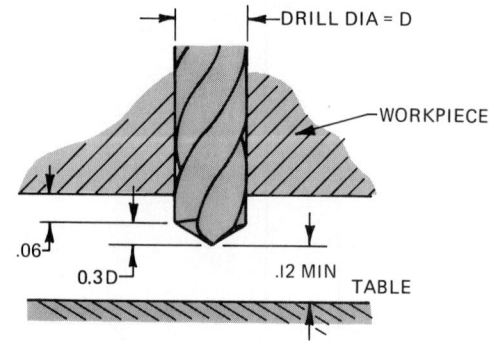

FIG. 17-2-5   Determining gage block height.

2. Any number of subcoordinate systems may be used to define features of a part as long as these systems can be related to the basic coordinate system of the given part.

3. Define part surfaces in relation to three mutually perpendicular reference planes. Establish these planes along part surfaces that parallel the machine axes if these axes can be predetermined.

4. Dimension the part precisely so that the physical shape can be readily determined. Dimension to points on the part surfaces.

5. Regular geometric contours, such as ellipses, parabolas, hyperbolas, etc., may be defined on the drawing by mathematical formulas. The NC machinery can easily be programmed to approximate these curves by linear interpolation, that is, as a series of short, straight lines whose endpoints are close enough together to ensure meeting the required tolerances for contour. In the case of arbitrary curves, the drawing should specify appropriate points on the curve by coordinate dimensions or a table of coordinates. Consideration should be given to the number of points needed to define the curve; however, one should keep in mind the fact that the tighter the tolerance or the smaller the radius of curvature, the closer together the points should be. Such terms as *blend smoothly* and *faired curve* are not used. Curves may also be defined by other coordinates, such as polar, spherical, or cylindrical, as applicable.

6. Changes in contour should be unambiguously defined with prime consideration for design intent.

7. Holes in a circular pattern should preferably be located with coordinate dimensions.

8. Where possible, express angular dimensions relative to the X axis in degrees and decimal parts of a degree.

9. Use plus and minus tolerances, not limit dimensions. Preferably, the tolerance should be equally divided bilaterally.

10. Positional tolerancing, form tolerancing, and datum referencing should be used where applicable. Datum features specified on the drawing in proper sequence will clearly indicate their usage for set-up.

11. Where profile tolerances are specified, the geometric boundary should be equally disposed bilaterally along the true profile. Avoid profile tolerances applied unilaterally along the true profile. Include no less than four defined points along the profile.

12. Tolerances are specified only on the basis of actual design requirement. The accuracy capability of NC equipment is not a basis for specifying more restrictive tolerances than are functionally required.

ASSIGNMENTS

See Assignments 5 and 6 for Unit 17-2 on pages 607–608.

ASSIGNMENTS FOR CHAPTER 17

ASSIGNMENTS FOR UNIT 17-1, TWO-AXIS CONTROL SYSTEMS

1. Prepare a chart listing the *X* and *Y* (coordinate) locations and the quadrant for the points *A* to *V* shown on Fig. 17-1-A. The grid is 10 × 10 to the centimeter or inch.

POINT	X AXIS	Y AXIS	QUADRANT
A			
B			
C			

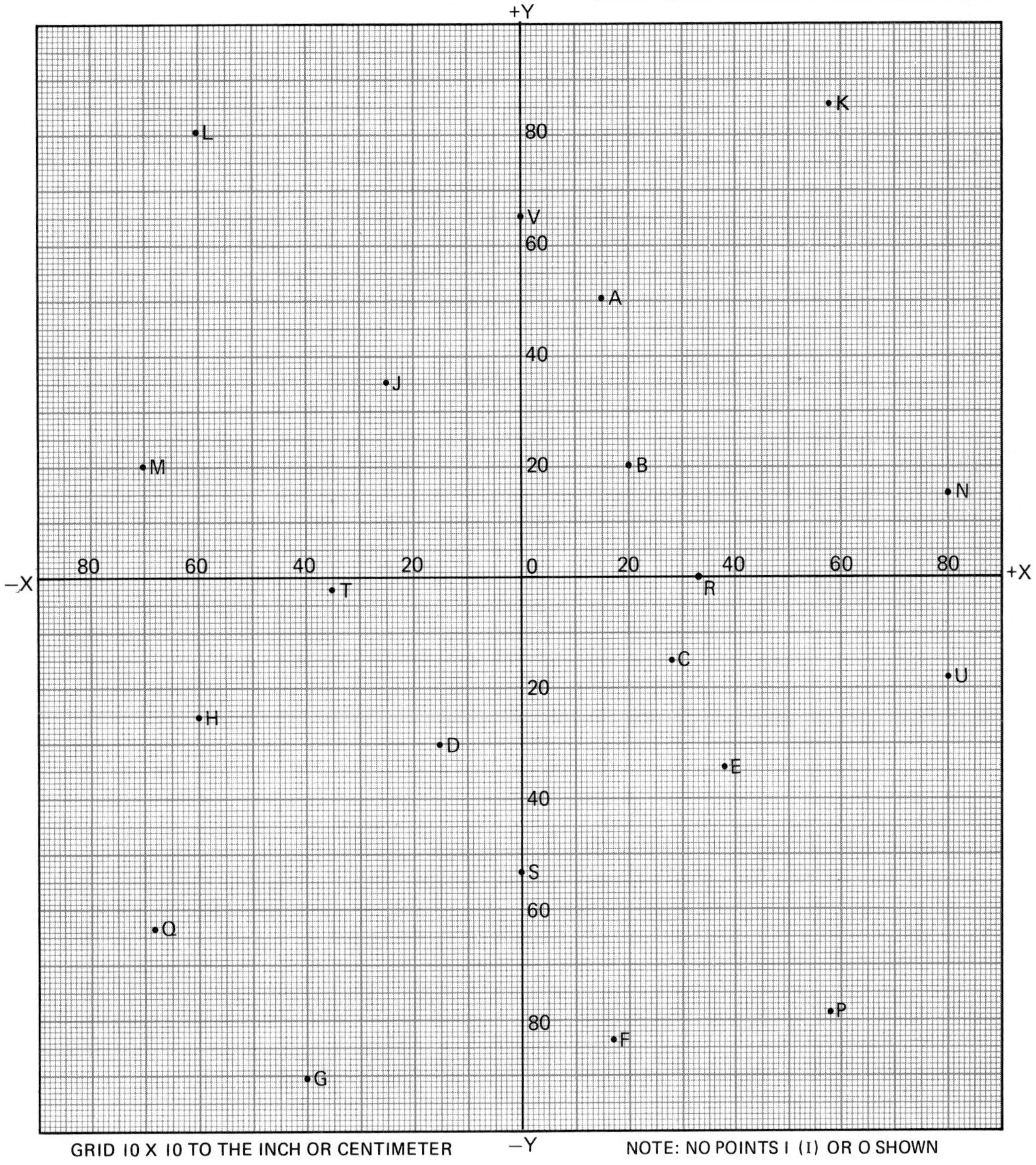

GRID 10 X 10 TO THE INCH OR CENTIMETER   NOTE: NO POINTS I (I) OR O SHOWN

FIG. 17-1-A   Chart.

2. Prepare two drawings of the cover plate shown in Fig. 17-1-B. Use point-to-point dimensioning on one drawing for the 10 holes; use coordinate dimensioning on the other drawing. Only the dimensions locating the holes need be shown. The radial and angular dimensions are to be replaced with coordinate dimensions and taken to three decimal places (inch). Below each drawing prepare a chart listing each hole and its *X* and *Y* coordinates. The letters shown at the holes indicate the sequence in which they are to be drilled. Scale 1:1.

3. Prepare two drawings of the cover plate shown in Fig. 17-1-C. Use point-to-point dimensioning on one drawing for the holes; use datum or coordinate dimensioning on the other drawing. Work with customary or metric dimensions as directed by your instructor. Only the dimensions locating the holes need be shown. Below each drawing prepare a chart listing each hole and its *X* and *Y* coordinates. The letters shown at the holes indicate the sequence in which they are to be drilled. Note the location of the origin. Scale 1:1.

FIG. 17-1-B   Cover plate.

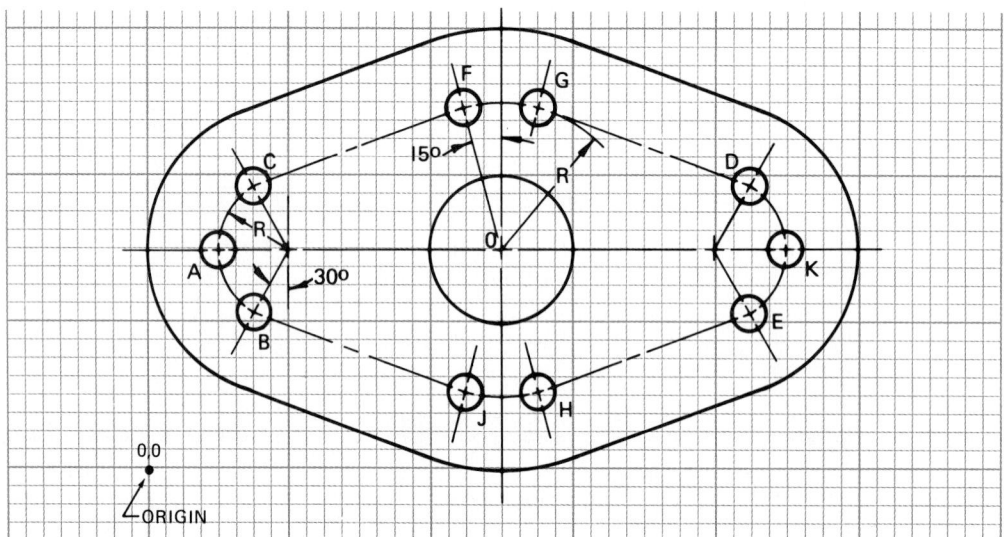

NOTE: GRID 10 X 10 TO THE INCH   NOTE: NO HOLE I (I) SHOWN

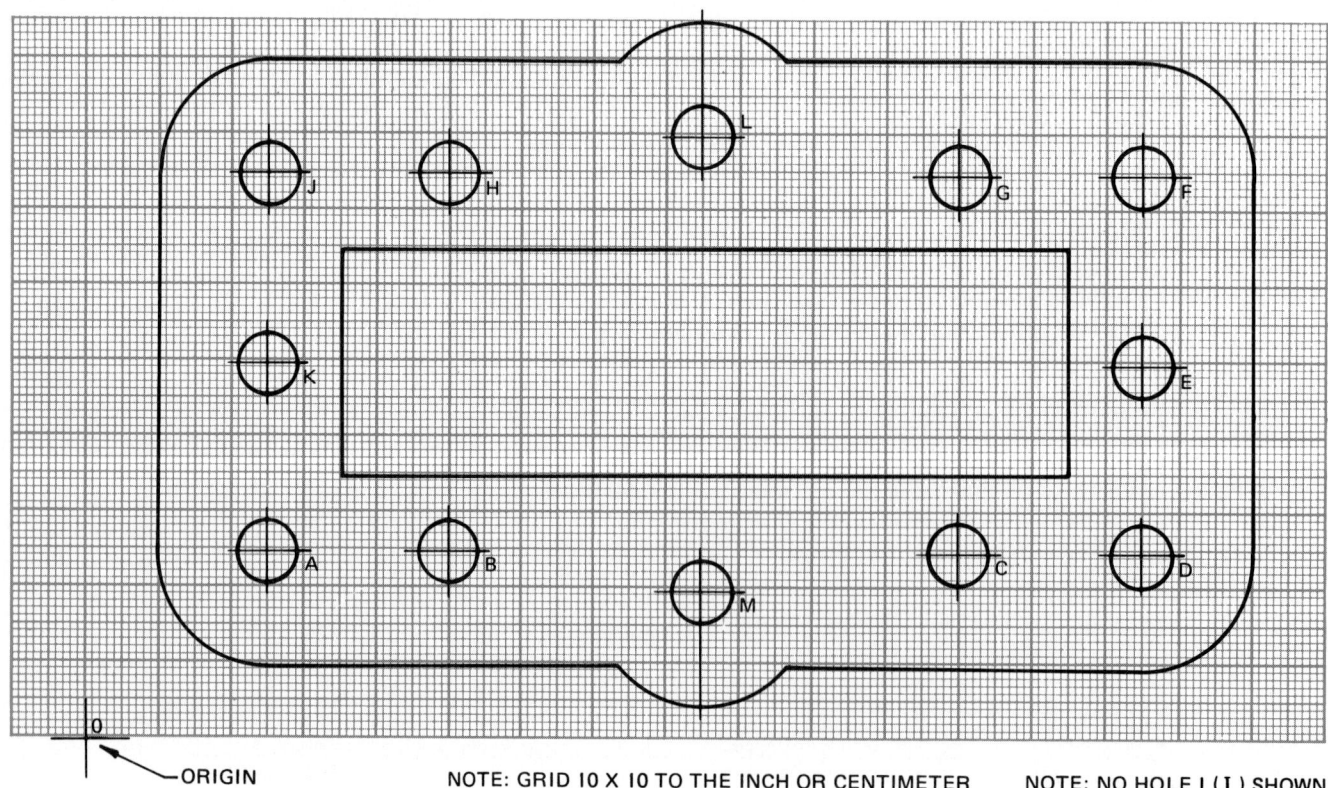

NOTE: GRID 10 X 10 TO THE INCH OR CENTIMETER   NOTE: NO HOLE I (I) SHOWN

FIG. 17-1-C   Cover plate.

FIG. 17-1-D   Terminal board.

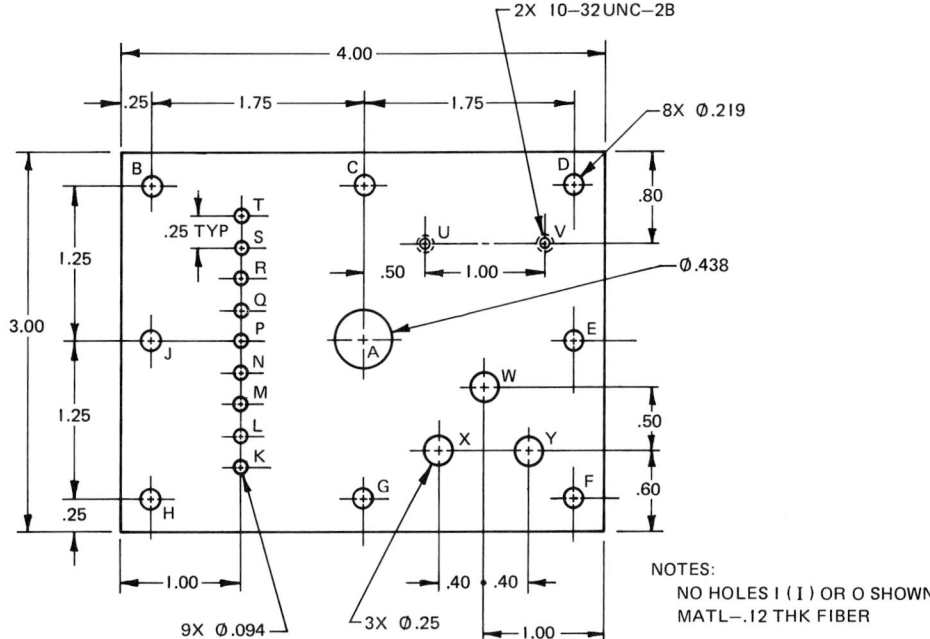

NOTES:
NO HOLES I (I) OR O SHOWN
MATL—.12 THK FIBER

4. Make a one-view drawing of the terminal board shown in Fig. 17-1-D. Point-to-point programming is to be used to locate each hole. Below the drawing prepare a chart listing each hole and its *X* and *Y* coordinates. The letters at the holes show the sequence in which they are to be drilled. Origin for the *X* and *Y* coordinates is the bottom left-hand corner of the part.

be shown. The depth of the tap drill is to be .10 in. below the last complete thread. Below the drawing prepare a chart listing each hole and its *X*, *Y*, and *Z* coordinates using point-to-point dimensioning. The letters at the holes show the sequence in which they are to be drilled. Calculating the *Z* coordinate is to be done in the same manner as for the part shown in Fig. 17-2-4. Scale 1:1.

Note: Programming will be for the tap-drill holes and the six through holes shown. Origin for the *X* and *Y* coordinates is the center of the end plate.

ASSIGNMENTS FOR UNIT 17-2, THREE-AXIS CONTROL SYSTEMS

5. Make a two-view drawing of the end plate shown in Fig. 17-2-A. Only the dimensions locating the holes need

FIG. 17-2-A   End plate.

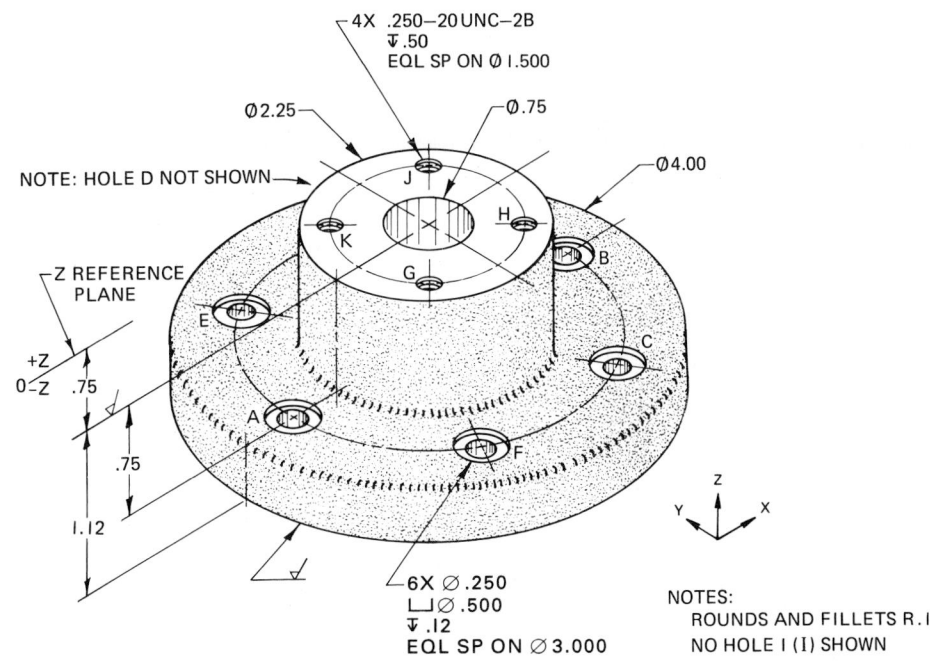

NOTES:
ROUNDS AND FILLETS R.12
NO HOLE I (I) SHOWN

6. Photocopy Fig. 17-2-B. On this photocopy locate points *P*1 to *P*10 on the coordinate chart 1. On chart 2 points *P*1 to *P*10 are located, but only one of their coordinates is known. Accurately lay out the missing coordinates on the chart and record their values in the table. Scale as shown.

CHART I

LOCATE POINTS PI TO PIO ON CHART

NOTE: ALL COORDINATES ARE +

POINT	X AXIS	Y AXIS	Z AXIS
PI	10	20	60
P2	20	70	70
P3	0	20	30
P4	100	60	75
P5	40	30	20
P6	40	60	10
P7	70	20	0
P8	50	50	50
P9	85	65	30
PIO	60	65	15

CHART 2

LOCATE COORDINATES AND RECORD IN TABLE

NOTE: ALL COORDINATES ARE +

POINT	X AXIS	Y AXIS	Z AXIS
PI	20		
P2			55
P3		60	
P4	30		
P5		30	
P6			0
P7			40
P8	65		
P9		30	
PIO			55

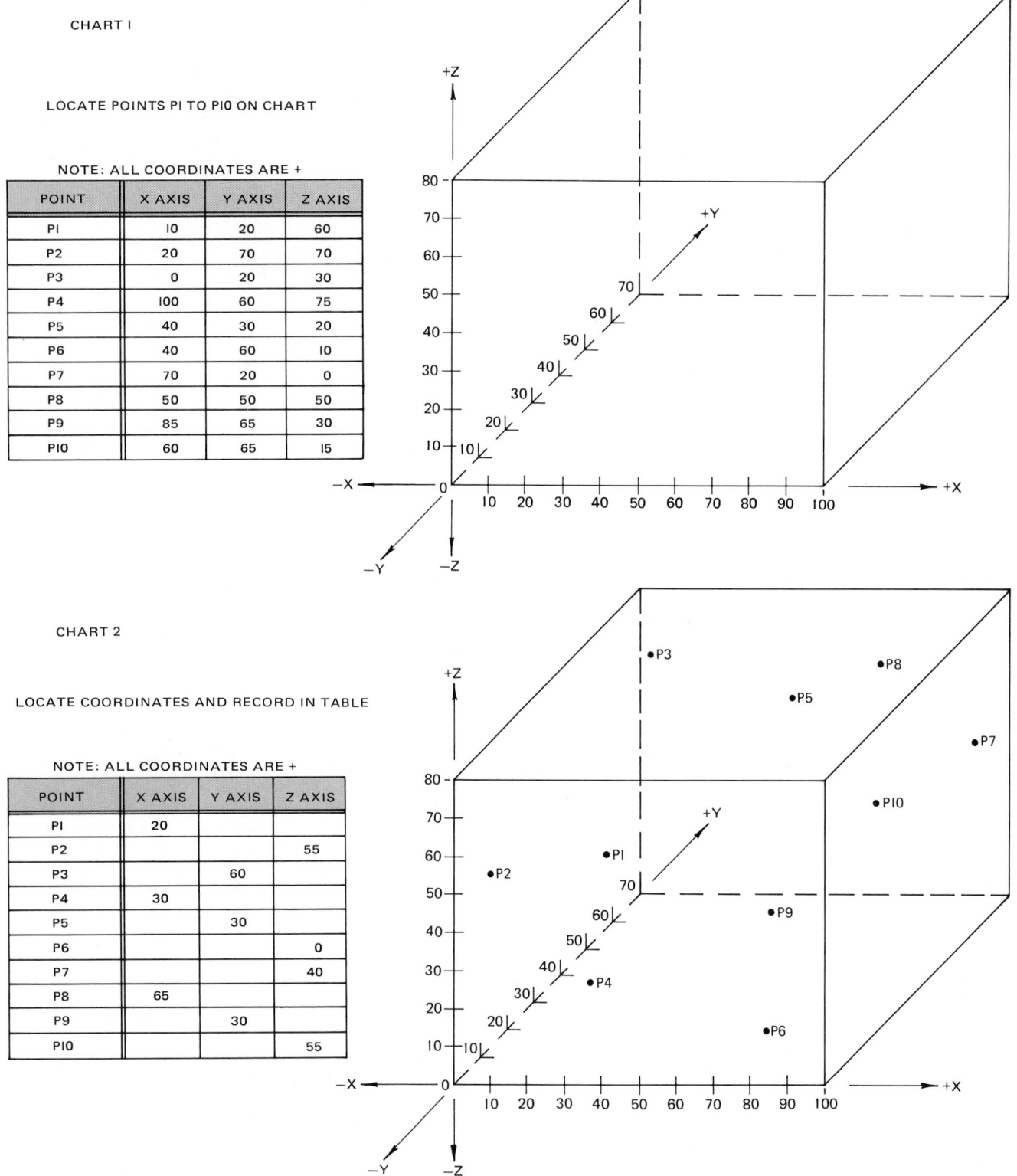

FIG. 17-2-B   Chart.

CHAPTER 18

WELDING DRAWINGS

Definitions

Field welds Welds not made in a shop or at the place of initial construction.

Weld symbol A symbol used to indicate the type of weld.

Welding symbol A method of representing the weld on drawings.

18-1 DESIGNING FOR WELDING

The primary importance of welding is to unite various pieces of metal so that they will operate as a unit structure to support the loads to be carried. In order to design such a structure, which will be both economical and efficient, the drafter must have a knowledge of the basic principles of welding practice and an understanding of the advantages and limitations of the process.

In order to produce an economical and pleasing design, the designer should endeavor to use the method of construction that is clearly the most advantageous for the application under consideration. This may mean a combination of welding and bolting or the incorporation of pressings, forgings, or even castings where appropriate. The possibility of using structural steel shapes and tubes should also be kept in mind (Figs. 18-1-1 and 18-1-2, pg. 610).

Welding Processes

Of more than 40 welding processes used in industry today, only a few are industrially important. Arc welding, gas welding, and resistance welding are the three most important types of welding.

The workpieces are melted along a common edge or surface so that their molten metal— and usually a filler metal also—is allowed to form a common pool or puddle. The pieces are fused when the puddle solidifies (Figs. 18-1-3 and 18-1-4 on pg. 610).

(A) CRANKS AND CRANKSHAFTS

(B) LINKS AND CLEVISES

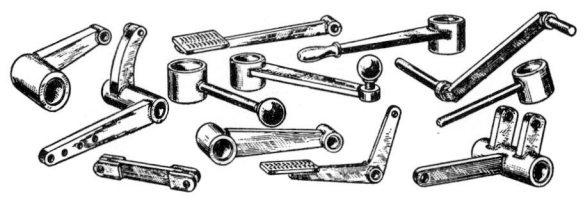

(C) WHEELS

(D) LEVERS

FIG. 18-1-1   A variety of weldments.
(James F. Lincoln Arc Welding Foundation)

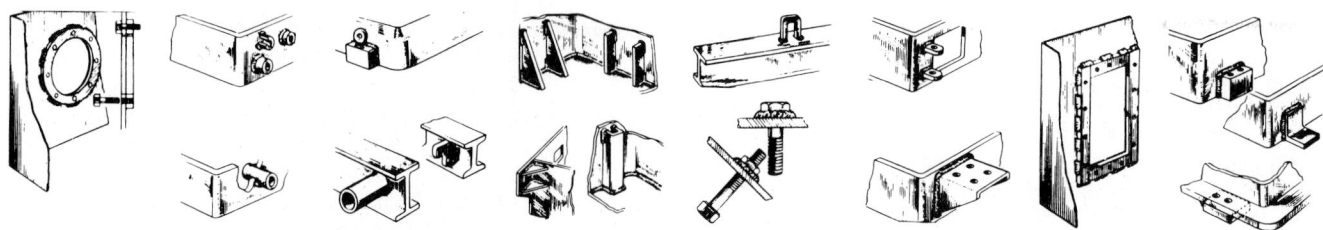

FIG. 18-1-2   Design ideas for fabricated parts. *(James F. Lincoln Arc Welding Foundation)*

FIG. 18-1-3   Preferred welding design.

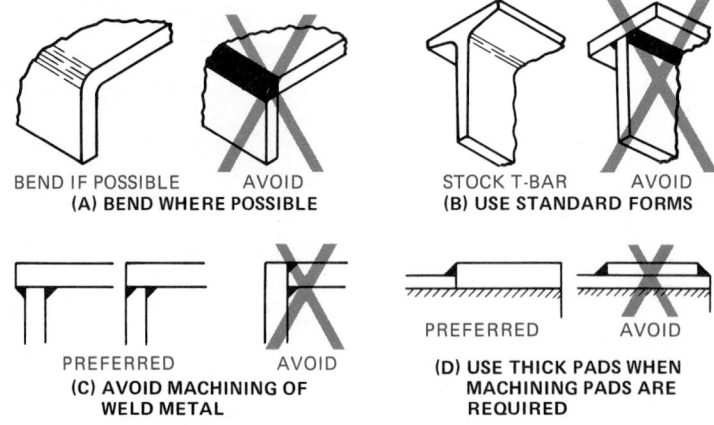

BEND IF POSSIBLE   AVOID
(A) BEND WHERE POSSIBLE

STOCK T-BAR   AVOID
(B) USE STANDARD FORMS

PREFERRED   AVOID
(C) AVOID MACHINING OF WELD METAL

PREFERRED   AVOID
(D) USE THICK PADS WHEN MACHINING PADS ARE REQUIRED

FIG. 18-1-4   Basic welding joints.

TYPE OF JOINT	APPLICABLE WELDS	
T-JOINT	FILLET PLUG SLOT SQUARE GROOVE BEVEL GROOVE	J-GROOVE FLARE BEVEL GROOVE SPOT PROJECTION SEAM
BUTT JOINT	SQUARE GROOVE V-GROOVE BEVEL GROOVE U-GROOVE	J-GROOVE FLARE V-GROOVE FLARE BEVEL GROOVE EDGE FLANGE
CORNER JOINT	FILLET SQUARE GROOVE V-GROOVE BEVEL GROOVE U-GROOVE J-GROOVE	FLARE V-GROOVE FLARE BEVEL GROOVE EDGE FLANGE CORNER FLANGE SPOT PROJECTION SEAM
LAP JOINT	FILLET PLUG SLOT BEVEL GROOVE	J-GROOVE FLARE BEVEL GROOVE SPOT PROJECTION SEAM
EDGE JOINT	PLUG SLOT SQUARE GROOVE BEVEL GROOVE V-GROOVE U-GROOVE J-GROOVE	EDGE FLANGE CORNER FLANGE SPOT PROJECTION SEAM EDGE

Gas welding, the most common form of which is oxyacetylene welding, gets its heat from the burning of flammable gases. This process is slow compared to other modern welding methods, so gas welding is normally confined to repair and maintenance work rather than major mass production (Fig. 18-1-5).

The major industrial welding process is arc welding, where heat is generated by an electric arc struck between a welding electrode, or rod, and the workpiece. The arc is quite hot, and melting and subsequent solidification of the weld metal occur very rapidly.

Resistance welding is also widely used, especially in mass-production work. As in arc welding, resistance welding employs electricity. But no arc is generated. Instead, heat is created from resistance losses as a high-amperage current is sent across a joint between two mating surfaces.

REFERENCES AND SOURCE MATERIAL

1. American Welding Society.
2. *Machine Design,* Fastening and joining reference issue.

ASSIGNMENT

See Assignment 1 for Unit 18-1 on pages 640–641.

18-2 WELDING SYMBOLS

The introduction of welding symbols enables the designer to indicate clearly the type and size of weld required to meet design requirements, and it is becoming increasingly important for the designer to specify the required type of weld correctly. Points that must be made clear are the type of weld, the joint preparation, the weld size, and the root opening (if any). These joints can be clearly specified on the drawing with welding symbols (Figs. 18-2-1 through 18-2-3, pp. 611–613).

METAL OR ALLOY	GAS	ARC
Aluminum —Commercially Pure —Al-Mn Alloy	 X X	 X X
Brass, Commercial	X	
Bronze, Commercial	X	
Copper (Deoxidized)	X	
Iron —Gray and Alloy —Malleable	 X	
Lead	X	
Magnesium Alloys	X	
Nickel and Nickel Alloys	X	X
Steels, Carbon —Low and Medium Carbon —High Carbon —Tool Steel	 X 	 X X
Steel, Cast	X	X
Steels, Stainless —Chromium —Chromium-Nickel	 X	 X X

FIG. 18-1-5   Weldability of various metals and alloys.

FIG. 18-2-1   Weld terminology.

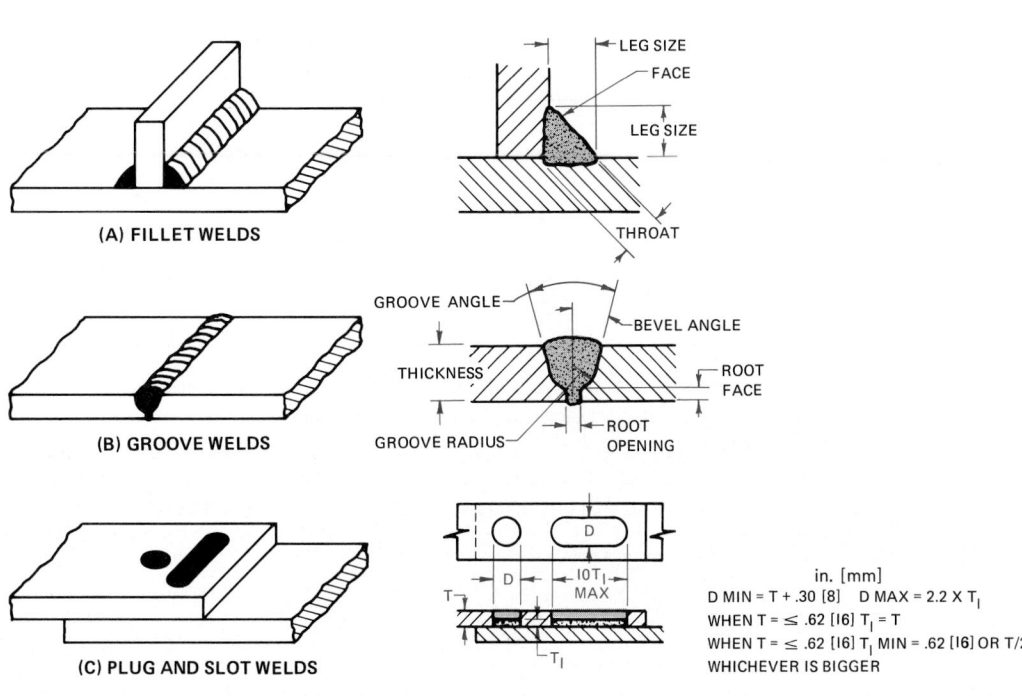

(A) FILLET WELDS

(B) GROOVE WELDS

(C) PLUG AND SLOT WELDS

in. [mm]
D MIN = T + .30 [8]   D MAX = 2.2 X T_l
WHEN T = ≤ .62 [16] T_l = T
WHEN T = ≤ .62 [16] T_l MIN = .62 [16] OR T/2
WHICHEVER IS BIGGER

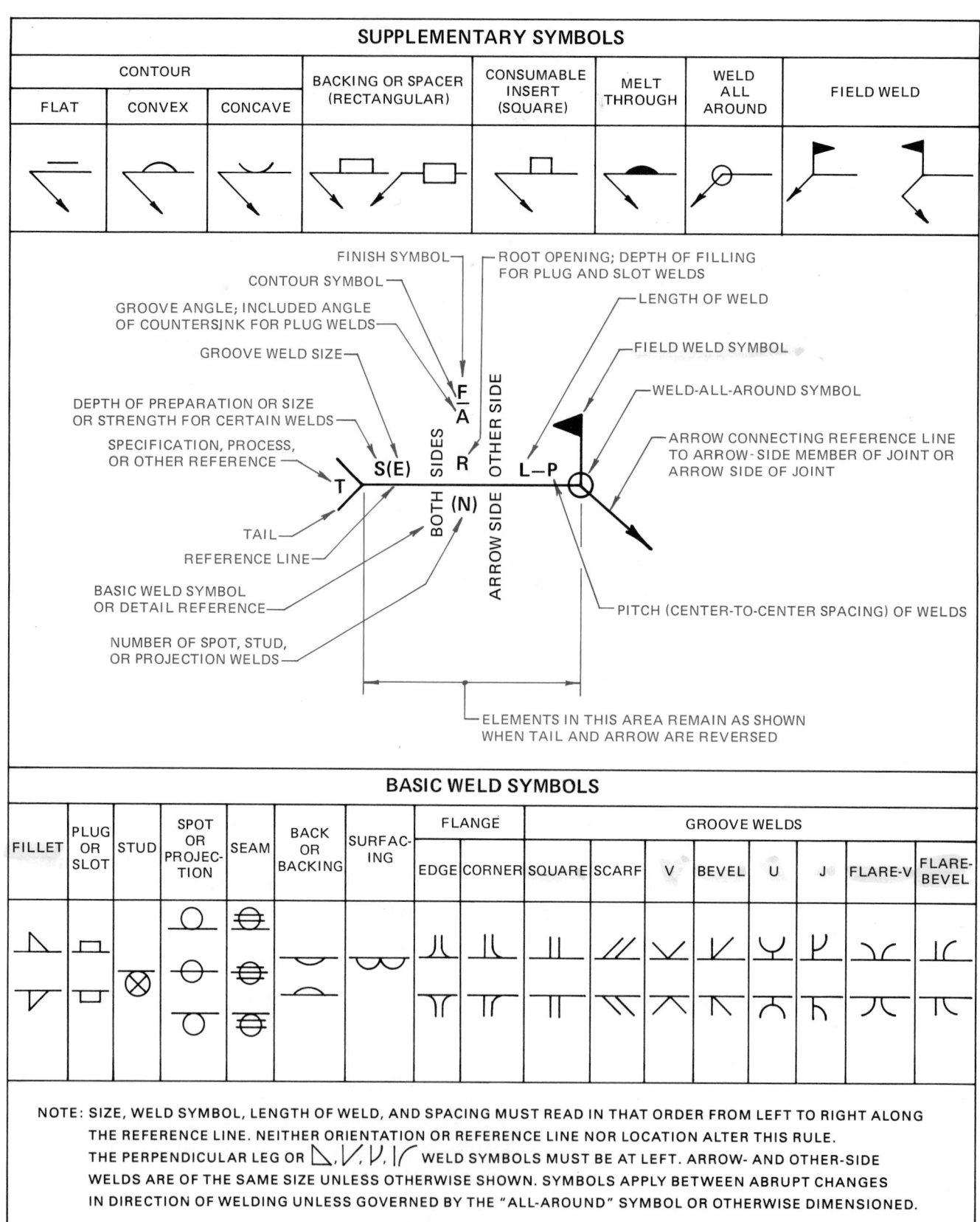

FIG. 18-2-2   Welding symbols.

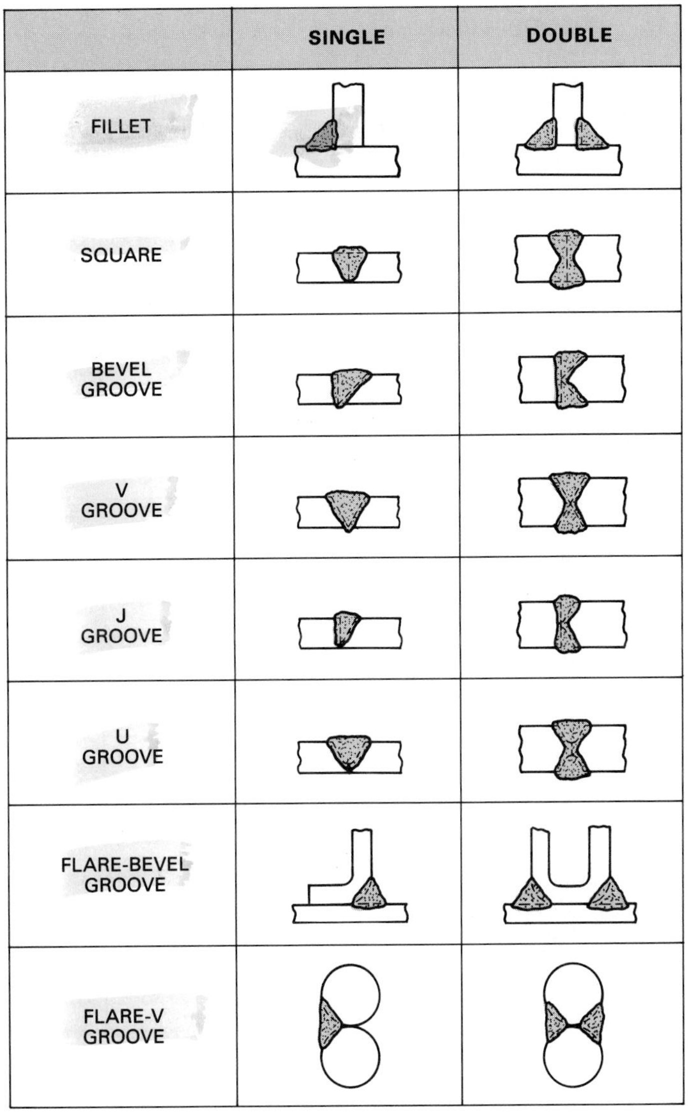

FIG. 18-2-3   Fillet and groove welds.

	SINGLE	DOUBLE
FILLET		
SQUARE		
BEVEL GROOVE		
V GROOVE		
J GROOVE		
U GROOVE		
FLARE-BEVEL GROOVE		
FLARE-V GROOVE		

Welding symbols are a shorthand language. They save time and money and, if used correctly, ensure understanding and accuracy. They need to be a universal language, and for this reason the symbols of the American Welding Society, already well established, have been adopted.

A distinction between the terms *weld symbol* and *welding symbol* should be understood. The weld symbol indicates the type of weld. The welding symbol is a method of representing the weld on drawings. It includes supplementary information and consists of the following eight elements. Not all elements need be used unless required for clarity.

1. Reference line
2. Arrow
3. Basic weld symbol
4. Dimensions and other data
5. Supplementary symbols
6. Finish symbols
7. Tail
8. Specification, process, or other reference

The tail of the symbol is used for designating the welding specifications, procedures, or other supplementary information to be used in the making of the weld (Fig. 18-2-4).

The use of letters can designate different welding and cutting processes (Figs. 18-2-5 and 18-2-6, pg. 614).

The use of the words *far side* and *near side* in the past has led to confusion because when joints are shown in section, all welds are equally distant from the reader, and the words *near* and *far* are meaningless. In the present system the joint is the basis of reference. Any joint, the welding of which is indicated by a symbol, will usually have an arrow side and another side. Accordingly, the words *arrow side, other side,* and *both sides*

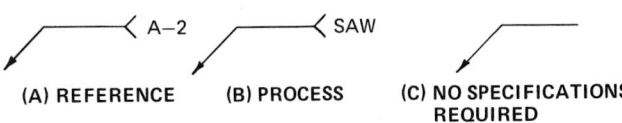

(A) REFERENCE   **(B) PROCESS**   **(C) NO SPECIFICATIONS REQUIRED**

FIG. 18-2-4   Location of reference and processes on welding symbols.

DESIGNATION	WELDING PROCESS
CAW	Carbon Arc Welding
CW	Cold Welding
DB	Dip Brazing
DFW	Diffusion Welding
EBW	Electron Beam Welding
ESW	Electroslag Welding
EXW	Explosion Welding
FB	Furnace Brazing
FCAW	Flux Cored Arc Welding
FOW	Forge Welding
FRW	Friction Welding
FW	Flash Welding
GMAW	Gas Metal Arc Welding
GTAW	Gas Tungsten Arc Welding
IB	Induction Brazing
IRB	Infrared Brazing
IW	Induction Welding
LBW	Laser Beam Welding
OAW	Oxyacetylene Welding
OHW	Oxyhydrogen Welding
PAW	Plasma Arc Welding
PEW	Percussion Welding
PGW	Pressure Gas Welding
PW	Projection Welding
RB	Resistance Brazing
RSEW	Resistance Seam Welding
RSW	Resistance Spot Welding
SAW	Submerged Arc Welding
SMAW	Shielded Metal Arc Welding
SW	Stud Welding
TB	Torch Brazing
TW	Thermit Welding
USW	Ultrasonic Welding
UW	Upset Welding

FIG. 18-2-5   Designation of welding processes by letters.

DESIGNATION	CUTTING PROCESS
AAC	Air-Carbon Arc Carbon
AC	Arc Cutting
AOC	Oxygen Arc Cutting
CAC	Carbon Arc Cutting
FOC	Chemical Flux Cutting
MAC	Metal Arc Cutting
OC	Oxygen Cutting
PAC	Plasma Arc Cutting
POC	Metal Powder Cutting

FIG. 18-2-6   Designation of cutting processes by letters.

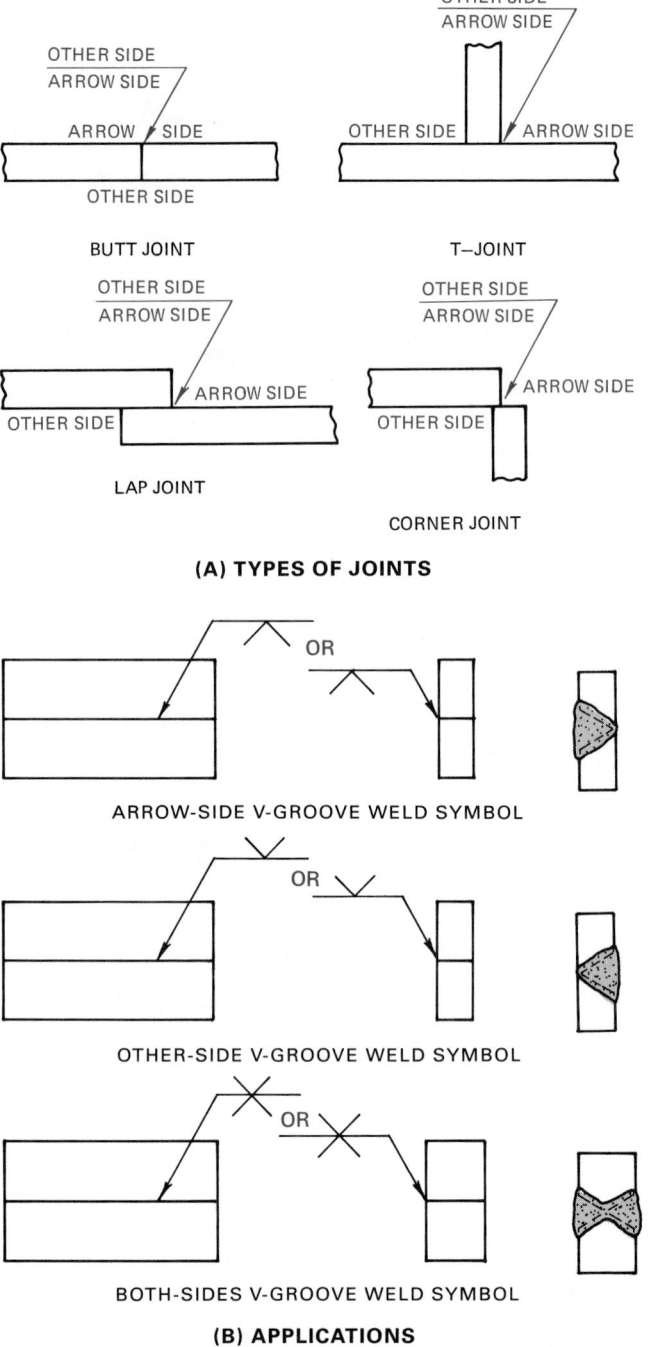

(A) TYPES OF JOINTS

ARROW-SIDE V-GROOVE WELD SYMBOL

OTHER-SIDE V-GROOVE WELD SYMBOL

BOTH-SIDES V-GROOVE WELD SYMBOL

(B) APPLICATIONS

FIG. 18-2-7   Arrow side and other side of joint.

are used here to locate the weld with respect to the joint (Fig. 18-2-7).

Location Significance of Arrow

1. In the case of fillet, groove, and flanged weld symbols, the arrow connects the welding symbol reference line to one side of the joint, and this side is considered the *arrow side* of the joint. The side opposite the arrow side of the joint is considered the *other side* of the joint.

2. When a joint is depicted by a single line on the drawing and the arrow of a welding symbol is directed to this line, the arrow side of the joint is considered the *near side* of the joint.

3. In the case of plug, slot, spot, projection, and seam weld symbols, the arrow connects the welding symbol reference line to the outer surface of one of the members of the joint at the center line of the desired weld. The member to which the arrow points is the *arrow-side* member. The remaining member of the joint is considered the *other-side* member.

Symbols with No Side Significance Some weld symbols have no arrow-side or other-side significance, although supplementary symbols used in conjunction with them may have such significance.

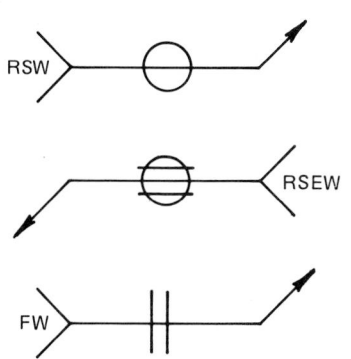

Orientation of Specific Weld Symbols Fillet, bevel-groove, J-groove, flare-bevel-groove, and corner-flange weld symbols are drawn with the perpendicular leg always to the left.

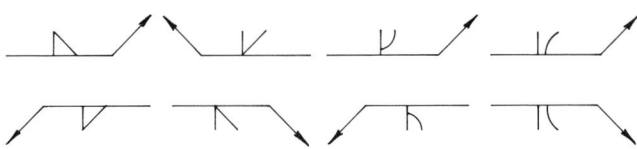

Break in Arrow When only one member of a joint is to be prepared, the arrow shall have a break and point toward that member (Fig. 18-2-8). If it is obvious which member is to be prepared, or there is no preference as to which member is to be prepared, the arrow need not be broken.

Location of Weld Symbol with Respect to Joint

1. Welds on the arrow side of the joint are shown by placing the weld symbol below the reference line.

2. Welds on the other side of the joint are shown by placing the weld symbol above the reference line.

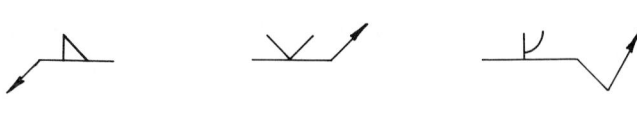

3. Welds on both sides of the joint are shown by placing the weld symbol on both sides of the reference line.

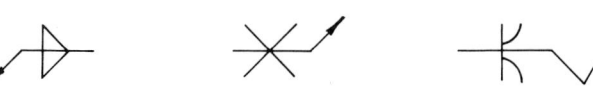

DRAWING CALLOUT   INTERPRETATION

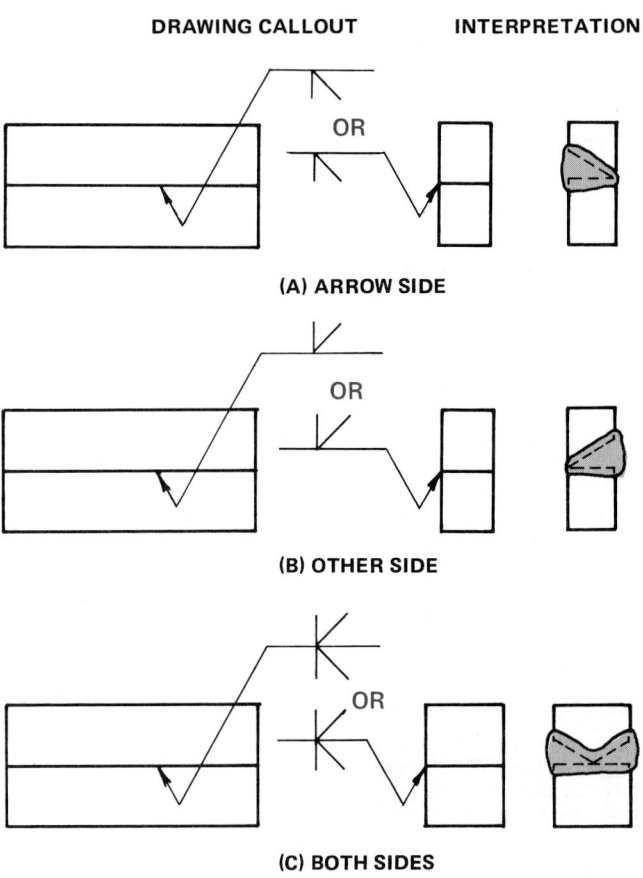

(A) ARROW SIDE

(B) OTHER SIDE

(C) BOTH SIDES

FIG. 18-2-8   Application of break in arrow of welding symbol.

Use of Field Weld Symbol

Field welds (welds not made in a shop or at the place of initial construction) are indicated by means of the field weld symbol placed at the intersection of the reference line and the arrow. The flag is placed above and at right angles to the reference line (Fig. 18-2-9).

Use of Weld-All-Around Symbol

A weld extending completely around a joint is indicated by means of a weld-all-around symbol placed at the intersection

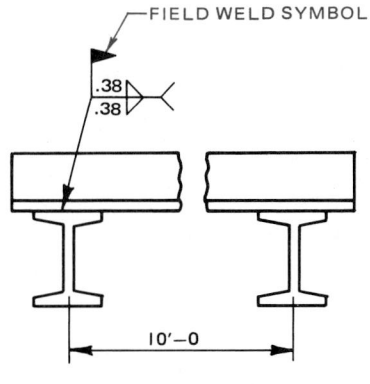

FIG. 18-2-9   Application of field weld symbol.

of the reference line and the arrow. (See Examples 1 and 2 in Fig. 18-2-10.)

Welds extending around the circumference of a pipe are excluded from the requirement regarding changes in direction and do not require the weld-all-around symbol to specify a continuous weld. (See Example 3 in Fig. 18-2-10.)

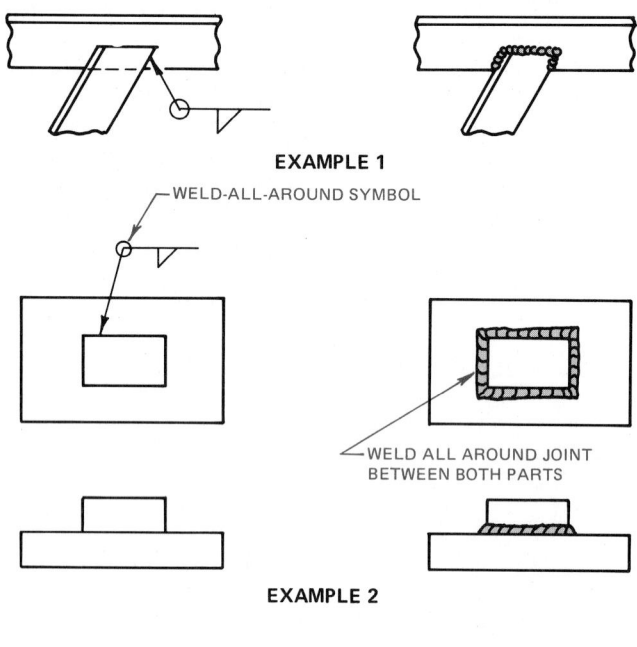

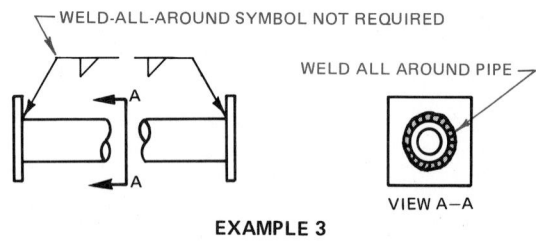

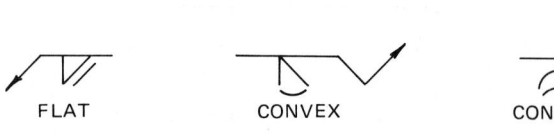

DRAWING CALLOUT   **DESIRED WELD**

FIG. 18-2-10   Application of weld-all-around.

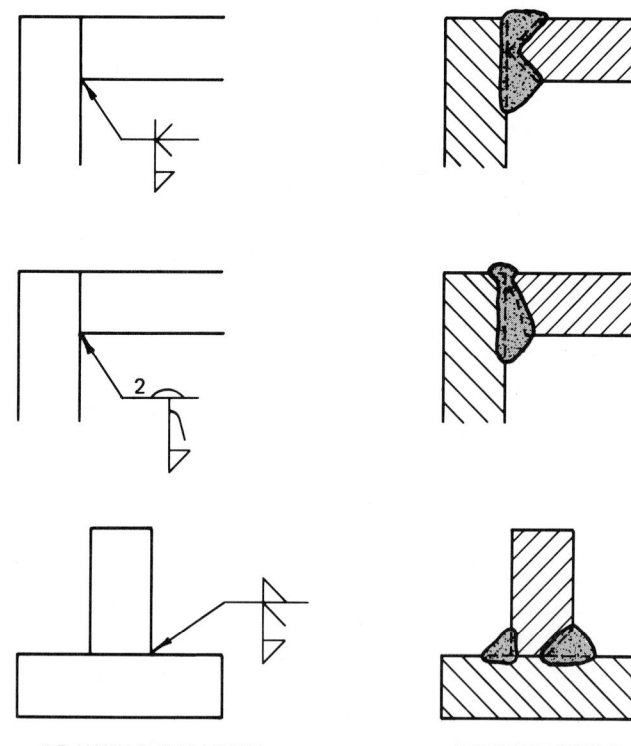

DRAWING CALLOUT   **DESIRED WELD**

FIG. 18-2-11   Combined welding symbols.

Combined Weld Symbols

For joints having more than one weld, a symbol is shown for each weld (Fig. 18-2-11).

Contours Obtained by Welding

Welds that are to be welded with approximately flush or convex faces without postweld finishing are specified by adding the flush or convex contour symbol to the welding symbol.

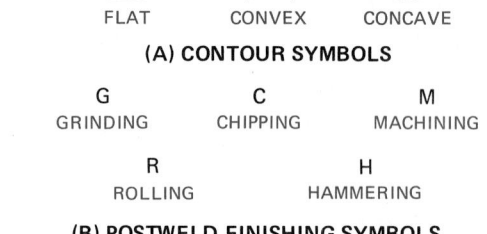

FLAT   CONVEX   CONCAVE
(A) CONTOUR SYMBOLS

G   C   M
GRINDING   CHIPPING   MACHINING

R   H
ROLLING   HAMMERING
(B) POSTWELD FINISHING SYMBOLS

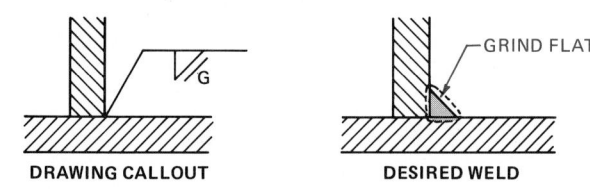

DRAWING CALLOUT   **DESIRED WELD**

(C) APPLICATION

FIG. 18-2-12   Finishing of welds.

Finishing of Welds

Finishing of welds, other than cleaning, is indicated where applicable by suitable contour symbols. Where postweld finishing of welds is required, the appropriate finishing symbol is added to the contour symbol (Fig. 18-2-12).

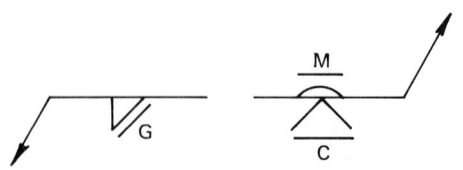

Multiple Reference Lines

Two or more reference lines may be used to indicate a sequence of operations. The first operation is specified on the reference line nearest the arrow. Subsequent operations are specified sequentially on other reference lines.

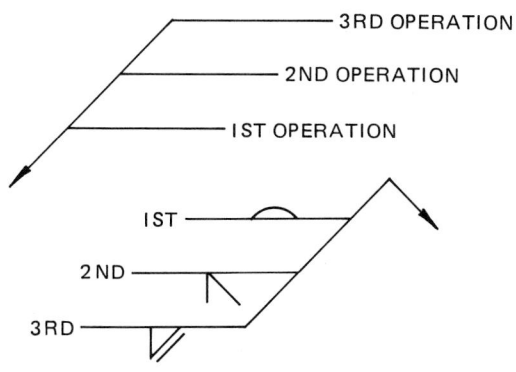

Tail of Welding Symbol

The welding and allied process to be used may be specified by placing the appropriate letter designations from Figs. 18-2-5 and 18-2-6 in the tail of the welding symbol.

The tail of additional reference lines may be used to specify data supplementary to welding symbol information. When no references are required, the tail may be omitted from the welding symbol.

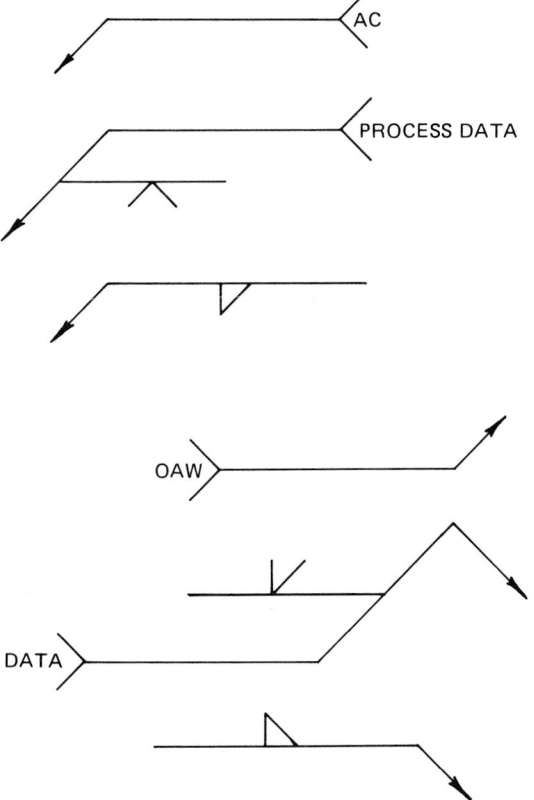

The Design of Welded Joints

Since loads are transferred from one member to another through the welds on a fabricated assembly, the type of joint and weld is specified by the designer. Figure 18-1-4 shows basic joint and weld types. Specifying the joint does not by itself describe the type of weld to be used. Several types of welds may be used for making a joint.

The fillet weld, requiring no groove penetration, is one of the most commonly used welds. Corner welds are also widely used in machine design. The corner-to-corner joint, shown in Fig. 18-2-13A, is difficult to assemble because neither plate can be supported by the other. The joint also requires a larger amount of weld than the other joints illustrated. The corner joint shown in Fig. 18-2-13B is easy to assemble and requires half the amount of weld metal as the joint in Fig. 18-2-13A. However, by using half the weld size, but placing two welds, one outside, as in Fig. 18-2-13C, it is possible to obtain the same total throat as with the first weld. Only half the weld metal is required.

With thick plates, a partial-penetration groove joint, as in Fig. 18-2-13D, is used. This requires beveling. For a deeper joint, a J preparation, as in Fig. 18-2-13E, may be used in preference to a bevel. The fillet weld in Fig. 18-2-13F is out of sight and makes a neat and economical corner.

The size of the weld should always be designed with reference to the size of the thinner member, as illustrated in Fig. 18-2-14. The joint cannot be made any stronger by using the

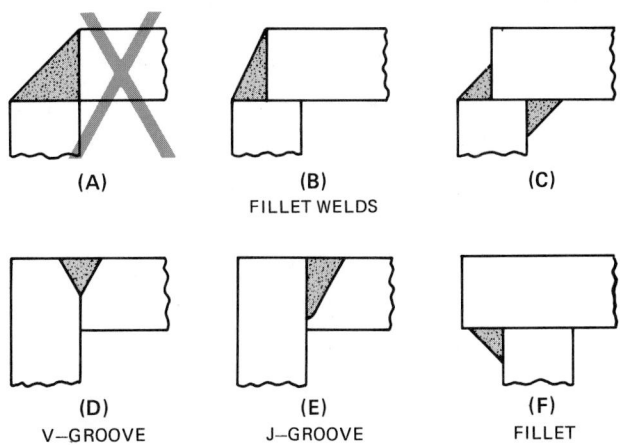

FIG. 18-2-13 Corner joints.

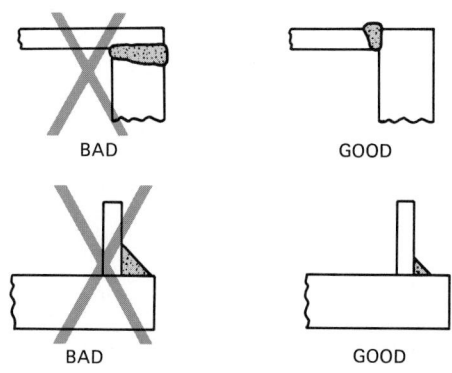

FIG. 18-2-14 Size of weld determined by thinner member.

thicker member for the weld size, and much more weld metal may be required.

The designer is frequently faced with the question of whether to use fillet or groove welds. Here cost becomes a major consideration. The fillet welds in Fig. 18-2-15A are easy to apply and require no special plate preparation.

In comparison, the double-bevel groove weld in Fig. 18-2-15B has about one-half the weld area of the fillet welds. However, it requires extra preparation and the use of smaller-diameter electrodes with lower welding currents to place the initial pass without burning through. As plate thickness increases, this initial low-deposition region becomes a less important factor, and the higher cost factor decreases in significance.

In Fig. 18-2-15C, it will be noted that the single-bevel groove weld requires about the same amount of weld metal as the fillet welds deposited in Fig. 18-2-15A. Thus, there is no apparent economic advantage. There are some disadvantages though. The single-bevel joint requires bevel preparation and initially a lower deposit rate at the root of the joint. From a design standpoint, however, it offers a direct transfer of force through the joint, which means that it is probably better under fatigue loading. Although the illustrated full-strength fillet welds, having leg sizes equal to 75 percent of the plate thickness, would be sufficient, some codes have lower allowable limits for fillet welds and may require a leg size equal to the plate thickness. In this case, the cost of the fillet-welded joint may exceed the cost of a single-bevel groove in thicker plates.

If the joint is so positioned that the weld can be made in a flat position, a single-bevel groove weld may be less expensive than if two fillet welds were specified. As can be seen in Fig. 18-2-16, one of the fillet welds would have to be made in the overhead position—a costly operation.

REFERENCES AND SOURCE MATERIAL

1. American Welding Society.
2. The Lincoln Electric Company.

ASSIGNMENTS

See Assignments 2 and 3 for Unit 18-2 on page 641.

18-3 FILLET WELDS

Fillet Weld Symbols

Figure 18-3-1 shows the fillet weld symbol and its relative position on the reference line. Figures 18-3-2 and 18-3-3 (on pp. 619–620) show applications of the fillet weld and appropriate symbols. In the illustrations that do not have figure numbers, the drawing callout is shown first (top or left side) followed by the interpretation.

ARROW SIDE   OTHER SIDE   BOTH SIDES

FIG. 18-3-1   Fillet weld symbol and its location significance.

1. Dimensions of fillet welds are shown on the same side of the reference line as the weld symbol and shown to the left of the weld symbol.

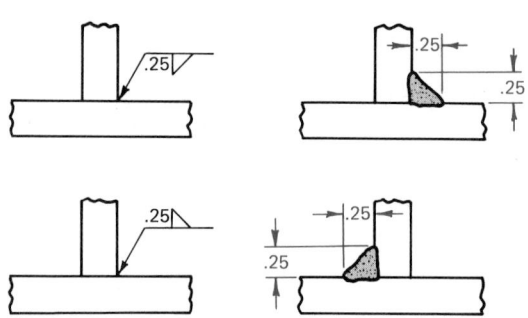

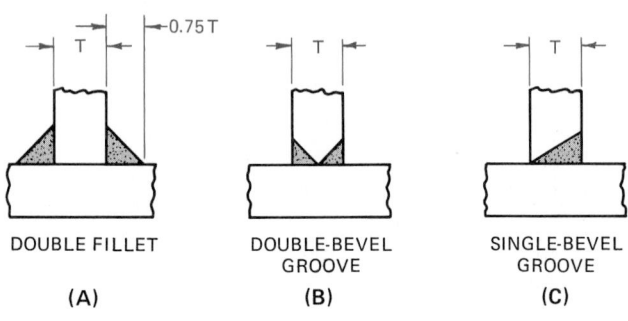

DOUBLE FILLET | DOUBLE-BEVEL GROOVE | SINGLE-BEVEL GROOVE
(A) | (B) | (C)

FIG. 18-2-15   Comparison between fillet and groove welds.

FIG. 18-2-16   In the flat position, a single-groove joint is less expensive than two fillet welds.

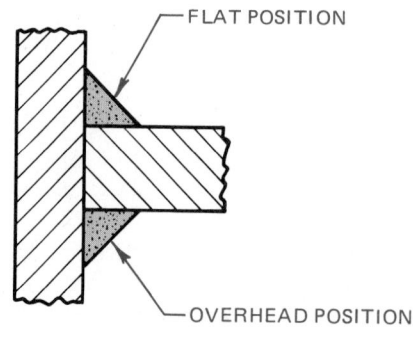

FLAT POSITION

OVERHEAD POSITION

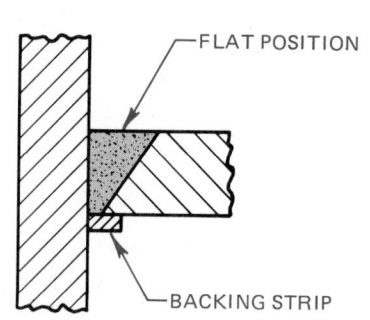

FLAT POSITION

BACKING STRIP

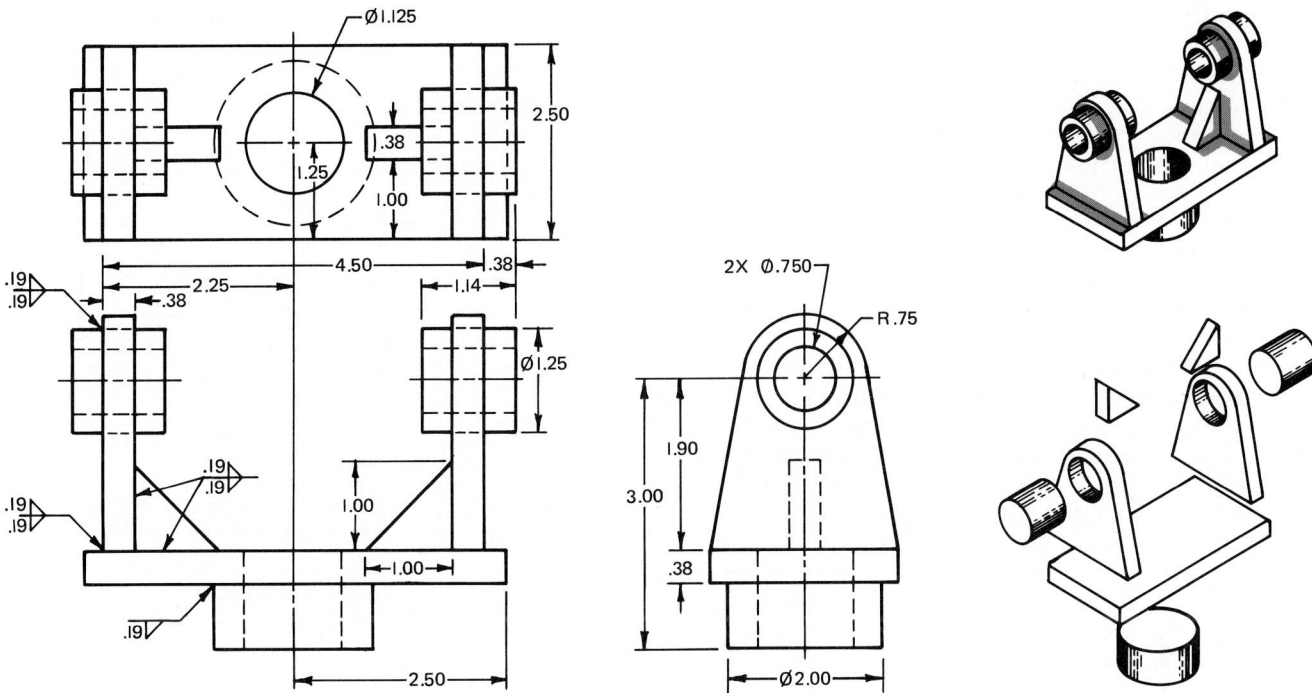

FIG. 18-3-2   Welded steel shaft support.

2. The dimensions of fillet welds on both sides of a joint are specified whether the dimensions are identical or different.

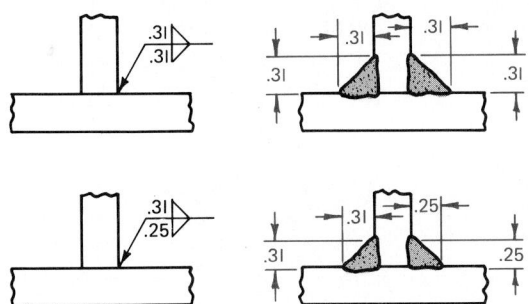

3. When there appears on a drawing a general note governing the dimension of fillet welds, such as ALL FILLET WELDS .25 IN. UNLESS OTHERWISE NOTED, and all the welds have dimensions governed by the note, the dimension need not be shown on the welding symbols.

4. When the dimensions of either arrow side or other side or both welds differ from the dimensions given in the general note, either or both welds are dimensioned.

5. The size of a fillet weld with unequal legs is shown to the left of the weld symbol. Weld orientation is not shown by the symbol. It is shown on the drawing when necessary.

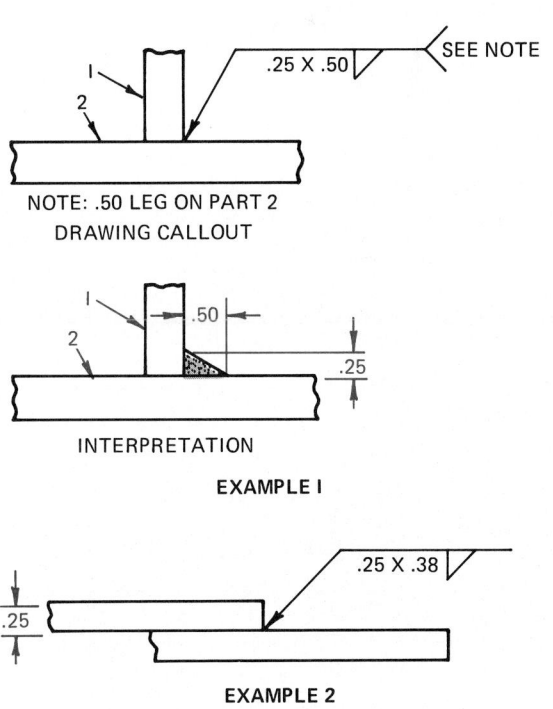

6. The length of a fillet weld, when specified on the welding symbol, is shown to the right of the weld symbol. (See figure at top of pg. 620.)

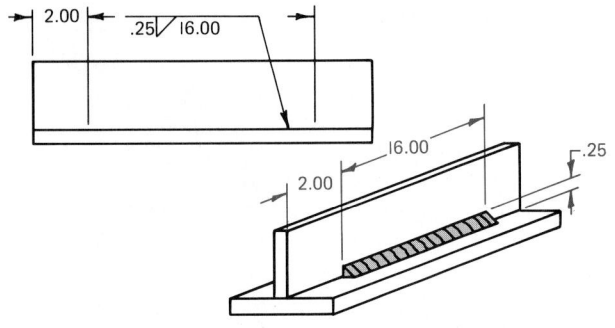

8. The pitch (center-to-center spacing) of intermittent fillet welding is shown as the distance between centers of increments on one side of the joint. It is shown to the right of the length dimension following a hyphen.

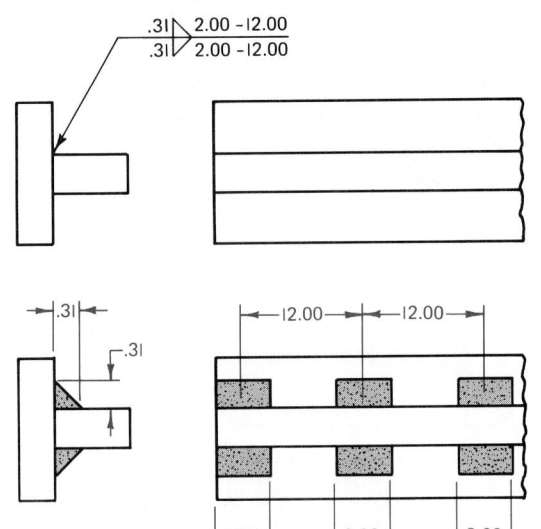

7. Specific lengths of fillet welds may be shown by symbols in conjunction with dimensions lines.

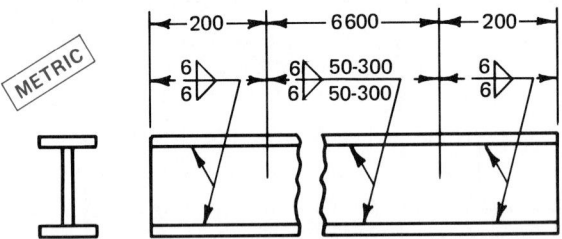

	DIMENSIONS IN INCHES				DIMENSIONS IN MILLIMETERS			
	Strength Design	Rigidity Design				Strength Design	Rigidity Design	
Plate Thickness	Full-strength Weld	50% of Full-strength Weld	33% of Full-strength Weld	Plate Thickness	Full-strength Weld	50% of Full-strength Weld	33% of Full-strength Weld
Up to .25	.12	.12	.12	Up to 6	3	3	3
.25	19	.19	.19	6	5	5	5
.31	.25	.19	.19	8	6	5	5
.38	.31	.19	.19	10	8	5	5
.44	.38	.19	.19	11	10	5	5
.50	.38	.19	.19	12	10	5	5
.56	.44	.25	.25	14	11	6	6
.62	.50	.25	.25	16	12	6	6
.75	.56	.31	.25	20	14	8	6
.88	.62	.38	.31	22	16	10	8
1.00	.62	.38	.31	25	16	10	8
1.12	.88	.44	.31	28	22	11	8
1.25	1.00	.50	.31	32	25	12	8
1.38	1.00	.50	.38	35	25	12	10
1.50	1.12	.56	.38	38	28	14	10

FIG. 18-3-3 Rule-of-thumb fillet weld sizes where the strength of the weld matches the plate.

9. Staggered intermittent fillet welds are shown with the weld symbols staggered.

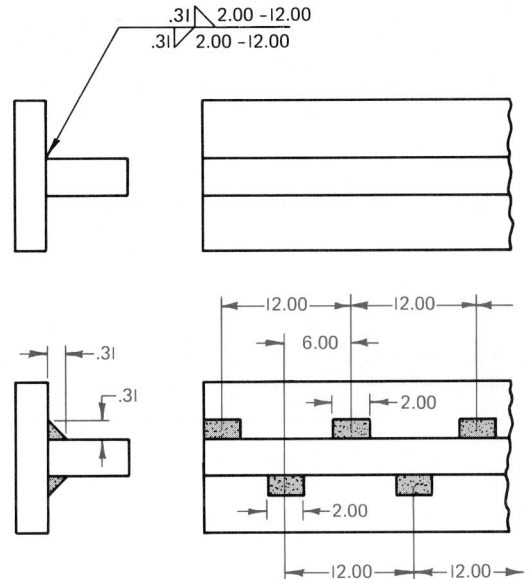

10. Fillet welds that are to be welded with approximately flat, convex, or concave faces without postweld finishing are specified by adding the flat, convex, or concave contour symbol to the weld symbol.

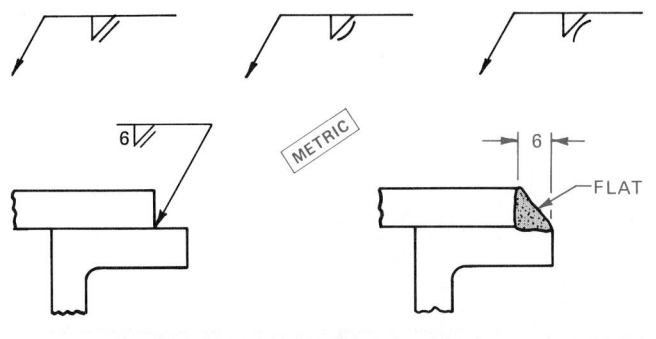

11. Fillet welds that are to be made flat-faced by mechanical means are shown by adding both the flush contour symbol and the user's standard finish symbol.

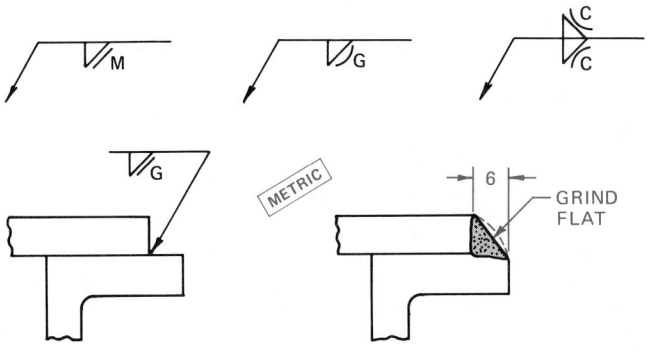

Size of Fillet Welds

Figure 18-3-3 gives the sizing of fillet welds for rigidity designs at various strengths and plate thicknesses, where the strength of the weld metal matches the plate.

In machine design work, where the primary design requirement is rigidity, members are often made with extra-heavy sections, so that improvement under load would be within very close tolerances. The question arises of how to determine the weld size for these types of rigidity designs.

A very practical method is to design the weld for the thinner plate, making it sufficient to carry one-third to one-half the carrying capacity of the plate. This means that if the plate were stressed one-third to one-half its usual value, the weld would be of sufficient size. Most rigidity designs are stressed much below these values. However, any reduction in weld size below one-third the full-strength value would give a weld too small in appearance for general acceptance.

EXAMPLE 1

What size fillet weld is required to match the strength of the fabricated design shown in Fig. 18-3-4A?

SOLUTION

With reference to Fig. 18-3-3, a full-strength weld is required. Thinner plate = .31 in. Fillet weld required = .25 in.

EXAMPLE 2

What size fillet weld is required to hold the rib to the plate shown in Fig. 18-3-4B? Weld design is for rigidity only, and only 33 percent of full-strength weld is required.

SOLUTION

Thinner plate = .31 in. With reference to Fig. 18-3-3, the weld size under rigidity design, 33 percent opposite .31 in., is .19 in.

Figure 18-3-5 (pg. 622) illustrates a cast part that has been redesigned using welded steel members of equal strength and rigidity. The thickness of the ribs and base were reduced by

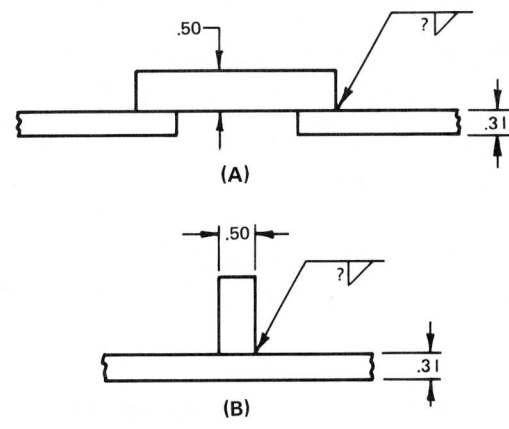

FIG. 18-3-4   Calculating fillet weld size.

approximately 50 percent due to the strength of the steel used. However, certain dimensions must not be altered as the welded steel design may be used as a replacement for the cast iron shaft support. The Ø1.75 in. is a good example of a dimension that should not be altered.

Figure 18-3-6 shows the application of the fillet welds for the shaft support shown in Fig. 18-3-5.

REFERENCES AND SOURCE MATERIAL

1. American Welding Society.
2. The Lincoln Electric Company.

ASSIGNMENTS

See Assignments 4 and 5 for Unit 18-3 on pages 642–644.

18-4   GROOVE WELDS

Use of Break in Arrow of Bevel and J-Groove Welding Symbols

When a bevel or J-groove weld symbol is used, the arrow points with a definite break toward the member that is to be chamfered. In cases where the member to be chamfered is

obvious, the break in the arrow may be omitted (Figs. 18-4-1 through 18-4-3, pp. 623–624).

Groove Weld Symbols

1. Dimensions of groove welds are shown on the same side of the reference line as the welding symbol.

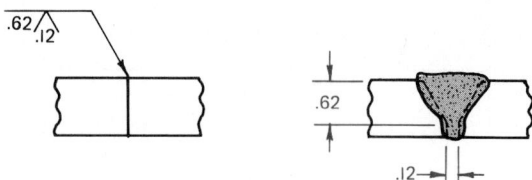

2. When both sides of a double-groove weld have the same dimensions, both sides are dimensioned; however, the root opening needs to appear only once.

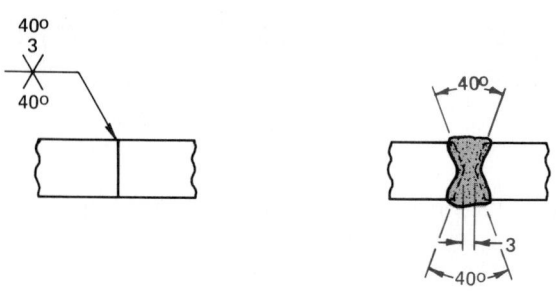

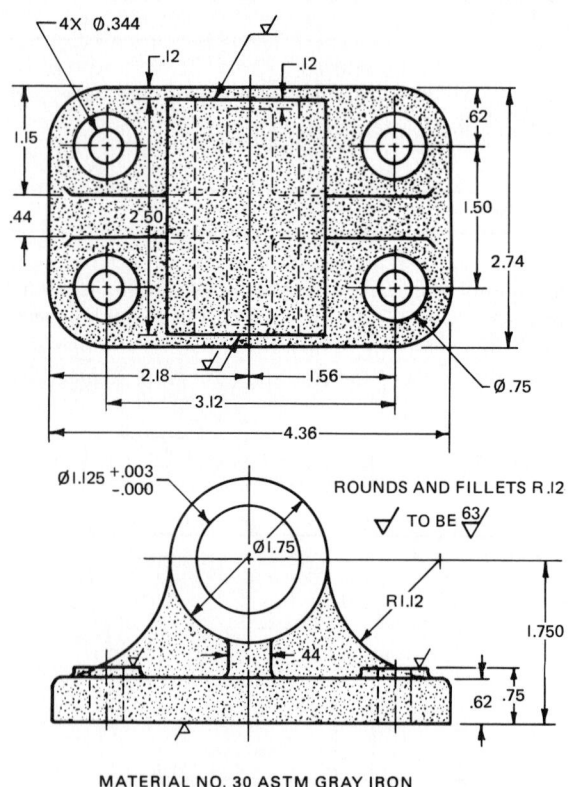

MATERIAL NO. 30 ASTM GRAY IRON

(A) CAST PART

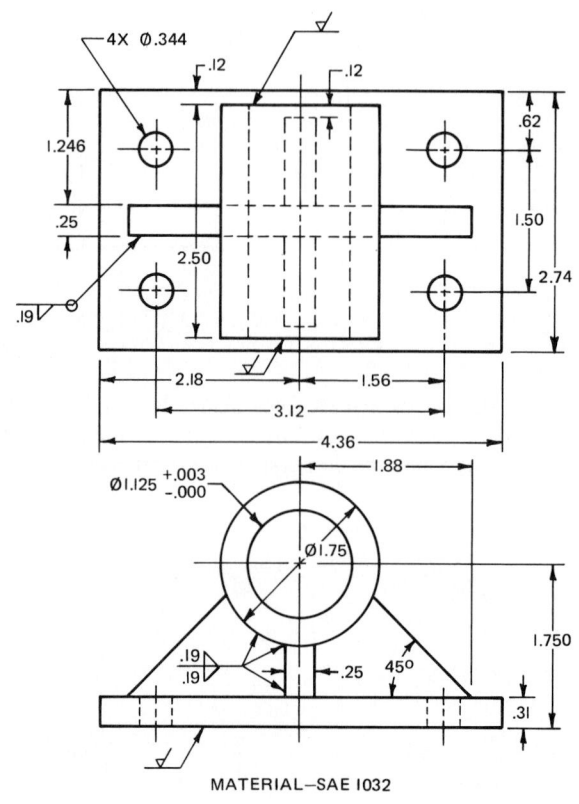

MATERIAL—SAE 1032

(B) WELDED PART

FIG. 18-3-5   Comparison of a cast shaft support with a welded steel shaft support.

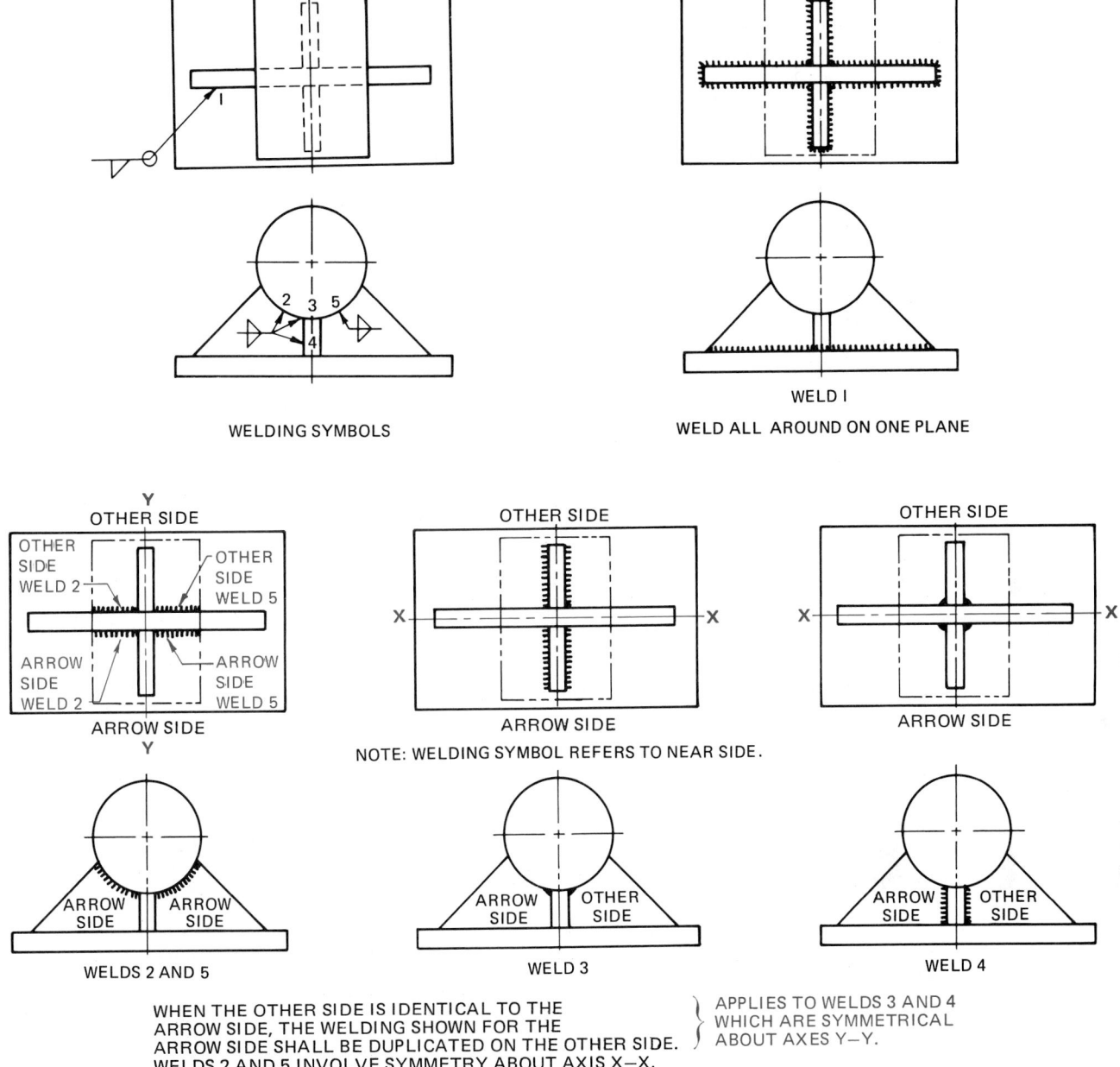

FIG. 18-3-6   Application of fillet weld for shaft support shown in Fig. 18-3-5B.

LOCATION SIGNIFICANCE	SQUARE	V	BEVEL	U	J	FLARE-V	FLARE-BEVEL
ARROW SIDE							
OTHER SIDE							
BOTH SIDES							

FIG. 18-4-1   Basic groove welding symbols and their location significance.

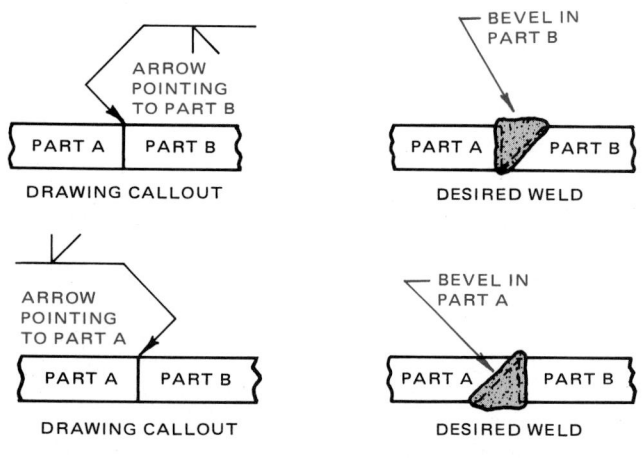

FIG. 18-4-2   Use of break in arrow.

3. When both sides of a double-groove weld differ in dimensions, both are dimensioned; however, the root opening needs to appear only once.

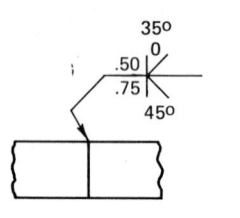

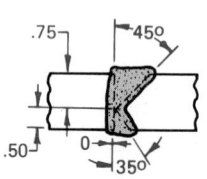

4. When there appears on the drawing a general note governing the dimensions of groove welds, such as ALL V-GROOVE WELDS ARE TO HAVE A 60° ANGLE UNLESS OTHERWISE NOTED, groove welds need not be dimensioned.

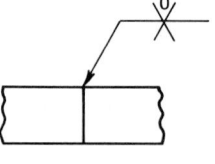

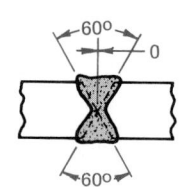

5. For bevel and groove welds, the arrow points with a definite break toward the member being beveled.

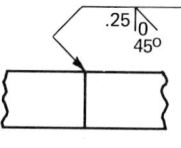

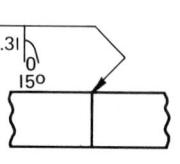

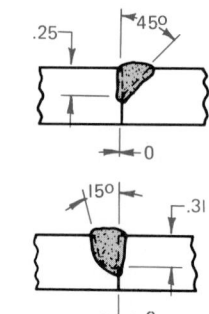

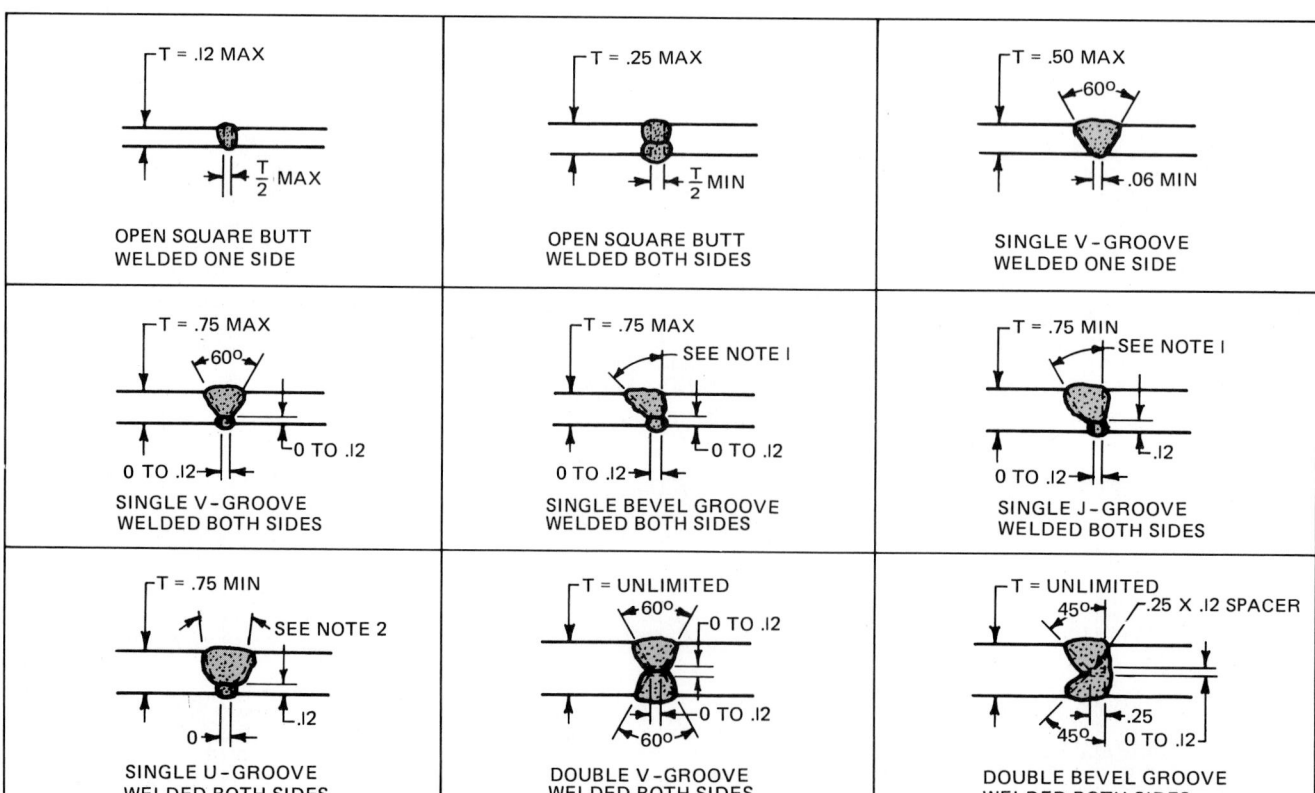

NOTE I: 45° ALL POSITIONS, 30° FLAT AND OVERHEAD ONLY.   NOTE 2: 45° ALL POSITIONS, 20° FLAT AND OVERHEAD ONLY.

FIG. 18-4-3   Spacing and material thickness for common butt joints.

6. When the dimensions of one or both welds differ from the dimensions given in the general note, both welds are dimensioned.

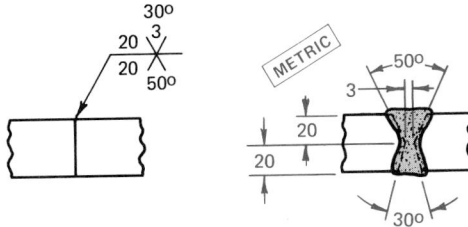

7. The size of groove welds is shown to the left of the weld symbol.

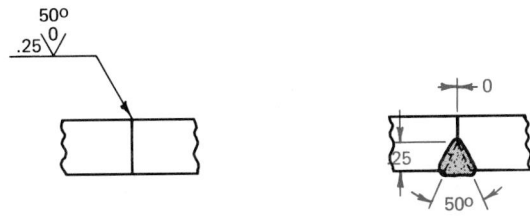

8. When the single-groove and symmetrical double-groove welds extend completely through the member or members being joined, the size of the weld need not be shown on the welding symbol.

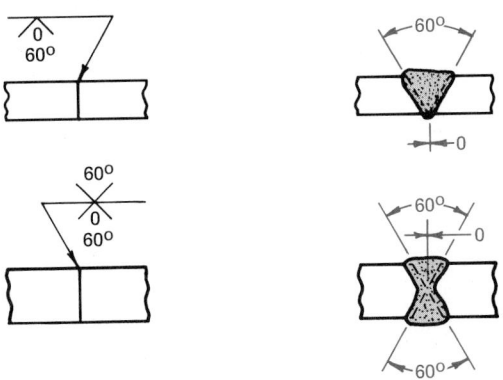

9. When the groove welds extend only partly through the member being joined, the size of the weld is shown on the welding symbol.

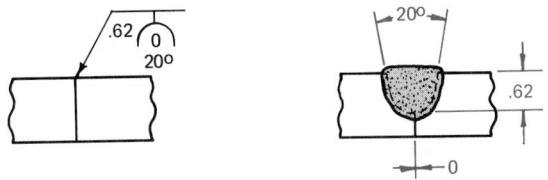

10. The depth of groove preparation and size of a groove weld (shown in parentheses at top of next column), when specified, are placed to the left of the weld symbol. Either or both may be shown (Fig. 18-4-4). Only the groove weld size is shown for square-groove welds.

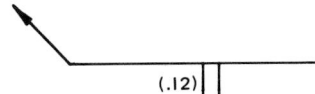

11. The size of flare-groove welds is considered as extending only to the tangent points. The extension beyond the point of tangency is treated as an edge or lap joint (Fig. 18-4-5, pg. 626).

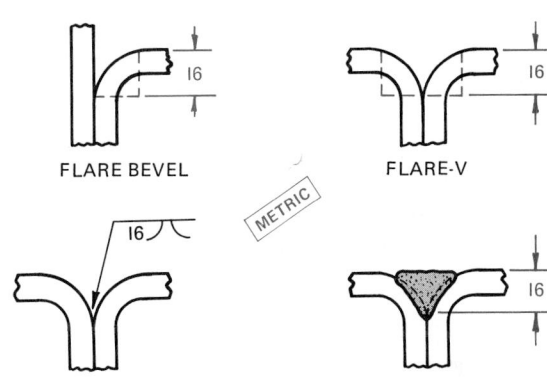

12. Root opening of groove welds is the user's standard unless otherwise indicated. Root opening of groove welds, when not the user's standard is shown inside the weld symbol.

13. Groove angle of groove welds is the user's standard, unless otherwise indicated. Groove angle of groove welds, when not the user's standard, is shown. (See figure at top of pg. 626.)

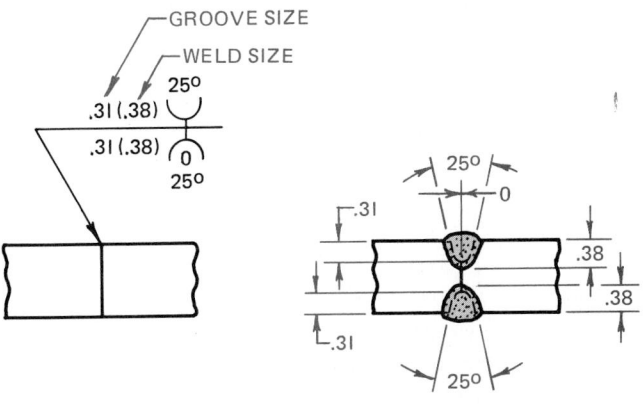

(A) DRAWING CALLOUT   **(B) INTERPRETATION**

FIG. 18-4-4   Groove weld symbol showing use of combined dimensions.

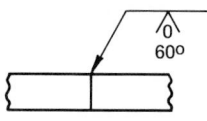
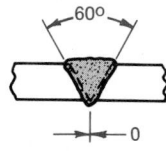
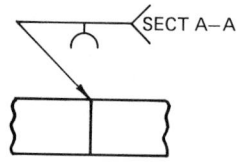
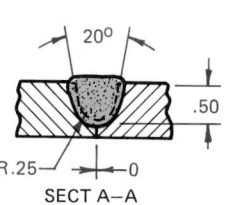

14. The groove radii and root faces of U- and J-groove welds are specified by a cross section, detail, or other data, with reference thereto in the tail of the welding symbol. (See figure at top of next column.)

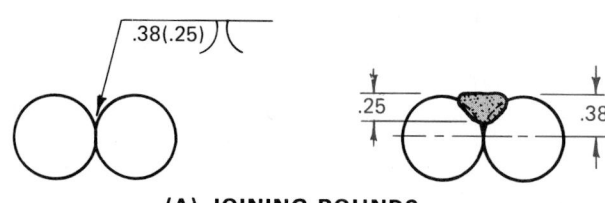

(A) JOINING ROUNDS

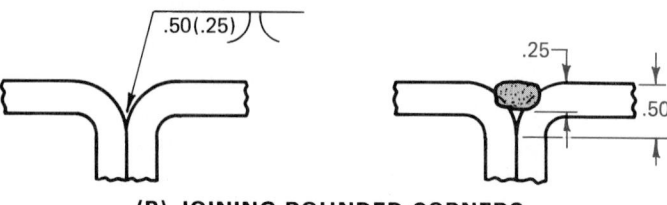

(B) JOINING ROUNDED CORNERS

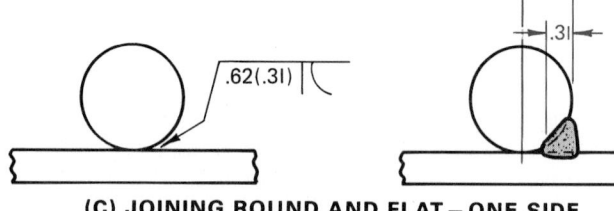

(C) JOINING ROUND AND FLAT — ONE SIDE

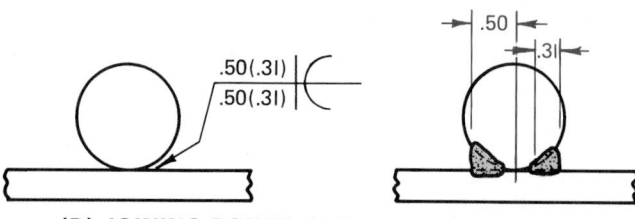

(D) JOINING ROUND AND FLAT — BOTH SIDES

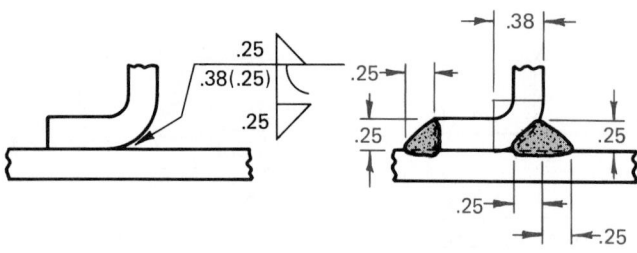

(E) COMBINED WELDS

FIG. 18-4-5   Application of flare-V and flare-bevel welds.

15. Groove welds that are to be welded with approximately flush or convex faces without postweld finishing are specified by adding the flush or convex contour symbol to the welding symbol.

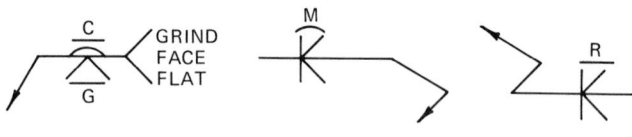

16. Groove welds whose faces are to be finished flush or convex by postweld finishing are specified by adding both the appropriate contour and finishing symbol to the welding symbol. Welds that require a flat but not flush surface require an explanatory note in the tail of the welding symbol.

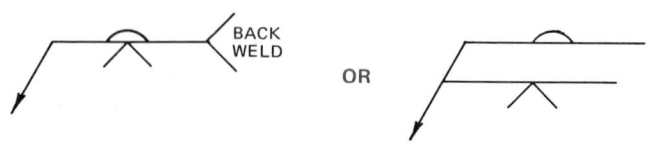

Back and Backing Welds

Back or backing weld symbols are used to indicate bead-type back or backing welds of single-groove welds.

The back and backing weld symbols are identical. The sequence of welding determines which designation applies. The back weld is made after the groove weld and the backing weld is made before the groove weld.

1. The back weld symbol is placed on the side of the reference line opposite a groove weld symbol. When a single reference line is used, BACK WELD is specified in the tail of the symbol. Alternately, if a multiple reference line is used, the back weld symbol is placed on a reference line subsequent to the reference line specifying the groove weld (Fig. 18-4-6A).

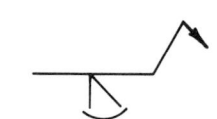

2. The backing weld symbol is placed on the side of the reference line opposite the groove weld symbol. When a

626

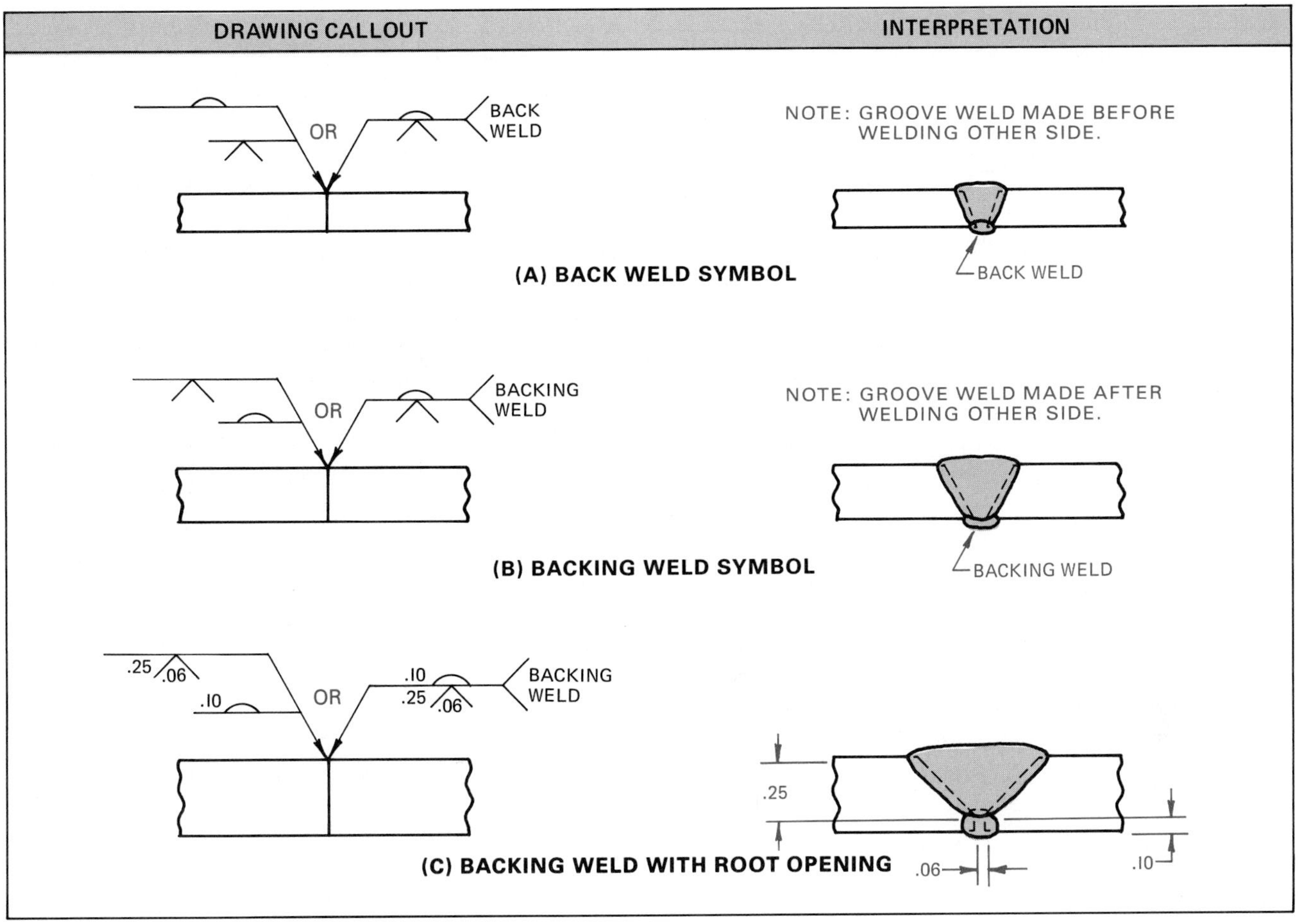

DRAWING CALLOUT	INTERPRETATION

(A) BACK WELD SYMBOL

NOTE: GROOVE WELD MADE BEFORE WELDING OTHER SIDE.

BACK WELD

(B) BACKING WELD SYMBOL

NOTE: GROOVE WELD MADE AFTER WELDING OTHER SIDE.

BACKING WELD

(C) BACKING WELD WITH ROOT OPENING

FIG. 18-4-6   Application of back and backing weld symbols.

single reference line is used, BACKING WELD is specified in the tail of the arrow. If a multiple reference line is used, the backing weld symbol is placed on a reference line prior to that specifying the groove weld (Figs. 18-4-6B and C).

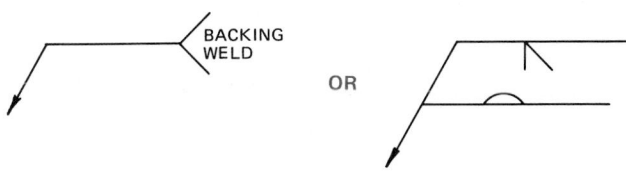

3. Back or backing welds may be welded with approximately flush or convex faces with or without postweld finishing.

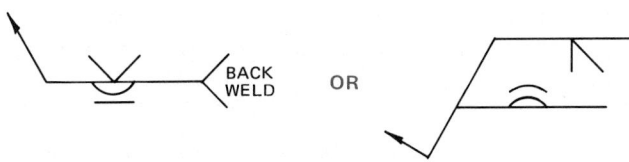

4. Back or backing welds may be finished approximately flush or convex by postweld finishing. Welds that require

a flat but not flush surface require an explanatory note in the tail of the symbol.

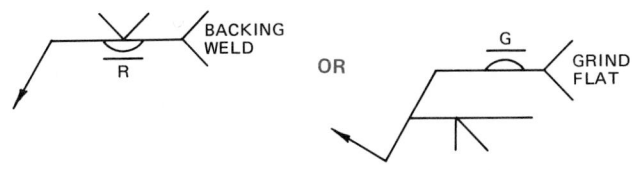

5. A joint with backing is specified by placing the backing symbol on the side of the reference line opposite the groove weld symbol. If the backing is to be removed after the welding, an R is placed in the backing symbol. Material and dimensions of backing are specified in the tail symbol or on the drawing (Fig. 18-4-7A, pg. 628).

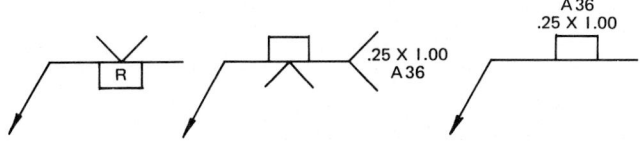

6. A joint with a required spacer is specified with the groove weld symbol modified to show a rectangle within

627

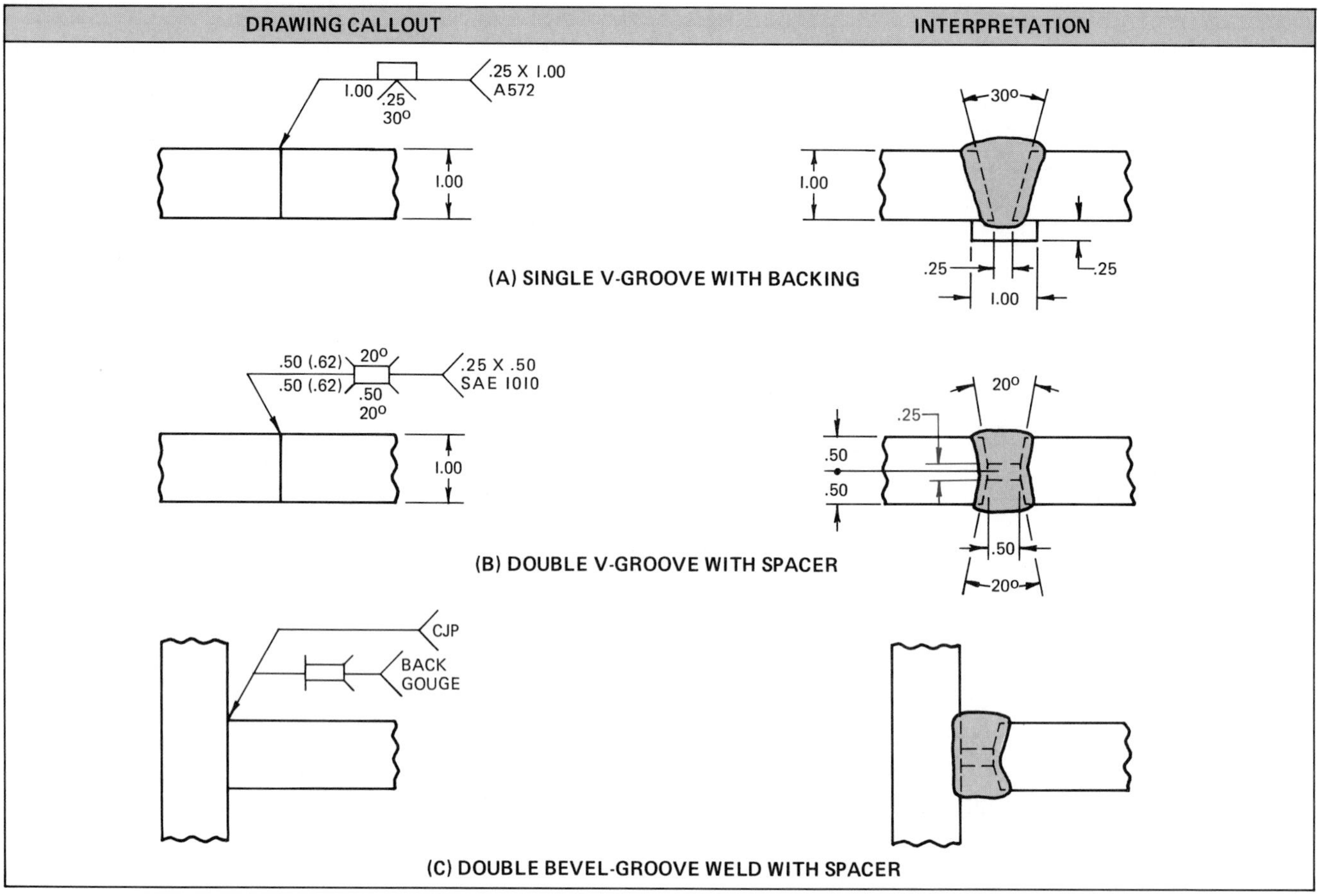

DRAWING CALLOUT	INTERPRETATION

.25 X 1.00
A 572
1.00   .25
30°

1.00

1.00

30°

1.00

.25   .25

1.00

(A) SINGLE V-GROOVE WITH BACKING

.50 (.62)   20°
.50 (.62)   .50
20°

.25 X .50
SAE 1010

1.00

20°

.25

.50

.50

.50

20°

(B) DOUBLE V-GROOVE WITH SPACER

CJP

BACK
GOUGE

(C) DOUBLE BEVEL-GROOVE WELD WITH SPACER

FIG. 18-4-7   Joints with backing and spacers.

it (Fig. 18-4-7B). In case of multiple reference lines, the rectangle need only appear on the reference line nearest the arrow (Fig. 18-4-7C). Material and dimensions of the spacer are specified in the tail of the symbol or on the drawing.

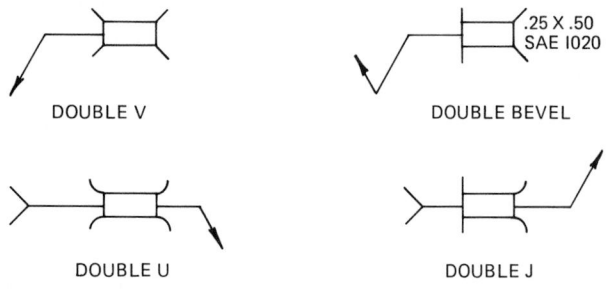

DOUBLE V   DOUBLE BEVEL

.25 X .50
SAE 1020

DOUBLE U   DOUBLE J

7. Consumable inserts are specified by placing the consumable insert symbol on the side of the reference line opposite the groove weld symbol. The AWS consumable insert class is placed in the tail of the symbol.

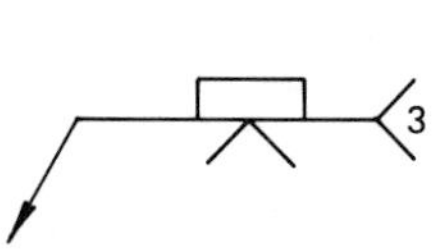

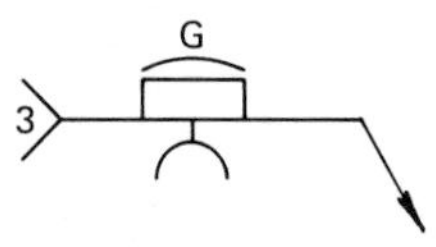

A joint requiring complete penetration involving back gouging may be specified using either a single or multiple reference line symbol. That welding symbol includes a reference to back gouging in its tail and (1) in the case of asymmetrical groove welds must show the depth of preparation from each side (Figs. 18-4-8A and B), together with groove angles and root openings, or (2) in the case of symmetrical groove welds, need not include any other information except the weld symbol (Fig. 18-4-8C), with groove angles and root opening.

Groove Joint Design

Figure 18-4-9 shows that the root opening R is the separation between the members to be formed. A root opening is used for electrode accessibility to the base or root of the joint. The smaller the angle of bevel, the larger the root opening must be to get good fusion at the root. If the root opening is too small, root fusion is more difficult to obtain and smaller electrodes must be used, thus slowing down the welding process.

Figure 18-4-10 shows how the root opening must be increased as the included angle of the bevel is decreased. Backup strips are used on larger root openings. All three preparations are acceptable; all are conducive to good welding procedure and good weld quality. Selection, therefore, is usually based on cost.

Root opening and joint preparation will directly affect weld cost, and selection should be made with this in mind. Joint

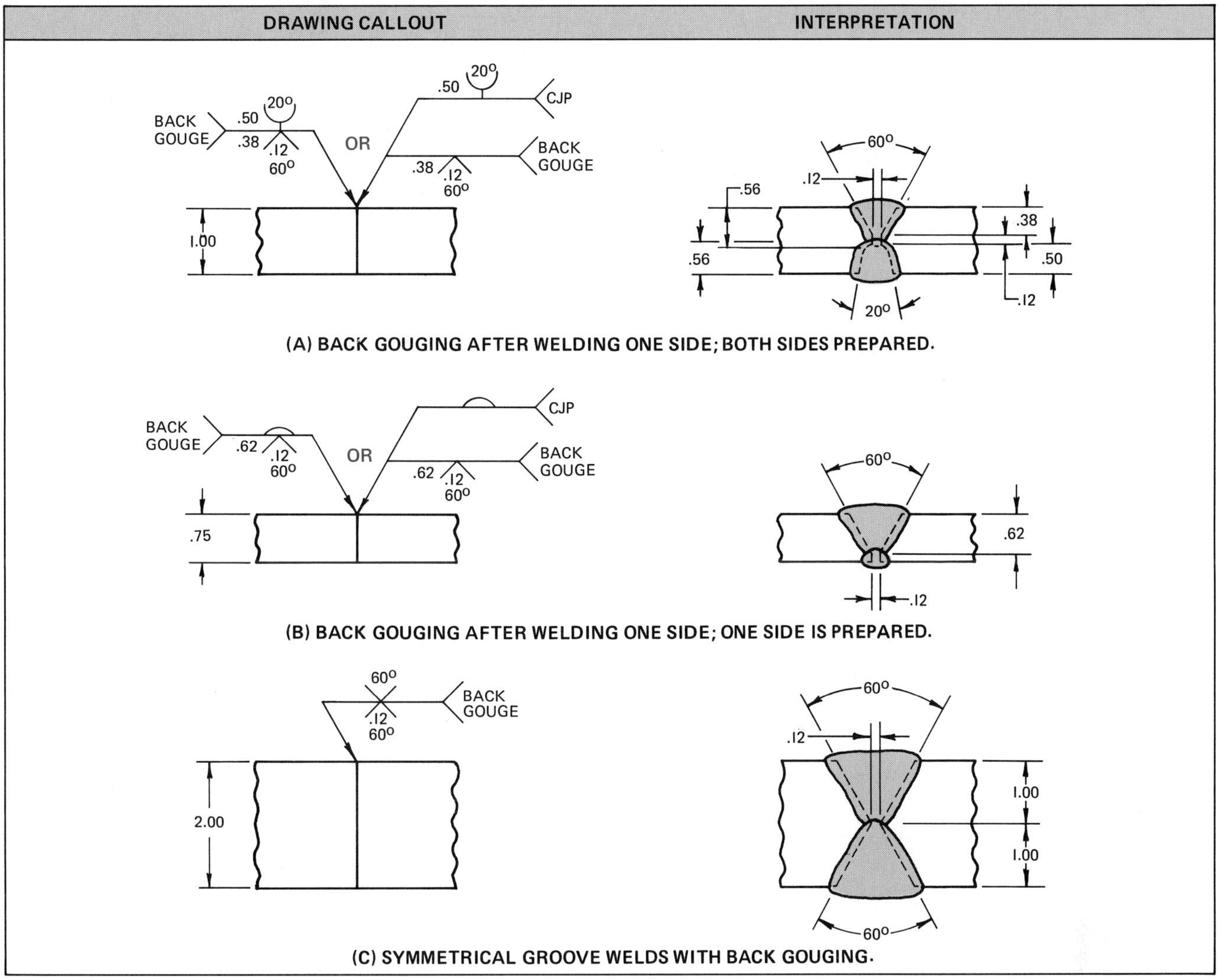

DRAWING CALLOUT	INTERPRETATION

(A) BACK GOUGING AFTER WELDING ONE SIDE; BOTH SIDES PREPARED.

(B) BACK GOUGING AFTER WELDING ONE SIDE; ONE SIDE IS PREPARED.

(C) SYMMETRICAL GROOVE WELDS WITH BACK GOUGING.

FIG. 18-4-8   Groove welds with back gouging.

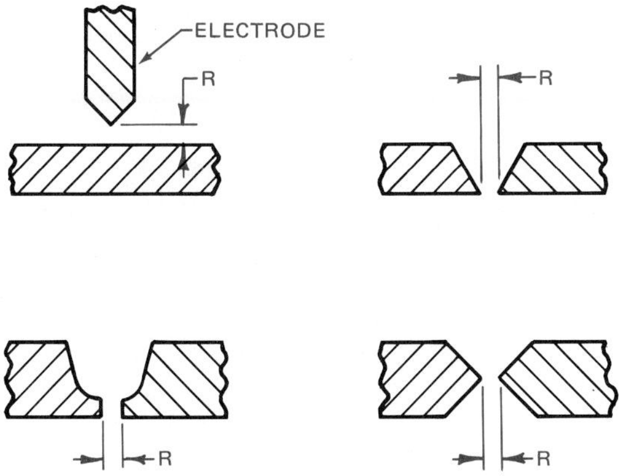

FIG. 18-4-9   Root openings.

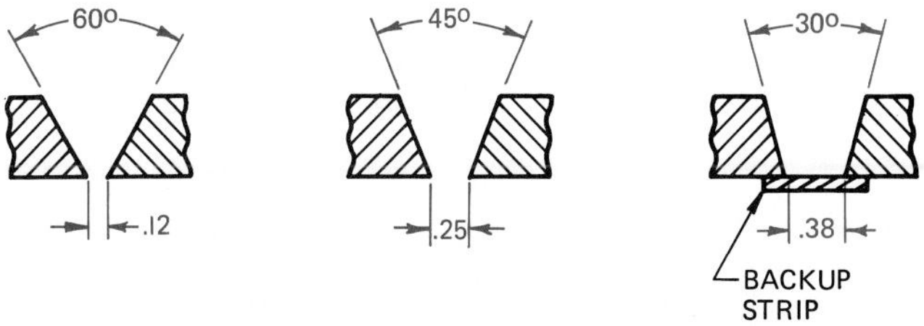

FIG. 18-4-10   Root openings increase as the angle decreases.

SINGLE V   DOUBLE V

FIG. 18-4-11   Single-V weld uses twice as much weld material as double-V weld.

preparation involves the work required on plate edges prior to welding and includes beveling and providing a root face.

Using a double-groove joint in preference to a single-groove, as in Fig. 18-4-11, halves the amount of welding. This reduces distortion and makes possible alternating the weld passes on each side of the joint, again reducing distortion.

REFERENCES AND SOURCE MATERIAL

1. American Welding Society.
2. Lincoln Electric Company.

ASSIGNMENTS

See Assignments 6 and 7 for Unit 18-4 on pages 645–647.

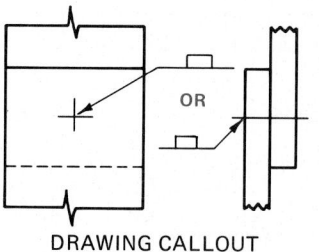
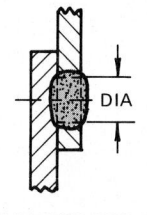

DRAWING CALLOUT   DESIRED WELD

18-5 OTHER BASIC WELDS

In order for engineers to keep abreast of national and international thinking and to reduce the complexity inherent in providing symbols for a variety of ways of making the same type of weld, new and revised weld symbols have been established (Fig. 18-5-1).

Plug Welds

1. Holes in the arrow-side member of a joint for plug welding are specified by placing the weld symbol below the reference line.

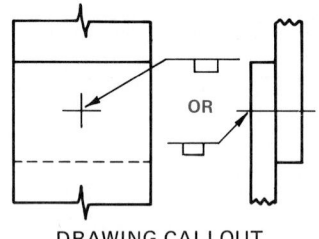
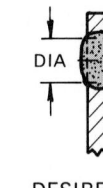

DRAWING CALLOUT   DESIRED WELD

2. Holes in the other-side member of a joint for plug welding are indicated by placing the weld symbol above the reference line. (See figure at top of next column.)

3. The size of a plug weld is shown on the same side and to the left of the weld symbol.

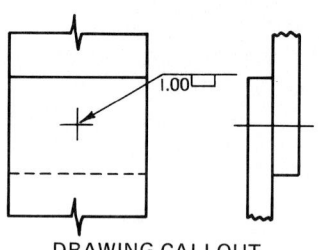
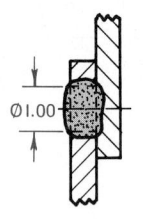

DRAWING CALLOUT   DESIRED WELD

4. The included angle of countersink of plug welds is the user's standard unless otherwise indicated. Included angle, when not the user's standard, is shown.

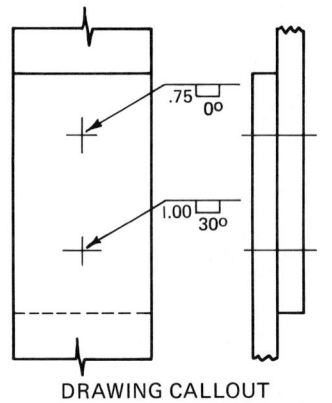
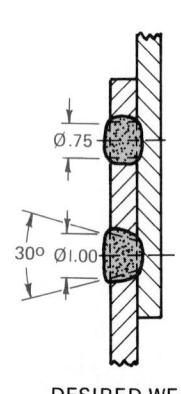

DRAWING CALLOUT   DESIRED WELD

LOCATION SIGNIFICANCE	PLUG OR SLOT	SPOT OR PROJECTION	SEAM	STUD	SURFACING	FLANGE	
						EDGE	CORNER
ARROW SIDE							
OTHER SIDE				NOT USED	NOT USED		
BOTH SIDES	NOT USED	NOT USED	NOT USED	NOT USED	NOT USED	NOT USED	NOT USED
NO ARROW-SIDE OR OTHER-SIDE SIGNIFICANCE	NOT USED			NOT USED	NOT USED	NOT USED	NOT USED

FIG. 18-5-1   Other basic welding symbols and their location significance.

5. The depth of filling of plug welds is complete unless otherwise indicated. When the depth of filling is less than complete, the depth of filling, in inches or millimeters, is shown inside the weld symbol.

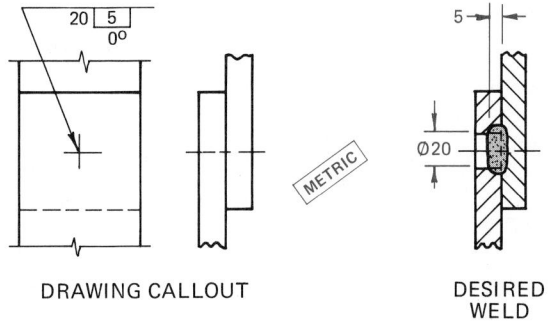

DRAWING CALLOUT   DESIRED WELD

6. Pitch (center-to-center spacing) of plug welds is shown to the right of the weld symbol.

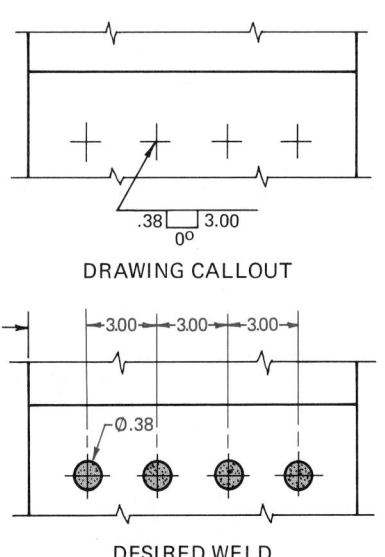

DRAWING CALLOUT

DESIRED WELD

7. Plug welds that are to be welded with approximately flush or convex faces without postweld finishing are specified by adding the flush or convex contour symbol to the weld symbol.

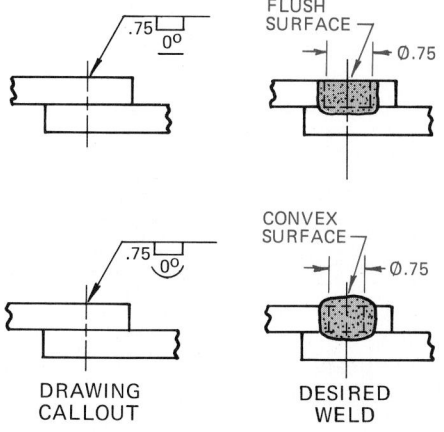

FLUSH SURFACE

CONVEX SURFACE

DRAWING CALLOUT   DESIRED WELD

8. Plug welds may be finished approximately flush or convex by postweld finishing. Welds that require a flat but not flush surface require an explanatory note in the tail of the symbol.

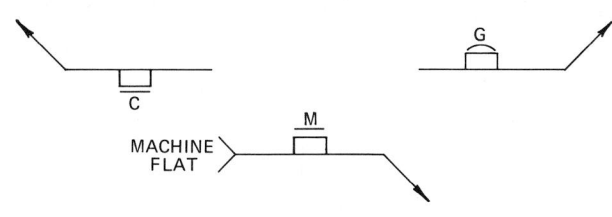

MACHINE FLAT

9. Plug welding symbols may be used to specify welding two or more members to another member. A section view of the joint shall be provided to clarify which members require preparation (Fig. 18-5-2, pg. 632).

Slot Welds

1. Slots in the arrow-side member of a joint for slot welding are indicated by placing the weld symbol below the reference line. Slot orientation must be shown on the drawing.

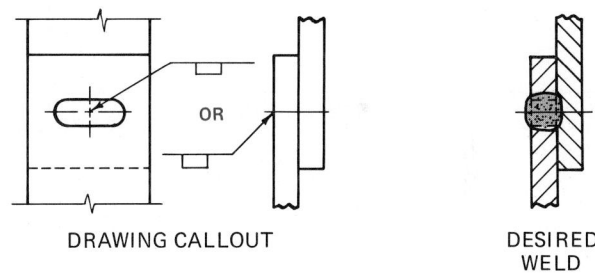

DRAWING CALLOUT   DESIRED WELD

2. Slots in the other-side member of a joint for slot welding are indicated by placing the weld symbol above the reference line.

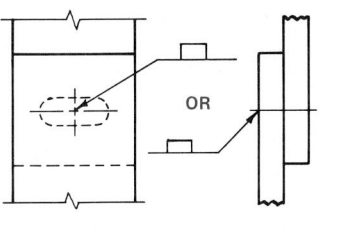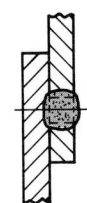

3. Depth of filling of slot welds is complete unless otherwise indicated. When the depth of filling is less than complete, the depth of filling, in inches or millimeters, is shown inside the welding symbol. (See figure at top of pg. 632.)

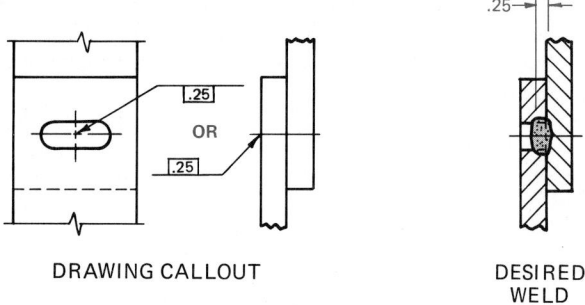

DRAWING CALLOUT   DESIRED
                                                    WELD

4. Length, width, spacing, included angle of countersink, orientation, and location of slot welds should be shown on
   the drawing or by a detail with reference to it on the welding symbol, observing the usual locational significance.
5. Slot welds may be welded with approximately flush or
   convex faces without postweld finishing.

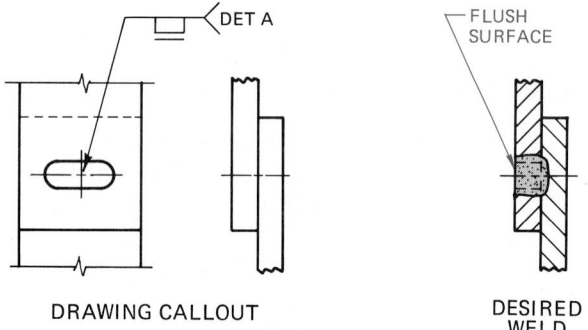

DRAWING CALLOUT   DESIRED
                                                    WELD

6. Slot welds may be finished approximately flush or
   convex by postweld finishing. Welds that require a flat
   but not flush surface require an explanatory note in the
   tail of the symbol (Fig. 18-5-3).

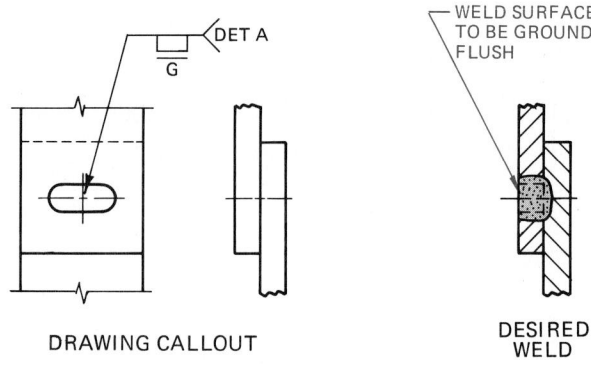

DRAWING CALLOUT   DESIRED
                                                    WELD

Spot Welds

The symbol for all spot welds is a circle, regardless of the
welding process used. There is no attempt to provide symbols
for different ways of making a spot weld, such as resistance,
arc, and electron beam welding (Fig. 18-5-4, pg. 634).

1. When the spot weld symbol is placed below the reference line, it indicates the arrow side. (See figure at top
   of next column.)

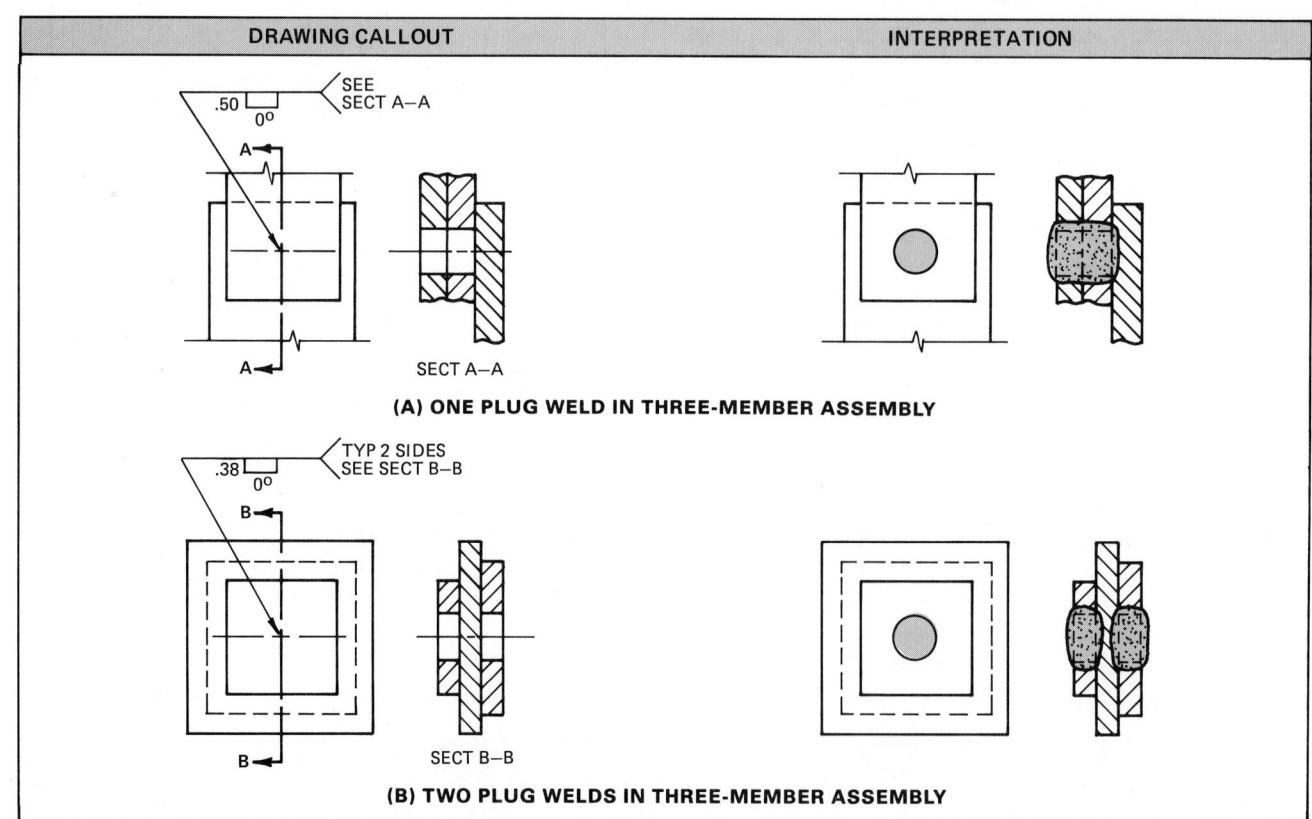

(A) ONE PLUG WELD IN THREE-MEMBER ASSEMBLY

(B) TWO PLUG WELDS IN THREE-MEMBER ASSEMBLY

FIG. 18-5-2   Plug welds for joints involving three or more members.

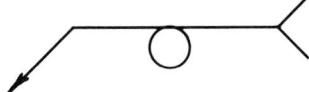

2. If the symbol is above the reference line, it indicates the other side.

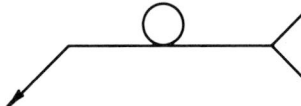

3. If the symbol is on the reference line, it indicates that there is no arrow or other side.

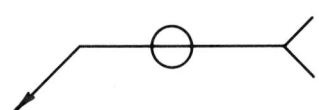

4. Dimensions of spot welds are shown on the same side of the reference line as the weld symbol, or on either side when the symbol is located astride the reference line and has no arrow-side or other-side significance. They are dimensioned by either the size or the strength. The size is designated as the diameter of the weld and is shown to the left of the weld symbol. The strength of the spot weld is designated in pounds (or newtons) per spot and is shown to the left of the weld symbol.

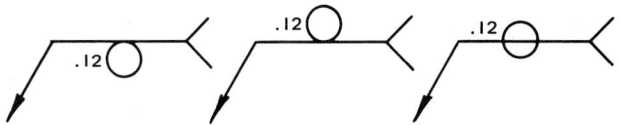

(A) SPECIFYING DIAMETER OF SPOT

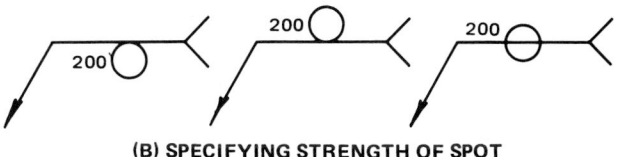

(B) SPECIFYING STRENGTH OF SPOT

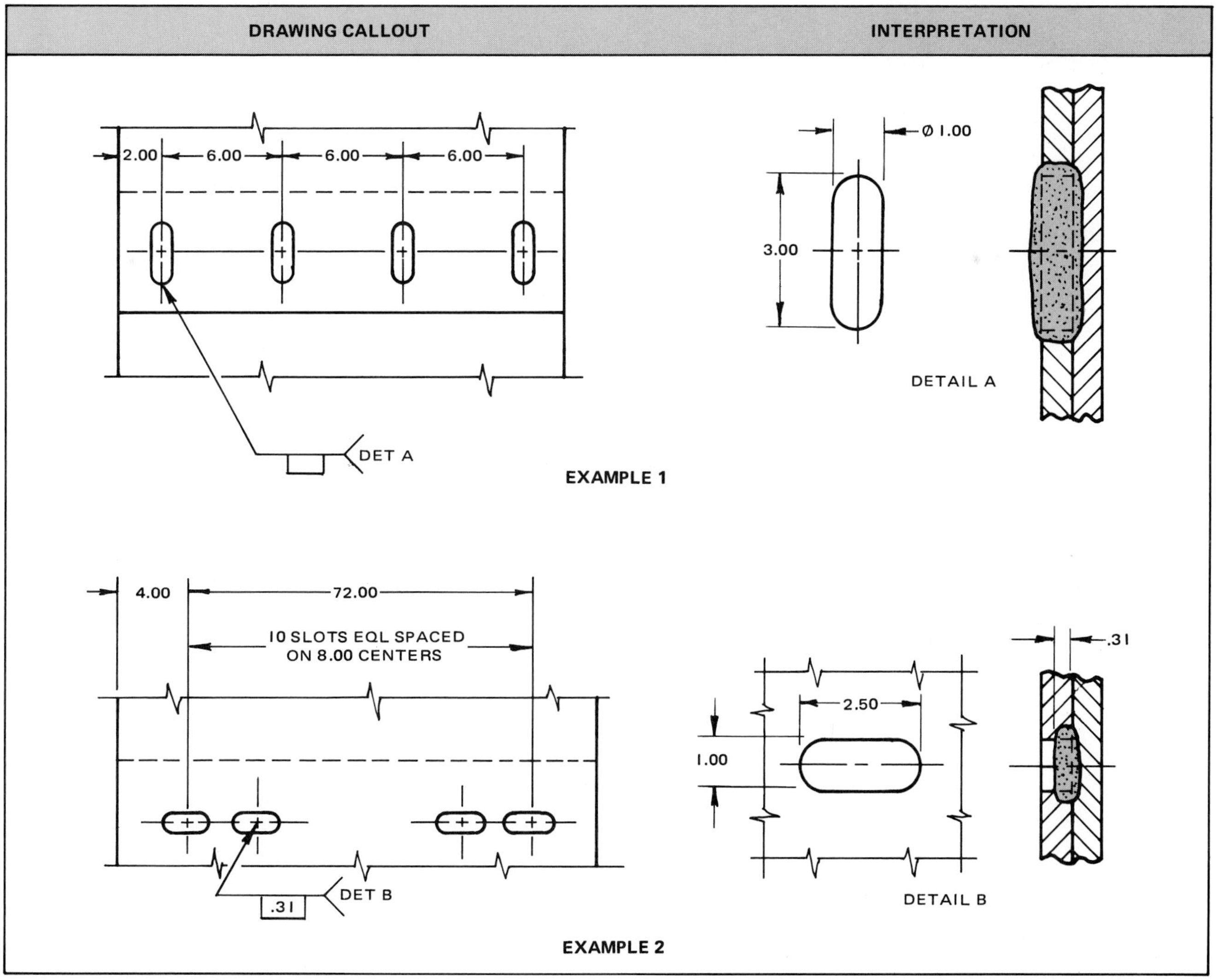

DRAWING CALLOUT	INTERPRETATION

EXAMPLE 1

2.00 — 6.00 — 6.00 — 6.00

DET A

∅ 1.00

3.00

DETAIL A

EXAMPLE 2

4.00 — 72.00

10 SLOTS EQL SPACED ON 8.00 CENTERS

.31   DET B

.31

2.50

1.00

DETAIL B

FIG. 18-5-3   Application of slot weld symbols.

FIG. 18-5-4   Application of spot weld symbols.

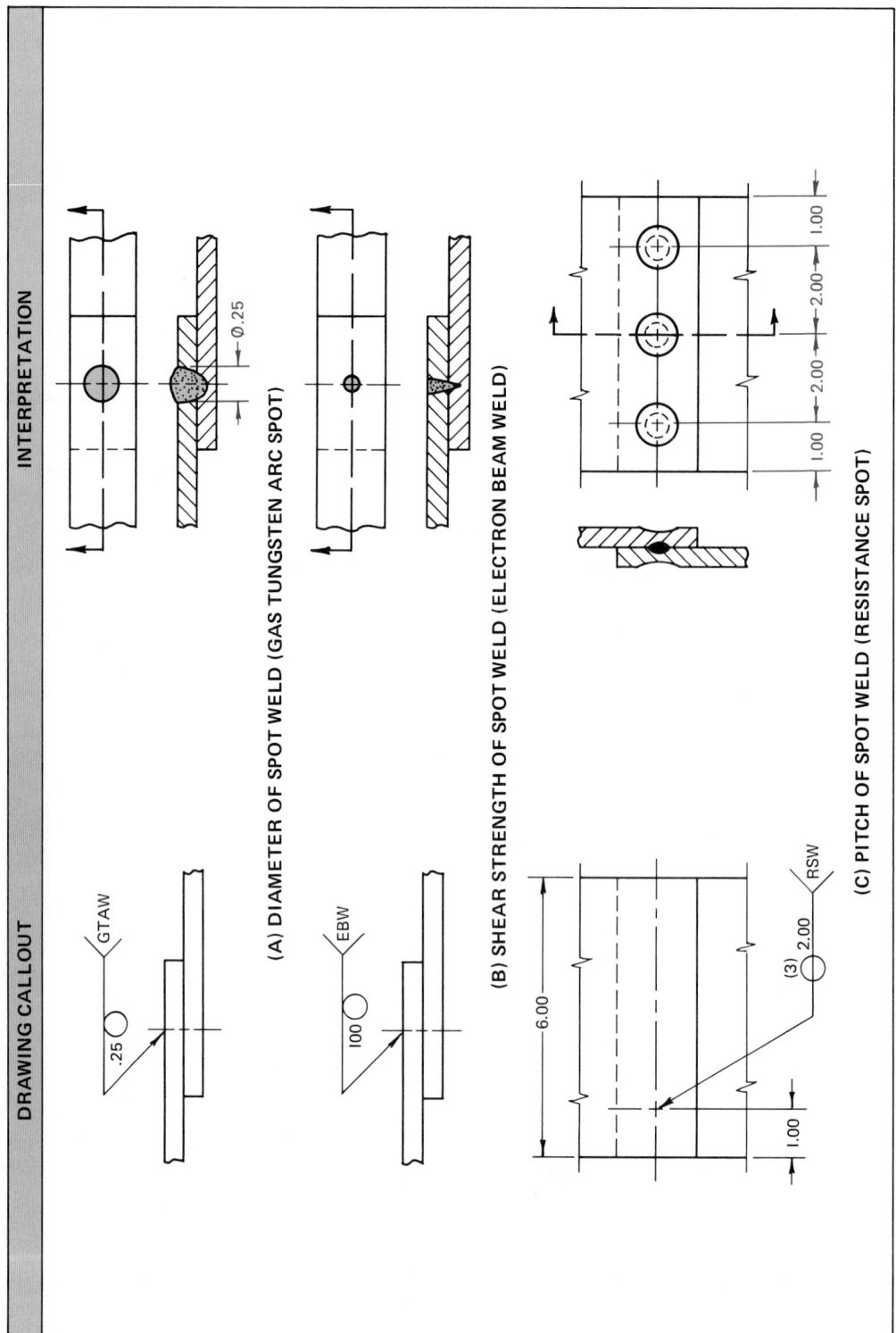

5. The pitch (center-to-center spacing) is shown to the right of the weld symbol.

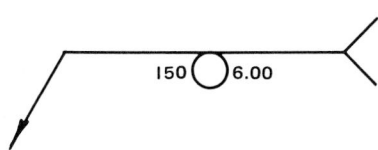

6. When spot welding extends less than the distance between abrupt changes in the direction of the welding or less than the full length of the joint, the extent is dimensioned.

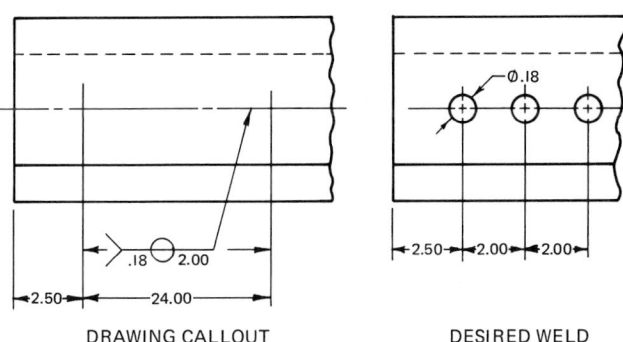

DRAWING CALLOUT   DESIRED WELD

7. When projection welding is used, the projection-welding process is referenced in the tail of the welding symbol. (See figure at top of next column.) The projection weld symbol is placed either above or below (not on) the reference line to designate in which member the embossment is placed (Fig. 18-5-5).

DRAWING CALLOUT   DESIRED WELD

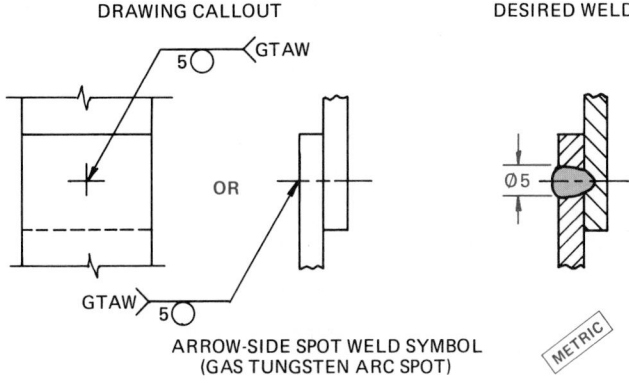

ARROW-SIDE SPOT WELD SYMBOL
(GAS TUNGSTEN ARC SPOT)

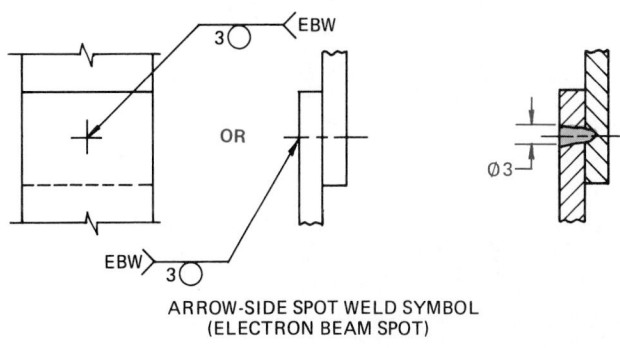

ARROW-SIDE SPOT WELD SYMBOL
(ELECTRON BEAM SPOT)

8. When a definite number of spot welds is desired in a joint, the number is specified in parentheses on the same side of the reference line as that of the weld symbol. The number may be above or below the weld symbol.

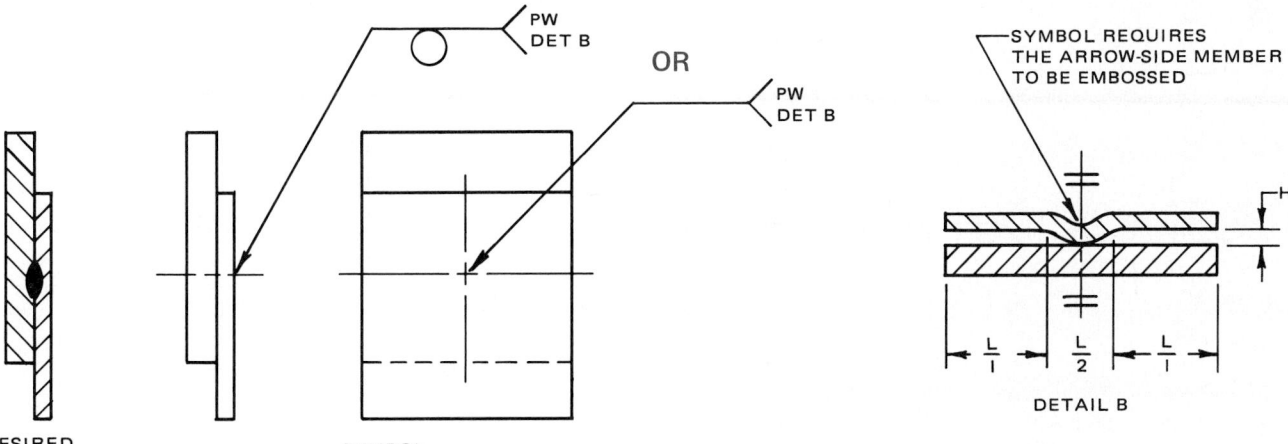

FIG. 18-5-5   Application of projection weld symbol.

9. A group of spot welds may be located on a drawing by intersecting center lines. The arrow points to at least one of the center lines passing through each weld symbol.

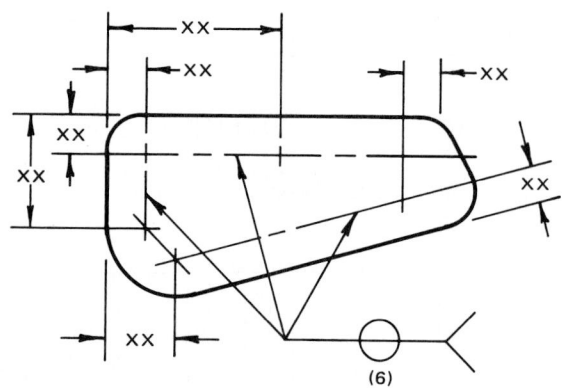

(6)

10. The exposed surface of either member of a spot-welded joint may be welded with approximately flush or convex faces without postweld finishing.

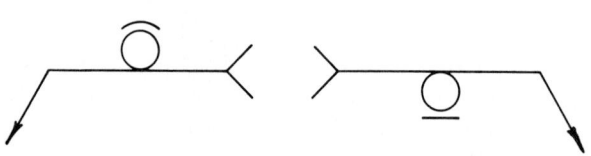

11. Spot welds may be finished approximately flush or convex by postweld finishing. Welds that require a flat but not flush surface require an explanatory note in the tail of the symbol.

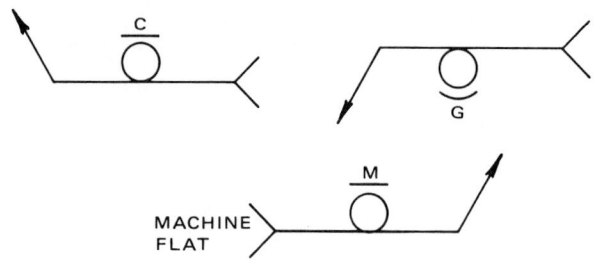

MACHINE
FLAT

Seam Welds

1. The symbol for all seam welds is a circle transversed by two horizontal parallel lines. This symbol is used for all seam welds regardless of the way they are made. The seam weld symbol is placed (1) below the reference line to indicate arrow side, (2) above the reference line to indicate the other side, and (3) on the reference line to indicate that there is no arrow or other side. (See figure at top of next column.)

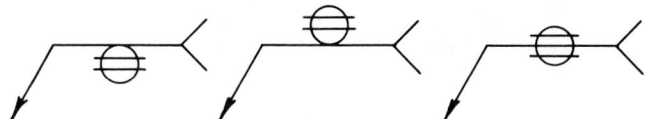

2. Dimensions of seam welds are shown on the same side of the reference line as the weld symbol. They are dimensioned by either size or strength. The size of seam welds is designated as the width of the weld and is shown to the left of the weld symbol. The strength of seam welds is designated in pounds per square inch (psi) or newtons per millimeter (N/mm) and is shown to the left of the weld symbol.

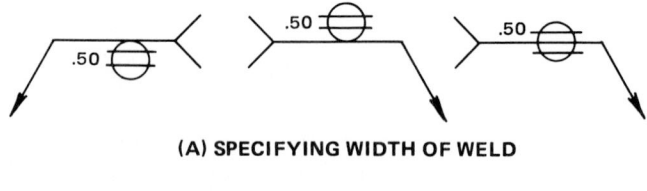

(A) SPECIFYING WIDTH OF WELD

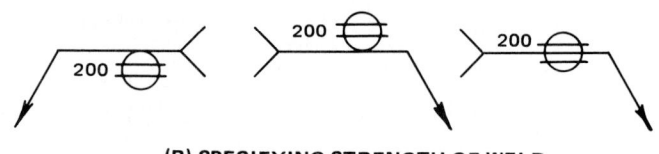

(B) SPECIFYING STRENGTH OF WELD

3. The length of a seam weld, when indicated on the welding symbol, is shown to the right of the weld symbol. When seam welding extends for the full distance between abrupt changes in the direction of the welding, no length dimension needs to be shown on the welding symbol. When a seam weld extends less than the full length of the joint, the extent of the weld should be shown.

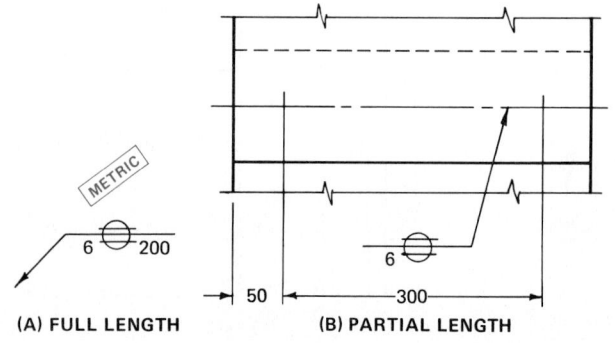

(A) FULL LENGTH   (B) PARTIAL LENGTH

4. The pitch of intermittent seam welds is shown as the distance between centers of the weld increments. The pitch is shown to the right of the length dimension. Unless

otherwise indicated, intermittent seam welds are interpreted as having the length and pitch measured parallel to the axis of the weld.

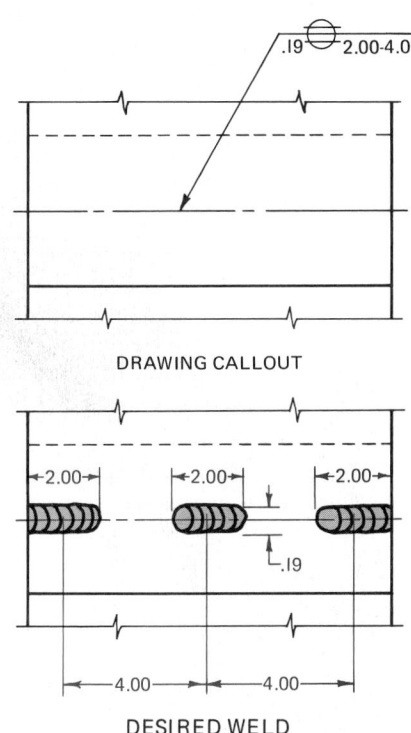

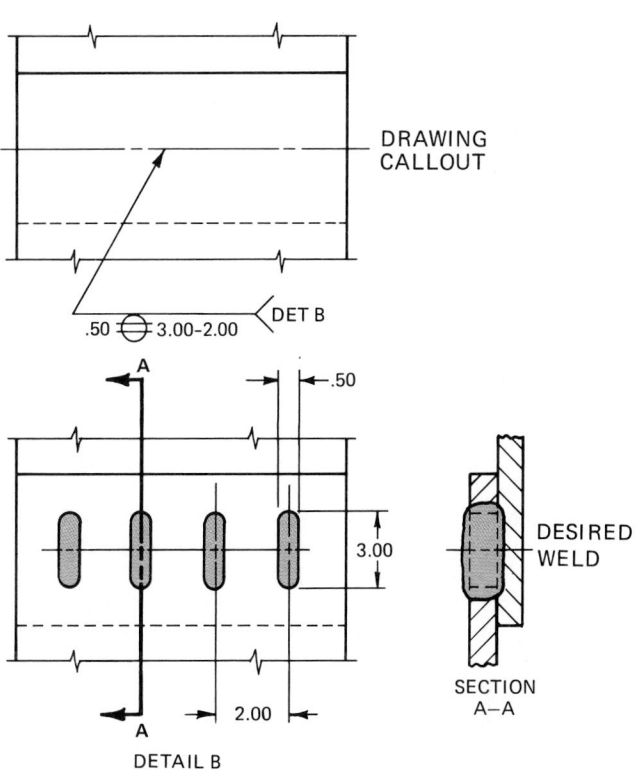

8. The seam weld process is referenced in the tail of the welding symbol.

5. The exposed surface of either member of a seam-welded joint may be welded with approximately flush or convex faces without postweld finishing.

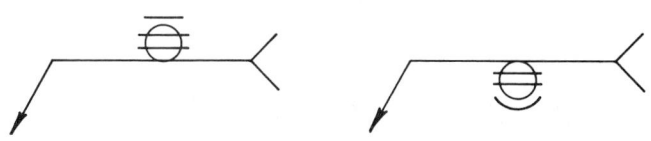

6. Seam welds may be finished approximately flush or convex by postweld finishing.

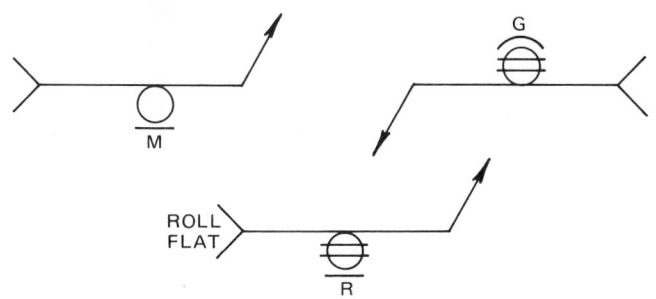

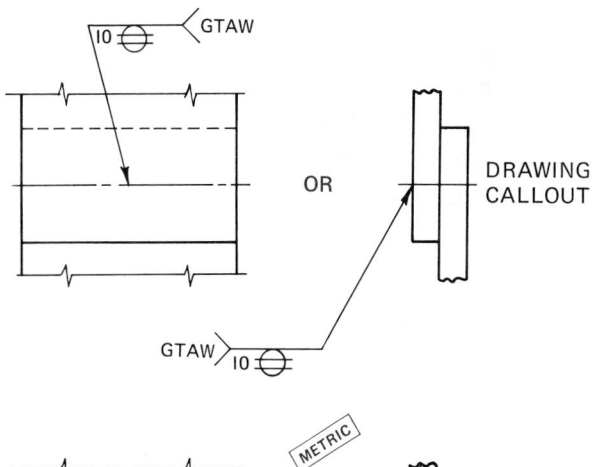

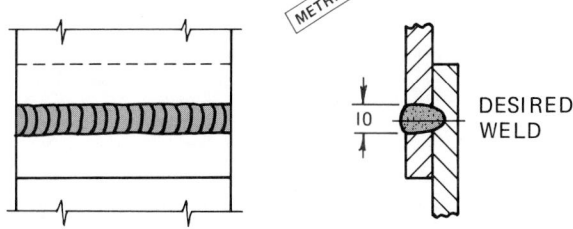

7. When the pitch or length of seam is not parallel to the axis of the weld, it must be shown on the drawing. (See figure at top of next column.)

Surfacing Welds

1. The surfacing weld symbol is used to indicate surfaces built up by welding. Surfaces built up by welding,

637

whether by single- or multiple-pass surfacing welds, are shown by the surfacing weld symbol. The surfacing weld symbol does not indicate the welding of a joint, and hence has no arrow- or other-side significance. This symbol is drawn below the reference line, and the arrow must point clearly to the surface on which the weld is to be deposited.

2. Dimensions used in conjunction with the surfacing weld symbol are shown on the same side of the reference line as the weld symbol. The size or thickness of the surface built up by welding is indicated by showing the minimum thickness of the weld deposit to the left of the weld symbol. When no specific thickness of weld deposit is desired, no size dimension need be shown on the welding symbol. When the entire area of a plane or curved surface is to be built up by welding, no dimension other than size (thickness of deposit) need be shown on the welding symbol.

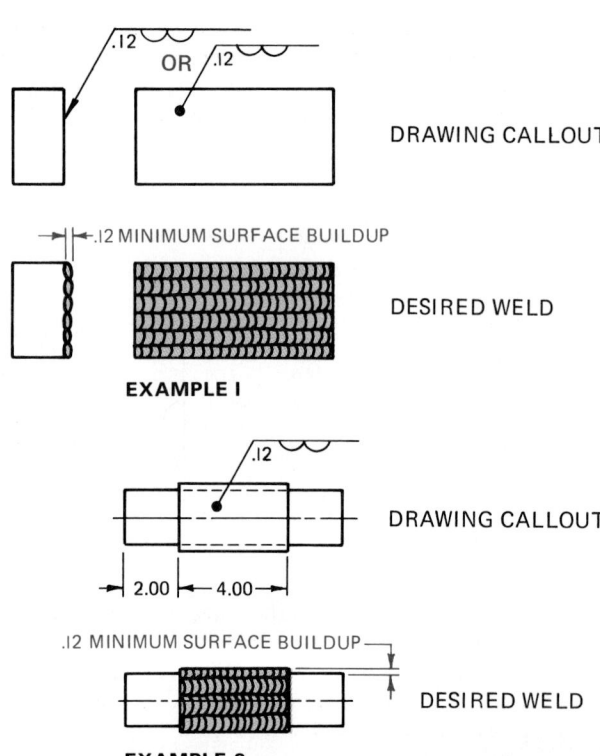

DRAWING CALLOUT

.12 MINIMUM SURFACE BUILDUP

DESIRED WELD

EXAMPLE 1

DRAWING CALLOUT

2.00   4.00

.12 MINIMUM SURFACE BUILDUP

DESIRED WELD

EXAMPLE 2

3. The direction of welding may be specified by a note in the tail of the welding symbol or indicated on the drawing.

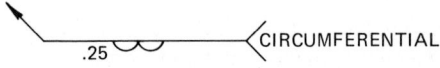

.25   CIRCUMFERENTIAL

4. Multiple-layer surfacing welds may be specified by using multiple reference lines with the required size (thickness) of each layer placed to the left of the weld symbol.

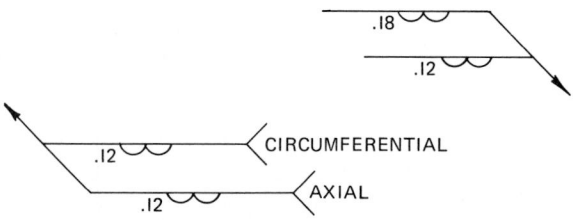

.18

.12

.12   CIRCUMFERENTIAL

.12   AXIAL

Flanged Welds

The following welding symbols are intended to be used for light-gage metal joints involving the flaring or flanging of the edges to be joined.

1. Edge-flange welds are shown by the edge-flange weld symbol. This symbol has no both-sides significance.

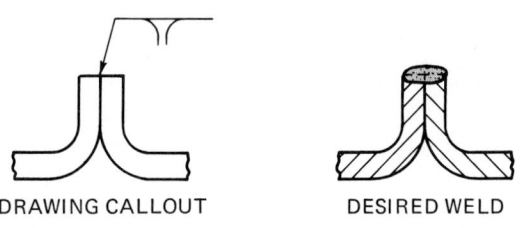

DRAWING CALLOUT   DESIRED WELD

2. Corner-flange welds are shown by the corner-flange weld symbol. This symbol has no both-sides significance.

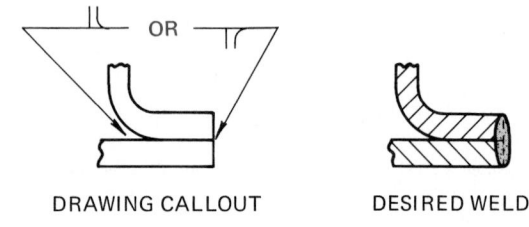

OR

DRAWING CALLOUT   DESIRED WELD

3. Corner-flange welds on joints not detailed on the drawing are specified by the corner-flange weld symbol. A broken arrow points to the member to be flanged. This symbol does not have both-sides significance.

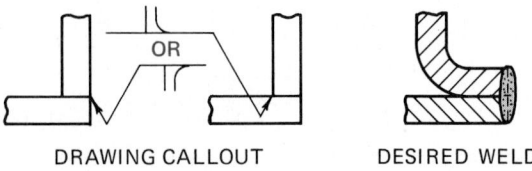

OR

DRAWING CALLOUT   DESIRED WELD

4. Edge-flange welds requiring complete joint penetration are specified by the edge-flange weld symbol with the

melt-through symbol placed on the opposite side of the reference line. The same welding symbol is used for joints either detailed or not detailed on the drawing.

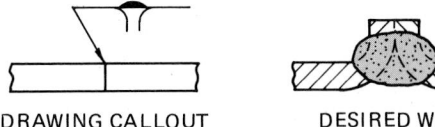

DRAWING CALLOUT   DESIRED WELD

5. Dimensions of flange welds are shown on the same side of the reference line as the weld symbol. The radius and the height above the point of tangency are indicated by showing both the radius and the height, separated by a plus mark, and placed to the left of the weld symbol. The radius and the height read in that order from left to right along the reference line.

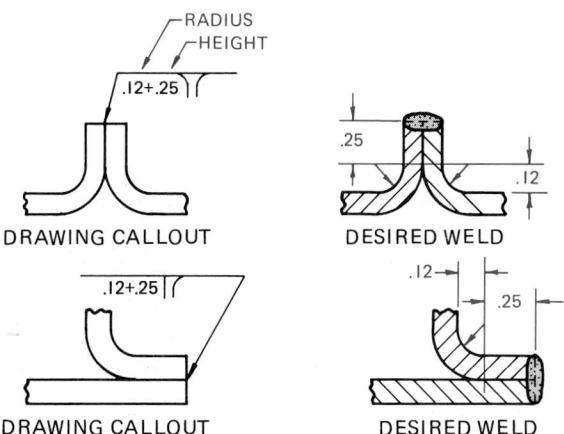

6. The size (thickness) of flange welds is shown by a dimension placed outward of the flanged dimensions.

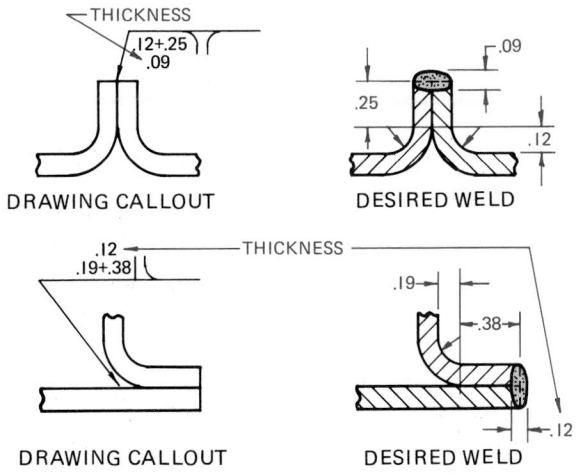

7. Root opening of flange welds is not shown on the welding symbol. If it is desired to specify this dimension, it is shown on the drawing.

8. For flange welds, when one or more pieces are inserted between the two outer pieces, the same welding symbol as for the two outer pieces is used regardless of the number of pieces inserted.

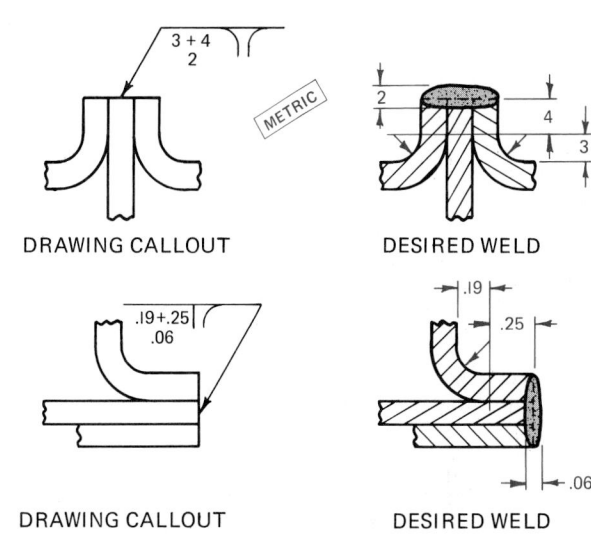

9. Corner-flange welds requiring complete joint penetration are specified by the corner-flange weld symbol with the melt-through symbol placed on the opposite side of the reference line. A broken arrow points to the member to be flanged.

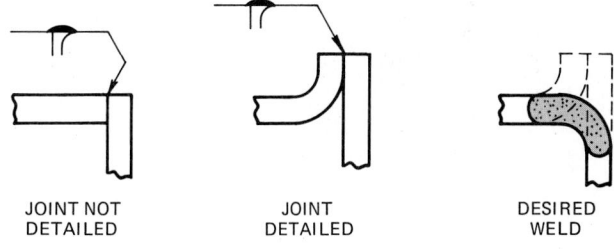

JOINT NOT   JOINT   DESIRED
DETAILED   DETAILED   WELD

Stud Welds

1. The stud weld symbol does not indicate the welding of a joint in the ordinary sense and therefore has no arrow- or other-side significance. The symbol is placed below the reference line and the arrow points clearly to the surface to which the stud is to be welded.

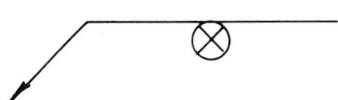

2. The required diameter of the stud is specified to the left of the weld symbol. (See figure at top of pg. 640.)

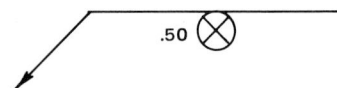

3. The pitch (center-to-center distance) of stud welds in a straight line, if specified, is placed to the right of the weld symbol. The spacing of stud welds in any configuration other than a straight line is dimensioned on the drawing.

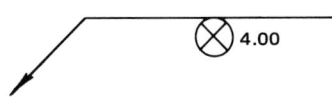

4. The number of stud welds is specified in parentheses below the stud weld symbol.

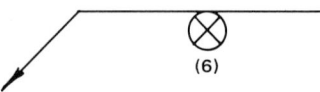

5. The location of the first and last stud weld in each single line is specified on the drawing. (See figure at top of next column.)

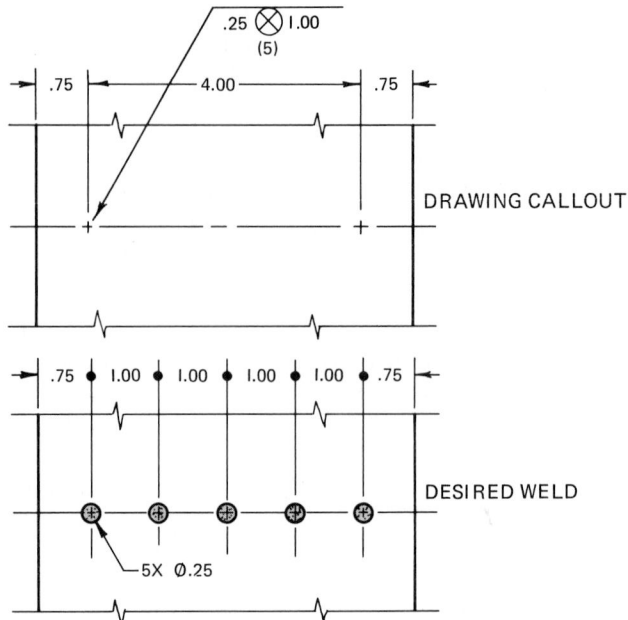

DRAWING CALLOUT

DESIRED WELD

REFERENCES AND SOURCE MATERIAL

1. American Welding Society.

ASSIGNMENTS

See Assignments 8 through 11 for Unit 18-5 on pages 648 through 651.

ASSIGNMENTS FOR CHAPTER 18

ASSIGNMENT FOR UNIT 18-1, DESIGNING FOR WELDING

1. Redesign one of the cast parts shown in Fig. 18-1-A or 18-1-B for fabrication by welding, using standard metal sizes and shapes. Make a detail assembly drawing. Welding symbols or sizes are not required. Include on the drawing an item list and identify each part on the assembly. Scale 1:1.

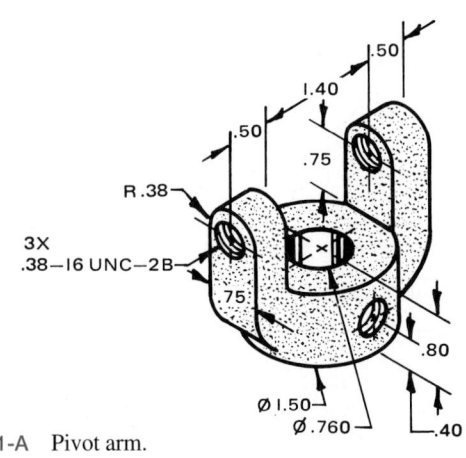

FIG. 18-1-A Pivot arm.

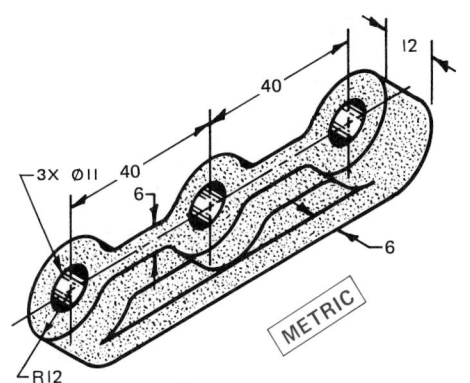

FIG. 18-1-B   Link.

ASSIGNMENTS FOR UNIT 18-2, WELDING SYMBOLS

2. Complete the enlarged views of the welded joints of the drawing callouts shown in Fig. 18-2-A. Use notes to explain any additional welding requirements.
3. Add the information shown above Fig. 18-2-B to the seven welding symbols shown in this assignment.

FIG. 18-2-A   Showing weld type and proportion on drawings.

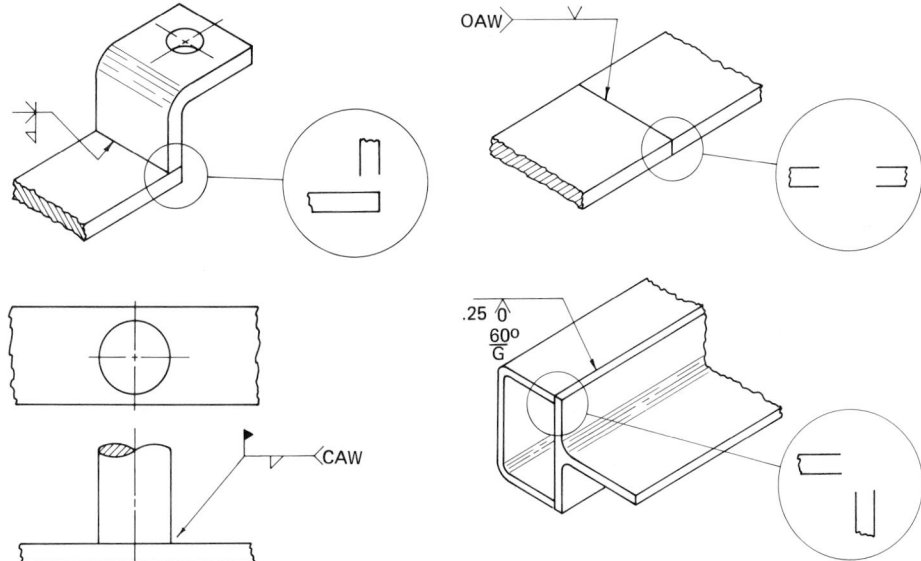

FIG. 18-2-B   Indicating welding symbols on drawings.

Weld	Welding Process Required	Type of Weld	Additional Requirements
1	Carbon Arc Welding	Bevel	
2	Oxyacetylene Welding	Durable Fillet	Both Sides Field Weld
3	Oxyacetylene Welding	Fillet	Both Sides
4	No Specifications Required	J-Groove	
5	Carbon Arc Welding	Fillet	All Around
6	Carbon Arc Welding	Fillet	All Around Field Weld
7	Gas Metal Arc Welding	Double V-Groove	

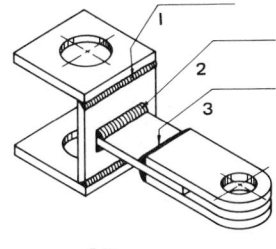

BEVEL AND
FILLET WELDS

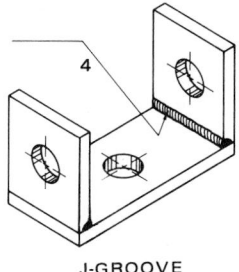

J-GROOVE
WELD

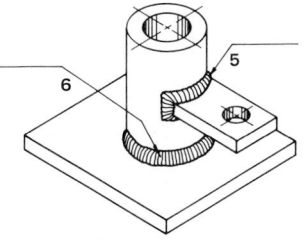

FILLET WELD
ALL AROUND

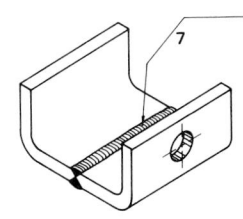

DOUBLE V-GROOVE
WELD

ASSIGNMENTS FOR UNIT 18-3, FILLET WELDS

4. Select one of the problems shown in Figs. 18-3-A through 18-3-D and make a working drawing complete with

dimensions and welding symbols. Include on the drawing an item list and identify each part of the assembly. Use full-strength welds. Scale 1:1.

FIG. 18-3-A   Slide bracket.

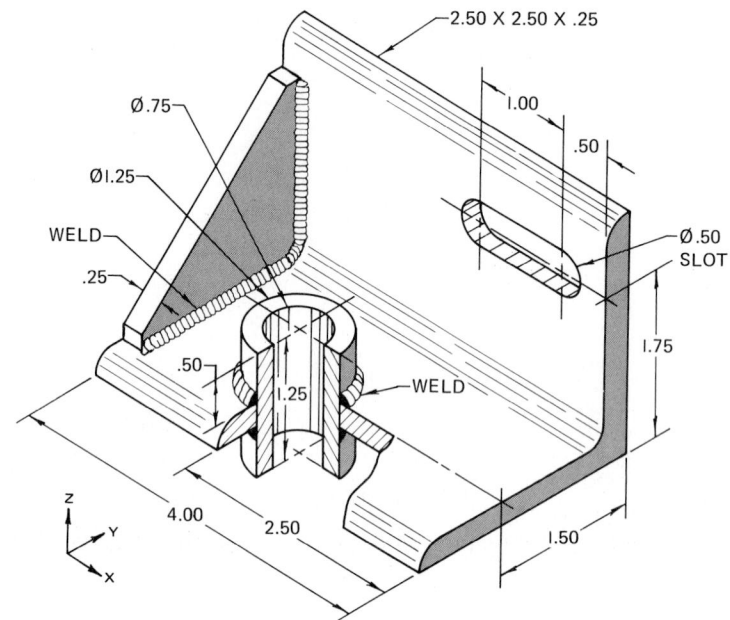

FIG. 18-3-B   Caster frame.

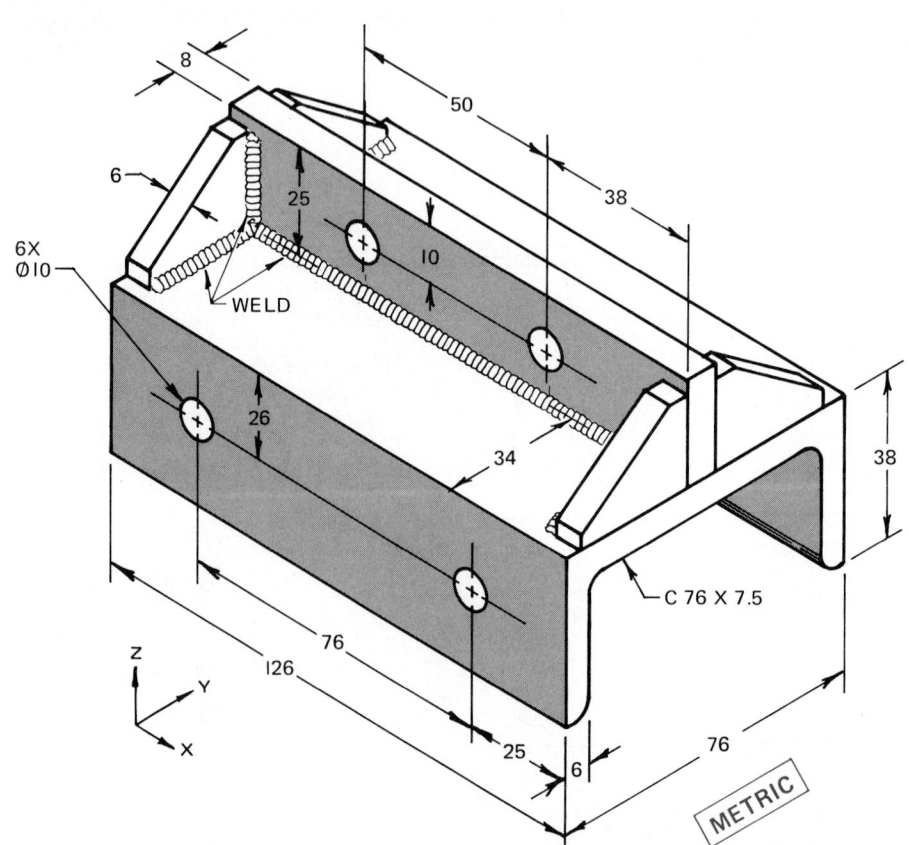

FIG. 18-3-C   Swing bracket.

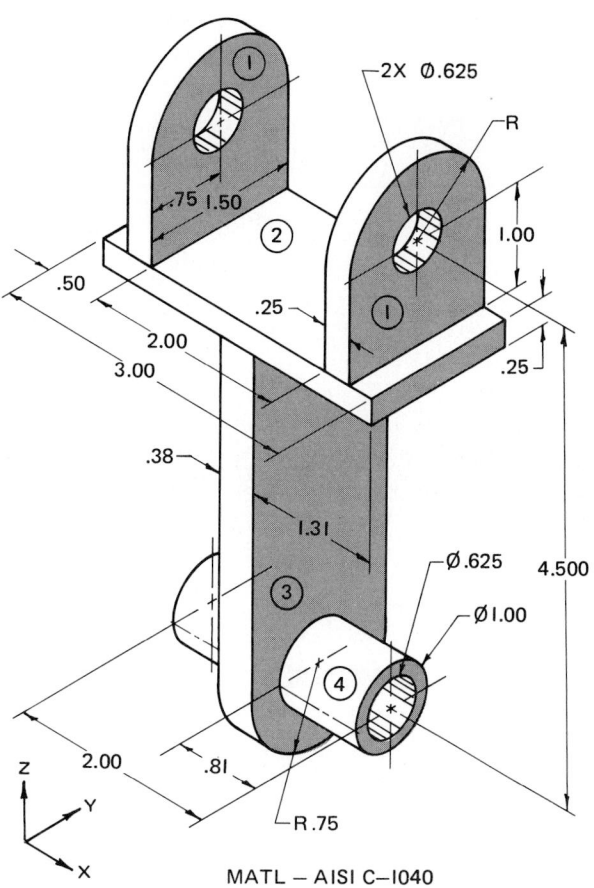

MATL — AISI C—1040

FIG. 18-3-D   Step bracket.

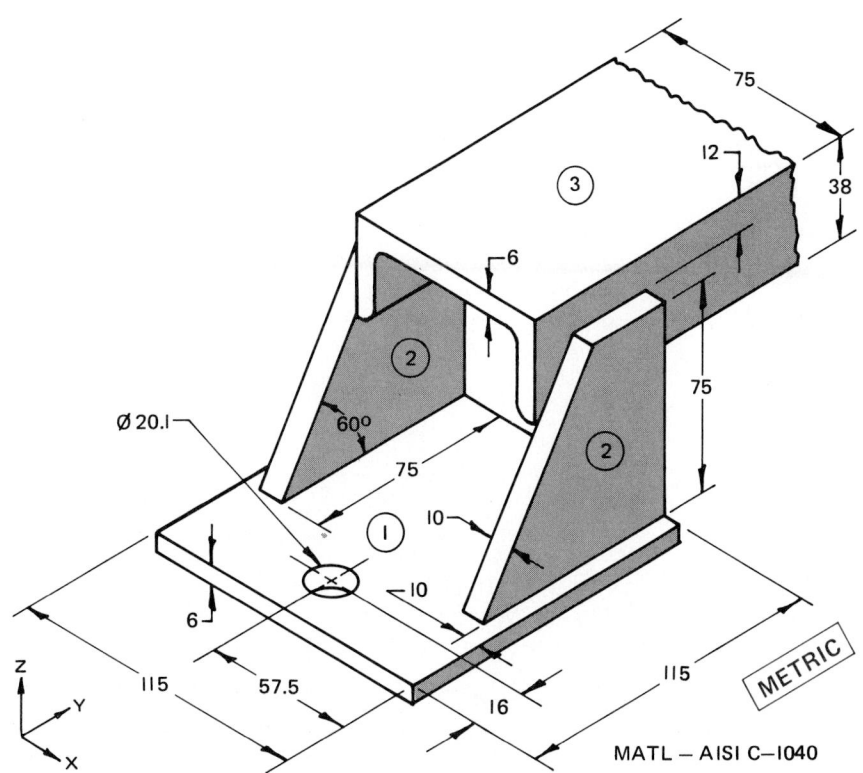

METRIC

MATL — AISI C—1040

5. With reference to Fig. 18-3-E, complete the welding symbols shown to the right of the desired welds.

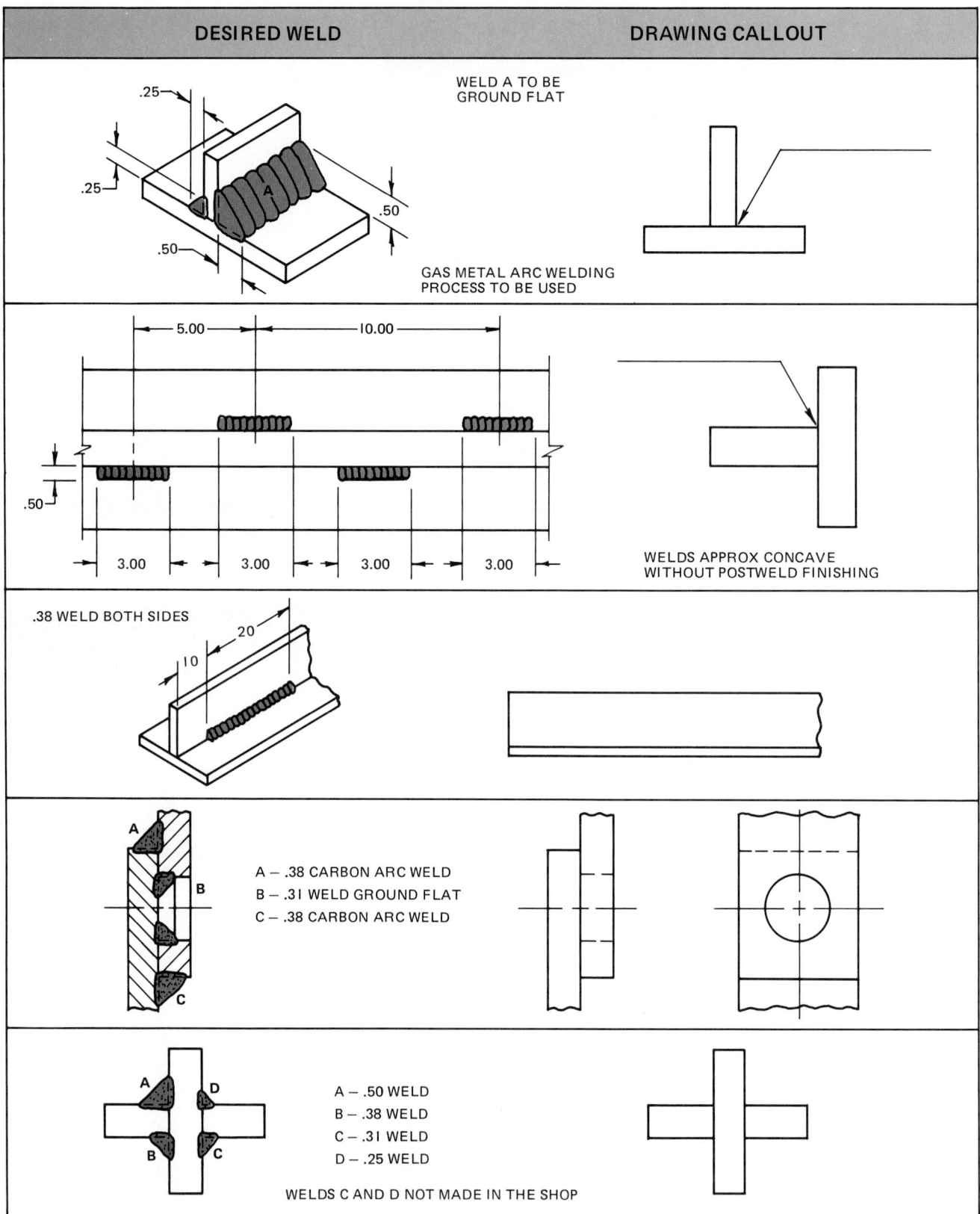

FIG. 18-3-E   Fillet weld symbols.

ASSIGNMENTS FOR UNIT 18-4, GROOVE WELDS

6. Select one of the problems shown in Figs. 18-4-A through 18-4-D (here and on pg. 646) and make a working drawing complete with dimensions and welding symbols.

Include on the drawing an item list and identify each part on the assembly. Use full-strength welds. Scale 1:1 for Figs. 18-4-A, B, and D. Scale 1:5 for Fig. 18-4-C. A comparison of a cast and welded steel part is shown in Fig. 18-3-5. For Fig. 18-4-A the size of the keyseat is to be selected from Table 21 of the Appendix.

FIG. 18-4-A   Swing bracket.

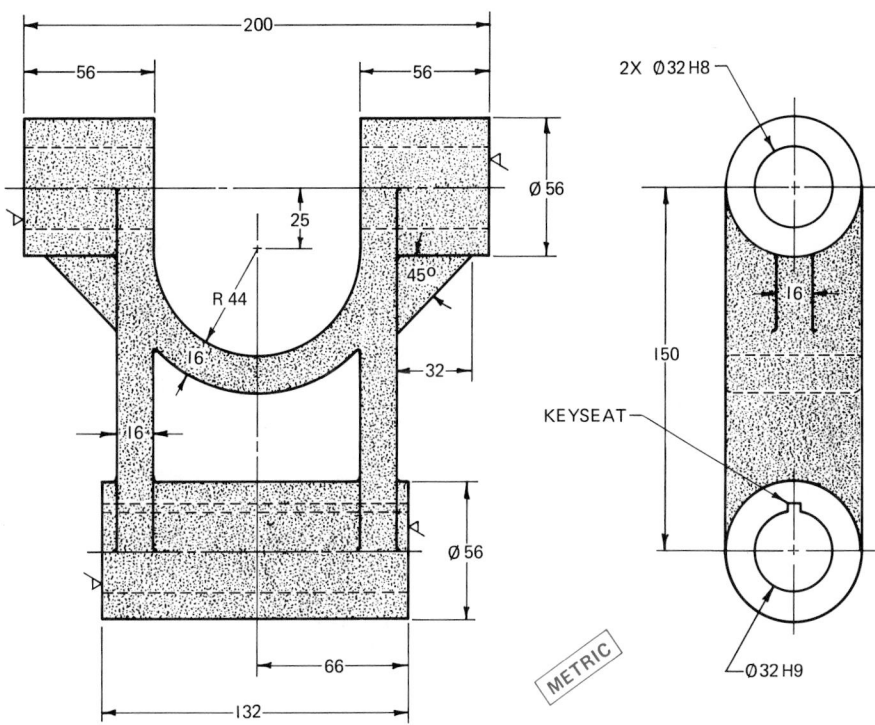

FIG. 18-4-B   Connecting link.

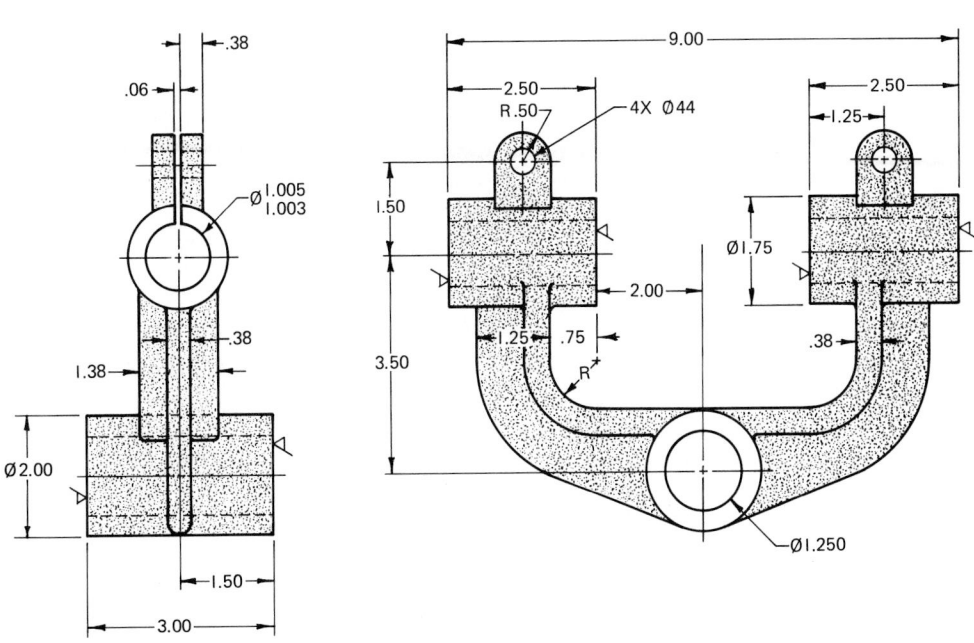

FIG. 18-4-C   Fan and motor base.

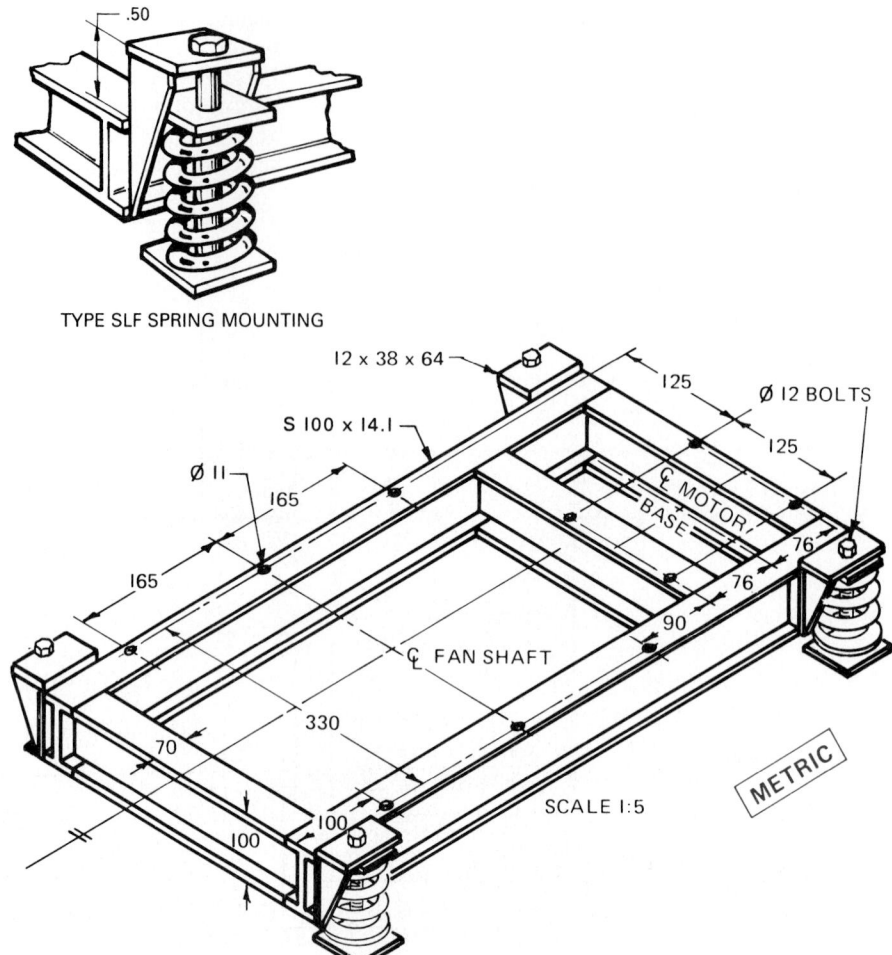

TYPE SLF SPRING MOUNTING

FIG. 18-4-D   Drill press base.

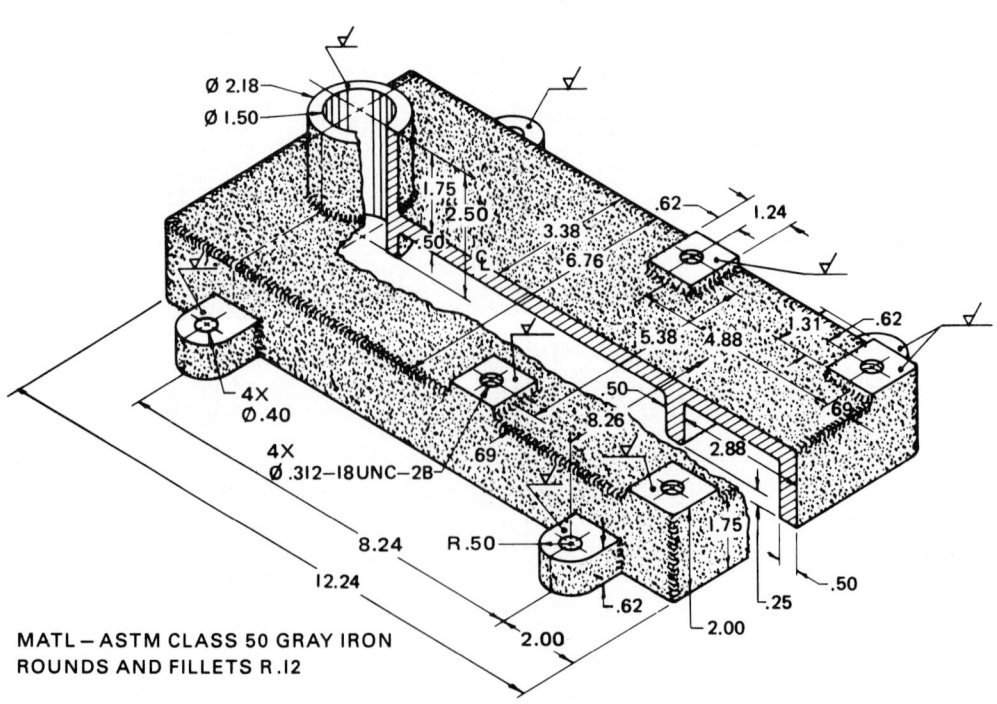

MATL — ASTM CLASS 50 GRAY IRON
ROUNDS AND FILLETS R.12

7. With reference to Fig. 18-4-E, prepare detailed sketches of the groove welds from the information shown.

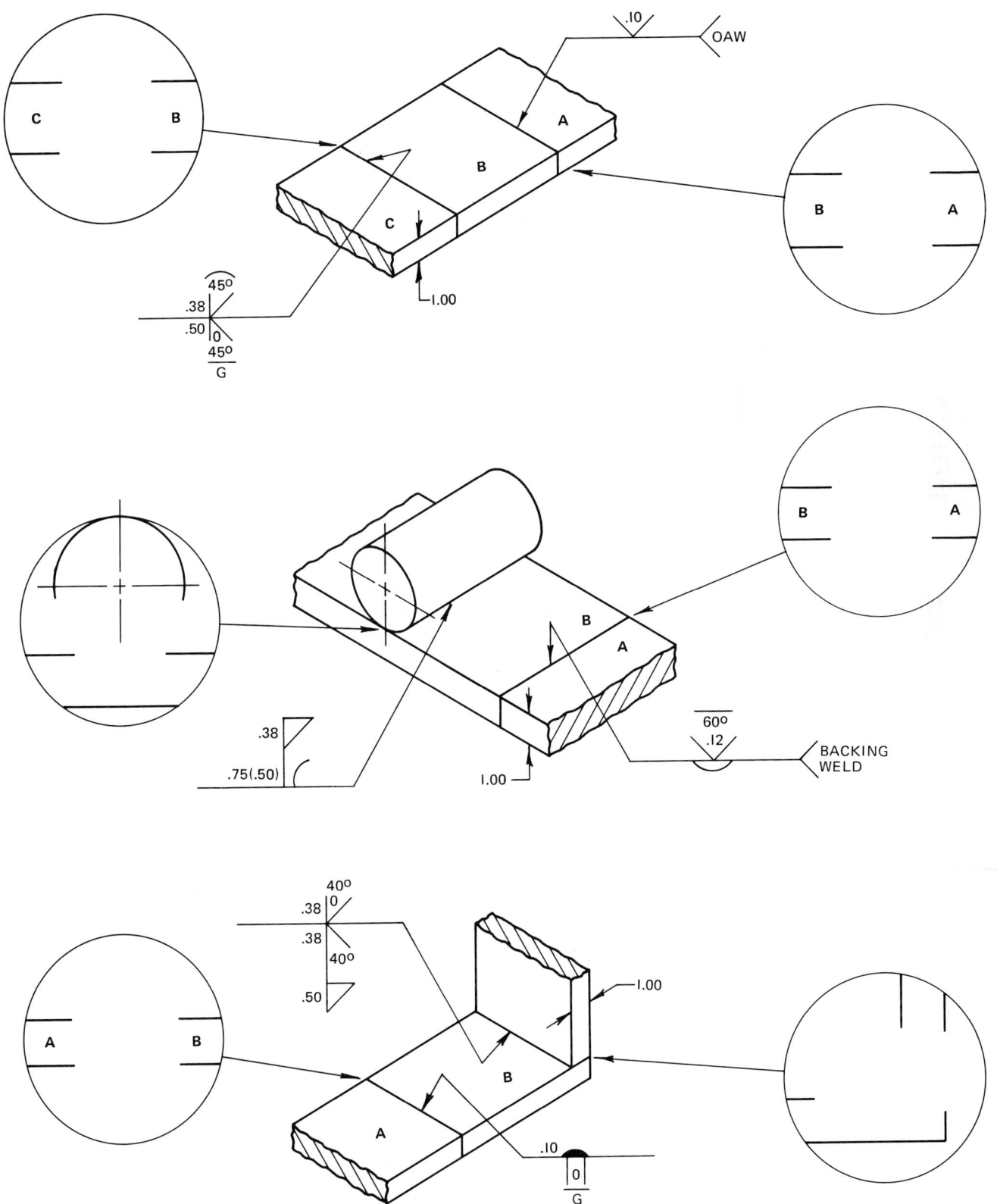

FIG. 18-4-E   Groove weld symbols.

ASSIGNMENTS FOR UNIT 18-5, OTHER BASIC WELDS

8. Prepare drawings of the parts and welds shown in Figs. 18-5-A through 18-5-C and add the weld-size dimensions.

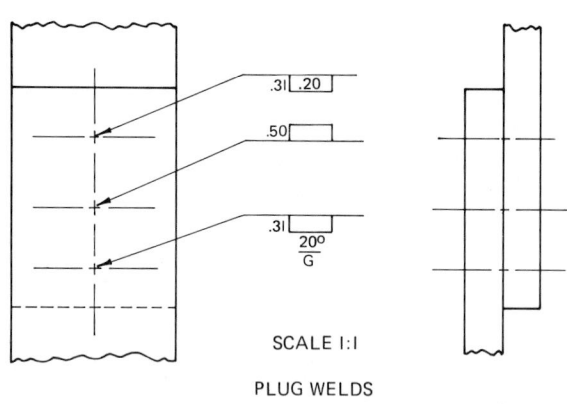

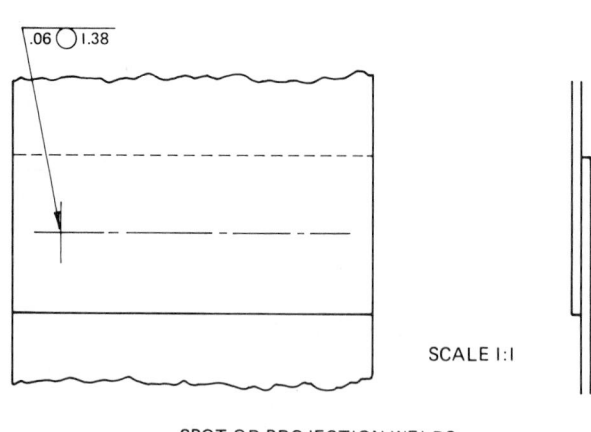

FIG. 18-5-A   Plug and spot welds.

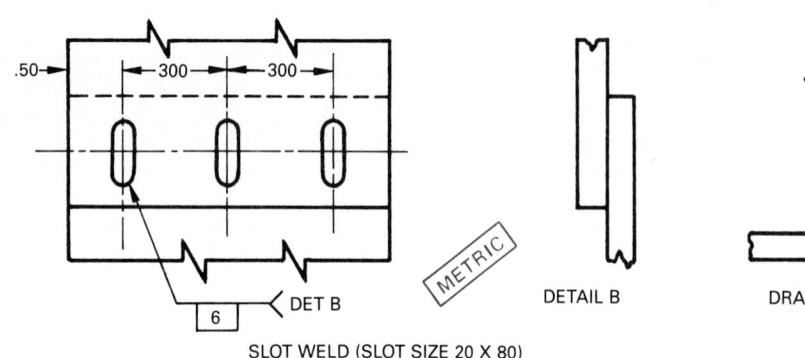

FIG. 18-5-B   Slot and flanged welds.

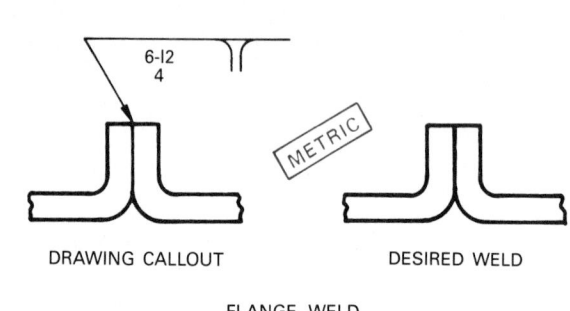

FIG. 18-5-C   Seam and surface welds.

9. With reference to Fig. 18-5-D, show the drawing callouts
   and sketch the welded assemblies shown.

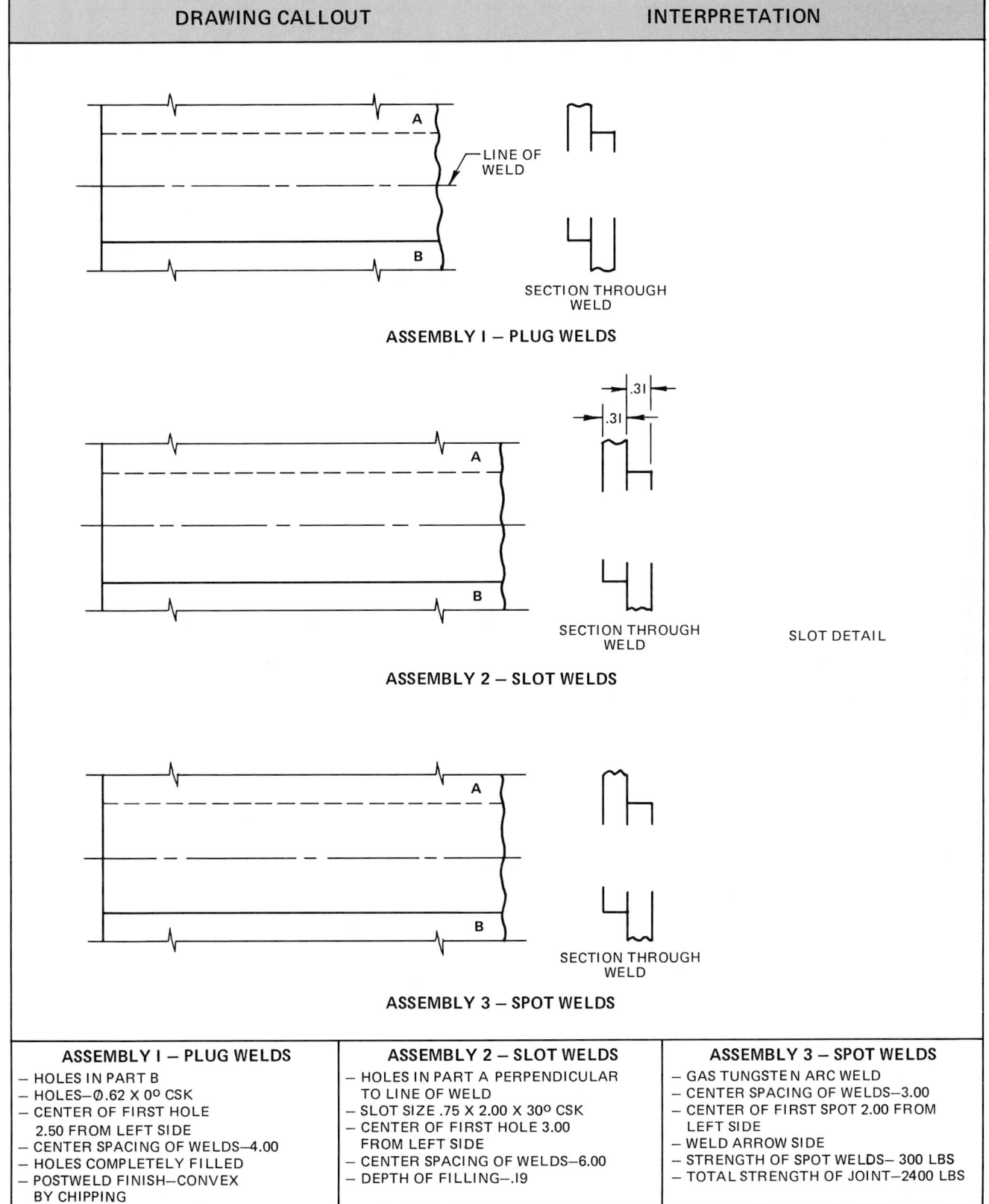

DRAWING CALLOUT	INTERPRETATION

ASSEMBLY I — PLUG WELDS

LINE OF WELD

SECTION THROUGH WELD

ASSEMBLY 2 — SLOT WELDS

.31

.31

SECTION THROUGH WELD

SLOT DETAIL

ASSEMBLY 3 — SPOT WELDS

SECTION THROUGH WELD

ASSEMBLY I — PLUG WELDS	ASSEMBLY 2 — SLOT WELDS	ASSEMBLY 3 — SPOT WELDS

– HOLES IN PART B
– HOLES—Ø.62 X 0º CSK
– CENTER OF FIRST HOLE
  2.50 FROM LEFT SIDE
– CENTER SPACING OF WELDS—4.00
– HOLES COMPLETELY FILLED
– POSTWELD FINISH—CONVEX
  BY CHIPPING
	– HOLES IN PART A PERPENDICULAR
	  TO LINE OF WELD
	– SLOT SIZE .75 X 2.00 X 30º CSK
	– CENTER OF FIRST HOLE 3.00
	  FROM LEFT SIDE
	– CENTER SPACING OF WELDS—6.00
	– DEPTH OF FILLING—.19
		– GAS TUNGSTEN ARC WELD
		– CENTER SPACING OF WELDS—3.00
		– CENTER OF FIRST SPOT 2.00 FROM
		  LEFT SIDE
		– WELD ARROW SIDE
		– STRENGTH OF SPOT WELDS— 300 LBS
		– TOTAL STRENGTH OF JOINT—2400 LBS

FIG. 18-5-D   Plug, slot, and spot welds.

10. With reference to Fig. 18-5-E, show the drawing callouts
    and sketch the welded assemblies shown.

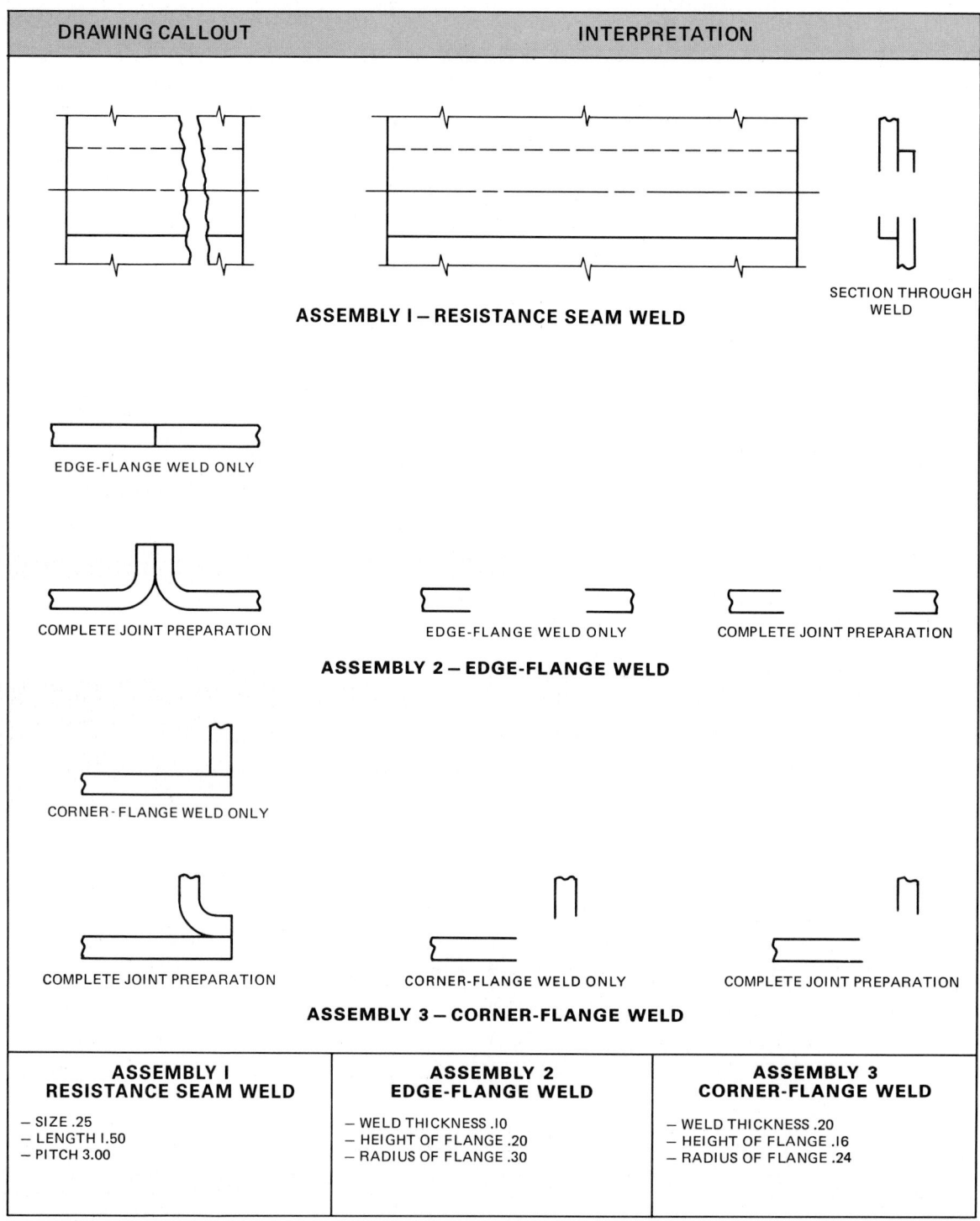

FIG. 18-5-E　Seam and flange welds.

11. Sketch the two assemblies shown in Fig. 18-5-F complete with welding symbols for the welding requirements shown.

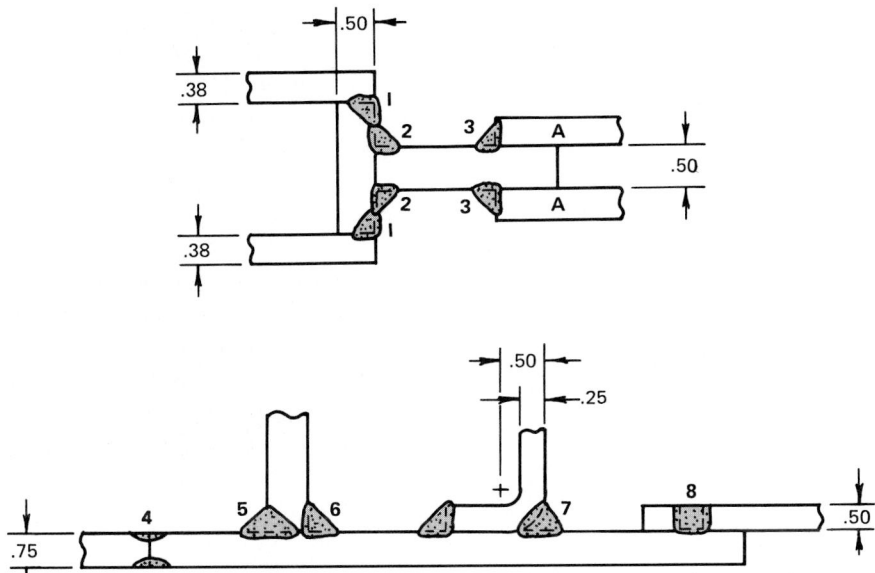

WELD NO.	WELDING PROCESS REQUIRED	TYPE OF WELD	SIZE	ADDITIONAL REQUIREMENTS
1	Carbon Arc	45° Bevel	.38	Convex contour by welding .31 groove preparation
2	Oxyacetylene	Fillet	.31	Staggered intermittent field welds–2.00 inches of weld on 4.00 inch centers
3	Carbon Arc	Fillet	.25	Weld on three sides of part A
4	Gas Metal Arc	Double V	.50 Top .25 Bottom	.12 gap, 90° angles, bottom weld to be ground flush
5	No Specifications	Combined – Sq Groove Fillet	.50 .38	Zero root opening
6	No Specifications	Fillet	.38	Intermittent 1.00 inch on 3.00 inch centers
7	No Specifications	Combined – Flare Bevel Fillets	.38 .25	
8		Plug	Ø.62	0° angle, grind flush, 4.00 inch pitch spacing

FIG. 18-5-F   Welding symbols.

CHAPTER 19

DESIGN CONCEPTS

Definitions

Assembly   A combination of two or more parts, joined by any of a number of different methods.

Brazing   The process of joining metallic parts by heating them at the junction points to a suitable temperature and using a nonferrous filler metal.

Crimping   Joining two or more metal pieces by folding over the metal of one part to squeeze or clinch the other part or parts.

Press fit   The assembling of a part, such as a shaft, into a hole that is slightly smaller in diameter.

Prototype   A sample which gives the designer the chance to see the newly-designed product as a three-dimensional object.

Riveting   Attaching parts of an assembly by using permanent fasteners.

Soft soldering   The process of joining metal parts by melting into their heated joints an alloy of nonferrous metal.

Subassembly   The attachment of component parts to facilitate the production of a larger assembly.

Welding   The process of joining metallic parts at their junction using heat, with or without pressure.

19-1 THE DESIGN PROCESS

The history of civilization is the story of the unique ability of men and women to use intelligence, imagination, and curiosity in the creation of tools and artifacts that ease the burden of physical labor.

Creativity has been defined as the exercise of imagination combined with knowledge and curiosity. Although more commonly associated with the arts—painting, sculpture, music, dance, literature—creativity is equally important in all fields of technology. The combined efforts of scientists, engineers, technicians, and skilled tradespersons have been largely responsible for the high living standards presently enjoyed by western civilization.

Technological designers must be creative within the limits of physical and scientific laws; artistic creativity has fewer restrictions. To be successful, a technical design must be functional, desirable, producible at a reasonable cost, and in many cases, visibly attractive and appealing. Like other abilities or talents, creativity is present in varying degrees in everyone and can be further developed with effort and practice.

The Design Process

The purpose of any design department is to create a product that not only will function efficiently but will also be a financial success. Although most designs are more complicated than the examples in this chapter, the main steps in designing a product follow a similar pattern.

In the design process, considerations should be given to each of the steps shown in Fig. 19-1-1. The design can be treated as a process in which the input is the problem and the output is the solution.

A detailer may work from a complete set of instructions, such as a complete assembly drawing, or may have a free hand in the design of the part. If a detailer extracts the information from a complete assembly drawing, many of the decisions have already been made by the designer or engineer. If the designer or engineer has made the decisions, there are several factors the detailer must consider before starting the detail drawing. Normally, the final design is a compromise of many factors.

The Engineering Approach to Successful Design

Each field of engineering has its techniques and rules—and its standards for the use of the construction materials peculiar to that field.

The steps from idea to production are based on logical and well-known design principles. These principles apply equally to the manufacture of gears, optical systems, industrial components,

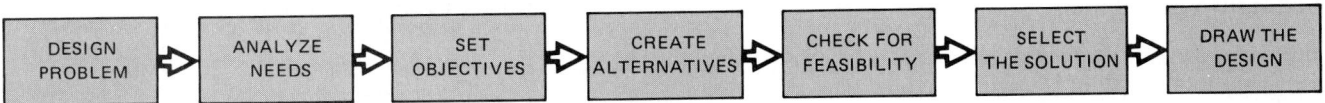

FIG. 19-1-1   Steps in the design process.

consumer articles, or rockets, regardless of the material of construction. These steps are not necessarily in order, but all are essential for a successful application.

Defining the End-Use Requirements

As an initial step, the product designer must anticipate the conditions of use and the performance requirements of the product. Consideration must be given to such things as environment, load, speed of production, life expectancy, optimum size, maintenance, shape, color, strength, and stiffness.

These end-use requirements can be ascertained through market analyses, surveys, examinations of similar products, testing, general experience, and frequently, material suppliers.

A clear definition of product requirements leads directly to choice of the construction material.

Selecting the Material

Material is a very important factor to consider in designing a part. Perhaps plastics are the better choice of material over wood or metal. Would one choice of metal be better than another? What elements will come into contact with water? If the part is to be immersed in an oil solution, will the choice of plastics be minimized or ruled out? Is strength a factor? If so, what materials will meet the stresses required? What material is in stock or easily obtainable? Is the material the correct choice if a plating or coating is required?

There are thousands of engineering materials available, yet no single one will exhibit all desired properties in their proper relationships. Therefore, a compromise among properties, cost, and manufacturing process determines the construction material. Even within one series, materials differ because of varying formulations. Just as steel compositions vary—tool steel and stainless steel, for example—so do the plastics.

The designer needs a firm set of properties and engineering data upon which to base the first design. The data can come from handbooks or, more likely, from the published literature provided by the manufacturers of materials. Chapter 12 provides the student with a variety of materials from which to choose.

Drafting the Preliminary Design

The designer blends the end-use requirements and the properties of the selected material into a preliminary design.

Engineering techniques and formulas are used to achieve the three requirements of design success:

- Economic feasibility
- Functional feasibility
- Attractive appearance

The production method to be used will often set limitations on design. The designer should be aware of the strengths and weaknesses of the method selected. The material supplier and the processor, with their experience in hundreds of applications, can assist here.

Prototyping the Design

The prototype is the opportunity for the designer to see the product as a three-dimensional object. This, too, is the first opportunity for checking the engineering design. The quality of the prototype is quite important. The method used in producing the prototype may not be the same as that planned for the final production line, but the design must be identical to that expected on the production line—otherwise tests may be misleading and analysis false.

If the search for the right material has been narrowed to only two or three, prototyping will help spotlight one.

Testing the Design

Every design should be given an actual or simulated service test while in the prototype stage to ensure that the obvious is not overlooked and that the not-so-obvious is taken into account. The end-use requirements dictate the design testing program. An engine part might be given temperature, vibration, and hydrocarbon-resistance tests; a luggage fixture might be subjected to abrasion tests; and a toaster knob might be checked for electrical and heat insulation. Other tests, such as field testing or consumer reactions, are part of the necessary procedure for completely evaluating any design.

Taking a Second Look

The second look at the design provides an answer to the basic question, "Is the product doing the right job at the right price?"

At this point, most products can be improved by redesigning for better production economies or for important functional or aesthetic changes. Weak sections can be strengthened, colors changed, and new features added. Substantial changes in design will require retesting. Now is the time to set up production. The first step is to write the specification.

Writing Meaningful Specifications

The purpose of the specification is to eliminate any variations in the product that will not satisfy the functional, aesthetic, or economic requirements. The specifications are a complete set of written requirements that the part must meet. The specifications for the part should include such things as the material of construction by brand and generic name, method of fabrication, dimensions, color, surface finish, packaging, printing, and every other detail of production to which there could be more than one possibility. See Fig. 12-4-5 for a typical specification report.

Setting Up Production

How many parts are required? When a large quantity is required, the number of methods of producing the parts increases. Perhaps a casting, a forging, or a stamping may be the most sound choice. If only a few parts are required, prefabricating or machining may be the better choice. See Chap. 13 for forming processes.

Production   Should the part be manufactured in the plant or sent out to be produced? In many cases company policy may be to produce the part within the plant. If this is the case, the production choices are limited to the methods available within the plant.

After the specification is written but before the production line can start, tooling must be designed, built, and integrated with the processing equipment. (In some cases, dies and molds can be started while testing is in progress.)

Production efficiency and economy can be realized through proper design of tools. The processor is an important source of aid in this area.

Time Factor   In some instances, such as when a breakdown of a machine is holding up production within the plant, the best method of producing a part may take second place if it involves too much time.

Workforce   This factor ties in with time. A machine breaks down at 2:00 P.M. It is essential that the machine be back in operation by 8:00 A.M. the following day. What personnel are available to produce the part, with overtime, or is there a night shift?

Controlling Quality

Good inspection practice requires a checklist to maintain a consistently good product. The inspection checklist, for the most part, will conform to the end-use requirements set forth in the specifications.

Here, too, it is beneficial to consult with the supplier or the molder, who knows the processing and finishing characteristics of the material chosen.

Part Specifications

All material applications start out as ideas in someone's mind. From this point the idea must be developed into a production item. The transition is accomplished in a series of logical steps. Ensuring the quality of the final production item starts with the writing of a set of specifications.

Remember, the specifications are a complete set of written requirements; the purpose is to ensure that the finished part will perform as intended. The scope of the specifications depends on the performance required of the part. In general, as specifications become more complex, the cost of the part increases.

Let's take a look at a typical, although hypothetical, part (shown in Fig. 19-1-2). Assume that it has been developed through the steps of the engineering approach and is ready for a meaningful specification.

Any good specifications should contain three basic portions: (1) the raw material, (2) the design of the part, and (3) the performance of the part in use.

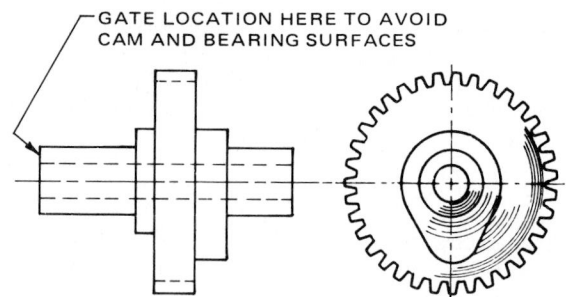

FIG. 19-1-2   Combined gear and cam.

Raw Material

The raw material for any part is selected for its physical properties, with due regard for economic and engineering requirements. The section of specifications dealing with raw material should be divided into two major parts: identification and quality.

Design of the Part

The second major portion of the specifications involves the design of the part. The design specification includes tolerances and surface finish. If the part is going to be cast, then parting lines, flash, gate location, and warpage must be considered.

Dimensional Tolerances   The dimensional tolerances should be as close as required for functioning. When tolerances become tighter than necessary, the cost of both tooling and fabrication rises very rapidly (Fig. 19-1-3). Such tight dimensional tolerances require very close control of the processing variables and necessitate extra inspection, which contributes to a high unit cost.

Other important considerations are:

1. Critical dimensions should be identified with specific tolerances; let overall tolerances control less important dimensions.
2. If parts are to be machined, allow generous tolerances in these areas.

Surface Finish   The type and degree of surface finish required should be clearly indicated. If a highly polished finish is necessary, it should be specified. On other surfaces, finish need not be covered except in general terms. Surfaces that must be clear of

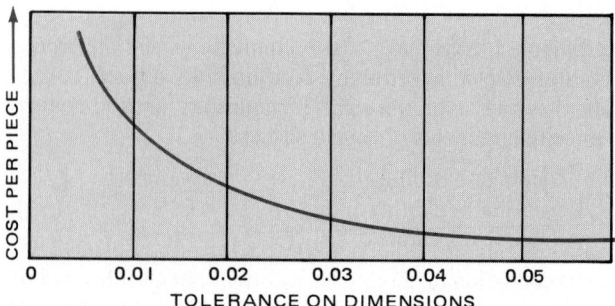

FIG. 19-1-3   Cost of part increases rapidly as tolerances get smaller.

imperfections, such as tool marks, sinks, blisters, and flow lines, should be clearly indicated.

Parting Line   The location of a parting line of a mold may be influenced by flash, part appearance, and structural design. In such cases, parting-line location should be specified on the drawings. Experienced processors can assist in altering part design to simplify tooling or molding.

The parting line in the gear-cam part could be located in several places. Placing the parting line on one end of the gear simplifies the tooling requirement.

Flash   Where flash is undesirable, the drawing should so indicate. The length of allowable flash may be given in a measurable dimension.

Gating   The gate should be located where it will cause the least difficulty. The specification on gating should define areas to avoid, such as cam or bearing surfaces. To permit maximum concentricity in the hypothetical part, the gate should be located as shown in Fig. 19-1-2.

Warpage   The allowable warpage of parts after molding should be specified, even though all dimensions may be met. When minimum warpage is desired, the cooling of parts in a jig or fixture may be necessary. It should be remembered that postmolding operations (annealing, moisture conditioning, etc.) may cause warpage.

Minimum warpage depends on a proper balance of several factors. These include uniform part thickness, location of knockout pins, and optimum molding conditions.

Performance Specifications

The third major portion of the specifications concerns the quality of the part. Earlier work on design, material selection, and end-use testing has given assurance of part performance. Performance specifications are concerned with two factors: the first is visual inspection, and the second is simulated service tests.

Visual Tests   This kind of test is concerned with color, weld lines, etc. The test is established by considering the effect of each of these on the final performance of the part. Not all the factors may be necessary for every part.

Simulated Service Test   The second test for a part should simulate the ultimate use for the part. Care must be taken to use a meaningful test. Excessive speeds, loads, or impacts well beyond ultimate requirements can frequently cause rejection of good parts.

A simulated service test on a part like the cam-gear assembly of Fig. 19-1-2 might consist of impact and/or torque loading.

Do's and Don'ts for Designers

Designers want reliably functioning parts that are dependably procurable at the lowest installed cost. They can best meet their needs by consulting with vendors and understanding custom-metal part manufacturing and pricing. Here is a checklist to use in the design of a part or assembly.

Don'ts

1. Don't specify tolerances tighter than essential for mechanism functioning.
2. Don't specify every dimension as mandatory; mark non-critical ones as reference only.
3. Don't specify material that is of a higher quality than necessary (too expensive) for the service.
4. Don't specify material that is available only on special purchase unless there is no alternative. If in doubt, ask your vendor.

Do's

1. Do leave adequate space for assembly, i.e., bolt clearance, finger grip, etc.
2. Do consider manufacturing economics.
3. Do consider utilizing stock items when you need only a small quantity of parts. Your savings in design time, procurement costs, and delivery time may be appreciable.
4. Do realize that for small quantities or one part, the cost of raw material is not important; material availability and minimum-quantity purchase restrictions are important.
5. Do realize that for large-quantity purchases, precise specification of raw material can be extremely important.
6. Do realize that the total cost of a custom part is not the purchase cost but the installed cost.
7. Do consider, in your product reliability, the relation between part cost, part reliability, and the cost of replacement of a broken part, including lost production time.

REFERENCES AND SOURCE MATERIAL

1. E.I. duPont de Nemours & Co.
2. The Wallace Barnes Co. Ltd.

ASSIGNMENTS

See Assignments 1 through 4 for Unit 19-1 on pages 664 to 665.

19-2 ASSEMBLY CONSIDERATIONS

An *assembly* is a combination of two or more parts that are joined by any of a number of different methods. A *subassembly* is made to facilitate the production of a larger assembly.

This unit describes various methods of attaching component parts to produce an assembly and some of the problems concerning assembly cost, tools, and practicability.

Although various examples of assembly methods and attachments are presented, they are not to be considered the only methods nor are they to have any preference over other means. Product design, volume, cost, and facilities are the determining factors influencing the need for an assembly.

The quality of the finished product as an assembly depends on effective attachment or fastening methods, regardless of the quality of the individual parts.

All assemblies, regardless of size, shape, or design, should be given the following considerations, since in many cases an analysis will dictate changes in design that will effect a cost savings.

Cost of Assembly

The following points should be carefully noted in determining the least expensive method of assembly.

Product Volume

Careful design will reduce costs of assembly at any volume level. However, many times the greatest savings are realized when the volume is high enough to justify the capital expenditure necessary for time-saving equipment that could not be justified at lower volumes. In many cases it can be readily demonstrated that the procurement of highly specialized machinery may be justified by the increased efficiency made possible by such equipment.

Product Design

The cost of any part or assembly is the responsibility of the design engineer and engineering management. To effectively control costs, it is essential that the design, fabrication, and assembly costs be continuously foremost in the minds of all responsible personnel. It is generally true that the simpler the design, the lower the cost of producing the finished product. When new designs or methods are being considered, all factors involved must be taken into account in calculating the savings or increased costs, including equipment obsolescence.

There is no general solution for any assembly problem. For example, an assembly may be made in one plant following a set sequence, while in another plant it may be made quite differently—the difference being due to established plant practices, equipment and facilities, tooling fixture design, volume, and labor costs.

Methods of assembly also have an important bearing on cost. Figure 19-2-1 shows a typical cost analysis covering seven possible methods for assembling a simple bracket to its carrying member, as illustrated in Fig. 19-2-2.

OPERATION OR MATERIAL	COST FACTOR	QTY & POS	COST COMPARISONS						
			SPOT WELDS	PROJ WELDS	RIVETS	ARC WELDS	BOLTS & NUTS	BOLTS & TAPPING PLATE	BLIND RIVETS
Method			1	2	3	4	5	6	7
Spot Welding	100	3B	300					200	
Projection Welding	100	3B		300					
Forming Weld Projection	89	3B		267					
Punching Hole	89	4A			356		356	356	356
Rivet	70	2A			140				
Driving Rivets	96	2A			192				192
Arc Welding (1 in.)	250	3C				750			
Bolt	115	2A					230	230	
Nut	106	2A					212		
Lockwasher	18	2A					36	36	
Assembling Bolts	136	2A					272	272	
Tapping Plate (Matl)	321	1A						321	
Drilling Hole	89	2A						178	
Tapping Hole	89	2A						178	
Blind Rivet	742	2A							1484
TOTAL COST			300	567	688	750	1106	1771	2032

The above table is for illustrative purposes only and its application should be adjusted to costs prevailing at the time of its use. Cost comparisons are based on spot welding as Unit 100. The table is not intended to indicate that the least costly method is the best; function and strength of assembly must also be considered.

FIG. 19-2-1   Assembly methods cost analysis with reference to Fig. 19-2-2.

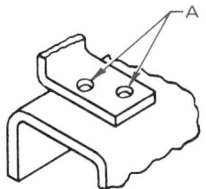 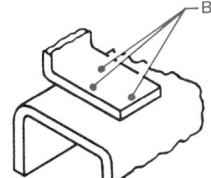 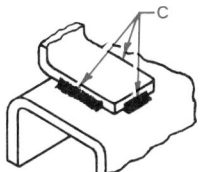 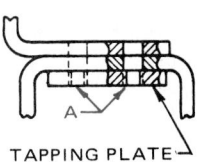

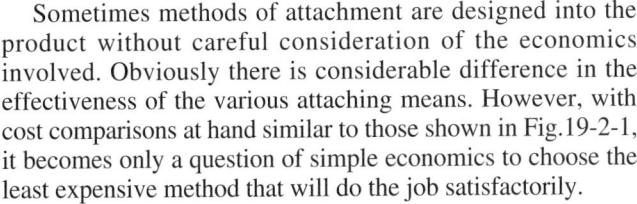

FIG. 19-2-2   Bracket assembly.

Sometimes methods of attachment are designed into the product without careful consideration of the economics involved. Obviously there is considerable difference in the effectiveness of the various attaching means. However, with cost comparisons at hand similar to those shown in Fig.19-2-1, it becomes only a question of simple economics to choose the least expensive method that will do the job satisfactorily.

Ease of Assembly

The cost of assembly labor and equipment, as well as space requirements for assembly operations and equipment, depends on the ease and speed with which the assembly can be made.

Intricate assemblies require careful, slow hand fitting, or expensive jigs and fixtures, or both. If the assembly is made from simple components that can be rapidly assembled, the cost will be lower.

Hole tolerances should be as liberal as is commensurate with the functional requirements of the assembly in order to facilitate the assembly operations. In many instances, a redesign may be justified to eliminate tight fits and unnecessarily close clearances that slow up the operation and make the assembly difficult. Not only should each subassembly be reviewed from this standpoint, but accessibility of all parts used to attach the subassembly to the main assembly should also be investigated to determine whether the tools normally used by the production department can reach the points of attachment.

Quality

Thought must be given to the finished appearance, the functional limitations, and the sales appeal of the completed product. Lack of attention to refinements in the assembly will otherwise completely offset the closest attention to the details.

Service

One of the most frequently heard criticisms of assembled products is the difficulty and cost of removing and replacing some minor part of the unit. Often the labor cost of replacing a bearing, gasket, or minor assembly exceeds the cost of the replaced parts. The cost of replacing parts can be minimized if consideration is given during the design period to providing for rapid and easy disassembly of functioning parts.

It must also be remembered that automobiles, trucks, industrial engines, airplanes, locomotives, etc., have to function in all types of weather, and parts may be subject to moisture and rust. Likewise, products made to handle corrosive vapors and liquids must be given special consideration, and the designer should use fastenings that will be least affected by such exposure.

Attachments

Attaching methods used in assemblies are broadly divided into three categories: permanent, semipermanent, and quickly detachable or connectable. Each has an important function in the assembly of component parts.

Permanent Attachments

Permanent attachments include welding, brazing, soldering, riveting, peening, staking, crimping, spinning, stapling, stitching, pressing, and shrinking. Welding is the most popular because of its satisfactory attachment and because it can be accomplished by many different processes.

Welding   *Welding* is the process of joining metallic parts by fusing them at their junction using heat, with or without pressure. For a more complete discussion, see Chap. 18 on welding. In considering resistance-welded attachments, select electrode shapes from a clearance standpoint (Fig. 19-2-3).

Brazing   *Brazing* is the process of joining metallic parts by heating them at the junction points to a suitable temperature and using a nonferrous filler metal that has a melting point below that of the base metals.

Soft Soldering   *Soft soldering* is the process of joining metal parts by melting into their heated joints an alloy of nonferrous metal. The silver brazing alloys, which are often called *hard solders,* have a much higher melting point and fall within the field of brazing.

Riveting   *Rivets* are a permanent type of fastener for attaching parts of an assembly. The most common types are solid, blind, tubular, and split.

Solid rivets are used in assemblies that are not intended to be taken apart.

Blind rivets are designed for use where it is impossible to have access to both ends of the rivet, such as riveting a bracket to a box section. In other applications, they may also be used in place of solid rivets; however, they are more costly than solid rivets. The cost of installation time plus the unit price of both methods should be considered before a process is chosen.

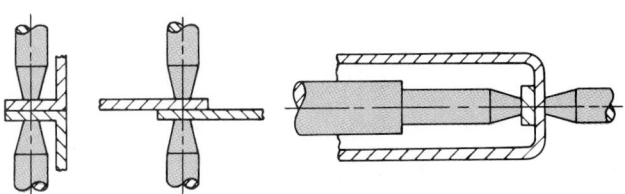

FIG. 19-2-3   Resistance welding.

Tubular rivets do not make joints as strong as solid rivets do, but they can be easily installed by either a spinning or a squeezing process. Although spinning is considered more desirable from a strength standpoint, the squeezing operation is more often used because of the simplicity of equipment.

Split rivets are somewhat limited in their applications. They are usually installed with the same type of squeezer equipment used for tubular rivets. An application of tubular and split rivets is shown in Fig. 19-2-4.

Impact riveting, known also as *peening,* is used to secure a shoulder pin or rivet in an assembly of two or more parts. Impact riveting can be used to advantage where stock thickness or hardness of parts varies; the operator can control the force and number of blows required to produce a secure assembly. In impact riveting, a round shaft is often swaged into contact with the sides of a hexagonal hole to solidly lock the pin and eliminate any possibility of rotation of the pin in the part (Fig. 19-2-5).

Spin riveting results in a better headbearing surface than impact or squeeze riveting and has less tendency to cause shaft distortion than squeeze riveting. However, it is usually slower in operation, and the tool cost is normally much higher. Spin riveting can be used to advantage where one of the assembled parts must be free to move.

Squeeze riveting can be used to advantage in fastening two or more parts where the holes for the rivet may be slightly mismatched, as well as in true-matched holes. Squeeze riveting of pins into parts tends to distort the pin below the riveting point and, by so doing, fills the holes in the plates even though they may be slightly mismatched. This method of riveting can be done with either hot or cold rivets (Fig. 19-2-6).

Crimping This method is used to secure two or more pieces of metal in an assembly by folding over the metal of one part to squeeze or clinch the other part or parts.

In crimping, the part must be designed to allow enough extruded metal on the crimped part to fold over in complete contact with its assembly mates but without excess metal that may be forced out from under the crimping punch. Successful crimping requires a die or tool designed for the specific crimping operation, as illustrated in Fig. 19-2-7. Crimping is less expensive than riveting or welding and can be used when the metal of one part is ductile enough to allow folding over without cracking.

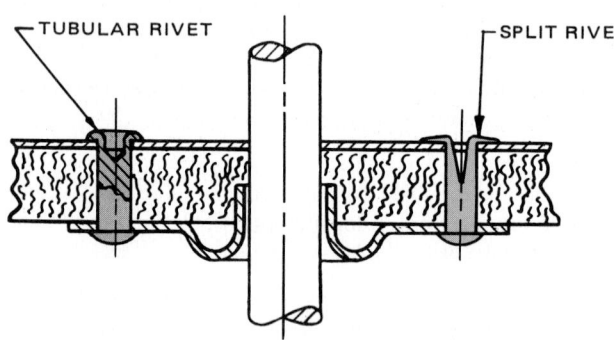

FIG. 19-2-4   Tubular and split rivet design.

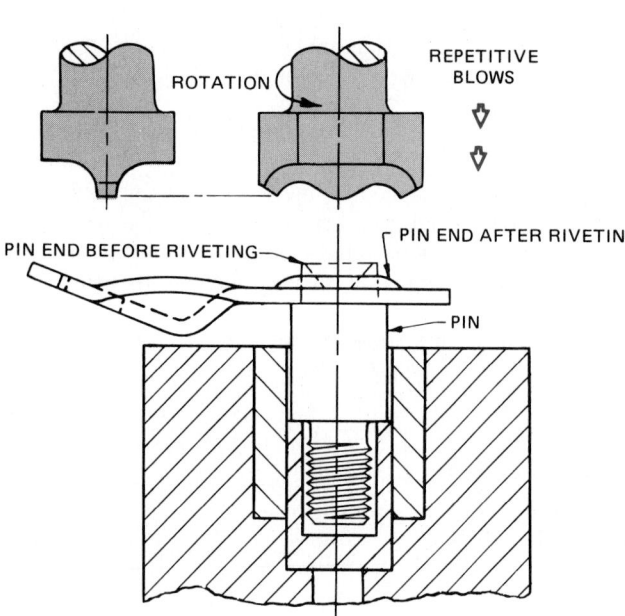

FIG. 19-2-5   Impact riveting.

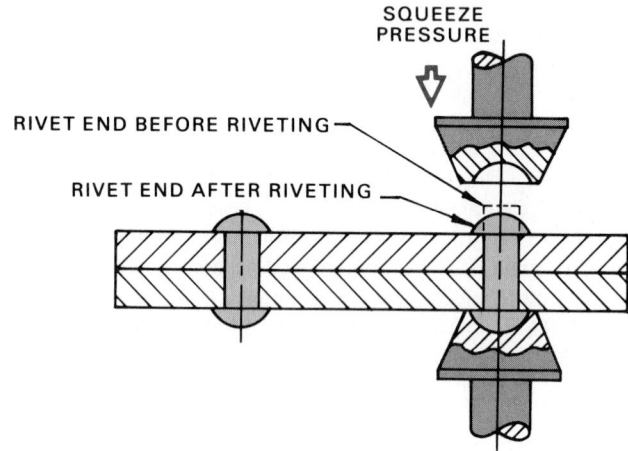

FIG. 19-2-6   Squeeze riveting.

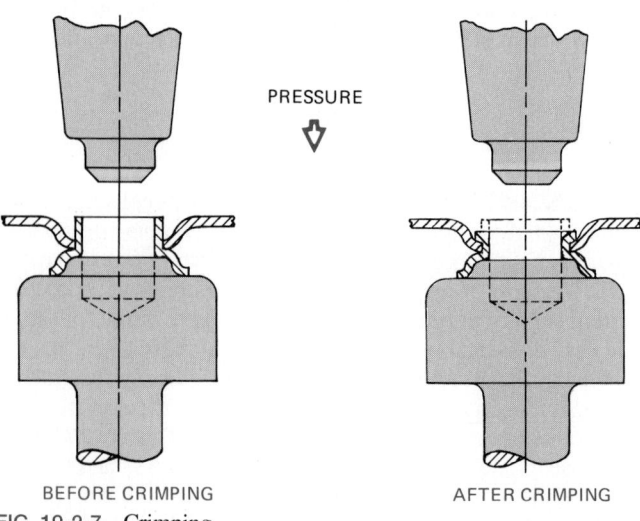

BEFORE CRIMPING   AFTER CRIMPING
FIG. 19-2-7   Crimping.

Stitching   This method is used to secure metal to metal, fabric to metal, rubber to metal, etc., as shown in Fig. 19-2-8.

Press Fit   The term *press fit* applies to the assembling of a part, such as a shaft, into a hole that is slightly smaller in diameter than the shaft (Fig. 19-2-9). The degree of interference depends on the size of the hole, the mass of material around the hole, and the kind and quality of material.

Shrink Fit   This method is a modification of a press fit, adapted particularly to large diameters. Diameters that would provide sufficient interference to hold the two parts together permanently could not be pressed together cold. In these cases, the ring is preheated, then slipped over the shaft or wheel and allowed to cool in place. As the ring cools, it shrinks to its normal diameter, thus producing a pressure on the shaft sufficient to hold it.

Cementing   Cementing or bonding with a suitable adhesive agent is another method used in production to make permanent or semipermanent assemblies.

Semipermanent Attachments

Semipermanent attachments include threaded fastenings, such as bolts, screws, studs, and nuts, as well as washers, nails, and pins. Many factors must be taken into consideration when a fastener selection is made, such as strength, appearance, permanence, corrosion resistance, materials to be joined, cost, assembling, and disassembling.

Bolts   The proper diameter for a bolt is usually determined by design requirement and controlled by the engineer or designer.

The factors that govern this decision are the strength requirement of the assembled unit and the material and heat treatment of the bolt. The type of head is also determined by design requirements such as unit pressure exerted by the bolt head, space limitations, and driving torque.

Hexagonal bolts are most commonly used. They have a washer face, or the underside of the head is chamfered. They may be used in a threaded hole or with a nut. A typical application of a hexagonal bolt is shown in Fig. 19-2-10.

Flanged hex-head bolts are usually specified where a bolt is to be used against a material that has a relatively low compressive strength, such as aluminum. The flanged head is also advantageous where an oversized hole or slotted hole is necessary.

Round-head bolts are made with variously shaped necks under the head for such specific purposes as embedding in wood or metal to prevent rotation or as a means of retention in thin metal, as shown in Fig. 19-2-11.

Square-head bolts are better adapted to heavy machinery, conveyors, and fixtures.

In selecting thread pitches, the bolt material strength and internal thread material strength must be considered, since a coarse pitch produces a stronger internal thread and a fine pitch produces a stronger external thread.

In general, coarse threads should be used in materials that have relatively low shear strength, such as castings and soft metals, and for applications requiring rapid assembly or disassembly. Fine threads should be used where fine adjustment is necessary and where thin walls may be encountered.

Studs   These are sometimes called *stud bolts*. Studs have threads on both ends, to be screwed permanently into a fixed

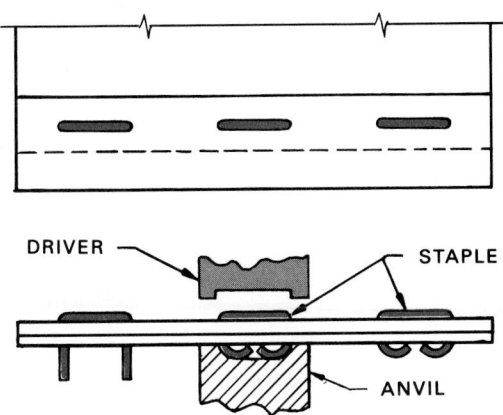

FIG. 19-2-8   Stitching.

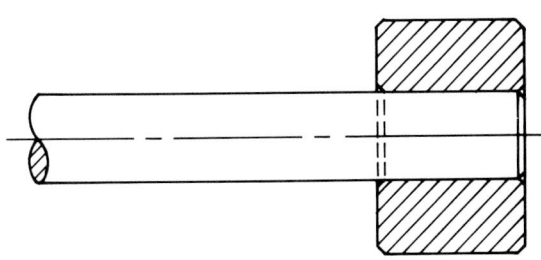

FIG. 19-2-9   Press fit.

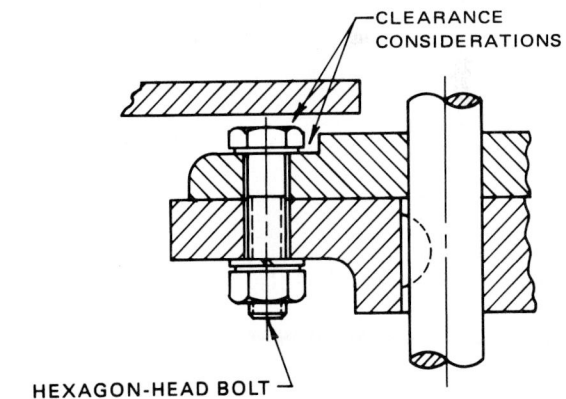

FIG. 19-2-10   Hexagon bolt application.

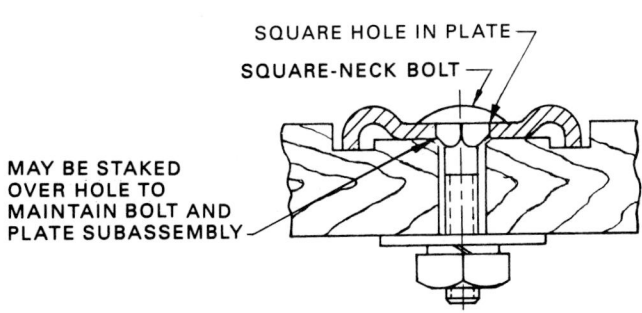

FIG. 19-2-11   Round-head square-neck bolt assembly.

part at one end and receive a nut on the exposed end. They are made of different materials depending on their use. Normally, they are made with coarse threads on the stud end and fine threads on the nut end, as shown in Fig. 19-2-12.

Machine Screws   Machine screws generally differ from bolts in range of diameters, head shapes, and driver provisions. Their use is restricted to light assemblies, such as instrument-panel mountings, moldings, and wire and pipe clips.

The *flat head* is used where a flush surface is required. The *oval head* is generally used for reasons of appearance. Other head types are used for functional reasons; for example, *pan* and *truss heads* are used to cover large clearance holes and elongated holes.

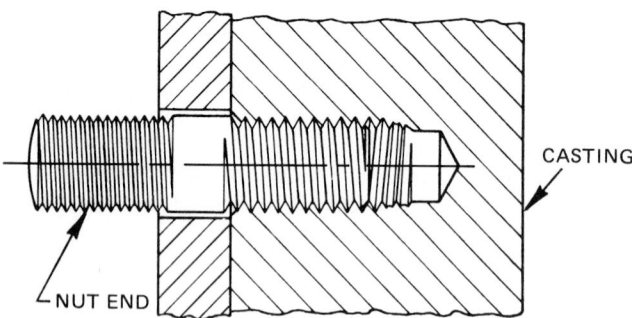

FIG. 19-2-12   Stud application.

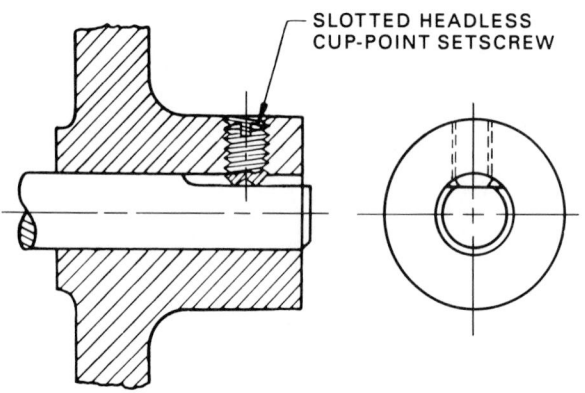

FIG. 19-2-13   Setscrew application.

FIG. 19-2-14   Screw-and-
washer
assemblies.

The *hexagonal heads* are preferable from a driving standpoint; however, they are not suitable for appearance in many locations. For appearance, the *cross-recess head* is popular.

Setscrews   Setscrews are used extensively in tools, jigs and fixtures, control knobs, hand wheels, cam levers, and collars. In order to avoid accidents to operators, setscrews with the end protruding above the hole should never be used on power-rotated or oscillating parts. A typical setscrew installation is shown in Fig. 19-2-13.

Screw-and-Washer Assemblies   A preassembled screw and washer comprise a unit assembly, as shown in Fig. 19-2-14. The washer is free to rotate relative to the screw and is held in place under the head of the screw by the threads, which are rolled after the washer is assembled. Screw-and-washer assemblies result in a labor savings, since only one part need be handled. In addition, they ensure that a washer will be included in the assembly. Procurement and stock control are also simplified. These are factors that bear consideration in specifying screws and washers for attachments and should be weighed against the added unit cost for screw-and-washer assemblies.

Drive Screws for Metal   Hardened metallic drive screws provide a permanent fastening for heavy sheet metal, castings, plastics, etc., and may be used in place of tapping screws or machine screws. Drive screws are hammered or otherwise forced into holes of suitable size. The unthreaded pilot guides the drive screw in straight, and the hardened spiral thread, which extends to the head, forms the required mating thread in the hole.

The thickness of metal into which the screw is to be driven must be at least approximately the same as the outside diameter of the drive screw to ensure adequate thread engagement. An advantage in using these screws in place of machine screws is the elimination of tapped holes; however, a pilot hole is necessary (Fig. 19-2-15).

Tapping Screws   These screws were developed primarily to eliminate tapping operations or nuts in certain assemblies of sheet-metal parts, plastics, and soft castings.

Nuts   Many types of nuts are available for specific requirements. It is desirable to minimize the use of special designs in favor of the more commonly used nuts.

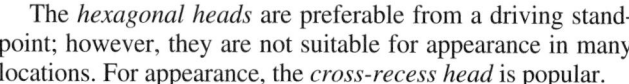
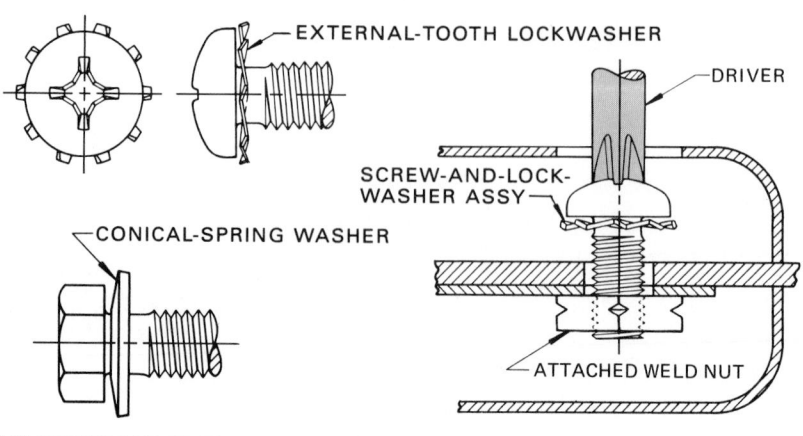

(A) SCREW AND WASHER ASSEMBLIES   (B) APPLICATION

Slotted nuts with cotter pins or wire can be used to help retain the nut on the bolt. *Jam nuts* are used where height is restricted or as a means of locking the working nut, if assembled as shown in Fig. 19-2-16A.

A *locknut* is a nut having a special means for gripping an externally threaded member so that relative back-off rotation between the nut and the companion member is impeded.

Prevailing-torque-type locknuts employ a self-contained locking feature, such as deformed or undersize threads, variable lead angle, plastic or fiber washers, or plug inserts. This type of nut resists screwing on, as well as unscrewing, and does not depend on bolt load for locking (Fig. 19-2-16B).

Free-running locknuts develop their locking action after the nut has been seated by reactive spring force against the threads or by friction against the bearing surface.

Spring nuts are made of thin spring metal and have arched prongs or formed embossments to fit a single lead of a mating screw thread. Spring nuts are used extensively for sheet-metal construction where relatively high torques and strength are not required.

Another type of spring nut is available that can be pushed on over rivets, tubing, nails, or other unthreaded parts and provides a positive bite that grips securely even on very smooth surfaces. Figure 19-2-16C shows a typical application.

Stamped nuts are usually fabricated from thin spring steel and have arched prongs formed to fit a single lead of a mating screw thread. They have the same functional usage as a spring nut, with the additional advantage of provisions for turning the nut (Fig. 19-2-16D).

Crown nuts are generally used where it is desirable to cover the end of the externally threaded part for purposes of appearance or protection from sharp edges.

Wing nuts, as the name implies, are provided with two wings to facilitate hand tightening and loosening. They are used where high torque is not required and where the nuts are to be disassembled and reassembled frequently.

Barrel and *sleeve nuts* are usually made to resemble a screw head at the outer or exposed end. They are used in assemblies where any other type of nut would present a less favorable appearance.

Clinch nuts were developed for sheetmetal assemblies where the nut is inaccessible for wrenching. They are provided on one side with a shoulder and smaller pilot, which is inserted into a preformed hole in the sheet metal, and are permanently attached by spinning or staking the portion of pilot extending through the hole (Fig. 19-2-17).

Weld nuts are similar to the clinch nuts in function. However, they are supplied with weld projections and are either

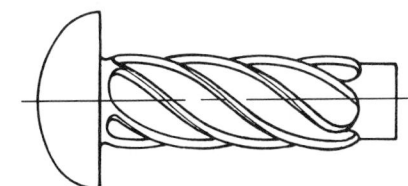

FIG. 19-2-15   Metallic drive screw.

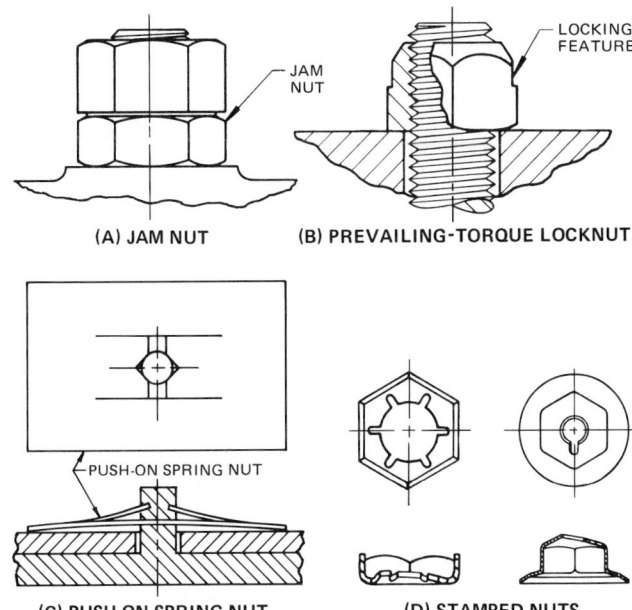

FIG. 19-2-16   Nut applications.

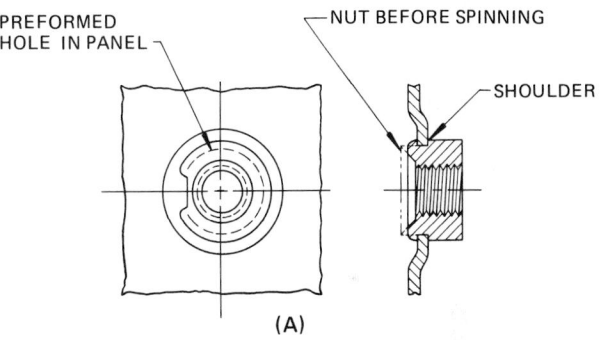

(A)

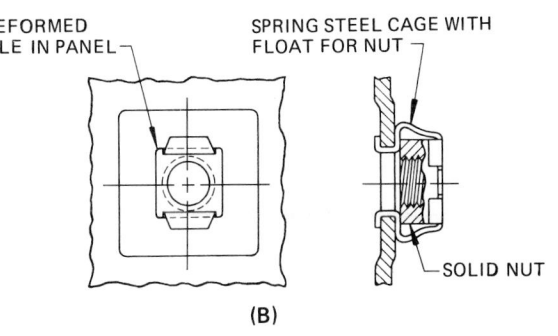

(B)

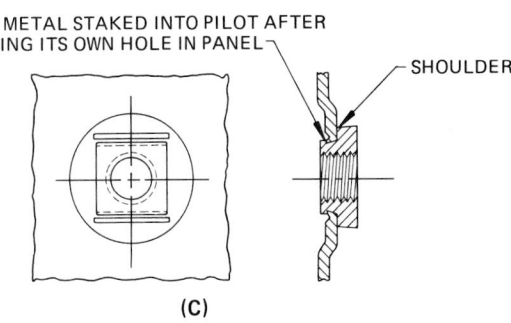

(C)

FIG. 19-2-17   Assembled clinch nuts.

661

spot- or projection-welded to the sheet metal instead of being peened.

Washers   The four basic types of washers are flat (plain) washers, conical-spring washers, helical-spring lock washers, and tooth lockwashers.

Flat washers are used under the head of a screw or bolt, or under a nut, for four principal purposes:

1. To spread the load over a greater area
2. To reduce frictional variations during assembly
3. To provide bearing surface over large clearance holes or slots
4. To prevent marring of parts during assembly

Conical-spring washers are made of steel that has been hardened and tempered. The relatively high supporting load and spring return make this washer effective where bolt tension may be lost because of such factors as thermal expansion or compression set of gaskets.

Helical-spring lockwashers are usually used as a hardened thrust washer or as a spacer.

Pins   Cotter pins, spring pins, groove pins, taper pins, and clevis pins are used to retain parts of an assembly in relative position.

Cotter pins are used for retaining slotted nuts, movable links or rods, etc., as shown in Fig. 19-2-18.

Groove pins are straight pins having longitudinal grooves rolled or pressed into the body that provide a reactive expansion effect when the pin is driven into a drilled hole.

Spring pins provide a spring effect that serves to retain the pin when it is driven into a drilled hole of diameter slightly smaller than that of the pin. These types of pins eliminate reaming or peening and can be disassembled a number of times without serious loss of holding power.

Taper pins serve the same functional purpose as groove pins. However, they require taper-reamed holes at assembly and are retained only by taper lock, which can totally disengage when minor displacement occurs. Assemblies using taper pins are more costly than those using groove pins.

The primary purpose of *clevis pins* is to attach clevises to rod ends and levers and to serve as bearings. They are held in place by cotter pins, as shown in Fig. 19-2-18.

Quickly Detachable Attachments   Quickly detachable attachments include clips or certain kinds of snap fasteners. These attaching means are convenient and time-saving by providing for quick assembly and disassembly.

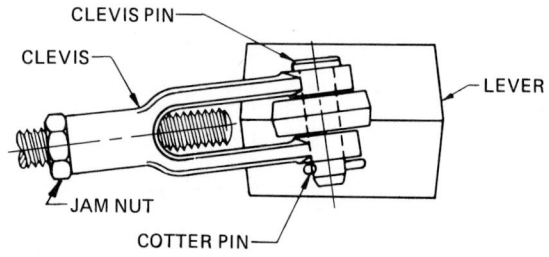

FIG. 19-2-18   Clevis-pin application.

Design Checklist

The following design checklist will serve as a helpful guide in reviewing a design.

1. Keep the number of separate pieces in a subassembly as low as practicable by combining single components into one assembly if functional requirements permit.
2. Check particularly left- and right-hand parts to determine whether they can be made identical and so avoid stocking an extra part.
3. Check advisability of using lock washers assembled to bolts as purchased assemblies.
4. Check to determine whether similar parts between models can be standardized.
5. If subassemblies require alignment, design the parts to secure such alignment without the use of special jigs by providing tabs, shoulders, notches, or contour locators.
6. Provide adequate clearances between parts for assembly tools.
7. Provide clearances between parts to allow for tolerance stackup variations.
8. Specify the simplest, most effective, and cheapest type of attachment practicable and commensurate with the functional requirements.
9. Avoid blind-riveting operations wherever possible.
10. Where rivets are used, provide sufficient clearances between parts and from flanges to permit the use of a standard riveting gun.
11. Avoid the use of slotted nuts and cotter pins wherever possible.
12. Standardize as far as practicable bolt and thread sizes. Hold the number of bolt lengths to a minimum and recheck frequently to reduce number of lengths in use.
13. Try to avoid riveting or welding operations in shop areas where this type of equipment is not normally used.

Design Approach to a Fabricated Structure

In designing machine frames and similar structures for fabrication by welding, the main considerations, apart from ensuring that the part will fulfill its intended function, are usually confined to the necessity for designing an article that will be pleasing in appearance and that can be economically produced. The drafter should try to avoid being unduly influenced by the design principles that have been developed for other methods of construction. For example, in designing machinery parts for fabrication, especially when they are intended to replace or supersede castings and forgings, it is generally essential for the drafter to avoid any tendency to design on the basis of making the weldment look like a casting or forging.

Weight (Mass) Saving

When castings are to be superseded by weldments, the higher labor costs of the weldment must be offset by simplifying the design and reducing the weight (mass). With a casting some extra thickness usually has been provided to allow for defective metal and maybe for shifting cores. With steel there is practically no risk of defective material so that this surplus can be eliminated. Moreover, since cast iron has less than half the tensile strength of steel, the weight (mass) of a steel part can be reduced proportionately. For example, for the same overall dimensions, because of the higher stresses that can be allowed, a steel section need be no more than half the thickness of a cast-iron one. This is shown in the drawing of a pump base in Fig. 19-2-19.

Conclusion

To a large extent, the ultimate cost of the job is usually the yardstick by which the advantages of any type of construction are measured. The drafter should, therefore, review those factors that contribute to the cost of a weldment. Although the cost of steel is low compared with that of cast iron or cast steel, and

FIG. 19-2-19   Design of pump base.

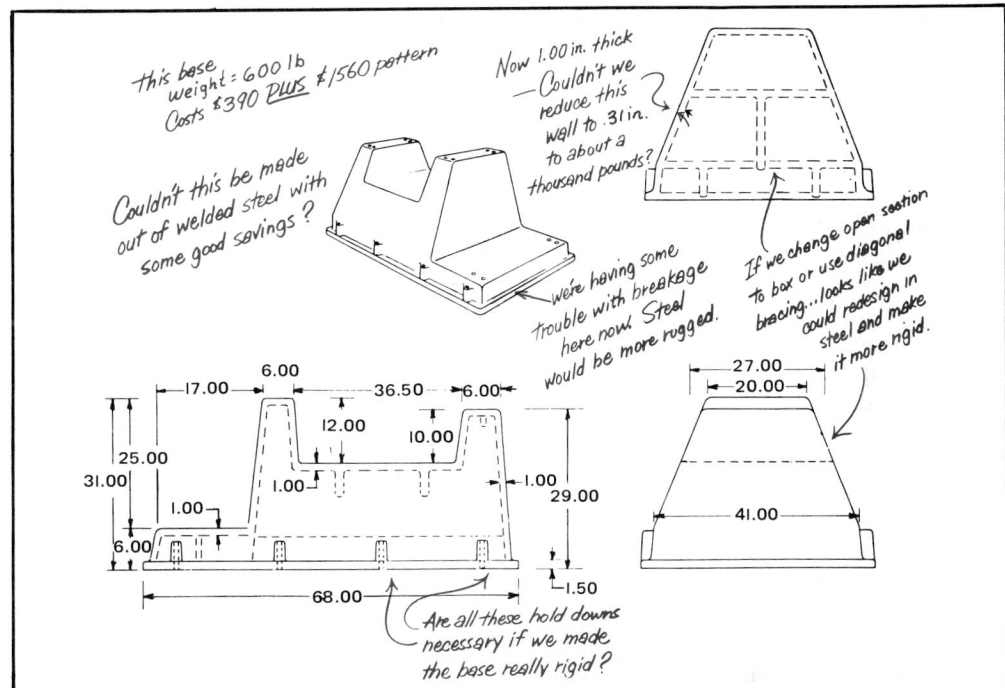

(A) ORIGINAL PUMP BASE SUBJECT OF COST STUDY

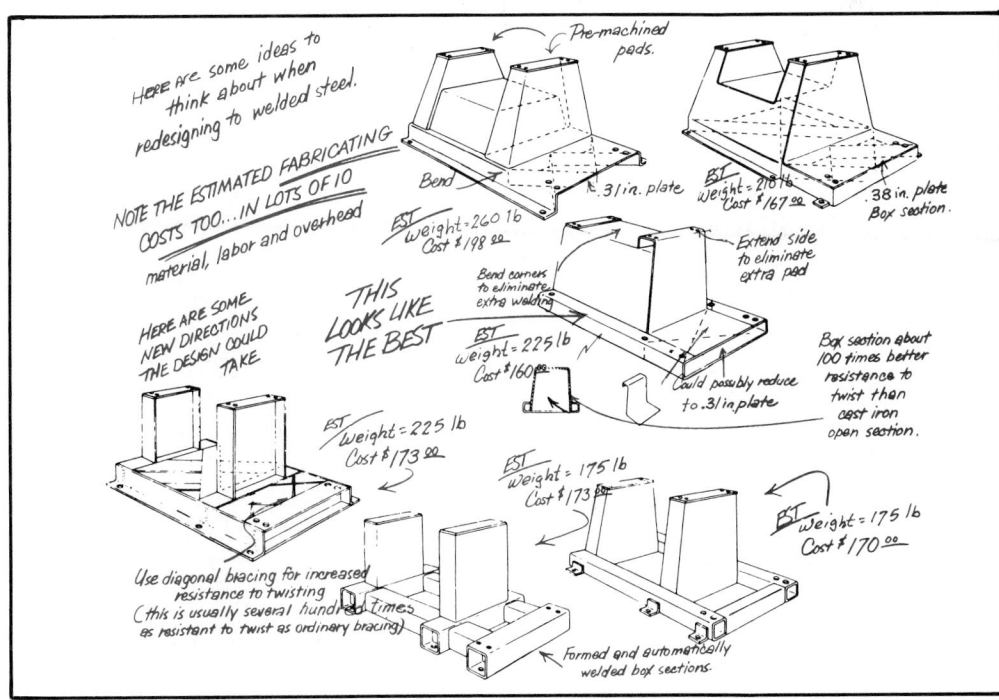

(B) ALTERNATIVE BASE DESIGNS

generally it is possible to use less metal in a weldment than in an equivalent casting, it is essential to remember that there are more operations involved in the production of a weldment than there are in a casting. The plate or section must be prepared for welding; the various components must then be assembled and fitted; finally, there is the actual welding, which may be followed by stress relieving.

Choice of Raw Materials

In this type of design, the drafter has a wide choice of raw materials, plates, structural shapes, forgings, tubes, castings, etc. Careful consideration of the function of the various components of the structure is desirable in order to enable the most suitable raw material to be selected to ensure efficiency, economy, and pleasing appearance. Steel plates will no doubt provide the basic element in the majority of cases, and by flame cutting there is no limit to the variety of shapes that can be produced.

Steel plate surfaces are usually flat and smooth enough to be used as seating or bolting surfaces without further machining. Moreover, where bearings in a plate are required or should a machined surface be considered desirable for the seating of bolt heads, collars, washers, etc., it is often not essential to weld on bosses such as would ordinarily be employed on a casting. If the plate is made a little thicker than normal, the machined areas can be spotfaced into the plate surface. The

spotfacing costs about the same as a boss machining operation, but the work in preparation and welding on of bosses is eliminated. See Fig. 19-2-20 for types of boss attachments.

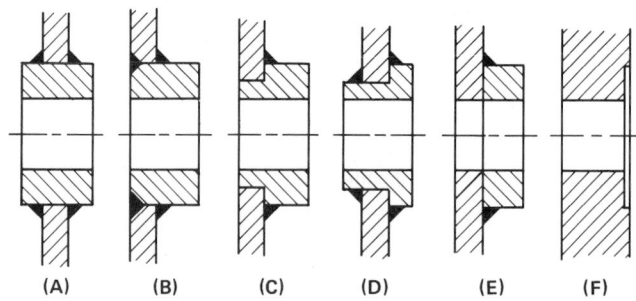

FIG. 19-2-20   Various designs for boss attachments.

REFERENCES AND SOURCE MATERIAL

1. General Motors Corp.

ASSIGNMENTS

See Assignments 5 through 10 for Unit 19-2 on page 665.

ASSIGNMENTS FOR CHAPTER 19

ASSIGNMENTS FOR UNIT 19-1, THE DESIGN PROCESS

1. Your drafting supervisor has assigned you the responsibility of designing an attractive single toggle-switch plate for use in kitchens and bathrooms. The toggle plate and clearance hole requirements are shown in Fig. 19-1-A. The production run will exceed 25,000 and four different color plates are required. Lay out the design of the plate and include on the drawing the production and specification data that you would submit with your design.

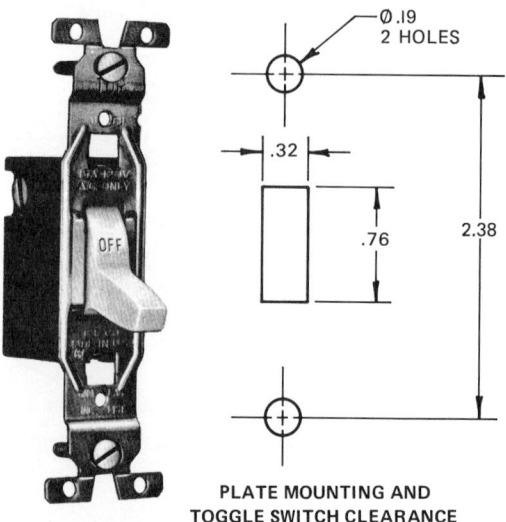

FIG. 19-1-A   Switch plate.

2.   Design a suitable handle and locking mechanism for a metal storage cabinet from the following data and the information shown in Fig. 19-1-B. The latch is to be released with a quarter turn (clockwise direction) of the square latch shaft. The shaft is .38 in. square and protrudes beyond the cover by 1.00 in. The lock and shaft are securely fastened to the inside face of the cover and as such do not require any support on the handle side of the cover plate. A .62 in. diameter hole is punched in the .062 in. thick cover plate. Quantity 5,000. Lay out the design of the handle and include on the drawing the production and specification data that you would submit with the design.

3.   Your drafting supervisor has asked you to design a table-mounted holder for a cassette microphone (Fig. 19-1-C). The material is to be of such quality as to have an attractive finish as well as support the mass of the microphone. A Ø20 mm tapered hole (large end) with a 3 mm wide slot on top is required for attaching the microphone to the holder. Lay out the holder and show the microphone in phantom lines.

4.   Design a personalized key holder for a recreation vehicle that is used occasionally. When not in use, it must hang on a wall or key rack. The material is to be lightweight and colorful.

ASSIGNMENTS FOR UNIT 19-2, ASSEMBLY CONSIDERATIONS

5.   Design a coat hanger support to be located above the backside window of an automobile.

6.   A .750 in. diameter steel rotary shaft must be supported at 6 ft. intervals along a ceiling. Light-duty journal bearings, 1.50 in. long, are recommended by the engineering department. A maximum of eight brackets are required. Purchase or design a suitable bracket to support the shaft.

7.   Same as Assignment 6 except the quantity required is 2,000.

8.   Two vertical .312 in. diameter electrical conductors are to be supported on the inside wall of an oil-filled metal tank. The vertical length of the conductors is 7 ft. The conductors, made of copper, have no insulation on them. The voltage they carry is such that there must be a minimum distance of 2.00 in. between conductors or any other metal when supported by nonconductive material, such as plastic, wood, etc. Although there are no external forces acting on these conductors, they should be supported every 24 in.

Prepare an assembly drawing showing the conductors, support, and tank wall. Include an item list. On a second sheet prepare the details of the parts required. Scale to suit.

9.   Design a container to store 8 or 9 tape cassettes and their containers. The measurements of the container are .70 × 2.70 × 4.30 in. It is to be installed on the bottom edge of an automobile or truck instrument panel.

10.   Design a napkin holder to sit on the table. The napkins can be stored either flat, 6.00 in. square, or folded in half, 3.00 × 6.00 in. Their position for dispensing is your choice as long as one napkin can be taken at a time. The holder must hold a minimum of 12 napkins and come in a variety of colors or a selection of wood grains to complement the modern kitchen.

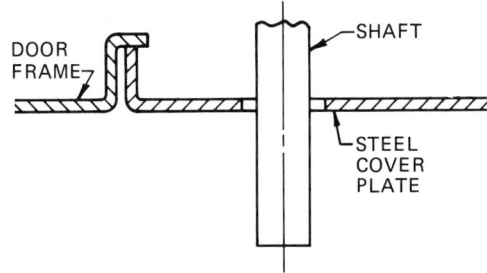

FIG. 19-1-B   Handle.

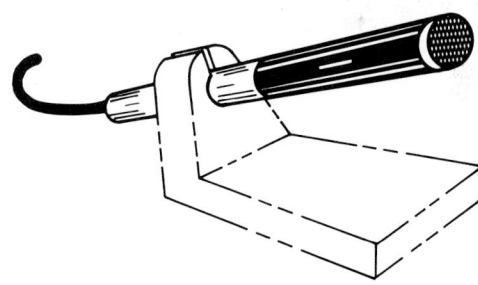

FIG. 19-1-C   Microphone holder.

POWER TRANSMISSIONS

CHAPTER 20

BELTS, CHAINS, AND GEARS

Definitions

Bevel gears   Gears that connect shafts whose axes intersect.

Belt drive   An endless flexible belt that connects two wheels or pulleys.

Chain drive   An endless chain whose links mesh with toothed wheels (sprockets).

Diametral pitch   In inch-size gears, the ratio of the number of teeth to a unit length of pitch diameter.

Idler pulleys   Grooved sheaves or flat pulleys in a drive system that do not serve to transmit power.

Involute   The specific form of the gear that best produces a constant angular velocity.

Miter gears   Bevel gears having the same diametral pitch or module, pressure angle, and number of teeth.

Module   In metric gears, the length of pitch diameter per tooth measured in millimeters.

Pinion   The smaller of two gears in mesh.

Pintle chains   Chains made up of individual cast links having a full, round barrel end with offset sidebars.

Pitch   The distance between centers of articulating joints.

Poly-V belts   Longitudinally grooved or serrated belts that use a flat belt as the tensile section and a series of adjacent V-shaped grooves for compression and tracking.

Positive-drive or **timing belts**   Belts that use a flat belt as the tensile section and a series of evenly spaced teeth on the bottom surface.

Rack   A straight bar having teeth that engage the teeth on a spur gear.

Ratio   In gears, the relationship between revolutions per minute, number of teeth, or pitch diameters.

Sheaves   The grooved wheels of pulleys.

Simple gear drive   A toothed driving wheel meshing with a similar driven wheel.

Spur gears   Gears that connect parallel shafts.

Worm gears   Gears that connect shafts whose axes do not intersect.

20-1 BELT DRIVES

The last 50 years have brought rubber-belt drives to a high state of technological refinement. The result is lighter, more compact drives capable of carrying higher loads at less cost.

Flat Belts

Flat-belt drives offer flexibility, shock absorption, efficient power transmission at high speeds, resistance to abrasive atmospheres, and comparatively low cost. The belts can operate on relatively small pulleys and can be spliced or connected for endless operation. However, because they require high tension, they also impose high bearing loads. They are sometimes noisier than other belt drives, will slip, and have comparatively low efficiency at moderate speeds (Fig. 20-1-1).

Flat belts for power transmission can be divided into three classes.

1. *Conventional:* plain flat belt without teeth, grooves, or serrations.
2. *Grooved or serrated:* basic flat belt modified to provide the advantages of another type of transmission product, e.g., V-belts.
3. *Positive drive:* basic flat belt modified to eliminate the need for frictional force for power transmission.

Conventional belts are available in two types: *reinforced*, which utilizes a tensile member to obtain strength, and *nonreinforced*, which depends upon the tensile strength of the basic material for its strength (Fig. 20-1-2A).

Longitudinally grooved or serrated belts use a flat belt as the tensile section and a series of adjacent V-shaped grooves for compression and tracking. These are generally known as *poly-V belts* (Fig. 20-1-2B).

Positive-drive belts use a flat belt as the tensile section and a series of evenly spaced teeth on the bottom surface. These teeth engage a similarly grooved pulley to achieve positive mesh. Positive-drive belts are also known as *timing belts* (Fig. 20-1-2C).

Conventional Flat Belts

Flat rubber belts were developed around the turn of the century primarily as a replacement for leather belts. With the advent of V-belts, fewer machines were designed to employ flat belts. Nevertheless, conventional flat belts are worthy of serious consideration in many applications. By being thin, flat belts are not subject to high centrifugal loads and thus can operate well over small pulleys at high speeds. This feature makes them well suited to miniature drives such as those used to power brushes in vacuum cleaners.

Conventional flat belts are available either as endless belts or as belting that can be spliced to make a needed length.

FIG. 20-1-1   Flat-belt drives.

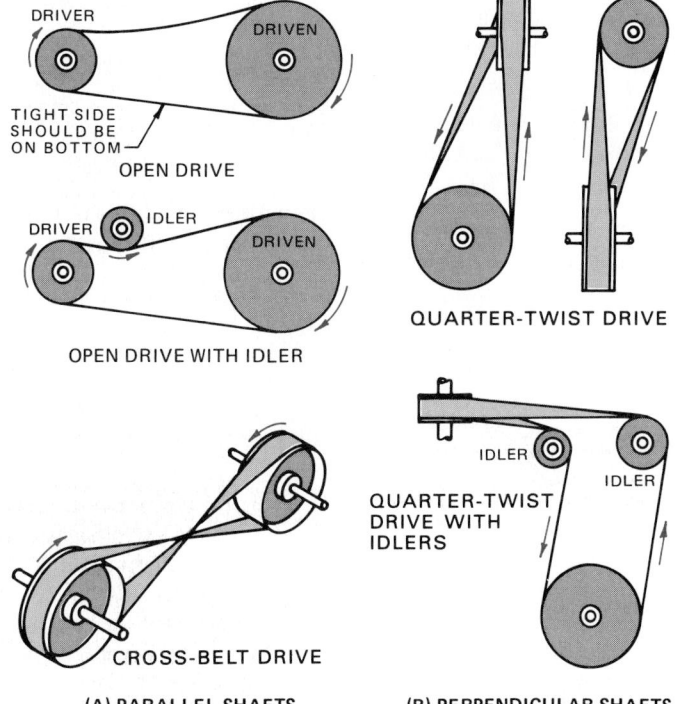

(A) PARALLEL SHAFTS   (B) PERPENDICULAR SHAFTS

(A) CONVENTIONAL   (B) GROOVED OR SERRATED   (C) POSITIVE DRIVE

FIG. 20-1-2   Flat belts.

Conventional belts are normally available in five basic materials:

1. Leather
2. Rubberized fabric or cord
3. Nonreinforced rubber or plastic
4. Reinforced leather
5. Fabric

Leather   Most leather belts are made of plies of belting bonded together. They provide excellent coefficient of friction, flexibility, and long life and are easily repaired. On the other hand, their initial cost is high, they must be cleaned, and they require belt dressing. They also stretch and shrink, depending on atmospheric conditions.

Rubberized Fabric or Cord   Many types and grains of rubberized belting are presently available. Almost all are moisture-, acid-, and alkali-resistant.

Rubberized Fabric   This is the least expensive type of flat belting. It is made up of plies of cotton or synthetic duck, impregnated with rubber.

Rubberized Cord   These belts consist of a series of plies of rubber-impregnated cords. They offer high tensile strength for a modest size and mass.

Nonreinforced Rubber or Plastic   For light-duty applications, flat belts are available in a number of unreinforced materials.

Rubber   Basically a simple strip of rubber, these belts are available in various compounds. They are designed specifically for low-horsepower (kW), low-speed drives. They are especially useful for fixed-center drives since they can be simply stretched into place over their pulleys.

Plastic   Unreinforced plastic belts transmit higher power loads than rubber belts. They are available in a number of plastic compounds.

Reinforced Leather   These belts consist of a plastic tensile member, generally reorientated nylon, and leather top and bottom layers.

Fabric   All-fabric belts may consist of a single piece of cotton or duck folded and sewn with rows of longitudinal stitches. Others are woven into endless forms.

The major advantage of all-fabric belts is their ability to track uniformly and to operate at high speeds. They are used typically in check-sorting machines.

Grooved Belts

These are basically flat belts with a longitudinally ribbed underside. The flat section of the belt serves as the load-carrying component, and the ribs provide traction in the sheave grooves.

This type of belt, although it bears a resemblance to the conventional V-belt, operates on a different principle. Rather than depending on wedging action to transmit power, it depends solely on friction between sheave and belt. Power capacity depends on belt width; only a single belt, with a varying number of ribs, is used for each drive.

Positive-Drive Belts

Another variation of the flat belt is the positive-drive belt, or timing belt. Basically a flat belt with a series of evenly spaced teeth on the inside circumference, it combines the advantages of the flat belt with the positive-grip features of chains and gears.

Positive-drive belts have many advantages. There is no slippage or speed variation, and a wide range of speed ratios is possible. Required belt tension is minimal, so that bearing loads are low.

These belts are not recommended where pulleys are misaligned.

Pulleys for Flat Belts

Different types of pulleys are used for flat, ribbed, and positive-drive belts.

Flat-Belt Pulleys   These are generally made of cast iron. However, they are also available in steel and in various rim and hub combinations. They may have solid, spoked, or split hubs as well as other modifications of the basic pulley.

Crowning   All power-transmission pulleys should be crowned or flanged (Fig. 20-1-3).

Other Types   Pulleys for ribbed and positive-drive belts are available in a variety of stock sizes and widths.

At least one pulley in a timing-belt drive must be flanged in order to keep the belt on the drive. For long-center drives, flanging both pulleys is recommended but not required. Idler pulleys should not be crowned.

V-Belts

V-belts remain the basic workhorse of industry, available from virtually every distributor and adaptable to practically any drive. They are presently available in a wide variety of standardized sizes and types, for transmitting almost any amount of load power.

Normally, V-belt drives operate best at belt speeds between 1,500 to 6,000 ft/min (8 to 30 m/s). For standard belts, ideal (peak-capacity) speed is approximately 4,500 ft/min (23 m/s). Narrow V-belts, however, will operate up to 10,000

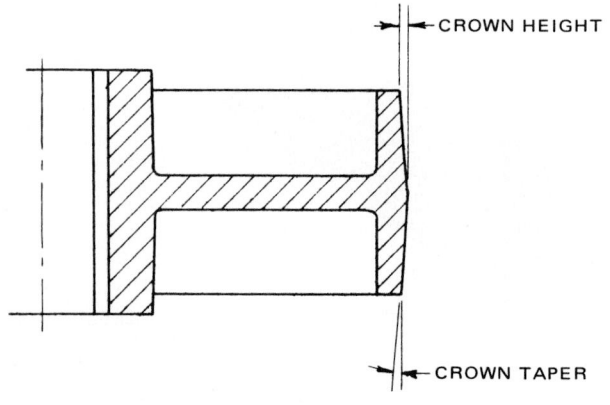

FIG. 20-1-3   Crown on pulley.

TYPE OF BELT*	MAXIMUM POWER		BELT SPEED FOR MAX. POWER		MAXIMUM SPEED		MAX. SPEED RATIO	SHOCK ABSORPTION
	(hp)	(kW)	(ft/min)	(m/s)	(ft/min)	(m/s)		
Constant-Speed								
Light-duty	7.5	5.6	3500	18	5000	25	8	Poor
Standard	350	260	4500	23	6000	30	7	Good
Super	500	375	5000	25	6000	30	7	Very good
Cogged	500	375	5000	25	6000	30	8	Very good
Steel cable	500	375	5000	25	8000	40	7	Poor
Narrow	270	200	7500	38	10000	50	7	Very good
Variable-Speed								
Conventional	300	225	—	—	6000	30	—	Good
Wide-range	75	55	—	—	6000	30	—	Good

*Stock items. Drives available to 1500 hp (1100 kW).

FIG. 20-1-4   V-belt characteristics.

ft/min (50 m/s). A summary of belt characteristics is given in Fig. 20-1-4.

Advantages   V-belt drives permit large speed ratios and provide long life (3 to 5 years). They are easily installed and removed, quiet, and low in maintenance. V-belts also provide shock absorption between driver and driven shafts.

Limitations   Because they are subject to a certain amount of creep and slip, V-belts should not be used where synchronous speeds are required.

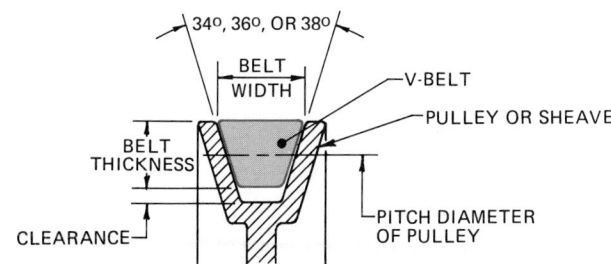

FIG. 20-1-5   V-belt and pulley.

Standard Dimensions

Cross Section   Industrial and agricultural V-belts are always made to standard cross sections (Fig. 20-1-5).

Industrial   These are made in two types: heavy-duty (conventional, narrow) and light-duty. Conventional belts are available in A, B, C, D, and E sections (Fig. 20-1-6). Narrow belts are made in 3V, 5V, and 8V sections. Light-duty belts come in 2L, 3L, 4L, and 5L sections.

Open-end belting is available in A, B, C, and D sections. Link-V belting, which is not covered by a standard, is made in A, B, C, D, and E sections, and in some sizes for low-horsepower (kilowatt) applications.

Agricultural   These belts are made in the same sections as conventional belts. They are designated HA, HB, HC, HD, and HE; in double-V sections HAA, HBB, HCC, and HDD are available.

Agricultural belts differ from industrial belts mainly in construction.

Automotive   Belts for automotive applications are made in six SAE-designated cross sections identified by the nominal top widths: .38, .50, .69, .75, .88, and 1.00 in. (10, 12, 17, 19, 22, and 25 mm).

FIG. 20-1-6   Industrial V-belts

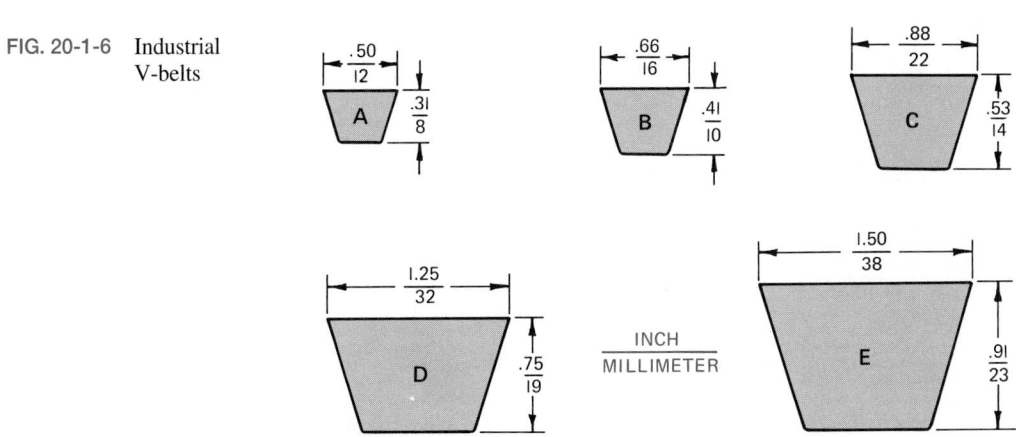

Length   Although endless V-belts can be manufactured in any length within a fairly wide range, manufacturers have standardized on certain lengths that are produced for stock.

Belt-Size Designation

For the different types of V-belts, the same basic method is used to designate belt size. Belt sizes are specified by a code designation consisting of symbols representing belt cross section followed by a designation of length. For conventional and light-duty belts, the length designation is in inches; for narrow belts the number represents tenths of an inch.

For example, a conventional V-belt designated B23 has a B cross section and a 23 in. standard length designation; a narrow belt designated 5V350 has a 5V cross section and a belt with a 35 in. effective outside length; and a light-duty V-belt designated 2L080 has a 2L cross section and an effective outside length of 80 in.

There are no standard methods for designating automotive belts. Variable-speed belts are designated by a code where the first two numbers denote the nominal belt width in sixteenths of an inch, the next two numbers denote the angle of the pulley groove, followed by the letter V, with numbers after that letter specifying length in tenths of an inch.

Basically, a V-belt consists of five component sections (Fig. 20-1-7):

1. Tensile members or load-carrying section
2. Low-durometer cushion section surrounding tensile members

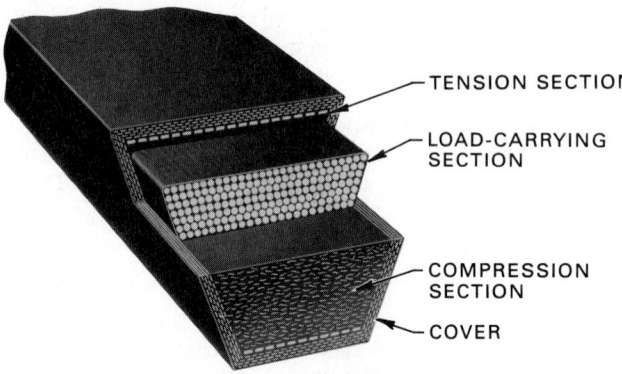

FIG. 20-1-7   Basic V-belt construction. *(American Beltrite Rubber Co.)*

FIG. 20-1-8   Common types of motor bases. *(T. B. Wood's Sons Co.)*

3. Flexible top section
4. Bottom compression section
5. Cover or jacket

Sheaves and Hubs

Most *sheaves* (the grooved wheels of pulleys) are made of cast iron, which is economical and stable and which provides long groove life. For light duty, sheaves may be of formed steel, cast iron, or plastic. Formed-steel sheaves are used primarily in automotive and agricultural applications. For special applications they may be made of steel or aluminum alloy. Typical V-belt applications are shown in Figs. 20-1-8 and 20-1-9.

Sheaves are made with either regular or deep grooves. A deep-groove sheave is generally used when the V-belt enters the sheave at an angle, for example, in a quarter-turn drive, on vertical-shaft drives, or wherever belt vibration may be a problem.

The Use of Idler Pulleys

Idler pulleys are grooved sheaves or flat pulleys that do not serve to transmit power. Usually they are used as belt tighteners when it is not possible to move either shaft for belt installation and take-up, as between two line shafts.

An inside idler pulley invariably decreases the arc of contact of the belts on each loaded sheave of the drive. It should be at least as large as the small loaded sheave and located, preferably, on the slack side of the drive (Fig. 20-1-10A).

A flat idler pulley, whether used inside or outside the drive, should be located as close as possible to the place where the belts leave the sheave. On the slack side of the drive, which is the preferred location, this means as close as possible to the driver sheave (Figs. 20-1-10A and B). On the tight side of the drive, this means as close as possible to the driven sheave (Figs. 20-1-10C and D).

How to Select A Light-Duty V-Belt Drive

The proper selection of V-belt drives for light machinery has been simplified and condensed into three steps. Complete selection involves the proper choice of:

1. V-pulley size for drive shaft and belt cross section
2. V-pulley size for driven shaft
3. Belt length for required center distance

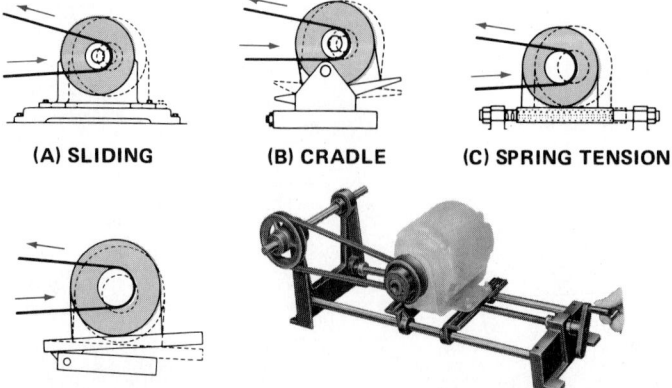

(A) SLIDING   **(B) CRADLE**   **(C) SPRING TENSION**

(D) PIVOTED   (E) APPLICATION OF A SLIDING MOTOR BASE

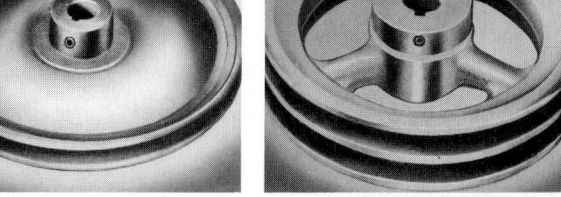

(A) SINGLE PULLEY   **(B) DOUBLE PULLEY**

(C) SINGLE DRIVE   **(D) MULTIPLE DRIVE**

FIG. 20-1-9   Single- and multiple-belt drives.

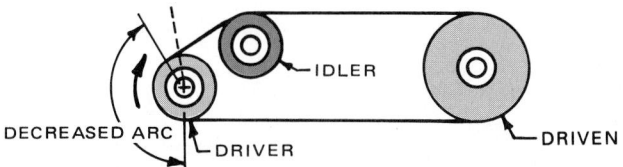

(A) INSIDE IDLER PULLEY, AT LEAST AS LARGE AS THE SMALL SHEAVE, ON THE SLACK SIDE OF THE DRIVE

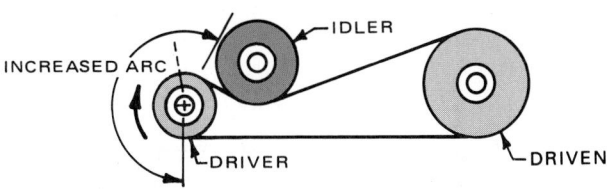

(B) OUTSIDE IDLER PULLEY, AT LEAST 1.3 LARGER THAN THE SMALL SHEAVE

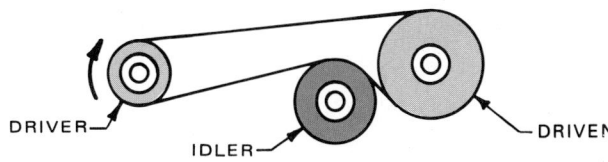

(C) OUTSIDE IDLER PULLEY ON THE TIGHT SIDE OF THE DRIVE

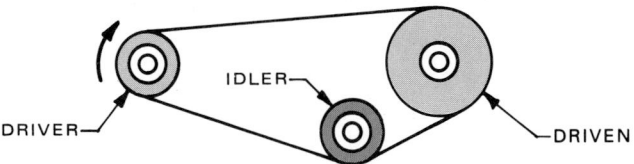

(D) INSIDE IDLER PULLEY ON THE TIGHT SIDE OF THE DRIVE

FIG. 20-1-10   Location of idler pulleys.

Proper duty classification helps to ensure maximum drive life. The following are typical duty classifications:

- *Light Duty*   Household washers, household ironers, dishwashers, fans and blowers, centrifugal pumps.
- *Normal Duty*   Oil burners, buffers, heating and ventilating fans, meat slicers, speed-up drives, drill presses, generators, power lawn mowers.
- *Heavy Duty*   Gasoline engine drives, metalworking machines, sanding machines, stokers, spray equipment, woodworking machines, lathes, industrial machines, refrigerators, compressors, piston or plunger pumps, grinders.

The horsepower (kilowatt) ratings listed in Fig. 20-1-11 (on pg. 674) are suitable for normal-duty applications. For light duty, multiply the normal-duty rating by 0.85. For heavy duty, multiply the normal rating by 1.2.

Three Easy Steps

Step 1:   Selecting Driver V-Pulley Diameter and Belt Cross Section   First, classify the application and apply the proper service factor, as explained above. Refer to Fig. 20-1-11 for driver V-pulley diameter and belt cross section.

Step 2:   Choosing Driven V-Pulley Diameter   Refer to Fig. 20-1-12 (pg. 675) for the speed of the motor. Locate desired driven speed in driver V-pulley column; read driven V-pulley diameter in the first column.

Step 3:   Finding Belt Length and Center Distance   Add the diameter of driver and driven V-pulley and refer to Fig. 20-1-13 (pg. 676). Locate the sum of V-pulley diameters at the top of the chart, read down to the required centers, and read the belt length in the belt length column.

Although the amount of stretch in V-belts is relatively small, some adjustment between centers of pulleys is necessary to compensate for stretch and side wear on the belts and sheaves.

To design a belt drive, the following information should be known:

1. The speed [revolutions per minute (r/min)] and horsepower (kilowatts) of the motor or driver unit
2. The speed (r/min) at which the driven shaft is to turn
3. The space available for the drive

EXAMPLE 1

A .5 hp, 1,750 r/min motor is to operate a drill press having a spindle speed of approximately 1,200 r/min. The center distance between the motor shaft and spindle is approximately 19.5 in. The type of drive required is V-belt.

SOLUTION

Since drill press operations come under the classification of normal duty, no adjustment needs to be made to the horsepower (kilowatt) rating.

Step 1: Selecting Driver V-Pulley Diameter and Belt Cross Section (Fig. 20-1-11)   Read down the extreme left column to the r/min figure nearest that of the speed of the motor, which

KILOWATT RATINGS (METRIC)

R/min of Small Pulley	Outside Diameter of Small V-Pulley—Millimeters														
	38	44	51	57	64	70	76	83	89	95	102	108	114	121	127
200											0.13	0.16	0.18	0.21	0.22
400			0.04	0.06	0.09	0.11	0.13	0.16	0.19	0.23	0.26	0.31	0.34	0.39	0.42
600	0.03	0.05	0.06	0.09	0.13	0.16	0.20	0.24	0.27	0.33	0.38	0.43	0.49	0.54	0.60
800	0.04	0.06	0.08	0.11	0.16	0.21	0.25	0.31	0.34	0.41	0.48	0.55	0.60	0.69	0.75
1000	0.04	0.07	0.09	0.13	0.19	0.25	0.31	0.36	0.41	0.48	0.56	0.64	0.74	0.82	0.90
1160	0.05	0.08	0.11	0.16	0.22	0.28	0.34	0.40	0.46	0.51	0.63	0.73	0.80	0.92	1.01
1400	0.06	0.09	0.13	0.17	0.25	0.32	0.40	0.48	0.55	0.63	0.72	0.82	0.93	1.06	1.16
1600	0.06	0.10	0.14	0.19	0.27	0.36	0.43	0.51	0.60	0.67	0.76	0.90	1.01	1.14	1.25
1750	0.06	0.11	0.15	0.19	0.28	0.38	0.47	0.55	0.63	0.72	0.81	0.93	1.07	1.20	1.33
2000	0.07	0.12	0.16	0.21	0.31	0.41	0.51	0.60	0.69	0.78	0.87	1.01	1.15	1.29	1.42
2200	0.07	0.13	0.18	0.23	0.33	0.43	0.54	0.64	0.74	0.84	0.94	1.05	1.20	1.34	1.48
2400	0.07	0.13	0.19	0.24	0.34	0.46	0.57	0.68	0.78	0.89	0.98	1.08	1.23	1.39	1.51
2600	0.07	0.14	0.19	0.26	0.35	0.48	0.59	0.72	0.81	0.93	1.03	1.10	1.26	1.41	1.56
2800	0.08	0.14	0.21	0.27	0.36	0.49	0.62	0.74	0.85	0.95	1.06	1.10	1.28	1.42	1.57
3000	0.08	0.16	0.22	0.29	0.37	0.51	0.63	0.76	0.88	0.98	1.09	1.10	1.26	1.41	1.55
3200	0.08	0.16	0.22	0.29	0.38	0.52	0.66	0.78	0.90	1.01	1.12	1.12	1.25	1.39	1.51
3450	0.09	0.16	0.24	0.31	0.38	0.53	0.67	0.80	0.92	1.03	1.13	1.13	1.20	1.33	1.45
3600	0.09	0.16	0.25	0.31	0.39	0.54	0.68	0.81	0.93	1.04	1.15	1.15	1.15	1.28	1.38
3800	0.09	0.16	0.25	0.31	0.39	0.54	0.69	0.81	0.93	1.05	1.15	1.15	1.15	1.19	1.28
4000	0.09	0.16	0.25	0.33	0.40	0.54	0.69	0.82	0.94	1.05	1.15	1.15	1.13	1.13	1.16

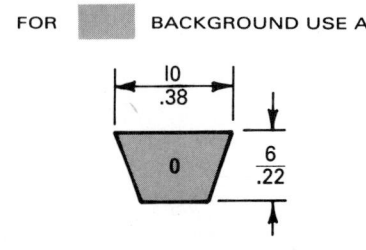

FOR [grey] BACKGROUND USE A

0 |← 10 / .38 →| 6 / .22

HORSEPOWER RATINGS (U.S. CUSTOMARY)

RPM of Small Pulley	Outside Diameter of Small V-Pulley—Inches														
	1.50	1.75	2.00	2.25	2.50	2.75	3.00	3.25	3.50	3.75	4.00	4.25	4.50	4.75	5.00
200	...	...	...	...	...	...	...	...	...	...	0.18	0.22	0.24	0.28	0.29
400	...	...	0.06	0.08	0.12	0.15	0.18	0.22	0.25	0.31	0.35	0.42	0.46	0.52	0.56
600	0.04	0.07	0.08	0.12	0.18	0.22	0.27	0.32	0.36	0.44	0.51	0.58	0.66	0.73	0.81
800	0.05	0.08	0.11	0.15	0.22	0.28	0.34	0.41	0.45	0.55	0.64	0.74	0.81	0.93	1.00
1000	0.06	0.10	0.12	0.18	0.26	0.33	0.42	0.48	0.55	0.64	0.75	0.86	0.99	1.10	1.21
1160	0.07	0.11	0.15	0.21	0.29	0.38	0.46	0.54	0.62	0.69	0.84	0.98	1.07	1.23	1.35
1400	0.08	0.12	0.17	0.23	0.33	0.43	0.53	0.64	0.74	0.84	0.96	1.10	1.25	1.42	1.55
1600	0.08	0.14	0.19	0.25	0.36	0.48	0.58	0.69	0.80	0.90	1.02	1.20	1.36	1.53	1.68
1750	0.08	0.15	0.20	0.25	0.38	0.51	0.63	0.74	0.85	0.96	1.08	1.25	1.43	1.61	1.78
2000	0.09	0.16	0.22	0.28	0.41	0.55	0.68	0.81	0.92	1.05	1.17	1.35	1.54	1.73	1.90
2200	0.09	0.17	0.24	0.31	0.44	0.58	0.72	0.86	0.99	1.12	1.25	1.41	1.61	1.80	1.99
2400	0.10	0.18	0.25	0.32	0.45	0.61	0.76	0.91	1.05	1.19	1.32	1.45	1.65	1.86	2.02
2600	0.10	0.19	0.26	0.35	0.47	0.64	0.79	0.96	1.09	1.24	1.38	1.48	1.69	1.89	2.09
2800	0.11	0.19	0.28	0.36	0.48	0.66	0.83	0.99	1.14	1.28	1.42	1.48	1.71	1.91	2.11
3000	0.11	0.21	0.29	0.39	0.49	0.68	0.85	1.02	1.18	1.32	1.46	1.48	1.69	1.89	2.08
3200	0.11	0.21	0.30	0.39	0.51	0.70	0.88	1.05	1.20	1.36	1.50	1.50	1.67	1.86	2.03
3450	0.12	0.22	0.32	0.41	0.51	0.71	0.90	1.07	1.23	1.38	1.52	1.52	1.61	1.78	1.94
3600	0.12	0.22	0.33	0.42	0.52	0.72	0.91	1.09	1.25	1.40	1.54	1.54	1.54	1.71	1.85
3800	0.12	0.22	0.33	0.42	0.52	0.72	0.92	1.09	1.25	1.41	1.54	1.54	1.54	1.59	1.72
4000	0.12	0.22	0.34	0.44	0.53	0.72	0.92	1.10	1.26	1.40	1.52	1.52	1.52	1.52	1.55

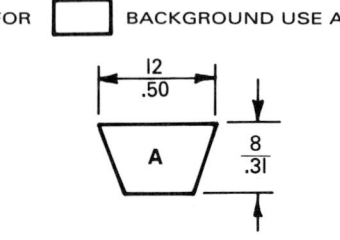

FOR [white] BACKGROUND USE A

A |← 12 / .50 →| 8 / .31

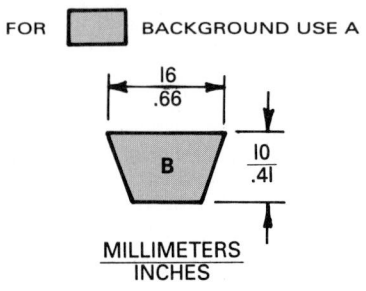

FOR [grey] BACKGROUND USE A

B |← 16 / .66 →| 10 / .41

MILLIMETERS / INCHES

NOTE:
THIS TABLE INCORPORATES A SERVICE FACTOR OF 1.3. FOR HEAVY DUTY, MULTIPLY NORMAL-DUTY RATING BY .85. FOR LIGHT DUTY, MULTIPLY NORMAL-DUTY RATING BY 1.20.

SIZES SHOWN ARE INCH SIZES SOFT CONVERTED TO MILLIMETERS.

FIG. 20-1-11   Calculating pulley diameter of drive shaft and belt cross section.
(T. B. Wood's Sons Co.)

is 1,750 r/min. Read across this line to the figure closest to the design horsepower or kilowatts of the drive. The closest horsepower rating is found to be .51. Read up from the .51 hp figure. The figure at the top of the column is the outside diameter of the motor pulley in inches. The .51 hp figure is in the white area. The reference at the side of the chart gives the size of the belt required.

Pulley size for motor = 2.75 in.

Belt section = .50 in. wide
× .31 in. thick

Step 2: Choosing Driven V-Pulley Diameter (Fig. 20-1-12)
Refer to the table for driven speeds for 1,750 r/min motors. Read across the top of the table to the figure nearest the small

DRIVEN SPEEDS FOR 1160 R/MIN MOTORS

DriveN V-Pulley OD in./mm	DriveR V-Pulley OD—in./mm												
	1.50 38	1.75 44	2.00 51	2.25 57	2.50 64	2.75 70	3.00 76	3.25 83	3.50 89	3.75 95	4.00 102	4.25 108	4.50 114
1.5/ 38	1160	1392	1625	1855	2085	2325	2550	2785	3015	3250	3480	3715	
2.0/ 51	829	995	1160	1325	1490	1658	1825	1988	2150	2315	2485	2650	
2.5/ 64	645	774	903	1031	1160	1290	1418	1546	1675	1805	1933	2032	2190
3.0/ 76	528	634	739	845	950	1057	1160	1266	1370	1475	1580	1685	1793
3.5/ 89	447	536	625	715	804	894	982	1071	1160	1248	1340	1428	1518
4.0/102	387	465	542	620	696	775	851	929	1008	1082	1160	1238	1315
4.5/114	341	409	477	545	614	682	750	819	886	955	1022	1091	1160
5.0/127	305	366	427	488	549	610	671	732	794	854	915	976	1039
5.5/140	277	332	381	442	497	553	608	663	718	774	829	884	939
6.0/152	253	302	353	404	454	505	555	605	655	706	756	806	857
7.0/178	215	258	301	344	388	430	474	516	560	602	648	688	732
8.0/203	187	224	262	297	337	374	411	449	486	524	561	599	636
10.0/254	149	179	208	238	268	298	328	357	387	417	446	477	506
12.0/305	123	148	173	197	222	247	272	296	321	346	370	395	420

DRIVEN SPEEDS FOR 1750 R/MIN MOTORS

DriveN V-Pulley OD in./mm	DriveR V-Pulley OD—in./mm												
	1.50 38	1.75 44	2.00 51	2.25 57	2.50 64	2.75 70	3.00 76	3.25 83	3.50 89	3.75 95	4.00 102	4.25 108	4.50 114
1.5/ 38	1750	2100	2450	2800	3150	3500	3850						
2.0/ 51	1250	1500	1750	2000	2250	2500	2750	3000	3250	3500	3750	4000	
2.5/ 64	974	1167	1360	1555	1750	1945	2140	2330	2530	2725	2915	3110	3305
3.0/ 76	797	955	1113	1272	1431	1590	1750	1910	2070	2225	2385	2545	2700
3.5/ 89	674	808	942	1077	1210	1346	1480	1615	1750	1885	2020	2155	2290
4.0/102	584	700	817	935	1050	1168	1283	1400	1518	1634	1750	1865	1985
4.5/114	516	618	720	824	926	1030	1131	1235	1339	1440	1543	1650	1750
5.0/127	462	554	646	737	830	922	1013	1105	1198	1290	1382	1473	1568
5.5/140	417	500	584	667	750	834	917	1000	1082	1167	1250	1333	1417
6.0/152	381	456	533	610	685	760	837	913	990	1065	1140	1217	1290
6.5/165	350	420	490	560	630	700	771	840	910	980	1050	1120	1190
7.0/178	324	389	454	518	584	648	713	778	843	907	973	1039	1102
8.0/203	282	339	394	451	507	564	620	676	734	789	845	902	959
9.0/229	250	300	350	400	450	500	550	600	650	700	750	800	850
10.0/254	224	270	315	360	405	450	495	540	585	630	675	720	765
11.0/279	203	244	285	326	366	407	448	488	530	570	610	652	692
12.0/305	186	224	261	298	336	373	410	446	485	522	560	596	634

DRIVEN SPEEDS FOR 3500 R/MIN MOTORS

DriveN V-Pulley OD in./mm	DriveR V-Pulley OD—in./mm												
	1.50 38	1.75 44	2.00 51	2.25 57	2.50 64	2.75 70	3.00 76	3.25 83	3.50 89	3.75 95	4.00 102	4.25 108	4.50 114
1.5/ 38	3500	4200	4900	5600	6300	7000	7700						
2.0/ 51	2500	3000	3500	4000	4500	5000	5500	6000	6500	7000	7500	8000	
2.5/ 64	1948	2324	2720	3110	3500	3890	4280	4660	5060	5450	5830	6220	6610
3.0/ 76	1594	1910	2236	2544	2862	3180	3500	3820	4140	4450	4770	5090	5400
3.5/ 89	1348	1616	1884	2154	2420	2692	2960	3230	3500	3770	4040	4310	4580
4.0/102	1168	1400	1634	1870	2030	2336	2566	2800	3036	3268	3500	3730	3970
4.5/114	1032	1236	1440	1648	1852	2060	2262	2470	2678	2880	3086	3300	3500
5.0/127	924	1108	1292	1474	1660	1844	2026	2210	2396	2580	2764	2946	3136
5.5/140	834	1000	1168	1334	1500	1668	1834	2000	2164	2334	2500	2666	2834
6.0/152	762	912	1066	1220	1370	1520	1774	1826	1980	2130	2280	2434	2580
6.5/165	700	840	980	1120	1260	1400	1542	1680	1820	1960	2100	2240	2380
7.0/178	648	778	908	1036	1168	1296	1426	1556	1686	1814	1946	2078	2204
8.0/203	564	678	788	902	1014	1128	1240	1352	1468	1578	1690	1804	1918
9.0/229	500	600	700	800	900	1000	1100	1200	1300	1400	1500	1600	1750
10.0/254	448	540	630	720	810	900	990	1080	1170	1260	1350	1440	1530
11.0/279	406	488	570	652	732	814	896	976	1060	1140	1220	1304	1384
12.0/305	372	448	522	596	672	746	820	892	970	1044	1120	1192	1268

FIG. 20-1-12 Calculating revolutions per minute and diameter of driven pulley.
(T. B. Wood's Sons Co.)

SUM OF BOTH V-BELT PULLEY DIAMETERS—DIMENSIONS IN INCHES

Installation Allowance	Take-Up	Belt Length	4	4.5	5	5.5	6	6.5	7	7.5	8	8.5	9	9.5	10	10.5	11	11.5	12	12.5	13	13.5	14	14.5	15	15.5	16
.50	.50	16	4.9	4.5	4.1																							
.62	.50	18	5.9	5.5	5.1	4.6																					
.62	.50	20	6.9	6.5	6.1	5.6	5.2																				
.62	.50	22	7.9	7.5	7.1	6.6	6.2	5.8																			
.62	.50	24	8.9	8.5	8.1	7.6	7.2	6.8	6.3	5.8																	
.62	.50	26	9.9	9.5	9.1	8.6	8.2	7.8	7.3	6.9	6.5																
.62	.50	28	10.9	10.5	10.1	9.6	9.2	8.8	8.4	7.9	7.6	7.1	6.6														
.62	.50	30	11.9	11.5	11.1	10.6	10.2	9.8	9.4	8.9	8.6	8.1	7.7	7.3													
.62	.50	32	12.9	12.5	12.1	11.6	11.2	10.8	10.4	10.0	9.6	9.1	8.7	8.4	8.0												
.62	.50	34	13.9	13.5	13.1	12.7	12.2	11.8	11.4	11.0	10.6	10.2	9.7	9.4	9.0	8.6											
.62	.50	36	14.9	14.5	14.1	13.7	13.2	12.8	12.4	12.0	11.6	11.2	10.7	10.4	10.0	9.6	9.0										
.62	.50	38	15.9	15.5	15.1	14.7	14.2	13.8	13.4	13.0	12.6	12.2	11.8	11.4	11.0	10.6	10.0	9.7	9.1								
.75	.50	40	16.9	16.5	16.1	15.7	15.3	14.8	14.4	14.0	13.6	13.2	12.8	12.4	12.0	11.6	11.1	10.7	10.1	9.8							
.75	.50	42	17.9	17.5	17.1	16.7	16.3	15.8	15.4	15.0	14.6	14.2	13.8	13.4	13.1	12.6	12.1	11.7	11.2	10.8	10.2						
.75	.50	44	18.9	18.5	18.1	17.7	17.3	16.8	16.4	16.0	15.6	15.2	14.8	14.4	14.1	13.6	13.1	12.8	12.2	11.9	11.2	10.9					
.75	.50	46	19.9	19.5	19.1	18.7	18.3	17.9	17.4	17.0	16.6	16.2	15.8	15.4	15.1	14.6	14.1	13.8	13.2	12.9	12.3	12.0	10.9	10.5			
.75	.50	48	20.9	20.5	20.1	19.7	19.3	18.9	18.4	18.0	17.7	17.2	16.8	16.4	16.1	15.6	15.1	14.8	14.3	13.9	13.3	13.0	12.0	11.6	11.3		
.75	.50	50	21.9	21.5	21.1	20.7	20.3	19.9	19.4	19.0	18.7	18.2	17.8	17.4	17.1	16.7	16.2	15.8	15.3	14.9	14.4	14.0	13.1	12.7	12.4	12.1	11.7
.75	.50	52	22.9	22.5	22.1	21.7	21.3	20.9	20.4	20.0	19.7	19.2	18.8	18.4	18.1	17.7	17.2	16.8	16.3	15.9	15.4	15.0	14.1	13.8	13.5	13.1	12.8
.75	.50	54	23.9	23.5	23.1	22.7	22.3	21.9	21.4	21.0	20.7	20.2	19.8	19.4	19.1	18.7	18.2	17.8	17.3	17.0	16.4	16.1	15.2	14.8	14.5	14.2	13.8
.75	.50	56	24.9	24.5	24.1	23.7	23.3	22.9	22.4	22.0	21.7	21.2	20.8	20.4	20.1	19.7	19.2	18.8	18.3	18.0	17.4	17.1	16.2	15.9	15.6	15.2	14.9
.75	.50	58	25.9	25.5	25.1	24.7	24.3	23.9	23.4	23.0	22.7	22.2	21.8	21.4	21.1	20.7	20.2	19.8	19.3	19.0	18.5	18.1	17.3	16.9	16.6	16.3	15.9
.88	.75	60	26.9	26.5	26.1	25.7	25.3	24.9	24.5	24.0	23.7	23.2	22.8	22.4	22.1	21.7	21.2	20.8	20.4	20.0	19.5	19.1	18.3	18.0	17.6	17.3	17.0
.88	.75	62	27.9	27.5	27.1	26.7	26.3	25.9	25.5	25.0	24.7	24.3	23.8	23.4	23.1	22.7	22.2	21.8	21.4	21.0	20.5	20.1	19.4	19.0	18.7	18.3	18.0
.88	.75	64	28.9	28.5	28.1	27.7	27.3	26.9	26.5	26.0	25.7	25.3	24.8	24.4	24.1	23.7	23.2	22.9	22.4	22.0	21.5	21.1	20.4	20.0	19.7	19.4	19.0
.88	.75	66	29.9	29.5	29.1	28.7	28.3	27.9	27.5	27.0	26.7	26.3	25.9	25.4	25.1	24.7	24.2	23.9	23.4	23.0	22.5	22.2	21.4	21.1	20.7	20.4	20.0
.88	.75	68	30.9	30.5	30.1	29.7	29.3	28.9	28.5	28.1	27.7	27.3	26.9	26.4	26.1	25.7	25.2	24.9	24.4	24.0	23.5	23.2	22.4	22.1	21.7	21.4	21.0
.88	.75	70	31.9	31.5	31.1	30.7	30.3	29.9	29.5	29.1	28.7	28.3	27.9	27.4	27.1	26.7	26.2	25.9	25.4	25.0	24.5	24.2	23.5	23.1	22.8	22.4	22.1
.88	.75	72	32.9	32.5	32.1	31.7	31.3	30.9	30.5	30.1	29.7	29.3	28.9	28.4	28.1	27.7	27.2	26.9	26.4	26.0	25.5	25.2	24.5	24.1	23.8	23.4	23.1
.88	.75	74	33.9	33.5	33.1	32.7	32.3	31.9	31.5	31.1	30.7	30.3	29.9	29.4	29.1	28.7	28.2	27.9	27.4	27.0	26.5	26.2	25.5	25.1	24.8	24.4	24.1
.88	.75	76	34.9	34.5	34.1	33.7	33.3	32.9	32.5	32.1	31.7	31.3	30.9	30.4	30.1	29.7	29.2	28.9	28.4	28.0	27.6	27.2	26.5	26.2	25.8	25.5	25.1
.88	.75	78	35.9	35.5	35.1	34.7	34.2	33.9	33.5	33.1	32.7	32.3	31.9	31.4	31.1	30.7	30.2	29.9	29.4	29.0	28.6	28.2	27.5	27.2	26.8	26.5	26.1
.88	.75	80	36.9	36.5	36.1	35.7	35.3	34.9	34.5	34.1	33.7	33.3	32.9	32.4	32.1	31.7	31.3	30.9	30.4	30.0	29.6	29.2	28.6	28.2	27.9	27.5	27.1
.88	.75	82	37.6	37.1	36.7	36.3	35.9	35.5	35.1	34.7	34.3	33.9	33.5	33.0	32.7	32.3	31.9	31.5	31.0	30.7	30.2	29.8	29.2	28.8	28.5	28.1	27.8
.88	.75	84	38.9	38.5	38.1	37.7	37.3	36.9	36.5	36.1	35.7	35.3	34.9	34.4	34.1	33.7	33.3	32.9	32.4	32.1	31.6	31.2	30.6	30.2	29.9	29.5	29.2

SUM OF BOTH V-BELT PULLEY DIAMETERS—DIMENSIONS IN MILLIMETERS (METRIC)

Installation Allowance	Take-Up	Belt Length	125	140	150	165	180	190	200	215	230	240	255	265	280	290	305	320	330	345	355	370
12	10	410	105																			
15	10	460	130	117																		
15	10	510	155	142	132																	
15	10	560	180	168	157	147																
15	10	610	205	193	183	173	160	147														
15	10	660	231	218	208	198	185	175	165													
15	10	710	257	269	234	224	213	201	193	180	168											
15	10	760	282	295	259	249	239	226	218	205	196	185										
15	10	810	308	295	284	274	264	254	244	231	221	213	203									
15	10	860	333	323	310	230	290	280	269	259	246	239	228	218								
15	10	910	358	348	335	325	315	305	295	284	272	264	254	244	229							
15	10	960	384	373	361	351	340	330	320	310	300	290	279	269	254	246	231					
20	10	1010	409	399	389	376	366	356	345	335	325	315	305	295	282	272	257	249				
20	10	1070	434	424	414	401	391	381	371	361	351	340	333	320	307	297	284	274	259			
20	10	1120	460	450	439	427	417	406	396	386	376	366	358	345	333	325	310	302	284	277		
20	10	1170	485	475	465	455	442	432	422	411	401	391	384	371	358	351	335	328	312	304	277	266
20	10	1220	511	500	490	480	467	457	450	437	427	417	409	396	384	376	363	353	338	330	304	295
20	10	1270	536	526	516	505	493	483	475	462	452	442	434	424	411	401	389	378	366	358	333	323
20	10	1320	513	551	541	531	518	508	500	488	478	467	460	450	437	427	414	404	391	381	358	351
20	10	1370	587	577	566	556	544	534	526	513	503	493	485	475	462	452	439	431	417	410	386	376
20	10	1420	612	602	592	582	569	559	551	538	528	518	511	500	488	478	465	457	442	434	411	404
20	10	1470	638	627	617	607	594	584	577	564	554	544	536	526	513	503	490	482	470	460	439	429
22	15	1520	663	653	643	632	622	610	602	589	579	569	561	551	538	528	518	508	495	485	465	457
22	15	1570	688	678	668	658	648	635	627	617	605	594	587	577	564	554	544	533	520	511	493	483
22	15	1630	714	704	693	683	673	660	653	643	630	620	612	602	589	582	569	559	546	536	518	508
22	15	1680	790	729	719	709	699	686	678	668	658	645	638	627	615	607	594	584	571	564	544	536
22	15	1730	765	754	744	734	724	714	704	693	683	671	663	653	640	632	620	610	597	589	569	561
22	15	1780	790	780	770	759	749	739	729	719	709	696	688	678	665	658	645	635	622	615	597	587
22	15	1830	815	805	798	785	765	765	754	744	734	721	714	704	691	683	671	660	647	640	622	612
22	15	1880	841	831	820	810	800	790	780	770	759	747	739	729	716	709	696	685	673	665	648	638
22	15	1930	866	886	846	836	826	815	805	795	785	772	765	754	742	734	721	711	701	691	673	665
22	15	1980	892	881	869	861	851	841	831	820	810	798	790	780	767	759	747	736	726	716	698	691
22	15	2030	917	907	897	886	876	866	856	846	836	823	815	805	795	785	772	762	752	742	726	716
22	15	2080	932	922	912	902	892	881	871	861	851	838	831	820	810	800	787	780	767	757	742	732
22	15	2130	968	958	947	937	927	917	907	897	886	874	866	856	846	836	823	815	803	792	777	767

FIG. 20-1-13   Determining V-belt length from pulley diameter and center distance. *(T. B. Wood's Sons Co.)*

pulley size. Column 2.75 corresponds exactly with the small pulley diameter. Read down this column to the figure nearest the desired speed (1,200 r/min) of the driven shaft. The nearest figure is 1,168. By reading to the left of this figure, the driven-pulley diameter is found to be 4.00 in.

Step 3: Finding Belt Length and Center Distance (Fig. 20-1-13)   Add the diameter of the pulleys and select the number in the top row that is nearest to this sum.

Motor pulley diameter = 2.75 in.

Spindle pulley diameter = 4.00 in.

Sum of diameter = 6.75 in.

The exact sum of the diameters is not shown on the top row; use 7.00 in. Read down this column to the figure which is closest to the desired center distance of 19.5 in. Use 19.4 in. since the approximate center distance required is 19.5 in. Follow along this line to the left to column Belt Length to obtain a belt length of 50 in.

REFERENCES AND SOURCE MATERIAL

1. *Machine Design,* Mechanical drives reference issue.
2. The Gates Rubber Co., Denver, CO.
3. T. B. Wood's Sons Co.

ASSIGNMENTS

See Assignments 1 and 2 for Unit 20-1 on pages 704–705.

20-2 CHAIN DRIVES

Nearly all types of power-transmission chains have two basic components: side bars or link plates, and pin and bushing joints. The chain articulates at each joint to operate around a toothed sprocket. The *pitch* of the chain is the distance between centers of the articulating joints.

Power-transmission chains have several advantages: relatively unrestricted shaft center distances, compactness, ease of assembly, elasticity in tension with no slip or creep, and ability to operate in relatively high-temperature atmosphere. A typical application can be seen in Fig. 20-2-1.

Basic Types

There are six major types of power-transmission chains, with numerous modifications and special shapes for specific applications. A seventh type, the bead chain, is often used for light-duty applications. Figure 20-2-2 shows basic characteristics of five of the major types.

Detachable

The malleable detachable chain is made in a range of sizes from .902 to 4.063 in. (23 to 103 mm) pitch and ultimate

FIG. 20-2-1   Chain drives.

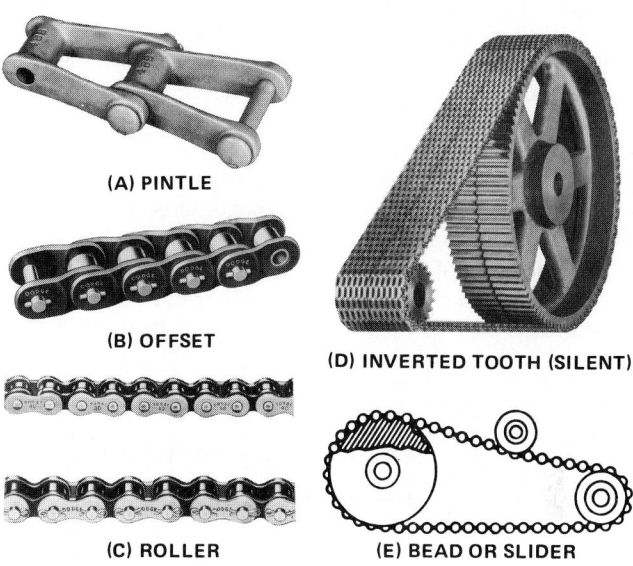

(A) PINTLE

(B) OFFSET

(C) ROLLER

(D) INVERTED TOOTH (SILENT)

(E) BEAD OR SLIDER

FIG. 20-2-2   Basic chain types.

strength from 700 to 17,000 lb/in^2 [5 to 110 megapascals (MPa)].

Of the same type is the steel detachable chain, covered by ANSI B29.6. This chain is made in sizes from .904 in. (23 mm) to just under 3.00 in. (76 mm) in pitch, with ultimate strength from 760 to 5,000 lb/in.^{2} (5 to 35 MPa).

The ends of the detachable link are referred to as the *bar end* and the *hook end*.

Pintle

For slightly higher speeds [to about 450 ft/min (2.2 m/s)] and heavier loads pintle chains are used. *Pintle chains* are made up of individual cast links having a full, round barrel end with offset sidebars. These links are intercoupled with steel pins. The ends of pintle chain links are referred to as the *barrel end* and *open end*.

Many of these chains have been designed to operate over sprockets intended for detachable chain. Therefore, chains range from just over 1.00 in. (25 mm) up to 6.00 in. (150 mm) in pitch with ultimate strengths from 3,600 to 30,000 lb/in.^{2} (25 to 200 MPa).

Offset-Sidebar

Steel offset-sidebar chains are used extensively as drive chains on construction machinery. They operate at speeds to 1,000 ft/min (5 m/s) and transmit loads to approximately 250 hp (185 kW).

Each link has two offset sidebars, one bushing, one roller, one pin, and if the chain is detachable, a cotter pin. Some offset-sidebar chains are made without rollers.

Roller

Transmission roller chain (Fig. 20-2-3) is available in pitches from .25 to 3.00 in. (6 to 75 mm). In the single-width roller, the ultimate strength ranges from 925 to 130,000 lb/in.^{2} (6 to 900 MPa). It is also available in multiple widths. Small-pitch sprockets can operate at speeds as high as 10,000 r/min, and 1,000 to 1,200 hp (750 to 900 kW) drives are not unusual.

These chains are assembled from roller links and pin links. If the chain is detachable, cotter pins are used in the chain pin holes.

ANSI B29.1 also covers a number of special types of roller chains. One is equipped with oil-impregnated, sintered, powdered metal bushings for self-lubrication. This chain handles lighter loads at reduced speeds and is limited in application because it does not use rollers. Instead, it uses bushings of the same outside diameter as normal rollers (Fig. 20-2-4).

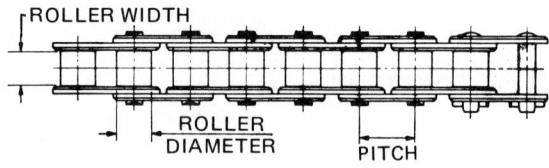

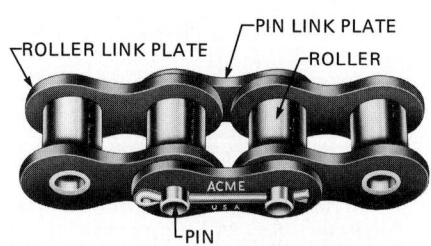

SINGLE STEEL   DOUBLE STEEL   DOUBLE CAST IRON

(B) SPROCKETS

FIG. 20-2-3   Roller chain terminology and sprockets.

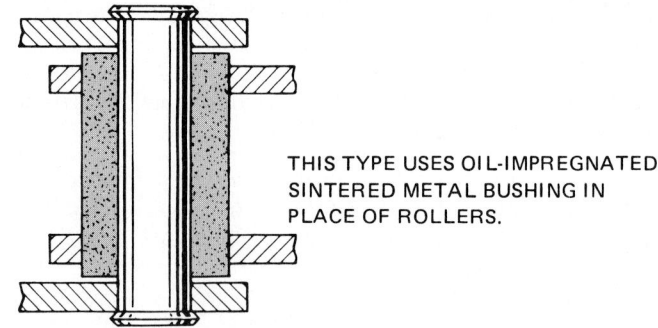

THIS TYPE USES OIL-IMPREGNATED SINTERED METAL BUSHING IN PLACE OF ROLLERS.

FIG. 20-2-4   Self-lubricating chain.

Another approach to self-lubrication has been the use of special roller chains with plastic sleeves between the chain rivets and bushings. The plastic reduces joint friction. Plastic chains are available for special applications.

Double-Pitch

These are basically the same as roller chains except that the pitch is twice as long. Roller and double-pitch chains have the same diameter pins and rollers, the same width rollers, and the same thickness of link plates.

Inverted-Tooth Silent

These are high-speed chains, used predominantly for prime-mover, power takeoff drives, such as on power cranes or shovels, machine tools, and pumps. Drives transmitting up to 1,200 hp (900 kW) are in use.

These chains are made up of a series of tooth links, alternately assembled with either pins or a combination of joint components in such a way that the joint articulates between adjoining pitches. Center-guide chain has guide links that engage a groove or grooves in the sprocket, while the side-guide chain has guides that engage the sides of the sprocket.

Bead or Slider

Bead chains are used as manually controlled or slow-speed drives in numerous products, such as television tuners, radio tuners, computing devices, time recorders, air conditioners, toys, display drives, ventilator controls, and venetian blinds.

Sprockets

Basic sprocket types used with precision steel roller chains conform to ANSI standards.

Used for mounting on flanges, hubs, or other devices, the plate sprocket is a flat, hubless sprocket.

Small- and medium-size hub sprockets are turned from bar stock or forgings or are made by welding a bar-stock hub to a hot-rolled plate. For small, low-load applications, only one hub extension may be needed. Large-diameter sprockets normally have two hub projections equidistant from the center plane of the sprocket.

Materials

Although normally machined from gray-iron castings, sprockets are also available in cast steel or welded hub construction.

Sprockets made of sintered powdered metal, and from nylon and other plastics, have become economical in large quantities. These sprockets offer many advantages. For example, plastic sprockets require minimum lubrication and are widely used where cleanliness is essential.

Design of Roller Chain Drives

The design of a roller chain drive consists primarily of the selection of the chain and sprocket sizes. It also includes the determination of chain length, center distance, method of lubrication, and in some cases, the arrangement of chain casings and idlers.

Unlike belt drives, which are based on lineal speeds in feet per minute or meters per second, the limiting factor of chain drives is based on the rotative speed, or revolutions per minute, of the smaller sprocket, which in most installations is the driven member.

Design of chain drives is based, not only on horsepower (kilowatts) and speeds, but on the following factors relative to broad service conditions:

1. Average horsepower (kilowatts) to be transmitted (Fig. 20-2-5).
2. The revolutions per minute of the driving and driven members.
3. Shaft diameter.
4. Permissible diameters of sprockets.
5. Load characteristics, whether smooth and steady, pulsating, heavy-starting, or subject to peaks.
6. Lubrication, whether periodic, occasional, or copious. Where chains are exposed to dust, dirt, or injurious foreign matter, chain cases should be used.
7. Life expectancy: the amount of service required, or total life. It is much better to "overchain" than to skimp on the size of the chain used.

In designing chain drives, it is of the utmost importance to consider and study the pitch or size of the chain used. The number of revolutions per minute and the size of the smaller or faster-moving sprocket determine the pitch of chain that should be used.

Smaller-pitch chains in single or multiple widths are adaptable for elevated-speed drives and also for any speed drives where smoother and quieter performance is essential.

Large-pitch chains are adaptable for slow- and medium-speed drives.

Multiple-width roller chains are becoming increasingly popular. They not only solve the problems of transmitting greater power at higher speeds, but because of their smoother action, they substantially reduce the noise factor (Fig. 20-2-6).

Chain fpm	Speed m/s	Power hp	kW	Type of Chain
350	1.0	20	15	Detachable
450	2.2	40	30	Pintle
1000	5	250	190	Offset-sidebar
2500	12.5	1500	1100	Roller
4000	20	2500	1850	Silent

FIG. 20-2-5   Tentative selection factors for chain drives.

Size of Sprockets   It is general practice to use a minimum-size sprocket of 17 teeth in order to obtain smooth operation at high speeds. Because of the lessening of tooth impact, 19- or 21-tooth sprockets should be considered from a standpoint of greater life expectancy and smoother operation. On slow-speed and special-purpose installations or where space limitations are involved, sprockets smaller than 17-tooth can be used. The normal maximum number of teeth is 120.

Ordinary practice indicates that the ratio of driver to driven sprockets should be no more than 6:1. The recommended chain wrap on the driver is 120°.

Center Distances   Center distances must be more than one-half the diameter of the smaller sprocket, plus one-half of the diameter of the larger sprocket; otherwise the sprocket teeth will touch. (When necessary, drives may be operated with a small amount of clearance between sprockets.) Best results are obtained by using a center distance of 30 to 50 times the pitch of the chain used. Eighty times the pitch is considered maximum.

Chain Tension   Chains should never run with both sides tight. Adjustable centers should be provided when possible to permit proper initial slack and to allow for periodic adjustment necessitated by natural chain wear. The chain sag should be equivalent to approximately 2 percent of the center distance (Fig. 20-2-7, pg. 680).

Idler sprockets should be used as a means of taking up chain slack where it is not possible to provide adjustable centers.

Chain Length   Chain length is a function of the number of teeth in both sprockets and of the center distance. In addition, the chain must consist of an integral number of pitches, with an even number preferable, in order to avoid the use of an offset link.

Chain Length Formula   For simplicity it is customary to compute the chain length in terms of chain pitches and then to

FIG. 20-2-6   Multiple-roller chain drive.

FIG. 20-2-7 Chain drives.

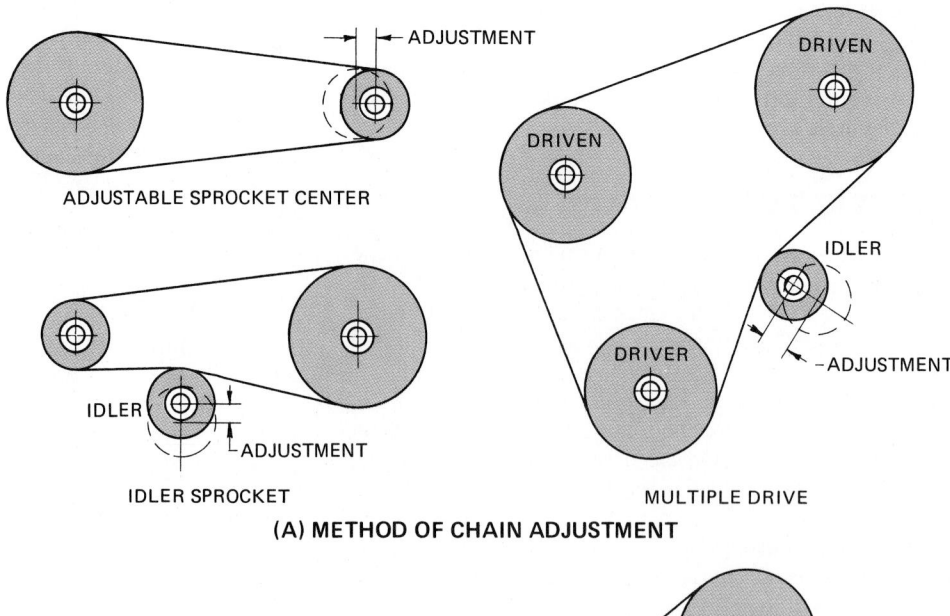

ADJUSTABLE SPROCKET CENTER

IDLER SPROCKET

MULTIPLE DRIVE

(A) METHOD OF CHAIN ADJUSTMENT

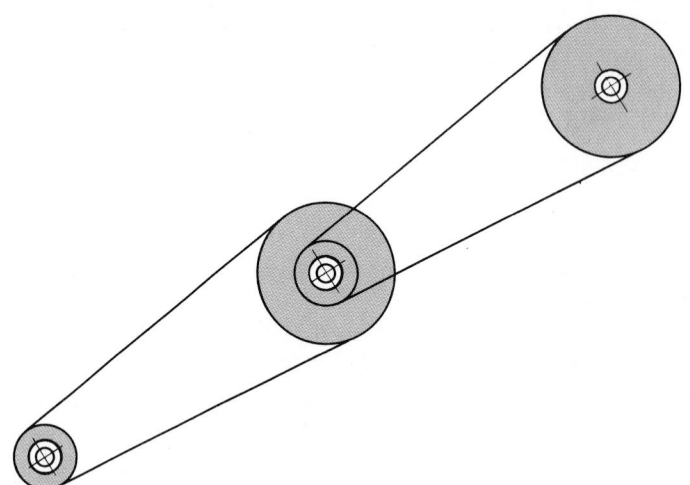

COUNTERSHAFT ADDED – ONE OR MORE DEPENDING ON DISTANCE

(B) CHAIN DRIVE WITH LONG CENTER DISTANCE

multiply the result by the chain pitch to obtain the length in inches (millimeters). The following formula is a quick and convenient method of finding the chain length in pitches (Fig. 20-2-8).

1. Divide center distance in inches (millimeters) by pitch of chain, obtaining C.
2. Add number of teeth in small sprocket to number of teeth in large sprocket, obtaining M.
3. Subtract number of teeth in small sprocket from number of teeth in large sprocket, obtaining value F to obtain the corresponding value of S.
4. Chain length in pitches equals

$$2C + \frac{M}{2} + \frac{S}{C}$$

A chain cannot contain the fractional part of a pitch. It is therefore necessary to increase the pitch to the next higher whole number, preferably an even number. The center distance must then be corrected.

5. Multiply number of pitches by chain pitch used in order to get chain length in inches (millimeters).

Drive Selection

The horsepower (kilowatt) ratings relate to the speed of the smaller sprocket, and drive selections are made on this basis, whether the drive is speed-reducing or speed-increasing. In making drive selections, consideration is given to the loads imposed on the chain by the type of input power and the type of equipment to be driven. Service factors are used to compensate for these loads, and the required horsepower (kilowatt) rating of the chain is determined by the following equation (Fig. 20-2-9):

Required horsepower (kilowatt) rating

$$= \frac{\text{hp (kW) to be transmitted} \times \text{service factor}}{\text{multiple–strand factor}}$$

Figures 20-2-10 to 20-2-13 (pp. 682–685) show the horsepower and kilowatt ratings for just a few of the many roller chains available. For additional information refer to manufacturers' catalogs.

The horsepower and kilowatt rating charts (Figs. 20-2-14 and 20-2-15, pp. 686–687) provide a quick means of determining the probable chain requirements.

Chain Drive Design

EXAMPLE 1

Select an electric motor-driven roller-chain drive to transmit 5 hp from a countershaft to the main shaft of a wire-drawing machine. The countershaft is 1.5 in. in diameter and operates at 1,200 r/min. The main shaft is also 1.5 in. and must operate between 378 and 382 r/min. Shaft centers, once established, are fixed and by initial calculations must be approximately 22.5 in. The load on the main shaft is uneven and presents peaks which place it in the heavy-shock load category.

SOLUTION

Step 1: Service Factor   The corresponding service factor from Fig. 20-2-9 for heavy-shock load and electric motor is 1.5.

STEPS	F	S	F	S	F	S	F	S	F	S	F	S
Step 1 Divide center distance, which is given in inches or millimeters, by pitch of chain used, obtaining *C*.	1	0.03	32	25.94	63	100.54	94	223.82	125	395.79	156	616.44
	2	0.10	33	27.58	64	103.75	95	228.61	126	402.14	157	624.37
	3	0.23	34	29.28	65	107.02	96	233.44	127	408.55	158	632.35
	4	0.41	35	31.03	66	110.34	97	238.33	128	415.01	159	640.38
	5	0.63	36	32.83	67	113.71	98	243.27	129	421.52	160	648.46
Step 2 Add number of teeth in smaller sprocket to number of teeth in larger sprocket, obtaining *M*.	6	0.91	37	34.68	68	117.13	99	248.26	130	428.08	161	656.59
	7	1.24	38	36.58	69	120.60	100	253.30	131	434.69	162	664.77
	8	1.62	39	38.53	70	124.12	101	258.39	132	441.36	163	673.00
	9	2.05	40	40.53	71	127.69	102	263.54	133	448.07	164	681.28
	10	2.53	41	42.58	72	131.31	103	268.73	134	454.83	165	689.62
Step 3 Subtract number of teeth in smaller sprocket from number of teeth in larger sprocket, which gives *F* in table. Use corresponding constant *S*.	11	3.06	42	44.68	73	134.99	104	273.97	135	461.64	166	698.00
	12	3.65	43	46.84	74	138.71	105	279.27	136	468.51	167	706.44
	13	4.28	44	49.04	75	142.48	106	284.67	137	475.42	168	714.92
	14	4.96	45	51.29	76	146.31	107	290.01	138	482.39	169	723.46
	15	5.70	46	53.60	77	150.18	108	295.45	139	489.41	170	732.05
	16	6.48	47	55.95	78	154.11	109	300.95	140	496.47	171	740.60
Step 4	17	7.32	48	58.36	79	158.09	110	306.50	141	503.59	172	749.37
	18	8.21	49	60.82	80	162.11	111	312.09	142	510.76	173	758.11
Chain length in pitches = $2C + \dfrac{M}{2} + \dfrac{S}{C}$	19	9.14	50	63.33	81	166.19	112	317.74	143	517.98	174	766.90
	20	10.13	51	65.88	82	170.32	113	323.44	144	525.25	175	775.74
	21	11.17	52	68.49	83	174.50	114	329.19	145	532.57	176	784.63
Step 5 Multiply number of pitches by chain pitch used in order to get chain length in inches or millimeters.	22	12.26	53	71.15	84	178.73	115	334.99	146	539.94	177	793.57
	23	13.40	54	73.86	85	183.01	116	340.84	147	547.36	178	802.57
	24	14.59	55	76.62	86	187.34	117	346.75	148	554.83	179	811.61
	25	15.83	56	79.44	87	191.73	118	352.70	149	562.36	180	820.70
	26	17.12	57	82.30	88	196.16	119	358.70	150	569.93	181	829.85
	27	18.47	58	85.21	89	200.64	120	364.76	151	577.56	182	839.04
	28	19.86	59	88.17	90	205.18	121	370.86	152	585.23	183	848.29
	29	21.30	60	91.19	91	209.76	122	377.02	153	592.96	184	857.58
	30	22.80	61	94.25	92	214.40	123	383.22	154	600.73	185	866.93
	31	24.34	62	97.37	93	219.08	124	389.48	155	608.56	—	—

FIG. 20-2-8   Determining chain length.

FIG. 20-2-9   Service- and multiple-strand factors for chain drives.

SERVICE FACTOR FOR SINGLE CHAINS HORSEPOWER OR KILOWATT RATINGS FOR MULTIPLE-STRAND CHAINS EQUAL SINGLE-STRAND RATINGS MULTIPLIED BY MULTIPLE-STRAND FACTOR				MULTIPLE-STRAND FACTOR	
	TYPE OF INPUT POWER				
TYPE OF DRIVEN LOAD	INTERNAL COMBUSTION ENGINE WITH HYDRAULIC DRIVE	ELECTRIC MOTOR OR TURBINE	INTERNAL COMBUSTION ENGINE WITH MECHANICAL DRIVE	NUMBER OF STRANDS	MULTIPLE-STRAND FACTOR
Smooth	1.0	1.0	1.2	2	1.7
Moderate shock	1.2	1.3	1.4	3	2.5
Heavy shock	1.4	1.5	1.7	4	3.3

No. of Teeth on Small Spkt.	6 PITCH	NO. 25 ASA STANDARD ROLLER CHAIN															
	Revolutions Per Minute of Small Sprocket																
	100	500	900	1200	1800	2500	3000	3500	4000	4500	5000	5500	6000	6500	7000	7500	8000
17	0.06	0.3	0.5	0.6	0.9	1.2	1.4	1.6	1.8	1.7	1.5	1.3	1.1	1.0	0.9	0.8	0.7
18	0.07	0.3	0.5	0.6	0.9	1.2	1.5	1.7	1.9	1.9	1.6	1.4	1.2	1.1	1.0	0.9	0.8
19	0.07	0.3	0.5	0.7	1.0	1.3	1.5	1.8	2.0	2.0	1.7	1.5	1.3	1.2	1.0	0.9	0.9
20	0.08	0.3	0.6	0.7	1.0	1.4	1.6	1.9	2.1	2.2	1.9	1.6	1.4	1.3	1.1	1.0	0.9
21	0.1	0.3	0.6	0.8	1.1	1.5	1.7	2.0	2.2	2.3	2.0	1.7	1.5	1.3	1.2	1.1	1.0
22	0.1	0.4	0.6	0.8	1.1	1.5	1.8	2.1	2.3	2.5	2.1	1.9	1.6	1.4	1.3	1.2	1.1
23	0.1	0.4	0.6	0.8	1.2	1.6	1.9	2.2	2.5	2.7	2.3	2.0	1.7	1.5	1.4	1.2	1.1
24	0.1	0.4	0.7	0.9	1.3	1.7	2.0	2.3	2.6	2.9	2.4	2.1	1.9	1.6	1.5	1.3	1.2
25	0.1	0.4	0.7	0.9	1.3	1.8	2.1	2.4	2.7	3.0	2.6	2.2	2.0	1.7	1.6	1.4	1.3
28	0.1	0.5	0.8	1.0	1.5	2.0	2.3	2.7	3.0	3.4	3.1	2.7	2.3	2.1	1.9	1.7	1.5
30	0.1	0.5	0.9	1.1	1.6	2.1	2.5	2.9	3.3	3.6	3.4	3.0	2.6	2.3	2.1	1.9	1.7
32	0.1	0.5	0.9	1.2	1.7	2.3	2.7	3.1	3.5	3.9	3.8	3.3	2.9	2.5	2.3	2.0	1.9
35	0.1	0.6	1.0	1.3	1.9	2.5	3.0	3.4	3.9	4.3	4.3	3.7	3.3	2.9	2.6	2.3	2.1
40	0.2	0.7	1.2	1.5	2.2	2.9	3.5	4.0	4.5	5.0	5.3	4.6	4.0	3.5	3.2	2.9	2.6
45	0.2	0.8	1.3	1.7	2.5	3.3	3.9	4.5	5.1	5.7	6.2	5.4	4.8	4.2	3.8	3.4	3.1

No. of Teeth on Small Spkt.	10 PITCH	NO. 35 ASA STANDARD ROLLER CHAIN															
	Revolutions Per Minute of Small Sprocket																
	100	500	900	1200	1800	2500	3000	3500	4000	4500	5000	5500	6000	6500	7000	7500	8000
17	0.2	0.9	1.6	2.1	2.9	4.0	4.2	3.3	2.7	2.3	2.0	1.7	1.5	1.3	1.2	1.1	1.0
18	0.2	1.0	1.7	2.2	3.1	4.2	4.6	3.6	3.0	2.5	2.1	1.8	1.6	1.4	1.3	1.2	1.1
19	0.2	1.1	1.8	2.3	3.3	4.5	5.0	3.9	3.1	2.7	2.3	2.0	1.8	1.6	1.4	1.3	1.1
20	0.3	1.1	1.9	2.4	3.5	4.7	5.4	4.3	3.5	2.9	2.5	2.2	2.0	1.7	1.5	1.4	1.2
21	0.3	1.2	2.0	2.6	3.7	5.0	5.8	4.6	3.7	3.1	2.7	2.3	2.0	1.8	1.6	1.5	1.3
22	0.3	1.2	2.1	2.7	3.9	5.2	6.2	4.9	4.0	3.4	2.9	2.5	2.2	1.9	1.7	1.6	1.4
23	0.3	1.3	2.2	2.8	4.1	5.5	6.5	5.2	4.3	3.6	3.1	2.7	2.3	2.1	1.9	1.7	1.5
24	0.3	1.4	2.3	3.0	4.3	5.8	6.8	5.6	4.6	3.8	3.3	2.8	2.5	2.2	2.0	1.8	1.6
25	0.3	1.4	2.4	3.1	4.5	6.0	7.1	5.9	4.9	4.1	3.5	3.0	2.6	2.3	2.1	1.9	1.7
28	0.4	1.6	2.7	3.5	5.1	6.8	8.0	7.0	5.8	4.8	4.1	3.6	3.1	2.8	2.5	2.2	2.0
30	0.4	1.7	2.9	3.8	5.4	7.3	8.7	7.8	6.4	5.4	4.6	4.0	3.5	3.1	2.8	2.5	2.3
32	0.4	1.8	3.1	4.1	5.8	7.8	9.3	8.6	7.0	5.9	5.0	4.4	3.8	3.4	3.0	2.7	2.5
35	0.5	2.0	3.4	4.5	6.4	8.7	10.2	9.8	8.1	6.8	5.8	5.0	4.4	4.0	3.5	3.1	2.8
40	0.6	2.3	4.0	5.2	7.4	10.0	11.8	12.0	9.8	8.3	7.0	6.1	5.4	4.8	4.3	3.8	3.5
45	0.7	2.7	4.5	5.9	8.4	11.3	13.4	14.3	11.8	9.8	8.4	7.3	6.4	5.7	5.1	4.6	

No. of Teeth on Small Spkt.	13 PITCH	NO. 40 ASA STANDARD ROLLER CHAIN															
	Revolutions Per Minute of Small Sprocket																
	50	200	400	600	900	1200	1800	2400	3000	3500	4000	4500	5000	5500	6000	6500	7000
17	0.3	0.9	1.8	2.6	3.7	4.8	6.7	4.3	3.1	2.5	2.0	1.7	1.4	1.3	1.1	1.0	0.9
18	0.3	1.0	1.9	2.7	4.0	5.1	7.3	4.7	3.4	2.7	2.2	1.8	1.6	1.4	1.2	1.1	0.9
19	0.3	1.1	2.0	2.9	4.1	5.4	7.8	5.1	3.8	2.9	2.4	2.0	1.7	1.5	1.3	1.1	1.0
20	0.3	1.1	2.1	3.0	4.4	5.7	8.3	5.5	4.0	3.1	2.6	2.1	1.8	1.6	1.4	1.2	1.1
21	0.3	1.2	2.3	3.2	4.7	6.1	8.7	6.0	4.3	3.4	2.8	2.3	2.0	1.7	1.5	1.3	1.2
22	0.4	1.3	2.4	3.4	4.9	6.4	9.2	6.4	4.6	3.6	3.0	2.5	2.1	1.8	1.6	1.4	1.3
23	0.4	1.3	2.5	3.6	5.1	6.7	9.6	6.8	4.9	3.9	3.2	2.7	2.3	2.0	1.7	1.5	1.4
24	0.4	1.4	2.6	3.7	5.4	7.0	10.1	7.3	5.2	4.1	3.4	2.8	2.4	2.1	1.8	1.6	1.5
25	0.4	1.5	2.7	3.9	5.6	7.3	10.5	7.8	5.5	4.4	3.6	3.0	2.6	2.2	2.0	1.7	
28	0.5	1.6	3.0	4.4	6.4	8.3	11.9	9.2	6.6	5.2	4.3	3.6	3.1	2.6	2.3	2.1	
30	0.5	1.8	3.3	4.8	6.9	8.9	12.8	10.1	7.3	5.8	4.8	3.1	3.4	2.9	2.6		
32	0.5	1.9	3.5	5.1	7.4	9.5	13.7	11.2	8.1	6.4	5.2	4.4	3.7	3.2	2.8		
35	0.6	2.1	3.9	5.6	8.1	10.5	15.1	12.8	9.2	7.3	6.0	5.0	4.3	3.7			
40	0.7	2.4	4.5	6.5	9.0	12.2	17.5	15.7	11.2	8.9	7.3	6.1	5.2				
45	0.7	2.7	5.1	7.4	10.6	13.8	19.8	18.7	13.4	10.6	8.8	7.3					

Dimensions in mm

FIG. 20-2-10 Kilowatt ratings for 6, 10, and 13 mm pitch single-strand roller chains.

No. of Teeth on Small Spkt.	16 PITCH					NO. 50 ASA STANDARD ROLLER CHAIN											
	Revolutions Per Minute of Small Sprocket																
	50	100	300	500	900	1200	1500	1800	2100	2400	2700	3000	3300	3500	4000	4500
17	0.54	1.00	2.7	4.2	7.2	9.3	10.7	8.0	6.3	5.2	4.3	4.0	3.2	3.0	2.4	2.0
18	0.57	1.07	2.9	4.5	7.7	10.0	11.6	8.7	6.9	5.7	4.7	4.0	3.6	3.2	2.7	2.2
19	0.60	1.13	3.0	4.8	8.1	10.5	12.6	9.4	7.5	6.1	5.1	4.4	3.9	3.4	2.9	2.4
20	0.64	1.2	3.2	5.1	8.7	11.2	13.6	10.2	8.1	6.6	5.6	4.8	4.1	3.8	3.1	2.6
21	0.67	1.3	3.4	5.3	9.1	11.8	14.4	11.0	8.7	7.1	6.0	5.1	4.4	4.1	3.3	3.0
22	0.71	1.3	3.6	5.6	9.5	12.4	15.1	11.7	9.3	7.6	6.4	5.5	4.8	4.3	3.6	3.0
23	0.75	1.4	3.7	5.9	10.0	13.0	15.9	12.6	10.0	8.1	7.0	5.9	5.1	4.7	3.8	3.2
24	0.78	1.5	3.9	6.1	10.5	13.6	16.6	13.4	10.6	8.7	7.2	6.2	5.4	5.0	4.0	3.4
25	0.81	1.5	4.1	6.4	11.0	14.2	17.4	14.2	11.3	9.3	7.8	6.7	5.8	5.2	4.3	3.6
28	0.90	1.7	4.6	7.3	12.4	16.1	19.6	17.0	13.4	11.0	9.1	7.9	6.9	6.2	5.1	0
30	0.99	1.8	5.0	7.8	13.4	17.3	21.2	18.7	14.8	12.2	10.2	8.8	7.6	7.0	5.7	
32	1.06	2.0	5.3	8.4	14.3	18.6	22.7	20.7	16.3	13.4	12.0	10.0	8.2	8.0	6.2	
35	1.17	2.2	5.9	9.2	15.8	20.4	25.0	23.7	18.7	15.3	13.0	11.0	10.0	9.0	7.1	
40	1.35	2.5	6.8	10.6	18.2	23.2	28.9	28.9	22.8	18.7	15.7	13.4	11.7	11.0	0	
45	1.54	2.9	7.7	12.2	20.7	26.9	32.7	34.4	27.2	22.3	19.0	16.0	14.0	0		

No. of Teeth on Small Spkt.	20 PITCH					NO. 60 ASA STANDARD ROLLER CHAIN											
	Revolutions Per Minute of Small Sprocket																
	50	100	200	500	700	900	1200	1400	1600	1800	2000	2200	2400	2600	2800	3000
17	0.9	1.7	3.2	7.3	10.0	12.5	16.2	13.6	11.0	9.3	7.9	6.9	6.0	5.3	4.8	4.3
18	1.0	1.8	3.4	7.8	10.5	13.3	17.2	14.8	12.0	10.1	8.6	7.5	6.5	5.8	5.2	4.7
19	1.0	1.9	3.6	8.3	11.2	14.0	18.2	16.0	13.1	11.0	9.3	8.1	7.1	6.3	5.7	5.1
20	1.1	2.1	3.8	8.8	11.9	14.9	19.2	17.3	14.1	11.9	10.1	8.7	7.7	6.8	6.1	5.5
21	1.2	2.2	4.0	9.2	12.5	15.7	20.3	18.6	15.1	12.8	10.8	9.4	8.3	7.3	6.6	5.9
22	1.2	2.3	4.2	9.7	13.1	16.5	21.3	20.0	16.3	13.8	11.6	10.1	8.9	7.8	7.0	6.3
23	1.3	2.4	4.5	10.1	13.7	17.3	22.4	21.3	17.4	14.7	12.5	10.7	9.5	8.4	7.5	6.8
24	1.3	2.5	4.7	10.6	14.4	18.1	23.4	22.7	18.5	15.6	13.2	11.5	10.1	9.0	8.0	7.2
25	1.4	2.6	4.9	11.1	15.1	19.0	24.6	24.2	19.7	16.7	14.1	12.2	10.8	9.5	8.5	7.7
28	1.6	3.0	5.5	12.6	17.0	21.4	27.7	28.7	23.3	19.7	16.8	14.5	12.7	11.3	10.1	9.1
30	1.7	3.2	6.0	13.6	18.4	23.1	30.0	31.8	25.9	21.8	18.6	16.0	14.1	12.5	11.2	10.1
32	1.8	3.4	6.4	14.5	19.7	24.7	32.0	35.0	28.5	24.0	20.4	17.0	15.6	13.8	12.3	11.1

No. of Teeth on Small Spkt.	25 PITCH					NO. 80 ASA STANDARD ROLLER CHAIN											
	Revolutions Per Minute of Small Sprocket																
	25	50	100	200	300	400	500	700	900	1000	1200	1400	1600	1800	2000	2200
17	1.2	2.1	4.0	7.5	10.8	14.0	17.1	23.1	29.0	28.0	21.3	16.9	13.9	11.6	9.9	8.6
18	1.2	2.3	4.3	8.0	11.5	14.8	18.2	24.6	30.8	30.6	23.3	18.5	15.1	12.7	10.8	9.4
19	1.3	2.4	4.5	8.4	12.2	15.7	19.2	26.1	32.7	33.2	25.3	20.1	16.4	13.7	11.7	10.1
20	1.4	2.6	4.8	9.0	12.8	16.6	20.4	27.6	34.6	35.9	27.3	21.6	17.8	14.8	12.7	11.0
21	1.4	2.7	5.0	9.4	13.6	17.6	21.5	29.1	36.5	38.6	29.4	23.3	19.1	16.0	13.7	11.9
22	1.5	2.8	5.3	9.9	14.2	18.5	22.6	30.6	38.3	41.4	31.5	25.0	20.4	17.2	14.6	12.7
23	1.6	3.0	5.6	10.4	14.9	19.4	23.6	32.1	40.2	44.2	33.6	26.7	21.9	18.4	15.7	13.6
24	1.7	3.1	5.8	10.9	15.7	20.3	24.8	33.6	42.1	46.3	35.9	28.4	23.3	19.5	16.6	14.5
25	1.7	3.3	6.1	11.3	16.3	21.2	26.0	35.1	44.0	48.4	38.1	30.3	24.8	20.7	17.8	15.4
28	2.0	3.7	6.9	12.8	18.5	23.9	29.3	39.7	49.7	54.7	45.2	35.9	29.4	24.6	21.0	18.2
30	2.1	4.0	7.4	13.8	19.9	25.8	31.6	42.7	53.6	58.9	50.1	39.8	32.5	27.3	23.3	18.3
32	2.3	4.3	8.0	14.8	21.3	27.7	33.8	45.8	57.4	63.2	55.2	43.8	35.9	30.1	25.7	

Dimensions in mm

FIG. 20-2-11   Kilowatt ratings for 16, 20, and 25 mm pitch single-strand roller chains.

No. of Teeth Small Spkt.	.25 PITCH		NO. 25 ASA STANDARD ROLLER CHAIN Revolutions Per Minute—Small Sprocket														
	100	500	900	1200	1800	2500	3000	3500	4000	4500	5000	5500	6000	6500	7000	7500	8000
17	.086	.37	.62	.81	1.16	1.56	1.84	2.11	2.38	2.28	1.95	1.69	1.48	1.31	1.18	1.06	0.96
18	.092	.39	.66	.86	1.23	1.66	1.95	2.25	2.53	2.49	2.12	1.84	1.62	1.43	1.28	1.16	1.05
19	.097	.41	.70	.91	1.31	1.76	2.07	2.38	2.69	2.70	2.30	2.00	1.75	1.55	1.39	1.25	1.14
20	.103	.44	.74	.96	1.38	1.86	2.19	2.52	2.84	2.91	2.49	2.16	1.89	1.68	1.50	1.35	1.23
21	.108	.46	.78	1.01	1.46	1.96	2.31	2.65	2.99	3.13	2.68	2.32	2.04	1.80	1.61	1.46	1.32
22	.114	.48	.82	1.06	1.53	2.06	2.43	2.79	3.15	3.36	2.87	2.49	2.18	1.93	1.73	1.56	1.42
23	.119	.51	.86	1.12	1.61	2.16	2.55	2.93	3.30	3.59	3.07	2.66	2.33	2.07	1.85	1.67	1.51
24	.125	.53	.90	1.17	1.69	2.26	2.67	3.07	3.46	3.83	3.27	2.83	2.48	2.20	1.97	1.78	1.61
25	.131	.56	.94	1.22	1.76	2.37	2.79	3.20	3.61	4.02	3.48	3.01	2.64	2.34	2.10	1.89	1.72
28	.148	.63	1.07	1.38	1.99	2.67	3.15	3.62	4.08	4.54	4.12	3.57	3.13	2.78	2.49	2.24	2.04
30	.159	.68	1.15	1.49	2.14	2.88	3.39	3.90	4.40	4.89	4.57	3.96	3.47	3.08	2.76	2.49	2.26
32	.170	.73	1.23	1.60	2.30	3.09	3.64	4.18	4.71	5.24	5.03	4.36	3.83	3.39	3.04	2.74	2.49
35	.188	.80	1.36	1.76	2.53	3.40	4.01	4.61	5.19	5.78	5.76	4.99	4.38	3.88	3.48	3.13	2.85
40	.217	.92	1.57	2.03	2.93	3.93	4.63	5.32	6.00	6.67	7.04	6.10	5.35	4.75	4.25	3.83	3.48
45	.246	.05	1.78	2.31	3.32	4.46	5.26	6.04	6.81	7.58	8.33	7.28	6.39	5.66	5.07	4.57	4.15

No. of Teeth Small Spkt.	.38 PITCH		NO. 35 ASA STANDARD ROLLER CHAIN Revolutions Per Minute—Small Sprocket														
	100	500	900	1200	1800	2500	3000	3500	4000	4500	5000	5500	6000	6500	7000	7500	8000
17	.29	1.25	2.12	2.75	3.95	5.31	5.63	4.47	3.66	3.06	2.62	2.27	1.99	1.77	1.58	1.42	1.29
18	.31	1.33	2.25	2.92	4.20	5.65	6.13	4.87	3.98	3.34	2.85	2.47	2.17	1.92	1.72	1.55	1.41
19	.33	1.41	2.39	3.10	4.46	5.99	6.65	5.28	4.42	3.62	3.09	2.68	2.35	2.09	1.87	1.68	1.53
20	.35	1.49	2.53	3.27	4.71	6.33	7.18	5.70	4.67	3.91	3.34	2.90	2.54	2.25	2.02	1.82	1.65
21	.37	1.57	2.66	3.45	4.97	6.68	7.73	6.13	5.02	4.21	3.59	3.11	2.73	2.42	2.17	1.96	1.77
22	.39	1.65	2.80	3.63	5.22	7.02	8.27	6.58	5.38	4.51	3.85	3.34	2.93	2.60	2.33	2.10	1.90
23	.41	1.73	2.94	3.81	5.48	7.37	8.68	7.03	5.75	4.82	4.12	3.57	3.13	2.78	2.49	2.24	2.03
24	.43	1.81	3.08	3.98	5.74	7.71	9.09	7.49	6.13	5.14	4.39	3.80	3.34	2.96	2.65	2.39	2.17
25	.44	1.89	3.21	4.16	6.00	8.06	9.50	7.97	6.52	5.47	4.67	4.05	3.55	3.15	2.82	2.54	2.31
28	.50	2.14	3.63	4.71	6.78	9.11	10.7	9.44	7.73	6.48	5.53	4.80	4.21	3.73	3.34	3.01	2.73
30	.54	2.31	3.91	5.07	7.30	9.81	11.6	10.5	8.57	7.18	6.14	5.32	4.67	4.14	3.70	3.34	3.03
32	.58	2.47	4.20	5.44	7.83	10.5	12.4	11.5	9.44	7.91	6.76	5.86	5.14	4.56	4.08	3.68	3.34
35	.64	2.72	4.62	5.99	8.63	11.6	13.7	13.2	10.8	9.06	7.73	6.70	5.88	5.22	4.67	4.21	3.82
40	.74	3.15	5.34	6.92	9.96	13.4	15.8	16.1	13.2	11.1	9.45	8.19	7.19	6.37	5.70	5.14	4.67
45	.84	3.57	6.06	7.85	11.3	15.2	17.9	19.2	15.8	13.2	11.3	9.77	8.57	7.60	6.80	6.14	0

No. of Teeth Small Spkt.	.50 PITCH		NO. 40 ASA STANDARD ROLLER CHAIN Revolutions Per Minute—Small Sprocket														
	50	200	400	600	900	1200	1800	2400	3000	3500	4000	4500	5000	5500	6000	6500	7000
17	.37	1.29	2.40	3.45	4.98	6.45	8.96	5.82	4.17	3.31	2.71	2.27	1.94	1.68	1.47	1.31	1.17
18	.39	1.37	2.55	3.68	5.30	6.86	9.76	6.34	4.54	3.60	2.95	2.47	2.11	1.83	1.60	1.42	1.27
19	.42	1.45	2.71	3.90	5.62	7.27	10.5	6.88	4.92	3.91	3.20	2.68	2.29	1.98	1.74	1.54	1.38
20	.44	1.53	2.86	4.12	5.94	7.69	11.1	7.43	5.31	4.22	3.45	2.89	2.47	2.14	1.88	1.67	1.49
21	.46	1.62	3.02	4.34	6.26	8.11	11.7	7.99	5.72	4.54	3.71	3.11	2.66	2.30	2.02	1.79	1.60
22	.49	1.70	3.17	4.57	6.58	8.52	12.3	8.57	6.13	4.87	3.98	3.34	2.85	2.47	2.17	1.92	1.72
23	.51	1.78	3.33	4.79	6.90	8.94	12.9	9.16	6.55	5.20	4.26	3.57	3.05	2.64	2.32	2.06	1.84
24	.54	1.87	3.48	5.02	7.23	9.36	13.5	9.76	6.99	5.54	4.54	3.80	3.25	2.81	2.47	2.19	1.96
25	.56	1.95	3.64	5.24	7.55	9.78	14.1	10.4	7.43	5.89	4.82	4.04	3.45	2.99	2.63	2.33	0
28	.63	2.20	4.11	5.93	8.54	11.1	15.9	12.3	8.80	6.99	5.72	4.79	4.09	3.55	3.11	2.76	0
30	.68	2.38	4.43	6.38	9.20	11.9	17.2	13.6	9.76	7.75	6.34	5.31	4.54	3.93	3.45	0	
32	.73	2.55	4.75	6.85	9.86	12.8	18.4	15.0	10.8	8.54	6.99	5.86	5.00	4.33	3.80	0	
35	.81	2.80	5.24	7.54	10.9	14.1	20.3	17.2	12.3	9.76	7.99	6.70	5.72	4.96	0		
40	.93	3.24	6.05	8.71	12.5	16.3	23.4	21.0	15.0	11.9	9.76	8.18	6.99	0			
45	1.06	3.68	6.87	9.89	14.2	18.5	26.6	25.1	17.9	14.2	11.7	9.76	0				

FIG. 20-2-12   Horsepower ratings for .25, .38, and .50 in. pitch single-strand roller chains. *(American Sprocket Chain Manufacturers Association)*

No. of Teeth Small Spkt.	.62 PITCH							NO. 50 ASA STANDARD ROLLER CHAIN Revolutions Per Minute—Small Sprocket								
	50	100	300	500	900	1200	1500	1800	2100	2400	2700	3000	3300	3500	4000	4500
17	.72	1.34	3.60	5.69	9.70	12.6	14.3	10.7	8.48	6.95	5.83	4.98	4.32	3.96	3.23	2.71
18	.77	1.43	3.83	6.05	10.3	13.4	15.6	11.7	9.24	7.58	6.35	5.42	4.70	4.31	3.52	2.95
19	.81	1.51	4.06	6.42	10.9	14.2	16.9	12.7	10.0	8.22	6.89	5.88	5.10	4.68	3.82	3.20
20	.86	1.60	4.30	6.78	11.6	15.0	18.2	13.7	10.8	8.87	7.44	6.35	5.51	5.05	4.12	3.45
21	.90	1.69	4.53	7.15	12.2	15.8	19.3	14.7	11.6	9.55	8.01	6.83	5.93	5.44	4.44	3.71
22	.95	1.77	4.76	7.52	12.8	16.6	20.3	15.8	12.5	10.2	8.59	7.33	6.36	5.83	4.76	3.98
23	1.00	1.86	5.00	7.89	13.4	17.4	21.3	16.9	13.3	10.9	9.18	7.83	6.79	6.23	5.08	4.26
24	1.04	1.95	5.23	8.26	14.1	18.3	22.3	18.0	14.2	11.7	9.78	8.34	7.24	6.64	5.42	4.54
25	1.09	2.04	5.47	8.63	14.7	19.1	23.3	19.1	15.1	12.4	10.4	8.88	7.70	7.06	5.76	4.83
28	1.20	2.30	6.18	9.76	16.6	21.6	26.3	22.7	17.9	14.7	12.3	10.5	9.13	8.37	6.83	0
30	1.33	2.42	6.66	10.5	17.9	23.2	28.4	25.1	19.9	16.3	13.7	11.7	10.1	9.28	7.57	0
32	1.42	2.66	7.14	11.3	19.2	24.9	30.4	27.7	21.9	18.0	15.0	12.9	11.1	10.2	8.34	0
35	1.57	2.93	7.86	12.4	21.2	27.4	33.5	31.7	25.1	20.5	17.2	14.7	12.8	11.7	9.55	0
40	1.81	3.38	9.08	14.3	24.4	31.1	38.7	38.7	30.6	25.1	21.0	18.0	15.6	14.3	0	
45	2.06	3.84	10.3	16.3	27.8	36.0	43.9	46.2	36.5	29.9	25.1	21.4	18.6	0		

No. of Teeth Small Spkt.	.75 PITCH							NO. 60 ASA STANDARD ROLLER CHAIN Revolutions Per Minute—Small Sprocket								
	50	100	200	500	700	900	1200	1400	1600	1800	2000	2200	2400	2600	2800	3000	3500
17	1.24	2.30	4.31	9.81	13.3	16.7	21.7	18.2	14.8	12.5	10.6	9.18	8.06	7.15	6.40	5.75	4.57
18	1.32	2.45	4.58	10.4	14.1	17.8	23.0	19.8	16.1	13.6	11.5	10.0	8.78	7.79	6.97	6.27	4.98
19	1.40	2.60	4.86	11.1	15.0	18.8	24.4	21.5	17.5	14.7	12.5	10.9	9.52	8.45	7.56	6.80	5.40
20	1.48	2.75	5.13	11.7	15.9	19.9	25.8	23.2	18.9	15.9	13.5	11.7	10.3	9.12	8.17	7.34	5.83
21	1.56	2.89	5.41	12.3	16.7	21.0	27.2	24.9	20.3	17.1	14.5	12.6	11.1	9.82	8.79	7.90	6.27
22	1.64	3.04	5.69	13.0	17.6	22.1	28.6	26.7	21.8	18.4	15.6	13.5	11.9	10.5	9.42	8.47	6.73
23	1.72	3.19	5.97	13.6	18.4	23.2	30.0	28.6	23.3	19.6	16.7	14.4	12.7	11.3	10.1	9.06	7.19
24	1.80	3.34	6.25	14.2	19.3	24.3	31.4	30.4	24.8	20.9	17.7	15.4	13.5	12.0	10.7	9.65	7.66
25	1.88	3.49	6.53	14.9	20.2	25.4	32.9	32.4	26.4	22.3	18.9	16.4	14.4	12.8	11.4	10.3	8.15
28	2.12	3.95	7.38	16.8	22.8	28.7	37.1	38.4	31.3	26.4	22.4	19.4	17.0	15.1	13.5	12.2	9.66
30	2.29	4.25	7.95	18.1	24.6	30.9	40.0	42.6	34.7	29.2	24.8	21.5	18.9	16.8	15.0	13.5	0
32	2.45	4.56	8.53	19.4	26.3	33.1	42.9	46.9	38.2	32.2	27.3	23.7	20.8	18.5	16.5	14.9	0

No. of Teeth Small Spkt.	1.00 PITCH							NO. 80 ASA STANDARD ROLLER CHAIN Revolutions Per Minute—Small Sprocket								
	25	50	100	200	300	400	500	700	900	1000	1200	1400	1600	1800	2000	2200	2400
17	1.55	2.88	5.38	10.0	14.5	18.7	22.9	31.0	38.9	37.6	28.6	22.7	18.6	15.6	13.3	11.5	10.1
18	1.64	3.07	5.72	10.7	15.4	19.9	24.4	33.0	41.3	41.0	31.2	24.8	20.3	17.0	14.5	12.6	11.0
19	1.74	3.25	6.07	11.3	16.3	21.1	25.8	35.0	43.8	44.5	33.9	26.9	22.0	18.4	15.7	13.6	12.0
20	1.84	3.44	6.42	12.0	17.2	22.3	27.3	37.0	46.4	48.1	36.6	29.0	23.8	19.9	17.0	14.7	12.9
21	1.94	3.62	6.76	12.6	18.2	23.6	28.8	39.0	48.9	51.7	39.4	31.2	25.6	21.4	18.3	15.9	13.9
22	2.04	3.81	7.11	13.3	19.1	24.8	30.3	41.0	51.4	55.5	42.2	33.5	27.4	23.0	19.6	17.0	14.9
23	2.14	4.00	7.46	13.9	20.0	26.0	31.7	43.0	53.9	59.2	45.1	35.8	29.3	24.6	21.0	18.2	15.9
24	2.24	4.19	7.81	14.6	21.0	27.2	33.3	45.0	56.4	62.0	48.1	38.1	31.2	26.2	22.3	19.4	17.0
25	2.34	4.38	8.17	15.2	21.9	28.4	34.8	47.0	59.0	64.9	51.1	40.6	33.2	27.8	23.8	20.6	8.34
28	2.65	4.94	9.23	17.2	24.8	32.1	39.3	53.2	66.6	73.3	60.6	48.1	39.4	33.0	28.2	24.4	0
30	2.85	5.33	9.94	18.5	26.7	34.6	42.3	57.3	71.8	78.9	67.2	53.3	43.6	36.6	31.2	24.5	0
32	3.06	5.71	10.7	19.9	28.6	37.1	45.3	61.4	77.0	84.7	74.0	58.7	48.1	40.3	34.4	0	

FIG. 20-2-13 Horsepower ratings for .62, .75, and 1.00 in. pitch single-strand roller chains. (*American Sprocket Chain Manufacturers Association*)

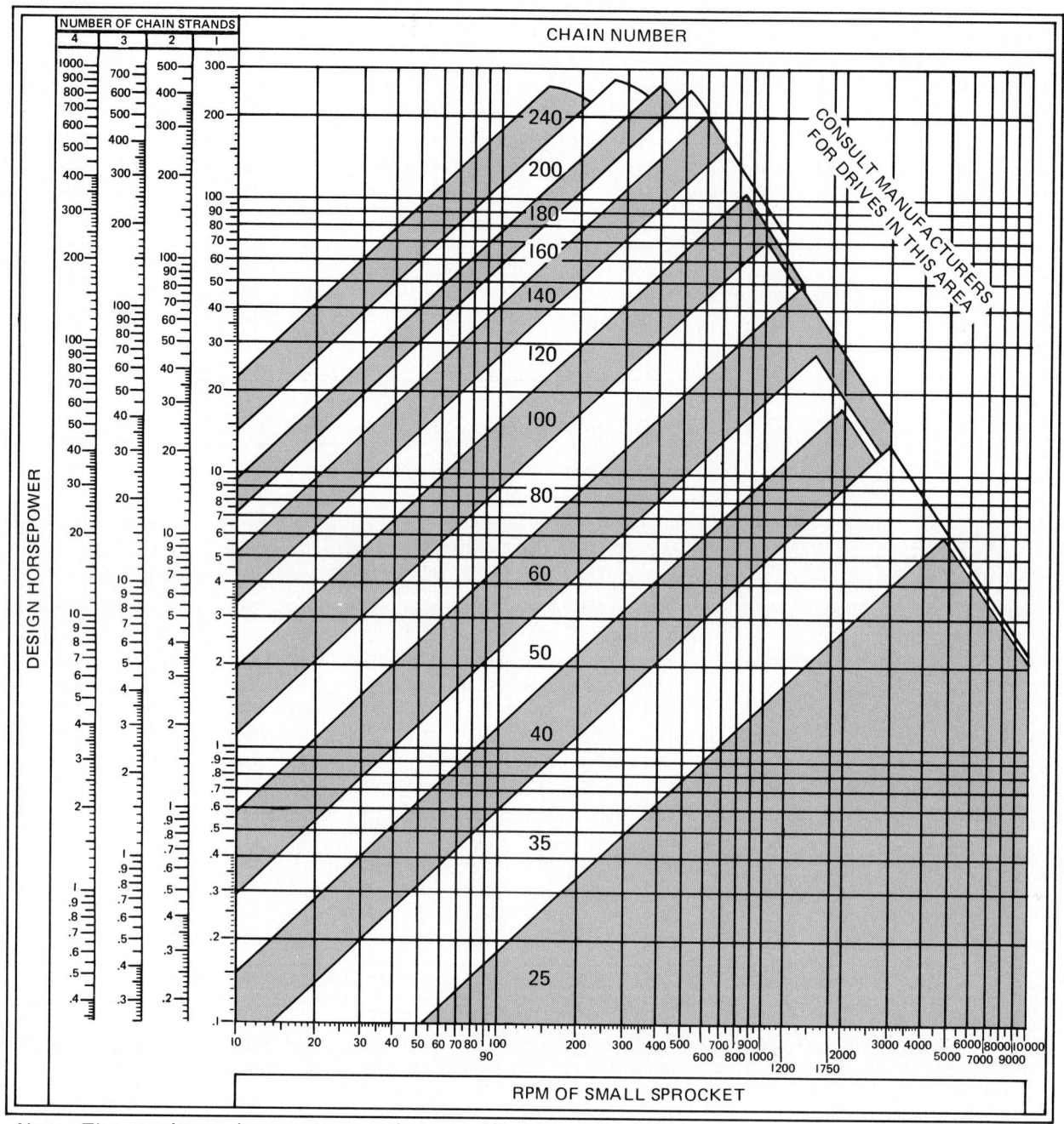

Note: The maximum horsepower rating specified in each of the strand columns is not limiting for chain drives. Consult chain manufacturers on those applications that are above the horsepower range of the chart.

FIG. 20-2-14   Horsepower rating chart.

Step 2: Design Horsepower   Design horsepower is 5 × 1.5 = 7.5 hp.

Step 3: Tentative Chain Selection   On the horsepower rating chart (Fig. 20-2-14) the suggested selection using a design of 7.5 hp and a 1,200-r/min sprocket is no. 40 (.50 in. pitch) chain.

If a multiple-strand chain has been selected, determine the required horsepower rating per strand from the following equation:

$$\text{Required horsepower rating} = \frac{\text{design hp}}{\text{multiple–strand factor}}$$

or refer to the right-hand columns shown in Fig. 20-2-9.

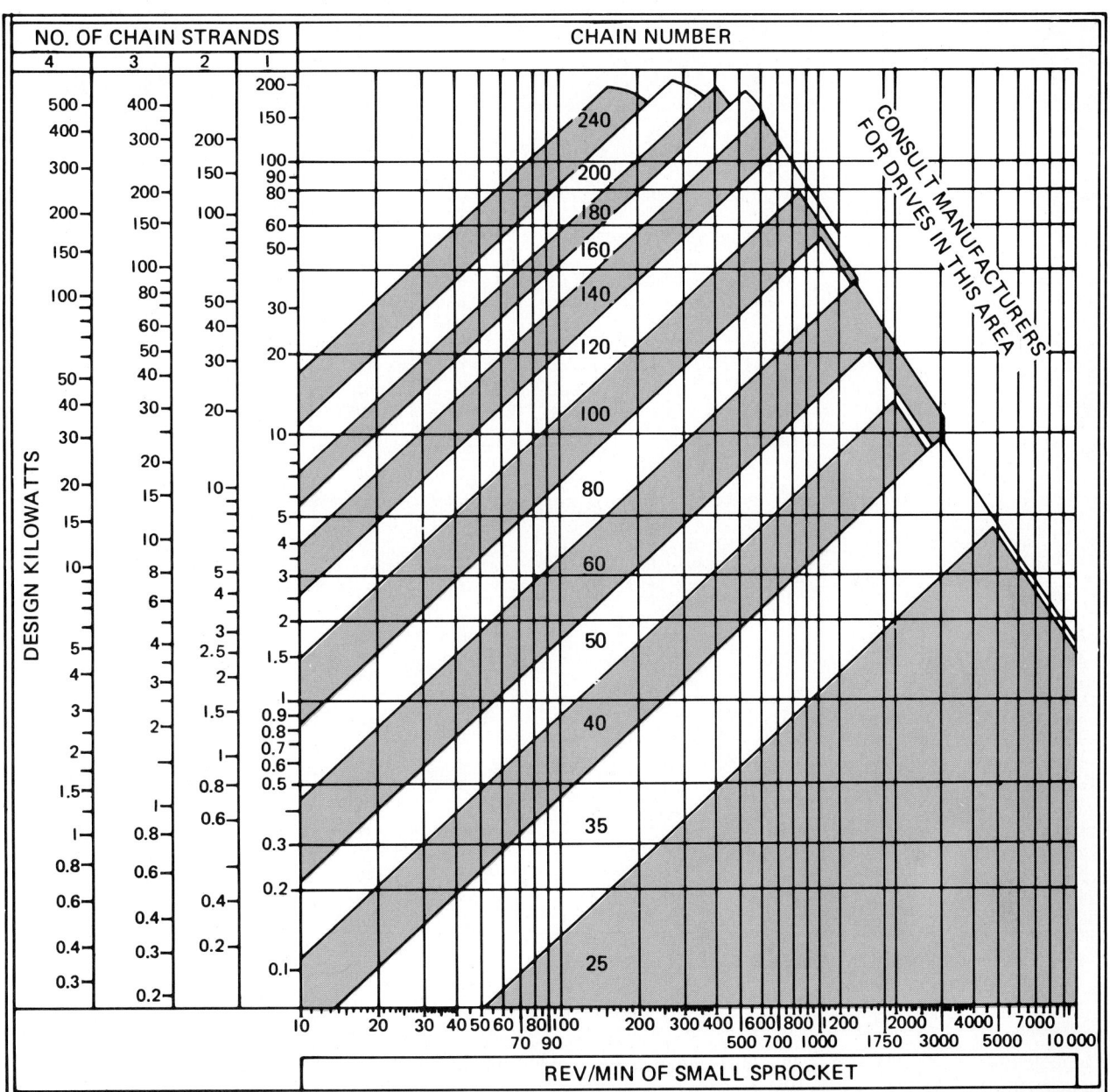

NOTE: THE MAXIMUM KILOWATT RATING SPECIFIED IN EACH OF THE STRAND COLUMNS IS NOT LIMITING FOR CHAIN DRIVES. CONSULT CHAIN MANUFACTURERS ON THOSE APPLICATIONS THAT ARE ABOVE THE KILOWATT RANGE OF THE CHART.

FIG. 20-2-15   Kilowatt rating chart.

Step 4: Final Selection of Chain and Small Sprocket   On the horsepower rating table for a no. 40 chain (Fig. 20-2-12) at 1,200 r/min, the computed design of 7.5 hp is realized with a 20-tooth sprocket. Follow down the column headed by the revolutions of the small sprocket (1,200 r/min) and find the nearest value to the design horsepower. Follow this line horizontally to the left to find the number of teeth for the small sprocket.

For intermediate speeds or sprocket sizes not tabulated, interpolate between the appropriate columns or lines. Check the maximum bore for the selected sprocket (Fig. 20-2-16, pg. 688). If the selected sprocket will not accommodate the shaft, use a larger sprocket or make a new sprocket and chain selection from

the rating table for the next larger chain number. In this problem, the 20-tooth sprocket will accommodate the 1.5 in. shaft.

Step 5: Selection of the Large Sprocket   Since the driver is to operate at 1,200 r/min and the driven at a minimum of 378 r/min, the speed ratio = 1,200/378 = 3.17:1 minimum. Therefore, the large sprocket should have 20 × 3.17 teeth = 63.4 teeth. Since the standard sprocket sizes near this number of teeth are either 60 or 70 teeth (Fig. 20-2-17, pg. 689), it may be more economical and time-saving to try to use a combination of standard sprockets. In rechecking the smaller sprocket, the 19-tooth sprocket would also be acceptable. This would

U.S. CUSTOMARY (INCH)										
No. of Teeth	.38 in. Pitch		.50 in. Pitch		.62 in. Pitch		.75 in. Pitch		1.00 in. Pitch	
	Maximum Bore	Maximum Hub Dia.	Maximum Bore	Maximum Hub Dia.	Maximum Bore	Maximum Hub Dia.	Maximum Bore	Maximum Hub Dia.	Maximum Bore	Maximum Hub Dia.
11	.59	.86	.78	1.17	.97	1.47	1.25	1.77	1.62	2.38
12	.62	.98	.88	1.33	1.16	1.67	1.28	2.02	1.78	2.70
13	.75	1.11	1.00	1.50	1.28	1.88	1.50	2.25	2.00	3.02
14	.84	1.23	1.16	1.66	1.31	2.08	1.75	2.50	2.28	3.34
15	.88	1.36	1.25	1.81	1.53	2.28	1.78	2.75	2.41	3.67
16	.97	1.47	1.28	1.98	1.69	2.48	1.97	2.98	2.72	3.98
17	1.09	1.59	1.38	2.14	1.78	2.69	2.22	3.22	2.81	4.31
18	1.22	1.72	1.53	2.30	1.88	2.89	2.28	3.47	3.12	4.64
19	1.25	1.84	1.69	2.45	2.06	3.08	2.44	3.70	3.31	4.95
20	1.28	1.95	1.78	2.62	2.25	3.28	2.69	3.95	3.50	5.28
21	1.31	2.08	1.78	2.78	2.28	3.48	2.81	4.19	3.75	5.59
22	1.44	2.20	1.94	2.94	2.44	3.69	2.94	4.44	3.88	5.92
23	1.56	2.31	2.09	3.09	2.62	3.89	3.12	4.67	4.19	6.23
24	1.69	2.44	2.25	3.27	2.81	4.08	3.25	4.91	4.56	6.56
25	1.75	2.56	2.28	3.42	2.84	4.28	3.38	5.16	4.69	6.88

METRIC (MILLIMETER)										
No. of Teeth	10 Pitch No. 25		13 Pitch No. 40		16 Pitch No. 50		20 Pitch No. 60		25 Pitch No. 80	
	Maximum Bore	Maximum Hub Dia.	Maximum Bore	Maximum Hub Dia.	Maximum Bore	Maximum Hub Dia.	Maximum Bore	Maximum Hub Dia.	Maximum Bore	Maximum Hub Dia.
11	15	22	20	30	25	37	32	45	41	50
12	16	25	22	34	29	42	33	51	45	69
13	20	28	25	38	33	47	38	57	51	76
14	22	32	30	42	33	53	45	64	58	85
15	22	35	32	46	39	58	45	70	61	93
16	25	38	32	50	43	63	50	76	69	101
17	28	40	35	54	45	68	56	82	71	109
18	31	44	39	58	48	73	58	88	79	118
19	32	47	43	62	52	78	62	94	84	126
20	33	50	45	67	57	83	68	100	89	134
21	34	53	45	71	58	88	71	106	95	142
22	36	56	49	75	62	94	75	111	98	150
23	40	59	53	79	67	99	79	119	106	158
24	43	62	57	83	71	104	83	124	116	167
25	45	65	58	87	72	109	86	131	119	175

FIG. 20-2-16   Maximum bore and hub diameter. *(American Sprocket Chain Manufacturers Association)*

require a large sprocket of 19×3.17 teeth = 60.2 teeth (use 60 teeth). Since the 19- and 60-tooth sprockets are acceptable and standard, it would be more economical to use this combination.

Step 6: Chain Length in Pitches   Since 19- and 60-tooth sprockets are to be placed on 22.5 in. centers, calculations are as follows to determine chain length:

$$\text{Chain length in pitches} = 2C + \frac{M}{2} + \frac{S}{C}$$

where C = center distance ÷ pitch
= 22.5 ÷ .5 = 45

M = total number of teeth on both sprockets
= 19 + 60 = 79

S = value obtained from table (Fig. 20-2-8)
F = 60 − 19 = 41
S = 42.58

Substituting values for C, M, and S, we get

$$\text{Chain length in pitches} = 2 \times 45 + \frac{79}{2} + \frac{42.58}{45} = 130.44$$

Since the chain is to couple to an even number of pitches, we will use 130 pitches because the leeway on the 22.5 in. centers is not critical.

Step 7: Chain Length in Inches (Millimeters)

$$\text{Length of chain} = \text{number of pitches} \times \text{pitch}$$
$$= 130 \times .5 = 65 \text{ in.}$$

For metric chain problems the same procedures are followed except that Figs. 20-2-10, 20-2-11, and 20-2-15 replace Figs. 20-2-12, 20-2-13, and 20-2-14.

NUMBER OF TEETH ON SPROCKET											
NO. 25		NO. 35		NO. 40		NO. 50		NO. 60		NO. 80	
9		9	48	8	39	9	39	9	34	9	34
10		10	54	9	40	10	40	10	35	10	35
11		11	60	10	41	11	41	11	36	11	36
12		12	70	11	42	12	42	12	37	12	37
13		13	72	12	43	13	43	13	38	13	38
14		14	80	13	44	14	44	14	39	14	39
15		15	84	14	45	15	45	15	40	15	40
16		16	96	15	46	16	46	16	41	16	41
17		17	112	16	47	17	47	17	42	17	42
18		18		17	48	18	48	18	43	18	43
19		19		18	49	19	49	19	44	19	44
20		20		19	50	20	50	20	45	20	45
21		21		20	51	21	51	21	46	21	46
22		22		21	52	22	52	22	47	22	47
24		23		22	53	23	53	23	48	23	48
25		24		23	54	24	54	24	49	24	54
26		25		24	55	25	55	25	50	25	60
28		26		25	56	26	56	26	51	26	
30		28		26	57	27	57	27	52	27	
32		30		27	58	28	58	28	53	28	
36		32		28	59	29	59	29	54	29	
40		35		30	60	30	60	30	60	30	
45		36		31	70	31	70	31	70	31	
48		40		32	72	32	72	32	72	32	
52		45		33	80	33	80	33	80	33	
60				34	84	34	84				84
				35	96	35	96				
				36	112	36	112				
				37		37					
				38		38					

FIG. 20-2-17   Stock sprockets.

ASSIGNMENTS

See Assignments 3 through 8 for Unit 20-2 on page 706.

20-3 GEAR DRIVES

The function of a gear is to transmit motion, rotating or reciprocating, from one machine part to another and where necessary to reduce or increase the revolutions of a shaft. *Gears* are rolling cylinders or cones having teeth on their contact surfaces to ensure positive motion (Figs. 20-3-1 and 20-3-2).

Gears are the most durable and rugged of all mechanical drives. For this reason, gears rather than belts or chains are used in automotive transmissions and most heavy-duty machine drives.

There are many kinds of gears, and they may be grouped according to the position of the shafts that they connect. *Spur gears* connect parallel shafts, *bevel gears* connect shafts whose axes intersect and *worm gears* connect shafts whose axes do not intersect. A spur gear with a rack converts rotary motion to reciprocating or linear motion. The smaller of two gears is known as the *pinion*.

FIG. 20-3-1   Gears. *(Boston Gear Works)*

MODULE	DIAMETRAL PITCH	PRESSURE ANGLE	
FOR METRIC SIZE GEARS	FOR INCH SIZE GEARS	14.5°	20°
6.35	4		
5.08	5		
4.23	6		
3.18	8		
2.54	10		
2.17	12		
1.59	16		
1.27	20		
1.06	24		
0.79	32		

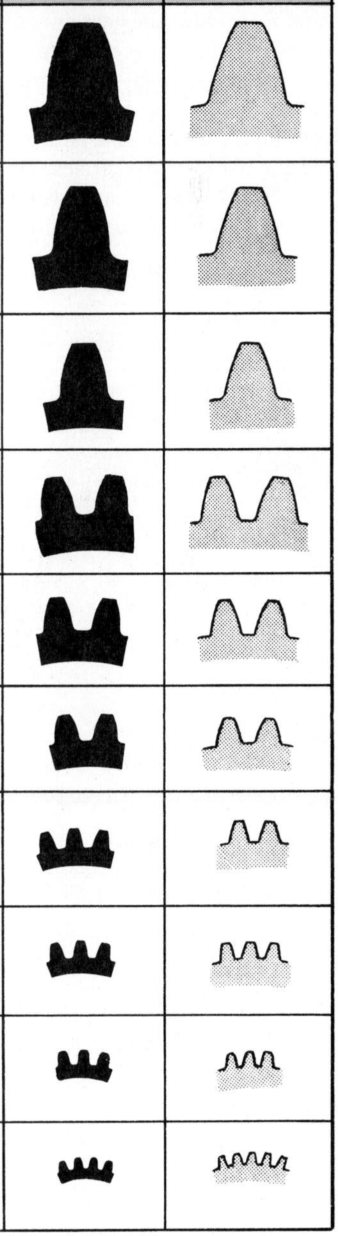

NOTE: MODULE SIZES SHOWN ARE CONVERTED INCH SIZES.

FIG. 20-3-2   Gear-teeth sizes.

Gear design is very complicated, dealing with such problems as strength, wear, and material selection. Normally a drafter selects a gear from a catalog. Most gears are made of cast iron or steel, but brass, bronze, and plastic are used when factors such as wear or noise must be considered.

Spur Gears

Spur gear proportions and the shape of gear teeth are standardized, and the definitions, symbols, and formulas are given in Figs. 20-3-3 to 20-3-6 (here and through pg. 692).

Gears are used to transmit motion and power at constant angular velocity. The specific form of the gear that best produces this constant angular velocity is the *involute.* Classically, the involute is described as the curve traced by a point on a taut string unwinding from a circle. This circle is called the *base circle.* Every involute gear has only one base circle from which all the involute surfaces of the gear teeth are generated. This base circle is not a physical part of the gear and cannot be measured directly. The contact between mating involutes takes place along a line that is always tangent to, and crosses between, the two base circles. This is the *line of action.*

The 14.5° pressure angle has been used for many years and remains useful for duplication or replacement of gearing.

Standard angles of 20 and 25° have become the standard for new gearing because of the smoother and quieter running characterics, greater load-carrying ability, and the fewer number of teeth affected by undercutting.

Standard spur gears having a 14.5° pressure angle should have a minimum of 16 teeth with at least 40 teeth in a mating pair. Gears with 20° pressure angle should have a minimum of 13 teeth with at least 26 teeth in a mating pair.

The formulas for the 14.5°-, 20°-, and the 25°-full-depth teeth are identical.

The 20° stub tooth differs from the 20° standard tooth depth. The stub tooth is shorter and stronger and therefore is preferred where maximum power transmission is required.

Drawing Gear Teeth

The teeth on a gear are not normally shown on the working drawings. Instead, they are represented by solid, broken, and hidden lines, which will be discussed under working drawings. However, presentation or display drawings normally require the teeth to be shown.

Since the exact form of an involute tooth would require too much time to draw, approximate methods are used. The two most common methods are shown in Fig. 20-3-6. To draw the

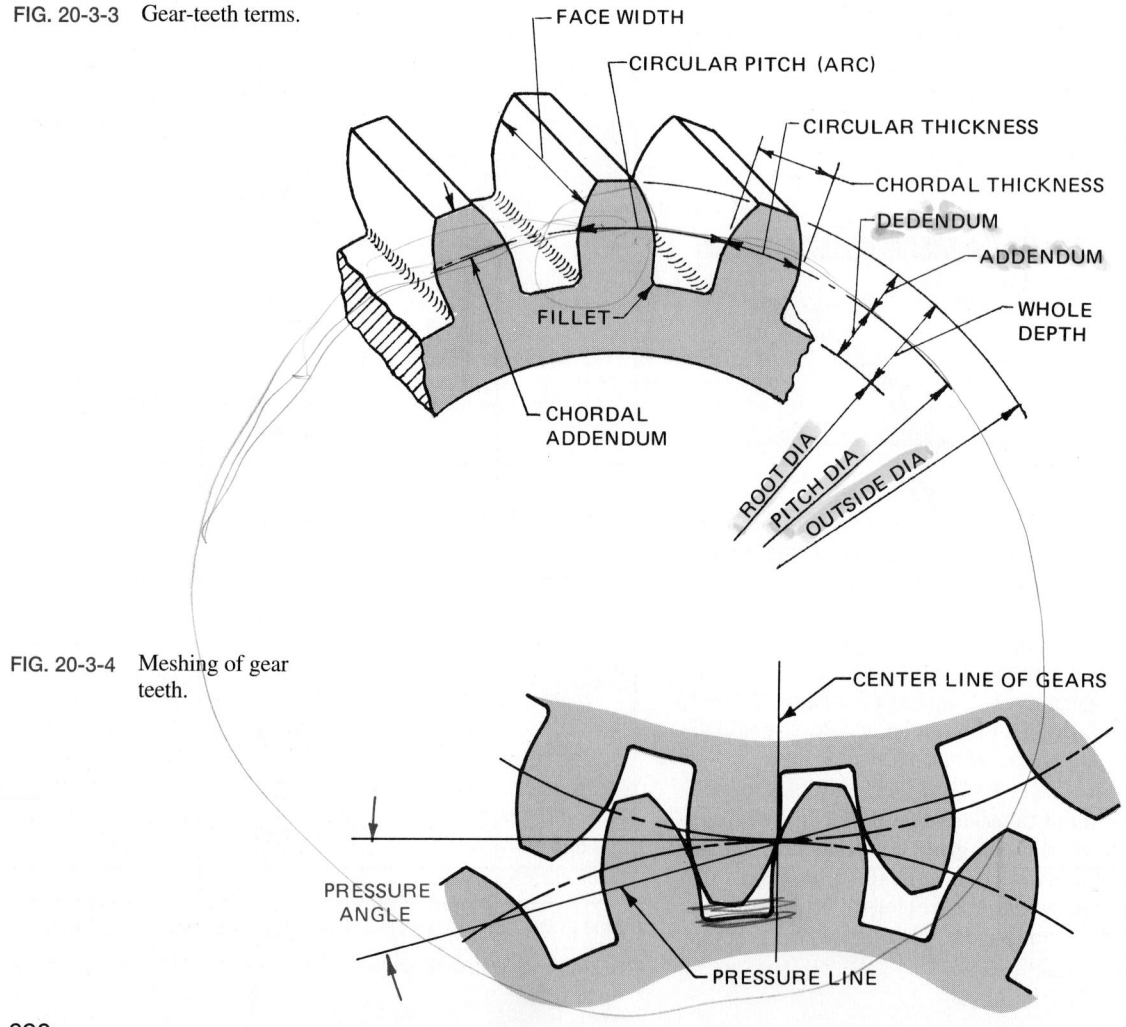

FIG. 20-3-3   Gear-teeth terms.

FIG. 20-3-4   Meshing of gear teeth.

TERM AND SYMBOL	DEFINITION	FORMULA	
		METRIC GEARS	INCH GEARS
Pitch diameter—PD	The diameter of an imaginary circle on which the gear tooth is designed	PD = MDL × N	PD = N ÷ DP
Number of teeth—N	The number of teeth on a gear	N = PD ÷ MDL	N = PD × DP
Module—MDL	The length of pitch diameter per tooth	MDL = PD ÷ N	
Diametral pitch—DP	A ratio equal to the number of teeth on a gear for every inch of pitch diameter	——	DP = N ÷ PD
Addendum—ADD	The radial distance from the pitch circle to the top of the tooth	14.5° or 20° ADD = MDL 20° stub ADD = 0.8 × MDL	14.5° or 20° ADD = 1 ÷ DP 20° stub ADD = 0.8 ÷ DP
Dedendum—DED	The radial distance from the pitch circle to the bottom of the tooth	14.5° or 20° DED = 1.157 × MDL 20° stub DED = MDL	14.5° or 20° DED = 1.157 ÷ DP 20° stub DED = 1 ÷ DP
Whole depth—WD	The overall height of the tooth	14.5° or 20° WD = 2.157 × MDL 20° stub WD = 1.8 × MDL	14.5° or 20° WD = 2.157 ÷ DP 20° stub WD = 1.8 ÷ DP
Clearance—CL	The radial distance between the bottom of one tooth and the top of the mating tooth	14.5° or 20° CL = 0.157 × MDL 20° stub CL = 0.2 × MDL	14.5° or 20° CL = 0.157 ÷ DP 20° stub CL = 0.2 ÷ DP
Outside diameter—OD	The overall diameter of the gear	14.5° or 20° OD = PD + 2ADD = PD + 2 MDL	14.5° or 20° OD = PD + 2ADD = (N + 2) ÷ DP
		20° stub OD = PD + 2ADD = PD + 1.6 MDL	20° stub OD = PD + 2ADD = (N + 1.6) ÷ DP
Root diameter—RD	The diameter at the bottom of the tooth	14.5° or 20° RD = PD − 2DED = PD − 2.314 MDL	14.5° or 20° RD = PD − 2DED = (N − 2.314) ÷ DP
		20° stub RD = PD − 2DED = PD − 2MDL	20° stub RD = PD − 2DED = (N − 2) ÷ DP
Base circle—BC	The circle from which the involute curve of the tooth is formed	BC = PD Cos PA	BC = PD Cos PA
Pressure angle—PA	The angle between the direction of pressure between contacting teeth and a line tangent to the pitch circle	14.5° or 20°	14.5° or 20°
Backlash	The clearance between the teeth of two meshing gears	——	——
Circular pitch—CP	The distance measured from the point of one tooth to the corresponding point on the adjacent tooth on the circumference of the pitch diameter	CP = 3.1416 PD ÷ N = 3.1416 MDL	CP = 3.1416 PD ÷ N = 3.1416 ÷ DP
Circular thickness—T	The thickness of a tooth or space measured on the circumference of the pitch diameter	T = 3.1416 PD ÷ 2N = 1.5708 PD ÷ N = 1.5708 MDL	T = 3.1416 PD ÷ 2N = 1.5708 ÷ DP
Chordal thickness—Tc	The thickness of a tooth or space measured along a chord on the circumference of the pitch diameter	Tc = PD sin (90° ÷ N)	Tc = PD sin (90° ÷ N)
Chordal addendum —ADDc	Chordal addendum, also known as Corrected addendum, is the perpendicular distance from chord to outside circumference of gear	ADDc = ADD + (T^2 ÷ 4PD)	ADDc = ADD + (T^2 ÷ 4PD)
Working depth —WKG DEPTH	The depth of engagement of two gears. The sum of two addendums	WKG DP = 2ADD	WKG DP = 2ADD

FIG. 20-3-5   Spur gear definitions and formulas.

teeth using the approximate representation of involute spur gear teeth, lay out the root, pitch, and outside circles. On the pitch circle mark off the circular thickness. Through the pitch point on the pitch circle draw the pressure line at an angle of 14.5° with the line tangent to the pitch circle for the 14.5° involute tooth (use 15° for convenience), 20° for the 20° involute tooth, or 25° for the 25° involute tooth. Draw the base circle tangent to this pressure line. With the compass set to a radius equal to one-eighth the pitch diameter and the compass point on the base circle, draw arcs passing through the circular thickness points established on the pitch diameter, starting at the base circle and ending at the top of the teeth. The part of the

tooth profile below the base circle is drawn as a radial line ending in a small fillet at the root circle.

For a closer approximation of the involute tooth profile, the Grant's odontograph method is used. Lay out the outside, pitch, root, and base circles and circular thickness in the same manner as used in the approximation method. The top portion of the tooth profile from point A to point B is drawn with the radius R, and the portion of the tooth profile from point B to point C is drawn with the radius r.

The values of the radii R and r are found by dividing the numbers found in the table by the diametral pitch for inch-size gears or by multiplying the numbers in the table by the module

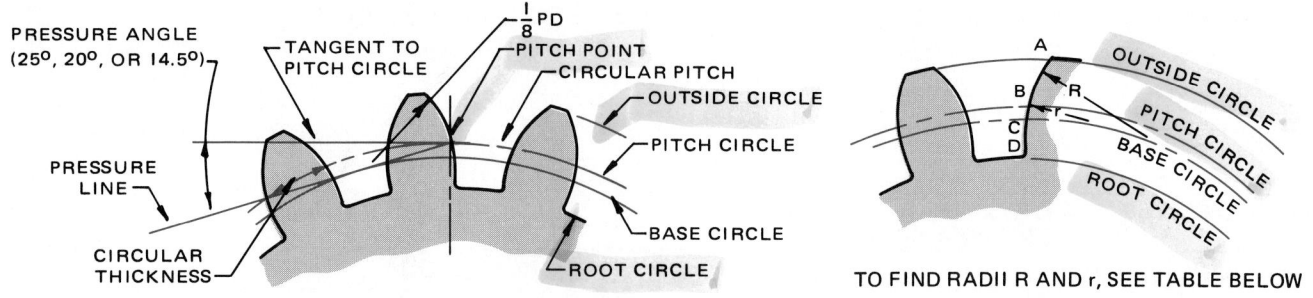

(A) APPROXIMATE REPRESENTATION OF INVOLUTE SPUR GEAR TEETH   **(B) GRANT'S ODONTOGRAPH REPRESENTATION**

INCH GEARS RADII IN INCHES			METRIC GEARS RADII IN MILLIMETERS	
RADIUS R DIVIDE NO. BY DP	RADIUS r DIVIDE NO. BY DP	NO. OF TEETH	RADIUS R MULTIPLY NO. BY MDL	RADIUS r MULTIPLY NO. BY MDL
2.51	0.96	12	2.51	0.96
2.62	1.09	13	2.62	1.09
2.72	1.22	14	2.72	1.22
2.82	1.34	15	2.82	1.34
2.92	1.46	16	2.92	1.46
3.02	1.58	17	3.02	1.58
3.12	1.69	18	3.12	1.69
3.22	1.79	19	3.22	1.79
3.32	1.89	20	3.32	1.89
3.41	1.98	21	3.41	1.98
3.49	2.06	22	3.49	2.06
3.57	2.15	23	3.57	2.15
3.64	2.24	24	3.64	2.24
3.71	2.33	25	3.71	2.33
3.78	2.42	26	3.78	2.42
3.85	2.50	27	3.85	2.50
3.92	2.59	28	3.92	2.59
3.99	2.67	29	3.99	2.67
4.06	2.76	30	4.06	2.76
4.13	2.85	31	4.13	2.85
4.20	2.93	32	4.20	2.93
4.27	3.01	33	4.27	3.01
4.33	3.09	34	4.33	3.09
4.39	3.16	35	4.39	3.16
4.20	3.23	36	4.20	3.23
4.45	4.20	37-40	4.45	4.20
4.63	4.63	41-45	4.63	4.63
5.06	5.06	46-51	5.06	5.06
5.74	5.74	52-60	5.74	5.74
6.52	6.52	61-70	6.52	6.52
7.72	7.72	71-90	7.72	7.72
9.78	9.78	91-120	9.78	9.78
13.38	13.38	121-180	13.38	13.38
21.62	21.62	181-360	21.62	21.62

FIG. 20-3-6   Methods of drawing involute spur gear teeth.

for metric-size gears. The lower portion of the tooth from points *C* to *D* is drawn as a radial line ending in a small fillet at the root circle.

If CAD is used, one tooth is drawn, and the remainder of the teeth may be created using the COPY/REPEAT or ARRAY command.

Working Drawings of Spur Gears

The working drawings of gears, which are normally cut from blanks, are not complicated. A sectional view is sufficient unless a front view is required to show web or arm details. Since the teeth are cut to shape by cutters, they need not be shown in the front view (Figs. 20-3-7 and 20-3-8).

ANSI recommends the use of phantom lines for the outside and root circles and a center line for the pitch circle. In the section view, the root and outside circles are shown as solid lines.

The dimensioning for the gear is divided into two groups, because the finishing of the gear blank and the cutting of the teeth are separate operations in the shop. The gear-blank dimensions are shown on the drawing while the gear tooth information is given in a table.

FIG. 20-3-7   Stock spur gear styles.

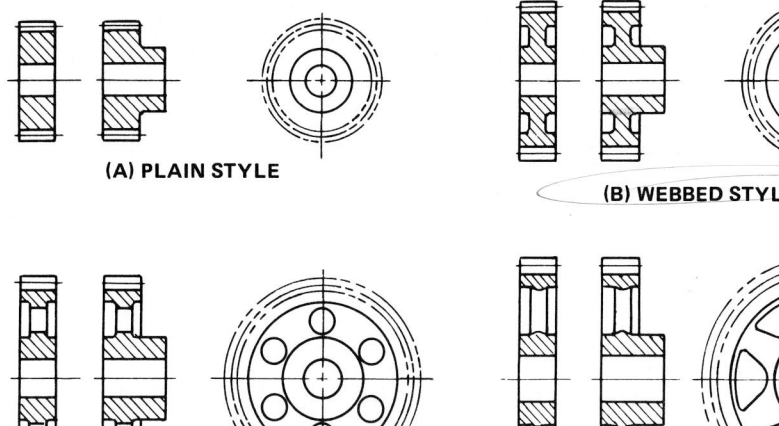

(A) PLAIN STYLE

(B) WEBBED STYLE

(C) WEBBED WITH CORED HOLES

(D) SPOKED STYLE

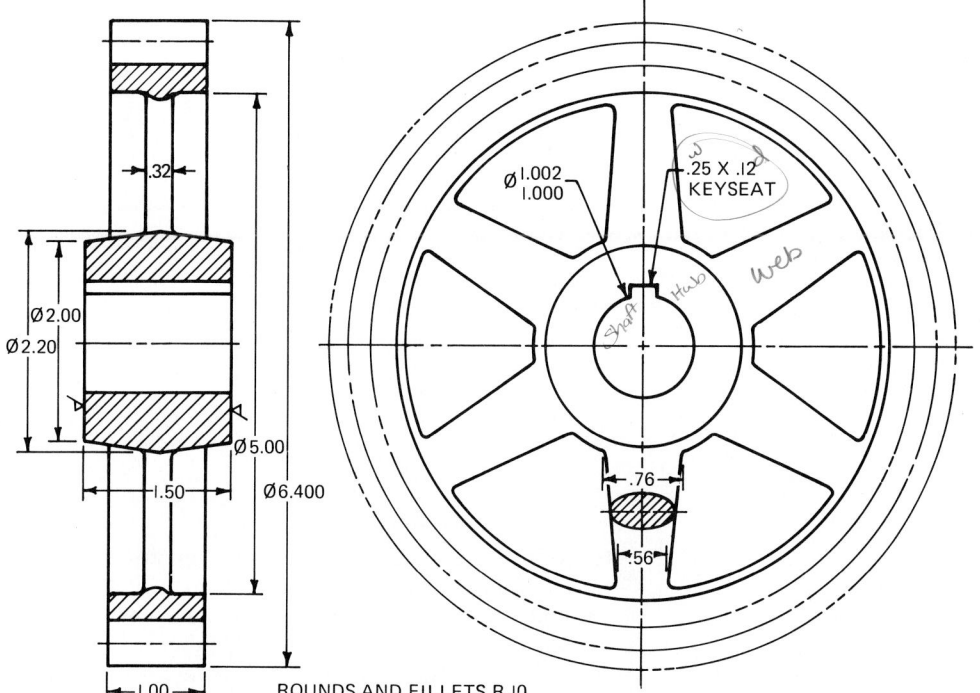

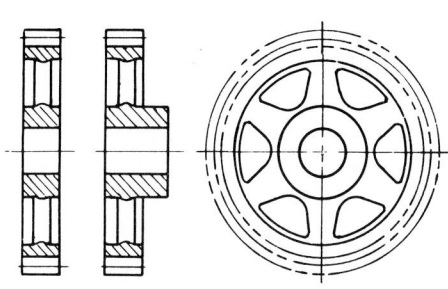

CUTTING DATA	
NUMBER OF TEETH	30
PITCH DIAMETER	6.000
DIAMETRAL PITCH	5
PRESSURE ANGLE	25°
WHOLE DEPTH	.431
CHORDAL ADDENDUM	.204
CHORDAL THICKNESS	.300
CIRCULAR THICKNESS	.314
WORKING DEPTH	.400

ROUNDS AND FILLETS R.10

FIG. 20-3-8   Working drawing of a spur gear.

The only differences in terminology between inch-size and metric-size gear drawings are the terms *diametral pitch* and *module.*

For inch-size gears, diametral pitch is used instead of module. The diametral pitch is a ratio of the number of teeth to a unit length of pitch diameter.

$$\text{Diametral pitch} = DP = \frac{N}{PD}$$

Module is the term used on metric gears. It is the length of pitch diameter per tooth measured in millimeters.

$$\text{Module} = MDL = \frac{PD}{N}$$

From these definitions it can be seen that the module is equal to the reciprocal of the diametral pitch and thus is not its metric dimensional equivalent. If the diametral pitch is known, the module can be obtained.

$$\text{Module} = 25.4 \div \text{diametral pitch}$$

Gears presently being used in North America are designed in the inch system and have a standard diametral pitch instead of a preferred standard module. Therefore it is recommended that the diametral pitch be referenced beneath the module when gears designed with standard inch pitches are used. For gears designed with standard modules, the diametral pitch need not be referenced on the gear drawing. The standard modules for metric gears are 0.8, 1, 1.25, 1.5, 2.25, 3, 4, 6, 7, 8, 9, 10, 12, and 16.

As metric gears were not readily available at the time of this writing, converted inch-size modules have been substituted to allow students to complete the drawing assignments (Fig. 20-3-2).

Spur Gear Calculations

Center Distance   The center distance between the two shaft centers is determined by adding the pitch diameter of the two gears and dividing the sum by 2.

EXAMPLE 1

A 12 DP, 36-tooth pinion mates with a 90-tooth gear. Find the center distance.

$$\begin{aligned}
\text{Pitch diameter} &= \text{number of teeth} \div \text{diametral pitch} \\
&= 36 \div 12 = 3.00 \text{ in. (pinion)} \\
&= 90 \div 12 = 7.50 \text{ in. (gear)}
\end{aligned}$$

$$\begin{aligned}
\text{Sum of the two pitch diameters} \\
= 3.00 \text{ in.} + 7.50 \text{ in.} = 10.50 \text{ in.}
\end{aligned}$$

$$\begin{aligned}
\text{Center distance} &= 1/2 \text{ sum of the two pitch diameters} \\
&= \frac{10.50}{2} = 5.25 \text{ in.}
\end{aligned}$$

EXAMPLE 2

A 3.18-module, 24-tooth pinion mates with a 96-tooth gear. Find the center distance.

$$\begin{aligned}
\text{Pitch diameter (PD)} &= \text{number of teeth} \times \text{module} \\
&= 24 \times 3.18 = \varnothing76.3 \text{ (pinion)} \\
&= 96 \times 3.18 = \varnothing305.2 \text{ (gear)}
\end{aligned}$$

$$\begin{aligned}
\text{Sum of the two pitch diameters} \\
= 76.3 + 305.2 = 381.5 \text{ mm}
\end{aligned}$$

$$\begin{aligned}
\text{Center distance} &= 1/2 \text{ sum of the two pitch diameters} \\
&= \frac{381.5}{2} = 190.75 \text{ mm}
\end{aligned}$$

Ratio   The *ratio* of gears is a relationship between any of the following:

1. Revolutions per minute of the gears
2. Number of teeth on the gears
3. Pitch diameter of the gears

The ratio is obtained by dividing the larger value of any of the three by the corresponding smaller value.

EXAMPLE 3

A gear rotates at 90 r/min and the pinion at 360 r/min.

$$\text{Ratio} = \frac{360}{90} = 4 \text{ or ratio} = 4{:}1$$

EXAMPLE 4

A gear has 72 teeth, the pinion 18 teeth.

$$\text{Ratio} = \frac{72}{18} = 4 \text{ or ratio} = 4{:}1$$

EXAMPLE 5

A gear with a pitch diameter of 8.500 in. meshes with a pinion having a pitch diameter of 2.125 in.

$$\text{Ratio} = \frac{PD \text{ of gear}}{PD \text{ of pinion}}$$

$$= \frac{8.500}{2.125} = 4$$

$$\text{or Ratio} = 4{:}1$$

Determining the Pitch Diameter and Outside Diameter
The pitch diameter of a gear can readily be found if the number of teeth and the diametral pitch or module are known. The outside diameter (OD) is equal to the pitch diameter plus two addendums. The addendum for a gear tooth other than a 20° stub tooth is equal to $1 \div DP$ (U.S. customary) or the module (metric).

EXAMPLE 6

A 25° spur gear has a diametral pitch of 5 and 40 teeth.

$$\text{Pitch diameter} = N \div DP$$
$$= 40 \div 5$$
$$= 8.00 \text{ in.}$$

$$OD = PD + 2\ ADD = 8.00 + 2\ \frac{1}{DP}$$
$$= 8.00 + \frac{2}{5}$$
$$= 8.40 \text{ in.}$$

EXAMPLE 7

A 20° spur gear has a module of 6.35 and 34 teeth.

$$\text{Pitch diameter} = N \times MDL$$
$$= 34 \times 6.35$$
$$= 216 \text{ mm}$$

$$OD = PD + 2\ ADD = 216 + 2(6.35)$$
$$= 228.7 \text{ mm}$$

ASSIGNMENTS

See Assignments 9 through 13 for Unit 20-3 on pages 706 through 708.

20-4 POWER-TRANSMITTING CAPACITY OF SPUR GEARS

Gear drives are required to operate under such a wide variety of conditions that it is very difficult and expensive to determine the best gear set for a particular application. The most economical procedure is to select standard stock gears with an adequate load rating for the application.

Approximate horsepower (kilowatt) ratings for spur gears of various sizes (numbers of teeth), at several operating speeds (revolutions per minute), are given in catalogs with the spur gear listings. Ratings for gear sizes and/or speeds not listed may be estimated from the values shown in Fig. 20-4-1 (pg. 696).

Pitch-line velocities exceeding 1,000 ft/min (5 m/s) for 14.5° PA (pressure angle), or 1,200 ft/min (6 m/s) for 20° PA, are not recommended for metallic spur gears. Ratings are listed for speeds below these limits.

The ratings given (or calculated) should be satisfactory for gears used under normal operating conditions, that is, when they are properly mounted and lubricated, carrying a smooth load (without shock) for 8 to 10 hours a day.

The charts shown in Fig. 20-4-1 indicate the approximate horsepower (kilowatt) ratings of 16- and 20-tooth steel spur gears of several tooth sizes operating at various speeds. They may be used to determine the approximate diametral pitch or module of a 16- or 20-tooth steel pinion that will carry the horsepower (kilowatts) required at the desired speed. The

intersection of the lines representing values of revolutions per minute and horsepower (kilowatts) indicates the approximate gear diametral pitch (module) required.

The number of teeth normally should not be less than 16 to 20 in a 14.5° pinion, nor less than 13 in a 20° pinion.

Ratings shown for spur gears in catalogs normally are for class 1 service. For other classes of service, the service factors in Fig. 20-4-2 (pg. 697) should be used.

Selecting the Spur Gear Drive

1. Determine the class of service.
2. Multiply the horsepower (kilowatts) required for the application by the service factor.
3. Select spur gear pinion with a catalog rating equal to or greater than the horsepower (kilowatts) determined in step 2.
4. Select driven spur gear with a catalog rating equal to or greater than the horsepower (kilowatts) determined in step 2.

EXAMPLE 1

Select a pair of 20° spur gears which will drive a machine at 150 r/min. Size of driving motor = 25 hp, 600 r/min. Service factor = 1.

SOLUTION

Since the service factor is 1, we do not need to increase or decrease the design horsepower. Refer to the charts in Fig. 20-4-2A, which show design data for 20° spur gears having 16 and 20 teeth. Selecting a pinion having 16 teeth and reading vertically on the column showing 600 r/min to horsepower rating of 25, we find that the required DP is 4. (Select the DP equal to or greater than the horsepower required.)

- Pinion: N = 16, DP = 4, PD = 16 ÷ 4 = 4.000 in.
- Ratio 4:1
- Gear: N = 16 × 4 = 64, DP = 4, PD = 64 ÷ 4 = 16.000 in.

EXAMPLE 2

A 5 hp, 1,200 r/min motor is used to drive a machine that runs 8 hours a day under moderate shock. If the machine is to run at 200 r/min and at the capacity of the motor, what spur gears would you select?

SOLUTION

The operating conditions of the machine are such that the machine fits into the service class 2 and requires a service factor of 1.2. Therefore

Horsepower required for design purposes = 5 hp × 1.2 = 6 hp

The pinion will be mounted on the motor and runs at 1,200 r/min. Selecting a 20° pinion having 16 teeth, refer to

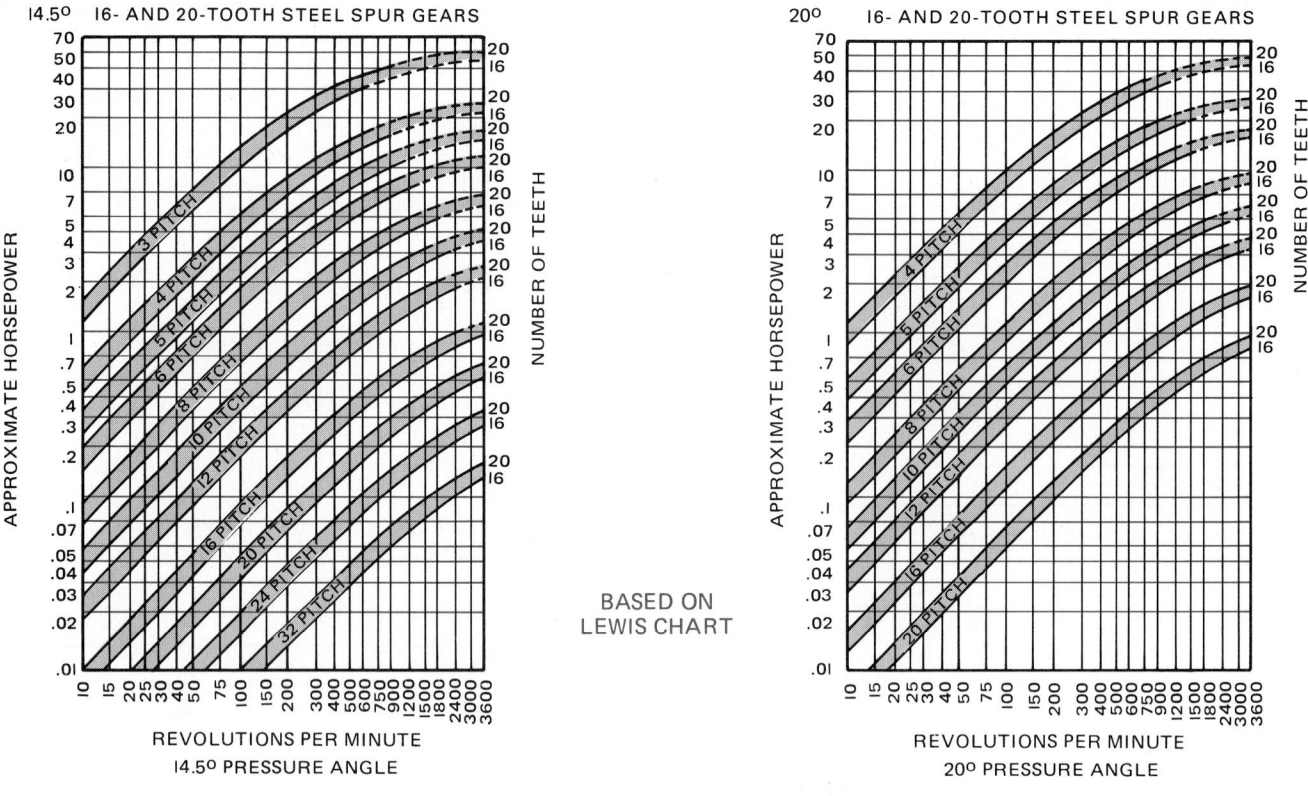

(A) DIAMETRAL PITCH SELECTION CHART

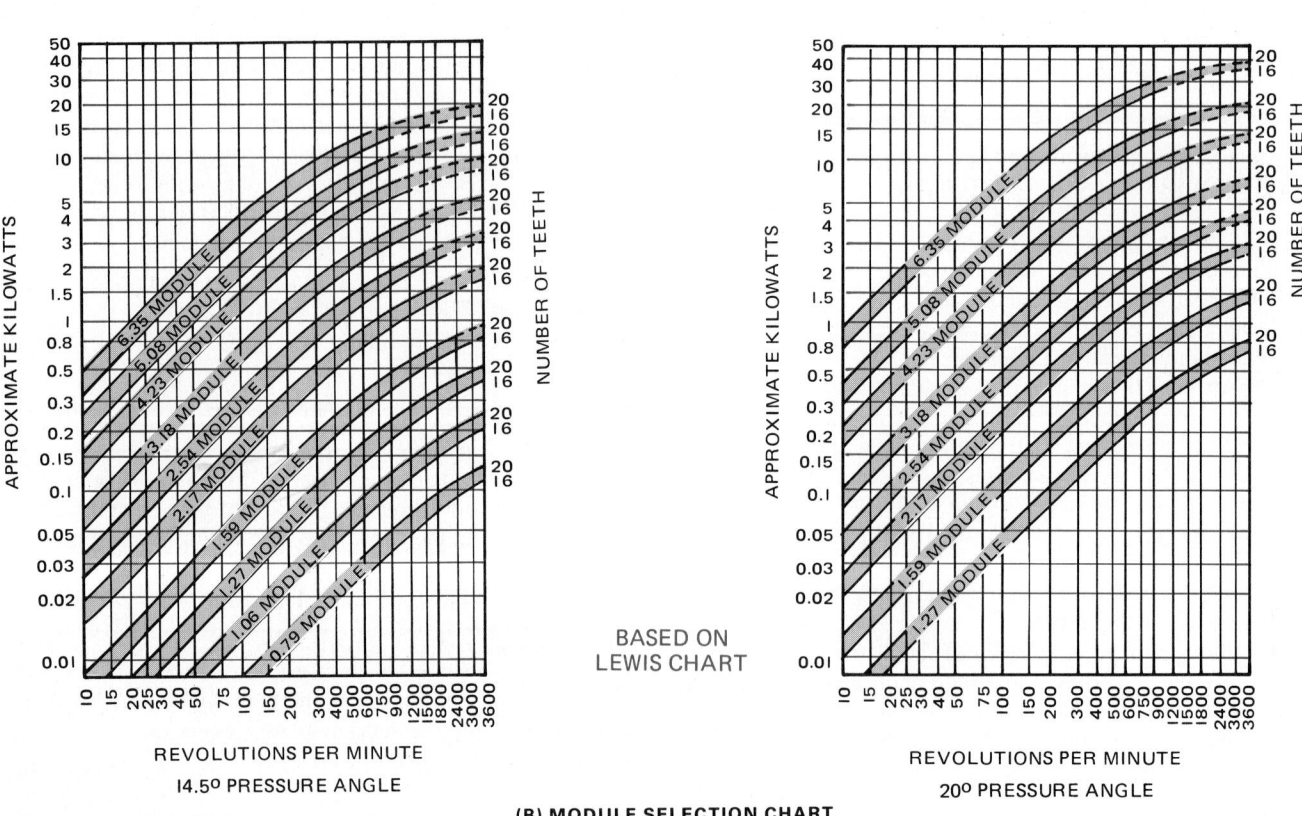

(B) MODULE SELECTION CHART

FIG. 20-4-1 Service class and factor for spur gears.

SERVICE CLASS	OPERATING CONDITIONS	SERVICE FACTOR
Class I	Continuous 8- to 10-hr per day duty, with smooth load (no shock)	1.0
Class II	Continuous 24-hr duty, with smooth load, or 8- to 10-hr per day, with moderate shock	1.2
Class III	Continuous 24-hr duty with moderate shock load	1.3
Class IV	Intermittent duty, not over 30-min per hr, with smooth load (no shock)	0.7
Class V	Hand operation, limited duty, with smooth load (no shock)	0.5

Heavy shock loads and/or severe conditions require the use of higher service factors. Such conditions may require factors of 1.5 to 2.0 or greater than required for Class I service.

FIG. 20-4-2 Service class and factor for spur gears.

Fig. 20-4-2A to find the required DP. Reading vertically on the 1,200 r/min line and horizontally at 6 hp we find that the required DP is 8.

- Pinion: N = 16, DP = 8, PD = 16 ÷ 8 = 2.000 in.
- Gear: N = $16 \times \frac{1200}{200}$ = 96, DP = 8, PD = 96 ÷ 8 = 12.000 in.

EXAMPLE 3

A 7.5 kW, 900 r/min motor is attached by means of 14.5° spur gears to a punch that operates 24 hours a day. The reduction in revolutions per minute is 4:1. Select a gear and pinion assuming that the punch is being operated at motor capacity.

SOLUTION

The operating conditions of the machine are such that the machine fits into the service class 3 and requires a service factor of 1.3. Therefore

Kilowatts required for design purposes = 7.5 × 1.3 = 9.75 kW

The pinion is mounted on the motor and runs at 900 r/min. Refer to Fig. 20-4-2B. Reading vertically on the 900 r/min line and horizontally at 9.75 kW, we may select either a pinion having a module of 5.08 and N of 20 or a module of 6.35 and N of 16. First, using a module of 5.08, we have

- Pinion: N = 20, MDL = 5.08, PD = 101.6 mm. Gear travels at 225 r/min, or one-fourth of pinion revolutions per minute.
- Gear: N = 4 × 20 = 80, MDL = 5.08, PD = 80 × 5.08 = 406.4 mm

Second, using a module of 6.35, we have

- Pinion: N = 16, MDL = 6.35, PD = 101.6 mm
- Gear: N = 64, MDL = 6.35, PD = 406.4 mm

Since both sets of gears are of the same diameter, the overall size is not a factor. Checking on the cost per set, we find that there would be considerable savings by selecting the gear and pinion having the module of 5.08. Since the extra strength of the gear set having a module of 6.35 is not required in this instance, we would recommend the gear and pinion having a module of 5.08.

REFERENCES AND SOURCE MATERIAL

1. Boston Gear Works.

ASSIGNMENTS

See Assignments 14 and 15 for Unit 20-4 on page 708.

20-5 RACK AND PINION

A *rack* is a straight bar having teeth that engage the teeth on a spur gear (Fig. 20-5-1). In theory, it is a spur gear having an infinite pitch diameter. Therefore all circular dimensions become linear. The addendum, dedendum, and tooth thickness are the same as the mating spur gear. To draw the teeth on a rack, lay out the addendum and dedendum distances from the pitch line. Divide the pitch line into lineal pitch distances equal in size to the circular pitch on the gear. Divide each of these spaces in half to get the lineal thickness. Through these points draw the tooth faces at angles of 14.5, 20, or 25° from the vertical lines. Darken in the top and bottom lines of the teeth and

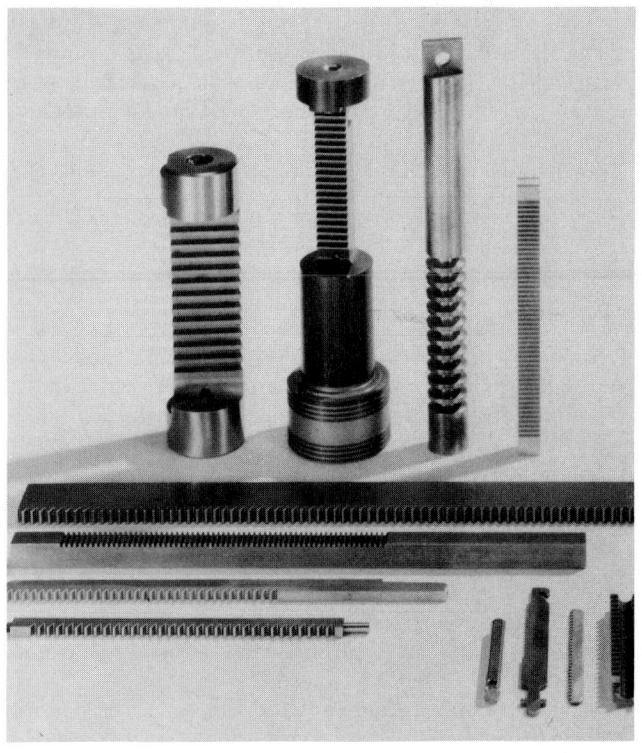

FIG. 20-5-1 Racks. *(Gear Specialties Inc.)*

FIG. 20-5-2   Rack and pinion.

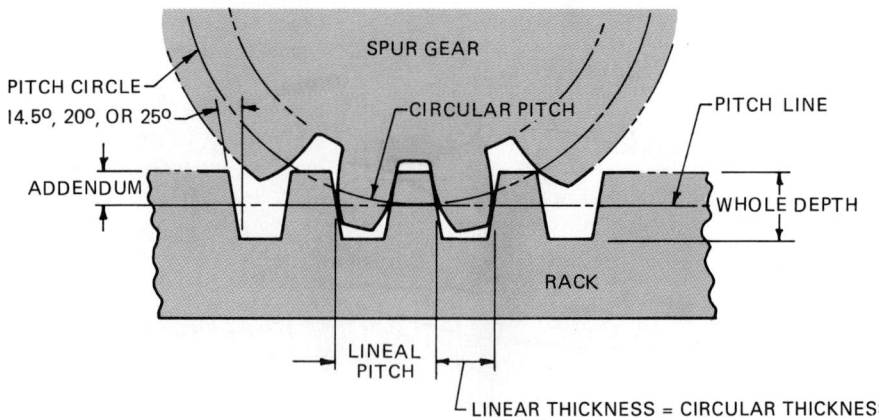

add the tooth fillets. The specifications for the teeth of the rack are given in the same manner as for spur gears (Fig. 20-5-2).

ASSIGNMENT ▓▓▓▓▓▓▓▓▓▓▓▓▓▓▓▓▓▓▓▓▓

See Assignment 16 for Unit 20-5 on pages 708–709.

20-6 BEVEL GEARS

Bevel gears are used to transmit power between two shafts whose axes intersect. The axes may intersect at any angle, but the most common is 90°. They are similar to rolling cones having the same apex. The teeth are the same shape as spur gear teeth but taper toward the cone apex. Therefore, many spur gear terms may apply to bevel gears. *Miter gears* are bevel gears having the same diametral pitch or module, pressure angle, and number of teeth.

Figures 20-6-1 and 20-6-2 show bevel gear definitions and formulas.

Working Drawings of Bevel Gears

The working drawings of bevel gears, like those of spur gears, give only the dimensions of the bevel gear blank. The cutting data for the teeth are given in a note or table. A single section view is normally used, unless a second view is required to show such details as spokes. Sometimes both the bevel gear and pinion are drawn together, showing their relationship. The dimensions and cutting data will depend on the method used in cutting the teeth, but the information shown in Fig. 20-6-3 is commonly used.

The actual gear teeth are often shown on assembly or display drawings. One of the most common conventions used for drawing the teeth is the Tredgold method, which is shown in Fig. 20-6-4 (pg. 700).

An arc whose radius is taken on the back cone is used as the pitch circle, and a tooth is developed using standard spur gear formulas. Tooth sizes taken on the OD and pitch diameter are

TERM	FORMULA
Addendum, dedendum, whole depth, pitch diameter, module, diametral pitch, number of teeth, circular pitch, chordal thickness, circular thickness	Same as for spur gears
Pitch cone radius	$\dfrac{PD}{2 \times \text{sin of pitch angle}}$
Pitch cone angle	Tan pitch angle $= \dfrac{PD \text{ of gear}}{PD \text{ of pinion}}$ $= \dfrac{N \text{ of gear}}{N \text{ of pinion}}$
Addendum angle	Tan addendum angle $= \dfrac{\text{Addendum}}{\text{Pitch cone radius}}$
Dedendum angle	Tan dedendum angle $= \dfrac{\text{Dedendum}}{\text{Pitch cone radius}}$
Face angle	Pitch cone angle plus addendum angle
Cutting angle	Pitch cone angle minus dedendum angle
Back angle	Same as pitch cone angle
Angular addendum	Cos of pitch cone angle × addendum
Outside diameter	Pitch diameter plus two angular addendums
Crown height	Divide ½ the outside diameter by the tangent of the face angle
Face width	1½ to 2½ times the circular pitch
Chordal addendum	Addendum + $\dfrac{\text{circular thickness}^2 \times \text{cos pitch cone angle}}{4PD}$

FIG. 20-6-1   Bevel gear formulas.

FIG. 20-6-2   Bevel gear nomenclature.

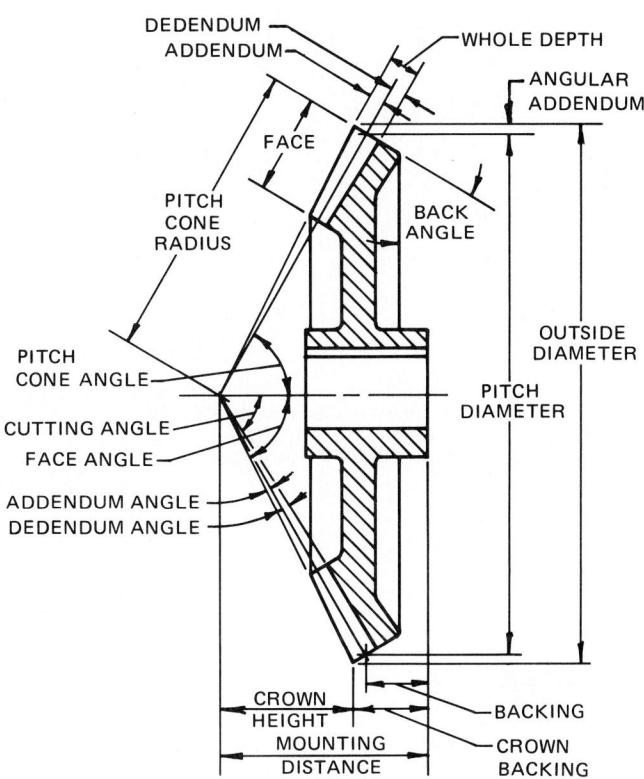

transferred to the front view, and the profiles for the teeth are drawn. Radial lines from these points are taken, and the small end of the tooth is developed. The teeth on the side or section view are now drawn by projecting the teeth from the front view. Cast iron is normally used for large gears and small gears that are not subject to heavy duty. Often a gear and pinion are made of different materials for efficiency and durability. The pinion is made of a stronger material because the teeth on the pinion come into contact more times than the teeth on the gear. Common combinations are steel and cast iron, and steel and bronze.

For hp or kW ratings of bevel gears, refer to manufacturers' catalogs.

ASSIGNMENTS

See Assignments 17 through 19 for Unit 20-6 on pages 708 to 709.

FIG. 20-6-3   Working drawing of a bevel gear.

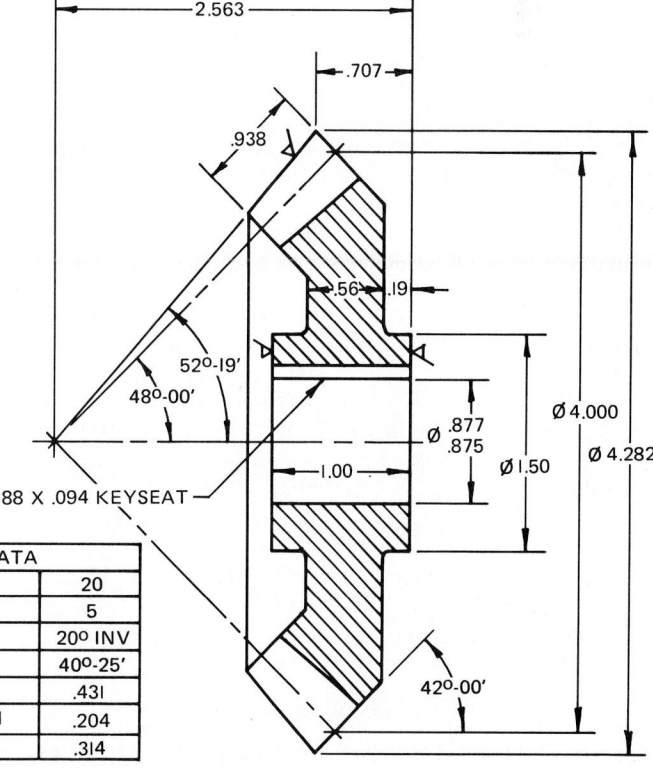

CUTTING DATA	
NO. OF TEETH	20
DIAMETRAL PITCH	5
TOOTH FORM	20° INV
CUTTING ANGLE	40°-25'
WHOLE DEPTH	.431
CHORDAL ADDENDUM	.204
CHORDAL THICKNESS	.314

FIG. 20-6-4   Bevel gear assembly or display drawing.

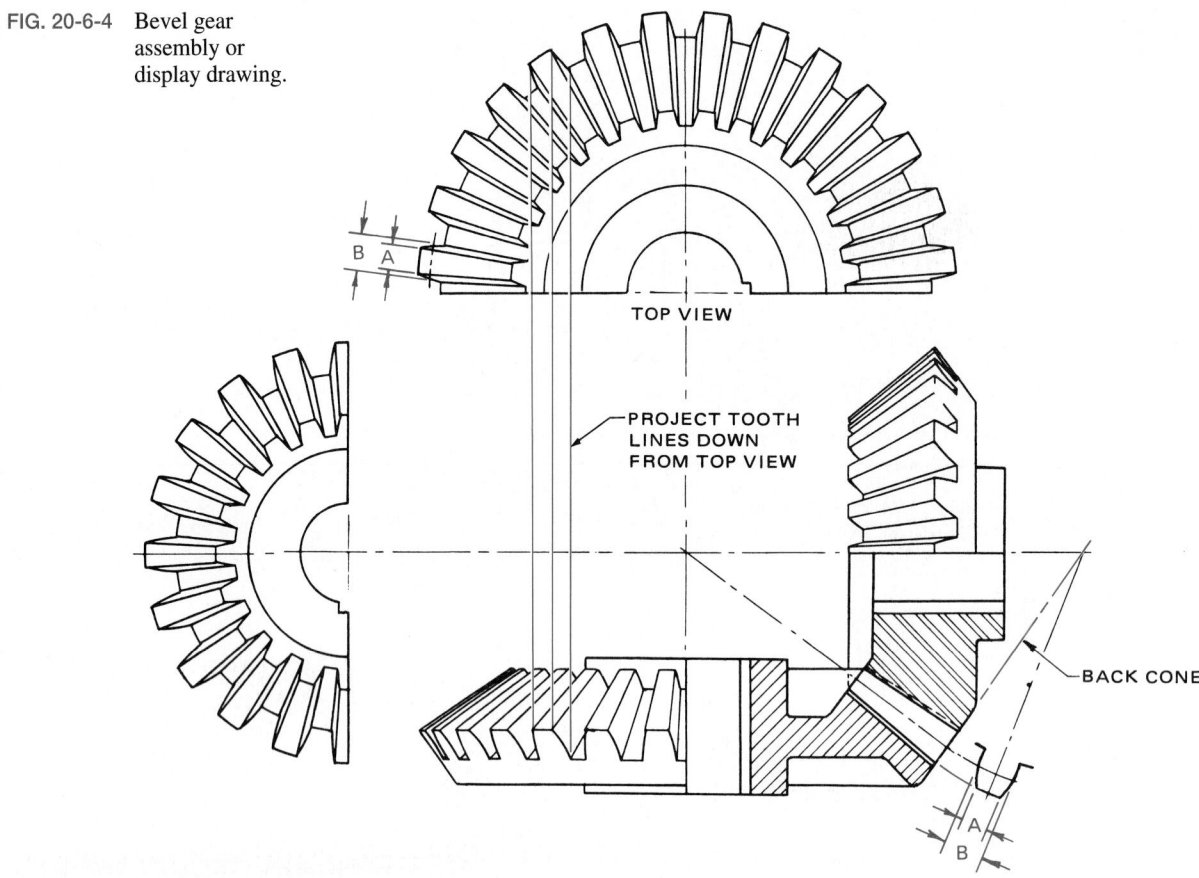

20-7 WORM AND WORM GEARS

Worm gears are used to transmit power between two shafts that are at right angles to each other and are nonintersecting. The teeth on the worm are similar to the teeth on the rack, and the teeth on the worm gear are curved to conform with the teeth on the worm. Thread terms such as *pitch* and *lead* are used on the worm.

Since a single-thread worm in one revolution advances the worm gear only one tooth and space, a large reduction in velocity is obtained. Another feature of worm gearing is the high mechanical advantage acquired. The ratio of worm gear speed to the worm speed is the ratio between the number of teeth on the worm gear and the number of threads on the worm. A worm gear with 33 teeth and a worm with a multiple thread of three has a ratio of 11:1.

About 50:1 is the maximum ratio recommended. Since a single-thread worm has a low lead (or helix) angle, it is inefficient and consequently not used to transmit power. The lead angle should be between 25 and 45° for efficiency in transmitting power; as a result, multithread worms are used. The number of threads on a worm may vary from one to eight.

Figures 20-7-1 to 20-7-5 (here and through pg. 702) supply data on worm gear drawings and formulas.

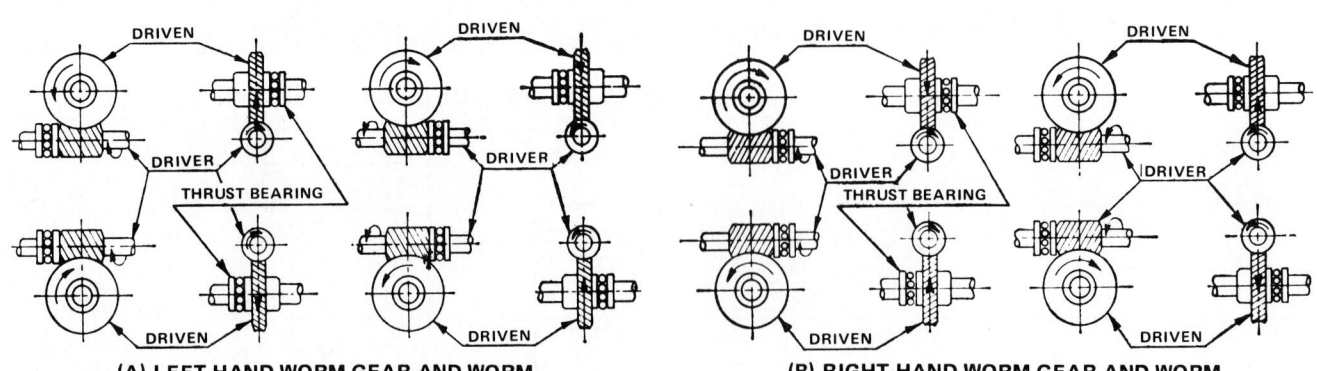

(A) LEFT-HAND WORM GEAR AND WORM   **(B) RIGHT-HAND WORM GEAR AND WORM**

FIG. 20-7-1   Location of bearings to absorb thrust load on worm and worm gear.

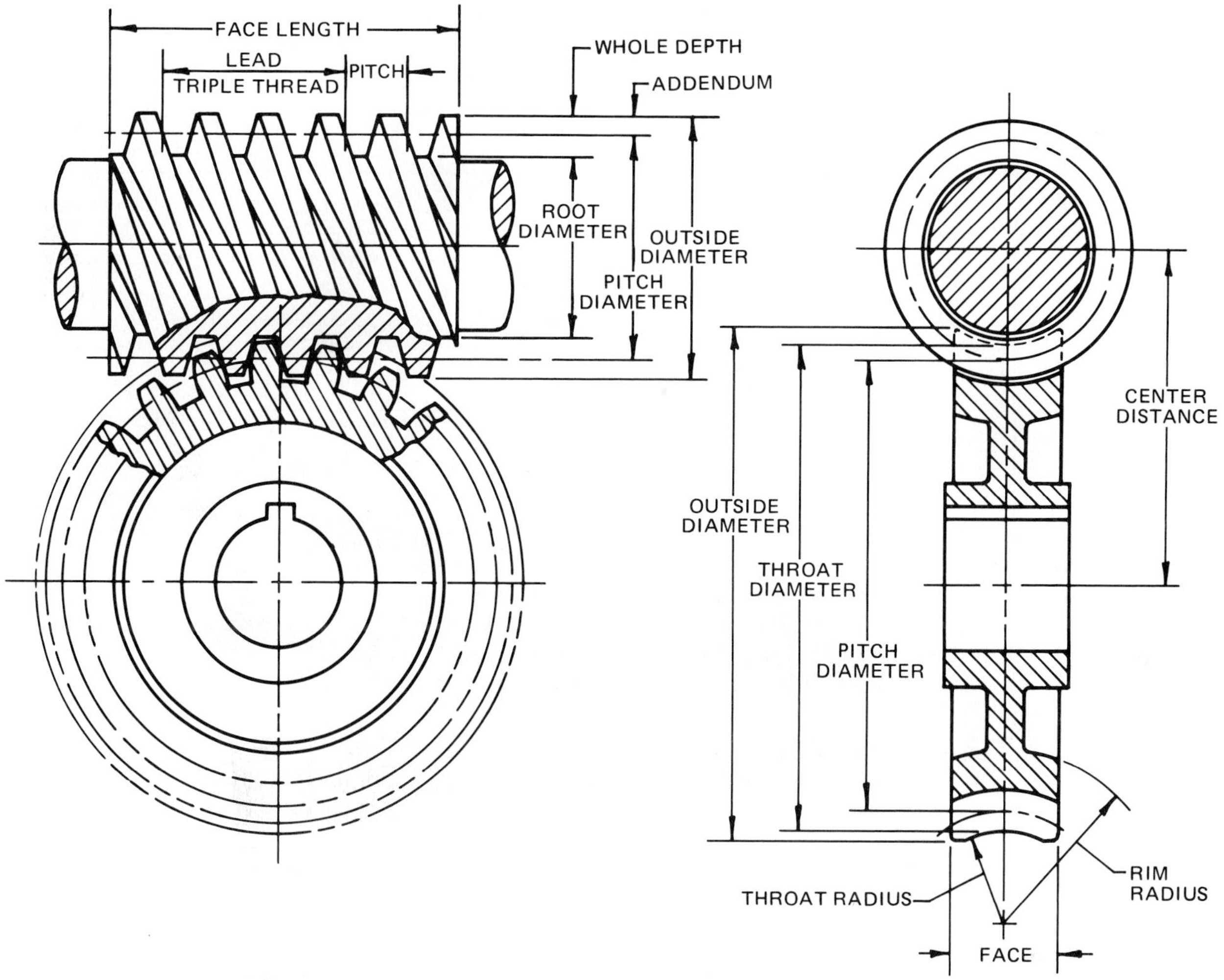

FACE LENGTH

LEAD

TRIPLE THREAD

PITCH

WHOLE DEPTH

ADDENDUM

ROOT DIAMETER

OUTSIDE DIAMETER

PITCH DIAMETER

CENTER DISTANCE

OUTSIDE DIAMETER

THROAT DIAMETER

PITCH DIAMETER

THROAT RADIUS

RIM RADIUS

FACE

FIG. 20-7-2   Worm and worm gear nomenclature.

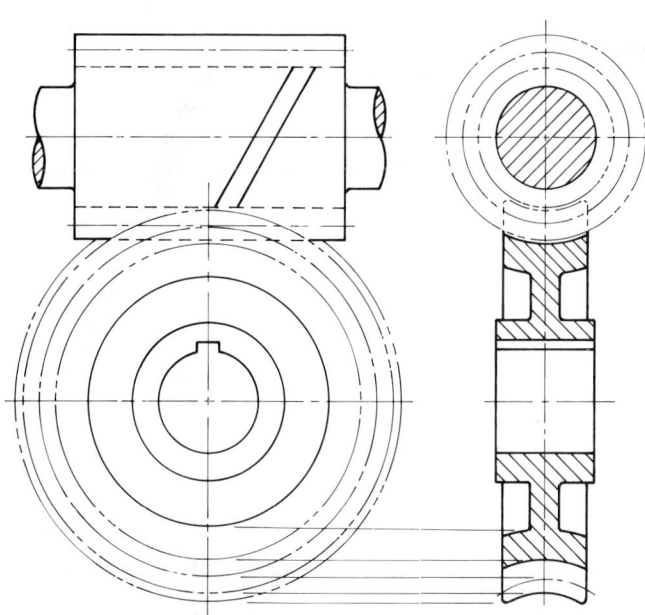

FIG. 20-7-3   Assembly drawing of a worm gear and worm.

THREADS OF LEFT HAND LEAN TO THE LEFT WHEN STANDING ON EITHER END

(A) LEFT-HAND WORM GEAR AND WORM

THREADS OF RIGHT HAND LEAN TO THE RIGHT WHEN STANDING ON EITHER END

(B) RIGHT-HAND WORM GEAR AND WORM

FIG. 20-7-4   Identifying right- and left-hand worms and worm gears.

Term	Symbol	Formula	Definition
Pitch diameter of worm	PDw	PDw = 2C − PDg	
Pitch diameter of gear	PDg	PDg = 2C − PDw or $\dfrac{NP}{\pi}$	
Pitch	P	$P = \dfrac{L}{T}$ $P = \dfrac{(2C - PDw) \times \pi}{N}$	The distance from one tooth to the corresponding point on the next tooth measured parallel to the worm axis. It is equal to the circular pitch on the worm gear
Lead	L	L = πPDg ÷ R L = P × T L = Tan La × πPDw	The distance the thread advances axially in one revolution of the worm
Threads	T	$T = \dfrac{L}{P}$	The number of threads or starts on worm; e.g., 2 for double thread, 3 for triple thread
Gear teeth	N	$N = \dfrac{\pi PDg}{P}$	Number of teeth on worm gear
Ratio	R	$R = \dfrac{N}{T}$	Divide number of gear teeth by number of worm threads
Center distance	C	$C = \dfrac{PDw + PDg}{2}$	
Addendum	ADD	ADD = 0.318P	Single and double threads
		ADD = 0.286P	Triple and quadruple threads
Whole depth	WD	WD = 0.686P	Single and double threads
		WD = 0.623P	Triple and quadruple threads
Outside diameter, worm	ODw	ODw = PDw + 2ADD	
Outside diameter, gear	ODg	ODg = TD + 0.4775P	Single and double threads
		ODg = TD + 0.3183P	Triple and quadruple threads
Throat diameter	TD	TD = PDg + 2ADD	
Face width, gear	F	F = 2.38P + .25 (inch) F = 2.38P + 6 (metric)	Single and double threads
		F = 2.15P + .20 (inch) F = 2.15P + 5 (metric)	Triple and quadruple threads
Face length, worm	FL	FL = 6 × P	
Lead angle	La	Tan La = $\dfrac{L}{PDw \times 3.1416}$	Divide lead by circumference of pitch diameter of worm. Quotient is tangent of lead angle.
Throat radius	R_t	$R_t = \dfrac{PDw}{2} - ADD$	Subtract addendum from half of pitch diameter of worm
Rim radius	R_r	$R_r = \dfrac{PDw}{2} + P$	

FIG. 20-7-5   Worm and worm gear formulas.

Working Drawings of Worm and Worm Gears

These are similar to working drawings of other gears. A one-view section drawing is normally used for the worm gear (Fig. 20-7-6). When a second view is required, the throat and root circles are shown as solid lines, and the outside circle is not shown on this view. As for the worm drawing, the root and the outside diameter are shown as solid lines, and a second view is not normally required.

When a worm and worm gear appear as an assembly drawing, both views are drawn and the conventional solid line for the OD of the worm and the throat diameter of the worm gear are shown as broken lines where the teeth mesh.

ASSIGNMENTS

See Assignments 20 through 22 for Unit 20-7 on pages 710 to 711.

FIG. 20-7-6   Working drawing
of a worm and
worm gear.

CUTTING DATA	
NO. OF TEETH	36
ADDENDUM	.159
WHOLE DEPTH	.343
NO. OF THREADS	2
PITCH (AXIAL)	.500
PRESSURE ANGLE	20°
LEAD ANGLE	7°-53'
LEAD—RH	

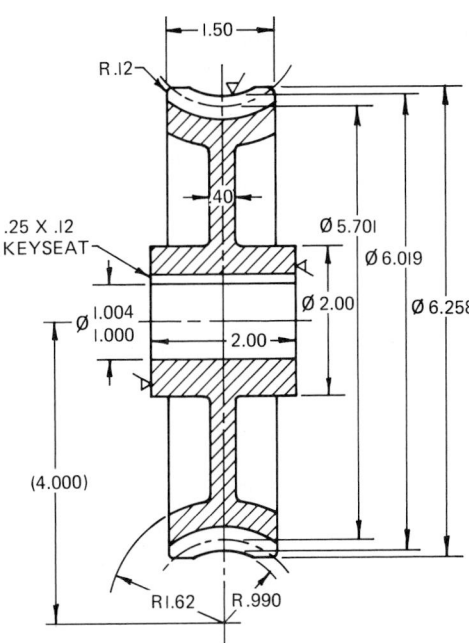

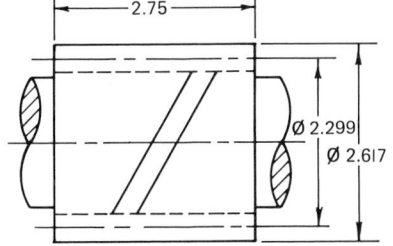

CUTTING DATA	
NO. OF THREADS	2
PITCH	.500
PRESSURE ANGLE	20°
LEAD ANGLE	7°-53'
LEAD—RH	
WHOLE DEPTH	.343
ADDENDUM	.159

20-8 COMPARISON OF CHAIN, GEAR, AND BELT DRIVES

Chains, gears, and belts are used for power drives between rotating shafts that cannot be directly coupled. In this unit the characteristics of these media are compared, and the conditions favorable to the use of each type of drive are discussed.

Chains

A *chain drive* consists of an endless chain whose links mesh with toothed wheels, called *sprockets*, which are keyed to the shafts of the driving and driven mechanisms.

Roller Chains   The unique feature of a roller chain is its freedom of joint action during its engagement with the sprocket. This is accomplished by articulation of the pins of the bushings, while the rollers turn on the outside of the bushings, thus eliminating rubbing action between the rollers and the sprocket teeth.

Silent Chains   Comparable ease of joint action occurs in the engagement of the silent chain with the sprocket.

Gears

A *simple gear drive* consists of a toothed driving wheel meshing with a similar driven wheel. Tooth forms are designed to ensure uniform angular rotation of the driven wheel during tooth engagement. Gears are available with precision-cut teeth or with unfinished teeth.

Belts

A *belt drive* consists of an endless flexible belt that connects two wheels or pulleys. Belt drives depend on friction between the belt and the pulley surfaces for the transmission of power.

In the case of V-belts, the friction for the transmission of the driving force is increased by wedging the belt into the grooves on the pulley.

V-belt drives are available in single or multiple strands for varying power-transmission requirements.

Another type of belt has shallow teeth molded on the inside of the driving face. The pulleys have teeth for engagement with the belt teeth.

Chain Drives Compared with Gear Drives

Advantages of Chains

Shaft center distances for chain drives are relatively unrestricted, whereas with gears, the center distance must be such that the pitch surfaces of the gears are tangent. This advantage often will result in a simpler, less costly, and more practical design.

Chains are easily installed. While all drive media require proper installation, the assembly tolerances for chain drives are not as restricted as those for gears. The resultant savings in the time of installation may be an important item in meeting the production schedule required of the driven machine.

The ease of chain installation is a definite advantage where later changes in design, such as speed ratio, capacity, and centers, are anticipated.

Advantages of Gears

When space limitations require the shortest possible distance between shaft centers, a gear drive is usually preferable to a chain drive.

The maximum speed ratio for satisfactory operation of a gear drive is usually greater than that for a chain drive.

Gears can be operated at higher rotative speeds than chain drives.

Chain Drives Compared with Belt Drives

Advantages of Chains

Chain drives do not slip or creep as do belt drives. As a result, chains maintain a positive speed ratio between the driving and the driven shafts, and they are more efficient since no power is lost because of slippage.

Chain drives are more compact than belt drives. For a given capacity, a chain will be narrower than a belt, and sprockets will be smaller in diameter than pulleys; thus the chain drive will occupy less overall space.

Chains are easy to install. A chain can be installed by wrapping it around the sprockets and then slipping the pins of a connecting link into position.

The required minimum arc of contact is smaller for chains than for belts. This advantage becomes more pronounced as the speed ratio increases and thus permits chain drives to operate on much shorter shaft center distances.

Where several shafts are to be driven from a single shaft, positive speed synchronism between the driven shafts is usually imperative. For such applications, chains are more suitable.

Chains do not deteriorate with age; nor are they affected by sun, oil, and grease. Chains can operate at higher temperatures. Chain drives are more practical for low speeds.

Chain elongation resulting from normal wear is a slow process; the chain therefore requires infrequent adjustment. Belt stretch, however, necessitates frequent tightening by shaft adjustment, by idlers, or by shortening the belt.

Advantages of Belts

Since no metal-to-metal contact occurs between a belt and pulleys, belts require no lubrication, although leather belts need periodic applications of belt dressing to preserve their flexibility.

Generally speaking, a belt drive operates with less noise than a chain drive.

Flat-belt drives can be used where extremely long center distances would make chain drives impractical.

In the extremely high-speed ranges, flat belts can be operated to better advantage than chains.

Conclusion

No one type of power drive is ideal for all types of service. This unit has discussed the relative merits of chain, gear, and belt drives, and should provide a guide to the selection of the best type for a given application.

REFERENCES AND SOURCE MATERIAL

1. American Chain Association.

ASSIGNMENTS

See Assignments 23 through 26 for Unit 20-8 on pages 712 to 713.

ASSIGNMENTS FOR CHAPTER 20

ASSIGNMENTS FOR UNIT 20-1, BELT DRIVES

1. *V-Belt Drive Problems.*   Do any three.
   (a) A .33 hp (0.25 kW), 1,750 r/min motor is to operate a furnace blower having a shaft speed of approximately 765 r/min. The center distance between the motor and blower shafts is approximately 13.5 in. (340 mm). Select a suitable V-belt.
   (b) A .5 hp (0.37 kW), 1,160 r/min motor is used to operate a drill press. The spindle speed is 520 r/min, ±5 r/min. The center distance between the motor and blower shafts is approximately 22 in. (550 mm). Select a suitable V-belt.
   (c) A 1.5 hp (1.1 kW), 1,750 r/min motor is to operate a band saw whose flywheel turns at approximately 800 r/min. A pulley attached to the flywheel shaft connects, by means of a V-belt, to the pulley on the motor shaft. Center-to-center distance of shafts is 13.5 in. (340 mm). Calculate the size of the V-belt required.
   (d) A .5 hp (0.37 kW), 1,750 r/min motor drives a power hacksaw. The shaft on the hacksaw is to run at approximately 750 r/min, and the center-to-center distance of the shafts is 15.5 in. (400 mm). Calculate the size of V-belt required.
   (e) A .75 hp (0.6 kW), 1,750 r/min motor is used to drive a punch machine whose flywheel turns at approximately 600 r/min. A pulley is attached to the flywheel shaft and connects to the motor pulley by means of a V-belt. Center-to-center distance is 17 in. (430 mm). Calculate the size of V-belt required.
2. *V-Belt Motor Drive.*   Lay out a .25 hp or 0.2 kW motor to drive shaft A between 815 and 835 r/min by means of a V-belt drive. Refer to Fig. 20-1-A or 20-1-B. Design for normal duty. Other details can be seen in Fig. 20-1-8. Draw top and front views. From manufacturers' catalogs select the belt and pulleys and call for them in an item list. Scale 1:4 inch or 1:5 metric.

HP	C	L
.12	10.75	4.62
.16	10.75	4.62
.25	11.25	5.12
.33	11.75	5.62
.50	12.62	6.50
.75	13.50	7.38

115 VOLT   1750 RPM

MOTOR DIMENSIONS

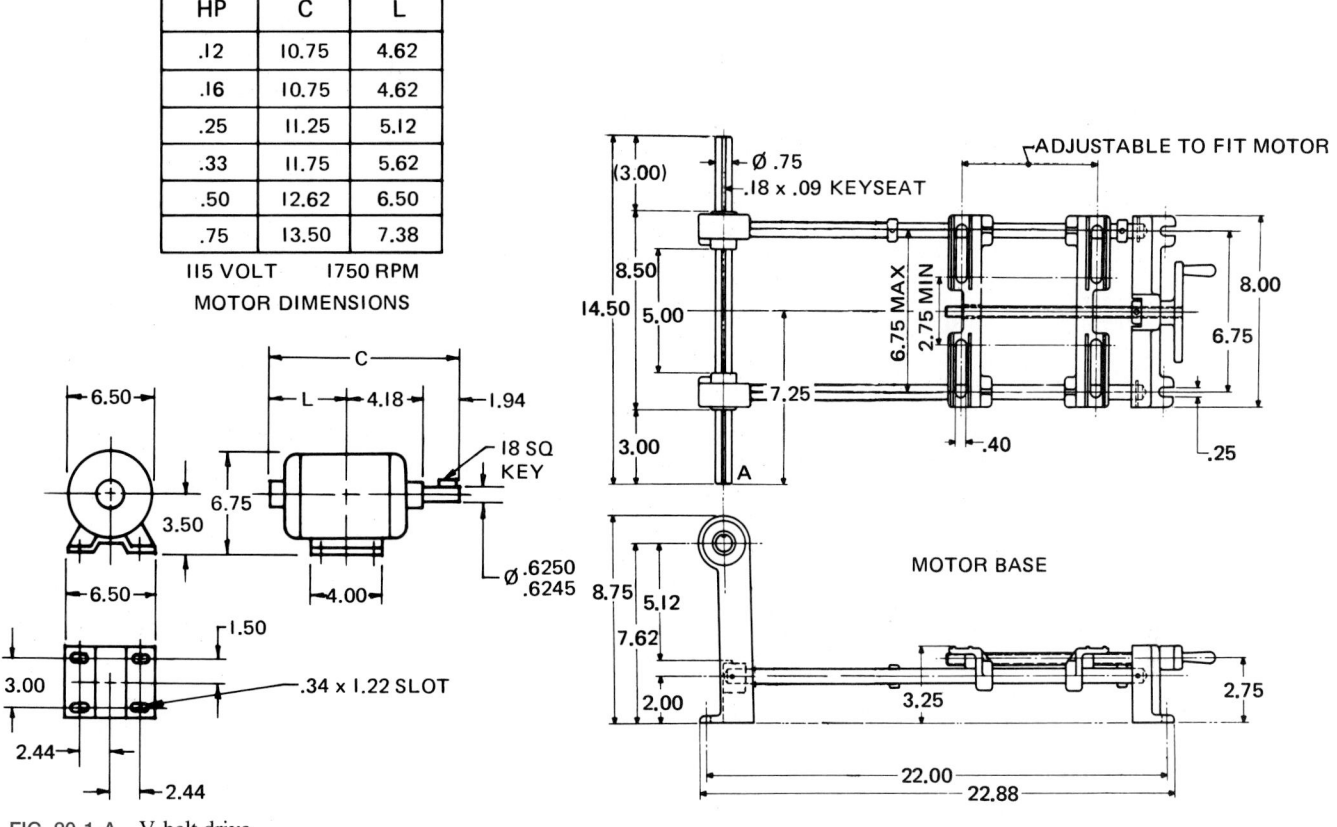

FIG. 20-1-A   V-belt drive.

kW	C	L
0.1	274	118
0.12	274	118
0.20	286	130
0.25	286	144
0.37	320	166
0.56	344	188

115 VOLT   1750 REV/MIN

MOTOR DIMENSIONS

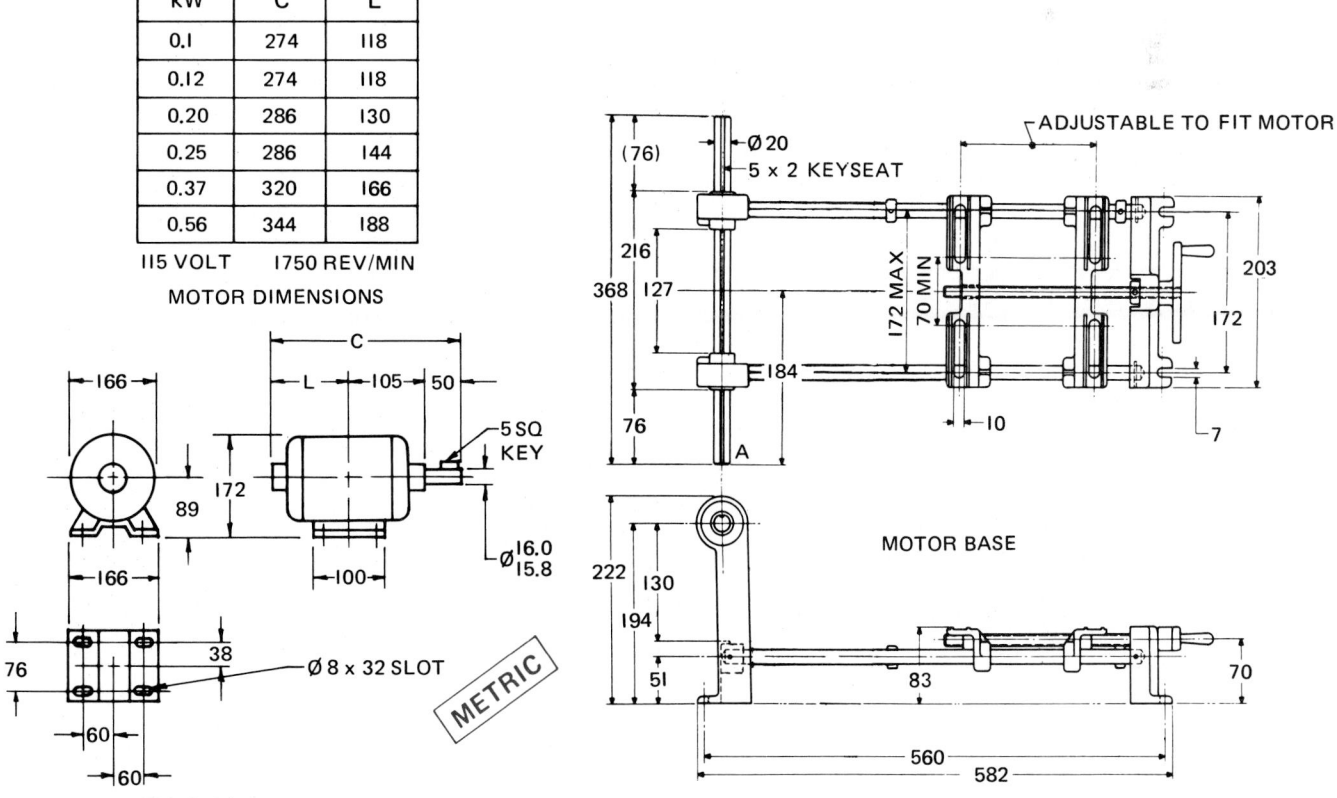

FIG. 20-1-B   V-belt drive.

ASSIGNMENTS FOR UNIT 20-2, CHAIN DRIVES

3. A tumbler barrel is to be driven at approximately 40 r/min by a speed reducer powered by a 5 hp (3.7 kW) electric motor. The reducer output speed is 100 r/min, and the output shaft is 1.75 in. (44 mm) in diameter. The shaft diameter of the tumbling barrel is 2 in. (50 mm). The shaft center distance is approximately 36 in. (900 mm). Select a single chain (heavy shock).

4. This is the same as Assignment 3 except that a double chain is to be used.

5. The head shaft of an apron conveyor, which handles rough castings from a shakeout, operates at 66 r/min and is driven by a gear motor whose output is 7.5 hp (5.6 kW) at 100 r/min. The head shaft has a 2 in. (50 mm) diameter, and the gear motor shaft has a 1.75 in. (44 mm) diameter. Shaft center distance should not exceed 42 in. (1055 mm). Select a multiple chain (moderate shock).

6. A gear-type lubrication pump located in the base of a large hydraulic press is to be driven at 860 r/min from a 1.25 in. (32 mm) diameter shaft operating at 1,000 r/min. The pump is rated at 3 hp (2.4 kW) and has a 1.375 in. (35 mm) diameter shaft. Shaft center distance must not be less than 10 in. (250 mm).

7. A 10 hp (7.5 kW), 480 r/min electric motor is to drive a line shaft, which is subject to light service, at 160 r/min. The motor shaft will be in approximately the same horizontal plane as the line shaft. The diameters of the motor shaft and line shaft are 1.69 and 1.75 in. (42 and 44 mm), respectively. A shaft distance of 48 to 60 in. (1,220 to 1,520 mm) will be acceptable. Select a triple chain.

8. A centrifugal fan is to be driven at 2,800 r/min by a 10 hp (7.5 kW) electric motor. The motor speed is 1,800 r/min, and the shaft has a 1.375 in. (35 mm) diameter. The fan shaft has a 1.25 in. (32 mm) diameter. The center distance is to be approximately 20 in. (500 mm). The overall drive must not exceed a 5 in. (125 mm) radius on the motor or a 3 in. (75 mm) radius at the fan.

ASSIGNMENTS FOR UNIT 20-3, GEAR DRIVES

9. Make working drawings for the two gears described in either Fig. 20-3-A or 20-3-B. Gear 1 will require one view

only and be drawn to scale 1:1. Gear 2 will require two views and will be drawn to scale 1:2. Select a proper key size and use your judgment for dimensions not given. Include with each gear drawing a cutting data block.

10. Prepare a working drawing of two gears in mesh from the information found in Fig. 20-3-C or 20-3-D. Show two views with three or four teeth shown in mesh. Add suitable keys and use your judgment for dimensions not given. Include cutting data for each gear. Scale 1:1.

GEAR #1	GEAR #2
Tooth form—14.5°	Tooth form—20°
PD—127	N—44
Module—6.35	Module—6.35
Face width—26	Face width—46
Web—10	Shaft—Ø45
Shaft—Ø28	Hub—Ø76 × 70.6 (total length)
Hub—Ø50 × 40 Lg	6 Spokes—16 Thk, 40 wide,
Matl—MI	tapered to 30 wide
	Matl—MI

METRIC

FIG. 20-3-B　Single spur gears.

PINION	GEAR
N—24	N—36
Shaft—Ø1.10	Face Width—1.10
Matl—Steel	Shaft—Ø1.25
	Web—.40
	Hub—Ø2.10 × 1.50 Lg
	Matl—MI

Tooth Form—14.5°
Center-to-center Distance—6.00

FIG. 20-3-C　Meshing spur gears.

GEAR #1	GEAR #2
Tooth form—14.5°	Tooth form—20°
PD—6.00	N—50
DP—5	DP—5
Face width—1.00	Face width—1.75
Web—.40	Shaft—Ø1.75
Shaft—Ø1.10	Hub—Ø3.00 × 2.75 (total length)
Hub—Ø1.90 × 1.50 Lg	6 Spokes—.60 Thk, 1.50 wide,
Matl—MI	tapered to 1.10 wide
	Matl—MI

FIG. 20-3-A　Single spur gears.

PINION	GEAR
N—16	N—24
Shaft—Ø30	Face Width—30
Matl—Steel	Shaft—Ø32
	Web—10
	Hub—Ø54 × 38 Lg
	Matl—MI

METRIC

Tooth Form—14.5°
Center-to-center Distance—127

FIG. 20-3-D　Meshing spur gears.

11. Complete the missing information on the gear-train problems shown in Fig. 20-3-E or 20-3-F.
12. Make a full section assembly drawing of the sliding-gear speed reducer shown in Fig. 20-3-G. Support the shafts on journal bearings. The gears are held to the shafts by setscrews and keys. Gears C and D are combined into one part and slide on the countershaft. Complete the chart on the figure showing the two speeds available (when gears C and E mesh and when gears F and D mesh) for the different motor inputs. Use your judgment for dimensions not shown. Scale 1:1.

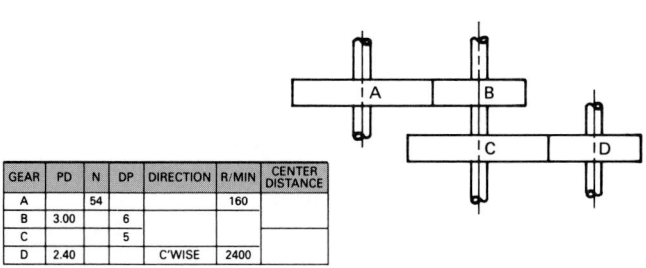

GEAR	PD	N	DP	DIRECTION	R/MIN	CENTER DISTANCE
A		54			160	
B	3.00		6			
C			5			
D	2.40			C'WISE	2400	

FIG. 20-3-E   Gear train calculations.

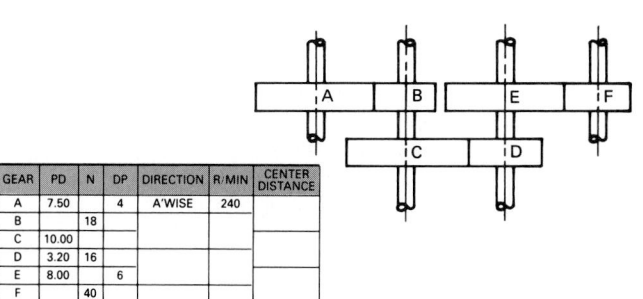

GEAR	PD	N	DP	DIRECTION	R/MIN	CENTER DISTANCE
A	7.50		4	A'WISE	240	
B		18				
C	10.00					
D	3.20	16				
E	8.00		6			
F		40				

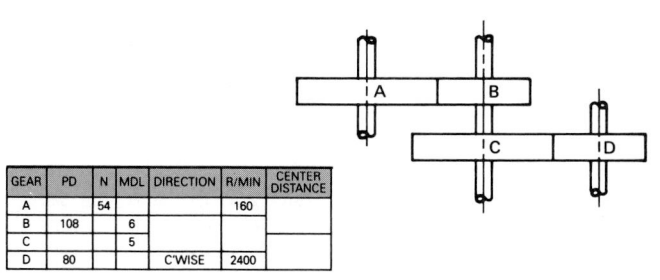

GEAR	PD	N	MDL	DIRECTION	R/MIN	CENTER DISTANCE
A		54			160	
B	108		6			
C			5			
D	80			C'WISE	2400	

FIG. 20-3-F   Gear train calculations.

GEAR	PD	N	MDL	DIRECTION	R/MIN	CENTER DISTANCE
A	182.88			A'WISE	240	
B		18	5.08			
C	203.52					
D	101.76	32				
E	203.04		4.23			
F		40				

INPUT R/MIN	OUTPUT R/MIN	GEARS IN CONTACT	COUNTER SHAFT R/MIN
1150		F & D	
		E & C	
1750		F & D	
		E & C	

GEAR	NUMBER OF TEETH
A	20
B	30
C	20
D	24
E	30
F	26

GEAR DATA:
    SHAFT DIA – 20
    FACE WIDTH – 15
    MODULE – 3.175

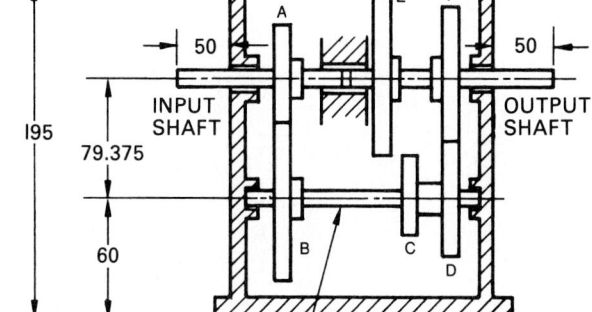

FIG. 20-3-G   Sliding-gear speed reducer.

13. Make a drawing of the speed reduction assembly shown in Fig. 20-3-H. The motor and worm gear reducer are mounted on a table. The coupling FC15 joins the two. A steel sprocket, mounted directly on the reducer shaft, is to move a chain at an approximate rate of 42 ft/hr. Call out on the assembly drawing the catalog numbers for the coupling and sprocket selected. Scale 1:2. Show the coupling and sprocket in full section.

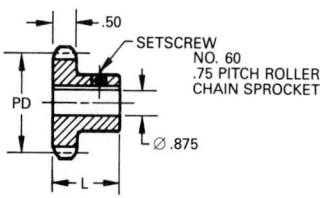

CATALOG NO.	PD	TEETH	HUB DIA	L
KS8	1.96	8	1.50	1.38
KS9	2.19	9	1.65	1.38
KSI0	2.43	10	1.94	1.38
KSII	2.66	11	2.10	1.25
KSI2	2.90	12	2.10	1.25

SPROCKET

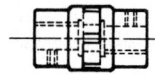

CATALOG NO.	HOLE DIA	HOLE LENGTH	O.D.	HUB DIA	HUB PROJ	LENGTH
FC 12	.50	.84	1.25	1.00	.60	2.30
FC 15	.50	1.00	1.50	1.25	.75	2.75
FC 20	.50	1.40	2.00	1.75	1.10	3.70

COUPLING

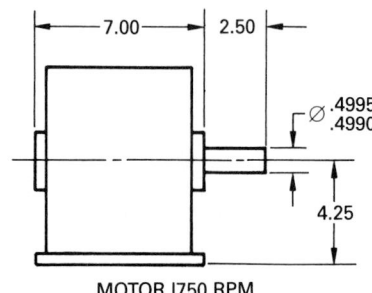

MOTOR 1750 RPM

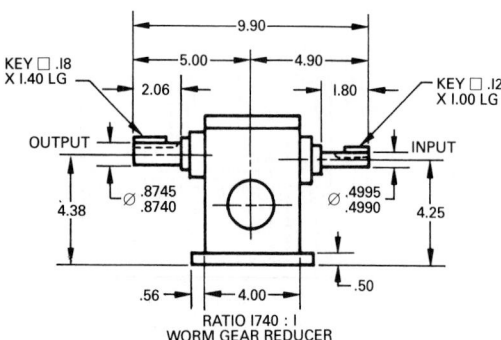

RATIO 1740 : 1
WORM GEAR REDUCER

FIG. 20-3-H   Speed-reducer assembly.

ASSIGNMENTS FOR UNIT 20-4, POWER-TRANSMITTING CAPACITY OF SPUR GEARS

14. Show your calculations for the designs of suitable pairs of 20° spur gears to operate the equipment described in a and b or c and d below.
    (a) A 1,200 r/min motor drives, by means of a spur gear and pinion, a machine rated at 8 hp and operating under moderate shock 12 hours a day. The reduction in r/min is 4:1. Select a suitable pair of spur gears to transmit the power required.
    (b) A punch press rated at 22 hp, 900 r/min is to be driven by a 30 hp, 1,200 r/min motor. The punch, which is subjected to moderate shock, will be in operation 16 hours a day. Select a suitable pair of spur gears to transmit the power required.
    (c) A 1,200 r/min motor drives, by means of a spur gear and pinion, a machine rated at 7.5 kW and operating under moderate shock 8 hours a day. The reduction in r/min is 3:1. Select a suitable pair of spur gears to transmit the power required.
    (d) An 1,800 r/min motor drives a machine which is rated at 2 kW and runs at 450 r/min under moderate shock 18 hours a day. Select a suitable pair of spur gears to transmit the power required.

15. Show your calculations for the design of suitable pairs of spur gears to operate the equipment described in a or b below.
    (a) A machine that works under smooth operating conditions is used twice a day for about 10 minutes. It is manually operated and is rated at 7 hp (6 kW) and runs at 800 r/min. Two motors are in stock at the plant: one rated at 7 hp (6 kW) and 1,200 r/min, the other at 5 hp (4 kW) and 750 r/min. Spur gears are to be used. Make a report to the plant engineer on the selection of motor and gears that you would recommend.
    (b) A 900 r/min motor drives an air compressor that operates between 15 to 20 minutes every hour. The compressor, which is rated at 5 kW (7.5 hp), runs at 600 r/min, under smooth operating conditions. Select a suitable pair of spur gears to transmit the power required.

ASSIGNMENT FOR UNIT 20-5, RACK AND PINION

16. On a B (A3) size sheet, make a working assembly drawing of one of the gear and racks shown in Fig. 20-5-A. Use your judgment for dimensions not given. Show four or five teeth in mesh. Scale 1:1.

ASSIGNMENTS FOR UNIT 20-6, BEVEL GEARS

17. Make a working drawing of one of the bevel gears from the data shown in Fig. 20-6-A. Use your judgment for dimensions not given. Scale 1:1.

18. Make an assembly working drawing of one of the gear assemblies from the data shown in Fig. 20-6-B. Add to the drawing the cutting data for the gears. Use your judgment for dimensions not given. Scale 1:1.

U.S. CUSTOMARY (in.)	METRIC (mm)
Gear—N—36 DP—5 Tooth form—14.5° Web—.50 Shaft—Ø1.25 Hub—Ø2.25 × 1.75 Lg Face width—1.25 Matl—MI Rack—Matl—Steel	Gear—N—30 MDL—5.08 Tooth form—14.5° Web—11 Shaft—Ø35 Hub—Ø58 × 45 Lg Face width—32 Matl—MI Rack—Matl—Steel

FIG. 20-5-A   Gear and rack assignments.

U.S. CUSTOMARY (in.)	METRIC (mm)
PD—4.500 Pitch cone angle—45° DP—4 Tooth form—14.5° Face width—1.25 Shaft—Ø1.00 Hub—Ø1.75 × 1.50 Lg Web thickness—.62	PD—114.3 Pitch cone angle—45° Module—6.35 Tooth form—14.5° Face width—32 Shaft—Ø24 Hub—Ø44 × 32 Lg Web thickness—16

FIG. 20-6-A   Single bevel gear.

19. The horizontal right-angle-drive bevel gear unit shown in Fig. 20-6-C has a ratio of 1:1. From the following information, make a one-view section assembly drawing of the top view with the cutting-plane line taken at the center of the shafts. Use your judgment for dimensions not given. Include an item list. Scale 1:1.

U.S. CUSTOMARY (in.)	METRIC (mm)
GEAR DATA	**GEAR DATA**
DP—4 Face width—1.10 N—22 Shaft—Ø1.00 Hub—Ø1.90 × 1.50 Lg Web—.75 Tooth form—14.5°	Module—6.35 Face width—30 N—22 Shaft—Ø25 Hub—Ø48 × 38 Lg Web—20 Tooth form—14.5°
PINION DATA	**PINION DATA**
N—14 Shaft—Ø.75 Hub—1.25 Lg Matl—Steel	N—14 Shaft—Ø20 Hub—32 Lg Matl—Steel

FIG. 20-6-B   Bevel gear assembly.

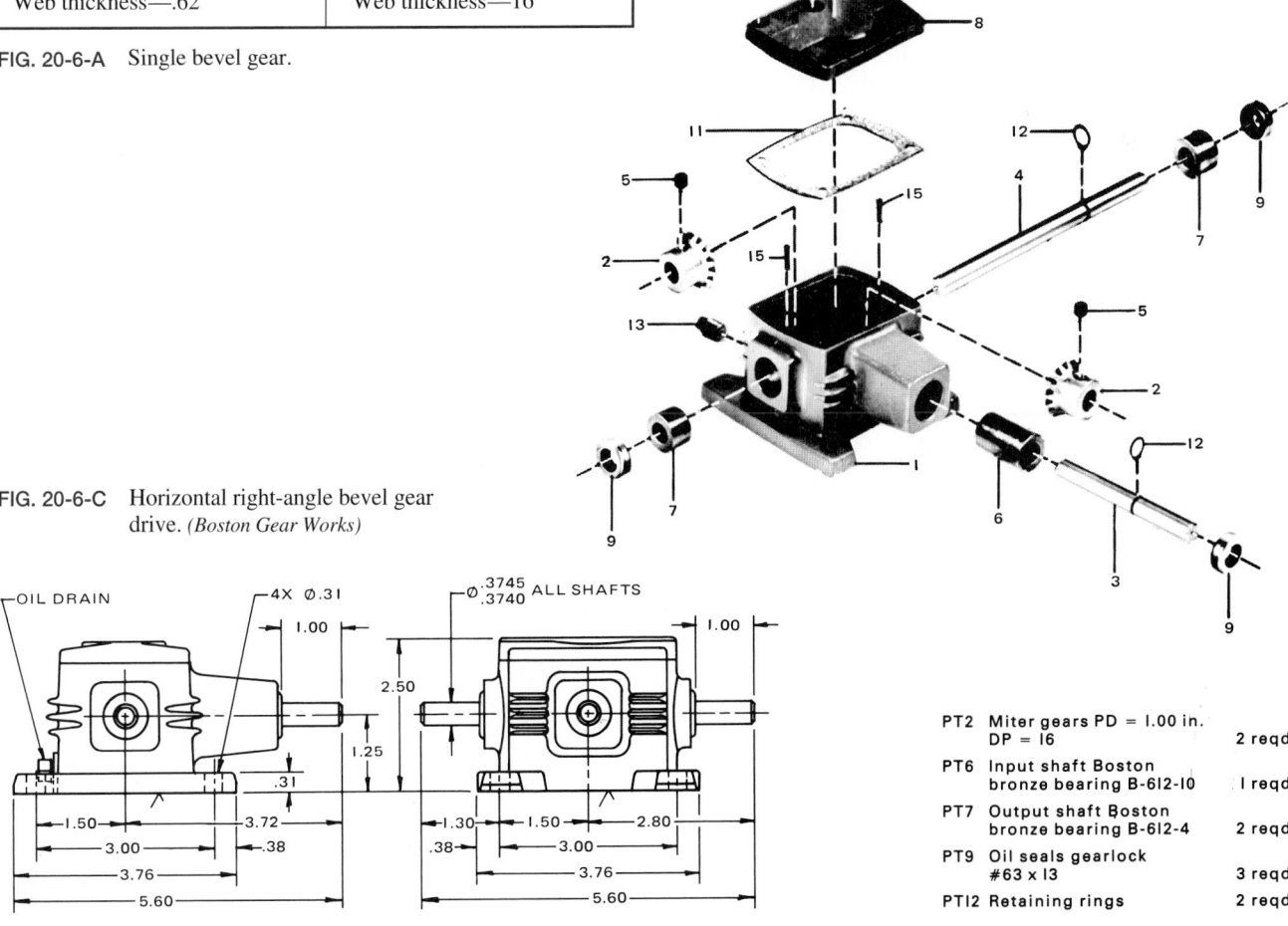

FIG. 20-6-C   Horizontal right-angle bevel gear drive. (Boston Gear Works)

PT2 Miter gears PD = 1.00 in. DP = 16   2 reqd
PT6 Input shaft Boston bronze bearing B-612-10   1 reqd
PT7 Output shaft Boston bronze bearing B-612-4   2 reqd
PT9 Oil seals gearlock #63 × 13   3 reqd
PT12 Retaining rings   2 reqd

709

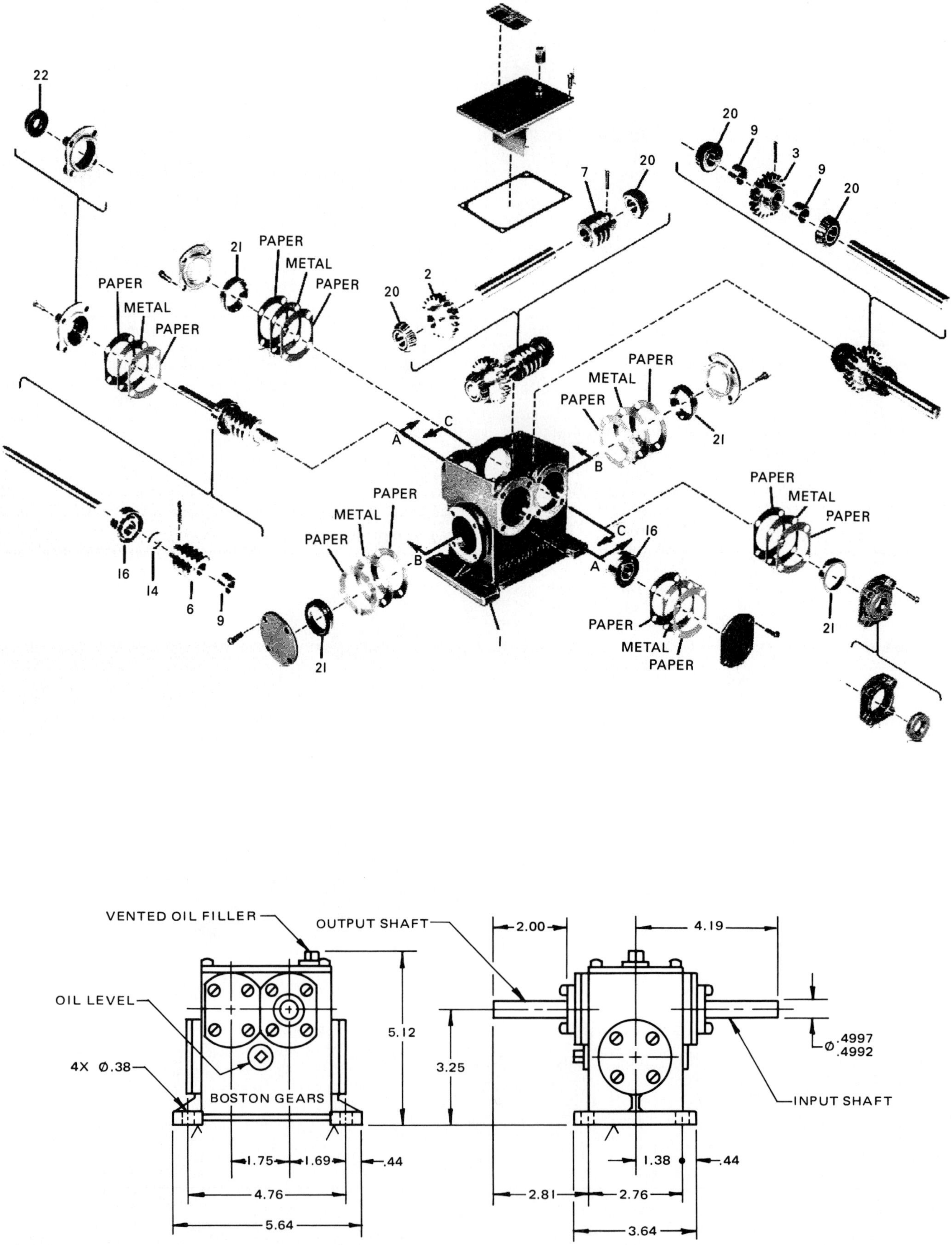

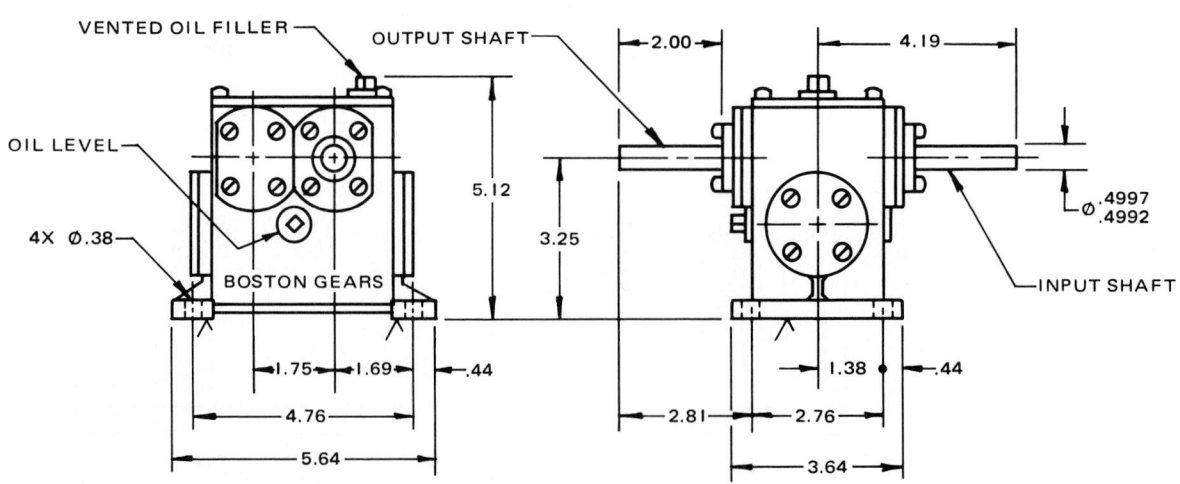

FIG. 20-7-A   Worm gear reducer. *(Boston Gear Works)*

ASSIGNMENTS FOR UNIT 20-7, WORM AND WORM GEARS

20. Make a working drawing of a worm and the mating worm gear from the data given in *a* or *b* below. Use your judgment for dimensions not given. Include on the drawing the cutting data. Scale 1:1.

   (*a*) A worm and worm gear have a pitch of .5236 in. The gear, made of cast iron, has 30 teeth; shaft dia = .88, hub dia = 1.75, hub length = 1.90, face width = 1.00, web thickness = .40. The worm, made of hardened steel, is 3.50 long on a Ø.88 shaft, pitch dia = 2.12, single thread, RH lead, pressure angle 14.5°.

   (*b*) A worm and worm gear have a pitch of 13.3 mm. The gear, made of cast iron, has 30 teeth, shaft dia = 22, hub dia = 44, hub length = 48, face width = 25, web thickness = 10. The worm, made of hardened steel, is 88 long on a Ø22 shaft, pitch dia = 54, single thread, RH lead, pressure angle 14.5°.

21. On a B (A3) size sheet, make a two-view detail assembly drawing of a worm and worm gear from the data given in *a* or *b* below. Use your judgment for dimensions not given. Include the cutting data on the drawing. Scale is half or 1:2.

   (*a*) A worm and worm gear have a pitch of .75 in. The gear, made of phosphor bronze, has 24 teeth, shaft dia = 1.25, hub dia = 2.25, hub length = 2.50, web thickness = .50. The worm, made of steel, has a pitch dia = 2.50, left-hand double thread, shaft dia = 1.00.

   (*b*) A worm and worm gear have a pitch of 19 mm. The gear, made of phosphor bronze, has 24 teeth, shaft dia = 32, hub dia = 58, hub length = 64, web thickness = 13. The worm, made of steel, has a pitch dia = 64, left-hand double thread, shaft dia = 26.

22. The horizontal parallel compound worm gear reducer shown in Figs. 20-7-A and 20-7-B has a ratio of 150:1. From the following information, draw three sectional assembly views through cutting-planes A-A, B-B, and C-C (right and left side views and a front view). Use your judgment for dimensions not shown. Note the flat face on the worm gears. Scale 1:1. C (A2) size paper.

PT 2	Worm gear XLB-2D	1 reqd
PT 3	Worm gear XLB-2C	1 reqd
PT 6	Worm XLB-6D	1 reqd
PT 7	Worm XLB-6C	1 reqd
PT 9	Spacer collar .52 ID × .75 OD	3 reqd
PT 14	Spacer collar .52 ID × .75 OD	1 reqd
PT 16	Ball bearing Nice #1616 NS	2 reqd
PT 20	Cone bearing Timken A4049 .50 ID × 1.38 OD × .44	4 reqd
PT 21	Bearing cup (part of PT 20)	4 reqd
PT 22	Oil seal .50 ID × 1.38 OD × .32	2 reqd

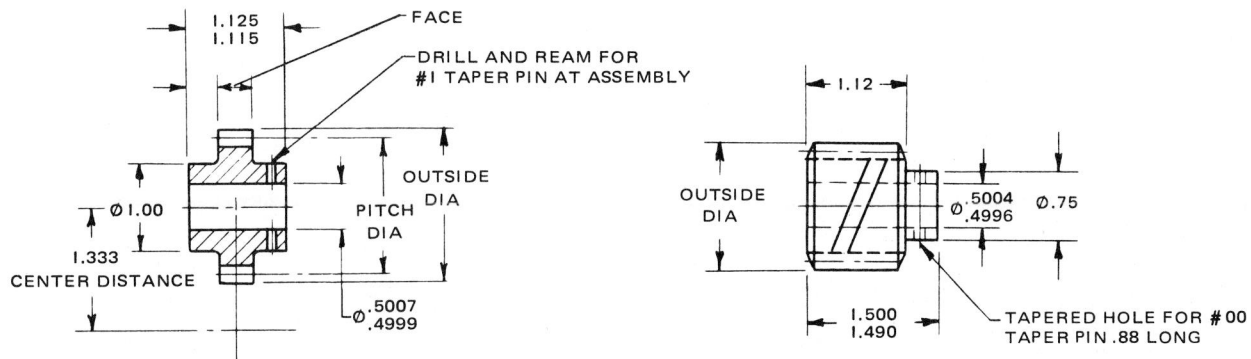

CATALOG NUMBER	RATIO	TEETH	PITCH DIA	OUTSIDE DIA	FACE
XLB-2E	4 TO 1	16	1.524	1.714	.38
XLB-2C	5 TO 1	20	1.667	1.833	.38
XLB-2A	10 TO 1	20	1.667	1.833	.38
XLB-2G	15 TO 1	30	1.875	2.000	.38
XLB-2B	20 TO 1	20	1.667	1.833	.38
XLB-2D	30 TO 1	30	1.875	2.000	.38

BRONZE WORM GEAR

CATALOG NUMBER	PITCH DIA	PITCH	NO. THREADS	LEAD ANGLE	OUTSIDE DIA
XLB-6E	1.143	3/10 C.P.	4	18° 29′	1.333
XLB-6C	1.000	12	4	18° 26′	1.166
XLB-6A	1.000	12	2	9° 28′	1.166
XLB-6G	.791	16	2	8° 59′	.916
XLB-6B	1.000	12	1	4° 46′	1.166
XLB-6D	.791	16	1	4° 31′	.916

STEEL WORMS

FIG. 20-7-B   14.5° worm gear set 1.333″ center distance. *(Boston Gear Works)*

711

ASSIGNMENTS FOR UNIT 20-8, COMPARISON OF CHAIN, GEAR, AND BELT DRIVES

23. Your supervisor has asked you to submit a report recommending the type of power transmission best suited for the machinery layouts shown in either Fig. 20-8-A or Fig. 20-8-B. Make a detailed report that specifies the power transmission parts required to properly operate the equipment.

24. Make a layout drawing showing a suitable mounting arrangement for the gear and bearing housing mounted on the mounting surface shown in Fig. 20-8-C. The design is for low-speed moderate use. Draw the gear and one shaft support in full section, and the pulley and the other shaft support in half-section. A partial top view is required to show the location of the mounting holes. Standard parts are to be used wherever possible and an item list is to be included on the drawing. Scale 1:1.

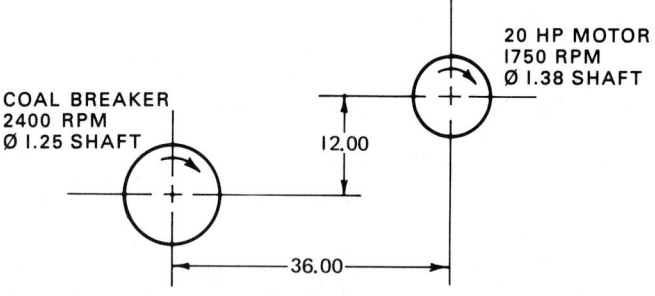

FIG. 20-8-A   Power transmission drive.

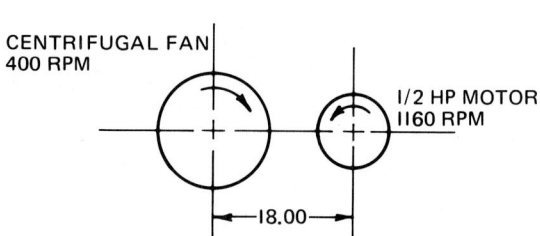

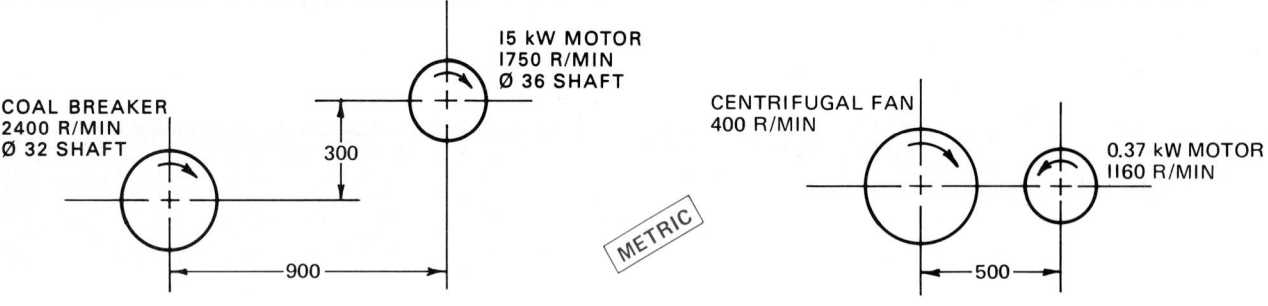

FIG. 20-8-B   Power transmission drive.

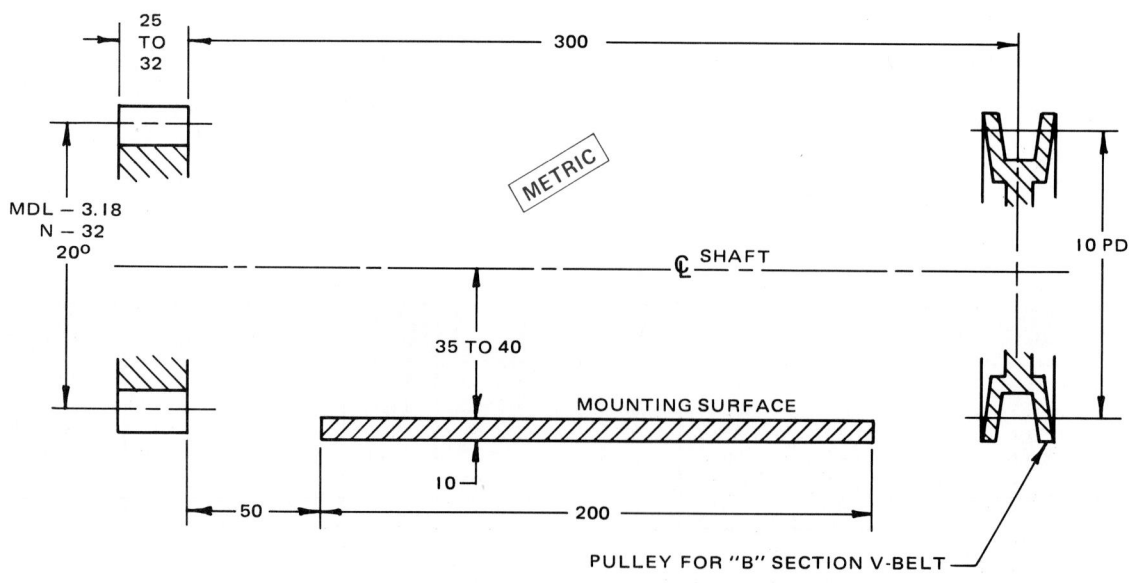

FIG. 20-8-C   Simple drive assembly using standard parts.

25. Make the necessary detail drawing for the assembly shown in Assignment 24.

26. The idler pulley shown in Fig. 20-8-D is fitted with a journal bearing and rotates freely on a Ø1.00 shaft. The shaft is mounted on a .38 steel bar, which in turn is fastened to a .12 in. steel wall. For adjusting the tension on the belt, the vertical position of the idler pulley can be positioned anywhere from .50 in. above to .50 in. below the 10.00 in. dimension shown on the drawing. Make the necessary assembly views to clearly show how this may be accomplished.

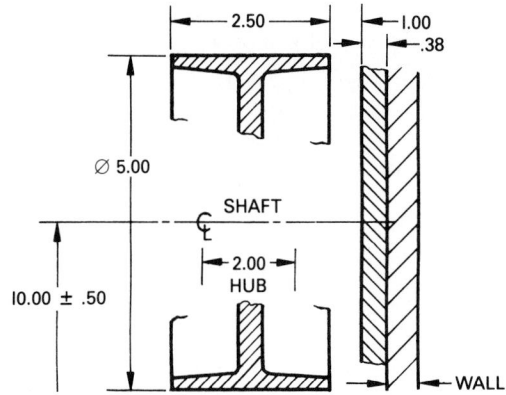

FIG. 20-8-D   Idler pulley assembly.

CHAPTER 21

COUPLINGS, BEARINGS, AND SEALS

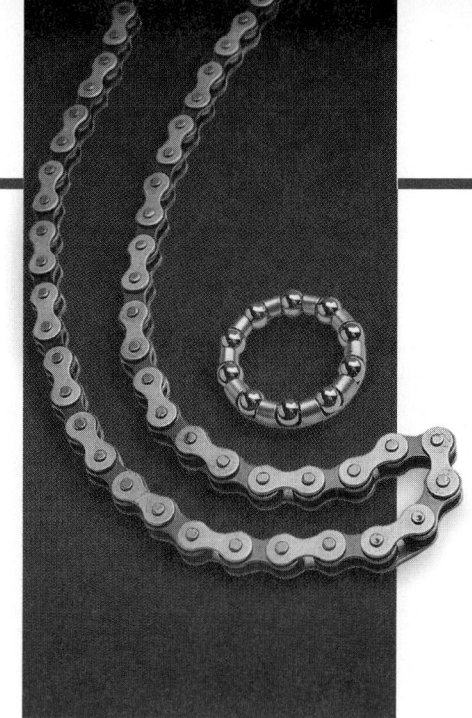

Definitions

Antifriction bearings   Bearings having minimal friction, such as ball, roller, and needle bearings.

Couplings   Devices used to join shafts.

Gasket   A device used to create and maintain a tight seal between separable members of a mechanical assembly.

Grease   A semisolid, combining a fluid with a thickening agent, used as a lubricant.

Hydrodynamic   Lubrication maintained by a squeezing or wedging of lubricant produced by the rolling action of the bearing itself.

Hydrostatic   Lubrication maintained by a pressurization system.

Journal   Any portion of a shaft supported by a bearing.

Lubricants   Materials used to reduce friction between rubbing surfaces, and as coolants to carry off heat in bearings.

Oils   Slippery hydrocarbon liquids used as lubricants.

Plain bearings   Bearings based on sliding action.

Premounted bearings   Preassembled units that consist of a bearing element and a housing.

Rolling-element bearings   Bearings based on rolling action.

Seals   In oil lubrication, devices that protect the bearing against contamination and retain the lubricant in the housing.

Sleeve   The general configuration of a bearing.

21-1 COUPLINGS AND FLEXIBLE SHAFTS

Couplings

Couplings, as the name implies, are used to couple or join shafts. There are two types of couplings: permanent couplings and clutches. Permanent couplings are not normally disconnected except for assembly or disassembly purposes, while clutches permit shafts to be connected or disconnected at will.

Permanent Couplings

Permanent couplings can be divided into three main categories: solid, flexible, and universal.

Solid Couplings   Solid couplings should be used only when driving and driven shafts are mounted on a common rigid base, so that shafts can be perfectly aligned and will stay that way in service. If two shafts are not in exact alignment and are connected by a rigid coupling, excess bearing wear may occur on the bearing supporting the shaft. The steel sleeve coupling and the flanged coupling shown in Fig. 21-1-1 are solid couplings.

Flexible Couplings   These are intended to compensate for unintentional misalignments or transient misalignments such as those caused by thermal expansion or vibration. They also prevent shock from being transferred from one shaft to another and are recommended where several power machines are connected on one shaft (Figs. 21-1-2 and 21-1-3).

There are many types of flexible couplings, but all are similar in operation. There are two hubs, one on each shaft,

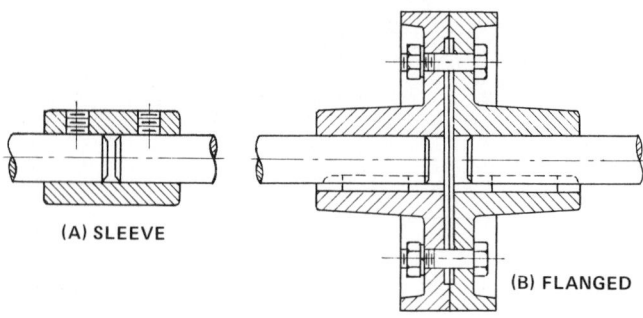

(A) SLEEVE

(B) FLANGED

FIG. 21-1-1   Solid couplings.

FIG. 21-1-2   Flexible coupling.

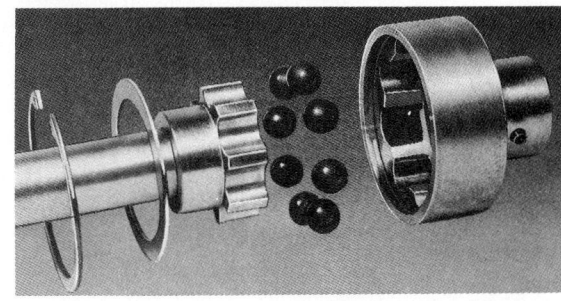

(A) ROLLER CHAIN   **(B) SILENT CHAIN**   **(C) MORFLEX**   **(D) EXPLODED ASSEMBLY OF A RUBBER BALL COUPLING**

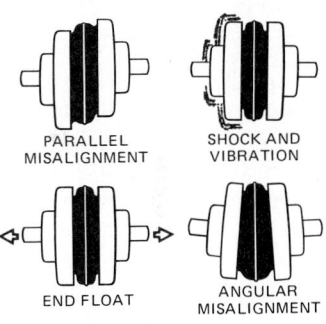

(E) APPLICATION   **(F) SURE-FLEX COUPLING**   **(G) 4-WAY FLEXACTION**

FIG. 21-1-3   Flexible couplings. *(A, B, and C, Morse Chain Co.; D, F, and G, Commonwealth Mfg.; E, T. B. Wood's Sons Co.)*

connected by an intermediate part, which may be flexible, floating, or both.

Flexible couplings may also be divided into three main categories: those that use mechanical movement, those that depend on the flexing of materials, and those that combine mechanical movement with flexing.

The table shown in Fig. 21-1-4 (pg. 716) lists the most common types of couplings and their main qualities. It should be used only as a guide, since special materials can considerably affect quality. Of course, there are exceptions to every rule. For most jobs, any one of several couplings may be suitable; cost determines the final selection.

To aid in selecting a coupling of the correct size, most manufacturers rate power transmitted in horsepower per 100 revolutions per minute or kW per 100 r/min and give maximum permissible revolutions per minute. The rating can be determined by the simple formula

$$\text{hp per 100 r/min} = \frac{\text{driving hp} \times 100 \times \text{service factor}}{\text{coupling r/min}}$$

or

$$\text{kilowatts per 100 r/min} = \frac{\text{driving kilowatts} \times 100 \times \text{service factor}}{\text{coupling r/min}}$$

The service factor depends on the source of driving power and the type of duty. For smooth power sources, such as an electric motor driving a smooth load like a centrifugal compressor, the factor is 1. It can be as high as 5 for reciprocating gasoline or diesel engines coupled to loads with cyclic torque variations, such as a single-cylinder compressor without a flywheel.

715

		CHAIN	STRAIGHT GEAR	CURVED GEAR	SLIDING INSERT	METAL BALL	METAL DISK	PLASTIC DISK	BELLOWS	HELICAL SPRING	RUBBER TIRE	PLASTIC INSERT	RUBBER INSERT	RUBBER BALL	RADIAL SPRING	GRID SPRING
Angular misalignment	-high							*	*	*	*			*	*	
	-average			*	*	*	*				*				*	*
	-low	*	*													
Out-of-parallel misalignment	-high							*			*			*	*	
	-average				*	*		*			*				*	*
	-low	*	*	*			*			*						
End float	-high		*	*										*	*	*
	-average	*			*	*	*	*	*	*	*	*	*			
	-low									*						
r/min	-high	*	*	*		*				*						
	-average				*		*	*	*	*	*	*	*	*	*	*
	-low															
Torsional resilience	-high										*			*	*	
	-average							*			*	*			*	*
	-low	*	*	*	*	*	*		*							
Lubrication required		*	*	*	*	*										

FIG. 21-1-4   Basic features of common flexible couplings.

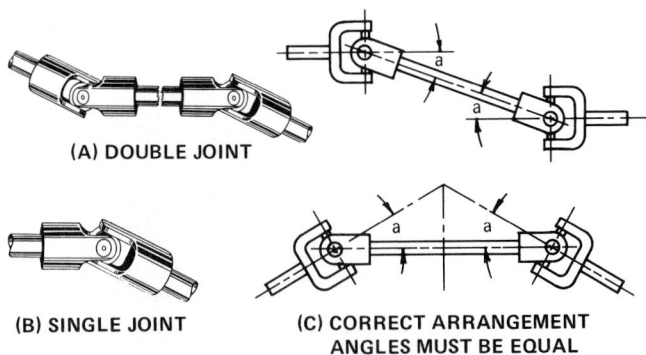

FIG. 21-1-5   Universal joints—Hook's type. *(Boston Gear Works)*

(A) DOUBLE JOINT

(B) SINGLE JOINT

(C) CORRECT ARRANGEMENT ANGLES MUST BE EQUAL

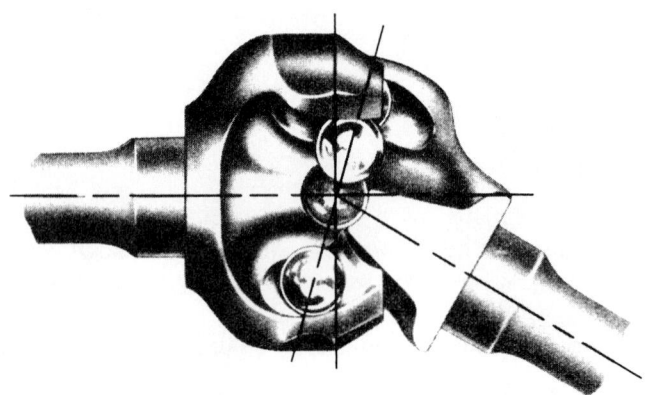

FIG. 21-1-6   Constant-velocity universal joint. *(The Bendix Corp.)*

Universal Couplings   Commonly called universal joints, universal couplings are for applications where angular displacement of shafts is a design requirement. It is easier to select universal couplings than flexible couplings because there are fewer types of them. Most common is the Hook's joint, which has a cross-type trunnion connected to driving and driven shafts by U-shaped end pieces (Fig. 21-1-5). Its main disadvantage is that because the trunnion is always at right angles to the driven shaft, it gives a sine-wave shaped variation in angular velocity between shafts. Other disadvantages are that it cannot compensate for out-of-parallel alignments and it does not compensate for changing distances between driving and driven points when the angle between shafts changes.

These disadvantages disappear when two universals are used, one with a sliding spline, as in automotive systems using the Hotchkiss drive. Here, the transmission and differential pinion shafts are parallel, so rotational fluctuations are canceled out. When two joints are used in this manner, the U-shaped fittings on the drive-shaft ends must be parallel, or else the rotational fluctuations will be increased instead of canceling out.

If constant velocity is essential with only one universal, a special constant-velocity universal must be used. Most of these have some type of ball drive, where the driving points of contact bisect the driving angle. They are more complex than the Hook's type and more expensive. The universal coupling shown in Fig. 21-1-6 is designed to transmit a constant velocity. The drive is through steel balls in races, designed so that the plane of contact between the balls and races always bisects the shaft angle. Flexible shafts also give constant velocity but are limited to transmitting relatively low power.

Flexible Shafts

Flexible shafts are used to transmit power around corners and at various angles when driving and driven elements are not aligned. Speedometers, tachometers, and indicating and recording instruments are typical applications.

Flexible shafts are constructed of helically wound wire and designed for transmission of rotary power and motion between two points located so that their relative positions preclude the use of solid shafts.

REFERENCES AND SOURCE MATERIAL

1. *Machine Design,* Mechanical drives reference issue.

ASSIGNMENTS

See Assignments 1 and 2 for Unit 21-1 on pages 736–737.

21-2   BEARINGS

Bearings permit smooth, low-friction movement between two surfaces. The movement can be either rotary (a shaft rotating within a mount) or linear (one surface moving along another).

Bearings can employ either a sliding or a rolling action. Bearings based on rolling action are called *rolling-element bearings.* Those based on sliding action are called *plain bearings.*

The basic principles of design and application of antifriction bearings were conceived many centuries ago. They originated for one purpose only—to lessen friction. Through the ages people wanted to move heavy objects across the earth's surface. As far back as 1100 B.C., we know that such friction was reduced by the insertion of rollers between the object and the surface over which it was being moved. The Assyrians and Babylonians used rollers to move enormous stones for their monuments and palaces. Down through history are recorded many similar examples of people's war on friction.

Plain Bearings

A plain bearing is any bearing that works by sliding action, with or without lubricant. This group encompasses essentially all types other than rolling-element bearings.

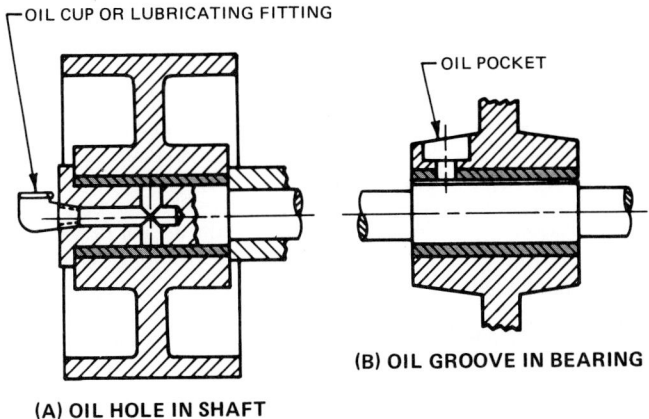

FIG. 21-2-1   Common methods of lubricating plain bearings.

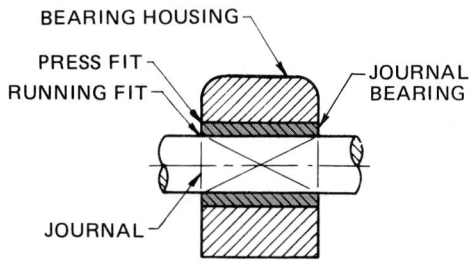

FIG. 21-2-2   Journal or sleeve bearing.

Plain bearings are often referred to as *sleeve bearings* or *thrust bearings,* terms that designate whether the bearing is loaded axially or radially.

Lubrication is critical to the operation of plain bearings, so their application and function are also often referred to according to the type of lubrication principle used. Thus, terms such as *hydrodynamic, fluid-film, hydrostatic, boundary-lubricated,* and *self-lubricated* are designations for particular types of plain bearings.

Although some materials have an inherent lubricity or can be lubricated by virtue of a film of slippery solid, most bearings operate with a fluid film—usually oil but sometimes a gas.

By far the largest number of bearings are oil-lubricated. The oil film can be maintained through pumping by a pressurization system, in which case the lubrication is termed *hydrostatic.* Or it can be maintained by a squeezing or wedging of lubricant produced by the rolling action of the bearing itself; this is termed *hydrodynamic* lubrication. The designs shown in Fig. 21-2-1 illustrate simple, effective arrangements for providing supplementary lubrication.

Bearing Types

Journal or Sleeve Bearings   These are cylindrical or ring-shaped bearings designed to carry radial loads (Fig. 21-2-2). The terms *sleeve* and *journal* are used more or less synonymously since *sleeve* refers to the general configuration while *journal* pertains to any portion of a shaft supported by a bearing. In another sense, however, the term *journal* may be reserved for two-piece bearings used to support the journals of an engine crankshaft.

The simplest and most widely used types of sleeve bearings are cast-bronze and porous-bronze (powdered-metal) cylindrical bearings. Cast-bronze bearings are oil- or grease-lubricated. Porous bearings are impregnated with oil and often have an oil reservoir in the housing.

Plastic bearings are being used increasingly in place of metal. Originally, plastic was used only in small, lightly loaded bearings where cost savings was the primary objective. More recently, plastics are being used because of functional advantages, including resistance to abrasion, and because they are available in large sizes.

Thrust Bearings   This type of bearing differs from a sleeve bearing in that loads are supported axially rather than radially (Fig. 21-2-3). Thin, disklike thrust bearings are called *thrust washers.*

FIG. 21-2-3   Thrust bearings.

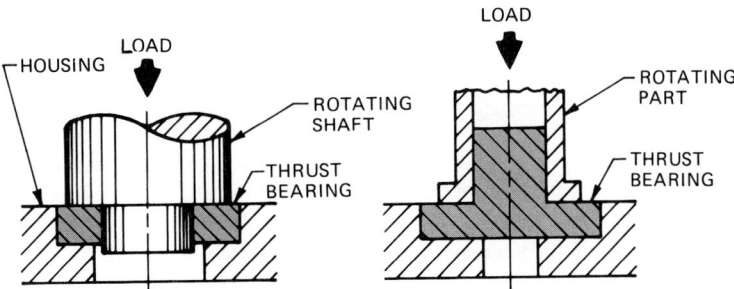

Bearing Materials

Babbitts   Tin and lead-base babbitts are among the most widely used bearing materials. They have an ability to embed dirt and have excellent compatibility properties under boundary-lubrication conditions.

In bushings for small motors and in automotive engine bearings, babbitt is generally used as a thin coating over a steel strip. For larger bearings in heavy-duty equipment, thick babbitt is cast on a rigid backing of steel or cast iron.

Bronzes and Copper Alloys   Dozens of copper alloys are available as bearing materials. Most of these can be grouped into four classes: copper-lead, lead-bronze, tin-bronze, and aluminum-bronze.

Aluminum   Aluminum bearing alloys have high wear resistance, load-carrying capacity, fatigue strength, and thermal conductivity; excellent corrosion resistance; and low cost. They are used extensively in connecting rods and main bearings in internal-combustion engines; in hydraulic gear pumps, in oil-well pumping equipment, in roll-neck bearings in steel mills; and in reciprocating compressors and aircraft equipment.

Porous Metals   Sintered-metal self-lubricating bearings, often called *powdered-metal bearings,* are simple and low in cost. They are widely used in home appliances, small motors, machine tools, business machines, and farm and construction equipment.

Common methods used when supplementary lubrication for oil-impregnated bearings is needed are shown in Fig. 21-2-4.

Plastics   Many bearings and bushings are being produced in a large variety of plastic materials. Many require no lubrication, and the high strength of modern plastics lends them to a variety of applications.

REFERENCES AND SOURCE MATERIAL

1. SKF Co. Ltd.
2. *Machine Design,* Mechanical drives reference issue.

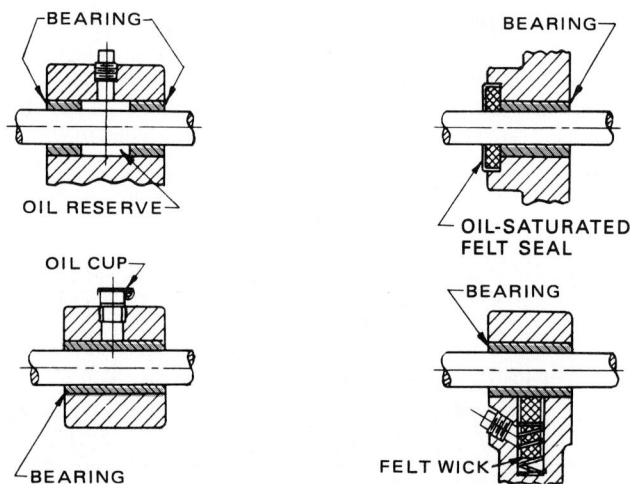

FIG. 21-2-4   Supplementary lubrication for oil-impregnated bearings.

ASSIGNMENTS

See Assignments 3 and 4 for Unit 21-2 on page 738.

21-3   ANTIFRICTION BEARINGS

Ball, roller, and needle bearings are classified as *antifriction bearings* since friction has been reduced to a minimum. They may be divided into two main groups: radial bearings and thrust bearings. Except for special designs, ball and roller bearings consist of two rings, a set of rolling elements, and a cage. The cage separates the rolling elements and spaces them evenly around the periphery (circumference of the circle). The nomenclature of an antifriction bearing is given in Fig. 21-3-1.

Bearing Loads

Radial Load   Loads acting perpendicular to the axis of the bearing are called *radial loads* (Fig. 21-3-2). Although radial bearings are designed primarily for straight radial service, they will withstand considerable thrust loads when deep ball tracks in the raceway are used.

Thrust Load   Loads applied parallel to the axis of the bearing are called *thrust loads.* Thrust bearings are not designed to carry radial loads.

Combination Radial and Thrust Loads   When loads are exerted both parallel and perpendicular to the axis of the bearings, a combination radial and thrust bearing is used. The load ratings listed in the manufacturer's catalogs for this type of bearing are for either pure thrust loads or a combination of both radial and thrust loads.

Ball Bearings

Ball bearings fall roughly into three classes: radial, thrust, and angular-contact. *Angular-contact bearings* are used for combined radial and thrust loads and where precise shaft location is needed. Uses of the other two types are described by their names: *radial bearings* for radial loads and *thrust bearings* for thrust loads (Fig. 21-3-3).

Radial Bearings

Deep-groove bearings are the most widely used ball bearings. In addition to radial loads, they can carry substantial thrust loads at high speeds, in either direction. They require careful alignment between shaft and housing.

Self-aligning bearings come in two types: internal and external. In internal bearings, the outer-ring ball groove is ground as a spherical surface. Externally self-aligning bearings have a spherical surface on the outside of the outer ring, which matches a concave spherical housing.

Double-row, deep-groove bearings embody the same principle of design as single-row bearings. Double-row bearings can be used where high radial and thrust rigidity is needed and space is limited. They are about 60 to 80 percent wider than

FIG. 21-3-1   Antifriction-
bearing
nomenclature.
(SKF Co.)

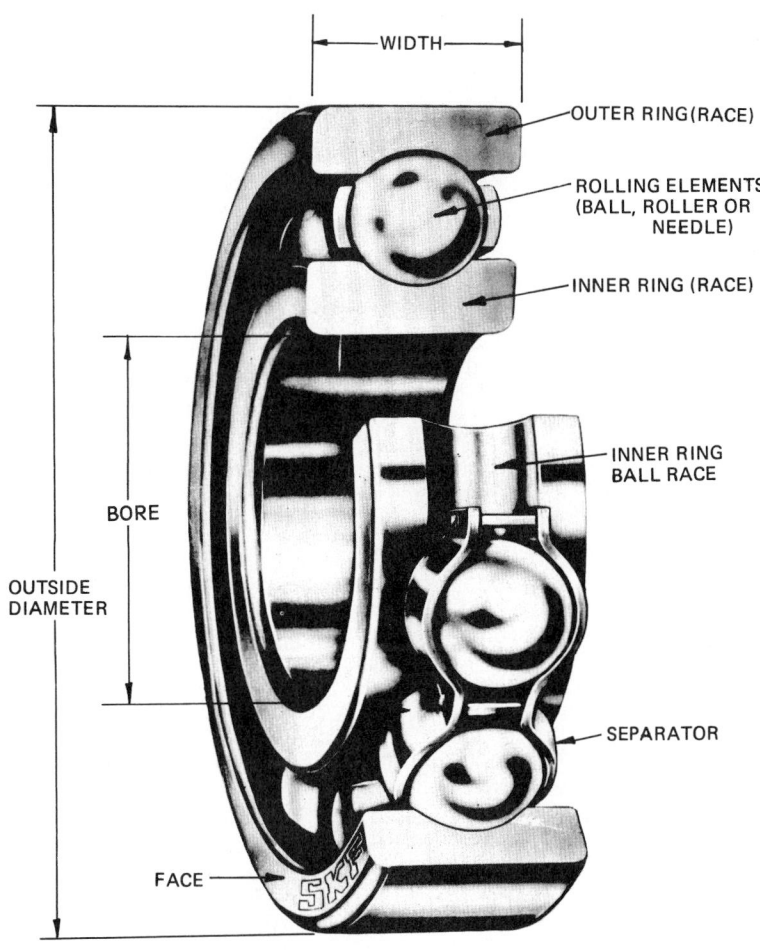

WIDTH

OUTER RING(RACE)

ROLLING ELEMENTS
(BALL, ROLLER OR
NEEDLE)

INNER RING (RACE)

INNER RING
BALL RACE

BORE

OUTSIDE
DIAMETER

SEPARATOR

FACE

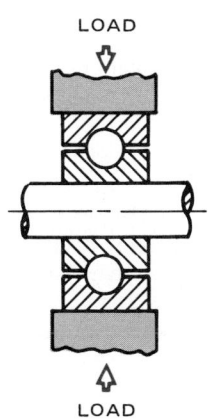

LOAD

LOAD

(A) RADIAL

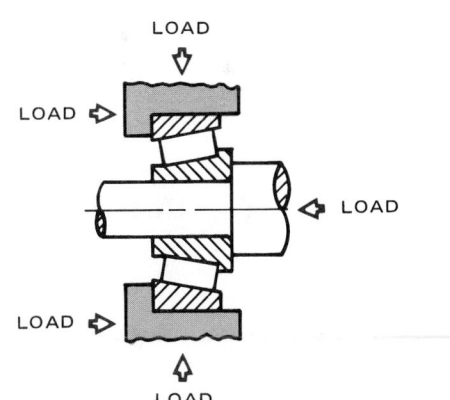

LOAD

LOAD

LOAD

LOAD

(B) THRUST

LOAD

LOAD

LOAD

LOAD

LOAD

(C) COMBINATION RADIAL AND THRUST

FIG. 21-3-2   Types of bearing loads.

DEEP-GROOVE

SELF-ALIGNING

DOUBLE-ROW

ANGULAR-CONTACT

THRUST

FIG. 21-3-3   Ball bearings. *(SKF Co.)*

719

comparable single-row, deep-groove bearings, and they have about 50 percent more radial capacity.

Angular-contact thrust bearings can support a heavy thrust load in one direction, combined with a moderate radial load. High shoulders on the inner and outer rings provide steep contact angles for high-thrust capacity and axial rigidity.

Thrust Bearings

In a sense, thrust bearings can be considered to be 90° angular-contact bearings. They support pure thrust loads at moderate speeds, but for practical purposes their radial load capacity is nil. Because they cannot support radial loads, ball thrust bearings must be used together with radial bearings.

Flat-race bearings consist of a pair of flat washers separated by the ball complement and a shaft-piloted retainer, so load capacity is limited. Contact stresses are high, and torque resistance is low.

One-directional, grooved-race bearings have grooved races very similar to those in radial bearings.

Two-directional, grooved-race bearings consist of two stationary races, one rotating race, and two ball complements.

Roller Bearings

The principal types of roller bearings are cylindrical, needle, tapered, and spherical. In general, they have higher load capacities than ball bearings of the same size and are widely used in heavy-duty, moderate-speed applications. However, except for cylindrical bearings, they have lower speed capabilities than ball bearings (Fig. 21-3-4).

Cylindrical Bearings

Cylindrical roller bearings have high radial capacity and provide accurate guidance to the rollers. Their low friction permits

(A) CYLINDRICAL   **(B) TAPERED**   **(C) SPHERICAL**

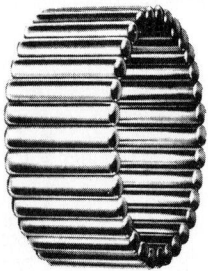 

LOOSE   CAGED
(D) NEEDLE

FIG. 21-3-4   Roller bearings. *(The Torrington Co. & Orange Roller Bearing Co.)*

operation at high speed, and thrust loads of some magnitude can be carried through the flange-roller end contacts.

Unlike ball bearings, cylindrical roller bearings are generally lubricated with oil; most of the oil serves as a coolant.

Needle Bearings

Needle bearings are roller bearings with rollers that have high length-to-diameter ratios. Compared with other roller bearings, needle bearings have much smaller rollers for a given bore size.

Loose-needle bearings are simply a full complement of needles in the annular space between two hardened machine components, which form the bearing raceways. They provide an effective and inexpensive bearing assembly with moderate speed capability, but they are sensitive to misalignment.

Caged assemblies are simply a roller complement with a retainer, placed between two hardened machine elements that act as raceways. Their speed capability is about three times higher than that of loose-needle bearings, but the smaller complement of needles reduces load capacity for the caged assemblies.

Thrust bearings are caged bearings with rollers assembled like the spokes of a wheel in a waferlike retainer.

Tapered Bearings

Tapered roller bearings are widely used in roll-neck applications in rolling mills, transmissions, gear reducers, geared shafting, steering mechanisms, and machine-tool spindles. Where speeds are low, grease lubrication suffices, but high speeds demand oil lubrication—and very high speeds demand special lubricating arrangements.

Spherical Bearings

Spherical roller bearings offer an unequaled combination of high load capacity, high tolerance to shock loads, and self-aligning ability, but they are speed-limited.

Single-row bearings are the most widely used tapered roller bearings. They have a high radial capacity and a thrust capacity about 60 percent of radial capacity.

Two-row bearings can replace two single-row bearings mounted back to back or face to face when the required capacity exceeds that of a single-row bearing.

Bearing Selection

Machine designers have a large variety of bearing types and sizes from which to choose. Each of these types has characteristics that make it best for a certain application. Although selection may sometimes present a complex problem requiring considerable experience, the following considerations are listed to serve as a general guide for conventional applications.

1. Ball bearings are normally the less-expensive choice in the smaller sizes and lighter loads, while roller bearings are less expensive for the larger sizes and heavier loads.
2. Roller bearings are more satisfactory under shock or impact loading than ball bearings.
3. If there is misalignment between housing and shaft, either a self-aligning ball or spherical roller bearing should be used.

4. Ball thrust bearings should be subjected only to pure thrust loads. At high speeds, a deep-groove or angular-contact ball bearing will usually be a better choice even for pure thrust loads.

5. Self-aligning ball bearings and cylindrical roller bearings have very low friction coefficients.

6. Deep-groove ball bearings are available with seals built into the bearings so that the bearing can be prelubricated and thus operate for long periods without attention.

Bearing Classifications

Because of standardization of boundary dimensions, it is possible to replace a bearing by another bearing produced by a different manufacturer without any modification to the existing assembly.

Ball and roller bearings are classified into various series: rigid ball journals, self-aligning ball journals, rigid roller journals, etc. Each series is subdivided into types—extra light, light, medium, and heavy—to meet varying load requirements. Each type is manufactured to a range of standard sizes, which are usually represented by the diameter of the bore. Therefore, when a bearing is ordered, the series, type, and size are specified.

Figure 21-3-5 shows a range of bearings to a common bore at A and to a common outside diameter at B. A selection can therefore be made for a given shaft size or for a given housing diameter, and the series selected will depend on the load that is applied to the bearing.

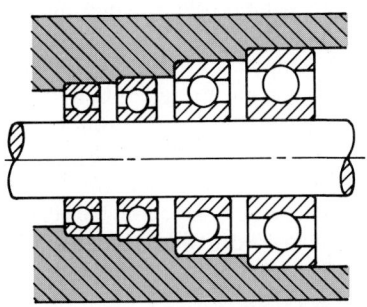

(A) COMMON BORE DIAMETER

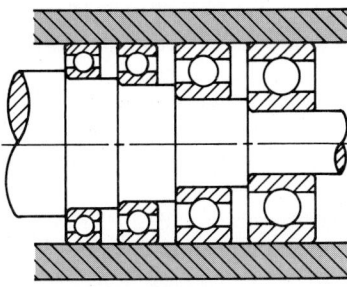

(B) COMMON OUTSIDE DIAMETER

FIG. 21-3-5   Standard bearing sizes.

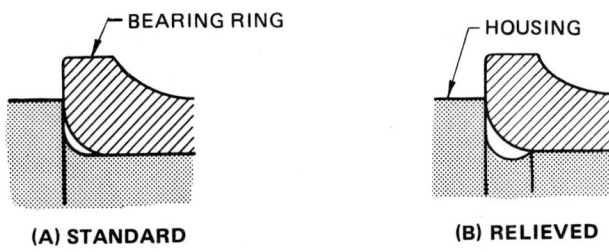

(A) STANDARD   **(B) RELIEVED**

FIG. 21-3-6   Correct shaft and housing fillet radii.

Shaft and Housing Fits

If a ball or roller bearing is to function satisfactorily, both the fit between the inner ring and the shaft and the fit between the outer ring and the housing must be suitable for the application. The desired fits can be obtained by selecting the proper tolerances for the shaft diameter and the housing bore.

Bearings may be mounted directly on the shaft or on tapered adapter sleeves. When the bearing is mounted directly on the shaft, the inner ring should be located against a shaft shoulder of proper height. This shoulder must be machined square with the bearing seat, and a shaft fillet should be used. The radius of the fillet must clear the corner radius of the inner ring (Fig. 21-3-6). This also applies when the outer ring is mounted in the housing.

To hold the bearing inner ring axially on the shaft, a locknut and lockwasher are commonly used (Fig. 21-3-7). Not only is this method effective and convenient, but nuts and washers

(A) LOCKWASHER AND LOCKNUT   **(B) ADAPTER SLEEVE**   **(C) WITHDRAWAL SLEEVE**

FIG. 21-3-7   Locking devices.

721

specially made for the purpose are also readily obtainable. A tab in the bore of the lockwasher engages a slot in the shaft, and one of the many tabs on the periphery of the washer is bent over into one of the slots in the nut OD.

Instead of a nut, a retaining ring fitted into a groove in the shaft can be used for simple bearing arrangements (Fig. 21-3-8).

If another machine component, such as a gear or pulley, is fitted alongside the bearing, the inner ring is often secured by means of a spacing sleeve. A sleeve is also frequently used for spacing the inner rings when the bearings are located reasonably close together.

Some bearings are merely mounted against a shoulder without other means of securing the inner ring axially. This is particularly the case where there are no axial forces tending to displace the bearings on the shaft. The housings for the two bearings are rigidly connected, and when thrust occurs, the bearing taking the load is pressed against its shoulder.

On long standard shafting it is impractical to apply bearings, with an interference fit, directly on the shaft. Therefore they are applied with tapered adapter sleeves. The outer surface of the sleeve is tapered to match the tapered bore of the bearing inner ring. This will provide the required tight fit between the inner ring and the shaft. The adapter sleeve is slotted to permit easy contraction and is threaded at the small end to fit a locknut. When the sleeve is drawn up tight between the bearing and the shaft, a press fit is provided at both the shaft and the inner ring.

If the operating conditions are such that the outer rings can be mounted with a push fit in the housing and *closed bearings* (bearings capable of carrying thrust load in either direction) are used, axial location may be controlled, as shown in Fig. 21-3-9A. The outer ring of the held bearing has a clearance of only .001 to .002 in. (0.05 to 0.1 mm) with the housing shoulders, while the floating bearings (Fig. 21-3-9B) have a free displacement axially in the housing.

One of the most critical factors affecting bearing operation is the mounting fit of the bearing on the shaft and in the housing. If there is any clearance or looseness between the shaft and the bore of the inner ring, the shaft, as it rotates, will roll along the bore of the inner ring. The rolling shaft in the bearing bore will cause the shaft to wear rapidly and become progressively looser. Soon it will become too sloppy for further operation. The best way to prevent this rolling action and wear is to press-fit the inner ring on the shaft.

Similar reasoning applies to a bearing subject to a load that rotates in space with the inner ring. Here, if the outer ring has a clearance in the housing bore, it will roll around the housing bore and wear loose. In this case it would be necessary to have the outer ring press-fit in the housing.

In all cases, it is necessary to press-fit the bearing ring that has relative rotation with respect to the direction of the radial load.

Seals for Grease Lubrication

In order that ball or roller bearings may operate properly, they must be protected against loss of lubricant and entrance of dirt and dust on the bearing surfaces. In its simplest and least space-requiring form, this is accomplished in some types of bearings by the use of a thin steel shield on one or both sides of the bearing, fastened in a groove in the outer ring and reaching almost to the inner rings, as illustrated in Fig. 21-3-10. All other types of bearings require a seal between the bearing housing and the shaft; the types and designs of seals are shown in Fig. 21-3-11. Other types of seals are explained in Units 21-5 and 21-6.

Seals for Oil Lubrication

With oil lubrication, the sealing devices have the double function of protecting the bearing against contamination and retaining the lubricant in the housing. Protection is obtained by means

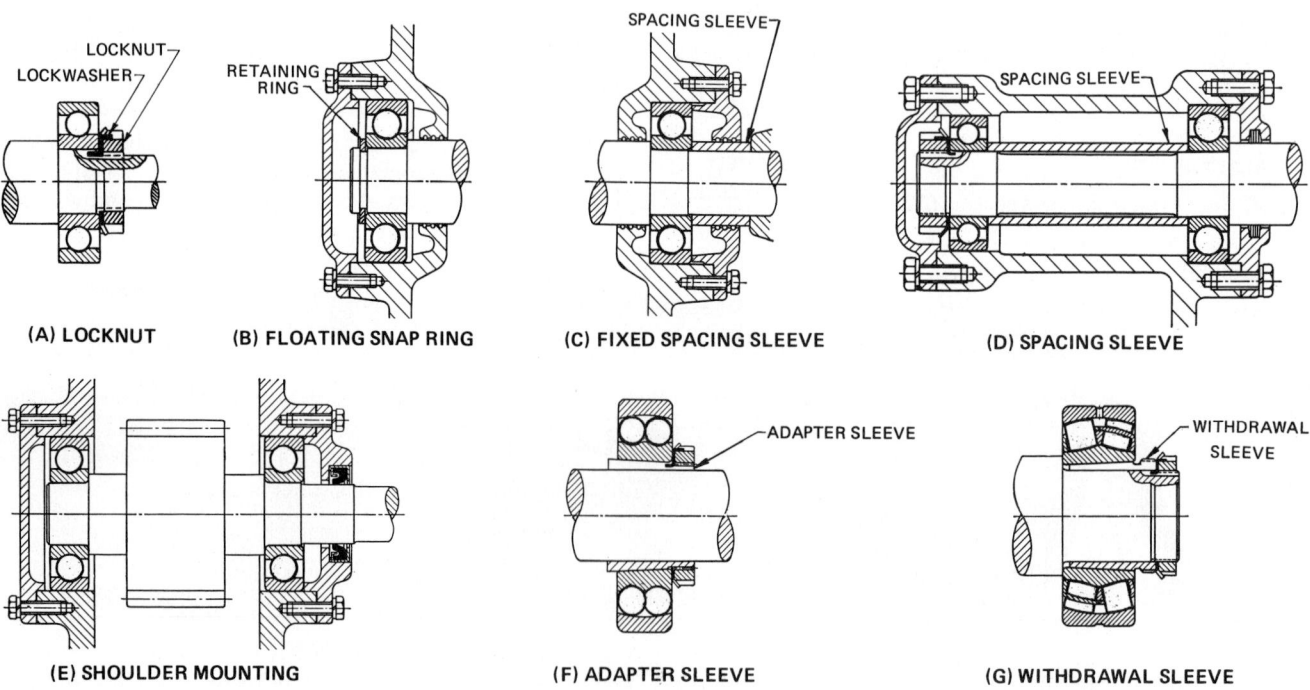

(A) LOCKNUT   (B) FLOATING SNAP RING   (C) FIXED SPACING SLEEVE   (D) SPACING SLEEVE

(E) SHOULDER MOUNTING   (F) ADAPTER SLEEVE   (G) WITHDRAWAL SLEEVE

FIG. 21-3-8   Axial mounting of inner rings.

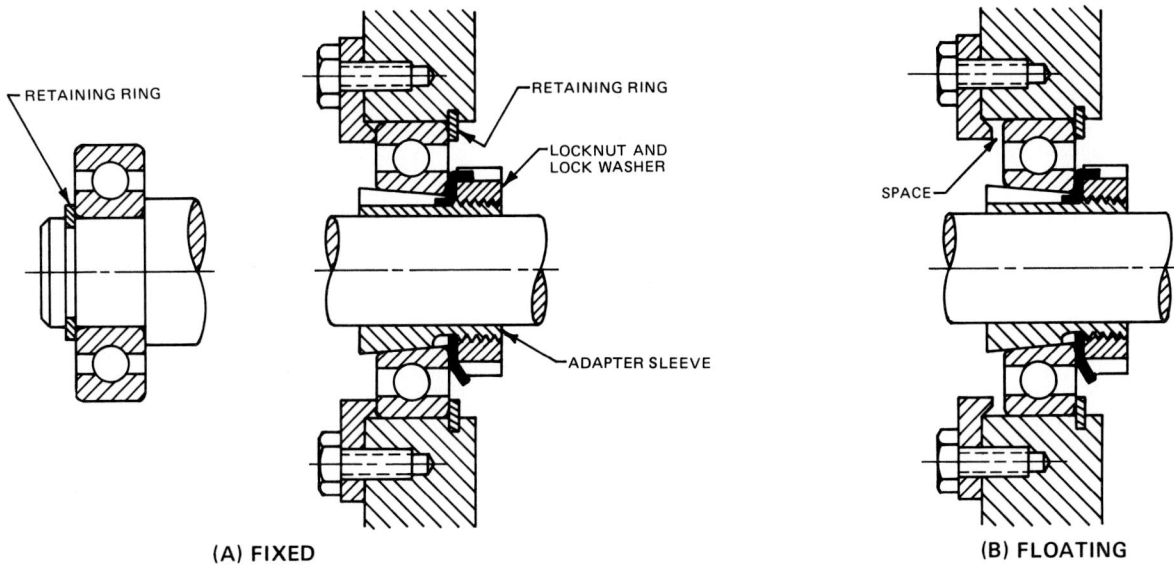

FIG. 21-3-9   Outer ring mountings.

of friction seals or flingers, as when grease lubrication is used. The essential feature for retaining the oil is a groove in the rotating shaft, or a rotating ring or collar from whose edges the oil is thrown by centrifugal force. The oil-groove seal shown in Fig. 21-3-12A (pg. 724) retains the oil effectively but should be used only in dry and dust-free places where there is little danger of contamination. Figure 21-3-12B shows examples of labyrinth seals, which retain the oil and protect against contamination.

Bearing Symbols

Simplified Representation   The simplified representation (general symbol) of rolling bearings (Fig. 21-3-13, pg. 724) should be used in all types of technical drawings, wherever it is not necessary to show the exact form or size of the rolling bearings or details of their inner design.

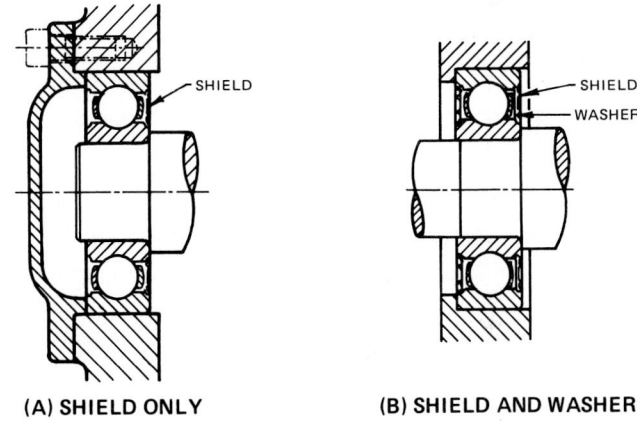

FIG. 21-3-10   Bearing seals for grease lubrication.

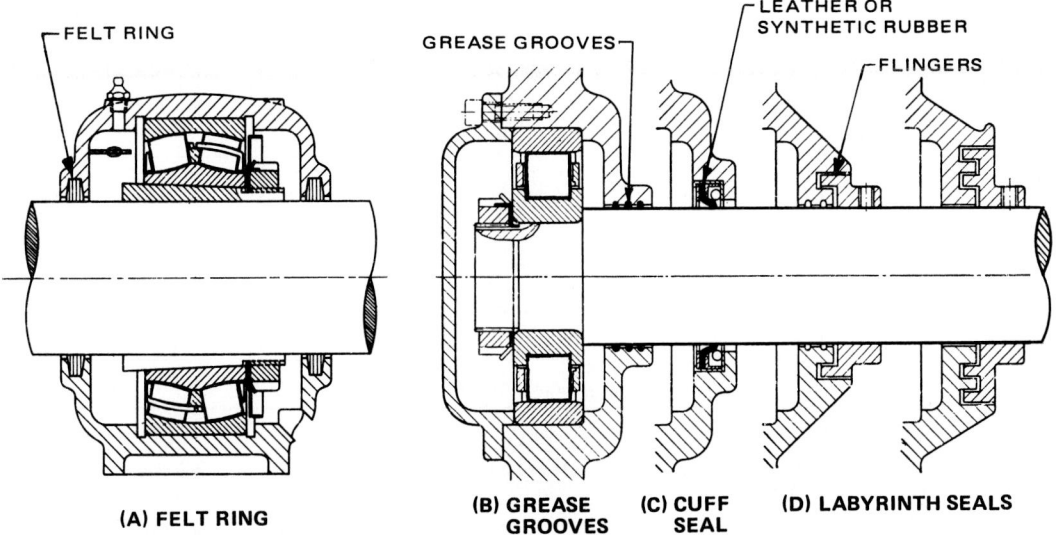

FIG. 21-3-11   Housing seals for grease lubrication.

FIG. 21-3-12   Housing seals for oil lubrication.

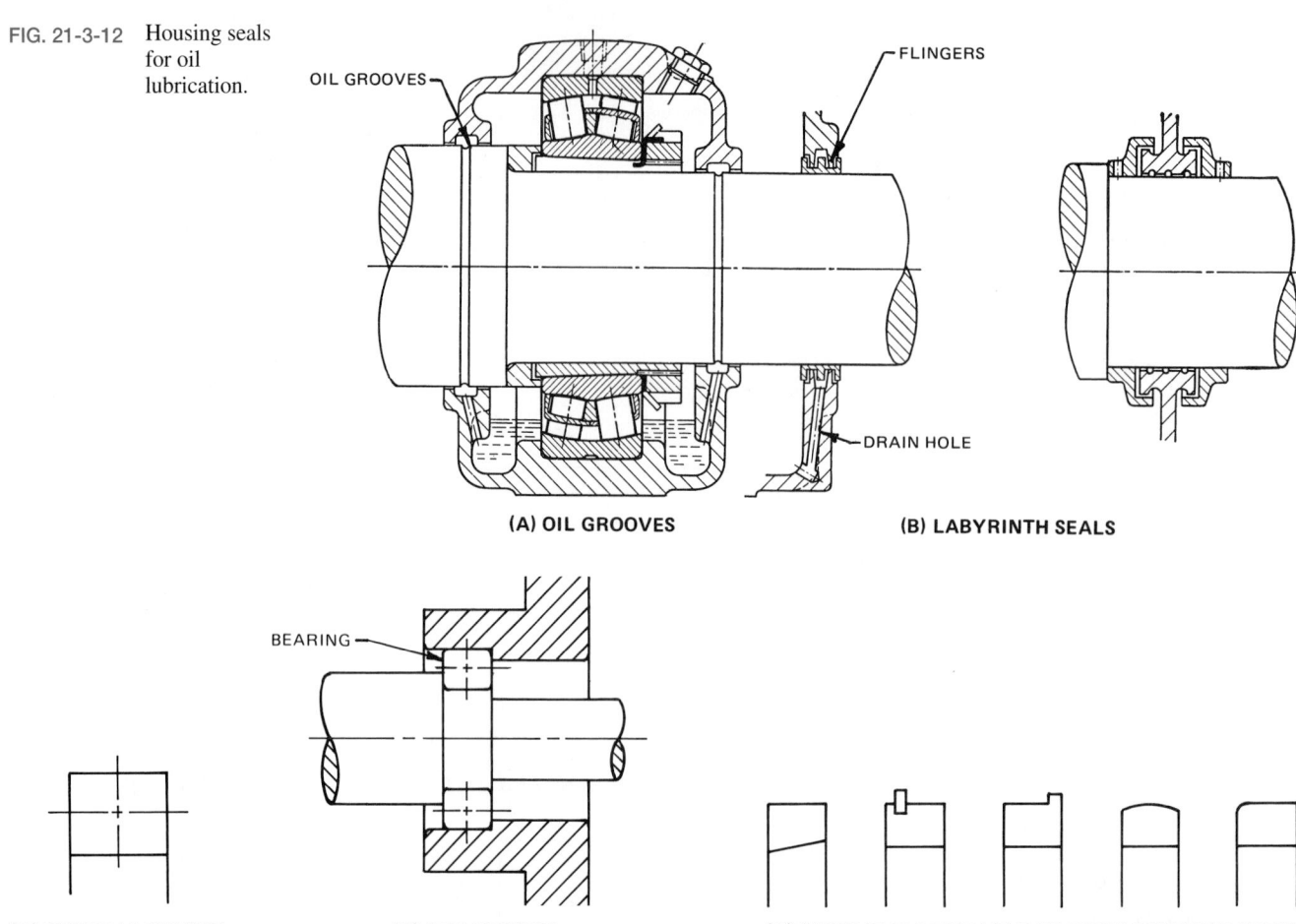

OIL GROOVES

FLINGERS

DRAIN HOLE

(A) OIL GROOVES

(B) LABYRINTH SEALS

BEARING

(A) GENERAL SYMBOL

(B) APPLICATION

(C) WHEN IT IS DESIRABLE TO SHOW CONTOUR FORM

FIG. 21-3-13   Simplified representation of ball and roller bearings.

BALL BEARINGS				ROLLER BEARINGS			THRUST BEARINGS		NEEDLE BEARINGS	
RADIAL DEEP-GROOVE	ANGULAR CONTACT	RADIAL DOUBLE ROW	SELF-ALIGNING DOUBLE ROW	CYLINDRICAL	SPHERICAL SELF-ALIGNING	TAPERED	BALL	ROLLER	RADIAL	AXIAL

(A) PICTORIAL

(B) SIMPLIFIED

(C) SCHEMATIC

FIG. 21-3-14   Representation of bearings on drawings.

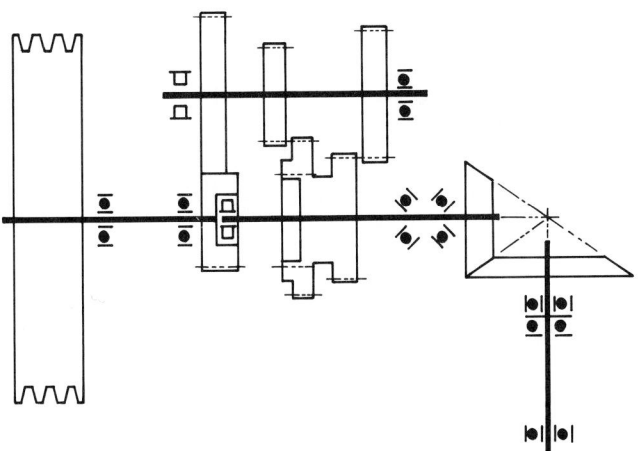

FIG. 21-3-15   Schematic representation of bearings.

FIG. 21-4-1   Premounted bearing unit.

Where it is desirable to show the functional principle of the set of rolling elements, symbols for the appropriate type of rolling element and raceway surface are added (Fig. 21-3-13C).

Pictorial Representation   Pictorial representation of bearings, as shown in Fig. 21-3-14A, is used chiefly in catalogs and magazines. It is not recommended for production drawings because of the extra drafting time required.

Schematic Representation   Designers and engineers frequently use schematic symbols in their initial design layout. The schematic diagrams of bearing types and their application are shown in Figs. 21-3-14C and 21-3-15.

REFERENCES AND SOURCE MATERIAL

1. *Machine Design,* Mechanical drives reference issue.
2. SKF Co. Ltd.

ASSIGNMENTS

See Assignments 5 through 11 for Unit 21-3 on pages 739 through 742.

21-4 PREMOUNTED BEARINGS

Premounted bearing units consist of a bearing element and a housing, usually assembled to permit convenient adaptation to a machinery frame. All components are incorporated within a single unit to ensure proper protection, lubrication, and operation of the bearing. Both plain and rolling-element bearing units are available in a variety of housing designs and for a wide range of shaft sizes, as shown in Figs. 21-4-1 and 21-4-2 (here and on pg. 726).

Provision for lubrication is made within the units, and sealing elements retain the lubricant and exclude foreign materials. Some types are prelubricated and sealed at the factory.

Rigid and Self-Aligning Types   Rigid premounted units require accurate alignment with the shaft.

Self-aligning units compensate for minor misalignment in mounting structures, shaft deflection, and changes that may occur after installation. Self-alignment in sleeve and in some rolling types is accomplished by the use of separate inner housings into which the bearing element is assembled.

Expansion and Nonexpansion Types   Expansion bearings permit axial shaft movement. The principal application for expansion units is in equipment where shafts become heated and increase in length at a greater rate than the structure on which the bearings are mounted.

Nonexpansion bearings restrict shaft movement relative to the mounting structure and keep shaft and attached components accurately positioned. These bearings also serve as thrust bearings within their capacity. Nonexpansion sleeve bearings usually require collars attached to the shaft at both ends of the housing.

Pillow blocks provide a convenient means of mounting shafts parallel to the surface of a supporting structure. Bolt holes are provided, usually elongated, to permit alignment, and dowel holes are sometimes predrilled for use in maintaining final position on the supporting member. Pillow blocks are available with rigid or self-aligning bearings of expansion or nonexpansion types and with either sleeve or rolling bearings. Housings are either split or solid.

REFERENCES AND SOURCE MATERIAL

1. *Machine Design,* Mechanical drives reference issue.

ASSIGNMENTS

See Assignments 12 and 13 for Unit 21-4 on pages 742–743.

21-5 LUBRICANTS AND RADIAL SEALS

Lubricants

There are two main reasons why lubricants are used in any bearing: (1) to reduce friction between rubbing surfaces, and

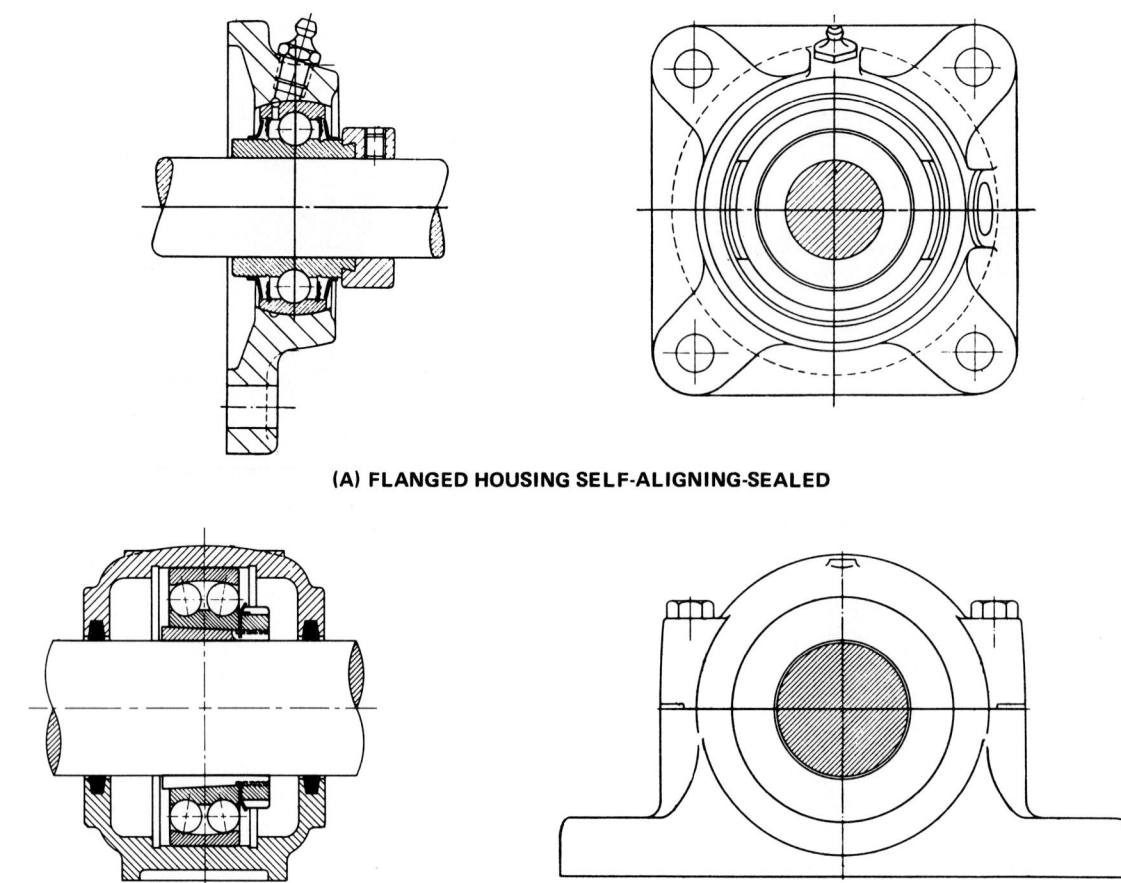

(A) FLANGED HOUSING SELF-ALIGNING-SEALED

(B) PILLOW BLOCK SELF-ALIGNING-SEALED

FIG. 21-4-2   Adjustable shaft support with ball bearings. *(Boston Gear Works)*

(2) as coolants to carry off heat which may be generated in bearings. Either or both of these functions may be required of a lubricant on a particular bearing.

As friction reducers, lubricants can be considered from two aspects. When a hydrodynamic bearing is started, for instance, metal-to-metal contact occurs. Here the actual oiliness of the lubricant lowers the coefficient of friction between the two sliding surfaces. In slider bearings operating on full fluid-film lubrication, the lubricant separates the two sliding surfaces completely, and shearing of the lubricant is substituted for sliding friction.

Any system of rolling elements, like a ball bearing, should theoretically reduce friction radically. If balls and rollers were perfectly smooth and inelastic, friction would be very low. But materials deform, and rolling elements slip under load. Also, uncaged balls or rollers tend to rub or slide against one another. When a separator or cage is present, the rolling elements slide against it, and the cage itself rubs against any guiding flange surfaces. Because of this sliding, lubrication is needed to minimize wear and friction.

Oils and Greases

Whether to use oil or grease (two of the three general types of lubricants) and what kind of oil or grease to use are questions that, for slider bearings, must usually be decided early in design phases, since bearing design depends on the lubricant and the type of lubrication selected.

Oils are slippery hydrocarbon liquids. *Grease* is a semisolid, combining a fluid lubricant with a thickening agent, usually a soap. In the past, the soaps in greases were considered storehouses for the oil. Pressure and temperature squeeze out the oil to lubricate bearing surfaces. This is probably only partly true. Soap molecules are attracted to metal surfaces.

Both oil and grease are used to lubricate rolling- and sliding-contact bearings. In fact, either type of lubricant can be used in some applications, but each type has peculiar qualities that equip it for certain types of applications.

Assets of Oil   Some of the advantages of oils are:

1. Oil is easier to drain and refill. This is important if lubricating intervals are close together. It is also easier to control the fill volume of the oil in the housing or reservoir.
2. An oil lubricant for a bearing might also be usable at many other points in the machine, even eliminating the need for a second grease-type lubricant.
3. Oil is more effective than grease in carrying heat away from bearing and housing surfaces. In addition, oils are available for a greater range of operating speeds and temperatures than greases.

4. Oil readily feeds into all areas of contact and can carry away dirt, water, and the products of wear.

Assets of Grease Some of the advantages of greases are:

1. Grease does not flow as readily as oil, so it can be more easily retained in a housing. Since grease is easily contained, leakproof designs are unnecessary.
2. Less maintenance is required. There is no oil level to maintain; regreasing is infrequent.
3. Grease has better sealing abilities than oil. This asset may help to keep dirt and moisture out of the housing.

Solid-Film Lubricants

Even the smoothest machined surfaces have microscopic roller-coaster profiles. When one such surface slides on another, the surface irregularities complicate lubrication. Under hydrodynamic conditions, a lubricating liquid may be interposed between the surfaces to prevent them from scraping one peak against another.

When a pure lubricant or a mixture of lubricants is applied as dry powder, grease, or an oil suspension, it is called a *solid lubricant*. When the lubricant is applied in a uniformly thin layer, confining a high concentration of lubricant to a given area, it is called a *bonded dry film*. Inorganic or organic binders and solvents provide the vehicle for these films.

Lubricating Devices

Available lubricating devices range from simple fittings to completely automatic systems. Lubricating devices may be classified as internal or external. The bearings they serve can be lubricated individually or as a group.

Individual bearing devices include oil cups, hydraulic grease fittings, and drip oilers. Group methods generally supply lubricant under pressure through a distribution system to a number of bearings.

Hand Lubrication *Hand lubrication* refers to the manual use of any portable or semiportable lubrication equipment for bearing-by-bearing application.

For hand lubrication from portable devices, accessibly located fittings must be provided. For oil-lubricated bearings, the simplest provision is a drilled hole into which fluid lubricant is dripped. To avoid plugging or contamination, a tube or cup with a spring-loaded hinged lid is usually installed.

Fittings must not only be accessible to the coupling on a portable lubricating device but for possible field installation of other types of lubrication equipment. Pipe-thread connections are the most universal.

Individual Bearings When life expectancy is satisfactory, certain bearings—prelubricated, permanently sealed bearings (rolling-element or plain), solid or dry-film lubricated plain bearings, and porous bushings—require no lubrication maintenance. But the capabilities of even these three types of bearings can be improved by adding internal reservoirs or external lubricating devices.

Internal Reservoirs The oldest method uses the direct contact of grease stored in the cavity of a rolling-element bearing or in the grooves of a plain bearing.

For oil lubrication, felt wicks or wool-waste packings are used to retain the oil and to transfer it to the moving surface by direct contact.

External Reservoirs These devices for individual bearings include drop-feed, constant-level, thermal-expansion, bottle-, wick-feed oilers, and pressure grease cups. Only the drop-feed or gravity oiler (Fig. 21-5-1) is in fairly common use.

Multiple Bearings A suitable housing or enclosure is required for all internal-reservoir methods for lubricating multiple bearings. This enclosure maintains proper lubricant level and prevents loss of lubricant from within the internal complex. The enclosure must also prevent the entrance of contaminants. Three common types of internal lubricating systems are shown in Fig. 21-5-2.

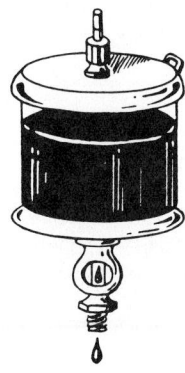

FIG. 21-5-1 External reservoir lubricating system.

(A) BATH LUBRICATION

(B) SPLASH LUBRICATION

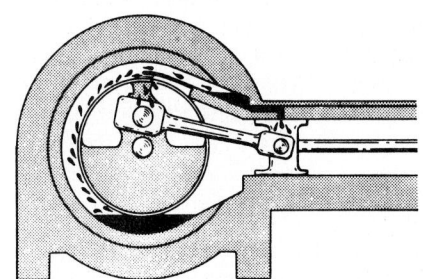

(C) SPLASH LUBRICATION WITH PRESSURE LUBRICATING SYSTEM

FIG. 21-5-2 Internal lubricating systems.

Grease and Oil Seals

Factors in the Selection of Oil Seals

Efficient oil seals are available today for every application. But if modern sealing techniques and advancements are to be utilized, oil seals must be selected, inspected, and installed correctly. There are a number of factors to be considered in carrying out these operations.

The environment in which the seal will operate is the most important factor in the selection of materials to be used for the seal. In particular, the user must consider:

1. Fluid to be sealed in
2. Fluid to be sealed out
3. Mean temperature of the environment
4. The shaft on which the sealing element runs
5. The sealing element designs for which tooling has been developed

The fluids to be sealed in are usually lubricants. The composition of fluid differs greatly, even within one classification.

The medium to be sealed out is usually air containing varying amounts of dust, gravel, water, slate, etc. Dirt can radically shorten seal life and often dictates the selection of the oil-seal compound. Dirt and dust get under the sealing lip and cause heavy wear and scoring of both the shaft and seal lip. Water and salt cause rusting of the shaft surface, with consequent pitting of the shaft and rapid wear of the seal lip.

The mean temperature of the environment has a radical effect on the seal life and strongly influences the choice of seal compound.

The shaft on which the seal must ride has some influence on the selection of the seal compound. If the shaft is hard and has a surface finish of better than 20 μin. (0.5 μm), any of the the compounds can be used. On rougher shafts, the acrylics and silicones wear too rapidly to be used.

Sealing Materials   The type of lubricant and the mean operating temperature usually govern the choice of the *elastomer* (any elastic substance resembling rubber) to be used for the seal compound. Since mean operating temperatures seldom exceed 220°F (105°C), nitrile rubber compounds are the most widely used sealing materials. They wear best, are easiest to mold, and are low in cost.

Silicone compounds are preferred for some applications. Not all silicone compounds are safe to use, however. Most will disintegrate rapidly in many automatic-transmission fluids and in some engine oils.

Fluorelastomer compounds, such as Viton, have a long life at very high temperatures in almost any lubricant. Their cost is high, however. They get quite stiff, but not brittle, at low temperatures.

Radial Seals

Typical seal designs being used today feature both single- and dual-lip sealing elements bonded securely to metal cases that add strength and rigidity to the seals. The bonding of the sealing element to the case eliminates internal leakage resulting from clamping.

For several years extensive tests and investigations have been conducted to evaluate the effect of various cross-sectional shapes for sealing elements. Several conditions must be satisfied in developing proper shapes, some of which conflict with others. The sealing element must be flexible enough to follow shaft runout but stiff enough to prevent collapse under operating conditions.

The combination of angles between the trim surface and the "approach angle" is critical. This is particularly true of the angle toward the oil. If an angle is too acute, an otherwise well-designed seal will perform poorly.

Installation   The bore must be round and smooth. It must have a proper lead-in chamfer, with a minimum of tool leads and marks, and no tool-return grooves. A bottom should be designed into the bore, and the bore should be concentric with the bearing retention surface.

The shaft should have a chamfer, and generally speaking, the surface should be approximately 20 μin. (0.5 μm) with above-C45 Rockwell hardness in abrasive applications and above-B80 Rockwell hardness when abrasive conditions are absent.

Felt Radial Seals

Felt is built-up fabric made by interlocking fibers through a suitable combination of mechanical work, chemical action, moisture, and heat, without spinning, weaving, or knitting. It may consist of one or more classes of fibers—wool, reprocessed wool, or reused wool—which are used alone or combined with animal, vegetable, and synthetic fibers.

Felt has long been used as an important material for sealing purposes. The main reasons are oil wicking, oil absorption, filtration, resiliency, low friction, polishing action, and cost (Fig. 21-5-3).

Radial Positive-Contact Seals

Radial positive-contact seals are dynamic rubbing seals. Operational effectiveness of a dynamic seal installation was once measured by an easy standard: If it did not leak too much too soon, it was a good seal. Today's operational concepts require sealing effectiveness with absolute minimal leakage over wide service parameters.

A radial positive-contact seal is a device that applies a sealing pressure to a mating cylindrical surface to retain fluids and, in some cases, to exclude foreign matter. Although this definition fits almost all dynamic-contact seals, including packings and felt rubbing seals, attention is given in this chapter to the types of seals more commonly known as oil seals or shaft seals (Fig. 21-5-4).

The rotating-shaft application of the radial seal is most common. However, the radial seal is also used where shaft motion is oscillating or reciprocating.

Among the factors that recommend a shaft seal over other possible sealing media are ease of installation and small space allocation necessary in design of equipment, relative low cost for high effectiveness, and ability to handle simultaneously a wide range of variables while providing a positive sealing effect throughout.

FIG. 21-5-3   Felt seal designs.

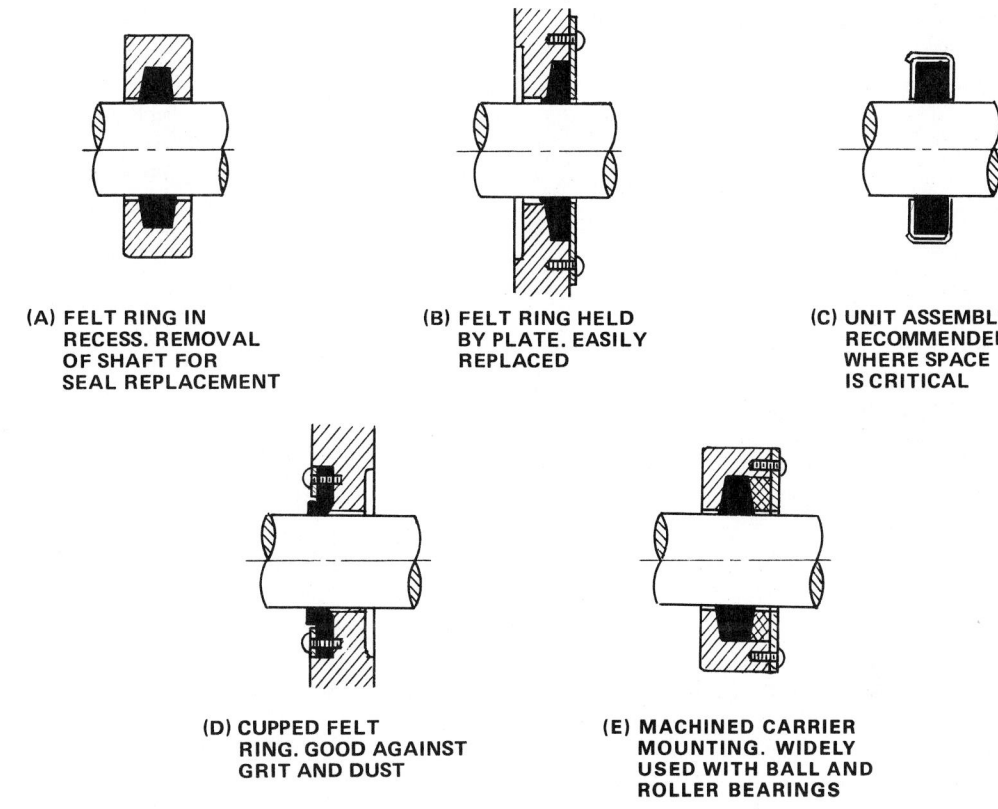

(A) FELT RING IN RECESS. REMOVAL OF SHAFT FOR SEAL REPLACEMENT

(B) FELT RING HELD BY PLATE. EASILY REPLACED

(C) UNIT ASSEMBLY RECOMMENDED WHERE SPACE IS CRITICAL

(D) CUPPED FELT RING. GOOD AGAINST GRIT AND DUST

(E) MACHINED CARRIER MOUNTING. WIDELY USED WITH BALL AND ROLLER BEARINGS

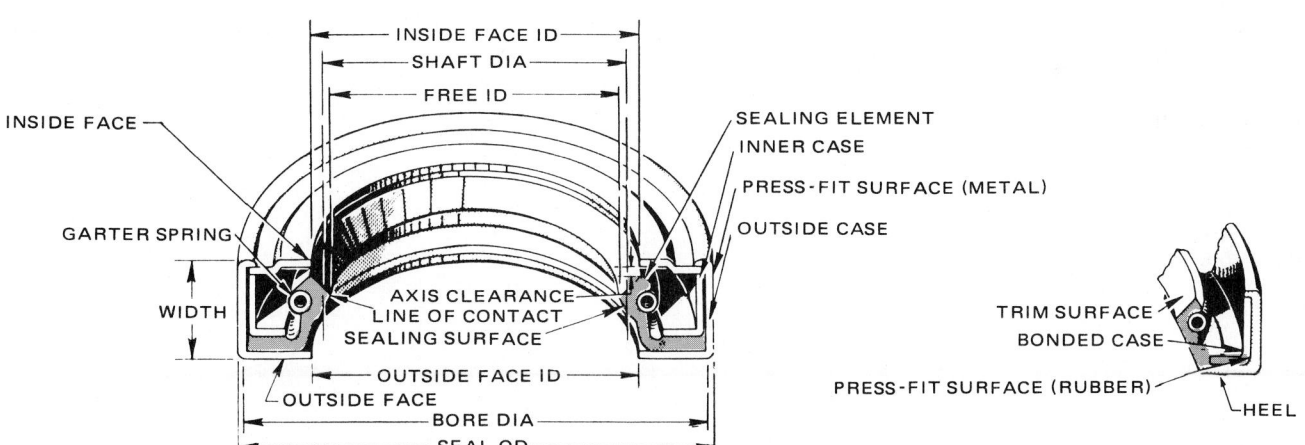

FIG. 21-5-4   Metal-cased seal nomenclature.

Types   Because of the variety of applications, the radial seal is manufactured in numerous types and sizes. They are usually categorized as:

1. Cased seals, where the leather or synthetic sealing element is retained in a precision-manufactured metal case
2. Bonded seals, with the synthetic element permanently bonded to a flat washer or to a formed-metal case

Seals of both categories can be provided with spring-tension elements, either garter-spring or finger type, for sealing low-viscosity fluids or where either shaft speed or eccentricity demands higher seal-contact pressures.

Clearance Seals

Clearance seals limit leakage by closely controlling the angular clearance between the rotating or reciprocating shaft and the relatively stationary housing. There are two basic clearance-seal types: labyrinth and bushing or ring. These seal types are employed when a small loss in efficiency because of leakage can be permitted. Clearance seals are used when pressure differentials are beyond the design limitations of contact seals (face and circumferential).

Labyrinths   Some advantages of labyrinth seals are reliability, simplicity, and flexibility in material selection. They are used mainly in heavy industrial, power, and aircraft applications where relatively high leakage rates may be tolerated and where design simplicity is an absolute necessity.

A labyrinth seal consists of one or more thin strips or knives, which are attached to either the stationary housing or the rotating shaft. A simple labyrinth is shown in Fig. 21-5-5A.

Bushing and Ring Seals   The bushing-type seal is a close-fitting stationary sleeve within which the shaft rotates. Leakage from a high-pressure station at one end of the bushing to a region of low pressure at the other end is controlled by the restricted clearance between shaft and bushing. Ideally, the bushing and shaft are perfectly concentric and no rubbing takes place (Fig. 21-5-5B).

Split-Ring Seals

Split rings are used for a large number of seal applications (Fig. 21-5-6). Expanding split rings (piston rings) are used in compressors, pumps, and internal-combustion engines. Applications for straight-cut and seal-joint rings are common in industrial and aerospace hydraulic and pneumatic cylinders (linear actuators), where the ruggedness of piston rings is advantageous and where various degrees of leakage can be tolerated.

Axial Mechanical Seals

By convention, the term *axial mechanical seal,* or *end-face seal,* designates a sealing device that forms a running seal between flat, precision-finished surfaces. Used for rotating shafts, the sealing surfaces usually are located in a plane at a right angle to the shaft. Forces that hold the rubbing faces in contact are parallel to the shaft.

Axial mechanical seals replace conventional stuffing boxes where a fluid must be contained in spite of a substantial pressure head. These seals have many advantages, such as:

1. Reduced friction and power losses
2. Elimination of wear on shaft or shaft sleeve
3. Zero or controlled leakage over a long service life
4. Relative insensitivity to shaft deflection or end play
5. Freedom from periodic maintenance

Axial mechanical seals do have disadvantages. As precision components they demand careful handling and installation.

While differing in design detail, all mechanical seals make use of the following elements:

1. Rotating seal rings
2. Stationary seal rings
3. Spring-loading devices
4. Static seals

The rotating seal ring and the stationary seal ring are spring-loaded together by the spring-loading apparatus, and sealing takes place on the surfaces of these two rings, which rub together. The static seal component of an axial mechanical seal stops leakage of fluid past the juncture of the rotating seal ring and the shaft.

Since the rotating seal ring is stationary with respect to the turning shaft, sealing at their junction point is accomplished easily through the use of gaskets, O-rings, V-rings, cups, and so forth.

End-Face Seals

The main advantage of an end-face seal is its low leakage rate. For example, the ratio of leakage between mechanical packings and end-face seals averages about 100:1. In addition, the end-face seal causes little wear of the sleeve or shaft on which it seals. Dynamic sealing is created on the seal faces in a perpendicular plane to the shaft.

FIG. 21-5-5   Clearance seals.

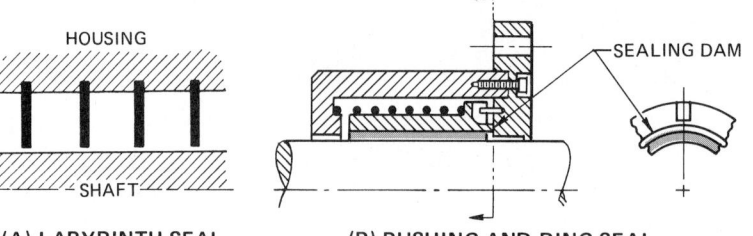

(A) LABYRINTH SEAL   **(B) BUSHING AND RING SEAL**

FIG. 21-5-6   Split-ring seals.

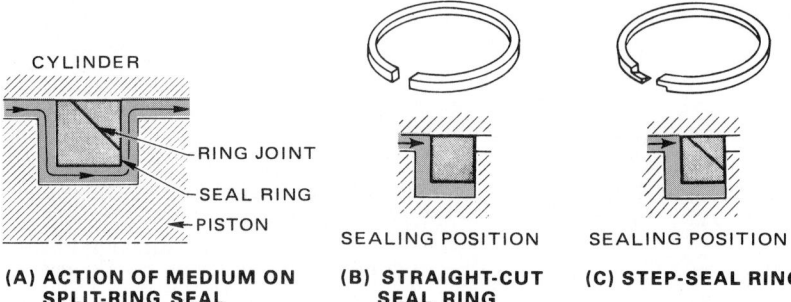

(A) ACTION OF MEDIUM ON SPLIT-RING SEAL   **(B) STRAIGHT-CUT SEAL RING**   **(C) STEP-SEAL RING**

The basic development of an end-face seal is shown in Fig. 21-5-7. A shaft with a simple O-ring as its sealing member is provided with a housing that incorporates one of the sealing faces. The housing encloses the O-rings and effects a preload on the shaft, thereby ensuring its sealing. A spring assembly is added to energize the end-face member axially, providing spring pressure against the end-face member to keep the faces together during periods of shutdown or lack of hydraulic pressure in the unit. To complete the basic seal, a stationary member is incorporated in the end cap of the unit.

A complete seal consists basically of two elements: the seal-head unit, which incorporates the housing, the end-face member, and the spring assembly; and the seal seat, which is the mating member that completes the precision-lapped face combination.

Shaft Sealing   Shaft sealing elements include the O-ring, V-ring, U-cup, wedge, and bellows (Fig. 21-5-8).

The first four of these elements constitute one category—the pusher-type seal. As the face wears, these sealing elements are pushed forward along the shaft to maintain the seal.

Pusher-Type Elements   For the case of the O-ring, the hydraulic pressure and a mechanical preloading factor provide the sealing effect.

For the V-ring, U-cup, and wedge, the sealing function is created by mechanical and hydraulic means. Mechanical preloading to the shaft is provided by spring action incorporated in the seal design and by hydraulic pressure in the stuffing box.

The V-ring and U-cup designs seal at the shaft surface and at the mating surface of the housing. The sealing action is obtained from both spring force and hydraulic pressure acting against the spreader element, which reacts against the wings of the seal, spreading it in both directions.

Bellows-Type Elements   The bellows-shaped sealing member differs from the pusher type in that it forms a static seal between itself and the shaft. Hence, all axial movement is taken up by bellows flexure.

Molded Packings

Molded packings are often called automatic, hydraulic, or mechanical packings. As a general group, these packings usually do not require any gland adjustment after installation. The fluid being sealed supplies the pressure needed to produce the force for sealing the packings against the wearing surface.

This general classification of packings can be subdivided into two categories: lip and squeeze types.

Lip-Type Packings   Lip-type packings of the flange, cup, U-cup, U-ring, and V-ring configurations are used almost exclusively for dynamic applications. Although rotary motions are encountered, the packings discussed here are used primarily for sealing during reciprocating motion. Hence, all the recommendations and designs mentioned apply to reciprocating service (Fig. 21-5-9).

Flanged Packings   The flange, sometimes called the *hat*, is the least popular of all the lip-type packings.

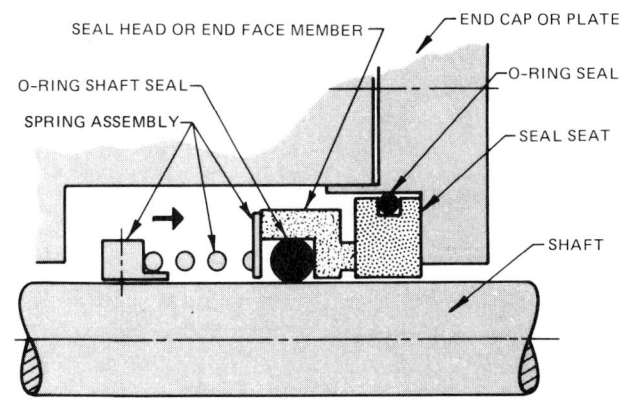

FIG. 21-5-7   Basic end-face seal design.

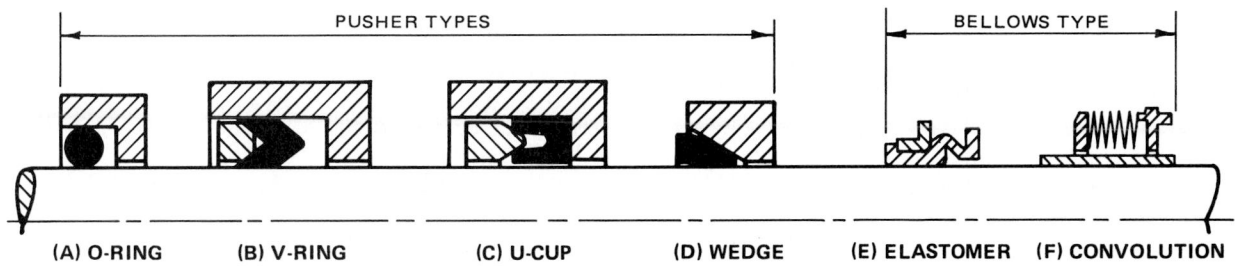

FIG. 21-5-8   Shaft seal configurations.

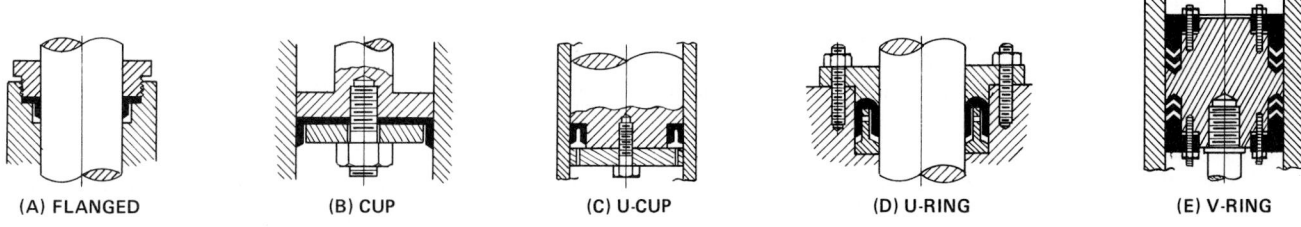

FIG. 21-5-9   Lip-type packings.

731

Cup Packings   Leather cup packings, one of the oldest types of lip or mechanical packings, are used in large volume for both hydraulic and pneumatic service at low and high pressures.

Squeeze-Type Packings   Squeeze-type molded packings are made in a variety of sizes and shapes (Fig. 21-5-10), but nearly all of them offer these advantages:

1. Low initial cost
2. Adaptability to limited space
3. Ease of installation
4. High efficiency
5. No need for adjustment
6. Tolerance to wide ranges of pressure, temperature, and fluids
7. Sealing in both directions
8. Relatively low friction

Squeeze-type packings are economical and easy to install and can be used whenever conditions permit. They are generally fitted to a rectangular groove, machined in a hydraulic or pneumatic mechanism. Nomenclature for the dimensions of a squeeze-packing seal groove is identified as that part which applies the "squeeze" to the cross section of the packing.

Commonly used squeeze-type packings are:

D-rings   D-shaped rings make good rod seals for reciprocating motion. They perform equally well in hydraulic or pneumatic applications.

Delta Rings   This triangle-shaped ring solves the twisting problem of O-rings, but since friction is greater, the expected life is relatively short. The delta ring has limited applications.

T-rings   The t-shaped ring is not susceptible to spiral failures. It can be used as a rod or piston seal for reciprocating motion, or it can be used for oscillating motion under low pressures.

Lobed Rings   This is a square-shaped ring with four rounded lobes. It can be used in conventional O-ring grooves for reciprocating, rotating, and oscillating motion. The lobed ring is superior to the O-ring in most rotating applications.

O-rings   The O-ring is the most common form of squeeze packing. It seals in both directions and has a low initial cost. O-ring seals work under the principle of controlled deformation. Some slight deformation is given the elastic O-ring in the form of diametral squeeze when it is installed (Fig. 21-5-11). But it is the pressure from the confined fluid that produces the deformation that causes the elastic O-ring to seal.

There are three types of applications for dynamic O-rings:

1. Reciprocating, where the sealing action is that of a piston ring or a seal around a piston rod.
2. Oscillating, where the seal rotates back and forth through a limited number of degrees or several complete turns. This may be combined with very short reciprocating strokes. The main difference between oscillation and rotation is the amount of motion involved.
3. Rotating, where a shaft turns inside the ID of the O-ring.

One of the ideal applications of an O-ring is as a piston seal in a hydraulic-actuating cylinder. Another common application uses the O-ring as a valve seat or as a valve stem packing.

Seal Symbols

The simplified representation of seals as shown in Fig. 21-5-12 is recommended for use on drawings, wherever it is not necessary to show the exact form and size of seals.

Where it is desirable to show the functional principle of the seal, symbols for the appropriate type of seal are added (Fig. 21-5-13).

REFERENCES AND SOURCE MATERIAL

1. *Machine Design,* Mechanical drives reference issue.

ASSIGNMENTS

See Assignments 14 through 17 for Unit 21-5 on pages 743 to 744.

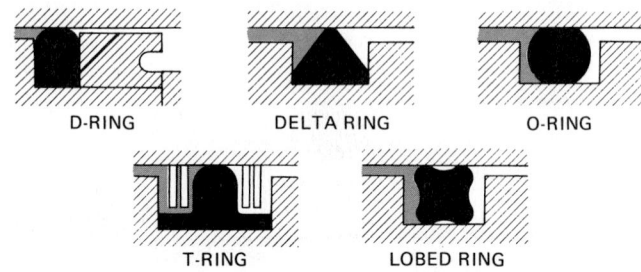

FIG. 21-5-10   Squeeze packings.

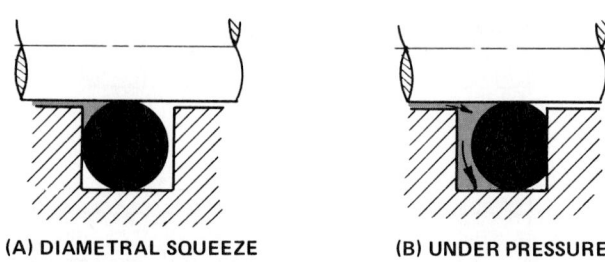

(A) DIAMETRAL SQUEEZE   **(B) UNDER PRESSURE**

O-RINGS ARE FITTED INTO RECTANGULAR GROOVES IN HYDRAULIC MECHANISMS, AND SEALED BY BEING FORCED, BY PRESSURE, INTO CREVICES.

FIG. 21-5-11   O-ring.

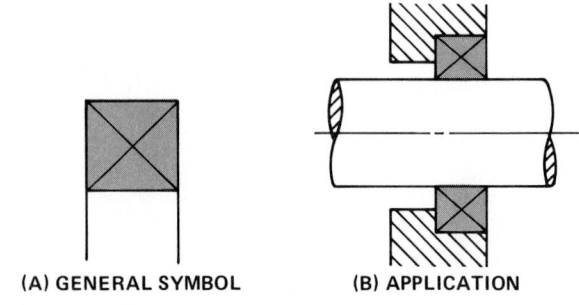

(A) GENERAL SYMBOL   **(B) APPLICATION**

FIG 21-5-12   Simplified representation of seals.

FIG. 21-5-13   Functional representation of seals.

	SHAFT SEALS		GROOVE SEALS
1	ONE TONGUE, LEFT SIDE, SEALING INSIDE.	1	SEALING OUTSIDE, LEFT.
2	ONE TONGUE, RIGHT SIDE, SEALING INSIDE.	2	SEALING OUTSIDE, RIGHT.
3	ONE TONGUE, LEFT SIDE, SEALING OUTSIDE.	3	SEALING OUTSIDE, LEFT, WITH BACKRING.
4	ONE TONGUE, RIGHT SIDE, SEALING OUTSIDE.	4	SEALING OUTSIDE, RIGHT, WITH BACKRING.
5	ONE TONGUE, LEFT SIDE, WITH DUST TONGUE, SEALING INSIDE.	5	SEALING INSIDE, LEFT.
6	ONE TONGUE, RIGHT SIDE, WITH DUST TONGUE, SEALING INSIDE.	6	SEALING INSIDE, RIGHT.
7	ONE TONGUE, RIGHT SIDE, WITH DUST TONGUE, SEALING OUTSIDE.	7	SEALING INSIDE, LEFT, WITH BACKRING.
8	ONE TONGUE, LEFT SIDE, WITH DUST TONGUE, SEALING OUTSIDE.	8	SEALING INSIDE, RIGHT, WITH BACKRING.
9	TONGUES, LEFT AND RIGHT, SEALING OUTSIDE.	9	SEALING INSIDE AND OUTSIDE.
10	TONGUES, LEFT AND RIGHT, SEALING INSIDE.	10	SEALING LEFT AND RIGHT.

21-6   STATIC SEALS AND SEALANTS

O-Ring Seals

All static O-ring seals are classified as gasket-type seals. Static O-ring seals are generally easier to design into a unit than dynamic O-ring seals. Wider tolerances and rougher surface finishes are allowed on metal mating members. The amount of squeeze applied to the O-ring cross section can also be increased. This type of nonmoving seal is used in flanges, flange fittings, flange unions and cylinder end caps, valve covers, plugs, etc.

Groove Design

A rectangular groove is the most common for O-rings used as flange gaskets.

The rectangular groove can be machined half in the face plate and half in the flange, or the entire groove can be cut in one member. In some flange-gasket designs, a triangular groove can be used to provide ease of machining and consequent reduced cost. Round-bottom grooves are also used.

Applications

Figure 21-6-1A (pg. 734) shows an application of an O-ring to a cylinder head cover. The higher the pressure, the tighter the seal. This design automatically preloads the O-ring in the groove. Cap screws or nuts are tightened only enough to maintain metal-to-metal surface contact. This type of installation will seal high pressures without the excessive bolt stress necessary with conventional gasketed joints.

Figure 21-6-1B illustrates two O-ring sizes used for sealing a rectangular pressure chamber. The outside O-ring is stretched in a groove and the cap screw heads are sealed by small O-rings in the counterbore. This design is a simple and effective method of sealing X-ray heads, gear pump end plates, and other applications without requiring lapped surfaces.

Figure 21-6-1C pictures an O-ring gasket used on a flange union. The O-ring makes a tight seal when the union is screwed down finger-tight. The round-bottom groove has the same diameter as the actual cross section of the O-ring. The O-ring protrudes .02 to .03 in. (0.39 to 0.79 mm) above the face of the flange. The volume of the groove is calculated to be the same as the minimum volume of the O-ring.

Figure 21-6-1D is a modification of the flange-type groove. This type of triangular groove is used where ease of machining and reduced cost are important. This type of design makes an effective seal. However, the O-ring is permanently deformed. Pressures are limited only by the clearance between the mating metal surfaces and the strength of the metal itself.

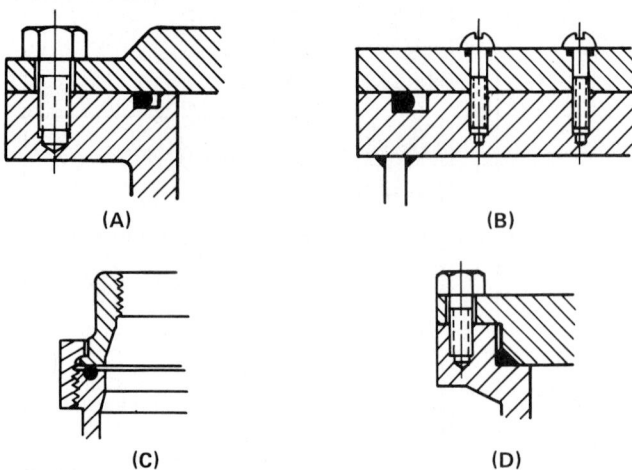

(A)   (B)

(C)   (D)

FIG. 21-6-1   Flanged-type static O-ring seal design.

Flat Nonmetallic Gaskets

A gasket creates and maintains a tight seal between separable members of a mechanical assembly. Although a seal may be obtained without a gasket, the gasket promotes an efficient initial seal and prolongs the useful life of an assembly.

Basic flange joints (Fig. 21-6-2) are suitable for all kinds of flat gaskets, plain or jacketed. For moderate pressures up to 200 lb/in.^{2} (1400 kPa), the simple flange joint is applicable.

Metal-to-metal joints are particularly suitable for truly compressible materials, such as cork composition and cork-and-rubber. These joints bear a great similarity to joints designed for rubber O-rings. Gasket design considerations are shown in Fig. 21-6-3.

Metallic Gaskets

Metallic gaskets are used for high pressures and for temperature extremes that cannot be handled by nonmetallic gaskets.

Solid metal gaskets usually require thick flanges. Thinner flanges can be used with metal O-rings. The rings are made of thin-walled metal tube, bent and welded to form a continuous circle.

Sealants

Sealants are used to exclude dust, dirt, moisture, and chemicals or to contain a liquid or gas. They can also protect against mechanical or chemical attack, exclude noise, improve appearance, and act as an adhesive.

Sealants are normally used for less severe conditons of temperature and pressure than gaskets. Sealants are categorized as hardening and nonhardening.

Hardening sealants may be either rigid or flexible, depending on their composition. Nonhardening types are characterized by plasticizers that come to the surface continually, so that the sealant stays "wet" after application.

The three basic sealing joints (Fig. 21-6-4, pg. 736) are:

Butt Joint   With a butt joint, several types of sealant may be used. If the thickness of the plate is sufficient, use sealant (A), or if the plates are thin, use bead sealant (B). Tape can also be used (C). If the joint moves due to dynamic loads or thermal expansion and contraction, a flexible sealant with good adhesion

FIG. 21-6-2   Flat gasket joints.

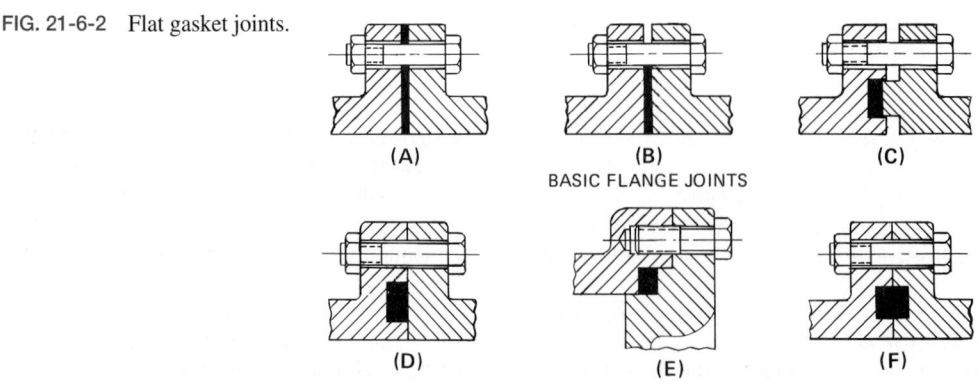

(A)   (B)   (C)

BASIC FLANGE JOINTS

(D)   (E)   (F)

METAL-TO-METAL JOINTS

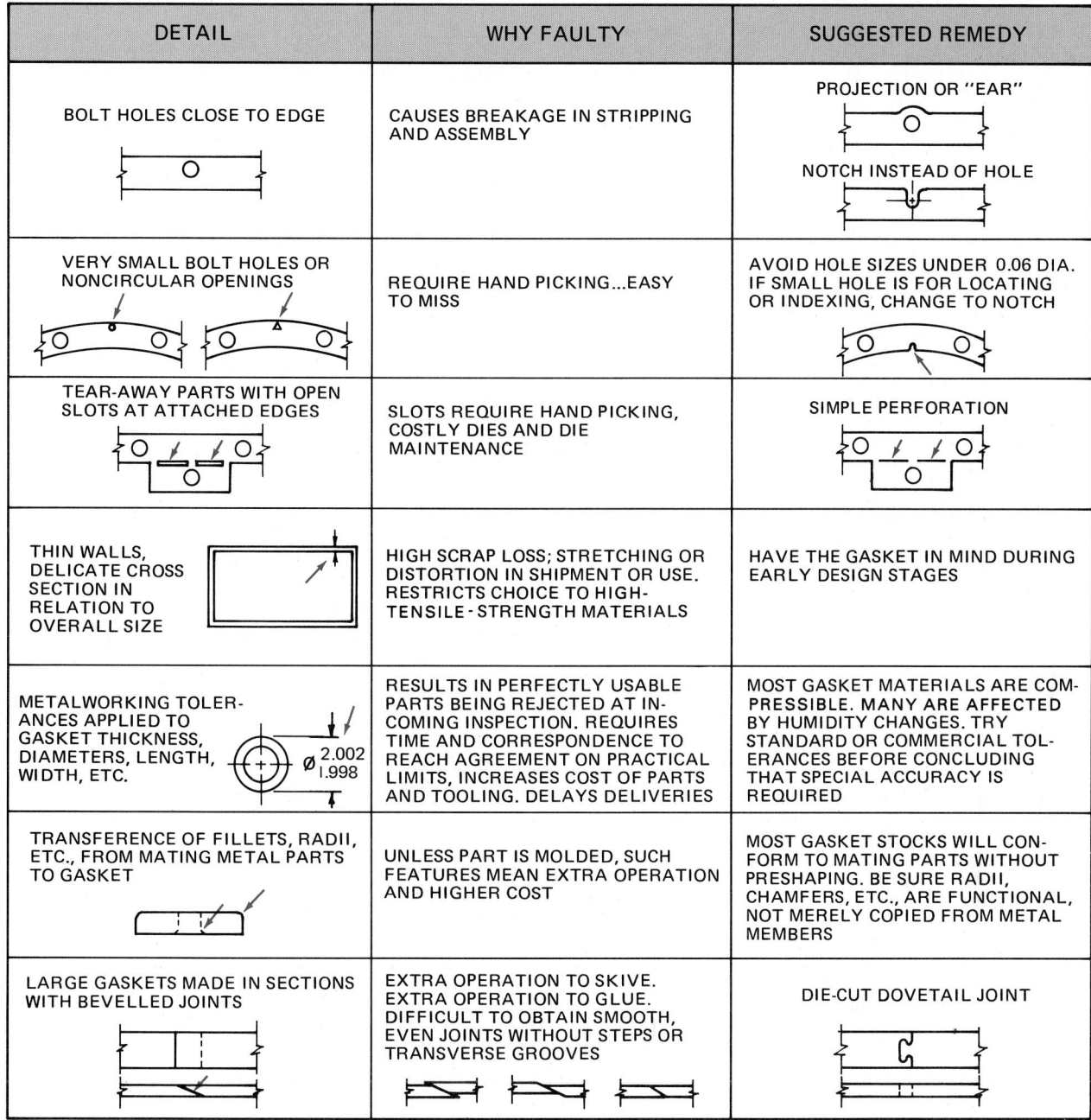

DETAIL	WHY FAULTY	SUGGESTED REMEDY
BOLT HOLES CLOSE TO EDGE	CAUSES BREAKAGE IN STRIPPING AND ASSEMBLY	PROJECTION OR "EAR" NOTCH INSTEAD OF HOLE
VERY SMALL BOLT HOLES OR NONCIRCULAR OPENINGS	REQUIRE HAND PICKING...EASY TO MISS	AVOID HOLE SIZES UNDER 0.06 DIA. IF SMALL HOLE IS FOR LOCATING OR INDEXING, CHANGE TO NOTCH
TEAR-AWAY PARTS WITH OPEN SLOTS AT ATTACHED EDGES	SLOTS REQUIRE HAND PICKING, COSTLY DIES AND DIE MAINTENANCE	SIMPLE PERFORATION
THIN WALLS, DELICATE CROSS SECTION IN RELATION TO OVERALL SIZE	HIGH SCRAP LOSS; STRETCHING OR DISTORTION IN SHIPMENT OR USE. RESTRICTS CHOICE TO HIGH-TENSILE-STRENGTH MATERIALS	HAVE THE GASKET IN MIND DURING EARLY DESIGN STAGES
METALWORKING TOLERANCES APPLIED TO GASKET THICKNESS, DIAMETERS, LENGTH, WIDTH, ETC. $\varnothing \begin{smallmatrix} 2.002 \\ 1.998 \end{smallmatrix}$	RESULTS IN PERFECTLY USABLE PARTS BEING REJECTED AT INCOMING INSPECTION. REQUIRES TIME AND CORRESPONDENCE TO REACH AGREEMENT ON PRACTICAL LIMITS, INCREASES COST OF PARTS AND TOOLING. DELAYS DELIVERIES	MOST GASKET MATERIALS ARE COMPRESSIBLE. MANY ARE AFFECTED BY HUMIDITY CHANGES. TRY STANDARD OR COMMERCIAL TOLERANCES BEFORE CONCLUDING THAT SPECIAL ACCURACY IS REQUIRED
TRANSFERENCE OF FILLETS, RADII, ETC., FROM MATING METAL PARTS TO GASKET	UNLESS PART IS MOLDED, SUCH FEATURES MEAN EXTRA OPERATION AND HIGHER COST	MOST GASKET STOCKS WILL CONFORM TO MATING PARTS WITHOUT PRESHAPING. BE SURE RADII, CHAMFERS, ETC., ARE FUNCTIONAL, NOT MERELY COPIED FROM METAL MEMBERS
LARGE GASKETS MADE IN SECTIONS WITH BEVELLED JOINTS	EXTRA OPERATION TO SKIVE. EXTRA OPERATION TO GLUE. DIFFICULT TO OBTAIN SMOOTH, EVEN JOINTS WITHOUT STEPS OR TRANSVERSE GROOVES	DIE-CUT DOVETAIL JOINT

FIG. 21-6-3   Common faults in gasket design and suggested remedies.

must be selected. If movement is anticipated, a flexible tape should be selected for the butt joint.

Lap Joint   To secure a lap joint, the sealant should be sandwiched between mating surfaces, and the seam should be riveted, bolted, or spot welded (A). Thick plates can be sealed with a bead of sealant (B), or tape can be used (C) if there is sufficient overlap to provide a surface to which the tape can adhere.

Angle Joint   As you can see, supported angle joints with a bead of sealant (C) or a sandwich seal (D) are superior to the butt joint shown in (A) or the sealant shown in (B).

Exclusion Seals

Exclusion seals are used to prevent the entry of foreign material into the moving parts of machinery (Fig. 21-6-5, pg. 736). This protection is necessary because foreign material contaminates the lubricant and accelerates wear and corrosion. Static joints are easily sealed by tight fits and gaskets. Sealing between parts having relative motion, such as between a housing and a moving shaft, is more difficult.

Sometimes seals designed only for inclusion are used to perform the inclusion and exclusion functions simultaneously. This is inadvisable, except under very light service conditions.

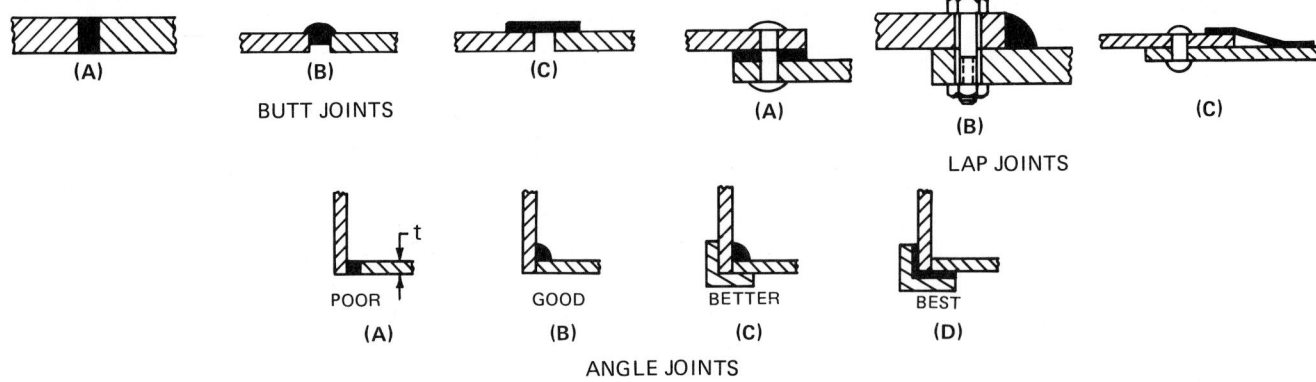

FIG. 21-6-4   Common methods for sealing joints.

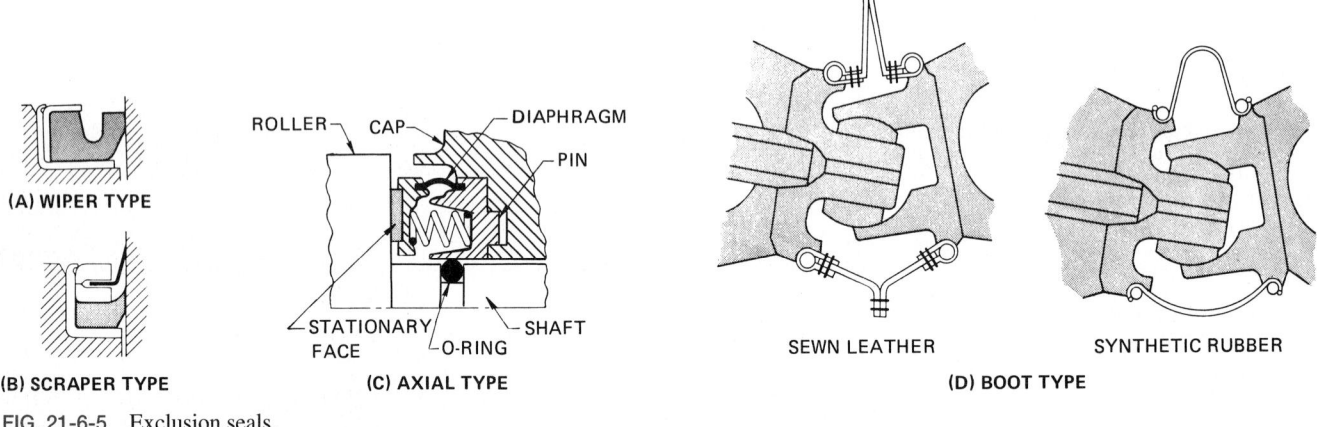

FIG. 21-6-5   Exclusion seals.

Inclusion seals usually do a poor exclusion job and are damaged by even small amounts of abrasive material.

Exclusion seals can be classified into four general groups: wiper, scraper, axial, and boot seals.

REFERENCES AND SOURCE MATERIAL

1. *Machine Design,* Mechanical drives reference issue.

ASSIGNMENTS ▦▦▦▦▦▦▦▦▦▦▦▦

See Assignments 18 and 19 for Unit 21-6 on pages 745–746.

ASSIGNMENTS FOR CHAPTER 21

ASSIGNMENTS FOR UNIT 21-1, COUPLINGS AND FLEXIBLE SHAFTS

1. Lay out the motor-to-gearbox drive unit shown in Fig. 21-1-A. A flexible coupling is required to connect the shafts, and the type of coupling specified is shown in the figure. Call for the correct-size coupling on the drawing. Scale is as specified.

2. Lay out the motor-to-pump drive assembly shown in Fig. 21-1-B. Flexible couplings are required to connect the shafts, and the type of coupling specified is shown in the figure. Call for the proper couplings on the drawing. Scale is as specified.

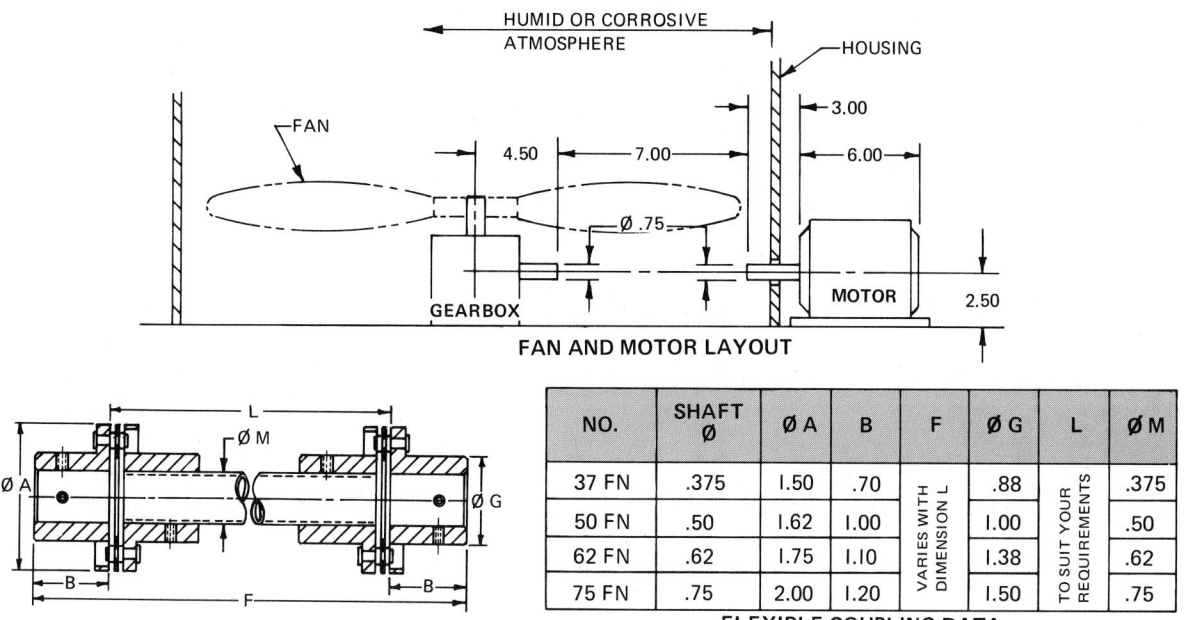

NO.	SHAFT Ø	Ø A	B	F	Ø G	L	Ø M
37 FN	.375	1.50	.70		.88		.375
50 FN	.50	1.62	1.00	VARIES WITH DIMENSION L	1.00	TO SUIT YOUR REQUIREMENTS	.50
62 FN	.62	1.75	1.10		1.38		.62
75 FN	.75	2.00	1.20		1.50		.75

FLEXIBLE COUPLING DATA

FIG. 21-1-A   Motor-to-gearbox drive.

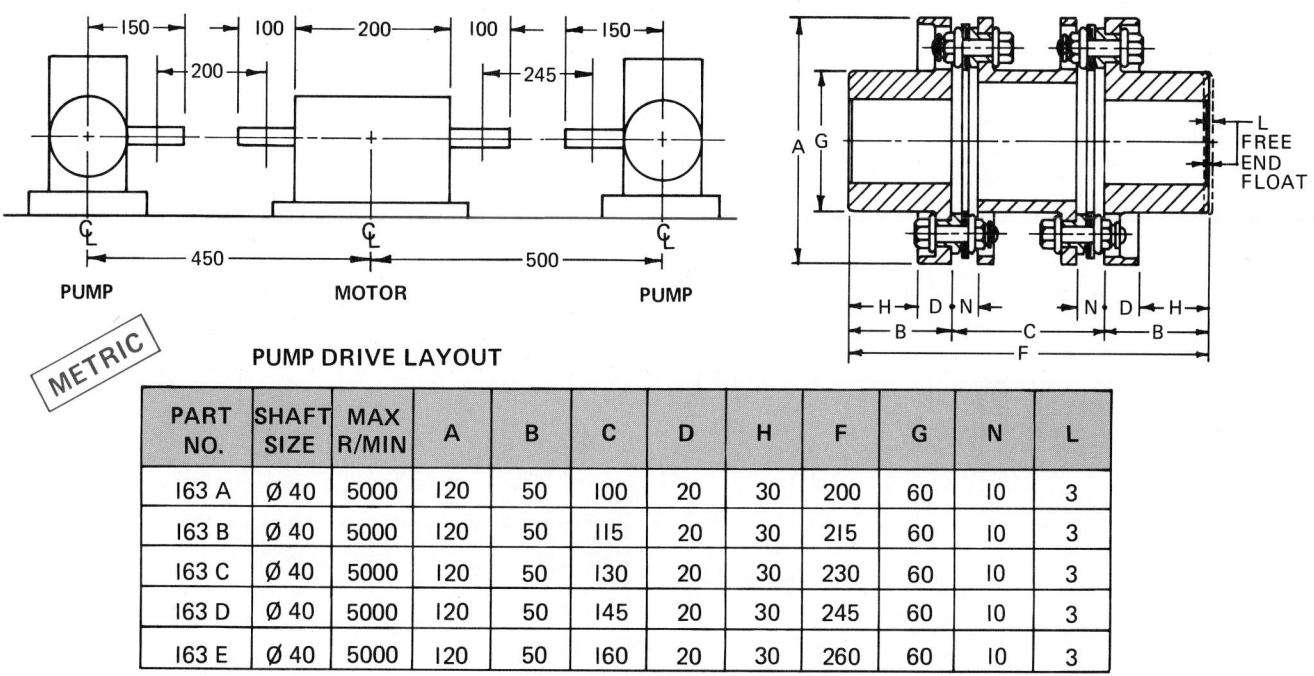

METRIC

PUMP DRIVE LAYOUT

PART NO.	SHAFT SIZE	MAX R/MIN	A	B	C	D	H	F	G	N	L
163 A	Ø 40	5000	120	50	100	20	30	200	60	10	3
163 B	Ø 40	5000	120	50	115	20	30	215	60	10	3
163 C	Ø 40	5000	120	50	130	20	30	230	60	10	3
163 D	Ø 40	5000	120	50	145	20	30	245	60	10	3
163 E	Ø 40	5000	120	50	160	20	30	260	60	10	3

COUPLING DATA

FIG. 21-1-B   Motor-to-pump drive.

ASSIGNMENTS FOR UNIT 21-2, BEARINGS

3. Complete the assembly drawings shown in Fig. 21-2-A or 21-2-B. The shaft, shown in the right view, is supported by two plain bearings press-fitted into the shaft support and lubricated by means of an oil fitting. Select suitable bolts and fasten the shaft support to the mounting plate.

   The shafts for the gearbox are supported by two plain bearings press-fitted into the gearbox. Setscrew collars, as shown in the Appendix, are mounted on the shafts to prevent lateral movement. Select suitable 16 DP (1.59 MDL), 20° spur gears from manufacturers' catalogs that revolve the smaller shaft four times as fast as the larger shaft. Lock the gears to the shafts using setscrews and flats on the shaft. Scale 1:1.

4. Complete the assembly drawings shown in Fig. 21-2-C or 21-2-D. For the assembly shown on the left, the largest portion of the shaft is positioned in the housing by a combination radial and thrust bearing. Also, complete the internal design of the housing in order to properly locate and support the bearing.

   The shaft shown in the assembly on the right rests on a thrust bearing. Complete the housing detail, and show the bearing in position.

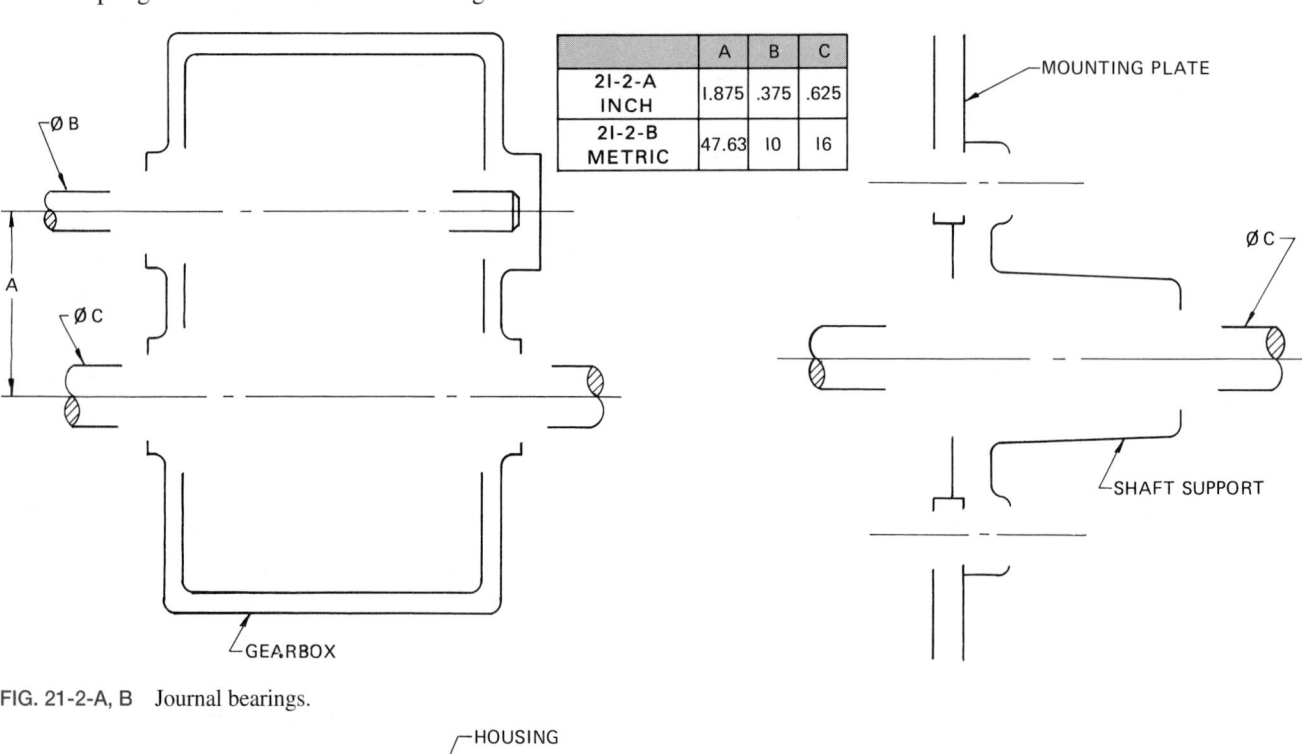

	A	B	C
21-2-A INCH	1.875	.375	.625
21-2-B METRIC	47.63	10	16

FIG. 21-2-A, B   Journal bearings.

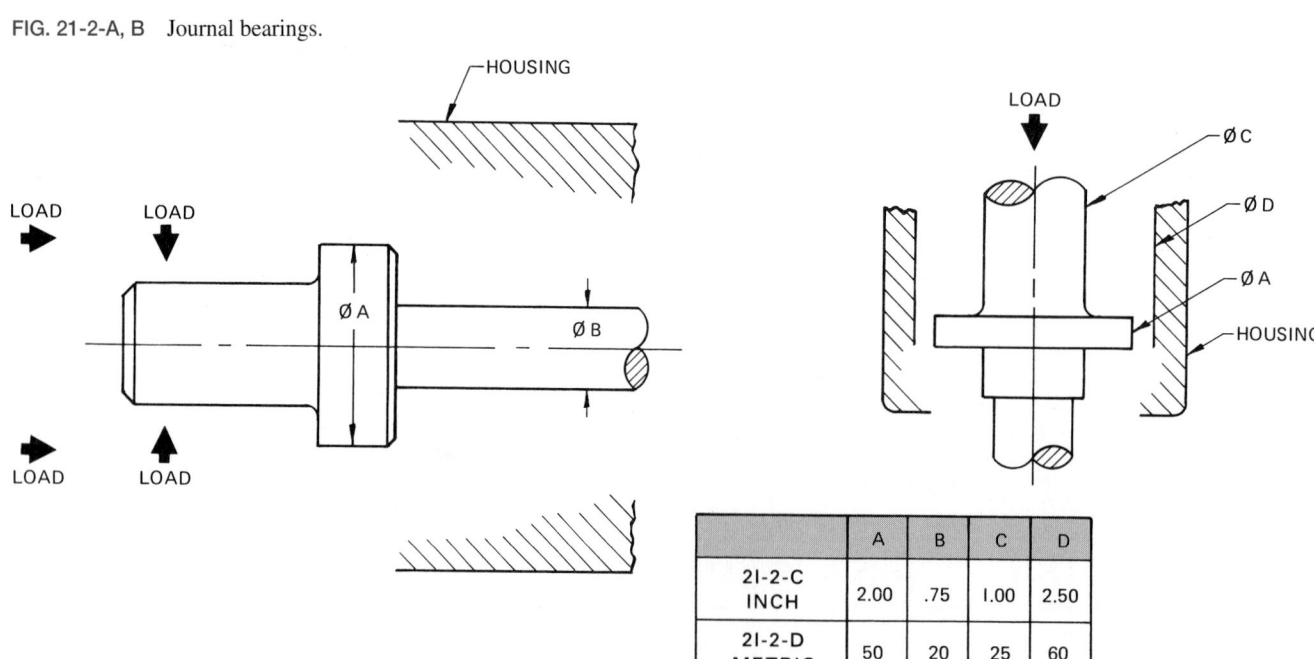

	A	B	C	D
21-2-C INCH	2.00	.75	1.00	2.50
21-2-D METRIC	50	20	25	60

FIG. 21-2-C, D   Thrust and journal bearings.

ASSIGNMENTS FOR UNIT 21-3, ANTIFRICTION BEARINGS

5. Complete the gearbox assembly drawing shown in Fig. 21-3-A or 21-3-B. Gears are mounted on shafts *A* and *B* and are positioned and held to the shafts by Woodruff keys and setscrews. The shafts are supported by radial ball bearings, which are positioned on the shafts with retaining rings. The bearings are to be positioned and held to the housing by internal shoulders on the castings and by cover plates bolted to the housing. Each shaft will have one floating and one fixed outer-ring mounting. The bearings will be purchased with seals on one side. From the information given, select suitable keys, bearings, retaining rings, and gears from the Appendix or manufacturers' catalogs. *Note:* Shaft *A* must be able to be removed from the housing with the gear in position. Scale 1:1.

6. Complete the gearbox assembly drawing shown in Fig. 21-3-C or 21-3-D. Gear 1 and the shaft are cast as a single unit. Gears 2, 3, 6, and 7 are fastened to their respective shafts by keys and are held in location by retaining rings. Gears 4, 5, and 8 are formed as one part which slides along the lay-shaft meshing with gear 3, 6, or 7. Retaining rings located at each end of this sliding-gear assembly locate it in the three positions, and a key locks the assembly to the shaft. Radial ball bearings are positioned at points *A* and *B* on each shaft. Each shaft will have one floating and one fixed outer-ring mounting. The gear end of the primary shaft must be designed to house bearing *A* of the main shaft. Refer to the manufacturers' catalogs or the Appendix for standard parts. Scale 1:1.

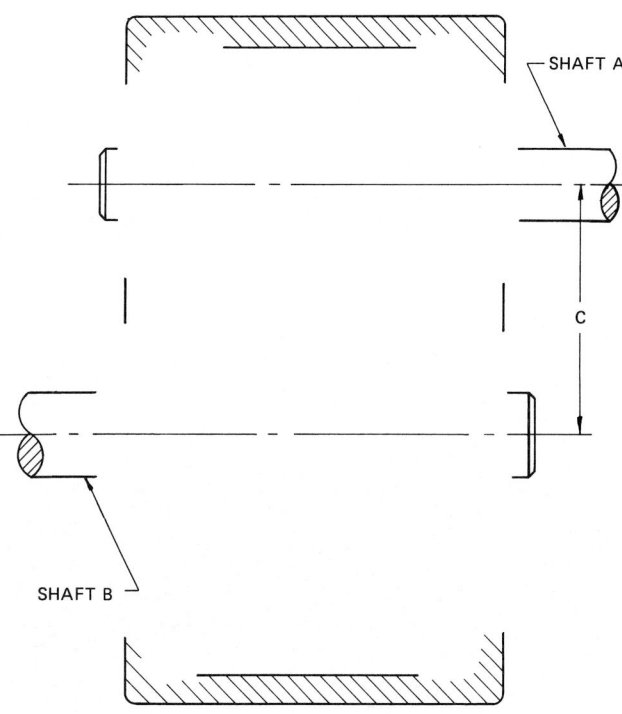

	Ø A	Ø B	C	GEAR DATA
21-3-A INCH	.750	1.000	3.000	DP 8 SHAFT A N=12 SHAFT B N=36
21-3-B METRIC	20	25	76.32	MODULE 3.18 SHAFT A N=12 SHAFT B N=36

FIG. 21-3-A, B   Ball bearings.

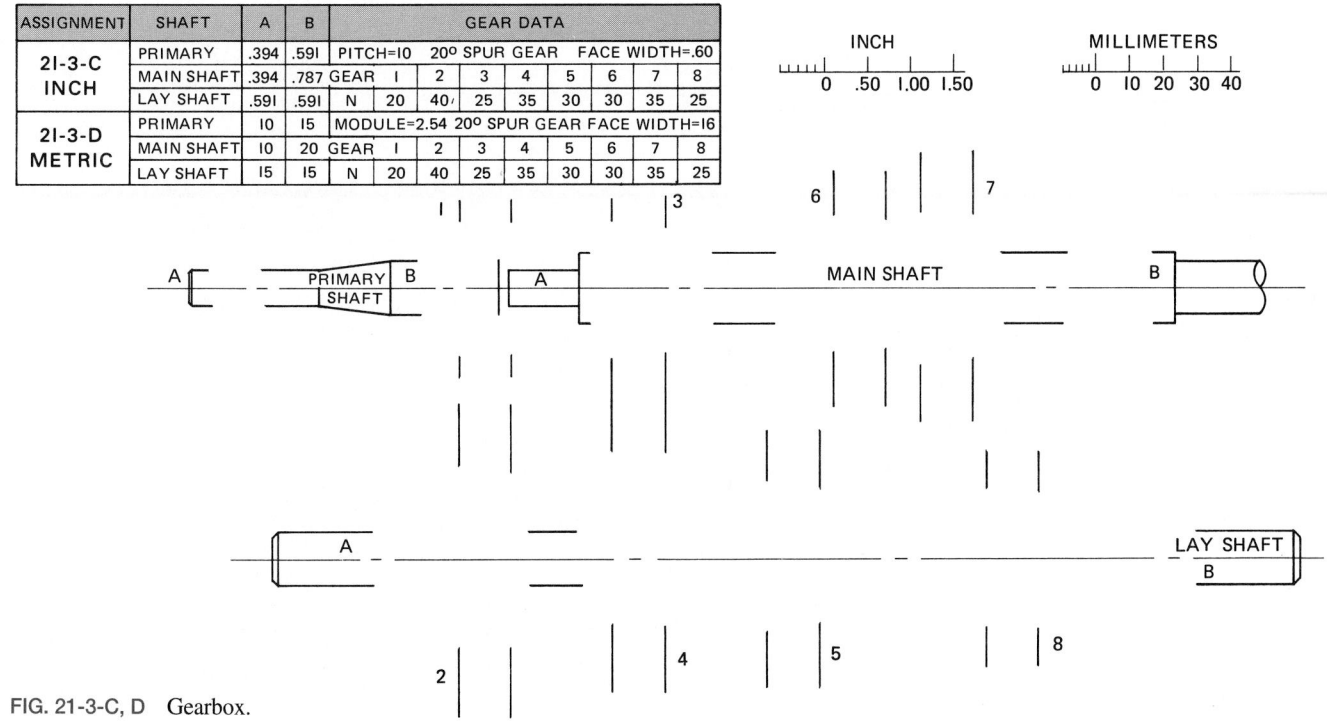

ASSIGNMENT	SHAFT	A	B	GEAR DATA								
21-3-C INCH	PRIMARY	.394	.591	PITCH=10	20º SPUR GEAR		FACE WIDTH=.60					
	MAIN SHAFT	.394	.787	GEAR	1	2	3	4	5	6	7	8
	LAY SHAFT	.591	.591	N	20	40	25	35	30	30	35	25
21-3-D METRIC	PRIMARY	10	15	MODULE=2.54	20º SPUR GEAR	FACE WIDTH=16						
	MAIN SHAFT	10	20	GEAR	1	2	3	4	5	6	7	8
	LAY SHAFT	15	15	N	20	40	25	35	30	30	35	25

FIG. 21-3-C, D   Gearbox.

7. Make a schematic representation of the bearings, gears, and belts, similar to Fig. 21-3-15, of one of the assemblies shown in Fig. 21-3-E or 21-3-F. Scale is to suit.

8. Prepare detail drawings of the bearing housing, belt pulley, detachable bushing shaft, and gear shown in Fig. 21-3-G. Include on your drawing an item list listing all of the parts.

FIG. 21-3-E   Lathe. *(Timken Roller Bearing Co.)*

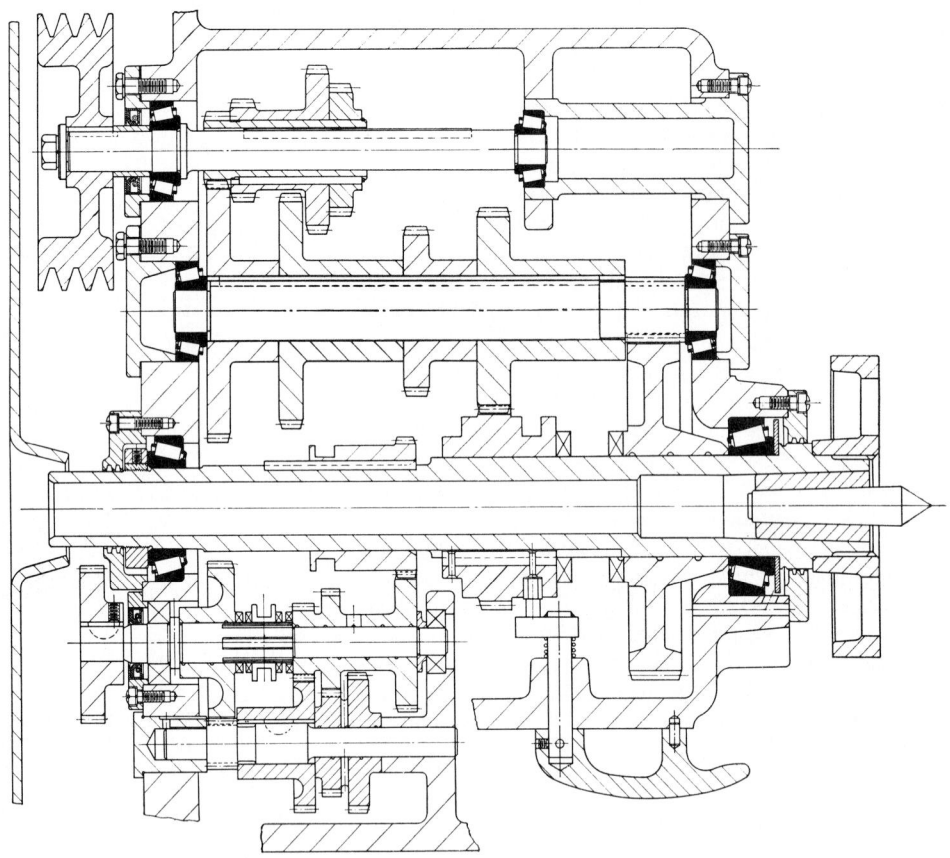

FIG. 21-3-F   Honing gearbox. *(Timken Roller Bearing Co.)*

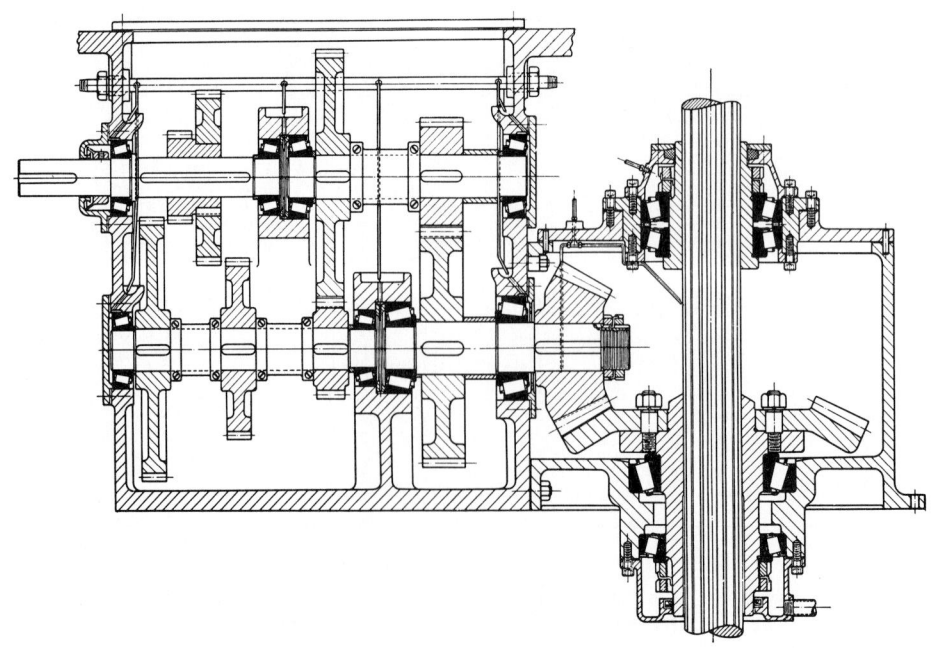

Sizes shown on the drawing are nominal. Select proper fit (shaft bases system) sizes using the tables in the Appendix.

9. The Ø1.000 in. shaft shown in Fig. 21-3-H is to be lubricated by means of an oil fitting positioned in the bracket and an oil groove in the journal bearing. The gears are locked to the shaft by square keys and locknuts. Gear data: Large gear: pressure angle = 20°, DP = 10, N = 48, Ø.875 shaft, face width = 1.00. Small gear: pressure angle = 20°, DP = 10, N = 24, Ø.750 shaft, face width = .90. Make a

one-view section assembly drawing of your own design from the information given. *Note:* Dimensions shown are nominal. Fits and clearances are to be selected from the Appendix.

10. This is the same as Assignment 9 except use retaining rings to hold the gears on the shaft and replace the one journal bearing with two standard journal bearings made of oil-impregnated material.

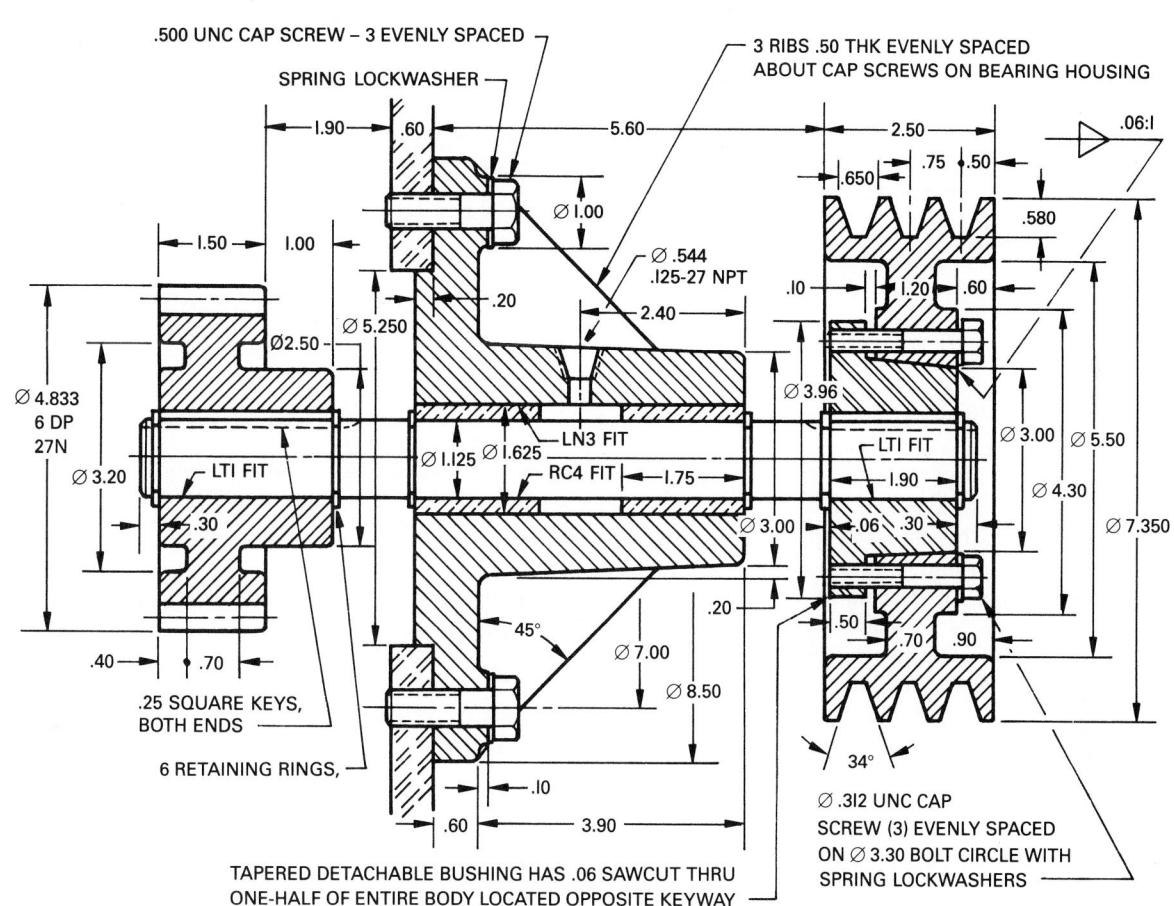

FIG. 21-3-G   Drive assembly.

FIG. 21-3-H   Gear drive assembly.

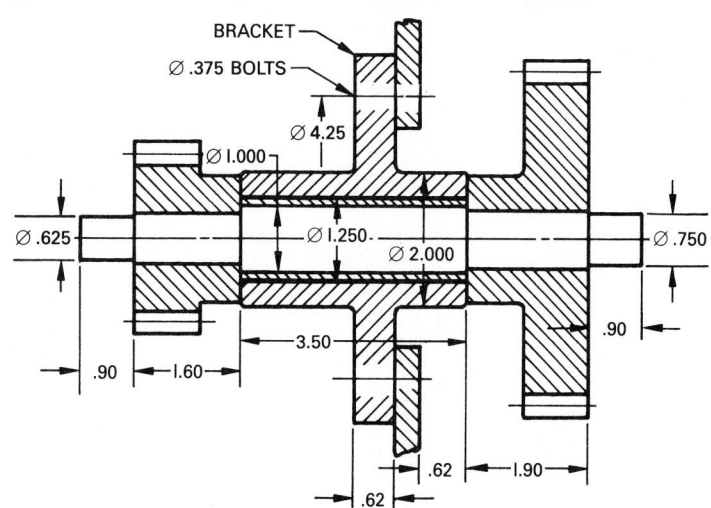

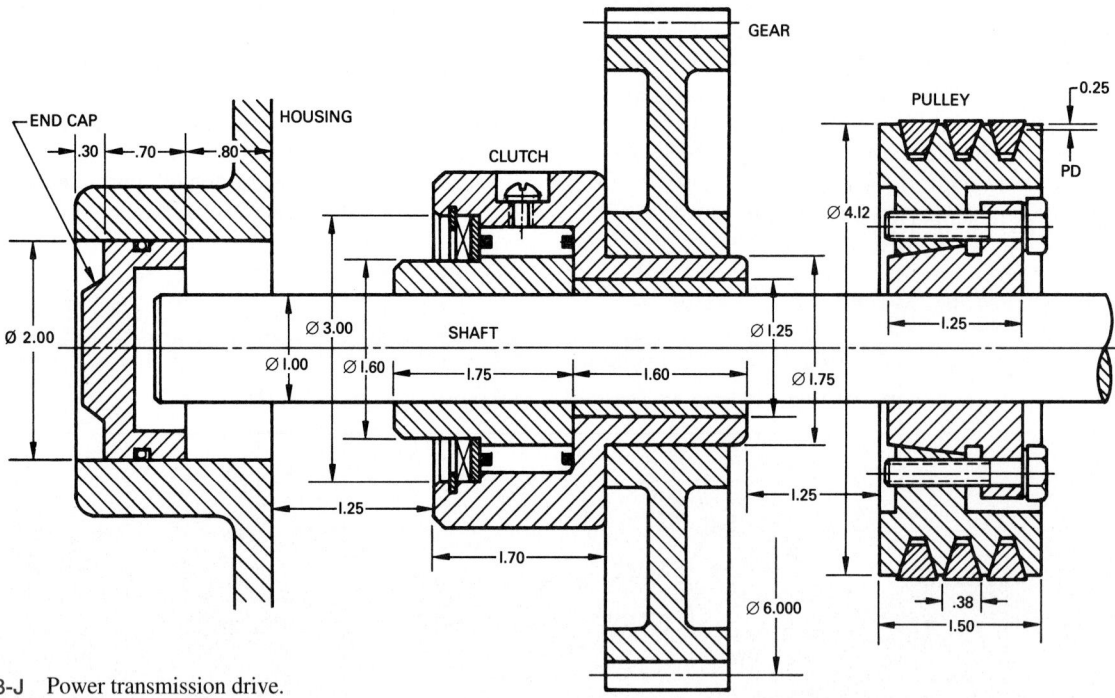

FIG. 21-3-J   Power transmission drive.

11. Design the power transmission drive shown in Fig. 21-3-J given the following information:

- The shaft is held by an SKF #6005 bearing (see Appendix) located in the housing. The end cap and a retaining ring hold the bearing in the housing. Two retaining rings hold the bearing in position on the shaft.
- The end cap is held in the housing by a retaining ring.
- The gear is held to the clutch by a key and retaining ring. A key locked in position by a setscrew holds the clutch in position on the shaft.
- The belt pulley assembly is held to the shaft by retaining rings and a square key.
- Dimensions shown are nominal sizes. Select proper fits using the tables in the Appendix.

ASSIGNMENTS FOR UNIT 21-4, PREMOUNTED BEARINGS

12. Make a one-view assembly drawing of the adjustable shaft support shown in Fig. 21-4-A. Show the bearing housing in its lowest position and a phantom outline of the bearing housing in its top position. Show only those dimensions that would be used for catalog purposes. Scale 1:1.

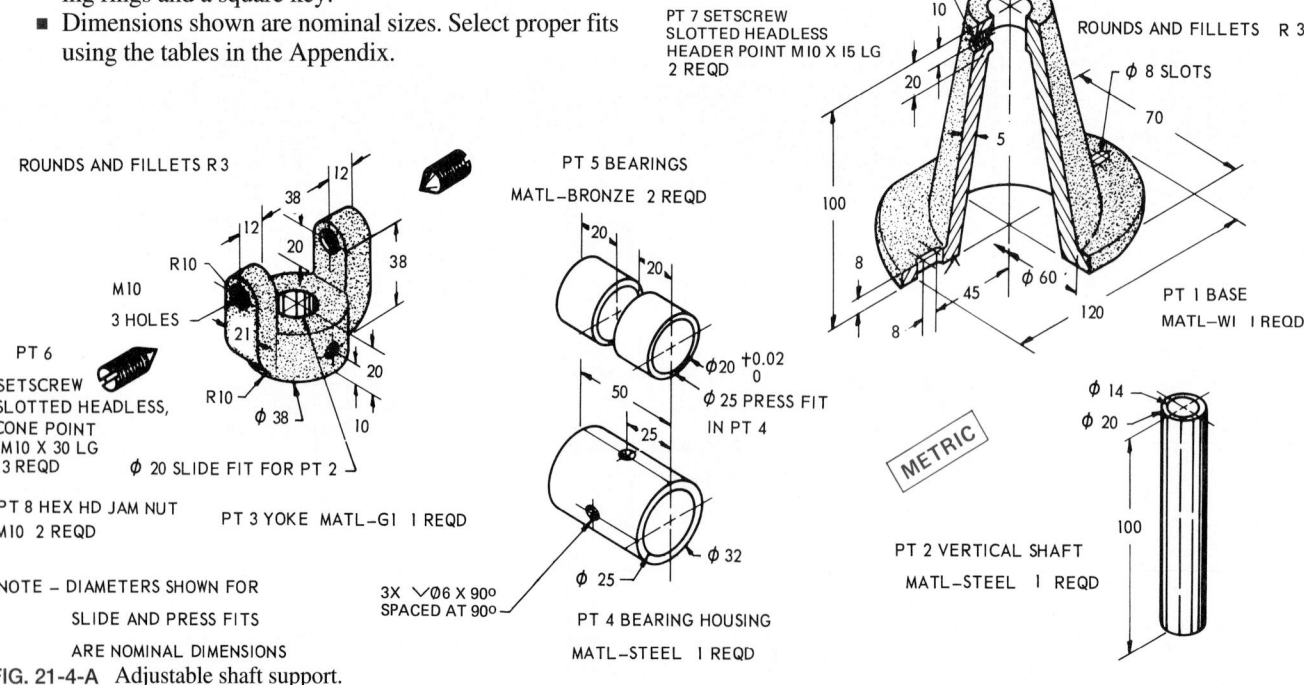

FIG. 21-4-A   Adjustable shaft support.

13. Make a two-view (front- and side-view) assembly drawing of the adjustable shaft support shown in Fig. 21-4-B.
Draw the front view in full section. Include on your drawing an item list. Scale 1:1.

ASSIGNMENTS FOR UNIT 21-5, LUBRICANTS AND RADIAL SEALS

14. Complete the two assemblies shown in Fig. 21-5-A or
21-5-B given the following information.
Radial Oil Seal Assembly.   The inner ring of the tapered
roller bearing is held laterally on the shaft by the shaft

shoulder. A cover plate which is bolted to the housing by
four socket-head cap screws has a stepped shoulder the
same diameter as that of the outside diameter of the bearing. This shoulder serves two purposes: It locks the outer
ring of the bearing in position, and it locates the cover plate
radially on the shaft. The cover plate has a recess to accommodate a metal-cased radial seal. The shaft diameter for the
oil seal should be slightly smaller than the diameter of
the shaft for the bearing. The outside face of the cover plate
and the oil seal should be flush. Scale 1:1.
Oil Ring Seal Assembly.   A magnetized ring press-fitted
into the housing firmly holds the mating ring on the shaft
element by magnetic force. The carbon ring in the face of

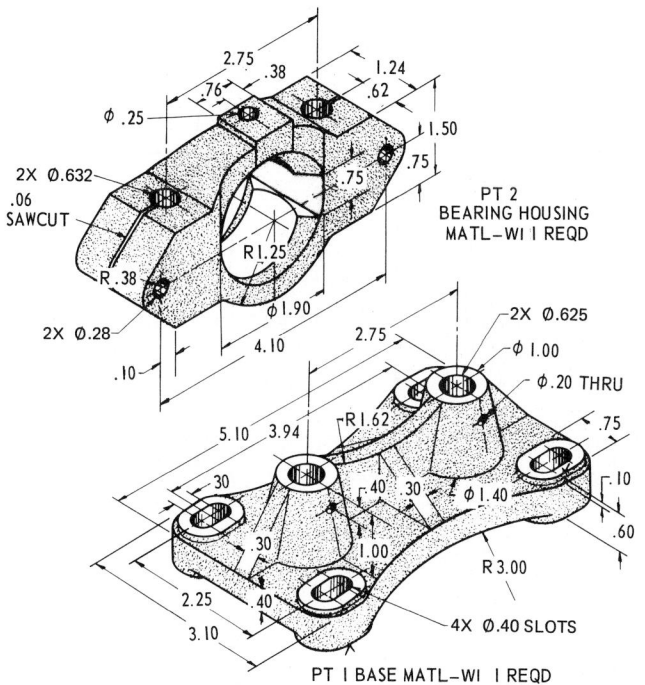

FIG. 21-4-B   Adjustable shaft support.

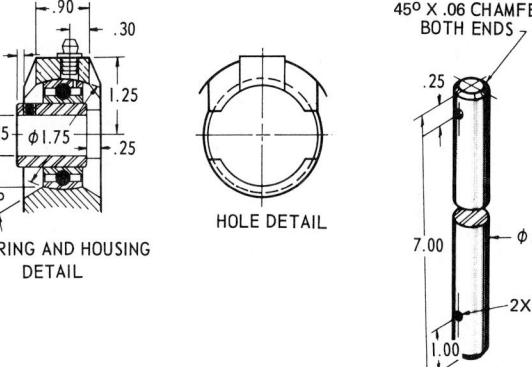

PT 4 BEARING – X 40I2
PT 5 BASE PIN MATL – .18 DRILL ROD X 1.25 LG, 2 REQD
PT 6 OIL CAP – XGF – D5
PT 7 BOLT, HEX HD REGULAR – .25 – 20 UNC X 1.62 LG, 2 REQD
PT 8 NUT, HD REGULAR – .25 – 20 UNC, 2 REQD

	A	B	C	D
2I–5–A INCH	.75	1.25	1.50	.75
2I–5–B METRIC	20	30	38	19

RADIAL OIL SEAL ASSEMBLY

OIL RING SEAL ASSEMBLY

FIG. 21-5-A, B   Oil seals.

the mating ring, in balanced contact with the lapped surface of the magnet, forms a permanent, self-adjusting face seal. O-rings in both the mating ring and the magnetic shaft element (between the element and housing) prevent leakage of confined fluids. Scale 2:1.

15. Complete the hydraulic cylinder assembly shown in Fig. 21-5-C. Add an item list calling out the standard sealing and fastener parts. All dimensions shown are nominal. Select all clearances, types of fits, O-rings, retaining rings, packings, seals, etc. The sealing and fastener requirements

are: at (A) bronze bushing with felt ring packing; at (B) two grooves on piston surface to accommodate split-ring seals; at (C) squeeze packing (O-ring) held in groove on cylinder head; at (D) three internal retaining rings held in one groove in housing to hold cylinder against step in housing. Scale 1:1.

16. Prepare detail drawings of the parts for the completed assembly in Assignment 15. Scale 1:1.

17. Prepare detail drawings for the parts shown in Fig. 21-5-D. Include on the drawing an item list. Scale 1:1.

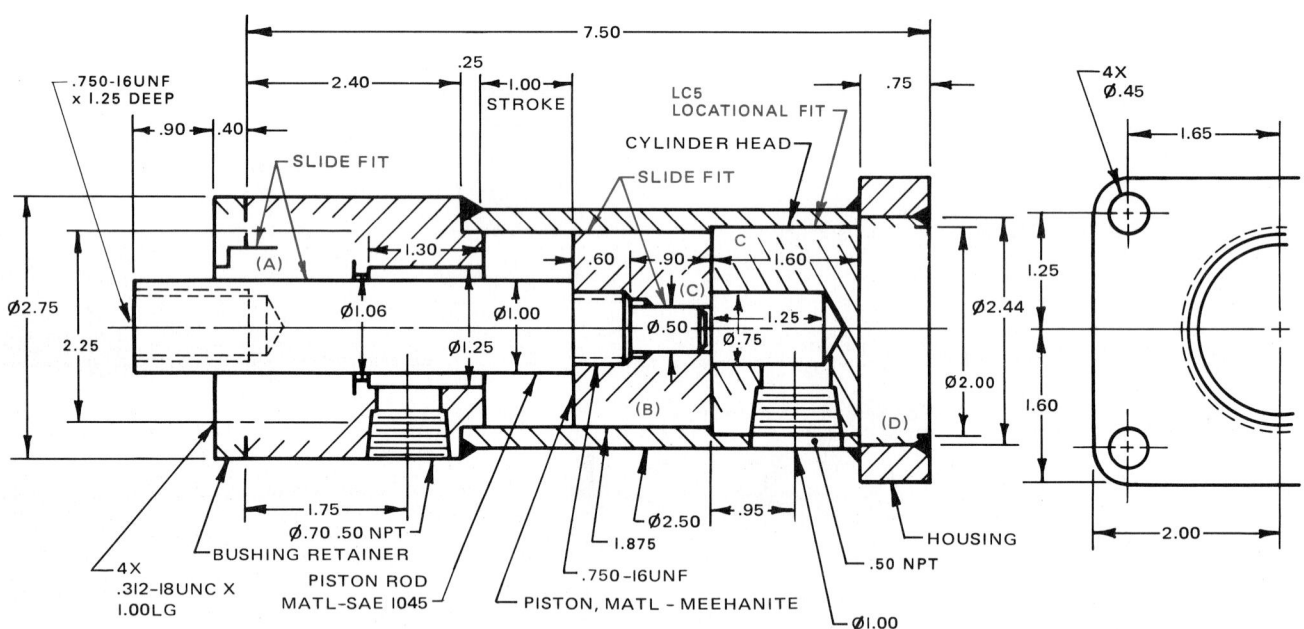

FIG. 21-5-C   Hydraulic cylinder.

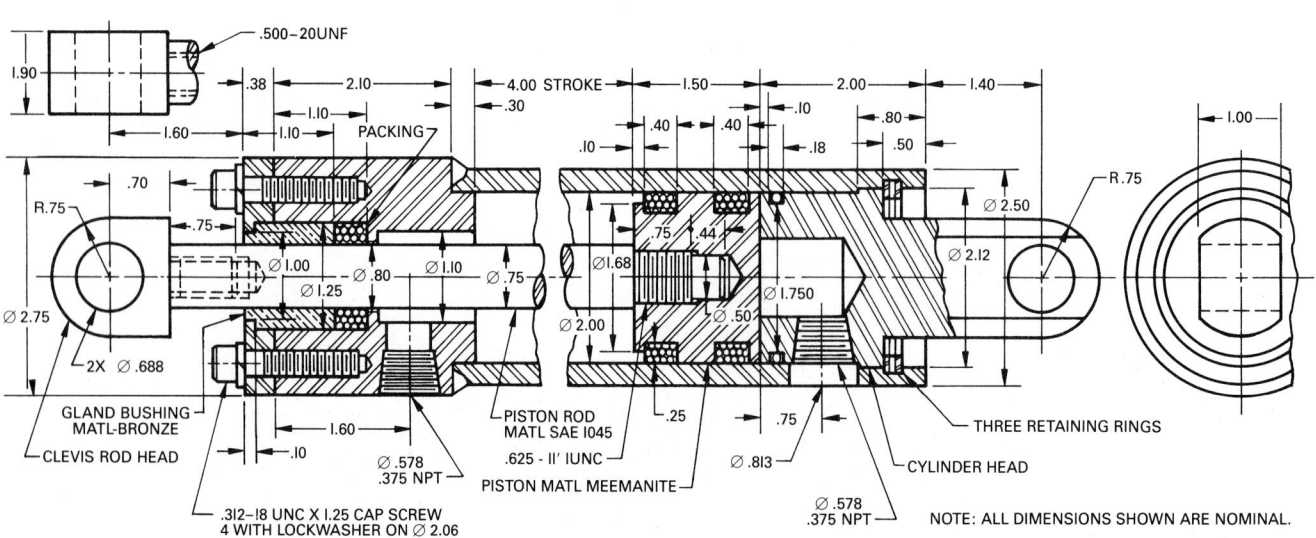

FIG. 21-5-D   Hydraulic cylinder.

ASSIGNMENTS FOR UNIT 21-6, STATIC SEALS AND SEALANTS

18. Complete the two assemblies shown in Fig. 21-6-A given the following information.
Flanged Pipe Coupling. The flanges are fastened by hex bolts, nuts, and lockwashers. Alignment is accomplished by a tongue-and-groove joint, similar to that

shown in Fig. 21-6-2C, and a gasket positioned in the groove provides the seal.
Cylinder Head Cap. The cylinder head cap is fastened to the cylinder head by four hex-head cap screws equally spaced. Locking is accomplished by lockwashers. Spotfacing on the cast head cap is required because of the rough finish of the casting. An O-ring provides the seal.

FIG. 21-6-A Flanged pipe coupling and cylinder head cap.

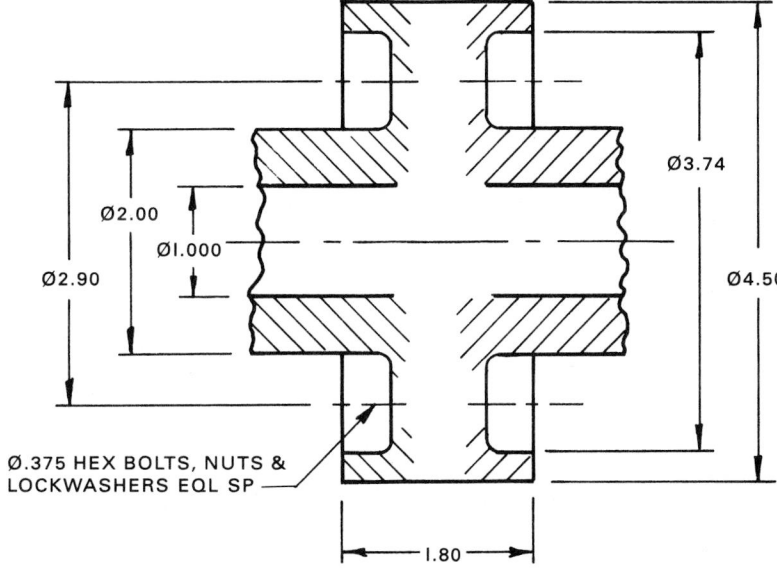

FLANGED PIPE COUPLING

Ø.375 HEX BOLTS, NUTS & LOCKWASHERS EQL SP

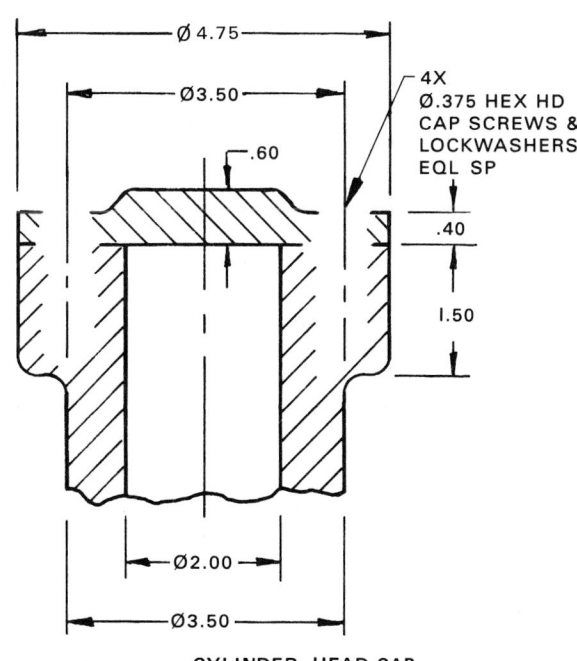

CYLINDER - HEAD CAP

19. Make a detail drawing of the gasket used with the domed
    cover shown in Fig. 21-6-B. Material is neoprene. Scale 1:2.

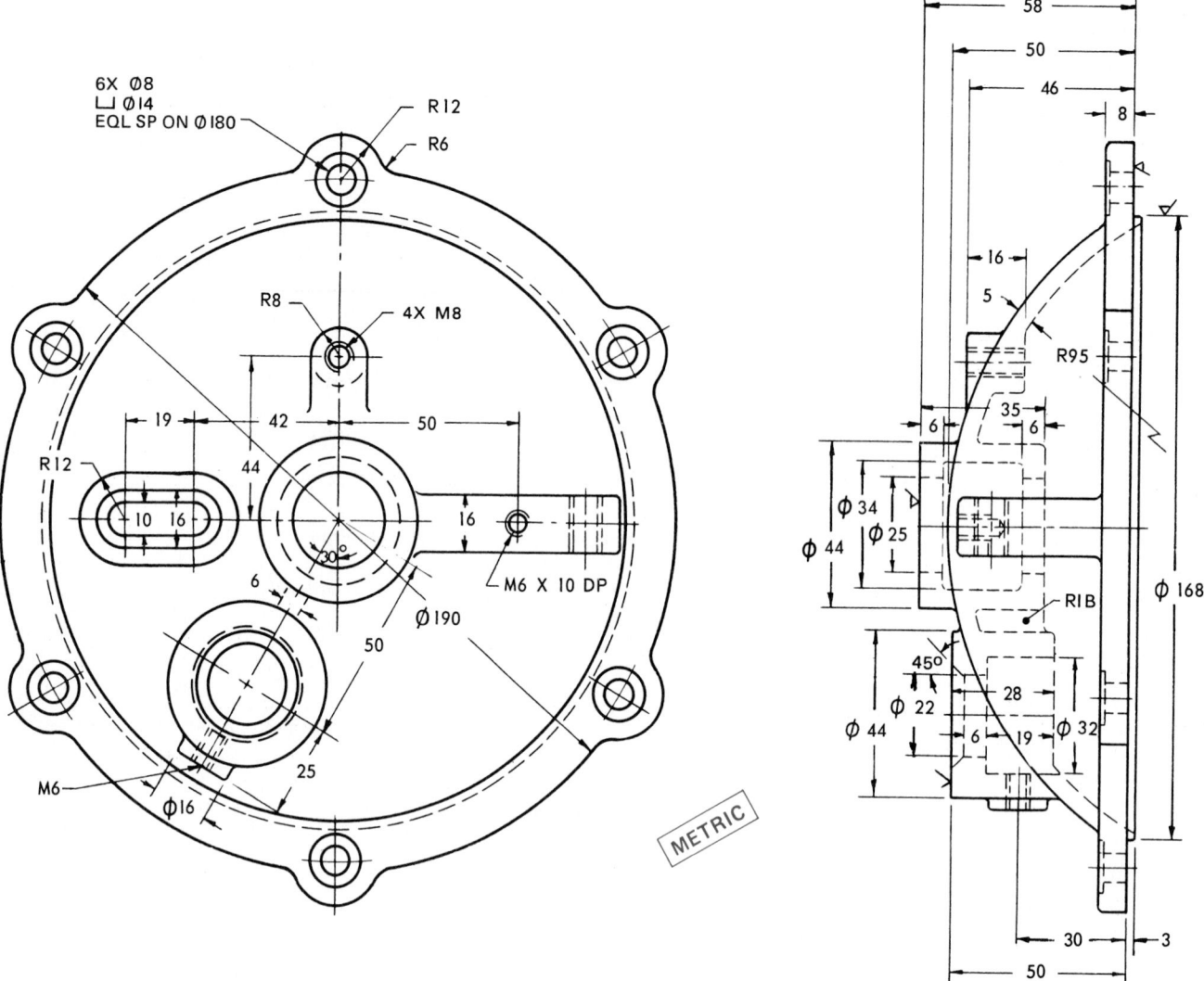

FIG. 21-6-B   Domed cover.

CHAPTER 22

CAMS, LINKAGES, AND ACTUATORS

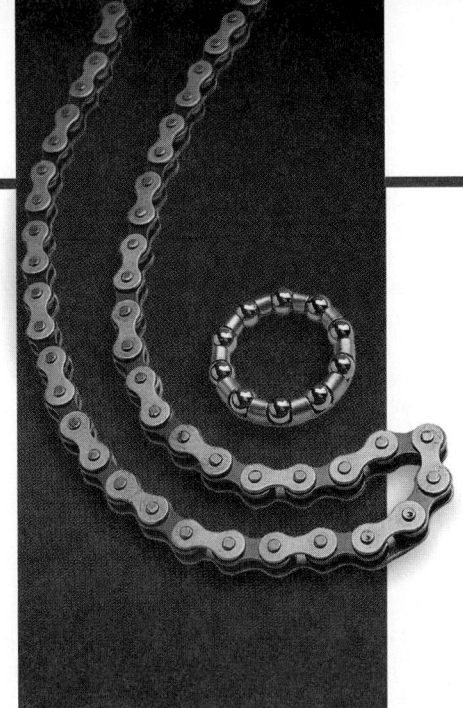

Definitions

Angular displacement   Displacement measured in degrees from a zero reference.

Bar linkages   Combinations of the crank, link, and sliding elements.

Cam   A machine element designed to generate a desired motion in a follower by means of direct contact.

Face cams   Cams in which the follower engages a groove on the face of the cam.

Harmonic motion or **crank motion**   Motion produced by a true eccentric cam operating against a flat follower whose surface is normal to the direction of linear displacement.

Indexing   The conversion of a constant-speed rotary-input motion to an intermittent rotary-output motion.

Linkage   Computing device made up of straight members joined together.

Linkage layout   The assignment of values to the various parameters in a linkage.

Locus   The path traced by a point as it moves in a linkage or mechanism according to certain controlled conditions.

Modified uniform motion   Straight-line motion that has been modified through the use of a radius to reduce the shock on the follower.

Number of stops   The number of times that the turret indexes from one station to the next and dwells for a specific period in one revolution.

Radial displacement   The distance from the cam center or as a displacement from the prime circle.

Straight-line mechanism   A linkage device used to guide a given point in an approximate straight line.

Straight-line motion   A type of motion used in connection with screw machines to control the feed of a cutting tool.

Uniform motion   A type of motion used when the follower is required to rise and drop at a uniform rate of speed.

Yoke-type followers   Followers in which the surface is flat or tangent to the curvature of the cam.

22-1 CAMS AND CAM MOTIONS

A *cam* is a machine element designed to generate a desired motion in a follower by means of direct contact. Cams are generally mounted on rotating shafts, although they can be used so that they remain stationary and the follower moves about them. Cams may also produce oscillating motion, or they may convert motions from one form to another.

The shape of a cam is always determined by the motion of the follower. The cam is actually the end product of a desired follower movement. From the standpoint of engineering alone, cams have many decided advantages over the fundamental kinematic four-bar linkages (Fig. 22-1-1). Once they are understood, cams are easier to design and the action produced by them can be more accurately forecast. For example, to cause the follower system to remain stationary during a portion of a

FIG. 22-1-1   Cam application. *(Manifold Machinery Co. Ltd.)*

cycle is very difficult when linkages are used. With a cam this is accomplished by a contour surface that runs concentric with the center of rotation. To produce a given motion, velocity, or acceleration during a specific portion of a cycle is very difficult to do with linkages, whereas it is comparatively easy with standard cam motions, especially when the design is achieved with the aid of a computer (Fig. 22-1-2).

By far the most popular types of cams are the OD or plate cam and the drum or cylinder cam. In the case of the OD cam, the body of the cam is usually shaped like a disk with the cam contour developed along its circumference. With these cams the line of action of a follower is usually perpendicular to the cam axis. With the drum cam, the cam track is normally machined around the circumference of the drum. In this type of cam the line of action is usually parallel to the cam axis. The level-winding mechanism on a fishing reel is an example of a drum cam. Other popular types of cams include the conjugate cam (multiple cams joined together); the face cam, in which the cam track is cut into the face of the disk; and the index cam, which is similar to a drum cam except that the motion of the follower passes in an arc over the cam itself.

As machine speeds increase, the need for properly designed quality cams becomes more evident. The essential specifications necessary to produce a cam of optimum quality are:

1. Proper dynamic design that considers the velocity, acceleration, and jerk characteristics of the follower system. These include vibration and shaft torque analysis.
2. Proper material selection that takes into account cost, wear, and surface stresses produced by the system.

Cam Nomenclature

Figures 22-1-3 and 22-1-4 illustrate the terminology associated with cams.

1. *Follower displacement* is usually defined as the position of the follower mechanism from a specific zero or rest position in relation to time or some fraction of the machine cycle (cam displacement) measured in degrees, inches, or millimeters.
2. *Cam displacement,* measured in degrees, inches or millimeters, is the cam motion measured from a specific zero or rest position and relates to the follower mechanism as defined above.
3. *Cam profile* is the actual working surface contour of the cam.
4. *Base circle* is the smallest circle drawn on the cam profile.
5. *Trace point* is the center line of the follower roller or its equivalent. When a flat follower is used, the cam profile is the envelope of successive positions of the flat follower.
6. *Pitch curve* is the locus of successive positions of the trace point as cam displacement takes place.
7. *Prime circle* is the smallest circle drawn on the pitch curve from the cam center. It is related to the base circle by the roller radius.
8. *Pressure angle* is the angle between the normal to the pitch curve and the instantaneous direction of motion of the follower.

(A) OD OR PLATE CAM

(B) BARREL (DRUM OR CYLINDER) CAM

(C) CONJUGATE CAM

(D) FACE CAM

(E) COMBINATION DRUM AND PLATE CAM

(F) INDEX CAM

FIG. 22-1-2   Common types of cams.

FIG. 22-1-3 Cam nomenclature.

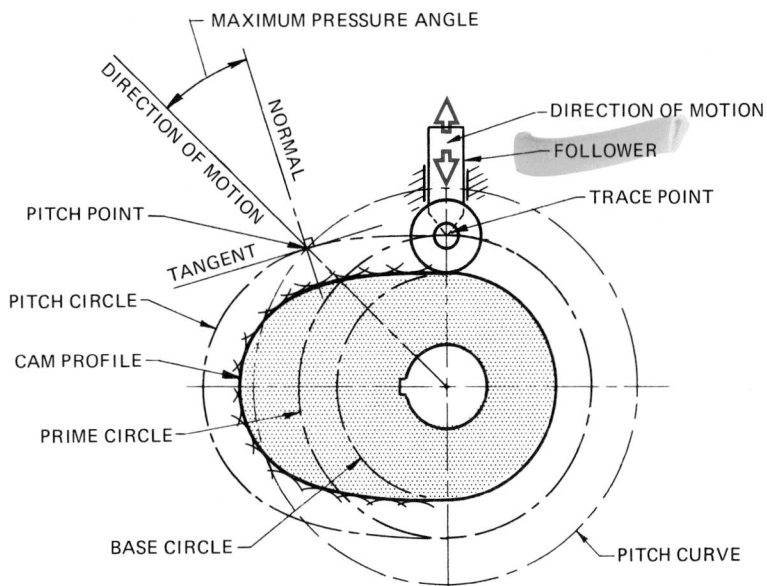

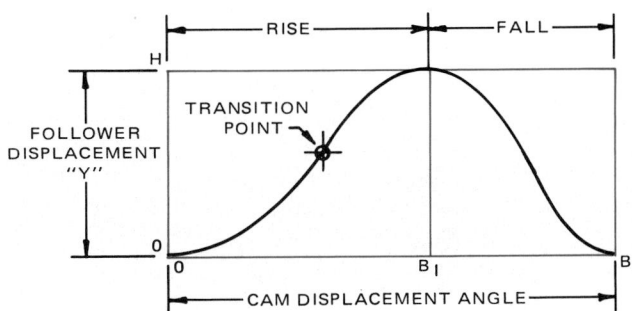

FIG. 22-1-4 Cam displacement diagram.

9. *Pitch point* is the position on the pitch curve where the pressure angle is maximum.
10. *Pitch circle* is the circle which passes through the pitch point.
11. *Transition point* is the position of maximum velocity where acceleration changes from plus to minus (force on follower changes direction). In a closed cam this is sometimes referred to as the *crossover point,* where, because of the reversing acceleration, the follower roller leaves one cam profile and crosses over to the opposite (or conjugate) one.

Figure 22-1-5 (pg. 750) illustrates typical cam and follower combinations used in machine design.

Cam Followers

The common types of cam followers are shown in Fig. 22-1-6 (pg. 750). The roller follower is more suitable where high speeds, heat (friction), and wear are factors (Fig. 22-1-7, pg. 751).

Cam Motions

In the early phases of the development of the cam mechanism, it is customary to work with only center lines to establish the desired motions. It is obvious that some data have been specified or determined from related parts of the design to establish

the cam and linkage requirements and to provide base points from which to start the cam linkage design. These data will usually be the motion requirements and timing relationships of a particular part of the machine, such as a feed slide, a folding mechanism, or a label applicator.

The choice of motion that the cam must produce will depend, first, on the cycle timing and, second, on the system or machine dynamics. For the purpose of showing cam layout techniques, cams producing the following motions will be discussed:

1. Uniform motion
2. Parabolic motion
3. Harmonic motion
4. Cycloidal motion
5. Modified trapezoidal motion
6. Modified sine motion
7. Synthesized, modified sine-harmonic motion

The first four are illustrated in Fig. 22-1-8 (pg. 751).

Uniform Motion (Constant-Velocity Motion)

Uniform motion is used when the follower is required to rise and drop at a uniform rate of speed. If a follower is to rise 1.50 in. in one-half of a revolution, or 180° of the cam, then for every 30° of cam rotation the follower would rise one-sixth of 1.50 in., or .25 in. This curve is referred to as a *straight-line motion,* and it is most commonly used in connection with screw machines to control the feed of a cutting tool. If it were used with a *dwell area* (where no rise or drop in the follower occurs) in a cam, there would be a jerk at the start and stop of the motion.

Since this kind of motion starts and ends abruptly, it is often modified slightly to reduce the shock on the follower. A radius is used at the beginning and end of the motion, and a line tangent to these arcs is drawn. The size of the radius varies between one-third and full-rise height depending on how sharp the rise is. This motion is known as *modified uniform motion.* Since this type of motion is not desirable for high speeds, motions that start and end slowly and reach their maximum speed in the center are used.

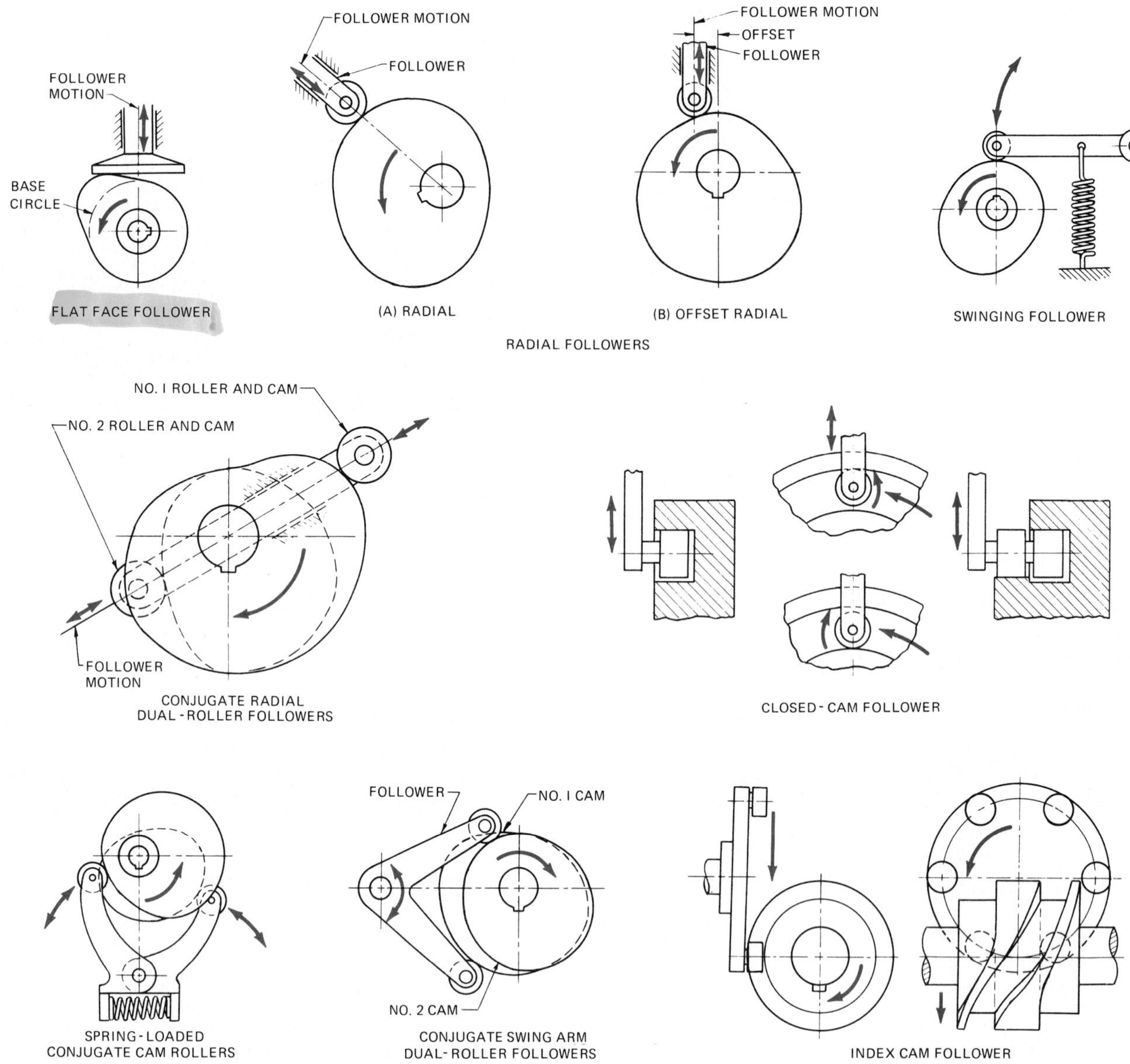

FOLLOWER
MOTION

BASE
CIRCLE

FLAT FACE FOLLOWER

FOLLOWER MOTION

FOLLOWER

(A) RADIAL

FOLLOWER MOTION
OFFSET
FOLLOWER

(B) OFFSET RADIAL

SWINGING FOLLOWER

RADIAL FOLLOWERS

NO. I ROLLER AND CAM

NO. 2 ROLLER AND CAM

FOLLOWER
MOTION

CONJUGATE RADIAL
DUAL-ROLLER FOLLOWERS

CLOSED-CAM FOLLOWER

SPRING-LOADED
CONJUGATE CAM ROLLERS

FOLLOWER

NO. I CAM

NO. 2 CAM

CONJUGATE SWING ARM
DUAL-ROLLER FOLLOWERS

INDEX CAM FOLLOWER

FIG. 22-1-5 Types of cam followers.

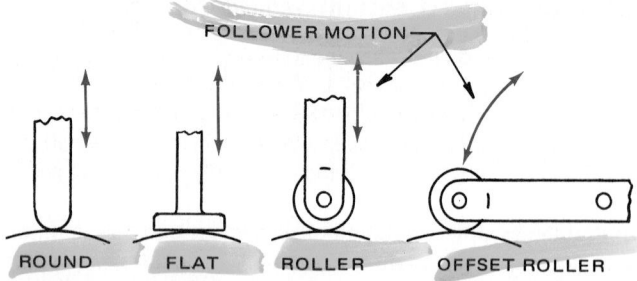

FOLLOWER MOTION

ROUND   FLAT   ROLLER   OFFSET ROLLER

FIG. 22-1-6 Cam roller followers.

Parabolic Motion

Parabolic motion, commonly referred to as *uniformly acceler-ated and retarded motion,* or *constant acceleration,* is described by a curve found by combining the cycloid and the constant-acceleration curve. The construction of the parabolic curve as shown in Fig. 22-1-8C is found in the same manner as detailed in Fig. 5-5-2.

When the uniformly accelerated and retarded method of construction is used for this motion, the divisions will increase and decrease by a ratio of 1:3:5:5:3:1. For instance, a follower is to rise 2.25 in. in 180°. Plotting points every 30° and using six proportional divisions of 1:3:5:5:3:1, we find in the first 30° the follower rises one-eighteenth of the total rise of 2.25 in., or .125 in.; in the next 30° the follower rises three-eighteenths of the rise of 2.25 in., or .375 in., and in the third 30° the follower rises five-eighteenths of the rise of 2.25 in., or .625 in.; the fourth, fifth, and last rises are .625, .375, and .125 in., respec-tively. This motion would produce a jerk if used in connection with a cam having a dwell.

Harmonic Motion

Harmonic motion, often referred to as *crank motion,* is pro-duced by a true eccentric operating against a flat follower whose surface is normal to the direction of linear displacement.

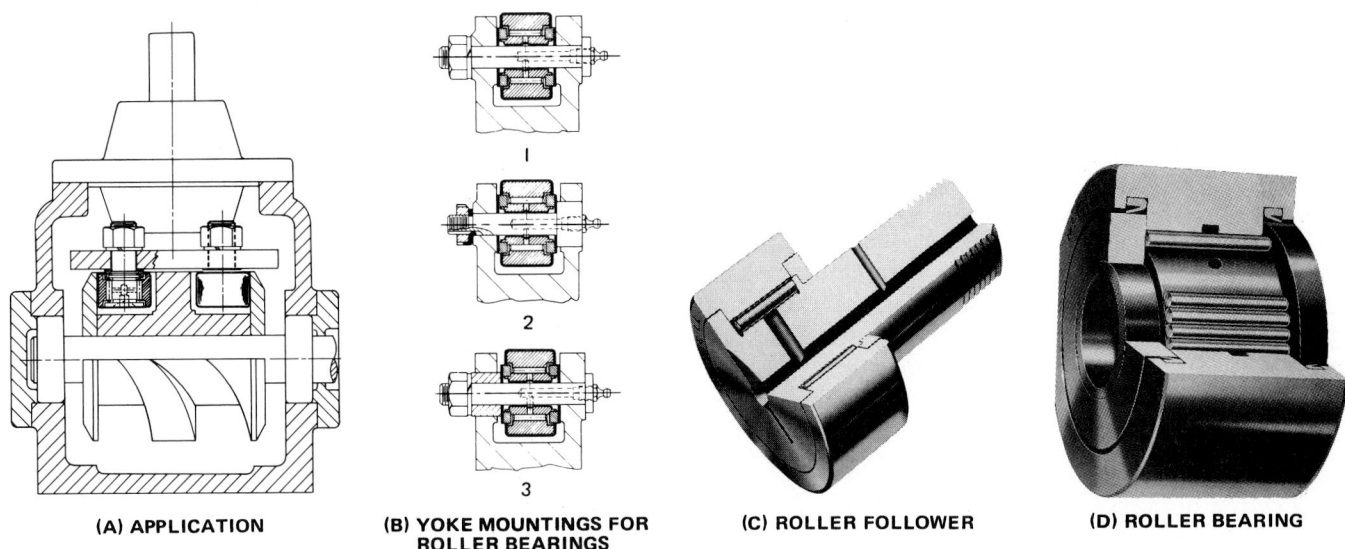

(A) APPLICATION

(B) YOKE MOUNTINGS FOR ROLLER BEARINGS

(C) ROLLER FOLLOWER

(D) ROLLER BEARING

FIG. 22-1-7   Typical cam and follower combination.

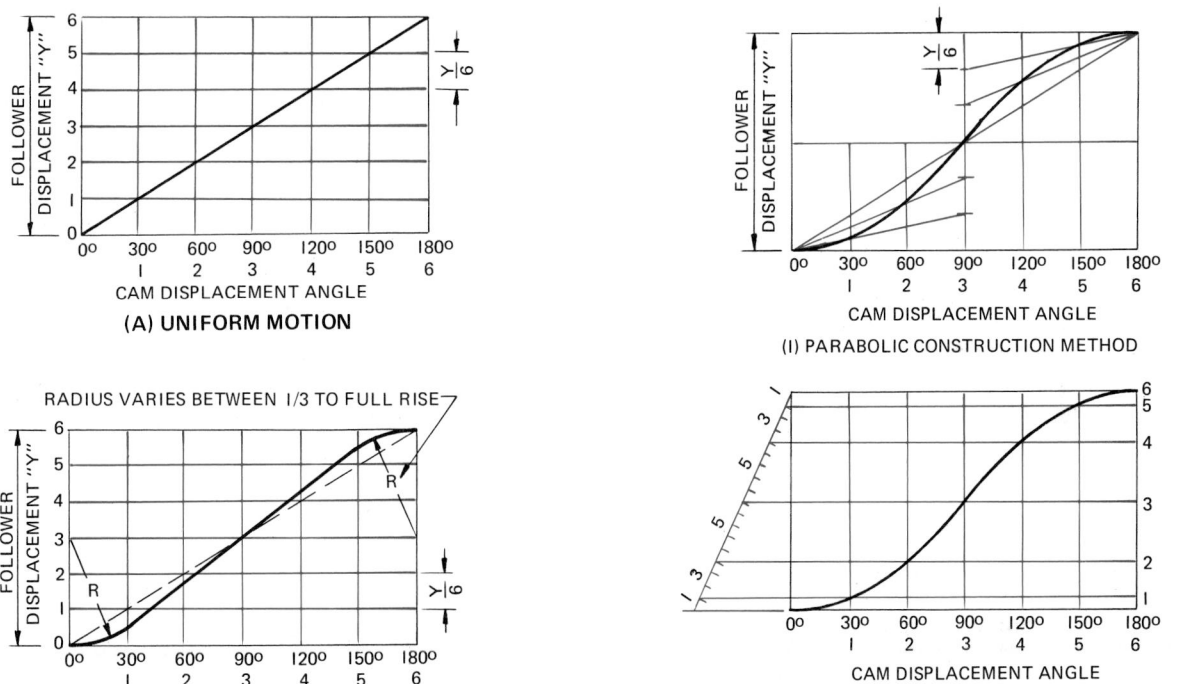

(A) UNIFORM MOTION

(I) PARABOLIC CONSTRUCTION METHOD

RADIUS VARIES BETWEEN 1/3 TO FULL RISE

(B) MODIFIED UNIFORM MOTION

(2) UNIFORMLY ACCELERATED AND RETARDED METHOD

(D) PARABOLIC MOTION

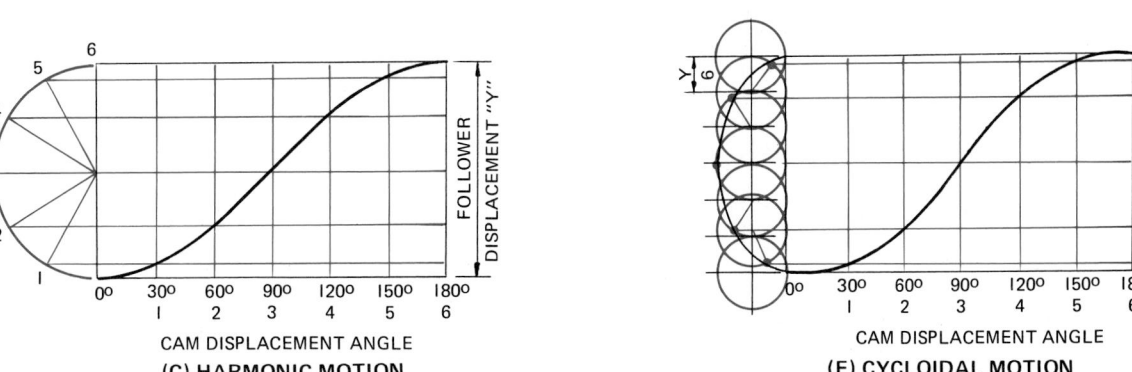

(C) HARMONIC MOTION

(E) CYCLOIDAL MOTION

FIG. 22-1-8   Cam motions.

Figure 22-1-9 illustrates this type of cam. However, it is more frequently necessary to produce a simple harmonic displacement with less than 360° of rotation of the cam, as illustrated in Fig. 22-1-10, and the ordinates for the cam pitch curve can then be determined as shown in Fig. 22-1-8C. It may be impossible to use a flat follower since the harmonic pitch curve usually has a reentrant, or reversing, curve and a flat follower would just "bridge" the hollow part. Since a roller follower is the most practical and reliable type, the development of the cam profile with this type of follower is shown. This motion would also produce a jerk if used in connection with a cam having a dwell.

To illustrate the effect of cam displacement for a given cam size and follower displacement on the pressure angle, the return, or fall, curve has been shown with a much larger angle.

Note that the maximum pressure angle has been considerably reduced.

Cycloidal Motion

Figure 21-1-8E illustrates the graphic method of laying out a cycloidal profile using a rolling circle, as shown on the left end of the illustration. The cycloidal curve, when generated accurately and used in a cam having a dwell, produces a very smooth, jerk-free motion. This curve is best suited for light loading at high speeds.

Modified Trapezoidal Motion

The modified trapezoid is made by combining the cycloid and the constant-acceleration curve. Manufacturing accuracy

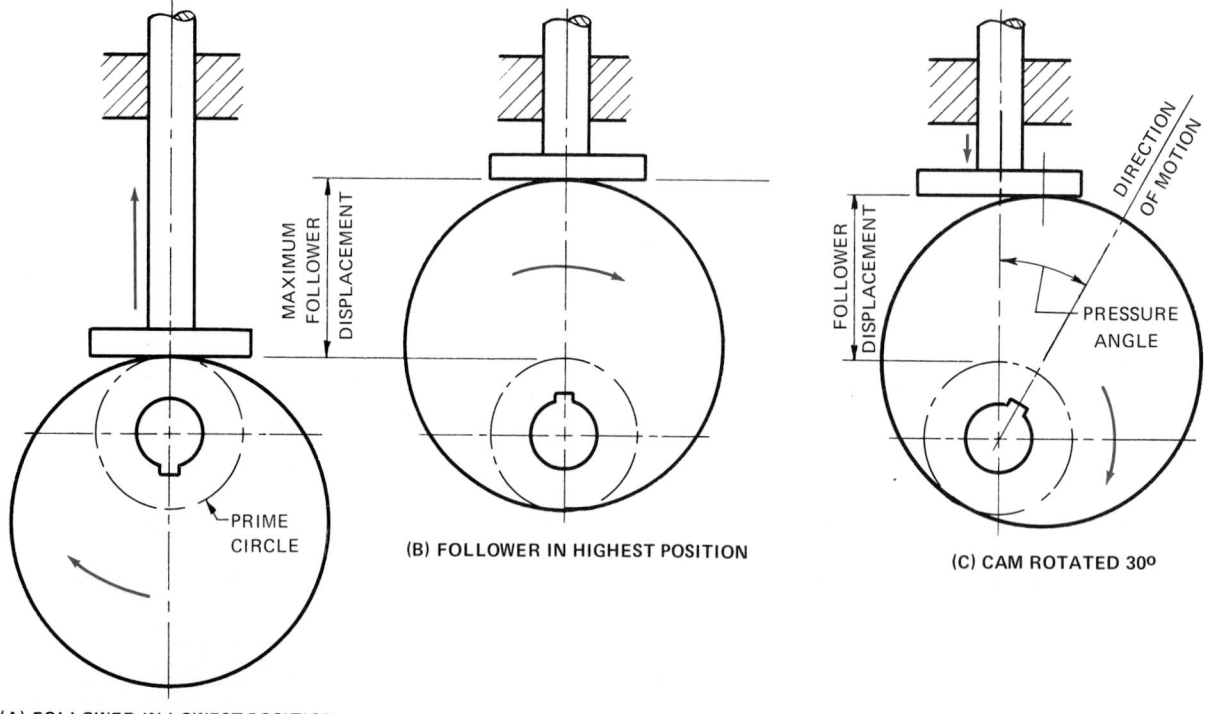

(A) **FOLLOWER IN LOWEST POSITION**

(B) **FOLLOWER IN HIGHEST POSITION**

(C) **CAM ROTATED 30°**

FIG. 22-1-9   Eccentric plate cam.

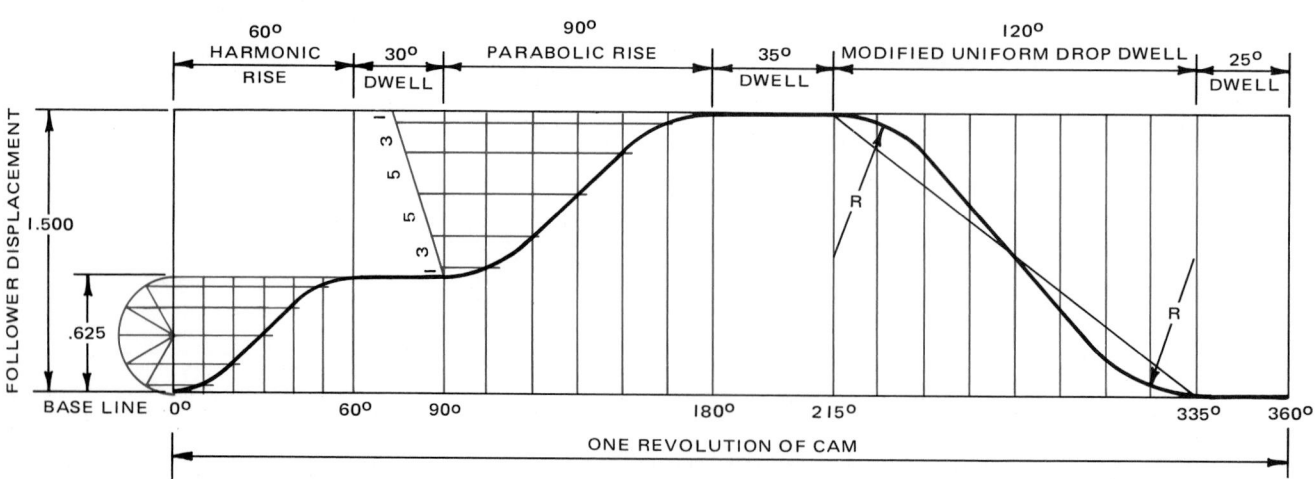

FIG. 22-1-10   Cam displacement diagram.

requirements are less critical with the modified trapezoid than with the cycloidal curve. An advantage over the cycloid is lower acceleration, which means lower forces on output members (the follower system). High inertias can be handled more satisfactorily with the modified trapezoid than with the cycloid. This curve also is jerk-free when used in a cam having a dwell (Fig. 22-1-11A).

Modified Sine-Curve Motion

The modified sine curve is a combination of cycloidal and harmonic curves. This curve will absorb more errors than the modified trapezoid or the cycloidal curve. The torque change from positive to negative is 0.2 in the modified trapezoid and 0.4 in the modified sine curve. This means that the modified sine curve can stand a more flexible, or elastic, input drive than the

modified trapezoid. This curve (modified sine) is ideal for high inertia, as well as for reasonably high speed (Fig. 22-1-11A).

Synthesized, Modified Sine-Harmonic Motion

Because of the complex makeup of the profiles of this curve, only the information shown in Fig. 22-1-11 is covered in this text.

Simplified Method for Laying Out Cam Motion

The method shown in Fig. 22-1-11B is a quick and accurate means of laying out a cam motion. The divisions shown on the lines in Fig. 22-1-11A are accurately divided into the proper divisions for the various cam motions. For example, it is required to construct a 2.00 in. parabolic rise in 120° of cam rotation.

FIG. 22-1-11   Simplified method of laying out a cam motion.

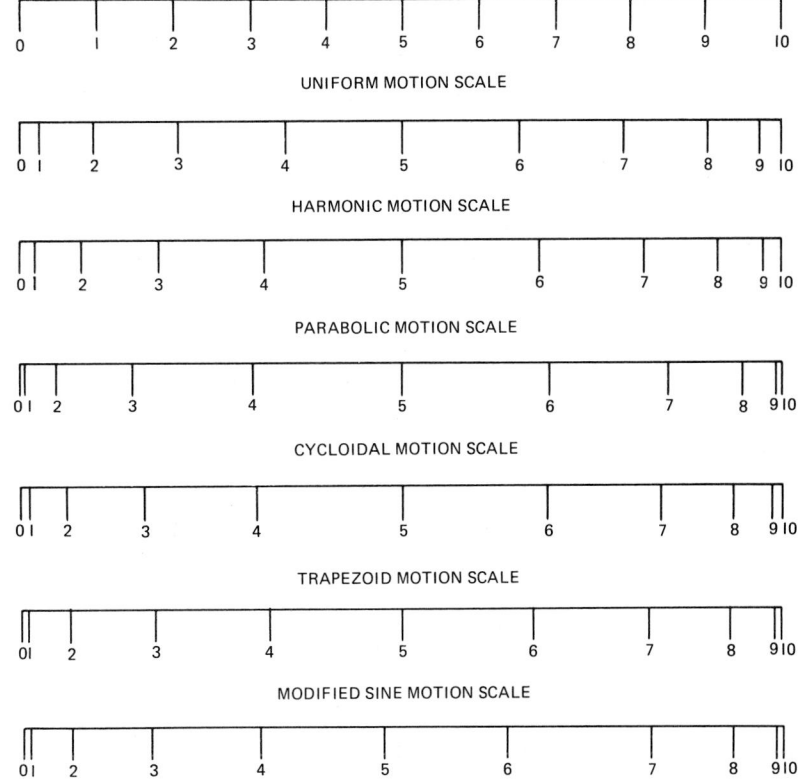

(A) COMMON CAM MOTION SCALES

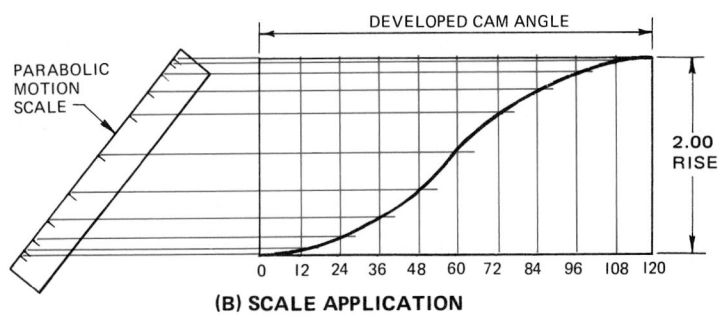

(B) SCALE APPLICATION

Method

1. Draw two parallel horizontal lines 2.00 in. apart representing the rise.
2. Select a suitable distance for the cam displacement and divide the 120° into 10 equal parts (12°, 24°, 36°, etc.).
3. Using the edge of a sheet of paper, mark on it the divisions for the parabolic motion shown in Fig. 22-1-11A.
4. Using this marked paper as the scale, lay the scale between the baseline and the top of the 2.00 in. rise, as shown in Fig. 22-1-11B, and transfer the points from the scale to the drawing.
5. Project these points horizontally to their respective cam divisions and draw the curve.

Cam Displacement Diagrams

In preparing cam drawings, a cam displacement diagram is drawn first to plot the motion of the follower. The curve on the drawing represents the path of the follower, not the face of the cam. The diagram can be any convenient length, but often it is drawn equal to the circumference of the base circle of the cam, and the height is drawn equal to the follower displacement. The lines drawn on the motion diagram are shown as radial lines on the cam drawing, and sizes are transferred from the motion diagram to the cam drawing. Figure 22-1-10 shows a cam

displacement diagram having three different types of motion plus three dwell periods. Most cam displacement diagrams have cam displacement angles of 360°.

REFERENCES AND SOURCE MATERIAL

1. Eonic Inc.
2. Commercial Cam and Machine Co.

ASSIGNMENT

See Assignment 1 for Unit 22-1 on pages 768–769.

22-2 PLATE CAMS

In preparing cam drawings, the radial ordinates should be laid out in the opposite direction to that in which the cam rotates.

In drawing plate cams, the prime circle is constructed first. This circle represents the face of a flat follower or the center line of a roller follower, whichever is used, in its lowest position. It also represents the baseline on the motion diagram.

One of the simplest cams to produce is the eccentric plate cam, as illustrated in Fig. 22-2-1. The shape of the cam is a

FIG. 22-2-1 Eccentric plate cam.

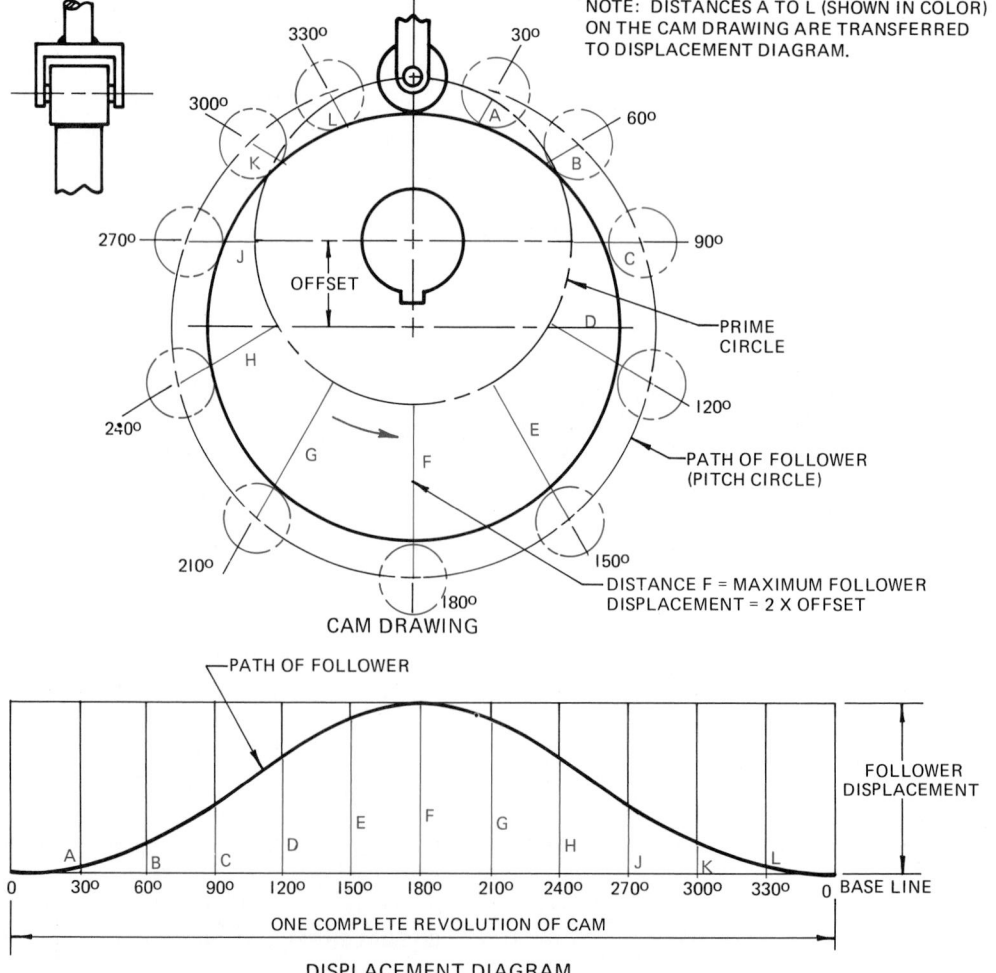

NOTE: DISTANCES A TO L (SHOWN IN COLOR) ON THE CAM DRAWING ARE TRANSFERRED TO DISPLACEMENT DIAGRAM.

perfect circle, and the offset distance for the camshaft is equal to one-half the follower displacement. Radial lines are marked off on the cam drawing, with the center of the shaft as center. The distance between the prime circle and the center of the roller follower on the radial lines is transposed to the displacement diagram. The length of the displacement diagram can be any convenient size, but often the circumference of the prime circle is chosen in order to keep it in the same scale as the follower displacement height. Since this type of cam does not provide a dwell period, it has limited applications.

Since most cams combine motions and dwells in their design, the drawing sequence is different than that for eccentric cams. The cam displacement diagram is drawn first. The ordinate lines constructed on the motion or displacement diagram are drawn on the cam drawing as radial lines, and the corresponding distances from the baseline to the motion curve are transposed to the cam drawing, locating the path of the follower. With cams using a roller follower, the roller diameter is then drawn in several positions along the path of the follower in order to construct the profile of the cam face. When the follower has a flat surface, the paths of the follower and the cam face are one.

Figure 22-2-2 shows a plate cam that produces a simple harmonic displacement with less than 360° of rotation of the cam. The ordinates for the cam pitch curve are constructed as shown in Fig. 22-1-8C. It may be impossible to use a flat follower since the harmonic pitch curve usually has a reentrant, or reversing, curve and a flat follower would just bridge the hollow part. Since a roller follower is the most practical and reliable type, the development of the cam profile with this type of follower is shown.

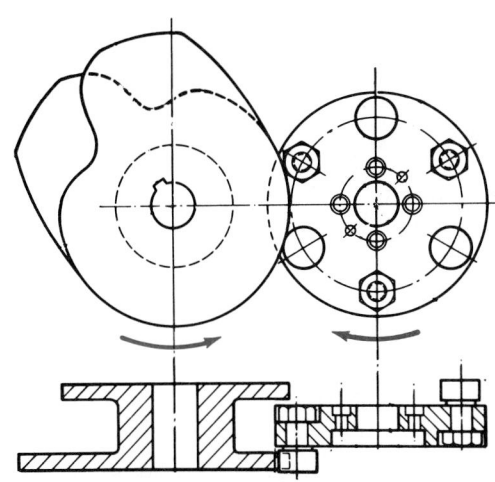

FIG. 22-2-3   Conjugate cam.

Conjugate Cams

Conjugate cams are used when a desired motion cannot be obtained with a single cam (Fig. 22-2-3). Many indexing mechanisms use conjugate cams to obtain the necessary indexing. A displacement diagram is required for each cam.

Timing Diagrams

A convenient method of relating the movement of various machine members that are activated by cams is by the use of a timing diagram. Figure 22-2-4 (pg. 756) shows the timing relationship for three cams. If displacements are plotted to scale, the diagram can be used for checking interferences. It can also be used for specifying the various types of transitions. If zero displacement is used to denote the prime circle radius, the timing diagram can be used by most manufacturers to produce finished cam data. The only additional data required would be a detailed drawing of the cam blank.

Dimensioning Cams

The old method of developing cam contours on the drawing board by layout has been outdated. In the past, a detailed cam was developed from an enlarged layout, using swung arcs and straight lines. However, it is difficult to make a quality cam that has been developed by this method.

To produce a master cam or a single cam, a table of cam radii with corresponding cam angles must be supplied. The cam is then cut on a milling machine, or some other suitable machine tool, by point settings. The result is a surface with a series of ridges, which must be filed down to a smooth profile. The cam radius, cutting radius, and frequency of machine setting determine the extent of filing and the final accuracy of the

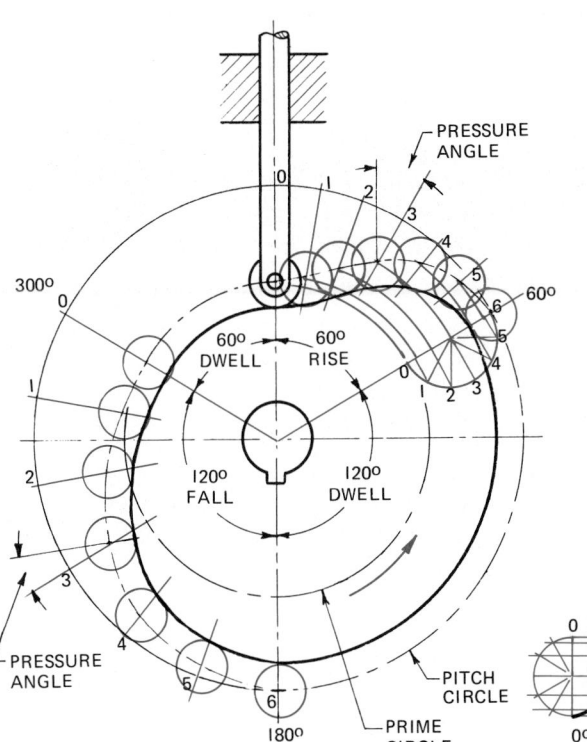

FIG. 22-2-2   Simple plate cam with harmonic motion.

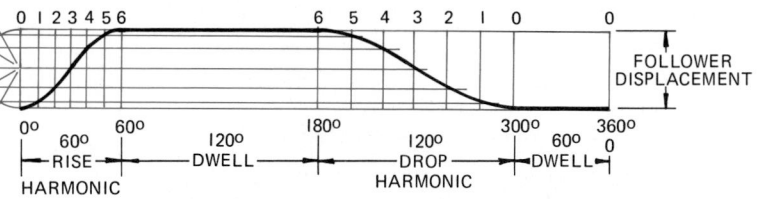

DISPLACEMENT DIAGRAM

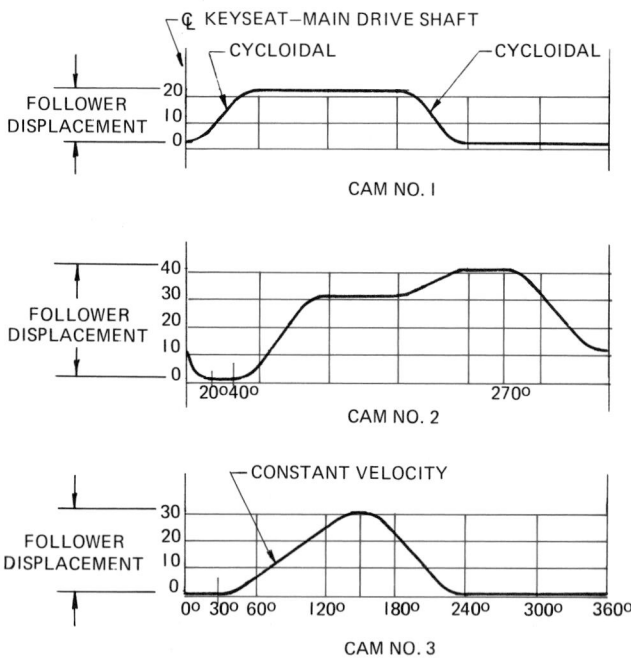

FIG. 22-2-4   Timing diagram.

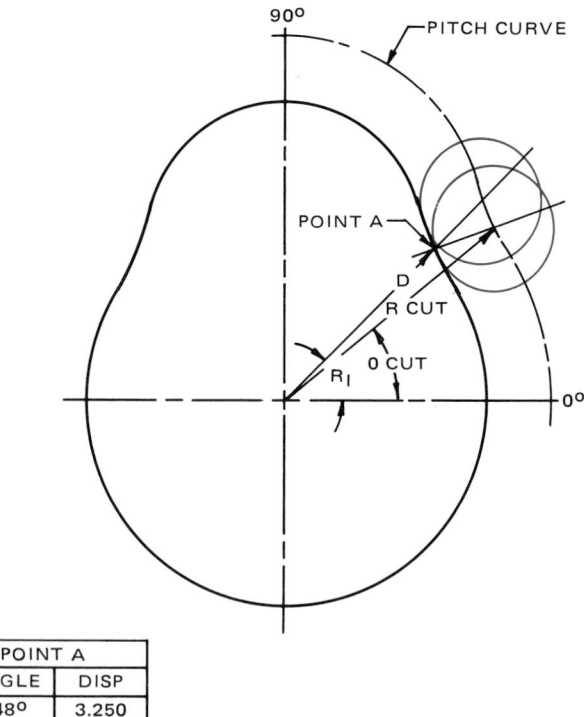

POINT A	
ANGLE	DISP
48°	3.250

FIG. 22-2-5   Dimensioning the cam profile.

profile. For accurate master cams, settings must often be in 0.5° increments, calculated to seconds. The preparation of this table may require the solution of six or eight equations for each of these machine settings.

If a cam has been developed by layouts and has a profile as shown in Fig. 22-2-5, it may appear that the easiest way to describe this contour is to scale the angle R_1 from 0° and scale the displacement D from the center of the cam to the profile surface. Admittedly, this method would define the surface or cam profile.

However, the only method of manufacturing that could be used would be to broach the cam with a very small point cutter. If a cutter radius is added to the displacement value and a cut is made, the adjacent contour is undercut. To properly cut the

point A and maintain the contour, a new set of coordinates must be established. These are shown as R_{cut} and O_{cut}. To produce such data becomes a laborious and expensive task.

In describing a profile, *always* dimension to the pitch curve produced by the follower center. This holds true whether the cam is developed by layout or analytically. The actual follower location requires two physical dimensions: radial displacement and angular displacement. *Radial displacement* is expressed as a distance from the cam center or as a displacement from the prime circle. *Angular displacement* is measured in degrees from some zero reference, such as a keyseat, a dowel hole, or a timing hole, as shown in Fig. 22-2-6.

The easiest method of presenting these data is in the tabular form, rather than dimensioning the detailed cam. Data should

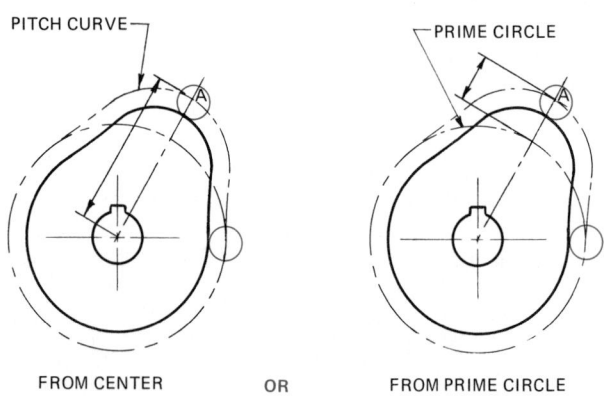

(A) DIMENSIONING RADIAL DISPLACEMENT

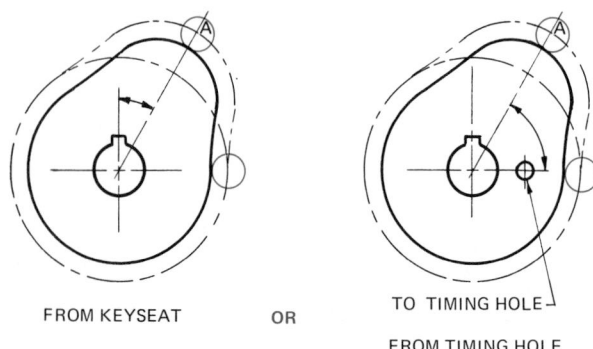

(B) DIMENSIONING ANGULAR DISPLACEMENT

FIG. 22-2-6   Dimensioning point A on the cam profile.

FIG. 22-2-7   Tolerancing polar data.

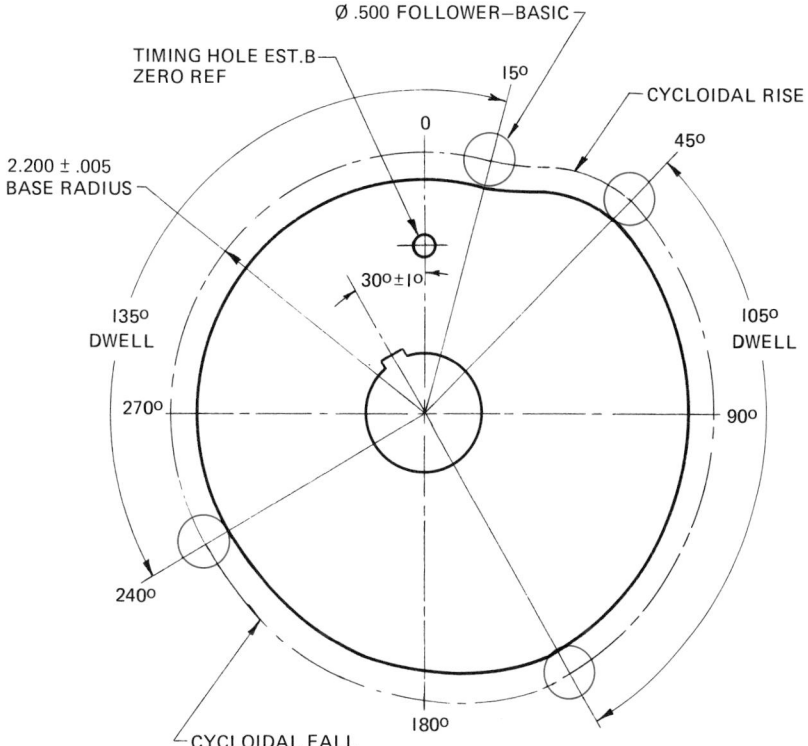

TOLERANCES:
1—TABULATED VALUES ARE TO THE CENTER OF A Ø .500 FOLLOWER.
2— ALL ANGLES ARE BASIC AND IN RELATION TO TIMING HOLE.
3—TOTAL TRANSITION DWELL TO DWELL MAY VARY ±.002 FROM TABULATED VALUES.
4—THE RELATIVE DIFFERENCE BETWEEN ANY TWO ADJACENT DEGREES MUST
    NOT EXCEED .0002 FROM THE DIFFERENCE ESTABLISHED BY THE TABULATED DATA.
5—DWELLS TO BE CONCENTRIC TO ID OF CAM WITHIN .0005 T.I.R. (TOTAL INDICATOR
    READING).

be given in at least 1° increments, although 0.5° increments are preferred. Increments of 0.5° allow the manufacturer to use discretion in selecting the intervals required to produce the finished cam.

Standard practice on tolerancing polar (angular) data is to hold the angles basic (zero tolerance) and to place all tolerance on the displacement value. The only angle that is toleranced is the one that relates the zero reference to some other point on the cam, such as a keyseat or dowel hole. Figure 22-2-7 shows this method, plus one way to tolerance the pitch curve contour.

Figure 22-2-8 (pg. 758) is an alternate method of establishing the pitch curve tolerances. Both of these examples contain three fundamental specifications:

1. Tolerance on basic cam size
2. Tolerance on total transition
3. Tolerance on the pitch curve over some increment of cam angle

It is the third specification that ensures smooth continuity of the cam surface.

By far the simplest method of describing the contour is by denoting the type of transition. In this case, the type of dynamic curve chosen is called out on the detailed print such as cycloidal, harmonic, modified sine, etc. Most cam manufacturers are capable of producing their own incremental data. In most cases,

the charge for this service is nominal. Figure 22-2-9 (pg. 758) is an illustration of this type of cam detail drawing.

Cam Size

Cam size depends primarily on three factors: the pressure angle, the curvature of profile, and the camshaft size. Secondary factors that affect size and design are cam-follower stresses, available cam material, and available space.

If a layout is made, such as shown in Fig. 22-2-10 (pg. 759), it becomes obvious that the maximum pressure angle for a given cam and follower displacement becomes smaller as the cam-pitch circle becomes larger. It is advisable to limit this maximum pressure angle to 30 or 35°.

Figure 22-2-10 also shows how cam curvature is related to cam size. For a given displacement H, cam rotation B, and roller radius R, the larger cam with pitch-curve radius Rp_2 has a much easier curve to manufacture. Note how the smaller cam with radius Rp_1 has much smaller radii of curvature near the high end of the displacement.

Figure 22-2-10 also shows the effect of roller diameter on the shape and accuracy of the cam profile. Always use the smallest possible roller consistent with the load it has to carry.

Another factor affecting cam size is the cutting away of a previously generated cam profile by virtue of too large a cam

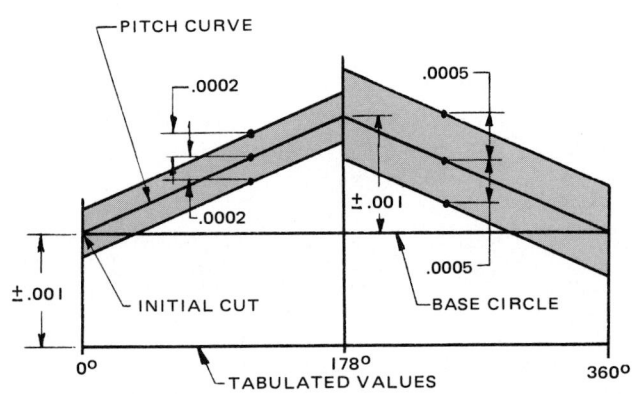

FIG. 22-2-8   Alternate method of establishing pitch curve tolerances.

Variations from the smooth curve representing the tabulated values shall not exceed the limits as follows:

A — Tabulated values are to the center of a .2500 ±.0002 diameter cutter.

B — The base circle, as established by the initial cut, shall establish a reference. This initial cut shall not vary by more than .001 from the tabulated values.

C — The values of all errors, in the interval from 0° through 178°, shall lie within bands allowing either cyclic or random variations not exceeding ± .0002 from the (straight) center line of the band. The center line shall start at the initial cut point and shall end at 178° within .001 from the initial cut point.

D — In the interval from 178° through 360°, all errors shall lie within bands allowing either cyclic or random variations not exceeding .0005 from the (straight) center line of the band. The center line of this interval shall start at the actual end point of the first interval center line at 178° and end at the initial cut position at 360°.

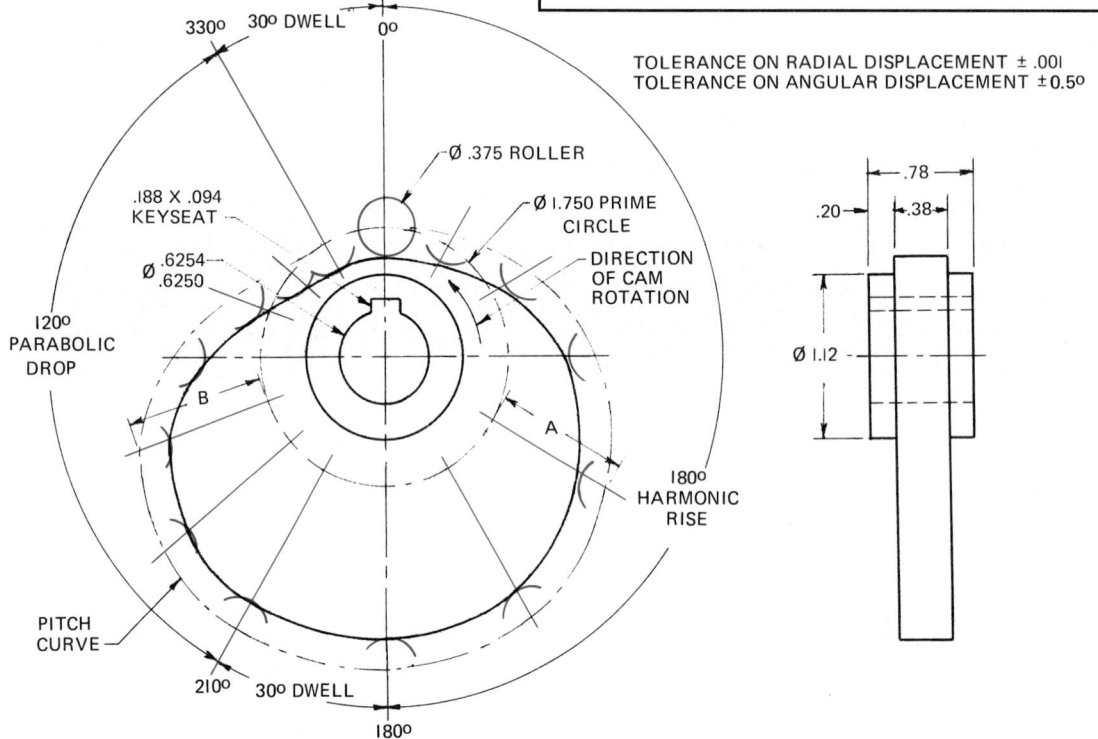

TOLERANCE ON RADIAL DISPLACEMENT ± .001
TOLERANCE ON ANGULAR DISPLACEMENT ±0.5°

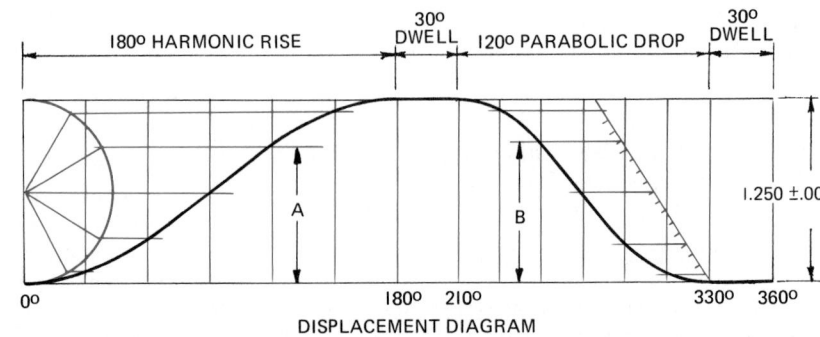

ANGULAR DISPLACE- MENT	RADIAL DISPLACEMENT FROM PRIME CIRCLE
0°	0
180°	1.250
210°	1.250
330°	0

DISPLACEMENT DIAGRAM

NOTE: ANGULAR AND RADIAL DISPLACEMENT DIMENSIONS FOR MOTIONS SUPPLIED BY CAM MANUFACTURER

FIG. 22-2-9   Plate cam drawing.

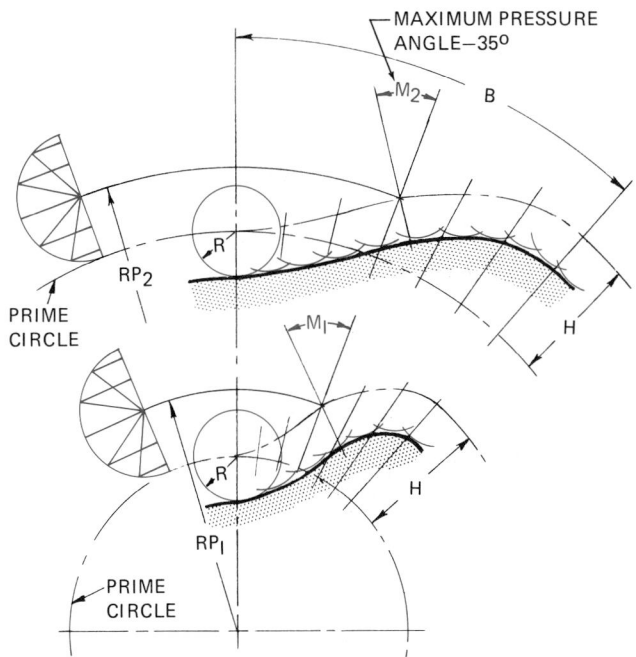

FIG. 22-2-10   Increasing the cam size decreases the pressure angle.

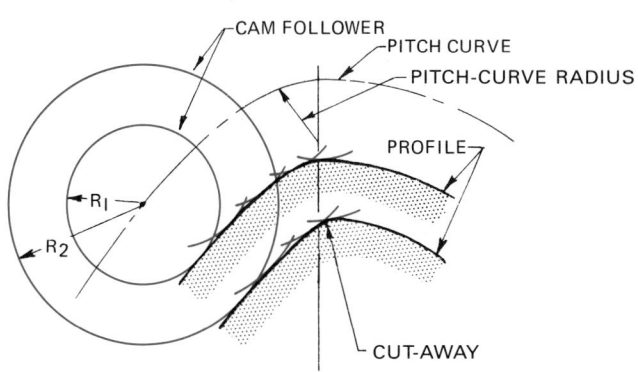

CAM FOLLOWER RADIUS MUST BE LESS THAN
PITCH-CURVE RADIUS AT ANY POINT TO AVOID
CUT-AWAY

FIG. 22-2-11   Factors affecting cam size.

follower. This is illustrated in Fig. 22-2-11. Basically, the cam-follower-roll radius must be less than the pitch-curve radii at any point along the pitch curve.

REFERENCES AND SOURCE MATERIAL

1. Eonic Inc.

ASSIGNMENTS

See Assignments 2 through 6 for Unit 22-2 on pages 768 through 771.

22-3   POSITIVE-MOTION CAMS

To ensure positive motion of the follower in both directions, positive-motion cams are employed. Two types are face cams

and cams with yoke-type followers. *Face cams* are similar to plate cams, except that the follower engages a groove on the face of the cam rather than on the outside edge of the cam. One disadvantage to this type of cam is that the outer edge of the cam groove tends to rotate the roller in the direction opposite to that of the inner edge, resulting in wear in both the cam and the roller. However, this is not serious at slow speeds. *Yoke-type followers* are used for operating light mechanisms. The follower surface is flat or tangent to the curvature of the cam. With this type of cam, only one-half of the cam displacement diagram need be drawn since the other half of the cam is identical to the first half (Figs. 22-3-1 and 22-3-2, here and on pg. 760).

ASSIGNMENTS

See Assignments 7 through 10 for Unit 22-3 on pages 769 to 770.

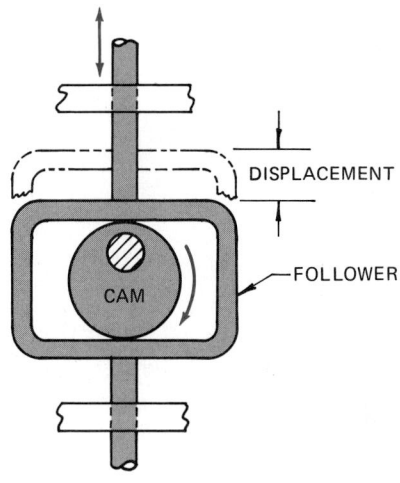

CAM – 360° CAM ROTATION

FIG. 22-3-1   Cam with yoke-type follower.

22-4   DRUM CAMS

The layout of a drum or cylinder cam starts, as with any cam, with the decision as to which profile and follower type will be used. Many cylinder cams are used with straight in-line followers so that the follower moves in a path parallel to the axis of the cam. The pitch surface is developed and shown as a rectangle (Fig. 22-4-1A, pg. 760), and the follower displacement is plotted with rectangular coordinates.

Theoretically, a tapered follower with its cone center on the cam axis should give the best results (Fig. 22-4-1B). Actually, straight rollers give excellent results as long as the roller length and diameter are not too large in relation to the cam cylinder diameter. Swinging followers are used on indexing-type cylindrical cams, as shown in Fig. 22-1-5.

For drum or cylindrical groove cams, the displacement diagram is replaced by the developed surface of the cam, as shown in Fig. 22-4-2 (pg. 761). The groove shown in the front view of the cam is found by projection.

759

FIG. 22-3-2   Face cam drawing.

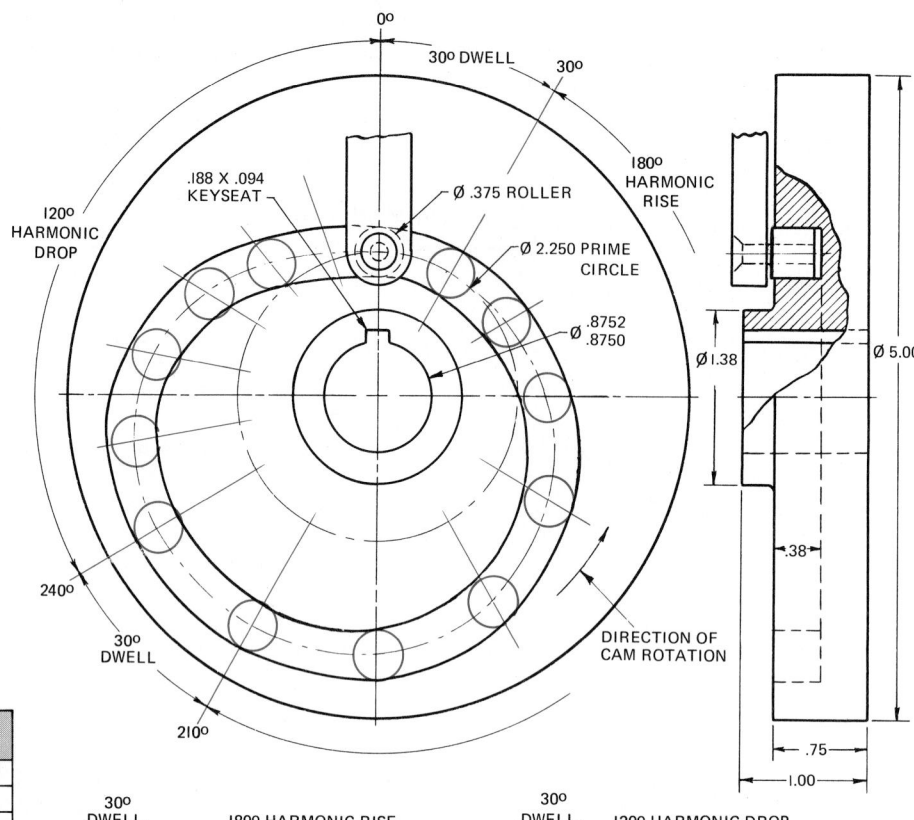

ANGULAR DISPLACEMENT FROM KEYSEAT	RADIAL DISPLACEMENT FROM PRIME CIRCLE
0°	0
30°	0
210°	.938
240°	.938

TOLERANCE ON RADIAL DISPLACEMENT ± .001
TOLERANCE ON ANGULAR DISPLACEMENT ± 0.5°
NOTE: ANGULAR AND RADIAL DISPLACEMENT
        DIMENSIONS FOR MOTIONS SUPPLIED
        BY CAM MANUFACTURER.

DISPLACEMENT DIAGRAM

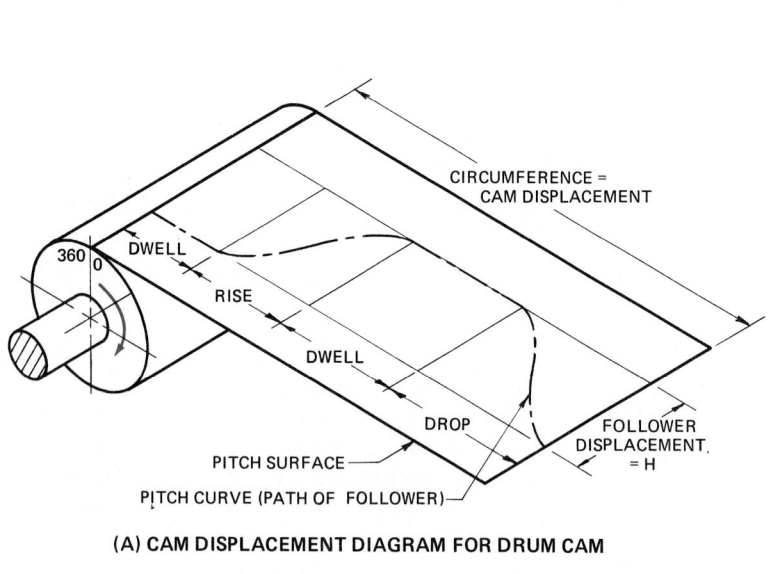

(A) CAM DISPLACEMENT DIAGRAM FOR DRUM CAM

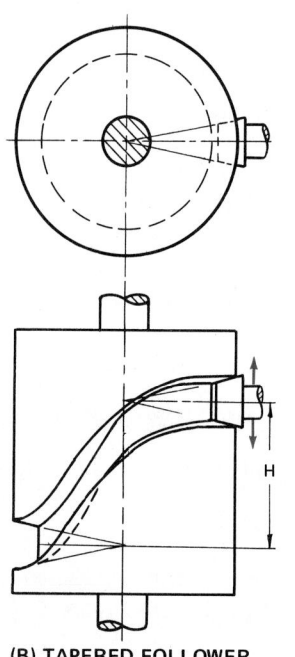

(B) TAPERED FOLLOWER

FIG. 22-4-1   Drum cam details.

Points from the developed surface of the cam and their corresponding points on the top view are projected to the front view, as shown by the letter *A* at position 210°.

ASSIGNMENTS ▧▧▧▧▧▧▧▧▧▧▧▧▧▧▧▧▧▧▧▧▧

See Assignments 11 and 12 for Unit 22-4 on pages 770–771.

22-5 INDEXING

Indexing is the conversion of a constant-speed rotary-input motion to an intermittent rotary-output motion. Press-feed tables, packaging machines, machine tools, switch gear, and feeding devices are but a few of the many machines found in industry that require indexing or intermittent motion.

In recent years advances in the design and manufacture of positive intermittent-motion devices have significantly

FIG. 22-4-2   Drum cam drawing.

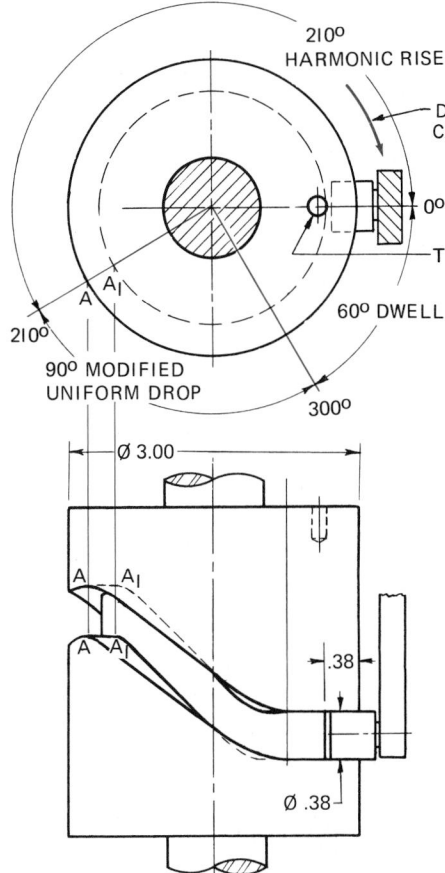

ANGULAR DISPLACEMENT FROM TIMING HOLE	DISPLACEMENT FROM BASE LINE
0°	0
210°	1.250
300°	0

TOLERANCE ON RADIAL DISPLACEMENT FROM BASELINE   ± .001

TOLERANCE ON ANGULAR DISPLACEMENT FROM BASELINE ±0.5°

NOTE: ANGULAR DISPLACEMENT AND DISPLACEMENT FROM BASELINE SUPPLIED BY CAM MANUFACTURER.

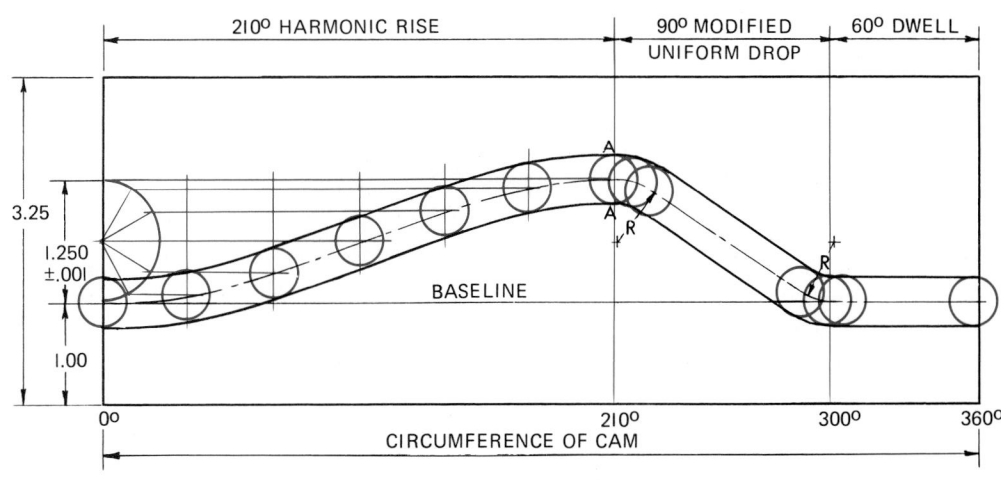

DISPLACEMENT DIAGRAM

improved the smoothness and speed of indexing motions possible with Geneva-type drives.

The manifold indexing mechanism (Fig. 22-5-1A) consists of two basic elements: a cam attached to the input shaft and a turret attached to the output shaft. The input and output shafts are at right angles but do not lie in the same plane. The cam is of concave globoidal form with a track that engages roller followers that project radially from the edge of the turret disk. Part of the cam track is straight, so that when the roller followers engage it, no movement of the turret can take place. The angle of the cam occupied by this part is called the *dwell angle*.

The remaining part of the cam track progresses along the cam axis in helical fashion, thus rotating the turret. Before a roller leaves one end of the cam track, another roller enters the other end to maintain continuity of movement. The angle of the cam occupied by this part of the track is called the *cam index angle*. Thus, one revolution of the cam represents one indexing cycle, during which the turret indexes from one station to the next and dwells for a specific period. The number of times that this takes place in one revolution of the turret is called the *number of stops*.

The tangent drive (Figs. 22-5-1C and 22-5-1E) consists of a constantly rotating driver and a driven wheel. The wheel may have four, five, six, or eight precision-machined radial slots. A matching cam follower, mounted on needle bearings on the driver, engages one of the slots on each revolution of the driver, thereby indexing the wheel. The concave section between the slots is precisely machined to mate with the locking hub of the driver to prevent movement of the wheel during dwell.

The tangent drive indexes over an angle equal to 360° divided by the number of slots or stations in its wheel. For example, each index of a four-station tangent drive is 90°; each index of a five-station drive is 72° (Fig. 22-5-2).

The time ratio of a tangent drive is expressed by the arc (in degrees) of each revolution of the driver that the wheel is being indexed and the arc of each revolution that the wheel is at rest or dwell. The time ratio refers to each revolution of the driver and, therefore, remains constant, regardless of the driver speed. The actual speed of indexing is a function of the driver speed and is directly proportional to it.

The indexing application shown in Fig. 22-5-1F employs an overrun clutch and a rack and gear. The input or driver shaft is connected to a rack that converts rotary motion into reciprocating motion. The gear, which is attached to an overrun clutch, rotates in both directions. The overrun clutch drives the shaft in one direction but overruns or freewheels on the shaft in the other direction, producing an intermittent rotary motion.

(A) BARREL CAM

(B) BARREL CAM

(C) 6-STATION DRIVE

(D) CONJUGATE CAMS

(E) 4-STATION DRIVE

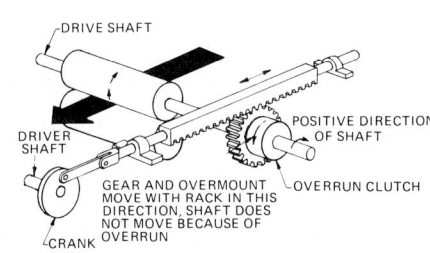

(F) OVERRUN CLUTCH, CRANK, AND GEAR AND RACK

FIG. 22-5-1   Indexing mechanisms. *(A, Manifold Machinery Co. Ltd.; B and D, Commercial Cam & Machine Co.; C and E, Geneva Motions Corp.)*

FIG. 22-5-2   Indexing drives.
(Geneva Motions Corp.)

DIMENSIONS		MODEL	A	B	C	D	DRIVER					WHEEL		
							E	F	G	H	K	L	M	N
4 STATIONS — STATIONS — TIME RATIO: 90° INDEX - 270° DWELL	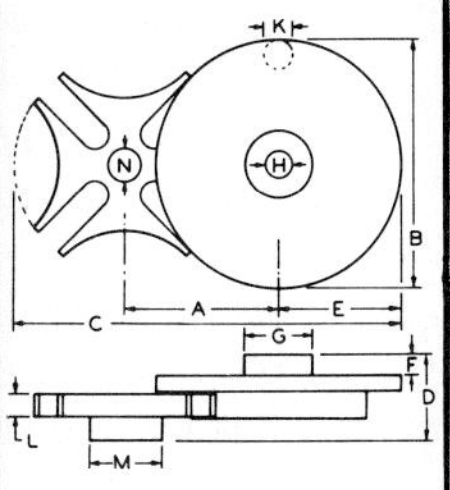 INCH	4S5	5.00	8.00	12.59	2.06	4.00	.50	1.50	.75	.88	.62	2.00	1.00
		4S5.5	5.50	8.62	13.75	2.06	4.40	.50	1.50	.75	.88	.62	2.00	1.00
		4S6	6.0	9.50	15.06	2.19	4.75	.50	1.50	.75	1.00	.75	2.00	1.00
	METRIC	4S75	75	124	194	48	62	12	40	16	16	12	45	20
		4S90	90	150	224	48	75	12	40	16	16	12	45	20
		4S100	100	164	256	50	82	12	40	20	20	16	48	22
		4S115	115	190	287	50	95	12	40	20	20	16	48	22
5 STATIONS — STATIONS — TIME RATIO: 108° INDEX - 252° DWELL	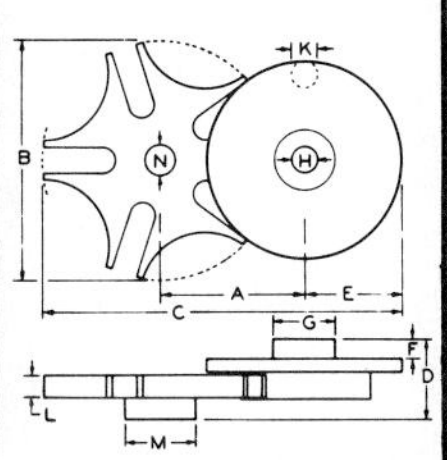 INCH	5S5	5.00	8.19	12.47	2.06	3.38	.50	1.50	.75	.88	.62	2.00	1.00
		5S5.5	5.50	9.00	13.69	2.06	3.69	.50	1.50	.75	.88	.62	2.00	1.00
		5S6	6.00	9.78	14.94	2.19	4.03	.50	1.50	.75	1.00	.75	2.00	1.00
	METRIC	5S75	75	124	190	48	53	12	40	16	16	12	45	20
		5S90	90	150	225	48	60	12	40	16	16	12	45	20
		5S100	100	164	252	50	70	12	40	20	20	16	48	22
		5S115	115	190	287	50	77	12	40	20	20	16	48	22
6 STATIONS — STATIONS — TIME RATIO: 120° INDEX - 240° DWELL	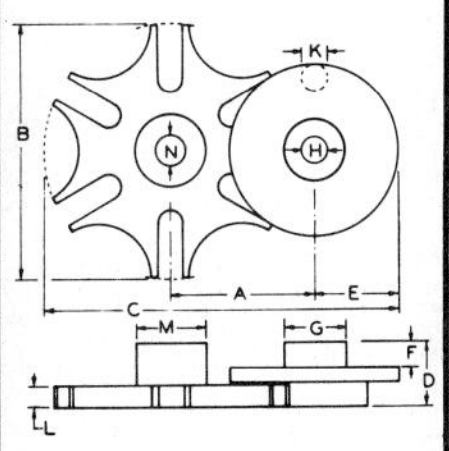 INCH	6S5	5.00	8.75	12.31	1.62	2.94	.50	1.50	.75	.88	.62	2.00	1.00
		6S5.5	5.50	9.62	13.50	1.62	3.14	.50	1.50	.75	.88	.62	2.00	1.00
		6S6	6.00	10.50	14.75	1.75	3.50	.50	1.50	.75	1.00	.75	2.00	1.00
	METRIC	6S75	75	134	188	35	46	12	40	16	16	12	45	20
		6S90	90	160	222	35	52	12	40	16	16	12	45	20
		6S100	100	180	250	38	60	12	40	20	20	16	48	22
		6S115	115	204	274	38	67	12	40	20	20	16	48	22
8 STATIONS — STATIONS — TIME RATIO: 135° INDEX - 225° DWELL	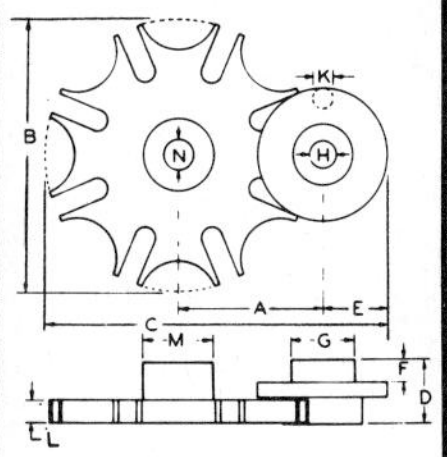 INCH	8S5	5.00	9.31	11.94	1.62	2.28	.50	1.50	.75	.75	.62	2.00	1.00
		8S5.5	5.50	10.22	13.12	1.62	2.50	.50	1.50	.75	.75	.62	2.00	1.00
		8S6	6.00	11.16	14.34	1.75	2.75	.50	1.50	.75	.75	.75	2.00	1.00
	METRIC	8S75	75	144	193	35	46	12	40	16	12	12	45	20
		8S90	90	170	215	35	40	12	40	16	12	12	45	20
		8S100	100	190	243	38	48	12	40	20	16	16	48	22
		8S115	115	214	274	38	52	12	40	20	16	16	48	22

REFERENCES AND SOURCE MATERIAL

1. Manifold Machinery Co. Ltd.
2. Geneva Motions Corp.

ASSIGNMENT

See Assignment 13 for Unit 22-5 on page 771.

22-6 LINKAGES

One of the ever-present problems in machine design is the mechanization of various interrelated motions. These motions are usually preassigned, often quite arbitrarily, and specify the relationships of moving parts or simply the end motion of a single part. Illustrations are present in all types of machinery. Typical examples are seen in textile machinery, packaging machinery, printing presses, valve mechanisms in steam locomotives, machine tools, automotive equipment, household articles, instruments, computing devices, and many other common mechanisms. Upon closer observation, it will be noticed that all these devices are simply combinations and arrangements of basic mechanical elements such as gear trains, cam actuators, cranks and links, sliders, bolts and pulleys, and other rotating and sliding parts. Combinations of the crank, link, and sliding elements are commonly termed *bar linkages*.

Locus of a Point

The *locus* of a point in a linkage or mechanism is the path traced by that point as it moves according to certain controlled conditions. The study of loci is important in machine design to determine:

1. The position of various links and joints in the cycle of the mechanism.
2. The relative speeds of different parts.
3. Forces exerted in the mechanism using applied machines in conjunction with diagrammatic layouts.

The designer is often called upon to make these diagrammatic layouts to assist in designing the most economical and space-saving container for linkages, as well as to ensure that parts of adjacent linkages will not foul one another at any point in the movement of the machine.

Cams Versus Linkages

The best-known solution for a function generator is the cam: flat cams for functions of single variables and barrel cams for functions of two variables.

As computing devices, linkage mechanisms enjoy a number of advantages over cams, with the one exception that the functions must be continuous. *Linkages* are essentially straight

members joined together. Only a small number of dimensions need to be held closely. The joints make use of standard bearings, and the links in effect form a solid chain and are not subject to undue acceleration limitations.

The harmonic transformer and the four-bar linkage shown in Fig. 22-6-1 are the two bar linkages most commonly used as function generators. The harmonic transformer consists of crank, connecting link, and slider. It may be driven from the crank end when a rotary input and linear output are desired or from the slide end when a linear input and rotary output are required. Two cranks and a connecting link form the four-bar linkage, whose input and output are both rotary. By assignment of correct values to the various parameters, these linkages will mechanize many single-variable functions. The selection of these values is termed a *linkage layout*. Typical linkage joints are shown in Fig. 22-6-2.

Straight-Line Mechanism

A *straight-line mechanism* is a linkage device used to guide a given point in an approximate straight line. Several such mechanisms use five or more links to produce exact straight-line

FIG. 22-6-1 Linkages used as function generators.

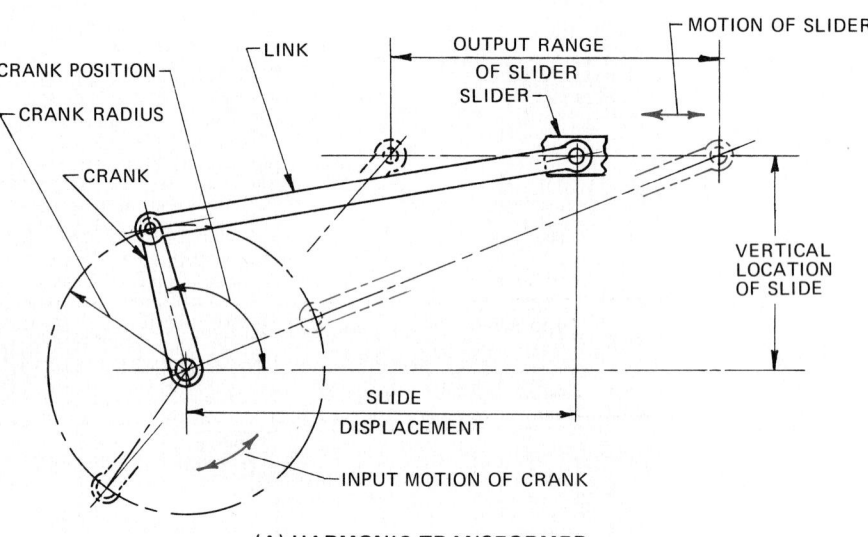

(A) HARMONIC TRANSFORMER

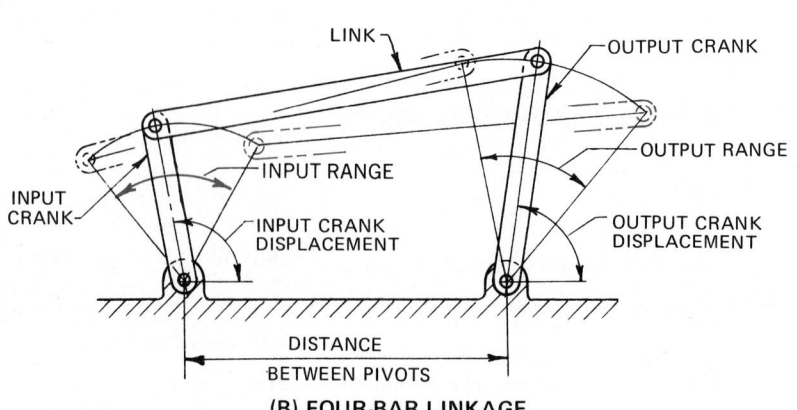

(B) FOUR-BAR LINKAGE

GEARS

Gears are an integral part of many products including automobile transmissions and heavy duty machinery. Gears can be designed manually or using a computer-aided design process.

COMPUTER-AIDED DRAFTING

Computer-aided drafting (CAD) has revolutionized the processes used in the design and manufacturing of products. Examples of CAD being used in various industries are shown here.

AIRBUS

300

0,0

ORIGIN

4

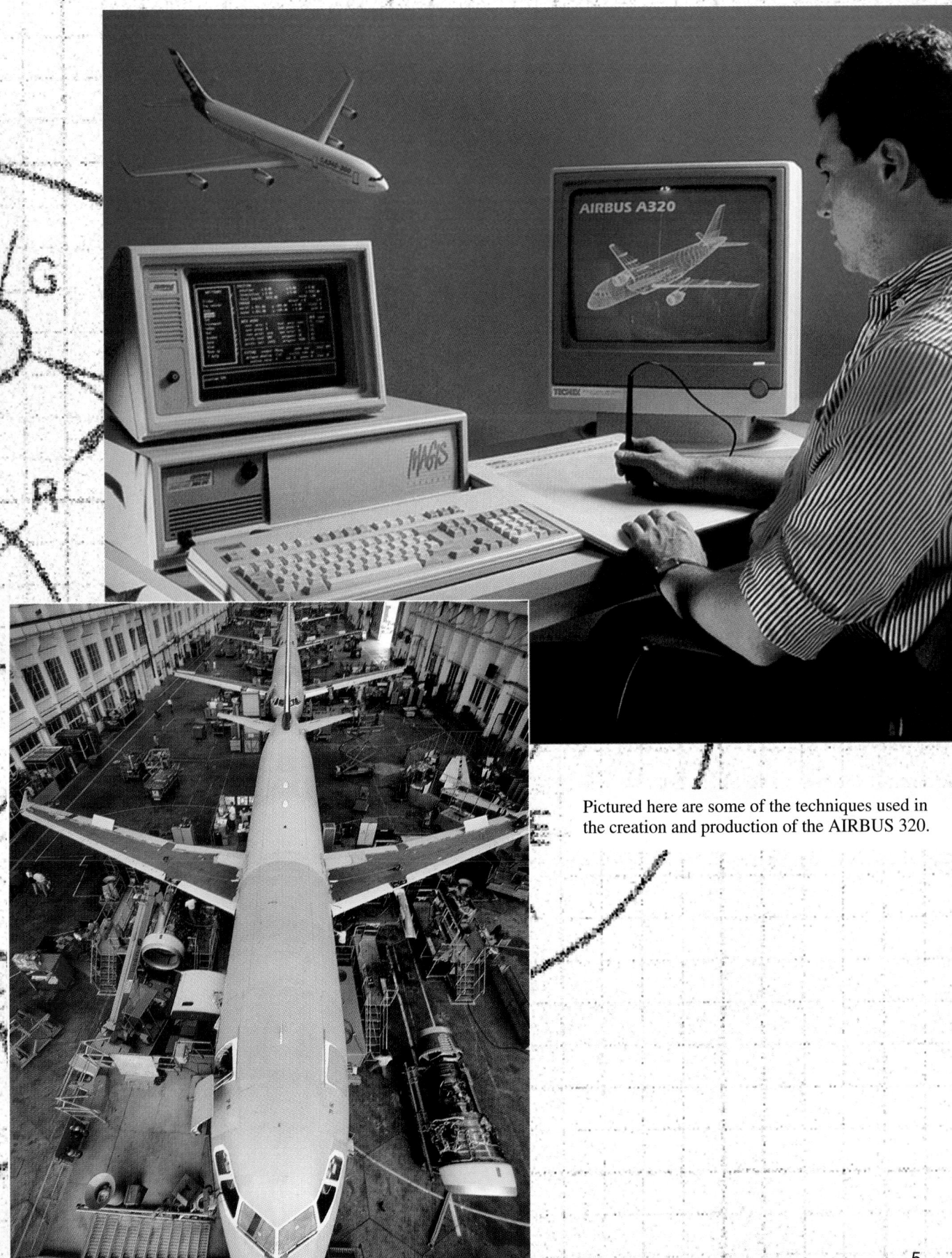

Pictured here are some of the techniques used in the creation and production of the AIRBUS 320.

CIVIL ENGINEERING

Civil engineering often uses the design team concept. These teams usually include the engineer, the architect, and the construction engineer.

CIRCUIT BOARD DESIGN

The increased use of electronics in modern products requires drafters to have a firm understanding of electronic drafting.

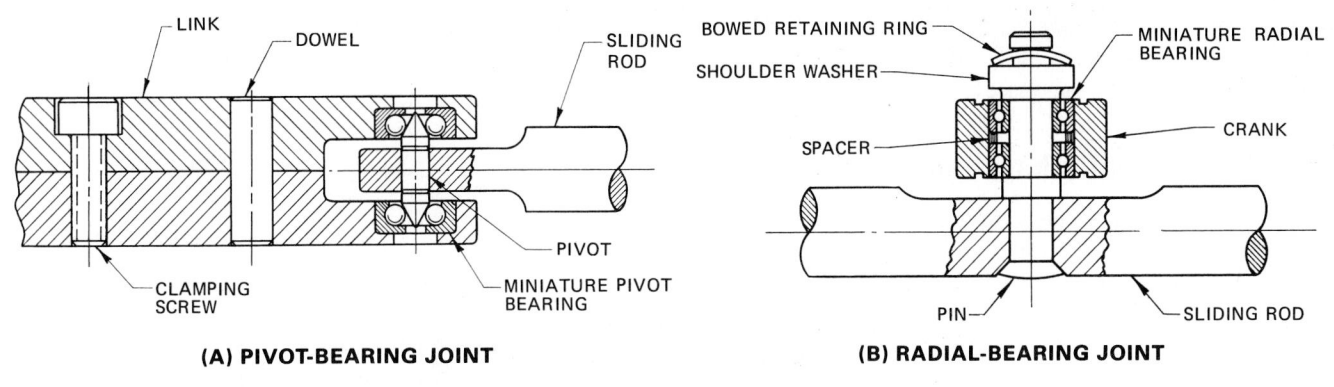

(A) PIVOT-BEARING JOINT

(B) RADIAL-BEARING JOINT

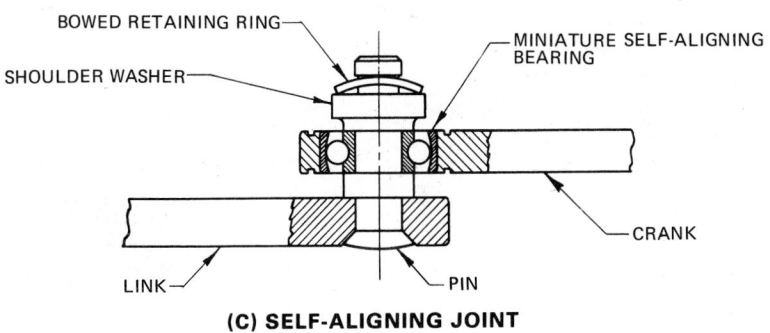

(C) SELF-ALIGNING JOINT

FIG. 22-6-2   Linkage joints.

FIG. 22-6-3   Terminology of a four-bar straight-line mechanism.

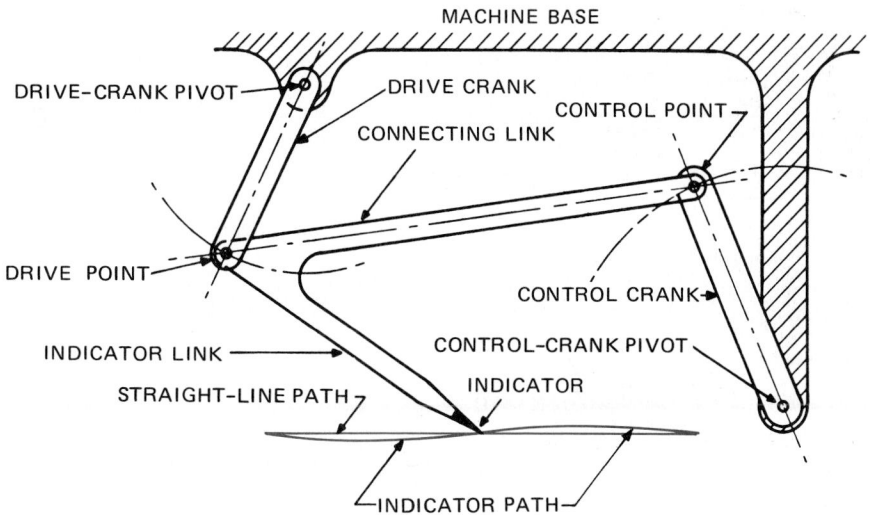

motion of a given point. A four-link (or four-bar) mechanism, using finite links, can only approximate a straight line.

The four elements of the four-bar linkage (Fig. 22-6-3) are:

1. A link, which will be caused to move so that one point on it, called the *indicator,* travels along the desired path. This link is composed of two rigidly connected parts: the *connecting link,* joining the drive point and the control point, and the *indicator link,* connecting the indicator to the drive point.

2. A *drive crank,* to which a turning torque is applied to move the mechanism and which is connected to the link at the drive point.

3. A *control crank,* which serves only to guide the link control point in the proper path.

4. The *base* of the machine, to which the two cranks are pivotally attached.

For identification purposes, *indicator path* is the term used to describe the approximate straight-line path through which the indicator travels, and *straight line* refers to the desired theoretical straight line. The indicator path and the straight line will coincide at three or four places.

Systems Having Linkages and Cams

A cam is of no value and can perform no useful function without a follower linkage. A simple follower is generally not thought of as a linkage since it is usually a slide or plunger, such as an automotive valve assembly in a simple L-head engine. A linkage is generally considered to be a group of levers and links (Fig. 22-6-4).

Figure 22-6-5 shows a typical cam linkage composed of a fairly heavy slide, a short link, and a bell crank. If the slide, which has the largest mass in the linkage, is to be moved with the most favorable accelerating forces, its displacement-to-time relationship must govern the shape of the cam profile. Since the bell crank and link swing about fixed and instantaneous centers during the stroke, the displacement increments of the slide and the cam profile do not bear the same relationship to

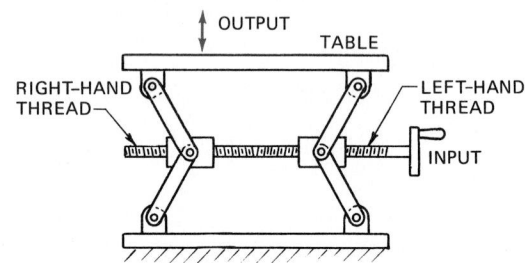

(A) CONVERTING ROTARY MOTION INTO RECIPROCATING MOTION

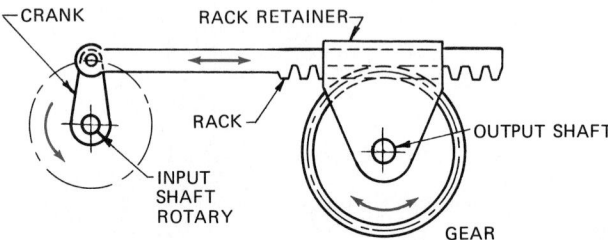

(B) CONVERTING ROTARY MOTION INTO OSCILLATING MOTION

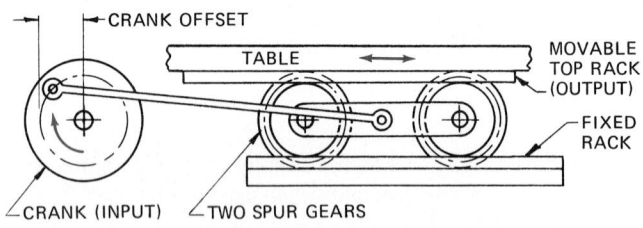

TABLE MOVES FOUR TIMES CRANK OFFSET

CRANK WITH GEARS AND RACK

(C) CONVERTING ROTARY MOTION INTO RECIPROCATING MOTION

FIG. 22-6-4   Converting rotary motion into oscillating or reciprocating motion.

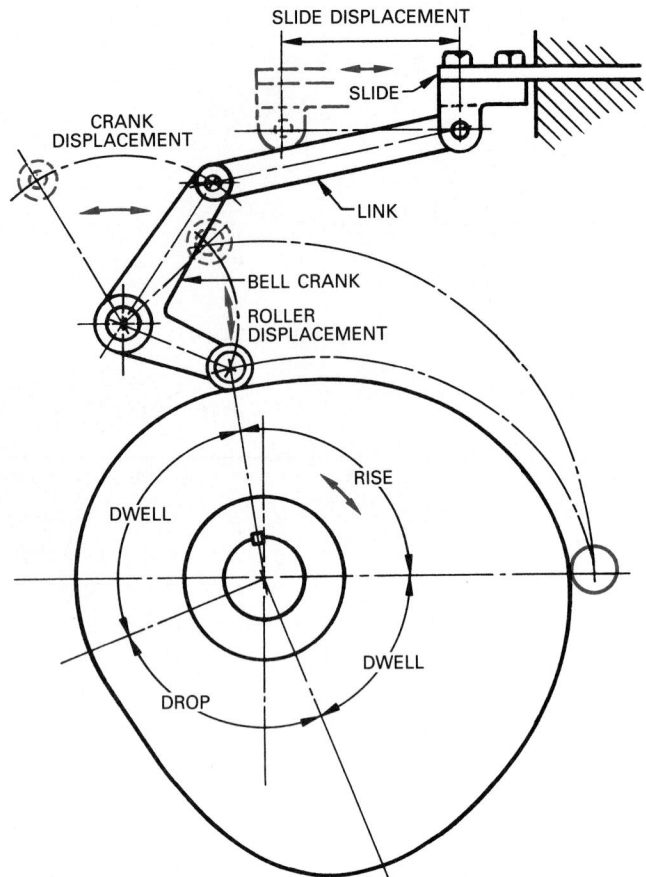

FIG. 22-6-5   Cam linkage.

the cam displacement. It is therefore necessary to plot the cam-pitch curve from the slide displacement at each increment of cam displacement. Figure 22-6-6 shows how this is done.

For extra accuracy and for high-speed machinery, these displacements should be calculated.

REFERENCES AND SOURCE MATERIAL

1. *Machine Design.*
2. Eonic Inc.

ASSIGNMENTS

See Assignments 14 through 17 for Unit 22-6 on pages 772 to 773.

22-7 RATCHET WHEELS

Ratchet wheels are used to transform reciprocating or oscillatory motion into intermittent motion, to transmit motion in one direction only, or as an indexing device. When a motion is to be transmitted at intervals rather than continuously and the loads are light, ratchets are ideal because of their low cost.

Common forms of ratchets and pawls are shown in Fig. 22-7-1. The teeth in the ratchet engage the teeth in the

FIG. 22-6-6 Plotting the pitch curve from the slide displacement for the cam shown in Fig. 22-6-5.

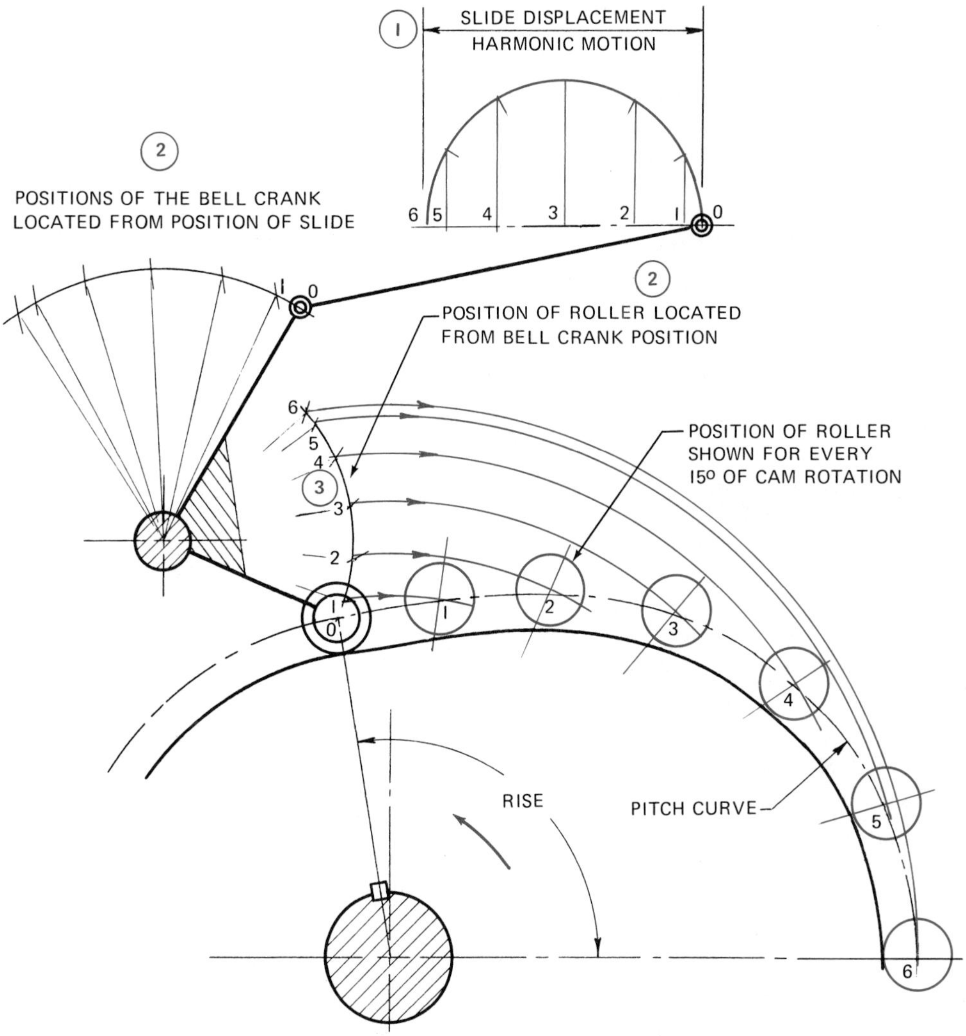

① SLIDE DISPLACEMENT HARMONIC MOTION

6 5 4 3 2 1 0

② POSITIONS OF THE BELL CRANK LOCATED FROM POSITION OF SLIDE

②

POSITION OF ROLLER LOCATED FROM BELL CRANK POSITION

POSITION OF ROLLER SHOWN FOR EVERY 15° OF CAM ROTATION

③

RISE

PITCH CURVE

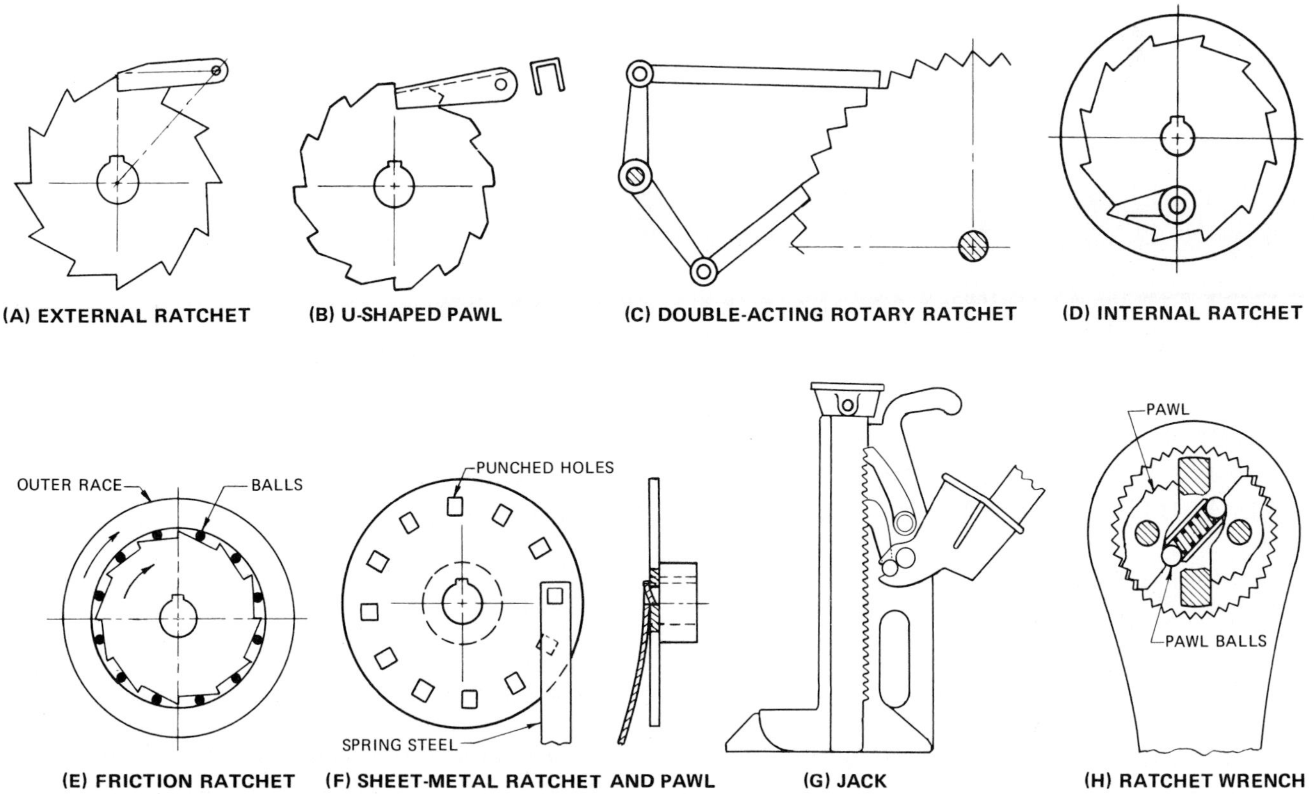

(A) EXTERNAL RATCHET

(B) U-SHAPED PAWL

(C) DOUBLE-ACTING ROTARY RATCHET

(D) INTERNAL RATCHET

OUTER RACE — BALLS

PUNCHED HOLES

PAWL

PAWL BALLS

SPRING STEEL

(E) FRICTION RATCHET

(F) SHEET-METAL RATCHET AND PAWL

(G) JACK

(H) RATCHET WRENCH

FIG. 22-7-1 Ratchet and pawl applications.

767

pawl, permitting rotation in one direction only. The pawl, which fits into the ratchet teeth as shown in Fig. 22-7-2, is pivoted at one end. A spring or counter-weight is normally used to maintain contact between the wheel and pawl. In Fig. 22-7-1G and H, lever or pawl balls are used to shift the pawl to the alternative position so that the ratchet will work in reverse.

In friction ratchets (Fig. 22-7-1E), balls are used between the ratchet and the follower. As the ratchet rotates in the direction of the arrow shown, the balls roll up on the high spots on the teeth, wedging the ratchet and outer race together. If the direction of the ratchet is reversed, the balls roll to the low points on the teeth and disengage the outer roller. This principle is used on overrunning clutches.

Ratchet wheels and pawls are also used widely to control drum rotation hoisting equipment.

In designing a ratchet wheel and pawl, lay out points *A, B,* and *C,* as shown in Fig. 22-7-2, on the same circle to ensure the smallest forces are acting on the system.

Another method to increase the number of stops made by the ratchet wheel without increasing the number of teeth is the use of multiple pawls (Fig. 22-7-3). Adding another pawl of different length doubles the number of indexing positions.

ASSIGNMENTS

See Assignments 18 through 20 for Unit 22-7 on page 773.

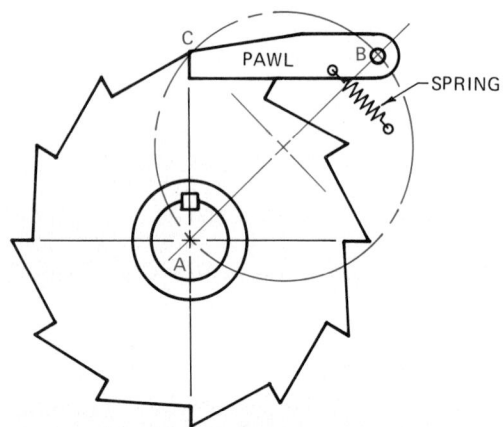

FIG. 22-7-2   Designing a ratchet wheel and pawl.

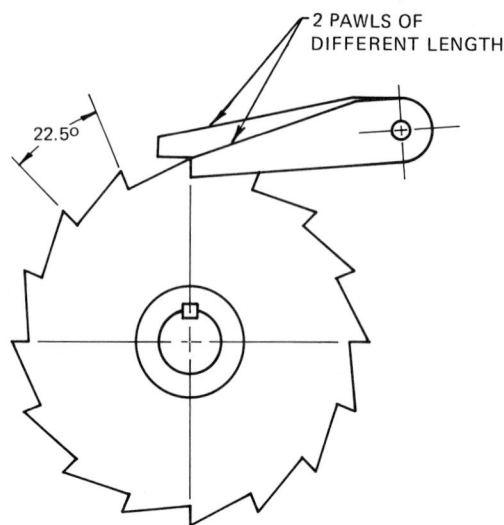

PAWLS LOCK RATCHET WHEEL EVERY 11.25°

FIG. 22-7-3   Ratchet with two pawls.

ASSIGNMENTS FOR CHAPTER 22

ASSIGNMENT FOR UNIT 22-1, CAMS AND CAM MOTIONS

1. Draw displacement diagrams for each of the two cams shown in Fig. 22-1-A or Fig. 22-1-B. Scale 1:1.

ASSIGNMENTS FOR UNIT 22-2, PLATE CAMS

2. Design a plate cam that will activate a Ø.50 in. roller follower as follows:
   - Rise 1.50 in. in 180° with harmonic motion
   - Dwell 30°
   - Drop 1.50 in. in 120° with modified uniform motion
   - Dwell for remainder

   Prime circle = Ø3.00 in., plate thickness = .50 in., shaft = Ø1.00 in., hub = Ø1.75 in. × 1.50 in. long, keyseat to suit.

CAM 1
—Rise 2.00 in. in 150° with harmonic motion
—Dwell for 45°
—Drop 2.00 in. in 120° with uniformly accelerated and retarded motion
—Dwell remainder
—Displacement diagram 2.00 in. high × 12.00 in. long

CAM 2
—Rise 1.50 in. in 120° with uniform motion
—Dwell for 60°
—Rise .50 in. in 60° with parabolic motion
—Drop 2.00 in. with harmonic motion for remainder
—Displacement diagram 2.00 in. high × 12.00 in. long

FIG. 22-1-A   Cam displacement assignments.

CAM 1
—Rise 50 mm in 120° with cycloidal motion
—Dwell for 60°
—Drop 10 mm in 90° with uniform motion
—Drop 40 mm in 90° with uniformly accelerated and retarded motion
—Displacement diagram 50 mm high × 300 mm long

CAM 2
—Rise 30 mm in 90° with modified uniform motion
—Dwell for 60°
—Rise 20 mm in 45° with harmonic motion
—Drop 50 mm in 120° with parabolic motion
—Dwell remainder
—Displacement diagram 50 mm high × 300 mm long

FIG. 22-1-B   Cam displacement assignments.

Add a chart to the drawing showing the angular and radial displacements every 15°, taking the radial measurements from the prime circle. Scale 1:1.

3. Design a plate cam that will activate a Ø10 mm roller follower as follows:
   ■ Rise 40 mm in 150° with uniformly accelerated and retarded motion
   ■ Dwell for 45°
   ■ Drop 40 mm in 120° with modified uniform motion
   ■ Dwell for remainder
   Prime circle = Ø70 mm, plate thickness = 10 mm, shaft = Ø26 mm, hub = Ø44 × 32 mm long, keyseat to suit. Add a chart to the drawing showing the angular and radial displacements every 15°, taking the radial measurements from the prime circle. Scale 1:1.

4. Lay out the parallel-drive indexing unit shown in Fig. 22-2-A (pg. 770) with the timing hole rotated to position *B*. Use your judgment for dimensions not shown. The angular and radial displacement values locate the center of the roller. Scale 1:2.

5. Layout the cam and follower shown in Fig. 22-2-B (pg. 770). Also show the arm and follower in phantom in their maximum position. On the cam displacement drawing plot both the path of the cam follower and the plunger it actuates.

6. Lay out the oscillating unit shown in Fig. 22-2-C (pg. 771) using sequence *F* and lever *L4* in a horizontal position in the dwell down position. Prime circle = Ø2.50 in., follower = Ø.75 in., rise = .625 in. Use your judgment for dimensions not shown. Scale 1:1.

ASSIGNMENTS FOR UNIT 22-3, POSITIVE-MOTION CAMS

7. Make a two-view drawing of a face cam from the following information:
   ■ Rise 1.20 in. in 150° with harmonic motion
   ■ Dwell 30°
   ■ Drop 1.20 in. in 120° with parabolic motion
   ■ Dwell for remainder
   Roller = Ø.50 in., prime circle = Ø3.00 in., OD of face cam = 6.50 in., cam thickness = 1.00 in., groove depth = .38 in., shaft = Ø1.00 in., hub = Ø1.75 in. × 1.50 in. long. Add a suitable keyseat. Prepare a chart showing angular and radial displacement for every 15°, taking the radial measurements from the prime circle. Scale 1:1.

8. Make a two-view drawing of a face cam from the following information:
   ■ Rise 24 mm in 120° with parabolic motion
   ■ Dwell 45°
   ■ Drop with cycloidal motion for the remainder
   Roller = Ø12 mm, prime circle = Ø80 mm, OD of face cam = 160 mm, cam thickness = 25 mm, groove depth = 12 mm, shaft = Ø24 mm, hub = Ø42 mm × 28 mm long. Add a suitable keyseat. Prepare a chart showing angular and radial displacement for every 15°, taking the radial measurements from the prime circle. Scale 1:1.

9. Make a two-view drawing of a yoke cam that will raise the yoke 1.40 in. The cam is an eccentric cam having a

FIG. 22-2-A Parallel-drive indexing unit. *(Commercial Cam & Machine Co.)*

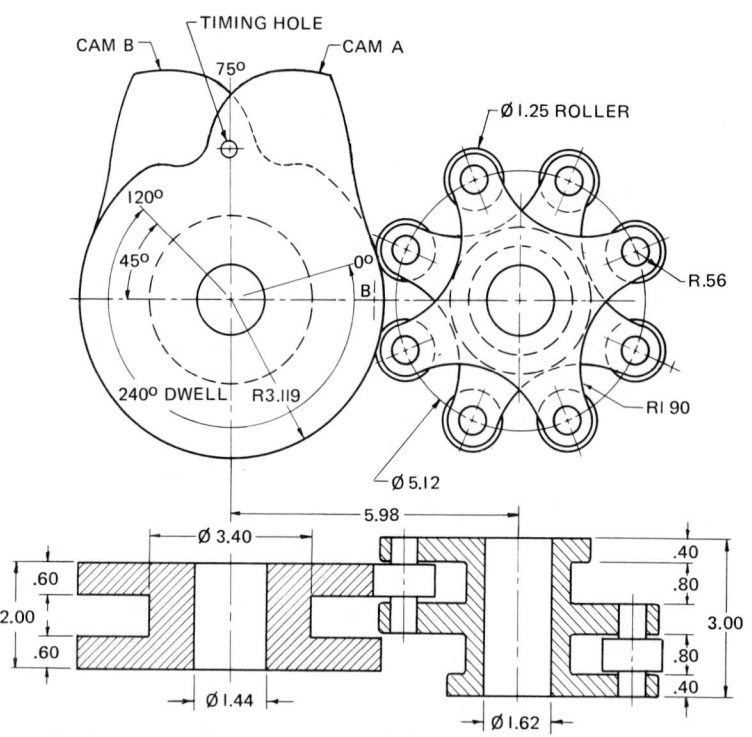

ANGULAR DISPLACEMENT AND RADIAL DISPLACEMENT FOR CAM A. CAM B IS OPPOSITE HAND TO CAM A. ANGULAR DIMENSIONS SHOWN ON DRAWING ARE FOR CAM A.									
0°	3.74	25°	4.08	50°	5.51	75°	4.86	95°	3.47
5°	3.75	30°	4.32	55°	5.46	80°	4.59	100°	3.61
10°	3.76	35°	4.64	60°	5.37	85°	4.26	105°	3.68
15°	3.80	40°	4.95	65°	5.24	90°	3.78	110°	3.73
20°	3.90	45°	5.24	70°	5.06	93°	3.41	115° AND 120°	3.74

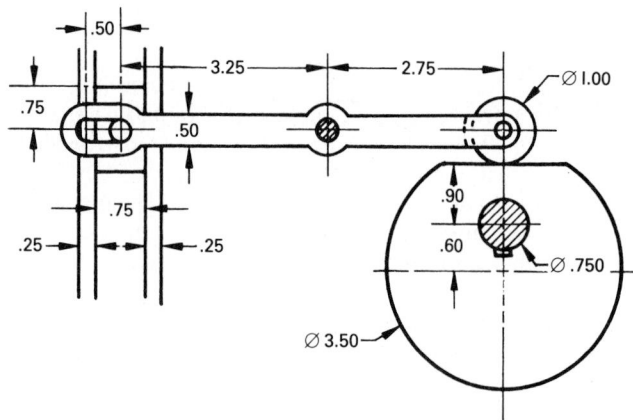

FIG. 22-2-B Eccentric cam.

dia. of 3.50 in. and a plate thickness of .75 in. Shaft = Ø1.06 in., hub = Ø1.75 in. × 1.10 in. long having the extension on one side only. The yoke is .50 in. thick and has a wall width of .75 in. A .25 in. × 1.25 in. steel guide bar is welded to the top and bottom of the yoke.

10. Make a two-view drawing of a yoke cam that will raise the yoke 35 mm. The cam is an eccentric cam having a dia. of 90 mm and a plate thickness of 20 mm. Shaft = Ø28 mm, hub = Ø44 mm × 30 mm long with the extension on one side only. The yoke is 14 mm thick and has a wall width of 20 mm. A 6 mm × 30 mm steel guide bar is welded to the top and bottom of the yoke.

ASSIGNMENTS FOR UNIT 22-4, DRUM CAMS

11. Make a two-view drawing of a drum cam from the following information:
    - Rise 1.50 in. in 120° with parabolic motion
    - Dwell for 60°
    - Drop 1.50 in. in 150° with modified sine motion
    - Dwell for remainder

    Roller follower = Ø.50 in., cam = Ø3.00 in. × 3.50 in. long, follower groove = .40 in. deep. Use your judgment for dimensions not given. Show the full development of the cam, which will serve as a motion diagram. Prepare a

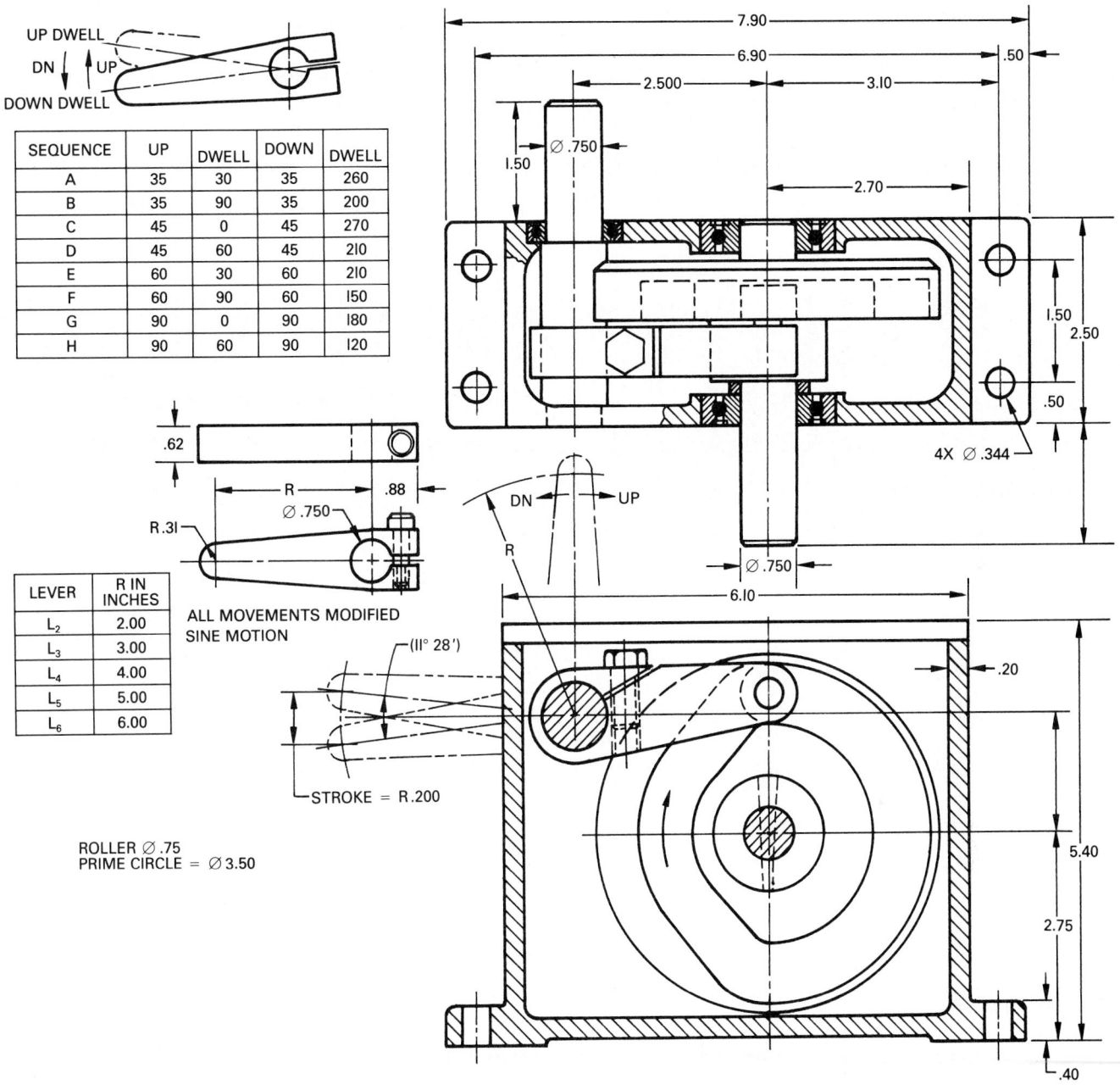

SEQUENCE	UP	DWELL	DOWN	DWELL
A	35	30	35	260
B	35	90	35	200
C	45	0	45	270
D	45	60	45	210
E	60	30	60	210
F	60	90	60	150
G	90	0	90	180
H	90	60	90	120

LEVER	R IN INCHES
L_2	2.00
L_3	3.00
L_4	4.00
L_5	5.00
L_6	6.00

ALL MOVEMENTS MODIFIED SINE MOTION

ROLLER ⌀.75
PRIME CIRCLE = ⌀3.50

STROKE = R.200

FIG. 22-2-C   Oscillating unit. *(Protected by patent—Stelron Cam Co.)*

chart showing the angular displacement from a timing hole located at 0°, and the displacement from the baseline for every 15°. Scale 1:1.

12. Make a two-view drawing of a drum cam from the following information:
  - Rise 32 mm in 150° with harmonic motion
  - Dwell for 45°
  - Drop 32 mm in 120° with trapezoid motion
  - Dwell for remainder

  Roller follower = ⌀14 mm, cam = ⌀70 mm × 64 mm long, follower groove = 10 mm deep. Use your judgment for dimensions not given. Show the full development of the cam, which will serve as a motion diagram. Prepare a

chart showing the angular displacement from a timing hole located at 0°, and the displacement from the baseline for every 15°. Scale 1:1.

ASSIGNMENT FOR UNIT 22-5, INDEXING

13. Make a two-view drawing of the indexing drive 6S5 or 6S75 shown in Fig. 22-5-2. Use your judgment for dimensions not shown. Draw an angular displacement diagram, plotting points every 5° on the index cycle. Add suitable keyseats. Scale 1:1.

ASSIGNMENTS FOR UNIT 22-6, LINKAGES

14. *Simple Crank Mechanism.* Lay out the two linkages shown in Fig. 22-6-A and plot the paths at 15° intervals of the points indicated. Scale 1:1.

15. Lay out the two linkages shown in Fig. 22-6-B and plot the paths at 15° intervals of point C in (A) and E in (B). Points C and E are located at midpoint of links. Scale 1:1.

16. Lay out the two linkages shown in Fig. 22-6-C. For the Peaucellier's mechanism plot the path taken by point C. Plot points by moving point D every .25 in., $OB = 4.50$, $DB = 1.75$ in. For the toggle linkage, plot the distance X for every 15° of rotation of point A. $AB = 2.50$ in., $BC =$

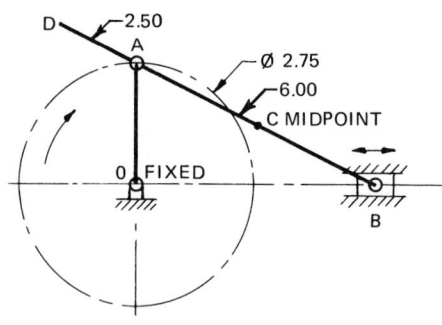

(A) SIMPLE CRANK

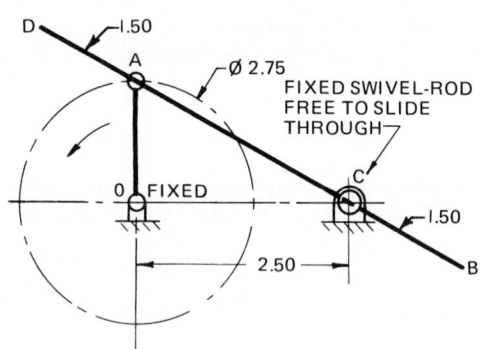

(B) CRANK WITH SLIDING ROD

FIG. 22-6-A Simple crank mechanisms.

FIG. 22-6-C Linkages.

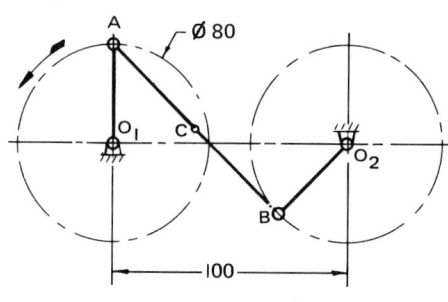

(A) CROSS-LINKED CRANK

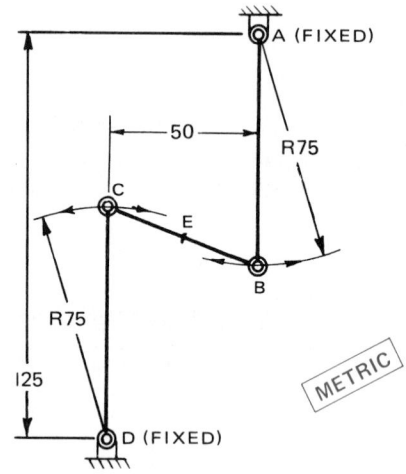

(B) WATT'S APPROXIMATE STRAIGHT-LINE MECHANISM

FIG. 22-6-B Linkage assignment.

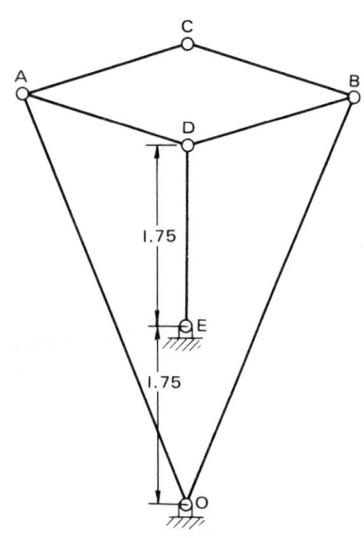

(A) PEAUCELLIER'S MECHANISM

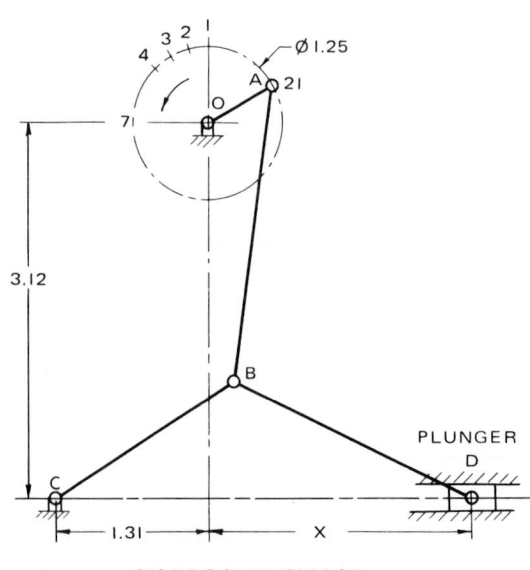

(B) TOGGLE LINKAGE

772

1.75 in., *BD* = 2.25 in. Scale 1:1. Motion displacement diagram size is 2.00 × 6.00 in.

17. Lay out the shaper shown in Fig. 22-6-D and complete the motion displacement diagram for the cutter travel for two complete strokes. Take positions every 30° of trunnion rotation starting at position 240°. Motion displacement diagram size is 4.00 × 9.00 in. Scale 1:4.

ASSIGNMENTS FOR UNIT 22-7, RATCHET WHEELS

18. Lay out a ratchet assembly using a U-shaped pawl with a ratchet wheel. The ratchet wheel is to have 24 teeth; OD of

Ø146 mm; hub Ø48 mm; shaft Ø32.5 mm; keyseat to suit; width of teeth 12 mm; depth of teeth 10 mm; and the hub is to extend 16 mm on one side. Scale 1:1. Show two views.

19. *Ratchet and Crank Mechanism.* Lay out a one-view drawing of the ratchet design shown in Fig. 22-7-A. Two pawls are used, a drive pawl as shown and a holding pawl held in position by a spring. Using crank rotation positions every 22.5°, plot the path of the end of the drive pawl. Use your judgment for dimensions not shown. Scale 1:1.

20. Lay out the pinion and pawl shown in Fig. 14-3-F but show the assembly in the reverse locking position. Scale 2:1.

FIG. 22-6-D   Shaper using Whitworth quick-return mechanism stroke.

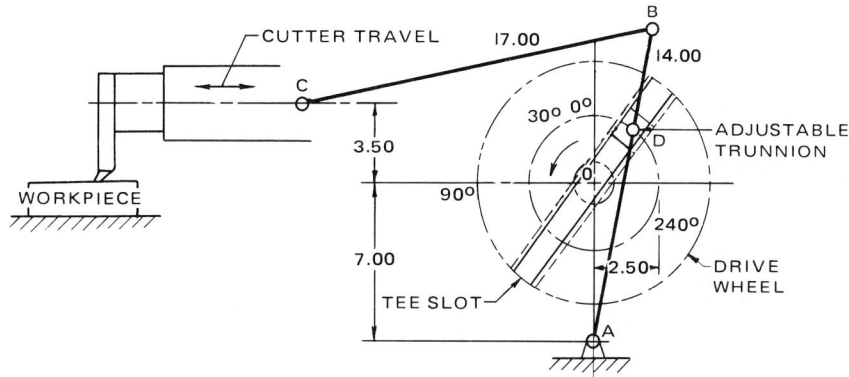

(A) SHAPER SHOWING QUICK-RETURN MECHANISM

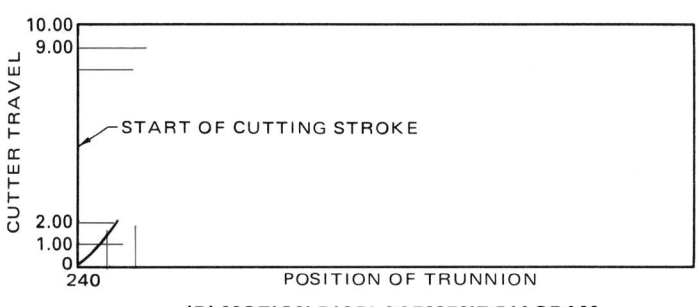

(B) MOTION DISPLACEMENT DIAGRAM

FIG. 22-7-A   Ratchet and crank mechanism.

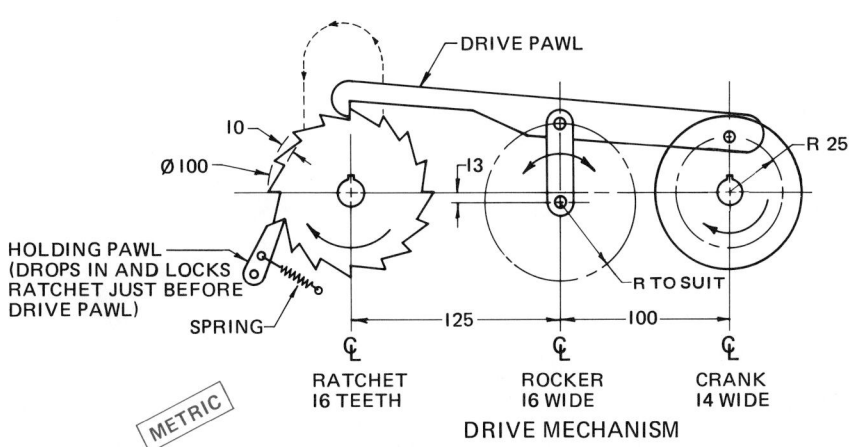

DRIVE MECHANISM

P A R T 5

SPECIAL FIELDS OF DRAFTING

CHAPTER 23

DEVELOPMENTS AND INTERSECTIONS

Definitions

Developable   Having the ability to have a thin sheet of flexible material wrapped smoothly about a surface.

Line of intersection   The line common to both surfaces when two surfaces meet.

Parallel-line development   A development technique used for an object having a single, curved surface.

Straight-line development   The development of an object that has surfaces on a flat plane of projection.

Surface development or **pattern drawing**   A layout used as a pattern for tracing out the developed shape on a flat material.

23-1 SURFACE DEVELOPMENTS

Many objects, such as cardboard and metal boxes, tin cans, funnels, cake pans, furnace pipes, elbows, ducts, and roof gutters, are made from flat sheet material that, when folded, formed, or rolled, will take the shape of an object. Since a definite shape and size are desired, a regular orthographic drawing of the object, as in Fig. 23-1-1, is made first; then a development drawing is made to show the complete surface or surfaces laid out in a flat plane.

Sheet-Metal Development

Surface development drawing is sometimes referred to as *pattern drawing,* because the layout, when made on heavy cardboard, thin metal, or wood, is used as a pattern for tracing out the developed shape on flat material. Such patterns are used extensively in sheet-metal shops.

When making a development drawing of an object that will be constructed of thin metal, such as a tin can or a dust pan,

the drafter must be concerned not only with the developed surfaces but with the joining of the edges of these surfaces and with exposed edges. An allowance must be made for the additional material necessary for such seams and edges. The drafter must also mark where the material is to be bent. The most common method of representing bend lines is shown in Fig. 23-1-2. If the finished drawing is not shown with the development drawing, instructions such as bend up 90°, bend down 180°, bend up 45° are shown beside each bend line. Figure 23-1-3 (pg. 778) shows a number of common methods for seaming and edging. Seams are used to join edges. The seams may be fastened together by lock seams, solder, rivets, adhesive, or welds. Exposed edges are folded or wired to give the edge added strength and to eliminate the sharp edge.

A surface is said to be *developable* if a thin sheet of flexible material, such as paper, can be wrapped smoothly about its surface. Objects that have plane, or flat surfaces or single-curved surfaces are developable, but if a surface is double-curved or warped, approximate methods must be used to develop the surface. The development of a spherical shape would thus be approximate, and the material would be stretched to compensate for small inaccuracies. For example, the coverings for a football or a basketball are made in segments, each segment cut to an approximate developed shape; the segments are then stretched and sewed together to give the desired shape.

Sheet-Metal Sizes

Metal thicknesses less than .25 in. (6 mm) are usually designated by a series of gage numbers, the more common gages being shown in Table 52 of the Appendix. Metal .25 in. and over is given in inch or millimeter sizes. In calling for the material size of sheet-metal developments, customary practice is to give the gage number, type of gage, and its inch or millimeter equivalent in parentheses, followed by the developed width and length (Fig. 23-1-4, pg. 778).

FIG. 23-1-1 Development of a rectangular box.

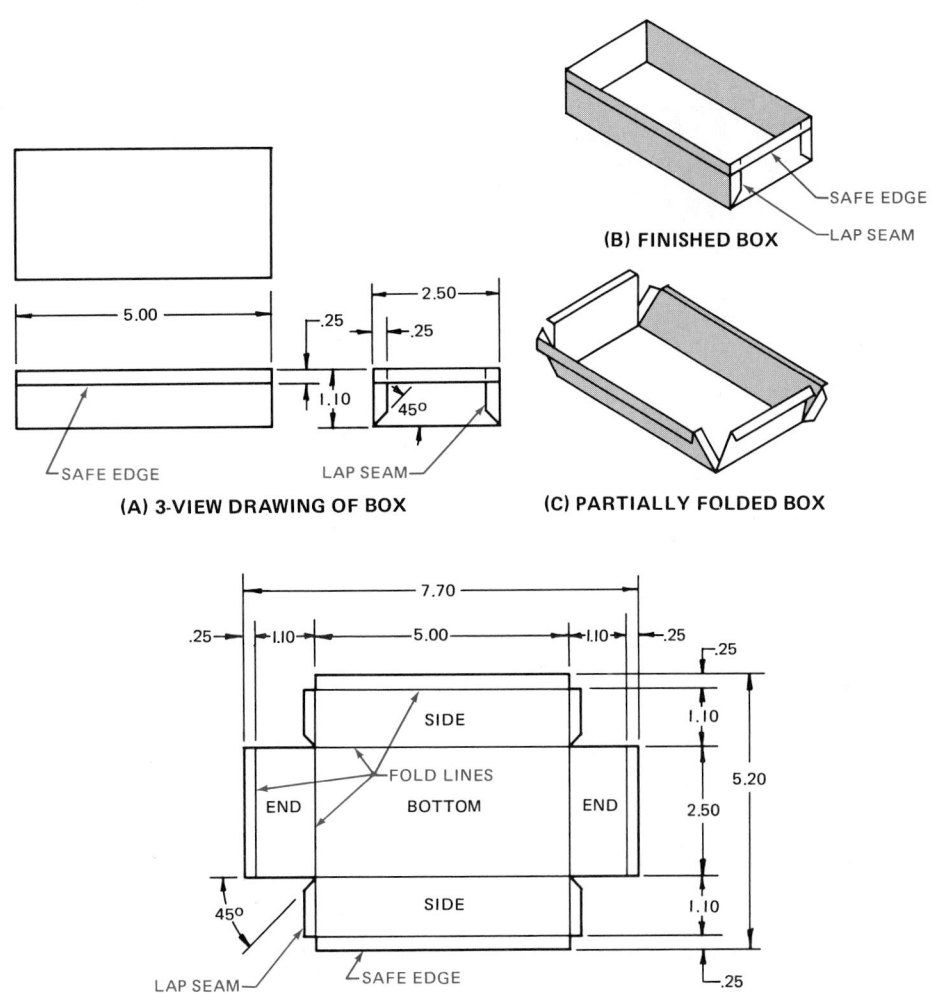

(A) 3-VIEW DRAWING OF BOX

(B) FINISHED BOX

(C) PARTIALLY FOLDED BOX

(D) DEVELOPMENT OF BOX

FIG. 23-1-2 Development drawing with a complete set of folding and assembly instructions.

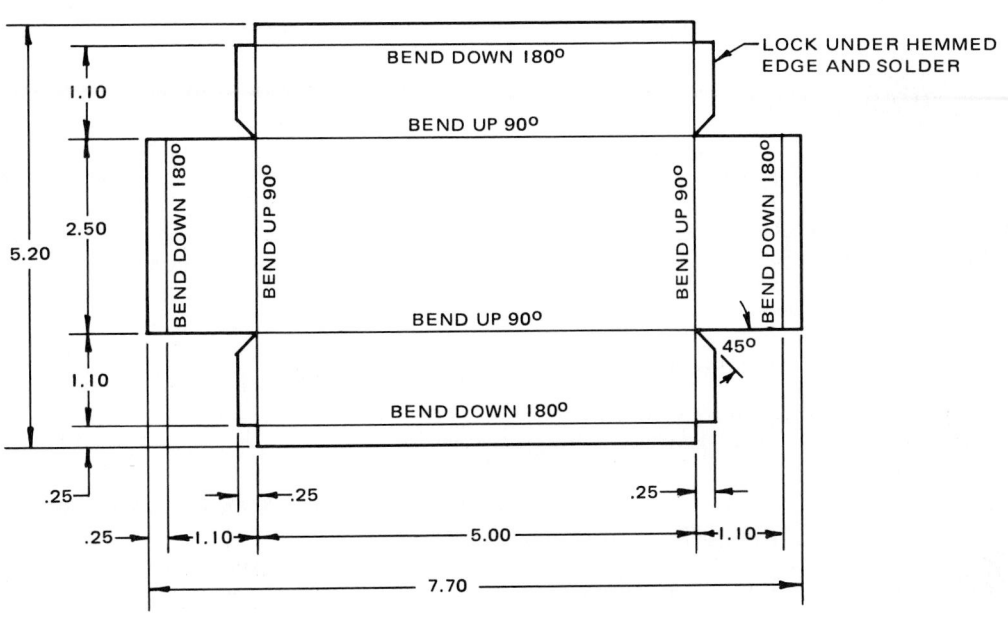

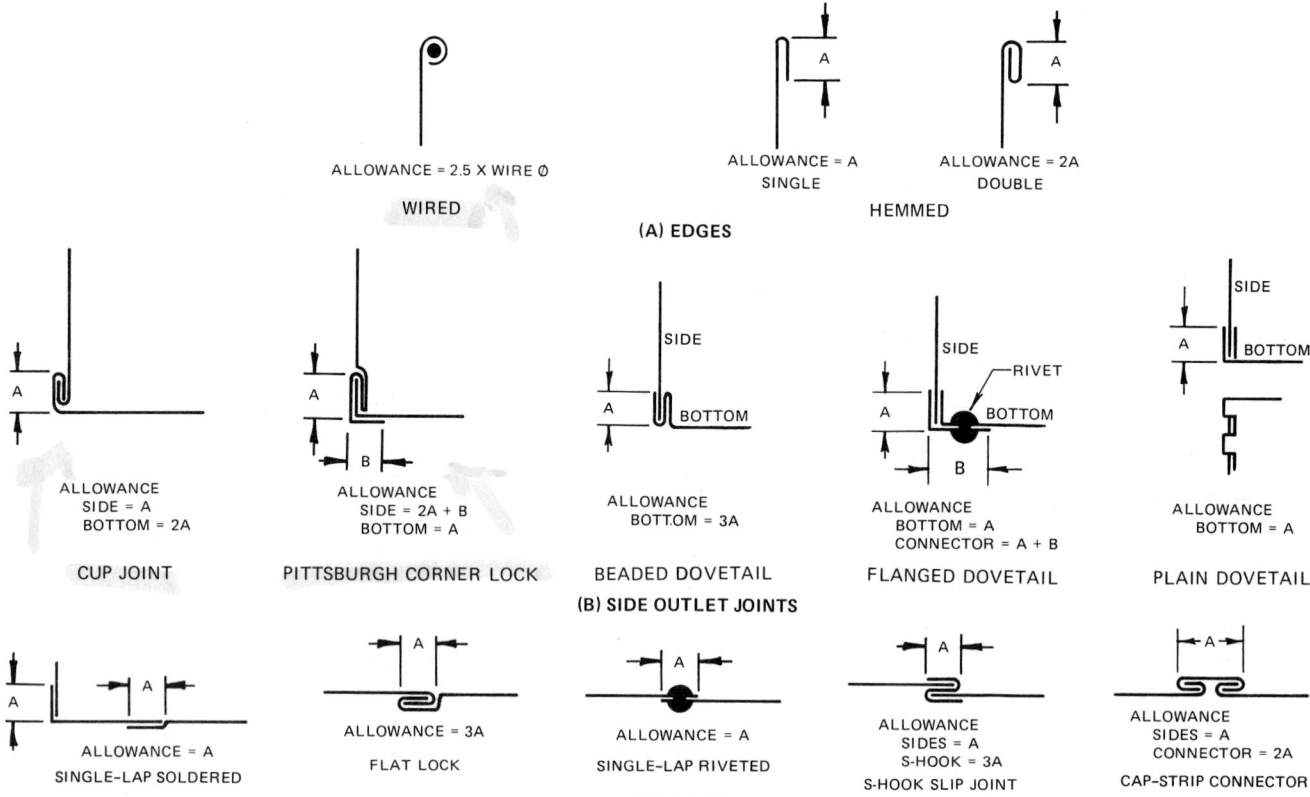

FIG. 23-1-3   Joints, seams, and edges.

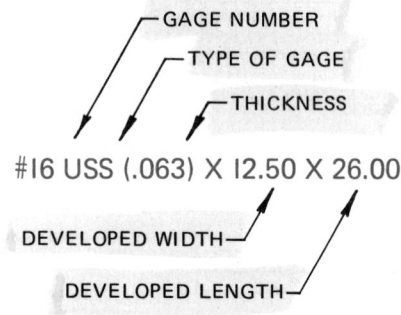

FIG. 23-1-4   Callout of sheet-metal sizes.

Straight-Line Development

Straight-line development is the term given to the development of an object that has surfaces on a flat plane of projection. The true size of each side of the object is known, and these sides can be laid out in successive order. Figure 23-1-1 shows the development of a simple rectangular box having a bottom and four sides. Note that in the development of the box an allowance is made for lap seams at the corners and for a folded edge. The fold lines are shown as thin, unbroken lines. Note also that all lines for each surface are straight.

ASSIGNMENT

See Assignment 1 for Unit 23-1 on page 799.

23-2   THE PACKAGING INDUSTRY

Packaging, which involves the principles of surface development, is one of the largest and most diversified industries in the world. Most products are packaged in metal, plastic, or cardboard containers. Many products, from candy-coated gum to large television sets, are packaged in cardboard containers (Figs. 23-2-1 and 23-2-2). Such containers, often referred to as "cartons," in many instances must be attractive as well as functional. They may be designed for sales appeal as well as for protection against contamination or damage from shipping and handling. They are also designed for temporary or permanent use.

Many cartons are printed, cut, creased, and sent to the customer in a flat position (Fig. 23-2-3). They take less space to store and ship and are readily assembled. Locking devices such as tabs hold each box together. Containers with tabs are used extensively by food chain operators, such as Dunkin Donuts, McDonald's, and Taco Bell. Unusual shapes, such as hexagons and octagons, as shown in Figs. 23-2-4 and 23-2-5 (pp. 779 to 780), are becoming popular because of their novel form.

ASSIGNMENTS

See Assignments 2 through 4 for Unit 23-2 on pages 800 to 801.

FIG. 23-2-1   Typical commercial containers. *(STUDIOHIO)*

FIG. 23-2-2   A familiar container made by cutting and folding a flat sheet. *(STUDIOHIO)*

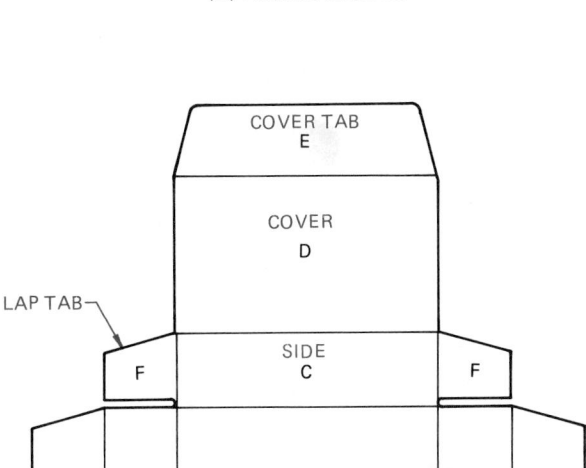

(A) FOLDED CARTON

(B) DEVELOPMENT OF CARTON

FIG. 23-2-3   Development of a one-piece carton with fold-down sides.

FIG. 23-2-4   Development of a truncated hexagon.

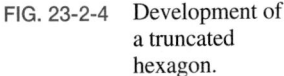

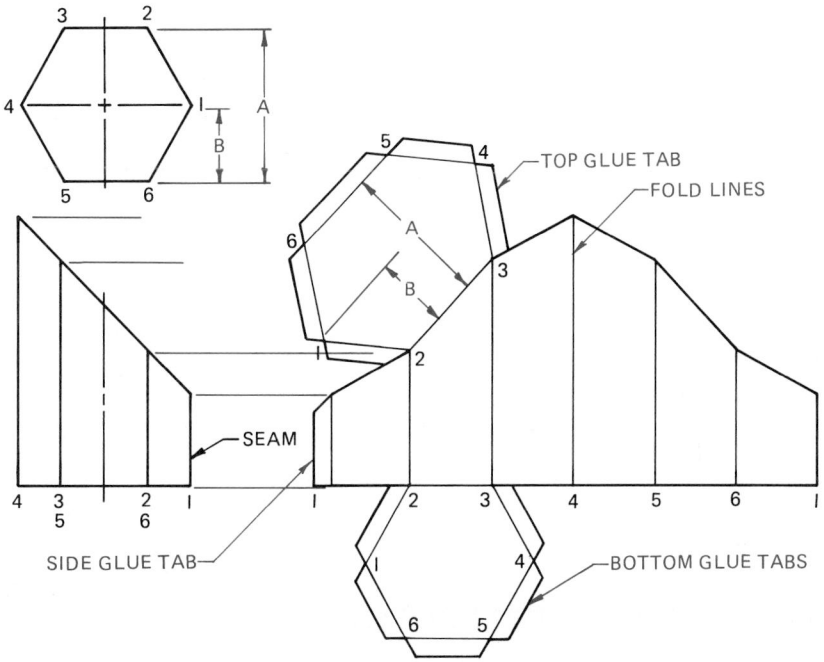

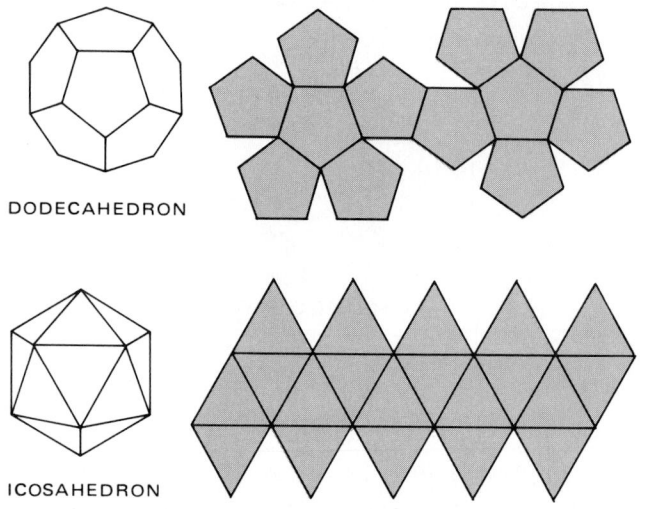

DODECAHEDRON

ICOSAHEDRON

FIG. 23-2-5   Twelve- and twenty-sided shapes.

23-3 RADIAL LINE DEVELOPMENT OF FLAT SURFACES

Development of a Right Pyramid with True Length of Edge Lines Shown   A right pyramid is a pyramid having all the lateral edges (from vertex to the base) of equal length (Fig. 23-3-1). Since the true length of the lateral edge is shown in the front view (line 0-1 or 0-3) and the top view shows the true lengths of the edges of the base (lines 1-2, 2-3, etc.), the development may be constructed as follows: with 0 as center (corresponding to the apex) and with a radius equal to the true length of the lateral edges (line 0-1 in the front view), draw an arc as shown. Drop a perpendicular from 0 to intersect the arc at point 3. With a radius equal to the length of the edges of the base (line 1-2 on the top view), start at point 3 and step off the distances 3-2, 2-1, 3-4, and 4-1 on the large arc. Join these points with straight lines. These points are then connected to point 0

FIG. 23-3-1   Development of a right pyramid with true length of edge lines shown.

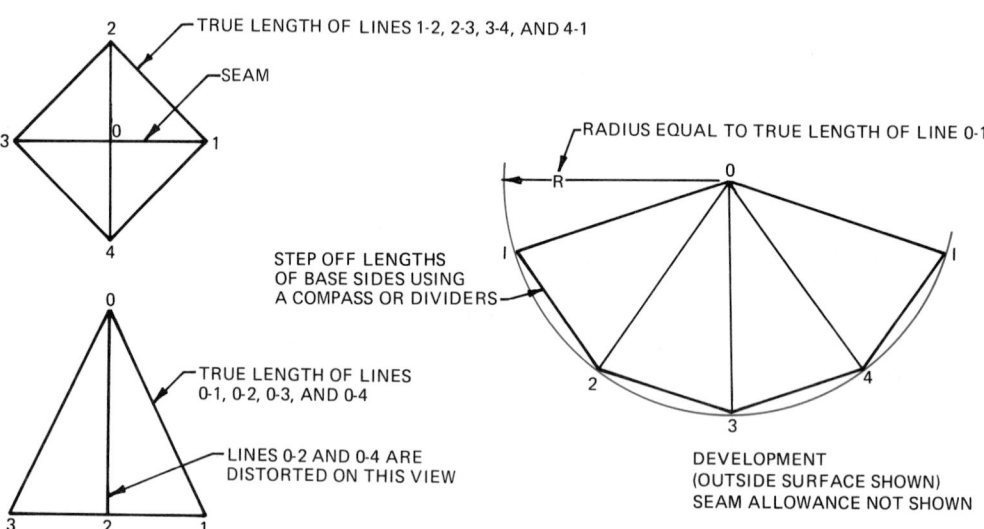

TRUE LENGTH OF LINES 1-2, 2-3, 3-4, AND 4-1

SEAM

STEP OFF LENGTHS OF BASE SIDES USING A COMPASS OR DIVIDERS

TRUE LENGTH OF LINES 0-1, 0-2, 0-3, AND 0-4

LINES 0-2 AND 0-4 ARE DISTORTED ON THIS VIEW

RADIUS EQUAL TO TRUE LENGTH OF LINE 0-1

DEVELOPMENT (OUTSIDE SURFACE SHOWN) SEAM ALLOWANCE NOT SHOWN

(A) DEVELOPMENT OF A PYRAMID

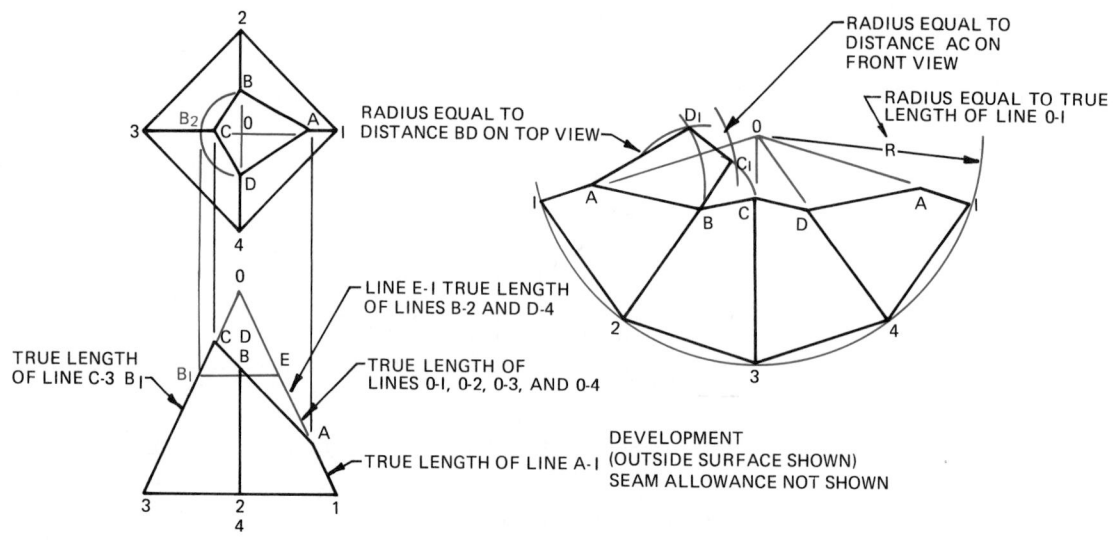

RADIUS EQUAL TO DISTANCE BD ON TOP VIEW

RADIUS EQUAL TO DISTANCE AC ON FRONT VIEW

RADIUS EQUAL TO TRUE LENGTH OF LINE 0-1

LINE E-1 TRUE LENGTH OF LINES B-2 AND D-4

TRUE LENGTH OF LINES 0-1, 0-2, 0-3, AND 0-4

TRUE LENGTH OF LINE C-3

TRUE LENGTH OF LINE A-1

DEVELOPMENT (OUTSIDE SURFACE SHOWN) SEAM ALLOWANCE NOT SHOWN

(B) DEVELOPMENT OF A TRUNCATED PYRAMID

by straight lines to complete the development. Lines 0-2, 0-3, and 0-4 are the lines on which the development is folded to shape the pyramid. The base and seam allowances have been omitted for clarity.

In developing a truncated pyramid of this type, the procedure is the same as described above, except that only a portion of lines 0-1, 0-2, 0-3, and 0-4 is required. The positions of points B and D in the top view are found by projecting lines horizontally from points B and D in the front view to intersect the true-length line 0-3 at B_1. Project a vertical line from point B_1 to intersect point B_2 in the top view. Rotate B_2 90° from point 0 to intersect line 0-2 at B and 0-4 at D. It will be noted that only lines A-1 and C-3 appear as their true length in the front view. The true length of lines B-2 and D-4 may be found by projecting a horizontal line from points B and D to point E on the true-length line 0-1.

To complete the development, step off distances 1-A on line 1-0, 2-B on line 2-0, 3-C on line 3-0, and 4-D on line 4-0. Join

points A, B, C, D, A with straight lines. The top surface of the truncated pyramid may be added to the development as follows: With A as center and with a radius equal to distance AC in the front view, swing an arc. With B as center and with a radius equal to line BC on the development, swing an arc intersecting the first arc at C_1. Join point B to point C_1 with a straight line. With A as center and with a radius equal to line AB in the development, swing an arc. With B as center and with a radius equal to distance BD in the top view, swing an arc intersecting the first arc at D_1. Join AD_1 and D_1C_1 with straight lines.

Development of a Right Pyramid with True Length of Edge Lines Not Shown In order to construct the development (Fig. 23-3-2), the true length of the edge lines 0-1, 0-2, etc., must first be found. The true length of the edge lines would be equal to the hypotenuse of a right triangle having one leg equal in length to the projected edge line in the top view and the

FIG. 23-3-2 Development of a right pyramid with true length of edge lines not shown.

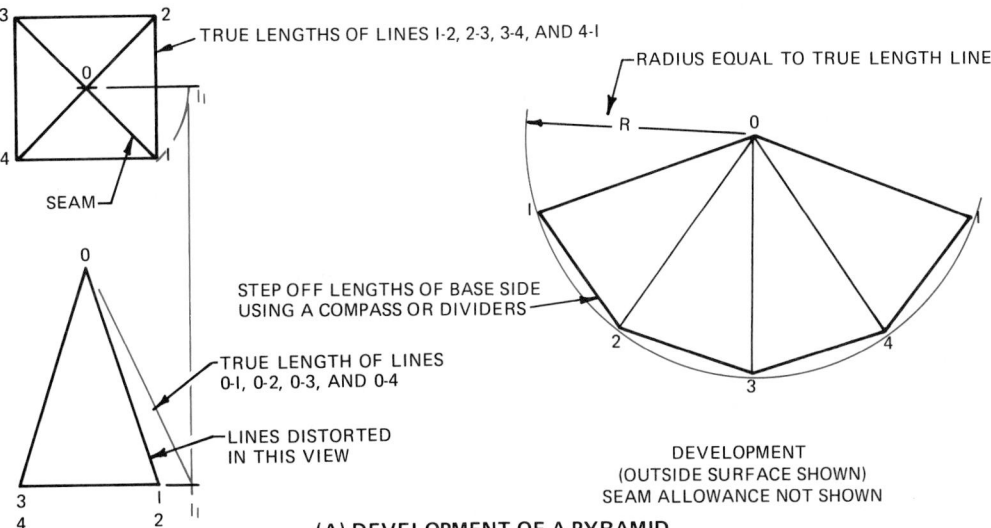

(A) DEVELOPMENT OF A PYRAMID

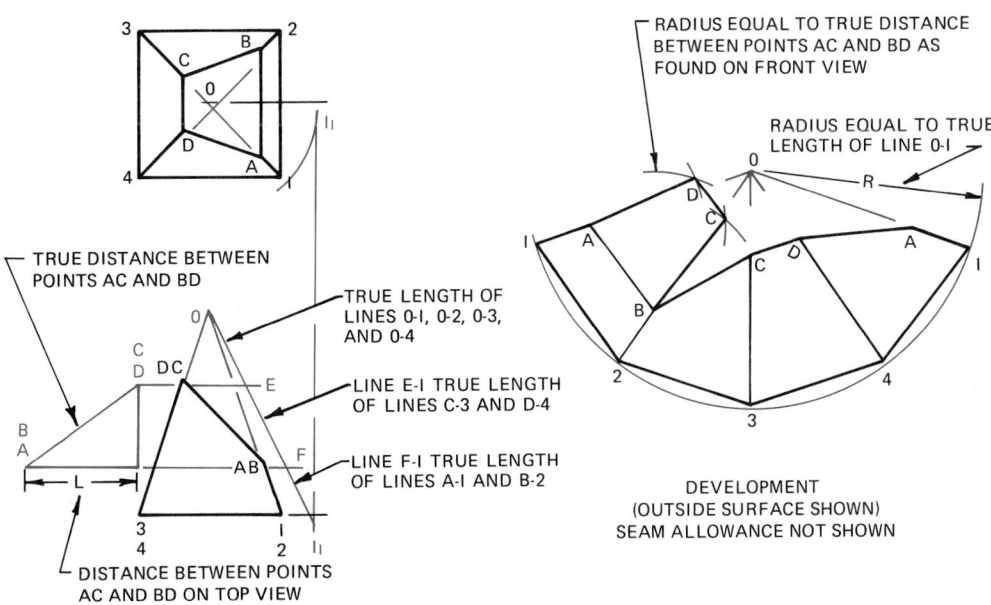

(B) DEVELOPMENT OF A TRUNCATED PYRAMID

781

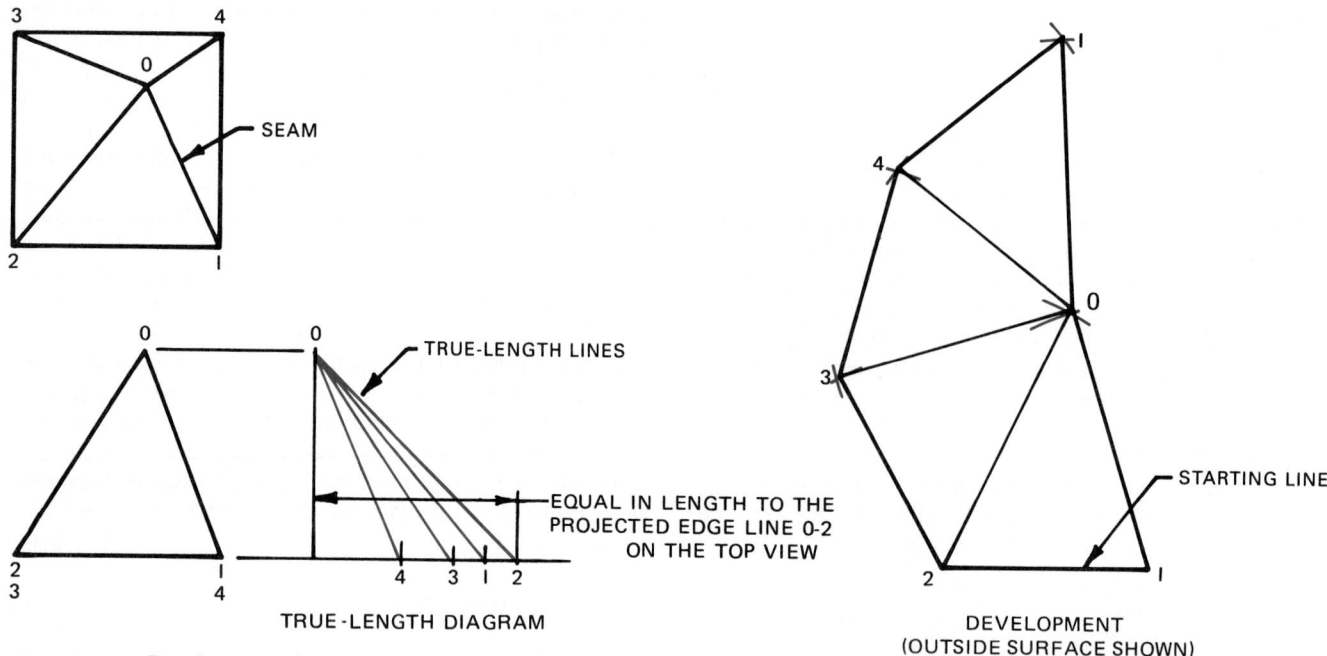

FIG. 23-3-3   Development of an oblique pyramid by triangulation.

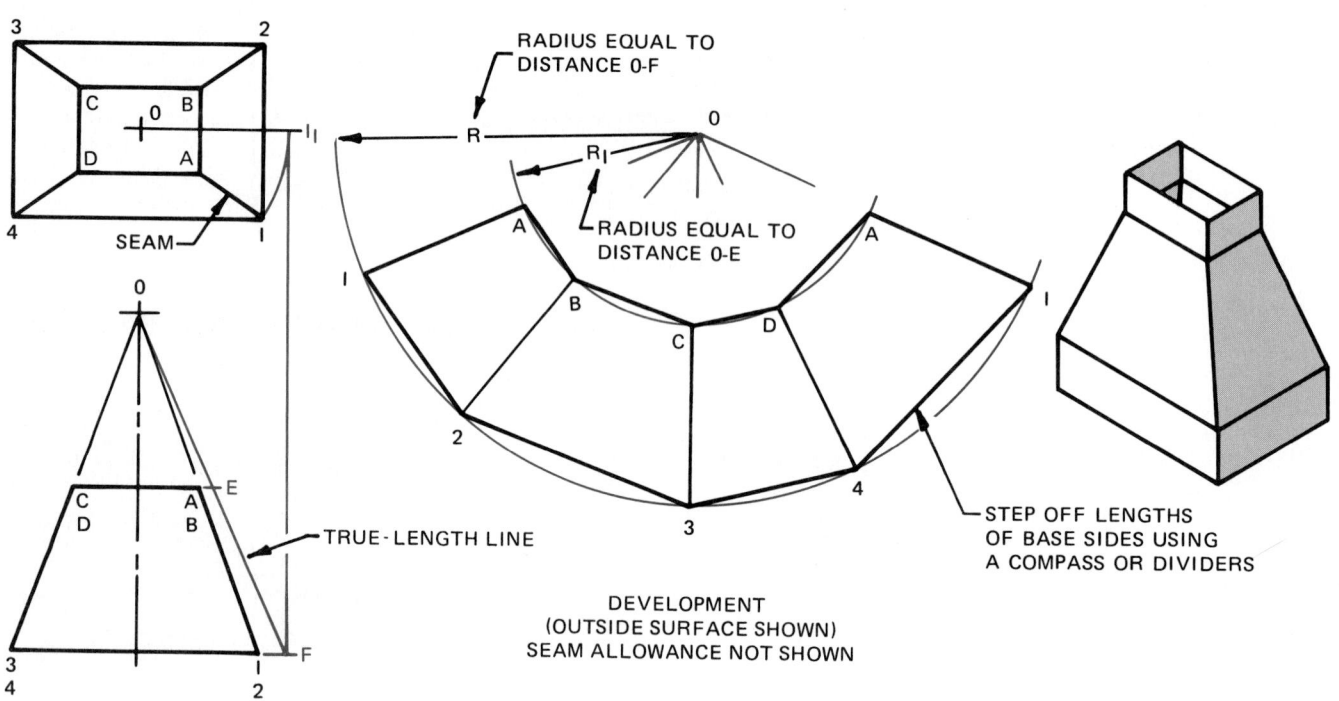

FIG. 23-3-4   Development of a transition piece.

other leg equal to the height of the projected edge line in the front view. Since only one true-length line is required, it may be developed directly on the front view rather than by making a separate true-length diagram. With 0 in the top view as center and radius equal to distance 0-1 in the top view, swing an arc from point 1 until it intersects the center line at point 1_1. Project a vertical line down to the front view, intersecting the baseline at point 1_1. Line $0\text{-}1_1$ is the true length of the edge lines. The

development may now be constructed in a similar manner to that outlined in the previous development.

In developing a truncated pyramid of this type, the procedure is the same except only the truncated edge lines are required. The true length of the truncated edge lines is required and may be found by projecting lines horizontally from points A, B, C, and D in the front view to intersect the true-length line $0\text{-}1_1$ at points F and E, respectively. Line $F\text{-}1_1$ is the true length of the truncated edge lines $A1$ and $B1$, and line $E1_1$ is the true length of the remaining truncated edge lines $C3$ and $D4$. The sides of the truncated pyramid may now be constructed in the development view. The top surface of the truncated pyramid may be added to the development as follows: With points A and B on the development as centers and with a radius equal in length to line BC on the development, swing light arcs. With a radius equal in length to the true distance between points A and C or B and D (this is found on the true-length diagram constructed to the left of the front view) and with center B, swing an arc intersecting the first arc at C. Repeat, using point A as center and intersecting the other arc at point C. Join points B, C, and A with straight lines to complete the top surface. The baseline and seam lines have been omitted for clarity.

Development of an Oblique Pyramid   Development of an oblique pyramid having all its lateral edges of unequal length is shown in Fig. 23-3-3. The true length of each of these edges must first be found as shown in the true-length diagram. The development may now be constructed as follows: Lay out baseline 1-2 in the development view equal in length to the baseline 1-2 found in the top view. With point 1 as center and radius equal in length to line 0-1 in the true-length diagram, swing an arc. With point 2 as center and radius equal in length to line 0-2 in the true-length diagram, swing an arc intersecting the first arc at 0. With point 0 as center and radius equal in length to line 0-3 in the true-length diagram, swing an arc. With point 2 as center and radius equal in length to baseline 2-3 found in the top view, swing an arc intersecting the first arc at point 3. Locate point 4 and point 1 in a similar manner, and join these points, as shown, with straight lines. The baseline and seam lines have been omitted on the development drawing.

Development of a Transition Piece   The development of the transition piece (Fig. 23-3-4) is created in a similar manner to that of the development of the right pyramid (Fig. 23-3-2).

ASSIGNMENT

See Assignment 5 for Unit 23-3 on pages 801–802.

23-4   PARALLEL LINE DEVELOPMENT OF CYLINDRICAL SURFACES

The lateral, or curved, surface of a cylindrically shaped object, such as a tin can, is developable since it has a single curved surface of one constant radius. The development technique used for such objects is called *parallel line development*. Figure 23-4-1 shows the development of the lateral surface of a simple hollow cylinder. The width of the development is equal to the height of the cylinder, and the length of the development is equal to the circumference of the cylinder plus the seam allowance. Figure 23-4-2 (pg. 784) shows the development of a cylinder with the top truncated at a 45° angle (one-half of a two-piece 90° elbow). Points of intersection are established to give the curved shape on the development. These points are derived from the intersection of a length location, representing a certain distance around the circumference from a starting point, and the height location at that same point on the circumference. The closer the points of intersection are to one another, the greater the accuracy of the development. An irregular curve is used to connect the points of intersection.

Figure 23-4-3 (pg. 784) shows the development of the surface of a cylinder with both the top and the bottom truncated at an angle of 22.5° (the center part of a three-piece elbow). It is normal practice in sheet-metal work to place the seam on the shortest side. In the development of elbows, however, this practice would result in considerable waste of material, as illustrated by Fig. 23-4-4A (pg. 784). To avoid this waste and to simplify cutting the pieces, the seams are alternately placed 180° apart, as illustrated by Fig. 23-4-4B for a two-piece elbow and by Fig. 23-4-4C for a three-piece elbow. Refer to Figs. 23-4-5 and 23-4-6 (pg. 785) for complete developments of two- and four-piece elbows.

ASSIGNMENTS

See Assignments 6 through 8 for Unit 23-4 on pages 802 to 803.

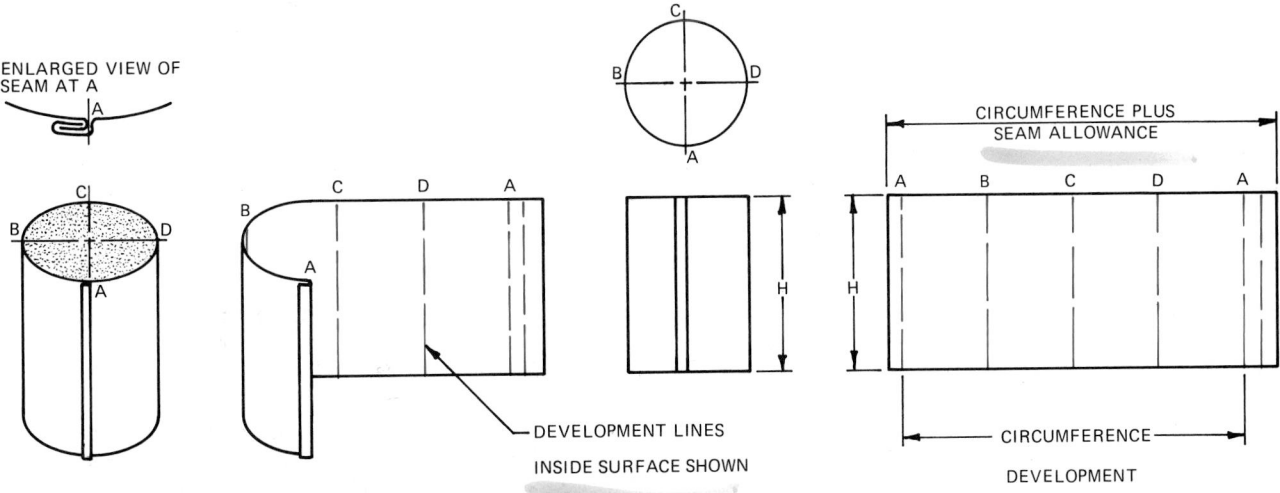

FIG. 23-4-1   Development of a cylinder.

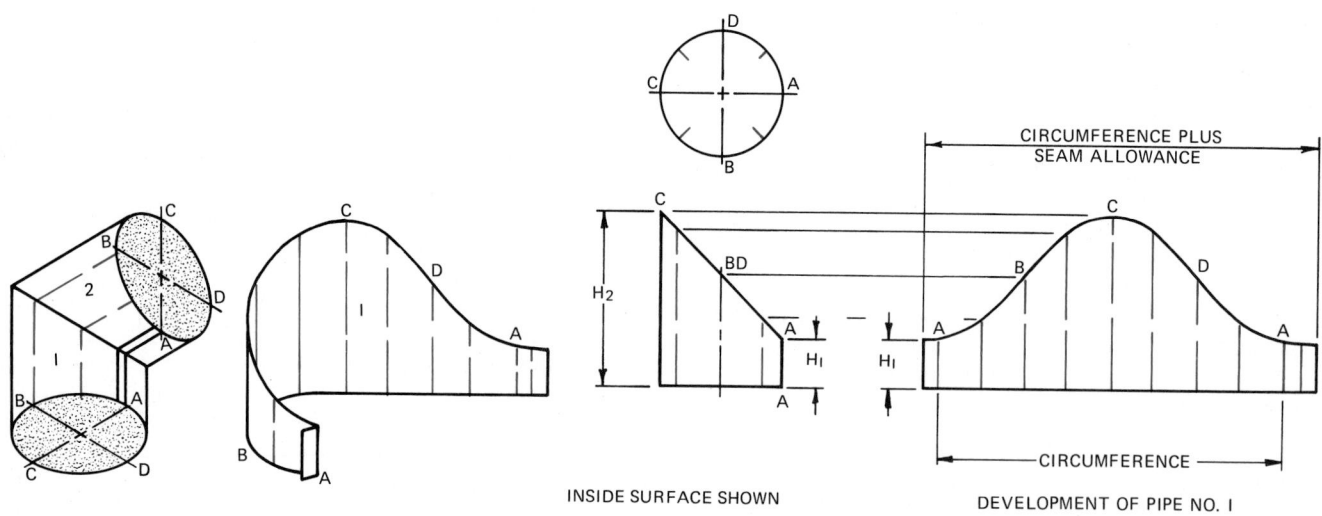

FIG. 23-4-2   Development of a truncated cylinder.

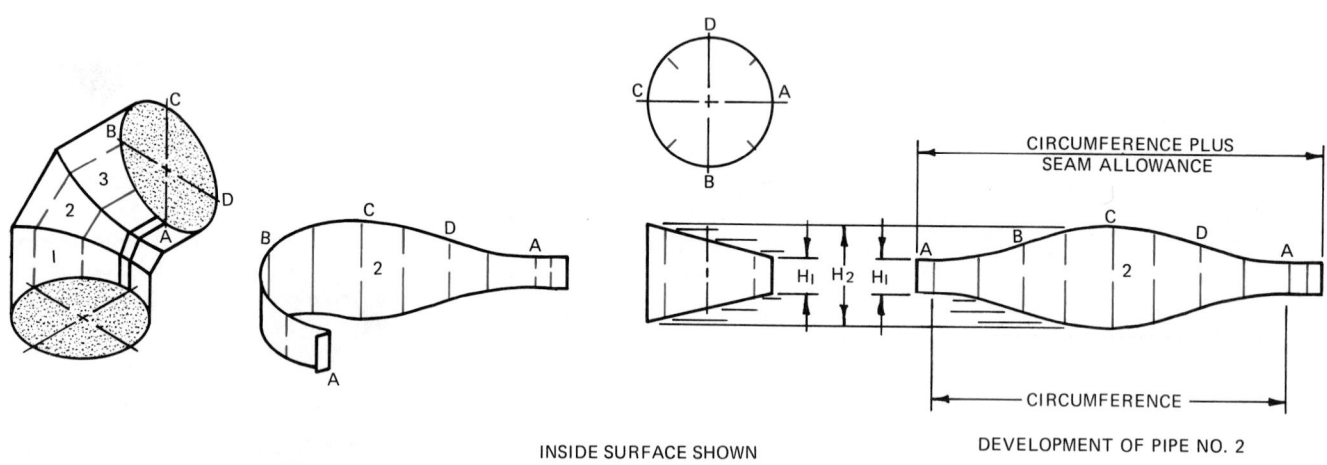

FIG. 23-4-3   Development of a cylinder with the top and bottom truncated.

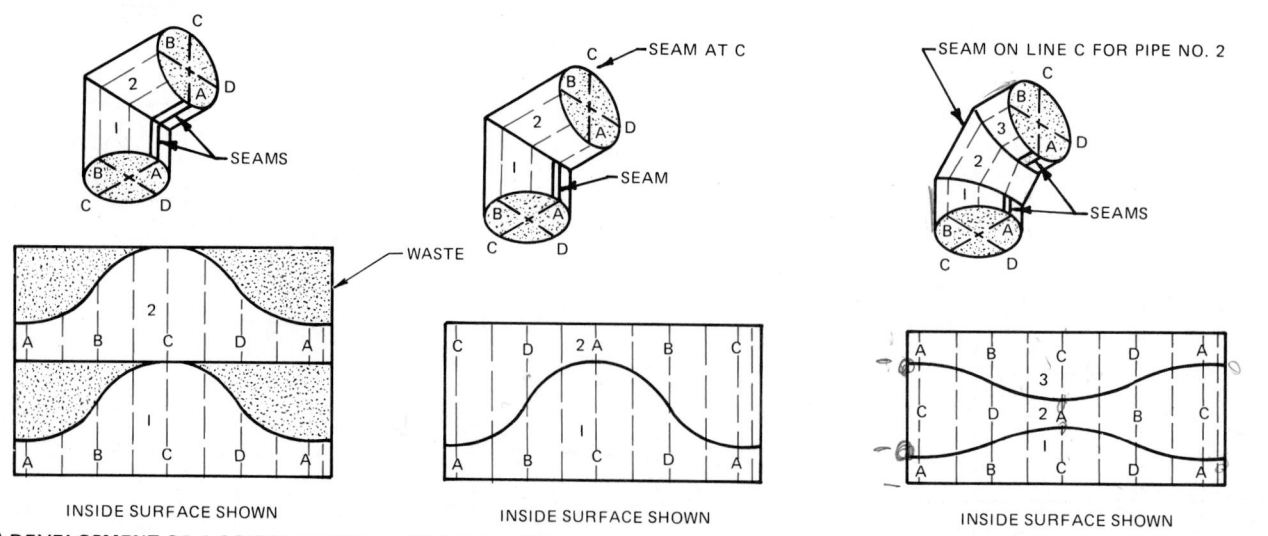

**(A) DEVELOPMENT OF A 2-PIECE ELBOW
WITH BOTH SEAMS ON LINE A**

**(B) DEVELOPMENT OF A 2-PIECE ELBOW
WITH SEAMS ON LINES A AND C**

**(C) DEVELOPMENT OF A 3-PIECE ELBOW WITH
SEAMS ALTERNATED ON LINES A AND C**

FIG. 23-4-4   Location of seams on elbows.

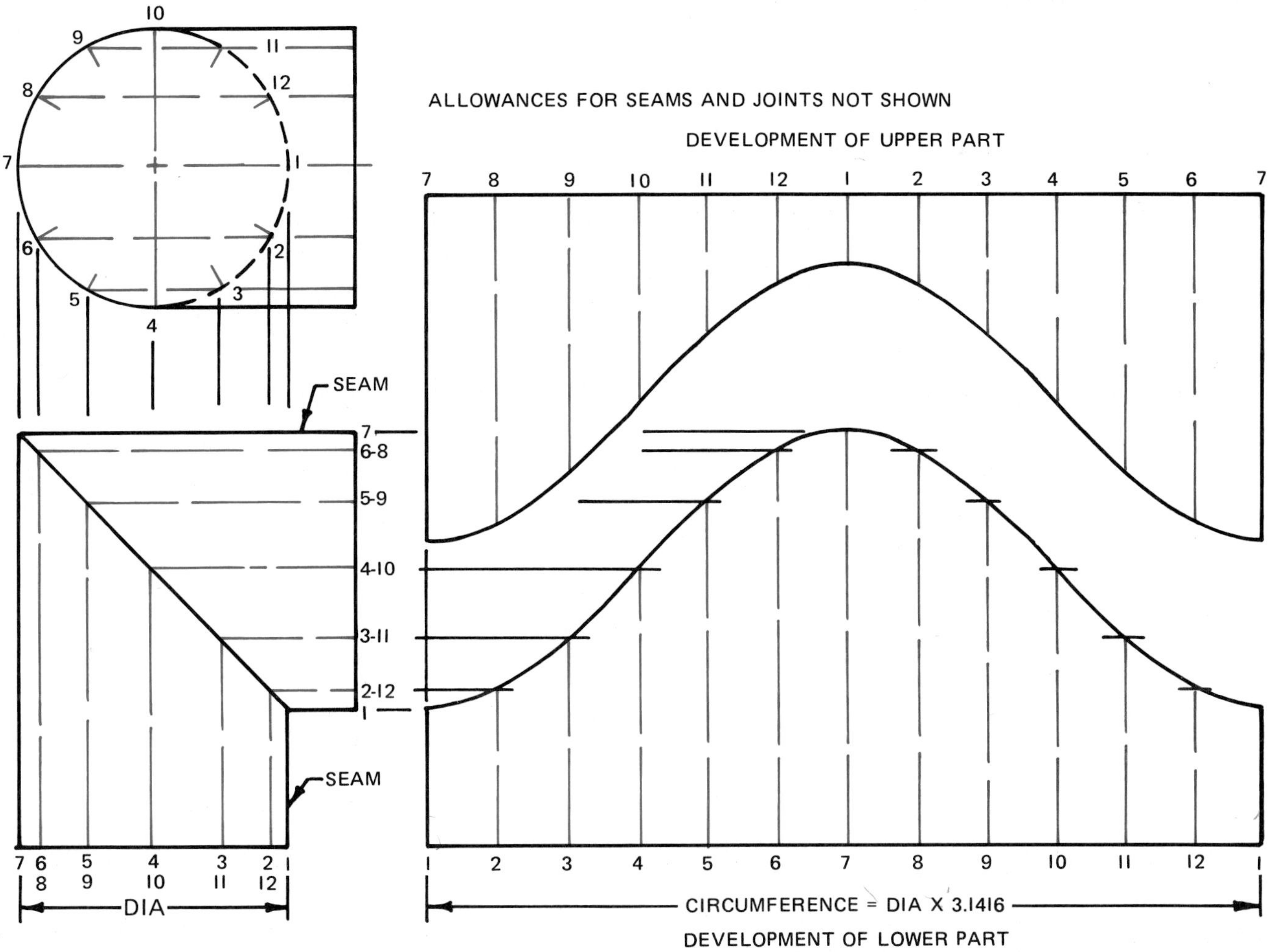

ALLOWANCES FOR SEAMS AND JOINTS NOT SHOWN

DEVELOPMENT OF UPPER PART

SEAM

SEAM

DIA

CIRCUMFERENCE = DIA X 3.1416

DEVELOPMENT OF LOWER PART

FIG. 23-4-5 Development of a two-piece elbow.

LEG

HEEL RADIUS

THROAT RADIUS

LEG

ALLOWANCES FOR SEAMS AND JOINTS NOT SHOWN

CIRCUMFERENCE

FIG. 23-4-6 Development of a four-piece elbow.

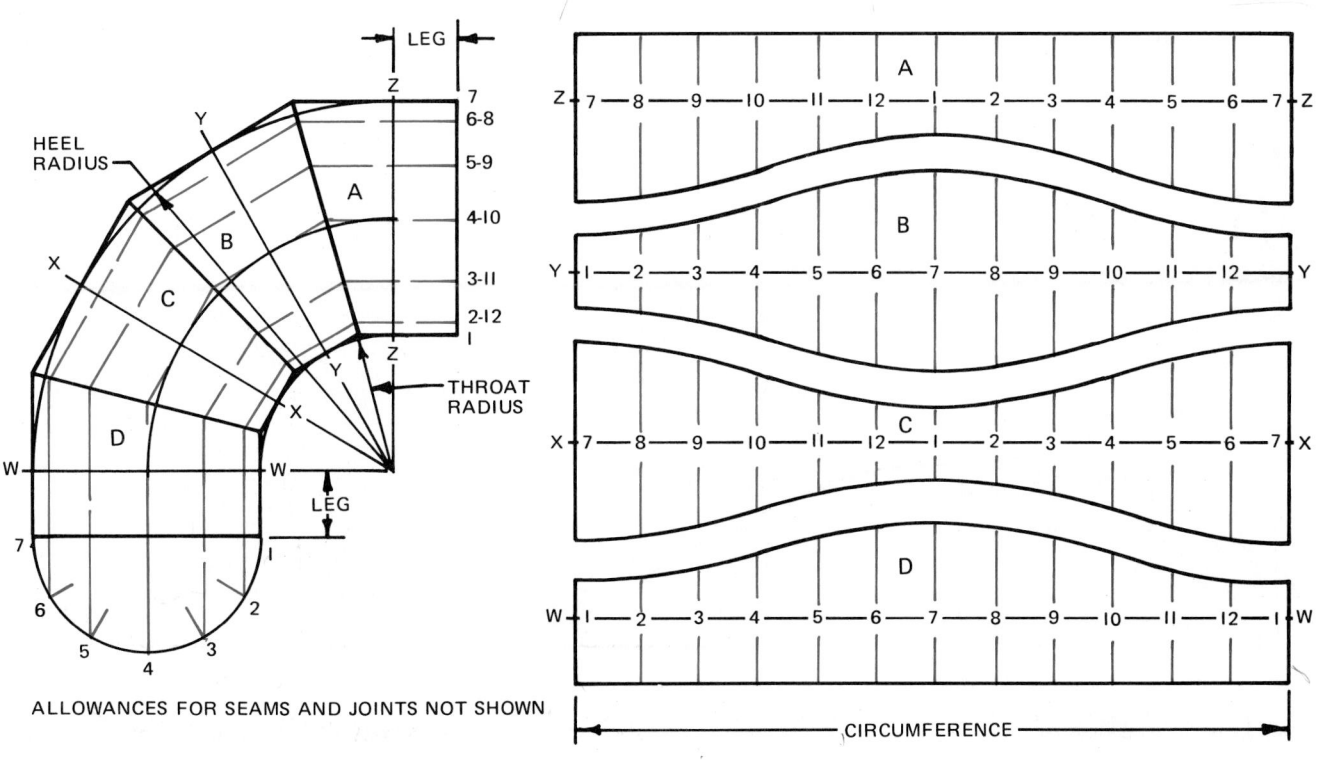

23-5 RADIAL LINE DEVELOPMENT OF CONICAL SURFACES

Development of a Cone   The surface of a cone is developable because a thin sheet of flexible material can be wrapped smoothly about it. The two dimensions necessary to make the development of the surface are the slant height of the cone and the circumference of its base. For a right circular cone (symmetrical about the vertical axis), the developed shape is a sector of a circle. The radius for this sector is the slant height of the cone, and the length around the perimeter of the sector is equal to the circumference of the base. The proportion of the height to the base diameter determines the size of the sector, as illustrated by Fig. 23-5-1A.

Figure 23-5-1B shows the steps in the development of a cone. The top view is divided into a convenient number of equal divisions, in this instance 12. The chordal distance between these points is used to step off the length of arc on the development. The radius R for the development is seen as the slant height in the front view. If a cone is truncated at an angle to the base, the inside shape on the development no longer has a constant radius; that is, it is an ellipse, which must be plotted by establishing points of intersection. The divisions made on the top view are projected down to the base of the cone in the front view. Element lines are drawn from these points to the apex of the cone. These element lines are seen in their true length only when the viewer is looking at right angles to them. Thus the points at which they cross the truncation line must be carried across, parallel to the base, to the outside element line, which is seen in its true length. The development is first made to represent the complete surface of the cone. Element lines are drawn from the step-off points about the circumference to the center point. True-length settings for each element line are taken from the front view and marked off on the corresponding element

lines in the development. An irregular curve is used to connect these points of intersection, giving the proper inside shape.

Development of a Truncated Cone   The development of a frustum of a cone is the development of a full cone less the development of the part removed, as shown in Fig. 23-5-2. Note that, at all times, the radius setting, either R_1 or R_2, is a slant height, a distance taken on the surface of the cone.

When the top of a cone is truncated at an angle to the base, the top surface will not be seen as a true circle. This shape must also be plotted by establishing points of intersection. True radius settings for each element line are taken from the front view and marked off on the corresponding element line in the top view. These points are connected with an irregular curve to give the correct oval shape for the top surface. If the development of the sloping top surface is required, an auxiliary view of this surface shows its true shape.

Development of an Oblique Cone   The development of an oblique cone is generally accomplished by the triangulation method (Fig. 23-5-3). The base of the cone is divided into a convenient number of equal parts and elements; 0-1, 0-2, etc., are drawn in the top view and projected down and drawn in the front view. The true lengths of the elements are not shown in either the top or front view but would be equal in length to the hypotenuse of a right-angle triangle having one leg equal in length to the projected element in the top view and the other leg equal to the height of the projected element in the front view.

When it is necessary to find the true length of a number of edges, or elements, a true-length diagram is drawn adjacent to the front view. This prevents the front view from being cluttered with lines.

Since the development of the oblique cone will be symmetrical, the starting line will be element 0-7. The development is constructed as follows: With 0 as center and radius equal to the true

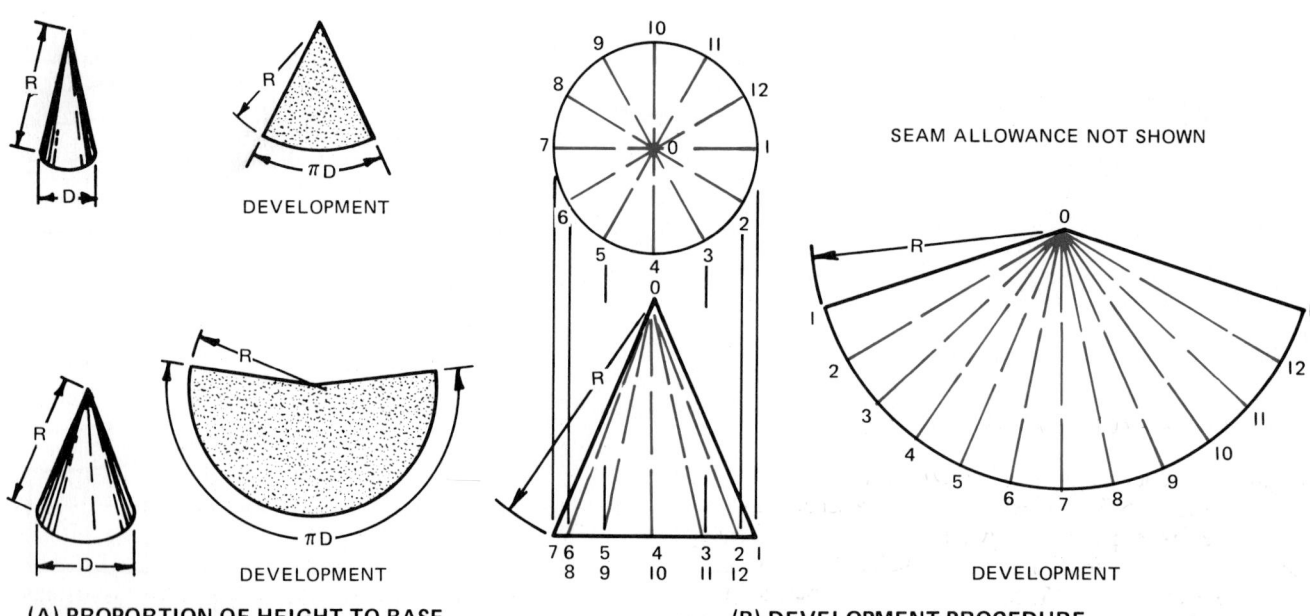

(A) PROPORTION OF HEIGHT TO BASE   **(B) DEVELOPMENT PROCEDURE**

FIG. 23-5-1   Development of a cone.

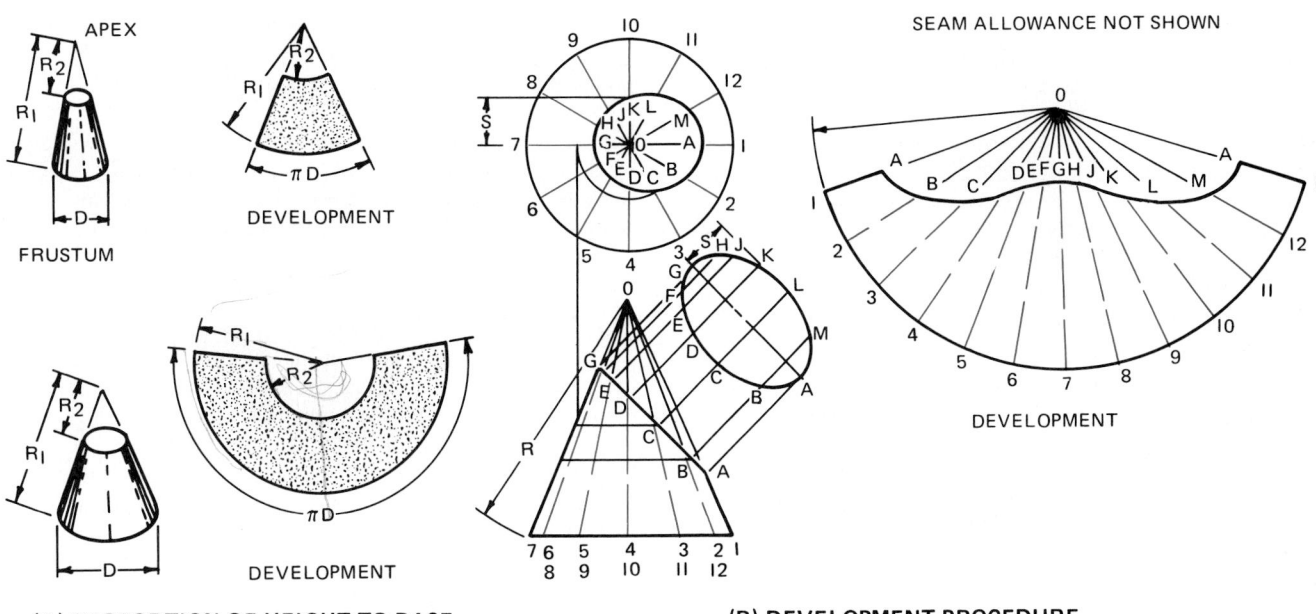

(A) PROPORTION OF HEIGHT TO BASE

(B) DEVELOPMENT PROCEDURE

FIG. 23-5-2 Development of a truncated cone.

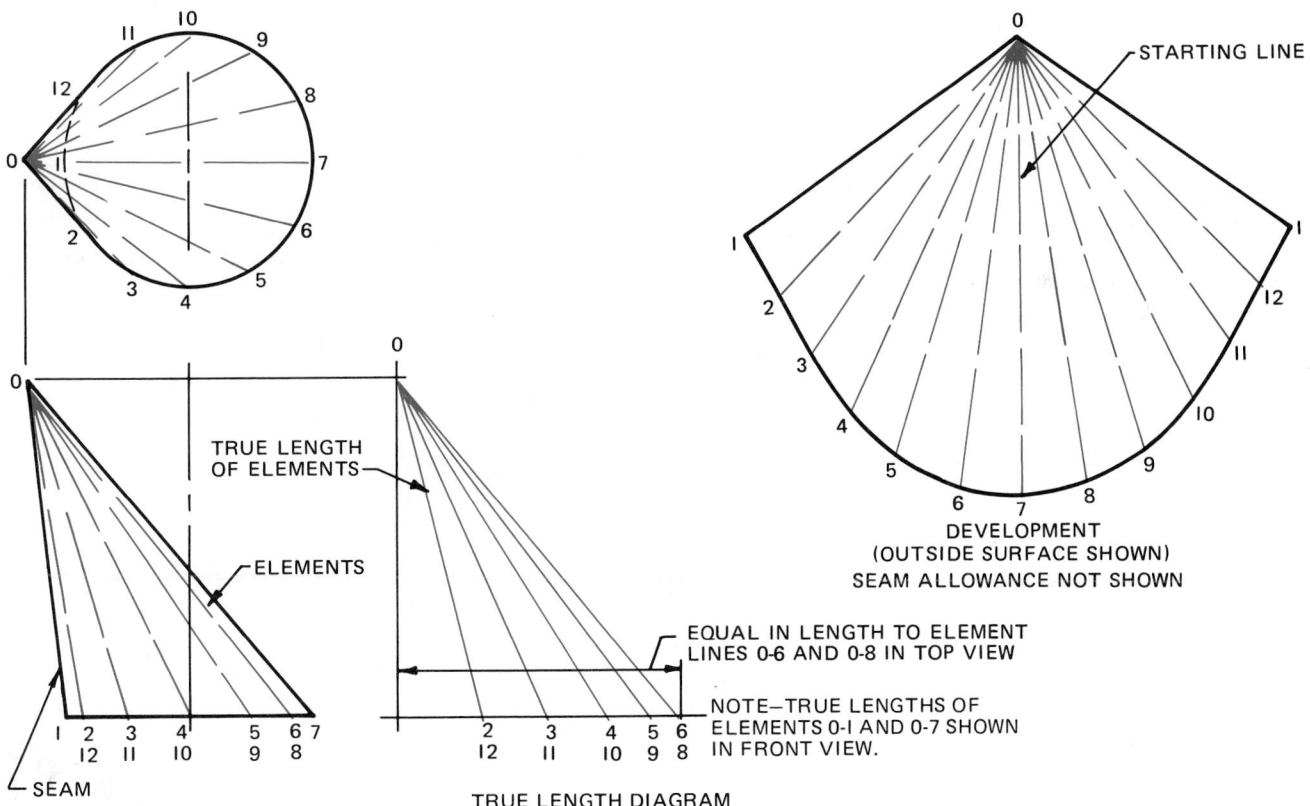

FIG. 23-5-3 Development of an oblique cone.

length of element 0-6, draw an arc. With 7 as center and radius equal to distance 6-7 in the top view, draw a second arc intersecting the first at point 6. Draw element 0-6 on the development. With 0 as center and the radius equal to the true length of element 0-5, draw an arc. With 6 as center and the radius equal to distance 5-6 in the top view, draw a second arc intersecting the first at point 5. Draw element 0-5 on the development. This is

repeated until all the element lines are located on the development view. No seam allowance is shown on the development.

ASSIGNMENT

See Assignment 9 for Unit 23-5 on pages 803–804.

FIG. 23-6-1 Forming a square-to-round transition piece.

23-6 DEVELOPMENT OF TRANSITION PIECES BY TRIANGULATION

Nondevelopable surfaces can be developed approximately by assuming the surface to be made from a series of triangular surfaces laid side by side to form the development. This form of development is known as triangulation (Figs. 23-6-1 and 23-6-2).

Development of a Transition Piece—Square to Round

The transition piece shown in Fig. 23-6-3 is used to connect round and square pipes. It can be seen from both the development and the pictorial drawings that the transition piece is made of four isosceles triangles whose bases connect with the square duct and four parts of an oblique cone having the circle as the base and the corners of the square pipe as the vertices. To make the development, a true-length diagram is drawn first. When the true length of line 1A is known, the four equal isosceles triangles can be developed. After the triangle G-2-3

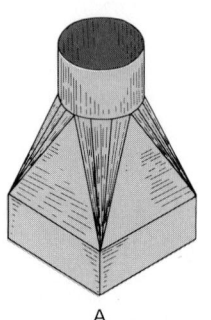

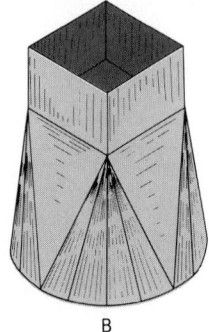

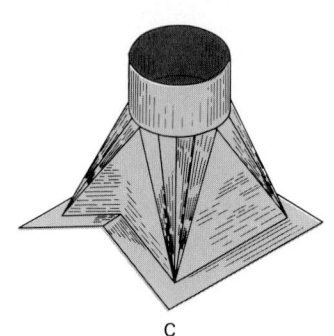

FIG. 23-6-2 Transition pieces.

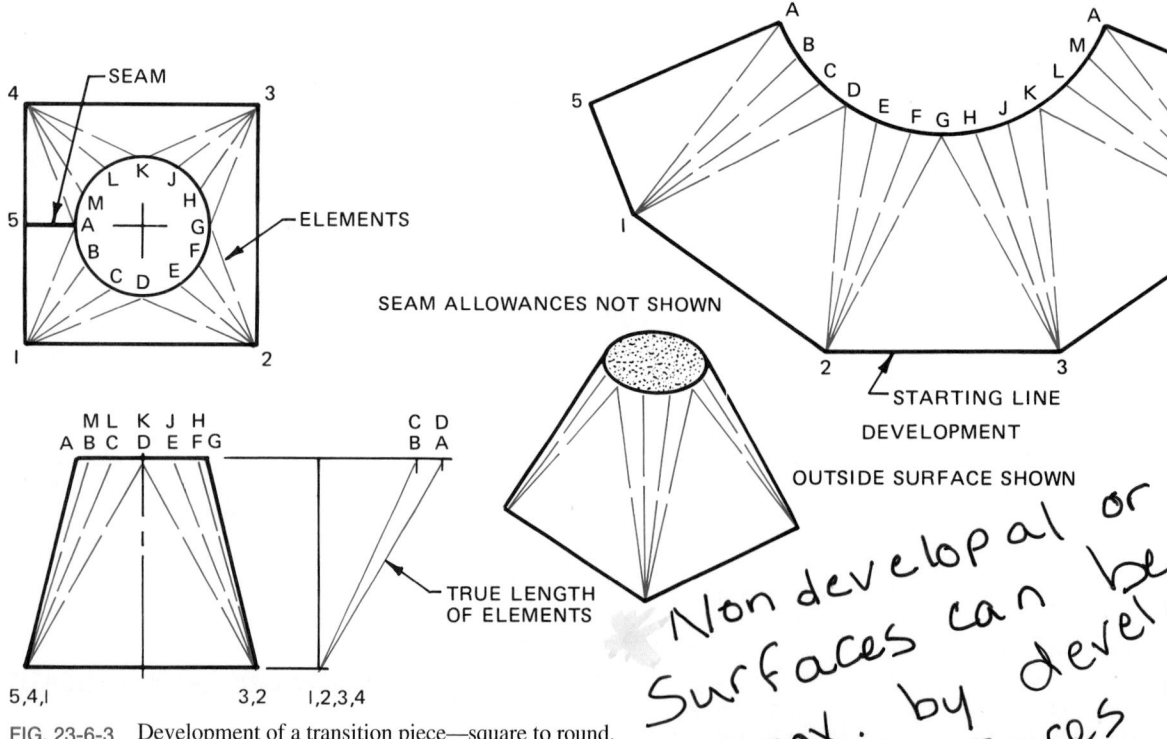

FIG. 23-6-3 Development of a transition piece—square to round.

Non developal or warped surfaces can be approx. by developing the surfaces as a series of triangular patches.

788

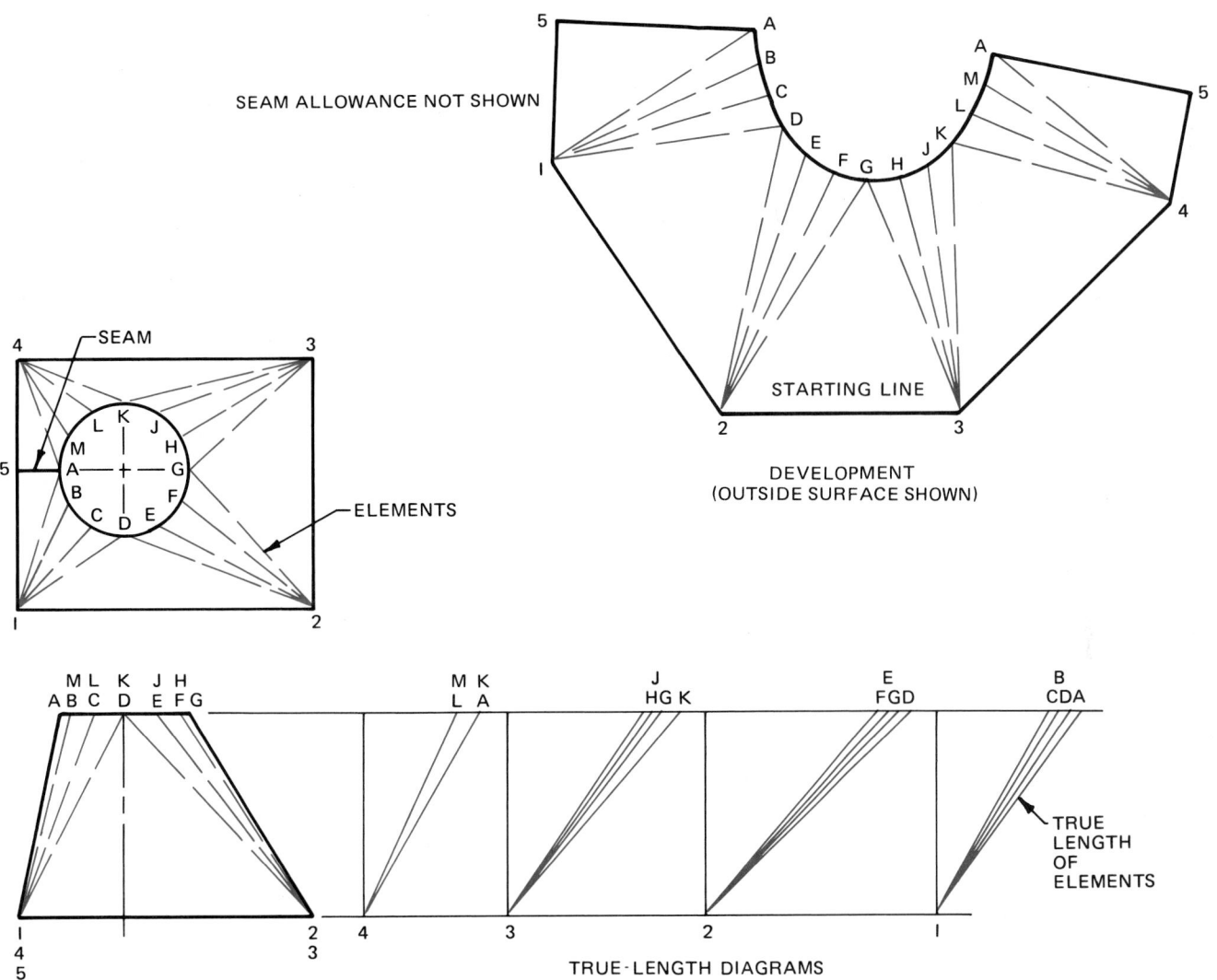

FIG. 23-6-4   Development of an offset transition piece—rectangular to round.

has been developed, the partial developments of the oblique cone are added until points *D* and *K* have been located. Next the isosceles triangles *D*-1-2 and *K*-3-4 are added, then the partial cones, and, last, half of the isosceles triangle placed at each side of the development.

Development of an Offset Transition Piece—Rectangular to Round

The development of the transition piece shown in Fig. 23-6-4 is constructed in the same manner as the one previously developed, except that all the elements are of different lengths. To avoid confusion, four true-length diagrams are drawn, and the true-length lines are clearly labeled.

Transition Piece Connecting Two Circular Pipes—Parallel Joints

The development of a transition piece connecting two circular pipes (Fig 23-6-5, pg. 790) is similar to the development of an oblique cone (Fig. 23-5-3), except that the cone is truncated. The apex of the cone, 0, is located by drawing the two given pipe diameters in their proper positions and extending

the radial lines 1-1_1 and 7-7_1 to intersect at point 0. First the development is made to represent the complete development of the cone, and then the top portion is removed. Radius settings for distances 0-2_1 and 0-3_1 on the development are taken from the true-length diagram.

Transition Piece Connecting Two Circular Pipes—Oblique Joints

When the joints between the pipe and transition piece are not perpendicular to the pipe axis, the transition piece may be developed as shown in Fig. 23-6-6 (pg. 790). Since the top and bottom of the transition piece will be elliptical in shape, a partial auxiliary view is required to find the true length of the chords between the end points of the elements. The development is then constructed in a manner similar to that outlined for Fig. 23-6-5.

ASSIGNMENT

See Assignment 10 for Unit 23-6 on pages 804–805.

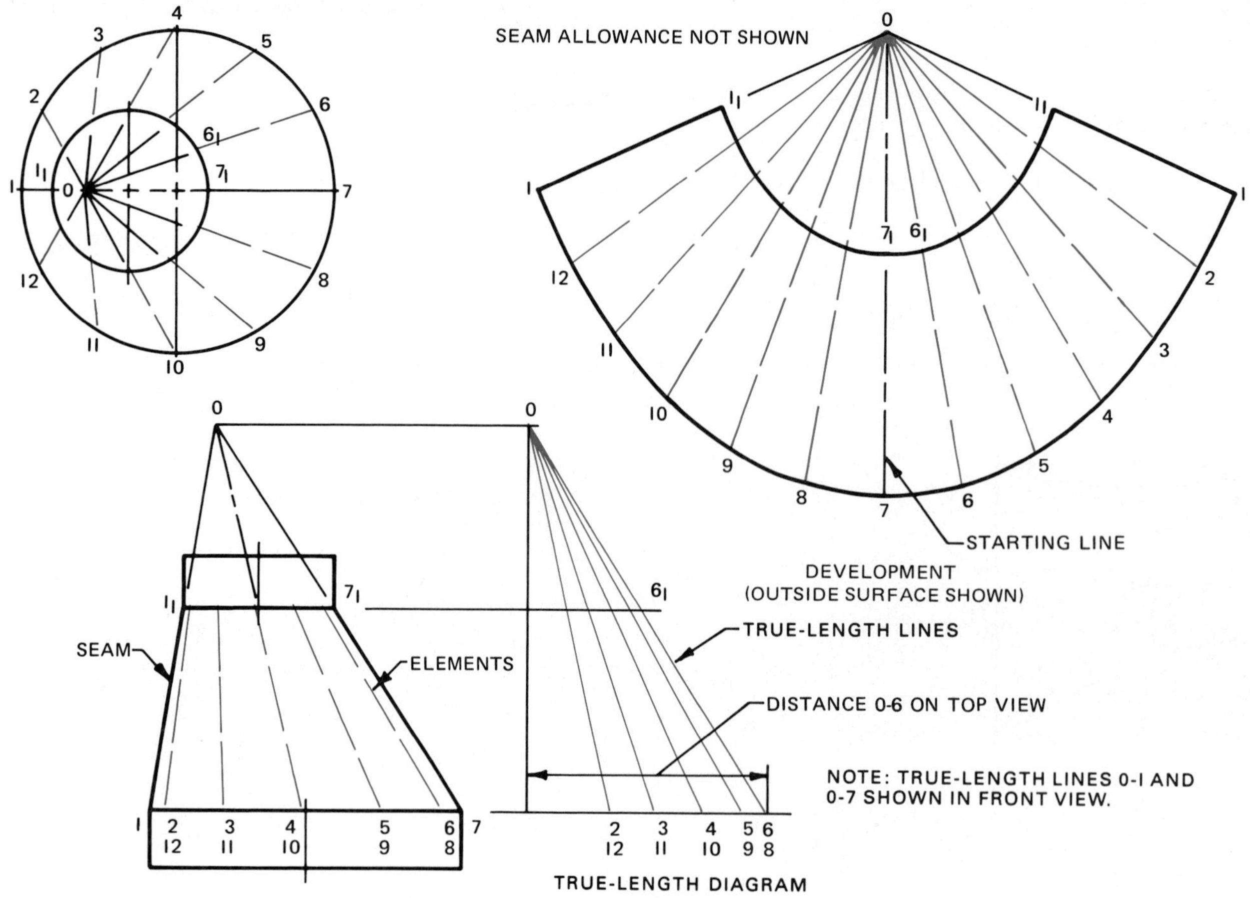

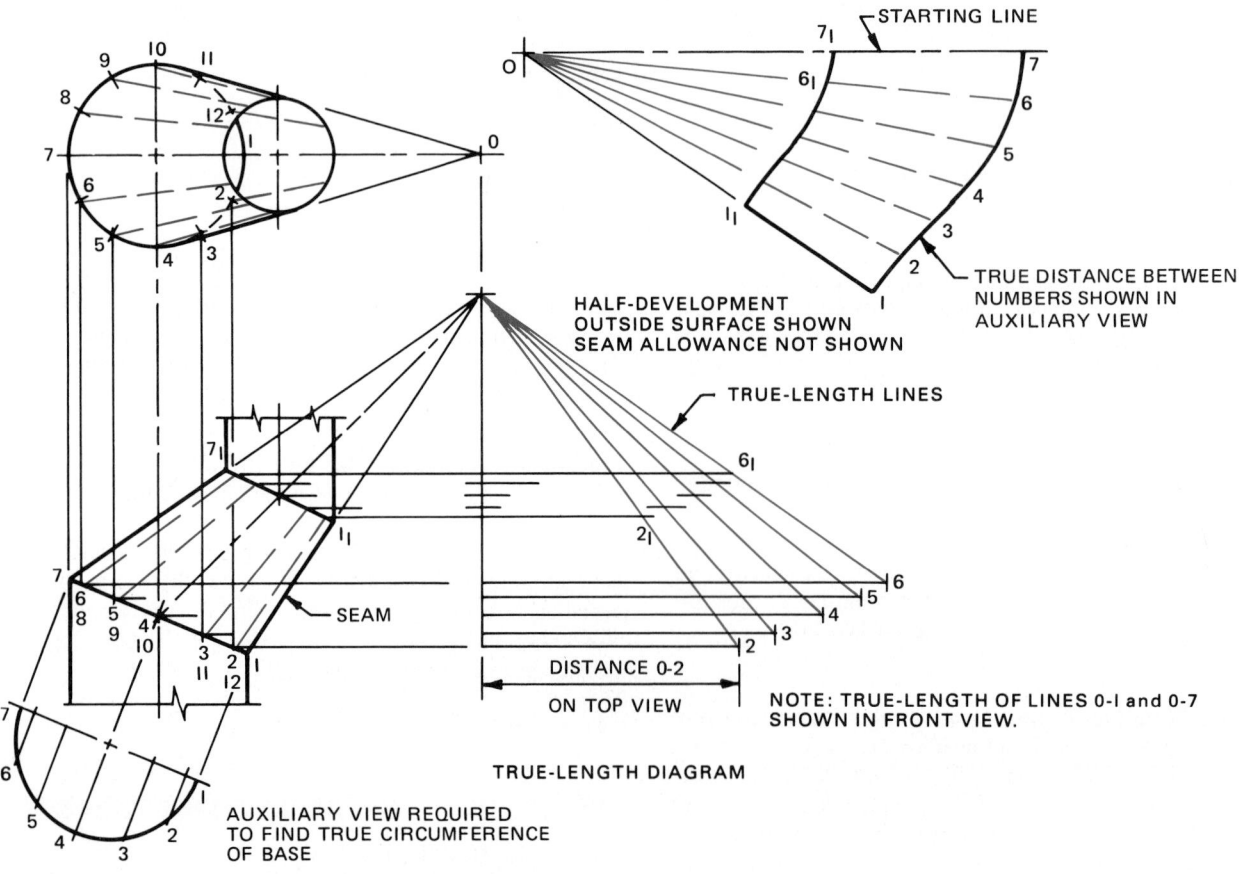

FIG. 23-6-5 Transition piece connecting two circular pipes—parallel joints.

FIG. 23-6-6 Transition piece connecting two circular pipes—oblique joints.

23-7 DEVELOPMENT OF A SPHERE

Since the surface of a sphere is double-curved, it is not developable. However, the surface may be approximately developed by either the gore or the zone method. In the gore method (Fig. 23-7-1) the surface is divided into a number of equal sections, each section being considered as a section of a cylinder.

Only one section need be developed, for it will serve as a pattern for the others. In the zone method (Fig. 23-7-2), the sphere is divided into horizontal zones and each zone is developed as a frustum of a cone.

ASSIGNMENTS

See Assignments 11 and 12 for Unit 23-7 on page 805.

FIG. 23-7-1   Development of a sphere—gore method.

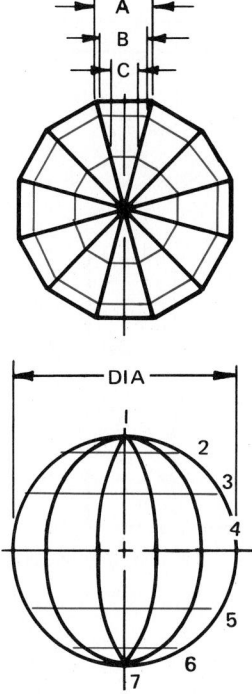

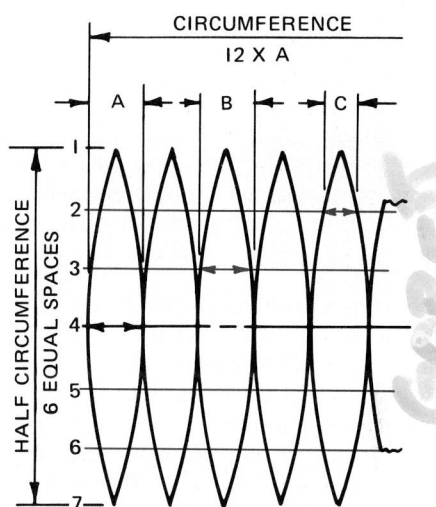

PARTIAL DEVELOPMENT
(OUTSIDE SURFACE SHOWN)
SEAM ALLOWANCE NOT SHOWN

FIG. 23-7-2   Development of a sphere—zone method.

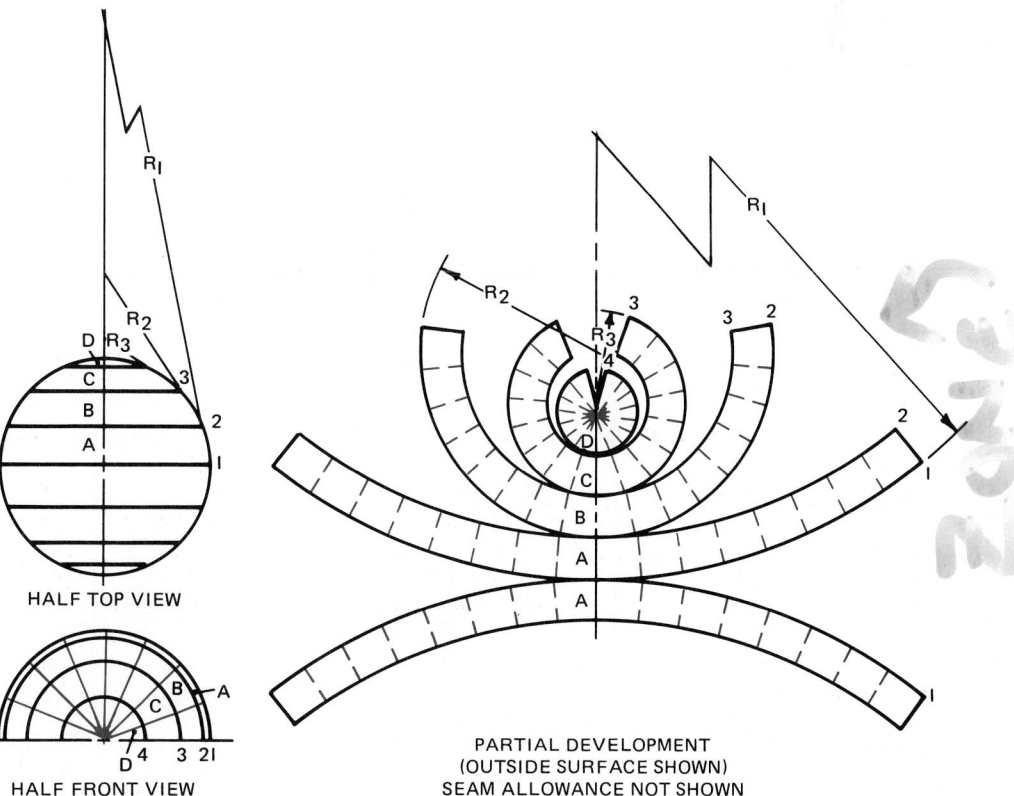

PARTIAL DEVELOPMENT
(OUTSIDE SURFACE SHOWN)
SEAM ALLOWANCE NOT SHOWN

791

23-8 INTERSECTION OF FLAT SURFACES—LINES PERPENDICULAR

Whenever two surfaces meet, there is a line common to both called the *line of intersection*. In making the orthographic drawing of objects that comprise two or more intersecting parts, the lines of intersection of these parts must be plotted on the orthographic views. Figures 23-8-1 and 23-8-2 illustrate this plotting technique for the intersection of flat-sided prisms. Figure 23-8-1 shows the development of the parts. A numbering technique is very valuable in plotting lines of intersection. In the illustrations shown, the lines of intersection appear in the front view. The end points for these lines are established by projecting the height position from the right side view to intersect the corresponding length position projected from the top view. When the prisms are flat-sided, the lines of intersection are straight, and the lines in the development are straight.

Intersecting Prisms—Triangle and Pyramid   In drawing the intersection of these two prisms, the points of intersection in the top view are found by projecting the points of intersection from the front view (Fig. 23-8-3). If a development of the pyramid is required, true-length lines, which do not appear in either the top or the front view, must be found. Since only a few true-length lines are unknown, line $0\text{-}D$ on the front view serves as a true-length diagram. Lines $0\text{-}2_1$, $0\text{-}1_1$, and $0\text{-}5_1$ on line $0\text{-}D$ are the true lengths of lines $0\text{-}2$, $0\text{-}1$, and $0\text{-}5$, respectively. Point 1 on surface $0\text{-}E\text{-}F$ and $0\text{-}B\text{-}C$ is located on the development as

follows: Draw a straight line through points 0 and 1 in the front view to intersect the base at point 7. Transfer distance $C7$ in the front view to the development view. Join points 0 and 7 with a straight line. With center 0 and radius equal to distance $0\text{-}1_1$ shown on the front view, swing an arc intersecting line $0\text{-}7$ at point 1.

Points on the development of the triangular prism are found by projecting lines from the top view to the development view

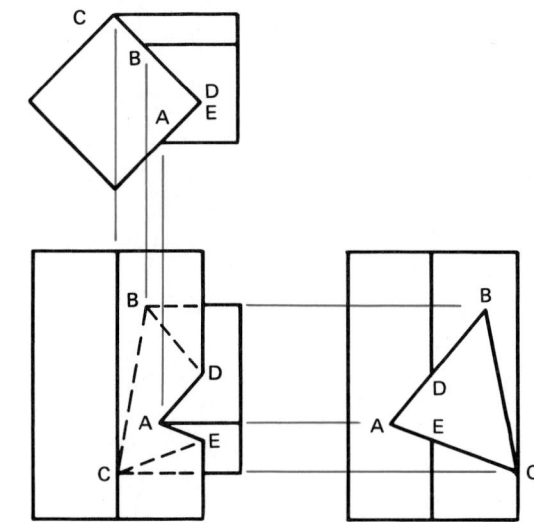

FIG. 23-8-2   Intersecting prisms at right angles.

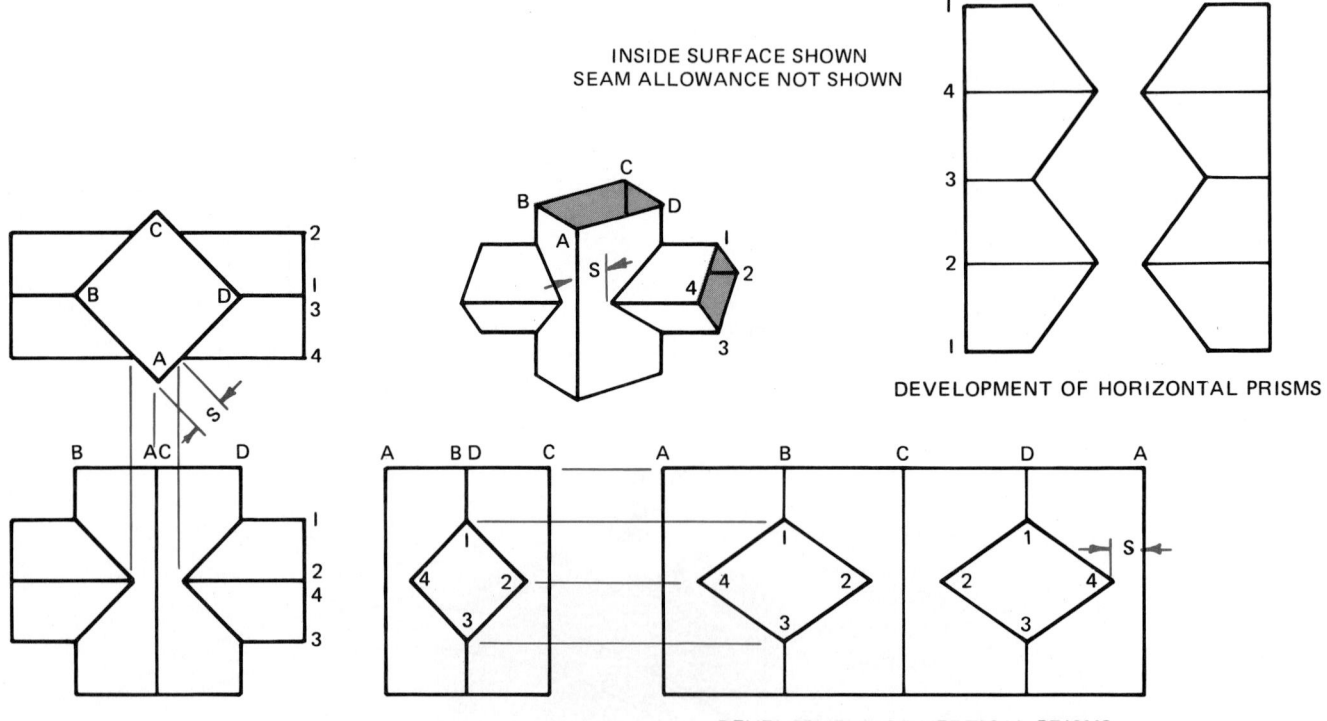

INSIDE SURFACE SHOWN
SEAM ALLOWANCE NOT SHOWN

DEVELOPMENT OF HORIZONTAL PRISMS

DEVELOPMENT OF VERTICAL PRISMS

FIG. 23-8-1   Plotting lines of intersection and making development drawings of intersecting prisms.

and transferring distances between points (numbers) and lines (letters) from the front view to corresponding points on the development.

Intersecting Prisms Not at Right Angles—Hexagon and Triangle An auxiliary view is required to locate points of intersection, such as points *A*, *B*, and *C*, as shown in Fig. 23-8-4. To complete the side view, ends of lines *D*, *E*, and *F* and points of intersection *A*, *B*, and *C* are projected from the top view. Distances between the lines of the hexagon and triangle are transferred from either the top view or the auxiliary view.

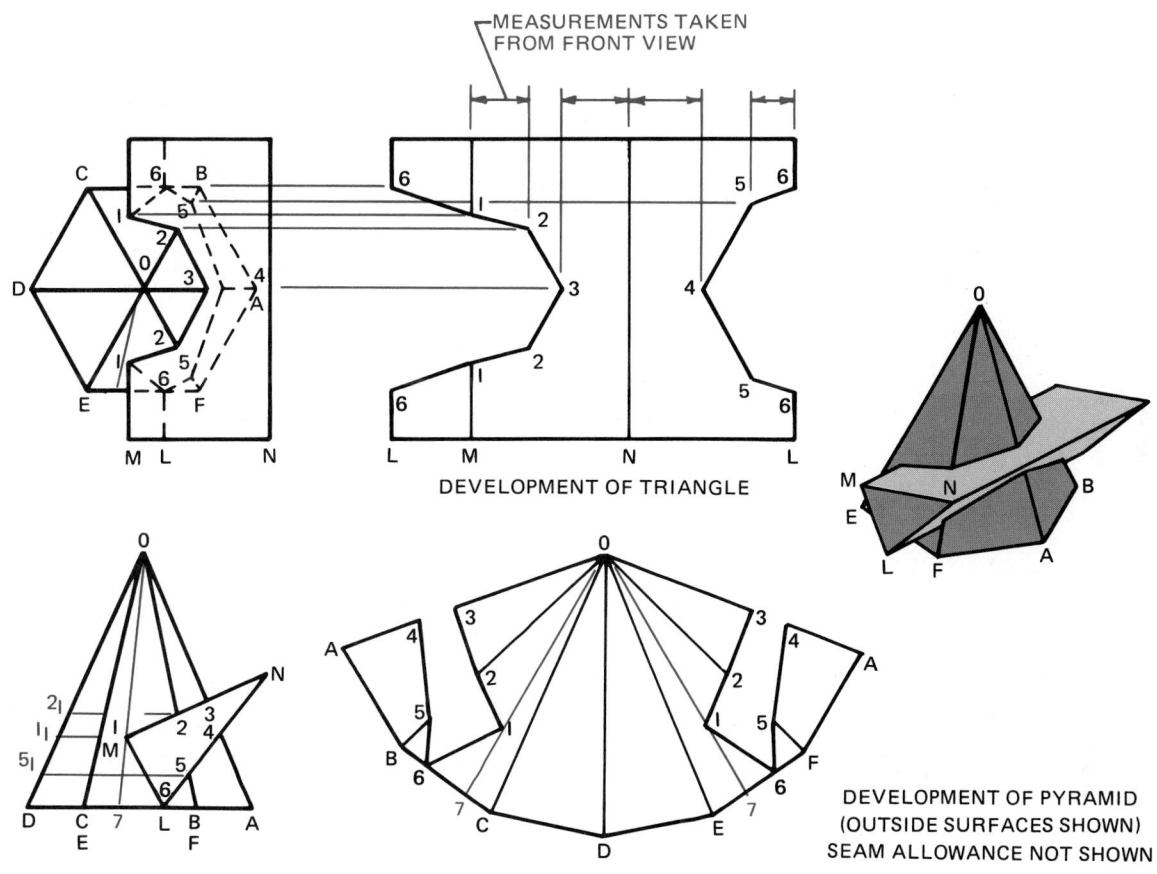

FIG. 23-8-3 Intersecting prisms—triangle and pyramid.

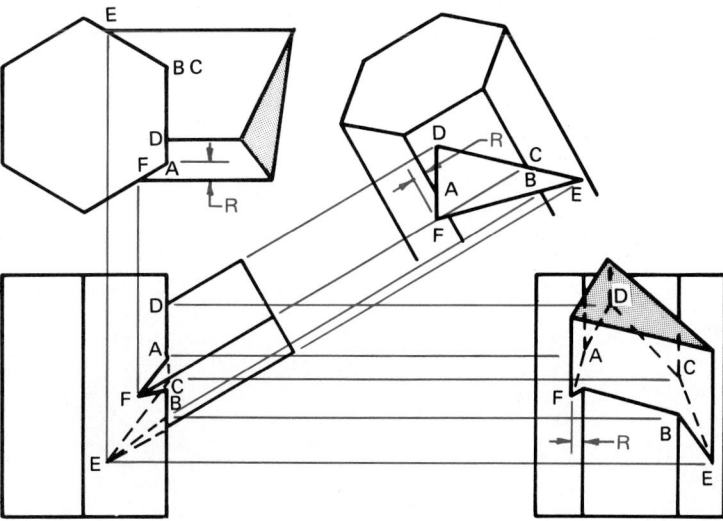

FIG. 23-8-4 Intersecting prisms not at right angles— hexagon and triangle.

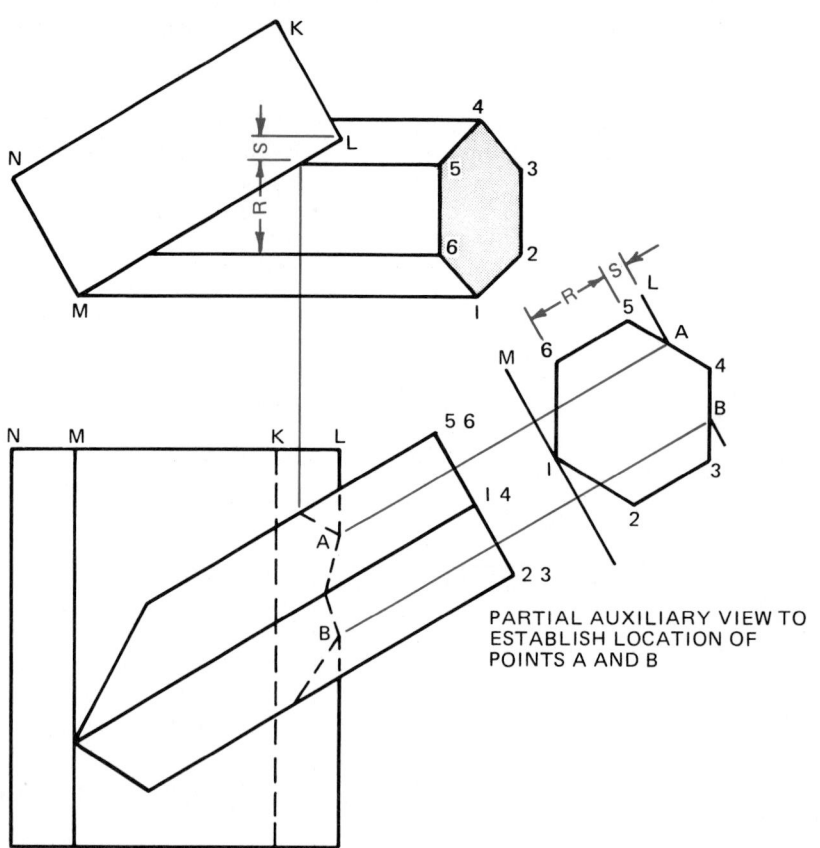

FIG. 23-8-5 Intersecting prisms not at right angles—hexagon and rectangle.

PARTIAL AUXILIARY VIEW TO
ESTABLISH LOCATION OF
POINTS A AND B

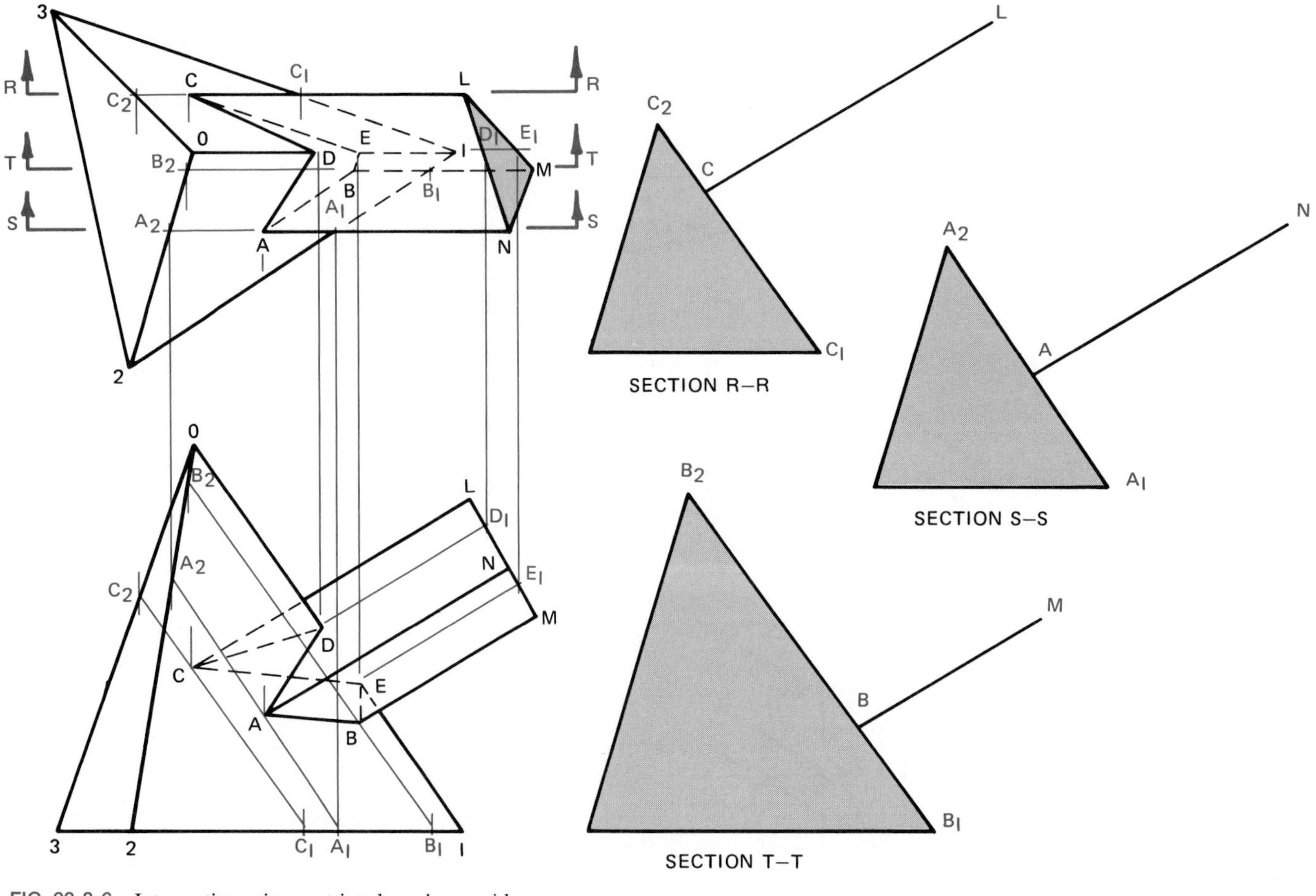

SECTION R–R

SECTION S–S

SECTION T–T

FIG. 23-8-6 Intersecting prisms—triangle and pyramid.

Intersecting Prisms Not at Right Angles—Hexagon and Rectangle   Often a partial auxiliary view is drawn to locate points of intersection, such as points *A* and *B* on line *L* in Fig. 23-8-5.

Intersecting Prisms—Triangle and Pyramid   Another method commonly used to locate points of intersection of lines and surfaces is the use of vertical cutting planes located on the edges piercing the surface (Fig. 23-8-6). Thus section *R-R* locates point *C*, section *S-S* locates point *A*, and section *T-T* locates point *B*. The sectional views shown are for illustrative purposes only and need not be drawn. To establish point *C*, extend line *LC* in the top view to intersect line 0-3 at C_2 and line 1-3 at C_1. Project vertical lines from C_1 and C_2 down the front view, locating points C_1 on baseline 1-3 and C_2 on line 0-3. Join points C_1 and C_2 at point *C*. Extend a vertical line up from point *C* to the top view, intersecting line C_2L at point *C*. Repeat for points *A* and *B*.

23-9   INTERSECTION OF CYLINDRICAL SURFACES

45° Reducing Tee   Figure 23-9-1 illustrates the intersection of a small pipe at an angle of 45° to a large pipe. The same techniques of plotting reference points are used as were previously described for a 90° reducing tee.

90° Reducing Tee   Figure 23-9-2 (pg. 796) illustrates the plotting technique for the intersection of cylinders. Because there are no edges on the cylinders, element or reference lines are established about the cylinders in their orthographic views. In the top view, the element lines for the small cylinder are drawn to touch the surface of the large cylinder; for example, line 2 touches at *T*. This point location is then projected down to the front view to intersect the corresponding element line, establishing the height at that point. The points of intersection thus established are connected by an irregular curve to produce the line of intersection. The same points of reference used to establish the line of intersection are used to draw the development.

ASSIGNMENT

See Assignment 13 for Unit 23-8 on pages 806–807.

ASSIGNMENT

See Assignment 14 for Unit 23-9 on pages 807–808.

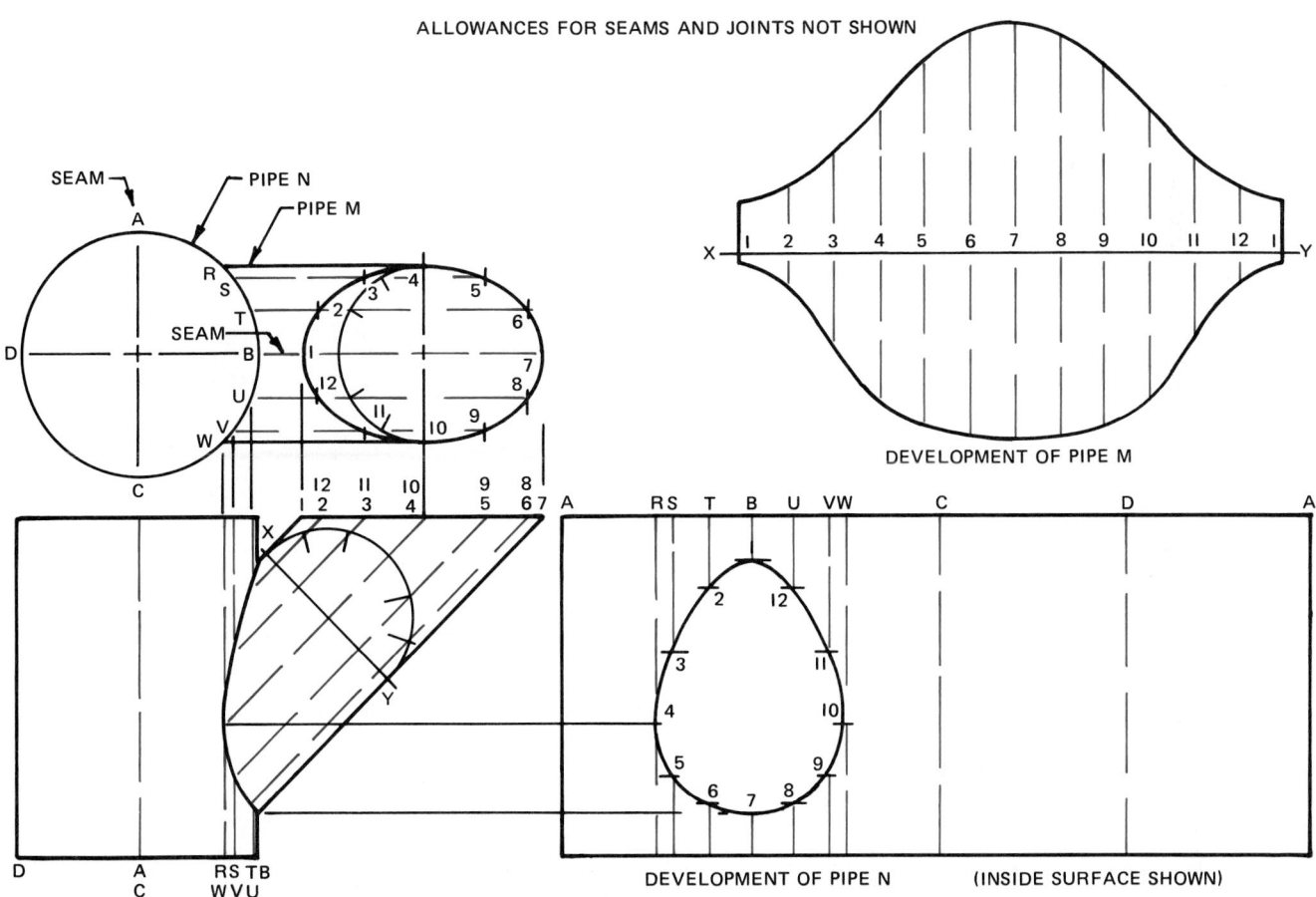

FIG. 23-9-1   Plotting lines of intersection and making development drawings for a 45° reducing tee.

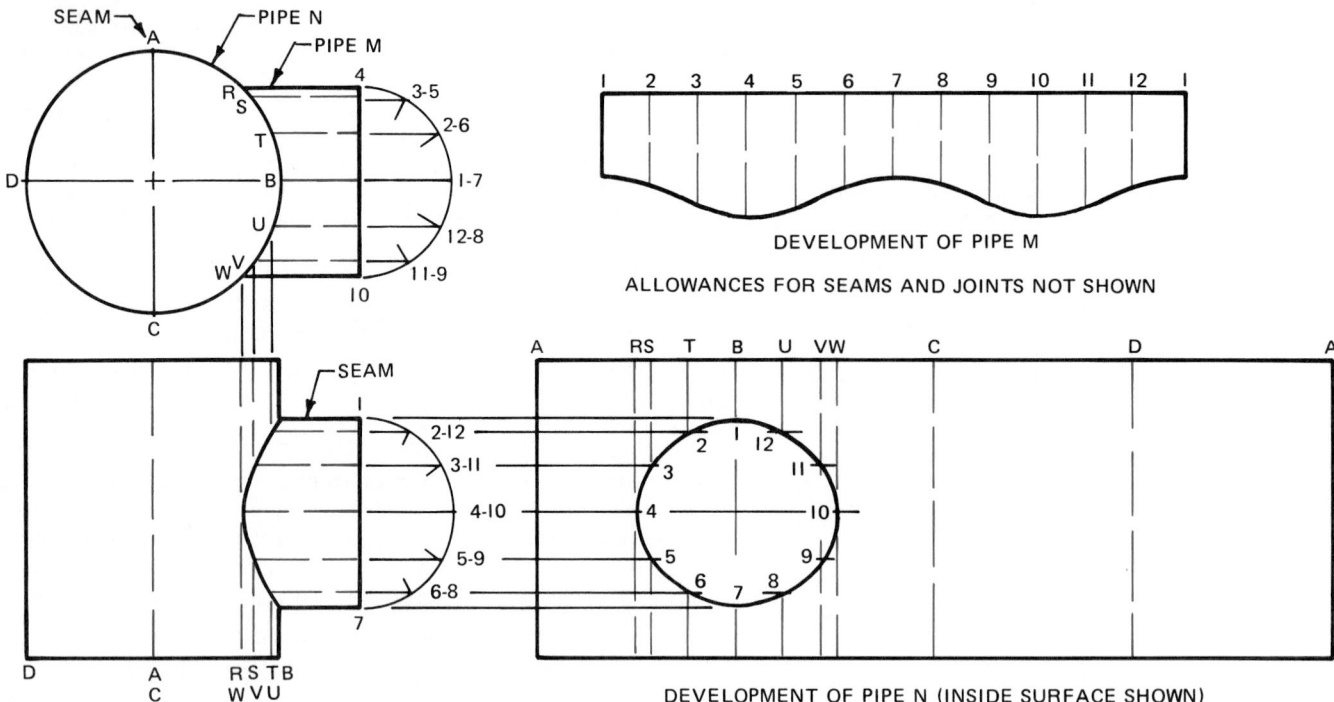

DEVELOPMENT OF PIPE M

ALLOWANCES FOR SEAMS AND JOINTS NOT SHOWN

DEVELOPMENT OF PIPE N (INSIDE SURFACE SHOWN)

FIG. 23-9-2   Plotting lines of intersection and making development drawings for a 90° reducing tee.

23-10   INTERSECTING PRISMS

Intersecting Prisms—Hexagon and Cone   The lines of intersection between the hexagon and cone (Fig. 23-10-1) are developed as follows. Divide each side of the hexagon into four parts (lines every 15°). Through point A in the top view, swing an arc intersecting horizontal center line 0-4 at A_1. From point A_1 drop a vertical line down to the front view, intersecting line 0-D at point A_1. Draw a light horizontal line through A_1 in the front view. Drop vertical lines down from point A in the top view, intersecting this line at point A. Repeat, locating point B. Point C may be located in the front view by extending a light horizontal line from point C on line 0-D. Join points A, B, and C with a line, which forms a hyperbolic curve.

Intersecting Prisms—Cone and Cylinder   The intersections of the cone and cylinder elements in the top view of Fig. 23-10-2 are first found and then projected down to the corresponding elements in the front view. A smooth curve is drawn through these points to produce the line of intersection.

Intersecting Prisms—Cone and Cylinder   The line of intersection between the cone and the cylinder in Fig. 23-10-3 is found by assuming the front view to have a series of horizontal cutting planes passing through points 2, 3, 4, 5, and 6. The cutting-plane line passes through the intersection of the cone and cylinder. Each point on the line of intersection is developed in a manner similar to that for Fig. 23-10-2.

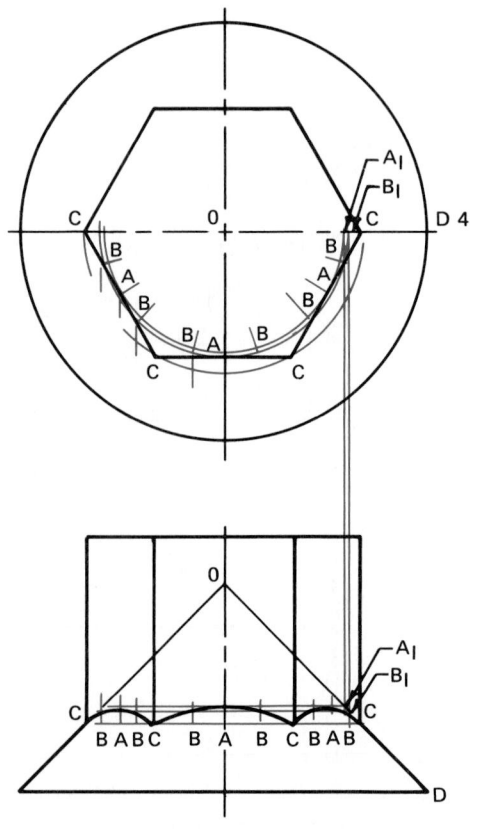

FIG. 23-10-1   Intersecting prisms—hexagon and cone.

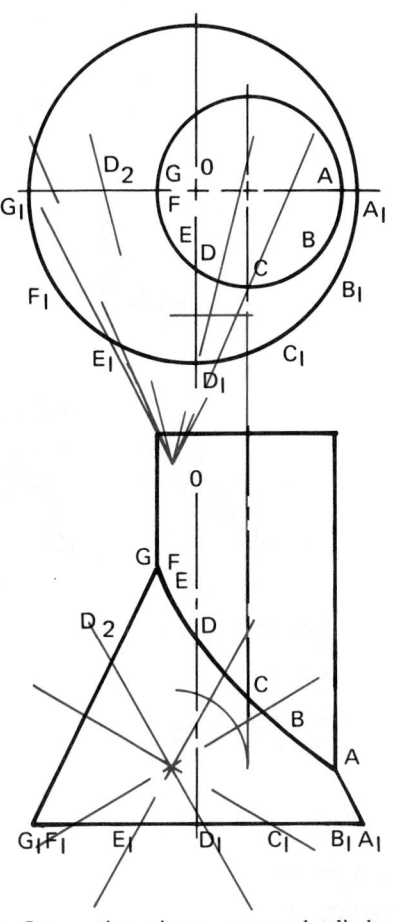

FIG. 23-10-2   Intersecting prisms—cone and cylinder.

FIG. 23-10-3   Intersecting prisms—cone and cylinder.

Intersecting Prisms—Pyramid and Cylinder In Fig. 23-10-4 (pg. 798) cutting plane lines taken horizontally through points 2, 3, 4, 5, and 6 in the front view are used to locate the lines of intersection. Point 5 on the line of intersection is located as follows: Draw a horizontal line through point 5 on the half-circle in the front view to intersect line 0-A at point 5_1. Extend a vertical line from point 5_1 to the top view to intersect line 0-A at point 5_1. Extend a line from 5_1 in the top view at an angle of 45° to intersect line 0-D. From this intersection, extend a line at an angle of 45° in the direction of line 0-C to intersect a horizontal line passing through point 5 on the half-circle in the top view. The intersection of these lines is point 5. To locate point 5 in the front view, drop a vertical line from point 5 in the top view to intersect the horizontal line passing through point 5 of the half-circle. Locate the other points in a similar manner, and connect them with a smooth, curved line.

Intersecting Prisms—Cone and Oblique Cylinder The cylinder shown in the auxiliary view in Fig. 23-10-5 (pg. 798) is divided into 12 equal parts. Element lines are drawn from the apex of the cone, point 0, through points 2 to 6 to the base of the cone, establishing points 2_1 to 6_1 inclusive. These points are projected to corresponding points in the front view and the element lines are drawn.

The element lines in the top view are located by projecting vertical lines up from points 2_1 to 6_1 in the front view to intersect the base circle in the top view. The lines of intersection are then found by projecting lines from the circle to meet their corresponding element lines.

ASSIGNMENT

See Assignment 15 for Unit 23-10 on page 808.

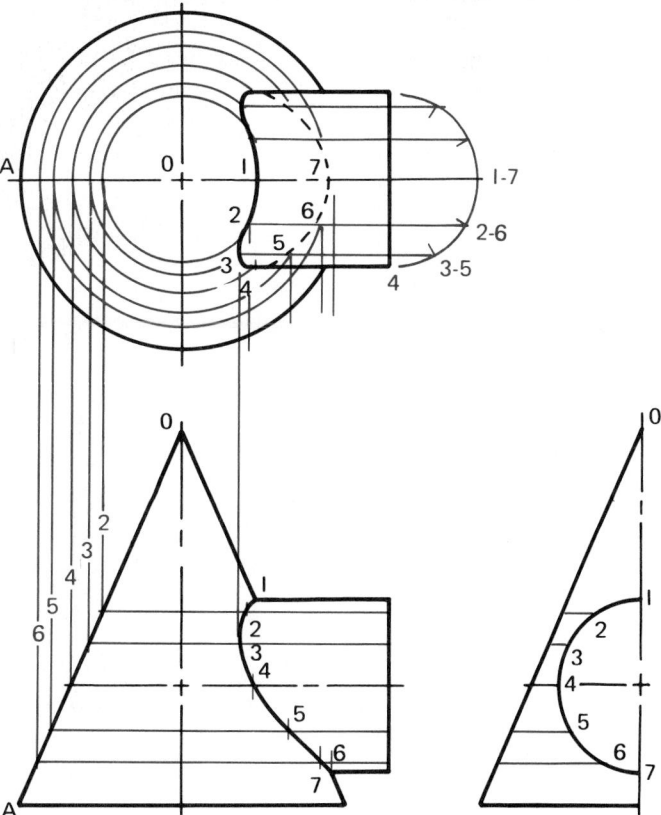

FIG. 23-10-4   Intersecting prisms—pyramid and cylinder.

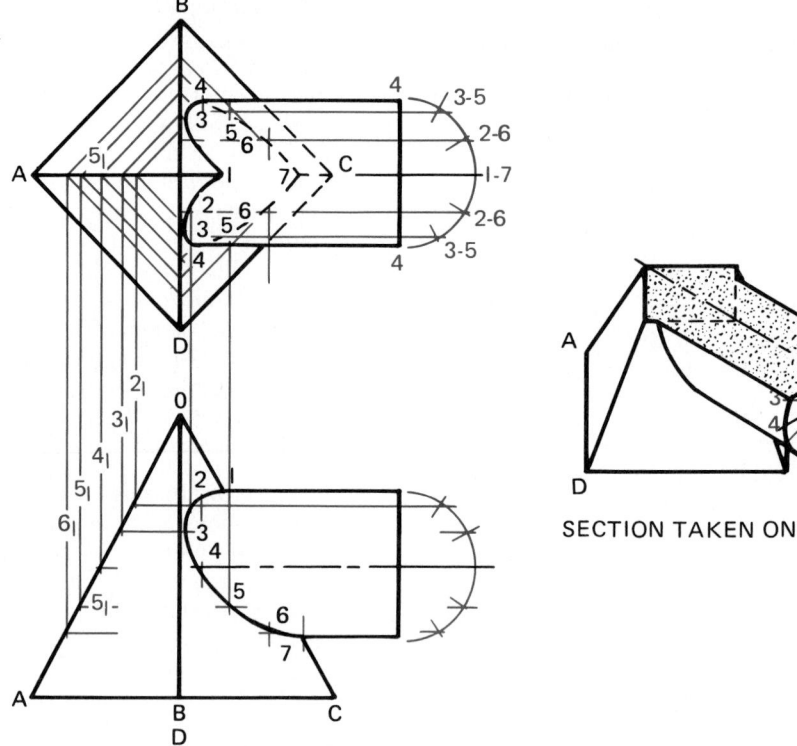

SECTION TAKEN ON PLANE 3

FIG. 23-10-5   Intersecting prisms—cone and oblique cylinder.

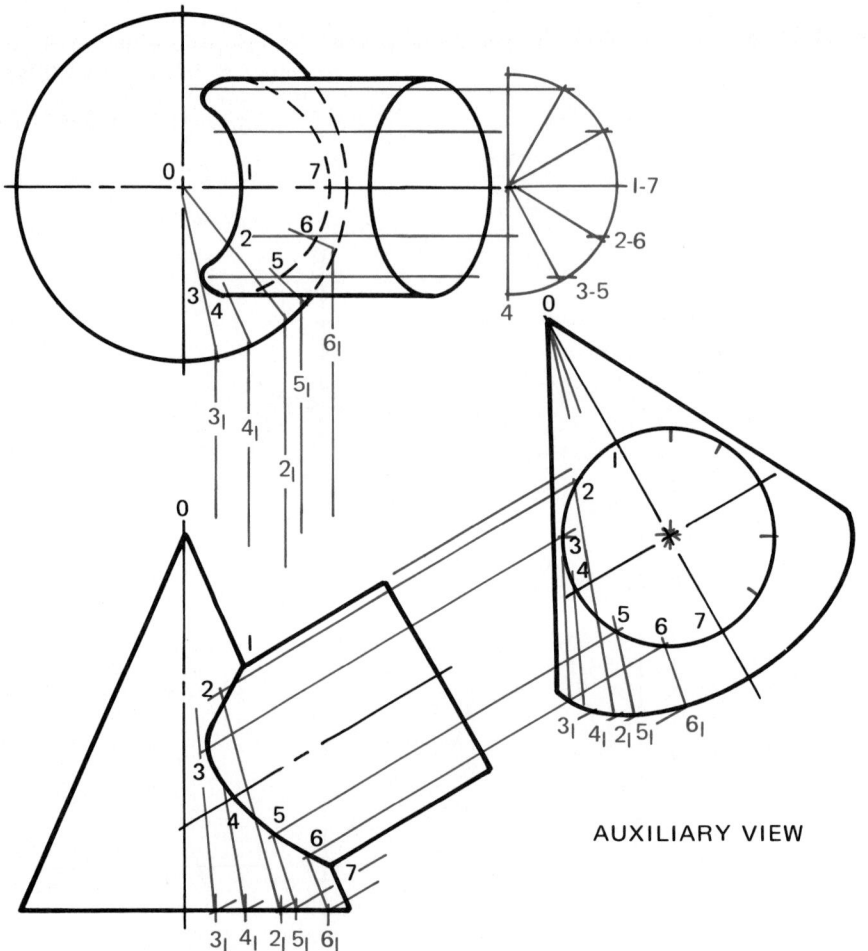

AUXILIARY VIEW

ASSIGNMENTS FOR CHAPTER 23

ASSIGNMENT FOR UNIT 23-1, SURFACE DEVELOPMENTS AND INTERSECTIONS

1. Make a development drawing complete with bending instructions of one of the parts shown in Figs. 23-1-A to 23-1-D. Dimension the development drawing, showing the distance between bend lines, and show the overall sizes. Scale 1:1.

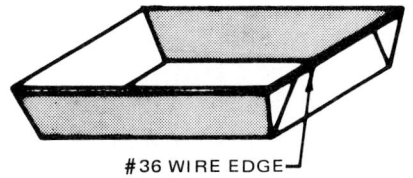

#36 WIRE EDGE

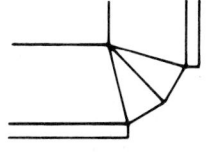

ENLARGED VIEW
OF ONE CORNER
OF DEVELOPMENT

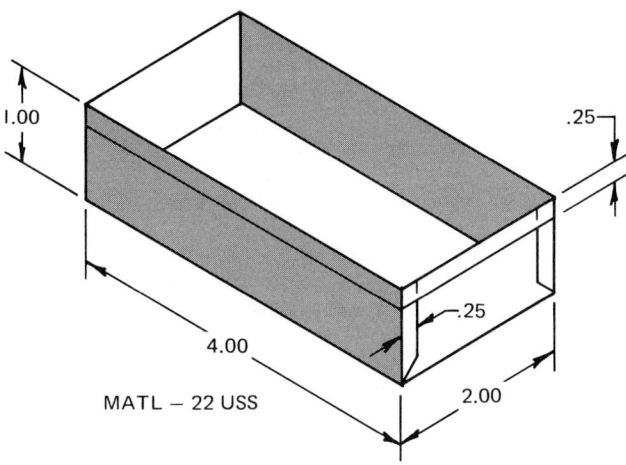

MATL — 22 USS

FIG. 23-1-A   Nail box.

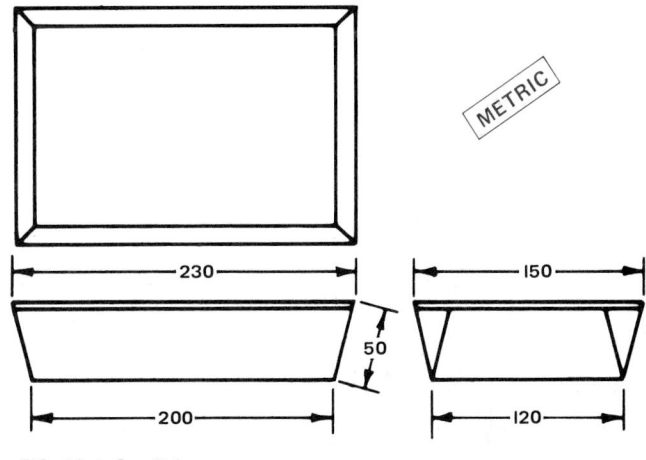

METRIC

FIG. 23-1-C   Cake tray.

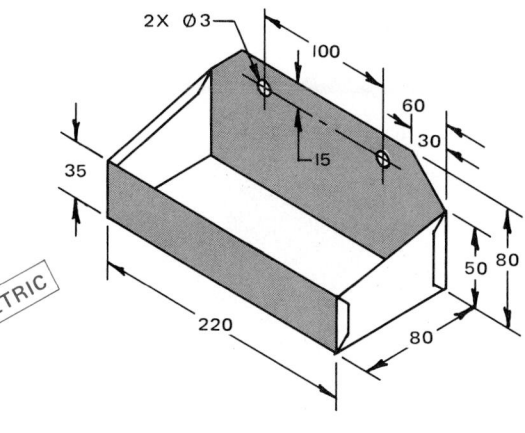

METRIC

FIG. 23-1-B   Wall tray.

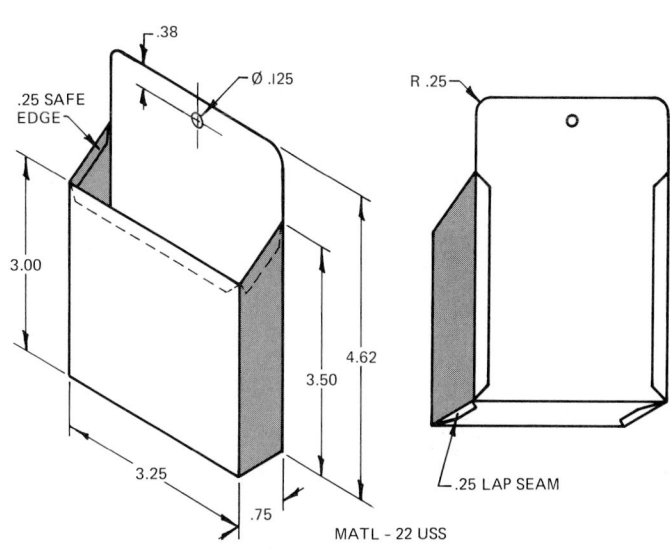

.25 SAFE EDGE

Ø .125

R .25

.25 LAP SEAM

MATL - 22 USS

FIG. 23-1-D   Memo pad holder.

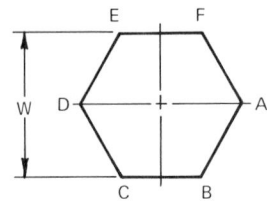

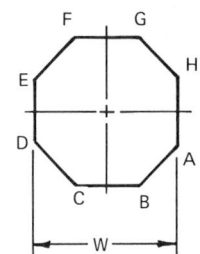

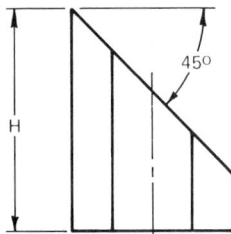

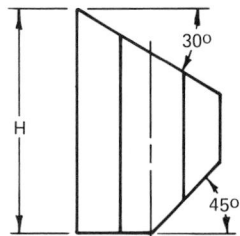

NOTES:
SEAM IS AT A.
TOP AND BOTTOM TO
HINGE AT C-D.

NOTES:
SEAM IS AT A.
TOP AND BOTTOM TO
HINGE AT D-E.

FIG. 23-2-A   Hexagon box.

FIG. 23-2-B   Octagon box.

ASSIGNMENTS FOR UNIT 23-2, THE PACKAGING INDUSTRY

2. Make a development drawing of one of the boxes shown in Figs. 23-2-A or 23-2-B. Scale 1:1. After the development drawing has been checked by your instructor, add suitable seams and joint allowances. Then cut out the development, score on the bend lines, and form and glue the box together.
   U.S. Customary   W = 1.75, H = 2.88. All seams .25 in. wide, placed on the inside and glued. Material is .02 in. cardboard.
   Metric   W = 44, H = 75. All seams 6 mm wide, placed on the inside and glued. Material is 0.5 mm cardboard.

3. Make a development drawing of either the pencil or swim goggle boxes shown in Figs. 23-2-C and 23-2-D. On the exterior surface of the box, lay out a design which has eye appeal (color can be used) and which contains in the design the name of the item being sold, a company name, a slogan, and any other feature you believe would improve the salability of the article. Cut out the development, score on the bend lines, and glue together. *Note:* With reference to the swim goggle box, the box is completely sealed and must be broken to remove the goggles. Scale 1:1.

4. Many containers are designed for a dual purpose. The main purpose is to accommodate the article being sold. The secondary purpose is to use the container as a novelty item after the article is removed. This next product is such a container. A company which produces cat and dog food wishes to have a container in the shape of a dodecahedron (12-sided) or icosahedron (20-sided) with illustrations of different animals on the sides. The container can be used

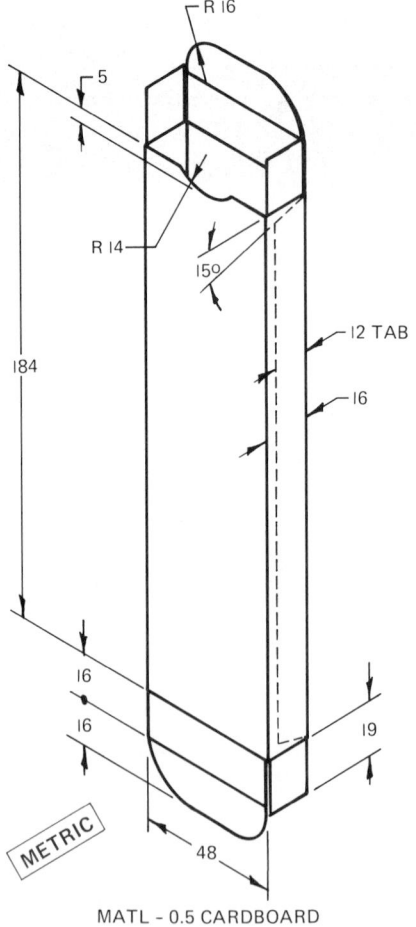

MATL - 0.5 CARDBOARD

FIG. 23-2-C   Pencil box.

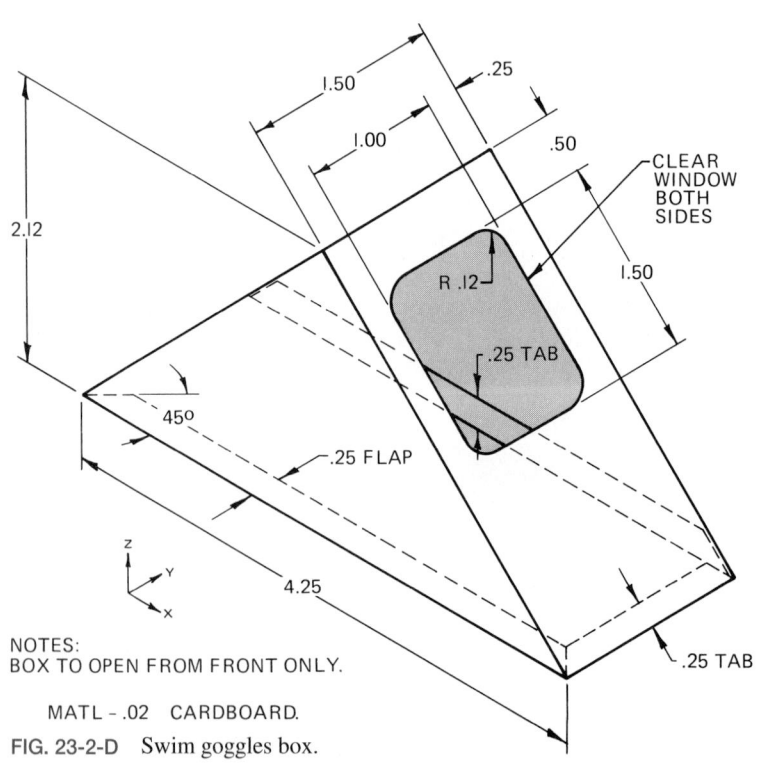

NOTES:
BOX TO OPEN FROM FRONT ONLY.

MATL - .02   CARDBOARD.

FIG. 23-2-D   Swim goggles box.

to hold small articles or as an art object (mobile) that can be hung from the ceiling.

Lay out one of the containers shown in Figs. 23-2-E and 23-2-F. One of the sides forms a lid. On the exterior surface of the container add a suitable design. Cut out the development, score on the bend lines, and glue together. Scale 1:1.

U.S. Customary   W = 1.75 in. All seams .20 in. wide, placed on the inside and glued. Material is .02 in. cardboard.

Metric   W = 45. All seams 6 mm wide, placed on the inside and glued. Material is 0.5 mm cardboard.

FIG. 23-2-E   Food container box.

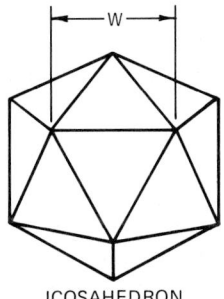

ICOSAHEDRON

FIG. 23-2-F   Food container box.

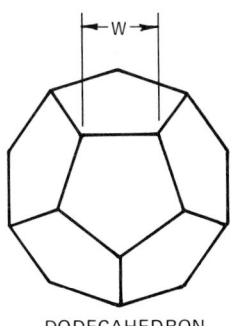

DODECAHEDRON

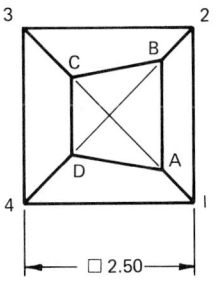

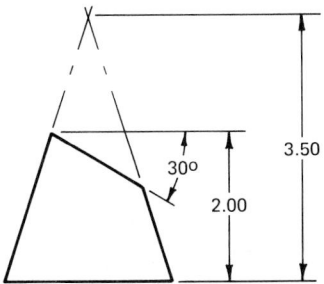

NOTES:
SEAM IS AT A–I.
TOP IS HINGED AT A–B.
BOTTOM IS HINGED AT I–2.

MATL – .0I CARDBOARD.

FIG. 23-3-A   Truncated concentric pyramid.

ASSIGNMENT FOR UNIT 23-3, RADIAL LINE DEVELOPMENT OF FLAT SURFACES

5. Make a development drawing of one of the concentric pyramids shown in Figs. 23-3-A to 23-3-E (here and on pg. 802). Add suitable seams. Scale 1:1.

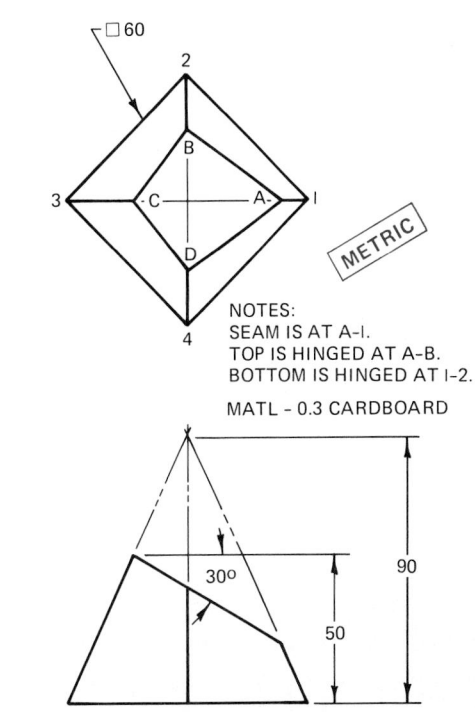

NOTES:
SEAM IS AT A–I.
TOP IS HINGED AT A–B.
BOTTOM IS HINGED AT I–2.

MATL – 0.3 CARDBOARD

FIG. 23-3-B   Truncated concentric pyramid.

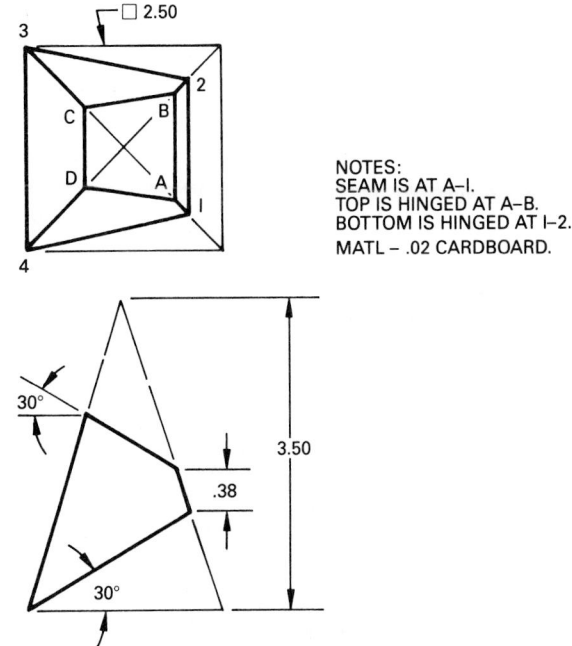

NOTES:
SEAM IS AT A–I.
TOP IS HINGED AT A–B.
BOTTOM IS HINGED AT I–2.

MATL – .02 CARDBOARD.

FIG. 23-3-C   Truncated concentric pyramid.

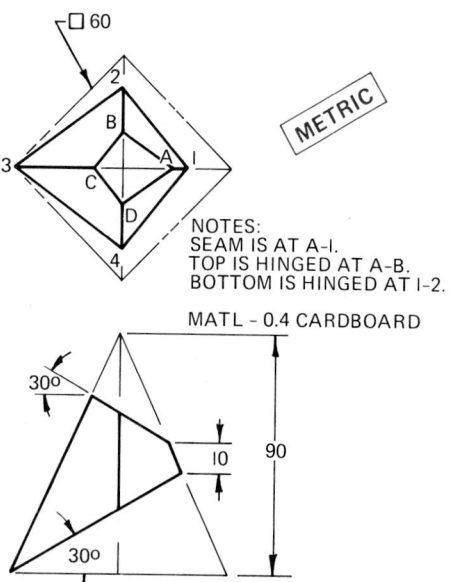

NOTES:
SEAM IS AT A–I.
TOP IS HINGED AT A–B.
BOTTOM IS HINGED AT I–2.

MATL – 0.4 CARDBOARD

FIG. 23-3-D   Truncated concentric pyramid.

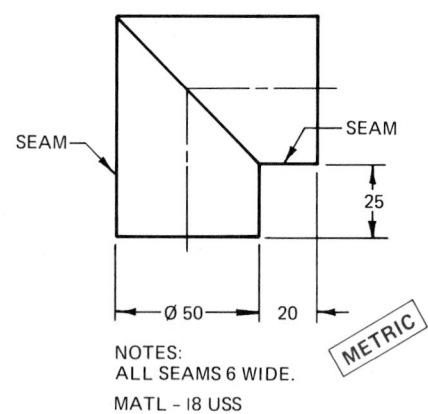

NOTES:
ALL SEAMS 6 WIDE.
MATL – 18 USS

FIG. 23-4-A   Two-piece elbow.

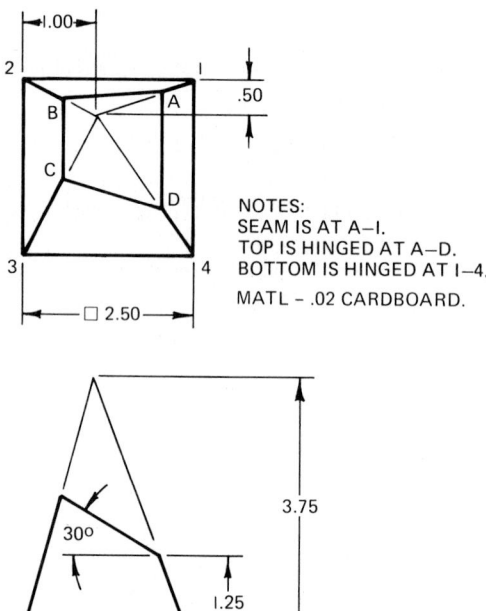

NOTES:
SEAM IS AT A–I.
TOP IS HINGED AT A–D.
BOTTOM IS HINGED AT I–4.

MATL – .02 CARDBOARD.

FIG. 23-3-E   Truncated concentric pyramid.

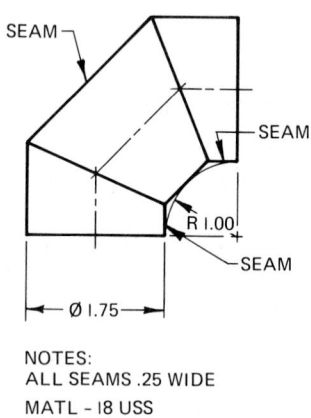

NOTES:
ALL SEAMS .25 WIDE
MATL – 18 USS

FIG. 23-4-B   Three-piece elbow.

ASSIGNMENTS FOR UNIT 23-4, PARALLEL LINE DEVELOPMENT

6. Make a two-view and a development drawing of one of the elbows shown in Figs. 23-4-A and 23-4-B. Add suitable seams. Scale 1:1.

7. Make a two-view and a development drawing of the four-piece elbow shown in Fig. 23-4-C. Add suitable seams. Scale 1:1.

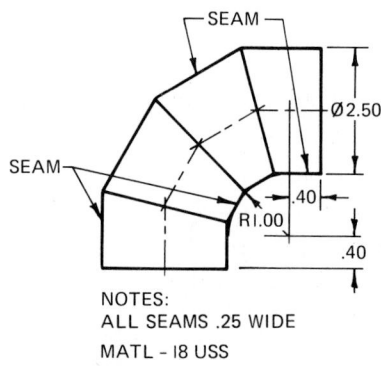

NOTES:
ALL SEAMS .25 WIDE
MATL – 18 USS

FIG. 23-4-C   Four-piece elbow.

FIG. 23-4-D   Sugar scoop.

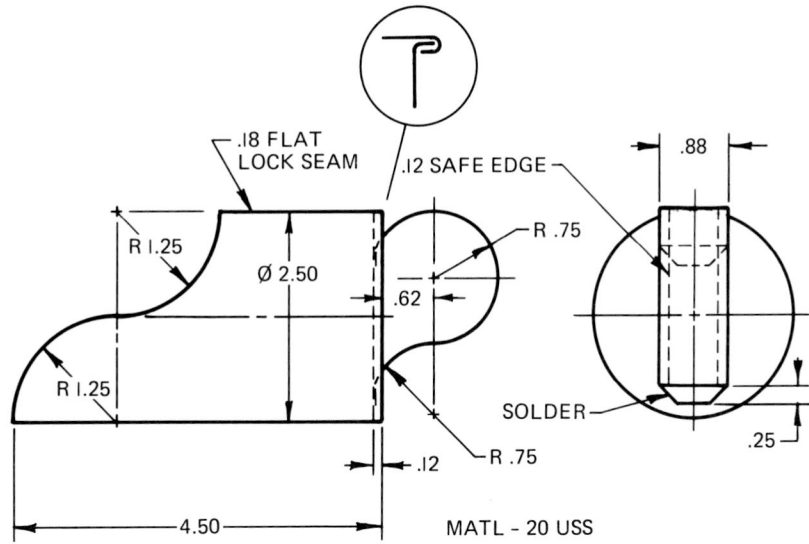

.18 FLAT LOCK SEAM
.12 SAFE EDGE
R .75
R 1.25
Ø 2.50
.62
R 1.25
.88
SOLDER
.25
.12
R .75
4.50
MATL – 20 USS

8. Make a two-view and a development drawing of one of the parts shown in Figs. 23-4-D or 23-4-E. Add suitable seams. Scale 1:1.

ASSIGNMENT FOR UNIT 23-5, RADIAL LINE DEVELOPMENT OF CONICAL SURFACES

9. Make a development drawing of one of the assembled parts shown in Figs. 23-5-A to 23-5-E (here and on pg. 804). Use your judgment for the other views required and add suitable seams. Scale is 1:1 for Figs. 23-5-A to 23-5-D, and 1:2 for Fig. 23-5-E.

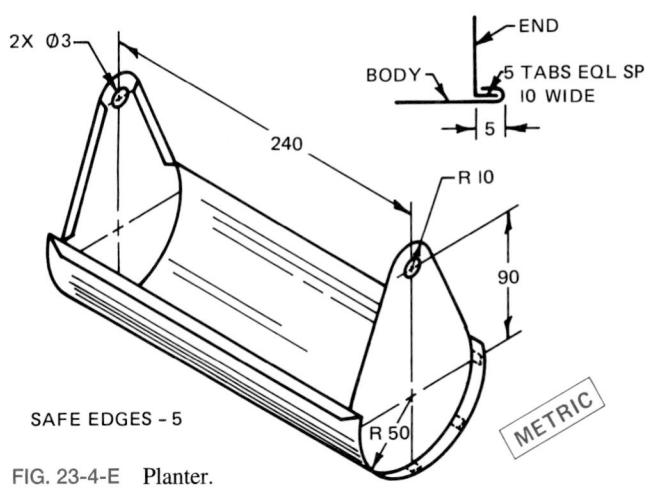

2X Ø3
240
R 10
90
END
BODY
5 TABS EQL SP
10 WIDE
5
SAFE EDGES – 5
R 50
METRIC

FIG. 23-4-E   Planter.

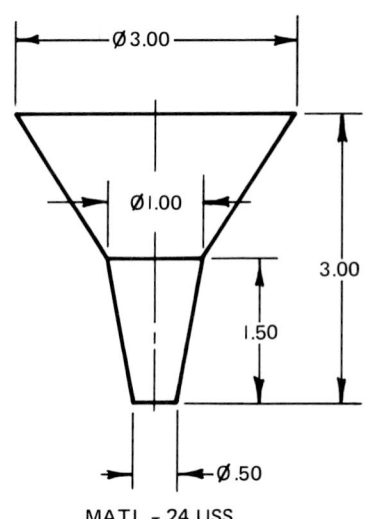

Ø 3.00
Ø 1.00
3.00
1.50
Ø .50
MATL – 24 USS
FIG. 23-5-A   Funnel.

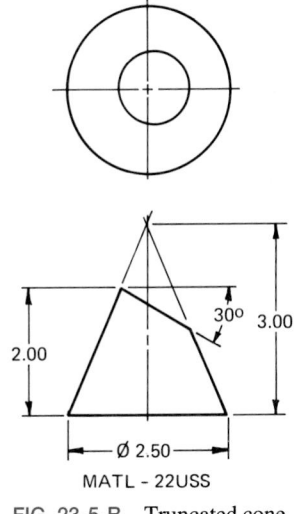

30°
3.00
2.00
Ø 2.50
MATL – 22USS
FIG. 23-5-B   Truncated cone.

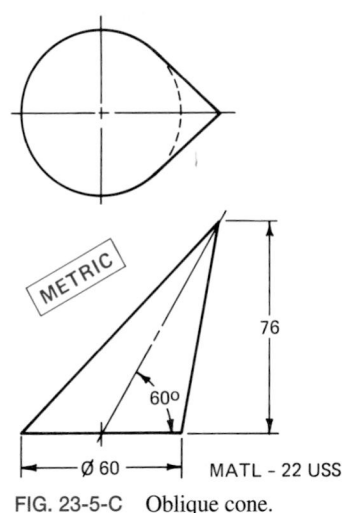

METRIC
76
60°
Ø 60   MATL – 22 USS
FIG. 23-5-C   Oblique cone.

803

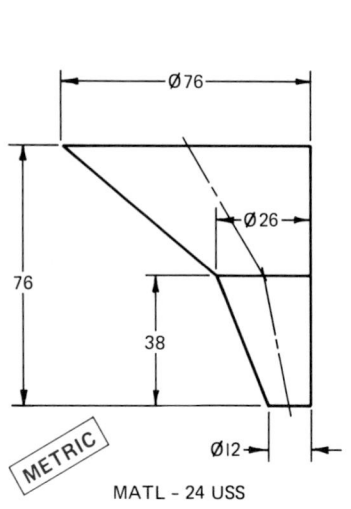

FIG. 23-5-D   Offset funnel.

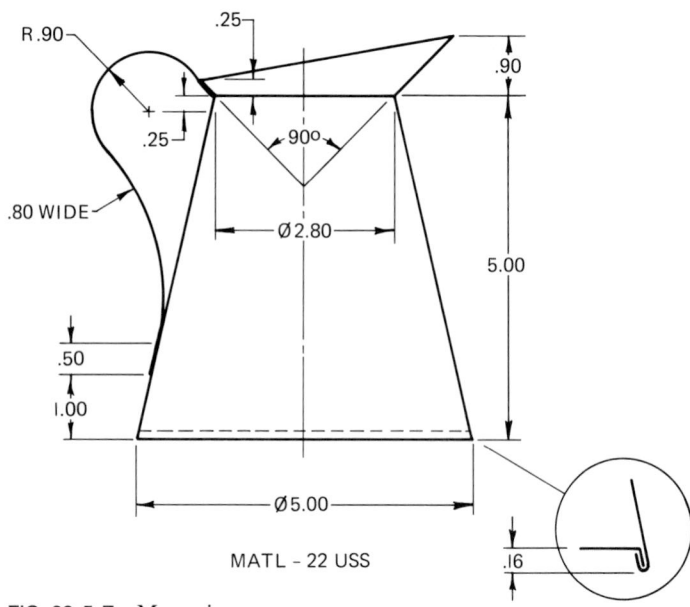

MATL – 22 USS

FIG. 23-5-E   Measuring can.

ASSIGNMENT FOR UNIT 23-6, DEVELOPMENT OF TRANSITION PIECES

10. Make a two-view drawing plus a development drawing of the transition piece of one of the parts shown in Figs.

23-6-A to 23-6-H. Use your judgment for size and location of seams. Scale 1:1.

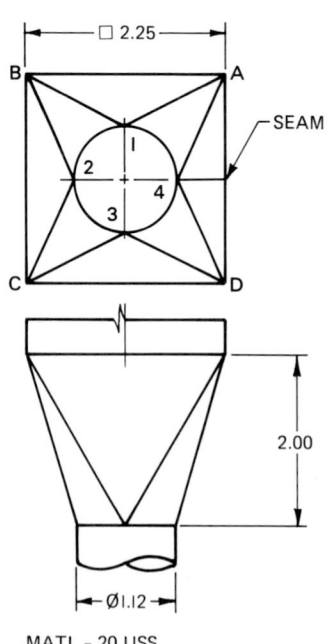

MATL – 20 USS

FIG. 23-6-A   Transition piece.

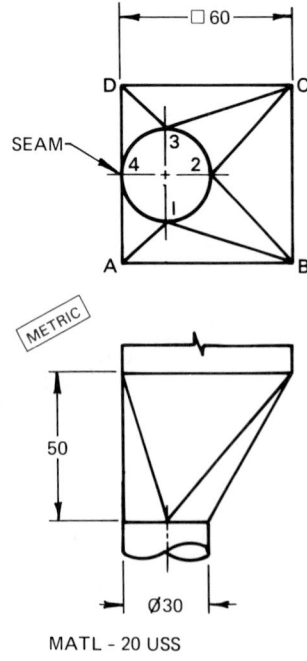

MATL – 20 USS

FIG. 23-6-B   Offset transition piece.

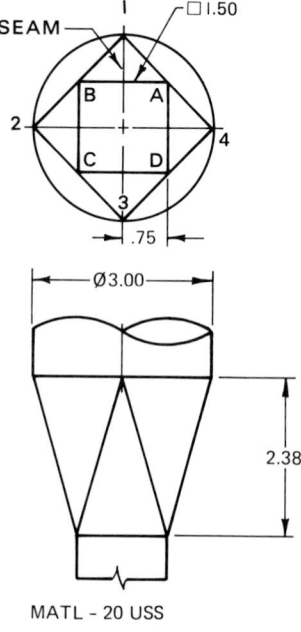

MATL – 20 USS

FIG. 23-6-C   Transition piece.

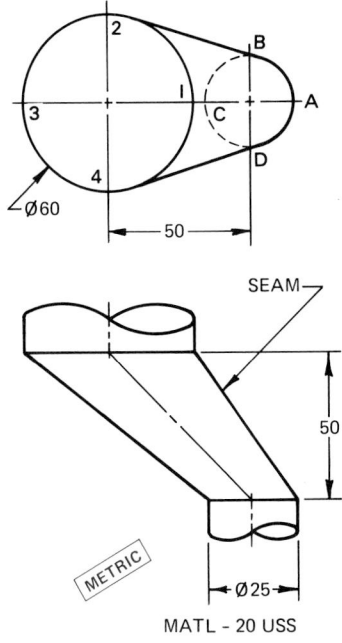

FIG. 23-6-D   Offset transition piece.

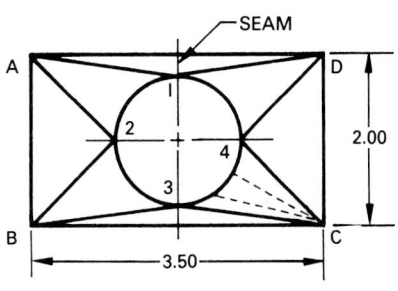

FIG. 23-6-E   Transition piece.

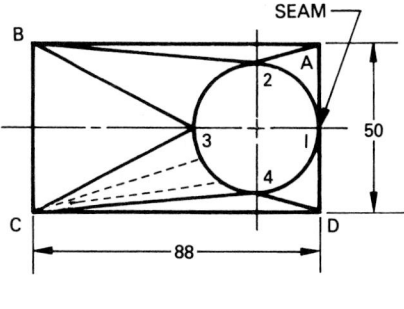

FIG. 23-6-F   Offset transition piece.

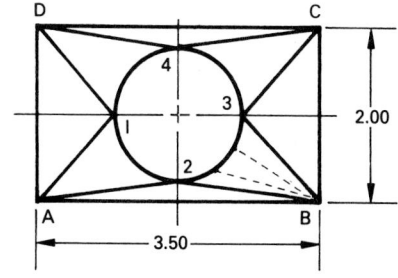

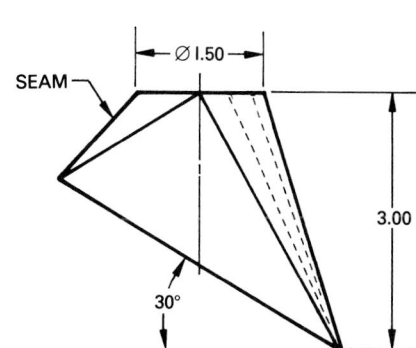

FIG. 23-6-G   Transition piece.

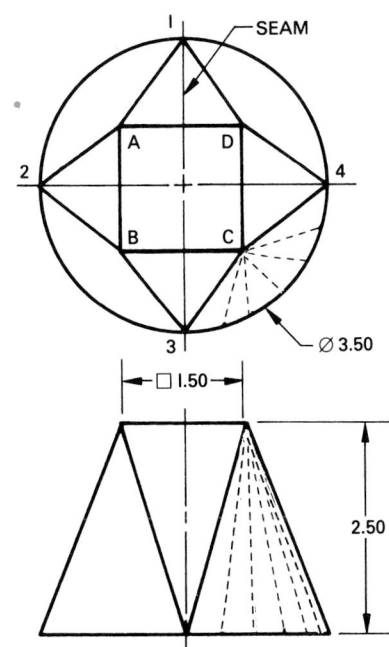

FIG. 23-6-H   Transition piece.

ASSIGNMENTS FOR UNIT 23-7, DEVELOPMENT OF A SPHERE

11. Make a development drawing of a ball. Use your judgment for the views required. Either the gore or zone method may be used. Scale 1:1. Ball diameter is 3.00 in. or 80 mm.

12. Make development drawings for the three parts of the thermos bottle liner shown in Fig. 23-7-A. Seam allowances need not be shown. Scale 1:2.

FIG. 23-7-A   Thermos bottle liner.

ASSIGNMENT FOR UNIT 23-8, INTERSECTION OF FLAT SURFACES— LINES PERPENDICULAR

13. Select one of the assembled parts shown in Figs. 23-8-A to 23-8-D. Complete the views and make development

drawings of the vertical prisms. The horizontal prisms go through the vertical prisms and the ends of the prisms are open. Do not show laps or seams. Scale 1:1.

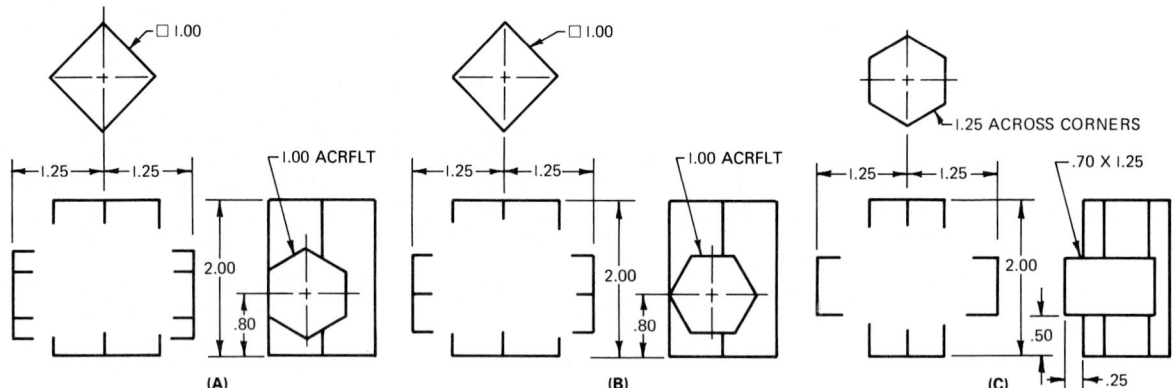

FIG. 23-8-A   Intersecting prisms.

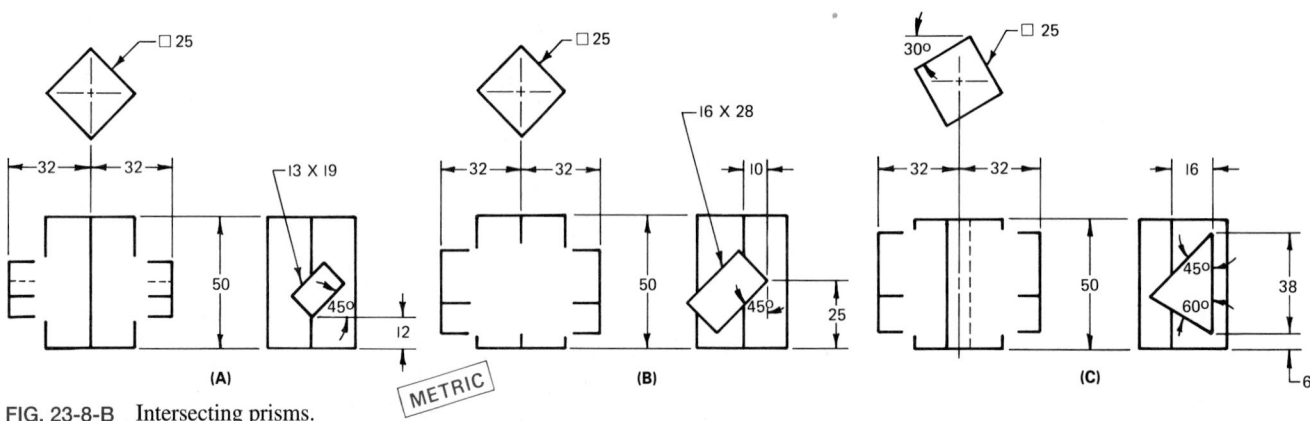

FIG. 23-8-B   Intersecting prisms.

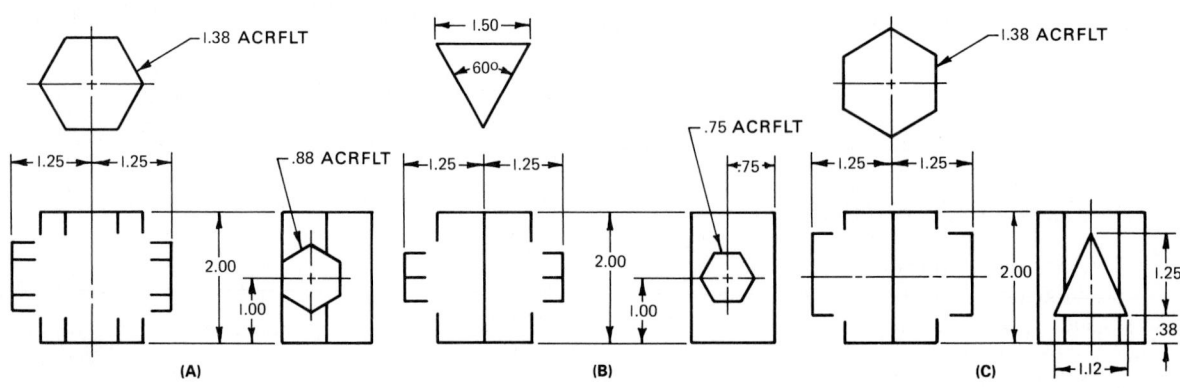

FIG. 23-8-C   Intersecting prisms.

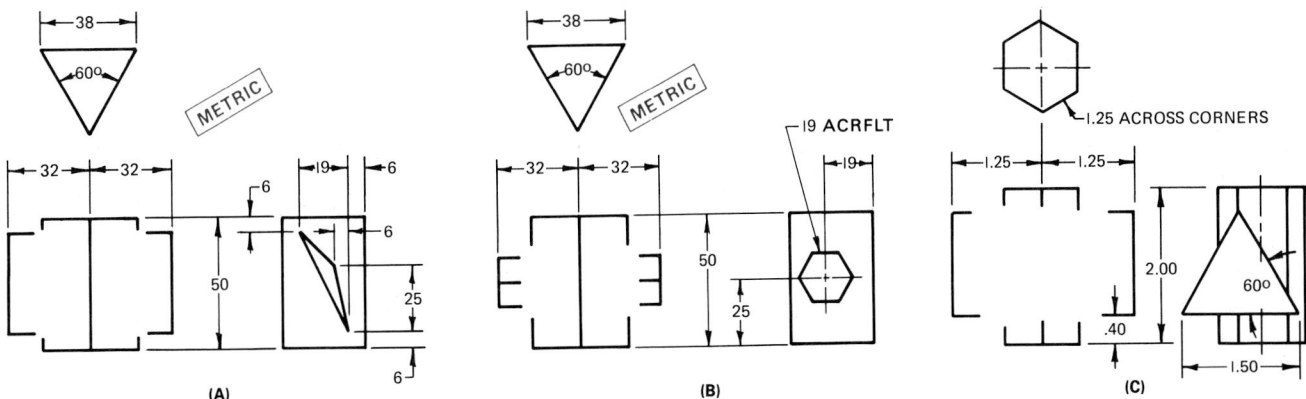

FIG. 23-8-D   Intersecting prisms.

ASSIGNMENT FOR UNIT 23-9, INTERSECTION OF CYLINDRICAL SURFACES

14. Select one of the intersections shown in each of Figs. 23-9-A to 23-9-C (here and on pg. 808). Draw all assemblies and complete the lines of intersection on all views. Do not show laps or seams. Scale 1:1.

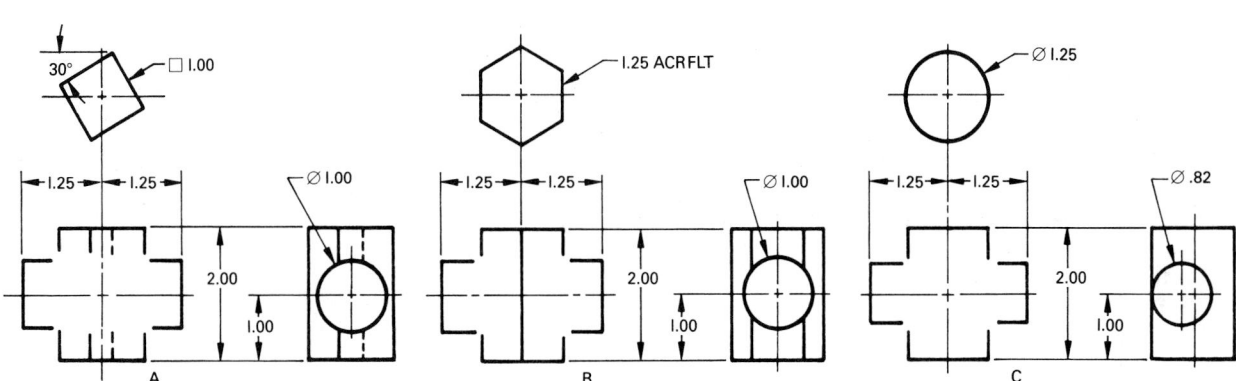

FIG. 23-9-A   Tees and laterals.

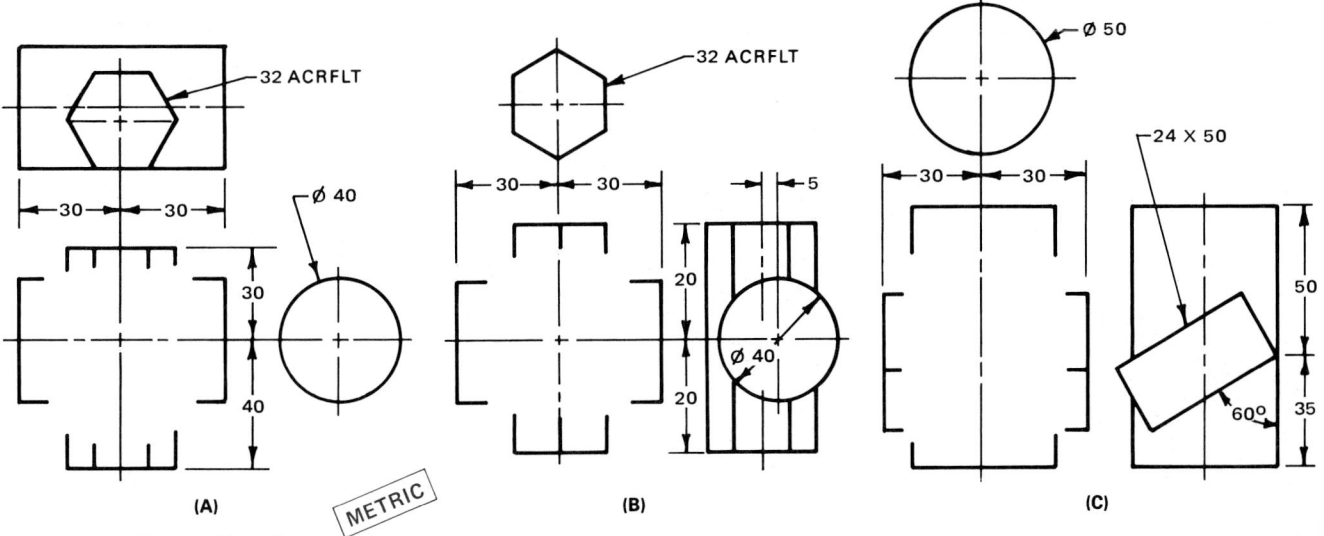

FIG. 23-9-B   Tees and laterals.

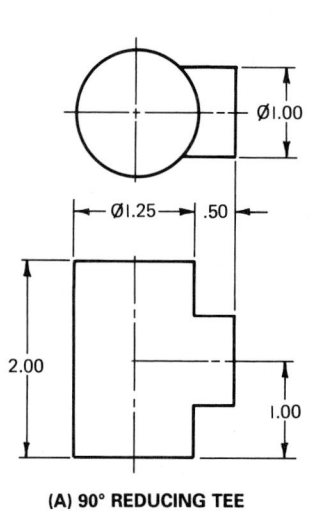

(A) 90° REDUCING TEE

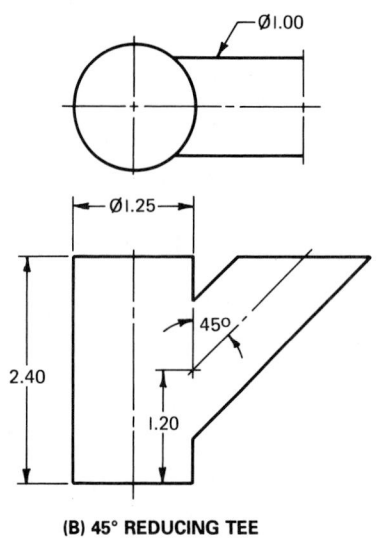

(B) 45° REDUCING TEE

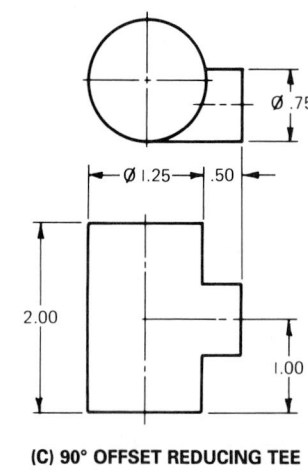

(C) 90° OFFSET REDUCING TEE

FIG. 23-9-C   Tees and laterals.

ASSIGNMENT FOR UNIT 23-10, INTERSECTING PRISMS

15. Select one of the intersections shown in Fig. 23-10-A. Complete the lines of intersection on the partially completed views, and make a development drawing of the conical prism or cone. For (A) and (C) the horizontal prisms pass through the vertical prisms. For (B) the Ø1.00 pipe passes through the cone. Scale 1:1.

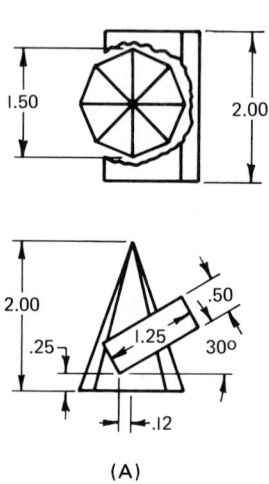

(A)

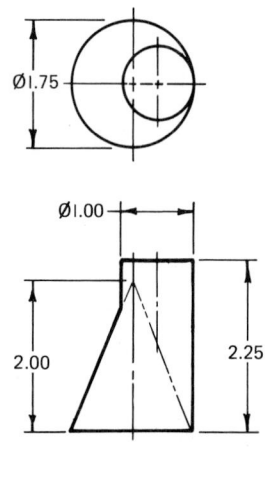

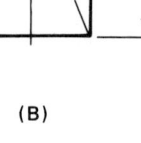

(B)

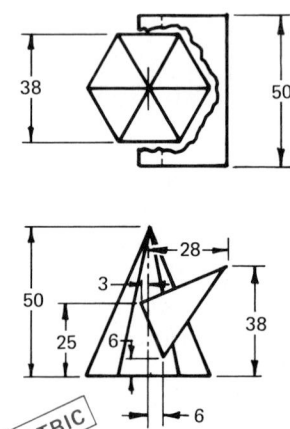

(C)

FIG. 23-10-A   Intersecting prisms.

CHAPTER 24

PIPE DRAWINGS

Definitions

Fittings Parts that are used to join pipe.

Reduced fittings Fittings used to connect different sizes of pipe.

Schedule number Numbers used by ANSI to designate any of 10 different pipe wall thicknesses.

Single-line piping drawings or **simplified representations** Drawings that use a single line to show the arrangement of the pipe and fittings.

Valves Devices used in piping systems to stop or regulate the flow of fluids and gases.

24-1 PIPES

One hundred years ago water was the only important fluid that was conveyed from one point to another in pipes. Today almost every conceivable fluid is handled in pipes during its production, processing, transportation, or utilization. The age of nuclear power and space flight has added fluids such as liquid metals, oxygen, and nitrogen to the list of more common fluids—oil, water, gases, and acids—that are being transported in pipes. Nor is the transportation of fluids the only phase of hydraulics that warrants attention now. Hydraulic and pneumatic mechanisms are used extensively for the control of machinery and numerous other types of equipment. Piping is also used as a structural element in columns and handrails. It is for these reasons that drafters and engineers should become familiar with pipe drawings.

Kinds of Pipes

Steel and Wrought-Iron Pipe

Steel or wrought-iron pipe carry water, steam, oil, and gas and are commonly used where high temperatures and pressures are encountered. Standard steel and cast-iron pipe is specified by the nominal diameter, which is always less than the actual inner diameter (ID) of the pipe. This pipe was available up to recent times in only three wall thicknesses—standard, extra-strong, and double extra-strong (Fig. 24-1-1). In order to use common fittings with these different wall thicknesses of pipe, the outer diameter (OD) of each remained the same, and the extra metal was added to the ID to increase the wall thickness of the extra-strong and double extra-strong pipe.

The nominal size of pipe is given in inch sizes, but the inside and outside diameters and wall thicknesses are given in millimeter sizes in the metric system.

Because of the demand for a greater variety of pipe for increased pressure and temperature uses, ANSI has now made available 10 different wall thicknesses of pipe, each designated by a schedule number. Standard pipe is now referred to as "schedule 40 pipe" and extra-strong pipe as "schedule 80." Pipe over 12 in. is referred to as "OD pipe," and the nominal size is the OD of the pipe.

Cast-Iron Pipe

Cast-iron pipe is often installed underground to carry water, gas, and sewage. It is also used for low-pressure steam connections. Cast-iron pipe joints are normally of the flanged type or the bell-and-spigot type.

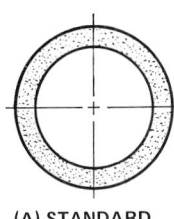 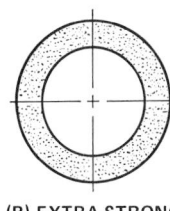 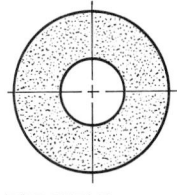

(A) STANDARD SCHEDULE 40   (B) EXTRA-STRONG SCHEDULE 80   (C) DOUBLE EXTRA-STRONG SCHEDULE 160

FIG. 24-1-1   A comparison between steel pipes.

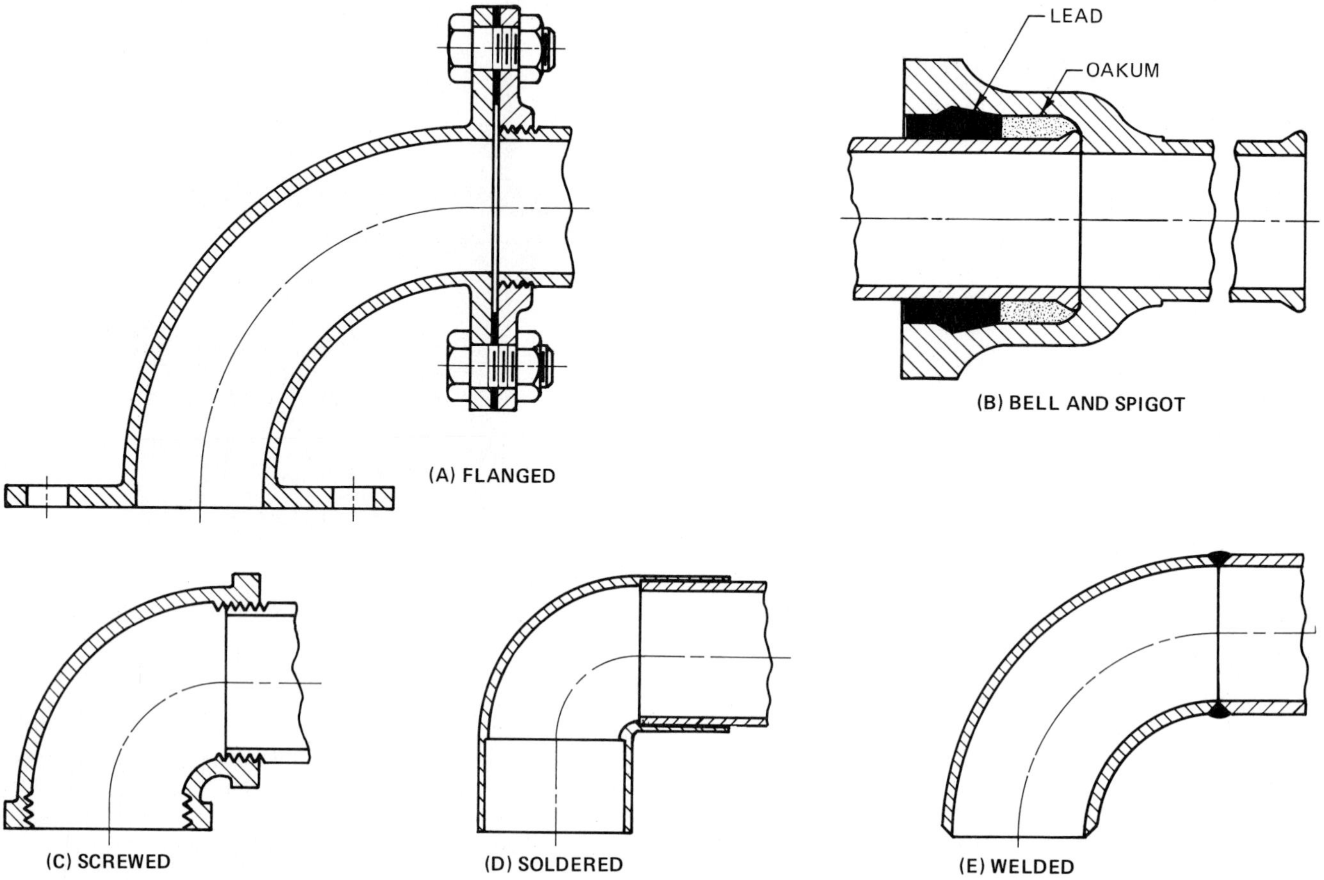

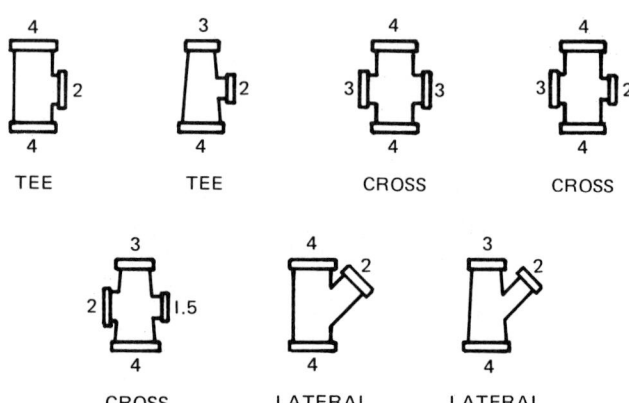

(A) FLANGED

(B) BELL AND SPIGOT

LEAD

OAKUM

(C) SCREWED

(D) SOLDERED

(E) WELDED

FIG. 24-1-2   Common types of pipe joints.

Seamless Brass and Copper Pipe

These types of pipe are used extensively in plumbing because of their ability to withstand corrosion. They have the same nominal diameter as steel or iron pipe, but they have thinner wall sections.

Copper Tubing

This pipe is used in plumbing and heating and where vibration and misalignment are factors, such as in automotive, hydraulic, and pneumatic designs.

Plastic Pipe

This pipe or tubing, because of its corrosion and chemical resistance, is used extensively in the chemical industry. It is flexible and readily installed but it is not recommended where heat or pressure is a factor.

Pipe Joints and Fittings

Parts that are used to join pipe are called *fittings*. They may be used to change size or direction and to join or provide branch connections. They fall into three general classes: screwed, welded, and flanged (Fig. 24-1-2). Other methods are used for cast-iron pipe and copper and plastic tubing.

TEE   TEE   CROSS   CROSS

CROSS   LATERAL   LATERAL

FIG. 24-1-3   Order of specifying the opening of reducing fittings.

Pipe fittings are specified by the nominal pipe size, the name of the fitting, and the material. Some fittings, such as tees, crosses, and elbows, are used to connect different sizes of pipe. These are called *reduced fittings,* and their nominal pipe sizes must be specified. The largest opening of the through run is given first, followed by the opposite end and the outlet. Figure 24-1-3 illustrates the method of designating sizes of reducing fittings.

Screwed Fittings

Screwed fittings, as shown in Fig. 24-1-4, are generally used on small pipe design of 2.50 in. or less. Common practice is to use a pipe compound (a mixture of lead and oil) on the threaded connection to provide a lubricant and to seal any irregularities.

The American standard pipe thread is of two types—tapered and straight. The tapered thread, which is the more common, employs a 1:16 taper on the diameter of both the external and internal threads (Fig. 24-1-5). This fixes the distance to which the pipe enters the fitting and ensures a tight joint. Straight threads are used for special applications, which are listed in the ANSI handbook.

Both the tapered and straight pipe threads have the same number of threads per inch of nominal pipe size, and a pipe with a tapered thread may thread into a fitting having a straight thread, resulting in a tight seal.

Tapered threads are designated on drawings as NPT (American Standard Pipe Taper Thread) and may be drawn with or without the taper, as shown in Fig. 24-1-6. When threads are drawn in tapered form, the taper is exaggerated. Straight pipe threads are designated on drawings as NPS (American Standard Pipe Straight Thread), and standard thread symbols are used. All pipe threads are assumed to be tapered unless otherwise specified.

Welded Fittings

Welded fittings are used where connections are to be permanent and on high-pressure and high-temperature lines. Other

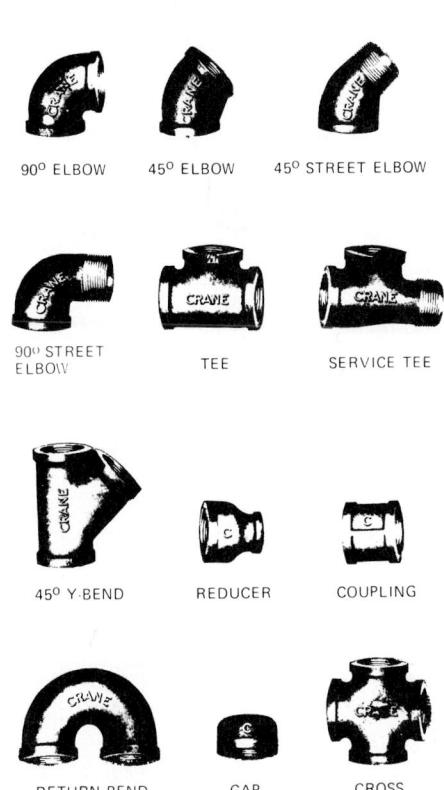

FIG. 24-1-4 Screwed fittings. (*Crane Canada Ltd.*)

90° ELBOW 45° ELBOW 45° STREET ELBOW

90° STREET ELBOW TEE SERVICE TEE

45° Y-BEND REDUCER COUPLING

RETURN BEND CAP CROSS

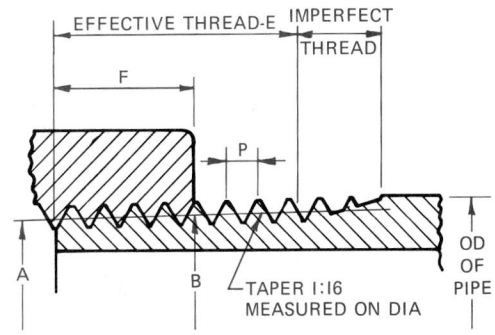

A = PITCH DIAMETER AT END OF PIPE D−(0.05D + 1.1)P
B = PITCH DIAMETER AT GAGING NOTCH (A + 0.0625F)
E = EFFECTIVE THREAD (0.8D + 6.8)P
F = NORMAL ENGAGEMENT BY HAND
P = PITCH
DEPTH OF THREAD = 0.8P

FIG. 24-1-5 American standard pipe thread.

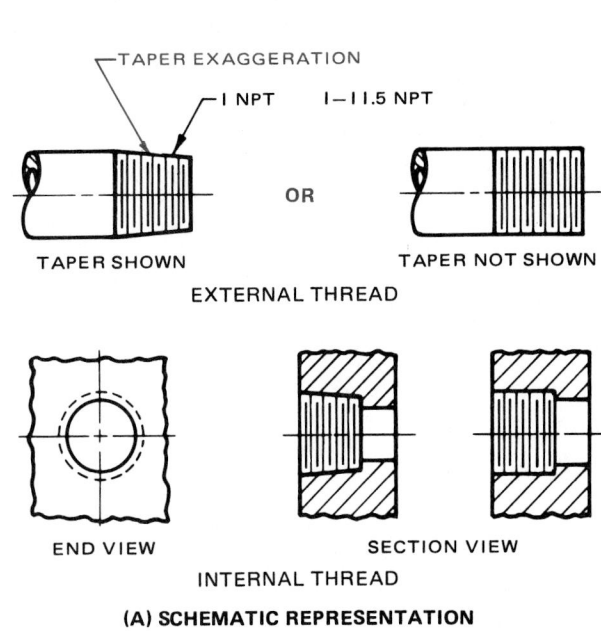

TAPER EXAGGERATION

1 NPT   1−11.5 NPT

OR

TAPER SHOWN   TAPER NOT SHOWN

EXTERNAL THREAD

END VIEW   SECTION VIEW

INTERNAL THREAD

(A) SCHEMATIC REPRESENTATION

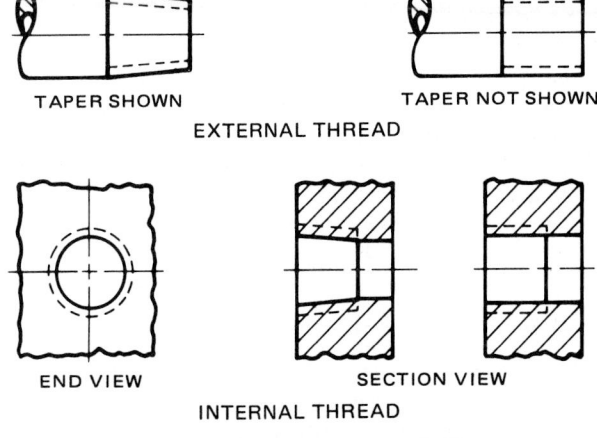

TAPER SHOWN   TAPER NOT SHOWN

EXTERNAL THREAD

END VIEW   SECTION VIEW

INTERNAL THREAD

(B) SIMPLIFIED REPRESENTATION

FIG. 24-1-6 Pipe thread conventions.

FIG. 24-1-7 Welded fittings.
(Tube Turns)

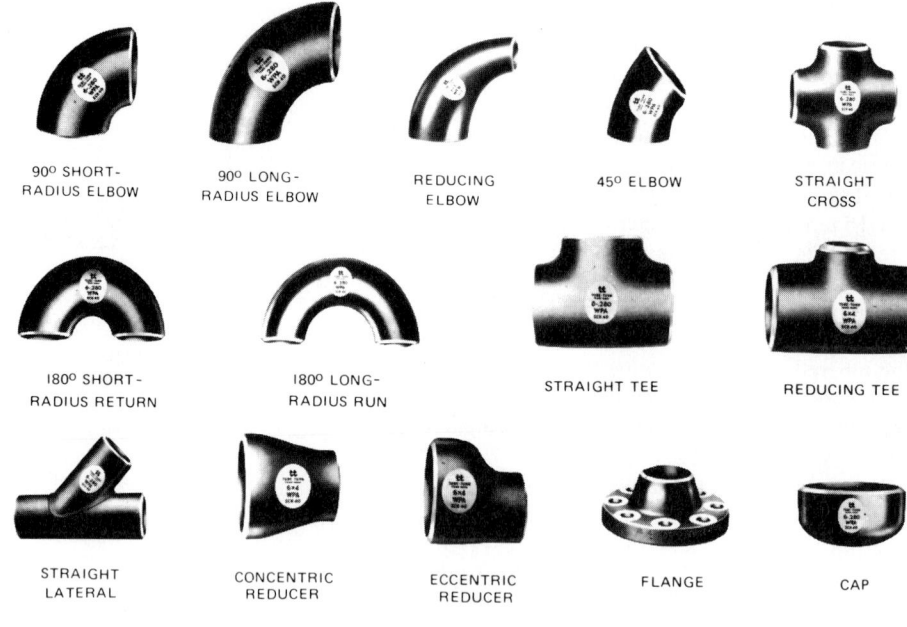

90° SHORT-RADIUS ELBOW   90° LONG-RADIUS ELBOW   REDUCING ELBOW   45° ELBOW   STRAIGHT CROSS

180° SHORT-RADIUS RETURN   180° LONG-RADIUS RUN   STRAIGHT TEE   REDUCING TEE

STRAIGHT LATERAL   CONCENTRIC REDUCER   ECCENTRIC REDUCER   FLANGE   CAP

FIG. 24-1-8 Flanged fittings.
(Crane Canada Ltd.)

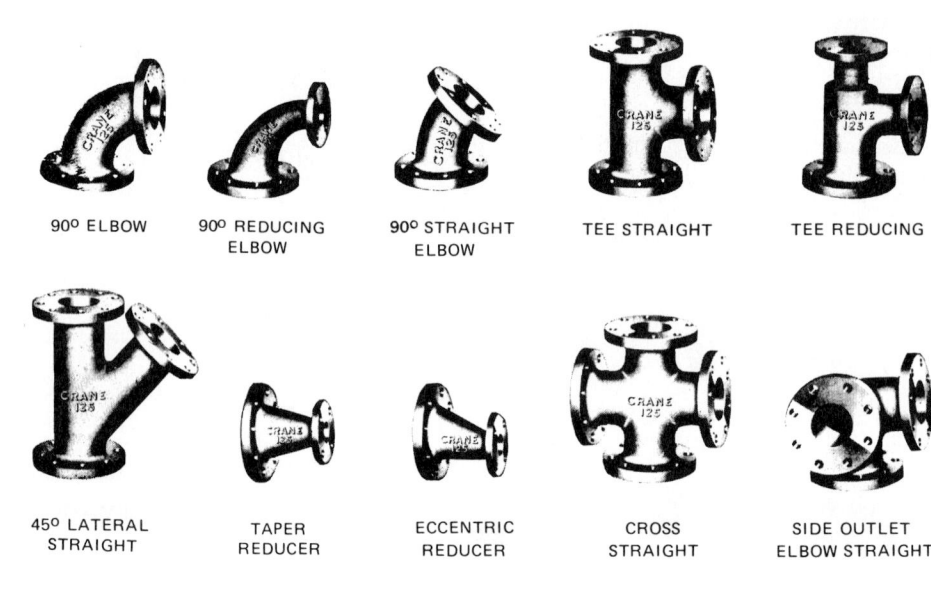

90° ELBOW   90° REDUCING ELBOW   90° STRAIGHT ELBOW   TEE STRAIGHT   TEE REDUCING

45° LATERAL STRAIGHT   TAPER REDUCER   ECCENTRIC REDUCER   CROSS STRAIGHT   SIDE OUTLET ELBOW STRAIGHT

advantages over flanged or screwed fittings are that welded pipes are easier to insulate, they may be placed closer together, and they are lighter in weight (mass). The ends of the pipe and pipe fittings are normally beveled, as shown in Fig. 24-1-7, to accommodate the weld. Joint rings may be used when welded pipe must be disassembled periodically.

Flanges

Flanges provide a quick means of disassembling pipe. Flanges are attached to the pipe ends by welding, screwing, or lapping. The flange faces are then drawn together by bolts, the size and spacing being determined by the size and working pressure of the joint (Figs. 24-1-8 and 24-1-9).

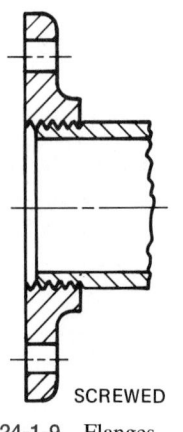
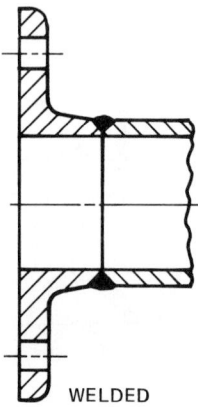

SCREWED   WELDED

FIG. 24-1-9 Flanges.

Valves

Valves are used in piping systems to stop or to regulate the flow of fluids and gases. A few of the more common types are described here.

Gate Valves

Gate valves are used to control the flow of liquids. The wedge, or gate, lifts to allow full, unobstructed flow and lowers to stop it completely (Fig. 24-1-10A). These valves are normally used where the operation is infrequent and they are not intended for throttling or close control.

Globe Valves

Globe valves are used to control the flow of liquids or gases. The design of the globe valve requires two changes in the direction of flow, which slightly reduces the pressure in the system. The globe valve shown in Fig. 24-1-10B is installed so that the pressure is on the disk, which assists the spring in the cap to make a tight closure. This type of valve is recommended for the control of air, steam, gas, or other compressibles where instantaneous on-and-off operation is essential. Figure 24-1-10C is recommended for the control of liquids, such as hot or cold water, gasoline, oil, or solvents, where the sudden closure of a valve might cause objectionable and destructive water hammer. The cap is fitted with a spring-loaded piston dashpot arrangement that retards closure times and helps eliminate shock.

Check Valves

As the name implies, check valves permit flow in one direction, but check all reverse flow. They are operated by the pressure and velocity of line flow alone, and they have no external means of control or operation (Fig. 24-1-10D).

Piping Drawings

The purpose of piping drawings is to show the size and location of pipes, fittings, and valves. Since these items may be purchased, a set of symbols has been developed to portray these features on a drawing.

There are two types of piping drawings in use—single-line and double-line drawings (Fig. 24-1-11, pg. 814). Double-line drawings take more time to draw and therefore are not recommended for production drawings. They are, however, suitable for catalogs and other applications where the visual appearance is more important than the extra drafting time taken to make the drawing.

Single-Line Drawings

Beyond dispute, single-line drawings, also known as *simplified representations,* of pipelines are able to provide substantial savings without loss of clarity or reduction of comprehensiveness of information. As such, the simplified method is used whenever possible.

Single-line piping drawings, as the name implies, use a single line to show the arrangement of the pipe and fittings. The center line of the pipe, regardless of pipe size, is drawn as a thick line to which the valve symbols are added. The size of the symbol is left to the direction of the drafter. When the pipelines carry different liquids, such as cold or hot water, a coded line symbol is often used.

Drawing Projection Two methods of projection are used—orthographic and pictorial. Orthographic projection, as shown

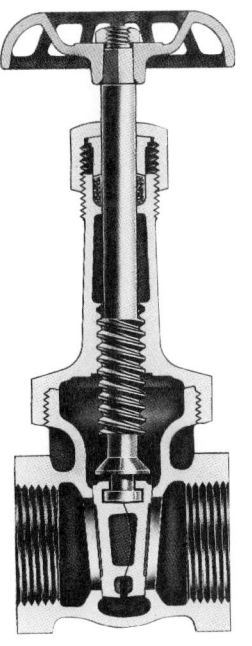

(A) GATE VALVE

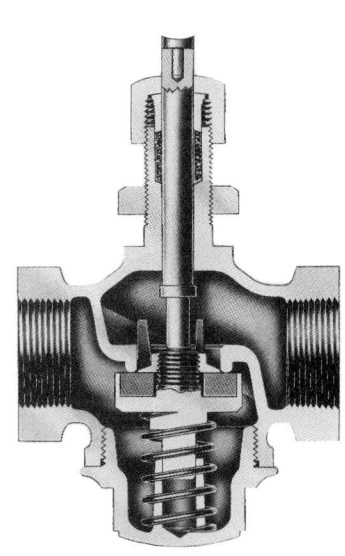

(B) GLOBE VALVE

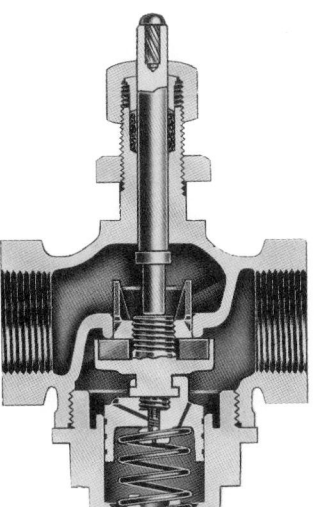

(C) GLOBE VALVE

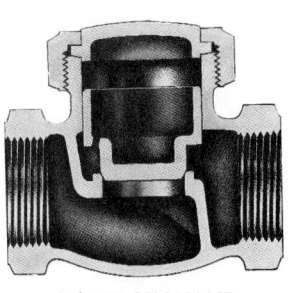

(D) CHECK VALVE

FIG. 24-1-10   Common valves. *(Jenkins Bros. Ltd.)*

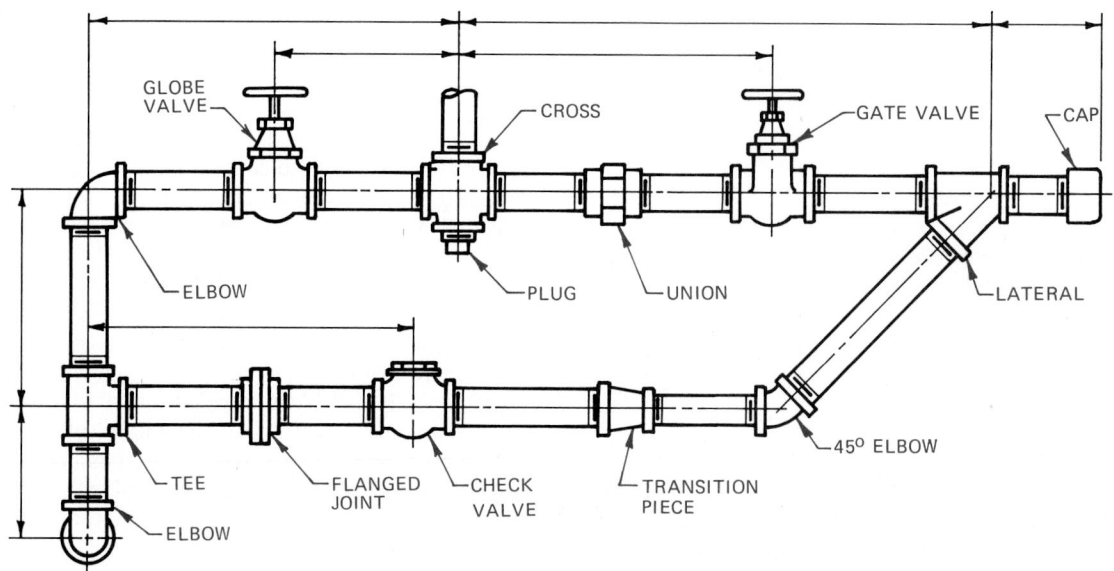

(A) DOUBLE-LINE DRAWING

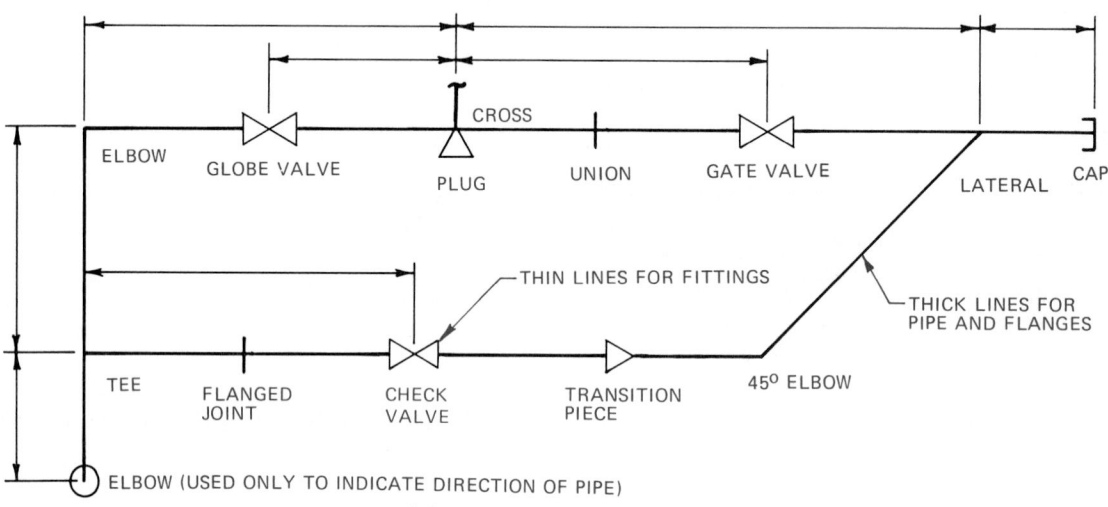

(B) SINGLE-LINE DRAWING

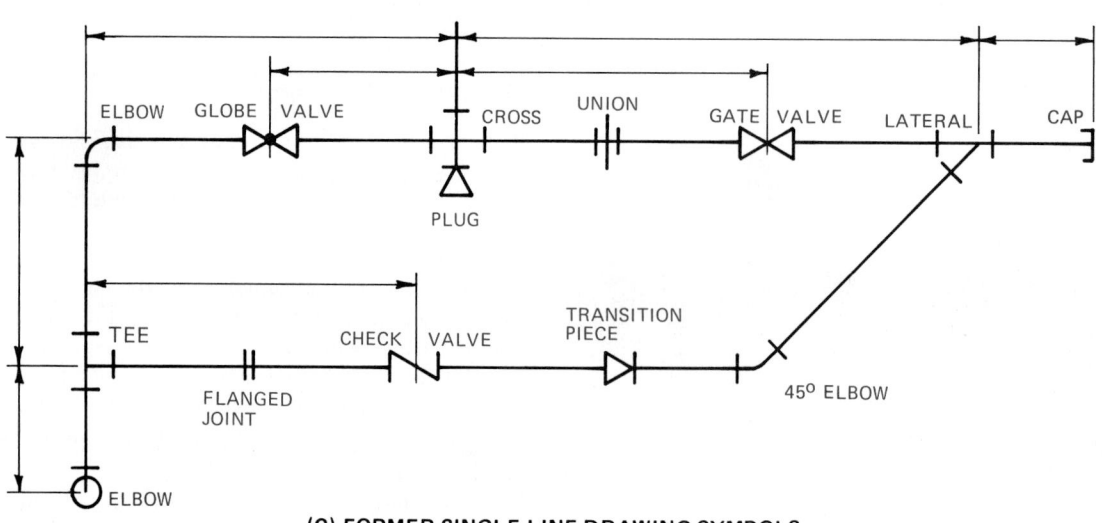

(C) FORMER SINGLE-LINE DRAWING SYMBOLS

FIG. 24-1-11   Pipe drawing symbols.

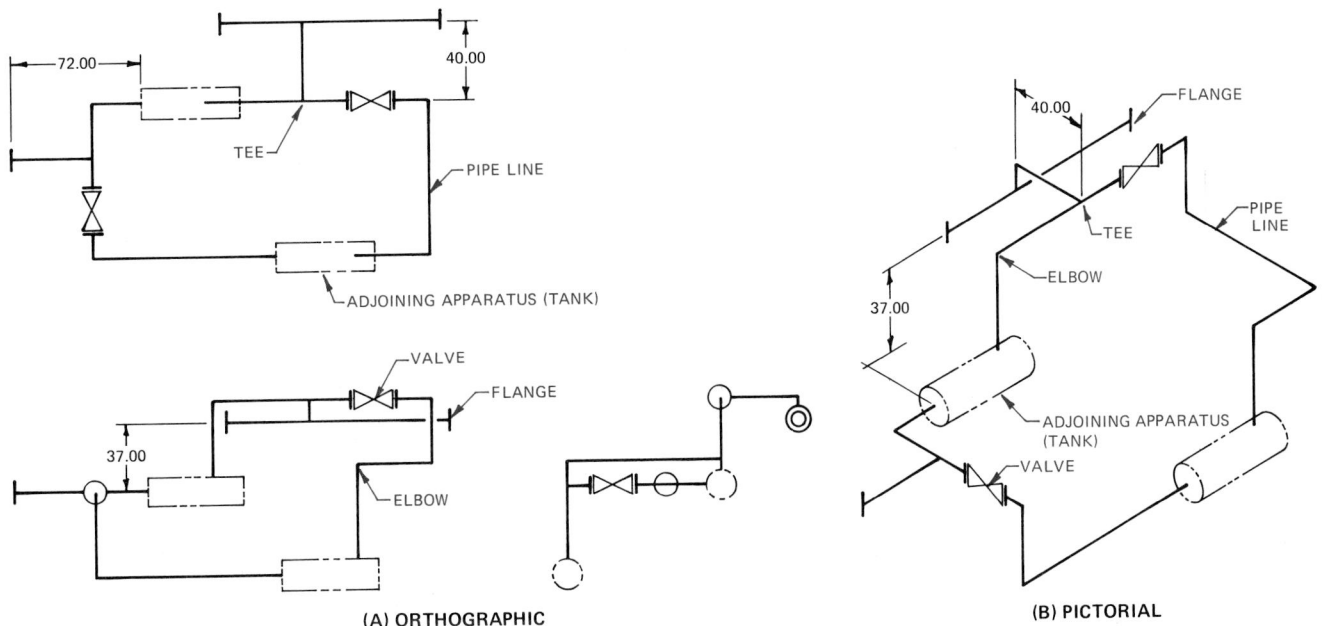

(A) ORTHOGRAPHIC

(B) PICTORIAL

FIG. 24-1-12 Single-line pipe drawings.

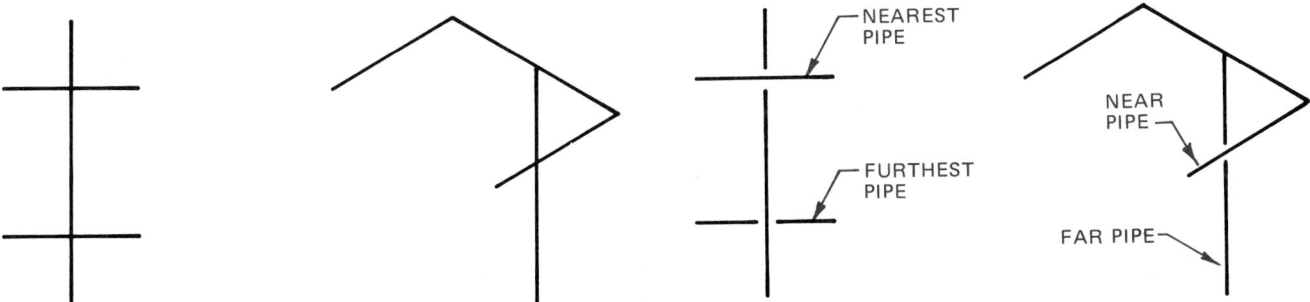

(A) CROSSING OF PIPE SHOWN WITHOUT INTERRUPTING THE PIPE PASSING BEHIND THE NEAREST PIPE

(B) USING AN INTERRUPTED LINE TO INDICATE PIPE FURTHEST AWAY

FIG. 24-1-13 Crossing of pipes.

in Fig. 24-1-12A, is recommended for the representation of single pipes, either straight or bent, in one plane only. However, this method is also used for more complicated piping.

Pictorial projection, as shown in Fig. 24-1-12B, is recommended for all pipes bent in more than one plane and for assembly and layout work, because the finished drawing is easier to understand.

Crossings Crossings or pipes without connections are normally depicted without interrupting the line representing the hidden line (Fig. 24-1-13). When it is desirable to show that one pipe must pass behind the other, the line representing the pipe farthest from the viewer will be shown with a break, or interruption, where the other pipe passes in front of it. For microform purposes, the break should not be less than 10 times the line thickness.

Connections Permanent connections or junctions, whether made by welding or other processes such as gluing and soldering, are to be shown on the drawing by a heavy dot

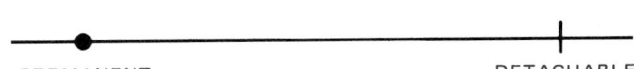

PERMANENT   DETACHABLE

FIG. 24-1-14 Pipe connections.

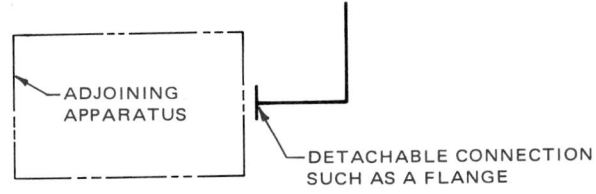

FIG. 24-1-15 Adjoining apparatus.

(Fig. 24-1-14). A general note or specification may describe the process used.

Detachable connections or junctions may be shown by a single thick line instead of a heavy dot, as shown in Figs. 24-1-14 and 24-1-15. The specifications, a general note, or the

815

item list will include the types of fittings, such as flanges, unions, or couplings, and whether the fittings are flanged or threaded.

Fittings If no specific symbols are standardized, fittings like tees, elbows, crosses, etc., are not specially drawn but are represented, like pipe, by a continuous line. The circular symbol for a tee or elbow may be used when it is necessary to indicate whether the piping is coming toward or going away from the viewer, as shown in Fig. 24-1-16. Elbows on isometric drawings may be shown without the radius. However, the change of direction that the piping takes should be quite clear if this method is used.

Adjoining Apparatus If needed, adjoining apparatus, such as tanks, machinery, etc., not belonging to the piping itself may be shown by an outline drawn with a thin phantom line (Fig. 24-1-15).

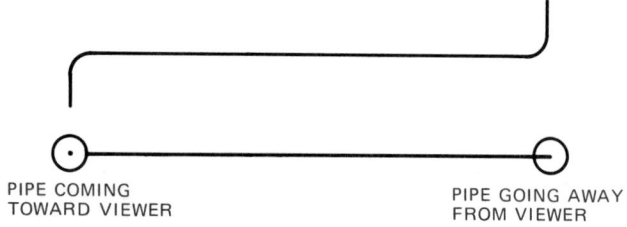

(A) PIPELINE WITHOUT FLANGES CONNECTED TO ENDS OF PIPE

PIPE COMING TOWARD VIEWER

PIPE GOING AWAY FROM VIEWER

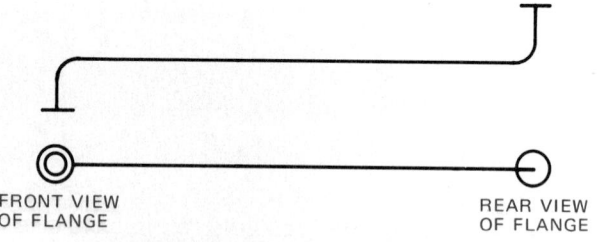

FRONT VIEW OF FLANGE

REAR VIEW OF FLANGE

(B) FLANGES CONNECTED TO ENDS OF PIPELINE

FIG. 24-1-16 Indicating ends of pipelines.

FIG. 24-1-17 Dimensioning piping.

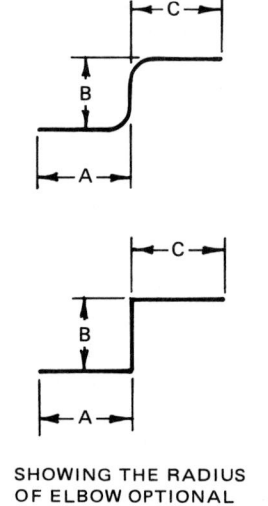

SHOWING THE RADIUS OF ELBOW OPTIONAL

(A) LINEAR DIMENSIONS

Dimensioning

- Dimensions for pipe and pipe fittings are always given from center to center of pipe and to the outer face of the pipe end or flange (Fig. 24-1-17).
- Pipe lengths are not normally shown on the drawings, but left to the pipe fitter.
- Pipe and fitting sizes and general notes are placed on the drawing beside the part concerned or, where space is restricted, indicated with a leader.
- An item list is usually provided with the drawing.
- Pipes with bends are dimensioned from vertex to vertex.
- Radii and angles of bends may be dimensioned as shown in Fig. 24-1-17B. Whenever possible, the smaller of the supplementary angles is to be specified.
- The outer diameter and wall thickness of the pipe may be specified on the line representing the pipe or elsewhere (item list, general note, specification, etc.).

Orthographic Piping Symbols

Pipe Symbols If flanges are not attached to the ends of the pipelines when orthographic projections are drawn in, pipeline symbols indicating the direction of the pipe are required. If the pipeline is coming toward the front (or viewer), it will be shown by two concentric circles, the smaller one being solid (Fig. 24-1-16A). If the pipeline is going toward the back (or away from the viewer), it will be shown by one circle. No extra lines are required on the other views.

Flange Symbols As shown in Fig. 24-1-16B, flanges are to be represented, irrespective of their type and sizes, by two concentric circles in the front view, by one circle in the rear view, and by a short stroke in the side view, while lines of equal thickness, as chosen for the representation of pipes, are used.

Valve Symbols Symbols representing valves are drawn with continuous thin lines (as opposed to thick lines for piping and flanges). The valve spindles should be shown only if it is necessary to define their positions. It will be assumed that unless

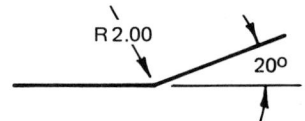

R 2.00   20°

(B) RADII AND ANGLES OF BENDS

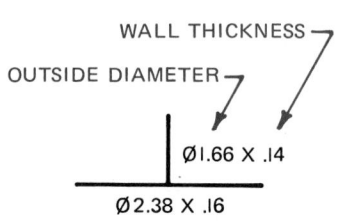

WALL THICKNESS

OUTSIDE DIAMETER

Ø1.66 X .14

Ø2.38 X .16

(C) PIPE SIZE INDICATED ON DRAWINGS

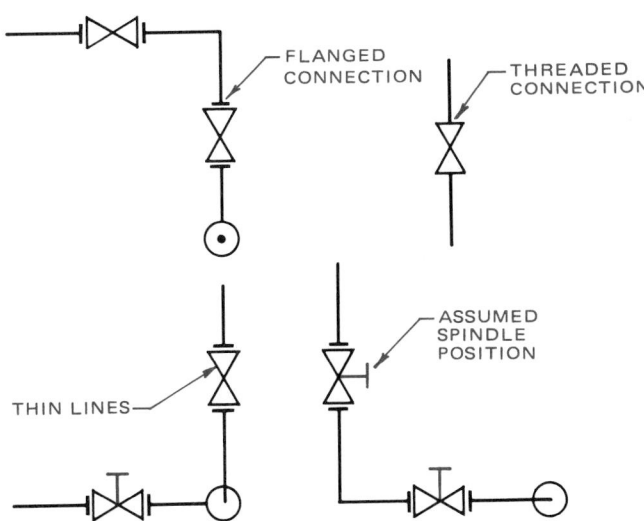

FLANGED CONNECTION

THREADED CONNECTION

ASSUMED SPINDLE POSITION

THIN LINES

NOTE: WHEN VALVE SPINDLES NOT SHOWN, IT WILL BE ASSUMED THAT THEY WILL BE IN THE POSITIONS INDICATED ABOVE.

FIG. 24-1-18   Valve symbols.

otherwise specified, the valve spindle is in the position shown in Fig. 24-1-18.

REFERENCES AND SOURCE MATERIAL

1. Crane Canada Ltd.
2. Jenkins Bros. Ltd.

ASSIGNMENTS

See Assignments 1 and 2 for Unit 24-1 on pages 822–823.

24-2   ISOMETRIC PROJECTION OF PIPING DRAWINGS

The scale of the drawing applies to the dimension taken along the coordinate axes (isometric axes). In drawing to the isometric projection method, the following rules should be observed:

- Parts of pipe that run parallel to the coordinate axes are drawn without any special indication of being parallel to the isometric axes (Figs. 24-2-1 and 24-2-2).
- With reference to calculations or programming for computer drafting, it will probably be necessary to indicate the X, Y, and Z axes (coordinates) on the drawing.

Flanges

Flanges are to be represented, irrespective of their type and size, by short lines of the same thickness as those chosen for the representation of the pipes (Fig. 24-2-3, pg. 818).

Flanges at the ends of vertical pipe parts should preferably be drawn to an angle of 30° to horizontal and flanges at the ends of horizontal pipe parts in a vertical direction.

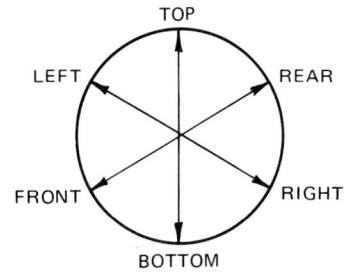

TOP

LEFT

REAR

FRONT

RIGHT

BOTTOM

(A) ISOMETRIC COORDINATE AXES

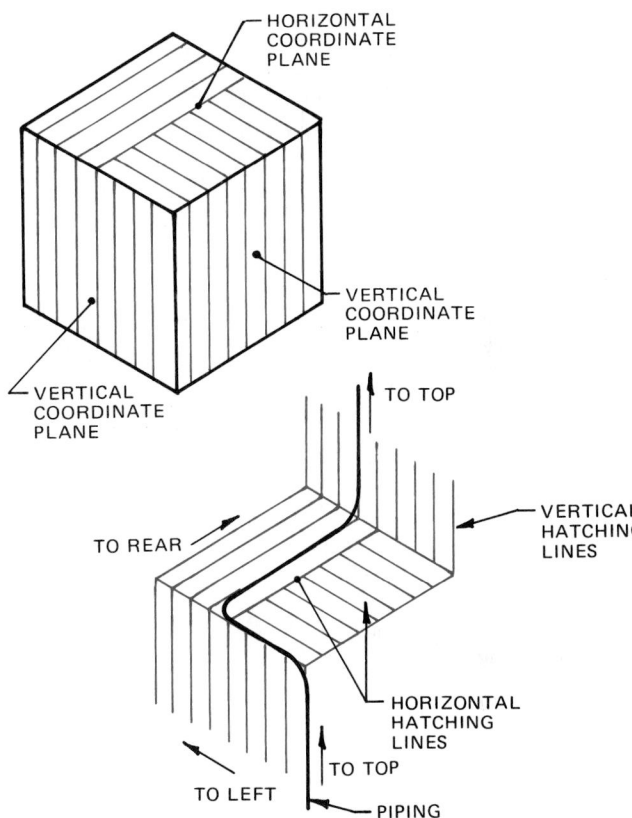

HORIZONTAL COORDINATE PLANE

VERTICAL COORDINATE PLANE

VERTICAL COORDINATE PLANE

TO TOP

TO REAR

VERTICAL HATCHING LINES

HORIZONTAL HATCHING LINES

TO TOP

TO LEFT

PIPING

(B) POSITIONING PIPING ON COORDINATE AXES

FIG. 24-2-1   Coordinate axes for piping drawings.

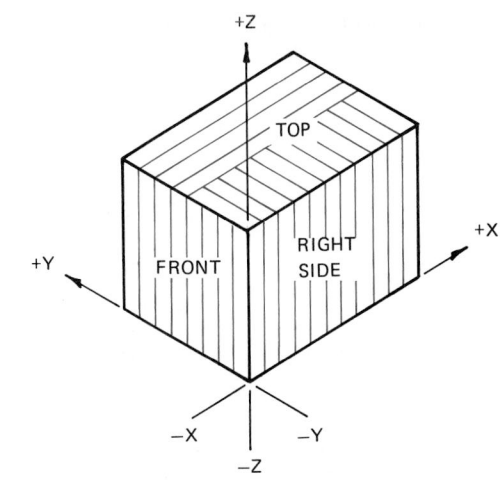

+Z

TOP

+X

RIGHT SIDE

+Y

FRONT

−X

−Y

−Z

FIG. 24-2-2   Coordinate axes for piping drawings.

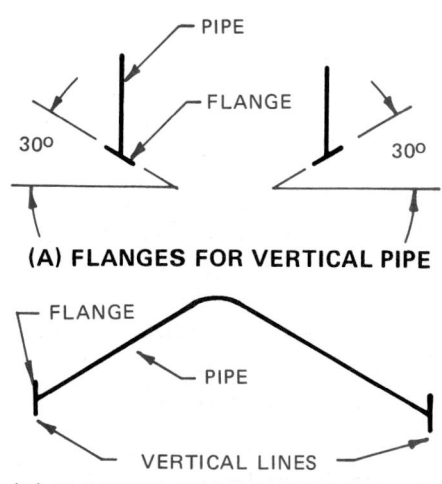

(A) FLANGES FOR VERTICAL PIPE

(B) FLANGES FOR HORIZONTAL PIPE

FIG. 24-2-3   Flange positioning for isometric drawings.

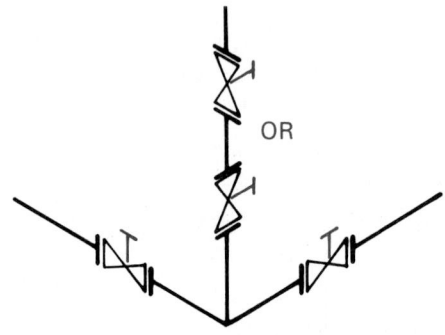

(A) VALVES WITH THREADED CONNECTIONS

(B) VALVES WITH FLANGE CONNECTIONS

FIG. 24-2-4   Valve symbols.

Valves

For isometric drawings it will be assumed that unless otherwise specified, the valve spindle is in the position shown in Fig. 24-2-4. Valve spindles should be drawn only if it is necessary to define their positions. Deviations from these positions can be described by specifying the angle to which the valve is rotated in a clockwise, or right-hand, direction when looking in the direction of the positive *X, Y,* or *Z* axes (Fig. 24-2-5).

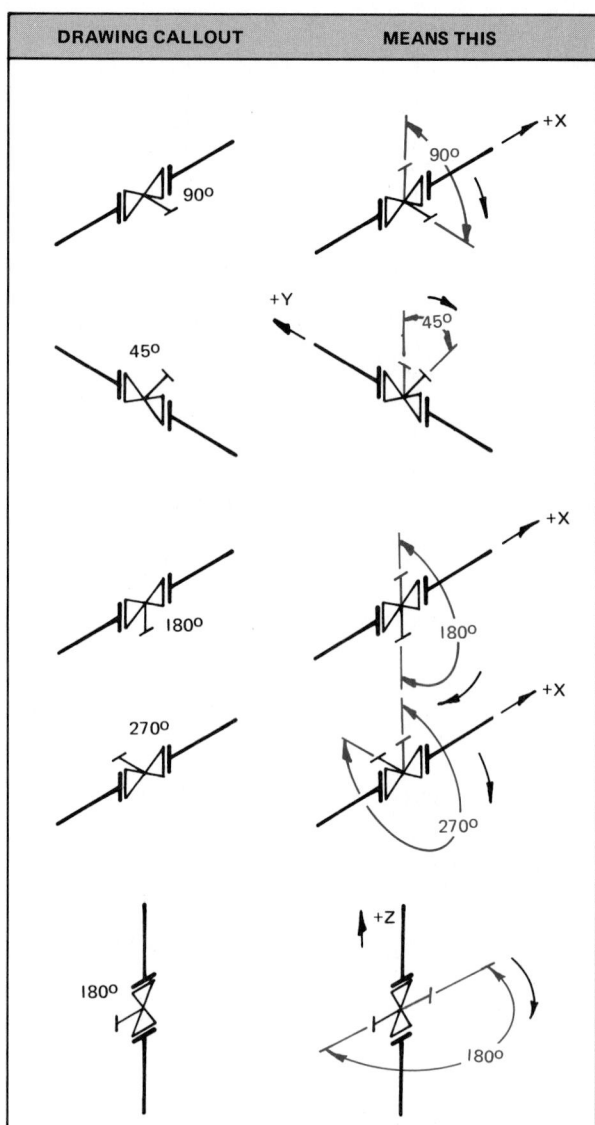

DRAWING CALLOUT	MEANS THIS

FIG. 24-2-5   Deviation from normal position of valve spindle.

Dimensioning

The preferred method of dimensioning isometric pipe drawings is the unidirectional system because of the ease in execution and reading (Fig. 24-2-6). An alternative to indicating the height of pipes is to use a level indicator symbol (Fig. 24-3-2).

REFERENCES AND SOURCE MATERIAL

1. Crane Canada Ltd.
2. Jenkins Bros. Ltd.

ASSIGNMENTS

See Assignments 3 and 4 for Unit 24-2 on pages 824–825.

FIG. 24-2-6   Unidirectional dimensioning.

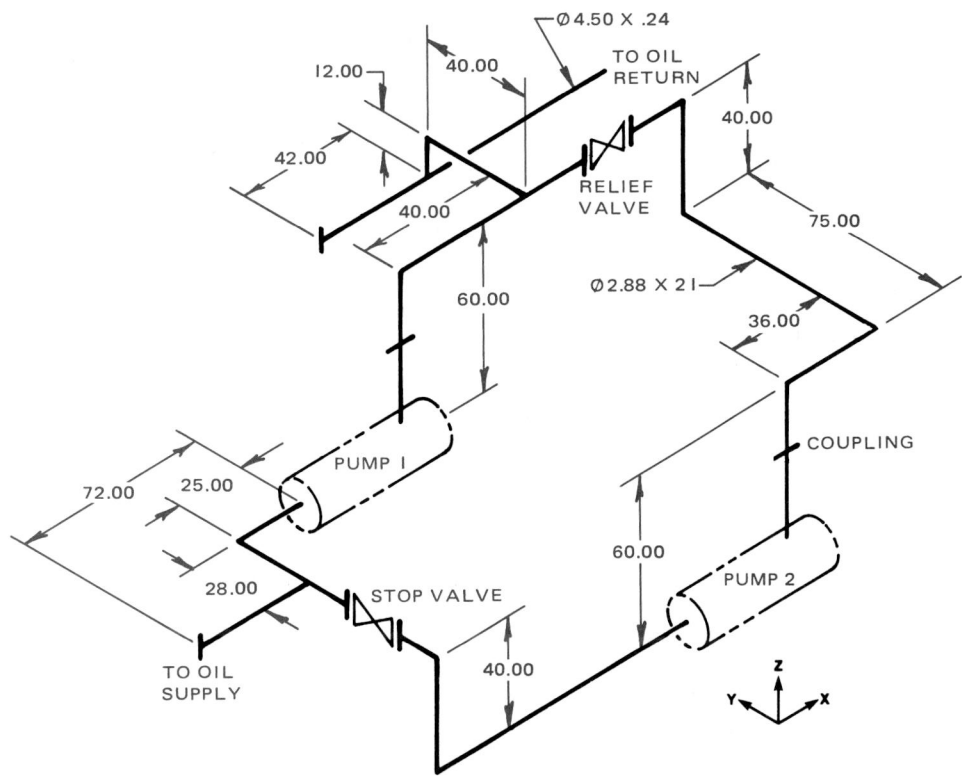

24-3 SUPPLEMENTARY PIPING INFORMATION

Direction of Flow   The direction of flow may be shown by an arrowhead on the line representing the piping, as shown in Fig. 24-3-1.

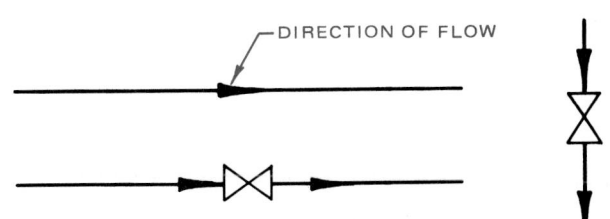

FIG. 24-3-1   Indicating direction of flow.

Level Indicators   Level indications in lieu of linear measurements may be used to show the height of pipelines and fittings. The preferred method of indicating levels is shown in Fig. 24-3-2.

Specifying Slope on Pipes   The direction of slope is shown by an arrow located above the pipe pointing from the higher to the lower level. The amount of slope may be specified by either a general note on the drawing or one of the methods shown in Fig. 24-3-3 (pg. 820). However, with long piping runs, it may be useful to specify the slope by reference to a datum and level indication as shown in Fig. 24-3-3C.

Support and Hangers   Support and hangers are to be represented by their appropriate symbols, as shown in Fig. 24-3-4A (pg. 820). The representation of repetitive accessories may be simplified, as shown in Fig. 24-3-4B.

FIG. 24-3-2   Level indicators.

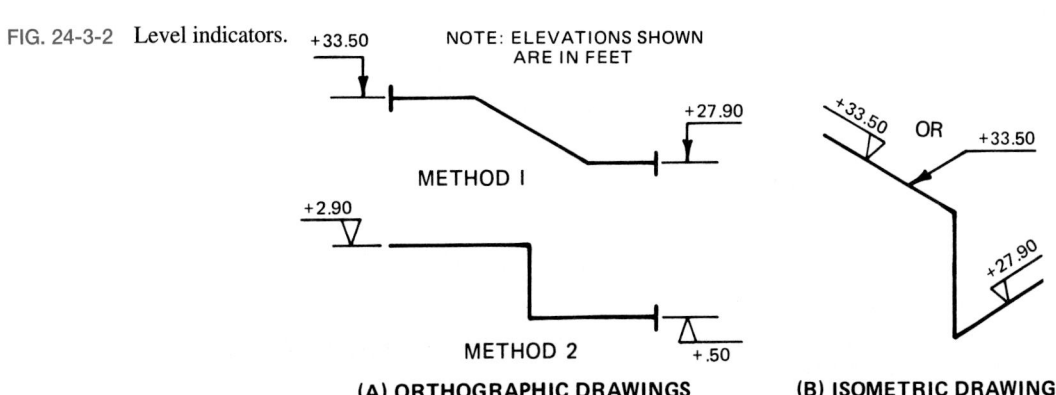

(A) ORTHOGRAPHIC DRAWINGS   **(B) ISOMETRIC DRAWINGS**

FIG. 24-3-3   Specifying slope of pipes.

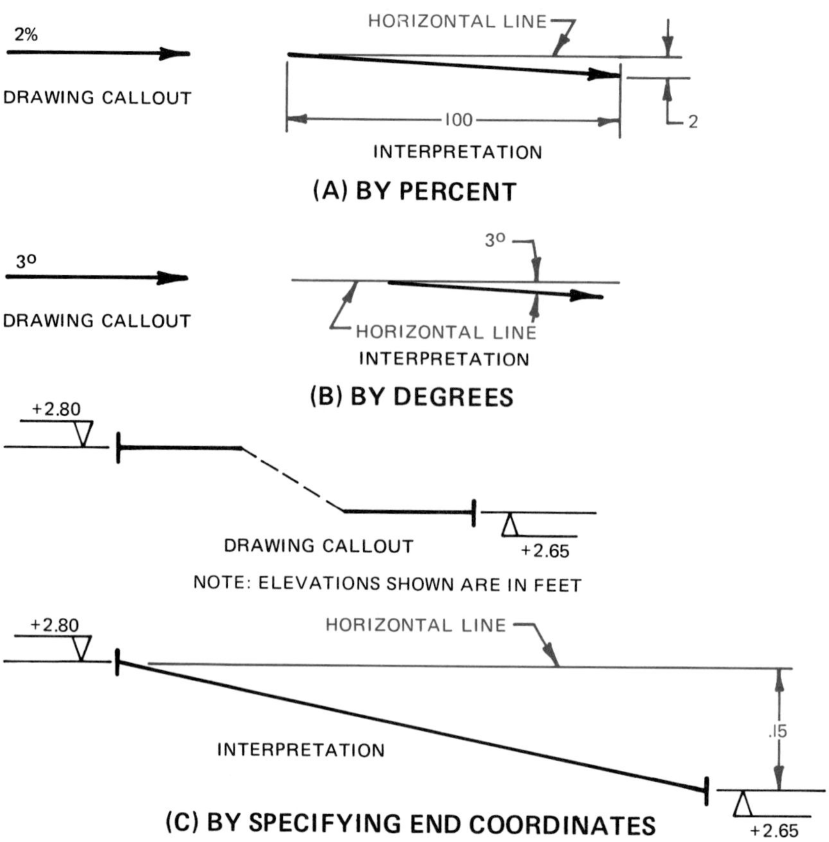

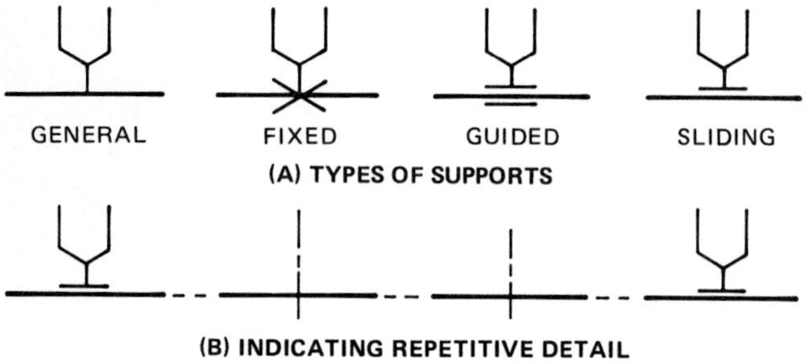

2%

DRAWING CALLOUT

HORIZONTAL LINE

⎯100⎯

⎣2

INTERPRETATION

(A) BY PERCENT

3°

DRAWING CALLOUT

3°

HORIZONTAL LINE

INTERPRETATION

(B) BY DEGREES

+2.80

+2.65

DRAWING CALLOUT

NOTE: ELEVATIONS SHOWN ARE IN FEET

+2.80

HORIZONTAL LINE

.15

+2.65

INTERPRETATION

(C) BY SPECIFYING END COORDINATES

GENERAL   FIXED   GUIDED   SLIDING

(A) TYPES OF SUPPORTS

(B) INDICATING REPETITIVE DETAIL

FIG. 24-3-4   Supports and hangers.

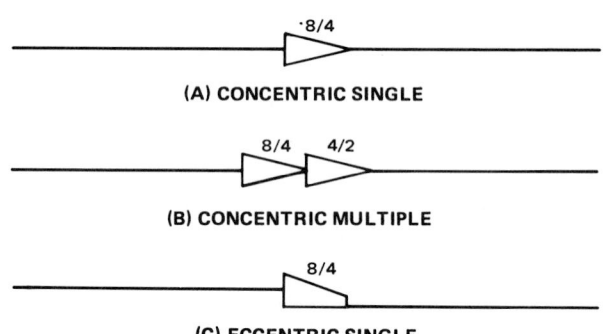

·8/4

(A) CONCENTRIC SINGLE

8/4   4/2

(B) CONCENTRIC MULTIPLE

8/4

(C) ECCENTRIC SINGLE

FIG. 24-3-5   Transition pieces.

Transition Pieces   Transition pieces for changing the cross section are indicated by the symbols shown in Fig. 24-3-5. The relevant nominal sizes are given above the symbols.

Pipe Runs Not Parallel with Coordinate Axes   Deviations from the directions of the coordinate axes are to be shown by means of hatched planes, as follows:

1. For a part of a pipe situated in a plane parallel to one of the vertical projection planes, vertical hatching lines are drawn to indicate the vertical projection plane (Fig. 24-3-6A).
2. For a part of a pipe situated in a plane parallel to the horizontal coordinate plane, horizontal hatching lines are drawn to indicate the horizontal projection plane (Fig. 24-3-6B).
3. For a part of a pipe not running parallel to any of the coordinate planes, both vertical and horizontal hatching lines are drawn to indicate the vertical and horizontal projection planes (Fig. 24-3-6C).

If desired, in addition to the coordinate planes, the prism of which the pipe part forms the diagonal may be shown in thin lines (Fig. 24-3-6D).

If such hatching is not convenient, for instance when using automated drafting equipment, it may be omitted but should be replaced with the thin-line rectangle or parallelepiped whose diagonal coincides with the pipe (Fig. 24-3-7). Application of projection planes is shown in Fig. 24-3-8.

REFERENCES AND SOURCE MATERIAL

1. Jenkins Bros. Ltd.

ASSIGNMENTS ▰▰▰▰▰▰▰▰▰▰▰▰▰▰▰▰

See Assignments 5 and 6 for Unit 24-3 on pages 826–827.

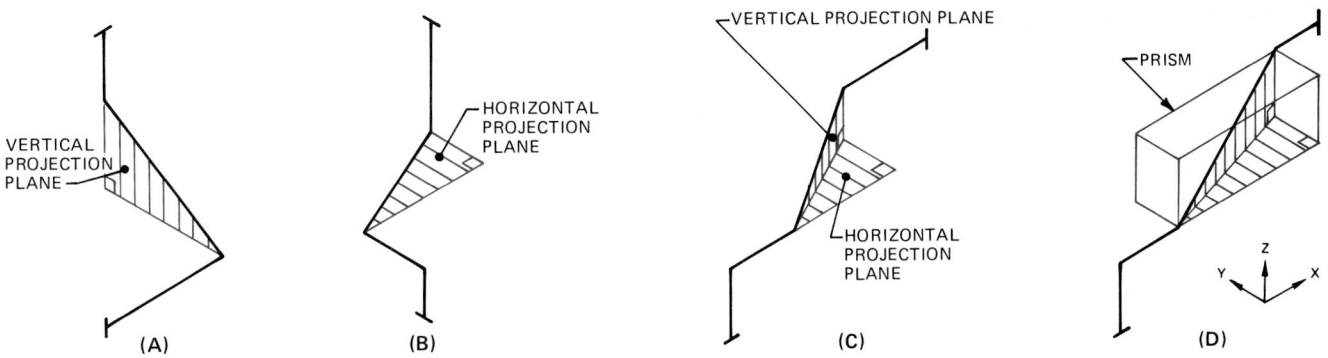

FIG. 24-3-6　Indication of pipe run not in the direction of the coordinate axes.

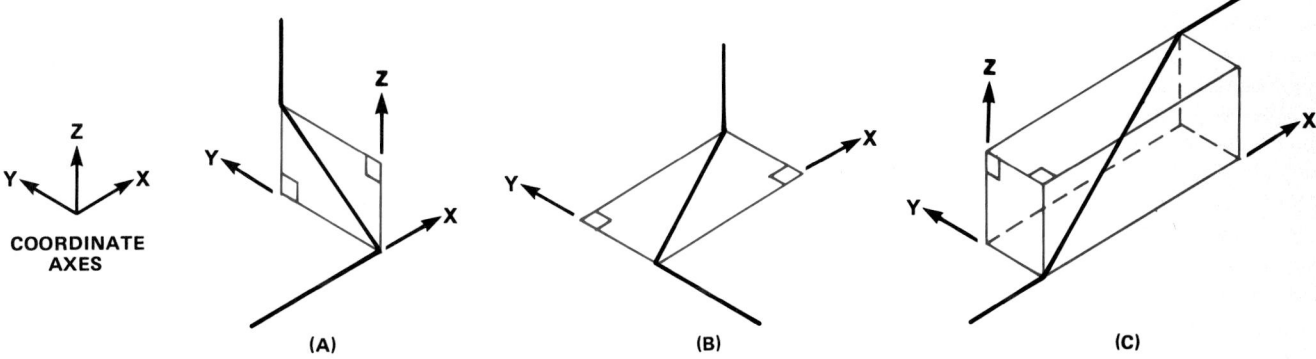

FIG. 24-3-7　Alternate method of indicating pipe run not in the direction of the coordinate axes.

(A) ISOMETRIC

NOTE—RADIUS TO INDICATE
DIRECTION OF RUN

(B) ORTHOGRAPHIC

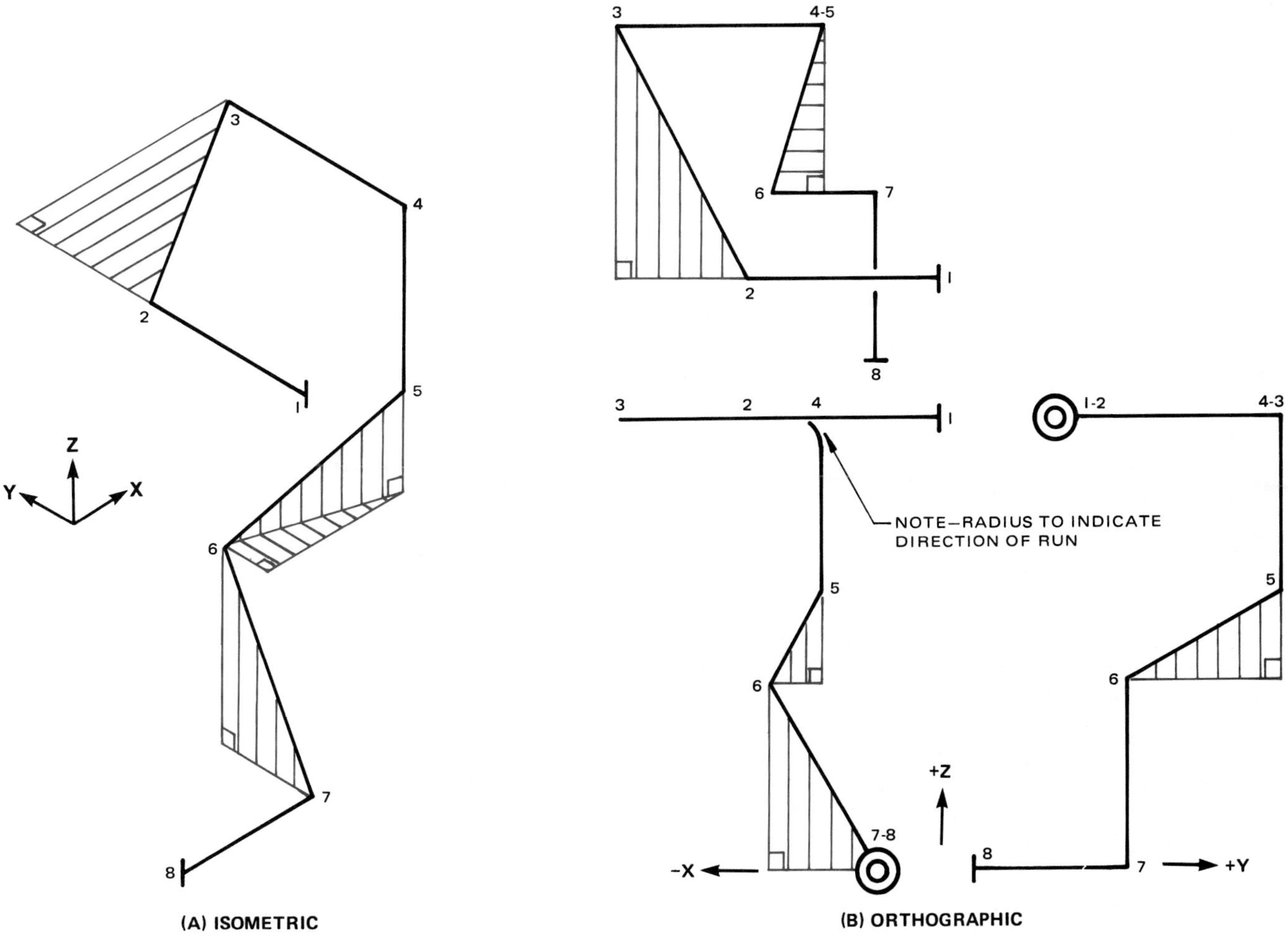

FIG. 24-3-8　Application of projection planes.

ASSIGNMENTS FOR CHAPTER 24

ASSIGNMENTS FOR UNIT 24-1, PIPES

1. Make a three-view drawing of the fuel oil supply system shown in Fig. 24-1-A. Include with the drawing an item list calling for all the pipe fittings and valves. The following valves are used: (1) relief valves, (2) globe valves, and (3) check valves. The numbers shown on the assignment correspond with the numbers listed above. Unions are used above the fuel oil pumps as detachable connections for ease of assembling and disassembling. The gages and temperature-sensing element have 1.00 in. pipe connections that necessitate the use of reducing tees. Scale is ½ in. = 1 ft. (U.S. customary) or 1:20 (metric)

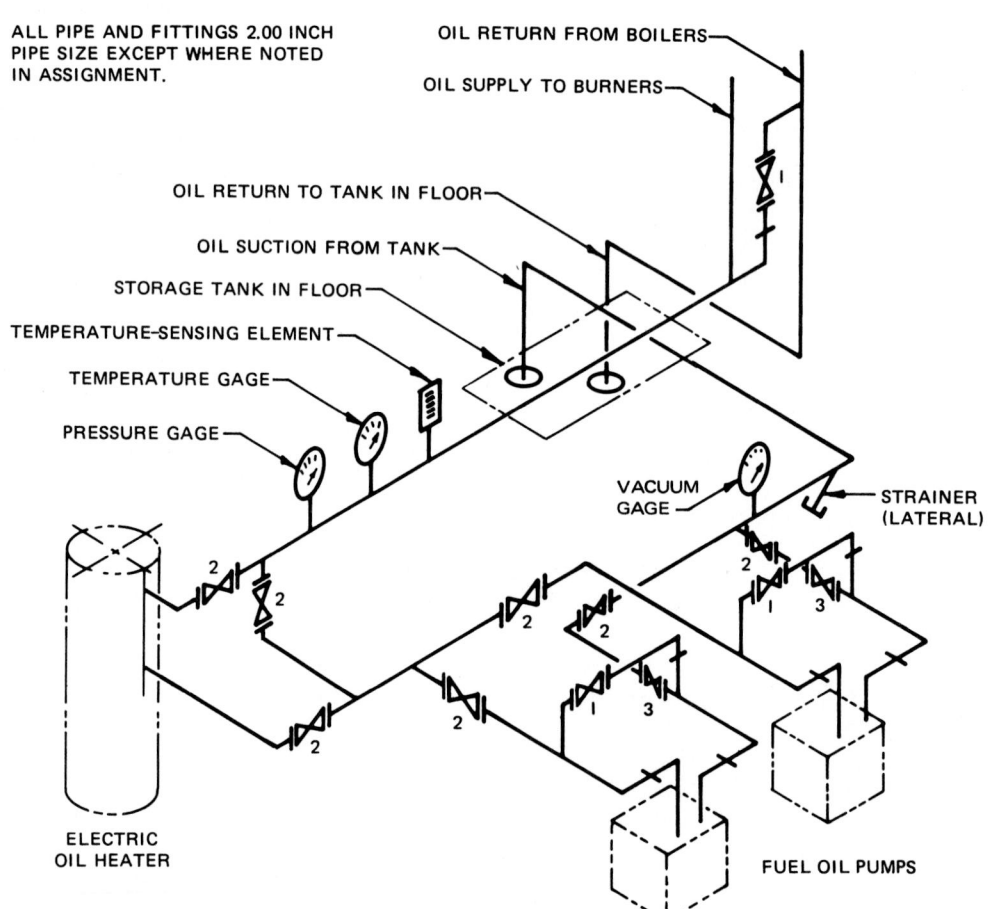

FIG. 24-1-A   Fuel oil supply system. *(Jenkins Bros. Ltd.)*

2. Heated tanks must be provided for the storage of industrial heating oil in most plants using this fuel for boiler furnaces generating heat or power or for processing furnaces.

To ensure uninterrupted service when cleaning, or in the event of a breakdown of one of the systems, a duplicate installation of tanks is shown in this layout. Since circulation must be provided to keep the oil fluid, a return line as well as a suction line from the tanks is shown. A valve is provided directly to the suction line. Connections are provided for both high and low suction. High suction guards against difficulties from sediment, while the low suction is necessary when the fuel oil supply is extremely low.

The free blow shown on the steam line connections to the heating coils in the tanks is important for testing for the presence of oil in the steam return line, since oil would indicate a leak.

Extra-heavy globe valves of the regrind-renew type are recommended on the oil lines to ensure maximum safety in the transmission of hazardous fluid and to meet code requirements. Outside screw and yoke gate valves are suggested because they show at a glance whether the valve is opened or closed.

Make a three-view drawing of the fuel oil storage connections with heating coil as shown in Fig. 24-1-B. Include with the drawing an item list calling for all the pipe fittings and valves. The oil lines are 1.50 in. pipe, and the steam line is 2.00 in. pipe. For the scale, see the drawing.

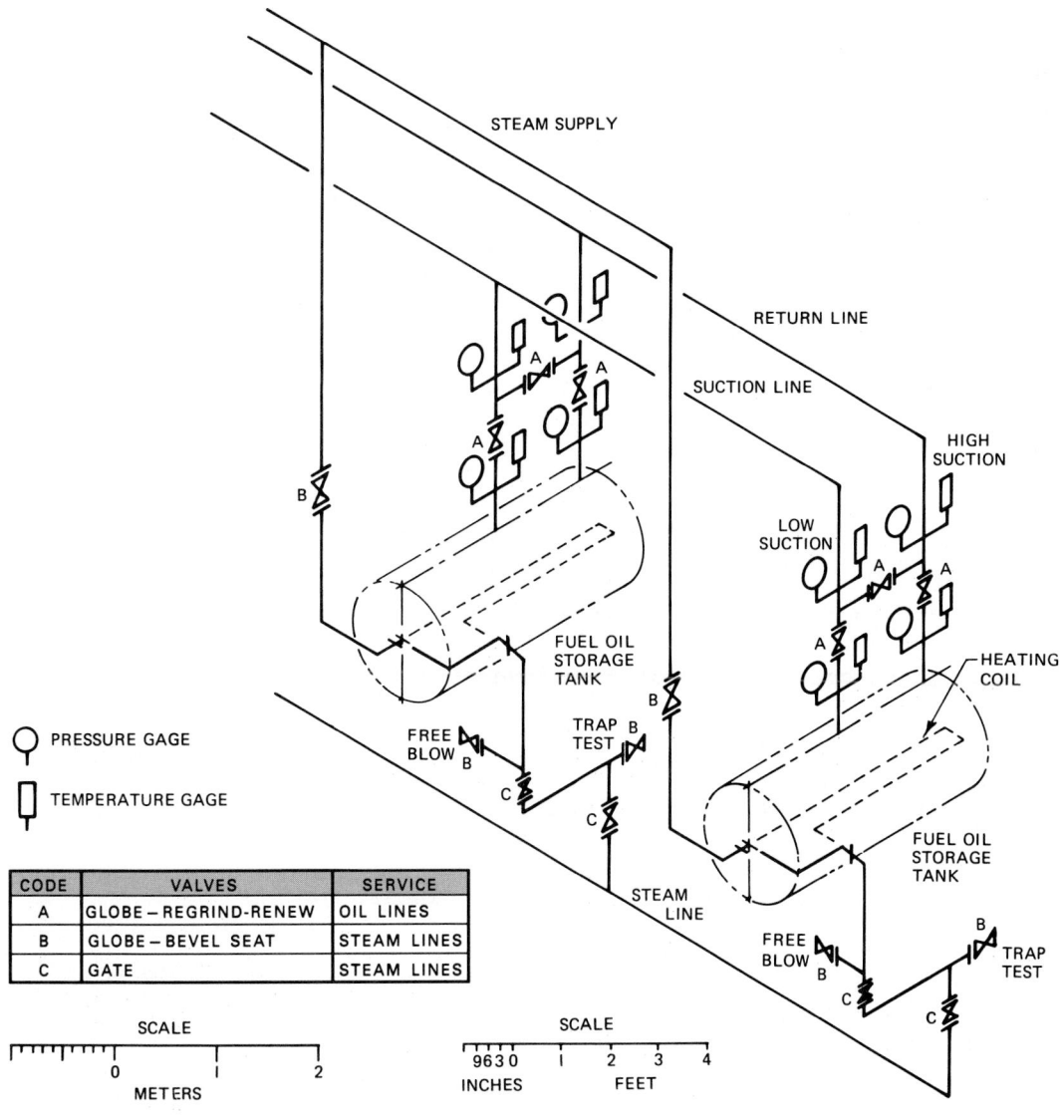

CODE	VALVES	SERVICE
A	GLOBE – REGRIND-RENEW	OIL LINES
B	GLOBE – BEVEL SEAT	STEAM LINES
C	GATE	STEAM LINES

FIG. 24-1-B   Fuel oil storage connections with heating coils. *(Jenkins Bros. Ltd.)*

ASSIGNMENTS FOR UNIT 24-2, ISOMETRIC PROJECTION OF PIPING DRAWINGS

3. For starting diesel engines, the most dependable and widely used method is an air system of the type illustrated in Fig. 24-2-A. With this starting system hooked up to diesel installations generating power and heat for such buildings as factories, hotels, large apartment houses, and stores, interruptions which might occur through failure of electric supply or storage cells are avoided.

Safety valves are provided for the compressor and the air storage tanks. Check valves are installed on the air storage tank feed lines and the compressor discharge lines to prevent accidental discharge of the tanks.

Piping is arranged so that the compressor will either fill the storage tanks and/or pump directly to the engines. Any of the three storage tanks may be used for starting, and pressure gages indicate their readiness. The engines are fitted with quick-opening valves to admit air quickly at full pressure and shut it off the instant rotation is obtained. A bronze globe valve is installed to permit complete shutdown of the engine for repairs and for regulation of the air flow. Drains are provided at low points to remove condensate from the air storage tanks, lines, and engine feeds.

Globe valves are recommended throughout this hookup except on the main shutoff lines where gate valves are used because of infrequent operation.

All valves connected to horizontal pipelines 6 ft (1800 mm) or higher above the floor will have their spindles located on the underside for ease of operation. Other horizontally positioned valves will have their spindles in the upright position.

All valves connected to vertical pipes will have their valve spindles oriented to the front of the drawing.

Flanges are to be located on the top pipeline near the three starting air tanks and near the air compressor for assembly and disassembly purposes. Flanges are located on the starting diesel engines.

Make an isometric piping drawing for the diesel engine starting system shown in Fig. 24-2-A. Scale is ¼ in. = 1 ft (U.S. customary) or 1:50 (metric). Include on your drawing an item list calling for the pipe fittings and valves. All the fittings are threaded, and 1.50 in. pipe is used throughout.

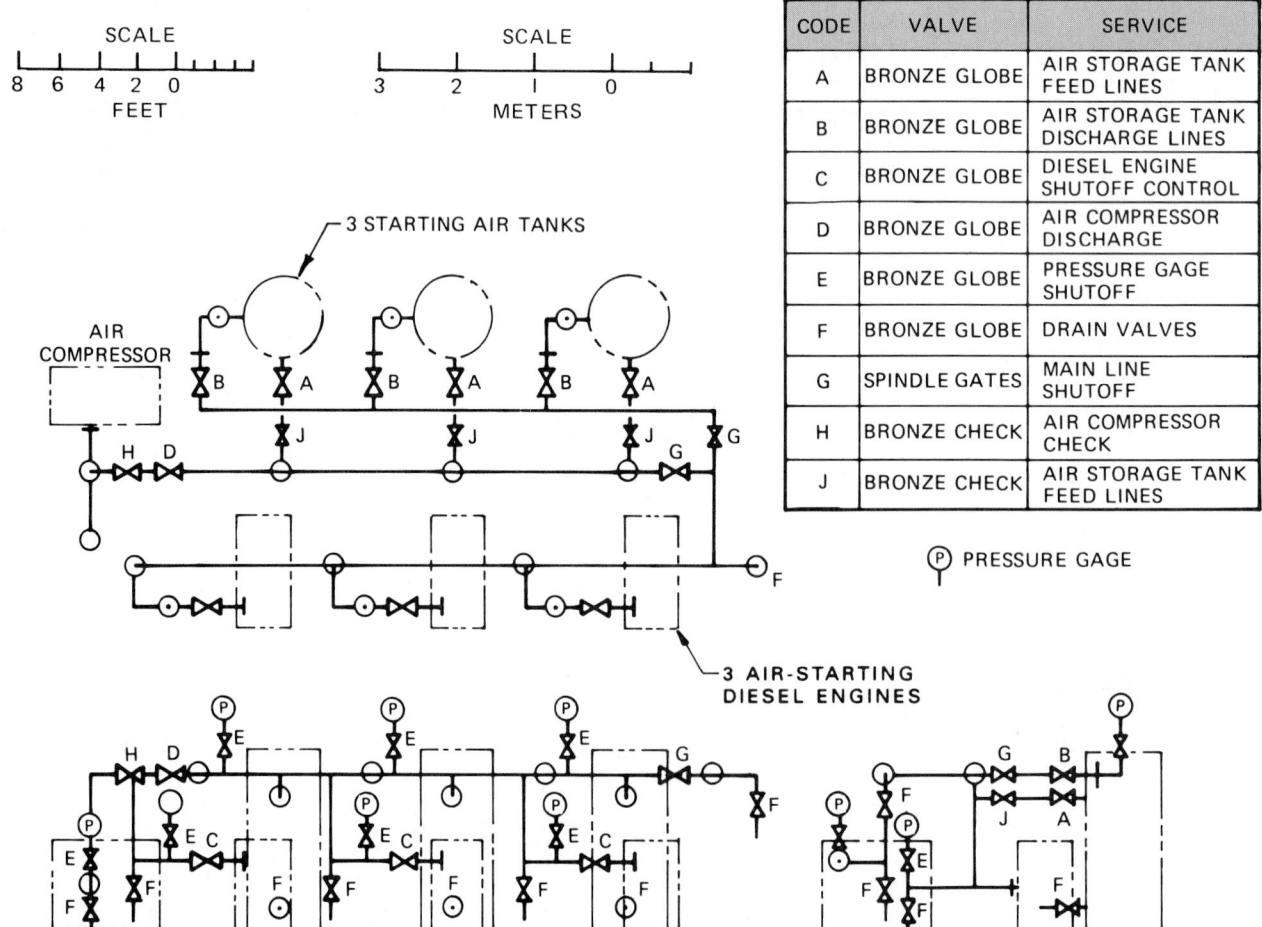

CODE	VALVE	SERVICE
A	BRONZE GLOBE	AIR STORAGE TANK FEED LINES
B	BRONZE GLOBE	AIR STORAGE TANK DISCHARGE LINES
C	BRONZE GLOBE	DIESEL ENGINE SHUTOFF CONTROL
D	BRONZE GLOBE	AIR COMPRESSOR DISCHARGE
E	BRONZE GLOBE	PRESSURE GAGE SHUTOFF
F	BRONZE GLOBE	DRAIN VALVES
G	SPINDLE GATES	MAIN LINE SHUTOFF
H	BRONZE CHECK	AIR COMPRESSOR CHECK
J	BRONZE CHECK	AIR STORAGE TANK FEED LINES

FIG. 24-2-A   Diesel engine air-starting system. *(Jenkins Bros. Ltd.)*

4. In the piping layout of a boiler room shown in Fig. 24-2-B, boilers 1, 3, and 5 are connected to supply the hot water to rooms located on the first floor. Check valves are placed on the cold-water lines adjacent to each boiler to prevent the hot water from backing into the cold-water lines. Gate valves are used to shut off the main hot- and cold-water supply lines. Globe valves are placed near the boilers on the hot-water lines. Flanged connections are used at the boilers for ease of assembly and disassembly. The size of pipe is indicated on the plane view, and the scale is shown on the drawing.

Make an isometric drawing of the piping layout of a boiler room. Include on the drawing an item list calling for the pipe fittings and valves. All fittings are of flanged type.

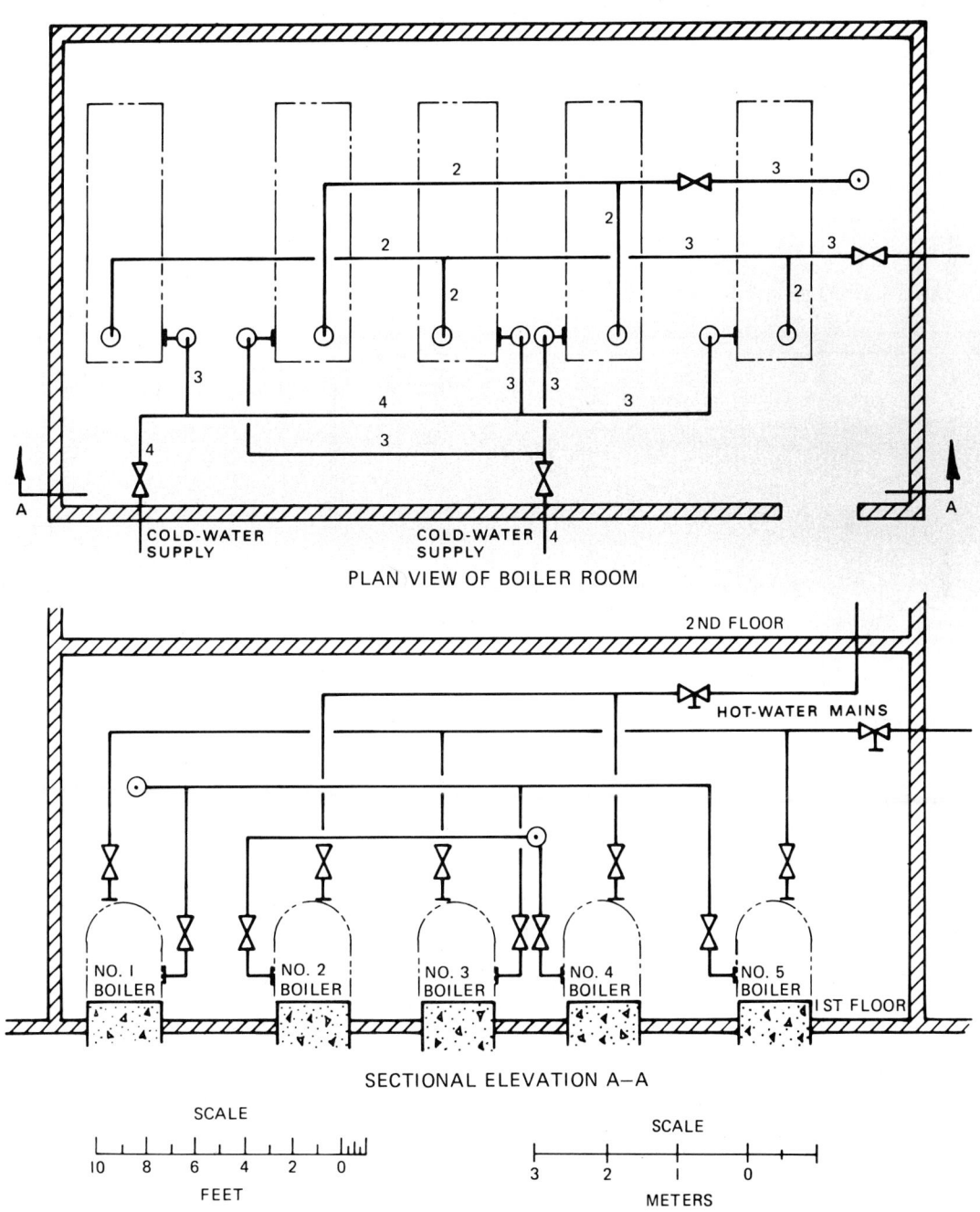

FIG. 24-2-B   Piping layout of boiler room. *(Jenkins Bros. Ltd.)*

ASSIGNMENTS FOR UNIT 24-3, SUPPLEMENTARY PIPING INFORMATION

5. The one-story, commercial building (Fig. 24-3-A) has been developed and steadily improved as a result of the movement of shopping centers to suburban areas. This type of building, which is multiplying rapidly, is constructed either with or without basement. It houses retail stores, service establishments, amusement centers, restaurants, and offices.

   Heat and plumbing services in such buildings are usually provided by the owner or operator, and for this reason, he or she might give careful consideration to low-cost and trouble-free installations. An oil- or gas-fired steam boiler with automatic control and a separate gas-fired heater for hot-water supply will generally meet these requirements.

   The two-pipe heating system, located in the basement in the installation illustrated, utilizes unit heaters with individual thermostatic controls. Valuable extra floor space is made available for tenants' use because the heaters are hung from the ceiling. Since the heaters in each store or each section of a store are automatically controlled, fuel savings are effected and even heating is ensured.

   Make an isometric drawing of the piping layout shown. Include with the drawing an item list calling for all the pipe fittings and valves. Pipe hangers are required for every 8 ft (2400 mm) of piping. Direction of flow, level indicators for horizontal piping using the basement floor as zero, indication of pipe runs not in the direction of the coordinate axes (see Fig. 24-3-6), and a drainage slope of 1:20 are to be shown on the drawing. Scale is ⅛ in. = 1 ft (U.S. customary) and 1:100 (metric). Use 1.50 in. pipe.

CODE	VALVE	SERVICE
A	BRONZE GATE	WATER SERVICE SHUTOFF
B	BRONZE GATE	DISTRIBUTION SHUTOFF
C	BRONZE GLOBE	WATER SUPPLY TO BOILER
D	BRONZE SWING CHECK	PREVENT BOILER BACKFLOW
E	BRONZE GLOBE	EMERGENCY BOILER FILL
F	BRONZE GLOBE	DRAINS SHUTOFF
G	BRONZE GLOBE	STEAM SUPPLY TO HEATERS
H	BRONZE GATE	CONDENSATE DRAIN SHUTOFFS
J	BRONZE GLOBE	STEAM MAIN TRAP CONNECTION
K	BRONZE GLOBE	TRAP BYPASS
L	BRONZE GATE	HOT WATER HEATER SHUTOFF
M	BRONZE SWING CHECK	PREVENT WATER HEATER BACKFLOW
N	BRONZE GATE	WATER SUPPLY SHUTOFFS
P	I.B.B.M. GATE	STEAM MAIN SHUTOFF
R	BRONZE GATE	RETURN SHUTOFF

1. FLOAT AND THERMO TRAP
2. WATER METER
3. HOT WATER TO STORE
4. COLD WATER TO STORE
5. DRAIN

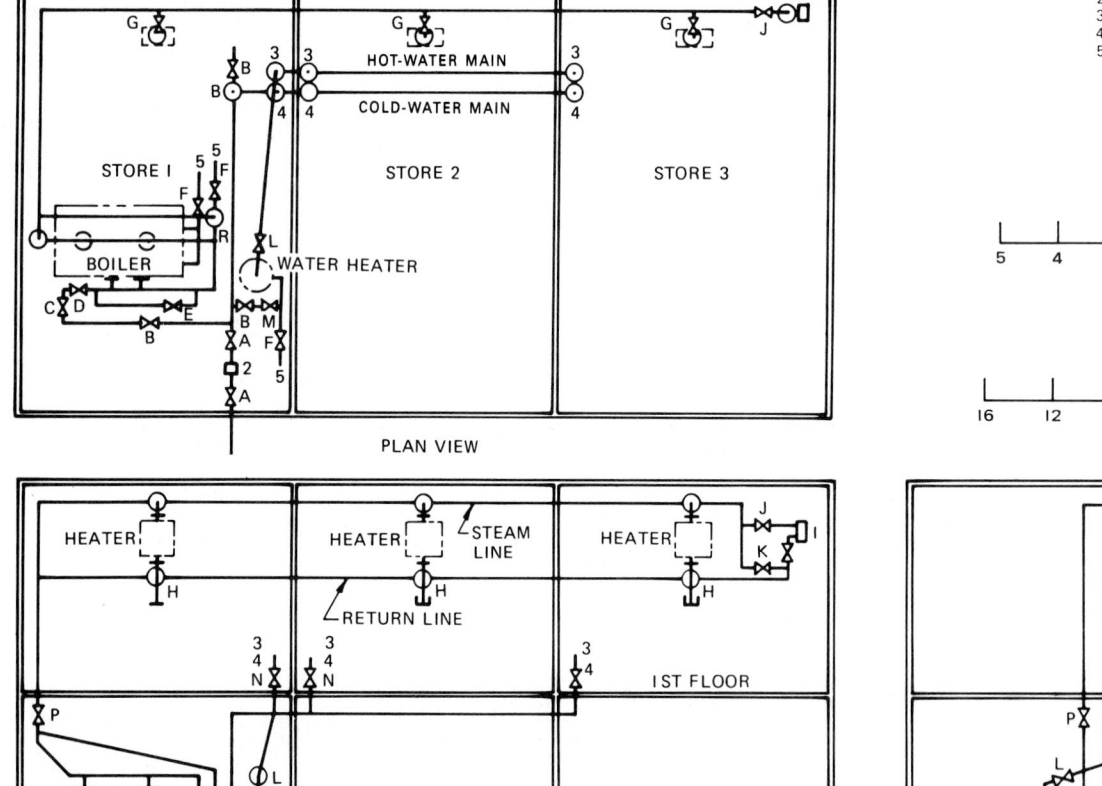

PLAN VIEW

FRONT ELEVATION

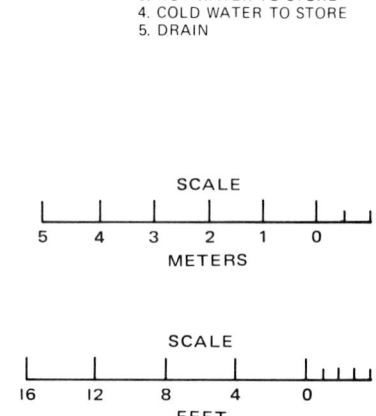

SCALE

5 4 3 2 1 0

METERS

SCALE

16 12 8 4 0

FEET

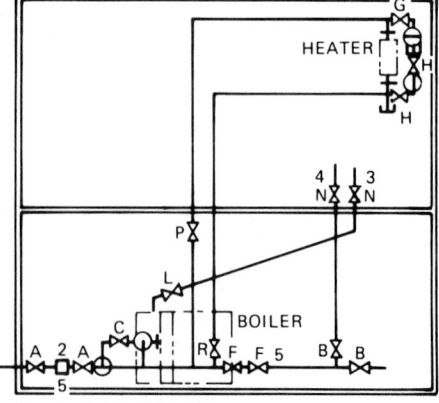

SIDE ELEVATION

FIG. 24-3-A Piping connections for plumbing and heating in a small building.
(*Jenkins Bros. Ltd.*)

6. Light and medium fuel oils, numbers 1, 2, 3, and 5 (cold), which do not require preheating can be handled in a layout such as shown in Fig. 24-3-B. Since the expense of a preheater installation can be eliminated, this relatively simple system is economical and easy to operate. Similar systems are often installed in hotels, apartment houses, office buildings, large residences, and small industrial plants.

Fuel oil, stored in an unheated tank, flows through a large-mesh, twin-type strainer to a motor-driven pump, which provides the necessary oil pressure for satisfactory operation. The oil then passes through a fine-mesh strainer, which removes any small particles that might clog the burner. Oil flow to the burner is controlled by a burner control valve, which opens or shuts according to the boiler pressure.

Although one fuel oil pump can adequately handle the maximum boiler demands, two are recommended to provide a second pump for standby service in case of breakdown. Each pump is provided with a pressure relief valve as a protection against excessive oil pressure, which might become high enough to cause leaks in the oil piping. Check valves in the relief lines prevent relieved oil from entering the idle standby pump.

Bronze valves are recommended throughout and must be of the appropriate pressure rating. The plug-type globe valve, recommended for the important individual burner shutoff, ensures positive tightness when closed and extremely close regulation of oil flow, both of which are essential to good oil burner operation.

The swing check valve indicated in this layout is exceptionally serviceable for the nonreturn control of steam, oil, water, and gas. It is generally used in connection with a gate valve, offering comparable full, free flow.

Make a single-line isometric drawing of the piping drawing shown. Design your own symbols for the indicators (gages, strainers, etc.) for which there are no standard symbols. The coding of these items should be shown clearly off the main drawing. Show the direction of flow and indicate on the drawing that all horizontal pipes require a slope of 1:20 for drainage purposes. Using the floor as zero elevation, scale the drawing and show by means of level indicator symbols the height of all horizontal pipelines. Include on your drawing an item list, listing all the valves and fittings. Scale is ¼ in. = 1 ft (U.S. customary) and 1:50 (metric).

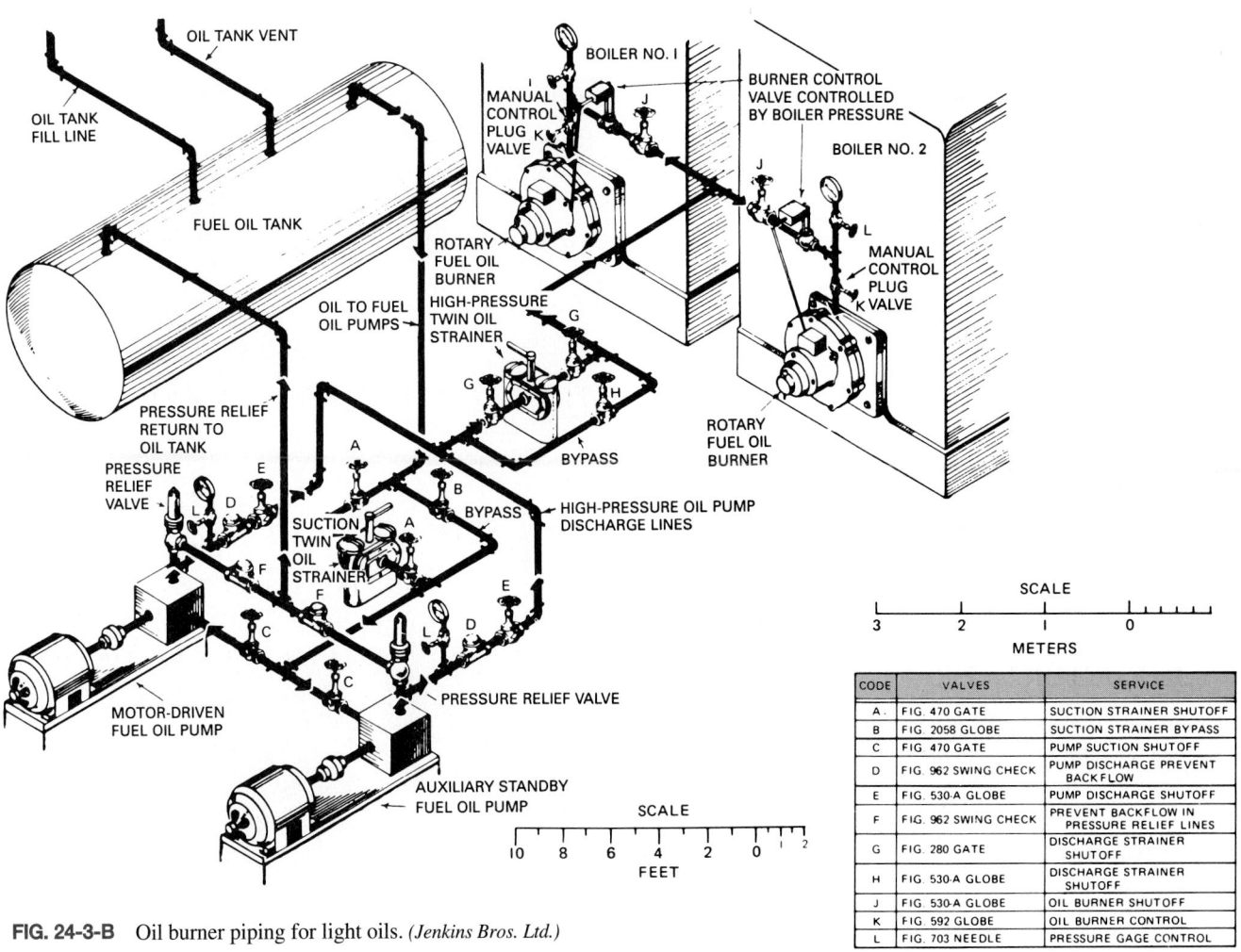

FIG. 24-3-B   Oil burner piping for light oils. *(Jenkins Bros. Ltd.)*

CODE	VALVES	SERVICE
A.	FIG. 470 GATE	SUCTION STRAINER SHUTOFF
B	FIG. 2058 GLOBE	SUCTION STRAINER BYPASS
C	FIG. 470 GATE	PUMP SUCTION SHUTOFF
D	FIG. 962 SWING CHECK	PUMP DISCHARGE PREVENT BACKFLOW
E	FIG. 530-A GLOBE	PUMP DISCHARGE SHUTOFF
F	FIG. 962 SWING CHECK	PREVENT BACKFLOW IN PRESSURE RELIEF LINES
G	FIG. 280 GATE	DISCHARGE STRAINER SHUTOFF
H	FIG. 530-A GLOBE	DISCHARGE STRAINER SHUTOFF
J	FIG. 530-A GLOBE	OIL BURNER SHUTOFF
K	FIG. 592 GLOBE	OIL BURNER CONTROL
L	FIG. 703 NEEDLE	PRESSURE GAGE CONTROL

CHAPTER 25

STRUCTURAL DRAFTING

Definitions

Bolt pitch   The distance between bolt holes.

Connection plate   The place or location of one member's attachment shape or plate, along with the means of fastening, to another's.

Framed type beam connections   Connections in which the beam is connected by means of fittings.

Levels   Detail dimensions that indicate elevation or height.

Mill tolerances   Permissible deviations from the published dimensions and contours brought about in the manufacture of structural steel.

Plain material   Steel, in any of its basic forms, before the fabricating shop begins working with it.

Seated beam connections   Connections in which the end of the beam rests on a ledge, or seat, which receives the load from the beam.

Shop drawings   Detail drawings done by the fabricator to depict individual building members.

Spread   The center-to-center distance between the open holes.

Tender   To quote a price for such activities as detailing, supplying, fabricating, etc. by steel fabricators.

25-1 STRUCTURAL DRAFTING

The training of the structural steel drafter is of vital importance to the engineering profession, the construction industry, and every structural steel fabricator.

The Building Process

The steps through which a building proceeds from conceptual planning to finished product are, generally speaking, as follows:

1. An owner with the appropriate financing establishes the requirements for a building to fulfill some particular function.
2. A design team (usually an architect and a structural engineer) studies the owner's needs in reference to set standards and conventions. The following factors may influence the preliminary design: available materials,

FIG. 25-1-1   Erecting fabricated steel supports for a building. *(Larry Hamill)*

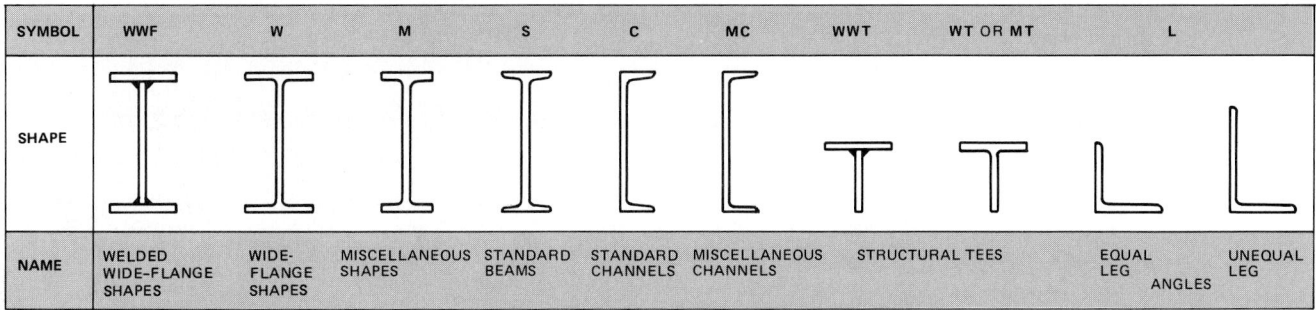

SYMBOL	WWF	W	M	S	C	MC	WWT	WT OR MT	L	
SHAPE										
NAME	WELDED WIDE–FLANGE SHAPES	WIDE–FLANGE SHAPES	MISCELLANEOUS SHAPES	STANDARD BEAMS	STANDARD CHANNELS	MISCELLANEOUS CHANNELS	STRUCTURAL TEES		EQUAL LEG	UNEQUAL LEG
										ANGLES

FIG. 25-1-2   Common structural steel shapes.

construction costs, building codes, zoning and health requirements, local bylaws, land condition, fire protection, finance, and setbacks.

3. With these parameters and the owner's requirements, the consultant or design team prepares sketches of the finished building, floor plans, and cost estimates, which are submitted to the owner for approval.

4. The design team then takes over the job. The structural design group designs the building frame, taking into consideration factors that influence the type and location of structural members.

5. When the structural arrangements have been finalized, layout drawings are made. These give distances from center line to center line, size and location of structural components, and other specifics of the design. When the layout drawings have been completed, checked, and approved, they are sent to steel fabricators for tendering. *Tendering* involves quoting a price (usually a price for detailing, supply, fabrication, and erection of the steel members).

6. When the contract has been received by the steel fabricator, he or she makes a list of material required so that the basic shapes can be ordered from the steel producer. The fabricator also begins to detail (draw the individual building members). These are referred to as *shop drawings.*

7. As the shop drawings are completed, they are sent to the shop in order that parts may be fabricated. It is usually during this period that the fabricator will make the erection drawings in conjunction with the structural design group.

8. As the steel is fabricated, it is either stored in the yard or sent to the construction site if it is required immediately.

9. At the site, the steel is erected using the erection drawings (Fig. 25-1-1).

Structural Steel—Plain Material

It is important to remember that the steel produced at the rolling mills and shipped to the fabricating shop comes in a wide variety of shapes (approximately 600) and forms. At this stage it is called *plain material.*

Many of these materials are shown in Fig. 25-1-2. They can be classified and designated as follows:

1. S shapes (formerly called standard beams or I beams) are rolled in many sizes 3 to 20 in. (75 to 500 mm).

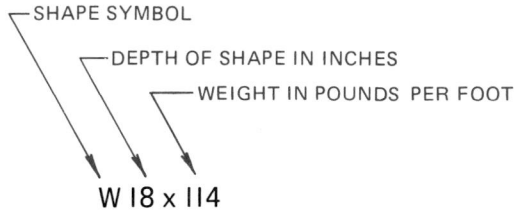

(A) INCH DESIGNATION

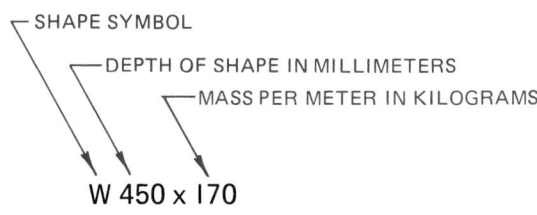

(B) METRIC DESIGNATION

FIG. 25-1-3   Structural steel callouts.

2. C shapes (formerly called standard channels) are available in sizes ranging from 3 to 18 in. (80 to 450 mm).

3. W shapes (formerly called wide-flange shapes) are welded wide-flange (WWF) beams and columns. W shapes are available in sizes ranging from 6 to 36 in. (150 to 900 mm). WWF shapes, sometimes referred to as H shapes, range in size from 14 to 48 in. (350 to 1200 mm).

4. M shapes (formerly called joists and light beams) are similar in contour to the W shapes. They are available in sizes ranging from 6 to 16 in. (150 to 400 mm).

5. Structural tees are produced by splitting S or W shapes, usually through the center of their webs, thus forming two T-shaped pieces from each beam.

6. L shapes, or angles, consisting of two legs set at right angles, are available in sizes ranging from 3 to 8 in. (75 to 200 mm).

7. Hollow structural sections (HSS) consist of round, square, and rectangular sections.

8. Plates, and round and rectangular bars.

When steel shapes are designated on drawings, a standard method of abbreviating should be followed that will identify the size and shape of the steel part (Figs. 25-1-3 and 25-1-4, here and on pg. 830). However, the method for calling for these

FIG. 25-1-4   Abbreviations for shapes, plates, bars, and tubes.

Shape	U.S. Customary Examples See Note 1		Metric Size Examples See Note 2
	New Designation	Old Designation	
Welded Wide-Flange Shapes (WWF Shapes)			
—Beams	WWF48 × 320	48WWF320	WWF1000 × 244
—Columns			WWF350 × 315
Wide-Flange Shapes (W Shapes)	W24 × 76	24WF76	W600 × 114
	W14 × 26	14B26	W160 × 18
Miscellaneous Shapes (M Shapes)	M8 × 18.5	8M18.5	M200 × 56
	M10 × 9	10JR9.0	M160 × 30
Standard Beams (S Shapes)	S24 × 100	24I100	S380 × 64
Standard Channels (C Shapes)	C12 × 20.7	12C20.7	C250 × 23
Structural Tees			
—cut from WWF Shapes	WWT24 × 160	ST24WWF160	WWT280 × 210
—cut from W Shapes	WT12 × 38	ST12WF38	WT130 × 16
—cut from M Shapes	MT4 × 9.25	ST4M9.25	MT100 × 14
Bearing Piles (HP Shapes)	HP14 × 73	14BP73	HP350 × 109
Angles (L Shapes)	L6 × 6 × .75	L6 × 6 × ¾	L75 × 75 × 6
(leg dimensions × thickness)	L6 × 4 × .62	16 × 4 × ⅝	L150 × 100 × 13
Plates (width × thickness)	20 × .50	20 × ½	500 × 12
Square Bar (side)	⌀ 1.00	Bar 1 ⌀	⌀ 25
Round Bar (diameter)	Ø1.25	Bar 1¼ Ø	Ø30
Flat Bar (width × thickness)	250 × .25	Bar 2½ × ¼	60 × 6
Round Pipe (type of pipe × OD × wall thickness)	12.75 OD × .375	12¾ × ⅜	XS 102 OD × 8
Square and Rectangular Hollow Structural Sections (outside dimensions × wall thickness)	HSS4 × 4 × .375	4 × 4RT × ⅜	HSS102 × 102 × 8
	HSS8 × 4 × .375	8 × 4RT × ⅜	
Steel Pipe Piles (OD × wall thickness)			320 OD × 6

Note 1—Values shown are nominal depth (inches) × weight per foot length (pounds).
Note 2—Values shown are nominal depth (millimeters) × mass per meter length (kilograms).
Note 3—Metric size examples shown are not necessarily the equivalents of the inch size examples shown.

standard shapes has changed over the last few years. When called upon to revise or modify existing drawings, the drafter must do so in the same convention used previously on the drawing. Therefore, it is important that the drafter not only have the most up-to-date knowledge but also be familiar with previous standards still in use on old drawings (Fig. 25-1-5).

The abbreviations shown are intended only for use on design drawings. When lists of materials are being prepared for ordering from the mills, the requirements of the respective

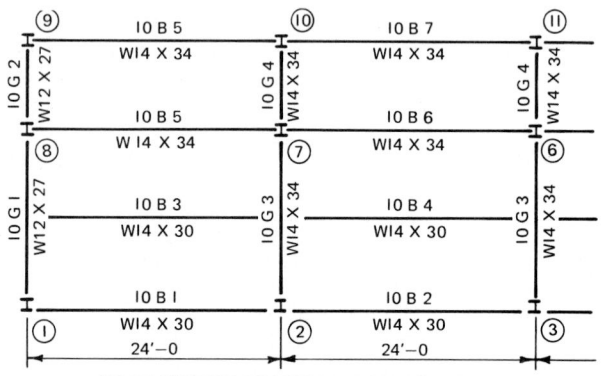

(B) BEAM DESIGNATION FROM 1972 ON

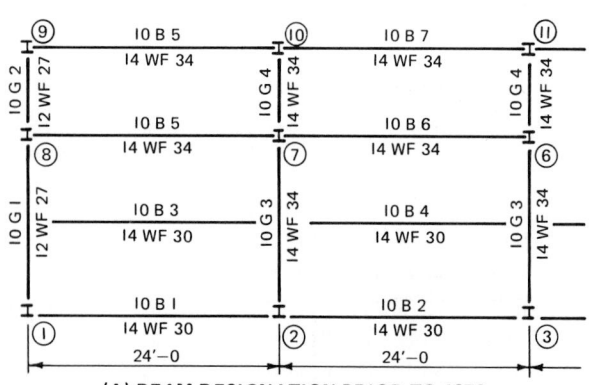

(A) BEAM DESIGNATION PRIOR TO 1972

FIG. 25-1-5   Building floor framing plan (partial view).

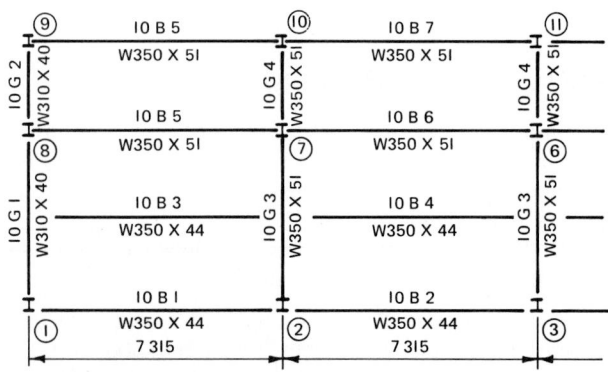

(C) METRIC BEAM DESIGNATION AND DIMENSIONING

mills from which the material is to be ordered should be observed.

Besides having to know the type of shapes available and their drawing designation, one must also be familiar with framing construction terms and where these shapes are used (Fig. 25-1-6).

All S, C, and MC shapes have a 16.67 percent slope on the inside faces of the flanges. This is equivalent to 9°28' or a bevel of 1:6. W-shaped beams and columns are rolled with parallel face flanges or with a 5 percent slope (2°51') on the inside of the flange (Fig. 25-1-7).

In structural steel shape tables, dimensions such as *K* and mean thickness of sloping flanges are given. Since the mean thickness of the sloping flange is given, these dimensions may also be used for all flange shapes.

If it is necessary to have the exact dimensions of a particular shape, they must be obtained from the individual mill's structural shape catalog. These catalogs also give the pertinent radial dimensions (Fig. 25-1-8, pg. 832).

It is customary, on details made to a scale of 1:8, 1:10, or smaller, for the curve indicating the toes of angles and of flanges, the interior fillets between legs of angles, and the interior fillets between web, or stem, and flanges to be omitted in the drawing. It is usual to exaggerate on detail drawings the thickness of the leg, stem, web, or flange.

Steel Grades

There are hundreds of grades of steel produced in mills today. However, only a few of those are suitable for structural applications. The most common structural grade used in the United States is ASTM A36. All structural members discussed in this chapter will be assumed to be fabricated from ASTM A36, while the bolts are made from A307 or A325 depending on the strength required.

Mill Tolerances

There are certain permissible deviations brought about in the manufacture of structural steel that the drafter should understand. These permissible deviations from the published dimensions and contours, as listed in the American Institute of Steel Construction (AISC) manual, in mill catalogs, and from the lengths specified by the purchaser, are referred to as *mill tolerances* (Figs. 25-1-9 through 25-1-11, pp. 833–834).

The factors that contribute to the necessity for a mill tolerance are as follows:

1. The high speed of the rolling operations required to prevent the metal from cooling before the process has been completed

1. Anchors or hangers for open-web steel joists
2. Anchors for structural steel
3. Bases of steel and iron for steel or iron columns
4. Beams, purlins, girts
5. Bearing plates for structural steel
6. Bracing for steel members or frames
7. Brackets attached to the steel frame
8. Columns, concrete-filled pipe, and struts
9. Conveyor structural steel framework
10. Steel joists, open-web steel joists, bracing, and accessories supplied with joists
11. Separators, angles, tees, clips, and other detail fittings
12. Floor and roof plates (raised pattern or plain (connected to steel frame)
13. Girders
14. Rivets and bolts
15. Headers or trimmers for support of open-web steel joists where such headers or trimmers frame into structural steel members
16. Light-gage cold-formed steel used to support floor and roofs
17. Lintels shown on the framing plans or otherwise scheduled
18. Shelf angles

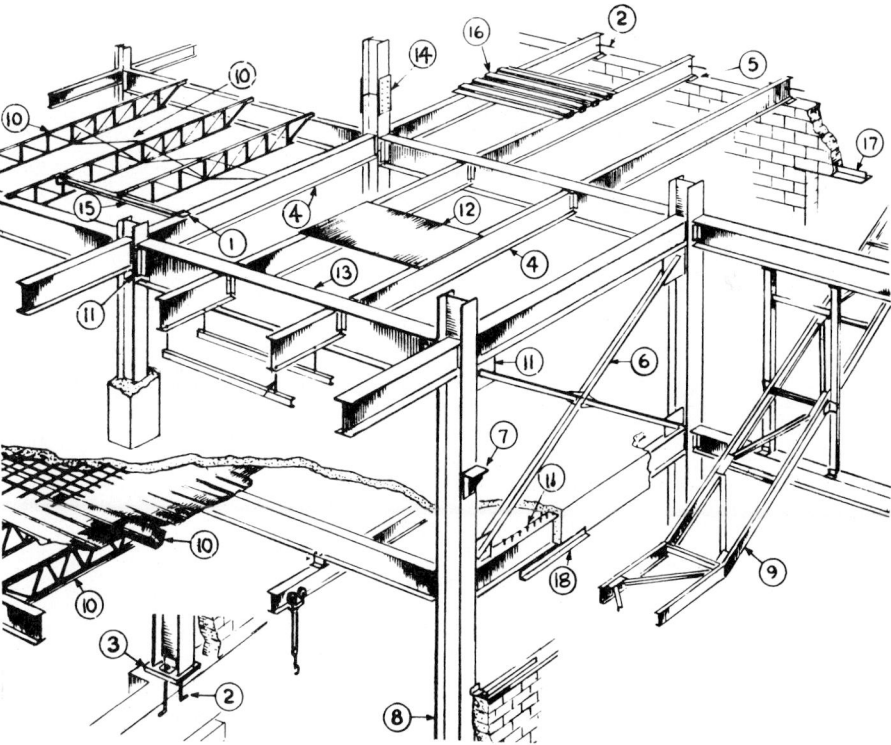

FIG. 25-1-6 Structural steel terms.

FIG. 25-1-7 Slopes and dimensions of flanges.

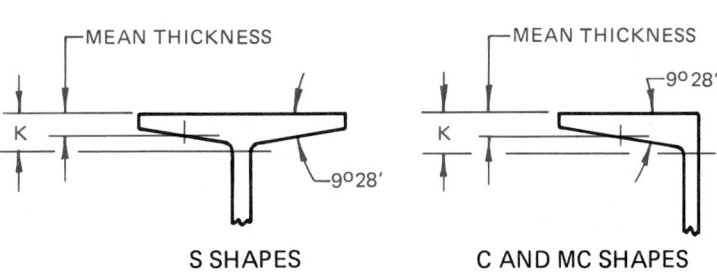

MEAN THICKNESS

MEAN THICKNESS

9° 28'

K

K

9°28'

S SHAPES

C AND MC SHAPES

		Flange		Web				
Inch Designation	Depth d	Width b	Thickness t	Thickness w	k	Metric Designation	Depth d	Width b
W24 × 94	24¼	9	⅞	½	1⅝	W610 × 140	617	230
W24 × 76	23⅞	9	11/16	7/16	1 7/16	W610 × 113	608	228
W18 × 114	18½	11⅞	1	⅝	1 11/16	W460 × 177	482	286
W18 × 105	18⅜	11¾	15/16	9/16	1⅝	W460 × 158	476	284
W18 × 96	18⅛	11¾	13/16	½	1½	W460 × 144	472	283
W18 × 60	18¼	7½	11/16	7/16	1 3/16	W460 × 89	463	192
W16 × 78	16⅜	8⅝	⅞	½	1⅝	W410 × 114	420	261
W16 × 40	16	7	½	5/16	1⅛	W410 × 60	407	178
W14 × 74	14¼	10⅛	13/16	7/16	1½	W360 × 110	357	255
W14 × 48	13¾	8	9/16	5/16	1¼	W360 × 72	350	204
W12 × 58	12¼	10	⅝	⅜	1⅜	W310 × 86	310	254
W12 × 50	12¼	8⅛	⅝	⅜	1⅜	W310 × 74	310	205
W12 × 36	12¼	6⅝	9/16	5/16	1 1/16	W310 × 52	317	167
W12 × 27	12	6½	⅜	¼	15/16	W310 × 39	310	165
W10 × 49	10	10	9/16	5/16	1⅛	W250 × 73	253	254
W10 × 33	9¾	8	7/16	5/16	1	W250 × 49	247	202
W10 × 25	10⅛	5¾	7/16	¼	1	W250 × 39	262	147
W10 × 15	10	4	¼	¼	13/16	W250 × 22	254	102
W8 × 35	8⅛	8	½	5/16	1	W200 × 52	206	204
W8 × 28	8	6½	7/16	5/16	15/16	W200 × 46	203	203
W8 × 20	8⅛	5¼	⅜	¼	⅞	W200 × 42	205	166
W8 × 15	8⅛	4	5/16	¼	13/16	W200 × 36	201	165
S24 × 100	24	7¼	⅞	¾	1¾	S610 × 149	610	184
S24 × 90	24	7⅛	⅞	⅝	1¾	S610 × 134	610	181
S20 × 95	20	7¼	15/16	13/16	1⅞	S510 × 141	508	183
S20 × 75	20	6⅜	13/16	⅝	1⅝	S510 × 112	508	162
S18 × 70	18	6¼	11/16	11/16	1½	S460 × 104	457	159
S15 × 50	15	5⅝	⅝	9/16	1⅜	S380 × 74	381	143
S12 × 50	12	5½	11/16	11/16	1 7/16	S310 × 74	305	139
S12 × 35	12	5⅛	9/16	7/16	1 3/16	S310 × 52	305	129
S10 × 35	10	5	½	⅝	1⅛	S250 × 52	254	126
S8 × 23	8	4⅛	7/16	7/16	1	S200 × 34	203	106
C15 × 50	15	3¾	⅝	11/16	1 7/16	C380 × 74	381	94
C15 × 40	15	3½	⅝	½	1 7/16	C380 × 60	381	89
C12 × 30	12	3⅛	½	½	1⅛	C310 × 45	305	80
C12 × 20.7	12	3	½	5/16	1⅛	C310 × 31	305	74
C10 × 30	10	3	7/16	11/16	1	C250 × 45	254	76
C10 × 20	10	2¾	7/16	⅜	1	C250 × 30	254	69

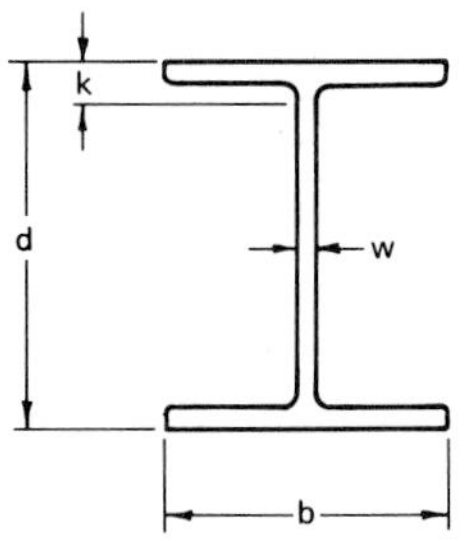

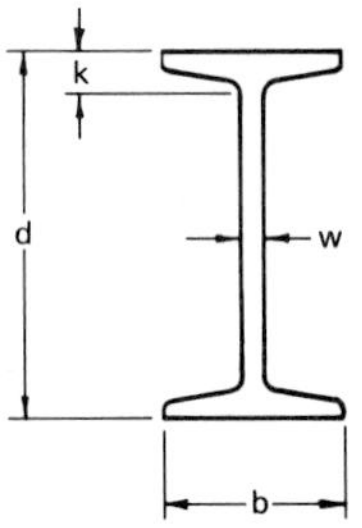

W SHAPES

S SHAPES

C SHAPES

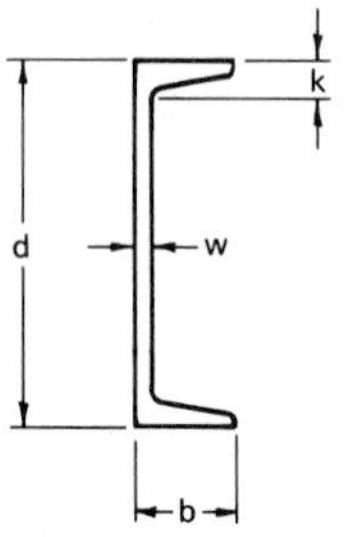

FIG. 25-1-8   Properties of common structural steel shapes.

2. The varying skill of operators in squeezing together the rolls for successive passes of the metal, particularly the final pass (Fig. 25-1-12, pg. 834)
3. The springing and wearing of the rolls, and other mechanical factors
4. The warping of the steel in the process of cooling
5. The subsequent shrinkage in the length of a shape that was cut while the metal was still hot

Under rolling tolerances (Fig. 25-1-9) note that the maximum overall depth *(C)* can be ¼ in. over the nominal depth. For example, a W24 × 94 beam (Fig. 25-1-8) is shown as having a 24¼ in. depth. However, its finished actual depth at *C*

after rolling could be 24½ in. The depth at center line *A* could be either 24⅜ in. (24¼ + ⅛) or 24⅛ in. (24¼ − ⅛). The width of flange *B* could be 9¼ in. (9 + ¼) or 8¹³⁄₁₆ in. (9 − ³⁄₁₆) in place of the 9 in. width.

Suppose the W24 × 94 is ordered cut to length from the mill as a 55-ft-long piece. It might be received by the fabricator, either with a length of 55'-¾ (55 ft + ¾ in.) or 54'-11⅝ (55 ft − ⅜ in.). The fabricators have standards for ordering the plain material that take into consideration these cutting tolerances.

Although this variation of length would not be tolerated in the shop, it is essential that the detailer be aware of its possible occurrence, so that when he or she specifies the required

ROLLING TOLERANCES (Inches)

Nominal Size	Depth A		Width of Flange B		Out of Square T or T₁	Web Off Center E	Maximum Depth of Any Cross Section C
	Over	Under	Over	Under	Max.	Max.	Over Nominal Size
12 in. and under	1/8	1/8	1/4	3/16	3/16	3/16	1/4
Over 12 in.	1/8	1/8	1/4	3/16	1/4	3/16	1/4

CUTTING TOLERANCES (Inches)

W Shapes Nominal Depth	Variation from Specified Length for Lengths Given									
	To 30 Incl		Over 30 to 40 Incl		Over 40 to 50 Incl		Over 50 to 65 Incl		Over 65	
	Over	Under	Over	Under	Over	Under	Over	Under	Over	Under
Beams 24 in. and under	3/8	3/8	1/2	3/8	5/8	3/8	3/4	3/8	1	3/8
Beams over 24 in. All Columns	1/2	1/2	5/8	1/2	3/4	1/2	1	1/2	1 1/8	1/2

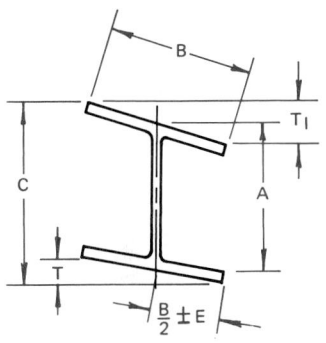

FIG. 25-1-9   Rolling and cutting tolerances for W shapes.

ROLLING TOLERANCES (Inches)

Shape	Nominal Specified Size	Depth A		Flange Width B		Out of Square T or T₁ B B	Out of Square C – D B
S	3 to 7 incl.	1/16	1/16	1/8	1/16	0.03	0.03
	Over 7 to 14 incl.	1/8	1/16	1/8	1/8	0.03	0.03
	Over 14 to 24 incl.	3/16	1/8	3/16	3/16	0.03	0.03
C	3 to 7 incl.	1/16	1/16	1/16	1/16	0.03	0.03
	Over 7 to 14 incl.	1/8	1/16	1/8	1/8	0.03	0.03
	Over 14	3/16	1/8	1/8	3/16	0.03	0.03

S SHAPES

CUTTING TOLERANCES (inches)

Shape	Variation from Specified Lengths									
	To 30 incl.		Over 30 to 40 Incl.		Over 40 to 50 Incl.		Over 50 to 65 Incl.		Over 65	
	Over	Under	Over	Under	Over	Under	Over	Under	Over	Under
S and C	1/2	1/4	3/4	1/4	1	1/4	1 1/8	1/4	1 1/4	1/4

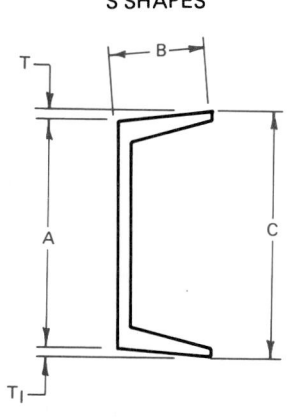

C SHAPES

FIG. 25-1-10   Rolling and cutting tolerances for S and C shapes.

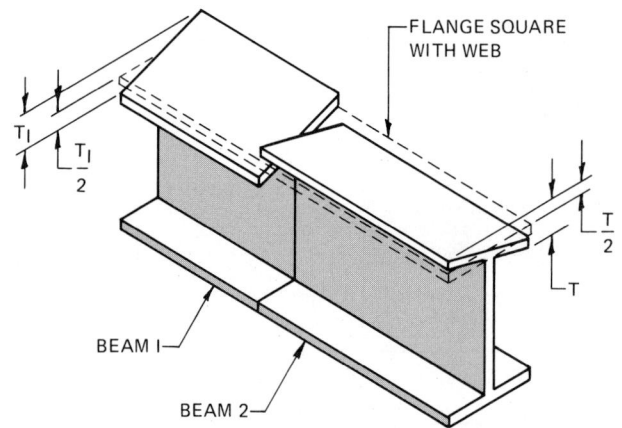

FIG. 25-1-11   Error between mating shapes.

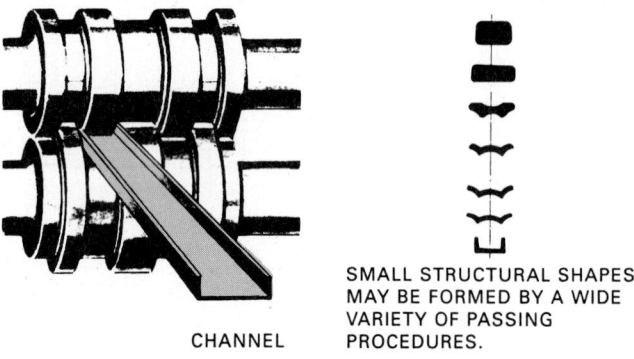

CHANNEL

SMALL STRUCTURAL SHAPES
MAY BE FORMED BY A WIDE
VARIETY OF PASSING
PROCEDURES.

FIG. 25-1-12   The making of a C shape.

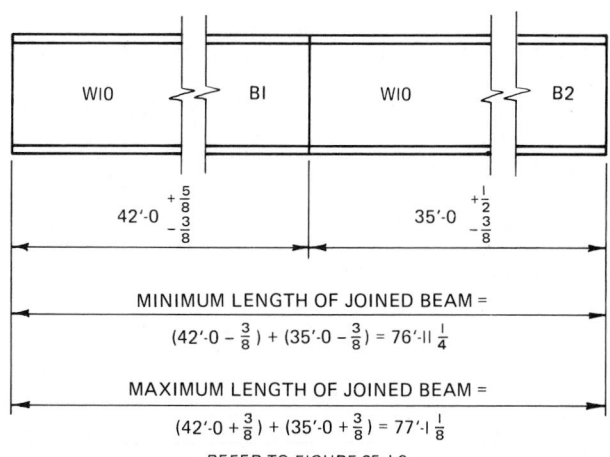

MINIMUM LENGTH OF JOINED BEAM =

$(42'-0 - \frac{3}{8}) + (35'-0 - \frac{3}{8}) = 76'-11\frac{1}{4}$

MAXIMUM LENGTH OF JOINED BEAM =

$(42'-0 + \frac{3}{8}) + (35'-0 + \frac{3}{8}) = 77'-1\frac{1}{8}$

REFER TO FIGURE 25-1-9

FIG. 25-1-13   Calculating minimum and maximum length of joined beams.

SCALES		PRINCIPAL DRAWINGS USED
INCH AND FOOT	MILLIMETER AND METER	
Full	1:1	Layout
3 = 1'-0	1:5	Layout
1½ = 1'-0	1:10	Layout or detail
¾ = 1'-0	1:20	Detail
⅜ = 1'-0		Erection or design
3/16 = 1'-0		Erection or design
3/32 = 1'-0		Erection or design
1 = 1'-0		Detail
½ = 1'-0		Detail or erection
¼ = 1'-0	1:50	Erection or design
⅛ = 1'-0	1:100	Erection or design

FIG. 25-1-14   Scales for structural drawings.

stock, the material obtained will fulfill the purpose for which it was ordered. Another example of mill tolerances is shown in Fig. 25-1-13.

Detailers can usually disregard mill tolerances when detailing light- and medium-mass trusses, standard beams, standard channels, struts, and most plate girders. But consideration must be given to the tolerances for all wide-flange beams and other heavy parts.

The handbooks prepared by AISC list all the available structural shapes, and their properties and dimensions. However, for the convenience of the student, all the dimensions required for examples and problems are reproduced in this text.

Structural Drawing Practices

Figure 25-1-14 is a table indicating the scale and the type of structural drafting in which the scale is most frequently used.

Dimensioning

The general practice in dimensioning structural drawings is to use the aligned method for dimensions and to place the dimensions above the dimension lines. Otherwise, the same general guidelines used in mechanical drafting will apply. All dimensions shown in this chapter will be in feet and inches or inches for U.S. customary. For metric, millimeters are used.

Dimensions should be arranged in a manner most convenient to all who must use the drawing. They should not crowd

the sketch and should cross the fewest possible number of other lines. The longest and overall dimensions should be farthest away from the views to which they apply. Dimensioning and descriptions of components (billing), in general, should be placed outside the picture. Dimensions should be given to the center lines of beams, to the backs of angles, and as explained later, to the backs of channels. They should be given to the top or bottom of beams and channels (whichever level is to be held), but never to both top and bottom, because of a possible overrun or underrun in the depth, resulting from rolling.

As shown in Fig. 25-1-15, when four or more equal spaces between bolts are required, it is recommended that the information be given as 4 @ 2 = 8 instead of repeating 2 four times. This reduces the possibility of error, both in reading the drawing and in layout of the work in the shop. Do not include in such an equation the distance locating the group itself from some reference point, even though the distance may happen to be the same as the increment of spacing.

Elevation detail dimensions, known as *levels,* are normally furnished by a note on the drawing. When it is desirable to show the level or vertical distance above some established reference point (usually ground level), the value is given in inches (or millimeters) and placed above the level symbol, as shown in Fig. 25-1-15. A plus or minus precedes the value, indicating that the level specified is higher or lower than the reference point.

Another dimensioning practice is to enclose bolt and hole sizes in diamond-shaped frames, as shown in Fig. 25-1-15. This helps to differentiate the circular sizes from the linear dimensions. Bolt symbols are shown in Fig. 25-1-16.

The neatness, and hence the legibility, of shop drawings is enhanced by lining up notes and dimensions that have the same common purpose. Thus if the 5 in. cut instructions (Fig. 25-1-15) were required at both ends of the beam, they would be shown in the same elevation, even though the dimension lines were not drawn from end to end of the sketch. Attention paid to these features, resulting in an orderly and systematic presentation of the necessary information, does much to enhance the finished appearance of a shop drawing.

REFERENCES AND SOURCE MATERIAL

1. American Institute of Steel Construction.
2. Canadian Institute of Steel Construction.

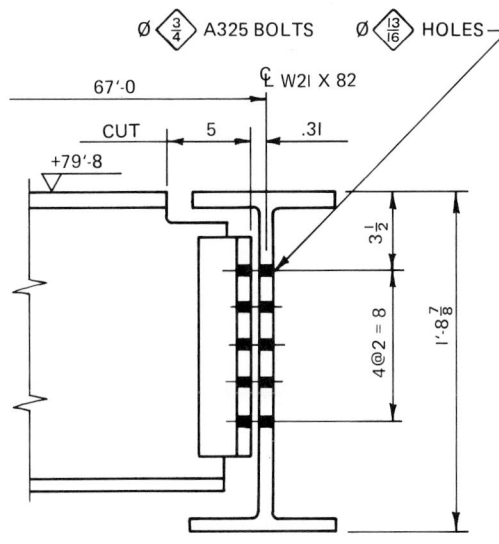

FIG. 25-1-15

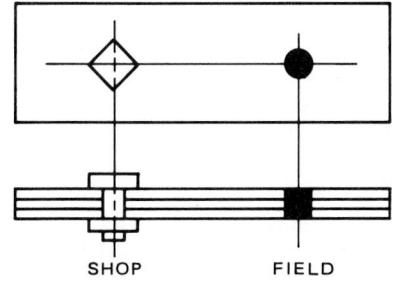

FIG. 25-1-16   Bolt symbols.

ASSIGNMENT

See Assignment 1 for Unit 25-1 on pages 852–853.

25-2 BEAMS

As a rule, each beam in a system of floor or roof framing makes a convenient erection unit. Hence, the required shop fabrication for each beam is shown on a shop drawing, which provides complete information for that beam. Such a drawing seldom pictures any part of the adjacent members to which this beam will later be joined in the field. However, in the preparation of the beam detail drawing, all the features that have a bearing on the later installation of the beam into its proper location in the frame, as indicated on the design drawing, must be investigated.

The location of the open holes to be provided in the beam for its field connection must match the location of similar holes in the supporting members. Proper clearances must be provided so that the beam can be swung into position after its supporting members have been erected. Any possible interference must be eliminated by cutting away the excess material.

The various fabricating-shop drafting rooms do not always agree among themselves on a standard way of making shop drawings. In this text, details will be presented in a manner that all shops could use.

The two principal kinds of beam connections most often used are the framed and the seated types. In the framed type, the beam is connected by means of fittings (usually a pair of short angles) attached to its web. With seated connections, the end of the beam rests on a ledge, or seat, which receives the load from the beam just as if the end of the beam rested upon a wall (Fig. 25-2-1).

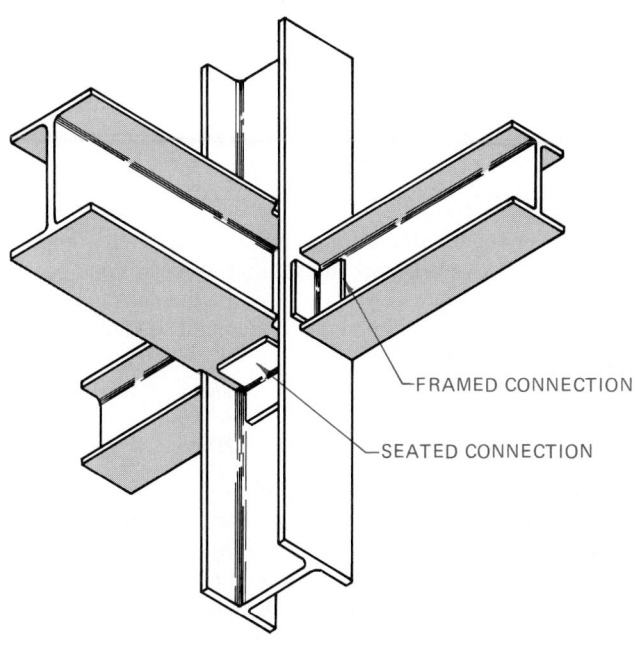

FIG. 25-2-1   Beam-to-column connections.

FIG. 25-2-2   Copes, blocks, and cuts.

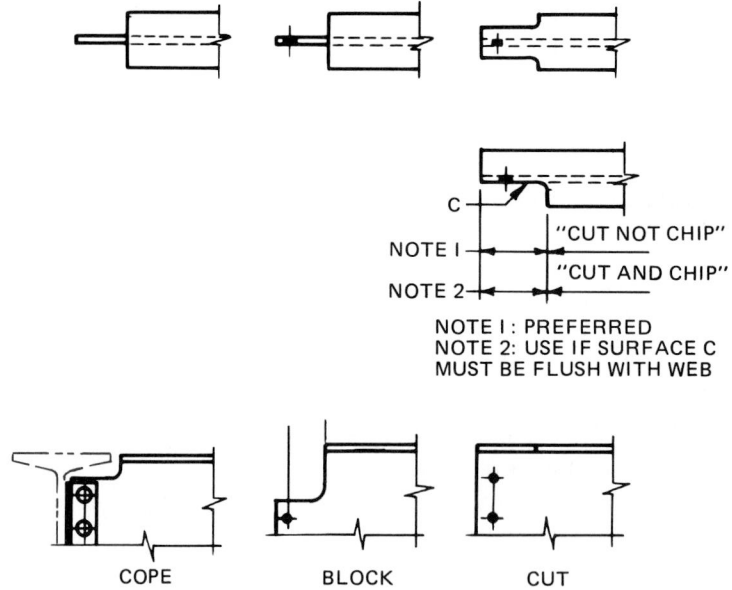

NOTE I : PREFERRED
NOTE 2: USE IF SURFACE C
MUST BE FLUSH WITH WEB

COPE   BLOCK   CUT

It should be noted that the depth of the beam, dimensions in relation to the depth of the beam, end connections, cuts, and spacing of holes are drawn to scale. Copes, blocks, and cuts are shown in Fig. 25-2-2. It is the practice of the structural detailer to draw the depth dimensions to scale so that the relation of detail is correct and so that the fabricator can interpret the relation of holes to bolts or holes more readily.

The length of beam and dimensions in relation to the length can be drawn to scale but are usually foreshortened. The reason that the length is usually foreshortened is that the scale length would, in most cases, take more space on a drawing than is economical and would have no practical value to the fabricator. However, foreshortening the length so much that the holes in the web or some of the detail will appear crowded or ambiguous should be avoided.

The structural detailer does not always draw to the exact scale, but exaggerates the drawing to clarify details. An example is the two lines that would show the thickness of the top or bottom flange of a beam.

Assembly Clearances

In order for members to assemble readily, clearances are required between beams and columns or beams and beams. It may also be necessary to cut or shape the ends of beams for mating parts to fit properly. The recommended clearances are shown in Fig. 25-2-3.

Simple Square-Framed Beams

The information—such as member length, size, and type, number of bolts, or type of fastener—required by the structural detailer is obtained from the design drawing. These drawings usually describe the type of construction, end loads or loads at support if not normal, type and size of bolts, member shape and size, and any other data that would be required by the detailer.

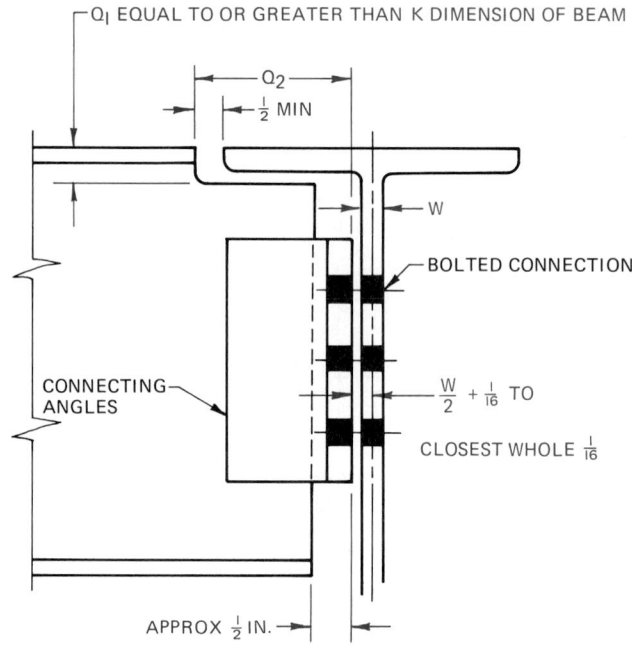

Q_1 EQUAL TO OR GREATER THAN K DIMENSION OF BEAM

Q_2

$\frac{1}{2}$ MIN

W

BOLTED CONNECTION

$\frac{W}{2} + \frac{1}{16}$ TO

CLOSEST WHOLE $\frac{1}{16}$

CONNECTING ANGLES

APPROX $\frac{1}{2}$ IN.

NOTE: PRACTICE IS TO MAKE Q_1 AND Q_2 DIMENSIONS MULTIPLES OF $\frac{1}{4}$ IN.

FIG. 25-2-3   Assembly clearance.

Figure 25-2-4 represents part of a design drawing for a steel-framed floor system as viewed from above. With its notes, it contains all the necessary information required by the shop detailer to detail the W18 × 60 beam, with the exception of connection-angle detail. Unless otherwise shown by dimensions or notes, members shown on the design drawing are presumed to be parallel or at right angles to one another, with their webs in a vertical plane, and to be in a level position from end to end. Elevation detail dimensions of beams are usually furnished

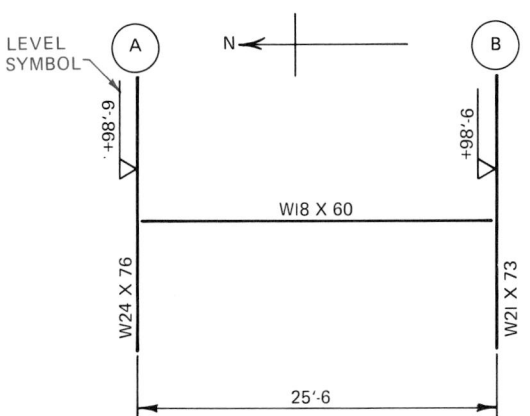

ELEVATION TOP OF STEEL SHOWN THUS: (+98'-6)

METHOD A

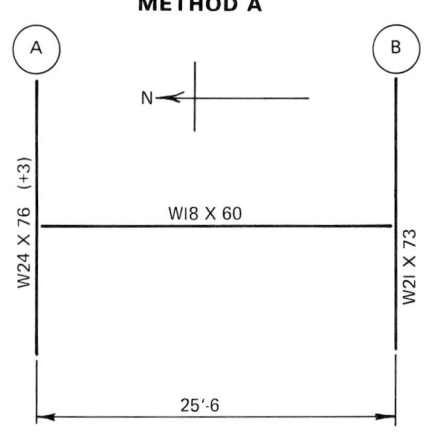

ELEVATION: ALL STEEL FLUSH, TOP AT ELEV. +98'-6

– CONNECTIONS: TWO ANGLES 4 X 3 X $\frac{1}{4}$ X 9
  AT EACH END OF BEAM

METHOD B

FIG. 25-2-4   Partial design drawing.

by a note on the drawing. In Fig. 25-2-4A the vertical distance, or elevation, is placed above the level symbol and is shown as +98'-6 for the W21 × 73 beam and +98'-9 for the W24 × 76 beam. It might have been given by a note reading ALL STEEL FLUSH, TOP AT ELEV. +98'-6, as shown by Fig. 25-2-4B. Note that the top elevation of W24 × 76 is designated by (+3), meaning that the top of the beam is 3 in. above the reference elevation of +98'-6, or (+98'-9) as presented in Fig. 25-2-4A.

Before starting the drawing, the detailer should first establish what the beam detail is going to look like. This is achieved by making sketches of the connections at both ends. The detailer first investigates the connection at one end, for this example the north end of the W18 × 60 of Fig. 25-2-4. A sketch, shown in Fig. 25-2-5A, is then made of the W18 framing into the W24. This section represents what would be seen if a viewer looked at the connection from the west side of the W18. A sketch is then made of the south end connection of W18. From the sketches the necessary detail requirements can be obtained. Of importance are the number of bolts or size of fillet welds required, and the size of the connecting angles. In this unit, the connecting-angle sizes are given. In Unit 25-3 the calculations for angle size and connections are covered in detail.

Note that the south end flange of the W18 × 60 is flush with that of the supporting W21 × 73 flange. Figure 25-2-5B is produced similarly to Fig. 25-2-5A for the north end except we find that the flange of the W18 will interfere with that of the W21. Thus it becomes necessary to notch out, or cope, the W18.

Some shops would not dimension such a cut but would give it a standard mark. Others would simply note on the drawing COPE TO W21 × 73 and let the shop work out its proper shape and size.

Note that the intersection of the horizontal and vertical is not a sharp corner (reentrant-cut) but is cut to a small radius to provide a fillet at this point. However, since the shop has been trained to provide these fillets, they are not usually shown on the detail drawing.

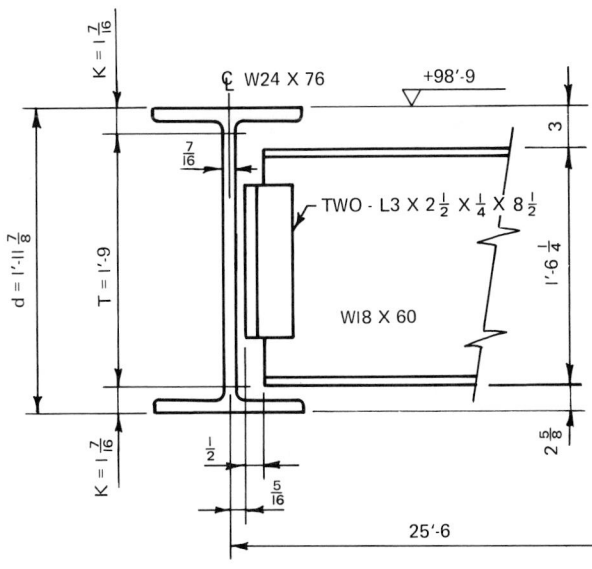

(A) NORTH-END BEAM CONNECTION

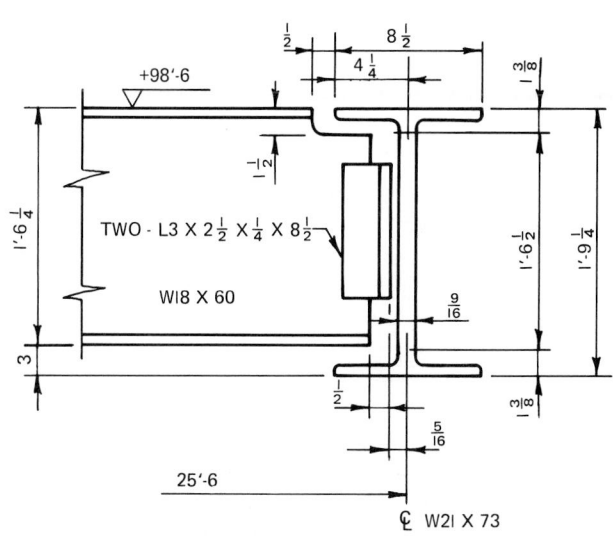

(B) SOUTH-END BEAM CONNECTION

FIG. 25-2-5   Detail of W18 × 60 beam connections.

Since the two beams are flush on top, the minimum depth of the cut Q is made equal to or greater than the K distance for the W21 × 73 beam.

K Distance for the W21 Beam = 1⅜   Following the guidelines shown in Fig. 25-2-3, the Q_1 dimension for this beam is 1½ in. The length of the cut, Q_2, as measured from the backs of the connecting angles, should allow for a minimum ½ in. clearance between the toe of the supporting beam and the flange of the supported beam. To determine the length of dimension Q_2, add ½ in. to half the flange width of the W21 beam. From this value subtract half the web thickness of the W21 beam and 1/16 in. The 1/16 in. dimension is the clearance allowed between the web face and the outer face of the connecting angles.

Therefore $Q_2 = ½ + 4⅛ - 7/32 - 1/16 = 4^{11}/_{32}$. As previously used for Q_1, the dimension Q_2 should be raised to the nearest length evenly divisible by ¼ in. Thus the 4 11/32 in. dimension (Q_2) is raised to 4½ in.

With the exception of the connecting-angle data, Fig. 25-2-6 is the completed shop drawing of the W18 × 60 beam. Note the following points.

1. The minus dimensions (−5/16), shown outside and opposite the dimension line for the back-to-back distance of the end connection angles (25'-5⅜), are the distances from the center lines of the supporting beams to the back of each connecting angle. For a beam framing to other shapes, the minus dimension (setback distance) is equal to half the web thickness of the supporting member plus 1/16 in., rounded off to the nearest 1/16 in.

2. The center-to-center distance of 25'-6 between the two supporting beams is shown for reference purposes.

3. The actual or ordered length of the W18 × 60 should be such that its ends are about ½ in. short of the backs of the connecting angles. This is to allow for inaccurate cutting to specified length at the mill or in the shop and will thereby eliminate any extra expense caused by recutting or trimming during fabrication.

4. No top or bottom views are necessary because no holes are required in either flange. In general, the shop should not be required to look at views that do not convey necessary instructions.

5. The end connection angles are shown, but not detailed. Information on detailing these angles is given in Unit 25-3. Note the end view of the angles is shown but the W18 beam is not drawn.

6. The complete beam is given a shipping or erection mark, B15, to identify it in the office, shop, and field. There are many systems currently in use for establishing the shipping mark. One of the most common methods is to use a capital letter followed by a sheet number. Each separate shipping piece, detailed on one sheet, has the same number preceded by a different letter. In this example the detail drawing of the W18 × 60 is the second sketch on sheet 15; the third would be C15, the fourth D15; and so on.

7. The connection angles are given assembly or template marks, usually lowercase letters. This is done for two reasons: (a) It saves the detailing of these angles again, when they are used on the same piece (as on the south end of the beam in this example) or on other beams on the same sheet. (b) The angles will be punched on a different machine from the one used for the beam. The assembly mark is a guarantee that the correct angle will be assembled on the correct beam. On the given detail, the material required to fabricate only one complete shipping piece is listed, or billed. When duplication of the shipping pieces is required, the shop multiplies the billing for one complete piece by the total number of assemblies required.

REFERENCES AND SOURCE MATERIAL

1. American Institute of Steel Construction.
2. Canadian Institute of Steel Construction.

ASSIGNMENTS

See Assignments 2 and 3 for Unit 25-2 on pages 852–854.

25-3 STANDARD CONNECTIONS

Standard framed-beam connections are used for framing structural steel. Since riveting is almost nonexistent in most fabricating plants today, rivets will not be considered in this context. Standard connection angles are shop-welded or bolted to the beam web and field-bolted to their supporting member. Only half-strength bolted connections of the friction type will be considered in this chapter.

When detailing individual members, the shop detailer should bear in mind that each individual member must be joined to other members. The place or location of one member's attachment shape or plate along with the means of fastening is called the *connection plate,* or the *connection* for short.

Bolted Connections

Bolts are placed on standard lines or gages. The distance between bolt holes is referred to as the *bolt pitch,* or *pitch.* The

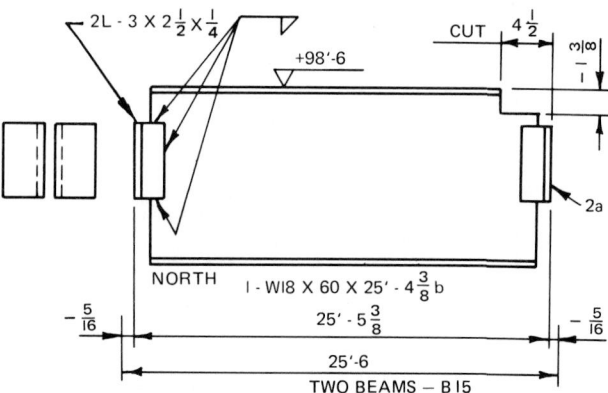

NOTE: DETAIL FOR THE CONNECTING ANGLES NOT SHOWN ON THIS DRAWING.

FIG. 25-2-6   Detail drawing of W18 × 60 beam.

gage and the pitch for a multiple bolt connection detail must be sufficiently large to allow for the wrench clearance when a bolt adjacent to a previously installed bolt or adjacent to another part of the shape being joined is tightened. Figure 25-3-1 shows the recommended gages to be used for structural shapes.

Of prime importance is consistency of detail; for example, gages on an individual member should not vary throughout the length of the member. If a connection plate, made from a sheared piece of steel, as shown in Fig. 25-3-2 (pg. 840), is to have three holes across its width of 10 in. for ⅝ in. bolts, the gage could be 3¾ in. with edge distances of 1¼ in. If the fasteners to be used are ¾ in. diameter bolts, the edge distance of 1¼ in. would have to be adjusted. The minimum distance from the center of a bolt hole to any edge should not be less than that

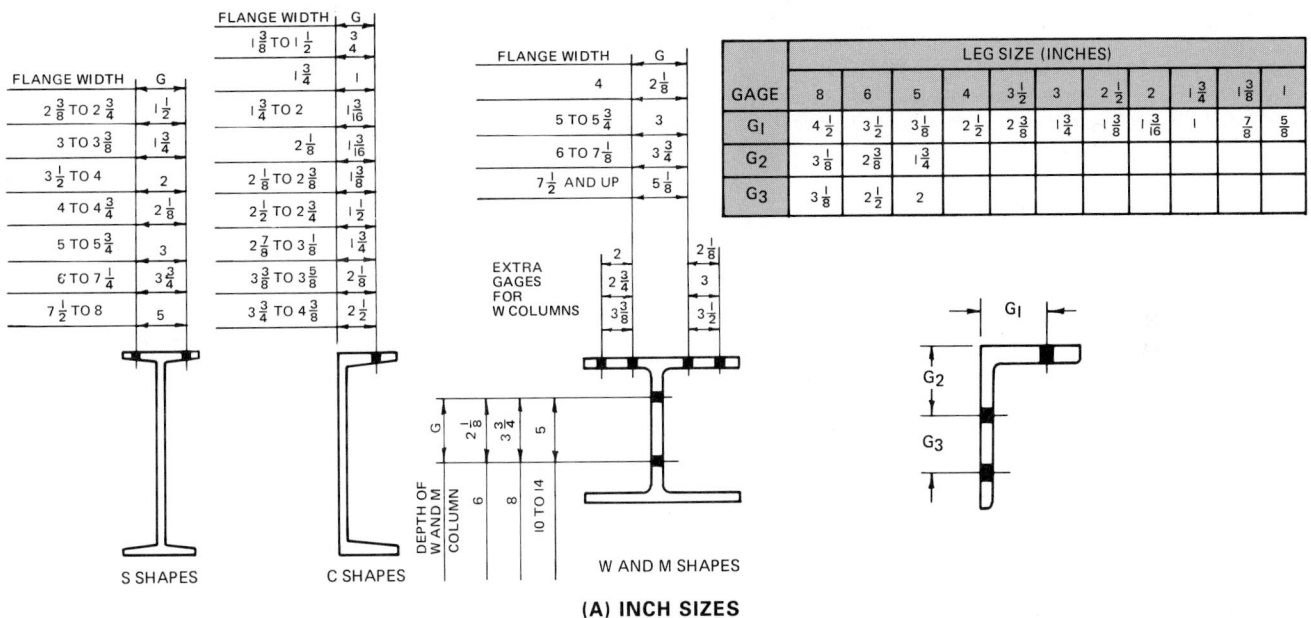

(A) INCH SIZES

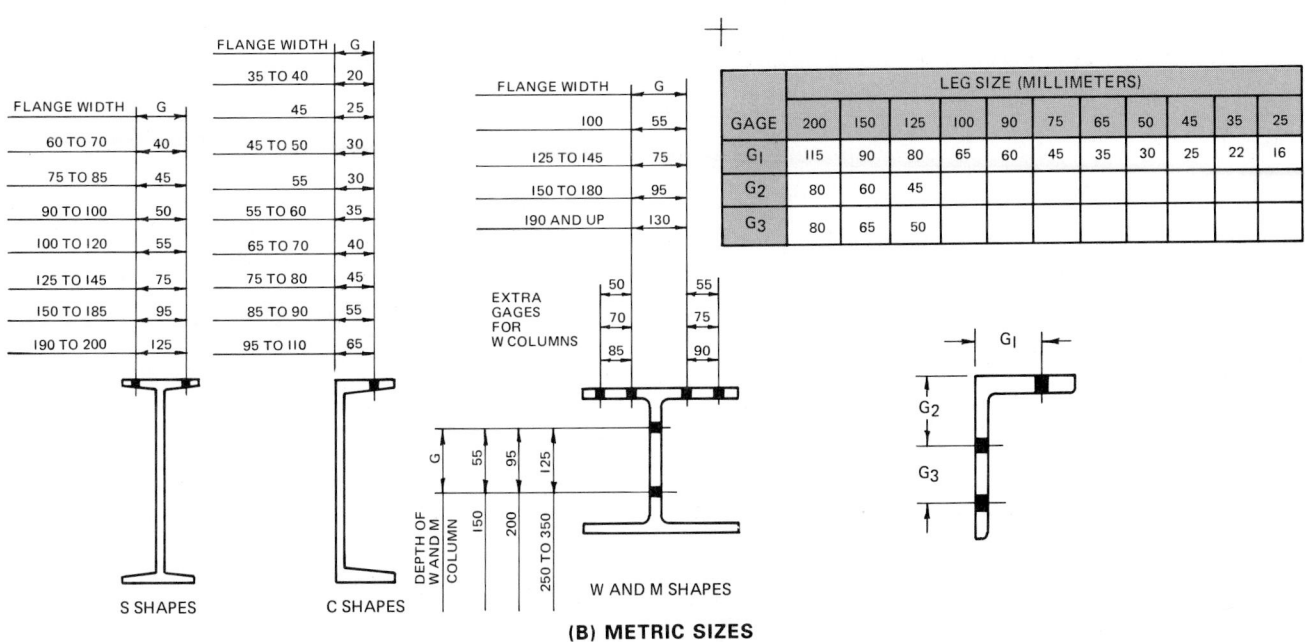

(B) METRIC SIZES

FIG. 25-3-1 Recommended gages.

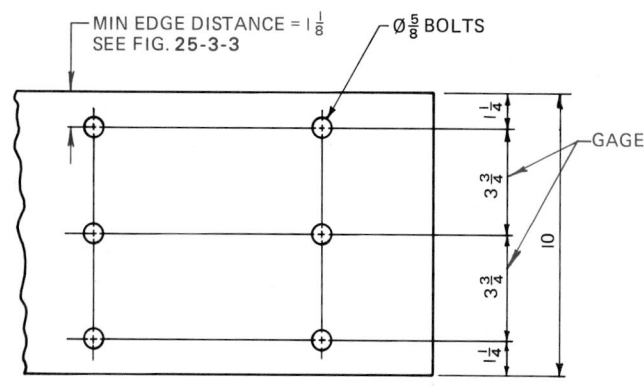

(A) $\frac{5}{8}$ - BOLTED CONNECTION

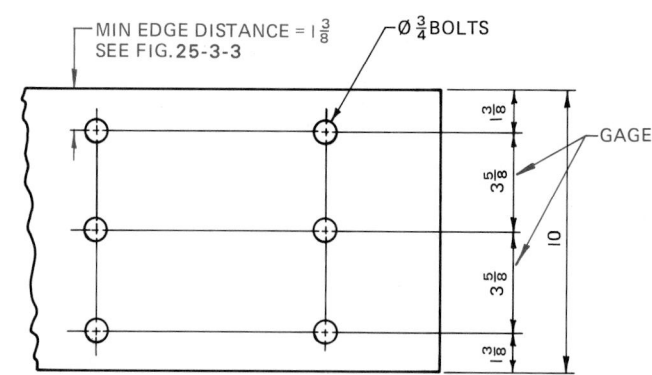

(B) $\frac{5}{8}$ - BOLTED CONNECTION

FIG. 25-3-2   Establishing gage sizes from edge distance.

	BOLT DIAMETER	AT SHEARED EDGE	AT ROLLED OR GAS-CUT EDGE
U.S. CUSTOMARY (INCHES)	½	1	¾
	⅝	1⅛	⅞
	¾	1⅜	1
	⅞	1½	1⅛
	1	1⅝	1³⁄₁₆
	1⅛	1¾	1⅜
	1¼	2	1½
	1⅜	2½	1¹³⁄₁₆
METRIC (mm)	14	26	20
	16	28	22
	20	34	26
	22	38	28
	24	42	30
	27	46	34
	30	52	38
	36	64	46

FIG. 25-3-3   Minimum edge distance for bolt holes.

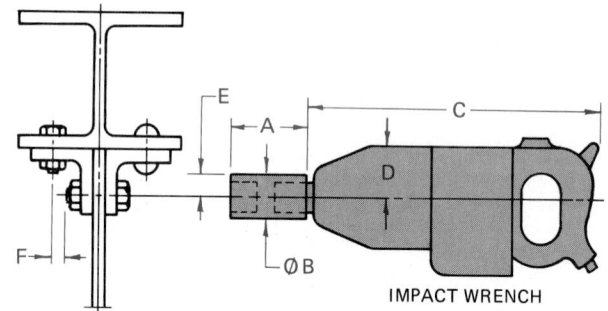

IMPACT WRENCH

		Sockets		Min. Clear.	
	Bolt Size	**A**	**B**	**E**	**F**
U.S. Customary (Inches)	⅝	2⅝	1¾		
	¾	3	2¼	1³⁄₃₂	1¼
	1	3½	2⅝	1⅛	1¼
	1¼	4	3⅛	1⁵⁄₁₆	1⁷⁄₁₆
	1⅜	4½	3¼	1⁹⁄₁₆	1¹¹⁄₁₆
Metric (mm)	16	70	55	28	32
	20	80	58	29	34
	24	90	65	33	36
	30	110	75	38	42
	36	130	85	43	48

		Bolt Size	C	D
U.S. Customary (Inches)	Light Wrenches	⅝ to 1	13¼ to 14	2
	Heavy Wrenches	1 to 1⅜	15¼ to 17¼	2
Metric (mm)	Light Wrenches	16 to 24	337 to 356	54
	Heavy Wrenches	24 to 36	375 to 438	64

FIG. 25-3-4   Minimum erection clearances.

given by Fig. 25-3-3. For the ¾ in. bolt, the minimum edge distance to the sheared edge is 1⅜ in. Since the plate is 10 in. wide and a minimum edge distance of 1⅜ in. is required, the gage required would be [10 − (2 × 1⅜)] ÷ 2, or 3⅝ in. If the connection plate had been an angle, the same reasoning as used above would still pertain to the connection detail. Another consideration is wrench clearance, which is illustrated in Fig. 25-3-4.

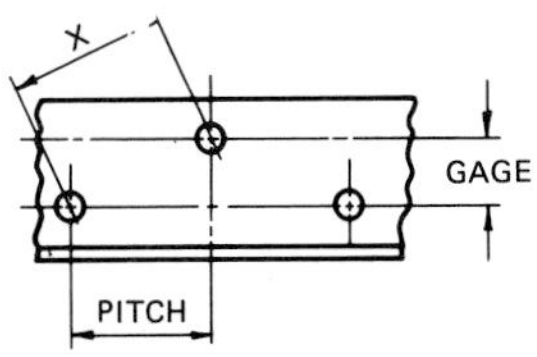
X
GAGE
PITCH

U.S. CUSTOMARY (in.)

Pitch (in.)	Gage (inches) 1	1¼	1½	1¾	2	2¼	2½	2¾	3	3¼	3½	3¾	4	4¼
¼	1.00	1.25	1.50	1.75	2.00	2.25	2.50	2.75	3.00	3.25	3.50	3.75	4.00	4.25
½	1.10	1.35	1.60	1.80	2.05	2.30	2.55	2.80	3.05	3.30	3.55	3.80	4.05	4.30
¾	1.25	1.45	1.70	1.90	2.15	2.35	2.60	2.85	3.10	3.35	3.60	3.85	4.10	4.35
1	1.40	1.60	1.80	2.00	2.25	2.45	2.70	2.95	3.15	3.40	3.65	3.90	4.15	4.40
1¼	1.60	1.75	1.95	2.15	2.35	2.55	2.80	3.00	3.25	3.50	3.70	3.95	4.20	4.45
1½	1.80	1.95	2.10	2.30	2.50	2.70	2.90	3.15	3.35	3.60	3.80	4.05	4.30	4.50
1¾	2.05	2.15	2.30	2.50	2.65	2.85	3.05	3.25	3.50	3.70	3.90	4.15	4.40	4.60
2	2.25	2.35	2.50	2.65	2.85	3.00	3.20	3.40	3.60	3.80	4.05	4.25	4.50	4.70
2¼	2.45	2.55	2.70	2.85	3.00	3.20	3.35	3.55	3.75	3.95	4.15	4.40	4.60	4.85
2½	2.70	2.80	2.90	3.05	3.20	3.35	3.55	3.70	3.90	4.10	4.30	4.50	4.75	4.95
2¾	2.95	3.00	3.15	3.25	3.40	3.55	3.70	3.90	4.10	4.25	4.45	4.65	4.85	5.10
3	3.15	3.25	3.35	3.50	3.60	3.75	3.90	4.10	4.25	4.40	4.60	4.80	5.00	5.20
3¼	3.40	3.50	3.60	3.70	3.80	3.95	4.10	4.25	4.40	4.60	4.80	5.00	5.20	5.35
3½	3.65	3.70	3.80	3.90	4.05	4.15	4.30	4.45	4.60	4.80	4.95	5.15	5.35	5.50
3¾	3.90	3.95	4.05	4.15	4.25	4.40	4.50	4.65	4.80	5.00	5.15	5.30	5.50	5.70

METRIC (mm)

Pitch (mm)	Gage (millimeters) 25	30	35	40	45	50	55	60	65	70	75	80	85	90	95	100	105	110
5	25	30	35	40	45	50	55	60	65	70	75	80	85	90	95	100	105	110
10	27	32	36	41	46	51	56	61	66	71	76	81	86	91	96	100	105	110
15	29	34	38	43	47	52	57	62	67	72	76	81	86	91	96	101	106	111
20	32	36	40	45	49	54	59	63	68	73	78	82	87	92	97	102	107	112
25	35	39	43	47	51	56	60	65	70	74	79	84	89	93	98	103	108	113
30	39	42	46	50	54	58	63	67	72	76	81	85	90	95	100	104	109	114
35	43	46	49	53	57	61	65	69	74	78	83	87	92	97	101	106	111	115
40	47	50	53	57	60	64	68	72	76	81	85	89	94	98	103	108	112	117
45	51	54	57	60	64	67	71	75	79	83	87	92	96	101	105	110	114	119
50	56	58	61	64	67	71	74	78	82	86	90	94	99	103	107	112	116	121
55	60	63	65	68	71	74	78	81	85	89	93	97	101	105	110	114	119	123
60	65	67	69	72	75	78	81	85	88	92	96	100	104	108	112	117	121	125
65	70	72	74	76	79	82	85	88	92	96	99	103	107	111	115	119	123	128
70	74	76	78	81	83	86	89	92	96	99	103	106	110	114	118	122	126	130
75	79	81	83	85	87	90	93	96	99	103	106	110	113	117	121	125	129	133
80	84	85	87	89	92	94	97	100	103	106	110	113	117	120	124	128	132	136
85	89	90	92	94	96	99	101	104	107	110	113	117	120	124	127	131	135	139
90	93	95	97	98	101	103	105	108	111	114	117	120	124	127	131	135	138	142

FIG. 25-3-5 Staggered fasteners.

Occasionally, a gage is too small for both holes to be placed adjacent to one another at right angles. When this happens, staggered centers are used, as illustrated in Fig. 25-3-5.

EXAMPLE 1

Given a flat bar 4½ in. in width, which is to have a double line of Ø⅝ in. bolts and a gage of 2 in., calculate the pitch of the bolts (Fig. 25-3-6).

SOLUTION

From Fig. 25-3-4 the clearance *E* required for a ⅝ in. bolt is $1\frac{3}{32}$. The minimum recommended distance between holes is 2*E*, or $2\frac{3}{16}$ in., which is greater than the gage of 2 in. Therefore, staggered holes will be required. To determine the minimum pitch, refer to Fig. 25-3-5. Read down the 2 in. gage column to find where the dimension $2\frac{3}{16}$ (2.18) will fall. If the dimension is not exact, use the next larger number (2.25). To the extreme left of the 2.25 value is the pitch required. For this problem, the pitch is 1.

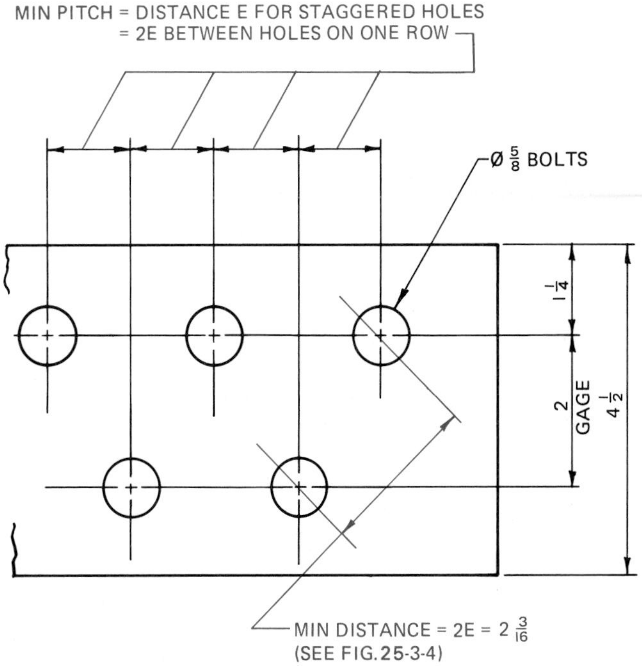
MIN PITCH = DISTANCE E FOR STAGGERED HOLES
= 2E BETWEEN HOLES ON ONE ROW
Ø 5/8 BOLTS
1 1/4
2
GAGE
4 1/2
MIN DISTANCE = 2E = 2 3/16
(SEE FIG.25-3-4)

FIG. 25-3-6 Gage and pitch layout.

EXAMPLE 2

Figure 25-3-7 shows a partial design drawing similar to Fig. 25-2-4 except that it includes information concerning the

connection. It represents part of a design drawing for a steel-framed floor system as viewed from above. With its notes, it contains all the necessary information required by the shop detailer to detail the W18 × 60 beam. With the exception of the connection-angle detail, all information pertaining to this beam was covered in Unit 25-2. Therefore, this unit will deal only with the connection-angle detail.

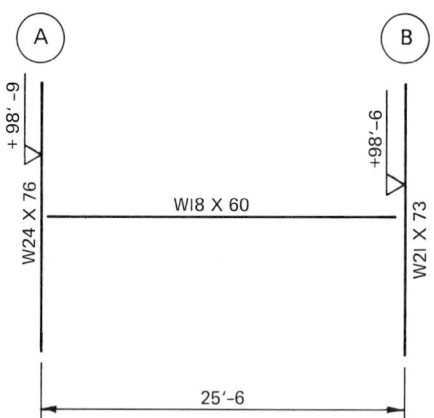

ELEVATION TOP OF STEEL SHOWN THUS (+98'-6)

NOTES:
ALL HOLES Ø $\frac{13}{16}$
ALL CONNECTIONS TO DEVELOP FULL LENGTH
    UNLESS OTHERWISE SPECIFIED
BOLTS: $\frac{3}{4}$ A325
CONNECTING ANGLES WELDED TO BEAM,
    BOLTED TO SUPPORT

FIG. 25-3-7 Partial design drawing.

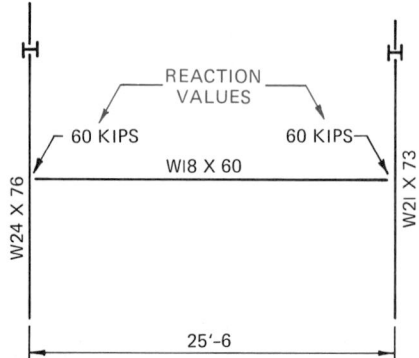

FIG. 25-3-8 Indicating beam reactions on drawings.

SOLUTION

The problem is to select a connection for the W18 beam that will be able to carry the reaction values. Often the reaction values are calculated and given on the design drawing, as shown in Fig. 25-3-8. If the reaction values are not given, the connection is designed to support half the total uniform load capacity. In this case, since the reaction values are not given and there is no indication on the design drawing for other than a uniformly distributed load when the building is complete, we consult Fig. 25-3-9, which states the maximum allowable load for a W18 × 60 beam having a span of 25'-6 lies between the 126 and 118 kip values (approximately 120 kips). One kip equals 1000 lb. Connection design load is one-half of this value or 60 kips.

Referring to Fig. 25-3-10 under the column Bolt Capacity in Kips Friction Connections, we find that four bolts are required to support this load. The length of the connecting angles shown for four bolts is 11½ in., for which the minimum and maximum depth recommendations for beams are 15 and 24 in. Since these limits bracket the actual depth of the W18 beam, the connection is acceptable.

Referring to the column Web Framing Leg with Welds, we find the maximum weld capacity of a four-bolt per vertical line angle with a ³⁄₁₆ in. weld is 140 kips for a 3 in. angle width and 134 kips for a 2½ in. angle width. Both are greater than the 60 kips allowable load and thus are acceptable. Therefore, the smaller angle width of 2½ in. is selected for the welded framing leg. Practice dictates that the angle thickness should be ¹⁄₁₆ in. greater than the weld size; hence the minimum required angle thickness is ³⁄₁₆ + ¹⁄₁₆ = ¼ in.

Another check for the minimum thickness of the angle and the minimum permissible web thickness for the beam must be made. In order to determine the minimum angle thickness, refer to the section Minimum Required Web Thickness and Angles Where Bolted of Fig. 25-3-10. Since connection angles

BEAM LOAD TABLE IN KIPS (1000 lb)								BEAM LOAD TABLE IN KILONEWTONS (kN)								
	W18 Beam				W12 Beam				W460 Beam				W310 Beam			
	Pounds per Foot				Pounds per Foot				Mass per Meter				Mass per Meter			
Span Feet	60	55	50	45	40	36	31	27	Span mm	89	82	74	67	60	52	45	39
16	177	160	145	130	81	74	63	55	5000	864	782	708	631	395	361	308	266
18	161	146	132	118	74	67	57	50	5500	785	711	644	573	359	328	280	242
20	148	134	121	108	67	62	53	46	6000	720	652	590	526	329	301	257	222
22	136	123	112	100	62	57	49	42	6500	665	601	545	485	304	278	237	205
24	126	115	104	92	58	53	45	39	7000	617	559	506	451	282	258	220	190
26	118	106	97	86	54	50	42	36	7500	576	521	472	420	263	241	205	177
28	111	100	91	81	51	46	40	34	8000	540	489	443	394	247	226	193	166
30	104	94	85	76	48	44	37	32	8500	508	460	417	371	232	212	181	157

FIG. 25-3-9 Beam load tables.

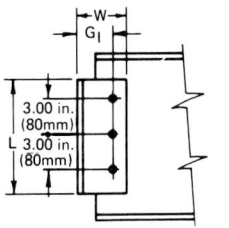
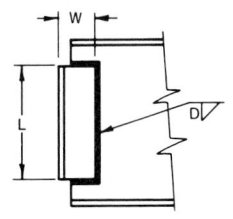

WEB FRAMING LEGS

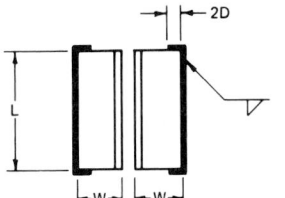
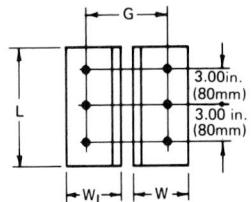

OUTSTANDING LEGS

EITHER LEG WITH 1.75 A325 BOLTS				WEB FRAMING LEG WITH E70XX WELDS				
				Weld Capacity In Kips				
				Fillet Size, D (in.)				
Angle Width and Gage	Bolts Per Vertical Line	Bolt Capacity In Kips Friction Connections	Bearing Connections	3/16	1/4	5/16		Angle Width (in.)
	2	38	47	81	108	135		
	3	54	70	114	152	190		
W = 3½	4	72	94	140	186	232		
G = 5½	5	90	118	165	220	275		W = 3
G₁ = 2¼	6	108	142	190	252	315		
	7	126	165	214	286	358		
	8	144	188	240	320	400		
	2	38	47	81	108	135		
	3	54	70	108	144	180		
W = 3	4	72	94	134	178	222		
G = 4	5	90	118	158	210	262		W = 2½
G₁ = 1¾	6	108	142	182	243	304		
	7	126	165	207	276	345		
	8	144	188	232	310	388		
Minimum Required Web Thickness and Angles Where Bolted (Inches)								
			.34	.27	.40	.54	.68	

EITHER LEG WITH M20-A325 BOLTS				WEB FRAMING LEG WITH E-480XX WELDS					
				Weld Capacity In Kilonewtons					
				Fillet Size, D (mm)					
Angle Width and Gage	Bolts Per Vertical Line	Bolt Capacity in Kilonewtons Friction Connections	Bearing Connections	4	6	8	10		Angle Width mm
	2	169	210	324	484	646	808		
	3	240	314	452	678	904	1130		
W = 90	4	320	419	551	826	1100	1380		
G = 130	5	400	524	649	974	1300	1620		W = 75
G₁ = 60	6	480	629	748	1120	1500	1870		
	7	560	734	846	1270	1690	2120		
	8	640	839	945	1420	1890	2360		
	2	169	210	330	484	646	808		
	3	240	314	428	641	855	1070		
W = 75	4	320	419	526	789	1050	1320		
G = 100	5	400	524	625	937	1250	1560		W = 65
G₁ = 45	6	480	629	723	1080	1450	1810		
	7	560	734	822	1230	1640	2050		
	8	640	839	920	1380	1840	2300		
Minimum Required Web Thickness and Angles Where Bolted									
FY = 300				8.6	6.9	10.3	13.8	17.2	

OUTSTANDING LEG WITH E70XX WELDS				LENGTH OF CONNECTING ANGLES			
	Weld Capacity in Kips						
	Fillet Size D (in.)			Bolts Per Vertical Line	Suggested Beam Depth Limits (in.)		Angle Length L (in.)
Angle Width (in.)	3/16	1/4	5/16		Min. Max.		
	24	32	40	2	8 – 12		5½
	56	74	92	3	12 – 18		8½
	94	126	158	4	15 – 24		11½
W = 3½	125	167	208	5	18 – 30		14½
	150	200	250	6	22 – 36		17½
	176	234	292	7	24 – 42		20½
	200	268	335	8	28 – 48		23½
	30	40	50	2	8 – 12		5½
	68	91	114	3	12 – 18		8½
	98	130	162	4	15 – 24		11½
W = 3	125	167	208	5	18 – 32		14½
	150	200	250	6	22 – 36		17½
	176	234	292	7	24 – 44		20½
	200	268	335	8	28 – 48		23½

OUTSTANDING LEG WITH E480XX WELDS					LENGTH OF CONNECTING ANGLES			
	Weld Capacity In Kilonewtons							
	Fillet Size D (mm)				Bolts Per Vertical Line	Suggested Beam Depth Limits (mm)		Angle Length L (mm)
Angle Width mm	4	6	8	10		Min. Max.		
	98	141	186	232	2	200 – 300		150
	228	329	438	546	3	300 – 450		230
	384	563	755	952	4	380 – 600		310
W = 90	490	743	1000	1263	5	450 – 800		390
	589	891	1200	1510	6	550 – 900		470
	687	1040	1400	1760	7	600 – 1100		550
	786	1190	1590	2000	8	700 – 1200		630
	125	180	238	297	2	200 – 300		150
	276	405	544	681	3	300 – 450		230
	386	581	789	1004	4	380 – 600		310
W = 75	490	743	1000	1263	5	450 – 800		390
	589	891	1200	1510	6	550 – 900		470
	687	1040	1400	1760	7	600 – 1100		550
	786	1190	1590	2000	8	700 – 1200		630
Minimum Required Web or Flange Thickness*					To be used for educational purposes only.			
FY = 300	3.4	5.2	6.9	8.6				

Note 1: Connection angles are assumed to be material with minimum yield strength of 44 000 psi (300 MPa).
Note 2: For connections with outstanding legs bolted, the minimum required thickness of the supporting material is one-half the thickness listed above, if beams are attached to one side of the supporting material.

*Thickness listed is for supporting material with beams attached to one side only. If beams are attached to both sides of the supporting material, use double the minimum thickness listed.

FIG. 25-3-10   Double-angle beam connections for .75 in. (M20)-A325 bolts and E480XX fillet welds.

are assumed to be material that has a yield strength of 44,000 lb/in² (see note 1 located below the table), the minimum web and angle thickness specified is .34 in. Note 2 states that if the connection is used for outstanding angle legs, the value as specified can be halved. Therefore, the minimum thickness of angle that can be used is .17 in. This is less than the ¼ in. required for welding, and as such, the ¼ in. angle thickness is selected. The web thickness of the W18 × 60 beam is 7/16 in., which is acceptable.

The next step is to select the gage (distance between bolt centers on the two outstanding legs). The recommended gage and angle widths shown are 5½ and 3½, or 4 and 3, respectively. Both the 3½ and 3 in. wide angles are acceptable as far as load capacities are concerned. The clearances shown in Fig. 25-3-11 (pg. 844) would be a factor in deciding to use either the 3 or 3½ in. wide outstanding lengths. In this problem, the 3 in. leg was chosen because the use of universal joints is now a widely accepted practice.

The G₁ distance of 1¾ in. shown in Fig. 25-3-1 should be used instead of the actual 1²⁵/₃₂ in. dimension shown in Fig. 25-3-11, since a hole clearance of 1/16 in. (13/16 in. diameter holes) is used for the Ø¾ bolts.

FIG. 25-3-11   How erection clearances control gage and connecting angle sizes.

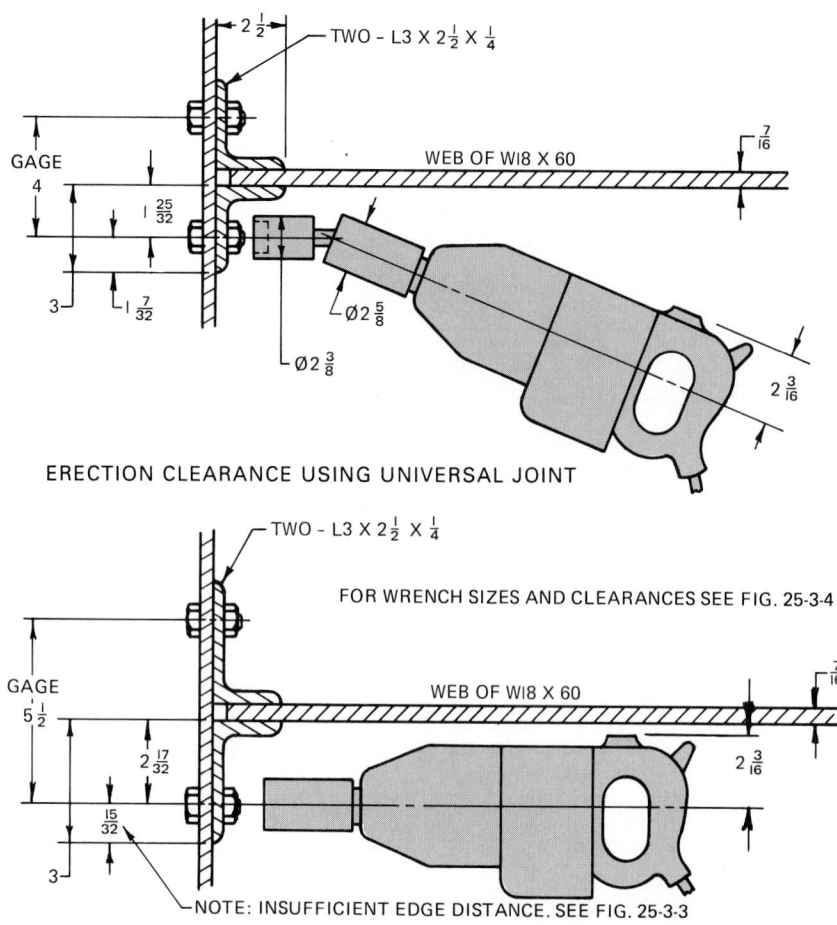

ERECTION CLEARANCE USING UNIVERSAL JOINT

FOR WRENCH SIZES AND CLEARANCES SEE FIG. 25-3-4

NOTE: INSUFFICIENT EDGE DISTANCE. SEE FIG. 25-3-3

ERECTION CLEARANCE WITHOUT USING UNIVERSAL JOINT

EXAMPLE 3

With reference to Fig. 25-3-12 select a connection for the W12 beam that will be able to carry the reaction value shown.

SOLUTION

The reaction value for the connection is 50 kips (50,000 lb). Referring to Fig. 25-3-10 under the column Bolt Capacity in Kips Friction Connections, we find that three bolts having a capacity of 54 kips are required to support this load. The length of the connecting angles shown for three bolts is 8½ in., for which the minimum and maximum depth recommendations for beams are 12 and 18 in. Since these limits bracket the actual depth of the W12 beam, the connection is acceptable.

Referring to the column Web Framing Leg with Welds in Fig. 25-3-10, we find the maximum weld capacity of three bolts per vertical line with a 3/16 in. weld is 114 kips for a 3 in. angle width. Practice dictates that the angle thickness should be 1/16 in. greater than the weld size; hence the minimum required angle thickness is 3/16 + 1/16 = 1/4 in.

Another check for the minimum thickness of the angle and the minimum permissible web thickness for the beam must be

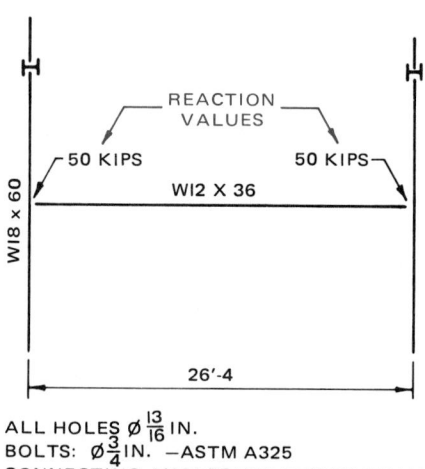

ALL HOLES Ø 13/16 IN.
BOLTS: Ø 3/4 IN. —ASTM A325
CONNECTING ANGLES WELDED TO BEAM, BOLTED TO SUPPORT.

FIG. 25-3-12   Partial design drawing.

made. In order to determine the minimum angle thickness, refer to the section Minimum Required Web Thickness and Angles Where Bolted. Since connection angles are assumed to be material that has a yield strength of 44,000 lb/in.[2] (see note 1 located below the table), the minimum web and angle thickness specified is .34. Note 2 states that if the connection is used for outstanding angle legs, the value specified can be halved. Therefore the minimum thickness of angle that can be used is .34 ÷ 2 = .17 in. This is less than the ¼ in. required for welding, and as such, the ¼ in. angle thickness is selected. The web thickness of the W12 × 36 beam is .31 in., which is acceptable.

The next step is to select the gage (distance between bolt centers on the two outstanding legs). The recommended gage and angle widths shown are 5½ and 3½ in., or 4 and 3 in., respectively. Both the 3½ and the 3 in. wide angle legs are acceptable as far as load capacities are concerned. Therefore the 3 in. wide leg is chosen.

The G_1 distance of 1¾ in. shown in Fig. 25-3-1 and the gage distance of 4 in. are selected since a hole clearance of ¹⁄₁₆ in. (¹³⁄₁₆ in. diameter holes) is used for the ¾ in. diameter bolts.

Some fabricators on occasion choose to use high-strength bolts in the shop to fasten the connection angles to the web of the beam. The combination of the shop and field high-strength bolts in one connection presents other problems, such as additional clearance requirements for entering and tightening the high-strength bolts. These requirements mean larger connection angles, larger gages, and larger spread. An alternate solution to the increased sizes would be to stagger the bolt centers of the outstanding leg. In order not to overcomplicate

the subject at this point, all connections discussed in this unit will be angles shop-welded to the web and field-bolted to the supporting members.

The location of the connection angles on the horizontal beam must now be set. Practice is to set the distance from the uppermost bolt hole to the top of the beam equal to the pitch distance. From the detailer's sketch, shown in Fig. 25-3-13, the complete detail drawing of the beam (Fig. 25-3-14) is made. Note the following points.

1. The end connection angles are completely detailed. Following their own individual shop standards, many shops would do this with less information.
2. The web leg is 2½ in.
3. The outstanding leg of each connection angle is 3 in. and the recommended gage for this leg size is 1¾ in.
4. The open holes, for connection to the web of the supporting beam, are spaced 4 in. apart, center to center. This distance is called the *spread*. It is $2 \times G_1$ + web thickness of the W18 × 60 beam = $2 \times 1¾ + ¹⁄₁₆ = 3¹⁵⁄₁₆$. Use 4 in.
5. The pitch, or distance, bolt to bolt, along any gage line is 3 in.
6. The end distance is equal to half of the remainder left after subtracting the sum of all bolt spaces from the length of the angles. In the case of the four-row connection, it equals $(11½ - 3 - 3 - 3) \div 2 = 1¼$ in.
7. Instead of noting them on each individual detail, as in Fig. 25-3-14, it is usual practice to call for the sizes of bolts and holes once on each sheet in a general note. Such a note covers all the bolts and holes on the sheet, with exceptions noted on the individual details where they occur.

REFERENCES AND SOURCE MATERIAL

1. American Institute of Steel Construction.
2. Canadian Institute of Steel Construction.

ASSIGNMENTS

See Assignments 4 and 5 for Unit 25-3 on pages 854–855.

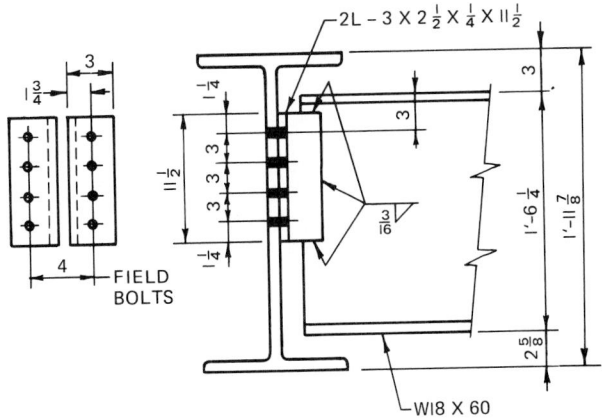

FIG. 25-3-13   Detail of north end of W18 × 60 beam from partial design drawing, Fig. 25-3-7.

FIG. 25-3-14   Complete beam detail of W18 × 60 from partial design drawing, Fig. 25-3-7.

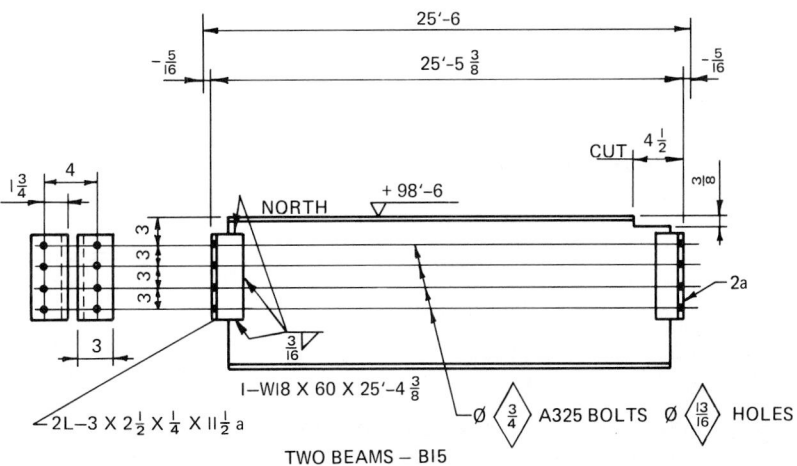

25-4 SECTIONING

In many instances the cross-hatching of sections may be omitted on shop drawings because its use is not needed to make the drawing any clearer. In structural detailing, the practice is to omit all lines on a drawing that serve no significant purpose.

Bottom Views

In shop detail drawings, the bottom flange or face of a shape is never viewed from below; that is, the drafter does not stand below the object and look squarely up at it. Rather, she or he cuts a section such as *A-A* in Fig. 25-4-1 and views the bottom flange by looking squarely down on the top side of it.

The reason for substituting a bottom section for a bottom view is to obtain a better correlation between it and the top view. For example, it will be more apparent whether a connection on one side of the top flange and a connection on the bottom flange are on the same side or opposite sides of the member. Note that the cross-hatching is omitted as previously mentioned.

Elimination of Top and Bottom Views

In the preceding examples, we found that top and bottom views were not necessary because no holes were required in either flange. Let us now look at an example where there are holes in the bottom and top flanges and see whether top and bottom views are required.

In Fig. 25-4-2, the detailing shows a top view and a picture of the bottom flange, taken as a section looking down. Note that the dash line in the top view and the solid lines in the bottom view that depict the web are not drawn continuously across the length of the member. Neither is the cut section of the web blackened or cross-hatched. Yet the drawing is complete, readable, and understandable to the fabricator. Remember, use as few lines as possible to describe the object and the shop fabrication.

Detailing in Fig. 25-4-3 eliminates these views. In order to eliminate views of the top and bottom flanges, instructions

(including necessary dimensions) for cutting these flanges at the right-hand end have been covered in the note on the web view concerning cutting. The transverse distance between gage lines on the flanges is covered by the note *GA* = 5. In both cases, symmetry about the center line of the beam web is understood. These notes must be explicit in showing what fabrication, if any, is required on each flange. Both methods of presentation are common practice.

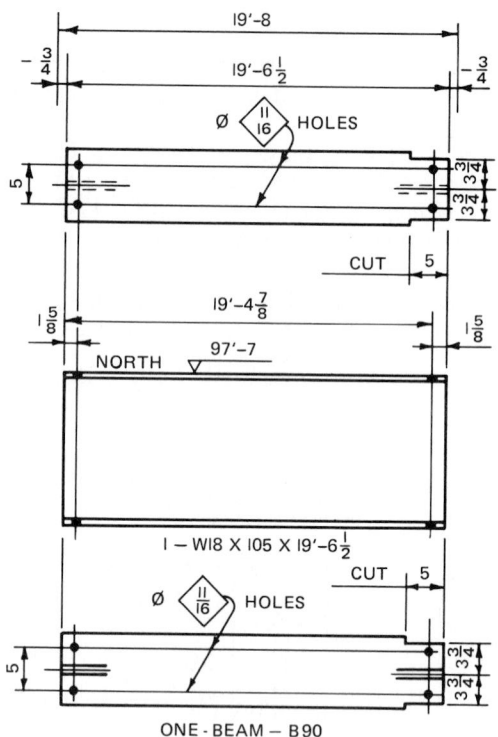

FIG. 25-4-2 Detail of beam.

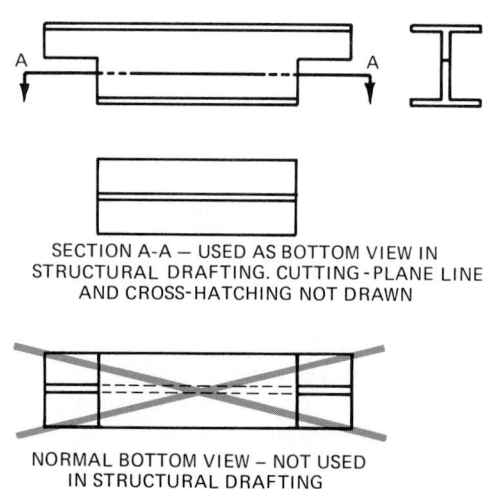

SECTION A-A — USED AS BOTTOM VIEW IN STRUCTURAL DRAFTING. CUTTING-PLANE LINE AND CROSS-HATCHING NOT DRAWN

NORMAL BOTTOM VIEW — NOT USED IN STRUCTURAL DRAFTING

FIG. 25-4-1 Bottom view in structural drafting.

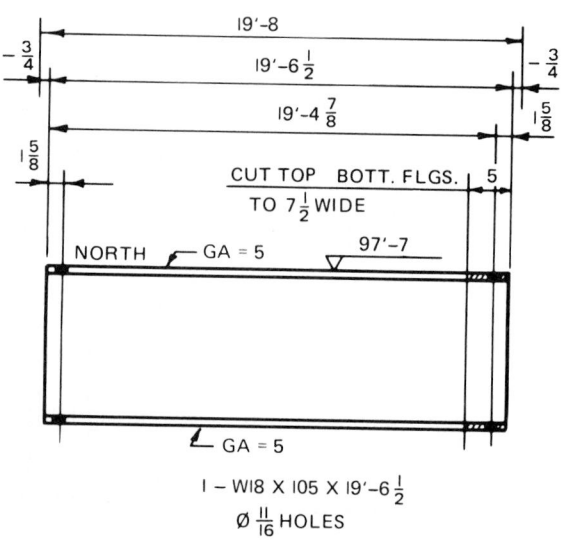

FIG. 25-4-3 Elimination of top and bottom views.

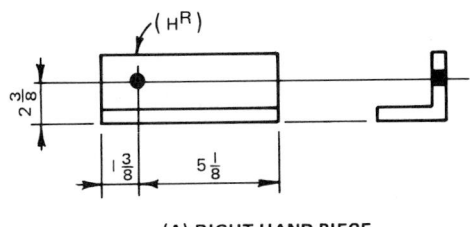

(A) RIGHT-HAND PIECE

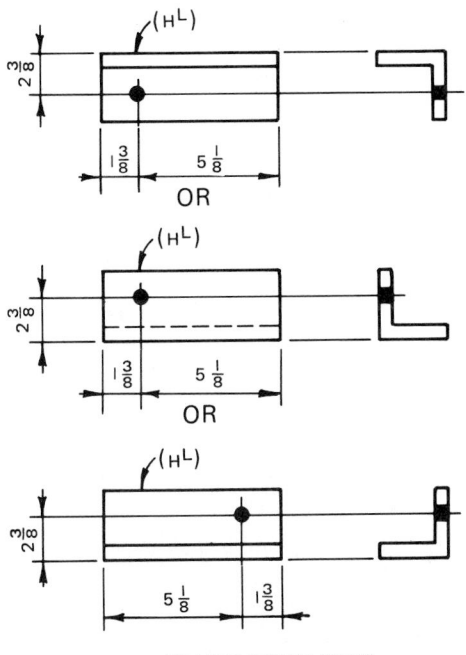

OR

OR

(B) LEFT DETAIL PIECE

FIG. 25-4-4   Right and left detail pieces.

Right- and Left-Hand Details

Very frequently detail material, such as connection angles and other fittings, is used under conditions where one piece must be the exact opposite of another. In such cases, both the right-hand (RH) and the left-hand (LH) pieces are fabricated from the same sketch, that is, from the detail of the RH piece.

The piece that is made like the drawing is identified by use of the letter R added to its assembly mark, thus: H^R for right. The one that is made opposite-hand has the letter L added to its assembly mark, thus: H^L for left. No assembly mark should be marked R unless there is an exact opposite, or LH, detail piece needed on the sheet, because all detail pieces are assumed to be RH as shown on the drawing unless otherwise noted. Likewise, no assembly mark should be marked L unless there is also a corresponding right.

If a drawing is placed in front of a mirror, the required RH detail would appear as represented by the drawing, and the required LH detail would appear as reflected in the mirror.

An understanding of rights and lefts, if not innate, may be acquired from Fig. 25-4-4. Note that the two end views of H^L of the drawing, even in their rotated position, still picture a fitting that is opposite-hand to H^R.

Pieces that in their assembled position may appear to be rights and lefts often really are alike. Thus the fittings shown on the left side of Fig. 25-3-14 can be turned upside down and used on the right side of the beam web.

When rights and lefts of whole shipping pieces are encountered, it is the practice in some fabricating shops to note the RH piece AS SHOWN and the LH piece OPPOSITE HAND in the required list. If two shipping pieces are involved, one the exact opposite of the other, the required listing under the single sketch might read (Fig. 25-4-5):

ONE CHANNEL A150^R—AS SHOWN
ONE CHANNEL A150^L—OPP HAND

FIG. 25-4-5   Right and left shipping pieces.

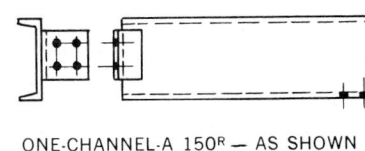
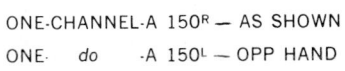
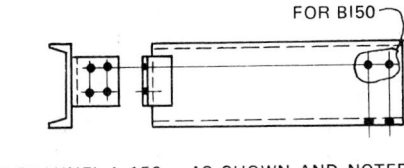

ONE-CHANNEL-A 150^R — AS SHOWN
ONE· *do* -A 150^L — OPP HAND

ONE-CHANNEL-A 150 — AS SHOWN AND NOTED
ONE· *do* -B 150 — OPP HAND AND NOTED

(A) IF THESE ARE THE DRAWINGS THE SHOP GETS TO WORK FROM

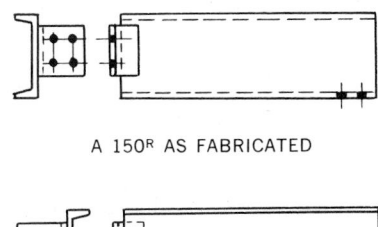

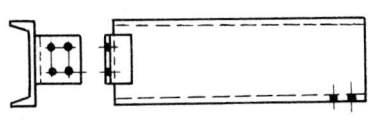

A 150^R AS FABRICATED

A 150 AS FABRICATED

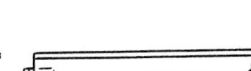
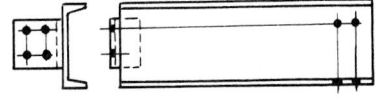

A 150^L AS FABRICATED

B 150 AS FABRICATED

(B) THIS IS WHAT THE SHOP FURNISHES (MAKES)

847

If two shipping pieces are involved, one practically but not exactly the opposite of the other, and they are detailed on the same sketch, the required listing under the single sketch would read:

ONE CHANNEL A150—AS SHOWN AND NOTED
ONE CHANNEL B150—OPP HAND AND NOTED

In the case of exact RH and LH shipping pieces, the shipping mark may be the same except for the R and L notation; in the case of combined but different RH and LH shipping pieces, the shipping marks are always different. In both cases only the RH or AS SHOWN shipping piece is detailed. It is the shop that does the reversing, according to the notation OPPOSITE HAND, in the required list.

Before an attempt is made to detail pieces involving combinations of rights and lefts, Fig. 25-4-5 should be studied. If differences between pieces are minor, it is common practice to combine the details of two or more different pieces in a single sketch by noting the differences, for example, in the case of the two web holes which are required in B150 but not required in A150.

REFERENCES AND SOURCE MATERIAL

1. American Institute of Steel Construction.

ASSIGNMENTS

See Assignments 6 and 7 for Unit 25-4 on page 856.

25-5 SEATED BEAM CONNECTIONS

Seated beam connections are used to connect beams to column webs or flanges. There are two types: unstiffened seat connections and stiffened seat connections. Only the unstiffened (angles) will be covered in this unit.

The following procedure is suggested for choosing a seated beam connection. Assume that a W12 × 27 beam is to be placed between two columns, as shown in Fig. 25-5-1. As in the example for the framed beam, Unit 25-3, the beam reactions must first be established. If they are not shown on the drawing, they must be calculated. To do this, the length of the beam must be known. This is found by subtracting half of the nominal depth of each of the two supporting columns from the center-to-center distance. The nominal depth of the columns can be found from Fig. 25-1-8. In this example, the length of the span is 16'-0 – 10 = 15'-2 in. From the beam load tables in Fig. 25-3-9 for a W12 × 27 beam having a span of 16 ft, the total allowable uniform load is 55 kips. The reaction at the end of the beam at each support is half the total load, or 27.5 kips. The length of the seated beam is now determined. From Fig. 25-1-8, the flange width *b* for the W12 × 27 beam is 6½ in. Note that Fig. 25-5-2 gives tables for a supporting angle length *L* of 6 and 8 in. Since the flange width of 6½ in. is greater than the *L* (length) of 6 in., the angle length of *L* = 8 in. for the seat angle will have to be used. The seat angle thickness must now be calculated. The web thickness of the W12 × 27 beam is ¼ in. Refer to Fig. 25-5-2, under the heading Outstanding Leg Capacity—kips, *L* = 8 in., and Beam Web Thickness = ¼ in. Read across until a leg capacity of 27½ kips or greater is found. This occurs at the kip value of 34 where the angle thickness is ⅝ in.

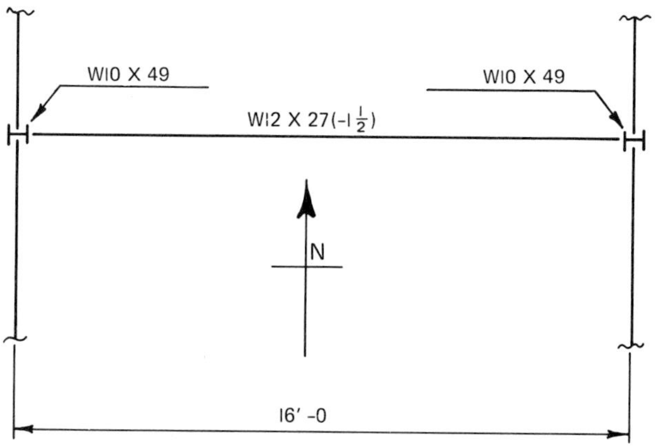

FIG. 25-5-1 Partial design drawing.

It is preferable for most fabricators to shop-weld the seat angle to the column, since the seat will provide support for the beam during erection. Under the heading Vertical Leg Weld Capacity, the angle thickness as previously determined was found to be ⅝ in.; therefore the maximum permissible weld size would be ⅝ – 1/16, or 9/16 in. (use ½ in.). From the table a ½ in. fillet weld will resist a force of 31 kips when a 4 × 4 angle size is used. To the right of the column, the angle thickness range of ⅜ to ⅝ in. is specified. The required angle thickness range as determined previously was ⅝ in.; therefore the dimensions for the seat angle are 4 × 4 × ⅝ × 8. The next step in the process is to make a sketch of the detail, as shown in Fig. 25-5-3.

The beam will be fastened to the seat angles using Ø⅝ in. A325 bolts. In order to determine their gage, reference should be made to Fig. 25-3-1. The recommended gage for a W12 × 27 that has a flange width of 6½ in. is 3¾ in. Also from Fig. 25-3-3, the minimum distance for a Ø⅝ in. bolt to a rolled edge is ⅞ in. Therefore the end distance (center of bolt to end of beam) is 4 – ⅞ – ½ = 2⅝. Use 2½ in.

A top, or cap, angle is used to provide lateral support at the top of the beam. Since it is not required to resist any movement at the end of the beam, this angle can be relatively small. For the top angle 4 × 4 × ¼ × 4 is recommended. In this example, no limitations have been specified by the design drawing as to the top angles, and therefore the angle can be placed as shown. However, if the top clearance had been critical, the angle could have been placed in the optional position on either side of the web, whichever provided the most convenient position for field erection purposes.

The top angle is welded to the column and bolted to the beam with two Ø⅝ bolts having a gage of 2⅛ in. as recommended in Fig. 25-3-1. The length of the beam required is equal to the center-to-center distance of the columns minus half of each of the column depths (or the column depth if both columns are the same) minus the ½ in. nominal clearance at each end. For this example, the length of the beam is 16'-0 – 10 – 2 (½) or 15'-1 in. The detail drawing of the W12 beam is shown in Fig. 25-5-4 (pg. 850).

REFERENCES AND SOURCE MATERIAL

1. American Institute of Steel Construction.

ASSIGNMENTS

See Assignments 8 through 11 for Unit 25-5 on pages 856 to 857.

SEATED BEAM CONNECTIONS*
METRIC BOLTS E480XX FILLET WELDS**
U.S. Customary (in.)

Outstanding Leg Capacity—Kips
(Based on 3½ or 4 in. Outstanding Leg)

Angle Length		L = 6 in.						L = 8 in.				
Angle Thickness	3/8	1/2	5/8	3/4	7/8	1	3/8	1/2	5/8	3/4	7/8	1
Beam Web Thickness 3/16	15	20	25	31	35	39	17	22	28	35	39	44
1/4	18	24	30	38	43	48	20	27	34	42	47	53
5/16	22	29	36	46	52	57	24	32	40	50	57	64
3/8	26	33	43	54	61	68	29	38	47	59	66	74
7/16	—	37	46	58	66	73	—	41	51	64	71	80
1/2	—	42	52	66	74	82	—	46	57	72	81	90
9/16	—	48	59	74	83	93	—	52	65	81	91	101

*To be used for educational purposes only.
**Welding resistances have been soft converted.

U.S. Customary (in.)
Vertical Leg Weld Capacity (Kips)

Fillet Weld Size D-E70XX Electrodes						†Angle Sizes	Angle Thickness
1/4	5/16	3/8	7/16	1/2	5/8		
22	26	28	29	31	32	4 x 4	3/8-5/8
34	40	43	45	48	50	5 x 3½	3/8-3/4
43	51	54	56	59	64	6 x 4	3/8-7/8
72	88	95	99	103	113	8 x 4	1/2-1

†Long vertical leg.

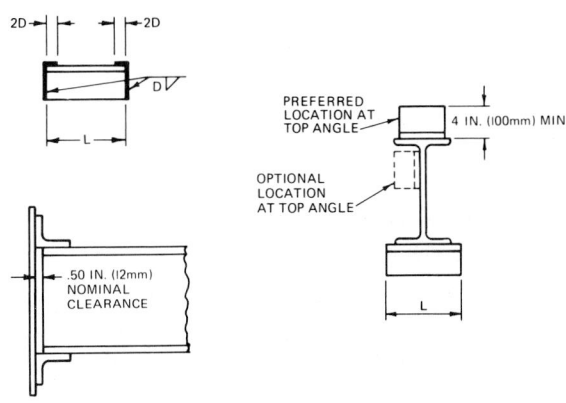

SEATED BEAM CONNECTIONS*
METRIC BOLTS E480XX FILLET WELDS**
Metric (mm)

Outstanding Leg Capacity (kN)
(Based on 90 or 100 mm Outstanding Leg)

Angle Length		L = 150 mm						L = 200 mm				
Angle Thickness	10	13	16	20	#25	##30	10	13	16	20	#25	##30
Beam Web Thickness 4	60.2	80.4	101	128	161	195	68.3	90.8	113	143	181	219
5	70.4	93.7	117	148	187	226	79.4	105	131	166	209	252
6	81.0	107	133	168	212	256	90.7	120	149	188	236	284
7	89.1	118	147	186	234	282	99.5	132	164	206	260	313
8	96.9	129	160	203	255	308	108	143	178	224	283	341
9	106	141	175	221	278	335	118	156	194	244	307	370
10	116	153	190	239	301	362	128	169	209	263	331	398
11		166	205	258	323	389		182	226	283	355	427
12		179	221	277	346	416		196	244	303	379	455

*To be used for educational purposes only.
**Welding resistances have been soft converted.

\#100 mm Outstanding Leg Only.
\##125 mm Outstanding Leg Only.

Metric (mm)
Vertical Leg Weld Capacity (kN)

Fillet Weld Size D—E480XX Electrodes						*Angle Sizes	Angle Thickness
6	8	10	12	14	16		
97	115	123	130	133	142	100 x 100	6:16
151	178	189	199	208	221	125 x 90	8:20
192	228	241	255	266	282	150 x 100	8:25
323	393	423	450	470	503	200 x 125	10:30

*Long vertical leg.

FIG. 25-5-2   Seated beam connections.

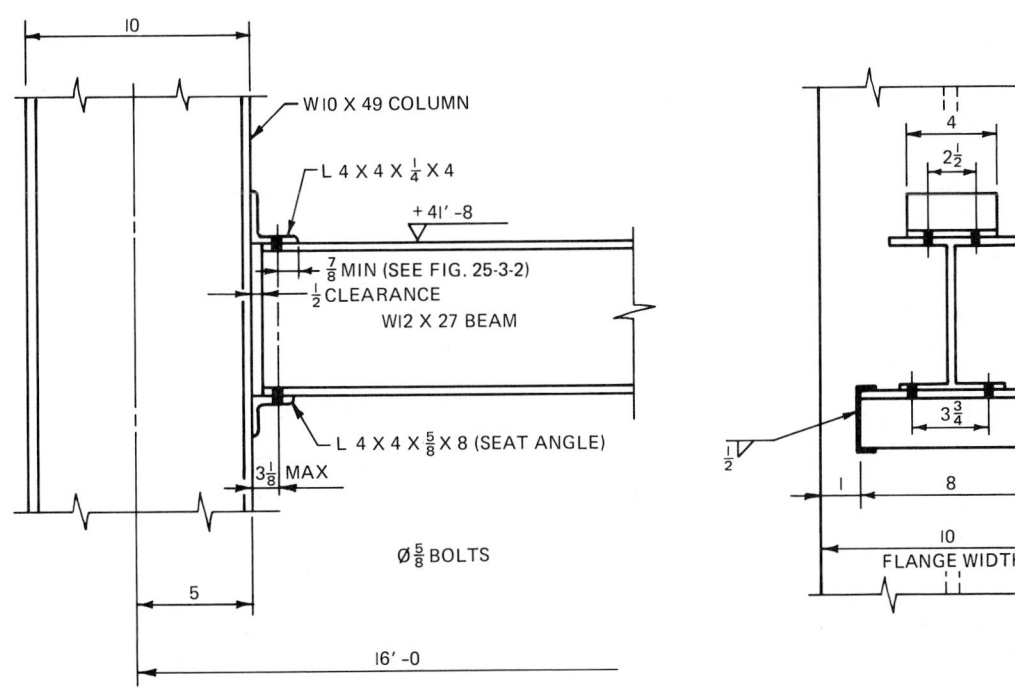

FIG. 25-5-3   Sketch of seated beam connection for partial design drawing, Fig. 25-5-1.

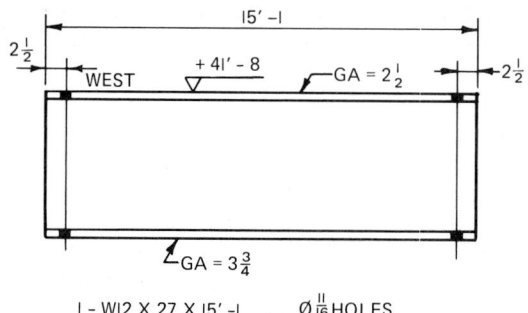

FIG. 25-5-4   Detail drawing of W12 beam shown in Fig. 25-5-1.

25-6 DIMENSIONING

Dimensioning techniques were discussed in Units 25-1 and 25-3. The following are additional items for consideration.

Note that longitudinal dimensions along the beams shown in Fig. 25-6-1 are given to the center line of the groups of open holes required for the field connections. This practice can serve two useful purposes. First, it simplifies the dimensioning work for the drafter and later for the checker, since the distances to the center lines of the beams are the dimensions given on the design drawing and the erection plan. Second, it is a convenience to the person laying out the beam details who first marks the locations of the center line for a group of holes on the beam and then centers a template at this point, by which he or she can center-punch the location of all the holes required in the group.

In the detail drawing (Fig. 25-6-1), the fabricator preferred to dimension to the center line of the channel webs rather than the backs of the channels. The direction in which the flanges of these channels are to be placed has been indicated on the drawing by the channel symbol, which has been drawn with the web parallel to the line showing the members. When they are installed, the flanges of these channels must point in the same direction as the flanges of the symbol.

In some shops, in addition to locating groups of holes as noted above, a method of using extension stub, or running dimensions, is employed. This consists of specifying the overall dimension from the left end of the beam to the center line of each group of holes, as shown in Fig. 25-6-2. Note that this practice was also followed in Fig. 25-6-1, to the first line of holes, but here it was done for reference only. In either case, it lessens the shop layout person's work by eliminating the need for calculations and therefore reduces the possibility of an error being made, especially in a shop that uses automated punching equipment.

If the fabricator had preferred to dimension to the backs of the channels, the dimensions would be as shown in Fig. 25-6-3. Dimensions and notes not shown are the same as in Fig. 25-6-1, except that no note is required to identify the dimension reference lines locating the groups of open holes in the beam web. The dimensions 2⅝ in. and 2⅞ in. to each side of the reference lines provide the clue that these open holes are to receive the connection for a channel. It would be understood that the back of the channel will be located toward the smaller of these two dimensions.

Another time-saving device commonly used in the drafting room when pieces are alike, except for some end or intermediate

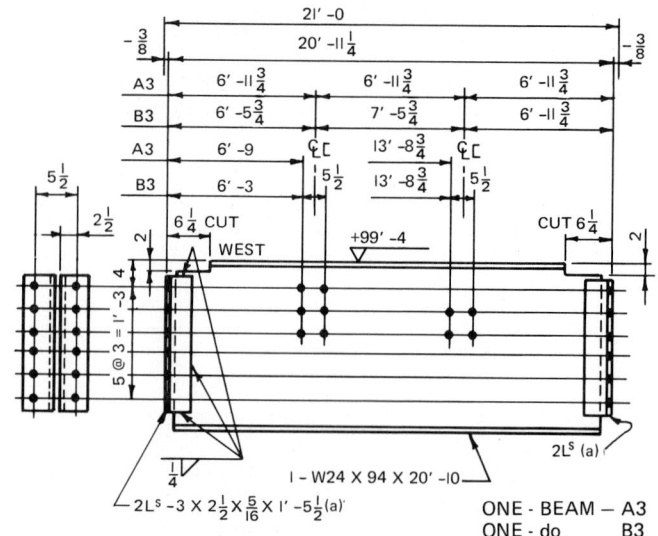

FIG. 25-6-1   Dimensioning to center line of channel webs.

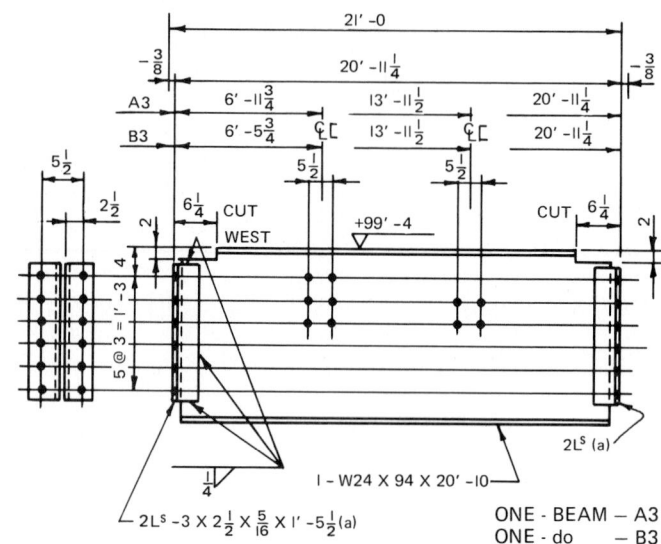

FIG. 25-6-2   Dimensioning from the left end of beam.

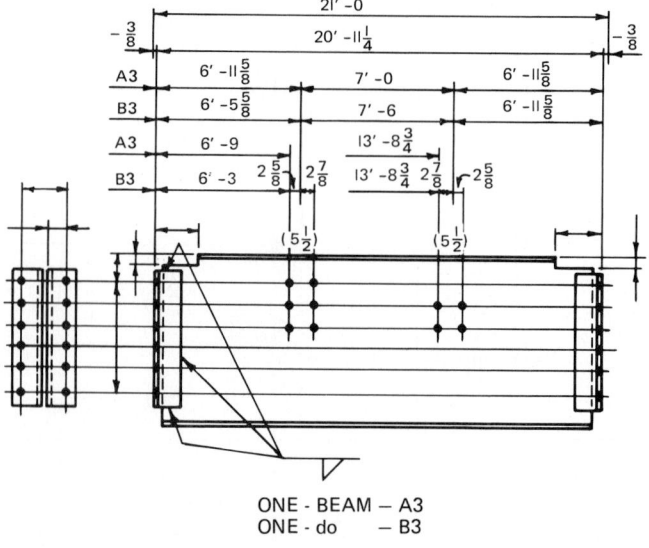

FIG. 25-6-3   Dimensioning to the backs of channels.

detail, is partial view detailing. Instead of completely redetailing each piece or trying to combine too many dissimilar details on one sketch by the use of notes indicating to which piece each detail applies, completely detail one piece first. Then, for the second piece, only the difference between it and the first piece needs to be detailed. This partial view is supplemented by a note stating that the parts not shown are THE SAME or possibly OPPOSITE HAND TO the first piece detailed. Figure 25-6-4 represents this practice.

Bills of Material

From the bills of material, or material bills, the workers in the yard where the structural shapes are stocked cut the material to length, cut the number of pieces shown on the bills, and send the material into the fabricating shop. The term *item list,* which means the same as bill of material, has not yet been adopted by the construction steel industry. From the bills of material the shipping department tallies the number of pieces to be shipped. Therefore it is extremely important that the drafter include in the bill of material all the material that is shown on the drawing.

The sample shop bill of material (Fig. 25-6-5) shows a typical form that is used in billing the material on shop drawings. The first items are the billing for beam A3 and beam B3 that are shown detailed in Fig. 25-6-1. Note that beam A3 is different from B3 only in the longitudinal spacing of holes in the web and that the material for the two beams is identical. When the material is the same, the billing of members on a shop drawing is grouped to avoid repetition.

Calculations of Weights (Masses)

After the billing operation, it may be necessary to figure the weight or mass of the material on the bill of material. The weight (mass) of the materials is very important in that the basis of payment for the fabricated steel may be a price per pound (kilogram). For that reason, the weight (mass) must be

FIG. 25-6-4   Partial view detailing.

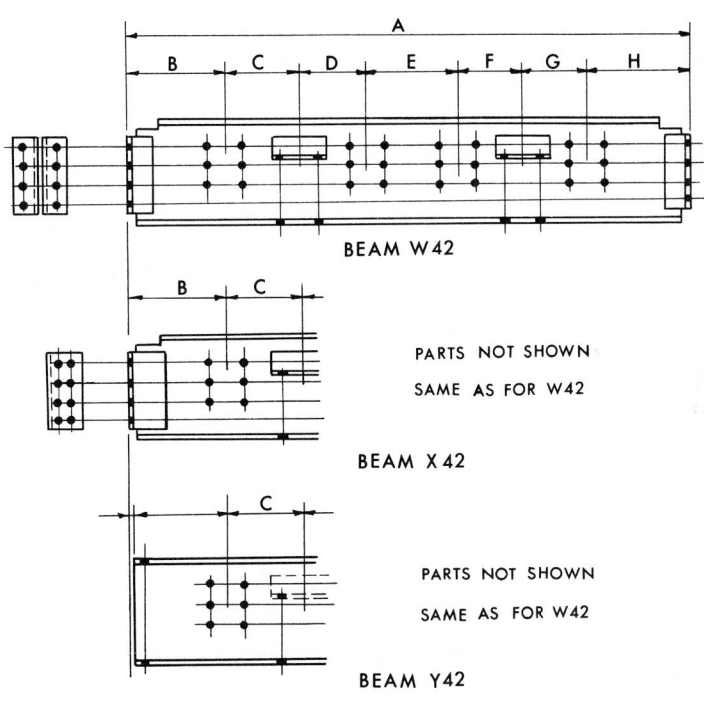

DATE TO P.A.   26

CONT. NO.

FG'D BY   DATE

BILL NO.   M 3a

CHK'D BY   DATE

CUSTOMER   INSP. AT MILLS   LUMP SUM POUND PRICE COST PLUS   MADE BY   DATE

DESCRIPTION   SPECN'S   KIND OF MAT'L   CHK'D BY   DATE

		BILL OF STRUCTURAL MATERIAL										SHIPPING LIST				
ITEM NO.	NO. PIECES	ASSEMBLY MARK	DESCR.	SIZE	LENGTH	P BILL REFERENCE	EST. WEIGHT				P. A.	NO. PCS.	DESCRIPTION	MARK		ACTUAL WEIGHT	SHIPPING RECORD
1	2		W	W24 X 36	20'-10		1	5	0	0		1	BEAM	A3			783
2	8	a	L	3 X 2½ X 5⁄16	1'-5½				6	6		1	BEAM	B3			783
3							1	5	6	6							
4																	
5																	

FIG. 25-6-5   Sample bill of materials.

accurate to the nearest pound (kilogram). Also, the shop uses the calculated weight (mass) of a member to avoid overloading the cranes or other transporting machinery. The shipping department uses the calculated weights (masses) for making up loads and as a basis of payment for shipping. The erection department is interested in weights (masses) of members to plan erection procedure and equipment. The weights (masses) of structural steel bolts and common structural steel shapes are found in the American and Canadian institutes of steel construction handbooks.

REFERENCES AND SOURCE MATERIAL

1. American Institute of Steel Construction.
2. Canadian Institute of Steel Construction.

ASSIGNMENTS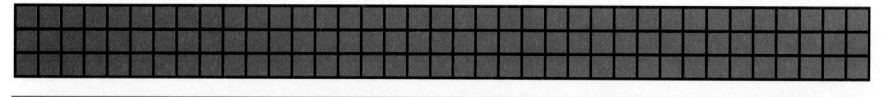

See Assignments 12 and 13 for Unit 25-6 on page 857.

ASSIGNMENTS FOR CHAPTER 25

ASSIGNMENT FOR UNIT 25-1, STRUCTURAL DRAFTING

1. Calculate the limits, tolerances, and sizes of the beams shown in either Fig. 25-1-A or Fig. 25-1-B. Refer to Fig. 25-1-8 and structural steel handbooks for sizes of structural steel shapes.

ASSIGNMENTS FOR UNIT 25-2, BEAMS

2. Make detail drawings of the two connections shown in either Fig. 25-2-A or Fig. 25-2-B. Refer to Fig. 25-1-8 and structural steel manuals. The connection angles are welded to the beam web, and the outstanding angles are bolted to the connecting beam. The bolts and holes need not be shown on these drawings. Scale 1:8 (U.S. customary) or 1:10 (metric).

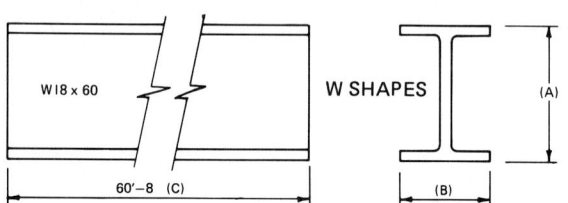

DIM	NOMINAL	TOLERANCE		LIMITS	
		+	−	UPPER	LOWER
A					
B					
C					

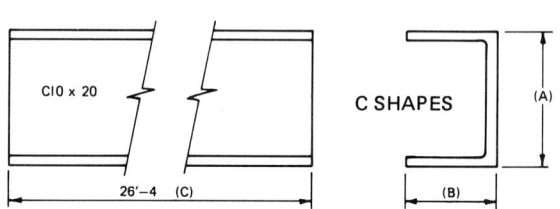

DIM	NOMINAL	TOLERANCE		LIMITS	
		+	−	UPPER	LOWER
A					
B					
C					

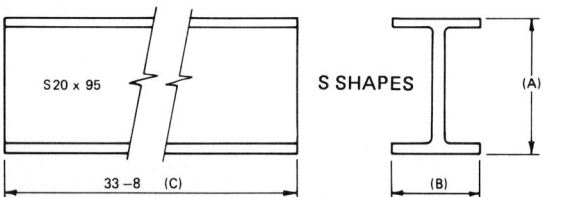

DIM	NOMINAL	TOLERANCE		LIMITS	
		+	−	UPPER	LOWER
A					
B					
C		+·	−		

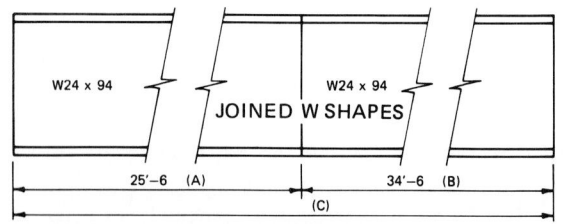

DIM	TOLERANCE		LIMITS	
	+	−	UPPER	LOWER
A				
B				
C				

FIG. 25-1-A   Beam sizes.

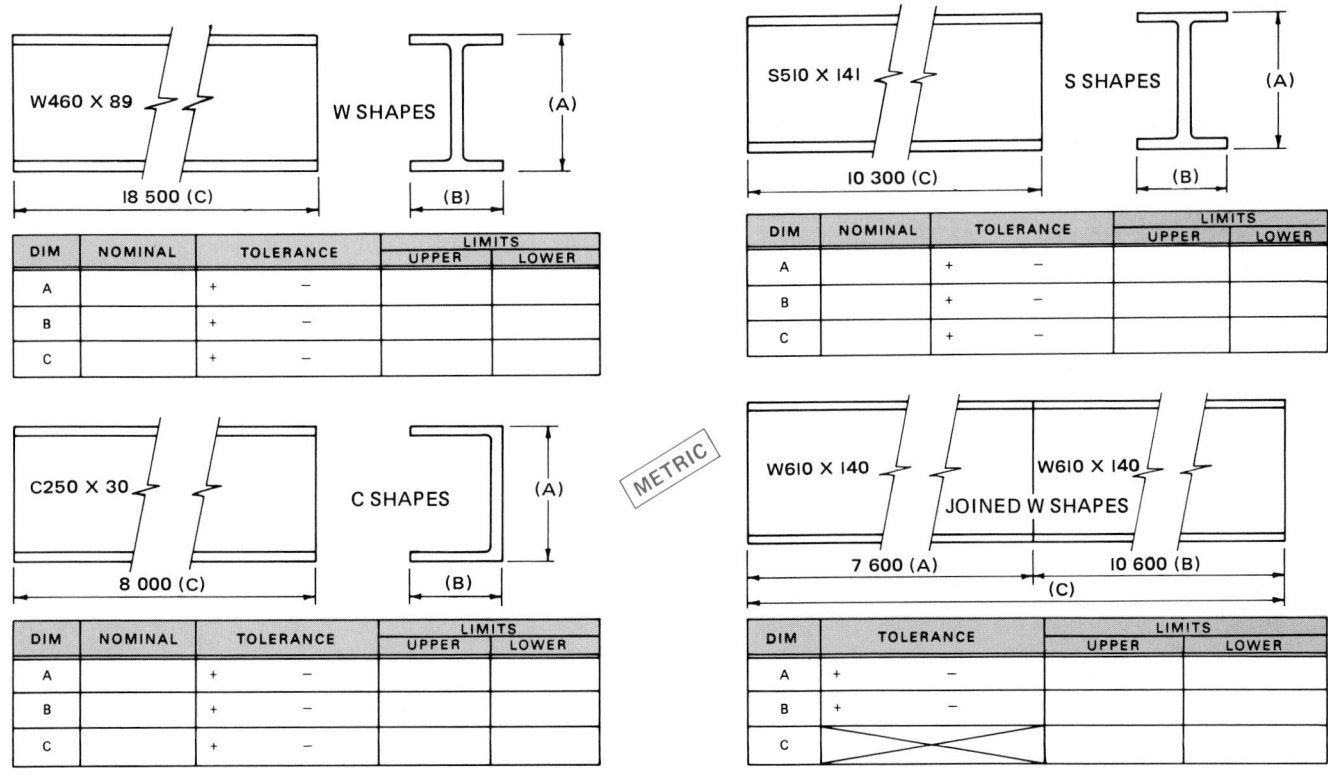

FIG. 25-1-B   Beam sizes.

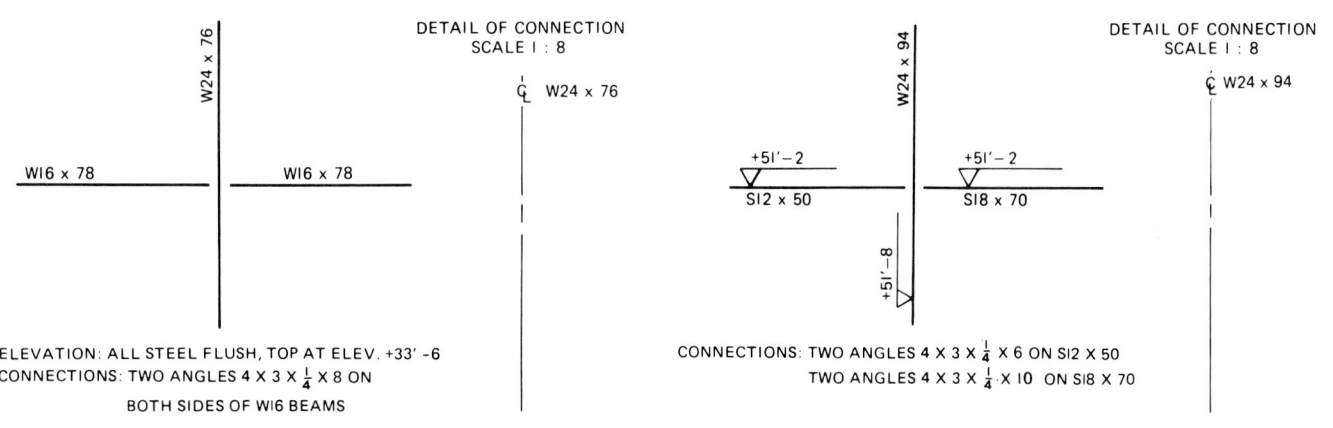

ELEVATION: ALL STEEL FLUSH, TOP AT ELEV. +33' -6
CONNECTIONS: TWO ANGLES 4 X 3 X $\frac{1}{4}$ X 8 ON
                BOTH SIDES OF W16 BEAMS

FIG. 25-2-A

CONNECTIONS: TWO ANGLES 4 X 3 X $\frac{1}{4}$ X 6 ON S12 X 50
                TWO ANGLES 4 X 3 X $\frac{1}{4}$ X 10 ON S18 X 70

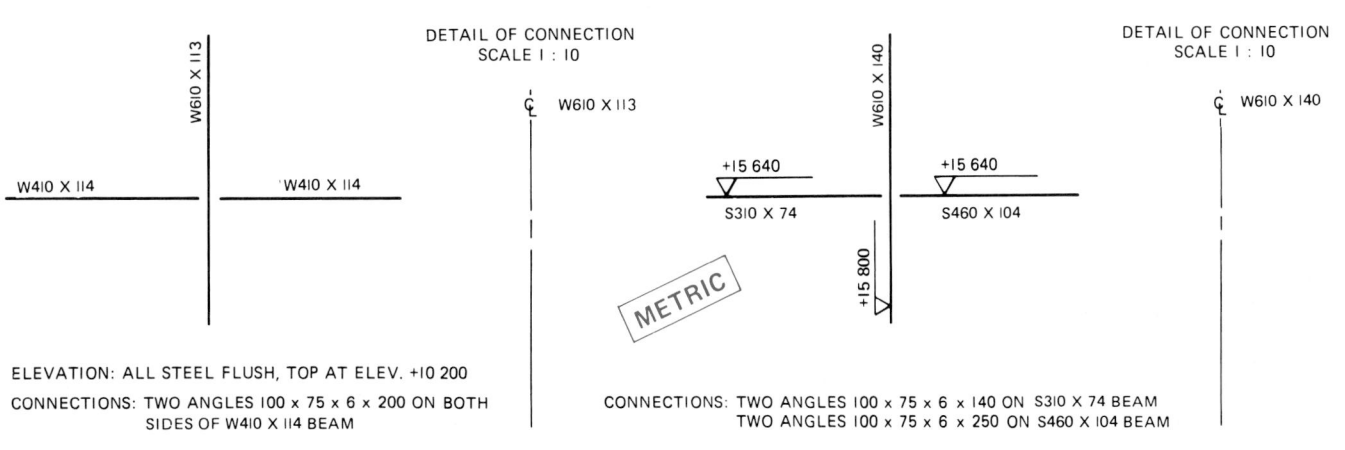

ELEVATION: ALL STEEL FLUSH, TOP AT ELEV. +10 200
CONNECTIONS: TWO ANGLES 100 x 75 x 6 x 200 ON BOTH
                SIDES OF W410 X 114 BEAM

CONNECTIONS: TWO ANGLES 100 x 75 x 6 x 140 ON S310 X 74 BEAM
                TWO ANGLES 100 x 75 x 6 x 250 ON S460 X 104 BEAM

FIG. 25-2-B   Beam and connection details.

3. Make sketches of the connections of both ends of the center beam shown in either Fig. 25-2-C or Fig. 25-2-D. After the beam connection sketches have been completed and approved by your instructor, prepare a working drawing of the beam from the sketches and information shown on the drawing. The bolt holes on the outstanding legs of the connection angles need not be shown. Use a conventional break to shorten the length of the beam.

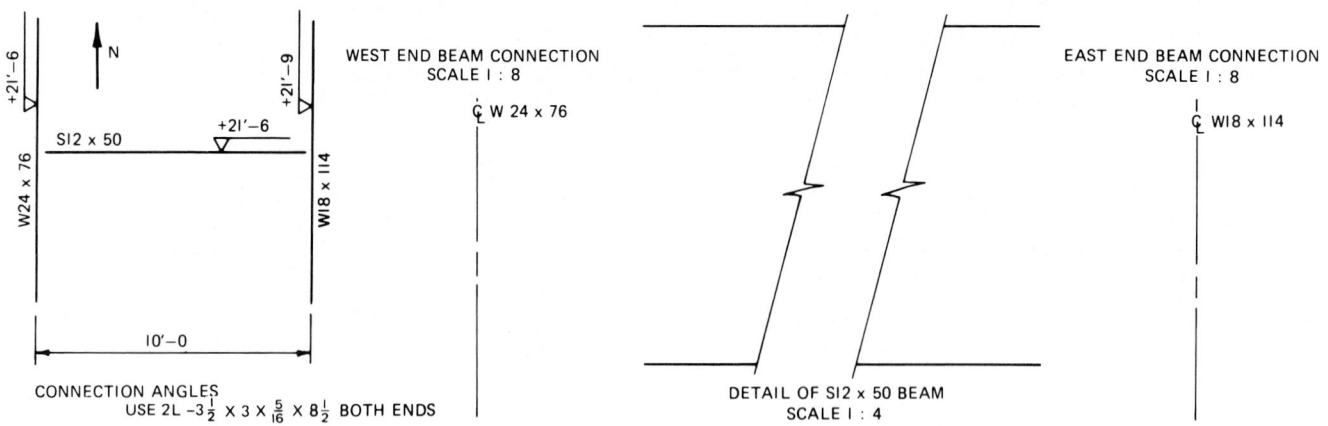

FIG. 25-2-C   Beam and connection details.

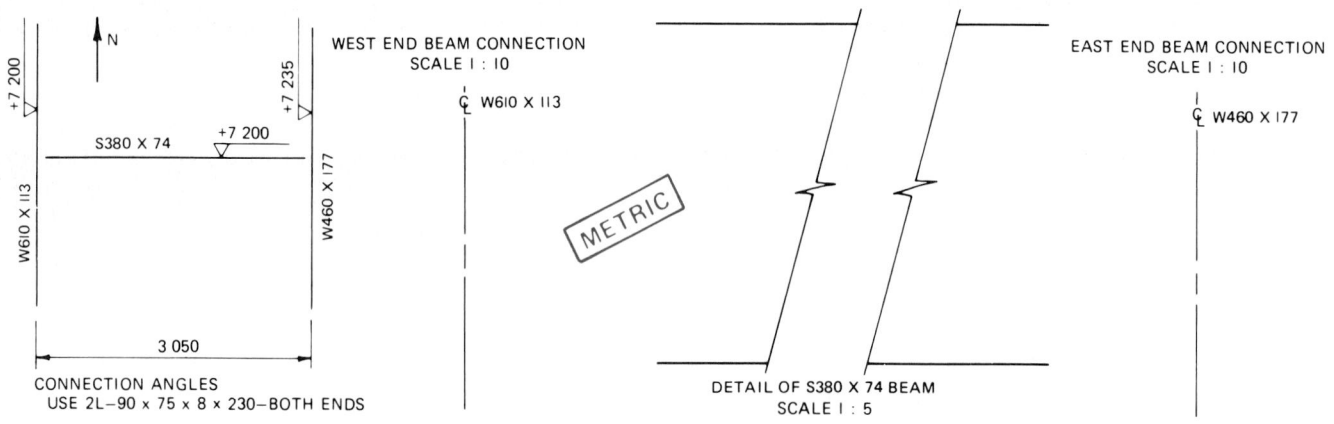

FIG. 25-2-D   Beam and connection details.

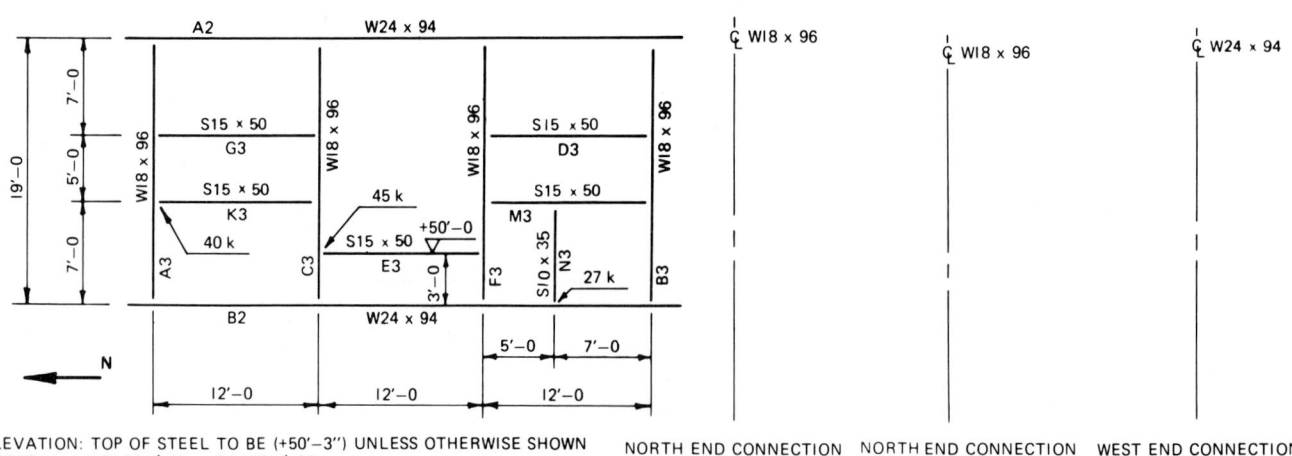

FIG. 25-3-A   Beam connections.

ASSIGNMENTS FOR UNIT 25-3, STANDARD CONNECTIONS

4. Sketch the following beam connections from the design sketch shown in Fig. 25-3-A or Fig. 25-3-B.
   (a) North end connection of beam E3.
   (b) South end connection of beam K3.
   (c) West end connection of beam N3.
   Scale 1:8 (U.S. customary) or 1:10 (metric).
5. Calculate the missing dimensions from the charts and drawings shown in Fig. 25-3-C.

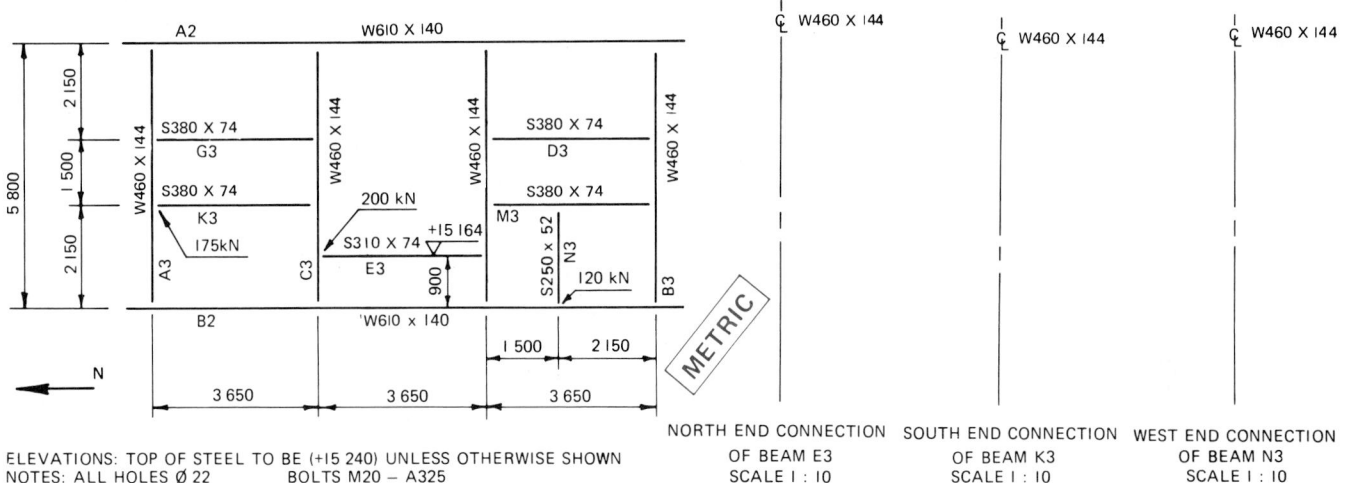

ELEVATIONS: TOP OF STEEL TO BE (+15 240) UNLESS OTHERWISE SHOWN
NOTES: ALL HOLES Ø 22   BOLTS M20 − A325

FIG. 25-3-B   Beam connections.

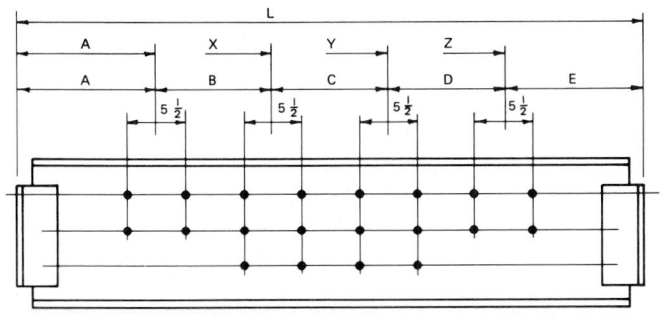

USING THE BEAM SHOWN ABOVE AND DIMENSIONS A−E CALCULATE DIMENSIONS L TO Z
AND COMPLETE THE CHART SHOWN BELOW.

	PROBLEM					
DIM	1	2	3	4	5	6
A	$5'-6\frac{3}{4}$	$3'-6$	$4'-1\frac{1}{2}$	$2'-10\frac{3}{4}$	$8'-0$	$4'-10$
B	$3'-6\frac{3}{8}$	$3'-5\frac{7}{8}$	$4'-7\frac{1}{4}$	$3'-10\frac{1}{2}$	$8'-9\frac{1}{2}$	$3'-11\frac{1}{2}$
C	$2'-7\frac{7}{8}$	$3'-6\frac{1}{2}$	$4'-1$	$4'-10$	$7'-9\frac{3}{4}$	$2'-11\frac{1}{2}$
D	$6'-10\frac{1}{2}$	$3'-6\frac{1}{4}$	$3'-9\frac{1}{4}$	$5'-9\frac{3}{4}$	$8'-7\frac{1}{2}$	$3'-11$
E	$3'-10\frac{1}{2}$	$3'-6\frac{1}{4}$	$4'-11\frac{3}{4}$	$6'-9\frac{1}{2}$	$7'-8$	$4'-11$
L						
X						
Y						
Z						

USING THE BEAM SHOWN ABOVE AND DIMENSIONS A−E CALCULATE DIMENSIONS L TO Z
AND COMPLETE THE CHART SHOWN BELOW.

	PROBLEM		
DIM	7	8	9
A	$3'-6\frac{1}{2}$	$4'-6$	$7'-7$
B	$6'-1\frac{1}{4}$	$9'-0\frac{1}{2}$	$13'-7\frac{1}{2}$
C	$8'-4\frac{3}{4}$	$14'-2\frac{1}{2}$	$19'-6\frac{3}{4}$
D	$11'-4$	$18'-3$	$27'-6\frac{1}{2}$
E	$13'-11$	$22'-6\frac{3}{4}$	$31'-9$
L			
V			
W			
X			
Y			
Z			

FIG. 25-3-C   Calculation of dimensions.

ASSIGNMENTS FOR UNIT 25-4, SECTIONING

6. Prepare one drawing for the two beams shown in Fig. 25-4-A (W16 × 40) or 25-4-B (W410 × 60). Eliminate the top and bottom views. Scale 1:8 (U.S. customary) or 1:10 (metric).
7. Make a complete working drawing of beam C6^R shown in Fig. 25-4-A or 25-4-B. Scale 1:8 (U.S. customary) or 1:10 (metric).

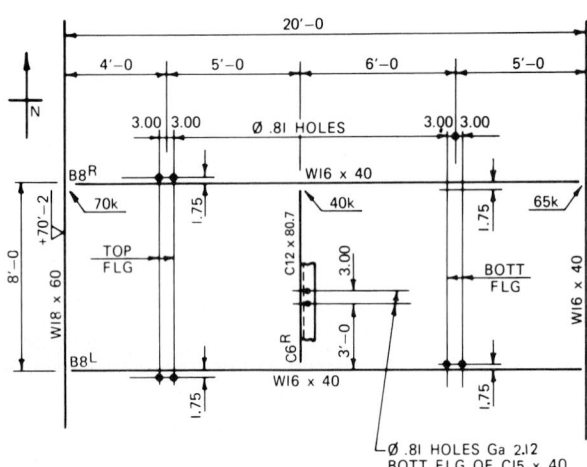

NOTES:
- TOP OF ALL MEMBERS AT ELEVATION +70'-3 EXCEPT WHERE NOTED
- SHOP CONNECTIONS WELDED
- FIELD CONNECTIONS .75–A325 FRICTION TYPE BOLTS
- USE DOUBLE-ANGLE BEAM CONNECTIONS

FIG. 25-4-A   One-view beam drawing.

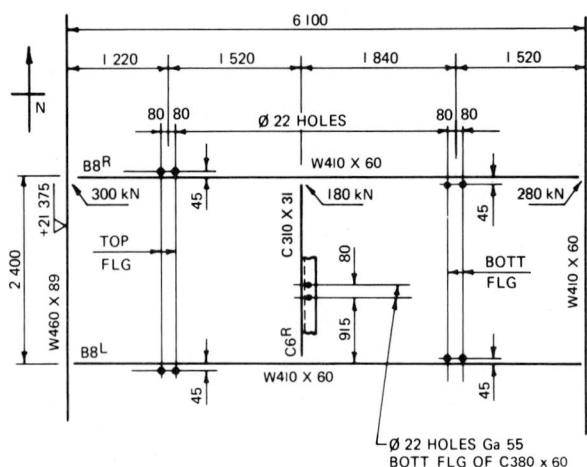

NOTES:
- TOP OF ALL MEMBERS AT ELEVATION +21 400 EXCEPT WHERE NOTED
- SHOP CONNECTIONS WELDED
- FIELD CONNECTIONS M20–A325 FRICTION TYPE BOLTS
- USE DOUBLE-ANGLE BEAM CONNECTIONS

METRIC

FIG. 25-4-B   One-view beam drawing.

ASSIGNMENTS FOR UNIT 25-5, SEATED BEAM CONNECTIONS

8. Make detailed sketches of the seated beam connections at both ends of beam A shown in Fig. 25-5-A. Scale 1:8.
9. Make a one-view detail drawing of beam A in Assignment 8. Scale is to suit.

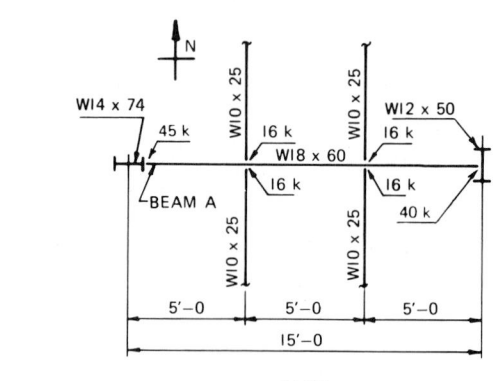

PLAN
TOP OF BEAM AT ELEV. +27'-3
Ø .75 A325 FRICTION BOLTS

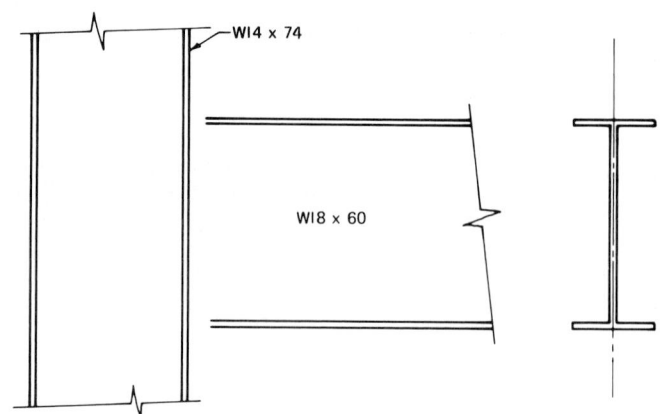

DETAIL OF WEST END OF W18 x 60 BEAM CONNECTION
SCALE 1 : 8

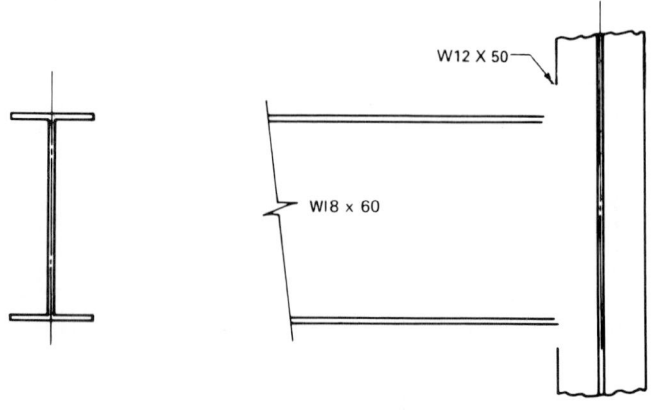

DETAIL OF EAST END OF W18 x 60 BEAM CONNECTION
SCALE 1 : 8

FIG. 25-5-A   Sketches of seated connections.

10. Make detailed sketches of the seated beam connection at both ends of beam A shown in Fig. 25-5-B. Scale 1:10.
11. Make a one-view detail drawing of beam A in Assignment 10. Scale is to suit.

ASSIGNMENTS FOR UNIT 25-6, DIMENSIONING

12. Prepare complete detail drawings of beams D3, E3, G3, K3, M3, N3, C3, and F3 shown in Fig. 25-6-A or 25-6-B. Dimension to the center line of channel webs. The connection angles are welded to the beams, and the outstanding angles are bolted to the connecting beams. Make sketches of the beam connections. Scale 1:8 (U.S. customary) or 1:10 (metric).
13. Prepare a bill of material and a shipping list for the beams in Assignment 12.

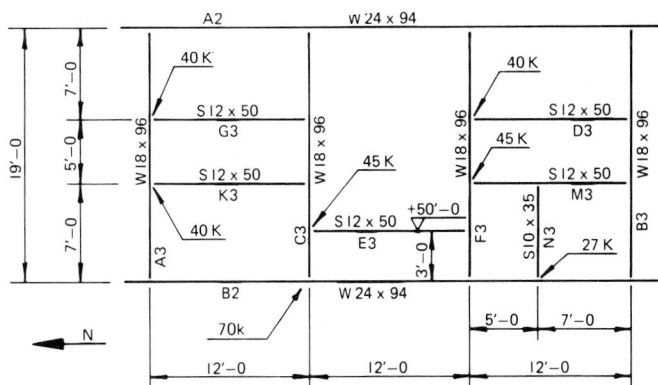

ELEVATION: TOP OF STEEL TO BE (+50'-3) UNLESS OTHERWISE SHOWN
NOTES: ALL HOLES Ø .81   BOLTS Ø .75–A325

FIG. 25-6-A   Detail drawings.

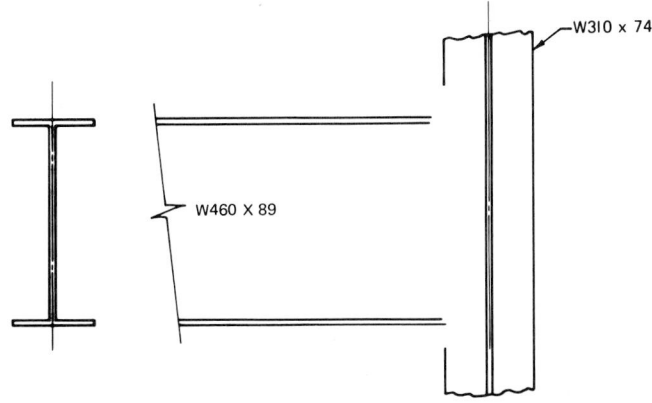

FIG. 25-5-B   Sketches of seated connections.

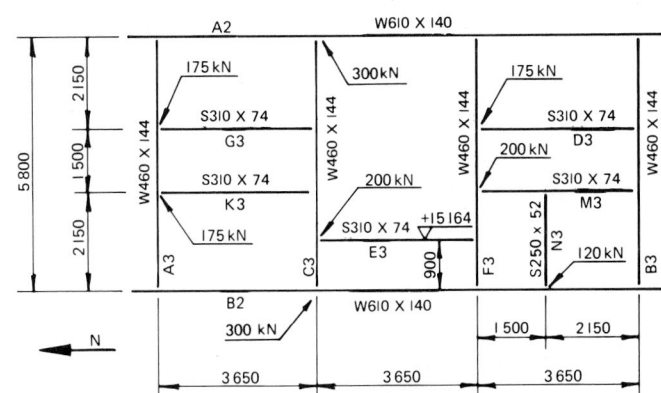

ELEVATIONS: TOP OF STEEL TO BE (+15 240) UNLESS OTHERWISE SHOWN
NOTES: ALL HOLES Ø 22   BOLTS M20 – A325

FIG. 25-6-B   Detail drawings.

CHAPTER 26

JIGS AND FIXTURES

Definitions

Drill bushings   Precision tools that guide cutting tools into precise locations in a workpiece.

Fixture   A device that supports, locates, and holds a workpiece securely in position while machining operations are being performed.

Jig   A device that holds the work and locates the path of the tool.

Jig body   The frame that holds the various parts of a jig assembly.

Liner   A device pressed into the jig plate or fixture both to provide precision mounting holes or correct slip fit for renewable bushings and to prevent wear of soft jig plates caused by frequent replacement of renewable bushings.

Locking pin   In jig design, a pin used to lock or hold the workpiece securely to the jig plate while the second or subsequent holes are being drilled.

Open jig or **plate jig** or **drill template**   A plate that has holes to guide the drill and locating pins that locate the workpiece on the jig.

Workpiece supports   Surfaces that support the workpiece in order to avoid distortion caused by either clamping or machining.

26-1 JIG AND FIXTURE DESIGN

With mass production and interchangeable assembly being used extensively in industry, it is imperative that components be machined and sized to identical standards. To do this, devices called jigs and fixtures are employed to hold and locate the work or to guide the tools while machining operations are being performed. Jigs and fixtures also cut down machining time, thus lowering production costs.

A *jig* is a device that holds the work and locates the path of the tool (Fig. 26-1-1). Usually, a jig may readily be moved

FIG. 26-1-1   Drill jig. *(Ex-Cell-O Corp. Ltd.)*

about or repositioned. An example would be a drill jig that may reposition the work several times when many holes are required in the workpiece, the drill being located each time by a drill bushing on the jig. Jigs are used extensively for drilling, reaming, tapping, and counterboring operations. A *fixture,* as the name implies, is fixed to the worktable of the machine and locates the work in an exact position relative to the cutting tool. The fixture does not guide the tool. Fixtures are often employed when milling, grinding, welding, and honing are required.

Jigs

There are two main types of jigs: those used for machining purposes and those used for assembly purposes.

858

When a jig is used in conjunction with a machine tool, its function is to locate the component, hold it firmly, and guide the cutting tool during its operation. The jig need not be secured to the machine. The term thus used normally refers to drilling, reaming, tapping, and boring operations (Fig. 26-1-2). The size of the jig in this case is limited by the proportions of the machine and the handling characteristic required of the jig. This type of jig is normally moved about frequently and is stored when not in use.

When a jig is used for assembly purposes, its function is to locate separate component parts and hold them rigidly in their correct positions relative to each other while they are being connected. These parts usually form large structural frameworks from which accurate locators are taken.

The Design of Jigs

The design of a jig is governed by five major factors:

1. The machining operation or operations involved
2. The number of parts to be produced
3. The degree of accuracy of the component
4. The stage of the component
5. Any other relevant factors, such as portability requirements and external locations

Machining Operation(s) Involved   As stated earlier, "jig" usually refers to a drilling, reaming, tapping, or boring device. More often than not, a jig may perform a combination of these functions—such as drill and ream, drill and tap, drill, ream and counterbore, etc. Drilling, reaming, and tapping jigs, and their combinations, are usually similar in construction since all these operations are performed on the one machine, the drill press.

Number of Parts to Be Produced   The number of parts to be produced has an important effect on design. For example, in very large-quantity production, the cost of an expensive clamping device may be recovered many times over, as a result of time saved through its use. Of course, in small-quantity production, the cost of the device might not be recovered; thus a cheaper device should be used.

Degree of Accuracy of the Component   It is logical, of course, that if a component is required to be very accurate, the tool producing it will have to be even more accurate.

Stage of the Component   The designer must know the stage of manufacturing of the component so that he or she can use any available machined faces for location purposes.

Any Other Relevant Factors   Sometimes it is necessary to bolt a jig to the table of the machine. For example, when a large hole is being drilled or reamed, the designer must know what facilities for clamping may be available on the machine.

Before designing a jig, the designer must have or be able to find all information such as that given above. The information is given to the designer in the form of a working drawing of the component, a process sheet showing the sequence of operations on the component, and general information usually available in the department.

Having progressed this far, there are several principles that must also be considered before the designer can finally decide on the design of the jig. The designer must consider:

1. The machine on which the operation is to be performed
2. Loading and unloading of the component: (a) clearances necessary for locating the part; (b) methods of guarding against improper loading
3. Rapid methods of clamping the work
4. Chip clearance and chip removal
5. Allowance for observation of operation where possible
6. Safety in operation

There are many other considerations, of course, but these are the major ones.

If the following questions are asked and answered before the jig design is started, much time and money (time *is* money) will be saved and trouble will be avoided.

Jigs in General

1. Can a component be inserted and withdrawn without difficulty?
2. Should the component be located to ensure symmetry or balance, i.e., optical balance or material balance?
3. Have the best points of location been chosen with regard to the accuracy of location and the function of the component?
4. Are hardened location points provided where necessary?
5. Can the locating points be adjusted where required to make allowance for wear of forging dies or patterns?
6. Are locations clear of flash or burrs?

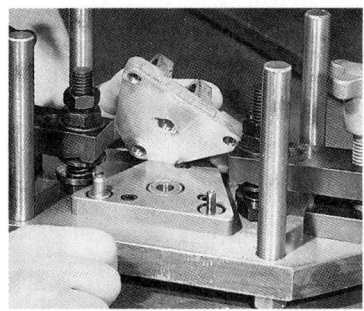

(A) LOADING THE WORKPIECE

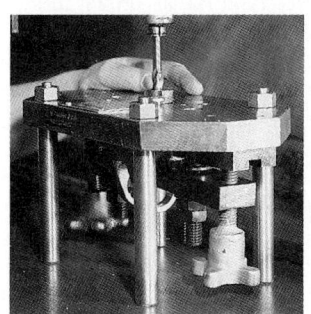

(B) TAPPING THE WORKPIECE

FIG. 26-1-2   Drilling and tapping jig. *(Northwestern Tools)*

7. Will the locating devices permit a commercial variation in the machining of the component without affecting the accuracy of location or causing it to bind in the jig?

8. Can the jig be easily cleared of metal shavings and grit, particularly on the locating faces?

9. Are all the clamps strong enough?

10. Will any clamp-operating lever or nut be in a dangerous position, i.e., near the cutter?

11. Are all clamps and clamping screws in the most accessible and natural positions?

12. Can wrenches be eliminated by the use of ball or eccentric levers?

13. Will one wrench fit all clamp-operating bolts and nuts?

14. Is the component well supported against the actions or pressure of the cutter?

15. Is the jig foolproof? That is, can the component, tools, or bushings, etc. be wrongly inserted or used?

16. Does the operator have an unobstructed view of the component, particularly at the points of location, clamping, and cut?

17. Is the jig as light as possible, consistent with the desired strength?

18. Can coolant, if used, reach the point of cut?

19. Have loose parts been eliminated wherever possible?

20. Have standard parts been used where circumstances allow?

21. Can the jig be designed to hold RH and LH or other similar or complementary components which may be required?

22. Where will burrs be formed, and is clearance for them arranged?

23. Are all corners and sharp edges that are likely to cut the operator shown well radiused on the drawing?

24. Are locating and other working faces and holes protected as far as possible from dirt and cuttings?

25. Will the jig as designed produce components within the required degree of accuracy?

Drilling and Boring Jigs

26. Do drills, tools, etc. enter the component at the face that directly adjoins the face of the component to which it fits?

27. Have all slip bushings necessary for reaming, spotfacing, tapping, counterboring, seating, etc. been arranged?

Milling Jigs

28. Have clamps, etc. been designed to permit the use of the smallest possible diameter of cutter(s)?

29. Will the cutter mandrel clear all parts of the jig when it passes over?

30. Have means for setting the cutter(s) in the correct position been provided?

Drill Jigs

Drill jigs are of two general types: open jigs and closed or box-type jigs.

Open Jigs

The simplest tool used to locate holes for drilling is the *open jig,* often referred to as the *plate jig* or *drill template.* It consists of a plate with holes to guide the drills, and it has locating pins that locate the workpiece on the jig; or the workpiece may be nested on the jig and then both are turned over for the drilling operation. Jigs of this type are usually without clamping devices. They are used where the cost of more elaborate tools would not be justified.

A separate base is often used with the template or top plate, thus forming the sandwich type of jig. The base may have holes or grooves to provide clearance for the end of the drill as it breaks through the work.

In the drill jig shown in Fig. 26-1-3, the component is not clamped into or onto the jig. The jig rests upon the component. Since the center-to-center distance between the holes is probably more critical and held to closer tolerances than the distance between the holes and the edge of the part, a locking pin is used to ensure the center-to-center hole accuracy. After the first hole is drilled, the locking pin is inserted into the drill jig and workpiece.

Drill Bushings

Drill bushings are precision tools that guide cutting tools such as drills and reamers into precise locations in a workpiece. Assembled in a jig or fixture, drill bushings are capable of producing duplicate parts to extremely close tolerances in regard to location and hole size.

A variety of bushings have been developed for a wide range of portable or machine drilling, reaming, and tapping operations. They include headless and head press-fit bushings, slip and fixed renewable bushings, headless and head liners, thin-wall bushings, and a number of embedment bushings for plastic or castable tooling, soft materials, and special applications.

Each type of bushing has its preferred use. Only proper selection can give the service that the manufacturer has built into the product. To select the proper bushing, it is necessary to consider not only the function of the jig, but the quantity of production. Life of the average bushing is no more than 5,000 to 10,000 pieces.

Press-Fit Bushings

Press-fit bushings are available in two basic styles: headless (type P) and headed (type H), as shown in Fig. 26-1-4. These bushings are permanently pressed into the jig plate or fixture. Press-fit bushings are recommended for use in limited production runs where replacement due to wear is not anticipated during the life of the tooling, and where a single operation, such as drilling only or reaming only, is performed. Headless press-fit bushings offer two advantages: They can be installed flush with the jig plate without counterboring the mounting hole, and they can be mounted closer together than headed bushings. However, where space permits, the use of headed press-fit bushings is preferable in any application where heavy axial loads may eventually force a headless bushing out of the jig plate. Typical sizes of standard press-fit drill jig bushings are shown in Table 78 in the Appendix.

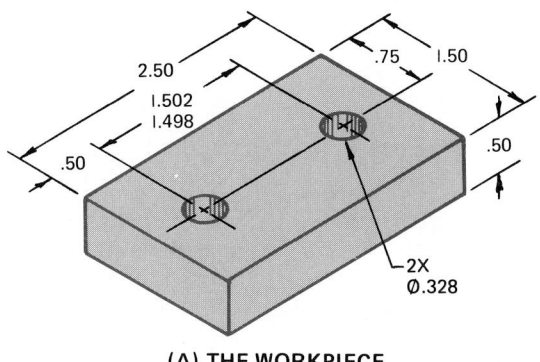

(A) THE WORKPIECE

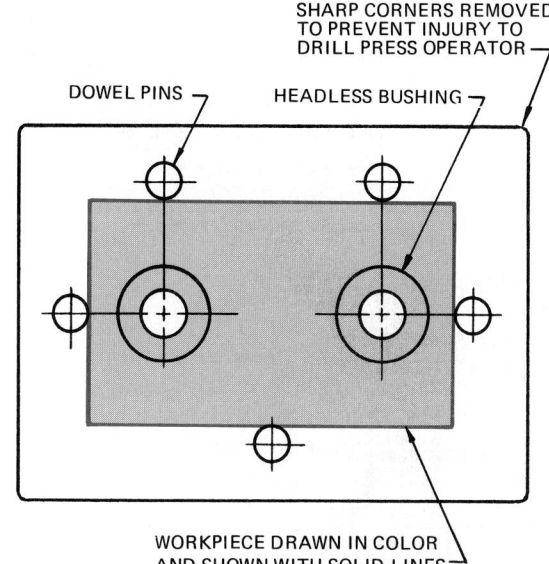

SHARP CORNERS REMOVED
TO PREVENT INJURY TO
DRILL PRESS OPERATOR

DOWEL PINS   HEADLESS BUSHING

WORKPIECE DRAWN IN COLOR
AND SHOWN WITH SOLID LINES

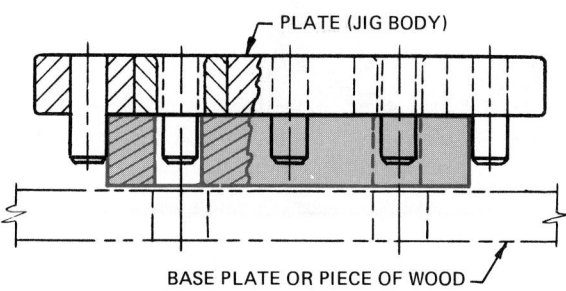

PLATE (JIG BODY)

BASE PLATE OR PIECE OF WOOD

**(B) PLACE JIG OVER WORKPIECE AND
DRILLING FIRST HOLE**

FIG. 26-1-3   Simple plate jig.

Installation

Chip Control   Sufficient clearance should be provided between the bushing and the workpiece to permit the removal of chips (Fig. 26-1-5, pg. 862). The exception to this rule occurs in drilling operations requiring maximum precision where the bushings should be in direct contact with the workpiece. However, suitable chip clearance should be provided in

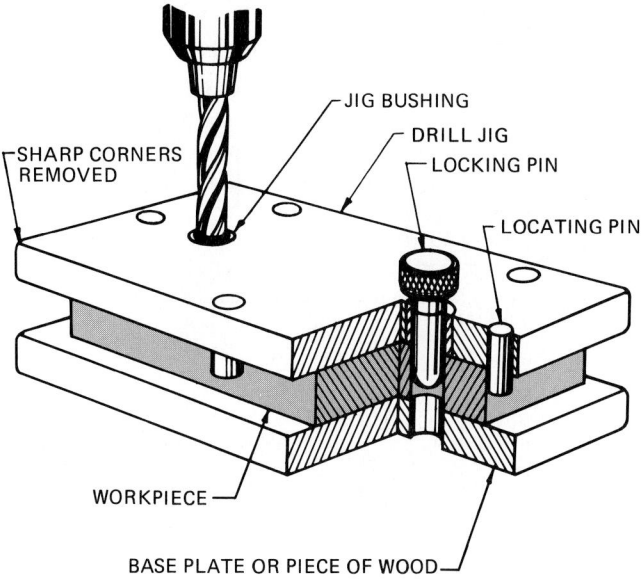

JIG BUSHING

DRILL JIG

LOCKING PIN

LOCATING PIN

SHARP CORNERS
REMOVED

WORKPIECE

BASE PLATE OR PIECE OF WOOD

**(C) ADD LOCKING PIN BEFORE STARTING
SECOND HOLE**

FIG. 26-1-3   Simple plate jig. (continued)

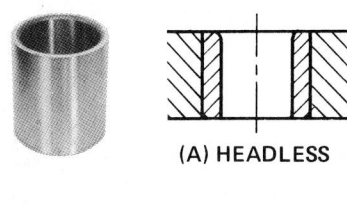

(A) HEADLESS

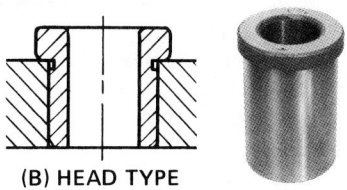

(B) HEAD TYPE

FIG. 26-1-4   Press-fit drill bushings. *(American Drill Bushing Co.)*

most applications because the abrasive action of metal particles will accelerate bushing wear.

Burr Clearance   Burr clearance should be provided between the bushing and the workpiece when wiry metals such as copper are drilled (Fig. 26-1-5B). Metals of this type tend to produce secondary burrs around the top of the drilled holes; the burrs act to lift the jig from the workpiece and to cause difficulty in the removal of workpieces from side-loaded jigs. The recommended burr clearance is one-half the bushing ID.

REFERENCES AND SOURCE MATERIAL

1. American Drill Bushing Co.

ASSIGNMENT

See Assignment 1 for Unit 26-1 on page 874.

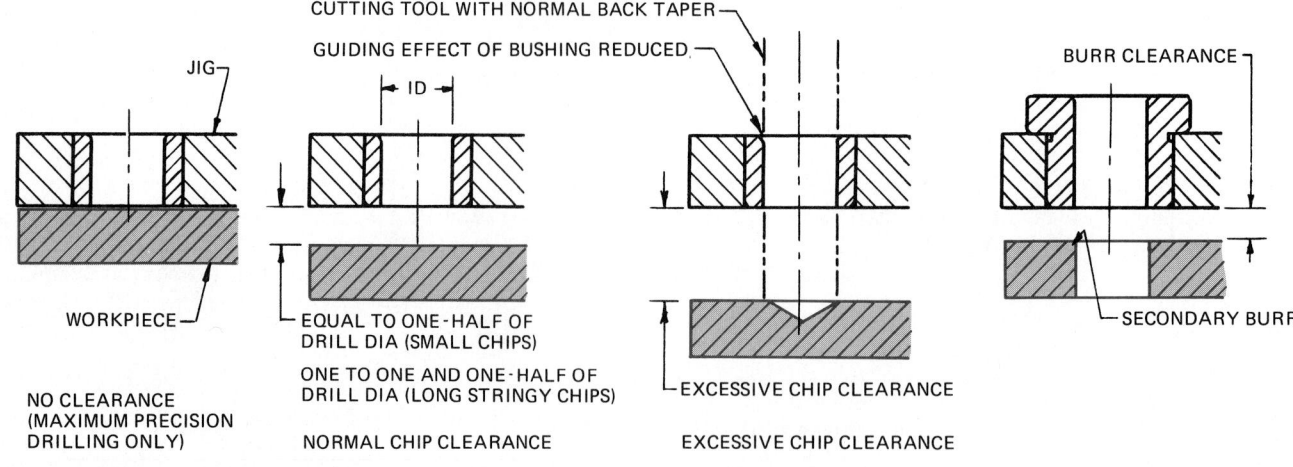

CUTTING TOOL WITH NORMAL BACK TAPER

GUIDING EFFECT OF BUSHING REDUCED

BURR CLEARANCE

JIG

ID

WORKPIECE

NO CLEARANCE
(MAXIMUM PRECISION
DRILLING ONLY)

EQUAL TO ONE-HALF OF
DRILL DIA (SMALL CHIPS)

ONE TO ONE AND ONE-HALF OF
DRILL DIA (LONG STRINGY CHIPS)

NORMAL CHIP CLEARANCE

EXCESSIVE CHIP CLEARANCE

EXCESSIVE CHIP CLEARANCE

SECONDARY BURR

(A) RECOMMENDED CLEARANCE BETWEEN WORKPIECE AND BUSHING

(B) BURR CLEARANCE

FIG. 26-1-5 Chip and burr clearance.

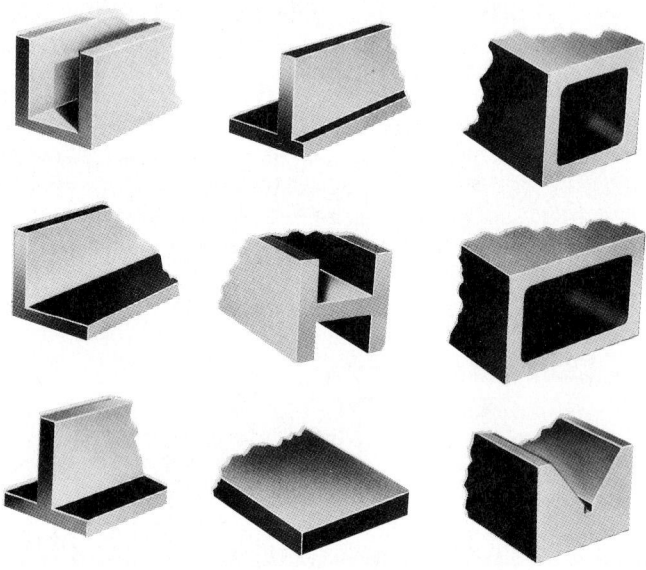

FIG. 26-2-1 Machined sections for construction of jigs and fixtures. *(Standard Parts Co.)*

FIG. 26-2-2 Typical jig feet.

26-2 DRILL JIG COMPONENTS

Jig Body

The frame that holds the various parts of a jig assembly is called the *jig body*. It may be in one piece or bolted or welded together. Rigid construction is necessary because of the accuracy required, yet the jig should be light enough to provide ease in handling. Sharp edges or burrs that might harm the operator should be removed. Supporting legs—a minimum of four being recommended—should be provided on the opposite side of each drilling surface. Standard shapes have been designed for jig bodies and are normally more economical than fabricating units in the shop (Figs. 26-2-1 and 26-2-2).

Cap Screws and Dowel Pins

The purpose of cap screws in jig design is to hold together fabricated parts. Dowel pins provide the necessary alignment between the parts, a minimum of two being recommended

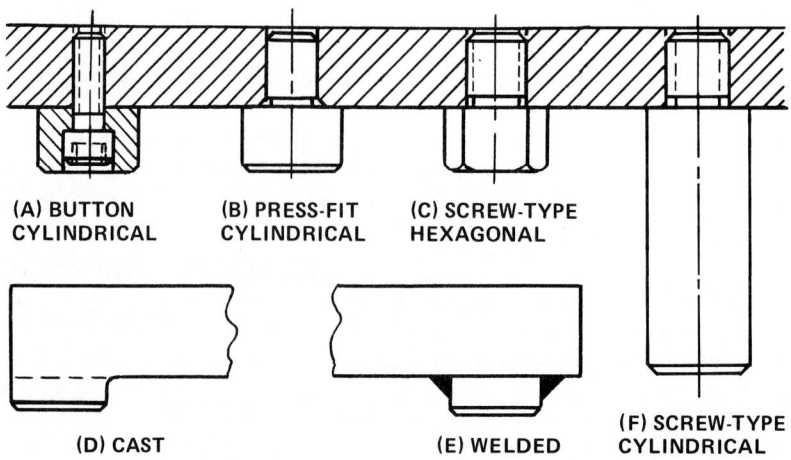

(A) BUTTON
CYLINDRICAL

(B) PRESS-FIT
CYLINDRICAL

(C) SCREW-TYPE
HEXAGONAL

(D) CAST

(E) WELDED

(F) SCREW-TYPE
CYLINDRICAL

(Fig. 26-2-3). Wherever possible, cap screws should be recessed and have socket fillister heads. This type of cap screw can be tightened, with a greater amount of pressure providing better holding power. When thin stock is to be fastened together and counterboring is not possible, a hexagon-head cap screw is used.

Dowel pins may be tapered or straight, the latter being used more frequently. A press fit into the two parts ensures the proper alignment required in jig design.

Locating Devices

The shape of the object determines the type of location best suited for the part. Pins, pads, and recesses are the more common methods used to locate the workpiece on the jig.

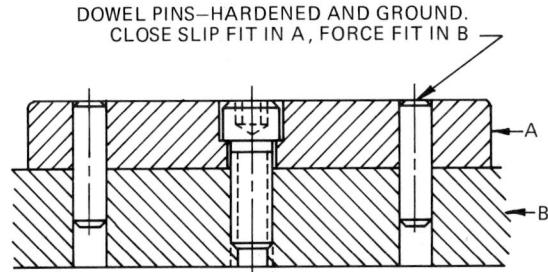

DOWEL PINS—HARDENED AND GROUND. CLOSE SLIP FIT IN A, FORCE FIT IN B

DOWEL PINS — USED TO ALIGN PARTS. MINIMUM OF 2 SOCKET SCREWS USED TO HOLD PARTS TOGETHER.

FIG. 26-2-3   Dowel pins and cap screws.

Internal Locating Devices   A machined recess in the jig plate (Fig. 26-2-4A) and a nesting ring (Fig. 26-2-4B) attached to the plate are two methods used to locate a part having a circular projection. The latter method is preferred because the part can be machined more readily and can be replaced when worn. Dowel pins—normally two—position the ring while fastening screws (the number being determined by the size of the ring) secure it to the plate. For small cylindrical extensions, a headless bushing mounted flush with the locating surface may be used, provided that the shoulder of the workpiece rests on the locating surface, as shown in Fig. 26-2-4C. An example of a drilling jig that has an internal locating device is shown in Fig. 26-2-5 (pg. 864).

External Locating Devices   Locating studs (Fig. 26-2-4D) provide an excellent means of locating workpieces with circular holes. When it is desirable to clamp the workpiece to the stud, the stud should be lengthened and fastened in place by a nut and washer (Fig. 26-2-4E). This secures the stud to the jig body and also provides for the interchanging of studs when necessary. Disk-type locators (Fig. 26-2-4F) are used when the locating diameter is over 2 in. (50mm). Dowels and fastening screws, the number determined by the size of the disk, locate and secure the disk to the plate.

Stops   When the workpiece cannot be located by recesses or projections as outlined above, locating stops are used. They are classified as either fixed or adjustable (Fig. 26-2-6, pg. 864).

The most common types of fixed stops are the stop pin, flattened shoulder plug, crowned shoulder plug, and stop pads. While stop pins (dowels) are the most economical, their main disadvantages are rapid wear and marring of the finished surface

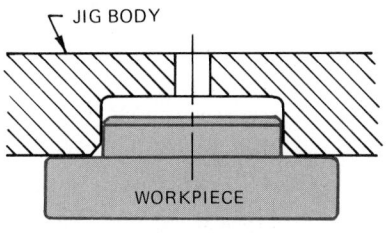

(A) MACHINED RECESS IN JIG BODY

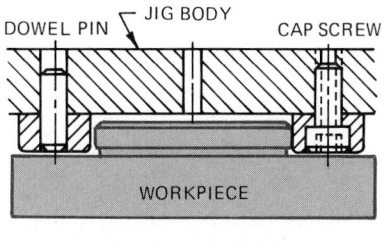

(B) NESTING RING

INTERNAL LOCATING DEVICES

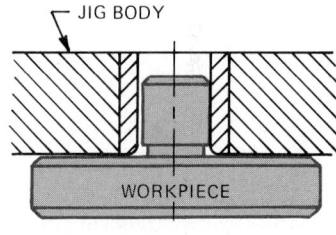

(C) HEADLESS BUSHING

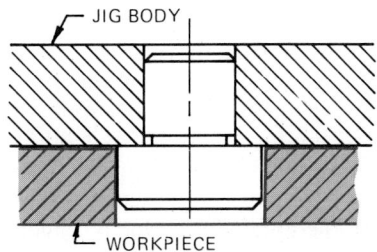

(D) STRAIGHT STUD—PRESS-FIT SHANK

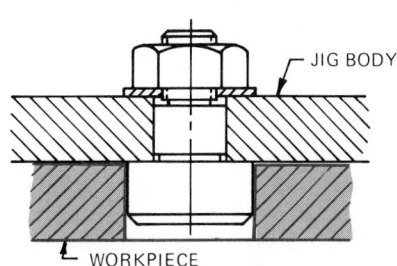

(E) STRAIGHT STUD—THREADED SHANK

EXTERNAL LOCATING DEVICES

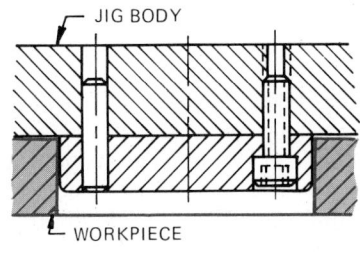

(F) DISK LOCATOR

FIG. 26-2-4   Common locating devices.

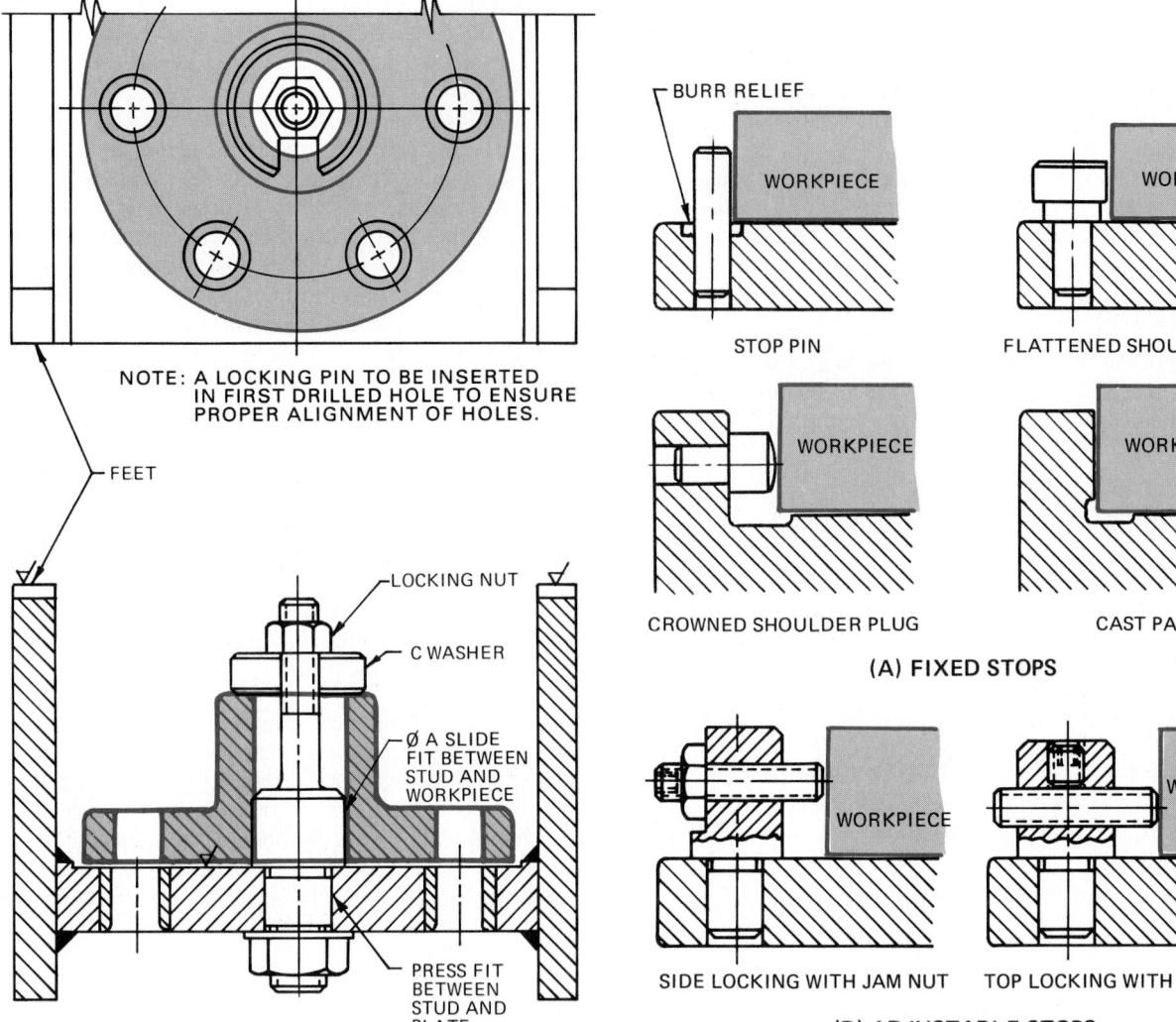

FIG. 26-2-5   Plate drill jig for drilling holes in flange.

FIG. 26-2-6   Fixed and adjustable stops.

of the workpiece. Shoulder plugs, with one side of the head flattened, provide a greater bearing surface and will not wear as readily.

Centralizers   Circular workpieces or flat workpieces with rounded or angled ends may be located or centered by centralizers, as shown in Figs. 26-2-7 and 26-2-8.

Workpiece Supports   The workpiece must be supported to avoid distortion caused by either clamping or machining (Fig. 26-2-9). The surfaces supporting the workpiece are called *workpiece supports* and are classified as either fixed or adjustable. They should be located, as nearly as possible, directly opposite the clamping force. It is recommended that four small work support areas be used in lieu of one large area, because the latter may produce a rocking condition.

The jig body with metal cut away and steel blocks, called *rest buttons,* are the more common types of fixed supports used.

Clamping Devices

The clamping components must be designed to securely hold the workpiece but not distort it, to be quickly and easily locked and unlocked, and to swing out of the way during loading and unloading. Some of the more common types of clamps are shown in Fig. 26-2-10 (pg. 866).

Screw clamps are commonly used because they do not tend to loosen under vibration and they provide adequate clamping force. One of the simplest types of screw clamps is the cone-point setscrew. The incline on the screw tends to push the workpiece against the locating pads as well as against the stops. It is best suited for clamping unfinished surfaces such as castings because the point of the setscrew will mar the workpiece surface. The toggle-head type of clamp provides a larger contact surface with the workpiece, thereby reducing the possibility of marring. It is also ideally suited for clamping workpieces having side drafts. Where only moderate clamping pressures are required, a knurled knob, lever nuts, or a thumbscrew may be used.

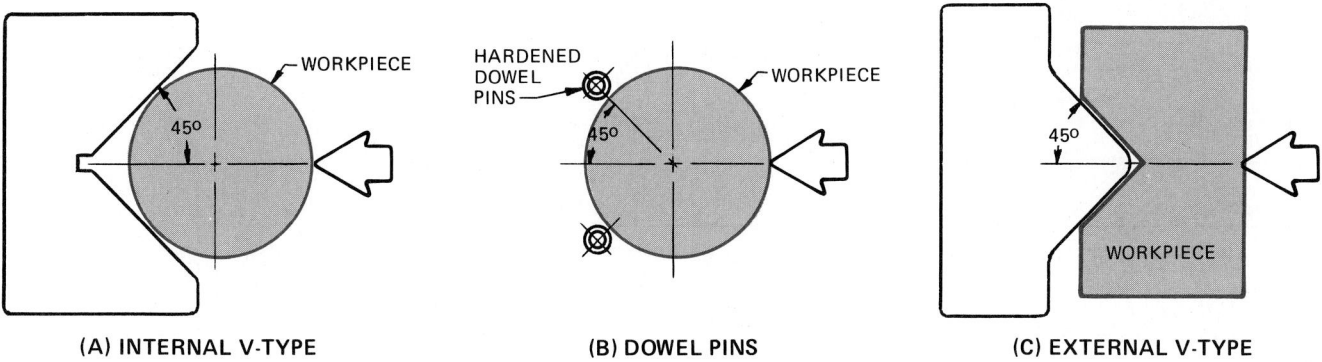

(A) INTERNAL V-TYPE **(B) DOWEL PINS** **(C) EXTERNAL V-TYPE**

FIG. 26-2-7 Centralizers.

FIG. 26-2-8 V-bushing drill jig. *(Acme Industrial Co.)*

FIG. 26-2-9 Workpiece supports.

The two-direction clamp provides both side and top clamping. As pressure is exerted on the end of the screw thread, the clamp is pivoted about the pivot pin, producing a downward pressure at the top of the workpiece.

Since screw-thread-type clamps are relatively slow, they are often used in combination with other devices to speed up the clamping and unclamping operation. The travel-cam lock assembly and the hinged cam assembly clamps are two such devices.

Locking Pins

A *locking pin* is used in jig design to lock or hold the workpiece securely to the jig plate while the second or subsequent holes are being drilled. After the first hole is drilled, the locking pin is inserted through the drill bushing into the drilled hole in the workpiece, locking the drill jig and workpiece together. When more than two holes are drilled, a second locking pin is used to maintain proper alignment. The use of a locking pin is illustrated in Fig. 26-1-3C.

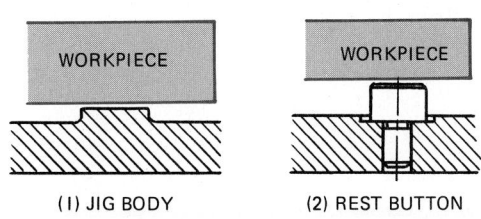

(I) JIG BODY (2) REST BUTTON

(A) FIXED WORKPIECE SUPPORTS

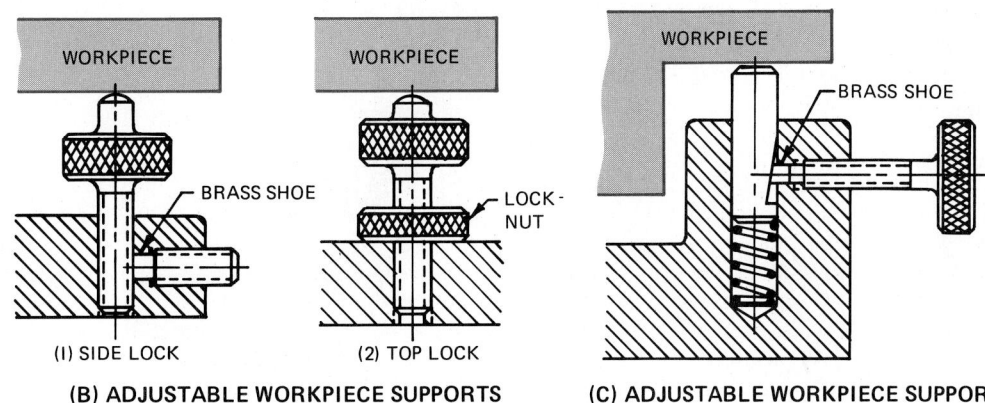

(I) SIDE LOCK (2) TOP LOCK

(B) ADJUSTABLE WORKPIECE SUPPORTS (JACK SCREWS)

(C) ADJUSTABLE WORKPIECE SUPPORT (JACK PIN)

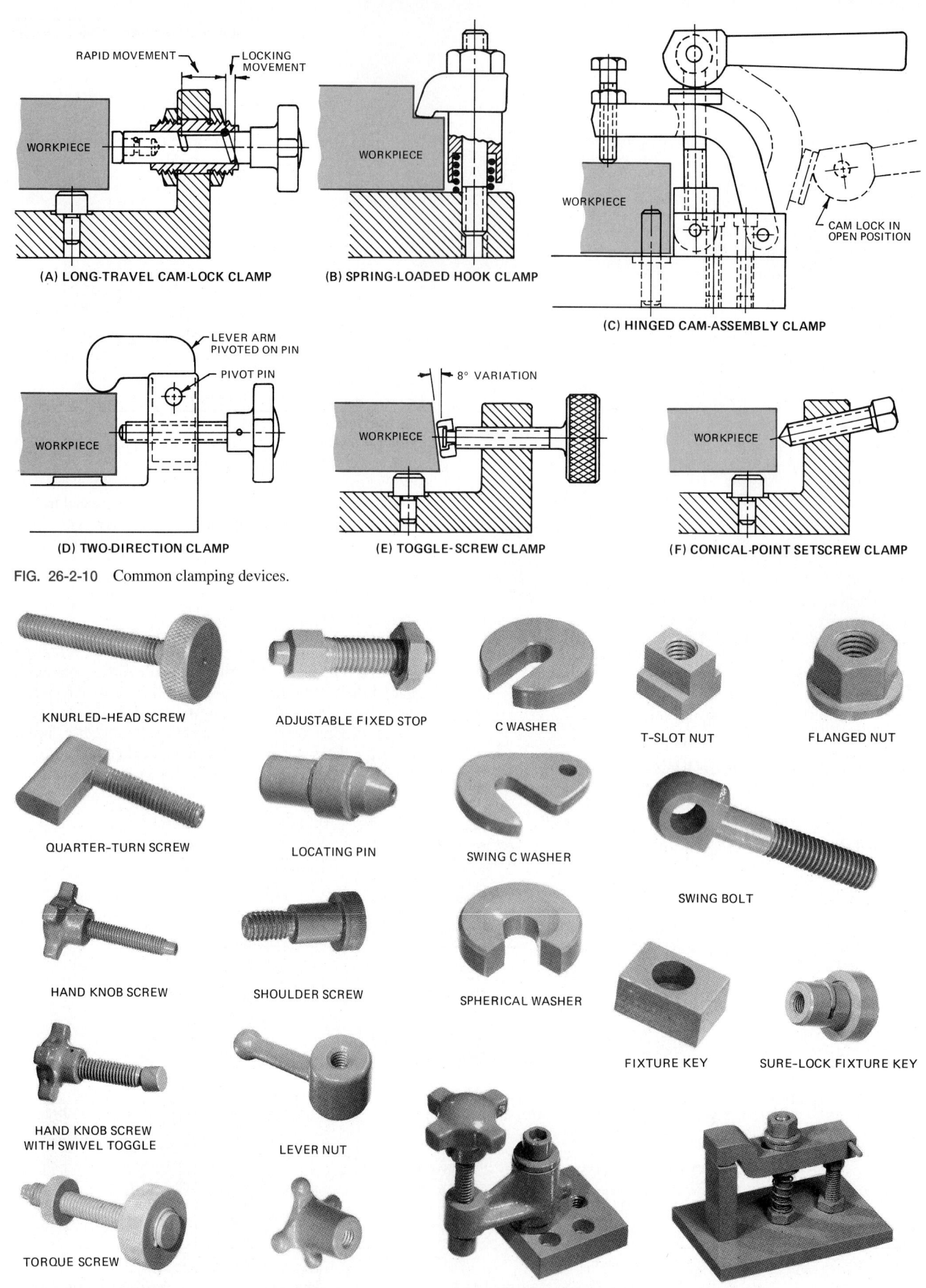

FIG. 26-2-10 Common clamping devices.

(A) LONG-TRAVEL CAM-LOCK CLAMP

RAPID MOVEMENT
LOCKING MOVEMENT
WORKPIECE

(B) SPRING-LOADED HOOK CLAMP

WORKPIECE

(C) HINGED CAM-ASSEMBLY CLAMP

WORKPIECE
CAM LOCK IN OPEN POSITION

(D) TWO-DIRECTION CLAMP

LEVER ARM PIVOTED ON PIN
PIVOT PIN
WORKPIECE

(E) TOGGLE-SCREW CLAMP

8° VARIATION
WORKPIECE

(F) CONICAL-POINT SETSCREW CLAMP

WORKPIECE

KNURLED-HEAD SCREW

ADJUSTABLE FIXED STOP

C WASHER

T-SLOT NUT

FLANGED NUT

QUARTER-TURN SCREW

LOCATING PIN

SWING C WASHER

SWING BOLT

HAND KNOB SCREW

SHOULDER SCREW

SPHERICAL WASHER

FIXTURE KEY

SURE-LOCK FIXTURE KEY

HAND KNOB SCREW WITH SWIVEL TOGGLE

LEVER NUT

TORQUE SCREW

BALL-HANDLE KNOB

SWING CLAMP

ADJUSTABLE GOOSE NECK CLAMP

FIG. 26-2-11 Standard jig and fixture parts. *(Standard Parts Co.)*

866

Miscellaneous Standard Parts

Figure 26-2-11 shows some of the more common standard jig components. The designer should, wherever possible, use standard parts in the design in order to simplify the work and reduce the manufacturing cost.

Design Examples

EXAMPLE 1

An alternative drill jig for the workpiece shown in Fig. 26-1-3 is shown in Fig. 26-2-12. This jig employs a lever arm and a knurled-head screw that apply pressure on two sides of the workpiece, forcing it against the locating pins. This jig not only locates but also holds the workpiece in position.

EXAMPLE 2

For drilling a series of bolt holes in a flange, a drill jig, as shown in Fig. 26-2-5, may be used. The base surface of the flange and the diameter A of the workpiece, which were previously machined, are used as locating surfaces. The workpiece slips over the stud and rests on the jig plate. The C washer is then inserted over the workpiece, and the locking nut is screwed down to securely clamp the parts together. The size of the locknut is selected to clear diameter A. The body of the

jig is designed to protect the threads on the stud from being damaged. The position shown is the loading and unloading position. For drilling, the jig must be inverted and the side walls, which act as feet, must be machined to level and true-up the jig. Notice that part of the sides has been machined away, leaving only the four small surfaces to act as jig feet.

EXAMPLE 3

The screw-latch clamp jig, similar to the one shown in Fig. 26-2-13, is frequently used because of its simple design and fast clamping action. All the parts shown, with the exception of the clamp plate, are standard items that can be purchased.

ASSIGNMENTS

See Assignments 2 through 4 for Unit 26-2 on pages 874 to 875.

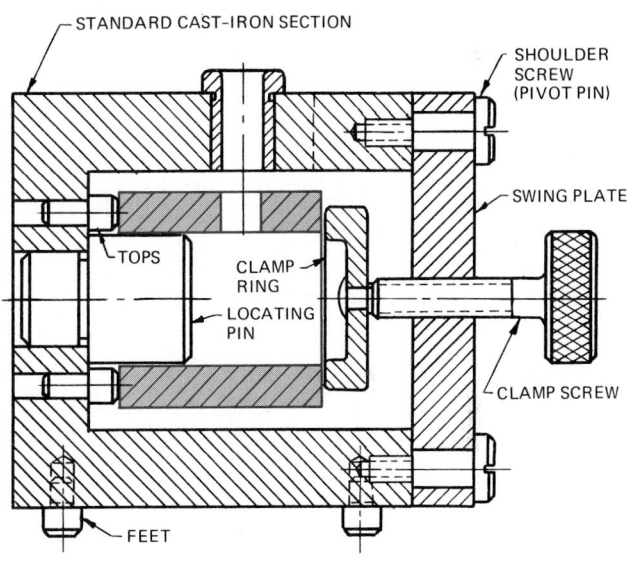

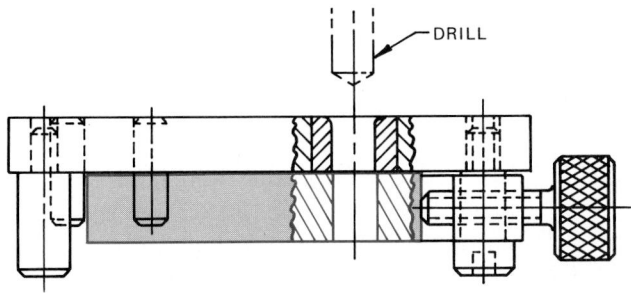

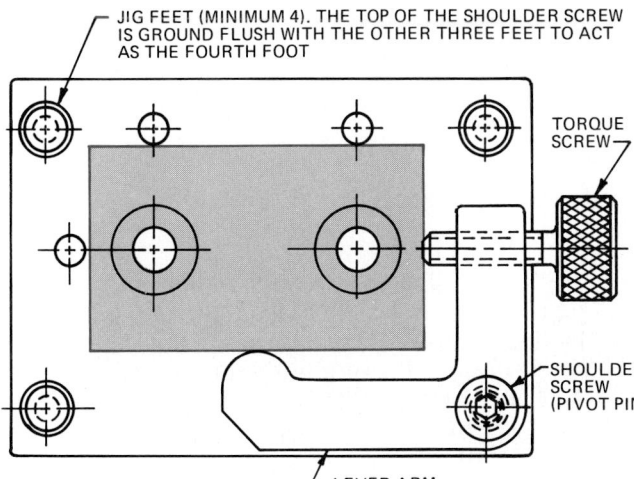

FIG. 26-2-12   Alternate plate jig for workpiece shown in Fig. 26-1-3.

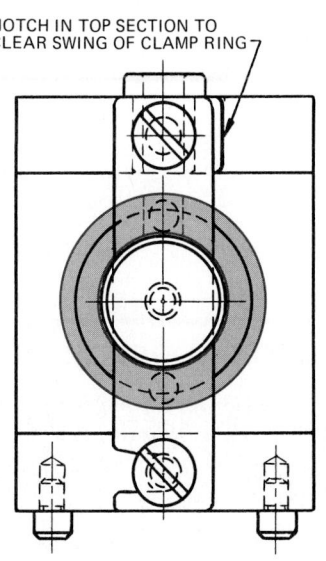

FIG. 26-2-13   Screw-latch clamp jig.

FIG. 26-3-1   Dimensioning jig drawings.

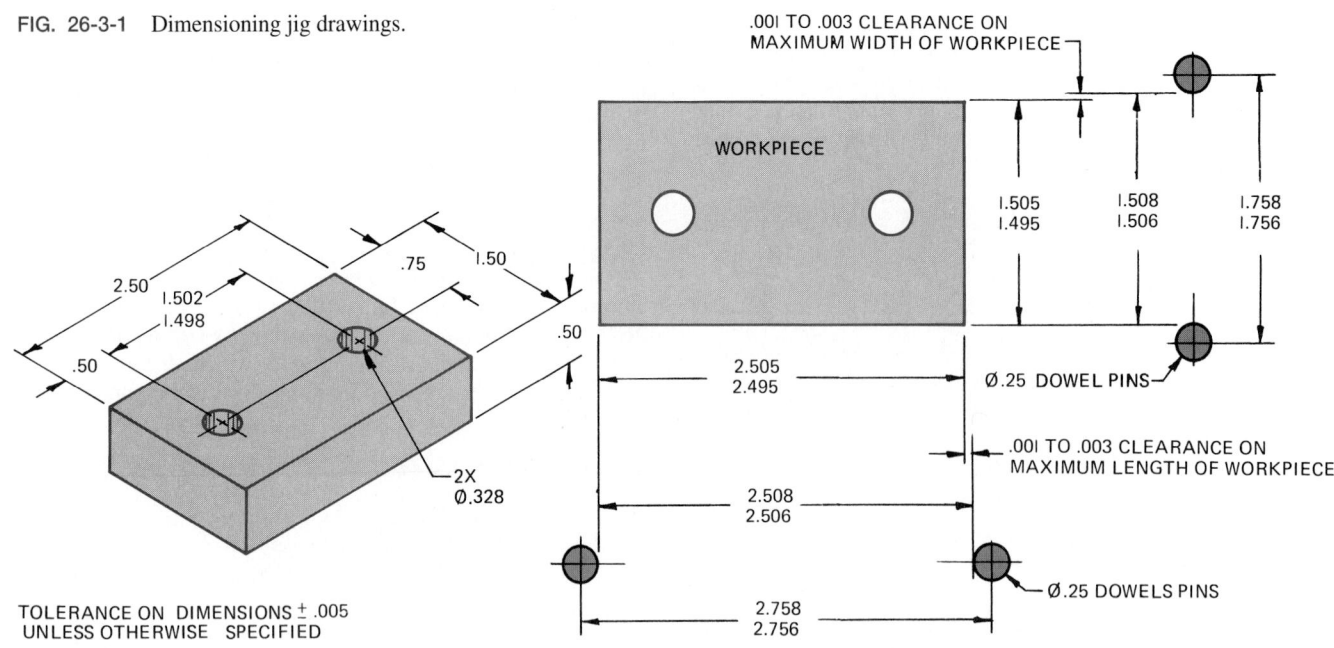

(A) CALCULATING DISTANCES BETWEEN DOWEL PINS

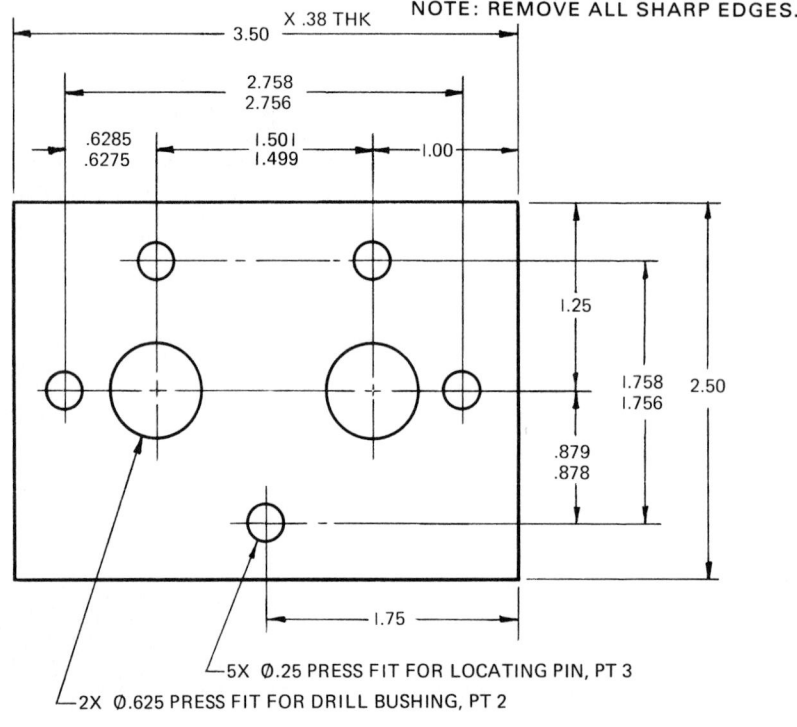

(B) DIMENSIONING JIG PLATE SHOWN IN FIGURE 26-1-3

26-3 DIMENSIONING JIG DRAWINGS

The finished detail drawing for the simple plate jig (Fig. 26-1-3) is shown in Fig. 26-3-1. The following is a brief explanation of why the various dimensions were chosen:

1. *Distance between holes.* The dimension between the Ø.328 in. holes on the workpiece is 1.498–1.502 in. Therefore the tolerance allowed on this dimension is

.004 in. The distance between the drill bushings on the jig plate must be kept to a closer tolerance because of bushing wear. A .002 in. tolerance was chosen for the center distance between the bushings, and the limits were placed midway between the workpiece limits. Thus the center-to-center distance was established at 1.499–1.501 in.

2. *Size of bushing holes.* In tool design it is general practice to show on the drawing only a note listing the nominal

FIG. 26-4-1   Vertical milling
fixture.
*(Ex-Cell-O
Corp. Ltd.)*

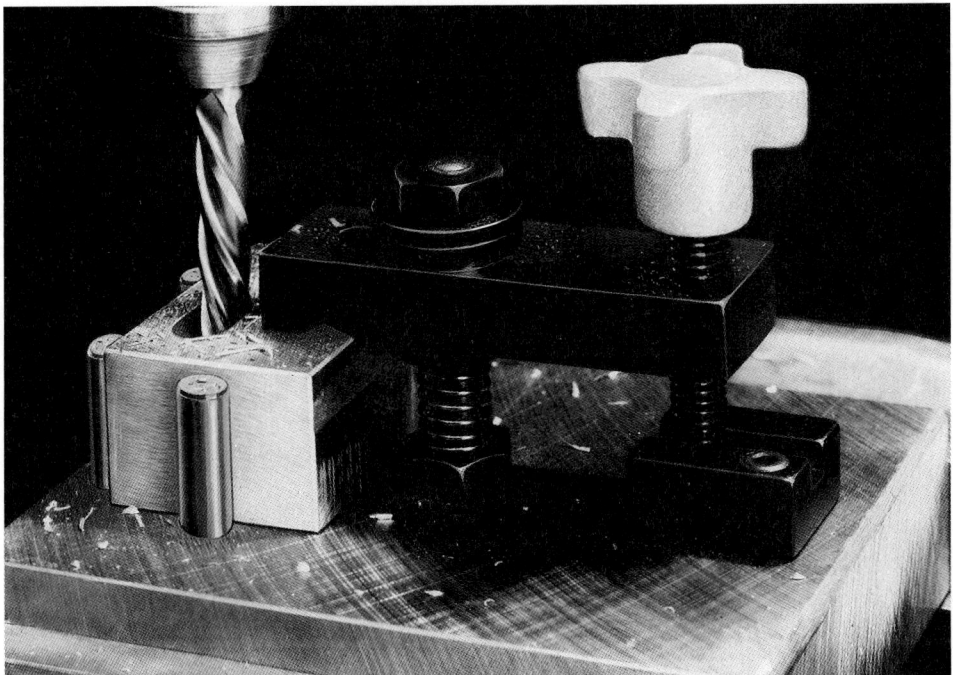

diameter of the hole and the part number of the mating part. It is the job of the machinist to select the proper diameters to ensure a press fit.

3. *Size of dowel pin holes.* Dowels are commercially available, at low cost, in a wide range of standard sizes. Standard commercial dowels are finished to .0002 in. (0.006 mm) larger than the nominal diameter with a tolerance of ± .0001 in. (0.003 mm). The size of the dowel pins used is Ø.25 (Ø.2501 – .2503) × .75 in. long. One end of the dowel pin has a chamfer to facilitate pressing it into the jig plate and to permit easy loading onto the workpiece.

   As mentioned earlier, a note calling for the nominal diameter and the part number of the mating part covers all the information required.

4. *Center distance between dowels.* Since the width and length of the workpiece are shown in nominal dimensions, the tolerance permitted on these dimensions is ±.005 in. Thus the size of the largest workpiece that would be permissible is 1.505 × 2.505 in. These are the workpiece sizes that are used in calculating the center-to-center distance between dowel pins. A clearance of .001 to .003 in. was decided on between the maximum workpiece size and the largest dowel pin. Thus the center-to-center distances between dowels were calculated to be 1.7563–1.7583 and 2.7563–2.7583 in.

5. *Center distance, bushing, and dowel.* The maximum limits were calculated by taking half the difference between the maximum limits of 2.758 and 1.501 in. for length and half the limit of 1.758 in. for width. An allowance of −.001 in. was given to these dimensions.

ASSIGNMENT

See Assignment 5 for Unit 26-3 on page 875.

26-4 FIXTURES

A fixture is a device that supports, locates, and holds a workpiece securely in position while machining operations are being performed. It should be noted that the accuracy of the machining depends on the quality of the machine and tools used.

Milling Fixtures

The most common type of fixture used is the milling fixture (Figs. 26-4-1 and 26-4-2). It may be clamped to the milling machine table or held in the milling machine vise. Before a milling fixture is designed, information such as the size and spacing of the T slots, crossfeed, and horizontal traverse of the table must be known. Most drafting offices have this

FIG. 26-4-2   Typical milling fixture. *(Standard Parts Co.)*

information tabulated in the form of a chart, and the designer may select the most suitable milling machine.

In laying out a fixture, the designer should check the drawing to see that no part of the fixture will interfere with the milling arbor or arbor supports. The standard practice of many designers is to show the cutter and arbor on the assembly drawing.

Fixture Components

Fixture Base

Fixture components and the workpiece are usually located on a base, which is securely fastened to the milling machine table with clamping lugs or slots (Fig. 26-4-3). The size of the lug opening corresponds to the T-slot width on the milling machine table. In addition, the base is usually provided with keys or tongues that sit on the table T slots, aligning the fixture so that the workpiece is perpendicular to the cutter arbor axis and parallel to the sides of the cutter. Standard fixture bases in a wide variety of sizes are at the disposal of the designer and are shown in Table 80 in the Appendix.

Clamps

In milling fixture design, forces resulting from the feed of the table and the rotation of the cutter are encountered. These forces are normally counteracted by the clamp forces. For this reason, fixture clamps must be of heavier design than jig clamps and must be properly located. See the examples shown in Figs. 26-4-4 and 26-4-5.

FIG. 26-4-3   Milling fixture base. *(Standard Parts Co.)*

FIG. 26-4-4   Toggle clamps.

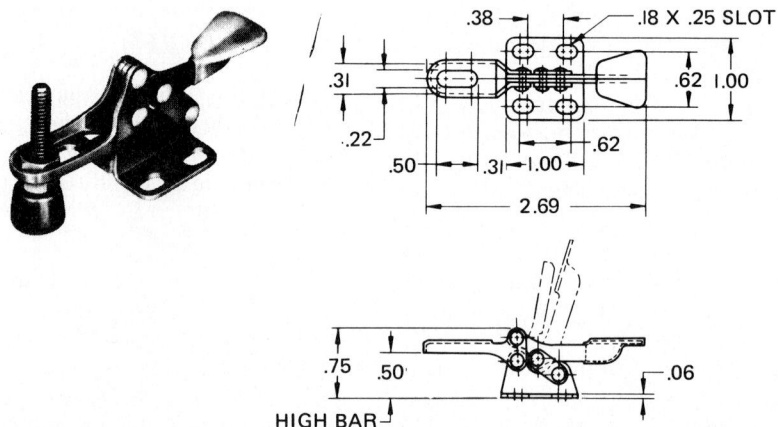

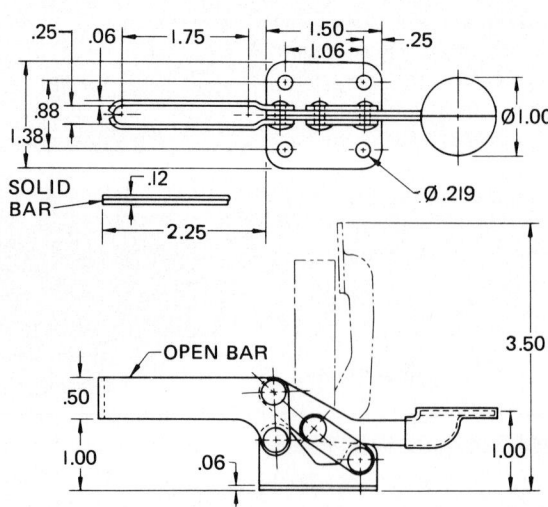

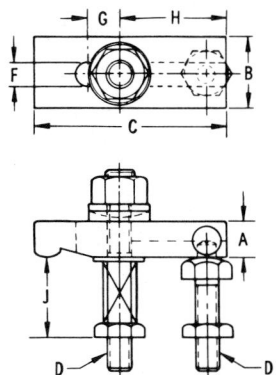
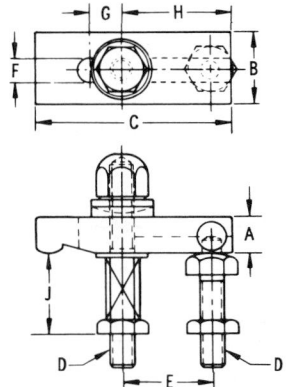
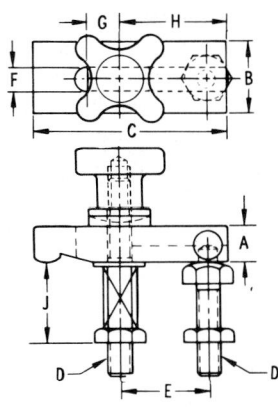

Clamp Assembly	Thread						Travel		Capacity
	A	B	C	D	E	F	G	H	J
20005M	10	16	50	M6	24	7	12	25	22
20010M	12	25	82	M8	32	9	46	38	28
20020M	16	32	100	M10	48	11	25	58	42
20030M	16	32	125	M12	48	13	38	58	42
20040M	22	38	125	M16	50	17	38	64	45
20050M	25	45	165	M20	66	21	45	82	50

FIG. 26-4-5   Strap clamp assemblies. *(American Drill Bushing Co.)*

FIG. 26-4-6   Application of holding components.

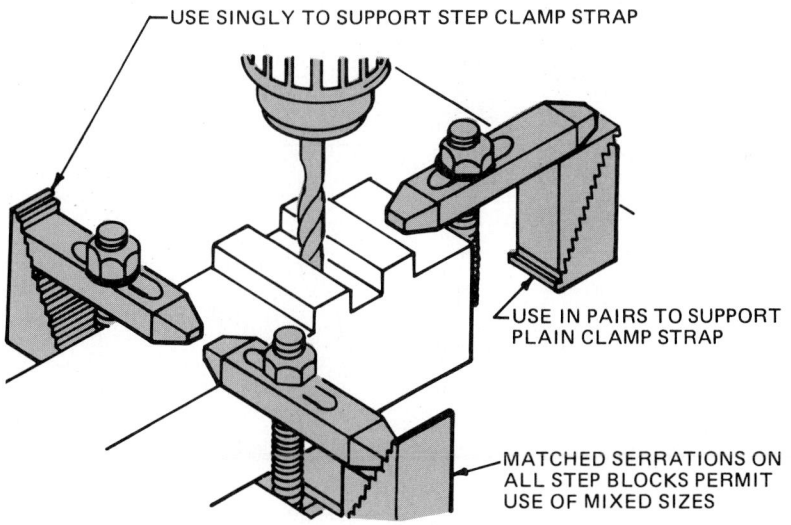

—USE SINGLY TO SUPPORT STEP CLAMP STRAP

—USE IN PAIRS TO SUPPORT PLAIN CLAMP STRAP

—MATCHED SERRATIONS ON ALL STEP BLOCKS PERMIT USE OF MIXED SIZES

Set Blocks

Cutter set blocks are mounted on the fixture to properly position the milling cutter in relation to the workpiece (Figs. 26-4-6 and 26-4-7, pg. 872). The locating surfaces of the set blocks are offset from the finished surfaces on the workpiece that are to be machined. Feeler gages the same thickness as the offset are placed on the located surfaces of the set block, and the position of the milling fixture is adjusted until the cutter touches the feeler gage. The space between the cutter and set block ensures clearance between the cutter and set block during the machining operation. Set blocks are normally fastened to the fixture body with cap screws and dowel pins.

Fixture Design Considerations

1. Is the fixture foolproof? Does the design permit only one way of loading?
2. Does the fixture permit rapid loading and unloading?
3. Is ample chip clearance provided?
4. Is the fixture kept as low as possible to avoid chatter and springing of the work?
5. Are the cutting forces taken on the base rather than on the clamp?
6. Does any part of the fixture interfere with the milling arbor or supports during the machining operation?
7. Are the clamps located in front of the workpiece?

FIG. 26-4-7   Holding
components.
*(American Drill
Bushing Co.)*

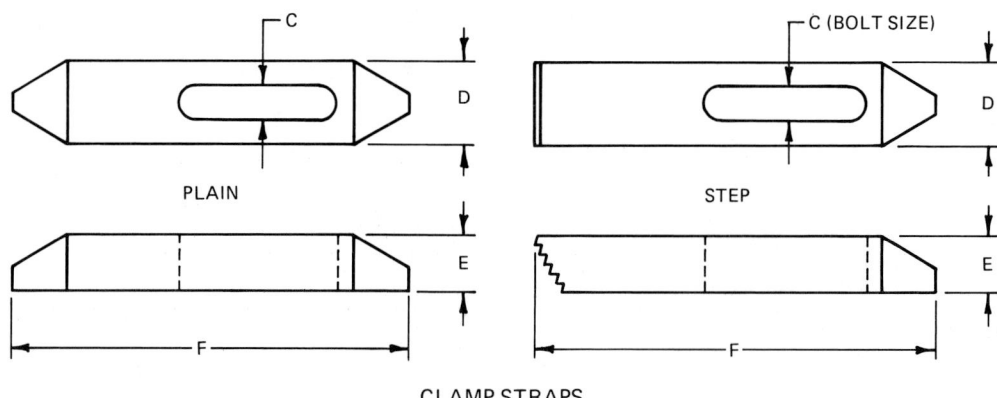

PLAIN   STEP

CLAMP STRAPS

	PART	C	D	E	F
INCH	A-930	.38 to .50	1.00	.50	2.50
	A-940	.50	1.19	.75	6.00
	A-950	.62	1.19	1.00	6.00
	A-960	.75	1.19	1.19	8.00
	A-970	1.00	2.00	1.38	10.00
MILLIMETER	A-930M	10 to 12	25	12	60
	A-940M	12	30	20	150
	A-950M	18	30	25	150
	A-960M	20	30	30	200
	A-970M	25	50	35	250

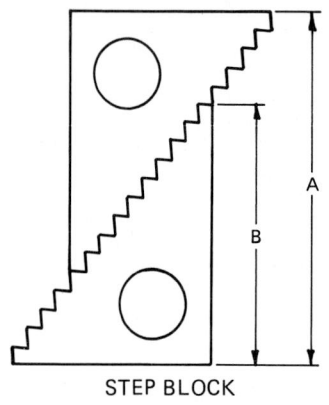

STEP BLOCK

	WIDTH	HEIGHT B	CAPACITY A
INCH	1.00	1.18	.75-1.50
	1.38	1.75	1.25-2.50
	1.75	3.50	2.50-6.00
mm	25	30	20-40
	35	45	30-60
	45	90	60-150

Sequence in Laying Out a Fixture

The following sequence is recommended in laying out a
fixture (Fig. 26-4-8):

1. Draw the necessary views of the workpiece. Leave sufficient room for drawing in the fixture details.
2. Draw the locating devices.
3. Draw the cutter and arbor.
4. Draw the clamping arrangement.
5. Draw the set blocks, if required.
6. Draw the fixture base and keying arrangements.

ASSIGNMENT

See Assignment 6 for Unit 26-4 on page 876.

FIG. 26-4-8 Sequence in
laying out a
fixture.
*(American Drill
Bushing Co.)*

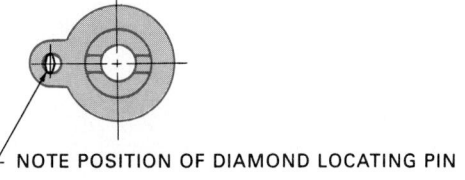

NOTE POSITION OF DIAMOND LOCATING PIN

SLOT TO BE MILLED

LOCATING PIN

(A) DRAW 3 VIEWS OF THE WORKPIECE AND ADD SUITABLE LOCATING DEVICES

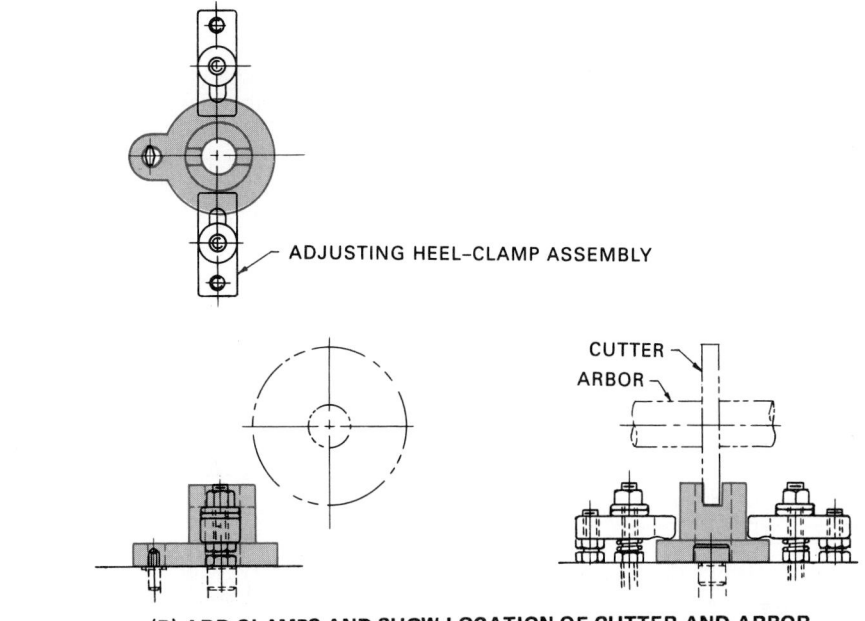

ADJUSTING HEEL-CLAMP ASSEMBLY

CUTTER

ARBOR

(B) ADD CLAMPS AND SHOW LOCATION OF CUTTER AND ARBOR

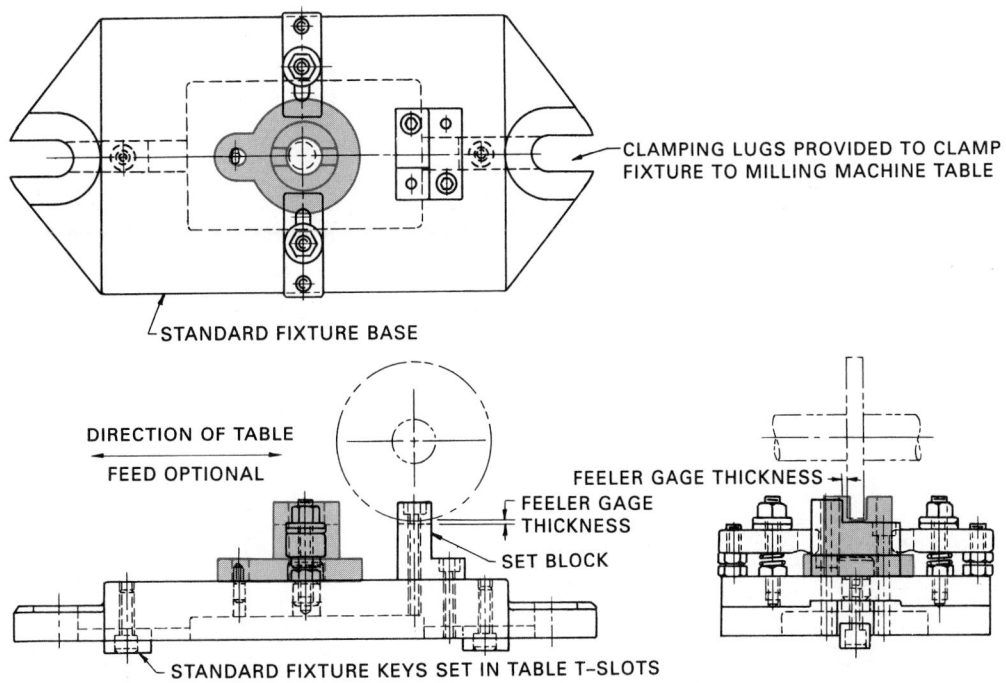

CLAMPING LUGS PROVIDED TO CLAMP
FIXTURE TO MILLING MACHINE TABLE

STANDARD FIXTURE BASE

DIRECTION OF TABLE
FEED OPTIONAL

FEELER GAGE THICKNESS

FEELER GAGE
THICKNESS

SET BLOCK

STANDARD FIXTURE KEYS SET IN TABLE T-SLOTS

(C) DRAW IN SET BLOCK AND BASE DETAIL

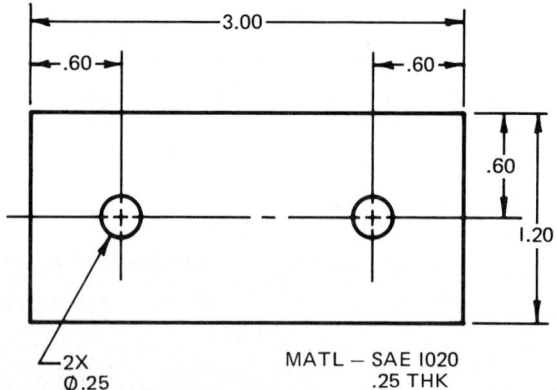

ASSIGNMENTS FOR CHAPTER 26

ASSIGNMENT FOR UNIT 26-1, JIG AND FIXTURE DESIGN

1. Design a simple plate jig for drilling the holes in one of the parts shown in Figs. 26-1-A through 26-1-C. Scale 1:1.

FIG. 26-1-A   Connector.

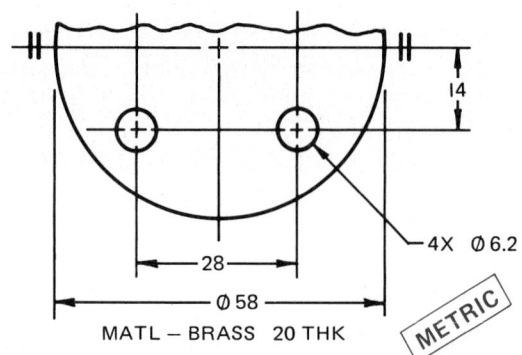

FIG. 26-1-B   Spacer.

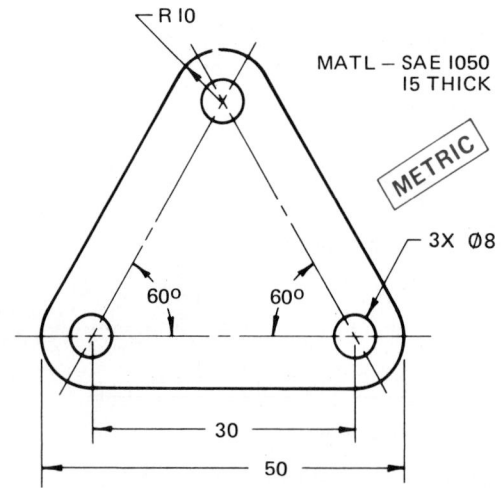

FIG. 26-1-C   Cover plate.

State the sequence of operations and the time at which locking pins are employed.

ASSIGNMENTS FOR UNIT 26-2, DRILL JIG COMPONENTS

2. Design a jig for drilling the small holes in the part in either Fig. 26-2-A or Fig. 26-2-B. The large center hole and finished base should be the features used for locating the part in the jig. A locking pin is recommended for alignment after the first hole is drilled. Standard components should be used wherever possible.

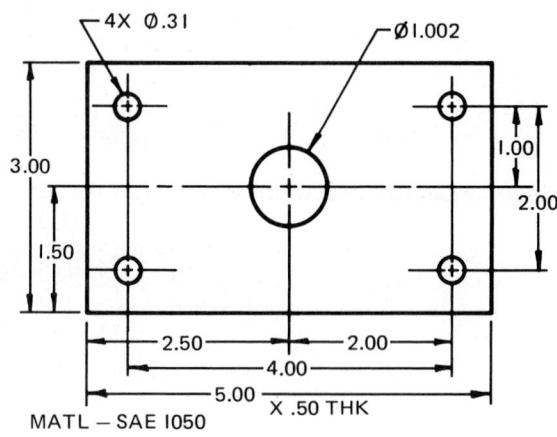

FIG. 26-2-A   Plate.

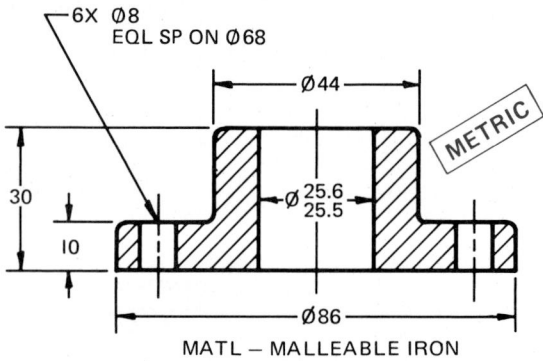

FIG. 26-2-B   Flanged bracket.

3. Design a drill jig for the smaller of the two holes shown in Fig. 26-2-C. The hole in the hub of the part and the finished surfaces should be the features used for locating the part in the jig. Standard components should be used wherever possible. Scale 1:1.

4. Design a drill jig for the Ø8 and Ø16.1 holes shown in Fig. 26-2-D. The hole in the hub of the part and the finished surfaces should be the features used for locating the part in the jig. A locking pin is optional, depending on the design. Standard components should be used wherever possible. Scale 1:1.

ASSIGNMENT FOR UNIT 26-3, DIMENSIONING JIG DRAWINGS

5. Design a simple plate jig for drilling the holes in one of the parts shown in Fig. 26-3-A or 26-3-B. The size of the dowel pins used in the design is .2502 ± .0001 in. or 6.006 ± 0.003 mm. After your overall design has been approved by your instructor, dimension the jig plate per procedures outlined in this unit. Scale 1:1.

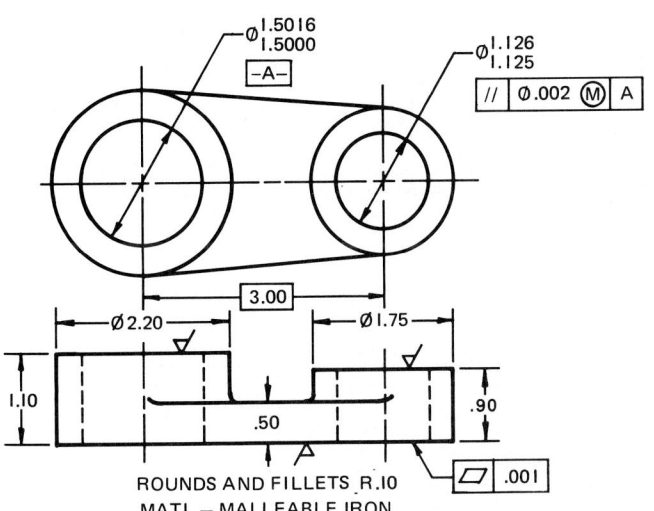

FIG. 26-2-C   Link.

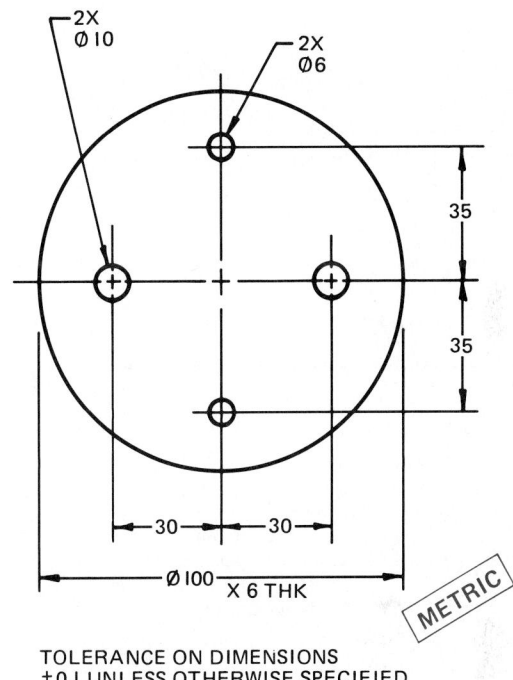

FIG. 26-3-A   Spacer.

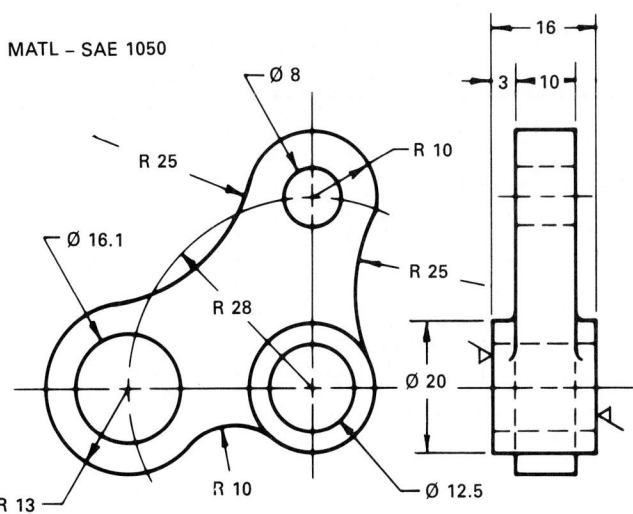

FIG. 26-2-D   Connector.

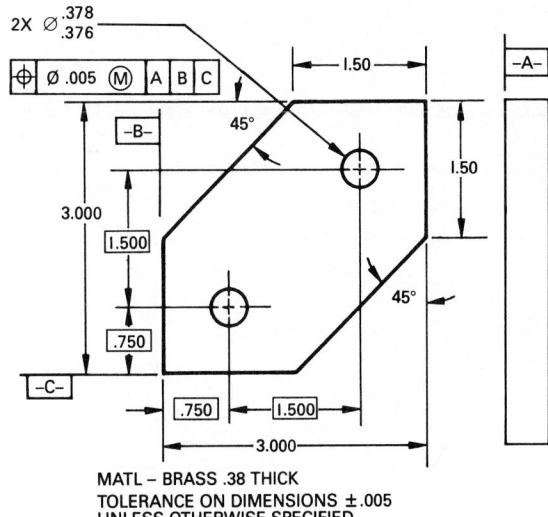

FIG. 26-3-B   Locking plate.

ASSIGNMENT FOR UNIT 26-4, FIXTURES

6. Design a simple milling fixture to mill out the two outside portions on the top of the part in Fig. 26-4-A or the slots shown in Figs. 26-4-B through 26-4-D. Use standard components and refer to manufacturers' catalogs wherever possible. Draw the workpiece in red. Scale to suit.

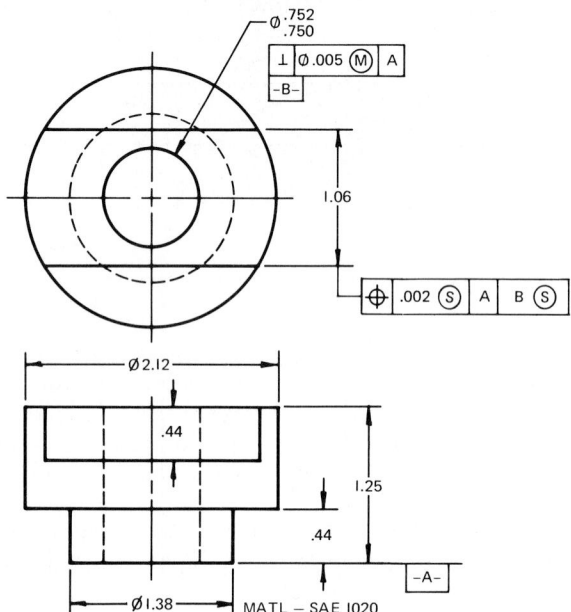

FIG. 26-4-A   Drive link.

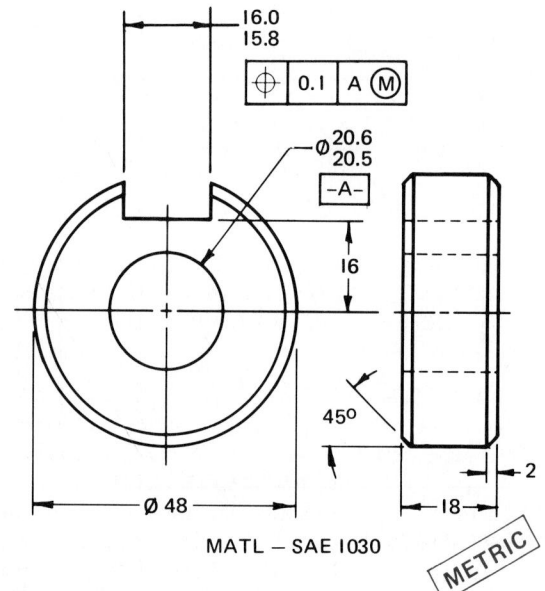

MATL – SAE 1030

METRIC

FIG. 26-4-B   Sleeve.

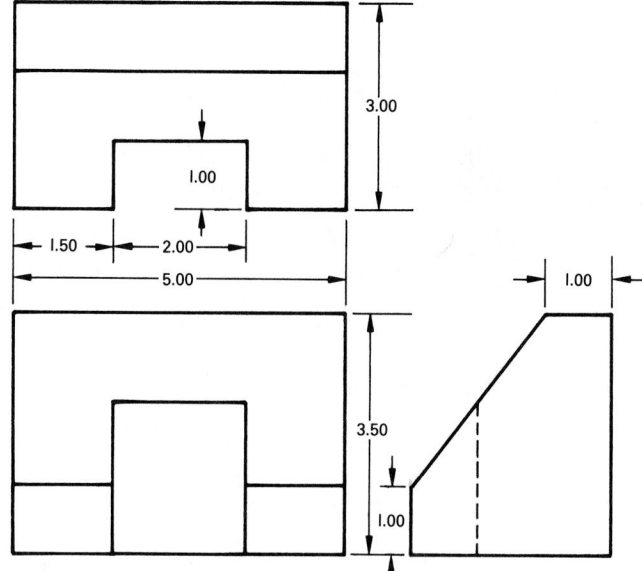

FIG. 26-4-C   Guide stand.

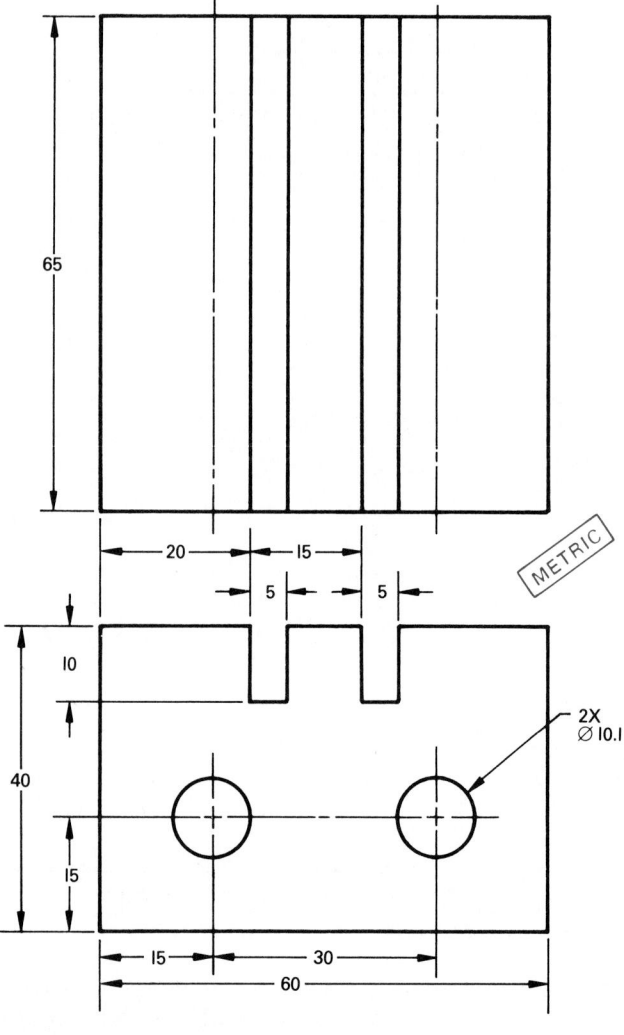

METRIC

FIG. 26-4-D   Locating guide.

CHAPTER 27

ELECTRICAL AND ELECTRONICS DRAWINGS

Definitions

Autoroute   The automatic routing of printed wiring traces on a board using a CAD system.

Block diagram   A diagram, consisting of a series of blocks and straight lines, representing a circuit.

Electrical drawings or **electrical diagrams**   Drawings used to show how to connect the wires and to explain how the circuits operate.

Highway   On a drawing, a thick line used to show several wires close together, as in a conduit, or held together by a harness.

Logic diagram   A diagram representing the logic elements and their implementations, but not necessarily expressing construction or engineering details.

Rat's nest   The first attempt at connecting the components on a schematic diagram.

Reference designation   A coding system used to indicate replaceable parts on a diagram.

Schematic capture   The drawing of a schematic diagram on a CAD system.

Schematic diagram or **elementary diagram**   Diagrams that show the electrical connections and function of a circuit using graphical symbols.

27-1 ELECTRICAL AND ELECTRONICS DRAWINGS

Mechanical drafters and technicians can no longer be isolated from electrical and electronics drawings. With the steady increase in automation and electronics equipment, they are now required to either produce or understand electrical and electronics drawings. In addition to the standard detail and assembly drawings used to manufacture and assemble electrical components, *electrical drawings,* also referred to as *electrical diagrams,* are used to show how to connect the wires and to explain how the circuits operate.

Although there are many types of electrical and electronics drawings, only the most widely used will be covered in this chapter. These are:

1. Schematic diagrams
2. Connection diagrams
3. Printed circuit (PC) drawings
4. Block and logic diagrams

Although electrical drawings for residential and commercial buildings are also widely used, this type of drawing should be dealt with in architectural texts.

Standardization

Since electrical and electronic drawings rely heavily on symbols to convey information to the person reading the drawing, it is important that the symbols be interpreted correctly. The American National Standards Institute (ANSI), the Institute of Electrical and Electronics Engineers (IEEE), and other groups have developed a number of standards to reduce the misinterpretation of information on these drawings. A few of these standards are:

ANSI/IEEE Std 315	Graphic Symbols for Electrical and Electronics Diagrams (ANSI Y32.2)
ANSI/IEEE Std 91	Graphic Symbols for Logic Diagrams (ANSI Y32.14)
ANSI Y14.15	Electrical and Electronics Diagrams
ANSI/IPC-D-275	Design Standard for Rigid Printed Boards and Rigid Printed Board Assemblies

877

The U.S. government has also developed a series of standards for use by military contractors. For example:

MIL-STD-681   Identification Coding and Application of Hook Up and Lead Wire

These and other standards should be referred to if applicable or required.

Using CAD for Electrical Drawings

Either manual drafting or CAD may be used to prepare electrical and electronics drawings. However, the advantages of using CAD far exceed those for manual drafting. CAD systems used for these drawings have all of the drawing functions of those used for mechanical drafting. They contain extensive libraries of electrical component symbols and are able to handle the complex processes of routing printed wiring traces on PC boards. It is important to note that for a beginning drafter, laying out PC boards manually is still probably the best way to understand the processes involved with this type of drawing. Even though a CAD system can greatly assist in the production of PC drawings, a drafter or CAD operator must still be able to use skill and judgment to evaluate the final results and revise, where needed, the computer's solution.

When electrical drawings are made using manual drafting techniques, the drafter usually is not directly involved with the actual manufacturing of the product. Using CAD, the drafter is not only creating a drawing, but is also creating data that will have a direct impact on other drawings and the processes used to build the product. For example, drawing a schematic diagram on a CAD system, sometimes referred to as *schematic capture,* not only shows the electrical function of the circuit, but also contains much other information about the components and their electrical connections. This information can be used to automatically route, or *autoroute,* the printed wiring traces on the board, or be transmitted to computer-aided manufacturing (CAM) systems to drill and machine the PC board. This shared CAD/CAM data can also be used by CAM systems which automatically place the electronic components on the board.

With the trend in electronics toward smaller and smaller packages, many printed wiring boards now utilize surface mount technology (SMT) and printed wiring traces as small as .010 in. wide. The small size of the components and the accuracy required in the placement of the components on the board require the use of robotics systems which can pick up the small components from the carrier cup, look at it and determine its position, and then correctly place it on the board. Systems with multiple robotics arms can place thousands of components in one hour.

CAD Graphics

Since most electrical and electronic drawings use the same component symbols arranged in different patterns, it has been a natural development to use computer graphics. Because of the repetitive nature of these drawings, CAD systems were one of the earliest applications for computer graphics. CAD systems have developed rapidly from mere symbol duplicating systems to highly complex systems which not only can be used to draw

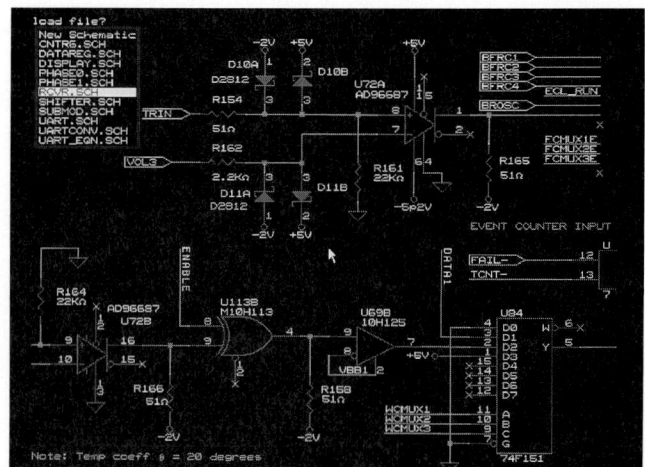

FIG. 27-1-1   Schematic diagram drawn on a CAD system. *(ORCAD)*

a schematic diagram, but can test and analyze the circuit and automatically produce the drawings needed to build a printed wiring board.

CAD software designed to produce electrical and electronic drawings often have symbol libraries containing thousands of different component symbols. A schematic diagram is drawn by retrieving the desired symbols from the library, positioning them on the diagram, and adding the connecting lines and text information. Figure 27-1-1 shows a portion of a schematic diagram drawn on a CAD system. Once the schematic diagram has been completed, the CAD system can be used to extract a list of all the electrical connections (net list) and a list of all of the components (material list). The net and material lists are important elements used in drawing a printed wiring board.

27-2 SCHEMATIC DIAGRAMS

Schematic diagrams, also known as *elementary diagrams,* show the electrical connections and function of a circuit using graphical symbols. They do not show the physical relationship of the components, nor mechanical connections. This type of drawing is mainly used by engineers and electronics technicians since they are interested mostly in the design and function of the equipment (Fig. 27-2-1).

Laying Out A Schematic Diagram

The connecting wires joining the electrical components are indicated on the diagram by straight horizontal or vertical lines. A line of medium thickness is recommended for general use on schematic diagrams. However, thicker lines may be used when emphasis of certain components or transmission paths is essential. Wire connections may be made at any convenient location on the diagram, and the connection normally is shown as a small, solid circle (dot). Where the connections can be shown as a single junction (which is preferred), the dot may be omitted. Figure 27-2-2 illustrates the "dot" and "no dot" methods of showing connections. Dots used for multiple junctions must be clearly visible, otherwise a connection could be mistaken for a

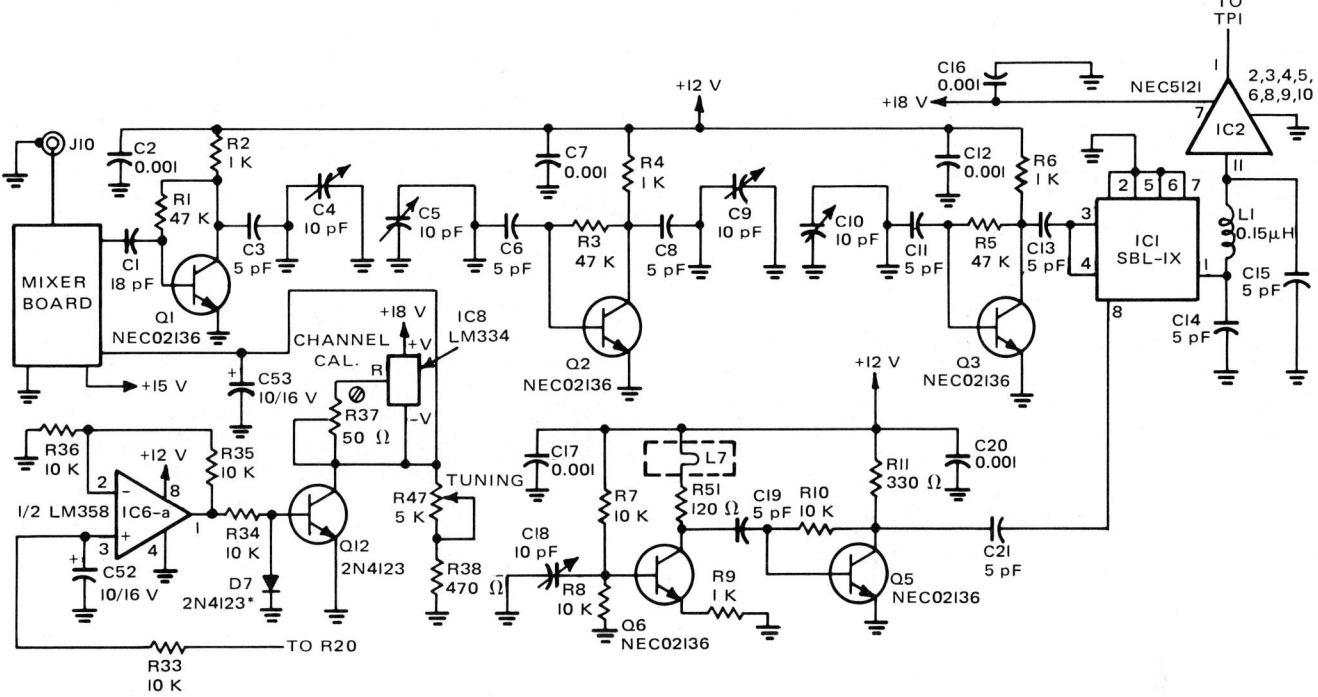

NOTE: ALL RESISTOR VALUES ARE IN OHMS (Ω).

FIG. 27-2-1 Partial schematic diagram of a receiver.

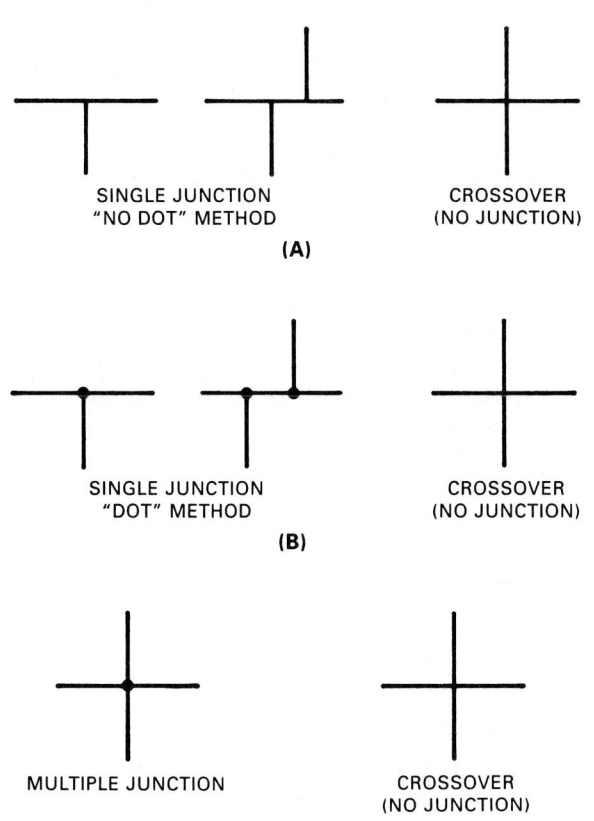

SINGLE JUNCTION
"NO DOT" METHOD

CROSSOVER
(NO JUNCTION)

(A)

SINGLE JUNCTION
"DOT" METHOD

CROSSOVER
(NO JUNCTION)

(B)

MULTIPLE JUNCTION

CROSSOVER
(NO JUNCTION)

(C)

FIG. 27-2-2 "Dot" and "no dot" methods of showing connections on a schematic diagram.

crossover. Ground symbols are used frequently on schematic diagrams instead of wire connections. The ground symbol is drawn so that the lines are horizontal and the symbol tapers towards the bottom of the drawing.

Graphical Symbols

Some of the standard symbols for components used for electrical schematic diagrams are shown in the Appendix. Although symbols may be drawn to any convenient size, electrical and electronic symbol templates are available for use by drafters and engineers.

Reference Designation

In conjunction with the graphical symbol all replaceable parts should be referenced with a letter and a number, called the *reference designation*. The letter identifies the class of component and is not an abbreviation for the component name. See Fig. 27-2-3 (pg. 880) for the letters used to identify commonly used components. Numbers within each component class should begin with 1 and run consecutively. Numbering (R1, R2, R3, etc.) within each class is normally done from left to right in a top to bottom order. The reference designation may be placed above, below, or on either side of the symbol (Fig. 27-2-4, pg. 880).

Numerical Values

Many components are also identified with a numerical value (resistance, capacitance, inductance, etc.) or a type number (2N4123, 1N914, etc.). These should always be placed below

Letter	Type of Component	Letter	Type of Component
U	Integrated Circuit Package	L	Inductor and Windings
Q	Transistor or Rectifier	HT	Electrical Headset
C	Capacitor	F	Fuse
R	Resistor	T	Transformer
A	Recording Unit	DS	Lamp
D or CR	Diode	S	Switch
J	Jack	BT	Battery
AR	Amplifier	LS	Speaker

FIG. 27-2-3   Item name designation letters shown on schematic diagrams.

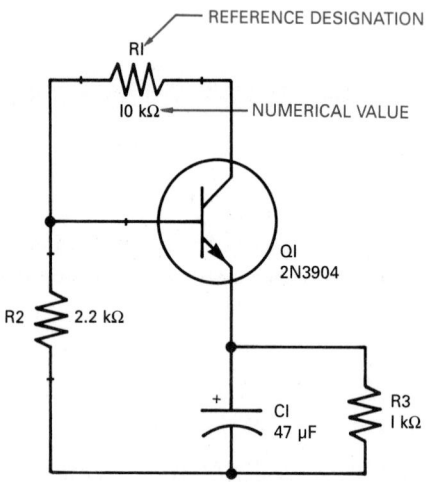

FIG. 27-2-4   Placement of reference designation and numerical value on graphical symbols.

the reference designation or on the opposite side of the symbol (Fig. 27-2-4). The numerical values often use multipliers to reduce the number of zeros in the value. Some of the standard multipliers and the symbols used to represent them on a drawing are:

MULTIPLIER	PREFIX	DRAWING SYMBOL
10^6	mega	M
10^3	kilo	k or K
10^{-3}	milli	m or MILLI
10^{-6}	micro	μ or U
10^{-12}	pico	p or P

A value of 23,000 ohms (23 kilo ohms) would be given on a drawing as 23K Ω. Likewise, a value of .47 picofarads would

be expressed as .47PF (Figs. 27-2-1 and 27-2-4). When many of the component values are in the same units, a note similar to the one shown below can be placed on the drawing, and the units part of the values omitted. For example:

UNLESS OTHERWISE SPECIFIED, CAPACITANCE IN MICROFARADS AND RESISTANCE IN OHMS

Integrated Circuit Symbols

One of the most recent far-reaching developments in the field of electronics is that of microelectronics, which started with the transistor and has evolved at a very rapid pace.

The symbols already discussed for schematic diagrams are meant to be used for individual parts. Integrated circuits (IC), which may contain hundreds of components and may be used as part of a larger circuit, require other symbols.

Two symbols are used to depict integrated circuits, the rectangle and the equilateral triangle, as shown in Fig. 27-2-5. The triangle symbol is used for ICs that are amplifiers. The symbol normally lists the manufacturer's number and function, and the pin numbers. Simplification of the overall diagram normally dictates the orientation of the rectangle.

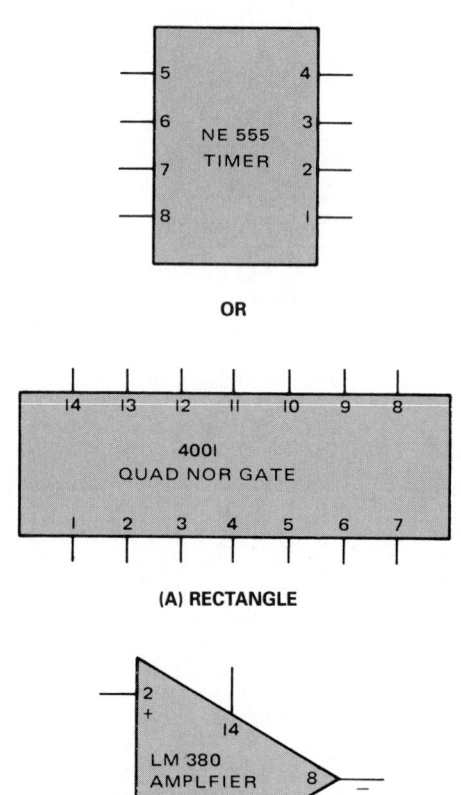

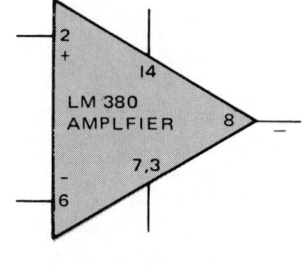

FIG. 27-2-5   Symbols for integrated circuits (composite assemblies).

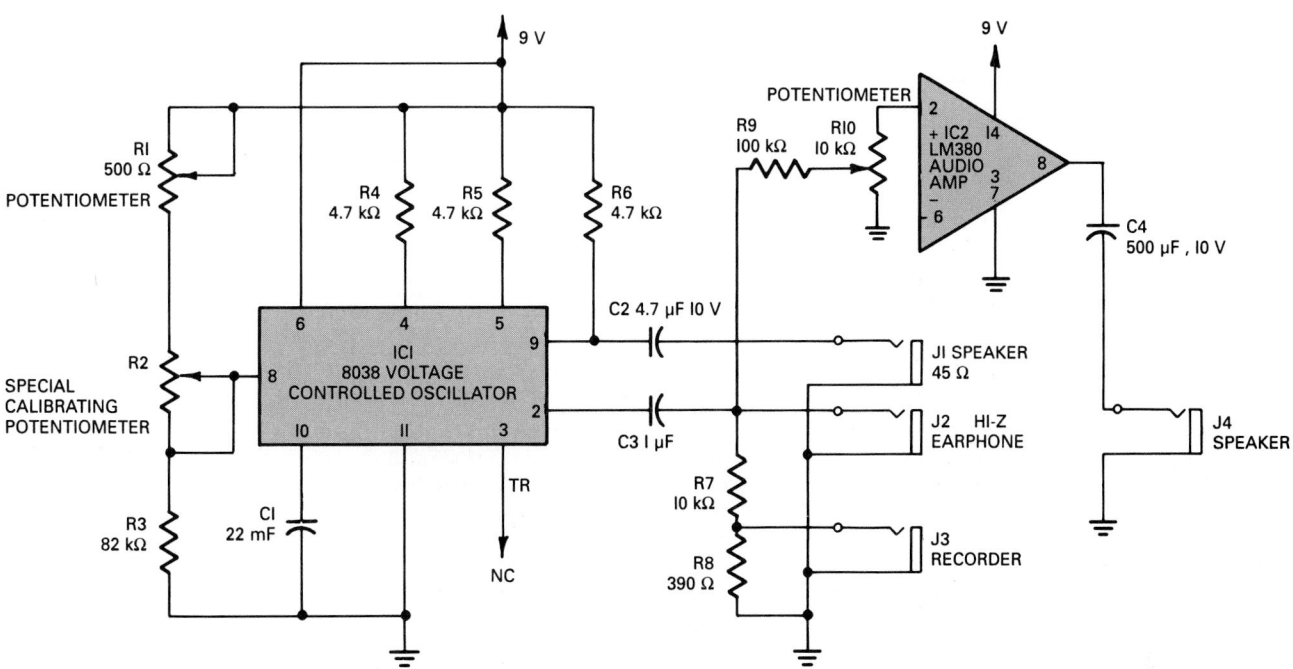

FIG. 27-2-6   Elementary diagram for a melody organ.

Although there are no particular sizes or rules governing the drawing of these symbols, generally they should be of one size, with a maximum of two sizes for one drawing. A typical schematic diagram incorporating ICs is shown in Fig. 27-2-6.

Basic Rules for Laying Out a Schematic Diagram

1. Keep lines to a minimum.
2. Avoid crossovers where possible.
3. Reposition, rotate, or invert component symbols to simplify the connections.
4. Maintain a uniform symbol size. The recommended size is approximately 1.5 times the size of those shown in ANSI/IEEE Std 315 (ANSI Y32.2).
5. Allow space for component identification (i.e., reference designation and component value).
6. Do not necessarily draw to scale.
7. Space symbols and lines so that the open areas appear balanced (no extremely large or extremely small open areas).
8. Align similar components where feasible to make a more pleasing and professional looking drawing.
9. Show switches in a position with no operating force applied, or indicate the switch position by a note. Relay contacts are shown in the de-energized or nonoperated condition.
10. When portions of a schematic diagram belong to separate printed circuit boards or subassemblies, show the grouping by enclosing each group of components with a phantom line (dashed line).
11. Use a standard grid spacing to simplify the positioning of the symbols and their connecting lines, thus saving many hours of drafting time. Most schematic symbols are designed so that the connection points are located on standard grid spacings (.100, .200, .250, etc.).

ASSIGNMENTS

See Assignments 1 through 5 for Unit 27-2 on pages 894 through 897.

27-3 WIRING (CONNECTION) DIAGRAMS

In this era of mass production of electronics equipment by nontechnical personnel, and with the publication of an increased number of repair manuals and building kits for the do-it-yourself enthusiast, a wiring diagram is required to show the proper electrical connections. Wiring diagrams supplement the assembly drawing. The assembly drawing shows how the components fit together, and the wiring diagram shows how to electrically connect the components.

As electrical and electronic symbols would be meaningless to many people using wiring diagrams, the components are often represented pictorially as shown in Fig. 27-3-1 (pg. 882). However, standard component symbols or simple geometric figures (circles, rectangles, etc.) are also used in diagrams intended for professional assemblers or repairers (Fig. 27-3-2, pg. 883).

In order to reduce assembly and repair time and lessen the chance of error when there are many connecting wires, color coding is often used. Each wire is covered with a different color of insulation. The color is indicated beside or on the wire on the diagram.

For relatively simple diagrams a separate line may be used to represent each wire. This type of diagram is known as a point-to-point diagram. Figure 27-3-1 illustrates a typical point-to-point diagram.

Often when there are several wires close together, as in a conduit, or held together by a harness, one thick line, called a

FIG. 27-3-1   Point-to-point connection diagram of a boat's electrical system.

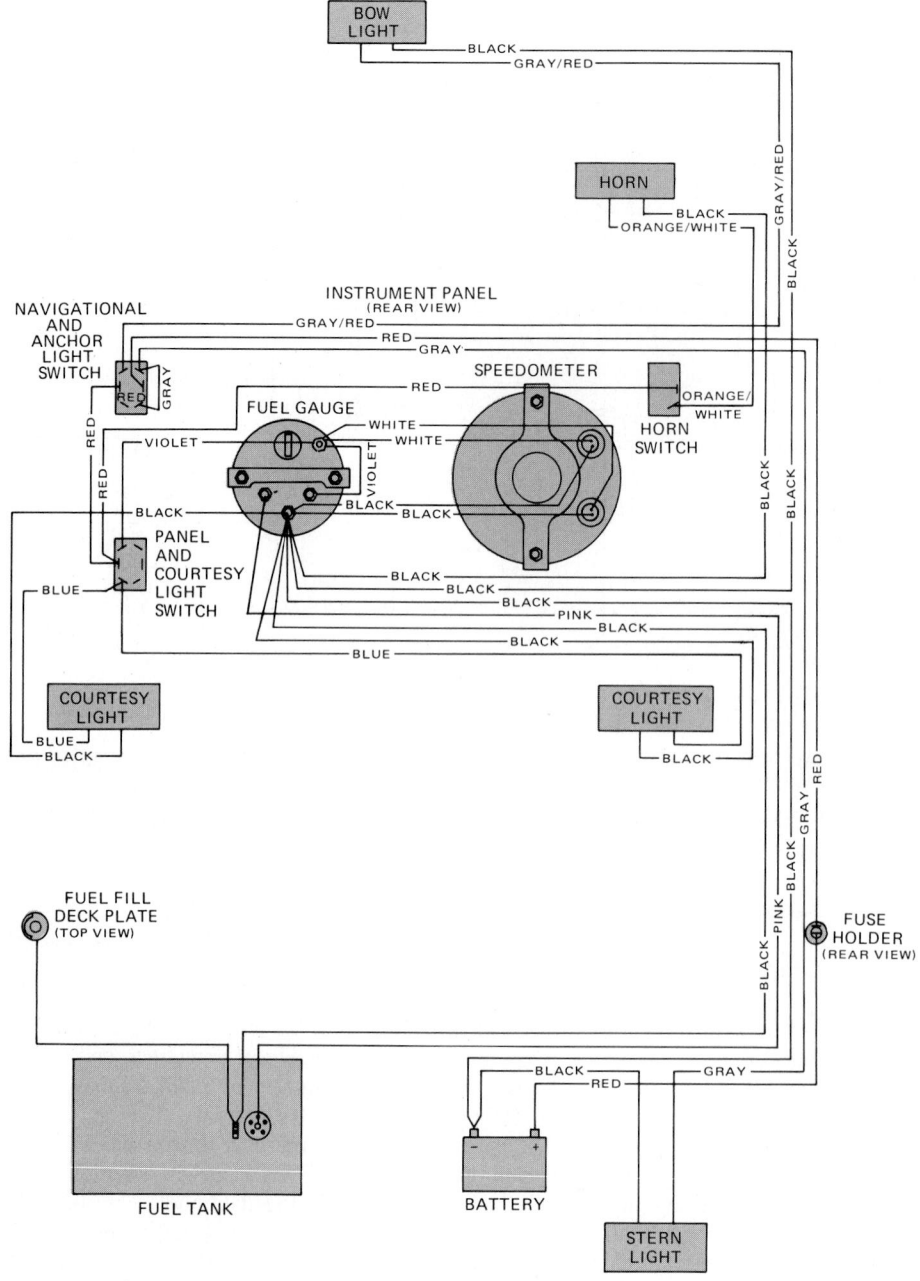

highway, is used instead of several separate lines. When clarity is required to show the direction a wire takes when it enters the highway, an arc or a 45° line is used to indicate the direction of travel. Several highways may be desirable on a drawing because of electrical, physical, or other factors. This type of drawing is called a highway-type wiring diagram (Fig. 27-3-2).

To minimize the cost of an electrical assembly, many of the wires are joined together prior to the final assembly. Each wire is cut to the required length, positioned on a board or jig, and taped together to form a wiring harness (Fig. 27-3-3). The length of each wire extending beyond the harness, its breakout point, and its color must be known.

Basic Rules for Laying Out A Wiring Diagram

1. Reference designations must be identical to those used on the corresponding schematic diagram. Where the component is inserted into a socket, the reference designation is prefixed with an "X" (XF1, XDS4, etc.).
2. Individual components are shown in the same general arrangement as they appear when viewed from the wiring direction. Size and spacing of components is mostly a matter of drawing convenience.

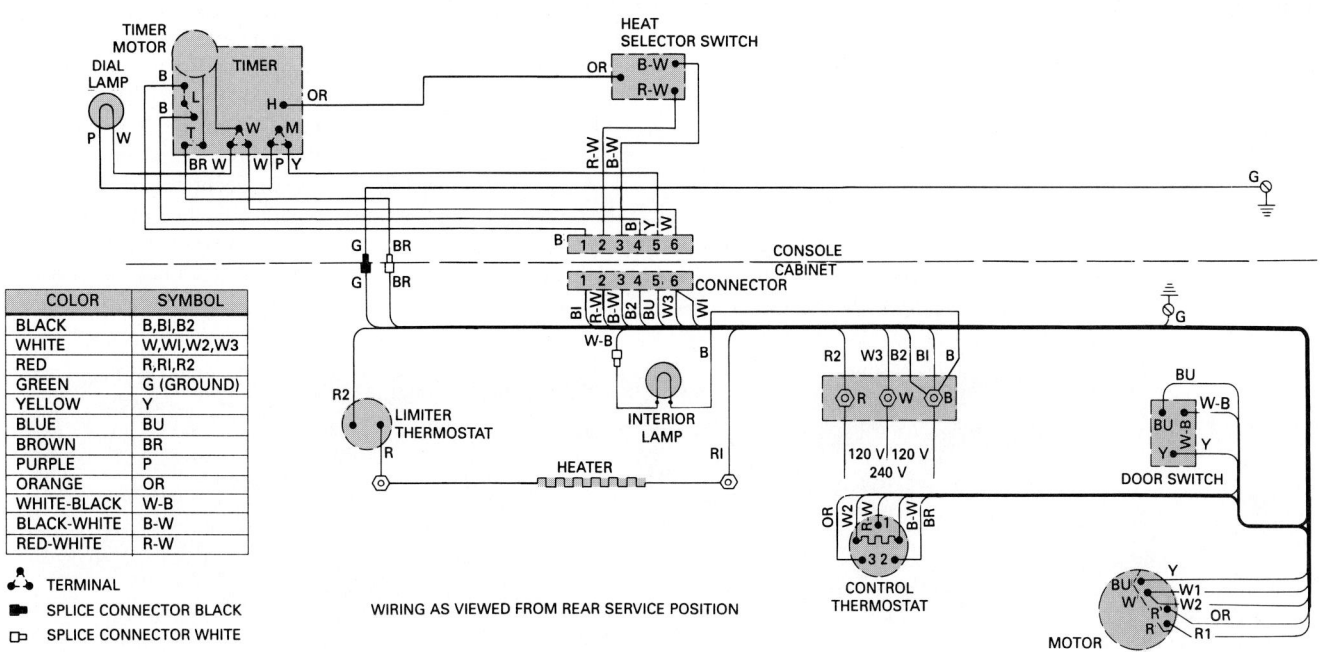

FIG. 27-3-2   Highway-type wiring diagram of a clothes dryer. *(Frigidaire)*

FIG. 27-3-3   Harness drawing.

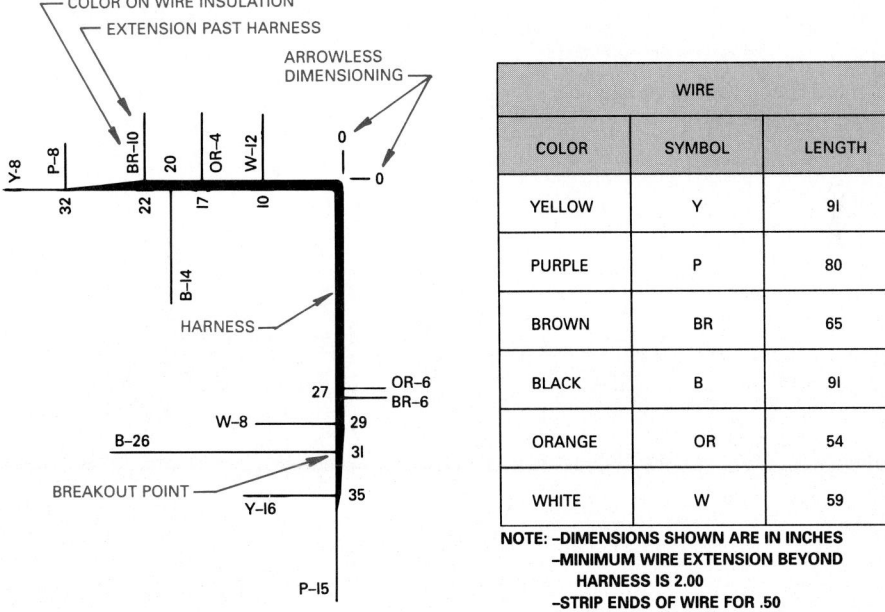

WIRE		
COLOR	SYMBOL	LENGTH
YELLOW	Y	91
PURPLE	P	80
BROWN	BR	65
BLACK	B	91
ORANGE	OR	54
WHITE	W	59

NOTE: –DIMENSIONS SHOWN ARE IN INCHES
–MINIMUM WIRE EXTENSION BEYOND
HARNESS IS 2.00
–STRIP ENDS OF WIRE FOR .50

3. Connecting wires are drawn as straight horizontal or vertical solid lines. Short angular lines or arcs may be used to clarify connections or to indicate directions. Note that the position of lines on the diagram do not normally indicate the actual wire position in the unit.
4. Connection lines should be uniformly spaced where possible.
5. Connection wires are usually identified by wire color, color code number, or a wire number.
6. Connection terminals should be identified exactly as shown on the parts. Components, such as batteries,

diodes, and transistors should have their polarity clearly identified.
7. The wiring diagram may be placed on the assembly drawing where practical.

ASSIGNMENTS ▐▐▐▐▐▐▐▐▐▐▐▐▐▐▐▐▐▐▐

See Assignments 6 through 9 for Unit 27-3 on pages 897 to 898.

27-4 PRINTED CIRCUITS

Practically all mass-produced electronics equipment (radios, televisions, etc.) are now using printed circuits (PC) instead of wire-lead connections. The advantages are many: uniformity in production, elimination of practically all assembly wiring errors, reduced cost, and miniaturization. Printed circuits or etched circuits are thin ribbons of copper or other metals formed on a plastic board and shaped to replace the wire leads formerly used to connect the components (Figs. 27-4-1 through

27-4-3). There are three types of PC boards: single-sided, with conductors on one side; double-sided, with conductors on both sides; and multilayer, with conductors at different levels. Only single-sided PC boards will be covered in this unit.

For some applications, manual PC drafting is still used but the trend is to computer graphics and photoplotting.

A printed circuit is made from an accurately drawn layout. The layout is normally drawn on a grid to an enlarged scale. A polyester film is laid over the layout and traced using pressure sensitive symbols and tapes. The symbols center the pads and holes in the layout, and are connected with tapes to form circuit

FIG. 27-4-1   Printed circuit used on the amplifier shown in Fig. 27-4-2. *(General Electric)*

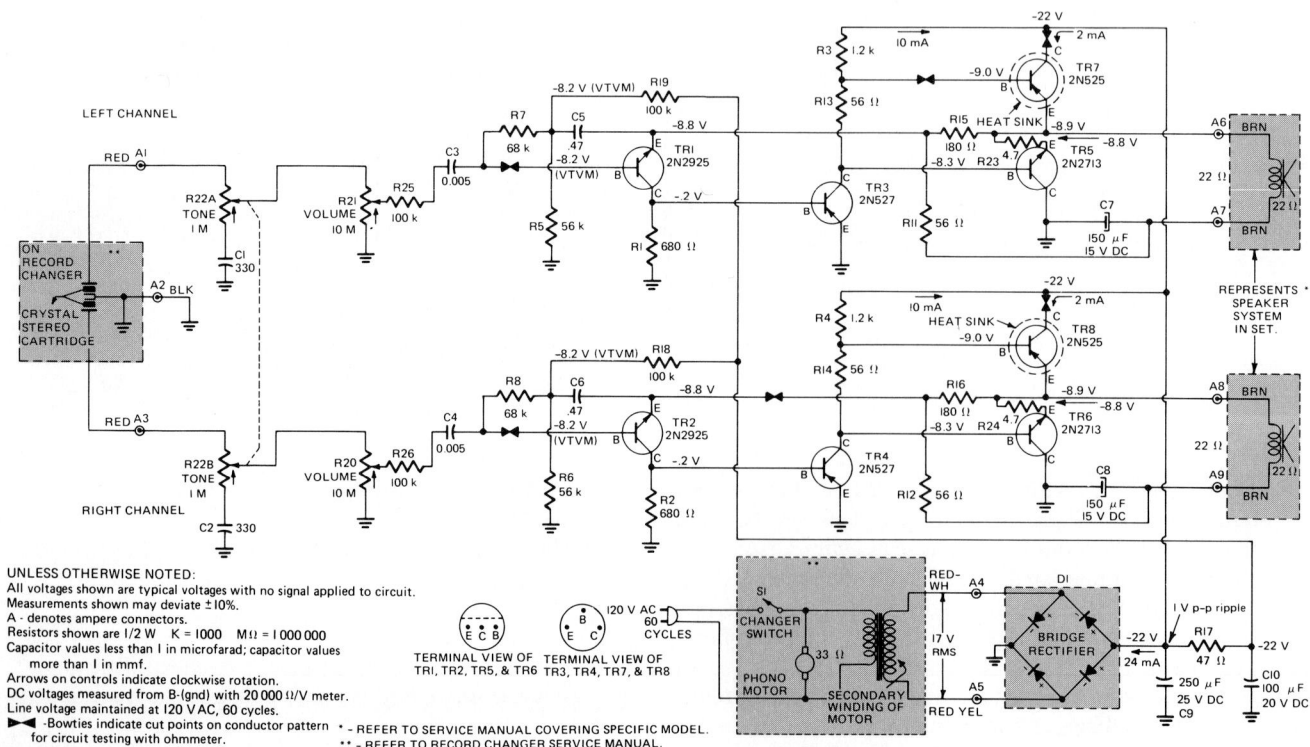

FIG. 27-4-2   Schematic diagram for a simple amplifier. *(General Electric)*

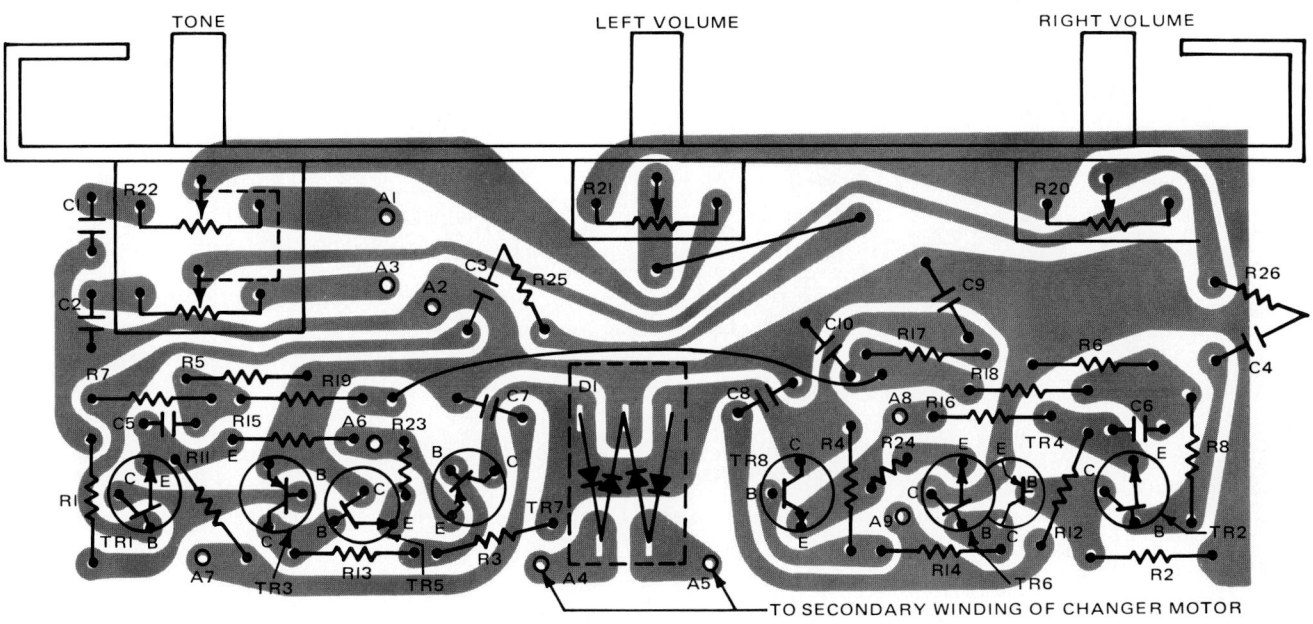

FIG. 27-4-3   Printed circuit and layout of components for amplifier
shown in Fig. 27-4-2. *(General Electric)*

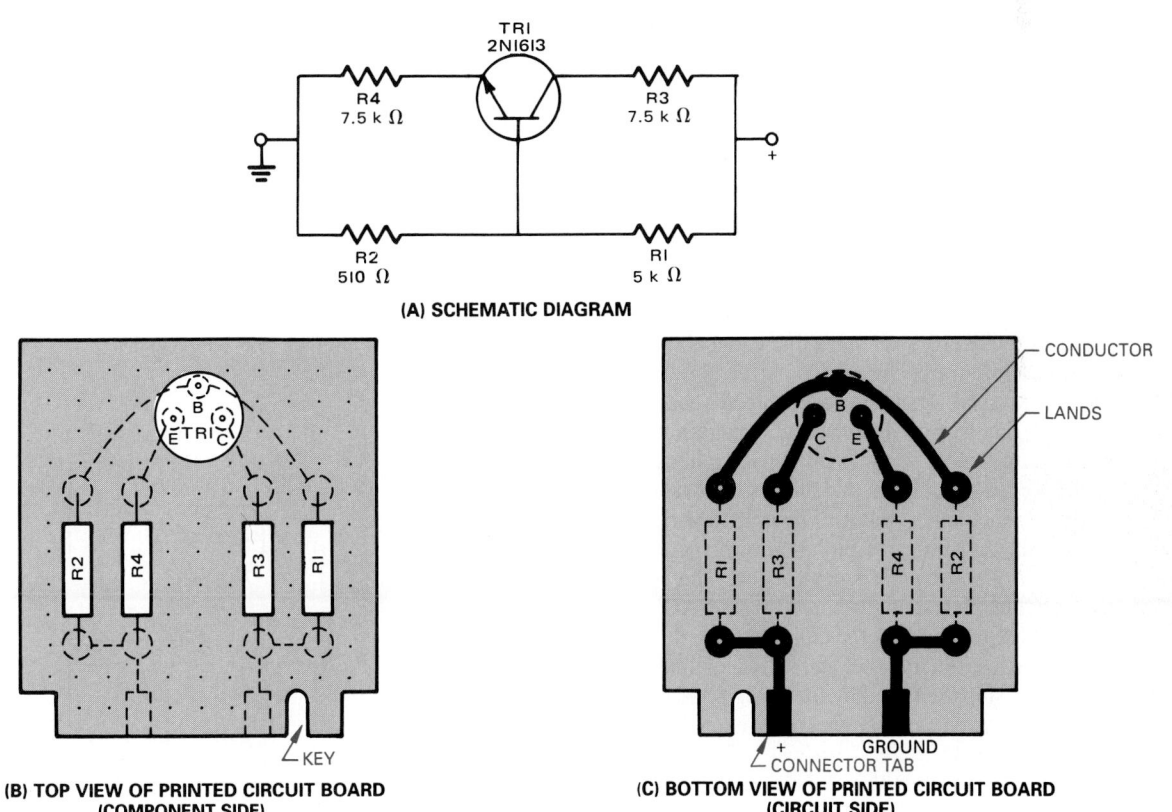

FIG. 27-4-4   Layout of a simple printed circuit board.

paths. The drawing is then photographically reduced to the desired size onto a copper foil which is bonded to a plastic board (Fig. 27-4-4). Holes are drilled at various locations in the board and leads from the components are inserted into circular conductors, called *lands,* located on the circuit side of the

board. When all of the components to be connected are in position, the circuit side of the board is then dipped into molten solder, making all of the soldered connections at one time.

Crossovers should be avoided whenever possible. Running the conductor around the land of the interference components,

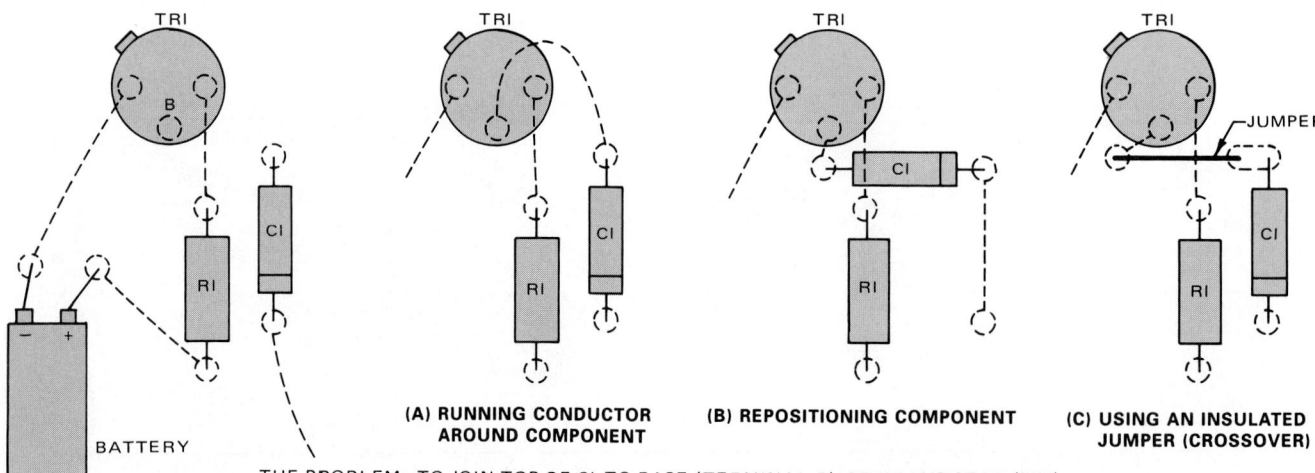

(A) RUNNING CONDUCTOR
AROUND COMPONENT

(B) REPOSITIONING COMPONENT

(C) USING AN INSULATED
JUMPER (CROSSOVER)

THE PROBLEM: TO JOIN TOP OF CI TO BASE (TERMINAL B) OF TRANSISTOR (TRI)

FIG. 27-4-5   Crossovers.

or repositioning the component as shown in Fig. 27-4-5A and B, should be given priority over crossovers. If, however, a crossover cannot be avoided when laying out a printed circuit, then a jumper wire (crossover) may be used, as shown in Fig. 27-4-5C. The jumper wire is treated in the same manner as other components and placed on the component side of the circuit board.

CAD for Printed Wiring Boards

After the schematic diagram is drawn, the printed wiring drawings are created in the following basic steps:

1. The board outline is drawn, showing the exact size and shape of the board. Any areas that cannot be used by the CAD system for placing components and running wiring are also marked. These areas are called keep-out areas.
2. The material list from the schematic diagram is then used to get symbols out of the printed wiring library for each of the components. Note that these symbols are different from the schematic symbols. The printed wiring symbols show the *physical* size and shape of the component instead of just the *electrical* function. The components can either be placed on the board by the CAD operator or automatically placed by the computer.
3. The schematic diagram net list, showing all of the wiring connections, is then used by the CAD system to connect all of the components. The first try at connecting the components is usually called a *rat's nest*. The rat's nest connections are just straight lines between connecting points without any effort to avoid crossing lines, pads, or components (Fig. 27-4-6). From the rat's nest the CAD operator can refine the arrangement of the components to reduce the number of crossing lines and to shorten the lines (Fig. 27-4-7). Because the rat's nest connections are much faster than the final autorouting, an experienced operator can often save hours of routing time by refining the component layout this way.
4. The final autorouting of the printed wiring connections must meet all of the design rules for the particular circuit. As each wire trace is routed, the computer checks

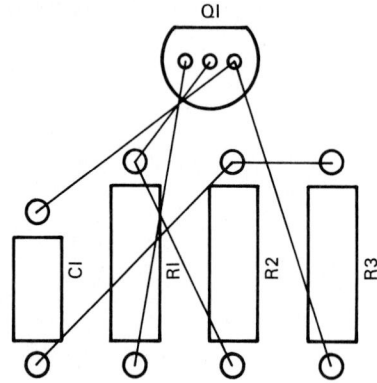

FIG. 27-4-6   First rat's nest connections.

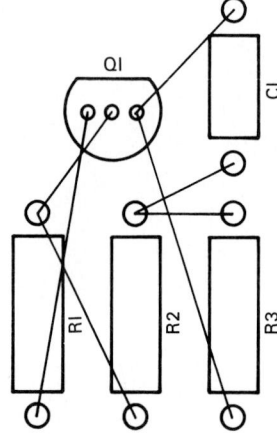

FIG. 27-4-7   Rat's nest after C1 has been moved to shorten connections and reduce crossings.

to make sure proper spacing is maintained to other traces and decides which side of the board (or layer) to run the trace to avoid crossing other wiring traces (Fig. 27-4-8). Depending upon the complexity of the circuit and how crowded the components are on the board, autorouting

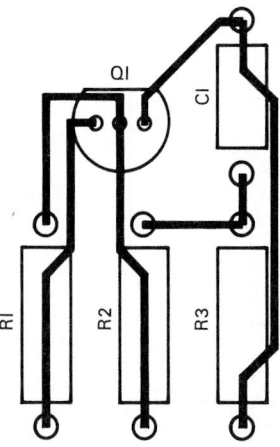

FIG. 27-4-8   Final autorouted connections.

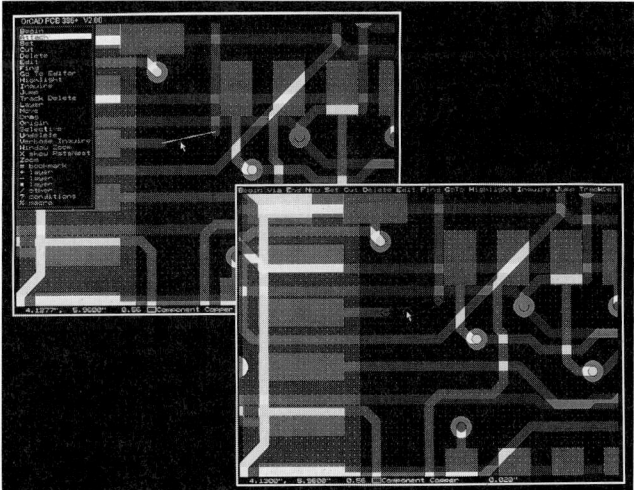

FIG. 27-4-9   Printed wiring as drawn on a CAD system. *(ORCAD)*

may take anywhere from a few minutes to several hours. Figure 27-4-9 shows printed wiring as drawn on a CAD system.

5. After the autorouting is completed, the traces can be either plotted and photographically reproduced on the board, or they can just be electronically transmitted to computers used especially for reproducing the circuit traces on the board.

Basic Rules for Laying Out A Printed Circuit

1. Establish the physical size of each component and where the leads are located. Physical requirements may determine the position of the components on the board. Transistor leads are shown in reverse when viewed from the component side of the circuit board. In most cases the emitter (E) is close to the tab on the transistor (Fig. 27-4-10). Check the orientation of components such as diodes and capacitors in Fig. 27-4-11 (pg. 888). Some capacitors are reversible, while others are not. The polarity is shown when the capacitor is not reversible.
2. Study the schematic diagram. What components have common connections?
3. Prepare a rough trial wiring diagram, with the components drawn to scale, prior to starting the final diagram. Remember, in most cases, an enlarged scale is desirable.
4. Locate the parts on the wiring layout, avoiding crossovers where possible.
5. Locate all holes at intersections of grid lines. Standard grid spacing should be used.
6. Keep conductors to a minimum length (Fig. 27-4-12, pg. 888), placed at least .10 in. (2.5 mm) from the edge of the circuit board, with a minimum width of .04 in. (1 mm), and spaced a minimum of .04 in. (1 mm) apart.
7. Allow a minimum center-to-center spacing for component holes equal to the component length plus .12 in. (3 mm).

FIG. 27-4-10   Location of leads on typical transistors.

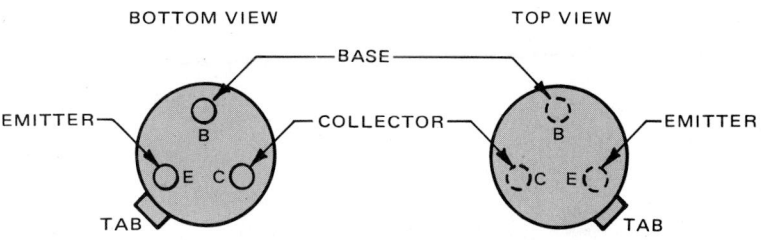

(A) METAL–CASE TYPE

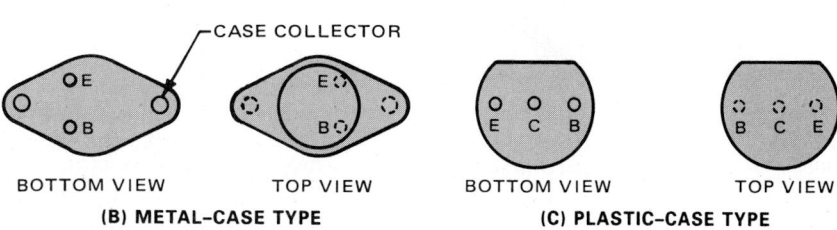

(B) METAL–CASE TYPE   **(C) PLASTIC–CASE TYPE**

FIG. 27-4-11   Orientation
of capacitors
and diodes.

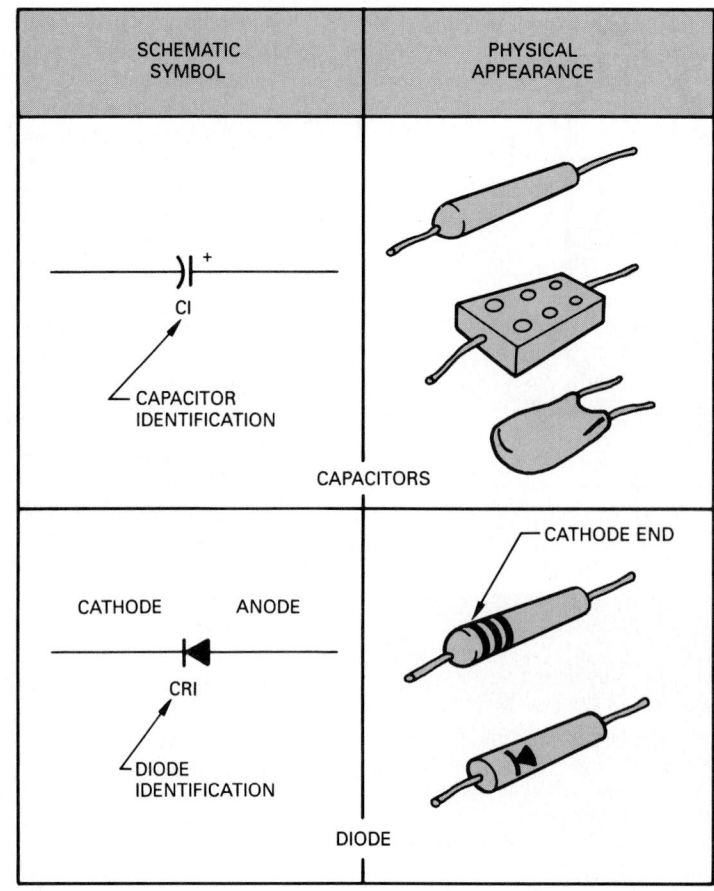

8. Locate long leads, such as grounds, preferably near or around the edge of the circuit board.
9. When using integrated circuits on PC boards, such as shown in Fig. 27-4-13, connect a socket base to hold the IC to the circuit board. Typical base sizes are shown in Fig. 27-4-14.

For more specific rules and standards, refer to ANSI/IPC-D-275.

ASSIGNMENTS

See Assignments 10 through 12 for Unit 27-4 on pages 898 through 900.

27-5   BLOCK AND LOGIC DIAGRAMS

Block Diagrams

Block diagrams in electrical and electronics drafting are used to simplify the understanding of circuits. Their utter simplicity enables you to tell at a glance the relative position and function of each part of the circuit. Block diagrams are used by

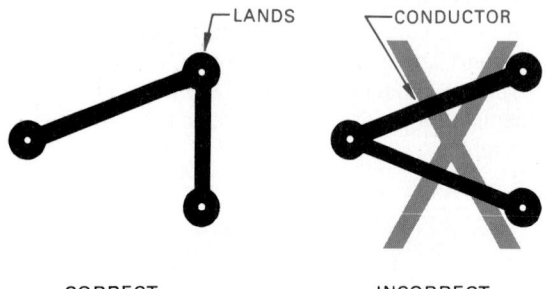

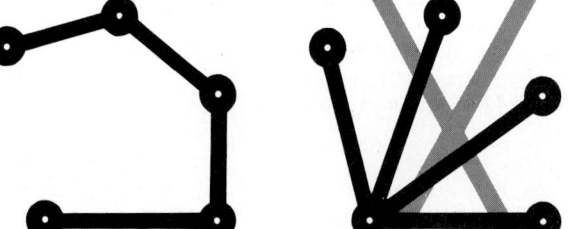

FIG. 27-4-12   Joining lands with conductors.

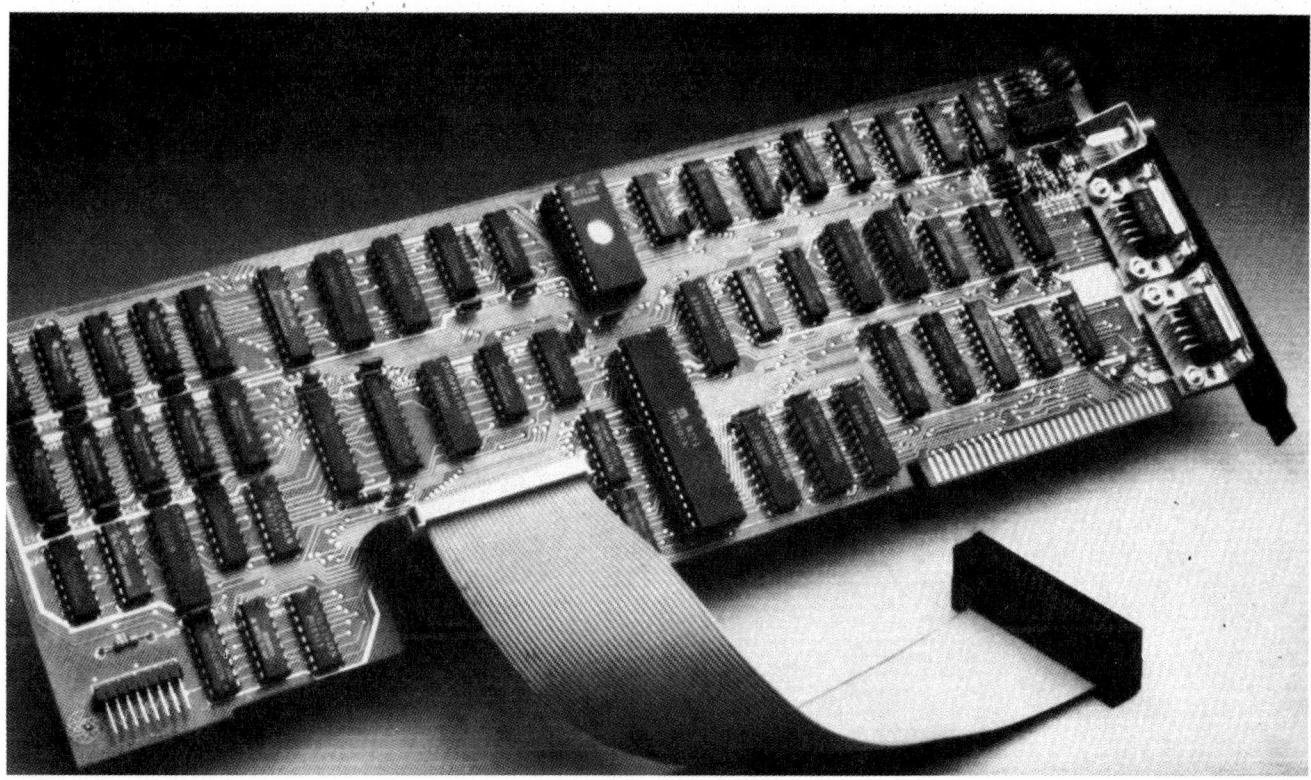

FIG. 27-4-13   A PC board with ICs. *(Quadram Corp.)*

FIG. 27-4-14   IC socket bases.

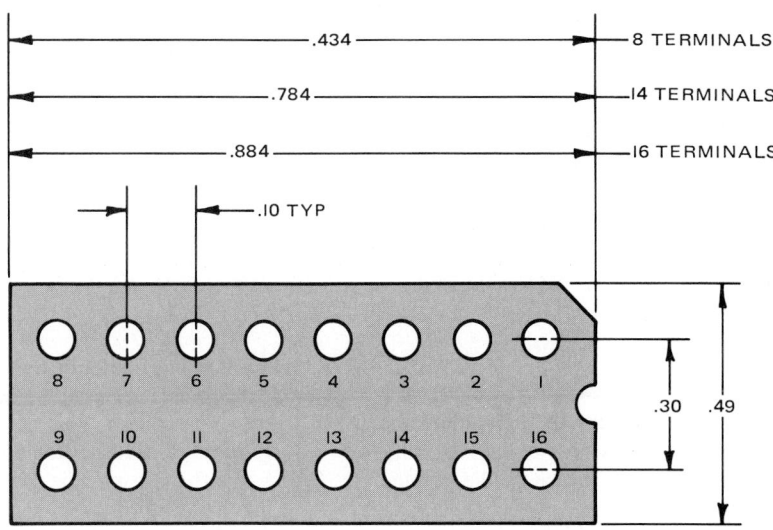

designers in the early stages of planning a project. You should note that a block diagram shows only the relationship between the components, and does not show the electrical connections.

A *block diagram,* as its name suggests, consists of a series of blocks (or boxes), connected by straight lines. The identities of the respective units are placed inside or adjacent to the blocks, in abbreviated form where necessary. Each block in the diagram represents a stage or subcircuit within a circuit. These blocks are usually drawn as squares, rectangles, or triangles and are uniform in size, shape, and spacing, regardless of the physical size they represent. Certain components such as antennas,

speakers, and microphones are shown by means of a symbol rather than a block (Fig. 27–5–1, pg. 890).

The blocks are joined by a single line, which indicates the signal path from block to block. The signal path is normally shown from left to right and the connecting line may be lighter or darker than the blocks, depending on which is being empha- sized. With arrows used on the connecting line to show the signal path, it is called a flow diagram. Block diagrams for alternative or future components are indicated by a broken line, the broken line being the same weight as the lines showing the solid blocks.

FIG. 27-5-1 Block diagram of a radio-phonograph combination.

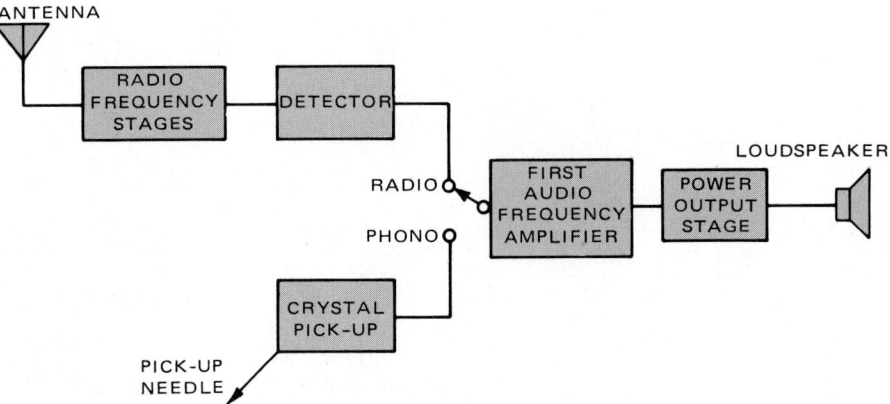

FIG. 27-5-2 Logic flow diagram.

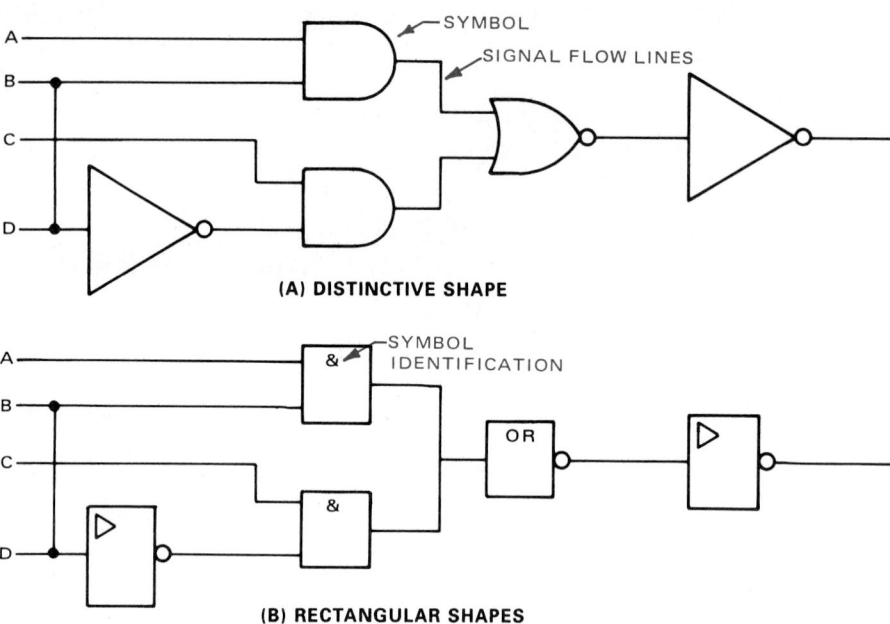

Logic Diagrams

A *logic diagram* is a diagram representing the logic elements and their implementations without necessarily expressing construction or engineering details. A logic symbol is the graphic representation of a logic function.

Computer design has been largely responsible for the steady growth of logic functions which can be performed by basic circuits (Fig. 27–5–2).

Graphic Symbols

The symbols representing logic functions shown in Fig. 27–5–2 are the symbols approved by ANSI/IEEE Std. 91 (ANSI Y32.14)—Graphic Symbols for Logic Diagrams. The distinctive symbols shown in Fig. 27–5–3 are the symbols more commonly used since they are easier to understand.

The following guidelines should be observed when using graphical symbols on logic diagrams.

1. In most cases, the meaning of a symbol is defined by its form. The size and line thickness does not affect the meaning of a symbol. In some cases, it may be desirable to use different sizes of symbols to emphasize certain aspects, or to facilitate the inclusion of additional information.
2. Logic symbols size should be governed by the space necessary for internal annotations and the length of the side needed to accommodate input and output lines and pin numbers at an acceptable spacing.
3. Graphic symbols may be drawn to any proportional size that suits a particular diagram, provided that the selection of size takes into account the anticipated reduction or enlargement. Standard proportions for these symbols are shown in Fig. 27-5-4.
4. It is preferred that all text be readable from the bottom, that is, right side up.

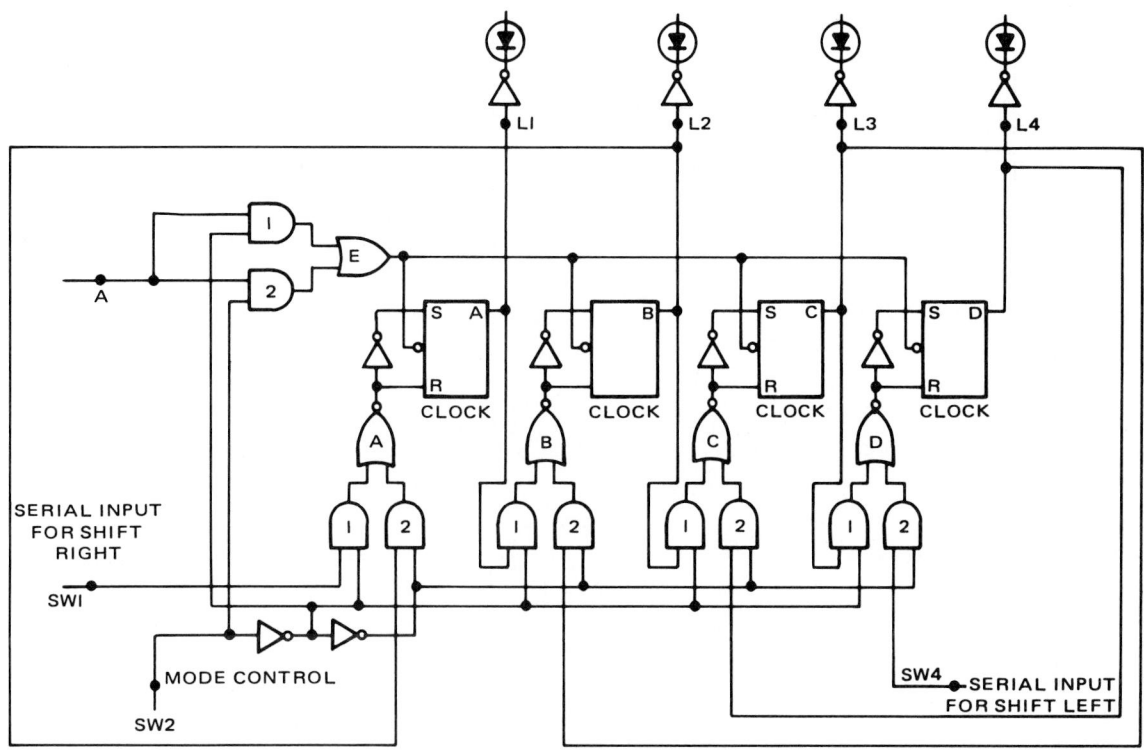

FIG. 27-5-3   Partial layout of shift-right, shift-left logic diagram. *(Heath Company)*

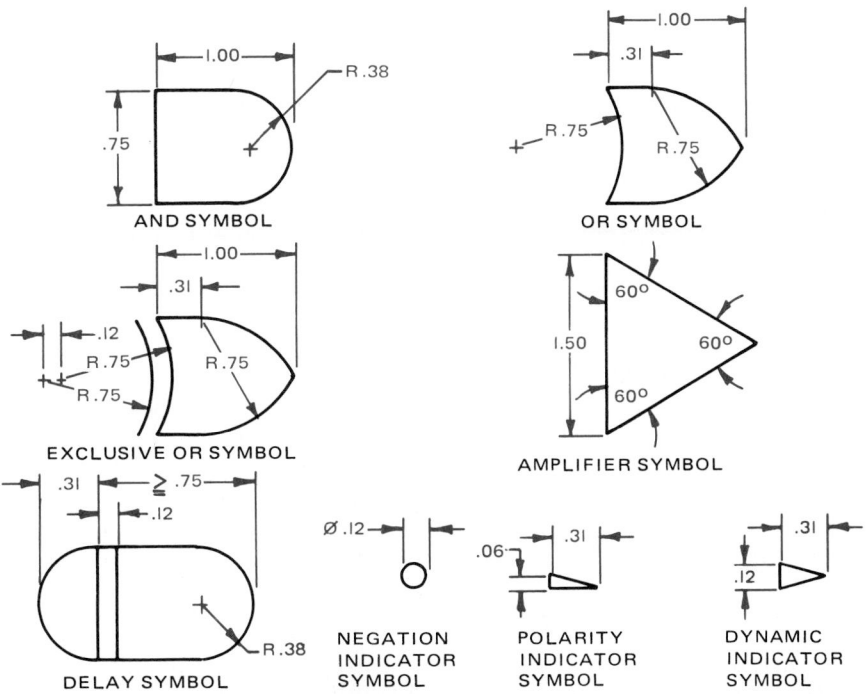

FIG. 27-5-4   Recommended symbol proportions for distinct-shape logic symbols.

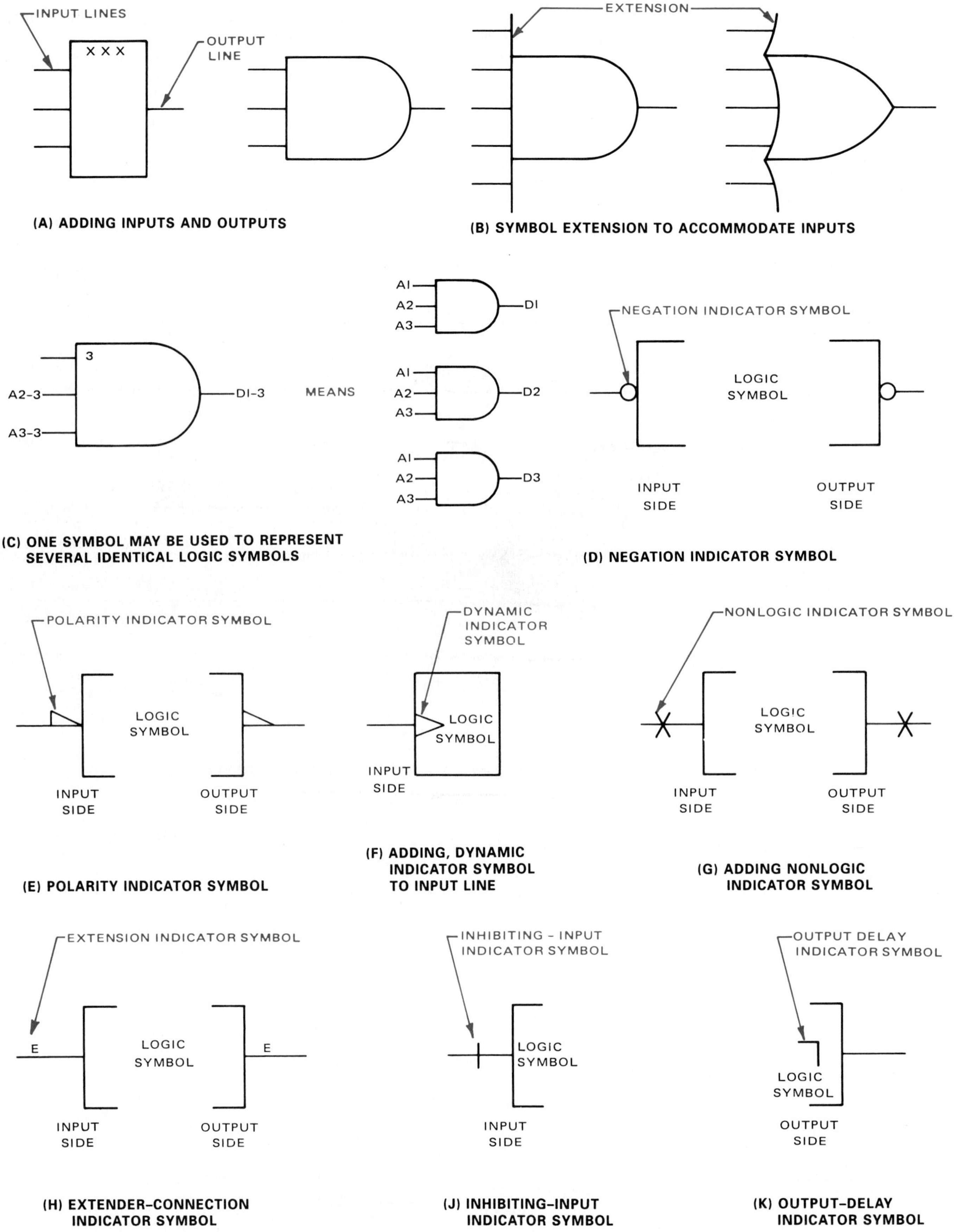

INPUT LINES
XXX
OUTPUT LINE
(A) ADDING INPUTS AND OUTPUTS
EXTENSION
(B) SYMBOL EXTENSION TO ACCOMMODATE INPUTS
3
A2-3
A3-3
DI-3
MEANS
AI
A2
A3
DI
D2
D3
(C) ONE SYMBOL MAY BE USED TO REPRESENT SEVERAL IDENTICAL LOGIC SYMBOLS
NEGATION INDICATOR SYMBOL
LOGIC SYMBOL
INPUT SIDE
OUTPUT SIDE
(D) NEGATION INDICATOR SYMBOL
POLARITY INDICATOR SYMBOL
LOGIC SYMBOL
INPUT SIDE
OUTPUT SIDE
(E) POLARITY INDICATOR SYMBOL
DYNAMIC INDICATOR SYMBOL
LOGIC SYMBOL
INPUT SIDE
(F) ADDING, DYNAMIC INDICATOR SYMBOL TO INPUT LINE
NONLOGIC INDICATOR SYMBOL
LOGIC SYMBOL
INPUT SIDE
OUTPUT SIDE
(G) ADDING NONLOGIC INDICATOR SYMBOL
EXTENSION INDICATOR SYMBOL
E
LOGIC SYMBOL
E
INPUT SIDE
OUTPUT SIDE
(H) EXTENDER-CONNECTION INDICATOR SYMBOL
INHIBITING - INPUT INDICATOR SYMBOL
LOGIC SYMBOL
INPUT SIDE
(J) INHIBITING-INPUT INDICATOR SYMBOL
OUTPUT DELAY INDICATOR SYMBOL
LOGIC SYMBOL
OUTPUT SIDE
(K) OUTPUT-DELAY INDICATOR SYMBOL

FIG. 27-5-5 Adding additional logic symbols to basic symbols.

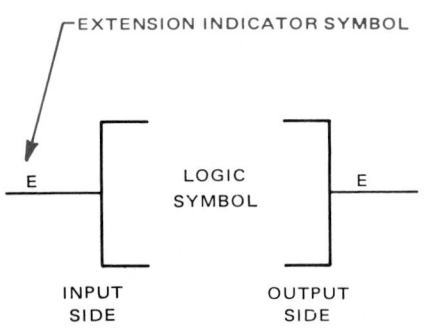

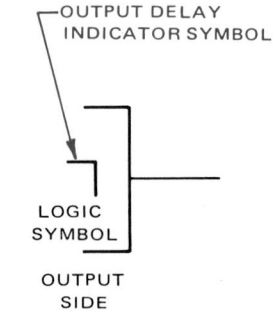

5. Input and output lines are preferably placed on opposite sides of the symbol and should join the outline of the symbol at right angles. Input lines are normally shown on the left side, output lines on the right side (Fig. 27-5-5).

6. Pin numbers, when used, are shown outside the graphic symbol (Fig. 27-5-6).

7. Recommended arrangements of internal information are shown in Fig. 27-5-7.

FIG. 27-5-6   Adding pin numbers to basic symbols.

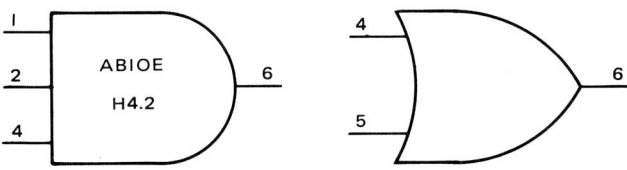

(A) DISTINCTIVE SHAPED SYMBOLS

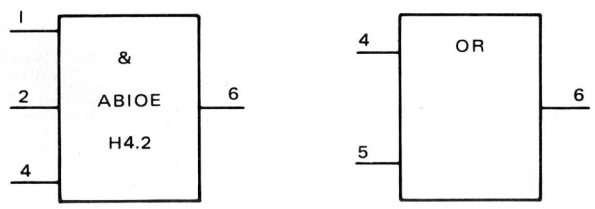

(B) RECTANGULAR SHAPED SYMBOLS

FIG. 27-5-7   Recommended arrangement of internal information.

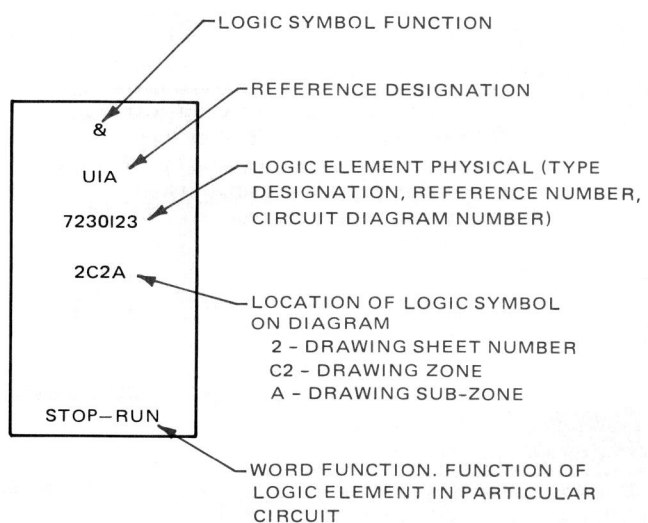

LOGIC SYMBOL FUNCTION

REFERENCE DESIGNATION

LOGIC ELEMENT PHYSICAL (TYPE DESIGNATION, REFERENCE NUMBER, CIRCUIT DIAGRAM NUMBER)

LOCATION OF LOGIC SYMBOL ON DIAGRAM
   2 – DRAWING SHEET NUMBER
   C2 – DRAWING ZONE
   A – DRAWING SUB-ZONE

WORD FUNCTION. FUNCTION OF LOGIC ELEMENT IN PARTICULAR CIRCUIT

EXAMPLE I

EXAMPLE 2

893

Basic Rules for Laying Out A Logic Diagram

1. The parts should be spaced to provide an even balance between blank spaces and lines.
2. The logic diagram should use a layout which follows the circuit, signal, or transmission path either from input to output, source to load, or in order of functional sequence.
3. The layout should be such that the principal flow of information is from left to right, or from top to bottom, of the diagram.
4. Functionally related symbols should be grouped and placed close to one another.

5. Lines should be drawn horizontally or vertically except in those cases where oblique lines aid in the clarity of the diagram.
6. Highways can be used to simplify logic diagrams where groups of similar signals are encountered.

ASSIGNMENTS

See Assignments 13 through 15 for Unit 27–5 on pages 901 to 902.

ASSIGNMENTS FOR CHAPTER 27

ASSIGNMENTS FOR UNIT 27-2, SCHEMATIC DIAGRAMS

1. Make a schematic diagram from one of the diagrams shown in Fig. 27-3-A or 27-3-B. For electrical symbols not shown in the Appendix, use the symbols shown on the assignment drawings. Where applicable, add reference designations and numerical values to the component symbols.
2. Make a diagram of one of the circuits shown in Figs. 27-2-A to 27-2-C. Refer to the Appendix for symbols. Where applicable, add reference designations and numerical values to the component symbols.
3. Make a schematic diagram of the radio amplifier shown in Fig. 27-2-D. Refer to the Appendix for symbols. Where applicable, add reference designations and numerical values to the component symbols.

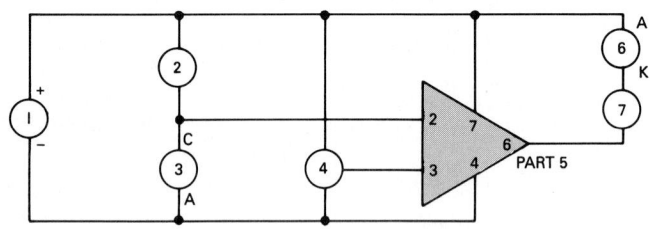

MATERIAL LIST

I	I2-V battery	5	74I operational
2	27 000-Ω resistor		amplifier (IC)
3	6.2-V photo diode	6	Semiconductor diode
4	100 000-Ω potentiometer	7	I000-Ω resistor
	(adjustable contact resistor)		

FIG. 27-2-A Low battery indicator.

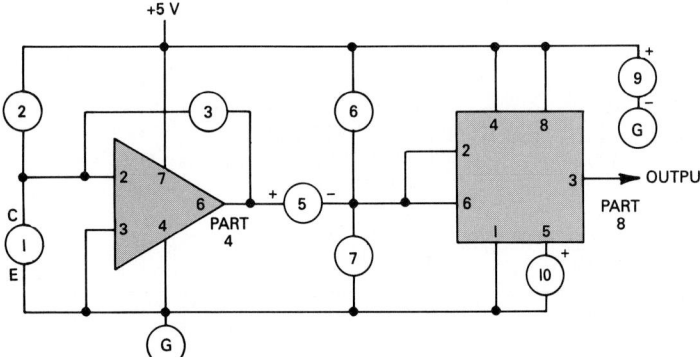

FIG. 27-2-B Fiber optic receiver.

MATERIAL LIST

I	Phototransistor	6	100 000 Ω resistor
	2 terminal (NPN-type)	7	100 000 Ω resistor
2	100 000-Ω resistor	8	555 timer IC
3	50 000-Ω variable	9	0.I μF capacitor
	resistor (rheostat)	I0	0.0I μF capacitor
4	741 operational amplifier IC	Ⓖ	Ground
5	0.I-μF capacitor		

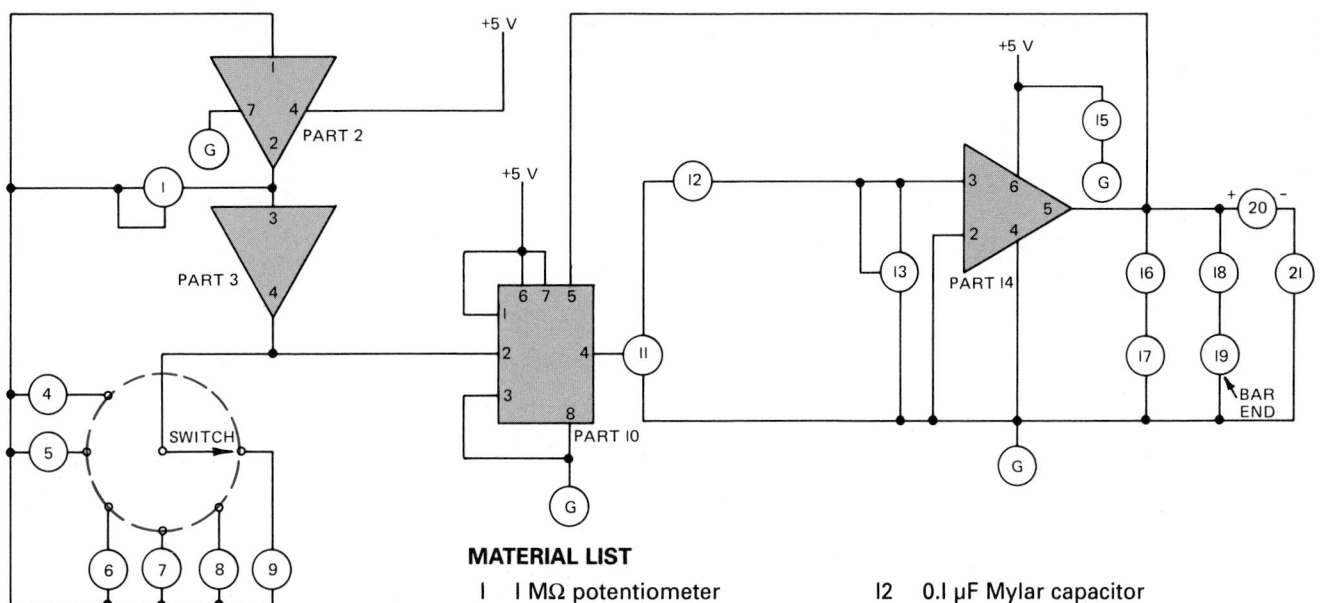

FIG. 27-2-C   Sound-effects generator.

MATERIAL LIST

I	I MΩ potentiometer	I2	0.I μF Mylar capacitor
2	Amplifier 4069 IC	I3	100 000 Ω potentiometer
3	Amplifier 4069 IC	I4	amplifier LM386 IC4
4	I μF capacitor	I5	0.I μF Mylar capacitor
5	0.47 μF capacitor	I6	0.047 μF Mylar capacitor
6	0.I μF capacitor	I7	I0 Ω resistor
7	0.05 μF capacitor	I8	I000 Ω resistor
8	500 μF capacitor	I9	Phototransistor-2 terminal (NPN type)
9	I00 μF capacitor	20	250 μF capacitor
I0	TL 507C IC (integrated circuit)	2I	8 Ω speaker
II	100 000 Ω potentiometer	Ⓖ	Ground

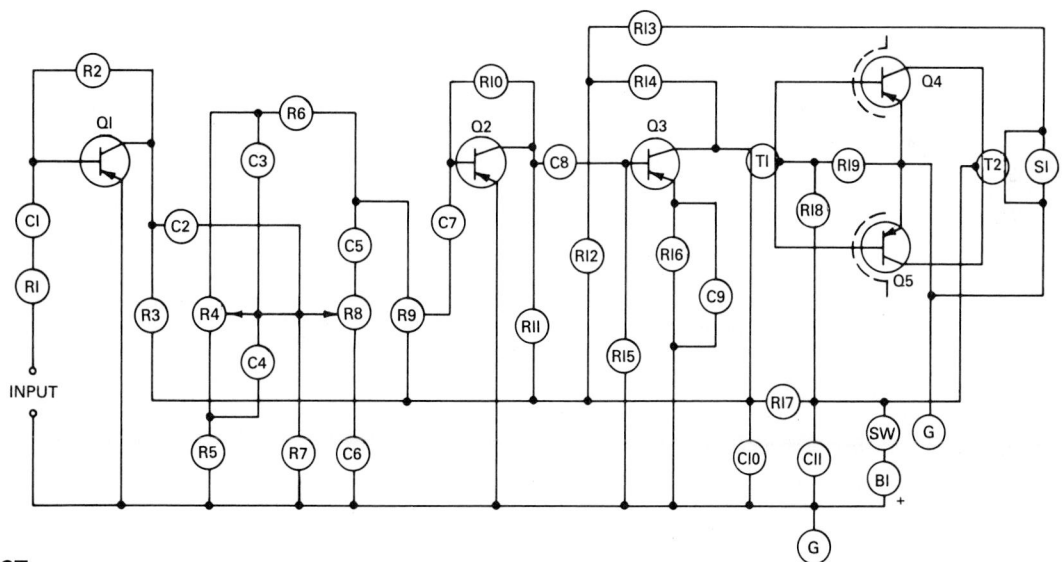

MATERIAL LIST

RI	10 000 Ω	R9	Adjustable Contact	CI	8 μF	TI	IK/IK C.T. transformer with magnetic core
R2	150 000 Ω		10 000 Ω	C2	.50 μF	T2	100 C.T./V.C. transformer with
R3	6800 Ω		Taper Audio V.C.	C3	.02 μF		adjustable magnetic core
R4	Adjustable Contact	RI0	220 000 Ω	C4	.20 μF	QI, Q2, Q3	GE2N323
	50 000 Ω	RII	2200 Ω	C5	.005 μF	Q4, Q5	GE2N321 (with clip-on heatsink)
	Bass Linear	RI2	4700 Ω	C6	.I0 μF	SI	speaker
R5	I000 Ω	RI3	33 000 Ω	C7	I0 μF	SW	switch
R6	I0 000 Ω	RI4	47 000 Ω	C8	I0 μF	BI	battery, 6V
R7	I00 000 Ω	RI5	1500 Ω	C9	50 μF	Ⓖ	Ground
R8	Adjustable Contact	RI6	330 Ω	CI0	50 μF		
	50 000 Ω	RI7	220 Ω	CII	50 μF		
	Treble Linear	RI8	I200 Ω				
		RI9	33 Ω				

FIG. 27-2-D   Five-transistor radio amplifier.

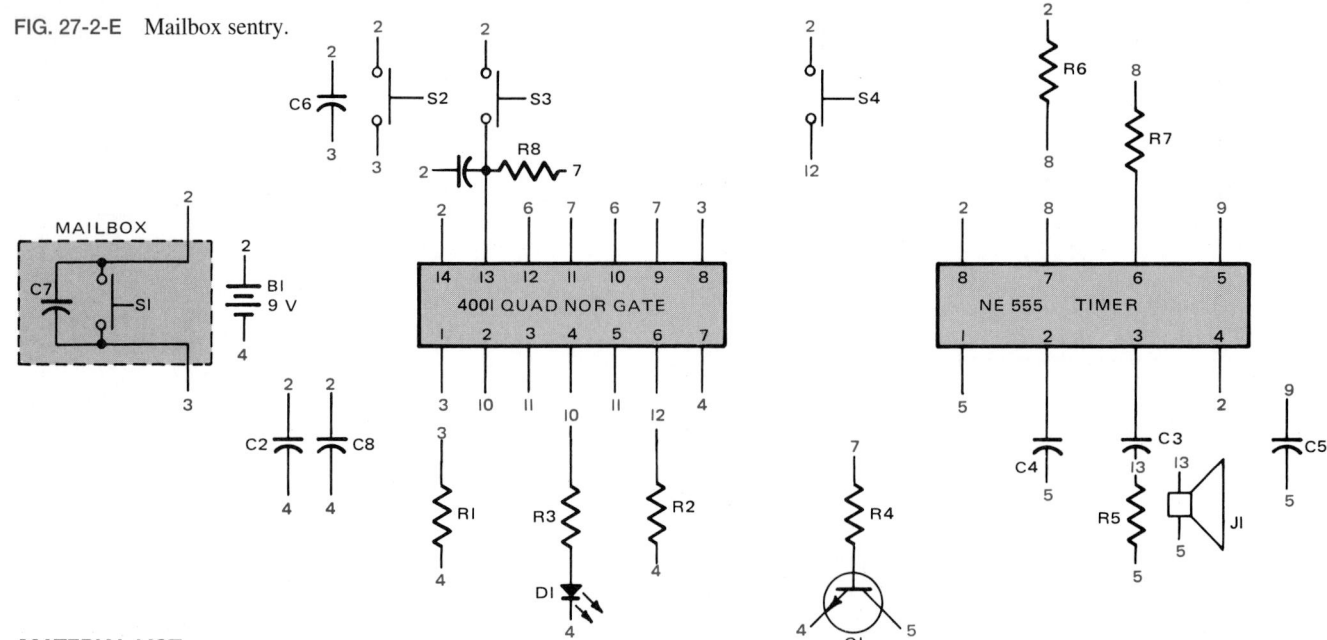

FIG. 27-2-E   Mailbox sentry.

MATERIAL LIST

BI   9-V battery
CI   I0-µF, 25-V electrolytic
C2, C3   I-I0-µF, 25-V electrolytic
C4   0.00I.-µF, 25-V ceramic disc capacitor
C5   0I.-µF, 25-V ceramic disc capacitor
C6, C7, C8   0.0I-µF, 25-V ceramic disc capacitor
IC   400I quad NOR gate
IC   555 timer

DI   Red light-emitting diode
QI   2N2222 NPN silicon transistor

The following are ¼-W, 10% resistors:
RI, R2   22 kΩ
R3   I kΩ
R4   4.7 kΩ
R5   I0 kΩ
R6   47 kΩ
R7   I00 kΩ
R8   2.2 MΩ

SI   Normally open microswitch, magnetic reed switch
S2 through S4   Normally open pushbutton switch, panel mount
JI   8-Ω, 2-in. speaker

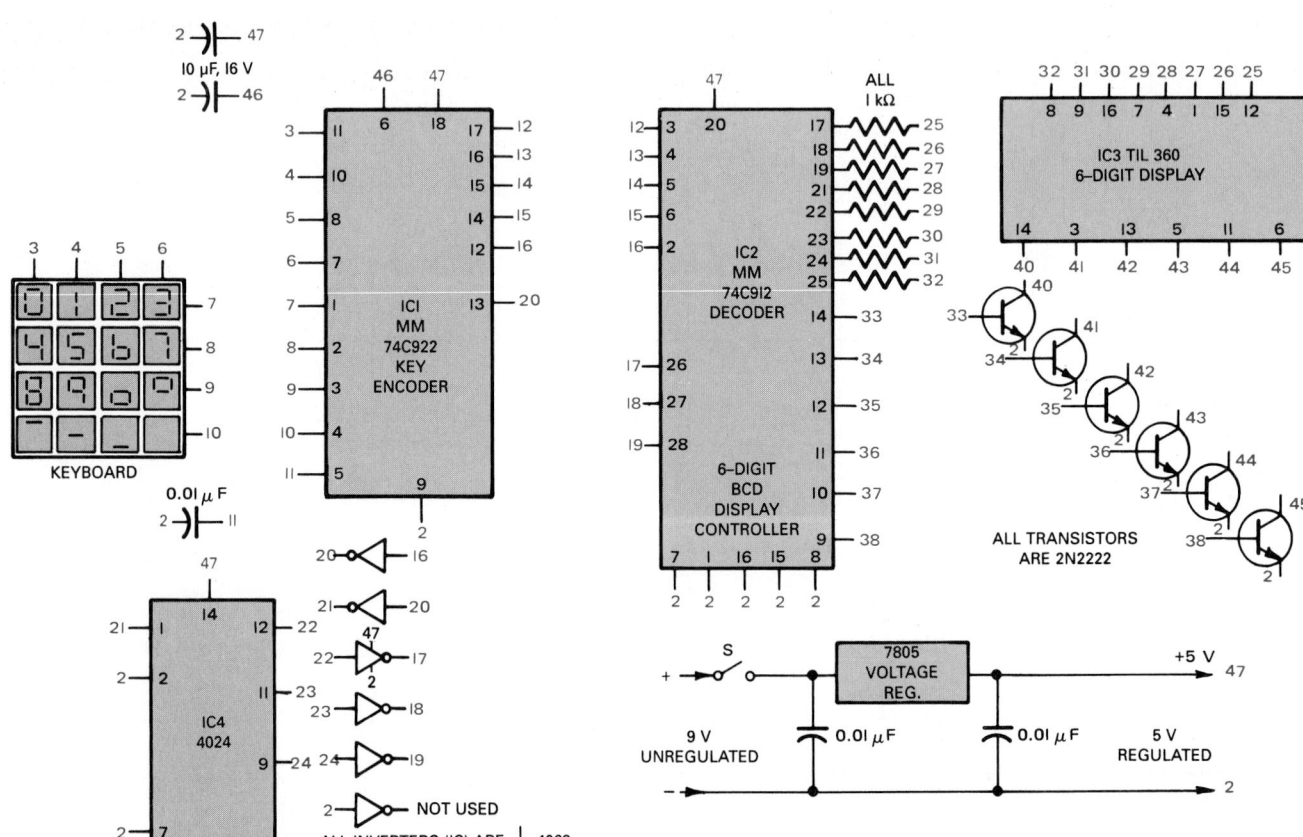

FIG. 27-2-F   Keyboard, display, and circuit using ICs.

4. Make a schematic diagram of the mailbox sentry shown in Fig. 27-2-E. Where applicable, add reference designations and numerical values to the component symbols. *Note:* Parts may be repositioned and similar numbers are electrically joined together.

5. Make a schematic diagram of the keyboard display circuit shown in Fig. 27-2-F. Where applicable, add reference designations and numerical values to the component symbols. *Note:* Parts may be repositioned, similar numbers are electrically joined together, number 2 is ground.

7. Make a point-to-point connection diagram of the clothes dryer shown in Fig. 27-3-2.

8. Make a point-to-point connection diagram of one of the appliances shown in Figs. 27-3-A or 27-3-B (pg. 898).

9. Prepare harness drawings for one of the appliances shown in Figs. 27-3-2, 27-3-A, or 27-3-B. Use the scale of 1:4 (U.S. Customary) or 1:5 (metric) to measure the cable lengths and positions, rounding off the measurements to the nearest .50 in. or 10 mm. Minimum lead extension extending from harness to be 2 in. or 50 mm. Strip the ends of wires for .50 in. (U.S. Customary) or 15 mm (metric). Use rectangular coordinate dimensioning.

ASSIGNMENTS FOR UNIT 27-3, WIRING (CONNECTION) DIAGRAMS

6. Make a highway-type connection diagram of the boat's electrical system shown in Fig. 27-3-1.

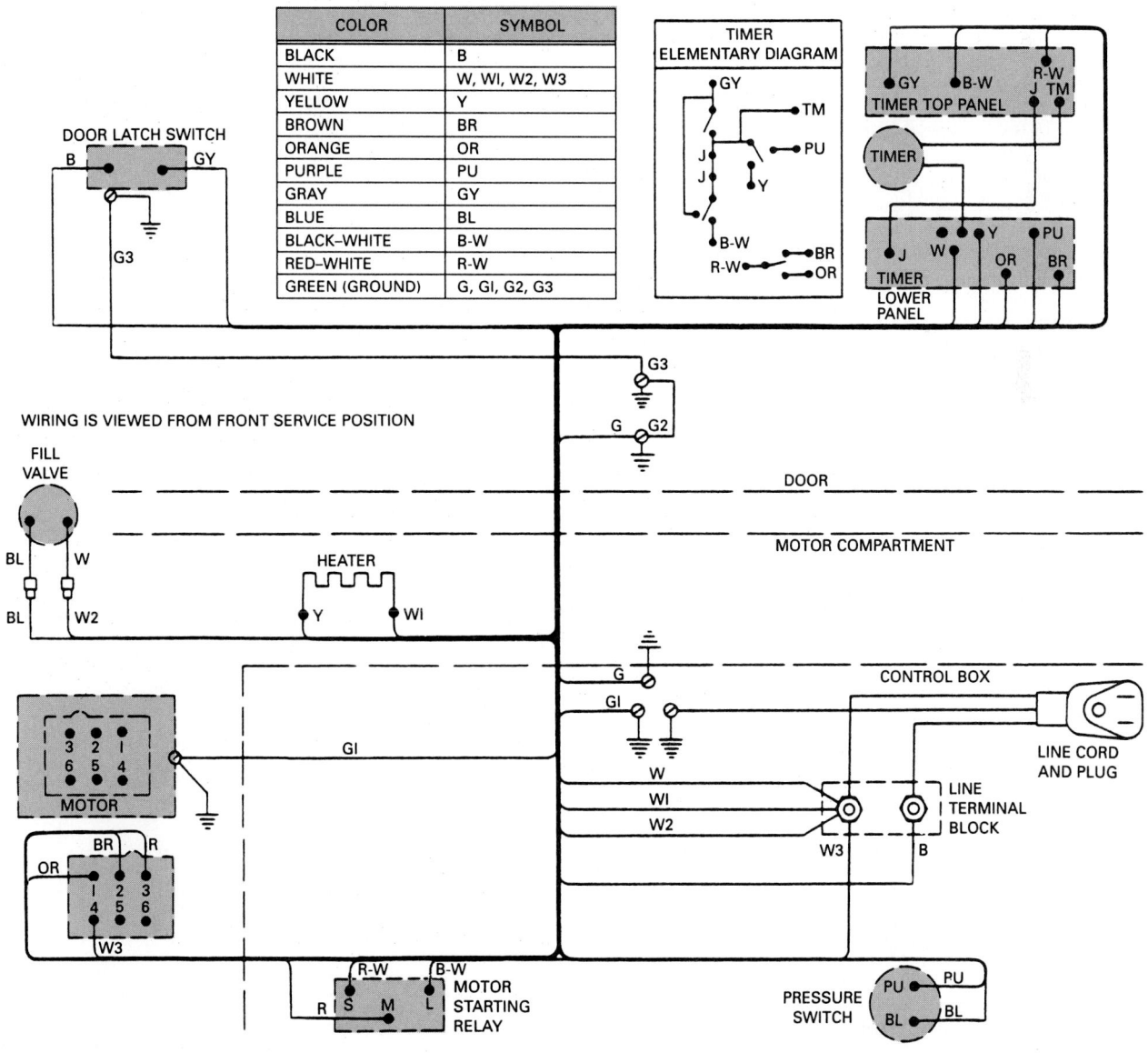

FIG. 27-3-A   Wiring diagram for a dishwasher. *(Frigidaire)*

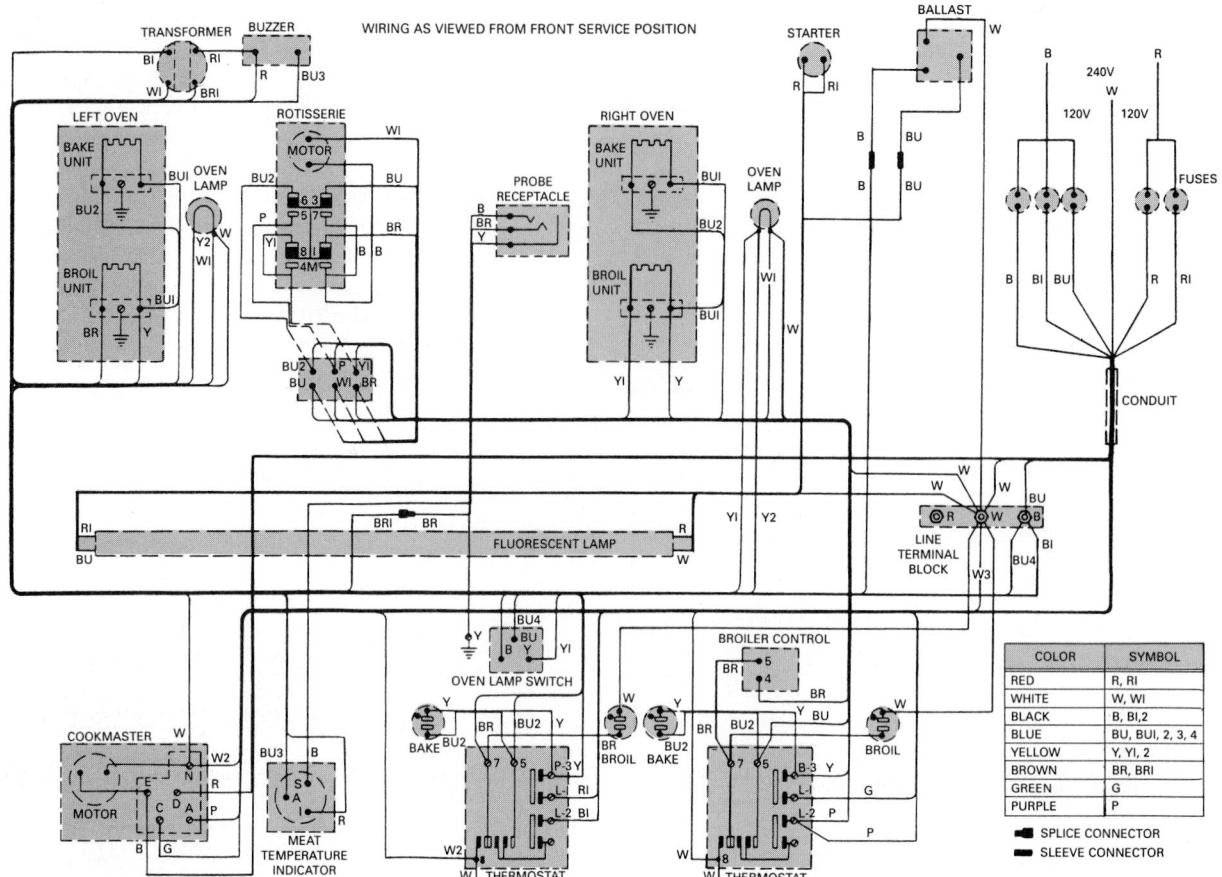

FIG. 27-3-B   Wiring diagram for a wall oven. *(Frigidaire)*

ASSIGNMENTS FOR UNIT 27-4, PRINTED CIRCUITS

10. Draw the top and bottom views of the circuit board shown in either Fig. 27-4-A or 27-4-B. Complete the conductor connections in the top view from the information shown in the schematic diagram. The grid lines shown on the printed circuit boards are .25 in. or 5 mm on centers. Conductor size and minimum clearance to be .06 in. or 1.5 mm. Land diameter is .12 in. or 3 mm. Scale 2:1.

11. Make a schematic diagram from the printed circuit and the component location diagram for one of the circuits shown in either Fig. 27-4-C or 27-4-D (pg. 900). Include on the drawing an item list calling for the capacitors, resistors, and transistors.

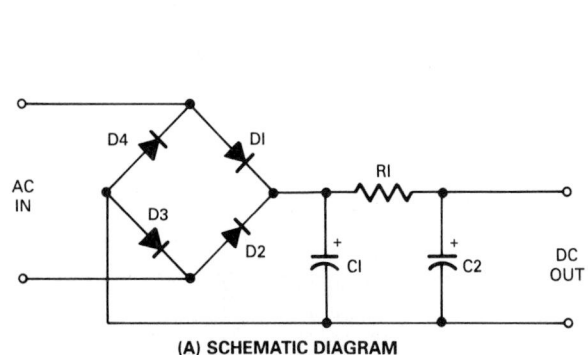

(A) SCHEMATIC DIAGRAM

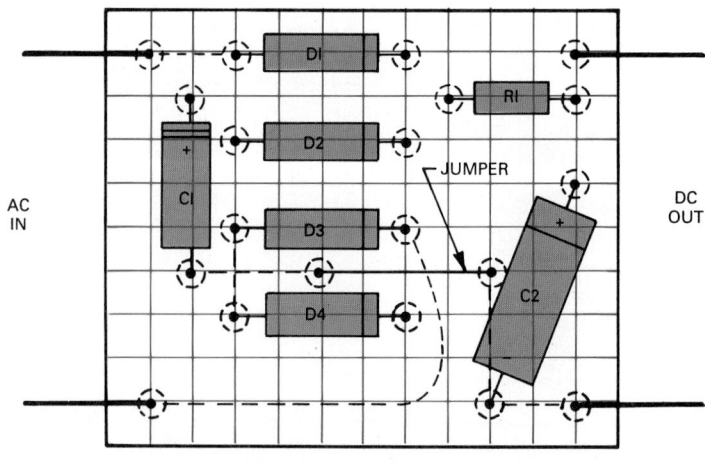

(B) PARTIALLY COMPLETED TOP VIEW (COMPONENT SIDE)

FIG. 27-4-A   Printed circuit board 1.

FIG. 27-4-B Printed circuit
board 2.

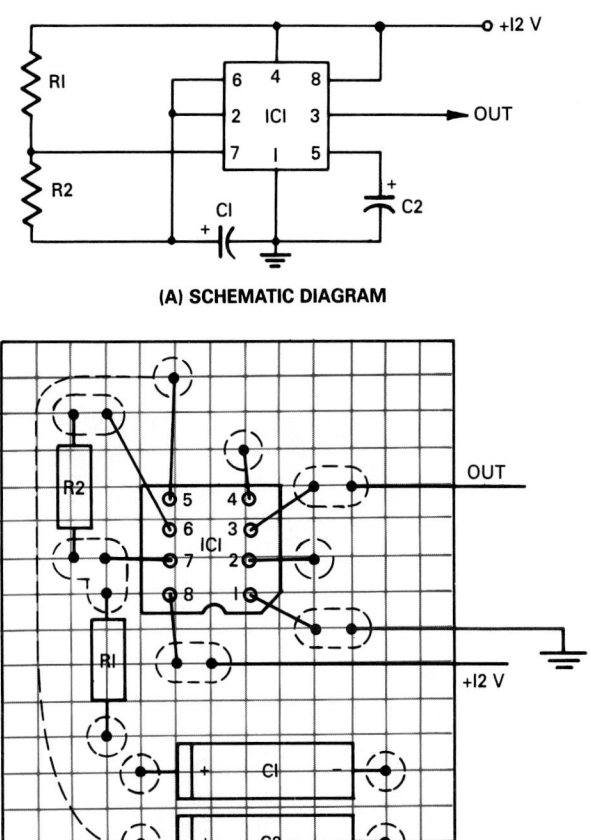

(A) SCHEMATIC DIAGRAM

(B) PARTIALLY COMPLETED TOP VIEW
(COMPONENT SIDE)

FIG. 27-4-C Preamplifier.
(General Electric)

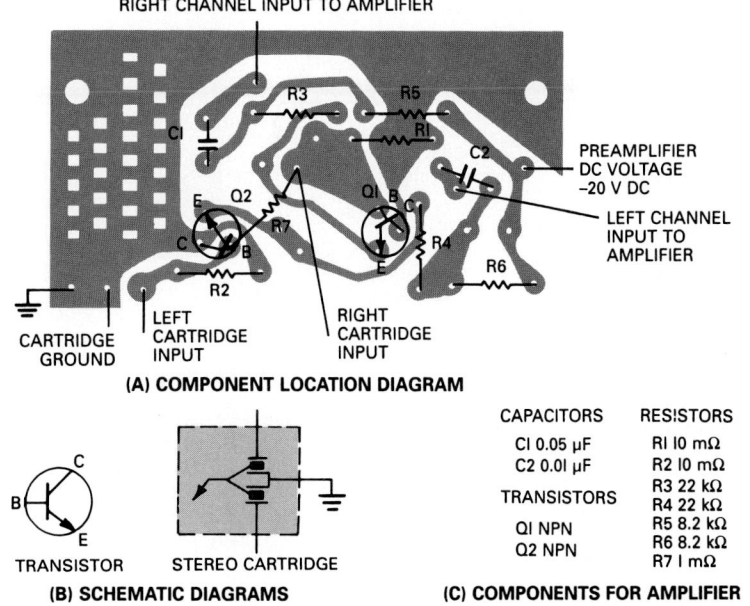

(A) COMPONENT LOCATION DIAGRAM

TRANSISTOR   STEREO CARTRIDGE
(B) SCHEMATIC DIAGRAMS

CAPACITORS	RESISTORS
CI 0.05 µF	RI I0 mΩ
C2 0.0I µF	R2 I0 mΩ
	R3 22 kΩ
TRANSISTORS	R4 22 kΩ
	R5 8.2 kΩ
QI NPN	R6 8.2 kΩ
Q2 NPN	R7 I mΩ

(C) COMPONENTS FOR AMPLIFIER

12. The three amplifiers in the schematic diagram shown in Fig. 27-4-E are to be replaced by the LM348 IC. Amplifier 4 (leads 12, 13, and 14) is not used. Redraw the schematic diagram.

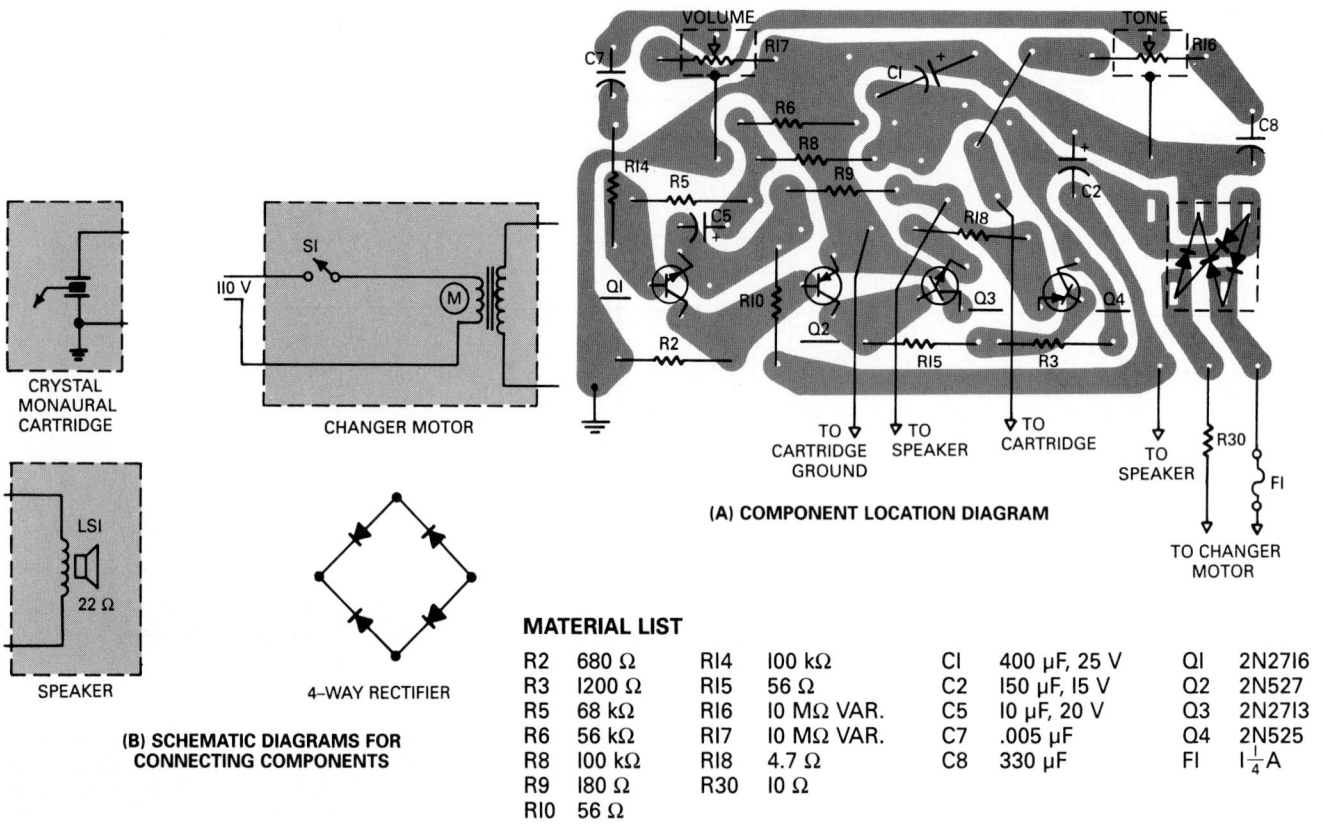

(A) COMPONENT LOCATION DIAGRAM

(B) SCHEMATIC DIAGRAMS FOR CONNECTING COMPONENTS

MATERIAL LIST

R2	680 Ω	RI4	100 kΩ	CI	400 μF, 25 V	QI	2N2716
R3	1200 Ω	RI5	56 Ω	C2	150 μF, 15 V	Q2	2N527
R5	68 kΩ	RI6	10 MΩ VAR.	C5	10 μF, 20 V	Q3	2N2713
R6	56 kΩ	RI7	10 MΩ VAR.	C7	.005 μF	Q4	2N525
R8	100 kΩ	RI8	4.7 Ω	C8	330 μF	FI	$1\frac{1}{4}$ A
R9	180 Ω	R30	10 Ω				
RI0	56 Ω						

(C) COMPONENTS

FIG. 27-4-D   Phonograph amplifier. *(General Electric)*

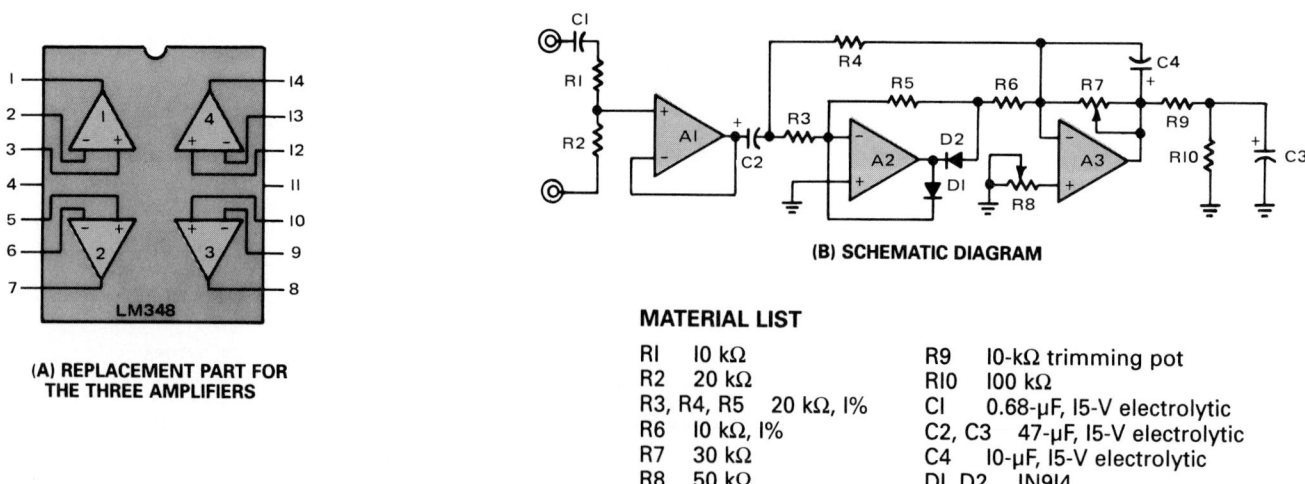

(A) REPLACEMENT PART FOR THE THREE AMPLIFIERS

(B) SCHEMATIC DIAGRAM

MATERIAL LIST

RI	10 kΩ	R9	10-kΩ trimming pot
R2	20 kΩ	RI0	100 kΩ
R3, R4, R5	20 kΩ, 1%	CI	0.68-μF, 15-V electrolytic
R6	10 kΩ, 1%	C2, C3	47-μF, 15-V electrolytic
R7	30 kΩ	C4	10-μF, 15-V electrolytic
R8	50 kΩ	DI, D2	1N914

(C) COMPONENTS

FIG. 27-4-E   Partial layout for an AC-DC converter.

ASSIGNMENTS FOR UNIT 27-5, BLOCK AND LOGIC DIAGRAMS

13. Prepare a block diagram of a home intercom system from Fig. 27-5-A and the following information:
    1. Four intercoms (located at the front door, the back door, the work shop, and the recreation room) connect to a multiplexer which in turn connects to two units, a speech synthesizer, and speech recognition hardware.
    2. The last two units in (1) connect (two-way flow) to the CPU module.
    3. Two sensors, one indoor and one outdoor, feed into the CPU module.
    4. The power to operate the system feeds into a line voltage monitor and battery backup system, which in turn is connected to the CPU module.
    5. The line voltage monitor and backup system also connects to an RCT unit, which in turn is connected (two-way flow) to the CPU module.

6. A 16-channel ac remote control transmitter links the power control receivers located throughout the house to the CPU module.

14. Prepare a block diagram with arrows showing direction flow of a hybrid TV set from the following information:
    1. Antenna connects to a TV tuner/mixer
    2. TV tuner/mixer connects to a broadband IF amplifier
    3. Broadband IF amplifier connects to two units: video bandpass filter and audio IF filter
    4. Video bandpass filter connects to detector
    5. Detector connects to video amplifier
    6. Video amplifier connects to CRT
    7. Audio IF filter connects to mixer/detector
    8. Mixer/detector connects the two amplifiers 5.5 MHz
    9. Each amplifier connects to an FM detector
    10. One FM detector connects to channel 1
    11. Other FM detector connects to channel 2

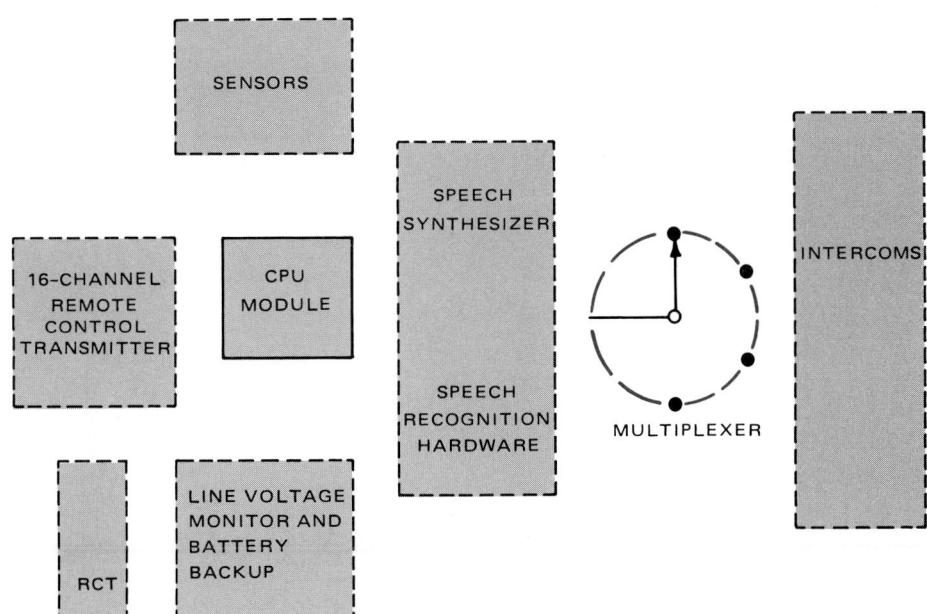

FIG. 27-5-A   Home intercom system.

15. Prepare a logic diagram from the information shown in Fig. 27-5-B. Convert to distinctive-shape logic symbols, and add the missing connections listed on the drawing.

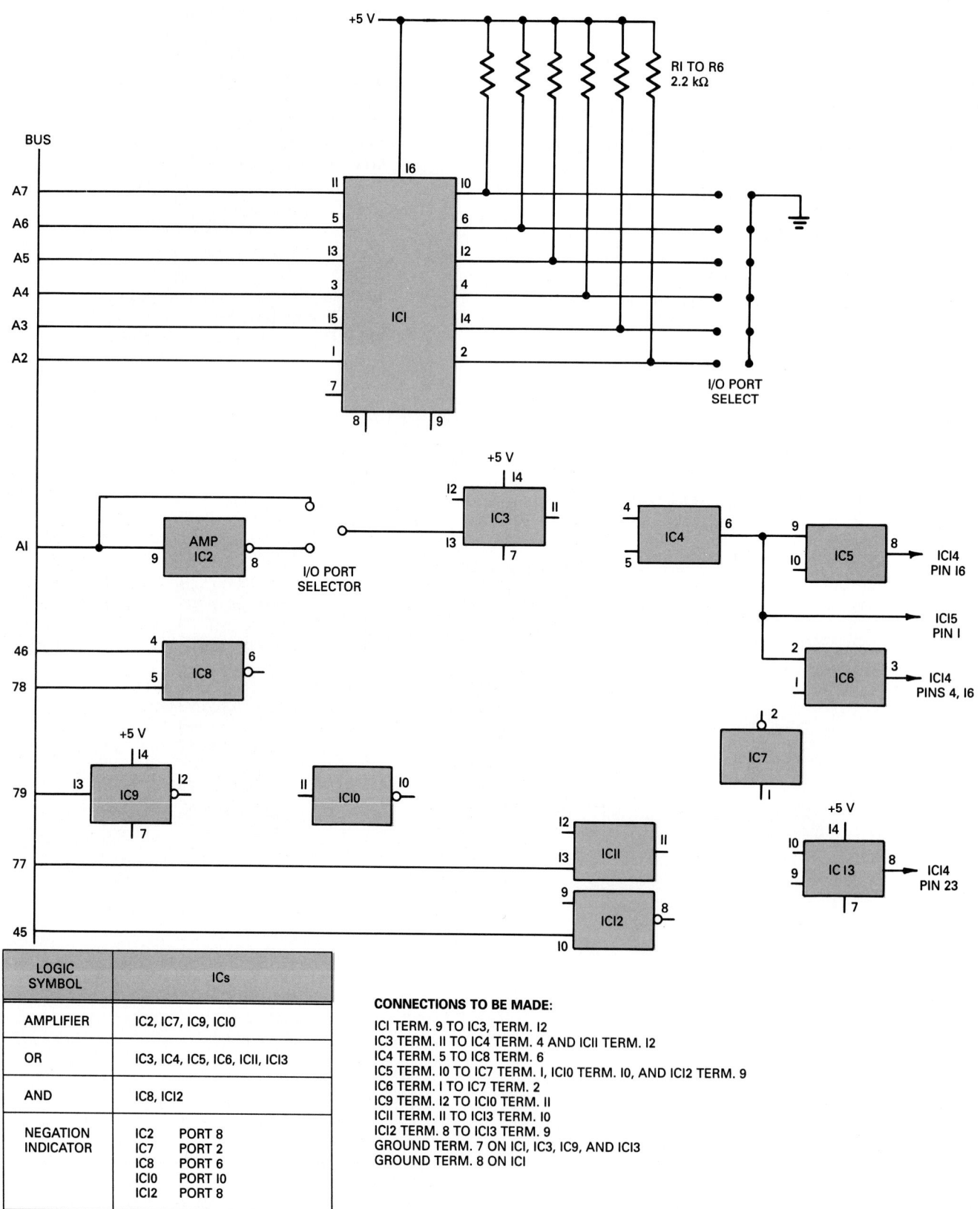

LOGIC SYMBOL	ICs	
AMPLIFIER	IC2, IC7, IC9, IC10	
OR	IC3, IC4, IC5, IC6, IC11, IC13	
AND	IC8, IC12	
NEGATION INDICATOR	IC2 IC7 IC8 IC10 IC12	PORT 8 PORT 2 PORT 6 PORT 10 PORT 8

CONNECTIONS TO BE MADE:

IC1 TERM. 9 TO IC3, TERM. 12
IC3 TERM. 11 TO IC4 TERM. 4 AND IC11 TERM. 12
IC4 TERM. 5 TO IC8 TERM. 6
IC5 TERM. 10 TO IC7 TERM. 1, IC10 TERM. 10, AND IC12 TERM. 9
IC6 TERM. 1 TO IC7 TERM. 2
IC9 TERM. 12 TO IC10 TERM. 11
IC11 TERM. 11 TO IC13 TERM. 10
IC12 TERM. 8 TO IC13 TERM. 9
GROUND TERM. 7 ON IC1, IC3, IC9, AND IC13
GROUND TERM. 8 ON IC1

FIG. 27-5-B   Partial logic diagram for a remote control housing wiring.

APPENDIX

STANDARD PARTS AND TECHNICAL DATA

ANSI	Y1.1	Abbreviations for Use on Drawings and in Text
ANSI	Y14.1	Drawing Sheet Size and Format
ANSI	Y14.2M	Line Conventions and Lettering
ANSI	Y14.3	Projections
ANSI	Y14.4	Pictorial Drawing
ANSI	Y14.5M	Dimensioning and Tolerancing for Engineering Drawings
ANSI	Y14.6	Screw Thread Representation
ANSI	Y14.7	Gears, Splines, and Serrations
ANSI	Y14.7.	Gear Drawing Standards—Part 1, for Spur, Helical, Double Helical and Rack
ANSI	Y14.9	Forgings
ANSI	Y14.10	Metal Stampings
ANSI	Y14.11	Plastics
ANSI	Y14.14	Mechanical Assemblies
ANSI	Y14.15	Electrical and Electronics Diagrams
ANSI	Y14.15A	Interconnection Diagrams
ANSI	Y14.17	Fluid Power Diagrams
ANSI	Y14.36	Surface Texture Symbols
ANSI	Y32.2	Graphic Symbols for Electrical and Electronics Diagrams
ANSI	Y32.9	Graphic Electrical Wiring Symbols for Architectural and Electrical Layout Drawings
ANSI	B1.1	Unified Screw Threads
ANSI	B4.2	Preferred Metric Limits and Fits
ANSI	B17.1	Keys and Keyseats
ANSI	B17.2	Woodruff Key and Keyslot Dimensions
ANSI	B18.2.1	Square and Hex Bolts and Screws
ANSI	B18.2.2	Square and Hex Nuts
ANSI	B18.3	Socket Cap, Shoulder, and Setscrews
ANSI	B18.6.2	Slotted-Head Cap Screws, Square-Head Setscrews, Slotted-Headless Setscrews
ANSI	B18.6.3	Machine Screws and Machine Screw Nuts
ANSI	B18.21.1	Lock Washers
ANSI	B27.2	Plain Washers
ANSI	B46.1	Surface Texture
ANSI	B94.6	Knurling
ANSI	B94.11M	Twist Drills
ANSI	Z210.1	Metric Practice

TABLE 1 ANSI publications of interest to drafters.

Fraction	Decimal	Fraction	Decimal	Fraction	Decimal	Fraction	Decimal
$\frac{1}{64}$	0.015625	$\frac{17}{64}$	0.265625	$\frac{33}{64}$	0.515625	$\frac{49}{64}$	0.765625
$\frac{1}{32}$	0.03125	$\frac{9}{32}$	0.28125	$\frac{17}{32}$	0.53125	$\frac{25}{32}$	0.78125
$\frac{3}{64}$	0.046875	$\frac{19}{64}$	0.296875	$\frac{35}{64}$	0.546875	$\frac{51}{64}$	0.796875
$\frac{1}{16}$	0.0625	$\frac{5}{16}$	0.3125	$\frac{9}{16}$	0.5625	$\frac{13}{16}$	0.8125
$\frac{5}{64}$	0.078125	$\frac{21}{64}$	0.328125	$\frac{37}{64}$	0.578125	$\frac{53}{64}$	0.828125
$\frac{3}{32}$	0.09375	$\frac{11}{32}$	0.34375	$\frac{19}{32}$	0.59375	$\frac{27}{32}$	0.84375
$\frac{7}{64}$	0.109375	$\frac{23}{64}$	0.359375	$\frac{39}{64}$	0.609375	$\frac{55}{64}$	0.859375
$\frac{1}{8}$	0.1250	$\frac{3}{8}$	0.3750	$\frac{5}{8}$	0.6250	$\frac{7}{8}$	0.8750
$\frac{9}{64}$	0.140625	$\frac{25}{64}$	0.390625	$\frac{41}{64}$	0.640625	$\frac{57}{64}$	0.890625
$\frac{5}{32}$	0.15625	$\frac{13}{32}$	0.40625	$\frac{21}{32}$	0.65625	$\frac{29}{32}$	0.90625
$\frac{11}{64}$	0.171875	$\frac{27}{64}$	0.421875	$\frac{43}{64}$	0.671875	$\frac{59}{64}$	0.921875
$\frac{3}{16}$	0.1875	$\frac{7}{16}$	0.4375	$\frac{11}{16}$	0.6875	$\frac{15}{16}$	0.9375
$\frac{13}{64}$	0.203125	$\frac{29}{64}$	0.453125	$\frac{45}{64}$	0.703125	$\frac{61}{64}$	0.953125
$\frac{7}{32}$	0.21875	$\frac{15}{32}$	0.46875	$\frac{23}{32}$	0.71875	$\frac{31}{32}$	0.96875
$\frac{15}{64}$	0.234375	$\frac{31}{64}$	0.484375	$\frac{47}{64}$	0.734375	$\frac{63}{64}$	0.984375
$\frac{1}{4}$	0.2500	$\frac{1}{2}$	0.5000	$\frac{3}{4}$	0.7500	1	1.0000

TABLE 2 Decimal equivalents of common inch fractions.

Quantity	Metric Unit	Symbol	Metric to Inch-Pound Unit	Inch-Pound to Metric Unit
Length	millimeter	mm	1 mm = 0.0394 in.	1 in. = 25.4 mm
	centimeter	cm	1 cm = 0.394 in.	1 ft. = 30.5 cm
	meter	m	1 m = 39.37 in. = 3.28 ft	1 yd. = 0.914 m = 914 mm
	kilometer	km	1 km = 0.62 mile	1 mile = 1.61 km
Area	square millimeter	mm²	1 mm² = 0.001 55 sq. in.	1 sq. in. = 6 452 mm²
	square centimeter	cm²	1 cm² = 0.155 sq. in.	1 sq. ft. = 0.093 m²
	square meter	m²	1 m² = 10.8 sq. ft. = 1.2 sq. yd.	1 sq. yd. = 0.836 m²
Mass	milligram	mg	1 g = 0.035 oz.	1 oz. = 28.3 g
	gram	g	1 kg = 2.205 lb.	1 lb. = 0.454 kg
	kilogram	kg	1 tonne = 1.102 tons	1 ton = 907.2 kg
	tonne	t		= 0.907 tonnes
Volume	cubic centimeter	cm³	1 mm³ = 0.000 061 cu. in.	1 fl. oz. = 28.4 cm³
	cubic meter	m³	1 cm³ = 0.061 cu. in.	1 cu. in. = 16.387 cm³
	milliliter	m	1 m³ = 35.3 cu ft. = 1.308 cu. yd.	1 cu. ft. = 0.028 m³
			1 ml = 0.035 fl. oz.	1 cu. yd. = 0.756 m³
Capacity	liter	L	U.S. Measure 1 pt. = 0.473 L 1 qt. = 0.946 L 1 gal = 3.785 L Imperial Measure 1 pt. = 0.568 L 1 qt. = 1.137 L 1 gal = 4.546 L	U.S. Measure 1 L = 2.113 pt. = 1.057 qt. = 0.264 gal. Imperial Measure 1 L = 1.76 pt. = 0.88 qt. = 0.22 gal.
Temperature	Celsius degree	°C	$°C = \frac{5}{9}(°F-32)$	$°F = \frac{9}{5} \times °C +32$
Force	newton	N	1 N = 0.225 lb (f)	1 lb (f) = 4.45N
	kilonewton	kN	1 kN = 0.225 kip (f) = 0.112 ton (f)	= 0.004 448 kN
Energy/Work	joule	J	1 J = 0.737 ft · lb	1 ft · lb = 1.355 J
	kilojoule	kJ	1 J = 0.948 Btu	1 Btu = 1.055 J
	megajoule	MJ	1 MJ = 0.278 kWh	1 kWh = 3.6 MJ
Power	kilowatt	kW	1 kW = 1.34 hp 1 W = 0.0226 ft · lb/min.	1 hp (550 ft · lb/s) = 0.746 kW 1 ft · lb/min = 44.2537 W
Pressure	kilopascal	kPa	1 kPa = 0.145 psi = 20.885 psf = 0.01 ton-force per sq. ft.	1 psi = 6.895 kPa 1 lb-force/sq. ft. = 47.88 Pa 1 ton-force/sq. ft. = 95.76 kPa
	*kilogram per square centimeter	kg/cm²	1 kg/cm² = 13.780 psi	
Torque	newton meter	N · m	1 N · m = 0.74 lb · ft	1 lb · ft = 1.36 N · m
	*kilogram meter	kg/m	1 kg/m = 7.24 lb · ft	1 lb · ft = 0.14 kg/m
	*kilogram per centimeter	kg/cm	1 kg/cm = 0.86 lb · in	1 lb · in = 1.2 kg/cm
Speed/Velocity	meters per second	m/s	1 m/s = 3.28 ft/s	1 ft/s = 0.305 m/s
	kilometers per hour	km/h	1 km/h = 0.62 mph	1 mph = 1.61 km/h

*Not SI units, but included here because they are employed on some of the gages and indicators currently in use in industry.

TABLE 3   Metric conversion tables.

Across Flats	ACRFLT	Machine Steel	MST
American National Standards Institute	ANSI	Machined	✓
And	&	Malleable Iron	MI
Angular	ANLR	Material	MATL
Approximate	APPROX	Maximum	MAX
Assembly	ASSY	Maximum Material Condition	Ⓜ or MMC
Between	↔	Meter	m
Bill of Material	B/M	Metric Thread	M
Bolt Circle	BC	Micrometer	μm
Brass	BR	Millimeter	mm
Brown and Sharpe Gage	B&S GA	Minimum	MIN
Bushing	BUSH	Module	MDL
Canada Standards Institute	CSI	Newton	N
Carbon Steel	CS	Nominal	NOM
Casting	CSTG	Not to Scale	x̲x̲
Cast Iron	CI	Number	NO
Center Line	CL or ℄	On Center	OC
Center to Center	C to C	Outside Diameter	OD
Centimeter	cm	Parallel	PAR
Chamfer	CHAM	Pascal	Pa
Circularity	CIR	Perpendicular	PERP
Cold-Rolled Steel	CRS	Pitch	P
Concentric	CONC	Pitch Circle	PC
Counterbore	⊔ or CBORE	Pitch Diameter	PD
Counterdrill	CDRILL	Plate	PL
Countersink	∨ or CSK	Radius	R
Cubic Centimeter	cm³	Reference or Reference Dimension	( ) or REF
Cubic Meter	m³	Regardless of Feature Size	* Ⓢ or RFS
Datum	DAT	Revolutions per Minute	rev/min
Degree (Angle)	° or DEG	Right Hand	RH
Depth	DP or ⌄̄	Root Diameter	RD
Diameter	∅ or DIA	Second (Arc)	(″)
Diametral Pitch	DP	Second (Time)	SEC
Dimension	DIM	Section	SECT
Drawing	DWG	Slotted	SLOT
Eccentric	ECC	Socket	SOCK
Equally Spaced	EQL SP	Spherical	SPHER
Figure	FIG	Spherical Diameter	S∅
Finish All Over	FAO	Spherical Radius	SR
Flat	FL	Spotface	⊔ or SFACE
Gage	GA	Square	□ or SQ
Gray Iron	GI	Square Centimeter	cm²
Head	HD	Square Meter	m²
Heat Treat	HT TR	Steel	STL
Heavy	HVY	Straight	STR
Hexagon	HEX	Symmetrical	⊣⊢ - - ⊣⊢ or SYM
Hydraulic	HYDR	Taper—Flat	◁
Inside Diameter	ID	—Conical	◁
International Organization for Standardization	ISO	Taper Pipe Thread	NPT
International Pipe Standard	IPS	Thread	THD
Kilogram	kg	Through	THRU
Kilometer	km	Tolerance	TOL
Least Material Condition	Ⓛ or LMC	True Profile	TP
Left Hand	LH	U.S. Gage	USG
Length	LG	Watt	W
Liter	L	Wrought Iron	WI
		Wrought Steel	WS

Datum symbol: [A] [A] or △ OR ▲

*Symbol no longer used in ASME Y14.5M-1994 standards.

TABLE 4 Abbreviations and symbols used on technical drawings.
(Note: See Tables 53 and 54 for geometric tolerancing symbols.)

ANGLE	SINE	COSINE	TAN	COTAN	ANGLE
0°	.0000	1.0000	.0000	θ	90°
1°	0.0175	0.9998	0.0175	57.290	89°
2°	0.0349	0.9994	0.0349	28.636	88°
3°	0.0523	0.9986	0.0524	19.081	87°
4°	0.0698	0.9976	0.0699	14.301	86°
5°	0.0872	0.9962	0.0875	11.430	85°
6°	0.1045	0.9945	0.1051	9.5144	84°
7°	0.1219	0.9925	0.1228	8.1443	83°
8°	0.1392	0.9903	0.1405	7.1154	82°
9°	0.1564	0.9877	0.1584	6.3138	81°
10°	0.1736	0.9848	0.1763	5.6713	80°
11°	0.1908	0.9816	0.1944	5.1446	79°
12°	0.2079	0.9781	0.2126	4.7046	78°
13°	0.2250	0.9744	0.2309	4.3315	77°
14°	0.2419	0.9703	0.2493	4.0108	76°
15°	0.2588	0.9659	0.2679	3.7321	75°
16°	0.2756	0.9613	0.2867	3.4874	74°
17°	0.2924	0.9563	0.3057	3.2709	73°
18°	0.3090	0.9511	0.3249	3.0777	72°
19°	0.3256	0.9455	0.3443	2.9042	71°
20°	0.3420	0.9397	0.3640	2.7475	70°
21°	0.3584	0.9336	0.3839	2.6051	69°
22°	0.3746	0.9272	0.4040	2.4751	68°
23°	0.3907	0.9205	0.4245	2.3559	67°
24°	0.4067	0.9135	0.4452	2.2460	66°
25°	0.4226	0.9063	0.4663	2.1445	65°
26°	0.4384	0.8988	0.4877	2.0503	64°
27°	0.4540	0.8910	0.5095	1.9626	63°
28°	0.4695	0.8829	0.5317	1.8807	62°
29°	0.4848	0.8746	0.5543	1.8040	61°
30°	0.5000	0.8660	0.5774	1.7321	60°
31°	0.5150	0.8572	0.6009	1.6643	59°
32°	0.5299	0.8480	0.6249	1.6003	58°
33°	0.5446	0.8387	0.6494	1.5399	57°
34°	0.5592	0.8290	0.6745	1.4826	56°
35°	0.5736	0.8192	0.7002	1.4281	55°
36°	0.5878	0.8090	0.7265	1.3764	54°
37°	0.6018	0.7986	0.7536	1.3270	53°
38°	0.6157	0.7880	0.7813	1.2799	52°
39°	0.6293	0.7771	0.8098	1.2349	51°
40°	0.6428	0.7660	0.8391	1.1918	50°
41°	0.6561	0.7547	0.8693	1.1504	49°
42°	0.6691	0.7431	0.9004	1.1106	48°
43°	0.6820	0.7314	0.9325	1.0724	47°
44°	0.6947	0.7193	0.9657	1.0355	46°
45°	0.7071	0.7071	0.0000	1.0000	45°
ANGLE	COSINE	SINE	COTAN	TAN	ANGLE

TABLE 5 Trigonometric functions.

NUMBER OR LETTER SIZE DRILL	SIZE		NUMBER OR LETTER SIZE DRILL	SIZE		NUMBER OR LETTER SIZE DRILL	SIZE		NUMBER OR LETTER SIZE DRILL	SIZE	
	mm	INCHES		mm	INCHES		mm	INCHES		mm	INCHES
80	0.343	.014	50	1.778	.070	20	4.089	.161	K	7.137	.281
79	0.368	.015	49	1.854	.073	19	4.216	.166	L	7.366	.290
78	0.406	.016	48	1.930	.076	18	4.305	.170	M	7.493	.295
77	0.457	.018	47	1.994	.079	17	4.394	.173	N	7.671	.302
76	0.508	.020	46	2.057	.081	16	4.496	.177	O	8.026	.316
75	0.533	.021	45	2.083	.082	15	4.572	.180	P	8.204	.323
74	0.572	.023	44	2.184	.086	14	4.623	.182	Q	8.433	.332
73	0.610	.024	43	2.261	.089	13	4.700	.185	R	8.611	.339
72	0.635	.025	42	2.375	.094	12	4.800	.189	S	8.839	.348
71	0.660	.026	41	2.438	.096	11	4.851	.191	T	9.093	.358
70	0.711	.028	40	2.489	.098	10	4.915	.194	U	9.347	.368
69	0.742	.029	39	2.527	.100	9	4.978	.196	V	9.576	.377
68	0.787	.031	38	2.578	.102	8	5.080	.199	W	9.804	.386
67	0.813	.032	37	2.642	.104	7	5.105	.201	X	10.084	.397
66	0.838	.033	36	2.705	.107	6	5.182	.204	Y	10.262	.404
65	0.889	.035	35	2.794	.110	5	5.220	.206	Z	10.490	.413
64	0.914	.036	34	2.819	.111	4	5.309	.209			
63	0.940	.037	33	2.870	.113	3	5.410	.213			
62	0.965	.038	32	2.946	.116	2	5.613	.221			
61	0.991	.039	31	3.048	.120	1	5.791	.228			
60	1.016	.040	30	3.264	.129	A	5.944	.234			
59	1.041	.041	29	3.354	.136	B	6.045	.238			
58	1.069	.042	28	3.569	.141	C	6.147	.242			
57	1.092	.043	27	3.658	.144	D	6.248	.246			
56	1.181	.047	26	3.734	.147	E	6.350	.250			
55	1.321	.052	25	3.797	.150	F	6.528	.257			
54	1.397	.055	24	3.861	.152	G	6.629	.261			
53	1.511	.060	23	3.912	.154	H	6.756	.266			
52	1.613	.064	22	3.988	.157	I	6.909	.272			
51	1.702	.067	21	4.039	.159	J	7.036	.277			

TABLE 6   Number and letter-size drills.

METRIC DRILL SIZES		Reference Decimal Equivalent (Inches)	METRIC DRILL SIZES		Reference Decimal Equivalent (Inches)	METRIC DRILL SIZES		Reference Decimal Equivalent (Inches)
Preferred	Available		Preferred	Available		Preferred	Available	
—	0.40	.0157	—	2.7	.1063	14	—	.5512
—	0.42	.0165	2.8	—	.1102	—	14.5	.5709
—	0.45	.0177	—	2.9	.1142	15	—	.5906
—	0.48	.0189	3.0	—	.1181	—	15.5	.6102
0.50	—	.0197	—	3.1	.1220	16	—	.6299
—	0.52	.0205	3.2	—	.1260	—	16.5	.6496
0.55	—	.0217	—	3.3	.1299	17	—	.6693
—	0.58	.0228	3.4	—	.1339	—	17.5	.6890
0.60	—	.0236	—	3.5	.1378	18	—	.7087
—	0.62	.0244	3.6	—	.1417	—	18.5	.7283
0.65	—	.0256	—	3.7	.1457	19	—	.7480
—	0.68	.0268	3.8	—	.1496	—	19.5	.7677
0.70	—	.0276	—	3.9	.1535	20	—	.7874
—	0.72	.0283	4.0	—	.1575	—	20.5	.8071
0.75	—	.0295	—	4.1	.1614	21	—	.8268
—	0.78	.0307	4.2	—	.1654	—	21.5	.8465
0.80	—	.0315	—	4.4	.1732	22	—	.8661
—	0.82	.0323	4.5	—	.1772	—	23	.9055
0.85	—	.0335	—	4.6	.1811	24	—	.9449
—	0.88	.0346	4.8	—	.1890	25	—	.9843
0.90	—	.0354	5.0	—	.1969	26	—	1.0236
—	0.92	.0362	—	5.2	.2047	—	27	1.0630
0.95	—	.0374	5.3	—	.2087	28	—	1.1024
—	0.98	.0386	—	5.4	.2126	—	29	1.1417
1.00	—	.0394	5.6	—	.2205	30	—	1.1811
—	1.03	.0406	—	5.8	.2283	—	31	1.2205
1.05	—	.0413	6.0	—	.2362	32	—	1.2598
—	1.08	.0425	—	6.2	.2441	—	33	1.2992
1.10	—	.0433	6.3	—	.2480	34	—	1.3386
—	1.15	.0453	—	6.5	.2559	—	35	1.3780
1.20	—	.0472	6.7	—	.2638	36	—	1.4173
1.25	—	.0492	—	6.8	.2677	—	37	1.4567
1.3	—	.0512	—	6.9	.2717	38	—	1.4361
—	1.35	.0531	7.1	—	.2795	—	39	1.5354
1.4	—	.0551	—	7.3	.2874	40	—	1.5748
—	1.45	.0571	7.5	—	.2953	—	41	1.6142
1.5	—	.0591	—	7.8	.3071	42	—	1.6535
—	1.55	.0610	8.0	—	.3150	—	43.5	1.7126
1.6	—	.0630	—	8.2	.3228	45	—	1.7717
—	1.65	.0650	8.5	—	.3346	—	46.5	1.8307
1.7	—	.0669	—	8.8	.3465	48	—	1.8898
—	1.75	.0689	9.0	—	.3543	50	—	1.9685
1.8	—	.0709	—	9.2	.3622	—	51.5	2.0276
—	1.85	.0728	9.5	—	.3740	53	—	2.0866
1.9	—	.0748	—	9.8	.3858	—	54	2.1260
—	1.95	.0768	10	—	.3937	56	—	2.2047
2.0	—	.0787	—	10.3	.4055	—	58	2.2835
—	2.05	.0807	10.5	—	.4134	60	—	2.3622
2.1	—	.0827	—	10.8	.4252			
—	2.15	.0846	11	—	.4331			
2.2	—	.0866	—	11.5	.4528			
—	2.3	.0906	12	—	.4724			
2.4	—	.0945	12.5	—	.4921			
2.5	—	.0984	13	—	.5118			
2.6	—	.1024	—	13.5	.5315			

TABLE 7   Metric twist drill sizes.

THREADS PER INCH AND TAP DRILL SIZES													
SIZE INCHES		Graded Pitch Series						Constant Pitch Series					
		Coarse UNC		Fine UNF		Extra Fine UNEF		8 UN		12 UN		16 UN	
Number or Fraction	Decimal	Threads per Inch	Tap Drill Dia.	Threads per Inch	Tap Drill Dia.	Threads per Inch	Tap Drill Dia.	Threads per Inch	Tap Drill Dia.	Threads per Inch	Tap Drill Dia.	Threads per Inch	Tap Drill Dia.
0	.060	—	—	80	$3/64$	—	—	—	—	—	—	—	—
2	.086	56	No. 50	64	No. 49	—	—	—	—	—	—	—	—
4	.112	40	No. 43	48	No. 42	—	—	—	—	—	—	—	—
5	.125	40	No. 38	44	No. 37	—	—	—	—	—	—	—	—
6	.138	32	No. 36	40	No. 33	—	—	—	—	—	—	—	—
8	.164	32	No. 29	36	No. 29	—	—	—	—	—	—	—	—
10	.190	24	No. 25	32	No. 21	—	—	—	—	—	—	—	—
$1/4$	.250	20	7	28	3	32	.219	—	—	—	—	—	—
$5/16$	.312	18	F	24	1	32	.281	—	—	—	—	—	—
$3/8$	.375	16	.312	24	Q	32	.344	—	—	—	—	UNC	—
$7/16$	.438	14	U	20	.391	28	Y	—	—	—	—	16	V
$1/2$	.500	13	.422	20	.453	28	.469	—	—	—	—	16	.438
$9/16$	.562	12	.484	18	.516	24	.516	—	—	UNC	—	16	.500
$5/8$	.625	11	.531	18	.578	24	.578	—	—	12	.547	16	.562
$3/4$	.750	10	.656	16	.688	20	.703	—	—	12	.672	UNF	—
$7/8$	.875	9	.766	14	.812	20	.828	—	—	12	.797	16	.812
1	1.000	8	.875	12	.922	20	.953	UNC	—	UNF	—	16	.938
$1 1/8$	1.125	7	.984	12	1.047	18	1.078	8	1.000	UNF	—	16	1.062
$1 1/4$	1.250	7	1.109	12	1.172	18	1.188	8	1.125	UNF	—	16	1.188
$1 3/8$	1.375	6	1.219	12	1.297	18	1.312	8	1.250	UNF	—	16	1.312
$1 1/2$	1.500	6	1.344	12	1.422	18	1.438	8	1.375	UNF	—	16	1.438
$1 5/8$	1.625	—	—	—	—	18	—	8	1.500	12	1.547	16	1.562
$1 3/4$	1.750	5	1.562	—	—	—	—	8	1.625	12	1.672	16	1.688
$1 7/8$	1.875	—	—	—	—	—	—	8	1.750	12	1.797	16	1.812
2	2.000	4.5	1.781	—	—	—	—	8	1.875	12	1.922	16	1.938
$2 1/4$	2.250	4.5	2.031	—	—	—	—	8	2.125	12	2.172	16	2.188
$2 1/2$	2.500	4	2.250	—	—	—	—	8	2.375	12	2.422	16	2.438
$2 3/4$	2.750	4	2.500	—	—	—	—	8	2.625	12	2.672	16	2.688
3	3.000	4	2.750	—	—	—	—	8	2.875	12	2.922	16	2.938
$3 1/4$	3.250	4	3.000	—	—	—	—	8	3.125	12	3.172	16	3.188
$3 1/2$	3.500	4	3.250	—	—	—	—	8	3.375	12	3.422	16	3.438
$3 3/4$	3.750	4	3.500	—	—	—	—	8	3.625	12	3.668	16	3.688
4	4.000	4	3.750	—	—	—	—	8	3.875	12	3.922	16	3.938

Note: The tap diameter sizes shown are nominal. The class and length of thread will govern the limits on the tapped hole size.

TABLE 8   Inch screw threads.

NOMINAL SIZE DIA (mm)	SERIES WITH GRADED PITCHES				SERIES WITH CONSTANT PITCHES																	
	COARSE		FINE		4		3		2		1.5		1.25		1		0.75		0.5		0.35	
Preferred	Thread Pitch	Tap Drill Size	Thread Pitch	Tap Drill Size	Thread Pitch	Tap Drill Size	Thread Pitch	Tap Drill Size	Thread Pitch	Tap Drill Size	Thread Pitch	Tap Drill Size	Thread Pitch	Tap Drill Size	Thread Pitch	Tap Drill Size	Thread Pitch	Tap Drill Size	Thread Pitch	Tap Drill Size	Thread Pitch	Tap Drill Size
1.6	0.35	1.25																				
1.8	0.35	1.45																				
2	0.4	1.6																				
2.2	0.45	1.75																				
2.5	0.45	2.05																			0.35	2.15
3	0.5	2.5																			0.35	2.65
3.5	0.6	2.9																			0.35	3.15
4	0.7	3.3																	0.5	3.5		
4.5	0.75	3.7																	0.5	4.0		
5	0.8	4.2																	0.5	4.5		
6	1	5.0															0.75	5.2				
8	1.25	6.7	1	7.0											1	7.0	0.75	7.2				
10	1.5	8.5	1.25	8.7									1.25	8.7	1	9.0	0.75	9.2				
12	1.75	10.2	1.25	10.8							1.5	10.5	1.25	10.7	1	11						
14	2	12	1.5	12.5							1.5	12.5	1.25	12.7	1	13						
16	2	14	1.5	14.5							1.5	14.5			1	15						
18	2.5	15.5	1.5	16.5					2	16	1.5	16.5			1	17						
20	2.5	17.5	1.5	18.5					2	18	1.5	18.5			1	19						
22	2.5	19.5	1.5	20.5					2	20	1.5	20.5			1	21						
24	3	21	2	22					2	22	1.5	22.5			1	23						
27	3	24	2	25					2	25	1.5	25.5			1	26						
30	3.5	26.5	2	28					2	28	1.5	28.5			1	29						
33	3.5	29.5	2	31					2	31	1.5	31.5										
36	4	32	3	33					2	34	1.5	34.5										
39	4	35	3	36					2	37	1.5	37.5										
42	4.5	37.5	3	39	4	38	3	39	2	40	1.5	40.5										
45	4.5	39	3	42	4	41	3	42	2	43	1.5	43.5										
48	5	43	3	45	4	44	3	45	2	46	1.5	46.5										

TABLE 9   Metric screw threads.

A-9

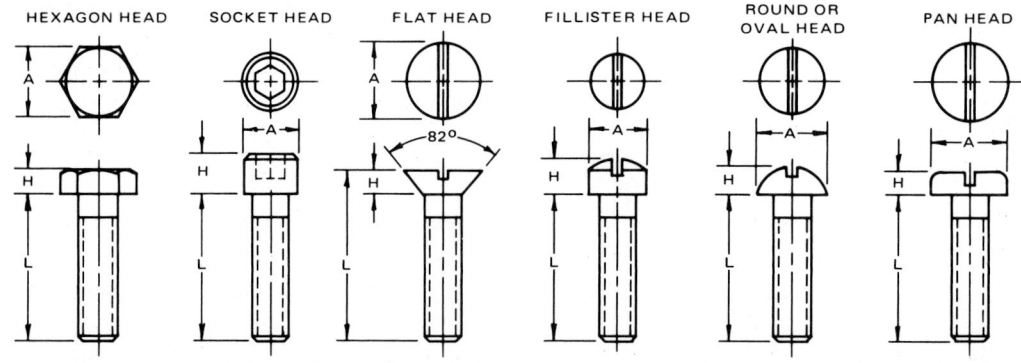

U.S. CUSTOMARY (INCHES)											METRIC (MILLIMETERS)										
Nominal Size	Hexagon Head		Socket Head		Flat Head		Fillister Head		Round or Oval Head		Nominal Size	Hexagon Head		Socket Head		Flat Head		Fillister Head		Pan Head	
	A	H	A	H	A	H	A	H	A	H		A	H	A	H	A	H	A	H	A	H
.250	.44	.17	.38	.25	.50	.14	.38	.24	.44	.19	M3	5.5	2	5.5	3	5.6	1.6	6	2.4	6	1.9
.312	.50	.22	.47	.31	.62	.18	.44	.30	.56	.25	4	7	2.8	7	4	7.5	2.2	8	3.1	8	2.5
.375	.56	.25	.56	.38	.75	.21	.56	.36	.62	.27	5	8.5	3.5	9	5	9.2	2.5	10	3.8	10	3.1
.438	.62	.30	.66	.44	.81	.21	.62	.37	.75	.33	6	10	4	10	6	11	3	12	4.6	12	3.8
.500	.75	.34	.75	.50	.88	.21	.75	.41	.81	.35	8	13	5.5	13	8	14.5	4	16	6	16	5
.625	.94	.42	.94	.62	1.12	.28	.88	.52	1.00	.44	10	17	7	16	10	18	5	20	7.5	20	6.2
.750	1.12	.50	1.12	.75	1.38	.35	1.00	.61	1.25	.55	12	19	8	18	12	23	6.4				
											16	24	10.5	24	16	29	8				
											20	30	13.1			35	9				

TABLE 10   Common machine and cap screws.

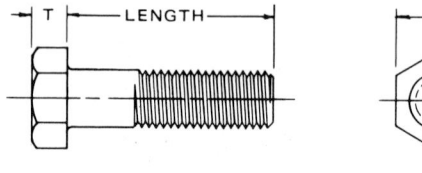

U.S. CUSTOMARY (INCHES)			METRIC (MILLIMETERS)		
Nominal Bolt Size	Width Across Flats F	Thickness T	Nominal Bolt Size and Thread Pitch	Width Across Flats F	Thickness T
.250	.438	.172	M5 x 0.8	8	3.9
.312	.500	.219	M6 x 1	10	4.7
.375	.562	.250	M8 x 1.25	13	5.7
.438	.625	.297			
.500	.750	.344	M10 x 1.5	15	6.8
.625	.938	.422	M12 x 1.75	18	8
.750	1.125	.500	M14 x 2	21	9.3
.875	1.312	.578	M16 x 2	24	10.5
1.000	1.500	.672	M20 x 2.5	30	13.1
1.125	1.688	.750	M24 x 3	36	15.6
1.250	1.875	.844	M30 x 3.5	46	19.5
1.375	2.062	.906	M36 x 4	55	23.4
1.500	2.250	1.000			

TABLE 11   Hexagon-head bolts and cap screws.

TABLE 12   Twelve-spline flange screws.

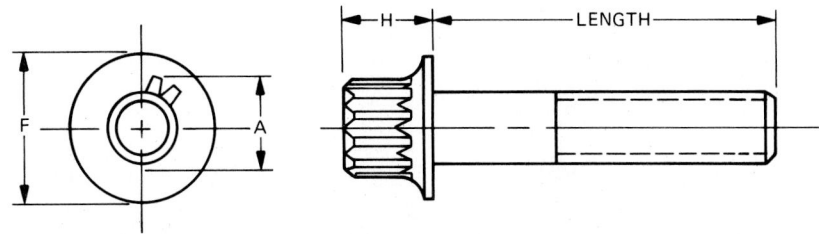

NOMINAL BOLT SIZE AND THREAD PITCH	HEAD SIZES		
	F	**A**	**H**
M5 x 0.8	9.4	5.9	5
M6 x 1	11.8	7.4	6.3
M8 x 1.25	15	9.4	8
M10 x 1.5	18.6	11.7	10
M12 x 1.75	22.8	14	12
M14 x 2	26.4	16.3	14
M16 x 2	30.3	18.7	16
M20 x 2.5	37.4	23.4	20

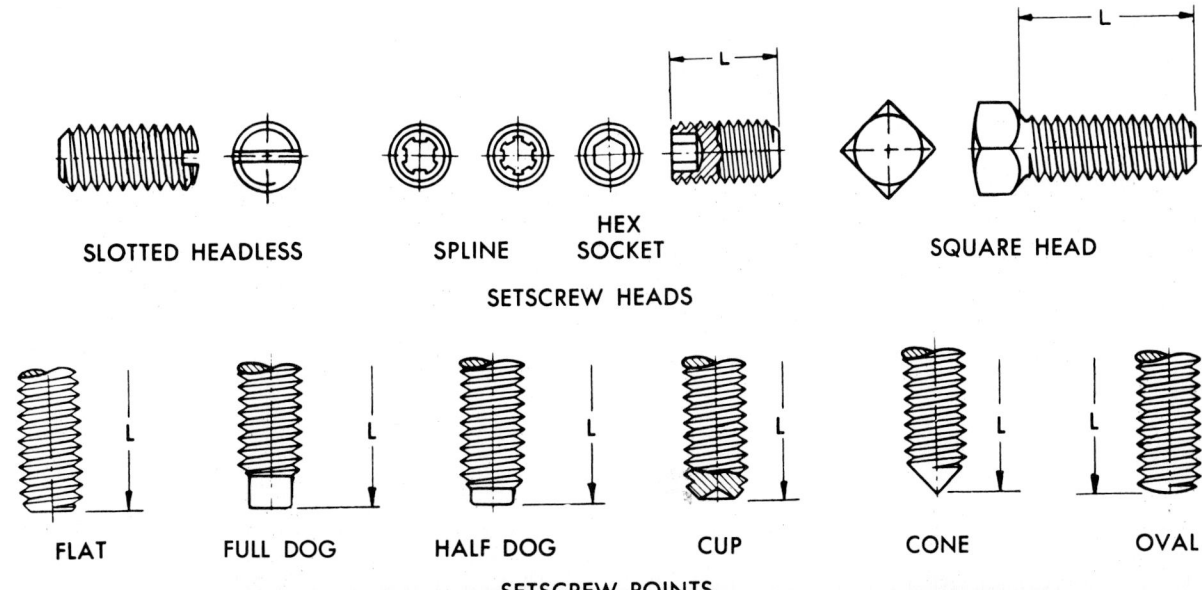

SLOTTED HEADLESS   SPLINE   HEX SOCKET   SQUARE HEAD

SETSCREW HEADS

FLAT   FULL DOG   HALF DOG   CUP   CONE   OVAL

SETSCREW POINTS

U.S. CUSTOMARY (INCHES)		METRIC (MILLIMETERS)	
Nominal Size	**Key Size**	**Nominal Size**	**Key Size**
.125	.06	M1.4	0.7
.138	.06	2	0.9
.164	.08	3	1.5
.190	.09	4	2
.250	.12	5	2
.312	.16	6	3
.375	.19	8	4
.500	.25	10	5
.625	.31	12	6
.750	.38	16	8

TABLE 13   Setscrews.

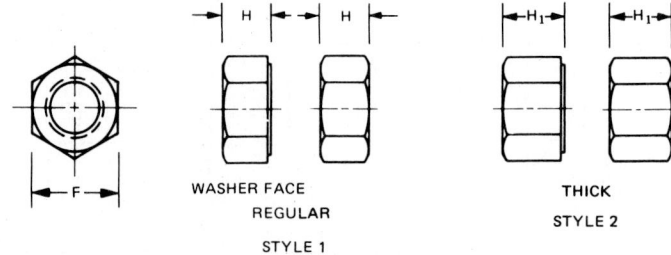

WASHER FACE
REGULAR
STYLE 1

THICK
STYLE 2

U.S. CUSTOMARY (INCHES)				METRIC (MILLIMETERS)			
	Distance Across Flats F	Thickness Max.			Distance Across Flats F	Thickness Max.	
Nominal Nut Size		Style 1 H	Style 2 H₁	Nominal Nut Size and Thread Pitch		Style 1 H	Style 2 H₁
.250	.438	.218	.281	M4 x 0.7	7	—	3.2
.312	.500	.266	.328	M5 x 0.8	8	4.5	5.3
.375	.562	.328	.406	M6 x 1	10	5.6	6.5
.438	.625	.375	.453	M8 x 1.25	13	6.6	7.8
.500	.750	.438	.562	M10 x 1.5	15	9	10.7
.562	.875	.484	.609	M12 x 1.75	18	10.7	12.8
.625	.938	.547	.719	M14 x 2	21	12.5	14.9
.750	1.125	.641	.812	M16 x 2	24	14.5	17.4
.875	1.312	.750	.906	M20 x 2.5	30	18.4	21.2
1.000	1.500	.859	1.000	M24 x 3	36	22	25.4
1.125	1.688	.969	1.156	M30 x 3.5	46	26.7	31
1.250	1.875	1.062	1.250	M36 x 4	55	32	37.6
1.375	2.062	1.172	1.375				
1.500	2.250	1.281	1.500				

TABLE 14   Hexagon-head nuts.

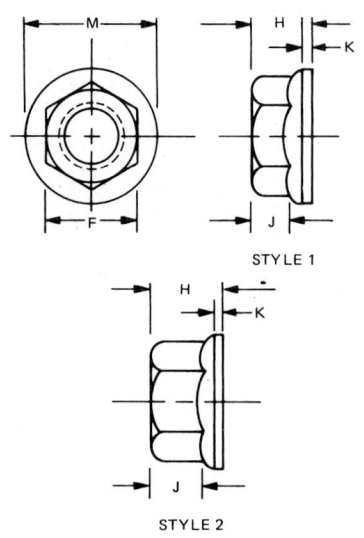

STYLE 1

STYLE 2

METRIC (MILLIMETERS)							
Nominal Nut Size and Thread Pitch	Width Across Flats F	Style 1				Style 2	
		H	J	K	M	H	J
M6 x 1	10	5.8	3	1	14.2	6.7	3.7
M8 x 1.25	13	6.8	3.7	1.3	17.6	8	4.5
M10 x 1.5	15	9.6	5.5	1.5	21.5	11.2	6.7
M12 x 1.75	18	11.6	6.7	2	25.6	13.5	8.2
M14 x 2	21	13.4	7.8	2.3	29.6	15.7	9.6
M16 x 2	24	15.9	9.5	2.5	34.2	18.4	11.7
M20 x 2.5	30	19.2	11.1	2.8	42.3	22	12.6

TABLE 15   Hex flange nuts.

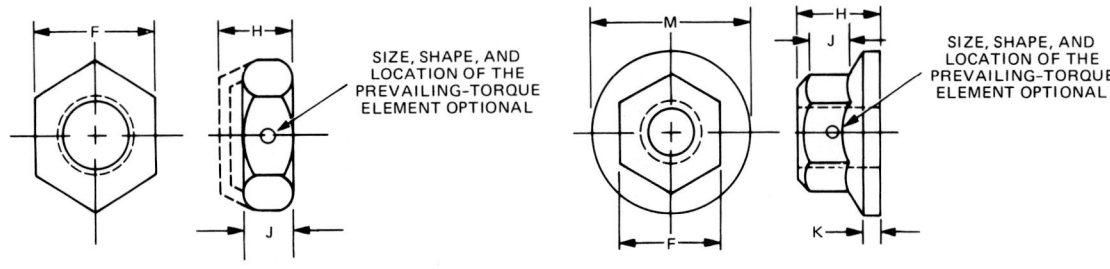

HEX NUTS   HEX FLANGE NUTS

NOMINAL NUT SIZE AND THREAD PITCH	WIDTH ACROSS FLATS F	HEX NUTS				HEX FLANGE NUTS					
		Style 1		Style 2		Style 1				Style 2	
		H max.	J max.	H max.	J max.	H	J	K	M	H	J
M5 × 0.8	8.0	6.1	2.3	7.6	2.9						
M6 × 1	10	7.6	3	8.8	3.7	7.6	3	1	14.2	8.8	3.7
M8 × 1.25	13	9.1	3.7	10.3	4.5	9.1	3.7	1.3	17.6	10.3	4.5
M10 × 1.5	15	12	5.5	14	6.7	12	5.5	1.5	21.5	14	6.7
M12 × 1.75	18	14.2	6.7	16.8	8.2	14.4	6.7	2	25.6	16.8	8.2
M14 × 2	21	16.5	7.8	18.9	9.6	16.6	7.8	2.3	29.6	18.9	9.6
M16 × 2	24	18.5	9.5	21.4	11.7	18.9	9.5	2.5	34.2	21.4	11.7
M20 × 2.5	30	23.4	11.1	26.5	12.6	23.4	11.1	2.8	42.3	26.5	12.6
M24 × 3	36	28	13.3	31.4	15.1						
M30 × 3.5	46	33.7	16.4	38	18.5						
M36 × 4	55	40	20.1	45.6	22.8						

TABLE 16 Prevailing-torque insert-type nuts.

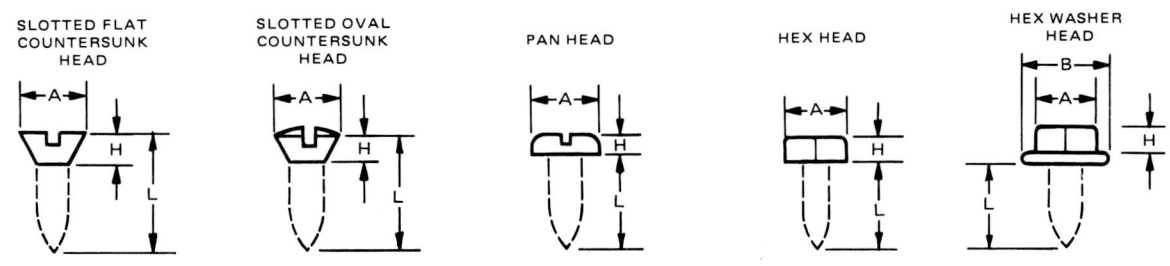

U.S. CUSTOMARY (INCHES)											METRIC (MILLIMETERS)											
NOMINAL SIZE		SLOTTED FLAT COUNTERSUNK HEAD		SLOTTED OVAL COUNTERSUNK HEAD		PAN HEAD		HEX HEAD		HEX WASHER HEAD			NOMINAL SIZE	SLOTTED FLAT COUNTERSUNK HEAD		SLOTTED OVAL COUNTERSUNK HEAD		PAN HEAD			HEX HEAD		HEX WASHER HEAD		
No.	DIA.	A	H	A	H	A	H	A	H	A	B	H		A	H	A	H	A	H	Slot Recess H	A	H	A	B	H
2	.086	.17	.05	.17	.05	.17	.05	.12	.05	.12	.17	.05	2	3.6	1.2	3.6	1.2	3.9	1.4	1.6	3.2	1.3	3.2	4.2	1.3
4	.112	.23	.07	.23	.07	.22	.07	.19	.06	.19	.24	.06	2.5	4.6	1.5	4.6	1.5	4.9	1.7	2	4	1.4	4	5.3	1.4
6	.138	.28	.08	.28	.08	.27	.08	.25	.09	.25	.33	.09	3	5.5	1.8	5.5	1.8	5.8	1.9	1.3	5	1.5	5	6.2	1.5
8	.164	.33	.10	.33	.10	.32	.10	.25	.11	.25	.35	.11	3.5	6.5	2.1	6.5	2.1	6.8	2.3	2.5	5.5	2.4	5.5	7.5	2.4
10	.190	.39	.17	.39	.17	.37	.11	.31	.12	.31	.41	.12	4	7.5	2.3	7.5	2.3	7.8	2.6	2.8	7	2.8	7	9.2	2.8
													5	9.5	2.9	9.5	2.9	9.8	3.1	3.5	8	3	8	10.5	3
													6	11.9	3.6	11.9	3.6	12	3.9	4.3	10	4.8	10	13.2	4.8
													8	15.2	4.4	15.2	4.4	15.6	5	5.6	13	5.8	13	17.2	5.8
													10	1.9	5.4	19	5.4	19.5	6.2	7	15	7.5	15	19.8	7.5
													12	22.9	6.4	22.9	6.4	23.4	7.5	8.3	18	9.5	18	23.8	9.5

TABLE 17 Tapping screws.

KIND OF MATERIAL	THREAD-FORMING								THREAD-CUTTING			SELF-DRILLING	
	Type A	Type B	Type AB	HEX HEAD B	SWAGE FORM*	SWAGE FORM* B	Type U	Type 21	Type F*	Type L	Type B-F*	DRIL-KWICK	TAPITS*
SHEET METAL .0156 to .0469in. thick (0.4 to 1.2mm) (Steel, Brass, Aluminum, Monel, etc.)	●	●	●	●	●	●		●				●	●
SHEET STAINLESS STEEL .0156 to .0469in thick (0.4 to 1.2mm)	●	●	●	●	●	●		●	●			●	●
SHEET METAL .20 to .50in. thick (1.2 to 5mm) (Steel, Brass, Aluminum, etc.)	●	●	●	●	●	●	●	●	●			●	
STRUCTURAL STEEL .20 to .50in. thick (1.2 to .5mm)				●	●		●		●				
CASTINGS (Aluminum, Magnesium, Zinc, Brass, Bronze, etc.)		●	●	●	●	●	●		●				
CASTINGS (Gray Iron, Malleable Iron, Steel, etc.)					●		●		●				
FORGINGS (Steel, Brass, Bronze, etc.)					●		●		●				
PLYWOOD, Resin Impregnated: Compreg, Pregwood, etc. NATURAL WOODS	●	●	●	●			●				●	●	●
ASBESTOS and other compositions: Ebony, Asbestos, Transite, Fiberglas, Insurok, etc.	●	●	●	●		●						●	●
PHENOL FORMALDEHYDE: Molded: Bakelite, Durez, etc. Cast: Catalin, Marblette, etc. Laminated: Formica, Textolite, etc.		●	●	●		●	●		●		●		
UREA FORMALDEHYDE: Molded: Plaskon, Beetle, etc. MELAMINE FORMALDEHYDE: Melantite, Melamac						●	●				●		
CELLULOSE ACETATES and NITRATES: Tenite, Lumarith, Plastacele Pyralin, Celanese, etc. ACRYLATE & STYRENE RESINS: Lucite, Plexiglas, Styron, etc.		●	●	●	●	●	●				●		
NYLON PLASTICS: Nylon, Zytel					●	●	●			●			

TABLE 18   Selector guide to thread-cutting screws.

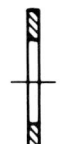

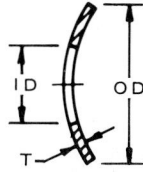

FLAT WASHER   LOCKWASHER   SPRING LOCKWASHER

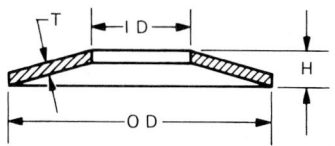

U.S. CUSTOMARY (INCHES)						
Bolt Size	Flat Washers Type A–N			Lockwashers Regular		
	ID	OD	Thick	ID	OD	Thick
#6	.156	.375	.049	.141	.250	.031
#8	.188	.438	.049	.168	.293	.040
#10	.219	.500	.049	.194	.334	.047
#12	.250	.562	.065	.221	.377	.056
.250	.281	.625	.065	.255	.489	.062
.312	.344	.688	.065	.318	.586	.078
.375	.406	.812	.065	.382	.683	.094
.438	.469	.922	.065	.446	.779	.109
.500	.531	1.062	.095	.509	.873	.125
.562	.594	1.156	.095	.572	.971	.141
.625	.656	1.312	.095	.636	1.079	.156
.750	.812	1.469	.134	.766	1.271	.188
.875	.938	1.750	.134	.890	1.464	.219
1.000	1.062	2.000	.134	1.017	1.661	.250
1.125	1.250	2.250	.134	1.144	1.853	.281
1.250	1.375	2.500	.165	1.271	2.045	.312
1.375	1.500	2.750	.165	1.398	2.239	.344
1.500	1.625	3.000	.165	1.525	2.430	.375

METRIC (MILLIMETERS)									
Bolt Size	Flat Washers			Lockwashers			Spring Lockwashers		
	ID	OD	Thick	ID	OD	Thick	ID	OD	Thick
2	2.2	5.5	0.5	2.1	3.3	0.5			
3	3.2	7	0.5	3.1	5.7	0.8			
4	4.3	9	0.8	4.1	7.1	0.9	4.2	8	0.3 / 0.4
5	5.3	11	1	5.1	8.7	1.2	5.2	10	0.4 / 0.5
6	6.4	12	1.5	6.1	11.1	1.6	6.2	12.5	0.5 / 0.5 / 0.7
7	7.4	14	1.5	7.1	12.1	1.6	7.2	14	0.5 / 0.8
8	8.4	17	2	8.2	14.2	2	8.2	16	0.6 / 0.9
10	10.5	21	2.5	10.2	17.2	2.2	10.2	20	0.8 / 1.1
12	13	24	2.5	12.3	20.2	2.5	12.2	25	0.9 / 1.5
14	15	28	2.5	14.2	23.2	3	14.2	28	1.0 / 1.5
16	17	30	3	16.2	26.2	3.5	16.3	31.5	1.2 / 1.7
18	19	34	3	18.2	28.2	3.5	18.3	35.5	1.2 / 2.0
20	21	36	3	20.2	32.2	4	20.4	40	1.5 / 2.25 / 1.75
22	23	39	4	22.5	34.5	4	22.4	45	2.5
24	25	44	4	24.5	38.5	5			
27	28	50	4	27.5	41.5	5			
30	31	56	4	30.5	46.5	6			

TABLE 19   Common washer sizes.

U.S. CUSTOMARY (INCHES)				METRIC (MILLIMETERS)			
OUTSIDE DIAMETER MAX.	INSIDE DIAMETER MIN.	STOCK THICKNESS T	H APPROX.	OUTSIDE DIAMETER MAX.	INSIDE DIAMETER MIN.	STOCK THICKNESS T	H APPROX.
.187	.093	.010	.015	4.8	2.4	0.16	0.33
.250	.125	.009	.017			0.25	0.38
		.013	.020	6.4	3.2	0.22	0.44
.281	.138	.010	.020			0.34	0.51
		.015	.023	7.9	4	0.27	0.55
.312	.156	.011	.022			0.42	0.64
		.017	.025	9.5	4.8	0.38	0.69
.343	.164	.013	.024			0.51	0.76
		.019	.028	12.7	6.5	0.46	0.86
.375	.190	.015	.027			0.64	0.97
		.020	.030			0.97	1.20
.500	.255	.018	.034	15.9	8.1	0.56	1.07
		.025	.038			0.81	1.22
.625	.317	.022	.042	19.1	9.7	0.71	1.3
		.032	.048			1.02	1.5
						1.42	1.8
.750	.380	.028	.051	22.2	11.2	0.79	1.5
		.040	.059			1.14	1.7
.875	.442	.031	.059	25.4	12.8	0.89	1.7
		.045	.067			1.27	1.9
						1.85	2.3
1.000	.505	.035	.067	28.6	14.4	0.97	1.9
		.050	.075			1.42	2.1
1.125	.567	.038	.073	31.8	16	1.02	2.1
		.056	.084			1.58	2.3
1.250	.630	.040	.082	34.9	17.6	1.12	2.2
		.062	.092			1.70	2.6
1.375	.692	.044	.088	38.1	19.2	1.14	2.4
		.067	.101			1.83	2.7
				44.5	22.4	1.45	2.9
						2.16	3.3
				50.8	25.4	1.65	3.3
						2.46	3.7
				63.5	31.8	2.03	4.1
						3.05	4.6

TABLE 20   Belleville washers. *(Wallace Barnes Co. Ltd.)*

U.S. CUSTOMARY (INCHES)						METRIC (MILLIMETERS)					
Diameter of Shaft		Square Key		Flat Key		Diameter of Shaft		Square Key		Flat Key	
		Nominal Size		Nominal Size				Nominal Size		Nominal Size	
From	To	W	H	W	H	Over	Up To	W	H	W	H
.500	.562	.125	.125	.125	.094	6	8	2	2		
.625	.875	.188	.188	.188	.125	8	10	3	3		
.938	1.250	.250	.250	.250	.188	10	12	4	4		
1.312	1.375	.312	.312	.312	.250	12	17	5	5		
1.438	1.750	.375	.375	.375	.250	17	22	6	6		
1.812	2.250	.500	.500	.500	.375	22	30	7	7	8	7
2.375	2.750	.625	.625			30	38	8	8	10	8
2.875	3.250	.750	.750			38	44	9	9	12	8
3.375	3.750	.875	.875			44	50	10	10	14	9
3.875	4.500	1.000	1.000			50	58	12	12	16	10

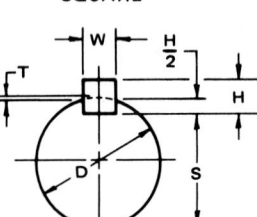
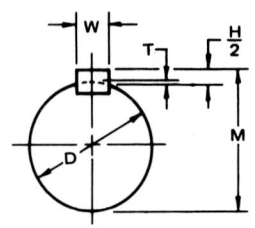

SQUARE   FLAT

C = ALLOWANCE FOR PARALLEL KEYS = .005 in. OR 0.12 mm

$$S = D - \frac{H}{2} - T = \frac{D - H + \sqrt{D^2 - W^2}}{2} \qquad T = \frac{D - \sqrt{D^2 - W^2}}{2}$$

$$M = D - T + \frac{H}{2} + C = \frac{D + H + \sqrt{D^2 - W^2}}{2} + C$$

W = NOMINAL KEY WIDTH (INCHES OR MILLIMETERS)

TABLE 21   Square and flat stock keys.

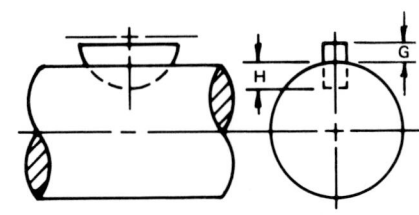

WOODRUFF KEYS

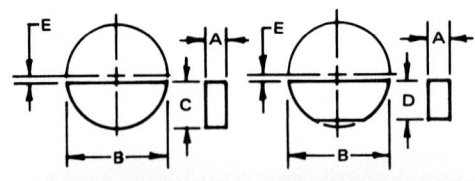

U.S. CUSTOMARY (INCHES)						METRIC (MILLIMETERS)					
Nominal Size	Key			Keyseats		Key No.	Nominal Size	Key			Keyseats	
A × B	E	C	D	H	G		A × B	E	C	D	H	G
.062 × .500	.047	.203	.194	.167	.037	204	1.6 × 12.7	1.5	5.1	4.8	4.24	0.94
.094 × .500	.047	.203	.194	.151	.053	304	2.4 × 12.7	1.3	5.1	4.8	3.84	1.35
.094 × .625	.062	.250	.240	.198	.053	305	2.4 × 15.9	1.5	6.4	6.1	5.03	1.35
.125 × .500	.049	.203	.194	.136	.069	404	3.2 × 12.7	1.3	5.1	4.8	3.45	1.75
.125 × .625	.062	.250	.240	.183	.069	405	3.2 × 15.9	1.5	6.4	6.1	4.65	1.75
.125 × .750	.062	.313	.303	.246	.069	406	3.2 × 19.1	1.5	7.9	7.6	6.25	1.75
.156 × .625	.062	.250	.240	.170	.084	505	4.0 × 15.9	1.5	6.4	6.1	4.32	2.13
.156 × .750	.062	.313	.303	.230	.084	506	4.0 × 19.1	1.5	7.9	7.6	5.84	2.13
.156 × .875	.062	.375	.365	.292	.084	507	4.0 × 22.2	1.5	9.7	9.1	7.42	2.13
.188 × .750	.062	.313	.303	.214	.100	606	4.8 × 19.1	1.5	7.9	7.6	5.44	2.54
.188 × .875	.062	.375	.365	.276	.100	607	4.8 × 22.2	1.5	9.7	9.1	7.01	2.54
.188 × 1.000	.062	.438	.428	.339	.100	608	4.8 × 25.4	1.5	11.2	10.9	8.61	2.54
.188 × 1.125	.078	.484	.475	.385	.100	609	4.8 × 28.6	2.0	12.2	11.9	9.78	2.54
.250 × .875	.062	.375	.365	.245	.131	807	6.4 × 22.2	1.5	9.7	9.1	6.22	3.33
.250 × 1.000	.062	.438	.428	.308	.131	808	6.4 × 25.4	1.5	11.2	10.9	7.82	3.33

NOTE: METRIC KEY SIZES WERE NOT AVAILABLE AT THE TIME OF PUBLICATION. SIZES SHOWN ARE INCH-DESIGNED KEY-SIZES SOFT CONVERTED TO MILLIMETERS. CONVERSION WAS NECESSARY TO ALLOW THE STUDENT TO COMPARE KEYS WITH SLOT SIZES GIVEN IN MILLIMETERS.

TABLE 22   Woodruff keys.

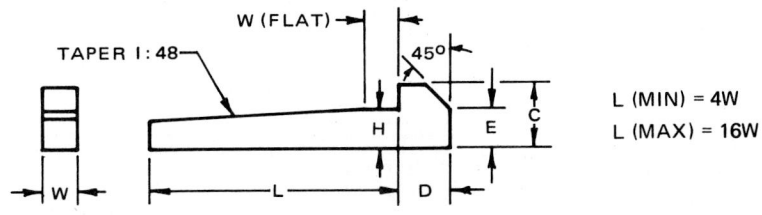

L (MIN) = 4W
L (MAX) = 16W

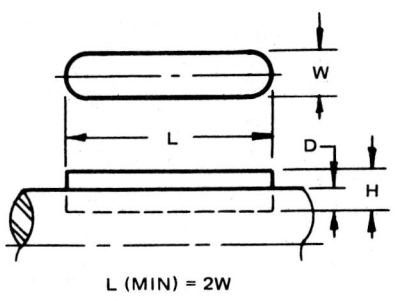

L (MIN) = 2W

Shaft Diameter	Square Type					Flat Type				
	W	H	C	D	E	W	H	C	D	E
.500–.562	.125	.125	.250	.219	.156	.125	.094	.188	.125	.125
.625–.875	.188	.188	.312	.281	.219	.188	.125	.250	.188	.156
.938–1.250	.250	.250	.438	.344	.344	.250	.188	.312	.250	.188
1.312–1.375	.312	.312	.562	.406	.406	.312	.250	.375	.312	.250
1.438–1.750	.375	.375	.688	.469	.469	.375	.250	.438	.375	.312
1.812–2.250	.500	.500	.875	.594	.625	.500	.375	.625	.500	.438
2.312–2.750	.625	.625	1.062	.719	.750	.625	.438	.750	.625	.500
2.875–3.250	.750	.750	1.250	.875	.875	.750	.500	.875	.750	.625

U.S. CUSTOMARY (INCHES)

METRIC (MILLIMETERS)

Shaft Diameter	Square Type					Flat Type				
	W	H	C	D	E	W	H	C	D	E
12–14	3.2	3.2	6.4	5.4	4	3.2	2.4	5	3.2	3.2
16–22	4.8	4.8	10	7	5.4	4.8	3.2	6.4	5	4
24–32	6.4	6.4	11	8.6	8.6	6.4	5	8	6.4	5
34–35	8	8	14	10	10	8	6.4	10	8	6.4
36–44	10	10	18	12	12	10	6.4	11	10	8
46–58	13	13	22	15	16	13	10	16	13	11
60–70	16	16	27	19	20	16	11	20	16	13
72–82	20	20	32	22	22	20	13	22	20	16

Note: Metric standards governing key sizes were not available at the time of publication. The sizes given in the above chart are "soft conversion" from current standards and are not representative of the precise metric key sizes which may be available in the future. Metric sizes are given only to allow the student to complete the drawing assignment.

TABLE 23   Square and flat gib-head keys.

U.S. CUSTOMARY (INCHES)

Key No.	L	W	H	D
2	.500	.094	.141	.094
4	.625	.094	.141	.094
6	.625	.156	.234	.156
8	.750	.156	.234	.156
10	.875	.156	.234	.156
12	.875	.234	.328	.219
14	1.00	.234	.328	.234
16	1.125	.188	.281	.188
18	1.125	.250	.375	.250
20	1.250	.219	.328	.219
22	1.375	.250	.375	.250
24	1.50	.250	.375	.250
26	2.00	.188	.281	.188
28	2.00	.312	.469	.312
30	3.00	.375	.562	.375
32	3.00	.500	.750	.500
34	3.00	.625	.938	.625

METRIC (MILLIMETERS)

Key No.	L	W	H	D
2	12	2.4	3.6	2.4
4	16	2.4	3.6	2.4
6	16	4	6	4
8	20	4	6	4
10	22	4	6	4
12	22	6	8.4	7
14	25	6	8.4	6
16	28	5	7	5
18	28	6.4	10	6.4
20	32	7	8	5
22	35	6.4	10	6.4
24	38	6.4	10	6.4
26	50	5	7	5
28	50	8	12	8
30	75	10	14	10
32	75	12	20	12
34	75	16	24	16

TABLE 24   Pratt and Whitney keys.

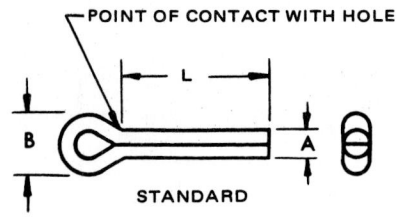

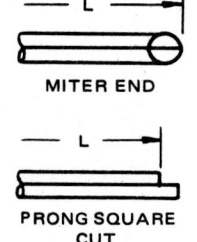

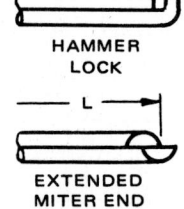

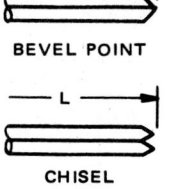

U.S. CUSTOMARY (INCHES)				METRIC (MILLIMETERS)			
Nominal Bolt or Thread Size Range	Nominal Cotter-Pin Size (A)	Cotter-Pin Hole	Min. End Clearance*	Nominal Bolt or Thread-Size Range	Nominal Cotter-Pin Size (A)	Cotter-Pin Hole	Min. End Clearance*
.125	.031	.047	.06	−2.5	0.6	0.8	1.5
.188	.047	.062	.08	2.5–3.5	0.8	1.0	2.0
.250	.062	.078	.11	3.5–4.5	1.0	1.2	2.0
.312	.078	.094	.11	4.5–5.5	1.2	1.4	2.5
.375	.094	.109	.14	5.5–7.0	1.6	1.8	2.5
.438	.109	.125	.14	7.0–9.0	2.0	2.2	3.0
.500	.125	.141	.18	9.0–11	2.5	2.8	3.5
.562	.141	.156	.25	11–14	3.2	3.6	5
.625	.156	.172	.40	14–20	4	4.5	6
1.000–1.125	.188	.203	.40	20–27	5	5.6	7
1.250–1.375	.219	.234	.46	27–39	6.3	6.7	10
1.500–1.625	.250	.266	.46	39–56	8.0	8.5	15
				56–80	10	10.5	20

*End of bolt to center of hole

TABLE 25   Cotter pins.

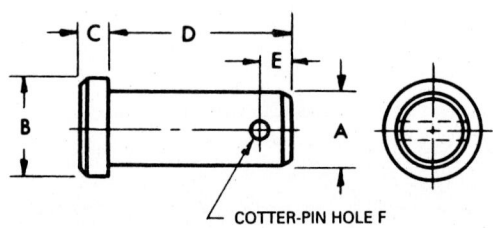

U.S. CUSTOMARY (INCHES)						METRIC (MILLIMETERS)					
Pin Dia.			Min.		Drill Size	Pin Dia.			Min.		Drill Size
A	B	C	D	E	F	A	B	C	D	E	F
.188	.31	.06	.59	.11	.078	4	6	1	16	2.2	1
.250	.38	.09	.80	.12	.078	6	10	2	20	3.2	1.6
.312	.44	.09	.97	.16	.109	8	14	3	24	3.5	2
.375	.50	.12	1.09	.16	.109	10	18	4	28	4.5	3.2
.500	.62	.16	1.42	.22	.141	12	20	4	36	5.5	3.2
.625	.81	.20	1.72	.25	.141	16	25	4.5	44	6	4
.750	.94	.25	2.05	.30	.172	20	30	5	52	8	5
1.000	1.19	.34	2.62	.36	.172	24	36	6	66	9	6.3

TABLE 26   Clevis pins.

U.S. CUSTOMARY (INCHES)

NUMBER	7/0	6/0	5/0	4/0	3/0	2/0	0	1	2	3	4	5	6	7	8	9
SIZE (LARGE END)	.062	.078	.094	.109	.125	.141	.152	.172	.193	.219	.250	.289	.314	.409	.492	.591
LENGTH .375	x	x														
.500	x	x	x	x	x	x	x									
.625	x	x	x	x	x	x	x									
.750		x	x	x	x	x	x	x	x	x						
.875						x	x	x	x	x						
1.000			x	x	x	x	x	x	x	x	x	x				
1.250						x	x	x	x	x	x	x	x			
1.500							x	x	x	x	x	x	x			
1.750								x	x	x	x	x	x			
2.000								x	x	x	x	x	x	x	x	
2.250									x	x	x	x	x	x	x	
2.500									x	x	x	x	x	x	x	
2.750										x	x	x	x	x	x	x

METRIC (MILLIMETERS)

SIZE (LARGE END)	1.6	2	2.4	2.8	3.2	3.6	4	4.4	4.9	5.6	6.4	7.4	8	10.4	12.5	15
LENGTH 10	x	x														
12	x	x	x	x	x	x	x									
16	x	x	x	x	x	x	x									
20		x	x	x	x	x	x	x	x	x						
22					x	x	x	x	x	x						
25			x	x	x	x	x	x	x	x	x	x				
30						x	x	x	x	x	x	x	x			
40							x	x	x	x	x	x	x			
45								x	x	x	x	x	x			
50								x	x	x	x	x	x	x	x	
55									x	x	x	x	x	x	x	
65									x	x	x	x	x	x	x	
70										x	x	x	x	x	x	x

TABLE 27 Taper pins.

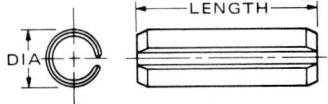

PIN DIAMETER (INCHES)

Length	.062	.094	.125	.156	.188	.250	.312
.250	x	x					
.375	x	x	x				
.500	x	x	x	x	x		
.625	x	x	x	x	x	x	
.750	x	x	x	x	x	x	x
.875	x	x	x	x	x	x	x
1.00	x	x	x	x	x	x	x
1.250		x	x	x	x	x	x
1.500		x	x	x	x	x	x
1.750			x	x	x	x	x
2.000		x	x	x	x	x	x
2.225			x	x	x	x	x
2.500				x	x	x	x
3.000						x	x
3.500						x	x

PIN DIAMETER (MILLIMETERS)

Length	1.5	2	2.5	3	4	5	6	8	10	12
5	x	x								
10	x	x	x	x						
15	x	x	x	x	x	x				
20	x	x	x	x	x	x	x			
25	x	x	x	x	x	x	x	x		
30		x	x	x	x	x	x	x	x	x
35		x	x	x	x	x	x	x	x	x
40		x	x	x	x	x	x	x	x	x
45				x	x	x	x	x	x	x
50				x	x	x	x	x	x	x
55					x	x	x	x	x	x
60					x	x	x	x	x	x
70							x	x	x	x
75							x	x	x	x
80							x	x	x	x

TABLE 28 Spring pins.

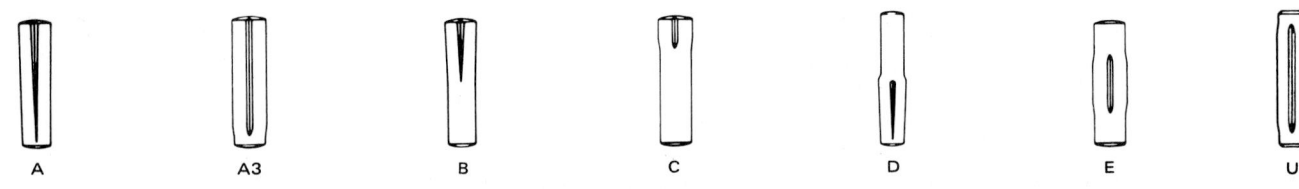

	U.S. CUSTOMARY (INCHES)								METRIC (MILLIMETERS)							
	PIN DIAMETER								PIN DIAMETER							
Length	.09	.125	.188	.250	.312	.375	.500	Length	2	3	4	5	6	8	10	12
.250	X	X						5	X	X	X					
.375	X	X	X					10	X	X	X	X	X			
.500	X	X	X	X				15	X	X	X	X	X	X		
.625	X	X	X	X	X			20	X	X	X	X	X	X	X	
.750	X	X	X	X	X	X		25	X	X	X	X	X	X	X	X
.875	X	X	X	X	X	X		30	X	X	X	X	X	X	X	X
1.000	X	X	X	X	X	X	X	35		X	X	X	X	X	X	X
1.250	X	X	X	X	X	X	X	40			X	X	X	X	X	X
1.500		X	X	X	X	X	X	45				X	X	X	X	X
1.750			X	X	X	X	X	50				X	X	X	X	X
2.000			X	X	X	X	X	55					X	X	X	X
2.250			X	X	X	X	X	60					X	X	X	X
2.275				X	X	X	X	65					X	X	X	X
3.000				X	X	X	X	70						X	X	X
								75						X	X	X

Note: Metric size pins were not available at the time of publication. Sizes were soft converted to allow students to complete drawing assignments.

TABLE 29 Groove pins.

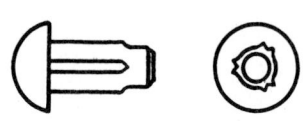

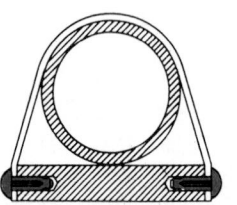

**WIDELY USED FOR
FASTENING BRACKETS**

**ATTACHING NAMEPLATES,
INSTRUCTION PANELS**

U.S. CUSTOMARY (INCHES)											METRIC (MILLIMETERS)									
STUD NUMBER	SHANK DIA.	DRILL SIZE	HEAD DIA.	STANDARD LENGTHS						STUD NUMBER	SHANK DIA.	DRILL SIZE	HEAD DIA.	STANDARD LENGTHS					
				.125	.188	.250	.312	.375	.500					4	6	8	10	12	14
0	.067	51	.130	●	●	●				0	1.7	1.7	3.3	●	●	●			
2	.086	44	.162	●	●	●				2	2.2	2.2	4.1	●	●	●			
4	.104	37	.211		●	●	●			4	2.6	2.6	5.4		●	●	●		
6	.120	31	.260			●	●	●		6	3.0	3.0	6.6			●	●	●	
7	.136	29	.309				●	●	●	7	3.4	3.4	7.8				●	●	●
8	.144	27	.309					●	●	8	3.8	3.8	7.8					●	●
10	.161	20	.359					●	●	10	4.1	4.1	9.1					●	●
12	.196	9	.408						●	12	5.0	5.0	10.4						●
14	.221	2	.457						●	14	5.6	5.6	11.6						●
16	.250	¼	.472						●	16	6.3	6.3	12						●

Note: Metric size studs were not available at the time of publication. Sizes were soft converted to allow students to complete drawing assignments.

TABLE 30 Grooved studs. *(Drive-Lok)*

Use these columns first to locate your correct

GRIP LENGTH

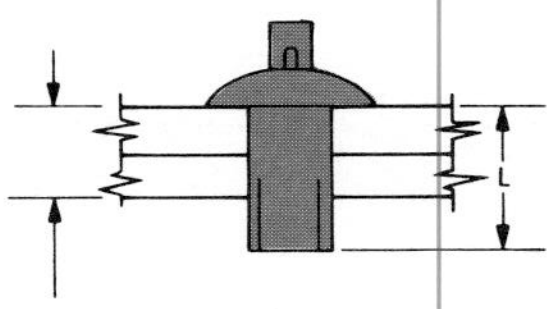

Grip = Total thickness of all sheets fastened together

.125in. (3mm) Dia — Part Numbers

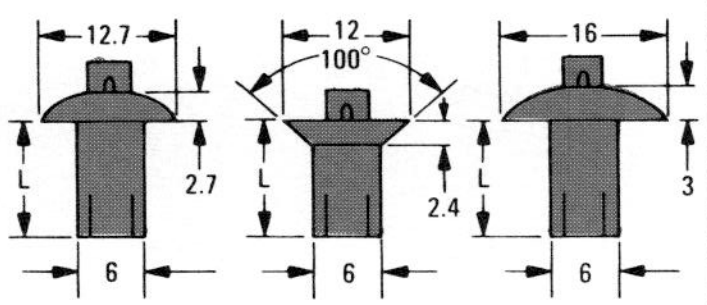

Minimum Grip	Nominal Grip	Maximum Grip	Length Under Head L	Universal Head	100° Csk Head	Full Brazier Head
1	2	3	5	✓	✓	✓
2	3	4	6	✓	✓	✓
4	5	6	8	✓	✓	✓
5	6	7	10	✓	✓	✓
7	8	9	12	✓	✓	✓
9	10	11	14	✓	✓	✓
11	12	13	16	✓		
13	14	15	18			
15	16	17	20			
19	20	21	24			
23	24	25	28			
27	28	29	32			

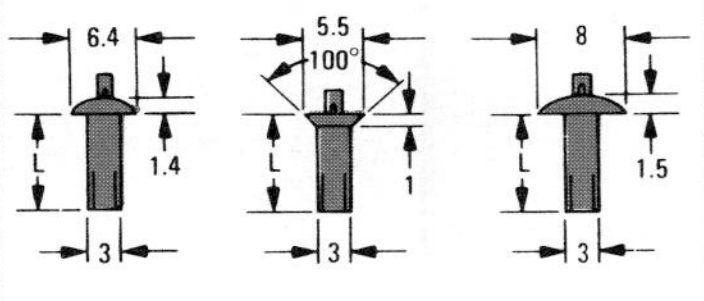

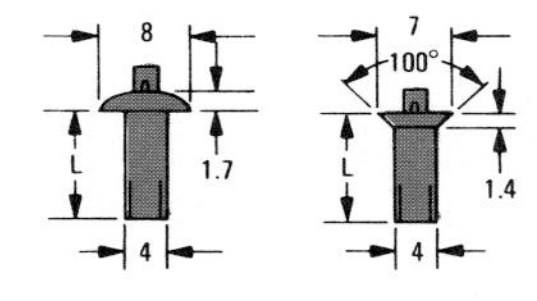

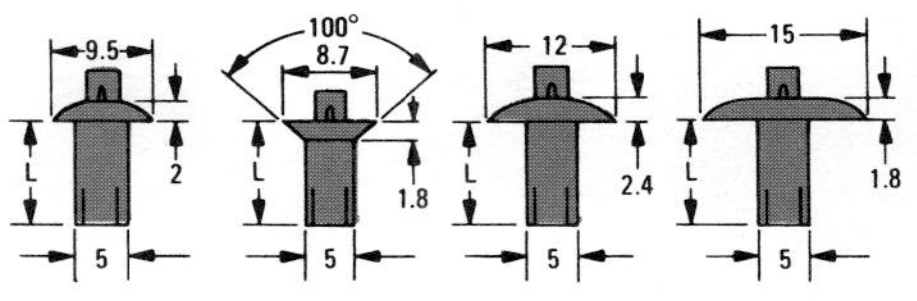

.156in. (4mm) Dia — Part Numbers

Length Under Head L	Universal Head	100° Csk Head
5	✓	
6	✓	✓
8	✓	✓
10	✓	✓
12	✓	✓
14	✓	✓
16	✓	✓
18	✓	✓
20	✓	✓
24		
28		
32		

.188in. (5mm) Dia — Part Numbers

Length Under Head L	Universal Head	100° Csk Head	Full Brazier Head	All Purpose Liner Head
5	✓		✓	✓
6	✓		✓	✓
8	✓	✓	✓	✓
10	✓	✓	✓	✓
12	✓	✓	✓	✓
14	✓	✓	✓	✓
16	✓	✓	✓	✓
18	✓	✓	✓	✓
20	✓	✓	✓	✓
24	✓		✓	
28	✓			
32	✓			

.250in. (6mm) Dia — Part Numbers

Length Under Head L	Universal Head	100° Csk Head	Full Brazier Head
5	✓	✓	✓
6	✓	✓	✓
8	✓	✓	✓
10	✓	✓	✓
12	✓	✓	✓
14	✓	✓	✓
16	✓	✓	✓
18	✓	✓	✓
20	✓	✓	✓
24	✓	✓	✓
28	✓	✓	✓
32	✓	✓	✓

Note: Metric drive rivets were not available at the time of publication. Sizes were soft converted to allow students to complete drawing assignments.

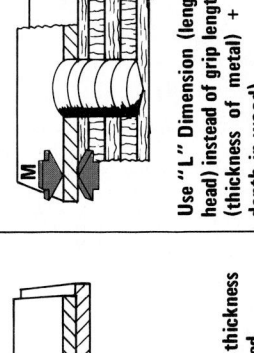

IN WOOD — Use "L" Dimension (length under head) instead of grip length. L = M (thickness of metal) + D (hole depth in wood).

IN METAL — GRIP — Grip Length = Total thickness of sheets to be fastened.

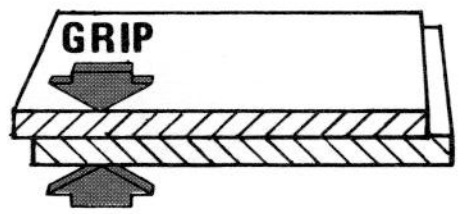

HIT THE PIN — Drive pin flush with rivet head

Expanding prongs clinch sheets tightly, eliminating gaps.

Metal and wood pulled tightly together. Nothing protrudes through wood.

TABLE 31   Aluminum drive rivets. (*Southco Lion Fasteners*)

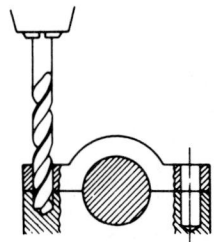

DRILL HOLE SLIGHTLY UNDERSIZE

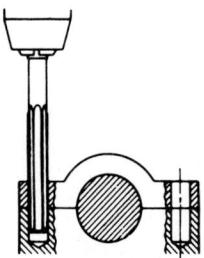

REAM FULL SIZE

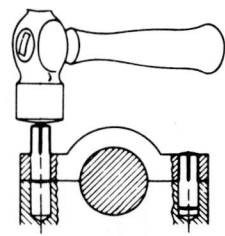

DRIVE OR PRESS LOK DOWELS INTO PLACE

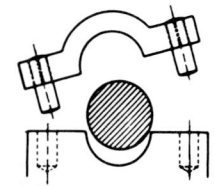

LOK DOWELS LOCK SECURELY AND PARTS SEPARATE EASILY

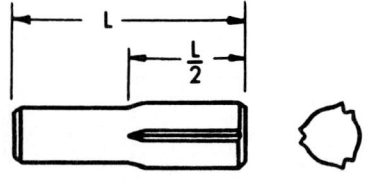

| | U.S. CUSTOMARY (INCHES) | | | | | | METRIC (MILLIMETERS) | | | | | |
| | DIAMETER | | | | | | | DIAMETER | | | | |
LENGTH	.125	.188	.250	.312	.375	.500	LENGTH	4	6	8	10	12	14
.375	•						10	•					
.500	•	•	•	•			12	•	•	•	•		
.625	•	•	•	•			16	•	•	•	•		
.750	•	•	•	•	•		20	•	•	•	•	•	
.875	•	•	•	•	•		22	•	•	•	•	•	
1.000	•	•	•	•	•	•	26	•	•	•	•	•	•
1.250		•	•	•	•	•	32		•	•	•	•	•
1.500			•	•	•	•	38			•	•	•	•
1.750				•	•	•	44				•	•	•
2.000					•	•	50					•	•

Note: Metric size dowels were not available at the time of publication. Sizes were soft converted to allow students to complete drawing assignments.

TABLE 32 Lok dowels. *(Drive-Lok)*

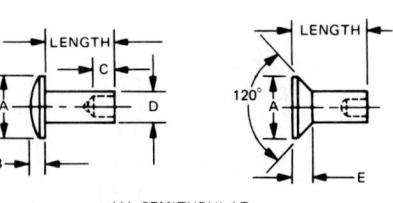

(A) SEMITUBULAR

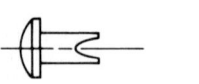 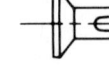

(B) SPLIT

	D	A	B	C	E	MIN. LENGTH
U.S. CUSTOMARY (INCHES)	.062	.125	.031	.062	—	.062
	.094	.156	.031	.062	.031	.078
	.109	.188	.031	.078	—	.078
	.125	.218	.047	.109	.049	.109
	.141	.250	.047	.125	.049	.125
	.188	.312	.062	.141	.062	.156
	.219	.438	.062	.188	.062	.188
	.250	.500	.078	.219	.094	.219
	.312	.562	.109	.250		.250
METRIC (MILLIMETERS)	1.5	2.8	0.4	1.2		1.6
	2.2	3.7	0.6	1.6	0.8	2.0
	2.5	4.7	0.7	2.0		2.0
	3.1	5.5	0.9	2.4	1.0	2.4
	3.6	5.9	1.0	3.2	1.2	3.2
	4.7	7.9	1.5	3.9	1.6	4.0
	5.4	11.1	1.7	4.8	1.8	4.8
	6.3	12.7	2.0	5.6	2.2	5.6
	7.7	14.3	2.4	6.2		6.4

TABLE 33 Semitubular and split rivets.

| U.S. CUSTOMARY (INCHES) | | | | METRIC (MILLIMETERS) | | | |
HOLE DIA.	PANEL RANGE	HEAD Dia.	Height	HOLE DIA.	PANEL RANGE	HEAD Dia.	Height
.125	.031–.140	.188	.047	3.18	0.8– 3.6	4.8	1.2
	.031–.125	.218	.062		0.8– 3.2	5.5	1.5
.156	.250–.375	.218	.047	4.01	5.9– 9.4	5.5	1.3
.188	.062–.156	.375	.125	4.75	1.6– 4.0	9.5	3.2
	.156–.281	.438	.094		4.0– 7.1	11.1	1.9
.219	.062–.125	.375	.094	5.54	1.6– 3.2	9.5	2.4
	.094–.312		.078		2.4– 8.0		2.0
.250	.094–.219	.625	.125	6.35	2.3– 5.6	16	3.2
	.125–.375	.750	.062		3.2– 9.5	19	1.3
.297	.140–.328	.500	.078	7.14	3.4– 8.1	12.3	1.9
.375	.250–.500	.438	.109	9.53	6.4–12.7	11.1	2.6
.500	.312–.375	.750	.109	12.7	8.1– 9.4	19	2.5

TABLE 34 Plastic rivets.

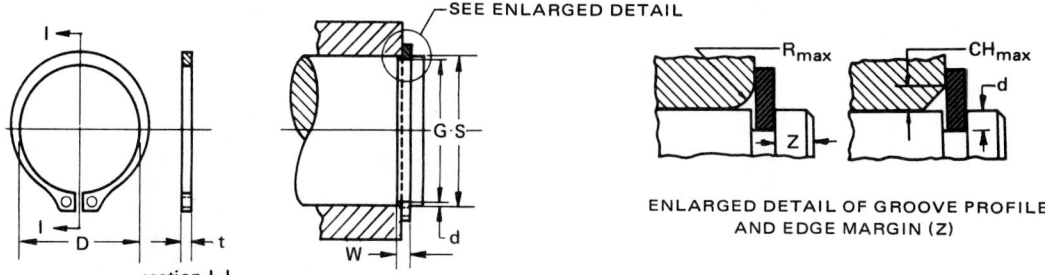

ENLARGED DETAIL OF GROOVE PROFILE
AND EDGE MARGIN (Z)

| | SHAFT DIA. | EXTERNAL SERIES | RETAINING RING DIMENSIONS | | GROOVE DIMENSIONS | | | | MAX. CORNER RADII AND CHAMFER OF RETAINED PARTS | | EDGE MARGIN | NOMINAL GROOVE DEPTH (REF) |
| | | | | | DIAMETER | | WIDTH | | | | | |
	S	Size—No.	D	t	G	Tol.	W	Tol.	R Max.	Ch. Max.	z	d
U.S. CUSTOMARY (INCHES)	.188	5100-18	.168	.015	.175	±.0015	.018	+.002	.014	.008	.018	.006
	.250	5100-25	.225	.025	.230	±.0015	.029	+.003	.018	.011	.030	.010
	.312	5100-31	.281	.025	.290	±.002	.029	+.003	.020	.012	.033	.011
	.375	5100-37	.338	.025	.352	±.002	.029	+.003	.026	.015	.036	.012
	.500	5100-50	.461	.035	.468	±.002	.039	+.003	.034	.020	.048	.016
	.625	5100-62	.579	.035	.588	±.003	.039	+.003	.041	.025	.055	.018
	.750	5100-75	.693	.042	.704	±.003	.046	+.003	.046	.027	.069	.023
	.875	5100-87	.810	.042	.821	±.003	.046	+.003	.051	.031	.081	.027
	1.000	5100-100	.925	.042	.940	±.003	.046	+.003	.057	.034	.090	.030
	1.125	5100-112	1.041	.050	1.059	±.004	.056	+.004	.063	.038	.099	.033
	1.250	5100-125	1.156	.050	1.176	±.004	.056	+.004	.068	.041	.111	.037
	1.375	5100-137	1.272	.050	1.291	±.004	.056	+.004	.072	.043	.126	.042
	1.500	5100-150	1.387	.050	1.406	±.004	.056	+.004	.079	.047	.141	.047
METRIC (MILLIMETERS)	4	M5100-4	3.6	0.25	3.80	−0.08	0.32	+0.05	0.35	0.25	0.3	0.10
	6	M5100-6	5.5	0.4	5.70	−0.08	0.5	+0.1	0.35	0.25	0.5	0.15
	8	M5100-8	7.2	0.6	7.50	−0.1	0.7	+0.15	0.5	0.35	0.8	0.25
	10	M5100-10	9.0	0.6	9.40	−0.1	0.7	+0.15	0.7	0.4	0.9	0.30
	12	M5100-12	10.9	0.6	11.35	−0.12	0.7	+0.15	0.8	0.45	1.0	0.33
	14	M5100-14	12.9	0.9	13.25	−0.12	1.0	+0.15	0.9	0.5	1.2	0.38
	16	M5100-16	14.7	0.9	15.10	−0.15	1.0	+0.15	1.1	0.6	1.4	0.45
	18	M5100-18	16.7	1.1	17.00	−0.15	1.2	+0.15	1.2	0.7	1.5	0.50
	20	M5100-20	18.4	1.1	18.85	−0.15	1.2	+0.15	1.2	0.7	1.7	0.58
	22	M5100-22	20.3	1.1	20.70	−0.15	1.2	+0.15	1.3	0.8	1.9	0.65
	24	M5100-24	22.2	1.1	22.60	−0.15	1.2	+0.15	1.4	0.8	2.1	0.70
	25	M5100-25	23.1	1.1	23.50	−0.15	1.2	+0.15	1.4	0.8	2.3	0.75
	30	M5100-30	27.9	1.3	28.35	−0.2	1.4	+0.15	1.6	1.0	2.5	0.83
	35	M5100-35	32.3	1.3	32.9	−0.2	1.4	+0.15	1.8	1.1	3.1	1.05
	40	M5100-40	36.8	1.6	37.7	−0.3	1.75	+0.2	2.1	1.2	3.4	1.15
	45	M5100-45	41.6	1.6	42.4	−0.3	1.75	+0.2	2.3	1.4	3.9	1.3
	50	M5100-50	46.2	1.6	47.2	−0.3	1.75	+0.2	2.4	1.4	4.2	1.4

TABLE 35   Retaining rings—external. (©1965, 1958 Waldes Koh-i-noor, Inc. Reprinted with permission.)

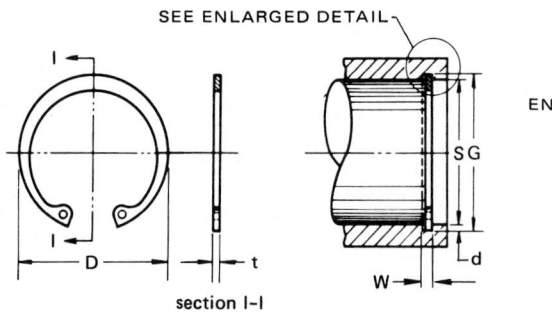

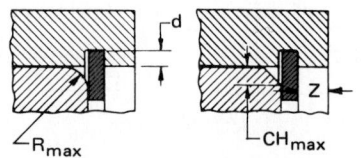

ENLARGED DETAIL OF GROOVE PROFILE
AND EDGE MARGIN (Z)

	HOUSING DIA.	INTERNAL SERIES	RETAINING RING DIMENSIONS		GROOVE DIMENSIONS				MAX. CORNER RADII AND CHAMFER OF RETAINED PARTS		EDGE MARGIN	NOMINAL GROOVE DEPTH
					DIAMETER		WIDTH					
	S	Size—No.	D	t	G	Tol.	W	Tol.	R Max.	Ch. Max.	z	d
U.S. CUSTOMARY (INCHES)	.250	N5000-25	.280	.015	.268	±.001	.018	+.002	.011	.008	.027	.009
	.312	N5000-31	.346	.015	.330	±.001	.018	+.002	.016	.013	.027	.009
	.375	N5000-37	.415	.025	.397	±.002	.029	+.003	.023	.018	.033	.011
	.500	N5000-50	.548	.035	.530	±.002	.039	+.003	.027	.021	.045	.015
	.625	N5000-62	.694	.035	.665	±.002	.039	+.003	.027	.021	.060	.020
	.750	N5000-75	.831	.035	.796	±.002	.039	+.003	.032	.025	.069	.023
	.875	N5000-87	.971	.042	.931	±.003	.046	+.003	.035	.028	.084	.028
	1.000	N5000-100	1.111	.042	1.066	±.003	.046	+.003	.042	.034	.099	.033
	1.125	N5000-112	1.249	.050	1.197	±.004	.056	+.004	.047	.036	.108	.036
	1.250	N5000-125	1.388	.050	1.330	±.004	.056	+.004	.048	.038	.120	.040
	1.375	N5000-137	1.526	.050	1.461	±.004	.056	+.004	.048	.038	.129	.043
	1.500	N5000-150	1.660	.050	1.594	±.004	.056	+.004	.048	.038	.141	.047
METRIC (MILLIMETERS)	8	MN5000-8	8.80	0.4	8.40	+0.6	0.5	+0.1	0.4	0.3	0.6	0.2
	10	MN5000-10	11.10	0.6	10.50	+0.1	0.7	+0.15	0.5	0.35	0.8	0.25
	12	MN5000-12	13.30	0.6	12.65	+0.1	0.7	+0.15	0.6	0.4	1.0	0.33
	14	MN5000-14	15.45	0.9	14.80	+0.1	1.0	+0.15	0.7	0.5	1.2	0.40
	16	MN5000-16	17.70	0.9	16.90	+0.1	1.0	+0.15	0.7	0.5	1.4	0.45
	18	MN5000-18	20.05	0.9	19.05	+0.1	1.0	+0.15	0.75	0.6	1.6	0.53
	20	MN5000-20	22.25	0.9	21.15	+0.15	1.0	+0.15	0.9	0.7	1.7	0.57
	22	MN5000-22	24.40	1.1	23.30	+0.15	1.2	+0.15	0.9	0.7	1.9	0.65
	24	MN5000-24	26.55	1.1	25.4	+0.15	1.2	+0.15	1.0	0.8	2.1	0.70
	25	MN5000-25	27.75	1.1	26.6	+0.15	1.2	+0.15	1.0	0.8	2.4	0.80
	30	MN5000-30	33.40	1.3	31.9	+0.2	1.4	+0.15	1.2	1.0	2.9	0.95
	35	MN5000-35	38.75	1.3	37.2	+0.2	1.4	+0.15	1.2	1.0	3.3	1.10
	40	MN5000-40	44.25	1.6	42.4	+0.2	1.75	+0.2	1.7	1.3	3.6	1.20
	45	MN5000-45	49.95	1.6	47.6	+0.2	1.75	+0.2	1.7	1.3	3.9	1.30
	50	MN5000-50	55.35	1.6	53.1	+0.2	1.75	+0.2	1.7	1.3	4.6	1.55

TABLE 36 Retaining rings—internal. *(©1965, 1958 Waldes Koh-i-noor, Inc. Reprinted with permission.)*

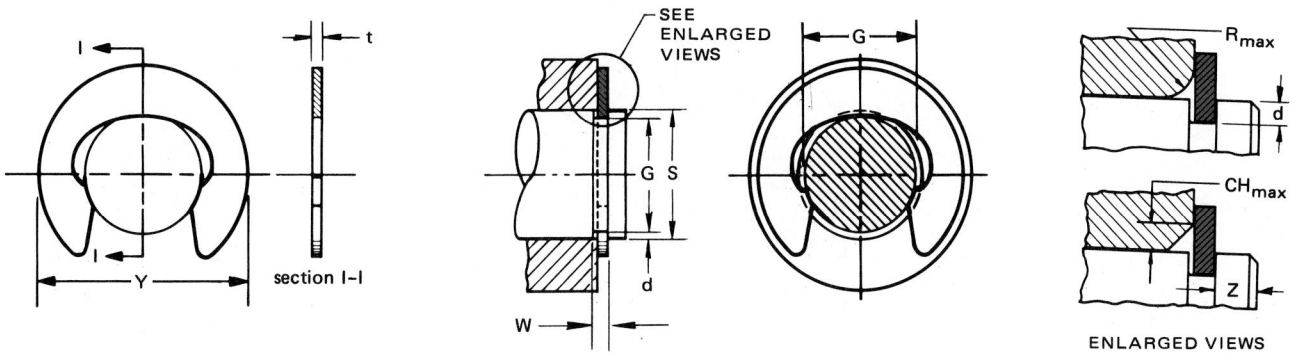

| | SHAFT DIA. | EXTERNAL SERIES 11-410 | RETAINING RING DIMENSIONS | | GROOVE DIMENSIONS | | | | MAXIMUM ALLOWABLE CORNER RADII AND CHAMFER OF RETAINED PARTS | | EDGE MARGIN | NOMINAL GROOVE DEPTH (REF) |
| | | | | | Diameter | | Width | | | | | |
	S	Size—No.	Y	t	G	Tol.	W	Tol.	R Max.	Ch. Max.	z	d
U.S. CUSTOMARY (INCHES)	.250	11-410-25	.311	.025	.222	−.004	.029	+.003	.023	.018	.030	.015
	.312	11-410-31	.376	.025	.278	−.004	.029	+.003	.024	.018	.036	.018
	.375	11-410-37	.448	.025	.337	−.004	.029	+.003	.026	.020	.040	.020
	.500	11-410-50	.581	.025	.453	−.006	.039	+.003	.030	.023	.050	.025
	.625	11-410-62	.715	.035	.566	−.006	.039	+.003	.033	.025	.062	.031
	.750	11-410-75	.845	.042	.679	−.006	.046	+.003	.036	.027	.074	.037
	.875	11-410-87	.987	.042	.792	−.006	.046	+.003	.040	.031	.086	.043
	1.000	11-410-100	1.127	.042	.903	−.006	.046	+.003	.046	.035	.100	.050
	1.125	11-410-112	1.267	.050	1.017	−.008	.056	+.004	.052	.040	.112	.056
	1.250	11-410-125	1.410	.050	1.130	−.008	.056	+.004	.057	.044	.124	.062
	1.375	11-410-137	1.550	.050	1.241	−.008	.056	+.004	.062	.048	.138	.069
	1.500	11-410-150	1.691	.050	1.354	−.008	.056	+.004	.069	.053	.150	.075
	1.750	11-410-175	1.975	.062	1.581	−.010	.068	+.004	.081	.062	.174	.087
	2.000	11-410-200	2.257	.062	1.805	−.010	.068	+.004	.091	.070	.200	.100
METRIC (MILLIMETERS)	8	M11-410-080	10	0.6	7	−0.1	0.7	+0.15	0.6	0.45	1.5	0.5
	10	M11-410-100	12.2	0.6	9	−0.1	0.7	+0.15	0.6	0.45	1.5	0.5
	12	M11-410-120	14.4	0.6	10.9	−0.1	0.7	+0.15	0.6	0.45	1.7	0.5
	14	M11-410-140	16.3	1	12.7	−0.1	1.1	+0.15	1	0.8	2	0.65
	16	M11-410-160	18.5	1	14.5	−0.1	1.1	+0.15	1	0.8	2.3	0.75
	18	M11-410-180	20.4	1.2	16.3	−0.1	1.3	+0.15	1.2	0.9	2.6	0.85
	20	M11-410-200	22.6	1.2	18.1	−0.2	1.3	+0.15	1.2	0.9	2.9	0.95
	22	M11-410-220	25	1.2	19.9	−0.2	1.3	+0.15	1.2	0.9	3.2	1.05
	24	M11-410-240	27.1	1.2	21.7	−0.2	1.3	+0.15	1.2	0.9	3.5	1.15
	25	M11-410-250	28.3	1.2	22.6	−0.2	1.3	+0.15	1.2	0.9	3.6	1.2
	30	M11-410-300	33.7	1.5	27	−0.2	1.3	+0.2	1.5	1.15	4.5	1.5
	35	M11-410-350	39.4	1.5	31.5	−0.25	1.6	+0.2	1.5	1.15	5.3	1.75
	40	M11-410-400	45	1.5	36	−0.25	1.6	+0.2	1.5	1.15	6	2
	45	M11-410-450	50.6	1.5	40.5	−0.25	1.6	+0.2	1.5	1.15	6.8	2.25
	50	M11-410-500	56.4	2	45	−0.25	2.2	+0.2	2	1.5	7.5	2.5

TABLE 37   Retaining rings—radial assembly. (©1965, 1958 Waldes
Koh-i-noor, Inc. Reprinted with permission.)

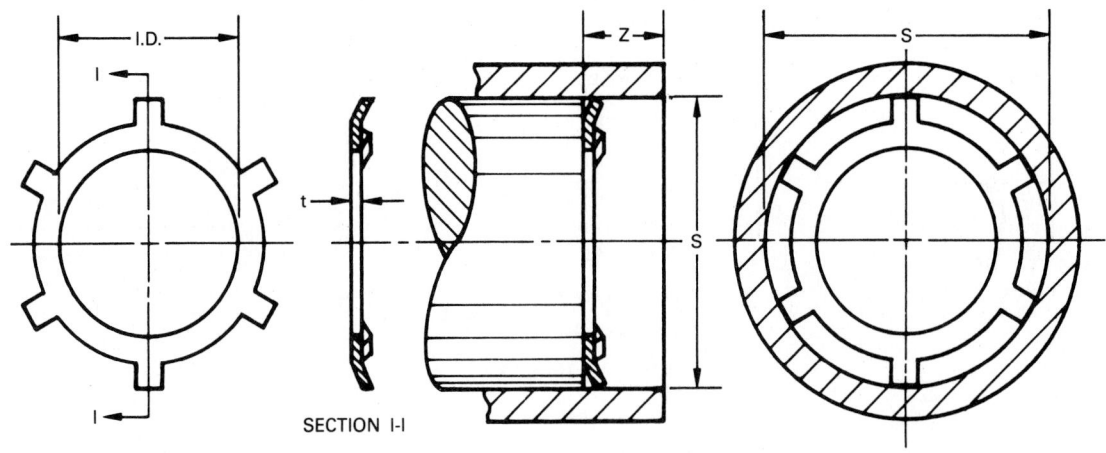

SECTION I-I

SHAFT DIAMETER		INTERVAL SERIES 5005	RING DIMENSIONS			APPLICATION DATA		
Decimal inch			OUTSIDE DIAMETER			THICKNESS		
S						.010 ± .001	.015 ± .002	
					No. of prongs	Allow. thrust load	Allow. thrust load	Z MIN
from	to	SIZE-NO.	OD	tol.				
.311	.313	5005-31	.136	± .005	5	80		.040
.374	.376	5005-37	.175		6	75		.040
.437	.439	5005-43	.237		6	70		.040
.498	.502	5005-50	.258		6	60		.040
.560	.564	5005-56	.312		6	50		.040
.623	.627	5005-62	.390	± .010	6	45		.040
.748	.752	5005-75	.500		8		75	.060
.873	.877	5005-87	.625		8		70	.060
.936	.940	5005-93	.687		10		70	.060
.998	1.002	5005-100	.750		10		75	.060
1.248	1.252	5005-125	.938		10		60	.060
1.371	1.379	Δ5015-137	1.050	± .015	16		150	.125
1.498	1.502	5005-150	1.188		12		60	.060
1.748	1.752	5005-175	1.438	± .010	12		55	.060
1.998	2.002	5005-200	1.600		14		55	.060

TABLE 38   Retaining rings—self-locking, internal. (©1965, 1958 Waldes Koh-i-noor, Inc. Reprinted with permission.)

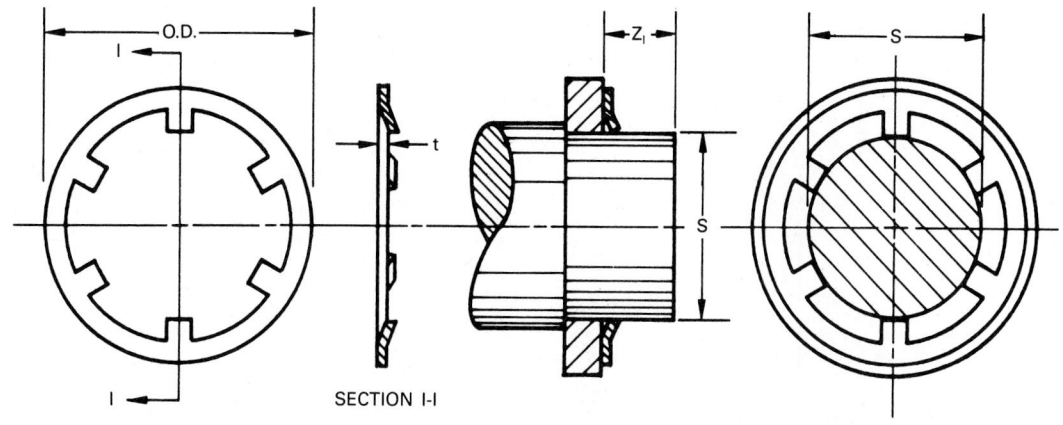

SECTION I-I

SHAFT DIAMETER		EXTERNAL SERIES 5105	RING DIMENSIONS			APPLICATION DATA		
							THICKNESS	
Decimal inch			OUTSIDE DIAMETER			.010	.015	
S					No. of prongs	Allow. thrust load	Allow. thrust load	Z MIN
from	to	SIZE-NO.	OD	tol.				
.093	.095	5105-9	.250		3	13		.040
.093	.095	Δ5505D-9	.240		3	10		.040
.124	.126	5105-12	.325		4	20		.040
.155	.157	5105-15	.356		4	25		.040
.187	.189	5105-18	.387		5	35		.040
.218	.220	5105-21	.418	± .005	6	35		.040
.239	.241	5105-24	.460		6		40	.060
.249	.251	5105-25	.450		6	40		.040
.311	.313	5105-31	.512		5	45		.040
.374	.376	5105-37	.575		6	45		.040
.437	.439	5105-43	.638		6		50	.060
.498	.502	5105-50	.750		6		50	.060
.560	.564	5105-56	.812		6		50	.060
.623	.627	5105-62	.875		7		50	.060
.637	.641	Δ5505-63	.875	± .010	8		50	.060
.748	.752	5105-75	1.000		8		55	.060
.873	.877	5105-87	1.125		10		60	.060
.998	1.002	5105-100	1.250		10		65	.060

TABLE 39 Retaining rings—self-locking, external. *(©1965, 1958 Waldes Koh-i-noor, Inc. Reprinted with permission.)*

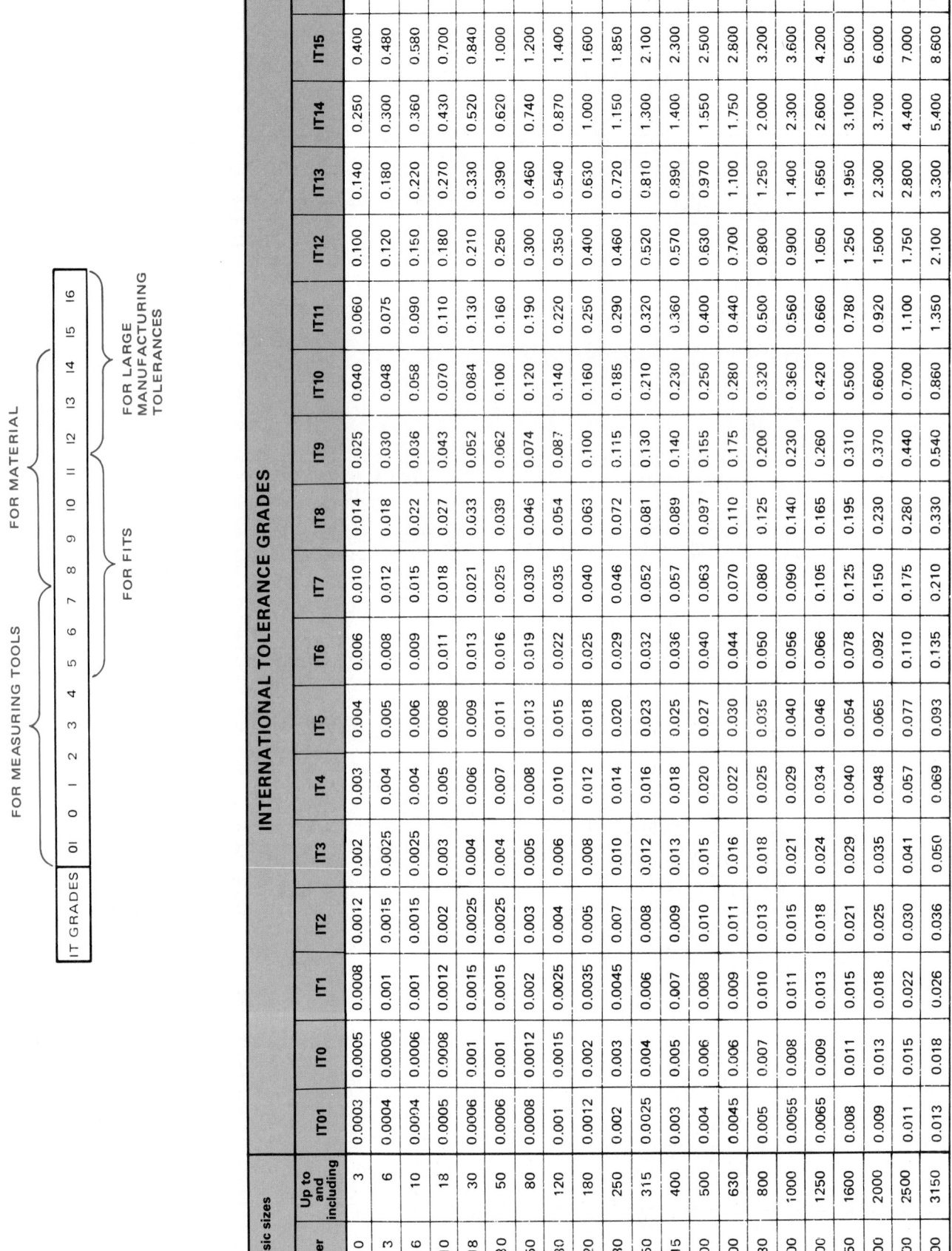

FOR MEASURING TOOLS — FOR MATERIAL

FOR FITS — FOR LARGE MANUFACTURING TOLERANCES

IT GRADES: 01 0 1 2 3 4 5 6 7 8 9 10 11 12 13 14 15 16

INTERNATIONAL TOLERANCE GRADES

Basic sizes																				
Over	Up to and including	IT01	IT0	IT1	IT2	IT3	IT4	IT5	IT6	IT7	IT8	IT9	IT10	IT11	IT12	IT13	IT14	IT15	IT16
0	3	0.0003	0.0005	0.0008	0.0012	0.002	0.003	0.004	0.006	0.010	0.014	0.025	0.040	0.060	0.100	0.140	0.250	0.400	0.600
3	6	0.0004	0.0006	0.001	0.0015	0.0025	0.004	0.005	0.008	0.012	0.018	0.030	0.048	0.075	0.120	0.180	0.300	0.480	0.750
6	10	0.0004	0.0006	0.001	0.0015	0.0025	0.004	0.006	0.009	0.015	0.022	0.036	0.058	0.090	0.150	0.220	0.360	0.580	0.900
10	18	0.0005	0.0008	0.0012	0.002	0.003	0.005	0.008	0.011	0.018	0.027	0.043	0.070	0.110	0.180	0.270	0.430	0.700	1.100
18	30	0.0006	0.001	0.0015	0.0025	0.004	0.006	0.009	0.013	0.021	0.033	0.052	0.084	0.130	0.210	0.330	0.520	0.840	1.300
30	50	0.0006	0.001	0.0015	0.0025	0.004	0.007	0.011	0.016	0.025	0.039	0.062	0.100	0.160	0.250	0.390	0.620	1.000	1.600
50	80	0.0008	0.0012	0.002	0.003	0.005	0.008	0.013	0.019	0.030	0.046	0.074	0.120	0.190	0.300	0.460	0.740	1.200	1.900
80	120	0.001	0.0015	0.0025	0.004	0.006	0.010	0.015	0.022	0.035	0.054	0.087	0.140	0.220	0.350	0.540	0.870	1.400	2.200
120	180	0.0012	0.002	0.0035	0.005	0.008	0.012	0.018	0.025	0.040	0.063	0.100	0.160	0.250	0.400	0.630	1.000	1.600	2.500
180	250	0.002	0.003	0.0045	0.007	0.010	0.014	0.020	0.029	0.046	0.072	0.115	0.185	0.290	0.460	0.720	1.150	1.850	2.900
250	315	0.0025	0.004	0.006	0.008	0.012	0.016	0.023	0.032	0.052	0.081	0.130	0.210	0.320	0.520	0.810	1.300	2.100	3.200
315	400	0.003	0.005	0.007	0.009	0.013	0.018	0.025	0.036	0.057	0.089	0.140	0.230	0.360	0.570	0.890	1.400	2.300	3.600
400	500	0.004	0.006	0.008	0.010	0.015	0.020	0.027	0.040	0.063	0.097	0.155	0.250	0.400	0.630	0.970	1.550	2.500	4.000
500	630	0.0045	0.006	0.009	0.011	0.016	0.022	0.030	0.044	0.070	0.110	0.175	0.280	0.440	0.700	1.100	1.750	2.800	4.400
630	800	0.005	0.007	0.010	0.013	0.018	0.025	0.035	0.050	0.080	0.125	0.200	0.320	0.500	0.800	1.250	2.000	3.200	5.000
800	1000	0.0055	0.008	0.011	0.015	0.021	0.029	0.040	0.056	0.090	0.140	0.230	0.360	0.560	0.900	1.400	2.300	3.600	5.600
1000	1250	0.0065	0.009	0.013	0.018	0.024	0.034	0.046	0.066	0.105	0.165	0.260	0.420	0.660	1.050	1.650	2.600	4.200	6.600
1250	1600	0.008	0.011	0.015	0.021	0.029	0.040	0.054	0.078	0.125	0.195	0.310	0.500	0.780	1.250	1.950	3.100	5.000	7.800
1600	2000	0.009	0.013	0.018	0.025	0.035	0.048	0.065	0.092	0.150	0.230	0.370	0.600	0.920	1.500	2.300	3.700	6.000	9.200
2000	2500	0.011	0.015	0.022	0.030	0.041	0.057	0.077	0.110	0.175	0.280	0.440	0.700	1.100	1.750	2.800	4.400	7.000	11.000
2500	3150	0.013	0.018	0.026	0.036	0.050	0.069	0.093	0.135	0.210	0.330	0.540	0.860	1.350	2.100	3.300	5.400	8.600	13.500

TABLE 40   International tolerance grades. (Values in millimeters.)

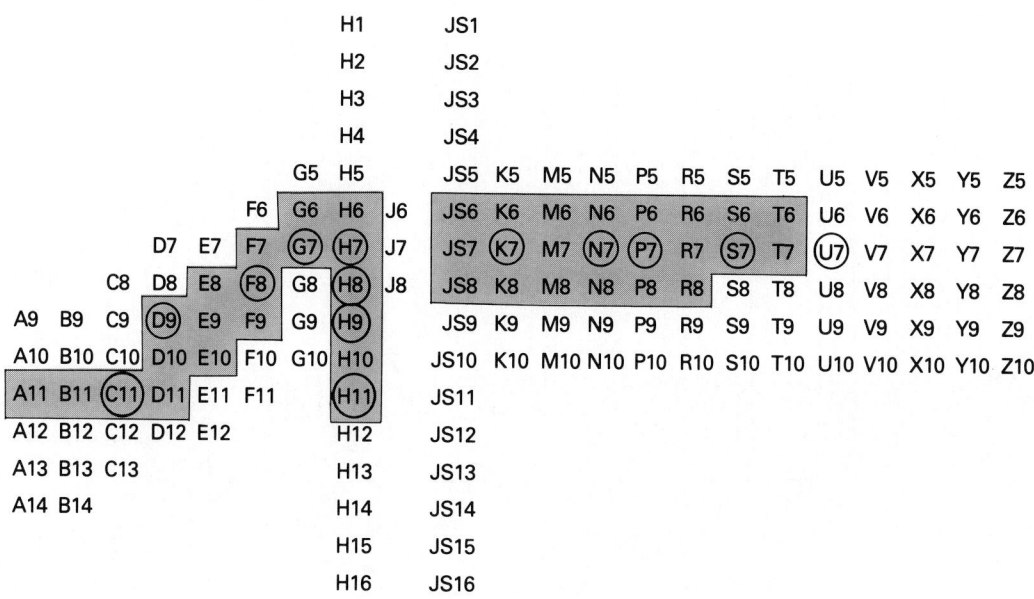

Legend: First choice tolerance zones encircled (ANSI B4.2 preferred)
Second choice tolerance zones framed (ISO 1829 selected)
Third choice tolerance zones open

TOLERANCE ZONES FOR INTERNAL DIMENSIONS (HOLES)

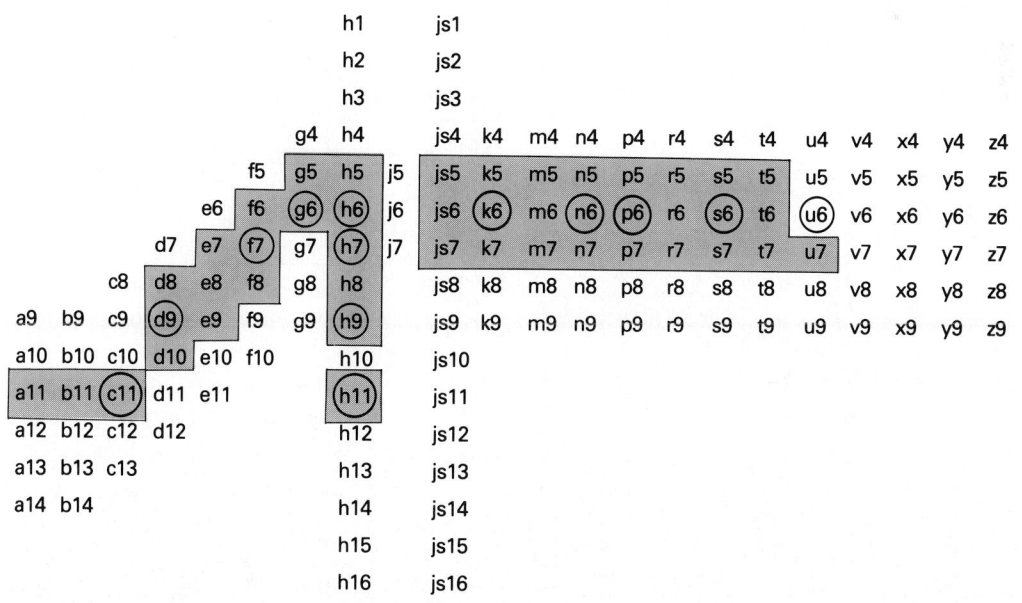

Legend: First choice tolerance zones encircled (ANSI B4.2 preferred)
Second choice tolerance zones framed (ISO 1829 selected)
Third choice tolerance zones open

TOLERANCE ZONES FOR EXTERNAL DIMENSIONS (SHAFTS)

TABLE 40   International tolerance grades. (Values in millimeters.)
(continued)

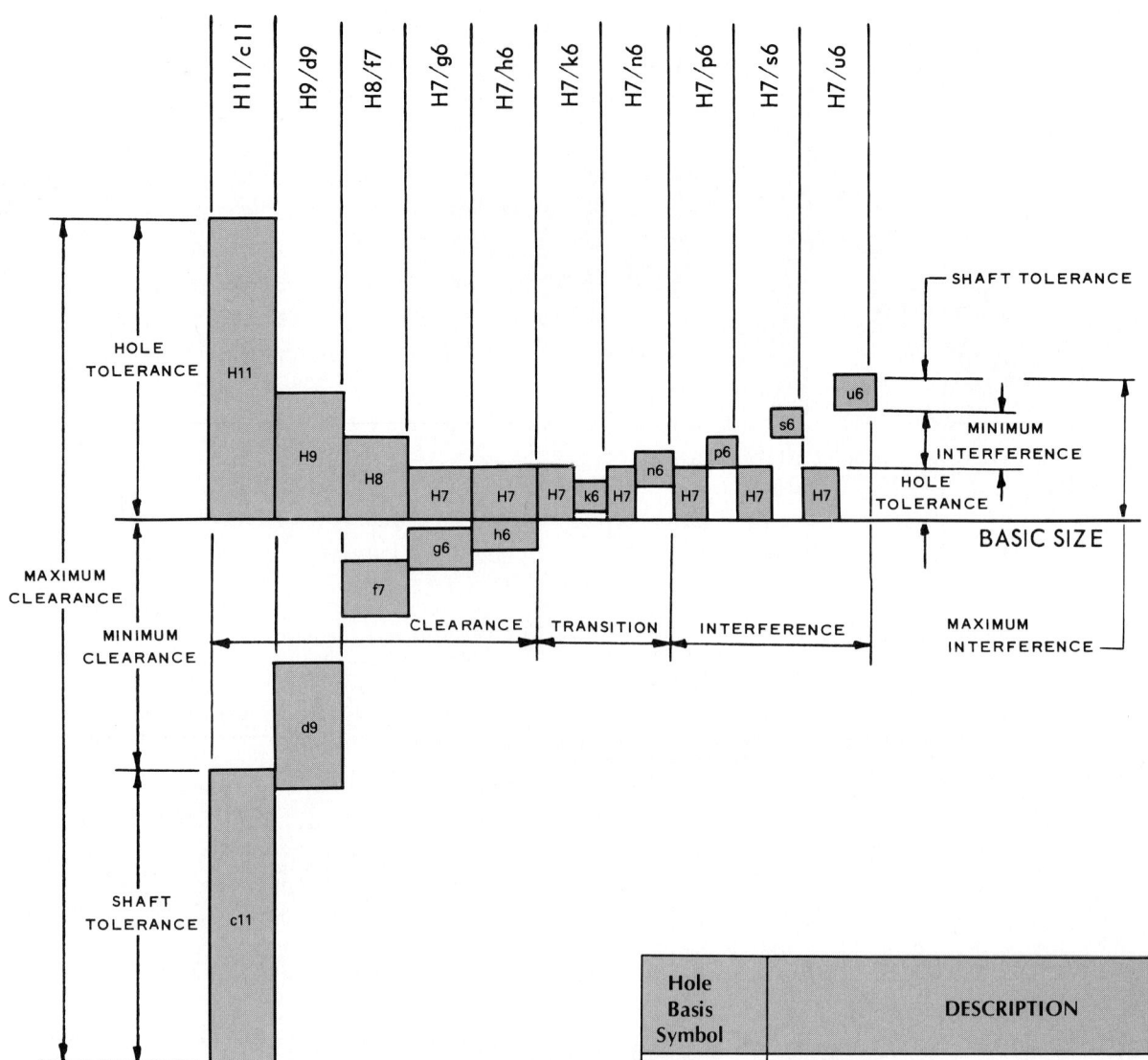

TABLE 41 Preferred hole basis fits description.

Hole Basis Symbol	DESCRIPTION
H11/c11	*Loose-running* fit for wide commercial tolerances or allowances on external members.
H9/d9	*Free-running* fit not for use where accuracy is essential, but good for large temperature variations, high running speeds, or heavy journal pressures.
H8/f7	*Close-running* fit for running on accurate machines and for accurate location at moderate speeds and journal pressures.
H7/g6	*Sliding* fit not intended to run freely, but to move and turn freely and locate accurately.

Hole Basis Symbol	DESCRIPTION
H7/h6	*Locational clearance* fit provides snug fit for locating stationary parts; but can be freely assembled and disassembled.
H7/k6	*Locational transition* fit for accurate location, a compromise between clearance and interference.
H7/n6	*Locational transition* fit for more accurate location where greater interference is permissible.
H7/p6	*Locational interference* fit for parts requiring rigidity and alignment with prime accuracy of location but without special bore pressure requirements.
H7/s6	*Medium drive* fit for ordinary steel parts or shrink fits on light sections, the tightest fit usable with cast iron.
H7/u6	*Force* fit suitable for parts which can be highly stressed or for shrink fits where the heavy pressing forces required are impractical.

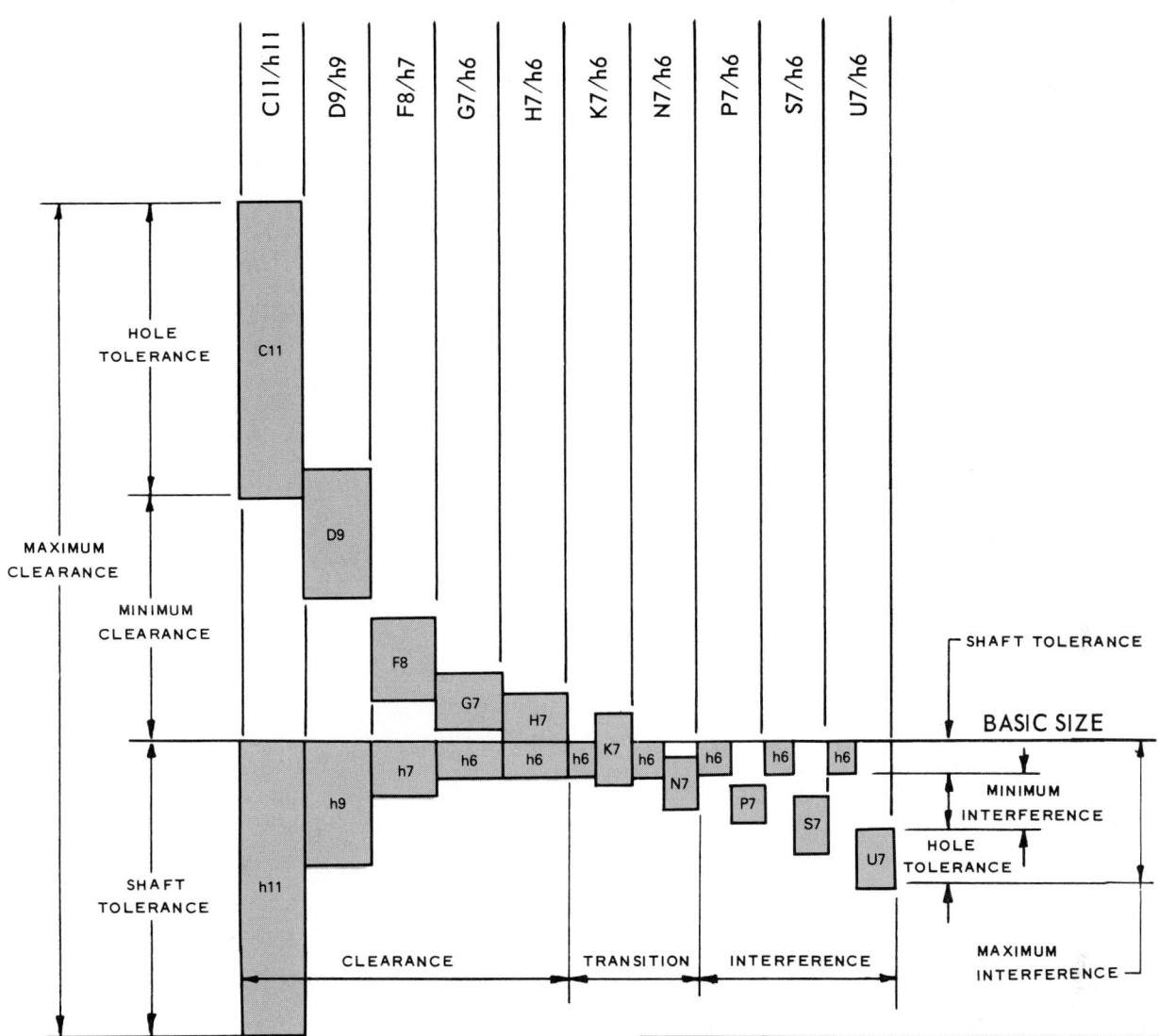

Shaft Basis Symbol	DESCRIPTION
C11/h11	*Loose-running* fit for wide commercial tolerances or allowances on external members.
D9/h9	*Free-running* fit not for use where accuracy is essential, but good for large temperature variations, high running speeds, or heavy journal pressures.
F8/h7	*Close-running* fit for running on accurate machines and for accurate location at moderate speeds and journal pressures.
G7/h6	*Sliding* fit not intended to run freely, but to move and turn freely and locate accurately.
H7/h6	*Locational clearance* fit provides snug fit for locating stationary parts; but can be freely assembled and disassembled.

Shaft Basis Symbol	DESCRIPTION
K7/h6	*Locational transition* fit for accurate location, a compromise between clearance and interference.
N7/h6	*Locational transition* fit for more accurate location where greater interference is permissible.
P7/h6	*Locational interference* fit for parts requiring rigidity and alignment with prime accuracy of location but without special bore pressure requirements.
S7/h6	*Medium drive* fit for ordinary steel parts or shrink fits on light sections, the tightest fit usable with cast iron.
U7/h6	*Force* fit suitable for parts which can be highly stressed or for shrink fits where the heavy pressing forces required are impractical.

TABLE 42   Preferred shaft basis fits description.

EXAMPLE: RC2 SLIDING FIT FOR A
Ø 1.50 NOMINAL HOLE DIAMETER — BASIC HOLE SYSTEM

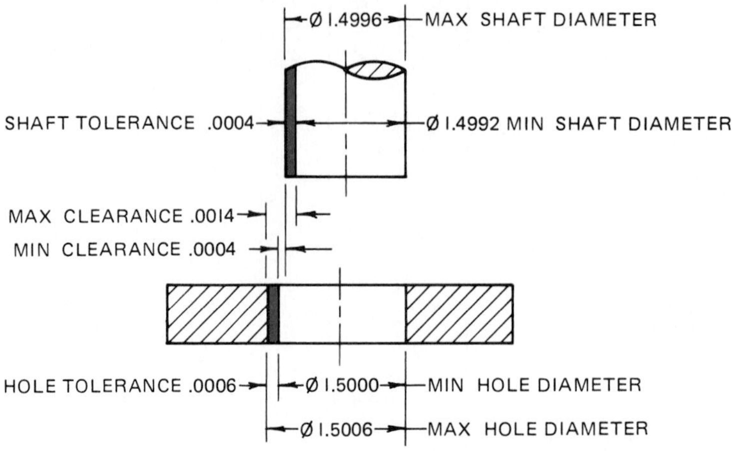

Ø 1.4996 — MAX SHAFT DIAMETER

SHAFT TOLERANCE .0004 — Ø 1.4992 MIN SHAFT DIAMETER

MAX CLEARANCE .0014

MIN CLEARANCE .0004

HOLE TOLERANCE .0006 — Ø 1.5000 — MIN HOLE DIAMETER

Ø 1.5006 — MAX HOLE DIAMETER

Nominal Size Range Inches		Class RC1 Precision Sliding			Class RC2 Sliding Fit			Class RC3 Precision Running			Class RC4 Close Running		
		Hole Tol. GR5	Minimum Clearance	Shaft Tol. GR4	Hole Tol. GR6	Minimum Clearance	Shaft Tol. GR5	Hole Tol. GR7	Minimum Clearance	Shaft Tol. GR6	Hole Tol. GR8	Minimum Clearance	Shaft Tol. GR7
Over	To	−0		+0	−0		+0	−0		+0	−0		+0
0	.12	+0.15	0.1	−0.12	+0.25	0.1	−0.15	+0.4	0.3	−0.25	+0.6	0.3	−0.4
.12	.24	+0.2	0.15	−0.15	+0.3	0.15	−0.2	+0.5	0.4	−0.3	+0.7	0.4	−0.5
.24	.40	+0.25	0.2	−0.15	+0.4	0.2	−0.25	+0.6	0.5	−0.4	+0.9	0.5	−0.6
.40	.71	+0.3	0.25	−0.2	+0.4	0.25	−0.3	+0.7	0.6	−0.4	+1.0	0.6	−0.7
.71	1.19	+0.4	0.3	−0.25	+0.5	0.3	−0.4	+0.8	0.8	−0.5	+1.2	0.8	−0.8
1.19	1.97	+0.4	0.4	−0.3	+0.6	0.4	−0.4	+1.0	1.0	−0.6	+1.6	1.0	−1.0
1.97	3.15	+0.5	0.4	−0.3	+0.7	0.4	−0.5	+1.2	1.2	−0.7	+1.8	1.2	−1.2
3.15	4.73	+0.6	0.5	−0.4	+0.9	0.5	−0.6	+1.4	1.4	−0.9	+2.2	1.4	−1.4
4.73	7.09	+0.7	0.6	−0.5	+1.0	0.6	−0.7	+1.6	1.6	−1.0	+2.5	1.6	−1.6
7.09	9.85	+0.8	0.6	−0.6	+1.2	0.6	−0.8	+1.8	2.0	−1.2	+2.8	2.0	−1.8
9.85	12.41	+0.9	0.8	−0.6	+1.2	0.8	−0.9	+2.0	2.5	−1.2	+3.0	2.5	−2.0
12.41	15.75	+1.0	1.0	−0.7	+1.4	1.0	−1.0	+2.2	3.0	−1.4	+3.5	3.0	−2.2

TABLE 43   Running and sliding fits. (Values in thousandths of an inch.)

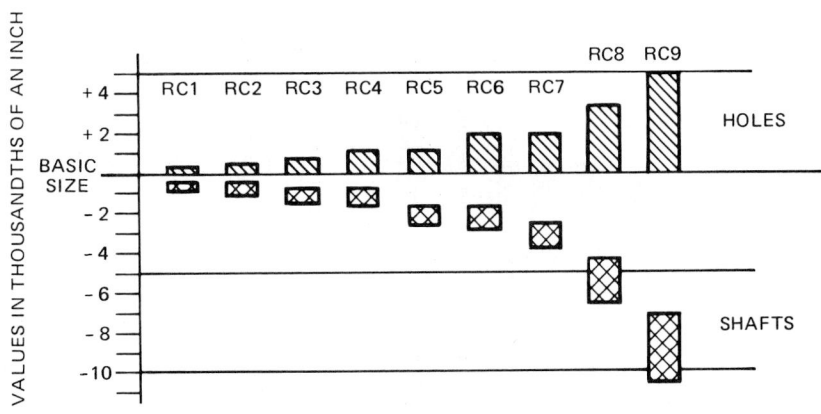

RUNNING AND SLIDING FITS
BASIC HOLE SYSTEM

Class RC5 Medium Running			Class RC6 Medium Running			Class RC7 Free Running			Class RC8 Loose Running			Class RC9 Loose Running		
Hole Tol. GR8	Minimum Clearance	Shaft Tol. GR7	Hole Tol. GR9	Minimum Clearance	Shaft Tol. GR8	Hole Tol. GR9	Minimum Clearance	Shaft Tol. GR8	Hole Tol. GR10	Minimum Clearance	Shaft Tol. GR9	Hole Tol. GR11	Minimum Clearance	Shaft Tol. GR10
−0		+0	−0		+0	−0		+0	−0		+0	−0		+0
+0.6	0.6	−0.4	+1.0	0.6	−0.6	+1.0	1.0	−0.6	+1.6	2.5	−1.0	+2.5	4.0	−1.6
+0.7	0.8	−0.5	+1.2	0.8	−0.7	+1.2	1.2	−0.7	+1.8	2.8	−1.2	+3.0	4.5	−1.8
+0.9	1.0	−0.6	+1.4	1.0	−0.9	+1.4	1.6	−0.9	+2.2	3.0	−1.4	+3.5	5.0	−2.2
+1.0	1.2	−0.7	+1.6	1.2	−1.0	+1.6	2.0	−1.0	+2.8	3.5	−1.6	+4.0	6.0	−2.8
+1.2	1.6	−0.8	+2.0	1.6	−1.2	+2.0	2.5	−1.2	+3.5	4.5	−2.0	+5.0	7.0	−3.5
+1.6	2.0	−1.0	+2.5	2.0	−1.6	+2.5	3.0	−1.6	+4.0	5.0	−2.5	+6.0	8.0	−4.0
+1.8	2.5	−1.2	+3.0	2.5	−1.8	+3.0	4.0	−1.8	+4.5	6.0	−3.0	+7.0	9.0	−4.5
+2.2	3.0	−1.4	+3.5	3.0	−2.2	+3.5	5.0	−2.2	+5.0	7.0	−3.5	+9.0	10.0	−5.0
+2.5	3.5	−1.6	+4.0	3.5	−2.5	+4.0	6.0	−2.5	+6.0	8.0	−4.0	+10.0	12.0	−6.0
+2.8	4.5	−1.8	+4.5	4.0	−2.8	+4.5	7.0	−2.8	+7.0	10.0	−4.5	+12.0	15.0	−7.0
+3.0	5.0	−2.0	+5.0	5.0	−3.0	+5.0	8.0	−3.0	+8.0	12.0	−5.0	+12.0	18.0	−8.0
+3.5	6.0	−2.2	+6.0	6.0	−3.5	+6.0	10.0	−3.5	+9.0	14.0	−6.0	+14.0	22.0	−9.0

TABLE 43   Running and sliding fits. (Values in thousandths of an inch.)
(continued)

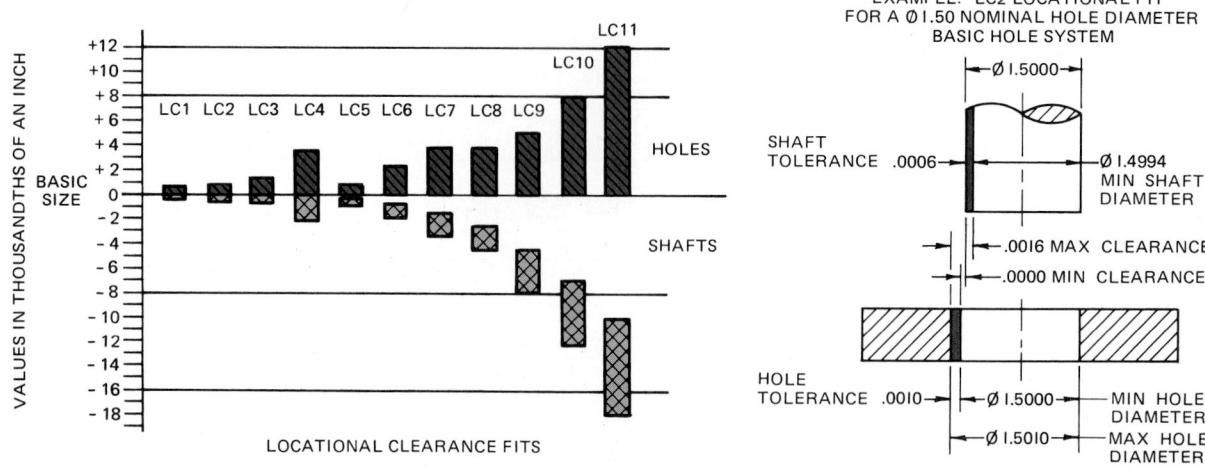

TABLE 44   Locational clearance fits. (Values in thousandths of an inch.)

Nominal Size Range Inches		Class LC1			Class LC2			Class LC3			Class LC4			Class LC5		
		Hole Tol. GR6	Minimum Clearance	Shaft Tol. GR5	Hole Tol. GR7	Minimum Clearance	Shaft Tol. GR6	Hole Tol. GR8	Minimum Clearance	Shaft Tol. GR7	Hole Tol. GR10	Minimum Clearance	Shaft Tol. GR9	Hole Tol. GR7	Minimum Clearance	Shaft Tol. GR6
Over	To	−0		+0	−0		+0	−0		+0	−0		+0	−0		+0
0	.12	+0.25	0	−0.15	+0.4	0	−0.25	+0.6	0	−0.4	+1.6	0	−1.0	+0.4	0.1	−0.25
.12	.24	+0.3	0	−0.2	+0.5	0	−0.3	+0.7	0	−0.5	+1.8	0	−1.2	+0.5	0.15	−0.3
.24	.40	+0.4	0	−0.25	+0.6	0	−0.4	+0.9	0	−0.6	+2.2	0	−1.4	+0.6	0.2	−0.4
.40	.71	+0.4	0	−0.3	+0.7	0	−0.4	+1.0	0	−0.7	+2.8	0	−1.6	+0.7	0.25	−0.4
.71	1.19	+0.5	0	−0.4	+0.8	0	−0.5	+1.2	0	−0.8	+3.5	0	−2.0	+0.8	0.3	−0.5
1.19	1.97	+0.6	0	−0.4	+1.0	0	−0.6	+1.6	0	−1.0	+4.0	0	−2.5	+1.0	0.4	−0.6
1.97	3.15	+0.7	0	−0.5	+1.2	0	−0.7	+1.8	0	−1.2	+4.5	0	−3.0	+1.2	0.4	−0.7
3.15	4.73	+0.9	0	−0.6	+1.4	0	−0.9	+2.2	0	−1.4	+5.0	0	−3.5	+1.4	0.5	−0.9
4.73	7.09	+1.0	0	−0.7	+1.6	0	−1.0	+2.5	0	−1.6	+6.0	0	−4.0	+1.6	0.6	−1.0
7.09	9.85	+1.2	0	−0.8	+1.8	0	−1.2	+2.8	0	−1.8	+7.0	0	−4.5	+1.8	0.6	−1.2
9.85	12.41	+1.2	0	−0.9	+2.0	0	−1.2	+3.0	0	−2.0	+8.0	0	−5.0	+2.0	0.7	−1.2
12.41	15.75	+1.4	0	−1.0	+2.2	0	−1.4	+3.5	0	−2.2	+9.0	0	−6.0	+2.2	0.7	−1.4

TABLE 44   Locational clearance fits. (Values in thousandths of an inch.)

Nominal Size Range Inches		Class LT1			Class LT2		
		Hole Tol. GR7	Maximum Interference	Shaft Tol. GR6	Hole Tol. GR8	Maximum Interference	Shaft Tol. GR7
Over	To	−0		+0	−0		+0
0	.12	+0.4	0.1	−0.25	+0.6	0.2	−0.4
.12	.24	+0.5	0.15	−0.3	+0.7	0.25	−0.5
.24	.40	+0.6	0.2	−0.4	+0.9	0.3	−0.6
.40	.71	+0.7	0.2	−0.4	+1.0	0.3	−0.7
.71	1.19	+0.8	0.25	−0.5	+1.2	0.4	−0.8
1.19	1.97	+1.0	0.3	−0.6	+1.6	0.5	−1.0
1.97	3.15	+1.2	0.3	−0.7	+1.8	0.6	−1.2
3.15	4.73	+1.4	0.4	−0.9	+2.2	0.7	−1.4
4.73	7.09	+1.6	0.5	−1.0	+2.5	0.8	−1.6
7.09	9.85	+1.8	0.6	−1.2	+2.8	0.9	−1.8
9.85	12.41	+2.0	0.6	−1.2	+3.0	1.0	−2.0
12.41	15.75	+2.2	0.7	−1.4	+3.5	1.0	−2.2

TABLE 45   Transition fits. (Values in thousandths of an inch.)

EXAMPLE: LT2 TRANSITION FIT
FOR A Ø1.50 NOMINAL HOLE DIAMETER
BASIC HOLE SYSTEM

VALUES IN THOUSANDTHS OF AN INCH

BASIC SIZE

LT1 LT2 LT3 LT4 LT5 LT6

= HOLE   = SHAFT

TRANSITION FITS

SHAFT TOLERANCE .0010 — Ø1.4995 MIN SHAFT DIAMETER
Ø1.5005
.0021 MAX. CLEARANCE
.0005 MAX INTEREFERNCE
HOLE TOLERANCE .0016 — Ø1.5000 — MIN HOLE DIAMETER
Ø1.5016 — MAX HOLE DIAMETER

Class LC6			Class LC7			Class LC8			Class LC9			Class LC10			Class LC11		
Hole Tol. GR9	Minimum Clearance	Shaft Tol. GR8	Hole Tol. GR10	Minimum Clearance	Shaft Tol. GR9	Hole Tol. GR10	Minimum Clearance	Shaft Tol. GR9	Hole Tol. GR11	Minimum Clearance	Shaft Tol. GR10	Hole Tol. GR12	Minimum Clearance	Shaft Tol. GR11	Hole Tol. GR13	Minimum Clearance	Shaft Tol. GR12
−0		+0	−0		+0	−0		+0	−0		+0	−0		+0	−0		+0
+1.0	0.3	−0.6	+1.6	0.6	−1.0	+1.6	1.0	−1.0	+2.5	2.5	−1.6	+4.0	4.0	−2.5	+6.0	5.0	−4.0
+1.2	0.4	−0.7	+1.8	0.8	−1.2	+1.8	1.2	−1.2	+3.0	2.8	−1.8	+5.0	4.5	−3.0	+7.0	6.0	−5.0
+1.4	0.5	−0.9	+2.2	1.0	−1.4	+2.2	1.6	−1.4	+3.5	3.0	−2.2	+6.0	5.0	−3.5	+9.0	7.0	−6.0
+1.6	0.6	−1.0	+2.8	1.2	−1.6	+2.8	2.0	−1.6	+4.0	3.5	−2.8	+7.0	6.0	−4.0	+10.0	8.0	−7.0
+2.0	0.8	−1.2	+3.5	1.6	−2.0	+3.5	2.5	−2.0	+5.0	4.5	−3.5	+8.0	7.0	−5.0	+12.0	10.0	−8.0
+2.5	1.0	−1.6	+4.0	2.0	−2.5	+4.0	3.6	−2.5	+6.0	5.0	−4.0	+10.0	8.0	−6.0	+16.0	12.0	−10.0
+3.0	1.2	−1.8	+4.5	2.5	−3.0	+4.5	4.0	−3.0	+7.0	6.0	−4.5	+12.0	10.0	−7.0	+18.0	14.0	−12.0
+3.5	1.4	−2.2	+5.0	3.0	−3.5	+5.0	5.0	−3.5	+9.0	7.0	−5.0	+14.0	11.0	−9.0	+22.0	16.0	−14.0
+4.0	1.6	−2.5	+6.0	3.5	−4.0	+6.0	6.0	−4.0	+10.0	8.0	−6.0	+16.0	12.0	−10.0	+25.0	18.0	−16.0
+4.5	2.0	−2.8	+7.0	4.0	−4.5	+7.0	7.0	−4.5	+12.0	10.0	−7.0	+18.0	16.0	−12.0	+28.0	22.0	−18.0
+5.0	2.2	−3.0	+8.0	4.5	−5.0	+8.0	7.0	−5.0	+12.0	12.0	−8.0	+20.0	20.0	−12.0	+30.0	28.0	−20.0
+6.0	2.5	−3.5	+9.0	5.0	−6.0	+9.0	8.0	−6.0	+14.0	14.0	−9.0	+22.0	22.0	−14.0	+35.0	30.0	−22.0

TABLE 44   Locational clearance fits. (Values in thousandths of an inch.)
(continued)

Class LT3			Class LT4			Class LT5			Class LT6		
Hole Tol. GR7	Maximum Interference	Shaft Tol. GR6	Hole Tol. GR8	Maximum Interference	Shaft Tol. GR7	Hole Tol. GR7	Maximum Interference	Shaft Tol. GR6	Hole Tol. GR8	Maximum Interference	Shaft Tol. GR7
−0		+0	−0		+0	−0		+0	−0		+0
+0.4	0.25	−0.25	+0.6	0.4	−0.4	+0.4	0.5	−0.25	+0.6	0.65	−0.4
+0.5	0.4	−0.3	+0.7	0.6	−0.5	+0.5	0.6	−0.3	+0.7	0.8	−0.5
+0.6	0.5	−0.4	+0.9	0.7	−0.6	+0.6	0.8	−0.4	+0.9	1.0	−0.6
+0.7	0.5	−0.4	+1.0	0.8	−0.7	+0.7	0.9	−0.4	+1.0	1.2	−0.7
+0.8	0.6	−0.5	+1.2	0.9	−0.8	+0.8	1.1	−0.5	+1.2	1.4	−0.8
+1.0	0.7	−0.6	+1.6	1.1	−1.0	+1.0	1.3	−0.6	+1.6	1.7	−1.0
+1.2	0.8	−0.7	+1.8	1.3	−1.2	+1.2	1.5	−0.7	+1.8	2.0	−1.2
+1.4	1.0	−0.9	+2.2	1.5	−1.4	+1.4	1.9	−0.9	+2.2	2.4	−1.4
+1.6	1.1	−1.0	+2.5	1.7	−1.6	+1.6	2.2	−1.0	+2.5	2.8	−1.6
+1.8	1.4	−1.2	+2.8	2.0	−1.8	+1.8	2.6	−1.2	+2.8	3.2	−1.8
+2.0	1.4	−1.2	+3.0	2.2	−2.0	+2.0	2.6	−1.2	+3.0	3.4	−2.0
+2.2	1.6	−1.4	+3.5	2.4	−2.2	+2.2	3.0	−1.4	+3.5	3.8	−2.2

TABLE 45   Transition fits. (Values in thousandths of an inch.)
(continued)

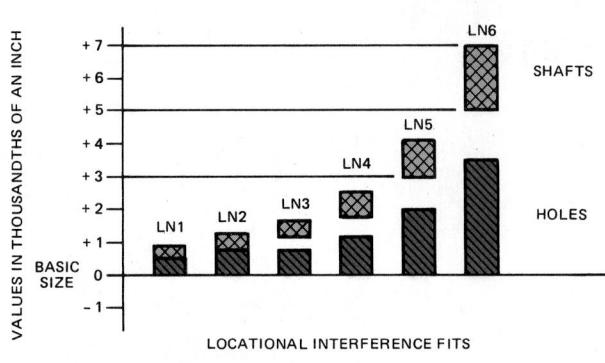

EXAMPLE: LN2 LOCATIONAL INTERFERENCE
FIT FOR A Ø1.50 NOMINAL HOLE DIAMETER
BASIC HOLE SYSTEM

Nominal Size Range Inches		Class LN1 Light Press Fit			Class LN2 Medium Press Fit		
		Hole Tol. GR6	Maximum Interference	Shaft Tol. GR5	Hole Tol. GR7	Maximum Interference	Shaft Tol. GR6
Over	To	−0		+0	−0		+0
0	.12	+0.25	0.4	−0.15	+0.4	0.65	−0.25
.12	.24	+0.3	0.5	−0.2	+0.5	0.8	−0.3
.24	.40	+0.4	0.65	−0.25	+0.6	1.0	−0.4
.40	.71	+0.4	0.7	−0.3	+0.7	1.1	−0.4
.71	1.19	+0.5	0.9	−0.4	+0.8	1.3	−0.5
1.19	1.97	+0.6	1.0	−0.4	+1.0	1.6	−0.6
1.97	3.15	+0.7	1.3	−0.5	+1.2	2.1	−0.7
3.15	4.73	+0.9	1.6	−0.6	+1.4	2.5	−0.9
4.73	7.09	+1.0	1.9	−0.7	+1.6	2.8	−1.0
7.09	9.85	+1.2	2.2	−0.8	+1.8	3.2	−1.2
9.85	12.41	+1.2	2.3	−0.9	+2.0	3.4	−1.2
12.41	15.75	+1.4	2.6	−1.0	+2.2	3.9	−1.4

TABLE 46   Locational interference fits. (Values in thousandths of an inch.)

Nominal Size Range Inches		Class FN1 Light Drive Fit			Class FN2 Medium Drive Fit		
		Hole Tol. GR6	Maximum Interference	Shaft Tol. GR5	Hole Tol. GR7	Maximum Interference	Shaft Tol. GR6
Over	To	−0		+0	−0		+0
0	.12	+0.25	0.5	−0.15	+0.4	0.85	−0.25
.12	.24	+0.3	0.6	−0.2	+0.5	1.0	−0.3
.24	.40	+0.4	0.75	−0.25	+0.6	1.4	−0.4
.40	.56	+0.4	0.8	−0.3	+0.7	1.6	−0.4
.56	.71	+0.4	0.9	−0.3	+0.7	1.6	−0.4
.71	.95	+0.5	1.1	−0.4	+0.8	1.9	−0.5
.95	1.19	+0.5	1.2	−0.4	+0.8	1.9	−0.5
1.19	1.58	+0.6	1.3	−0.4	+1.0	2.4	−0.6
1.58	1.97	+0.6	1.4	−0.4	+1.0	2.4	−0.6
1.97	2.56	+0.7	1.8	−0.5	+1.2	2.7	−0.7
2.56	3.15	+0.7	1.9	−0.5	+1.2	2.9	−0.7
3.15	3.94	+0.9	2.4	−0.6	+1.4	3.7	−0.9

TABLE 47   Force and shrink fits. (Values in thousandths of an inch.)

FORCE AND SHRINK FITS

EXAMPLE: FN2 MEDIUM DRIVE FIT FOR A Ø1.50 NOMINAL HOLE DIAMETER BASIC HOLE SYSTEM

	Class LN3 Heavy Press Fit			Class LN4			Class LN5			Class LN6		
	Hole Tol. GR7	Maximum Interference	Shaft Tol. GR6	Hole Tol. GR8	Maximum Interference	Shaft Tol. GR7	Hole Tol. GR9	Maximum Interference	Shaft Tol. GR8	Hole Tol. GR10	Maximum Interference	Shaft Tol. GR9
	−0		+0	−0		+0	−0		+0	−0		+0
	+0.4	0.75	−0.25	+0.6	1.2	−0.4	+1.0	1.8	−0.6	+1.6	3.0	−1.0
	+0.5	0.9	−0.3	+0.7	1.5	−0.5	+1.2	2.3	−0.7	+1.8	3.6	−1.2
	+0.6	1.2	−0.4	+0.9	1.8	−0.6	+1.4	2.8	−0.9	+2.2	4.4	−1.4
	+0.7	1.4	−0.4	+1.0	2.2	−0.7	+1.6	3.4	−1.0	+2.8	5.6	−1.6
	+0.8	1.7	−0.5	+1.2	2.6	−0.8	+2.0	4.2	−1.2	+3.5	7.0	−2.0
	+1.0	2.0	−0.6	+1.6	3.4	−1.0	+2.5	5.3	−1.6	+4.0	8.5	−2.5
	+1.2	2.3	−0.7	+1.8	4.0	−1.2	+3.0	6.3	−1.8	+4.5	10.0	−3.0
	+1.4	2.9	−0.9	+2.2	4.8	−1.4	+4.0	7.7	−2.2	+5.0	11.5	−3.5
	+1.6	3.5	−1.0	+2.5	5.6	−1.6	+4.5	8.7	−2.5	+6.0	13.5	−4.0
	+1.8	4.2	−1.2	+2.8	6.6	−1.8	+5.0	10.3	−2.8	+7.0	16.5	−4.5
	+2.0	4.7	−1.2	+3.0	7.5	−2.0	+6.0	12.0	−3.0	+8.0	19	−5.0
	+2.2	5.9	−1.4	+3.5	8.7	−2.2	+6.0	14.5	−3.5	+9.0	23	−6.0

TABLE 46 Locational interference fits. (Values in thousandths of an inch.)
(continued)

Class FN3 Heavy Drive Fit			Class FN4 Shrink Fit			FN5 Heavy Shrink Fit		
Hole Tol. GR7	Maximum Interference	Shaft Tol. GR6	Hole Tol. GR7	Maximum Interference	Shaft Tol. GR6	Hole Tol. GR8	Maximum Interference	Shaft Tol. GR7
−0		+0	−0		+0	−0		+0
			+0.4	0.95	−0.25	+0.6	1.3	−0.4
			+0.5	1.2	−0.3	+0.7	1.7	−0.5
			+0.6	1.6	−0.4	+0.9	2.0	−0.6
			+0.7	1.8	−0.4	+1.0	2.3	−0.7
			+0.7	1.8	−0.4	+1.0	2.5	−0.7
			+0.8	2.1	−0.5	+1.2	3.0	−0.8
+0.8	2.1	−0.5	+0.8	2.3	−0.5	+1.2	3.3	−0.8
+1.0	2.6	−0.6	+1.0	3.1	−0.6	+1.6	4.0	−1.0
+1.0	2.8	−0.6	+1.0	3.4	−0.6	+1.6	5.0	−1.0
+1.2	3.2	−0.7	+1.2	4.2	−0.7	+1.8	6.2	−1.2
+1.2	3.7	−0.7	+1.2	4.7	−0.7	+1.8	7.2	−1.2
+1.4	4.4	−0.9	+1.4	5.9	−0.9	+2.2	8.4	−1.4

TABLE 47 Force and shrink fits. (Values in thousandths of an inch.)
(continued)

Example ⌀ 50H9/d9

Hole Size ⌀ 50.062 / 50.000

Shaft Size ⌀ 49.920 / 49.858

Clearance Max. 0.204 / Min. 0.080

Example ⌀ 70H7/g6

Hole Size ⌀ 70.030 / 70.000

Shaft Size ⌀ 69.990 / 69.971

Clearance Max. 0.059 / Min. 0.010

		PREFERRED HOLE BASIS CLEARANCE FITS														
UP TO AND INCLUDING		LOOSE RUNNING			FREE RUNNING			CLOSE RUNNING			SLIDING			LOCATIONAL CLEARANCE		
		Hole H11	Shaft c11	Fit	Hole H9	Shaft d9	Fit	Hole H8	Shaft f7	Fit	Hole H7	Shaft g6	Fit	Hole H7	Shaft h6	Fit
1	MAX	1.060	0.940	0.180	1.025	0.980	0.070	1.014	0.994	0.030	1.010	0.998	0.018	1.010	1.000	0.016
	MIN	1.000	0.880	0.060	1.000	0.955	0.020	1.000	0.984	0.006	1.000	0.992	0.002	1.000	0.994	0.000
1.2	MAX	1.260	1.140	0.180	1.225	1.180	0.070	1.214	1.194	0.030	1.210	1.198	0.018	1.210	1.200	0.016
	MIN	1.200	1.080	0.060	1.200	1.155	0.020	1.200	1.184	0.006	1.200	1.192	0.002	1.200	1.194	0.000
1.6	MAX	1.660	1.540	0.180	1.625	1.580	0.070	1.614	1.594	0.030	1.610	1.598	0.018	1.610	1.600	0.016
	MIN	1.600	1.480	0.060	1.600	1.555	0.020	1.600	1.584	0.006	1.600	1.592	0.002	1.600	1.594	0.000
2	MAX	2.060	1.940	0.180	2.025	1.980	0.070	2.014	1.994	0.030	2.010	1.998	0.018	2.010	2.000	0.016
	MIN	2.000	1.880	0.060	2.000	1.955	0.020	2.000	1.984	0.006	2.000	1.992	0.002	2.000	1.994	0.000
2.5	MAX	2.560	2.440	0.180	2.525	2.480	0.070	2.514	2.494	0.030	2.510	2.498	0.018	2.510	2.500	0.016
	MIN	2.500	2.380	0.060	2.500	2.455	0.020	2.500	2.484	0.006	2.500	2.492	0.002	2.500	2.494	0.000
3	MAX	3.060	2.940	0.180	3.025	2.980	0.070	3.014	2.994	0.030	3.010	2.998	0.018	3.010	3.000	0.016
	MIN	3.000	2.880	0.060	3.000	2.955	0.020	3.000	2.984	0.006	3.000	2.992	0.002	3.000	2.994	0.000
4	MAX	4.075	3.930	0.220	4.030	3.970	0.090	4.018	3.990	0.040	4.012	3.996	0.024	4.012	4.000	0.020
	MIN	4.000	3.855	0.070	4.000	3.940	0.030	4.000	3.978	0.010	4.000	3.988	0.004	4.000	3.992	0.000
5	MAX	5.075	4.930	0.220	5.030	4.970	0.090	5.018	4.990	0.040	5.012	4.996	0.024	5.012	5.000	0.020
	MIN	5.000	4.855	0.070	5.000	4.940	0.030	5.000	4.978	0.010	5.000	4.988	0.004	5.000	4.992	0.000
6	MAX	6.075	5.930	0.220	6.030	5.970	0.090	6.018	5.990	0.040	6.012	5.996	0.024	6.012	6.000	0.020
	MIN	6.000	5.855	0.070	6.000	5.940	0.030	6.000	5.978	0.010	6.000	5.988	0.004	6.000	5.992	0.000
8	MAX	8.090	7.920	0.260	8.036	7.960	0.112	8.022	7.987	0.050	8.015	7.995	0.029	8.015	8.000	0.024
	MIN	8.000	7.830	0.080	8.000	7.924	0.040	8.000	7.972	0.013	8.000	7.986	0.006	8.000	7.991	0.000
10	MAX	10.090	9.920	0.260	10.036	9.960	0.112	10.022	9.987	0.050	10.015	9.995	0.029	10.015	10.000	0.024
	MIN	10.000	9.830	0.080	10.000	9.924	0.040	10.000	9.972	0.013	10.000	9.986	0.005	10.000	9.991	0.000
12	MAX	12.110	11.905	0.315	12.043	11.950	0.136	12.027	11.984	0.061	12.018	11.994	0.035	12.018	12.000	0.029
	MIN	12.000	11.795	0.095	12.000	11.907	0.050	12.000	11.966	0.016	12.000	11.983	0.006	12.000	11.989	0.000
16	MAX	16.110	15.905	0.315	16.043	15.950	0.136	16.027	15.984	0.061	16.018	15.994	0.035	16.018	16.000	0.029
	MIN	16.000	15.795	0.095	16.000	15.907	0.050	16.000	15.966	0.016	16.000	15.983	0.006	16.000	15.989	0.000
20	MAX	20.130	19.890	0.370	20.052	19.935	0.169	20.033	19.980	0.074	20.021	19.993	0.041	20.021	20.000	0.034
	MIN	20.000	19.760	0.110	20.000	19.883	0.065	20.000	19.959	0.020	20.000	19.980	0.007	20.000	19.987	0.000
25	MAX	25.130	24.890	0.370	25.052	24.935	0.169	25.033	24.980	0.074	25.021	24.993	0.042	25.021	25.000	0.034
	MIN	25.000	24.760	0.110	25.000	24.883	0.065	25.000	24.959	0.020	25.000	24.980	0.007	25.000	24.987	0.000
30	MAX	30.130	29.890	0.370	30.052	29.935	0.169	30.033	29.980	0.074	30.021	29.993	0.041	30.021	30.000	0.034
	MIN	30.000	29.760	0.110	30.000	29.883	0.065	30.000	29.959	0.020	30.000	29.980	0.007	30.000	29.987	0.000
40	MAX	40.160	39.880	0.440	40.062	39.920	0.204	40.039	39.975	0.089	40.025	39.991	0.050	40.025	40.000	0.041
	MIN	40.000	39.720	0.120	40.000	39.858	0.080	40.000	39.950	0.025	40.000	39.975	0.009	40.000	39.984	0.000
50	MAX	50.160	49.870	0.450	50.062	49.920	0.204	50.039	49.975	0.089	50.025	49.991	0.050	50.025	50.000	0.041
	MIN	50.000	49.710	0.130	50.000	49.858	0.080	50.000	49.950	0.025	50.000	49.975	0.009	50.000	49.984	0.000
60	MAX	60.190	59.860	0.520	60.074	59.900	0.248	60.046	59.970	0.106	60.030	59.990	0.059	60.030	60.000	0.049
	MIN	60.000	59.670	0.140	60.000	59.826	0.100	60.000	59.940	0.030	60.000	59.971	0.010	60.000	59.981	0.000
80	MAX	80.190	79.850	0.530	80.074	79.900	0.248	80.046	79.970	0.106	80.030	79.990	0.059	80.030	80.000	0.049
	MIN	80.000	79.660	0.150	80.000	79.826	0.100	80.000	79.940	0.030	80.000	79.971	0.010	80.000	79.981	0.000
100	MAX	100.220	99.830	0.610	100.087	99.880	0.294	100.054	99.964	0.125	100.035	99.988	0.069	100.035	100.000	0.057
	MIN	100.000	99.610	0.170	100.000	99.793	0.120	100.000	99.929	0.036	100.000	99.966	0.012	100.000	99.978	0.000
120	MAX	120.220	119.820	0.620	120.087	119.880	0.294	120.054	119.964	0.125	120.035	119.988	0.069	120.035	120.000	0.057
	MIN	120.000	119.600	0.180	120.000	119.793	0.120	120.000	119.929	0.036	120.000	119.966	0.012	120.000	119.978	0.000
160	MAX	160.250	159.790	0.710	160.100	159.855	0.345	160.063	159.957	0.146	160.040	159.986	0.079	160.040	160.000	0.065
	MIN	160.000	159.540	0.210	160.000	159.755	0.145	160.000	159.917	0.043	160.000	159.961	0.014	160.000	159.975	0.000

TABLE 48   Preferred hole basis fits. (Dimensions in millimeters.)

Example Ø 10H7/n6

Hole Size Ø $\begin{array}{l}10.015\\10.000\end{array}$

Shaft Size Ø $\begin{array}{l}10.019\\10.010\end{array}$

Max. Clearance 0.005

Max. Interference 0.019

Example Ø 35H7/u6

Hole Size Ø $\begin{array}{l}35.025\\35.000\end{array}$

Shaft Size Ø $\begin{array}{l}35.076\\35.060\end{array}$

Min. Interference 0.035

Max. Interference 0.076

PREFERRED HOLE BASIS TRANSITION AND INTERFERENCE FITS															
UP TO AND INCLUDING	LOCATIONAL TRANSN.			LOCATIONAL TRANSN.			LOCATIONAL INTERF.			MEDIUM DRIVE			FORCE		
	Hole H7	Shaft k6	Fit	Hole H7	Shaft n6	Fit	Hole H7	Shaft p6	Fit	Hole H7	Shaft s6	Fit	Hole H7	Shaft u6	Fit
1 MAX	1.010	1.006	0.010	1.010	1.010	0.006	1.010	1.012	0.004	1.010	1.020	−0.004	1.010	1.024	−0.008
MIN	1.000	1.000	−0.006	1.000	1.004	−0.010	1.000	1.006	−0.012	1.000	1.014	−0.020	1.000	1.018	−0.024
1.2 MAX	1.210	1.206	0.010	1.210	1.210	0.006	1.210	1.212	0.004	1.210	1.220	−0.004	1.210	1.224	−0.008
MIN	1.200	1.200	−0.006	1.200	1.204	−0.010	1.200	1.206	−0.012	1.200	1.214	−0.020	1.200	1.218	−0.024
1.6 MAX	1.610	1.606	0.010	1.610	1.610	0.006	1.610	1.612	0.004	1.610	1.620	−0.004	1.610	1.624	−0.008
MIN	1.600	1.600	−0.006	1.600	1.604	−0.010	1.600	1.606	−0.012	1.600	1.614	−0.020	1.600	1.618	−0.024
2 MAX	2.010	2.006	0.010	2.010	2.010	0.006	2.010	2.012	0.004	2.010	2.020	−0.004	2.010	2.024	−0.008
MIN	2.000	2.000	−0.006	2.000	2.004	−0.010	2.000	2.006	−0.012	2.000	2.014	−0.020	2.000	2.018	−0.024
2.5 MAX	2.510	2.506	0.010	2.510	2.510	0.006	2.510	2.512	0.004	2.510	2.520	−0.004	2.510	2.524	−0.008
MIN	2.500	2.500	−0.006	2.500	2.504	−0.010	2.500	2.506	−0.012	2.500	2.514	−0.020	2.500	2.518	−0.024
3 MAX	3.010	3.006	0.010	3.010	3.010	0.006	3.010	3.012	0.004	3.010	3.020	−0.004	3.010	3.024	−0.008
MIN	3.000	3.000	−0.006	3.000	3.004	−0.010	3.000	3.006	−0.012	3.000	3.014	−0.020	3.000	3.018	−0.024
4 MAX	4.012	4.009	0.011	4.012	4.016	0.004	4.012	4.020	0.000	4.012	4.027	−0.007	4.012	4.031	−0.011
MIN	4.000	4.001	−0.009	4.000	4.008	−0.016	4.000	4.012	−0.020	4.000	4.019	−0.027	4.000	4.023	−0.031
5 MAX	5.012	5.009	0.011	5.012	5.016	0.004	5.012	5.020	0.000	5.012	5.027	−0.007	5.012	5.031	−0.011
MIN	5.000	5.001	−0.009	5.000	5.008	−0.016	5.000	5.012	−0.020	5.000	5.019	−0.027	5.000	5.023	−0.031
6 MAX	6.012	6.009	0.011	6.012	6.016	0.004	6.012	6.020	0.000	6.012	6.027	−0.007	6.012	6.031	−0.011
MIN	6.000	6.001	−0.009	6.000	6.008	−0.016	6.000	6.012	−0.020	6.000	6.019	−0.027	6.000	6.023	−0.031
8 MAX	8.015	8.010	0.014	8.015	8.019	0.005	8.015	8.024	0.000	8.015	8.032	−0.008	8.015	8.037	−0.013
MIN	8.000	8.001	−0.010	8.000	8.010	−0.019	8.000	8.015	−0.024	8.000	8.023	−0.032	8.000	8.028	−0.037
10 MAX	10.015	10.010	0.014	10.015	10.019	0.005	10.015	10.024	0.000	10.015	10.032	−0.008	10.015	10.037	−0.013
MIN	10.000	10.001	−0.010	10.000	10.010	−0.019	10.000	10.015	−0.024	10.000	10.023	−0.032	10.000	10.028	−0.037
12 MAX	12.018	12.012	0.017	12.018	12.023	0.006	12.018	12.029	0.000	12.018	12.039	−0.010	12.018	12.044	−0.015
MIN	12.000	12.001	−0.012	12.000	12.012	−0.023	12.000	12.018	−0.029	12.000	12.028	−0.039	12.000	12.033	−0.044
16 MAX	16.018	16.012	0.017	16.018	16.023	0.006	16.018	16.029	0.000	16.018	16.039	−0.010	16.018	16.044	−0.015
MIN	16.000	16.001	−0.012	16.000	16.012	−0.023	16.000	16.018	−0.029	16.000	16.028	−0.039	16.000	16.033	−0.044
20 MAX	20.021	20.015	0.019	20.021	20.028	0.006	20.021	20.035	−0.001	20.021	20.048	−0.014	20.021	20.054	−0.020
MIN	20.000	20.002	−0.015	20.000	20.015	−0.028	20.000	20.022	−0.035	20.000	20.035	−0.048	20.000	20.041	−0.054
25 MAX	25.021	25.015	0.019	25.021	25.028	0.006	25.021	25.035	−0.001	25.021	25.048	−0.014	25.021	25.061	−0.027
MIN	25.000	25.002	−0.015	25.000	25.015	−0.028	25.000	25.022	−0.035	25.000	25.035	−0.048	25.000	25.048	−0.061
30 MAX	30.021	30.015	0.019	30.021	30.028	0.006	30.021	30.035	−0.001	30.021	30.048	−0.014	30.021	30.061	−0.027
MIN	30.000	30.002	−0.015	30.000	30.015	−0.028	30.000	30.022	−0.035	30.000	30.035	−0.048	30.000	30.048	−0.061
40 MAX	40.025	40.018	0.023	40.025	40.033	0.008	40.025	40.042	−0.001	40.025	40.059	−0.018	40.025	40.076	−0.035
MIN	40.000	40.002	−0.018	40.000	40.017	−0.033	40.000	40.026	−0.042	40.000	40.043	−0.059	40.000	40.060	−0.076
50 MAX	50.025	50.018	0.023	50.025	50.033	0.008	50.025	50.042	−0.001	50.025	50.059	−0.018	50.025	50.086	−0.045
MIN	50.000	50.002	−0.018	50.000	50.017	−0.033	50.000	50.026	−0.042	50.000	50.043	−0.059	50.000	50.070	−0.086
60 MAX	60.030	60.021	0.028	60.030	60.039	0.010	60.030	60.051	−0.002	60.030	60.072	−0.023	60.030	60.106	−0.057
MIN	60.000	60.002	−0.021	60.000	60.020	−0.039	60.000	60.032	−0.051	60.000	60.053	−0.072	60.000	60.087	−0.106
80 MAX	80.030	80.021	0.028	80.030	80.039	0.010	80.030	80.051	−0.002	80.030	80.078	−0.029	80.030	80.121	−0.072
MIN	80.000	80.002	−0.021	80.000	80.020	−0.039	80.000	80.032	−0.051	80.000	80.059	−0.078	80.000	80.102	−0.121
100 MAX	100.035	100.025	0.032	100.035	100.045	0.012	100.035	100.059	−0.002	100.035	100.093	−0.036	100.035	100.146	−0.089
MIN	100.000	100.003	−0.025	100.000	100.023	−0.045	100.000	100.037	−0.059	100.000	100.071	−0.093	100.000	100.124	−0.146
120 MAX	120.035	120.025	0.032	120.035	120.045	0.012	120.035	120.059	−0.002	120.035	120.101	−0.044	120.035	120.166	−0.109
MIN	120.000	120.003	−0.025	120.000	120.023	−0.045	120.000	120.037	−0.059	120.000	120.079	−0.101	120.000	120.144	−0.166
160 MAX	160.040	160.028	0.037	160.045	160.052	0.013	160.040	160.068	−0.003	160.040	160.125	−0.060	160.040	160.215	−0.150
MIN	160.000	160.003	−0.028	160.000	160.027	−0.052	160.000	160.043	−0.068	160.000	160.000	−0.125	160.000	160.190	−0.215

TABLE 48 Preferred hole basis fits. (Dimensions in millimeters.)
(continued)

Example Ø 100C11/h11

Hole Size Ø $\begin{array}{l}100.390\\100.170\end{array}$

Shaft Size Ø $\begin{array}{l}100.000\\99.780\end{array}$

Clearance $\begin{array}{l}\text{Max. }0.610\\\text{Min. }0.170\end{array}$

Example Ø 11H7/h6

Hole Size Ø $\begin{array}{l}11.018\\11.000\end{array}$

Shaft Size Ø $\begin{array}{l}11.000\\10.989\end{array}$

Clearance $\begin{array}{l}\text{Max. }0.029\\\text{Min. }0.000\end{array}$

PREFERRED SHAFT BASIS CLEARANCE FITS																	
UP TO AND INCLUDING		LOOSE RUNNING			FREE RUNNING			CLOSE RUNNING			SLIDING			LOCATIONAL CLEARANCE		
		Hole C11	Shaft h11	Fit	Hole D9	Shaft h9	Fit	Hole F8	Shaft h7	Fit	Hole G7	Shaft h6	Fit	Hole H7	Shaft h6	Fit
1	MAX	1.120	1.000	0.180	1.045	1.000	0.070	1.020	1.000	0.030	1.012	1.000	0.018	1.010	1.000	0.016
	MIN	1.060	0.940	0.060	1.020	0.975	0.020	1.006	0.990	0.006	1.002	0.994	0.002	1.000	0.994	0.000
1.2	MAX	1.320	1.200	0.180	1.245	1.200	0.070	1.220	1.200	0.030	1.212	1.200	0.018	1.210	1.200	0.016
	MIN	1.260	1.140	0.060	1.220	1.175	0.020	1.206	1.190	0.006	1.202	1.194	0.002	1.200	1.194	0.000
1.6	MAX	1.720	1.600	0.180	1.645	1.600	0.070	1.620	1.600	0.030	1.612	1.600	0.018	1.610	1.600	0.016
	MIN	1.660	1.540	0.060	1.620	1.575	0.020	1.606	1.590	0.006	1.602	1.594	0.002	1.600	1.594	0.000
2	MAX	2.120	2.000	0.180	2.045	2.000	0.070	2.020	2.000	0.030	2.012	2.000	0.018	2.010	2.000	0.016
	MIN	2.060	1.940	0.060	2.020	1.975	0.020	2.006	1.990	0.006	2.002	1.994	0.002	2.000	1.994	0.000
2.5	MAX	2.620	2.500	0.180	2.545	2.500	0.070	2.520	2.500	0.030	2.512	2.500	0.018	2.510	2.500	0.016
	MIN	2.560	2.440	0.060	2.520	2.475	0.020	2.506	2.490	0.006	2.502	2.494	0.002	2.500	2.494	0.000
3	MAX	3.120	3.000	0.180	3.045	3.000	0.070	3.020	3.000	0.030	3.012	3.000	0.018	3.010	3.000	0.016
	MIN	3.060	2.940	0.060	3.020	2.975	0.020	3.006	2.990	0.006	3.002	2.994	0.002	3.000	2.994	0.000
4	MAX	4.145	4.000	0.220	4.060	4.000	0.090	4.028	4.000	0.040	4.016	4.000	0.024	4.012	4.000	0.020
	MIN	4.070	3.925	0.070	4.030	3.970	0.030	4.010	3.988	0.010	4.004	3.992	0.004	4.000	3.992	0.000
5	MAX	5.145	5.000	0.220	5.060	5.000	0.090	5.028	5.000	0.040	5.016	5.000	0.024	5.012	5.000	0.020
	MIN	5.070	4.925	0.070	5.030	4.970	0.030	5.010	4.988	0.010	5.004	4.992	0.004	5.000	4.992	0.000
6	MAX	6.145	6.000	0.220	6.060	6.000	0.090	6.028	6.000	0.040	6.016	6.000	0.024	6.012	6.000	0.020
	MIN	6.070	5.925	0.070	6.030	5.970	0.030	6.010	5.988	0.010	6.004	5.992	0.004	6.000	5.992	0.000
8	MAX	8.170	8.000	0.260	8.076	8.000	0.112	8.035	8.000	0.050	8.020	8.000	0.029	8.015	8.000	0.024
	MIN	8.080	7.910	0.080	8.040	7.964	0.040	8.013	7.985	0.013	8.005	7.991	0.005	8.000	7.991	0.000
10	MAX	10.170	10.000	0.260	10.076	10.000	0.112	10.035	10.000	0.050	10.020	10.000	0.029	10.015	10.000	0.024
	MIN	10.080	9.910	0.080	10.040	9.964	0.040	10.013	9.985	0.013	10.005	9.991	0.005	10.000	9.991	0.000
12	MAX	12.205	12.000	0.315	12.093	12.000	0.136	12.043	12.000	0.061	12.024	12.000	0.035	12.018	12.000	0.029
	MIN	12.095	11.890	0.095	12.050	11.957	0.050	12.016	11.982	0.016	12.006	11.989	0.006	12.000	11.989	0.000
16	MAX	16.205	16.000	0.315	16.093	16.000	0.136	16.043	16.000	0.061	16.024	16.000	0.035	16.018	16.000	0.029
	MIN	16.095	15.890	0.095	16.050	15.957	0.050	16.016	15.982	0.016	16.006	15.989	0.006	16.000	15.989	0.000
20	MAX	20.240	20.000	0.370	20.117	20.000	0.169	20.053	20.000	0.074	20.028	20.000	0.041	20.021	20.000	0.034
	MIN	20.110	19.870	0.110	20.065	19.948	0.065	20.020	19.979	0.020	20.007	19.987	0.007	20.000	19.987	0.000
25	MAX	25.240	25.000	0.370	25.117	25.000	0.169	25.053	25.000	0.074	25.028	25.000	0.041	25.021	25.000	0.034
	MIN	25.110	24.870	0.110	25.065	24.948	0.065	25.020	24.979	0.020	25.007	24.987	0.007	25.000	24.987	0.000
30	MAX	30.240	30.000	0.370	30.117	30.000	0.169	30.053	30.000	0.074	30.028	30.000	0.041	30.021	30.000	0.034
	MIN	30.110	29.870	0.110	30.065	29.948	0.065	30.020	29.979	0.020	30.007	29.987	0.007	30.000	29.987	0.000
40	MAX	40.280	40.000	0.440	40.142	40.000	0.204	40.064	40.000	0.089	40.034	40.000	0.050	40.025	40.000	0.041
	MIN	40.120	39.840	0.120	40.080	39.938	0.080	40.025	39.975	0.025	40.009	39.984	0.009	40.000	39.984	0.000
50	MAX	50.290	50.000	0.450	50.142	50.000	0.204	50.064	50.000	0.089	50.034	50.000	0.050	50.025	50.000	0.041
	MIN	50.130	49.840	0.130	50.080	49.938	0.080	50.025	49.975	0.025	50.009	49.984	0.009	50.000	49.984	0.000
60	MAX	60.330	60.000	0.520	60.174	60.000	0.248	60.076	60.000	0.106	60.040	60.000	0.059	60.030	60.000	0.049
	MIN	60.140	59.810	0.140	60.100	59.926	0.100	60.030	59.970	0.030	60.010	59.981	0.010	60.000	59.981	0.000
80	MAX	80.340	80.000	0.530	80.174	80.000	0.248	80.076	80.000	0.106	80.040	80.000	0.059	80.030	80.000	0.049
	MIN	80.150	79.810	0.150	80.100	79.926	0.100	80.030	79.970	0.030	80.010	79.981	0.010	80.000	79.981	0.000
100	MAX	100.390	100.000	0.610	100.207	100.000	0.294	100.090	100.000	0.125	100.047	100.000	0.069	100.035	100.000	0.057
	MIN	100.170	99.780	0.170	100.120	99.913	0.120	100.036	99.965	0.036	100.012	99.978	0.012	100.000	99.978	0.000
120	MAX	120.400	120.000	0.620	120.207	120.000	0.294	120.090	120.000	0.125	120.047	120.000	0.069	120.035	120.000	0.057
	MIN	120.180	119.780	0.180	120.120	119.913	0.120	120.036	119.965	0.036	120.012	119.978	0.012	120.000	119.978	0.000
160	MAX	160.460	160.000	0.710	160.245	160.000	0.345	160.106	160.000	0.146	160.054	160.000	0.079	160.040	160.000	0.065
	MIN	160.210	159.750	0.210	160.145	159.900	0.145	160.043	159.960	0.043	160.014	159.975	0.014	160.000	159.975	0.000

TABLE 49 Preferred shaft basis fits. (Dimensions in millimeters.)

Example Ø 16N7/h6

Hole Size Ø $\frac{15.995}{15.977}$

Shaft Size Ø $\frac{16.000}{15.989}$

Max. Clearance 0.006

Max. Interference 0.023

Example Ø 45U7/h6

Hole Size Ø $\frac{44.939}{44.914}$

Shaft Size Ø $\frac{45.000}{44.984}$

Min. Interference 0.045

Max. Interference 0.086

		PREFERRED SHAFT BASIS TRANSITION AND INTERFERENCE FITS															
UP TO AND INCLUDING		LOCATIONAL TRANSN.			LOCATIONAL TRANSN.			LOCATIONAL INTERF.			MEDIUM DRIVE			FORCE		
		Hole K7	Shaft h6	Fit	Hole N7	Shaft h6	Fit	Hole P7	Shaft h6	Fit	Hole S7	Shaft h6	Fit	Hole U7	Shaft h6	Fit
1	MAX	1.000	1.000	0.006	0.996	1.000	0.002	0.994	1.000	0.000	0.986	1.000	−0.008	0.982	1.000	−0.012
	MIN	0.990	0.994	−0.010	0.986	0.994	−0.014	0.984	0.994	−0.016	0.976	0.994	−0.024	0.972	0.994	−0.028
1.2	MAX	1.200	1.200	0.006	1.196	1.200	0.002	1.194	1.200	0.000	1.186	1.200	−0.008	1.182	1.200	−0.012
	MIN	1.190	1.194	−0.010	1.186	1.194	−0.014	1.184	1.194	−0.016	1.176	1.194	−0.024	1.172	1.194	−0.028
1.6	MAX	1.600	1.600	0.006	1.596	1.600	0.002	1.594	1.600	0.000	1.586	1.600	−0.008	1.582	1.600	−0.012
	MIN	1.590	1.594	−0.010	1.586	1.594	−0.014	1.584	1.594	−0.016	1.576	1.594	−0.024	1.572	1.594	−0.028
2	MAX	2.000	2.000	0.006	1.996	2.000	0.002	1.994	2.000	0.000	1.986	2.000	−0.008	1.982	2.000	−0.012
	MIN	1.990	1.994	−0.010	1.986	1.994	−0.014	1.984	1.994	−0.016	1.976	1.994	−0.024	1.972	1.994	−0.028
2.5	MAX	2.500	2.500	0.006	2.496	2.500	0.002	2.494	2.500	0.000	2.486	2.500	−0.008	2.482	2.500	−0.012
	MIN	2.490	2.494	−0.010	2.486	2.494	−0.014	2.484	2.494	−0.016	2.476	2.494	−0.024	2.472	2.494	−0.028
3	MAX	3.000	3.000	0.006	2.996	3.000	0.002	2.994	3.000	0.000	2.986	3.000	−0.008	2.982	3.000	−0.012
	MIN	2.990	2.994	−0.010	2.986	2.994	−0.014	2.984	2.994	−0.016	2.976	2.994	−0.024	2.972	2.994	−0.028
4	MAX	4.003	4.000	0.011	3.996	4.000	0.004	3.992	4.000	0.000	3.985	4.000	−0.007	3.981	4.000	−0.011
	MIN	3.991	3.992	−0.009	3.984	3.992	−0.016	3.980	3.992	−0.020	3.973	3.992	−0.027	3.969	3.992	−0.031
5	MAX	5.003	5.000	0.011	4.996	5.000	0.004	4.992	5.000	0.000	4.985	5.000	−0.007	4.981	5.000	−0.011
	MIN	4.991	4.992	−0.009	4.984	4.992	−0.016	4.980	4.992	−0.020	4.973	4.992	−0.027	4.969	4.992	−0.031
6	MAX	6.003	6.000	0.011	5.996	6.000	0.004	5.992	6.000	0.000	5.985	6.000	−0.007	5.981	6.000	−0.011
	MIN	5.991	5.992	−0.009	5.984	5.992	−0.016	5.980	5.992	−0.020	5.973	5.992	−0.027	5.969	5.992	−0.031
8	MAX	8.005	8.000	0.014	7.996	8.000	0.005	7.991	8.000	0.000	7.983	8.000	−0.008	7.978	8.000	−0.013
	MIN	7.990	7.991	−0.010	7.981	7.991	−0.019	7.976	7.991	−0.024	7.968	7.991	−0.032	7.963	7.991	−0.037
10	MAX	10.005	10.000	0.014	9.996	10.000	0.005	9.991	10.000	0.000	9.983	10.000	−0.008	9.978	10.000	−0.013
	MIN	9.990	9.991	−0.010	9.981	9.991	−0.019	9.976	9.991	−0.024	9.968	9.991	−0.032	9.963	9.991	−0.037
12	MAX	12.006	12.000	0.017	11.995	12.000	0.006	11.989	12.000	0.000	11.979	12.000	−0.010	11.974	12.000	−0.015
	MIN	11.988	11.989	−0.012	11.977	11.989	−0.023	11.971	11.989	−0.029	11.961	11.989	−0.039	11.950	11.989	−0.044
16	MAX	16.006	16.000	0.017	15.995	16.000	0.006	15.989	16.000	0.000	15.979	16.000	−0.010	15.974	16.000	−0.015
	MIN	15.988	15.989	−0.012	15.977	15.989	−0.023	15.971	15.989	−0.029	15.961	15.989	−0.039	15.956	15.989	−0.044
20	MAX	20.006	20.000	0.019	19.993	20.000	0.006	19.986	20.000	−0.001	19.973	20.000	−0.014	19.967	20.000	−0.020
	MIN	19.985	19.987	−0.015	19.972	19.987	−0.028	19.965	19.987	−0.035	19.952	19.987	−0.048	19.946	19.987	−0.054
25	MAX	25.006	25.000	0.019	24.993	25.000	0.006	24.986	25.000	−0.001	24.973	25.000	−0.014	24.960	25.000	−0.027
	MIN	24.985	24.987	−0.015	24.972	24.987	−0.028	24.965	24.987	−0.035	24.952	24.987	−0.048	24.939	24.987	−0.061
30	MAX	30.006	30.000	0.019	29.993	30.000	0.006	29.986	30.000	−0.001	29.973	30.000	−0.014	29.960	30.000	−0.027
	MIN	29.985	29.987	−0.015	29.972	29.987	−0.028	29.965	29.987	−0.035	29.952	29.987	−0.048	29.939	29.987	−0.061
40	MAX	40.007	40.000	0.023	39.992	40.000	0.008	39.983	40.000	−0.001	39.966	40.000	−0.018	39.949	40.000	−0.035
	MIN	39.982	39.984	−0.018	39.967	39.984	−0.033	39.958	39.984	−0.042	39.941	39.984	−0.059	39.924	39.984	−0.076
50	MAX	50.007	50.000	0.023	49.992	50.000	0.008	49.983	50.000	−0.001	49.966	50.000	−0.018	49.939	50.000	−0.045
	MIN	49.982	49.984	−0.018	49.967	49.984	−0.033	49.958	49.984	−0.042	49.941	49.984	−0.059	49.914	49.984	−0.086
60	MAX	60.009	60.000	0.028	59.991	60.000	0.010	59.979	60.000	−0.002	59.958	60.000	−0.023	59.924	60.000	−0.057
	MIN	59.979	59.981	−0.021	59.961	59.981	−0.039	59.949	59.981	−0.051	59.928	59.981	−0.072	59.894	59.981	−0.106
80	MAX	80.009	80.000	0.028	79.991	80.000	0.010	79.979	80.000	−0.002	79.952	80.000	−0.029	79.909	80.000	−0.072
	MIN	79.979	79.981	−0.021	79.961	79.981	−0.039	79.949	79.981	−0.051	79.922	79.981	−0.078	79.879	79.981	−0.121
100	MAX	100.010	100.000	0.032	99.990	100.000	0.012	99.976	100.000	−0.002	99.942	100.000	−0.036	99.889	100.000	−0.089
	MIN	99.975	99.978	−0.025	99.955	99.978	−0.045	99.941	99.978	−0.059	99.907	99.978	−0.093	99.854	99.978	−0.146
120	MAX	120.010	120.000	0.032	119.990	120.000	0.012	119.976	120.000	−0.002	119.934	120.000	−0.044	119.869	120.000	−0.109
	MIN	119.975	119.978	−0.025	119.955	119.978	−0.045	119.941	119.978	−0.059	119.899	119.978	−0.101	119.834	119.978	−0.166
160	MAX	160.012	160.000	0.037	159.988	160.000	0.013	159.972	160.000	−0.003	159.915	160.000	−0.060	159.825	160.000	−0.150
	MIN	159.972	159.975	−0.028	159.948	159.975	−0.052	159.932	159.975	−0.068	159.875	159.975	−0.125	159.785	159.975	−0.215

TABLE 49   Preferred shift basis fits. (Dimensions in millimeters.)
(continued)

MORSE TAPERS		
	TAPER	
No. of Taper	inches per Foot	mm per 100 mm
0	.625	5.21
1	.599	4.99
2	.599	4.99
3	.602	5.02
4	.623	5.19
5	.631	5.26
6	.626	5.22
7	.624	5.20

BROWN AND SHARPE TAPERS		
	TAPER	
No. of Taper	inches per Foot	mm per 100 mm
1	.502	4.18
2	.502	4.18
3	.502	4.18
4	.502	4.18
5	.502	4.18
6	.503	4.19
7	.502	4.18
8	.501	4.18
9	.501	4.18
10	.516	4.3
11	.501	4.18
12	.500	4.17
13	.500	4.17
14	.500	4.17
15	.500	4.17
16	.500	4.17

TABLE 50   Machine tapers.

PRODUCT	OUTSTANDING FEATURES	APPLICATION METHOD	COLOR
1357	A high performance adhesive with long bonding range, excellent initial strength. Meets specification requirements of MMM-A-121 (supersedes MIL-A-1154 C), MIL-A-5092 B, Type II, and MIL-A-21366. Bonds rubber, cloth, wood, foamed glass, paper honeycomb, decorative plastic laminates. Also used with metal-to-metal for bonds of moderate strength.	Spray or Brush	Gray/ Green or Olive
2210	Fast drying, exhibits aggressive tack that allows coated surfaces to knit easily under hand roller pressure. Excellent water and oil resistance. Meets specification requirements of MMM-A-121 (supersedes MIL-A-1154 C), MIL-A-21366, and MMM-A-00130a. Bonds a wide range of materials including rubber, leather, cloth, aluminum, wood, hardboard. Used extensively for bonding decorative plastic laminates.	Brush, Roller, or Trowel	Yellow
2215	Fast drying and has a rapid rate of strength build-up. Its aggressive tack permits adhesive coated surfaces to bond easily with moderate pressure. Bonds decorative plastic laminates to metal, wood, and particle board. Also used for general bonding of rubber, leather, cloth, aluminum, wood, hardboard, etc.	Spray	Light Yellow
2218	Offers rapid strength build-up, high-ultimate strength. Has a high softening point and excellent resistance to plastic flow. Adhesion to steel is especially good. Meets specification requirements of MMM-A-121 (supersedes MIL-A-1154 C). Bonds high density decorative plastic laminates to metal or wood. Widely used to fabricate honeycomb and sandwich-type building panels with various face sheets, including porcelain enamel steel.	Spray	Green
2226	Water dispersed, has high immediate bond strength, long bonding range. Changes color from blue to green while drying. Used to bond foamed plastics, plastic laminate, wood, rubber, plywood, wallboard, wood veneer, plaster, and canvas to themselves and to each other.	Spray or Brush	Wet: Lt. Blue Dry: Green
4420	High performance, fast-drying adhesive designed for application by pressure curtain coating and mechanical roll coating. Bonds decorative plastic laminates to plywood or particle board and is suitable for conveyor line production of laminated panels of various types such as aluminum to wood or hardboard.	Roll Coating	Yellow
4488	Lower viscosity version of Cement 4420 for use specifically with flow-over or Weir-type curtain coaters. Bonds decorative plastic laminates to plywood or particle board and is suitable for conveyor line production of laminated panels of various types, such as aluminum to wood or hardboard.	Curtain and Roll Coating	Yellow
4518	Designed for spray application with automatic or production line equipment. Dries very fast, requires pressure from a niproll (rotary press) or platen press to assure proper bonding. Bonds decorative plastic laminates to plywood or particle board on both flat work and postforming. Also used for conveyor line production of laminated panels of various types such as aluminum to wood or hardboard. Meets requirements of MIL-A-5092 B, Type II.	Spray	Green
4729	Similar to Cement 2218 except that it is formulated with a nonflammable solvent. Requires force drying to prevent blushing.	Spray	Red
5034	Water dispersed, has high immediate bond strength and long bonding range. Bonds foamed plastics, plastic laminate, wood, rubber, plywood, wallboard, wood veneer, plaster, and canvas to themselves and to each other.	Spray or Brush	Neutral

TABLE 51   Physical properties and application data of adhesives. *(3M Co.)*

NORTH AMERICAN GAGES													EUROPEAN GAGES								
Ferrous metals, such as galvanized steel, tin plate						Nonferrous metals, such as copper, brass, aluminum			Steel and iron wire and bare copper piano wire			Galvanized steel, tin plate, copper, strip steel and steel, copper and aluminum tubes						Nonferrous		
U.S. Standard (USS)			U.S. Standard (Revised) Formerly Manufactures Standard			American Standard or Brown and Sharpe (B & S)			United States Steel Wire Gage			Birmingham (BWG)			New Birmingham (BG)			Imperial Wire Gage Imperial Standard (SWG)		
Gage	in.	mm	Gage	in.	mm	Gage	in.	mm	Gage	in.	mm	Gage	in.	mm	Gage	in.	mm	Gage	in.	mm
			3	.240	6.01	3	.229	5.83												
4	.234	5.95	4	.224	5.70	4	.204	5.19	4	.225	5.72	4	.238	6.05	4	.250	6.35	4	.232	5.89
5	.219	5.56	5	.209	5.31	5	.182	4.62	5	.207	5.26	5	.220	5.59	5	.223	5.65	5	.212	5.39
6	.203	5.16	6	.194	4.94	6	.162	4.12	6	.192	4.88	6	.203	5.16	6	.198	5.03	6	.192	4.88
7	.188	4.76	7	.179	4.55	7	.144	3.67	7	.177	4.50	7	.180	4.57	7	.176	4.48	7	.176	4.47
8	.172	4.37	8	.164	4.18	8	.129	3.26	8	.162	4.11	8	.165	4.19	8	.157	3.99	8	.160	4.06
9	.156	3.97	9	.149	3.80	9	.114	2.91	9	.148	3.77	9	.148	3.76	9	.140	3.55	9	.144	3.66
10	.141	3.57	10	.135	3.42	10	.102	2.59	10	.135	3.43	10	.134	3.40	10	.125	3.18	10	.128	3.25
11	.125	3.18	11	.120	3.04	11	.091	2.30	11	.121	3.06	11	.120	3.05	11	.111	2.83	11	.116	2.95
12	.109	2.78	12	.105	2.66	12	.081	2.05	12	.106	2.68	12	.109	2.77	12	.099	2.52	12	.104	2.64
13	.094	2.38	13	.090	2.78	13	.072	1.83	13	.092	2.32	13	.095	2.41	13	.088	2.24	13	.092	2.34
14	.078	1.98	14	.075	1.90	14	.064	1.63	14	.080	2.03	14	.083	2.11	14	.079	1.99	14	.080	2.03
15	.070	1.79	15	.067	1.71	15	.057	1.45	15	.072	1.83	15	.072	1.83	15	.070	1.78	15	.072	1.83
16	.063	1.59	16	.060	1.52	16	.051	1.29	16	.063	1.63	16	.065	1.65	16	.063	1.59	16	.064	1.63
17	.056	1.43	17	.054	1.37	17	.045	1.15	17	.054	1.37	17	.058	1.47	17	.056	1.41	17	.056	1.42
18	.050	1.27	18	.048	1.21	18	.040	1.02	18	.048	1.21	18	.049	1.25	18	.050	2.58	18	.048	1.22
19	.044	1.11	19	.042	1.06	19	.036	0.91	19	.041	1.04	19	.042	1.07	19	.044	1.19	19	.040	1.02
20	.038	0.95	20	.036	0.91	20	.032	0.81	20	.035	0.88	20	.035	0.89	20	.039	1.00	20	.036	0.91
21	.034	0.87	21	.033	0.84	21	.029	0.72	21	.032	0.81	21	.032	0.81	21	.035	0.89	21	.032	0.81
22	.031	0.79	22	.030	0.76	22	.025	0.65	22	.029	0.73	22	.028	0.71	22	.031	0.79	22	.028	0.71
23	.028	0.71	23	.027	0.68	23	.023	0.57	23	.026	0.66	23	.025	0.64	23	.028	0.71	23	.024	0.61
24	.025	0.64	24	.024	0.61	24	.020	0.51	24	.023	0.58	24	.022	0.56	24	.025	0.63	24	.022	0.56
25	.022	0.56	25	.021	0.53	25	.018	0.46	25	.020	0.52	25	.020	0.51	25	.022	0.56	25	.020	0.51
26	.019	0.48	26	.018	0.46	26	.016	0.40	26	.018	0.46	26	.018	0.46	26	.020	0.50	26	.018	0.46
27	.017	0.44	27	.016	0.42	27	.014	0.36	27	.017	0.44	27	.016	0.41	27	.017	0.44	27	.016	0.42
28	.016	0.40	28	.015	0.38	28	.013	0.32	28	.016	0.41	28	.014	0.36	28	.016	0.40	28	.015	0.38
29	.014	0.36	29	.014	0.34	29	.011	0.29	29	.015	0.38	29	.013	0.33	29	.014	0.35	29	.014	0.35
30	.013	0.32	30	.012	0.31	30	.010	0.25	30	.014	0.36	30	.012	0.31	30	.012	0.31	30	.012	0.32
31	.011	0.28	31	.011	0.27	31	.009	0.23	31	.013	0.34	31	.010	0.25	31	.011	0.28			
32	.010	0.26	32	.010	0.25	32	.008	0.20	32	.013	0.33	32	.009	0.23				32	.011	0.27
33	.009	0.24	33	.009	0.23	33	.007	0.18	33	.012	0.30	33	.008	0.20	33	.009	0.22	33	.010	0.25
34	.009	0.22	34	.008	0.21	34	.006	0.16	34	.010	0.26	34	.007	0.18	34	.008	0.20	34	.009	0.23
									35	.010	0.24	35	.005	0.13	35	.007	0.18	35	.008	0.21
36	.007	0.18	36	.007	0.17	36	.005	0.13	36	.009	0.23	36	.004	0.10	36	.006	0.16			
									37	.008	0.22							37	.007	0.17
38	.006	0.16	38	.006	0.15	38	.004	0.10	38	.008	0.20				38	.005	0.12	38	.006	0.15
									39	.008	0.19									
									40	.007	0.18				40	.004	0.10	40	.005	0.12
									41	.007	0.17							42	.004	0.10

NOTE: METRIC STANDARDS GOVERNING GAGE SIZES WERE NOT AVAILABLE AT THE TIME OF PUBLICATION. THE SIZES GIVEN IN THE ABOVE CHART ARE SOFT CONVERSION FROM CURRENT INCH STANDARDS AND ARE NOT MEANT TO BE REPRESENTATIVE OF THE PRECISE METRIC GAGE SIZES WHICH MAY BE AVAILABLE IN THE FUTURE. CONVERSIONS ARE GIVEN ONLY TO ALLOW THE STUDENT TO COMPARE GAGE SIZES READILY WITH THE METRIC DRILL SIZES.

TABLE 52   Wire and sheet-metal gages and thicknesses.

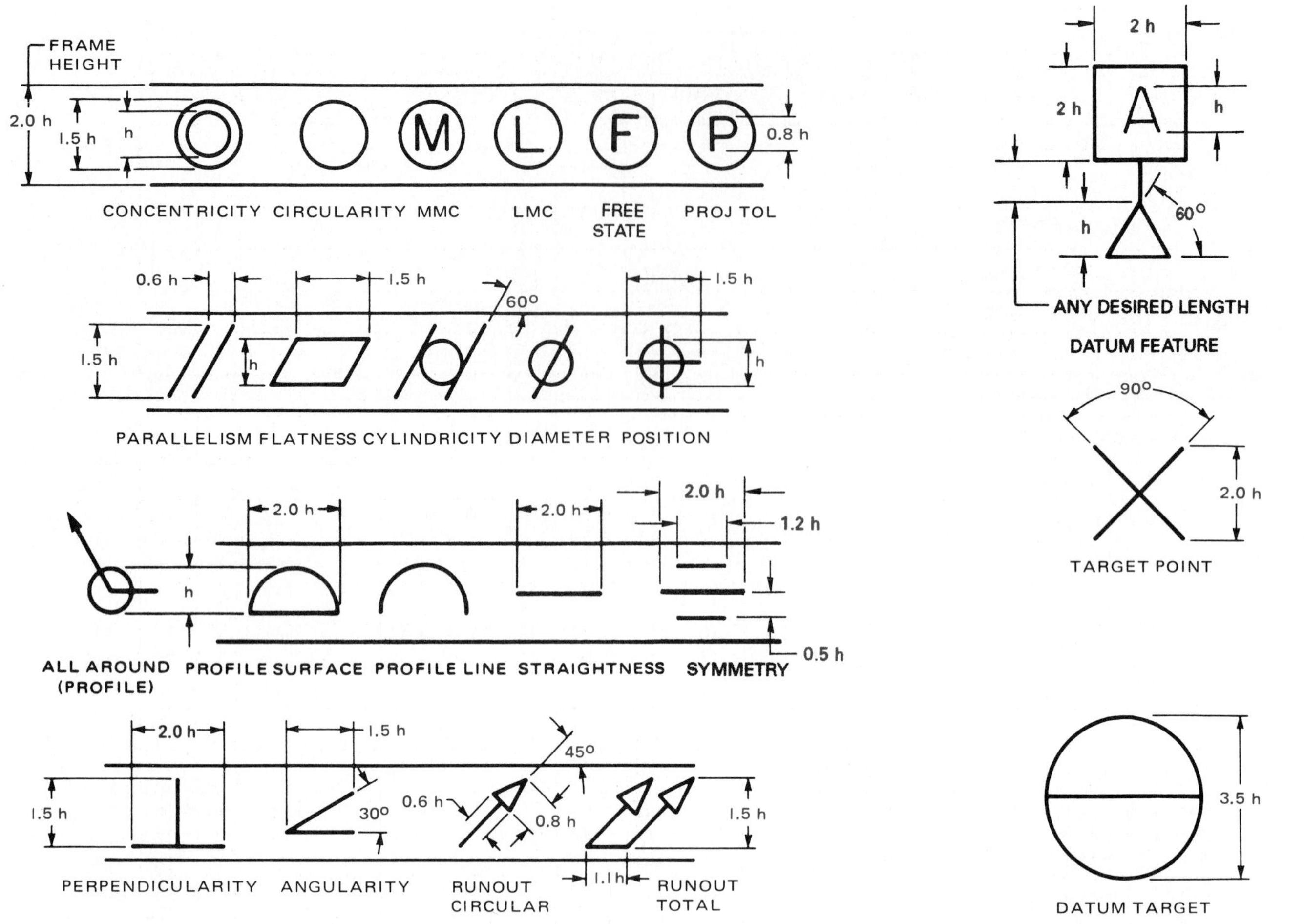
FRAME HEIGHT
2.0 h
1.5 h
h
0.8 h
CONCENTRICITY
CIRCULARITY
MMC
LMC
FREE STATE
PROJ TOL
M
L
F
P
0.6 h
1.5 h
60°
1.5 h
h
h
1.5 h
PARALLELISM
FLATNESS
CYLINDRICITY
DIAMETER
POSITION
2.0 h
2.0 h
2.0 h
1.2 h
h
0.5 h
ALL AROUND (PROFILE)
PROFILE SURFACE
PROFILE LINE
STRAIGHTNESS
SYMMETRY
2.0 h
1.5 h
45°
1.5 h
30°
0.6 h
0.8 h
1.5 h
1.1 h
PERPENDICULARITY
ANGULARITY
RUNOUT CIRCULAR
RUNOUT TOTAL
2 h
2 h
A
h
60°
h
ANY DESIRED LENGTH
DATUM FEATURE
90°
2.0 h
TARGET POINT
3.5 h
DATUM TARGET

TABLE 53 Form and proportion of geometric tolerancing symbols.

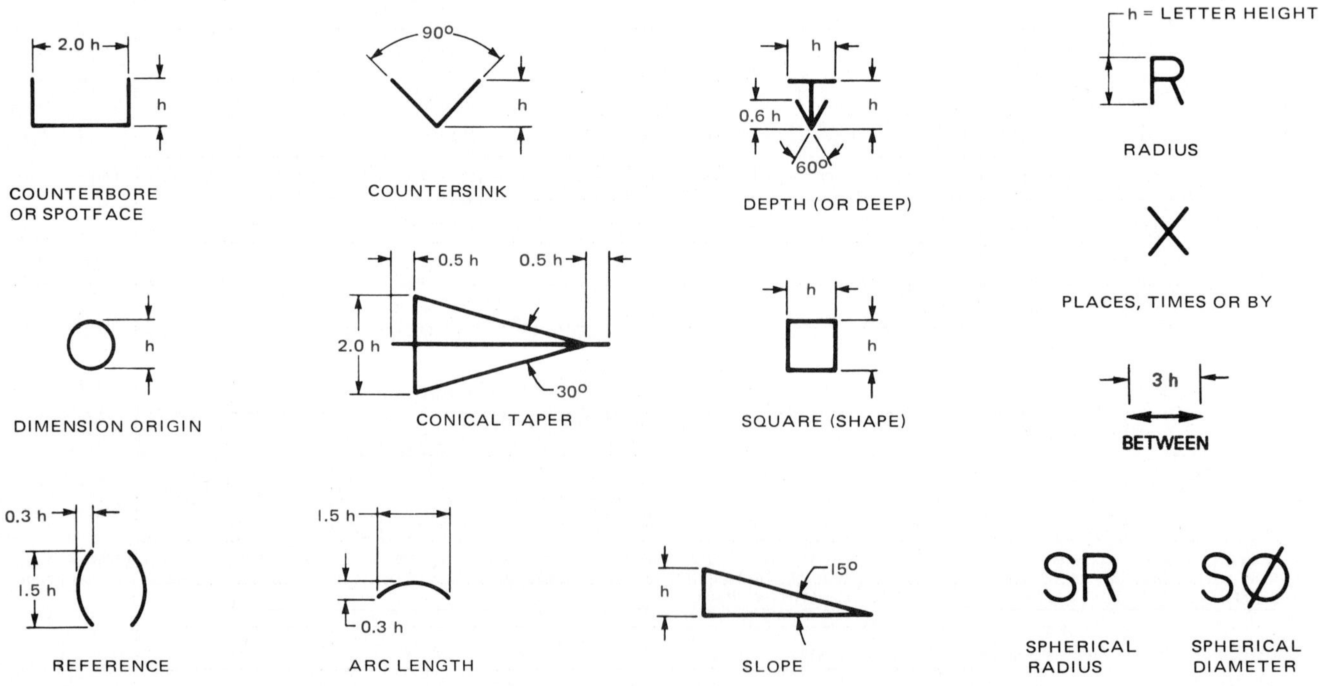
2.0 h
h
COUNTERBORE OR SPOTFACE
90°
h
COUNTERSINK
h
0.6 h
h
60°
DEPTH (OR DEEP)
h = LETTER HEIGHT
R
RADIUS
X
PLACES, TIMES OR BY
h
DIMENSION ORIGIN
0.5 h
0.5 h
2.0 h
30°
CONICAL TAPER
h
h
SQUARE (SHAPE)
3 h
BETWEEN
0.3 h
1.5 h
REFERENCE
1.5 h
0.3 h
ARC LENGTH
15°
h
SLOPE
SR
SØ
SPHERICAL RADIUS
SPHERICAL DIAMETER

TABLE 54 Form and proportion of dimensioning symbols.

SYMBOL FOR:	ASME Y14.5M–1994	ISO	CAN/CSA–B78.2–M91
STRAIGHTNESS	—	—	—
FLATNESS	▱	▱	▱
CIRCULARITY	◯	◯	◯
CYLINDRICITY	⌭	⌭	⌭
PROFILE OF A LINE	⌒	⌒	⌒
PROFILE OF A SURFACE	⌓	⌓	⌓
ALL AROUND–PROFILE	⌖⊖	⌖⊖	⌖⊖
ANGULARITY	∠	∠	∠
PERPENDICULARITY	⊥	⊥	⊥
PARALLELISM	//	//	//
POSITION	⊕	⊕	⊕
CONCENTRICITY/COAXIALITY	◎	◎	◎
SYMMETRY	≡	≡	≡
CIRCULAR RUNOUT	*↗	↗	*↗
TOTAL RUNOUT	*↗↗	↗↗	*↗↗
AT MAXIMUM MATERIAL CONDITION	Ⓜ	Ⓜ	Ⓜ
AT LEAST MATERIAL CONDITION	Ⓛ	Ⓛ	Ⓛ (PROPOSED)
REGARDLESS OF FEATURE SIZE	NONE	NONE	NONE
PROJECTED TOLERANCE ZONE	Ⓟ	Ⓟ	Ⓟ
DIAMETER	⌀	⌀	⌀
BASIC DIMENSION	50	50	50
REFERENCE DIMENSION	(50)	(50)	(50)
DATUM FEATURE	▨Ⓐ	* OR * ▨Ⓐ	▨*Ⓐ
DATUM TARGET	Ⓐ1⌀6	Ⓐ1⌀6	Ⓐ1⌀6
TARGET POINT	✕	✕	✕
DIMENSION ORIGIN	⊕→	⊕→	⊕→
FEATURE CONTROL FRAME	⊕⌀0.5ⓂA B C	⊕⌀0.5ⓂA B C	⊕⌀0.5ⓂA B C
CONICAL TAPER	▷	▷	▷
SLOPE	◹	◹	◹
COUNTERBORE/SPOTFACE	⌴	⌴ (PROPOSED)	⌴
COUNTERSINK	⌵	⌵ (PROPOSED)	⌵
DEPTH/DEEP	⤓	⤓ (PROPOSED)	⤓
SQUARE (SHAPE)	□	□	□
DIMENSION NOT TO SCALE	15	15	15
NUMBER OF TIMES/PLACES	8X	8X	8X
ARC LENGTH	1̄0̄5̄	1̄0̄5̄	1̄0̄5̄
RADIUS	R	R	R
SPHERICAL RADIUS	SR	SR	SR
SPHERICAL DIAMETER	S⌀	S⌀	S⌀
BETWEEN	←→	NONE	←→ (PROPOSED)

* MAY BE FILLED IN

Table 55 Comparison of ASME (ANSI), ISO, and CSA symbols

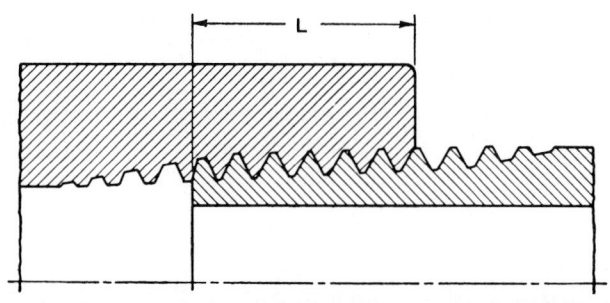

	Nominal Pipe Size Inches	Outside Diameter	Threads Per Inch	Wall Thickness			Approx. Distance Pipe Enters Fitting L	Weight (lbs/ft)		
				Sched. 40 (Standard)	Sched. 80 (Extra Strong)	Sched. 160		Sched. 40 (Standard)	Sched. 80 (Extra Strong)	Sched. 160
U.S. CUSTOMARY (INCHES)	⅛ (.125)	.405	27	.068	.095	—	.188	.24	.31	—
	¼ (.250)	.540	18	.088	.119	—	.281	.42	.54	—
	⅜ (.375)	.675	18	.091	.126	—	.297	.57	.74	—
	½ (.500)	.840	14	.109	.147	.188	.375	.85	1.09	1.31
	¾ (.750)	1.050	14	.113	.154	.219	.406	1.13	1.47	1.94
	1.00	1.315	11.50	.133	.179	.250	.500	1.68	2.17	2.84
	1.25	1.660	11.50	.140	.191	.250	.549	2.27	3.00	3.76
	1.50	1.900	11.50	.145	.200	.281	.562	2.72	3.63	4.86
	2	2.375	11.50	.154	.218	.344	.578	3.65	5.02	7.46
	2.5	2.875	8	.203	.276	.375	.875	5.79	7.66	10.01
	3	3.500	8	.216	.300	.438	.938	7.58	10.25	14.31
	3.5	4.000	8	.226	.318	—	1.000	9.11	12.51	—
	4	4.500	8	.237	.337	.531	1.062	10.79	14.98	22.52
	5	5.563	8	.258	.375	.625	1.156	14.62	20.78	32.96
	6	6.625	8	.280	.432	.719	1.250	18.97	28.57	45.34
	8	8.625	8	.322	.500	.906	1.469	28.55	43.39	74.71

	Nominal Pipe Size Inches	Outside Diameter mm	Threads Per Inch	Wall Thickness			Approx. Distance Pipe Enters Fitting L	Mass (kg/m)		
				Sched. 40 (Standard)	Sched. 80 (Extra Strong)	Sched. 160		Sched. 40 (Standard)	Sched. 80 (Extra Strong)	Sched. 160
METRIC (MILLIMETERS)	⅛ (.125)	10.3	27	1.7	2.4	—	5	0.36	0.46	—
	¼ (.250)	13.7	18	2.2	3.0	—	7	0.63	0.80	—
	⅜ (.375)	17.1	18	2.3	3.2	—	8	0.85	1.10	—
	½ (.500)	21.3	14	2.8	3.7	4.8	10	1.26	1.62	1.95
	¾ (.750)	26.7	14	2.9	3.9	5.6	11	1.68	2.19	2.89
	1.00	33.4	11.50	3.4	4.6	6.4	13	2.50	3.23	4.23
	1.25	42.1	11.50	3.6	4.9	6.4	14	3.38	4.46	5.60
	1.50	48.3	11.50	3.7	5.1	7.1	14	4.05	5.40	7.23
	2.00	60.3	11.50	3.9	5.5	8.7	15	5.43	7.47	11.10
	2.50	73	8	5.2	7.0	9.5	22	8.62	11.40	14.90
	3.00	88.9	8	5.5	7.6	11.1	24	11.28	15.25	21.30
	3.50	101.6	8	5.7	8.1	—	25	13.56	18.62	—
	4.00	114.3	8	6.0	8.6	13.5	27	16.06	22.30	33.51
	5.00	141.3	8	6.6	9.5	15.9	29	21.76	30.92	49.05
	6.00	168.3	8	7.1	11.0	18.3	32	28.23	42.52	67.47
	8.00	219	8	8.2	12.7	23.0	38	42.49	64.57	111.18

TABLE 56 American standard wrought steel pipe.

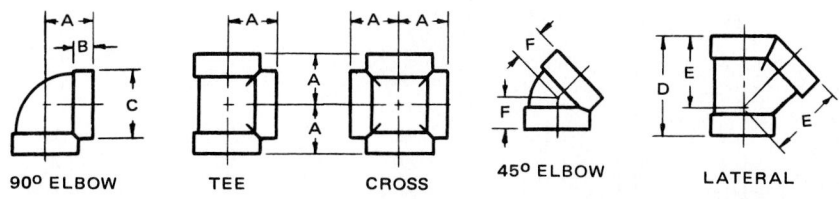

Nominal Pipe Size in Inches	U.S. CUSTOMARY (INCHES)						METRIC (MILLIMETERS)					
	A	Min B	Min C	D	E	F	A	Min B	Min C	D	E	F
.25	.81	.38	.93	—	—	.73	21	10	24	—	—	19
.375	.95	.44	1.12	—	—	.80	24	11	28	—	—	20
.50	1.12	.50	1.34	2.50	1.87	.88	28	13	34	64	47	22
.75	1.31	.56	1.63	3.00	2.25	.98	33	14	41	76	57	25
1.00	1.50	.62	1.95	3.50	2.75	1.12	38	16	50	89	70	28
1.25	1.75	.69	2.39	4.25	3.25	1.29	44	18	61	108	83	33
1.50	1.94	.75	2.68	4.87	3.81	1.43	49	19	68	124	97	36
2.00	2.25	.84	3.28	5.75	4.25	1.68	57	21	83	146	108	43
2.50	2.70	.94	3.86	6.75	5.18	1.95	69	24	98	171	132	50
3.00	3.08	1.00	4.62	7.87	6.12	2.17	78	25	117	200	155	55
3.50	3.42	1.06	5.20	8.87	6.87	2.39	87	27	132	225	174	61
4.00	3.79	1.12	5.79	9.75	7.62	2.61	96	28	147	248	194	66
5.00	4.50	1.18	7.05	11.62	9.25	3.05	114	30	179	295	235	77
6.00	5.13	1.28	8.28	13.43	10.75	3.46	130	33	210	341	273	88
8.00	6.56	1.47	10.63	16.94	13.63	4.28	167	37	270	430	346	109
10.00	8.08	1.68	13.12	20.69	16.75	5.16	205	43	333	613	425	131

TABLE 57   American standard (125 lb) cast-iron screwed-pipe fittings.

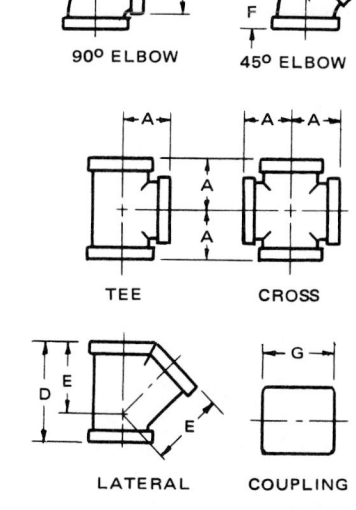

Nominal Pipe Size in Inches	U.S. CUSTOMARY (INCHES)							METRIC (MILLIMETERS)						
	A	B	C	D	E	F	G	A	B	C	D	E	F	G
.125	.69	.20	.69	—	—	—	.96	18	5.0	18	—	—	—	24
.250	.81	.22	.84	—	—	.73	1.06	21	5	21	—	—	19	27
.375	.95	.23	1.02	1.93	1.43	.80	1.16	24	6	26	49	36	20	29
.500	1.12	.25	1.20	2.32	1.71	.88	1.34	28	6	30	59	43	22	34
.750	1.31	.27	1.46	2.77	2.05	.98	1.52	33	7	37	70	52	25	39
1.00	1.50	.30	1.77	3.28	2.43	1.12	1.67	38	8	45	83	62	28	42
1.25	1.75	.34	2.15	3.94	2.92	1.29	1.93	44	9	55	100	74	33	49
1.50	1.94	.37	2.43	4.38	3.28	1.43	2.15	49	9	62	111	83	36	55
2.00	2.25	.42	2.96	5.17	3.93	1.68	2.53	57	11	75	131	100	43	64
2.50	2.70	.48	3.59	6.25	4.73	1.95	2.88	69	12	91	159	120	50	73
3.00	3.08	.55	4.29	7.26	5.55	2.17	3.18	78	14	109	184	141	55	81
3.50	3.42	.60	4.84	—	—	2.39	3.43	87	15	123	—	—	61	87
4.00	3.79	.66	5.40	8.98		2.61	3.69	96	17	137	228	177	66	94
5.00	4.50	.78	6.58		6.97	3.05	—	114	20	167	—	—	77	—
6.00	5.13	.90	7.77			3.46	—	130	23	197	—	—	88	—

TABLE 58   American standard (150 lb) malleable-iron screwed-pipe fittings.

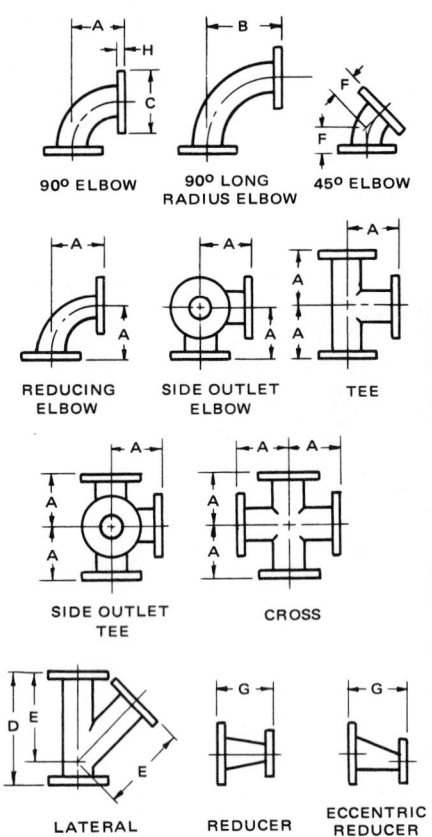

NOMINAL PIPE SIZE IN INCHES		A	B	C	D	E	F	G	H
U.S. CUSTOMARY (INCHES)	1.50	4.00	6.00	5.00	9.00	7.00	2.25	—	.56
	2.00	4.50	6.50	6.00	10.50	8.00	2.50	5.00	.62
	2.50	5.00	7.00	7.00	12.00	9.50	3.00	5.50	.69
	3.00	5.50	7.75	7.50	13.00	10.00	3.00	6.00	.75
	3.50	6.00	8.50	8.50	14.50	11.50	3.50	6.50	.81
	4.00	6.50	9.00	9.00	15.00	12.00	4.00	7.00	.94
	5.00	7.50	10.25	10.00	17.00	13.50	4.50	8.00	.94
	6.00	8.00	11.50	11.00	18.00	14.50	5.00	9.00	1.00
	8.00	9.00	14.00	13.50	22.00	17.50	5.50	11.00	1.12
	10.00	11.00	16.50	16.00	25.50	20.50	6.50	12.00	1.19
METRIC (MILLIMETERS)	1.50	102	152	127	229	178	57	—	14
	2.00	114	165	152	267	203	64	127	16
	2.50	127	178	178	305	241	76	140	18
	3.00	140	197	190	330	254	76	152	19
	3.50	153	216	216	368	292	89	165	21
	4.00	165	229	229	381	305	102	178	24
	5.00	190	260	254	432	343	114	203	24
	6.00	203	292	280	457	368	127	229	25
	8.00	229	356	343	559	445	140	279	28
	10.00	279	419	406	648	521	165	305	30

TABLE 59   American standard flanged fittings.

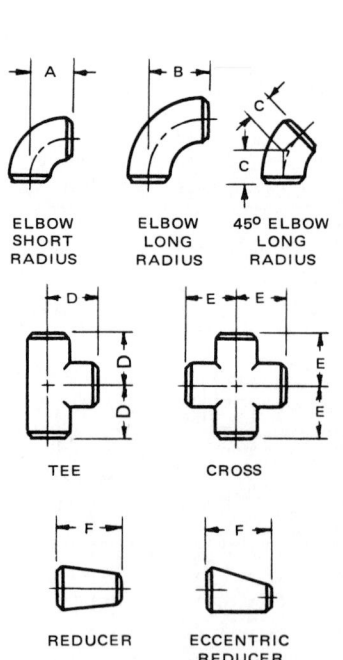

NOMINAL PIPE SIZE IN INCHES		A	B	C	D	E	F
U.S. CUSTOMARY (INCHES)	1.50	1.50	2.25	1.12	2.25	2.25	2.50
	2.00	2.00	3.00	1.38	2.50	2.50	3.00
	2.50	2.50	3.75	1.75	3.00	3.00	3.50
	3.00	3.00	4.50	2.00	3.38	3.38	3.50
	3.50	3.50	5.25	2.25	3.75	3.75	4.00
	4.00	4.00	6.00	2.50	4.12	4.12	4.00
	5.00	5.00	7.50	3.12	4.89	4.89	5.00
	6.00	6.00	9.00	3.75	5.62	5.62	5.50
	8.00	8.00	12.00	5.00	7.00	7.00	6.00
	10.00	10.00	15.00	6.25	8.50	8.50	7.00
METRIC (MILLIMETERS)	1.50	38	57	28	57	57	64
	2.00	51	76	35	64	64	76
	2.50	64	95	44	76	76	89
	3.00	76	114	51	86	86	89
	3.50	89	133	57	95	95	102
	4.00	102	152	64	105	105	102
	5.00	127	190	79	124	124	127
	6.00	152	229	95	143	143	140
	8.00	203	305	127	178	178	152
	10.00	254	381	159	216	216	178

TABLE 60   American standard steel butt-welding fittings.

Nominal Pipe Size in Inches	Lift Check			Swing Check		
	A		B	C		D
	Screwed	Flanged		Screwed	Flanged	
U.S. CUSTOMARY (INCHES)						
2.00	6.50	—	3.50	6.50	8.00	4.25
2.50	7.00	—	4.25	7.00	8.50	4.81
3.00	8.00	9.50	5.00	8.00	9.50	5.06
3.50	—	—	—	9.00	10.50	5.81
4.00	10.00	11.50	6.25	10.00	11.50	6.19
5.00	—	13.00	7.00	11.25	13.00	7.19
6.00	—	14.00	8.25	12.50	14.00	7.50
8.00	—	—	—	—	19.50	10.19
10.00	—	—	—	—	24.50	12.12
METRIC (MILLIMETERS)						
2.00	165	—	89	165	203	108
2.50	178	—	108	178	216	122
3.00	203	241	127	203	241	129
3.50	—	—	—	229	267	148
4.00	254	292	159	254	292	157
5.00	—	330	178	286	330	183
6.00	—	356	210	318	356	190
8.00	—	—	—	—	495	259
10.00	—	—	—	—	622	308

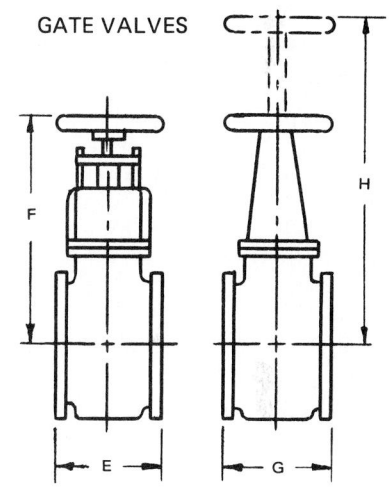

CHECK VALVES

LIFT   SWING

Nominal Pipe Size in Inches	Nonrising Spindle			Rising Spindle		
	E		F	G		H
	Screwed	Flanged		Screwed	Flanged	
U.S. CUSTOMARY (INCHES)						
2.00	4.75	7.00	10.50	4.75	7.00	13.12
2.50	5.50	7.50	11.19	5.50	7.50	14.50
3.00	6.00	8.00	12.62	6.00	8.00	16.62
3.50	6.62	8.50	13.31	6.62	8.50	18.44
4.00	7.12	9.00	15.25	7.12	9.00	21.06
5.00	8.12	10.00	17.88	8.12	10.00	25
6.00	9.00	10.50	20.19	9.00	10.50	29.25
8.00	10.00	11.50	24.00	10.00	11.50	37.25
10.00	—	13.00	28.19	—	13.00	44.12
METRIC (MILLIMETERS)						
2.00	121	178	267	121	178	333
2.50	140	190	284	140	190	368
3.00	152	203	321	152	203	422
3.50	168	216	338	168	216	468
4.00	181	229	387	181	229	535
5.00	206	254	454	206	254	635
6.00	229	267	513	229	267	743
8.00	254	292	610	254	292	946
10.00	—	330	716	—	330	1121

GATE VALVES

Nominal Pipe Size in Inches	Globe			Angle		
	J		K	L		M
	Screwed	Flanged		Screwed	Flanged	
U.S. CUSTOMARY (INCHES)						
2.00	4.75	7.00	9.44	3.50	3.88	10.38
2.50	5.50	7.50	11.06	3.88	4.50	12.06
3.00	6.00	8.00	12.38	4.69	4.62	12.69
3.50	6.62	8.50	13.19	5.00	5.38	13.62
4.00	7.12	9.00	15.25	6.00	5.88	14.88
5.00	8.12	10.00	17.25	6.31	6.50	17.69
6.00	9.00	10.50	18.81	8.00	8.00	19.06
8.00	10.00	11.50	22.12	—	9.25	22.75
10.00	—	13.00	24.75	—	10.62	24.94
METRIC (MILLIMETERS)						
2.00	121	178	240	89	98	264
2.50	140	190	281	98	114	281
3.00	152	203	314	119	118	346
3.50	168	216	335	127	136	348
4.00	181	229	387	152	149	378
5.00	206	254	438	160	165	449
6.00	229	267	478	203	203	484
8.00	254	292	562	—	235	578
10.00	—	330	629	—	270	633

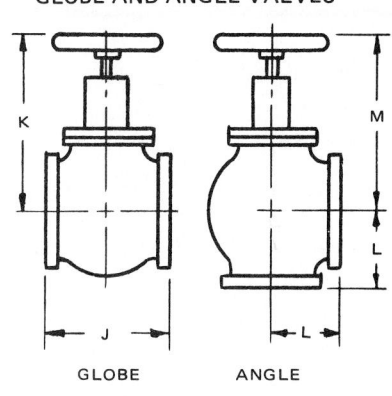

GLOBE AND ANGLE VALVES

GLOBE   ANGLE

Dimensions taken from manufacturer's catalogs for drawing purposes.

TABLE 61   Common valves.

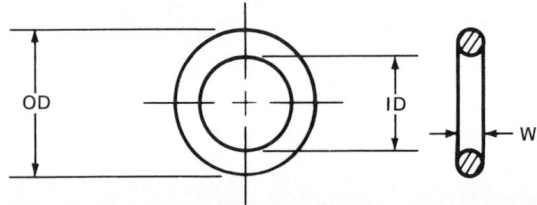

U.S. CUSTOMARY (INCHES)							METRIC (MILLIMETERS)					
W = .071	W = .087	W = .125	W = .174	W = .209	W = .256		W = 1.5	W = 2	W = 3	W = 4	W = 5	W = 6
ID	OD	OD	OD	OD	OD	OD	ID	OD	OD	OD	OD	OD	OD
.124	x						3	6					
.158	x						4	7					
.132	x						5	8					
.248	x						6	9					
.280	x						7	10					
.295	x	x	x				8	11	12	14			
.354	x	x	x				9	12	13	15			
.394	x	x	x	x	x	x	10	13	14	16	18	20	
.465	x	x	x	x	x	x	12	15	16	18	20	22	24
.551	x	x	x	x	x	x	14	17	18	20	22	24	26
.591	x	x	x	x	x	x	15	18	19	21	23	25	27
.622	x	x	x	x	x	x	16	19	20	22	24	26	28
.630	x	x	x	x	x	x	18	21	22	24	26	28	30
.662	x	x	x	x	x	x	20	23	24	26	28	30	32
.669	x	x	x	x	x	x	22	25	26	28	30	32	34
.787	x	x	x	x	x	x	24	27	28	30	32	34	36
.850	x	x	x	x	x	x	25	28	29	31	33	35	37
1.024	x	x	x	x	x	x	26	29	30	32	34	36	38
1.102	x	x	x	x	x	x	28	31	32	34	36	38	40
1.181	x	x	x	x	x	x	30	33	34	36	38	40	42
1.220	x	x	x	x	x	x	32	35	36	38	40	42	44
1.299	x	x	x	x	x	x	34	37	38	40	42	44	46
1.319	x	x	x	x	x	x	35	38	39	41	43	45	47
1.496	x	x	x	x	x	x	36	39	40	42	44	46	48
1.524	x	x	x	x	x	x	38	41	42	44	46	48	50
1.575	x	x	x	x	x	x	40	43	44	46	48	50	52
1.654	x	x	x	x	x	x	42	45	46	48	50	52	54
1.732	x	x	x	x	x	x	44	47	48	50	52	54	56
1.772	x	x	x	x	x	x	45	48	49	51	53	55	57
1.811	x	x	x	x	x	x	46	49	50	52	54	56	58
1.890	x	x	x	x	x	x	48	51	52	54	56	58	60
2.008	x	x	x	x	x	x	50	53	54	56	58	60	62
2.087	x	x	x	x	x	x	52	55	56	58	60	62	64
2.165	x	x	x	x	x	x	54	57	58	60	62	64	66
2.205	x	x	x	x	x	x	55	58	59	61	63	65	67
2.244	x	x	x	x	x	x	56	59	60	62	64	66	68
2.323	x	x	x	x	x	x	58	61	62	64	66	68	70
2.402	x	x	x	x	x	x	60	63	64	66	68	70	72
2.639	x	x	x	x	x	x	65	68	69	71	73	75	77
2.835	x	x	x	x	x	x	70	73	74	76	78	80	82
3.031	x	x	x	x	x	x	75	78	79	81	83	85	87
3.228	x	x	x	x	x	x	80	83	84	86	88	90	92

TABLE 62   O-rings.

Inside Dia		Outside Dia		Width	
in.	mm	in.	mm	in.	mm
.375	10	.753	19	.25	6
.375	10	.840	21	.31	8
.438	11	1.003	26	.31	8
.438	11	1.128	28	.31	8
.500	12	1.003	26	.31	8
.500	12	1.128	28	.31	8
.500	12	1.254	32	.38	10
.562	14	1.003	26	.31	8
.562	14	1.128	28	.31	8
.625	16	1.250	32	.38	10
.625	16	1.128	28	.31	8
.625	16	1.250	32	.38	10
.688	18	1.379	35	.38	10
.688	18	1.128	28	.31	8
.688	18	1.254	32	.38	10
.750	20	1.379	35	.38	10
.750	20	1.254	32	.38	10
.750	20	1.379	35		10
.750	20	1.503	38	.38	10
.8125	21	1.756	44	.44	12
.8125	21	1.254	32	.38	10
.8125	21	1.379	35	.38	10
.8125	21	1.503	38	.38	10
.875	22	1.756	44	.44	12
.875	22	1.379	35	.38	10
.875	22	1.503	38	.38	10
.875	22	1.628	42	.44	12
.938	24	1.756	44	.44	12
.938	24	1.503	38	.38	10
.938	24	1.628	42	.38	10
1.000	25	1.756	44	.44	12
1.000	25	1.503	38	.38	10
1.000	25	1.756	44	.44	12
1.000	25	1.878	48	.44	12
1.000	25	2.004	50	.44	12
1.062	26	1.503	38	.38	10
	26	1.628	42	.44	12
	26	1.756	44	.44	12
1.125	28	1.628	42	.44	12
	28	1.756	44	.44	12
	28	1.987	50	.50	12
1.188	30	1.832	46	.44	12
	30	1.987	50	.50	12
	30	2.254	58	.50	12
1.25	32	1.756	44	.44	12
	32	1.878	48	.44	12
	32	2.066	52	.50	12
1.312	34	2.060	52	.44	12
	34	2.254	58	.50	12
	34	2.378	60	.50	12
1.375	35	2.066	52	.44	12
	35	2.254	58	.50	12
	35	2.441	62	.50	12
1.438	36	2.254	58	.50	12
	36	2.506	64	.50	12
	36	2.627	66	.50	12
1.500	38	2.254	58	.38	10
	38	2.410	62	.50	12
	38	2.720	70	.50	12
1.562	40	2.441	62	.50	12
	40	2.690	68	.50	12
	40	2.879	74	.50	12
1.625	42	2.441	62	.38	10
	42	2.879	74	.38	10
	42	2.627	66	.50	12
	42	2.879	74	.50	12
	42	3.066	78	.50	12
1.75	44	2.254	58	.50	12
	44	2.441	62	.50	12
	44	2.506	64	.50	12

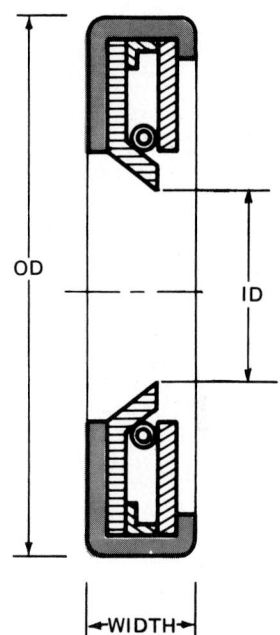

TABLE 63   Oil seals.

Shaft Dia		Outside Dia		Width		Size of Set Screw	
in.	mm	in.	mm	in.	mm	in.	mm
.375	10	.75	20	.40	10	.250	M6
.500	12	1.00	25	.44	10	.250	M6
.625	16	1.10	28	.50	12	.3125	M8
.750	20	1.20	30	.56	14	.3125	M8
.875	22	1.50	40	.56	14	.3125	M8
1.000	24	1.60	40	.60	16	.3125	M8

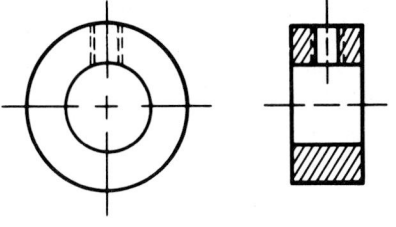

TABLE 64   Setscrew collars.

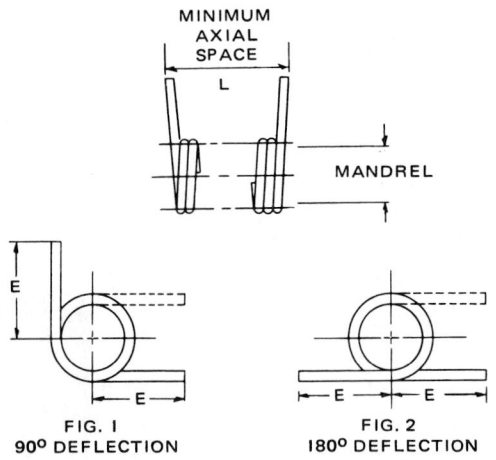

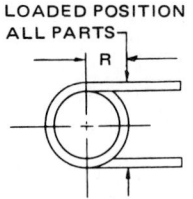

FIGURES SHOW SPRINGS
WOUND LEFT-HAND

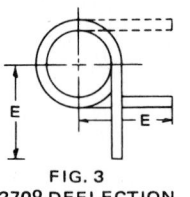

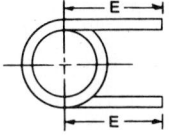

| FIG. I | FIG. 2 | FIG. 3 | FIG. 4 |
| 90° DEFLECTION | 180° DEFLECTION | 270° DEFLECTION | 360° DEFLECTION |

U.S. CUSTOMARY (INCHES)								METRIC (MILLIMETERS)							
Wire Dia	Out-side Dia	Pos. of Ends Fig.	Def, Deg.	Torque in · lb	Radius R	E	Min. Axial Space L	Wire Dia	Out-side Dia	Pos. of Ends Fig.	Def, Deg.	Torque N · m	Radius R	E	Min. Axial Space L
.014	.124	1	90	.070	.250	.50	.067	0.35	3.2	1	90	0.008	6.4	13	1.7
	.133	2	180		.250	.50	.105		3.4	2	180		6.4	13	2.7
	.124	3	270		.250	.50	.158		3.2	3	270		6.4	13	4.0
	.194	2	180		.375	.75	.077		4.9	2	180		9.5	19	2.0
	.201	3	270		.375	.75	.102		5.1	3	270		9.5	19	2.6
	.204	4	360		.375	.75	.126		5.2	4	360		9.5	19	3.2
.017	.160	1	90	.117	.250	.50	.081	0.43	4.1	1	90	0.013	6.4	13	2.1
	.172	2	180		.250	.50	.127		4.4	2	180		6.4	13	3.2
	.160	3	270		.250	.50	.192		4.1	3	270		6.4	13	4.9
	.249	2	180		.375	.75	.094		6.3	2	180		9.5	19	2.4
	.259	3	270		.375	.75	.123		6.6	3	270		9.5	19	3.1
	.235	4	360		.375	.75	.170		6.0	4	360		9.5	19	4.3
.020	.191	1	90	.187	.375	.75	.095	0.51	4.9	1	90	0.021	9.5	19	2.4
	.179	2	180		.375	.75	.170		4.6	2	180		9.5	19	4.3
	.175	3	270		.375	.75	.245		4.5	3	270		9.5	19	6.2
	.242	2	180		.500	1.00	.130		6.2	2	180		12.7	25	3.3
	.268	3	270		.500	1.00	.165		6.8	3	270		12.7	25	4.2
	.254	4	360		.500	1.00	.250		6.5	4	360		12.7	25	6.4
.023	.204	1	90	.308	.375	.75	.109	0.59	5.2	1	90	0.035	9.5	19	2.8
	.191	2	180		.375	.75	.196		4.9	2	180		9.5	19	5.0
	.187	3	270		.375	.75	.282		4.8	3	270		9.5	19	7.2
	.259	2	180		.500	1.00	.150		6.6	2	180		12.7	25	3.8
	.251	3	270		.500	1.00	.213		6.4	3	270		12.7	25	5.4
	.271	4	360		.500	1.00	.253		6.9	4	360		12.7	25	6.5
.028	.267	1	90	.515	.500	1.00	.133	0.71	6.8	1	90	0.058	12.7	25	3.4
	.249	2	180		.500	1.00	.238		6.3	2	180		12.7	25	6.0
	.245	3	270		.500	1.00	.344		6.2	3	270		12.7	25	8.8
	.340	2	180		.500	1.00	.182		8.7	2	180		12.7	25	4.6
	.329	3	270		.500	1.00	.259		8.4	3	270		12.7	25	6.6
	.355	4	360		.500	1.00	.308		9.0	4	360		12.7	25	7.9

TABLE 65 Torsion springs. *(Wallace Barnes Co. Ltd.)*

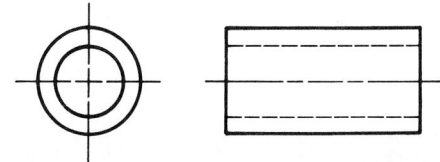

Inside Dia in.	mm	Outside Dia in.	mm	15 .50	20 .75	25 1.00	30 1.25	35 1.50	40 1.75	50 2.00	60 2.50	75 3.00
.375	10	.625	16	•	•	•	•					
		.750	20	•	•	•	•					
.500	12	.625	16	•	•	•	•	•				
		.750	20	•	•	•	•	•	•	•		
.625	16	.875	22	•	•	•	•	•	•	•		
		1.000	25	•	•	•	•	•	•	•		
.750	20	1.000	25	•	•	•	•	•	•	•		
		1.125	28	•	•	•	•	•	•	•		
.875	22	1.125	28		•	•	•	•	•	•	•	
		1.250	32		•	•	•	•	•	•	•	
1.000	25	1.250	32			•	•	•	•	•	•	
		1.375	35			•	•	•	•	•	•	•

TABLE 66 Standard plain (journal) bearings.

Series	Bearing Number	Bore d in.	mm	Outside Diameter D in.	mm	Width W in.	mm	Basic Load Rating Kips	kN
Light	7205B	0.984	25	2.047	52	0.591	15	2.5	11.4
	7206B	1.181	30	2.441	62	0.630	16	3.4	15.6
	7207B	1.378	35	2.835	72	0.670	17	4.6	20.8
	7208B	1.575	40	3.150	80	0.709	18	5.5	24.5
	7209B	1.772	45	3.347	85	0.748	19	6.2	27.5
	7210B	1.969	50	3.543	90	0.787	20	6.4	28.5
	7211B	2.165	55	3.937	100	0.827	21	8.2	36
	7212B	2.362	60	4.331	110	0.866	22	9.7	43
Medium	7304B	0.787	20	2.047	52	0.591	15	3.1	13.4
	7305B	0.984	25	2.441	62	0.670	17	4.3	19
	7306B	1.181	30	2.835	72	0.748	19	5.4	24
	7307B	1.378	35	3.150	80	0.827	21	6.3	28
	7308B	1.575	40	3.543	90	0.906	23	7.8	34.5
	7309B	1.772	45	3.937	100	0.984	25	10	45
	7310B	1.969	50	4.331	110	1.063	27	11.6	52
	7311B	2.165	55	4.724	120	1.142	29	13.4	61
	7312B	2.362	60	5.118	130	1.221	31	15.6	69.5
Heavy	7405B	0.984	25	3.150	80	0.827	21	6.8	30.5
	7406B	1.181	30	3.543	90	0.906	23	8.3	36.5
	7407B	1.378	35	3.937	100	0.984	25	10.4	46.5
	7408B	1.575	40	4.331	110	1.063	27	12	54
	7409B	1.772	45	4.724	120	1.142	29	14.6	65.5
	7410B	1.969	50	5.118	130	1.221	31	16.6	73.5
	7411B	2.165	55	5.512	140	1.299	33	19	85
	7412B	2.362	60	5.906	150	1.378	35	20.4	91.5

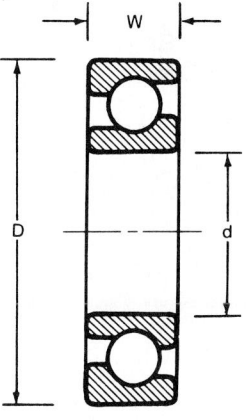

TABLE 67 Angular contact ball bearings. *(SKF Co. Ltd.)*

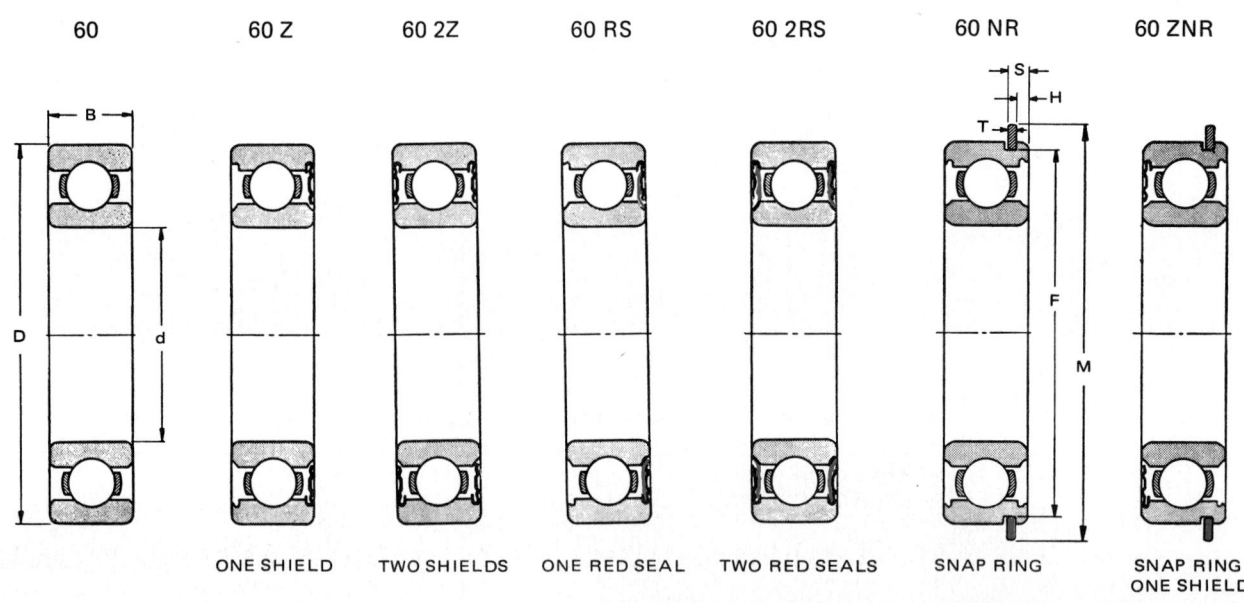

60	60 Z	60 2Z	60 RS	60 2RS	60 NR	60 ZNR
	ONE SHIELD	TWO SHIELDS	ONE RED SEAL	TWO RED SEALS	SNAP RING	SNAP RING ONE SHIELD

						Nominal Bearing Dimensions									
						d		D		B		F	M	S	H	T
Bearing Number						in.	mm	in.	mm	in.	mm	Millimeters				
6000	6000 Z	6000 2Z	6000 RS	6000 2RS			.375	10	1.031	26	.312	8				
6001	6001 Z	6001 2Z	6001 RS	6001 2RS			.500	12	1.125	28	.312	8				
6002	6002 Z	6002 2Z	6002 RS	6002 2RS	6002 NR	6002 ZNR	.625	15	1.250	32	.375	9	30	36.5	3	2	1
6003	6003 Z	6003 2Z	6003 RS	6003 2RS			.750	17	1.656	35	.500	10				
6004	6004 Z	6004 2Z	6004 RS	6004 2RS			.875	20	1.750	42	.500	2				
6005	6005 Z	6005 2Z	6005 RS	6005 2RS			1.000	25	1.875	47	.500	12				
6006	6006 Z	6006 2Z	6006 RS	6006 2RS			1.1875	30	2.156	55	.531	13				
6007	6007 Z	6007 2Z	6007 RS	6007 2RS			1.375	35	2.500	62	.562	14				
6008	6008 Z	6008 2Z	6008 RS	6008 2RS			1.500	40	2.688	68	.594	15				
6009							1.750	45	3.000	75	.625	16				
6010	6010 Z	6010 2Z	6010 RS	6010 2RS	6010 NR	6010 ZNR	2.000	50	3.125	80	.625	16	76.8	86	4	2.5	1.6
6011	6011 Z	6011 2Z	6011 RS	6011 2RS	6011 NR	6011 ZNR	2.250	55	3.562	90	.719	18	86.8	96.5	5	3	2.4
6012							2.375	60	3.750	95	.719	18				
6013	6013 Z	6013 2Z			6013 NR	6013 ZNR	2.500	65	4.000	100	.719	18	96.8	106	5	3	2.4
6014							2.750	70	4.375	110	.781	20				
6015	6015 Z	6015 2Z			6015 NR	6015 ZNR	3.000	75	4.500	115	.781	20	112	121	5	3	2.4

Note: When shield is required on same side as snap ring, order as ZNBR.
       Inch size dimensions shown are for inch size bearings.

TABLE 68   Radial ball bearings. *(SKF Co. Ltd.)*

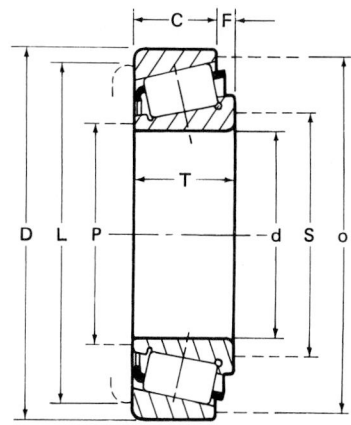

Bearing Number	Nominal External Sizes						U.S. CUSTOMARY (in.)						METRIC (mm)					
	Bore		Outside Dia		Width													
	in.	mm	in.	mm	in.	mm	F	C	S	P	L	O	F	C	S	P	L	O
4059	.590	15	1.375	35	.430	11	.089	.344	.781	.719	1.094	1.250	2	9	20	18	28	32
03062	.625	16	1.625	41	.562	14	.125	.438	.844	.781	1.312	1.469	3	11	21	20	33	37
5062	.625	16	1.850	46	.566	14	.129	.438	.938	.844	1.562	1.688	3	11	24	21	40	43
1749	.688	17	1.570	40	.545	14	.125	.420	.906	.781	1.312	1.438	3	11	23	20	37	36
9195	.695	17.6	1.938	49	.906	20	.219	.562	1.062	.969	1.625	1.750	6	14	27	24	41	44
9070	.695	17.6	1.938	49	.906	23	.219	.688	1.062	.969	1.562	1.750	6	17	27	24	39	44
1774	.748	19	2.240	57	.753	20	.138	.625	1.062	.969	1.875	2.031	4	16	27	24	48	51
1949	.750	20	1.781	45	.610	16	.135	.475	1.000	.922	1.500	1.625	4	12	25	59	38	41
9067	.750	20	1.938	50	.835	21	.148	.688	1.000	.922	1.531	1.750	4	17	25	23	39	44
9078	.750	20	1.938	50	.906	23	.219	.562	1.000	.922	1.625	1.750	6	17	25	23	39	44
7075	.750	20	2.000	51	.591	15	.091	.375	1.000	.938	1.719	1.844	3	12	25	24	44	47
7087	.857	22	1.969	50	.531	14	.156	.375	1.125	1.062	1.719	1.844	4	10	28	27	44	47
1380	.875	22	2.125	54	.762	20	.200	.562	1.125	1.094	1.750	1.906	5	15	28	28	44	48
1280	.875	22	2.250	54	.875	22	.188	.688	1.125	1.094	1.875	2.062	5	17	28	28	48	52
4643	1.000	25	1.980	48	.560	14	.140	.420	1.219	1.156	1.719	1.844	4	10	33	29	44	47
1780	1.000	25	2.240	57	.762	20	.781	.625	1.188	1.172	1.844	2.031	4	16	30	30	48	52

TABLE 69   Tapered roller bearings. *(SKF Co. Ltd.)*

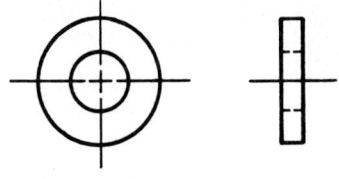

Inside Dia		Thickness		Outside Dia					
in.	mm	in.	mm	in.	mm	in.	mm	in.	mm
.500	12	.12 .18	3 5	.80	20	1.10	25	1.25	30
.625	16	.12 .18	3 5	1.25	30	1.50	40	1.75	46
.750	20	.12 .18	3 5	1.30	34	1.60	40	2.00	50
.875	22	.12 .18	3 5	1.30	34	1.60	40	2.00	50
1.000	25	.12 .18 .25	3 5 6	1.60	40	2.00	50	2.25	60

TABLE 70   Thrust plain bearings.

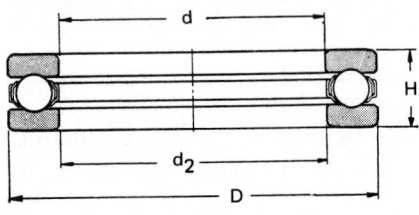

BEARING NUMBER	Nominal Bearing Dimensions								Max. Fillet Radius* in.	Shoulder Diameter in.	
	d		d₂		D		H			Shaft Min.	Housing Max.
	in.	mm	in.	mm	in.	mm	in.	mm			
51100	.3937	10	.402	10.2	.9449	24	.354	9	.012	.750	.563
51101	.4724	12	.480	12.2	1.0236	26	.354	9	.012	.844	.656
51102	.5906	15	.598	15.2	1.1024	28	.354	9	.012	.938	.750
51103	.6693	17	.677	17.2	1.1811	30	.354	9	.012	1.031	.844
51104	.7874	20	.795	20.2	1.3780	35	.394	10	.012	1.188	.969
51105	.9843	25	.992	25.2	1.6535	42	.433	11	.024	1.438	1.219
51106	1.1811	30	1.189	30.2	1.8504	47	.433	11	.024	1.563	1.469
51107X	1.3780	35	1.386	35.2	2.0472	52	.472	12	.024	1.813	1.656
51108	1.5748	40	1.583	40.2	2.3622	60	.512	13	.024	2.063	1.875
51109	1.7717	45	1.780	45.2	2.5591	65	.551	14	.024	2.281	2.063
51110	1.9685	50	1.976	50.2	2.7559	70	.551	14	.024	2.500	2.250
51111	2.1654	55	2.173	55.2	3.0709	78	.630	16	.024	2.750	2.500
51112	2.3622	60	2.370	60.2	3.3465	85	.669	17	.039	3.000	2.719
51113	2.5591	65	2.567	65.2	3.5433	90	.709	18	.039	3.188	2.875
51114	2.7559	70	2.764	70.2	3.7402	95	.709	18	.039	3.375	3.063
51115	2.9528	75	2.961	75.2	3.9370	100	.748	19	.039	3.625	3.250
51116	3.1496	80	3.157	80.2	4.1339	105	.748	19	.039	3.813	3.438
51117	3.3465	85	3.354	85.2	4.3307	110	.748	19	.039	4.000	3.625

* The maximum fillet on the shaft or in the housing, which will be cleared by the bearing corner.

TABLE 71   Thrust roller bearings. *(SKF Co. Ltd.)*

STEEL AND IRON SPUR GEARS
14.5° PRESSURE ANGLE
(Will not operate with 20° Spurs)
Actual Tooth Size

SPUR GEARS STEEL AND IRON
20° PRESSURE ANGLE
(Will not operate with 14.5° Spurs)
Actual Tooth Size

Style	No. of Teeth	Inch Sizes 8 Pitch		1.25 Face		Metric Sizes (mm) 3.18 Module		30 Face	
		Pitch Dia	Hole	Hub Dia	Hub Proj	Pitch Dia	Hole	Hub Dia	Hub Proj
Steel Plain	12	1.500	.750	1.12	.75	38.2	20	28	20
	14	1.750	.750	1.38	.75	44.5	20	35	20
	15	1.875	.875	1.50	.75	47.7	22	40	20
	16	2.000	.875	1.62	.75	50.9	22	40	20
	18	2.250	.875	1.88	.75	57.2	22	48	20
	20	2.500	.875	2.12	.75	63.6	22	54	20
	22	2.750	.875	2.38	.75	70.0	22	60	20
Cast Iron — Web	24	3.000	.875	2.12	1.00	76.3	22	54	25
	28	3.500	.875	2.25	1.00	89.0	22	56	25
	30	3.750	.875	2.25	1.00	95.4	22	56	25
	32	4.000	1.000	2.25	1.00	101.8	25	56	25
	36	4.500	1.000	2.50	1.00	114.5	25	64	25
	40	5.000	1.000	2.50	1.00	127.2	25	64	25
	42	5.250	1.000	2.50	1.00	133.6	25	64	25
	44	5.500	1.000	2.50	1.00	139.9	25	64	25
	48	6.000	1.000	2.50	1.00	152.6	25	64	25
	54	6.750	1.000	2.50	1.00	171.7	25	64	25
Cast Iron — Spoke	56	7.000	1.000	2.50	1.00	178.1	25	64	25
	60	7.500	1.000	2.50	1.00	190.8	25	64	25
	64	8.000	1.000	2.50	1.00	203.5	25	64	25
	72	9.000	1.000	2.50	1.00	229.0	25	64	25
	80	10.000	1.125	3.00	1.12	254.4	28	76	28
	84	10.500	1.125	3.00	1.12	267.1	28	76	28
	88	11.000	1.125	3.00	1.12	279.8	28	76	28
	96	12.000	1.125	3.00	1.12	305.3	28	76	28

Note: Metric size gears were not available at time of publication. Values were soft converted for problem solving only.

TABLE 72 Eight-pitch (3.18 module) spur-gear data. *(Boston Gear Works)*

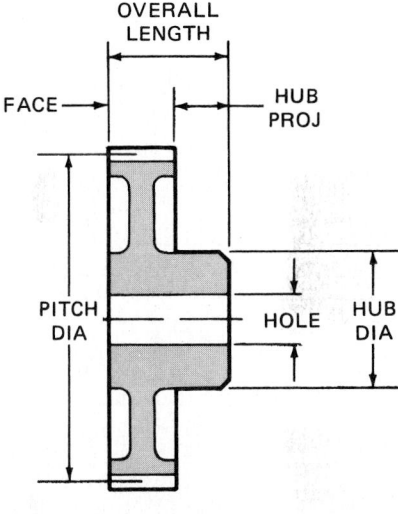

3.18 Module (8 Pitch)

Style	No. of Teeth	Inch Sizes 6 Pitch		1.50 Face		Metric Sizes (mm) 4.23 Module		40 Face	
		Pitch Dia	Hole	Hub Dia	Hub Proj	Pitch Dia	Hole	Hub Dia	Hub Proj
Steel Plain	12	2.000	1.00	1.50	.88	50.8	25	38	22
	14	2.333	1.00	1.81	.88	59.2	25	46	22
	15	2.500	1.00	2.00	.88	63.5	25	50	22
	16	2.667	1.00	2.16	.88	67.7	25	55	22
	18	3.000	1.00	2.50	.88	76.1	25	64	22
	20	3.333	1.00	2.84	.88	84.6	25	72	22
	21	3.500	1.00	3.00	.88	88.8	25	76	22
Cast Iron — Web	24	4.000	1.12	2.50	1.00	101.5	28	64	25
	27	4.500	1.12	2.50	1.00	114.2	28	64	25
	30	5.000	1.12	2.50	1.00	126.9	28	64	25
	32	5.333	1.12	2.50	1.00	135.4	28	64	25
	33	5.500	1.12	2.50	1.00	139.6	28	64	25
	36	6.000	1.12	2.50	1.00	152.3	28	64	25
	40	6.667	1.12	2.50	1.00	169.2	28	64	25
	42	7.000	1.12	2.50	1.00	177.7	28	64	25
Cast Iron — Spoke	48	8.000	1.12	2.50	1.00	203.0	28	64	25
	54	9.000	1.12	2.50	1.00	228.4	28	64	25
	60	10.000	1.25	3.00	1.25	253.8	30	76	30
	64	10.667	1.25	3.00	1.25	270.7	30	76	30
	66	11.000	1.25	3.00	1.25	279.2	30	76	30
	72	12.000	1.25	3.00	1.25	304.6	30	76	30
	84	14.000	1.25	3.25	1.25	355.3	30	82	30

Note: Metric size gears were not available at time of publication. Values were soft converted for problem solving only.

TABLE 73 Six-pitch (4.23 module) spur-gear data. *(Boston Gear Works)*

SPUR GEARS STEEL AND IRON
14.5° PRESSURE ANGLE
(Will not operate with 20° Spurs)
Actual Tooth Size

SPUR GEARS STEEL AND IRON
20° PRESSURE ANGLE
(Will not operate with 14.5° Spurs)
Actual Tooth Size

Style	No. of Teeth	Inch Sizes 5 Pitch 2.00 Face				Metric Sizes (mm) 5.08 Module 50 Face			
		Pitch Dia	Hole	Hub Dia	Hub Proj.	Pitch Dia	Hole	Hub Dia	Hub Proj.
Steel Plain	12	2.400	1.06	1.78	.88	61.0	26	45	22
	14	2.800	1.06	2.18	.88	71.1	26	55	22
	15	3.000	1.06	2.38	.88	76.2	26	60	22
	16	3.200	1.06	2.59	.88	81.3	26	65	22
	18	3.600	1.06	3.00	.88	91.4	26	75	22
	20	4.000	1.06	3.38	.88	101.6	26	85	22
Cast Iron — Web	24	4.800	1.06	3.00	1.25	121.9	26	75	30
	25	5.000	1.06	3.00	1.25	127.0	26	75	30
	30	6.000	1.06	3.00	1.25	152.4	26	75	30
	35	7.000	1.18	3.00	1.25	177.8	30	75	30
	40	8.000	1.18	3.00	1.25	203.2	30	75	30
	45	9.000	1.18	3.00	1.25	228.6	30	75	30
Cast Iron — Spoke	50	10.000	1.18	3.50	1.25	254.0	30	90	30
	55	11.000	1.18	3.50	1.25	279.4	30	90	30
	60	12.000	1.18	3.50	1.25	304.8	30	90	30
	70	14.000	1.18	3.50	1.25	355.6	30	90	30
	80	16.000	1.18	3.50	1.25	406.4	30	90	30
	90	18.000	1.18	3.50	1.25	457.2	30	90	30
	100	20.000	1.31	3.75	1.50	508.0	32	95	38
	110	22.000	1.31	3.75	1.50	558.8	32	95	38
	120	24.000	1.31	4.00	1.50	609.6	32	100	38

Note: Metric size gears were not available at the time of publication. Values were soft converted for problem solving only.

TABLE 74   Five-pitch (5.08 module) spur-gear data. *(Boston Gear Works)*

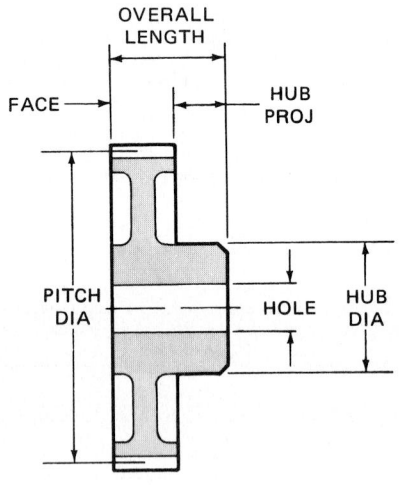

5.08 Module (5 Pitch)

Style	No. of Teeth	Inch Sizes 4 Pitch 2.75 Face				Metric Sizes (mm) 6.35 Module 70 Face			
		Pitch Dia	Hole	Hub Dia	Hub Proj.	Pitch Dia	Hole	Hub Dia	Hub Proj.
Steel Plain	12	3.000	1.12	2.25	.88	76.2	28	58	22
	14	3.500	1.12	2.75	.88	88.9	28	70	22
	15	3.750	1.12	3.00	.88	95.3	28	76	22
	16	4.000	1.12	3.25	.88	101.6	28	82	22
	18	4.500	1.12	3.75	.88	114.3	28	96	22
	20	5.000	1.12	4.25	.88	127.0	28	108	22
	22	5.500	1.12	4.75	.88	139.7	28	120	22
Cast Iron — Web	24	6.000	1.12	3.50	1.50	152.4	28	90	38
	28	7.000	1.25	3.50	1.50	177.8	30	90	38
	30	7.500	1.25	3.50	1.50	190.5	30	90	38
	32	8.000	1.25	3.50	1.50	203.2	30	90	38
	36	9.000	1.25	3.50	1.50	228.6	30	90	38
	40	10.000	1.25	4.00	1.50	254.0	30	100	38
	42	10.500	1.25	4.00	1.50	266.7	30	100	38
	44	11.000	1.25	4.00	1.50	279.4	30	100	38
Cast Iron — Spoke	48	12.000	1.25	4.00	1.50	304.8	30	100	38
	54	13.500	1.25	4.00	1.50	342.9	30	100	38
	56	14.000	1.25	4.00	1.50	355.6	30	100	38
	60	15.000	1.25	4.00	1.50	381.0	30	100	38
	64	16.000	1.25	4.00	1.50	406.4	30	100	38
	72	18.000	1.25	4.00	1.50	457.2	30	100	38
	80	20.000	1.38	4.50	1.50	508.0	35	115	38

Note: Metric size gears were not available at the time of publication. Values were soft converted for problem solving only.

TABLE 75   Four-pitch (6.35 module) spur-gear data. *(Boston Gear Works)*

20° PRESSURE ANGLE—STRAIGHT TOOTH

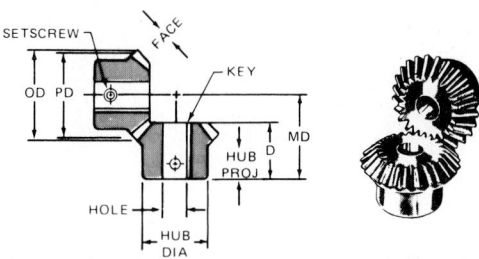

Module (Pitch)	Teeth	U.S. CUSTOMARY (in.)									METRIC (mm)									
		Pitch Dia	Face	OD Approx.	D	MD	Hub Dia	Hub Prov. Approx.	Hole	Keyseat	Pitch Dia	Face	OD Approx.	D	MD	Hub Dia	Hub Prov. Approx.	Hole	Keyseat
2.54 (10)	20	2.000	.44	2.15	1.36	2.00	1.62	.81	.500	.12 × .06	50	11.2	54	34	50	42	20	12	3 × 1.5
									.625	.18 × .09								16	5 × 2.5
									.750	.18 × .09								20	5 × 2.5
2.54 (10)	25	2.500	.55	2.65	1.62	2.44	2.00	.94	.750	.18 × .09	62	14	68	41	62	50	24	20	5 × 2.5
									.875	.18 × .09								22	5 × 2.5
									1.000	.25 × .12								25	6 × 3
3.18 (8)	24	3.000	.64	3.18	1.58	2.56	1.75	.81	.750	.18 × .09	76	16	80	40	65	45	20	20	5 × 2.5
					1.76	2.75	2.25	1.06	1.000	.25 × .12				45	70	58	27	25	6 × 3
					1.76	2.75	2.50	1.12	1.250	.25 × .12				45	70	64	28	30	6 × 3
3.18 (8)	28	3.500	.75	3.68	2.09	3.25	2.50	1.25	1.000		88	16	93	53	82	64	32	25	
									1.188	.25 × .12								30	6 × 3
									1.250									32	
4.23 (6)	24	4.000	.86	4.24	2.31	3.62	3.00	1.31	1.25	.25 × .12	100	22	108	58	92	76	33	32	6 × 3
									1.50	.38 × .18								38	10 × 5
	27	4.500	.96	4.74	2.62	4.12	3.25	1.50	1.50	.38 × .18	114	24	120	66	104	82	38	38	10 × 5
5.08 (5)	25	5.000	1.10	5.29	3.00	4.62	3.50	1.75	1.38	.31 × .16	128	28	134	76	117	88	44	35	8 × 4
									1.50	.38 × .18								38	10 × 5
									1.75	.38 × .18								44	10 × 5

NOTE: METRIC SIZE GEARS WERE NOT AVAILABLE AT TIME OF PUBLICATION. VALUES WERE SOFT CONVERTED FOR PROBLEM SOLVING ONLY.

TABLE 76   Miter gears. *(Boston Gear Works)*

20° PRESSURE ANGLE

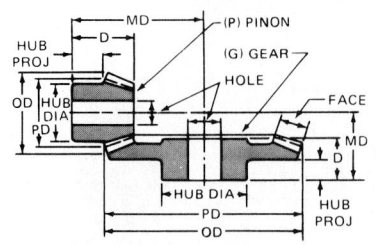 

Module (Pitch)	Teeth	U.S. CUSTOMARY (in.)								METRIC (mm)							
		Pitch Dia	Face	Hole	D	MD	Hub Dia	Hub Proj. Approx.	OD Approx.	Pitch Dia	Face	Hole	D	MD	Hub Dia	Hub Proj. Approx.	OD Approx.
2.54 (10)	50	5.000	.70	.750	1.30	2.62	2.00	1.00	5.06	127	18	20	33.0	66.6	50	25	128.5
	25	2.500			1.55	3.38	2.00	.75	2.75	63.5			39.4	85.8	50	20	69.8
	60	6.000	.78	1.00	1.86	2.75	3.00	1.38	6.04	152.4	20	25	47.2	69.8	76	35	153.4
	20	2.000		.875	2.16	4.38	1.75	1.30	2.27	50.8		22	54.9	111.2	45	33	57.6
	60	6.000	.72	.875	1.62	2.25	2.50	1.12	6.03	152.4	18.3	22	41.1	57.2	64	28	153.2
	15	1.500		.625	1.60	3.88	1.44	.84	1.78	38.1		16	40.6	98.6	36	22	45.2
3.18 (8)	40	5.000	.82	1.000	1.84	2.88	3.00	1.25	5.08	127.2	20.8	25	46.7	67.6	76	32	129
	20	2.500			2.28	4.00	2.12	1.40	2.81	63.6			57.9	101.6	54	35	71.4
	48	6.000	.84	.875	1.62	2.38	2.75	1.00	6.05	152.6	21.4	22	41.2	60.4	70	25	153.7
	16	2.000		.750	2.08	4.25	1.75	1.88	2.35	50.9		20	52.8	108	45	48	59.7
	64	8.000	.84	1.000	1.88	2.75	2.75	1.25	8.04	203.5	21.4	25	47.7	69.8	70	32	204.2
	16	2.000		.875	2.09	5.25	1.88	1.22	2.36	50.9		22	53.1	133.4	48	30	60
4.23 (6)	40	5.000	.82	1.000	1.84	2.88	3.00	1.25	5.08	169.2	20.8	25	46.7	67.6	76	32	129
	20	2.500			2.28	4.00	2.12	1.40	2.81	84.6			57.9	101.6	54	35	71.4
	48	6.000	.84	.875	1.62	2.38	2.75	1.00	6.05	203	21.4	22	41.1	60.4	70	25	153.7
	16	2.000		.750	2.08	4.25	1.75	1.18	2.35	67.7		20	52.8	108	45	30	59.7
	64	8.000	.84	1.000	1.88	2.75	2.75	1.25	8.04	270.2	21.4	25	47.7	69.8	70	32	204.2
	16	2.000		.875	2.09	5.25	1.88	1.22	2.36	67.7		22	53.1	133.4	48	30	60
	36	6.000	1.06	1.125	2.25	3.50	3.25	1.50	6.10	152.3	27	28	57.2	88.9	82	38	155
	18	3.000			2.76	4.75	2.50	1.60	3.41	76.1			70.1	120.6	64	40	86.6
	45	7.500	1.06	1.125	2.12	3.00	3.25	1.25	7.57	190.4	27	28	53.8	76.2	82	32	192.3
	15	2.500		.875	2.56	5.25	2.12	1.44	2.95	63.4		22	65.0	133.4	54	36	75

NOTE: METRIC SIZE GEARS WERE NOT AVAILABLE AT TIME OF PUBLICATION. VALUES WERE SOFT CONVERTED FOR PROBLEM SOLVING ONLY.

TABLE 77   Bevel gears. (*Boston Gear Works*)

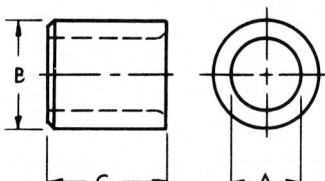

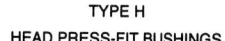

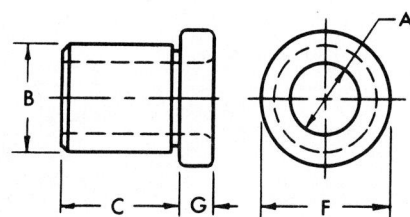

Inside Dia (A)		Outside Dia (B)		Bushing Length C											Technical Data	
From	To	Max.	Min.	.25	.31	.38	.50	.75	1.00	1.38	1.75	2.12	2.50	3.00	F	G
.125	.194	.3141	.3138	•	•	•	•	•	•	•	•				.42	.12
.188	.257	.4078	.4075	•	•	•	•	•	•	•	•	•			.50	.16
.188	.316	.5017	.5014	•	•	•	•	•	•	•	•	•			.60	.22
.312	.438	.6267	.6264	•	•	•	•	•	•	•	•	•			.80	.22
.312	.531	.7518	.7515	•	•	•	•	•	•	•	•	•	•		.92	.22
.500	.656	.8768	.8765		•	•	•	•	•	•	•	•	•	•	1.10	.25
.500	.766	1.0018	1.0015				•	•	•	•	•	•	•	•	1.24	.32
.625	1.031	1.3772	1.3768					•	•	•	•	•	•	•	1.60	.38
1.000	1.390	1.7523	1.7519					•	•	•	•	•	•	•	1.98	.38

U.S. CUSTOMARY (in.)

From	To	Max.	Min.	6	8	10	12	16	20	25	30	35	40	45	F	G
3	5	7.978	7.971	•	•	•	•	•	•	•	•	•	•		10	3
5	6.5	10.358	10.351	•	•	•	•	•	•	•	•	•	•		12	4
5	8	12.972	12.736	•	•	•	•	•	•	•	•	•	•		15	5
8	11	15.918	15.911	•	•	•	•	•	•	•	•	•	•		20	6
8	13.5	19.096	19.088	•	•	•	•	•	•	•	•	•	•	•	24	6
12	16	22.271	22.263		•	•	•	•	•	•	•	•	•	•	28	6
12	20	25.446	25.438			•	•	•	•	•	•	•	•	•	32	8
16	26	34.981	34.971			•	•	•	•	•	•	•	•	•	40	10
25	35	44.508	44.498			•	•	•	•	•	•	•	•	•	50	10

METRIC (mm)

NOTE: METRIC SIZE BUSHINGS WERE NOT AVAILABLE AT TIME OF PUBLICATION. VALUES SHOWN WERE SOFT CONVERTED FOR PROBLEM SOLVING ONLY.

TABLE 78   Press-fit drill jig bushings. *(American Drill Bushing Co.)*

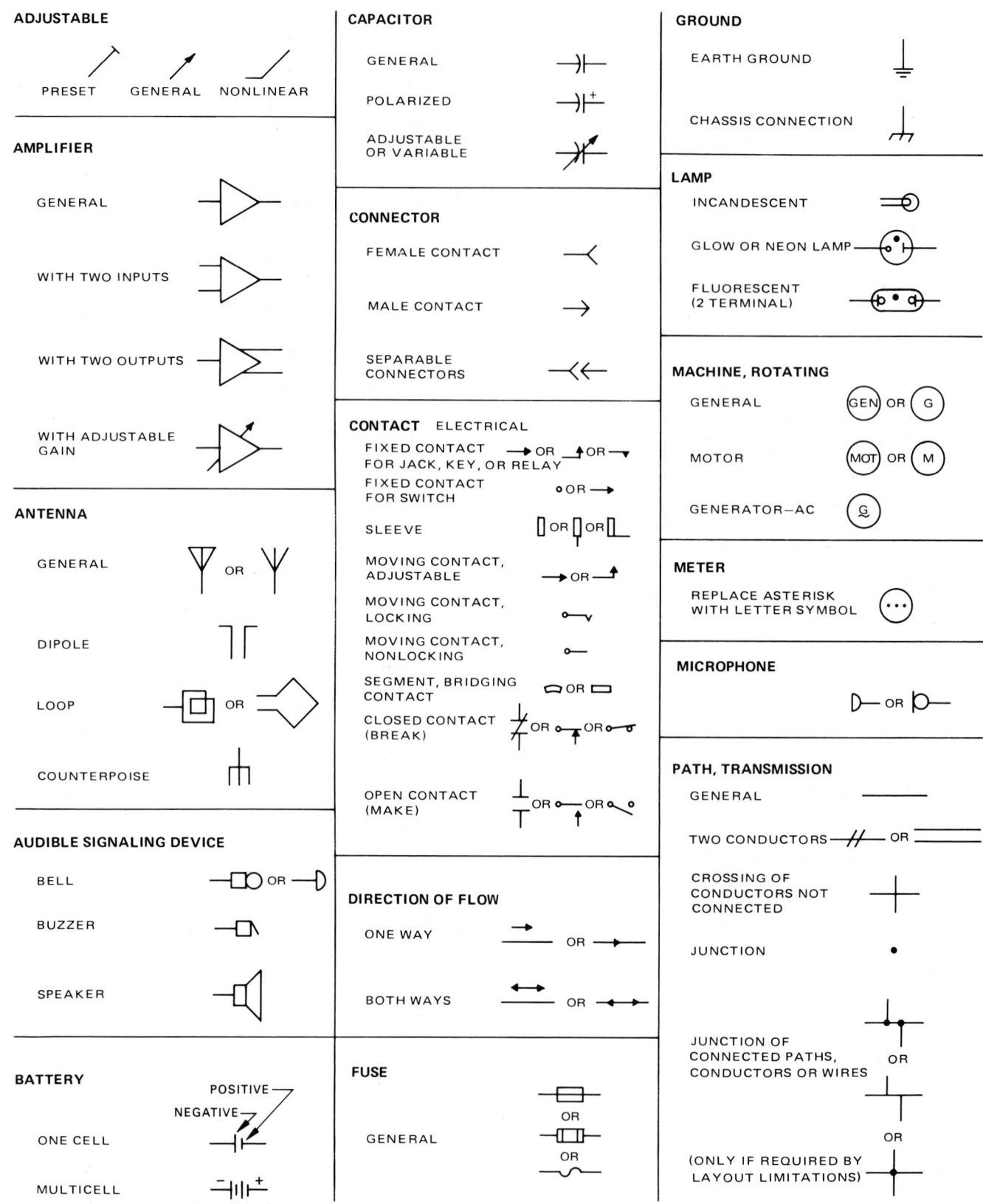

TABLE 79 Graphic symbols for electrical and electronics diagrams.

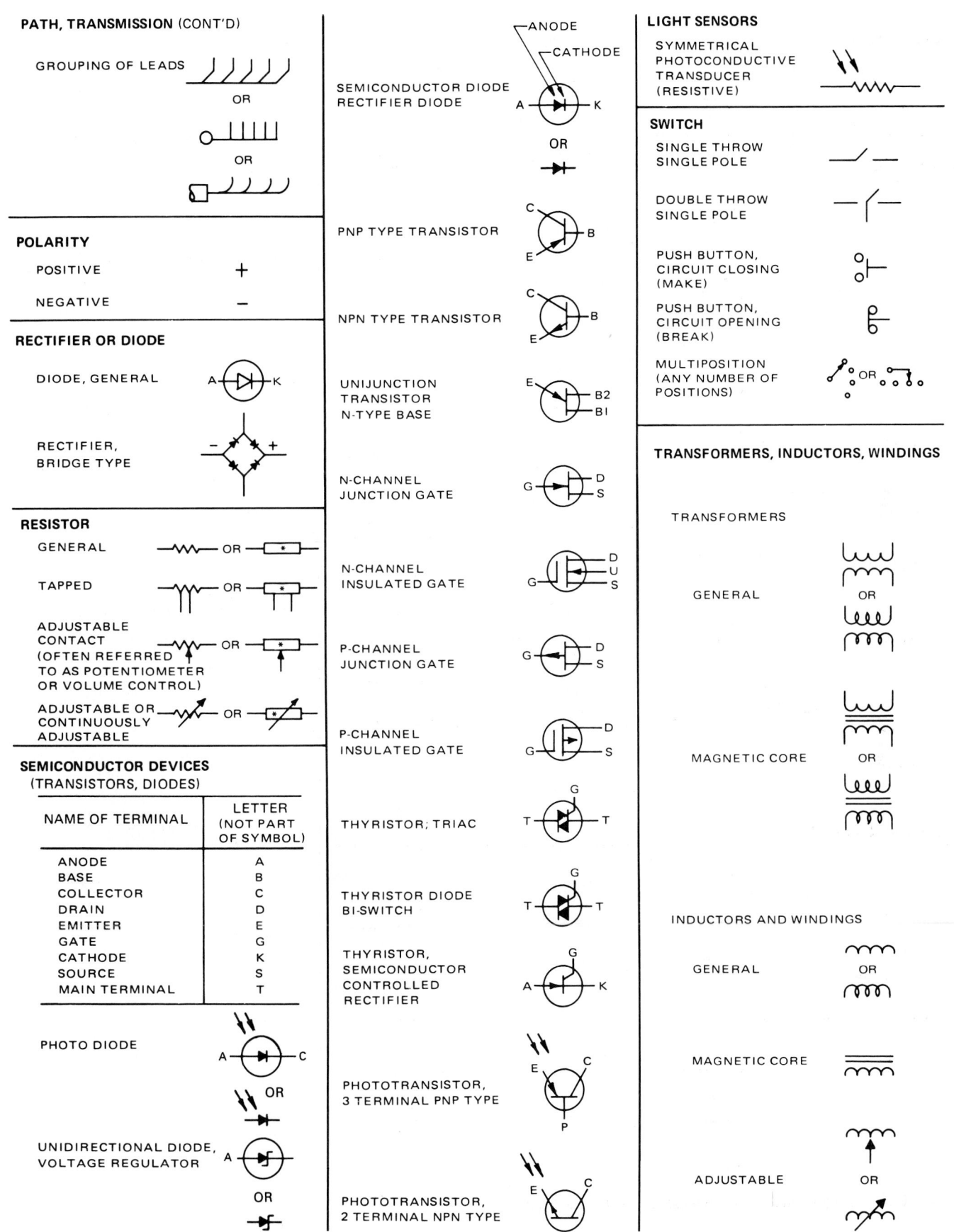

PATH, TRANSMISSION (CONT'D)

GROUPING OF LEADS

OR

OR

POLARITY

POSITIVE $+$

NEGATIVE $-$

RECTIFIER OR DIODE

DIODE, GENERAL A⟶K

RECTIFIER, BRIDGE TYPE

RESISTOR

GENERAL — OR — *

TAPPED — OR — *

ADJUSTABLE CONTACT (OFTEN REFERRED TO AS POTENTIOMETER OR VOLUME CONTROL) — OR — *

ADJUSTABLE OR CONTINUOUSLY ADJUSTABLE — OR — *

SEMICONDUCTOR DEVICES
(TRANSISTORS, DIODES)

NAME OF TERMINAL	LETTER (NOT PART OF SYMBOL)
ANODE	A
BASE	B
COLLECTOR	C
DRAIN	D
EMITTER	E
GATE	G
CATHODE	K
SOURCE	S
MAIN TERMINAL	T

PHOTO DIODE A⟶C

OR

UNIDIRECTIONAL DIODE, VOLTAGE REGULATOR A

OR

SEMICONDUCTOR DIODE RECTIFIER DIODE — ANODE — CATHODE A — K

OR

PNP TYPE TRANSISTOR C B E

NPN TYPE TRANSISTOR C B E

UNIJUNCTION TRANSISTOR N-TYPE BASE E B2 B1

N-CHANNEL JUNCTION GATE G D S

N-CHANNEL INSULATED GATE G D U S

P-CHANNEL JUNCTION GATE G D S

P-CHANNEL INSULATED GATE G D S

THYRISTOR; TRIAC T G T

THYRISTOR DIODE BI-SWITCH T G T

THYRISTOR, SEMICONDUCTOR CONTROLLED RECTIFIER A G K

PHOTOTRANSISTOR, 3 TERMINAL PNP TYPE E C P

PHOTOTRANSISTOR, 2 TERMINAL NPN TYPE E C

LIGHT SENSORS

SYMMETRICAL PHOTOCONDUCTIVE TRANSDUCER (RESISTIVE)

SWITCH

SINGLE THROW SINGLE POLE

DOUBLE THROW SINGLE POLE

PUSH BUTTON, CIRCUIT CLOSING (MAKE)

PUSH BUTTON, CIRCUIT OPENING (BREAK)

MULTIPOSITION (ANY NUMBER OF POSITIONS) OR

TRANSFORMERS, INDUCTORS, WINDINGS

TRANSFORMERS

GENERAL OR

MAGNETIC CORE OR

INDUCTORS AND WINDINGS

GENERAL OR

MAGNETIC CORE

ADJUSTABLE OR

TABLE 79 Graphic symbols for electrical and electronics diagrams. (continued)

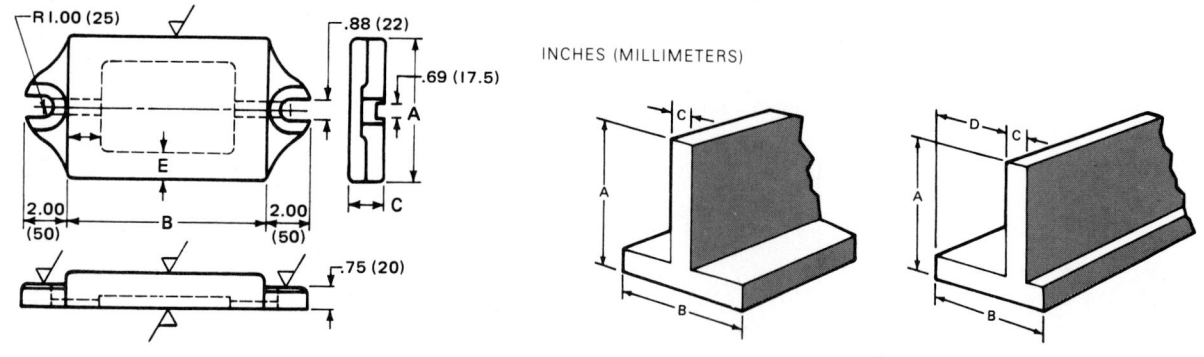

INCHES (MILLIMETERS)

	A	B	C	D	E	Cat. No.	A-B	C	Cat. No.	A-B	C	D
U.S. Customary (in.)	6.00	9.00	1.25	1.88	1.35	ET-300	3.00	.62	UT-300	3.00	.62	1.75
	8.00	12.00	1.38	2.00	1.62	ET-400	4.00	.75	UT-400	4.00	.75	2.62
	10.00	15.00	1.50	2.00	1.75	ET-500	5.00	.75	UT-500	5.00	.75	3.25
	12.00	18.00	1.75	2.25	2.00	ET-600	6.00	1.00	UT-600	6.00	1.00	3.50
						ET-800	8.00	1.25	UT-800	8.00	1.25	4.88
Metric (mm)	150	225	30	45	35	ET-75M	75	15	UT-75M	75	15	45
	200	300	35	50	40	ET-100M	100	20	UT-100M	100	20	65
	250	375	40	50	45	ET-125M	125	20	UT-125M	125	20	85
	300	450	45	55	50	ET-150M	150	25	UT-150M	150	25	90
						ET-200M	200	30	UT-200M	200	30	120

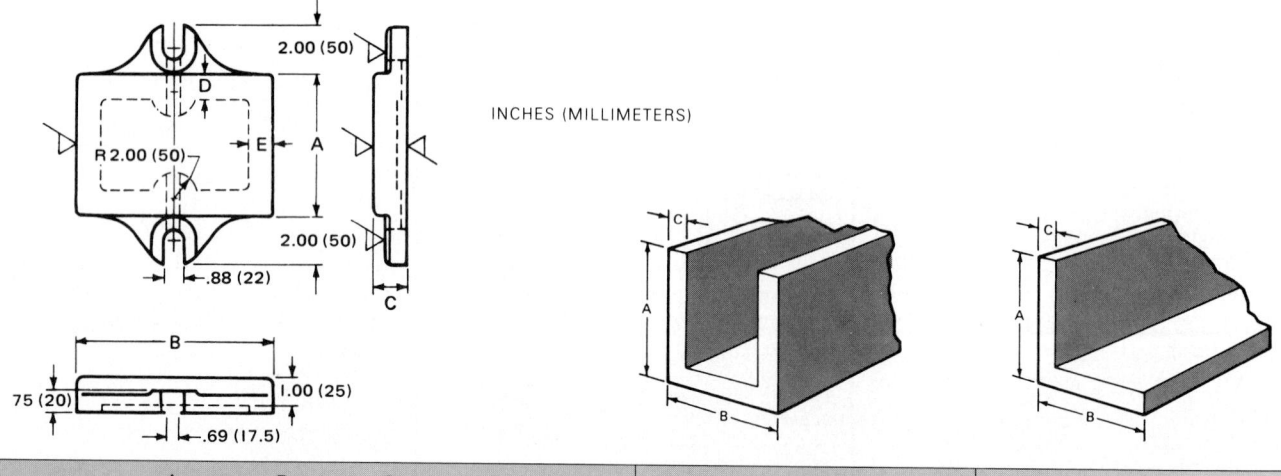

INCHES (MILLIMETERS)

	A	B	C	D	E	Cat. No.	A-B	C	Cat. No.	A-B	C
U.S. Customary (in.)	6.00	9.00	1.25	1.25	1.25	U-300	3.00	.62	L-300	3.00	.62
	8.00	12.00	1.38	1.25	1.25	U-400	4.00	.75	L-400	4.00	.75
	10.00	15.00	1.50	1.38	1.38	U-500	5.00	.75	L-500	5.00	.75
						U-600	6.00	1.00	L-600	6.00	1.00
						U-800	8.00	1.25	L-800	8.00	1.25
Metric (mm)	150	225	30	30	30	U-75M	75	15	L-75M	75	15
	200	300	35	30	30	U-100M	100	20	L-100M	100	20
	250	375	40	35	35	U-125M	125	20	L-125M	125	20
						U-150M	150	25	L-150M	150	25
						U-200M	200	30	L-200M	200	30

TABLE 80   Fixture bases and microsections.

INDEX

PHOTO CREDITS

Table of Contents: pg. iii, Doug Martin; pg. iv, (b), courtesy Mayline Company; pg. iv, (t), Bruno de Houques/TSW; pg. v, Autocad/MacIntosh; pg. vi, Glencoe file photo; pg. vii, Glencoe file photo; pg. viii, STUDIOHIO; pg. ix, Doug Martin; pg. x, courtesy of International Business Machines Corporation; pg. xi, Autocad/MacIntosh; pg. xii, Doug Martin; pg. xii, Doug Martin; pg. xiv, STUDIOHIO; pg. xv, Doug Martin.

Part Openers: pg. 1, © 1994 Comstock; pg. 259, Charly Franklin/FPG International; pg. 379, Andrew Sacks/Tony Stone Images; pg. 667, Earl Zubkoff Photography; pg. 775, Martin Rogers/Tony Stone Images.

Chapter Openers: pp. 2, 22, 35, 45, 76, 92, 136, 171, 227, 260, 292, 327, 347, 380, 441, 481, 598, 609, 652, 668, 714, 747, 776, 809, 828, 858, 877, Aaron Haupt.

Insert A: pg. 1, (t, c), John Terence Turner/FPG International; pg. 1, (b), E. Alan McGee/FPG International; pg, 2, (t, bl), TSW; pg. 2, (b), Tom Tracy/TSW; pg. 2, (c), Jim Cambon/TSW; pg. 3, (b), John Garrett/TSW; pg. 3, (t), John Garrett/TSW; pg. 4, (c), Charles Thatcher/TSW; pg. 4, (t), Bob Thamason/TSW; pg. 5, (c), Bob Thamason/TSW; pg. 4, (b), Lester Lefkowitz/TSW; pg. 5, (b), Lester Lefkowitz/TSW; pg. 5, (t), Paul Chesley/TSW; pg. 6, (b), Andy Sacks/TSW; pg. 6, (t,c), Andrew Sacks/TSW; pg. 7, (b), Sylvain Coffie/TSW; pg. 7, (t), Andrew Sacks/TSW; pg. 8, (b), Andrew Sacks/TSW.

Insert B: pg. 1, (b), Rosenfeld Images/The Stock Market; pg. 1, (t, c), Jim Pickerell/Westlight; pg. 2, (b), Charles Thatcher/TSW; pg. 2, (c), Jim Pickerell/Westlight; pg. 2, (t), Daniel Bosler/TSW; pg. 3, (b), Jim Cambon/TSW; pg. 3, (t), Andrew Sacks/TSW; pg. 4, (tr), Tom Sheppard/TSW; pg. 4, (b), Lonnie Duka/TSW; pg. 4, (bl), Bruno de Houques/TSW; pg. 4, (t), Bruno de Houques/TSW; pg. 5, (t), Bruno de Houques/TSW; pg. 6, (b), Charles Thatcher/TSW; pg. 6, (c), Greg Pease/TSW; pg. 6, (t), Bruce Ando/TSW; pg. 7, (b), Franz Edson/TSW; pg. 7, (c), Poulides/Thatcher/TSW; pg. 8, (bl), John McDermott/TSW; pg. 8, (br), K. Love/Custom Medical Stock Photo; pg. 8, (t, c), Doug Martin; pg. 8, (t, c), Pete Seaward/TSW.